Japanese Scientific and Technical Literature

Japanese Scientific and Technical Literature

A Subject Guide

ROBERT W. GIBSON, JR.
AND
BARBARA K. KUNKEL

Greenwood Press
Westport, Connecticut

Library of Congress Cataloging in Publication Data

Gibson, Robert W 1923-
Japanese scientific and technical literature.

Bibliography: p.
Includes index.
1. Scientific literature—Japan—Bibliography.
2. Technical literature—Japan—Bibliography. 3. Science—Information services—Japan. 4. Technology—Information services—Japan. I. Kunkel, Barbara K., 1948- . II. Title.
Z699.5.S3G5 016.5 80-39693
ISBN 0-313-22929-5 (lib. bdg.)

Library of Congress Catalog Card Number: 80-39693
ISBN: 0-313-22929-5

First published in 1981

Greenwood Press
A division of Congressional Information Service, Inc.
88 Post Road West, Westport, Connecticut 06881

Printed in the United States of America

10 9 8 7 6 5 4 3 2 1

CONTENTS

PREFACE

Japanese goods are competing very successfully in a number of product areas in Western world markets. The creation of these products is supported by advances in Japanese science and technology. These advances are, in turn, supported by research in scientific and technical areas.

The study of the flow of information is of interest to librarians and information scientists. Librarians at the General Motors Research Laboratories became interested in the two-way flow of information about Japanese scientific and technological research between Japan and the West. One result of this interest is this study.

The authors wish to express their appreciation to Dr. Paul F. Chenea, Vice President, General Motors Research Laboratories, and to Mr. Richard J. McDonald, Manager, Administrative Services, General Motors Research Laboratories, for their support during the project study period. The authors also gratefully acknowledge Dr. Theodore F. Welch, Assistant University Librarian for Development and Lecturer in Japanese at Northwestern University, and Naomi Fukuda, Assistant Head (retired) of the Asia Library and Curator of the Japanese Collection at the University of Michigan, for their assistance in the preparation of this study. In addition, two members of the Library staff deserve special recognition for their outstanding efforts as a part of the study group—they are Richard Stoey and Donald Sunday.

The authors sincerely hope that the publication of this study will shed light on the exchange of information between Japan and the West and provide the impetus for more extensive sharing of research information and scientific cooperation.

Robert W. Gibson, Jr.
Barbara K. Kunkel

INTRODUCTION

For centuries the Japanese have been absorbing and improving foreign research efforts. Because of Japan's geographic isolation, however, much of their innovative research has been unknown to the Western world. For example, from mathematical treatises imported from China, the seventeenth century Japanese understood the basics of addition, subtraction, multiplication, and division, which were performed on the abacus (in Japan, called a *soroban*). The solution of simple equations, quadratic equations, and simultaneous equations, and the Pythagorean theorem were also known to them. Building upon this base borrowed from China, a mathematical genius named Seki Takakazu (1642-1708) and one of his disciples developed the basic elements of differential and integral calculus at approximately the same time that Newton and Leibniz were inventing calculus in Europe.

In the postwar world, Japan has once again been an extensive borrower of foreign science and technology. In the years following World War II, Japan made a concentrated national effort to apply science and technology to achieve economic growth. These years were marked by active efforts to use foreign technology to fill the gaps which separated Japan from technologically advanced countries. As Japanese economic capabilities were strengthened, foreign technology was absorbed and improved. The Japanese are now capable of developing and utilizing their own technological innovations. Industrial and economic activities in Japan have increased to such a level that industrial production capacity has surpassed that of all Western European countries. Japan is first in world production of cameras, two-wheeled vehicles, synthetic fibers, and shipbuilding. It is second in world production of passenger cars and third in production of steel. Japan is now the second economic power in the free world and has earned an important place in the world economic structure.

Illustrative of Japan's prominent position in world scientific affairs is the fact that investment in research in Japan is steadily increasing (figure I-1). When research and development expenditures in Japan are plotted against those of other major industrialized nations, this increasing investment is obvious (figure I-2). A distribution chart of expenditures by organization shows that Japanese corporations account for the major portion, followed by universities and research institutes (figure I-3). It should be noted here that over 70 percent of Japan's total research and development funds come from the private sector, whereas in the United States only about one-third of

Figure I-1. Japan's Research and Development Expenditures.

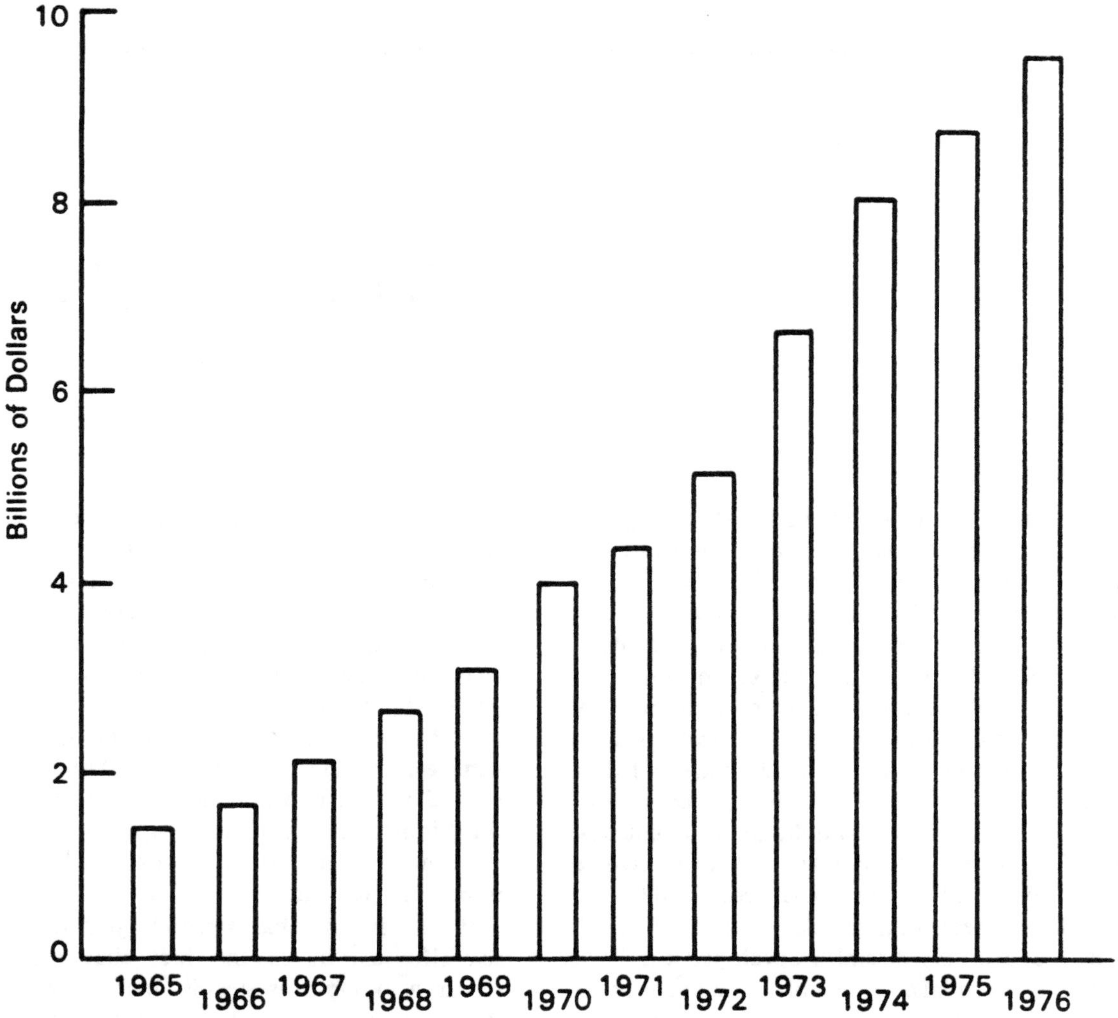

From: Japan Science and Technology Agency, "Outline of the White Paper on Science and Technology." January 1977. p. 48. and Japan Economic Yearbook 1978/79, p.62.

Figure I-2. Research and Development Expenditures in Major Industrialized Nations.

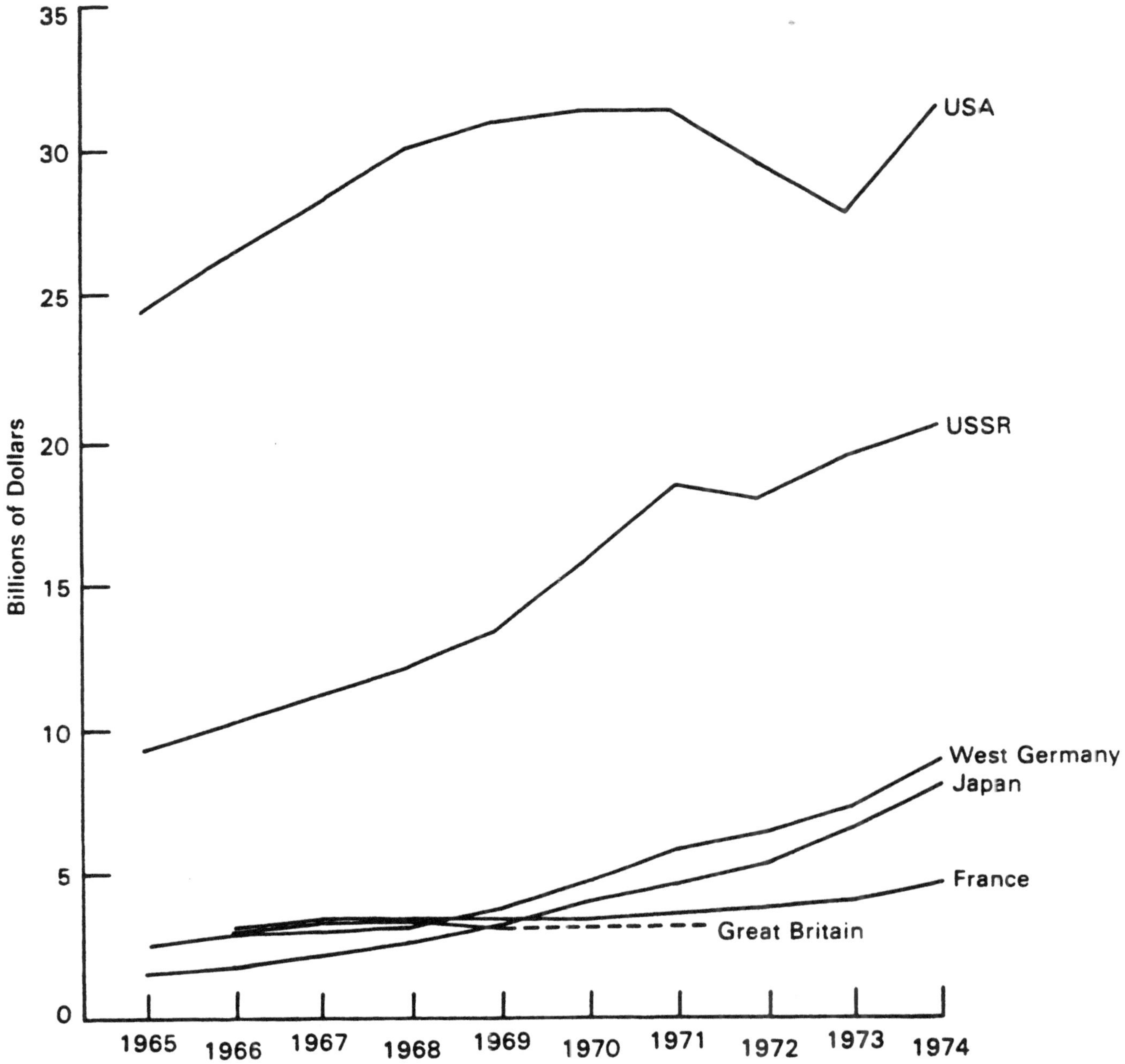

From: Japan Science and Technology Agency, "Outline of the White Paper on Science and Technology." January 1977. p. 49.

Figure I-3. Research and Development Expenditures by Organization.

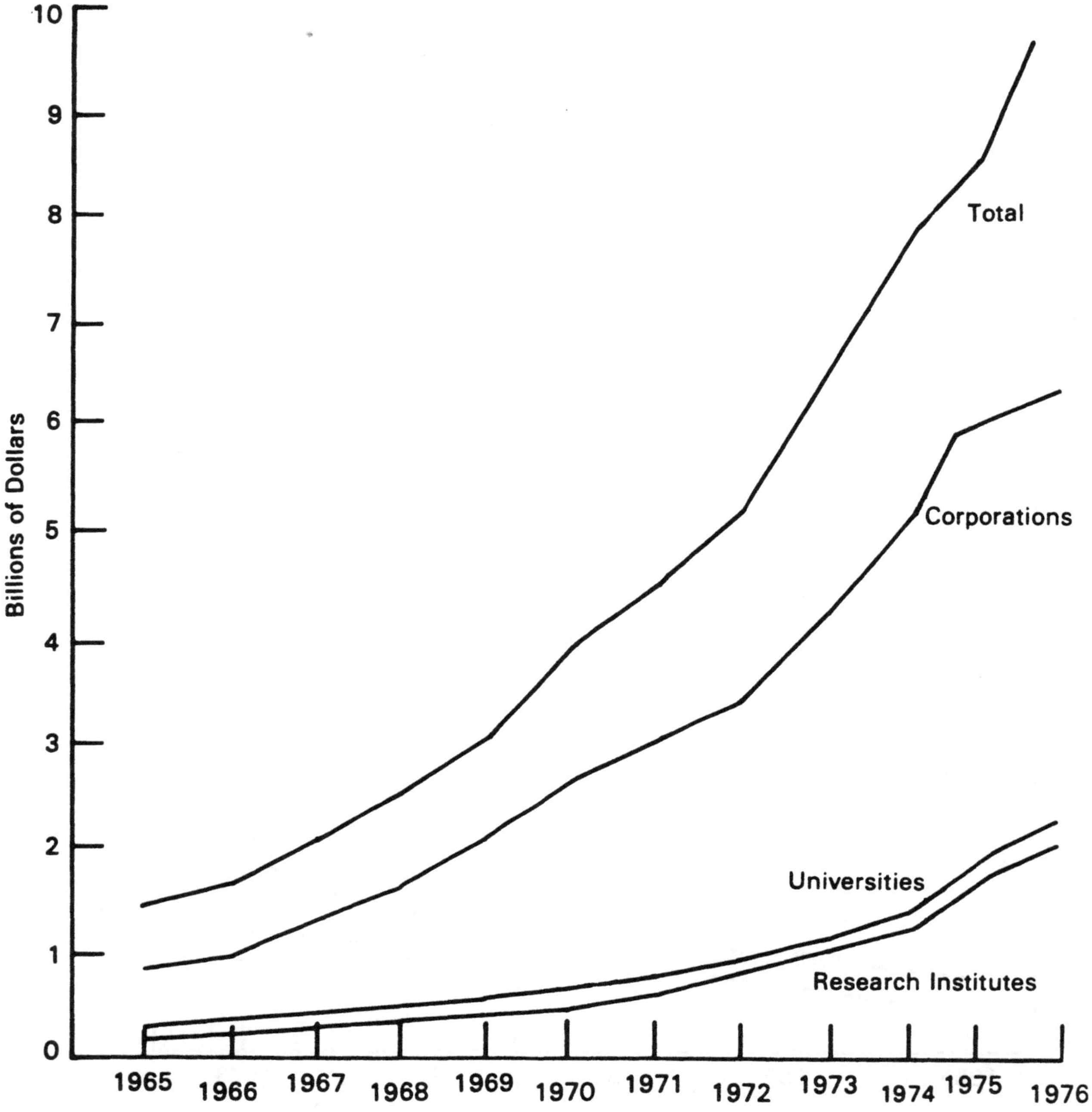

From: Japan Science and Technology Agency, "Outline of the White Paper on Science and Technology." January 1977. p. 51 and Japan Economic Yearbook 1978/79, p. 61, 63.

research and development funding comes from industry, the rest coming through government financing of defense, space, and nuclear projects. Thus Japanese industry has the comparative advantage of directing more research and development efforts toward consumer and industrial markets, which improves its commercial competitiveness both internally and abroad. A percentage distribution chart of the type of research being conducted (figure I-4) shows expenditures of 15 percent for basic research, 21.7 percent for applied research, and 63.3 percent for development for fiscal year 1974. As in most industrialized countries, development accounts for the largest portion of the total research and development expenditures in Japan.

The number of researchers in Japan has also grown fairly steadily in recent years (figure I-5), with over 255,000 people now involved in this effort. This research labor force is the third largest in the world (behind only the United States and the Soviet Union).

It follows that a national research effort of this size would be accompanied by a documentation effort of equal magnitude. A special project group was established in the General Motors Research Laboratories Library to attempt to define the dimensions of Japanese research documentation. Their study focused on these questions:

Figure I-4.
Percentage Distribution of Research and Development Expenditures by Type.

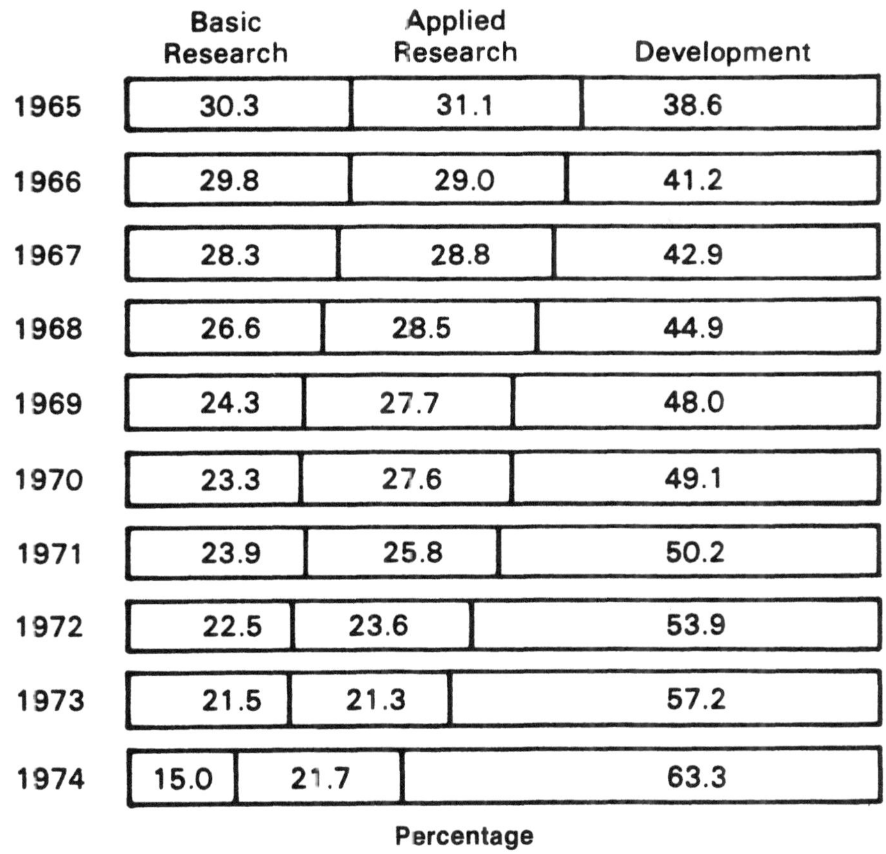

From: Japan Science and Technology Agency, "Outline of the White Paper on Science and Technology." January 1977. p. 53.

Figure I-5. Number of Researchers in Japan.

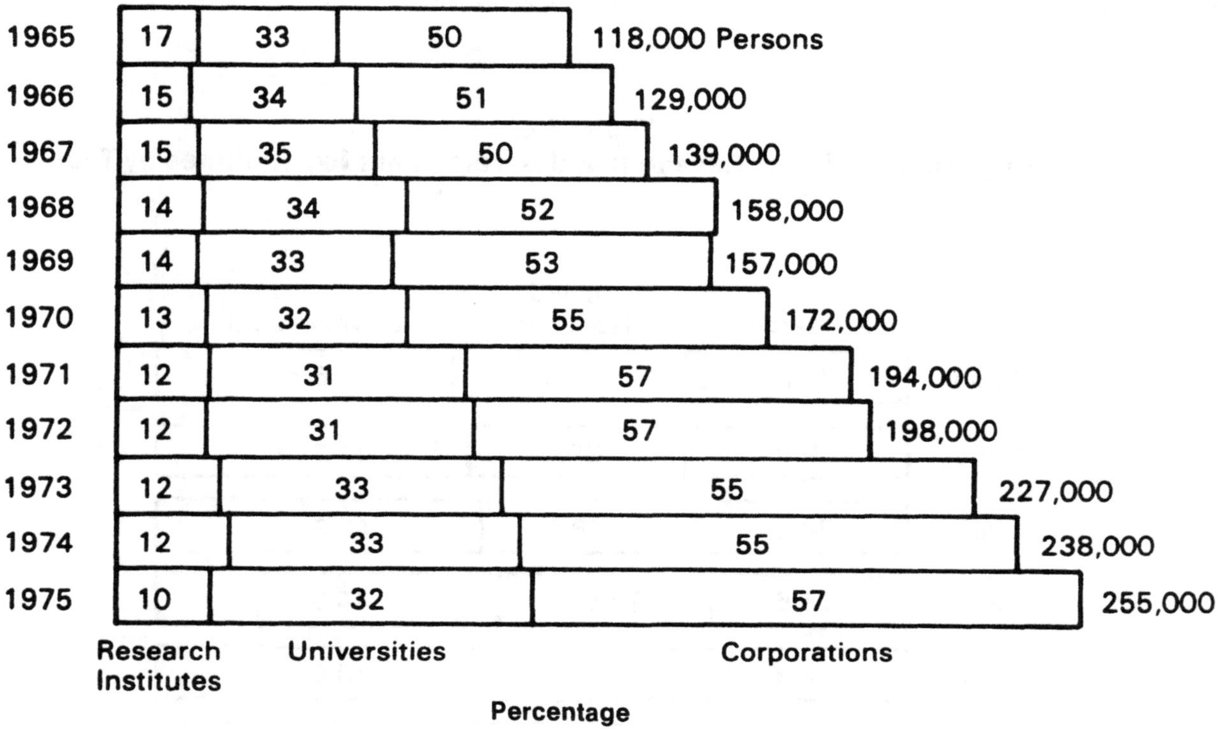

From: Japan Science and Technology Agency, "Outline of the White Paper on Science and Technology." January 1977. p. 55.

What is the size and scope of Japan's scientific and technical documentation effort?

How much of the Japanese scientific and technical literature is accessible in the West?

Does the accessible portion of this literature document the "best" research being done in Japan?

What problems does the language barrier pose for Western researchers in the areas of accessing and using Japanese publications?

Bibliography

Japan Economic Yearbook 1978/79. Tokyo: Oriental Economist, 1979.

Japan Science and Technology Agency. *Outline of the White Paper on Science and Technology*. January 1977.

Japan Science and Technology Agency. *Prospect of Japan's Science and Technology for the Next 30 Years: A Selection from a Survey on the Forecast of Technological Development*. February 1977.

Peck, Morton J. "Technology." In *Asia's New Giant; How the Japanese Economy Works*, edited by Hugh Patrick and Henry Rosovsky, pp. 525-585. Washington, D.C.: Brookings Institution, 1976.

ABOUT THE AUTHORS

ROBERT W. GIBSON is Head of the Library Department, General Motors Research Laboratories, General Motors Technical Center, Warren, Michigan.

BARBARA K. KUNKEL was formerly a member of the same department.

Part 1

ANALYSIS OF INFORMATION ACTIVITIES AND BIBLIOGRAPHIC CONTROL IN JAPAN: SCIENTIFIC AND TECHNICAL LITERATURE

chapter one

PRE-STUDY EVALUATION

The initial step in this project involved personal visits and telephone contacts within the United States and Canada to ascertain what, if anything, other individuals or groups knew about the publication and dissemination of Japanese scientific and technical literature. There was no intention of re-plowing the same field if previous studies existed in this area.

Nearly 100 people were contacted in five major groups—individuals employed by the United States government or quasi-governmental agencies; individuals employed by foreign governments (primarily Canadian); librarians or information specialists employed by private industry in the United States, especially those from companies with Japanese connections or from companies competing directly with the Japanese; university or special librarians whose libraries maintain Japanese literature collections; and representatives of major Western world indexing and abstracting services.

The first group contacted were individuals employed by the United States government or quasi-governmental agencies. Some of those surveyed included personnel from the Departments of State, Defense, Commerce, and the Interior; the National Academy of Sciences; the National Science Foundation; agencies such as the Office of Naval Research, the Foreign Technology section of the United States Air Force, the Foreign Technology and Science Division of the United States Army, the Defense Documentation Center, the National Technical Information Service, the National Aeronautics and Space Administration; and the Japanese Section of the Asian Division, the Science and Technology Division, and the National Referral Center, all of the Library of Congress.

Extensive efforts have been made by the scientific community over the years since Sputnik to assess Russian technical publications. It would seem logical that a similar effort might be in progress concerning the Japanese technical output, especially with the general lack of Japanese language competence in the United States. However, these contacts indicated that they were aware of no efforts currently being mounted to analyze or evaluate Japanese scientific and technical literature. Several individuals noted that while there are cooperative exchange agreements between the United States and Japan, a relatively small portion of the Japanese technical literature is received in the United States. Although there is a large body of Japanese scientific and technical literature, there appears to be no central gathering source in Japan. The Japanese seem to have difficulty accessing their own materials. It is nearly impossible for United States organizations to gain access to

Japanese literature unless they have Japanese sources or contacts for materials. It was also noted that scientific and technical material in Japan seems to be oriented more toward academic interests than toward industrial interests. In addition, much communication between Japanese researchers appears to be transmitted on an informal level rather than through the literature.

Another group contacted included individuals employed by foreign governments and included the Library of Parliament (Canada), the National Research Council of Canada, and the Scientific Advisor of the Japanese Embassy in Washington, D.C. The Canadian contacts noted that while literature is exchanged between Japan and Canada, no efforts are currently being made to analyze or evaluate this material. A Japanese contact noted that great emphasis is placed on those involved in research in Japan knowing intimately who their colleagues are and what they are doing. Much of the research done in Japan is government sponsored and, as such, is treated as being for the use of the Japanese government. Although most Japanese scientific and technical literature is currently published in Japanese, there is a trend toward a higher percentage of English language publishing.

Another large group contacted was made up of librarians or information specialists employed by private industry in the United States, especially individuals from organizations who are meeting the Japanese head on in the market place. They were asked what efforts were being made by their organizations to access and disseminate Japanese scientific and technical publications. By and large, none of the libraries or information centers in the various companies attempts to index or analyze Japanese publications. All of them subscribe to a few Japanese periodicals (usually ten to twelve titles written either in English or with English summaries) and fill interlibrary loan and translation requests from their researchers. Most of the contacts seemed to feel that, although a great deal of research was being published in Japan, the language barrier formed too much of a hurdle for them to cover more than a very selected area. Several of the companies have subsidiaries in Japan and receive some literature and information from them. Access to Japanese patents also poses a problem for many of these corporations, since there are a large number of patents and access is difficult.

In addition to these contacts, several university, public, and special librarians whose libraries maintain Japanese scientific and technical literature collections were contacted. Those librarians in libraries with major Japanese collections do not attempt any analysis of that literature. In fact, few good collections of Japanese scientific and technical literature were located. One librarian did mention, however, that there appears to be a trend toward higher quality in scientific and technical literature received from Japan.

Individuals affiliated with indexing and abstracting services stressed that Western researchers do not know about all of the Japanese scientific and technical literature, and noted that it is not safe to assume that the "best" scientific and technical research information is translated into English. This latter assumption is fairly commonly held by individuals working in scientific areas. Although a significant portion of Japanese periodicals are now published in English, the corollary of the "best" is not necessarily true.

The indexing and abstracting services are aware that an invisible college does exist among Japanese researchers, but are of the opinion that there is little communication between those involved in research at different universities or companies. It is difficult for Western indexing and abstracting services to do justice to Japanese scientific and technical literature because of the time and difficulties involved in translation of the Japanese language. The Japanese would like to be able to publish more in English, but they also encounter problems in translation.

The general impressions received from all of these contacts indicate that, although a vast amount of scientific and technical literature is being published, there is not a great deal of effort in the United States (or Western world) to attain access to it, mainly because of the language problem. It should not be assumed that Western sources cover any significant amount of this literature, nor that the small portion that is covered documents the "best" Japanese research efforts.

After several weeks of visits and communications, it appeared that no one individual or group in this country was very well versed in the Japanese scientific and technical literature. Several of those contacted had positive, but unproven, ideas about the state of publishing in Japan, but subsequent study would reveal how little was really known about the problem. It must be noted here that the above conclusions reflect the thoughts and impressions of those people contacted and are not necessarily in agreement with the findings of this report.

chapter two

METHODOLOGY

After the pre-study visits and consultations were completed, it was acknowledged that additional study was needed to fully understand the information flow of Japanese scientific and technical literature, both within Japan and with other countries. This further study took a two-pronged approach: one portion was devoted to examining the existing Japanese information network; the second portion involved an extensive analysis of Japanese scientific and technical publications.

Most of the information about the organizations and plans that have shaped the Japanese information network was found in publications and documents acquired during the pre-study evaluation. The information which is detailed in chapter 3 is a combination of the best and most current information available from the documents located.

The methodology for the second portion of the study (analysis of Japanese scientific and technical publications) was vastly different. Three types of scientific and technical publications were studied: monographs, periodicals, and technical reports.

Monographs tend to play the secondary role of summarizing research in the areas of science and technology; accordingly, information on this type of publication was emphasized less than the information on periodicals.

It was necessary to do original work for the analysis of the Japanese scientific and technical periodicals since no comprehensive studies were found to have been done. The groundwork was done by analyzing the 7,087 scientific and technical periodical titles listed in the National Diet Library's 1974 publication entitled *Directory of Japanese Scientific Periodicals*. This book is written in English and, although it was somewhat laborious, an analysis of the titles with complete bibliographic information (i.e. Japanese title, translated title, publisher, beginning date, language of publication, and subject area designation) was not too difficult.

After this preliminary analysis was completed, the study group decided that verification of the comprehensiveness of the Japanese periodicals contained in the study was necessary. A visit was made to the Japanese Section of the Asian Department of the Library of Congress, where the study group learned that the list was not comprehensive and that only by including additional publications in the study would any degree of comprehensiveness be approached. Several of the newly referred publications were published only in the Japanese language. Figure 2-1 details a list of all the publications covered and analyzed.

Figure 2-1.

SOURCES FOR JAPANESE SCIENTIFIC & TECHNICAL PERIODICAL TITLES

Directory of Japanese Scientific Periodicals
(National Diet Library, 1974)

Wazasshi Mokuroku 1975 (Catalogue of Japanese Periodicals)
(National Diet Library, 1976)

Wazasshi Mokuroku Tsuiroku, January 1976-December 1977 (Catalogue of Japanese Periodicals)
(National Diet Library, 1978)

Ulrich's International Periodical Directory, 17th ed., 1977-1978

Catalogue of Japanese Periodicals '78
(Maruzen Company, Ltd., 1978)

Current Japanese Periodicals for '78
(Japan Publications Trading Company, Ltd., 1978)

Shuppan Nenkan 1978 (Publishing Yearbook 1978)

Chikuji Kankobutsu Shozo Mokuroku (Index to Periodical Holdings, as of September 30, 1976)
(Japan Information Center of Science and Technology, 1976)

Japanese periodicals indexed by various Western indexing and abstracting publications. (also see Figure 2-3)

As no one in the study group could read Japanese characters, the services of five graduate students from the University of Michigan fluent in technical Japanese were obtained. They identified the scientific and technical Japanese periodicals contained in the additional sources, compared these with the titles already identified, and combined all the titles to form as comprehensive a listing as possible of current Japanese scientific and technical periodicals, with as much bibliographic information as was available. Titles which had ceased publication as of August 1978 were dropped, and titles which appeared in more than one source were combined to avoid duplicate listings. The only exception to this method was for titles which are issued in both Japanese and International editions, in which case both edition titles were included. Supplemental issues to periodicals which are issued regularly and have their own numbering scheme were also retained.

After a monumental effort, a list of 9,116 titles was prepared and classified by issuing institution, language of publication, subject area, and by whether or not they are indexed by either Japanese or Western indexing services. This final compilation was then used for comparisons and cross-tabulation. The results are discussed in chapter 4.

Information from an article entitled "Bibliographical Analysis of Japanese Scientific Periodicals," which was published in the *National Diet Library Newsletter*, no. 46, March 1977, was used to categorize the majority of the titles by issuing institution. Titles which were included in that study but which have subsequently ceased publication were deleted. Additional titles which were not included in that study were classified according to the following breakdown: universities, research institutes affiliated with universities, national and local organizations (which includes governmental agencies and other groups, either national or local in scope, which do not fall into any other category, such as a prefectural public health liaison council), national and local research institutes (which includes, for the most part, government research institutes), public corporations (which are government-owned enterprises such as Japan National Railways, the telephone company, and the tobacco monopoly), other research institutes, learned societies, other corporations (which includes industry-wide associations, such as the Japan Automobile Manufacturers Association), private enterprises, private research institutes (which are research institutes sponsored by private enterprises), and mass media and publishing companies.

An exhaustive effort was made to ascertain the language of publication of each title and to include that information for analysis and evaluation. The titles were divided into two groups: (1) those that were written partially or fully in Western languages (includes titles whose text is completely in Western languages—for the most part English, titles whose text is written in both Japanese and Western languages, and those titles whose text is written in Japanese with Western-language summaries) and (2) those whose text is in Japanese only (periodicals that have a Western-language title but whose text is completely in Japanese were also included in this group).

A modified Universal Decimal Classification (UDC) scheme, which is used by the National Diet Library, was employed to categorize each title by subject area. This scheme for the areas of science and technology is shown in figure 2-2. In this system, the natural sciences correspond to the pure sciences used in Western classification schemes and the applied sciences correspond to the area known as technology in Western schemes.

Since one of the primary interests of the study was to determine the flow of scientific and technical information contained in these periodicals, a plan was formulated to explore that area. It can be hypothesized that Western researchers and scholars working in the areas of science and technology gain access to foreign literature through citations in index and abstract journals. Thus, the major portion of knowledge about foreign research is gleaned from references in these index and abstract periodicals. This hypothesis can be carried one step further to say that information contained in foreign periodicals which are not covered by indexes and abstracts is virtually inaccessible to Western researchers. In this study, an attempt was made to identify those Japanese periodical titles which are covered by Western indexing and abstracting services to ascertain the amount of Japanese scientific and technical periodical literature which is accessible to Western researchers. Time constraints did not permit an exhaustive perusal of Western indexes and abstracts, but those included in the study cover the majority of the English-language index and abstract journals which are international in scope.

The titles of the publications of these services are shown here (figure 2-3) with the number of Japanese periodical titles they survey and a comparison of that number to their total periodical coverage. The titles abstracted by any one abstracting and indexing service are directly proportional to the available in-house Japanese-language capability, usually very limited.

Figure 2-2.

JAPANESE SCIENTIFIC AND TECHNICAL PERIODICALS SUBJECT CATEGORIES

Natural Sciences

Natural sciences in general
Mathematics
Astronomy, surveying, geodesy
Physics
Chemistry
Mineralogical sciences & crystallography
Geophysics
General geology, rocks, minerals
Oceanography, surface waters, hydrology
Meteorology, climatology
Biosciences (including paleontology)
Anthropology, ethnology
Botany
Zoology
Geography, charts
Archaeology

Medical Sciences

Medical sciences in general
Anatomy
Physiology
Health, preventive medicine, environmental pollution
Accidents-prevention, protection, safety, fires
Pharmacy, pharmacology, therapeutics
Disease, pathology
Surgery, ophthalmology
Gynecology, obstetrics

Agricultural Sciences

Agriculture, forestry, stockbreeding, animal produce, fisheries
Plant injuries, diseases, pests, crop damage and protection
Field crops, industrial crops, fruit growing
Arboriculture, trees, shrubs
Horticulture, gardens, gardening
Animal husbandry, stockbreeding, animal produce, dairy and other animal produce, comparative pathology, veterinary medicine
Insect and reptile management, breeding industries, sericulture
Fisheries

Applied Sciences

Applied sciences, medicine, technology, agriculture in general
Inventions, patents
Engineering & technology (includes information science)
Mechanical engineering & machinery
Nuclear technology
Electrical engineering (includes communication)
Mining and mineral dressing
Military & naval engineering & technology
Civil, structural engineering
Railway, highway engineering
Hydraulic engineering, natural waters, and channels
Public health engineering
Land and road vehicles (not railway)
Shipbuilding, ships, boats
Aeronautical engineering
Domestic science, household economy, housekeeping
Management
Typography and printing
Transport
Chemical industry, chemical technology (includes oils, fats & petroleum)
Beverages, stimulants, food industry, preservation
Glass and ceramic industries
Metallurgy (including workshop practice generally)
Industries & crafts based on processible materials
Papers and pulp industry
Textile and cordage industry
Macromolecular materials, rubber & plastics
Precision mechanisms, apparatus & machines, instruments, computer
Building industry, physical planning, landscape, architecture
The Arts, recreation, entertainment, sport, etc.
Photography and cinematography

Figure 2-3.

JAPANESE PERIODICAL TITLES SURVEYED BY VARIOUS WESTERN INDEXING AND ABSTRACTING PUBLICATIONS

Title	Number of Japanese Titles	Total Number of Titles Indexed
API Abstracts	7 (4.5%)	154
Applied Mechanics Reviews	57 (5.4%)	1060
APTIC	27 (46.6%)	58
Automobile Abstracts	16 (4.6%)	348
Bibliography of Agriculture	486 (7.5%)	6500
Biological Abstracts	430 (5.2%)	8312
Ceramic Abstracts	9 (2.3%)	400
Chemical Abstracts	1401 (10.0%)	14000
Copper Extracts	37 (8.6%)	432
Engineering Index	105 (5.2%)	2022
Excerpta Medica	226 (5.0%)	4536
Index Medicus	102 (3.4%)	3000
INSPEC (includes Computer & Control Abstracts, Electrical & Electronics Abstracts, and Physics Abstracts)	185 (9.3%)	2000
International Aerospace Abstracts	37 (3.6%)	1026
ISMEC Bulletin	10 (3.4%)	294
Mathematical Reviews	122 (8.0%)	1524
Metals Abstracts	40 (5.7%)	701
Pollution Abstracts	7 (2.5%)	278
RAPRA Abstracts	9 (3.4%)	264
Science Citation Index	80 (2.9%)	2717
World Aluminum Abstracts	112 (10.3%)	1089
TOTAL	3505 (6.9%)	50715

Number of Japanese titles (excluding duplicates) = 1729

The two services with the highest percentages are utilizing abstracting support from technical abstractors in Japan.

Those Japanese scientific and technical periodical titles which are covered in the only Japanese indexing and abstracting service which could be identified were also included in the study as an indication of the Japanese scientific and technical periodicals accessible to Japanese researchers. The National Diet Library does prepare a Japanese Periodical Index quite similar in format to the H. W. Wilson Company's *Readers' Guide to Periodical Literature.* However, the depth of indexing is only to verv broad categories, not unlike the breakdown shown in figure 2-2. Thus the study group did not include this material as indexed because it would prove nearly impossible to retrieve by a subject approach.

The study group found that very little information is available on publication of technical reports in Japan—mainly because there is no central gathering point (such as the National Technical Information Service [NTIS] in the United States) and also because many technical reports in Japan are proprietary, either to the sponsoring industry or to the government. The information that was located, primarily through secondary sources, is included in chapter 4.

chapter three

SCIENTIFIC AND TECHNICAL INFORMATION ORGANIZATIONS IN JAPAN

Japan's Developing National Information System

Since World War II, the national government of Japan has taken an active interest in reorganizing technical information production and dissemination activities.

In 1947, the Scientific Research Organization Renovation Committee was appointed, whose deliberations resulted in the formation of two organizations in 1949. These two were the Science Council of Japan, formed to represent Japanese scientists, and the Scientific and Technical Administration Commission, formed to serve as an intermediary between the Science Council of Japan and the government and to coordinate information activities among various government agencies.

The Science Council of Japan was created by legislation to promote the development of science, to facilitate its permeation into government, industry, and national life, and to represent Japanese scientists, both within their own country and on the international level. This step was taken on the conviction that science provides the basis of a nation's culture and that the development and promotion of scientific activities could contribute to the rehabilitation of Japan and to the advancement of science in the world.

The council encompasses all fields of science, including the humanities, social sciences, and natural sciences, and consists of seven divisions:

(1) literature, philosophy, pedagogy, psychology, sociology, and history

(2) law and political science

(3) economics, commerce, and business administration

(4) pure science

(5) engineering

(6) agriculture

(7) medicine, dentistry, and pharmacology

The tasks of the Science Council of Japan are to discuss important matters concerning science, and to coordinate scientific investigations. The council may submit reports and recommendations to the government. Other activities of the council include the maintenance of contacts with learned societies, research institutions, and researchers, the organization of lecture meetings and symposia, and the compilation and publication of journals, annual reports, and bulletins.

In 1956, the Science and Technology Agency, which was set up within the prime minister's office, superseded the Scientific and Technical Administration Commission. The Council for Science and Technology came into being in 1959 as an advisory group to the prime minister offering opinions on all scientific and technical policies. It is presided over by the prime minister, and membership consists of the minister of finance, minister of education, director-general of the Economic Planning Board, director-general of the Science and Technology Agency, president of the Science Council of Japan, and five persons of learning and experience appointed for a three-year term by the prime minister with the approval of the National Diet.

According to law, the prime minister must consult the Council for Science and Technology when coordination among the various ministries of the government is necessary in relation to the following matters: (1) establishment of basic and overall policies for science and technology, (2) the setting up of long-range and comprehensive research objectives, (3) establishment of policies for encouraging important research programs, and (4) inquiries to, or reports or recommendations from, the Science Council of Japan.

In response to an inquiry from the prime minister, the council submitted a report in October 1969, which discussed basic policies on effective use of scientific and technological information. In this report, a plan for a National Information System for Science and Technology (NIST) was proposed. This NIST Plan was designed to strengthen the functions of existing information services and to establish new centers (i.e., regional service centers, specialized centers, data centers, and a general center as a nucleus) as well as to coordinate these services to create an efficient information network.

The Science and Technology Agency was given the responsibility for promoting the NIST plan. In accordance with this responsibility, a Committee for NIST Implementation, composed of information specialists, was appointed and began work. In February 1973, a Consultative Committee for Scientific and Technological Information was formed to act as a private advisory body to the minister in charge of the Science and Technology Agency. The responsibility of this committee was to prepare a plan to make NIST more effective. In August 1974, the committee submitted a report entitled "Program of the National Information System for Science and Technology." According to this report, the NIST plan involves five basic areas:

(1) information activities should be comprehensive and involve long-range planning

(2) NIST should be implemented as a national project and include cooperation between all related organizations

(3) various user needs should be met exactly

(4) close coordination with other information systems should be planned

(5) encouragement should be given to cooperation with international information organizations

To implement the plan, a Central Coordination Organ (CCO) would be established to coordinate the activities of the centers which contribute to the program. These associated centers would be:

General Information Center—This center would be responsible for acquiring and processing the world's scientific and technological literature and for photocopying and translating this material. Abstracting, indexing, and disseminating this information are also functions of the center. The functions of this center form the core of NIST, and the Japan Information Center of Science and Technology (JICST), which will be discussed in depth later in this report, is being upgraded to fill these requirements.

Specialized Information Centers—These centers would collect and process information in the specific subject fields that they serve. It is also expected that they would perform technological forecasting. Abstracting, indexing, and compiling information in machine-readable form would enable the centers to utilize computers to serve the information needs of the users in their subject fields. The NIST plan calls for the establishment of more than twenty such centers.

Data Centers—Numerical data derived from various experiments and observations are vital to scientific research. This data, stored in national experiment stations or observatories, would be systematically arranged and classified under such categories as observation of nature, biological data, physical and chemical properties data, and engineering data. Several data centers are already in operation and will be further integrated into the NIST program. They are: National Institute of Environmental Pollution Research, National Research Centre for Disaster Prevention, Japan Atomic Energy Research Institute, National Space Development Agency, Japan Marine Science and Technology Centre, and the Small Business Promotion Corporation. Other centers which are planned include: Life-Science Research Promotion Centre, Central Medical Treatment Information Centre, National Institute of Hygienic Sciences, Agricultural Information Centre, Ministry of Posts and Telecommunications, Earth Science and Energy, and Engineering Information.

Regional Service Centers—These centers would be responsible for disseminating various types of information including National Information Systems (NATIS) activities. They will also attempt to improve availability of local information. Three categories of centers will be included: bloc-unit regional centers (eventually in ten major cities, each encompassing seven prefectures), prefectural-unit regional centers (to provide information and technical guidance), and special area service centers (in such places as industrial complexes).

The Clearinghouse—This center would collect and disseminate information on research topics, ongoing research, research centers, and researchers. It would also provide guidance to the information centers. The center's main function would be to act as a referral center for scientific and technical information.

Central Depository—Information collected by the various centers which could not be stored by them due to space limitations would be sent to the depository. The depository would then store these materials if they were judged as being worthy of long-range preservation. This depository would be a part of JICST's responsibilities.

Other Organizations—The secondary information provided by the above components of the NIST implementation plan would have great impact on reference services provided by libraries. Training facilities for librarians and information specialists fall in this category and would also be a part of the NIST program.

Figure 3-1 illustrates the interrelationships between the various facets of the NIST plan. The NIST plan is quite comprehensive in scope and once fully implemented should provide Japanese researchers and scientists with the beginnings of an organized and efficient method of accessing information. It is important to understand the functions and activities of the Japan Information Center of Science and Technology (JICST) because it is in the process of being upgraded to form a keystone in the NIST Program.

Japan Information Center of Science and Technology (JICST)

ORGANIZATION AND ACTIVITIES OF JICST

The Japan Information Center of Science and Technology (JICST) is a semigovernmental organization attached to the Office of the Prime Minister of Japan and serves as a core for activities related to scientific and technological information in Japan. It was created under the provisions of the Japan Information Center of Science and Technology Law (Act No. 84 of 1957) and went into operation on August 16, 1957.

The purpose and activities of JICST are:

> To collect worldwide scientific information comprehensively and to process it systematically;
>
> To disseminate information rapidly and appropriately to organizations and individuals, regularly and upon request; and
>
> To offer other far-reaching services to encourage the scientific documentation efforts of organizations and to assist them in the solution of complex problems.*

In August 1957, JICST began with a total budget of 80 million yen, of which 40 million yen were contributed by the government and 40 million yen were contributed by private sources. In fiscal year 1976, the JICST budget called for total capital expenditures of 952 million yen. From fiscal year 1958 through fiscal year 1977, JICST has received 8.24 billion yen in total funds. Budgetary control over JICST is exercised by the Science and Technology Agency. Figure 3-2 details the organizational structure, officers, and staff at JICST.

JICST SERVICES

Collecting Information

JICST collects information related to science and technology in the form of domestic and foreign journals, technical reports, conference presentations (mostly foreign meetings), and patent literature from more than fifty countries.

*Japan Information Center of Science and Technology, *JICST 1977: Science Information Service Activities*, 1977, p. 1.

Figure 3-1. Organization of NIST Elements.

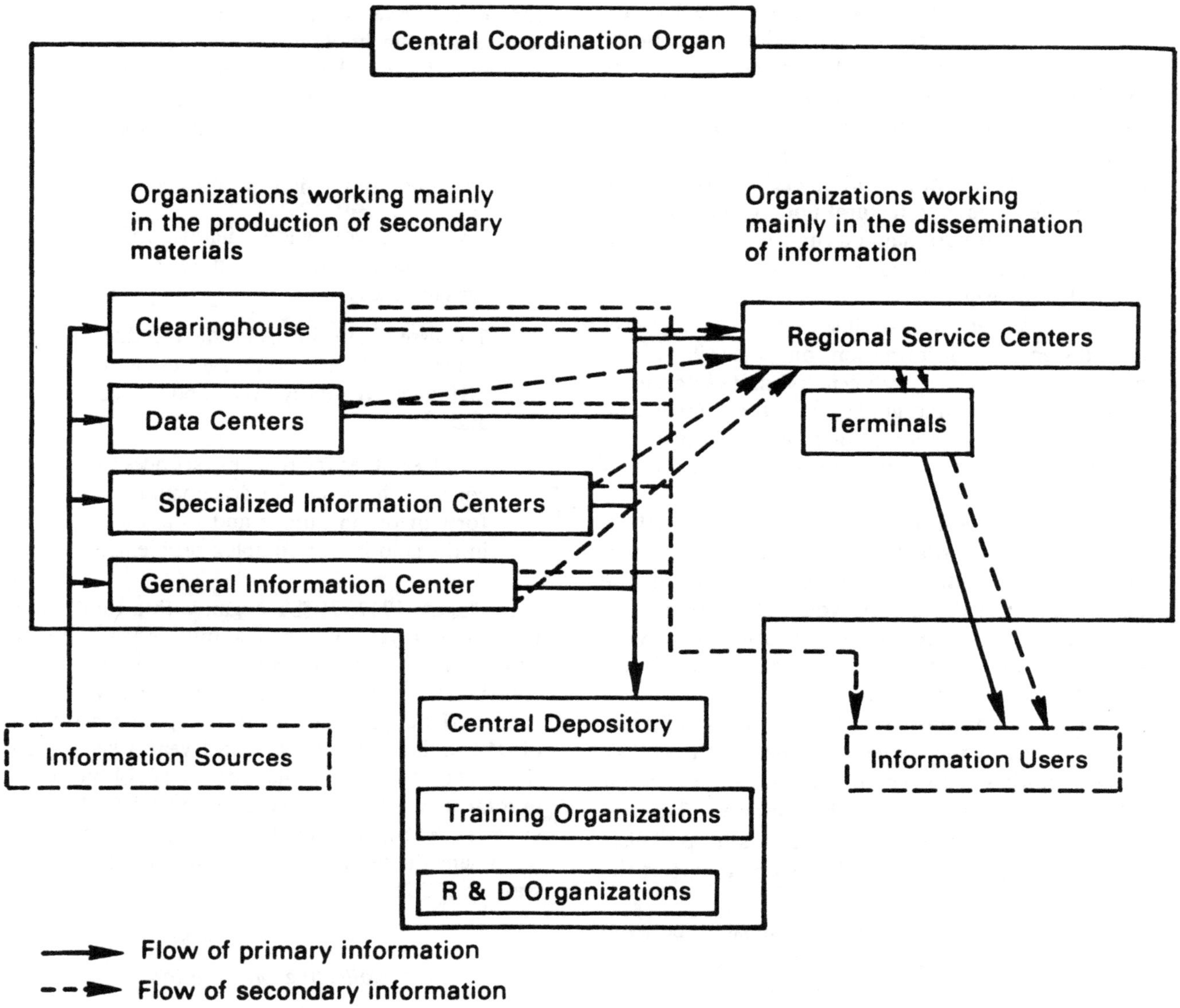

From: Suzuki, Herihachiro, "National plans for library/information services and the National Diet Library," *Library and Information Science, No. 14, 1976, p. 8.*

Figure 3-2. JICST Organizational Structure.

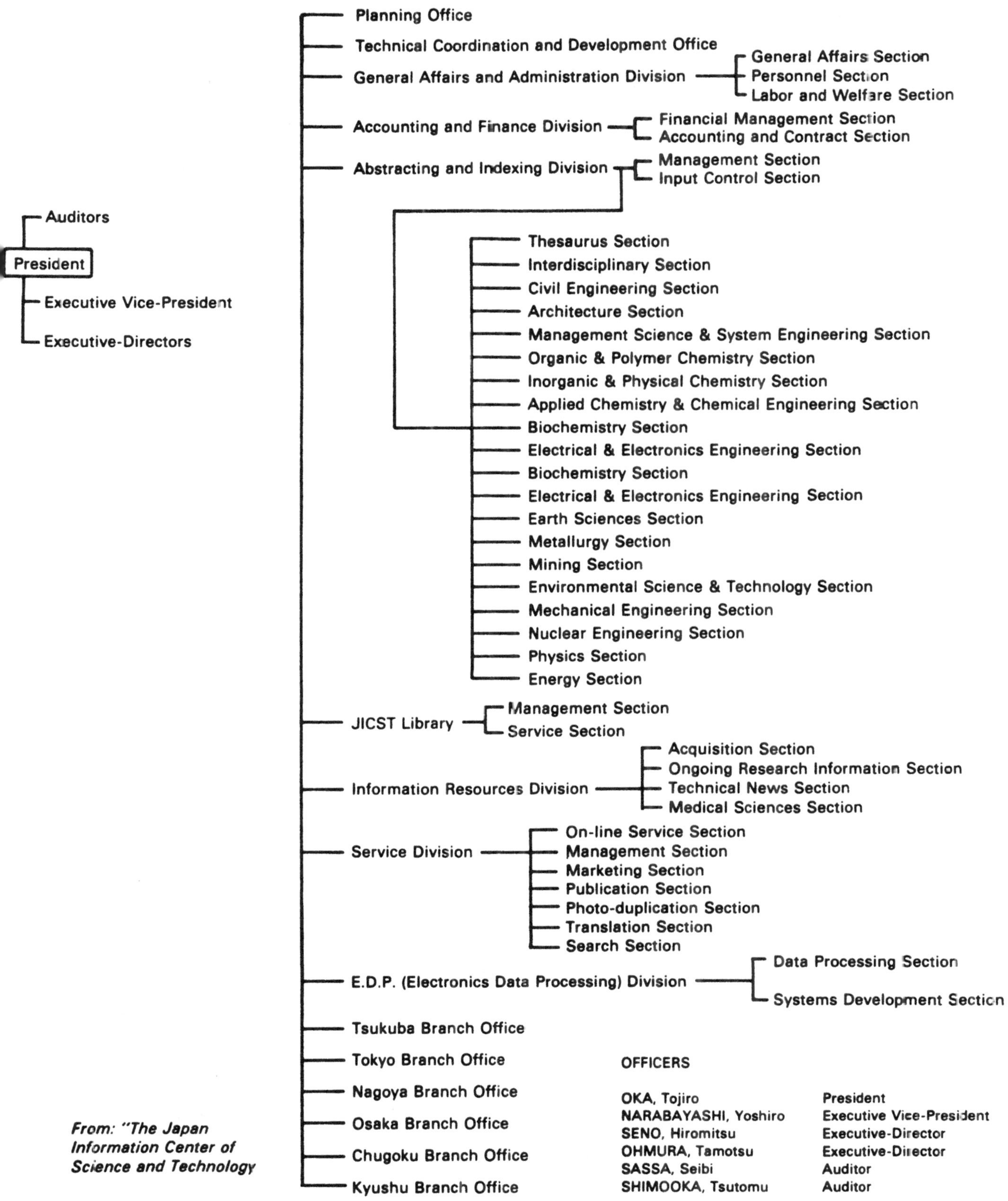

OFFICERS

OKA, Tojiro	President
NARABAYASHI, Yoshiro	Executive Vice-President
SENO, Hiromitsu	Executive-Director
OHMURA, Tamotsu	Executive-Director
SASSA, Seibi	Auditor
SHIMOOKA, Tsutomu	Auditor

From: "The Japan Information Center of Science and Technology

Types of materials collected and present totals are as follows:

TYPE OF MATERIAL	*NUMBER OF ITEMS*
Foreign journals	6,310 journals
Domestic journals	3,300 journals
Technical reports	12,500 reports
Conference presentations	300 items
Patent disclosures	49,000 items
Patent announcements	40 types

Figure 3-3 illustrates the breakdown of JICST-collected journals by country, and figure 3-4 shows the breakdown of JICST-collected journals by subject area.

Information Processing

JICST realized the need to use computers shortly after their inception and acquired a small computer in 1961. In order to process the abstracted and indexed information, JICST was forced to develop computer equipment capable of handling Chinese characters (*kanji*), Japanese syllabary letters (*kana*), Roman, Russian, and Greek alphabets, as well as numerals and symbols.

Personnel to operate and input into these systems include about 100 information officers with scientific and technical backgrounds, plus clerical staff. In addition, more than 5,000 scientists and engineers who are actively engaged throughout Japan in scientific and industrial research are registered by JICST to extract, process, and abstract more than 370,000 items per year. The average length of each abstract is 300 Japanese characters (*kanji* and *kana*), which is equivalent to about 100 words in English.

The JICST system operates in both batch and online modes. The batch system is divided into the following subsystems:

(1) Journal article file construction system

(2) Journal article editing system

(3) Word control system

(4) Materials control system

(5) Search system

The three basic purposes of the system are: (1) to provide a total system involving input and output to minimize manual handling; (2) to facilitate versatile utilization of input data; and (3) to use the Japanese language in nonromanized form using both *kanji* (Chinese characters adopted by the Japanese) and *kana* (made up of *hiragana*—used to form grammatical endings to words and *katakana*—used to write foreign words for which the Japanese have no equivalent word) in addition to alphanumeric coverage.

The following projects are now in operation or in the planning stages at JICST:

(1) Automatic editing and photocomposition of the *Current Bibliography on Science and Technology* is designed to prepare original plates for printing output for issues and annual indexes of the publication. The abstracts in Japanese are punched onto paper tape, made machine readable with *kanji* teletypewriters, and transferred onto magnetic tapes. These tapes are then utilized as master tapes for the use of the entire information processing system at JICST. These master tapes contain the entire texts of abstracts, translated titles, original titles, keywords, classification codes, bibliographical notes, and everything that is necessary for the publication of abstract journals and indexes and for retrieval services. This project is now operational and at the full level of performance will process approximately thirty thousand abstracts per month.

(2) A terminology control system is designed for keyword control and for preparation of a Japanese thesaurus. This system is aimed at integrating the scientific terminology of various languages and their thesaurus interrelationships, the provision of terminology translation and compilation, and the revision of thesaurus organization. This system is designed to increase JICST's efficiency in retrieving and indexing material in the Japanese language.

(3) An information retrieval system using magnetic tape and a Selective Dissemination of Information (SDI) service are being developed which will be available in *kanji* or in binary code decimal form. It will be an online real-time system for retrieval which will connect not only to domestic users, but will also serve as the basis for a national system of technical information exchange abroad.

(4) Computer technology will be applied to the source document control problem to facilitate identification of the ever growing library collection. This control system will eventually be used to compile a national union catalog of scientific and technical periodicals. In addition, JICST is working on an international serials catalog which, when completed in the future, will bring together all domestic data on scientific and technical journals. A hard line has been drawn between scholarly research documents and popular scientific magazines. JICST has already compiled a bibliographic master file and a library

Figure 3-3 JICST Collection of Journals by Country.

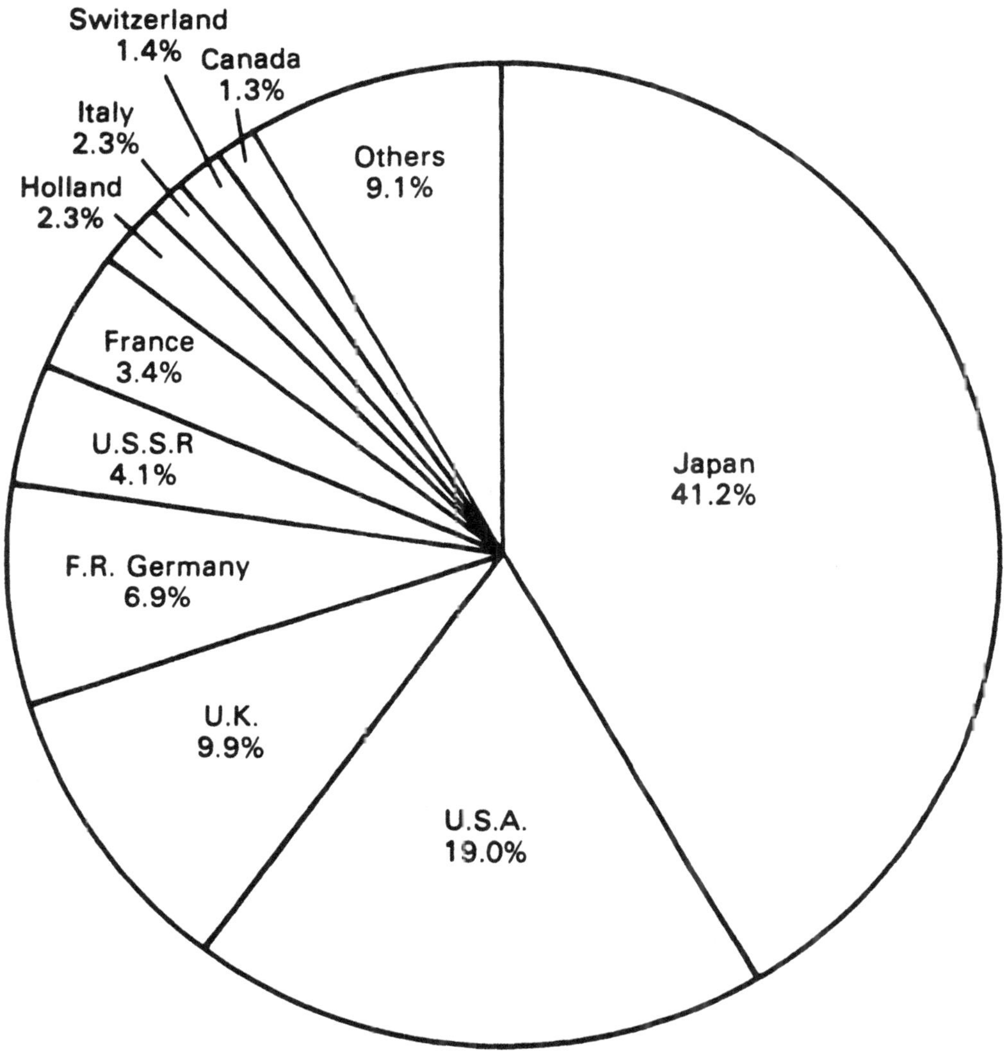

From: "Japan Information Center of Science and Technology (JICST)", National Diet Library Newsletter, No. 53, July 1979. p. 3.

Figure 3-4. JICST Collection of Journals by Subject Area.

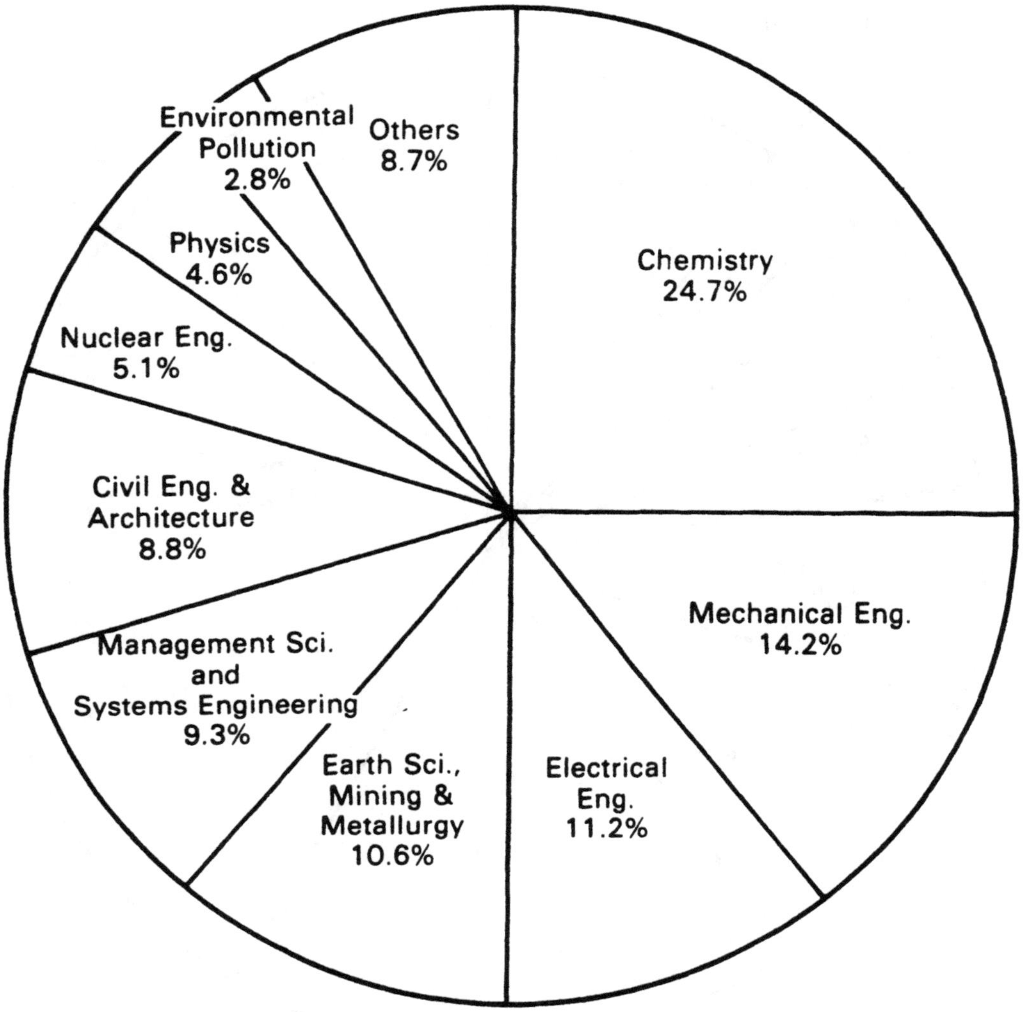

From: "Japan Information Center of Science and Technology (JICST)", National Diet Library Newsletter, No. 53, July 1979. p. 3.

routine system as well as designed data elements to be included on the master tapes and layout format for input card and paper tapes. Since the JICST collection now contains over nine thousand journals, twelve thousand technical reports (only a small part of the known documents in the technical fields are acquired), and forty-nine thousand patent specifications, this source document control system will be designed to accommodate a vast library collection with an anticipated growth rate of 10 percent per year.

(5) The computer system will also be used to provide business office automation procedures such as control of copying, receipt of orders for publications, shipping, and maintenance of payroll.

Figure 3-5 is a flow chart of the JICST information processing system.

Information Service

A flow chart of the activities of the Information Service is shown in figure 3-6. The various sections within the Information Service which perform these activities are described in detail below.

PUBLICATION SERVICE

JICST produces a number of publications. They are:

Current Bibliography on Science and Technology (*Kagaku Gijutsu Bunken Sokuho*)—a compilation of Japanese-language abstracts from Japanese and foreign literature. It is published in eleven different series, which are:

Chemistry and Chemical Industry	
Foreign	Three times per month
Japanese (Chemical Abstract of Japan)	Monthly
Civil Engineering and Architecture	Semimonthly
Earth Science, Mining, and Metallurgy	Semimonthly
Electronics and Electrical Engineering	Semimonthly
Energy	Monthly
Environmental Pollution	Monthly
Management Science and Systems Engineering	Monthly
Mechanical Engineering	Semimonthly
Nuclear Engineering	Monthly
Pure and Applied Physics	Semimonthly

Current Foreign Patent Report (Chemical edition) (*Gaikoku Tokkyo Sokuho·Kagaku-hen*)—a weekly abstracting journal that covers chemical patents from the United States, Great Britain, and West Germany. Translated versions of the patents are available in this publication.

Foreign Technology Highlights (Kaigai Gijutsu Hairaito)—a monthly publication that offers news items on new technology, introduction of new products, and trends in foreign science and technology of interest to persons active in science and technology and related fields. About 5,200 titles of articles from periodicals published in forty-eight countries provide the input for this publication.

Information on Technology for Food Industry (*Shokuhin Kogyo Gijutsu Joho*)—a bimonthly, not-for-sale publication, available at JICST offices or the Food Industry Center. It provides scientific and technical information of interest to the food industry.

Guide to the Public Laboratories and Research Organizations in Japan (Kokyo Shiken Kenkyu Kikan Annai)—an annual publication that lists research projects being planned or currently conducted by public research institutions. Three indexes are available: subject index, investigator index, and organization index. Subjects are limited to physical science and technology, medical and agricultural science. Lists of reports which are difficult to obtain are also included.

JICST Unified Classification Table for Science and Engineering (*JICST Rikogaku Toitsu Bunruihyo*)—a table that lists the major and minor subject headings used in the various series of *Current Bibliography on Science and Technology.*

Classification Table for Current Bibliography on Science and Technology (*Kagaku Gijutsu Bunken Sokuho Henbetsu Bunruihyo*)—a table that gives related terms for the subheadings in the classification scheme and assigns UDC (Universal Decimal Classification) reference numbers to aid in conducting searches on particular subjects in the various series of *Current Bibliography on Science and Technology*. It contains roughly 2,300 terms with each item having an average of 1.2 classification codes.

JICST Thesaurus of Scientific and Technical Terms (*JICST Kagaku Gijutsu Yoho Shisorasu*)—a comprehensive thesaurus containing 35,572 terms, 30,609 terms as descriptors, and 4,963 terms as nondescriptors covering scientific and technical fields (excluding medicine, agriculture, and biology). The terms selected are based on those used in the indexing

Figure 3-5. The JICST Information Processing System.

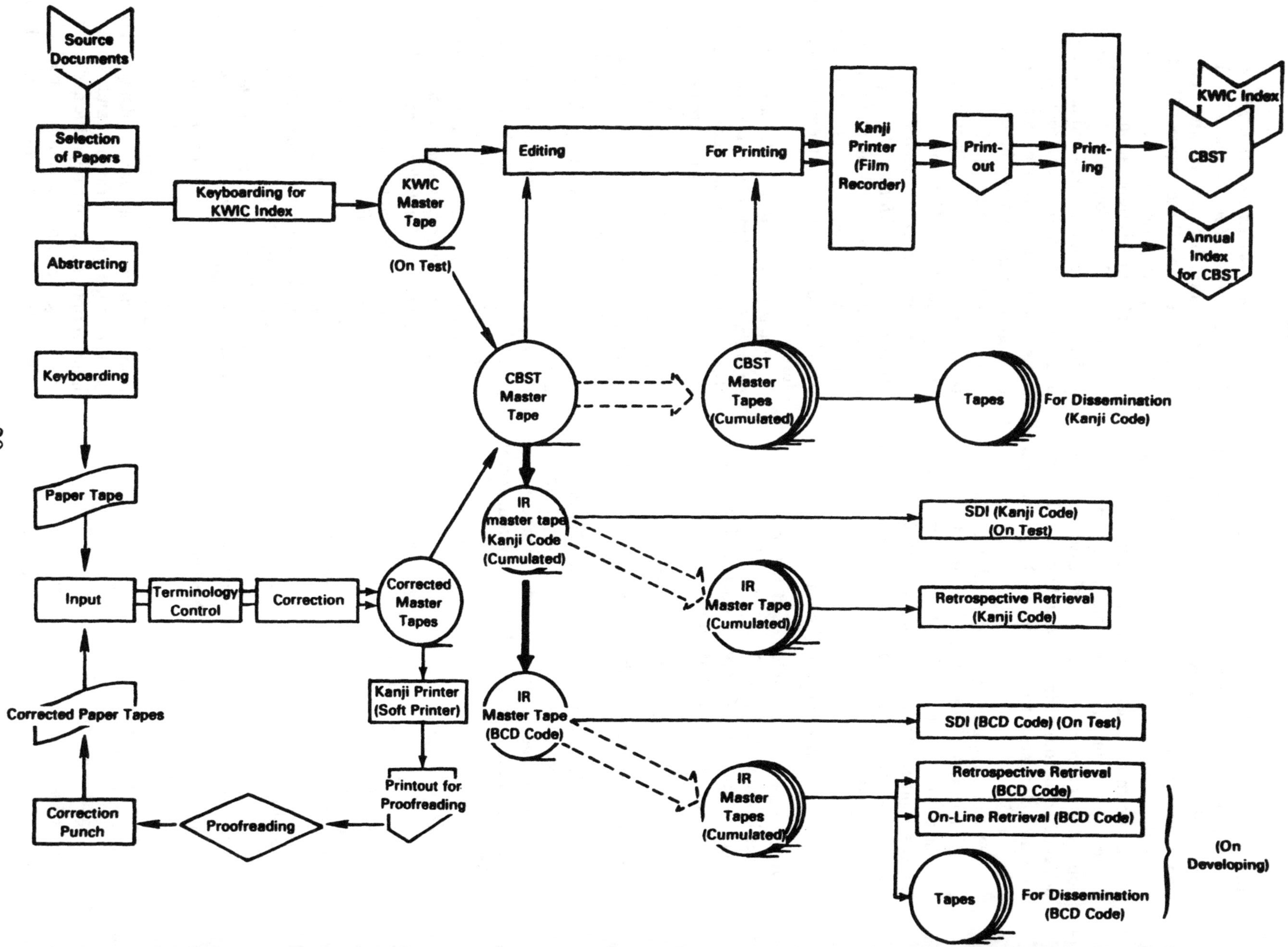

Figure 3-6. The Activities of the JICST Information Service.

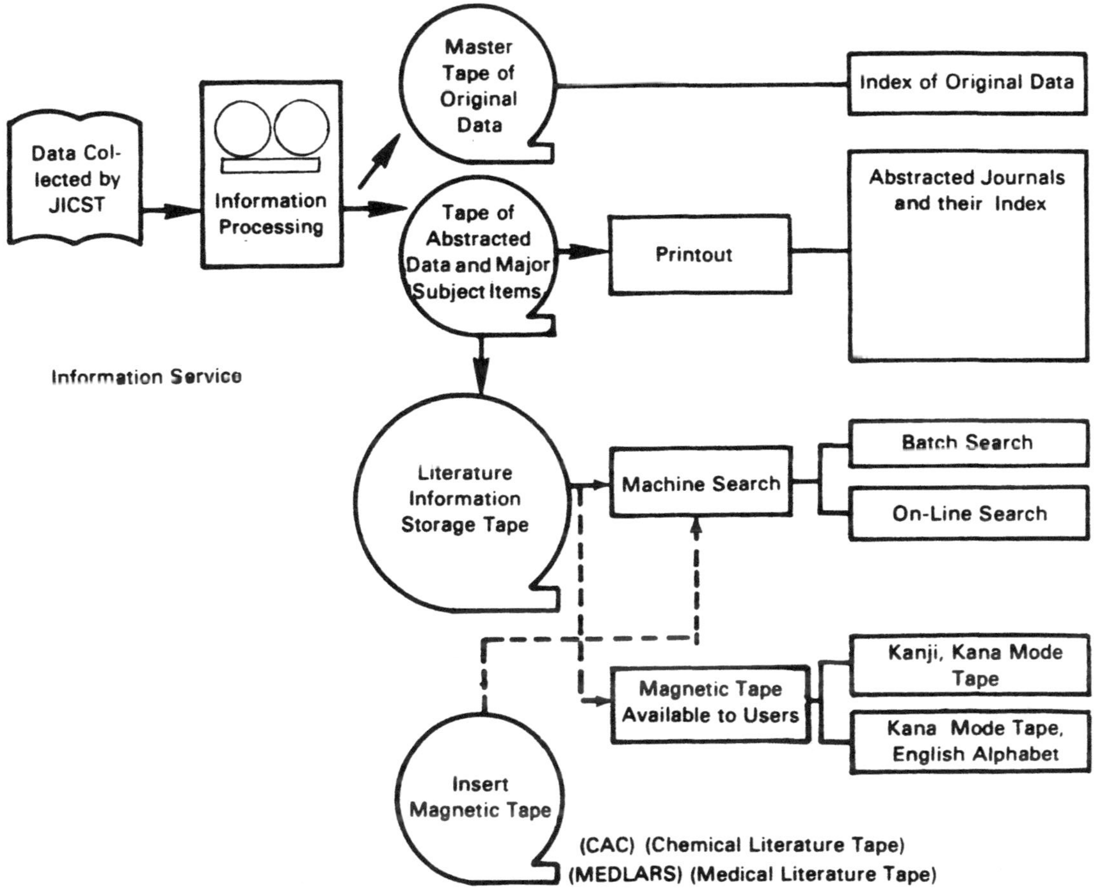

From: Japan Information Center for Science and Technology. ***JICST Yoran 1978.*** *Tokyo, 1978 p. 15.*

of the *Current Bibliography on Science and Technology*, with some additional terms from other sources. The thesaurus is updated annually and consists of two parts: the main body and the subject category index. The main language of the thesaurus is Japanese, but some foreign language terms have also been added. Each item which is indexed in *Current Bibliography on Science and Technology* is assigned 5 to 6 thesaurus terms to facilitate searching.

Information and Documentation (*Joho Kanri*)—a monthly professional journal containing current information and articles useful for specialists involved in information and information science service activities.

Information Science and Technology Research Conference Proceedings (Joho Kagaku Gijutsu Kenkyushu Kaihatsuhyo Ronbunshu)—an annual publication covering original papers and discussions from an all-Japan research conference, sponsored by JICST, and held in the fall of each year from 1964 to date.

Foreign Technical Information for Small and Medium-Sized Enterprises (*Chusho Kigyo Kaigai Gijutsu Joho*)—a bimonthly publication prepared under the auspices of the Small and Medium Enterprises Agency.

Metal Finishing Technology Abstracts (*Kinzoku Hyomen Shori Gijutsu Bunkenshu*)—a semiannual publication covering the area of metal finishing.

Handbook of Science and Technology Information (*Kagaku Gijutsu Joho Handobukku*)—an annual publication on science and technology information.

Index to Japanese Patents (*Nippon Tokkyo Sakuin*)—an annual list indexed by patent applicant and subject.

CUSTOM SERVICES

JICST performs certain customized services to satisfy the specialized requests of their users. Particular attention is given to maintaining secrecy as far as the user is concerned, and anyone can use these services with complete anonymity.

Reproduction Service—orders are accepted for items which are held in repositories in Japan and abroad, as well as items stored by JICST.

Translation Service—translation services are offered for translating foreign language material into Japanese and vice versa. A network of more than 2,500 specialists in science and technology is maintained by JICST for this purpose.

Survey Service—a survey service is offered to search the most recent or retrospective information on special topics, or patents, in the field of science and technology.

Other Custom Services—other services provide notices which report the issuance of specially prepared abstracts, the production of current bibliographic cards, and reproduction of journals' tables of contents.

The following chart shows the trends in the custom services over the past years:

YEAR	*REPRODUCTIONS*	*TRANSLATIONS*	*SURVEYS*
1958	23,504	577	594
1971	392,278	6,122	1,816
1972	381,500	6,237	1,597
1973	408,050	6,454	1,508
1974	374,049	5,822	1,471

TAPE LITERATURE SEARCH SERVICE

This service provides a compilation of the *Current Bibliography on Science and Technology* recorded on magnetic tape. These tapes are available in two configurations—one in English and *kana* (syllabary), and the other in *kanji* (characters)—and are offered on the basis of lease agreements.

MACHINE SEARCH SERVICE

At the present time in Japan, the following large-scale online systems are available:

- Japan National Railway's Reservation System
- Nationwide Banking Data Communication System
- Automobile Registration Verification System
- Securities Information System
- University of Tokyo On-Line Information Retrieval System (TOOL-IR)
- JICST On-Line Information System (JOIS)

The JICST On-Line Information System (JOIS) is an online system in which searches can be executed immediately. It is a conversational type system, and an optional trial-and-error method is possible. The operational procedures are very simple, making the system fairly easy to use. The system contains five online files which can be searched:

- JICST Science and Technology file
- Chemical Abstracts Condensates file
- MEDLARS file
- TOXLINE
- Clearing file

Figure 3-7 gives an overview of these files.

The JICST Science and Technology file covers the entire field of science and technology, is compiled by JICST, and corresponds to the hard copy of *Current Bibliography on Science and Technology* (except for the section on medicine and agriculture in the Environmental Pollution series). The items contained in the file originate from some eight thousand journals, technical reports, and industrial information published both in Japan and abroad. The file also contains old chemical patents up to 1975.

The publication scheme of *Current Bibliography on Science and Technology* allows for the inclusion of items in the online file about a month earlier than by other channels. The items are included in the online file even before the compilations of *Current Bibliography on Science and Technology* are published. A sample search from the JICST Science and Technology file is shown in figure 3-8.

The Chemical Abstracts Condensates file corresponds to the hard copy of *Chemical Abstracts* (excluding the abstracts), published by the Chemical Abstracts Service (United States), and covers the entire field of chemistry. The compiled input of approximately fourteen thousand indexed items consists of journal articles, conference papers, academic and governmental publications, reports, and patents from over 125 countries. The patents which are included cover approximately 15 percent of all granted patents.

The MEDLARS file is compiled by the United States National Library of Medicine, corresponds to the hard copy of *Index Medicus*, and covers the fields of medicine and pharmacy, as well as related fields of biology, biochemistry and pharmacology.

The Clearing file contains references to ongoing research projects being carried out at national research institutions, semigovernmental research institutions, private research institutions affiliated with universities, public research laboratories, and nonprofit research institutions. The information on the projects is assigned keywords and classification codes for ease in searching.

JICST offers free training sessions at various terminal sites to teach interested persons the operation and use of JOIS. A textbook, *Handbook of JOIS Use*, is also provided to users. The JOIS system is searched using keywords, classification codes, and names.

The system is presently available to users through terminals installed in the JICST head office, as well as in branch offices in Tokyo, Osaka, and Nagoya. The files are presently available under the following time schedule:

	MON.	*TUE.*	*WED.*	*THU.*	*FRI.*
9:45 a.m.-Noon	MED	JICST	CAC	JICST	CAC
1:00 p.m.-4:15 p.m.	CAC	MED	JICST	MED	JICST

The current cost for using the system is 200 yen/minute for use of the terminal (online time). The offline service charges include an offline handling charge of 500 yen/request, plus a charge of 15 yen/item. In addition, a royalty of 10 yen/item is charged for responses from the Chemical Abstracts Condensates File.

In July, 1978, JICST leased-line, dial-up access became available. The number of calls has greatly increased with the implementation of this service. As of April, 1979, there are about 270 terminals connected to the JICST online system forming a network from Sapporo to Hiroshima, and it is expected to extend to Kyushu in the near future.

Figure 3-9 is a map showing present and future locations of terminal sites, figure 3-10 shows JICST's plan for online retrieval service operations through 1985, and figure 3-11 shows the linkage between the JICST services to give an overall picture of JICST operation.

TRAINING COURSES IN INFORMATION CONTROL TECHNOLOGY

Training courses and lectures on the basics and general features, including methods of information control by computerized information retrieval, are given in Tokyo, Osaka, and Nagoya, as well as on a nationwide basis. Top specialists throughout the country in this field gather at the Information Science and Technology Conference.

Figure 3-7

JOIS Files

NAME OF FILE	CUMULATIVE PERIOD	NUMBER OF ITEMS	SUBJECT AREA	INFORMATION SOURCE	MOD
JICST Science and Technology file	April 1975–to date	About 370,000 items per year	All science and technology	Corresponds to *Current Bibliography on Science and Technology,* published by JICST	Kanji
Chemical Abstracts Condensates	January 1974–to date	About 390,000 items per year	Chemistry and chemical engineering	Corresponds to *Chemical Abstracts,* published by Chemical Abstracts Service, U.S.	English numbers
MEDLARS	January 1974–to date	About 280,000 items per year	Medicine and pharmacology	Corresponds to *Index Medicus,* published by U.S. National Library of Medicine	English numbers
TOXLINE	January 1974–to date	About 90,000 items per year	Toxicology	Corresponds to the U.S. National Library of Medicine's Toxline file	English numbers
Clearing file	1976–to date	About 10,000 items per year	All science and technology	On-going research from about 400 research institutes	

Each file includes the pertinent bibliographic citation, keywords, and classifications, but contains no summaries. In addition, all entries are in the Roman alphabet and Kana (syllabary).

FROM: Japan Information Center of Science and Technology. JICST Yōran (JICST Directory). Tokyo, JICST, 1977, p. 20, and Japan Information Center of Science and Technology. *JICST – 20 Years of Progress.* Tokyo, JICST, 1977. p. 12 and Uchilda, Hisako, "Information system, data bases, and on-line services of the Japan Information Center of Science and Technology (JICST)", *Journal of Chemical Information and Computer Science,* V. 19, no. 4 (1979) p. 199.

Figure 3-8

Sample JOIS Printout

```
IN JOIS,      ,NAKAI,NAKAI
    S:
       JICST ｵﾝﾗｲﾝ ｹﾝｻｸ ｻｰﾋﾞｽ
       NO      ﾃﾞｰﾀﾍﾞｰｽ     ﾁｸｾｷ ﾊﾝｲ                       ｹﾝｽｳ         ｼｭｳﾘｮｳ ﾖﾃｲ
        2 --- CAC          (V.080N.01-V.087N.19)  1,467,651 ｹﾝ       12:00
        3 --- MEDLARS      ( 1974.01 - 1977.11 )    965,229 ｹﾝ       12:00
       ﾃﾞｰﾀﾍﾞｰｽ ﾉ ﾊﾞﾝｺﾞｳ ﾊ ?    2.3,ﾅｼ=N
    U: 2
       ﾃﾞｰﾀﾍﾞｰｽ CAC
       ﾊﾞﾝｺﾞｳ    ﾁｸｾｷ ﾊﾝｲ                     ｹﾝｽｳ
        S --- (V.087N.19)                 7,126 ｹﾝ
        0 --- (V.086N.01-V.087N.19)     351,732 ｹﾝ
        1 --- (V.084N.01-V.085N.26)     390,735 ｹﾝ
        2 --- (V.082N.01-V.083N.26)     391,883 ｹﾝ
        3 --- (V.080N.01-V.081N.26)     333,301 ｹﾝ
    S: ﾃﾞｰﾀﾍﾞｰｽ ﾉ ﾊﾝｲ ﾊ ?    S,ﾊﾞﾝｺﾞｳｰﾊﾞﾝｺﾞｳ
    U: 3
    S: ｻｰﾋﾞｽ ｵ ｶｲｼ ｼﾏｽ    1977.12.07    10:42:10
       ﾃﾞｰﾀﾍﾞｰｽ    CAC          (V.080N.01-V.081N.26)         333,301 ｹﾝ
 1] U: LIBRARY
    S:  ..    26 ｹﾝ
 2] U: N = JPN
    S: ﾀｸﾞ ｶﾞ ｱﾘﾏｾﾝ
 2] U: L = EN
    S: ﾀｸﾞ ｶﾞ ｱﾘﾏｾﾝ
 2] U: ¥D
    S: ｼｭﾂﾘｮｸ ｵﾜﾘﾏｼﾀ
 2] U: ¥Q

***************************************************************************
*      ﾃﾞｰﾀﾍﾞｰｽ    ｼﾂﾓﾝ ｼﾞｶﾝ    ｶｲﾄｳ ｹﾝｽｳ  ¥OFF ｶｲｽｳ    ¥OFF ｹﾝｽｳ     ﾘｮｳｷﾝ      *
*      CAC         00:06:00            0 ｹﾝ       0 ｶｲ          0 ｹﾝ   ¥1,200      *
*                                                                                  *
*                                                           ﾙｲｹｲ     ¥1,200      *
***************************************************************************

    S: ﾂｷﾞ ﾉ ｼﾂﾓﾝ ｶﾞ ｱﾘﾏｽｶ ?     ﾊｲ=Y,ｲｲｴ=N
    U: N

***************************************************************************
*      ﾃﾞｰﾀﾍﾞｰｽ    ｼﾂﾓﾝ ｼﾞｶﾝ    ｶｲﾄｳ ｹﾝｽｳ  ¥OFF ｶｲｽｳ    ¥OFF ｹﾝｽｳ     ﾘｮｳｷﾝ      *
*      CAC         00:06:00            0 ｹﾝ       0 ｶｲ          0 ｹﾝ   ¥1,200      *
*                                                                                  *
*      ｶｲﾜ ｼﾞｶﾝ    00:06:00                                 ｺﾞｳｹｲ    ¥1,200      *
***************************************************************************

    S: ｶｲﾜ ｵ ｼｭｳﾘｮｳ ｼﾏｽ   1977.12.07    10:48:24
---------------------------------------------------------------------------
```

Figure 3-9. Present and Future JOIS Terminal Locations.

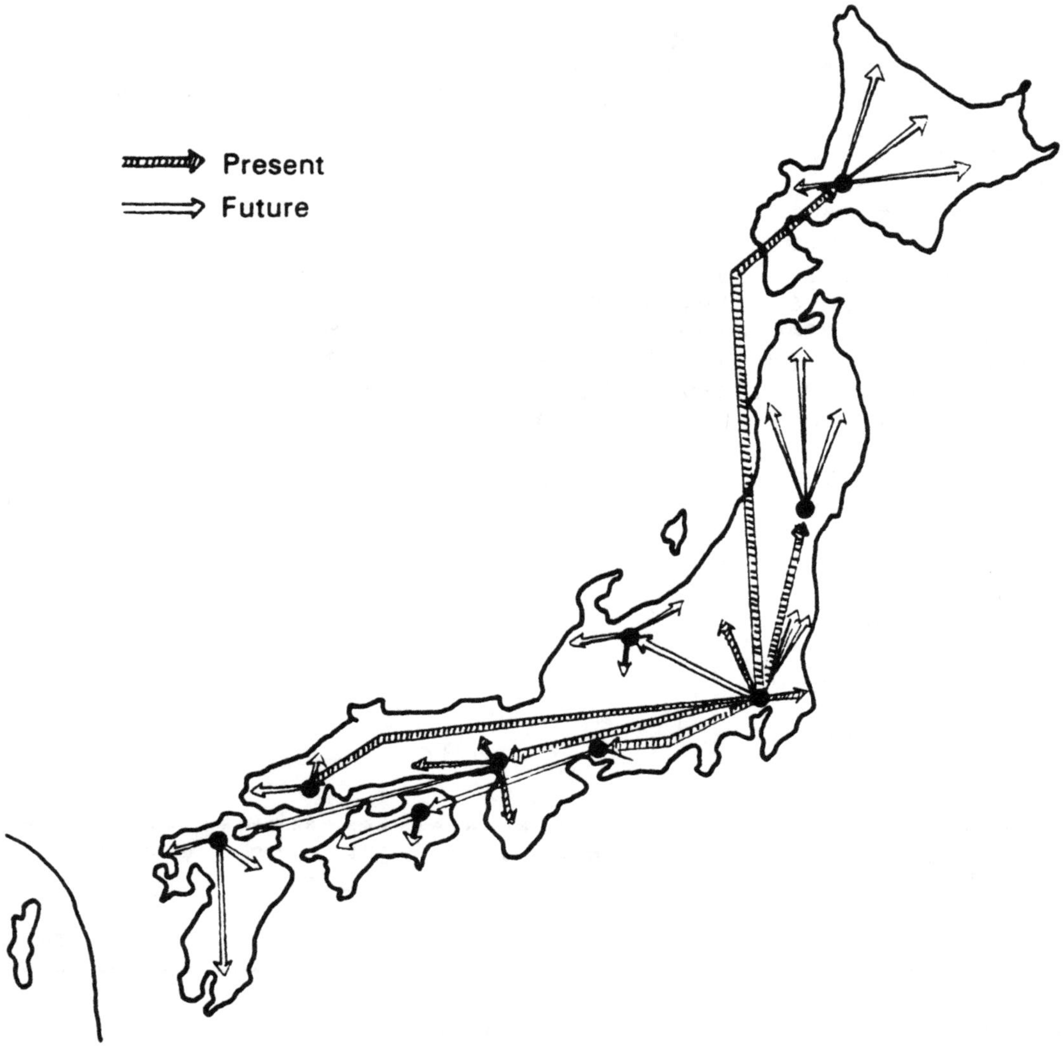

From: *Uchida, Hisako, "Information System, Data Bases, and On-Line Services of the Japan Information Center of Science and Technology (JICST)," Journal of Chemical Information and Computer Sciences, vol. 19, no. 4, 1979. p. 200.*

Figure 3-10. JICST Plan for Online Retrieval Operations Through 1985

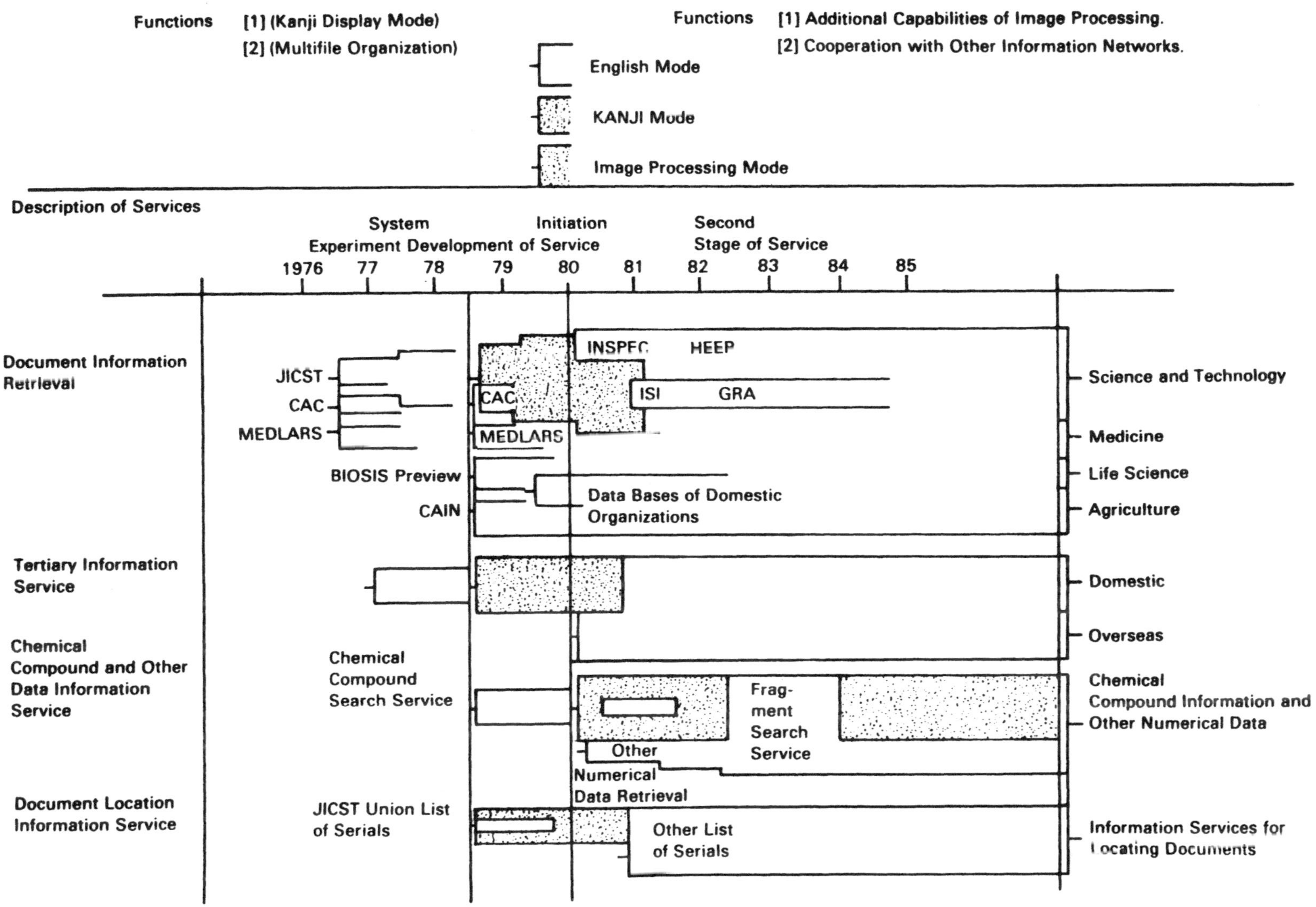

Figure 3-11. Linkage Between JICST Services.

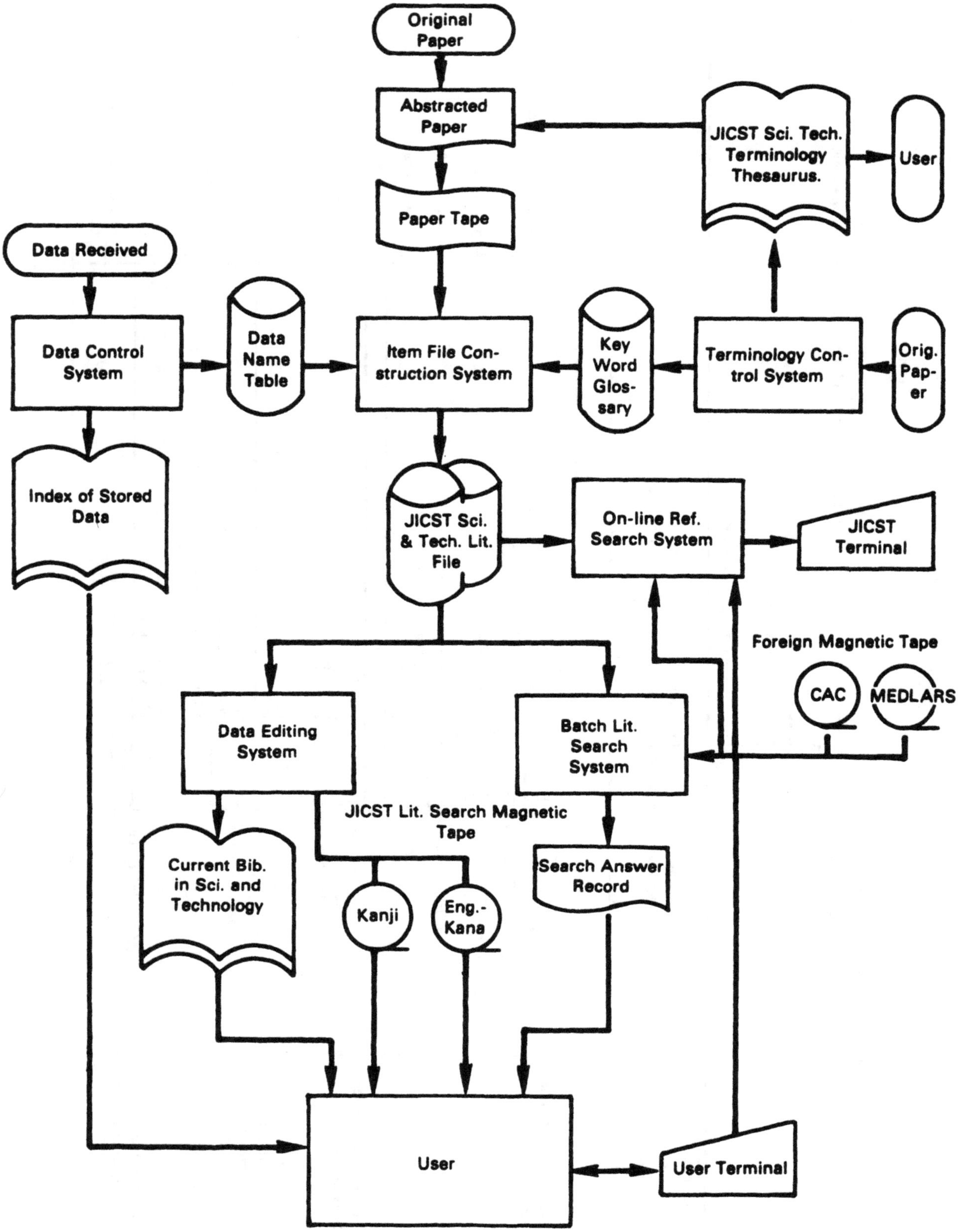

From: Japan Information Center for Science and Technology, <u>JICST 1976.</u> Tokyo, JICST, 1976. p. 18.

READER SERVICE

The materials which have been acquired by JICST can be examined in the reading room at the JICST headquarters. There is also an instant reproduction service. Reading rooms are also located in the Tokyo, Osaka, Nagoya, Kyushu, and Shikoku branch offices, where JICST publications, various handbooks, and reference materials can be reviewed.

CLEARINGHOUSE

The Clearinghouse serves as a guide to specialists or agencies seeking specialized information, as well as methods of information acquisition on scientific and technical subjects. This is also the service in JICST which is responsible for the *Guide to Public Laboratories and Research Organizations in Japan.*

Bibliography

JAPAN'S DEVELOPING NATIONAL INFORMATION SYSTEM

Aines, Andrew A. and Day, Melvin S. "National Planning of Information Services." In *Annual Review of Information Science and Technology*, vol. 10, edited by Carlos S. Cuadra, Ann W. Luke, and Jessica L. Harris, pp. 22-42. New York: American Society for Information Science, 1975.

"Current Development of NIST Programs." *USACO Reporter*, vol. 1, no. 2, July 1976.

Japan Science and Technology, *Program of the National Information System for Science and Technology (NIST)*, November 1975.

Kobayashi, Yutaka. "Development of a Comprehensive Network for Scientific and Technical Information in Japan," *Library Trends* 17 (January 1969): 258-266.

Koizumi, Akira. "The Present and Future of Library Networking in Japan." In *Japanese and U.S. Research Libraries at the Turning Point; Proceedings of the Third Japan-U.S. Conference on Libraries and Information Science in Higher Education*, Kyoto, Japan, October 28-31, 1975, pp. 25-42. Metuchen, New Jersey: Scarecrow Press, 1977.

"Scientific and Technological Information Activities in Japan." *National Diet Library Newsletter*, no. 42 (November 1975), pp. 1-12.

Suzuki, Heihachiro. "National Plans for Library/Information Services and the National Diet Library." *Library and Information Science*, no. 14 (1976), pp. 1-14.

Yoshida, Tomizo. "Organization of Scientific Activities." In *Science in Japan*, edited by Arthur H. Livermore, pp. 1-24. Washington, D.C.: American Association for the Advancement of Science, 1965.

JAPAN INFORMATION CENTER OF SCIENCE AND TECHNOLOGY

Japan Information Center of Science and Technology. *Guide to the Electronic Computer System*. Tokyo JICST, January 1978.

Japan Information Center of Science and Technology. *JICST*. Tokyo: JICST, no date.

Japan Information Center of Science and Technology. *JICST—20 Years of Progress*. Tokyo: JICST, 1977.

Japan Information Center of Science and Technology. *JICST 1976*. Tokyo: JICST, 1976.

Japan Information Center of Science and Technology. *JICST Service Guide 1976*. Tokyo: JICST, 1976.

Japan Information Center of Science and Technology. *JICST Service Guide 1977*. Tokyo: JICST, 1977.

Japan Information Center of Science and Technology. *JICST Yoran* (JICST Directory). Tokyo: JICST, 1977.

Japan Information Center of Science and Technology. *JICST Yoran* (JICST Directory). Tokyo: JICST, 1978.

Japan Information Center of Science and Technology. *JOIS Online Information Search Service*. Tokyo: JICST, 1978.

Japan Information Center of Science and Technology. *JOIS Online Reference Service; User's Guide*. Tokyo: JICST, [1977?]

Japan Information Center of Science and Technology. *JOIS On-Line Search Service*. Tokyo: JICST, [1976?]

"Japanese Information Center of Science and Technology (JICST)." In *Encyclopedia of Information Systems and Services*, 2nd edition, edited by Anthony T. Kruzas, p. 308. Ann Arbor, Anthony T. Kruzas Associates, 1974.

"Japan Information Center of Science and Technology." *National Diet Library Newsletter*, no. 53 (July 1979), pp. 1-8.

Japan Information Center of Science and Technology. *Sabisu Gaido*. Tokyo: JICST, 1978.

Japan Information Center of Science and Technology. *Thesaurus of Scientific and Technical Terms*. Tokyo: JICST, 1978.

"JICST's Plan for 1977-1978." *USACO Reporter*, vol. 1, no. 3 (May 1977).

Sasaki, Chikashi. "Japan Information Center of Science and Technology." In *University and Research Libraries in Japan and the United States; proceedings of the First Japan-U.S. Conference on Libraries and Information Science in Higher Education*, Tokyo, May 15-19, 1969, edited by Thomas R. Buckman, Yukihisa Suzuki, and Warren T. Tsuneishi, pp. 245-249. Chicago: American Library Association, 1972.

Tokizane, Soichi. *On-Line Retrieval Systems in Japan*. September 1977.

Welch, Theodore F. *Toshokan: Libraries in Japanese Society*. Chicago: American Library Association, 1976.

Uchida, Hisako, "Information System, Data Bases, and On-Line Services of the Japan Information Center of Science and Technology (JICST)," *Journal of Chemical Information and Computer Sciences*, vol. 19, no. 4, 1979, pp. 199-201.

chapter four

ANALYSIS OF JAPANESE SCIENTIFIC AND TECHNICAL PUBLICATIONS

Three types of publications (monographs, periodicals, and technical reports) were studied in an attempt to define the dimensions of the Japanese research documentation effort, and to gain an understanding of the size and scope of this effort. When the study was begun, two major difficulties quickly surfaced.

The first of these is the vastness of the Japanese publishing effort. One author noted that "being allowed to report on the Japan publishing scene is a bewildering honor—something like owning a small trenching tool and laying hands on the blueprint for the Grand Canyon" (Kern, 1971, p. 172). Japan boasts the third largest publishing conglomeration in the world, with over 3,000 publishers distributing their publications in over 23,000 retail outlets. To further illustrate, in 1977 alone, the Japanese issued over 960 million copies of books with a total value of over 874 billion yen (approximately 4.37 billion dollars). They also produced over 2.5 billion copies of magazines with a sale price of over 741 billion yen (approximately 3.7 billion dollars). One can easily realize that evaluating this literature is no small task.

The second difficulty is the complexity of the Japanese language. Japanese is written in a combination of three orthographies (Chinese and two syllabic scripts) and has a syntactic structure radically different from English. Although the prewar number of Chinese-Japanese characters in use was far more than are currently allowed by law, the number of characters needed to produce an average edition of a Japanese newspaper is estimated at about 4,000. The Japanese government has designated a list of standard characters which number 1,850. Characters in actual use, however, exceed this number due to the fact that personal, place and other proper names were not included in the standard list. These 1,850 characters alone require a tremendous effort to learn and master. The Japanese had been seriously handicapped in the utilization of computers to mechanize the processing of natural language information because there was no keyboard to input the material. One of the first priorities of the JICST computerization efforts was the development of such a keyboard. Even so, the large number of characters in current use eliminates the possibility of a single key for each character (space for such machines is at a premium). Thus, the JICST keyboard relies on striking several keys to produce some of the necessary characters.

This language barrier may be one reason why Western researchers have not fully recognized the extent and quality of Japanese research efforts, a situation very similar to Western perception of Russian scientific literature prior to Sputnik.

Monographs

Since the introduction of moveable type printing presses from Korea in the 1590s, the Japanese have been avid book buyers. Underlying the Japanese public's appetite for books in the twentieth century has been the achievement of near universal literacy soon after 1900. Traditionally, the Japanese have relied on their personal libraries rather than utilizing the public libraries. Since libraries in Japan play a relatively minor role, the Japanese purchase, rather than borrow, their reading material. In fact, there appears to be a bookstore on almost every other corner. The bookstore is far more accessible than the libraries, which would account for the fact that the Japanese buy more books per capita than any other nation.

Approximately 1,037,640,000 volumes were printed in 1978, with a value of approximately 962,888,410,000 yen (or about 4.3 billion dollars). Over 674 million of these volumes were actually sold, which amounts to an average of about six books purchased by every man, woman and child in Japan.

The year 1978 was also a record year for new titles published—26,906. This compares with a total of 33,841 new titles published in the United States during the same period. Figure 4-1 shows a comparison of the new titles published in the United States and Japan in 1977 by subject.

However, monographs tend to play the secondary role of summarizing research in the areas of science and technology. Since timeliness of information was an important objective of this study, much more emphasis was placed on analysis and evaluation of periodical literature. In the area of the natural sciences, there were 2,082 new titles published in Japan and 4,660 new titles published in the United States. In the area of engineering, there were 2,439 new titles published in Japan and 1,614 new titles published in the United States. These figures tend to verify the theory that Japan places a greater emphasis on research in the applied sciences than on research in the natural sciences.

Periodicals

The results of an analysis and evaluation conducted on the 9,116 current Japanese scientific and technical periodical titles follow:

ANALYSIS BY ISSUING INSTITUTION

Figure 4-2 shows a breakdown of the total 9,116 Japanese scientific and technical periodicals by issuing institution. Learned societies publish the largest portion of the titles, with national and local research institutes, national and local organizations, and universities following. These four "publishers" account for almost two-thirds of the scientific and technical periodicals issued in Japan.

ANALYSIS OF TOTAL NUMBER OF JAPANESE SCIENTIFIC AND TECHNICAL PERIODICAL TITLES SURVEYED

Figure 4-3 shows a breakdown of the total number of Japanese scientific and technical periodical titles by language. It clearly shows that fewer than one-quarter of the total number are written either fully or partially in Western languages, with three-quarters being published solely in Japanese, not nearly as high a percentage as we would have predicted at the beginning of the study.

The 9,116 total titles were also divided into the four broad subject categories shown in figure 2-2. The results of this breakdown are shown in figure 4-4. The applied sciences account for the largest portion of the literature (44.7 percent), followed by the agricultural sciences (20.1 percent), the medical sciences (18.4 percent), and last, the natural sciences (16.8 percent). This would be a logical distribution according to the hypothesis presented in the discussion of Japanese monographs that stated that the Japanese place a greater emphasis on applied research than on natural research.

These titles were then evaluated by language of publication within each of these four broad subject areas, and a breakdown of the results is shown in figure 4-5. In the natural sciences, there are a sizeable number of periodicals which are written either fully or partially in Western languages (38.1 percent). In the area of medical sciences, a little more than 30 percent of the periodical literature is published utilizing Western languages. Applied sciences follow with 20 percent in Western languages, and agricultural sciences show the smallest portion in Western languages (18 percent). It is logical that the majority of the Japanese scientific and technical periodicals would be written in Japanese, but the minority number of periodicals issued in Western languages is significant. However, although these resources present no language problem, further analysis will show that much of this literature is not actually accessible to Western researchers.

Figure 4-1. Comparison of New Book Titles Published in the United States and Japan in 1978.

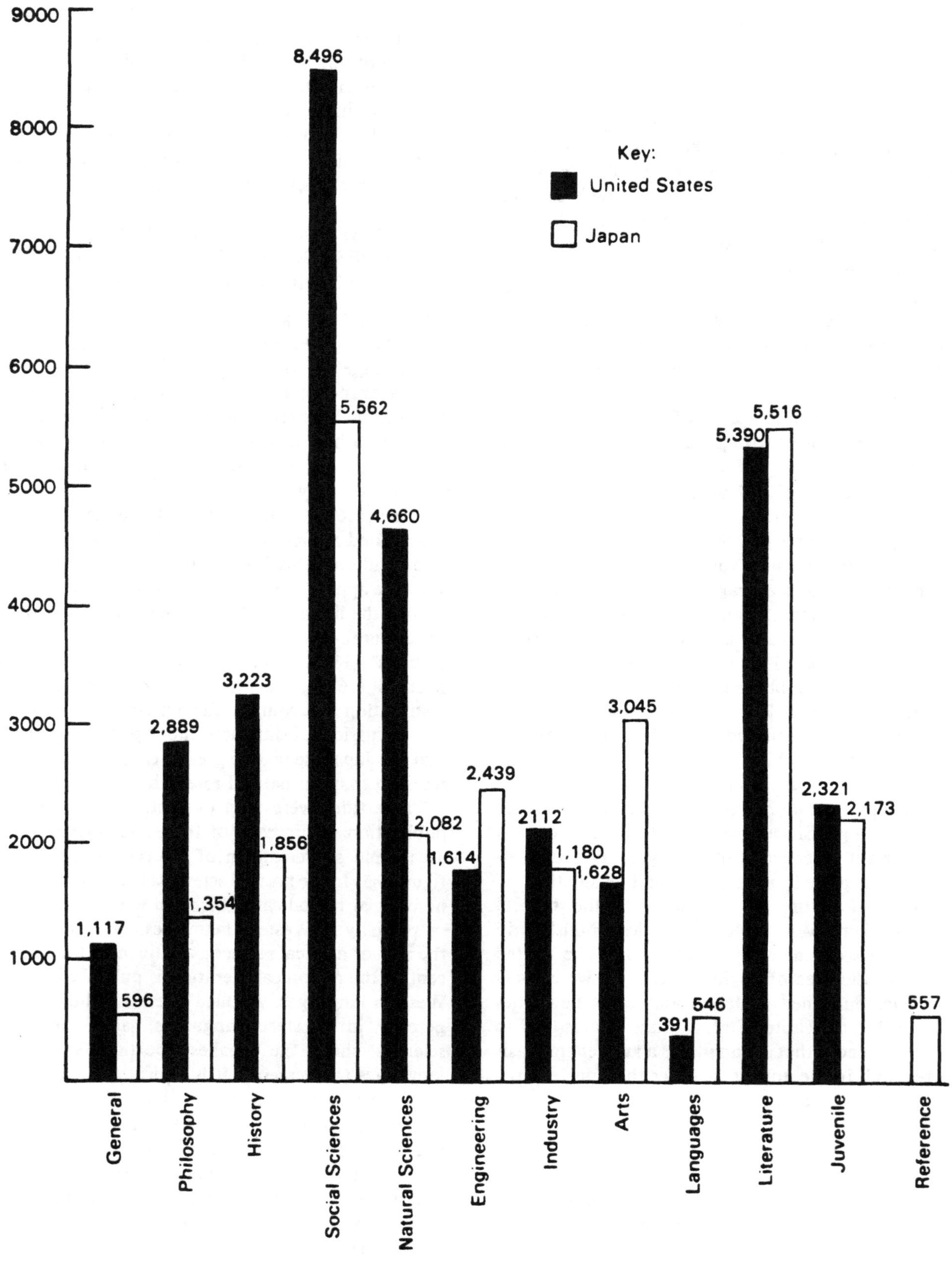

Figure 4-2. Total Japanese Scientific and Technical Periodicals by Issuing Institution.

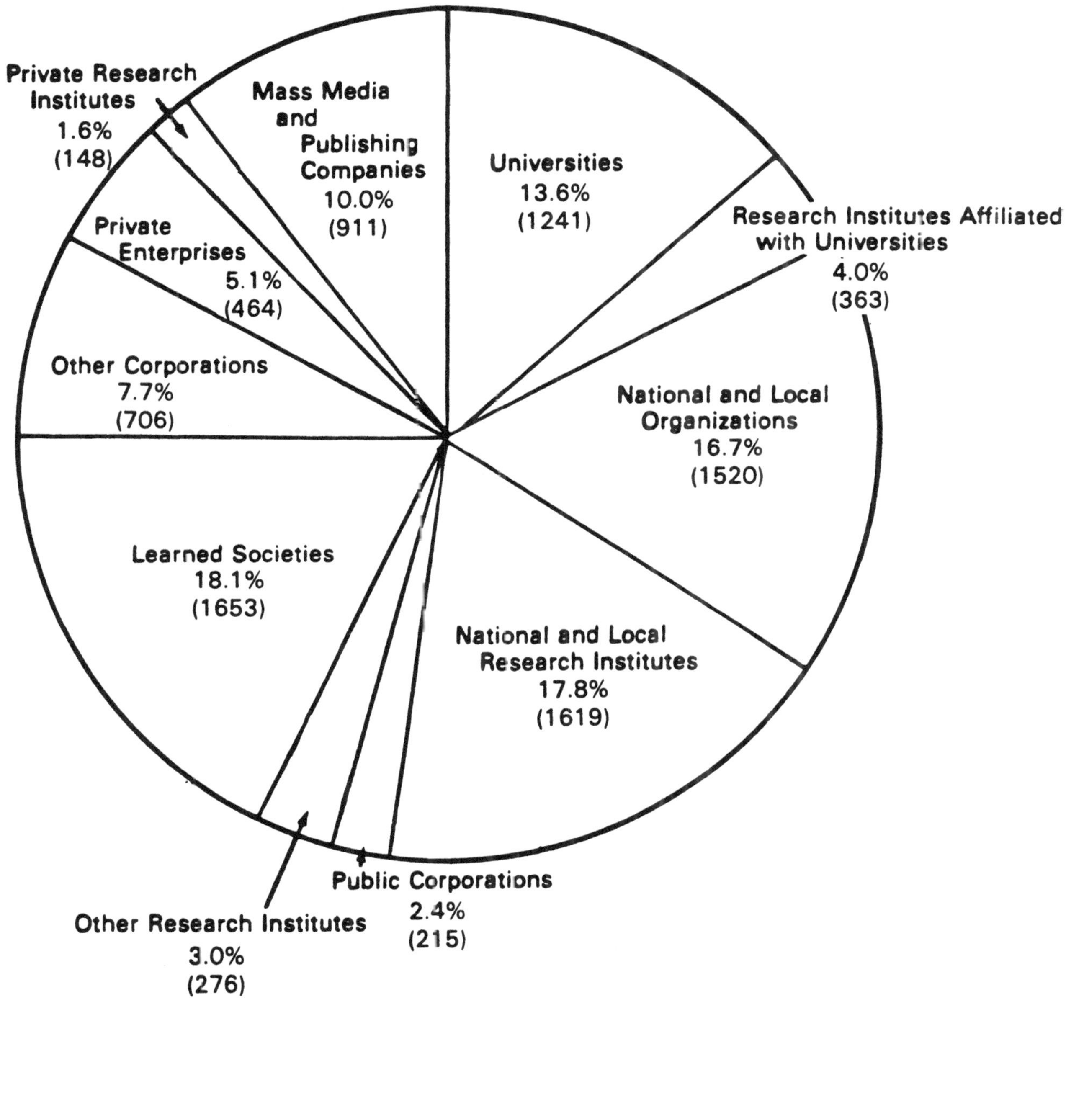

Figure 4-3.
Total Japanese Scientific and Technical Periodicals
by Language

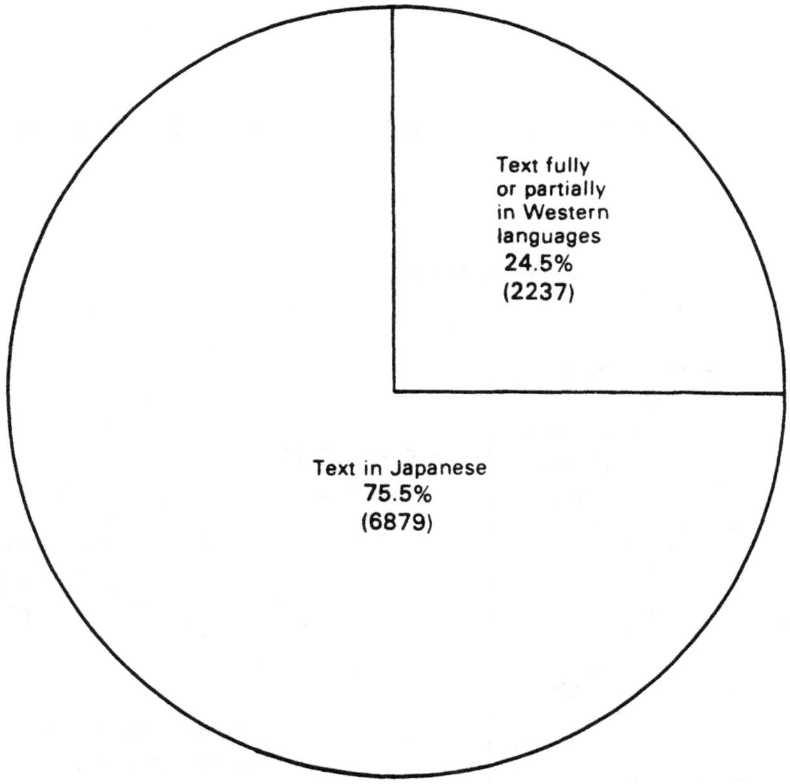

Figure 4-4.
Total Japanese Scientific and Technical Periodicals
by Subject

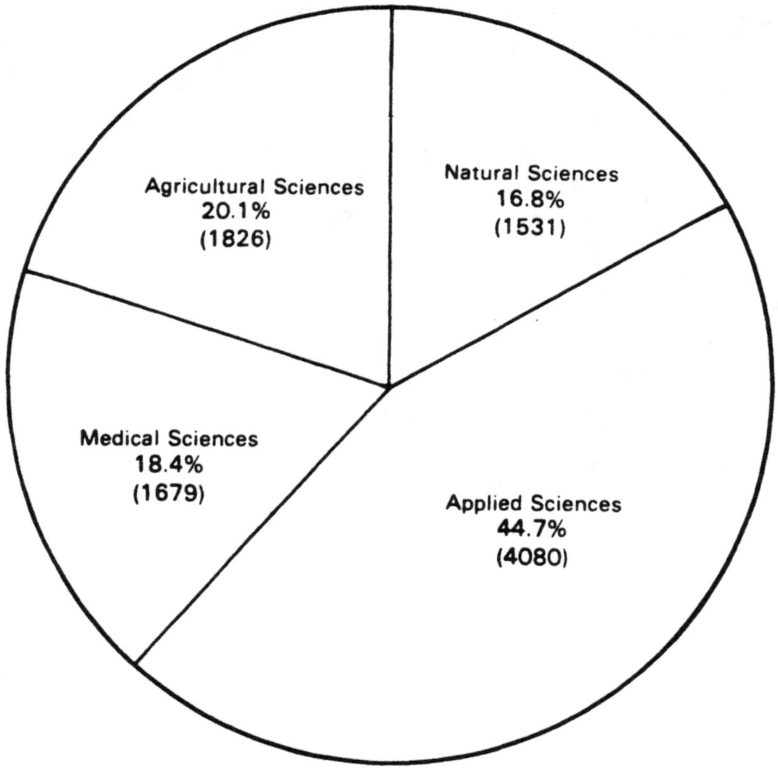

Figure 4-5.

Total Japanese Scientific and Technical Periodicals by Subject and Language

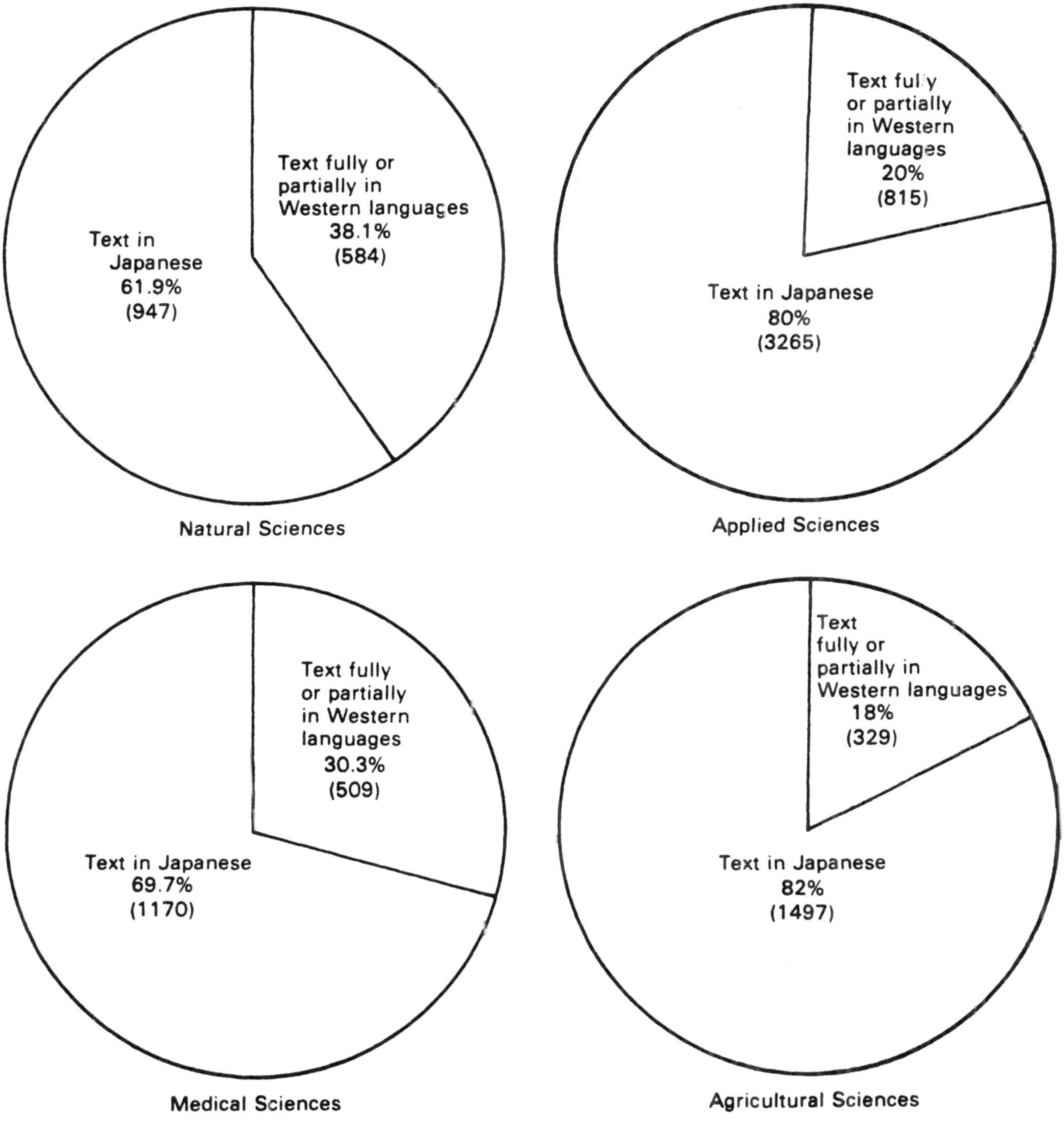

JAPANESE SCIENTIFIC AND TECHNICAL PERIODICALS WHICH ARE INDEXED AND NOT INDEXED BY WESTERN SOURCES

This type of evaluation is very important if the term "indexed" is equated with "accessible to Western researchers" and the term "not indexed" is equated with "inaccessible to Western researchers." If these terms are kept in mind as the following analysis and evaluation is reviewed, it takes on a greater significance in light of the potential importance of this literature to Western researchers and scientists.

Figure 4-6 shows the number and percentage of the total 9,116 Japanese scientific and technical periodicals which are indexed by Western sources and those which are not indexed by Western sources. The Western abstracting and indexing sources used in the survey are those shown in figure 2-3. Only 19 percent of the Japanese scientific and technical periodical literature is covered by Western indexes, while 81 percent is not covered by any of the Western indexing or abstracting services included in the study. These figures further indicate that only a small part of the Japanese scientific and technical periodical literature is available to Western researchers for their perusal, and the vast majority of the information is virtually unknown to scientists and engineers in the West.

One would conjecture that the majority of the periodicals which are covered in Western indexing and abstracting sources would be written either fully or partially in Western languages. Western-language publications pose no translation problem for the Western indexing and abstracting services and thus are much easier to include than publications written in Japanese. Figure 4-7 bears this out, showing that 71 percent of the Japanese scientific and technical periodicals which are indexed or abstracted by Western sources are written either fully or partially in Western languages. The 29 percent of the periodical literature which is written in Japanese and indexed or abstracted by Western sources is mainly processed by those indexing and abstracting services which either employ Japanese translators on their staffs or receive input directly from Japanese indexers and abstractors (Chemical Abstracts, INSPEC, and World Aluminum Abstracts). Thus the language problem forms a significant barrier for inclusion of Japanese-language material in Western sources.

Figure 4-8 illustrates the broad subject areas of the Japanese scientific and technical periodicals which are indexed by Western sources. The chart shows that the largest portion of the indexed literature falls into the area of applied sciences, with medical sciences, natural sciences, and agricultural sciences following in descending order.

The information shown in figure 4-9 further substantiates the findings illustrated in figure 4-7 by demonstrating that within each of the four broad subject areas, most of the periodicals which are indexed or abstracted in Western sources are written either fully or partially in Western languages.

When analyzing Japanese scientific and technical periodicals which are not indexed or abstracted in Western sources, it is easy to conjecture the inverse of the supposition for indexed material—that is, that Japanese scientific and technical periodicals which are not covered in Western indexes and abstracts would be written in Japanese. Figure 4-10 bears out that conjecture. It shows that 86.3 percent of the periodical literature which is not covered by Western sources is written in Japanese. It is interesting, however, that within this set of not indexed periodicals are 1,008 titles (13.7 percent) which are written fully or partially in Western languages. Language of publication would present no barrier to the inclusion of these titles, yet they are not indexed or abstracted by the sources included in the study.

Figure 4-11 shows a subject breakdown of the Japanese scientific and technical periodical titles which are not indexed in Western sources. Almost one-half of the unindexed periodical titles fall into the area of applied sciences, one-fifth fall into the area of agricultural sciences, and roughly one-sixth fall into the areas of medical sciences and natural sciences.

Figure 4-12 illustrates a language breakdown for the Japanese scientific and technical periodicals which are not indexed by Western sources for each of the four broad subject areas. In each area, it is obvious that the largest portion of the unindexed periodicals are written in Japanese. The chart, however, does illustrate the point raised in the discussion of the unindexed periodicals by language (illustrated in figure 4-10) that there are a number of scientific and technical periodicals written either fully or partially in Western languages which are not covered in Western sources. This is particularly apparent in the area of natural sciences, where almost one-quarter of the unindexed periodicals is written in Western languages. No obvious explanation is apparent for the exclusion of this literature from Western sources. Since language does not present a problem in this case, there must be some other reason

Figure 4-6.
Japanese Scientific and Technical Periodicals
Indexed and Not Indexed by Western Sources

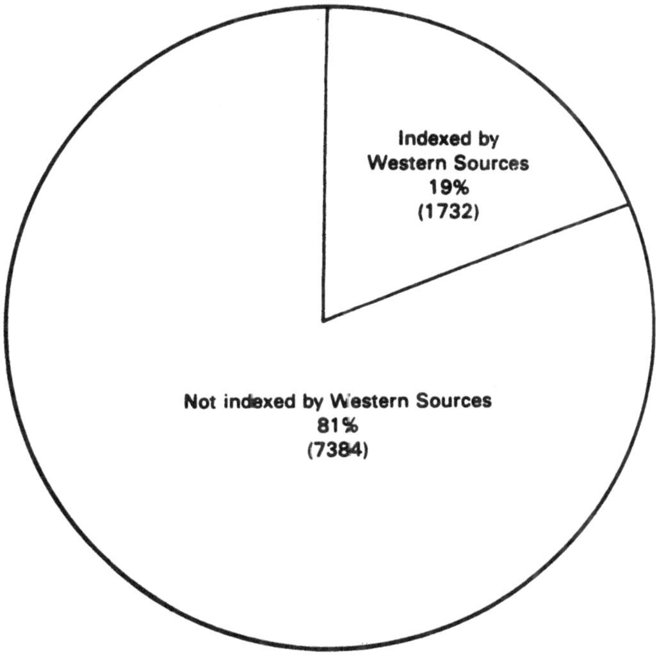

Figure 4-7.
Japanese Scientific and Technical Periodicals
Indexed by Western Sources
by Language

Figure 4-8.
Japanese Scientific and Technical Periodicals
Indexed by Western Sources
by Subject

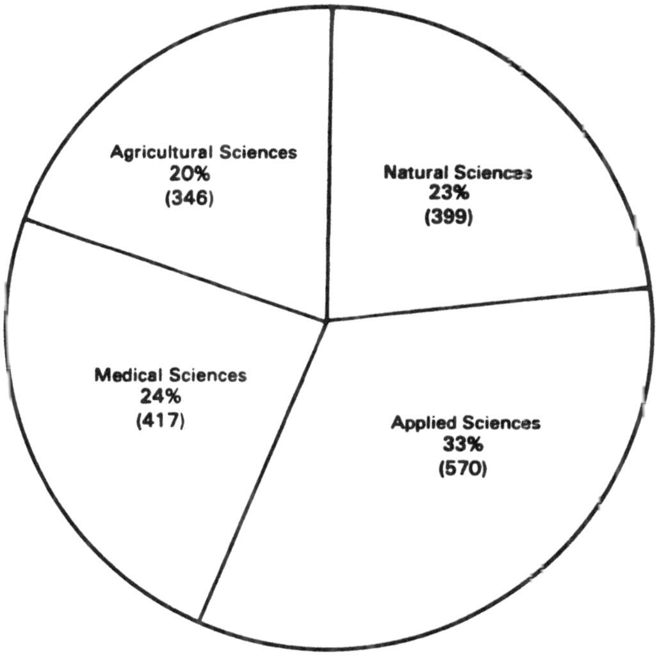

Figure 4-9.
Japanese Scientific and Technical Periodicals
Indexed by Western Sources
by Subject and Language

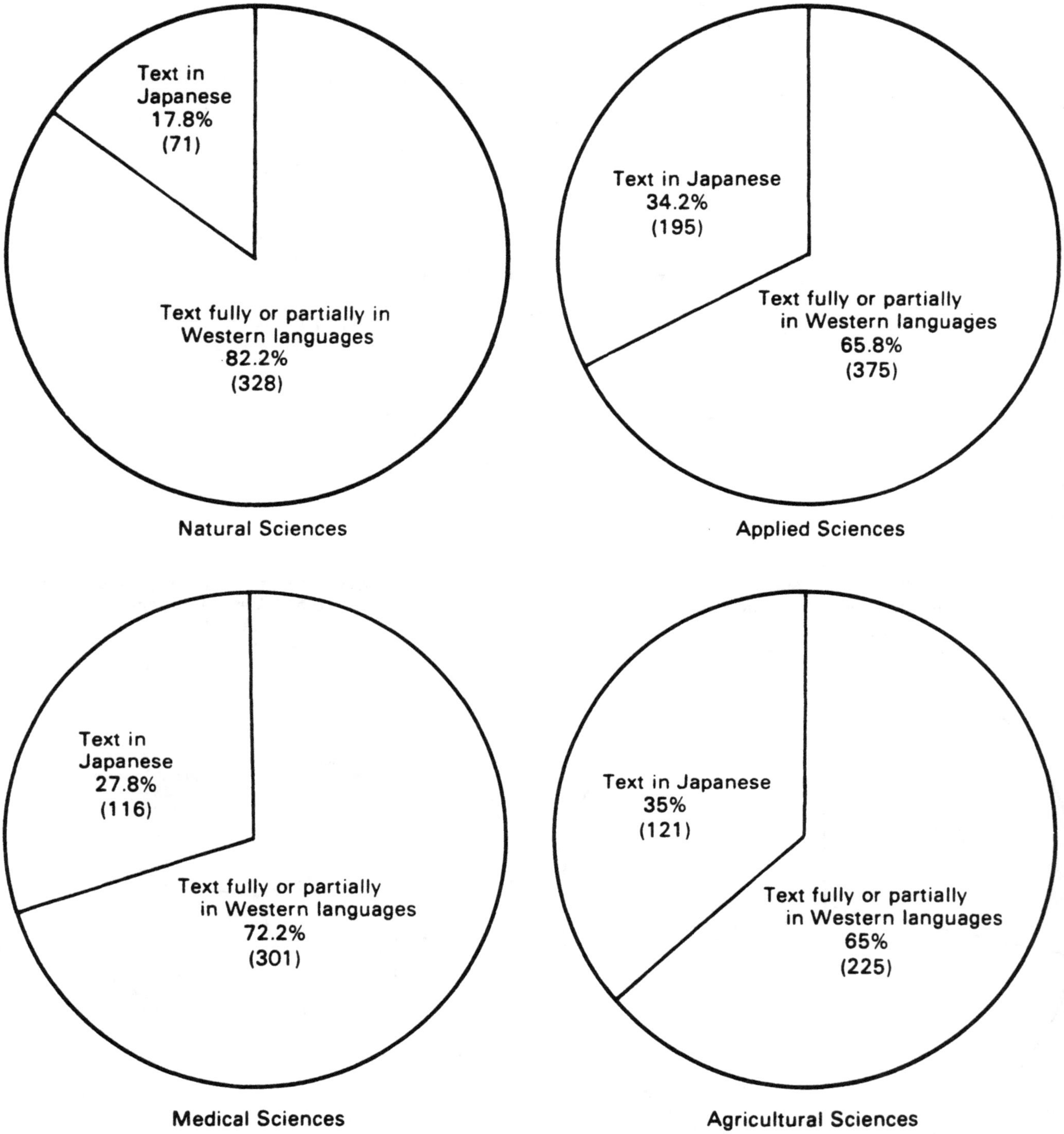

Figure 4-10.

Japanese Scientific and Technical Periodicals
Not Indexed by Western Sources
by Language

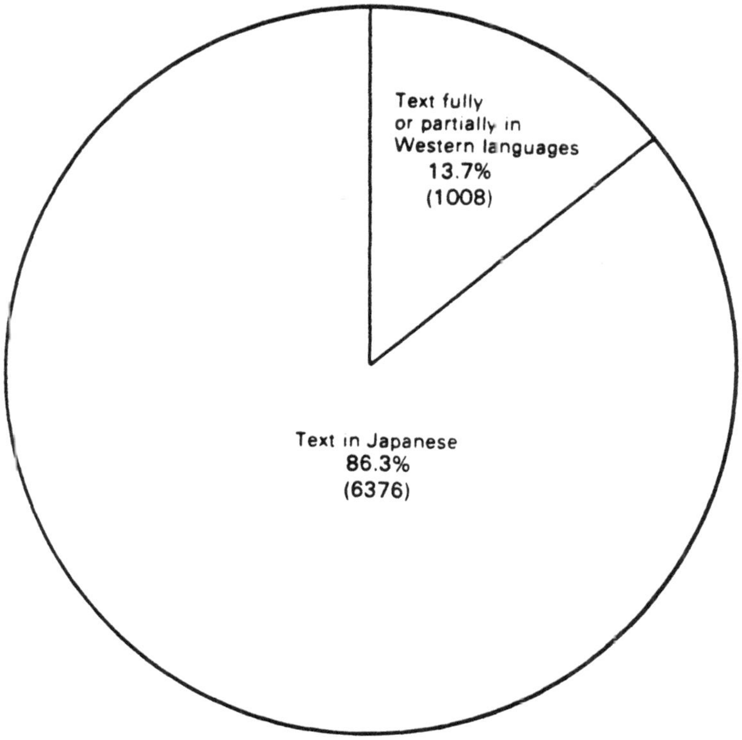

Figure 4-11.

Japanese Scientific and Technical Periodicals
Not Indexed by Western Sources
by Subject

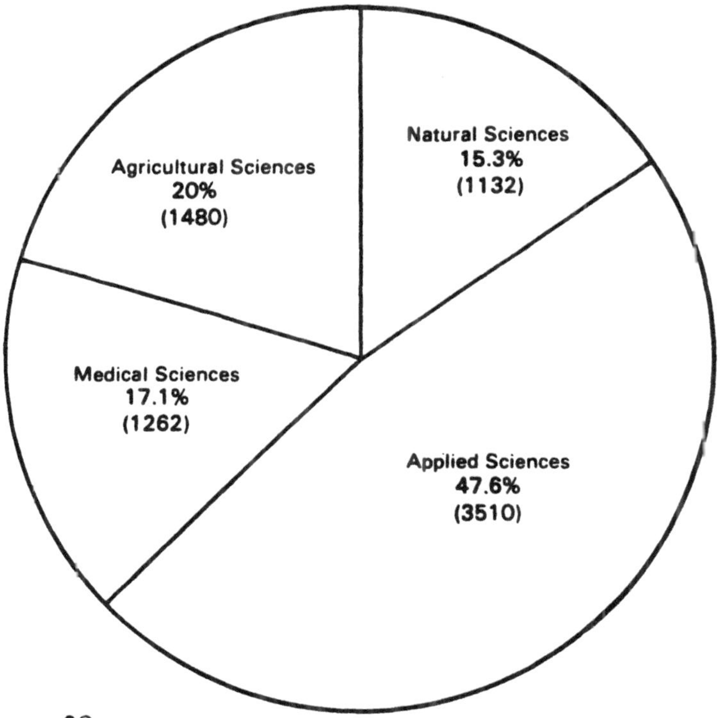

Figure 4-12.
Japanese Scientific and Technical Periodicals
Not Indexed by Western Sources
by Subject and Language

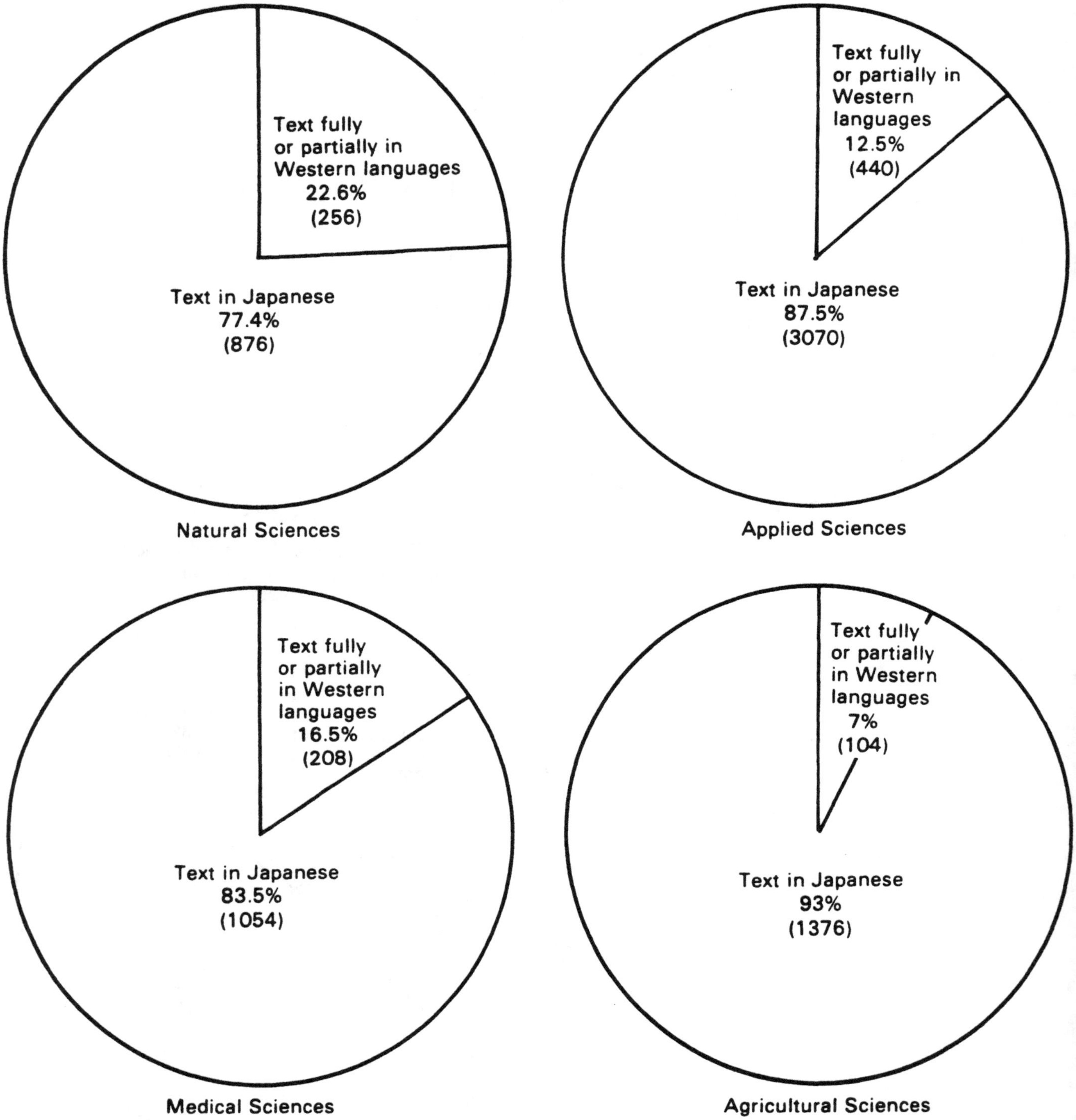

why these periodicals are not surveyed by Western indexes and abstracts. Availability? Knowledge of their existence?

JAPANESE SCIENTIFIC AND TECHNICAL PERIODICALS INDEXED BY JAPANESE SOURCES

The Japan Center of Science and Technology (JICST), the only Japanese indexing and abstracting source similar in format and depth of indexing to the Western indexing sources, was used as a base point for the study. Since JICST appears to be the most comprehensive indexing and abstracting service for scientific and technical periodicals in Japan, their coverage was analyzed to give an indication of the information from periodical literature which is available to Japanese researchers.

Although JICST lists 3,036 periodical titles in their holdings list, 457 of these titles were either not within the scope of this study (many were in the area of economics) or had ceased publication since the issuance of the holdings list. Therefore, these 457 titles were dropped, and 2,579 periodical titles were analyzed by subject and language.

One would conjecture that the majority of Japanese scientific and technical periodicals indexed and abstracted by a Japanese indexing and abstracting service would be written in Japanese. This, however, is not the case. Figure 4-13 shows that although 60.0 percent of the titles covered by JICST are published in Japanese, 40.0 percent of the scientific and technical periodicals covered by JICST are published either fully or partially in Western languages.

This information raised an interesting question: Why is a Japanese indexing and abstracting service covering such a high percentage of periodical literature that is written in a language foreign to them? Could this be construed as meaning that much of the "best" research being conducted in Japan (assuming that JICST has a selection policy which chooses only the most valuable research publications for inclusion in their system) is being published in Western languages? Or does it mean that it is easier for the Japanese to include a portion of the Western material—process it into the JICST data base—than it is to make use of Western indexing services?

A breakdown of the Japanese scientific and technical periodicals which are indexed in Japan indicates (figure 4-14) that the largest portion of material covered is in the area of applied sciences (59.4 percent), with the areas of natural sciences, medical sciences, and agricultural sciences following. This result would be expected in light of the theory that greater emphasis is placed on the applied sciences in Japan than on the other subject areas.

When the Japanese scientific and technical periodicals which are indexed by JICST are analyzed by language within each of the four subject areas (figure 4-15), the results are very interesting. In the areas of natural sciences, medical sciences and agricultural sciences, the majority of the periodicals covered are written either fully or partially in Western languages. This further supports the question raised in the discussion of Japanese-indexed periodicals by language, which asks why this portion of Western-language periodicals is so high. However, in the area of applied sciences, the portion which is written fully or partially in Western languages (29.3 percent) is much smaller than the portion written in Japanese (70.7 percent).

Is it possible that this means that much of the "best" research in the applied sciences area is being documented only in Japanese? As this is the area in which the Japanese seem to place their greatest emphasis, it is important to ascertain the language of publication of this "best" research, as it has direct bearing on whether Western researchers are aware of it.

JAPANESE SCIENTIFIC AND TECHNICAL PERIODICALS INDEXED AND NOT INDEXED BY JAPANESE OR WESTERN SOURCES

It is important to combine the findings of the previous two sections (Japanese scientific and technical periodicals indexed and not indexed by Western sources, and Japanese scientific and technical periodicals indexed and not indexed by Japanese sources) to determine the total amount of the Japanese periodical literature in the fields of science and technology which is being covered or not covered in the world.

Figure 4-16 shows an overall comparison of the total number of indexed titles and the total number of titles which are not indexed. When the Japanese and Western source statistics are combined, 35.8 percent of the Japanese scientific and technical periodicals fall into the indexed area and 64.2 percent fall into the unindexed area. If the term "accessible to Japanese or Western researchers" is substituted for "indexed" and the term "not accessible to Japanese or Western researchers" is substituted for "not indexed," this chart shows that only about 35 percent of the Japanese scientific and technical information which is documented in the periodical literature is accessible to researchers in Japan and the

Figure 4-13.
Japanese Scientific and Technical Periodicals Indexed by Japanese Sources (JICST) by Language

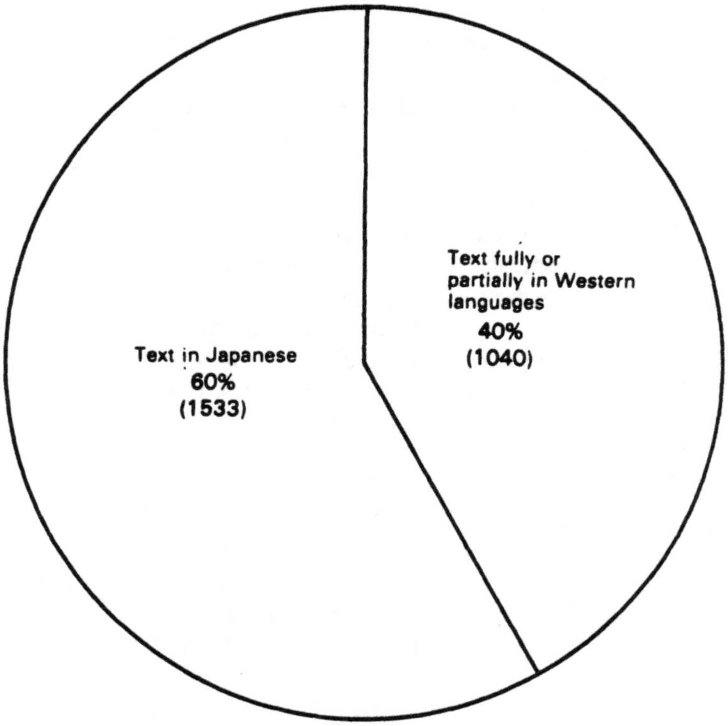

Figure 4-14.
Japanese Scientific and Technical Periodicals Indexed by Japanese Sources (JICST) by Subject

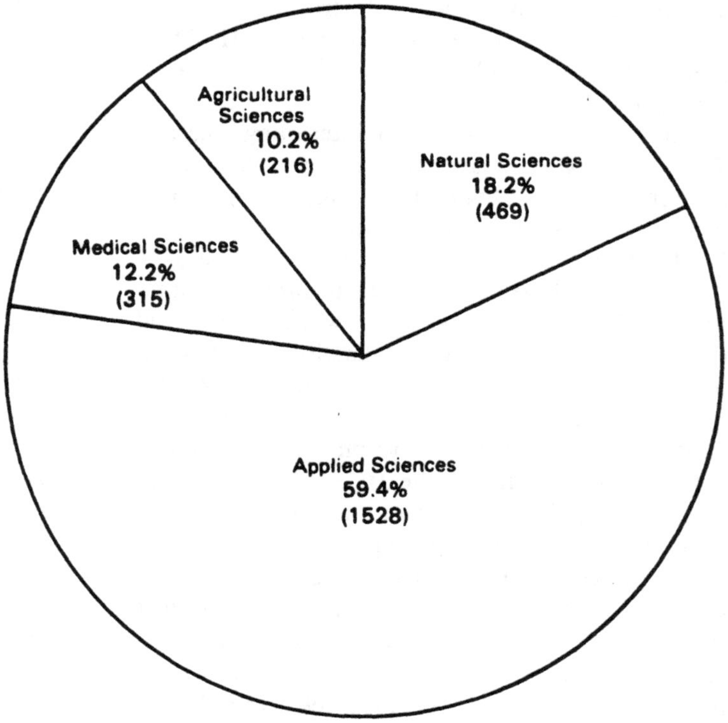

Figure 4-15.
Japanese Scientific and Technical Periodicals Indexed by Japanese Sources (JICST) by Subject and Language

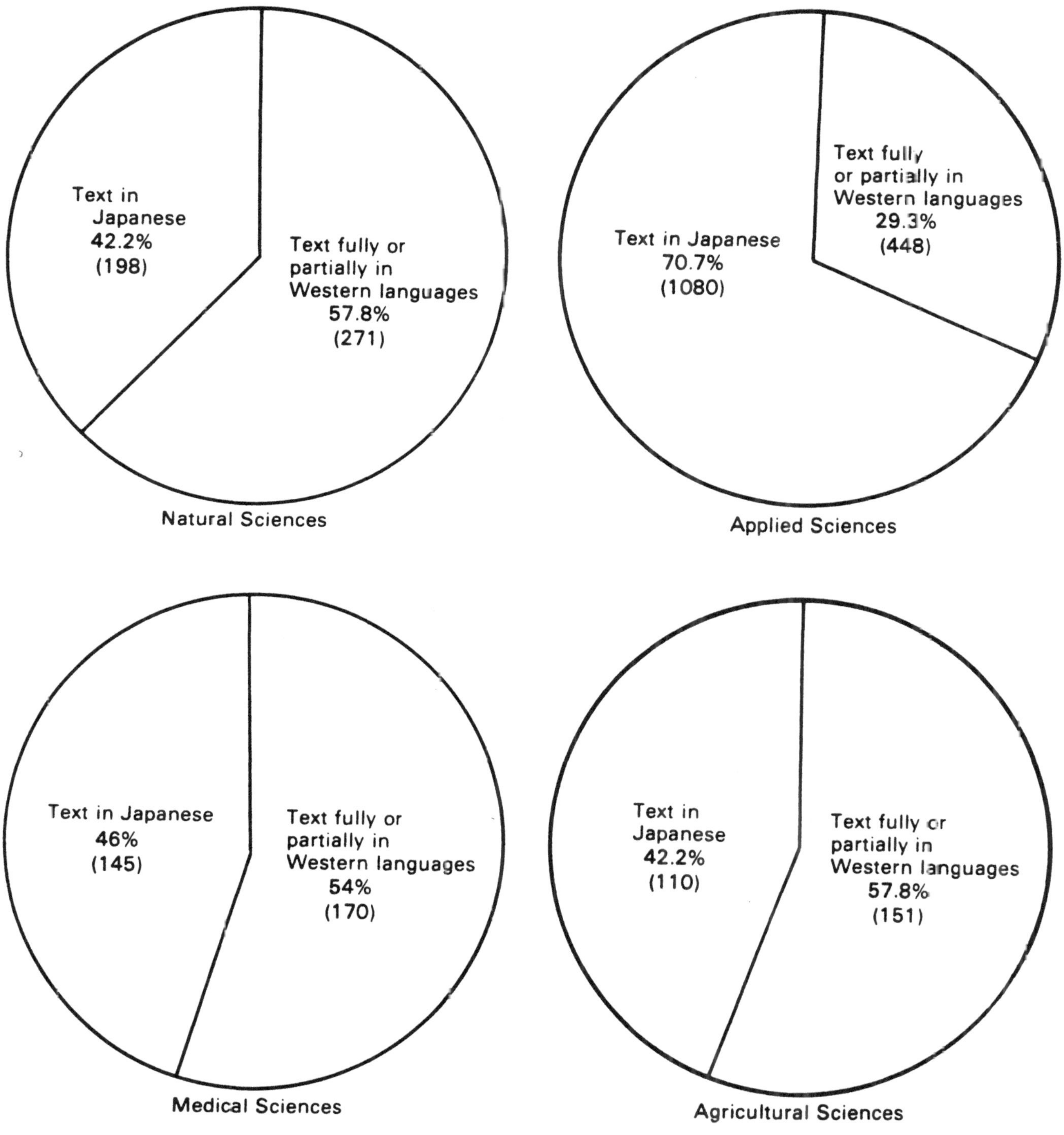

Figure 4-16.
Japanese Scientific and Technical Periodicals Indexed and Not Indexed by Japanese or Western Sources

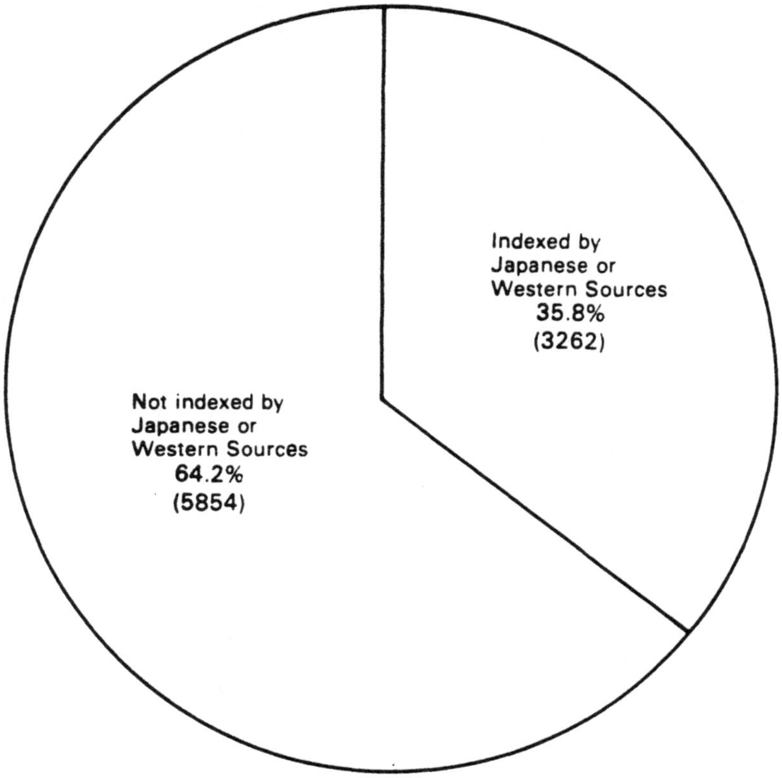

West, and almost 65 percent of the Japanese scientific and technical information which appears in the periodical literature is virtually inaccessible by researchers in either Japan or the West.

Figure 4-17 shows a language breakdown of the Japanese scientific and technical periodicals which are indexed by Japanese or Western sources. If this chart is compared to the chart illustrating the language breakdown of periodicals indexed by Western sources (figure 4-7), one can see that although there is an overlap, for example, Western-language titles which are covered by both the Japanese and Western sources, the coverage of the Japanese-language periodicals increases from 29 percent covered by Western sources to 54.4 percent covered by Japanese or Western sources. This increase is explained by the number of Japanese-language periodicals which are indexed by Japanese sources. Even more interesting is the fact that almost one-half of the Japanese scientific and technical periodicals which are covered by Japanese or Western indexing and abstracting sources are written either fully or partially in Western languages.

The addition of the scientific and technical periodicals indexed by Japanese sources to those indexed by Western sources indicates that the number of titles which fall into the applied sciences area greatly increases. Figure 4-18 shows that over one-half of all indexed and abstracted information falls into that category. Coverage of the natural sciences area is the next largest area, with the medical sciences and the agricultural sciences following.

Figure 4-19, which shows language of publication within each of the four subject areas, further substantiates the findings discovered in figure 4-17. There is one variation, however, to the statement made in the discussion of the language breakdown of the Japanese scientific and technical periodicals

Figure 4-17.
Japanese Scientific and Technical Periodicals
Indexed by Japanese or Western Sources
by Language

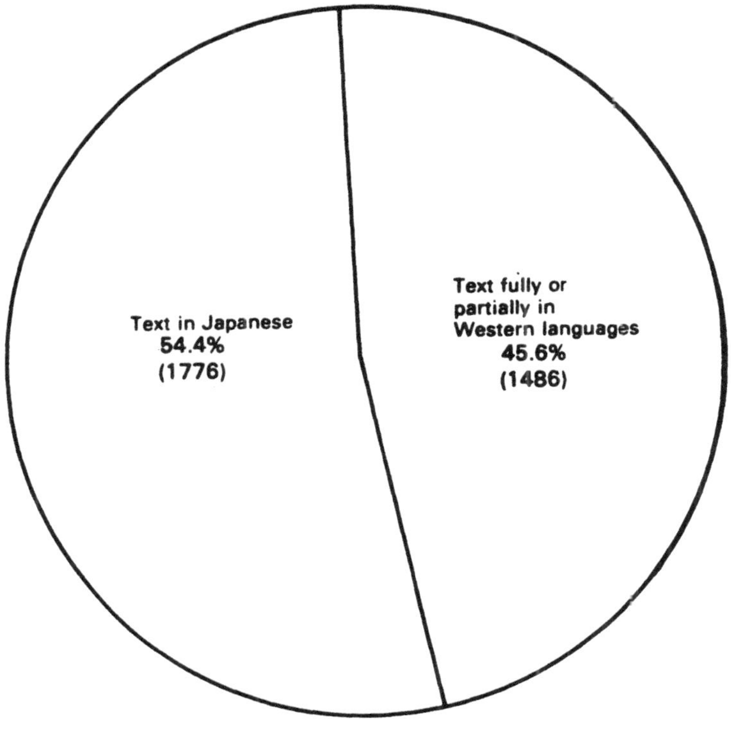

Figure 4-18.
Japanese Scientific and Technical Periodicals
Indexed by Japanese or Western Sources
by Subject

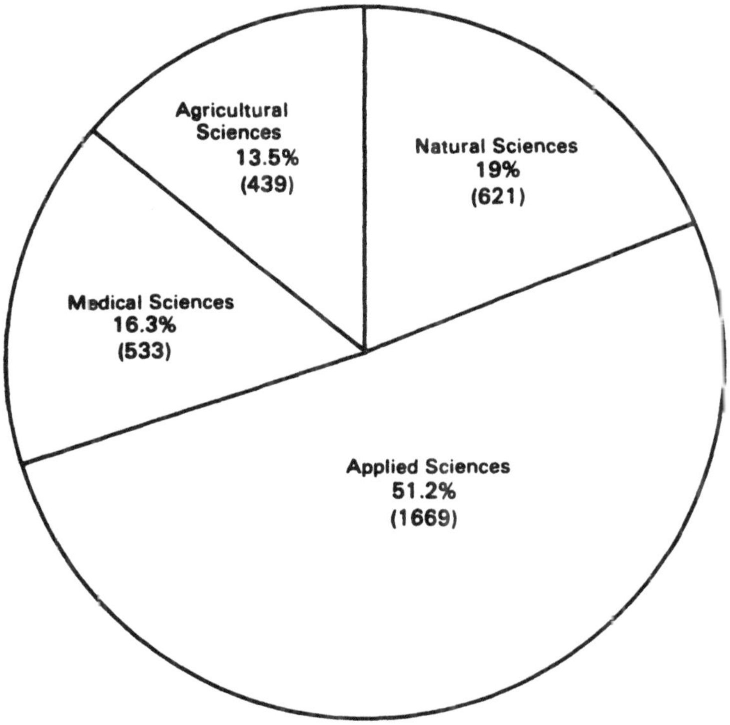

Figure 4-19.
Japanese Scientific and Technical Periodicals
Indexed by Japanese or Western Sources
by Subject and Language

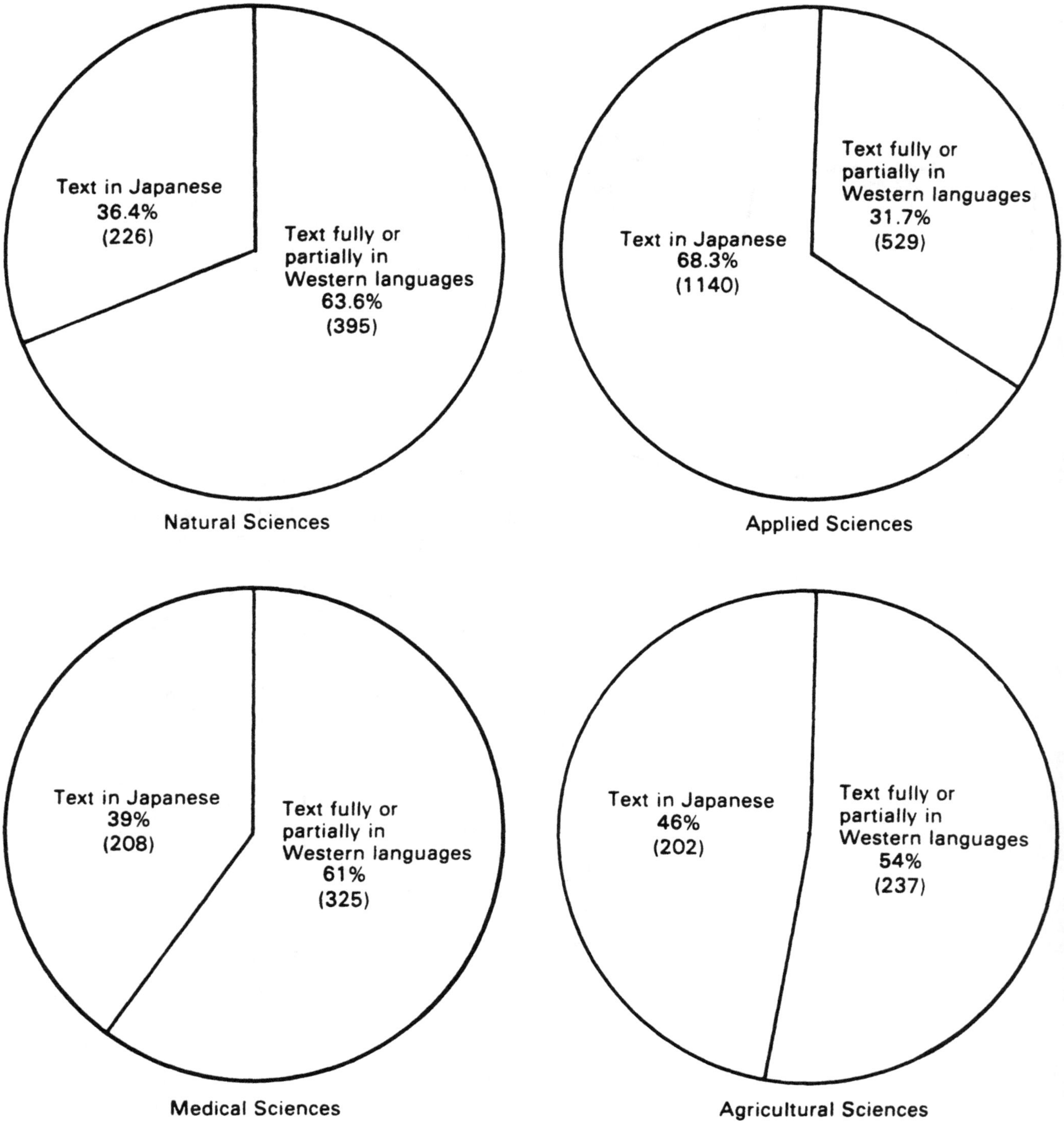

indexed by Japanese or Western sources, and that is in the applied sciences field. Although in the natural, medical, and agricultural sciences the majority of the indexed material is written fully or partially in Western languages, in the area of applied sciences, the majority of the indexed information is in Japanese. This figure is obviously affected by the relatively high amount of Japanese-language material which is indexed by Japanese sources in this subject area. It also tends to make one suspect that much "good" research in the applied sciences area is being published only in the Japanese language.

After analyzing the Japanese scientific and technical periodical titles which are not indexed by Japanese or Western sources and therefore are virtually inaccessible to Western or Japanese researchers, it was found that the vast majority of this literature (87.2 percent) was published only in Japanese. Figure 4-20 also shows, however, that there is a body of some 751 periodicals, which form 12.8 percent of this unindexed portion of the literature, that is written either fully or partially in Western languages. Subsequent charts will show where this unindexed Western-language material falls in the subject breakdowns.

Figure 4-21 shows a subject breakdown of the unindexed Japanese scientific and technical periodical titles. The largest portion is in the applied sciences, followed by the agricultural, medical, and natural sciences.

Figure 4-22 shows a breakdown of the language of publication within each of the four subject areas for Japanese scientific and technical periodicals not indexed by Japanese or Western sources. In the natural sciences, 20.8 percent of the unindexed periodicals are in Western languages. This is followed by 16.1 percent of the medical science periodicals which are in Western languages and unindexed, 11.9 percent of the applied sciences, and 6.6 percent of the agricultural sciences. An interesting question is why this Western-language material, especially in the natural and medical sciences, is not being covered by indexes and abstracts in Japan or in the West.

Figure 4-23 was designed to better illustrate the breakdown of Japanese scientific and technical periodicals which are indexed and not indexed by Japanese and Western sources. There are 690 titles (7.6 percent) which are indexed by Western sources only and 1,041 titles (11.4 percent) which are indexed by *both* Japanese and Western sources. These two figures combined indicate the total Japanese scientific and technical periodicals which are indexed by Western sources (1,731 titles, forming 19 percent of the total). There are 1,531 titles (16.8 percent) which are indexed by Japanese sources only. This figure, combined with the previously totaled Western sources, produces a total of 3,262 indexed titles, which form 35.8 percent of the total number of periodicals. This total of indexed items, subtracted from the 9,116 titles included in the survey, leaves 5,854 titles, or 64.2 percent, which are not indexed by Japanese or Western sources.

JAPANESE SCIENTIFIC AND TECHNICAL PERIODICALS BY LANGUAGE WHICH ARE INDEXED OR NOT INDEXED BY WESTERN SOURCES

Figure 4-24 shows that out of all the Japanese scientific and technical periodicals which are published either fully or partially in Western languages, approximately one-half are indexed by Western sources and approximately one-half are not indexed by Western sources. Since language is no barrier, one could conjecture that there is some other reason, budget or space constraints, for example, why Western sources are not covering these 1,000-plus titles. When the periodicals which are written in Japanese are evaluated, only 7.3 percent are indexed by Western sources, leaving 92.7 percent of the Japanese-language scientific and technical periodicals virtually inaccessible to Western researchers.

JAPANESE SCIENTIFIC AND TECHNICAL PERIODICALS BY LANGUAGE WHICH ARE INDEXED AND NOT INDEXED BY JAPANESE OR WESTERN SOURCES

Figure 4-25 illustrates the indexed and unindexed portions of each language breakdown which are indexed or not indexed by Japanese or Western sources. Of the total Japanese scientific and technical periodicals which are published either fully or partially in Western languages, 1,486 titles (66.4 percent) are indexed by Japanese or Western sources, but almost one-third of the Western language periodicals (751 titles, which form 33.6 percent) are not indexed either in Japan or in the West. Of those periodicals which are published in Japanese, about one-quarter of the titles are indexed by Japanese or Western sources, but almost three-quarters of the Japanese-language scientific and technical periodicals are not indexed in Japan or in the West.

Figure 4-20.
Japanese Scientific and Technical Periodicals Not Indexed by Japanese or Western Sources by Language

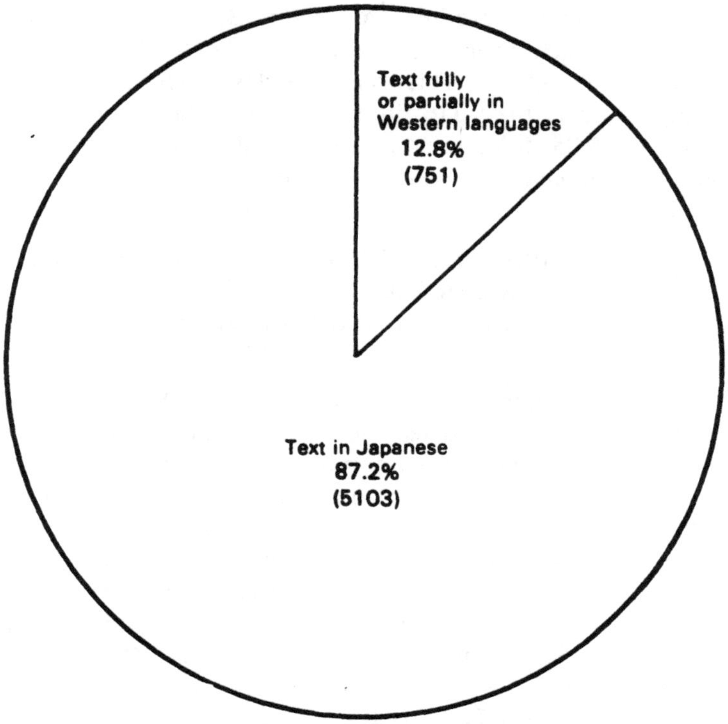

Figure 4-21.
Japanese Scientific and Technical Periodicals Not Indexed by Japanese or Western Sources by Subject

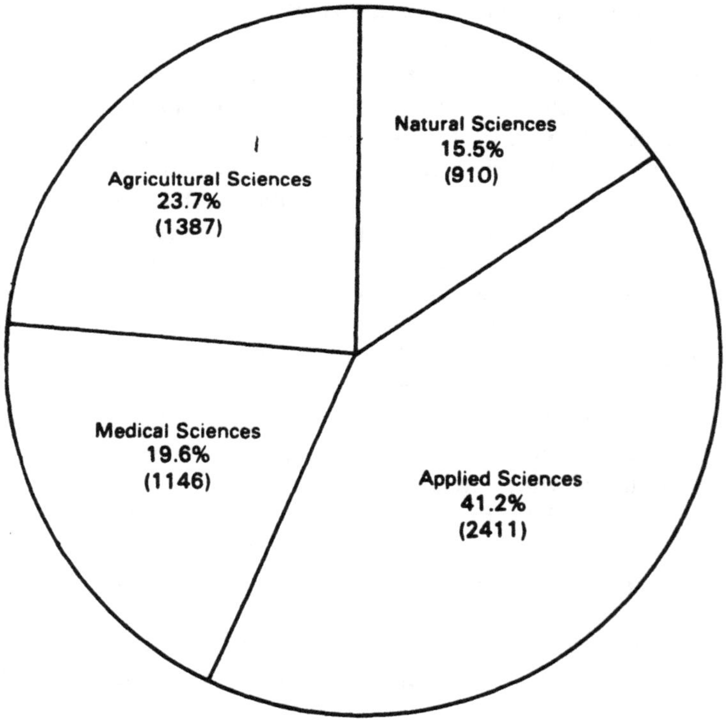

Figure 4-22.
Japanese Scientific and Technical Periodicals
Not Indexed by Japanese or Western Sources
by Subject and Language

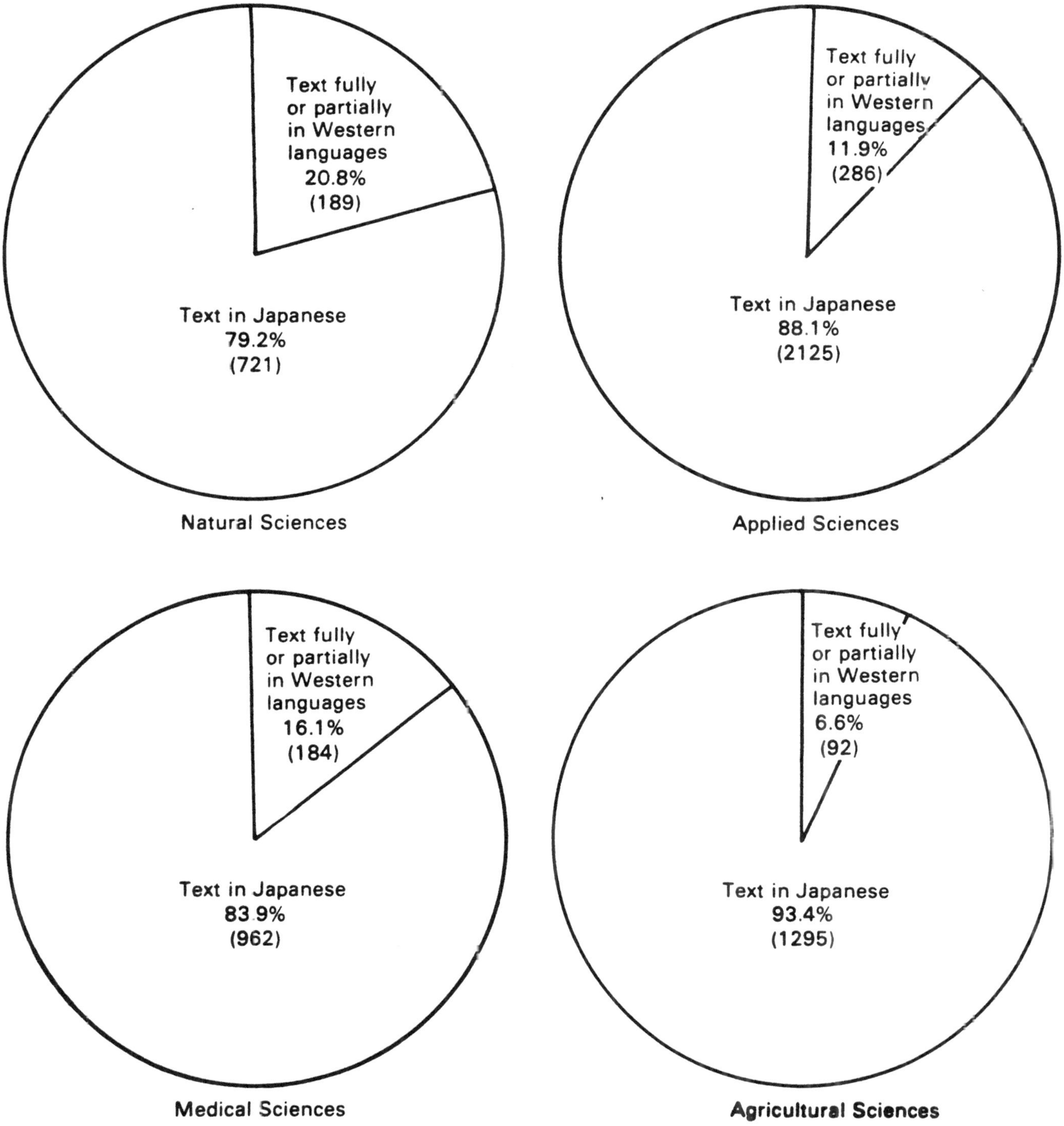

Figure 4-23
Japanese Scientific and Technical Periodicals
Indexed and Not Indexed
by Japanese and Western Sources

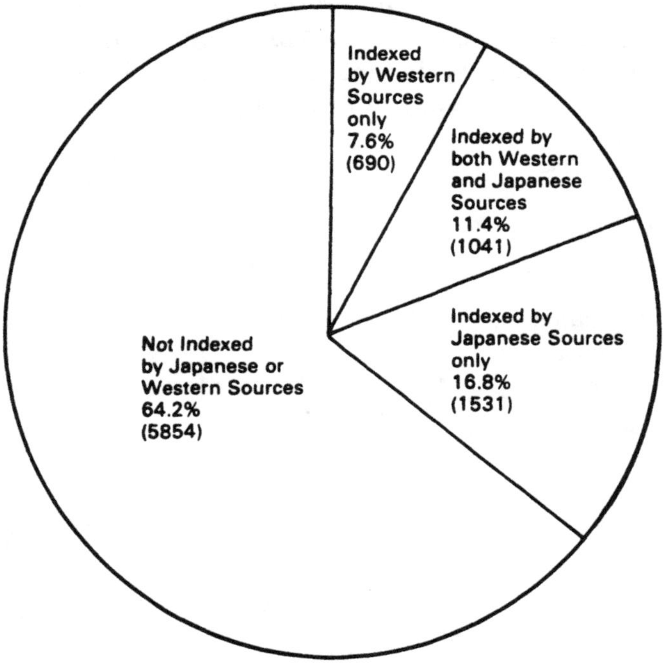

Figure 4-24.
Japanese Scientific and Technical Periodicals
by Language
Indexed or Not Indexed by Western Sources

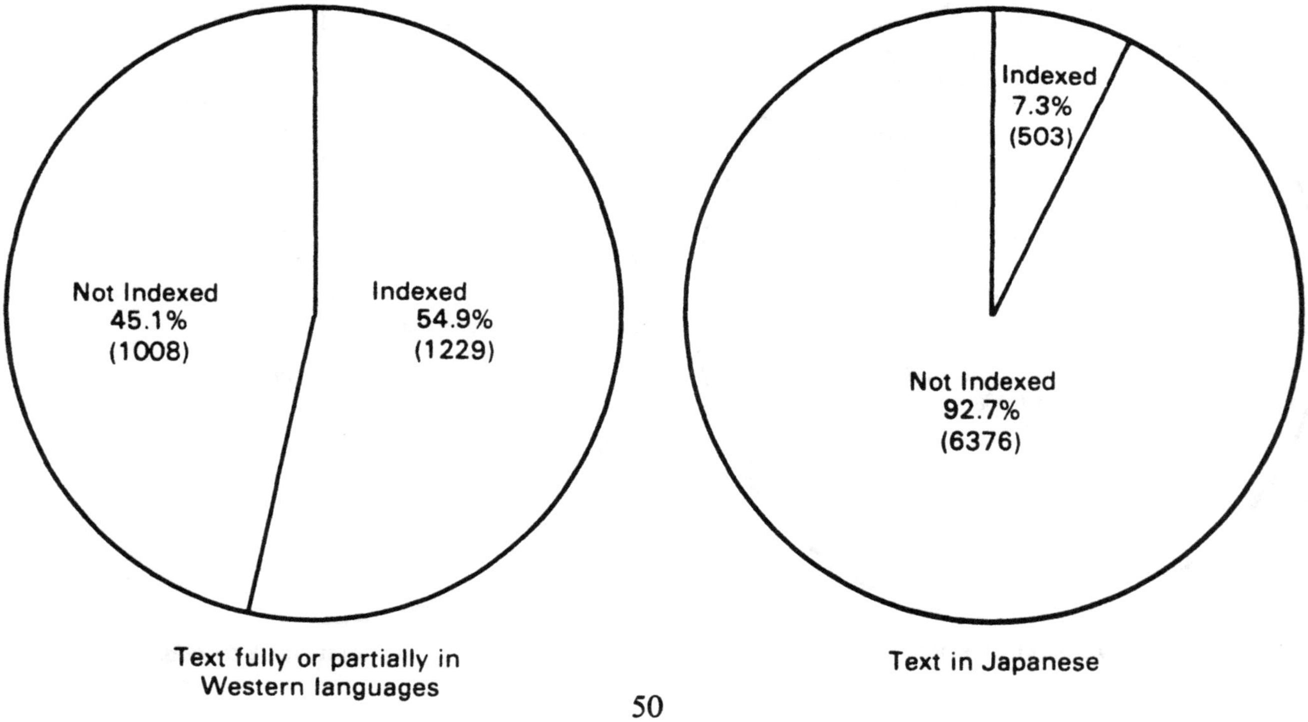

Figure 4-25.
Japanese Scientific and Technical Periodicals
by Language
Indexed and Not Indexed by Japanese or Western Sources

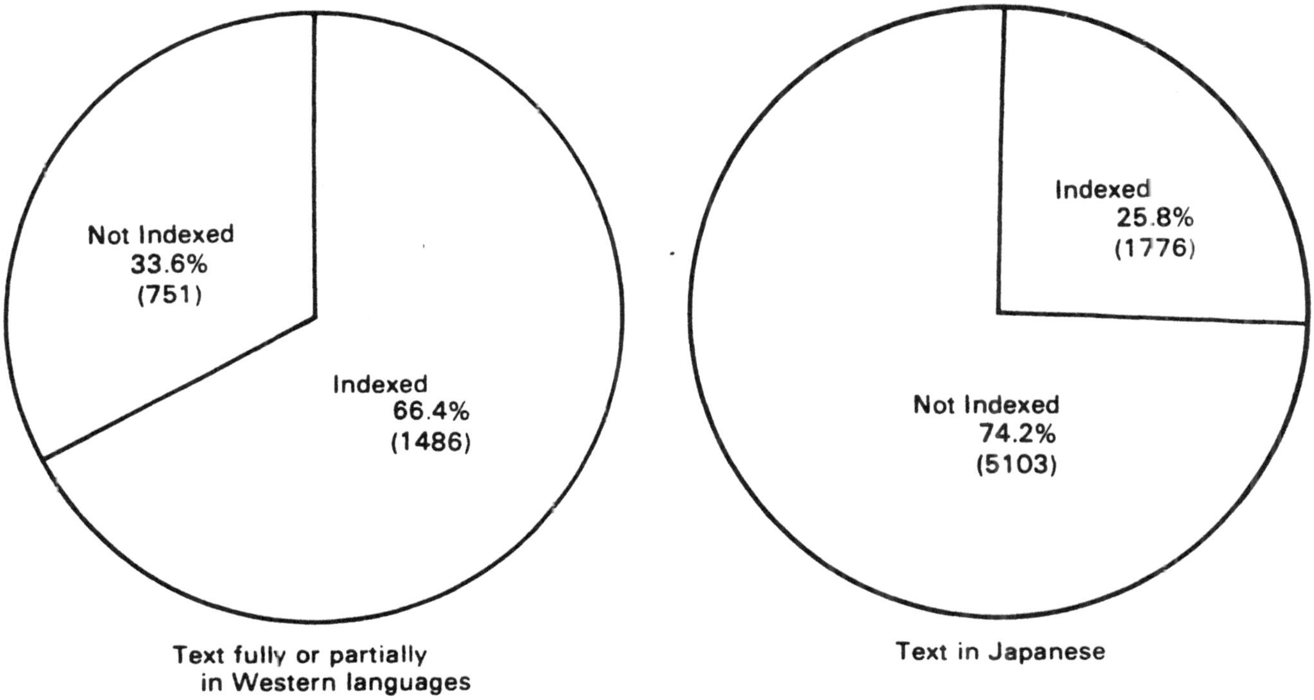

JAPANESE SCIENTIFIC AND TECHNICAL PERIODICALS BY SUBJECT WHICH ARE INDEXED OR NOT INDEXED BY WESTERN SOURCES

Figure 4-26 shows the breakdown of the information covered by the Western sources by subject. In the natural sciences, approximately one-quarter of the Japanese scientific and technical periodicals are indexed, leaving almost three-quarters which are not indexed in Western sources. In the applied sciences, only 14 percent of the Japanese scientific and technical periodicals are indexed, with 86 percent not covered by Western sources. In the medical sciences, one-quarter of the titles are indexed, with three-quarters not indexed in Western sources. In the agricultural sciences, less than one-fifth of the Japanese scientific and technical periodicals are indexed in Western sources, with over four-fifths being unindexed. If one subscribes to the theory that "indexed" means "accessible" and "not indexed" means "inaccessible," then very little of the Japanese research which is documented in Japanese scientific and technical periodicals is accessible in the West.

JAPANESE SCIENTIFIC AND TECHNICAL PERIODICALS BY SUBJECT WHICH ARE INDEXED OR NOT INDEXED BY JAPANESE OR WESTERN SOURCES

When the total indexed and unindexed Japanese scientific and technical periodicals are evaluated by subject area (figure 4-27), the unindexed portion of the periodical literature is greater in every subject area. In each subject area, the portion of the Japanese scientific and technical periodical literature which is not indexed by sources in either Japan or the West exceeds 50 percent, and in the case of the agricultural sciences, the unindexed portion of the literature exceeds 75 percent.

JAPANESE SCIENTIFIC AND TECHNICAL PERIODICALS BY SUBJECT WHICH ARE INDEXED OR NOT INDEXED BY WESTERN SOURCES—WITH A LANGUAGE BREAKDOWN

Figure 4-28 was created to put the information shown on the previous charts in perspective. This chart shows that for each of the four subject areas, the largest portion of the periodical literature which is covered by Western sources is in Western languages. The largest portion of the unindexed literature is in Japanese. In each area, however, there is a sizeable portion of the unindexed portion which is in Western languages.

JAPANESE SCIENTIFIC AND TECHNICAL PERIODICALS BY SUBJECT WHICH ARE INDEXED OR NOT INDEXED BY JAPANESE OR WESTERN SOURCES—WITH A LANGUAGE BREAKDOWN

Figure 4-29 shows a subject breakdown of the indexed and unindexed Japanese scientific and technical periodicals combining all the indexing and abstracting sources (Japanese and Western). It also shows a language breakdown within each indexed and unindexed portion of the literature of a subject area. In all four subject areas, the unindexed portion exceeds 50 percent, and in the agricultural sciences, it exceeds 75 percent. Again, there is a sizeable portion of the literature in Western languages which is not indexed by any source.

Figures 4-28 and 4-29 should serve to shatter any myths which are held by Western researchers (or Japanese researchers, for that matter) that most of the Japanese scientific and technical periodical literature is accessible to them. If Western researchers restrict themselves only to the Japanese periodical literature which is written in Western languages (which is possibly typical), they will only be seeing the tip of the iceberg of information which is being published in any subject area of science and technology in Japan.

Technical Reports

Analysis of technical report literature is important because this form of publication often represents the most up-to-date information on current research.

In Japan, technical reports appear to be published mainly as products of research going on at research institutes. Since there is no central source for accessing and disseminating technical reports in Japan, and no abstracting and indexing services which comprehensively cover them, analysis of the quality and quantity of this type of Japanese scientific and technical literature is virtually impossible.

However, since most of the technical report literature appears to be published by research institutes, a look at the number and type of research institutes may provide an indication of the volume of the report publication effort.

In 1976, there were about 1,220 research institutes which were affiliated with universities. The most recent information on other types of research institutes was published in 1965 (Yoshida, pp. 4-17). At that time, about 60 government research institutes which were involved in basic and applied research under the administration or supervision of agencies within the various government ministries were identified. There

Figure 4-26.
Japanese Scientific and Technical Periodicals
by Subject
Indexed or Not Indexed by Western Sources

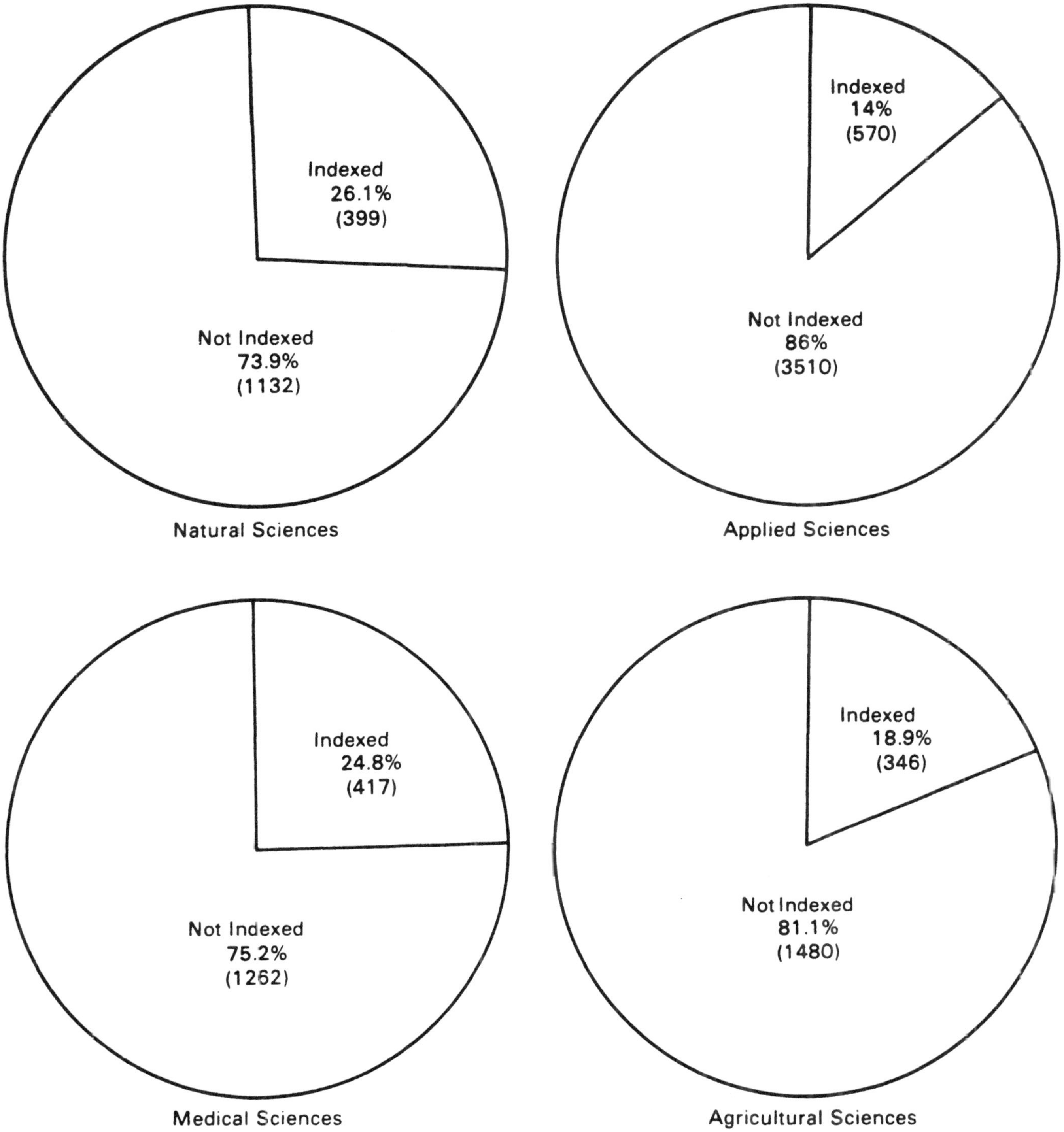

Figure 4-27.
Japanese Scientific and Technical Periodicals
by Subject
Indexed or Not Indexed by Japanese or Western Sources

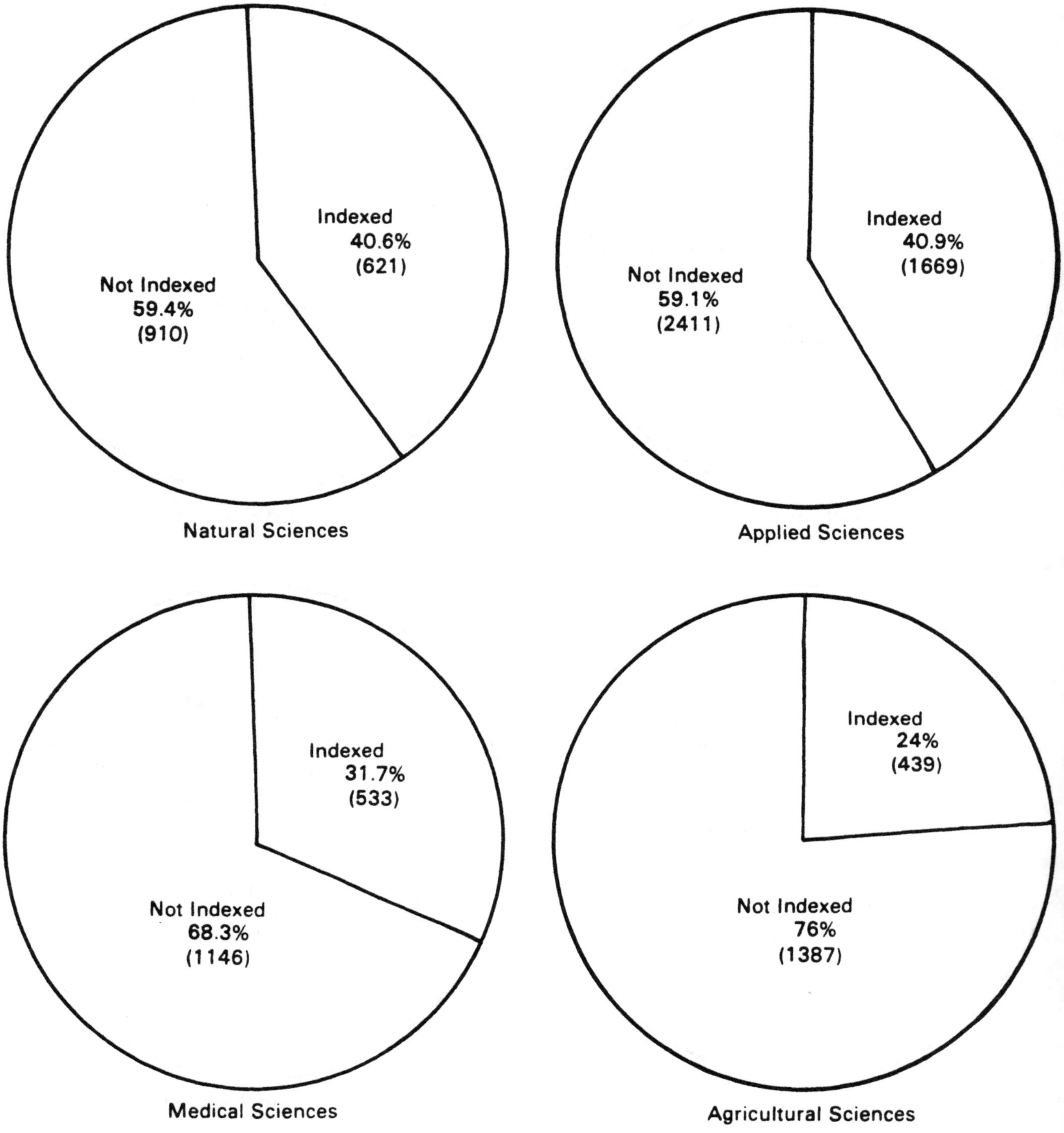

Figure 4-28.
Japanese Scientific and Technical Periodicals
by Subject
Indexed or Not Indexed by Western Sources
by Language

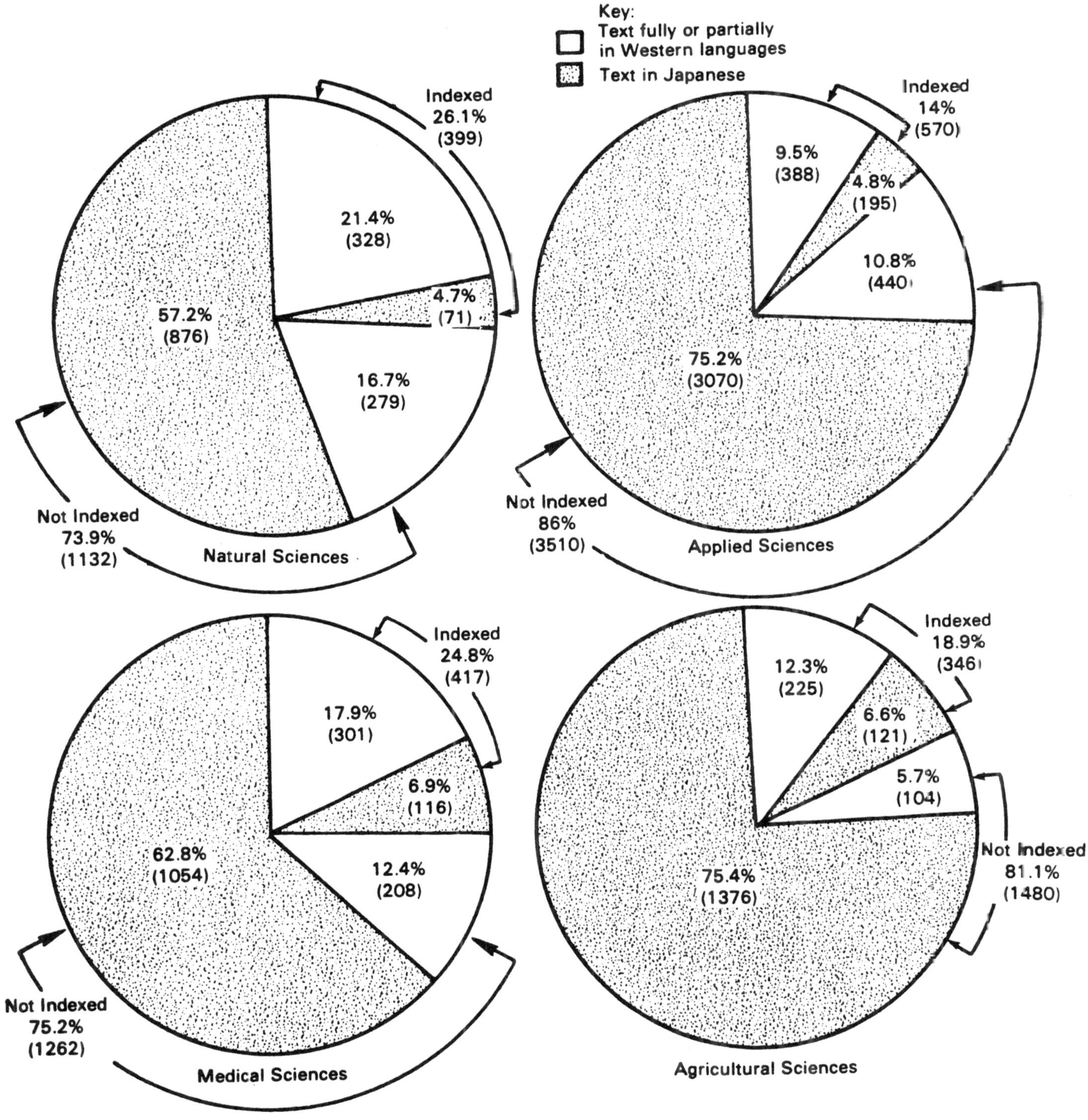

Figure 4-29.

Japanese Scientific and Technical Periodicals by Subject Indexed or Not Indexed by Japanese or Western Sources by Language

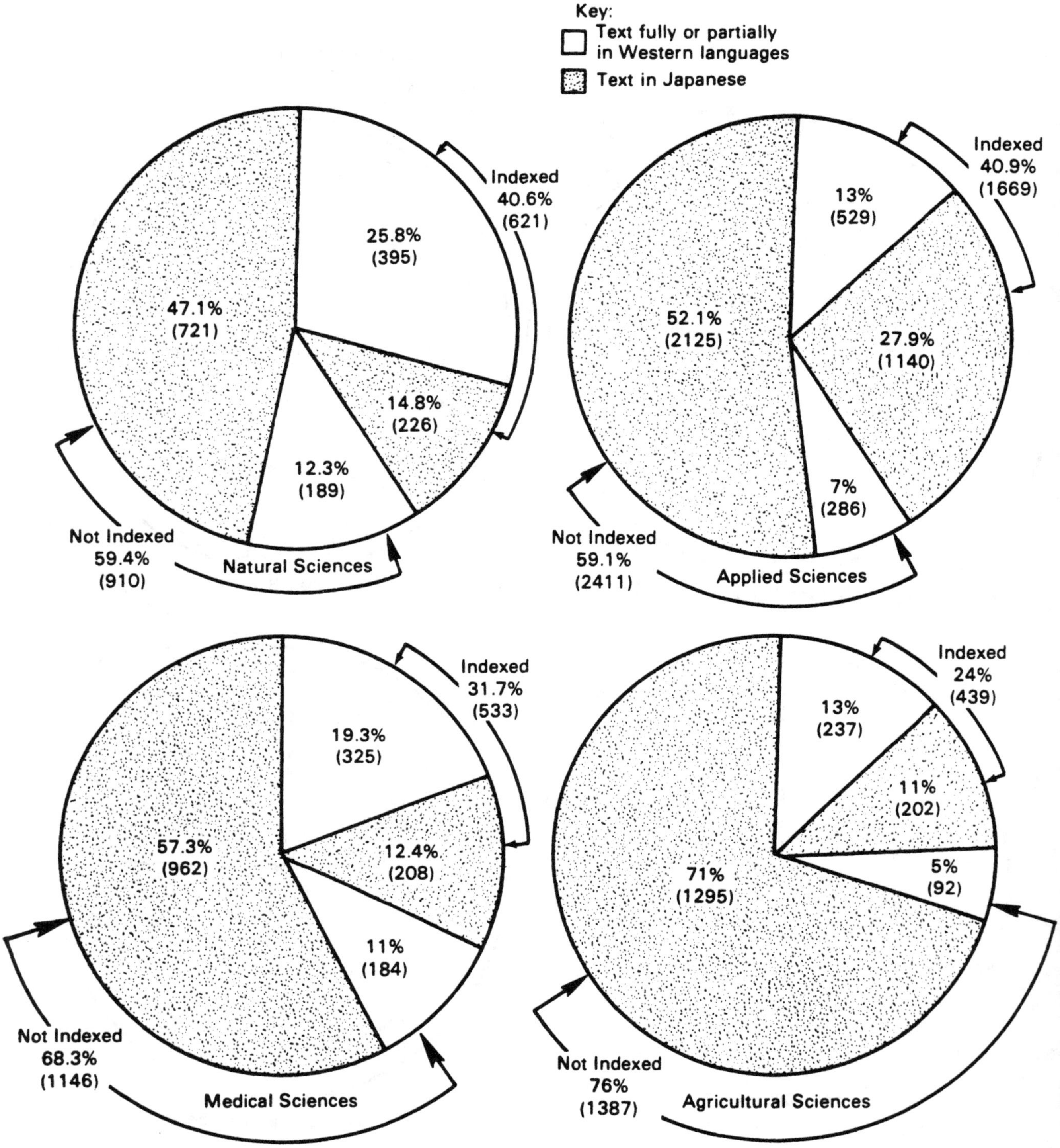

were also 448 prefectural or municipal research institutes and 134 research institutes under the management of nonprofit corporations or private individuals. Thus, one might estimate that there are currently over 2,000 research institutes throughout Japan, most of which probably publish technical reports.

Information on the actual volume of the technical reports published by these institutes is sketchy. The National Diet Library has identified some ninety technical report series within their holdings which represent the output of sixty-one Japanese research institutes. Since this number covers only report series and not individual reports, and covers only sixty-one out of a probable two thousand institutes, it does not provide a very good indication of the total publishing effort in this field.

From this information, one can only assume that there is a large amount of technical report literature being produced in Japan, but the actual volume is unknown.

The question of quality of this type of publication also needs to be explored. An indication of the value of technical reports can be found in the results of a two-year study carried out under the auspices of the National University Libraries Council beginning in 1974. A total of 4,131 questionnaires returned by researchers at five universities in the Kansai area were studied. These researchers were asked to evaluate fifteen various types of publications with respect to their value. Their responses were then used to derive a weighted value indicating their importance to researchers. The results, for those involved in scientific research, showed the following types of publications and their weighted values: periodicals (96.1), books (77.6), personal communication (45.3), reprints (42.2), and technical reports (40.2), followed by other types of publications with lesser values. Since technical reports were ranked fifth in this study, this would seem to indicate that the material contained in them is valuable and is being used by Japanese researchers.

In another part of the study, researchers indicated that they feel that technical reports have a relatively brief life span and are used mainly as sources of current information. Scientists also indicated that technical reports are extremely difficult to obtain, even after they are identified.

In summary, then, one could conclude that there is a large volume of valuable technical report literature being published in Japan, but access to this material is extremely difficult, even for researchers in Japan.

Informal Channels of Communication

It is apparent from personal interviews and articles surveyed that there is an informal communication network between researchers and scientists in Japan. It also appears that Japanese researchers are aware of what their colleagues are doing in fields of advanced research. Thus, the analysis of the accessibility of research to Japanese researchers as shown in the section of the report on periodical analysis is not the whole story. Even though the findings reveal that Japanese researchers are not able to access a large amount of the Japanese scientific and technical literature through the secondary sources, these researchers seem to have partially filled this gap through informal communications.

It is very difficult to analyze the effect of informal communication channels between Japan and the West, since little work has been done in this area. One would tend to believe, however, that this communication network is limited due primarily to the language barrier. Further research in this area is needed before definite conclusions can be drawn about the volume of information on Japanese scientific and technical research that is communicated to the West through informal channels.

Bibliography

MONOGRAPHS

The Bowker Annual of Library & Book Trade Information. 24th ed. New York: R. R. Bowker Company, 1979.

"Japan: Book Publishing is an Industry, not a Business," *Wilson Library Bulletin*, vol. 50, no. 10 (June 1976), p. 769.

Kern, Lawrence E. "Japan: The World's Third Largest Publishing Conglomeration." *Publisher's Weekly*, vol. 200, pt. 1 (September 27, 1971), pp. 172-174.

Lottman, Herbert R. "Publishing in Japan." *Publisher's Weekly*, vol. 208, pt. 2 (December 29, 1975), pp. 34-48.

Malcolm, Andrew H. "Japan's Reading Craze at a Peak in Recessing." *New York Times*, March 26, 1976, p. 10, column 1.

Shuppan Nenkan 1979. Tokyo: Suppan Nyusu-sha, 1979.

PERIODICALS

"Bibliographical Analysis on Japanese Scientific Periodicals (1)." *National Diet Library Newsletter*, no. 45 (March 1977), pp. 10-20.

"Bibliographical Analysis on Japanese Scientific Periodicals (2)." *National Diet Library Newsletter*, no. 47 (July 1977), pp. 17-21.

Catalogue of Japanese Periodicals '78. Tokyo: Maruzen Company, Ltd., 1978.

Chikuji Kankobutsu Shozo Mokuroku, September 30, 1976 (Index to Periodical Holdings, as of September 30, 1976), Tokyo: Japan Information Center of Science and Technology, 1976.

Current Japanese Periodicals for '78. Tokyo: Japan Publications Trading Company, Ltd., 1978.

Directory of Japanese Scientific Periodicals. Tokyo: National Diet Library, 1974.

Shuppan Nenkan 1978. (Publishing Yearbook 1978). Tokyo: Shuppan Nyusu-sha, 1978.

Sources of Serials. New York: R. R. Bowker, 1977.

Ulrich's International Periodical Directory. 17th ed., 1977-1978. New York: R. R. Bowker, 1977.

Wazasshi Mokuroku 1975 (Catalogue of Japanese Periodicals). Tokyo: National Diet Library, 1976.

Wazasshi Mokuroku Tsuiroku, January 1976-December 1977 (Catalogue of Japanese Periodicals: Supplement). Tokyo: National Diet Library, 1978.

TECHNICAL REPORTS

Fry, Bernard, and Kortendick, Rev. James J., Jr., eds. *The Production and Use of Technical Reports*. Washington, D.C.: Catholic University of America Press, 1955.

Hirase, Susume and Hayashi, Ryohei. "Research Scholars in their Quest for Information: A Study Based on Survey Data." In *Japanese and U.S. Research Libraries at the Turning Point; Proceedings of the Third Japan-U.S. Conference on Libraries and Information Science in Higher Education*, Kyoto, Japan, October 28-31, 1975, edited by Robert D. Stevens, Raynard C. Swank, and Theodore F. Welch, pp. 161–170. Metuchen, N.J.: Scarecrow Press, 1977.

Suzuki, Heihachiro, "National Plans for Library/Information Services and the National Diet Library," *Library and Information Science*, no. 14 (1976), pp. 1-14.

Teramura, Yuiko, "Technical Reports in Japan," *National Diet Library Newsletter*, no. 47 (July 1977), pp. 1-14.

Yoshida, Tomizo, "Organization of Scientific Activities." In *Science in Japan*, edited by Arthur H. Livermore, pp. 1–24. Washington, D.C.: American Association for the Advancement of Science, 1965.

chapter five

VERIFICATION OF RESULTS

After the two portions of the study, the dimensions of the Japanese scientific and technical information network and the scientific and technical publishing effort, were completed, the study group sought on-the-scene verification of the results of their research. A trip was made to Japan to determine how Japanese information specialists (both librarians and potential literature consumers) evaluated the survey findings and to fill possible gaps in the information.

The first stop was the National Diet Library where the major points of the study were presented to not only Mr. Kishida, Librarian, and Mr. Sakai, Deputy Librarian, but also subsequently to the division chiefs of the various departments in the library. The National Diet Library is Japan's largest library, the major producer of national bibliographical records. NDL was modeled after the United States Library of Congress and observed its thirtieth anniversary in November, 1979. The National Diet Library renders services to the Diet, to the Executive and Judicial branches of the government, and to the general public, by acquiring, processing, preserving, and offering for use books and other library materials. There is a staff of over 850 people. Its collection includes 1,970,000 Japanese and 630,000 foreign books and 13,000 Japanese and 11,000 foreign serial titles, including newspapers.

These briefings resulted in a request by the division chiefs that they be allowed some time to study a copy of the report and that an additional meeting be scheduled with them. After some discussion, this group did agree that the statistics which appeared in the report were valid. There were some comments concerning the validity of the inclusion of some of the periodical titles which were listed. The feeling of this group was that some publications were probably unworthy of indexing and abstracting, as they were more popular in nature and not really addressed to the scientific community. This may be true, as our source list did not always make such an evaluation. However, good, technical, and often unique information occasionally appears in nontechnical periodicals.

In our attempt to explain our inclusion policy, we tried to build on the cliché that "one man's meat is another man's poison" by describing some of our experiences a few years ago in the automotive safety information center, which is a part of the General Motors Research Library. One of the questions that was asked of the center concerned seat belts for pregnant women. An extensive search in this area located only a few articles which had appeared in the *Ladies Home Journal*. This publication is obviously not one that normally would be scanned and indexed in a technical information center. We discovered that the same situation was true in many areas of automotive safety information.

Some concern was expressed about the fact that only the JICST indexing effort was included in our study. An attempt was made to indicate that our study group had been aware of the NDL effort in periodical literature control, but that they had investigated the material and decided that the broad categorization of subjects used in NDL indexing would preclude in-depth searching of the individual items. A major difference in definition between the West and Japan evolved from this discussion. The Western approach towards retrieval indexing goes to a greater depth and specificity than the broad categorical division used at NDL. This difference probably exists because so much of the Western indexing is coupled with computerization. As was discussed in the chapter on methodology, the only major, publicly available in-depth indexing in Japan is that being performed at JICST.

The discussions with personnel from the JICST operation were important, as a major portion of the survey was based on the JICST indexing effort and the list of journals that were processed by JICST personnel. Although the listing that was used in our survey, supplied by JICST, included some 3,000 periodical titles, budgetary restrictions have caused the JICST indexing effort to actually cover no more than 2,000 titles. Also, some of the titles that comprised the listing were really outside the specific fields that JICST is currently covering.

Most of the JICST input is from the periodical literature. However, there was an indication that JICST felt that the portion of the information covered was probably the core information of the 9,000-plus titles in our survey. JICST does not receive many conference papers and, of those received, only about 10 percent are abstracted. Few conference papers other than foreign are included in the JICST data base. It was learned that the Ministry of Education is compiling a scientific and technical periodical union list which includes both JICST and the National Diet Library holdings.

In addition to JICST services, many companies in Japan have enhanced their coverage of the literature by creating their own computerized systems to access their own company publications as well as foreign data bases of special interest to them, such as INSPEC or SAE papers. However, little evidence of intercompany networking of these company-originated data bases was uncovered.

Contact with governmental agencies such as MITI (Ministry of International Trade and Industry) indicated that there is no government-wide information data base or control system to which the technical staff contributes nor from which they may query information. However, a visit to the Institute for the Dissemination and Research of Government Data was suggested. This institute is a nonprofit organization, which is recognized by the Office of the Prime Minister but is not a governmentally supported agency. They do attempt to collect a variety of government publications. It was started in 1971 and has been built on direct contacts by institute staff members with individuals they may know in various governmental offices. The collection ranges from mimeographed budget proposals to annual reports. Although some scientific publications are included, it is not at all exhaustive. Unfortunately, little subject searching is possible although the material is broadly classified. Very little awareness of the total amount of report literature was visible in Japan, and nothing was available similar to the United States NTIS organization.

Most of the government documents are in the public domain with the exception of overseas trade policy statements and internal scientific and technical reports. Governmentally sponsored contract report literature is sizable in the science and technology field; however, up to 50 percent is classified and accordingly not generally available.

In discussions with Japanese scientific and technical personnel, as well as with librarians and information people, variations of three questions were posed. (1) In their research did they rely on Western or Japanese sources of information? (2) Were they "satisfied" with the information available? (3) Given the low priority of Japanese effort expended on indexing, did they think that in-depth indexing was important to accessing the information that they required? As could be expected, the answers were extremely varied. On the first question, the answer showed a 30 percent to 70 percent reliance on Western sources. The overall aggregate of responses indicated that although there was relatively little in-depth indexing in Japan, those interviewed were grateful that the West was making indexing tools available and that there undoubtedly should be more coverage in Japan of domestic information. When individually pressed, these individuals did agree that if the "invisible college" let them down it was more than likely that important information would be missed.

The exchange of information with Japanese technical individuals would indicate that rather than rely on a library to provide necessary information, Japanese researchers probably use the library as only one node of several which the individual engineers and scientists would query when trying to locate needed information. Among researchers there does exist an "invisible college," but its effectiveness only extends to the point of public competitiveness. Most of the technical people with whom contact was made indicated that it would be beneficial to the total research effort if the "invisible college" were replaced or supported by a more formal information exchange.

Information centers, organizations for acquiring, processing, disseminating, and evaluating technical information, rarely exist in Japan except at the largest companies such as Mitsubishi, Toshiba, Hitachi, Toyota, and Nissan, where centers are possible due to larger staff involvements with documentation effort. There appears to be very little information flow from any one of these centers to another due to competitive proprietary restrictions.

There is some exchange of information within industries through the industrial trade associations such as the Japanese Society of Automotive Engineers and the Pharmaceutical Society of Japan. The exchange of information is especially enhanced when there are strong chairmen of association committees.

In addition to visiting governmental agencies and corporations, part of the time spent in Japan was involved with seminars presented at some of the American Information Centers in Kyoto, Osaka, and Nagoya. Over 350 librarians from public, university, and special libraries discussed similarities and dissimilarities between American and Japanese approaches to librarianship. There appears to be a difference in priority in the libraries of Japan and the Western world, with the Western libraries emphasizing service and the Japanese emphasizing collection building. This difference in approach may have a bearing on the relative lack of importance placed on Japanese indexing and abstracting efforts.

In the individual contacts that were made with various librarians throughout Japan, the importance of more attention being given in Japan to abstracting and indexing of the scientific and technical literature, rather than spending huge additional sums of money and manpower to increase the number of Japanese articles published in English, was stressed. Several times statements were heard that more of the Western world should learn Japanese, but since this move did not seem to be occurring, the Japanese felt compelled to increase English language publications. Furthermore, an attempt was made to express the opinion that if a wider, more complete coverage of the total Japanese publishing effort could be included in additional indexing services, greater overall worldwide use would be made of Japanese scientific and technical periodicals. More importantly, the increased bibliographic control of Japanese scientific and technical information would aid greatly in improving the intra-Japanese exchange of information. Access to actual documents retrieved from this additional indexing is difficult, since interlibrary lending in Japan is not well developed. While it is quite true that many of our scientists are frightened off by non-English articles, it should be suggested that if there is knowledge of potentially important information, users will request translation of either the whole or pertinent portions of the material.

The visits verified that most of the information in the report was quite correct. Those interviewed expressed surprise that foreigners would make such an intensive survey of Japanese scientific and technical output, especially as the Japanese had not made such a study of their own publishing. Although a vast amount of scientific and technical literature is being published in Japan, there has not been, with few exceptions, a comparable effort, either internally or externally, toward providing bibliographical access to it.

It was interesting that the November 1978 issue of the *National Diet Library Newsletter*, in a report of the Council on the NDL Collection of Scientific and Technological Materials, noted that "only a part of scientific research information of Japan is provided through international information services, and a larger part of it still remains unknown abroad. In order to improve this situation, NDL should investigate and work out an innovative measure in collaboration with the institutions concerned."

As we move toward a global village (as Marshall McLuhan called it), it is vital that librarians and information specialists in the United States and Japan develop more and better communication channels for the effective exchange of information.

chapter six

CONCLUSIONS

The complexity and magnitude of the Japanese scientific and technical publishing effort are such that a definitive study is nearly impossible. However, some salient points did emerge during the course of this study, and this, combined with some additional information, was sufficient to draw the conclusions which follow.

(1) Japan exhibits strong vertical organization—not only in its hierarchy but in its social structure as well. Since World War II, Japan has placed extremely strong emphasis on a high priority for economic survival. Continued adherence to this channeling of the country's money and effort has left few resources for areas of lower priority, such as bibliographic control of periodical literature.

(2) The Japanese have developed a comprehensive national plan whose purpose is to ensure a smooth flow of scientific and technical information, thus creating a solid base for Japanese activities in the area of science and technology.

(3) The action agency for this national plan is the Japan Information Center of Science and Technology (JICST). JICST's mission is to collect, process, store, and disseminate scientific and technical information. JICST is currently being upgraded to implement the national plan, which will result in an integrated Japanese information network.

(4) The volume of Japanese scientific and technical publications (including monographs, periodicals, and technical reports) shows that the information being produced and disseminated in Japan is of equal magnitude to that being published and disseminated by other countries where science and technology play an important role, but identification of, and access to, this material is often difficult.

(5) If one accepts the hypothesis that Japanese scientific and technical periodical literature is virtually inaccessible to Western researchers and scientists unless it is indexed or abstracted by a Western indexing and abstracting service, and applies the hypothesis to the results of the study, one arrives at some rather startling conclusions:

(a) Less than 20 percent of the Japanese scientific and technical periodical literature is indexed by the Western indexes and abstracts;

(b) About 25 percent of the Japanese scientific and technical periodical literature is written fully or partially in Western languages, and only about one-half of that literature is accessible to Western researchers and scientists through the indexes and abstracts;

(c) The majority (75 percent) of Japanese scientific and technical periodical literature is written in Japanese, and of that literature, only 7 percent is accessible to Western researchers and scientists through the indexes and abstracts;

(d) The larger portion (44.7 percent) of the scientific and technical periodical literature is published in the applied sciences, with 19.9 percent of the periodicals being published in the agricultural sciences, 18.3 percent in the medical sciences, and 16.9 percent in the natural sciences. In the applied sciences, only 14 percent of that literature is accessible to Western research scientists through the indexes and abstracts.

(6) Informal channels of communication seem to be used extensively among researchers within Japan. The amount of information which is communicated between Japanese researchers and Western researchers is unknown, since little research seems to have been done in this area.

(7) The availability of Japanese periodicals in this country appears to be meager for most scientific and technical disciplines. For example, during the course of the study, a list of seventy-two titles in the subject area of land vehicles was investigated for bibliographic control. Of the seventy-two titles, twenty-four (or 35 percent) were indexed by either Western or Japanese sources. Twelve percent of the titles contained articles or summaries in Western languages, while 88 percent were in Japanese. It is interesting that the statistics for the subsection on land vehicles followed the percentages for the total titles surveyed. It is extremely important to mention that of the seventy-two titles, the study group was able to locate only four in United States libraries. A further investigation, unfortunately, revealed that the true holdings of these four are in no way complete—a most distressing revelation.

(8) There does not seem to be a concerted effort to hide Japanese developments. Much has been written which was obvious from the monograph statistics as well as the large number of scientific and technical periodicals located by the study.

(9) The few Japanese industrial concerns that were visited give the indication that although each is concerned in varying degrees about technical literature, each has gone its own way toward attempting to solve the information problem. Some of the larger industrial concerns are beginning to build in-house data banks.

(10) The first results of technical research sponsored by private industry or the government are generally written as reports which are not generally available.

(11) The low level of Japanese bibliographical control of the scientific and technical literature is probably caused, to some degree, by the low level of interlibrary loans.

(12) It appears, from this study, that Western researchers and scientists are severely handicapped in accessing Japanese scientific and technical literature. Most of the information flow between the Western world and Japan has been one way (to the Japanese), undoubtedly caused in part by the great language barrier. As a result, Japanese specialists in various subject disciplines are currently far more aware of worldwide developments than are Western scientists of Japanese developments. This handicap must somehow be overcome if Western researchers are to cooperate with the Japanese researchers in areas of technological advancement.

The results of this study indicate directions that might be taken to ensure a greater degree of utilization of Japanese scientific and technical results by Western researchers. Some of these might include:

(1) Governmentally supported efforts to survey published Japanese scientific and technical information similar to that inaugurated for Russian literature after the advent of Sputnik.

(2) Encouragement in implementation of cooperative programs between Western indexing and abstracting services and JICST.

(3) Increased dialogues between Western and Japanese librarians, similar to those reported in the proceedings of the Japan-United States Conferences on Libraries and Information Science in Higher Education, by emphasizing especially the need for greater Japanese emphasis on bibliographical control as well as more effort towards indexing and abstracting journal articles and reports.

(4) A concerted effort by the major Western academic and corporate libraries to improve their holdings of Japanese scientific and technical periodicals.

In the preindustrial world of mathematician Seki Takakuzu's time, Europe's ignorance of his mathematical achievements was of no great consequence. In today's highly competitive world, however, we can ill afford to ignore advanced scientific and technical research being done anywhere.

This study does not attempt to be the final word on Japanese scientific and technical literature. Many questions have been left unanswered, but perhaps it will provide the impetus for further research which is necessary before a true picture of the quality and quantity of scientific and technical literature produced in Japan can be drawn.

Part 2

SUBJECT GUIDE TO JAPANESE SCIENTIFIC AND TECHNICAL JOURNALS

INTRODUCTION

Introduction

This subject guide covers 9,116 titles in science and technology published in Japan. The comprehensive listing was compiled using the *Directory of Japanese Scientific Periodicals*, published by the National Diet Library in 1974; *Wazasshi Mokuroku 1975* (Catalogue of Japanese Periodicals), published by the National Diet Library in 1976; *Wazasshi Mokuroku Tsuiroku, January 1976-December 1977* (Catalogue of Japanese Periodicals: Supplement), published by the National Diet Library in 1978; *Ulrich's International Periodical Directory*, 17th edition, 1977-1978; Catalogue *of Japanese Periodicals '78*, published by Maruzen Company, Ltd., *Current Japanese Periodicals for '78*, published by the Japan Publications Trading Company, Ltd.; *Shuppan Nenkan 1978* (publishing Yearbook 1978); *Chikuji Kankobutsu Shozo Mokuroku*, September 30, 1976 (index to Periodical Holdings, as of September 30, 1976), published by the Japan Information Center of Science and Technology in 1976; and from various Western indexing and abstracting services which include Japanese periodicals in their coverage.

A modified Universal Decimal Classification (UDC) scheme, which is used by the National Diet Library, was employed to categorize each title by subject area. This breakdown for the areas of science and technology was shown as figure 3-3, chapter 3. In this system, the natural sciences correspond to the pure sciences used in Western classification schemes, and the applied sciences correspond to the area known as technology in Western schemes.

Arrangement of Entries

The titles in this directory are categorized by the modified Universal Decimal Classification scheme, and under each classification the titles are arranged alphabetically. Articles, prepositions, and conjunctions in English titles are not used in the arrangement either in the main part or in the index.

Description of Entry

Each entry gives the information (if available) in the following order:

1. Romanized Japanese title.
2. Title in English.
   The title without parentheses indicates the authorized English title appearing in the journal itself. The English title if in parentheses is not given in the journal but is a tentative one given by its publisher or by the translators.
3. Publisher and/or editing body in romanized Japanese.

4. English translation of publisher and/or editing body.
5. Address of the publisher.
6. Language of publication.
7. Codes used to indicate abstracting and indexing services which index each title:

ABBREVIATION	*TITLE*
API	API Abstracts
APP	Applied Mechanics Reviews
APT	APTIC
AUT	Automobile Abstracts
BIB	Bibliography of Agriculture
BIO	Biological Abstracts
CER	Ceramic Abstracts
CHE	Chemical Abstracts
COP	Copper Extracts
ENG	Engineering Index
EXC	Excerpta Medica
IND	Index Medicus
INS	INSPEC (includes Computer & Control Abstracts, Electrical & Electronic Abstracts, and Physics Abstracts)
IAA	International Aerospace Abstracts
ISM	ISMEC Bulletin
JICST	Japan Information Center of Science and Technology
MAT	Mathematical Reviews
MET	Metals Abstracts
POL	Pollution Abstracts
RAP	RAPRA Abstracts
SCI	Science Citation Index
WAA	World Aluminum Abstracts

NATURAL SCIENCES

Natural Sciences in General

ACADEMIA 1
ZENKOKU NIHON GAKUSHI KAI
JAPANESE

AICHI KYOIKU DAIGAKU KENKYU HOKOKU, SHIZENKAGAKU 2
(BULLETIN OF AICHI UNIVERSITY OF EDUCATION, NATURAL SCIENCE)
AICHI KYOIKU DAIGAKU (AICHI UNIVERSITY OF EDUCATION)
1, HIROSAWA, IGAYA-CHO, KARIYA-SHI, 448
JAPANESE AND ENGLISH
INDEXED: CHE, JICST

AICHI SHUKUTOKU DAIGAKU RONSHU 3
(BULLETIN OF AICHI SHUKUTOKU COLLEGE)
AICHI SHUKUTOKU DAIGAKU (AICHI SHUKUTOKU COLLEGE)
JAPANESE

AICHI-KEN KAGAKU KYOIKU SENTA KENKYU HOKOKUSHO 4
(RESEARCH REPORT OF THE SCIENCE EDUCATION CENTER, AICHI PREFECTURE)
AICHI-KEN KAGAKU KYOIKU SENTA
1-8, SHIRAKABE-MACHI, HIGASHI-KU, NAGOYA-SHI 461
JAPANESE

AICHI-KEN KAGAKU KYOIKU SENTA KENKYU KIYO 5
(BULLETIN OF THE SCIENCE EDUCATION CENTER, AICHI PREFECTURE)
AICHI-KEN KAGAKU KYOIKU SENTA
1-8, SHIRAKABE-MACHI, HIGASHI-KU, NAGOYA-SHI 461
JAPANESE

AIDEA 6
(IDEA)
JAPANESE
INDEXED: JICST

AIZU TANKI DAIGAKU GAKUHO - SHIZEN KAGAKU-HEN 7
(GAZETTE OF AIZU JUNIOR COLLEGE - NATURAL SCIENCE)
FUKUSHIMA-KENRITSU AIZU TANKI DAIGAKU
(FUKUSHIMA PREFECTURAL AIZU JUNIOR COLLEGE)
JAPANESE

AJIA DAIGAKU KYOYOBU KIYO 8
(BULLETIN OF THE ASIA UNIVERSITY, GENERAL EDUCATION DEPARTMENT)
AJIA DIAGAKU (ASIA UNIVERSITY)
JAPANESE

AKADEMIA JINBUN-SHIZENKAGAKU-HEN, HOKEN TAIIKU-HEN NANZAN DAIGAKU 9
(ACADEMIA; JOURNAL OF THE NANZAN ACADEMIC SOCIETY - HUMANITIES, NATURAL SCIENCES, PHYSICAL EDUCATION)
JAPANESE

AKITA DAIGAKU KYOIKU GAKUBU KENKYU KIYO, SHIZENKAGAKU 10
(MEMOIRS OF THE FACULTY OF EDUCATION, AKITA UNIVERSITY, NATURAL SCIENCE)
AKITA DAIGAKU KYOIKUGAKUBU KENKYU KIYO HENSHU IINKAI (MEMOIRS EDITORIAL COMMITTEE OF THE FACULTY OF EDUCATION, AKITA UNIVERSITY)
TEGATA GAKUEN-MACHI, AKITA-SHI 010
JAPANESE AND ENGLISH
INDEXED: CHE, JICST

AKIYOSHI-DAI KAGAKU HAKUBUTSUKAN HOKOKU 11
(BULLETIN OF THE AKIYOSHI-DAI SCIENCE MUSEUM)
AKIYOSHI-DAI KAGAKU HAKUBUTSUKAN (AKIYOSHI-DAI SCIENCE MUSEUM)
AKIYOSHI-DAI, AKIYOSHI-CHO, MIYA-GUN, YAMAGUCHI 754-05
JAPANESE AND ENGLISH

ANATA NO KAGAKU GIJUTSU 12
(YOUR SCIENCE AND TECHNOLOGY)
KAGAKU GIJUTSUCHOHEN NIHON KAGAKU GIJUTSU SHINKO AIDAN (JAPAN FOUNDATION FOR THE PROMOTION OF SCIENCE AND TECHNOLOGY)
JAPANESE

ANNALS OF THE JAPAN ASSOCIATION FOR PHILOSOPHY OF SCIENCE 13
KAGAKU KISORON GAKKAI (JAPAN ASSOCIATION FOR PHILOSOPHY OF SCIENCE)
C/O TOKEI SURI KENKYUSHO, 6-7, 4-CHOME, MINAMIAZABU, MINATO-KU, TOKYO 106
ENGLISH
INDEXED: BIO, MAT

ANNUAL REPORT OF SCIENCE COUNCIL OF JAPAN 14
NIHON GAKUJUTSU KAIGI (SCIENCE COUNCIL OF JAPAN)
22-34, 7-CHOME, ROPPONGI, MINATO-KU, TOKYO 106
ENGLISH

APUROCHI 15
(APPROACH)
JAPANESE
INDEXED: JICST

ASAHIKAWA DAIGAKU KIYO 16
(JOURNAL OF THE ASAHIKAWA UNIVERSITY)
KITA NIHON GAKUIN DAIGAKU (KITA NIHON GAKUIN UNIVERSITY)
JAPANESE

ASHIYA DAIGAKU RONSO 17
(ASHIYA UNIVERSITY REVIEW)
ASHIYA DAIGAKU (ASHIYA UNIVERSITY)
JAPANESE

BEPPU DAIGAKU KIYO 18
(BULLETIN OF BEPPU UNIVERSITY)
JAPANESE

BOEI DAIGAKKO KYOKAN KENKYU YOROKU 19
(DIGEST OF RESEARCH BY FACULTY MEMBERS, DEFENSE ACADEMY)
BOEI DAIGAKKO (DEFENSE ACADEMY)
10-20, 1-CHOME, HASHIRIMIZU, YOKOSUKA-SHI 239
JAPANESE

BOEI DAIGAKKO RIKOGAKU KENKYU HOKOKU 20
(REPORT OF THE DEFENSE ACADEMY SCIENTIFIC AND ENGINEERING)
BOEI DAIGAKKO (DEFENSE ACADEMY)
10-20, 1-CHOME, HASHIRIMIZU, YOKOSUKA-SHI 239
JAPANESE, WITH EUROPEAN SUMMARY
INDEXED: JICST

BULLETIN OF THE FACULTY OF TECHNOLOGY, KANAZAWA UNIVERSITY 21
KANAZAWA UNIVERSITY, FACULTY OF TECHNOLOGY
1-1 MARUNOUCHI, KANAZAWA
JAPANESE, WITH ENGLISH SUMMARY

BULLETIN OF THE KYUSHU INSTITUTE OF TECHNOLOGY, MATHEMATICS, NATURAL SCIENCE 22
KYUSHU KOGYO DAIGAKU (KYUSHU INSTITUTE OF TECHNOLOGY)
SENSUI-CHO, TOBATA-KU, KITAKYUSHU-SHI 804
ENGLISH
INDEXED: CHE, INS, MAT, JICST

BUREN 23
(BRAIN)
JAPANESE
INDEXED: JICST

CHIBA DAIGAKU BUARI GAKUBU KIYO. SHIZEN KAGAKU 24
(BULLETIN OF CHIDA UNIVERSITY, FACULTY OF ARTS AND SCIENCES. NATURAL SCIENCES)
JAPANESE
INDEXED: JICST

CHIBA DAIGAKU GAKUHO 25
(REPORT OF THE CHIBA UNIVERSITY)
JAPANESE

CHIBA DAIGAKU KYOYOBU KENKYU HOKOKU, B 26
(JOURNAL OF THE COLLEGE OF ARTS AND SCIENCES)
CHIBA DAIGAKU KYOYOBU
1-33, YAYOI-CHO, CHIBA-SHI 280
JAPANESE AND ENGLISH

CHIBA-KENRITSU KAZUSA HAKUBUTSUKANPO 27
(REPORT OF CHIBA PREFECTURAL KAZUSA MUSEUM)
CHIBA-KENRITSU KAZUSA HAKUBUTSUKAN
352-3, MORISAKI, OTA, KISARAZU-SHI 292
JAPANESE

CHICHIBU SHIZENKAGAKU HAKUBUTSUKAN KENKYU HOKOKU 28
(BULLETIN OF THE CHICHIBU MUSEUM OF NATURAL HISTORY)
CHICHIBU SHIZENKAGAKU HAKUBUTSUKAN (CHICHIBU MUSEUM OF NATURAL HISTORY)
1417, NAGATORO, NAGATORO-MACHI, CHICHIBU-GUN, SAITAMA 369-13
JAPANESE

CHIIKI TO KAGAKUSHA 29
(LOCALITIES AND SCIENTISTS)
NIHON KAGAKUSHA KAIGI HIROSHIMA-KEN SHIBU
JAPANESE

CHOSEN GAKUJUTSU TSUHO 30
(KOREAN SCIENCE INFORMATION)
ZAI NIHON CHOSENJIN KAGAKUSHA KYOKAI (ASSOCIATION OF KOREAN SCIENTISTS IN JAPAN)
C/O CHOSEN SYUPPAN KAIKAN, 33-14, 4-CHOME, HAKUSAN, BUNKYO-KU, TOKYO 112
JAPANESE
INDEXED: JICST

CHOSEN SHOGAKKAI GAKUJUTSU RONBUNSHU 31
(SCIENCE REPORT OF KOREAN SCHOLARSHIP SOCIETY)
CHOSEN SHOGAKKAI (KOREAN ASSOCIATION OF SCHOLARSHIP)
8-1, 1-CHOME, NISHI SIINJUKU, SHINJUKU-KU, TOKYO 160
JAPANESE AND ENGLISH

CHUKYO DAIGAKU KYOYO RONSO 32
(CHUKYO UNIVERSITY BULLETIN)
CHUKYO DAIGAKU GAKUJUTSU KENKYUKAI (THE CHUKYO UNIVERSITY ACADEMIC RESEARCH SOCIETY
JAPANESE

CHUKYO JOSHI DAIGAKU KIYO 33
(MEMOIRS OF CHUKYO WOMEN'S UNIVERSITY)
CHUKYO JOSHI DAIGAKU (CHUKYO WOMEN'S UNIVERSITY)
YOKONE-MACHI, OSAKA-SHI 474
JAPANESE

CHUO DAIGAKU DAIGAKUIN KENKYU NENPO 34
(ANNUAL RESEARCH REPORT OF THE CHUO UNIVERSITY GRADUATE SCHOOL)
CHUO UNIVERSITY GRADUATE SCHOOL
JAPANESE

CHUO DAIGAKU GAKUHO 35
(REPORT OF THE CHUO UNIVERSITY)
CHUO UNIVERSITY
JAPANESE

CHUO DAIGAKU RIKOGAKUBU KIYO 36
(BULLETIN OF THE FACULTY OF SCIENCE AND TECHNOLOGY, CHUO UNIVERSITY)
CHUO DAIGAKU RIKOGAKUBU (FACULTY OF SCIENCE AND TECHNOLOGY, CHUO UNIVERSITY)
13-27, 1-CHOME, KASUGA, BUNKYO-KU, TOKYO 112
JAPANESE, WITH ENGLISH SUMMARY
INDEXED: CHE, JICST

CHUO GAKUIN DAIGAKU RONSO 37
(REVIEW OF THE CHUO GAKUIN UNIVERSITY)
CHUO GAKUIN UNIVERSITY
JAPANESE

CHUO GAKUJUTSU KENKYUJO KIYO 38
(BULLETIN OF THE CHUO ACADEMIC RESEARCH INSTITUTE)
CHUO ACADEMIC RESEARCH INSTITUTE
JAPANESE

DAITO BUNKA DAIGAKU KIYO SHAKAI SHIZEN KAGAKU 39
(BULLETIN OF DAITO BUNKA UNIVERSITY, SOCIAL AND NATURAL SCIENCES)
JAPANESE

DOHO DAIGAKU RONSO 40
(DOHO UNIVERSITY REVIEW)
DOHO GAKKAI (DOHO LEARNED SOCIETY)
JAPANESE

DOKKYO DAIGAKU KYOYO SHOGAKU KENKYU 41
(DOKKYO UNIVERSITY GENERAL EDUCATION RESEARCH)
DOKKYO DAIGAKU GAKUJUTSU KENKYUKAI (DOKKYO UNIVERSITY ACADEMIC RESEARCH ASSOCIATION)
JAPANESE

EHIME DAIGAKU KIYO, DAI-2-BU, SHIZENKAGAKU, A-SHIRIZU, SUGAKU, BUTSURIGAKU 42
(MEMOIRS OF THE EHIME UNIVERSITY, SECTION 2, NATURAL SCIENCE, SERIES A, MATHEMATICS, PHYSICS)
EHIME DAIGAKU (EHIME UNIVERSITY)
SANBAN, BUNKYO-CHO, MATSUYAMA-SHI 790
JAPANESE AND EUROPEAN
INDEXED: MAT, JICST

EHIME DAIGAKU KYOYOBU KIYO 43
(BULLETIN OF EHIME UNIVERSITY, DEPARTMENT OF LIBERAL ARTS)
EHIME DAIGAKU (EHIME UNIVERSITY)
JAPANESE

FERISU JOGAKUIN DAIGAKU KIYO 44
(BULLETIN OF FERISU JOGAKUIN UNIVERSITY)
JAPANESE

FUJI DAIGAKU KIYO 45
(REVIEW OF FUJI COLLEGE)
FUJI DAIGAKU GAKUJUTSU KENKYUKAI (FUJI UNIVERSITY, ARTS AND SCIENCE RESEARCH ASSOCIATION)
JAPANESE

FUJI JOSHI DAIGAKU, FUJI JOSHI TANKI DAIGAKU KIYO 46
(RESEARCH BULLETIN OF FUJI WOMEN'S COLLEGE, FUJI WOMEN'S JUNIOR COLLEGE)
FUJI JOSHI DAIGAKU, FUJI JOSHI TANKI DAIGAKU (FUJI WOMEN'S COLLEGE, FUJI JUNIOR WOMEN'S COLLEGE)
NISHI-2-CHOME, KITA-16-JO, KITA-KU, SAPPORO 065
JAPANESE AND EUROPEAN
INDEXED: BIO

FUJI RONSO 47
(FUJI RESEARCH REPORTS)
FUJI TANKI DAIGAKU GAKUJUTSU KENKYUKAI (FUJI JUNIOR COLLEGE SCHOLARLY RESEARCH SOCIETY)
JAPANESE

FUKUI DAIGAKU KYOIKU GAKUBU KIYO, DAI-2-BU, SHIZENKAGAKU 48
(MEMOIRS OF THE FACULTY OF EDUCATION, FUKUI UNIVERSITY, SERIES 2, NATURAL SCIENCE)
FUKUI DAIGAKU KYOIKU GAKUBU (FACULTY OF EDUCATION, FUKUI UNIVERSITY)
9-1, 3-CHOME, BUNKYO, FUKUI-SHI 910
ENGLISH

FUKUOKA DAIGAKU KENKYUJOHO 49
(BULLETIN OF THE FUKUOKA UNIVERSITY RESEARCH INSTITUTE)
JAPANESE

FUKUOKA DAIGAKU KENKYUJOHO BESSATSU 50
(BULLETIN OF THE FUKUOKA UNIVERSITY RESEARCH INSTITUTE: SPECIAL ISSUE)
FUKUOKA DAIGAKU KENKYUJO
JAPANESE

FUKUOKA DAIGAKU RIGAKU SHUHO 51
(FUKUOKA UNIVERSITY SCIENCE REPORTS)
FUKUOKA DAIGAKU KENKYUJO (FUKUOKA UNIVERSITY RESEARCH INSTITUTE)
JAPANESE

FUKUOKA KYOIKU DAIGAKU KIYO, DAI-3-BUNSATSU, RIKA-HEN 52
(BULLETIN OF FUKUOKA UNIVERSITY OF EDUCATION, PART 3, NATURAL SCIENCE)
FUKUOKA KYOIKU DAIGAKU (FUKUOKA UNIVERSITY OF EDUCATION)
729, AKAMA, MUNAKATA-MACHI, MUNAKATA-GUN, FUKUOKA 811-41
JAPANESE AND ENGLISH
INDEXED: BIO, CHE, MAT, JICST

FUKUSHI DAIGAKU HYORON 53
(FUKUSHI UNIVERSITY REVIEW)
FUKUSHI DAIGAKU HYORON HENSHU IINKAN (THE REVIEW EDITING COMMITTEE, FUKUSHI UNIVERSITY)
JAPANESE

FUKUSHIMA DAIGAKU KYOIKUGAKUBU RIKA HOKOKU 54
(SCIENCE REPORTS OF THE FACULTY OF EDUCATION, FUKUSHIMA UNIVERSITY)
FUKUSHIMA DAIGAKU KYOIKUGAKUBU (FACULTY OF EDUCATION, FUKUSHIMA UNIVERSITY)
HAMADA-CHO, FUKUSHIMA-SHI 960
JAPANESE AND ENGLISH
INDEXED: CHE, JICST

FUKUYAMA DAIGAKU KYOYOBU KIYO 55
(BULLETIN OF FUKUYAMA UNIVERSITY, DEPARTMENT OF LIBERAL ARTS)
FUKUYAMA DAIGAKU KYOYOBU (FUKUYAMA UNIVERSITY, DEPARTMENT OF ARTS AND SCIENCES)
JAPANESE

GAKUJUTSU GEPPO 56
(JAPANESE SCIENTIFIC MONTHLY)
NIHON GAKUJUTSU SHINKOKAI (JAPAN SOCIETY FOR THE PROMOTION OF SCIENCE)
3-1, 5-CHOME, KOJI-MACHI, CHIYODA-KU, TOKYO 102
JAPANESE
INDEXED: JICST

GENDAI KAGAKU RONSO 57
(REVIEW OF CONTEMPORARY SCIENCE)
HAKUTEISHA
JAPANESE

GIFU DAIGAKU KYOIKUGAKUBU KENKYU HOKOKU, SHIZENKAGAKU 58
(SCIENCE REPORT OF THE FACULTY OF EDUCATION, GIFU UNIVERSITY, NATURAL SCIENCE)
GIFU DAIGAKU KYOIKUGAKUBU (FACULTY OF EDUCATION, GIFU UNIVERSITY)
NAGARA, GIFU-SHI 500
JAPANESE AND ENGLISH
INDEXED: BIB, CHE, MAT, JICST

GIFU DAIGAKU KYOYOBU KENKYU HOKOKU 59
(BULLETIN OF THE FACULTY OF GENERAL EDUCATION, GIFU UNIVERSITY)
GIFU DAIGAKU KYOYOBU (FACULTY OF GENERAL EDUCATION, GIFU UNIVERSITY)
1466, NAGARA JONOUCHI, GIFU-SHI 500
JAPANESE AND EUROPEAN

GUNMA DAIGAKU KYOIKUGAKUBU KIYO, SHIZENKAGAKU-HEN 60
(SCIENCE REPORT OF THE GUNMA UNIVERSITY, NATURAL SCIENCE)
GUNMA DAIGAKU KYOIKUGAKUBU (FACULTY OF EDUCATION, GUNMA UNIVERSITY)
1375, ARAMAKI-CHO, MAEBASHI-SHI 371
JAPANESE AND EUROPEAN
INDEXED: BIO, CHE, JICST

GUNMA DAIGAKU KYOYOBU KIYO 61
(GUNMA JOURNAL OF LIBERAL ARTS AND SCIENCE)
GUNMA DAIGAKU KYOYOBU (FACULTY OF GENERAL STUDY, GUNMA UNIVERSITY)
1375, ARAMAKI-CHO, MAEBASHI-SHI 371
JAPANESE AND ENGLISH
INDEXED: CHE, JICST

HAKUBUTSUKAN KENKYU 62
(MUSEUM STUDIES)
NIHON HAKUBUTSUKAN KYOKAI (JAPANESE ASSOCIATION OF MUSEUMS)
C/O URAGAMI-TENSHUDO DAIICHI BLDG., 10-1, 1-CHOME, NIHONBASHI KAYABA-CHO, CHUO-KU, TOKYO 103
JAPANESE

HANAZONO DAIGAKU KENKYU KIYO 63
(RESEARCH BULLETIN OF HANAZONO UNIVERSITY)
JAPANESE

HANNAN RONSHU 64
(PAPERS OF HANNAN UNIVERSITY)
HANNAN DAIGAKU (HANNAN UNIVERSITY)
JAPANESE

HIGASHI NIHON GAKUEN DAIGAKU KYOYOBU RONSHU 65
(PAPERS OF FACULTY OF GENERAL EDUCATION HIGASHI NIHON GAKUEN UNIVERSITY)
JAPANESE

HIGASHIYAMA GAKUEN KENKYU KIYO 66
(RESEARCH BULLETIN OF HIGASHIYAMA GAKUEN)
HIGASHIYAMA GAKUEN GAKUGEI KENKYUJO
JAPANESE

HIROSAKI DAIGAKU KYOIKUGAKUBU KIYO 67
(BULLETIN OF THE FACULTY OF EDUCATION, HIROSAKI UNIVERSITY)
HIROSAKI DAIGAKU KYOIKUGAKUBU (FACULTY OF EDUCATION HIROSAKI UNIVERSITY)
1, BUNKYO-CHO, HIROSAKI-SHI 036
JAPANESE AND ENGLISH

HIROSHIMA DAIGAKU KYOIKU GAKUBU KIYO. 2 68
(BULLETIN OF HIROSHIMA UNIVERSITY, FACULTY OF EDUCATION. 2)
JAPANESE
INDEXED: JICST

HIROSHIMA DAIGAKU KYOIKUGAKUBU KIYO, DAI-3-BU, SHIZENKAGAKU 69
(BULLETIN OF THE FACULTY OF EDUCATION, HIROSHIMA UNIVERSITY, PART 3, SCIENCE AND TECHNICS)
HIROSHIMA DAIGAKU KYOIKUGAKUBU (FACULTY OF EDUCATION HIROSHIMA UNIVERSITY
HIGASHISENDA-CHO, HIROSHIMA-SHI 730
JAPANESE AND ENGLISH
INDEXED: BIO, JICST

HIROSHIMA DAIGAKU KYOYOBU KIYO, III, SHIZENKAGAKU 70
(STUDIES IN NATURAL SCIENCES, MEMOIRS OF THE FACULTY OF GENERAL EDUCATION, HIROSHIMA UNIVERSITY, III)
HIROSHIMA DAIGAKU KYOYOBU (FACULTY OF GENERAL EDUCATION, HIROSHIMA UNIVERSITY)
1-89, 1-CHOME, HIGASHISENDA-CHO, HIROSHIMA-SHI 730
JAPANESE AND ENGLISH

HIROSHIMA JOGAKUIN DAIGAKU RONSHU 71
(ESSAYS AND STUDIES BY THE FACULTY OF HIROSHIMA JOGAKUIN COLLEGE)
HIROSHIMA JOGAKUIN DAIGAKU (HIROSHIMA JOGAKUIN COLLEGE)
13-1, 4-CHOME, USHIDA-HIGASHI, HIROSHIMA-SHI 730
JAPANESE, WITH EUROPEAN SUMMARY
INDEXED: BIO, CHE

HIROSHIMA KEIZAI DAIGAKU KENKYU RONSHU: JINBUN SHAKAI 72
(RESEARCH PAPERS OF HIROSHIMA ECONOMICS UNIVERSITY HUMANITIES SOCIETY, NATURAL SCIENCES)
HIROSHIMA KEIZAI DAIGAKU KEIZAI GAKKAI (ECONOMICS UNIVERSITY, ECONOMICS LEARNED SOCIETY)
JAPANESE

HITOTSUBASHI JOURNAL OF ARTS AND SCIENCES 73
HITOTSUBASHI DAIGAKU HITOTSUBASHI GAKKAI (THE HITOTSUBASHI ACADEMY HITOTSUBASHI UNIVERSITY)
2-1, NAKA, KUNITACHI-SHI, TOKYO 186
ENGLISH
INDEXED: CHE, MAT

HITOTSUBASHI KENKYU 74
(HITOTSUBASHI RESEARCH)
HITOTSUBASHI DAIGAKU DAIGAKUIN KENKYU HENSHU IINKAI (HITOTSUBASHI UNIVERSITY. GRADUATE SCHOOL)
JAPANESE

HITOTSUBASHI RONSO 75
(PAPERS OF HITOTSUBASHI UNIVERSITY)
NIHON HYORONSHA
JAPANESE
INDEXED: JICST

HOKKAI GAKUEN DAIGAKU GAKUEN RONSHU 76
(RESEARCH PAPERS OF HOKKAI GAKUEN UNIVERSITY)
HOKKAI GAKUEN DAIGAKU GAKUJUTSU KENKYUKAI
JAPANESE
INDEXED: JICST

HOKKAIDO KAGAKU KENKYUHI JIYU KADAI NI YORU KENKYU HOKOKUSHO 77
(REPORT OF FREE-THEME RESEARCHES WITH HOKKAIDO SCIENCE RESEARCH FUNDING)
HOKKAIDO (HOKKAIDO PREFECTURAL GOVERNMENT)
NISHI-6-CHOME, KITA-3-JO, CHUO-KU, SAPPOROSHI 060
JAPANESE

HOKKAIDO KAITAKU KINENKAN KENKYU NENPO 78
(ANNUAL REPORT OF THE HISTORICAL MUSEUM OF HOKKAIDO)
HOKKAIDO KAITAKU KINENKAN (HISTORICAL MUSEUM OF HOKKAIDO)
SHIMONOPPORO, ATSUBESHI-MACHI, SHIRAISHI-KU, SAPPORO-SHI 061-01
JAPANESE, WITH ENGLISH SUMMARY

HOKKAIDO KOMAZAWA DAIGAKU KENKYU KIYO 79
(RESEARCH REPORT OF HOKKAIDO KOMAZAWA UNIVERSITY)
KOMAZAWA DAIGAKU HOKKAIDO KYOYOBU (KOMAZAWA UNIVERSITY HOKKAIDO, FACULTY OF ARTS AND SCIENCES)
JAPANESE

HOKKAIDO KYOIKU DAIGAKU KIYO, DAI-2-BU, A, SUGAKU, BUTSURIGAKU, KAGAKU, KOGAKU-HEN 80
(JOURNAL OF HOKKAIDO UNIVERSITY OF EDUCATION, SECTION 2, A, MATHEMATICS, PHYSICS, CHEMISTRY, ENGINEERING)
HOKKAIDO KYOIKU DAIGAKU (HOKKAIDO UNIVERSITY OF EDUCATION)
NISHI-13-CHOME, MINAMI-24-JO, CHUO-KU,SAPPORO-SHI 060
JAPANESE AND EUROPEAN
INDEXED: CHE, MAT, JICST

HOKKAIDO KYOIKU DAIGAKU KIYO, DAI-2-BU, B, SEIBUTSUGAKU, CHIGAKU, NOGAKU-HEN 81
(JOURNAL OF HOKKAIDO UNIVERSITY OF EDUCATION, SECTION 2, B, BIOLOGY, GEOLOGY, AGRICULTURE SCIENCE)
HOKKAIDO KYOIKU DAIGAKU (HOKKAIDO UNIVERSITY OF EDUCATION)
NISHI-13 CHOME, MINAMI-24-JO, CHUO-KU,SAPPORO-SHI 060
JAPANESE AND EUROPEAN
INDEXED: BIO, CHE, JICST

HOKKAIDO KYOIKU DAIGAKU TAISETSUZAN KAGAKU KENKYUJO HOKOKU 82
(REPORTS OF HOKKAIDO EDUCATION UNIVERSITY AND TAISETSUZAN SCIENCE RESEARCH INSTITUTE)
JAPANESE
INDEXED: JICST

HOSEI DAIGAKU KYOYOBU KIYO 83
(BULLETIN OF HOSEI UNIVERSITY, DEPARTMENT OF LIBERAL ARTS)
JAPANESE

HOZON KAGAKU 84
(SCIENCE FOR CONSERVATION)
TOKYO KOKURITSU BUNKAZAI KENKYUSHO HOZON KAGAKU-BU (DEPARTMENT OF CONSERVATION SCIENCE, TOKYO NATIONAL RESEARCH INSTITUTE OF CULTURAL PROPERTIES)
UENO KOEN, TAITO-KU, TOKYO 110
JAPANESE, WITH ENGLISH SUMMARY

HOZON KAGAKU KENKYUSHITSU KIYO 85
(REPORT OF CONSERVATORY SCIENCE RESEARCH ROOM)
GANKOJI BUKKYO MINZOKU SHIRYO KENKYUJO (GANKOJI BUDDHIST FOLKLORE MATERIALS RESEARCH INSTITUTE)
JAPANESE

IBARAKI DAIGAKU KYOIKU GAKUBU KIYO 86
(BULLETIN OF IBARAKI UNIVERSITY FACULTY OF EDUCATION)
JAPANESE
INDEXED: JICST

IBARAKI DAIGAKU KYOYOBU KIYO 87
(BULLETIN, COLLEGE OF GENERAL EDUCATION, IBARAKI UNIVERSITY)
IBARAKI DAIGAKU KYOYOBU (COLLEGE OF GENERAL EDUCATION, IBARAKI UNIVERSITY)
2127, WATASATO-CHO, MITO-SHI 310
JAPANESE

IBARAKI-KIRISUTOKYO DAIGAKU KIYO 88
(BULLETIN OF THE IBARAKI CHRISTIAN COLLEGE)
JAPANESE

ICHIMURA GAKUEN TANKI DAIGAKU SHIZEN KAGAKU KENKYUKAI KAISHI 89
(JOURNAL OF THE NATURAL SCIENTIFIC SOCIETY OF ICHIMURA GAKUEN JUNIOR COLLEGE)
ICHIMURA GAKUEN TANKI DAIGAKU SHIZEN KAGAKU KENKYUKAI (THE NATURAL SCIENTIFIC SOCIETY OF ICHIMURA GAKUEN JUNIOR COLLEGE
61-1, UCHIKUBO OTSU, INUYAMA-SHI 484
JAPANESE, WITH ENGLISH SUMMARY

ISHIKAWA-KON HAKUSAN SHIZEN HOGO SENTA KENKYU HOKOKU 90
(RESEARCH REPORT OF THE ISHIKAWA PREFECTURE HAKUSAN NATURE CONSERVATION CENTER)
JAPANESE

IWATE DAIGAKU KYOIKUGAKUBU KENKYU NENPO 91
(ANNUAL REPORT OF THE FACULTY OF EDUCATION UNIVERSITY OF IWATE)
IWATE DAIGAKU KYOIKUGAKUBU (COLLEGE OF EDUCATION, UNIVERSITY OF IWATE)
18-33, 3-CHOME, UEDA, MORIOKA-SHI 020
JAPANESE AND ENGLISH
INDEXED: BIO, JICST

IWATE IKA DAIGAKU KYOYOBU KENKYU NENPO 92
(ANNUAL RESEARCH REPORT OF THE IWATE MEDICAL COLLEGE, ARTS AND SCIENCE FACULTY)
IWATE IKA DAIGAKU KYOYOBU (LIBERAL ARTS FACULTY OF THE IWATE MEDICAL COLLEGE)
JAPANESE

JAPANESE PERIODICALS INDEX. NATURAL SCIENCES. PART 1: SCIENCE AND TECHNOLOGY 93
JAPANESE
INDEXED: JICST

JAPANESE STUDIES IN THE HISTORY OF SCIENCE 94
TOKYO KOGYO DAIGAKU, OOKAYAMA, MEGURO-KU, TOKYO 152
ENGLISH
INDEXED: CHE

JOBU DAIGAKU RONSHU 95
(PAPERS OF THE JOBU UNIVERSITY)
JOBU DAIGAKU GAKKAI (LEARNED SOCIETY OF THE JOBU UNIVERSITY)
JAPANESE
INDEXED:

JOURNAL OF THE FACULTY OF SCIENCE, SHINSHU UNIVERSITY 96
SHINSHU DAIGAKU RIGAKUBU (THE FACULTY OF LIBERAL ARTS AND SCIENCE, SHINSHU UNIVERSITY)
1-1, 3-CHOME, ASAHI, MATSUMOTO-SHI 390
JAPANESE
INDEXED: BIO, CHE, MAT, JICST

JOURNAL OF THE FACULTY OF TEXTILE SCIENCE AND TECHNOLOGY, SHINSHU UNIVERSITY, SERIES F, PHYSICS AND MATHEMATICS 97
SHINSHU DAIGAKU SEN'IGAKUBU (THE FACULTY OF TEXTILE SCIENCE AND TECHNOLOGY, SHINSHU UNIVERSITY)
15-1, 3-CHOME, TOKIDA, UEDA-SHI 386
EUROPEAN

JOURNAL OF SCIENCE OF THE HIROSHIMA UNIVERSITY SERIES A, PHYSICS AND CHEMISTRY 98
HIROSHIMA DAIGAKU RIGAKUBU (FACULTY OF SCIENCE, HIROSHIMA UNIVERSITY)
1-89, 1-CHOME, HIGASHISENDA-CHO, HIROSHIMA-SHI
ENGLISH
INDEXED: CHE, INS, JICST

JOURNAL OF THE YOKOHAMA CITY UNIVERSITY. SERIES C: NATURAL SCIENCE 99
JAPANESE
INDEXED: JICST

JUNTENDO DAIGAKU BUNRIGAKU KIYO 100
(ARTS AND SCIENCES BULLETIN OF JUNTENDO UNIVERSITY)
JUNTENDO UNIVERSITY
JAPANESE

KAGAKU (IWANAMI SHOTEN) 101
(SCIENCE)
IWANAMI SHOTEN
5-5, 2-CHOME, HITOTSUBASHI, CHIYODA-KU, TOKYO 101
JAPANESE
INDEXED: CHE, INS, JICST

KAGAKU ASAHI 102
(SCIENTIFIC ASAHI)
ASAHI SHINBUNSHA
2-3, YURAKU-CHO, CHIYODA-KU, TOKYO 100
JAPANESE
INDEXED: CHE, JICST

KAGAKU BUNKEN BESSATSU 103
(SCIENCE BIBLIOGRAPHY SUPPLEMENT)
KAGAKU BUNKENSHA
JAPANESE

KAGAKU BUNKEN TOKUBETSUGO 104
(SCIENCE BIBLIOGRAPHY SPECIAL NUMBER)
KAGAKU BUNKENSHA
JAPANESE

KAGAKU GIJUSTU BUNKEN JOHO 105
(INFORMATION SERVICE (BIBLIOGRAPHIC INFORMATION ON SCIENCE AND TECHNOLOGY))
FUKUOKA-KEN BUNKA KAIKON (FUKUOKA PREFUCTURE CULTURAL HALL)
JAPANESE

KAGAKU GIJUTSU BENKEN SABISU 106
(SCIENCE AND TECHNOLOGY INFORMATION SERVICE)
KOKURITSU KOKKAI TOSHOKAN (NATIONAL DIET LIBRARY)
10-1, 1-CHOME, NAGATA-CHO, CHIYODA-KU, TOKYO 100
JAPANESE
INDEXED: JICST

KAGAKU GIJUTSU BUNKEN TOYAMA 107
(TOYAMA SCIENCE AND TECHNICAL DOCUMENTS)
KAGAKU GIJUTSU BUNKEN RIYO SHINKOKAI (THE SOCIETY FOR SCIENCE AND TECHNICAL DOCUMENTS UTILIZATION PROMOTION)
C/O TOYAMA-KENRITSU TOSHOKAN, 206-3, CHAYA-MACHI, TOYAMA-SHI 930-01
JAPANESE

KAGAKU GIJUTSU CHOSA 108
(PROGRESS IN SCIENCE AND TECHNOLOGY)
KAGAKU GIJUTSU-CHO KEIKAKU-KYOKU (PLANNING BUREAU, SCIENCE AND TECHNOLOGY AGENCY)
2-1,2-CHOME, KASUMIGASEKI, CHIYODA-KU, TOKYO 100
JAPANESE
INDEXED: JICST

KAGAKU GIJUTSU KANKEI ZASSHI KIJI SAKUIN 109
(PERIODICALS INDEX OF SCIENCE AND INDUSTRY)
FUKUOKA-KEN BUNKA KAIKAN
2-1, 5-CHOME, TENJIN, CHUO-KU, FUKUOKA-SHI 810
JAPANESE

KAGAKU GIJUTSU KENKYU CHOSA HOKOKU 110
(REPORT ON THE SURVEY OF RESEARCH AND DEVELOPMENT IN JAPAN)
SORI-FU TOKEI-KYOKU (BUREAU OF STATISTICS PRIME MINISTER'S OFFICE) NIPPON TOKEI KYOKAI (JAPAN STATISTICAL ASSOCIATION)
95, WAKAMATSU-CHO, SHINJUKU-KU, TOKYO 162
JAPANESE
INDEXED: JICST

KAGAKU GIJUTSU-CHO GEPPO 111
(BULLETIN OF THE SCIENCE AND TECHNOLOGY AGENCY)
KAGAKU GIJUTSU-CHO (THE SCIENCE AND TECHNOLOGY AGENCY)
2, AKASAKA AOI-CHO, MINATO-KU, TOKYO 107
JAPANESE
INDEXED: JICST

KAGAKU GIJUTSU-CHO MUKI ZAISHITSU KENKYUJO KENKYU HOKOKUSHO 112
(RESEARCH REPORTS OF SCIENCE AND TECHNOLOGY AGENCY, NATIONAL INSTITUTE IN RESEARCHES IN INORGANIC MATERIALS)
JAPANESE
INDEXED: JICST

KAGAKU GIJUTSU-CHO MUKI ZAISHITSU KENKYUJO KENKYU RENBUNSHU 113
(RESEARCH PAPERS OF THE SCIENCE AND TECHNOLOGY AGENCY, NATIONAL INSTITUTE FOR RESEARCHES IN INORGANIC MATERIALS)
JAPANESE
INDEXED: JICST

KAGAKU GIJUTSU-CHO NENPO 114
(ANNUAL REPORT OF THE SCIENCE AND TECHNOLOGY AGENCY)
KAGAKU GIJUTSU-CHO SHINKO-KYOKU (PROMOTION BUREAU, THE SCIENCE AND TECHNOLOGY AGENCY) OKURA-SHO INSATSU-KYOKU
2, AOI-CHO, AKASAKA, MINATO-KU, TOKYO 107
JAPANESE
INDEXED: JICST

KAGAKU GIJUTSU-CHO NO GENKYO 115
(REVIEW OF THE SCIENCE AND TECHNOLOGY AGENCY)
KAGAKU GIJUTSU-CHO SHINKO-KYOKU (PROMOTION BUREAU, THE SCIENCE AND TECHNOLOGY AGENCY)
2-1, 2-CHOME, KASUMIGASEKI, CHIYODA-KU, TOKYO 100
JAPANESE

KAGAKU GIJUTSU-CHO SHIGEN CHOSAJO SHIRYO 116
(MATERIALS OF SCIENCE AND TECHNOLOGY AGENCY)
NATIONAL INSTITUTE OF RESOURCES
JAPANESE
INDEXED: JICST

KAGAKU GIJUTSU-CHO SHIGEN CHOSAKAI HOKOKU 117
(REPORT OF THE RESOURCES COUNCIL, THE SCIENCE AND TECHNOLOGY AGENCY)
KAGAKU GIJUTSU-CHO SHIGEN CHOSAKAI (THE SCIENCE AND TECHNOLOGY AGENCY, THE RESOURCES COUNCIL)
2-1, 2-CHOME, KASUMIGASEKI, CHIYODA-KU, TOKYO 100
JAPANESE
INDEXED: JICST

KAGAKU GIJUTSU-CHO SHIGEN CHOSAKAI KANKOKU 118
(RECOMMENDATION OF THE SCIENCE AND TECHNOLOGY AGENCY)
NATIONAL INSTITUTE OF RESOURCES
JAPANESE
INDEXED: JICST

KAGAKU GIJUTSU-CHO SHIGEN-KYOKU SHIGEN TOKEI-KA SHIRYO 119
(MATERIALS OF THE SCIENCE AND TECHNOLOGY AGENCY)
NATURAL RESOURCE BUREAU, NATURAL RESOURCE STATISTICS OFFICE
JAPANESE
INDEXED: JICST

KAGAKU GIJUTSU-CHO SHIGEN-KYOKU SHIRYO 120
(MATERIALS OF SCIENCE AND TECHNOLOGY AGENCY)
BUREAU OF NATURAL RESOURCES
JAPANESE
INDEXED: JICST

KAGAKU KISORON KENKYU 121
(JOURNAL OF THE JAPAN ASSOCIATION FOR PHILOSOPHY OF SCIENCE)
KAGAKU KISORON GAKKAI (THE JAPAN ASSOCIATION FOR PHILOSOPHY OF SCIENCE)
C/O TOKEI SURI KENKYUSHO, 6-7, 4-CHOME, MINAMIAZABU, MINATO-KU, TOKYO 106
JAPANESE
INDEXED: INS, JICST

KAGAKU NO JIKKEN 122
(SCIENCE THROUGH EXPERIMENTS)
KYORITSU SHUPPAN, K. K.
6-19, 4-CHOME, KOHINATA, BUNKYO-KU, TOKYO 112
JAPANESE
INDEXED: CHE

KAGAKU SHINBUN 123
(THE SCIENCE NEWS)
KAGAKU SHINBUN-SHA
8-1, 1-CHOME, HAMAMATSU-CHO, MINATO-KU, TOKYO 105
JAPANESE

KAGAKU TETSUGAKU 124
(PHILOSOPHY OF SCIENCE)
NIHON KAGAKU TETSUGAKKAI (PHILOSOPHY OF SCIENCE SOCIETY, JAPAN) RISOSHA
46, AKAGIMOTO-MACHI, SHINJUKU-KU, TOKYO 162
JAPANESE

KAGAKU TO GAIKO 125
(SCIENCE AND FOREIGN POLICY)
GAIMU-SHO KOKUSAI RENGO-KYOKU, KAGAKU-KA (SCIENTIFIC AFFAIRS DIVISION, UNITED NATIONS BUREAU, MINISTRY OF FOREIGN AFFAIRS)
2-1, 2-CHOME, KASUMIGASEKI, CHIYODA-KU, TOKYO 100
JAPANESE
INDEXED: JICST

KAGAKU TO GIJUTSU NO HIROBA 126
(SCIENTIST AND ENGINEERS)
KAGAKU TO GIJUTSU NO HIROBA NO KAI
36, BANSHU-CHO, SHINJUKU-KU, TOKYO 160
JAPANESE

KAGAKU TO KOGYO (OSAKA KOKEN KYOKAI) 127
(SCIENCE AND INDUSTRY)
OSAKA KOKEN KYOKAI (THE OSAKA SOCIETY OF INDUSTRIAL RESEARCH)
C/O OSAKA-SHIRITSU KOGYO KENKYUSHO, 38, KITAOGI-CHO, KITA-KU, OSAKA-SHI 530
JAPANESE
INDEXED: BIB, CHE, JICST

KAGAKU YOMIURI 128
(SCIENCE YOMIURI)
JAPANESE
INDEXED: JICST

KAGAKU/NINGEN 129
(SCIENCE/MAN)
KANTO GAKUIN DAIGAKU KOGAKU BU KYOYO KA (UNIVERISTY ENGINEERING FACULTY EDUCATION DEPARTMENT)
JAPANESE

KAGAKUSHI KENKYU 130
(JOURNAL OF HISTORY OF SCIENCE, JAPAN)
NIHON KAGAKUSHI GAKKAI (THE HISTORY OF SCIENCE SOCIETY OF JAPAN)
5-5, 2-CHOME, HITOTSUBASHI, CHIYODA-KU, TOKYO 101
JAPANESE, WITH EUROPEAN SUMMARY
INDEXED: CHE, JICST

KAGAKUSHI SHUKAN 131
(STUDIES IN THE HISTORY OF SCIENCE AND TECHNOLOGY)
TOKYO KOGYO DAIGAKU KAGAKUSHI GIJUTSUSHI DANWAKAI (MEETING OF THE HISTORY OF SCIENCE AND TECHNOLOGY, TOKYO INSTITUTE OF TECHNOLOGY)
KOGYO DAIGAKU JINMONKEI KAGAKUGAIRON KENKYUSHITSU, 12-1, 2-CHOME, OOKAYAMA,MEGURO-KU, TOKYO 152
JAPANESE

KAGAWA DAIGAKU KYOIKUGAKUBU KENKYU HOKOKU, DAI-II-BU 132
(MEMOIRS OF THE FACULTY OF EDUCATION, KAGAWA UNIVERSITY, PART II)
KAGAWA DAIGAKU KYOIKUGAKUBU (THE FACULTY OF EDUCATION KAGAWA UNIVERSITY)
1-1, SAIWAI-CHO, TAKAMATSU-SHI 760
JAPANESE, WITH ENGLISH SUMMARY
INDEXED: JICST

KAGOSHIMA DAIGAKU GAKUHO GOGAI 133
(KAGOSHIMA UNIVERSITY GAZETTE EXTRA)
JAPANESE

KAGOSHIMA DAIGAKU KYOIKU GAKUBU KENKYU KIYO. JINBUN SHAKAI KAGAKU-HEN 134
(RESEARCH BULLETIN OF KAGOSHIMA UNIVERSITY, FACULTY OF EDUCATION)
JAPANESE
INDEXED: JICST

KAGOSHIMA DAIGAKU KYOIKUGAKUBU KENKYU KIYO, SHIZENKAGAKU-HEN 135
(BULLETIN OF THE FACULTY OF EDUCATION, UNIVERSITY OF KAGOSHIMA, NATURAL SCIENCE)
KAGOSHIMA DAIGAKU KYOIKUGAKUBU (THE FACULTY OF EDUCATION UNIVERSITY OF KAGOSHIMA)
251, KAMOIKE-CHO, KAGOSHIMA-SHI 890
JAPANESE AND ENGLISH
INDEXED: MAT, JICST

KAGOSHIMA DAIGAKU NANKAKEN SHIRYO SENTA KOKOKU 136
(REPORT OF THE RESEARCH AND INFORMATION CENTER FOR THE SOUTHERN SCIENCE, KAGOSHIMA UNIVERSITY)
KAGOSHIMA DAIGAKU NANPO KAGAKU KENKYU SHIRYO SENTA
201, KAMONOIKE-CHO, KAGOSHIMA-SHI 890
JAPANESE

KAGOSHIMA DAIGAKU RIGAKUBU KIYO, CHIGAKU, SEIBUTSUGAKU 137
(REPORTS OF THE FACULTY OF SCIENCE, KAGOSHIMA UNIVERSITY, EARTH SCIENCES AND BIOLOGY)
KAGOSHIMA DAIGAKU RIGAKUBU (FACULTY OF SCIENCE, KAGOSHIMA UNIVERSITY)
21-35, 2-CHOME, KORIMOTO, KAGOSHIMA-SHI 890
JAPANESE AND ENGLISH
INDEXED: BID, CHE, JICST

KAGOSHIMA DAIGAKU RIGAKUBU KIYO, SUGAKU, BUTSURIGAKU, KAGAKU 138
(REPORTS OF THE FACULTY OF SCIENCE, KAGOSHIMA UNIVERSITY, MATHEMATICS, PHYSICS AND CHEMISTRY)
KAGOSHIMA DAIGAKU RIGAKUBU (FACULTY OF SCIENCE, KAGOSHIMA UNIVERSITY)
201, KAMOIKE-CHO, KAGOSHIMA-SHI 890
JAPANESE AND ENGLISH
INDEXED: BID, CHE, MAT, JICST

KAGOSHIMA DAIGAKU RIKA HOKOKU 139
(SCIENCE REPORTS OF KAGOSHIMA UNIVERSITY)
KAGOSHIMA DAIGAKU KYOYO-BU (COLLEGE OF LIBERAL ARTS, KAGOSHIMA UNIVERSITY)
201, KAMOIKE-CHO, KAGOSHIMA-SHI 890
JAPANESE AND EUROPEAN
INDEXED: CHE, MAT, JICST

KAGOSHIMA-KENRITSU TANKI DAIGAKU KIYO, SHIZEN KAGAKU-HEN 140
(BULLETIN OF KAGOSHIMA PREFECTURAL JUNIOR COLLEGE, NATURAL SCIENCE)
KAGOSHIMA-KENRITSU TANKI DAIGAKU (KAGOSHIMA PREFECTURAL JUNIOR COLLEGE)
44, SHIMOISHIKI-CHO, KAGOSHIMA-SHI 890
JAPANESE, WITH ENGLISH SUMMARY

KAIGAI CHOSADAN KOKOKU 141
(REPORT ON FOREIGN SCIENTIFIC INFORMATION)
KAGAKU GIJUTSU-CHO (SCIENCE AND TECHNOLOGY AGENCY)
2-1, 2-CHOME, KASUMIGASEKI, CHIYODA-KU, TOKYO 100
JAPANESE

KAIGAI KAGAKU GIJUTSU JOHO SHIRYO 142
(SCIENTIFIC AND TECHNICAL INFORMATION IN FOREIGN COUNTRIES)
KAGAKU GIJUTSU-CHO KEIKAKU-KYOKU (PLANNING BUREAU, SCIENCE AND TECHNOLOGY AGENCY)
2-1, 2-CHOME, KASUMIGASEKI, CHIYODA-KU, TOKYO 100
JAPANESE

KAIGAI KAGAKU GIJUTSU SHIRYO GEPPO 143
(MONTHLY LIST OF FOREIGN SCIENTIFIC AND TECHNICAL PUBLICATIONS)
KOKURITSU KOKKAI TOSHOKAN SANKOSHOSHI-BU (REFERENCE AND BIBLIOGRAPHY DIVISION, NATIONAL DIET LIBRARY)
10-1, 1-CHOME, NAGATA-CHO, CHIYODA-KU, TOKYO 100
ENGLISH
INDEXED: JICST

KAKYOKYO NYUSU 144
(SCIENCE EDUCATION COOPERATIVE NEWS)
KAGAKU KYOIKU KENKYU KYOGI KAI LINKAI (COMMITTEE OF THE SCIENCE EDUCATION RESEARCH CONFERENCE)
JAPANESE

KANAGAWA-KEN HAKUBUTSUKAN KYOKAI KAIHO 145
(KANAGAWA-KEN MUSEUM GAZETTE)
KANAGAWA-KEN HAKUBUTSUKAN KYOKAI (MUSEUMS ASSOCIATION OF KANAGAWA PREFECTURE)
C/O ENOSHIMA SUIZOKUKAN 17-25, 2-CHOME, KATASEKAIGAN, FUJISAWA-SHI 251
JAPANESE

KANAGAWA-KENRITSU GAIGO TANKI DAIGAKU KIYO, SHIZEN KAGAKU-HEN 146
(BULLETIN OF THE COLLEGE OF FOREIGN STUDIES, YOKOHAMA, NATURAL SCIENCE)
KANAGAWA-KENRITSU TANKI DAIGAKU (COLLEGE OF FOREIGN STUDIES, YOKOHAMA)
800-1, OKAMURA-CHO, ISOGO-KU, YOKOHAMA-SHI 235
JAPANESE AND ENGLISH
INDEXED: BIO

KANAGAWA-KENRITSU HAKUBUTSUKAN CHOSA KENKYU HOKOKU, SHIZEN KAGAKU 147
(RESEARCH REPORT OF THE KANAGAWA PREFECTURAL MUSEUM, NATURAL HISTORY)
KANAGAWA-KENRITSU HAKUBUTSUKAN (THE KANAGAWA PREFECTURAL MUSEUM)
5-60, MINAMI NAKA-DORI, NAKA-KU, YOKOHAMA-SHI 231
JAPANESE

KANAGAWA-KENRITSU HAKUBUTSUKAN KENKYUHOKOKU, SHIZEN KAGAKU 148
(BULLETIN OF THE KANAGAWA PREFECTURAL MUSEUM)
KANAGAWA-KENRITSU HAKUBUTSUKAN (KANAGAWA PREFECTURAL MUSEUM)
5-60, MINAMI NAKA-DORI, NAKA-KU, YOKOHAMA-SHI 231
JAPANESE AND ENGLISH
INDEXED: JICST

KANAZAWA DAIGAKU KYOIKUGAKUBU KIYO, SHIZENKAGAKU-HEN 149
(BULLETIN OF THE FACULTY OF EDUCATION KANAZAWA UNIVERSITY, SERIES NATURAL SCIENCE)
KANAZAWA DAIGAKU KYOIKUGAKUBU (THE FACULTY OF EDUCATION, KANAZAWA UNIVERSITY)
1-1, MARUNOUCHI, KANAZAWA-SHI 920
JAPANESE AND ENGLISH
INDEXED: JICST

KANAZAWA DAIGAKU KYOYOBU RONSHU, SHIZENKAGAKU-HEN 150
(THE ANNALS OF SCIENCE, COLLEGE OF LIBERAL ARTS, KANAZAWA UNIVERSITY)
KANAZAWA DAIGAKU KYOYOBU (COLLEGE OF LIBERAL ARTS, KANAZAWA UNIVERSITY)
1-1, MARUNOUCHI, KANAZAWA-SHI 920
JAPANESE AND EUROPEAN
INDEXED: BIO, JICST

KANKYO 151
(ENVIORNMENT)
NIHON KANKYO CHOSA KAI (JAPAN ENVIRONMENT INVESTIGATION COMMITTEE)
JAPANESE

KANKYO KAGAKU SOGO KENKYUJO NENPO 152
(ANNUAL OF THE ENVIRONMENTAL SCIENCE INTEGRATED RESEARCH INSTITUTE)
JAPANESE

KANSAI GAKUIN DAIGAKU RONKO 153
(PAPERS OF KANSAI GAKUIN UNIVERSITY)
KANSAI GAKUIN UNIVERSITY
JAPANESE

KANSAI IKA DAIGAKU KYOYOBU KIYO 154
(BULLETIN OF THE FACULTY OF LIBERAL ARTS, KANSAI MEDICAL COLLEGE)
KANSAI IKA DAIGAKU KYOYOBU (FACULTY OF LIBERAL ARTS, KANSAI MEDICAL COLLEGE)
18-89, UYAMA HIGASHI-MACHI, HIRAKATA-SHI 573
JAPANESE

KANSEI GAKUIN UNIVERSITY ANNUAL STUDIES 155
KANSEI GAKUIN DAIGAKU (KANSEI GAKUIN UNIVERSITY)
UEGAHARA ICHIBAN-CHO, NISHINOMIYA-SHI 662
EUROPEAN

KEIO GIJUKU DAIGAKU HIYOSHI RONBUNSHU, SHIZEN KAGAKU-HEN 156
(HIYOSHI SCIENCE REVIEW)
KEIO GIJUKU DAIGAKU HIYOSHI KENKYUSHITSU (HIYOSHI SCIENCE SOCIETY, KEIO UNIVERSITY)
665, HIYOSHI-CHO, KOHOKU-KU, YOKOHAMA-SHI 223
JAPANESE, WITH ENGLISH SUMMARY

KEIO GIJUKU DAIGAKUHO 157
(KEIO GIJUKU UNIVERSITY REPORT)
KEIO GIJUKU DAIGAKU (KEIO GIJUKU UNIVERSITY)
JAPANESE

KIHJO GAKUIN DAIGAKU ROHSHU 158
(PAPERS OF KIHJO GAKUIN UNIVERSITY)
KIHJO GAKUIN DAIGAKU (KIHJO GAKUIN UNIVERSITY)
JAPANESE

KIKAN KAGAKU TO SHISO 159
(SCIENCE AND PHILOSOPHY)
SHIN NIHON SHUPPANSHA
13-14, 2-CHOME, FUJIMI, CHIYODA-KU, TOKYO 102
JAPANESE

KINKI DAIGAKU KYOYO BU KENKYU KIYO 160
(BULLETIN OF THE SCHOOL OF ARTS AND SCIENCES, KINKI UNIVERSITY)
KINKI DAIGAKU KYOYO BU (SCHOOL OF ARTS AND SCIENCES, KINKI UNIVERSITY)
JAPANESE

KINKI DAIGAKU RIKOGAKUBU KENKYU HOKOKU 161
(JOURNAL OF THE FACULTY OF SCIENCE AND TECHNOLOGY, KINKI UNIVERSITY)
KINKI DAIGAKU RIKOGAKUBU (FACULTY OF SCIENCE AND TECHNOLOGY, KINKI UNIVERSITY)
321, KOWAKAE, HIGASHIOSAKA-SHI 577
JAPANESE, WITH EUROPEAN SUMMARY
INDEXED: CHE, JICST

KITASATO DAIGAKU KYOYOBU KIYO 162
(KITASATO UNIVERSITY JOURNAL OF LIBERAL ARTS AND SCIENCES)
JAPANESE

KOBE DAIGAKU KYOIKUGAKUBU KENKYU SHUROKU 163
(BULLETIN OF THE FACULTY OF EDUCATION, KOBE UNIVERSITY)
KOBE DAIGAKU KYOIKUGAKUBU (THE FACULTY OF EDUCATION, KOBE UNIVERSITY)
3-11, TSURUKABUTO, NADA-KU, KOBE-SHI 657
JAPANESE
INDEXED: CHE, JICST

KOBE GAKUIN DAIGAKIN KIYO 164
(BULLETIN OF KOBE GAKUIN UNIVERSITY)
KOBE GAKUIN DAIGAKU KYOYO BU
JAPANESE

KOBE JOGAKUIN DAIGAKU RONSHU 165
(PAPERS OF KOBE WOMEN'S GAKUIN UNIVERSITY)
JAPANESE
INDEXED: JICST

KOBE JOSHI TANKI DAIGAKU GAKKAI RONKO, RIKA-HEN 166
(PROCEEDINGS OF THE SOCIETY OF THE KOBE WOMEN'S JUNIOR COLLEGE)
KOBE JOSHI TANKI DAIGAKU
2-71, NAKAYAMATE-DORI, IKUTA-KU, KOBE-SHI 650
JAPANESE

KOCHI DAIGAKU GAKUJUTSU KENKYU HOKOKU, SHIZENKAGAKU 167
(RESEARCH REPORTS OF THE KOCHI UNIVERSITY, NATURAL SCIENCE)
KOCHI DAIGAKU (KOCHI UNIVERSITY)
1000, ASAKURA, KOCHI-SHI 780
JAPANESE AND EUROPEAN
INDEXED: BIB, CHE, JICST

KOCHI DAIGAKU KYOIKUGAKUBU KENKYU HOKOKU, DAI-3-BU, SHIZEN 168
(BULLETIN OF THE FACULTY OF EDUCATION, KOCHI UNIVERSITY)
KOCHI DAIGAKU KYOIKUGAKUBU (THE FACULTY OF EDUCATION KOCHI UNIVERSITY)
1000, ASAKURA, KOCHI-SHI 780
JAPANESE AND ENGLISH

KOCHI GAKUEN TANKI DAIGAKU KIYO 169
(BULLETIN OF KOCHI GAKUEN JUNIOR COLLEGE)
KOCHI GAKUEN TANKI DAIGAKU (KOCHI GAKUEN JUNIOR COLLEGE)
292-26, JINGAMORI, ASAHITENJIN-CHO, KOCHI-SHI 780
JAPANESE AND ENGLISH

KOCHI JOSHI DAIGAKU KIYO, SHIZENKAGAKU-HEN 170
(BULLETIN OF KICHI WOMEN'S UNIVERSITY, NATIONAL SCIENCE SERIES)
KOCHI JOSHI DAIGAKU
5-15, EIKOKUJI-CHO, KOCHI-SHI 780
JAPANESE
INDEXED: CHE, JICST

KOGAKKAN DAIGAKU KIYO 171
(BULLETIN OF KOGAKKAN UNIVERSITY)
KOGAKKAN UNIVERSITY
JAPANESE

KOGAKUIN DAIGAKU KENKYU RONSO 172
(KOGAKUIN UNIVERSITY RESEARCH REVIEW)
KOGAKUIN DAIGAKU (KOGAKUIN UNIVERSITY)
JAPANESE

KOKUGAKUIN DAIGAKU DAIGAKUIN KIYO 173
(BULLETIN OF KOKUGAKUIN UNIVERSITY, GRADUATE SCHOOL)
KOKUGAKUIN UNIVERSITY, GRADUATE SCHOOL
JAPANESE

KOKUGAKUIN DAIGAKU HAKUBUTSU KANGAKU KIYO 174
(MUSEUM STUDIES BULLETIN OF KOKUGAKUIN UNIVERSITY)
KOKUGAKUIN DAIGAKU HAKUBUTSU KANGAKU KOZA
JAPANESE

KOKUGAKUIN DAIGAKU KIYO 175
(BULLETIN OF KOKUGAKUIN UNIVERSITY)
JAPANESE

KOKUGAKUIN ZASSHI 176
(KOKUGAKUIN MAGAZINE)
KOKUGAKUIN DAIGAKU
JAPANESE

KOKURITSU HAKUBUTSUKAN NYUSU 177
(TOKYO NATIONAL MUSEUM NEWS)
TOKYO KOKURITSU HABUBUTSUKAN (TOKYO NATIONAL MUSEUM)
13-9, UENOKOEN, TAITO-KU, TOKYO 110
JAPANESE

KOKURITSU KAGAKU HAKUBUTSUKAN NYUSU 178
(NATIONAL SCIENCE MUSEUM NEWS)
JAPANESE

KOKURITSU KAGAKU HAKUBUTSUKAN SENPO 179
(MEMOIRS OF THE NATIONAL SCIENCE MUSEUM)
KOKURITSU KAGAKU HAKUBUTSUKAN (NATIONAL SCIENCE MUSEUM)
7-20, UENO-KOEN, TAITO-KU, TOKYO 110
JAPANESE AND ENGLISH
INDEXED: BIB, BIO, CHE, JICST

KOKUSAI DAIGAKU KIYO 180
(BULLETIN OF KOKUSAI UNIVERSITY)
KOKUSAI DAIGAKU GAKKAI (KOKUSAI UNIVERSITY LEARNED SOCIETY)
JAPANESE

KOKYO SHIKEN KENKYU KIKAN ANNAI 181
(DIRECTORY OF PUBLIC RESEARCH INSTITUTE)
NIHON KAGAKU GIJUTSU JOHO SENTA (THE JAPAN INFORMATION CENTER OF SCIENCE AND TECHNOLOGY)
1-1, 2-CHOME, NAGATA-CHO, CHIYODA-KU, TOKYO 100
JAPANESE

KONAN DAIGAKU KIYO RIGAKU-HEN 182
(MEMOIRS OF THE KONAN UNIVERSITY SCIENCE SERIES)
KONAN DAIGAKU (KONAN UNIVERSITY)
OKAMOTO, HIGASHINADA-KU, KOBE-SHI 658
JAPANESE AND ENGLISH
INDEXED: BIO, CHE, JICST

KONAN JOSHI DAIGAKU KENKYU KIYO 183
(KONAN WOMEN'S COLLEGE RESEARCHES)
KONAN JOSHI DAIGAKU (KONAN WOMEN'S COLLEGE)
2-23, 6-CHOME, MORIKITA-CHO, HIGASHIHADA-KU, KOBE-SHI 658
JAPANESE AND EUROPEAN
INDEXED: CHE, JICST

KUMAMOTO DAIGAKU KYOIKUGAKUBU KIYO, DAI-1-BUNSATSU, SHIZENKAGAKU 184
(MEMOIRS OF THE FACULTY OF EDUCATION, KUMAMOTO UNIVERSITY, NATURAL SCIENCE)
KUMAMOTO DAIGAKU KYOIKUGAKUBU (THE FACULTY OF EDUCATION, KUMAMOTO UNIVERSITY)
40-1, 2-CHOME, KUROKAMI, KUMAMOTO-SHI 860
JAPANESE AND EUROPEAN
INDEXED: CHE, MAT, JICST

KUMAMOTO DAIGAKU KYOYOBU KIYO, SHIZEN KAGAKU HEN 185
(BULLETIN OF THE KYOTO UNIVERSITY GENERAL EDUCATION FACULTY, NATURAL SCIENCE EDITION)
JAPANESE

KUMAMOTO DAIGAKU RIGAKUBU KENKYU GYOSEKI 186
(RESEARCH ACCOMPLISHMENTS OF KUMAMOTO UNIVERISTY FACULTY OF SCIENCE)
JAPANESE

KUNI NO SHIKEN KENKYU GYOMU KEIKAKU 187
(NATIONAL RESEARCH PROJECTS)
KAGAKU GIJUTSU-CHO KENKYU CHOSEI-KYOKU (RESEARCH COORDINATION BUREAU, SCIENCE AND TECHNOLOGY AGENCY)
2-1, 2-CHOME, KASUMIGASEKI, CHIYODA-KU, TOKYO 100
JAPANESE

KURUME DAIGAKU RONSO 188
(KURUME UNIVERSITY JOURNAL)
KURUME DAIGAKU SHOGAKUBU, IGAKUBU SHINGAKUKATEI (FACULTY OF COMMERCE AND PREMEDICAL COURSE, KURUME UNIVERSITY)
1635, MII-MACHI, KURUME-SHI 830
JAPANESE AND ENGLISH
INDEXED: CHE

KUSHIRO RONSHU 189
(REPORTS OF HOKKAIDO UNIVERSITY OF EDUCATION, KUSHIRO)
HOKKAIDO KYOIKU DAIGAKU KUSHIRO BUNKO
15-55, 1-CHOME, SHIROYAMA, KUSHIRO-SHI 085
JAPANESE

KUSHIRO-SHIRITSU KYODO HAKUBUTSUKANHO 190
(SCIENCE REPORT OF THE KUSHIRO MUNICIPAL MUSEUM)
KUSHIRO-SHIRITSU KYODO HAKUBUTSUKAN (THE KUSHIRO MUNICIPAL MUSEUM)
10-35, 1-CHOME, TSURUGATAI, KUSHIRO-SHI 085
JAPANESE

KYODO TO HAKUBUTSUKAN 191
(NATIVE LAND AND MUSEUM)
TOTTORI-KENRITSU HAKUBUTSUKAN (TOTTORI PREFECTURAL MUSEUM)
2-124, HIGASHI-MACHI, TOTTORI-SHI 680
JAPANESE

KYORIN DAIGAKU IGAKUBU KYOYOKALEI KENKYUHOKOKU 192
(KYORIN JOURNAL OF ARTS AND SCIENCE, KYORIN UNIVERSITY SCHOOL OF MEDICINE)
KYORIN DAIGAKU IGAKUBU (KYORIN UNIVERSITY SCHOOL OF MEDICINE)
JAPANESE

KYOTO GAKUEN DAIGAKU RONSHU 193
(COLLECTED PAPERS OF KYOTO GAKUEN UNIVERSITY)
KYOTO GAKUEN DAIGAKU GAKKAI (LEARNED SOCIETY OF KYOTO GAKUEN UNIVERSITY)
JAPANESE

KYOTO KOKURITSU HAKUBUTSUKAN NENPO 194
(ANNUAL REPORT OF THE KYOTO NATIONAL MUSEUM)
KYOTO KOKURITSU HAKUBUTSUKAN (KYOTO NATIONAL MUSEUM)
527, CHAYA-MACHI, HIGASHIYAMA-KU, KYOTO-SHI 605
JAPANESE

KYOTO KYOIKU DAIGAKU KIYO, B, SHIZENKAGAKU 195
(BULLETIN OF KYOTO UNIVERSITY OF EDUCATION, SERIES B, MATHEMATICS AND NATURAL SCIENCE)
KYOTO KYOIKU DAIGAKU (KYOTO UNIVERSITY OF EDUCATION)
1, FUKAKUSA FUJINOMORI-CHO, FUSHIMI-KU, KYOTO 612
JAPANESE AND EUROPEAN
INDEXED: BIO, CHE, MAT, JICST

KYOTO-FURITSU DAIGAKU GAKUJUTSU HOKOKU,RIGAKU, SEIKATSUKAGAKU 196
(THE SCIENTIFIC REPORT OF THE KYOTO PREFECTURAL UNIVERSITY, NATURAL SCIENCE AND LIVING SCIENCE)
KYOTO-FURITSU DAIGAKU (KYOTO PREFECTURAL UNIVERSITY)
SHIMOKAMO HANGI-CHO, SAKYO-KU, KYOTO-SHI 605
JAPANESE AND ENGLISH
INDEXED: CHE, JICST

KYUSHU JOSHI DAIGAKU KIYO 197
(BULLETIN OF KYUSHU WOMEN'S UNIVERSITY)
KYUSHU JOSHI DAIGAKU
1, ORIO JIYUGAOKA, YAHATANISHI-KU,
KITAKYUSHU-SHI 807
JAPANESE

KYUSHU OTANI KENKYU KIYO 198
(KYUSHU OTANI RESEARCH BULLETIN)
KYUSHU OTANI GAKKAI (KYUSHU OTANI SOCIETY)
JAPANESE

KYUSHU SANGYO DAIGAKU KYOYOBU KIYO 199
(BULLETIN OF THE FACULTY OF GENERAL EDUCATION,
KYUSHU INDUSTRIAL UNIVERSITY)
JAPANESE

KYUSHU TOKAI DAIGAKU KIYO 200
(PROCEEDINGS OF KYUSHU TOKAI UNIVERSITY)
JAPANESE

MEIJI DAIGAKU KAGAKU GIJUTSU KENKYUSHO KIYO 201
(MEMOIRS OF THE INSTITUTE OF SCIENCES AND
TECHNOLOGY, MEIJI UNIVERSITY)
MEIJI DAIGAKU KAGAKU GIJUTSU KENKYUSHO
(INSTITUTE OF SCIENCES AND TECHNOLOGY, MEIJI
UNIVERSITY)
5158, IKUTA, TAMA-KU, KAWASAKI-SHI 214
JAPANESE AND ENGLISH
INDEXED: CHE, JICST

MEIJI DAIGAKU KAGAKU GIJUTSU KENKYUSHO NENPO 202
(ANNUAL REPORTS OF THE INSTITUTE OF SCIENCES AND
TECHNOLOGY, MEIJI UNIVERSITY)
MEIJI DAIGAKU KAGAKU GIJUTSU KENKYUSHO
5158, IKUTA, TAMA-KU, KAWASAKI-SHI 214
JAPANESE
INDEXED: CHE, JICST

MEIJI DAIGAKU KYOYO RONSHU 203
(PAPERS OF THE MEIJI UNIVERSITY ARTS AND SCIENCES)
MEIJI DAIGAKU KYOYO RONSHU KANKOKAI
(MEIJI UNIVERSITY ARTS AND SCIENCE PUBLISHING
ASSOCIATION)
JAPANESE

MEIJO DAIGAKU RIKOGAKUBU KENKYU HOKOKU 204
(REPORTS OF THE FACULTY OF SCIENCE AND TECHNOLOGY,
MEIJO UNIVERSITY)
MEIJO DAIGAKU RIKOGAKUBU (FACULTY OF SCIENCE AND
TECHNOLOGY, MEIJO UNIVERSITY)
69, YAGOTO-URAYAMA, TEMPAKU-CHO, SHOWA-KU,
NAGOYA-SHI 468
JAPANESE AND ENGLISH
INDEXED: CHE, JICST

MEISEI DAIGAKU KENKYU KIYO RIKOGAKUBU 205
(RESEARCH BULLETIN OF MEISEI UNIVERSITY - PHYSICAL
SCIENCE AND ENGINEERING)
MEISEI DAIGAKU (MEISEI UNIVERSITY)
337, HODOKUBO, HINO-SHI 191
JAPANESE AND ENGLISH
INDEXED: CHE, MAT

MEMOIRS OF THE DEFENSE ACADEMY, MATHEMATICS, 206
PHYSICS, CHEMISTRY AND ENGINEERING
BOEI DAIGAKKO (DEFENSE ACADEMY)
10-20, 1-CHOME, HASHIRIMIZU, YOKOSUKA-SHI 239
EUROPEAN
INDEXED: CHE, ENG, INS, MAT, JICST

MEMOIRS OF THE FACULTY OF SCIENCE, KYOTO UNIVERSITY 207
SERIES OF PHYSICS, ASTROPHYSICS, GEOPHYSICS
AND CHEMISTRY
KYOTO DAIGAKU RIGAKUBU (FACULTY OF SCIENCE,
KYOTO UNIVERSITY)
KITASHIRAKAWA OIWAKE-CHO, SAKYO-KU, KYOTO-SHI 606
ENGLISH
INDEXED: CHE, INS, JICST

MEMOIRS OF THE SCHOOLS OF SCIENCE AND ENGINEERING, 208
WASEDA UNIVERSITY
WASEDA DAIGAKU RIKOGAKUBU (THE SCHOOL OF SCIENCE
AND ENGINEERING, WASDEA UNIVERSITY)
4-170, NISHIOKUBO, SHINJUKU-KU, TOKYO 160
ENGLISH
INDEXED: CHE, MAT, JICST

MINAMI KYUSHU DAIGAKU ENGEIGAKUBU KENKYU HOKOKU, 209
SHIZEN KAGAKU, JINBUN SHAKAI KAGAKU
(BULLETIN OF THE FACULTY OF HORTICULTURE, MINAMI
KYUSHU UNIVERSITY NATURAL SCIENCE, CULTURAL
SCIENCE, SOCIAL SCIENCE)
MINAMI KYUSHU DAIGAKU ENGEIGAKUBU
HIBARIGAOKA, TAKANABE-CHO, KOYU-GUN,
MIYAZAKI 884
JAPANESE

MISCELLANEOUS REPORTS OF THE NATIONAL SCIENCE 210
MUSEUM
JAPANESE
INDEXED: JICST

MITSUBISHI SOGO KENKYUJO SHOHO 211
(REPORT OF THE MITSUBISHI GENERAL RESEARCH
INSTITUTE)
JAPANESE
INDEXED: JICST

MIYAGI KYOIKU DAIGAKU KIYO 212
(BULLETIN OF MIYAGI EDUCATION COLLEGE)
MIYAGI KIYOKU DAIGAKU
JAPANESE

MIYAGI KYOIKU DAIGAKU RIKA KYOIKU KENKYU 213
SHISETSU NENPO
(ANNUAL REPORT FROM THE RESEARCH INSTITUTE FOR
SCIENCE EDUCATION, MIYAGI UNIVERSITY OF
EDUCATION)
MIYAGI KYOIKU DAIGAKU KYOIKUGAKUBU FUZOKU RIKA
KYOIKU KENKYU SHISETSU (RESEARCH INSTITUTE OF
SCIENCE EDUCATION, FACULTY OF SCIENCE, MIYAGI
UNIVERSITY OF EDUCATION
AOBA, ARAMAKI, SENDAI-SHI 980
JAPANESE AND ENGLISH

MIYAZAKI DAIGAKU KYOIKUGAKUBU KIYO, 214
SHIZENKAGAKU
(MEMOIRS OF THE FACULTY OF EDUCATION, MIYAZAKI
UNIVERSITY, NATURAL SCIENCE)
MIYAZAKI DAIGAKU KYOIKUGAKUBU (FACULTY OF THE
EDUCATION, MIYAZAKI UNIVERSITY)
130, NISHIMARUYAMA-CHO, MIYAZAKI-SHI 880
JAPANESE AND ENGLISH
INDEXED: CHE, MAT

MOMOYAMA GAKUIN DAIGAKU SOGO KENKYUJOHO 215
(REPORT OF THE MOMOYAMA GAKUIN UNIVERSITY
COLLECTIVE RESEARCH INSTITUTE)
MOMOYAMA GAKUIN UNIVERSITY COLLECTIVE
RESEARCH INSTITUTE
JAPANESE

MONBU-SHO KAGAKU KENKYUHI NI YORU KENKYU HOKOKU 216
SHUROKU, RIGAKU-HEN 1
(REPORTS ON RESEARCHES BY GRANTEES MINISTRY OF
EDUCATION, SCIENCE PART 1)
NIHON GAKUJUTSU SHINKOKAI (JAPAN SOCIETY FOR THE
PROMOTION OF SCIENCE)
3-1, 5-CHOME, KOJI-MACHI, CHIYODA-KU, TOKYO 102
JAPANESE

MONBU-SHO KAGAKU KENKYUHI NI YORU KENKYU HOKOKU 217
SHUROKU, RIGAKU-HEN 2
(REPORTS ON RESEARCHES BY GRANTEES MINISTRY OF
EDUCATION, SCIENCE, PART 2)
NIHON GAKUJUTSU SHINKOKAI (JAPAN SOCIETY FOR THE
PROMOTION OF SCIENCE)
3-1, 5-CHOME, KOJI-MACHI, CHIYODA-KU, TOKYO 102
JAPANESE

MUSASHINO DAIGAKU RONSHU 218
(PAPERS OF THE MUSASHINO UNIVERSITY)
MUSASHINO UNIVERSITY
JAPANESE

NAGANO-KEN TANKI DAIGAKU KIYO 219
(THE JOURNAL OF NAGANO PREFECTURE JUNIOR COLLEGE)
NAGANO-KEN TANKI DAIGAKU (NAGANO PREFECTURE
JUNIOR COLLEGE)
49-7, 8-CHOME, MIWA, NAGANO-SHI 380
JAPANESE
INDEXED: JICST

NAGAOKA-SHIRITSU KAGAKU HAKUBUTSUKAN KENKYU 220
HOKOKUSHO
(BULLETIN OF THE NAGAOKA MUNICIPAL SCIENCE MUSEUM)
NAGAOKA-SHIRITSU KAGAKU HAKUBUTSUKAN (NAGAOKA
MUNICIPAL SCIENCE MUSEUM)
C/O YUKYUZAN KOEN, OYAMA-MACHI, NAGAOKA-SHI 940
JAPANESE AND ENGLISH

NAGASAKI DAIGAKU GAKUHO 221
(ACADEMIC REPORT OF NAGASAHI UNIVERSITY)
JAPANESE

NAGASAKI DAIGAKU KYOIKUGAKUBU, SHIZENKAGAKU 222
KENKYU HOKOKU
(SCIENCE BULLETIN OF THE FACULTY OF EDUCATION,
NAGASAKI UNIVERSITY)
NAGASAKI DAIGAKU KYOIKUGAKUBU (FACULTY OF
EDUCATION, NAGASAKI UNIVERSITY)
1-14, BUNKYO-MACHI, NAGASAKI-SHI 852
JAPANESE AND ENGLISH
INDEXED: CHE, MAT, JICST

NAGASAKI DAIGAKU KYOYOBU KIYO, SHIZENKAGAKU 223
(BULLETIN OF FACULTY OF LIBERAL ARTS, NAGASAKI
UNIVERSITY , NATURAL SCIENCE)
NAGASAKI DAIGAKU KYOYOBU (FACULTY OF LIBERAL ARTS,
NAGASAKI UNIVERSITY)
1-14, BUNKYO-MACHI, NAGASAKI-SHI 852
JAPANESE AND ENGLISH
INDEXED: JICST

NAGASAKI-SHIRITSU HAKUBUTSUKAN KANPO 224
(BULLETIN OF THE NAGASAKI CITY MUSEUM)
NAGASAKI-SHIRITSU HAKUBUTSUKAN (NAGASAKI CITY MUSEUM)
C/O KOKUSAI BUNKA KAIKAN, 7-8, HIRANOMACHI, NAGASAKI-SHI 850
JAPANESE

NAGOYA DAIGAKU KYOYOBU KIYO, SHIZEN KAGAKU, SHINRIGAKU, HOKEN TAIIKUGAKU 225
(RESEARCH BULLETIN, NATURAL SCIENCE, PSYCHOLOGY, AND HEALTH AND PHYSICAL EDUCATION, DEPARTMENT OF GENERAL EDUCATION, NAGOYA UNIVERSITY)
NAGOYA DAIGAKU KYOYOBU
FURO-CHO, CHIKUSA-KU, NAGOYA-SHI 466
JAPANESE AND ENGLISH
INDEXED: CHE, JICST

NAGOYA GAKUIN DAIGAKU RONSHU, JINBUN, SHIZEN KAGAKU-HEN 226
(NAGOYA GAKUIN UNIVERSITY REVIEW, HUMANITIES, NATURAL SCIENCE)
NAGOYA GAKUIN DAIGAKU SANGYO KAGAKU KENKYUSHO (NAGOYA GAKUIN UNIVERSITY, INSTITUTE OF INDUSTRIAL SCIENCE)
1350, KAMISHINANO-CHO, SETO-SHI 480-12
JAPANESE AND ENGLISH
INDEXED: JICST

NAGOYA GAKUIN RANSO 227
(COLLECTED PAPERS OF NAGOYA ACADEMY)
JAPANESE

NAGOYA-SHIRITSU DAIGAKU KYOYOBU KIYO, SHIZENKAGAKU-HEN 228
(BULLETIN OF THE DEPARTMENT OF GENERAL EDUCATION, NAGOYA CITY UNIVERSITY, NATURAL SCIENCE SECTION)
NAGOYA-SHIRITSU DAIGAKU (NAGOYA CITY UNIVERSITY)
1-1, YAMANOHATA, MIZUHO-CHO, MIZUHO-KU, NAGOYA-SHI 467
JAPANESE AND EUROPEAN
INDEXED: BID, CHE, JICST

NANZAN RONSHU 229
(NANZAN PAPERS)
NANZAN DAIGAKU DAIGAKUIN KENKYUKAI (NANZAN UNIVERSITY, GRADUATE SCHOOL RESEARCH SOCIETY)
JAPANESE

NARA DAIGAKU KIYO 230
(BULLETIN OF NARA UNIVERSITY)
JAPANESE

NARA JOSHI DAIGAKU KIYO 231
(BULLETIN OF NARA WOMEN'S UNIVERSITY)
JAPANESE

NARA KYOIKU DAIGAKU KIYO, SHIZENKAGAKU 232
(BULLETIN OF NARA UNIVERSITY OF EDUCATION, NATURAL SCIENCE)
NARA KYOIKU DAIGAKU (THE NARA UNIVERSITY OF EDUCATION)
TAKABATAKE-CHO, NARA-SHI 630
JAPANESE AND ENGLISH
INDEXED: BID, CHE, MAT, JICST

NATURAL SCIENCE REPORT OF THE OCHANOMIZU UNIVERSITY 233
OCHANOMIZU JOSHI DAIGAKU (OCHANOMIZU UNIVERSITY)
1-1, 2-CHOME, OTSUKA, BUNKYO-KU, TOKYO 112
EUROPEAN
INDEXED: APP, CHE, MAT, JICST

NAYORO JOSHI TANKI DAIGAKU GAKUJUTSU KENKYU HOKOKU 234
(ACADEMIC RESEARCH REPORT OF NAYORO WOMEN'S JUNIOR COLLEGE)
JAPANESE

NICHIFUTSU RIKOKA KAISHI 235
(BULLETIN DE LA SOCIETE FRANCO-JAPONAISE DES SCIENCE PURES ET APPLIQUEES)
NICHIFUTSU RIKOKAKAI
C/O NICHIFUTSU, KAIKAN, 2-3
KANDA SURUGADAI, CHIYODA-KY, TOKYO 101
JAPANESE AND EUROPEAN

NIHON DAIGAKU BUNRIGAKUBU KENKYU NENPO 236
(ANNUAL REPORT OF THE RESEARCHES, NIHON UNIVERSITY)
NIHON DAIGAKU BUNRIGAKUBU (MISHIMA) (MISHIMA COLLEGE OF HUMANITIES AND SCIENCES, NIHON UNIVERSITY)
2-CHOME, BUNKYO-CHO, MISHIMA-SHI 411
JAPANESE AND ENGLISH

NIHON DAIGAKU BUNRIGAKUBU SHIZEN KAGAKU KENKYUJO KENKYU HOKOKU 237
(RESEARCH REPORT OF NIHON UNIVERSITY DEPARTMENT OF HUMANITIES, LITERATURE AND SCIENCE, NATURAL SCIENCE RESEARCH INSTITUTE)
JAPANESE
INDEXED: JICST

NIHON DAIGAKU BUNRIGAKUBU SHIZEN KAGAKU KENKYUJO KENKYU KIYO 238
(RESEARCH BULLETIN OF NATURAL SCIENCE RESEARCH INSTITUTE, DEPARTMENT OF HUMANITIES AND SCIENCES, NIHON UNIVERSITY)
JAPANESE

NIHON DAIGAKU SHIZEN KAGAKU KENKYUSHO KENKYU KIYO 239
(PROCEEDINGS OF THE INSTITUTE OF NATURAL SCIENCES, NIHON UNIVERSITY)
NIHON DAIGAKU SHIZEN KAGAKU KENKYUSHO (INSTITUTE OF NATURAL SCIENCES, NIHON UNIVERSITY)
25-40, 3-CHOME, SAKURAJOSUI, SETAGAYAKU, TOKYO 156
JAPANESE AND EUROPEAN

NIHON DAIGAKU TOYAMA GAKUEN KIYO 240
(THE BULLETIN OF TOYAMA GAKUEN, NIHON UNIVERSITY)
JAPANESE

NIHON FUJIN KAGAKUSHA NO KAI NYUSU 241
(NEWS OF THE SOCIETY OF JAPANESE WOMEN SCIENTIST)
NIHON FUJIN KAGAKUSHA NO KAI (SOCIETY OF JAPANESE WOMEN SCIENTIST)
C/O GAN KENKYUSHO, 37-1, 1-CHOME, KAMIIKEBUKURO, TOSHIMA-KU, TOKYO 170
JAPANESE

NIHON FUKUSHI DAIGAKU KENKYA KIYO 242
(RESEARCH BULLETIN OF NIHON FUKUSHI UNIVERSITY)
NIHON FUKUSHI DAIGAKU (NIHON FUKUSHI UNIVERSITY)
JAPANESE

NIHON GAKUJUTSU KAIGI 243
(ANNUAL REPORT OF THE SCIENCE COUNCIL OF JAPAN)
JAPANESE
INDEXED: BIB

NIHON GAKUJUTSU KAIGI GEPPO 244
(MONTHLY REPORT OF THE SCIENCE COUNCIL OF JAPAN)
NIHON GAKUJUTSU KAIGI JIMU-KYOKU (SCIENCE COUNCIL OF JAPAN)
22-34, 7-CHOME, ROPPONGI, MINATO-KU, TOKYO 106
JAPANESE
INDEXED: JICST

NIHON GAKUSHIIN KIYO 245
(BULLETIN OF NIHON GAKUSHIIN)
JAPANESE

NIHON KAGAKUSHA TSUSHIN 246
(JAPAN SCIENTISTS NEWS)
NIHON KAGAKUSHA KAIGI (SCIENTISTS COUNCIL OF JAPAN)
JAPANESE

NIHON NO KAGAKU TO GIJUTSU 247
(JAPAN SCIENCE AND TECHNOLOGY)
NIHON KAGAKU GIJUTSU SHINKO ZAIDAN (JAPAN SCIENCE FOUNDATION)
2-1, KITANOMARU KOEN, CHIYODA-KU, TOKYO 102
JAPANESE
INDEXED: APT, JICST

NIHON NO KAGAKUSHA 248
(JOURNAL OF JAPANESE SCIENTISTS)
NIHON KAGAKUSHA KAIGI (JAPAN SCIENTISTS ASSOCIATION)
9-16, 1-CHOME, YUSHIMA, BUNKYO-KU, TOKYO 113
JAPANESE

NIHON SHIZEN HOGO KYOKAI CHOSA HOKOKU 249
(REPORTS OF THE NATURE CONSERVATION SOCIETY OF JAPAN)
NIHON SHIZEN HOGO KYOKAI (NATURE CONSERVATION SOCIETY OF JAPAN)
C/O TORANOMON DENKI BUILDING, 15, SHIBA NISHIKUBO AKEFUNE-CHO, MINATO-KU, TOKYO 105
JAPANESE

NIIGATA DAIGAKU GAKUHO 250
(ACADEMIC REPORT OF NIIGATA UNIVERSITY)
JAPANESE

NIIGATA DAIGAKU KYOIKUGAKUBU KIYO, SHIZENKAGAKU-HEN 251
(MEMOIRS OF THE FACULTY OF EDUCATION, NIIGATA UNIVERSITY)
NIIGATA DAIGAKU KYOIKUGAKUBU (THE FACULTY OF EDUCATION, NIIGATA UNIVERSITY)
746, NIBAN-CHO, ASAHIMACHI-DORI, NIIGATA-SHI 951
JAPANESE AND ENGLISH
INDEXED: CHE, JICST

NIIGATA DAIGAKU KYOIKUGAKUBU NAGAOKA BUNKO KENKYU KIYO 252
(RESEARCH BULLETIN OF THE NIIGATA UNIVERSITY SCHOOL OF EDUCATION NAGAOKA BRANCH SCHOOL)
JAPANESE

NIIGATA DAIGAKU KYOIKUGAKUBU TAKADA BUNKO KENKYU KIYO 253
(MEMOIRS OF TAKADA BRANCH, FACULTY OF EDUCATION, UNIVERSITY OF NIIGATA)
NIIGATA DAIGAKU KYOIKUGAKUBU TAKADA BUNKO (TAKADA BRANCH, FACULTY OF EDUCATION, UNIVERSITY OF NIIGATA)
1, SAIJO-CHO, TAKADA-SHI 943)
JAPANESE AND ENGLISH

NIIGATA DAIGAKU KYOYOBU KENKYU KIYO 254
(JOURNAL OF GENERAL EDUCATION DEPARTMENT)
NIIGATA DAIGAKU KYOYOBU (NIIGATA UNIVERSITY)
8050, IGARASHI NINO-MACHI, NIIGATA-SHI 950-21
JAPANESE AND ENGLISH

NIIGATA-KENRITSU KYOIKU SENTA KENKYU SHUROKU: RIKA KENKYU HEN, DAI 1 255
(COLLECTED RESEARCH PAPERS OF THE NIIGATA PREFECTURAL EDUCATION CENTER: SCIENTIFIC RESEARCH EDITION, NUMBER 1)
JAPANESE

NIIGATA-KENRITSU KYOIKU SENTA KENKYU SHUROKU: RIKA KENKYU HEN, DAI 2 256
(COLLECTED RESEARCH PAPERS OF THE NIIGATA PREFECTURAL EDUCATION CENTER: SCIENTIFIC RESEARCH EDITION, NUMBER 2)
JAPANESE

NIPPON SHIKA DAIGAKU SHINGAKU KATEI KIYO 257
(BULLETIN OF NIPPON DENTAL COLLEGE, GENERAL EDUCATION)
NIPPON SHIKA DAIGAKU SHINGAKU KATEI
9-20, 1-CHOME, FUJIMI, CHIYODA-KU, TOKYO 102
JAPANESE AND ENGLISH
INDEXED: CHE

NISHO GAKUSHA DAIGAKU RONSHU 258
(NISHO GAKUSHO UNIVERSITY PAPERS)
JAPANESE

NITCHO KAGAKU GIJUTSU 259
(REVIEW OF THE JAPAN-KOREA CO-OPERATIVE COMMITTEE FOR SCIENCE AND TECHNOLOGY)
NITCHO KAGAKU GIJUTSU KYORYOKU IINKAI
C/O MATSUSHIMA BUILDING, 1-17, KANDA JINBO-CHO, CHIYODA-KU, TOKYO 101
JAPANESE

NOTORUDAMU JOSHI DAIGAKU KENKYU KIYO 260
(MEMOIRS OF NOTRE DAME WOMEN'S COLLEGE)
NOTORUDAMU JOSHI DAIGAKU
1, SHIMOKAMO MINAMINONOGAMI-CHO, SAKYO-KU, KYOTO-SHI 606
JAPANESE AND ENGLISH

OBIRIN RONSHU IPPAN KYOYO-HEN 261
(OBIRIN REVIEW: GENERAL ARTS AND SCIENCES)
OBIRIN DAIGAKU TANKI DAIGAKU
(OBIRIN UNIVERSITY AND JUNIOR COLLEGE)
JAPANESE

OECD KAGAKU GIJUTSU JOHO SHIRYO 262
(OECD SCIENCE AND TECHNOLOGY INFORMATION MATERIALS)
JAPANESE
INDEXED: JICST

OITA DAIGAKU KYOIKUGAKUBU KENKYU KIYO, SHIZENKAGAKU 263
(RESEARCH BULLETIN OF THE FACULTY OF EDUCATION, OITA UNIVERSITY, NATURAL SCIENCE)
OITA DAIGAKU KYOIKUGAKUBU (FACULTY OF EDUCATION, OITA UNIVERSITY)
DANOHARA, OITA-SHI 870-11
JAPANESE AND EUROPEAN
INDEXED: CHE, MAT, JICST

OKAYAMA DAIGAKU KYOIKUGAKUBU KENKYU SHUROKU 264
(BULLETIN OF THE SCHOOL OF EDUCATION, OKAYAMA UNIVERSITY)
OKAYAMA DAIGAKU KYOIKUGAKUBU (THE SCHOOL OF EDUCATION, OKAYAMA UNIVERSITY)
TSUSHIMA, OKAYAMA-SHI 700
JAPANESE AND EUROPEAN

OKAYAMA DAIGAKU KYOYOGAKUBU KIYO 265
(BULLETIN OF THE FACULTY OF GENERAL EDUCATION, OKAYAMA UNIVERSITY)
JAPANESE

OKAYAMA DAIGAKU YOGO KYOYU YOSEIJO KIYO 266
(MEMOIRS OF THE TRAINING SCHOOL FOR THE NURSE-TEACHER, OKAYAMA UNIVERSITY)
OKAYAMA DAIGAKU YOGO KYOYU YOSEIJO
TSUSHIMA, OKAYAMA-SHI 700
JAPANESE AND ENGLISH

OKAYAMA RIKA DAIGAKU KIYO 267
(BULLETIN OF THE OKAYAMA COLLEGE OF SCIENCE)
OKAYAMA RIKA DAIGAKU (THE OKAYAMA COLLEGE OF SCIENCE)
957-10, SHUKU, OKAYAMA-SHI 700
JAPANESE AND EUROPEAN
INDEXED: JICST

OKAYAMA-KEN KYOIKU SENTA JOHO 268
(REPORT OF THE OKAYAMA PREFECTURAL EDUCATION CENTER)
JAPANESE

OKAYAMA-KEN KYOIKU SENTA JOHO: TOKUSHU 269
(REPORT OF THE OKAYAMA PREFECTURAL EDUCATION CENTER: SPECIAL ISSUE)
JAPANESE

OKIDAI RONSO 270
(COLLECTED TREATISES OF OKINAWA UNIVERSITY)
OKINAWA UNIVERSITY
JAPANESE

OKINAWA KOKUSAI DAIGAKU KYOYOBU KIYO 271
(JOURNAL OF THE GENERAL EDUCATION DEPARTMENT, OKINAWA KOKUSAI UNIVERSITY)
JAPANESE

OKINAWA-KENRITSU KYOIKU SENTA KIYO 272
(BULLETIN OF OKINAWA PREFECTURAL EDUCATION CENTER)
OKINAWA-KENRITSU KYOIKU SENTA
JAPANESE

OKUSHU DAIGAKU KIYO 273
(BULLETIN OF OKUSHU UNIVERSITY)
OKUSHU DAIGAKU GAKUJUTSU KENKYUKAI
(OKUSHU UNIVERSITY LEARNED RESEARCH SOCIETY)
JAPANESE

OSAKA DAIGAKU TEION SENTA DAYORI 274
(OSAKA UNIVERSITY LOW TEMPERATURE CENTER LETTER)
JAPANESE

OSAKA DENKI TSUSHIN DAIGAKU KENKYU RONSHU, SHIZENKAGAKU-HEN 275
(REPORT OF OSAKA ELECTRO-COMMUNICATION UNIVERSITY)
OSAKA DENKI TSUSHIN DAIGAKU (OSAKA ELECTRO-COMMUNICATION UNIVERSITY)
18-8, HATSU-CHO, NEYAGAWA-SHI 572
JAPANESE AND ENGLISH
INDEXED: JICST

OSAKA GAKUIN DAIGAKU JINMON SHIZEN RONSHU 276
(BULLETIN OF THE CULTURAL AND NATURAL SCIENCES, OSAKA GAKUIN UNIVERSITY)
OSAKA GAKUIN DAIGAKU JINMON SHIZEN GAKKAI
(SOCIETY FOR THE STUDY OF CULTURAL AND NATURAL SCIENCE, OSAKA GAKUIN UNIVERSITY)
JAPANESE

OSAKA GAKUIN DAIGAKU RONSO 277
(TRANSACTIONS OF OSAKA GAKUIN UNIVERSITY)
JAPANESE

OSAKA KOGYO DAIGAKU KIYO, RIKO-HEN 278
(MEMOIRS OF THE OSAKA INSTITUTE OF TECHNOLOGY, SCIENCE AND TECHNOLOGY)
OSAKA KOGYO DAIGAKU (OSAKA INSTITUTE OF TECHNOLOGY)
16-1, 5-CHOME, OMIYA, ASAHI-KU, OSAKA-SHI 535
JAPANESE AND ENGLISH
INDEXED: CHE, MAT, JICST

OSAKA KYOIKU DAIGAKU KIYO, 3, SHIZENKAGAKU 279
(MEMOIRS OF OSAKA KYOIKU UNIVERSITY, III, NATURAL SCIENCE AND APPLIED SCIENCE)
OSAKA KYOIKU DAIGAKU (OSAKA KYOIKU UNIVERSITY)
43, MINAMI KAWAHORI-CHO, TENNOJI-KU, OSAKA-SHI 543
JAPANESE AND EUROPEAN
INDEXED: BIO, MAT, JICST

OSAKA SANGYO DAIGAKU RONSHU, SHIZENKAGAKU-HEN 280
(JOURNAL OF OSAKA INDUSTRIAL UNIVERSITY, NATURAL SCIENCE)
OSAKA SANGYO DAIGAKU GAKKAI
1-1, 3-CHOME, NAKAGAKI-UCHI, DAITO-SHI 574
JAPANESE AND ENGLISH
INDEXED: CHE, JICST

OSAKA TO KAGAKU KYOIKU 281
(OSAKA AND SCIENCE EDUCATION)
OSAKA-FU KAGAKU KYOIKU SENTA (OSAKA SCIENCE EDUCATION CENTER)
JAPANESE

OSAKA-FURITSU DAIGAKU GAKUHO 282
(REPORT OF OSAKA MUNICIPAL UNIVERSITY)
JAPANESE

OSAKA-SHIRITSU SHIZEN-SHI HAKUBUTSUKAN KENKYU HOKOKU 283
(RESEARCH BULLETIN OF THE OSAKA MUSEUM OF NATURAL HISTORY)
JAPANESE

OSAKA-SHIRITSU SHIZENKAGAKU HAKUBUTSUKAN KANPO 284
(ANNUAL REPORT OF THE OSAKA MUSEUM OF NATURAL HISTORY)
OSAKA-SHIRITSU SHIZENKAGAKU HAKUBUTSUKAN (OSAKA MUSEUM OF NATURAL HISTORY)
C/O NAGAI KOEN, HIGASHINAGAI-CHO, HIGASHISUMIYOSHI-KU, OSAKA-SHI 546
JAPANESE

OSAKA-SHIRITSU SHIZENKAGAKU HAKUBUTSUKAN KENKYU HOKOKU 285
(BULLETIN OF THE OSAKA MUSEUM OF NATURAL HISTORY)
OSAKA-SHIRITSU SHIZENKAGAKU HAKUBUTSUKAN (OSAKA MUSEUM OF NATURAL HISTORY)
C/O NAGAI KOEN, HIGASHINAGAI-CHO, HIGASHISUMIYOSHI-KU, OSAKA-SHI 546
JAPANESE AND ENGLISH
INDEXED: BIO

OTANI DAIGAKU KENKYU NENPO 286
(OTANI UNIVERSITY RESEARCH ANNUAL)
OTANI GAKKAI (OTANI LEARNED SOCIETY)
JAPANESE

PROCEEDINGS OF THE FACULTY OF SCIENCE OF TOKAI UNIVERSITY 287
TOKAI DAIGAKU RIGAKUBU (FACULTY OF SCIENCE, TOKAI UNIVERSITY)
C/O TOKAI BUILDING, 27-4, 3-CHOME, SHINJUKU-KU, TOKYO 160
ENGLISH
INDEXED: MAT, JICST

PROCEEDINGS OF THE JAPAN ACADEMY 288
NIPPON GAKUSHIIN (THE JAPAN ACADEMY)
UENO KOEN, TAITO-KU, TOKYO 110
EUROPEAN
INDEXED: BIB, BIO, CHE, INS, MAT, MET, SCI, WAA, JICS

RANGAKU SHIRYO KENKYUKAI KENKYU HOKOKU 289
(REPORTS FROM THE JAPANESE SOCIETY FOR THE STUDY OF DUTCH LEARNING)
RANGAKU SHIRYO KENKYUKAI (SOCIETY FOR THE STUDY OF MATERIALS ON DUTCH LEARNING)
C/O OGATA IGAKU KAGAKU KENKYUSHO, 11-4, 1-CHOME, HIGASHIKANDA, CHIYODA-KU, TOKYO 101
JAPANESE

REITAKU DAIGAKU KIYO 290
(BULLETIN OF REITAKU UNIVERSITY)
JAPANESE
INDEXED: JICST

REPORT OF THE RESEARCH INSTITUTE OF SCIENCE AND TECHNOLOGY, NIHON UNIVERSITY 291
NIHON DAIGAKU RIKOGAKU KENKYUSHO (THE RESEARCH INSTITUTE OF SCIENCE AND TECHNOLOGY, NIHON UNIVERSITY)
1-8, KANDA SURUGADAI, CHIYODA-KU, TOKYO 101
ENGLISH
INDEXED: APP, CHE, JICST

REPORT OF SCIENCE AND ENGINEERING RESEARCH LABORATORY 292
WASEDA DAIGAKU RIKOGAKU KENKYUSHO (SCIENCE AND ENGINEERING RESEARCH LABORATORY, WASEDA UNIVERSITY)
17, KIKUI-CHO, SHINJUKU-KU, TOKYO 162
ENGLISH

REPORTS OF FACULTY OF SCIENCE, SHIZUOKA UNIVERSITY 293
SHIZUOKA DAIGAKU RIGAKUBU (FACULTY OF SCIENCE, SHIZUOKA UNIVERSITY)
836, OYA, SHIZUOKA-SHI 420
EUROPEAN
INDEXED: BIO, CHE, MAT, JICST

RIKA 294
(SCIENCE)
NIHON RIKA KYOIKU KYOKAI (JAPAN SCIENCE EDUCATION ISSUE)
JAPANESE

RIKA NENPYO 295
(SCIENCE ALMANAC)
TOKYO TENMONDAI (TOKYO ASTRONOMICAL OBSERVATORY)
3-10, 2-CHOME, NIHONBASHI, CHUO-KU, TOKYO 103
JAPANESE

RIKAGAKKAISHI 296
(JOURNAL OF PHYSICS, CHEMISTRY AND EARTH SCIENCE)
TOYAMA-KEN RIKA GAKKAI
C/O TOYAMA-KEN KAGAKU KYOIKU CENTER, 2238, GOFUKU, TOYAMA-SHI 930
JAPANESE

RIKAGAKU KENKYUSHO HOKOKU 297
(REPORTS OF THE INSTITUTE OF PHYSICAL AND CHEMICAL RESEARCH)
RIKAGAKU KENKYUSHO (THE INSTITUTE OF PHYSICAL AND CHEMICAL RESEARCH)
2-1, HIROSAWA, WAKO-SHI 351
JAPANESE, WITH ENGLISH SUMMARY
INDEXED: CHE, INS, WAA, JICST

RIKAGAKU KENKYUSHO KENKYU NENPO 298
(ANNUAL REPORTS OF RESEARCH ACTIVITIES OF THE INSTITUTE OF PHYSICAL AND CHEMICAL RESEARCH)
RIKAGAKU KENKYUSHO (THE INSTITUTE OF PHYSICAL AND CHEMICAL RESEARCH)
2-1, HIROSAWA, WAKO-SHI 351
JAPANESE
INDEXED: CHE, JICST

RIKAGAKU KENKYUSHO KENKYU RONBUNSHU 299
(COLLECTED PAPERS OF INSTITUTE OF PHYSICAL AND CHEMICAL RESEARCH)
RIKAGAKU KENKYUSHO (INSTITUTE OF PHYSICAL AND CHEMICAL RESEARCH)
2-1, HIROSAWA, WAKO-SHI 351
JAPANESE AND ENGLISH
INDEXED: JICST

RIKKYO DAIGAKU KENKYU HOKOKU, SHIZENKAGAKU 300
(ST. PAUL'S REVIEW OF SCIENCE)
RIKKYO DAIGAKU IPPAN KYOIKUBU (COLLEGE OF GENERAL EDUCATION, RIKKYO UNIVERSITY)
3, NISHI IKEBUKURO, TOSHIMA-KU, TOKYO 171
JAPANESE AND ENGLISH

RISSHO DAIGAKU KYOYOBU KIYO 301
(BULLETIN OF THE RISSHO UNIVERSITY GENERAL EDUCATION DEPARTMENT)
JAPANESE

RISSHO JOSHI DAIGAKU KIYO 302
(BULLETIN OF THE RISSHO WOMEN'S UNIVERSITY)
JAPANESE

RITSUMEIKAN DAIGAKU DAIGAKUIN RONSHU 303
(COLLECTED PAPERS OF THE RITSUMEIKAN UNIVERSITY GRADUATE SCHOOL)
RITSUMEIKAN DAIGAKU DAIGAKUINSEI KYOGIKAI (RITSUMEIKAN UNIVERSITY GRADUATE STUDENTS COUNCIL)
JAPANESE

RITSUMEIKAN DAIGAKU RIKOGAKU KENKYUSHO KIYO 304
(MEMOIRS OF THE RESEARCH INSTITUTE OF SCIENCE AND ENGINEERING, RITSUMEIKAN UNIVERSITY)
RITSUMEIKAN DAIGAKU RIKOGAKU KENKYUSHO (THE RESEARCH INSTITUTE OF SCIENCE AND ENGINEERING, RITSUMEIKAN UNIVERSITY)
TOJIIN KITA-MACHI, KITA-KU, KYOTO-SHI 603
JAPANESE AND EUROPEAN
INDEXED: CHE, INS, JICST

ROKKODAI RONSHU 305
(COLLECTED PAPERS OF ROKKO OBSERVATORY)
KOBE DAIGAKU DAIGAKUIN KENKYU KAI (KOBE UNIVERSITY GRADUATE SCHOOL RESEARCH SOCIETY)
JAPANESE

RYUKYU DAIGAKU RIKO GAKUBU KIYO: RIGAKU-HEN 306
(BULLETIN OF THE RYUKYU UNIVERSITY FACULTY OF SCIENCE AND ENGINEERING: SCIENCE EDITION)
JAPANESE

RYUTANI DAIGAKU RONSHU 307
(COLLECTED PAPERS OF RYUTANI UNIVERSITY)
RYUTANI GAKKAI (THE RYUTANI LEARNED SOCIETY)
JAPANESE

SADO HAKUBUTSUKAN KANPO 308
(BULLETIN OF SADO MUSEUM)
SADO HAKUBUTSUKAN (THE SADO MUSEUM)
NAKAE, YAWATA, SAWADA-CHO, SADO-GUN, NIIGATA 952-13
JAPANESE

SAGA DAIGAKU KYOIKUGAKUBU KENKYU RONBUNSHU 309
(REPORT OF THE FACULTY OF EDUCATION, SAGA UNIVERSITY)
SAGA DAIGAKU KYOIKUGAKUBU (FACULTY OF EDUCATION, SAGA UNIVERSITY)
1, HONJO-MACHI, SAGA-SHI 840
JAPANESE AND ENGLISH

SAGA DAIGAKU KYOYOBU KENKYU KIYO 310
(JOURNAL OF THE FACULTY OF LIBERAL ARTS, SAGA UNIVERSITY)
SAGA DAIGAKU KYOYOBU (THE FACULTY OF LIBERAL ARTS, SAGA UNIVERSITY)
1, HONJO-MACHI, SAGA-SHI 840
JAPANESE AND ENGLISH

SAGA DAIGAKU RIKOGAKUBU SHUHO 311
(COLLECTIVE REPORTS OF THE FACULTY OF SCIENCE AND ENGINEERING, SAGA UNIVERSITY)
JAPANESE

SAGAMI JOSHI DAIGAKU KIYO 312
(JOURNAL OF SAGAMI WOMEN'S UNIVERSITY)
SAGAMI JOSHI DAIGAKU (SAGAMI WOMEN'S UNIVERSITY)
1-1, 2-CHOME, BUNKYO, SAGAMIHARA-SHI 228
JAPANESE
INDEXED: JICST

SAIENSU 313
(SCIENCE)
NIHON KEIZAI SHINBUNSHA
9-5, 1-CHOME, OTE-MACHI, CHIYODA-KU, TOKYO 100
JAPANESE
INDEXED: JICST

SAIENSU EKO 314
(SCIENCE ECHO)
GAKUSHU KENKYUSHA (LEARNING RESEARCH PUBLISHING COMPANY)
JAPANESE

SAITAMA DAIGAKU KIYO 315
(BULLETIN OF THE SAITAMA UNIVERSITY)
SAITAMA DAIGAKU KYOYO GAKUBU (SAITAMA UNIVERSITY, DEPARTMENT OF ARTS AND SCIENCES
JAPANESE

SAITAMA DAIGAKU KIYO, SHIZENKAGAKU-HEN 316
(JOURNAL OF SAITAMA UNIVERSITY, NATURAL SCIENCE)
SAITAMA DAIGAKU KYOYOBU (COLLEGE OF LIBERAL ART, SAITAMA UNIVERISTY)
225, SHIMOOKUBO, URAWA-SHI 338
JAPANESE AND ENGLISH
INDEXED: CHE, JICST

SAITAMA DAIGAKU KIYO. KYOIKU GAKUBU. SUGAKU, SHIZENKAGAKU 317
(JOURNAL OF SAITAMA UNIVERSITY. MATHEMATICS AND NATURAL SCIENCE)
JAPANESE

SAITAMA DAIGAKU RIKOGAKUBU JIBAN SUIRI SHISETSU NENPO 318
(ANNUAL REPORT OF HYDROSCIENCE AND GEOTECHNOLOGY LABORATORY, FACULTY OF SCIENCE AND ENGINEERING, SAITAMA UNIVERSITY)
JAPANESE
INDEXED: JICST

SAITO HOONKAI MUSEUM RESEARCH BULLETIN 319
SAITO HOONKAI (SAITO GRATIUDE FOUNDATION)
2-20, HON-CHO, SENDAI-SHI 980
JAPANESE AND ENGLISH
INDEXED: BIO

SAN'IN BUNKA KENKYU KIYO, SHIZEN KAGAKU-HEN 320
(MEMOIRS OF NATURAL AND CULTURAL RESEARCHES OF THE SAN'IN REGION)
SHIMANE DAIGAKU (SHIMANE UNIVERSITY)
1060, NISHIKAWAZU-MACHI, MATSUE-SHI 690
JAPANESE AND ENGLISH
INDEXED: BIO

SCIENCE REPORTS OF THE COLLEGE OF GENERAL EDUCATION, OSAKA UNIVERSITY 321
OSAKA DAIGAKU KYOYOBU (COLLEGE OF GENERAL EDUCATION, OSAKA UNIVERSITY)
1-1, MACHIKANEYAMA-CHO, TOYONAKA-SHI 560
EUROPEAN
INDEXED: BIO, CHE, INS, MAT, JICST

SCIENCE REPORTS OF THE HIROSAKI UNIVERSITY 322
HIROSAKI DAIGAKU RIGAKUBU (FACULTY OF SCIENCE, KANAZAWA UNIVERSITY)
3-BUNKYO-CHO, HIROSAKI-SHI 036
JAPANESE, WITH EUROPEAN SUMMARY
INDEXED: CHE, MAT, JICST

SCIENCE REPORTS OF KANAZAWA UNIVERSITY 323
KANAZAWA DAIGAKU RIGAKUBU (THE FACULTY OF SCIENCE, KANAZAWA UNIVERSITY
1-1, MARUNOUCHI, KANAZAWA-SHI 920
EUROPEAN
INDEXED: CHE, MAT, JICST

SCIENCE REPORTS OF THE RESEARCH INSTITUTE TOHOKU UNIVERSITY, SERIES A, PHYSICS, CHEMISTRY AND METALLURGY 324
TOHOKU DAIGAKU KENKYUSHO RENGOKAI (THE ASSOCIATION OF THE RESEARCH INSTITUTE TOHOKU UNIVERSITY)
C/O TOHOKU DAIGAKU KINZOKU ZAIRYO KENKYUSHO, 1-1, 2-CHOME, KATAHIRA, SENDAI-SHI 980
EUROPEAN
INDEXED: CHE, ENG, INS, MET, SCI, WAA, JICST

SCIENCE REPORTS OF THE SAITAMA UNIVERSITY, SERIES A, MATHEMATICS, PHYSICS AND CHEMISTRY 325
SAITAMA DAIGAKU (SAITAMA UNIVERSITY)
255, SHIMOOKUBO, URAWA-SHI 338
ENGLISH
INDEXED: CHE, MAT, JICST

SCIENCE REPORTS OF THE TOHOKU UNIVERSITY, SERIES I, PHYSICS, CHEMISTRY, ASTRONAUTICS 326
TOHOKU DAIGAKU (TOHOKU UNIVERSITY)
1-1, 2-CHOME, KATAHIRA, SENDAI-SHI 980
ENGLISH
INDEXED: APP, BIB, CHE, IAA, INS. JICST

SCIENCE REPORTS OF THE TOKYO KYOIKU DAIGAKU, SECTION A, MATHEMATICS AND PHYSICS 327
TOKYO KYOIKU DAIGAKU RIGAKUBU (FACULTY OF SCIENCE, TOKYO UNIVERSITY OF EDUCATION)
29-1, 3-CHOME, OTSUKA, BUNKYO-KU, TOKYO 112
EUROPEAN
INDEXED: CHE, MAT

SCIENCE REPORTS OF THE YOKOHAMA NATIONAL UNIVERSITY, SECTION 1, MATHEMATICS, PHYSICS, CHEMISTRY 328
YOKOHAMA KOKURITSU DAIGAKU KYOIKUGAKUBU (FACULTY OF EDUCATION, YOKOHAMA NATIONAL UNIVERSITY)
SHIMIZUGAOKA, MINAMI-KU, YOKOHAMA-SHI 232
EUROPEAN
INDEXED: CHE, MAT, JICST

SCIENTIFIC PAPERS OF THE COLLEGE OF GENERAL EDUCATION, UNIVERSITY OF TOKYO 329
TOKYO DAIGAKU KYOYOGAKUBU (THE COLLEGE OF GENERAL EDUCATION, UNIVERSITY OF TOKYO)
865, KOMABA-CHO, MEGURO-KU, TOKYO 153
EUROPEAN
INDEXED: APP, BIB, BIO, CHE, INS, MAT, JICST

SCIENTIFIC PAPERS OF THE INSTITUTE OF PHYSICAL AND CHEMICAL RESEARCH 330
RIKAGAKU KENKYUSHO (INSTITUTE OF PHYSICAL AND CHEMICAL RESEARCH)
2-1, HIROSAWA, WAKO-SHI 351
ENGLISH
INDEXED: APP, BIB, CHE, ENG, INS, JICST

SEI MARIANNA IKA DAIGAKU KIYO, IPPAN KYOIKU 331
(BULLETIN OF THE ST. MARIANNA UNIVERSITY, SCHOOL OF MEDICINE (GENERAL EDUCATION))
SEI MARIANNA IKA DAIGAKU
2095, SUGAO, TAKATSU-KU, KAWASAKI-SHI 213
JAPANESE AND ENGLISH
INDEXED: CHE

SEIKEI DAIGAKU IPPAN KENKYU HOKOKU 332
(GENERAL RESEARCH REPORT OF SEIKEI UNIVERSITY)
JAPANESE

SEINAN GAKUIN DAIGAKU BUNRI RONSHU 333
(LITERATURE AND SCIENCE REPORTS OF SEINAN GAKUIN UNIVERSITY)
SEINAN GAKUIN DAIGAKU GAKUJUTSU KENKYUSHO
2-92, 6-CHOME, NISHIJIN, NISHI-KU, FUKUOKA-SHI 814
JAPANESE

SEINAN GAKUIN DAIGAKU GAKUJUTSU KENKYUJO KIYO 334
(BULLETIN OF SEINAN GAKUIN ACADEMIC RESEARCH INSTITUTE)
JAPANESE

SEISHIN JOSHI DAIGAKU RONSO 335
(COLLECTED PAPERS OF SACRED HEART WOMEN'S UNIVERSITY)
JAPANESE

SENDAI DAIGAKU KIYO 336
(BULLETIN OF SENDAI UNIVERSITY)
SENDAI DAIGAKU (SENDAI UNIVERSITY)
JAPANESE

SENDAI-SHI KAGAKUKAN JIHO 337
(REVIEW OF SENDAI CITY SCIENCE MUSEUM)
SENDAI-SHI KAGAKUKAN
1-2, 2-CHOME, ICHIBAN-CHO, SENDAI-SHI 980
JAPANESE

SENSHU SHIZEN KAGAKU KIYO 338
(BULLETIN OF THE ASSOCIATION OF NATURAL SCIENCE, SENSHU UNIVERSITY)
SENSHU DAIGAKU SHIZEN KAGAKU KENKYUKAI
(SENSHU UNIVERSITY NATURAL SCIENCE RESEARCH SOCIETY)
JAPANESE

SENSOKU RONSO 339
(COLLECTED PAPERS OF SENSOKU UNIVERSITY AND JUNIOR COLLEGE)
SENSOKU GAKUEN DAIGAKU.TANKI DAIGAKU
(SENSOKU GAKUEN UNIVERSITY AND JUNIOR COLLEGE)
JAPANESE

SHIGA DAIGAKU KYOIKUGAKUBU KIYO, SHIZENKAGAKU 340
(MEMOIRS OF THE FACULTY OF EDUCATION, SHIGA UNIVERSITY, NATURAL SCIENCE)
SHIGA DAIGAKU KYOIKUGAKUBU (THE FACULTY OF EDUCATION, SHIGA UNIVERSITY)
ISHIYAMA HIRATSU-CHO, OTSU-SHI 520
JAPANESE AND ENGLISH
INDEXED: BIO, CHE, JICST

SHIGA-KENRITSU TANKI DAIGAKU GAKUJUTSU ZASSHI 341
(SCIENTIFIC REPORTS OF SHIGA PREFECTURAL JUNIOR COLLEGE)
SHIGA-KENRITSU TANKI DAIGAKU (SHIGA PREFECTURAL JUNIOR COLLEGE)
HASSAKA-CHO, HIKONE-SHI 522
JAPANESE AND EUROPEAN
INDEXED: CHE, JICST

SHIMANE DAIGAKU KYOIKUGAKUBU KIYO, SHIZENKAGAKU 342
(MEMOIRS OF THE FACULTY OF EDUCATION, SHIMANE UNIVERSITY, NATURAL SCIENCE)
SHIMANE DAIGAKU KYOIKUGAKUBU (FACULTY OF EDUCATION, SHIMANE UNIVERSITY)
1060, NISHIKAWATSU-MACHI, MATSUE-SHI 690
JAPANESE AND EUROPEAN
INDEXED: BIO, CHE, MAT, JICST

SHINSHU DAIGAKU KYOIKUGAKUBU KIYO 343
(JOURNAL OF THE FACULTY OF EDUCATION, SHINSHU UNIVERSITY)
SHINSHU DAIGAKU KYOIKUGAKUBU (FACULTY OF EDUCATION, SHINSHU UNIVERSITY)
NISHINAGANO, NAGANO-SHI 380
JAPANESE AND ENGLISH
INDEXED: JICST

SHINSHU DAIGAKU KYOYOBU KIYO DAI-NIBU SHIZEN KAGAKU 344
(BULLETIN OF THE ARTS AND SCIENCE FACULTY OF SHINSHU UNIVERSITY, PART TWO, NATURAL SCIENCE)
JAPANESE

SHIRITSU ASAHIKAWA KYODO HAKUBUTSUKAN KENKYU HOKOKU 345
(SCIENCE REPORT OF THE ASAHIKAWA MUSEUM)
SHIRITSU ASAHIKAWA KYODO HAKUBUTSUKAN (THE ASAHIKAWA MUSEUM)
1-JO 1-CHOME, 4-KU, ASAHIKAWA-SHI 070
JAPANESE, WITH ENGLISH SUMMARY

SHIRITSU NAGOYA KAGAKUKAN JIGYO GAIYO 346
(ANNUAL REPORT OF NAGOYA MUNICIPAL SCIENCE MUSEUM)
SHIRITSU NAGOYA KAGAKUKAN (NAGOYA MUNICIPAL SCIENCE MUSEUM)
17-22, 2-CHOME, SAKAE, NAKA-KU, NAGOYA-SHI 460
JAPANESE

SHIRITSU NAGOYA KAGAKUKAN KIYO 347
(SCIENCE REPORT OF NAGOYA MUNICIPAL SCIENCE MUSEUM)
SHIRITSU NAGOYA KAGAKUKAN (NAGOYA MUNICIPAL SCIENCE MUSEUM)
17-22, 2-CHOME, SAKAE, NAKA-KU, NAGOYA-SHI 460
JAPANESE

SHIRITSU NAGOYA KAGAKUKAN HYUSU 348
(NAGOYA MUNICIPAL SCIENCE MUSEUM MONTHLY)
SHIRITSU NAGOYA KAGAKUKAN (NAGOYA MUNICIPAL SCIENCE MUSEUM)
17-22, 2-CHOME, SAKAE, NAKA-KU, NAGOYA-SHI 460
JAPANESE

SHIZEN 349
(NATURE)
CHUO KORONSHA
2-1, KYOBASHI, CHUO-KU, TOKYO 104
JAPANESE
INDEXED: JICST

SHIZEN HOGO 350
(CONSERVATION OF NATURE)
NIHON SHIZEN HOGO KYOKAI (NATURE CONSERVATION SOCIETY OF JAPAN)
15, SHIBA NISHIKUBO AKEFUNE-CHO, MINATO-KU, TOKYO 105
JAPANESE
INDEXED: JICST

SHIZEN KAGAKU KENKYU 351
(STUDIES IN SCIENCES)
HITOTSUBASHI GAKKAI (THE HITOTSUBASHI ACADEMY)
C/O HITOTSUBASHI DAIGAKU, 2-1, NAKA, KUNITACHI-SHI, TOKYO 186
JAPANESE
INDEXED: JICST

SHIZEN KAGAKU RONSO 352
(BULLETIN OF THE SOCIETY OF NATURAL SCIENCE AND PHYSICAL EDUCATION, WOMEN'S UNIVERSITY OF KYOTO)
KYOTO JOSHI DAIGAKU SHIZEN KAGAKU HOKEN TAIIKUGAKKAI (THE SOCIETY OF NATURAL SCIENCE AND PHYSICAL EDUCATION WOMEN'S UNIVERSITY
17, KITAHIYOSHI-CHO, IMAGUMANO, HIGASHI-KU, KYOTO-SHI 605
JAPANESE, WITH ENGLISH SUMMARY

SHIZEN KAGAKU TO HAKUBUTSUKAN, KAGAKUNOME 353
(NATURAL SCIENCE AND MUSEUMS)
KAGAKU HAKUBUTSUKAN KOENKAI (ASSOCIATION FOR PROMOTION OF SCIENCE MUSEUM IN JAPAN)
7, UENOKOEN, TAITO-KU, TOKYO 110
JAPANESE
INDEXED: JICST

SHIZEN SAIGAI KAGAKU SHIRYO KAISEKI KENKYU 354
(ANALYSIS OF SCIENTIFIC DATA ON NATURAL DISASTERS)
SHIZEN SAIGAI KAGAKU SHIRYO SHUSHU KAISEKI SOGO KENKYU HAN (GENERAL RESEARCH UNIT FOR ANALYSIS OF COLLECTED SCIENTIFIC DATA ON NATURAL DISASTERS)
JAPANESE

SHIZEN TO SHAKAI, HOKURIKU 355
(NATURE AND SOCIETY, HOKURIKU)
TOYAMA CHIGAKKAI, ISHIKAWA CHIRIGAKKAI, FUKUI CHIRIGAKKAI (TOYAMA GEOGRAPHICAL SOCIETY, ISHIKAWA GEOGRAPHICAL SOCIETY, FUKUI GEOGRAPHICAL SOCIETY)
C/O TOYAMA DAIGAKU CHIGAKU KENKYUSHITSU, GOFUKU, TOYAMA-SHI 930
JAPANESE

SHIZENSHI KENKYU 356
(OCCASIONAL PAPERS FROM THE OSAKA MUSEUM OF NATURAL HISTORY)
OSAKA-SHIRITSU SHIZENSHI HAKUBUTSUKAN (OSAKA MUSEUM OF NATURAL HISTORY)
C/O NAGAI KOEN, HIGASHINAGAI-CHO, HIGASHISUMIYOSHI-KU, OSAKA-SHI 546
JAPANESE AND ENGLISH
INDEXED: BIO

SHIZUOKA DAIGAKU KYOIKUGAKUBU KENKYU HOKOKU, SHIZENKAGAKU-HEN 357
(BULLETIN, FACULTY OF EDUCATION SHIZUOKA UNIVERSITY, NATURAL SCIENCE)
SHIZUOKA DAIGAKU KYOIKUGAKUBU (FACULTY OF EDUCATION SHIZUOKA UNIVERSITY)
836, OYA, SHIZUOKA-SHI 420
JAPANESE AND EUROPEAN
INDEXED: JICST

SHIZUOKA DAIGAKU KYOYOBU KENKYU HOKOKU, SHIZEN KAGAKU HEN 358
(RESEARCH REPORT OF THE SHIZUOKA UNIVERSITY ARTS AND SCIENCES FACULTY, NATURAL SCIENCE EDITION)
JAPANESE

SHIZUOKA YAKKA DAIGAKU IPPAN KYOYO KENKYU KIYO 359
(REPORT OF GENERAL EDUCATION STUDY OF SHIZUOKA MEDICAL COLLEGE)
SHIZUOKA YAKKA DAIGAKU KYOYO KENKYUSHITSU
2-1, 2-CHOME, OJIKA, SHIZUOKA-SHI 420
JAPANESE AND ENGLISH

SHOWA DAIGAKU KYOYOBU KIYO 360
(BULLETIN OF THE ARTS AND SCIENCE FACULTY OF SHOWA UNIVERSITY)
JAPANESE

SHUDAI RONSO 361
(PAPERS OF SHUDO UNIVERSITY)
HIROSHIMA SHUDO DAIGAKU DAIGAKUIN KENKYUKAI
HIROSHIMA SHUDO UNIVERSITY, GRADUATE SCHOOL, RESEARCH SOCIETY
JAPANESE

SOFIA 362
(SOPHIA)
JOCHI DAIGAKU (SOPHIA UNIVERSITY)
JAPANESE

SONODA GAKUEN JOSHI DAIGAKU RONBUNSHU 363
(PAPERS OF THE SONODA GAKUEN WOMEN'S UNIVERSITY)
JAPANESE
INDEXED: JICST

SUZUGAMINE JOSHI TANDAI KENKYUSHUHO SHIZEN KAGAKU 364
(BULLETIN OF THE SUZUGAMINE WOMEN'S COLLEGE NATURAL SCIENCE)
SUZUGAMINE JOSHI TANKI DAIGAKU (SUZUGAMINE WOMEN'S COLLEGE)
6-18, 4-CHOME, INOKUCHI, HIROSHIMA-SHI 733
JAPANESE AND ENGLISH
INDEXED: BIO

TAISHO DAIGAKU DAIGAKUIN KENKYU RONSHU 365
(RESEARCH PAPERS OF THE GRADUATE SCHOOL, TAISHO UNIVERSITY)
JAPANESE

TAISHO DAIGAKU GAKUHO 366
(TAISHO UNIVERSITY GAZETTE)
TAISHO DAIGAKU SHUPPANBU (TAISHO UNIVERSITY PUBLICATIONS)
JAPANESE

TAKADA GAKUHO 367
(TAKADA LEARNED SOCIETY GAZETTE)
TAKADA GAKKAI (TAKADA LEARNED SOCIETY)
JAPANESE

TAKUSHOKU DAIGAKU DAIGAKUIN KENKYU NENPO 368
(ANNUAL ACADEMIC RESEARCH REPORT OF THE TAKUSHOKU UNIVERSITY, GRADUATE SCHOOL)
TAKUSHOKU UNIVERSITY GRADUATE SCHOOL
JAPANESE

TAKUSHOKU DAIGAKU RONSHU 369
(PAPERS OF THE TAKUSHOKU UNIVERSITY)
JAPANESE

TENNEN KINENBUTSU SHOHO 370
(REPORTS ON THE NATURAL MONUMENTS IN CHIBA PREFECTURE CHIBA-KEN KYOIKU IINKAI)
CHIBA-KEN KYOIKU IINKAI
1-1, ICHIBA-CHO, CHIBA-SHI 280-91
JAPANESE

TENRI DAIGAKU GAKUHO, JINBUN, SHAKAI, SHIZENHEN 371
(BULLETIN OF TENRI UNIVERSITY, HUMAN SCIENCES, SOCIAL SCIENCES, NATURAL SCIENCES)
TENRI DAIGAKU GAKUJUTSU KENKYUKAI (ACADEMIC RESEARCH ASSOCIATION TENRI UNIVERSITY)
SOMANOUCHI-CHO, TENRI-SHI 632
JAPANESE AND EUROPEAN

TENRI DAIGAKU OYASATO KENKYUSHO KENKYU HOKOKU 372
(RESEARCH REPORT OF THE TENRI UNIVERSITY, OYASATO RESEARCH INSTITUTE)
JAPANESE

TEZUKAYAMA DAIGAKU KIYO 373
(BULLETIN OF THE TEZUKAYAMA UNIVERSITY)
JAPANESE

TEZUKAYAMA DAIGAKU RONSHU 374
(PAPERS OF THE TEZUKAYAMA UNIVERSITY)
JAPANESE

TEZUKAYAMA GAKUIN DAIGAKU KENKYU RONSHU 375
(RESEARCH PAPERS OF THE TEZUKAYAMA GAKUIN UNIVERSITY)
JAPANESE

TEZUKAYAMA GAKUIN TANKI DAIGAKU KENKYU NENPO 376
(TEZUKAYAMA GAKUIN JUNIOR COLLEGE ANNUAL REPORT OF SCIENTIFIC STUDIES)
TEZUKAYAMA GAKUIN TANKI DAIGAKU
3, TEZUKAYAMA NAKA, SUMIYOSHI-KU, OSAKA-SHI 558
JAPANESE AND ENGLISH
INDEXED: JICST

TEZUKAYAMA TANKI DAIGAKU KIYO, SHIZEN KAGAKU 377
(JOURNAL OF THE TEZUKAYAMA JUNIOR COLLEGE. NATURAL SCIENCE)
JAPANESE

TOA DAIGAKU GAKUJUTSU KENKYUJO KENKYU RONSO 378
(RESEARCH PAPERS OF TOA UNIVERSITY SCIENTIFIC RESEARCH CENTER)
JAPANESE

TOHOKU DAIGAKU GAKUHO 379
(REPORT OF TOHOKU UNIVERSITY)
JAPANESE

TOHOKU DAIGAKU KAGAKU KEISOKU KENKYUJO NENJI YORAN 380
(ANNUAL SURVEY OF SCIENTIFIC INSTRUMENTATION INSTITUTE OF TOHOKU UNIVERSITY)
JAPANESE
INDEXED: JICST

TOHOKU FUKUSHI DAIGAKU KIYO 381
(BULLETIN OF TOHOKU FUKUSHI UNIVERSITY)
JAPANESE

TOHOKU KOGYO DAIGAKU KIYO, B: RIKAKEI-HEN 382
(BULLETIN OF THE TOHOKU INSTITUTE OF TECHNOLOGY SECTION B)
TOHOKU KOGYO DAIGAKU (TOHOKU INSTITUTE OF TECHNOLOGY)
19, KOSHIJI, NAGA-MACHI, SENDAI-SHI 982
JAPANESE AND ENGLISH

TOKAI CHIKU KAGAKU SHISETSU NYUSU 383
(TOKAI REGION SCIENCE FACILITIES NEWS)
TOKAI CHIKU KAGAKU SHISETSU KYOGIKAI
(COUNCIL OF THE TOKAI REGION SCIENCE FACILITIES)
JAPANESE

TOKAI DAIGAKU KIYO KYOYO GAKUBU 384
(JOURNAL OF THE FACULTY OF LIBERAL ARTS, TOKAI UNIVERSITY)
TOKAI DAIGAKU SHUPPANKAI
27-4, 3-CHOME, SHINJUKU, SHINJUKU-KU, TOKYO 160
JAPANESE AND ENGLISH

TOKAI DAIGAKU KIYO NUMAZU KYOYOBU 385
(BULLETIN OF TOKAI UNIVERSITY, NUMAZU FACULTY OF GENERAL EDUCATION)
TOKAI DAIGAKU NUMAZU KYOYOBU KIYO IINKAI
JAPANESE

TOKAI DAIGAKU TANKI DAIGAKUBU KIYO 386
(PROCEEDINGS OF THE JUNIOR COLLEGE OF TOKAI UNIVERSITY)
TOKAI DAIGAKU TANKI DAIGAKUBU (JUNIOR COLLEGE OF TOKAI UNIVERSITY) AND TOKAI DAIGAKU SHUPPANKAI
C/O TOKAI BUILDING, 27-4, 3-CHOME, SHINJUKU, SHINJUKU-KU, TOKYO 160
JAPANESE AND ENGLISH

TOKUSHIMA BUNRI DAIGAKU, TOKUSHIMA BUNRI DAIGAKU KENKYU KIYO 387
(TECHNICAL BULLETIN OF TOKUSHIMA UNIVERSITY OF ARTS AND SCIENCE)
TOKUSHIMA BUNRI DAIGAKU (TOKUSHIMA UNIVERSITY OF ARTS AND SCIENCE)
YAMASHIRO-CHO, TOKUSHIMA-SHI 770
JAPANESE AND ENGLISH

TOKUSHIMA DAIGAKU GAKUGEI KIYO, SHIZENKAGAKU 388
(JOURNAL OF GAKUGEI, TOKUSHIMA UNIVERSITY, NATURAL SCIENCE)
TOKUSHIMA DAIGAKU KYOIKU GAKUBU (FACULTY OF EDUCATION TOKUSHIMA UNIVERSITY)
1-1, MINAMI JOSANJIMA-CHO, TOKUSHIMA-SHI 770
JAPANESE AND EUROPEAN
INDEXED: BIO, CHE, JICST

TOKUSHIMA DAIGAKU KYOYOBU KIYO, SHIZENKAGAKU 389
(JOURNAL OF SCIENCE, COLLEGE OF GENERAL EDUCATION, UNIVERSITY OF TOKUSHIMA)
TOKUSHIMA DAIGAKU KYOYOBU (COLLEGE OF GENERAL EDUCATION UNIVERSITY OF TOKUSHIMA)
1-1, MINAMI JOSANJIMA-CHO, TOKUSHIMA-SHI 770
JAPANESE AND ENGLISH
INDEXED: JICST

TOKUYAMA DAIGAKU RONSO 390
(PAPERS AT TOKUYAMA UNIVERSITY)
JAPANESE

TOKYO DAIGAKU SOGO KENKYU SHIRYOKAN GYOSEKISHU 391
(THE UNIVERSITY MUSEUM UNIVERSITY OF TOKYO COLLECTED REPRINTS)
TOKYO DAIGAKU SOGO KENKYU SHIRYOKAN (UNIVERSITY MUSEUM UNIVERSITY OF TOKYO)
3-1, 7-CHOME, HONGO, BUNKYO-KU, TOKYO 113
JAPANESE AND ENGLISH

TOKYO GAKUGEI DAIGAKU KIYO, 4, SUGAKU, SHIZENKAGAKU 392
(BULLETIN OF TOKYO GAKUGEI UNIVERSITY, SERIES 4, MATHEMATICS AND NATURAL SCIENCES)
TOKYO GAKUGEI DAIGAKU (TOKYO GAKUGEI UNIVERSITY)
1-1, 4-CHOME, NUKUI KITA-MACHI, KOGANEI-SHI, TOKYO 184
JAPANESE AND ENGLISH
INDEXED: CHE, INS, MAT, JICST

TOKYO JOSHI DAIGAKU RONSHU, KAGAKU BUMON HOKOKU 393
(SCIENCE REPORTS OF TOKYO WOMEN'S CHRISTIAN COLLEGE)
TOKYO JOSHI DAIGAKU GAKKAI (THE ACADEMIC SOCIETY OF TOKYO WOMEN'S CHRISTIAN COLLEGE)
2-CHOME, ZENPUKUJI, SUGINAMI-KU, TOKYO 167
JAPANESE AND ENGLISH
INDEXED: MAT, JICST

TOKYO KEIZAI DAIGAKU JINBUN SHIZEN KAGAKU RONSHU 394
(TOKYO ECONOMICS UNIVERSITY HUMANITIES AND NATURAL SCIENCE PAPERS)
JAPANESE

TOKYO KOKURITSU HAKUBUTSUKAN KIYO 395
(PROCEEDINGS OF THE TOKYO NATIONAL MUSEUM)
TOKYO KOKURITSU HAKUBUTSUKAN (TOKYO NATIONAL MUSEUM)
13-9, UENO KOEN, TAITO-KU, TOKYO 110
JAPANESE, WITH ENGLISH SUMMARY

TOKYO KOKURITSU HAKUBUTSUKAN NENPO 396
(TOKYO NATIONAL MUSEUM ANNUAL REPORT)
TOKYO KOKURITSU HAKUBUTSUKAN (TOKYO NATIONAL MUSEUM)
13-9, UENO KOEN, TAITO-KU, TOKYO 110
JAPANESE

TOKYO KOKURITSU HAKUBUTSUKAN ZUHAN MOKUROKU 397
(ILLUSTRATED CATALOGUES OF TOKYO NATIONAL MUSEUM)
TOKYO KOKURITSU HAKUBUTSUKAN (TOKYO NATIONAL MUSEUM)
13-9, UENO KOEN, TAITO-KU, TOKYO 110
JAPANESE

TOKYO NOGYO DAIGAKU IPPAN KYOIKU GAKUJUTSU SHUHO 398
(JOURNAL OF GENERAL EDUCATION, TOKYO NOGYO DAIGAKU)
TOKYO NOGYO DAIGAKU GAKUJUTSU ZASSHI HENSHU IINKAI (EDITORIAL BOARD FOR ACADEMIC JOURNALS, TOKYO UNIVERSITY OF AGRICULTURE) AND TOKYO NOGYO DAIGAKU (TOKYO UNIVERSITY OF AGRICULTURE)
1-1, 1-CHOME, SAKURAGAOKA, SETAGAYA-KU, TOKYO 156
JAPANESE AND ENGLISH
INDEXED: CHE

TOKYO NOKO DAIGAKU, IPPAN KYOIKUBU KIYO 399
(BULLETIN OF THE GENERAL EDUCATION FACULTY, TOKYO AGRICULTURAL AND INDUSTRIAL COLLEGE)
JAPANESE

TOKYO RIKA DAIGAKU KENKYU RONBUNSHU 400
(REPORTS OF SCIENCE UNIVERSITY OF TOKYO)
TOKYO RIKA DAIGAKU (SCIENCE UNIVERSITY OF TOKYO)
1-3, KAGURAZAKA, SHINJUKU-KU, TOKYO 162
JAPANESE AND EUROPEAN
INDEXED: JICST

TOKYO SHOSEN DAIGAKU KENKYU HOKOKU, SHIZENKAGAKU 401
(JOURNAL OF THE TOKYO UNIVERSITY OF MERCANTILE MARINE, NATURAL SCIENCE)
TOKYO SHOSEN DAIGAKU (TOKYO UNIVERSITY OF MERCANTILE MARINE)
1-6, 2-CHOME, ETCHUJIMA, KOTO-KU, TOKYO 135
JAPANESE, WITH EUROPEAN SUMMARY
INDEXED: CHE, JICST

TOKYO YAKKA DAIGAKU, IPPAN KYOIKU KENKYU KIYO 402
(BULLETIN, TOKYO COLLEGE OF PHARMACY, LIBERAL ARTS AND SCIENCES)
JAPANESE

TOKYO-TO TAKAO SHIZEN KAGAKU HAKUBUTSUKAN KENKYU HOKOKU 403
(RESEARCH REPORT OF TOKYO MUNICIPAL TAKAO NATIONAL SCIENCE MUSEUM)
JAPANESE

TOKYO-TORITSU DAIGAKU GAKUBO 404
(REPORT OF THE TOKYO METROPOLITAN COLLEGE)
JAPANESE

TO'OKU GIJUKU KENKYU KIYO 405
(RESEARCH BULLETIN OF THE TO'OKU GIJUKU)
JAPANESE

TORE KAGAKU SHINKOKAI JIGYO HOKOKUSHO, I 406
(ANNUAL REPORT, TORAY SCIENCE FOUNDATION, PART I)
TORE KAGAKU SHINKOKAI (TORAY SCIENCE FOUNDATION)
2-2, NIHONBASHI MURO-MACHI, CHUO-KU, TOKYO 103
JAPANESE AND ENGLISH

TORE KAGAKU SHINKOKAI JIGYO HOKOKUSHO, II 407
(ANNUAL REPORT, TORAY SCIENCE FOUNDATION, PART 2)
TORE KAGAKU SHINKOKAI (TORAY SCIENCE FOUNDATION)
2-2, NIHONBASHI MURO-MACHI, CHUO-KU, TOKYO 103
JAPANESE AND ENGLISH

TORE KAGAKU SHINKOKAI KAGAKU KOENKAI KIROKU 408
(PROCEEDINGS OF THE SYMPOSIUM OF TORAY SCIENCE FOUNDATION)
TORE KAGAKU SHINKOKAI (TORAY SCIENCE FOUNDATION)
2-2, NIHONBASHI MURO-MACHI, CHUO-KU, TOKYO 103
JAPANESE

TOTTORI DAIGAKU KYOIKUGAKUBU KENKYU HOKOKU, SHIZENKAGAKU 409
(THE JOURNAL OF THE FACULTY OF EDUCATION, NATURAL SCIENCE (TOTTORI UNIVERSITY))
TOTTORI DAIGAKU KYOIKUGAKUBU (THE FACULTY OF EDUCATION, TOTTORI UNIVERSITY)
1-1, KOYAMA-CHO, TOTTORI-SHI 680
JAPANESE AND EUROPEAN
INDEXED: BIB, BIO, CHE, COP, JICST

TOTTORI DAIGAKU KYOYOBU KIYO 410
(BULLETIN OF THE FACULTY OF ARTS AND SCIENCES OF TOTTORI UNIVERSITY)
JAPANESE

TOTTORI-KENRITSU KAGAKU HAKUBUTSUKAN KENKYU HOKOKU 411
(BULLETIN OF THE TOTTORI SCIENCE MUSEUM)
TOTTORI-KENRITSU KAGAKU HAKUBUTSUKAN (TOTTORI PREFECTURAL SCIENCE MUSEUM)
2-121, HIGASHI-MACHI, TOTTORI-SHI 680
JAPANESE, WITH ENGLISH SUMMARY

TOWA DAIGAKU KIYO 412
(BULLETIN OF THE TOWA UNIVERSITY)
TOWA UNIVERSITY
JAPANESE

TOYAMA DAIGAKU KYOIKUGAKUBU KIYO, B RIKA-KEI 413
(MEMOIRS OF THE FACULTY OF EDUCATION, TOYAMA UNIVERSITY, B SCIENCE)
TOYAMA DAIGAKU KYOIKUGAKUBU (FACULTY OF EDUCATION, TOYAMA UNIVERSITY)
3190 GOFUKU, TOYAMA-SHI 930
JAPANESE AND ENGLISH
INDEXED: CHE, JICST

TOYAMA DAIGAKU KYOYOBU KIYO, SHIZENKAGAKU-HEN 414
(MEMOIRS OF THE FACULTY OF GENERAL EDUCATION, TOYAMA UNIVERSITY, NATURAL SCIENCE)
TOYAMA DAIGAKU KYOYOBU
3190, GOFUKU, TOYAMA-SHI 930
JAPANESE AND ENGLISH

TOYO DAIGAKU DAIGAKUIN KIYO 415
(BULLETIN OF THE TOYO UNIVERSITY GRADUATE SCHOOL)
JAPANESE

TOYO DAIGAKU KIYO, KYOYOKATEI-HEN, SHIZENKAGAKU 416
(JOURNAL OF THE TOYO UNIVERSITY, GENERAL EDUCATION, NATURAL SCIENCE)
TOYO DAIGAKU
5-28, HAKUSAN, BUNKYO-KU, TOKYO 112
JAPANESE AND ENGLISH
INDEXED: CHE

TOYODA KENKYU HOKOKU 417
(REPORTS OF TOYODA PHYSICAL AND CHEMICAL RESEARCH INSTITUTE)
TOYODA RIKAGAKU KENKYUSHO (TOYODA PHYSICAL AND CHEMICAL RESEARCH INSTITUTE)
2-12, HISAKATA, SHOWA-KU, NAGOYA-SHI 468
JAPANESE, WITH ENGLISH SUMMARY
INDEXED: CHE, JICST

UNIVERSITY MUSEUM, UNIVERSITY OF TOKYO, BULLETIN 418
TOKYO DAIGAKU SOGO KENKYU SHIRYOKAN (THE UNIVERSITY MUSEUM, THE UNIVERSITY OF TOKYO)
3-1, 7-CHOME, HONGO, BUNKYO-KU, TOKYO 113
ENGLISH

UTSUNOMIYA DAIGAKU GAKUGEI GAKUBU KIYO. 1 419
(BULLETIN OF UTSUNOMIYA UNIVERSITY FACULTY OF ARTS AND SCIENCE)
JAPANESE
INDEXED: JICST

UTSUNOMIYA DAIGAKU KYOIKUGAKUBU KIYO, DAI-2-BU 420
(BULLETIN OF THE FACULTY OF EDUCATION UTSONOMIYA UNIVERSITY, SECTION 2)
UTSUNOMIYA DAIGAKU KYOIKUGAKUBU (THE FACULTY OF EDUCATION UTSUNOMIYA UNIVERSITY)
350, MINE-MACHI, UTSUNOMIYA-SHI 320
JAPANESE AND ENGLISH
INDEXED: BIO, MAT, JICST

UTSUNOMIYA DAIGAKU KYOYOBU KENKYU HOKOKU, DAI-ICHI BU 421
(BULLETIN OF THE FACULTY OF GENERAL EDUCATION, UTSUNOMIYA UNIVERSITY, SECTION 1)
JAPANESE

UTSUNOMIYA DAIGAKU KYOYOBU KENKYU HOKOKU, DAI-2-BU 422
(BULLETIN OF THE FACULTY OF GENERAL EDUCATION, UTSUNOMIYA UNIVERSITY, SECTION 2)
UTSUNOMIYA DAIGAKU KYOYOBU
350, MINE-MACHI, UTSUNOMIYA-SHI 320
JAPANESE AND ENGLISH
INDEXED: BIO

WAKASUGI KENKYUJO KIYO 423
(ESSAYS AND STUDIES (WAKASUGI RESEARCH INSTITUTE))
JAPANESE

WAKAYAMA DAIGAKU KYOIKUGAKUBU KIYO, SHIZENKAGAKU 424
(BULLETIN OF THE FACULTY OF EDUCATION WAKAYAMA UNIVERSITY, NATURAL SCIENCE)
WAKAYAMA DAIGAKU KYOIKUGAKUBU (THE FACULTY OF EDUCATION WAKAYAMA UNIVERSITY)
1-1, MASAGO-CHO, WAKAYAMA-SHI 640
JAPANESE AND ENGLISH
INDEXED: BIO, CHE, JICST

WASEDA DAIGAKU DAIGAKUIN RIKOGAKU KENKYU IHO 425
(SYNOPSES OF ENGINEERING PAPERS GRADUATE SCHOOL OF SCIENCE AND ENGINEERING, WASEDA UNIVERSITY)
WASEDA DAIGAKU DAIGAKUIN RIKOGAKU KENKYUKA (GRADUATE SCHOOL OF SCIENCE AND ENGINEERING, WASEDA UNIVERSITY)
4-170, NISHIOKUBO, SHINJUKU-KU, TOKYO 160
JAPANESE AND ENGLISH
INDEXED: JICST

WASEDA DAIGAKU KYOIKUGAKUBU GAKUJUTSU KENKYU, SEIBUTSU, CHIGAKU-HEN 426
(THE SCIENTIFIC RESEARCHES)
WASEDA DAIGAKU KYOIKUKAI
1-647, TOTSUKA, SHINJUKU-KU, TOKYO 160
JAPANESE
INDEXED: JICST

WASEDA DAIGAKU RIKOGAKU KENKYUSHO HOKOKU 427
(BULLETIN OF SCIENCE AND ENGINEERING RESEARCH LABORATORY, WASEDA UNIVERSITY)
WASEDA DAIGAKU RIKOGAKU KENKYUSHO (SCIENCE AND ENGINEERING RESEARCH LABORATORY, WASEDA UNIVERSITY)
17, KIKUI-CHO, SHINJUKU-KU, TOKYO 162
JAPANESE AND EUROPEAN
INDEXED: CHE, MAT, JICST

WASEDA DAIGAKU SHISUTEMU KAGAKU KENKYUJO KIYO 428
(BULLETIN OF THE SYSTEM SCIENCE INSTITUTE, WASEDA UNIVERITY)
JAPANESE
INDEXED: JICST

WASEDA JINBUN SHIZEN KAGAKU KENKYU 429
(WASEDA LIBERAL ARTS AND NATURAL SCIENCE RESEARCH)
WASEDA DAIGAKU SHAKAIKA GAKUBU GAKKAI (WASEDA UNIVERSITY DEPARTMENT OF SOCIOLOGY SOCIETY
JAPANESE

YAMA TO HAKUBUTSUKAN 430
(MOUNTAIN AND MUSEUM)
OMACHI SANGAKU HAKUBUTSUKAN (OMACHI ALPINE MUSEUM)
OMACHI, OMACHI-SHI 380
JAPANESE

YAMAGATA DAIGAKU KIYO, SHIZENKAGAKU 431
(BULLETIN OF THE YAMAGATA UNIVERSITY, NATURAL SCIENCE)
YAMAGATA DAIGAKU (YAMAGATA UNIVERSITY)
KOSHIRAKAWA-MACHI, YAMAGATA-SHI 990
JAPANESE AND ENGLISH
INDEXED: CHE, INS, JICST

YAMAGUCHI DAIGAKU KYOIKUGAKUBU KENKYU RONSO, SHIZENKAGAKU 432
(BULLETIN OF THE FACULTY OF EDUCATION, YAMAGUCHI UNIVERSITY)
YAMAGUCHI DAIGAKU KYOIKUGAKUBU (FACULTY OF EDUCATION YAMAGUCHI UNIVERSITY)
1677-1, YOSHIDA, YAMAGUCHI-SHI 753
JAPANESE AND EUROPEAN
INDEXED: JICST

YAMAGUCHI DAIGAKU KYOYOBU KIYO 433
(BULLETIN OF YAMAGUCHI UNIVERSITY, FACULTY OF GENERAL EDUCATION)
JAPANESE
INDEXED: JICST

YAMAGUCHI DAIGAKU RIKA HOKOKU 434
(THE SCIENCE REPORT OF THE YAMAGUCHI UNIVERSITY)
YAMAGUCHI DAIGAKU BUNRIGAKUBU RIGAKKAI (THE SCIENTIFIC SOCIETY OF THE YAMAGUCHI UNIVERSITY FACULTY OF LITERATURE AND SCIENCE, YAMAGUCHI UNIVERSITY
JAPANESE AND ENGLISH
INDEXED: JICST

YAMAGUCHI JOSHI DAIGAKU KENKYU HOKOKU; DAI-2 BU, SHIZEN KAGAKU 435
(BULLETIN OF YAMAGUCHI WOMEN'S UNIVERSITY. SECTION 2: NATURAL SCIENCE.)
YAMAGUCHI WOMEN'S UNIVERSITY
JAPANESE

YAMAGUCHI JOSHI TANKI DAIGAKU KENKYU HOKOKU DAI-2-BU SHIZENKAGAKU 436
(BULLETIN OF YAMAGUCHI WOMEN'S JUNIOR COLLEGE, SECTION II, NATURAL SCIENCE)
YAMAGUCHI JOSHI TANKI DAIGAKU (YAMAGUCHI JOSHI WOMEN'S JUNIOR COLLEGE)
2-1, 3-CHOME, SAKURABATAKE, YAMAGUCHI-SHI 753
JAPANESE AND ENGLISH

YAMAGUCHI-KEN NO SHIZEN 437
(NATURAL HISTORY OF YAMAGUCHI-PREFECTURE)
YAMAGUCHI-KENRITSU YAMAGUCHI HAKUBUTSUKAN (YAMAGUCHI PREFECTURAL YAMAGUCHI MUSEUM)
8-2, KASUGA-CHO, YAMAGUCHI-SHI 753
JAPANESE

YAMAGUCHI-KENRITSU YAMAGUCHI HAKUBUTSUKAN KENKYU HOKOKU 438
(BULLETIN OF THE YAMAGUCHI PREFECTURAL YAMAGUCHI MUSEUM)
YAMAGUCHI-KENRITSU YAMAGUCHI HABUBUTSUKAN (YAMAGUCHI PREFECTURAL YAMAGUCHI MUSEUM)
8-2, KASUGA-CHO, YAMAGUCHI-SHI 753
ENGLISH

YAMANASHI DAIGAKU KYOIKUGAKUBU KENKYU HOKOKU 439
(MEMOIRS OF THE FACULTY OF LIBERAL ARTS AND EDUCATION, YAMANASHI UNIVERSITY)
YAMANASHI DAIGAKU KYOIKUGAKUBU (FACULTY OF EDUCATION, YAMANASHI UNIVERSITY)
4-CHOME, TAKEDA, KOFU-SHI 400
JAPANESE AND ENGLISH
INDEXED: JICST

YAMANASHI DAIGAKU KYOIKUGAKUBU KIYO 440
(BULLETIN, FACULTY OF EDUCATION, YAMANASHI UNIVERSITY)
YAMANASHI DAIGAKU KYOIKUGAKUBU (FACULTY OF EDUCATION, YAMANASHI UNIVERSITY)
4-CHOME, TAKEDA, KOFU-SHI 400
JAPANESE AND ENGLISH

YASHIRO GAKUEN DAIGAKU KIYO 441
(BULLETIN OF YASHIRO GAKUEN UNIVERSITY)
YASHIRO GAKUEN DAIGAKU GAKUJUTSN KENKYU KAI (YASHIRO GAKUEN UNIVERSITY ARTS AND SCIENCE RESEARCH SOCIETY)
JAPANESE

YASUDA GAKUEN KENKYU KIYO 442
(RESEARCH BULLETIN OF YASUDA GAKUEN)
JAPANESE

YOKOHAMA KOKURITSU DAIGAKU RIKA KIYO, DAI-2-RUI, SEIBUTSUGAKU, CHIGAKU 443
(SCIENCE REPORTS OF THE YOKOHAMA NATIONAL UNIVERSITY SECTION 2. BIOLOGICAL AND GEOLOGICAL SCIENCES)
YOKOHAMA KOKURITSU DAIGAKU (YOKOHAMA NATIONAL UNIVERSITY)
41, SHIMIZUGAOKA, MINAMI-KU, YOKOHAMA-SHI 232
JAPANESE AND EUROPEAN
INDEXED: CHE, MET, WAA, JICST

YOKOHAMA-SHIRITSU DAIGAKU KIYO, KAGAKU BUTSURI-HEN 444
(JOURNAL OF YOKOHAMA CITY UNIVERSITY, CHEMICAL AND PHYSICAL SERIES)
YOKOHAMA-SHIRITSU DAIGAKU (YOKOHAMA CITY UNIVERSITY)
4646, MUTSUURA-MACHI, KANAZAWA-KU, YOKOHAMA-SHI 236
JAPANESE AND ENGLISH

YOKOHAMA-SHIRITSU DAIGAKU RONSO, SHIZENKAGAKU KEIRETSU 445
(THE BULLETIN OF YOKOHAMA CITY UNIVERSITY, NATURAL SCIENCE)
YOKOHAMA-SHIRITSU DAIGAKU GAKUJUTSU KENKYUKAI (ARTS AND SCIENCE SOCIETY OF YOKOHAMA CITY UNIVERSITY)
JAPANESE AND ENGLISH
INDEXED: JICST

YOKOSUKA-SHI HAKUBUTSUKAN KENKYU HOKOKU SHIZENKAGAKU 446
(BULLETIN OF THE YOKOSUKA CITY MUSEUM)
YOKOSUKA-SHI HAKUBUTSUKAN (YOKOSUKA CITY MUSEUM)
95, FUKADADAI, YOKOSUKA-SHI 238
JAPANESE AND ENGLISH

YOKOSUKA-SHI HAKUBUTSUKAN ZAPPO 447
(ANNUAL REPORT OF THE YOKOSUKA CITY MUSEUM)
YOKOSUKA-SHI HAKUBUTSUKAN (YOKOSUKA CITY MUSEUM)
95, FUKADADAI, YOKISUKA-SHI 238
JAPANESE
INDEXED: BIO

ZASSHI KIJI SAKUIN, KAGAKU GIJUTSU-HEN 448
(JAPANESE PERIODICALS INDEX, SCIENCE AND TECHNOLOGY)
KOKURITSU KOKKAI TOSHOKAN (NATIONAL DIET LIBRARY)
10-1, 1-CHOME, NAGATA-CHO, CHIYODA-KU, TOKYO 100
JAPANESE AND EUROPEAN
INDEXED: BIB, JICST

Mathematics

ANNALS OF THE INSTITUTE OF STATISTICAL MATHEMATICS 449
TOKEI SURI KENKYUSHO (INSTITUTE OF STATISTICAL MATHEMATICS)
6-7, 4-CHOME, MINAMI AZABU, MINATO-KU, TOKYO 106
EUROPEAN
INDEXED: MAT, SCI, JICST

ATARASHII SANSU KENKYU 450
(RESEARCH ON NEW MATHEMATICS)
SHIN SANSU KYOIKU KENKYUKAI-HEN TOYOKAN SHUPPANSHA (NEW MATHEMATICS EDUCATIONAL RESEARCH COMMITTEE TOYOKAN PUBLISHING COMPANY)
JAPANESE

BULLETIN OF THE FACULTY OF SCIENCE, IBARAKI UNIVERSITY, SERIES A, MATHEMATICS 451
IBARAKI DAIGAKU RIGAKUBU SUGAKU KYOSHITSU (DEPARTMENT OF MATHEMATICS, FACULTY OF SCIENCE, IBARAKI UNIVERSITY)
1-1, 2-CHOME, BUNKYO, MITO-SHI 310
ENGLISH
INDEXED: MAT

BULLETIN OF MATHEMATICAL STATISTICS 452
KYUSHU DAIGAKU TOKEI KAGAKU KENKYUKAI (RESEARCH ASSOCIATES OF STATISTICAL SCIENCE, KYUSHU UNIVERSITY)
C/O KYUSHU DAIGAKU RIGAKUBU, HAKOZAKI-CHO, HIGASHI-KU, FUKUOKA-SHI 812
ENGLISH
INDEXED: MAT, SCI

COLLECTED PAPERS OF THE MATHEMATICAL SOCIETY OF WAKAYAMA UNIVERSITY 453
JAPANESE
INDEXED: MAT

COMMENTARII MATHEMATICI UNIVERSITATIS SANCTI PAULI 454
RIKKYO DAIGAKU RIGAKUBU SUGAKU KYOSHITSU (INSTITUTE OF MATHEMATICS, FACULTY OF SCIENCE, RIKKYO UNIVERSITY) AND KINOKUNIYA SHOTEN (KINOKUNIYA BOOKSTORE COMPANY, LIMITED)
17-7, 3-CHOME, SHINJUKU, SHINJUKU-KU, TOKYO 160
EUROPEAN
INDEXED: MAT, JICST

EPOCH MAKING 455
HENMI KEISANJAKU K.K.
4-4, KANDA SURUGADAI, CHIYODA-KU, TOKYO 101
JAPANESE

FUNKCIALAJ EKVACIOJ 456
KYORITSU SHUPPAN K.K.
6-19, 4-CHOME, KOHINATA, BUNKYO-KU, TOKYO 112
EUROPEAN

GEKKAN SHUZANKAI 457
(ABACUS MONTHLY)
SHUZAN PURINTO KYOZAIBU
20-2, 4-CHOME, TAISHIDO, SETAGAYA-KU, TOKYO 154
JAPANESE

GENDAI SUGAKU 458
(MODERN MATHEMATICS)
GENDAI SUGAKUSHA
SHIN YANAGINOBANBA NIOMON KUDARU, SAKYO-KU, KYOTO-SHI 606
JAPANESE

HIROSHIMA MATHEMATICAL JOURNAL 459
HIROSHIMA DAIGAKU RIGAKUBU SUGAKU KYOSHITSU (DEPARTMENT OF MATHEMATICS, FACULTY OF SCIENCE, HIROSHIMA UNIVERSITY)
EUROPEAN
INDEXED: MAT, JICST

HOKKAIDO MATHEMATICAL JOURNAL 460
HOKKAIDO DAIGAKU RIGAKUBU (THE FACULTY OF SCIENCE HOKKAIDO UNIVERSITY)
NISHI-8-CHOME, KITA-10-JO, KITA-KU, SAPPORO-SHI 060
ENGLISH
INDEXED: MAT, JICST

JAPANESE JOURNAL OF MATHEMATICS 461
NIHON GAKUJUTSU KAIGI (SCIENCE COUNCIL OF JAPAN) AND KINOKUNIYA SHOTEN
17-7, 3-CHOME, SHINJUKU, SHINJUKU-KU, TOKYO 160
ENGLISH
INDEXED: MAT

JOURNAL OF THE FACULTY OF SCIENCE, UNIVERSITY OF TOKYO, SECTION 1. A, MATHEMATICS 462
TOKYO DAIGAKU (UNIVERSITY OF TOKYO)
3-1, 7-CHOME, HONGO, BUNKYO-KU, TOKYO 113
ENGLISH
INDEXED: CHE, MAT, JICST

JOURNAL OF THE MATHEMATICAL SOCIETY OF JAPAN 463
NIHON SUGAKKAI (MATHEMATICAL SOCIETY OF JAPAN)
C/O TOKYO DAIGAKU RIGAKUBU, 3-1, 7-CHOME, HONGO, BUNKYO-KU, TOKYO 113
EUROPEAN
INDEXED: MAT, SCI, JICST

JOURNAL OF MATHEMATICS OF KYOTO UNIVERSITY 464
KYOTO DAIGAKU RIGAKUBU SUGAKU KYOSHITSU (DEPARTMENT OF MATHEMATICS, KYOTO UNIVERSITY)
KITASHIRAKAWA, SAKYO-KU, KYOTO-SHI 606
EUROPEAN
INDEXED: MAT

JOURNAL OF MATHEMATICS, TOKUSHIMA UNIVERSITY 465
TOKUSHIMA DAIGAKU (TOKUSHIMA UNIVERSITY)
2-6, SHINZO-MACHI, TOKUSHIMA-SHI 770
ENGLISH
INDEXED: MAT

KEIRYO KOKUGOGAKU 466
(MATHEMATICAL LINGUISTICS)
JAPANESE
INDEXED: MAT, JICST

KODAI MATHEMATICAL SEMINAR REPORTS 467
TOKYO KOGYO DAIGAKU SUGAKU KYOSHITSU (DEPARTMENT OF MATHEMATICS, TOKYO INSTITUTE OF TECHNOLOGY)
OOKAYAMA, MEGURO-KU, TOKYO 152
ENGLISH
INDEXED: MAT

KUMAMOTO JOURNAL OF SCIENCE, MATHEMATICS 468
KUMAMOTO DAIGAKU RIGAKUBU SUGAKU KYOSHITSU (FACULTY OF SCIENCE, KUMAMOTO UNIVERSITY)
39-1, 2-CHOME, KUROKAMI, KUMAMOTO-SHI 860
ENGLISH
INDEXED: INS, MAT, JICST

KYOTO DAIGAKU SURI KAISEKI KENKYUSHO KOKYUROKU 469
(PAPERS OF RESEARCH INSTITUTE FOR MATHEMATICAL SCIENCES, KYOTO UNIVERSITY)
KYOTO DAIGAKU SURI KAISEKI KENKYUSHO (RESEARCH INSTITUTE FOR MATHEMATICAL SCIENCES, KYOTO UNIVERSITY)
KITA SHIRAKAWA OIWAKE-CHO, SAKYO-KU, KYOTO-SHI 606
JAPANESE AND ENGLISH
INDEXED: MAT

KYUSHU DAIGAKU KYOYOBU SUGAKU ZASSHI 470
(MATHEMATICAL REPORTS OF COLLEGE OF GENERAL EDUCATION, KYUSHU UNIVERSITY)
KYUSHU DAIGAKU KYOYOBU SUGAKU KYOSHITSU
4116, HAKOZAKI, HIGAHI-KU, FUKUOKA-SHI 812
JAPANESE AND ENGLISH
INDEXED: MAT

MATHEMATICA JAPONICAE 471
OSAKA-FURITSU DAIGAKU SUGAKU KYOSHITSU (DEPARTMENT OF MATHEMATICS, UNIVERSITY OF OSAKA PREFECTURE)
MOZU UME-MACHI, SAKAI-SHI 591
EUROPEAN
INDEXED: MAT

MATHEMATICAL JOURNAL OF OKAYAMA UNIVERSITY 472
OKAYAMA DAIGAKU RIGAKUBU SUGAKU KYOSHITSU (DEPARTMENT OF MATHEMATICS, FACULTY OF SCIENCE, OKAYAMA UNIVERSITY)
TSUSHIMA, OKAYAMA-SHI 700
EUROPEAN
INDEXED: MAT

MATHEMATICS SEMINAR NOTES. KOBE UNIVERSITY 473
JAPANESE AND ENGLISH
INDEXED: MAT

MEMOIRS OF THE FACULTY OF SCIENCE, KYUSHU UNIVERSITY, SERIES A, MATHEMATICS 474
KYUSHU DAIGAKU RIGAKUBU (FACULTY OF SCIENCE, KYUSHU UNIVERSITY)
3576, HAKOZAKI, HIGASHI-KU, FUKUOKA-SHI 812
ENGLISH
INDEXED: MAT, MET, WAA

MEMOIRS OF NUMERICAL MATEHMATICS 475
ENGLISH
INDEXED: MAT

MUSCIDAE 476
NIHON SEIBUTSU-CHIRI GAKKAI (JAPAN BIOLOGICAL-GEOGRAPHY SOCIETY)
ENGLISH
INDEXED: BIB

NAGOYA MATHEMATICAL JOURNAL 477
NAGOYA DAIGAKU RIGAKUBU, SUGAKU KYOSHITSU (DEPARTMENT OF MATHEMATICS, FACULTY OF SCIENCE, NAGOYA UNIVERSITY)
FURO-CHO, CHIKUSA-KU, NAGOYA-SHI 464
EUROPEAN
INDEXED: MAT, SCI, JICST

NIHON SHUZAN 478
(ABACUS OF JAPAN)
NIHON SHUZAN RENMEI (THE LEAGUE OF JAPAN ABACUS ASSOCIATION)
C/O NIHON SHOKO KAIGISHO, 3-2, MARUNOUCHI, CHIYODA-KU, TOKYO 100
JAPANESE

NIHON SUGAKU KYOIKU GAKKAISHI 479
(JOURNAL OF SOCIETY OF MATHEMATICAL EDUCATION)
NIHON SUGAKU KYOIKU GAKKAI (JAPAN SOCIETY OF MATHEMATICAL EDUCATION)
C/O TOKYO KYOIKU DAIGAKU, 3-29, OTSUKA, BUNKYO-KU, TOKYO 112
JAPANESE

NIHON TOKEI GAKKAISHI 480
(JOURNAL OF THE JAPAN STATISTICAL SOCIETY)
NIHON TOKEI GAKKAI (THE JAPAN STATISTICAL SOCIETY)
C/O TOKEI SURI KENKYUSHO, 6-7, 4-CHOME, MINAMI AZABU, MINATO-KU, TOKYO 106
JAPANESE AND ENGLISH
INDEXED: JICST

NIHON TOKEI KENKYUJOHO 481
(BULLETIN OF JAPAN STATISTICS RESEARCH INSTITUTE)
HOSEI DAIGAKU NIHON TOKEI KENKYUJO (HOSEI UNIVERSITY, JAPAN STATISTICS RESEARCH INSTITUTE)
JAPANESE

NIKKAGIREN SUGAKU KEIKAKU SHINPOJUMU HOBUN SHIRIZU 482
(JUSE SYMPOSIUM ON MATHEMATICAL PLANNING SERIES)
NIHON KAGAKU GIJUTSU RENMEI
10-11, 5-CHOME, SENDAGAYA, SHIBUYA-KU, TOKYO 151
JAPANESE
INDEXED: JICST

OSAKA JOURNAL OF MATHEMATICS 483
OSAKA DAIGAKU RIGAKUBU, OSAKA-SHIRITSU DAIGAKU RIGAKUBU (DEPARTMENT OF MATHEMATICS, OSAKA UNIVERSITY AND OSAKA CITY UNIVERSITY)
1-1, MACHIKANEYAMA-CHO, TOYONAKA-SHI 560
EUROPEAN
INDEXED: MAT, JICST

PUBLICATIONS OF THE MATHEMATICAL SOCIETY OF JAPAN 484
NIHON SUGAKKAI (MATHEMATICAL SOCIETY OF JAPAN)
C/O TOKYO DAIGAKU RIGAKUBU, 3-1, 7-CHOME, BUNKYO-KU, TOKYO 113
ENGLISH
INDEXED: MAT

PUBLICATIONS OF THE RESEARCH INSTITUTE FOR MATHEMATICAL SCIENCES 485
KYOTO DAIGAKU SURI KAISEKI KENKYUSHO (RESEARCH INSTITUTE FOR MATHEMATICAL SCIENCES, KYOTO UNIVERSITY)
KITASHIRAKAWA OIWAKE-CHO, SAKYO-KU, KYOTO-SHI 606
ENGLISH

RAAG RESEARCH NOTES 486
OYO KIKAGAKU KENKYU KYOKAI (RESEARCH ASSOCIATES OF APPLIED GEOMETRY)
HONGO, BUNKYO-KU, TOKYO 113
JAPANESE, WITH ENGLISH SUMMARY
INDEXED: JICST

REPORTS OF THE FACULTY OF SCIENCE AND ENGINEERING. SAGA UNIVERSITY. 487
JAPANESE
INDEXED: MAT

REPORTS OF STATISTICAL APPLICATION RESEARCH 488
NIPPON KAGAKU GIJUTSU RENMEI (UNION OF JAPANESE SCIENTISTS AND ENGINEERS)
10-11, 5-CHOME, SENDAGAYA, SHIBUYA-KU TOKYO 151
ENGLISH
INDEXED: BIB, CHE, INS, MAT

RESEARCH ASSOCIATION OF APPLIED GEOMETRY. RESEARCH NOTES AND MEMORANDA OF APPLIED GEOMETRY IN POST-RAAG 489
JAPANESE
INDEXED: MAT

SANSU TO SUGAKU 490
(ARITHMETIC AND MATHEMATICS)
KYOIKU SOGO KENKYUJO (RESEARCH INSTITUTE FOR EDUCATIONAL COORDINATION)
JAPANESE

SANSUKA SHIDO SHIRYO 491
(MATERIALS FOR THE DEVELOPMENT OF MATHEMATICS)
MEIJI TOSHO SHUPPAN
(MEIKU BOOKS PUBLISHING COMPANY)
1
JAPANESE

SCIENCE REPORTS OF NIIGATA UNIVERSITY SERIES A, MATHEMATICS 492
NIIGATA DAIGAKU RIGAKUBU (THE FACULTY OF SCIENCE, NIIGATA UNIVERSITY)
8050, IGARASHI NI-NO-CHO, NIIGATA-SHI 950-21
EUROPEAN
INDEXED: CHE, MAT, JICST

SOOCHOW JOURNAL OF MATHEMATICAL AND NATURAL SCIENCES 493
JAPANESE
INDEXED: MAT

SUGAKU 494
(MATHEMATICS)
NIHON SUGAKKAI (MATHEMATICAL SOCIETY OF JAPAN)
IWANAMI SHOTEN
5-5, 2-CHOME, HITOTSUBASHI, CHIYODA-KU, TOKYO 101
JAPANESE
INDEXED: JICST

SUGAKU KYOIKU KENKYU 495
(OSAKA JOURNAL OF MATHEMATICS EDUCATION)
OSAKA KYOIKU DAIGAKU SUGAKU KYOSHITSU (OSAKA EDUCATION UNIVERSITY, MATHEMATICS DEPARTMENT)
JAPANESE

SUGAKU KYOSHITSU 496
(MATHEMATICAL STUDY ROOM)
KOKUDOSHA
17-6, 1-CHOME, MEJIRODAI, BUNKYO-KU, TOKYO 112
JAPANESE

SUGAKU NO AYUMI 497
(THE COURSE OF MATHEMATICS)
SUGAKU NO AYUMI KANKOKAI
C/O TOKYO DAIGAKU RIGAKUBU SUGAKU KYOSHITSU, 3-1, 7-CHOME, HONGO, BUNKYO-KU, TOKYO 113
JAPANESE

SUGAKU SEMINA 498
(MATHEMATICS SEMINAR)
NIHON HYORONSHA
14, SUGA-CHO, SHINJUKU-KU, TOKYO 160
JAPANESE
INDEXED: JICST

SUGAKUSHI KENKYU 499
(JOURNAL OF HISTORY OF MATHEMATICS OF JAPAN)
NIHON SUGAKUSHI GAKKAI (THE HISTORY OF MATHEMATICS SOCIETY OF JAPAN)
C/O FUJI TANKI DAIGAKU, 7-18, 1-CHOME, SHIMOOCHIAI, SHINJUKU-KU, TOKYO 161
JAPANESE
INDEXED: MAT

SUGEI PAZURU 500
(FASCINATING PUZZLES)
SUGEI PAZURU AIKOKAI (FASCINATING PUZZLES CLUB)
C/O MR. AKIO SUZUKI, 31, 1-CHOME, ISHIGAMIDO-CHO, HIGASHI-KU, NAGOYA-SHI 461
JAPANESE

SUKEN KENKYU RIPOTO 501
(RESEARCH REPORTS, INSTITUTE OF STATISTICAL MATHEMATICS)
TOKEI SURI KENKYUSHO (INSTITUTE OF STATISTICAL MATHEMATICS)
6-7, 4-CHOME, MINAMI AZABU, MINATO-KU, TOKYO 106
JAPANESE
INDEXED: JICST

SURI KAGAKU 502
(MATHEMATICAL SCIENCES)
DAIYAMONDOSHA (DAIAMOND PUBLICATIONS)
4-2, 1-CHOME, KASUMIGASEKI, CHIYODA-KU, TOKYO 100
JAPANESE
INDEXED: JICST

SURI KAISEKI KENKYUJO TAYORI 503
(MATHEMATICAL ANALYSIS RESEARCH INSTITUTE LETTER)
KYOTO DAIGAKU SURI KAISEKI KENKYUJO (KYOTO UNIVERSITY MATHEMATICAL ANALYSIS RESEARCH INSTITUTE)
JAPANESE

T.R.U. MATHEMATICS 504
ENGLISH
INDEXED: MAT

TENSOR 505
TENSORU GAKKAI (THE TENSOR SOCIETY)
C/O SAGAMI KOGYO DAIGAKU, 1-25, 1-CHOME NISHIKAIGAN, TSUJIDO, FUJISAWA-SHI 251
EUROPEAN
INDEXED: MAT

TOHOKU MATHEMATICAL JOURNAL 506
TOHOKU DAIGAKU (THE TOHOKU UNIVERSITY)
1-1, 2-CHOME, KATAHIRA, SENDAI-SHI 980
EUROPEAN
INDEXED: MAT

TOKEI 507
(STATISTICS)
NIHON TOKEI KYOKAI (JAPAN STATISTICAL ASSOCIATION)
95, WAKAMATSU-CHO, SHINJUKU-KU, TOKYO 162
JAPANESE

TOKEI KENKYU TSUSHIN 508
(INFORMATION PAMPHLET IN STATISTICS)
TOKEI SURI KENKYUSHO (INSTITUTE OF STATISTICAL MATHEMATICS)
6-7, 4-CHOME, MINAMI AZABU, MINATO-KU, TOKYO 106
JAPANESE AND EUROPEAN
INDEXED: JICST

TOKEI SURI KENKYUSHO IHO 509
(PROCEEDINGS OF THE INSTITUTE OF STATISTICAL MATHEMATICS)
TOKEI SURI KENKYUSHO (INSTITUTE OF STATISTICAL MATHEMATICS)
6-7, 4-CHOME, MINAMI AZABU, MINATO-KU, TOKYO 106
JAPANESE, WITH EUROPEAN SUMMARY
INDEXED: MAT, JICST

TOKEI SURI KENKYUSHO NENPO 510
(ANNUAL REPORT OF THE INSTITUTE OF STATISTICAL MATHEMATICS)
TOKEI SURI KENKYUSHO (THE INSTITUTE OF STATISTICAL MATHEMATICS)
6-7, 4-CHOME, MINAMI AZABU, MINATO-KU, TOKYO 106
JAPANESE
INDEXED: JICST

TOKEI-KYOKU KENKYU IHO 511
(RESEARCH MEMOIR OF THE BUREAU OF STATISTICS)
SORI-FU TOKEI-KYOKU (BUREAU OF STATISTICS OFFICE OF THE PRIME MINISTER JAPAN)
95, WAKAMATSU-CHO, SHINJUKU-KU, TOKYO 162
JAPANESE, WITH ENGLISH SUMMARY
INDEXED: JICST

TOKEIGAKU 512
(STATISTICS)
KEIZAI TOKEI KENKYUKAI (ECONOMIC STATISTICS RESEARCH ASSOCIATION)
JAPANESE

TOKEIGAKU KENKYU 513
(STUDIES IN STATISTICS)
TOKYO JOSHI DAIGAKU BUNRIGAKUBU SUGAKU KENKYUSHITSU (MATHEMATICAL INSTITUTE, FACULTY OF HUMANITIES AND SCIENCE TOKYO WOMEN'S CHRISTIAN COLLEGE)
6-1, 2-CHOME, ZENPUKUJI, SUGINAMI-KU, TOKYO 167
JAPANESE

TSUDA JUKU DAIGAKU KIYO 514
(JOURNAL OF TSUDA COLLEGE)
JAPANESE
INDEXED: MAT

WASEDA DAIGAKU KYOIKUGAKUBU, GAKUJUTSU KENKYU, SUGAKU-HEN 515
(THE SCIENTIFIC RESEARCH, MATHEMATICS, FACULTY OF EDUCATION, WASEDA UNIVERSITY)
WASEDA DAIGAKU KYOIKUKAI
1-647, TOTSUKA, SHINJUKU-KU, TOKYO 160
JAPANESE

YOKOHAMA MATHEMATICAL JOURNAL 516
YOKOHAMA-SHIRITSU DAIGAKU, SUGAKU KYOSHITSU (DEPARTMENT OF MATHEMATICS YOKOHAMA CITY UNIVERSITY)
4646, MUTSUURA-CHO, KANAZAWA-KU, YOKOHAMA-SHI 236
ENGLISH
INDEXED: MAT, JICST

ZUGAKU KENKYU 517
(JOURNAL OF GRAPHIC SCIENCE JAPAN)
NIHON ZUGAKU KENKYUKAI (JAPAN SOCIETY FOR GRAPHIC SCIENCE)
C/O TOKYO DAIGAKU KYOYO GAKUBU ZUGAKU KYOSHITSU, 8-1, 3-CHOME, KOMABA, MEGURO-KU, TOKYO 153
JAPANESE

Astronomy, Surveying, Geodesy

ABSOLUTE INTENSITY OF MAIN AURORAL EMISSIONS OF SYOWA STATION IN 1967 518
JAPANESE
INDEXED: JICST

ANNALS OF THE TOKYO ASTRONOMICAL OBSERVATORY, UNIVERSITY OF TOKYO, SECOND SERIES 519
TOKYO TENMONDAI (TOKYO ASTRONOMICAL OBSERVATORY)
OSAWA, MITAKA-SHI, TOKYO 181
ENGLISH

BULLETINS TIME SERVICE OF THE MIZUSAWA OBSERVATORY 520
JAPANESE
INDEXED: JICST

CONTRIBUTIONS FROM THE DEPARTMENT OF ASTRONOMY, UNIVERSITY OF KYOTO 521
ENGLISH

ITTO SUIJUNTEN KENSOKU SEIKA SHUROKU 522
(LAND SURVEY RESULTS FROM THE FIRST BENCH MARK)
KENSETSU-SHO KOKUDO CHIRI-IN SOKUCHI-BU (GEODETIC DIVISION, GEOPHYSICAL SURVEY INSTITUTE, MINISTRY OF CONSTRUCTION)
24-13, 3-CHOME, HIGASHIYAMA, MEGURO-KU, TOKYO 153
JAPANESE

KOKYO SOKURYO NO KIROKU 523
(RECORDS OF THE LAND SURVEY)
KENSETSU-SHO KOKUDO CHIRI-IN (GEOGRAPHICAL SURVEY OFFICE, MINISTRY OF CONSTRUCTION)
24-13, 3-CHOME, HIGASHIYAMA, MEGURO-KU, TOKYO 153
JAPANESE

NIHON TENMON KENKYUKAI HOBUN 524
(MEMOIRS OF THE JAPAN ASTRONOMICAL STUDY ASSOCIATION)
NIHON TENMON KENKYUKAI (JAPAN ASTRONOMICAL STUDY ASSOCIATION)
59, IZUMI, YUGAWARA-MACHI, ASHIGARASHIMO-GUN, KANAGAWA 259-03
JAPANESE
INDEXED: JICST

NIHON TENMON KENKYUKAI KAIHO 525
(BULLETIN OF JAPAN ASTRONOMICAL STUDY ASSOCIATION)
NIHON TENMON KENKYUKAI (JAPAN ASTRONOMICAL STUDY ASSOCIATION)
59, IZUMI, YUGAWARA-MACHI, ASHIGARASHIMO-GUN, KANAGAWA 259-03
JAPANESE

NIHON TENMON KENKYUKAI KANSOKU GEPPO 526
(MONTHLY BULLETIN OF THE JAPAN ASTRONOMICAL ASSOCIATION)
NIHON TENMON KENKYUKAI (THE JAPAN ASTRONOMICAL STUDY ASSOCIATION
59, IZUMI, YUGAWARA-MACHI, ASHIGARASHIMO-GUN, KANAGAWA 259-03
JAPANESE

PUBLICATIONS OF THE ASTRONOMICAL SOCIETY OF JAPAN 527
NIHON TENMON GAKKAI (ASTRONOMICAL SOCIETY OF JAPAN)
C/O TOKYO TENMONDAI, 21-1, 2-CHOME, OSAWA, MITAKA-SHI, TOKYO 181
EUROPEAN
INDEXED: CHE, IAA, INS, JICST

PURANETARIUMU 528
(PLANETARIUM)
ZENKOKU PURANETARIUMU RENRAKU KYOGIKAI (NATIONWIDE PLANETARIUM LIAISON COUNCIL)
JAPANESE

REKISHO NENPYO 529
(METEOROLOGICAL ALMANAC)
NIHON TENMON GAKKAI (ASTRONOMICAL SOCIETY OF JAPAN)
21-1, 2-CHOME, OSAWA, MITAKA-SHI, TOKYO 181
JAPANESE

SEIKEI KISHO TENMON KANSOKUSHO HOKOKU 530
(REPORT OF SEIKEI METEOROLOGICAL AND ASTRONOMICAL OBSERVATORY)
SEIKEI GAKUEN (SEIKEI GAKUEN UNIVERSITY)
10-13, 3-CHOME, KICHIJOJI KITAMACHI, MUSASHINO-SHI, TOKYO 180
JAPANESE

SENDAI ASTRONOMIAJ REPORTOJ 531
TOHOKU DAIGAKU RIGAKUBU (FACULTY OF SCIENCE, TOHOKU UNIVERSITY)
1-1, 2-CHOME, KATAHIRA, SENDAI-SHI 980
JAPANESE AND EUROPEAN
INDEXED: JICST

SHASHIN SOKURYO 532
(JOURNAL OF THE JAPAN SOCIETY OF PHOTOGRAMMETRY)
NIHON SHASHIN SOKURYO GAKKAI (THE JAPAN SOCIETY OF PHOTOGRAMMETRY)
C/O NISHIMAKI BUILDING, 4-CHOME, ROKUBAN-CHO, CHIYODA-KU, TOKYO 102
JAPANESE AND ENGLISH
INDEXED: JICST

SOKUCHI GAKKAISHI 533
(JOURNAL OF THE GEODETIC SOCIETY OF JAPAN)
NIHON SOKUCHI GAKKAI (THE GEODETIC SOCIETY OF JAPAN)
C/O KENSETSU-SHO, KOKUDO-CHIRI-IN, 24-13, 3-CHOME, HIGASHIYAMA, MEGURO-KU, TOKYO 153
JAPANESE, WITH ENGLISH SUMMARY
INDEXED: JICST

SOKURYO 534
(MONTHLY ON SURVEYING)
NIHON SOKURYO KYOKAI (JAPANESE ASSOCIATION OF SURVEYORS)
3-4, 1-CHOME, KOISHIKAWA, BUNKYO-KU, TOKYO 112
JAPANESE
INDEXED: JICST

SOKURYOGYO 535
(MONTHLY JOURNAL OF THE SURVEYING ENTERPRISE ASSOCIATION OF JAPAN)
ZENKOKU SOKURYOGYO KYOKAI (SURVEYING ENTERPRISE ASSOCIATION OF JAPAN)
1-433, NISHI OKUBO, SHINJUKU-KU, TOKYO 160
JAPANESE
INDEXED: JICST

SUIROBU KANSOKU HOKOKU TENMON SOKUCHI-HEN 536
(SURVEY REPORT OF THE HYDROGRAPHIC DEPARTMENT, ASTRONOMICAL SURVEYING)
KAIJO HOAN-CHO (MARITIME SAFETY AGENCY)
1-3, 2-CHOME, KASUMIGASEKI, CHIYODA-KU, TOKYO 100
ENGLISH

TENKAI 537
(HEAVENS)
TOA TENMON GAKKAI (THE ORIENTAL ASTRONOMICAL ASSOCIATION)
C/O YAMAMOTO TENMONDAI, KIRYUMACHI, KAMITANAKAMI, OTSU-SHI 520-21
JAPANESE

TENMON GAIDO 538
(AMATEUR ASTRONOMER)
SEIBUNDO SHINKOSH
1-5, KANDA NISHIKI-CHO, CHIYODA-KU, TOKYO 101
JAPANESE

TENMON GEPPO 539
(ASTRONOMICAL HERALD)
NIHON TENMON GAKKAI (ASTRONOMICAL SOCIETY OF JAPAN)
C/L TOKYO TENMONDAI, 21-1, 2-CHOME, OSAWA, MITAKA SHI, TOKYO 181
JAPANESE
INDEXED: CHE, JICST

TENMON KANSOKU NENPYO 540
(NAUTICAL ALMANAC)
CHIJIN SHOKAN
1-10, 1-CHOME, KORAKU, BUNKYO-KU, TOKYO 112
JAPANESE

TENMON TO KISHO 541
(ASTRONOMY AND METEOROLOGY)
CHIGIN SHOKAN
1-10, 1-CHOME, KORAKU, BUNKYO-KU,TOKYO 112
JAPANESE
INDEXED: JICST

TENSOKU RYAKUREKI 542
(ABRIDGED NAUTICAL ALMANAC)
KAIJO HOAN-CHO (MARITIME SAFETY AGENCY)
1-3, 2-CHOME, KASUMIGASEKI, CHIYODA-KU, TOKYO 100
JAPANESE

TENSOKUREKI 543
(NAUTICAL ALMANAC)
KAIJO HOAN-CHO (MARITIME SAFETY AGENCY)
1-3, 2-CHOME, KASUMIGASEKI, CHIYODA-KU, TOKYO 100
JAPANESE

TENTAI ICHIHYO 544
(JAPANESE EPHEMERIS)
KAIJO HOAN-CHO (MARITIME SAFETY AGENCY)
1-3, 2-CHOME, KASUMIGASEKI, CHIYODA-KU, TOKYO 100
JAPANESE

TOKYO ASTRONOMICAL BULLETIN, SERIES 2 545
TOKYO TENMONDAI (TOKYO ASTRONOMICAL OBSERVATORY)
OSAWA, MITAKA-SHI, TOKYO 181
JAPANESE
INDEXED: IAA, INS

TOKYO TENMONDAIHO 546
(TOKYO ASTRONOMICAL OBSERVATORY REPORT)
TOKYO TENMONDAI (TOKYO ASTRONOMICAL OBSERVATORY)
OSAWA, MITAKA-SHI, TOKYO 181
JAPANESE
INDEXED: IAA, INS

UCHU KAGAKU GIJUTSU CHOSA 547
(RESEARCH ON SPACE SCIENCES AND TECHNOLOGY)
KAGAKUGIJUTSU-CHO KENKYU CHOSEI-KYOKU (RESEARCH
CODDINATION BUREAU, SCIENCE AND TECHNOLOGY AGENCY)
2-1, 2-CHOME, KASUMIGASEKI, CHIYODA-KU, TOKYO 100
JAPANESE
INDEXED: JICST

UCHU KAGAKUKAN NYUSU 548
(MT. IKOMA SPACE SCIENCE MUSEUM NEWS)
IKOMAYAMA UCHU KAGAKUKAN (MT. IKOMA SPACE SCIENCE
MUSEUM)
2312-1, NAHATA-CHO, IKOMA-SHI 630-02
JAPANESE

YAMAMOTO SOKUHO 549
(YAMAMOTO CIRCULAR)
YAMAMOTO TENMONDAI (THE YAMAMOTO OBSERVATORY)
289, KAMITANAKAMI KIRYU-MACHI, OTSU-SHI 520-21
JAPANESE

Physics

ANNUAL REVIEW, INSTITUTE OF PLASMA PHYSICS 550
NAGOYA DAIGAKU PURAZUMA KENKYUSHO (INSTITUTE OF PLASMA PHYSICS, NAGOYA UNIVERSITY)
FURO-CHO, CHIKUSA-KU, NAGOYA-SHI 464
ENGLISH
INDEXED: JICST

BUNKO KENKYU 551
(JOURNAL OF THE SPECTROSCOPICAL SOCIETY OF JAPAN)
NIHON BUNKO GAKKAI (THE SPECTROSCOPICAL SOCIETY OF JAPAN)
25-2, 3-CHOME, HYAKUNIN-CHO, SHINJUKU-KU TOKYO 160
JAPANESE AND ENGLISH
INDEXED: CHE, INS, MET, WAS, JICST

BUSSEI KENKYU 552
(MATERIAL SCIENCE)
BUSSEI KENKYU KANKOKAI
C/O KYOTO DAIGAKU, YUKAWA KINENKAN, KITASHIRAKAWA OIWAKE-CHO, SAKYO-KU, KYOTO-SHI 606
JAPANESE
INDEXED: CHE, JICST

BUSSEIKEN DAYORI 553
(COMMUNICATIONS OF THE INSTITUTE OF SOLID STATE PHYSICS)
TOKYO DAIGAKU BUSSEI KENKYUSHO
7-22-1, ROPPONGI TOKYO 106
JAPANESE
INDEXED: CHE

BUTSURI KYOIKU 554
(JOURNAL OF THE PHYSICS, EDUCATION SOCIETY OF JAPAN)
NIHON BUTSURI KYOIKU GAKKAI (THE PHYSICS EDUCATION SOCIETY OF JAPAN)
C/O TOKYO KYOIKU DAIGAKU RIGAKUBU, 29-1, 3-CHOME, OTSUKA, BUNKYO-KU, TOKYO 112
JAPANESE
INDEXED: JICST

BUTSURIGAKU RONBUN SENSHU 555
(SERIES OF SELECTED PAPERS IN PHYSICS)
NIHON BUTSURI GAKKAI (PHYSICAL SOCIETY OF JAPAN)
C/O KIKAI SHINKO KAIKAN, 5-8, 3-CHOME, SHIBA-KOEN, MINATO-KU, TOKYO 105
ENGLISH

BUTSURIGAKUSHI KENKYU 556
(STUDY OF PHYSICAL HISTORY)
BUTSURIGAKU-SHI KENKYU KANKOKAI (EDITORIAL OFFICE OF STUDY OF PHYSICAL HISTORY)
C/O TOKAI DAIGAKU OBUTSU KENKYUSHITSU, 2-28, TOMIGAYA, SHIBUYA-KU, TOKYO 151
JAPANESE

CHO-TEION GIJUTSU 557
(CRYOGENIC ENGINEERING)
JAPAN INDUSTRIAL PUBLISHING
10-6, 3-CHOME, AZABU-JUBAN, MINATO-KU, TOKYO 106
JAPANESE
INDEXED: JICST

CONTRIBUTIONS FROM THE INSTITUTE OF LOW TEMPERATURE SCIENCE, SERIES A, PHYSICS 558
HOKKAIDO DAIGAKU TEION KAGAKU KENKYUSHO (THE INSTITUTE OF LOW TEMPERATURE SCIENCE, HOKKAIDO UNIVERSITY)
NISHI-8-CHOME, KITA-19-JO, KITA-KU, SAPPORO-SHI 060
ENGLISH
INDEXED: JICST

CRL HOKOKU 559
(CRL REPORT, JAPANESE EDITION)
TOKYO DAIGAKU UCHUSEN KANSOKUJO (COSMIC RAY LABORATORY, TOKYO UNIVERSITY)
2-1, 3-CHOME, MIDORI-CHO, TANASHI-SHI, TOKYO 188
JAPANESE

CRL REPORT 560
TOKYO DAIGAKU UCHUSEN KANSOKUJO (COSMIC RAY LABORATORY, TOKYO UNIVERSITY)
2-1, 3-CHOME, MIDORI-CHO, TANASHI-SHI, TOKYO 188
ENGLISH

HANDOTAI KENKYU 561
(SEMICONDUCTOR RESEARCH)
HANDOTAI KENKYU SHINKOKAI (SEMICONDUCTOR RESEARCH INSTITUTE)
KAWAUCHI, SENDAI-SHI 980
JAPANESE AND ENGLISH

HANDOTAI KENKYUSHO HOKOKU 562
(REPORTS OF SEMICONDUCTOR RESEARCH FOUNDATION)
HENDOTAI KENKYU SHINKOKAI (SEMICONDUCTOR RESEARCH FOUNDATION)
KAWAUCHI-SENDAI-SHI 980
JAPANESE
INDEXED: JICST

HIRUZEN KENKYUJO KENKYU HOKOKU 563
(BULLETIN OF THE HIRUZEN RESEARCH INSTITUTE)
OKAYAMA RIKA DAIGAKU HIRUZEN KENKYUJO (OKAYAMA SCHOOL OF PHYSICS, HIRUZEN RESEARCH INSTITUTE)
JAPANESE

JAPANESE JOURNAL OF APPLIED PHYSICS 564
OYO BUTSURIGAKU OBUNSHI KANKOKAI (PUBLICATION OFFICE OF JAPANESE JOURNAL OF APPLIED PHYSICS)
24-8, 4-CHOME, SHINBASHI, MINATO-KU, TOKYO 105
ENGLISH
INDEXED: CHE, COP, ENG, IAA, INS, MET, SCI, WAA, JICST

JASCO REPORT 565
NIHON BUNKO KOGYO K.K. (JAPAN SPECTROSCOPIC COMPANY, LIMITED)
2967-5, ISHIKAWA-CHO, HACHIOJI-SHI, TOKYO 192
JAPANESE
INDEXED: JICST

JOURNAL OF THE FACULTY OF SCIENCE, HOKKAIDO UNIVERSITY, SERIES 2, PHYSICS 566
HOKKAIDO DAIGAKU RIGAKUBU (FACULTY OF SCIENCE, HOKKAIDO UNIVERSITY)
NISHI 8-CHOME, KITA 10-JO, KITA-KU, SAPPORO-SHI 060
ENGLISH
INDEXED: JICST

JOURNAL OF THE PHYSICAL SOCIETY OF JAPAN 567
NIHON BUTSURI GAKKAI (PHYSICAL SOCIETY OF JAPAN)
KIKAI SHINKO BUILDING, 5-8, 3-CHOME, SHIBAKOEN, MINATO-KU, TOKYO 105
EUROPEAN
INDEXED: APP, CHE, COP, ENG, IAA, INS, MAT, MET, SCI, WAA, JICST

KAGAKU GIJUTSU BUNKEN SOKUHO, BUTSURI, OYOBUTSURI-HEN 568
(CURRENT BIBLIOGRAPHY ON SCIENCE AND TECHNOLOGY, SERIES PURE AND APPLIED PHYSICS)
NIHON KAGAKU GIJUTSU JOHO SENTA (JAPAN INFORMATION CENTER OF SCIENCE AND TECHNOLOGY)
1-1, 2-CHOME, NAGATA-CHO- CHIYODA-KU, TOKYO 100
JAPANESE
INDEXED: JICST

KEK 569
MONBU-SHO KO-ENERUGI BUTSURIGAKU KENKYUSHO (NATIONAL LABORATORY FOR HIGH ENERGY PHYSICS)
MAENO, OHO-MACHI, TSUKUBA-GUN, IBARAKI 300-32
ENGLISH
INDEXED: CHE

KEK ACCELERATORS 570
MONBU-SHO KO-ENERGUGI BUTSURIGAKU KENKYUSHO (NATIONAL LABORATORY FOR HIGH ENERGY PHYSICS)
MAENO, OHO-MACHI, TSUKUBA-GUN, IBARAKI 300-32
JAPANESE AND ENGLISH

KEK ANNUAL REPORT 571
MONBU-SHO KO-ENERUGI BUTSURIGAKU KENKYUSHO (NATIONAL LABORATORY FOR HIGH ENERGY PHYSICS)
MAENO, OHO-MACHI, TSUKUBA-GUN, IBARAKI 300-32
ENGLISH

KEK EXPERIMENTAL FACILITIES 572
MONBU-SHO KO-ENERUGI BUTSURIGAKU KENKYUSHO (NATIONAL LABORATORY FOR HIGH ENERGY PHYSICS)
MAENO, OHO-MACHI, TSUKUBA-GUN, IBARAKI 300-32
JAPANESE AND ENGLISH

KEK PREPRINT 573
MONBU-SHO KO-ENERUGI BUTSURIGAKU KENKYUSHO (NATIONAL LABORATORY FOR HIGH ENERGY PHYSICS)
MAENO-OHO-MACHI, TSUKUBA-GUN, IBARAKI 300-32
ENGLISH

KINZOKU BUTSURI SEMINA 574
(METAL PHYSICS SEMINAR)
AGNE GIJUTSU CENTER
KITAMURA BUILDING, 5-1-25, MINAMIAOYAMA, MINATO-KU
TOKYO 107
JAPANESE
INDEXED: CHE, JICST

KOENERUGIKEN GEPPO 575
(MONTHLY BULLETIN OF THE NATIONAL LABORATORY FOR
HIGH ENERGY PHYSICS)
MONBU-SHO KO-ENERUGI BUTSURIGAKU KENKYUSHO
(NATIONAL LABORATORY FOR HIGH ENERGY PHYSICS)
MAENO, OHO-MACHI, TSUKUBA-GUN, IBARAKI 300-32
JAPANESE

KOGAKU 576
(JAPANESE JOURNAL OF OPTICS)
OYO BUTSURIGAKKAI KOGAKU KONWAKAI (OPTICS
DIVISION, JAPAN SOCIETY OF APPLIED PHYSICS)
1-5, 21-CHOME, SHIBAKOEN, MINATO-KU TOKYO 105
JAPANESE
INDEXED: JICST

KOON GAKKAI SHI 577
(JOURNAL OF THE HIGH TEMPERATURE SOCIETY)
JAPANESE

KOTAI BUTSURI 578
(SOLID STATE PHYSICS)
SAIENSU PRESUSHA (SCIENCE PRESS COMPANY)
C/O KOJI-MACHI MANSHON, 5-3, KOJI-MACHI,
CHIYODA-KU, TOKYO 102
JAPANESE
INDEXED: CHE, INS, JICST

MEIDAI UCHUSEN KENKYUSHITSU KIJI 579
(PROCEEDINGS OF THE COSMIC-RAY RESEARCH LABORATORY
OF NAGOYA UNIVERSITY)
NAGOYA DAIGAKU RIGAKUBU UCHUSEN BOENKYO KENKYU
SHISETSU (COSMIC RAY RESEARCH LABORATORY AND
DEPARTMENT OF PHYSICS, NAGOYA UNIVERSITY)
FURO-CHO, CHIKUSA-KU, NAGOYA-SHI 464
JAPANESE

MEMOIRS OF THE FACULTY OF SCIENCES, KYUSHU 580
UNIVERSITY, SERIES B, PHYSICS
KYUSHU DAIGAKU RIGAKUBU (FACULTY OF SCIENCE,
KYUSHU UNIVERSITY)
3576, HAKOZAKI, HIGAHI-KU, FUKUOKA-SHI 812
ENGLISH
INDEXED: CHE, INS, WAA, JICST

NAGOYA DAIGAKU PURAZUMA KENKYUSHO SHIRYO, GIJUTSU 581
HOKOKU
(INSTITUTE OF PLASMA PHYSICS, DATA AND TECHNICAL
BOOK)
NAGOYA DAIGAKU PURAZUMA KENKYUSHO (INSTITUTE OF
PLASMA PHYSICS, NAGOYA UNIVERSITY)
FURO-CHO, CHIKUSA-KU, NAGOYA-SHI 464
JAPANESE
INDEXED: JICST

NIHON BUTSURI GAKKAI BUNKAKAI YOKOSHU 582
(PREPRINTS OF THE SYMPOSIUM OF THE PHYSICAL SOCIETY
OF JAPAN)
NIHON BUTSURI GAKKAI (PHYSICAL SOCIETY OF JAPAN)
5-8, 3-CHOME, SHIBAKOEN, MINATO-KU, TOKYO 105
JAPANESE
INDEXED: JICST

NIHON BUTSURI GAKKAISHI 583
(PROCEEDINGS OF THE PHYSICAL SOCIETY OF JAPAN)
NIHON BUTSURI GAKKAI (PHYSICAL SOCIETY OF JAPAN)
KIKAI SHINKO BUILDING, 5-8, 3-CHOME, SHIBAKOEN,
MINATO-KU, TOKYO 105
JAPANESE
INDEXED: CHE, JICST

NIHON ISOTOPE KAIGI HOBUNSHU 584
(PAPERS OF THE JAPAN ISOTOPE CONFERENCE)
JAPANESE

NIHON ONKYO GAKKAI KENKYU HAPPYOKAI KOEN RONBUNSHU 585
(REPORT OF THE MEETING, ACOUSTICAL SOCIETY OF JAPAN)
NIHON ONKYO GAKKAI (ACOUSTICAL SOCIETY OF JAPAN)
C/O IKEDA BUILDING, 7-7, 2-CHOME, YOYOGI,
SHIBUYA-KU, TOKYO 151
JAPANESE

NIHON ONKYO GAKKAISHI 586
(JOURNAL OF THE ACOUSTICAL SOCIETY OF JAPAN)
NOHON ONKYO GAKKAI (ACOUSTICAL SOCIETY OF JAPAN)
C/O IKEDA BUILDING, 7-7, 2-CHOME, YOYOGI,
SHIBUYA-KU, TOKYO 151
JAPANESE, WITH EUROPEAN SUMMARY
INDEXED: CHE, INS, JICST

NIPPON BUTSURI GAKKAI NENKAI YOKOSHU 587
(PREPRINTS OF THE ANNUAL MEETING, JAPAN PHYSICS
SOCIETY)
JAPANESE
INDEXED: JICST

NIPPON BUTSURI GAKKAI. AKI NO BUNKAKAI YOKOSHU 588
(PREPRINTS OF THE AUTUMN SECTIONAL MEETING, JAPAN
PHYSICS SOCIETY)
JAPANESE
INDEXED: JICST

ORYOKU TO HIZUMI 589
(JAPANESE JOURNAL OF EXPERIMENTAL MECHANICS)
CHUO KAGAKUSHA
20-8, 1-CHOME, MUKOGAOKA, BUNKYO-KU, TOKYO 113
JAPANESE
INDEXED: JICST

OYO BUTSURI 590
(APPLIED PHYSICS)
OYO BUTSURI GAKKAI (THE JAPAN SOCIETY OF APPLIED
PHYSICS)
5-8 3-CHOME, SHIBAKOEN, MINATO-KU, TOKYO 105
JAPANESE
INDEXED: CHE, INS, JICST

OYO BUTSURI GAKKAI GAKUJUTSU KOENKAI KOEN YOKOSHU 591
(PREPRINTS OF THE SYMPOSIUM OF THE JAPAN SOCIETY
OF APPLIED PHYSICS)
OYO BUTSURI GAKKAI (JAPAN SOCIETY OF APPLIED
PHYSICS)
5-8, 3-CHOME, SHIBAKOEN, MINATO-KU, TOKYO 105
JAPANESE

OYO BUTSURI GAKKAI KANKEI RENGO KOENKAI KOEN 592
YOKOSHU
(PREPRINTS OF THE JOINT CONGRESS OF THE JAPAN
SOCIETY OF APPLIED PHYSICS)
OYO BUTSURI GAKKAI (JAPAN SOCIETY OF APPLIED
PHYSICS)
5-8, 3-CHOME, SHIBAKOEN, MINATO-KU, TOKYO 105
JAPANESE

OYO BUTSURI GAKKAI KOGAKU KONDANKAI MAEZURI 593
(OPTICS MEETING PREPRINTS, APPLIED PHYSICS SOCIETY)
JAPANESE
INDEXED: JICST

OYO BUTSURI GAKKAI KOGAKU KONWAKAI TOKI KOSHUKAI 594
TEKISUTO
(OPTICS MEETING, WINTER CLASS TEXT, APPLIED PHYSICS
SOCIETY)
JAPANESE
INDEXED: JICST

OYO RIKIGAKU KENKYUSHO SHOHO 595
(BULLETIN OF RESEARCH FOR APPLIED MECHANICS)
KYUSHU DAIGAKU OYO RIKIGAKU KENKYUSHO (RESEARCH
INSTITUTE FOR APPLIED MECHANICS, KYUSHU UNIVERSITY)
HAKOZAKI, HIGASHI-KU, FUKUOKA-SHI 812
JAPANESE, WITH ENGLISH SUMMARY
INDEXED: JICST

PHYSICS REPORTS OF THE KUMAMOTO UNIVERSITY 596
DEPARTMENT OF PHYSICS, KUMAMOTO UNIVERSITY
774-6, OMI-CHO KUMAMOTO
ENGLISH
INDEXED: CHE

PROCEEDINGS OF JAPAN NATIONAL CONGRESS FOR APPLIED 597
MECHANICS
NIHON GAKUJUTSU KAIGI RIKIGAKU KENKYU RENRAKU
IINKAI (JAPAN NATIONAL COMMITTEE FOR THEORETICAL AND
APPLIED MECHANICS)
C/O KIYA BUILDING, 20-8, 1-CHOME, MUKOGAOKA,
BUNKYO-KU, TOKYO 113
ENGLISH
INDEXED: APP

PROGRESS OF THEORETICAL PHYSICS 598
KISO BUTSURIGAKU KENKYUSHO RIRON
BUTSURIGAKU KANKOKAI
C/O KYOTO DAIGAKU KISOBUTSURIGAKU KENKYUSHO,
KITASHIRAKAWA OIWAKE-CHO, SAKYO-KU, KYOTO-SHI 606
ENGLISH
INDEXED: CHE, IAA, INS, MAT, MET, SCI, JICST

REPORTS OF THE INSTITUTE OF HIGH SPEED MECHANICS, 599
TOHOKU UNIVERSITY
TOHOKU DAIGAKU KOSOKU RIKIGAKU KENKYUSHO (THE
INSTITUTE OF HIGH SPEED MECHANICS,
TOHOKU UNIVERSITY)
1-1, 2-CHOME, KATAHIRA, SENDAI-SHI 980
ENGLISH
INDEXED: IAA, INS, JICST

REPORTS ON PROGRESS IN POLYMER PHYSICS IN JAPAN 600
GAKUJUTSU BUNKEN FUKYUKAI (ASSOCIATION FOR SCIENCE
DOCUMENTS INFORMATION)
C/O TOKYO KOGYO DAIGAKU, OOKAYAMA, MEGURO-KU,
TOKYO 152
ENGLISH
INDEXED: CHE, JICST

REPORTS OF RESEARCH INSTITUTE FOR APPLIED MECHANICS 601
KYUSHU DAIGAKU OYORIKIGAKU KENKYUSHO (RESEARCH
INSTITUTE FOR APPLIED MECHANICS, KYUSHU UNIVERSITY)
HAKOZAKI, HIGAHI-KU, FUKUOKA-SHI 812
ENGLISH
INDEXED: APP, CHE, ENG, INS, JICST

RESEARCH REPORT, INSTITUTE OF PLASMA PHYSICS, 602
NAGOYA UNIVERSITY (IPPJ)
NAGOYA DAIGAKU PURAZUMA KENKYUSHO (INSTITUTE OF
PLASMA PHYSICS, NAGOYA UNIVERSITY)
FURO-CHO, CHIKUSA-KU, NAGOYA-SHI 464
ENGLISH
INDEXED: CHE, JICST

RIFP REPORT 603
KYOTO DAIGAKU KISO BUTSURIGAKU KENKYUSHO (RESEARCH
INSTITUTE FOR FUNDAMENTAL PHYSICS, KYOTO UNIVERSITY)
KITASHIRAKAWA OIWAKE-CHO, SAKYO-KU, KYOTO-SHI 606
ENGLISH

RIKIGAKU KENKYUJO HOKOKU 604
(REPORT OF THE APPLIED MECHANICS INSTITUTE)
JAPANESE

RRK 605
HIROSHIMA DAIGAKU RIRON BUTSURIGAKU KENKYUSHO
(RESEARCH INSTITUTE FOR THEORETICAL PHYSICS,
HIROSHIMA UNIVERSITY)
1294, MATOBA, TAKEHARA-CHO, TAKEHARA-SHI 725
ENGLISH

RYUHYO KENKYU SHISETSU KENKYU HOKOKU SHI 606
(COLLECTED RESEARCH REPORTS OF THE DRIFTING ICE
RESEARCH FACILITY)
HOKKAIDO DAIGAKU TEIONKAGAKU KENKYUJO RYUHO KENKYU
SHISETSU (HOKKAIDO UNIVERSITY LOW TEMPERATURE
SCIENCE RESEARCH INSTITUTE, DRIFTING ICE RESEARCH
FACILITY)
JAPANESE

SCIENCE OF LIGHT 607
TOKYO KYOIKU DAIGAKU KOGAKU KENKYUSHO (THE
INSTITUTE FOR OPTICAL RESEARCH, TOKYO KYOIKU
UNIVERSITY)
22-17, 3-CHOME, HYAKUNIN-CHO, SHINJUKU-KU,
TOKYO 160
ENGLISH
INDEXED: CHE, INS, SCI, JICST

SCIENCE REPORTS OF NIIGATA UNIVERSITY, SERIES B, 608
PHYSICS
NIIGATA DAIGAKU RIGAKUBU
(FACULTY OF SCIENCE, NIIGATA UNIVERSITY)
8050, IGARASHI NINO-MACHI, NIIGATA-SHI 950-21
EUROPEAN

SCIENTIFIC REPORTS OF THE RESEARCH INSTITUTE FOR 609
THEORETICAL PHYSICS, HIROSHIMA UNIVERSITY
HIROSHIMA DAIGAKU RIRONBUTSURIGAKU KENKYUSHO
(RESEARCH INSTITUTE FOR THEORETICAL PHYSICS,
HIROSHIMA UNIVERSITY)
ENGLISH
INDEXED: MAT

SHINKU 610
(JOURNAL OF THE VACUUM SOCIETY OF JAPAN)
NIHON SHINKU KYOKAI (THE VACUUM SOCIETY OF JAPAN)
5-8, 3-CHOME, SHIBAKOEN, MINATO-KU, TOKYO 105
JAPANESE, WITH ENGLISH SUMMARY
INDEXED: CHE, INS, JICST

SHINPEN BUTSURIGAKU SENSHU 611
(SERIES OF SELECTED PAPERS IN PHYSICS, NEW SERIES)
NIHON BUTSURI GAKKAI (PHYSICAL SOCIETY OF JAPAN)
C/O KIKAI SHINKO KAIKAN, 5-8, 3-CHOME, SHIBAKOEN,
MINATO-KU, TOKYO 105
ENGLISH

SORYUSHIRON KENKYU 612
(STUDY OF ELEMENTARY PARTICLES)
RIRON BUTSURIGAKU KANKOKAI (PUBLICATION OFFICE
OF PROGRESS OF THEORETICAL PHYSICS)
C/O KYOTO DAIGAKU KISO BUTSURIGAKU KENKYUSHO,
KITASHIRAKAWA OIWAKE-CHO, SAKYO-KU,
KYOTO-SHI 606
JAPANESE
INDEXED: CHE, JICST

SUPPLEMENT OF THE PROGRESS OF THEORETICAL PHYSICS 613
KISOBUTSURIGAKU KENKYUSHO RIRON BUTSURIGAKU
KANKOKAI
C/O KYOTO DAIGAKU KISOBUTSURIGAKU KENKYUSHO,
KITASHIRAKAWA OIWAKE-CHO, SAKYO-KU, KYOTO-SHI 606
EUROPEAN
INDEXED: INS

TECHNICAL REPORT OF THE INSTITUTE OF PLASMA PHYSICS, 614
NAGOYA UNIVERSITY (IPPJ-T)
NAGOYA DAIGAKU PURAZUMA KENKYUSHO (INSTITUTE OF
PLASMA PHYSICS, NAGOYA UNIVERSITY)
FURO-CHO, CHIKUSA-KU, NAGOYA-SHI 464
ENGLISH
INDEXED: JICST

TECHNICAL REPORT OF ISSP, SERIES A 615
TOKYO DAIGAKU BUSSEI KENKYUSHO (INSTITUTE FOR
SOLID STATE PHYSICS, UNIVERSITY OF TOKYO)
22-1, 7-CHOME, ROPPONGI, MINATO-KU, TOKYO 106
ENGLISH
INDEXED: JICST

TECHNICAL REPORT OF ISSP, SERIES B 616
TOKYO DAIGAKU BUSSEI KENKYUSHO (INSTITUTE FOR
SOLID STATE PHYSICS, UNIVERSITY OF TOKYO)
22-1, 7-CHOME, ROPPONGI, MINATO-KU, TOKYO 106
ENGLISH
INDEXED: JICST

TEION KAGAKU, BUTSURI-HEN 617
(LOW TEMPERATURE SCIENCE, SERIES A)
HOKKAIDO DAIGAKU TEION KAGAKU KENKYUSHO (THE
INSTITUTE OF LOW TEMPERATURE SCIENCE, HOKKAIDO
UNIVERSITY)
NISHI 8-CHOME, KITA 19-JO, KITA-KU, SAPPORO-SHI 060
JAPANESE, WITH ENGLISH SUMMARY
INDEXED: CHE, INS, JICST

TEION KAGAKU, BUTSURI-HEN SHIRYOSHU 618
(LOW TEMPERATURE SCIENCE, SERIES A, DATA REPORT)
HOKKAIDO DAIGAKU TEION KAGAKU KENKYUSHO (INSTITUTE
OF LOW TEMPERATURE SCIENCE, HOKKAIDO UNIVERSITY)
NISHI 8-CHOME, KITA 19-JO, KITA-KU, SAPPORO-SHI 060
JAPANESE

TOHOKU DAIGAKU KOSOKU RIKIGAKU KENKYUSHO HOKOKU 619
(REPORT OF THE INSTITUTE OF HIGH SPEED MECHANICS,
TOHOKU UNIVERSITY)
TOHOKU DAIGAKU KOSOKU RIKIGAKU KENKYUSHO (THE
INSTITUTE OF HIGH SPEED MECHANICS, TOHOKU
UNIVERSITY)
1-1, 2-CHOME, KATAHIRA, SENDAI-SHI 980
JAPANESE
INDEXED: CHE, ENG, MAT, JICST

TOKYO DAIGAKU RIGAKUBU BUTSURIGAKU KYOSHITSU KENJI 620
HOKOKU
(ANNUAL REPORTS OF THE DEPARTMENT OF PHYSICS,
FACULTY OF SCIENCE, UNIVERSITY OF TOKYO)
TOKYO DAIGAKU, RIGAKUBU BUTSURI KYOSHITSU
(DEPARTMENT OF PHYSICS, FACULTY OF SCIENCE,
UNIVERSITY OF TOKYO)
3-1, 7-CHOME, HONGO, BUNKYO-KU TOKYO 113
JAPANESE
INDEXED: JICST

UCHUSEN KENKYU 621
(STUDIES OF COSMIC-RAY)
UCHUSEN KENKYU HENSHU-GAKARI (EDITORIAL BOARD
OF STUDIES OF COSMIC-RAY)
C/O TOKYO DAIGAKU GENSHIKAKU KENKYUSHO TOSHOKAN,
2-1, 3-CHOME, MIDORI-CHO, TANASHI-SHI, TOKYO 188
JAPANESE AND EUROPEAN
INDEXED: CHE, JICST

UCHUSEN KENKYUSHITSU HOKOKU 622
(REPORT OF THE COSMIC-RAY LABORATORY)
RIKAGAKU KENKYUSHO UCHUSEN KENKYUSHITSU (COSMIC RAY LABORATORY, INSTITUTE OF PHYSICAL AND CHEMICAL RESEARCH)
7-13, 1-CHOME, KAGA, ITABASHI-KU, TOKYO 173
JAPANESE

Chemistry

AIONIKUSU 623
(IONICS)
JAPANESE
INDEXED: JICST

AMIRAZE SHINPOJIUMU 624
(PROCEEDINGS OF THE SYMPOSIUM ON AMYLASE)
AMIRAZE KENKYUKAI
(AMYLASE RESEARCH GROUP)
JAPANESE

BESSATSU KAGAKU 625
(CHEMISTRY (SPECIAL ISSUE))
KAGAKU DOJIN (STAFF OF KAGAKU)
JAPANESE

BULLETIN OF THE CHEMICAL SOCIETY OF JAPAN 626
NIPPON KAGAKKAI (THE CHEMICAL SOCIETY OF JAPAN)
1-5, KANDA-SURUGADAI, CHIYODA-KU, TOKYO 101
ENGLISH
INDEXED: API, CER, CHE, EXC, SCI, JICST

BULLETIN OF THE INSTITUTE FOR CHEMICAL RESEARCH, 627
KYOTO UNIVERSITY
KYOTO DAIGAKU KAGAKU KENKYUSHO (THE INSTITUTE FOR CHEMICAL RESEARCH, KYOTO UNIVERSITY)
GOKASHO, UJI-SHI 611
ENGLISH
INDEXED: BIB, BIO, CHE, INS, JICST

BUNSEKI 628
(ANALYSIS)
JAPANESE
INDEXED: JICST

BUNSEKI KAGAKU 629
(JAPAN ANALYST)
NIHON BUNSEKI KAGAKKAI (THE JAPAN SOCIETY FOR ANALYTICAL CHEMISTRY)
C/O KOGYO GIJUTSU-IN TOKYO KOGYO SHIKENJO, 1-5, 1-CHOME, HON-MACHI, SHIBUYA-KU, TOKYO 151
JAPANESE, WITH ENGLISH SUMMARY
INDEXED: CHE, JICST

BUNSEKI KAGAKU SHINPO SOSETSU 630
(ANNUAL REVIEW OF JAPAN ANALYST)
NIHON BUNSEKI KAGAKKAI (THE JAPAN SOCIETY FOR ANALYTICAL CHEMISTRY)
C/O KOGYO GIJUTSU-IN TOKYO KOGYO SHIKENJO, 1-5, 1-CHOME, HON-MACHI, SHIBUYA-KU, TOKYO 151
JAPANESE
INDEXED: CHE

BUNSEKI KIKI 631
(ANALYSIS AND INSTRUMENT)
SANGYO KAIHATSUSHA
11-10, 1-CHOME, KANDA, CHIYODA-KU, TOKYO 100
JAPANESE
INDEXED: CHE, JICST

BUNSHIKEN RETAZU 632
(MOLECULAR RESEARCH LETTERS)
BUNSHI KAGAKU KENKYUJO
(MOLECULAR SCIENCE RESEARCH INSTITUTE)
JAPANESE

BUNSHIKEN RIFURETTO 633
(MOLECULAR RESEARCH LEAFLET)
BUNSHI KAGAKU KENKYUJO (MOLECULAR SCIENCE RESEARCH INSTITUTE)
JAPANESE

CHEMISTRY LETTERS 634
NIHON KAGAKKAI (THE CHEMICAL SOCIETY OF JAPAN)
1-5, KANDA SURUGADAI, CHIYODA-KU, TOKYO 101
JAPANESE AND ENGLISH
INDEXED: BIB, CHE, SCI, JICST

CHIBA DAIGAKU SEIBUTSU KASSEI KENKYUSHO HOKOKU 635
(ANNUAL REPORT OF RESEARCH INSTITUTE FOR CHEMOBIODYNAMICS, CHIBA UNIVERSITY)
CHIBA DAIGAKU SEIBUTSU KASSEI KENKYUSHO
OKUBO-CHO NARASHINO
JAPANESE AND ENGLISH
INDEXED: BIB, BIO, CHE, JICST

CHOSEN NO GENDAI KAGAKU GIJUTSU 636
(CONTEMPORARY KOREAN SCIENCE AND TECHNOLOGY)
JAPANESE
INDEXED: JICST

DENKI KAGAKU NO SHINPO 637
(ADVANCES IN ELECTROCHEMISTRY)
DENKI KAGAKU KYOKAI (ELECTROCHEMICAL SOCIETY OF JAPAN)
12-1, 1-CHOME, YURAKU-CHO, CHIYODA-KU, TOKYO 100)
JAPANESE

DENKI KAGAKU OYOBI KOGYO BUTSURI KAGAKU 638
(ELECTRO-CHEMISTRY AND INDUSTRIAL PHYSICAL-CHEMISTRY)
DENKI KAGAKU KYOKAI (ELECTRO-CHEMICAL SOCIETY OF JAPAN
1-11, YURAKU-CHO, CHIYODA-KU, TOKYO 100
JAPANESE AND ENGLISH
INDEXED: CHE, JICST

EHIME DAIGAKU KIYO DAI-2-BU SHIZENKAGAKU, C-SHIRIZU 639
KAGAKU
(MEMOIRS OF THE EHIME UNIVERSITY, SECTION 2, NATURAL SCIENCE, SERIES C, CHEMISTRY)
EHIME DAIGAKU
3-BAN, BUNKYO-CHO, MATSUYAMA-SHI 790
JAPANESE AND ENGLISH
INDEXED: CHE

FUSSO KAGAKU TORONKAI KOEN YOKOSHU 640
(PREPRINTS OF THE FORUM ON FLUORINE CHEMISTRY)
JAPANESE
INDEXED: JICST

GENDAI KAGAKU 641
(MODERN CHEMISTRY)
GENDAI KAGAKU HENSYU GURUPU AND TOKYO KAGAKU DOJIN
36-7, 3-CHOME, SENGOKU, BUNKYO-KU, TOKYO 112
JAPANESE
INDEXED: CHE, JICST

GIJUTSU KENKYU-KAI HOKOKU 642
(REPORT OF THE TECHNICAL RESEARCH GROUP)
BUNSHI KAGAKU KENKYUJO KANRI-BU GIJUTSU-KA
(MOLECULAR SCIENCE RESEARCH INSTITUTE, ADMINISTRATIVE DIVISION, TECHNICAL SECTION)
JAPANESE

GURUKURON-SAN SHINPOJIUMU HOKOKUSHU 643
(GLUDURONIC ACID SYMPOSIUM, REPORT)
JAPANESE
INDEXED: JICST

HETEROCYCLES 644
SENDAI FUKUSOKAN KAGAKU KENKYUSHO (THE SENDAI INSTITUTE OF HETEROCYCLIC CHEMISTRY)
C/O TOHOKU DAIGAKU YAKUGAKUBU, AOBA, ARAMAKI, SENDAI-SHI 980
ENGLISH
INDEXED: CHE, SCI

HYOMEN (HYOMEN DANWAKAI) 645
(SURFACE)
HYOMEN DANWAKAI KOROIDO KONWAKAI (MEETING OF SURFACE, MEETING OF COLLOID CHEMISTRY)
C/O SANKYO BUILDING, 3-20, KANDA NISHIKI-CHO, CHIYODA-KU, TOKYO 101
JAPANESE
INDEXED: CHE, JICST

HYOMEN (IBARAKI DAIGAKU) 646
(SURFACE)
IBARAKI DAIGAKU KOGAKUBU HYOMEN SHORI KENKYUSHITSU (SURFACE TREATMENT LABORATORY, DEPARTMENT OF ENGINEERING, IBARAKI UNIVERSITY)
12-1, 4-CHOME, NAKA NARISAWA-CHO, HITACHI-SHI 316
JAPANESE AND ENGLISH
INDEXED: JICST

JOURNAL OF THE FACULTY OF TEXTILE SCIENCE AND 647
TECHNOLOGY, SHINSHU UNIVERSITY, SERIES C, CHEMISTRY
SHINSHU DAIGAKU SEN'IGAKUBU (THE FACULTY OF TEXTILE SCIENCE & TECHNOLOGY, SHINSHU UNIVERSITY)
15-1, 3-CHOME, TOKIDA, UEDA-SHI 386
ENGLISH
INDEXED: CHE, JICST

JOURNAL OF THE RESEARCH INSTITUTE FOR CATALYSIS, HOKKAIDO UNIVERSITY 648
HOKKAIDO DAIGAKU SHOKUBAI KENKYUSHO (THE RESEARCH INSTITUTE FOR CATALYSIS, HOKKAIDO UNIVERSITY)
NISHI-10-CHOME, KITA-11-JO, KITA-KU, SAPPORO-SHI 060
EUROPEAN
INDEXED: API, CHE, JICST

KAGAKU 649
(CHEMISTRY)
KAGAKU DOJIN
YANAGINOBABA OIKE-SAGARU, NAKAGYO-KU, KYOTO-SHI 604
JAPANESE
INDEXED: CHE, JICST

KAGAKU GIJUTSU BUNKEN SOKUHO KAGAKU, KAGAKU KOGYO-HEN (KOKUNAI-HEN) 650
(CURRENT BIBIOGRAPHY ON SCIENCE AND TECHNOLOGY, SERIES ON CHEMISTRY AND CHEMICAL ENGINEERING (JAPANESE))
NIHON KAGAKU GIJUTSU JOHO SENTA (JAPAN INFORMATION CENTER OF SCIENCE AND TECHNOLOGY)
5-2, 2-CHOME, NAGATA-CHO, CHIYODA-KU, TOKYO 100
JAPANESE
INDEXED: JICST

KAGAKU GIJUTSU BUNKEN SOKUHO KAGAKU, KAGAKU KOGYO-HEN (GAIKOKU-HEN) 651
(CURRENT BIBLIOGRAPHY ON SCIENCE AND TECHNOLOGY, SERIES ON CHEMISTRY AND CHEMICAL ENGINEERING (FOREIGN))
NIHON KAGAKU GIJUTSU JOHO SENTA (JAPAN INFORMATION CENTER OF SCIENCE AND TECHNOLOGY)
5-2, 2-CHOME, NAGATA-CHO, CHIYODA-KU, TOKYO 100
JAPANESE
INDEXED: CHE, JICST

KAGAKU KYOIKU 652
(CHEMICAL EDUCATION)
NIPPON KAGAKKAI (THE CHEMICAL SOCIETY OF JAPAN)
1-5, KANDA SURUGADAI, CHIYODA-KU, TOKYO 101
JAPANESE
INDEXED: CHE, JICST

KAGAKU NO RYOIKI 653
(JOURNAL OF JAPANESE CHEMISTRY)
NANKODO
42-6, 3-CHOME, HONGO, BUNKYO-KU, TOKYO 113
JAPANESE
INDEXED: CHE, JICST

KAGAKU NO RYOIKI ZOKAN 654
(JOURNAL OF JAPANESE CHEMISTRY, SPECIAL ISSUE)
NANKODO
42-6, 3-CHOME, HONGO, BUNKYO-KU TOKYO 113
JAPANESE
INDEXED: CHE, JICST

KAGAKU SHOHO 655
(CHEMICAL ABSTRACTS)
KAGAKU JOHO KYOGIKAI
C/O NIHON KAGAKUKAI, 1-5, KANDA SURUGADAI, CHIYODA-KU, TOKYO 101
JAPANESE AND ENGLISH
INDEXED: JICST

KAGAKU SOSETSU 656
(CHEMISTRY IN GENERAL)
TOKYO DAIGAKU SHUPPANKAI
3-1, 7-CHOME, HONGO, BUNKYO-KU, TOKYO 113
JAPANESE AND ENGLISH
INDEXED: CHE

KAGAKU TO YAKUGAKU NO KYOSHITSU 657
(CHEMISTRY AND PHARMACY, QUARTERLY JOURNAL)
HIROKAWA SHOTEN
27-14, 3-CHOME, HONGO, BUNKYO-KU, TOKYO 113-91
JAPANESE
INDEXED: CHE, JICST

KAGAKU ZOKAN 658
(CHEMISTRY (SUPPLEMENT ISSUE))
KAGAKU DONIN (BY THE STAFF)
JAPANESE

KAGAKUSHI KENKYU 659
(RESEARCH ON THE HISTORY OF CHEMISTRY)
KAGAKUSHI KENKYUKAI (ASSOCIATION FOR RESEARCH ON THE HISTORY OF CHEMISTRY)
JAPANESE

KANZEI CHUO BUNSEKISHOHO 660
(REPORTS OF THE CENTRAL CUSTOMS LABORATORY)
OKURA-SHO KANZEI CHUO BUNSEKISHO (ANALYZING CENTRE FOR CUSTOMS, MINISTRY OF FINANCE)
531, IWASE, MATSUDO-SHI 271
JAPANESE
INDEXED: CHE, JICST

KIKAN REOROJI 661
(RHEOLOGY)
SOGO KAGAKU KENKYUSHO
6-13, 4-CHOME, KUDAN MINAMI, CHIYODA-KU, TOKYO 102
JAPANESE
INDEXED: JICST

KYORITSU KAGAKU RAIBURARI 662
(KYORITSU CHEMISTRY LIBRARY)
KYORITSU SHUPPAN COMPANY, LIMITED
4-6-19, KOHINATA, BUNKYO-KU, TOKYO
JAPANESE
INDEXED: CHE

MEMOIRS OF THE FACULTY OF SCIENCE, KYUSHU UNIVERSITY, SERIES C, CHEMISTRY 663
KYUSHU DAIGAKU RIGAKUBU (FACULTY OF SCIENCE, KYUSHU UNIVERSITY)
3576, HAKOJAKI, HIGASHI-KU, FUKUOKA-SHI 812
ENGLISH
INDEXED: BIB, CHE, INS, WAA, JICST

MONBU-SHO KAGAKU KENKYUHI NI YORU KENKYU HOKOKU SHUROKU, KAGAKU-HEN 664
(REPORTS ON RESEARCHES BY GRANTEES MINISTRY OF EDUCATION, CHEMISTRY)
NIHON GAKUJUTSU SHINKOKAI (JAPAN SOCIETY FOR THE PROMOTION OF SCIENCE)
3-1, 5-CHOME, KOJI-MACHI, CHIYODA-KU, TOKYO 102
JAPANESE
INDEXED: CHE

NIHON KAGAKU SORAN 665
(CHEMICAL ABSTRACTS OF JAPAN)
JAPANESE

NIIGATA RIKAGAKU 666
(JOURNAL OF PHYSICS AND CHEMISTRY OF NIIGATA)
NIIGATA-KEN RIKAGAKU GIJUTSU SHOKUIN KYOGIKAI (NIIGATA PREFECTURAL COUNCIL OF PHYSICAL AND CHEMICAL ENGINEERS)
JAPANESE

NIPPON KAGAKKAI TENNEN YUKI KAGOBUTSU TORONKAI KOEN YOSHISHU 667
(SYMPOSIUM ON THE CHEMISTRY OF NATURAL PRODUCTS, SYMPOSIUM PAPERS)
TENNEN YUKI KAGOBUTSU TORONKAI JUNBI IINKAI (ORGANIZING COMMITTEE OF SYMPOSIUM ON THE CHEMISTRY OF NATURAL PRODUCTS)
C/O TOKYO DAIGAKU RIGAKUBU, 3-1, 7-CHOME, HONGO, BUNKYO-KU, TOKYO 113
JAPANESE

NIPPON KAGAKKAISHI 668
(JOURNAL OF THE CHEMICAL SOCIETY OF JAPAN)
NIPPON KAGAKKAI (THE CHEMICAL SOCIETY OF JAPAN)
1-5, KANDA SURUGADAI, CHIYODA-KU, TOKYO 101
JAPANESE, WITH ENGLISH SUMMARY
INDEXED: API, BIB, CER, CHE, WAA, JICST

NIPPON KAGAKUKAI KOEN YOKOSHU 669
(PREPRINTS OF LECTURES AT THE JAPAN CHEMICAL SOCIETY)
JAPANESE
INDEXED: JICST

NIPPON KAGAKUKAI. KOSOKU HANNO TORONKAI KOEN YOKOSHU 670
(PREPRINTS OF THE SYMPOSIUM PAPERS ON HIGH SPEED REACTIONS, JAPAN CHEMICAL SOCIETY)
JAPANESE
INDEXED: JICST

NIPPON KAGAKUKAI. NMR TORONKAI KOEN YOKOSHU 671
(PREPRINTS OF THE SYMPOSIUM PAPERS ON NMR, JAPAN CHEMICAL SOCIETY)
JAPANESE
INDEXED: JICST

OTO KEMIKARU 672
(AUTO CHEMICALS)
JIDOSHA SANGYO TSUSHINSHA (AUTOMOBILE INDUSTRY NEWS COMPANY)
JAPANESE

PEPTIDE INFORMATION 673
ENGLISH

PORAROGURAFI 674
(REVIEW OF POLAROGRAPHY)
NIHON PORAROGURAFU GAKKAI (THE POLAROGRAPHIC SOCIETY OF JAPAN)
C/O KYOTO DAIGAKU NOGAKUBU NOGEIKAGAKU KYOSHITSU, KITASHIRAKAWA, SAKYO-KU, KYOTO-SHI 606
JAPANESE AND EUROPEAN
INDEXED: BIB, CHE, JICST

REPORTS OF THE RESEARCH LABORATORY FOR SURFACE SCIENCE, OKAYAMA UNIVERSITY 675
OKAYAMA DAIGAKU RIGAKUBU KAIMEN-KAGAKU KENKYU SHISETSU (LABORATORY FOR SURFACE SCIENCE, OKAYAMA UNIVERSITY)
EUROPEAN
INDEXED: CHE, ENG. INS, JICST

REVIEW OF PHYSICAL CHEMISTRY OF JAPAN 676
NIHON BUTSURI KAGAKU KENKYUKAI (THE PHYSICO-CHEMICAL SOCIETY OF JAPAN)
C/O KYOTO DAIGAKU RIGAKUBU KAGAKU KYOSHITSU BUTSURI KAGAKU KENKYUSHITSU, KITASHIRAKAWA OIWAKE-CHO, SAKYO-KU, KYOTO-SHI 606
ENGLISH
INDEXED: BIB, CHE, INS, SCI, JICST

RIKAGAKU KENKYUJO NYUSU 677
(INSTITUTE OF PHYSICAL AND CHEMICAL RESEARCH NEWS)
JAPANESE

SCIENCE REPORTS OF NIIGATA UNIVERSITY, SERIES C, CHEMISTRY 678
NIIGATA DAIGAKU RIGAKUBU (THE FACULTY OF SCIENCE NIIGATA UNIVERSITY)
8050, IGARASHI NINO-CHO, NIIGATA-SHI 950-21
EUROPEAN
INDEXED: CHE, JICST

SCIENCE REPORTS OF THE SOCIETY FOR THE RESEARCH OF PHYSICS-CHEMISTRY 679
KINKI DAIGAKU RIGAKUBU FUSE
ENGLISH
INDEXED: CHE

SHITSURYO BUNSEKI 680
(MASS SPECTROSCOPY)
SHITSURYO BUNSEKI GAKKAI (SOCIETY OF MASS SPECTROSCOPY, JAPAN)
C/O OSAKA FURITSU DAIGAKU KYOYOGAKUBU, 4-804, MOZU UME-CHO, SAKAI-SHI 591
JAPANESE AND EUROPEAN
INDEXED: CHE, INS, JICST

SHOKUBAI 681
(CATALYST)
SHOKUBAI GAKKAI (CATALYSIS SOCIETY OF JAPAN)
C/O TOKYO KOGYO DAIGAKU, 12-1, 2-CHOME, OOKAYAMA, MEGURO-KU, TOKYO 152
JAPANESE
INDEXED: CHE, JICST

SHOKUBAI TORONKAI KOEN YOKOSHU 682
(PREPRINTS OF PAPERS OF THE SYMPOSIUM ON CATALYSTS)
JAPANESE
INDEXED: JICST

TEIKOKU GAKUEN KIYO 683
(MEMOIRS OF THE TEIKOKU WOMEN'S UNIVERSITY)
TEIKOKU WOMEN'S UNIVERSITY
173, TODA-CHO, 6-CHOME, MORIGUCHI, OSAKA 570
JAPANESE AND ENGLISH
INDEXED: CHE

YAMAGUCHI DAIGAKU IGAKUBU TANPAKU KAGAKU KENKYUKAI NENPO 684
(ANNUAL OF YAMAGUCHI UNIVERSITY, SCHOOL OF MEDICINE, PROTEIN CHEMISTRY RESEARCH SOCIETY)
JAPANESE

YAMAZAKI KAGAKU KENKYUJO HOKOKU 685
(REPORT OF THE YAMAZAKI CHEMISTRY INSTITUTE)
JAPANESE
INDEXED: JICST

YOYUEN KAGAKU TORONKAI KOEN YOSHISHU 686
(SUMMARY PAPERS AT SYMPOSIUM ON MOLTEN SALT)
DENKI KAGAKU KYOKAI YOYUEN IINKAI (MOLTEN SALT COMMITTEE, ELECTRO-CHEMISTRY ASSOCIATION)
C/O KYOTO DAIGAKU KOGAKUBU, YOSHIDA HON-CHO, SAKYO-KU, KYOTO-SHI 606
JAPANESE
INDEXED: JICST

Mineralogical Sciences and Crystallography

COLLECTION OF REPORTS OF THE SYNTHETIC CRYSTAL RESEARCH LABORATORY, FACULTY OF ENGINEERING, NAGOYA UNIVERSITY 687
NAGOYA DAIGAKU KOGAKUBU JINKO KETSHO KENKYU SHISETSU (SYNTHETIC CRYSTAL RESEARCH LABORATORY, FACULTY OF ENGINEERING, NAGOYA UNIVERSITY)
FURO-CHO, CHIKUSA-KU, NAGOYA-SHI 464
JAPANESE AND ENGLISH

EKISHO TORONKAI KOEN YOKOSHU (PREPRINTS OF PAPERS AT THE SYMPOSIUM ON LIQUID CRYSTALS) 688
JAPANESE
INDEXED: JICST

GANSEKI KOBUTSU KOSHO GAKKAISHI (JOURNAL OF THE JAPANESE ASSOCIATION OF MINERALOGISTS, PETROIOGISTS AND ECONOMIC GEOLOGISTS) 689
NIHON GANSEKI KOBUTSU KOSHO GAKKAI (JAPANESE ASSOCIATION OF MINERALOGIST, PETROLOGISTS AND ECONOMIC GEOLOGISTS)
C/O TOHOKU DAIGAKU RIGAKUBU, AOBAYAMA, SENDAI-SHI 980
JAPANESE AND EUROPEAN
INDEXED: CHE, JICST

HOSEKI GAKKAISHI (JOURNAL OF THE GEMMOLOGICAL SOCIETY) 690
GEMMOLOGICAL SOCIETY OF JAPAN
C/O INSTITUTE OF MINERALOGY, PETROLOGY, ECONOMIC GEOLOGY, TOHOKU UNIVERSITY SENDAI 980
JAPANESE, WITH ENGLISH SUMMARY
INDEXED: CHE

INSEKI TO INTETSU (METEORIC STONE AND METEORIC IRON) 691
NIHON INSEKI KENKYUKAI
59, IZUMI, YUGAWARA-CHO, ASHIGARA SHIMOGUN KANAGAWA
JAPANESE
INDEXED: CHE, JICST

KOBUTSUGAKU ZASSHI (JOURNAL OF THE MINERALOGICAL SOCIETY OF JAPAN) 692
NIHON KOBUTSU GAKKAI (THE MINERAL SOCIETY OF JAPAN)
C/O OSAKA DAIGAKU KYOYOBU, 1-1, MACHIKANEYAMA-CHO, TOYONAKA-SHI 560
JAPANESE
INDEXED: CHE, COP, JICST

MINERALOGICAL JOURNAL 693
NIHON KOBUTSU GAKKAI (THE MINERALOGICAL SOCIETY OF JAPAN)
C/O OSAKA DAIGAKU KYOYOBU, 1-1, MACHIKANEYAMA-CHO, TOYONAKA-SHI 560
ENGLISH
INDEXED: CHE, JICST

NIHON KESSHO GAKKAISHI (JOURNAL OF THE CRYSTALLOGRAPHIC SOCIETY OF JAPAN) 694
NIHON KESSHO GAKKAI (CRYSTALLOGRAPHIC SOCIETY OF JAPAN)
C/O NIHON GAKKAI JIMUSENTA, 4-16, 2-CHOME, YAYOI, BUNKYO-KU, TOKYO 113
JAPANESE
INDEXED: CHE, WAA, JICST

NIHON KESSHO SEICHO GAKKAISHI (JOURNAL OF THE JAPANESE ASSOCIATION OF CRYSTAL GROWTH) 695
JAPANESE ASSOCIATION OF CRYSTAL GROWTH
JAPANESE

NIPPON KESSHO SEICHO GAKKAISHI (JOURNAL OF THE JAPANESE ASSOCIATION OF CRYSTAL GROWTH) 696
JAPANESE
INDEXED: JICST

SCIENCE REPORT OF THE TOHOKU UNIVERSITY, SERIES III, MINERALOGY, PETROLOGY AND ECONOMIC GEOLOGY 697
TOHOKU DAIGAKU (TOHOKU UNIVERSITY)
1-1, 2-CHOME, KATAHIRA, SENDAI-SHI 980
ENGLISH
INDEXED: CHE, JICST

Geophysics

ANNUAL REPORT OF THE GEOPHYSICAL OBSERVATIONS MADE AT THE INTERNATIONAL LATITUDE OBSERVATORY OF MIZUSAWA 698
MONBU-SHO MIZUSAWA IDO KANSOKUSHO (INTERNATIONAL LATITUDE OBSERVATORY OF MIZUSAWA, MINISTRY OF EDUCATION
2-12, HOSHIGAOKA-MACHI, MIZUSAWA-SHI 023
ENGLISH
INDEXED: JICST

ANNUAL REPORT OF THE INTERNATIONAL POLAR MOTION SERVICE 699
KOKUSAI KYOKU-UNDO KANSOKU JIGYO CHUO-KYOKU (CENTRAL BUREAU OF THE INTERNATIONAL POLAR MOTION SERVICE)
C/O IDO KANSOKUSHO, 2-12, HOSHIGAOKAMACHI, MIZUSAWA-SHI 023
ENGLISH
INDEXED: JICST

ANNUAL REPORT OF THE SEISMOLOGICAL OBSERVATIONS MADE AT THE INTERNATIONAL LATITUDE OBSERVATORY OF MIZUSAWA 700
MONBU-SHO MIZUSAWA IDO KANSOKUSHO (THE INTERNATIONAL LATITUDE OBSERVATORY OF MIZUSAWA, MINISTRY OF EDUCATION)
2-12, HOSHIGAOKA-MACHI, MIZUSAWA-SHI 023
ENGLISH

BULLETIN OF THE INTERNATIONAL INSTITUTE OF SEISMOLOGY AND EARTHQUAKE ENGINEERING 701
KENSETSU-SHO KENCHIKU KENKYUSHO KOKUSAI JISHIN KOGAKUBU (INTERNATIONAL INSTITUTE OF SEISMOLOGY AND EARTHQUAKE ENGINEERING, PUBLIC WORKS RESEARCH INSTITUTE, MINISTRY OF CONSTRUCTION)
28-8, 3-CHOME, HYAKUNIN-CHO, SHINJUKU-KU, TOKYO 160
ENGLISH
INDEXED: JICST

BULLETIN OF THE NIIGATA AIRGLOW OBSERVATORY 702
NIIGATA DAIGAKU RIGAKUBU (FACULTY OF SCIENCE, NIIGATA UNIVERSITY)
8050, IGARASHI NINO-CHO, NIIGATA-SHI 950-21
ENGLISH

BULLETIN OF THE URAKAWA SEISMOLOGICAL OBSERVATORY 703
HOKKAIDO DAIGAKU RIGAKUBU FUZOKU URAKAWA JISHIN KANAOKUSHO (URAKAWA SEISMOLOGICAL OBSERVATORY, FACULTY OF SCIENCE, HOKKAIDO UNIVERSITY)
956-2, KAMIKINEUSU, URAKAWAMACHI, URAKAWA-GUN, HOKKAIDO 057-01
ENGLISH

BULLETIN OF VOLCANIC ERUPTIONS 704
NIHON KAZAN GAKKAI (VOLCANOLOGICAL SOCIETY OF JAPAN)
C/O TOKYO DAIGAKU JISHIN KENKYUSHO, 1-1, 1-CHOME, YAYOI, BUNKYO-KU TOKYO 113
ENGLISH

BULLETINS: TIME SERVICE OF THE MIZUSAWA OBSERVATORY 705
MONBU-SHO IDO KANSOKUSHO (INTERNATIONAL LATITUDE OBSERVATORY OF MIZUSAWA, MINISTRY OF EDUCATION)
2-12, HOSHIGAOKA-MACHI. MIZUSAWA-SHI 023
ENGLISH

CHIGAKU KENKYU HOKOKU 706
(BULLETIN OF EARTH SCIENCE RESEARCH)
JAPANESE
INDEXED: JICST

CHIJIKI KANSOKUJO YOHO 707
(MEMOIRS OF THE KAKIOKA MAGNETIC OBSERVATORY)
KISHO-CHO KAKIOKA CHIJIKI KANSOKUJO (KAKIOKA MAGNETIC OBSERVATORY, JAPAN METEOROLOGICAL AGENCY)
KAKIOKA, YASATO-CHO, NIIHARI-GUN, IBARAKI 315-01
JAPANESE AND ENGLISH

CHIKYU KAGAKU 708
(EARTH CHEMISTRY)
NIHON CHIKYU KAGAKKAI (GEOCHEMICAL SOCIETY OF JAPAN)
C/O KISHO KENKYUSHO, CHIKYU KAGAKU KENKYUBU, 4-35 KOENJIKITA, SUGINAMI-KU, TOKYO 166
JAPANESE, WITH ENGLISH SUMMARY

CHIKYU KAGAKU 709
(EARTH SCIENCE)
CHIGAKU DANTAI KENKYUKAI (THE ASSOCIATION FOR GEOLOGICAL COLLABORATION IN JAPAN)
32-12, 2-CHOME, MINAMIIKEBUKURO, TOSHIMA-KU, TOKYO 171
JAPANESE AND EUROPEAN
INDEXED: CHE, JICST

CHINETSU 710
(JOURNAL OF THE JAPAN GEOTHERMAL ENERGY ASSOCIATION)
NIHON CHINETSU CHOSAKAI (JAPAN GEOTHERMAL ENERGY ASSOCIATION)
1-3, YURAKU-CHO, CHIYODA-KU, TOKYO 100
JAPANESE, WITH ENGLISH SUMMARY
INDEXED: CHE, JICST

CHINETSU, BESSATSU 711
(SUBTERRANEAN HEAT (SPECIAL ISSUE))
NIHON CHINETSU CHOSAKAI (JAPAN SUBTERREANEAN HEAT RESEARCH ASSOCIATION)
JAPANESE

CHISHO, SHIZEN-HEN 712
(TERRESTRIAL PHENOMENA, SCIENCE EDITION)
HOSEI DAIGAKU DAICHIBU CHIRI GAKU KENKYUKAI (HOSEI UNIVERSITY, SECTION 1, RESEARCH ACADEMY OF GEOGRAPHY)
JAPANESE

CONTRIBUTIONS OF THE GEOPHYSICAL INSTITUTE, KYOTO UNIVERSITY 713
KYOTO DAIGAKU RIGAKUBU CHIKYU BUTSURIGAKU KYOSHITSU (GEOPHYSICAL INSTITUTE, KYOTO UNIVERSITY)
KITASHIRAKAWA OIWAKE-CHO, SAKYO-KU, KYOTO-SHI 606
ENGLISH
INDEXED: CHE, JICST

COSMIC RAY INTENSITY DURING THE POST INTERNATIONAL GEOPHYSICAL COOPERATION 714
JAPANESE
INDEXED: JICST

GEOCHEMICAL JOURNAL 715
NIHON CHIKYU KAGAKKAI (THE GEOCHEMICAL SOCIETY OF JAPAN)
C/O KISHO KENKYUSHO CHIKYU KAGAKU KENKYUBU, 35-8, 4-CHOME, KOENJIKITA, SUGINAMI-KU, TOKYO 166
EUROPEAN
INDEXED: BIO, CHE, IAA, JICST

GEOPHYSICAL MAGAZINE 716
KISHO-CHO (JAPAN METEOROLOGICAL AGENCY)
3-4, 1-CHOME, OTE-MACHI, CHIYODA-KU, TOKYO 100
ENGLISH
INDEXED: APP, CHE, JICST

GEOPHYSICAL NOTES 717
JAPANESE
INDEXED: JICST

HOKKAIDO DAIGAKU CHIKYU-BUTSURIGAKU KENKYU HOKOKU 718
(GEOPHYSICAL BULLETIN OF THE HOKKAIDO UNIVERSITY)
HOKKAIDO DAIGAKU RIGAKUBU CHIKYU BUTSURIGAKU KYOSHITSU (DEPARTMENT OF GEOPHYSICS, FACULTY OF SCIENCE, HOKKAIDO UNIVERSITY)
NISHI-9-CHOME, KITA-11-JO, KITA-KU, SAPPORO-SHI 060
JAPANESE, WITH EUROPEAN SUMMARY
INDEXED: JICST

IDO KANSOKUSHO IHO 719
(PROCEEDINGS OF THE INTERNATIONAL LATITUDE OBSERVATORY OF MIZUSAWA)
MONBU-SHO IDO KANSOKUSHO (INTERNATIONAL LATITUDE OBSERVATORY OF MIZUSAWA, MINISTRY OF EDUCATION)
2-12, HOSHIGAOKA-MACHI, MIZUSAWA-SHI 023
JAPANESE, WITH ENGLISH SUMMARY
INDEXED: JICST

IISEE LECTURE NOTES 720
KOKUSAI JISHIN KOGAKU KENSHUSHO (INTERNATIONAL INSTITUTE OF SEISMOLOGY AND EARTHQUAKE ENGINEERING)
28-8, 3-CHOME, HYAKUNIN-CHO, SHINJUKU-KU, TOKYO 160
ENGLISH

IQSY DATA ON ATMOSPHERICS, WHISTLERS, VLF EMISSIONS AND SOLAR RADIO EMISSIONS 721
JAPANESE
INDEXED: JICST

IQSY DATA OF IONOSPHERIC ABSORPTION AT KOKUBUNJI 722
(INTERNATIONAL YEARS OF THE QUIET SUN)
JAPANESE
INDEXED: JICST

IQSY DATA OF IONOSPHERIC DRIFTS AT YAMAGAWA IN JAPAN 723
JAPANESE
INDEXED: JICST

JAPANESE ANTARCTIC RESEARCH 724
NIHON GAKUJUTSU KAIGI NANKYOKU TOKUBETSU IINKAI (JAPANESE NATIONAL COMMITTEE ON ANTARCTIC RESEARCH, SCIENCE COUNCIL OF JAPAN)
22-24, 7-CHOME, ROPPONGI, MINATO-KU, TOKYO 106
ENGLISH

JAPANESE CONTRIBUTION TO THE INTERNATIONAL GEOPHYSICAL YEAR AND THE INTERNATIONAL GEOPHYSICAL COOPERATION 725
JAPANESE
INDEXED: JICST

JAPANESE CONTRIBUTION TO THE INTERNATIONAL YEARS OF THE QUIET SUN 726
JAPANESE
INDEXED: JICST

JARE DATA REPORTS 727
KOKURITSU KAGAKU HAKUBUTSUKAN KYOKUCHI KENKYU SENTA (POLAR RESEARCH CENTER, NATIONAL SCIENCE MUSEUM)
9-10, 1-CHOME, KAGA, ITABASHI-KU, TOKYO 173
ENGLISH
INDEXED: CHE

JARE SCIENTIFIC REPORT, SERIES A, AERONOMY 728
KOKURITSU KAGAKU HAKUBUTSUKAN KYOKUCHI KENKYU SENTA (POLAR RESEARCH CENTER, NATIONAL SCIENCE MUSEUM)
9-10, 1-CHOME, KAGA, ITABASHI-KU, TOKYO 173
EUROPEAN

JARE SCIENTIFIC REPORT, SERIES F, LOGISTICS 729
KOKURITSU KAGAKU HAKUBUTSUKAN KYOKUCHI KENKYU SENTA (POLAR RESEARCH CENTER, NATIONAL SCIENCE MUSEUM)
9-10, 1-CHOME, KAGA, ITABASHI-KU, TOKYO 173
EUROPEAN

JIBAN SHINDO SHINPOJIUMU SHIRYOSHU 730
(MATERIALS FROM THE SYMPOSIUM ON GROUND TREMORS)
JAPANESE
INDEXED: JICST

JISHIN 731
(JOURNAL OF THE SEISMOLOGICAL SOCIETY OF JAPAN)
JISHIN GAKKAI (THE SEISMOLOGICAL SOCIETY OF JAPAN)
C/O TOKYO DAIGAKU JISHIN KENKYUSHO, 1-1, 1-CHOME, YAYOI, BUNKYO-KU, TOKYO 113
JAPANESE AND ENGLISH
INDEXED: CHE, JICST

JISHIN GEPPO 732
(SEISMOLOGICAL BULLETIN OF THE JAPAN METEOROLOGICAL AGENCY)
KISHO-CHO (JAPAN METEOROLOGICAL AGENCY)
3-4, 1-CHOME, OTE-MACHI, CHIYODA-KU, TOKYO 100
ENGLISH

JISHIN KANSOKU GEPPO 733
(MONTHLY REPORT ON SEISMOLOGY (EARTHQUAKE SURVEY))
KAGAKU GIJUTSUCHO KOKURITSU BOSAI KAGAKU GIJUTSU SENTA DAI-NI KENKYUBU (SCIENCE AND TECHNOLOGY AGENCY, NATIONAL DISASTER PREVENTION CENTER RESEARCH DEPARTMENT 2)
JAPANESE

JISHIN KANSOKU HOKOKU 734
(SEISMOLOGICAL BULLETIN)
KYOTO DAIGAKU RIGAKUBU FUZOKU ABUYAMA JISHIN KANSOKUSHO (ABUYAMA SEISMOLOGICAL OBSERVATORY, KYOTO UNIVERSITY)
NASAHARA, TAKATSUKI-SHI 569
ENGLISH

JISHIN KANSOKU HOKOKU 735
SERIES 2
(SEISMOLOGICAL BULLETIN (SERIES 2))
KISHOCHO JISHIN KANSOKUJO (METEOROLIGICAL AGENCY, SEISMOLOGICAL OBSERVATORY)
JAPANESE

JISHIN TO YOCHI 736
(EARTHQUAKES AND FORECASTING)
NIHON JISHIN YOCHI KURABU (JAPAN EARTHQUAKE FORECASTING CLUB)
JAPANESE

JISHIN YOCHI KANSOKU SENTA SHIRYO 737
(REPORT OF THE EARTHQUAKE PREDICTION OBSERVATION CENTER)
TOKYO DAIGAKU JISHIN KENKYUSHO (EARTHQUAKE RESEARCH INSTITUTE, UNIVERSITY OF TOKYO)
1-1, 1-CHOME, YAYOI, BUNKYO-KU, TOKYO 113
JAPANESE, WITH ENGLISH SUMMARY
INDEXED: JICST

JISHIN YOCHI RENTAKUKAI KAIHO 738
(BULLETIN OF THE LIAISON COMMITTEE OF EARTHQUAKE FORECAST)
KENSETSU-SHO KOKUDO CHIRI-IN (GEOGRAPHICAL SURVEY INSTITUTE, MINISTRY OF CONSTRUCTION)
24-13, 3-CHOME, HIGASHIYAMA, MEGURO-KU, TOKYO 153
JAPANESE

JISUBERI 739
(LANDSLIP)
JAPANESE
INDEXED: JICST

JISUBERI GIJUTSU 740
(LANDSLIP TECHNOLOGY)
JAPANESE
INDEXED: JICST

JOURNAL OF EARTH SCIENCE, NAGOYA UNIVERSITY 741
NAGOYA DAIGAKU RIGAKUBU CHIKYUKAGAKU KYOSHITSU (DEPARTMENT OF EARTH SCIENCES, FACULTY OF SCIENCE, NAGOYA UNIVERSITY)
FURO-CHO, CHIKUSA-KU, NAGOYA-SHI 464
EUROPEAN
INDEXED: CHE, JICST

JOURNAL OF THE FACULTY OF SCIENCE, HOKKAIDO UNIVERSITY, SERIES 7, GEOPHYSICS 742
HOKKAIDO DAIGAKU RIGAKUBU (THE FACULTY OF SCIENCE HOKKAIDO UNIVERSITY)
NISHI 8-CHOME, KITA-10-JO, KITA-KU, SAPPORO-SHI 060
ENGLISH
INDEXED: CHE, JICST

JOURNAL OF GEOMAGNETISM AND GEOELECTRICITY 743
NIHON CHIKYU DENKI JIKI GAKKAI (THE SOCIETY OF TERRESTRIAL MAGNETISM AND ELECTRICITY OF JAPAN)
C/O TOKYO DAIGAKU RIGAKUBU CHIKYU BUTSURIGAKU KYOSHITSU, 11-16, 2-CHOME, YOYOI, BUNKYO-KU, TOKYO 113
EUROPEAN
INDEXED: CHE, IAA, INS, JICST

JOURNAL OF PHYSICS OF THE EARTH 744
JISHIN GAKKAI, NIHON KAZAN GAKKAI (THE SEISMOLOGICAL SOCIETY OF JAPAN, THE VOLCANOLOGICAL SOCIETY OF JAPAN)
3-1, 7-CHOME, HONGO, BUNKYO-KU, TOKYO 113
ENGLISH
INDEXED: APP, CHE, INS, JICST

KAZAN 745
(BULLETIN OF THE VOLCANOLOGICAL SOCIETY OF JAPAN)
NIHON KAZAN GAKKAI (THE VOLCANOLOGICAL SOCIETY OF JAPAN)
C/O TOKYO DAIGAKU JISHIN KENKYUSHO, 1-1, 1-CHOME, YAYOI, BUNKYO-KU, TOKYO 113
JAPANESE, WITH ENGLISH SUMMARY
INDEXED: CHE, JICST

KAZAN FUNKA YOCHI RENRAKUKAI KAIHO 746
(REPORT OF THE VOLCANIC ERUPTION FORECASTING LIAISON COUNCIL)
KISHO-CHO (METEOROLOGICAL AGENCY)
JAPANESE

KAZAN HOKOKU 747
(VOLCANOLOGICAL BULLETIN OF THE JAPAN METEOROLOGICAL AGENCY)
KISHO-CHO (JAPAN METEOROLOGICAL AGENCY)
3-4, 1-CHOME, OTE-MACHI, CHIYODA-KU, TOKYO 100
JAPANESE AND ENGLISH

KENSHIN JIHO 748
(QUARTERLY JOURNAL OF SEISMOLOGY)
KISHO-CHO (JAPAN METEOROLOGICAL AGENCY)
3-4, 1-CHOME, OTE-MACHI, CHIYODA-KU, TOKYO 100
JAPANESE, WITH ENGLISH SUMMARY

KOKUSAI CHIKYU KANSOKU SHIRYO-SHITSU SHIRYO MOKUROKU 749
(INTERNATIONAL EARTH OBERVATION DATA ROOM, DATA INDEX)
JAPANESE
INDEXED: JICST

KOKUSAI JISHIN OYOBI JISHIN KOGAKU KENSHU NENPO 750
(INTERNATIONAL STUDY REPORT ON THE EARTHQUAKE AND EARTHQUAKE ENGINEERING))
KENSETSU-SHO KENCHIKU KENKYUSHO (BUILDING RESEARCH INSTITUTE, MINISTRY OF CONSTRUCTION)
28-8, 3-CHOME, HYAKUNIN-CHO, SHINJUKU-KU, TOKYO 160
JAPANESE

KYOKUCHI 751
(POLAR NEWS)
NIHON KYOKUCHI KENKYU SHINKOKAI (JAPAN POLAR RESEARCH ASSOCIATION)
4-2, 3-CHOME, KASUMIGASEKI, CHIYODA-KU, TOKYO 100
JAPANESE
INDEXED: CHE

KYOKUCHIKEN NYUSU 752
(POLAR AREA RESEARCH NEWS)
KOKURITSU KYOKUCHI KENKYUJO (NATIONAL POLAR AREA RESEARCH INSTITUTE)
JAPANESE

KYOSHIN SOKUHO 753
(STRONG MOTION EARTHQUAKE REPORT)
KOKURITSU BOSAI KAGAKU GIJUTSU SENTA (NATIONAL RESEARCH CENTER FOR DISASTER PREVENTION)
15-1, 6-CHOME, GINZA, CHUO-KU, TOKYO 104
JAPANESE

KYOTO DAIGAKU RIGAKUBU CHIKYU BUTSURIGAKU KENKYUSHISETSU HOKOKU 754
(REPORTS OF GEOPHYSICAL RESEARCH STATION, KYOTO UNIVERSITY)
KYOTO DAIGAKU RIGAKUBU FUZOKU CHIKYU BUTSURIGAKU KENKYUSHISETSU (GEOPHYSICAL RESEARCH STATION, KYOTO UNIVERSITY)
3088-7, NOGUCHIBARA, BEPPU-SHI 874
JAPANESE AND ENGLISH
INDEXED: CHE

KYUSHU DAIGAKU RIGAKUBU SHIMABARA KAZAN KANSOKUSHO KENKYU HOKOKU 755
(SCIENCE REPORTS OF THE SHIMABARA VOLCANO OBSERVATORY, THE FACULTY OF SCIENCE, KYUSHU UNIVERSITY)
KYUSHU DAIGAKU RIGAKUBU (FACULTY OF SCIENCE, KYUSHU UNIVERSITY)
3576 HAKOZAKI, HIGASHI-KU, FUKUOKA-SHI 812
JAPANESE
INDEXED: CHE

MATSUSHIRO GUNPATSU JISHIN SHIRYO HOKOKU 756
(REPORT ON DATA OF EARTHQUAKES ORIGINATING IN THE MATSUSHIRO DISTRICT)
JAPANESE
INDEXED: JICST

MATSUSHIRO JISHIN SENTA KENKYU HOKOKU 757
(REPORT OF COOPERATIVE RESEARCH FOR MATSUSHIRO EARTHQUAKE SWARM)
MATSUSHIRO JISHIN SENTA (RESEARCH CENTER OF MATSUSHIRO EARTHQUAKE SWARM)
3511, NISHIJO, MATSUSHIRO-MACHI, NAGANO-SHI 381-12
JAPANESE
INDEXED: JICST

MATSUSHIRO-GUN HATSUJISHIN SHIRYO GEPPO 758
(ANNUAL REPORT OF THE MATSUYO EARTHQUAKE OCCURRENCES DATA)
MATSUYO JISHIN SENTA (MATSUYO EARTHQUAKE CENTER)
JAPANESE

MEMOIRS OF NATIONAL INSTITUTE OF POLAR RESEARCH, SPECIAL ISSUE (JAPAN) 759
NATIONAL INSTITUTE OF POLAR RESEARCH
9-10, KAGA 1-CHOME, ITATASHI-KU TOKYO
ENGLISH
INDEXED: CHE

MONTHLY NOTES OF THE INTERNATIONAL POLAR MOTION SERVICE 760
KOKUSAI KYOKU-UNDO KANSOKU JIGYO CHUO-KYOKU (CENTRAL BUREAU OF THE INTERNATIONAL POLAR MOTION SERVICE)
C/O IDO KANSOKUSHO, 2-12, HOSHIGAOKAMACHI, MIZUSAWA-SHI 023
ENGLISH
INDEXED: JICST

NANKYOKU SHIRYO 761
(ANTARCTIC RECORD)
KOKURITSU KAGAKU HAKUBUTSUKAN, KYOKUCHI KENKYU SENTA (POLAR RESEARCH CENTER, NATIONAL SCIENCE MUSEUM)
9-10, 1-CHOME, KAGA, ITABASHI-KU, TOKYO 173
JAPANESE AND ENGLISH
INDEXED: BIO, CHE

NIHON CHIKYU KAGAKKAI NYUSU 762
(JAPAN GEOCHEMISTRY SOCIETY NEWS)
JAPANESE

NIIGATA DAIGAKU RIGAKUBU JIBAN SAIGAI KENKYU SHISETSU KENKYU NENPO 763
(ANNUAL REPORT OF THE RESEARCH LABORATORY OF GROUND FAILURE (THE JIBANSAIGAI-KEN) A)
JAPANESE
INDEXED: CHE, JICST

NIPPON JISHIN KOGAKU SHINPOJIUMU KOENSHU 764
(PAPERS FROM THE JAPAN SEISMOLOGICAL ENGINEERING SYMPOSIUM)
JAPANESE
INDEXED: JICST

OSAKA KANKU JISHIN GEPPO 765
(MONTHLY REPORT OF EARTHQUAKES, OSAKA DISTRICT METEOROLOGICAL OBSERVATORY)
OSAKA KANKU KISHODAI (THE OSAKA DISTRICT METEOROLOGICAL OBSERVATORY)
6-25, HOENSAKA-CHO, HIGASHI-KU, OSAKA-SHI 540
JAPANESE
INDEXED: JICST

POST-IGC DATA ON ATMOSPHERICS, WHISTLERS AND SOLAR RADIO EMISSIONS 766
JAPANESE
INDEXED: JICST

POST-IGS 1961 DATA ON UPPER AIR (RADIOSONDE) OBSERVATION DEW POINT TEMPERATURE ATMOSPHERIC ELECTRICITY DURING WORLD METEOROLOGICAL INTERVALS 767
JAPANESE
INDEXED: JICST

PROCEEDINGS OF THE UNITED STATES-JAPAN CONFERENCE ON RESEARCH RELATED TO EARTHQUAKE PREDICTION PROBLEMS 768
JAPANESE
INDEXED: JICST

PUBLICATIONS OF THE INTERNATIONAL LATITUDE OBSERVATORY OF MIZUSAWA 769
MONBU-SHO IDO KANSOKUSHO (INTERNATIONAL LATITUDE OBSERVATORY OF MIZUSAWA MINISTRY OF EDUCATION)
2-12, HOSHIGAOKA-MACHI, MIZUSAWA-SHI 023
ENGLISH
INDEXED: JICST

REPORT OF THE GEOMAGNETIC AND GEOELECTRIC OBSERVATIONS 770
KISHO-CHO KAKIOKA CHIJIKI KANSOKUJO (KAKIOKA MAGNETIC OBSERVATORY JAPAN METEOROLOGICAL AGENCY)
KAKIOKA, YASATO-CHO, NIIHARI-GUN, IBARAKI 315-01
ENGLISH

REPORT ON THE GEOMAGNETIC OBSERVATION DURING THE POST-INTERNATIONAL GEOPHYSICAL COOPERATION 1961 771
JAPANESE
INDEXED: JICST

REPORT OF THE KAKIOKA MAGNETIC OBSERVATORY, GEOMAGNETISM, KAKIOKA 772
KISHO-CHO CHIJIKI KANSOKUJO
KAKIOKA, YASATO-CHO, NIIHARI-GUN, IBARAKI 315-01
ENGLISH

REPORT OF THE KAKIOKA MAGNETIC OBSERVATORY, GEOMAGNETISM, KANOYA 773
KISHO-CHO KAKIOKA CHIJIKI KANSOKUJO
KAKIOKA, YASATO-CHO, NIIHARI-GUN, IBARAKI 315-01
ENGLISH

REPORT OF THE KAKIOKA MAGNETIC OBSERVATORY, GEOMAGNETISM, MEMANBETSU 774
KISHO-CHO KAKIOKA CHIJIKI KANSOKUJO
595, KAKIOKA, YASATO-CHO, NIIHARI-GUN, IBARAKI 351-01
ENGLISH

ROCK MAGNETISM AND PALEOGEOPHYSICS 775
ROCK MAGNETISM AND PALEOPHYSICS RESEARCH GROUP IN JAPAN
C/O TOKYO DAIGAKU RIGAKUBU CHIKYU BUTSURIGAKU KYOSHITSU, 3-1, 7-CHOME, HONGO, BUNKYO-KU, TOKYO 113
ENGLISH

SAKURAJIMA BAKUHATSU SOKUHO 776
(CURRENT REPORT OF VOLCANIC ACTIVITY ON SAKURAJIMA (AN ISLAND))
KAGOSHIMA CHIHO KISHODAI (KAGOSHIMA REGIONAL METEOROLOGICAL OBSERVATORY) (WEATHER STATION)
JAPANESE

SCIENCE REPORTS OF THE TOHOKU UNIVERSITY, SERIES 5, GEOPHYSICS 777
TOHOKU DAIGAKU RIGAKUBU CHIKYU BUTSURIGAKKA (GEOPHYSICAL INSTITUTE, THE FACULTY OF SCIENCE, TOHOKU UNIVERSITY)
AOBA, ARAMAKI, SENDAI-SHI 980
ENGLISH
INDEXED: APP, CHE, JICST

SEISMOLOGICAL BULLETIN 778
KISHO-CHO JISHIN KANSOKUJO (SEISMOLOGICAL OBSERVATORY, JAPAN METEOROLOGICAL AGENCY)
3511, NISHIJO, MATSUSHIRO-MACHI, NAGANO-SHI 381-12
ENGLISH

SEISMOLOGICAL BULLETIN OF KOCHI EARTHQUAKE OBSERVATORY 779
KOCHI EARTHQUAKE OBSERVATORY, FACULTY OF LIBERAL ARTS AND SCIENCE, KOCHI UNIVERSITY
ASAKURA, KOCHI-SHI 780
ENGLISH

SEISMOLOGICAL BULLETIN OF NATIONAL RESEARCH CENTER FOR DISASTER PREVENTION 780
KOKURITSU BOSAI KAGAKU GIJUTSU SENTA (NATIONAL RESEARCH CENTER FOR DISASTER PREVENTION)
15-1, 6-CHOME, GINZA, CHUO-KU, TOKYO 104
ENGLISH

STRONG-MOTION EARTHQUAKE RECORDS IN JAPAN 781
KAGAKU GIJUTSU-CHO, KOKURITSU BOSAI KAGAKU GIJUTSU SENTA (NATIONAL RESEARCH CENTER FOR DISASTER PREVENTION, SCIENCE AND TECHNOLOGY AGENCY)
15-1, 6-CHOME, GINZA, CHUO-KU, TOKYO 104
ENGLISH

SUIRO-BU KANSOKU HOKOKU, CHIJIKI-HEN 782
(DATA REPORT OF HYDROGRAPHIC OBSERVATIONS, SERIES OF GEOMAGNETISM)
KAIJO HOAN-CHO (MARITIME SAFETY AGENCY)
1-3, 2-CHOME, KASUMIGASEKI, CHIYODA-KU, TOKYO 100
ENGLISH

SUIRO-BU KANSOKU HOKOKU, ZENKOKU JIKI SOKURYOHEN 783
(DATA REPORT OF HYDROGRAPHIC OBSERVATIONS, SERIES OF MAGNETIC SURVEY)
KAIJO HOAN-CHO SUIRO-BU (HYDROGRAPHIC DEPARTMENT, MARITIME SAFETY AGENCY)
1-3, 2-CHOME, KASUMIGASEKI, CHIYODA-KU, TOKYO 100
ENGLISH

TOKYO DAIGAKU JISHIN KENKYUSHO IHO 784
(BULLETIN OF THE EARTHQUAKE RESEARCH INSTITUTE, UNIVERSITY OF TOKYO)
TOKYO DAIGAKU JISHIN KENKYUSHO
1-1, YAYOI 1-CHOME, BUNKYO-KU TOKYO
JAPANESE, WITH ENGLISH SUMMARY
INDEXED: APP, CHE, INS, JICST

TOKYO DAIGAKU JISHIN KENKYUSHO KENKYU SOKUHO 785
(SPECIAL BULLETIN OF THE EARTHQUAKE RESEARCH INSTITUTE, UNIVERSITY OF TOKYO)
TOKYO DAIGAKU JISHIN KENKYUSHO (EARTHQUAKE RESEARCH INSTITUTE, UNIVERSITY OF TOKYO)
1-1, 1-CHOME, YAYOI, BUNKYO-KU, TOKYO 113
JAPANESE, WITH ENGLISH SUMMARY

X-RAY KYOKKO X-RAY 786
(AURORA POLARIS, X-RAY)
KYOKKO KENKYUSHO (KYOKKO RESEARCH LABORATORY)
14-1, SAIWAI-CHO, CHIGASAKI-SHI 253
JAPANESE

General Geology, Rocks, Minerals

AISEKIKAI 787
(ROCK WORLD)
JUSEKISHA TREES AND ROCKS COMPANY
JAPANESE

ANNUAL REPORT OF THE INSTITUTE OF GEOSCIENCE, THE UNIVERSITY OF TSUKUBA 788
JAPANESE

BUTSURI TANKO 789
(GEOPHYSICAL EXPLORATION)
BUTSURI TANKO GIJUTSU KYOKAI (SOCIETY OF EXPLORATION GEOPHYSICS OF JAPAN)
C/O CHISHITSU CHOSASHO, 135, HISAMOTO, TAKATSU-KU, KAWASAKI-SHI 213
JAPANESE
INDEXED: CHE, JICST

BUTSURI TANKO CHOSA KENKYU ICHIRAN 790
(SURVEY LIST OF GEOPHYSICAL PROSPECTING)
TSUSHO SANGYO-SHO KOGYO GIJUTSU-IN CHISHITSU CHOSASHO (GEOLOGICAL SURVEY OF JAPAN AGENCY OF INDUSTRIAL SCIENCE AND TECHNOLOGY, MINISTRY OF INTERNATIONAL TRADE AND INDUSTRY)
135, HISAMOTO, TAKATSU-KU, KAWASAKI-SHI 213
JAPANESE
INDEXED: JICST

BUTSURI TANSA 791
(GEOPHYSICAL PROSPECTS)
NIHON BUTSURI TANKO K.K. (NIHON GEOPHYSICAL PROSPECT COMPANY, LIMITED)
2-1, 8-CHOME, NAKAMAGOME, OTA-KU, TOKYO 144
JAPANESE

CHIBA-KEN NO JIBAN CHINKA 792
(RESEARCHES ON LANDSUBSIDENCE IN CHIBA PREFECTURE)
CHIBA-KEN KOGAI KENKYUSHO (CHIBA PREFECTURAL RESEARCH INSTITUTE FOR ENVIRONMENTAL POLLUTION)
5-1, 3-CHOME, INAGE KAIGAN, CHIBA-SHI 280
JAPANESE

CHIGAKU BUNKEN SOKUHO 793
(CURRENT REPORT ON EARTH SCIENCE LITERATURE)
KOGYO GIJUTSUIN CHISHITSU CHOSASHO SHIRYOSHITSU-HEN CHIGAKU BUNKEN SENTA (GEOLOGICAL SURVEY DATA ROOM, INSTITUTE OF INDUSTRY AND TECHNOLOGY)
JAPANESE

CHIGAKU KENKYU 794
(JOURNAL OF THE STUDY OF GEOSCIENCE OF JAPAN)
NIPPON CHIGAKU KENKYUKAI (THE SOCIETY FOR THE STUDY OF GEOSCIENCE OF JAPAN)
C/O NIHON CHIGAKU KENKYU KAIKAN, KARASUMA DEMIZU NISHI-IRU, KAMIGYO-KU, KYOTO-SHI 602
JAPANESE AND ENGLISH
INDEXED: CHE, JICST

CHIGAKU KYOIKU 795
(EDUCATION OF EARTH SCIENCE)
NIHON CHIGAKU KYOIKU GAKKAI (JAPAN SOCIETY OF EARTH SCIENCE EDUCATION)
C/O TOKYO GAKUGEI DAIGAKU CHIGAKU KYOSHITSU, NUKUIKITA-MACHI, KOGANEI-SHI, TOKYO 184
JAPANESE

CHIGAKU ZASSHI 796
(JOURNAL OF GEOGRAPHY)
TOKYO CHIGAKU KYOKAI (TOKYO GEOGRAPHICAL SOCIETY)
12-2, NIBAN-CHO, CHIYODA-KU, TOKYO 102
JAPANESE, WITH EUROPEAN SUMMARY
INDEXED: CHE, JICST

CHISHITSU BUNKEN MOKUROKU 797
(BIBLIOGRAPHY FOR GEOLOGY OF JAPAN)
TSUSHO SANGYO-SHO KOGYO GIJUTSU-IN CHISHITSU CHOSASHO (GEOLOGICAL SURVEY OF JAPAN, AGENCY OF INDUSTRIAL SCIENCE AND TECHNOLOGY, MINISTRY OF INTERNATIONAL TRADE AND INDUSTRY)
135, HISAMOTO, TAKATSU-KU, KAWASAKI-SHI 213
JAPANESE

CHISHITSU CHOSASHO GEPPO 798
(BULLETIN OF THE GEOLOGICAL SURVEY OF JAPAN)
TSUSHO SANGYO-SHO KOGYO GIJUTSU-IN CHISHITSU CHOSASHO (GEOLOGICAL SURVEY OF JAPAN, AGENCY OF INDUSTRIAL SCIENCE AND TECHNOLOGY, MINISTRY OF INTERNATIONAL TRADE AND INDUSTRY)
135, HISAMOTO, TAKATSU-KU, KAWASAKI-SHI 213
JAPANESE AND ENGLISH
INDEXED: BIO, CHE, JICST

CHISHITSU CHOSASHO HOKOKU 799
(REPORTS, GEOLOGICAL SURVEY OF JAPAN)
TSUSHO SANGYO-SHO KOGYO GIJUTSU-IN CHISHITSU CHOSASHO (GEOLOGICAL SURVEY OF JAPAN, AGENCY OF INDUSTRIAL SCIENCE AND TECHNOLOGY, MINISTRY OF INTERNATIONAL TRADE AND INDUSTRY)
135, HISAMOTO, TAKATSU-KU, KAWASAKI-SHI 213
JAPANESE AND ENGLISH
INDEXED: BIO, CHE, JICST

CHISHITSU CHOSASHO KAGAKU BUNSEKI-HO 800
(CHEMICAL ANALYTICAL METHOD FOR GEOLOGY, GEOLOGICAL SURVEY OF JAPAN)
TSUSHO SANGYO-SHO KOGYO GIJUTSU-IN CHISHITSU CHOSASHO (GEOLOGICAL SURVEY OF JAPAN, AGENCY OF INDUSTRIAL SCIENCE AND TECHNOLOGY, MINISTRY OF INTERNATIONAL TRADE AND INDUSTRY)
135, HISAMOTO, TAKATSU-KU, KAWASAKI-SHI 213
JAPANESE

CHISHITSU CHOSASHO NENPO 801
(ANNUAL REPORT OF GEOLOGICAL SURVEY OF JAPAN)
TSUSHO SANGYO-SHO KOGYO GIJUTSU-IN CHISHITSU CHOSASHO (GEOLOGICAL SURVEY OF JAPAN, AGENCY OF INDUSTRIAL SCIENCE AND TECHNOLOGY, MINISTRY OF INTERNATIONAL TRADE AND INDUSTRY)
135, HISAMOTO, TAKATSU-KU, KAWASAKI-SHI 213
JAPANESE
INDEXED: JICST

CHISHITSU CHOSASHO TOKUBETSU HOKOKU 802
(SPECIAL REPORT, GEOLOGICAL SURVEY OF JAPAN)
TSUSHO SANGYO-SHO KOGYO GIJUTSU-IN CHISHITSU CHOSASHO (GEOLOGICAL SURVEY OF JAPAN, AGENCY OF INDUSTRIAL SCIENCE AND TECHNOLOGY, MINISTRY OF INTERNATIONAL TRADE AND INDUSTRY)
135, HISAMOTO, TAKATSU-KU, KAWASAKI-SHI 213
JAPANESE AND ENGLISH
INDEXED: JICST

CHISHITSU GAKURONSHU 803
(MEMOIRS OF THE GEOLOGICAL SOCIETY OF JAPAN)
NIHON CHISHITSU GAKKAI (GEOLOGICAL SOCIETY OF JAPAN)
JAPANESE

CHISHITSU KOGAKU 804
(GEOTECHNICS)
NIHON BUTSURI TANKO K. K. CHISHITSU KOGAKU KENKYUSHO (NIHON GEOPHYSICAL PROSPECTING COMPANY, LIMITED, THE GEOTECHNICAL RESEARCH OFFICE)
2-21, 2-CHOME, NAKAMAGOME, OTA-KU, TOKYO 143
JAPANESE, WITH ENGLISH SUMMARY
INDEXED: INS, JICST

CHISHITSU NYUSU 805
(MONTHLY REVIEW OF GEOLOGY)
TSUSHO SANGYO-SHO KOGYO GIJUTSU-IN CHISHITSU CHOSASHO (GEOLOGICAL SURVEY OF JAPAN, AGENCY OF INDUSTRIAL SCIENCE AND TECHNOLOGY, MINISTRY OF INTERNATIONAL TRADE AND INDUSTRY)
2-12, 4-CHOME, KUDANMINAMI, CHIYODA-KU, TOKYO 102
JAPANESE
INDEXED: CHE, WAA, JICST

CHISHITSUGAKU ZASSHI 806
(JOURNAL OF THE GEOLOGICAL SOCIETY OF JAPAN)
NIHON CHISHITSU GAKKAI (GEOLOGICAL SOCIETY OF JAPAN)
C/O TOKYO DAIGAKU RIGAKUBU CHISHITSUGAKU KYOSHITSU, 3-1, 7-CHOME, HONGO, BUNKYO-KU, TOKYO 113
JAPANESE AND ENGLISH
INDEXED: BIO, CHE, JICST

CHOSA KENKYU HOKOKU KOEN YOSHIROKU 807
(ABSTRACTS OF REPORTS AND PAPERS FROM THE HOKKAIDO BRANCH, GEOLOGICAL SURVEY OF JAPAN)
TSUSHO SANGYO-SHO KOGYO GIJUTSU-IN CHISHITSU CHOSASHO HOKKAIDO SHISHO (HOKKAIDO BRANCH, GEOLOGICAL SURVEY OF JAPAN, AGENCY OF INDUSTRIAL SCIENCE AND TECHNOLOGY)
NISHI-18-CHOME, MINAMI-1-JO, CHUO-KU, SAPPORO-SHI 060
JAPANESE

CHUO ONSEN KENKYUSHO NENPO 808
(ANNUAL REPORT OF THE HOT SPRING RESEARCH CENTERS INSTITUTE)
CHUO ONSEN KENKYUSHO (HOT SPRING RESEARCH CENTER)
42-10, 3-CHOME, TAKADA, TOSHIMA-KU, TOKYO 171
JAPANESE AND ENGLISH

COLLECTED PAPERS ON EARTH SCIENCES 809
NAGOYA DAIGAKU RIGAKUBU CHIKYU KAGAKU KYOSHITSU
(DEPARTMENT OF EARTH SCIENCES, NAGOYA UNIVERSITY)
FURO-CHO, CHIKUSA-KU, NAGOYA-SHI 464
ENGLISH
INDEXED: JICST

DAI-4-KI KENKYU 810
(QUATERNARY RESEARCH)
NIHON DAI-4-KI GAKKAI (JAPAN ASSOCIATION FOR QUATERNARY RESEARCH)
C/O NIHON GAKKAI JIMU SENTA, 4-16, 2-CHOME, YAYOI, BUNKYO-KU, TOKYO 113
JAPANESE, WITH EUROPEAN SUMMARY
INDEXED: JICST

DOKUTSUGAKU ZASSHI 811
(JOURNAL OF THE SPELEOLOGICAL SOCIETY OF JAPAN)
NIHAN DOKUTSU GAKKAI (THE SPELEOLOGICAL SOCIETY OF JAPAN)
JAPANESE

EHIME DAIGAKU KIYO, DAI-2-BU, SHIZENKAGAKU, D-SHIRIZU, CHIGAKU 812
(MEMOIRS OF THE EHIME UNIVERSITY, SECTION 2, NATURAL SCIENCE, SERIES D, EARTH SCIENCE)
EHIME DAIGAKU (EHIME UNIVERSITY)
3, BUNKYO-CHO, MATSUYAMA-SHI 790
JAPANESE AND ENGLISH

EHIME NO CHIGAKU 813
(EHIME EARTH SCIENCE)
EHIME CHIGAKKAI (EHIME EARTH SCIENCE ASSOCIATION)
C/O MR. NAGAI, 5-29, YUWATARI-CHO, MATSUYAMASHI 790
JAPANESE

FUKUSHIMA-KEN NO ONSEN 814
(HOT SPRINGS IN FUKUSHIMA)
FUKUSHIMA-KEN KOSEIBU (FUKUSHIMA PREFECTURAL GOVERNMENT)
2-16, SUGITSUMA-CHO, FUKUSHIMA-SHI 960
JAPANESE

GEOLOGY AND MINERAL RESOURCES OF JAPAN 815
TSUSHO SANGYO-SHO KOGYO GIJUTSU-IN CHISHITSU CHOSAJO (GEOLOGICAL SURVEY OF JAPAN, AGENCY OF INDUSTRIAL SCIENCE AND TECHNOLOGY, MINISTRY OF INTERNATIONAL TRADE AND INDUSTRY)
135, HISAMOTO, TAKATSU-KU, KAWASAKI-SHI 213
ENGLISH

HAKONE OWAKUDANI, OSAWA JISUBERI CHOSA HOKOKU 816
(INVESTIGATIONS OF LANDSLIDING AT OWAKUDANI, HAKONE AND OSAWA, MT. FUJI)
KANAGAWA-KEN DOBOKUBU (PUBLIC WORKS BUREAU, KANAGAWA PREFECTURE)
1, NIHON-ODORI, NAKA-KU, YOKOHOMA-SHI 231
JAPANESE

HIROSHIMA CHIGAKKAIHO 817
(ANNUAL BULLETIN OF HIROSHIMA EARTH SCIENCE ASSOCIATION)
HIROSHIMA CHIGAKUKAI
C/O HIROSHIMA DAIGAKU RIGAKUBU CHISHITSUGAKU KOBUTSUGAKU KYOSHITSU, 1-89, 1-CHOME HIGASHISENDA-CHO, HIROSHIMA-SHI 730
JAPANESE

HIROSHIMA DAIGAKU CHIGAKU KENKYU HOKOKU 818
(GEOLOGICAL REPORT OF THE HIROSHIMA UNIVERSITY)
HIROSHIMA DAIGAKU RIGAKUBU CHISHITSUGAKU KOBUTSUGAKU KYOSHITSU (INSTITUTE OF GEOLOGY AND MINERALOGY, FACULTY OF SCIENCE, HIROSHIMA UNIVERSITY
1-89, 1-CHOME, HIGASHISENDA-CHO, HIROSHIMASHI 730
JAPANESE, WITH ENGLISH SUMMARY
INDEXED: BIO, CHE, JICST

ISHI TO KAWA 819
(NEWS OF THE LIAISON COMMITTEE FOR THE EARTH SCIENCE EDUCATION, ISHIKAWA PREFECTURE)
ISHIKAWA-KEN CHIGAKU KYOIKU RENRAKUKAI
C/O ISHIKAWA-KEN KYOIKU SENTA, TAKAO-CHO, KANAZAWA-SHI 921
JAPANESE

JAPAN CAVING 820
C/O EHIME DAIGAKU RIGAKUBU CHISHITSUGAKU KYOSHITSU, BUNKYO-CHO, MATSUYAMA-SHI 790
JAPANESE

JAPANESE JOURNAL OF GEOLOGY AND GEOGRAPHY 821
NIHON GAKUJUTSU KAIGI (SCIENCE COUNCIL OF JAPAN)
22-34, 7-CHOME, ROPPONGI, MINATO-KU, TOKYO 106
ENGLISH
INDEXED: JICST

JARE SCIENTIFIC REPORTS, SERIES C, EARTH SCIENCE, GEOLOGY, GEOGRAPHY, SEISMOLOGY, GEODECY AND GEOCHEMISTRY 822
KOKURITSU KAGAKU HAKUBUTSUKAN KYOKUCHI KENKYU SENTA
(POLAR RESEARCH CENTER, NATIONAL SCIENCE MUSEUM)
9-10, 1-CHOME, KAGA, ITABASHI-KU, TOKYO 173
EUROPEAN

JIBAN CHINKA 823
(REPORT ON GROUND SUBSIDENCE)
OSAKA-SHI KANKYO HOKEN-KYOKU KANKYO-BU
1-4, NAKANOSHIMA, KITA-KU, OSAKA-SHI 530
JAPANESE

JOURNAL OF THE FACULTY OF SCIENCE, HOKKAIDO UNIVERSITY, SERIES 4, GEOLOGY AND MINERALOGY 824
HOKKAIDO DAIGAKU RIGAKUBU (THE FACULTY OF SCIENCE HOKKAIDO UNIVERSITY)
NISHI-8-CHOME, KITA-10-JO, KITA-KU, SAPPORO-SHI 060
ENGLISH
INDEXED: BIO, CHE, JICST

JOURNAL OF THE FACULTY OF SCIENCE, UNIVERSITY OF TOKYO, SECTION II, GEOLOGY, MINERALOGY, GEOGRAPHY, GEOPHYSICS 825
TOKYO DAIGAKU (UNIVERSITY OF TOKYO)
3-1, 7-CHOME, HONGO, BUNKYO-KU, TOKYO 113
ENGLISH
INDEXED: BIO, CHE, JICST

JOURNAL OF GEOSCIENCES, OSAKA CITY UNIVERSITY 826
OSAKA-SHIRITSU DAIGAKU RIGAKUBU CHIGAKU KYOSHITSU
(DEPARTMENT OF GEOSCIENCES, FACULTY OF SCIENCES, OSAKA CITY UNIVERSITY)
459, SUGIMOTO-CHO, SUMIYOSHI-KU, OSAKA-SHI 558
ENGLISH
INDEXED: BIO, CHE, JICST

JOURNAL OF SCIENCE OF THE HIROSHIMA UNIVERSITY, SERIES C, GEOLOGY AND MINERALOGY 827
HIROSHIMA DAIGAKU (HIROSHIMA UNIVERSITY)
1-89, 1-CHOME, HIGASHISENDA-CHO, HIROSHIMASHI 730
ENGLISH
INDEXED: BIO, CHE, JICST

KAIGAI CHISHITSU KIHO 828
(QUARTERLY REVIEW OF GEOLOGY, ABROAD)
TSUSHO SANGYO-SHO KOGYO GIJUTSU-IN CHISHITSU CHOSAJO (GEOLOGICAL SURVEY OF JAPAN, AGENCY OF INDUSTRIAL SCIENCE AND TECHNOLOGY, MINISTRY OF INTERNATIONAL TRADE AND INDUSTRY)
135, HISAMOTO, TAKATSU-KU, KAWASAKI-SHI 213
JAPANESE

KAIGAI CHISHITSU KOZO CHOSA HOKOKUSHO 829
(SURVEY REPORT OF OVERSEAS GEOLOGICAL STRUCTURE)
KINZOKU KOGYO JIGYODAN (METAL MINING AGENCY OF JAPAN)
C/O DAI-11 MORI BUILDING, SHIBA NISHIKUBO AKEFUNE-CHO, MINATO-KU, TOKYO 105
JAPANESE

KANAGAWA CHIGAKU 830
(EARTH SCIENCE OF KANAGAWA)
KANAGAWA CHIGAKKAI (THE EARTH SCIENCE ASSOCIATION OF KANAGAWA)
41, SHIMIZUGAOKA, MINAMI-KU, YOKOHAMA-SHI 232
JAPANESE

KANAGAWA-KEN ONSEN KENKYUSHO HOKOKU 831
(BULLETIN OF THE HOT SPRING RESEARCH INSTITUTE, KANAGAWA PREFECTURE)
KANAGAWA-KEN ONSEN KENKYUSHO (HOT SPRING RESEARCH INSTITUTE OF KANAGAWA PREFECTURE)
997, YUMOTO, HAKONE-MACHI, ASHIGARASHIMO-GUN, KANAGAWA 250-03
JAPANESE, WITH ENGLISH SUMMARY
INDEXED: JICST

KUMAMOTO DAIGAKU RIGAKUBU CHIGAKU KENKYU HOKOKU 832
(GEOLOGICAL REPORTS OF THE KUMAMOTO UNIVERSITY)
KUMAMOTO DAIGAKU RIGAKUBU CHIGAKU KYOSHITSU
(DEPARTMENT OF GEOLOGY, FACULTY OF SCIENCE, KUMAMOTO UNIVERSITY)
39-1, 2-CHOME, KUROKAMI, KUMAMOTO-SHI 860
JAPANESE

KUMAMOTO JOURNAL OF SCIENCE, GEOLOGY 833
KUMAMOTO DAIGAKU RIGAKUBU CHIGAKU KYOSHITSU
(DEPARTMENT OF GEOLOGY, FACULTY OF SCIENCE,
KUMAMOTO UNIVERSITY)
39-1, 2-CHOME, KUROKAMI, KUMAMOTO-SHI 860
ENGLISH
INDEXED: BIO, CHE, JICST

KUMAMOTO-KEN KOSENSHI 834
(MINERAL SPRINGS IN KUMAMOTO)
KUMAMOTO-KEN EISEI KOGAI KENKYUSHO (KUMAMOTO
PREFECTURAL INSTITUTE FOR PUBLIC HEALTH AND
ENVIRONMENTAL POLLUTION)
4-33, MINAMISENDANBATA-CHO, KUMAMOTO-SHI 860
JAPANESE

KYUSHU DAIGAKU KYOYOBU CHIGAKU KENKYU HOKOKU 835
(REPORTS ON EARTH SCIENCE)
KYUSHU DAIGAKU KYOYOBU CHIGAKU KYOSHITSU
(INSTITUTE OF EARTH SCIENCE, DEPARTMENT OF GENERAL
EDUCATION, KYUSHU UNIVERSITY)
JAPANESE AND EUROPEAN
INDEXED: CHE, JICST

KYUSHU DAIGAKU RIGAKUBU KENKYU HOKOKU, 836
CHISHITSUGAKU NO BU
(SCIENCE REPORTS OF THE FACULTY OF SCIENCE,
KYUSHU UNIVERSITY, GEOLOGY)
KYUSHU DAIGAKU RIGAKUBU (FACULTY OF SCIENCE,
KYUSHU UNIVERSITY)
3576, HAKOZAKI, HIGASHI-KU, FUKUOKA-SHI 812
JAPANESE, WITH ENGLISH SUMMARY
INDEXED: JICST

MEMOIRS OF THE FACULTY OF SCIENCE, KYOTO 837
UNIVERSITY, SERIES OF GEOLOGY AND MINERALOGY
KYOTO DAIGAKU RIGAKUBU (FACULTY OF SCIENCE,
KYOTO UNIVERSITY)
KITASHIRAKAWA OIWAKE-CHO, SAKYO-KU, KYOTO-SHI 606
ENGLISH
INDEXED: JICST

MEMOIRS OF THE FACULTY OF SCIENCE, KYUSHU 838
UNIVERSITY, SERIES D, GEOLOGY
KYUSHU DAIGAKU RIGAKUBU (FACULTY OF SCIENCE
KYUSHU UNIVERSITY)
3576, HAKOZAKI, HIGASHI-KU, FUKUOKA-SHI 812
ENGLISH
INDEXED: BIO, CHE, WAA, JICST

NAGANO-KEN CHIGAKKAIHO 839
(REVIEW OF NAGANO EARTH SCIENCE ASSOCIATION)
NAGANO-KEN CHIGAKKAI
C/O SHINSHU DAIGAKU RIGAKUBU CHISHITSUGAKU
KYOSHITSU, 1-1, 3-CHOME, ASAHI, MATSUMOTO-SHI 390
JAPANESE

NAGASAKI-KEN CHIGAKKAISHI 840
(JOURNAL OF NAGASAKI EARTH SCIENCE ASSOCIATION)
NAGASAKI-KEN CHIGAKKAI
C/O NAGASAKI DAIGAKU KYOIKUGAKUBU CHIGAKU
KYOSHITSU, 1-14, BUNKYO-CHO, NAGASAKI-SHI 852
JAPANESE

NATIONAL SCIENCE MUSEUM BULLETIN. SERIES C: GEOLOGY 841
KOKURITSU KAGAKU HAKUBUTSUKAN (NATIONAL SCIENCE
MUSEUM)
7-20 UENO PARK, DAITO-KU, TOKYO 110, JAPAN
ENGLISH
INDEXED: BIO, CHE, JICST

NIHON CHIKASUI GAKKAI KAISHI 842
(JOURNAL OF THE JAPANESE ASSOCIATION OF
GROUNDWATER HYDROLOGY)
JAPANESE ASSOCIATION OF GROUNDWATER HYDROLOGY
C/O CHISHITSU CHOSAJO MIZUSHIGEN-KA, 135, HISAMOTO,
TAKATSU-KU, KAWASAKI-SHI 213
JAPANESE AND ENGLISH
INDEXED: EXC

NIHON DAIGAKU BUNRIGAKUBU SHIZEN KAGAKU KENKYUJO 843
KENKYU KIYO: OYO CHIGAKU
(RESEARCH BULLETIN OF NATURAL SCIENCE RESEARCH
INSTITUTE, DEPARTMENT OF HUMANITIES AND SCIENCES,
NIHON UNIVERSITY: APPLIED EARTH SCIENCE)
JAPANESE

NIHON ONSEN KYOKAISHI 844
(BULLETIN OF THE JAPAN HOT SPRING ASSOCIATION)
NIHON ONSEN KYOKAI (JAPAN HOT SPRING ASSOCIATION)
8-3, 1-CHOME, MARUNOUCHI, CHIYODA-KU, TOKYO 100
JAPANESE

NIIGATA DAIGAKU RIGAKUBU CHISHITSU KOBUTSUGAKU 845
KYOSHITSU KENKYU HOKOKU
(BULLETIN OF THE DEPARTMENT OF GEOLOGY & MINEROLOGY,
FACULTY OF SCIENCE, NIIGATA UNIVERSITY)
JAPANESE
INDEXED: JICST

OITA-KEN ONSEN CHOSA KENKYUKAI HOKOKU 846
(ANNUAL BULLETIN OF THE HOT SPRING RESEARCH
ASSOCIATION OF OITA PREFECTURE)
OITA-KEN KANKYO HOKEN-BU
1-1, 3-CHOME, OTE-MACHI, OITA-SHI 870
JAPANESE
INDEXED: JICST

OKAYAMA DAIGAKU ONSEN KENKYUSHO HOKOKU 847
(PAPERS OF THE INSTITUTE FOR THERMAL SPRING
RESEARCH, OKAYAMA UNIVERSITY)
OKAYAMA DAIGAKU ONSEN KENKYUSHO (THE INSTITUTE FOR
THERMAL SPRING RESEARCH, OKAYAMA UNIVERSITY)
827, YAMADA, MIASA-CHO, TOHAKU-GUN, TOTTORI 682-02
JAPANESE AND EUROPEAN
INDEXED: CHE, JICST

ONSEN KAGAKU 848
(JOURNAL OF THE BALNEOLOGICAL SOCIETY OF JAPAN)
NIPPON ONSEN KAGAKKAI (BALNEOLOGICAL SOCIETY OF
JAPAN)
C/O TOKYO-TORITSU DAIGAKU RIGAKUBU KAGAKU KYOSHITSU,
2-1, FUKAZAWA, SETAGAYA-KU, TOKYO 158
JAPANESE AND EUROPEAN
INDEXED: JICST

ONSEN KOGAKKAISHI 849
(JOURNAL OF THE SOCIETY OF ENGINEERS FOR MINERAL
SPRINGS, JAPAN)
ONSEN KOGAKKAI (THE SOCIETY OF ENGINEERS FOR
MINERAL SPRINGS, JAPAN)
C/O CHUO ONSEN KENKYUSHO, 42-10, 3-CHOME, TAKADA,
TOSHIMA-KU, TOKYO 171
JAPANESE, WITH EUROPEAN SUMMARY
INDEXED: CHE, JICST

OYO CHISHITSU 850
(ENGINEERING GEOLOGY)
NIHON OYO CHISHITSU GAKKAI (SOCIETY OF ENGINEERING
GEOLOGY OF JAPAN)
C/O CHISHITSU CHOSAJO, 135, HISAMOTO, TAKATSU-KU,
KAWASAKI-SHI 213
JAPANESE AND ENGLISH

PEDOROJISUTO 851
(PEDOLOGIST)
PEDOROJISUTO KONDANKAI (GROUP OF JAPANESE
PEDOLOGISTS)
C/O NORIN-SHO NOGYO GIJUTSU KENKYUSHO, 1-7, 2-CHOME,
NISHIGAHARA, KITA-KU, TOKYO 114
JAPANESE
INDEXED: BIB, CHE, JICST

SCIENCE REPORTS OF NIIGATA UNIVERSITY, SERIES E, 852
GEOLOGY AND MINERALOGY
NIIGATA DAIGAKU RIGAKUBU (FACULTY OF SCIENCE,
NIIGATA UNIVERSITY
2-8050, IGARAHI, NIIGATA-SHI 950-21
ENGLISH
INDEXED: BIO, CHE

SCIENCE REPORTS OF THE TOHOKU UNIVERSITY, SERIES 2, 853
GEOLOGY
TOHOKU DAIGAKU (TOHOKU UNIVERSITY)
C/O TOHOKU DAIGAKU RIGAKUBU CHISHITSUGAKU
KOSEIBUTSUGAKU KYOSHITSU, AOBAYAMA, SENDAI-SHI 980
ENGLISH
INDEXED: CHE, JICST

SCIENCE REPORTS OF THE TOKYO KYOIKU DAIGAKU, 854
SECTION C, GEOGRAPHY, GEOLOGY AND MINERALOGY
TOKYO KYOIKU DAIGAKU RIGAKUBU CHIGAKUKA
(GEOGRAPHICAL INSTITUTE AND GEOLOGICAL AND
MINERALOGICAL INSTITUTE, TOKYO UNIVERSITY OF
EDUCATION)
29-1, 3-CHOME, OTSUKA, BUNKYO-KU, TOKYO 112
EUROPEAN
INDEXED: BIB, BIO, CHE, JICST

SHIGEN CHOSAKAI GEPPO 855
(MONTHLY REPORT ON NATURAL RESOURCES RESEARCH)
KAGAKU GIJUTSUCHO KEIKAKUKYOKU (SCIENCE AND
TECHNOLOGY AGENCY, PLANNING BUREAU)
JAPANESE

SHIGEN TOKEI GEPPO TSU SHO SANGYO DAIJIN KANBO 856
CHOSA TOKEIBU-HEN
(MONTHLY REPORT ON NATURAL RESOURCES STATISTICS
MINISTRY OF INTERNAL TRADE AND INDUSTRY, MINISTER'S
SECRETARIAT, RESEARCH AND STATISTICS DEPARTMENT)
TSUSAN TOKEI KYOKAI (MINISTRY OF INTERNATIONAL
TRADE & INDUSTRY STATISTICS ASSOCIATION
JAPANESE

SHIZUOKA CHIGAKU 857
(EARTH SCIENCE OF SHIZUOKA)
SHIZUOKA-KEN CHIGAKUKAI (THE EARTH SCIENCE ASSOCIATION OF SHIZUOKA PREFECTURE)
C/O SHIZUOKA DAIGAKU KYOIKUGAKUBU CHIGAKU KYOSHITSU, 836, OYA, SHIZUOKA-SHI 422
JAPANESE

SHIZUOKA DAIGAKU CHIGAKU KENKYU HOKOKU 858
(CONTRIBUTIONS FROM GEOLOGICAL INSTITUTE, SHIZUOKA UNIVERSITY)
SHIZUOKA DAIGAKU RIGAKUBU CHIGAKU KYOSHITSU (GEOLOGICAL INSTITUTE FACULTY OF SCIENCE, SHIZUOKA UNIVERSITY)
836, OYA, SHIZUOKA-SHI 422
JAPANESE
INDEXED: JICST

SHIZUOKA DAIGAKU CHIKYU KAGAKU KENKYU HOKOKU 859
(BULLETIN OF EARTH SCIENCE RESEARCH, SHIZUOKA UNIVERSITY)
SHIZUOKA DAIGAKU
JAPANESE

SPECIAL PAPER OF PALAEONTOLOGICAL SOCIETY OF JAPAN 860
ENGLISH

TAIRA CHIGAKU DOKOKAI KAIHO 861
(ANNUAL BULLETIN OF TAIRA EARTH SCIENCE ASSOCIATION)
TAIRA CHIGAKU DOKOKAI
1, TANAKA, TAIRA MAKUNOUCHI, IWAKI-SHI 970
JAPANESE

TAISEKIGAKU KENKYUKAIHO 862
(NEWS LETTER OF THE JAPANESE ASSOCIATION OF SEDIMENTALOGISTS)
TAISEKIGAKU KENKYUKAI
C/O CHISHITSU CHOSASHO, 135, HISAMOTO-CHO, TAKATSU-KU, KAWASAKI-SHI 213
JAPANESE AND ENGLISH

TOHOKU DAIGAKU RIGAKUBU CHISHITSUGAKU KOSEIBUTSUGAKU KYOSHITSU KENKYU HOBUN HOKOKU 863
(CONTRIBUTIONS FROM THE INSTITUTE OF GEOLOGY AND PALEONTOLOGY, TOHOKU UNIVERSITY)
TOHOKU DAIGAKU (TOHOKU UNIVERSITY)
C/O TOHOKU DAIGAKU RIGAKUBU CHISHITSUGAKU KOSEIBUTSUGAKU KYOSHITSU, AOBAYAMA, SENDAI-SHI 980
JAPANESE, WITH ENGLISH SUMMARY
INDEXED: BIO, CHE, JICST

TSUKUMO CHIGAKU 864
(TSUKUMO EARTH SCIENCE)
KYOTO DAIGAKU KYOYOBU, CHIGAKU KYOSHITSU (INSTITUTE OF EARTH SCIENCE, KYOTO UNIVERSITY)
YOSHIDA NIHONMATSU-CHO, SAKYO-KU, KYOTO-SHI 606
JAPANESE
INDEXED: CHE

Oceanography, Surface Waters, Hydrology

AICHI-KEN RYOSUI NENPO 865
(ANNUAL REPORT OF THE QUANTITY OF WATER, AICHI PREFECTURE)
AICHI-KEN DOBOKU-BU (PUBLIC WORKS BUREAU, AICHI PREFECTURE)
1-2, 3-CHOME, SANNOMARU, NAKA-KU, NAGOYA-SHI 460
JAPANESE

ANNUAL REPORT OF HYDROLOGICAL RESEARCH IN THE AREA OF LAKE BIWA 866
APPLIED HYDROLOGY SECTION, DISASTER PREVENTION RESEARCH INSTITUTE, KYOTO UNIVERSITY
GOKASHO, UJI-SHI 611
ENGLISH
INDEXED: JICST

BULLETIN OF MARINE ENGINEERING SOCIETY IN JAPAN 867
ENGLISH

BULLETIN OF THE MARINE PARK RESEARCH STATIONS 868
ENGLISH

BULLETIN OF THE OCEAN RESEARCH INSTITUTE OF TOKYO 869
TOKYO DAIGAKU KAIYO KENKYUSHO (OCEAN RESEARCH INSTITUTE, UNIVERSITY OF TOKYO)
15-1, 1-CHOME, MINAMIDAI, NAKANO-KU, TOKYO 164
ENGLISH
INDEXED: BIO

BULLETIN OF THE PLANKTON SOCIETY OF JAPAN 870
PLANKTON SOCIETY OF JAPAN
C/O PLANKTON LABORATORY, OCEAN RESEARCH INSTITUTE, UNIVERSITY OF TOKYO, MINAMIDAI, NAKANO-KU, TOKYO 309
JAPANESE AND ENGLISH
INDEXED: BIO

CATALOGUE OF OCEANOGRAPHIC DATA (NON-JAPANESE DATA) 871
KAIJO HOAN-CHO KAIYO SHIRYO SENTA (JAPANESE OCEANOGRAPHIC DATA CENTER, HYDROGRAPHIC DEPARTMENT, MARITIME SAFETY AGENCY)
3-1, 5-CHOME, CHUO-KU, TOKYO 104
ENGLISH

CHOIHYO 872
(TIDE TABLE (METEOROLOGICAL AGENCY))
KISHO-CHO (METEOROLOGICAL AGENCY)
3-4, 1-CHOME, OTE-MACHI, CHIYODA-KU TOKYO 100
JAPANESE

CHOSEKI KANSOKU 873
(TIDAL OBSERVATIONS)
KISHO-CHO (METEOROLOGICAL AGENCY)
3-4, 1-CHOME, OTE-MACHI, CHIYODA-KU, TOKYO 100
JAPANESE AND ENGLISH

CHOSEKIHYO 874
(TIDE TABLES (MARITIME SAFETY AGENCY))
KAIJO HOAN-CHO (MARITIME SAFETY AGENCY)
1-3, 2-CHOME, KASUMIGASEKI, CHIYODA-KU, TOKYO 100
JAPANESE

CSK ATLAS 875
KAIJO HOAN-CHO SUIRO-BU KAIYO SHIRYO SENTA (JAPAN OCEANOGRAPHIC DATA CENTER, HYDROGRAPHIC DEPARTMENT, MARITIME SAFETY AGENCY)
ENGLISH

CSK NEWSLETTER 876
KAIJO HOAN-CHO SUIRO-BU KAIYO SHIRYO SENTA (JAPAN OCEANOGRAPHIC DATA CENTER, HYDROGRAPHIC DEPARTMENT, MARITIME SAFETY AGENCY)
3-1, 5-CHOME, TSUKIJI, CHUO-KU, TOKYO 104
ENGLISH
INDEXED: JICST

DATA OF OCEANOGRAPHIC OBSERVATIONS AND EXPLORATORY FISHINGS 877
JAPANESE
INDEXED: JICST

DATA REPORT OF CSK 878
KAIJO HOAN-CHO SUIRO-BU KAIYO SHIRYO SENTA (JAPAN OCEANOGRAPHIC DATA CENTER, HYDROGRAPHIC DEPARTMENT, MARITIME SAFETY AGENCY)
3-1, 5-CHOME, TSUKIJI, CHUO-KU,TOKYO 104
ENGLISH
INDEXED: JICST

FUKUE CHOSEKIHYO 879
(TIDE TABLES OF FUKUE)
FUKUE SOKKOSHO
KIBA-MACHI, FUKUE-SHI 853
JAPANESE

GEKKAN INSUPESU 880
(INNER SPACE)
KAIYO KAIHATSU KENKYUSHO
C/O CHISAN MANSHON, 33-16, 2-CHOME, JINGUMAE, SHIBUYA-KU, TOKYO 150
JAPANESE

HAIDORORОJI 881
(HYDROLOGY)
HYDROLOGY DANWAKAI (RESEARCH GROUP FOR HYDROLOGY)
C/O TOKYO KYOIKU DAIGAKU RIGAKUBU CHIRIGAKU KYOSHITSU, 29-1, 3-CHOME, OTSUKA, BUNKYO-KU, TOKYO 112
JAPANESE, WITH ENGLISH SUMMARY

HIRATO CHOIHYO 882
(TIDE TABLES OF HIRATO)
HIRATO SOKKOSHO KISHO DOKOKAI
321, DAIZENHARA, IWANOUE-CHO, HIRATO-SHI 859-51
JAPANESE

JARE SCIENTIFIC REPORTS, SERIES D, OCEANOGRAPHY 883
KOKURITSU KAGAKU HAKUBUTSUKAN, KYOKUCHI KENKYU SENTA (POLAR RESEARCH CENTER, NATIONAL SCIENCE MUSEUM)
9-10, 1-CHOME, KAGA, ITABASHI-KU, TOKYO 173
EUROPEAN

JODC NYUSU 884
(JODC NEWS)
KAIJO HOAN-CHO SUIRO-BU KAIYO SHIRYO SENTA (JAPAN OCEANOGRAPHIC DATA CENTER, HYDROGRAPHIC DEPARTMENT, MARITIME SAFETY AGENCY)
3-1, 5-CHOME, TSUKIJI, CHUO-KU, TOKYO 104
JAPANESE

KAICHU KOEN JOHO 885
(MARINE PARKS JOURNAL)
KAICHU KOEN SENTA
C/O TORANOMON DENKI BUILDING 15, SHIBA NISHIKUBO AKEFUNE-CHO, MINATO-KU, TOKYO 105
JAPANESE

KAIGAN SHOKO KENCHI SENTA CHOI NENPO 886
(ANNUAL TIDAL OBSERVATIONS OF THE COAST RISE AND FALL SURVEY CENTER)
KENSETSU-SHO KOKUDO CHIRI-IN SOKUCHI-BU (GEODETIC DIVISION GEOGRAPHICAL SURVEY INSTITUTE, MINISTRY OF CONSTRUCTION
24-13, 3-CHOME, HIGASHIYAMA, MEGURO-KU, TOKYO 153
JAPANESE

KAIYO CHISHITSU 887
(JOURNAL OF MARINE GEOLOGY)
KAIYO CHISHITSU KENKYUKAI
C/O TOKYO DAIGAKU KAIYO KENKYUSHO, 15-1, 1-CHOME, MINAMIDAI, NAKANO-KU, TOKYO 164
JAPANESE AND ENGLISH

KAIYO CHOSA GEPPO 888
(MONTHLY OCEANOGRAPHIC OBSERVATIONS)
AOMORI-KEN SUISAN SHIKENJO (AOMORI FISHERIES EXPERIMENT STATION)
196, AJIGASAWA HON-CHO, NISHI-GUN, AOMORI 038-27
JAPANESE

KAIYO KAGAKU 889
(MARINE SCIENCES MONTHLY)
KAIYO SHUPPAN K. K.
1-19, 3-CHOME, MITA, MINATO-KU, TOKYO 108
JAPANESE
INDEXED: JICST

KAIYO KAGAKU GIJUTSU SENTA NENPO 890
(ANNUAL REPORT OF JAPAN MARINE SCIENCE AND TECHNOLOGY CENTER)
KAIYO KAGAKU GIJUTSU SENTA
2-15, NATSUSHIMA, YOKOSUKA-SHI 237
JAPANESE

KAIYO KAIHATSU 891
(OCEAN EXPLOITATION)
JAPAN INDUSTRIAL PUBLISHING
10-6, 3-CHOME, AZABU JUBAN, MINATO-KU, TOKYO 106
JAPANESE
INDEXED: JICST

KAIYO KAIHATSU KANKEI SHIRYO MOKUROKU 892
(CATOLOG OF MATERIALS RELATED TO OCEAN DEVELOPMENT)
NIHON KIKAI KOGYO RENGO KAI (JAPANESE MACHINE INDUSTRY ASSOCIATION/FEDERATION)
JAPANESE

KAIYO KAIHATSU NO GENJO TO TENBO 893
(MARINE DEVELOPMENT, IT'S SITUATION AND PROSPECT)
KAGAKU GIJUTSU-CHO (SCIENCE AND TECHNOLOGY AGENCY)
2, AKASAKA AOI-CHO, MINATO-KU, TOKYO 107
JAPANESE

KAIYO SANGYO KENKYU SHIRYO 894
(OCEAN INDUSTRY RESEARCH MATERIALS)
KAIYO SANGYO KENKYU KAI (OCEAN INDUSTRY RESEARCH SOCIETY)
JAPANESE

KAIYOHAKU DAYORI 895
(LETTER OF THE OCEANIC MUSEUM)
OKINAWA-KEN OKINAWA KOKUSAI KAIYO HAKURANKAI KYORYOKU KYOKU (OKINAWA PREFECTURE OKINAWA INTERNATIONAL OCEANIC EXHIBITION COOPERATIVE BUREAU)
JAPANESE

KAIYOHAKU NYUSU 896
(NEWS OF THE INTERNATIONAL OCEAN EXPOSITION OKINAWA, 1975)
OKINAWA KOKUSAI KAIYO HAKURANKAI KYOKAI (JAPAN ASSOCIATION FOR THE INTERNATIONAL OCEAN EXPOSITION OKINAWA, 1975)
C/O SHUWA DAI-2 TORANOMON BUILDING, 20, NISHIKUBO SAKURAGAWA-CHO, SHIBA, MINATO-KU, TOKYO 105
JAPANESE

KENCHO KIROKU 897
(TIDAL RECORD)
KENSETSU-SHO KOKUDO CHIRI-IN SOKUCHI-BU (TEODETIC DIVISION GEOGRAPHICAL SURVEY INSTITUTE, MINISTRY OF CONSTRUCTION)
24-13, 3-CHOME, HIGASHIYAMA, MEGURO-KU, TOKYO 153
JAPANESE

KIKAN KAIYO JIHO 898
(QUARTERLY REPORT ON THE OCEANS)
NIHO KAIYO KYOKAI (JAPAN OCEANS SOCIETY)
JAPANESE

KOGYO JIJI TSUSHIN KAIYO KAIHATSUBAN 899
(INDUSTRIAL NEWS, MARINE DEVELOPMENT EDITION)
KOGYO JIJI TSUSHINSHA (THE INDUSTRIAL NEWS COMPANY)
3-10, KANDA OGAWA-CHO, CHIYODA-KU, TOKYO 101
JAPANESE

KUMAMOTO-KEN CHOSEKIHYO 900
(TIDE TABLES, KUMAMOTO PREFECTURE)
KUMAMOTO CHIHO KISHODAI (KUMAMOTO LOCAL METEOROLOGICAL OBSERVATORY)
12-20, 2-CHOME, KYO-MACHI, KUMAMOTO-SHI 860
JAPANESE

KUROSHIO KYODO CHOSA SANKO SHIRYO 901
(JOINT SURVEY REFERENCE DATA ON THE KUROSHIO CURRENT)
JAPANESE
INDEXED: JICST

KYOTO UNIVERSITY. MISAKI MARINE BIOLOGICAL INSTITUTE. BULLETIN. 1962. 902
JAPANESE

KYUSHU CHOSEKIHYO 903
(TIDE TABLES, KYUSHU PREFECTURE)
KAIJO HOAN-CHO SUIRO-BU (HYDROGRAPHIC DEPARTMENT, MARITIME SAFETY AGENCY)
23-7, 5-CHOME, SHINBASHI, MINATO-KU, TOKYO 105
JAPANESE

MARIN PABIRION 904
(MARINE PAVILION)
KUSHIMOTO KAICHU KOEN SENTA
ARITA, KUSHIMOTO-CHO, NISHIMURO-GUN, WAKAYAMA 649-34
JAPANESE

NAIKAI CHOSEKIHYO 905
(INLAND SEA TIDE TABLE)
KAIJO HOAN-CHO SUIRO-BU (HYDROGRAPHIC DEPARTMENT, MARITIME SAFETY AGENCY)
23-7, 5-CHOME, SHINBASHI, MINATO-KU, TOKYO 105
JAPANESE

NATSUSHIMA 906
(JAPAN MARINE SCIENCE AND TECHNOLOGY CENTER)
KAIYO KAGAKU GIJUTSU SENTA (JAPAN MARINE SCIENCE AND TECHNOLOGY CENTER)
2-15, NATSUSHIMA-MACHI, YOKOSUKA-SHI 237
JAPANESE

NIHON KAIGAKKAISHI 907
(BULLETIN OF THE JAPAN SEA SCIENCES SOCIETY)
JAPANESE

NIHON KAIIKI KENKYUSHO HOKOKU 908
(BULLETIN OF THE JAPAN SEA RESEARCH INSTITUTE, KANAZAWA UNIVERSITY)
KANAZAWA DAIGAKU NIHON KAIIKI KENKYUSHO (JAPAN SEA RESEARCH INSTITUTE, KANAZAWA UNIVERSITY)
1-1, MARUNOUCHI, KANAZAWA-SHI 920
JAPANESE, WITH ENGLISH SUMMARY
INDEXED: JICST

NIHON KAISUI GAKKAISHI 909
(BULLETIN OF THE SOCIETY OF SEA WATER SCIENCE, JAPAN)
NIHON KAISUI GAKKAI (SOCIETY OF SEA WATER SCIENCE)
C/O NIHON GAKKAI JIMU SENTA, 4-16, 2-CHOME, YAYOI, BUNKYO-KU, TOKYO 113
JAPANESE, WITH ENGLISH SUMMARY
INDEXED: CHE, JICST

NIHON KAIYO GAKKAISHI 910
(JOURNAL OF THE OCEANOGRAPHICAL SOCIETY OF JAPAN)
NIHON KAIYO GAKKAI (OCEANOGRAPHICAL SOCIETY OF JAPAN)
C/O TOKYO DAIGAKU KAIYO KENKYUSHO, 15-1, 1-CHOME, MINAMIDAI, NAKANO-KU, TOKYO 164
JAPANESE AND ENGLISH
INDEXED: BIO, CHE, JICST

NIHON PURANKUTON GAKKAIHO 911
(BULLETIN OF THE PLANKTON SOCIETY OF JAPAN)
JAPANESE

OCEAN AGE 912
C/O DAI-1 TOEI BUILDING, 11-2, KANDA MITOSHIRO-CHO, CHIYODA-KU, TOKYO 101
JAPANESE
INDEXED: JICST

OCEANOGRAPHICAL MAGAZINE 913
KISHO-CHO (METEOROLOGICAL AGENCY)
3-4, 1-CHOME, OTE-MACHI, CHIYODA-KU, TOKYO 100
ENGLISH
INDEXED: APP, CHE

OKINAWA-KEN MIZU CHOSA HOKOKUSHO 914
(WATER SURVEY, OKINAWA PREFECTURE)
OKINAWA-KEN (OKINAWA PREFECTURAL GOVERNMENT)
2-14, 1-CHOME, IZUMIZAKI, NAHA-SHI 900
JAPANESE

PACIFIC GEOLOGY 915
TOKAI DAIGAKU SHUPPANKAI
27-4, 3-CHOME, SHINJUKU, SHINJUKU-KU, TOKYO 160
ENGLISH
INDEXED: CHE, JICST

PRELIMINARY REPORT OF THE HAKUHO-MARU CRUISE KH 916
TOKYO DAIGAKU KAIYO KENKYUSHO (OCEAN RESEARCH INSTITUTE, UNIVERSITY OF TOKYO)
15-1, 1-CHOME, MINAMIDAI, NAKANO-KU, TOKYO 164
ENGLISH

RECORDS OF OCEANOGRAPHIC WORKS IN JAPAN 917
NIHON GAKUJUTSU KAIGI (SCIENCE COUNCIL OF JAPAN)
22-34, 7-CHOME, ROPPONGI, MINATO-KU, TOKYO 106
ENGLISH

RIKUSUIGAKU ZASSHI 918
(JAPANESE JOURNAL OF LIMNOLOGY)
NIHON RIKUSUI GAKKAI (JAPANESE SOCIETY OF LIMNOLOGY)
C/O KYOTO DAIGAKU, OTSU RINKO JIKKENJO, SHIMOSAKAHON-MACHI, OTSU-SHI 520-01
JAPANESE AND ENGLISH
INDEXED: BIB, BIO, EXC, JICST

RYURYO NENPO 919
(FLUX STATISTICS)
KENSETSU-SHO KASEN-KYOKU (RIVER BUREAU, MINISTRY OF CONSTRUCTION)
C/O ZENKOKU CHOSONKAIKAN HOKAN 508, 11-35, 1-CHOME, NAGATO-CHO, CHIYODA-KU, TOKYO 100
JAPANESE

SETO MARINE BIOLOGICAL LABORATORY. CONTRIBUTIONS 920
SETO MARINE BIOLOGICAL LABORATORY. KYOTO UNIVERSITY
SIRAHAMA, WAKAYAMA-KEN 649-22, JAPAN
JAPANESE AND ENGLISH

SETO NAIKAI CHOSEKIHYO 921
(TIDE TABLES OF THE SETO INLAND SEA)
KAIJO HOAN-CHO SUIRO-BU (HYDROGRAPHIC DEPARTMENT, MARITIME SAFETY AGENCY) AND HEIWA NO UMI KYOKAI
23-7, 5-CHOME, SHINBASHI, MINATO-KU, TOKYO 105
JAPANESE

SHIGA DAIGAKU KOSHO KENKYUJO RONBUNSHU 922
(REPORT OF THE LIMNOLOGICAL LABORATORY OF SHIGA UNIVERSITY)
SHIGA DAIGAKU KYOIKU GAKUBU (SHIGA UNIVERSITY, DEPARTMENT OF EDUCATION)
JAPANESE

SHINRIN CHISUI SHIKEN IHO 923
(JOURNAL OF THE FOREST HYDROLOGICAL EXPERIMENTS)
JAPAN RINGYO SHIKENJO (NORINSHO RINGYO SHIKENJO)
JAPANESE
INDEXED: BIB

SUIKEN KAGAKU KENKYUJO NENPO 924
(ANNUAL REPORT OF THE WATER RESEARCH INSTITUTE, NAGOYA UNIVERSITY)
NAGOYA DAIGAKU SUIKEN KAGAKU KENKYUJO (WATER RESEARCH INSTITUTE, NAGOYA UNIVERSITY)
FURO-CHO, CHIKUSA-KU, NAGOYA 464
JAPANESE

SUION NO KENKYU 925
(WATER TEMPERATURE RESEARCH)
SUION CHOSAKAI
25-2, 3-CHOME, HYAKUNIN-CHO, SHINJUKU-KU, TOKYO 160
JAPANESE
INDEXED: JICST

SUIRI KOENKAI KOENSHU 926
(PAPERS OF THE ANNUAL MEETING ON HYDROLOGY)
DOBOKU GAKKAI (JAPAN SOCIETY OF CIVIL ENGINEERS)
1-CHOME, YOTSUYA, SHINJUKU-KU, TOKYO 160
JAPANESE
INDEXED: JICST

SUIRI NENPYO 927
(ANNUAL REPORT OF HYDROLOGIC OBSERVATIONS)
KENSETSU-SHO KINKI CHIHO KENSETSU-KYOKU
OTEMAENO-CHO, HIGASHI-KU, OSAKA-SHI 540
JAPANESE

SUIRO YOHO 928
(HYDROGRAPHIC BULLETIN)
KAIJO HOAN-CHO SUIRO-BU (HYDROGRAPHIC DEPARTMENT, MARITIME SAFETY AGENCY)
1-3, 2-CHOME, KASUMIGASEKI, CHIYODA-KU, TOKYO 100
JAPANESE
INDEXED: JICST

SUIRO-BU KANKOKU HOKOKU, CHOSEKI-HEN 929
(DATA REPORT OF HYDROGRAPHIC OBSERVATIONS, SERIES OF TIDE)
KAIJO HOAN-CHO (MARITIME SAFETY AGENCY)
1-3, 2-CHOME, KASUMIGASEKI, CHIYODA-KU, TOKYO 104
ENGLISH

SUIRO-BU KANSOKU HOKOKU, KAIYO-HEN 930
(DATA REPORT OF HYDROGRAPHIC OBSERVATIONS, SERIES OF OCEANOGRAPHY)
KAIJO HOAN-CHO (MARITIME SAFETY AGENCY)
1-3, 2-CHOME, KASUMIGASEKI, CHIYODA-KU, TOKYO 100
ENGLISH

SUIRO-BU KENKYU HOKOKU 931
(REPORT OF HYDROGRAPHIC RESEARCH)
KAIJO HOAN-CHO (MARITIME SAFETY AGENCY)
1-3, 2-CHOME, KASUMIGASEKI, CHIYODA-KU, TOKYO 100
JAPANESE AND ENGLISH

SUISHITSU NENPYO 932
(WATER QUALITY STATISTICS)
KENSETSU-SHO KASEN-KYOKU (RIVER BUREAU, MINISTRY OF CONSTRUCTION)
C/O CHIYODA DAI-1 BUILDING, 1-27, KANDA NISHIKI-CHO, CHIYODA-KU, TOKYO 101
JAPANESE

TANSUI MORUI KENKYU 933
(STUDIES OF FRESHWATER ALGAE)
TANSUI MORUI KENKYUKAI (FRESHWATER ALGAE RESEARCH ASSOCIATION)
JAPANESE

TOKAI DAIGAKU KAIYO KAGAKU HAKUBUTSUKAN NENPO 934
(ANNUAL REPORT OF THE MARINE SCIENCE MUSEUM, TOKAI UNIVERSITY)
JAPANESE

TOKAI DAIGAKU KAIYOGAKUBU GYOSEKISHU 935
(COLLECTED RESEARCH OF THE FACULTY OF MARINE SCIENCE AND TECHNOLOGY, TOKAI UNIVERSITY)
JAPANESE

TOKAI DAIGAKU KIYO, KAIYOGAKUBU 936
(JOURNAL OF THE FACULTY OF MARINE SCIENCE AND TECHNOLOGY)
TOKAI DAIGAKU KAIYO GAKUBU (FACULTY OF MARINE SCIENCE AND TECHNOLOGY, TOKAI UNIVERSITY) AND TOKAI DAIGAKU SHUPPANKAI (TOKAI UNIVERSITY PRESS)
C/O TOKAI BUILDING, 27-4, 3-CHOME, SHINJUKU, SHINJUKU-KU, TOKYO 160
JAPANESE AND ENGLISH
INDEXED: BIO

TOKYO DAIGAKU KAIYO KENKYUSHO IHO 937
(BULLETIN OF THE OCEAN INSTITUTE, UNIVERSITY OF TOKYO)
TOKYO DAIGAKU KAIYO KENKYUJO (OCEAN RESEARCH INSTITUTE, UNIVERSITY TOKYO)
15-1, 1-CHOME, MINAMI-DAI, NAKANO-KU, TOKYO 164
ENGLISH

TOKYOWAN CHORYU CHOSA HOKOKUSHO 938
(TIDAL CURRENT SURVEY OF TOKYO BAY)
UNYU-SHO DAI-2 KOWAN KENSETSU-KYOKU YOKOHAMA CHOSA SEKKEI JIMUSHO (INVESTIGATION & DESIGN OFFICE, YOKOHAMA, SECOND DISTRICT CONSTRUCTION BUREAU, MINISTRY OF TRANSPORTATION)
2-5, 1-CHOME, TAKASHIMA-CHO, NISHI-KU, YOKOHAMA-SHI 220
JAPANESE

TSUKUBA DAIGAKU SUIRI JIKKEN SENTA HOKOKU 939
(BULLETIN OF THE ENVIRONMENTAL RESEARCH CENTER, UNIVERSITY OF TSUKUBA)
JAPANESE

UMI 940
(SEA)
NICHIFUTSU KAIYO GAKKAI (LA SOCIETE FRANCO-JAPONEISE D'OCEANOGRAPHIE)
C/O NICHIFUTSU KAIKAN, 2-3, KANDA SURUGADAI, CHIYODA-KU, TOKYO 101
EUROPEAN
INDEXED: BIO, CHE

UMI TO GIJUTSU 941
(SEA AND TECHNOLOGY)
KAIYO KAIHATSU GIJUTSU KENKYUJO (OCEAN TECHNICAL DEVELOPMENT RESEARCH INSTITUTE)
JAPANESE

Meteorology, Climatology

ABASHIRI CHIHO NOGYO KISHO GEPPO (KISHO NO BU) 942
(MONTHLY REPORT OF AGRICULTURAL METEOROLOGY, ABASHIRI DISTRICT (METEOROLOGY SECTION))
ABASHIRI CHIHO KISHODAI (ABASHIRI LOCAL METEOROLOGICAL OBSERVATORY)
64, DAI-MACHI, ABASHIRI-SHI 093
JAPANESE

ABASHIRI CHIHO NOGYO KISHO SHUHO 943
(WEEKLY REPORT OF AGRICULTURAL METEOROLOGY, ABASHIRI DISTRICT))
ABASHIRI CHIHO KISHODAI (ABASHIRI LOCAL METEOROLOGICAL OBSERVATORY)
64, DAI-MACHI, ABASHIRI-SHI 093
JAPANESE

AEROLOGICAL DATA OF JAPAN 944
KISHO-CHO (METEOROLOGICAL AGENCY)
3-4, 1-CHOME, OTE-MACHI, CHIYODA-KU, TOKYO 100
ENGLISH
INDEXED: JICST

AICHI-KEN KISHO GEPPO 945
(MONTHLY REPORT OF METEOROLOGY, AICHI PREFECTURE)
NAGOYA CHIHO KISHODAI (NAGOYA LOCAL METEOROLOGICAL OBSERVATORY)
2-18, HIYORI-CHO, CHIKUSA-KU, NAGOYA-SHI 464
JAPANESE

AKITA-KEN KISHO GEPPO 946
(MONTHLY REPORT OF METEOROLOGY, AKITA PREFECTURE)
AKITA CHIHO KISHODAI (AKITA LOCAL METEOROLOGICAL OBSERVATORY)
78-4, YABASE, AKITA-SHI 010
JAPANESE

AMTEX NYUSU 947
(AMTEX NEWS)
AMTEX RENRAKU KYOGIKAI (KISHO KENKYUSHO UCHI) (AMTEX LIASON COUNCIL (IN THE METEOROLOGICAL RESEARCH INSTITUTE)
JAPANESE AND ENGLISH

AMTEX REPORT 948
AMTEX RENRAKU KYOGIKAI
35-8, 4-CHOME, KOENJI-KITA, SUGINAMI-KU, TOKYO 166
ENGLISH

ANNUAL REPORT OF THE METEOROLOGICAL OBSERVATIONS MADE AT THE INTERNATIONAL LATITUDE OBSERVATORY OF MIZUSAWA 949
MONBU-SHO IDO KANSOKUSHO (THE INTERNATIONAL LATITUDE OBSERVATORY OF MIZUSAWA, MINISTRY OF EDUCATION)
2-12, HOSHIGAOKA-CHO, MIZUSAWA-SHI 023
ENGLISH
INDEXED: JICST

AOMORI-KEN KISHO GEPPO 950
(MONTHLY REPORT OF METEOROLOGY, AOMORI PREFECTURE)
AOMORI CHIHO KISHODAI (AOMORI LOCAL METEOROLOGICAL OBSERVATORY)
155, TSUKUDA, MATSUMORI, AOMORI-SHI 030
JAPANESE

AOMORI-KEN NOGYO KISHO GEPPO 951
(AOMORI PREFECTURAL AGRICULTURAL WEATHER MONTHLY)
AOMORI CHIHO KISHODAI AOMORI-KEN (AOMORI PREFECTURE: AOMORI REGIONAL WEATHER STATION)
JAPANESE

BOSAI KISHO 952
(METEOROLOGY FOR THE PREVENTION OF DISASTER, MIYAZAKI METEOROLOGICAL COUNCIL FOR THE PREVENTION OF DISASTER)
MIYAZAKI-KEN BOSAI KISHO RENRAKU KYOGIKAI
C/O MIYAZAKI CHIHO KISHODAI, WACHI KAWARA-CHO, MIYAZAKI-SHI 880
JAPANESE

BOSAI SHIRYO SHIRIZU 953
(DISASTER PREVENTION SERIES)
FUKUOKA KANKU KISHODAI (FUKUOKA DISTRICT METEOROLOGICAL OBSERVATORY)
2-36, 1-CHOME, OHORI, CHUO-KU, FUKUOKA-SHI 810
JAPANESE

BULLETIN OF THE AEROLOGICAL OBSERVATORY 954
KOSOKISHODAI
TATENO, YATABE-MACHI, TSUKUBA-GUN, IBARAKI 300-21
ENGLISH
INDEXED: JICST

CHIBA-KEN KISHO GEPPO 955
(MONTHLY REPORT OF METEOROLOGY, CHIBA PREFECTURE)
CHOSHI CHIHO KISHODAI (CHOSHI LOCAL METEOROLOGICAL OBSERVATORY)
31, MINAMI-CHO, CHOSHI-SHI 288
JAPANESE

CHIBA-KEN KISHO NENPO 956
(ANNUAL REPORT OF METEOROLOGY, CHIBA PREFECTURE)
CHOSHI CHIHO KISHODAI (CHOSHI LOCAL METEOROLOGICAL OBSERVATORY)
31, MINAMI-CHO CHOSHI-SHI 288
JAPANESE

CHUGOKU CHIHO DENRYOKU KISHO GAIHO 957
(REPORT OF THE POWER AND WEATHER, CHUGOKU DISTRICT)
HIROSHIMA CHIHO KISHODAI (HIROSHIMA LOCAL METEOROLOGICAL OBSERVATORY)
40-1, 1-CHOME, EBAMINAMI, HIROSHIMA-SHI 733
JAPANESE

CHUGOKU CHIHO RAIU GAIHO 958
(REPORT OF THE THUNDERSTORM, CHUGOKU DISTRICT)
HIROSHIMA CHIHO KISHODAI (HIROSHIMA LOCAL METEOROLOGICAL OBSERVATORY)
40-1, 1-CHOME, EBAMINAMI, HIROSHIMA-SHI 733
JAPANESE

DENRYOKU KISHO RENRAKUKAI IHO (DENRYOKU TO KISHO) 959
(YEARLY REPORT OF THE POWER AND WEATHER COORDINATING COMMITTEE)
DENRYOKU KISHO RENRAKUKAI
C/O NIHON KISHO KYOKAI, TONEN BUILDING, 17-6, 3-CHOME, KANDA NISHIKI-CHO, CHIYODA-KU, TOKYO 101
JAPANESE

EHIME-KEN KISHO GEPPO 960
(MONTHLY REPORT OF METEOROLOGY, EHIME PREFECTURE)
MATSUYAMA CHIHO KISHODAI (MATSUYAMA LOCAL METEOROLOGICAL OBSERVATORY)
102, KITAMOCHIDA-CHO, MATSUYAMA-SHI 790
JAPANESE

EHIME-KEN KISHO NENPO 961
(ANNUAL REPORT OF METEOROLOGY, EHIME PREFECTURE)
MATSUYAMA CHIHO KISHODAI (MATSUYAMA LOCAL METEOROLOGICAL OBSERVATORY)
102, KITAMOCHIDA-CHO, MATSUYAMA-SHI 790
JAPANESE

FUKUI DAIGAKU SEKISETSU KENKYUSHITSU KENKYU HOKOKU 962
(RESEARCH REPORT, THE LABORATORY OF SNOW AND ICE, FUKUI UNIVERSITY)
FUKUI DAIGAKU SEKISETSU KENKYUSHITSU
9-1, 3-CHOME, BUNKYO, FUKUI-SHI 910
JAPANESE, WITH ENGLISH SUMMARY

FUKUI-KEN KISHO NENPO 963
(ANNUAL REPORT OF METEOROLOGY, FUKUI PREFECTURE)
FUKUI CHIHO KISHODAI (FUKUI LOCAL METEOROLOGICAL OBSERVATORY)
5-2, 2-CHOME, TOYOSHIMA, FUKUI-SHI 910
JAPANESE

FUKUI-KEN NOGYO KISHO GEPPO 964
(MONTHLY REPORT OF AGRICULTURAL METEOROLOGY, FUKUI PREFECTURE)
FUKUI CHIHO KISHODAI (FUKUI LOCAL METEOROLOGICAL OBSERVATORY)
5-2, 2-CHOME, TOYOSHIMA, FUKUI-SHI 910
JAPANESE

FUKUI-KEN NOGYO KISHO NENPO 965
(ANNUAL REPORT OF AGRICULTURAL METEOROLOGY, FUKUI PREFECTURE)
FUKUI CHIHO KISHODAI (FUKUI LOCAL METEOROLOGICAL OBSERVATORY)
5-2, 2-CHOME, TOYOSHIMA, FUKUI-SHI 910
JAPANESE

FUKUOKA KANKU KISHO KENKYUKAISHI 966
(PROCEEDINGS OF THE FUKUOKA DISTRICT MEETING FOR THE STUDY OF METEOROLOGY)
FUKUOKA KANKU KISHODAI (FUKUOKA DISTRICT METEOROLOGICAL OBSERVATORY)
2-36, 1-CHOME, OHORI, CHUO-KU, FUKUOKA-SHI 810
JAPANESE

FUKUOKA KANKU KISHODAI GIJUTSU TSUSHIN 967
(FUKUOKA DISTRICT METEOROLOGICAL OBSERVATORY TECHNICAL TIMES)
FUKUOKA KANKU KISHODAI
2-36, 1-CHOME, OHORI, CHUO-KU, FUKUOKA-SHI 810
JAPANESE

FUKUOKA KANKU KISHODAI YOHO 968
(MEMOIRS OF THE FUKUOKA DISTRICT METEOROLOGICAL OBSERVATORY)
FUKUOKA KANKU KISHODAI
2-36, 1-CHOME, OHORI, CHUO-KU, FUKUOKA-SHI 810
JAPANESE
INDEXED: JICST

FUKUOKA-KEN KISHO GEPPO 969
(MONTHLY REPORT OF METEOROLOGY, FUKUOKA PREFECTURE)
FUKUOKA KANKU KISHODAI (FUKUOKA DISTRICT METEOROLOGICAL OBSERVATORY)
2-36, 1-CHOME, OHORI, CHUO-KU, FUKUOKA-SHI 810
JAPANESE

FUKUSHIMA-KEN KISHO GEPPO 970
(MONTHLY REPORT OF METEOROLOGY, FUKUSHIMA PREFECTURE)
FUKUSHIMA CHIHO KISHODAI (FUKUSHIMA LOCAL METEOROLOGICAL OBSERVATORY)
9, 1-CHOME, MATSUMOTO-CHO, FUKUSHIMA-SHI 960
JAPANESE

FUNE TO KAIJO KISHO 971
(SHIP AND MARITIME METEOROLOGY)
KISHO-CHO (METEOROLOGICAL AGENCY)
3-4, 1-CHOME, OTE-MACHI, CHIYODA-KU, TOKYO 100
JAPANESE

GIFU-KEN KISHO GEPPO 972
(MONTHLY REPORT OF METEOROLOGY, GIFU PREFECTURE)
GIFU CHIHO KISHODAI (GIFU LOCAL METEOROLOGICAL OBSERVATORY)
6, NINOMARU, KANO, GIFU-SHI 500
JAPANESE

GIJUTSU JIHO, SAPPORO KANKU KISHODAI 973
(TECHNICAL BULLETIN OF SAPPORO DISTRICT METEOROLOGICAL OBSERVATORY)
SAPPORO KANKU KISHODAI
NISHI-18-CHOME, KITA-2-JO, CHUO-KU, SAPPORO-SHI 060
JAPANESE

GIJUTSU NOTO 974
(TECHNICAL NOTES)
OKINAWA KISHODAI (OKINAWA WEATHER OBSERVATORY)
JAPANESE

GIJUTSU TSUSHIN 975
(TECHNICAL DISPATCH)
FUKUOKA KANKU KISHODAI (FUKUOKA DISTRICT METEOROLOGICAL OBSERVATORY)
JAPANESE

GUNMA-KEN KISHO GEPPO 976
(MONTHLY REPORT OF METEOROLOGY, GUNMA PREFECTURE)
MAEBASHI CHIHO KISHODAI (MAEBASHI LOCAL METEOROLOGICAL OBSERVATORY)
20-12, 3-CHOME, SHOWA-CHO, MAEBASHI-SHI 371
JAPANESE

GUNMA-KEN KISHO NENPO 977
(ANNUAL REPORT OF METEOROLOGY, GUNMA PREFECTURE)
MAEBASHI CHIHO KISHODAI (MAEBASHI LOCAL METEOROLOGICAL OBSERVATORY)
20-12, 3-CHOME, SHOWA-CHO, MAEBASHI-SHI 371
JAPANESE

HAKODATE KAIYO KISHODAI KAIJO KISHO HOKOKU 978
(MARINE METEOROLOGICAL REPORT OF THE HAKODATE MARINE OBSERVATORY)
HAKODATE KAIYO KISHODAI (THE HAKODATE MARINE OBSERVATORY)
181, AKAGAWADORI-CHO, KAMEDA-SHI 040
JAPANESE
INDEXED: JICST

HAKODATE KAIYO KISHODAI YOHO 979
(BULLETIN OF THE HAKODATE MARINE OBSERVATORY)
HAKODATE KAIYO KISHODAI (THE HAKODATE MARINE OBSERVATORY)
181, AKAGAWADORI-CHO, KAMEDA-SHI 040
JAPANESE AND ENGLISH
INDEXED: JICST

HIROSHIMA-KEN KISHO GEPPO 980
(MONTHLY REPORT OF METEOROLOGY, HIROSHIMA PREFECTURE)
HIROSHIMA CHIHO KISHODAI (HIROSHIMA LOCAL METEOROLOGICAL OBSERVATORY)
40-1, 1-CHOME, EBA-MINAMI, HIROSHIMA-SHI 733
JAPANESE
INDEXED: JICST

HIROSHIMA-KEN KISHO NENPO 981
(ANNUAL REPORT OF METEOROLOGY, HIROSHIMA PREFECTURE)
HIROSHIMA CHIHO KISHODAI (HIROSHIMA LOCAL METEOROLOGICAL OBSERVATORY)
40-1, 1-CHOME EBAMINAMI, HIROSHIMA-SHI 733
JAPANESE
INDEXED: JICST

HOKKAIDO KAIHATSU-KYOKU SAROBETSU SOGO CHOSA CHUKAN HOKOKU KISHO BUMON 982
(INTERIM REPORT OF THE SAROBETSU GENERAL SURVEY, HOKKAIDO DEVELOPMENT BUREAU)
JAPANESE
INDEXED: JICST

HOKKAIDO KISHO GEPPO 983
(MONTHLY REPORT OF METEOROLOGY, HOKKAIDO)
SAPPORO KANKU KISHODAI (SAPPORO DISTRICT METEOROLOGICAL OBSERVATORY)
NISHI-18-CHOME, KITA-2-JO, CHUO-KU,SAPPORO-SHI 060
JAPANESE

HOKKAIDO NO KISHO 984
(METEOROLOGICAL DATA OF HOKKAIDO)
NIPPON KISHO KYOKAI HOKKAIDO HONBU (HOKKAIDO OFFICE, JAPAN WEATHER ASSOCIATION)
NISHI-18-CHOME, KITA-2-JO, CHUO-KU, SAPPORO-SHI 060
JAPANESE

HOKUBU KANKU KISHO KENKYU KAISHI 985
(JOURNAL OF THE NORTHERN DISTRICT METEOROLOGICAL RESEARCH SOCIETY)
JAPANESE
INDEXED: JICST

HOKUBU-KANKU KISHO SHOKUIN RONBUNSHU 986
(SCIENTIFIC PAPERS OF THE STAFF MEMBERS OF THE NORTHERN DISTRICT, SAPPORO DISTRICT METEOROLOGICAL OBSERVATORY)
NICHI-18-CHOME, KITA-2-JO, SAPPORO-SHI 060
JAPANESE AND ENGLISH
INDEXED: JICST

HYOGO-KEN KISHO GEPPO 987
(MONTHLY REPORT OF METEOROLOGY, HYOGO PREFECTURE)
KOBE KAIYO KISHODAI (KOBE MARINE OBSERVATORY)
178, 7-CHOME, NAKA YAMATE-DORI IKUTA-KU, KOBE-SHI 650
JAPANESE

HYOGO-KEN KISHO NENPO 988
(ANNUAL REPORT OF METEOROLOGY, HYOGO PREFECTURE)
KOBE KAIYO KISHODAI (KOBE MARINE OBSERVATORY)
178, 7-CHOME, NAKA YAMATE-DORI, IKUTA-KU, KOBE-SHI 650
JAPANESE

IBARAKI-KEN KISHO GEPPO 989
(MONTHLY REPORT OF METEOROLOGY, IBARAKI PREFECTURE)
MITO CHIHO KISHODAI (MITO LOCAL METEOROLOGICAL OBSERVATORY)
4-6, 1-CHOME, KANAMACHI, MITO-SHI 310
JAPANESE

IGC DATA OF IONOSPHERIC ABSORPTION AT KOKUBUNJI 990
JAPANESE
INDEXED: JICST

IJO GENSHO CHOSA HOKOKU (FUKUOKA KANKU KISHODAI) 991
(UNUSUAL METEOROLOGICAL PHENOMENA REPORT, FUKUOKA DISTRICT METEOROLOGICAL OBSERVATORY)
FUKUOKA KANKU KISHODAI (FUKUOKA DISTRICT METEOROLOGICAL OBSERVATORY)
2-36, 1-CHOME, OHORI, CHUO-KU, FUKUOKA-SHI 810
JAPANESE

IJO GENSHO CHOSA HOKOKU (OSAKA KANKU KISHODAI) 992
(UNUSUAL METEOROLOGICAL PHENOMENA REPORT, OSAKA DISTRICT METEOROLOGICAL OBSERVATORY))
OSAKA KANKU KISHODAI (OSAKA DISTRICT METEOROLOGICAL OBSERVATORY)
25, 6-CHOME, HOENZAKA-MACHI, HIGASHI-KU, OSAKA-SHI 540
JAPANESE

IJO GENSHO CHOSA HOKOKU (SAPPORO KANKU KISHODAI) 993
(UNUSUAL METEOROLOGICAL PHENOMENA REPORT, SAPPORO DISTRICT METEOROLOGICAL OBSERVATORY)
SAPPORO KANKU KISHODAI (SAPPORO DISTRICT METEOROLOGICAL OBSERVATORY)
NISHI 18-CHOME, KITA 2-JO, CHUO-KU, SAPPORO-SHI 060
JAPANESE

IJO GENSHO CHOSA HOKOKU (SENDAI KANKU KISHODAI) 994
(UNUSUAL METEOROLOGICAL PHENOMENA REPORT, SENDAI DISTRICT METEOROLOGICAL OBSERVATORY)
SENDAI KANKU KISHODAI (SENDAI DISTRICT METEOROLOGICAL OBSERVATORY)
1, TEPPO-CHO, SENDAI-SHI 983
JAPANESE

IJO GENSHO CHOSA HOKOKU (TOKYO KANKU KISHODAI) 995
(UNUSUAL METEOROLOGICAL PHENOMENA REPORT, TOKYO DISTRICT METEOROLOGICAL OBSERVATORY)
TOKYO KANKU KISHODAI (TOKYO DISTRICT METEOROLOGICAL OBSERVATORY)
3-4, 1-CHOME OTE-MACHI, CHIYODA-KU, TOKYO 100
JAPANESE

IJO KISHO CHOSA HOKOKU (TOKYO KANKU KISHODAI) 996
(UNUSUAL METEOROLOGICAL REPORT, TOKYO DISTRICT METEOROLOGICAL OBSERVATORY))
TOKYO KANKU KISHODAI (TOKYO DISTRICT METEOROLOGICAL OBSERVATORY)
3-4, 1-CHOME, OTE-MACHI, CHIYODA-KU, TOKYO 100
JAPANESE

IJO KISHO HOKOKU (CHOSHI CHIHO KISHODAI) 997
(UNUSUAL METEOROLOGICAL REPORT, CHIBA PREFECTURE)
CHOSHI CHIHO KISHODAI (CHOSHI LOCAL METEOROLOGICAL OBSERVATORY)
31, MINAMI-CHO, CHOSHI-SHI 288
JAPANESE

IJO KISHO HOKOKU (FUKUOKA KANKU KISHODAI) 998
(UNUSUAL METEOROLOGICAL REPORT, FUKUOKA DISTRICT METEOROLOGICAL OBSERVATORY)
FUKUOKA KANKU KISHODAI (FUKUOKA DISTRICT METEOROLOGICAL OBSERVATORY)
2-36, 1-CHOME, OHORI, CHUO-KU, FUKUOKA-SHI 810
JAPANESE

IJO KISHO HOKOKU (HIROSHIMA CHIHO KISHODAI) 999
(UNUSUAL METEOROLOGICAL REPORT, HIROSHIMA DISTRICT METEOROLOGICAL OBSERVATORY)
HIROSHIMA CHIHO KISHODAI (HIROSHIMA LOCAL METEOROLOGICAL OBSERVATORY)
40-1, 1-CHOME, EBAMINAMI, HIROSHIMA-SHI 733
JAPANESE

IJO KISHO HOKOKU (KISHO-CHO) 1000
(UNUSUAL METEOROLOGICAL REPORT, JAPAN METEOROLOGICAL AGENCY)
KISHO-CHO YOHO-BU (JAPAN METEOROLOGICAL AGENCY)
3-4, 1-CHOME, OTE-MACHI, CHIYODA-KU, TOKYO 100
JAPANESE

IJO KISHO HOKOKU (KOFU CHIHO KISHODAI) 1001
(UNUSUAL METEOROLOGICAL REPORT, KOFU LOCAL METEOROLOGICAL OBSERVATORY)
KOFU CHIHO KISHODAI (KOFU LOCAL METEROLOGICAL OBSERVATORY)
7-29, 4-CHOME, IIDA, KOFU-SHI 400
JAPANESE

IJO KISHO HOKOKU (OSAKA KANKU KISHODAI) 1002
(UNUSUAL METEOROLOGICAL REPORT, OSAKA DISTRICT METEOROLOGICAL OBSERVATORY)
OSAKA KANKU KISHODAI (OSAKA DISTRICT METEOROLOGICAL OBSERVATORY)
25, 6-CHOME, HOENZAKA-MACHI, HIGASHI-KUR, OSAKA-SHI 540
JAPANESE
INDEXED: JICST

IJO KISHO HOKOKU (SENDAI KANKU KISHODAI) 1003
(UNUSUAL METEOROLOGICAL REPORT, SENDAI DISTRICT METEOROLOGICAL OBSERVATORY)
SENDAI KANKU KISHODAI (SENDAI DISTRICT METEOROLOGICAL OBSERVATORY)
1, TEPPO-CHO, SENDAI-SHI 893
JAPANESE

IJO KISHO HOKOKU (TOKYO KANKU KISHODAI) 1004
(UNUSUAL METEOROLOGICAL REPORT, TOKYO DISTRICT METEOROLOGICAL OBSERVATORY)
TOKYO KANKU KISHODAI (TOKYO DISTRICT METEOROLOGICAL OBSERVATORY)
3-4, 1-CHOME, OTE-MACHIE, CHIYODA-KU, TOKYO 100
JAPANESE

IJO TENKO KANSHI HOKOKU 1005
(REPORT ON OBSERVATIONS OF ABNORMAL WEATHER CONDITIONS)
KISHOCHO YOJOBU (METEOROLOGICAL AGENCY, FORECAST DEPARTMENT)
JAPANESE

IONOSPHERIC DATA AT SYOWA BASE (ANTARCTICA) 1006
YUSEI-SHO DENPA KENKYUSHO (RADIO RESEARCH LABORATORY, MINISTRY OF POST AND TELECOMMUNICATIONS)
2-1, 4-CHOME, NUKUI KITA-MACHI, KOGANEI-SHI, TOKYO 184
JAPANESE

IONOSPHERIC DATA IN JAPAN 1007
YUSEI-SHO DENPA KENKYUSHO (RADIO RESEARCH LABORATORY, MINISTRY OF POST AND TELECOMMUNICATIONS)
2-1, 4-CHOME, NUKUI KITA-MACHI, KOGANEI-SHI, TOKYO 184
ENGLISH

IRAKO SOKKOSHO KISHO GEPPO 1008
(MONTHLY REPORT OF METEOROLOGY, IRAKO WEATHER STATION)
IRAKO SOKKOSHO
56-2, KINGOROSAKA, FUKUE, ATSUMI-CHO, ATSUMI-GUN, AICHI 441-36
JAPANESE

IRAKO SOKKOSHO KISHO NENPO 1009
(ANNUAL REPORT OF METEOROLOGY, IRAKO WEATHER STATION)
IRAKO SOKKOSHO
56-2, KINGOROSAKA, FUKUE, ATSUMI-CHO, ATSUMI-GUN, AICHI 441-36
JAPANESE

ISHIKAWA-KEN NOGYO KISHO GEPPO 1010
(MONTHLY REPORT OF AGRICULTURAL METEOROLOGY, ISHIKAWA PREFECTURE)
KANAZAWA CHIHO KISHODAI (KANAZAWA LOCAL METEOROLOGICAL OBSERVATORY)
32-25, 1-CHOME, YAYOI, KANAZAWA-SHI 921
JAPANESE

ISHIKAWA-KEN NOGYO KISHO NENPO 1011
(ANNUAL REPORT OF AGRICULTURAL METEOROLOGY, ISHIKAWA PREFECTURE)
KANAZAWA CHIHO KISHODAI (KANAZAWA LOCAL METEOROLOGICAL OBSERVATORY)
39-25, 1-CHOME, YAYOI, KANAZAWA-SHI 921
JAPANESE

IWATE-KEN KISHO GEPPO 1012
(MONTHLY REPORT OF METEOROLOGY, IWATE PREFECTURE)
MORIOKA CHIHO KISHODAI (MORIOKA LOCAL METEOROLOGICAL OBSERVATORY)
7-60, SANNO-CHO, MORIOKA-SHI 020
JAPANESE

IWATE-KEN NOGYO KISHO GEPPO 1013
(IWATE PREFECTURE AGRICULTURAL WEATHER MONTHLY)
IWATE KEN, MORIOKA CHIHO KISHODAI (MORIOKA WEATHER STATION, IWATE PREFECTURE)
JAPANESE

JAPANESE PROGRESS IN CLIMATOLOGY 1014
TOKYO KYOIKU DAIGAKU KIKOGAKU DANWAKAI (JAPANESE CLIMATOLOGICAL SEMINAR, TOKYO UNIVERSITY OF EDUCATION)
29-1, 3-CHOME, OTSUKA, BUNKYO-KU, TOKYO 112
ENGLISH
INDEXED: BIB

JARE SCIENTIFIC REPORTS, SERIES B, METEOROLOGY 1015
KOKURITSU KAGAKU HAKUBUTSUKAN, KYOKUCHI KENKYU SENTA (POLAR RESEARCH CENTER, NATIONAL SCIENCE MUSEUM)
9-10, 1-CHOME, KAGA, ITABAHI-KU, TOKYO 173
EUROPEAN

JOURNAL OF THE METEOROLOGICAL SOCIETY OF JAPAN 1016
NIHON KISHO GAKKAI (METEOROLOGICAL SOCIETY OF JAPAN)
C/O KISHO-CHO, 3-4, 1-CHOME, OTE-MACHI, CHIYODA-KU, TOKYO 100
ENGLISH
INDEXED: APP, CHE, JICST

KAGAWA-KEN KISHO GEPPO 1017
(MONTHLY REPORT OF METEOROLOGY, KAGAWA PREFECTURE)
TAKAMATSU CHIHO KISHODAI (TAKAMATSU LOCAL METEOROLOGICAL OBSERVATORY)
1277-1, FUSEISHI-CHO, TAKAMATSU-SHI 760
JAPANESE
INDEXED: JICST

KAGAWA-KEN KISHO NENPO 1018
(ANNUAL REPORT OF METEOROLOGY, KAGAWA PREFECTURE)
TAKAMATSU CHIHO KISHODAI (TAKAMATSU LOCAL METEOROLOGICAL OBSERVATORY)
1277-1, FUSEISHI-CHO, TAKAMATSU-SHI 760
JAPANESE
INDEXED: JICST

KAGOSHIMA-KEN NOGYO KISHO GEPPO 1019
(MONTHLY REPORT OF AGRICULTURAL METEOROLOGY, KAGOSHIMA PREFECTURE)
KAGOSHIMA CHIHO KISHODAI (KAGOSHIMA LOCAL METEOROLOGICAL OBSERVATORY)
24-13, 1-CHOME, ARATA, KAGOSHIMA-SHI 890
JAPANESE

KAIJO KISHO GAIHO 1020
(MARINE METEOROLOGICAL REPORT)
MAIZURU KAIYO KISHODAI (MAIZURU MARINE OBSERVATORY)
KITASUI, MAIZURU-SHI 625
JAPANESE

KAISHO TO KISHO 1021
(OCEANOGRAPHY & METEOROLOGY)
NAGASAKI KAIYO KISHODAI (NAGASAKI MARINE OBSERVATORY)
5, MINAMIYAMATE-MACHI, NAGASAKI-SHI 850
JAPANESE AND ENGLISH

KAIYO SOKUHO (HAKODATE KAIYO KISHODAI) 1022
(OCEANOGRAPHIC OBSERVATION REPORT, HAKODATE MARINE OBSERVATORY)
HAKODATE KAIYO KISHODAI
181, AKAGAWADORI-CHO, KAMEDA-SHI 040
JAPANESE

KAIYO SOKUHO (KOBE KAIYO KISHODAI) 1023
(OCEANOGRAPHIC OBSERVATION REPORT, KOBE MARINE OBSERVATORY)
KOBE KAIYO KISHODAI (KOBE MARINE OBSERVATORY)
178, 7-CHOME, NAKAYAMATE-DORI, IKUTA-KU, KOBE-SHI 650
JAPANESE

KAIYO SOKUHO (MAIZURU KAIYO KISHODAI) 1024
(OCEANOGRAPHIC OBSERVATION REPORT, MAIZURU MARINE OBSERVATORY)
MAIZURU KAIYO KISHODAI (MAIZURU MARINE OBSERVATORY)
KITASUI, MAIZURU-SHI 625
JAPANESE

KAIYO SOKUHO (NAGASAKI KAIYO KISHODAI) 1025
(OCEANOGRAPHIC PROMPT REPORT, NAGASAKI MARINE OBSERVATORY)
NAGASAKI KAIYO KISHODAI (NAGASAKI MARINE OBSERVATORY)
5, MINAMIYAMATE-MACHI, NAGASAKI-SHI 850
JAPANESE

KANAGAWA-KEN KISHO GEPPO 1026
(MONTHLY REPORT OF METEOROLOGY, KANAGAWA PREFECTURE)
YOKOHAMA CHIHO KISHODAI (YOKOHAMA LOCAL METEOROLOGICAL OBSERVATORY)
99, YAMATE-MACHI, NAKA-KU, YOKOHAMA-SHI 231
JAPANESE

KANSOKUBU JIHO 1027
(BULLETIN OF THE OBSERVATION DEPARTMENT)
KISHO-CHO (METEOROLOGICAL AGENCY)
3-4, 1-CHOME, OTE-MACHI, CHIYODA-KU, TOKYO 100
JAPANESE

KANSOKUSHO KISHO NENPO-ZENKOKU KISHO-HYO 1028
(ANNUAL REPORT OF CLIMATOLOGICAL STATIONS)
KISHO-CHO (METEOROLOGICAL AGENCY)
3-4, 1-CHOME, OTE-MACHI, CHIYODA-KU, TOKYO 100
JAPANESE AND ENGLISH

KARUIZAWA KISHO NENPO 1029
(ANNUAL REPORT OF METEOROLOGY, KARUIZAWA WEATHER STATION)
KARUIZAWA SOKKOJO (KARUIZAWA WEATHER STATION)
1151-2, MICHINOUE, NABERIBASHI, OIWAKE, KARUIZAWA-MACHI, KITASAKU-GUN, NAGANO 389-01
JAPANESE

KIKOGAKU KENKYU 1030
(CLIMATOLOGICAL NOTES)
HOSEI DAIGAKU CHIRIGAKU KYOSHITSU
17-1, 2-CHOME, FUJIMI, CHIYODA-KU, TOKYO 102
ENGLISH

KINKI CHUGOKU SHIKOKU CHIKU KISHO KENKYUKAISHI 1031
(JOURNAL OF METEOROLOGICAL SOCIETY OF KINKI, CHUGOKU, SHIKOKU DISTRICT)
OSAKA KANKU KISHODAI (OSAKA DISTRICT METEOROLOGICAL OBSERVATORY)
25, 6-CHOME, HOENZAKA-CHO, HIGASHI-KU, OSAKA-SHI 540
JAPANESE
INDEXED: JICST

KISETSU YOHO SHIRYO 1032
(QUARTERLY FORECASTING DATA)
KISHO-CHO (METEOROLOGICAL AGENCY)
3-4, 1-CHOME, OTE-MACHI, CHIYODA-KU, TOKYO 100
JAPANESE

KISHO 1033
(WEATHER)
NIHON KISHO KYOKAI (JAPAN METEOROLOGICAL ASSOCIATION)
C/O KISHO-CHO, 3-4, 1-CHOME, OTE-MACHI, CHIYODA-KU, TOKYO 100
JAPANESE
INDEXED: JICST

KISHO DAIGAKKO KENKYU SHUHO 1034
(MEMOIRS ON GEOPHYSICS AND GEOPHYSICAL EDUCATION)
KISHO DAIGAKKO (METEOROLOGICAL COLLEGE)
7-481, ASAHI-CHO, KASHIWA-SHI 277
JAPANESE AND ENGLISH

KISHO GIJUTSU NOTO 1035
(METEOROLOGICAL TECHNICAL NOTE)
NIHON KISHO GAKKAI KANSAI SHIBU (KANSAI BRANCH, METEOROLOGICAL SOCIETY OF JAPAN)
6-25, HOENZAKA-CHO, HIGASHI-KU, OSAKA-SHI 540
JAPANESE

KISHO KENKYU NOTO 1036
(METEOROLOGICAL STUDY NOTE)
NIHIN KISHO GAKKAI (METEOROLOGICAL SOCIETY OF JAPAN)
C/O KISHO-CHO, 3-4, 1-CHOME, OTE-MACHI, CHIYODA-KU TOKYO 100
JAPANESE
INDEXED: CHE, JICST

KISHO KENKYUSHO KENKYU HOKOKU 1037
(PAPERS IN METEOROLOGY AND GEOPHYSICS)
KISHO-CHO KISHO KENKYUSHO (METEOROLOGICAL RESEARCH INSTITUTE METEOROLOGICAL AGENCY)
35-8, 4-CHOME, KOENJI-KITA, SUGINAMI-KU, TOKYO 166
JAPANESE AND ENGLISH

KISHO KENKYUSHO NENPO 1038
(ANNUAL REPORT OF METEOROLOGICAL RESEARCH INSTITUTE)
KISHO-CHO KISHO KENKYUSHO (METEOROLOGICAL RESEARCH INSTITUTE METEOROLOGICAL AGENCY)
35-8, 4-CHOME, KOENJI-KITA, SUGINAMI-KU, TOKYO 166
JAPANESE

KISHO KENKYUSHO NYUSU 1039
(METEOROLOGICAL RESEARCH INSTITUTE NEWS)
KISHO-CHO KISHO KENKYUSHO (METEOROLOGICAL RESEARCH INSTITUTE, METEOROLOGICAL AGENCY)
35-8, 4-CHOME, KOENJI-KITA, SUGINAMI-KU, TOKYO 166
JAPANESE

KISHO NENPO (NIHON SENBAI KOSHA) 1040
(ANNUAL REPORT OF THE METEOROLOGY, JAPAN TABACCO & SALT PUBLIC CORPORATION))
NIHON SENBAI KOSHA
2-1, AKASAKA AOI-CHO, MINATO-KU, TOKYO 107
JAPANESE

KISHO SHUSHI 1041
(JOURNAL OF THE METEOROLOGICAL SOCIETY OF JAPAN)
NIHON KISHO GAKKAI (METEOROLOGICAL SOCIETY OF JAPAN)
3-4, 1-CHOME, OTE-MACHI, CHIYODA-KU, TOKYO 100
ENGLISH

KISHO TO NOSAGYO 1042
(METEOROLOGY AND AGRICULTURE)
KUMAMOTO CHIHO KISHODAI (KUMAMOTO LOCAL METEOROLOGICAL OBSERVATORY)
6, SUIZENJI, KUMAMOTO-SHI 862
JAPANESE

KISHO YORAN 1043
(METEOROLOGICAL SURVEY)
JAPANESE
INDEXED: JICST

KISHO-CHO GEPPO, ZENKOKU KISHOHYO 1044
(MONTHLY REPORT OF THE JAPAN METEOROLOGICAL AGENCY)
KISHO-CHO (METEOROLOGICAL AGENCY)
3-4, 1-CHOME, OTE-MACHI, CHIYODA-KU, TOKYO 100
JAPANESE
INDEXED: JICST

KISHO-CHO GIJUTSU HOKOKU 1045
(TECHNICAL REPORT OF THE JAPAN METEOROLOGICAL AGENCY)
KISHO-CHO (METEOROLOGICAL AGENCY)
3-4, 1-CHOME, OTE-MACHI, CHIYODA-KU, TOKYO 100
JAPANESE AND ENGLISH
INDEXED: CHE, JICST

KISHO-CHO KANSOKU GIJUTSU SHIRYO 1046
(TECHNICAL DATA SERIES, JAPAN METEOROLOGICAL AGENCY)
KISHO-CHO (METEOROLOGICAL AGENCY)
3-4, 1-CHOME, OTE-MACHI, CHIYODA-KU, TOKYO 100
JAPANESE AND ENGLISH

KISHO-CHO KENKYU JIHO 1047
(JOURNAL OF METEOROLOGICAL RESEARCH)
KISHO-CHO (METEOROLOGICAL AGENCY)
3-4, 1-CHOME, OTE-MACHI, CHIYODA-KU, TOKYO 100
JAPANESE, WITH ENGLISH SUMMARY
INDEXED: CHE, JICST

KISHO-CHO NENPO, ZENKOKU KISHOHYO DAI-1-BU 1048
(ANNUAL REPORT OF THE JAPAN METEOROLOGICAL AGENCY,
METEOROLOGICAL OBSERVATIONS, PART 1)
KISHO-CHO (METEOROLOGICAL AGENCY)
3-4, 1-CHOME, OTE-MACHI, CHIYODA-KU, TOKYO 100
JAPANESE

KISHO-CHO NENPO, ZENKOKU KISHOHYO DAI-2-BU 1049
(ANNUAL REPORT OF THE JAPAN METEOROLGOCIAL AGENCY,
METEOROLOGICAL OBSERVATIONS, PART 2)
KISHO-CHO, (METEOROLOGICAL AGENCY)
3-4, 1-CHOME, OTE-MACHI, CHIYODA-KU, TOKYO 100
JAPANESE AND ENGLISH

KISHO-CHO TOSHO GEPPO 1050
(METEOROLOGICAL AGENCY LIBRARY MONTHLY)
METEOROLOGICAL AGENCY
JAPANESE
INDEXED: JICST

KOBE KAIYO KISHODAI IHO 1051
(BULLETIN OF THE KOBE MARINE OBSERVATORY)
KOBE KAIYO KISHODAI (KOBE MARINE OBSERVATORY)
7-178, NAKAYAMATE-DORI, IKUTA-KU, KOBE-SHI 650
JAPANESE, WITH ENGLISH SUMMARY
INDEXED: CHE, JICST

KOCHI-KEN KISHO GEPPO 1052
(MONTHLY REPORT OF METEOROLOGY, KOCHI PREFECTURE)
KOCHI CHIHO KISHODAI (KOCHI DISTRICT
METEOROLOGICAL OBSERVATORY)
7-3, 1-CHOME, HIJIMA-CHO, KOCHI-SHI 780
JAPANESE

KOCHI-KEN KISHO NENPO 1053
(ANNUAL REPORT OF METEOROLOGY, KOCHI PREFECTURE)
KOCHI CHIHO KISHODAI (KOCHI LOCAL METEOROLOGICAL
OBSERVATORY)
7-3, 1-CHOME, HIJIMA-CHO, KOCHI-SHI 780
JAPANESE

KOKU KISHO KANSOKU SHISHIN 1054
(OBSERVATION MANUAL FOR THE AERONAUTICAL METEOROLOGY)
KISHO-CHO KANSOKU-BU (FORECAST DEPARTMENT,
METEOROLOGICAL AGENCY)
3-4, 1-CHOME, OTE-MACHI, CHIYODA-KU, TOKYO 100
JAPANESE

KOKU KISHO NOTO 1055
(ABSTRACTS IN AVIATION METEOROLOGY)
KISHO-CHO (METEOROLOGICAL AGENCY)
3-4, 1-CHOME, OTE-MACHI, CHIYODA-KU, TOKYO 100
JAPANESE

KONGETSU NO TENKO TO NOSAGYO (MATSUE CHIHO KISHODAI) 1056
(MONTHLY NEWS OF THE WEATHER AND AGRICULTURE,
MATSUE LOCAL METEOROLOGICAL OBSERVATORY))
MATSUE CHIHO KISHODAI
1281, NISHITSUDA-MACHI, MATSUE-SHI 690
JAPANESE

KONGETSU NO TENKO TO NOSAGYO (MIYAZAKI CHIHO 1057
KISHODAI)
(MONTHLY NEWS OF THE WEATHER AND AGRICULTURE,
MIYAZAKI LOCAL METEOROLOGICAL OBSERVATORY)
MIYAZAKI CHIHO KISHODAI
331-2, WACHIKAWARA-CHO, MIYAZAKI-SHI 880
JAPANESE

KONSHU NO TENKI TO NOSAGYO (MATSUE CHIHO KISHODAI) 1058
(WEEKLY NEWS OF THE WEATHER AND AGRICULTURE, MATSUE
LOCAL METEOROLOGICAL OBSERVATORY)
MATSUE CHIHO KISODAI
1282, NISHITSUDA-MACHI, MATSUE-SHI 690
JAPANESE

KOSO KISHODAI IHO 1059
(JOURNAL OF THE AEROLOGICAL OBSERVATORY AT TATENO)
KOSO KISHODAI (AEROLOGICAL OBSERVATORY)
TATENO, YATABE-MACHI, TSUKUBA-GUN, IBARAKI 300-21
JAPANESE, WITH ENGLISH SUMMARY
INDEXED: JICST

KOTOSHI NO TENKO TO NOSAKUGYO 1060
(THIS YEAR'S CLIMATE AND FARMING INDUSTRY)
MATSUE CHIHO KISHODAI (MATSUE REGIONAL WEATHER
STATION)
JAPANESE

KOZUI 1061
(FLOODS)
KENSETSU-SHO KANTO CHIHO KENSETSU-KYOKU (KANTO
REGIONAL CONSTRUCTION BUREAU, MINISTRY OF
CONSTRUCTION)
3-1, 1-CHOME, OTE-MACHI, CHIYODA-KU, TOKYO 100
JAPANESE
INDEXED: JICST

KOZUI SOKUHO 1062
(FLOOD REPORT)
KENSETSU-SHO KANTO CHIHO KENSETSU-KYOKU
(KANTO REGIONAL CONSTRUCTION BUREAU, MINISTRY OF
CONSTRUCTION)
3-1, 1-CHOME, OTE-MACHI, CHIYODA-KU, TOKYO 100
JAPANESE

KUMAMOTO-KEN KISHO GEPPO 1063
(MONTHLY REPORT OF METEOROLOGY, KUMAMOTO PREFECTURE)
KUMAMOTO CHIHO KISHODAI (KUMAMOTO LOCAL
METEOROLOGICAL OBSERVATORY)
12-20, 2-CHOME, KYO-MACHI, KUMAMOTO-SHI 860
JAPANESE

KUMAMOTO-KEN NOGYO KISHO GEPPO 1064
(KUMAMOTO PREFECTURE AGRICULTURAL METEOROLOGY
MONTHLY REPORT)
KUMAMOTO-KEN, KUMAMOTO CHIHO KISHODAI (KUMAMOTO
PREFECTURE, KUMAMOTO REGIONAL WEATHER STATION)
JAPANESE

KYOTO-FU KISHO GEPPO 1065
(MONTHLY REPORT OF METEOROLOGY, KYOTO PREFECTURE)
KYOTO CHIHO KISHODAI (KYOTO LOCAL METEOROLOGICAL
OBSERVATORY)
38, NISHINOKYO KASADONO-MACHI, NAKAGYO-KU,
KYOTO-SHI 604
JAPANESE

KYOTO-FU KISHO NENPO 1066
(ANNUAL REPORT OF METEOROLOGY, KYOTO PREFECTURE)
KYOTO CHIHO KISHODAI (KYOTO LOCAL METEOROLOGICAL
OBSERVATORY)
38, NISHINOKYO KASANDONO-MACHI, NAKAGYO-KU,
KYOTO-SHI 604
JAPANESE

KYUSHU CHIHO DENRYOKU KISHO GAIHO 1067
(KYUSHU REGIONAL ELECTRICAL POWER METEOROLOGICAL
SUMMARY)
FUKUOKA KANKU KISHODAI, DENRYOKU KISHO RENRAKUKAI
KYUSHU CHIHO IINKAI (ELECTRICAL POWER METEOROLOGICAL
LIASON ASSOCIATION, KYUSHU REGIONAL COMMITTEE)
JAPANESE

MAIZURU KAIYO KISHODAI YOHO 1068
(BULLETIN OF THE MAIZURU MARINE OBSERVATORY)
MAIZURU KAIYO KISHODAI (MAIZURU MARINE OBSERVATORY)
KITASUI- MAIZURU-SHI 625
JAPANESE AND ENGLISH

MEAN MAPS, LONG RANGE WEATHER FORECASTING 1069
KISHO-CHO (METEOROLOGICAL AGENCY)
3-4, 1-CHOME, OTE-MACHI, CHIYODA,KU, TOKYO 100
ENGLISH

MIE-KEN KISHO GEPPO 1070
(MONTHLY REPORT OF METEOROLOGY, MIE PREFECTURE)
TSU CHIHO KISHODAI (TSU LOCAL METEOROLOGICAL
OBSERVATORY) AND NIPPON KISHO KYOKAI TSU SHIBU
(TSU OFFICE JAPAN WEATHER ASSOCIATION)
1376-1, MATSUMURA, TSUOKI, TSU-SHI 514
JAPANESE

MIYAGI-KEN KISHO GEPPO 1071
(MONTHLY REPORT OF METEOROLOGY, MIYAGI PREFECTURE)
SENDAI KANKU KISHODAI (SENDAI DISTRICT METEOROLOGICAL OBSERVATORY)
1, TEPPO-MACHI, SENDAI-SHI 983
JAPANESE

MIYAZAKI-KEN KISHO GEPPO 1072
(MONTHLY REPORT OF METEOROLOGY, MIYAZAKI PREFECTURE)
MIYAZAKI CHIHO KISHODAI (MIYAZAKI LOCAL METEOROLOGICAL OBSERVATORY)
331-2, WACHIKAWARA-CHO, MIYAZAKI-SHI 880
JAPANESE

MIYAZAKI-KEN NOGHO KISHO GEPPO 1073
(MONTHLY REPORT OF AGRICULTURAL METEOROLOGY, MIYAZAKI PREFECTURE)
MIYAZAKI CHIHO KISHODAI (MIYAZAKI LOCAL METEOROLOGICAL OBSERVATORY)
331-2 WASHIKAWARA-CHO, MIYAZAKI-SHI 880
JAPANESE

MIYAZAKI-KEN NOGYO KISHO NENPO 1074
(ANNUAL REPORT OF AGRICULTURAL METEOROLOGY, MIYAZAKI PREFECTURE)
MIYAZAKI CHIHO KISHODAI (MIYAZAKI LOCAL METEOROLOGICAL OBSERVATORY)
331-2 WACHIKAWARA-CHO, MIYAZAKI-SHI 880
JAPANESE

MIYAZAKI-KEN NOGYO KISHO SHUHO 1075
(WEEKLY REPORT OF AGRICULTURAL METEOROLOGY, MIYAZAKI PREFECTURE)
MIYAZAKI CHIHO KISHODAI (MIYAZAKI LOCAL METEOROLOGICAL OBSERVATORY)
331-2, WACHIKAWARA-CHO, MIYAZAKI-SHI 880
JAPANESE

NAGANO-KEN KISHO NENPO 1076
(ANNUAL REPORT OF METEOROLOGY, NAGANO PREFECTURE)
NAGANO CHIHO KISHODAI (NAGANO LOCAL METEOROLOGICAL OBSERVATORY)
2417, HAKOSHIMIZU, NAGANO-SHI 380
JAPANESE

NAGANO-KEN NOGYO KISHO GEPPO 1077
(MONTHLY REPORT OF AGRICULTURAL METEOROLOGY, NAGANO PREFECTURE)
NAGANO CHIHO KISHODAI (NAGANO LOCAL METEOROLOGICAL OBSERVATORY)
2417, HAKOSHIMIZU, NAGANO-SHI 380
JAPANESE

NAGASAKI-KEN KISHO GEPPO 1078
(MONTHLY REPORT OF METEOROLOGY, NAGASAKI PREFECTURE)
NAGASAKI KAIYO KISHODAI (NAGASAKI MARINE OBSERVATORY)
5, MINAMI YAMATE-MACHI, NAGASAKI-SHI 850
JAPANESE

NAGOYA KOKU KISHO HYO 1079
(NAGOYA AVIATION WEATHER TABLE)
NOGAOYA KOKU SOKKOJO (NAGOYA AVIATION WEATHER STATION)
JAPANESE

NARA-KEN KISHO GEPPO 1080
(MONTHLY REPORT OF METEOROLOGY, NARA PREFECTURE)
NARA CHIHO KISHODAI (NARA LOCAL METEOROLOGICAL OBSERVATORY)
7-CHOME, HANDA HIRAKI-CHO, NARA-SHI 630
JAPANESE

NARA-KEN KISHO NENPO 1081
(ANNUAL REPORT OF METEOROLOGY, NARA PREFECTURE)
NARA CHIHO KISHODAI (NARA LOCAL METEOROLOGICAL OBSERVATORY)
7-CHOME, HANDA HIRAKI-CHO, NARA-SHI 630
JAPANESE

NIIGATA-KEN KISHO NENPO 1082
(ANNUAL REPORT OF METEOROLOGY, NIIGATA PREFECTURE)
NIIGATA CHIHO KISHODAI (NIIGATA LOCAL METEOROLOGICAL OBSERVATORY)
4-1, 4-CHOME, SAIWAINISHI, NIIGATA-SHI 950
JAPANESE

NIIGATA-KEN NOGYO KISHO GEPPO 1083
(MONTHLY REPORT OF AGRICULTURAL METEOROLOGY, NIIGATA PREFECTURE)
NIIGATA CHIHO KISHODAI (NIIGATA LOCAL METEORLOGICAL OBSERVATORY)
4-1, 4-CHOME, SAIWAINISHI, NIIGATA-SHI 950
JAPANESE

NISHI NIHON KAIKYO JUNPO 1084
(TEN-DAY MARINE REPORT OF THE EAST CHINA SEA)
NAGASAKI KAIYO KISHODAI (NAGASAKI MARINE OBSERVATORY)
5, MINAMI YAMATE-MACHI, NAGASAKI-SHI 850
JAPANESE

NISHI-NIHON KISHO GEPPO 1085
(WESTERN JAPAN METEOROLOGICAL MONTHLY)
FUKUOKA KANKU KISHODAI (FUKUOKA DISTRICT METEOROLOGICAL OBSERVATORY)
JAPANESE

NOGYO KISHO 1086
(JOURNAL OF AGRICULTURAL METEOROLOGY)
NIHON NOGYO KISHO GAKKAI (SOCIETY OF AGRICULTURAL METEOROLOGY OF JAPAN)
C/O NOGYO GIJUTSU KENKYUSHO KISHO-KA, 2-1, NISHIGAHARA, KITA-KU, TOKYO 114
JAPANESE AND ENGLISH
INDEXED: BIB, CHE

NOGYO KISHO KENKYU SHUROKU 1087
(COLLECTED PAPERS ON AGRICULTURAL METEOROLOGY)
NORINSHO NOGYO GIJUTSU KENKYUJO KISHOKA (MINISTRY OF AGRICULTURE AND FORESTRY, ACRICULTURAL TECHNOLOGY RESEARCH INSTITUTE
JAPANESE

NOGYO KISHO NENPO (KISHO-CHO) 1088
(ANNUAL REPORT OF AGRICULTURAL METEOROLOGY, JAPAN METEOROLOGICAL AGENCY)
KISHO-CHO (METEOROLOGICAL AGENCY)
3-4, 1-CHOME, OTE-MACHI, CHIYODA-KU, TOKYO 100
JAPANESE

NOGYO KISHO NENPO (KUMAMOTO-KEN) 1089
KUMAMOTO CHIHO KISHODAI (KUMAMOTO LOCAL METEOROLOGICAL OBSERVATORY)
18-1, 6-CHOME, SUIZENJI, KUMAMOTO-SHI 862
JAPANESE

NOGYO KISHO SOKUHO 1090
(CURRENT REPORT ON AGRICULTURAL METEOROLOGY)
MATSUE CHIHO KISHODAI (MATSUE REGIONAL METEOROLOGICAL OBSERVATORY)
JAPANESE

OAME YOSO SHIRYO 1091
(DATA FOR THE FORECAST OF HEAVY RAINFALLS)
KISHO-CHO (THE JAPAN METEOROLOGICAL AGENCY)
3-4, 1-CHOME, OTE-MACHI, CHIYODA-KU, TOKYO 100
JAPANESE AND ENGLISH

OITA-KEN KISHO GEPPO 1092
(MONTHLY REPORT OF METEOROLOGY, OITA PREFECTURE)
OITA CHIHO KISHODAI (OITA LOCAL METEOROLOGICAL OBSERVATORY)
1-38, 3-CHOME, NAGAHAMA-CHO, OITA-SHI 870
JAPANESE

OKAYAMA-KEN KISHO GEPPO 1093
(MONTHLY REPORT OF METEOROLOGY, OKAYAMA PREFECTURE)
OKAYAMA CHIHO KISHODAI (OKAYAMA LOCAL METEOROLOGICAL OBSERVATORY)
1380-3, TSUSHIMA, OKAYAMA-SHI 700
JAPANESE

OKAYAMA-KEN KISHO NENPO 1094
(ANNUAL REPORT OF METEOROLOGY, OKAYAMA PREFECTURE)
OKAYAMA CHIHO KISHODAI (OKAYAMA LOCAL METEOROLOGICAL OBSERVATORY)
1380-3, TSUSHIMA, OKAYAMA-SHI 700
JAPANESE

OKAYAMA-KEN NO KISHOREKI 1095
(METEOROLOGICAL CALENDER OF OKAYAMA-KEN)
NIHON KISHO KYOKAI OKAYAMA SHIBU (OKAYAMA BRANCH OF JAPAN WEATHER ASSOCIATION)
C/O OKAYAMA CHIHO KISHODAI, 1380, TSUSHIMA, OKAYAMA-SHI 700
JAPANESE

OKINAWA GIJUTSU NOTO 1096
(TECHNICAL NOTES OF OKINAWA METEOROLOGICAL OBSERVATORY)
OKINAWA KISHODAI (OKINAWA METEOROLOGICAL OBSERVATORY)
1126, AMEKU, NAHA-SHI 900
JAPANESE

OKINAWA KANKU KISHO KENKYU KAISHI 1097
(COLLECTED PAPERS OF OKINAWA METEOROLOGICAL OBSERVATORY)
OKINAWA KISHODAI (OKINAWA METEOROLOGICAL OBSERVATORY)
1126, AMEKU, NAHA-SHI 900
JAPANESE

OKINAWA KANNAI KISHO KENKYU KAISHI 1098
(JOURNAL OF METEOROLOGICAL RESEARCH IN OKINAWA)
OKINAWA KISHODAI (OKINAWA METEOROLOGICAL OBSERVATORY)
JAPANESE

OKINAWA KISHO GEPPO 1099
(MONTHLY METEOROLOGICAL REPORT FOR OKINAWA)
OKINAWA KISHO DAI
(OKINAWA METEOROLOGICAL OBSERVATORY)
JAPANESE

OSAKA KANKU FUKEN KISHO KENKYU KAISHI 1100
(PROCEEDINGS OF THE METEOROLOGICAL RESEARCH SOCIETY OF THE OSAKA REGION)
OSAKA KANKU KISHODAI
JAPANESE

OSAKA KANKU KISHO KENKYUKAISHI 1101
(MAGAZINE OF OSAKA DISTRICT METEOROLOGICAL SOCIETY)
OSAKA KANKU KISHODAI (OSAKA DISTRICT METEOROLOGICAL SOCIETY)
6-25, HOENZAKA-MACHI, HIGASHI-KU, OSAKA-SHI 540
JAPANESE

OSAKA KANKU KISHODAI GIJUTSU JOHO 1102
(OSAKA DISTRICT METEOROLOGICAL OBSERVATORY TECHNICAL INFORMATION)
JAPANESE

OSAKA KANKU KISHODAI TOKUBETSU CHOSA HOKOKU 1103
(SPECIAL REPORT OF OSAKA DISTRICT METEOROLOGICAL OBSERVATORY)
OSAKA KANKU KISHODAI (OSAKA DISTRICT METEOROLOGICAL SOCIETY)
6-25, HOENZAKA-MACHI HIGAHI-KU, OSAKA-SHI 540
JAPANESE

OSAKA-FU KISHO GEPPO 1104
(MONTHLY REPORT OF METEOROLOGY, OSAKA PREFECTURE)
OSAKA KANKU KISHODAI (OSAKA DISTRICT METEOROLOGICAL SOCIETY)
6-25, HOENZAKA-MACHI, HIGASHI-KU, OSAKA-SHI 540
JAPANESE
INDEXED: JICST

OSAKA-FU KISHO NENPO 1105
(ANNUAL REPORT OF METEOROLOGY, OSAKA PREFECTURE)
OSAKA KANKU KISHODAI (OSAKA DISTRICT OF METEOROLOGICAL OBSERVATORY)
6-25, HOENZAKA-MACHI, HIGASHI-KU, OSAKA-SHI 540
JAPANESE

OSHIMA KANNAI NOGYO KISHO GEPPO 1106
(MONTHLY REPORT OF AGRICULTURAL METEOROLOGY, OSHIMA DISTRICT METEOROLOGICAL OBSERVATORY)
HAKODATE KAIYO KISHODAI (THE HAKODATE MARINE OBSERVATORY)
181, AKAGAWADORI-CHO, KAMEDA-SHI 040
JAPANESE

P. R. METEO (KISHO KYOKAI DAYORI) 1107
(P. R. METEO. (METEOROLOGICAL ASSOCIATION))
NIHON KISHO KYOKAI (METEOROLOGICAL ASSOCIATION)
C/O KISHO-CHO, 3-4, 1-CHOME, OTE-MACHI, CHIYODA-KU, TOKYO 100
JAPANESE

PAPERS IN METEOROLOGY AND GEOPHYSICS 1108
KISHO-CHO KISHO KENKYUSHO (METEOROLOGICAL RESEARCH INSTITUTE, JAPAN METEOROLOGICAL AGENCY)
35-8, 4-CHOME, KOENJI-KITA, SUGINAMI-KU, TOKYO 166
JAPANESE AND ENGLISH
INDEXED: APP, CHE, JICST

PROCEEDINGS OF THE RESEARCH INSTITUTE OF ATMOSPHERICS, NAGOYA UNIVERSITY 1109
NAGOYA DAIGAKU KUDEN KENKYUSHO (RESEARCH INSTITUTE OF ATMOSPHERICS, NAGOYA UNIVERSITY)
3-13, HONOHARA, TOYOKAWA-SHI 442
ENGLISH
INDEXED: INS, JICST

REPORT OF IONOSPHERE AND SPACE RESEARCH IN JAPAN 1110
NIHON GAKUJUTSU KAIGI DENRISO KENKYU RENRAKU SHOIINKAI (INSTITUTE OF SPACE AND AERONAUTICAL SCIENCE, UNIVERSITY OF TOKYO)
6-1, 4-CHOME, KOMABA, MEGURO-KU, TOKYO 153
ENGLISH
INDEXED: APP, CHE, INS, JICST

RESULTS OF MARINE METEOROLOGICAL AND OCEANOGRAPHICAL OBSERVATIONS 1111
KISHO-CHO (JAPAN METEOROLOGICAL AGENCY)
3-4, 1-CHOME, OTE-MACHI, CHIYODA-KU, TOKYO 100
JAPANESE AND ENGLISH
INDEXED: APP

RYUKYU KISHO CHOSA HOKOKU 1112
(RYUKYUAN METEOROLOGICAL SURVEY REPORT)
RYUKYU SEIFU KISHO CHO (RYUKYU GOVERNMENT METEOROLOGICAL BUREAU)
JAPANESE

SAGA-KEN KISHO GEPPO 1113
(MONTHLY REPORT OF METEOROLOGY, SAGA PREFECTURE)
SAGA CHIHO KISHODAI (SAGA LOCAL METEOROLOGICAL OBSERVATORY)
8-14, 2-CHOME, JONAI, SAGA-SHI 840
JAPANESE

SAITAMA-KEN KISHO GEPPO 1114
(MONTHLY REPORT OF METEOROLOGY, SAITAMA PREFECTURE)
KUMAGAYA CHIHO KISHODAI (KUMAGAYA LOCAL METEOROLOGICAL OBSERVATORY)
1192-4, KUMAGAYA, KUMAGAYA-SHI 360
JAPANESE

SAITAMA-KEN KISHO NENPO 1115
(ANNUAL REPORT OF METEOROLOGY, SAITAMA PREFECTURE)
KUMAGAYA CHIHO KISHODAI (KUMAGAYA LOCAL METEOROLOGICAL OBSERVATORY)
1192-4, KUMAGAYA, KUMAGAYA-SHI 360
JAPANESE

SEISAN KISHO RYUSU 1116
(SEISAN (P.N.) WEATHER NEWS)
KAGAWA-KEN BOSAI KISHO RENRAKUKAI SEISAN SHIBU (KAGAWA PREFECTURE DISASTER PREVENTION WEATHER LIAISON ASSOCIATION, SEISAN (P.N.) BRANCH)
JAPANESE

SENDAI KANKU GIJUTSU SHIRYO 1117
(TECHNICAL DATA OF SENDAI DISTRICT METEOROLOGICAL OBSERVATORY)
SENDAI KANKU KISHODAI (SENDAI DISTRICT METEOROLOGICAL OBSERVATORY)
1, TEPPO-MACHI, SENDAI-SHI 983
JAPANESE

SEPPYO 1118
(JOURNAL OF THE JAPANESE SOCIETY OF SNOW AND ICE)
NIHO SEPPYO GAKKAI (JAPANESE SOCIETY OF SNOW AND ICE)
C/O KANKYO JOHO KAGAKU SENTA, SOGO DAI-3 BLDG., 1-6, KOJI MACHI, CHIYODA-KU, TOKYO 102
JAPANESE, WITH EUROPEAN SUMMARY
INDEXED: JICST

SHIGA-KEN KISHO GEPPO 1119
(MONTHLY REPORT OF METEOROLOGY, SHIGA PREFECTURE)
HIKONE CHIHO KISHODAI (HIKONE LOCAL METEOROLOGICAL OBSERVATORY)
5-25, 2-CHOME, SHIRO-MACHI, HIKONE-SHI 522
JAPANESE

SHIGA-KEN KISHO NENPO 1120
(ANNUAL REPORT OF METEOROLOGY, SHIGA PREFECTURE)
HIKONE CHIHO KISHODAI (HIKONE LOCAL METEOROLOGICAL OBSERVATORY)
5-25, 2-CHOME, SHIRO-MACHI, HIKONE-SHI 522
JAPANESE

SHIKOKU URYO GEPPO 1121
(MONTHLY PRECIPITATION REPORT IN SHIKOKU DISTRICT)
TAKAMATSU CHIHO KISHODAI (TAKAMATSU LOCAL METEOROLOGICAL OBSERVATORY)
1277-1, FUSEISHI-CHO, TAKAMATSU-SHI 760
JAPANESE

SHIMANE-KEN KISHO NENPO 1122
(ANNUAL REPORT OF METEOROLOGY, SHIMANE PREFECTURE)
MATSUE CHIHO KISHODAI (MATSUE LOCAL METEOROLOGICAL OBSERVATORY)
1282, NISHI TSUDA-MACHI, MATSUE-SHI 690
JAPANESE

SHIMANE-KEN NOGYO KISHO GEPPO 1123
(MONTHLY REPORT OF AGRICULTURAL METEOROLOGY, SHIMANE PREFECTURE)
MATSUE CHIHO KISHODAI (MATSUE LOCAL METEOROLOGICAL OBSERVATORY)
1282, NISHI TSUDA-MACHI, MATSUE-SHI 690
JAPANESE

SHIZUOKA-KEN KISHO GEPPO 1124
(MONTHLY REPORT OF METEOROLOGY, SHIZUOKA PREFECTURE)
SHIZUOKA CHIHO KISHODAI (SHIZUOKA LOCAL METEOROLOGICAL OBSERVATORY)
1-5, 2-CHOME, MAGARIGANE, SHIZUOKA-SHI 420
JAPANESE

SOKKO JIHO 1125
(WEATHER SERVICE BULLETIN)
KISHO-CHO (METEOROLOGICAL AGENCY)
3-4, 1-CHOME, OTE-MACHI, CHIYODA-KU, TOKYO 100
JAPANESE

SOYA CHIHO NOGYO KISHO GEPPO 1126
(MONTHLY REPORT OF AGRICULTURAL METEOROLOGY, SOYA DISTRICT METEOROLOGICAL OBSERVATORY)
WAKKANAI CHIHO KISHODAI (WAKKANI LOCAL METEOROLOGICAL OBSERVATORY)
C/O WAKKANAI KOWAN GODOCHOSHA, KAIUN-CHO, WAKKANAI-SHI 097
JAPANESE

SUIGAI TOKEI 1127
(STATISTICS OF FLOOD)
KENSETSU-SHO KASEN-KYOKU (RIVER BUREAU, MINISTRY OF CONSTRUCTION)
1-3, 2-CHOME, KASUMIGASEKI, CHIYODA-KU, TOKYO 100
JAPANESE

SUIMON KISHO 1128
(HYDROMETEOROLOGY)
OITA-KEN SUIMON KISHO RENRAKU KYOGIKAI (THE OITA PREFECTURE HYDROMETEOROLOGICAL LIAISON CONFERENCE)
C/O OITA-KEN MIZUSHIGEN TAISAKU JIMUKYOKU, 1, 3-CHOME, OTE-MACHI, OITA-SHI 870
JAPANESE
INDEXED: JICST

TADOTSU KISHO NENPO 1129
(ANNUAL REPORT OF METEOROLOGY, TADOTSU WEATHER STATION)
TAKAMATSU CHIHO KISHODAI, TADOTSU SOKKOSHO
251, HIGASHI SHIN-MACHI, TADOTSU-CHO, TADO-GUN, KAGAWA 764
JAPANESE

TAIFU KEIROZU 1130
(TRAJECTORIES OF TROPICAL CYCLONES)
KISHO-CHO (METEOROLOGICAL AGENCY)
3-4, 1-CHOME, OTE-MACHI, CHIYODA-KU, TOKYO 100
JAPANESE

TAIKI 1131
(ATMOSPHERE)
KISHO KENKYUSHO (METEOROLOGICAL RESEARCH INSTITUTE, METEOROLOGICAL AGENCY)
35-8, 4-CHOME, KOENJI KITA, SUGINAMI-KU, TOKYO 166
JAPANESE

TENKI 1132
(WEATHER, METEOROLOGICAL SOCIETY OF JAPAN)
NIHON KISHO GAKKAI (METEOROLOGICAL SOCIETY OF JAPAN)
C/O KISHO-CHO, 3-4, 1-CHOME, OTE-MACHI, CHIYODA-KU, TOKYO 100
JAPANESE
INDEXED: JICST

TENKI 1133
(WEATHER)
TENKI HAKKOSHO (TENKI PUBLISHERS)
JAPANESE

TENKIZU 1134
(DAILY WEATHER MAPS)
KISHO-CHO (METEOROLOGICAL AGENCY)
3-4, 1-CHOME, OTE-MACHI, CHIYODA-KU, TOKYO 100
JAPANESE AND ENGLISH

TOCHIGI-KEN KISHO GEPPO 1135
(MONTHLY REPORT OF METEOROLOGY, TOCHIGI PREFECTURE)
UTSUNOMIYA CHIHO KISHODAI (UTSUNOMIYA LOCAL METEOROLOGICAL OBSERVATORY)
1-7, AKEBONO-CHO, UTSUNOMIYA-SHI 320
JAPANESE

TOHOKU CHIHO CHOKI YOHO SOKUHO 1136
(BULLETIN OF LONG-RANGE WEATHER FORECASTING OF TOHOKU DISTRICT)
SENDAI KANKU KISHODAI (SENDAI DISTRICT METEOROLOGICAL OBSERVATORY)
1, TEPPO-MACHI, SENDAI-SHI 983
JAPANESE

TOHOKU CHIHO KISHO KENKYUKAI, NIHON KISHO GAKKAI TOHOKU SHIBU KOENKAI SHIRYO 1137
(BULLETIN OF TOHOKU DISTRICT METEOROLOGICAL RESEARCH GROUP)
SENDAI KANKU KISHODAI (SENDAI DISTRICT METEOROLOGICAL OBSERVATORY)
1, TEPPO-MACHI, SENDAI-SHI 983
JAPANESE

TOHOKU CHIHO KISHO KENKYUKAISHI 1138
(JOURNAL OF METEOROLOGICAL RESEARCH OF TOHOKU DISTRICT)
SENDAI KANKU KISHODAI (SENDAI DISTRICT METEOROLOGICAL OBSERVATORY)
1, TEPPO-MACHI, SENDAI-SHI 983
JAPANESE AND ENGLISH
INDEXED: JICST

TOHOKU CHIHO RAIU GAIHO 1139
(REPORT OF THE THUNDERSTORM, TOHOKU DISTRICT)
SENDAI KANKU KISHODAI (SENDAI DISTRICT METEOROLOGICAL OBSERVATORY)
1, TEPPO-MACHI, SENDAI-SHI 983
JAPANESE

TOHOKU GIJUTSU DAYORI 1140
(TECHNICAL NEWS OF TOHOKU DISTRICT METEOROLOGICAL OBSERVATORY)
SENDAI KANKU KISHODAI
1, TEPPO-MACHI, SENDAI-SHI 983
JAPANESE

TOHOKU NO NOGYO KISHO 1141
(BULLETIN OF THE AGRICULTURAL METEOROLOGY OF TOHOKU DISTRICT OF JAPAN)
NIHON NOGYO KISHO GAKKAI TOHOKU SHIBU (THE TOHOKU BRANCH OF THE SOCIETY OF AGRICULTURAL METEOROLOGY OF JAPAN)
C/O NORIN-SHO TOHOKU NOGYO SHIKENJO, 4, SHIMOKURIYAGAWA AKAHIRA, MORIOKA-SHI 020-01
JAPANESE

TOKACHI KISHO GEPPO 1142
(MONTHLY REPORT OF METEOROLOGY, TOKACHI)
NIHON KISHO KYOKAI OBIHIRO SHIBU (THE OBIHIRO BRANCH, JAPAN METEOROLOGICAL ASSOCIATION)
9-CHOME, HIGASHI-5-JO, OBIHIRO-SHI 080
JAPANESE

TOKAN GIJUTSU NYUSU 1143
(TECHNICAL INFORMATION NEWS OF THE TOKYO DISTRICT METEOROLOGICAL OBSERVATORY)
TOKYO KANKU KISHODAI
3-4, 1-CHOME, OTE-MACHI, CHIYODA-KU, TOKYO 100
JAPANESE

TOKUSHIMA-KEN KISHO GEPPO 1144
(MONTHLY REPORT OF METEOROLOGY, TOKUSHIMA PREFECTURE)
TOKUSHIMA CHIHO KISHODAI (TOKUSHIMA LOCAL METEOROLOGICAL OBSERVATORY)
3-36, 2-CHOME, YAMATO-CHO, TOKUSHIMA-SHI 770
JAPANESE

TOKUSHIMA-KEN KISHO NENPO 1145
(ANNUAL REPORT OF METEOROLOGY, TOKUSHIMA PREFECTURE)
TOKUSHIMA CHIHO KISHODAI (TOKUSHIMA LOCAL METEOROLOGICAL OBSERVATORY)
3-36, 2-CHOME, YAMATO-CHO, TOKUSHIMA-SHI 776
JAPANESE

TOKYO JOURNAL OF CLIMATOLOGY 1146
TOKYO KYOIKU DAIGAKU KIKOGAKU KYOSHITSU (LABORATORY OF CLIMATOLOGY, TOKYO UNIVERSITY OF EDUCATION)
29-1, 3-CHOME, OTSUKA, BUNKYO-KU, TOKYO 112
EUROPEAN

TOKYO KANKU CHIHO KISHO KENKYUKAISHI 1147
(GEOPHYSICAL NOTES OF THE TOKYO DISTRICT METEOROLOGICAL OBSERVATORY)
TOKYO KANKU KISHODAI
3-4, 1-CHOME, OTE-MACHI, CHIYODA-KU, TOKYO 100
JAPANESE

TOKYO-TO KISHO GEPPO 1148
(MONTHLY REPORT OF METEOROLOGY, TOKYO)
TOKYO KANKU KISHODAI (TOKYO DISTRICT METEOROLOGICAL OBSERVATORY)
3-4, 1-CHOME, OTE-MACHI, CHIYODA-KU, TOKYO 100
JAPANESE

TOKYO-TO KISHO NENPO 1149
(ANNUAL REPORT OF METEOROLOGY, TOKYO)
TOKYO KANKU KISHODAI (TOKYO DISTRICT METEOROLOGICAL OBSERVATORY)
3-4, 1-CHOME, OTE-MACHI, CHIYODA-KU, TOKYO 100
JAPANESE

TOTTORI-KEN KISHO GEPPO 1150
(MONTHLY REPORT OF METEOROLOGY, TOTTORI PREFECTURE)
TOTTORI CHIHO KISHODAI (TOTTORI LOCAL METEOROLOGICAL OBSERVATORY)
1640, KOYAMA-CHO, TOTTORI-SHI 680
JAPANESE

TOTTORI-KEN KISHO NENPO 1151
(ANNUAL REPORT OF METEOROLOGY, TOTTORI PREFECTURE)
TOTTORI CHIHO KISHODAI (TOTTORI LOCAL METEOROLOGICAL OBSERVATORY)
1640, KOYAMA-CHO, TOTTORI-SHI 680
JAPANESE

TOTTORI-KEN NOGYO KISHO HAN NENPO 1152
(SEMI-ANNUAL AGRICULTURAL WEATHER REPORT FOR TOTTORI PREFECTURE)
TOTTORI-KEN, TOTTORI CHIHO KISHODAI (TOTTORI PREFECTURE, TOTTORI REGIONAL WEATHER STATION)
JAPANESE

TOTTORI-KEN NOGYO KISHO SHUHO 1153
(WEEKLY REPORT OF AGRICULTURAL METEOROLOGY, TOTTORI PREFECTURE)
TOTTORI CHIHO KISHODAI (TOTTORI LOCAL METEOROLOGICAL OBSERVATORY)
1640, KOYAMA-CHO, TOTTORI-SHI 680
JAPANESE

TOYAMA-KEN NOGYO KISHO GEPPO 1154
(MONTHLY REPORT OF AGRICULTURAL METEOROLOGY, TOYAMA PREFECTURE)
TOYAMA CHIHO KISHODAI (TOYAMA LOCAL METEOROLOGICAL OBSERVATORY)
2415, ISHIZAKA, TOYAMA-SHI 930
JAPANESE

TOYAMA-KEN NOGYO KISHO NENPO 1155
(ANNUAL REPORT OF AGRICULTURAL METEOROLOGY, TOYAMA PREFECTURE)
TOYAMA CHIHO KISHODAI (TOYAMA LOCAL METEOROLOGICAL OBSERVATORY)
2415, ISHIZAKA, TOYAMA-SHI 930
JAPANESE

UMI NO KISHO 1156
(MARINE METEOROLOGY)
KAIYO KISHO GAKKAI (OCEANOGRAPHICAL AND METEOROLOGICAL SOCIETY)
C/O KOBE KAIYO KOSHODAI, 7 NAKAYAMATE-DORI, IKUTA-KU, KOBE-SHI 650
JAPANESE

UMI TO SORA 1157
(SEA AND SKY)
KAIYO KISHO GAKKAI (OCEANOGRAPHICAL AND METEOROLOGICAL SOCIETY)
C/O KOBE KAIYO KISHODAI, 7 NAKAYAMATE-DORI, IKUTA-KU, KOBE-SHI 650
JAPANESE AND ENGLISH
INDEXED: POL, JICST

URYO NENPYO (KENSETSU-SHO) 1158
(RAINFALL STATISTICS)
KENSETSU-SHO KASEN-KYOKU (RIVER BUREAU, MINISTRY OF CONSTRUCTION) AND NIHON KASEN KYOKAI
C/O ZENKOKU CHOSON KAIKAN, 11-35, 1-CHOME, NAGATA-CHO, CHIYODA-KU, TOKYO 100
JAPANESE

URYO NENPYO (TOKYO-TO) 1159
(RAINFALL STATISTICS, TOKYO)
TOKYO-TO KENSETSU-KYOKU KASEN-BU (RIVER DIVISION, CONSTRUCTION BUREAU, TOKYO METROPOLITAN GOVERNMENT)
5-1, 3-CHOME, MARUNOUCHI, CHIYODA-KU, TOKYO 100
JAPANESE

UZU 1160
(VORTEX)
KISHO-CHO YOHO-BU (FORECAST DEPARTMENT, METEOROLOGICAL AGENCY)
3-4, 1-CHOME, OTE-MACHI, CHIYODA-KU, TOKYO 100
JAPANESE

WAKAYAMA-KEN KISHO GEPPO 1161
(MONTHLY REPORT OF METEOROLOGY, WAKAYAMA PREFECTURE)
WAKAYAMA CHIHO KISHODAI (WAKAYAMA LOCAL METEOROLOGICAL OBSERVATORY)
4, ONOSHIBA-CHO, WAKAYAMA-SHI 640
JAPANESE

WAKAYAMA-KEN KISHO NENPO 1162
(ANNUAL REPORT OF METEOROLOGY, WAKAYAMA PREFECTURE)
WAKAYAMA CHIHO KISHODAI (WAKAYAMA LOCAL METEOROLOGICAL OBSERVATORY)
4, ONOSHIBA-CHO, WAKAYAMA-SHI 640
JAPANESE

WMO KAIYO KIKO GAIYO 1163
(MARINE METEOROLOGICAL OBSERVATION REPORT RESPOND TO WMO ADVICE)
KISHO-CHO (METEOROLOGICAL AGENCY)
3-4, 1-CHOME, OTE-MACHI, CHIYODA-KU, TOKYO 100
JAPANESE

YAMAGATA-KEN KISHO GEPPO 1164
(MONTHLY REPORT OF METEOROLOGY, YAMAGATA PREFECTURE)
YAMAGATA CHIHO KISHODAI (YAMAGATA LOCAL METEOROLOGICAL OBSERVATORY)
5-77, 1-CHOME, MIDORI-MACHI, YAMAGATA-SHI 990
JAPANESE

YAMAGATA-KEN SAII NENPYO 1165
(CHRONOLOGICAL TABLE OF NATURAL CALAMITY IN YAMAGATA PREFECTURE)
YAMAGATA CHIHO KISHODAI (YAMAGATA LOCAL METEOROLOGICAL OBSERVATORY)
5-77, 1-CHOME, MIDORI-MACHI, YAMAGATA-SHI 990
JAPANESE

YAMAGUCHI CHIHO CHOSEKIHYO 1166
(TIDE TABLES, YAMAGUCHI PREFECTURE)
YAMAGUCHI SOKKOSHO (YAMAGUCHI METEOROLOGICAL OBSERVATORY)
2-1, SUFU-MACHI, YAMAGUCHI-SHI 753
JAPANESE

YAMAGUCHI-KEN KISHO GEPPO 1167
(MONTHLY REPORT OF METEOROLOGY, YAMAGUCHI PREFECTURE)
SHIMONOSEKI CHIHO KISHODAI (SHIMONOSEKI LOCAL METEOROLOGICAL OBSERVATORY)
15-19, MEICHI-CHO, SHIMONOSEKI-SHI 751
JAPANESE

YAMANASHI-KEN KISHO GEPPO 1168
(MONTHLY REPORT OF METEOROLOGY, YAMANASHI PREFECTURE)
KOFU CHIHO KISHODAI (KOFU LOCAL METEOROLOGICAL OBSERVATORY)
7-29, 4-CHOME, IIDA-CHO, KOFU-SHI 400
JAPANESE

YAMANASHI-KEN KISHO NENPO 1169
(ANNUAL REPORT OF METEOROLOGY, YAMANASHI PREFECTURE)
KOFU CHIHO KISHODAI (KOFU LOCAL METEOROLOGICAL OBSERVATORY)
7-29, 4-CHOME, IIDA-CHO, KOFU-SHI 400
JAPANESE

YAMANASHI-KEN NO KISHO 1170
(METEOROLOGICAL DATA OF YAMANASHI PREFECTURE)
KOFU CHIHO KISHODAI (KOFU LOCAL METEOROLOGICAL OBSERVATORY)
7-29, 4-CHOME, IIDA-CHO, KOFU-SHI 400
JAPANESE

ZENKOKU IJO KISHO GAIKYO 1171
(UNUSUAL METEOROLOGICAL REPORT, ALL JAPAN)
KISHO-CHO YOHO-BU (FORECAST DEPARTMENT, METEOROLOGICAL AGENCY)
3-4, 1-CHOME, OTE-MACHI, CHIYODA-KU, TOKYO 100
JAPANESE

ZENKOKU KAIKYO JUNPO 1172
(TEN DAY MARINE REPORT)
KISHO-CHO (METEOROLOGICAL AGENCY)
3-4, 1-CHOME, OTE-MACHI, CHIYODA-KU, TOKYO 100
JAPANESE

Biosciences (Including Paleontology)

ADVANCES IN BIOPHYSICS 1173
BIOPHYSICAL SOCIETY OF JAPAN AND UNIVERSITY OF TOKYO PRESS
ENGLISH
INDEXED: BIO, CHE, IND

AMINO ACID AND NUCLEIC ACID 1174
AMINOSAN KAKUSAN SHUDANKAI (ASSOCIATION OF AMINO ACID & NUCLEIC ACID) AND KYOWA HAKKO KOGYO K.K. TOKYO KENKYUSHO (TOKYO RESEARCH LABORATORY KYOWA HAKKO)
6-6, 3-CHOME, ASAHI-MACHI, MACHIDA-SHI, TOKYO 194
JAPANESE AND ENGLISH
INDEXED: BIO, CHE

ANNUAL REPORT OF BIOLOGICAL WORKS, FACULTY OF SCIENCE, OSAKA UNIVERSITY 1175
JAPANESE
INDEXED: JICST

ANNUAL REPORT, MITSUBISHI KASEI SEIMEI KAGAKU KENKYUSHO 1176
11, MINAMIOTANI, MACHIDA-SHI, TOKYO 194
ENGLISH

ANNUAL REPORT OF THE NATIONAL INSTITUTE OF GENETICS, JAPAN 1177
MONBU-SHO KOKURITSU IDENGAKU KENKYUSHO (NATIONAL INSTITUTE OF GENETICS, JAPAN)
1111, YATA, MISHIMA-SHI 411
ENGLISH
INDEXED: BIB, BIO

ANNUAL REPORT OF THE SADO MARINE BIOLOGICAL STATION, NIIGATA UNIVERSITY 1178
NIIGATA DAIGAKU RIGAKUBU (FACULTY OF SCIENCE, NIIGATA UNIVERSITY)
2-8050, IGARASHI, NIIGATA-SHI 950-21
ENGLISH
INDEXED: BIO

AOMORI-KEN SEIBUTSU GAKKAISHI 1179
(JOURNAL OF AOMORI-KEN BIOLOGICAL SOCIETY)
AOMORI-KEN SEIBUTSU GAKKAI (THE AOMORI-KEN BIOLOGICAL SOCIETY)
C/O HIROSAKI DAIGAKU RIGAKUBU SEIBUTSUGAKU KYOSHITSU, 3, BUNKYO-CHO, HIROSAKI-SHI 036
JAPANESE

ASHI NO HAZUE 1180
(NEWS OF THE CIRCLE INTERESTING IN LIVING THINGS)
IKIMONO SHUMI NO KAI (THE CIRCLE INTERESTING IN LIVING THINGS)
MR. Y. TSUTSUI, 9-7, 4-CHOME, SENRIYAMANISHI, SUITA-SHI 565
JAPANESE

BAIOTORONIKUSU 1181
(BIOTRONICS)
KYUSHU DAIGAKU SEIBUTSU KANKYO CHOSETSU SENTA (BIOTRON INSTITUTE, KYUSHU UNIVERSITY)
HAKOZAKI, HIGASHI-KU, FUKUOKA-SHI 812
JAPANESE AND ENGLISH

BIOLOGICAL JOURNAL OF OKAYAMA UNIVERSITY 1182
OKAYAMA DAIGAKU RIGAKUBU SEIBUTSUGAKU KYOSHITSU (DEPARTMENT OF BIOLOGY, FACULTY OF SCIENCE, OKAYAMA UNIVERSITY)
TSUSHIMA, OKAYAMA-SHI 700
ENGLISH
INDEXED: BIB, BIO, CHE, JICST

BLAKISTON 1183
AOMORI-KENRITSU HIROSAKI KOTOGAKKO SEIBUTSU-CLUB HIROSAKI
JAPANESE
INDEXED: CHE

BULLETIN OF THE MARINE BIOLOGICAL STATION OF ASAMUSHI, TOHOKU UNIVERSITY 1184
TOHOKU DAIGAKU RIGAKUBU ASAMUSHI RINKAI JIKKENSHO (MARINE BIOLOGICAL STATION, FACULTY OF SCIENCE, TOHOKU UNIVERSITY)
9, SAKAMOTO, ASAMUSHI, AOMORI-SHI 039-34
ENGLISH
INDEXED: BIO, CHE, JICST

CALANUS 1185
KUMAMOTO DAIGAKU RIGAKUBU FUZOKU AITSU RINKAI JIKKENSHO (AITSU MARINE BIOLOGICAL STATION, KUMAMOTO UNIVERSITY)
AITSU, MATSUSHIMA-MACHI, AMAKUSA-GUN, KUMAMOTO 861-61
JAPANESE AND ENGLISH

CHIBA SEIBUTSUSHI 1186
(JOURNAL OF THE BIOLOGICAL SOCIETY OF CHIBA PREFECTURE)
CHIBA-KEN SEIBUTSU GAKKAI (BIOLOGICAL SOCIETY OF CHIBA PREFECTURE)
C/O CHIBA DAIGAKU RIGAKUBU SEIBUTSU GAKKA, YAYOI-CHO, CHIBA-SHI 274
JAPANESE, WITH ENGLISH SUMMARY

CHIKUHO HAKUBUTSU 1187
(NATURHISTORICA CHIKUHOANA)
CHIKUHO HAKUBUTSU KENKYUKAI (THE SOCIETY OF NATURAL HISTORY OF CHIKUHO)
C/O FUKUOKA-SHIRITSU TOSHOKAN, 2-58, NISHI-CHO, IZUKA-SHI 820
JAPANESE

CIS: CHROMOSOME INFORMATION SERVICE 1188
SENSHOKUTAI GAKKAI (SOCIETY OF CHROMOSOME RESEARCH)
C/O KOKUSAI KIRISTOKYO DAIGAKU, 10-2, 3-CHOME, OSAWA, MITAKA-SHI, TOKYO 181
ENGLISH
INDEXED: BIB, BIO, JICST

CONTRIBUTIONS FROM THE INSTITUTE OF LOW TEMPERATURE SCIENCE, SERIES B, BIOLOGY 1189
HOKKAIDO DAIGAKU TEION KAGAKU KENKYUSHO (INSTITUTE OF LOW TEMPERATURE SCIENCE, HOKKAIDO UNIVERSITY)
NISHI-8-CHOME, KITA-19-JO, KITA-KU, SAPPORO-SHI 060
ENGLISH
INDEXED: BIO, CHE, JICST

CONTRIBUTIONS FROM THE MISAKI MARINE BIOLOGICAL STATION 1190
TOKYO DAIGAKU RIGAKUBU FUZOKU RINKAI JIKKENSHO (MARINE BIOLOGICAL STATION, FACULTY OF SCIENCE, UNIVERSITY OF TOKYO)
JAPANESE AND ENGLISH

CONTRIBUTIONS FROM OTSU HYDROBIOLOGICAL STATION, UNIVERSITY OF KYOTO 1191
KYOTO DAIGAKU RIGAKUBU FUZOKU OTSU RINKO JIKKENSHO (OTSU HYDROBIOLOGICAL STATION, UNIVERSITY OF KYOTO)
SHIMOSAKA HON-MACHI, OTSU-SHI 520-01
ENGLISH
INDEXED: BIO

CYTOLOGIA 1192
KOKUSAI SAIBO GAKKAI (INTERNATIONAL ASSOCIATION OF CYTOLOGY)
C/O TOYO BUNKO, 28-21, 2-CHOME, HONKOMAGOME, BUNKYO-KU, TOKYO 113
ENGLISH
INDEXED: BIB, BIO, CHE, JICST

DEVELOPMENT, GROWTH AND DIFFERENTIATION 1193
NIHON HASSEI SEIBUTSU GAKKAI (JAPANESE SOCIETY OF DEVELOPMENTAL BIOLOGISTS)
C/O NAGOYA DAIGAKU RIGAKUBU SEIBUTSUGAKU KYOSHITSU, FURO-CHO, CHIKUSA-KU, NAGOYA-SHI 464
ENGLISH
INDEXED: BIB, BIO, CHE, EXC, JICST

EBINO KOGEN YAGAI SEIBUTSU JIKKENSHITSU KENKYU GYOSEKI 1194
(REPORTS FROM THE EBINO BIOLOGICAL LABORATORY, KYUSHU UNIVERSITY)
KYUSHU DAIGAKU RIGAKUBU SEIBUTSUGAKU KYOSHITSU (KYUSHU UNIVERSITY DEPARTMENT OF SCIENCE, BIOLOGY SECTION)
JAPANESE

EHIME DAIGAKU KIYO, DAI-2-BU SHIZENKAGAKU, B-SHIRIZU, SEIBUTSUGAKU 1195
(MEMOIRS OF THE EHIME UNIVERSITY, SECTION 2, NATURAL SCIENCE, SERIES B, BIOLOGY)
EHIME DAIGAKU (EHIME UNIVERSITY)
3, BUNKYO-CHO, MATSUYAMA-SHI 790
JAPANESE AND ENGLISH

GEKKAN SAIBO 1196
(CELL)
NYU SAIENSUSHA
3-21, KANDA NISHIKI-CHO, CHIYODA-KU, TOKYO 101
JAPANESE
INDEXED: JICST

GEKKAN SOSHIKI BAIYO 1197
NYU SAIENSUSHA (NEW SCIENCE PUBLISHING COMPANY)
JAPANESE

GUNMA SEIBUTSU 1198
(BULLETIN OF GUNMA BIOLOGICAL EDUCATION SOCIETY)
GUNMA-KEN KOTO GAKKO KYOIKU KENKYU KAI SEIBUTSU BU KAI (GUNMA PREFECTURE, HIGH SCHOOL EDUCATIONAL RESEARCH ASSOCIATION, BIOLOGY DIVISION)
JAPANESE

HIROSAKI DAIGAKU RIGAKUBU FUZOKU FUKAURA RINKAI JISSHUSHO HOKOKU 1199
(REPORT OF THE FUKAURA MARINE BIOLOGICAL LABORATORY)
HIROSAKI DAIGAKU RIGAKUBU (FACULTY OF SCIENCE, HIROSAKI UNIVERSITY)
3, BUKYO-CHO, HIROSAKI-SHI 036
JAPANESE AND ENGLISH

HIROSHIMA DAIGAKU SEIBUTSU GAKKAISHI 1200
(BULLETIN OF THE BIOLOGICAL SOCIETY OF HIROSHIMA UNIVERSITY)
HIROSHIMA DAIGAKU SEIBUTSU GAKKAI (BIOLOGICAL SOCIETY OF HIROSHIMA UNIVERSITY)
C/O HIROSHIMA DAIGAKU RIGAKUBU, HIGASHI SENDA-CHO, HIROSHIMA-SHI 730
JAPANESE, WITH ENGLISH SUMMARY

IDEN 1201
(HEREDITY)
SHOKABO
8-1, YONBAN-CHO, CHIYODA-KU, TOKYO 102
JAPANESE
INDEXED: BIB, JICST

IDENGAKU ZASSHI 1202
(JAPANESE JOURNAL OF GENETICS)
NIHON IDEN GAKKAI (GENETICS SOCIETY OF JAPAN)
C/O KOKURITSU IDENGAKU KENKYUSHO, 1111, YATA, MISHIMA-SHI 411
JAPANESE AND EUROPEAN
INDEXED: BIB, BIO, CHE, EXC, JICST

JARE SCIENTIFIC REPORTS, SERIES E, BIOLOGY 1203
KOKURITSU KAGAKU HAKUBUTSUKAN KYOKUCHI KENKYU SENTA (POLAR RESEARCH CENTER, NATIONAL SCIENCE MUSEUM)
9-10, 1-CHOME, KAGA, ITABASHI-KU, TOKYO 173
EUROPEAN
INDEXED: BIO

JIBP SYNTHESIS 1204
ENGLISH
INDEXED: BIB

JOURNAL OF BIOCHEMISTRY 1205
NIHON SEIKAGAKKAI (JAPANESE BIOCHEMICAL SOCIETY)
C/O GAKKAI CENTER BUILDING, 4-16, 2-CHOME, YAYOI, BUNKYO-KU, TOKYO 113
ENGLISH
INDEXED: BIB, BIO, CHE, EXC, IND, SCI, JICST

JOURNAL OF BIOLOGY, OSAKA CITY UNIVERSITY 1206
ENGLISH

JOURNAL OF THE FACULTY OF TEXTILE SCIENCE AND TECHNOLOGY, SHINSHU UNIVERSITY, SERIES A, BIOLOGY 1207
SHINSHU DAIGAKU SEN'IGAKUBU (FACULTY OF TEXTILE SCIENCE & TECHNOLOGY, SHINSHU UNIVERSITY)
15-1, 3-CHOME, TOKIDA, UEDA-SHI 386
JAPANESE AND ENGLISH
INDEXED: BIB, BIO, JICST

JOURNAL OF GENERAL AND APPLIED MICROBIOLOGY 1208
OYO BISEIBUTSUGAKU KENKYU SHOREIKAI (MICROBIOLOGY RESEARCH FOUNDATION)
C/O GAKKAI SENTA BUILDING, 4-16, 2-CHOME, YAYOI, BUNKYO-KU, TOKYO 113
ENGLISH
INDEXED: BIB, BIO, CHE, EXC, JICST

KAGAKU TO SEIBUTSU 1209
(CHEMISTRY AND BIOLOGY)
NIHON NOGEI KAGAKUKAI (AGRICULTURAL CHEMICAL SOCIETY OF JAPAN)
3-1, 7-CHOME, HONGO, BUNKYO-KU, TOKYO 113
JAPANESE
INDEXED: CHE, JICST

KANAZAWA DAIGAKU RIGAKUBU NOTO RINKAI JIKKENJO NENPO 1210
(ANNUAL REPORT OF THE NOTO MARINE LABORATORY, FACULTY OF SCIENCE, KANAZAWA UNIVERSITY)
KANAZAWA DAIGAKU RIGAKUBU FUZOKU NOTO RINKAI JIKKENJO (NOTO MARINE LABORATORY, FACULTY OF SCIENCE, UNIVERSITY OF KANAZAWA)
OGI, UCHIURA-CHO, SUZU-GUN ISHIKAWA 927-05
JAPANESE AND EUROPEAN

KASEKI 1211
(FOSSILS)
NIHON KOSEIBUTSU GAKKAI (PALAEONTOLOGICAL SOCIETY OF JAPAN)
C/O NIHON GAKKAI JIMU SENTA, 4-16, 2-CHOME, YAYOI, BUNKYO-KU, TOKYO 113
JAPANESE

KASEKI KENKYUKAI KAISHI 1212
(BULLETIN OF THE FOSSIL RESEARCH SOCIETY)
JAPANESE

KEITO SEIBUTSU 1213
(STRAIN BIOLOGY)
GAKKAI SENTA (ACADEMIC CENTER)
JAPANESE

KIHARA SEIBUTSUGAKU KENKYUJO SEIKEN JIHO 1214
(BIOLOGICAL RESEARCH REVIEW OF THE KIHARA BIOLOGICAL RESEARCH INSTITUTE)
JAPANESE

KISHU SEIBUTSU 1215
(KISHU BIOLOGY)
WAKAYAMA-KEN SEIBUTSU DOKOKAI (WAKAYAMA PREFECTURE BIOLOGY ASSOCIATION)
JAPANESE

KOGOSEI KANKEI BUNKEN SHOROKUSHU 1216
(PHOTOSYNTHESIS ABSTRACTS)
BUSSHITSU TAISHA KENKYUKAI
C/O TOKYO DAIGAKU RIGAKUBU SEIBUTSU KAGAKU KYOSHITSU, 3-1, 7-CHOME, HONGO, BUNKYO-KU, TOKYO 113
EUROPEAN

KOGYO GIJUTSU-IN BISEIBUTSU KOGYO GIJUTSU KENKYUJO KENKYU HOKOKU 1217
(RESEARCH REPORT OF THE AGENCY OF INDUSTRIAL SCIENCE AND TECHNOLOGY)
MICROBIOLOGICAL INDUSTRIAL TECHNICAL RESEARCH INSTITUTE
JAPANESE
INDEXED: JICST

KOKURITSU IDENGAKU KENKYUSHO NENPO 1218
(ANNUAL REPORT OF THE NATIONAL INSTITUTE OF GENETICS, JAPAN)
MONBU-SHO, KOKURITSU IDENGAKU KENKYUSHO (NATIONAL INSTITUTE OF GENETICS)
1111, YATA, MISHIMA-SHI 411
JAPANESE
INDEXED: JICST

KUMAMOTO DAIGAKU RIGAKUBU FUZOKU AITSU RINKAI JIKKENSHO GYOSEKISHU 1219
(AITSU MARINE BIOLOGICAL STATION, KUMAMOTO UNIVERSITY, COLLECTED REPRINTS)
KUMAMOTO DAIGAKU RIGAKUBU FUZOKU AITSU RINKAI JIKKENSHO (AITSU MARINE BIOLOGICAL STATION, KUMAMOTO UNIVERSITY)
AITSU, MATSUSHIMA-MACHI, AMAKUSA-GUN, KUMAMOTO 861-61
JAPANESE AND ENGLISH

KUMAMOTO JOURNAL OF SCIENCE, BIOLOGY 1220
KUMAMOTO DAIGAKU RIGAKUBU SEIBUTSUGAKU KYOSHITSU (DEPARTMENT OF BIOLOGY, FACULTY OF SCIENCE, KUMAMOTO UNIVERSITY)
39-1, 2-CHOME, KUROKAMI, KUMAMOTO-SHI 860
EUROPEAN
INDEXED: BIB, BIO, CHE, JICST

KUROKO 1221
(KUROKO)
HIRADOJIMA SEIBUTSU KENKYUKAI (SOCIETY FOR THE STUDY OF THE BIOLOGY, HIRADO ISLANDS)
C/O SHIJIKI SHOGAKKO, SHIJIKI-CHO, HIRADO-SHI 859-56
JAPANESE

KYOTO DAIGAKU RIGAKUBU OTSU RINKO JIKKENSHO HOBUN GYOSEKI 1222
(CONTRIBUTIONS FROM OTSU HYDROBIOLOGICAL STATION, UNIVERSITY OF KYOTO)
KYOTO DAIGAKU RIGAKUBU FUZOKU OTSU RINKO JIKKENSHO (OTSU HYDROBIOLOGICAL STATION, UNIVERSITY OF KYOTO)
SHIMOSAKA HON-MACHI, OTSU-SHI 520-01
JAPANESE, WITH ENGLISH SUMMARY

MEMOIRS OF THE FACULTY OF SCIENCE, KYOTO UNIVERSITY, SERIES OF BIOLOGY 1223
KYOTO DAIGAKU RIGAKUBU (FACULTY OF SCIENCE, KYOTO UNIVERSITY)
KITASHIRAKAWA OIWAKE-CHO, SAKYO-KU, KYOTO-SHI 606
EUROPEAN
INDEXED: BIB, BIO, JICST

MEMOIRS OF THE INSTITUTE FOR PROTEIN RESEARCH, OSAKA UNIVERSITY 1224
OSAKA DAIGAKU TANPAKUSHITSU KENKYUSHO (INSTITUTE FOR PROTEIN RESEARCH, OSAKA UNIVERSITY)
5311, YAMADAKAMI, SUITA-SHI 565
ENGLISH
INDEXED: CHE, JICST

MICHURIN SEIBUTSUGAKU KENKYU 1225
(JAPANESE JOURNAL OF MICHURIN BIOLOGY)
NIHON NOGYO SEIBUTSUGAKU KENKYUKAI
C/O TOKYO NOKO DAIGAKU IPPAN KYOIKUBU, SEIBUTSU KENKYUSHITSU, 5-8, 3-CHOME, SAIWAI-CHO, FUCHU-SHI, TOKYO 183
JAPANESE, WITH ENGLISH SUMMARY
INDEXED: CHE

MICROBIOASSAY 1226
ENGLISH

MIZUNAMI-SHI KASEKI HAKUBUTSUKAN KENKYU HOKOKU 1227
(BULLETIN OF THE MIZUNAMI FOSSIL MUSEUM)
MIZUNAMI FOSSIL MUSEUM
JAPANESE

MUKAISHIMA RINKAI JIKKENJO GYOSEKI 1228
(CONTRIBUTIONS FROM THE MUKAISHIMA MARINE BIOLOGICAL STATION)
HIROSHIMA DAIGAKU MUKAISHIMA RINKAI JIKKENJO (MUKAISHIMA MARINE BIOLOGICAL STATION, HIROSHIMA UNIVERSITY)
JAPANESE AND ENGLISH

NANKI SEIBUTSU 1229
(JOURNAL OF THE NANKI BIOLOGICAL SOCIETY)
NANKI SEIBUTSU DOKOKAI (THE NANKI BIOLOGICAL SOCIETY)
C/O KYOTO DAIGAKU SETO RINKAI JIKKENJO, SHIRAHAMA-CHO, NISHIMURO-GUN, WAKAYAMA 649-22
JAPANESE

NANKI SEIBUTSU DOKOKAI KAIHO 1230
(NEWS OF THE NANKI BIOLOGICAL SOCIETY)
NANKI SEIBUTSU DOKOKAI (THE NANKI BIOLOGICAL SOCIETY)
C/O KYOTO DAIGAKU SETO RINKAI JIKKENJO, SHIRAHAMA-CHO, NISHIMURO-GUN, WAKAYAMA 649-22
JAPANESE

NARA JOSHI DAIGAKU SEIBUTSU GAKKAISHI 1231
(BIOLOGICAL JOURNAL OF NARA WOMEN'S UNIVERSITY)
NARA JOSHI DAIGAKU SEIBUTSU GAKKAI (BIOLOGICAL SOCIETY OF NARA WOMEN'S UNIVERSITY)
C/O NARA JOSHI DAIGAKU, KITAUOYANISHIMACHI, NARA-SHI 630
JAPANESE, WITH EUROPEAN SUMMARY
INDEXED: BIO, CHE

NATURE STUDY 1232
OSAKA-SHIRITSU SHIZENSHI HAKUBUTSUKAN TOMONOKAI
C/O OSAKA-SHIRITSU SHIZENSHI HAKUBUTSUKAN, NAGAI KOEN, NAGAI-CHO, HIGASHISUMIYOSHI-KU, OSAKA-SHI 546
JAPANESE

NIBS BULLETIN OF BIOLOGICAL RESEARCH 1233
NIHON SEIBUTSU KAGAKU KENKYUSHO (JAPAN INSTITUTE FOR BIOLOGICAL SCIENCE)
28-19, 3-CHOME, AKEBONO-CHO, TACHIKAWA-SHI, TOKYO 190
ENGLISH
INDEXED: CHE

NIHON DAIGAKU NOJUIGAKUBU DOSHOKUBUTSU KENKYUKAIHO 1234
(REPORT OF THE SOCIETY FOR THE STUDY OF ANIMALS AND PLANTS)
NIHON DAIGAKU NOJUIGAKUBU DOSHOKUBUTSU KENKYUKAI (COLLEGE OF AGRICULTURE VETERINERY MEDICINE, NIHON UNIVERSITY SOCIETY FOR THE ANIMALS AND PLANTS)
C/O NIHON DAIGAKU NAJUI GAKUBU, 34-1, 3-CHOME, SHIMOUMA-CHO, SETAGAYA-KU, TOKYO 154
JAPANESE

NIHON KOSEIBUTSU GAKKAI HOKOKU KIJI 1235
(TRANSACTIONS AND PROCEEDINGS OF THE PALAEONTOLOGICAL SOCIETY OF JAPAN)
NIHON KOSEIBUTSU GAKKAI
C/O NIHON GAKKAI JIMU SENTA, 4-16, 2-CHOME, YAYOI, BUNKYO-KU, TOKYO 113
ENGLISH
INDEXED: BIO, JICST

NIHON SEIBUTSU CHIRI GAKKAI TSUSHIN 1236
(INFORMATION NEWS OF THE BIOGEOGRAPHICAL SOCIETY OF JAPAN)
NIHON SEIBUTSU CHIRIGAKKAI
C/O TOKYO SUISAN DAIGAKU SHIRYOKAN, 5-7, 4-CHOME, KONAN, MINATO-KU, TOKYO 108
JAPANESE

NIHON SEIBUTSU CHIRI GAKKAIHO 1237
(BULLETIN OF THE BIOGEOGRAPHICAL SOCIETY OF JAPAN)
NIHON SEIBUTSU CHIRI GAKKAI (THE BIOGEOGRAPHICAL SOCIETY OF JAPAN)
C/O TOKYO SUISAN DAIGAKU SHIRYOKAN, 5-7, 4-CHOME, KONAN, MINATO-KU, TOKYO 108
JAPANESE AND ENGLISH
INDEXED: BIO

NIHON SEITAI GAKKAISHI 1238
(JAPANESE JOURNAL OF ECOLOGY)
NIHON SEITAI GAKKAI (THE ECOLOGICAL SOCIETY OF JAPAN)
C/O TOHOKU DAIGAKU RIGAKUBU SEIBUTSUGAKU KYOSHITSU, AOBA, ARAMAKI, SENDAI-SHI 980
JAPANESE AND EUROPEAN
INDEXED: BIB, BIO, CHE

NIHON SHIZEN HOGO KYOKAI SHIRYO 1239
(RESEARCH MATERIALS OF THE NATURE CONSERVATION SOCIETY OF JAPAN)
NIHON SHIZEN HOGO KYOKAI
15, SHIBA NISHIKUBO AKEFUNE-CHO, MINATO-KU TOKYO 105
JAPANESE

NISSEIKEN DAYORI 1240
(JOURNAL OF THE NIPPON INSTITUTE OF BIOLOGICAL SCIENCES)
NIPPON SEIBUTSU KAGAKU KENKYUSHO (NIPPON INSTITUTE FOR BIOLOGICAL SCIENCE)
28-19, 3-CHOME, AKEBONO-CHO, TACHIKAWASHI, TOKYO 190
JAPANESE

OBIHIRO CHIKUSAN DAIGAKU SEIBUTSU KENKYUSHO KENKYU HOKOKU 1241
(BULLETIN OF THE BIOLOGICAL INSTITUTE, OBIHIRO ZOOTECHNICAL UNIVERSITY)
OBIHIRO CHIKUSAN DAIGAKU SEIBUTSU KENKYUSHO (BIOLOGICAL INSTITUTE, OBIHIRO ZOOTECHNICAL UNIVERSITY)
JAPANESE AND ENGLISH

OKAYAMA DAIGAKU TAMANO RINKAI JIKKENSHO GYOSEKI 1242
(CONTRIBUTIONS FROM THE TAMANO MARINE LBAORATORY, OKAYAMA UNIVERSITY)
OKAYAMA DAIGAKU RIGAKUBU FUZOKU RINKAI JIKKENSHO (TAMANO MARINE LABORATORY, OKAYAMA UNIVERSITY)
3-3, 2-CHOME, SHIBUKAWA, TAMANO-SHI 706
JAPANESE AND EUROPEAN

PUBLICATIONS FROM THE AKKESHI MARINE BIOLOGICAL STATION 1243
HOKKAIDO DAIGAKU RIGAKUBU FUZOKU RINKAI JIKKENSHO
AIKAPP, AKKESHI-CHO, AKKESHI-GUN, HOKKAIDO 088-12
EUROPEAN

PUBLICATIONS FROM THE AMAKUSA MARINE BIOLOGICAL LABORATORY, KYUSHU UNIVERSITY 1244
KYUSHU DAIGAKU RIGAKUBU FUZOKU AMAKUSA RINKAI JIKKENSHO (AMAKUSA MARINE BIOLOGICAL LABORATORY, KYUSHU UNIVERSITY)
2231, TOMIOKA, REIHOKU-CHO, AMAKUSA-GUN, KUMAMOTO 863-25
ENGLISH
INDEXED: BIO

PUBLICATIONS OF THE SETO MARINE BIOLOGICAL LABORATORY 1245
KYOTO DAIGAKU RIGAKUBU FUZOKU SETO RINKAI JIKKENSHO (SETO MARINE BIOLOGICAL LABORATORY, FACULTY OF SCIENCE, UNIVERSITY OF KYOTO)
SHIRAHAMA-CHO, NISHIMURO-GUN, WAKAYAMA 649-22
ENGLISH
INDEXED: BIO, CHE

REPORT FROM THE INSTITUTE OF APPLIED MICROBIOLOGY, UNIVERSITY OF TOKYO 1246
TOKYO DAIGAKU OYO BISEIBUTSU KENKYUSHO (INSTITUTE OF APPLIED MICROBIOLOGY, UNIVERSITY OF TOKYO)
1-1, 1-CHOME, YAYOI, BUNKYO-KU, TOKYO 113
EUROPEAN
INDEXED: BIB, CHE, JICST

REPORTS OF THE USA MARINE BIOLOGICAL STATION 1247
KOCHI DAIGAKU BUNRIGAKUBU FUZOKU USA RINKAI JIKKENSHO (USA MARINE BIOLOGICAL STATION, KOCHI UNIVERSITY)
JAPANESE AND ENGLISH
INDEXED: BIB, BIO

SAISHU TO SHIIKU 1248
(COLLECTING AND BREEDING)
SAISHU TO SHIIKU NO KAI (THE SOCIETY OF COLLECTING AND BREEDING)
C/O TOKYO NOGYO DAIGAKU IKUSHUGAKU KENKYUSHO, 4-28, 2-CHOME, KAMIYOGA, SETAGAYA-KU, TOKYO 158
JAPANESE AND ENGLISH
INDEXED: BIB

SAROBETSU SOGO CHOSA CHUKAN HOKOKUSHO. SEIBUTSU BUMON 1249
(INTERIM REPORT OF THE SAROBETSU GENERAL SURVEY. BIOLOGICAL SECTION)
JAPANESE
INDEXED: JICST

SCIENCE REPORTS OF NIIGATA UNIVERSITY, SERIES D, BIOLOGY 1250
NIIGATA DAIGAKU RIGAKUBU (FACULTY OF SCIENCE, NIIGATA UNIVERSITY)
8050, IGARASHININO-CHO, NIIGATA-SHI 950-21
ENGLISH
INDEXED: BIO, CHE, JICST

SCIENCE REPORTS OF THE SAITAMA UNIVERSITY, SERIES B: BIOLOGY AND EARTH SCIENCES 1251
SAITAMA DAIGAKU (SAITAMA UNIVERSITY)
255, SHIMOOKUBO, URAWA-SHI 338
ENGLISH
INDEXED: JICST

SCIENCE REPORTS OF THE TOHOKU UNIVERSITY (SERIES IV: BIOLOGY) 1252
TOHOKU DAIGAKU (TOHOKU UNIVERISTY)
1-1, 2-CHOME, KATAHIRA, SENDAI-SHI 980
ENGLISH
INDEXED: BIO, CHE, EXC, JICST

SCIENCE REPORTS OF THE TOKYO KYOIKU DAIGAKU, SECTION B, ZOOLOGY AND BOTANY 1253
TOKYO KYOIKU DAIGAKU (TOKYO UNIVERSITY OF EDUCATION)
29-1, 3-CHOME, OTSUKA, BUNKYO-KU, TOKYO 112
ENGLISH
INDEXED: BIO, CHE

SEIBUTSU BUTSURI 1254
(BIOPHYSICS)
NIHON SEIBUTSU BUTSURI GAKKAI (BIOPHYSICAL SOCIETY OF JAPAN)
87, TANAKA MONZEN-CHO, SAKYO-KU, KYOTO-SHI 606
JAPANESE
INDEXED: CHE, JICST

SEIBUTSU BUTSURI KAGAKU 1255
(PHYSICO-CHEMICAL BIOLOGY)
DENKI EIDO GAKKAI (SOCIETY OF ELECTROPHORESIS)
C/O TOKYO IDA SHIKA DAIGAKU KYOYOBU, 8-30, 2-CHOME, KOHODAI, ICHIKAWA-SHI 272
JAPANESE AND ENGLISH
INDEXED: BIO, CHE, JICST

SEIBUTSU FUKUOKA 1256
(BIOLOGIA FUKUOKA)
FUKUOKA-KEN KOTO GAKKO SEIBUTSU KENKYUBUKAI (BIOLOGICAL SOCIETY OF FUKUOKA-KEN HIGH SCHOOL TEACHERS)
C/O KENRITSU SHUKENKAN KOKO, 1-10, 6-CHOME, NISHIARAI, NISHI-KU, FUKUOKA-SHI 814
JAPANESE

SEIBUTSU KAGAKU 1257
(BIOLOGICAL SCIENCE)
NIHON SEIBUTSU KAGAKUSHA KYOKAI (JAPAN SOCIETY OF BIOLOGICAL SCIENTIST)
AND IWANAMI SHOTEN
2-3, HITOTSUBASHI, CHIYODA-KU, TOKYO 101
JAPANESE
INDEXED: BIB, CHE, JICST

SEIBUTSU KAGAKU NYUSU 1258
(BIOLOGICAL SCIENCE NEWS)
SEIBUTSU KAGAKU NYUSU HENSHU IINKAI
C/O GAKKAISHI KANKO SENTA, 4-16, 2-CHOME, YAYOI, BUNKYO-KU, TOKYO 113
JAPANESE

SEIBUTSU KANKYO CHOSETSU 1259
(ENVIRONMENT CONTROL IN BIOLOGY)
NIHON SEIBUTSU KENKYO CHOSETSU GAKKAI (JAPANESE SOCIETY OF ENVIRONMENT CONTROL IN BIOLOGY)
C/O KYUSHU DAIGAKU SEIBUTSU KANKYO CHOSETSU SENTA, HAKOZAKI, HIGASHI-KU, FUKUOKA-SHI 812
JAPANESE AND ENGLISH
INDEXED: BIB

SEIBUTSU KYOIKU 1260
(JOURNAL OF THE BIOLOGICAL EDUCATION)
NIHON SEIBUTSU KYOIKU GAKKAI (SOCIETY OF THE BIOLOGICAL EDUCATION OF JAPAN)
8-10, 4-CHOME, KOHINATA, BUNKYO-KU, TOKYO 112
JAPANESE

SEIBUTSUGAKUSHI KENKYU 1261
(JAPANESE JOURNAL OF THE HISTORY OF BIOLOGY)
NIHON KAGAKUSHI GAKKAI SEIBUTSUGAKUSHI BUNKAKAI (BIOLOGICAL HISTORY UNIT OF THE HISTORY OF SCIENCE SOCIETY OF JAPAN)
C/O TOKYO KOGYO DAIGAKU KOGAKUBU KAGAKU GAIRON KENKYUSHITSU, OOKAYAMA, MEGURO-KU, TOKYO 152
JAPANESE

SEIKAGAKU 1262
(JOURNAL OF JAPANESE BIOCHEMICAL SOCIETY)
NIHON SEIKAGAKKAI (JAPANESE BIOCHEMICAL SOCIETY)
GAKKAI SENTA BUILDING, 4-16, 2-CHOME, YAYOI, BUNKYO-KU, TOKYO 113
JAPANESE
INDEXED: BIO, CHE, IND, JICST

SEIKAGAKU JIKKEN KOZA 1263
(LECTURES ON EXPERIMENTS IN BIOCHEMISTRY)
NIHON SEIKAGAKUKAI
JAPANESE
INDEXED: BIB

SEIKEN JIHO 1264
(REPORT OF THE KIHARA INSTITUTE FOR BIOLOGICAL RESEARCH)
KIHARA SEIBUTSUGAKU KENKYUSHO
122-23, 3-CHOME, MUTSUKAWA-CHO, MINAMI-KU, YOKOHAMA-SHI 233
JAPANESE AND EUROPEAN
INDEXED: BIB, BIO, CHE, JICST

SEITAIGAKU KENKYU 1265
(ECOLOGICAL REVIEW)
TOHOKU DAIGAKU HAKKODASAN SHOKUBUTSU JIKKENSHO (MT. HAKKODA BOTANICAL LABORATORY)
C/O TOHOKU DAIGAKU RIGAKUBU SEIBUTSUGAKU KYOSHITSU, AOBA, ARAMAKI, SENDAI-SHI 980
EUROPEAN
INDEXED: BIB, BIO, JICST

SENSHOKUTAI 1266
(KROMOSOMO)
SENSHOKATAI GAKKAI (SOCIETY OF CHROMOSOME RESEARCH)
10-2, 3-CHOME, OSAWA, MITAKA-SHI, TOKYO 181
JAPANESE AND ENGLISH
INDEXED: BIB, BIO, CHE, JICST

SERIES OF STOCK CULTURE IN BIOLOGICAL FIELD 1267
KEIGAKU PUBLISHING COMPANY
ENGLISH
INDEXED: BIB

SHIGA SHIZEN KYOIKU KENKYU SHISETSU KENKYU GYOSEKI 1268
(BULLETIN OF INSTITUTE OF NATURAL EDUCATION IN SHIGA HEIGHTS)
SHINSHU DAIGAKU KYOIKUGAKUBU FUZOKU SHIGA SHIZEN KYOIKU KENKYU SHISETSU (INSTITUTE OF NATURAL EDUCATION IN SHIGA HEIGHTS, COLLEGE OF EDUCATION, SHINSHU UNIVERSITY)
7148, SHIGAKOGEN, HIRAO, YAMANOUCHI-MACHI, SHIMOTAKAI-GUN, NAGANO 381-04
JAPANESE, WITH EUROPEAN SUMMARY
INDEXED: BIO

SHIMA MARINRANDO KENKYU HOKOKU 1269
(RESEARCH REPORT OF SHIMA MARINELAND)
SHIMA MARINRANDO
JAPANESE

SHINSHU DAIGAKU YAGAI SEIBUTSU KENKYUKAI KAIHO 1270
(JOURNAL OF FIELD WORKS SOCIETY OF SHINSHU UNIVERSITY)
SHINSHU DAIGAKU NOGAKUBU YAGAI SEIBUTSU KENKYUKAI
8304, MINAMIMINOWA-MURA, KAMIINAGUN, NAGANO 399-45
JAPANESE

SIEBOLDIA ACTA BIOLOGICA 1271
KYUSHU DAIGAKU KYOYOBU SEIBUTSUGAKU KYOSHITSU (BIOLOGICAL LABORATORY, COLLEGE OF GENERAL EDUCATION, KYUSHU UNIVERSITY)
4-CHOME, ROPPONMATSU, CHUO-KU, FUKUOKA-SHI 810
JAPANESE AND ENGLISH

SPECIAL PUBLICATIONS FROM THE SETO MARINE BIOLOGICAL LABORATORY 1272
KYOTO DAIGAKU RIGAKUBU FUZOKU SETO RINKAI JIKKENJO (SETO MARINE BIOLOGICAL LABORATORY, FACULTY OF SCIENCE, KYOTO UNIVERSITY)
SHIRAHAMA-CHO, NISHIMURO-GUN, WAKAYAMA 649-22
ENGLISH

SUGADAIRA KOGEN SEIBUTSU JIKKENJO KENKYU HOKOKU 1273
(RESEARCH REPORT OF THE SUGADAIRA PLATEAU BIOLOGICAL LABORATORY)
TOKYO KYOIKU DAIGAKU (TOKYO UNIVERSITY OF EDUCATION, FACULTY OF SCIENCE, SUGADAIRA PLATEAU BIOLOGICAL LABORATORY)
JAPANESE

SUGASHIMA RINKAI JIKKENJO GYOSEKISHU 1274
(CONTRIBUTIONS FROM SUGASHIMA MARINE BIOLOGICAL STATION)
NAGOYA DAIGAKU RIGAKUBU FUZOKU SUGASHIMA RINKAI JIKKENJO (SUGASHIMA MARINE BIOLOGICAL STATION, FACULTY OF SCIENCE, NAGOYA UNIVERSITY)
SUGASHIMA-CHO, TOBA-SHI 517
JAPANESE AND ENGLISH

TANPAKUSHITSU KAKUSAN KOSO 1275
(PROTEIN, NUCLEIC ACID, ENZYME)
KYORITSU SHUPPAN K.K.
6-19, 4-CHOME, KOHINATA, BUNKYO-KU, TOKYO 112
JAPANESE
INDEXED: CHE, JICST

TEION KAGAKU, SEIBUTSU-HEN 1276
(LOW TEMPERATURE SCIENCE, SERIES B)
HOKKAIDO DAIGAKU TEION KAGAKU KENKYUSHO (INSTITUTE OF LOW TEMPERATURE SCIENCE, HOKKAIDO UNIVERSITY)
NISHI-8-CHOME, KITA-19-JO, KITA-KU, SAPPORO-SHI 060
JAPANESE, WITH ENGLISH SUMMARY
INDEXED: BIB, BIO, CHE, JICST

TOKYO KYOIKU DAIGAKU SHIMODA RINKAI JIKKENSHO GYOSEKI 1277
(CONTRIBUTIONS FROM THE SHIMODA MARINE BIOLOGICAL STATION)
TOKYO KYOIKU DAIGAKU SHIMODA RINKAI JIKKENJO (SHIMODA MARINE BIOLOGICAL STATION, TOKYO UNIVERSITY OF EDUCATION)
10-1, 5-CHOME, SHIMODA-SHI 415
ENGLISH

TOKYO KYOIKU DAIGAKU SUGADAIRA KOGEN SEIBUTSU JIKKENSHO KENKYU HOKOKU 1278
(BULLETIN OF THE SUGADAIRA BIOLOGICAL LABORATORY)
TOKYO KYOIKU DAIGAKU FUZOKU SUGADAIRA KOGEN SEIBUTSU JIKKENSHO (SUGADAIRA BIOLOGICAL LABORATORY OF TOKYO UNIVERSITY OF EDUCATION)
SUGADAIRA, SANADA-MACHI, CHIISAGATA-GUN, NAGANO 386-22
JAPANESE AND ENGLISH
INDEXED: BIO

TOKYO NOKO DAIGAKU KOTANPAKUSHITSU KIYO KENKYU SHISETSU NENPO 1279
(ANNUAL OF THE RESEARCH FACILITIES TO UTILIZE SOLID PROTEINS)
TOKYO NOKO DAIGAKU (TOKYO AGRICULTURE & INDUSTRY UNIVERSITY
JAPANESE

TOKYOWAN NO SEIBUTSUSO CHOSA HOKOKU 1280
(SURVEY REPORT OF THE BIOLOGICAL CONDITION IN THE TOKYO BAY)
KANAGAWA-KEN KOGAI TAISAKU JIMU-KYOKU (COUNTERPLANNING OFFICE FOR THE PUBLIC NUISANCE, KANAGAWA PREFECTURE)
1, NIHON-ODORI, NAKA-KU, YOKOHAMA-SHI 232
JAPANESE

TOSEIKEN KAISHI 1281
(JOURNAL OF BTAT)
TOKYO-TO SEIBUTSU KYOIKU KENKYUKAI (BIOLOGY TEACHER'S ASSOCIATION OF TOKYO)
C/O TOKYO-TORITSU NERIMA KOTO GAKKO, 28-25, 4-CHOME, KASUGA-CHO, NERIMA-KU, TOKYO 176
JAPANESE, WITH ENGLISH SUMMARY

WASEDA SEIBUTSU 1282
(WASEDA BIOLOGY)
WASEDA DAIGAKU SEIBUTSU DOKOKAI (WASEDA UNIVERSITY BIOLOGICAL SOCIETY)
C/O WASEDA DAIGAKU IPPAN KYOIKU SEIBUTSUGAKU KYOSHITSU, 1-647, TOTSUKA-CHO, SHINJUKU-KU, TOKYO 160
JAPANESE

WATAKUSHITACHI NO SHIZEN 1283
(NATURE)
NIHON CHORUI HOGO RENMEI (THE JAPANESE ASSOCIATION FOR PRESERVATION OF BIRDS)
8-20, NANPEIDAI-CHO, SHIBUYA-KU, TOKYO 150
JAPANESE

YOKOHAMA-SHIRITSU DAIGAKU KIYO, SEIBUTSUGAKU-HEN 1284
(JOURNAL OF THE YOKOHAMA CITY UNIVERSITY, BIOLOGY SERIES)
YOKOHAMA-SHIRITSU DAIGAKU SEIBUTSUGAKU KYOSHITSU (YOKOHAMA CITY UNIVERSITY)
4646, MUTSUURA-MACHI, KANAZAWA-KU, YOKOHAMASHI 236
JAPANESE AND ENGLISH

YOSHINOGAWA NO SEIBUTSU SEISANRYOKU NO KENKYU 1285
(RESEARCH REPORT OF THE BIOLOGICAL PRODUCTIVITY IN THE YOSHINO RIVER)
NARA JOSHI DAIGAKU RIGAKUBU DOBUTSU KYOSHITSU (INSTITUTE OF ZOOLOGY, FACULTY OF SCIENCE, NARA WOMEN'S UNIVERSITY)
KITAUOYANISHI-MACHI, NARA-SHI 630
JAPANESE
INDEXED: JICST

YUKI KAGOBUTSU GOSEIHO 1286
(METHODS OF SYNTHESIZING ORGANIC CHEMICAL COMPOUNDS)
JAPANESE
INDEXED: JICST

Anthropology, Ethnology

BULLETIN OF THE NATIONAL SCIENCE MUSEUM. SERIES D. ANTHROPOLOGY 1287
ENGLISH
INDEXED: BIO, JICST

JINRUI IDENGAKU ZASSHI 1288
(JAPANESE JOURNAL OF HUMAN GENETICS)
NIHON JINRUI IDEN GAKKAI (JAPAN SOCIETY OF HUMAN GENETICS)
C/O TOKYO IKA SHIKA DAIGAKU JINRUI IDENGAKU KENKYUSHITSU, 1-5, YUSHIMA, BUNKYO-KU, TOKYO 113
JAPANESE AND EUROPEAN
INDEXED: BIO, CHE, EXC

JINRUI KAGAKU 1289
(ANTHROPOLOGICAL SCIENCE)
KYUGAKA RENGOKAI (KYUGAKU FEDERATION)
JAPANESE

JINRUIGAKU KENKYUJO KIYO 1290
(BULLETIN OF THE ANTHROPOLOGICAL INSTITUTE)
NANZAN DAIGAKU JINRUIGAKU KENKYUSHO (NANZAN UNIVERSITY ANTHROPOLOGICAL RESEARCH INSTITUTE)
JAPANESE

JINRUIGAKU SHUHO 1291
(ANTHROPOLOGICAL REPORTS)
OSAKA-SHIRITSU DAIGAKU IGAKUBU KAIBOGAKU KYOSHITSU
4-54, 1-CHOME, ASAHI-CHO, ABENO-KU, OSAKA-SHI 545
JAPANESE, WITH ENGLISH SUMMARY

JINRUIGAKU ZASSHI 1292
(JOURNAL OF THE ANTHROPOLOGICAL SOCIETY OF NIPPON)
NIPPON JINRUI GAKKAI (ANTHROPOLOGICAL SOCIETY OF NIPPON)
C/O TOKYO DAIGAKU RIGAKUBU JINRUIGAKU KYOSHITSU, 3-1, 7-CHOME, HONGO, BUNKYO-KU, TOKYO 113
JAPANESE AND EUROPEAN
INDEXED: BIO, EXC, JICST

JOURNAL OF THE FACULTY OF SCIENCE, UNIVERSITY OF TOKYO, SECTION 5, ANTHROPOLOGY 1293
TOKYO DAIGAKU (UNIVERSITY OF TOKYO)
3-1, 7-CHOME, HONGO, BUNKYO-KU, TOKYO 113
ENGLISH
INDEXED: BIO, CHE

KANATO KAMATO 1294
WASEDA DAIGAKU JINRUIGAKUKAI (WASEDA UNIVERSITY ANTHROPOLOGY ASSOCIATION)
JAPANESE

KIKAN JINRUIGAKU 1295
(QUARTERLY JOURNAL OF ANTHROPOLOGY)
KYOTO DAIGAKU JINRUIGAKU KENKYUKAI (SOCIETY OF ANTHROPOLOGY OF KYOTO UNIVERSITY) AND SHAKAI SHISOSHA
25-21, 1-CHOME, HONGO, BUNKYO-KU, TOKYO 113
JAPANESE

KOKURITSU MINZOKUGAKU HAKUBUTSUKAN KENKYU KIYO 1296
(BULLETIN OF THE NATIONAL MUSEUM OF ETHNOLOGY)
NATIONAL MUSEUM OF ETHNOLOGY
JAPANESE

MINZOKU KAGAKU 1297
(RACIAL SCIENCE)
MINAOKU KAGAKU KENKYUSHO (RACIAL SCIENCE RESEARCH INSTITUTE)
1-67, KOTAKE-CHO, NERIMA-KU, TOKYO 176
JAPANESE

MINZOKUGAKU KENKYU 1298
(JAPANESE JOURNAL OF ETHNOLOGY)
NIHON MINZOKU GAKKAI (THE JAPANESE SOCIETY OF ETHNOLOGY)
1-17, 3-CHOME, HIGASHI-CHO, HOYA-SHI, TOKYO 188
JAPANESE, WITH EUROPEAN SUMMARY

NANZAN DAIGAKU JINRUIGAKU KENKYUSHO KIYO 1299
(BULLETIN OF THE ANTHROPOLOGICAL INSTITUTE OF NANZAN UNIVERSITY)
NANZAN DAIGAKU JINRUIGAKU KENKYUSHO (THE ANTHROPOLOGICAL INSTITUTE OF NANZAN UNIVERSITY)
18, YAMAZATO-CHO, SHOWA-KU, NAGOYA-SHI 466
JAPANESE AND ENGLISH

Botany

BOTANICAL MAGAZINE, TOKYO 1300
NIHON SHOKUBUTSU GAKKAI (BOTANICAL SOCIETY OF JAPAN)
C/O TOYO BUNKO, 28-21, 2-CHOME, HONKOMAGOME, BUNKYO-KU, TOKYO 113
ENGLISH
INDEXED: BIB, BIC, CHE, JICST

BOTANY 1301
KUMAMOTO KINEN SHOKUBUTSU SAISHUKAI
C/O KUMAMOTO HAKUBUTSUKAN, 7-17, HANAHATA-CHO, KUMAMOTO-SHI 860
JAPANESE

BULLETIN OF THE NATIONAL SCIENCE MUSEUM. SERIES B. BOTANY. 1302
ENGLISH
INDEXED: BIB, BIO, CHE, JICST

FLORA JAPONICA EXSICCATA 1303
(JAPANESE DRIED FLORA)
KOKURITSU KAGAKU HAKUBUTSU-KAN
JAPANESE AND ENGLISH
INDEXED: BIB

FUJI TAKERUI SHOKUBUTSUEN HOKOKU 1304
(RESEARCH OF THE FUJI BAMBOO GARDEN)
NIHON TAKE SASA NO KAI (JAPANESE BAMBOO SOCIETY)
C/O SHIZEN SHINDO HOMBU, 1285, KAMADO, GOTENBA-SHI 412
JAPANESE, WITH EUROPEAN SUMMARY
INDEXED: JICST

GINKGOANA; CONTRIBUTIONS TO THE FLORA OF ASIA AND THE PACIFIC REGION 1305
ACADEMIA SCIENTIFIC BOOKS, INC.
ENGLISH
INDEXED: BIB

HATTORI SHOKUBUTSU KENKYUSHO HOKOKU 1306
(JOURNAL OF THE HATTORI BOTANICAL LABORATORY)
HATTORI SHOKUBUTSU KENKYUSHO (HATTORI BOTANICAL LABORATORY)
3888, HON-MACHI, NICHINAN-SHI 889-25
EUROPEAN
INDEXED: BIB, CHE

HIKOBIA (JOURNAL OF HIROSHIMA BOTANICAL CLUB) 1307
HIKOBIAKAI (HIROSHIMA BOTANICAL CLUB)
C/O HIROSHIMA DAIGAKU RIGAKUBU, SHOKUBUTSUGAKU KYOSHITSU, HIGASHISENDA-CHO, HIROSHIMA-SHI 730
JAPANESE AND EUROPEAN
INDEXED: BIO, CHE

HOKKAIDO DAIGAKU RIGAKUBU KAISO KENKYUSHO HOKOKU 1308
(REPORT OF THE INSTITUTE OF ALGOLOGICAL RESEARCH, FACULTY OF SCIENCE, HOKKAIDO UNIVERSITY)
HOKKAIDO DAIGAKU RIGAKUBU FUZOKU KAISO KENKYU SHISETSU (THE INSTITUTE OF ALGOLOGICAL RESEARCH, FACULTY OF SCIENCE, HOKKAIDO UNIVERISTY)
1-13, BOKOIMINAMI-MACHI, MURORAN-SHI 051
JAPANESE

HOKKAIDO DAIGAKU SHOKUBUTSUEN KENKYU HOKOKU 1309
(SAPPORO BULLETIN OF THE BOTANIC GARDEN, HOKKAIDO UNIVERSITY)
HOKKAIDO DAIGAKU NOGAKUBU FUZOKU SHOKUBUTSUEN (BOTANIC GARDEN, FACULTY OF AGRICULTURE, HOKKAIDO UNIVERSITY)
NISHI-8-CHOME, KITA-3-JO, CHUO-KU, SAPPORO-SHI 060
JAPANESE AND ENGLISH

HOKURIKU NO SHOKUBUTSU 1310
(JOURNAL OF GEOBOTANY)
HOKURIKU NO SHOKUBUTSU NO KAI
C/O MR. NOBUO SATOMI, 4-361, HISAYASU, KANAZAWA-SHI 921
JAPANESE AND ENGLISH
INDEXED: BIB

HOKURIKU NO SHOKUBUTSU 1311
(HOKURIKU JOURNAL OF BOTANY)
JAPANESE

JAPANESE JOURNAL OF BOTANY 1312
NIHON GAKUJUTSU KAIGI (SCIENCE COUNCIL OF JAPAN)
22-34, 7-CHOME, ROPPONGI, MINATO-KU, TOKYO 106
ENGLISH
INDEXED: BIB

JOURNAL OF THE FACULTY OF SCIENCE, HOKKAIDO UNIVERSITY, SERIES 5, BOTANY 1313
HOKKAIDO DAIGAKU RIGAKUBU (FACULTY OF SCIENCE, HOKKAIDO UNIVERSITY)
NISHI-8-CHOME, KITA-10-JO, KITA-KU, SAPPORO-SHI 060
ENGLISH
INDEXED: BIB, BIO, CHE, JICST

JOURNAL OF THE FACULTY OF SCIENCE, UNIVERSITY OF TOKYO, SECTION III, BOTANY 1314
TOKYO DAIGAKU (UNIVERSITY OF TOKYO)
3-1, 7-CHOME, HONGO, BUNKYO-KU, TOKYO 113
ENGLISH
INDEXED: BIO, CHE

JOURNAL OF SCIENCE OF THE HIROSHIMA UNIVERSITY, SERIES B, DIVISION 2, BOTANY 1315
HIROSHIMA DAIGAKU (HIROSHIMA UNIVERSITY)
HIGASHISENDA-CHO, HIROSHIMA-SHI 730
EUROPEAN
INDEXED: BIB, BIO, CHE

KANAZAWA DAIGAKU RIGAKUBU FUZOKU SHOKUBUTSUEN NENPO 1316
(ANNUAL REPORT OF BOTANIC GARDEN, FACULTY OF SCIENCE, KANAZAWA UNIVERSITY)
KANAZAWA DAIGAKU RIGAKUBU FUZOKU SHOKUBUTSUEN (BOTANIC GARDEN, FACULTY OF SCIENCE, KANAZAWA UNIVERSITY)
1, MARUNOUCHI, KANAZAWA-SHI 920
JAPANESE AND ENGLISH

KINKI SHOKUBUTSU DOKOKAI KAISHI 1317
(BULLETIN OF THE KINKI BOTANCIAL SOCIETY)
KINKI SHOKUBUTSU DOKOKAI
C/O SENBA CHUGAKKO, 1-7, ITOYA-CHO, HIGASHI-KU, OSAKA-SHI 540
JAPANESE, WITH ENGLISH SUMMARY

KONGETSU NO NOYAKU 1318
(AGRICULTURAL CHEMICALS MONTHLY)
JAPANESE

KYOTO-FUFURITSU SHOKYBUTSUEN HOKOKU 1319
(REPORT OF THE KYOTO PREFECTURAL BOTANICAL GARDENS)
JAPANESE

NAGANO-KEN SHOKUBUTSU KENKYUKAISHI 1320
(BULLETIN OF THE BOTANICAL SOCIETY OF NAGANO PREFECTURE)
NAGANO-KEN SHOKUBUTSU KENKYUKAI (BOTANICAL SOCIETY OF NAGANO PREFECTURE)
JAPANESE

NIHON KAFUN GAKKAI KAISHI 1321
(JAPANESE JOURNAL OF PALYNOLOGY)
NIHON KAFUN GAKKAI (PALYNOLOGICAL SOCIETY OF JAPAN)
C/O SHIZUOKA DAIGAKU RIGAKUBU SEIBUTSUGAKU KYOSHITSU, 836, OYA, SHIZUOKA-SHI 420
JAPANESE AND ENGLISH
INDEXED: CHE

NIHON KIN GAKKAI KAIHO 1322
(TRANSACTIONS OF THE MYCOLOGICAL SOCIETY OF JAPAN)
NIHON KIN GAKKAI (MYCOLOGICAL SOCIETY OF JAPAN)
C/O KOKURITSU KAGAKU HAKUBUTSUKAN BUNKAN, 23-1, 3-CHOME, HYAKUNIN-CHO, SHINJUKU-KU, TOKYO 160
JAPANESE AND ENGLISH
INDEXED: BIB, BIO, CHE

NIHON SHIDA NO KAI KAIHO 1323
(JOURNAL OF THE NIPPON FERNIST CLUB)
NIHON SHIDA NO KAI (NIPPON FERNIST CLUB)
C/O MR. SATORU KURATA, 33-3, 4-CHOME, IBEBUKURO, TOSHIMA-KU, TOKYO 171
JAPANESE

NIHON SHOKUBUTSU BUNRUI GAKKAI KAIHO 1324
(PROCEEDINGS OF THE JAPAN SOCIETY OF PLANT TAXONOMISTS)
NIHON SHOKUBUTSU BUNRUI GAKKAI (JAPAN SOCIETY OF PLANT TAXONOMISTS)
C/O KOKURITSU KAGAKU HAKUBUTSUKAN BUNKAN, 23-1, 3-CHOME, HYAKUNIN-CHO, SHINJUKU-KU, TOKYO 160
JAPANESE

NIHON SHOKUBUTSU SEIRI GAKKAI TSUSHIN 1325
(NEWS OF THE JAPANESE SOCIETY OF PLANT PHYSIOLOGISTS)
NIHON SHOKUBUTSU SEIRI GAKKAI
C/O NIHON GAKKAI JIMU SENTA, 4-16, 2-CHOME, YAYOI, BUNKYO-KU, TOKYO 113
JAPANESE

NIHON SHOKUBUTSUEN KYOKAI KAIHO 1326
(JOURNAL OF THE JAPAN BOTANICAL GARDEN ASSOCIATION)
JAPAN BOTANICAL GARDEN ASSOCIATION
JAPANESE

OYO SHOKUBUTSU SHAKAIGAKU KENKYU 1327
(APPLIED PHYTOSOCIOLOGY)
OYO SHOKUBUTSU SHAKAIGAKU KENKYUKAI (THE JAPANESE SOCIETY FOR APPLIED PHYTOSOCIOLOGY)
C/O TOKYO DAIGAKU NOGAKUBU RYOKUCHIGAKU KYOSHITSU, 1-1, 1-CHOME, YAYOI, BUNKYO-KU, TOKYO 113
JAPANESE, WITH ENGLISH SUMMARY

PLANT AND CELL PHYSIOLOGY 1328
NIHON SHOKUBUTSU SEIRI GAKKAI (JAPANESE SOCIETY OF PLANT PHYSIOLOGISTS)
SHIMODACHIYURI-DORI OGAWA HIGASHIIRU, KAMIGYO-KU, KYOTO-SHI 602
ENGLISH
INDEXED: BIB, BIO, CHE, JICST

REPORTS FROM THE MICROALGAE RESEARCH INSTITUTE OF JAPAN 1329
ENGLISH

SCIENTIFIC PAPERS OF THE INSTITUTE OF ALGOLOGICAL RESEARCH, FACULTY OF SCIENCE, HOKKAIDO UNIVERSITY 1330
HOKKAIDO DAIGAKU RIGAKUBU (FACULTY OF SCIENCE, HOKKAIDO UNIVERSITY)
NISHI-8-CHOME, KITA-10-JO, KITA-KU, SAPPORO-SHI 060
ENGLISH
INDEXED: BIO

SENTAI CHII ZAPPO 1331
(MISCELLANEA BRYOLOGICA ET LICHENOLOGICA)
HATTORI SHOKUBUTSU KENKYUSHO (HATTORI BOTANICAL LABORATORY)
3888, HON-MACHI, NICHINAN-SHI 889-25
JAPANESE, WITH EUROPEAN SUMMARY
INDEXED: BIB, BIO, CHE

SHIZEN KYOIKUEN HOKOKU 1332
(MISCELLANEOUS REPORTS OF THE NATIONAL PARK FOR NATURE STUDY)
KOKURITSU KAGAKU HAKUBUTSUKAN FUZOKU SHIZEN KYOIKUEN (NATIONAL SCIENCE MUSEUM AFFILIATION, NATIONAL PARK OF NATURE STUDY)
JAPANESE

SHOKUBUTSU BUNRUI CHIRI 1333
(ACTA PHYTOTAXOMINICA ET GEOBOTANICA)
SHOKUBUTSU BUNRUI CHIRI GAKKAI (PHYTOGEOGRAPHICAL SOCIETY)
C/O KYOTO DAIGAKU RIGAKUBU SHOKUBUTSUGAKU KYOSHITSU, KITASHIRAKAWA OIWAKE-CHO, SAKYO-KU, KYOTO-SHI 606
JAPANESE AND ENGLISH
INDEXED: BIB, BIO, CHE,

SHOKUBUTSU KENKYU ZASSHI 1334
(JOURNAL OF JAPANESE BOTANY)
TSUMURA KENKYUSHO (TSUMURA LABORATORY)
1421, IZUMI, KOMAE-SHI, TOKYO 182
JAPANESE AND ENGLISH
INDEXED: BIB, BIO, CHE, JICST

SHOKUBUTSU NO KAGAKU CHOSETSU 1335
(CHEMICAL REGULATION OF PLANTS)
SHOKUBUTSU KAGAKU CHOSETSU KENKYUKAI (THE SOCIETY FOR CHEMICAL REGULATION OF PLANTS)
C/O TOKYO DAIGAKU NOGAKUBU NOGEI KAGAKUKA, 1-1, 1-CHOME, YAYOI, BUNKYO-KU, TOKYO 113
JAPANESE

SHOKUBUTSU NO TOMO 1336
(PLANT'S COMPANION)
NIHON SHOKUBUTSU TOMONOKAI
7-12, 2-CHOME, KINUTA, SETAGAYA-KU, TOKYO 157
JAPANESE

SHOKUBUTSU SHUNJU 1337
(BOTANY SPRING & AUTUMN)
JAPANESE

SHOKUBUTSU TO BUNKA 1338
(PLANTS AND CULTURE)
YASAKA SHOBO
1-56, KANDA JINBO-CHO, CHIYODA-KU, TOKYO 101
JAPANESE

SHOKUBUTSU TO SHIZEN 1339
(NATURE AND PLANTS)
NYU SAIENSUSHA
3-21, KANDA NISHIKI-CHO, CHIYODA-KU, TOKYO 101
JAPANESE
INDEXED: CHE

SHOKUCHU SHOKUBUTSU KENKYUKAI KAISHI 1340
(QUARTERLIES OF INSPECTIVOROUS PLANT SOCIETY)
SHOKUCHU SHOKUBUTSU KENKYUKAI (INSPECTIVOROUS PLANT SOCIETY)
C/O NIHON SHIKA DAIGAKU SEIBUTSUGAKU KYOSHITSU, 9-20, 1-CHOME, FUJIMI-CHO, CHIYODA-KU, TOKYO 102
JAPANESE AND EUROPEAN
INDEXED: JICST

SORUI 1341
(BULLETIN OF JAPANESE SOCIETY OF PHYCOLOGY)
NIPPON SORUI GAKKAI (JAPANESE SOCIETY OF PHYCOLOGY)
C/O HOKKAIDO DAIGAKU RIGAKUBU FUZOKU KAISO KENKYU SHISETSU, 1-13, BOKOIMINAMI-MACHI, MURORAN-SHI 051
JAPANESE AND ENGLISH
INDEXED: BIO, CHE

SORUI SHOKUSEI TO SUISHITSU ODAKU 1342
(ALGAE FLORA AND WATER POLLUTION)
KANAGAWA-KEN KOGAI TAISAKU JIMU-KYOKU (COUNTERPLANNING OFFICE FOR THE PUBLIC NUISANCE, KANAGAWA PREFECTURE)
1, NIHON-ODORI, NAKA-KU, YOKOHAMA-SHI 232
JAPANESE

WHEAT INFORMATION SERVICE 1343
KIHARA SEIBUTSUGAKU KENKYUSHO (KIHARA INSTITUTE FOR BIOLOGICAL RESEARCH)
122-23, 3-CHOME, MUTSUGAWA, MINAMI-KU, YOKOHAMA-SHI 233
ENGLISH
INDEXED: BIB, BIO

YAKKOSO TSUCHITORIMOCHI NO TOMO 1344
(NEWS OF THE SOCIETY FOR THE RESERVATION OF YAKKOSO & TSUCHITORIMOCHI)
YAKKOSO TSUCHITORIMOCHI O HOGOSURU KAI HONBU
19-4, 3-CHOME, MEJIRODAI, BUNKYO-KU, TOKYO 112
JAPANESE

YASO TOMONOKAI 1345
(NEWS OF THE SOCIETY OF THE WILD GRASS)
YASO TOMONOKAI
18-3, 4-CHOME, SHIROGANEDAI, MINATO-KU, TOKYO 108
JAPANESE

Zoology

ACTA ARACHNOLOGICA 1346
TOA KUMO GAKKAI (ARACHNOLOGICAL SOCIETY OF EAST ASIA)
C/O OTEMON GAKUIN DAIGAKU, 230, AI, IBARAKI-SHI 567
JAPANESE AND ENGLISH
INDEXED: BIB, BIO

ACTA SERIOLOGIAI 1347
JAPANESE AND ENGLISH
INDEXED: EXC

AIKYU NO TOMO 1348
(FRIENDS OF AIKYU)
AIKYU NO TOMOSHA
JAPANESE

ANIMA 1349
HEIBONSHA
4-1, YONBAN-CHO, CHIYODA-KU, TOKYO 102
JAPANESE

ANNOTATIONES ZOOLOGICAE JAPONENSES 1350
NIHON DOBUTSU GAKKAI (ZOOLOGICAL SOCIETY OF JAPAN)
C/O TOYO BUNKO, 28-21, 2-CHOME, HONKOMAGOME, BUNKYO-KU, TOKYO 113
EUROPEAN
INDEXED: BIB, BIO, CHE, JICST

APPLIED ENTOMOLOGY AND ZOOLOGY 1351
NIHON OYO DOBUTSU KONCHU GAKKAI (JAPANESE SOCIETY OF APPLIED ENTOMOLOGY AND ZOOLOGY)
C/O NORIN-SHO NOGYO GIJUTSU KENKYUSHO, 2-1, NISHIGAHARA, KITA-KU, TOKYO 114
EUROPEAN
INDEXED: BIB, BIO, CHE, JICST

ASHIOTO 1352
(NEWS OF THE PHOENIX NATURAL ZOOLOGICAL GARDEN)
FENIKKUSU SHIZEN DOBUTSUEN
3083, HAMAYAMA, SHIOJI, MIYAZAKI-SHI 880-01
JAPANESE

ATYPUS 1353
TOA KUMO GAKKAI (ARACHNOLOGICAL SOCIETY OF EAST ASIA)
C/O OTEMON GAKUIN DAIGAKU, 230, AI, IBARAKI-SHI 567
JAPANESE

BADO RAIFU 1354
(BIRD LIFE)
AICHO NO TOMOSHA
4-4, DEN'ENCHOFUMINAMI, OTA-KU, TOKYO 145
JAPANESE

BULLETIN OF THE NATIONAL SCIENCE MUSEUM. SERIES A. ZOOLOGY. 1355
ENGLISH
INDEXED: BIB, BIO, CHE, JICST

CHIROPTERA 1356
KOMORI KENKYUKAI (KOMARI RESEARCH COMMITTEE)
JAPANESE AND ENGLISH

CHO TO GA 1357
(TRANSACTIONS OF THE LEPIDOPTEROLOGICAL SOCIETY OF JAPAN)
NIPPON RINSHI GAKKAI (THE LEPIDOPTEROLOGICAL SOCIETY OF JAPAN)
C/O OGATA BYOIN, 3-18, IMABASHI, HIGASHI-KU, OSAKA-SHI 541
JAPANESE AND ENGLISH

CHOJU TOKEI 1358
(ANNUAL STATISTICS OF BIRDS AND ANIMALS)
KANKYO-CHO SHIZENHOGO-KYOKU (NATURE CONSERVATION BUREAU, ENVIRONMENT AGENCY)
1-1, 3-CHOME, KASUMIGASEKI, CHIYODA-KU, TOKYO 100
JAPANESE

DEDENMUSHI 1359
(NEWS OF GIFU RESEARCH ASSOCIATION OF CONCHOLOGY)
GIFU-KEN KAIRUI KENKYUKAI
C/O MR. J. MIYAZAKI, KOMENO, KASAMATSU-CHO, HAJIMA-GUN, GIFU 501-61
JAPANESE

DOBUTSU BUNRUI GAKKAI KAIHO 1360
(CIRCULAR OF THE JAPANESE SOCIETY OF SYSTEMATIC ZOOLOGY)
DOBUTSU BUNRUI GAKKAI (JAPANESE SOCIETY OF SYSTEMATIC ZOOLOGY)
C/O KOKURITSU KAGAKU HAKUBUTSUKAN BUNKAN, 23-1, 3-CHOME, HYAKUNIN-CHO, SHINJUKU-KU, TOKYO 160
JAPANESE

DOBUTSU BUNRUI GAKKAISHI 1361
(PROCEEDINGS OF THE JAPANESE SOCIETY OF SYSTEMATIC ZOOLOGY)
DOBUTSU BUNRUI GAKKAI (JAPANESE SOCIETY OF SYSTEMATIC ZOOLOGY)
C/O KOKURITSU KAGAKU HAKUBUTSUKAN BUNKAN, 23-1, 3-CHOME, HYAKUNIN-CHO, SHINJUKU-KU, TOKYO 160
JAPANESE AND ENGLISH

DOBUTSU SHINRIGAKU NENPO 1362
(ANNUAL OF ANIMAL PSYCHOLOGY)
NIHON DOBUTSU SHINRI GAKKAI (JAPANESE SOCIETY FOR ANIMAL PSYCHOLOGY)
C/O TOKYO DAIGAKU BUNGAKUBU SHINRIGAKU KENKYUSHITSU, 3-1, 7-CHOME, HONGO, BUNKYO-KU, TOKYO 113
JAPANESE AND ENGLISH

DOBUTSU TO DOBUTSUEN 1363
(ANIMALS AND ZOOS)
TOKYO DOBUTSUEN KYOKAI (TOKYO ZOOLOGICAL PARK SOCIETY)
C/O UENO ZOOLOGICAL GARDENS, 9-83, UENO KOEN, TAITO-KU, TOKYO 110
JAPANESE AND ENGLISH

DOBUTSU TO SHIZEN 1364
(NATURE AND ANIMALS)
NYU SAIENSUSHA
C/O HISATOMO BUILDING, 3-21, KANDA NISHIKI-CHO, CHIYODA-KU, TOKYO 101
JAPANESE

DOBUTSUEN SUIZOKKAN ZASSHI 1365
(JOURNAL OF THE JAPANESE ASSOCIATION OF ZOOLOGICAL GARDENS AND AQUARIUMS)
DOBUTSU SUIZOKKAN ZASSHI HENSHU IINKAI AND NIHON DOBUTSUEN SUIZOKKAN KYOKAI
C/O UENO DOBUTSUEN, 9-83, UENO KOEN, TAITO-KU, TOKYO 110
JAPANESE

DOBUTSUGAKU ZASSHI 1366
(ZOOLOGICAL MAGAZINE)
NIHON DOBUTSU GAKKAI (ZOOLOGICAL SOCIETY OF JAPAN)
C/O TOYO BUNKO, 28-21, 2-CHOME, HONKOMAGOME, BUNKYO-KU, TOKYO 113
JAPANESE AND EUROPEAN
INDEXED: BIB, BIO, CHE, JICST

ESAKIA; OCCASIONAL PAPERS OF THE HIKOSAN BIOLOGICAL LABORATORY IN ENTOMOLOGY 1367
KYUSHU DAIGAKU NOGAKUBU FUZOKU HIKOSAN SEIBUTSUGAKU JIKKENSHO (HIKOSAN BIOLOGICAL LABORATORY, FACULTY OF AGRICULTURE, KYUSHU UNIVERSITY)
EIHIKOSAN, SOEDA-CHO, TAGAWA-GUN, FUKUOKA 824-07
EUROPEAN
INDEXED: BIB

EZOSHIRO 1368
(JOURNAL OF THE HAKODATE ENTOMOLOGICAL ASSOCIATION)
HAKODATE KONCHU DOKOKAI (HAKODATE ENTOMOLOGICAL ASSOCIATION)
C/O SHIRITSU HAKODATE HAKUBUTSUKAN GORYOKAKU BUNKAN, GORYOKAKUKOEN, HAKODATE-SHI 040
JAPANESE

FREUNDE 1369
BAIERU YAKUHIN K.K.
2-10, HON-CHO, HIGASHI-KU, OSAKA-SHI 541
JAPANESE

GARUI TSUSHIN 1370
(JAPAN HETEROCERISTS' JOURNAL)
NIHON GARUI GAKKAI (JAPAN HETEROCERISTS' SOCIETY)
C/O KOKURITSU KAGAKU HAKUBUTSUKAN, UENO KOEN, TAITO-KU, TOKYO 110
JAPANESE, WITH ENGLISH SUMMARY

GEIKEN TSUSHIN 1371
(WHALES RESEARCH INSTITUTE NEWS)
GEIRUI KANKYUSHO (WHALES RESEARCH INSTITUTE)
3-1, 1-CHOME, ETCHUJIME, KOTO-KU, TOKYO 135
JAPANESE

GENSEI 1372
(JOURNAL OF THE ENTOMOLOGICAL SOCIETY OF KOCHI)
KOCHI KONCHU KENKYUKAI (ENTOMOLOGICAL SOCIETY OF KOCHI)
C/O MR. MIROYASU OKAMOTO, 215, ROKUSENJI-MACHI, KOCHI-SHI 780
JAPANESE

GYORUIGAKU ZASSHI 1373
(JAPANESE JOURNAL OF ICHTHYOLOGY)
NIHON GYORUI GAKKAI (JAPANESE ASSOCIATION OF ICHTHYOLOGY)
C/O TOKYO SUISAN DAIGAKU GYORUIGAKU KOZA, 5-7, 4-CHOME, KONAN, MINATO-KU, TOKYO 108
JAPANESE AND EUROPEAN
INDEXED: BIB, CHE

HACHU-RYOSEIRUIGAKU ZASSHI 1374
(JAPANESE JOURNAL OF HERPETOLOGY)
NIHON HACHU RYOSEIRUI GAKKAI (HERPETOLOGICAL SOCIETY OF JAPAN)
C/O KYOTO KYOIKU DAIGAKU SEIBUTSUGAKU KYOSHITSU, 1, FUKAKUSA FUJINOMORI-CHO, FUSHIMI-KU, KYOTO-SHI 612
JAPANESE AND ENGLISH
INDEXED: BIO

HONYU DOBUTSUGAKU ZASSHI 1375
(JOURNAL OF THE MAMMALOGICAL SOCIETY OF JAPAN)
NIHON HONYU DOBUTSU GAKKAI (MAMMALOGICAL SOCIETY OF JAPAN)
C/O KOKURITSU KAGAKU HAKUBUTSUKAN BUNKAN, 23-1, 3-CHOME, HYAKUNIN-CHO, SHINJUKU-KU, TOKYO 160
JAPANESE, WITH EUROPEAN SUMMARY
INDEXED: BIO

HONYURUI KAGAKU 1376
(MAMMALIAN SCIENCE)
HONYURUI KENKYU GURUPU HOKU-DAI NOGAKUBU OYO DOBUTSUGAKU KYOSHITSUNAI (MAMMALIAN SCIENCE GROUP HOKKAIDO UNIVERSITY, DEPARTMENT OF AGRICULTURE)
JAPANESE

INSECTA MATSUMURANA (JOURNAL OF THE FACULTY OF AGRICULTURE, HOKKAIDO UNIVERSITY, SERIES ON ENTOMOLOGY) 1377
HOKKAIDO DAIGAKU NOGAKUBU KONCHUGAKU KYOSHITSU (ENTOMOLOGICAL INSTITUTE, FACULTY OF AGRICULTURE, HOKKAIDO UNIVERSITY)
NISHI-9-CHOME, KITA-9-JO, KITA-KU, SAPPORO-SHI 060
JAPANESE AND ENGLISH
INDEXED: BIO, CHE

INSEKUTARYUMU 1378
(INSECTARIUM)
TOKYO DOBUTSUEN KYOKAI (TOKYO ZOOLOGICAL PARK SOCIETY)
C/O UENO ZOOLOGICAL GARDENS, 9-83, UENO KOEN, TAITO-KU, TOKYO 110
JAPANESE

JAPANESE JOURNAL OF ZOOLOGY 1379
NIHON GAKUJUTSU KAIGI (SCIENCE COUNCIL OF JAPAN)
22-34, 7-CHOME, ROPPONGI, MINATO-KU, TOKYO 106
ENGLISH
INDEXED: BIB

JOURNAL OF THE FACULTY OF SCIENCE, UNIVERSITY OF TOKYO, SECTION 4, ZOOLOGY 1380
TOKYO DAIGAKU (UNIVERSITY OF TOKYO)
3-1, 7-CHOME, HONGO, BUNKYO-KU, TOKYO 113
ENGLISH
INDEXED: BIB, BIO, CHE, JICST

JOURNAL OF THE FACULTY OF SCIENCE, UNIVERSITY OF TOKYO, SERIES 6, ZOOLOGY 1381
HOKKAIDO DAIGAKU RIGAKUBU (FACULTY OF SCIENCE, HOKKAIDO UNIVERSITY)
NISHI-8-CHOME, KITA-10-JO, KITA-KU, SAPPORO-SHI 060
ENGLISH
INDEXED: BIB, BIO, CHE, JICST

JOURNAL OF SCIENCE OF THE HIROSHIMA UNIVERSITY, SERIES B, DIVISION 1, ZOOLOGY 1382
HIROSHIMA DAIGAKU (HIROSHIMA UNIVERSITY)
HIGASHISENDA-CHO, HIROSHIMA-SHI 730
ENGLISH
INDEXED: BIB, BIO, CHE

KAGOSHIMA DAIGAKU HABU KENKYUKAI HOKOKUSHU 1383
(TRIMERESURUS RESEARCH, KAGOSHIMA UNIVERSITY)
KAGOSHIMA DAIGAKU IGAKUBU (FACULTY OF MEDICINE, KAGOSHIMA UNIVERSITY)
7-82, SHIROYAMA-CHO, KAGOSHIMA-SHI 892
JAPANESE
INDEXED: CHE

KAIRUIGAKU ZASSHI 1384
(VENUS (THE JAPANESE JOURNAL OF MALACOLOGY))
NIHON KAIRUI GAKKAI (MALACOLOGICAL SOCIETY OF JAPAN)
C/O KOKURITSU KAGAKU HAKUBUTSUKAN BUNKAN, 23-1, 3-CHOME, HYAKUNIN-CHO, SHINJUKU-KU, TOKYO 160
JAPANESE AND ENGLISH
INDEXED: BIO

KAKOCHO 1385
(JOURNAL OF THE SOCIETY OF NAGOYA ENTOMOLOGISTS)
NAGOYA KONCHU DOKOKAI
C/O MR. ANDO, 5-24, 1-CHOME, OTOWA, ICHINOMIYA-SHI 491
JAPANESE

KEIKYU ABURATSUBO MARIN PAKU SUIZOKUKAN NENPO 1386
(ANNUAL REPORT OF THE KEIKYU ABURATSUBO MARINE PARK AQUARIUM)
KEIKYU ABURATSUBO MARIN PAKU (KEIKYU ABURATSUBO MARINE PARK AQUARIUM)
1082, KOAJIRO JONOUCHI, MISAKI-CHO, MIURA-SHI 238-02
JAPANESE

KOKAKURUI NO KENKYU 1387
(RESEARCHES ON CRUSTACEA)
NIHON KOKAKURUI GAKKAI (CARCINOLOGICAL SOCIETY OF JAPAN)
C/O ODAWARA KOKAKURUI HAKUBUTSUKAN, 3-11, AZABU JUBAN, MINATO-KU, TOKYO 106
JAPANESE AND ENGLISH

KONCHU 1388
(ENTOMOLOGY)
NIHON KONCHU GAKKAI (ENTOMOLOGICAL SOCIETY OF JAPAN)
C/O KOKURITSU YOBO EISEI KENKYUSHO EISEI KONCHU-BU, 10-35, 2-CHOME, KAMIOSAKI, SHINAGAWA-KU, TOKYO 141
JAPANESE AND EUROPEAN
INDEXED: BIB, BIO, CHE

KONCHU TO SHIZEN 1389
(NATURE AND INSECTS)
NYU SAIENSUSHA (NEW SCIENCE PUBLISHING COMPANY)
3-21, KANDA NISHIKI-CHO, CHIYODA-KU, TOKYO 101
JAPANESE
INDEXED: CHE

KONCHUGAKU HYORON 1390
(ENTOMOLOGICAL REVIEW OF JAPAN)
NIPPON KOCHU GAKKAI (JAPAN COLEOPTEROLOGICAL SOCIETY)
C/O DR. MASAO HAYASHI, NO. 199, TAKAAI JUTAKU, 3-1, NISHITAKAAI-CHO, HIGASHISUMIYOSHI-KU, OSAKA-SHI 546
JAPANESE AND EUROPEAN
INDEXED: BIB

KYOTO UNIVERSITY AFRICAN STUDIES 1391
KYOTO DAIGAKU JINBUN KAGAKU KENKYUSHO (KYOTO UNIVERSITY, RESEARCH INSTITUTE FOR HUMANISTIC STUDIES) AND KYOTO DAIGAKU AFURIKA RUIJIN'EN GAKUJUTSU CHOSATAI (COMMITTEE OF KYOTA UNIVERSITY)
C/O KYOTO DAIGAKU JINBUN KAGAKU KENKYUSHO, 47, HIGASHIOGURA-CHO, KITASHIRAKAWA, SAKYO-KU, KYOTO-SHI 606
JAPANESE
INDEXED: BIO

MEETING REPORTS OF THE CHUGOKU SHIKOKU SECTION OF THE ZOOLOGICAL SOCIETY OF JAPAN 1392
NIHON DOBOTSU GAKKAI CHUGOKU SHIKOKU SHIBU (CHUGOKU SHIKOKU SECTION OF THE ZOOLOGICAL SOCIETY OF JAPAN)
C/O HIROSHIMA DAIGAKU RIGAKUBU, HIGASHISENDA-CHO, HIROSHIMA-SHI 730
JAPANESE

MONKI 1393
(MONKEY)
NIHON MONKI SENTA (JAPAN MONKEY CENTER)
26, INUYAMA KANRIN, INUYAMA-SHI 484
JAPANESE

MUSHI 1394
(INSECTS)
FUKUOKA MUSHI NO KAI
C/O KYUSHU DAIGAKU NOGAKUBU KONCHUGAKU KYOSHITSU, 3575-1, HAKOZAKI, HIGASHI-KU, FUKUOKA-SHI 812
ENGLISH
INDEXED: BIB, BIO

NAPI NEWS 1395
NAGOYA KONCHU DOKOKAI (SOCIETY OF NAGOYA ENTOMOLOGISTS)
C/O MR. ANDO, 5-24, 1-CHOME, OTOWA, ICHINOMIYA-CHI 491
JAPANESE

NIHON DOBUTSU BUNRUI 1396
(JAPAN FAUNA CLASSIFICATION)
SANSEIDO
JAPANESE
INDEXED: BIB

NIHON OYO DOBUTSU KONCHU GAKKAI CHUGOKU-SHIBU KAIHO 1397
(BULLETIN OF THE JAPAN SOCIETY FOR APPLIED ZOOLOGY AND ENTOMOLOGY, CHUGOKU REGIONAL OFFICE)
JAPANESE

NIHON OYO DOBUTSU KONCHU GAKKAISHI 1398
(JAPANESE JOURNAL OF APPLIED ENTOMOLOGY AND ZOOLOGY)
NIHON OYO DOBUTSU KONCHU GAKKAI (JAPANESE SOCIETY OF APPLIED ENTOMOLOGY AND ZOOLOGY)
C/O NORIN-SHO NOGYO GIJUTSU KENKYUSHO, 2-1, NISHIGAHARA, KITA-KU, TOKYO 114
JAPANESE AND ENGLISH
INDEXED: BIB, BIO, CHE, JICST

NIHON SENCHU KENKYUKAI SHI 1399
(JAPANESE JOURNAL OF NEMATOLOGY)
NIHON SENCHU KENKYUKAI
JAPANESE
INDEXED: BIB

NIPPON DOBUTSUEN SUIZOKUKAN KYOKAI GEPPO 1400
(MONTHLY REPORT OF JAPANESE ASSOCIATION OF ZOOLOGICAL GARDENS AND AQUARIUMS)
NIPPON DOBUTSUEN SUIZOKUKAN KYOKAI (JAPANESE ASSOCIATION OF ZOOLOGICAL GARDENS AND AQUARIUMS)
C/O UENO DOBUTSUEN, 9-83, UENO-KOEN, TAITO-KU, TOKYO 110
JAPANESE

NIPPON DOBUTSUEN SUIZOKUKAN NENPO 1401
(ANNUAL REPORT OF THE JAPANESE ASSOCIATION OF ZOOLOGICAL GARDENS AND AQUARIUMS)
NIPPON DOBUTSUEN SUIZOKUKAN KYOKAI
C/O UENO DOBUTSUEN, 9-83, UENO-KOEN, TAITO-KU, TOKYO 110
JAPANESE

NIPPONIA NIPPON 1402
DOBUTSU NO KAGAKU KENKYUKAI
24-16, 5-CHOME, NAKANO, NAKANO-KU, TOKYO 164
JAPANESE

NYU ENTOMOROJISUTO 1403
(NEW ENTOMOLOGIST)
SHINSHU KONCHU GAKKAI (ENTOMOLOGICAL SOCIETY OF SHINSHU)
C/O SHINSHU DAIGAKU SEN'IGAKUBU SEIBUTSUGAKU KYOSHITSU, 15-1, 3-CHOME, TOKIDA, UEDA-SHI 386
JAPANESE, WITH EUROPEAN SUMMARY
INDEXED: BIB, BIO

OHMU 1404
KYUSHU DAIGAKU NOGAKUBU DOBUTSUGAKU KYOSHITSU (ZOOLOGICAL LABORATORY, FACULTY OF AGRICULTURE, KYUSHU UNIVERSITY)
3575-1, HAKOZAKI, HIGASHI-KU, FUKUOKA-SHI 812
ENGLISH

OKERA 1405
(NEWS OF THE SOCIETY OF IBARAKI ENTOMOLOGISTS)
IBARAKI KONCHU DOKOKAI
C/O MR. NOZAKI, 65-9, ISHINAZAKA-MACHI, HITACHI-SHI 319-12
JAPANESE

PRIMATES: JOURNAL OF PRIMATOLOGY 1406
NIHON MONKI SENTA (JAPAN MONKEY CENTER)
26, INUYAMA KANRIN, INUYAMA-SHI 484
ENGLISH
INDEXED: BIO, CHE

REICHORUI KENKYUSHO NENPO 1407
(ANNUAL REPORTS OF THE PRIMATE RESEARCH INSTITUTE, KYOTO UNIVERSITY)
KYOTO DAIGAKU REICHORUI KENKYUSHO (PRIMATE RESEARCH INSTITUTE, KYOTO UNIVERSITY)
INUYAMA KANRIN, INUYAMA-SHI 484
JAPANESE

RESEARCHES ON POPULATION ECOLOGY 1408
KOTAIGUN KEITAIGAKKAI (SOCIETY OF POPULATION ECOLOGY)
C/O KYOTO DAIGAKU NOGAKUBU KONCHUGAKU KENKYUSHITSU, KITASHIRAKAWA, SAKYO-KU, KYOTO-SHI 606
ENGLISH
INDEXED: BIB, BIO

SCIENTIFIC REPORT OF THE LABORATORY FOR AMPHIBIAN BIOLOGY 1409
HIROSHIMA DAIGAKU (HIROSHIMA UNIVERSITY)
1-89, 1-CHOME, HIGASHISENDA-CHO, HIROSHIMA-SHI 730
ENGLISH

SCIENTIFIC REPORTS OF THE WHALES RESEARCH INSTITUTE 1410
GEIRUI KENKYUSHO (WHALES RESEARCH INSTITUTE)
3-1, 1-CHOME, ETCHUJIMA, KOTO-KU, TOKYO 135
ENGLISH
INDEXED: BIO, CHE

SNAKE 1411
NIHON HEBIZOKU GAKUJUTSU KENKYUSHO (JAPAN SNAKE INSTITUTE)
YABUZUKA HON-CHO, NITTA-GUN, GUNMA 379-23
JAPANESE AND ENGLISH
INDEXED: BIO, CHE

SURUGA NO KONCHU 1412
(INSECTS IN SURUGA)
SHIZUOKA KONCHU DOKOKAI (ENTOMOLOGICAL SOCIETY OF SHIZUOKA)
C/O MR. M. TAKAHASHI, 13-11, 5-CHOME, KITAANDO, SHIZUOKA-SHI 420
JAPANESE

TINEA 1413
NIHON GARUI GAKKAI (JAPAN HETEROCERISTS' SOCIETY)
C/O KOKURITSU KAGAKU HAKUBUTSUKAN, UENO KOEN, TAITO-KU, TOKYO 143
JAPANESE AND EUROPEAN

TOA KUMO GAKKAI KAIMU HOKOKU 1414
(REPORT ON THE ASSOCIATION'S AFFAIRS, SOCIETY FOR EAST ASIAN ARACHNOLOGY)
JAPANESE

TOHOKU KONCHU 1415
(TOHOKU ENTOMOLOGY)
NIHON KONCHU GAKKAI TOHOKU SHIBU (JAPAN ENTOMOLOGY LEARNED SOCIETY, TOHOKU BRANCH)
JAPANESE

TOKYO-TO TAMA DOBUTSU KOEN JIGYO GAIYO 1416
(ANNUAL REPORT OF TAMA ZOOLOGICAL PARK, TOKYO)
TOKYO-TO TAMA DOBUTSU KOEN (TAMA ZOOLOGICAL PARK)
300, HODOKUBO, HINO-SHI, TOKYO 191
JAPANESE

TOMBA 1417
(ACTA ODONATOLOGICA)
ENGLISH

TORI 1418
(BULLETIN OF THE ORNITHOLOGICAL SOCIETY OF JAPAN)
NIHON CHO GAKKAI (ORNITHOLOGICAL SOCIETY OF JAPAN)
C/O YAMASHINA CHORUI KENKYUSHO, 8-20, NANPEIDAI, SHIBUYA-KU, TOKYO 150
JAPANESE AND ENGLISH
INDEXED: BIO

TRANSACTIONS OF THE SHIKOKU ENTOMOLOGICAL SOCIETY 1419
EHIME DAIGAKU NOGAKUBU KONCHUGAKU KENKYUSHITSU (ENTOMOLOGICAL LABORATORY, COLLEGE OF AGRICULTURE, EHIME UNIVERSITY) AND SHIKOKU KONCHU GAKKAI (SHIKOKU ENTOMOLOGICAL SOCIETY)
C/O EHIME DAIGAKU NOGAKUBU KONCHUGAKU KENKYUSHITSU, 5-7, 3-CHOME, TARUMI, MATSUYAMA-SHI 790
EUROPEAN
INDEXED: BIB, BIO

YACHO 1420
(WILD BIRD)
NIHON YACHO NO KAI (WILD BIRD SOCIETY OF JAPAN)
1-4, 1-CHOME, SHIBUYA, SHIBUYA-KU, TOKYO 150
JAPANESE

YADORIGA 1421
(JOURNAL OF THE LEPIDOPTEROLOGICAL SOCIETY OF JAPAN)
NIPPON RINSHI GAKKAI (LEPIDOPTEROLOGICAL SOCIETY OF JAPAN)
C/O OGATA BYOIN, 3-18, IMABASHI, HIGASHI-KU, OSAKA-SHI 541
JAPANESE

YAMASHINA CHORUI KENKYUSHO HOKOKU 1422
(MISCELLANEOUS REPORTS OF THE YAMASHINA INSTITUTE FOR ORNITHOLOGY)
YAMASHINA CHORUI KENKYUSHO (YAMASHINA INSTITUTE FOR ORNITHOLOGY)
8-20, NANPEIDAI, SHIBUYA-KU, TOKYO 150
JAPANESE AND ENGLISH
INDEXED: BIO, CHE

YASEI SEIBUTSU 1423
(WILDLIFE)
SEKAI YASEI SEIBUTSU KIKIN NIHON IINKAI (INTERNATIONAL FOUNDATION FOR WILDLIFE, JAPAN COUNCIL)
JAPANESE

YOHO KENKYUJO KIKANSHI 1424
(BEE SCIENCE)
JAPANESE

Geography, Charts

AKITA CHIRI 1425
(AKITA GEOGRAPHY)
AKITA GEOGRAPHICAL SOCIETY
JAPANESE

BOSO KENKYU - CHIBA CHIRI GAKKISHI 1426
(BOSO RESEARCH - JOURNAL OF THE CHIBA GEOGRAPHICAL SOCIETY)
JAPANESE

BULLETIN OF THE DEPARTMENT OF GEOGRAPHY, UNIVERSITY OF TOKYO 1427
TOKYO DAIGAKU RIGAKUBU CHIRIGAKU KYOSHITSU
(DEPARTMENT OF GEOGRAPHY, UNIVERSITY OF TOKYO)
3-1, 7-CHOME, HONGO, BUNKYO-KU, TOKYO 113
ENGLISH

BULLETIN OF THE GEOGRAPHICAL SURVEY INSTITUTE 1428
KENSETSU-SHO KOKUDO CHIRI-IN (GEOGRAPHICAL SURVEY INSTITUTE, MINISTRY OF CONSTRUCTION)
24-13, 3-CHOME, HIGASHIYAMA, MEGURO-KU, TOKYO 153
ENGLISH
INDEXED: BIB, JICST

CHIRI 1429
(GEOGRAPHY)
KOKON SHOIN
2-10, KANDA SURUGAKAI, CHIYODA-KU, TOKYO 101
JAPANESE
INDEXED: JICST

CHIRI CHOSAJO HOKOKU 1430
(BULLETIN OF THE GEOGRAPHIC SURVEY BUREAU)
JAPANESE

CHIRI KAGAKU 1431
(GEOGRAPHICAL SCIENCE)
CHIRI KAGAKU GAKKAI (HIROSHIMA GEOGRAPHICAL ASSOCIATION)
C/O HIROSHIMA DAIGAKU BUNGAKUBU CHIRI KYOSHITSU, HIGASHI SENDA-CHO, HIROSHIMASHI 730
JAPANESE AND EUROPEAN

CHIRI KYOIKU KENKYU NENPO 1432
(ANNUAL REPORT OF RESEARCH ON GEOGRAPHY EDUCATION)
JAPANESE

CHIRI SHISO 1433
(GEOGRAPHICAL DOCUMENTS)
NIHON DAIGAKU CHIRI GAKKAI (NIHON UNIVERSITY, ACADEMY OF GEOGRAPHY)
JAPANESE

CHIRIGAKU HOKOKU 1434
(GEOGRAPHICAL REPORT)
AICHI KYOIKU DAIGAKU CHIRI GAKKAI (GEOGRAPHICAL SOCIETY OF AICHI KYOIKU UNIVERSITY)
C/O AICHI KYOIKU DAIGAKU CHIRIGAKU KYOSHITSU, 1, HIROSAWA, IGAYA-CHO, KARIYA-SHI 448
JAPANESE
INDEXED: JICST

CHIRIGAKU HYORON 1435
(GEOGRAPHICAL REVIEW OF JAPAN)
NIPPON CHIRI GAKKAI (ASSOCIATION OF JAPANESE GEOGRAPHERS)
C/O GAKKAI SENTA BUILDING, 4-16, 2-CHOME, YAYOI, BUNKYO-KU, TOKYO 113
JAPANESE AND EUROPEAN

CHIRIGAKU KENKYU 1436
(GEOGRAPHICAL STUDY)
KAGAWA DAIGAKU CHIRIGAKU KENKYUKAI (GEOGRAPHICAL SOCIETY, KAGAWA UNIVERSITY)
1, SAIWAI-CHO, TAKAMATSU-SHI 760
JAPANESE

CHIRIGAKUHO 1437
(JOURNAL OF GEOGRAPHY, OSAKA KYOIKU UNIVERSITY)
OSAKA KYOIKU DAIGAKU CHIRIGAKU KYOSHITSU
(DEPARTMENT OF GEOGRAPHY, OSAKA KYOIKU UNIVERSITY)
43, MINAMI KAWAHORI-CHO, TENNOJI-KU, OSAKA-SHI 543
JAPANESE

CHIZU 1438
(MAP (JOURNAL OF THE JAPAN CARTOGRAPHERS ASSOCIATION))
NIHON KOKUSAI CHIZU GAKKAI (JAPAN CARTOGRAPHERS ASSOCIATION)
C/O KENSETSU-SHO KOKUDO CHIRI-IN, 24-13, 3-CHOME, HIGASHIYAMA, MEGURO-KU, TOKYO 153
JAPANESE, WITH EUROPEAN SUMMARY

CHIZU NO TOMO 1439
(MAP'S COMPANION)
CHIZU KYOKAI
C/O SHINSEN BUILDING, 8-2, SHINSEN-CHO, SHIBUYA-KU, TOKYO 150
JAPANESE

CHIZU SENTA NYUSU 1440
(JAPAN MAP CENTER NEWS)
NIHON CHIZU SENTA (JAPAN MAP CENTER)
10-12, 2-CHOME, DOGENZAKA, SHIBUYA-KU, TOKYO 150
JAPANESE

CHIZU TO SOKURYO 1441
(MAPPING AND SURVEYING)
SOKURYOSHA TOMO NO KAI
(SURVEYERS COMPANION ASSOCIATION)
JAPANESE

CONTOUR 1442
KYOTO DAIGAKU CHIRI DOKOKAI (KYOTO UNIVERSITY GEOGRAPHY ASSOCIATION)
JAPANESE AND ENGLISH

GAKUGEI CHIRI 1443
(GAKUGEI GEOGRAPHY)
TOKYO GAKUGEI DAIGAKU CHIRI GAKKAI (TOKYO GAKUGEI UNIVERSITY GEOGRAPHY ASSOCIATION)
JAPANESE

GEOGRAPHICAL REPORTS OF TOKYO METROPOLITAN UNIVERSITY 1444
TOKYO-TORITSU DAIGAKU RIGAKUBU CHIRIGAKU KYOSHITSU
(DEPARTMENT OF GEOGRAPHY, TOKYO METROPOLITAN UNIVERSITY)
1-1, 2-CHOME, FUKAZAWA, SETAGAYA-KU, TOKYO 158
EUROPEAN

GIFU CHIRI 1445
(GIFU GEOGRAPHY)
GIFU CHIRI GAKKAI
JAPANESE

GUNMA DAIGAKU CHIRI GAKKAI RONBUNSHU 1446
(PAPERS OF THE GUNMA UNIVERSITY GEOGRAPHY SOCIETY)
JAPANESE

HIROSHIMA DAIGAKU BUNGAKUBU CHIRIGAKU KENKYUSHITSU KENKYU HOKOKU 1447
(RESEARCH REPORT OF GEOGRAPHY RESEARCH DEPARTMENT, FACULTY OF LITERATURE, HIROSHIMA UNIVERSITY)
DAIGAKU BUNGAKUBU CHIRIGAKU KYOSHITSU
JAPANESE

HOSEI CHIRI 1448
(HOSEI GEOGRAPHY)
HOSEI DAIGAKU CHIRI GAKKAI (HOSEI UNIVERSITY GEOGRAPHICAL SOCIETY)
JAPANESE

HOSEI DAIGAKU CHIRIGAKU SHUHO 1449
(GEOGRAPHICAL PAPERS OF HOSEI UNIVERSITY)
HOSEI DAIGAKU CHIRIGAKU SHUHO KANKOKAI
JAPANESE

KAGOSHIMA CHIRI GAKKAI KIYO 1450
(MEMOIRS OF KAGOSHIMA GEOGRAPHICAL SOCIETY)
KAGOSHIMA CHIRI GAKKAI
C/O KAGOSHIMA DAIGAKU KYOIKUGAKUBU CHIRIGAKU KYOSHITSU, 251, KAMOIKE-CHO, KAGOSHIMA-SHI 890
JAPANESE
INDEXED: JICST

KOKUDO CHIRI-IN JIHO 1451
(GSI JOURNAL)
KENSETSU-SHO KOKUDO CHIRI-IN (GEOGRAPHICAL SURVEY INSTITUTE, MINISTRY OF CONSTRUCTION)
24-13, 3-CHOME, HIGASHIYAMA, MEGURO-KU, TOKYO 153
JAPANESE
INDEXED: JICST

KOMAZAWA CHIRI 1452
(KOMAZAWA GEOGRAPHY)
KOMAZAWA DAIGAKU CHIRI GAKKAI (KOMAZAWA UNIVERSITY GEOGRAPHIC SOCIETY)
JAPANESE

KOMAZAWA DAIGAKU DAIGAKUIN CHIRIGAKU KENKYU 1453
(GEOGRAPHICAL RESEARCH OF THE GRADUATE SCHOOL, KOMAZAWA UNIVERSITY)
KOMAZAWA DAIGAKU DAIGAKUIN CHIRIGAKU KENKYUKAI
(KOMAZAWA UNIVERSITY GRADUATE SCHOOL GEOGRAPHICAL RESEARCH ASSOCIATION)
JAPANESE

NIHON DAIGAKU BUNRIGAKUBU SHIZENKAGAKU KENKYUSHO KENKYU KIYO, CHIRI 1454
(PROCEEDINGS OF THE INSTITUTE OF NATURAL SCIENCES, NIHON UNIVERSITY, GEOGRAPHY)
NIHON DAIGAKU BUNRIGAKUBU SHIZENKAGAKU KENKYUSHO CHIRIGAKU KYOSHITSU (DEPARTMENT OF THE INSTITUTE OF NATURAL SCIENCES, NIHON UNIVERSITY)
C/O NIHON DAIGAKU BUNRIGAKUBU, 25-40, 3-CHOME, SAKURAJOSUI, SETAGAYA-KU, TOKYO 156
JAPANESE, WITH EUROPEAN SUMMARY
INDEXED: CHE

OCHANOMIZU CHIRI 1455
(OCHANOMIZU GEOGRAPHY)
OCHANOMIZU JOSHI DAIGAKU CHIRIGAKU KYOSHITSU (SCHOOL OF GEOGRAPHY, OCHANOMIZU WOMEN'S UNIVERSITY)
1-1, 2-CHOME, OTSUKA, BUNKYO-KU, TOKYO 112
JAPANESE

OKAYAMA DAIGAKU CHIRIGAKU KENKYU HOKOKU 1456
(OKAYAMA UNIVERSITY PUBLICATIONS IN GEOGRAPHY (CITY AND COUNTRY))
OKAYAMA DAIGAKU HOBUN BUNGAKUBU CHIRIGAKU KYOSHITSU
JAPANESE

PAIONIA 1457
(PIONEER)
KANSAI CHIRIGAKU KENKYUKAI (KANSAI GEOGRAPHICAL RESEARCH SOCIETY)
JAPANESE

SAITAMA CHIRI 1458
(SAITAMA GEOGRAPHY)
SAITAMA CHIRI GAKKAI (SAITAMA GEOGRAPHICAL SOCIETY)
JAPANESE

SCIENCE REPORT OF THE TOHOKU UNIVERSITY, SERIES 7: GEOGRAPHY 1459
TOHOKU DAIGAKU RIGAKUBU (FACULTY OF SCIENCE, TOHOKU UNIVERSITY)
AOBAYAMA, SENDAI-SHI 980
ENGLISH
INDEXED: BIB, JICST

SENSHU DAIGAKU CHIRIGAKU KENKYUKAI KIYO 1460
(BULLETIN OF THE SENSHU UNIVERSITY GEOGRAPHIC RESEARCH SOCIETY)
JAPANESE

SHUDAI CHIRI 1461
(BULLETIN OF THE GEOGRAPHY, AKITA UNIVERSITY)
AKITA DAIGAKU KYOIKUGAKUBU CHIRIGAKU KENKYUKAI (FACULTY OF EDUCATION, AKITA UNIVERSITY)
1-1, TEGATA GAKUEN-CHO, AKITA-SHI 010
JAPANESE

TOHOKU CHIRI 1462
(ANNALS OF THE TOHOKU GEOGRAPHICAL ASSOCIATION)
TOHOKU CHIRI GAKKAI (TOHOKU GEOGRAPHICAL ASSOCIATION)
C/O TOHOKU DAIGAKU RIGAKUBU CHIRIGAKU KYOSHITSU, AOBAYAMA, SENDAI-SHI 980
JAPANESE, WITH EUROPEAN SUMMARY

TOHOKU GAKUIN DAIGAKU RONSO, REKISHI, CHIRIGAKU 1463
(PAPERS OF TOHOKU GAKUIN UNIVERSITY, HISTORY, GEOGRAPHY)
TOHOKU GAKUIN DAIGAKU BUNKEI HOGAKU
JAPANESE

WASEDA DAIGAKU KYOIKU GAKUBU GAKUJUTSU KENKYU. CHIRIGAKU. REKISHIGAKU. SHAKAI KAGAKU-HEN 1464
(WASEDA UNIVERSITY SCHOOL OF EDUCATION. ARTS AND SCIENCE RESEARCH. GEOGRAPHY, HISTORY, SOCIAL SCIENCES)
WASEDA DAIGAKU KYOIKUKAI (WASEDA UNIVERSITY EDUCATIONAL SOCIETY)
JAPANESE

YAMANASHI CHIRI 1465
(YAMANASHI GEOGRAPHY)
YAMANASHI CHIRI GAKKAI (YAMANASHI GEOGRAPHICAL SOCIETY)
JAPANESE

Archaeology

AIZU TO KOKO 1466
(AIZU AND ARCHAEOLOGY)
AIZU KOKO GAKKAI (AIZU ARCHAEOLOGICAL SOCIETY)
JAPANESE

AWAJI KOKO KENKYUKAISHI 1467
(RECORDS OF THE AWAJI ARCHAEOLOGICAL RESEARCH SOCIETY)
AWAJI ARCHAEOLOGICAL RESEARCH SOCIETY
JAPANESE

DOSHISHA KOKO 1468
(DOSHISHA UNIVERSITY ARCHAEOLOGY)
DOSHISHA DAIGAKU KOKOGAKU KENKYUKAI
JAPANESE

EHIME KOKOGAKU 1469
(EHIME ARCHAEOLOGY)
EHIME KOKOGAKU (EHIME ARCHAEOLOGICAL SOCIETY)
JAPANESE

FUKUSHIMA DAIGAKU KOKOGAKU KENKYUKAI GEPPO 1470
(MONTHLY REPORT OF THE FUKUSHIMA UNIVERSITY ARCHAEOLOGICAL SOCIETY)
JAPANESE

FUKUSHIMA DAIGAKU KOKOGAKU KENKYUKAI KENKYU KIYO 1471
(RESEARCH BULLETIN OF THE FUKUSHIMA UNIVERSITY ARCHAEOLOGICAL RESEARCH SOCIETY)
JAPANESE

FUKUSHIMA DAIGAKU KOKOGAKU KENKYUKAI KENKYU KIYO: FUKUSEIHAN 1472
(RESEARCH BULLETIN OF THE FUKUSHIMA UNIVERSITY ARCHAEOLOGICAL RESEARCH SOCIETY: REPRINT EDITION)
JAPANESE

FUKUSHIMA KOKO 1473
(FUKUSHIMA ARCHAEOLOGY)
FUKUSHIMA-KEN KOKO GAKKAI
JAPANESE

GEKKAN KOKOGAKU JANARU 1474
(MONTHLY ARCHAEOLOGICAL JOURNAL)
NYU SAIENSU SHA (NEW SCIENCE COMPANY)
JAPANESE

HAYAMI KOKO 1475
(HAYAMI ARCHAEOLOGY)
KYUSHU SENSHI KENKYUKAI (KYUSHU PREHISTORIC RESEARCH SOCIETY)
JAPANESE

HOKKAIDO KOKOGAKU 1476
(HOKKAIDO ARCHAEOLOGY)
HOKKAIDO KOKOGAKU KAI (HOKKAIDO ARCHAEOLOGICAL SOCIETY)
JAPANESE

HOSHO 1477
(SOARING PHOENIX)
SAITAMA DAIGAKU KOKOGAKU KENKYU KAI (SAITAMA UNIVERSITY ARCHAEOLOGY RESEARCH SOCIETY)
JAPANESE

INDO KOKO KENKYU 1478
(INDIAN ARCHAEOLOGICAL RESEARCH)
INDO KOKO KENKYUKAI (SOCIETY FOR ARCHAEOLOGICAL RESEARCH IN INDIA)
JAPANESE

ISHIKAWA KOKOGAKU KENKYUKAI KAISHI 1479
(BULLETIN OF THE ISHIKAWA ARCHAEOLOGICAL ASSOCIATION)
ISHIKAWA KOKOGAKU KENKYUKAI
C/O MR. TAKAHORI, 66, HAKAWAYOKE-MACHI, KANAZAWA-SHI 920
JAPANESE

JODAI BUNKA 1480
KOKUGAKUIN DAIGAKU KOKO GAKKAI (KOKUGAKUIN UNIVERSITY ARCHAEOLOGICAL SOCIETY)
JAPANESE

JOSHO KOKO 1481
(JOSHO ARCHAEOLOGY)
JOSHO KOKOJUKU KENKYUKAI
JAPANESE

KAGAWA-KEN KOKO GAKKAI HOKOKU 1482
(REPORT OF THE KAGAWA ANTIQUARIAN INVESTIGATIONS)
KAGAWA-KEN KOKO GAKKAI (KAGAWA ANTIQUARIAN INVESTIGATIONS)
C/O SANUKINOMIYA SHAMUSHO, ZENTSUJI-SHI 765
JAPANESE

KAGOSHIMA DAIGAKU KOKOGAKU KENKYUKAI KIYO 1483
(BULLETIN OF THE KAGOSHIMA UNIVERSITY ARCHAEOLOGICAL RESEARCH SOCIETY)
JAPANESE

KAI KOKO 1484
(KAI ARCHAEOLOGY)
YAMANASHI-KEN KOKOGAKUSHI SHIRYOSHITSU (YAMANOSHI PREFECTURE ARCHAEOLOGY AND HISTORY DOCUMENTS OFFICE)
JAPANESE

KAMAKURA GAKUEN KOKOGAKUBU BUHO 1485
(KAMAKURA ACADEMY DEPARTMENT OF ARCHAEOLOGY BULLETIN)
JAPANESE

KANSAI DAIGAKU KOKOGAKU KENKYU NENPO 1486
(ANNUAL OF KANSAI UNIVERSITY ARCHAEOLOGY RESEARCH)
KANSAI DAIGAKU KOKOGAKU KENKYUKAI (KANSAI UNIVERSITY ARCHAEOLOGY RESEARCH SOCIETY)
JAPANESE

KEIO GIJUKU DAIGAKU KOKOGAKU KENKYUKAI HOKOKU 1487
(REPORT OF ARCHAEOLOGICAL RESEARCH ASSOCIATION, KEIO GIJUKU UNIVERSITY)
JAPANESE

KINRIN 1488
WASEDA DAIGAKU KOKOGAKU KENKYUKAI (WASEDA UNIVERSITY ARCHAEOLOGICAL SOCIETY)
JAPANESE

KODAI 1489
(ANCIENT TIMES)
WASEDA DAIGAKU KOKO GAKKAI
JAPANESE

KOKO 1490
(ARCHAEOLOGY)
FUKUSHIMA-KENRITSU BANJO KOTO GAKKO SHIGAKU KENKYU KU (FUKUSHIMA PREFECTURAL BANJO HIGH SCHOOL, HISTORICAL RESEARCH CLUB)
JAPANESE

KOKOGAKU KENKYU 1491
(ARCHAEOLOGICAL STUDY)
KOKOGAKU KENKYUKAI (SOCIETY OF ARCHAEOLOGICAL STUDY OF JAPAN)
C/O OKAYAMA DAIGAKU HOBUNGAKUBU KOKOGAKU KENKYUSHITSU, TSUSHIMA, OKAYAMA-SHI 700
JAPANESE

KOKOGAKU RONKO 1492
(REPORT OF ARCHAEOLOGICAL CONTROVERSIES)
NARA-KEN KASHIWARA KOKOGAKU KENKYUJO (NARA PREFECTURE, KASHIWARA ARCHAEOLOGICAL RESEARCH INSTITUTE)
JAPANESE

KOKOGAKU RONKYU 1493
(ARCHAEOLOGICAL DISCUSSION)
YAMANASHI SHIGAKU KENKYUKAI (YAMANASHI HISTORICAL RESEARCH SOCIETY)
JAPANESE

KOKOGAKU RONSO 1494
(REVUE D'ARCHEOLOGIE)
BEPPU DAIGAKU KOKOGAKU KENKYUKAI (BEPPU UNIVERSITY ARCHAEOLOGICAL RESEARCH GROUP)
JAPANESE

KOKOGAKU SHUKAN. FUKUSEIHAN 1495
(COLLECTED PUBLICATIONS ON ARCHAEOLOGY. REPRINT)
KOKO GAKKAI (ARCHAEOLOGICAL SOCIETY)
JAPANESE

KOKOGAKU ZASSHI 1496
(JOURNAL OF THE ARCHAEOLOGICAL SOCIETY OF NIPPON)
NIHON KOKO GAKKAI (ARCHAEOLOGICAL SOCIETY OF NIPPON)
C/O TOKYO KOKURITSU HAKUBUTSUKAN, 13-9, UENO KOEN, TAITO-KU, TOKYO 110
JAPANESE AND ENGLISH

KUMANORO KOKO 1497
(KUMANORO ARCHAEOLOGY)
NANKI KOKO DOKOKAI (NANKI ARCHAEOLOGICAL ASSOCIATION)
JAPANESE

KURASHIKI KOKOKAN KENKYU SHUHO 1498
(KURASHIKI ARCHAEOLOGICAL MUSEUM, COLLECTED RESEARCH REPORTS)
JAPANESE

MIYAKYO KOKO 1499
(MIYAKYO ARCHAEOLOGY)
MIYAGI KYOIKU DAIGAKU KOKOGAKU KENKYUKAI (MIYAGI EDUCATION COLLEGE ARCHAEOLOGICAL SOCIETY)
JAPANESE

MOGURA 1500
(MOLE)
KAGOSHIMA KENRITSU SHUSSUI KOTO GAKKO KOKOGAKUBU (KAGOSHIMA PREFECTURE HIGH SCHOOL ARCHAEOLOGY DEPARTMENT)
JAPANESE

MORIMACHI KOKO 1501
(MORIMACHI ARCHAEOLOGY)
MORIMACHI KOKOGAKU KENKYUKAI (MORIMACHI ARCHAEOLOGY RESEARCH SOCIETY)
JAPANESE

NAGANO-KEN KOKO GAKKAISHI 1502
(JOURNAL OF THE NAGANO PREFECTURE ANTHROPOLOGICAL SOCIETY))
NAGANO-KEN KOKO GAKKAI (NAGANO PREFECTURE ANTHROPOLOGICAL SOCIETY)
C/O K. KASAI, 437, SHIMONOHARA TANAKA-CHO, SHIMO SUWA-MACHI, SUWA-GUN, NAGANO 393
JAPANESE

NEW WAVE ARCHAEOLOGY 1503
ATARASHII KOKOGAKU NOLTA
JAPANESE AND ENGLISH

OKIKOKUDAI KOKO 1504
(ARCHAEOLOGICAL PAPERS)
OKINAWA KOKUSAI DAIGAKU BUNGAKUBU KOKOGAKU KENKYU SHITSU
JAPANESE

OKITAMA KOKO 1505
(OKITAMA ARCHAEOLOGY)
OKITAMA KOKO GAKKAI (OKITAMA ARCHAEOLOGICAL SOCIETY)
JAPANESE

OMI 1506
(OMI)
OMI KOKOGAKU KENKYUKAI (OMI ARCHAEOLOGICAL RESEARCH ASSOCIATION)
JAPANESE

ONIHANAYAMA 1507
TAKU KOKOGAKKAI
JAPANESE

OTSUKA KOKO 1508
(JOURNAL OF OTSUKA ARCHAEOLOGICAL SOCIETY)
OTSUKA KOKO GAKKAI (OTSUKA ARCHAEOLOGICAL SOCIETY)
JAPANESE

RISSHO KOKO 1509
(RISSHO UNIVERSITY ARCHAEOLOGICAL RESEARCH)
RISSHO DAIGAKU KOKOGAKU KENKYUKAI (RISSHO UNIVERSITY ARCHAEOLOGICAL RESEARCH ASSOCIATION)
JAPANESE

RYUTANI DAIGAKU BUNGAKUBU KOKOGAKU SHIRYO SHITSU KENKYU HOKOKU 1510
(RESEARCH REPORT OF THE RYUTANI UNIVERSITY FACULTY OF LITERATURE ARCHAEOLOGOCAL MATERIALS OFFICE)
JAPANESE

SADO KOKO REKISHI 1511
(SADO ARCHAEOLOGY AND HISTORY)
SADO KOKO REKISHI GAKKAI (SADO ARCHAEOLOGOCAL AND HISTORICAL SOCIETY)
JAPANESE

SAITAMA KENKYU 1512
(SAITAMA (ARCHAEOLOGICAL) RESEARCH)
SAITAMAKEN KOKO GAKKAI (ARCHAEOLOGICAL SOCIETY OF SAITAMA PREFECTURE)
JAPANESE

SAITAMA KOKO 1513
(SAITAMA ARCHAEOLOGY)
SAITAMA KOKO GAKKAI (SAITAMA ARCHAEOLOGICAL SOCIETY)
JAPANESE

SEIRYO 1514
KASHIWARA KOKOGAKU KENKYUJO (KASHIWARA ARCHAEOLOGICAL RESEARCH INSTITUTE)
JAPANESE

SENSHI 1515
KOMAZAWA DAIGAKU KOKOGAKU KENKYUSHITSU
JAPANESE

SHIKAOIMACHI KOKOGAKU KENKYU HOKOKU 1516
(REPORT OF SHIKAOIMACHI ARCHAEOLOGICAL RESEARCH)
SHIKAOIMACHI KOKOGAKU KENKYUKAI, SHIAOIMACHI KYOIKU IINKAI
JAPANESE

SHIMINE 1517
SHINJIN KOKOGAKU SHIMINE JINKAI
JAPANESE

SHIMOTSUKE KOKOGAKU 1518
(SHIMOTSUKE ARCHAEOLOGY)
SHIMOTSUKE KOKOGAKU KENKYUKAI (SHIMOTSUKE ARCHAEOLOGICAL RESEARCH ASSOCIATION)
JAPANESE

SHINANO KOKO 1519
(SHINAMO ARCHAEOLOGY)
NAGANO KOKO GAKKAI (NAGANO ARCHAEOLOGICAL STUDY GROUP)
JAPANESE

SHINOBU KOKO 1520
(SHINOBU ARCHAEOLOGY)
SHINOBU KOKO GAKKAI (SHINOBU ARCHAEOLOGICAL STUDY GROUP)
JAPANESE

SUGATA KOKO 1521
(SUGATA ARCHAEOLOGY)
SHIMANE DAIGAKU KOKOGAKU KENKYUKAI (SHIMANE UNIVERSITY ARCHAEOLOGICAL RESEARCH SOCIETY)
JAPANESE

SUNPU KOKO 1522
(SUNPU ARCHAEOLOGY)
SUNPU KOKO GAKKAI (SUNPU ARCHAEOLOGICAL SOCIETY)
JAPANESE

SUWA KOKO GAKU: FUKUSEIHAN 1523
(SUWA ARCHAEOLOGY: REPRINT EDITION)
TAKIZAWA KEIZO
JAPANESE

TAMA KOKO 1524
(TAMA ARCHAEOLOGY)
TAMA KOKOGAKU KENKYUKAI
JAPANESE

TEZUKAYAMA DAIGAKU KOKOGAKU JISSHUSHITSU DAYORI 1525
(LETTERS OF THE TEZUKAYAMA UNIVERSITY ARCHAEOLOGICAL PRACTICE ROOM)
JAPANESE

TEZUKAYAMA KOKOGAKU 1526
(TEZUKAYAMA ARCHAEOLOGY)
TEZUKAYAMA DAIGAKU KOKOGAKU KENKYUSHITSU (TEZUKAYAMA UNIVERSITY ARCHAEOLOGICAL RESEARCH ROOM)
JAPANESE

TOKACHI KOKO 1527
(TOKACHI ARCHAEOLOGY)
TOKACHI KAWA RYUIKI-SHI KENKYUKAI (RESEARCH ASSOCIATION FOR THE HISTORY OF THE TOKACHI RIVER VALLEY)
JAPANESE

TOZO KAWARA HAN 1528
(EXCAVATED TILES)
GIFU SEINEN KEKOGAKU KENKYUKAI (GIFU YOUTH ARCHAEOLOGICAL RESEARCH SOCIETY)
JAPANESE

URAWA KOKO GAKKAI KENKYU CHOSA HOKOKUSHO 1529
(REPORT OF THE URAWA ARCHAEOLOGICAL RESEARCH SURVEY)
JAPANESE

YOKOSUKA KOKOGAKKAI NENPO 1530
(ANNUAL REPORT OF YOKOSUKA ARCHAEOLOGICAL SOCIETY)
JAPANESE

YUSEI KOKO 1531
(POSTAL SERVICE ARCHAEOLOGY)
YUSEI KOKOGAKKAI (POSTAL SERVICE ARCHAEOLOGICAL SOCIETY)
JAPANESE

MEDICAL SCIENCES

Medical Sciences in General

ABSTRACTS OF PAPERS OF THE JOURNAL OF JAPANESE SOCIETY OF INTERNAL MEDICINE 1532
JAPANESE SOCIETY OF INTERNAL MEDICINE
5, 2-CHOME, KANDA-SURUGADI, CHIYODA-KU, TOKYO 309
JAPANESE AND ENGLISH
INDEXED: BIO

ACTA MEDICA ET BIOLOGICA 1533
NIIGATA DAIGAKU IGAKUBU (NIIGATA UNIVERSITY, SCHOOL OF MEDICINE)
ICHIBAN-CHO, ASAHICHO-DORI, NIIGATA-SHI 951
ENGLISH
INDEXED: BIO, CHE, EXC, JICST

ACTA MEDICA NAGASAKIENSIA 1534
NAGASAKI DAIGAKU IGAKUBU (NAGASAKI UNIVERSITY, SCHOOL OF MEDICINE)
12-4, SAKAMOTO-MACHI, NAGASAKI-SHI 852
EUROPEAN
INDEXED: CHE, EXC

ACTA MEDICA OKAYAMA 1535
OKAYAMA DAIGAKU IGAKUBU (OKAYAMA UNIVERSITY, MEDICAL SCHOOL)
5-1, 2-CHOME, SHIKATA-CHO, OKAYAMA-SHI 700
ENGLISH
INDEXED: BIO, CHE, EXC, SCI

ACTA MEDICA UNIVERSITATIS KAGOSHIMAENSIS 1536
KAGOSHIMA DAIGAKU IGAKUBU (FACULTY OF MEDICINE, KAGOSHIMA UNIVERSITY)
7-82, SHIROYAMA-CHO, KAGOSHIMA-SHI 892
EUROPEAN
INDEXED: BIO, CHE, EXC, JICST

ACTA SCHOLAE MEDICINALIS UNIVERSITATIS IN KYOTO 1537
KYOTO DAIGAKU IGAKUBU
SAKYO-KU, KYOTO
ENGLISH
INDEXED: CHE, JICST

AICHI IKA DAIGAKU IGAKKAI ZASSHI 1538
(JOURNAL OF THE AICHI MEDICAL UNIVERSITY ASSOCIATION)
AICHI IKA DAIGAKU IGAKKAI
21, YAZAKOKARIMATA, NAGAKUTA-CHO, AICHI-GUN, AICHI 480-11
JAPANESE, WITH ENGLISH SUMMARY
INDEXED: CHE, EXC

AICHI-KEN SHINSHIN SHOGAI-SHA KORONI HATTATSU SHOGAI KENKYUSHO NENPO 1539
(ANNUAL REPORT OF THE RESEARCH INSTITUTE OF HANDICAP TO GROWTH, AICHI PREFECTURAL MENTALLY AND PHYSICALLY HANDICAPPED PERSONS COLONY)
AICHI-KEN SHINSHIN SHOGAISHA KORONI HATTATSU SHOGAI KENKYUSHO
713-1, KAMIYA-CHO, KASUGAI-SHI 480-03
JAPANESE

AINO BYOIN IGAKU ZASSHI 1540
(MEDICAL JOURNAL OF AINO HOSPITAL)
KOSHOKAI AINO BYOIN
381, TAKADA, OTANOUCHI, IBARAGI-SHI 567
JAPANESE AND ENGLISH

AKITA-KEN ISHIKAI ZASSHI 1541
(AKITA MEDICAL JOURNAL)
AKITA-KEN ISHIKAI (AKITA MEDICAL ASSOCIATION)
1-40, SENSHU MEITOKU-MACHI, AKITA-SHI 010
JAPANESE

AKITA-KEN NOSON IGAKKAI ZASSHI 1542
(AKITA JOURNAL OF RURAL MEDICINE)
AKITA-KEN NOSON IGAKKAI (THE AKITA ASSOCIATION OF RURAL MEDICINE)
C/O AKITA-KEN KOSEI NOGYO KYODO KUMIAI RENGOKAI, 3, OMACHI, AKITA-SHI 010
JAPANESE

ANNUAL REPORT OF THE INSTITUTE FOR VIRUS RESEARCH, KYOTO UNIVERSITY 1543
KYOTO DAIGAKU UIRUSU KENKYUSHO (THE INSTITUTE FOR VIRUS RESEARCH, KYOTO UNIVERSITY)
SHOGOIN KAWAHARA-MACHI, SAKYO-KU, KYOTO-SHI 606
ENGLISH
INDEXED: JICST

ANNUAL REPORTS OF THE KIZAZATO INSTITUTE 1544
ENGLISH

AOMORI-KEN ISHIKAIHO 1545
(JOURNAL OF AOMORI MEDICAL ASSOCIATION)
AOMORI-KEN ISHIKAI (AOMORI MEDICAL ASSOCIATION)
2-8, SHIN-MACHI, AOMORI-SHI 030
JAPANESE

AOMORI-KENRITSU CHUO BYOIN ISHI 1546
(MEDICAL JOURNAL OF AOMORI PREFECTURAL CENTRAL HOSPITAL)
AOMORI-KENRITSU CHUO BYOIN (AOMORI PREFECTURAL CENTRAL HOSPITAL) AND AOMORI-KENRITSU CHUO BYOIN KYOSAIKAI
2-24, 1-CHOME, NAGASHIMA, AOMORI-SHI 030
JAPANESE
INDEXED: CHE

ASAHIGAWASO KENKYU NENPO 1547
(ANNUAL REPORT OF ASAHIGAWASO)
ASAHIGAWASO
GIONCHISAKI, OKAYAMA-SHI 703
JAPANESE

ASAHIKAWA-SHIRITSU BYOIN ISHI 1548
(JOURNAL OF ASAHIKAWA MUNICIPAL HOSPITAL)
ASAHIKAWA-SHIRITSU BYOIN
1, KINSEI-CHO, ASAHIKAWA-SHI 070
JAPANESE

ASIAN MEDICAL JOURNAL 1549
NIHON ISHIKAI (JAPAN MEDICAL ASSOCIATION) AND NIHON GAKUJUTSU SHUPPANSHA (JAPAN MEDICAL PUBLISHERS, INC.)
ZENKYOREN BUILDING, 7-9, 2-CHOME, HIRAKAWA-CHO, CHIYODA-KU, TOKYO 102
ENGLISH
INDEXED: CHE, EXC

ATARASHII ISHI 1550
(NEW-MEDICAL PRACTITIONER)
SHIN NIHON ISHI KYOKAI (NEW JAPAN MEDICAL ASSOCIATION)
HIDAKA BUILDING, 10-2, 1-CHOME, NISHI IKEBUKURO, TOSHIMA-KU, TOKYO 171
JAPANESE

AYURUBEDA KENKYU 1551
(ALPHA BETA RESEARCH)
AYURUBEDA KENKYU JUNBIKAI
JAPANESE

BAIOFIDOBAKKU KENKYU 1552
(JAPANESE JOURNAL OF BIOFEEDBACK RESEARCH)
BAIOFIDOBAKKU KENKYUKAI, JOCHI DAIGAKU BUNGAKUBU, SHINRIGAKU KENKYU SHITSUNAI (BIOFEEDBACK RESEARCH SOCIETY, SOPHIA UNIVERSITY, DEPARTMENT OF LITERATURE, PSYCHOLOGY RESEARCH DIVISION)
JAPANESE

BEHAVIORMETRIKA 1553
ENGLISH

BIKEN JOURNAL 1554
OSAKA DAIGAKU BISEIBUTSUBYO KENKYUSHO (RESEARCH INSTITUTE FOR MICROBIAL DISEASES, OSAKA UNIVERSITY)
YAMADA-KAMI, OAZA, SUITA-SHI 565
EUROPEAN
INDEXED: BIO, CHE, EXC, IND, SCI, JICST

BIYO NO IGAKU 1555
(MEDICAL SCIENCE FOR BEAUTY ART)
NIHON BIYO IGAKU KENKYUKAI (JAPAN ASSOCIATION OF MEDICAL SCIENCE FOR BEAUTY ART)
12-5, 1-CHOME, SHINBASHI, MINATO-KU, TOKYO 105
JAPANESE

BOEI IKA DAIGAKKO ZASSHI 1556
(JOURNAL OF THE NATIONAL DEFENSE MEDICAL COLLEGE)
JAPANESE

BULLETIN OF THE INSTITUTE OF CONSTITUTIONAL MEDICINE, KUMAMOTO UNIVERSITY (SUPPLEMENT) 1557
KUMAMOTO DAIGAKU TAISHITSU IGAKU KENKYUSHO (INSTITUTE OF CONSTITUTIONAL MEDICINE, KUMAMOTO UNIVERSITY)
24-1, 4-CHOME, KUHONJI, KUMAMOTO-SHI 862
ENGLISH

BULLETIN OF THE OSAKA MEDICAL SCHOOL 1558
OSAKA IKA DAIGAKU (OSAKA MEDICAL COLLEGE)
DAIGAKU-MACHI, TAKATSUKI-SHI 569
EUROPEAN
INDEXED: BIO, CHE, EXC, IND, JICST

BULLETIN OF TOKYO MEDICAL AND DENTAL UNIVERSITY 1559
TOKYO IKA SHIKA DAIGAKU (TOKYO MEDICAL AND DENTAL UNIVERSITY)
5-45, 1-CHOME, YUSHIMA, BUNKYO-KU, TOKYO 113
EUROPEAN
INDEXED: BIO, CHE, EXC, IND, JICST

BULLETIN OF THE YAMAGUCHI MEDICAL SCHOOL 1560
YAMAGUCHI DAIGAKU IGAKUBU (YAMAGUCHI UNIVERSITY, SCHOOL OF MEDICINE)
KOGUSHI, UBE-SHI 755
EUROPEAN
INDEXED: CHE, EXC

BYOIN 1561
(JOURNAL OF JAPAN HOSPITAL ASSOCIATION)
IGAKU SHOIN
29-11, 5-CHOME, HONGO, BUNKYO-KU, TOKYO 113
JAPANESE

BYOIN KANRI 1562
(HOSPITAL ADMINISTRATION)
NIHON BYOIN KANRI GAKKAI (THE JAPANESE SOCIETY ON HOSPITAL ADMINISTRATION)
C/O BYOIN KANRI KENKYUSHO, 1-CHOME, TOYAMA-CHO, SHINJUKU-KU, TOKYO 162
JAPANESE, WITH EUROPEAN SUMMARY

BYOIN KANRI KENKYUJO KIYO 1563
(BULLETIN OF THE NATIONAL INSTITUTE OF HOSPITAL ADMINISTRATION)
NATIONAL INSTITUTE OF HOSPITAL ADMINISTRATION
JAPANESE

BYOIN KANRI KENKYUSHO NENPO 1564
(ANNUAL REPORT OF THE NATIONAL INSTITUTE OF HOSPITAL ADMIMISTRATION)
KOSEI-SHO BYOIN KANRI KENKYUSHO (NATIONAL INSTITUTE OF HOSPITAL ADMINISTRATION, MINISTRY OF HEALTH AND WELFARE)
1-CHOME, TOYAMA-CHO, SHINJUKU-KU, TOKYO 162
JAPANESE

BYOIN SETSUBI 1565
(MAGAZINE OF HOSPITAL EQUIPMENT)
NIHON BYOIN SETSUBI KYOKAI (JAPAN HOSPITAL EQUIPMENT ASSOCIATION)
C/O BYOIN KANRI KENKYUSHO, 1-CHOME, TOYAMA-CHO, SHINJUKU-KU, TOKYO 162
JAPANESE

BYOIN SHINBUN 1566
(HOSPITALS NEWS)
BYOIN SHINBUNSHA
4-1, 2-CHOME, HIRAKAWA-CHO, CHIYODA-KU, TOKYO 102
JAPANESE

CHIBA IGAKKAI ZASSHI 1567
(JOURNAL OF CHIBA MEDICAL SOCIETY)
CHIBA IGAKKAI (CHIBA MEDICAL SOCIETY)
C/O CHIBA DAIGAKU IGAKUBU, 8-1, 1-CHOME, INOHAMA, CHIBA-SHI 280
JAPANESE AND ENGLISH
INDEXED: BIO, CHE, EXC, JICST

CHIBA-KEN IRYO GENKYO CHOSA 1568
(MEDICAL CARE IN CHIBA PREFECTURE)
CHIBA-KEN ISHIKAI
204, SHINMEI-CHO, CHIBA-SHI 280
JAPANESE

CHIBA-KEN ISHIKAIHO 1569
(JOURNAL OF THE CHIBA MEDICAL ASSOCIATION)
CHIBA-KEN ISHIKAI (THE CHIBA MEDICAL ASSOCIATION)
C/O EISEI KAIKAN, 204, SHINMEI-CHO, CHIBA-SHI 280
JAPANESE

CHIIKI IRYO 1570
(MEDICAL TREATMENT IN LOCAL DISTRICT)
KOKUMIN KENKYO HOKEN SHINRYO SHISETSU IGAKKAI (MEDICAL SOCIETY OF THE NATIONAL HEALTH INSURANCE CLINICAL FACILITIES)
9-13, 2-CHOME, MOTOAZABU, MINATO-KU, TOKYO 106
JAPANESE

CHIIKI IRYO ZOKANGO 1571
(MEDICAL TREATMENT IN LOCAL DISTRICT. SPECIAL ISSUE)
KOKUMIN KENKO HOKEN SHINRYO SHISETSU IGAKKAI (MEDICAL SOCIETY OF THE NATIONAL HEALTH INSURANCE CLINICAL FACILITIES)
9-13, 2-CHOME, MOTOAZABU, MINATO-KU, TOKYO 106
JAPANESE

CHUBU IJI SHINPO 1572
(CHUBU MEDICAL NEWS)
CHUBU IJI SHINPOSHA
1298, MINAMIGUMI, YAICHISHOKU FUTABA-CHO,
GIFU-SHI 500
JAPANESE

CHUGOKU SHIKOKU ISHIKAI RENGOKAI IGAKKAI SHINPOJUMU KOENSHU 1573
(PROCEEDING OF THE SYMPOSIUM, THE FEDERATION OF THE CHUGOKU AND SHIKOKU MEDICAL SOCIETY)
YAMAGUCHI-KEN ISHIKAI (THE YAMAGUCHI MEDICAL SOCIETY)
2-33, 3-CHOME, YUDA ONSEN, YAMAGUCHI-SHI 753
JAPANESE

CHUO BYOIN NENPO 1574
(ANNUAL REPORT OF THE CENTRAL HOSPITAL, AICHI PREFECTURAL MENTALLY AND PHYSICALLY HANDICAPPED PERSONS COLONY))
AICHI-KEN SHINSHIN SHOGAISHA KORONI
713-1, KAIDO, KAMIYA-CHO, KASUGAISHI 480-03
JAPANESE

CIEA PRECLINICAL REPORTS 1575
KAWASAKI
JAPANESE AND ENGLISH
INDEXED: EXC

COMPUTERS IN MEDICINE 1576
KOKUSAI IGAKU JOHO SENTA (INTERNATIONAL MEDICAL INFORMATION CENTER)
C/O NOGUCHI HAUSU 306, 26, DAIKYO-CHO, SHINJUKU-KU, TOKYO 160
JAPANESE AND ENGLISH

DEVELOPING ECONOMIES 1577
TOKYO
JAPANESE AND ENGLISH
INDEXED: EXC

DOAI IGAKU ZASSHI 1578
(MEDICAL JOURNAL OF THE FRATERNITY MEMORIAL HOSPITAL)
DOAI KINEN BYOIN (FRATERNITY MEMORIAL HOSPITAL)
1-11, 2-CHOME, YOKOAMI, SUMIDA-KU, TOKYO 130
JAPANESE
INDEXED: JICST

DOCTOR 1579
YAKUJI NYUSUSHA
C/O INDO BUILDING, DOSHO-MACHI, HIGASHI-KU, OSAKA-SHI 541
JAPANESE

DOKKYA JOURNAL OF MEDICAL SCIENCES 1580
ENGLISH

DOKUTA SARON 1581
(DOCTOR SALON LIBRARY)
TOTSU
6-3, 2-CHOME, SARUGAKU-CHO, CHIYODA-KU, TOKYO 101
JAPANESE

DONAN IGAKKAISHI 1582
(JOURNAL OF THE MEDICAL ASSOCIATION OF SOUTH HOKKAIDO)
DONAN IGAKKAI (THE MEDICAL ASSOCIATION OF SOUTH HOKKAIDO)
33-19, MOTO-MACHI, HAKODATE-SHI 040
JAPANESE

DOSHISHA SHINRI 1583
(DOSHISHA PSYCHOLOGICAL REVIEW)
DOSHISHA DAIGAKU SHINRIGAKU KENKYUKAI (DOSHISHA UNIVERSITY RESEARCH ASSOCIATION OF PSYCHOLOGY)
KAWASHIMA UMEZONO-CHO, UKYO-KU, KYOTO-SHI 615
JAPANESE AND ENGLISH

EHIME-KENRITSU BYOIN GAKKAI KAISHI 1584
(EHIME JOURNAL OF MEDICINE)
EHIME-KENRITSU BYOIN (EHIME PREFECTURAL HOSPITAL) AND EHIME-KEN KOEI KIGYO KANRI-KYOKU
4-2, 4-CHOME, ICHIBAN-CHO MATSUYAMA-SHI 790
JAPANESE

FUKUI-KEN ISHI-KAI DAYORI 1585
(FUKUI PREFECTURE MEDICAL SOCIETY LETTER)
JAPANESE

FUKUOKA DAIGAKU IGAKU KIYO 1586
(MEDICAL BULLETIN OF FUKUOKA UNIVERSITY)
FUKUOKA DAIGAKU KENKYUJO
JAPANESE

FUKUOKA IGAKU ZASSHI 1587
(FUKUOKA ACTA MEDICA)
FUKUOKA IGAKKAI (FUKUOKA MEDICAL ASSOCIATION)
C/O KYUSHU DAIGAKU IGAKUBU, KATAKASU, HIGASHI-KU, FUKUOKA-SHI 812
JAPANESE AND EUROPEAN
INDEXED: CHE, EXC, IND, JICST

FUKUOKA-KENRITSU BYOIN NENPO 1588
(ANNUAL REPORT OF FUKUOKA PREFECTURAL HOSPITAL)
FUKUOKA-KEN EISEI-BU
1-1, 1-CHOME, TENJIN-CHO, CHUO-KU, FUKUOKA-SHI 810
JAPANESE

FUKUSHIMA IGAKU ZASSHI 1589
(FUKUSHIMA MEDICAL JOURNAL)
FUKUSHIMA IGAKU ZASSHI KANKOKAI (SOCIETY FOR PUBLICATION OF THE FUKUSHIMA MEDICAL JOURNAL)
5-75, SUGITSUMA-CHO, FUKUSHIMA-SHI 960
JAPANESE AND ENGLISH
INDEXED: CHE, EXC, JICST

FUKUSHIMA JOURNAL OF MEDICAL SCIENCE 1590
FUKUSHIMA IGAKU ZASSHI KANKOKAI (SOCIETY FOR PUBLICATION OF THE FUKUSHIMA MEDICAL JOURNAL)
5-75, SUGITSUMA-CHO, FUKUSHIMA-SHI 960
ENGLISH
INDEXED: BIO, CHE, EXC, IND, JICST

GAIRAI KANJA OYOBI BYORUIBETSU NYU (TAI) IN KANJA TOKEI 1591
(ANNUAL STATISTICS ON PATIENTS IN NIIGATA HOSPITAL, NIIGATA PREFECTURAL CANCER CENTER)
NIIGATA-KENRITSU GAN SENTA NIIGATA BYOIN
2-10, KAWAGISHI-CHO, NIIGATA-SHI 951
JAPANESE

GEKKAN HOKEN SHINRYO 1592
(MONTHLY REVIEW OF INSURANCE TREATMENT)
IGAKU TSUSHINSHA
3-13, KANDA NISHIKI-CHO, CHIYODA-KU, TOKYO 101
JAPANESE
INDEXED: JICST

GEKKAN ISHIYAKU SHINGAKU 1593
(ADVANCED MEDICAL, DENTAL, & PHARMACOLOGICAL STUDIES MONTHLY)
GENBUNSHA (GENBUN PUBLISHING COMPANY)
JAPANESE

GEKKAN MEDIKARU SAIENSU 1594
(MONTHLY MEDICAL SCIENCE)
UMEZAKI
JAPANESE

GENBAKU SHOGAI CHOSA IINKAI GYOSEKI HOKOKUSHO 1595
(ABCC TECHNICAL REPORT SERIES)
KOSEI-SHO KOKURITSU YOBO EISEI KENKYUSHO GENBAKU SHOGAI CHOSA IINKAI (ATOMIC BOMB CASUALTY COMMISSION, THE NATIONAL INSTITUTE OF HEALTH)
10-35, 2-CHOME, KAMIOOSAKI, SHINAGAWAKU, TOKYO 141
JAPANESE, WITH ENGLISH SUMMARY

GENBAKU SHOGAI CHOSA IINKAI NENPO 1596
(ATOMIC BOMB CASUALTY COMMISSION, ANNUAL REPORT)
KOSEI-SHO KOKURITSU YOBO EISEI KENKYUSHO GENBAKU SHOGAI CHOSA IINKAI (ATOMIC BOMB CASUALTY COMMISSION, THE NATIONAL INSTITUTE OF HEALTH)
10-35, 2-CHOME, KAMIOOSAKI, SHINAGAWAKU, TOKYO 141
JAPANESE AND ENGLISH

GENDAI IGAKU 1597
(CURRENT MEDICINE)
AICHI-KEN ISHIKAI (AICHI MEDICAL ASSOCIATION)
14-28, 4-CHOME, SAKAE, NAKA-KU, NAGOYA-SHI 460
JAPANESE AND ENGLISH
INDEXED: CHE

GENDAI IRYO 1598
(MODERN MEDICAL CARE)
GENDAI IRYOSHA
1-21, KANDA NISHIKI-CHO, CHIYODA-KU, TOKYO 101
JAPANESE

GENDAI NO RINSHO 1599
(MODERN CLINICS)
GENDAI NO RINSHOSHA
2, YOSHIDA KONOE-CHO, SAKYO-KU, KYOTO-SHI 606
JAPANESE

GENDAI NO SHINRYO 1600
(CLINICAL MEDICINE)
KINDAI IGAKUSHA
C/O KAWAZU BUILDING, 20-17, 1-CHOME, JINNAN, SHIBUYA-KU, TOKYO 150
JAPANESE, WITH ENGLISH SUMMARY

GIFU DAIGAKU IGAKUBU KIYO 1601
(ACTA SCHOLAE MEDICINALIS UNIVERSITATIS IN GIFU)
GIFU DAIGAKU IGAKUBU (GIFU UNIVERSITY, SCHOOL OF MEDICINE)
40, TSUKASA-MACHI, GIFU-SHI 500
JAPANESE AND ENGLISH
INDEXED: CHE, EXC, JICST

GIFU-KENRITSU GERO ONSEN BYOIN, ONSEN IGAKU KENKYUSHO NENPO 1602
(ANNUAL REPORT OF THE INSTITUTE OF MEDICAL SCIENCE BY HOT SPRINGS, GIFU PREFECTURAL GERO HOT SPRINGS HOSPITAL)
GIFU-KENRITSU GERO ONSEN BYOIN
1162, YUNOSHIMA, GERO-MACHI, MASHITA-GUN, GIFU 509-22
JAPANESE

GINKYO GAKUEN KIYO 1603
(BULLETIN OF GINKYO COLLEGE OF MEDICAL TECHNOLOGY)
GINKYO GAKUEN (GINKYO COLLEGE OF MEDICAL TECHNOLOGY)
JAPANESE

GUNMA IGAKU 1604
(JOURNAL OF GUNMA MEDICAL ASSOCIATION)
GUNMA-KEN IGAKKAI
7-4, 1-CHOME, CHIYODA-MACHI, MAEBASHI-SHI 371
JAPANESE

GUNMA REPORTS ON MEDICAL SCIENCES 1605
GUNMA DAIGAKU IGAKUBU (GUNMA UNIVERSITY, SCHOOL OF MEDICINE)
39-22, SHOWA-MACHI, MAEBASHI-SHI 371
ENGLISH
INDEXED: CHE, JICST

HAMAMATSU IKA DAIGAKU GAKUHO 1606
(ACADEMIC REPORT OF THE HAMAMATSU MEDICAL UNIVERSITY)
HAMAMATSU MEDICAL UNIVERSITY
JAPANESE

HATTATSU NINGENGAKU KENKYU 1607
(RESEARCH ON DEVELOPMENTAL HUMAN STUDIES)
JAPANESE

HIROSAKI DAIGAKU IRYO GIJUTSU TANKI DAIGAKUBU KIYO 1608
(BULLETIN OF THE HIROSAKI UNIVERSITY, MEDICAL TECHNOLOGY JUNIOR COLLEGE DEPARTMENT)
HIROSAKI UNIVERSITY
JAPANESE

HIROSAKI IGAKU 1609
(HIROSAKI MEDICAL JOURNAL)
HIROSAKI DAIGAKU IGAKUBU (HIROSAKI UNIVERSITY, SCHOOL OF MEDICINE)
5, ZAIFU-CHO, HIROSAKI-SHI 036
JAPANESE AND ENGLISH
INDEXED: CHE, EXC, JICST

HIROSHIMA DAIGAKU GENBAKU HOSHANO IGAKU KENKYUSHO NENPO 1610
(PROCEEDINGS OF THE RESEARCH INSTITUTE FOR NUCLEAR MEDICINE AND BIOLOGY, HIROSHIMA UNIVERSITY)
HIROSHIMA DAIGAKU GENBAKU HOSHANO KENKYUSHO (RESEARCH INSTITUTE FOR NUCLEAR MEDICINE AND BIOLOGY, HIROSHIMA UNIVERSITY)
KASUMI, HIROSHIMA-SHI 734
JAPANESE AND EUROPEAN
INDEXED: CHE

HIROSHIMA DAIGAKU IGAKU ZASSHI 1611
(MEDICAL JOURNAL OF HIROSHIMA UNIVERSITY)
HIROSHIMA DAIGAKU IGAKUBU (HIROSHIMA UNIVERSITY, SCHOOL OF MEDICINE)
C/O HIROSHIMA DAIGAKU IGAKU TOSHOKAN, 2-3, 1-CHOME, KASUMI, HIROSHIMA-SHI 734
JAPANESE AND ENGLISH
INDEXED: BIO, CHE, IND, SCI, JICST

HIROSHIMA IGAKU 1612
(JOURNAL OF HIROSHIMA MEDICAL ASSOCIATION)
HIROSHIMA IGAKKAI (HIROSHIMA MEDICAL ASSOCIATION)
1-1, 1-CHOME, KANNONHON-MACHI, HIROSHIMA-SHI 733
JAPANESE
INDEXED: CHE

HIROSHIMA JOURNAL OF MEDICAL SCIENCES 1613
HIROSHIMA DAIGAKU IGAKUBU (HIROSHIMA UNIVERSITY, SCHOOL OF MEDICINE)
2-3, 1-CHOME, KASUMI, HIROSHIMA-SHI 734
ENGLISH
INDEXED: EXC, JICST

HIROSHIMA-KENRITSU BYOIN NENPO 1614
(MEDICAL JOURNAL OF HIROSHIMA PREFECTURAL HOSPITAL)
HIROSHIMA-KENRITSU BYOIN
5-54, 1-CHOME, KANDA, UJINA, HIROSHIMA-SHI 734
JAPANESE, WITH ENGLISH SUMMARY

HITACHI IGAKKAISHI 1615
(HITACHI MEDICAL JOURNAL)
ODAIRA KINEN TOKYO KITACHI BYOIN
5-7, 3-CHOME, YUSHIMA, BUNKYO-KU, TOKYO 113
JAPANESE

HOIGAKU NO JISSAI TO KENKYU 1616
(CASE STUDY OF THE MEDICAL JURISPRUDENCE)
C/O TOHOKU DAIGAKU IGAKUBU HOIGAKU KYOSHITSU,
2-1, SEIRYO-MACHI, SENDAI-SHI 980
JAPANESE

HOKKAIDO IGAKU ZASSHI 1617
(HOKKAIDO JOURNAL OF MEDICAL SCIENCE)
HOKKAIDO IGAKKAI (HOKKAIDO MEDICAL SOCIETY)
C/O HOKKAIDO DAIGAKU IGAKUBU, NISHI-7-CHOME, KITA-15-JO, KITA-KU, SAPPORO-SHI 060
JAPANESE AND ENGLISH
INDEXED: BIO, CHE, IND

HOKKAIDO IHO 1618
(INFORMATION BULLETIN OF HOKKAIDO MEDICAL ASSOCIATION)
HOKKAIDO ISHIKAI
C/O ESUKO BUILDING, NISHI-11-CHOME, ODORI, CHUO-KU, SAPPORO-SHI 060
JAPANESE

HOKKAIDO NO IRYO 1619
(MEDICAL CARE IN HOKKAIDO)
HOKKAIDO ISHIKAI (HOKKAIDO MEDICAL ASSOCIATION)
NISHI-6-CHOME, ODORI, CHUO-KU, SAPPORO-SHI 060
JAPANESE

HOKKAIDO UNIVERSITY MEDICAL LIBRARY SERIES 1620
HOKKAIDO UNIVERSITY, SCHOOL OF MEDICINE, MEDICAL LIBRARY
WEST 7 NORTH 15, SAPPORO 060
ENGLISH
INDEXED: BIO, CHE

HOKKAIDORITSU BYOIN NENPO 1621
(ANNUAL REPORT OF HOKKAIDO HOSPITAL)
HOKKAIDO EISEI-BU
NISHI-6-CHOME, KITA-3-JO, CHUO-KU, SAPPOROSHI 060
JAPANESE

HONPO IGAKU KENKYU GENKYO 1622
(STATUS OF MEDICAL RESEARCH IN JAPAN)
NIHON ISHIKAI (THE JAPAN MEDICAL ASSOCIATION)
2-5, KANDA SURUGADAI, CHIYODA-KU, TOKYO 101
JAPANESE
INDEXED: JICST

HYOGO IKA DAIGAKU IGAKKAI ZASSHI 1623
(HYOGO MEDICAL UNIVERSITY ACADEMIC MAGAZINE)
HYOGO MEDICAL UNIVERSITY
JAPANESE

HYOGO IKA DAIGAKUSHI 1624
(JOURNAL OF HYOGO COLLEGE OF MEDICINE)
HYOGO COLLEGE OF MEDICINE
JAPANESE

HYOGO-KEN ISHIKAIHO 1625
(MONTHLY REPORT OF HYOGO MEDICAL ASSOCIATION)
HYOGO-KEN ISHIKAI
57-4, 4-CHOME, SHIMOYAMATE-DORI, IKUTA-KU, KOBE-SHI 650
JAPANESE

IDO NO NIPPON 1626
(MEDICAL WAY IN JAPAN)
IDO NO NIPPONSHA
1-45, HON-CHO, OPPAMA, YOKOSUKA-SHI 237
JAPANESE

IGAKU BUNKEN GEPPO 1627
(MONTHLY MEDICAL BIBLIOGRAPHY)
TOKYO JOSHI IKA DAIGAKU TOSHOKAN (LIBRARY OF THE TOKYO WOMEN'S MEDICAL COLLEGE)
JAPANESE

IGAKU CHUO ZASSHI 1628
(JAPAN CENTRA REVUO MEDICINA)
IGAKU CHUO ZASSHI KANKOKAI (JAPANESE MEDICAL ABSTRACTING SOCIETY)
5-18, 2-CHOME, TAKAIDO HIGASHI, SUGINAMI-KU, TOKYO 168
JAPANESE
INDEXED: JICST

IGAKU HYORON 1629
(JAPANA MEDICINA REVUO)
SHIN NIHON ISHI KYOKAI (NEW JAPANESE DOCTORS ASSOCIATION)
C/O HIDAKA BUILDING, 10-2, 1-CHOME, NISHI IKEBUKURO, TOSHIMA-KU, TOKYO 171
JAPANESE

IGAKU KENKYU 1630
(ACTA MEDICINA)
DAIDO GAKKAN SHUPPANBU
C/O KYUSHU DAIGAKU IGAKUBU, HOIGAKU KYOSHITSU, KATAKASU, HIGASHI-KU, FUKUOKA-SHI 812
JAPANESE AND ENGLISH
INDEXED: BIO, CHE, EXC

IGAKU KYOIKU 1631
(MEDICAL EDUCATION)
NIHON IGAKU KYOIKU GAKKAI (JAPAN ASSOCIATION OF MEDICAL EDUCATION)
39-5, 2-CHOME, HONGO, BUNKYO-KU, TOKYO 113
JAPANESE

IGAKU KYOIKU BUNKEN SOKUHO 1632
(PROMPT NEWS OF LITERATURES ON MEDICAL EDUCATION)
KOKUSAI IGAKU JOHO SENTA (INTERNATIONAL MEDICAL INFORMATION CENTER)
30, DAIKYO-MACHI, SHINJUKU-KU, TOKYO 160
JAPANESE AND ENGLISH

IGAKU NO AYUMI 1633
(STRIDES OF MEDICINE)
ISHIYAKU SHUPPAN K.K. (ISHIYAKU PUBLISHERS, INCORPORATED)
7-10, 1-CHOME, HONKOMAGOME, BUNKYO-KU, TOKYO 113
JAPANESE AND EUROPEAN
INDEXED: CHE

IGAKU SENWAKU 1634
(FRAMEWORK OF MEDICAL CHOICES)
NIHON IGAKU BUNKA HOZONKAI (SOCIETY FOR THE PRESERVATION OF JAPANESE MEDICAL CULTURE)
JAPANESE

IGAKU TO FUKUIN 1635
(MEDICINE AND GOSPEL)
NIHON KIRISUTOSHA IKA RENMEI
531-23, 1-CHOME, TOTSUKA, SHINJUKU-KU, TOKYO 160
JAPANESE

IGAKU TO IRYO (GEKKAN) 1636
(MEDICAL SCIENCE AND MEDICAL CARE (MONTHLY))
NIPPON IGAKU KYOKAI (JAPAN MEDICAL ASSOCIATION)
7-2-8 HONGO, BUNKYO-KU TOKYO 113
JAPANESE

IGAKU TO SEIBUTSUGAKU 1637
(MEDICINE AND BIOLOGY)
IGAKU SEIBUTSUGAKU SOKUHOKAI
C/O OGATA IGAKU KAGAKU KENKYUSHO, 11-4, 1-CHOME, HIGASHI KANDA, CHIYODA-KU, TOKYO 101
JAPANESE
INDEXED: CHE, JICST

IGAKU TOSHOKAN 1638
(MEDICAL LIBRARIES)
NIHON IGAKU TOSHOKAN KYOKAI (JAPAN MEDICAL LIBRARY ASSOCIATION)
4-16, 2-CHOME, YAYOI, BUNKYO-KU, TOKYO 113
JAPANESE
INDEXED: CHE, JICST

IGAKUSHI KENKYU 1639
(STUDIUM HISTORIAE MEDICAE)
IGAKUSHI KENKYUKAI (COLLEQUIM STUDIUM HISTORIAE MEDICAL)
C/O OSAKA DAIGAKU IGAKUBU EISEIGAKU KYOSHITSU, 33, JOAN-CHO, KITA-KU, OSAKA-SHI 530
JAPANESE

IGAKUSHI KENKYU: TSUSHIN 1640
(RESEARCH ON MEDICAL HISTORY: CORRESPONDENCE)
IGAKUSHI KENKYU KAI (MEDICAL HISTORY RESEARCH ASSOCIATION)
JAPANESE

IGAKUSHI TSUSHIN 1641
(MEDICAL HISTORY TIMES)
IGAKUSHI KENKYUKAI KANTO CHIHOKAI (COLLEQUIM STUDIUM HISTORIAE MEDICAE) AND IGAKUSHI KENKYUKAI
C/O OSAKA DAIGAKU IGAKUBU EISEIGAKU KYOSHITSU, 33, JOAN-CHO, KITA-KU, TOKYO 530
JAPANESE

IJI JOHO KURIPPINGU 1642
(MEDICAL REPORT CLIPPINGS)
IJI JOHO KENKYUKAI (MEDICAL REPORTS RESEARCH GROUP)
JAPANESE

IJI KAI 1643
(MEDICAL WORLD)
IJI DOGYOKAI (MEDICAL GUILD)
JAPANESE

IJI KENKYUSHO CHOSA KENKYU HOKOKU 1644
(RESEARCH REPORTS OF MEDICAL LABORATORY)
YUSEI-SHO IJI KENKYUSHO (THE MEDICAL LABORATORY OF HEALTH DIVISION, MINISTRY OF POST AND TELECOMMUNICATIONS)
3-2, 1-CHOME, KASUMIGASEKI, CHIYODA-KU, TOKYO 100
JAPANESE

IJI KOHO 1645
(MEDICAL JOURNAL)
NIHON IJI SHINBUNSHA-HEN NIHON BYOIU KANRI KYOKAI (JAPAN MEDICAL NEWSPAPER COMPANY EDITOR, JAPAN HOSPITAL MANAGEMENT ASSOCIATION)
JAPANESE

IJI NIPPO 1646
(DAILY NEWS OF MEDICINE)
IJI NIPPOSHA
3-6, AWAJI-CHO, HIGASHI-KU, OSAKA-SHI 541
JAPANESE

IJI SHINBUN 1647
(MEDICAL AFFAIRS NEWSPAPER)
NIHON BYOIN KANRI KYOKAI (JAPAN ASSOCIATION FOR HOSPITAL ADMINISTRATION)
JAPANESE

IJIN YAKUJIN 1648
(BAN'YU PHARMACEUTICAL NEWS)
KYOWA KIKAKU, IJIN YAKUJINSHA
9-3, 2-CHOME, SHIMOMEGURO, MEGURO-KU, TOKYO 153
JAPANESE

IMIC NYUSU 1649
(IMIC NEWS)
KOKUSAI IGAKU JOHO SENTA (INTERNATIONAL MEDICAL INFORMATION CENTER)
26, DAIKYO-CHO, SHINJUKU-KU, TOKYO 160
JAPANESE

INSATSU-KYOKU IHO 1650
(MEDICAL JOURNAL OF THE GOVERNMENT PRINTING BUREAU)
INSATSU-KYOKU IGAKKAI (THE MEDICAL ASSOCIATION OF THE GOVERNMENT PRINTING BUREAU)
C/O OKURA-SHO INSATSU-KYOKU TOKYO BYOIN, 3-6, 2-CHOME, NISHIGAHARA, KITA-KU, TOKYO 114
JAPANESE

INTERNATIONAL MEDICAL NEWS 1651
NIHON KOKUSAI IGAKU KYOKAI (INTERNATIONAL MEDICAL SOCIETY OF JAPAN)
C/O MARUNOUCHI NOMURA BUILDING, 1-1, 2-CHOME, OTE-MACHI, CHIYODA-KU, TOKYO 100
JAPANESE AND EUROPEAN

IRYO 1652
(MEDICAL TREATMENT)
IRYO DOKOKAI (MEDICAL TREATMENT ASSOCIATION)
C/O KOSEI-SHO IMU-KYOKU 2-2, 1-CHOME, KASUMIGASEKI, CHIYODA-KU, TOKYO 100
JAPANESE, WITH ENGLISH SUMMARY
INDEXED: CHE, JICST

IRYO KAI 1653
(MEDICAL TREATMENT WORLD)
RINSHO TSUSHIN SHA (CLINICAL REPORTS COMPANY)
JAPANESE

IRYO MEDICAL JOURNAL OF THE NATIONAL HOSPITAL SANATORIUMS 1654
TOKYO
JAPANESE AND ENGLISH
INDEXED: EXC

IRYO NO HIROBA 1655
(MEDICAL PLAZA)
KOSEI KYOSAIKAI
21, HONSHIO-CHO, SHINJUKU-KU, TOKYO 160
JAPANESE

IRYO SHISETSU CHOSA, BYOIN HOKOKU 1656
(SURVEY OF MEDICAL FACILITIES, REPORT OF HOSPITAL STATISTICS)
KOSEI-SHO DAIJIN KANBO TOKEI CHOSA-BU (THE DIVISION OF HEALTH AND WELFARE STATISTICS, MINISTER'S SECRETARIAT, MINISTRY OF HEALTH AND WELFARE)
42, ICHIGAYA HONMURA-CHO, SHINJUKU-KU, TOKYO 162
JAPANESE

IRYO SHISETSU CHOSA, ISHI, SHIKA ISHI, YAKUZAISHI CHOSA, BYOIN HOKOKU (SAGA-KEN) 1657
(SURVEY OF MEDICAL FACILITIES, DOCTORS, DENTISTS, PHARMACEUTICS AND HOSPITALS, SAGA PREFECTURE)
SAGA-KEN
1-59, 1-CHOME, JONAI, SAGA-SHI 840
JAPANESE

IRYO SHISETSU CHOSA, SOKUHO 1658
(PROMPT REPORT OF MEDICAL FACILITIES SURVEY)
KOSEI-SHO DAIJIN KANBO TOKEI CHOSA-BU (HEALTH AND WELFARE STATISTICS DEPARMENT, MINISTER'S SECRETARIAT, MINISTRY OF HEALTH AND WEALTH)
42, ICHIGAYA HONMURA-CHO, SHINJUKU-KU, TOKYO 162
JAPANESE

IRYO TO FUKUSHI 1659
(MEDICAL CARE AND WELFARE)
NIHON IRYO SHAKAI JIGYO KYOKAI HEN, KAWASHIMA SHOTEN (JAPAN MEDICAL CARE ASSOCIATION EDITION, KAWASHIMA BOOKSTORE)
JAPANESE

IRYOHIN NEWS 1660
NAKAYAMASHIKI SANGYO K. K.
6-48, MINAMI-SENJU, ARAKAWA-KU, TOKYO 116
JAPANESE

ISHI, SHIKAISHI, YAKUZAISHI CHOSA 1661
(SURVEY OF DOCTORS, DENTISTS AND PHARMACEUTISTS)
KOSEI-SHO DAIJIN KANBO TOKEI CHOSA-BU (HEALTH AND WELFARE STATISTICS DEPARTMENT, MINISTER'S SECRETARIAT MINISTRY OF HEALTH AND WELFARE)
42, ICHIGAYA HONMURA-CHO, SHINJUKU-KU, TOKYO 162
JAPANESE

ISHIYAKU SHINPO 1662
(NEWS FOR DOCTOR, DENTIST AND PHARMACEUTIST)
TOKYO ISHIYAKU SHUPPANSHA
1-3, KANDA OGAWA-MACHI, CHIYODA-KU, TOKYO 101
JAPANESE

ITAN 1663
(HISTORY OF MEDICINE)
NIHON ISHI GAKKAI KANSAI SHIBU (KANSAI BRANCH OF THE JAPAN SOCIETY OF MEDICAL HISTORY)
6-18, SEIMEI-DORI, ABENO-KU, OSAKA-SHI 545
JAPANESE

IWACHIDORI 1664
(ROCK PLOVER, PUBLICATION OF THE OGASA MEDICAL ASSOCIATION)
OGASA ISHIKAI (OGASA PHYSICIANS SOCIETY)
JAPANESE

IWATE IGAKU ZASSHI 1665
(JOURNAL OF THE IWATE MEDICAL ASSOCIATION)
IWATE IGAKKAI (IWATE MEDICAL ASSOCIATION)
C/O IWATE IKA DAIGAKU, 19-1, UCHIMARU, MORIOKA-SHI 020
JAPANESE AND ENGLISH
INDEXED: BIO, CHE

IWATE-KENRITSU BYOIN IGAKKAI ZASSHI 1666
(JOURNAL OF THE MEDICAL ASSOCIATION OF IWATE PREFECTURAL HOSPITAL)
1-1, 1-CHOME, HONCHO-DORI, MORIOKA-SHI 020
JAPANESE

IWATE-KENRITSU BYOINTO NENPO 1667
(ANNUAL REPORT ON IWATE PREFECTURAL HOSPITAL AND OTHER HOSPITALS)
IWATE-KEN IRYO-KYOKU
10-2, UCHIMARU, MORIOKA-SHI 020
JAPANESE

IZUMI 1668
(FOUNTAIN, FUJISAWA MEDICAL NEWS)
IZUMISHA (IZUMI AND COMPANY, LIMITED)
25-1, 3-CHOME, IMABASHI, HIGASHI-KU, OSAKA-SHI 541
JAPANESE
INDEXED: CHE

JAPAN INTERNATIONAL MEDICAL TRIBUNE 1669
MEDIKARU TORIBYUN NIHON SHISHA
7-1, 2-CHOME, HATCHOBORI, CHUO-KU, TOKYO 104
JAPANESE

JAPAN MEDICAL GAZETTE 1670
ENGLISH

JAPAN PHARMACEUTICAL LETTER 1671
ENGLISH

JAPANESE JOURNAL OF EXPERIMENTAL MEDICINE 1672
TOKYO DAIGAKU IKAGAKU KENKYUSHO (INSTITUTE OF MEDICAL SCIENCE, UNIVERSITY OF TOKYO)
6-1, 4-CHOME, SHIROGANEDAI, MINATO-KU, TOKYO 108
EUROPEAN
INDEXED: BIO, CHE, EXC, IND, SCI, JICST

JAPANESE JOURNAL OF MEDICAL SCIENCE AND BIOLOGY 1673
KOKURITSU YOBO EISEI KENKYUSHO FUZOKU TOSHOKAN (LIBRARY J.J.M.S.B. EDITORIAL BOARD)
10-35, 2-CHOME, KAMIOSAKI, SHINAGAWA-KU, TOKYO 141
ENGLISH
INDEXED: BIB, BIO, CHE, EXC, IND, SCI, JICST

JAPANESE JOURNAL OF MEDICINE 1674
NIHON NAIKA GAKKAI (JAPANESE SOCIETY OF INTERNAL MEDICINE)
NIHON SHINPAN HONSHA BUILDING, 5F, 33-5, 3-CHOME, HONGO, BUNKYO-KU, TOKYO 113
ENGLISH
INDEXED: BIO, CHE, EXC

JAPANESE JOURNAL OF MICROBIOLOGY 1675
NIHON SAIKIN GAKKAI, NIHON UIRUSU GAKKAI (JAPANESE SOCIETY FOR BACTERIOLOGY, SOCIETY OF JAPANESE VIROLOGISTS) AND IGAKU SHOIN (IGAKU SHOIN LIMITED)
29-11, 5-CHOME, HONGO, BUNKYO-KU, TOKYO 113
ENGLISH
INDEXED: BIO, EXC, IND, JICST

JAPANESE JOURNAL OF TROPICAL MEDICINE 1676
NIHON NETTAI IGAKKAI (JAPAN SOCIETY OF TROPICAL MEDICINE)
C/O TOKYO DAIGAKU IKAGAKU KENKYUSHO, SHIROGANEDAI, MINATO-KU, TOKYO 108
ENGLISH

JAPANESE PERIODICALS INDEX. NATURAL SCIENCES, PART 2: MEDICAL SCIENCES 1677
JAPANESE
INDEXED: JICST

JAPANESE PSYCHOLOGICAL RESEARCH 1678
NIHON SHINRI GAKKAI (JAPANESE PSYCHOLOGICAL ASSOCIATION)
802, BUNKYO CENTER BUILDING, 37-13, 4-CHOME, HONGO, BUNKYO-KU, TOKYO 113
EUROPEAN
INDEXED: BIO

JICHI IKA DAIGAKU NENPO 1679
(ANNUAL REPORT OF THE JICHI MEDICAL COLLEGE)
JAPANESE

JIKEIDAI MEDICAL JOURNAL 1680
TOKYO JIKEIKAI IKA DAIGAKU (JIKEI UNIVERSITY, SCHOOL OF MEDICINE)
25-8, 3-CHOME, NISHI SHINBASHI, MINATO-KU, TOKYO 105
ENGLISH
INDEXED: BIO, CHE, EXC, JICST

JIKKEN DOBUTSU 1681
(BULLETIN OF THE EXPERIMENTAL ANIMALS)
NIHON JIKKEN DOBUTSU KENKYUKAI (JAPAN EXPERIMENTAL ANIMAL RESEARCH ASSOCIATION)
C/O TOKYO DAIGAKU IKAGAKU KENKYUSHO, 6-1, 4-CHOME, SHIROGANEDAI, MINATO-KU, TOKYO 108
JAPANESE AND ENGLISH
INDEXED: BIO, CHE, IND

JIKKEN DOBUTSU DANWAKAI SHOROKUSHU 1682
(ABSTRACTS OF THE SYMPOSIUM OF THE EXPERIMENTAL ANIMALS RESEARCH SOCIETY)
NIHON JIKKEN DOBUTSU KENKYUKAI (EXPERIMENTAL ANIMALS RESEARCH SOCIETY OF JAPAN)
C/O TOKYO DAIGAKU IKAGAKU KENKYUSHO, 6-1, 4-CHOME, SHIROGANEDAI, MINATO-KU, TOKYO 108
JAPANESE

JINSEN IGAKU 1683
(JOURNAL OF JINSEN MEDICAL SCIENCES)
JINSEN IGAKU HENSHUBU
2-7, DAIGAKU-CHO, TAKATSUKI-SHI, OSAKA-FU
JAPANESE AND ENGLISH
INDEXED: BIO, CHE

JOURNAL OF APPLIED MEDICINE 1684
NIHON OYO IGAKKAI (THE SOCIETY OF APPLIED MEDICINE)
108, SHIMOGAMO MIYAZAKI-CHO, SAKYO-KU, KYOTO-SHI 606
JAPANESE AND ENGLISH

JOURNAL OF THE WAKAYAMA MEDICAL SOCIETY 1685
WAKAYAMA
JAPANESE AND ENGLISH
INDEXED: EXC

JUNTENDO DAIGAKU GAKUSEI KENKYU HOKOKUSHU 1686
(RESEARCH REPORT OF JUNTENDO UNIVERSITY STUDENTS)
JUNTENDO DAIGAKU
1-2, 2-CHOME, HONGO, BUNKYO-KU, TOKYO 113
JAPANESE

JUNTENDO IGAKU 1687
(JUNTENDO MEDICAL JOURNAL)
JUNTENDO IGAKKAI (JUNTENDO MEDICAL SOCIETY)
1-1, 2-CHOME, HONGO, BUNKYO-KU, TOKYO 113
JAPANESE AND ENGLISH
INDEXED: CHE, JICST

JUSHO SHINSHIN SHOGAI NI KANSURU KENKYU HOKOKUSHO 1688
(RESEARCH REPORT ON SERIOUS HANDICAPS IN MENTAL AND PHYSICAL)
TOKYO-TO EISEI-KYOKU BYOIN KANRI-BU (DIVISION OF HOSPITAL ADMINISTRATION, SANITARY BUREAU, TOKYO)
7-1, 5-CHOME, HIROO, SHIBUYA-KU, TOKYO 150
JAPANESE

JUZEN IGAKKAI ZASSHI 1689
(JOURNAL OF THE JUZEN MEDICAL SOCIETY)
KANAZAWA DAIGAKU JUZEN IGAKKAI (JUZEN MEDICAL SOCIETY OF KANAZAWA UNIVERSITY)
13-1, TAKARA-MACHI, KANAZAWA-SHI 920
JAPANESE, WITH ENGLISH SUMMARY
INDEXED: CHE

KAGAKU KEISATSU KENKYUSHO HOKOKU, BOHAN SHONEN-HEN 1690
(REPORTS OF THE NATIONAL RESEARCH INSTITUTE OF POLICE SCIENCE, RESEARCH ON PREVENTION OF CRIME AND DELINQUENCY)
KEISATSU-CHO, KAGAKU KEISATSU KENKYUSHO (NATIONAL RESEARCH INSTITUTE OF POLICE SCIENCE, NATIONAL POLICE AGENCY)
6, SANBAN-CHO, CHIYODA-KU, TOKYO 102
JAPANESE, WITH EUROPEAN SUMMARY
INDEXED: CHE, JICST

KAGAMI 1691
(MIRROR)
NIHON JIBI-INKOKA IKAI RENGOKAI (NIHON OTORHINOLARYNGOLOGICAL MEDICAL FEDERATION)
JAPANESE

KAGAWA-KEN ISHIKAISHI 1692
(JOURNAL OF KAGAWA MEDICAL ASSOCIATION)
KAGAWA-KEN ISHIKAI (KAGAWA MEDICAL ASSOCIATION)
4-15, 5-CHOME, BANCHO, TAKAMATSU-SHI 760
JAPANESE

KAGOSHIMA DAIGAKU IGAKU ZASSHI 1693
(MEDICAL JOURNAL OF KAGOSHIMA UNIVERSITY)
KAGOSHIMA DAIGAKU IGAKU ZASSHI HENSHU-BU (EDITORIAL OFFICE OF MEDICAL JOURNAL OF KAGOSHIMA UNIVERSITY)
C/O KAGOSHIMA DAIGAKU IGAKUBU, 7-82, SHIROYAMA-MACHI, KAGOSHIMA-SHI 892
JAPANESE, WITH EUROPEAN SUMMARY
INDEXED: BIO, CHE, JICST

KAGOSHIMA NETTAI IGAKU KENKYUKAI IHO 1694
(BULLETIN OF KAGOSHIMA TROPICAL MEDICINE, FACULTY OF MEDICINE, KAGOSHIMA UNIVERSITY)
KAGOSHIMA DAIGAKU IGAKUBU FUZOKU NETTAI IGAKU KENKYU SHISETSU (RESEARCH INSTITUTE OF TROPICAL MEDICINE, KAGOSHIMA UNIVERSITY)
3-1, HIRATO-CHO, NAZE-SHI 894
JAPANESE AND ENGLISH

KAGOSHIMA-KENRITSU BYOIN NENPO 1695
(ANNUAL REPORT OF KAGOSHIMA PREFECTURAL HOSPITAL)
KAGOSHIMA-KEN EISEI-BU
14-50, YAMASHITA-CHO, KAGOSHIMA-SHI 892
JAPANESE

KAIJO IGAKU KENKYU 1696
(JOURNAL OF DISEASES OF SEAMAN)
SEN'IN HOKENKAI (SEAMAN'S INSURANCE ASSOCIATION)
5-6, 1-CHOME, SHIBUYA, SHIBUYA-KU, TOKYO 150
JAPANESE

KANAGAWA-KENRITSU BYOIN NENPO 1697
(ANNUAL REPORT OF KANAGAWA PREFECTURAL HOSPITAL)
KANAGAWA-KEN EISEI-BU
1, NIHON-ODORI, NAKA-KU, YOKOHAMA-SHI 231
JAPANESE

KANAZAWA DAIGAKU JUZEN IGAKUKAI ZASSHI 1698
(BULLETIN OF THE JUZEN MEDICAL SOCIETY, KANAZAWA UNIVERSITY)
JAPANESE

KANAZAWA IKA DAIGAKU ZASSHI 1699
(JOURNAL OF KANAZAWA MEDICAL UNIVERISTY)
KANAZAWA IKA DAIGAKU IGAKU KAI (KANAZAWA MEDICAL UNIVERSITY MEDICAL SOCIETY)
JAPANESE

KANGO GAKUSEI 1700
(NURSE STUDENTS)
MEDIKARU FURENDOSHA
1-32, 4-CHOME, KUDANKITA, CHIYODA-KU, TOKYO 102
JAPANESE

KANJA CHOSA 1701
(PATIENT SURVEY)
KOSEI-SHO DAIJIN KANBO TOKEI CHOSA-BU (HEALTH AND WELFARE STATISTICS DEPARTMENT, MINISTER'S SECRETARIAT, MINISTRY OF HEALTH AND WELFARE)
42, ICHIGAYA HONMURA-CHO, SHINJUKU-KU, TOKYO 162
JAPANESE

KANPO KENKYU 1702
(CHINESE HERBAL MEDICINE RESEARCH)
GEKKAN KANPO KENKYU (CHINESE HERBAL MEDICINE RESEARCH MONTHLY)
JAPANESE

KANSAI DENRYOKU BYOIN IGAKU ZASSHI 1703
(ANNALS OF KANSAI DENRYOKU HOSPITAL)
KANSAI DENRYOKU BYOIN
3-1, 3-CHOME, DOJIMA HAMADORI, FUKUSHIMA-KU, OSAKA-SHI 553
JAPANESE, WITH ENGLISH SUMMARY

KANSAI IKA DAIGAKU ZASSHI 1704
(JOURNAL OF THE KANSAI MEDICAL SCHOOL)
KANSAI IKA DAIGAKU IGAKKAI (SOCIETY OF KANSAI MEDICAL SCHOOL)
1, FUMIZONO-CHO, MORIGUCHI-SHI 570
JAPANESE, WITH EUROPEAN SUMMARY
INDEXED: BIO, CHE, EXC

KANTO TEISHIN BYOIN KIYO 1705
(REPORT OF KANTO COMMUNICATION HOSPITAL)
KANTO TEISHIN BYOIN
9-22, 5-CHOME, HIGASHI-GOTANDA, SHINAGAWA-KU, TOKYO 141
JAPANESE

KARADA NO KAGAKU 1706
(POPULAR MEDICINE)
NIHON HYORONSHA
14, SUGA-CHO, SHINJUKU-KU, TOKYO 160
JAPANESE

KARUTE 1707
(DIE ALZTLICHE KARTE)
NIHON KEMIFA K. K.
12-12, 2-CHOME, HONKOMAGOME, BUNKYO-KU, TOKYO 113
JAPANESE

KAWASAKI BYOIN IGAKU ZASSHI 1708
(MEDICAL JOURNAL OF KAWASAKI HOSPITAL)
KAWASAKI IKA DAIGAKU FUZOKU KAWASAKI BYOIN
2-1, NAKAYAMASHITA, OKAYAMA-SHI 700
JAPANESE AND ENGLISH

KAWASAKI IGAKKAISHI 1709
(KAWASAKI MEDICAL JOURNAL)
KAWASAKI IKA DAIGAKU (KAWASAKI MEDICAL UNIVERSITY)
JAPANESE

KAWASAKI MEDICAL JOURNAL 1710
KAWASAKI MEDICAL SCHOOL
KURASHIKI 701-01
ENGLISH
INDEXED: BIO, CHE, EXC

KEIO IGAKU 1711
(JOURNAL OF THE KEIO MEDICAL SOCIETY)
KEIO IGAKKAI (KEIO MEDICAL SOCIETY)
C/O KEIO GIJUKU DAIGAKU IGAKUBU, 35, SHINANO-MACHI, SHINJUKU-KU, TOKYO 160
JAPANESE, WITH EUROPEAN SUMMARY
INDEXED: CHE, JICST

KEIO JOURNAL OF MEDICINE 1712
KEIO GIJUKU DAIGAKU IGAKUBU (SCHOOL OF MEDICINE, KEIO UNIVERSITY)
35, SHINANO-MACHI, SHINJUKU-KU, TOKYO 160
ENGLISH
INDEXED: CHE, EXC, IND, JICST

KENKO HOKEN BYOIN BYOREKI NENPO 1713
(ANNUAL REPORT ON THE CASE HISTORY OF HEALTH INSURANCE HOSPITALS)
ZENKOKU SHAKAI HOKEN KYOKAI RENGOKAI (ALL JAPAN FEDERATION OF SOCIAL INSURANCE ASSOCIATION)
10-24, 2-CHOME, HIGASHI GOTANDA, SHINAGAWA-KU, TOKYO 141
JAPANESE

KENKO TO SHIZEN 1714
(HEALTH AND NATURE)
KEN'YUKAN
3-3, TOYOTAMA MINAMI, NERIMA-KU, TOKYO 176
JAPANESE

KENMIN SHOBYO CHOSA (IWATE-KEN) 1715
(SURVEY ON THE RESIDENTS' WOUND AND DISEASE IN IWATE PREFECTURE)
IWATE-KEN
10-1, UCHIMARU, MORIOKA-SHI 020
JAPANESE

KENSA TO GIJUTSU 1716
(MODERN MEDICAL LABORATORY)
IGAKU SHOIN
29-11, 5-CHOME, HONGO, BUNKYO-KU, TOKYO 113
JAPANESE
INDEXED: CHE

KIKAN HOKEN SHINRYO 1717
(INSURANCE EXAMINATION QUARTERLY)
IGAKU TSUSHINSHA
JAPANESE

KINKI DAIGAKU IGAKU ZASSHI 1718
(MEDICAL JOURNAL OF KINKI UNIVERSITY)
KINKI DAIGAKU IGAKKAI (KINKI UNIVERSITY MEDICAL SOCIETY)
JAPANESE

KITAKANTO IGAKU 1719
(KITAKANTO MEDICAL JOURNAL)
KITAKANTO IGAKKAI (KITAKANTO MEDICAL SOCIETY)
C/O GUNMA DAIGAKU IGAKUBU, 39-22, 3-CHOME, SHOWA-MACHI, MAEBASHI-SHI 371
JAPANESE, WITH ENGLISH SUMMARY
INDEXED: CHE, EXC, JICST

KITANO BYOIN KIYO 1720
(KITANO HOSPITAL JOURNAL OF MEDICINE)
TAZUKE KOFUKAI IGAKU KENKYUSHO (TAZUKE KOFUKAI MEDICAL RESEARCH INSTITUTE)
C/O KITANO BYOIN, NISHI OGI-CHO, KITA-KU,
JAPANESE AND EUROPEAN
INDEXED: CHE, EXC

KITAZATO ARCHIVES OF EXPERIMENTAL MEDICINE 1721
KITAZATO JIKKEN IGAKU HENSHU-BU (KITAZATO INSTITUTE)
C/O KITAZATO KENKYUSHO, 9-1, 5-CHOME, SHIROGANE, MINATO-KU, TOKYO 108
ENGLISH
INDEXED: BIO, CHE, EXC

KITAZATO IGAKU 1722
(KITAZATO MEDICINE)
KITAZATO IGAKKAI (KITAZATO MEDICAL SOCIETY)
1, ASAMIZODAI, SAGAMIHARA-SHI 228
JAPANESE AND ENGLISH
INDEXED: CHE,

KITAZATO MEDIKARU NYUSU 1723
(MONTHLY THE KITAZATO MEDICAL NEWS)
KITAZATO MEDIKARU NYUSU HENSHU-BU (KITAZATO INSTITUTE)
C/O KITAZATO KENKYUSHO, 9-1, 5-CHOME, SHIROGANE, MINATO-KU, TOKYO 108
JAPANESE

KOBE DAIGAKU IGAKUBU KIYO 1724
(MEDICAL JOURNAL OF KOBE UNIVERSITY)
KOBE DAIGAKU IGAKUBU KIYO HENSHU IINKAI (EDITORIAL BOARD OF THE MEDICAL JOURNAL OF KOBE UNIVERSITY) AND KOBE DAIGAKU IGAKKAI (MEDICAL SOCIETY OF KOBE UNIVERSITY)
12-1, 7-CHOME, KUSUNOKI-CHO, IKUTA-KU, KOBE-SHI 650
JAPANESE, WITH ENGLISH SUMMARY
INDEXED: BIO, CHE, JICST

KOBE JOURNAL OF MEDICAL SCIENCE 1725
KOBE DAIGAKU IGAKUBU (KOBE UNIVERSITY SCHOOL OF MEDICINE)
12-1, 7-CHOME, KUSUNOKI-CHO, IKUTA-KU, KOBE-SHI 650
ENGLISH
INDEXED: BIO, CHE, EXC, IND, JICST

KOBE-SHIRITSU BYOIN KIYO 1726
(KOBE MUNICIPAL HOSPITAL BULLETIN)
KOBE-SHI EISEI-KYOKU (SANITARY BUREAU, KOBE MUNICIPAL OFFICE)
6-7, KANO-CHO, IKUTA-KU, KOBE-SHI 650
JAPANESE

KOKU IGAKU JIKKENTAI HOKOKU 1727
(REPORTS OF AEROMEDICAL LABORATORY)
BOEI-CHO KOKU IGAKU JIKKENTAI (AEROMEDICAL LABORATORY, JAPAN AIR SELF DEFENSE FORCE)
2-10, 1-CHOME, SAKAE-CHO, TACHIKAWA-SHI 190
JAPANESE
INDEXED: IAA, JICST

KOKU IGAKU JIKKENTAI NENPO 1728
(ANNUAL REPORT OF AEROMEDICAL LABORATORY)
KOKU IGAKU JIKKENTAI (AEROMEDICAL LABORATORY, JAPAN AIR SELF DEFENSE FORCE)
2-10, 1-CHOME, SAKAE-CHO, TACHIKAWA-SHI 190
JAPANESE
INDEXED: JICST

KOKU JIETAI KOKU IGAKU SHIRYO 1729
(AVIATION MEDICINE)
BOEI-CHO KOKU BAKURYO KANBU (AIR STAFF OFFICE, DEFENSE AGENCY)
7-45, 9-CHOME, AKASAKA, MINATO-KU, TOKYO 107
JAPANESE

KOKU UCHU IGAKU SHINRIGAKU 1730
(JAPANESE JOURNAL OF AEROSPACE MEDICINE AND PSYCHOLOGY)
NIHON KOKU UCHU IGAKU SHINRIGAKKAI (JAPAN AEROSPACE MEDICAL AND PSYCHOLOGICAL SOCIETY)
C/O TOKYO JIKEIKAI IKA DAIGAKU UCHU IGAKU KENKYUSHO, 25-8, 3-CHOME, NISHI-SHINBASHI, MINATO-KU, TOKYO 105
JAPANESE AND ENGLISH

KOKUMIN IRYO 1731
(MODERN MEDICAL TREATMENT)
KONDO SHINRYOSHO
ZENKOKU IRYO MONDAI KONDANKAI
2-129, 2-CHOME, TOTSUKA, SHINJUKU-KU, TOKYO 160
JAPANESE

KOKUNAI IGAKU ZASSHI KIJI SAKUIN 1732
(INDEX OF JAPANESE MEDICAL PERIODICALS)
JAPAN MEDIKARU SABISU (JAPAN MEDICAL SERVICE)
2-1, 1-CHOME, KUDAN-KITA, CHIYODA-KU, TOKYO 102
JAPANESE

KOKURA KINEN BYOIN KIYO 1733
(KOKURA MEMORIAL HOSPITAL JOURNAL OF MEDICINE)
KOKURA KINEN BYOIN (KOKURA MEMORIAL HOSPITAL)
1-1, KIFUNE-MACHI, KOKURA-KU, KITAKYUSHU-SHI 802
JAPANESE AND EUROPEAN
INDEXED: JICST

KOKURITSU BYOIN NENPO 1734
(ANNUAL REPORT OF NATIONAL HOSPITALS, JAPAN)
KOSEI-SHO IMU-KYOKU (MEDICAL AFFAIRS BUREAU, MINISTRY OF HEALTH AND WELFARE)
2-2, 1-CHOME, KASUMIGASEKI, CHIYODA-KU, TOKYO 100
JAPANESE

KOKURITSU DAIGAKU BYOIN NENPO 1735
(ANNUAL REPORT OF THE NATIONAL UNIVERSITY HOSPITAL)
MONBU-SHO DAIGAKU GAKUJUTSU-KYOKU (HIGHER EDUCATION AND SCIENCE BUREAU, MINISTRY OF EDUCATION)
2-2, 3-CHOME, KASUMIGASEKI, CHIYODA-KU, TOKYO 100
JAPANESE

KOKURITSU RYOYOJO NENPO 1736
(ANNUAL REPORT OF THE NATIONAL SANATORIUM)
KOSEI-SHO IMU-KYOKU (MEDICAL AFFAIRS BUREAU, MINISTRY OF HEALTH AND WELFARE)
2-2, 1-CHOME, KASUMIGASEKI, CHIYODA-KU, TOKYO 100
JAPANESE

KOHNICHI NO IGAKU 1737
(MEDIZIN VON HEUTE)
BAIERU YAKUHIN K. K. (BAYER YAKUHIN, LIMITED)
10-1, HON-MACHI, HIGASHI-KU, OSAKA-SHI 541
JAPANESE

KOTSU IGAKU 1738
(JOURNAL OF TRANSPORTATION MEDICINE)
NIHON KOTSU IGAKKAI (JAPANESE ASSOCIATION OF TRANSPORTATION MEDICINE)
C/O CHUO TETSUDO BYOIN, 1-3, 2-CHOME, YOYOGI, SHIBUYA-KU, TOKYO 151
JAPANESE, WITH EUROPEAN SUMMARY
INDEXED: CHE, JICST

KOWA IHO 1739
(KOWA MEDICAL MAGAZINE)
KOWA K. K. (KOWA COMPANY, LIMITED)
3-3, NIHONBASHI HON-CHO, CHUO-KU, TOKYO 103
JAPANESE

KOZAN IGAKU KENKYUKAI KAISHI 1740
(REPORT OF MEDICAL RESEARCH SOCIETY FOR MINING AND SMELTING INDUSTRIES)
KOZAN IGAKU KENKYUKAI
C/O NIHON KOGYO KYOKAI, 2-3, UCHISAIWAI-CHO, CHIYODA-KU, TOKYO 100
JAPANESE

KUMAMOTO DAIGAKU TAISHITSU IGAKU KENKYUSHO HOKOKU 1741
(REPORT OF INSTITUTE OF CONSTITUTIONAL MEDICINE, KUMAMOTO UNIVERSITY)
KUMAMOTO DAIGAKU TAISHITSU IGAKU KENKYUSHO (INSTITUTE OF CONSTITUTIONAL MEDICINE, KUMAMOTO UNIVERSITY)
24-1, 4-CHOME, KUHONJI, KUMAMOTO-SHO 862
JAPANESE
INDEXED: CHE

KUMAMOTO IGAKKAI ZASSHI 1742
(JOURNAL OF THE KUMAMOTO MEDICAL SOCIETY)
KUMAMOTO IGAKKAI (KUMAMOTO MEDICAL SOCIETY)
C/O KUMAMOTO DAIGAKU IGAKUBU, 2-1, 2-CHOME, HONJO, KUMAMOTO-SHI 860
JAPANESE, WITH ENGLISH SUMMARY
INDEXED: CHE, JICST

KUMAMOTO MEDICAL JOURNAL 1743
KUMAMOTO DAIGAKU IGAKUBU (KUMAMOTO UNIVERSITY MEDICAL SCHOOL)
2-1, 2-CHOME, HONJO, KUMAMOTO-SHI 860
ENGLISH
INDEXED: BIO, CHE, EXC, IND, JICST

KURASHI TO KENKO 1744
(MODERN LIVING AND HEALTH)
HOKEN DOJINSHA
3-6, 1-CHOME, SARUGAKU-CHO, CHIYODA-KU, TOKYO 101
JAPANESE
INDEXED: JICST

KURASHIKI CHUO BYOIN NENPO 1745
(JAHRESBERICHT DES KURASHIKI-ZENTRALHOSPITALS)
KURASHIKI CHUO BYOIN (KURASHIKI CENTRAL HOSPITAL)
1-1, 1-CHOME, MIWA, KURASHIKI-SHI 710
JAPANESE, WITH EUROPEAN SUMMARY
INDEXED: CHE, JICST

KURUME IGAKKAI ZASSHI 1746
(JOURNAL OF THE KURUME MEDICAL ASSOCIATION)
KURUME IGAKKAI (KURUME MEDICAL ASSOCIATION)
C/O KURUME DAIGAKU IGAKUBU, 67, ASAHI-MACHI, KURUME-SHI 830
JAPANESE, WITH EUROPEAN SUMMARY
INDEXED: CHE, JICST

KURUME MEDICAL JOURNAL 1747
KURUME DAIGAKU IGAKUBU (KURUME UNIVERSITY SCHOOL OF MEDICINE)
67, ASAHI-MACHI, KURUME-SHI 830
ENGLISH
INDEXED: BIO, CHE, EXC, JICST

KYOIKU TO IGAKU 1748
(EDUCATION AND MEDICINE)
KYOIKU TO IGAKU NO KAI (ASSOCIATION FOR THE RESEARCH OF EDUCATION AND MEDICINE)
KEIO TSUSHIN K. K.
19-30, 2-CHOME, MITA, MINATO-KU, TOKYO 108
JAPANESE

KYORIN IGAKKAI ZASSHI 1749
(JOURNAL OF THE KYORIN MEDICAL SOCIETY)
KYORIN IGAKKAI (KYORIN MEDICAL SOCIETY)
20-2, 6-CHOME, SHINKAWA, MITAKA-SHI, TOKYO 181
JAPANESE, WITH ENGLISH SUMMARY
INDEXED: CHE

KYOSAI IHO 1750
(MEDICAL JOURNAL OF MUTUAL AID ASSOCIATION)
KOKKA KOMUIN KYOSAI KUMIAI RENGOKAI (FEDERATION OF MUTUAL AID ASSOCIATION)
2-CHOME, AKASAKA AOI-CHO, MINATO-KU, TOKYO 107
JAPANESE AND ENGLISH
INDEXED: CHE, EXC, JICST

KYOSEI IGAKU 1751
(JOURNAL OF CORRECTIONAL MEDICINE)
NIHON KYOSEI IGAKKAI (JAPANESE ASSOCIATION OF CORRECTIONAL MEDICINE)
C/O HOMU-SHO DAI-1 BEKKAN, 1-1, 1-CHOME, KASUMIGASEKI, CHIYODA-KU, TOKYO 100
JAPANESE AND ENGLISH
INDEXED: JICST

KYOTO IGAKKAI ZASSHI 1752
(JOURNAL OF THE KYOTO MEDICAL ASSOCIATION)
KYOTO-FU ISHI-KAI
MATSUBARA SAGARU, ONMAE-DORI, NAKAGYO-KU, KYOTO-SHI 604
JAPANESE
INDEXED: CHE

KYOTO IHO 1753
(KYOTO MEDICAL JOURNAL)
KYOTO IHOSHA
MATSUBARA SAGARU, ONMAI-DORI, NAKAGYO-KU, KYOTO-SHI 604
JAPANESE

KYOTO-FURITSU IKA DAIGAKU ZASSHI 1754
(JOURNAL OF KYOTO PREFECTURAL UNIVERSITY OF MEDICINE)
KYOTO-FURITSU IKA DAIGAKU IGAKKAI (MEDICAL SOCIETY IN KYOTO PREFECTURAL UNIVERSITY OF MEDICINE)
HIROKOJI, KAWARA-MACHI, KAMIGYO-KU, KYOTO-SHI 602
JAPANESE AND ENGLISH
INDEXED: CHE, JICST

KYOTO-SHIRITSU BYOINHO 1755
(KYOTO PRIVATE HOSPITAL REPORT)
KYOTO SHIRITSU BYOIHHOSHA (KYOTO PRIVATE HOSPITAL REPORT COMPANY)
JAPANESE

KYUDAI IHO 1756
(KYUSHU UNIVERSITY MEDICAL BULLETIN)
KYUSHU DAIGAKU IGAKUBU DOSOKAI (KYUSHU UNIVERSITY MEDICAL SCHOOL ALUMNI ASSOCIATION)
JAPANESE

KYUSHU TETSUDO IGAKKAISHI 1757
(JOURNAL OF TRANSPORTATION MEDICINE)
NIHON KOTSU IGAKKAI KYUSHU CHIHOKAI
C/O MOJI TETSUDO BYOIN, 1-1, 2-CHOME, TAKADA, MOJI-KU, KITAKYUSHU-SHI 800
JAPANESE
INDEXED: EXC

LIBRARY SYSTEM 1758
KEIO GIJUKU DAIGAKU IGAKU JOHO SENTA (MEDICAL LIBRARY AND INFORMATION CENTER, KEIO UNIVERSITY)
35, SHINANO-MACHI, SHINJUKU-KU, TOKYO 16[illegible]
JAPANESE
INDEXED: JICST

MAINICHI RAIFU 1759
(MAINICHI LIFE)
MAINICHI SHINBUNSHA
1-1, HITOTSUBASHI, CHIYODA-KU, TOKYO 100
JAPANESE

MATSUNAGA KENKYUSHO KIYO 1760
(JOURNAL OF MATSUNAGA LABORATORY)
MATSUNAGA KENKYUSHO
3-420, MIYOSHI-CHO, FUKUYAMA-SHI 720
JAPANESE

MATSUYAMA SEKIJUJI BYOIN IGAKU ZASSHI 1761
(HOSPITAL JOURNAL OF MEDICINE)
MATSUYAMA RED CROSS
JAPANESE

MBS MEDIKARU BIJINESU SEKURETARI 1762
(MBS MEDICAL BUSINESS SECRETARY)
NIHON IRYO JIMU KANRISHI KYOKAI
C/O DAI-2 NISHIWAKI BUILDING, 55-8, 1-CHOME,
YOYOGI, SHIBUYA-KU, TOKYO 151
JAPANESE

ME JOHO 1763
(MEDICAL AND BIOLOGICAL ENGINEERING INFORMATION)
IRYO GIJUTSU KENKYU KAIHATSU ZAIDAN (FOUNDATION
FOR RESEARCH ON MEDICAL AND BIOLOGICAL ENGINEERING)
9-11, 1-CHOME, NISHI SHINBASHI, MINATO-KU, TOKYO 105
JAPANESE
INDEXED: JICST

MEDICAL JOURNAL OF OSAKA UNIVERSITY 1764
OSAKA DAIGAKU IGAKUBU (OSAKA UNIVERSITY MEDICAL
SCHOOL)
33, JOAN-CHO, KITA-KU, OSAKA-SHI 530
EUROPEAN
INDEXED: EXC, JICST

MEDICAL JOURNAL OF SHINSHU UNIVERSITY 1765
SHINSHU DAIGAKU IGAKUBU (THE FACULTY OF MEDICINE,
SHINSHU UNIVERSITY)
1-1, 3-CHOME, ASAHI-MACHI, MATSUMOTO-SHI 390
EUROPEAN
INDEXED: JICST

MEDICAL TECHNOLOGY 1766
ISHIYAKU SHUPPAN K.K. (ISHIYAKU PUBLISHERS,
INCORPORATED)
7-10, 1-CHOME, HONKOMAGOME, BUNKYO-KU, TOKYO 113
JAPANESE
INDEXED: CHE

MEDICAMENT NEWS 1767
RAIFU SAIENSU (LIFE SCIENCE)
C/O ITO BUILDING, 42-14, UDAGAWA-CHO, SHIBUYA-KU,
TOKYO 150
JAPANESE

MEDICINA 1768
IGAKU SHOIN
29-11, 5-CHOME, HONGO, BUNKYO-KU, TOKYO 113
JAPANESE

MEDICO-PHILOSOPHIA 1769
IGAKU KENKYUSHA
5-13, 1-CHOME, TOMIGAYA, SHIBUYA-KU, TOKYO 151
JAPANESE

MEDIKARU BYU 1770
(MEDICAL VIEW)
MEJIKARU BYUSHA (MEDICAL VIEW COMPANY,
LIMITED)
C/O KAGAYA SUZUKI BUILDING, 2-4, KOJI-MACHI,
CHIYODA-KU, TOKYO 102
JAPANESE

MEDIKARU NYUSU 1771
(MEDICAL NEWS)
DAINIHON SEIYAKU K. K. (DAINIHON PHARMACEUTICAL
COMPANY, LIMITED)
2-7, NIHONBASHI HON-CHO, CHUO-KU, TOKYO 103
JAPANESE

MEDIKKUSU 1772
(MEDIX)
HITACHI MEJIKO (HITACHI MEDICO)
JAPANESE

MEIDAI BUN'IN NENPO 1773
(ANNUAL REPORT OF THE NAGOYA UNIVERSITY BRANCH
HOSPITAL)
NAGOYA DAIGAKU IGAKUBU FUZOKU BYOIN BUN'IN
(NAGOYA UNIVERSITY, BRANCH HOSPITAL)
2-1, HIGASHIMONZEN-CHO, HIGASHI-KU, NAGOYA-SHI 461
JAPANESE AND EUROPEAN

MEIDAI IGAKUBU GAKUYU JIHO 1774
(ALUMNI REVIEW OF THE MEIJI UNIVERSITY, DEPARTMENT OF
MEDICINE)
MEIDAI IGAKUBU GAKUYUKAI (ALUMNI OF THE MEIJI
UNIVERSITY, DEPARTMENT OF MEDICINE)
JAPANESE

MIE IGAKU 1775
(JOURNAL OF MEDICINE, MIE PREFECTURAL UNIVERSITY)
MIE IGAKKAI, MIE-KENRITSU DAIGAKU IGAKUBU (MIE
MEDICAL ASSOCIATION, MIE PREFECTURAL UNIVERSITY,
SCHOOL OF MEDICINE)
JAPANESE
INDEXED: CHE

MIE MEDICAL JOURNAL 1776
MIE DAIGAKU IGAKUBU (MIE UNIVERSITY, SCHOOL OF
MEDICINE)
2-174, EDOBASHI, TSU-SHI 514
EUROPEAN
INDEXED: BIO, CHE, EXC, JICST

MINAMI BYOIN IGAKU ZASSHI 1777
(MEDICAL JOURNAL OF KYOTO MINAMI HOSPITAL)
MINAMI BYOIN (MINAMI HOSPITAL)
MINAMI-NAKANO-CHO, NISHI-7-JO, SHIMOGYO-KU,
KYOTO-SHI 600
JAPANESE

MINAMI OSAKA BYOIN IGAKU ZASSHI 1778
(MEDICAL JOURNAL OF MINAMI OSAKA HOSPITAL)
MINAMI OSAKA BYOIN (MINAMI OSAKA HOSPITAL)
1-60, HIGASHIKAGAYA-CHO, SUMIYOSHI-KU, OSAKA-SHI 558
JAPANESE AND ENGLISH
INDEXED: CHE, JICST

MIN'IREN IRYO 1779
(QUARTERLY REVIEW OF THE FEDERATION OF ALL JAPAN
DEMOCRATIC MEDICAL TREATMENT ORGANS)
ZEN NIHON MINSHU IRYO KIKAN RENGOKAI
C/O SHINJUKU NOKYO KAIKAN, 5-5, 2-CHOME, YOYOGI,
SHIBUYA-KU, TOKYO 151
JAPANESE

MIN'IREN SHINBUN 1780
(NEWS OF THE FEDERATION OF ALL JAPAN DEMOCRATIC
MEDICAL TREATMENT ORGANS)
ZEN NIHON MINSHU IRYO KIKAN RENGOKAI
C/O SHINJUKU NOKYO KAIKAN, 5-5, 1-CHOME, YOYOGI,
SHIBUYA-KU 151
JAPANESE

MINZOKU EISEI 1781
(HUMAN ECOLOGY AND RACE HYGIENE)
NIHON MINZOKU EISEI GAKKAI (JAPANESE SOCIETY OF
RACE HYGIENE)
C/O TOKYO IKA SHIKA DAIGAKU IGAKUBU NOSON KOSEI
IGAKU KENKYU SHISETSU, 5-45, 1-CHOME, YUSHIMA,
BUNKYO-KU, TOKYO 113
JAPANESE, WITH ENGLISH SUMMARY
INDEXED: CHE

MITSUKOSHI KOSEI JIGYODAN KENKYU NENPO 1782
(RESEARCH REPORT OF MITSUKOSHI WELFARE WORKS
FOUNDATION)
MITSUKOSHI KOSEI JIGYODAN (MITSUKOSHI WELFARE WORKS
FOUNDATION)
24-3, 3-CHOME, SHINJUKU, SHINJUKU-KU, TOKYO 160
JAPANESE AND ENGLISH

MIYAZAKI IKA DAIGAKU KIYO 1783
(BULLETIN OF THE MEDICAL COLLEGE OF MIYAZAKI)
MEDICAL COLLEGE OF MIYAZAKI
JAPANESE

MODAN MEDIA 1784
(MODERN MEDIA)
EIKEN KAGAKU K. K. (EIKEN CHEMICAL COMPANY)
33-8, 1-CHOME, HONGO, BUNKYO-KU, TOKYO 113
JAPANESE

MODERN MEDICINE OF JAPAN 1785
ASAHI SHINBUNSHA
2-3, YURAKU-CHO, CHIYODA-KU, TOKYO 100
JAPANESE

MONBU-SHO KAGAKU KENKYUHI NI YORU KENKYU HOKOKU 1786
SHUROKU, IGAKU OYOBI YAKUGAKU-HEN I
(REPORTS ON RESEARCHES BY GRANTEES, MINISTRY OF
EDUCATION, MEDICAL AND PHARMACEUTICAL SCIENCES,
PART 1)
NIHON GAKUJUTSU SHINKOKAI (JAPAN SOCIETY FOR THE
PROMOTION OF SCIENCE)
3-1, 5-CHOME, KOJI-MACHI, CHIYODA-KU, TOKYO 102
JAPANESE

MONBU-SHO KAGAKU KENKYUHI NI YORU KENKYU HOKOKU 1787
SHUROKU, IGAKU OYOBI YAKUGAKU-HEN II
(REPORTS ON RESEARCHES BY GRANTEES, MINISTRY OF
EDUCATION, MEDICAL AND PHARMACEUTICAL SCIENCES,
PART 2)
NIHON GAKUJUTSU SHINKOKAI (JAPAN SOCIETY FOR THE
PROMOTION OF SCIENCE)
3-1, 5-CHOME, KOJI-MACHI, CHIYODA-KU, TOKYO 102
JAPANESE

MURORAN SEITETSUJO BYOIN ISHI 1788
(MEDICAL RECORD OF MURORAN IRON WORKS HOSPITAL)
SHIN NIHON SEITETSU MURORAN SEITETSUJO BYOIN
1-45, CHIRIBETSU-MACHI, MURORAN-SHI 050
JAPANESE

NAGASAKI IGAKKAI ZASSHI 1789
(NAGASAKI MEDICAL JOURNAL)
NAGASAKI IGAKKAI (NAGASAKI MEDICAL ASSOCIATION)
C/O NAGASAKI DAIGAKU IGAKUBU, 12-4, SAKAMOTO-MACHI, NAGASAKI-SHI 852
JAPANESE AND EUROPEAN
INDEXED: CHE, BIO, JICST

NAGASAKI-KEN IRYO TOKEI 1790
(ANNUAL STATISTICS OF MEDICAL TREATMENT, NAGASAKI PREFECTURE)
NAGASAKI-KEN HOKEN-BU
2-13, EDO-MACHI, NAGASAKI-SHI 850
JAPANESE

NAGOYA IGAKU 1791
(JOURNAL OF THE NAGOYA MEDICAL ASSOCIATION)
NAGOYA IGAKKAI (NAGOYA MEDICAL ASSOCIATION)
65, TSURUMAI-CHO, SHOWA-KU, NAGOYA-SHI 466
JAPANESE
INDEXED: CHE, EXC, JICST

NAGOYA JOURNAL OF MEDICAL SCIENCE 1792
NAGOYA DAIGAKU IGAKUBU (NAGOYA UNIVERSITY, SCHOOL OF MEDICINE)
65, TSURUMAI-CHO, SHOWA-KU, NAGOYA-SHI 466
EUROPEAN
INDEXED: BIO, CHE, EXC, IND, JICST

NAGOYA MEDICAL JOURNAL 1793
NAGOYA-SHIRITSU DAIGAKU IGAKUBU (NAGOYA CITY UNIVERSITY, MEDICAL SCHOOL)
MIZUHO-CHO, MIZUHO-KU, NAGOYA-SHI 467
EUROPEAN
INDEXED: CHE, EXC, JICST

NAGOYA-SHIRITSU DAIGAKU IGAKKAI ZASSHI 1794
(JOURNAL OF THE NAGOYA CITY UNIVERSITY, MEDICAL ASSOCIATION)
NAGOYA-SHIRITSU DAIGAKU IGAKKAI (NAGOYA CITY UNIVERSITY, MEDICAL ASSOCIATION)
KAWASUMI, MIZUHO-CHO, MIZUHO-KU, NAGOYA-SHI 467
JAPANESE
INDEXED: CHE, EXC, JICST

NARA IGAKU ZASSHI 1795
(JOURNAL OF NARA MEDICAL ASSOCIATION)
NARA IGAKKAI (NARA MEDICAL ASSOCIATION)
C/O NARA-KENRITSU DAIGAKU, SHIJO-MACHI, KASHIHARA-SHI 634
JAPANESE AND EUROPEAN
INDEXED: BIO, CHE, EXC, JICST

NETTAI 1796
(TROPICAL MEDICINE)
NIHON NETTAI IGAKU KYOKAI (JAPAN ASSOCIATION FOR TROPICAL MEDICINE)
TOGIN BUILDING, 916, 4-2, 1-CHOME, MARUNOUCHI, CHIYODA-KU, TOKYO 100
JAPANESE AND ENGLISH

NETTAI IGAKU 1797
(TROPICAL MEDICINE)
NAGASAKI DAIGAKU NETTAI IGAKU KENKYUSHO (INSTITUTE FOR TROPICAL MEDICINE, NAGASAKI UNIVERSITY)
12-4, SAKAMOTO-CHO, NAGASAKI-SHI 852
JAPANESE AND ENGLISH
INDEXED: BIO, CHE, EXC

NICHIDAI IGAKU ZASSHI 1798
(MEDICAL JOURNAL OF NIHON UNIVERSITY)
NIHON DAIGAKU IGAKKAI (NIHON UNIVERSITY, MEDICAL ASSOCIATION)
30-1, OYAGUCHI KAMI-CHO, ITABASHI-KU, TOKYO 173
JAPANESE, WITH ENGLISH SUMMARY
INDEXED: BIO, CHE, JICST

NICHIDAI KOKU KAGAKU 1799
(NIHON UNIVERSITY JOURNAL OF ORAL SCIENCE)
NIHON DAIGAKU MATSUDO SHIKA DAIGAKU (NIHON UNIVERSITY MATSUDO DENTAL UNIVERSITY)
JAPANESE

NICHIDOKU IHO 1800
(JAPAN-GERMANY MEDICAL REPORTS)
NICHIDOKU YAKUHIN K.K. (MEDICAL-SCIENTIFIC DEPARTMENT, NICHIDOKU YAKUHIN COMPANY, LIMITED)
2-112, NISHIMIYAHARA-CHO, HIGASHI YODOGAWA-KU, OSAKA-SHI 532
JAPANESE
INDEXED: CHE

NICHIFUTSU IGAKU 1801
(BULLETIN MEDICAL FRANCO-JAPONAIS)
NICHIFUTSU IKAKAI (JAPANO-FRENCH MEDICAL SOCIETY)
2-3, KANDA SURUGADAI, CHIYODA-KU, TOKYO 101
JAPANESE AND EUROPEAN

NICHII NYUSU 1802
(JAPAN MEDICAL ASSOCIATION NEWS)
NIHON ISHIKAI (JAPAN MEDICAL ASSOCIATION)
2-5, KANDA SURUGADAI, CHIYODA-KU, TOKYO 101
JAPANESE

NIHON BYOIN KYOKAI ZASSHI 1803
(MONTHLY JOURNAL OF THE HOSPITAL ASSOCIATION OF JAPAN)
NIHON BYOIN KYOKAI (THE HOSPITAL ASSOCIATION OF JAPAN)
C/O BANCHO KYOSAI KAIKAN, 2, NIBAN-CHO, CHIYODA-KU, TOKYO 102
JAPANESE

NIHON DAIGAKU SHIGAKUBU DOSOKAI KAISHI 1804
(ALUMNI BULLETIN OF THE NIHON UNIVERSITY DENTAL SCHOOL)
JAPANESE

NIHON GAKUJUTSU KAIGI DAI-7-BU SHUHO 1805
(REPORT OF THE 7TH DIVISION, SCIENCE COUNCIL OF JAPAN)
NIHON GAKUJUTSU KAIGI (SCIENCE COUNCIL OF JAPAN)
22-34, 7-CHOME, ROPPONGI, MINATO-KU, TOKYO 106
JAPANESE

NIHON HOIGAKU ZASSHI 1806
(JAPANESE JOURNAL OF LEGAL MEDICINE)
NIHON HOI GAKKAI (MEDICO-LEGAL SOCIETY OF JAPAN)
C/O TOKYO DAIGAKU IGAKUBU, HOIGAKU KYOSHITSU, 3-1, 7-CHOME, HONGO, BUNKYO-KU, TOKYO 113
JAPANESE AND ENGLISH
INDEXED: CHE, EXC

NIHON HOKEN IGAKKAISHI 1807
(JOURNAL OF THE ASSOCIATION OF LIFE INSURANCE MEDICINE OF JAPAN)
NIHON HOKEN IGAKKAI (THE ASSOCIATION OF LIFE INSURANCE MEDICINE OF JAPAN)
C/O SHIN KOKUSAI BUILDING, SEIMEI HOKEN KYOKAI, 3-4, MARUNOUCHI, CHIYODA-KU, TOKYO 100
JAPANESE

NIHON HOSENKINKENKYUKAI KAIHO 1808
(ACTINOMYCETOLOGIST)
NIHON HOSENKINKENKYUKAI (SOCIETY FOR ACTINOMYCETES, JAPAN)
C/O CHIBA DAIGAKU SEIBUTSU KASSEI KENKYUSHO, 9-1, 3-CHOME, IZUMI-CHO, NARASHINO-SHI 275
JAPANESE AND ENGLISH

NIHON IGAKKAI SOKAI KAISHI 1809
(PROCEEDINGS OF THE SYMPOSIUM OF JAPANESE ASSOCIATION OF MEDICAL SCIENCES)
NIHON IGAKKAI (JAPANESE ASSOCIATION OF MEDICAL SCIENCES)
2-5, KANDA SURUGADAI, CHIYODA-KU TOKYO 101
JAPANESE

NIHON IJI SHINBUN 1810
(JAPAN MEDICAL NEWS)
NIHON IJI SHINBUNSHA
IZUMI KAIKAN, NIHONBASHI KAYABA-CHO, CHUO-KU, TOKYO 103
JAPANESE

NIHON IJI SHINPO 1811
(JAPAN MEDICAL JOURNAL)
NIHON IJI SHINPOSHA
2-9, KANDA SURUGADAI, CHIYODA-KU, TOKYO 101
JAPANESE

NIHON IRYO EISEI SHINBUN 1812
(JAPAN MEDICAL TREATMENT SANITARY NEWS)
NIHON IRYO EISEI SHINBUNSHA
2-20, JUNKEI-CHO, MINAMI-KU, OSAKA-SHI 542
JAPANESE

NIHON ISHI GAKKAI KAIHO 1813
(TRANSACTIONS OF THE JAPAN MEDICAL HISTORY SOCIETY)
JAPANESE

NIHON ISHIGAKU ZASSHI 1814
(JOURNAL OF THE JAPANESE SOCIETY OF MEDICAL HISTORY)
NIHON ISHI GAKKAI (JAPANESE SOCIETY OF MEDICAL HISTORY)
C/O JUNTENDO DAIGAKU IGAKUBU ISHIGAKU KENKYUSHITSU, 1-1, 2-CHOME, HONGO, BUNKYO-KU, TOKYO 113
JAPANESE, WITH ENGLISH SUMMARY

NIHON ISHIKAI ZASSHI 1815
(JOURNAL OF THE JAPAN MEDICAL ASSOCIATION)
NIHON ISHIKAI (JAPAN MEDICAL ASSOCIATION)
2-5, KANDA SURUGADAI, CHIYODA-KU, TOKYO 101
JAPANESE
INDEXED: CHE

NIHON NETTAI IGAKKAI ZASSHI 1816
(JAPANESE JOURNAL OF TROPICAL MEDICINE AND HYGIENE)
NIHON NETTAI IGAKKAI (JAPAN SOCIETY OF TROPICAL MEDICINE)
C/O NAGASAKI DAIGAKU NETTAI IGAKU KENKYUSHO, 12-4, SAKAMOTO-MACHI, NAGASAKI-SHI 852
JAPANESE AND ENGLISH
INDEXED: CHE

NIHON NOSON IGAKKAI ZASSHI 1817
(JOURNAL OF THE JAPANESE ASSOCIATION OF RURAL MEDICINE)
NIHON NOSON IGAKKAI (JAPANESE ASSOCIATION OF RURAL MEDICINE)
C/O NOKYO BUILDING, 8-3, 1-CHOME, OTE-MACHI, CHIYODA-KU, TOKYO 100
JAPANESE, WITH ENGLISH SUMMARY

NIHON NOSON IGAKU KENKYUJO NENPO 1818
(ANNUAL REPORT OF JAPAN INSTITUTE OF RURAL MEDICINE)
NIHON NOSON IGAKU KENKYUJO (JAPAN INSTITUTE OF RURAL MEDICINE)
JAPANESE

NIHON OYO SHINRI GAKKAI KAIHO 1819
(BULLETIN OF THE JAPAN SOCIETY FOR APPLIED PSYCHOLOGY)
NIHON OYO SHINRI GAKKAI
C/O KEIO GIJUKU DAIGAKU GAKUSEI SODANSHITSU, 15-45, 2-CHOME, MITA, MINATO-KU, TOKYO 108
JAPANESE

NIHON OYO SHINRI GAKKAI TAIKAI RONBUNSHU 1820
(PROCEEDINGS OF THE SYMPOSIUM OF JAPAN SOCIETY FOR APPLIED PSYCHOLOGY)
NIHON OYO SHINRI GAKKAI
C/O KEIO GIJUKU DAIGAKU GAKUSEI SODANSHITSU, 15-45, 2-CHOME, MITA, MINATO-KU, TOKYO 108
JAPANESE

NIHON P.S. GAKKAI SHI 1821
(JOURNAL OF MIE JAPAN P.S. SOCIETY)
NIHON P.S. GAKKAI
JAPANESE

NIHON SAIKINGAKU ZASSHI 1822
(JAPANESE JOURNAL OF BACTERIOLOGY)
NIHON SAIKIN GAKKAI (JAPANESE SOCIETY FOR BACTERIOLOGY)
C/O NIHON GAKKAI JIMU SENTA, 4-16, 2-CHOME, YAYOI, BUNKYO-KU, TOKYO 113
JAPANESE, WITH EUROPEAN SUMMARY
INDEXED: CHE, BIO, EXC, IND, JICST

NIHON SEIKISHO GAKKAI ZASSHI 1823
(JAPANESE JOURNAL OF BIOMETEOROLOGY)
NIHON SEIKISHO GAKKAI (JAPANESE SOCIETY OF BIOMETEOROLOGY)
C/O TOKYO IKA SHIKA DAIGAKU IGAKUBU EISEIGAKU KYOSHITSU, 5-45, 1-CHOME, YUSHIMA, BUNKYO-KU, TOKYO 113
JAPANESE AND ENGLISH

NIHON SHAKAI HOKEN IGAKKAI ENZETSUSHU 1824
(PROCEEDINGS OF THE SYMPOSIUM OF THE JAPAN SOCIETY OF SOCIAL INSURANCE MEDICINE)
NIHON SHAKAI HOKEN IGAKKAI
1-1, 8-CHOME, ICHIJO, NAKANOSHIMA, TOYOHIRA-KU, SAPPORO-SHI 062
JAPANESE

NIHON SHINRI GAKKAI TAIKAI HAPPYO RONBUNSHU 1825
(RESEARCH PAPERS AT THE ANNUAL MEETING OF JAPAN SOCIETY OF PSYCHOLOGY)
NIHON SHINRI GAKKAI
37-13, 4-CHOME, HONGO, BUNKYO-KU, TOKYO 113
JAPANESE

NIHON SOGO AIIKU KENKYUSHO KIYO 1826
(REPORT OF THE STUDIES OF NIPPON AIIKU RESEARCH INSTITUTE)
NIHON SOGO AIIKU KENKYUSHO
6-8, 5-CHOME, MINAMI AZABU, MINATO-KU, TOKYO 106
JAPANESE

NIHON TOSHI IGAKKAISHI 1827
(JOURNAL OF THE JAPANESE SOCIETY OF URBAN MEDICINE)
NIHON TOSHI IGAKKAI (JAPANESE SOCIETY OF URBAN MEDICINE)
C/O TOKYO-TORITSU EBARA BYOIN, 5-10, 4-CHOME, HIGASHI YUKIGAYA, OTA-KU, TOKYO 145
JAPANESE

NIHON TOYO IGAKKAI KAIHO 1828
(NEWS OF THE SOCIETY FOR ORIENTAL MEDICINE IN JAPAN)
NIHON TOYO IGAKKAI (SOCIETY FOR ORIENTAL MEDICINE IN JAPAN)
C/O CHUJOTO BUILDING, 3-8, NIHONBASHI-DORI, CHUO-KU, TOKYO 103
JAPANESE

NIHON TOYO IGAKKAISHI 1829
(JOURNAL OF SOCIETY FOR ORIENTAL MEDICINE IN JAPAN)
NIHON TOYO IGAKKAI (SOCIETY FOR ORIENTAL MEDICINE IN JAPAN)
C/O CHUJOTO BUILDING, 3-8, NIHONBASHI-DORI, CHUO-KU, TOKYO 103
JAPANESE AND ENGLISH

NIHON UIRUSU GAKKAI KAIHO 1830
(BULLETIN OF THE SOCIETY OF JAPANESE VIROLOGISTS)
NIHON UIRUSU GAKKAI
C/O NIHON GAKKAI JIMU SENTA, 4-16, 2-CHOME, YAYOI, BUNKYO-KU, TOKYO 113
JAPANESE

NIHON UNIVERSITY JOURNAL OF MEDICINE 1831
NIHON DAIGAKU IGAKUBU (NIHON UNIVERSITY, SCHOOL OF MEDICINE)
30, OYAGUCHI KAMI-MACHI, ITABASHI-KU, TOKYO 173
ENGLISH
INDEXED: EXC, JICST

NIIGATA IGAKKAI ZASSHI 1832
(NIIGATA MEDICAL JOURNAL)
NIIGATA IGAKKAI (NIIGATA MEDICAL SOCIETY)
C/O NIIGATA DAIGAKU FUZOKU TOSHOKAN ASAHI-MACHI BUNKAN, ASAHI-MACHI, NIIGATA-SHI 950
JAPANESE
INDEXED: CHE, JICST

NIIGATA-KEN ISHIKAIHO 1833
(JOURNAL OF NIIGATA MEDICAL ASSOCIATION)
NIIGATA-KEN ISHIKAI
36, NIBAN-CHO, IGAKU-CHO, NIIGATA-SHI 950
JAPANESE

NIIGATA-KENRITSU BYOIN IGAKKAISHI 1834
(ANNUAL REPORT OF MEDICAL ASSOCIATION OF NIIGATA PREFECTURAL HOSPITAL)
NIIGATA-KENRITSU CHUO BYOIN (NIIGATA PREFECTURAL CENTRAL HOSPITAL)
3-20, OTE-MACHI, JOETSU-SHI 943
JAPANESE

NIIGATA-KENRITSU BYOIN NENPO 1835
(ANNUAL REPORT OF NIIGATA PREFECTURAL HOSPITAL)
NIIGATA-KEN BYOIN-KYOKU
602, ICHIBAN-CHO, GAKKO-DORI, NIIGATA-SHI 951
JAPANESE

NIKKEI MEDIKARU 1836
(NIKKEI MEDICAL)
NIKKEI MAGUROUHIRUSHA (NIKKEI MCGRAW-HILL, INCORPORATED)
C/O NIKKEI BEKKAN, 1-2, 2-CHOME, UCHI-KANDA, CHIYODA-KU, TOKYO 101
JAPANESE

NINGEN IGAKU 1837
(HUMAN MEDICINE)
NINGEN IGAKUSHA
16-23, 5-CHOME, SENRIOKA, SETTSU-SHI 564
JAPANESE

NINGEN NO IGAKU 1838
(HOLISTIC MEDICINE)
JITCHI IKA NO TAME NO KAI (MEDICAL PRACTITIONER'S SOCIETY)
C/O KYOWA KIKAKU, 2-20, SHINBASHI, MINATO-KU, TOKYO 105
JAPANESE AND ENGLISH

NIPPON IKA DAIGAKU ZASSHI 1839
(JOURNAL OF NIPPON MEDICAL SCHOOL)
NIPPON IKA DAIGAKU IGAKKAI (MEDICAL ASSOCIATION OF NIPPON MEDICAL SCHOOL)
1-5, 1-CHOME, SENDAGI, BUNKYO-KU, TOKYO 113
JAPANESE AND ENGLISH
INDEXED: CHE, JICST

NIPPON RINSHO DENSHI KENBIKYO GAKKAISHI 1840
(JOURNAL OF CLINICAL ELECTRON MICROSCOPY)
NIPPON RINSHO DENSHI KENBIKYO GAKKAI (CLINICAL
ELECTRON MICROSCOPY SOCIETY OF JAPAN)
AND KISSEI YAKUHIN KOGYO KAISHA (KISSEI
PHARMACEUTICAL COMPANY)
2-5, NIHONBASHI HON-CHO, CHUO-KU, TOKYO 103
JAPANESE AND ENGLISH
INDEXED: CHE

NISHI IGAKU 1841
(NISHI SYSTEM OF MEDICAL SCIENCE)
NISHI-KAI
SEIBU KANKO KAIKAN, 12, GOBAN-CHO, CHIYODA-KU,
TOKYO 102
JAPANESE

NISSEI BYOIN IGAKU ZASSHI 1842
(JOURNAL OF THE NISSEI HOSPITAL)
NISSEI BYOIN (NISSEI HOSPITAL)
4-11, ITACHIBORI MINAMI-DORI, NISHI-KU, OSAKA-SHI
550
JAPANESE, WITH ENGLISH SUMMARY

NISSEKI IGAKU 1843
(JAPANESE RED CROSS MEDICAL JOURNAL)
OSAKA SEKIJUJI BYOIN ISHIYAKUGAKU KENKYUKAN
(MEDICAL, DENTAL AND PHARMACEUTICAL INSTITUTE,
OSAKA RED CROSS HOSPITAL)
50, FUDEGASAKI-MACHI, TENNOJI-KU, OSAKA-SHI 543
JAPANESE AND ENGLISH

OCHANOMIZU IGAKU ZASSHI 1844
(OCHANOMIZU MEDICAL JOURNAL)
OCHANOMIZU IGAKKAI (OCHANOMIZU MEDICAL ASSOCIATION)
C/O TOKYO IKASHIKA DAIGAKU IGAKUBU, 5-45, 1-CHOME,
YUSHIMA, BUNKYO-KU, TOKYO 113
JAPANESE
INDEXED: CHE, JICST

OGATA IGAKU KAGAKU KENKYUSHO KENKYU HOKOKU 1845
(RESEARCH BULLETIN OF THE OGATA INSTITUTE FOR
MEDICAL AND CHEMICAL RESEARCH)
OGATA INSTITUTE FOR MEDICAL AND CHEMICAL RESEARCH
2-3-5 NIHONBASHI BAKURO-CHO, CHUO-KU, TOKYO 103
JAPANESE
INDEXED: CHE

OGATA IGAKU KENKYUSHO NENPO 1846
(ANNUAL REPORT OF OGATA INSTITUTE FOR MEDICAL AND
CHEMICAL RESEARCH)
OGATA IGAKU KAGAKU KENKYUSHO (OGATA INSTITUTE FOR
MEDICAL AND CHEMICAL RESEARCH)
11-4, 1-CHOME, HIGASHI KANDA, CHIYODA-KU, TOKYO 101
JAPANESE

OHARA SOGO BYOIN NENPO 1847
(ANNUAL REPORT OF OHARA GENERAL HOSPITAL)
OHARA SOGO BYOIN KENKYUSHO (LABORATORY OF OHARA
GENERAL HOSPITAL)
6-11, O-MACHI, FUKUSHIMA-SHI 960
JAPANESE, WITH EUROPEAN SUMMARY

OITA-KEN ISHIKAI KAIHO 1848
(MONTHLY BULLETIN OF OITA MEDICAL ASSOCIATION)
OITA-KEN ISHI-KAI
6-23, NIAGE-CHO, OITA-SHI 870
JAPANESE

OKAYAMA IGAKKAI ZASSHI 1849
(JOURNAL OF MEDICAL ASSOCIATION OF OKAYAMA)
OKAYAMA IGAKKAI (MEDICAL ASSOCIATION OF OKAYAMA)
C/O OKAYAMA DAIGAKU IGAKUBU, 5-1, 2-CHOME,
SHIKATA-CHO, OKAYAMA-SHI 700
JAPANESE, WITH EUROPEAN SUMMARY
INDEXED: CHE

OKAYAMA SAISEIKAI SOGO BYOIN ZASSHI 1850
(JOURNAL OF OKAYAMA SAISEIKAI GENERAL HOSPITAL)
OKAYAMA SAISEIKAI SOGO BYOIN (OKAYAMA SAISEIKAI
GENERAL HOSPITAL)
17-18, 1-CHOME, IFUKU-CHO, OKAYAMA-SHI 700
JAPANESE AND ENGLISH

OKINAWA IGAKKAI ZASSHI 1851
(OKINAWA MEDICAL JOURNAL)
OKINAWA IGAKKAI (OKINAWA MEDICAL ASSOCIATION)
C/O OKINAWA-KEN ISHI KAIKAN, 26-1, HIGASHI-MACHI,
NAHA-SHI 900
JAPANESE

OMORI ISHIKAI KAIHO 1852
(REPORT OF THE OMORI MEDICAL DOCTORS ASSOCIATION)
JAPANESE

OSAKA CITY MEDICAL JOURNAL 1853
OSAKA-SHI IGAKKAI (OSAKA CITY MEDICAL CENTER)
4-54, 1-CHOME, ASAHI-MACHI, ABENO-KU, OSAKA-SHI 545
EUROPEAN
INDEXED: BIO, CHE, IND, JICST

OSAKA DAIGAKU IGAKU ZASSHI 1854
(MEDICAL JOURNAL OF OSAKA UNIVERSITY)
OSAKA DAIGAKU IGAKUBU
33, JOAN-CHO, KITA-KU, OSAKA-SHI 530
JAPANESE
INDEXED: BIO, CHE, IND, JICST

OSAKA DAIGAKU IRYO GIJUTSU TANKI DAIGAKUBU 1855
KENKYU KIYO: SHIZEN KAGAKU, IRYO KAGAKU-HEN
(STUDIES IN NATURAL SCIENCE AND HEALTH TECHNOLOGY)
JAPANESE

OSAKA IKA DAIGAKU ZASSHI 1856
(JOURNAL OF OSAKA MEDICAL COLLEGE)
OSAKA IKA DAIGAKU IGAKKAI
2-7, DAIGAKU-MACHI, TAKATSUKI-SHI 569
JAPANESE
INDEXED: CHE, JICST

OSAKA-SHIRITSU DAIGAKU IGAKU ZASSHI 1857
(JOURNAL OF THE OSAKA CITY MEDICAL CENTER)
OSAKA-SHI IGAKKAI (OSAKA CITY MEDICAL CENTER)
C/O OSAKA-SHIRITSU DAIGAKU IGAKUBU, 4-54, 1-CHOME,
ASAHI-MACHI, ABENO-KU, OSAKA-SHI 545
JAPANESE
INDEXED: EXC, JICST

OSAKASHI IGAKKAI ZASSHI 1858
(JOURNAL OF THE OSAKA MEDICAL ASSOCIATION)
OSAKASHI IGAKKAI
JAPANESE

OTARU-SHI IJI KENKYU KAISHI 1859
(OTARU MEDICAL JOURNAL)
OTARU-SHI IJI KENKYUKAI
C/O SHIRITSU OTARU BYOIN, 1, RYOTOKU-MACHI, OTARU
JAPANESE
INDEXED: CHE, JICST

PSYCHOLOGIA (AN INTERNATIONAL JOURNAL OF PSYCHOLOGY 1860
IN THE ORIENT)
PUSHIKORIOGIA-KAI (POSYCHOLOGIA SOCIETY)
C/O KYOTO DAIGAKU KYOIKUGAKUBU, KYOIKU SHINRIGAKU
KYOSHITSU, YOSHIDA HON-CHO, SAKYO-KU, KYOTO-SHI 606
EUROPEAN

REPORTS OF THE NUKADA INSTITUTE 1861
NUKADA IGAKU SEIBUTSUGAKU KENKYUSHO (THE NUKADA
INSTITUTE FOR MEDICAL AND BIOLOGICAL RESEARCH)
5-18, INAGE-CHO, CHIBA-SHI 280
ENGLISH

RIKKYO DAIGAKU SHINRIGAKKA KENKYU NENPO 1862
(ANNUAL RESEARCH REPORT OF RIKKYO UNIVERSITY
PSYCHOLOGY DEPARTMENT)
RIKKYO DAIGAKU BUNGAKU-BU SHINRIGAKU KENKYUSHITSU
(RIKKYO UNIVERSITY SCHOOL OF LITERATURE,
PSYCHOLOGY LABORATORY)
JAPANESE

RIKUJO JIEITAI SAPPORO-CHIKU BYOIN KENKYU NENPO 1863
(ANNUAL RESEARCH REPORT, SAPPORO DISTRICT HOSPITAL,
GROUND SELF DEFENSE FORCE)
RIKUJO JIEITAI SAPPORO-CHIKU BYOIN
12-CHOME, 1-JO, HIRAGISHI, TOYOHIRA-KU,
SAPPORO-SHI 062
JAPANESE AND ENGLISH

RINSHO KENKYU KIYO 1864
(CLINICAL RESEARCH BULLETIN)
TOKYO-TO EISEI KYOKU (TOKYO METROPOLITAN GOVERNMENT,
BUREAU OF HEALTH)
JAPANESE

RINSHO KENSA 1865
(JOURNAL OF MEDICAL TECHNOLOGY)
IGAKU SHOIN
29-11, 5-CHOME, HONGO, BUNKYO-KU, TOKYO 113
JAPANESE
INDEXED: CHE

RINSHO KENSA SENTA DAYORI 1866
(CLINICAL EXAMINATION CENTER LETTER)
NAGOYA ISHI KYODO KUMIAI, NAGOYA RINSHO KENSA
SENTA (NAGOYA MEDICAL DOCTOR'S COOPERATIVE
ASSOCIATION, NAGOYA CLINICAL EXAMINATION CENTER)
JAPANESE

RINSHO TO UIRUSU 1867
(CLINICAL VIROLOGY)
KINDAI SHUPPAN
5-3, MARUYAMA-CHO, SHIBUYA-KU, TOKYO 150
JAPANESE

RISSHO DAIGAKU TETSUGAKU. SHINRIGAKKAI KIYO 1868
(BULLETIN OF THE RISSHO UNIVERSITY PHILOSOPHICAL AND PSYCHOLOGICAL SOCIETY)
JAPANESE

RYOIKU NO MADO 1869
(NEWS OF THE JAPAN WELFARE FOUNDATION FOR MENTALLY AND PHYSICALLY HANDICAPPED CHILDREN)
ZENKOKU SHINSHIN SHOGAIZI FUKUSHI ZAIDAN
43, TOYAMA-CHO, SHINJIKU-KU, TOKYO 162
JAPANESE

SAIMINGAKU KENKYU 1870
(JAPANESE JOURNAL OF HYPNOSIS)
NIHON SAIMIN IGAKU SHINRI GAKKAI (JAPANESE SOCIETY OF HYPNOSIS) AND KAWASHIMA SHOTEN
15-17, 7-CHOME, NISHISHINJUKU, SHINJUKU-KU, TOKYO 160
JAPANESE

SAISEI 1871
(WELFARE)
SAISEIKAI (SOCIAL WELFARE ORGANIZATION IMPERIAL GIFT FOUNDATION INCORPORATED SAISEIKAI)
4-36, 1-CHOME, MITA, MINATO-KU, TOKYO 108
JAPANESE AND ENGLISH

SAISHIN IGAKU 1872
(MEDICINE TODAY)
SAISHIN IGAKUSHA
37, CHAYA-MACHI, KITA-KU, OSAKA-SHI 530
JAPANESE
INDEXED: CHE

SAITAMA IKA DAIGAKU ZASSHI 1873
(JOURNAL OF THE SAITAMA MEDICAL SCHOOL)
SAITAMA IKA DAIGAKU IGAKKAI (THE MEDICAL RESEARCH SOCIETY OF THE SAITAMA MEDICAL SCHOOL)
JAPANESE

SAN'EI REPORTO, IGAKU KANKEI 1874
(SAN'EI REPORT OF MEDICINE)
SAN'EI SOKKI K. K. (SAN'EI INSTRUMENT COMPANY, LIMITED)
1-57, TENJIN-CHO, KODAIRA-SHI, TOKYO 187
JAPANESE

SANGYO IGAKU 1875
(JAPANESE JOURNAL OF INDUSTRIAL HEALTH)
NIHON SANGYO EISEI GAKKAI (JAPAN ASSOCIATION OF INDUSTRIAL HEALTH)
C/O KOSHU EISEI BUILDING, 29-8, 1-CHOME, SHINJUKU, SHINJUKU-KU, TOKYO 160
JAPANESE, WITH EUROPEAN SUMMARY
INDEXED: CHE, EXC, JICST

SAPPORO IGAKU ZASSHI 1876
(SAPPORO MEDICAL JOURNAL)
SAPPORO IKA DAIGAKU (SAPPORO MEDICAL COLLEGE)
NISHI-17-CHOME, MINAMI-1-JO, CHUO-KU, SAPPORO-SHI 060
JAPANESE AND ENGLISH
INDEXED: CHE, EXC, JICST

SAPPORO IKA DAIGAKU IGAKU SHINGAKU KATEI KIYO 1877
(JOURNAL OF PRE-MEDICAL COURSE, SAPPORO MEDICAL COLLEGE)
SAPPORO IKA DAIGAKU (SAPPORO MEDICAL COLLEGE)
NISHI-17-CHOME, MINAMI-1-JO, CHUO-KU, SAPPORO-SHI
JAPANESE AND ENGLISH
INDEXED: BIO, JICST

SEI MARIA IGAKU 1878
(ST. MARY'S MEDICAL JOURNAL)
SEI MARIA BYOIN (ST. MARY'S HOSPITAL)
422, TSUBUKUHON-MACHI, KURUME-SHI 830
JAPANESE

SEI MARIANNA IKA DAIGAKU ZASSHI 1879
(ST. MARIANNA MEDICAL JOURNAL)
SEI MARIANNA IKA DAIGAKU
2095, SUGAO, TAKATSU-KU, KAWASAKI-SHI 213
JAPANESE, WITH ENGLISH SUMMARY

SEIMEI TO KIKETSU 1880
(MIND BODY AND LIFE)
SHIN KETSUEKI GAKKAI (SOCIETY OF NEO-HAEMATOLOGY) AND SEIMEI MONDAI KONWAKAI
4-26, NAKA MONZEN-CHO, KAGAMIHARA-SHI 504
JAPANESE

SEITAI JOHO KAGAKU KENKYU 1881
(BIOMEDICAL INFORMATION SCIENCE)
KOBE DAIGAKU IGAKUBU DAINI SEIRIGAKU KYOSHITSU SEITAI JOHO SHORI KENKYU GURUPU (BIOMEDICAL INFORMATION HANDLING RESEARCH GROUP, PHYSIOLOGY SECTION #2, KOBE UNIVERSITY MEDICAL SCHOOL)
JAPANESE

SHAKAI HOKEN IGAKU ZASSHI 1882
(JOURNAL OF HEALTH INSURANCE MEDICINE)
ZENKOKU SHAKAI HOKEN KYOKAI RENGOKAI (ALL JAPAN FEDERATION OF SOCIAL INSURANCE ASSOCIATION)
10-24, 2-CHOME, HIGASHI GOTANDA, SHINAGAWA-KU, TOKYO 141
JAPANESE

SHAKAI HOKEN NARUWA SOGO BYOIN NENPO 1883
(ANNUAL REPORT OF SOCIAL INSURANCE NARUWA GENERAL HOSPITAL)
SHAKAI HOKEN NARUWA SOGO BYOIN
3-1, OHI-MACHI, KANAZAWA-SHI 920
JAPANESE

SHAKAI IGAKU KENKYU 1884
(KYOTO JOURNAL OF SOCIAL MEDICINE)
KYOTO SHAKAI IGAKU KENKYUKAI (KYOTO INSTITUTE OF SOCIAL MEDICINE)
C/O KYOTO DAIGAKU IGAKUBU KOSHU EISEIGAKU KYOSHITSU, YOSHIDA KONOE-CHO, SAKYO-KU, KYOTO-SHI 606
JAPANESE

SHAKAI IGAKU KENKYUKAI SOKAI KIRYOKU 1885
(MINUTES OF THE GENERAL MEETING OF THE SOCIAL MEDICINE RESEARCH ASSOCIATION)
JAPANESE

SHAKAI IRYO CHOSA HOKOKU 1886
(RESEARCH REPORT OF SOCIAL MEDICINE)
KOSEI-SHO DAIJIN KANBO TOKEI CHOSA-BU (HEALTH AND WELFARE STATISTICS DIVISION, MINISTER'S SECRETARIAT, MINISTRY OF HEALTH AND WELFARE)
42, ICHIGAYA HONMURA-CHO, SHINJUKU-KU, TOKYO 162
JAPANESE

SHIGA-KEN KANJA JITTAI CHOSA 1887
(PATIENT SURVEY IN SHIGA PREFECTURE)
SHIGA-KEN KOSEI-BU
1-1, 4-CHOME, KYO-MACHI, OTSU-SHI 520
JAPANESE

SHIGA-KENRITSU SEIJINBYO SENTA NENPO 1888
(ANNUAL REPORT OF THE CENTER FOR ADULT DISEASES, SHIGA PREFECTURE)
SHIGA-KENRITSU SEIJINBYO SENTA
328-1, MORIYAMA-CHO, MORIYAMA-SHI 524
JAPANESE

SHIKOKU IGAKU ZASSHI 1889
(SHIKOKU ACTA MEDICA)
TOKUSHIMA DAIGAKU IGAKUBU (TOKUSHIMA UNIVERSITY, SCHOOL OF MEDICINE) AND TOKUSHIMA IGAKKAI (TOKUSHIMA MEDICAL ASSOCIATION)
C/O TOKUSHIMA DAIGAKU IGAKUBU, 3, KURATE-MACHI, TOKUSHIMA-SHI 770
JAPANESE AND ENGLISH
INDEXED: CHE, EXC, JICST

SHIMANE IGAKU 1890
(JOURNAL OF THE SHIMANE MEDICAL ASSOCIATION)
SHIMANE-KEN ISHIKAI (THE SHIMANE MEDICAL ASSOCIATION)
8, SODESHI-CHO, MATSUE-SHI 690
JAPANESE

SHIMANE-KEN KANJA CHOSA 1891
(PATIENT SURVEY IN SHIMANE PREFECTURE)
SHIMANE-KEN
1, TONO-MACHI, MATSUE-SHI 690
JAPANESE

SHINDAN TO CHIRYO 1892
(DIAGNOSIS AND TREATMENT)
SHINDAN TO SHIRYOSHA
406, MARUNOUCHI BUILDING, MARUNOUCHI, CHIYODA-KU, TOKYO 100
JAPANESE AND ENGLISH
INDEXED: CHE

SHIN'EN KENKYU 1893
(SHIN'EN RESEARCH)
AKITA KEIZAI DAIGAKU SATO SHIN'EN KENKYUKAI
JAPANESE

SHINRI SOKUTEI JONARO 1894
(PSYCHOLOGICAL MEASUREMENT JOURNAL)
NIHON SHINRI TEKISEI KENKYUSHO
(JAPAN PSYCHOLOGICAL APTITUDE RESEARCH INSTITUTE)
JAPANESE

SHINRIGAKU HYORON 1895
(QUARTERLY REVIEW OF PSYCHOLOGY)
SHINRIGAKU HYORON KANKOKAI
C/O KYOTO DAIGAKU SHINRIGAKU KYOSHITSU, YOSHIDA, SAKYO-KU, KYOTO-SHI 606
JAPANESE, WITH ENGLISH SUMMARY

SHINRIGAKU KENKYU 1896
(JAPANESE JOURNAL OF PSYCHOLOGY)
NIHON SHINRI GAKKAI (THE JAPANESE PSYCHOLOGICAL ASSOCIATION)
802, BUNKYO SENTA BUILDING, 37-13, 4-CHOME, HONGO, BUNKYO-KU, TOKYO 113
JAPANESE, WITH EUROPEAN SUMMARY

SHINRYO TECHO 1897
(MEDICAL CARE NOTE)
NIHON KAYAKU K. K. JIGYO-BU
15-5, ICHIBAN-CHO, CHIYODA-KU, TOKYO 102
JAPANESE

SHINSHU DAIGAKU IGAKUBU GYOSEKI ROKU 1898
(RECORD OF FINDINGS, DEPARTMENT OF MEDICINE OF THE SHINSHU UNIVERSITY)
SHINSHU UNIVERSITY, DEPARTMENT OF MEDICINE
JAPANESE

SHINSHU DAIGAKU IRYO GIJUTSU TANKI DAIGAKUBU KIYO 1899
(TREATISES AND STUDIES, COLLEGE OF MEDICAL TECHNOLOGY AND NURSING, SHINSHU UNIVERSITY)
JAPANESE

SHINSHU IGAKU ZASSHI 1900
(SHINSHU MEDICAL JOURNAL)
NAGANO-KEN IGAKKAI (NAGANO MEDICAL SOCIETY)
C/O SHINSHU DAIGAKU IGAKUBU, 1-1, 3-CHOME, ASAHI-MACHI, MATSUMOTO-SHI 390
JAPANESE
INDEXED: BIO, CHE, EXC, JICST

SHIRITSU EBETSU SOGO BYOIN ISHI 1901
(JOURNAL OF EBETSU MUNICIPAL GENERAL HOSPITAL)
EBETSU MUNICIPAL GENERAL HOSPITAL
JAPANESE

SHIRITSU OMACHI BYOIN IHO 1902
(ANNALS OF THE OMACHI HOSPITAL)
SHIRITSU OMACHI SOGO BYOIN
313C, OMACHI, OMACHI-SHI 398
JAPANESE

SHIRITSU SAPPORO BYOIN ISHI 1903
(JOURNAL OF SAPPORO MUNICIPAL GENERAL HOSPITAL (ACTA MEDICA NOSOCOMI SAPPORO))
SHIRITSU SAPPORO BYOIN (SAPPORO MUNICIPAL HOSPITAL)
NISHI-9-CHOME, KITA-1-JO, CHUO-KU, SAPPOROSHI 060
JAPANESE
INDEXED: JICST

SHIZUOKA-KENRITSU BYOIN NENPO 1904
(ANNUAL REPORT OF THE SHIZUOKA CENTRAL PREFECTURAL HOSPITAL)
SHIZUOKA-KEN EISEI-BU
9-6, OITE-CHO, SHIZUOKA-SHI 420
JAPANESE

SHIZUOKA-KENRITSU CHUO BYOIN IGAKU ZASSHI 1905
(SHIZUOKA JOURNAL OF MEDICINE)
SHIZUOKA-KENRITSU CHUO BYOIN (SHIZUOKA CENTRAL PREFECTURAL HOSPITAL)
6-1, 3-CHOME, TAKASHO, SHIZUOKA-SHI 420
JAPANESE AND ENGLISH

SHOIN 1906
(MUSIC OF THE PINES)
SHOINKAI JIMUKYOKU JICHI IKA DAIGAKU KYOMUKANAI (SHOIN SOCIETY, BUSINESS OFFICE JICHI MEDICAL UNIVERSITY, DEPARTMENT OF INSTRUCTION)
JAPANESE

SHOJINKAISHI 1907
(MEMOIRS OF SHOJINKAI)
SHOJINKAI
2-16, KANESHITA-CHO, MORIGUCHI-SHI 570
JAPANESE

SHONAI IGAKKAI ZASSHI 1908
(JOURNAL OF THE SHONAI MEDICAL ASSOCIATION)
SHONAI IGAKKAI (THE SHONAI MEDICAL ASSOCIATION)
C/O MR. S. URAMOTO, 7-3, 7-CHOME, SEIJO, SETAGAYA-KU, TOKYO 157
JAPANESE

SHOWA IGAKKAI ZASSHI 1909
(JOURNAL OF THE SHOWA MEDICAL ASSOCIATION)
SHOWA DAIGAKU, SHOWA IGAKKAI (SHOWA UNIVERSITY AND THE SHOWA MEDICAL ASSOCIATION)
5-8, 1-CHOME, HATANODAI, SHINAGAWA-KU, TOKYO 142
JAPANESE, WITH EUROPEAN SUMMARY
INDEXED: CHE, EXC,

SHUJII 1910
(SIE SIND DER HARS ARZT)
SHOJU KURABU
29-8, 1-CHOME, SHINJUKU, SHINJUKU-KU, TOKYO 160
JAPANESE

SOBIETO SHINRIGAKU KENKYU 1911
(PSYCHOLOGY IN THE SOVIET UNION)
SOBIETO SHINRIGAKU KENKYUKAI (JAPANESE ASSOCIATION FOR THE RESEARCH OF SOVIET PSYCHOLOGY)
C/O TOKYO GAKUGEI DAIGAKU SHINRIGAKU KENKYUSHITSU, NUKUIKITA-MACHI, KOGANEI-SHI, TOKYO 184
JAPANESE

SUMITOMO BYOIN IGAKU ZASSHI 1912
(MEDICAL JOURNAL OF SUMITOMO HOSPITAL)
JAPANESE

TEISHIN IGAKU 1913
(MEDICAL JOURNAL FOR COMMUNICATION)
TEISHIN IGAKU KYOKAI (TEISHIN MEDICAL SOCIETY) AND HOKEN DOJINSHA
3-6, 1-CHOME, SARUGAKU-CHO, CHIYODA-KU, TOKYO 101
JAPANESE
INDEXED: CHE

TOA IHO 1914
(EAST ASIA MEDICAL)
TOA IGAKUSHA (EAST ASIA MEDICINE)
6-7, 5-CHOME, KYODO, SETAGAYA-KU, TOKYO 156
JAPANESE

TOCHIGI-KEN IGAKKAISHI 1915
(JOURNAL OF TOCHIGI MEDICAL SOCIETY)
TOCHIGI-KEN IGAKKAI
372, TOMATSURIMOTO-MACHI, UTSUNOMIYA-SHI 320
JAPANESE

TOHO IGAKKAI ZASSHI 1916
(JOURNAL OF THE MEDICAL SOCIETY OF TOHO UNIVERSITY)
TOHO DAIGAKU IGAKUBU IGAKKAI (MEDICAL SOCIETY OF TOHO UNIVERSITY)
21-16, 5-CHOME, OMORI NISHI, OTA-KU, TOKYO 143
JAPANESE AND ENGLISH
INDEXED: CHE, EXC, JICST

TOHOKU IGAKU ZASSHI 1917
(TOKOKU MEDICAL JOURNAL)
TOHOKU IGAKKAI (TOHOKU MEDICAL SOCIETY)
C/O TOHOKU DAIGAKU IGAKUBU, 2-1, SEIRYO-MACHI, SENDAI-SHI 980
JAPANESE AND ENGLISH
INDEXED: BIO, CHE, EXC, JICST

TOHOKU JOURNAL OF EXPERIMENTAL MEDICINE 1918
TOHOKU JANARU HENSHU IINKAI (EDITORIAL BOARD OF THE TOHOKU JOURNAL OF EXPERIMENTAL MEDICINE) AND TOHOKU DAIGAKU IGAKUBU (TOHOKU UNIVERSITY, SCHOOL OF MEDICINE)
2-1, SEIRYO-MACHI, SENDAI-SHI 980
ENGLISH
INDEXED: BIO, CHE, EXC, IND, SCI, JICST

TOHOKU SHINRIGAKU KENKYU 1919
(TOHOKU UNIVERSITY PSYCHOLOGICAL JOURNAL)
TOHOKU SHINRI GAKKAI (TOHOKU SOCIETY OF PSYCHOLOGY)
C/O TOHOKU DAIGAKU BUNGAKUBU SHINRIGAKU KENKYSUSHITSU, 1-1, 2-CHOME, KATAHIRA, SENDAI-SHI 980
JAPANESE

TOKUSHIMA JOURNAL OF EXPERIMENTAL MEDICINE 1920
TOKUSHIMA DAIGAKU IGAKUBU (SCHOOL OF MEDICINE, TOHOKU UNIVERSITY)
18-15, 3-CHOME, KURAMOTO-CHO, TOKUSHIMA-SHI 770
ENGLISH
INDEXED: BIO, CHE, EXC, IND, JICST

TOKYO IGAKU 1921
(TOKYO JOURNAL OF MEDICAL SCIENCES)
TOKYO IGAKKAI (TOKYO SOCIETY OF MEDICAL SCIENCES)
3-1, 7-CHOME, HONGO, BUNKYO-KU, TOKYO 113
JAPANESE
INDEXED: CHE, JICST

TOKYO IKA DAIGAKU KIYO 1922
(BULLETIN OF TOKYO MEDICAL COLLEGE)
JAPANESE

TOKYO IKA DAIGAKU ZASSHI 1923
(JOURNAL OF TOKYO MEDICAL COLLEGE)
TOKYO IKA DAIGAKU IGAKKAI (MEDICAL SOCIETY OF TOKYO MEDICAL COLLEGE)
1-412, HIGASHI OKUBO, SHINJUKU-KU, TOKYO 160
JAPANESE AND ENGLISH
INDEXED: CHE, EXC, JICST

TOKYO IKA SHIKA DAIGAKU IYO KIZAI KENKYUJO HOKOKU 1924
(REPORT OF THE INSTITUTE ON TOOLS AND MATERIALS FOR MEDICAL USE, TOKYO MEDICAL AND DENTAL UNIVERSITY)
JAPANESE
INDEXED: JICST

TOKYO IKA SHIKA DAIGAKU KYOYOBU KENKYU KIYO 1925
(BULLETIN OF THE DEPARTMENT OF GENERAL EDUCATION, TOKYO MEDICAL AND DENTAL UNIVERSITY)
TOKYO IKA SHIKA DAIGAKU KYOYOBU
8-30, 2-CHOME, KONODAI, ICHIKAWA-SHI 272
JAPANESE AND ENGLISH

TOKYO IKA SHIKA DAIGAKU SHIKA ZAIRYO KENKYUJO HOKOKU 1926
(REPORT OF THE INSTITUTE ON DENTAL MATERIALS, TOKYO MEDICAL AND DENTAL UNIVERSITY)
JAPANESE
INDEXED: JICST

TOKYO JIKEIKAI IKA DAIGAKU FUZOKU KENKYUSHITSU NENPO 1927
(ANNUAL REPORT OF THE ATTACHED LABORATORY, TOKYO JIKEIKAI MEDICAL COLLEGE)
TOKYO JIKEIKAI IKA DAIGAKU FUZOKU KENKYUSHITSU
25-8, 3-CHOME, NISHISHINBASHI, MINATO-KU, TOKYO 105
JAPANESE

TOKYO JIKEIKAI IKA DAIGAKU ZASSHI 1928
(TOKYO JIKEIKAI MEDICAL JOURNAL)
TOKYO JIKEIKAI IKA DAIGAKU (TOKYO JIKEIKAI MEDICAL COLLEGE) AND TOKYO JIKEIKAI IKA DAIGAKU SEIIKAI (TOKYO JIKEIKAI MEDICAL COLLEGE SEIIKAI)
25-8, 3-CHOME, NISHISHINBASHI, MINATO-KU, TOKYO 105
JAPANESE, WITH ENGLISH SUMMARY
INDEXED: CHE, EXC, JICST

TOKYO JOSHI IKA DAIGAKU ZASSHI 1929
(JOURNAL OF TOKYO WOMEN'S MEDICAL COLLEGE)
TOKYO JOSHI IKA DAIGAKU GAKKAI (SOCIETY OF TOKYO WOMEN'S MEDICAL COLLEGE)
C/O TOKYO JOSHI IKA DAIGAKU TOSHOKAN, 10, KAWADA-CHO, SHINJUKU-KU, TOKYO 160
JAPANESE, WITH ENGLISH SUMMARY
INDEXED: CHE, EXC, JICST

TOKYO KYOIKU DAIGAKU FUZOKU ROGAKKO KIYO 1930
(BULLETIN OF THE NATIONAL SCHOOL FOR THE DEAF, TOKYO UNIVERSITY OF EDUCATION)
JAPANESE

TOKYO-TO ISHIKAI ZASSHI 1931
(JOURNAL OF THE TOKYO MEDICAL ASSOCIATION)
TOKYO-TO ISHIKAI (THE TOKYO MEDICAL ASSOCIATION
2-5, KANDA SURUGADAI, CHIYODA-KU, TOKYO 101
JAPANESE

TOKYO-TO KANSATSU IMUIN KIYO 1932
(MEMOIRS OF TOKYO MEDICAL EXAMINER'S OFFICE)
TOKYO-TO KANSATSU IMUIN
21-18, 4-CHOME, OTSUKA, BUNKYO-KU, TOKYO 112
JAPANESE

TOKYO-TO SEISHIN IGAKU SOGO KENKYUJO TOSHO DAYORI 1933
(TOKYO METROPOLITAN PSYCHOLOGICAL MEDICINE GENERAL RESEARCH LIBRARY LETTER)
JAPANESE

TOKYO-TO SEISHIN IGAKU SOGO KENKYUSHO KENKYU KIYO 1934
(RESEARCH BULLETIN OF THE TOKYO PSYCHIATRIC GENERAL RESEARCH INSTITUTE)
JAPANESE

TOKYO-TO YOIKUIN GEPPO 1935
(JOURNAL OF TOKYO METROPOLITAN WELFARE HOME)
TOKYO-TO YOIKUIN (TOKYO METROPOLITAN WELFARE HOME)
35-1, SAKAE-CHO, ITABASHI-KU, TOKYO 173
JAPANESE

TOKYO-TORITSU MATSUZAWA BYOIN JIGYO GAIYO 1936
(ABSTRACT REPORT OF TOKYO MUNICIPAL MATSUZAWA HOSPITAL)
TOKYO-TORITSU MATSUZAWA BYOIN
1-1, 2-CHOME, KAMI-KITAZAWA, SETAGAYA-KU TOKYO 156
JAPANESE

TORITSU BYOSAN'IN RINSHO KENKYU HOKOKUSHO 1937
(CLINICAL STUDY REPORT OF THE MUNICIPAL HOSPITALS AND MATERNITY HOSPITALS)
TOKYO-TO EISEI-KYOKU BYOIN KANRI-BU (HOSPITAL ADMINISTRATION DIVISION, SANITARY BUREAU, TOKYO)
7-1, 5-CHOME, HIROO, SHIBUYA-KU, TOKYO 150
JAPANESE

TOSHIBA MEDIKARU NYUSU 1938
(TOSHIBA MEDICAL NEWS)
TOSHIBA MEDIKARU K.K. (TOSHIBA MEDICAL SYSTEMS COMPANY, LIMITED)
26-5, 3-CHOME, HONGO, BUNKYO-KU, TOKYO 113
JAPANESE

TOTTORI DAIGAKU IRYO GIJUTSU TANKI DAIGAKU-BU KENKYU HOKOKU 1939
(RESEARCH REPORT OF TORRORI JUNIOR MEDICAL TECHNOLOGY COLLEGE)
TOTTORI DAIGAKU IRYO GIJUTSU TANKI DAIGAKUBU (TOTTORI UNIVERSITY - MEDICAL TECHNOLOGY JUNIOR COLLEGE)
JAPANESE

TOTTORI-KEN KANJA CHOSA 1940
(PATIENT SURVEY IN TOTTORI PREFECTURE)
TOTTORI-KEN
1-220, HIGASHI-CHO, TOTTORI-SHI 680
JAPANESE

TOYAMA-KEN I-TSUHO 1941
(JOURNAL OF TOYAMA MEDICAL ASSOCIATION)
TOYAMA-KEN ISHIKAI (TOYAMA MEDICAL ASSOCIATION)
8-6, 2-CHOME, SOGAWA, TOYAMA-SHI 930
JAPANESE

TOYO IGAKU 1942
(EASTERN MEDICINE)
SHIZENSHA MIDORI SHOBO HATSUBAI (NATURE FIRM MIDORI PUBLISHING COMPANY SALES)
JAPANESE

TOYO IJUTSU 1943
(ORIENTAL MEDICINE)
SHINKYU SHINKOKAI
MODOROGAIKE-HAN, KITA-KU, KYOTO-SHI 603
JAPANESE

TRANSACTIONS OF THE SOCIETY FOR ACTINOMYCETES, JAPAN 1944
NIHON HOSENKIN KENKYUKAI (SOCIETY FOR ACTINOMYCETES, JAPAN)
C/O CHIBA DAIGAKU SEIBUTSU KASSEI KENKYUSHO 9-1, 3-CHOME, IZUMI-CHO, NARASHINO-SHI 275
JAPANESE AND ENGLISH

UIRUSU 1945
(VIRUS)
NIHON UIRUSU GAKKAI (SOCIETY OF JAPANESE VIROLOGISTS)
C/O TOKYO DAIGAKU IKAGAKU KENKYUSHO, 6-1, 4-CHOME, SHIROGANE, MINATO-KU, TOKYO 108
JAPANESE, WITH EUROPEAN SUMMARY
INDEXED: CHE, EXC

UIRUSUGAKU NO SHINTEN 1946
(PROGRESS IN VIRUS RESEARCH)
KYOTO DAIGAKU UIRUSU KENKYUSHO (INSTITUTE FOR VIRUS RESEARCH, KYOTO UNIVERSITY)
SHOGOIN KAWAHARA-CHO, SAKYO-KU, KYOTO-SHI 606
JAPANESE
INDEXED: CHE

UNDO CHINO SHOGAI KENKYU 1947
(REPORT ON A STUDY OF CHILDREN WITH MOTOR AND/OR INTELLIGENCE DISABILITIES)
TOKYO KYOIKU DAIGAKU KYOIKUGAKUBU TOKUSHU KYOIKU GAKKA (SCHOOL OF EDUCATION FOR THE HANDICAPPED, FACULTY OF EDUCATION, TOKYO UNIVERSITY OF EDUCATION)
29-1, 3-CHOME, OTSUKA, BUNKYO-KU, TOKYO 112
JAPANESE

WAKAYAMA IGAKU 1948
(WAKAYAMA MEDICINE)
WAKAYAMA IGAKKAI (WAKAYAMA MEDICAL SOCIETY)
C/O WAKAYAMA-KENRITSU IKA DAIGAKU, 9, KYUBAN-CHO, WAKAYAMA-SHI 640
JAPANESE, WITH EUROPEAN SUMMARY
INDEXED: CHE, JICST

WAKAYAMA MEDICAL REPORTS 1949
WAKAYAMA-KENRITSU IKA DAIGAKU (WAKAYAMA MEDICAL COLLEGE)
9, KYUBAN-CHO, WAKAYAMA-SHI 640
EUROPEAN
INDEXED: BIO, CHE, EXC

WAKAYAMA-KENRITSU IKA DAIGAKU SHINGAKU KATEI KIYO 1950
(BULLETIN OF WAKAYAMA MEDICAL COLLEGE, PRE-MEDICAL COURSE)
WAKAYAMA MEDICA COLLEGE
JAPANESE

WASEDA DAIGAKU KYOIKU GAKUBU GAKUJUTSU KENKYU 1951
KYOIKU SHAKAI KYOIKU SHINRI ITAIIKU-HEN
(WASEDA UNIVERSITY, SCHOOL OF EDUCATION, ARTS & SCIENCE RESEARCH - EDUCATION, SOCIAL EDUCATION, PSYCHOLOGY, PHYSICAL EDUCATION)
WASEDA DAIGAKU KYOIKU KAI (WASEDA UNIVERSITY EDUCATION SOCIETY)
JAPANESE

WASEDA SHINRIGAKU NENPO 1952
(WASEDA PSYCHOLOGICAL REPORTS)
WASEDA DAIGAKU BUNGAKUBU SHINRIGAKKAI (WASEDA UNIVERSITY, DEPARTMENT OF LITERATURE, ACADEMY OF PSYCHOLOGY)
JAPANESE

YAMAGATA-KEN ISHIKAI KAIHO 1953
(JOURNAL OF THE YAMAGATA MEDICAL ASSOCIATION)
YAMAGATA-KEN ISHIKAI (YAMAGATA MEDICAL ASSOCIATION)
9-19, 2-CHOME, KASUMI-CHO, YAMAGATA-SHI, YAMAGATA 990
JAPANESE

YAMAGATA-KENRITSU BYOIN IGAKU ZASSHI 1954
(YAMAGATA JOURNAL OF MEDICINE)
YAMAGATA-KENRITSU CHUO BYOIN (YAMAGATA PREFECTURAL CENTRAL HOSPITAL)
YAMAGATA-KEN BYOISHI HENSHU IINKAI
7-17, SAKURA-CHO, YAMAGATA-SHI 990
JAPANESE

YAMAGATA-SHIRITSU BYOIN SAISEIKAN IGAKU ZASSHI 1955
(MEDICAL MAGAZINE OF THE YAMAGATA MUNICIPAL HOSPITAL, SAISEIKAN)
YAMAGATA MUNICIPAL HOSPITAL, SAISEIKAN
JAPANESE

YAMAGUCHI DAIGAKU KYOIKU GAKUBU KENKYU RONSO 1956
DAISANBU GEINO, TAIIKU, KYOIKU, SHINRI
(RESEARCH PAPERS OF YAMAGUCHI UNIVERSITY, SCHOOL OF EDUCATION PART 3, PERFORMING ARTS, PHYSICAL EDUCATION, PHYSIOLOGY)
JAPANESE

YAMAGUCHI IGAKU 1957
(YAMAGUCHI MEDICINE)
YAMAGUCHI DAIGAKU IGAKKAI (YAMAGUCHI UNIVERSITY, MEDICAL ASSOCIATION)
KOGUSHI, UBE-SHI 755
JAPANESE
INDEXED: CHE

YAMAGUCHI MEDICAL JOURNAL 1958
YAMAGUCHI
JAPANESE AND ENGLISH
INDEXED: EXC

YAMAGUCHI SANGYO IGAKU NENPO 1959
(ANNUAL REPORT, SOCIETY OF YAMAGUCHI INDUSTRIAL HEALTH)
YAMAGUCHI SANGYO IGAKKAI (SOCIETY OF YAMAGUCHI INDUSTRIAL HEALTH)
C/O YAMAGUCHI DAIGAKU IGAKUBU KOSHU EISEI KYOSHITSU, 1144, KOGUSHI, UBE-SHI 755
JAPANESE
INDEXED: JICST

YAMAGUCHI-KEN IGAKU KAISHI 1960
(MEMOIRS OF YAMAGUCHI MEDICAL SOCIETY)
YAMAGUCHI-KEN ISHIKAI
2-33, 3-CHOME, YUDAONSEN, YAMAGUCHI-SHI 753
JAPANESE

YAMAGUCHI-KEN ISHI KAIHO 1961
(YAMAGUCHI PREFECTURE PHYSICIAN'S ASSOCIATION REPORT)
JAPANESE

YOKOHAMA IGAKU 1962
(YOKOHAMA MEDICAL JOURNAL)
YOKOHAMA SHIDAI IGAKKAI (MEDICAL ASSOCIATION OF YOKOHAMA MUNICIPAL UNIVERSITY)
2-33, URAFUNE-CHO, MINAMI-KU, YOKOHAMA-SHI 232
JAPANESE, WITH EUROPEAN SUMMARY
INDEXED: CHE, JICST

YOKOHAMA MEDICAL BULLETIN 1963
YOKOHAMA-SHIRITSU DAIGAKU IGAKUBU (YOKOHAMA CITY UNIVERSITY, SCHOOL OF MEDICINE)
2-33, URAFUNE-CHO, MINAMI-KU, YOKOHAMA-SHI 232
ENGLISH
INDEXED: BIO, CHE, EXC, JICST

YOKOHAMA-SHIRITSU DAIGAKU KIYO, TAIRYOKU IGAKU-HEN 1964
(JOURNAL OF YOKOHAMA CITY UNIVERSITY, SERIES OF SPORT SCIENCES AND MEDICINE)
YOKOHAMA-SHIRITSU DAIGAKU (YOKOHAMA CITY UNIVERSITY)
4646, MUTSUURA-MACHI, KANAZAWA-KU, YOKOHAMASHI 236
JAPANESE AND ENGLISH

YONAGO ACTA MEDICA 1965
TOTTORI DAIGAKU IGAKUBU (TOTTORI UNIVERSITY, SCHOOL OF MEDICINE)
86, NISHI-MACHI, YONAGO-SHI 683
ENGLISH
INDEXED: CHE, EXC, IND, SCI, JICST

YONAGO IGAKU ZASSHI 1966
(JOURNAL OF THE YONAGO MEDICAL ASSOCIATION)
YONAGO IGAKKAI (YONAGO MEDICAL ASSOCIATION)
C/O TOTTORI DAIGAKU IGAKUBU, 86, NISHI-MACHI, YONAGO-SHI 683
JAPANESE, WITH ENGLISH SUMMARY
INDEXED: CHE

ZENKOKU HOKEN'I SHINBUN 1967
(ALL JAPAN SOCIAL INSURANCE NEWS)
ZENKOKU HOKEN'I DANTAI RENGOKAI (FEDERATION OF ALL JAPAN SOCIAL INSURANCE MEDICAL DOCTOR'S UNIONS)
C/O SHINJUKU NOKYO KAIKAN, 5-5, 2-CHOME, YOYOGI, SHIBUYA-KU, TOKYO 151
JAPANESE, WITH ENGLISH SUMMARY

ZENNIPPON BYOIN 1968
(ALL NIPPON HOSPITAL)
ZENNIPPON BYOIN KYOKAI (ALL NIPPON HOSPITAL ASSOCIATION)
1-9, 3-CHOME, KAMIHON-CHO, TENNOJI-KU, OSAKA-SHI 543
JAPANESE

ZENNIPPON BYOIN KANRIGAKKAI ZASSHI 1969
(JOURNAL OF ALL NIPPON HOSPITAL MANAGEMENT ASSOCIATION)
ZENNIPPON BYOIN KYOKAI
1-9, 3-CHOME, KAMIHON-CHO, TENNOJI-KU, OSAKA-SHI 543
JAPANESE

Anatomy

ACTA HISTOCHEMICA ET CYTOCHEMICA 1970
NIHON SOSHIKI SAIBO KAGAKKAI (JAPAN SOCIETY OF HISTOCHEMISTRY AND CYTOCHEMISTRY)
C/O KYOTO DAIGAKU KEKKAKU KYOBU SHIKKAN KENKYUSHO SAIBO KAGAKUBU, 53, SHOGOIN KAWARA-CHO, SAKYO-KU, KYOTO-SHI 615
JAPANESE AND ENGLISH
INDEXED: BIB, BIO, CHE, SCI

CONNECTIVE TISSUE 1971
TOKYO
JAPANESE AND ENGLISH
INDEXED: EXC

EXPERIMENTAL ANIMALS 1972
TOKYO
JAPANESE AND ENGLISH
INDEXED: EXC

HIROSHIMA DAIGAKU IGAKUBU KAIBOGAKU DAI-1-KOZA GYOSEKISHU 1973
(STUDIES FROM THE DEPARTMENT OF ANATOMY, HIROSHIMA UNIVERSITY SCHOOL OF MEDICINE)
HIROSHIMA DAIGAKU IGAKUBU KAIBOGAKU KYOSHITSU DAI-1-KOZA (DEPARTMENT OF ANATOMY HIROSHIMA UNIVERSITY SCHOOL OF MEDICINE)
2-3, 1-CHOME, KASUMI HIROSHIMA-SHI 734
JAPANESE, WITH EUROPEAN SUMMARY

KAIBOGAKU ZASSHI 1974
(ACTA ANATOMICA NIPPONICA)
NIHON KAIBO GAKKAI (SOCIETY ANATOMICA NIPPONICA)
C/O TOKYO DAIGAKU IGAKUBU, 3-1, 7-CHOME, BUNKYO-KU, TOKYO 113
JAPANESE AND ENGLISH
INDEXED: CHE, EXC

KANAZAWA DAIGAKU IGAKUBU KAIBOGAKU KYOSHITSU GYOSEKI 1975
(REPORT OF THE DEPARTMENT OF ANATOMY, KANAZAWA UNIVERSITY)
KANAZAWA DAIGAKU IGAKUBU KAIBOGAKU DAI-3-KOZA (SCHOOL OF MEDICINE, KANAZAWA UNIVERSITY)
13-1, TAKARA-MACHI, KANAZAWA-SHI 920
ENGLISH

KOKO KAIBO KENKYU 1976
(ORAL CAVITY ANATOMICAL RESEARCH)
TOKYO SHIKA DAIGAKU KOKO GAKKO SHITSU (TOKYO DENTAL UNIVERSITY, ANATOMICAL STUDIES CLASSROOM)
JAPANESE

KOKU KAIBO KENKYU 1977
(STUDIES ON ORAL ANATOMY)
TOKYO SHIKA DAIGAKU KAIBOGAKU KYOSHITSU (ANATOMY DIVISION, TOKYO DENTAL UNIVERSITY)
9-18, 2-CHOME, MISAKI-CHO, CHIYODA-KU, TOKYO 101
JAPANESE, WITH ENGLISH SUMMARY

MAKU 1978
(MEMBRANE)
KITAMI SHOBO COMPANY, LIMITED
14-26, KITAMI 7-CHOME, SETAGAYA-KU, TOKYO 157
JAPANESE, WITH ENGLISH SUMMARY
INDEXED: BIO, CHE, EXC, JICST

MIE DAIGAKU IGAKUBU KAIBOGAKU KYOSHITSU GYOSEKISHU 1979
(REPORTS FROM THE DEPARTMENT OF ANATOMY, MIE UNIVERSITY SCHOOL MEDICINE)
MIE UNIVERSITY, SCHOOL OF MEDICINE, DEPARTMENT OF ANATOMY
JAPANESE

MIYADAKE GAIKOTSU KAIBO 1980
(MIYADAKE SKELETAL DISSECTION)
JAPANESE

MYAKKANGAKU 1981
(JOURNAL OF JAPANESE COLLEGE OF ANGIOLOGY)
NIHON MYAKKAN GAKKAI (JAPANESE COLLEGE OF ANGIOLOGY)
C/O TOKYO DAIGAKU IGAKUBU DAI-1-GEKA KYOSHITSU, 3-1, 7-CHOME, HINGO, BUNKYO-KU, TOKYO 113
JAPANESE AND ENGLISH

NIHON HEIKATSUKIN GAKKAI ZASSHI 1982
(JAPANESE JOURNAL OF SMOOTH MUSCLE RESEARCH)
NIHON HEIKATSUKIN GAKKAI (JAPAN SOCIETY OF SMOOTH MUSCLE RESEARCH)
C/O NARA-KENRITSU IKA DAIGAKU DAI-1-GEKAGAKU KYOSHITSU, 840, SHIJO-CHO, KASHIHARA-SHI 634
JAPANESE, WITH ENGLISH SUMMARY
INDEXED: BIO, CHE, EXC, IND

NIHON SOSHIKI SAIBO KAGAKKAI KOEN SHOROKU 1983
(ABSTRACTS OF THE SYMPOSIUM OF THE JAPAN SOCIETY OF HISTOCHEMISTRY AND CYTOCHEMISTRY)
NIHON SOSHIKI SAIBO KAGAKKAI (JAPAN SOCIETY OF HISTOCHEMISTRY AND CYTOCHEMISTRY)
YOSHIDA KONOE-CHO, SAKYO-KU, KYOTO-SHI 606
JAPANESE

NIHON SOSHIKIGAKU KIROKU 1984
(ARCHIVUM HISTOLOGICUM JAPONICUM)
NIHON SOSHIKI KAGAKKAI (SOCIETATIS HISTOCHEMICAE JAPONICAE)
5-1, 2-CHOME, SHIKADA-CHO, OKAYAMA-SHI 700
JAPANESE, WITH ENGLISH SUMMARY
INDEXED: BIO, CHE, EXC, IND, SCI
JICST

NOVA ANGIOLOGICAE 1985
(NOVA ANGIOROJI)
NIHON MYAKKAN GAKKAI (JAPANESE COLLEGE OF ANGIOLOGY), HEKISUTO JAPAN K. K. (HOECHST JAPAN LIMITED)
10-33, 4-CHOME, AKASAKA, MINATO-KU, TOKYO 107
JAPANESE AND ENGLISH

OKAJIMAS FOLIA ANATOMICA JAPONICA 1986
OKAJIMA FORIA ANATOMIKA YAPONIKA HENSHUBU (EDITORIAL BOARD OF THE OKAJIMA FOLIA ANATOMICA JAPONICA)
C/O KEIO DAIGAKU IGAKUBU KAIBOGAKU KYOSHITSU, 35, SHINANO-MACHI, SHINJUKU-KU, TOKYO 160
EUROPEAN
INDEXED: BIB, BIO, CHE, EXC, JICST

SEITAI KEISOKU RONBUNSHU 1987
(BULLETIN OF HUMAN BODY MEASUREMENT)
SEI MARIANNA IKA DAIGAKU (ST. MARIANNA UNIVERSITY SCHOOL OF MEDICINE) AND SEI MARIANNA IKA DAIGAKU OGAWARA KENKYUSHITSU
2095, SUGAO, TAKATSU-KU, KAWASAKI-SHI 213
JAPANESE AND EUROPEAN
INDEXED: BIO

SHOKAKI NAISHIKYO NO SHINPO 1988
(PROGRESS OF DIGESTIVE ENDOSCOPY)
KYOWA KIKAKU TSUSHIN (KYOWA PLANNING NEWS)
JAPANESE

TISSUE CULTURE STUDIES IN JAPAN, THE ANNUAL BIBLIOGRAPHY 1989
NIHON SOSHIKI BAIYO GAKKAI (THE JAPANESE TISSUE CULTURE ASSOCIATION)
C/O TOKYO DAIGAKU IGAKUBU KAIBOGAKU KYOSHITU, 3-1, 7-CHOME, HONGO, BUNKYO-KU, TOKYO 113
EUROPEAN

TOTTORI DAIGAKU KAIBOGAKU KYOSHITSU GYOSEKISHU 1990
(ARBEITEN AUS DEM ANATOMISCHEN INSTITUTE DER UNIVERSITY TOTTORI)
TOTTORI DAIGAKU KAIBOGAKU KYOSHITSU (DEPARTMENT OF ANATOMY, TOTTORI UNIVERSITY, SCHOOL OF MEDICINE)
86, NISHI-MACHIE, YONAGO-SHI 683
EUROPEAN

Physiology

AKATSUKI NO HOSHI EISEN KIYO 1991
(BULLETIN OF THE AKATSUKI NO HOSHI, SHOOL OF NUTRITION)
FUKUYAMA AKATSUKI NO HOSHI EIYO SENMON GAKKO (FUKUYAMA CITY, AKATSUKI NO HOSHI SCHOOL OF NUTRITION)
JAPANESE

AMINOSAN 1992
(AMINO ACIDS)
TANABE AMINOSAN KENKYU KIKIN (TANABE AMINO ACID RESEARCH FOUNDATION)
3-33, HIRANO-MACHI, HIGASHI-KU, OSAKA-SHI 541
JAPANESE
INDEXED: CHE

ANNUAL BULLETIN (RESEARCH INSTITUTE OF LOGOPEDICS AND PHONIATRICS, UNIVERSITY OF TOKYO) 1993
TOKYO DAIGAKU IGAKUBU ONSEI GENGO IGAKU KENKYU SHISETSU (RESEARCH INSTITUTE OF LOGOPEDICS AND PHONIATRICS, FACULTY OF MEDICINE, UNIVERSITY OF TOKYO)
3-1, 7-CHOME, HONGO, BUNKYO-KU, TOKYO 113
ENGLISH

ANNUAL REPORT OF THE NATIONAL INSTITUTE OF NUTRITION 1994
KOSEI-SHO KOKURITSU EIYO KENKYUSHO (NATIONAL INSTITUTE OF NUTRITION, MINISTRY OF HEALTH AND WELFARE)
1, TOYAMA-CHO, SHINJUKU-KU, TOKYO 162
JAPANESE
INDEXED: BIO, CHE, JICST

ANNUAL REPORT OF THE TANABE AMINO ACID RESEARCH FOUNDATION 1995
TANABE AMINOSAN KENKYU KIKIN (TANABE AMINO ACID RESEARCH FOUNDATION)
3-33, HIRANO-CHO, HIGASHI-KU, OSAKA-SHI 541
ENGLISH

AUDIOLOGY JAPAN 1996
NIHON OJIOROJI GAKKAI (JAPAN AUDIOLOGICAL SOCIETY)
C/O TOKYO DAIGAKU IGAKUBU JIBIKA, 3-1, 7-CHOME, HONGO, BUNKYO-KU, TOKYO
JAPANESE, WITH ENGLISH SUMMARY
INDEXED: EXC

BITAMIN 1997
(VITAMIN)
NIPPON BITAMIN GAKKAI (VITAMIN SOCIETY OF JAPAN)
C/O NIPPON ITARIA KYOTO KAIKAN, 4, YOSHIDA USHINOMIYA-CHO, SAKYO-KU, KYOTO-SHI 606
JAPANESE, WITH ENGLISH SUMMARY
INDEXED: BIB, BIO, CHE, JICST

BITAMIN B KENKYU IINKAI KIJI 1998
(REPORT OF THE COMMITTEE TO STUDY VITAMIN B)
JAPANESE
INDEXED: JICST

DAEKISEN SHINPOJUMU 1999
(SYMPOSIUM FOR THE SALIVARY GLAND)
NIHON DAEKISEN KENKYUKAI (JAPAN SALIVARY GLAND SOCIETY)
C/O CHIBA DAIGAKU IGAKUBU BYORIGAKU KYOSHITSU, 8-1, 1-CHOME, INOHANA-CHO, CHIBA-SHI 280
JAPANESE AND EUROPEAN

DENPUNTO GIJUTSU KENKYU KAIHO 2000
(REPORT OF 'STARCH SUGAR' TECHNOLOGY RESEARCH SOCIETY)
JAPANESE
INDEXED: JICST

EIYO NIPPON 2001
(NUTRITION OF JAPAN)
NIPPON EIYO SHIKAI (JAPAN DIETETIC ASSOCIATION CORPORATION)
C/O OTSUKA BUILDING, 16-6, 1-CHOME, KITA-OTSUKA, TOSHIMA-KU, TOKYO 170
JAPANESE, WITH ENGLISH SUMMARY

EIYO, SHOKURYO, HAKKO, SUISHITSU GAIKOKU BUNKEN 2002
KAISHI
(BULLETIN OF LITERATURE ON FOOD, WATER, NUTRITION AND FERMENTATION)
GAIKOKU BUNKENKAI (SOCIETY FOR INTRODUCING FOREIGN LITERATURE)
C/O OSAKA-SHIRITSU EISEI KENKYUSHO, 38, KITAOGI-MACHI, KITA-KU, OSAKA-SHI 530
JAPANESE AND EUROPEAN
INDEXED: JICST

EIYO TO RYORI 2003
(NUTRITION AND COOKERY)
JOSHI EIYO DAIGAKU SHUPPANBU (WOMEN'S NUTRITION COLLEGE PUBLISHING BUREAU)
JAPANESE

EIYO TO RYORI NO KENKYU 2004
(RESEARCH ON NUTRITION AND COOKERY)
YAMANASHI GAKUIN TANKI DAIGAKU SHOKUMOTSU EIYOKA (YAMANASHI ACADEMY JUNIOR COLLEGE, DEPARTMENT OF FOOD AND NUTRITION)
JAPANESE

EIYO TO SHOKURYO 2005
(FOOD AND NUTRITION)
NIHON EIYO, SHOKURYO GAKKAI (JAPANESE SOCIETY OF FOOD AND NUTRITION)
C/O NIPPON GAKKAI JIMU SENTA, 4-16, 2-CHOME, YAYOI, BUNKYO-KU, TOKYO 113
JAPANESE, WITH EUROPEAN SUMMARY
INDEXED: BIB, CHE, JICST

EIYOGAKU ZASSHI 2006
(JAPANESE JOURNAL OF NUTRITION)
KOKUMIN EIYO SHINKOKAI (PROMOTION OF NATIONAL NUTRITION SOCIETY) AND DAIICHI SHUPPAN K.K.
1-39, KANDA JINBO-CHO, CHIYODA-KU, TOKYO 101
JAPANESE, WITH ENGLISH SUMMARY
INDEXED: BIB, CHE, JICST

ELECTROPHYSIOLOGY 2007
SHOWA DAIGAKU IGAKUBU DAI-1-SEIRIGAKU KYOSHITSU (DEPARTMENT OF PHYSIOLOGY, SCHOOL OF MEDICINE, SHOWA UNIVERSITY)
5-8, 1-CHOME, HATANODAI, SHINAGAWA-KU, TOKYO 142
ENGLISH
INDEXED: CHE

ENDOCRINOLOGICA JAPONICA 2008
NIHON NAIBUNPI GAKKAI (ENDOCRINOLOGICAL SOCIETY OF JAPAN)
C/O TOKYO JIKEIKAI IKA DAIGAKU DAI-2-KAIBOGAKU KYOSHITSU, 25-8, 3-CHOME, NISHISHINBASHI MINATO-KU, TOKYO 105
JAPANESE
INDEXED: BIO, CHE, EXC, JICST

FUKUYAMA AKATSUKI NO HOSHI GAKUIN EIYO SENMON GAKKO 2009
KENKYU KIYO
(RESEARCH BULLETIN OF THE FUKUYAMA AKATSUKI NO HOSHI ACADEMY, SCHOOL OF NUTRITION)
JAPANESE

GAKKO KYUSHOKU 2010
(SCHOOL LUNCH)
ZENKOKU GAKKO KYUSHOKU KYOKAI (ALL JAPAN SCHOOL LUNCH ASSOCIATION)
2-32, 1-CHOME, FUJIMI, CHIYODA-KU, TOKYO 102
JAPANESE

GAKKO KYUSHOKU KOHO 2011
(SCHOOL LUNCH NEWS)
NIHON GAKKO KYUSHOKUKAI (JAPAN SCHOOL LUNCH CORPORATION)
15, SHIBA NISHIKUBO AKEFUNE-CHO, MINATO-KU, TOKYO 105
JAPANESE

GUNMA SYMPOSIA ON ENDOCRINOLOGY 2012
GUNMA DAIGAKU NAIBUNPI KENKYUSHO (INSTITUTE OF ENDOCRINOLOGY, GUNMA UNIVERSITY)
39-15, 3-CHOME, SHOWA-MACHI, MAEBASHI-SHI 371
ENGLISH
INDEXED: BIO, CHE, EXC

HISSU AMINOSAN KENKYU 2013
(REPORTS OF THE RESEARCH COMMITTEE OF ESSENTIAL AMINO ACIDS (JAPAN))
HISSU AMINOSAN KENKYU IINKAI (RESEARCH COMMITTEE OF ESSENTIAL AMINO ACIDS (JAPAN)
C/O KOKURITSU KOSHU EISEIIN EIYO SEIKAGAKU-BU, 6-1, 4-CHOME, SHIBA SHIROGANEDAI, MINATO-KU, TOKYO 108
JAPANESE, WITH ENGLISH SUMMARY

HOPUKINZU KENKYU 2014
(HOPKINS RESEARCH)
NIHON HOPUKINZU KYOKAI TOKYO BUKAI (JAPAN HOPKINS ASSOCIATION, TOKYO DIVISION)
JAPANESE

HORUMON TO RINSHO 2015
(CLINICAL ENDOCRINOLOGY)
IGAKU NO SEKAI-SHA
5-1, 2-CHOME, AKASAKA, MINATO-KU, TOKYO 107
JAPANESE
INDEXED: CHE

JAPANESE JOURNAL OF PHYSIOLOGY 2016
NIHON SEIRI GAKKAI (PHYSIOLOGICAL SOCIETY OF JAPAN) AND TODAI SHUPPANKAI (UNIVERSITY OF TOKYO PRESS)
3-1, 7-CHOME, HONGO, BUNKYO-KU, TOKYO 113
ENGLISH
INDEXED: BIB, BIO, CHE, EXC, IND, IAA

JOSHI EIYO DAIGAKU KIYO 2017
(JOURNAL OF KAGAWA NUTRITION COLLEGE)
JOSHI EIYO DAIGAKU
24-3, 3-CHOME, KOMAGOME, TOSHIMA-KU, TOKYO 170
JAPANESE AND ENGLISH
INDEXED: CHE, JICST

JOURNAL OF NUTRITIONAL SCIENCE AND VITAMINOLOGY 2018
NIPPON BITAMIN GAKKAI (VITAMIN SOCIETY OF JAPAN)
C/O NIPPON ITARIA KYOTO KAIKAN, 4, YOSHIDA USHINOMIYA-CHO, SAKYO-KU, KYOTO-SHI 606
ENGLISH
INDEXED: BIB, BIO, CHE, EXC, JICST

KANAGAWA-KENRITSU EIYO TANKI DAIGAKU KIYO 2019
(JOURNAL OF KANAGAWA PREFECTURAL JUNIOR COLLEGE OF NUTRITION)
KANAGAWA-KENRITSU EIYO TANKI DAIGAKU (KANAGAWA PREFECTURAL JUNIOR COLLEGE OF NUTRITION)
399, SAKURAGAOKA, HODOGAYA-KU, YOKOHAMA-SHI 240
JAPANESE, WITH ENGLISH SUMMARY
INDEXED: CHE

KENMIN EIYO NO GENJO (CHIBA-KEN) 2020
(NUTRITIONAL STATISTICS IN CHIBA PREFECTURE))
CHIBA-KEN EISEI-BU (CHIBA PREFECTURAL GOVERNMENT)
1-1, ICHIBA-CHO, CHIBA-SHI 280-91
JAPANESE

KETSUEKI TO MYAKKAN 2021
(BLOOD ANGIOLOGY)
KAGAKU HYORONSHA
2-10, KANDA TSUKASA-CHO, CHIYODA-KU, TOKYO 101
JAPANESE

KOKURITSU EIYO KENKYUSHO KENKYU HOKOKU 2022
(RESEARCH REPORT OF THE NATIONAL INSTITUTE OF NUTRITION)
KOSEI-SHO KOKURITSU EIYO KENKYUSHO (NATIONAL INSTITUTE OF NUTRITION, MINISTRY OF HEALTH AND WELFARE)
1, TOYAMA-CHO, SHINJUKU-KU, TOKYO 162
JAPANESE
INDEXED: CHE, JICST

KOKYU TO JUNKAN 2023
(RESPIRATION AND CIRCULATION)
IGAKU SHOIN
29-11, 5-CHOME, HONGO, BUNKYO-KU, TOKYO 113
JAPANESE AND ENGLISH
INDEXED: BIO, CHE

KOTSU TAISHA 2024
(BONE METABOLISM)
KOTSU TAISHA KENKYUKAI (JAPANESE SOCIETY FOR BONE METABOLISM RESEARCH)
13-10, 2-CHOME, HONGO, BUNKYO-KU, TOKYO 113
JAPANESE AND ENGLISH

KYOSEN SOKUHO 2025
(BIBLIOGRAPHY OF THYNUS GLAND AND IMMUNOLOGY)
KOKUSAI IGAKU JOHO SENTA (INTERNATIONAL MEDICAL INFORMATION CENTER)
30, DAIKYO-MACHI, SHINJUKU-KU, TOKYO 160
JAPANESE AND ENGLISH

KYUSHOKU KANRI 2026
(MASS FEEDING ADMINISTRATION)
NIHON KYUSHOKU GIJUTSU KYOKAI (JAPANESE SOCIETY OF MASS FEEDING TECHNICS)
51, CHAYA-MACHI, KITA-KU, OSAKA-SHI 530
JAPANESE

KYUSHU KETSUEKI KENKYU DOKOKAISHI 2027
(JOURNAL OF KYUSHU HEMATOLOGICAL SOCIETY)
KYUSHU KETSUEKI KENKYU DOKOKAI (KYUSHU HEMATOLOGICAL SOCIETY) AND KYUSHU DAIGAKU SHIGAKUBU KOKU BYORIGAKU KYOSHITSU (DEPARTMENT OF ORAL PATHOLOGY FACULTY OF DENISTRY, KYUSHU UNIVERSITY)
C/O KYUSHU DAIGAKU SHIGAKUBU KOKO BYORIGAKU KYOSHITSU, KATAKASU, FUKUOKA-SHI 812
JAPANESE AND ENGLISH
INDEXED: BIO, EXC

MASS FEEDING 2028
SHIBATA SHOTEN
33-5, 3-CHOME, HONGO, BUNKYO-KU, TOKYO 113
JAPANESE

NAGOYA EIYO TANKI DAIGAKU DAIGAKU KIYO 2029
(JOURNAL OF NAGOYA NUTRITION COLLEGE)
NAGOYA EIYO TANKI DAIGAKU (NAGOYA NUTRITION JUNIOR COLLEGE)
JAPANESE

NIHON JUNKANKIGAKUSHI 2030
(JAPANESE CIRCULATION JOURNAL)
NIHON JUNKANKI GAKKAI (JAPANESE CIRCULATION SOCIETY)
C/O KYOTO DAIGAKU IGAKUBU NAIKA DAI-3-KOZA, SHOGOIN KAWARA-CHO, SAKYO-KU, KYOTO-SHI 606
JAPANESE AND ENGLISH
INDEXED: CHE, EXC, JICST

NIHON KETSUEKI GAKKAI SOKAI PUROGURAMU SHOROKUSHU 2031
(ABSTRACTS AND PROGRAM OF THE SYMPOSIUM OF JAPAN HEMATOLOGICAL SOCIETY)
NIHON KETSUEKI GAKKAI (JAPAN HEMATOLOGICAL SOCIETY)
C/O KOKURITSU GAN SENTA, 1-1, 5-CHOME, TSUKIJI, CHUO-KU, TOKYO 104
JAPANESE

NIHON KETSUEKI GAKKAI ZASSHI 2032
(ACTA HAEMATOLOGICA JAPONICA)
NIHON KETSUEKI GAKKAI (JAPAN HAEMATOLOGICAL SOCIETY)
C/O KYOTO DAIGAKU IGAKUBU NAIKA DAI-1-KOZA, SHOGOIN KAWARA-CHO, SAKYO-KU, KYOTO-SHI 606
JAPANESE AND EUROPEAN
INDEXED: BIO, CHE, EXC, IND, JICST

NIHON NAIBUNPI GAKKAI TOBU BUKAISHI 2033
(ENDOCRINOLOGICA JAPONICA)
NIHON NAIBUNPI GAKKAI OBUN-SHI HENSHU IINKAI (EDITORIAL BOARD OF ENDOCRINOLOGICA JAPONICA)
TOKYO JIKEIKAI IKA DAIGAKU DAI-2 KAIBO KYOSHITSU
25-8, 3-CHOME, NISHISHINBASHI, MINATO-KU, TOKYO 105
ENGLISH

NIHON NAIBUNPI GAKKAI ZASSHI 2034
(FOLIA ENDOCRINOLOGICA JAPONICA)
NIHON NAIBUNPI GAKKAI (JAPAN ENDOCRINOLOGICAL SOCIETY)
C/O KYOTO-FURITSU IKA DAIGAKU DAI-1 NAIKA KYOSHITSU, HIROKOJIAGARU, KAWARA-MACHI-DORI, KAMIGYO-KU, KYOTO-SHI 602
JAPANESE, WITH EUROPEAN SUMMARY
INDEXED: BIO, CHE, EXC, JICST

NIHON RINSHO TAISHA GAKKAI KIROKU 2035
(PROCEEDINGS OF THE JAPAN SOCIETY OF CLINICAL BIOCHEMISTRY AND METABOLISM)
NIHON RINSHO TAISHA GAKKAI (JAPANESE SOCIETY OF CLINICAL BIOCHEMISTRY AND METABOLISM)
C/O TOKYO DAIGAKU IGAKUBU, 3-1, 7-CHOME, HONGO, BUNKYO-KU, TOKYO 113
JAPANESE
INDEXED: CHE

NIHON SEIRIGAKU ZASSHI 2036
(JOURNAL OF THE PHYSIOLOGICAL SOCIETY OF JAPAN)
NIHON SEIRI GAKKAI (PHYSIOLOGICAL SOCIETY OF JAPAN)
C/O TOYO BUNKO, 28-21, 2-CHOME, HONKOMAGOME, BUNKYO-KU, TOKYO 113
JAPANESE, WITH EUROPEAN SUMMARY
INDEXED: CHE, BIB, BIO, EXC, JICST

ONSEI GAKKAI KAIHO 2037
(BULLETIN, PHONETIC SOCIETY OF JAPAN)
NIHON ONSEI GAKKAI (PHONETIC SOCIETY OF JAPAN)
13-2, 2-CHOME, DAITA, SETAGAYA-KU, TOKYO 154
JAPANESE AND ENGLISH

ONSEI GENGO IGAKU 2038
(JAPAN JOURNAL OF LOGOPEDICS AND PHONIATRICS)
NIHON ONSEI GENGO IGAKKAI (THE JAPAN SOCIETY OF LOGOPEDICS AND PHONIATRICS)
C/O TOKYO DAIGAKU IGAKUBU ONSEI GENGO IGAKU KENKYUSHISETSU, 3-1, 7-CHOME, HONGO, BUNKYO-KU TOKYO 113
JAPANESE, WITH ENGLISH SUMMARY

ONSEI KAGAKU KENKYU 2039
(STUDIA PHONOLOGICA)
KYOTO DAIGAKU ONSEI KAGAKU SOGO KENKYU BUKAI (INSTITUTE FOR PHONETIC SCIENCES, KYOTO UNIVERSITY)
YOSHIDA NIHONMATSU-CHO, SAKYO-KU, KYOTO-SHI 606
ENGLISH

OSAKA SHIKA DAIGAKU SEIRIGAKU KYOSHITSU RONBUNSHU 2040
(COLLECTED PAPERS FROM INSTITUTE OF PHYSIOLOGY, OSAKA DENTAL COLLEGE)
OSAKA SHIKA DAIGAKU SEIRIGAKU KYOSHITSU
1-47, KYOBASHI, HIGASHI-KU, OSAKA-SHI 540
JAPANESE AND ENGLISH

PROCEEDINGS OF ANNUAL MEETING OF JAPAN ENDOCRINOLOGICAL SOCIETY 2041
NIHON NAIBUNPI GAKKAI (JAPAN ENDOCRINOLOGICAL SOCIETY)
C/O KYOTO FURITSU IKA DAIGAKU DAI-1- NAIKA KYOSHITSU HIROKOJI, KAWARAMACHI-DORI, KAMIGYO-KU, KYOTO-SHI 602
ENGLISH

RINSHO EIYO 2042
(JOURNAL OF CLINICAL NUTRITION)
ISHIYAKU SHUPPAN K.K. (ISHIYAKU PUBLISHERS INCORPORATED)
7-10, 1-CHOME, HONKOMAGOME, BUNKYO-KU, TOKYO 113
JAPANESE
INDEXED: CHE, JICST

RINSHO KAGAKU SHINPOJUMU 2043
(PROCEEDING OF THE SYMPOSIUM ON CHEMICAL PHYSIOLOGY AND PATHOLOGY)
IKAGAKU SHIMPOJUMU JIMU-KYOKU (JAPANESE SOCIETY OF CLINICAL CHEMISTRY)
C/O OSAKA DAIGAKU FUZOKU BYOIN YAMAMURA NAIKA, 3, DOJIMA HAMA-DORI, FUKUSHIMA-KU, OSAKA-SHI 553
JAPANESE, WITH EUROPEAN SUMMARY
INDEXED: BIO, CHE

RINSHO KETSUEKI 2044
(JAPANESE JOURNAL OF CLINICAL HEMATOLOGY)
NIHON RINSHO KETSUEKI GAKKAI (THE JAPANESE SOCIETY OF CLINICAL HEMATOLOGY)
C/O NIHON KETSUEKI KENSA KIKAI KENTEI KYOKAI, 1-464, HIGASHIOKUBO, SHINJUKU-KU, TOKYO 160
JAPANESE, WITH EUROPEAN SUMMARY

RINSHO SEIRI 2045
(CLINICAL PHYSIOLOGY)
SHINKO KOEKI K. K. ISHO SHUPPAN-BU (PUBLICATION DEPARTMENT OF MEDICAL BOOKS, SHINKO TRADING COMPANY, LIMITED)
C/O FUKUYOSHI-CHO BUILDING, 2-6, 2-CHOME, ROPPONGI, MINATO-KU, TOKYO 106
JAPANESE

SEICHO 2046
(JOURNAL OF GROWTH)
SEICHO DANWAKAI
C/O AICHI GAKUIN DAIGAKU SHIGAKUBU DAI-1 KAIBOGAKU KYOSHITSU, 2-11, SUEMORI-DORI, CHIKUSA-KU, NAGOYA-SHI 464
JAPANESE, WITH ENGLISH SUMMARY
INDEXED: CHE

SEIRI SEITAI 2047
(PHYSIOLOGY AND ECOLOGY)
SEIRI SEITAI KANKOKAI
KYOTO
JAPANESE, WITH ENGLISH SUMMARY
INDEXED: CHE

SEITAI NO KAGAKU 2048
(SCIENCE OF THE HUMAN BODY)
IGAKU SHOIN (IGAKU SHOIN LIMITED)
29-11, 5-CHOME, HONGO, BUNKYO-KU, TOKYO 113
JAPANESE
INDEXED: CHE, JICST

SEITOKU EIYO TANKI DAIGAKU KIYO 2049
(MEMOIRS OF SEITOKU JUNIOR COLLEGE OF NUTRITION)
SEITOKU EIYO TANKI DAIGAKU (SEITOKU JUNIOR COLLEGE OF NUTRITION)
4-6, 1-CHOME, NISHISHINKOIWA, KATSUSHIKAKU, TOKYO 124
JAPANESE AND ENGLISH

SHIN EIYO 2050
(NEW NUTRITION)
SANRO
47-1, KITAYAMABUSHI-CHO, SHINJUKU-KU, TOKYO 162
JAPANESE

SHOKU SEIKATSU 2051
(FOOD LIFE)
KOKUMIN EIYO KYOKAI (THE ASSOCIATION OF NATIONAL NUTRITION)
C/O NIHON KOSHU EISEI KYOKAI BUILDING, 78, HANAZONO-CHO, SHINJUKU-KU, TOKYO 160
JAPANESE

TABEMONO TO KENKO 2052
(FOOD AND HEALTH)
YUKI NOGYO KENKYUKAI (ORGANIC AGRICULTURE RESEARCH SOCIETY)
JAPANESE

TAISHA 2053
(METABOLISM AND DISEASE)
NAKAYAMA SHOTEN
2-24, KANDA JINBO-CHO, CHIYODA-KU, TOKYO 101
JAPANESE
INDEXED: CHE

TAISHA IJO CHIRYO KENKYU KIKIN KENKYU GYOSEKISHU 2054
(COLLECTED RESEARCH FINDINGS OF THE FOUNDATION FOR RESEARCH ON THE TREATMENT OF ABNORMAL METABOLISM)
FOUNDATION FOR RESEARCH ON THE TREATMENT OF ABNORMAL METABOLISM
JAPANESE

TOSHITSU NO TAISHA 2055
(CARBOHYDRATES AND METABOLISM)
TOSHITSU TAISHA KENKYU KAIHEN OTSUKA SEIYAKU (CARBOHYDRATE AND METABOLISM RESEARCH COMMITTEE OTSUKA PHARMACEUTICAL COMPANY)
JAPANESE

YAMAGUCHI DAIGAKU TANPAKU KENKYUKAI NENPO 2056
(ANNUAL REPORT OF THE SOCIETY FOR PROTEIN CHEMISTRY, YAMAGUCHI UNIVERSITY)
YAMAGUCHI DAIGAKU IGAKUBU SEIKAGAKU KYOSHITSU (DEPARTMENT OF BIOCHEMISTRY, SCHOOL OF MEDICINE, YAMAGUCHI UNIVERSITY), YAMAGUCHI DAIGAKU TANPAKU KENKYUKAI (SOCIETY FOR PROTEIN CHEMISTRY)
C/O YAMAGUCHI DAIGAKU IGAKUBU, OGUSHI, UBE-SHI 755
JAPANESE

ZESSHOKU KENKYU 2057
(LASTING RESEARCH)
ZESSHOKU KENKYU KAI
JAPANESE

Health, Preventive Medicine, Environmental Pollution

AICHI-KEN EISEI KENKYUSHOHO 2058
(REPORT OF AICHI PREFECTURAL INSTITUTE OF PUBLIC HEALTH)
AICHI-KEN EISEI KENKYUSHO (AICHI PREFECTURAL INSTITUTE OF PUBLIC HEALTH)
7-6, NAGARE, TSUJI-MACHI, KITA-KU, NAGOYA-SHI 462
JAPANESE
INDEXED: CHE, JICST

AICHI-KEN KOGAI CHOSA SENTA SHOHO 2059
(ANNUAL REPORT OF AICHI PREFECTURAL RESEARCH CENTER FOR ENVIRONMENTAL PROTECTION)
AICHI-KEN KOGAI CHOSA SENTA
1-2, 3-CHOME, SANNOMARU, NAKA-KU, NAGOYA-SHI 460
JAPANESE
INDEXED: CHE

AKITA-KEN EISEI KAGAKU KENKYUSHOHO 2060
(REPORT OF THE AKITA PREFECTURE INSTITUTE OF PUBLIC HEALTH)
AKITA-KEN EISEI KAGAKU KENKYUSHO (AKITA PREFECTURE INSTITUTE OF PUBLIC HEALTH)
1-40, SENSHUMEITOKU-CHO, AKITA-SHI 010
JAPANESE AND ENGLISH

AKITA-KEN KOGAI GIJUTSU SENTA SHOHO 2061
(REPORT OF AKITA PREFECTURAL RESEARCH INSTITUTE FOR ENVIRONMENTAL PROTECTION)
AKITA-KEN KOGAI GIJUTSU SENTA (AKITA PREFECTURAL RESEARCH INSTITUTE FOR ENVIRONMENTAL PROTECTION)
4-82, 1-CHOME, BARAJIMA, AKITA-SHI 010
JAPANESE

AKUSHU NO KENKYU 2062
(ODOR CONTROL ASSOCIATION JOURNAL)
AKUSHU KOGAI KENKYUKAI (ODOR CONTROL ASSOCIATION OF JAPAN)
2-12, 1-CHOME, MUKOGAOKA, BUNKYO-KU, TOKYO 112
JAPANESE
INDEXED: APT, JICST

ANNUAL REPORT OF THE RESEARCH INSTITUTE OF ENVIRONMENTAL MEDICINE, NAGOYA UNIVERSITY 2063
NAGOYA DAIGAKU KANKYO IGAKU KENKYUSHO (RESEARCH INSTITUTE OF ENVIRONMENTAL MEDICINE, NAGOYA UNIVERSITY)
FURO-CHO, CHIKUSA-KU, NAGOYA-SHI 464
ENGLISH
INDEXED: CHE, JICST

ANNUAL REPORT OF THE TOKYO METROPOLITAN RESEARCH INSTITUTE FOR ENVIRONMENTAL PROTECTION 2064
TOKYO-TO KOGAI KENKYUSHO
2-5, YURAKU-CHO, CHIYODA-KU, TOKYO 100
ENGLISH
INDEXED: CHE

AO TO MIDORI 2065
(BLUE SKY AND TREE PLANTING)
AO TO MIDORI HENSHU IINKAI AND KAEDE SHUPPANSHA
OZAQ, OZAWA BUILDING, 1-10, WAKABA, SHINJUKU-KU, TOKYO 160
JAPANESE

AOMORI-KEN EISEI KENKYUSHO SHOHO 2066
(ANNUAL REPORT OF AOMORI PREFECTURE INSTITUTE OF PUBLIC HEALTH)
AOMORI-KEN EISEI KENKYUSHO (AOMORI PREFECTURE INSTITUTE OF PUBLIC HEALTH)
25, SAWADA, TSUKURIMICHI, AOMORI-SHI 030
JAPANESE

AOMORI-KEN KOGAI SENTA SHOHO 2067
(REPORT OF AOMORI PREFECTURE POLLUTION CENTER)
JAPANESE

ARUKORU KENKYU 2068
(JAPANESE JOURNAL OF STUDIES ON ALCOHOL)
NIHON ARUKORU IGAKKAI (JAPANESE MEDICAL SOCIETY OF ALCOHOL STUDIES)
C/O KYOTO FURITSU IKA DAIGAKU HOIGAKU KYOSHITSU, HIROKOJI, KAWARA-MACHI, KAMIGYO-KU, KYOTO-SHI 602
JAPANESE, WITH ENGLISH SUMMARY
INDEXED: CHE

ATOMU FUKUSHIMA 2069
(FUKUSHIMA ATOM)
FUKUSHIMA-KEN SEIKATSU KANKYO BU KANKO SEIBIKA (ENVIRONMENTAL EQUIPMENT OFFICE, LIVING ENVIRONMENT DIVISION, FUKUSHIMA PREFECTURE)
JAPANESE

BCG MEN'EKIRYOHO KENKYUKAI KAISHI 2070
(BCG RESEARCH ASSOCIATION OF IMMUNOLOGY)
JAPANESE AND ENGLISH

BOEI EISEI 2071
(NATIONAL DEFENSE MEDICAL JOURNAL)
BOEI EISEI KYOKAI (NATIONAL DEFENSE MEDICAL SOCIETY)
C/O BOEI-CHO, EISEI-KYOKU, EISEI-KA, 7-45, 9-CHOME, AKASAKA, MINATO-KU, TOKYO 107
JAPANESE AND ENGLISH
INDEXED: CHE

BOSEI EISEI 2072
(JAPANESE JOURNAL OF MATERNAL HEALTH)
NIHON BOSEI EISEI GAKKAI (JAPAN SOCIETY OF MATERNAL HEALTH)
C/O SHIMADA BUILDING, 2-17, 3-CHOME, NIHONBASHI, CHUO-KU, TOKYO
JAPANESE

BOSHI EISEI NO GENKYO (NIIGATA-KEN) 2073
(ANNUAL REPORT ON MATERNAL AND CHILD HEALTH, NIIGATA PREFECTURE))
NIIGATA-KEN (NIIGATA PREFECTURAL GOVERNMENT)
602, ICHIBAN-CHO, GAKKO-DORI, NIIGATA-SHI 951
JAPANESE

BOSHI EISEI NO OMONARU TOKEI 2074
(STATISTICS RELATING TO MATERNAL AND CHILD HEALTH IN JAPAN)
KOSEI-SHO JIDO KATEI-KYOKU (CHILDREN AND FAMILIES BUREAU, MINISTRY OF HEALTH AND WELFARE)
2-2, 1-CHOME, KASUMIGASEKI, CHIYODA-KU, TOKYO 100
JAPANESE

BOSHI HOKEN KAZOKU KEIKAKU NENPO 2075
(MATERNAL AND CHILD WELFARE AND FAMILY PLANNING, ANNUAL REPORT)
NIHON KAZOKU KEIKAKU KYOKAI (JAPAN FAMILY PLANNING ASSOCIATION)
1-2, ICHIGAYA SADOHARA-CHO, SHINJUKU-KU, TOKYO 162
JAPANESE

CHIBA DAIGAKU KANKYO KAGAKU KENKYU HOKOKU 2076
(REPORT OF ENVIRONMENTAL RESEARCH ORGANIZATION OF CHIBA UNIVERSITY)
JAPANESE

CHIBA-KEN EISEI KENKYUSHO NENPO 2077
(ANNUAL REPORT OF PUBLIC HEALTH LABORATORY, CHIBA PREFECTURE)
CHIBA-KEN EISEI KENKYUSHO
204, SHINMEI-CHO, CHIBA-SHI 280
JAPANESE

CHIBA-KEN KOGAI KENKUJO JIBAN CHINKA KENKYU JIGYO HOKOKU 2078
(PROJECT RESEARCH REPORT, CHIBA PREFECTURAL RESEARCH INSTITUTE FOR ENVIRONMENTAL POLLUTION)
JAPANESE
INDEXED: JICST

CHIBA-KEN KOGAI KENKYUJO KENKYU HOKOKU 2079
(BULLETIN OF CHIBA PREFECTURAL RESEARCH INSTITUTE FOR ENVIRONMENTAL POLLUTION)
JAPANESE
INDEXED: JICST

CHIBA-KEN KOGAI KENKYUSHO NENPO 2080
(ANNUAL REPORT OF THE PREFECTURAL RESEARCH INSTITUTE FOR ENVIRONMENTAL POLLUTION, CHIBA)
CHIBA-KEN KOGAI KENKYUSHO (PREFECTURAL RESEARCH INSTITUTE FOR ENVIRONMENTAL POLLUTION)
254, IWASAKI, ICHIHARA-SHI 290
JAPANESE

CHIBA-SHI TAIKI OSEN SHOSA HOKOKU 2081
(INVESTIGATIONS ON AIR POLLUTION, CHIBA CITY)
CHIBA-SHI SHIMIN SEIKATSU-BU
1-1, CHIBA-KO, CHIBA-SHI 280
JAPANESE

CHUDOKUKENHO 2082
(REPORT OF THE TOXICOLOGY INSTITUTE, KUMAMOTO UNIVERSITY, MEDICAL SCHOOL)
KUMAMOTO DAIGAKU IGAKUBU FUZOKU CHUDOKU KENKYU SHISETSU (TOXICOLOGY INSTITUTE, KUMAMOTO UNIVERSITY MEDICAL SCHOOL)
2-1, 2-CHOME, HONSHO, KUMAMOTO-SHI 860
JAPANESE AND ENGLISH

CHUO HOKEN KANRISHOHO 2083
(HEALTH CONTROL)
NIHON KOKUYU TETSUDO CHUO HOKEN KANRISHO (CENTRAL HEALTH INSTITUTE, JAPAN NATIONAL RAILWAYS)
2-1, YOYOGI, SHIBUYA-KU, TOKYO 151
JAPANESE AND ENGLISH

DENRYOKU TO KOGAI 2084
(ELECTRIC POWER AND POLLUTION)
DENRYOKU KOGAI KENKYUKAI (ELECTRIC POWER POLLUTION RESEARCH ASSOCIATION)
JAPANESE

DENSENBYO OYOBI SHOKUCHUDOKU TOKEI 2085
(STATISTICS OF COMMUNICABLE DISEASE AND FOOD POISONING)
KOSEI-SHO DAIJIN KANBO TOKEI CHOSA-BU (HEALTH AND WELFARE STATISTICS DIVISION, MINISTER'S SECRETARIAT, MINISTRY OF HEALTH AND WELFARE
42, ICHIGAYA HONMURA-CHO, SHINJUKU-KU, TOKYO 162
JAPANESE

DRUG ACTION AND DRUG RESISTANCE IN BACTERIA 2086
UNIVERSITY OF TOKYO PRESS AND
UNIVERSITY PARK PRESS
7-3-1 HONGO, BUNKYO-KU, TOKYO
ENGLISH
INDEXED: CHE

EHIME NO KOGAI 2087
(ENVIRONMENTAL POLLUTION IN EHIME PREFECTURE)
EHIME-KEN (EHIME PREFECTURAL GOVERNMENT)
4-2, 4-CHOME, ICHIBAN-CHO, MATSUYAMA-SHI 790
JAPANESE

EHIME-KENRITSU EISEI KENKYUSHO NENPO 2088
(ANNUAL REPORT OF EHIME PREFECTURAL HYGIENIC LABORATORY)
EHIME-KENRITSU EISEI KENKYUSHO (EHIME PREFECTURAL INSTITUTE OF PUBLIC HEALTH)
234, 8-CHOME, SANBAN-CHO, MATSUYAMA-SHI 790
JAPANESE
INDEXED: CHE

EIKEN NYUSU (YAMAGATA-KEN) 2089
(NEWS BULLETIN OF THE YAMAGATA PREFECTURAL INSTITUTE OF PUBLIC HEALTH)
YAMAGATA-KEN EISEI KENKYUSHO (YAMAGATA PREFECTURAL INSTITUTE OF PUBLIC HEALTH)
6-6, 1-CHOME, TOKA-MACHI, YAMAGATA-SHI 990
JAPANESE

EISEI DOBUTSU 2090
(JAPANESE JOURNAL OF SANITARY ZOOLOGY)
NIHON EISEI DOBUTSU GAKKAI (JAPAN SOCIETY OF SANITARY ZOOLOGY)
C/O NIHON KANKYO EISEI SENTA, 198-3, YOTSUYA-KAMI-MACHI, SAIWAI-KU, KAWASAKI-SHI 210
JAPANESE AND EUROPEAN
INDEXED: BIB, BIO, CHE

EISEI GAKKO KIJI 2091
(MEDICAL SCHOOL REVIEW, GROUND SELF DEFENSE FORCE)
RIKUJO JIEITAI MISHUKU SHUSHINKAI
2-24, 1-CHOME, IKEJIRI, SETAGAYA-KU, TOKYO 154
JAPANESE

EISEI GYOSEI GYOMU HOKOKU 2092
(ADMINISTRATIVE REPORT ON PUBLIC HEALTH)
KOSEI-SHO DAIJIN KANBO TOKEI CHOSA-BU (HEALTH AND WELFARE STATISTICS DIVISION, MINISTER'S SECRETARIAT, MINISTRY OF HEALTH AND WELFARE
42, ICHIGAYA HONMURA-CHO, SHINJUKU-KU, TOKYO 162
JAPANESE

EISEI KENSA 2093
(JAPAN JOURNAL OF MEDICAL TECHNOLOGY)
NIHON EISEI KENSA GISHIKAI (ASSOCIATION OF MEDICAL TECHNOLOGIST)
C/O KITASATO DAIGAKU, 9-1, 5-CHOME, SHIROGANE, MINATO-KU, TOKYO 108
JAPANESE

EISEI NENPO (AICHI-KEN) 2094
(ANNUAL REPORT OF PUBLIC HEALTH, AICHI PREFECTURE))
AICHI-KEN EISEI-BU (PUBLIC HEALTH DEPARTMENT, AICHI PREFECTURAL GOVERNMENT)
1-2, 3-CHOME, SANNO-MARU, NAKA-KU, NAGOYA-SHI 460
JAPANESE

EISEI NENPO (IWATE-KEN) 2095
(ANNUAL REPORT OF PUBLIC HEALTH, IWATE PREFECTURE)
IWATE-KEN
10-1, UCHIMARU, MORIOKA-SHI 020
JAPANESE

EISEI NENPO (KANAZAWA-SHI) 2096
(ANNUAL REPORT OF PUBLIC HEALTH, KANAZAWA CITY))
KANAZAWA-SHI KOSEI-BU
1-1, 1-CHOME, HIROSAKA, KANAZAWA-SHI 920
JAPANESE

EISEI NENPO (KUMAMOTO-KEN) 2097
(ANNUAL REPORT OF PUBLIC HEALTH, KUMAMOTO PREFECTURE)
KUMAMOTO-KEN EISEI-BU (PUBLIC HEALTH DEPARTMENT, KUMAMOTO PREFECTURAL GOVERNMENT)
18-1, 6-CHOME, SUIZENJI, KUMAMOTO-SHI 862
JAPANESE

EISEI NENPO (NAGASAKI-KEN) 2098
(ANNUAL REPORT OF PUBLIC HEALTH, NAGASAKI PREFECTURE)
NAGASAKI-KEN HOKEN-BU
2-13, EDO-MACHI, NAGASAKI-SHI 850
JAPANESE

EISEI NENPO (NAGOYA-SHI 2099
(ANNUAL REPORT OF PUBLIC HEALTH, NAGOYA CITY)
NAGOYA-SHI EISEI-KYOKU
1-1, 3-CHOME, SANNOMARU, NAKA-KU, NAGOYA-SHI 460
JAPANESE

EISEI NENPO (NIIGATA-KEN) 2100
(ANNUAL REPORT OF PUBLIC HEALTH, NIIGATA PREFECTURE)
NIIGATA-KEN EISEI-BU (HEALTH DEPARTMENT, NIIGATA PREFECTURAL GOVERNMENT)
1-5290, GAKKO-MACHI, NIIGATA-SHI 950
JAPANESE

EISEI NENPO (OSAKA-FU) 2101
(ANNUAL REPORT OF PUBLIC HEALTH, OSAKA PREFECTURE)
OSAKA-FU EISEI-BU (HEALTH DEPARTMENT, OSAKA PREFECTURAL GOVERNMENT)
OTEMAENO-CHO, HIGASHI-KU, OSAKA-SHI 540
JAPANESE

EISEI NENPO (SHIZUOKA-KEN) 2102
(ANNUAL REPORT OF PUBLIC HEALTH, SHIZUOKA PREFECTURE)
SHIZUOKA-KEN (SHIZUOKA PREFECTURAL GOVERNMENT)
9-6, OTE-MACHI, SHIZUOKA-SHI 420
JAPANESE
INDEXED: JICST

EISEI NENPO (TOCHIGI-KEN) 2103
(ANNUAL REPORT OF PUBLIC HEALTH, TOCHIGI PREFECTURE)
TOCHIGI-KEN (TOCHIGI PREFECTURAL GOVERNMENT)
504, HANAWADA-CHO, UTSUNOMIYA-SHI 320
JAPANESE

EISEI NENPO (TOKYO-TO) 2104
(ANNUAL REPORT OF PUBLIC HEALTH, TOKYO)
TOKYO-TO EISEI-KYOKU SOMU-BU (BUREAU OF PUBLIC HEALTH, TOKYO METROPOLITAN GOVERNMENT)
5-1, 3-CHOME, MARUNOUCHI, CHIYODA-KU, TOKYO 100
JAPANESE

EISEI NENPO (YOKOHAMA-SHI) 2105
(ANNUAL REPORT OF PUBLIC HEALTH, YOKOHAMA CITY)
YOKOHAMA-SHI EISEI-KYOKU (HEALTH BUREAU, YOKOHAMA CITY OFFICE)
1-1, MINATO-CHO, NAKA-KU, YOKOHAMA-SHI 231
JAPANESE

EISEI NENPO, KYOIKU SHIRYO-HEN (KANAGAWA-KEN) 2106
(ANNUAL REPORT OF PUBLIC HEALTH, KANAGAWA PREFECTURE SERIES ON EDUCATION)
KANAGAWA-KEN EISEI-BU (PUBLIC HEALTH DEPARTMENT, KANAGAWA PREFECTURAL GOVERNMENT)
1, NIHON-ODORI, NAKA-KU, YOKOHAMA-SHI 231
JAPANESE

EISEI SHIKENJO HOKOKU 2107
(BULLETIN OF NATIONAL INSTITUTE OF HYGIENIC SCIENCE)
KOKURITSU EISEI SHIKENJO (NATIONAL INSTITUTE OF HYGIENIC SCIENCE)
18-1, 1-CHOME, KAMIYOGA, SETAGAYA-KU, TOKYO 158
JAPANESE, WITH EUROPEAN SUMMARY
INDEXED: BIB, CHE, EXC, JICST

EISEI TOKEI NENPO (FUKUI-KEN) 2108
(ANNUAL REPORT OF PUBLIC HEALTH, FUKUI PREFECTURE)
FUKUI-KEN KOSEI-BU (PUBLIC WELFARE DEPARTMENT, FUKUI PREFECTURAL GOVERNMENT)
101, GOHONMARU, FUKUI-SHI 910
JAPANESE

EISEI TOKEI NENPO (HOKKAIDO) 2109
(ANNUAL REPORT OF PUBLIC HEALTH, HOKKAIDO)
HOKKAIDO EISEI-BU (PUBLIC HEALTH DEPARTMENT, HOKKAIDO PREFECTURAL GOVERNMENT)
NISHI-6-CHOME, KITA-3-JO, CHUO-KU, SAPPORO-SHI 060
JAPANESE

EISEI TOKEI NENPO (IBARAKI-KEN) 2110
(ANNUAL REPORT OF PUBLIC HEALTH, IBARAKI PREFECTURE)
IBARAKI-KEN EISEI-BU (PUBLIC HEALTH DEPARTMENT, IBARAKI PREFECTURAL GOVERNMENT)
5-38, 1-CHOME, SANNOMARU, MITO-SHI 310
JAPANESE

EISEI TOKEI NENPO (ISHIKAWA-KEN) 2111
(ANNUAL REPORT OF PUBLIC HEALTH, ISHIKAWA PREFECTURE)
ISHIKAWA-KEN KOSEI-BU (PUBLIC HEALTH DEPARTMENT, ISHIKAWA PREFECTURAL GOVERNMENT)
1-1, 2-CHOME, HIROSAKA, KANAZAWA-SHI 920
JAPANESE

EISEI TOKEI NENPO (KAGOSHIMA-KEN) 2112
(ANNUAL REPORT OF PUBLIC HEALTH, KAGOSHIMA PREFECTURE)
KAGOSHIMA-KEN (KAGOSHIMA PREFECTURAL GOVERNMENT)
14-50, YAMASHITA-CHO, KAGOSHIMA-SHI 892
JAPANESE

EISEI TOKEI NENPO (MIYAGI-KEN) 2113
(ANNUAL REPORT OF PUBLIC HEALTH, MIYAGI PREFECTURE)
MIYAGI-KEN EISEI-BU (SANITARY DEPARTMENT, MIYAGI PREFECTURAL GOVERNMENT)
27, KOTODAI-DORI, SENDAI-SHI 980
JAPANESE

EISEI TOKEI NENPO (MIYAZAKI-KEN) 2114
(ANNUAL REPORT OF PUBLIC HEALTH, MIYAZAKI PREFECTURE)
MIYAZAKI-KEN KANKYO HOKEN-BU (PUBLIC HEALTH DEPARTMENT, MIYAZAKI PREFECTURAL GOVERNMENT)
10-1, 2-CHOME, TACHIBANA-DORI HIGASHI, MIYAZAKI-SHI 880
JAPANESE

EISEI TOKEI NENPO (OKAYAMA-KEN) 2115
(ANNUAL REPORT OF PUBLIC HEALTH, OKAYAMA PREFECTURE)
OKAYAMA-KEN EISEI-BU (PUBLIC HEALTH DEPARTMENT, OKAYAMA PREFECTURAL GOVERNMENT)
4-6, 2-CHOME, UCHIYAMASHITA, OKAYAMA-SHI, 700-91
JAPANESE

EISEI TOKEI NENPO (OSAKA-SHI) 2116
(ANNUAL REPORT OF PUBLIC HEALTH, OSAKA CITY)
OSAKA-SHI KANKYO HOKEN-KYOKU KANRI-BU (OSAKA PREFECTURAL GOVERNMENT)
3-2, SENBA-CHUO, HIGASHI-KU, OSAKA-SHI 540
JAPANESE

EISEI TOKEI NENPO (SHIGA-KEN) 2117
(ANNUAL REPORT OF PUBLIC HEALTH, SHIGA PREFECTURE)
SHIGA-KEN (SHIGA PREFECTURAL GOVERNMENT)
1-1, 4-CHOME, KYO-MACHI, OTSU-SHI 520
JAPANESE

EISEI TOKEI NENPO (TOKUSHIMA-KEN) 2118
(ANNUAL REPORT OF PUBLIC HEALTH, TOKUSHIMA PREFECTURE)
TOKUSHIMA-KEN (TOKUSHIMA PREFECTURAL GOVERNMENT)
1-1, BANDAI-CHO, TOKUSHIMA-SHI 770
JAPANESE

EISEI TOKEI NENPO (TOTTORI-KEN) 2119
(ANNUAL REPORT OF PUBLIC HEALTH, TOTTORI PREFECTURE)
TOTTORI-KEN KOSEI-BU (WELFARE DEPARTMENT, TOTTORI PREFECTURAL GOVERNMENT)
1-220, HIGASHI-CHO, TOTTORI-SHI 680
JAPANESE

EISEI TOKEI NENPO (TOYAMA-KEN) 2120
(ANNUAL REPORT OF PUBLIC HEALTH, TOYAMA PREFECTURE)
TOYAMA-KEN KOSEI-BU (WELFARE DEPARTMENT, TOYAMA PREFECTURAL GOVERNMENT)
1-7, SHINSOGAWA, TOYAMA-SHI 930
JAPANESE

EISEI TOKEI NENPO (YAMAGATA-KEN) 2121
(ANNUAL REPORT OF PUBLIC HEALTH, YAMAGATA PREFECTURE)
YAMAGATA-KEN EISEI-BU (HEALTH DIVISION, YAMAGATA PREFECTURAL GOVERNMENT)
4-51, 3-CHOME, HATAGO-MACHI, YAMAGATA-SHI 990
JAPANESE

EISEI TOKEI NENPO (YAMAGUCHI-KEN) 2122
(ANNUAL REPORT OF PUBLIC HEALTH, YAMAGUCHI PREFECTURE)
YAMAGUCHI-KEN EISEI-BU (PUBLIC HEALTH DEPARTMENT, YAMAGUCHI PREFECTURE)
1-1, TAKI-MACHI, YAMAGUCHI-SHI 753
JAPANESE

EISEI TOKEI NENPO (YAMANASHI-KEN) 2123
(ANNUAL REPORT OF PUBLIC HEALTH, YAMANASHI PREFECTURAL))
YAMANASHI-KEN KOSEI-BU (WELFARE DEPARTMENT, YAMANASHI PREFECTURAL GOVERNMENT)
6-1, 1-CHOME, MARUNOUCHI, KOFU-SHI 400
JAPANESE

EISEI TOKEI NENPO GYOMU TOKEIHEN (NAGASAKI-KEN) 2124
(ANNUAL REPORT OF PUBLIC HEALTH, NAGASAKI PREFECTURE, SERIES ON OPERATIONAL STATISTICS)
NAGASAKI-KEN EISEI-BU (PUBLIC HEALTH DEPARTMENT, NAGASAKI PREFECTURAL GOVERNMENT)
2-13, EDO-MACHI, NAGASAKI-SHI 850
JAPANESE

EISEI TOKEI NENPO JINKO DOTAIHEN (NAGASAKI-KEN) 2125
(ANNUAL REPORT OF PUBLIC HEALTH, NAGASAKI PREFECTURE, SERIES ON VITAL STATISTICS)
NAGASAKI-KEN EISEI-BU (PUBLIC HEALTH DEPARTMENT, NAGASAKI PREFECTURAL GOVERNMENT)
2-13, EDO-MACHI, NAGASAKI-SHI 850
JAPANESE

EISEI TOKEISHO (KANAGAWA-KEN) 2126
(ANNUAL REPORT OF PUBLIC HEALTH, KANAGAWA PREFECTURE)
KANAGAWA-KEN EISEI-BU (PUBLIC HEALTH DEPARTMENT, KANAGAWA PREFECTURAL GOVERNMENT)
1, NIHON-ODORI, NAKA-KU, YOKOHAMA-SHI 232
JAPANESE

EISEI TOKEISHO (SHIMANE-KEN) 2127
(ANNUAL REPORT OF PUBLIC HEALTH, SHIMANE PREFECTURE)
SHIMANE-KEN KANKYO HOKEN-BU (PUBLIC HEALTH DEPARTMENT, SHIMANE PREFECTURAL GOVERNMENT)
1, TONO-MACHI, MATSUE-SHI 690
JAPANESE

ENBI SHOKUHIN EISEI KYOGIKAI GIJUTSU SANKO SHIRYO 2128
(JAPAN HYGENIC PVC ASSOCIATION (JHPA) TECHNICAL REPORT)
ENBI SHOKUHIN EISEI KYOGIKAI (JAPAN HYGENIC PVC ASSOCIATION)
C/O MIKI TORANOMON BUILDING, 11, SHIBA NISHIKUBO AKEFUNE-CHO, MINATO-KU, TOKYO 105
JAPANESE

ENBI SHOKUHIN EISEI KYOGIKAI KAIHO 2129
(JAPAN HYGENIC PVC ASSOCIATION (JPHA) KAIHO)
ENBI SHOKUHIN EISEI KYOIKAI (JAPAN HYGENIC PVC ASSOCIATION)
C/O MIKI TORANOMON BUILDING, 11, SHIBA NISHIKUBO AKEFUNE-CHO, MINATO-KU, TOKYO 105
JAPANESE

ENERUGI TO KOGAI 2130
(ENERGY AND POLLUTION)
ENERUGI JANARUSHA
C/O IWAI BUILDING, 1-20, YOTSUYA, SHINJUKU-KU, TOKYO 160
JAPANESE

EPI; ENVIRONMENTAL PROTECTION & INDUSTRY 2131
JAPANESE
INDEXED: JICST

FUJIMURA GAKUEN TOKYO JOSHI TAIIKU DAIGAKU KIYO 2132
(BULLETIN OF FUJIMURA ACADEMY, TOKYO WOMEN'S PHYSICAL EDUCATION COLLEGE)
TOKYO JOSHI TAIIKU DAIGAKU, TOKYO JOSHI TAIIKU TANKI DAIGAKU (TOKYO WOMEN'S PHYSICAL EDUCATION COLLEGE AND TOKYO WOMEN'S PHYSICAL EDUCATION JUNIOR COLLEGE)
JAPANESE

FUKUI DAIGAKU KYOIKUGAKUBU KIYO, DAI-6-BU, TAIIKUGAKU 2133
(MEMOIRS OF THE DEPARTMENT OF EDUCATION, FUKUI UNIVERSITY, PHYSICAL EDUCATION)
FUKUI DAIGAKU KYOIKUGAKUBU (FACULTY OF EDUCATION, FUKUI UNIVERSITY)
9-1, 3-CHOME, BUNKYO, FUKUI-SHI 910
JAPANESE

FUKUI-KEN EISEI KENKYUSHO CHOSA KENKYU HOKOKU 2134
(RESEARCH REPORT OF FUKUI PREFECTURAL INSTITUTE OF PUBLIC HEALTH)
FUKUI-KEN EISEI KENKYUSHO (FUKUI PREFECTURAL INSTITUTE OF PUBLIC HEALTH)
39-4, HARAME-CHO, FUKUI-SHI 910
JAPANESE

FUKUI-KEN EISEI KENKYUSHO KENKYU HOKOKU 2135
(SCIENCE REPORT OF FUKUI PREFECTURAL INSTITUTE OF PUBLIC HEALTH)
FUKUI-KEN EISEI KENKYUSHO (FUKUI PREFECTURAL INSTITUTE OF PUBLIC HEALTH)
39-4, HARAME-CHO, FUKUI-SHI 910
JAPANESE

FUKUI-KEN EISEI KENKYUSHO SHOHO 2136
(ANNUAL REPORT OF FUKUI PREFECTURAL INSTITUTE OF PUBLIC HEALTH)
FUKUI-KEN EISEI KENKYUSHO (FUKUI PREFECTURAL INSTITUTE OF PUBLIC HEALTH)
39-4, HARAME-CHO, FUKUI-SHI 910
JAPANESE

FUKUI-KEN KOGAI SENTA SHOHO 2137
(REPORT OF THE PUBLIC HAZARDS CENTER, FUKUI PREFECTURE)
FUKUI-KEN KOGAI SENTA
39-4, HARAME-CHO, FUKU-SHI 910
JAPANESE

FUKUOKA DAIGAKU TAIIKUGAKU KENKYU 2138
(FUKUOKA UNIVERSITY REVIEW OF PHYSICAL EDUCATION)
FUKUOKA DAIGAKU KENKYUJO (FUKUOKA UNIVERSITY, RESEARCH INSTITUTE)
JAPANESE

FUKUOKA-KEN EISEI KENKYUSHO KENKYU HOKOKU 2139
(REPORT OF FUKUOKA PREFECTURAL INSTITUTE OF PUBLIC HEALTH)
FUKUOKA-KEN EISEI KENKYUSHO (FUKUOKA PREFECTURAL INSTITUTE OF PUBLIC HEALTH)
39, MUKAEDA, MUKAISANO, DAZAIFU-CHO, CHIKUSHI-GUN, FUKUOKA 818-01
JAPANESE, WITH ENGLISH SUMMARY

FUKUOKA-KEN EISEI KENKYUSHO SHOHO 2140
(ANNUAL REPORT OF FUKUOKA PREFECTURAL INSTITUTE OF PUBLIC HEALTH)
FUKUOKA-KEN EISEI KENKYUSHO (FUKUOKA PREFECTURAL INSTITUTE OF PUBLIC HEALTH)
39, MUKAEDA, MUKAISANO, DAZAIFU-CHO, CHIKUSHI-GUN, FUKUOKA 818-01
JAPANESE

FUKUSHIMA-KEN EISEI KOGAI KENKYUSHO KENKYU HOKOKU 2141
(REPORT OF THE FUKUSHIMA PREFECTURAL INSTITUTE OF PUBLIC HEALTH AND SAFETY)
FUKUSHIMA-KEN EISEI KOGAI KENKYUSHO
15-4 HOKIDA AZA MITOUCHI, FUKUSHIMA
JAPANESE
INDEXED: CHE

FUMIN TO KENKO 2142
(HEALTH OF PEOPLES OF OSAKA PREFECTURE)
OSAKA-FU EISEI-BU (PUBLIC HEALTH DIVISION, OSAKA PREFECTURAL GOVERNMENT)
OTEMAENO-CHO, HIGASHI-KU, OSAKA-SHI 540
JAPANESE

FUNABASHI-SHI KOGAI CHOSA HOKOKUSHO 2143
(POLLUTION SURVEY REPORT OF FUNABASHI CITY)
FUNABASHI-SHI KOGAI TAISAKU-BU (POLLUTION CONTROL BUREAU, FUNABASHI CITY)
10-25, 2-CHOME, MINATO-MACHI, FUNABASHI-SHI 273
JAPANESE

GAKKO HOKEN 2144
(SCHOOL HEALTH)
NIHON GAKKO HOKENKAI
20, SHIBA NISHIKUBO AKEFUNE-CHO, MINATO-KU, TOKYO 106
JAPANESE

GAKKO HOKEN CHOSA KEKKA HOKOKU (SAITAMA-KEN) 2145
(ANNUAL STATISTICS OF SCHOOL HEALTH, SAITAMA PREFECTURE)
SAITAMA-KEN KIKABU-BU
15-1, 3-CHOME, TAKASAGO-CHO, URAWA-SHI 336
JAPANESE

GAKKO HOKEN KAGAKU 2146
(SCHOOL HEALTH SCIENCE)
OSAKA-SHI GAKKO YAKUZAISHIKAI (SCHOOL PHARMACIST ASSOCIATION, OSAKA) AND HOKEN KAGAKUSHA
3-17, KATSUYAMA-DORI, TENNOJI-KU, OSAKA-SHI 543
JAPANESE

GAKKO HOKEN KAIHO 2147
(BULLETIN FOR PRESERVATION OF SCHOOL HEALTH)
FUKUOKA-KEN GAKKO HOKENKAI
6-29, NISHINAKASU, CHUO-KU, FUKUOKA-SHI 810
JAPANESE

GAKKO HOKEN KENKYU 2148
(SCHOOL HEALTH STUDY)
HOKEN KENKYUSHA
21-4, 3-CHOME, MEJIRODAI, BUNKYO-KU, TOKYO 112
JAPANESE

GAKKO HOKEN TOKEI CHOSA HOKOKU (MIYAGI-KEN) 2149
(ANNUAL STATISTICS OF SCHOOL HEALTH AND HYGIENE, MIYAGI PREFECTURE)
MIYAGI-KEN SOMU-BU CHOSA-KA (RESEARCH SECTION, GENERAL AFFAIRS DIVISION, MIYAGI PREFECTURAL GOVERNMENT)
8-1, 3-CHOME, HON-CHO, SENDAI-SHI 980
JAPANESE

GAKKO HOKEN TOKEI CHOSA HOKOKUSHO (MONBU-SHO) 2150
(ANNUAL STATISTICS OF SCHOOL HEALTH AND HYGIENE, MINISTRY OF EDUCATION)
MONBU-SHO DAIJIN KANBO TOKEI-KA (STATISTICS SECTION, MINISTER'S SECRETARIAT, MINISTRY OF EDUCATION)
2-2, 3-CHOME, KASUMIGASEKI, CHIYODA-KU, TOKYO 100
JAPANESE

GAKKO HOKEN TOKEI CHOSA KEKKA HOKOKU (HIROSHIMA-KEN) 2151
(ANNUAL STATISTICS OF SCHOOL HEALTH AND HYGIENE, HIROSHIMA PREFECTURE)
HIROSHIMA-KEN KIKAKU-BU TOKEI-KA
10-52, MOTO-MACHI, HIROSHIMA-SHI 730
JAPANESE

GAKKO KIHON CHOSA GAKKO HOKEN TOKEI CHOSA KEKKA HOKOKUSHO (GUNMA-KEN) 2152
(ANNUAL STATISTICS OF SCHOOL HEALTH AND HYGIENE, GUNMA PREFECTURE)
GUNMA-KEN KIKAKU-BU TOKEI-KA (STATISTICS SECTION, PLANNING DIVISION, GUNMA PREFECTURAL GOVERNMENT)
1-1, 1-CHOME, OTE-MACHI, MAEBASHI-SHI 371
JAPANESE

GAKKO KIHON CHOSA GAKKO HOKEN TOKEI CHOSA KEKKA HOKOKUSHO (SHIMANE-KEN) 2153
(ANNUAL STATISTICS OF SCHOOL HEALTH AND HYGIENE, SHIMANE PREFECTURE)
SHIMANE-KEN SOMU-BU TOKEI-KA (STATISTICS SECTION, ADMINISTRATIVE DIVISION, SHIMANE PREFECTURAL GOVERNMENT)
1, TONO-MACHI, MATSUE-SHI 690
JAPANESE

GAKKO TAIIKU 2154
(SCHOOL PHYSICAL EDUCATION)
NIHON TAIIKUSHA
16-10, 3-CHOME, HONGO, BUNKYO-KU, TOKYO 113
JAPANESE

GEKKAN KENKO IGAKU 2155
(HEALTH MEDICINE MONTHLY)
KENKO IGAKUSHA (HEALTH MEDICINE PUBLISHER)
JAPANESE

GEKKAN KOGAI O NOGASUNA 2156
(MONTHLY, DON'T LET POLLUTION GO)
HON KOGAI YUSHUTSU TSUHO SENTA (ANTI-POLLUTION EXPORT BULLETIN CENTER)
JAPANESE

GIFU-KEN EISEI KENKYUSHO NENPO 2157
(ANNUAL REPORT OF GIFU PREFECTURAL INSTITUTE OF PUBLIC HEALTH)
GIFU-KEN EISEI KENKYUSHO
NAGAMORI NOISHIKI, GIFU-SHI 500
JAPANESE

GIFU-KEN EISEI KENKYUSHO SHOHO 2158
(SCIENCE REPORT OF GIFU PREFECTURAL INSTITUTE OF PUBLIC HEALTH)
GIFU-KEN EISEI KENKYUSHO (GIFU PREFECTURAL INSTITUTE OF PUBLIC HEALTH)
NAGAMORI NOISHIKI, GIFU-SHI 500
JAPANESE, WITH ENGLISH SUMMARY

GIFU-KEN KOGAI KENKYUSHO NENPO 2159
(ANNUAL REPORT OF GIFU PREFECTURAL INSTITUTE OF ENVIRONMENTAL POLLUTION)
GIFU-KEN KOGAI KENKYUSHO (GIFU PREFECTURAL INSTITUTE OF ENVIRONMENTAL POLLUTION)
NAGAMORI NOISHIKI, GIFU-SHI 500
JAPANESE
INDEXED: CHE

GUNMA-KEN EISEI KENKYUSHO, GUNMA-KEN KOGAI KENKYU SENTA NENPO 2160
(ANNUAL REPORT OF GUNMA INSTITUTE OF PUBLIC HEALTH, AND GUNMA RESEARCH CENTER FOR ENVIRONMENTAL SCIENCE)
GUNMA-KEN EISEI KENKYUSHO, GUNMA-KEN KOGAI KENKYU SENTA (GUNMA INSTITUTE OF PUBLIC HEALTH, GUNMA RESEARCH CENTER FOR ENVIRONMENTAL SCIENCE)
21-19, 3-CHOME, IWAGAMI-CHO, MAEBASHI-SHI 371
JAPANESE

HATARAKU HITO NO KOKORO NO KENKO 2161
(MENTAL HEALTH OF THE WORKING CLASS)
NIHON SEISHIN EISEI RENMEI (JAPAN MENTAL HEALTH FEDERATION)
JAPANESE

HERUSU 2162
(HEALTH)
SHIZEN KAN HERUSU HENSHU-BU (NATURE HALL 'HEALTH' EDITORIAL DEPARTMENT)
JAPANESE

HIRATSUKA HOKENJO NENPO 2163
(HIRATSUKA HEALTH CENTER REPORT)
KANAGAWA-KEN HIRASUKA HOKENJO
6-21, TOYOHARA-SHO, HIRATSUKA-SHI 254
JAPANESE

HIROSAKI DAIGAKU IGAKUBU EISEIGAKU KYOSHITSU GYOSEKISHU 2164
(TECHNICAL RESEARCHES OF SCHOOL OF HYGIENICS, MEDICAL DEPARTMENT, HIROSAKI UNIVERSITY)
HIROSAKI DAIGAKU IGAKUBU EISEIGAKU KYOSHITSU
5, ZAIFU-CHO, HIROSAKI-SHI 036
JAPANESE

HIROSHIMA-KEN EISEI KENKYUSHO KOGAI KENKYUSHO KENKYU HOKOKU 2165
(BULLETIN OF THE HIROSHIMA PREFECTURAL INSTITUTE OF PUBLIC HEALTH AND ENVIRONMENTAL SCIENCE)
HIROSHIMA-KEN EISEI KENKYUSHO (HIROSHIMA PREFECTURAL INSTITUTE OF PUBLIC HEALTH)
5-70, 1-CHOME, UJINA KANDA, HIROSHIMA-SHI 734
JAPANESE, WITH ENGLISH SUMMARY
INDEXED: CHE, JICST

HIROSHIMA-KEN EISEI KENKYUSHO NENPO 2166
(ANNUAL REPORT OF THE LABORATORY OF PUBLIC HEALTH, HIROSHIMA PREFECTURE)
HIROSHIMA-KEN EISEI KENKYUSHO (LABORATORY OF PUBLIC HEALTH, HIROSHIMA PREFECTURE)
5-70, 1-CHOME, UJINA KANDA, HIROSHIMA-SHI 734
JAPANESE

HOKEN BUTSURI 2167
(JOURNAL OF THE JAPAN HEALTH PHYSICS SOCIETY)
NIHON HOKEN BUTSURI KYOGIKAI (JAPAN HEALTH PHYSICS SOCIETY)
C/O GAKKAI SENTA BUILDING, 4-16, 2-CHOME, YAYOI, BUNKYO-KU, TOKYO 113
JAPANESE
INDEXED: CHE, JICST

HOKEN DOJIN, KURASHI TO KENKO 2168
(LIVING AND HEALTH)
HOKEN DOJINSHA
3-6, 1-CHOME, SARUGAKU-CHO, CHIYODA-KU, TOKYO 101
JAPANESE

HOKEN EISEI KISO CHOSA GAIKYO 2169
(SUMMARY REPORT OF HEALTH AND HYGIENE RESEARCH)
HOSEI-SHO DAIJIN KANBO TOKEI CHOSA-BU
42, ICHIGAYA HONMURA-CHO, SHINJUKU-KU, TOKYO 162
JAPANESE

HOKEN EISEI KISO CHOSA HOKOKU 2170
(HEALTH AND HYGIENE RESEARCH REPORT)
KOSEI-SHO DAIJIN KANBO TOKEI CHOSA-BU (HEALTH AND WELFARE STATISTICS DIVISION, MINISTRY OF HEALTH AND WELFARE)
42, ICHIGAYA HONMURA-CHO, SHINJUKU-KU, TOKYO 162
JAPANESE

HOKEN NO KAGAKU 2171
(HEALTH CARE)
HOKEN KAGAKU KENKYUKAI (HEALTH SCIENCE RESEARCH ASSOCIATION) AND KYORIN SHOIN
2-1, 4-CHOME, YUSHIMA, BUNKYO-KU, TOKYO 113
JAPANESE

HOKEN SANGYO JIHO 2172
(HEALTH AND INDUSTRY)
HOKEN SANGYO JIHOSHA
14-5, 3-CHOME, MINAMIKEBUKURO, TOSHIMA-KU, TOKYO 171
JAPANESE

HOKEN TAIIKU KYOSHITSU 2173
(HEALTH AND PHYSICAL EDUCATION)
TAISHUKAN SHOTEN
3-24, KANDA NISHIKI-CHO, CHIYODA-KU, TOKYO 101
JAPANESE

HOKEN TSUSHIN 2174
(PUBLIC HEALTH REVIEW)
HOKEN TSUSHINSHA
22, KITASHIRAKAWA NISHISENOUCHI-CHO, SAKYO-KU, KYOTO-SHI 606
JAPANESE

HOKENFU ZASSHI 2175
(JAPANESE JOURNAL FOR PUBLIC HEALTH NURSE)
IGAKU SHOIN
29-11, 5-CHOME, HONGO, BUNKYO-KU, TOKYO 113
JAPANESE

HOKENJO UN'EI HOKOKU 2176
(ANNUAL REPORT OF HEALTH CENTERS ACTIVITIES)
KOSEI-SHO DAIJIN KANBO TOKEI CHOSA-BU (HEALTH AND WELFARE STATISTICS DIVISION, MINISTER'S SECRETARIAT, MINISTRY OF HEALTH AND WELFARE)
42, ICHIGAYA HONMURA-CHO, SHINJUKU-KU, TOKYO 162
JAPANESE

HOKKAIDO KAN'EI DAYORI 2177
(HOKKAIDO ENVIRONMENTAL PROTECTION LETTER)
HOKKAIDO EISEIBU SHOKUHIN EISEIKA (HOKKAIDO SANITATION DEPARTMENT, FOOD PRESERVATION SECTION)
JAPANESE

HOKKAIDO KOGAI BOSHI KENKYU JOHO 2178
(REPORT OF HOKKAIDO RESEARCH INSTITUTE FOR ENVIRONMENTAL POLLUTION)
JAPANESE

HOKKAIDO NO TAIKI OSEN CHOSA SOKUTEI KEKKA HOKOKU 2179
(RESULT OF THE INVESTIGATION OF AIR POLLUTION IN HOKKAIDO)
HOKKAIDO EISEI-BU (PUBLIC HEALTH DIVISION, HOKKAIDO)
NISHI-6-CHOME, KITA-3-JO, CHUO-KU, SAPPORO-SHI 060
JAPANESE

HOKKAIDORITSU EISEI KENKYUSHOHO 2180
(REPORT OF THE HOKKAIDO INSTITUTE OF PUBLIC HEALTH)
HOKKAIDORITSU EISEI KENKYUSHO (HOKKAIDO INSTITUTE OF PUBLIC HEALTH)
NISHI-12-CHOME, KITA-19-JO, KITA-KU, SAPPORO-SHI 060
JAPANESE AND ENGLISH
INDEXED: BIO, CHE, JICST

HOPPO SANGYO EISEI 2181
(JOURNAL OF NORTHERN OCCUPATIONAL HEALTH)
HOPPO SANGYO EISEI KYOKAI
C/O IWAMIZAWA ROSAI BYOIN, 18, TONEBETSU-CHO, IWAMIZAWA-SHI 068
JAPANESE, WITH ENGLISH SUMMARY

HYOGO-KEN EISEI KENKYUSHO KENKYU HOKOKU 2182
(BULLETIN OF THE PUBLIC HEALTH INSTITUTE OF HYOGO PREFECTURE)
HYOGO-KEN EISEI KENKYUSHO (PUBLIC HEALTH INSTITUTE OF HYOGO PREFECTURE)
2-1, ARATA-CHO, HYOGO-KU, KOBE-SHI 652
JAPANESE AND ENGLISH
INDEXED: CHE

HYOGO-KEN KOGAI KENKYUSHO KENKYU HOKOKU 2183
(REPORT OF THE ENVIRONMENTAL SCIENCE INSTITUTE OF HYOGO PREFECTURE)
HYOGO-KEN KOGAI KENKYUSHO
2-1, ARATA-CHO, HYOGO-KU, KOBE-SHI 652
JAPANESE, WITH ENGLISH SUMMARY
INDEXED: CHE, JICST

IBARAKI-KEN EISEI KENKYUSHO NENPO 2184
(ANNUAL REPORT OF IBARAKI PREFECTURAL INSTITUTE OF HEALTH)
IBARAKI-KEN EISEI KENKYUSHO (IBARAKI PREFECTURAL INSTITUTE OF HEALTH)
4-1, ATAGO-CHO, MITO-SHI 310
JAPANESE
INDEXED: CHE

IBARAKI-KEN KOGAI GIJUTSU SENTA NENPO 2185
(ANNUAL REPORT OF THE ENVIRONMNTAL POLLUTION RESEARCH CENTER, IBARAKI PREFECTURE)
IBARAKI-KEN KOGAI GIJUTSU SENTA (ENVIRONMENTAL POLLUTION RESEARCH CENTER, IBARAKI PREFECTURE)
4043, ISHIKAWA-CHO, MITO-SHI 310
JAPANESE
INDEXED: CHE

IKUYUKAI KOGAI KENKYUSHO SHOHO 2186
(SCIENCE REPORT OF IKUYUKAI RESEARCH LABORATORY FOR ENVIRONMENTAL POLLUTION)
HOSEI DAIGAKU DAI-2 KOTO GAKKO (THE 2ND HIGH SCHOOL, HOSEI UNIVERSITY)
164, KIZUKI, NAKAHARA-KU, KAWASAKI-SHI 211
JAPANESE

INDUSTRIAL HEALTH 2187
RODO-SHO RODO EISEI KENKYUSHO (NATIONAL INSTITUTE OF INDUSTRIAL HEALTH)
2051, KIZUKI SUMIYOSHI-CHO, KAWASAKI-SHI 211
ENGLISH
INDEXED: BIO, CHE, EXC, JICST

INFOMEISHON, TOKYO-TO KOGAI KENKYUSHO SHIRYOSHITSU GEPPO 2188
(MONTHLY NEWS OF INFORMATION OFFICE OF THE TOKYO METROPOLITAN RESEARCH INSTITUTE FOR ENVIRONMENTAL PROTECTION)
TOKYO-TO KOGAI KENKYUSHO
2-5, YURAKU-CHO, CHIYODA-KU, TOKYO 100
JAPANESE

ISHIKAWA-KEN EISEI KOGAI KENKYUSHO NENPO 2189
(ANNUAL REPORT OF PUBLIC HEALTH AND PUBLIC NUISANCE LABORATORY, ISHIKAWA PREFECTURE)
ISHIKAWA-KEN EISEI KOGAI KENKYUSHO
2-251, MINMA, KANAZAWA-SHI 921
JAPANESE

IWAKUNI-SHI NO KOGAI GAIKYO 2190
(ENVIRONMENTAL POLLUTION IN IWAKUNI CITY)
IWAKUNI-SHI (IWAKUNI MUNICIPAL GOVERNMENT)
14-51, 1-CHOME, IMAZU-CHO, IWAKUNI-SHI 740
JAPANESE

IWATE NO HOKEN 2191
(HEALTH IN IWATE PREFECTURE)
IWATE-KEN KOKUMIN KENKO HOKEN DANTAI RENGOKAI
C/O IWATE-KEN KOKUHO KAIKAN, 7-30, 3-CHOME, OSAWAKAWARA, MORIOKA-SHI 020
JAPANESE

IWATE-KEN EISEI KENKYUSHO NENPO 2192
(ANNUAL REPORT OF THE IWATE INSTITUTE OF PUBLIC HEALTH)
IWATE-KEN EISEI KENKYUSHO (IWATE PREFECTURAL INSTITUTE OF PUBLIC HEALTH)
15-25, UCHIMARU, MORIOKA-SHI 020
JAPANESE

JAPANESE JOURNAL OF PHYSICAL FITNESS AND SPORTS MEDICINE 2193
JAPANESE AND ENGLISH
INDEXED: EXC

JINKO DOTAI TOKEI 2194
(VITAL STATISTICS OF POPULATION, ANNUAL REPORT)
KOSEI-SHO DAIJIN KANBO TOKEI CHOSA-BU (HEALTH AND WELFARE STATISTICS DEPARTMENT, MINISTER'S SECRETARIAT, MINISTRY OF HEALTH AND WELFARE)
2-2, 1-CHOME, KASUMIGASEKI, CHIYODA-KU, TOKYO 100
JAPANESE

JINKO DOTAI TOKEI GEPPO 2195
(VITAL STATISTICS OF POPULATION, MONTHLY REPORT)
KOSEI-SHO DAIJIN KANBO TOKEI CHOSA-BU (HEALTH AND STATISTICS DEPARTMENT, MINISTER'S SECRETARIAT, MINISTRY OF HEALTH AND WELFARE)
42, ICHIGAYA HONMURA-CHO, SHINJUKU-KU, TOKYO 162
JAPANESE

JINKO MONDAI KENKYU 2196
(JOURNAL OF POPULATION PROBLEMS)
KOSEI-SHO JINKO MONDAI KENKYUSHO (INSTITUTE OF POPULATION PROBLEMS, MINISTRY OF HEALTH AND WELFARE)
2-2, 1-CHOME, KASUMIGASEKI, CHIYODA-KU, TOKYO 100
JAPANESE, WITH ENGLISH SUMMARY

JINKO MONDAI KENKYUSHO JIGYO HOKOKUSHO 2197
(ANNUAL REPORT OF THE ACTIVITIES OF THE INSTITUTE OF POPULATION PROBLEMS)
KOSEI-SHO JINKO MONDAI KENKYUSHO (INSTITUTE OF POPULATION PROBLEMS, MINISTRY OF HEALTH AND WELFARE)
2-2, 1-CHOME, KASUMIGASEKI, CHIYODA-KU, TOKYO 100
JAPANESE

JINKO MONDAI KENKYUSHO KENKYU SHIRYO 2198
(RESEARCH SERIES OF THE INSTITUTE OF POPULATION PROBLEMS)
KOSEI-SHO JINKO MONDAI KENKYUSHO (INSTITUTE OF POPULATION PROBLEMS, MINISTRY OF HEALTH AND WELFARE)
2-2, 1-CHOME, KASUMIGASEKI, CHIYODA-KU, TOKYO 100
JAPANESE

JINKO MONDAI KENKYUSHO NENPO 2199
(ANNUAL REPORTS OF THE INSTITUTE OF POPULATION PROBLEMS)
KOSEI-SHO JINKO MONDAI KENKYUSHO (INSTITUTE OF POPULATION PROBLEMS, MINISTRY OF HEALTH AND WELFARE)
2-2, 1-CHOME, KASUMIGASEKI, CHIYODA-KU, TOKYO 100
JAPANESE, WITH ENGLISH SUMMARY
INDEXED: BIO

JINKO SUIKAI GEPPO 2200
(MONTHLY REPORT ON CURRENT POPULATION ESTIMATES)
SORI-FU TOKEI-KYOKU (BUREAU OF STATISTICS, PRIME MINISTER'S OFFICE)
95 WAKAMATSU-CHO, SHINJUKU-KU, TOKYO 162
JAPANESE AND ENGLISH

JISHU KOZA 2201
(STUDY SERIES ON ENVIRONMENTAL POLLUTION)
JISHU KOZA HENSHUSHITSU AND AKI SHOBO
2-9, KANDA JINBO-CHO, CHIYODA-KU, TOKYO 101
JAPANESE

JOSHI TAIIKU 2202
(WOMAN'S PHYSICAL EDUCATION)
NIHON TAIIKU RENMEI (JAPANESE UNION OF WOMAN'S PHYSICAL EDUCATION)
C/O OMYC, 3-1, YOYOGI KAMIZONO-CHO, SHIBUYA-KU, TOKYO 151
JAPANESE

JOURNAL OF ANTIBIOTICS 2203
NIHON KOSEIBUSSHITSU GAKUJITSU KYOGIKAI (JAPAN ANTIBIOTICS RESEARCH ASSOCIATION)
TOKYO
ENGLISH
INDEXED: BIB, BIO

JOURNAL OF SCIENCE OF LABOUR, PART 2 2204
RODO-SHO RODO KAGAKU KENKYUSHO (INSTITUTE FOR SCIENCE OF LABOUR)
1544, SUGAO, TAKATSU-KU, KAWASAKI-SHI 213
ENGLISH
INDEXED: JICST

JUNTENDO DAIGAKU HOKEN TAIIKU KIYO 2205
(JUNTENDO UNIVERSITY BULLETIN OF HEALTH PHYSICAL EDUCATION)
JUNTENDO DAIGAKU TAIIKU GAKUBU (SCHOOL OF PHYSICAL EDUCATION, JUNTENDO UNIVERSITY)
4-54, 5-CHOME, FUJISAKI, NAGASHINO-SHI 275
JAPANESE AND ENGLISH

KAGAKU GIJUTSU BUNKEN SOKUHO, KANKYO KOGAI-HEN 2206
(CURRENT BIBLIOGRAPHY ON SCIENCE AND TECHNOLOGY, SERIES ON ENVIRONMENTAL POLLUTION)
NIHON KAGAKU GIJUTSU HOJO SENTA (JAPAN INFORMATION CENTER OF SCIENCE AND TECHNOLOGY)
5-2, 2-CHOME, NAGATA-CHO, CHIYODA-KU, TOKYO 100
JAPANESE
INDEXED: JICST

KAGAWA NO EISEI 2207
(ANNUAL REPORT OF PUBLIC HEALTH, KAGAWA PREFECTURE)
KAGAWA-KEN (KAGAWA PREFECTURAL GOVERNMENT)
1-10, 4-CHOME, BAN-CHO, TAKAMATSU-SHI 760
JAPANESE

KAGAWA-KEN EISEI KOGAI KENKYUSHO NENPO 2208
(ANNUAL REPORT OF KAGAWA PREFECTURAL RESEARCH INSTITUTE OF HEALTH)
KAGAWA-KEN EISEI KOGAI KENKYUSHO (KAGAWA PREFECTURAL RESEARCH INSTITUTE OF HEALTH)
4-15, 5-CHOME, BAN-CHO, TAKAMATSU-SHI 760
JAPANESE

KAGAWA-KEN TOKUSHU KYOIKU KENKYUKAI. ONISHI SEISHIN EISEI KENKYUJO KENKYU KIYO 2209
(RESEARCH BULLETIN OF THE KAGAWA PREFECTURE SPECIAL EDUCATION RESEARCH ASSOCIATION AND THE ONISHI MENTAL HYGIENE RESEARCH INSTITUTE)
JAPANESE

KAGOSHIMA DAIGAKU TAIIKUKA HOKOKU 2210
(PHYSICAL EDUCATION REPORTS OF THE KAGOSHIMA UNIVERSITY)
KAGOSHIMA DAIGAKU KYOYOBU (COLLEGE OF LIBERAL ARTS, KAGOSHIMA UNIVERSITY)
201, KAMOIKE-CHO, KAGOSHIMA-SHI 890
JAPANESE

KAGOSHIMA-KEN KOGAI EISEI KENKYUSHOHO 2211
(ANNUAL REPORT OF KAGOSHIMA PREFECTURAL INSTITUTE OF PUBLIC HEALTH AND ENVIRONMENTAL POLLUTION)
KAGOSHIMA-KEN KOGAI EISEI KENKYUSHO
1-24, SHIROYAMA-CHO, KAGOSHIMA-SHI 892
JAPANESE AND ENGLISH

KAGOSHIMA-KEN NO KOGAI 2212
(ENVIRONMENTAL POLLUTION IN KAGOSHIMA PREFECTURE)
KAGOSHIMA-KEN (KAGOSHIMA PREFECTURAL GOVERNMENT)
14-50, YAMASHITA-CHO, KAGOSHIMA-SHI 892
JAPANESE

KANAGAWA NO KOGAI, GENJO TO TAISAKU NO ARAMASHI 2213
(ENVIRONMENTAL POLLUTION IN KANAGAWA PREFECTURE, IT'S SITUATION AND COUNTERPLAN)
KANAGAWA-KEN KOGAI TAISAKU JIMU-KYOKU (POLLUTION CONTROL BUREAU, KANAGAWA PREFECTURE)
1, NIHON-ODORI, NAKA-KU, YOKOHAMA-SHI 231
JAPANESE

KANAGAWA-KEN EISEI KENKYUSHO KENKYU HOKOKU 2214
(BULLETIN OF THE KANAGAWA PREFECTURAL PUBLIC HEALTH LABORATORY)
KANAGAWA-KEN EISEI KENKYUSHO (KANAGAWA PREFECTURAL PUBLIC HEALTH LABORATORY)
52, NAKAO-CHO, ASAHI-KU, YOKOHAMA-SHI 241
JAPANESE, WITH ENGLISH SUMMARY
INDEXED: CHE

KANAGAWA-KEN EISEI KENKYUSHO NEMPO 2215
(ANNUAL REPORT OF THE KANAGAWA PREFECTURE PUBLIC HEALTH LABORATORY)
KANAGAWA-KEN EISEI KENKYUSHO
52-2, NAKAO-CHO, HODOGAYA, YOKOHAMA
JAPANESE
INDEXED: CHE, JICST

KANAGAWA-KEN KOGAI SENTA NENPO 2216
(ANNUAL REPORT OF KANAGAWA PREFECTURAL ENVIRONMENTAL CENTER)
KANAGAWA-KEN KOGAI SENTA
87-1, 1-CHOME, FUTAMATAGAWA, ASAHI-KU, YOKOHAMA-SHI 241
JAPANESE, WITH ENGLISH SUMMARY

KANAGAWA-KEN KOSHU EISEI GAKKAISHI 2217
(BULLETIN OF THE KANAGAWA PREFECTURE PUBLIC HEALTH ASSOCIATION)
KANAGAWA-KEN KOSHU EISEI KYOKAI (KANAGAWA PREFECTURE PUBLIC HEALTH ASSOCIATION)
C/O KANAGAWA-KEN EISEI-BU, 1-1, NIHON-ODORI, NAKA-KU, YOKOHAMA-SHI 231
JAPANESE

KANAGAWA-KEN TAIKI OSEN CHOSA KENKYU HOKOKU 2218
(TECHNICAL REPORT ON AIR POLLUTION IN YOKOHAMA-KAWASAKI INDUSTRIAL AREA)
KANAGAWA-KEN KOGAI TAISAKU JIMU-KYOKU (POLLUTION CONTROL BUREAU, KANAGAWA PREFECTURE)
1, NIHON-ODORI, NAKA-KU, YOKOHAMA-SHI 231
JAPANESE
INDEXED: APT, JICST

KANKOCHO KOGAI SENMON SHIRYO 2219
(REPORT OF PUBLIC NUISANCE, GOVERNMENT AND PUBLIC OFFICE)
KOGAI KENKYU TAISAKU SENTA (RESEARCH CENTER FOR PUBLIC NUISANCE)
10, SAN'EI-CHO, SHINJUKU-KU, TOKYO 160-91
JAPANESE
INDEXED: APT

KANKYO 2220
(ENVIRONMENT)
GYOSEI
JAPANESE

KANKYO EISEI 2221
(ENVIRONMENTAL SANITATION)
KANKYO EISEI KENKYUKAI (ENVIRONMENTAL SANITATION RESEARCH ASSOCIATION)
3-3, KANDA JINBO-CHO, CHIYODA-KU, TOKYO 101
JAPANESE

KANKYO GIJUTSU 2222
(ENVIRONMENTAL CONSERVATION ENGINEERING)
KANKYO GIJUTSU KENKYUKAI
2-93, KYOMACHIBORI, NISHI-KU, OSAKA-SHI 550
JAPANESE
INDEXED: APT, JICST

KANKYO HAKAI 2223
(JOURNAL OF ENVIRONMENTAL DISRUPTION)
KOGAI MONDAI KENKYUKAI
C/O YOTSUYA MANSHON, SAN'EI-CHO, SHINJUKU-KU, TOKYO 160
JAPANESE
INDEXED: JICST

KANKYO HOKEN REPOTO 2224
(ENVIRONMENT AND PUBLIC HEALTH REPORT)
NIHON KOSHU EISEI KYOKAI (JAPAN PUBLIC HEALTH ASSOCIATION)
29-8, 1-CHOME, SHINJUKU, SHINJUKU-KU, TOKYO 160
JAPANESE

KANKYO HOZEN TAIMUZU 2225
(ENVIRONMENTAL PROTECTION TIMES)
NIHON KANKYU HOZEN KYOKAI
6-4, 1-CHOME, KUDANKITA, CHIYODA-KU, TOKYO 102
JAPANESE

KANKYO IGAKU KENKYUSHO NENPO 2226
(ANNUAL REPORT OF THE RESEARCH INSTITUTE OF ENVIRONMENTAL MEDICINE, NAGOYA UNIVERSITY (JAPANESE EDITION))
NAGOYA DAIGAKU KANKYO IGAKU KENKYUSHO (RESEARCH INSTITUTE OF ENVIRONMENTAL MEDICINE, NAGOYA UNIVERSITY)
FURO-CHO, CHIKUSA-KU, NAGOYA-SHI 464
JAPANESE

KANKYO JOHO KAGAKU 2227
(ENVIRONMENTAL INFORMATION SCIENCE)
KANKYO JOHO KAGAKU SENTA (ENVIRONMENTAL INFORMATION SCIENCE CENTER)
C/O NISHIMAKI BUILDING, 4, ROKUBAN-CHO, CHIYODA-KU, TOKYO 102
JAPANESE, WITH ENGLISH SUMMARY
INDEXED: JICST

KANKYO JOHOBU NYUSU 2228
(ENVIRONMENTAL INFORMATION SECTION NEWS)
KOKU-RITSU KOGAI KENKYUJO KANKYO JOHOBU (ENVIRONMENTAL INFORMATION SECTION, NATIONAL POLLUTION RESEARCH INSTITUTE)
JAPANESE

KANKYO KAGAKU KENKYU 2229
(STUDIES IN ENVIRONMENTAL SCIENCES)
HIROSHIMA DAIGAKU SOGO KAGAKU-BU (GENERAL SCIENCES DEPARTMENT, HIROSHIMA UNIVERSITY)
JAPANESE

KANKYO KENKYU 2230
(QUARTERLY REVIEW ON ENVIRONMENT)
KOGAI CHOSA SENTA
C/O HITACHI SEISAKUJO, SHIN-MARUNOUCHI BUILDING, 5-1, 1-CHOME, MARUNOUCHI, CHIYODA-KU, TOKYO 100
JAPANESE
INDEXED: JICST

KANKYO KOGAI BUNKENSHU 2231
(BIBLIOGRAPHY OF ENVIRONMENTAL POLLUTION)
NIHON KAGAKU GIJUTSU JOHO SENTA (JAPAN INFORMATION CENTER OF SCIENCE AND TECHNOLOGY)
5-2, 2-CHOME, NAGATA-CHO, CHIYODA-KU, TOKYO 100
JAPANESE
INDEXED: JICST

KANKYO KOGAI KANKEI INDEKKUSU 2232
(ENVIRONMENT AND POLLUTION INDEX)
NIHON KAIHATSU GINKO CHUO SHIRYOSHITSU (DATA CENTER, JAPAN DEVELOPMENT BANK)
1-9, OTE-MACHI, CHIYODA-KU, TOKYO 100
JAPANESE

KANKYO KOGAI SHINBUN 2233
(ENVIRONMENTAL POLLUTION NEWS)
KANKYO KOGAI SHINBUNSHA
C/O EGAWA BUILDING, 3, FUNA-CHO, SHINJUKU-KU, TOKYO 160
JAPANESE

KANKYO OYOBI SHOKUHINTO NO HOSHANO OSEN NI TSUITE 2234
(MONITORING OF ENVIRONMENTAL RADIOACTIVITY IN JAPAN)
NAIKAKU HOSHANO TAISAKU HONBU (RADIOACTIVITY COUNTERMEASURE HEADQUARTERS OF THE CABINET)
2-1, 2-CHOME, KASUMIGASEKI, CHIYODA-KU, TOKYO 100
JAPANESE

KANKYO SOZO 2235
(ENVIRONMENTAL CREATION)
J.I.P.
3-23, 1-CHOME, MEGURO, MEGURO-KU, TOKYO 153
JAPANESE
INDEXED: APT, JICST

KATEI TO KENKO 2236
(DAILY LIFE AND HEALTH)
ZENKOKU HOKENSHOCHOKAI
5-2, 10-CHOME, SHIMOTAKAIDO, SUGINAMI-KU, TOKYO 167
JAPANESE

KEN'EKI EISEI 2237
(QUARANTINE HYGIENE)
NIHON KEN'EKI EISEI KYOKAI (JAPAN QUARANTINE HYGIENE ASSOCIATION)
JAPANESE

KEN'EKI GYOMU NENPO 2238
(ANNUAL REPORT OF QUARANTINE AFFAIRS)
KOSEI-SHO KOSHU EISEI-KYOKU (PUBLIC HEALTH BUREAU, MINISTRY OF HEALTH AND WELFARE)
2-2, 1-CHOME, KASUMIGASEKI- CHIYODA-KU, TOKYO 100
JAPANESE

KENKO GAIDO 2239
(GUIDE FOR HEALTH)
NIHON KOSHU EISEI KYOKAI (JAPAN PUBLIC HEALTH ASSOCIATION)
29-8, 1-CHOME, SHINJUKU, SHINJUKU-KU, TOKYO 160
JAPANESE

KENKO KAIGI 2240
(HEALTH ASSEMBLY)
IRYO TOSHO SHUPPANSHA
22-20, 1-CHOME, KOTO-CHO, NAKANO-KU, TOKYO 165
JAPANESE

KENKO KAIHATSU 2241
(HEALTH DEVELOPMENT)
YODOGAWA ZENRINKAN (YODOGAWA GOOD NEIGHBOR HALL)
JAPANESE

KENKO KANRI 2242
(HEALTH CONTROL)
HOKEN BUNKASHA
5-2, 3-CHOME, OTSUKA, BUNKYO-KU, TOKYO 112
JAPANESE
INDEXED: CHE, JICST

KENKO KYOIKU 2243
(PUBLIC HEALTH EDUCATION)
KAWAI SEIYAKU K. K.
51-8, 2-CHOME, ARAI, NAKANO-KU, TOKYO 165
JAPANESE

KENKO KYOSHITSU 2244
(HEALTH STUDY)
HIGASHIYAMA SHOBO
2, NISHINOKYO FIJNOKI-CHO, NAGAGYO-KU, KYOTO-SHI 604
JAPANESE
INDEXED: JICST

KENKO NA KURASHI 2245
(LONGER AND HEALTHIER LIFE)
KENKO NA KURASHI NO KAI
1-2, 5-CHOME, HONGO, BUNKYO-KU, TOKYO 113
JAPANESE

KENKO NIHON 2246
(HEALTHY JAPAN)
KENKO NIHONSHA
13-5, 4-CHOME, AKASAKA, MINATO-KU, TOKYO 107
JAPANESE

KENKO TO CHOJU 2247
(HEALTH AND LONG LIFE)
KENKO HODOKAI
10-14, SAKAE-MACHI, KITA-KU, TOKYO 114
JAPANESE
INDEXED: JICST

KENKO TO EIYO 2248
(HEALTH AND NUTRITION)
KENKO TO EIYOSHA
33-13, 3-CHOME, YAGUCHI, OTA-KU, TOKYO 144
JAPANESE

KENKO TO TAIRYOKU 2249
(HEALTH AND PHYSICAL STRENGTH)
MONBU-SHO TAIIKU-KYOKU (PHYSICAL EDUCATION BUREAU, MINISTRY OF EDUCATION)
1-17, 2-CHOME, MINAMIAYOAMA, MINATO-KU, TOKYO 107
JAPANESE

KITAKYUSHU-SHI KANKYO EISEI KENKYU SHOHO 2250
(REPORT OF THE KITAKYUSHU ENVIRONMENTAL HEALTH RESEARCH INSTITUTE)
KITAKYUSHU ENVIRONMENTAL HEALTH RESEARCH INSTITUTE
JAPANESE

KITAKYUSHU-SHI NO KOGAI 2251
(ENVIORNMENTAL POLLUTION IN KITAKYUSHU CITY)
KITAKYUSHU-SHI KOGAI TAISAKU-KYOKU (POLLUTION CONTROL BUREAU, KITAKYUSHU CITY)
1-1, JONAI, KOKURA-KU, KITAKYUSHU-SHI 803
JAPANESE

KOBE-SHI KANKYO HOKEN KENKYUSHO KENKYU HOKOKU 2252
(ANNUAL REPORT OF PUBLIC HEALTH RESEARCH INSTITUTE OF KOBE CITY)
KOBE-SHI KANKYO HOKEN KENKYUSHO
1-5, KANO-CHO, IKUTA-KU, KOBE-SHI 650
JAPANESE AND ENGLISH

KOBE-SHI TAIKI OSEN CHOSA KENKYU HOKOKU 2253
(SURVEY REPORT ON THE AIR POLLUTION IN KOBE CITY)
KOBE-SHI KOGAI TAISAKU-BU (PUBLIC NUISANCE BUREAU, KOBE CITY)
6-7, KANO-CHO, IKUTA-KU, KOBE-SHI 653
JAPANESE

KOCHI-KEN EISEI KENKYUSHO KENKYU HOKOKU 2254
(RESEARCH REPORT OF THE KOCHI PREFECTURAL PUBLIC HEALTH LABORATORY))
KOCHI-KEN EISEI KENKYUSHO (KOCHI PREFECTURAL PUBLIC HEALTH LABORATORY)
4-1, 2-CHOME, MARUNOUCHI, KOCHI-SHI 780
JAPANESE

KOGAI 2255
(POLLUTION CONTROL)
TSUSHO SANGYO-SHO KOGYO GIJUTSU-IN KOGAI SHIGEN KENKYUSHO (NATIONAL RESEARCH INSTITUTE FOR POLLUTION AND RESOURCES, AGENCY OF INDUSTRIAL SCIENCE AND TECHNOLOGY)
37-11, 2-CHOME, OGIKUBO, SUGINAMI-KU, TOKYO 167
JAPANESE, WITH EUROPEAN SUMMARY
INDEXED: APT, CHE, JICST

KOGAI BOSHI JIGYODAN NYUSU 2256
(POLLUTION PREVENTION CORPORATION NEWS)
POLLUTION PREVENTION CORPORATION
JAPANESE

KOGAI BOSHI KANRISHA 2257
(POLLUTION PREVENTION MANAGER)
KOGAI MONDAI KENKYUJO (POLLUTION PROBLEMS RESEARCH INSTITUTE)
JAPANESE

KOGAI BOSHI NETSUKANRI HOKKAIDO 2258
(ENERGY AND POLLUTION CONTROL)
HOKKAIDO KOGAIBOSHI NETSUKANRI KYOKAI (HOKKAIDO ASSOCIATION FOR ENERGY AND POLLUTION CONTROL, JAPAN)
NISHI-7-CHOME, MINAMI-15-JO, CHUO-KU, SAPPORO-SHI 064
JAPANESE
INDEXED: JICST

KOGAI CHOSA HOKOKUSHO (TOKYO-TO, MEGURO-KU) 2259
(REPORT OF ENVIRONMENTAL POLLUTION IN MEGURO WARD, TOKYO)
TOKYO-TO MEGURO-KU KENCHIKU KOGAI-BU
4-5, 2-CHOME, CHUO-MACHI, MEGURO-KU, TOKYO 152
JAPANESE

KOGAI GEPPO 2260
(MONTHLY REPORT ON POLLUTION CONTROL)
TOKYO-TO KOGAI-KYOKU (BUREAU OF ENVIRONMENTAL PROTECTION, TOKYO)
8-1, 3-CHOME, MARUNOUCHI, CHIYODA-KU, TOKYO 100
JAPANESE

KOGAI JIREI OYOBI SHORI JOKYO 2261
(POLLUTION CONTROL, CASE STUDY)
CHIBA-KEN EISEI-BU (BUREAU OF PUBLIC HEALTH, CHIBA PREFECTURE)
1-1, ICHIBA-CHO, CHIBA-SHI 280-91
JAPANESE

KOGAI JOHO 2262
(MONTHLY REVIEW OF ENVIRONMENTAL POLLUTION)
KOGAI JOHOSHA
31, ICHIGAYA SANAI-CHO, SHINJUKU-KU, TOKYO 162
JAPANESE

KOGAI JOHO 2263
(POLLUTION INFORMATION)
TOKYO-TO KOGAIKYOKU SOMUBU CHOSEIKA
(TOKYO MUNICIPAL POLLUTION BUREAU, GENERAL AFFAIRS DIVISION, REGULATION SECTION)
JAPANESE

KOGAI KENKYU 2264
(RESEARCH ON ENVIRONMENTAL DISRUPTION TOWARD INTERDISCIPLINARY COOPERATION)
TOKEI KENKYUKAI AND IWANAMI SHOTEN
5-5, 2-CHOME, HITOTSUBASHI, CHIYODA-KU, TOKYO 101
JAPANESE
INDEXED: BIB, JICST

KOGAI MONDAI JOHO 2265
(WEEKLY NEWS OF POLLUTION)
NIHON BIJINESU K. K.
1-1, 1-CHOME, JINGUMAE, SHIBUYA-KU, TOKYO 150
JAPANESE

KOGAI NO GENJO TO TAISAKU (FUKUOKA-SHI) 2266
(ENVIRONMENTAL POLLUTION, PRESENT SITUATION AND COUNTERPLAN, FUKUOKA CITY)
FUKUOKA-SHI EISEI-KYOKU (PUBLIC HEALTH BUREAU, FUKUOKA MUNICIPAL GOVERNMENT)
1-1, 1-CHOME, TENJIN, CHUO-KU, FUKUOKA-SHI 810
JAPANESE

KOGAI NO GENJO TO TAISAKU (KYOTO-SHI) 2267
(ENVIRONMENTAL POLLUTION, PRESENT SITUATION AND COUNTERPLAN, KYOTO CITY)
KYOTO-SHI KOGAI TAISAKUSHITSU (KYOTO MUNICIPAL GOVERNMENT)
OIKE, TERAMACHI-DORI, NAKAKYO-KU, KYOTO-SHI 604
JAPANESE

KOGAI NO GENKYO TO TAISAKU (IWATE-KEN) 2268
(ENVIRONMENTAL POLLUTION, PRESENT SITUATION AND COUNTERPLAN, IWATE PREFECTURE)
IWATE-KEN (IWATE PREFECTURAL GOVERNMENT)
10-1, UCHIMARU, MORIOKA-SHI 020
JAPANESE

KOGAI NO GENKYO TO TAISAKU (OSAKA-SHI) 2269
(ENVIRONMENTAL POLLUTION, PRESENT SITUATION AND COUNTERPLAN, OSAKA CITY)
OSAKA-SHI (OSAKA MUNICIPAL GOVERNMENT)
1-4, NAKANOSHIMA, KITA-KU, OSAKA-SHI 530
JAPANESE

KOGAI NO GENKYO TO TAISAKU (SAGA-KEN) 2270
(ENVIRONMENTAL POLLUTION, PRESENT SITUATION AND COUNTERPLAN, SAGA PREFECTURE)
SAGA-KEN (SAGA PREFECTURAL GOVERNMENT)
1-59, 1-CHOME, JONAI, SAGA-SHI 840
JAPANESE

KOGAI NO JOKYO NI KANSURU NENJI HOKOKU 2271
(GUNMA-KEN)
(ANNUAL REPORT ON THE ENVIRONMENTAL POLLUTION, GUNMA PREFECTURE)
GUNMA-KEN (GUNMA PREFECTURAL GOVERNMENT)
1-1, 1-CHOME, OTE-MACHI, MAEBASHI-SHI 371
JAPANESE

KOGAI NO JOKYO NI KANSURU NENJI HOKOKU (KANKYO-CHO) 2272
(ANNUAL REPORT ON THE EVNIRONMENTAL POLLUTION, ENVIRONMENT AGENCY)
KANKYO-CHO (ENVIRONMENT AGENCY)
1-1, 3-CHOME, KASUMIGASEKI, CHIYODA-KU, TOKYO 100
JAPANESE

KOGAI NO JOKYO NI KANSURU NENJI HOKOKU (TOCHIGI-KEN) 2273
(ANNUAL REPORT ON THE EVNIRONMENTAL POLLUTION, TOCHIGI PREFECTURE)
TOCHIGI-KEN (TOCHIGI PREFECTURAL GOVERNMENT)
504, HANAWADA-CHO, UTSUNOMIYA-SHI 320
JAPANESE

KOGAI SHIGEN KENKYUSHO HOKOKU 2274
(SCIENCE REPORT OF THE NATIONAL RESEARCH INSTITUTE FOR POLLUTION AND RESOURCES)
TSUSHO SANGYO-SHO KOGYO GIJUTSU-IN KOGAI SHIGEN KENKYUSHO
188, KOTOBUKI-CHO, KAWAGUCHI-SHI 332
JAPANESE, WITH ENGLISH SUMMARY
INDEXED: CHE, JICST

KOGAI SHIGEN KENKYUSHO IHO 2275
(BULLETIN OF THE NATIONAL RESEARCH INSTITUTE FOR POLLUTION AND RESOURCES)
TSUSHO SANGYO-SHO KOGYO GIJUTSU-IN KOGAI SHIGEN KENKYUSHO
188, KOTOBUKI-CHO, KAWAGUCHI-SHI 332
JAPANESE AND ENGLISH
INDEXED: CHE, JICST

KOGAI SHIGEN KENKYUSHO NENPO 2276
(ANNUAL REPORT OF NATIONAL RESEARCH INSTITUTE FOR POLLUTION AND RESOURCES)
TSUSHO SANGYO-SHO KOGYO GIJUTSU-IN KOGAI SHIGEN KENKYUSHO
188, KOTOBUKI-CHO, KAWAGUCHI-SHI 332
JAPANESE
INDEXED: JICST

KOGAI SHIGENKEN NYUSU 2277
(NATIONAL RESEARCH INSTITUTE FOR POLLUTION AND RESOURCES NEWS)
TSUSHO SANGYO-SHO KOGYO GIJUTSU-IN KOGAI SHIGEN KENKYUSHO
188, KOTOBUKI-CHO, KAWAGUCHI-SHI 332
JAPANESE
INDEXED: JICST

KOGAI SHIRYO 2278
(DATA OF ENVIRONMENTAL POLLUTION)
YOKOHAMA-SHI KOGAI TAISAKU-KYOKU (POLLUTION CONTROL BUREAU, YOKOHAMA CITY)
1-1, MINATO-MACHI, NAKA-KU, YOKOHAMA-SHI 231
JAPANESE

KOGAI SOKUTEI CHOSA NO GAIYO 2279
(OUTLINE OF MEASUREMENT AND INVESTIGATION OF ENVIRONMENTAL POLLUTION)
KOMAE-SHI SHIMIN-BU (KOMAE CITY)
1585, IZUMI, KOMAE-SHI 182
JAPANESE

KOGAI TO TAISAKU 2280
(JOURNAL OF PUBLIC NUISANCE)
KOGAI TAISAKU GIJUTSU DOYUKAI
1-244, 9-CHOME, AKASAKA, MINATO-KU, TOKYO 107
JAPANESE
INDEXED: CHE, JICST

KOJO KOGAI CHOSA HOKOKUSHO 2281
(SURVEY REPORT ON THE POLLUTION CAUSED BY INDUSTRY)
KASUGAI-SHI (KASUGAI CITY)
5800-1, 5-CHOME, MIZUHO-DORI, KASUGAISHI 486
JAPANESE

KOKU KOGAI, KENKYU TO TAISAKU 2282
(AIRCRAFT NUISANCE; RESEARCH AND CONTROL)
JAPANESE
INDEXED: JICST

KOKUMIN EISEI NO DOKO 2283
(NATIONAL PUBLIC HEALTH, IT'S TREND)
KOSEI TOKEI KYOKAI (HEALTH AND WELFARE STATISTICS SOCIETY)
13-14, 5-CHOME, ROPPONGI, MINATO-KU, TOKYO 106
JAPANESE

KOKUMIN KENKO CHOSA 2284
(NATIONAL HEALTH SURVEY)
(KOSEI-SHO DAIJIN KANBO TOKEI CHOSA-BU (HEALTH AND WELFARE STATISTICS DIVISION, MINISTER'S SECRETARIAT, MINISTRY OF HEALTH AND WELFARE)
42, ICHIGAYA HONMURA-CHO, SHINJUKU-KU, TOKYO 162
JAPANESE

KOKURITSU EISEI SHIKENJO CHOSA GEPPO 2285
(MONTHLY REPORT OF THE NATIONAL INSTITUTE OF HYGIENIC SCIENCES)
KOKURITSU EISEI SHIKENJO
18-1, 1-CHOME, KAMIYOGA, SETAGAYA-KU, TOKYO 158
JAPANESE

KOKURITSU KOSHU EISEI-IN KENKYU HOKOKU 2286
(BULLETIN OF THE INSTITUTE OF PUBLIC HEALTH)
KOKURITSU KOSHU EISEI-IN (INSTITUTE OF PUBLIC HEALTH)
6-1, 4-CHOME, SHIROGANEDAI, MINATO-KU, TOKYO 108
JAPANESE AND ENGLISH
INDEXED: BIB, CHE, EXC

KOKURITSU KOSHU EISEI-IN NENPO 2287
(ANNUAL REPORT OF THE INSTITUTE OF PUBLIC HEALTH)
KOKURITSU KOSHU EISEIIN (THE INSTITUTE OF PUBLIC HEALTH)
6-1, 4-CHOME, SHIROGANEDAI, MINATO-KU, TOKYO 108
JAPANESE

KOKURITSU YOBO EISEI KENKYUSHO NENPO 2288
(ANNUAL REPORT OF THE NATIONAL INSTITUTE OF HEALTH)
KOKURITSU YOBO EISEI KENKYUSHO (NATIONAL INSTITUTE OF HEALTH, MINISTRY OF HEALTH AND WELFARE)
10-35, 2-CHOME, KAMIOSAKI, SHINAGAWA-KU, TOKYO 141
JAPANESE AND ENGLISH

KOSEI 2289
(PUBLIC WELFARE)
KOSEI MONDAI KENKYUKAI
1-18, NIHONBASHI KAYABA-CHO, CHUO-KU, TOKYO 103
JAPANESE

KOSEI NO SHIHYO 2290
(HEALTH AND WELFARE STATISTICS)
KOSEI TOKEI KYOKAI (HEALTH AND WELFARE STATISTICS SOCIETY)
13-14, 5-CHOME, ROPPONGI, MINATO-KU, TOKYO 106
JAPANESE

KOSEI TOKEI CHIIKI KEIKO SEIMITSU CHOSA GAIKYO 2291
(OUTLINE OF THE MINUTE INVESTIGATION FOR THE AREA CHARACTERISTICS SHOWN BY HEALTH STATISTICS)
KOSEI-SHO DAIJIN KANBO TOKEI CHOSA-BU (HEALTH AND WELFARE STATISTICS DEPARTMENT, MINISTER'S SECRETARIAT, MINISTRY OF HEALTH AND WELFARE)
42, ICHIGAYA HONMURA-CHO, SHINJUKU-KU, TOKYO 162
JAPANESE

KOSHU EISEI 2292
(JOURNAL OF PUBLIC HEALTH PRACTICE)
IGAKU SHOIN
29-11, 5-CHOME, HONGO, BUNKYO-KU, TOKYO 113
JAPANESE
INDEXED: CHE, JICST

KOSHU EISEI JOHO 2293
(PUBLIC HEALTH REVIEW)
NIHON KOSHU EISEI KYOKAI (JAPANESE SOCIETY OF PUBLIC HEALTH)
29-8, 1-CHOME, SHINJUKU, SHINJUKU-KU, TOKYO 160
JAPANESE
INDEXED: APT

KOSHU EISEI NI KANSURU KENKYU HOKOKU 2294
(RESEARCH REPORT ON PUBLIC HEALTH)
TOKYO-TO EISEI-KYOKU SOMU-BU (GENERAL AFFAIRS DIVISION, BUREAU OF PUBLIC HEALTH, TOKYO)
5-1, 3-CHOME, MARUNOUCHI, CHIYODA-KU, TOKYO 100
JAPANESE
INDEXED: JICST

KOTAI HAIBUTSU 2295
(SOLID WASTE)
KOTAI HAIBUTSU HENSHU IIHKAI
KOGYO SHUPPANSHA
12-2, 2-CHOME, MISAKI-CHO, CHIYODA-KU, TOKYO 101
JAPANESE
INDEXED: JICST

KUMAMOTO-KEN EISEI KOGAI KENKYUSHOHO 2296
(ANNUAL REPORT OF KUMAMOTO RESEARCH INSTITUTE FOR PUBLIC HEALTH AND ENVIRONMENTAL POLLUTION)
KUMAMOTO-KEN EISEI KOGAI KENKYUSHO
4-3, MINAMI SENDANBATA-MACHI, KUMAMOTO-SHI 860
JAPANESE

KYO 2297
NIPPON SHINYAKU COMPANY, LIMITED
NISHIOJI 8-JO SAGARU, MINAMI-KU, KYOTO 601
JAPANESE
INDEXED: CHE

KYOTO KOSHU EISEI GAKKAISHI 2298
(MEMOIRS OF KYOTO SOCIETY OF PUBLIC HEALTH)
KYOTO KOSHU EISEI KYOKAI
C/O KYOTO-FU KOSHU EISEI-BU, YABUNOUCHI-CHO, SHIMOTACHIYURI-DORI SHIN-MACHI NISHIIRU, KAMIGYO-KU KYOTOSHI 602
JAPANESE

KYOTO-FU EISEI KENKYUSHO NENPO, KYOTO-FU KOGAI KENKYUSHO NENPO 2299
(ANNUAL REPORT OF KYOTO PREFECTURAL INSTITUTE OF PUBLIC HEALTH, ANNUAL REPORT OF KYOTO PREFECTURAL INSTITUTE OF ENVIRONMENTAL POLLUTION)
KYOTO-FU EISEI KENKYUSHO, KYOTO-FU KOGAI KENKYUSHO
HIGASHIOJI 5-JO AGARU, HIGASHIYAMA-KU, KYOTOSHI 605
JAPANESE AND ENGLISH

KYOTO-FU NI OKERU KOGAI NO GENKYO TO TAISAKU 2300
(ENVIRONMENTAL POLLUTION IN KYOTO PREFECTURE, IT'S SITUATION AND COUNTER PLAN)
KYOTO-FU (KYOTO PREFECTURAL GOVERNMENT)
YABUNOUCHI-CHO, SHIMOTACHIYURI-DORI SHIN-MACHI NISHIIRU, KAMIGYO-KU, KYOTO-SHI 602
JAPANESE

KYOTO-FU NO JINKO 2301
(POPULATION OF KYOTO PREFECTURE)
KYOTO-FU SOMU-BU (GENERAL AFFAIRS DIVISION, KYOTO PREFECTURAL GOVERNMENT)
YABUNOUCHI-CHO, SHIMOTACHIYURI-DORI SHIN-MACHI NISHIIRU, KAMIGYO-KU, KYOTO-SHI 602
JAPANESE

KYOTO-SHI EISEI KENKYUSHO NENPO 2302
(ANNUAL REPORTS OF KYOTO CITY RESEARCH INSTITUTE OF PUBLIC HEALTH)
KYOTO-SHI EISEI KENKYU-SHO (KYOTO CITY RESEARCH INSTITUTE OF PUBLIC HEALTH)
1-2, MIBU-HIGASHI TAKADA-CHO, NAKAGYO-KU, KYOTO-SHI 604
JAPANESE

KYOTO-SHI EISEI TOKEI NENPO 2303
(ANNUAL REPORT, HEALTH STATISTICS OF KYOTO CITY)
HYOTO-SHI EISEI-KYOKU (PUBLIC HEALTH BUREAU, KYOTO MUNICIPAL GOVERNMENT)
488, KAMI-HONNOJIMAE-CHO, TERAMACHIDORI OSHIKOJI-SAGARU, NAKAGYO-KU, KYOTO-SHI 604
JAPANESE

KYOTO-SHI EISEI-KYOKU JIGYO SEISEKI 2304
(STATISTICS ON THE ACTIVITIES OF PUBLIC HEALTH BUREAU, KYOTO-CITY)
KYOTO-SHI EISEI-KYOKU (PUBLIC HEALTH BUREAU, KYOTO MUNCIPAL GOVERNMENT)
488, KAMI-HONNOJIMAE-CHO, TERAMACHIDORI OSHIKOJI-SAGARU, NAKAGYO-KU, KYOTO-SHI 604
JAPANESE

KYOTO-SHI EISEI-KYOKU JIGYO TOKEI 2305
(ANNUAL REPORT OF PUBLIC HEALTH BUREAU, KYOTO CITY)
KYOTO-SHI EISEI-KYOKU (PUBLIC HEALTH BUREAU, KYOTO MUNICIPAL GOVERNMENT)
488, HONNOJIMAE-CHO, OSHIKOJI-SAGARU, TERAMACHI-DORI, NAKAGYO-KU, KYOTO-SHI 604
JAPANESE

KYUSHU DAIGAKU IRYO GIJUTSU TANKI DAIGAKUBU KIYO 2306
(MEMOIRS OF KYUSHU UNIVERSITY OF HEALTH SCIENCE)
JAPANESE

KYUSHU DAIGAKU TAIIKUGAKU KENKYU 2307
(KYUSHU UNIVERSITY, RESEARCH OF PHYSICAL EDUCATION)
KYUSHU DAIGAKU KYOYO-BU HOKEN TAIIKUGAKU KYOSHITSU (THE PHYSICAL EDUCATION LABORATORY, GENERAL EDUCATION DEPARTMENT, KYUSHU UNIVERSITY)
2-1, 4-CHOME, ROPPONMATSU, CHUO-KU, FUKUOKA-SHI 810
JAPANESE, WITH ENGLISH SUMMARY

MEGURO-KU NO KOGAI 2308
(ENVIRONMENTAL POLLUTION IN MEGURO WARD)
TOKYO-TO MEGURO-KU (MEGURO WARD, TOKYO)
4-5, 2-CHOME, CHUO-MACHI, MEGURO-KU, TOKYO 152
JAPANESE

MICROBIOLOGY AND IMMUNOLOGY 2309
CENTER FOR ACADEMIC PUBLICATIONS
4-16, YAYOI 2-CHOME, BUNKYO-KU, TOKYO 113
ENGLISH
INDEXED: BIB, CHE, EXC

MIE DAIGAKU KANKYO KAGAKU KENKYU KIYO 2310
(REPORT OF ENVIRONMENTAL SCIENCE OF MIE UNIVERSITY)
MIE DAIGAKU KANKYO KAGAKU SOGO KENKYU SHITSU (MIE UNIVERSITY ENVIRONMENTAL SCIENCE, COLLECTIVE RESEARCH OFFICE)
JAPANESE

MIE-KEN EISEI KENKYUSHO NENPO 2311
(ANNUAL REPORT OF MIE PREFECTURAL INSTITUTE OF PUBLIC HEALTH)
MIE-KEN EISEI KENKYUSHO
1-172, SAKAE-MACHI, TSU-SHI 514
JAPANESE AND ENGLISH

MINATO-KU NO KOGAI 2312
(ENVIRONMENTAL POLLUTION IN MINATO WARD)
TOKYO-TO MINATO-KU (MINATO WARD, TOKYO)
5-25, 1-CHOME, SHIBA KOEN, MINATO-KU, TOKYO 105
JAPANESE

MIYAGI-KEN EISEI KENKYUSHO HOKOKU, GYOSEKISHU 2313
(ANNUAL REPORT OF MIYAGI PREFECTURAL INSTITUTE OF PUBLIC HEALTH)
MIYAGI-KEN EISEI KENKYUSHO (MIYAGI PREFECTURAL INSTITUTE OF PUBLIC HEALTH)
27, MASUE, HARANO-MACHI, ODAWARA, SENDAI-SHI 983
JAPANESE
INDEXED: JICST

MIYAGI-KEN EISEI KENKYUSHO JIGYO HOKOKU 2314
(ANNUAL REPORT OF THE ACTIVITIES OF MIYAGI PREFECTURAL INSTITUTE OF PUBLIC HEALTH)
MIYAGI-KEN EISEI KENKYUSHO
27, MASUE, HARANO-MACHI, ODAWARA, SENDAI-SHI 983
JAPANESE

MIYAGI-KEN KOGAI GIJUTSU SENTA HOKOKU 2315
(ANNUAL REPORT OF MIYAGI TECHNOLOGY CENTER FOR POLLUTION)
MIYAGI-KEN KOGAI GIJUTSU SENTA
8-1, 3-CHOME, HON-CHO, SENDAI-SHI 980
JAPANESE

MIYAGI-KEN KOSHU EISEI GAKKAI KAISHI 2316
(MEMOIRS OF MIYAGI PREFECTURE PUBLIC HEALTH ASSOCIATION)
MIYAGI-KEN KOSHU EISEI GAKKAI
C/O MIYAGI-KEN KOSHU EISEI KYOKAI, 6-9, 1-CHOME, TORI-MACHI, SENDAI-SHI 980
JAPANESE

MIYAZAKI-KEN EISEI KENKYUSHOHO 2317
(ANNUAL REPORT OF MIYAZAKI PREFECTURAL PUBLIC HEALTH LABOROATORY)
MIYAZAKI-KEN EISEI KENKYUSHO (MIYAZAKI PREFECTURAL PUBLIC HEALTH LABORATORY)
5-30, KITATAKAMATSU-CHO, MIYAZAKI-SHI 880
JAPANESE

MIYAZAKI-KEN NO EISEI 2318
(PUBLIC HEALTH IN MIYAZAKI PREFECTURE)
MIYAZAKI-KEN KANKYO HOKEN-BU (MIYAZAKI PREFECTURAL GOVERNMENT)
10-1, 2-CHOME, TACHIBANADORI HIGASHI, MIYAZAKI-SHI 880
JAPANESE

MUKOGAWA JOSHI DAIGAKU KIYO; TAIIKUGAKU-HEN 2319
(BULLETIN OF THE MUKOGAWA WOMEN'S UNIVERSITY PHYSICAL EDUCATION EDITION)
MUKOGAWA WOMEN'S UNIVERSITY
JAPANESE

NAGANO-KEN EISEI KOGAI KENKYUSHO CHOSA KENKYU HOKOKU 2320
(BULLETIN OF NAGANO RESEARCH INSTITUTE FOR HEALTH AND POLLUTION)
NAGANO-KEN EISEI KOGAI KENKYUSHO (NAGANO RESEARCH INSTITUTE FOR HEALTH AND POLLUTION)
1978, KOMEMURA, AMORI, NAGANO-SHI 380
JAPANESE

NAGASAKI-KEN EISEI KENKYUSHOHO 2321
(ANNUAL REPORT OF NAGASAKI PREFECTURAL INSTITUTE OF PUBLIC HEALTH)
NAGASAKI-KEN EISEI KOGAI KENKYUSHO
32-31, NAMEISHI-MACHI, NAGASAKI-SHI 852
JAPANESE

NAGOYA-SHI EISEI KENKYUSHOHO 2322
(ANNUAL REPORT OF NAGOYA CITY HEALTH RESEARCH INSTITUTE)
NAGOYA-SHI EISEI KENKYUSHO
1-11, HAGIYAMA-CHO, MIZUHO-KU, NAGOYA-SHI 467
JAPANESE
INDEXED: JICST

NAGOYA-SHI KOGAI KENKYU SHOHO 2323
(REPORT OF ENVIRONMENTAL POLLUTION RESEARCH INSTITUTE, CITY OF NAGOYA)
JAPANESE
INDEXED: JICST

NAKANO-KU NO KOGAI CHOSA HOKOKUSHO 2324
(REPORT OF ENVIRONMENTAL POLLUTION IN NAKANO WARD, TOKYO)
TOKYO-TO NAKANO-KU KANKYO-BU (NAKANO WARD, TOKYO)
8-1, 4-CHOME, NAKANO, NAKANO-KU, TOKYO 164
JAPANESE

NAOETSU CHIIKI NO TAIKI OSEN CHOSA HOKOKUSHO 2325
(SURVEY REPORT ON THE AIR POLLUTION IN NAOETSU AREA, NIIGATA)
NIIGATA-KEN KIKAKU-BU (NIIGATA PREFECTURAL GOVERNMENT)
ICHIBAN-CHO, GAKKO-DURI, NIIGATA-SHI 951
JAPANESE

NARA-KEN EISEI KENKYUSHO NENPO 2326
(ANNUAL REPORT OF THE INSTITUTE OF PUBLIC HEALTH, NARA PREFECTURE)
NARA-KEN EISEI KENKYUSHO
57-6, OMORI-CHO, NARA-SHI 630
JAPANESE

NIHON EISEIGAKU ZASSHI 2327
(JAPAN JOURNAL OF HYGIENE)
NIHON EISEI GAKKAI (JAPANESE SOCIETY FOR HYGIENE)
C/O KYOTO DAIGAKU IGAKUBU EISEIGAKU, KYOSHITSU YOSHIDA KONOE-CHO, SAKYO-KU, KYOTO-SHI 606
JAPANESE AND ENGLISH
INDEXED: CHE, JICST

NIHON JIDOKA KENSHIN GAKKAI KAISHI 2328
(RECORD OF THE JAPAN AUTOMATED PHYSICAL EXAMINATION STUDY SOCIETY)
JAPAN AUTOMATED PHYSICAL EXAMINATION STUDY SOCIETY
JAPANESE

NIHON JINKO GAKKAI KAIHO 2329
(BULLETIN OF THE POPULATION ASSOCIATION OF JAPAN)
NIHON JINKO GAKKAI (THE POPULATION ASSOCIATION JAPAN)
C/O KOSEI-SHO JINKO MONDAI KENKYUSHO, 2-2, 1-CHOME, KASUMIGASEKI, CHIYODA-KU, TOKYO 100
JAPANESE

NIHON JOSHI TAIIKU DAIGAKU KIYO 2330
(ANNUAL REPORT OF PHYSICAL EDUCATION)
NIHON JOSHI TAIIKU DAIGAKU (JAPAN WOMEN'S COLLEGE OF PHYSICAL EDUCATION)
7-1, 8-CHOME, KITAKARASUYAMA, SETAGAYA-KU, TOKYO 154
JAPANESE, WITH ENGLISH SUMMARY

NIHON KANKYO EISEI SENTA SHOHO 2331
(REPORT OF THE JAPANESE ENVIRONMENT AND HEALTH CENTER)
JAPANESE ENVIRONMENT AND HEALTH CENTER
JAPANESE

NIHON KOSHU EISEI GAKKAI KINKI CHIHOKAI KOEN YOSHISHU 2332
(ABSTRACT OF SPEECHES AT KINKI DISTRICT MEETING OF JAPANESE SOCIETY OF PUBLIC HEALTH)
NIHON KOSHU EISEI GAKKAI KINKI CHIHOKAI JIMUKYOKU
C/O KYOTO-FU EISEI-BU, YABUNOUCHI-CHO, SHIMOTACHIYURI-DORI SHIN-MACHI NISHIIRU, KAMIGYO-KU, KYOTO-SHI 602
JAPANESE

NIHON KOSHU EISEI ZASSHI 2333
(JAPANESE JOURNAL OF PUBLIC HEALTH)
NIHON KOSHU EISEI GAKKAI (JAPANESE SOCIETY OF PUBLIC HEALTH)
29-8, 1-CHOME, SHINJUKU, SHINJUKU-KU, TOKYO 160
JAPANESE AND ENGLISH
INDEXED: JICST

NIHON SUISHORI SEIBUTSU GAKKAISHI 2334
(JOURNAL OF THE JAPAN WATER TREATMENT BIOLOGICAL SOCIETY)
JAPANESE

NIHON TAIIKU DAIGAKU KIYO 2335
(BULLETIN OF THE JAPAN PHYSICAL EDUCATION UNIVERSITY)
JAPANESE

NIHON TAIIKU DAIGAKU TAIIKU KENKYUJO SHOHO 2336
(REPORT OF THE PHYSICAL EDUCATION RESEARCH INSTITUTE, JAPAN PHYSICAL EDUCATION UNIVERSITY)
JAPANESE

NIIGATA-KEN EISEI KENKYUSHO GYOSEKI HOKOKU 2337
(REPORT OF THE PUBLIC HEALTH LABORATORY, NIIGATA PREFECTURE)
NIIGATA-KEN EISEI KENKYUSHO (NIIGATA PREFECTURE PUBLIC HEALTH LABORATORY)
ICHIBANBORI-DORI-CHO, NIIGATA-SHI 951
JAPANESE

NIIGATA-KEN EISEI KENKYUSHO, NIIGATA-KEN KOGAI KENKYUSHO NENPO 2338
(ANNUAL REPORT OF PUBLIC HEALTH LABORATORY, ENVIRONMENTAL POLLUTION LABORATORY, NIIGATA PREFECTURE)
NIIGATA-KEN EISEI KENKYUSHO, NIIGATA-KEN KOGAI KENKYUSHO (PUBLIC HEALTH LABORATORY, ENVIRONMENTAL POLLUTION LABORATORY, NIIGATA PREFECTURE)
5294, ICHIBANBORI-DORI-CHO, NIIGATA-SHI 951
JAPANESE

NIIGATA-KEN EISEI KENKYUSHO, NIIGATA-KEN KOGAI KENKYUSHO SHUDANKAI SHOROKU 2339
(SUMMARY OF THE MEETING OF THE PUBLIC HEALTH LABORATORY, NIIGATA PREFECTURE AND ENVIRONMENTAL POLLUTION LABORATORY, NIIGATA PREFECTURE))
NIIGATA-KEN EISEI KENKYUSHO, NIIGATA-KEN KOGAI KENKYUSHO (PUBLIC HEALTH LABORATORY, ENVIRONMENTAL POLLUTION LABORATORY, NIIGATA PREFECTURE)
5924, ICHIBANBORI-DORI-CHO, NIIGATA-SHI 951
JAPANESE

OITA KOGYO DAIGAKU KOGAI KENKYUJO HO 2340
(REPORTS OF THE ENVIRONMENTAL SCIENCE AND TECHNOLOGY LABORATORY, OITA INSTITUTE OF TECHNOLOGY)
JAPANESE

OKAYAMA-KEN EISEI KENKYUSHO NENPO 2341
(ANNUAL REPORT OF HYGIENIC LABORATORY OF OKAYAMA PREFECTURE)
OKAYAMA-KEN EISEI KENKYUSHO (HYGIENIC LABORATORY OF OKAYAMA PREFECTURE)
1-17, 1-CHOME, FURUGYO-CHO, OKAYAMA-SHI 703
JAPANESE AND ENGLISH

OKINAWA-KEN KOGAI EISEI KENKYUSHOHO 2342
(REPORT OF OKINAWA PREFECTURAL INSTITUTE OF PUBLIC HEALTH)
OKINAWA-KEN KOGAI EISEI KENKYUSHO (OKINAWA PREFECTURAL INSTITUTE OF PUBLIC HEALTH)
2-1, 1-CHOME, KUMOJI, NAHA-SHI 900
JAPANESE AND ENGLISH

OSAKA NO SORA 2343
(SKY OVER THE CITY OF OSAKA)
OSAKA-FU KOGYO KYOKAI (OSAKA INDUSTRIAL ASSOCIATION)
C/O OSAKA-FU SHOKO KAIKAN, 5-16, MINAMIHON-CHO, HIGASHI-KU, OSAKA-SHI 541
JAPANESE

OSAKA TAIIKU DAIGAKU KIYO 2344
(BULLETIN OF OSAKA PHYSICAL EDUCATION UNIVERSITY)
JAPANESE

OSAKA-FU KOGAI KANSHI SENTA CHOSASHITSU KENKYU HOKOKU 2345
(PROCEEDINGS OF THE RESEARCH SECTION, OSAKA ENVIRONMENTAL POLLUTION CONTROL CENTER)
OSAKA-FU KOGAI KANSHI SENTA (OSAKA ENVIRONMENTAL POLLUTION CONTROL CENTER)
3-62, 1-CHOME, NAKAMICHI, HIGASHINARI-KU, OSAKA-SHI 537
JAPANESE AND ENGLISH

OSAKA-FURITSU KOSHU EISEI KENKYUJO KOSHU EISEIBU HOKOKU 2346
(REPORT OF THE PUBLIC HEALTH DIVISION, OSAKA MUNICIPAL PUBLIC HEALTH RESEARCH INSTITUTE)
JAPANESE

OSAKA-FURITSU KOSHU EISEI KENKYUSHO KENKYU HOKOKU, KANKYO EISEI-HEN 2347
(PROCEEDINGS OF THE OSAKA PREFECTURAL INSTITUTE OF PUBLIC HEALTH, EDUCATION OF ENVIRONMENT HYGIENE)
OSAKA-FURITSU KOSHU EISEI KENKYUSHO
3-69, 1-CHOME, NAKAMICHI, HIGASHINARI-KU, OSAKA-SHI 537
JAPANESE

OSAKA-FURITSU KOSHU EISEI KENKYUSHO KENKYU HOKOKU, KOSHU EISEI-HEN 2348
(PROCEEDINGS OF THE OSAKA PREFECTURAL INSTITUTE OF PUBLIC HEALTH, EDUCATION OF AIR AND PUBLIC HEALTH)
OSAKA-FURITSU KOSHU EISEI KENKYUSHO
3-59, 1-CHOME, NAKAMICHI, HIGASHINARI-KU, OSAKA-SHI 537
JAPANESE
INDEXED: JICST

OSAKA-FURITSU KOSHU EISEI KENKYUSHO KENKYU HOKOKU, RODO EISEI-HEN 2349
(PROCEEDINGS OF THE OSAKA PREFECTURAL INSTITUTE OF PUBLIC HEALTH, EDUCATION OF INDUSTRIAL HEALTH)
OSAKA-FURITSU KOSHU EISEI KENKYUSHO
3-69, 1-CHOME, NAKAMICHI, HIGASHINARI-KU, OSAKA-SHI 537
JAPANESE
INDEXED: JICST

OSAKA-FURITSU KOSHU EISEI KENKYUSHO KENKYU HOKOKU, SHOKUHIN EISEI-HEN 2350
(PROCEEDINGS OF THE OSAKA PREFECTURAL INSTITUTE OF PUBLIC HEALTH, EDUCATION OF FOOD SANITATION)
OSAKA-FURITSU KOSHU EISEI KENKYUSHO
3-69, 1-CHOME, NAKAMICHI, HIGASHINARI-KU OSAKA-SHI 537
JAPANESE

OSAKA-FURITSU KOSHU EISEI KENKYUSHO NENPO 2351
(ANNUAL REPORT OF THE OSAKA PREFECTURAL INSTITUTE OF PUBLIC HEALTH)
OSAKA-FURITSU KOSHU EISEI KENKYUSHO
3-69, 1-CHOME, NAKAMICHI, HIGASHINARI-KU, OSAKA-SHI 537
JAPANESE AND ENGLISH

OSAKA-SHI CHUO TAIIKUKAN TAIIKU IJI NENPO 2352
(ANNUAL REPORT ON HEALTH AND PHYSICAL EDUCATION, CENTRAL GYMNASIUM OF OSAKA CITY)
OSAKA-SHI CHUO TAIIKUKAN TAIIKU IJI SODANSHO
6, HOENSAKA-MACHI, HIGASHI-KU, OSAKA-SHI 540
JAPANESE

OSAKA-SHI NO TAIIKU TO KOSEI 2353
(PHYSICAL EDUCATION AND HEALTH IN THE CITY OF OSAKA)
OSAKA-SHI TAIIKU KOSEI KYOKAI
(PHYSICAL EDUCATION AND HEALTH ASSOCIATION OF OSAKA)
JAPANESE

OSAKA-SHIRITSU DAIGAKU HOKEN TAIIKUGAKU KENKYU HOKOKU 2354
(ANNALS OF PHYSICAL EDUCATION, OSAKA CITY UNIVERSITY)
OSAKA-SHIRITSU DAIGAKU KYOYO-BU HOKEN TAIIKU-KA
(THE HEALTH AND PHYSICAL EDUCATION LABORATORY, OSAKA CITY UNIVERSITY)
191, SUGIMOTO-CHO, SUMIYOSHI-KU, OSAKA-SHI 558
JAPANESE AND ENGLISH

OSAKA-SHIRITSU DAIGAKU HOKEN TAIIKUGAKU KENKYU KIYO 2355
(ANNALS OF PHYSICAL EDUCATION)
OSAKA-SHIRITSU DAIGAKU KYOYO-BU HOKEN TAIIKU KENKYUSHITSU (THE HEALTH AND PHYSICAL EDUCATION LABORATORY, OSAKA CITY UNIVERSITY)
191, SUGIMOTO-CHO, SUMIYOSHI-KU, OSAKA-SHI 558
JAPANESE

OSAKA-SHIRITSU EISEI KENKYUSHO KENKYU HOKOKU 2356
(REPORT OF THE OSAKA CITY INSTITUTE OF HYGIENE)
OSAKA-SHIRITSU EIDEI KENKYUSHO (OSAKA CITY INSTITUTE OF HYGIENE)
38, KITAOGI-MACHI, KITA-KU, OSAKA-SHI 530
JAPANESE
INDEXED: JICST

PORYUSHON REBYU 2357
(POLLUTION REVIEW)
KOGAI KENKYUKAI
TEKINTO/KOGYO JOHO KENKYUKAI
1-5, KANDA SURUGADAI, CHIYODA-KU, TOKYO 101
JAPANESE AND ENGLISH

PPM 2358
NIHON KOGYO SHINBUNSHA
7-2, 1-CHOME, OTE-MACHI, CHIYODA-KU, TOKYO 100
JAPANESE
INDEXED: APT, CHE, JICST

PUBLIC CLEANSING SERVICE IN TOKYO 2359
TOKYO-TO SEISO-KYOKU KIKAKU-BU (DIVISION OF PLANNING, BUREAU OF PUBLIC CLEANSING, TOKYO)
8-1, 3-CHOME, MARUNOUCHI, CHIYODA-KU, TOKYO 100
ENGLISH

RIYO TENBO 2360
(BARBER AND BEAUTY SHOP VIEW)
ZENKOKU RIYO KANKYO EISEI DOGYO KUMIAI RENGO KAI (NATIONAL BARBER AND BEAUTY SHOP ENVIRONMENTAL HYGIENE, NATIONAL BARBER AND BEAUTY SHOP INDUSTRY FEDERATION)
JAPANESE

RODO EISEI 2361
(JAPANESE JOURNAL OF INDUSTRIAL HYGIENE)
CHUO RODO SAIGAI BOSHI KYOKAI (JAPAN INDUSTRIAL SAFETY ASSOCIATION)
35-1, 5-CHOME, SHIBA, MINATO-KU, TOKYO 108
JAPANESE
INDEXED: CHE, JICST

RODO EISEI KENKYUSHO NENPO 2362
(ANNUAL REPORT OF THE NATIONAL INSTITUTE OF INDUSTRIAL HEALTH)
RODO-SHO RODO EISEI KENKYUSHO (THE NATIONAL INSTITUTE OF INDUSTRIAL HEALTH)
2051, KIZUKI SUMIYOSHI-CHO, NAKAHARA-KU, KAWASAKI-SHI 211
JAPANESE
INDEXED: JICST

RODO EISEI KOGAKU 2363
(INDUSTRIAL SANITARY ENGINEERING)
NIHON RODO EISEI KOGAKKAI
2051, KIZUKI SUMIYOSHI-CHO, NAKAHARA-KU, KAWASAKI-SHI 211
JAPANESE

RODO KAGAKU 2364
(JOURNAL OF SCIENCE OF LABOUR)
RODO KAGAKU KENKYUSHO (INSTITUTE FOR SCIENCE OF LABOUR)
1544, SUGAO, TAKATSU-KU, KAWASAKI-SHI 213
JAPANESE AND ENGLISH
INDEXED: CHE, JICST

RODO KAGAKU KENKYUSHO NENPO 2365
(ANNUAL REPORT OF THE INSTITUTE OF SCIENCE OF LABOUR)
RODO KAGAKU KENKYUSHO (THE INSTITUTE FOR SCIENCE OF LABOUR)
1544, SUGAO, TAKATSU-KU, KAWASAKI-SHI 213
JAPANESE
INDEXED: JICST

RODO NO KAGAKU 2366
(DIGEST OF SCIENCE OF LABOUR)
RODO KAGAKU KENKYUSHO (THE INSTITUTE FOR SCIENCE OF LABOUR)
1544, SUGAO, TAKATSU-KU, KAWASAKI-SHI 213
JAPANESE
INDEXED: JICST

SAGA-KEN EISEI KENKYUSHOHO 2367
(ANNUAL REPORT OF THE SAGA PREFECTURAL INSTITUTE OF PUBLIC HEALTH)
SAGA-KEN EISEI KENKYUSHO
12-13, 1-CHOME, MIZUGAE, SAGA-SHI 840
JAPANESE

SAIGAI IGAKU 2368
(JAPANESE JOURNAL OF ACCIDENTAL MEDICINE)
KANEHARA SHUPPAN K. K.
31-14, 2-CHOME, YUSHIMA, BUNKYO-KU, TOKYO 113
JAPANESE

SAIKIN NO OMONA JINKO TOKEI 2369
(LATEST DEMOGRAPHIC STATISTICS)
KOSEI-SHO JINKO MONDAI KENKYUSHO (INSTITUTE OF POPULATION PROBLEMS, MINISTRY OF HEALTH AND WELFARE)
2-2, 1-CHOME, KASUMIGASEKI, CHIYODA-KU, TOKYO 100
JAPANESE

SAITAMA KOGAI MANSURI 2370
(SAITAMA ENVIRONMENTAL POLLUTION MONTHLY)
SAITAMA-KEN KANKYO-BU (ENVIRONMENTAL POLLUTION DIVISION, SAITAMA PREFECTURE)
3-15, TAKASAGO, URAWA-SHI 336
JAPANESE

SAKAI-SHI TAIKI OSEN CHOSA HOKOKU 2371
(REPORT OF AIR POLLUTION SURVEY IN SAKAI CITY)
SAKAI-SHI KOGAI TAISAKU-BU (PUBLIC NUISANCE DIVISION, SAKAI MUNICIPAL GOVERNMENT)
3-1, MINAMIKAWARA-MACHI, SAKAI-SHI 590
JAPANESE

SANGYO HAIKIBUTSU JITTAI CHOSA HOKOKUSHO 2372
(GUNMA-KEN)
(SURVEY REPORT ON INDUSTRIAL WASTE, GUNMA PREFECTURE)
GUNMA-KEN (GUNMA PREFECTURAL GOVERNMENT)
1-1, 1-CHOME, OTE-MACHI, MAEBASHI-SHI 371
JAPANESE

SANGYO HAIKIBUTSU JITTAI CHOSA KEKKA HOKOKUSHO 2373
(OSAKA-SHI)
(SURVEY REPORT ON INDUSTRIAL WASTE, OSAKA CITY))
OSAKA-SHI KANKYO HOKEN-KYOKU (OSAKA MUNICIPAL GOVERNMENT)
1-4, NAKANOSHIMA, KITA-KU, OSAKA-SHI 530
JAPANESE

SANGYO HAISUI TAISAKU SEMINA 2374
(SEMINAR ON COUNTER MEASURES TO INDUSTRIAL DRAINAGE)
JAPANESE
INDEXED: JICST

SANGYO KOGAI 2375
(PUBLIC NUISANCE)
SANGYO KOGAI BOSHI KYOKAI (PUBLIC NUISANCE CONTROL ASSOCIATION)
4-2, 2-CHOME, NAGATA-CHO, CHIYODA-KU, TOKYO 100
JAPANESE
INDEXED: APT, JICST

SANGYO KOGAI BOSHI GIJUTSU 2376
(INDUSTRIAL POLLUTION CONTROL TECHNOLOGY)
TSUSHO SANGYO-SHO KOGYO GIJUTSU-IN (AGENCY OF INDUSTRIAL SCIENCE AND TECHNOLGOY, MINISTRY OF INTERNATIONAL TRADE AND INDUSTRY)
3-1, 1-CHOME, KASUMIGASEKI, CHIYODA-KU, TOKYO 100
JAPANESE

SANGYO KOGAI NYUSU 2377
(INDUSTRIAL PUBLIC NUISANCE NEWS)
SANGYO KOGAI CHOSA KENKYUSHO
3-5, 1-CHOME, KANDA JINBO-CHO, CHIYODA-KU, TOKYO 101
JAPANESE

SANGYO TO KANKYO 2378
(INDUSTRY AND ENVIRONMENT)
TSUSAN SHIRYO CHOSAKAI
TAKATO BUILDING, 3-3, KANDA KAJI-CHO, CHIYODA-KU, TOKYO 101
JAPANESE

SAPPORO-SHI NO KOGAI NO GENKYO TO TAISAKU 2379
(ENVIRONMENTAL POLLUTION, IT'S PRESENT CONDITION AND COUNTERPLAN, SAPPORO CITY)
SAPPORO-SHI KANKYU-KYOKU KOGAI-BU (POLLUTION CONTROL DIVISION, ENVIRONMENT BUREAU, SAPPORO CITY)
NISHI-2-CHOME, KITA-1-JO, CHUO-KU, SAPPORO-SHI 060
JAPANESE

SEIKATSU EISEI 2380
(JOURNAL OF URBAN LIVING AND HEALTH ASSOCIATION)
OSAKA SEIKATSU EISEI KYOKAI (URBAN LIVING AND HEALTH ASSOCIATION)
C/O OSAKA-SHIRITSU EISEI KENKYUSHO, 38, KITAOGI-MACHI KITA-KU, OSAKA-SHI 530
JAPANESE
INDEXED: JICST

SEIKATSU TO KANKYO 2381
(LIFE AND ENVIRONMENT)
NIHON KANKYO EISEI SENTA (JAPAN ENVIRONMENTAL SANITATION CENTER)
C/O NIHON KANKYO EISEI SENTA, 198-3, YOTSUYA KAMI-CHO, KAWASAKI-KU, KAWASAKI-SHI 210
JAPANESE
INDEXED: APT, JICST

SENDAI-SHI EISEI SHIKENSHOHO 2382
(REPORT OF THE SENDAI MUNICIPAL INSTITUTE OF PUBLIC HEALTH)
SENDAI-SHI EISEI SHIKENSHO
59-7, HIGASHI-KYUBAN-CHO, SENDAI-SHI 980
JAPANESE AND ENGLISH

SHIGA-KENRITSU EISEI KENKYU SHOHO 2383
(REPORT OF SHIGA (PREFECTURE) HEALTH RESEARCH INSTITUTE)
SHIGA (PREFECTURE) HEALTH RESEARCH INSTITUE
JAPANESE

SHIGA-KENRITSU EISEI KOGAI KENKYUSHOHO 2384
(REPORT OF THE SHIGA PREFECTURAL INSTITUTE OF PUBLIC HEALTH)
SHIGA-KENRITSU EISEI KOGAI KENKYUSHO (SHIGA PREFECTURAL INSTITUTE OF PUBLIC HEALTH)
13-45, GOTENHAMA, OTSU-SHI 520
JAPANESE
INDEXED: CHE

SHIKOKU KOSHU EISEI GAKKAI ZASSHI 2385
(JOURNAL OF SHIKOKU PUBLIC HEALTH SOCIETY)
TOKUSHIMA DAIGAKU IGAKUBU KOSHU EISEIGAKU KYOSHITSU (DEPARTMENT OF PUBLIC HEALTH, SCHOOL OF MEDICINE, TOKUSHIMA UNIVERSITY)
3-15, KURAMOTO-CHO, TOKUSHIMA-SHI 770
JAPANESE

SHIMANE-KEN EISEI KOGAI KENKYUSHO GYOSEKI HOKOKU 2386
(RESEARCH REPORT OF SHIMANE PREFECTURAL PUBLIC HEALTH LABORATORY)
SHIMANE-KEN EISEI KOGAI KENKYUSHO
420, TAIRIN-CHO, MATSUE-SHI 690
JAPANESE

SHIMANE-KEN EISEI KOGAI KENKYUSHO NENPO 2387
(ANNUAL REPORT OF SHIMAME PREFECTURAL PUBLIC HEALTH LABORATORY)
SHIMANE-KEN EISEI KOGAI KENKYUSHO
420, TAIRIN-CHO, MATSUE-SHI 690
JAPANESE

SHIN TAIIKU 2388
(PHYSICAL TRAINING)
SHIN TAIIKUSHA
18-4, 2-CHOME, KOHINATA, BUNKYO-KU, TOKYO 112
JAPANESE

SHINAGAWA-KU NO KOGAI 2389
(ENVIRONMENTAL POLLUTION IN SHINAGAWA WARD)
TOKYO-TO SHINAGAWA-KU (SHINAGAWA WARD, TOKYO)
1-36, 2-CHOME, HIRO-MACHI, SHINAGAWA-KU, TOKYO 140
JAPANESE

SHIZUOKA-KEN EISEI KENKYUSHO NENPO 2390
(ANNUAL REPORT OF SHIZUOKA PUBLIC HEALTH LABORATORY)
SHIZUOKA-KEN EISEI KENKYUSHO
6-2, 3-CHOME, TAKASHO, SHIZUOKA-SHI 420
JAPANESE

SHIZUOKA-KEN KOGAI BOSHI SENTA NENPO 2391
(ANNUAL REPORT OF ENVIRONMENTAL POLLUTION CONTROL AND RESEARCH CENTER OF SHIZUOKA PREFECTURE)
SHIZUOKA-KEN KOGAI BOSHI SENTA (ENVIRONMENTAL POLLUTION CONTROL & RESEARCH CENTER OF SHIZUOKA PREFECTURE)
6-2, 3-CHOME, TAKASHO, SHIZUOKA-SHI 420
JAPANESE

SHOKUCHUDOKU TOKEIHYO 2392
(STATISTICS ON FOOD POISONING)
TOKYO-TO EISEI-KYOKU KANKYO EISEI-BU (BUREAU OF PUBLIC HEALTH, TOKYO)
5-1, 3-CHOME, MARUNOUCHI, CHIYODA-KU, TOKYO 100
JAPANESE

SHOKUHIN EISEI 2393
(FOOD HYGIENE)
NIHON SHOKUHIN EISEI KYOKAI (JAPAN FOOD HYGIENE ASSOCIATION)
6-1, 2-CHOME, JINGUMAE, SHIBUYA-KU, TOKYO 150
JAPANESE

SHOKUHIN EISEI KANKEI JIGYO HOKOKU 2394
(RESEARCH REPORT ON FOOD SANITATION)
TOKYO-TO EISEI-KYOKU KANKYO EISEI-BU (BUREAU OF PUBLIC HEALTH, TOKYO)
5-1, 3-CHOME, MARUNOUCHI, CHIYODA-KU, TOKYO 100
JAPANESE

SHOKUHIN EISEI KENKYU 2395
(FOOD HYGIENE STUDY)
KOSEI-SHO SHOKUHIN EISEI-KA NYUMIKU EISEI-KA SHOKUHIN KAGAKU-KA (FOOD HYGIENE SECTION, ENVIRONMENTAL SANITATION BUREAU, MINISTRY OF HEALTH AND WELFARE)
JAPANESE
INDEXED: JICST

SHOKUHIN EISEIGAKU ZASSHI 2396
(JOURNAL OF THE FOOD HYGIENIC SOCIETY OF JAPAN)
NIHON SHOKUHIN EISEI GAKKAI (JAPAN FOOD HYGIENIC SOCIETY)
C/O KOKURITSU EISEI SHIKENJO, 18-1, 1-CHOME, KAMIYOGA, SETAGAYA-KU, TOKYO 158
JAPANESE AND ENGLISH
INDEXED: BIB, CHE, EXC, JICST

SHOKUIN KOSEI JIHO 2397
(REVIEW OF PERSONNEL WELFARE)
YOKOHAMA-SHI SOMUKYOKU ROMUKA, YOKOHAMA-SHI KENKO HOKEN KUMIAI, YOKOHAMA-SHI SHOKUIN KOSEI KAI (YOKOHAMA MUNICIPAL GENERAL AFFAIRS BUREAU, LABOR SECTION; YOKOHAMA MUNICIPAL HEALTH INSURANCE LEAGUE)
JAPANESE

SHONI HOKEN SHIRIZU 2398
(CHILD HEALTH SERIES)
NIHON SHONI HOKEN KYOKAI (THE JAPANESE SOCIETY OF CHILD HEALTH, TOKYO)
C/O HOKEN KAIKAN, 1-2, SADOHARACHO, ICHIGAYA, SHINJUKU-KU, TOKYO 162
JAPANESE

SODAI, KYOIKU, TAIIKU, REPOTO 2399
(WASEDA UNIVERSITY SCIENCE REPORT ON EDUCATION AND PHYSICAL EDUCATION)
WASEDA DAIGAKU KYOIKU GAKUBU KYOIKUGAKKA TAIIKU SENSHU TAIIKUGAKU JIKKENSHITSU
1-647, TOTSUKA SHINJUKU-KU, TOKYO 160
JAPANESE

SODAN'IN NO AYUMI 2400
(PREVENTION OF NARCOTIC POISOINING)
KOSEI-SHO YAKUMU-KYOKU (PHARMACEUTICAL AND SUPPLY BUREAU, MINISTRY OF HEALTH AND WELFARE)
2-2, 1-CHOME, KASUMIGASEKI, CHIYODA-KU, TOKYO 100
JAPANESE

SOKEN SEIKATSU 2401
(HEALTHY LIFE)
SOKEN KENKYUSHA
35-8, 1-CHOME, MATSUGAOKA, NAKANO-KU, TOKYO 165
JAPANESE

SOON SHINDO TAISAKU 2402
(COUNTERPLAN FOR THE PREVENTION OF NOISE AND OSCILLATION)
KANAGAWA-KEN KOGAI SENTA (KANAGAWA PREFECTURAL ENVIRONMENTAL CENTER)
87-1, 1-CHOME, FUTAMATAGAWA, ASAHI-KU, YOKOHAMA-SHI 241
JAPANESE

SUGINAMI-KU KOGAI HOKOKUSHO 2403
(SURVEY REPORT ON ENVIRONMENTAL POLLUTION IN SUGINAMI WARD, TOKYO)
TOKYO-TO SUGINAMI-KU KANKYO-BU (SUGINAMI WARD, TOKYO)
15-1, 1-CHOME, ASAGAYAMINAMI, SUGINAMI-KU, TOKYO 166
JAPANESE

SUMITOMO SANGYO EISEI 2404
(SUMITOMO BULLETIN OF INDUSTRIAL HEALTH)
SUMITOMO BYOIN SANGYO EISEI KENKYUSHITSU (JURIDICAL FOUNDATION SUMITOMO HOSPITAL, INSTITUTE OF INDUSTRIAL HEALTH)
5-15, NAKANOSHIMA, KITA-KU, OSAKA-SHI 530
JAPANESE AND ENGLISH
INDEXED: BIO, EXC

TAIIKU KENKYU 2405
(PHYSICAL EDUCATION STUDY)
NIHON TAIIKU GAKKAI KANAGAWASHIBU (THE JAPANESE SOCIETY OF PHYSICAL EDUCATION)
C/O YOKOHAMA KOKURITSU DAIGAKU KYOIKUGAKUBU, HOKEN TAIIKUGAKU KYOSHITSU, 41, SHIMIZUGAOKA, MINAMIKU, YOKOHAMA-SHI 232
JAPANESE

TAIIKU KENKYU 2406
(RESEARCH IN PHYSICAL EDUCATION)
CHUO DAIGAKU HOKEN TAIIKU IINKAI (CHUO UNIVERSITY) COMMITTEE ON HEALTH AND PHYSICAL EDUCATION)
JAPANESE

TAIIKU KENKYUSHO KIYO 2407
(BULLETIN OF THE INSTITUTE OF PHYSICAL EDUCATION, KEIO UNIVERSITY)
KEIO GIJUKU DAIGAKU TAIIKU KENKYUSHO (INSTITUTE OF PHYSICAL EDUCATION, KEIO UNIVERSITY)
665, HIYOSHI-MACHI, KOHOKU-KU, YOKOHAMA-SHI 223
JAPANESE AND ENGLISH

TAIIKU NO KAGAKU 2408
(JOURNAL OF HEALTH, PHYSICAL EDUCATION AND RECREATION)
NIHON TAIIKU GAKKAI (THE JAPANESE SOCIETY OF PHYSICAL EDUCATION)
2-1, 4-CHOME, YUSHIMA, BUNKYO-KU, TOKYO 113
JAPANESE

TAIIKU. SUPOTSU REKURIESHON 2409
(PHYSICAL EDUCATION, SPORTS, AND RECREATION)
ZENKOKU DAIGAKU TAIIKU RENJO KIKANSHI HENSHU IINKAI (JOURNAL EDITING COMMITTEE OF THE NATIONAL UNIVERSITIES' PHYSICAL EDUCATION FEDERATION)
JAPANESE

TAIIKUGAKU KENKYU 2410
(RESEARCH JOURNAL OF PHYSICAL EDUCATION)
NIHON TAIIKU GAKKAI (THE JAPANESE SOCIETY OF PHYSICAL EDUCATION)
C/O TOKYO DAIGAKU KYOIKUGAKUBU TAIIKUGAKU KENKYUSHITSU, 3-1, 7-CHOME, HONGO, BUNKYO-KU TOKYO 113
JAPANESE, WITH ENGLISH SUMMARY

TAIIKUGAKU KENKYU SHUROKU 2411
(COLLECTED RESEARCH ON PHYSICAL EDUCATION)
NIHON TAIIKU GAKKAI HYOGO SHIBU (JAPAN PHYSICAL EDUCATION SOCIETY, HYOGO BRANCH)
JAPANESE

TAIIKUGAKU KIYO 2412
(PROCEEDINGS OF THE DEPARTMENT OF PHYSICAL EDUCATION, COLLEGE OF GENERAL EDUCATION, UNIVERISTY OF TOKYO)
8-1, 3-CHOME, KOMABA, MEGURO-KU, TOKYO 153
JAPANESE AND ENGLISH

TAIKI OSEN CHOSA HOKOKUSHO (CHIBA-KEN) 2413
(SURVEY REPORT ON AIR POLLUTION, CHIBA PREFECTURE)
CHIBA-KEN EISEI-BU (CHIBA PREFECTURAL GOVERNMENT)
1-1, ICHIBA-CHO, CHIBA-SHI 280-91
JAPANESE

TAIKI OSEN CHOSA HOKOKUSHO (SAITAMA-KEN) 2414
(SURVEY REPORT ON AIR POLLUTION, SAITAMA PREFECTURE)
SAITAMA-KEN KENMIN SEIKATSU-BU (SAITAMA PREFECTURAL GOVERNMENT)
15-1, 3-CHOME, TAKASAGO, URAWA-SHI 336
JAPANESE

TAIKI OSEN KENKYU 2415
(JOURNAL OF JAPAN SOCIETY OF AIR POLLUTION)
TAIKI OSEN KENKYU ZENKOKU KYOGIKAI (JAPAN SOCIETY OF AIR POLLUTION)
6-1, 4-CHOME, SHIROGANEDAI, MINATO-KU, TOKYO 108
JAPANESE, WITH ENGLISH SUMMARY
INDEXED: BIB, CHE, JICST

TAIKI OSEN NI KANSURU KENKYU CHOSA 2416
(INVESTIGATIONS ON AIR POLLUTION)
SENDAI-SHI EISEI SHIKENJO (SENDAI MUNICIPAL INSTITUTE OF PUBLIC HEALTH)
59-7, HIGASHIKYUBAN-CHO, SENDAI-SHI 980
JAPANESE

TAIKI OSEN NYUSU 2417
(AIR POLLUTION NEWS)
TAIKI OSEN KENKYU ZENKOKU KYOGIKAI (JAPAN SOCIETY OF AIR POLLUTION)
6-1, 4-CHOME, SHIROGANEDAI, MINATO-KU, TOKYO 108
JAPANESE
INDEXED: CHE

TAIKI OSEN OYOBI SUISHITSU ODAKU NO JINTAI NI ATAERU EIKYO CHOSA KENKYU 2418
(RESEARCH REPORT ON THE INFLUENCE UPON HUMAN BODY BROUGHT BY AIR AND WATER POLLUTION)
CHIBA-KEN EISEI-BU (PUBLIC HEALTH DIVISION, CHIBA PREFECTURAL GOVERNMENT)
1-1, ICHIBA-CHO, CHIBA-SHI 280-91
JAPANESE

TAIKI OSEN SOKUTEI KEKKA HOKOKUSHO 2419
(RESULT OF AIR POLLUTION MEASUREMENT)
SAITAMA-KEN KOGAI SENTA (POLLUTION CONTROL CENTER OF SAITAMA PREFECTURE)
639-1, KAMI-OKUBO-HIGASHI, URAWA-SHI 338
JAPANESE

TAIKI, SOON CHOSA HOKOKUSHO (KUMAMOTO-KEN) 2420
(SURVEY REPORT ON AIR AND NOISE, KUMAMOTO PREFECTURE)
KUMAMOTO-KEN KIKAKU-BU
18-1, 6-CHOME, SUIZENJI, KUMAMOTO-SHI 862
JAPANESE

TAIRYOKU KAGAKU 2421
(JAPANESE JOURNAL OF PHYSICAL FITNESS)
NIHON TAIRYOKU IGAKKAI (JAPANESE PHYSICAL FITNESS SOCIETY)
C/O TOKYO JIKEIKAI IKA DAIGAKU DAI-1 SEIRIGAKU KYOSHITSU, 25-8, 3-CHOME, NISHI-SHINBASHI, MINATO-KU TOKYO 105
JAPANESE

TAIRYOKU KENKYU (TAIRYOKU IGAKU KENKYU HOKOKU) 2422
(BULLETIN OF THE PHYSICAL FITNESS RESEARCH INSTITUTE)
MEIJI SEIMEI KOSEI JIGYODAN (MEIJI LIFE FOUNDATION OF HEALTH AND WELFARE)
1-18, 1-CHOME, SHIROGANEDAI, MINATO-KU, TOKYO 108
JAPANESE AND ENGLISH
INDEXED: EXC

TAIRYOKU UNDO NORYOKU CHOSA HOKOKUSHO 2423
(RESEARCH REPORT OF PHYSICAL STRENGTH AND MOTOR ABILITY)
MOMBU-SHO TAIIKU-KYOKU (PHYSICAL EDUCATION BUREAU, MINSTRY OF EDUCATION)
2-2, 3-CHOME, KASUMIGASEKI, CHIYODA-KU, TOKYO 100
JAPANESE

TEKKO RODO EISEI 2424
(JOURNAL OF LABOUR HYGIENE IN IRON AND STEEL INDUSTRY)
NIHON TEKKO RENMEI EISEI IINKAI (COMMITTEE ON INDUSTRIAL HEALTH OF JAPAN IRON AND STEEL FEDERATION)
NIHON TEKKO RENMEI (JAPAN IRON AND STEEL FEDERATION)
C/O KEIDANREN KAIKAN, 9-4, 1-CHOME, OTE-MACHI, CHIYODA-KU, TOKYO 100
JAPANESE, WITH ENGLISH SUMMARY
INDEXED: CHE, JICST

TENRI DAIGAKU GAKUHO; TAIIKU-HEN 2425
(BULLETIN OF TENRI UNIVERSITY; PHYSICAL EDUCATION)
TENRI DAIGAKU JINBUN GAKKAI
JAPANESE

TETSUDO RODO KAGAKU 2426
(BULLETIN OF RAILWAY LABOUR SCIENCE RESEARCH INSTITUTE)
NIHON KOKUYU TETSUDO TETSUDO RODO KAGAKU KENKYUSHO (RAILWAY LABOUR SCIENCE RESEARCH INSTITUTE)
5-6, 2-CHOME, IZUMI-MACHI, KOKUBUNJI-SHI, TOKYO 185
JAPANESE
INDEXED: JICST

TOCHIGI-KEN EISEI KENKYUSHO GYOSEKI HOKOKU 2427
(REPORT OF THE TOCHIGI PREFECTURAL INSTITUTE OF HEALTH)
TOCHIGI-KEN EISEI KENKYUSHO (TOCHIGI PREFECTURAL INSTITUTE OF HEALTH)
927-47, 2-CHOME, MIYAHARA-CHO, UTSUNOMIYA-SHI 320
JAPANESE

TOKAI-SHI NO KOGAI GAIKYO 2428
(GENERAL SURVEY ON POLLUTION, TOKAI CITY)
TOKAI-SHI KIKAKU-BU
10-4, MINAMINOYAMA, NAWA-MACHI, TOKAI-SHI 476
JAPANESE

TOKEI KARA MITA AICHI-KEN NO EISEI SHIHYO 2429
(PUBLIC HEALTH INDICES FROM STATISTICAL DATA)
AICHI-KEN EISEI-BU (PUBLIC HEALTH DEPARTMENT, AICHI PREFECTURAL GOVERNMENT)
1-2, 3-CHOME, SANNOMARU, NAKA-KU, NAGOYA-SHI 460
JAPANESE

TOKUSHIMA DAIGAKU KYOYOBU KIYO, HOKEN TAIIKU 2430
(JOURNAL OF HEALTH AND PHYSICAL EDUCATION, COLLEGE OF GENERAL EDUCATION, UNIVERSITY OF TOKUSHIMA)
TOKUSHIMA DAIGAKU KYOYO-BU (COLLEGE OF GENERAL EDUCATION, UNIVERSITY OF TOKUSHIMA)
1, MINAMIJO SANJIMA-CHO, TOKUSHIMA-SHI 770
JAPANESE

TOKUSHIMA-KEN EISEI KENKYUSHO NENPO 2431
(ANNUAL REPORT OF THE TOKUSHIMA PREFECTURE INSTITUTE OF PUBLIC HEALTH)
TOKUSHIMA-KEN EISEI KENKYUSHO (TOKUSHIMA PREFECTURE INSTITUTE OF PUBLIC HEALTH)
3, SHINKURA-CHO, TOKUSHIMA-SHI 770
JAPANESE
INDEXED: JICST

TOKYO KYOIKU DAIGAKU TAIIKU GAKUBU KIYO 2432
(BULLETIN OF THE FACULTY OF PHYSICAL EDUCATION, TOKYO UNIVERSITY OF EDUCATION)
TOKYO KYOIKU DAIGAKU TAIIKU GAKUBU (THE FACULTY OF PHYSICAL EDUCATION, TOKYO UNIVERSITY OF EDUCATION)
40-1, 1-CHOME, NISHIHARA, SHIBUYA-KU, TOKYO 151
JAPANESE, WITH EUROPEAN SUMMARY

TOKYO-TO AKASAKA HOKENJO JIGYO GAIYO 2433
(REPORT OF AKASAKA HEALTH CENTER, TOKYO)
TOKYO-TO AKASAKA HOKENJO
5-15, 1-CHOME, MINAMIAOYAMA, MINATO-KU, TOKYO 160
JAPANESE

TOKYO-TO ASAKUSA HOKENJO JIGYO GAIYO 2434
(REPORT OF ASAKUSA HEALTH CENTER, TOKYO)
TOKYO-TO ASAKUSA HOKENJO
14-16, 1-CHOME, HANAKAWADO, TAITO-KU, TOKYO 100
JAPANESE

TOKYO-TO EBARA HOKENJO JIGYO GAIYO 2435
(REPORT OF EBARA HEALTH CENTER, TOKYO)
TOKYO-TO EBARA HOKENJO
2-9, 1-CHOME, NISHI-NAKANOBU, SHINAGAWA-KU TOKYO 142
JAPANESE

TOKYO-TO EDOGAWA HOKENJO JIGYO GAIYO 2436
(REPORT OF EDOGAWA HEALTH CENTER, TOKYO)
TOKYO-TO EDOGAWA HOKENJO
24-19, 4-CHOME, CHUO, EDOGAWA-KU, TOKYO 132
JAPANESE

TOKYO-TO EISEI-KYOKU GAKKAISHI 2437
(REPORTS OF TOKYO PUBLIC HEALTH STUDIES)
TOKYO-TO EISEI-KYOKU SOMU-BU TOKEICHOSA-KA
(GENERAL AFFAIRS, DIVISION, BUREAU OF PUBLIC HEALTH, TOKYO METROPOLITAN GOVERNMENT)
5-1, 3-CHOME, MARUNOUCHI, CHIYODA-KU, TOKYO 100
JAPANESE

TOKYO-TO EISEI-KYOKU GEPPO 2438
(MONTHLY REPORT OF THE BUREAU OF PUBLIC HEALTH, TOKYO METROPOLITAN GOVERNMENT)
TOKYO-TO EISEI-KYOKU SOMU-BU (GENERAL AFFAIRS DIVISION, BUREAU OF PUBLIC HEALTH, TOKYO METROPOLITAN GOVERNMENT)
5-1, 3-CHOME, MARUNOUCHI, CHIYODA-KU, TOKYO 100
JAPANESE

TOKYO-TO EISEI-KYOKU JIGYO GAIYO 2439
(OUTLINE PROJECT OF BUREAU OF PUBLIC HEALTH, TOKYO METROPOLITAN GOVERNMENT)
TOKYO-TO EISEI-KYOKU SOMU-BU (GENERAL AFFAIRS DIVISION, BUREAU OF PUBLIC HEALTH, TOKYO METROPOLITAN GOVERNMENT)
5-1, 3-CHOME, MARUNOUCHI, CHIYODA-KU, TOKYO 100
JAPANESE

TOKYO-TO FUCHU HOKENJO JIGYO GAIYO 2440
(REPORT OF FUCHU HEALTH CENTER, TOKYO)
TOKYO-TO FUCHU HOKENJO
8876, BABA-HIGASHI, FUCHU-SHI, TOKYO 183
JAPANESE

TOKYO-TO FUKAGAWA HOKENJO JIGYO GAIYO 2441
(REPORT OF FUKAGAWA HEALTH CENTER, TOKYO)
TOKYO-TO FUKAGAWA HOKENJO
9-7, 3-CHOME, SHIRAKAWA, KOTO-KU, TOKYO 135
JAPANESE

TOKYO-TO HACHIOJI HOKENJO JIGYO GAIYO 2442
(REPORT OF HACHIOJI HEALTH CENTER, TOKYO)
TOKYO-TO HACHIOJI HOKENJO
13-18, ASAHI-CHO, HACHIOJI-SHI, TOKYO 192
JAPANESE

TOKYO-TO HIMON'YA HOKENJO JIGYO GAIYO 2443
(REPORT OF HIMON'YA HEALTH CENTER, TOKYO)
TOKYO-TO HIMON'YA HOKENJO
15-3, 5-CHOME, HIMON'YA, MEGURO-KU, TOKYO 152
JAPANESE

TOKYO-TO HONGO HOKENJO JIGYO GAIYO 2444
(REPORT OF HONGO HEALTH CENTER, TOKYO)
TOKYO-TO HONGO HOKENJO
20-18, 5-CHOME, SENDAGI, BUNKYO-KU, TOKYO 113
JAPANESE

TOKYO-TO ITABASHI HIGASHI HOKENJO HOKENJO DAYORI 2445
(NEWS OF ITABASHI HIGASHI HEALTH CENTER, TOKYO)
TOKYO-TO ITABASHI HIGASHI HOKENJO
38-2, OYAMA-CHO, ITABASHI-KU, TOKYO 173
JAPANESE

TOKYO-TO ITABASHI-HIGASHI HOKENJO JIGYO GAIYO 2446
(REPORT OF ITABASHI-HIGASHI HEALTH CENTER, TOKYO)
TOKYO-TO ITABASHI, HIGASHI HOKENJO
8-2, OYAMA-CHO, ITABASHI-KU, TOKYO 173
JAPANESE

TOKYO-TO ITSUKAICHI HOKENJO JIGYO GAIYO 2447
(REPORT OF ITSUKAICHI HEALTH CENTER, TOKYO)
TOKYO-TO ITSUKAICHI HOKENJO
978, ITSUKAICHI, ITSUKAICHI-MACHI, NISHITAMA-GUN, TOKYO 190-01
JAPANESE

TOKYO-TO KANDA HOKENJO JIGYO GAIYO 2448
(REPORT OF KANDA HEALTH CENTER, TOKYO)
TOKYO-TO KANDA HOKENJO
3-10, KANDA NISHIKI-CHO, CHIYODA-KU, TOKYO 101
JAPANESE

TOKYO-TO KATSUSHIKA-KITA HOKENJO JIGYO GAIYO 2449
(REPORT OF KATSUSHIKA-KITA HEALTH CENTER, TOKYO)
TOKYO-TO KATSUSHIKA-KITA HOKENJO
18-9, 4-CHOME, KANAMACHI, KATSUSHIKA-KU, TOKYO 125
JAPANESE

TOKYO-TO KOGAI KENKYU SHIRYO 2450
(RESEARCH MATERIALS ON ENVIRONMENTAL PROTECTION, TOKYO)
TOKYO-TO KOGAI KENKYUSHO (THE TOKYO METROPOLITAN RESEARCH INSTITUTE FOR ENVIRONMENTAL PROTECTION)
2-5, YURAKU-CHO, CHIYODA-KU, TOKYO 100
JAPANESE
INDEXED: JICST

TOKYO-TO KOGAI KENKYUSHO NENPO 2451
(ANNUAL REPORT OF THE TOKYO METROPOLITAN RESEARCH INSTITUTE FOR ENVIRONMENTAL PROTECTION)
TOKYO-TO KOGAI KENKYUSHO (TOKYO METROPOLITAN RESEARCH INSTITUTE FOR ENVIRONMENTAL PROTECTION)
2-5, YURAKU-CHO, CHIYODA-KU, TOKYO 100
JAPANESE, WITH ENGLISH SUMMARY
INDEXED: JICST

TOKYO-TO KOGAI-KYOKU JIGYO GAIYO 2452
(REPORT OF BUREAU OF ENVIRONMENTAL PROTECTION, TOKYO)
TOKYO-TO KOGAI-KYOKU
5-1, 3-CHOME, MARUNOUCHI, CHIYODA-KU, TOKYO 100
JAPANESE

TOKYO-TO KOISHIKAWA HOKENJO JIGYO GAIYO 2453
(REPORT OF KOISHIKAWA HEALTH CENTER, TOKYO)
TOKYO-TO KOISHIKAWA HOKENJO
9-21, 1-CHOME, KASUGA, BUNKYO-KU, TOKYO 112
JAPANESE

TOKYO-TO MEGURO HOKENJO JIGYO GAIYO 2454
(REPORT OF MEGURO HEALTH CENTER, TOKYO)
TOKYO-TO MEGURO HOKENJO
13-22, 3-CHOME, NAKA-MEGURO, MEGURO-KU, TOKYO 153
JAPANESE

TOKYO-TO MUSASHI CHOFU HOKENJO JIGYO GAIYO 2455
(REPORT OF MUSASHI CHOFU HEALTH CENTER, TOKYO)
TOKYO-TO MUSASHI CHOFU HOKENJO
46-1, 5-CHOME, FUDA, CHOFU-SHI, TOKYO 183
JAPANESE

TOKYO-TO NIHONBASHI HOKENJO JIGYO GAIYO 2456
(REPORT OF NIHONBASHI HEALTH CENTER, TOKYO)
TOKYO-TO NIHONBASHI HOKENJO
1-2, NIHONBASHI HORIDOME-CHO, CHUO-KU, TOKYO 103
JAPANESE

TOKYO-TO SEISO-KYOKU GEPPO 2457
(MONTHLY REPORT OF THE BUREAU OF PUBLIC CLEANSING)
TOKYO-TO SEISO-KYOKU KIKAKU-BU (DIVISION OF PLANNING, BUREAU OF PUBLIC CLEANSING, TOKYO)
8-1, 3-CHOME, MARUNOUCHI, CHIYODA-KU, TOKYO 100
JAPANESE

TOKYO-TO SEISO-KYOKU JIGYO GAIYO 2458
(REPORT OF BUREAU OF PUBLIC CLEANSING, TOKYO)
TOKYO-TO SEISO-KYOKU KIKAKU-BU
8-1, 3-CHOME, MARUNOUCHI, CHIYODA-KU, TOKYO 100
JAPANESE

TOKYO-TO SHIMURA HOKENJO JIGYO GAIYO 2459
(REPORT OF SHIMURA HEALTH CENTER, TOKYO)
TOKYO-TO SHIMURA HOKENJO
8-1, 2-CHOME, SAKASHITA, ITABASHI-KU, TOKYO 174
JAPANESE

TOKYO-TO SHINJUKU HOKENJO JIGYO GAIYO 2460
(REPORT OF SHINJUKU HEALTH CENTER, TOKYO)
TOKYO-TO SHINJUKU HOKENJO
5-8, 7-CHOME, NISHI-SHINJUKU, SHINJUKU-KU, TOKYO 160
JAPANESE

TOKYO-TO SHITAYA HOKENJO JIGYO GAIYO 2461
(REPORT OF SHITAYA HEALTH CENTER, TOKYO)
TOKYO-TO SHITAYA HOKENJO
2-11, 1-CHOME, SHITAYA, TAITO-KU, TOKYO 110
JAPANESE

TOKYO-TO SHODOKUSHO JIGYO GAIYO 2462
(REPORT OF DISINFECTING CENTER, TOKYO)
TOKYO-TO SHODOKUSHO
8-5, 1-CHOME, EDAKAWA, KOTO-KU, TOKYO 135
JAPANESE

TOKYO-TO TAMAGAWA HOKENJO JIGYO GAIYO 2463
(REPORT OF TAMAGAWA HEALTH CENTER, TOKYO)
TOKYO-TO TAMAGAWA HOKENJO
19-18, 4-CHOME, TODOROKI, SETAGAYA-KU, TOKYO 158
JAPANESE

TOKYO-TO USHIGOME HOKENJO JIGYO GAIYO 2464
(REPORT OF USHIGOME HEALTH CENTER, TOKYO)
TOKYO-TO USHIGOME HOKENJO
50, BENTEN-CHO, SHINJUKU-KU, TOKYO 123
JAPANESE

TOKYO-TO YOTSUYA HOKENJO JIGYO GAIYO 2465
(REPORT OF YOTSUYA HEALTH CENTER, TOKYO)
TOKYO-TO YOTSUYA HOKENJO
4-17, YOTSUYA, SHINJUKU-KU, TOKYO 160
JAPANESE

TOKYO-TO YUKIGAYA HOKENJO JIGYO GAIYO 2466
(REPORT OF YUKIGAYA HEALTH CENTER, TOKYO)
TOKYO-TO YUKIGAYA HOKENSHO
1-1, YUKIGAYA OTSUKA-MACHI, OTA-KU, TOKYO 140
JAPANESE

TOKYO-TORITSU BOSHI HOKENIN JIGYO GAIYO 2467
(REPORT OF TOKYO METROPOLITAN MOTHER AND CHILD HEALTH CENTER)
TOKYO-TORITSU BOSHI HOKENIN
27-33, 1-CHOME, SAKURA, SETAGAYA-KU, TOKYO 156
JAPANESE

TOKYO-TORITSU EISEI KENKYUSHO JIGYO GAIYO 2468
(REPORT OF TOKYO METROPOLITAN RESEARCH LABORATORY OF PUBLIC HEALTH)
TOKYO-TORITSU EISEI KENKYUSHO
24-1, 3-CHOME, HYAKUNIN-CHO, SHINJUKU-KU, TOKYO 160
JAPANESE

TOKYO-TORITSU EISEI KENKYUSHO KENKYU NENPO 2469
(ANNUAL REPORT OF TOKYO METROPOLITAN RESEARCH LABORATORY OF PUBLIC HEALTH)
TOKYO-TORITSU EISEI KENKYUSHO (TOKYO METROPOLITAN RESEARCH LABORATORY OF PUBLIC HEALTH)
24-1, 3-CHOME, HYAKUNIN-CHO, SHINJUKU-KU, TOKYO 160
JAPANESE AND ENGLISH
INDEXED: CHE, JICST

TOMAKOMAI-SHI NI OKERU KOGAI CHOSA NO GENKYO 2470
(SURVEY REPORT ON ENVIRONMENTAL POLLUTION IN TOMAKOMAI CITY)
TOMAKOMAI-SHI SHIMIN-BU (TOMAKOMAI MUNICIPAL GOVERNMENT)
5-6, 4-CHOME, ASAHI-CHO, TOMAKOMAI-SHI 053
JAPANESE

TOMIN O KOGAI KARA BOEI SURU KEIKAKU 2471
(DEFENSE PLAN FOR THE CITIZEN FROM ENVIRONMENTAL POLLUTION)
TOKYO-TO KOGAI-KYOKU (BUREAU OF ENVIRONMENTAL PROTECTION, TOKYO)
5-1, 3-CHOME, MARUNOUCHI, CHIYODA-KU, TOKYO 100
JAPANESE

TOSHI TO HAIKIBUTSU 2472
(JOURNAL OF SOLID WASTES)
KANKYO SANGYO SHINBUNSHA
11-15, 2-CHOME, SHIBADAIMON, MINATO-KU, TOKYO 105
JAPANESE
INDEXED: APT, JICST

TOTTORI-KEN EISEI KENKYUSHOHO 2473
(BULLETIN OF TOTTORI PREFECTURE PUBLIC HEALTH LABORATORY)
TOTTORI-KEN EISEI KENKYUSHO (TOTTORI PREFECTURE PUBLIC HEALTH LABORATORY)
2-470, MATSUNAMI-CHO, TOTTORI-SHI 680
JAPANESE

TOTTORI-KEN EISEI TOKEISHO 2474
(ANNUAL REPORT OF PUBLIC HEALTH IN TOTTORI PREFECTURE)
TOTTORI-KEN KOSEI-BU (PUBLIC HEALTH DIVISION, TOTTORI PREFECTURAL GOVERNMENT)
1-22, HIGASHI-MACHI, TOTTORI-SHI 680
JAPANESE

TOYAMA-KEN KOGAI SENTA HOKOKU 2475
(REPORT OF TOYAMA ENVIRONMENTAL POLLUTION CONTROL CENTER)
TOYAMA-KEN KOGAI SENTA (TOYAMA ENVIRONMENTAL POLLUTION CONTROL CENTER)
TAIKOYAMA, KOSUGI-CHO, IMIZU-GUN TOYAMA 939-03
JAPANESE

TSURUMI DAIGAKU KIYO DAI-3-BU HOKEN, HOKEN SHIKA-HEN 2476
(BULLETIN OF TSURUMI COLLEGE, PART 3, STUDIES IN HEALTH AND DENTAL HYGIENE)
TSURUMI DAIGAKU
84, TSURUMI-CHO, TSURUMI-KU, YOKOHAMA-SHI 230
JAPANESE

UBE-SHI NO KOGAI 2477
(ENVIRONMENTAL POLLUTION IN UBE CITY)
UBE-SHI SHIMIN-BU (UBE MUNICIPAL GOVERNMENT)
7-1, 1-CHOME, TOKIWA-CHO, UBE-SHI 755
JAPANESE

WAKAYAMA-KEN EISEI KENKYUSHO NENPO 2478
(ANNUAL REPORT OF WAKAYAMA PREFECTURE INSTITUTE OF PUBLIC HEALTH)
WAKAYAMA-KEN EISEI KENKYUSHO (WAKAYAMA PREFECTURE INSTITUTE OF PUBLIC HEALTH)
271-2, MINATO HIGASHINOTSUBO, WAKAYAMA-SHI 640
JAPANESE

WAKUCHIN KA KENKYU VAKZIN 2479
(VACCINE)
WAKUCHINKA KENKYU KAI
JAPANESE

WASEDA DAIGAKU KYOIKUGAKU-BU, TAIIKUGAKU JIKKENSHITSU KIYO 2480
(BULLETIN OF EXPERIMENT LABORATORY OF PHYSICAL EDUCATION, STUDIES, SCHOOL OF EDUCATION, WASEDA UNIVERSITY)
WASEDA UNIVERSITY, SCHOOL OF EDUCATION
JAPANESE

YAKURI TO CHIRYO 2481
(PHARMACOLOGY AND THERAPEUTICS)
RAIFU SAIENSU PUBLISHING COMPANY
5-3-3 YAESU, CHUO-KU, TOKYO 104
JAPANESE

YAMAGATA EIKEN NYUSU 2482
(NEWS OF YAMAGATA PREFECTURAL INSTITUTE OF PUBLIC HEALTH)
YAMAGATA-KEN EISEI KENKYUSHO
6-6, 1-CHOME, TOKA-MACHI, YAMAGATA-SHI 990
JAPANESE

YAMAGATA-KEN EISEI KENKYUSHOHO 2483
(REPORT OF YAMAGATA PREFECTURAL INSTITUTE OF PUBLIC HEALTH)
YAMAGATA-KEN EISEI KENKYUSHO
6-6, 1-CHOME, TOKA-MACHI, YAMAGATA-SHI 990
JAPANESE AND ENGLISH

YAMAGATA-KEN NO KOGAI 2484
(ENVIRONMENTAL POLLUTION IN YAMAGATA PREFECTURE)
YAMAGATA-KEN GYOSEI SOGO TAISAKUSHITSU (YAMAGATA PREFECTURAL GOVERNMENT)
4-51, 3-CHOME, HATAGO-CHO, YAMAGATA-SHI 990
JAPANESE

YAMAGUCHI-KEN EISEI KENKYUSHO GYOSEKI HOKOKU 2485
(REPORT OF YAMAGUCHI PREFECTURAL RESEARCH INSTITUTE OF HEALTH)
YAMAGUCHI-KEN EISEI KENKYUSHO (YAMAGUCHI PREFECTURAL RESEARCH INSTITUTE OF HEALTH)
5-67, 2-CHOME, AOI, YAMAGUCHI-SHI 753
JAPANESE

YAMAGUCHI-KEN EISEI KENKYUSHO NENPO 2486
(ANNUAL REPORT OF THE YAMAGUCHI PREFECTURAL RESEARCH INSTITUTE OF HEALTH)
YAMAGUCHI-KEN EISEI KENKYUSHO (YAMAGUCHI PREFECTURAL RESEARCH INSTITUTE OF HEALTH)
5-67, 2-CHOME, AOI, YAMAGUCHI-SHI 753
JAPANESE

YAMAGUCHI-KEN NO EISEI 2487
(ANNUAL REPORT OF PUBLIC HEALTH IN YAMAGUCHI PREFECTURE)
YAMAGUCHI-KEN EISEI-BU
1-1, TAKI-MACHI, YAMAGUCHI-SHI 753
JAPANESE

YAMAGUCHI-KEN SEISHIN EISEI SENTA SHOHO 2488
(REPORT OF THE YAMAGUCHI PREFECTURE CENTER FOR MENTAL HEALTH)
YAMAGUCHI PREFECTURE CENTER FOR MENTAL HEALTH
JAPANESE

YAMANASHI-KENRITSU EISEI KENKYUSHO NENPO 2489
(ANNUAL REPORT OF YAMANASHI PREFECTURAL INSTITUTE FOR PUBLIC HEALTH)
7-31, 1-CHOME, FUJIMI, KOFU-SHI 400
JAPANESE AND ENGLISH

YOKOHAMA KOKURITSU DAIGAKU KANKYO KAGAKU KENKYU SENTA KIYO 2490
(BULLETIN INSTITUTE OF ENVIRONMENTAL SCIENCE AND TECHNOLOGY. YOKOHAMA NATIONAL UNIVERSITY)
JAPANESE

YOKOHAMA-SHI EISEI KENKYUSHO NENPO 2491
(ANNUAL REPORT OF YOKOHAMA CITY INSTITUTE OF HEALTH)
YOKOHAMA-SHI EISEI KENKYUSHO (YOKOHAMA CITY INSTITUTE OF HEALTH)
2-17, 1-CHOME, TAKIGASHIRA, ISOGO-KU, YOKOHAMA-SHI 235
JAPANESE, WITH ENGLISH SUMMARY

YOKOHAMA-SHI TAIKI OSEN CHOSA HOKOKU 2492
(SURVEY REPORT OF AIR POLLUTION OF YOKOHAMA CITY)
YOKOHAMA-SHI KOGAI TAISAKU-KYOKU (ENVIRONMENTAL PROTECTION BUREAU, YOKOHAMA CITY)
1-1, MINATO-CHO, NAKA-KU, YOKOHAMA-SHI 231
JAPANESE

YUSEI HOGO TOKEISHO 2493
(STATISTICAL REPORT OF EUGENIC PROTECTION)
KOSEI-SHO DAIJIN KANBO TOKEI CHOSA-BU (HEALTH AND WELFARE STATISTICS DIVISION, MINISTER'S SECRETARIAT, MINISTRY OF HEALTH AND WELFARE)
42, ICHIGAYA HONMURA-CHO, SHINJUKU-KU, TOKYO 162
JAPANESE
INDEXED: JICST

ZENKOKU KOGAIKEN KAISHI 2494
(JOURNAL OF ENVIRONMENTAL LABORATORIES ASSOCIATION)
JAPANESE

Accidents — Prevention, Protection, Safety, Fires

ANZEN 2495
(SAFETY)
CHUO RODO SAIGAI BOSHI KYOKAI (JAPAN INDUSTRIAL SAFETY ASSOCIATION)
35-1, 5-CHOME, SHIBA, MINATO-KU, TOKYO 108
JAPANESE
INDEXED: JICST

ANZEN EISEI 2496
(SAFETY AND HYGIENE)
NIHON ANZEN EISEI KYOKAI (THE JAPAN SAFETY AND HYGIENE COUNCIL)
1-1, KANAYA-CHO, KITA-KU, OSAKA-SHI 530
JAPANESE
INDEXED: JICST

ANZEN EISEI NO HIROBA 2497
(SAFETY PLAZA)
CHUO RODO SAIGAI BOSHI KYOKAI (JAPAN INDUSTRIAL ASSOCIATION)
35-1, 5-CHOME, SHIBA, MINATO-KU, TOKYO 108
JAPANESE

ANZEN EISEI TAIMUZU 2498
(SAFETY AND HYGIENE TIMES)
NIHON RODO BUNKA KYOKAI
25-47, 3-CHOME, NISHI-SHINBASHI, MINATO-KU, TOKYO 105
JAPANESE

ANZEN GEPPO 2499
(SAFETY MONTHLY)
KAIJO JIEITAI KOKU SHUDAN (MARITIME SELF DEFENSE FORCE, AIR DIVISION)
JAPANESE

ANZEN KANRI 2500
(SAFETY MANAGEMENT)
NIHON KOKUYU TETSUDO SHOKUIN KYOKU
(JAPAN NATIONAL RAILWAY, PERSONNEL BUREAU)
JAPANESE

ANZEN KOGAKU 2501
(SAFETY ENGINEERING)
ANZEN KOGAKU HENSHU IINKAI AND
ANZEN KOGAKU KYOKAI (JAPAN SOCIETY FOR SAFETY ENGINEERING)
DAIWA GINKO BUILDING, 3F, 4-47, ONOE-CHO, NAKA-KU, YOKOHAMA-SHI 231
JAPANESE, WITH ENGLISH SUMMARY
INDEXED: APT, JICST

BOKA 2502
(FIRE PREVENTION)
NIHON BOKA KYOKAI (JAPAN FIRE PREVENTION ASSOCIATION)
C/O SHOBO-CHO YOBO-KA, 1-2, 2-CHOME, KASUMIGASEKI, CHIYODA-KU, TOKYO 100
JAPANESE

BOKA KANRI 2503
(FIRE PREVENTION MANAGEMENT)
KOJIMACHI BOKA KANRISHA KENKYU KAI (FIRE PREVENTION SUPERVISOR'S RESEARCH SOCIETY)
JAPANESE

BOKA KANRI 2504
(FIRE PREVENTION MANAGEMENT)
SHIMOYA BOKA KANRISHA KENKYUKAI (SHIMOYA FIRE PREVENTION SUPERVISORS RESEARCH SOCIETY)
JAPANESE

BOSAI (SHIGA-KEN BOSAI KISHO RENRAKUKAI) 2505
(PREVENTION OF DISASTER)
HIKONE CHIHO KISHODAI (HIKONE METEOROLOGICAL OBSERVATORY)
SHIGA-KEN BOSAI KISHO RENRAKUKAI
C/O HIKONE CHIHO KISHODAI, 5-25, 2-CHOME, SHIRO-MACHI, HIKONE-SHI 522
JAPANESE
INDEXED: JICST

BOSAI (TOKYO RENGO BOKA KYOKAI) 2506
(DISASTER PREVENTION)
TOKYO RENGO BOKA KYOKAI (TOKYO CONSOLIDATED FIRE PREVENTION ASSOCIATION)
C/O TOKYO SHOBO-CHO, 11-39, 1-CHOME, NAGATA-CHO, CHIYODA-KU, TOKYO 100
JAPANESE AND ENGLISH

BOSAI (ZENKOKU BOSAI KYOKAI) 2507
(DISASTER PREVENTION)
ZENKOKU BOSAI KYOKAI (JAPAN DISASTER PREVENTION ASSOCIATION)
C/O SHOYU KAIKAN, 3-3, 3-CHOME, KASUMIGASEKI, CHIYODA-KU, TOKYO 100
JAPANESE

BOSAI KAGAKU GIJUTSU 2508
(BULLETIN OF THE NATIONAL RESEARCH CENTER FOR DISASTER PREVENTION)
KAGAKU GIJUTSU-CHO KOKURITSU BOSAI KAGAKU GIJUTSU SENTA (NATIONAL RESEARCH CENTER FOR DISASTER PREVENTION, SCIENCE AND TECHNOLOGY AGENCY)
15-1, 6-CHOME, GINZA, CHUO-KU, TOKYO 104
JAPANESE
INDEXED: JICST

BOSAI KAGAKU GIJUTSU INFOMESHON SABISU 2509
(DISASTER PREVENTION SCIENTIFIC AND TECHNICAL INFORMATION SERVICE)
KAGAKU GIJUTSU-CHO KOKURITSU BOSAI KAGAKU GIJUTSU SENTA KIKAKUKA SHIRYO CHOSA SHITSU (SCIENCE AND TECHNOLOGY AGENCY, NATIONAL DISASTER DEFENSE SCIENCE AND TECHNOLOGY CENTER PLANNING DIVISION RESEARCH)
JAPANESE

BOSAI KAGAKU GIJUTSU KENKYU SHIRYO 2510
(REVIEW OF RESEARCH FOR DISASTER PREVENTION)
KAGAKU GIJUTSU-CHO KOKURITSU BOSAI KAGAKU GIJUTSU SENTA (NATIONAL RESEARCH CENTER FOR DISASTER PREVENTION, SCIENCE AND TECHNOLOGY AGENCY)
15-1, 6-CHOME, GINZA, CHUO-KU, TOKYO 104
JAPANESE AND ENGLISH
INDEXED: JICST

BOSAI KAGAKU GIJUTSU SOGO KENKYU HOKOKU 2511
(REPORTS OF COOPERATIVE RESEARCH FOR DISASTER PREVENTION)
KAGAKU GIJUTSU-CHO KOKURITSU BOSAI KAGAKU GIJUTSU SENTA (NATIONAL RESEARCH CENTER FOR DISASTER PREVENTION, SCIENCE AND TECHNOLOGY AGENCY)
15-1, 6-CHOME, GINZA, CHUO-KU, TOKYO 104
JAPANESE, WITH ENGLISH SUMMARY
INDEXED: JICST

BOSAI KAGAKU GIJUTSU SOGO KENKYU SOKUHO 2512
(NOTES OF COOPERATIVE RESEARCH FOR DISASTER PREVENTION)
KAGAKU GIJUTSU-CHO KOKURITSU BOSAI KAGAKU GIJUTSU SENTA (NATIONAL RESEARCH CENTER FOR DISASTER PREVENTION, SCIENCE AND TECHNOLOGY AGENCY)
15-1, 6-CHOME, GINZA, CHUO-KU, TOKYO 104
JAPANESE, WITH ENGLISH SUMMARY
INDEXED: JICST

BULLETIN OF THE DISASTER PREVENTION RESEARCH INSTITUTE 2513
KYOTO DAIGAKU BOSAI KENKYUSHO (DISASTER PREVENTION RESEARCH INSTITUTE, KYOTO UNIVERSITY)
GOKASHO, UJI-SHI 611
ENGLISH
INDEXED: CHE

GAKKO ANZEN 2514
(SCHOOL SAFETY)
NIHON GAKKO ANZENKAI YAMAGUCHI-KEN SHIBU, YAMAGUCHI-KEN KYOKU IINKAI (JAPAN SCHOOL SAFETY COMMITTEE, YAMAGUCHI PREFECTURE CHAPTER, YAMAGUCHI EDUCATION COMMITTEE
JAPANESE

GAKKO ANZEN 2515
(SCHOOL SAFETY)
NIHON GAKKO ANZEN KAI (JAPAN SCHOOL SAFETY SOCIETY)
JAPANESE

GAKKO ANZEN NAGANO 2516
(SCHOOL SAFETY, NAGANO)
NIHON GAKKO ANZENKAI NAGANO-KEN SHIBU (JAPAN SCHOOL SAFETY COMMITTEE NAGANO PREFECTURE BRANCH)
JAPANESE

GURAFU BOKA KANRI 2517
(GRAPH: FIRE-PREVENTION CONTROL)
ZENKOKU KAJO HOREI SHUPPAN K.K.
17, SAN'EI-CHO, SHINJUKU-KU, TOKYO 160
JAPANESE

HAIEN JIKKEN HOKOKUSHO 2518
(EXPERIMENTAL REPORT OF SMOKE EXHAUST)
NIHON KASAI GAKKAI (JAPAN FIRE DISASTER SOCIETY)
C/O GAKKAI SENTA BUILDING, 4-16, 2-CHOME, YAYOI, BUNKYO-KU, TOKYO 113
JAPANESE

HANZAIGAKU ZASSHI 2519
(ACTA CRIMINOLOGIAE ET MEDICINE LEGALLS JAPONICA)
NIHON HANZAI GAKKAI (THE JAPANESE ASSOCIATION OF CRIMINOLOGY)
C/O TOKYO IKA SHIKA DAIGAKU, 5-45, 1-CHOME, YUSHIMA, BUNKYO-KU, TOKYO 113
JAPANESE AND EUROPEAN

HOSU NYUSU 2520
(MAINTENANCE NEWS)
TOKYO KASAI HOCHI SETSUBI HOSHU KYOKAI
81-3, YARAI-CHO, SHINJUKU-KU, TOKYO 162
JAPANESE

IWATE-KEN SHOBO NENPO 2521
(ANNUAL REPORT OF THE FIRE PREVENTION IN IWATE PREFECTURE)
IWATE-KEN SOMU-BU (GENERAL AFFAIRS DIVISION, IWATE PREFECTURAL GOVERNMENT)
10-1, UCHIMARU, MORIOKA-SHI 020
JAPANESE

JITSUMU TENBO 2522
(BUSINESS PRACTICE VIEW)
JAPANESE
INDEXED: JICST

KAGAKU KEISATSU KENKYUSHO HOKOKU 2523
(REPORTS OF THE NATIONAL RESEARCH INSTITUTE OF POLICE SCIENCE)
KEISATSU-CHO KAGAKU KEISATSU KENKYUSHO (NATIONAL RESEARCH INSTITUTE OF POLICE SCIENCE, NATIONAL POLICE AGENCY)
6, SANBAN-CHO, CHIYODA-KU, TOKYO 102
JAPANESE, WITH ENGLISH SUMMARY
INDEXED: JICST

KAGAKU KEISATSU KENKYUSHO NENPO 2524
(NATIONAL RESEARCH INSTITUTE OF POLICE SCIENCE, ANNUAL REPORT)
KEISATSU-CHO KAGAKU KEISATSU KENKYUSHO (NATIONAL RESEARCH INSTITUTE OF POLICE SCIENCE, NATIONAL POLICE AGENCY)
6, SANBAN-CHO, CHIYODA-KU, TOKYO 102
JAPANESE
INDEXED: JICST

KAGAKU KEISATSU KENKYUSHO SHIRYO 2525
(DATA OF NATIONAL RESEARCH INSTITUTE OF POLICE SCIENCE)
KEISATSU-CHO KAGAKU KEISATSU KENKYUSHO (NATIONAL RESEARCH INSTITUTE OF POLICE SCIENCE, NATIONAL POLICE AGENCY)
6, SANBAN-CHO, CHIYODA-KU, TOKYO 102
JAPANESE

KAIHAN TOKEI GEPPO 2526
(MONTHLY STATISTICS ON MARINE DISASTERS)
UN'YU-SHO DAIJIN KANBO TOKEI CHOSA-BU TOKEI DAI-1-KA (STATISTICS AND SURVEY DIVISION, MINISTER'S SECRETARIAT, MINISTRY OF TRANSPORTATION)
1-3, 2-CHOME, KASUMIGASEKI, CHIYODA-KU, TOKYO 100
JAPANESE

KAIHAN TOKEI NENPO 2527
(ANNUAL REPORT OF SHIPWRECK STATISTICS)
UN'YU-SHO DAIJIN KANBO JOHO KANRI-BU (DEPARTMENT OF RESEARCH AND DATA PROCESSING, MINISTER'S SECRETARIAT, MINISTRY OF TRANSPORTATION)
1-3, 2-CHOME, KASUMIGASEKI, CHIYODA-KU, TOKYO 100
JAPANESE

KASAI (NIHON KASAIGAKKAI-SHI) 2528
(JAPANESE ASSOCIATION OF FIRE SCIENCE AND ENGINEERING)
NIHON KASAI GAKKAI (FIRE PREVENTION SOCIETY OF JAPAN) AND KOGYO CHOSAKAI
4-16, 2-CHOME, YAYOI, BUNKYO-KU, TOKYO 113
JAPANESE
INDEXED: JICST

KASAI HOCHIKI 2529
(THE FIRE ALARM SYSTEMS)
NIHON KASAI HOCHIKI KOGYOKAI (THE ASSOCIATION OF FIRE ALARMS OF JAPAN)
C/O TANAKA BUILDING, 3-4, 5-CHOME, SOTOKANDA, CHIYODA-KU, TOKYO 101
JAPANESE

KASAI KYUKYU TOKEI 2530
(STATISTICS ON FIRE RESCUE)
KYOTO-SHI SHOBO-KYOKU
101, HIGASHIIRU ENOKI-CHO, OSHIKOJI-DORI, TERA-MACHI, NAKAGYO-KU, KYOTO-SHI 604
JAPANESE

KASAI NENPO 2531
(FIRE DISASTER ANNUAL REPORT)
SHOBO-CHO (FIRE DEFENSE AGENCY)
1-2, 2-CHOME, KASUMIGASEKI, CHIYODA-KU, TOKYO 100
JAPANESE

KENTEI JIHO 2532
(EXAMINATION REPORT)
NIHON SHOBO KENTEI KYOKAI (JAPAN FIRE PREVENTION EXAMINATION ASSOCIATION)
JAPANESE

KINDAI SHOBO 2533
(MODERN FIRE SERVICE)
ZENKOKU KAJO HOREI SHUPPAN K.K.
17, SAN'EI-CHO, SHINJUKU-KU, TOKYO 160
JAPANESE

KOBE-SHI SHOBO HAKUSHO 2534
(FIRE ANNUAL REPORT, KOBE CITY)
KOBE-SHI SHOBO-KYOKU (FIRE DEFENSE BUREAU, KOBE CITY)
7, 6-CHOME, KANO-CHO, IKUTA-KU, KOBE-SHI 650
JAPANESE

KOKURITSU BOSAI KAGAKU GIJUTSU SENTA GIJUTSU JOHO 2535
(TECHNICAL INFORMATION OF NATIONAL RESEARCH CENTER FOR DISASTER PREVENTION)
KOKURITSU BOSAI KAGAKU GIJUTSU SENTA (NATIONAL RESEARCH CENTER FOR DISASTER PREVENTION)
15-1, 6-CHOME, GINZA, CHUO-KU, TOKYO 104
JAPANESE

KOKURITSU BOSAI KAGAKU GIJUTSU SENTA KENKYU SOKUHO 2536
(RESEARCH NOTES OF THE NATIONAL RESEARCH CENTER FOR DISASTER PREVENTION)
KAGAKU GIJUTSU-CHO KOKURITSU BOSAI KAGAKU GIJUTSU SENTA (NATIONAL RESEARCH CENTER FOR DISASTER PREVENTION, SCIENCE AND TECHNOLOGY AGENCY)
15-1, 6-CHOME, GINZA, CHUO-KU, TOKYO 104
JAPANESE AND ENGLISH
INDEXED: JICST

KYOTO DAIGAKU BOSAI KENKYUSHO NENPO 2537
(DISASTER PREVENTION RESEARCH INSTITUTE ANNUALS)
KYOTO DAIGAKU BOSAI KENKYUSHO (DISASTER PREVENTION RESEARCH INSTITUTE, KYOTO UNIVERSITY)
GOKASHO, UJI-SHI 611
JAPANESE
INDEXED: JICST

KYOTO SHOBO 2538
(NEWS OF KYOTO FIRE PREVENTION ASSOCIATION)
KYOTO SHOBO HENSHU IINKAI
C/O KYOTO-SHI SHOBO GAKKO, FUKAKUSA ECHIGO YASHIKI-CHO, FUSHIMI-KU, KYOTO-SHI 612
JAPANESE

NIHON BOSAI KYODO TSUSHIN 2539
(CORRESPONDENCE OF DISASTER PREVENTION IN JAPAN)
NIHON BOSAI KYODO TSUSHIN-SHA
33-4, 3-CHOME, YUSHIMA, BUNKYO-KU, TOKYO 113
JAPANESE

NIHON KASAI GAKKAI RONBUNSHU 2540
(BULLETIN OF THE FIRE PREVENTION SOCIETY OF JAPAN)
NIHON KASAI GAKKAI (JAPANESE ASSOCIATION OF FIRE SCIENCE AND ENGINEERING)
JAPAN ACADEMIC SOCIETIES' CENTER 4-16, 2-CHOME, YAYOI, BUNKYO-KU, TOKYO 113
JAPANESE, WITH ENGLISH SUMMARY
INDEXED: JICST

NIHON SAIGAI IGAKKAI KAISHI 2541
(JOURNAL OF JAPAN ACCIDENT MEDICAL ASSOCIATION)
NIHON SAIGAI IGAKKAI (JAPAN ACCIDENT MEDICAL ASSOCIATION)
C/O TOKYO ROSAI BYOIN, 13-21, 4-CHOME, OMORIMINAMI, OTA-KU, TOKYO 143
JAPANESE
INDEXED: JICST

NIHON SAIGAI IGAKKAI NITTEI OYOBI ENZETSU NAIYO SHOROKU 2542
(JAPANESE DISASTER MEDICAL ASSOCIATION, THE DAY'S PROGRAM AND THE SUMMARIES OF CONTENTS OF THE LECTURE)
NIHON SAIGAI IGAKKAI
C/O KYUSHU ROSAI BYOIN, 522-2, KUZUHARA, MINAMI-KU, KOKURA, KITA-KYUSHU-SHI 800-02
JAPANESE

NIHON SHOBO 2543
(JAPAN FIRE DEFENSE)
NIHON SHOBO KYOKAI (JAPAN FIRE DEFENSE ASSOCIATION)
18, SHIBA NISHIKUBO AKEFUNE-CHO, MINATO-KU, TOKYO 105
JAPANESE

NIHON SHOBO KENTEI KYOKAI CHOSA KENKYU SHIRYO 2544
(SURVEY RESEARCH DATA OF THE JAPAN FIRE PREVENTION INSPECTION ASSOCIATION)
JAPAN FIRE PREVENTION INSPECTION ASSOCIATION
JAPANESE

NIHON SHOBO SHINBUN 2545
(JAPAN FIRE NEWS)
NIHON SHOBO SHINBUNSHA
18, 2-CHOME, MIYAMAE-CHO, NAKA-KU, NAGOYA-SHI 460
JAPANESE

NOMI ARAMU 2546
(NOHMI ALARM)
NOMI BOSAI K.K.
7-3, 4-CHOME, KUDAN MINAMI, CHIYODA-KU, TOKYO 102
JAPANESE

OSAKA-FU SHOHO TOKEI 2547
(ANNUAL REPORT OF THE FIRE PREVENTION IN OSAKA PREFECTURE)
OSAKA-FU SEIKATSU KANKYO-BU (LIFE ENVIRONMENT DEPARTMENT, OSAKA PREFECTURAL GOVERNMENT)
OTEMAENO-CHO, HIGASHI-KU, OSAKA-SHI 540
JAPANESE

OSAKA-SHI SHOBO NENPO 2548
(ANNUAL REPORT OF THE FIRE PREVENTION IN OSAKA CITY)
OSAKA-SHI SHOBO-KYOKU (FIRE DEFENSE BUREAU, OSAKA MUNICIPAL GOVERNMENT)
1-12, NISHINAGAHORI-KITADORI, NISHI-KU, OSAKA-SHI 550
JAPANESE

RODO ANZEN EISEI KOHO 2549
(LABOR SAFETY AND HEALTH INFORMATION)
RODO KIJUN CHOSA KAI (LABOR STANDARDS RESEARCH ASSOCIATION)
JAPANESE

RODO SAIGAI DOKO CHOSA MAIGETSU CHOSA HOKOKU 2550
(MONTHLY SURVEY REPORT ON THE TREND OF LABOUR ACCIDENT)
RODO-SHO DAIJIN KANBO TOKEI CHOSA-BU (STATISTICS AND INFORMATION DEPARTMENT, MINISTER'S SECRETARIAT, MINISTRY OF LABOUR)
3-1, 1-CHOME, OTE-MACHI, CHIYODA-KU, TOKYO 100
JAPANESE
INDEXED: JICST

SAIGAI KAGAKU KENKYUSHO KENKYU HOKOKUSHO 2551
(REPORTS OF THE INSTITUTE OF CALAMITY SCIENCE)
SAIGAI KAGAKU KENKYUSHO (INSTITUTE OF CALAMITY RESEARCH)
C/O OSAKA DAIGAKU KOGYO KAIKAN, 125-18, 5-CHOME, FUJISHIRODAI, SUITA-SHI 565
JAPANESE

SAIGAI KAGAKU SOGO SHINPOJUMU KOENKAI RONBUNSHU 2552
(PROCEEDINGS OF THE GENERAL SYMPOSIUM ON DISASTER SCIENCE)
MONBU-SHO TOKUBETSU KENKYU SAIGAI KAGAKU SOGO KENKYU-HAN
2-2, 3-CHOME, KASUMIGASEKI, CHIYODA-KU, TOKYO 100
JAPANESE
INDEXED: JICST

SAIGAI TOKEI 2553
(DISASTER STATISTICS)
KENSETSU-SHO KASEN-KYOKU (RIVER BUREAU, MINISTRY OF CONSTRUCTION)
1-2, KASUMIGASEKI, CHIYODA-KU, TOKYO
JAPANESE

SANGYO ANZEN KENKYUSHO ANZEN SHIRYO 2554
(SAFETY DOCUMENT OF THE RESEARCH INSTITUTE OF INDUSTRIAL SAFETY)
RODO-SHO SANGYO ANZEN KENKYUSHO (THE RESEARCH INSTITUTE OF INDUSTRIAL SAFETY, MINISTRY OF LABOUR)
35-1, 5-CHOME, SHIBA, MINATO-KU, TOKYO 108
JAPANESE

SANGYO ANZEN KENKYUSHO GIJUTSU SHIRYO (RIIS-TN) 2555
(TECHNICAL NOTE OF THE RESEARCH INSTITUTE OF INDUSTRIAL SAFETY)
RODO-SHO SANGYO ANZEN KENKYUSHO (RESEARCH INSTITUTE OF INDUSTRIAL SAFETY, MINISTRY OF LABOUR)
35-1, 5-CHOME, SHIBA, MINATO-KU, TOKYO 108
JAPANESE
INDEXED: JICST

SANGYO ANZEN KENKYUSHO GIJUTSU SHISHIN (RIIS-TR) 2556
(TECHNICAL RECOMMENDATION OF THE RESEARCH INSTITUTE OF INDUSTRIAL SAFETY)
RODO-SHO SANGYO ANZEN KENKYUSHO (RESEARCH INSTITUTE OF INDUSTRIAL SAFETY)
35-1, 5-CHOME, SHIBA, MINATO-KU, TOKYO 108
JAPANESE
INDEXED: JICST

SANGYO ANZEN KENKYUSHO KENKYU HOKOKU 2557
(BULLETIN OF THE RESEARCH INSTITUTE OF INDUSTRIAL SAFETY)
RODO-SHO SANGYO ANZEN KENKYUSHO (THE RESEARCH INSTITUTE OF INDUSTRIAL SAFETY, MINISTRY OF LABOUR)
35-1, 5-CHOME, SHIBA, MINATO-KU, TOKYO 108
JAPANESE, WITH ENGLISH SUMMARY
INDEXED: JICST

SANGYO ANZEN KENKYUSHO NENPO 2558
(ANNUAL REPORT OF THE RESEARCH INSTITUTE OF INDUSTRIAL SAFETY)
RODO-SHO SANGYO ANZEN KENKYUSHO (THE RESEARCH INSTITUTE OF INDUSTRIAL SAFETY, MINISTRY OF LABOUR)
35-1, 5-CHOME, SHIBA, MINATO-KU, TOKYO 108
JAPANESE

SEIFUTI DAIJESTO 2559
(SAFETY DIGEST)
NIHON HOAN YOHIN KYOKAI (JAPAN SAFETY APPLIANCES ASSOCIATION)
C/O KYOIKU BUILDING, 6-16, 4-CHOME, KOHINATA, BUNKYO-KU, TOKYO 112
JAPANESE

SEN'IN TO SAIGAI BOSHI 2560
(SHIP CREWMEN AND ACCIDENT PREVENTION)
SEN'IN SAIGAI BOSHI KYOKAI (SHIP CREWMEN'S ACCIDENT PREVENTION ASSOCIATION)
JAPANESE

SETSUGAI JIKKEN KENKYUJO HOKOKU 2561
(REPORT OF THE EXPERIMENTAL STATION ON SNOW DISASTERS)
JAPANESE
INDEXED: JICST

SHOBO BOSAI NENPO (FUKUSHIMA-KEN) 2562
(ANNUAL REPORT OF FIRE AND DISASTER PREVENTION, FUKUSHIMA PREFECTURE))
FUKUSHIMA-KEN SEIKATSU KANKYO-BU
2-16, SUGITSUMA-CHO, FUKUSHIMA-SHI 960
JAPANESE

SHOBO BOSAI NENPO (KYOTO-FU) 2563
(ANNUAL REPORT OF FIRE AND DISASTER PREVENTION, KYOTO PREFECTURE)
KYOTO-FU KIKAKU KANRI-BU (PLANNING AND ADMINISTRATION DIVISION, KYOTO PREFECTURAL GOVERNMENT)
YABMOUCHI-MACHI, SHIMOTACHIYURIDORI KAMAZA HIGASHIIRU, KAMIGYO-KU, KYOTO-SHI 602
JAPANESE

SHOBO BOSAI NO GENKYO (AOMORI-KEN) 2564
(ANNUAL REPORT OF FIRE AND DISASTER PREVENTION, AOMORI PREFECTURE)
AOMORI-KEN SOMU-BU (ADMINISTRATIVE DIVISION, AOMORI PREFECTURAL GOVERNMENT)
1-1, 1-CHOME, NAGASHIMA, AOMORI-SHI 030
JAPANESE

SHOBO KAGAKU KENKYUSHOHO 2565
(REPORT OF FIRE SCIENCE LABORATORY)
TOKYO SHOBO-CHO SHOBO KAGAKU KENKYUSHO (FIRE SCIENCE LABORATORY TOKYO FIRE DEPARTMENT)
13-20, 1-CHOME, HATAGAYA, SHIBUYA-KU, TOKYO 151
JAPANESE

SHOBO KENKYUSHITSU NENPO 2566
(ANNUAL REPORT OF FIRE RESEARCH INSTITUTE IN NAGOYA MUNICIPAL GOVERNMENT)
NAGOYA-SHI SHOBO-KYOKU (FIRE DEFENSE BUREAU, NAGOYA MUNICIPAL GOVERNMENT)
2280-12, NAGAHAZAMA, SHIMOSHIDAMI, MORIYAMA-KU, NAGOYA-SHI 463
JAPANESE

SHOBO KENKYUSHO GIJUTSU SHIRYO 2567
(TECHNICAL MEMORANDUM OF FIRE RESEARCH INSTITUTE, FIRE DEFENSE AGENCY)
SHOBO-CHO SHOBO KENKYUSHO (FIRE RESEARCH INSTITUTE, FIRE DEFENSE AGENCY)
14-1, 3-CHOME, NAKAHARA, MITAKA-SHI, TOKYO 181
JAPANESE

SHOBO KENKYUSHO HOKOKU 2568
(REPORT OF FIRE RESEARCH INSTITUTE, FIRE DEFENSE AGENCY)
SHOBO-CHO SHOBO KENKYUSHO (FIRE RESEARCH INSTITUTE, FIRE DEFENSE AGENCY)
14-1, 3-CHOME, NAKAHARA, MITAKA-SHI, TOKYO 181
JAPANESE, WITH ENGLISH SUMMARY
INDEXED: JICST

SHOBO KENKYUSHO NENPO 2569
(ANNUAL REPORT OF FIRE RESEARCH INSTITUTE, FIRE DEFENSE AGENCY)
SHOBO-CHO SHOBO KENKYUSHO (FIRE RESEARCH INSTITUTE, FIRE DEFENSE AGENCY)
14-1, 3-CHOME, NAKAHARA, MITAKA-SHI, TOKYO 181
JAPANESE
INDEXED: JICST

SHOBO KENSHU 2570
(FIRE PREVENTION STUDY)
SHOBO DAIGAKKO
(FIRE PREVENTION COLLEGE)
JAPANESE

SHOBO NENPO (KANAGAWA-KEN) 2571
(ANNUAL REPORT OF THE FIRE PREVENTION IN KANAGAWA PREFECTURE)
KANAGAWA-KEN KIKAKU CHOSA-BU (PLANNING AND SURVEY DIVISION, KANAGAWA PREFECTURAL GOVERNMENT)
1, NIHON-ODORI, NAKA-KU, YOKOHAMA-SHI 231
JAPANESE

SHOBO NENPO (NAGOYA-SHI) 2572
(ANNUAL REPORT OF THE FIRE PREVENTION IN NAGOYA CITY)
NAGOYA-SHI SHOBO-KYOKU (FIRE DEFENSE BUREAU, NAGOYA MUNICIPAL GOVERNMENT)
1-1, 3-CHOME, SANNOMARU, NAKA-KU, NAGOYA-SHI 460
JAPANESE

SHOBO NENPO (SHOBO-CHO) 2573
(ANNUAL REPORT OF FIRE PREVENTION, FIRE DEFENSE AGENCY)
SHOBO-CHO (FIRE DEFENSE AGENCY)
1-2, 2-CHOME, KASUMIGASEKI, CHIYODA-KU, TOKYO 100
JAPANESE

SHOBO NENPO (TOKYO-TO) 2574
(ANNUAL REPORT OF FIRE PREVENTION IN TOKYO METROPOLIS)
TOKYO-TO SOMU-KYOKU GYOSEI-BU (ADMINISTRATIVE BUREAU, TOKYO METROPOLITAN GOVERNMENT)
5-1, 3-CHOME, MARUNOUCHI, CHIYODA-KU, TOKYO 100
JAPANESE

SHOBO NENPO (YOKOHAMA-SHI) 2575
(ANNUAL REPORT OF FIRE PREVENTION IN YOKOHAMA CITY)
YOKOHAMA-SHI SHOBO-KYOKU (THE FIRE DEFENSE BUREAU, YOKOHAMA MUNICIPAL GOVERNMENT)
1-15, KAWABATA-CHO, HODOGAYA-KU, YOKOHAMA-SHI 240
JAPANESE

SHOBO TSUSHIN 2576
(INFORMATION OF FIRE PROTECTION)
SHOBO TSUSHINSHA (FIRE PROTECTION INFORMATION PUBLISHING COMPANY)
JAPANESE

SHOKAKI 2577
(FIRE EXTINGUISHER)
NIHON SHOKAKI KOGYOKAI (JAPAN FIRE EXTINGUISHER MANUFACTURERS' ASSOCIATION)
25-7, 3-CHOME, ASAKUSABASHI, TAITO-KU, TOKYO 187
JAPANESE

SHOKEN SHUHO 2578
(BULLETIN OF THE FIRE SCIENCE LABORATORY)
SHOBO-CHO SHOBO KENKYUSHO (FIRE RESEARCH INSTITUTE, FIRE DEFENSE AGENCY)
14-1, 3-CHOME, NAKAHARA, MITAKA-SHI, TOKYO 181
JAPANESE

SHUYO KAIUNKOKU KAINAN TOKEI 2579
(MARINE DISASTER'S STATISTICS OF MAJOR MARINE TRANSPORTATION COUNTRIES)
KAIJO HOAN-CHO KEIBI KYUNAN-BU (GUARD AND RESCUE DEPARTMENT, MARITIME SAFETY AGENCY)
1-3, 2-CHOME, KASUMIGASEKI, CHIYODA-KU, TOKYO 100
JAPANESE

SHUYO SAIGAI CHOSA 2580
(DISASTER RECORD)
KOKURITSU BOSAI KAGAKU GIJUTSU SENTA (NATIONAL RESEARCH CENTER FOR DISASTER PREVENTION)
15-1, 6-CHOME, GINZA, CHUO-KU, TOKYO 104
JAPANESE

SHUYO SAIGAI KIROKU 2581
(RECORD OF THE MAJOR DISASTERS)
KENSETSU-SHO KASEN-KYOKU (RIVER BUREAU, MINISTRY OF CONSTRUCTION)
1-3, 2-CHOME, KASUMIGASEKI, CHIYODA-KU, TOKYO 100
JAPANESE

TOKYO SHOBO 2582
(NEWS OF TOKYO FIRE PREVENTION ASSOCIATION)
TOKYO SHOBO KYOKAI (TOKYO FIRE PREVENTION ASSOCIATION)
C/O TOKYO SHOBO-CHO, 11-39, 1-CHOME, NAGATA-CHO, CHIYODA-KU, TOKYO 100
JAPANESE

TOKYO SHOBO-CHO TOKEISHO 2583
(ANNUAL STATISTICS OF FIRE PREVENTION, TOKYO FIRE DEPARTMENT)
TOKYO SHOBO-CHO (TOKYO FIRE DEPARTMENT)
11-39, 1-CHOME, NAGATA-CHO, CHIYODA-KU, TOKYO 100
JAPANESE

TOKYO-TO NO SAIGAI 2584
(DISASTER IN TOKYO METROPOLIS)
TOKYO-TO SOMU-KYOKU (GENERAL AFFAIRS BUREAU, TOKYO METROPOLITAN GOVERNMENT)
5-1, 3-CHOME, MARUNOUCHI, CHIYODA-KU, TOKYO 100
JAPANESE

UMI TO ANZEN 2585
(SEA AND SAFETY)
NIHON KAINAN BOSHI KYOKAI (THE JAPAN ASSOCIATION FOR PREVENTING SEA CASUALITIES
C/O SENPAKU SHINKO BUILDING, 35, SHIBA KOTOHIRA-CHO, MINATO-KU, TOKYO 105
JAPANESE

YAMAGATA-KEN SHOBO NENPO 2586
(ANNUAL REPORT OF FIRE PREVENTION IN YAMAGATA PREFECTURE)
YAMAGATA-KEN SEIKATSU KANKYO-BU (LIVING AND ENVIRONMENTAL DEPARTMENT, YAMAGATA PREFECTURAL GOVERNMENT)
4-51, 3-CHOME, HATAGO-CHO, YAMAGATA-SHI 990
JAPANESE

YOBO JIHO 2587
(ACCIDENT PREVENTION JOURNAL)
NIHON SONGAI HOKEN KYOKAI (THE MARINE AND FIRE INSURANCE ASSOCIATION OF JAPAN)
9, 2-CHOME, KANDA AWAJI-CHO, CHIYODA-KU, TOKYO 101
JAPANESE

YOKKAICHI-SHI CHIIKI BOSAI KEIKAKU 2588
(REGIONAL PLANNING ON FIRE PREVENTION IN YOKKAICHI CITY)
YOKKAICHI-SHI BOSAI KAIGI
1-5, SUWA-CHO, YOKKAICHI-SHI 510
JAPANESE

YOKYUJO KAINAN TOKEI 2589
(STATISTICS ON MARINE DISASTER REQUESTED RESCUE)
KAIJO HOAN-CHO KEIBI KYUNAN-BU (GUARD AND RESCUE DEPARTMENT, MARITIME SAFETY AGENCY)
1-3, 2-CHOME, KASUMIGASEKI, CHIYODA-KU, TOKYO 100
JAPANESE

YUKI 2590
(NEWS OF KOBE FIRE PREVENTION ASSOCIATION)
KOBE-SHI SHOBO-KYOKU (FIRE DEFENSE BUREAU, KOBE MUNICIPAL GOVERNMENT)
KOBE-SHI BOKA KYOKAI RENRAKU KYOGIKAI
7, 6-CHOME, KANO-CHO, IKUTA-KU, KOBE-SHI 650
JAPANESE

ZENKOKU SHOBO 2591
(FIRE PREVENTION IN JAPAN)
ZENKOKU KAJOHOREI SHUPPAN K.K.
17, SAN'EI-CHO, SHINJUKU-KU, TOKYO 160
JAPANESE

ZENKOKU SHOBOCHOKAI KAIHO 2592
(NATIONAL FIRE CHIEFS ASSOCIATION, CONFERENCE REPORT
(CONFERENCE REPORT OF N-F-C-A))
ZENKOKU SHOBOCHOKAI JIMUKYOKU (NATIONAL FIRE
CHIEF'S ASSOCIATION ADMINISTRATIVE BUREAU)
JAPANESE

Pharmacy, Pharmacology, Therapeutics

AICHI KYOIKU DAIGAKU FUZOKU YOGO GAKKO KENKYU KIYO 2593
(BULLETIN OF THE NURSING SCHOOL, AICHI COLLEGE OF EDUCATION)
AICHI DAIGAKU FUZOKU YOGO GAKKO (AICHI COLLEGE OF EDUCATION, NURSING SCHOOL)
JAPANESE

AICHI-KENRITSU KANGO TANKI DAIGAKU ZASSHI 2594
(JOURNAL OF AICHI PREFECTURAL COLLEGE OF NURSING)
AICHI-KENRITSU KANGO TANKI DAIGAKU ZASSHI HENSHU IINKAI
AICHI-KENRITSU KANGO TANKI DAIGAKU TOSHOKAN
TOGOKU, KAMI-SHIDAMI, MORIYAMA-KU, NAGOYA-SHI 463
JAPANESE

AIGO 2595
(CARE)
NIHON SEISHIN HAKUJAKUSHA AIGO KYOKAI (JAPAN'S SOCIETY OF CARE FOR THE MENTALLY HANDICAPPED)
JAPANESE

AIGO NYUSU 2596
(CARE NEWS)
NIHON SEISHIN HAKUJAKUSHA AIGO KYOKAI (JAPAN'S SOCIETY OF CARE FOR THE MENTALLY HANDICAPPED)
JAPANESE

ANNUAL REPORT, NATIONAL INSTITUTE OF RADIOLOGICAL SCIENCES 2597
HOSHASEN IGAKU SOGO KENKYUSHO (NATIONAL INSTITUTE OF RADIOLOGICAL SCIENCES)
9-1, 4-CHOME, ANAGAWA, CHIBA-SHI 280
ENGLISH
INDEXED: JICST

BOKIN BOBAI 2598
(JOURNAL OF ANTIBACTERIAL AND ANTIFUNGAL AGENTS, JAPAN)
NIHON BOKIN BOBAI GAKKAI (JAPAN SOCIETY FOR ANTIBACTERIAL AND ANTIFUNGAL AGENTS)
JAPANESE

BULLETIN OF PHARMACEUTICAL RESEARCH INSTITUTE 2599
YAKUGAKU KENKYUSHO (PHARMACEUTICAL RESEARCH INSTITUTE)
C/O OSAKA IKA DAIGAKU, 2-7, DAIGAKU-CHO, TAKATSUKI-SHI 569
JAPANESE AND ENGLISH
INDEXED: BIO, JICST

BULLETIN OF THE TOKYO METROPOLITAN REHABILITATION CENTER FOR THE PHYSICALLY AND MENTALLY HANDICAPPED 2600
ENGLISH

BYOIN YAKUHO 2601
(HOSPITAL PHARMACEUTICAL REPORT)
JAPANESE
INDEXED: JICST

CHEMICAL AND PHARMACEUTICAL BULLETIN 2602
NIHON YAKUGAKKAI (PHARMACEUTICAL SOCIETY OF JAPAN)
12-15, 2-CHOME, SHIBUYA, SHIBUYA-KU, TOKYO 150
ENGLISH
INDEXED: BIB, BIO, CHE, EXC, IND, SCI, JICST

CHEMOTHERAPY 2603
NIHON KAGAKU RYOHO GAKKAI (JAPAN SOCIETY OF CHEMOTHERAPY)
20-8, 2-CHOME, KAMIOSAKI, SHINAGAWA-KU, TOKYO 141
JAPANESE AND ENGLISH
INDEXED: CHE, EXC, JICST

CHIKEN ISHIYAKU JOHO 2604
(INVESTIGATIONAL DRUG INFORMATION, MEDICAL AND DENTAL)
IJI SHUPPANSHA
4-11, NIHONBASHI HON-CHO, CHUO-KU, TOKYO 103
JAPANESE AND ENGLISH

CHIKEN'YAKU JOHO, GAIKOKU 2605
(INVESTIGATIONAL DRUG REPORT, WORLDWIDE)
IYAKUHIN SANGYO KENKYUSHO (DRUG BUSINESS RESEARCH AND COMPANY, LIMITED)
25-2, 1-CHOME, KOISHIKAWA, BUNKYO-KU, TOKYO 112
JAPANESE AND ENGLISH

CHIKEN'YAKU JOHO, KOKUNAI 2606
(INVESTIGATIONAL DRUG REPORT, INTERNAL)
IYAKUHIN SANGYO KENKYUSHO (DRUG BUSINESS RESEARCH CONSULTANTS)
25-2, 1-CHOME, KOISHIKAWA, BUNKYO-KU, TOKYO 112
JAPANESE

CHIRYO 2607
(JOURNAL OF THERAPY)
NANZANDO
1-11, 4-CHOME, YUSHIMA, BUNKYO-KU, TOKYO 113-91
JAPANESE
INDEXED: CHE

CHIRYO YAKUHO 2608
(THERAPEUTIC PHARMACEUTICAL REPORT)
JAPANESE
INDEXED: JICST

CHOONPA IGAKU 2609
(JAPANESE JOURNAL OF MEDICAL ULTRASONICS)
NIHON CHOONPA IGAKKAI (MEDICAL ULTRASONICS ASSOCIATION OF JAPAN)
JAPANESE

CHUGAI IYAKU 2610
(CHUGAI MEDICAL NEWS)
CHUGAI SEIYAKU K.K. (CHUGAI PHARMACEUTICAL COMPANY, LIMITED)
10-6, 1-CHOME, IWAMOTO-CHO, CHIYODA-KU, TOKYO 101
JAPANESE

CLINICIAN 2611
EIZAI K.K.
6-10, 4-CHOME, KOISHIKAWA, BUNKYO-KU, TOKYO 112
JAPANESE

CONTENTS 2612
NIHON IYAKU JOHO SENTA (JAPAN PHARMACEUTICAL INFORMATION CENTER)
12-15, 2-CHOME, SHIBUYA, SHIBUYA-KU, TOKYO 150
JAPANESE
INDEXED: JICST

DAIICHI YAKKA DAIGAKU KENKYU NENPO 2613
(THE ANNUAL REPORT OF THE DAIICHI COLLEGE OF PHARMACEUTICAL SCIENCES)
DAIICHI YAKKA DAIGAKU (THE DAIICHI COLLEGE OF PHARMACEUTICAL SCIENCES)
93, TAMAGAWA-CHO, MINAMI-KU, FUKUOKA-SHI 815
JAPANESE, WITH ENGLISH SUMMARY

DBC NYUSU 2614
(DBC NEWS)
DOJIN YAKKAGAKU KENKYUSHO (DOJINDO COMPANY, LIMITED, RESEARCH LABORATORY)
2681, MURAZOE, KENGUN-MACHI, KUMAMOTO-SHI 862
JAPANESE AND ENGLISH

DENSHI IGAKU 2615
(MEDICAL ELECTRONICS)
DENPA JIKKENSHA
4-2, 1-CHOME, EBISU-MINAMI, SHIBUYA-KU, TOKYO 150
JAPANESE
INDEXED: JICST

DOJIN NYUSU 2616
(DOJIN NEWS)
DOJIN YAKU KAGAKU KENKYUJO (DOJIN PHARMACEUTICAL RESEARCH INSTITUTE)
JAPANESE

DORAGGU BIJINESU 2617
(DRUG BUSINESS)
IYAKUHIN SANGYO KENKYUJO (DRUG INDUSTRY RESEARCH INSTITUTE)
JAPANESE

DOTAITO NYUSU RETA 2618
(DOTITE NEWS LETTER)
DOJIN YAKKAGAKU KENKYUSHO (DOJINDO COMPANY, LIMITED, RESEARCH LABORATORY)
2861, MURAZOE, KENGUN-MACHI, KUMAMOTO-SHI 862
JAPANESE

DOTAITO SEIKAGAKU NYUSU 2619
(DOTITE BIOCHEMICAL NEWS)
DOJIN YAKKAGAKU KENKYUSHO (DOJINDO COMPANY, LIMITED, RESEARCH LABORATORY)
2861, MURAZOE, KENGUN-MACHI, KUMAMOTO-SHI 862
JAPANESE

EHIME DAIGAKU KYOIKU GAKUBU SHOGAINI KYOIKU KENKYUSHITSU KENKYU KIYO 2620
(BULLETIN OF THE INSTITUTE FOR THE EDUCATION OF HANDICAPPED CHILDREN, FACULTY OF EDUCATION, EHIME UNIVERSITY.)
JAPANESE

EISEI KAGAKU 2621
(JOURNAL OF HYGIENIC CHEMISTRY)
NIHON YAKUGAKKAI (PHARMACEUTICAL SOCIETY OF JAPAN)
12-15, 2-CHOME, SHIBUYA, SHIBUYA-KU, TOKYO 150
JAPANESE AND ENGLISH
INDEXED: CHE, EXC, JICST

EREKUTORONIKUSU NO RINSHO 2622
(ELECTRONICS IN MEDICAL CLINICS)
MEDIKARU EREKUTORO TAIMUSUSHA (MEDICAL ELECTRO TIMES COMPANY)
39-4, 3-CHOME, HONGO, BUNKYO-KU, TOKYO 113
JAPANESE

FARUMASHIA 2623
(PHARMACY)
NIHON YAKUGAKKAI (PHARMACEUTICAL SOCIETY OF JAPAN)
12-15, 2-CHOME, SHIBUYA, SHIBUYA-KU, TOKYO 150
JAPANESE
INDEXED: CHE, JICST

FUJISAWA YAKUHIN KOGYO CHUO KENKYU SHOHO 2624
(REPORT OF THE FUJISAWA MEDICAL SUPPLIES CENTRAL RESEARCH INSTITUTE)
JAPANESE
INDEXED: JICST

GAIKOKU IYAKU JOHO 2625
(INTERNATIONAL DRUG REPORT)
IYAKUHIN SANGYO KENKYUSHO (DRUG BUSINESS RESEARCH AND COMPANY, LIMITED)
25-2, 1-CHOME, KOISHIKAWA, BUNKYO-KU, TOKYO 112
JAPANESE

GEKKAN SHIN IRYO 2626
(NEW MEDICAL TREATMENT MONTHLY)
EMUI SHINKO KYOKAI (SOCIETY FOR THE ADVANCEMENT OF M.E.)
JAPANESE

GEKKAN YAKUJI 2627
(PHARMACEUTICAL MONTHLY)
YAKUJI KANKYUKAI (PHARMACEUTICAL RESEARCH ASSOCIATION) AND YAKUGYO JIHOSHA
C/O INAOKA BUILDING, 2-36, KANDA JINBO-CHO, CHIYODA-KU, TOKYO 101
JAPANESE
INDEXED: CHE, JICST

GEKKAN YAMABIKO (SHUGAI KENKYU) 2628
(PREVENTION OF ALCOHOLISM MAGAZINE)
NIHON SHUGAI BOSHI KYOKAI (JAPAN SOCIETY FOR PREVENTION OF ALCOHOLISM)
20-14, 7-CHOME, NISHISHINJUKU SHINJUKU-KU, TOKYO 160
JAPANESE

GIFU YAKKA DAIGAKU KIYO 2629
(ANNUAL PROCEEDINGS OF GIFU COLLEGE OF PHARMACY)
GIFU YAKKA DAIGAKU (GIFU COLLEGE OF PHARMACY)
492-36, MITAHORA, GIFU-SHI, GIFU 502
JAPANESE AND ENGLISH
INDEXED: BIO, CHE, EXC, JICST

GINKAI 2630
(NEWS LETTER, SENJU PHARMACEUTICAL COMPANY, LIMITED)
SENJU SEIYAKU K.K. (SENJU PHARMACEUTICAL COMPANY, LIMITED)
3-28, HIRANO-MACHI, HIGASHI-KU, OSAKA-SHI 541
JAPANESE

HIROSHIMA-KEN BYOIN YAKUZAISHIKAI KAISHI 2631
(JOURNAL OF PHARMACEUTICAL ASSOCIATION OF HIROSHIMA PREFECTURAL HOSPITAL)
HIROSHIMA-KEN BYOIN YAKUZAISHIKAI
2-3, 1-CHOME, KASUMI-CHO, HIROSHIMA-SHI 734
JAPANESE

HOIKEN SHINPOJUMU RONBUNSHU 2632
(PROCEEDINGS OF THE SYMPOSIUM, NATIONAL INSTITUTE OF RADIOLOGICAL SCIENCES)
HOSHASEN IGAKU SOGO KENKYUSHO (NATIONAL INSTITUTE OF RADIOLOGICAL SCIENCES)
9-1, 4-CHOME, ANAGAWA, CHIBA-SHI 280
JAPANESE
INDEXED: JICST

HOKKAIDO RIHABIRITESHON GAKKAI ZASSHI 2633
(HOKKAIDO REHABILITATION)
HOKKAIDO RIHABIRITESHON GAKKAI (HOKKAIDO REHABILITATION ASSOCIATION)
C/O SAPPORO IKA DAIGAKU SEIKEI GEKA, MINAMI ICHIJO NISHI-16-CHOME, SAPPORO-SHI 060
JAPANESE

HOSHASEN IGAKU SOGO KENKYUJO TOKUBETSU KENKYU 2634
(SPECIAL RESEARCH RADIATION MEDICINE GENERAL RESEARCH INSTITUTE)
JAPANESE
INDEXED: JICST

HOSHASEN IGAKU SOGO KENKYUSHO NENPO 2635
(ANNUAL REPORT, NATIONAL INSTITUTE OF RADIOLOGICAL SCIENCES, JAPANESE EDITION)
HOSHASEN IGAKU SOGO KENKYUSHO (NATIONAL INSTITUTE OF RADIOLOGICAL SCIENCES)
9-1, 4-CHOME, ANAGAWA, CHIBA-SHI 280
JAPANESE
INDEXED: JICST

HOSHASEN KAGAKU 2636
(RADIOLOGICAL SCIENCES)
HOSHASEN IGAKU SOGO KENKYUSHO (NATIONAL INSTITUTE OF RADIOLOGY SCIENCE) AND JITSUGYO KOHOSHA
2-12, 4-CHOME, KUDAN-MINAMI, CHIYODA-KU, TOKYO 102
JAPANESE
INDEXED: CHE, JICST

HOSHI YAKKA DAIGAKU KIYO 2637
(PROCEEDINGS OF THE HOSHI COLLEGE OF PHARMACY)
HOSHI YAKKA DAIGAKU (HOSHI COLLEGE OF PHARMACY)
4-41, 2-CHOME, EBARA, SHINAGAWA-KU, TOKYO 142
JAPANESE AND ENGLISH
INDEXED: CHE, EXC, JICST

HYOYAKU KAI 2638
(HYOGO PREFECTURE PHARMACISTS ASSOCIATION)
HYOGO-KEN YAHNZAISHI KAI (HYOGO PREFECTURE PHARMACISTS ASSOCIATION)
JAPANESE

HYOYAKUKAI 2639
(JOURNAL OF PHARMACEUTICAL SOCIETY OF HYOGO)
HYOGO-KEN YAKUZAISHIKAI, HYOYAKUKAI HENSU JIMU-KYOKU (PHARMACEUTICAL SOCIETY OF HYOGO)
39-14, 6-CHOME, SHIMOYAMATE-DORI, IKUTA-KU, KOBE-SHI 150
JAPANESE
INDEXED: CHE, JICST

IGAKU TO KIKAI 2640
(MEDICINE AND INSTRUMENTS)
IGAKU JANARU-SHA
1-14, SUEYOSHIBASHI-DORI, MINAMI-KU, OSAKA-SHI 542
JAPANESE

IKAKIKAIGAKU ZASSHI 2641
(JOURNAL OF JAPANESE MEDICAL INSTRUMENTS)
NIHON IKAKIKAI GAKKAI (MEDICAL INSTRUMENTS SOCIETY OF JAPAN)
39-15, 3-CHOME, HONGO, BUNKYO-KU, TOKYO 113
JAPANESE, WITH ENGLISH SUMMARY
INDEXED: EXC, JICST

INTANASHONARU NASHINGU REBYU 2642
(INTERNATIONAL NURSING REVIEW)
NIHON KANGO KYOKAI SHUPPANKAI (JAPAN NURSING ASSOCIATION PUBLISHING GROUP)
JAPANESE

IRYO SANGYO KIJI DAIJESUTO 2643
(MEDICAL TREATMENT INDUSTRY NEWS DIGEST)
IYAKUHIN SANGYO KENKYUSHO (MEDICAL SUPPLIES INDUSTRY RESEARCH INSTITUTE)
JAPANESE

IRYO SHINSEIHIN GEPPO 2644
(NEW MEDICAL TREATMENT PRODUCTS MONTHLY)
IRYOHIN SANGYO KONKYUSHO (MEDICAL TREATMENT PRODUCTS INDUSTRY RESEARCH BULLETIN)
JAPANESE

ISHIZU SHIYAKU JIHO 2645
(ISHIZU REAGENT NEWS)
ISHIZU SEIYAKU K.K. (ISHIZU PHARMACEUTICAL COMPANY, LIMITED)
2-26, DOSHO-MACHI, HIGASHI-KU, OSAKA-SHI 541
JAPANESE

IYAKU JANARU 2646
(MEDICINE AND DRUG JOURNAL)
IYAKU JANARUSHA
C/O KANDA BUILDING, 2-2, KANDA JINBO-CHO, CHIYODA-KU, TOKYO 101
JAPANESE

IYAKU NO MON 2647
(GATEWAY TO MEDICINE)
IYAKU NO MONSHA
3, 3-CHOME, NIHONBASHI HON-CHO, CHUO-KU, TOKYO 103
JAPANESE
INDEXED: CHE

IYAKU TOKKYO SHOROKU 2648
(ABSTRACTS OF PHARMACEUTICAL PATENTS)
NIHON PATERITTO K.K.
YAKUGYO JIHOSHA
C/O INAOKA BUILDING, 2-36, KANDA JINBO-CHO, CHIYODA-KU, TOKYO 101
JAPANESE

IYAKUHIN KENKYU 2649
(STUDY OF MEDICAL SUPPLIES)
NIHON KOTEISHO KYOKAI (SOCIETY OF JAPANESE PHARMACOPOEIA)
12-15, 2-CHOME, SHIBUYA, SHIBUYA-KU, TOKYO 150
JAPANESE, WITH ENGLISH SUMMARY
INDEXED: CHE

IYAKUHIN SANGYO NENPO 2650
(ANNUAL REPORT OF MEDICAL SUPPLIES INDUSTRY)
IYAKUHIN SANGYO KENKYUSHO (DRUG BUSINESS RESEARCH AND COMPANY, LIMITED)
21-11, 2-CHOME, HAKUSAN, BUNKYO-KU, TOKYO 112
JAPANESE

IYAKUHIN SOGO SAYO KENKYU 2651
(RESEARCH ON DRUG INTERACTIONS)
IYAKUHIN SOGO SAYO KENYUKAI, TOHOKU DAIGAKU YAKUGAKU-BU YOKUHIN SAYOGAKU KYOSHITSU (TOHOKU UNIVERSITY, SCHOOL OF PHARMACY, DRUG INTERACTION RESEARCH GROUP)
JAPANESE

IYO DENSHI 2652
(MEDICAL ELECTRONICS)
DENSHI KIKAI KOGYOKAI, IYO DENSHI KIKI GIJUTSU IINKAI AND DENSHI KEISOKU SHUPPANSHA
6-9, 2-CHOME, FUJIMI, CHIYODA-KU, TOKYO 102
JAPANESE

IYO DENSHI TO SEITAI KOGAKU (NIHON M-E GAKKAI ZASSHI) 2653
(JAPANESE JOURNAL OF MEDICAL ELECTRONICS AND BIOLOGICAL ENGINEERING)
NIHON M-E GAKKAI (JAPAN SOCIETY OF MEDICAL ELECTRONICS AND BIOLOGICAL ENGINEERING) AND KORONASHA (CORONA PUBLISHING COMPANY, LIMITED)
46-10, 4-CHOME, SENGOKU, BUNKYO-KU, TOKYO 112
JAPANESE
INDEXED: EXC, IND, INS, JICST

IYO KIZAI KENKYUSHO HOKOKU 2654
(REPORTS OF THE INSTITUTE FOR MEDICAL AND DENTAL ENGINEERING, TOKYO MEDICAL AND DENTAL UNIVERSITY)
TOKYO IKA SHIKA DAIGAKU IYO KIZAI KENKYUSHO
3, YUSHIMA, BUNKYO-KU, TOKYO 113
JAPANESE AND ENGLISH
INDEXED: CHE

IYO KOSO 2655
(JOURNAL OF MEDICAL ENZYMOLOGY)
MEJIKARU BYU SHA (MEDICAL VIEW COMPANY)
JAPANESE

JAPAN MEDICAL NEWS 2656
NIHON IYAKUHIN YUSHUTSU KUMIAI (JAPAN PHARMACEUTICAL, MEDICAL AND DENTAL SUPPLY EXPORTERS' ASSOCIATION)
3-6, 4-CHOME, NIHONBASHI ASSOCIATION HON-CHO, CHUO-KU, TOKYO 103
ENGLISH

JAPAN PHARMACEUTICAL ABSTRACTS 2657
ENGLISH

JAPANESE JOURNAL OF ANTIBIOTICS 2658
NIHON KOSEI BUSSHITSU GAKUJUTSU KYOGIKAI (JAPAN ANTIBIOTICS RESEARCH ASSOCIATION)
20-8, 2-CHOME, KAMIOSAKI, SHINAGAWA-KU, TOKYO 141
JAPANESE AND ENGLISH
INDEXED: BIB, CHE, EXC, IND

JAPANESE JOURNAL OF PHARMACOLOGY 2659
NIHON YAKURI GAKKAI (JAPANESE PHARMACOLOGICAL SOCIETY)
C/O KYOTO DAIGAKU IGAKUBU YAKURIGAKU KYOSHITSU, YOSHIDA KONOE-CHO, SAKYO-KU, KYOTO-SHI 606
ENGLISH
INDEXED: BIO, CHE, EXC, IND, JICST

JAPANESE MEDICAL ULTRASONICS 2660
NIHON CHOONPA IGAKKAI (JAPAN SOCIETY OF ULTRASONICS IN MEDICINE)
C/O JUNTENDO DAIGAKU IGAKUBU CHOONPA SENTA, 1-1, 2-CHOME, HONGO, BUNKYO-KU, TOKYO 113
ENGLISH

JIKKEN CHIRYO 2661
(EXPERIMENTAL THERAPY)
TAKEDA YAKUHIN KOGYO K.K. (TAKEDA PHARMACEUTICAL INDUSTRIES COMPANY, LIMITED)
2-27, DOSHO-MACHI, HIGASHI-KU, OSAKA-SHI 541
JAPANESE
INDEXED: CHE, JICST

JOHO (NIHON IYAKU JOHO SENTA) 2662
(INFORMATION, JPIC)
NIHON IYAKU JOHO SENTA (JAPAN PHARMACEUTICAL INFORMATION CENTER)
12-15, 2-CHOME, SHIBUYA, SHIBUYA-KU, TOKYO 150
JAPANESE

JOURNAL OF ANTIBIOTICS (INTERNATIONAL EDITION) 2663
NIHON KOSEI BUSSHITSU GAKUJUTSU KYOGIKAI (JAPAN ANTIBIOTICS RESEARCH ASSOCIATION)
20-8, 2-CHOME, KAMIOSAKI, SHINAGAWA-KU, TOKYO 141
ENGLISH
INDEXED: BIO, CHE, EXC, IND, SCI, JICST

JOURNAL OF RADIATION RESEARCH 2664
NIHON HOSHASEN EIKYO GAKKAI (JAPAN RADIATION RESEARCH SOCIETY) AND HOSHASEN IGAKU SOGO KENKYUSHO (NATIONAL INSTITUTE OF RADIOLOGICAL SCIENCES)
9-1, 4-CHOME, ANAGAWA, CHIBA-SHI 280
ENGLISH
INDEXED: BIB, BIO, CHE, EXC, INS, POL, JICST

KAGAKU RYOHO KENKYUSHO IHO 2665
(STUDIES FROM THE CHEMOTHERAPEUTIC INSTITUTE FOR MEDICAL RESEARCH)
KAGAKU RYOHO KENKYUSHO (CHEMOTHERAPEUTIC INSTITUTE FOR MEDICAL RESEARCH)
1-14, 6-CHOME, KONODAI, ICHIKAWA-SHI 272
JAPANESE
INDEXED: CHE, JICST

KAKU IGAKU (NIHON KAKU IGAKKAI KIKANSHI) 2666
(JAPANESE JOURNAL OF NUCLEAR MEDICINE)
NIHON KAKU IGAKKAI (JAPANESE SOCIETY OF NUCLEAR MEDICINE)
C/O 2ND INTERNAL MEDICINE, UNIVERSITY OF TOKYO, 3-1, 7-CHOME, HONGO, BUNKYO-KU, TOKYO 113
JAPANESE AND ENGLISH
INDEXED: CHE, EXC, JICST

KANAZAWA IRIGAKU SOSHO 2667
(JOURNAL OF RADIOLOGY AND PHYSICAL THERAPY, UNIVERSITY OF KANAZAWA)
KANAZAWA DAIGAKU IGAKUBU HOSHASEN IGAKU KYOSHITSU (DEPARTMENT OF RADIOLOGY, SCHOOL OF MEDICINE, KANAZAWA UNIVERSITY)
13-1, TAKARA-MACHI, KANAZAWA-SHI 920
JAPANESE, WITH ENGLISH SUMMARY
INDEXED: EXC

KANGO 2668
(JAPANESE JOURNAL OF NURSING)
NIHON KANGO KYOKAI (JAPAN NURSING ASSOCIATION)
8-2, 5-CHOME, JINGUMAE, SHIBUYA-KU, TOKYO 150
JAPANESE

KANGO GIJUTSU 2669
(JAPANESE JOURNAL OF NURSING ART)
MEJIKARU FURENDO-SHA (MEDICAL FRIEND COMPANY, LIMITED)
1-32, 4-CHOME, KUDANKITA, CHIYODA-KU, TOKYO 102
JAPANESE

KANGO JISSEN NO KAGAKU 2670
(SCIENCE OF PRACTICAL NURSING)
KANGO NO KAGAKUSHA (NURSING SCIENCE COMPANY)
JAPANESE

KANGO KENKYU 2671
(JAPANESE JOURNAL OF NURSING RESEARCH)
IGAKU SHOIN
29-11, 5-CHOME, HONGO, BUNKYO-KU, TOKYO 113
JAPANESE

KANGO KYOIKU 2672
(JAPANESE JOURNAL OF NURSING EDUCATION)
IGAKU SHOIN
29-11, 5-CHOME, HONGO, BUNKYO-KU, TOKYO 113
JAPANESE

KANGO TENBO 2673
(JAPANESE JOURNAL OF NURSING SCIENCE)
MEJIKARU FURENDO SHA (MEDICAL FRIEND COMPANY, LIMITED)
JAPANESE

KANGOGAKU ZASSHI 2674
(JAPANESE JOURNAL OF NURSING)
IGAKU SHOIN
29-11, 5-CHOME, HONGO, BUNKYO-KU, TOKYO 113
JAPANESE

KANKO SHIKISO 2675
(MEDICAL RESEARCHES FOR PHOTOSENSITIZING DYES)
KANKO SHIKISO KENKYUKAI (PHOTOSENSITIZING DYES RESEARCH SOCIETY)
C/O KYOTO DAIGAKU IGAKUBU BYORIGAKU KYOSHITSU, YOSHIDA KONOE-CHO, SAKYO-KU, KYOTO-SHI 606
JAPANESE, WITH EUROPEAN SUMMARY
INDEXED: CHE, JICST

KANPO NO RINSHO 2676
(PRACTICAL KANPO)
TOA IGAKU KYOKAI (ASSOCIATION OF EAST-ASIAN MEDICINE)
2-20, SHIN'OGAWA-MACHI, SHINJUKU-KU, TOKYO 162
JAPANESE
INDEXED: JICST

KEIRAKU CHIRYO 2677
(KEIRAKU)
KEIRAKU CHIRYO KENKYUKAI
27-19, 2-CHOME, KOMAZAWA, SETAGAYA-KU, TOKYO 154
JAPANESE

KEIRAKU SHINRYO 2678
(ACUPUNCTURE)
TOYO HARI IGAKKAI
C/O FUJI KOPU, 26, DAIKYO-CHO, SHINJUKU-KU, TOKYO 160
JAPANESE

KENKYU YAKUBUTSU JOHO 2679
(DRUG RESEARCH REPORT)
IYAKUHIN SANGYO KENKYUSHO (MEDICAL SUPPLIES (DRUGS) INDUSTRY RESEARCH INSTITUTE)
JAPANESE

KETSUEKI JIGYO NO GENJO (NIIGATA-KEN) 2680
(SITUATION OF BLOOD ENTERPRISE, NIIGATA PREFECTURE)
NIIGATA-KEN EISEI-BU (NIIGATA PREFECTURAL GOVERMENT)
602, ICHIBAN-CHO, GAKKO-DORI, NIIGATA-SHI 951
JAPANESE

KIKAN BYOIN SEISHIN IGAKU 2681
(QUARTERLY (JOURNAL OF) INSTITUTIONAL PSYCHOLOGICAL MEDICINE)
BYOIN SEISHIN IGAKUKAI (SOCIETY FOR INSTITUTIONAL PSYCHOLOGICAL MEDICINE)
JAPANESE

KIKAN SEISHIN RYOHO 2682
(PSYCHOTHERAPY QUARTERLY) JAPANESE JOURNAL OF PSYCHOTHERAPY)
KONGO SHUPPAN (KONGO PUBLISHING COMPANY)
JAPANESE

KITAZATO DAIGAKU YAKUGAKUBU KENKYU GYOSEKISHU 2683
(ARCHIVES OF THE FACULTY OF PHARMACY, KITAZATO UNIVERSITY)
KITAZATO DAIGAKU YAKUGAKUBU
9-1, 5-CHOME, SHIROGANE, MINATO-KU, TOKYO 108
JAPANESE AND ENGLISH

KODOMO IRYO SENTA IGAKUSHI 2684
(JOURNAL OF THE CHILDREN'S MEDICAL TREATMENT CENTER)
KANAGAWA-KENRITSU KODOMO IRYO SENTA (KANAGAWA PREFECTURAL CHILDREN'S MEDICAL (TREATMENT) CENTER.)
JAPANESE

KOKURITSU TOKUSHU KYOIKU SOGO KENKYUSHO KENKYU KIYO 2685
(BULLETIN OF THE NATIONAL INSTITUTE FOR SPECIAL EDUCATION)
JAPANESE

KOKUSAI RIHABIRITESHON NYUSU 2686
(INTERNATIONAL REHABILITATION REVIEW)
NIHON SHOGAISHA RIHABIRITESHON KYOKAI
13-15, 3-CHOME, HIGASHIIKEBUKURO, TOSHIMA-KU, TOKYO 170
JAPANESE

KOMAZAWA SHINRIGAKU KENKYU 2687
(KOMAZAWA PSYCHOLOGICAL RESEARCH)
KOMAZAWA SHINRIGAKU KENKYUKAI (KONAZAWA PSYCHOLOGICAL RESEARCH ASSOCIATION)
JAPANESE

KOWA YAKUHO 2688
(KOWA PHARMACEUTICAL MAGAZINE)
KOWA K.K. (KOWA COMPANY, LIMITED)
3-3, NIHONBASHI HON-CHO, CHUO-KU, TOKYO 103
JAPANESE

KURINIKA 2689
(CLINICA)
KINDAI IGAKUSHA (MODERN MEDICAL COMPANY)
JAPANESE

KUSURI NO CHISHIKI 2690
(PHARMACEUTICAL REVIEW)
SUMITOMO KOGYO K.K. IYAKU JIGYO-BU (PHARMACEUTICAL DIVISION, SUMITOMO CHEMICAL COMPANY, LIMITED)
2-40, DOSHO-MACHI, HIGASHI-KU, OSAKA-SHI 541
JAPANESE
INDEXED: JICST

KUSURI NO HIROBA 2691
(PHARMACEUTICAL SPACE)
KUSURI O KANSHI SURU KOKUMIN UNDO NO KAI
23-11, 5-CHOME, MINAMITANAKA, NERIMA-KU, TOKYO 177
JAPANESE

KYOIKUGAKU SHINRIGAKU KENKYU KIYO 2692
(RESEARCH BULLETIN ON EDUCATION AND PSYCHOLOGY)
GIFU DAIGAKU KYOIKUGAKU-BU KYOIKU KENKYU SHITSU (GIFU UNIVERSITY SCHOOL OF EDUCATION, EDUCATIONAL RESEARCH OFFICE)
JAPANESE

KYORITSU YAKKA DAIGAKU KENKYU NENPO 2693
(ANNUAL REPORT OF THE KYORITSU COLLEGE OF PHARMACY)
KYORITSU YAKKA DAIGAKU (KYORITSU COLLEGE OF PHARMACY)
5-30, 1-CHOME, SHIBA KOEN, MINATO-KU, TOKYO 105
JAPANESE, WITH ENGLISH SUMMARY
INDEXED: CHE, JICST

KYOYAKUKAISHI 2694
(JOURNAL OF KYOYAKU SOCIETY)
KYOYKUKAI HONBU
C/O KYOTO YAKKA DAIGAKU, 5, NAKAUCHI-CHO, YAMASHINA GORYO, HIGASHIYAMA-KU, KYOTO-SHI 607
JAPANESE

KYUKYU IGAKU 2695
(JAPANESE JOURNAL OF ACUTE MEDICINE)
HERUSU SHUPPAN (HERUSU PUBLISHING)
JAPANESE

KYUSHU YAKUGAKKAI KAIHO 2696
(JOURNAL OF THE KYUSHU PHARMACEUTICAL SOCIETY)
KYUSHU YAMAGUCHI YAKUGAKKAI (KYUSHU YAMAGUCHI PHARMACEUTICAL SOCIETY)
C/O KYUSHU DAIGAKU IGAKUBU FUZOKU BYOIN YAKUZAIBU, 1276, KATAKASU, HIGASHI-KU, FUKUOKA-SHI 812
JAPANESE, WITH ENGLISH SUMMARY
INDEXED: CHE

MEDI CONTENTS 2697
IYAKUHIN SANGYO KENKYUSHO (DRUG PRODUCTION RESEARCH SUBSTITUTE)
JAPANESE AND ENGLISH

MEDICAL CORNER 2698
MEIJI SEIKA GAKUJUTSU-BU
2-8, KYOBASHI, CHUO-KU, TOKYO 104
JAPANESE

MEDICAL DIGEST 2699
DAIICHI SEIYAKU K.K. GAKUJUTSU-BU
14-10, 3-CHOME, NIHONBASHI, CHUO-KU, TOKYO 103
JAPANESE
INDEXED: CHE

MEDICAL NEWS 2700
KYOWA KIKAKU (KYOWA KIKAKU COMPANY, LIMITED)
DAI-NIPPON SEIYAKU K.K. (DAI-NIPPON PHARMACEUTICAL COMPANY, LIMITED)
2-7, NIHONBASHI HON-CHO, CHUO-KU, TOKYO 103
JAPANESE

MEDICAL PHARMACY 2701
DAIICHI SEIYAKU
14-10, 3-CHOME, NIHONBASHI, CHUO-KU, TOKYO 103
JAPANESE
INDEXED: CHE

MEDIYA SAKURU 2702
(MEDIA CIRCLE)
NISSUI SEIYAKU K.K. (NISSUI PHARMACEUTICAL COMPANY, LIMITED)
5-11, 2-CHOME, KOMAGOME, TOSHIMA-KU, TOKYO 170
JAPANESE

MEIJI SEIKA KENKYU NENPO 2703
(SCIENTIFIC REPORTS OF MEIJI SEIKA KAISHA)
MEIJI SEIKA K.K. CHUO KENKYUSHO (MEIJI SEIKA KAISHA LIMITED, RESEARCH LABORATORIES)
MOROOKA-CHO, KOHOKU-KU, YOKOHAMA-SHI 222
JAPANESE AND ENGLISH
INDEXED: BIO, CHE, JICST

MEIJI YAKUKA DAIGAKU KENKYU KIYO 2704
(BULLETIN OF MEIJI PHARMACEUTICAL COLLEGE)
JAPANESE
INDEXED: JICST

MEMOIRS OF THE FACULTY OF PHARMACEUTICAL SCIENCES OSAKA UNIVERSITY 2705
OSAKA DAIGAKU YAKUGAKUBU (FACULTY OF PHARMACEUTICAL SCIENCES, OSAKA UNIVERSITY)
1-1, 6-CHOME, TONEYAMA, TOYONAKA-SHI 560
JAPANESE AND ENGLISH

MINOPHAGEN MEDICAL REVIEW 2706
MINOPHAGEN SEIYAKU HONPO
10-22, 8-CHOME, AKASAKA, MINATO-KU, TOKYO 107
JAPANESE

MOCHIDA SEIYAKU K.K. RINSHO BUNKENSHU 2707
(CLINICAL DOCUMENTS OF MOCHIDA PHARMACEUTICAL COMPANY, LIMITED)
MOCHIDA SEIYAKU K.K. (MOCHIDA PHARMACEUTICAL COMPANY, LIMITED)
1-1, YOTSUYA, SHINJUKU-KU, TOKYO 160
JAPANESE

MUKOGAWA JOSHI DAIGAKU KIYO, YAKUGAKU-HEN 2708
(BULLETIN OF MUKOGAWA WOMEN'S UNIVERSITY, PHARMACEUTICAL SCIENCE))
MUKOGAWA JOSHI DAIGAKU (MUKOGAWA WOMEN'S UNIVERSITY)
6-46, IKEBIRAKI-CHO, NISHINOMIYA-SHI 663
JAPANESE, WITH ENGLISH SUMMARY
INDEXED: JICST

NAGOYA-SHI KOSEIIN KIYO 2709
(NAGOYA GERIATRIC MEDICINE)
JAPANESE

NAGOYA-SHIRITSU DAIGAKU YAKUGAKUBU KENKYU NENPO 2710
(ANNUAL REPORT OF THE FACULTY OF PHARMACEUTICAL SCIENCES, NAGOYA CITY UNIVERSITY)
NAGOYA-SHIRITSU DAIGAKU YAKUGAKUBU (THE FACULTY OF PHARMACEUTICAL SCIENCES, NAGOYA CITY UNIVERSITY)
3-1, TANABE-DORI, MIZUHO-KU, NAGOYA-SHI 467
JAPANESE, WITH EUROPEAN SUMMARY
INDEXED: BIO, CHE, JICST

NARA-KEN YAKUJI NENPO 2711
(PHARMACEUTICAL REPORT OF NARA PREFECTURE)
NARA-KEN KOSEI-BU (NARA PREFECTURAL GOVERNMENT)
NOBORI OJI-CHO, NARA-SHI 630
JAPANESE

NARA-KEN YAKUJI SHIDOSHO HOKOKU 2712
(REPORT OF PHARMACEUTICAL GUIDANCE DIVISION, NARA PREFECTURE)
NARA-KEN YAKUJI SHIDOSHO
605-10, GOSE, GOSE-SHI 639-22
JAPANESE

NASU SUTESHON 2713
(JOURNAL OF NURSING ADMINISTRATION FOR PATIENTS)
IGAKU SHOIN
29-11, 5-CHOME, HONGO, BUNKYO-KU, TOKYO 113
JAPANESE

NICHIYAKU FAMASHISUTO 2714
(JAPAN PHARMACIST)
NICHIYAKU FAMASHISUTO SHUPPANSHA
8-7, 6-CHOME, GINZA, CHUO-KU, TOKYO 104
JAPANESE

NICHIYAKU SHINBUN 2715
(WEEKLY NEWS OF DRUG INDUSTRY)
NICHIYAKU SHINBUNSHA
10-18, 8-CHOME, NISHIGOTANDA, SHINAGAWA-KU, TOKYO 141
JAPANESE

NIHON CHOONPA IGAKKAI KENKYU HAPPOKAI KOEN RONBUNSHU 2716
(PROCEEDINGS OF THE JAPAN SOCIETY OF ULTRASONICS IN MEDICINE)
NIHON CHOONPA IGAKKAI (JAPAN SOCIETY OF ULTRASONICS IN MEDICINE)
C/O JUNTENDO DAIGAKU IGAKUBU CHOONPA SENTA, 2-1, HONGO, BUNKYO-KU, TOKYO 113
JAPANESE

NIHON CHOONPA IGAKKAI KOEN RONBUNSHU 2717
(COLLECTED LECTURES AND ESSAYS OF THE JAPAN MEDICAL ULTRASONICS SOCIETY)
JAPANESE

NIHON DAIGAKU YAKUGAKU KENKYU HOKOKU 2718
(PHARMACEUTICAL BULLETIN OF NIHON UNIVERSITY)
NIHON DAIGAKU RIKOGAKUBU YAKUGAKKA (DEPARTMENT OF PHARMACY, COLLEGE OF SCIENCE AND TECHNOLOGY, NIHON UNIVERSITY)
1-8, KANDA SURUGADAI, CHIYODA-KU, TOKYO 101
JAPANESE AND ENGLISH
INDEXED: CHE

NIHON HOSHASEN EIKYO GAKKAI NYUSU 2719
(NEWS OF THE JAPAN RADIATION RESEARCH SOCIETY)
NIHON HOSHASEN EIKYO GAKKAI (JAPAN RADIATION RESEARCH SOCIETY)
C/O HOSHASEN IGAKU SOGO KENKYUSHO, 9-1, 4-CHOME, ANAGAWA, CHIBA-SHI 280
JAPANESE

NIHON HOSHASEN GIJUTSU GAKKAI SOKAI SHOROKUSHU 2720
(ABSTRACTS OF THE SYMPOSIUM OF JAPANESE SOCIETY OF RADIOLOGICAL TECHNOLOGY)
NIHON HOSHASEN GIJUTSU GAKKAI (JAPANESE SOCIETY OF RADIOLOGICAL TECHNOLOGY)
C/O NIJO PURAZA, 88, NISHINOKYO KITATSUBOI-CHO, NAKAGYO-KU, KYOTO-SHI 604
JAPANESE
INDEXED: JICST

NIHON HOSHASEN GIJUTSU GAKKAI ZASSHI 2721
(JAPANESE JOURNAL OF RADIOLOGICAL TECHNOLOGY)
NIHON HOSHASEN GIJUTSU GAKKAI (JAPANESE SOCIETY OF RADIOLOGICAL TECHNOLOGY)
C/O NIJO PURAZA, 88, NISHINOKYO KITATSUBOI-CHO, NAKAGYO-KU, KYOTO-SHI 604
JAPANESE, WITH EUROPEAN SUMMARY
INDEXED: CHE, JICST

NIHON HOSHASEN GIJUTSU GAKKAI ZASSHI, TOKUSHU 2722
(JOURNAL OF THE JAPAN RADIATION TECHNOLOGY SOCIETY, SPECIAL EDITION)
JAPANESE

NIHON HOSHASEN GISHIKAI ZASSHI 2723
(JOURNAL OF JAPAN RADIOLOGICAL TECHNICIANS ASSOCIATIONS)
NIHON HOSHASEN GISHIKAI (JAPAN RADIOLOGICAL TECHNICIANS' ASSOCIATION)
14-18-402, 5-CHOME, MINAMISHINAGAWA, SHINAGAWA-KU, TOKYO 140
JAPANESE

NIHON IGAKU HOSHASEN GAKKAI ZASSHI 2724
(NIPPON ACTA RADIOLOGICA, THE OFFICIAL JOURNAL OF JAPAN RADIOLOGICAL SOCIETY)
NIHON IGAKU HOSHASEN GAKKAI (JAPAN RADIOLOGICAL SOCIETY)
ROOM NO. 301 AKAMON ABITASHION, 29-13, 5-CHOME, HONGO, BUNKYO-KU, TOKYO 113
JAPANESE AND ENGLISH
INDEXED: CHE, EXC, JICST

NIHON IKA KIKAI SHINBUN 2725
(JAPAN MEDICAL INSTRUMENTS NEWS)
NIHON IKA KIKAI SHINBUNSHA
15-3, MIYAMOTO-CHO, ITABASHI-KU, TOKYO 174
JAPANESE

NIHON IYAKUHIN FUKUSAYO BUNKEN SHOROKUSHU 2726
(SUMMARY OF THE JAPAN-PHARMACEUTICS PHYSICAL REACTIONS BIBLIOGRAPHY)
NIHON IYAKU JOHO SENTA, YAKUJI NIPPOSHA (HATSUBAI) (JAPAN PHARMACEUTICAL INFORMATION CENTER, PHARMACEUTICS DAILY)
JAPANESE

NIHON KAGAKU RYOHO GAKKAI SOKAI SHOROKU 2727
(ABSTRACTS OF THE SYMPOSIUM OF THE JAPAN SOCIETY OF CHEMOTHERAPY)
NIHON KAGAKU RYOHO GAKKAI (JAPAN SOCIETY OF CHEMOTHERAPY)
20-8, 2-CHOME, KAMIOSAKI, SHINAGAWA-KU, TOKYO 141
JAPANESE

NIHON KANKO SHIKISO KENKYUSHO HOKOKU 2728
(BULLETIN OF JAPANESE RESEARCH INSTITUTE FOR PHOTOSENSITIZING DYES)
NIHON KANKO SHIKISO KENKYUSHO (JAPANESE RESEARCH INSTITUTE FOR PHOTOSENSITIZING DYES COMPANY, LIMITED)
2-3, 1-CHOME, SHIMOISHII, OKAYAMA-SHI 700
JAPANESE AND ENGLISH
INDEXED: CHE, JICST

NIHON M-E GAKKAI TAIKAI YOKOSHU 2729
(PREPRINTS OF THE SYMPOSIUM OF THE JAPAN SOCIETY OF MEDICAL ELECTRONICS AND BIOLOGICAL ENGINEERING)
NIHON M-E GAKKAI (JAPAN SOCIETY OF MEDICAL ELECTRONICS AND BIOLOGICAL ENGINEERING)
5-8, 3-CHOME, SHIBA KOEN, MINATO-KU, TOKYO 105
JAPANESE

NIHON MUKI YAKUHIN KYOKUI KAIHO 2730
(PROCEEDINGS OF THE JAPAN INORGANIC PHARMACEUTICAL SOCIETY)
JAPANESE

NIHON NAISHIKYO GAKKAISHI 2731
(GASTROENTEROLOGICAL ENDOSCOPY)
NIHON SHOKAKI NAISHIKYO GAKKAI (JAPAN GASTRO-ENTEROLOGICAL ENDOSCOPY SOCIETY)
3-4, KANDA OGAWA-MACHI, CHIYODA-KU, TOKYO 101
JAPANESE AND ENGLISH

NIHON ONSEN KIKO BUTSURI IGAKKAI ZASSHI 2732
(JOURNAL OF THE JAPANESE ASSOCIATION OF PHYSICAL MEDICINE BALNEOLOGY AND CLIMATOLOGY)
NIHON ONSEN KIKO BUTSURI IGAKKAI (JAPANESE ASSOCIATION OF PHYSICAL MEDICINE BALNEOLOGY AND CLIMATOLOGY)
C/O TOKYO DAIGAKU IGAKUBU BUTSURYO NAIKA, 3-1, 7-CHOME, HONGO, BUNKYO-KU, TOKYO 113
JAPANESE, WITH EUROPEAN SUMMARY
INDEXED: CHE, EXC, JICST

NIHON RYODORAKU JIRITSU SHINKEI ZASSHI 2733
(JAPANESE JOURNAL OF RYODORAKU, AUTONOMIC NERVOUS SYSTEM)
NIHON RYODORAKU JIRITSU SHINKEI GAKKAI
1-4, NAKATSU-HONODORI, OYODO-KU, OSAKA-SHI 532
JAPANESE
INDEXED: JICST

NIHON SHINKYU CHIRYO GAKKAISHI 2734
(JOURNAL OF THE JAPAN ACUPUNCTURE AND MOXIBUSTION SOCIETY)
NIHON SHINKYU CHIRYO GAKKAI (JAPAN ACUPUNCTURE AND MOXIBUSTION SOCIETY)
8-7, UDAGAWA-CHO, SHIBUYA-KU, TOKYO 150
JAPANESE

NIHON SHINKYU SHINPO 2735
(CURRENT REPORT ON JAPANESE ACUPUNCTURE)
NIHON SHINKYU SHIKAI
(JAPAN ACUPUNCTURE SOCIETY)
JAPANESE

NIHON YAKUGYO SHINBUN 2736
(JAPAN DRUG NEWS)
NIHON YAKUGYO SHINBUNSHA
C/O SHOEI BUILDING, 1-12, AWAJI-CHO, HIGASHI-KU, OSAKA-SHI 541
JAPANESE

NIHON YAKURIGAKU ZASSHI 2737
(FOLIA PHARMACOLOGICA JAPONICA)
NIHON YAKURI GAKKAI (JAPANESE PHARMACOLOGICAL SOCIETY) AND KYOTO DAIGAKU IGAKUBU YAKURIGAKU KYOSHITSU (C/O DEPARTMENT OF PHARMACOLOGY, FACULTY OF MEDICINE, KYOTO UNIVERSITY)
YOSHIDA KONOE-CHO, SAKYO-KU, KYOTO-SHI 606
JAPANESE, WITH ENGLISH SUMMARY
INDEXED: BIO, CHE, EXC, JICST

NIHON YAKUZAISHIKAI ZASSHI 2738
(JOURNAL OF THE JAPAN PHARMACEUTICAL ASSOCIATION)
NIHON YAKUZAISHIKAI (JAPAN PHARMACEUTICAL ASSOCIATION)
C/O YAKUGAKU KAIKAN, 12-15, 2-CHOME, SHIBUYA, SHIBUYA-KU, TOKYO 150
JAPANESE
INDEXED: CHE

NIHON YUKETSU GAKKAI SOKAI GAKKAI ENZETSU SHOROKU 2739
(SUMMARIES OF REPORTS, GENERAL CONGRESS OF JAPAN SOCIETY OF BLOOD TRANSFUSION)
NIHON YUKETSU GAKKAI (JAPAN SOCIETY OF BLOOD TRANSFUSION)
C/O TOKYO JOSHI IKA DAIGAKU BYOIN YUKETSUBU, 10, ICHIGAYA KAWADA-CHO, SHINJUKU-KU, TOKYO 162
JAPANESE

NIHON YUKETSU GAKKAI ZASSHI 2740
(JOURNAL OF THE JAPAN SOCIETY OF BLOOD TRANSFUSION)
NIHON YUKETSU GAKKAI (JAPAN SOCIETY OF BLOOD TRANSFUSION)
C/O TOKYO JOSHI IKA DAIGAKU BYOIN YUKETSUBU, 10, ICHIGAYA KAWADA-CHO, SHINJUKU-KU, TOKYO 162
JAPANESE, WITH ENGLISH SUMMARY
INDEXED: CHE

NIKKAN YAKUGYO 2741
(DAILY NEWS OF DRUG INDUSTRY)
YAKUGYO JIHOSHA
C/O INAOKA BUILDING, 2-36, KANDA JINBO-CHO, CHIYODA-KU, TOKYO 101
JAPANESE

NIPPON KAGAKU RYOHO GAKKAI SOKAI PUROGURAMU 2742
(PROGRAM OF THE GENERAL MEETING OF THE JAPANESE SOCIETY OF CHEMICAL MEDICAL TREATMENT)
JAPANESE
INDEXED: JICST

NIRS-M 2743
HOSHASEN IGAKU SOGO KENKYUSHO
9-1, 4-CHOME, ANAGAWA, CHIBA-SHI 280
JAPANESE

NIRS-R 2744
HOSHASEN IGAKU SOGO KENKYUSHO
9-1, 4-CHOME, ANAGAWA, CHIBA-SHI 280
JAPANESE

ONKEN KIYO 2745
(BULLETIN OF THE INSTITUTE OF BALNEOTHERAPEUTICS)
KYUSHU DAIGAKU ONSEN CHIRYOGAKU KENKYUSHO (INSTITUTE OF BALNEOTHERAPEUTICS, KYUSHU UNIVERSITY)
TSURUMI HARA, BEPPU-SHI 874
JAPANESE, WITH ENGLISH SUMMARY
INDEXED: CHE, JICST

OSAKA NO BYOJAKU KYOJAKU KYOIKU 2746
(EDUCATION FOR THE MENTALLY AND PHYSICALLY HANDICAPPED)
OSAKA BYOJAKU KYOJAKU KYOIKU KENKYUKAI (OSAKA SOCIETY FOR RESEARCH ON THE MENTALLY AND PHYSICALLY HANDICAPPED)
JAPANESE

OSAKA-FU YAKU ZASSHI 2747
(JOURNAL OF THE OSAKA PHARMACEUTICAL ASSOCIATION)
OSAKA-FU YAKU ZASSHI HENSHU-KYOKU
12, MINAMIOGI-MACHI, KITA-KU, OSAKA-SHI 530
JAPANESE

OSAKA-FURITSU KOSHU EISEI KENKYUSHO KENKYU HOKOKU 2748
(YAKUJI SHIDO-HEN)
(BULLETIN OF OSAKA PREFECTURAL INSTITUTE OF PUBLIC HEALTH, SERIES OF PHARMACEUTICAL GUIDANCE)
OSAKA-FURITSU KOSHU EISEI KENKYUSHO
3-69, 1-CHOME, NAKAMICHI, HIGASHINARI-KU, OSAKA-SHI 537
JAPANESE
INDEXED: CHE

OTSUKA YAKUHO 2749
(JOURNAL OF THE OTSUKA PHARMACEUTICAL FACTORY)
OTSUKA SEIYAKU KOJO (OTSUKA PHARMACEUTICAL FACTORY)
TATEIWA, MUYA-CHO, NARUTO-SHI 772
JAPANESE

OYO YAKURI 2750
(PHARMACOMETRICS)
OYO YAKURI KENKYUKAI
AND KINDAI IGAKUSHA
C/O KAWATSU BUILDING, 20-17, 1-CHOME, JINNAN,
SHIBUYA-KU, TOKYO 150
JAPANESE AND ENGLISH
INDEXED: EXC, JICST

PHARMACEUTICAL DAILY NEWS 2751
YAKUGYO JIHOSHA (YAKUGYO JIHO COMPANY)
2-36, KANDA JINBO-CHO, CHIYODA-KU, TOKYO 101
ENGLISH

PURASUMIN BUNKENSHU 2752
(BIBLIOGRAPHY OF PLASMIN)
KOKUSAI IGAKU JOHO SENTA
DAIICHI SEIYAKU K.K.
14-10, 3-CHOME, NIHONBASHI EDOBASHI, CHUO-KU,
TOKYO 103
JAPANESE AND ENGLISH

RIGAKU RYOHO TO SAGYO RYOHO 2753
(JAPANESE JOURNAL OF PHYSICAL THERAPY AND
OCCUPATIONAL THERAPY)
IGAKU SHOIN
29-11, 5-CHOME, HONGO, BUNKYO-KU, TOKYO 113
JAPANESE

RIHABIRI TECHO SSK 2754
(REHABILITATION NOTEBOOK SSK)
SHINTAI SHOGAISHA DANTAI TEIKI KANKOBUTSU KYOKAI
(PHYSICALLY HANDICAPPED GROUP PERIODICAL PUBLISHING
ASSOCIATION)
JAPANESE

RIHABIRITESHON 2755
(REHABILITATION)
ZEN KOKUTETSU SHINSHOSHA KYOKAI (THE HANDICAPPED
PERSONS ASSOCIATION OF ALL JAPAN NATIONAL RAILWAY)
5-1, KOJI-MACHI, CHIYODA-KU, TOKYO 102
JAPANESE

RIHABIRITESHON IGAKU 2756
(JAPANESE JOURNAL OF REHABILITATION MEDICINE)
NIHON RIHABIRITESHON IGAKKAI AND IGAKU SHOIN
29-11, 5-CHOME, HONGO, BUNKYO-KU, TOKYO 113
JAPANESE, WITH ENGLISH SUMMARY

RIHABIRITESHON KENKYU 2757
(STUDY OF CURRENT REHABILITATION)
NIHON SHOGAISHA RIHABIRITESHON KYOKAI
13-15, 3-CHOME, HIGASHIKEBUKURO, TOSHIMA-KU,
TOKYO 170
JAPANESE

RIHABIRITESHON SHINRIGAKU KENKYU 2758
(JAPANESE JOURNAL OF REHABILITATION PSYCHOLOGY)
SHINRI RIHABIRITESHON KENKYUSHO HENSHU-BU
C/O YASURAGISO, YASUKOGEN FUKUSHI-MURA, YASU-MACHI,
ASAKURA-GUN, FUKUOKA 838-02
JAPANESE

RINSHO HOSHASEN 2759
(JAPANESE JOURNAL OF CLINICAL RADIOLOGY)
KANEHARA SHUPPAN K.K. (KANEHARA SHUPPAN COMPANY,
LIMITED)
31-14, 2-CHOME, YUSHIMA, BUNKYO-KU, TOKYO 113-91
JAPANESE
INDEXED: JICST

RINSHO KANGO 2760
(JAPANESE JOURNAL OF CLINICAL NURSING)
HERUSU SHUPPAN (HEALTH PUBLISHING)
JAPANESE

RINSHO KOGAKU 2761
(JOURNAL OF CLINICAL ENGINEERING)
TOA IYO DENSHI K.K. (TOA MEDICAL ELECTRONICS
COMPANY, LIMITED)
8-2, KAKOGAWA KOGYO DANCHI, KITANO, NOGUCHI-CHO,
KAKOGAWA-SHI 675
JAPANESE

RINSHO NO AYUMI 2762
(CLINICAL STEP)
TANABE SEIYAKU K.K.
3-21, DOSHO-MACHI, HIGASHI-KU, OSAKA-SHI 541
JAPANESE

RINSHO RIGAKU RYOHO 2763
(JOURNAL OF JAPANESE PHYSICAL THERAPY ASSOCIATION)
NIHON RIGAKU RYOHOSHI KYOKAI (JAPAN PHYSICAL
THERAPISTS ASSOCIATION)
JAPANESE

RINSHO YAKURI 2764
(JAPANESE JOURNAL OF CLINICAL PHARMACOLOGY)
RINSHO YAKURIGAKU KENKYUKAI (JAPANESE ASSOCIATION
OF CLINICAL PHARMACOLOGY)
C/O JITCHUKEN IGAKU KENKYUSHO, 1433, NOGAWA,
TAKATSU-KU, KAWASAKI-SHI 211
JAPANESE AND ENGLISH
INDEXED: CHE

RINSHOI NO HOSHASEN 2765
(CLINICAL RADIATION)
OCHAHOMIZU JIMUKYOKU (C.R.C.-CLINICAL RADIATION
CENTER) AND OCHA NO MIZU (C.R.C.-BUSINESS OFFICE)
JAPANESE

RYOIKU 2766
(REHABILITATION, JAPANESE SOCIETY FOR CRIPPLED
CHILDREN)
NIHON SHITAI FUJIYUJI KYOKAI (JAPANESE SOCIETY
FOR CRIPPLED CHILDREN)
13-15, 3-CHOME, HIGASHIIKEBUKURO, TOSHIMA-KU,
TOKYO 170
JAPANESE

SAGA-KEN YAKUGYO SHIDOSHO HOKOKU 2767
(REPORT OF THE SAGA PREFECTURE PHARMACEUTICAL
RESEARCH INSTITUTE)
SAGA-KEN YAKUGYO SHIDOSHO (SAGA PREFECTURE
PHARMACEUTICAL RESEARCH INSTITUTE)
SHUKU-MACHI, TOSU-SHI 841
JAPANESE

SANKYO KENKYUSHO NENPO 2768
(ANNUAL REPORT OF SANKYO RESEARCH LABORATORY)
SANKYO K.K. CHUO KENKYUSHO (RESEARCH LABORATORY,
SANKYO COMPANY, LIMITED)
2-58, 1-CHOME, HIRO-MACHI, SHINAGAWA-KU, TOKYO 140
JAPANESE AND ENGLISH
INDEXED: BIO, CHE, JICST

SEIROKA KANGO DAIGAKU KIYO 2769
(BULLETIN OF ST. LUKE'S COLLEGE OF NURSING)
SEIROKA KANGO DAIGAKU (ST. LUKE'S COLLEGE OF
NURSING)
10-1, AKASHI-CHO, CHUO-KU, TOKYO 104
JAPANESE

SEISHIN YAKURYO KIKIN KENKYU NENPO 2770
(ANNUAL REPORT OF THE PHARMACOPSYCHIATRY RESEARCH
FOUNDATION)
SEISHIN SHINKEI-KEI YAKUBUTSU CHIRYO KENKYU KIKIN
(PHARMACOPSYCHIATRY RESEARCH FOUNDATION)
3-35, HIRANO-MACHI, HIGASHI-KU, OSAKA-SHI 541
JAPANESE

SHIGA-KEN YAKUJI SHIDOSHO SHOHO 2771
(REPORT OF THE SHIGA PREFECTURE PHARMACEUTICAL
RESEARCH INSTITUTE)
SHIGA-KEN YAKUJI SHIDOSHO
200-2, OHARAICHIBA, KOGA-MACHI, KOGA-GUN,
SHIGA 520-34
JAPANESE

SHINRYO TO SHIN'YAKU 2772
(MEDICAL CONSULTATION AND NEW REMEDIES)
IJI SHUPPAN-SHA
C/O KYODO BUILDING, 2-5, NIHONBASHI HON-CHO,
CHUO-KU, TOKYO 103
JAPANESE

SHIN'YAKU JOHO 2773
(NEW DRUG REPORT)
IYAKUHIN SANGYO KENKYUSHO (DRUG BUSINESS RESEARCH
AND COMPANY, LIMITED)
25-2, 1-CHOME, KOISHIKAWA, BUNKYO-KU, TOKYO 112
JAPANESE

SHIN'YAKU TO CHIRYO 2774
(NEW REMEDIES AND THERAPY)
YAMANOUCHI SEIYAKU K.K. (YAMANOUCHI PHARMACEUTICAL
COMPANY, LIMITED)
2-5, NIHONBASHI HON-CHO, CHUO-KU, TOKYO 103
JAPANESE
INDEXED: CHE

SHIN'YAKU TO RINSHO 2775
(JOURNAL OF NEW REMEDIES AND CLINIC)
SHIN'YAKU TO RINSHOSHA
16-32, 4-CHOME, FUJIMI-DAI, NERIMA-KU, TOKYO 177
JAPANESE
INDEXED: CHE, JICST

SHIONOGI KENKYUSHO NENPO 2776
(ANNUAL REPORT OF SHIONOGI RESEARCH LABORATORY)
SHIONOGI KENKYUSHO (SHIONOGI RESEARCH LABORATORY, SHIONOGI COMPANY, LIMITED)
2-47, SAGISUKAMI-DORI, FUKUSHIMA-KU, OSAKA-SHI 553
JAPANESE AND ENGLISH
INDEXED: BIO, CHE, EXC, JICST

SHITAI FUJIJU KYOIKU 2777
(JAPANESE JOURNAL OF EDUCATION OF THE HANDICAPPED)
NIHON SHITAI FUJIYUJI KYOKAI (JAPANESE SOCIETY FOR CRIPPLED CHILDREN)
3-13-15 HIGASHI IKEBUKURO, TOSHIMA-KU, TOKYO 170
JAPANESE

SHOGAISHA MONDAI KENKYU 2778
(RESEARCH ON PROBLEMS OF THE HANDICAPPED)
ZENKOKU SHOGAISHA MONDAI KENKYU KAI (NATIONAL ASSOCIATION FOR RESEARCH ON THE PROBLEMS OF THE HANDICAPPED)
JAPANESE

SHOWA DAIGAKU YAKUGAKUBU KENKYU GYOSEKISHU 2779
(ARCHIVES OF THE FACULTY OF PHARMACY, SHOWA UNIVERSITY)
SHOWA DAIGAKU YAKUGAKUBU
5-8, 1-CHOME, HATANODAI, SHINAGAWA-KU, TOKYO 142
JAPANESE AND EUROPEAN

SHOWA YAKKA DAIGAKU KIYO 2780
(ANNUAL REPORT OF THE SHOWA PHARMACEUTICAL COLLEGE)
SHOWA YAKKA DAIGAKU (SHOWA PHARMACEUTICAL COLLEGE)
1-8, 5-CHOME, TSURUMAKI, SETAGAYA-KU, TOKYO 154
JAPANESE
INDEXED: JICST

SHOYAKUGAKU ZASSHI 2781
(JAPANESE JOURNAL OF PHARMACOGNOSY)
NIHON SHOYAKU GAKKAI (JAPANESE SOCIETY OF PHARMACOGNOSY)
C/O KYOTO DAIGAKU YAKUGAKUBU, YOSHIDA SHIMODACHI-CHO, KYOTO-SHI 606
JAPANESE AND ENGLISH
INDEXED: BIO, CHE, JICST

SOGO KANGO 2782
(COMPREHENSIVE NURSING, QUARTERLY)
GENDAISHA (GENDAISHA PUBLISHING COMPANY, LIMITED)
C/O SANKO BUILDING, 130, WASEDA TSURUMAKI-CHO, SHINJUKU-KU, TOKYO 162
JAPANESE

SOGO RIHABIRITESHON 2783
(SOGO REHABILITATION)
IGAKU SHOIN
29-11, 5-CHOME, HONGO, BUNKYO-KU, TOKYO 113
JAPANESE
INDEXED: JICST

STETHOSCOPE 2784
SANKYO K.K. (SANKYO COMPANY, LIMITED)
7-12, 2-CHOME, GINZA, CHUO-KU, TOKYO 104
JAPANESE

TAISHO SEIYAKU K.K. KENKYU HOKOKU 2785
(REPORT OF TAISHO RESEARCH LABORATORY)
TAISHO SEIYAKU K.K. KENKYU-BU AND TAISHO SEIYAKU K.K. (TAISHO PHARMACEUTICAL COMPANY, LIMITED)
34-1, 3-CHOME, TAKADA, TOSHIMA-KU, TOKYO 164
JAPANESE AND ENGLISH

TAKEDA KENKYUSHOHO 2786
(JOURNAL OF THE TAKEDA RESEARCH LABORATORIES)
TAKEDA YAKUHIN KOGYO K.K. CHUO KENKYUSHO (CENTRAL RESEARCH DIVISION, TAKEDA CHEMICAL INDUSTRIES, LIMITED)
4-54, JUSO-NISHINO-CHO, HIGASHIYODOGAWA-KU, OSAKA-SHI 532
JAPANESE AND ENGLISH
INDEXED: BIB, BIO, CHE, JICST

TAKEDA YAKUHO 2787
(TAKEDA DRUG NEWS)
TAKEDA YAKUHIN KOGYO K.K. (TAKEDA CHEMICAL INDUSTRIES, LIMITED)
2-27, DOSHO-MACHI, HIGASHI-KU, OSAKA-SHI 541
JAPANESE

TAKEDA-KAI KAISHI TAKEDA YAKUHO 2788
(JOURNAL OF TAKEDA, TAKEDA DRUG NEWS)
TAKEDA YAKUHIN KOGYO K.K. (TAKEDA CHEMICAL INDUSTRIES, LIMITED)
2-27, DOSHO-MACHI, HIGASHI-KU, OSAKA-SHI 541
JAPANESE

TANABE SEIYAKU KENKYU HOBUN 2789
(PUBLICATIONS OF RESEARCH LABORATORIES OF TANABE SEIYAKU COMPANY, LIMITED)
TANABE SEIYAKU (PHARMACEUTICAL INDUSTRY) COMPANY, LIMITED
JAPANESE

TANABE SEIYAKU KENKYU NENPO 2790
(ANNUAL REPORT OF TANABE SEIYAKU COMPANY, LIMITED)
TANABE SEIYAKU K.K. KENKYUSHO (RESEARCH LABORATORY OF TANABE SEIYAKU COMPANY, LIMITED)
962, KASHIMA-CHO, HIGASHIYODOGAWA-KU, OSAKA-SHI 532
JAPANESE AND ENGLISH
INDEXED: CHE, JICST

TETSUDO YAKUGAKU KENKYU NENPO 2791
(ANNUAL REPORT OF RAILROAD PHARMACY)
TETSUDO YAKUGAKU KENKYUKAI
C/O CHUO TETSUDO BYOIN YAKUZAI-KA, 1-3, 2-CHOME YOYOGI, SHIBUYA-KU, TOKYO 151
JAPANESE
INDEXED: JICST

TOHOKU YAKKA DAIGAKU KENKYU NENPO 2792
(ANNUAL REPORT OF TOHOKU COLLEGE OF PHARMACY)
TOHOKU YAKKA DAIGAKU (TOHOKU COLLEGE OF PHARMACY)
4-1, 4-CHOME, KOMATSUSHIMA, SENDAI-SHI 983
JAPANESE AND EUROPEAN
INDEXED: CHE, JICST

TOKYO TANABE QUARTERLY 2793
TOKYO TANABE SEIYAKU K.K. (TOKYO TANABE SEIYAKU COMPANY, LIMITED)
7-3, 2-CHOME, NIHONBASHI HON-CHO, CHUO-KU, TOKYO 103
JAPANESE

TOKYO YAKKA DAIGAKU KENKYU NENPO 2794
(ANNUAL REPORT OF THE TOKYO COLLEGE OF PHARMACY)
TOKYO YAKKA DAIGAKU (TOKYO COLLEGE OF PHARMACY)
20-1, 3-CHOME, KITA SHINJUKU, SHINJUKU-KU, TOKYO 160
JAPANESE AND EUROPEAN
INDEXED: CHE, JICST

TOKYO-TO NO KENKETSU JIJO 2795
(SITUATION OF OFFERING BLOOD IN TOKYO)
TOKYO-TO EISEI-KYOKU
5-1, 3-CHOME, MARUNOUCHI, CHIYODA-KU, TOKYO 100
JAPANESE

TOKYO-TO YAKUZAISHIKAISHI 2796
(JOURNAL OF TOKYO PHARMACEUTICAL ASSOCIATION)
TOKYO-TO YAKUZAISHIKAI
1-21, KANDA NISHIKI-CHO, CHIYODA-KU, TOKYO 101
JAPANESE

TOYAKUKO KAIHO 2797
(PHARMACEUTICAL INDUSTRY)
TOKYO IYAKUHIN KOGYO KYOKAI (PHARMACEUTICAL MANUFACTURER'S ASSOCIATION OF TOKYO)
2-9, NIHONBASHI HON-CHO, CHUO-KU, TOKYO 103
JAPANESE
INDEXED: JICST

TOYAMA-KEN YAKUJI KENKYUSHO SHOHO 2798
(ANNUAL REPORT OF TOYAMA PREFECTURAL PHARMACEUTICAL RESEARCH INSTITUTE))
TOYAMA-KEN YAKUJI KENKYUSHO (TOYAMA PREFECTURAL PHARMACEUTICAL RESEARCH INSTITUTE)
4-2, 1-CHOME, CHITOSE-CHO, TOYAMA-SHI 930
JAPANESE

WAKAN'YAKU 2799
(ORIENTAL MEDICINES)
WAKAN'YAKU DOKOKAI
UCHIDA WAKAN'YAKU K.K.
4-13, NIHONBASHI HON-CHO, CHUO-KU, TOKYO 103
JAPANESE

WAKAN'YAKU SHINPOJUMU KIROKU 2800
(PROCEEDINGS OF THE SYMPOSIUM ON WAKAN'YAKU)
TOYAMA DAIGAKU YAKUGAKUBU WAKAN'YAKU KENKYU SHISETSU (DRUG RESEARCH INSTITUTE, FACULTY OF PHARMACEUTICAL SCIENCES, TOYAMA UNIVERSITY)
3190, GOFUKU, TOYAMA-SHI 980
JAPANESE, WITH ENGLISH SUMMARY

WAKAYAMA-KEN YAKUJI SHIDOSHO HOKOKU 2801
(REPORT OF THE WAKAYAMA PREFECTURAL LABORATORY OF PHARMACOLOGY)
WAKAYAMA-KEN YAKUJI SHIDOSHO (WAKAYAMA PREFECTURAL LABORATORY OF PHARMACOLOGY)
1-1, KOMATSUBARA-DORI, WAKAYAMA-SHI 640
JAPANESE

WAKO JUN'YAKU JIHO 2802
(WAKO PHARMACY TIMES)
WAKO JUN'YAKU KOGYO K.K. (WAKO PURE CHEMICAL INDUSTRIES, LIMITED)
3-10, DOSHO-MACHI, HIGASHI-KU, OSAKA-SHI 541
JAPANESE

WAKSMAN FOUNDATION OF JAPAN, INCORPORATED REPORT 2803
NIHON WAKKUSUMAN ZAIDAN (WAKSMAN FOUNDATION OF JAPAN, INCORPORATED)
C/O KEIO DAIGAKU IGAKUBU, 30-8, DAIKYO-MACHI, SHINJUKU-KU, TOKYO 160
ENGLISH

X-REI JANARU 2804
(X-RAY JOURNAL)
X-REI JANARU
3-12, 4-CHOME, TSUKIJI, CHUO-KU, TOKYO 104
JAPANESE

YAKKO 2805
(DRUG POT)
KANAGAWA-KEN YAKUZAISHIKAI
14-11, NISHI-MACHI, ISOGO-KU, YOKOHAMA-SHI 235
JAPANESE

YAKKYOKU 2806
(JOURNAL OF PRACTICAL PHARMACY)
NANZANDO
1-11, 4-CHOME, YUSHIMA, BUNKYO-KU, TOKYO 113-91
JAPANESE
INDEXED: CHE

YAKKYOKU NO RYOIKI 2807
(SPHERE OF PHARMACY)
JAPANESE
INDEXED: JICST

YAKKYOKU NO TOMO 2808
(PHARMACY COMPANION)
YAMANOUCHI SEIYAKU K.K. (YAMANOUCHI PHARMACEUTICAL COMPANY, LIMITED)
2-5, NIHONBASHI HON-CHO, CHUO-KU, TOKYO 103
JAPANESE

YAKKYOKU SHINBUN 2809
(DRUGGIST AND PHARMACIST NEWS)
YAKKYOKU SHINBUNSHA
4-8, NIHONBASHI HON-CHO, CHUO-KU, TOKYO 103
JAPANESE

YAKUBUTSU RYOHO 2810
(MEDICAL TREATMENT)
IJI NIPPOSHA
C/O FUNABA BUILDING, 3-6, AWAJI-CHO, HIGASHI-KU, OSAKA-SHI 541
JAPANESE AND ENGLISH

YAKUGAKU TOSHOKAN 2811
(PHARMACEUTICAL LIBRARY BULLETIN)
NIPPON YAKUGAKU TOSHOKAN KYOGIKAI (JAPAN PHARMACEUTICAL LIBRARY ASSOCIATION)
3-1, 7-CHOME, HONGO, BUNKYO-KU, TOKYO 113
JAPANESE
INDEXED: JICST

YAKUGAKU ZASSHI 2812
(JOURNAL OF THE PHARMACEUTICAL SOCIETY OF JAPAN)
NIHON YAKUGAKKAI (PHARMACEUTICAL SOCIETY OF JAPAN)
12-15, 2-CHOME, SHIBUYA, SHIBUYA-KU, TOKYO 150
JAPANESE AND ENGLISH
INDEXED: BIB, BIO, EXC, JICST

YAKUGYO JIHO 2813
(DRUG INDUSTRY TIMES)
YAKUGYO JIHOSHA
2-36, KANDA JINBO-CHO, CHIYODA-KU, TOKYO 101
JAPANESE

YAKUGYO SHIMBUN 2814
(JAPANESE WEEKLY ON PHARMACY AND CHEMISTRY)
1-13-12 KIKAWA NISHINO-CHO, HIGOSHI YODOGAWAKU, OSAKA, JAPAN
JAPANESE

YAKUJI KOGYO SEISAN DOTAI TOKEI GEPPO 2815
(MONTHLY STATISTICS OF THE PRODUCTION OF PHARMACEUTICAL INDUSTRY)
KOSEI-SHO YAKUMU-KYOKU (PHARMACEUTICAL AND SUPPLY BUREAU, MINISTRY OF HEALTH AND WELFARE)
YAKUGYO KEIZAI KENKYUSHO
C/O TOKYO KURABU BUILDING, 2-6, 3-CHOME, KASUMIGASEKI, CHIYODA-KU, TOKYO 100
JAPANESE

YAKUJI KOGYO SEISAN DOTAI TOKEI NENPO 2816
(ANNUAL STATISTICS OF PRODUCTION OF PHARMACEUTICAL INDUSTRY)
KOSEI-SHO YAKUMU-KYOKU (PHARMACEUTICAL AND SUPPLY BUREAU, MINISTRY OF HEALTH AND WELFARE)
YAKUGYO KEIZAI KENKYUSHO
C/O TOKYO KURABU BUILDING, 2-6, 3-CHOME KASUMIGASEKI, CHIYODA-KU, TOKYO 100
JAPANESE

YAKUJI NIPPO 2817
(PHARMACEUTICAL NEWS)
YAKUJI NIPPOSHA (YAKUJI NIPPO LIMITED)
1-11, KANDA IZUMI-CHO, CHIYODA-KU, TOKYO 101
JAPANESE

YAKUJI NYUSU 2818
(NEWS OF PHARMACEUTICAL INDUSTRY)
YAKUJI NYUSUSHA
C/O INDO BUILDING, DOSHO-MACHI, HIGASHI-KU, OSAKA-SHI 541
JAPANESE

YAKUJI SHINPO 2819
(DRUG WEEKLY)
YAKUJI SHINPOSHA
3-4, NIHONBASHI HON-CHO, CHUO-KU, TOKYO 103
JAPANESE AND ENGLISH

YAKUMU KOHO 2820
(BULLETIN OF PHARMACEUTICAL AND SUPPLY BUREAU)
KOSEI-SYO YAKUMU-KYOKU (PHARMACEUTICAL AND SUPPLY BUREAU, MINISTRY OF HEALTH AND WELFARE)
YAKUMU KOHOSHA
26-8, 5-CHOME, ASAGAYAKITA, SUGINAMI-KU, TOKYO 166
JAPANESE

YAKUNICHI SHINBUN 2821
(DRUG NEWS)
YAKUNICHI SHINBUNSHA
C/O DENKI BUILDING, SAKURABASHI-DORI, TOYAMA-SHI 930
JAPANESE

YAKUSHIGAKU ZASSHI 2822
(JOURNAL OF JAPANESE HISTORY OF PHARMACY)
NIHON YAKUSHIGAKKAI (JAPANESE SOCIETY OF HISTORY OF PHARMACY)
C/O NIHON DAIGAKU RIKO GAKUBU YAKUGAKKA SHOYAKUGAKU KYOSHITSU, 1-8, KANDA SURUGADAI, CHIYODA-KU, TOKYO 101
JAPANESE, WITH ENGLISH SUMMARY

YAKUZAIGAKU 2823
(ARCHIVES OF PRACTICAL PHARMACY)
NIHON YAKUGAKKAI YAKUZAIGAKU HENSHU IINKAI (PHARMACEUTICAL SOCIETY OF JAPAN)
AND YAKUJI NIPPOSHA (YAKUJI NIPPO LIMITED)
1-11, KANDA IZUMI-CHO, CHIYODA-KU, TOKYO 101
JAPANESE, WITH ENGLISH SUMMARY
INDEXED: CHE, JICST

YAMANOUCHI SEIYAKU KENKYU HOKOKU 2824
(REPORT OF YAMANOUCHI CENTRAL RESEARCH LABORATORIES)
YAMANOUCHI SEIYAKU K.K. (YAMANOUCHI PHARMACEUTICAL COMPANY, LIMITED)
1-8, 1-CHOME, AZUSAWA, ITABASHI-KU, TOKYO 174
JAPANESE AND ENGLISH
INDEXED: CHE

YOSHITOMI SEIYAKU KENKYUSHOHO 2825
(REPORT OF YOSHITOMI RESEARCH LABORATORY)
YOSHITOMI SEIYAKU K.K. KENKYUSHO (YOSHITOMI RESEARCH LABORATORY)
955, KOIWAI, YOSHITOMI-MACHI, CHIKUJO-GUN, FUKUOKA 871
JAPANESE

ZENKOKU KANGO KYOIKUGAKKAISHI 2826
(JOURNAL OF NATIONAL NURSING EDUCATION SOCIETY)
ZENKOKU KANGO KYOIKU KENKYUKAI
C/O SEIBO JOSHI TANKIDAIGAKU, 16-11, 4-CHOME SHIMOOCHIAI, SHINJUKU-KU, TOKYO 160
JAPANESE

Disease, Pathology

ACTA DERMATOLOGICA-KYOTO 2827
KYOTO DAIGAKU IGAKUBU HIFUKA KYOSHITSU (DEPARTMENT OF DERMATOLOGY, FACULTY OF MEDICINE, KYOTO UNIVERSITY)
SHOGOIN KAWARA-MACHI, SAKYO-KU, KYOTO-SHI 606
ENGLISH

ACTA PAEDIATRICA JAPONICA 2828
NIHON SHONIKA GAKKAI (SOCIETAS PAEDIATRICA JAPONICA)
29-8, 1-CHOME, SHINJUKU, SHINJUKU-KU, TOKYO 160
EUROPEAN
INDEXED: CHE, EXC

ACTA PATHOLOGICA JAPONICA 2829
NIHON BYORI GAKKAI (JAPANESE PATHOLOGICAL SOCIETY)
C/O TOKYO DAIGAKU IGAKUBU BYORIGAKU KYOSHITSU, 3-1, 7-CHOME, HONGO, BUNKYO-KU, TOKYO 113
EUROPEAN
INDEXED: BIO, CHE, EXC, IND, SCI

ACTA TUBERCULOSEA JAPONICA 2830
(OFFICIAL ORGAN OF THE TUBERCULOSIS RESEARCH INSTUTUTE, KYOTO UNIVERSITY)
JAPANESE
INDEXED: JICST

AICHIGAKUIN DAIGAKU SHIGAKKAISHI 2831
(AICHI-GAKUIN JOURNAL OF DENTAL SCIENCE)
AICHIGAKUIN DAIGAKU SHIGAKKAI (AICHIGAKUIN UNIVERSITY)
2-11, SUEMORI-DORI, CHIKUSA-KU, NAGOYA-SHI 464
JAPANESE, WITH EUROPEAN SUMMARY
INDEXED: BIO

AKITA-KEN NOKEKKAN KENKYU SENTA KENKYU RONBUN SHOROKUSHU 2832
(ABSTRACTS OF PAPERS FROM THE RESEARCH INSTITUTE OF BRAIN AND BLOOD VESSELS, AKITA)
AKITA-KEN NOKEKKAN KENKYU SENTA (RESEARCH INSTITUTE OF BRAIN AND BLOOD VESSELS, AKITA)
6-10, SENSHU KUBOTA-MACHI, AKITA-SHI 010
JAPANESE

ANESTHESIOLOGY NEWS 2833
HEKISUTO JAPAN K.K.
10-13, 8-CHOME, AKASAKA, MINATO-KU, TOKYO 107
JAPANESE

ANNUAL PUBLICATIONS OF THE SOCIETY OF NIPPON DENTAL COLLEGE 2834
NIHON SHIKA DAIGAKU SHIGAKKAI (THE SOCIETY OF NIPPON DENTAL COLLEGE)
C/O NIHON SHIKA DAIGAKU, 9-20, 1-CHOME, FUJIMI, CHIYODA-KU, TOKYO 102
ENGLISH

ANNUAL REPORT OF THE CENTER FOR ADULT DISEASES 2835
OSAKA-FURITSU SEIJINBYO SENTA (CENTER FOR ADULT DISEASES AND OSAKA ASSOCIATION FOR PREVENTION OF ADULT DISEASES)
1, MORI-MACHI MINAMI, HIGASHINARI-KU, OSAKA-SHI 537
ENGLISH
INDEXED: BIO, CHE, JICST

ANNUAL REPORT. INSTITUTE FOR CANCER RESEARCH, OSAKA UNIVERSITY MEDICAL SCHOOL 2836
OSAKA DAIGAKU IGAKUBU FUZOKU GAN KENKYU SHISETSU (INSTITUTE FOR CANCER RESEARCH, OSAKA UNIVERSITY MEDICAL SCHOOL)
3-12, DOJIMAHAMA-DORI, FUKUSHIMA-KU, OSAKA-SHI 553
ENGLISH

ANNUAL REPORT. JUNTENDO UNIVERSITY, TOKYO MEDICAL ULTRASONICS RESEARCH CENTER 2837
JUNTENDO UNIVERSITY SCHOOL OF MEDICINE, MEDICAL ULTRASONICS RESEARCH CENTER
HONGO, TOKYO, JAPAN
ENGLISH

ARERUGI 2838
(JAPANESE JOURNAL OF ALLERGOLOGY)
NIHON ALLERGY GAKKAI (JAPANESE SOCIETY OF ALLERGY)
C/O NIHON IKA DAIGAKU BISEIBUTSU MEN'EKIGAKU KYOSHITSU, 1-5, 1-CHOME, SENDAGI, BUNKYO-KU, TOKYO 113
JAPANESE AND ENGLISH
INDEXED: CHE, EXC

ASAHI SEIMEI SEIJINBYO KENKYUSHO NENPO 2839
(ANNUAL REPORT OF THE INSTITUTE FOR ADULT DISEASES, ASAHI LIFE FOUNDATION)
ASAHI SEIMEI SEIJINBYO KENKYUSHO (INSTITUTE FOR ADULT DISESASES, ASAHI LIFE FOUNDATION)
9-14, 1-CHOME, NISHISHINJUKU, SHINJUKU-KU, TOKYO 161
JAPANESE AND ENGLISH

AURIS, NASUS, LARYNX 2840
KOKUSAI JIBI INKOKAGAKU SHINKOKAI (SOCIETY FOR PROMOTION OF INTERNATIONAL OTORHINOLARYNGOLOGY)
4-16, YAYOI, BUNKYO-KU, TOKYO 113
ENGLISH
INDEXED: CHE, EXC

BIBLIOGRAPHY ON STOMACH CANCER 2841
ENGLISH

BULLETIN OF THE HEART INSTITUTE JAPAN 2842
NIHON SHINZO KETSUATSU KENKYUSHO (THE HEART INSTITUTE JAPAN) AND TOKYO JOSHI IKA DAIGAKU (TOKYO WOMEN'S MEDICAL COLLEGE)
10, ICHIGAYA KAWADA-CHO, SHINJUKU-KU, TOKYO 162
ENGLISH
INDEXED: BIO, CHE

BULLETIN OF KANAGAWA DENTAL COLLEGE 2843
ENGLISH

BULLETIN OF TOKYO DENTAL COLLEGE 2844
TOKYO SHIKA DAIGAKU (TOKYO DENTAL COLLEGE)
9-18, 2-CHOME, MISAKI-CHO, CHIYODA-KU, TOKYO 101
ENGLISH
INDEXED: BIO, CHE

CANCER RESEARCH INSTITUTE REPORT 2845
KANAGAWA DAIGAKU GAN KENKYUJO (KANAGAWA UNIVERSITY, CANCER RESEARCH INSTITUTE)
ENGLISH

CELL STRUCTURE AND FUNCTION 2846
JAPAN SOCIETY FOR CELL BIOLOGY, DEPARTMENT OF PATHOLOGY, OKAYAMA UNIVERSITY MEDICAL SCHOOL
2-5-1, SHIKATA-CHO, OKAYAMA 700
ENGLISH
INDEXED: BIB, CHE, EXC

CHIBA-KEN NO KEKKAKU 2847
(DATA ON TUBERCULOSIS IN CHIBA PREFECTURE)
CHIBA-KEN EISEI-BU (HEALTH DEPARTMENT, CHIBA PREFECTURAL GOVERNMENT)
1-1, ICHIBA-CHO, CHIBA-SHI 280-91
JAPANESE

CHIRYO KYOIKU KENKYU 2848
(JAPANESE JOURNAL OF MEDI COPEDAGOGY)
CHIRYO KYOIKU KENKYUKAI
JAPANESE

CHOKAKU GENGO SHOGAI 2849
(HEARING, SPEECH DISORDERS)
CHOKAKU GENGO SHOGAI KANKOKAI
JAPANESE

CHOKAKU SHOGAI 2850
(AUDITORY DISORDERS)
RO KYOIKU KENKYUKAI (DEAF EDUCATION RESEARCH ASSOCIATION)
C/O TOKYO KYOIKU DAIGAKU KONODAI BUNKO, 2-1, 2-CHOME, KONODAI, ICHIKAWA-SHI 272
JAPANESE

CIRCULATION NEWS 2851
HEKISUTO JAPAN K.K.
10-13, 8-CHOME, AKASAKA, MINATO-KU, TOKYO 107
JAPANESE

CLINICAL REPORT 2852
KINDAI IGAKUSHA
20-17, 1-CHOME, JINNAN, SHIBUYA-KU, TOKYO 150
JAPANESE AND ENGLISH

COLLECTED PAPERS FROM THE NATIONAL CANCER CENTER RESEARCH INSTITUTE 2853
KOKURITSU GAN SENTA TOSHOKAN (NATIONAL CANCER CENTER LIBRARY)
KOKURITSU GAN SENTA (NATIONAL CANCER CENTER)
1-1, 5-CHOME, TSUKIJI, CHUO-KU, TOKYO 104
ENGLISH

DE: JOURNAL OF DENTAL ENGINEERING 2854
SHIKA RIKO GAKKAI
C/O TOKYO IKA SHIKA DAIGAKU, 5-45, 1-CHOME, YUSHIMA, BUNKYO-KU, TOKYO 113
JAPANESE
INDEXED: CHE, JICST

DENTAL DIAMOND 2855
DENTARU DAIYAMONDO SHA
(DENTAL DIAMOND COMPANY)
JAPANESE AND ENGLISH

DENTARU MIRA 2856
(DENTAL MIRROR)
IGAKU SHOBO
9, DAIKYO-CHO, SHINJUKU-KU, TOKYO 160
JAPANESE

DENTISTRY IN JAPAN 2857
(NIHON NO SHIKA IRYO)
NIHON SHIKA IGAKKAI (JAPANESE ASSOCIATION FOR DENTAL SCIENCE)
NIHON SHIKA ISHIKAI (JAPAN DENTAL ASSOCIATION)
1-20, 4-CHOME, KUDAN KITA, CHIYODA-KU, TOKYO 102
EUROPEAN

DIABETES NEWS 2858
HEKISUTO JAPAN K.K.
10-13, 8-CHOME, AKASAKA, MINATO-KU, TOKYO 107
JAPANESE

DOMYAKU KOKA 2859
(JOURNAL OF JAPAN ARTERIOSCLEROSIS SOCIETY)
JAPAN ARTERIOSCLEROSIS SOCIETY
JAPANESE

EQUILIBRIUM RESEARCH 2860
NIHON HEIKO SHINKEI KAGAKU KAI (JAPAN SOCIETY FOR STUDY OF NEUROLOGICAL STABILITY)
JAPANESE AND ENGLISH

FOLIA PSYCHIATRICA ET NEUROLOGICA JAPONICA 2861
NIPPON SEISHIN SHINKEI GAKKAI (JAPANESE SOCIETY FOR PSYCHIATRY AND NEUROLOGY)
C/O TOYO BUNKO 28-21, 2-CHOME, HONKOMAGOME, BUNKYO-KU, TOKYO 113
ENGLISH
INDEXED: BIO, CHE, EXC

FUKUJUJI 2862
(RED DOUBLE-BARRED CROSS)
KEKKAKU YOBOKAI (ANTI-TUBERCULOSIS ASSOCIATION)
3-12, 1-CHOME, MISAKI-CHO, CHIYODA-KU, TOKYO 101
JAPANESE

FUKUOKA SHIKA DAIGABU GAKKAI ZASSHI 2863
(FUKUOKA DENTAL COLLEGE SOCIETY JOURNAL)
JAPANESE

FUKUOKA-KEN SHIKA ISHIKAISHI 2864
(JOURNAL OF FUKUOKA PREFECTURE DENTAL ASSOCIATION)
FUKUOKA-KEN SHIKA ISHIKAI (FUKUOKA PREFECTURE DENTAL ASSOCIATION)
12-43, 1-CHOME, DAIMYO, CHUO-KU, FUKUOKA-SHI 810
JAPANESE

GAKU KOKU SOGO KENKYU SHISETSU HOKOKUSHO 2865
(REPORT OF INSTITUTE OF STOMATOGNATHIC SCIENCE, FACULTY OF DENTISTRY, TOKYO MEDICAL AND DENTAL UNIVERSITY)
TOKYO IKA SHIKA DAIGAKU SHIGAKUBU GAKU KOKU SOGO KENKYU SHISETSU (INSTITUTE OF STOMATOGNATHIC SCIENCE, FACULTY OF DENTISTRY, TOKYO MEDICAL AND DENTAL UNIVERSITY)
5-45, 1-CHOME, YUSHIMA, BUNKYO-KU, TOKYO 113
JAPANESE AND ENGLISH

GAN NO RINSHO 2866
(JAPANESE JOURNAL OF CANCER CLINICS)
ISHIYAKU SHUPPAN K.K. (ISHIYAKU PUBLISHERS, INC.
7-10, 1-CHOME, HONKOMAGOME, BUNKYO-KU, TOKYO 113
JAPANESE, WITH ENGLISH SUMMARY
INDEXED: CHE

GAN TO KAGAKU RYOHO 2867
(CANCER AND CHEMOTHERAPHY)
GAN TO KOGAKU RYOHO-SHA (CANCER AND CHEMOTHERAPHY COMPANY)
JAPANESE

GANN (THE JAPANESE JOURNAL OF CANCER RESEARCH) 2868
NIHON GAN GAKKAI (THE JAPANESE CANCER ASSOCIATION)
37-1, 1-CHOME, KAMIIKEBUKURO, TOSHIMA-KU, TOKYO 170
ENGLISH
INDEXED: BIO, CHE, EXC, IND, SCI, JICST

GANN MONOGRAPHS ON CANCER RESEARCH 2869
JAPANESE AND ENGLISH
INDEXED: BIO, EXC

GASTROENTEROLOGIA JAPONICA 2870
NIHON SHOKAKIBYO GAKKAI (JAPANESE SOCIETY OF GASTROENTEROLOGY)
4-12, 7-CHOME, GINZA, CHUO-KU, TOKYO 104
ENGLISH
INDEXED: BIO, CHE, EXC, IND

GERIATRIC MEDICINE 2871
RAIFU SAIENSU (LIFE SCIENCE COMPANY, LIMITED)
C/O ITO BUILDING, 42-14, UDAGAWA-CHO, SHIBUYA-KU, TOKYO 150
JAPANESE

GIFU SEISHIN EISEI 2872
(JOURNAL OF MENTAL HEALTH ASSOCIATION GIFU)
GIFU-KEN SEISHIN EISEI KYOKAI
C/O GIFU-KEN EISEI-BU, YABUTA, GIFU-SHI 500
JAPANESE

GIFU SHIKA GAKKAI ZASSHI 2873
(JOURNAL OF GIFU DENTAL SOCIETY)
JAPANESE

GIFU-KEN SEISHIN EISEI SENTA SHOHO 2874
(REPORT OF GIFU PREFECTURAL MENTAL HEALTH CENTER)
GIFU-KEN SEISHIN EISEI SENTA
C/O GIFU SOGO CHOSHA, 1, TSUKASA-CHO, GIFU-SHI 500
JAPANESE

GUNMA-KEN SHONI HOKENKAI KAIHO 2875
(JOURNAL OF CHILD HEALTH ASSOCIATION, GUNMA PREFECTURE)
GUNMA-KEN SHONI HOKENKAI
1-1, 1-CHOME, OTE-MACHI, MAEBASHI-SHI 371
JAPANESE

HAI TO SHIN 2876
(LUNG AND HEART)
KOKUSEIDO SHUPPAN K.K. (KOKUSEIDO PUBLISHING COMPANY, LIMITED)
23-5, 3-CHOME, HONGO, BUNKYO-KU, TOKYO 113
JAPANESE, WITH ENGLISH SUMMARY

HAIGAN 2877
(LUNG CANCER)
NIHON HAIGAN GAKKAI (JAPAN LUNG CANCER SOCIETY)
C/O CHIBA DAIGAKU IGAKUBU HAIGAN KENKYUSHO, 8-1, 1-CHOME, INOHANA, CHIBA-SHI 280
JAPANESE, WITH ENGLISH SUMMARY

HARIMA (TONYOBYO KYOKAIHO) 2878
(JOURNAL OF DIABETES ASSOCIATION)
HARIMA TONYOBYO KYOKAI
C/O SHINKI BUS K.K., 202, EKIMAE-MACHI, HIMEJI-SHI 670
JAPANESE

HIFU 2879
(SKIN RESEARCH)
NIHON HIFUKAGAKKAI OSAKA CHIHOKAI (OSAKA DERMATOLOGICAL ASSOCIATION)
C/O OSAKA DAIGAKU HIFUKA KYOSHITSU, 3-1, DOJIMA HAMADORI, FUKUSHIMA-KU, OSAKA-SHI 553
JAPANESE AND ENGLISH
INDEXED: BIO, CHE

HIFUKA KIYO 2880
(ACTA DERMATOLOGICA)
HIFUKA KIYO HENSHU-BU (EDITORIAL OFFICE OF ACTA DERMATOLOGICA)
C/O KYOTO DAIGAKU IGAKUBU HIFUKA KYOSHITSU, 53, SHOGOIN KAWARA-MACHI, SAKYO-KU, KYOTO-SHI 606
JAPANESE AND EUROPEAN
INDEXED: CHE, EXC, JICST

HIFUKA NO RINSHO 2881
(JOURNAL OF CLINICAL DERMATOLOGY)
KANEHARA SHUPPAN K.K.
31-14, 2-CHOME, HONGO, YUSHIMA, BUNKYO-KU, TOKYO 113-91
JAPANESE

HINYOKIKA KIYO 2882
(ACTA UROLOGICA JAPONICA)
KYOTO DAIGAKU IGAKUBU HINYOKIKAGAKU KYOSHITSU (DEPARTMENT OF UROLOGY, FACULTY OF MEDICINE, KYOTO UNIVERSITY)
53, SHOGOIN KAWARA-MACHI, SAKYO-KU, KYOTO-SHI 606
JAPANESE, WITH ENGLISH SUMMARY
INDEXED: CHE, EXC, JICST

HIROSHIMA DAIGAKU SHIGAKU ZASSHI 2883
(JOURNAL OF HIROSHIMA UNIVERSITY DENTAL SOCIETY)
HIROSHIMA DAIGAKU SHIGAKKAI (HIROSHIMA UNIVERSITY DENTAL SOCIETY)
2-3, 1-CHOME, KASUMI, HIROSHIMA-SHI 734
JAPANESE
INDEXED: CHE, JICST

HOKENFU NO KEKKAKU TENBO 2884
(REVIEW OF TUBERCULOSIS FOR PUBLIC HEALTH NURSE)
KEKKAKU YOBOKAI (JAPAN ANTI-TUBERCULOSIS ASSOCIATION)
3-12, 1-CHOME, MISAKI-CHO, CHIYODA-KU, TOKYO 101
JAPANESE

HOKKAIDO SHIKA ISHIKAISHI, DOSHIKAI TSUSHIN 2885
(JOURNAL OF THE HOKKAIDO DENTAL ASSOCIATION)
HOKKAIDO SHIKA ISHIKAI (HOKKAIDO DENTAL ASSOCIATION)
7-2, ODORI NISHI, CHUO-KU, SAPPORO-SHI 060
JAPANESE
INDEXED: CHE

HOKURIKU MASUIGAKU ZASSHI 2886
(HOKURIKU JOURNAL OF ANESTHESIOLOGY)
NIHON MASUIGAKKAI HOKURIKU CHIHOKAI (HOKURIKU BRANCH, JAPAN ANESTHESIOLOGY ASSOCIATION)
13-1, TAKARA-MACHI, KANAZAWA-SHI 920
JAPANESE AND ENGLISH
INDEXED: CHE, EXC

HOTETSU RINSHO 2887
(PRACTICE IN PROSTHODONTICS)
ISHIYAKU SHUPPAN K.K. (ISHIYAKU PUBLISHERS, INCORPORATED)
7-10, 1-CHOME, HONKOMAGOME, BUNKYO-KU, TOKYO 113
JAPANESE

HYOGO-KEN GAN SENTA NENPO 2888
(ANNUAL OF THE HYOGO (PREFECTURAL) CANCER CENTER 'HOSPITAL')
JAPANESE
INDEXED: JICST

HYOGO-KENRITSU BYOIN GAN SENTA KIYO 2889
(BULLETIN OF HYOGO CANCER HOSPITAL)
HYOGO CANCER HOSPITAL
JAPANESE

I TO CHO 2890
(STOMACH AND INTESTINE)
SOKI IGAN KENKYUKAI (RESEARCH SOCIETY FOR EARLY GASTRIC CANCER) AND IGAKU SHOIN
29-11, 5-CHOME, HONGO, BUNKYO-KU, TOKYO 113
JAPANESE, WITH ENGLISH SUMMARY
INDEXED: BIO, EXC

IBUNPI KENKYUKAISHI KENKYUKAI KIROKU 2891
(PROCEEDINGS OF THE SYMPOSIUM OF THE STOMACH SECRETION SOCIETY))
IBUMPI KENKYUKAI
C/O KAGOSHIMA DAIGAKU IGAKUBU SHUYO KENKYU SHISETSU, 7-82, SHIROYAMA-MACHI, KAGOSHIMA-SHI 892
JAPANESE

IGAN TO SHUDAN KENSHIN 2892
(GASTRIC CANCER)
NIHON I SHUDAN KENSHIN GAKKI (JAPANESE SOCIETY OF GASTRIC MASS SURVEY)
C/O NIHON DAIGAKU BYOIN ARIGA NAIKA KYOSHITSU, 1-8, KANDA SURUGADAI, CHIYODA-KU, TOKYO 101
JAPANESE

IMMUNOLOGY REVIEW 2893
HEKISUTO JAPAN K.K.
10-13, 8-CHOME, AKASAKA, MINATO-KU, TOKYO 107
JAPANESE

IWATE IKA DAIGAKU SHIGAKU ZASSHI 2894
(DENTAL JOURNAL OF IWATE MEDICAL UNIVERSITY)
IWATE IKA DAIGAKU SHIGAKUKAI (IWATE MEDICAL UNIVERSITY DENTAL SOCIETY)
JAPANESE

JAPANESE HEART JOURNAL 2895
TOKYO DAIGAKU IGAKUBU DAI-2-NAIKA (THE 2ND DEPARTMENT OF INTERNAL MEDICINE, FACULTY OF MEDICINE, UNIVERSITY OF TOKYO)
3-1, 7-CHOME, HONGO, BUNKYO-KU, TOKYO 113
ENGLISH
INDEXED: BIO, CHE, EXC, IND, SCI

JAPANESE JOURNAL OF CLINICAL ONCOLOGY 2896
GAN KENKYU SHINKOKAI (FOUNDATION OF CLINICAL ONCOLOGY)
C/O KOKURITSU GAN SENTA, 1-1, 5-CHOME, TSUKIJI, CHUO-KU, TOKYO 104
ENGLISH
INDEXED: BIO, CHE, EXC

JAPANESE JOURNAL OF TUBERCULOSIS AND CHEST DISEASES 2897
KOKURITSU DAIGAKU KEKKAKU KENKYUSHO DANWAKAI (THE ASSOCIATION OF THE RESEARCH INSTITUTE FOR TUBERCULOSIS OF NATIONAL UNIVERSITIES IN JAPAN)
C/O TOKYO DAIGAKU IKAGAKU KENKYUSHO, 6-1, 4-CHOME, SHIBA SHIROGANEDAI, MINATO-KU, TOKYO 108
ENGLISH
INDEXED: BIO, CHE, EXC, JICST

JIBI INKOKA 2898
(OTOLARYNGOLOGY)
IGAKU SHOIN
29-11, 5-CHOME, HONGO, BUNKYO-KU, TOKYO 113
JAPANESE, WITH ENGLISH SUMMARY
INDEXED: EXC

JIBI INKOKA RINSHO 2899
(PRACTICA OTOLOGICA KYOTO)
JIBI INKOKA RINSHOKAI (ASSOCITION OF OTO-RHINO-LARYNGOLOGICAL CLINIC)
C/O KYOTO DAIGAKU IGAKUBU JIBI INKO-KAGAKU KYOSHITSU, 53, SHOGOIN KAWARA-MACHI, SAKYO-KU, KYOTO-SHI 606
JAPANESE
INDEXED: CHE, JICST

JIBI INKOKA TENBO 2900
(OTO-RHINO-LARYNGOLOGY)
JIBI INKOKA TENBOKAI
C/O TOKYO JIKEIKAI IKA DAIGAKU JUBI INKOKA KYOSHITSU, 25-8, 3-CHOME, NISHI SHINBASHI, MINATO-KU, TOKYO 105
JAPANESE AND ENGLISH

JIBI TO RINSHO 2901
(OTOLOGIA FUKUOKA)
JIBI TO RINSHOKAI
C/O KYUSHU DAIGAKU IGAKUBU JIBI INKO-KAGAKU KYOSHITSU, 1276, KATAKASU, HIGASHI-KU, KUFUOKA-SHI 812
JAPANESE AND EUROPEAN
INDEXED: CHE, EXC

JIDO SEISHIN IGAKU TO SONO KINSETSU RYOIKI 2902
(JAPANESE JOURNAL OF CHILD PSYCHIATRY)
JIDO SEISHIN IGAKU TO SONO KINSETSU RYOIKI HENSHU-BU (EDITORIAL OFFICE OF JAPANESE JOURNAL OF CHILD PSYCHIATRY)
C/O KYOTO DAIGAKU IGAKUBU FUZOKU BYOIN SEISHIN IGAKU KYOSHITSU, 53, SHOGOIN KAWARA-MACHI, SAKYO-KU KYOTO-SHI 606
JAPANESE AND ENGLISH

JIKEIKAI SEISHIN IGAKU KENKYUSHO GYOSEKI 2903
(ARBEITEN AUS DEM ZIKEI INSTITUTE FUR PSYCHIATRIE)
JIKEIKAI SEISHIN IGAKU KENKYUSHO (ZIKEI INSTITUTE FUR PSYCHIATRIE)
100, URAYASU-HONMACHI, OKAYAMA-SHI 702
JAPANESE AND ENGLISH

JIN TO YUSEKI 2904
(KIDNEY AND DIALYSIS)
TOKYO IGAKUSHA
(TOKYO MEDICAL COMPANY)
JAPANESE

JIRITSU SHINKEI ZASSHI 2905
(JOURNAL OF AUTONOMIC NERVOUS SYSTEM)
NIHON SHINKYU GAKKAI (JAPANESE INSERT BURNING SOCIETY)
C/O OSAKA DAIGAKU IGAKUBU DAI-2 SEIRIGAKU KYOSHITSU, 2-7, DAIGAKU-MACHI, TAKATSUKI-SHI 569
JAPANESE

JOKEN HANSHA 2906
(CONDITIONED REFLEX)
KOBE IKA DAIGAKU
ROKKODAI-CHO, NADU-KU, KOBE
JAPANESE
INDEXED: CHE

JOSAI SHIKA DAIGAKU KIYO 2907
(BULLETIN OF THE JOSAI DENTAL UNIVERSITY)
JOSAI SHIKA DAIGAKU
1-1, KEYAKIDAI, SAKATO-CHO, IRIMA-GUN, SAITAMA 350-02
JAPANESE, WITH ENGLISH SUMMARY
INDEXED: CHE

JOURNAL OF CANCER AND VIROLOGY 2908
ENGLISH

JOURNAL OF CARDIOGRAPHY 2909
MEDIKARU EREKUTORO TAIMUSU SHA (MEDICAL ELECTRO TIMES COMPANY)
JAPANESE AND ENGLISH

JOURNAL OF DERMATOLOGY 2910
JAPANESE DERMATOLOGICAL ASSOCIATION
TAI-SEI-BIRU, 3-14,10 HONGO BUNKYO-KU, TOKYO
ENGLISH
INDEXED: BIO, CHE, EXC, IND

JOURNAL OF NIHON UNIVERSITY, SCHOOL OF DENTISTRY 2911
NIHON DAIGAKU SHIGAKUBU (NIHON UNIVERSITY, SCHOOL OF DENTISTRY)
8-13, 1-CHOME, KANDA SURUGADAI, CHIYODA-KU, TOKYO 101
EUROPEAN
INDEXED: CHE

JOURNAL OF OSAKA DENTAL UNIVERSITY 2912
OSAKA SHIKA DAIGAKU (OSAKA DENTAL UNIVERSITY)
1-47, KYOBASHI, HIGASHI-KU, OSAKA-SHI 540
ENGLISH
INDEXED: EXC, JICST

JOURNAL OF OSAKA UNIVERSITY DENTAL SCHOOL 2913
OSAKA DAIGAKU SHIGAKUBU (OSAKA UNIVERSITY DENTAL SCHOOL)
32, JOAN-CHO, KITA-KU, OSAKA-SHI 530
EUROPEAN
INDEXED: CHE, EXC

KAGOSHIMA DAIGAKU IGAKUBU GAN KENKYU NENPO 2914
(ANNUAL REPORT OF THE CANCER RESEARCH, KAGOSHIMA UNIVERSITY)
KAGOSHIMA DAIGAKU IGAKUBU FUZOKU SHUYO KENKYU SHISETSU (INSTITUTE OF CANCER RESEARCH, FACULTY OF MEDICINE, KAGOSHIMA UNIVERSITY)
7-82, SHIROYAMA-MACHI, KAGOSHIMA-SHI 892
JAPANESE

KANAGAWA SHIGAKU 2915
(JOURNAL OF THE KANAGAWA ODONTOLOGICAL SOCIETY)
KANAGAWA SHIKA DAIGAKU GAKKAI
82, INEOKA-CHO, YOKOSUKA-SHI 238
JAPANESE, WITH ENGLISH SUMMARY

KANAGAWA-KENRITSU SEIJINBYO SENTA NENPO 2916
(ANNUAL REPORT OF THE CENTER FOR ADULT DISEASES, KANAGAWA PREFECTURE)
KANAGAWA-KENRITSU SEIJINBYO SENTA
54-2, NAKAO-CHO, HODOGAYA-KU, YOKOHAMA-SHI 240
JAPANESE

KANSEN ENSHO MEN'EKI 2917
(INFECTION, INFLAMMATION AND IMMUNITY)
IYAKU NO MONSHA
3-3, NIHONBASHI HON-CHO, CHUO-KU, TOKYO 103
JAPANESE
INDEXED: CHE

KANSENSHOGAKU ZASSHI 2918
(JOURNAL OF THE JAPANESE ASSOCIATION FOR INFECTIOUS DISEASES)
NIHON DENSENBYO GAKKAI (JAPANESE ASSOCIATION FOR INFECTIOUS DISEASES)
20-12, 3-CHOME, KOMAGOME, TOSHIMA-KU, TOKYO 170
JAPANESE, WITH ENGLISH SUMMARY
INDEXED: CHE, EXC, IND

KANZO 2919
(ACTA HEPATOLOGICA JAPONICA)
NIHON KANZO GAKKAI (JAPAN ASSOCIATION FOR THE STUDY OF THE LIVER)
C/O TOYO BUNKO, 28-2, 2-CHOME, HON KOMAGOME, BUNKYO-KU, TOKYO 113
JAPANESE, WITH EUROPEAN SUMMARY

KEKKAKU 2920
(BULLETIN OF THE JAPANESE SOCIETY OF TUBERCULOSIS)
NIHON KEKKAKUBYO GAKKAI (JAPANESE SOCIETY FOR TUBERCULOSIS)
3-12, 1-CHOME, MISAKI-CHO, CHIYODA-KU, TOKYO 101
JAPANESE, WITH EUROPEAN SUMMARY
INDEXED: CHE, EXC

KEKKAKU NENPO 2921
(TUBERCULOSIS YEARBOOK)
KEKKAKU YOBOKAI (JAPAN ANTI-TUBERCULOSIS ASSOCIATION)
3-12, 1-CHOME, MISAKI-CHO, CHIYODA-KU, TOKYO 101
JAPANESE

KEKKAKU NO KENKYU 2922
(TUBERCULOSIS RESEARCH)
HOKKAIDO DAIGAKU KEKKAKU KENKYUSHO (RESEARCH INSTITUTE FOR TUBERCULOSIS, HOKKAIDO UNIVERSITY)
NISHI-6-CHOME, KITA-11-JO, KITA-KU, SAPPORO-SHI 060
JAPANESE
INDEXED: BIO, CHE, JICST

KEKKAKU NO TOKEI 2923
(TUBERCULOSIS STATISTICS)
KOSEI-SHO KOSHU EISEI-KYOKU (PUBLIC HEALTH BUREAU, MINISTRY OF HEALTH AND WELFARE)
3-12, 1-CHOME, MISAKI-CHO, CHIYODA-KU, TOKYO 101
JAPANESE

KEKKAKU OYOBI KOKYUKI SHIKKAN BUNKEN NO SHOROKU SOKUHO 2924
(ABSTRACTS OF THE CURRENT LITERATURE ON TB AND RESPIRATORY DISEASES)
KEKKAKU YOBOKAI (JAPAN ANTI-TUBERCULOSIS ASSOCIATION)
3-12, 1-CHOME, MISAKI-CHO, CHIYODA-KU, TOKYO 101
JAPANESE

KENRITSU GAN SENTA NIIGATA BYOIN GYOSEKISHU 2925
(LIST OF PUBLISHED PAPERS IN PREFECTURAL CANCER CENTER, NIIGATA HOSPITAL)
KENRITSU GAN SENTA NIIGATA BYOIN (PREFECTURAL CANCER CENTER, NIIGATA HOSPITAL)
2-10, KAWAGISHI-CHO, NIIGATA-SHI 951
JAPANESE

KENRITSU GAN SENTA NIIGATA BYOIN ISHI 2926
(JOURNAL OF CANCER CENTER, NIIGATA HOSPITAL)
KENRITSU GAN SENTA NIIGATA BYOIN (PREFECTURAL CANCER CENTER, NIIGATA HOSPITAL)
2-10, KAWAGISHI-CHO, NIIGATA-SHI 951
JAPANESE AND ENGLISH

KETSUGO SOSHIKI 2927
(CONNECTIVE TISSUE)
JAPANESE SOCIETY FOR CONNECTIVE TISSUE RESEARCH
C/O DEPARTMENT OF PATHOLOGY, TOKYO MEDICAL COLLEGE, 412, 1-CHOME, HIGASHIOKUBO, SHINJUKU-KU, TOKYO
JAPANESE AND ENGLISH
INDEXED: CHE

KISEICHUGAKU ZASSHI 2928
(JAPANESE JOURNAL OF PARASITOLOGY)
NIHON KISEICHU GAKKAI (JAPANESE SOCIETY OF PARASITOLOGY)
C/O SCHOOL OF MEDICINE, KEIO UNIVERSITY, SHINANO-MACHI, SHINJUKU-KU, TOKYO 160
JAPANESE AND ENGLISH
INDEXED: BIB, BIO, CHE, EXC

KITATAMA 2929
(BULLETIN OF THE KITATAMA DENTIST ASSOCIATION)
KITATAMA SHIKA ISHIKAI (KITATAMA DENTIST ASSOCIATION)
1-22, 5-CHOME, HON-CHO, KOGANEI-SHI, TOKYO 184
JAPANESE

KODO KEIRYOGAKU 2930
(JAPANESE JOURNAL OF BEHAVIORMETRICS)
NIHON KODO KEIRYO GAKKAI (JAPAN BEHAVIORMETRICS ACADEMY) AND TOKYO DAIGAKU IGAKUBU YAKUGAKU KYOSHITSUNAI (TOKYO UNIVERSITY, DEPARTMENT OF MEDICINE, EPIDEMIOLOGY SECTION)
JAPANESE

KOKO EISEI GAKKAI ZASSHI 2931
(ORAL HYGIENE ACADEMY MAGAZINE)
ORAL HYGIENE ACADEMY
JAPANESE

KOKOBYO GAKKAI ZASSHI 2932
(ORAL DISEASE ACADEMY MAGAZINE)
ORAL DISEASE ACADEMY
JAPANESE

KOKORO TO SHAKAI 2933
(MENTAL HEALTH)
NIHON SEISHIN EISEIKAI (JAPAN MENTAL HEALTH SOCIETY) AND SEISHIN SHOBO
91, BENTEN-CHO, SHINJUKU-KU, TOKYO 162
JAPANESE

KOKU EISEI GAKKAI ZASSHI 2934
(JOURNAL OF DENTAL HEALTH)
KOKU EISEI GAKKAI (JAPANESE SOCIETY FOR DENTAL HEALTH)
C/O NIHON SHIKA DAIGAKU EISEIGAKU KYOSHITSU, 9-20, 1-CHOME, FUJIMI-CHO, CHIYODA-KU, TOKYO 102
JAPANESE, WITH ENGLISH SUMMARY
INDEXED: CHE

KOKUBYO GAKKAI ZASSHI 2935
(JOURNAL OF THE JAPAN STOMATOLOGICAL SOCIETY)
KOKUBYO GAKKAI (JAPAN STOMATOLOGICAL SOCIETY)
C/O TOKYO IKASHIKA DAIGAKU, 5-45, 1-CHOME, BUNKYO-KU, TOKYO 113
JAPANESE AND EUROPEAN
INDEXED: CHE, JICST

KOKUNAI IGAN BUNKEN SOKUHO 2936
(ABSTRACTS ON GASTRIC CANCER LITERATURE IN JAPAN)
WHO KOKUSAI IGAN JOHO SENTA (INTERNATIONAL INFORMATION CENTER FOR STOMACHIC CANCER, WHO)
1-1, 5-CHOME, TSUKIJI, CHUO-KU, TOKYO 104
JAPANESE

KOKUNAI IGAN BUNKEN SOKUHO: NENKAN RUISEKI-HAN 2937
(CURRENT REPORT OF DOMESTIC STOMACH CANCER BIBLIOGRAPHY: ANNUAL CUMMULATIVE EDITION)
WHO KOKUSAI IGAN JOHO SENTA (WHO INTERNATIONAL STOMACH CANCER REPORT CENTER)
JAPANESE

KOKURITSU CHORYOKU GENGO SHOGAI SENTA 2938
(NATIONAL HEARING AND SPEECH IMPAIRMENT CENTER)
NATIONAL HEARING AND SPEECH IMPAIRMENT CENTER
JAPANESE

KOKURITSU GAN SENTA NENPO 2939
(ANNUAL REPORT OF NATIONAL CANCER CENTER, JAPAN)
KOKURITSU GAN SENTA (NATIONAL CANCER CENTER, JAPAN)
1-1, 5-CHOME, TSUKIJI, CHUO-KU, TOKYO 104
JAPANESE

KOKURITSU KEKKAKU RYOYOSHO KYODO KENKYU NENPO 2940
(ANNUAL REPORT ON THE JOINT STUDIES BY NATIONAL TUBERCULOSIS SANATORIA)
KOSEI-SHO IMU-KYOKU KOKURITSU RYOYOSHO CHUO KENKYU KYOGIKAI (CENTRAL RESEARCH COMMITTEE FOR NATIONAL SANATORIUM, MEDICAL AFFAIRS BUREAU, MINISTRY OF HEALTH AND WELFARE)
1-1, 3-CHOME, TAKEOKA, KIYOSE-SHI, TOKYO 180-04
JAPANESE

KOKURITSU TAMA KENKYUSHO NENPO 2941
(ANNUAL REPORT OF THE NATIONAL INSTITUTE FOR LEPROSY RESEARCH, MINISTRY OF HEALTH AND WELFARE)
KOSEI-SHO KOKURITSU TAMA KENKYUSHO (NATIONAL INSTITUTE FOR LEPROSY RESEARCH, MINISTRY OF HEALTH AND WELFARE)
4-155, AOBA-CHO, HIGASHIMURAYAMA-SHI, TOKYO 189
JAPANESE
INDEXED: JICST

KOKUSAI SHIKA JANARA 2942
(INTERNATIONAL JOURNAL OF DENTAL MEDICINE)
KOKUSAI ISHO SHUPPAN (INTERNATIONAL MEDICAL PUBLISHING)
JAPANESE

KONO RINSHO IGAKU KENKYUSHO-SHI 2943
(REPORT OF KONO CLINICAL MEDICINE RESEARCH INSTITUTE)
KONO RINSHO IGAKU KENKYUSHO (KONO CLINICAL MEDICINE RESEARCH INSTITUTE)
28-15, 1-CHOME, KITASHINAGAWA, SHINAGAWA-KU, TOKYO 140
JAPANESE, WITH ENGLISH SUMMARY

KOSANKINBYO KENKYU ZASSHI, KYOBU SHIKKAN KENKYU ZASSHI 2944
(JOURNAL OF TUBERCULOSIS AND LEPROSY)
KOSEIKAI
4-12, HIROSE-MACHI, SENDAI-SHI 980
JAPANESE
INDEXED: CHE, JICST

KOSEISHO GAN KENKYU HOKOKUSHU 2945
(ANNUAL REPORT ON CANCER RESEARCH, MINISTRY OF HEALTH AND WELFARE)
KOKURITSU GAN SENTA, UN'EIBU, KIKAKU SHITSU (NATIONAL CANCER CENTER, ADMINISTRATION DEPARTMENT, PLANNING OFFICE)
JAPANESE

KUCHI 2946
(MOUTH)
KORIYAMA SHIKA ISHIKAI (KORIYAMA DENTAL SOCIETY)
C/O KORIYAMA HOKENJO, 1-1, 1-CHOME, HAYAMA, KORIYAMA-SHI 963
JAPANESE

KUINTO ESSENSU 2947
(QUINTESSENCE INTERNATIONAL, JOURNAL OF PRACTICAL DENTISTRY)
ISHIYAKU SUPPAN
7-10, 1-CHOME, HONKOMAGOME, BUNKYO-KU, TOKYO 113
JAPANESE

KUMAMOTO DOMONKAISHI 2948
(KUMAMOTO JOURNAL OF PEDIATRICS)
KUMAMOTO DAIGAKU IGAKUBU SHONIKA DOMONKAI (ASSOCIATION OF PEDIATRICS, FACULTY OF MEDICINE, KUMAMOTO UNIVERSITY)
1-1, 1-CHOME, HONJO, KUMAMOTO-SHI 860
JAPANESE

KYOTO DAIGAKU KEKKAKU KYOBU SHIKKAN KENKYUSHO KIYO 2949
(BULLETIN OF THE CHEST DISEASE RESEARCH INSTITUTE, KYOTO UNIVERSITY)
KYOTO DAIGAKU KEKKAKU KYOBU SHIKKAN KENKYUSHO (CHEST DISEASE RESEARCH INSTITUTE, KYOTO UNIVERSITY)
53, SHOGOIN KAWARA-CHO, SAKYO-KU, KYOTO-SHI 606
JAPANESE AND ENGLISH
INDEXED: BIO, CHE, EXC, IND, JICST

KYOTO DAIGAKU KOKUKAGAKU KIYO 2950
(BULLETIN OF STOMATOLOGY, KYOTO UNIVERSITY)
KYOTO DAIGAKU IGAKUBU KOKUGEKAGAKU KYOSHITSU (DEPARTMENT OF STOMATOLOGY, FACULTY OF MEDICINE, KYOTO UNIVERSITY)
53, SHOGOIN KAWARA-CHO, SAKYO-KU, KYOTO-SHI 606
JAPANESE AND ENGLISH
INDEXED: BIO

KYUSHU DAIGAKU KYOBU SHIKKAN KENKYU SHISETSU KIYO 2951
(BULLETIN OF THE RESEARCH INSTITUTE FOR DISEASE OF THE CHEST, KYUSHU UNIVERSITY)
KYUSHU DAIGAKU KYOBU SHIKKAN KENKYU SHISETSU (RESEARCH INSTITUTE FOR DISEASE OF THE CHEST, KYUSHU UNIVERSITY)
1752, MEINOHAMA, NISHI-KU, FUKUOKA-SHI 814
JAPANESE, WITH EUROPEAN SUMMARY
INDEXED: CHE, JICST

KYUSHU SAIBOSHIN KENKYUKAI ZASSHI 2952
(JOURNAL OF THE KYUSHU SOCIETY OF CYTOLOGICAL DIAGNOSIS)
KYUSHU SAIBOSHIN KENKYUKAI (KYUSHU SOCIETY OF CYTOLOGICAL DIAGNOSIS)
C/O KYUSHU DAIGAKU BYOIN 1276, KATAKASU, HIGASHI-KU, FUKUOKA-SHI 812
JAPANESE

KYUSHU SHIKA DAIGAKU SHINGAKU KATEI KENKYU KIYO 2953
(RESEARCH BULLETIN OF THE KYUSHU DENTAL UNIVERSITY PRE-DENTAL DIVISION)
KYUSHU SHIKA DAIGAKU SHINGAKU KATEI (PRE-DENTAL DIVISION, KYUSHU DENTAL UNIVERSITYT)
JAPANESE

KYUSHU SHIKA GAKKAI ZASSHI 2954
(JOURNAL OF THE KYUSHU DENTAL SOCIETY)
KYUSHU SHIKA GAKKAI (KYUSHU DENTAL SOCIETY)
C/O KYUSHU SHIKA DAIGAKU, 6-1, 2-CHOME, MANAZURU, KOKURA-KU, KITA KYUSHU-SHI 803
JAPANESE AND ENGLISH
INDEXED: CHE, JICST

KYUSHU SHINKEI SEISHIN IGAKU 2955
(KYUSHU NEURO-PSYCHIATRY)
NISHI NIHON SHINKEI SEISHIN GAKKAI (WEST-JAPANESE ASSOCIATION OF NEURO-PSYCHIATRY)
C/O KYUSHU DAIGAKU IGAKUBU SHINKEI SEISHIN IGAKU KYOSHITSU, KATAKASU, HIGASHI-KU, FUKUOKA-SHI 812
JAPANESE, WITH EUROPEAN SUMMARY
INDEXED: CHE

MASUI 2956
(JAPANESE JOURNAL OF ANESTHESIOLOGY)
KOKUSEIDO SHUPPAN K.K. (KOKUSEIDO PUBLISHING COMPANY, LIMITED)
23-5, 3-CHOME, HONGO, BUNKYO-KU, TOKYO 113
JAPANESE, WITH ENGLISH SUMMARY
INDEXED: CHE, EXC, JICST

MASUI TO SOSEI 2957
(HIROSHIMA JOURNAL OF ANESTHESIA)
HIROSHIMA MASUI IGAKKAI (HIROSHIMA ANESTHESIA ASSOCIATION)
C/O HIROSHIMA DAIGAKU IGAKUBU MASUIGAKU KYOSHITSU, 2-3, 1-CHOME, KASUMI-CHO, HIROSHIMA-SHI 734
JAPANESE AND ENGLISH
INDEXED: CHE, EXC, JICST

MATSUMOTO SHIGAKU 2958
(JOURNAL OF THE MATSUMOTO DENTAL COLLEGE SOCIETY)
MATSUMOTO SHIKA DAIGAKU GAKKAI (MATSUMOTO DENTAL COLLEGE SOCIETY)
JAPANESE

MATSUMOTO SHIKA DAIGAKU KENKYU KAISHI 2959
(RECORDS OF THE MATSUMOTO DENTAL UNIVERSITY RESEARCH)
MATSUMOTO DENTAL UNIVERSITY
JAPANESE

MATSUMOTO SHIKA DAIGAKU KIYO 2960
(BULLETIN OF MATSUMOTO DENTAL COLLEGE)
MATSUMOTO SHIKA DAIGAKU
HIROOKAGO, SHIOJIRI-SHI 399-07
JAPANESE AND ENGLISH

MEDICAL ULTRASONICS 2961
ENGLISH

MEGURO KISEICHUKAN KENKYU HOKOKU 2962
(RESEARCH BULLETIN OF THE MEGURO PARASITOLOGICAL MUSEUM)
MEGURO KISEICHUKAN (MEGURO PARASITOLOGICAL MUSEUM)
1-1, 4-CHOME, SHIMOMEGURO, MEGURO-KU, TOKYO 153
ENGLISH
INDEXED: BIO

MEGURO KISEICHUKAN NYUSU 2963
(MEGURO PARASITOLOGICAL MUSEUM NEWS)
MEGURO KISEICHUKAN (MEGURO PARASITOLOGICAL MUSEUM)
1-1, 4-CHOME, SHIMOMEGURO, MEGURO-KU, TOKYO 153
JAPANESE AND ENGLISH

MICHINOKU SHIGAKKAI ZASSHI 2964
(JOURNAL OF MICHINOKU DENTAL SOCIETY)
MICHINOKU SHIGAKKAI (MICHINOKU DENTAL SOCIETY)
C/O TOHOKU DAIGAKU SHIGAKUBU, 4-1, SEIRYO-CHO, SENDAI-SHI 980
JAPANESE AND ENGLISH

MIE SEISHIN IRYO 2965
(JOURNAL OF MIE PSYCHIATRIST ASSOCIATION)
MIE SEISHIN'IKAI
C/O MIE-KENRITSU TAKAJAYA BYOIN, 2251-1, KOMORI-CHO, TAKAJAYA, TSU-SHI 514
JAPANESE

NAIKA 2966
(INTERNAL MEDICINE)
NANKODO
42-6, 3-CHOME, HONGO, BUNKYO-KU, TOKYO 113
JAPANESE
INDEXED: CHE

NAIKA HOKAN 2967
(JAPANESE ARCHIVES OF INTERNAL MEDICINE)
NAIKA HOKAN KANKOKAI (SOCIETY FOR PUBLICATION OF JAPANESE ARCHIVES OF INTERNAL MEDICINE)
C/O KYOTO DAIGAKU IGAKUBU NAIKAGAKU KYOSHITSU, 53, SHOGOIN KAWARA-CHO, KYOTO-SHI 606
JAPANESE, WITH ENGLISH SUMMARY
INDEXED: CHE, EXC, JICST

NICHIDAI KURINIKKU 2968
(CLINICAL JOURNAL OF NIHON UNIVERSITY)
NIHON DAIGAKU IGAKUBU FUZOKU ITABASHI BYOIN (ITABASHI HOSPITAL, FACULTY OF MEDICINE, NIHON UNIVERSITY)
30-1, OYAGUCHI KAMI-MACHI, ITABASHI-KU, TOKYO 173
JAPANESE AND ENGLISH

NICHIDAI SHIGAKU 2969
(NIHON UNIVERSITY DENTAL JOURNAL)
NIHON DAIGAKU SHIGAKKAI (SOCIETY OF DENTISTRY, NIHON UNIVERSITY)
8-13, 1-CHOME, KANDA SURUGADAI, CHIYODA-KU, TOKYO 101
JAPANESE
INDEXED: JICST

NIHON BYORI GAKKAI KAISHI 2970
(TRANSACTIONES SOCIETATIS PATHOLOGICAE JAPONICAE)
NIHON BYORI GAKKAI (JAPANESE PATHOLOGICAL SOCIETY)
C/O TOKYO DAIGAKU IGAKUBU BYORIGAKU KYOSHITSU, HONGO, BUNKYO-KU, TOKYO 113
JAPANESE

NIHON BYORIGAKKAI KAIHO 2971
(TRANSACTIONS SOCIETY PATHOLOGICAL JAPONICAE)
JAPANESE AND ENGLISH

NIHON DAICHO KOMONBYO GAKKAI ZASSHI 2972
(JOURNAL OF THE JAPAN SOCIETY OF COLO-PROCTOLOGY)
NIHON DAICHO KOMONBYO GAKKAI (JAPAN SOCIETY OF COLO-PROCTOLOGY)
C/O TOHO DAIGAKU IGAKUBU DAI-1 GEKA KYOSHITSU, 11-1, 6-CHOME, OMORI NISHI, OTA-KU, TOKYO 143
JAPANESE
INDEXED: CHE, EXC

NIHON GAN CHIRYO GAKKAISHI 2973
(JOURNAL OF JAPAN SOCIETY FOR CANCER THERAPY)
NIHON GAN CHIRYO GAKKAI (JAPAN SOCIETY FOR CANCER THERAPY)
53, SHOGOIN KAWARA-MACHI, SAKYO-KU, KYOTO-SHI 606
JAPANESE AND ENGLISH
INDEXED: BIO, EXC, IND

NIHON GANGAKKAI SOKAI KIJI 2974
(ARTICLES OF THE GENERAL MEETING OF THE JAPAN CANCER SOCIETY)
JAPANESE

NIHON GODO GANGAKKAIGI ENDAISHU 2975
(LIST OF PAPERS OF JAPANESE JOINT CONGRESS ON CANCER)
NIHON GAN GAKKAI, NIHON GAN CHIRYO GAKKAI, NIHON HAIGAN GAKKAI (JAPANESE CANCER ASSOCIATION, JAPANESE SOCIETY FOR CANCER THERAPY, JAPAN SOCIETY OF LUNG CANCER)
C/O GAN KENKYUSHO, 37-1, 1-CHOME, KAMIIKEBUKURO, TOSHIMA-KU, TOKYO 171
JAPANESE

NIHON HIFUKA GAKKAI ZASSHI 2976
(JAPANESE JOURNAL OF DERMATOLOGY)
NIHON HIFUKA GAKKAI (JAPANESE DERMATOLOGICAL ASSOCIATION)
C/O TAISEI BUILDING, 14-10, 3-CHOME, HONGO, BUNKYO-KU, TOKYO 113
JAPANESE AND ENGLISH
INDEXED: BIO, CHE, IND

NIHON HINYOKIKA GAKKAI ZASSHI 2977
(JAPANESE JOURNAL OF UROLOGY)
NIHON HINYOKIKA GAKKAI (JAPANESE UROLOGICAL ASSOCIATION)
C/O TOKYO DAIGAKU IGAKUBU HINYOKIKA KYOSHITSU, 3-1, 7-CHOME, BUNKYO-KU, TOKYO 113
JAPANESE AND ENGLISH
INDEXED: CHE, EXC, IND

NIHON HOTETSU SHIKA GAKKAI ZASSHI 2978
(JOURNAL OF JAPAN PROSTHODONTIC SOCIETY)
NIHON HOTETSU SHIKA GAKKAI (JAPAN PROSTHODONTIC SOCIETY)
C/O KOKU HOKEN KYOKAI, 38-6, 1-CHOME, KOMAGOME, TOSHIMA-KU, TOKYO 170
JAPANESE
INDEXED: JICST

NIHON JIBI INKOKA GAKKAI KAIHO 2979
(JOURNAL OF THE OTO-RHINO-LARYNGOLOGICAL SOCIETY OF JAPAN)
NIHON JIBI INKOKA GAKKAI (OTO-RHINO-LARYNGOLOGICAL SOCIETY OF JAPAN)
23-14-807, 3-CHOME, TAKANAWA, MINATO-KU, TOKYO 108
JAPANESE AND ENGLISH
INDEXED: CHE, EXC

NIHON JIDO SEISHIN IGAKKAI SOKAI 2980
(PROCEEDINGS OF THE GENERAL MEETING OF THE CHILD PSYCHIATRY SOCIETY OF JAPAN)
NIHON JIDO SEISHIN IGAKKAI
C/O KYOTO DAIGAKU IGAKUBU FUZOKU BYOIN, 53, SHOGOIN KAWARA-CHO, SAKYO-KU, KYOTO-SHI 605
JAPANESE

NIHON JINZO GAKKAISHI 2981
(JAPANESE JOURNAL OF NEPHROLOGY)
NIHON JINZO GAKKAI (JAPANESE SOCIETY OF NEPHROLOGY)
C/O NIHON DAIGAKU IGAKUBU DAI-2 NAIKA, 30, OYAGUCHI KAMI-CHO, ITABASHI-KU, TOKYO 173
JAPANESE AND ENGLISH

NIHON KIKAN SHOKUDOKA GAKKAI KAIHO 2982
(JOURNAL OF THE JAPAN BRONCHO-ESOPHAGOLOGICAL SOCIETY)
NIHON KIKAN SHOKUDOKA GAKKAI (JAPAN BRONCHO-ESOPHAGOLOGICAL SOCIETY)
C/O KEIO DAIGAKU IGAKUBU, 35, SHINANO-MACHI, SHINJUKU-KU, TOKYO 160
JAPANESE, WITH ENGLISH SUMMARY
INDEXED: BIO, EXC

NIHON KOKUKA GAKKAI ZASSHI 2983
(JOURNAL OF THE JAPANESE STOMATOLOGICAL SOCIETY)
NIHON KOKUKA GAKKAI (JAPANESE STOMATOLOGICAL SOCIETY)
C/O TOKYO DAIGAKU IGAKUBU KOKU GEKAGAKU KYOSHITSU, 3-1, 7-CHOME, HONGO, BUNKYO-KU, TOKYO 113
JAPANESE
INDEXED: CHE, JICST

NIHON KYOBU RINSHO 2984
(JAPANESE JOURNAL OF CHEST DISEASES)
KOKUSEIDO SHUPPAN K.K. (KOKUSEIDO PUBLISHING COMPANY, LIMITED)
23-5, 3-CHOME, HONGO, BUNKYO-KU, TOKYO 113
JAPANESE, WITH ENGLISH SUMMARY
INDEXED: APT

NIHON KYOBU SHIKKAN GAKKAI ZASSHI 2985
(JAPANESE JOURNAL OF THORACIC DISEASES)
NIHON KYOBU SHIKKAN GAKKAI (JAPAN SOCIETY OF CHEST DISEASES)
C/O TOYO BUNKO, 28-21, 2-CHOME, KOMAGOME, BUNKYO-KU, TOKYO 112
JAPANESE AND ENGLISH

NIHON KYOSEI SHIKA GAKKAI ZASSHI 2986
(JOURNAL OF JAPAN ORTHODONTIC SOCIETY)
NIHON KYOSEI SHIKA GAKKAI (JAPAN ORTHODONTIC SOCIETY)
C/O TOKYO IKA SHIKA DAIGAKU SHIGAKUBU KYOSEIGAKU KYOSHITSU, YUSHIMA, BUNKYO-KU, TOKYO 113
JAPANESE, WITH EUROPEAN SUMMARY
INDEXED: JICST

NIHON MONAI-KEI GAKKAI KAISHI 2987
(JOURNAL OF THE JAPAN SOCIETY OF THE RETICULOENDTHELIAL SYSTEM)
NIHON MONAI-KEI GAKKAI (JAPAN SOCIETY OF THE RETICULOENDTHELIAL SYSTEM)
C/O AICHI-KEN GAN SENTA KENKYUSHO, KANOKODEN, CHIKUSA-KU, NAGOYA-SHI 464
JAPANESE

NIHON NAIKA GAKKAI ZASSHI 2988
(JOURNAL OF THE JAPANESE SOCIETY OF INTERNAL MEDICINE)
NIHON NAIKA GAKKAI (JAPANESE SOCIETY OF INTERNAL MEDICINE)
33-5, 3-CHOME, HONGO, BUNKYO-KU, TOKYO 113
JAPANESE
INDEXED: CHE, IND, JICST

NIHON RIKA RYOJISHI KYOKAI NYUSU 2989
(JAPANESE PHYSICAL THERAPY ASSOCIATION NEWS)
JAPANESE PHYSICAL THERAPY ASSOCIATION
JAPANESE

NIHON RINSHO 2990
(JAPANESE JOURNAL OF CLINICAL MEDICINE)
NIHON RINSHOSHA
1, 3-CHOME, DOSHO-MACHI, HIGASHI-KU, OSAKA-SHI 541
JAPANESE
INDEXED: CHE, IND

NIHON RINSHO BYORI DOGAKUIN TSUSHIN 2991
(REPORT OF THE JAPAN CLINICAL PATHOLOGY ACADEMY)
JAPAN CLINICAL PATHOLOGY ACADEMY
JAPANESE

NIHON RONEN IGAKKAI ZASSHI 2992
(JAPANESE JOURNAL OF GERIATRICS)
NIHON RONEN IGAKKAI (JAPAN GERIATRICS SOCIETY)
C/O TOKYO DAIGAKU IGAKUBU, 3-1, 7-CHOME, BUNKYO-KU, TOKYO 113
JAPANESE, WITH ENGLISH SUMMARY
INDEXED: EXC

NIHON RYUMACHI GAKKAI SOKAI ENDAI NAIYO SHOROKU 2993
(ABSTRACTS OF THE SYMPOSIUM OF THE JAPAN RHEUMATISM ASSOCIATION)
NIHON RYUMACHI GAKKAI (THE JAPAN RHEUMATISM ASSOCIATION)
C/O KYOTO FURITSU IKA DAIGAKU SEIKEI GEKAGAKU KYOSHITSU, HIROKOJI AGARU, KAWARACHO-DORI, KAMIGYO-KU, KYOTO-SHI 602
JAPANESE

NIHON SEISHIN BYOIN KYOKAI GEPPO 2994
(MONTHLY BULLETIN OF THE JAPAN MENTAL HOSPITAL ASSOCIATION)
NIHON SEISHIN BYOIN KYOKAI (JAPAN MENTAL HOSPITAL ASSOCIATION)
C/O SHINBASHI EKIMAE BUILDING, 2-20, SHINBASHI, MINATO-KU, TOKYO 105
JAPANESE

NIHON SEISHIN EISEIKAI NYUSU RETA 2995
(NEWS LETTER, JAPAN ASSOCIATION OF MENTAL HEALTH)
NIHON SEISHIN EISEIKAI (JAPAN ASSOCIATION OF MENTAL HEALTH)
C/O SHINKEI KENKYUSHO, 91, BENTENCHO, SHINJUKU-KU, TOKYO 162
JAPANESE

NIHON SEISHIN HAKUJAKU KENKYU KYOKAI KAISHI 2996
(JAPANESE JOURNAL FOR THE SCIENTIFIC STUDY OF MENTAL DEFICIENCY)
JAPANESE

NIHON SEISHIN SHINKEI GAKKAI SOKAI SHOROKUSHU 2997
(ABSTRACTS OF THE SYMPOSIUM OF THE JAPANESE SOCIETY OF PHYSICHIATRY AND NEUROLOGY)
NIHON SEISHIN SHINKEI GAKKAI (JAPANESE SOCIETY OF PSYCHIATRY AND NEUROLOGY)
C/O TOKYO DAIGAKU IGAKUBU SEISHIN IGAKU KYOSHITSU, 3-1, 5-CHOME, HONGO BUNKYO-KU, TOKYO 113
JAPANESE

NIHON SHIGI 2998
(JAPAN DENTAL TECHNICIAN)
NIHON SHIKA GIKOSHIKAI (JAPAN DENTAL TECHNICIANS ASSOCIATION)
JAPANESE

NIHON SHIKA DAIGAKU KIYO 2999
(BULLETIN OF JAPAN DENTAL UNIVERSITY)
NIHON SHIKA DAIGAKU SHINGAKU KATEI
JAPANESE

NIHON SHIKA DAIGAKU SHIGAKKAI KAIHO 3000
(JOURNAL OF THE DENTAL SOCIETY, JAPAN DENTAL UNIVERSITY)
JAPANESE

NIHON SHIKA HOSHASEN GAKKAI SOKAI KOEN SHOROKUSHU 3001
(ABSTRACTS OF THE SYMPOSIUM OF THE JAPANESE SOCIETY OF DENTAL RADIOLOGY)
NIHON SHIKA HOSHASEN GAKKAI (JAPANESE SOCIETY OF DENTAL RADIOLOGY)
C/O NIHON DAIGAKU SHIGAKUBU HOSHASEN KYOSHITSU, 1-8, KANDA SURUGADAI, CHIYODA-KU, TOKYO 101
JAPANESE

NIHON SHIKA HOZONGAKU ZASSHI 3002
(JAPANESE JOURNAL OF CONSERVATIVE DENTISTRY)
NIHON SHIKA HOZON GAKKAI (JAPANESE SOCIETY OF CONSERVATIVE DENTISTRY)
C/O TOKYO SHIKA DAIGAKU HOZONGAKU KYOSHITSU, 9-18, 2-CHOME, MISAKI-CHO, CHIYODA-KU, TOKYO 101
JAPANESE, WITH ENGLISH SUMMARY

NIHON SHIKA HYORON 3003
(NIPPON DENTAL REVIEW)
NIHON SHIKA HYORONSHA
C/O MORITA BUILDING, 3-6, 3-CHOME, GINZA NISHI, CHUO-KU, TOKYO 104
JAPANESE

NIHON SHIKA IGAKKAI KAIHO 3004
(JOURNAL OF THE JAPANESE ASSOCIATION FOR DENTAL SCIENCE)
JAPANESE

NIHON SHIKA ISHIGAKKAI KAISHI 3005
(JOURNAL OF JAPAN DENTAL HISTORY ASSOCIATION)
NIHON SHIKA ISHIGAKKAI
C/O NIHON DAIGAKU MATSUDO SHIKA DAIGAKU SHIGAKUSHI KENKYUSHITSU, 870-1, 2-CHOME, NISHI, SAKAE-CHO, MATSUDO-SHI 271
JAPANESE

NIHON SHIKA ISHIKAI ZASSHI 3006
(JOURNAL OF THE JAPAN DENTAL ASSOCIATION)
NIHON SHIKA ISHIKAI (JAPAN DENTAL ASSOCIATION)
1-20, 4-CHOME, KUDAN KITA, CHIYODA-KU, TOKYO 102
JAPANESE
INDEXED: CHE

NIHON SHIKA SHINBUN 3007
(JAPAN DENTAL TIMES)
RINSHO TSUSHINSHA
C/O KUDANSHITA BUILDING, 3-4, KANDA JINBO-CHO, CHIYODA-KU, TOKYO 100
JAPANESE

NIHON SHIKA ZAIRYO KIKAI GAKKAI ZASSHI 3008
(JOURNAL OF THE JAPAN RESEARCH SOCIETY OF DENTAL MATERIALS AND APPLICANCES)
NIHON SHIKA ZAIRYO KIKAI GAKKAI (JAPAN RESEARCH SOCIETY OF DENTAL MATERIALS & APPLICANCES) AND TOKYO IKA SHIKA DAIGAKU SHIKA RIKOGAKU KYOSHITSU (DEPARTMENT OF DENTAL MATERIALS)
1-5, YUSHIMA, BUNKYO-KU, TOKYO 113
JAPANESE, WITH ENGLISH SUMMARY
INDEXED: CHE

NIHON SHINSEIJI GAKKAI ZASSHI 3009
(ACTA NEONATOLOGICA JAPONICA)
NIHON SHINSEIJI GAKKAI (JAPAN SOCIETY OF NEONATOLOGY)
C/O NIHON DAIGAKU IGAKUBU SHONIKAGAKU KYOSHITSU, 30-1, OYAGUCHI-KAMI-MACHI, ITABASHI-KU, TOKYO 173
JAPANESE AND ENGLISH
INDEXED: EXC

NIHON SHISHUBYO GAKKAI KAISHI 3010
(JOURNAL OF THE JAPANESE ORGANIZATION FOR RESEARCH OF PERIODONTOLOGY)
NIHON SHISHUBYO GAKKAI
C/O DEPARTMENT OF PERIODONTOLOGY, SCHOOL OF DENTISTRY, AICHI GAKUIN UNIVERSITY, 2-11, SUEMORIDORI, CHIKUSA-KU, NAGOYA-SHI 464
JAPANESE, WITH ENGLISH SUMMARY
INDEXED: CHE

NIHON SHOKAKIBYO GAKKAI ZASSHI 3011
(JOURNAL OF THE JAPANESE SOCIETY OF GASTROENTEROLOGY)
NIHON SHOKAKIBYO GAKKAI (JAPANESE SOCIETY OF GASTROENTEROLOGY)
4-12, 7-CHOME, GINZANISHI, CHUO-KU, TOKYO 104
JAPANESE
INDEXED: CHE, JICST

NIHON SHONIKA GAKKAI ZASSHI 3012
(ACTA PAEDIATRICA JAPONICA)
NIHON SHONIKA GAKKAI (JAPAN PEDIATRIC SOCIETY)
C/O KOSHU EISEI KYOKAI BUILDING, 29-8, 1-CHOME, SHINJUKU, SHINJUKU-KU, TOKYO 160
JAPANESE
INDEXED: EXC, JICST

NIHON TAISHITSUGAKU ZASSHI 3013
(JAPANESE JOURNAL OF CONSTITUTIONAL MEDICINE)
NIHON TAISHITSU KENKYUKAI (JAPAN CONSTITUTIONAL MEDICAL RESEARCH ASSOCIATION)
C/O KUMAMOTO DAIGAKU TAISHITSU IGAKU KENKYUSHO, 24-1, 4-CHOME, KUHONJI, KUMAMOTO-SHI 862
JAPANESE AND EUROPEAN
INDEXED: CHE, JICST

NIIGATA DAIGAKU NO KENKYUJO GYOSEKISHU 3014
(REPORT OF ACHIEVEMENTS OF THE NIGATA UNIVERSITY BRAIN RESEARCH INSTITUTE)
NIIGATA UNIVERSITY BRAIN RESEARCH INSTITUTE
JAPANESE

NIIGATA SHIGAKKAI ZASSHI 3015
(NIIGATA DENTAL JOURNAL)
NIIGATA SHIGAKKAI (NIIGATA DENTAL ASSOCIATION)
2-5274, GAKKO-MACHI, NIIGATA-SHI 951
JAPANESE

NIIGATA-KENRITSU GAN SENTA NIIGATA BYOIN GYOSEKISHU 3016
(COLLECTED ACHIEVEMENTS OF THE NIIGATA PREFECTURAL CANCER CENTER, NIIGATA HOSPITAL)
KENRITSU GAN SENTA NIIGATA BYOIN-HENSHU IINKAI HEN (PREFECTURAL CANCER CENTER, NIIGATA HOSPITAL EDITORIAL COMMITTEE)
JAPANESE

NIPPON DENTAL COLLEGE ANNUAL PUBLICATIONS 3017
JAPANESE
INDEXED: JICST

NIPPON RINSHO SAIBO GAKKAI ZASSHI 3018
(JOURNAL OF JAPANESE SOCIETY OF CLINICAL CYTOLOGY)
NIPPON RINSHO SAIBO GAKKAI (JAPANESE SOCIETY OF CLINICAL CYTOLOGY)
C/O GAN KENKYUSHO FUZOKU BYOIN FUJINKA, 37-1, 1-CHOME, KAMIIKEBUKURO, TOSHIMA-KU, TOKYO 170
JAPANESE, WITH ENGLISH SUMMARY

NIPPON SHIKA MASUI GAKKAI ZASSHI 3019
(JOURNAL OF JAPANESE DENTAL ANESTHESIA SOCIETY)
NIPPON SHIKA MASUI GAKKAI (JAPANESE SOCIETY OF DENTAL ANESTHESIA)
C/O TOKYO IKA SHIKA DAIGAKU SHIKA MASUIGAKU KYOSHITSU, 5-45, 1-CHOME, YUSHIMA, BUNKYO-KU, TOKYO 113
JAPANESE AND ENGLISH
INDEXED: CHE

NISHI NIHON HIFUKA 3020
(NISHI NIHON JOURNAL OF DERMATOLOGY)
NISHI NIHON HIFUKA HENSHU-BU
C/O KYUSHU DAIGAKU IGAKUBU HIFUKA KYOSHITSU, 1276, KATAKASU, HIGASHI-KU, FUKUOKA 812
JAPANESE, WITH ENGLISH SUMMARY
INDEXED: JICST

NISHI NIHON HINYOKIKA 3021
(NISHI NIHON JOURNAL OF UROLOGY)
NIHON HINYOKIKA GAKKAI NISHI NIHON RENGO CHIHOKAI (NISHI NIHON SECTION OF JAPANESE UROLOGICAL ASSOCIATION)
C/O KYUSHU DAIGAKU IGAKUBU HINYOKIKA, 1276, KATAKASU, HIGASHI-KU, FUKUOKA 812
JAPANESE, WITH ENGLISH SUMMARY
INDEXED: BIO, JICST

NISSHI KOHO 3022
(NEWS OF THE JAPAN DENTAL ASSOCIATION)
NIHON SHIKA ISHIKAI (JAPAN DENTAL ASSOCIATION)
1-20, 4-CHOME, KUDAN KITA, CHIYODA-KU, TOKYO 102
JAPANESE

NO KENKYUKAI KAISHI 3023
(JAPANESE JOURNAL OF THE NEUROSCIENCES RESEARCH ASSOCIATION)
JAPANESE

NO KENKYUSHO GYOSEKISHU (NIIGATA DAIGAKU) 3024
(RESEARCH RESULTS OF THE BRAIN RESEARCH INSTITUTE, NIIGATA UNIVERSITY)
NIIGATA DAIGAKU NO KENKYUSHO (BRAIN RESEARCH INSTITUTE, NIIGATA UNIVERSITY)
757, ICHIBAN-CHO, ASAHICHO-DORI, NIIGATA-SHI 951
JAPANESE

NO SHINKEI GEKA 3025
(NEUROLOGICAL SURGERY)
IGAKU SHOIN
29-11, 5-CHOME, HONGO, BUNKYO-KU, TOKYO 113
JAPANESE, WITH ENGLISH SUMMARY
INDEXED: EXC

NO TO HATTATSU 3026
(BRAIN AND DEVELOPMENT)
NIHON SHONI SHINKEIGAKU KENKYUKAI (JAPANESE SOCIETY OF PEDIATRIC NEUROLOGY) AND SHINDAN TO CHIRYOSHA
C/O 4TH FLOOR, MARUNOUCHI BUILDING, MARUNOUCHI, CHIYODA-KU, TOKYO 100
JAPANESE, WITH ENGLISH SUMMARY
INDEXED: CHE, EXC

NO TO SHINKEI 3027
(BRAIN AND NERVE)
IGAKU SHOIN
29-11, 5-CHOME, HONGO, BUNKYO-KU, TOKYO 113
JAPANESE, WITH ENGLISH SUMMARY
INDEXED: CHE, EXC

NOHA TO KINDENZU 3028
(JAPANESE JOURNAL OF EEG AND ENG)
NIHON NOHA KINDENZU GAKKAI
C/O TOKYO BUNKO, 28-21, 2-CHOME, HONKOMAGOME, BUNKYO-KU, TOKYO 113
JAPANESE

NOSOTCHU NO KENKYU 3029
(STUDIES ON CEREBRAL APOPLEXY)
NOKEKKAN KENKYUSHO (INSTITUTE OF BRAIN AND BLOOD VESSELS)
6-23, OTE-MACHI, ISEZAKI-SHI 372
JAPANESE AND ENGLISH

OKAYAMA-KEN SHIKA ISHIKAIHO 3030
(REPORT OF OKAYAMA DENTAL ASSOCIATION)
OKAYAMA-KEN SHIKA ISHIKAI (OKAYAMA DENTAL ASSOCIATION)
1-5, ISHIZEKI-CHO, OKAYAMA-SHI 700
JAPANESE

ONISHI SEISHIN EISEI KENKYUSHO KIYO 3031
(REVIEW OF ONISHI MENTAL HYGIENE LABORATORY)
ONISHI SEISHIN EISEI KENKYUSHO (ONISHI MENTAL HYGIENE LABORATORY)
336, KAMITENJIN-CHO, TAKAMATSU-SHI 760
JAPANESE

OSAKA DAIGAKU SHIGAKU ZASSHI 3032
(JOURNAL OF OSAKA UNIVERSITY DENTAL SOCIETY)
OSAKA DAIGAKU SHIGAKKAI (OSAKA UNIVERSITY DENTAL SOCIETY)
32, JOAN-CHO, KITA-KU, OSAKA-SHI 530
JAPANESE, WITH EUROPEAN SUMMARY
INDEXED: CHE, EXC

OSAKA SEISHIN EISEI 3033
(OSAKA MENTAL HEALTH)
OSAKA SEISHIN EISEI KYOGIKAI (OSAKA MENTAL HEALTH ASSOCIATION)
C/O OSAKA-FU SEISHIN EISEI SODANSHO, 3-69, 1-CHOME, NAKAMICHI, HIGASHINARI-KU, OSAKA-SHI 537
JAPANESE

OSAKA-FURITSU KOSHU EISEI KENKYUSHO KENKYU HOKOKU, SEISHIN EISEI-HEN 3034
(PROCEEDINGS OF THE OSAKA PREFECTURE PUBLIC HEALTH INSTITUTE)
OSAKA-FURITSU KOSHU EISEI KENKYUSHO (OSAKA PREFECTURE PUBLIC HEALTH INSTITUTE)
3-69, 1-CHOME, NAKAMICHI, HIGASHINARI-KU, OSAKA-SHI 537
JAPANESE
INDEXED: JICST

PROCEEDINGS OF THE CONGRESS OF THE JAPAN SOCIETY FOR CANCER THERAPY 3035
NIHON GAN CHIRYO GAKKAI (JAPAN SOCIETY FOR CANCER THERAPY)
53, SHOGOIN KAWARA-CHO, SAKYO-KU, KYOTO-SHI 606
EUROPEAN

PROCEEDINGS OF THE JAPANESE CANCER ASSOCIATION 3036
NIHON GANGAKKAI (JAPANESE CANCER ASSOCIATION)
37-1, 1-CHOME, KAMI IKEBUKURO, TOSHIMA-KU, TOKYO 170
JAPANESE
INDEXED: JICST

PROGRESS OF MEDICAL PARASITOLOGY IN JAPAN 3037
MEGURO KISEICHU-KAN (MEGURO PARASITOLOGICAL MUSEUM)
3-557, SHIMOMEGURO, MEGURO-KU, TOKYO 153
ENGLISH
INDEXED: BIO

RECENT ADVANCES IN RES RESEARCH 3038
NIHON MONAIKEI GAKKAI (JAPAN SOCIETY OF THE RETICULOENDOTHELIAL SYSTEM) AND BUNKODO
2-7, 7-CHOME, HONGO, BUNKYO-KU, TOKYO 113
ENGLISH
INDEXED: BIO, CHE

REPORTS ON MEDICAL RESEARCH PROBLEMS OF THE JAPAN ANTI-TUBERCULOSIS ASSOCIATION 3039
KEKKAKU YOBOKAI (JAPAN ANTI-TUBERCULOSIS ASSOCIATION)
3-12, 1-CHOME, MISAKI-CHO, CHIYODA-KU, TOKYO 101
ENGLISH
INDEXED: CHE, JICST

REPURA 3040
(LA LEPRO)
NIHON RAI GAKKAI (JAPANESE LEPROSY ASSOCIATION)
C/O KOKURITSU TAMA KENKYUSHO, 4-1455, AOBA-CHO, HIGASHIMURAYAMA-SHI, TOKYO 189
JAPANESE AND ENGLISH
INDEXED: CHE, EXC, JICST

RINSHO BYORI 3041
(JAPANESE JOURNAL OF CLINICAL PATHOLOGY)
NIHON RINSHO BYORI GAKKAI (JAPAN SOCIETY OF CLINICAL PATHOLOGY) AND RINSHO BYORI KANKO-KAI
C/O KURAMATA BUILDING, 5-4, 4-CHOME, YUSHIMA, BUNKYO-KU, TOKYO 113
JAPANESE
INDEXED: CHE, IND

RINSHO HIFUKA 3042
(JAPANESE JOURNAL OF CLINICAL DERMATOLOGY)
IGAKU SHOIN
29-11, 5-CHOME, HONGO, BUNKYO-KU, TOKYO 113
JAPANESE

RINSHO HINYOKIKA 3043
(JAPANESE JOURNAL OF CLINICAL UROLOGY)
IGAKU SHOIN
29-11, 5-CHOME, HONGO, BUNKYO-KU, TOKYO 113
JAPANESE
INDEXED: EXC

RINSHO HYOKA 3044
(CLINICAL EVALUATION)
RINSHO HYOKA KANKOKAI (CLINICAL EVALUATION PUBLISHING ASSOCIATION)
JAPANESE

RINSHO IGAKU SHINBUN 3045
(NEWS OF CLINICAL MEDICINE)
RINSHO IGAKU SHINBUNSHA
1-13, 2-CHOME, NISHITOKOROZAWA, TOKOROZAWA-SHI 359
JAPANESE

RINSHO KAGAKU 3046
(JAPANESE JOURNAL OF CLINICAL CHEMISTRY)
NIHON RINSHO KAGAKU KENKYUKAI (JAPAN CLINICAL CHEMISTRY RESEARCH ASSOCIATION)
JAPANESE

RINSHO KAGAKU 3047
(JOURNAL OF CLINICAL SCIENCE)
RINSHO KAGAKUSHA
C/O SEKAI HOKEN TSUSHINSHA, 33, 3-CHOME, HIRANO-MACHI, HIGASHI-KU, OSAKA-SHI 541
JAPANESE
INDEXED: CHE

RINSHO KOKYU SEIRI 3048
(CLINIC OF RESPIRATORY PHYSIOLOGY)
MEDIKARU JANARUSHA
TAITO FAIZA K.K. TOKYO EIGYOSHO JONAN BUNSHITSU
C/O DAI-1 HANADA BUILDING, NISHIGOTANDA, SHINAGAWA-KU, TOKYO 141
JAPANESE

RINSHO MEN'EKI 3049
(CLINICAL IMMUNOLOGY)
RINSHO MEN'EKI HENSHU IINKAI
TOKYO IGAKUSHA
13-10, 2-CHOME, HONGO, BUNKYO-KU, TOKYO 113
JAPANESE

RINSHO NOHA 3050
(CLINICAL ELECTROENCEPHALOGRAPHY)
NAGAI SHOTEN
3-111, KAMIFUKUSHIMAKITA, FUKUSHIMA-KU, OSAKA-SHI 553
JAPANESE

RINSHO SEIJINBYO 3051
(JOURNAL OF ADULT DISEASES)
TOKYO IGAKUSHA
13-10, 2-CHOME, HONGO, BUNKYO-KU, TOKYO 113
JAPANESE

RINSHO SEISHIN IGAKU 3052
(JAPANESE JOURNAL OF CLINICAL PSYCHIATRY)
KOKUSAI ISHO SHUPPAN
C/O ISHII BUILDING, 5-8, 2-CHOME, HONGO, BUNKYO-KU, TOKYO 113
JAPANESE

RINSHO SHIKA 3053
(FOLIA ODONTOLOGICA PRACTICA)
RINSHO SHIKASHA
TAKOYAKUSHI SAKAI-MACHI HIGASHI, NAKAGYO-KU, KYOTO-SHI 604
JAPANESE
INDEXED: CHE, JICST

RINSHO SHINKEIGAKU 3054
(CLINICAL NEUROLOGY)
NIHON SHINKEI GAKKAI (JAPANESE SOCIETY OF NEUROLOGY)
C/O AKAMON BUILDING, 26-4, 5-CHOME, HONGO, BUNKYO-KU, TOKYO 113
JAPANESE AND ENGLISH

RINSHO SHIN'ONZU 3055
(CARDIOVASCULAR SOUND BULLETIN)
RINSHO SHIN'ONZU KENKYUKAI (JAPANESE SOCIETY OF CARDIOVASULAR SOUND)
MEDIKARU EREKUTORO TAIMUSUSHA (MEDICAL ELECTRO TIMES COMPANY, LIMITED)
C/O FUKUDA DENSHI BUILDING, 39-4, 3-CHOME, BUNKYO-KU, TOKYO 113
JAPANESE AND ENGLISH

RINSHO SHINRIGAKU KENKYU 3056
(STUDIES ON CLINICAL PSYCOLOGY)
NIHON RINSHO SHINRIGAKKAI (JAPANESE SOCIETY OF CLINICAL PSYCOLOGY)
C/O KOKURITSU SEISHIN EISEI KENKYUSHO, 7-3, 1-CHOME, KONODAI, ICHIKAWA-SHI 272
JAPANESE

RINSHO SHONI IGAKU 3057
(JOURNAL OF CLINICAL PEDIATRICS)
SHONI AIIKU KYOKAI (ASSOCIATION FOR THE CARE OF THE CHILD)
C/O HOKKAIDO DAIGAKU IGAKUBU SHONIKAGAKU KYOSHITSU, KITA-14-JO, NISHI-5-CHOME, KITA-KU, SAPPORO-SHI 060
JAPANESE, WITH ENGLISH SUMMARY
INDEXED: CHE, EXC, JICST

RINSHO TO KENKYU 3058
(JAPANESE JOURNAL OF CLINICAL AND EXPERIMENTAL MEDICINE)
DAIDO GAKKAN SHUPPAN-BU
C/O KYUSHU DAIGAKU IGAKUBU HOIGAKU KYOSHITSU, KATAKASU, HIGASHI-KU, FUKUOKA-SHI 812
JAPANESE AND ENGLISH
INDEXED: CHE

RINSHO TO SAIKIN 3059
(CLINICAL BACTERIOLOGY)
KINDAI SHUPPAN (MODERN PUBLISHING)
JAPANESE

RONENGAKU BUNKENSHU 3060
(BIBLIOGRAPHY OF GERONTOLOGY AND GERIATRICS)
YOKUFUKAI BYOIN
12-1, 1-CHOME, TAKAIDO NISHI, SUGINAMI-KU, TOKYO 168
JAPANESE AND ENGLISH

RYUMACHI 3061
(OFFICIAL JOURNAL OF THE JAPAN RHEUMATISM ASSOCIATION)
NIHON RYUMACHI KYOKAI (JAPAN RHEUMATISM ASSOCIATION)
3-1, 7-CHOME, HONGO, BUNKYO-KU, TOKYO 113
JAPANESE AND ENGLISH
INDEXED: EXC

SAIBOKAKU BYORIGAKU ZASSHI 3062
(JOURNAL OF KARYOPATHOLOGY)
SAIBOKAKU BYORIGAKU ZASSHI HENSHUBU (EDITORIAL OFFICE OF JOURNAL OF KARYOPATHOLOGY)
C/O OKAYAMA DAIGAKU IGAKUBU BYORIGAKU KYOSHITSU, 5-1, 2-CHOME, SHIKADA-MACHI, OKAYAMA-SHI 700
JAPANESE, WITH ENGLISH SUMMARY
INDEXED: CHE

SEIJINBYO 3063
(ADULT DISEASES)
OSAKA-FURITSU SEIJINBYO SENTA (CENTER FOR ADULT DISEASES, OSAKA PREFECTURE)
OSAKA SEIJINBYO YOBO KYOKAI
C/O OSAKA-FU EISEI-BU, 2, OTEMAENO-CHO, OSAKA-SHI 540
JAPANESE

SEISHIN BUNSEKI 3064
(TOKYO JOURNAL OF PSYCHOANALYSIS)
TOKYO SEISHIN BUNSEKIGAKU KENKYUSHO (THE TOKYO INSTITUTE FOR PSYCHOANALYSIS)
281, 1-KU, NISHI NASUNO-MACHI, TOCHIGI 329-27
JAPANESE AND ENGLISH

SEISHIN BUNSEKI KENKYU 3065
(JAPANESE JOURNAL OF PSYCHOANALYSIS)
NIHON SEISHIN BUNSEKI GAKKAI (JAPAN PSYCHOANALYTICAL ASSOCIATION)
C/O KEIO GIJUKU DAIGAKU IGAKUBU SHINKEIKA KYOSHITSU, 35, SHINANO-MACHI, SHINJUKU-KU, TOKYO 160
JAPANESE

SEISHIN BYOIN KANRI KENKYUKAISHI 3066
(JOURNAL OF MENTAL HOSPITAL CONTROL RESEARCH ASSOCIATION)
NIHON SEISHIN BYOIN KYOKAI (JAPAN MENTAL HOSPITAL ASSOCIATION)
2-20, SHINBASHI, MINATO-KU, TOKYO 105
JAPANESE

SEISHIN BYOINCHO KENSHUKAISHI 3067
(REPORT OF MENTAL HOSPITAL DIRECTORS' TRAINING ASSEMBLY)
NIHON SEISHIN BYOIN KYOKAI (JAPAN MENTAL HOSPITAL ASSOCIATION)
2-20, SHINBASHI, MINATO-KU, TOKYO 105
JAPANESE

SEISHIN EISEI JITTAI CHOSA (KANAGAWA-KEN) 3068
(STATISTICAL REPORT OF MENTAL HEALTH, KANAGAWA PREFECTURE)
KANAGAWA-KEN
1, NIHON ODORI, NAKA-KU, YOKOHAMA-SHI 231
JAPANESE

SEISHIN EISEI KANRI KENKYU 3069
(MENTAL HEALTH CARE)
SEISHIN EISEI KANRI KENKYUKAI
C/O OSAKA DAIGAKU BYOIN, SEISHINKA, DOJIMA HAMA-DORI FUKUSHIMA-KU, OSAKA-SHI 553
JAPANESE

SEISHIN EISEI KENKYU 3070
(JOURNAL OF MENTAL HEALTH)
KOSEI-SHO KOKURITSU SEISHIN EISEI KENKYUSHO (NATIONAL INSTITUTE OF MENTAL HEALTH, MINISTRY OF HEALTH AND WELFARE)
7-3, 1-CHOME, KONODAI-MACHI, ICHIKAWA-SHI 272
JAPANESE, WITH EUROPEAN SUMMARY
INDEXED: BIO

SEISHIN EISEI SENTA NENPO 3071
(ANNUAL REPORT OF THE HOKKAIDO MENTAL HEALTH CENTER)
SEISHIN EISEI SENTA (HOKKAIDO MENTAL HEALTH CENTER)
818, SHIRAISHI HON-DORI, SHIRAISHI-KU, SAPPORO-SHI 062
JAPANESE

SEISHIN EISEI SHIRYO 3072
(ANNUAL REPORT ON MENTAL HEALTH)
KOSEI-SHO KOKURITSU SEISHIN EISEI KENKYUSHO (NATIONAL INSTITUTE OF MENTAL HEALTH, MINISTRY OF HEALTH AND WELFARE)
7-3, 1-CHOME, KONODAI-MACHI, TACHIKAWA-SHI 272
JAPANESE

SEISHIN HAKUJAKU NO KENKYU -- SHISETU NI OKERU 3073
(RESEARCH ON THE FEEBLE-MINDED -- THE INSTITUTIONALIZED)
NIHON SEISHIN HAKUJAKUSHA AIGO KYOKAI (JAPAN ASSOCIATION FOR THE CARE OF THE FEEBLE-MINDED)
JAPANESE

SEISHIN HAKUJAKUJI KENKYU 3074
(STUDIES ON FEEBLE-MINDED CHILDREN)
ZEN NIHON TOKUSHU KYOIKU KENKYU RENMEI
NIHON BUNKA KAGAKUSHA
15-17, 6-CHOME, HONKOMAGOME, BUNKYO-KU, TOKYO 113
JAPANESE

SEISHIN HOKEN 3075
(MENTAL HEALTH)
NISHI NIHON SEISHIN HOKEN GAKKAI (KYUSHU ASSOCIATION OF MENTAL HEALTH)
C/O KYUSHU DAIGAKU IGAKUBU SHINKEI SEISHIN IGAKU KYOSHITSU, KATAKASU, HIGASHI-KU, FUKUOKA-SHI 812
JAPANESE

SEISHIN IGAKU 3076
(CLINICAL PSYCHIATRY)
IGAKU SHOIN
29-11, 5-CHOME, HONGO, BUNKYO-KU, TOKYO 113
JAPANESE

SEISHIN IGAKU KENKYUSHO GYOSEKISHU 3077
(BULLETIN OF THE SEISHIN IGAKU INSTITUTE)
SEISHIN IGAKU KENKYUSHO (INSTITUTE OF PSYCHIATRY, TOKYO)
11-11, 4-CHOME, KOMONE, ITABASHI-KU, TOKYO 174
JAPANESE AND EUROPEAN
INDEXED: BIO, EXC, JICST

SEISHIN SHINKEIGAKU ZASSHI 3078
(PSYCHIATRIA ET NEUROLOGIA JAPONICA)
NIHON SEISHIN SHINKEI GAKKAI (JAPANESE SOCIETY OF NEUROLOGY AND PSYCHIATRY)
C/O TOYO BUNKO, 28-21, 2-CHOME, HONKOMAGOME, BUNKYO-KU, TOKYO 113
JAPANESE, WITH EUROPEAN SUMMARY
INDEXED: CHE, EXC

SEISHIN SHINTAI IGAKU 3079
(JOURNAL OF JAPANESE PSYCHOSOMATIC SOCIETY)
NIHON SEISHIN SHINTAI IGAKKAI (JAPANESE PSYCHOSOMATIC SOCIETY) AND IGAKU SHOIN
29-11, 5-CHOME, HONGO, BUNKYO-KU, TOKYO 113
JAPANESE AND ENGLISH
INDEXED: BIO

SENTEN IJO 3080
(CONGENITAL ANOMALIES)
NIHON SENTEN IJO GAKKAI (CONGENITAL ANOMALIES RESEARCH ASSOCIATION OF JAPAN)
C/O KYOTO DAIGAKU IGAKUBU KAIBOGAKU KYOSHITSU, YOSHIDA KONOE-CHO, SAKYO-KU, KYOTO-SHI 606
JAPANESE AND ENGLISH
INDEXED: BIO, EXC, JICST

SEXUAL MEDICINE 3081
MEDEIKARA TORIBIYUN NIHON SHISHA (MEDICAL TRIBUNE, JAPAN BRANCH OFFICE)
JAPANESE AND ENGLISH

SHIGAKU 3082
(ODONTOLOGY)
NIHON SHIKA DAIGAKU SHIGAKKAI (SOCIETY OF NIPPON DENTAL COLLEGE)
9-20, 1-CHOME, FUJIMI, CHIYODA-KU, TOKYO 102
JAPANESE AND EUROPEAN
INDEXED: BIO, CHE, JICST

SHIKA GAKUHO 3083
(JOURNAL OF THE TOKYO DENTAL COLLEGE SOCIETY)
TOKYO SHIKA DAIGAKU GAKKAI (TOKYO DENTAL COLLEGE SOCIETY)
9-18, 2-CHOME, MISAKI-CHO, CHIYODA-KU, TOKYO 101
JAPANESE, WITH ENGLISH SUMMARY
INDEXED: CHE, EXC

SHIKA GIKO 3084
(JOURNAL OF DENTAL TECHNICS)
ISHIYAKU SHUPPAN
7-10, 1-CHOME, HONKOMAGOME, BUNKYO-KU, TOKYO 113
JAPANESE
INDEXED: CHE

SHIKA HOSHASEN 3085
(DENTAL RADIOLOGY)
NIHON SHIKA HOSHASEN GAKKAI (JAPANESE SOCIETY OF DENTAL RADIOLOGY)
C/O NIHON DAIGAKU SHIGAKUBU HOSHASENGAKU KYOSHITSU, 1-8, KANDA SURUGADAI, CHIYODA-KU, TOKYO 101
JAPANESE, WITH EUROPEAN SUMMARY

SHIKA IGAKU 3086
(JOURNAL OF THE OSAKA ODONTOLOGICAL SOCIETY)
OSAKA SHIKA GAKKAI (OSAKA ODONTOLOGICAL SOCIETY)
1-47, KYOBASHI, HIGASHI-KU, OSAKA-SHI 540
JAPANESE, WITH EUROPEAN SUMMARY
INDEXED: CHE, JICST

SHIKA JIHO 3087
(JOURNAL OF DENTISTRY)
SHIKA JIHOSHA
43-7, 1-CHOME, ZOSHIGAYA, TOSHIMA-KU, TOKYO 171
JAPANESE
INDEXED: JICST

SHIKA KISO IGAKKAI ZASSHI 3088
(JAPANESE JOURNAL OF ORAL BIOLOGY)
SHIKA KISO IGAKKAI (JAPANESE ASSOCIATION FOR ORAL BIOLOGY)
C/O NIHON DAIGAKU SHIGAKUBU, 1-8, KANDA SURUGADAI, CHIYODA-KU, TOKYO 101
JAPANESE

SHIKA PEN KURABU 3089
(JOURNAL OF DENTAL PEN CLUB)
SHIKA PEN KURABUSHA
8-12, 1-CHOME, KAMIMEGURO, MEGURO-KU, TOKYO 153
JAPANESE

SHIKA RIKOGAKU ZASSHI 3090
(JOURNAL OF THE JAPAN SOCIETY FOR DENTAL APPARATUS AND MATERIALS)
SHIKA RIKO GAKKAI (JAPAN SOCIETY FOR DENTAL APPARATUS AND MATERIALS)
C/O TOKYO IKA SHIKA DAIGAKU, 5-45, 1-CHOME, YUSHIMA, BUNKYO-KU, TOKYO 113
JAPANESE, WITH EUROPEAN SUMMARY
INDEXED: JICST

SHIKA SARON 3091
(DENTAL NEWS AND RECREATION)
TOKYO SHIKA SARONSHA
34-5, 5-CHOME, KOENJIMINAMI, SUGINAMI-KU, TOKYO 166
JAPANESE

SHIKAI GEPPO 3092
(HYOGO DENTISTS' ASSOCIATION MONTHLY)
HYOGO-KEN SHIKA ISHIKAI (HYOGO DENTISTS' ASSOCIATION)
5-41, YAMAMOTODORI, IKUTA-KU, KOBE-SHI 650
JAPANESE

SHIKAI HOCHI 3093
(DENTISTRY NEWS)
SHIKAI HOCHI SHINBUNSHA
3-2, HIMURO-CHO, MINAMI-KU, NAGOYA-SHI 457
JAPANESE

SHIKAI KOHO 3094
(DENTAL BULLETIN)
SHIKAI KOHOSHA
4-29, CHIHAYA-CHO, TOSHIMA-KU, TOKYO 171
JAPANESE

SHIKAI TENBO 3095
(DENTAL OUTLOOK)
ISHIYAKU SHUPPAN K.K. (ISHIYAKU PUBLISHERS, INC.)
7-10, 1-CHOME, HONKOMAGOME, BUNKYO-KU, TOKYO 113
JAPANESE
INDEXED: CHE

SHINKEI GEKA 3096
(NEUROLOGICAL MEDICINE)
KAGAKU HYORON SHA (SCIENCE REVIEW COMPANY)
JAPANESE

SHINKEI JOHO KENKYUSHITSU NENPO 3097
(BULLETIN OF NEUROINFORMATION LABORATORY)
NAGASAKI DAIGAKU IGAKUBA SHINKEI JOHO KENKYUSHITSA (NAGASAKI UNIVERSITY, MEDICAL COLLEGE, NEUROLOGICAL INFORMATION RESEARCH ROOM)
JAPANESE

SHINKEI KENKYU NO SHINPO 3098
(ADVANCES IN NEUROLOGICAL SCIENCES)
IGAKU SHOIN
29-11, 5-CHOME, HONGO, BUNKYO-KU, TOKYO 113
JAPANESE, WITH EUROPEAN SUMMARY
INDEXED: BIO, CHE, EXC

SHINKEI KENKYUSHO GYOSEKISHU 3099
(RESEARCH RESULTS OF NEUROPSYCHIATRIC RESEARCH INSTITUTE)
SHINKEI KENKYUSHO (NEUROPSYCHIATRIC RESEARCH INSTITUTE)
91, BENTEN-CHO, SHINJUKU-KU, TOKYO 162
JAPANESE AND ENGLISH

SHINKIN TO SHINKINSHO 3100
(JAPANESE JOURNAL OF MEDICAL MYCOLOGY)
NIHON ISHINKIN GAKKAI (JAPANESE SOCIETY FOR MEDICAL MYCOLOGY)
C/O JUNTENDO DAIGAKU IGAKUBU, 1-1, 2-CHOME, HONGO, BUNKYO-KU, TOKYO 113
JAPANESE AND EUROPEAN
INDEXED: JICST

SHINZO 3101
(HEART)
SHINZO HENSHU IINKAI
IJI TSUSHINSHA
4-15, 4-CHOME, AKASAKA, MINATO-KU, TOKYO 107
JAPANESE

SHIYO 3102
(JOURNAL OF THE TOKYO DENTAL ASSOCIATION)
TOKYOTO SHIKAISHIKAI (TOKYO MUNICIPAL MEDICAL AND DENTAL ASSOCIATION)
JAPANESE

SHIZUOKA-KEN SEISHIN EISEI SENTA SHOHO 3103
(REPORT OF MENTAL HEALTH CENTER, SHIZUOKA PREFECTURE)
SHIZUOKA-KEN SEISHIN EISEI SENTA
594-5, MAGARIKANE, SHIZUOKA-SHI 420
JAPANESE

SHONI GEKA NAIKA 3104
(JAPANESE JOURNAL OF PEDIATRIC SURGERY AND MEDICINE)
TOKYO IGAKUSHA
25-12, 3-CHOME, HONGO, BUNKYO-KU, TOKYO 113
JAPANESE AND ENGLISH

SHONI HOKEN KENKYU 3105
(JOURNAL OF CHILD HEALTH)
NIHON SHONI HOKEN KYOKAI (JAPANESE SOCIETY OF CHILD HEALTH)
1-2, ICHIGAYA SATOHARA-CHO, SHINJUKU-KU, TOKYO 162
JAPANESE

SHONI IGAKU 3106
(PEDIATRICS)
IGAKU SHOIN (MEDICAL ACADEMY)
JAPANESE

SHONI NO HOKEN 3107
(CHILDREN'S HEALTH)
OSAKA-SHIRITSU SHONI HOKEN SENTA (OSAKA MUNICIPAL CHILDREN'S HEALTH CENTER)
JAPANESE

SHONI NO SEISHIN TO SHINKEI 3108
(PSYCHIATRIA ET NEUROLOGIA PEDIATRICA JAPONICA)
SHONI SEISHIN SHINKEIGAKU KENKYUKAI (SOCIETAS PSYCHIATRIA ET NEUROLOGIA PEDIATRICA JAPONICA)
38-3, 1-CHOME, HIGASHI NAKANO, NAKANO-KU, TOKYO 164
JAPANESE AND ENGLISH

SHONI SHIKAGAKU ZASSHI 3109
(JAPANESE JOURNAL OF PEDODONTICS)
NIHON SHONI SHIKA GAKKAI (JAPAN SOCIETY OF PEDODONTICS)
C/O TOKYO SHIKA DAIGAKU SHONI SHIKAGAKU KYOSHITSU, 9-18, 2-CHOME, MISAKI-CHO, CHIYODA-KU, TOKYO 101
JAPANESE, WITH EUROPEAN SUMMARY

SHONIKA 3110
(PEDIATRICS)
KANEHARA SHUPPAN K.K.
31-14, 2-CHOME, YUSHIMA, BUNKYO-KU, TOKYO 113-91
JAPANESE
INDEXED: APT

SHONIKA KIYO 3111
(ANNALES PAEDIATRICI JAPONICI)
SHONIKA KIYO KANKOKAI (SOCIETY FOR THE PUBLICATION OF ANNALES PAEDIATRICI JAPONICI)
C/O KYOTO-DAIGAKU IGAKUBU SHONIKAGAKU KYOSHITSU, 53, SHOGOIN KAWARA-CHO, SAKYO-KU, KYOTO-SHI 606
JAPANESE AND EUROPEAN
INDEXED: CHE, EXC

SHONIKA RINSHO 3112
(JAPANESE JOURNAL OF PEDIATRICS)
NIHON SHONI IJI SHUPPANSHA
C/O WAKODO BUILDING, 2-16, KANDA KAJI-CHO, CHIYODA-KU, TOKYO 101
JAPANESE

SHONIKA SHINRYO 3113
(JOURNAL OF PEDIATRIC PRACTICE)
SHINDAN TO CHIRYOSHA
C/O MARU BUILDING, 406, MARUNOUCHI, CHIYODA-KU, TOKYO 100
JAPANESE AND EUROPEAN
INDEXED: APT

SOGO NYUYOJI KENKYU 3114
(GENERAL INFANT RESEARCH)
DOBUN SHOIN (DOBUN PUBLISHERS)
JAPANESE

SOGO RINSHO 3115
(CLINIC ALL-ROUND)
NAGAI SHOTEN
3-111, KITA, KAMIFUKUSHIMA, FUKUSHIMA-KU, OSAKA-SHI 553
JAPANESE
INDEXED: CHE

TAI GAN KYOKAIHO 3116
(BULLETIN OF JAPAN ANTI-CANCER ASSOCIATION)
NIHON TAI GAN KYOKAI
C/O ASAHI BUILDING, 6-7, 6-CHOME, GINZA, CHUO-KU, TOKYO 104
JAPANESE

TAISHABYO 3117
(METABOLIC DISEASE)
IYAKU NO MONSHA
3-3, NIHONBASHI HON-CHO, CHUO-KU, TOKYO 113
JAPANESE

TAISHITSU IGAKU KENKYUSHO HOKOKU 3118
(BULLETIN OF THE INSTITUTE OF CONSTITUTIONAL MEDICINE)
KUMAMOTO DAIGAKU TAISHITSU IGAKU KENKYUSHO (INSTITUTE OF CONSTITUTIONAL MEDICINE, KUMAMOTO UNIVERSITY)
24-1, 4-CHOME, KUHONJI, KUMAMOTO-SHI 862
JAPANESE, WITH EUROPEAN SUMMARY
INDEXED: JICST

TOHOKU DAIGAKA SHIGAKU KENKYU 3119
(TOHOKU UNIVERSITY DENTAL RESEARCH)
TOHOKU DAIGAKU SHIGAKUBU (TOHOKU UNIVERSITY DENTAL FACULTY)
JAPANESE

TOHOKU NO KORONI 3120
(JOURNAL OF THE TOHOKU SOCIETY FOR MICROBIOLOGY)
NIHON SAIKIN GAKKAI TOHOKU SHIBU (TOHOKU SOCIETY FOR MICROBIOLOGY)
C/O MIYAGI EISEI KENKYUSHO, 27, MASUE, ODAWARA, HARA-MACHI, SENDAI-SHI 983
JAPANESE

TOHOKU PHYCHOLOGICA FOLIA 3121
TOHOKU DAIGAKU (TOHOKU UNIVERSITY)
1-1, 2-CHOME, KATAHIRA-CHO, SENDAI-SHI 980
EUROPEAN

TOHOKU SHIKA DAIGAKU GAKKAISHI 3122
(JOURNAL OF TOHOKU DENTAL UNIVERSITY)
JAPANESE

TOJIBIKAIHO 3123
(TOKYO EAR AND NOSE MEDICAL SOCIETY REPORT)
TOKYOTO JIBIINKO KA IKAI JIMU-KYOKU (BUSINESS OFFICE OF THE TOKYO OTOLARYNGOLOGICAL SOCIETY)
JAPANESE

TOKAI JOURNAL OF EXPERIMENTAL AND CLINICAL MEDICINE 3124
TOKAI UNIVERSITY PRESS
SHINJUKU TOKAI BUILDING, 3-27-4 SHINJUKU, SHINJUKU-KU, TOKYO 160
ENGLISH
INDEXED: BIO, CHE

TOKAI RYUMACHI 3125
(OFFICIAL JOURNAL OF THE TOKAI RHEUMATISM ASSOCIATION)
NIHON RYUMACHI GAKKAI TOKAI CHIHOKAI
C/O KOKURITSU NAGOYA BYOIN, 1-1, 4-CHOME, SANNOMARU, NAKA-KU, NAGOYA-SHI 460
JAPANESE

TOKYO-TO GAN KENSHIN SENTA NENPO 3126
(ANNUAL REPORT OF TOKYO METROPOLITAN CANCER DETECTION CENTER)
TOKYO-TO GAN KENSHIN SENTA (TOKYO METROPOLITAN CANCER DETECTION CENTER)
2-5, KANDA SURUGADAI, CHIYODA-KU, TOKYO 102
JAPANESE

TOKYO-TO ROJIN SOGO KENKYUSHO NENPO 3127
(ANNUAL REPORT OF THE TOKYO METROPOLITAN INSTITUTE OF GERONTOLOGY)
TOKYO-TO ROJIN SOGO KENKYUSHO (TOKYO METROPOLITAN INSTITUTE OF GERONTOLOGY)
35-2, SAKAE-CHO, ITABASHI-KU, TOKYO 173
JAPANESE

TOKYO-TO SHINKEI KAGAKU SOGO KENKYUSHO NENPO 3128
(ANNUAL REPORT OF THE TOKYO INSTITUTE OF NEUROSCIENCES)
TOKYO-TO SHINKEI KAGAKU SOGO KENKYUSHO (TOKYO METROPOLITAN INSTITUTE FOR NEUROSCIENCES)
2-6, MUSASHIDAI, FUCHU-SHI, TOKYO 183
JAPANESE

TONYOBYO 3129
(JOURNAL OF THE JAPAN DIABETIC SOCIETY)
NIHON TONYOBYO GAKKAI (JAPAN DIABETIC SOCIETY)
C/O TOYO BUNKO, 28-21, 2-CHOME, HONKOMAGOME, BUNKYO-KU, TOKYO 113
JAPANESE, WITH EUROPEAN SUMMARY
INDEXED: CHE, EXC

TONYOBYO NO SHIMPO 3130
(ADVANCES IN DIABETES)
SHINDAN TO CHIRYOSHA
406 MARUNOUCHI BUILDING, CHIYODA-KU, TOKYO
JAPANESE
INDEXED: CHE

TONYOBYO TO TAISHA 3131
(DIABETES JOURNAL)
DEIABETEISU JANARU SHA (DIABETES JOURNAL COMPANY)
JAPANESE

TOYO IGAKU TO PEIN KURINIKKU 3132
(ORIENTAL MEDICINE AND THE PAIN CLINIC)
OSAKA IKA DAIGAKU MASUIGAKU KYOSHITSU
2-9, DAIGAKU-MACHI, TAKATSUKI-SHI 569
JAPANESE

TOYU 3133
(DIABETES COMPANION)
TONYOYU NO KAI
(DIABETES COMPANION ASSOCIATION)
JAPANESE

TSURUMI SHIGAKU 3134
(TSURUMI UNIVERSITY DENTAL JOURNAL)
TSURUMI DAIGAKU SHIGAKKAI (TSURUMI UNIVERSITY DENTAL SOCIETY)
JAPANESE

TUBERCULOSIS, LEPROSY AND CANCER 3135
TOHOKU DAIGAKU KOSANKINBYO KENKYUSHO (RESEARCH INSTITUTE FOR TUBERCULOSIS, LEPROSY AND CANCER, TOHOKU UNIVERSITY)
4-12, HIROSE-MACHI, SENDAI-SHI 980
ENGLISH
INDEXED: BIO, CHE, EXC, IND, INS, JICST

TUMOR RESEARCH, EXPERIMENTAL AND CLINICAL 3136
SAPPORO IKA DAIGAKU FUZOKU GAN KENKYUSHO (CANCER RESEARCH INSTITUTE, SAPPORO MEDICAL COLLEGE)
NISHI-17-CHOME, MINAMI-1-JO, CHUO-KU, SAPPORO-SHI 060
ENGLISH
INDEXED: BIO, CHE, EXC

VD - JAPANESE JOURNAL OF VENEREAL DISEASES 3137
NIHON SEIBYO YOBO KYOKAI (JAPANESE ASSOCIATION FOR PREVENTION OF VENEREAL DISEASE)
14-10, 3-CHOME, HONGO, BUNKYO-KU, TOKYO 113
JAPANESE

WAY 3138
DENTARU RISACHI INTANASHONARU (DENTAL RESEARCH INTERNATIONAL)
JAPANESE AND ENGLISH

YOKUFUKAI CHOSA KENKYU KIYO 3139
(ACTA GERONTOLOGICA JAPONICA)
YOKUFUKAI BYOIN (YOKUFUKAI GERIATRIC HOSPITAL) AND YOKUFUKAI
12-1, 1-CHOME, TAKAIDONISHI, SUGINAMI-KU, TOKYO 168
JAPANESE AND EUROPEAN
INDEXED: CHE, EXC, JICST

ZENKOKU IGAN TOROKU CHOSA HOKOKU 3140
(DATA ON JAPANESE REGISTERED GASTRIC CANCER)
IGAN KENKYUKAI, KOKURITSU GANSENTA (JAPANESE RESEARCH SOCIETY FOR GASTRIC CANCER, NATIONAL CANCER CENTER)
1-1,5-CHOME, TSUKIJI, CHUO-KU, TOKYO 104
JAPANESE

Surgery, Ophthalmology

CHUBU NIPPON SEIKEI GEKA SAIGAI GEKA GAKKAI ZASSHI 3141
(CENTRAL JAPAN JOURNAL OF ORTHOPAEDIC AND TRAUMATIC SURGERY)
CHUBU NIHON SEIKEI GEKA SAIGAI GEKA GAKKAI (CENTRAL JAPAN ORTHOPAEDIC AND TRAUMATIC SURGICAL SOCIETY) AND KYOTO DAIGAKU IGAKUBU SEIKEI GEKA KYOSHITSU (DEPARTMENT OF ORTHOPAEDICS SURGERY, KYOTO UNIVERSITY)
SHOGOIN KAWARA-CHO, SAKYO-KU, KYOTO-SHI 606
JAPANESE

GAISHO 3142
(TRAUMATOLOGY)
KAGAKU HYORONSHA
2-10, KANDA TSUKASA-CHO, CHIYODA-KU, TOKYO 101
JAPANESE, WITH ENGLISH SUMMARY

GANKA 3143
(OPHTHALMOLOGY)
KANEHARA SHUPPAN K.K.
31-14, 2-CHOME, YUSHIMA BUNKYO-KU, TOKYO 113-91
JAPANESE

GANKA RINSHO IHO 3144
(JAPANESE REVIEW OF CLINICAL OPHTHALMOLOGY)
GANKA RINSHO IHOKAI
12-27, 3-CHOME, KUGAYAMA, SUGINAMI-KU, TOKYO 168
JAPANESE

GEKA 3145
(SURGERY)
NANKODO
42-6, 3-CHOME, HONGO, BUNKYO-KU, TOKYO 113
JAPANESE

GEKA CHIRYO 3146
(SURGICAL THERAPY)
NAGAI SHOTEN
3-111, KAMI FUKUSHIMA KITA, FUKUSHIMA-KU, OSAKA-SHI 553
JAPANESE
INDEXED: CHE

GEKA SHINRYO 3147
(SURGICAL DIAGNOSIS AND TREATMENT)
SHINDAN TO CHIRYOSHA
4-406, MARUNOUCHI BUILDING, CHIYODA-KU, TOKYO 100
JAPANESE
INDEXED: BIO, CHE

GEKKAN GEKA SHOREI 3148
(JOURNAL OF SURGICAL CASE REPORT)
JAPANESE

HOKKAIDO GEKA ZASSHI 3149
(HOKKAIDO JOURNAL OF SURGERY)
HOKKAIDO GEKA GAKKAI (THE HOKKAIDO SURGICAL SOCIETY)
C/O HOKKAIDO DAIGAKU IGAKUBU, NISHI-7-CHOME, KITA-15-JO, KITA-KU, SAPPORO-SHI 060
JAPANESE, WITH ENGLISH SUMMARY

HOKKAIDO SEIKEI SAIGAI GEKA ZASSHI 3150
(HOKKAIDO JOURNAL OF ORTHOPEDIC AND TRAUMATIC SURGERY)
HOKKAIDO SEIKEI SAIGAI GEKA GAKKAI (HOKKAIDO ORTHOPEDIC AND TRAUMATIC SOCIETY)
C/O HOKKAIDO DAIGAKU IGAKUBU SEIKEI GEKA KYOSHITSU, NISHI-5-CHOME, KITA-14-JO, KITA-KU, SAPPORO-SHI 060
JAPANESE AND ENGLISH
INDEXED: EXC

HYOGO-KEN GEKA SEIKEI GEKA IKAI 3151
(HYOGO ASSOCIATION OF SURGEONS AND ORTHOPEDIC SURGEONS)
HYOGO-KEN GEKA SEIKEI GEKAIKAI (HYOGO ASSOCIATION OF SURGEONS AND ORTHOPEDIC SURGEONS)
1-34, 4-CHOME, KITANAGASA-DORI, IKUTA-KU, KOBE-SHI 650
JAPANESE

HYOGO-KEN ZENGEKA IKAI KAISHI 3152
(JOURNAL OF HYOGO ASSOCIATION OF SURGEONS)
HYOGO ASSOCIATION OF SURGEONS
JAPANESE

ISHOKU 3153
(TRANSPLANTATION)
NIHON ISHOKU GAKKAI (JAPAN TRANSPLANTATION MEDICAL ASSOCIATION)
C/O KYOTO FURITSU IKA DAIGAKU, HIROKOJI, KAWARACHO-DORI, KAMIGYO-KU, KYOTO-SHI 602
JAPANESE AND ENGLISH

JAPANESE JOURNAL OF OPHTHALMOLOGY 3154
TOKYO DAIGAKU IGAKUBU GANKAGAKU KYOSHITSU (DEPARTMENT OF OPHTHALMOLOGY, SCHOOL OF MEDICINE, UNIVERSITY OF TOKYO)
3-1, 7-CHOME, BUNKYO-KU, TOKYO 113
ENGLISH
INDEXED: CHE, EXC

JAPANESE JOURNAL OF SURGERY 3155
JAPANESE AND ENGLISH
INDEXED: EXC

JINKO ZOKI 3156
(ARTIFICIAL ORGANS)
NIHON ZINKO ZOKI GAKKAI (THE JAPANESE SOCIETY FOR ARTIFICIAL ORGANS)
C/O TOKYO DAIGAKU IGAKUBU DAI-2 GEKA KYOSHITSU, 3-1, 7-CHOME, HONGO, BUNKYO-KU, TOKYO 113
JAPANESE

JUTSUGO TAISHA KENKYUKAISHI 3157
(JOURNAL OF POST-OPERATION METABOLISM RESEARCH ASSOCIATION)
JUTSUGO TAISHA KENKYUKAI
C/O TODAI BUN'IN GEKA, 28-6, 3-CHOME, HYLGADAI, BUNKYO-KU, TOKYO 112
JAPANESE

KEISEI GEKA 3158
(JAPANESE JOURNAL OF PLASTIC AND RECONSTRUCTIVE SURGERY)
KOKUSEIDO SHUPPAN K.K. (KOKUSEIDO PUBLISHING COMPANY, LIMITED)
23-5, 3-CHOME, HONGO, BUNKYO-KU, TOKYO 113
JAPANESE, WITH ENGLISH SUMMARY
INDEXED: EXC

KOKURITSU TOKYO SHIRYOKU SHOGAI SENTA KENKYU HOKOKUSHO 3159
(RESEARCH REPORT OF THE NATIONAL TOKYO EYESIGHT HANDICAP CENTER)
NATIONAL TOKYO EYESIGHT HANDICAP CENTER
JAPANESE

KYOBU GEKA 3160
(JAPANESE JOURNAL OF THORACIC SURGERY)
NANKODO
42-6, 3-CHOME, HONGO, BUNKYO-KU, TOKYO 113
JAPANESE AND ENGLISH
INDEXED: EXC

NEUROLOGIA MEDICO-CHIRURGICA 3161
NIHON NOSHINKEI GEKA GAKKAI (JAPAN NEUROSURGICAL SOCIETY)
C/O TOKYO DAIGAKU IGAKUBU FUZOKU BYOIN NOSHINKEI GEKA, 3-1, 7-CHOME, HONGO, BUNKYO-KU, TOKYO 113
EUROPEAN
INDEXED: EXC

NIHON GANKA GAKKAI ZASSHI 3162
(ACTA SOCIETATIS OPHTHALMOLOGICAE JAPONICAE)
NIHON GANKA GAKKAI (JAPANESE OPHTALMOLOGICAL SOCIETY)
C/O NIHON ISHIKAIKAN, 2-5, KANDASURUGADAI, CHIYODA-KU, TOKYO 101
JAPANESE, WITH ENGLISH SUMMARY
INDEXED: CHE, EXC

NIHON GEKA GAKKAI ZASSHI 3163
(JOURNAL OF JAPAN SURGICAL SOCIETY)
NIHON GEKA GAKKAI (JAPAN SURGICAL SOCIETY)
2-5, KANDA SURUGADAI, CHIYODA-KU, TOKYO 101
JAPANESE AND ENGLISH
INDEXED: CHE, EXC

NIHON GEKA HOKAN 3164
(ARCHIV FUR JAPANISCHE CHIRURGI)
NIHON GEKA HOKAN HENSHUSHITSU (EDITORIAL BOARD ARCHIV FUR JAPANISCHE CHIRUGI)
C/O KYOTO DAIGAKU IGAKUBU GEKA SEIKEI GEKA KYOSHITSU, 54, SHOGOIN KAWARA-CHO, SAKYO-KU, KYOTO-SHI 606
JAPANESE AND ENGLISH

NIHON GEKAKEI GAKKAI RENGOKAISHI 3165
(RECORDS OF THE JAPAN SURGICAL SOCIETIES FEDERATION)
JAPAN SURGICAL SOCIETIES FEDERATION
JAPANESE

NIHON ISHOKU GAKKAI ZASSHI 3166
(JAPANESE JOURNAL OF TRANSPLANTATION)
NIHON ISHOKU GAKKAI
C/O KYOTO-FURITSU IKA DAIGAKU, HIROKOJI, KAWARA-CHO-DORI, KAMIGYO-KU, KYOTO-SHI 602
JAPANESE

NIHON JAKUSHI SHASHI KENKYUKAI ZASSHI 3167
(JOURNAL OF JAPANESE ASSOCIATION OF STRABISMUS AND AMBLYOPIA)
GANKA RINSHO IHOKAI (CLINICAL OPHTHAMOLOGY MEDICAL INFORMATION SOCIETY)
JAPANESE

NIHON JAKUSHI SHASHI KENKYUKAIHO 3168
(JOURNAL OF JAPAN ORTHOPTIC COUNCIL)
NIHON JAKUSHI SHASHI KENKYUKAI (JAPAN ORTHOPTIC COUNCIL)
C/O TEIKYO DAIGAKU IGAKUBU GANKA KYOSHITSU, 11-1, 2-CHOME, KAGA, ITABASHI-KU, TOKYO 173
JAPANESE

NIHON KONTAKUTO RENZU GAKKAI KAISHI 3169
(JOURNAL OF JAPAN CONTACT LENS SOCIETY)
NIHON GANKA KIYOKAI (FOLIA OPHTALMOLOGICA JAPONICA EDITORIAL ASSOCIATION)
C/O OSAKA DAIGAKU IGAKUBU FUZOKU BYOIN OSAKA AIBANKU, 3-CHOME, DOJIMA HAMA-DORI, FUKUSHIMA-KU, OSAKA-SHI 553
JAPANESE AND ENGLISH
INDEXED: CHE, EXC

NIHON KONTAKUTORENZU GAKKAI KAIHO 3170
(BULLETIN OF THE JAPAN CONTACT LENS ASSOCIATION)
JAPANESE

NIHON KYOBU GEKA GAKKAI ZASSHI 3171
(JOURNAL OF JAPANESE ASSOCIATION FOR THORACIC SURGERY)
NIHON KYOBU GEKA GAKKAI (JAPANESE ASSOCIATION FOR THORACIC SURGERY)
C/O NIHON ISHI KAIKAN, 2-5, SURUGADAI, CHIYODA-KU, TOKYO 101
JAPANESE AND ENGLISH
INDEXED: CHE, EXC

NIHON NO GANKA 3172
(OPHTHALMIC PRACTICE)
NIHON GANKAIKAI (JAPANESE ASSOCIATION OF OPHTHALMOLOGISTS)
C/O HIOKI BUILDING, 1, SUGA-CHO, SHINJUKU-KU, TOKYO 160
JAPANESE

NIHON RINSHO GEKA IGAKKAI ZASSHI 3173
(JOURNAL OF THE JAPANESE PRACTICAL SURGEON SOCIETY)
NIHON RINSHO GEKA IGAKKAI (JAPANESE PRACTICAL SURGEON SOCIETY)
C/O NIHON IKA DAIGAKU DAI-1-BYOIN DAI-2-GEKA KYOSHITSU, 5-5, 3-CHOME, IIDABASHI, CHIYODA-KU, TOKYO 102
JAPANESE AND ENGLISH

NIHON SHINO KUNRENSHI KYOKAI KAIHO 3174
(JOURNAL OF JAPANESE ASSOCIATION OF CERTIFIED ORTHOPTIST)
JAPANESE ASSOCIATION OF CERTIFIED ORTHOPTIST
JAPANESE

NIHON SHOKAKI GEKAGAKKAI ZASSHI 3175
(JAPANESE JOURNAL OF GASTROENTEROLOGICAL SURGERY)
NIHON SHOKAKI GEKA GAKKAI
C/O TOKYO JOSHI IDAI FUZOKU SHOKAKIBYO SENTA, 10, KAWADA-CHO, SHINJUKU-KU, TOKYO 162
JAPANESE
INDEXED: EXC

NIHON SHONI GEKA GAKKAI ZASSHI 3176
(JOURNAL OF THE JAPANESE SOCIETY OF PEDIATRIC SURGEONS)
NIHON SHONI GEKA GAKKAI (JAPANESE SOCIETY OF PEDIATRIC SURGEONS)
C/O JUNTENDO DAIGAKU IGAKUBU SHONI GEKA KYOSHITSU NO. 315, 1-1, 2-CHOME, HONGO, BUNKYO-KU, TOKYO 113
JAPANESE

NIPPON KOKU GEKA GAKKAI ZASSHI 3177
(JAPANESE JOURNAL OF ORAL SURGERY)
NIPPON KOKU GEKA GAKKAI (JAPANESE SOCIETY OF ORAL SURGEONS)
C/O TOKYO IKA DAIGAKU KOKU GEKAGAKU KYOSHITSU, 7-1, 6-CHOME, NISHISHINJUKU, SHINJUKU-KU, TOKYO 160
JAPANESE

NIPPON SEIKEI GEKA GAKKAI ZASSHI 3178
(JOURNAL OF THE JAPANESE ORTHOPAEDIC ASSOCIATION)
NIPPON SEIKEI GEKA GAKKAI (JAPANESE ORTHOPAEDIC ASSOCIATION)
C/O TOKYO DAIGAKU IGAKUBU SEIKEI GEKA KYOSHITSU, 3-1, 7-CHOME, HONGO, BUNKYO-KU, TOKYO 113
JAPANESE AND ENGLISH
INDEXED: CHE, EXC

RINSHO GANKA 3179
(JAPANESE JOURNAL OF CLINICAL OPHTHALMOLOGY)
IGAKU SHOIN
29-11, 5-CHOME, HONGO, BUNKYO-KU, TOKYO 113
JAPANESE, WITH ENGLISH SUMMARY
INDEXED: CHE, EXC

RINSHO GEKA 3180
(CLINICAL SURGERY)
IGAKU SHOIN
29-11, 5-CHOME, HONGO, BUNKYO-KU, TOKYO 113
JAPANESE
INDEXED: CHE

RINSHO SEIKEI GEKA 3181
(CLINICAL ORTHOPEDIC SURGERY)
IGAKU SHOIN
29-11, 5-CHOME, HONGO, BUNKYO-KU, TOKYO 113
JAPANESE

SAITAMA-KEN NOSHINKEI GEKA IKAI KAIHO 3182
(REPORT OF THE NEUROSURICAL ASSOCIATION OF SAITAMA PREFECTURE)
SAITAMA-KEN ISHIKAI (SAITAMA PREFECTURE MEDICAL ASSOCIATION)
JAPANESE

SEIKEI GEKA 3183
(ORTHOPEDIC SURGERY)
NANKODO
42-6, 3-CHOME, HONGO, BUNKYO-KU, TOKYO 113
JAPANESE AND ENGLISH

SEIKEI GEKA TO SAIGAI GEKA 3184
(ORTHOPEDICS AND TRAUMATOLOGY)
NISHI NIHON SEIKEI SAIGAI GEKA GAKKAI (WEST-JAPANESE SOCIETY OF ORTHOPEDICS AND TRAUMATOLOGY)
KYUSHU DAIGAKU IGAKUBU SEIKEI GEKA KYOSHITSU (DEPARTMENT OF ORTHOPEDICS SURGERY FACULTY)
1276, KATAKASU, HIGASHI-KU, FUKUOKA-SHI 812
JAPANESE AND ENGLISH

SHIKAKU SHOGAI KENKYU 3185
(VISION IMPAIRMENT RESEARCH)
NIHON RAITOHAUSU SHIKAKU KEKKAN KENKYU SHITSU (JAPAN LIGHTHOUSE VISION DEFECT RESEARCH ROOM)
JAPANESE

SHONI GEKA 3186
(JAPANESE JOURNAL OF PEDIATRIC SURGERY)
TOKYO IGAKUSHA (TOKYO MEDICAL PUBLISHERS)
JAPANESE

SHUJUTSU 3187
(OPERATION)
KANEHARA SHUPPAN K.K. (KANEHARA SHUPPAN COMPANY, LIMITED)
31-14, 2-CHOME, YUSHIMA, BUNKYO-KU, TOKYO 113-91
JAPANESE

TOHOKU SEIKEI SAIGAI GEKA KIYO 3188
(TOHOKU ARCHIVO POR ORTHOPEDIA KEJ AKCIDENTA HIRURGIO)
TOHOKU SEIKEI SAIGAI GEKA GAKKAI (TOHOKU SOCIETE DE ORTOPEDIA KAJ AKCIDENTA HIRURGIO)
TOHOKU DAIGAKU SEIKEI GEKA KYOSHITSU (DEPARTMENT OF ORTHOPEDIC SURGERY, TOHOKU UNIVERSITY)
1-1, SEIRYO-MACHI, SENDAI-SHI 980
JAPANESE

TORANSU PURANTO 3189
(TRANSPLANT)
JINZO ISHOKU FUKYU KAI
20-15, 2-CHOME, SHINBASHI, MINATO-KU, TOKYO 105
JAPANESE
INDEXED: CHE

Gynecology, Obstetrics

ACTA OBSTETRICA ET GYNAECOLOGICA JAPONICA 3190
NIHON SANKA FUJINKA GAKKAI (JAPAN SOCIETY OF OBSTETRICS AND GYNECOLOGY)
29-8, 1-CHOME, SHINJUKU, SHINJU-KU, TOKYO 160
ENGLISH
INDEXED: EXC

BUNBEN TO MASUI 3191
(JOURNAL OF OBSTETRICS AND ANESTHESIA)
MUTSU BUNBEN KENKYUKAI (INSTITUTE OF OBSTETRICAL ANESTHESIA) AND KOKUSEIDO SHUPPAN K.K.
23-5, 3-CHOME, HONGO, BUNKYO-KU, TOKYO 113
JAPANESE
INDEXED: EXC

JOSANPU 3192
(MIDWIVES)
NIHON JOSANPUKAI SHUPPAN-BU (JAPAN MIDWIVES ASSOCIATION)
8-21, 1-CHOME, FUJIMI, CHIYODA-KU, TOKYO 102
JAPANESE

JOSANPU ZASSHI 3193
(JAPANESE JOURNAL FOR MIDWIFE)
IGAKU SHOIN
29-11, 5-CHOME, HONGO, BUNKYO-KU, TOKYO 113
JAPANESE

NAGANO-KEN SANKA FUJINKAIKAIHO 3194
(BULLETIN OF GYNECOLOGICAL SOCIETY OF NAGANO)
NAGANO-KEN SANKA FUJINKAIKAI (GYNECOLOGICAL SOCIETY OF NAGANO)
C/O SHINSHU DAIGAKU IGAKUBU SANFUJINKA KYOSHITSU, 3-1, ASAHI, MATSUMOTO-SHI 390
JAPANESE

NICHIBO IHO 3195
(REPORT OF THE JAPAN MEDICAL ASSOCIATION FOR THE PROTECTION OF PREGNANT MOTHERS)
NIHON BOSEI HOGO IKYOKAI (JAPAN MEDICAL ASSOCIATION FOR THE PROTECTION OF PREGNANT MOTHERS)
JAPANESE

NIHON FUNIN GAKKAI ZASSHI 3196
(JAPANESE JOURNAL OF FERTILITY AND STERILITY)
NIHON FUNIN GAKKAI (JAPANESE SOCIETY OF FERTILITY AND STERILITY)
5-22, 7-CHOME, OMORINISHI, OTA-KU, TOKYO 143
JAPANESE AND ENGLISH
INDEXED: CHE, EXC

NIHON SANKA FUJINKA GAKKAI CHUGOKU SHIKOKU GODO CHIHO BUKAI ZASSHI 3197
(CHUGOKU AND SHIKOKU DISTRICTS JOURNAL OF THE JAPAN SOCIETY OF OBSTETRICS AND GYNECOLOGY)
NIHON SANKA FUJINKA GAKKAI CHUGOKU SHIKOKU CHIHO BUKAI (CHUGOKU SHIKOKU DISTRICTS BRANCH OF JAPAN SOCIETY OF OBSTETRICS AND GYNECOLOGY)
C/O OKAYAMA DAIGAKU IGAKUBU SANFUJINKA KYOSHITSU, 5-1, 2-CHOME, SHIKATA-MACHI, OKAYAMA-SHI 700
JAPANESE

NIHON SANKA FUJINKA GAKKAI KANTO RENGOKAI CHIHO BUKAI KAIHO 3198
(KANTO DISTRICTS JOURNAL OF THE JAPAN SOCIETY OF OBSTETRICS AND GYNECOLOGY)
NIHON SANKA FUJINKA GAKKAI KANTO RENGOKAI CHIHOBUKAI
29-8, 1-CHOME, SHINJUKU, SHINJUKU-KU, TOKYO 160
JAPANESE

NIHON SANKA FUJINKA GAKKAI KUMAMOTO CHIHO BUKAI ZASSHI 3199
(KUMAMOTO DISTRICTS JOURNAL OF THE JAPAN SOCIETY OF OBSTETRICS AND GYNECOLOGY)
NIHON SANKA FUJINKA GAKKAI KUMAMOTO CHIHO BUKAI (KUMAMOTO DISTRICTS BRANCH OF JAPAN SOCIETY OF OBSTETRICS AND GYNECOLOGY)
C/O KUMAMOTO DAIGAKU IGAKUBU SANFUJINKAGAKU KYOSHITSU, 1-1, 1-CHOME, HONJO KUMAMOTO-SHI, 860
JAPANESE

NIHON SANKA FUJINKA GAKKAI NIIGATA CHIHO BUKAI KAISHI 3200
(NIIGATA DISTRICTS JOURNAL OF THE JAPAN SOCIETY OF OBSTETRICS AND GYNECOLOGY)
NIPPON SANFUJINKA GAKKAI NIIGATA CHIHO BUKAI NIPPON BOSEI HOGOIKYOKAI NIIGATA SHIBU
1-754, ASAHIMACHI-DORI, NIIGATA-SHI 951
JAPANESE

NIHON SANKA FUJINKA GAKKAI SOKAI GAKUJUTSU KOENSHU 3201
(PROCEEDINGS OF THE SYMPOSIUM OF THE JAPANESE OBSTETRICAL AND GYNECOLOGICAL SOCIETY)
NIHON SANKA FUJINKA GAKKAI (THE JAPANESE OBSTETRICAL AND GYNECOLOGICAL SOCIETY)
78, HANAZONO-CHO, SHINJUKU-KU, TOKYO 160
JAPANESE AND ENGLISH

NIHON SANKA FUJINKA GAKKAI TOKYO CHIHO BUKAI KAIHO 3202
(TOKYO DISTRICTS JOURNAL OF THE JAPAN SOCIETY OF OBSTETRICS AND GYNECOLOGY)
NIHON SANKA FUJINKA GAKKAI TOKYO CHIHO BUKAI (THE TOKYO DISTRICTS, JAPAN SOCIETY OF OBSTETRICS AND GYNECOLOGY)
C/O NIHON KOSHU EISEI KYOKAI BUILDING, 29-8, 1-CHOME SHINJUKU, SHINJUKU-KU, TOKYO 160
JAPANESE

NIHON SANKA FUJINKA GAKKAI ZASSHI 3203
(JOURNAL OF THE JAPAN SOCIETY OF OBSTETRICS AND GYNECOLOGY, JAPANESE EDITION)
NIHON SANKA FUJINKA GAKKAI (JAPAN SOCIETY OF OBSTETRICS AND GYNECOLOGY)
29-8, 1-CHOME, SHINJUKU, SHINJUKU-KU, TOKYO 160
JAPANESE
INDEXED: CHE, JICST

RINSHO FUJINKA SANKA 3204
(CLINICAL GYNECOLOGY AND OBSTETRICS)
IGAKU SHOIN
29-11, 5-CHOME, HONGO, BUNKYO-KU, TOKYO 113
JAPANESE

SANFUJINKA CHIRYO 3205
(OBSTETRICAL AND GYNECOLOGICAL THERAPY)
NAGAI SHOTEN
3-111, KAMI-FUKUSHIMA-KITA, FUKUSHIMA-KU, OSAKA-SHI 553
JAPANESE
INDEXED: CHE

SANFUJINKA NO JISSAI 3206
(JAPANESE JOURNAL OF PRACTICAL GYNECOLOGY AND OBSTETRICS)
KANEHARA SHUPPAN K.K.
31-14, 2-CHOME, YUSHIMA, BUNKYO-KU, TOKYO 113-91
JAPANESE

SANFUJINKA NO SEKAI 3207
(WORLD OF OBSTETRICS AND GYNECOLOGY)
IGAKU NO SEKAISHA
5-1, 2-CHOME, AKASAKA MINATO-KU, TOKYO 107
JAPANESE

SANFUJINKA NO SHINPO 3208
(ADVANCES IN OBSTETRICS AND GYNECOLOGY)
KINKI SANKA FUJINKA GAKKAI (OBSTETRICAL AND GYNECOLOGICAL SOCIETY OF KINKI DISTRICT)
C/O KYOTO-FU ISHI KAIKAN, KYOTO SANFUJINKAIKAI, GOZENDORI MATSUBARA SAGARU, NAKAGYO-KU, KYOTO-SHI 604
JAPANESE
INDEXED: CHE

SANKA TO FUJINKA 3209
(OBSTETRICS AND GYNECOLOGY)
SHINDAN TO CHIRYOSHA
C/O MARU BUILDING, MARUNOUCHI, CHIYODA-KU, TOKYO 100
JAPANESE

SHUSANKI IGAKU 3210
(PERINATAL MEDICINE)
TOKYO IGAKUSHA
25-12, 3-CHOME, HONGO, BUNKYO-KU, TOKYO 113
JAPANESE

AGRICULTURAL SCIENCES

Agricultural Sciences in General

ABSTRACT OF STATISTICS ON AGRICULTURE, FORESTRY AND FISHERIES JAPAN 3211
NORIN-SHO NORIN KEIZAI-KYOKU TOKEI JOHO-BU (STATISTICS AND INFORMATION DIVISION, AGRICULTURE AND FORESTRY ECONOMIC BUREAU, MINISTRY OF AGRICULTURE AND FORESTRY)
C/O OTORI BUILDING, 11-4, 2-CHOME, MEGURO, MEGURO-KU, TOKYO 153
ENGLISH

AFF NORIN-SHO KOHO 3212
NORIN-SHO (MINISTRY OF AGRICULTURE AND FORESTRY) AND NORIN KOSAIKAI
2-1, 1-CHOME, KASUMIGASEKI, CHIYODA-KU, TOKYO 100
JAPANESE
INDEXED: BIB

AGRICULTURAL AND BIOLOGICAL CHEMISTRY 3213
NIHON NOGEIKAGAKKAI (AGRICULTURAL CHEMICAL SOCIETY OF JAPAN)
4-16, 2-CHOME, YAYOI, BUNKYO-KU, TOKYO 113
ENGLISH
INDEXED: BIB, BIO, CHE, EXC, SCI, JICST

AICHI NORIN SUISAN TOKEI NENPO 3214
(STATISTICAL YEARBOOK OF AGRICULTURE, FORESTRY AND FISHERY, AICHI)
NORIN-SHO TOKAI NOSEI-KYOKU TOKEI JOHO-BU (STATISTICS AND INFORMATION DIVISION, TOKAI AGRICULTURAL ADMINISTRATION BUREAU, MINISTRY OF AGRICULTURE AND FORESTRY)
2-2, 1-CHOME, SANNOMARU, NAKA-KU, NAGOYA-SHI 460
JAPANESE

AICHI-KEN NOGYO SHIKENJO NENPO 3215
(ANNUAL REPORT OF AICHI AGRICULTURAL EXPERIMENT STATION)
AICHI-KEN NOGYO SHIKENJO AND AICHI-KEN NOGYO SHIKENJO NEMPO
IKEURA-CHO ANJO
JAPANESE
INDEXED: CHE

AICHI-KEN NOGYO SOGO SHIKENJO KENKYU HOKOKU, A, SAKUMOTSU 3216
(RESEARCH BULLETIN OF THE AICHI-KEN AGRICULTURAL RESEARCH CENTER, SERIES A)
AICHI-KEN NOGYO SOGO SHIKENJO (AICHI-KEN AGRICULTURAL RESEARCH CENTER)
NAGAKUDE-CHO, AICHI-GUN, AICHI 480-11
JAPANESE, WITH ENGLISH SUMMARY
INDEXED: CHE

AICHI-KEN NOGYO SOGO SHIKENJO NENPO 3217
(ANNUAL REPORT OF THE AICHI-KEN AGRICULTURAL RESEARCH CENTER)
AICHI-KEN NOGYO SOGO SHIKENJO (AICHI-KEN AGRICULTURAL RESEARCH CENTER)
NAGAKUDE-CHO, AICHI-GUN, AICHI 480-11
JAPANESE
INDEXED: BIB

AINO 3218
(AGRICULTURE (LOVE OF AGRICULTURE))
ZENKOKU AINOKAI HONBU (ALL JAPAN AGRICULTURAL SOCIETY, MAIN OFFICE)
JAPANESE

AJIA NOGYO 3219
(AGRICULTURE, ASIA)
AJIA NOGYO KORYU KONWAKAI (ASSOCIATION OF AGRICULTURAL RELATIONS IN ASIA)
11, ICHIGAYA FUNAKAWARA-CHO, SHINJUKU-KU, TOKYO 162
JAPANESE AND EUROPEAN
INDEXED: CHE

AKITA NORIN SUISAN TOKEI NENPO 3220
(STATISTICAL YEARBOOK OF AGRICULTURE, FORESTRY AND FISHERY, AKITA)
NORIN-SHO TOHOKU NOSEI-KYOKU AKITA TOKEI JOHO JIMU-SHO (AKITA STATISTICS AND INFORMATION OFFICE, TOHOKU AGRICULTURAL ADMINISTRATION BUREAU, MINISTRY OF AGRICULTURE AND FORESTRY)
1-3, 7-CHOME, SANNO, AKITA-SHI 010
JAPANESE

AKITA-KEN NOGYO KIHON CHOSA KEKKA HOKOKUSHO 3221
(FUNDAMENTAL RESEARCH REPORT OF AGRICULTURE, AKITA PREFECTURE)
AKITA-KEN SOMU-BU
1-1, 4-CHOME, SANNO, AKITA-SHI 010
JAPANESE

AKITA-KEN NOGYO NENJI TOKEISHO 3222
(ANNUAL STATISTICS OF AGRICULTURE, AKITA PREFECTURE)
NORIN-SHO TOHOKU NOSEI-KYOKU (TOHOKU AGRICULTURAL ADMINISTRATION BUREAU, MINISTRY OF AGRICULTURE AND FORESTRY)
3-1, 3-CHOME, HON-MACHI, SENDAI-SHI 980
JAPANESE

AKITA-KEN NOGYO SHIKENJO KENKYU HOKOKU 3223
(BULLETIN OF THE AKITA AGRICULTURAL EXPERIMENT STATION)
AKITA-KEN NOGYO SHIKENJO (AKITA AGRICULTURAL EXPERIMENT STATION)
NIIDA, AKITA-SHI 010-14
JAPANESE
INDEXED: CHE

AKITA-KEN NOGYO SHIKENJO KENKYU JIHO 3224
(REPORT OF AKITA AGRICULTURAL EXPERIMENT STATION)
AKITA-KEN NOGYO SHIKENJO (AKITA AGRICULTURAL EXPERIMENT STATION)
NIIDA, AKITA-SHI 010-14
JAPANESE

AKITA-KEN NOGYO SHIKENJO KENKYU JOHO 3225
(RESEARCH INFORMATION OF AKITA AGRICULTURAL EXPERIMENT STATION)
AKITA-KEN NOGYO SHIKENJO
NIIDA, AKITA-SHI 010-14
JAPANESE

AKITA-KEN NOGYO SHIKENJO NENPO 3226
(ANNUAL REPORT OF AKITA AGRICULTURAL EXPERIMENT STATION)
AKITA-KEN NOGYO SHIKENJO
NIIDA, AKITA-SHI 010-14
JAPANESE

AKITA-KENRITSU NOGYO TANKI DAIGAKU KENKYU HOKOKU 3227
(RESEARCH REPORT OF THE AKITA PREFECTURAL COLLEGE OF AGRICULTURE)
AKITA PREFECTURAL COLLEGE OF AGRICULTURE
OHGATA-MURA, AKITA-KEN 010-04 AKITA
JAPANESE
INDEXED: CHE

AMA; AGRICULTURAL MECHANIZATION IN ASIA 3228
NOKISANGYO CHOSA KENKYUSHO (INSTITUTE OF AGRICULTURAL MACHINE INDUSTRY)
C/O SHIN NORIN BUILDING, 2-7, KANDA NISHIKI-CHO, CHIYODA-KU, TOKYO 101
ENGLISH
INDEXED: BIB, JICST

ANNUAL REPORT OF TOKYO UNIVERSITY OF AGRICULTURE AND TECHNOLOGY 3229
TOKYO NOKO DAIGAKU FUZOKU TOSHOKAN (LIBRARY, TOKYO UNIVERSITY OF AGRICULTURE AND TECHNOLOGY)
5-8, 3-CHOME, SAIWAI-CHO, FUCHU-SHI, TOKYO 183
JAPANESE AND EUROPEAN
INDEXED: BIB, JICST

AOMORI NOGYO 3230
(AGRICULTURE IN AOMORI)
AOMORI-KEN NOGYO KAIRYO FUKYUKAI (AOMORI PREFECTURE AGRICULTURAL IMPROVEMENT ASSOCIATION)
C/O AOMORI KENCHO NORIN-BU NOMU-KA, 1, NAGASHIMA, AOMORI-SHI 030
JAPANESE

AOMORI NORIN SUISAN TOKEI NENPO 3231
(STATISTICAL YEARBOOK OF AGRICULTURE, FORESTRY AND FISHERY, AOMORI)
NORIN-SHO TOHOKU NOSEI-KYOKU AOMORI TOKEI JIMUSHO (AOMORI STATISTICS AND INFORMATION OFFICE, TOHOKU AGRICULTURAL ADMINISTRATION BUREAU, MINISTRY OF AGRICULTURE AND FORESTRY)
20-35, 3-CHOME, CHUO, AOMORI-SHI 030
JAPANESE

AOMORI NOSHI GEPPO 3232
(MONTHLY REPORT OF AOMORI PREFECTURE AGRICULTURAL EXPERIMENT STATION)
AOMORI-KEN NOGYO SHIKENJO (AOMORI PREFECTURE AGRICULTURAL EXPERIMENT STATION)
1, SUNAMORI, KUROISHI, KUROISHI-SHI 036-03
JAPANESE

AOMORI-KEN NOGYO SHIKENJO JIGYO GAIYO 3233
(OUTLINE OF THE WORKS OF AOMORI PREFECTURE AGRICULTURAL EXPERIMENT STATION)
AOMORI-KEN NOGYO SHIKENJO
1, SUNAMORI, KUROISHI, KUROISHI-SHI 036-03
JAPANESE

AOMORI-KEN NOGYO SHIKENJO JIGYO SEISEKISHO 3234
(REPORT OF AOMORI PREFECTURE AGRICULTURAL EXPERIMENT STATION)
AOMORI-KEN NOGYO SHIKENJO
1, SUNAMORI, KUROISHI, KUROISHI-SHI 036-03
JAPANESE

AOMORI-KEN NOGYO SHIKENJO KENKYU HOKOKU 3235
(BULLETIN OF THE AOMORI AGRICULTURAL EXPERIMENT STATION)
AOMORI-KEN NOGYO SHIKENJO (AOMORI PREFECTURE AGRICULTURAL EXPERIMENT STATION)
1, SUNAMORI, KUROISHI, KUROISHI-SHI 036-03
JAPANESE, WITH ENGLISH SUMMARY
INDEXED: BIO

AOMORI-KEN NOGYO SHIKENJO NENPO 3236
(ANNUAL BULLETIN OF THE AOMORI PREFECTURE AGRICULTURAL EXPERIMENT STATION)
AOMORI-KEN NOGYO SHIKENJO (AOMORI PREFECTURE AGRICULTURAL EXPERIMENT STATION)
1, SUNAMORI, KUROISHI, KUROISHI-SHI 036-03
JAPANESE

AOMORI-KEN NOGYO SHIKENJO SHIKEN SEISEKI GAIYOSHU 3237
(EXPERIMENTAL RESULT OF AOMORI PREFECTURE AGRICULTURAL EXPERIMENT STATION)
AOMORI-KEN NOGYO SHIKENJO
1, SUNAMORI, KUROISHI, KUROISHI-SHI 036-03
JAPANESE

AOMORI-KEN NORIN DOBOKU GIJUTSU SHIKENSHITSU HOKOKU 3238
(REPORT OF AGRICULTURAL ENGINEERING LABORATORY, AOMORI PREFECTURE)
AOMORI-KEN NORIN DOBOKU GIJUTSU SHIKENSHITSU
20-41, 3-CHOME, CHUO, AOMORI-SHI 030
JAPANESE

ATARASHII GIJUTSU 3239
(NEW AGRICULTURAL TECHNOLOGY)
NORIN-SHO NORIN SUISAN GIJUTSU KAIGI
2-1, 1-CHOME, KASUMIGASEKI, CHIYODA-KU, TOKYO 100
JAPANESE

AURORA 3240
SHINNOKAI
HON-CHO, ONO-CHO, KAMEDA-GUN, HOKKAIDO 041-12
JAPANESE

BERICHTE DES OHARA INSTITUTE FUR LANDWIRTSCHAFTLICHE BIOLOGIE, OKAYAMA UNIVERSITY 3241
OKAYAMA DAIGAKU NOGYO SEIBUTSU KENKYUSHO (DAS OHARA INSTITUTE FUR LANDWIRTSCHAFTLICHE BIOLOGIE, OKAYAMA UNIVERSITY)
20-1, 2-CHOME, CHUO, KURASHIKI-SHI 710
ENGLISH
INDEXED: BIB, BIO, CHE, JICST

BIBLIOGRAPHY OF AGRICULTURAL SCIENCES IN JAPAN 3242
NIHON DOKUMENTESHON KYOKAI (JAPAN DOCUMENTATION SOCIETY)
5-7, 2-CHOME, KOISHIKAWA, BUNKYO-KU, TOKYO 112
ENGLISH
INDEXED: JICST

BULLETIN OF THE UNIVERSITY OF OSAKA PREFECTURE, SECTION B, AGRICULTURE AND BIOLOGY 3243
OSAKA-FURITSU DAIGAKU (UNIVERSITY OF OSAKA PREFECTURE)
4-804, MOZUUME-MACHI, SAKAI-SHI 591
JAPANESE AND EUROPEAN
INDEXED: BIB, CHE, JICST

CHIBA NO RINGYO 3244
(FOREST INDUSTRY IN CHIBA)
CHIBA-KEN NORIN-BU RINMU-KA (CHIBA PREFECTURE AGRICULTURE AND FORESTRY DEPARTMENT, FOREST ADMINISTRATION SECTION)
JAPANESE

CHIBA NORIN SUISAN TOKEI NENPO 3245
(STATISTICAL YEARBOOK OF AGRICULTURE, FORESTRY AND FISHERY, CHIBA)
NORIN-SHO KANTO NOSEI-KYOKU CHIBA TOKEI JOHO JIMUSHO (CHIBA STATISTICS AND INFORMATION OFFICE, KANTO AGRICULTURAL ADMINISTRATION BUREAU, MINISTRY OF AGRICULTURE AND FORESTRY)
1-4, 5-CHOME, TODOROKI-CHO, CHIBA-SHI 280
JAPANESE

CHIBA-KEN NOGYO SHIKENJO GYOMU NENPO 3246
(ANNUAL REPORT OF THE CHIBA PREFECTURE AGRICULTURAL EXPERIMENT STATION)
CHIBA-KEN NOGYO SHIKENJO (CHIBA PREFECTURE AGRICULTURAL EXPERIMENT STATION)
808, DAIZENNO-CHO, CHIBA-SHI 280-02
JAPANESE

CHIBA-KEN NOGYO SHIKENJO KENKYU HOKOKU 3247
(BULLETIN OF THE CHIBA PREFECTURE AGRICULTURAL EXPERIMENT STATION)
CHIBA-KEN NOGYO SHIKENJO (CHIBA PREFECTURE AGRICULTURAL EXPERIMENT STATION)
808, DAIZENNO-CHO, CHIBA-SHI 280-02
JAPANESE
INDEXED: CHE, JICST

CHIBA-KEN NOGYO SHIKENJO TOKUBETSU HOKOKU 3248
(SPECIAL BULLETIN OF THE CHIBA-KEN AGRICULTURAL EXPERIMENT STATION)
CHIBA-KEN NOGYO SHIKENJO
808, DAIZENNO-CHO, CHIBA-SHI 280-02
JAPANESE, WITH ENGLISH SUMMARY

CHIBA-KEN NORINGYO NO DOKO 3249
(ANNUAL REPORT OF AGRICULTURE AND FORESTRY IN CHIBA PREFECTURE)
CHIBA-KEN NORIN-BU
1-1, ICHIBA-MACHI, CHIBA-SHI 280-91
JAPANESE

CHOSA SHIRYO 3250
(RESEARCH DATA)
JAPANESE
INDEXED: BIB

CHUGOKU CHIKI KYODO KENKYU SEIKA SHUROKU 3251
(REPORT OF THE COOPERATIVE RESEARCH IN CHUGOKU REGION)
CHUGOKU NOGYO SHIKENJO (CHUGOKU NATIONAL AGRICULTURAL EXPERIMENT STATION)
HIGASHI FUKATSU-MACHI, FUKUYAMA-SHI 720
JAPANESE, WITH ENGLISH SUMMARY
INDEXED: CHE, JICST

CHUGOKU NOGYO KENKYU 3252
(CHUGOKU AGRICULTURAL RESEARCH)
CHUGOKU NOGYO KENKYU KYOGIKAI (CHUGOKU AGRICULTURAL RESEARCH ASSOCIATION)
C/O NORIN-SHO CHUGOKU NOGYO SHIKENJO, HIGASHI FUKATSU-MACHI, FUKUYAMA-SHI 720
JAPANESE
INDEXED: CHE, JICST

CHUGOKU NOGYO SHIKENJO HOKOKU, A, SAKUMOTSU-BU 3253
(BULLETIN OF THE CHUGOKU NATIONAL AGRICULTURAL EXPERIMENT STATION, SERIES A)
CHUGOKU NOGYO SHIKENJO (CHUGOKU NATIONAL AGRICULTURAL EXPERIMENT STATION)
HIGASHI FUKATSU-MACHI, FUKUYAMA-SHI 720
JAPANESE, WITH ENGLISH SUMMARY
INDEXED: BIB, CHE, JICST

CHUGOKU NOGYO SHIKENJO HOKOKU, C, NOGYO KEIEI-BU 3254
(BULLETIN OF THE CHUGOKU NATIONAL AGRICULTURAL EXPERIMENT STATION, SERIES C)
CHUGOKU NOGYO SHIKENJO (CHUGOKU NATIONAL AGRICULTURAL EXPERIMENT STATION)
HIGASHI FUKATSU-MACHI, FUKUYAMA-SHI 720
JAPANESE AND ENGLISH
INDEXED: BIB, JICST

CHUGOKU NOGYO SHIKENJO HOKOKU, D, GIJUTSU RENRAKUSHITSU, KAKUBU KYOTSU 3255
(BULLETIN OF THE CHUGOKU NATIONAL AGRICULTURAL EXPERIMENT STATION, SERIES D)
CHUGOKU NOGYO SHIKENJO (CHUGOKU NATIONAL AGRICULTURAL EXPERIMENT STATION)
HIGASHI FUKATSU-MACHI, FUKUYAMA-SHI 720
JAPANESE AND ENGLISH
INDEXED: JICST

CHUGOKU NOGYO SHIKENJO HOKOKU, E, KANKYO-BU 3256
(BULLETIN OF THE CHUGOKU NATIONAL AGRICULTURAL EXPERIMENT STATION, SERIES E, ENVIRONMENT STATION)
CHUGOKU NOGYO SHIKENJO (CHUGOKU NATIONAL AGRICULTURAL EXPERIMENT STATION)
HIGASHI FUKATSU-MACHI, FUKUYAMA-SHI 720
JAPANESE AND ENGLISH
INDEXED: BIB, CHE, EXC, JICST

CHUGOKU SHIKOKU NO NOSEI 3257
(AGRICULTURAL ADMINISTRATION, CHUGOKU SHIKOKU DISTRICT)
NORIN-SHO CHUGOKU SHIKOKU NOSEI-KYOKU (CHUGOKU SHIKOKU AGRICULTURAL ADMINISTRATION BUREAU, MINISTRY OF AGRICULTURE AND FORESTRY)
9-24, TENJIN-CHO, OKAYAMA-SHI 700
JAPANESE

CHUGOKU SHIKOKU NOGYO JOSEI HOKOKU 3258
(ANNUAL REVIEW OF AGRICULTURE, CHUGOKU SHIKOKU DISTRICT)
NORIN-SHO CHUGOKU SHIKOKU NOSEI-KYOKU (CHUGOKU SHIKOKU AGRICULTURAL ADMINISTRATION BUREAU, MINISTRY OF AGRICULTURE AND FORESTRY)
9-24, TENJIN-CHO, OKAYAMA-SHI 700
JAPANESE

DOJO JOKA SHISUTEMU 3259
(SOIL PURIFICATION SYSTEMS)
MOKAN JOKA KENKYUKAI NIHON DOJO JOKA JIGYO KYOKAI (CAPILLARY PURIFICATION RESEARCH SOCIETY, SOIL PURIFICATION ENTERPRISES ASSOCIATION)
JAPANESE

DOJO NO BUTSURISEI 3260
(SOIL PHYSICAL CONDITIONS AND PLANT GROWTH, JAPAN)
DOJO BUTSURI KENKYUKAI
C/O HOKKAIDO NOGYO SHIKENJO, 1, HITSUJIGAOKA, TOYOHIRA-KU, SAPPORO-SHI 061-01
JAPANESE
INDEXED: JICST

EHIME DAIGAKU NOGAKUBU KIYO 3261
(MEMOIRS OF THE COLLEGE OF AGRICULTURE, EHIME UNIVERSITY)
EHIME DAIGAKU NOGAKUBU (COLLEGE OF AGRICULTURE, EHIME UNIVERSITY)
5-7, 3-CHOME, TARUMI-CHO, MATSUYAMA-SHI 790
JAPANESE AND ENGLISH
INDEXED: BIB, BIO, CHE, JICST

EHIME DAIGAKU SOGO NOGAKU KENKYU IHO 3262
(RESEARCH BULLETIN FOR AGRICULTURE OF THE EHIME UNIVERSITY)
EHIME DAIGAKU NOGAKUBU KEIEI NOGAKKA (INSTITUTE OF FARM BUSINESS OF EHIME UNIVERSITY)
5-7, 3-CHOME, TARUMI-CHO, MATSUYAMA-SHI 790
JAPANESE
INDEXED: JICST

EHIME NORIN SUISAN TOKEI NENPO 3263
(STATISTICAL YEARBOOK OF AGRICULTURE, FORESTRY AND FISHERY, EHIME)
NORIN-SHO CHUGOKU SHIKOKU NOSEI-KYOKU EHIME TOKEI JOHO JIMUSHO (EHIME STATISTICS AND INFORMATION OFFICE, CHUGOKU SHIKOKU AGRICULTURAL ADMINISTRATION BUREAU, MINISTRY OF AGRICULTURE AND FORESTRY)
1-5, DOGO ICHIMAN, MATSUYAMA-SHI 790
JAPANESE

EHIME-KEN NOGYO SHIKENJO KENKYU HOKOKU 3264
(BULLETIN OF THE EHIME AGRICULTURAL EXPERIMENT STATION)
EHIME-KEN NOGYO SHIKENJO (EHIME AGRICULTURAL EXPERIMENT STATION)
1-2, DOGO ICHIMAN, MATSUYAMA-SHI 790
JAPANESE

EHIME-KENRITSU SAIJO NOGYO KOTOGAKKO KENKYU KIYO 3265
(RESEARCH BULLETIN OF EHIME PREFECTURAL SAIJO AGRICULTURAL HIGH SCHOOL)
EHIME PREFECTURAL SAIJO AGRICULTURAL HIGH SCHOOL
JAPANESE

EHIME-KENRITSU SAIJO NOGYO KOTOGAKKO SEITOKAI KENKYU HOKOKU 3266
(RESEARCH REPORT OF THE EHIME PREFECTURAL SAIJO AGRICULTURAL HIGH SCHOOL STUDENTS ASSOCIATION)
SAIJO NOGYO KOTOGAKKO (SAIJO AGRICULTURAL HIGH SCHOOL)
JAPANESE

ENGEI SHIN CHISHIKI, YASAI HEN 3267
(NEW KNOWLEDGE ABOUT GARDENING, VEGETABLES)
JAPANESE

ENSHURIN (TOKYO DAIGAKU NOGAKUBU) 3268
(EXPERIMENTAL FORESTS)
JAPANESE
INDEXED: JICST

EXPERIMENTAL REPORTS OF EQUINE HEALTH LABORATORY 3269
JAPANESE AND ENGLISH
INDEXED: BIB

FARMING JAPAN 3270
NORIN SHIJO KENKYUSHO (AGRICULTURE AND FORESTRY MARKET RESEARCH INSTITUTE)
6-6, 3-CHOME, IIDABASHI, CHIYODA-KU, TOKYO 102
ENGLISH
INDEXED: BIB

FUKUGO HIRYO 3271
(COMPLEX FERTILIZER)
HIRYO KYOKAI SHINBUN-BU (NEWSPAPER DIVISION, JAPAN FERTILIZER ASSOCIATION)
2-3, KYOBASHI, CHUO-KU, TOKYO 104
JAPANESE

FUKUI NORIN SUISAN TOKEI NENPO 3272
(STATISTICAL YEARBOOK OF AGRICULTURE, FORESTRY AND FISHERY, FUKUI)
NORIN-SHO HOKURIKU NOSEI-KYOKU FUKUI TOKEI JOHO JIMUSHO (FUKUI STATISTICS AND INFORMATION OFFICE, HOKURIKU AGRICULTURAL ADMINISTRATION BUREAU, MINISTRY OF AGRICULTURE AND FORESTRY)
14-15, 3-CHOME, HINODE, FUKUI-SHI 910
JAPANESE

FUKUI-KEN NO NOGYO 3273
(CENSUS OF AGRICULTURE, FUKUI PREFECTURE)
FUKUI-KEN (FUKUI PREFECTURAL GOVERNMENT)
17-1, 3-CHOME, OTE, FUKUI-SHI 910
JAPANESE

FUKUI-KEN NOGYO SHIKENJO CHIKUSAN SHIKENJO ENGEI SENTA KENKYU SOKUHO 3274
(CURRENT REPORT ON RESEARCH OF THE FUKUI PREFECTURE AGRICULTURAL EXPERIMENT STATION, ANIMAL HUSBANDRY EXPERIMENT STATION AND HORTICULTURAL CENTER)
FUKUI-KEN NOGYO SHIKENJO
(FUKUI PREFECTURE AGRICULTURAL EXPERIMENT STATION)
JAPANESE

FUKUI-KEN NOGYO SHIKENJO GYOMU NENPO 3275
(OPERATION YEARBOOK OF FUKUI AGRICULTURAL EXPERIMENT STATION)
FUKUI-KEN NOGYO SHIKENJO
RYO-MACHI, FUKUI-SHI 910
JAPANESE

FUKUI-KEN NOGYO SHIKENJO HOKOKU 3276
(BULLETIN OF THE FUKUI AGRICULTURAL EXPERIMENT STATION)
FUKUI-KEN NOGYO SHIKENJO (FUKUI PREFECTURE AGRICULTURAL EXPERIMENT STATION)
RYO-MACHI, FUKUI-SHI 910
JAPANESE, WITH ENGLISH SUMMARY
INDEXED: CHE

FUKUI-KEN NOGYO SHIKENJO TOKUBETSU HOKOKU 3277
(SPECIAL REPORT OF THE FUKUI AGRICULTURAL EXPERIMENT STATION)
FUKUI AGRICULTURAL EXPERIMENT STATION
52-22, RYO-MACHI, FUKUI
JAPANESE AND ENGLISH
INDEXED: CHE

FUKUI-KEN NORIN SUISAN TOKEI NENPO 3278
(CENSUS OF AGRICULTURE, FORESTRY AND FISHERY IN FUKUI PREFECTURE)
FUKUI-KEN
17-1, 3-CHOME, OTE, FUKUI-SHI 910
JAPANESE

FUKUOKA HISHIRYO KENSAJO JIGYO HOKOKU, HIRYO NO BU 3279
(ANNUAL REPORT OF THE FUKUOKA FERTILIZER AND FEED INSPECTION STATION, FERTILIZER SECTION)
NORIN-SHO FUKUOKA HISHIRYO KENSAJO (FUKUOKA FERTILIZER AND FEED INSPECTION STATION, MINISTRY OF AGRICULTURE AND FORESTRY)
1-1, TEHAYA, HIGASHI-KU, FUKUOKA-SHI 813
JAPANESE

FUKUOKA NORIN SUISAN TOKEI NENPO 3280
(ANNUAL STATISTICS OF AGRICULTURE, FORESTRY AND FISHERY, FUKUOKA)
NORIN-SHO KYUSHU NOSEI-KYOKU (KYUSHU AGRICULTURAL ADMINISTRATION BUREAU, MINISTRY OF AGRICULTURE AND FORESTRY)
1-2, NINOMARU, KUMAMOTO-SHI 860
JAPANESE

FUKUOKA NORIN SUISAN TOKEIHYO 3281
(STATISTICAL YEARBOOK OF AGRICULTURE, FORESTRY AND FISHERY, FUKUOKA)
NORIN-SHO KYUSHU NOSEI-KYOKU FUKUOKA TOKEI JOHO JIMUSHO (FUKUOKA STATISTICS AND INFORMATION OFFICE, KYUSHU AGRICULTURAL ADMINISTRATION BUREAU, MINISTRY OF AGRICULTURE AND FORESTRY)
JAPANESE

FUKUOKA-KEN NO NOGYO 3282
(CENSUS OF AGRICULTURE, FUKUOKA PREFECTURE)
FUKUOKA-KEN KIKAKU KAIHATSU-BU (PLANNING AND DEVELOPMENT DIVISION, FUKUOKA PREFECTURAL GOVERNMENT)
1-1, 1-CHOME, TENJIN, CHUO-KU, FUKUOKA-SHI 810
JAPANESE

FUKUOKA-KEN NOGYO NO DOKO 3283
(TENDENCY OF AGRICULTURE, FUKUOKA PREFECTURE)
FUKUOKA-KEN (FUKUOKA PREFECTURAL GOVERNMENT)
1-1, 1-CHOME, TENJIN, CHUO-KU, FUKUOKA-SHI 810
JAPANESE

FUKUOKA-KENRITSU NOGYO SHIKENJO GYOMU NENPO 3284
(OPERATION YEARBOOK OF FUKUOKA AGRICULTURAL EXPERIMENT STATION)
FUKUOKA-KENRITSU NOGYO SHIKENJO, FUKUOKA-KENRITSU NOGYO KOSHUJO
KAMIKOGA, CHIKUSHINO-SHI 818
JAPANESE

FUKUOKA-KENRITSU NOGYO SHIKENJO KENKYU HOKOKU 3285
(RESEARCH REPORT OF THE FUKUOKA AGRICULTURAL EXPERIMENT STATION)
FUKUOKA-KENRITSU NOGYO SHIKENJO (AGRICULTURAL EXPERIMENT STATION OF FUKUOKA PREFECTURE)
KAMIKOGA, CHIKUSHINO-SHI 818
JAPANESE

FUKUSHIMA NORIN SUISAN TOKEI NENPO 3286
(STATISTICAL YEARBOOK OF AGRICULTURE, FORESTRY AND FISHERY, FUKUSHIMA)
NORIN-SHO TOHOKU NOSEI-KYOKU FUKUSHIMA TOKEI JOHO JIMUSHO (FUKUSHIMA STATISTICS AND INFORMATION OFFICE, TOHOKU AGRICULTURAL ADMINISTRATION BUREAU, MINISTRY OF AGRICULTURE AND FORESTRY)
4-15, YAMASHITA-CHO, FUKUSHIMA-SHI 960
JAPANESE

FUKUSHIMA-KEN NO NOGYO (NOGYO KIHON CHOSA KEKKA) 3287
(REPORT ON THE RESULT OF AGRICULTURAL BASIC SURVEY, FUKUSHIMA PREFECTURE)
FUKUSHIMA-KEN KIKAKU KAIHATSU-BU (PLANNING AND DEVELOPMENT DIVISION, FUKUSHIMA PREFECTURAL GOVERNMENT)
2-16, SUGIZUMA-CHO, FUKUSHIMA-SHI 960
JAPANESE

FUKUSHIMA-KEN NOGYO NO UGOKI 3288
(TENDENCY OF AGRICULTURE, FUKUSHIMA PREFECTURE)
NORIN-SHO TOHOKU NOSEI-KYOKU (TOHOKU AGRICULTURAL ADMINISTRATION BUREAU, MINISTRY OF AGRICULTURE AND FORESTRY)
3-1, 3-CHOME, HON-MACHI, SENDAI-SHI 980
JAPANESE

FUKUSHIMA-KEN NOGYO SHIKENJO KENKYU HOKOKU 3289
(BULLETIN OF THE FUKUSHIMA AGRICULTURAL EXPERIMENT STATION)
FUKUSHIMA-KEN NOGYO SHIKENJO (FUKUSHIMA AGRICULTURAL EXPERIMENT STATION)
20, WAKAMIYAMAE, TOMITA-CHO, KORIYAMA-SHI 963
JAPANESE

GAMMA FIELD SYMPOSIA 3290
NORIN-SHO HOSHASEN IKUSHUJO (INSTITUTE OF RADIATION BREEDING, MINISTRY OF AGRICULTURE AND FORESTRY)
OMIYA-MACHI, NAKA-GUN, IBARAKI 319-22
ENGLISH
INDEXED: CHE, BIB, JICST

GENDAI NOGYO 3291
(PRESENT-DAY AGRICULTURE)
NOSAN GYOSON BUNKA KYOKAI
6-1, 7-CHOME, AKASAKA, MINATO-KU, TOKYO 107
JAPANESE

GENDAI RINGYO 3292
(MODERN FOREST INDUSTRY)
ZENKOKU RINGYO KAIRYO FUKYU KYOKAI (NATIONAL FOREST INDUSTRY IMPROVEMENT PROMOTION ASSOCIATION)
JAPANESE

GIFU DAIGAKU NOGAKUBU KENKYU HOKOKU 3293
(RESEARCH BULLETIN OF THE FACULTY OF AGRICULTURE, GIFU UNIVERSITY)
GIFU DAIGAKU NOGAKUBU (FACULTY OF AGRICULTURE, GIFU UNIVERSITY)
3-1, NAKA MONZEN-CHO, KAGAMIHARA-SHI 504
JAPANESE AND ENGLISH
INDEXED: BIB, CHE, JICST

GIFU NORIN SUISAN TOKEI NENPO 3294
(STATISTICAL YEARBOOK OF AGRICULTURE, FORESTRY AND FISHERY, GIFU)
NORIN-SHO TOKAI NOSEI-KYOKU GIFU TOKEI JOHO JIMUSHO (GIFU STATISTICS AND INFORMATION OFFICE, TOKAI AGRICULTURAL ADMINISTRATION BUREAU, MINISTRY OF AGRICULTURE AND FORESTRY)
5-13, KINRYU-CHO, GIFU-SHI 500
JAPANESE

GIFU-KEN KOREICHI NOGYO SHIKENJO HOKOKU 3295
(REPORT OF THE GIFU PREFECTURE HIGH ALTITUDE AGRICULTURE EXPERIMENT STATION)
GIFU-KEN KOREICHI NOGYO SHIKENJO
216-2, MUKAI-MACHI, FURUKAWA-MACHI, YOSHIKI-GUN, GIFU 509-42
JAPANESE

GIFU-KEN NOGYO SHIKENJO HOKOKU 3296
(RESEARCH BULLETIN OF THE GIFU AGRICULTURAL EXPERIMENT STATION)
GIFU-KEN NOGYO SHIKENJO (GIFU PREFECTURE AGRICULTURAL EXPERIMENT STATION)
729, MATAMARU GONGEN, GIFU-SHI 501-11
JAPANESE

GIFU-KEN NOGYO SHIKENJO SHIKEN KENKYU NO SEIKA 3297
(REPORT OF GIFU PREFECTURE AGRICULTURAL EXPERIMENT STATION)
GIFU-KEN NOGYO SHIKENJO (GIFU PREFECTURE AGRICULTURAL EXPERIMENT STATION)
729, MATAMARU GONGEN, GIFU-SHI 501-11
JAPANESE

GOJO NOYU 3298
(AGRICULTURE PROPAGATION)
MIYAGI-KEN NOGYO FUKYU KYOKAI (MIYAGI PREFECTURE PROPAGATION ASSOCIATION)
C/O MIYAGI-KEN NOGYO FUKYU-KA, 8-1, 3-CHOME, HON-MACHI, SENDAI-SHI 980
JAPANESE

GUNMA NOGYO TO SEIKATSU 3299
(AGRICULTURE AND FARMERS' LIFE IN GUNMA PREFECTURE)
GUNMA-KEN NOGYO KAIRYO KYOKAI (GUNMA ASSOCIATION FOR AGRICULTURAL IMPROVEMENT)
C/O KENCHO, 1-1, 1-CHOME, OTE-CHO, MAEBASHI-SHI 371
JAPANESE

GUNMA NORIN SUISAN TOKEI NENPO 3300
(STATISTICAL YEARBOOK OF AGRICULTURE, FORESTRY AND FISHERY, GUNMA)
NORIN-SHO KANTO NOSEI-KYOKU GUNMA TOKEI JOHO JIMUSHO (GUNMA STATISTICS AND INFORMATION OFFICE, KANTO AGRICULTURAL ADMINISTRATION BUREAU, MINISTRY OF AGRICULTURE AND FORESTRY)
12-10, 3-CHOME, SHOWA-CHO, MAEBASHI-SHI 371
JAPANESE

GUNMA-KEN NOGYO SHIKENJO HOKOKU 3301
(BULLETIN OF THE GUNMA AGRICULTURAL EXPERIMENT STATION)
GUNMA-KEN NOGYO SHIKENJO (GUNMA AGRICULTURAL EXPERIMENT STATION)
1251, EGI-MACHI, MAEBASHI-SHI 371
JAPANESE
INDEXED: CHE

GUNMA-KEN NOGYO SHIKENJO TOKUBETSU HOKOKU 3302
(BULLETIN OF THE GUNMA AGRICULTURAL EXPERIMENT STATION, SPECIAL ISSUE)
GUNMA-KEN NOGYO SHIKENJO (GUNMA AGRICULTURAL EXPERIMENT STATION)
1251, EGI-MACHI, MAEBASHI-SHI 371
JAPANESE

GUNMA-KEN SHICHOSONBETSU NORIN TOKEISHO 3303
(STATISTICS OF AGRICULTURE AND FORESTRY BY TOWNS, GUNMA PREFECTURE)
NORIN-SHO KANTO NOSEI-KYOKU GUNMA TOKEI JOHO JIMUSHO (GUNMA STATISTICS AND INFORMATION OFFICE, KANTO AGRICULTURAL ADMINISTRATION BUREAU, MINISTRY OF AGRICULTURE AND FORESTRY)
12-10, 3-CHOME, SHOWA-MACHI, MAEBASHI-SHI 371
JAPANESE

GURAFU KINDAI NOGYO 3304
JAPANESE
INDEXED: BIB

HATACHI NOGYO 3305
(NEW DEVELOPMENT ON AGRICULTURE)
HATACHI NOGYO KENKYUKAI (FIELD AGRICULTURE SOCIETY)
8-8, 2-CHOME, NISHI SHINBASHI, MINATO-KU, TOKYO 105
JAPANESE

HIKEN KAIHO 3306
(PUBLICATION OF THE FERTILIZER CONTROL OFFICIALS)
NIHON HIRYO KENTEI KYOKAI (FOUNDATIONAL JAPAN ANALYSIS AND INSPECTION ASSOCIATION FOR FERTILIZER AND FOOD STUFF)
12-15, 1-CHOME, FUKUZUMI, KOTO-KU, TOKYO 135
JAPANESE
INDEXED: CHE

HIROSAKI DAIGAKU NOGAKUBU GAKUJUTSU HOKOKU 3307
(BULLETIN OF THE FACULTY OF AGRICULTURE, HIROSAKI UNIVERSITY)
HIROSAKI DAIGAKU NOGAKUBU (FACULTY OF AGRICULTURE HIROSAKI UNIVERSITY)
3, BUNKYO-CHO, HIROSAKI-SHI 036
JAPANESE AND ENGLISH
INDEXED: BIB, CHE, JICST

HIROSHIMA NOGYO TANKI DAIGAKU KENKYU HOKOKU 3308
(BULLETIN OF THE HIROSHIMA AGRICULTURAL COLLEGE)
HIROSHIMA NOGYO TANKI DAIGAKU (HIROSHIMA AGRICULTURAL COLLEGE)
SAIIJO-CHO, KAMO-GUN, HIROSHIMA 724-06
JAPANESE, WITH EUROPEAN SUMMARY
INDEXED: BIB, BIO, CHE, JICST

HIROSHIMA NORIN SUISAN TOKEI NENPO 3309
(STATISTICAL YEARBOOK OF AGRICULTURE, FORESTRY AND FISHERY, HIROSHIMA)
NORIN-SHO CHUGOKU SHIKOKU NOSEI-KYOKU HIROSHIMA TOKEI JOHO JIMUSHO (HIROSHIMA STATISTICS AND INFORMATION OFFICE, CHUGOKU SHIKOKU AGRICULTURAL ADMINISTRATION BUREAU)
C/O HIROSHIMA GODOCHOSHA, 6-30, KAMI HATCHOBORI, HIROSHIMA-SHI 730
JAPANESE

HIROSHIMA-KENRITSU NOGYO SHIKENJO CHOSA SHIKEN HOKOKU 3310
(SURVEY AND RESEARCH REPORTS OF HIROSHIMA PREFECTURAL AGRICULTURAL EXPERIMENT STATION))
HIROSHIMO-KENRITSU NOGYO SHIKENJO
HACHIHONMATSU-MACHI, HIGASHIHIROSHIMA-SHI 739-01
JAPANESE

HIROSHIMA-KENRITSU NOGYO SHIKENJO GYOMU NENPO 3311
(ANNUAL REPORT OF HIROSHIMA PREFECTURAL AGRICULTURAL EXPERIMENT STATION)
HIROSHIMA-KENRITSU NOGYO SHIKENJO (HIROSHIMA PREFECTURAL AGRICULTURAL EXPERIMENT STATION)
HACHIHONMATSU-MACHI, HIGASHIHIROSHIMA-SHI 739-01
JAPANESE

HIROSHIMA-KENRITSU NOGYO SHIKENJO HOKOKU 3312
(BULLETIN OF HIROSHIMA PREFECTURAL AGRICULTURAL EXPERIMENT STATION)
HIROSHIMA-KENRITSU NOGYO SHIKENJO (HIROSHIMA PREFECTURAL AGRICULTURAL EXPERIMENT STATION)
HACHIHONMATSU-MACHI, HIGASHIHIROSHIMA-SHI 739-01
JAPANESE, WITH ENGLISH SUMMARY
INDEXED: BIB, CHE

HIROSHIMA-KENRITSU NOGYO SHIKENJO KENKYU GYOSEKISHU 3313
(REPORT OF HIROSHIMA PREFECTURAL AGRICULTURAL EXPERIMENT STATION)
HIROSHIMA-KENRITSU NOGYO SHIKENJO
HACHIHONMATSU-MACHI, HIGASHIHIROSHIMA-SHI 739-01
JAPANESE

HIRYO JIHO 3314
(MANURIAL TIMES)
NORIN-SHO NOSAN ENGEI-KYOKU (AGRICULTURE PRODUCTION BUREAU, MINISTRY OF AGRICULTURE AND FORESTRY)
NORIN KYOKAI
2-1, 1-CHOME, KASUMIGASEKI, CHIYODA-KU, TOKYO 100
JAPANESE

HIRYO JIKKEN TENJIHO SEISEKISHO 3315
(ANNUAL RECORD OF THE PILOT FIELD FOR FERTILIZER TESTING, ISHIKAWA PREFECTURE)
ISHIKAWA-KEN NOGYO SEISAN SHIZAITO JIKKEN TENJIHO SETCHI JIMUKYOKU
NAKABAYASHI, NONOICHI-MACHI, ISHIKAWA-GUN, ISHIKAWA 921
JAPANESE

HIRYO KANKEI TOKEI SHIHYO 3316
(STATISTICS INDEX FOR FERTILIZER)
NORIN-SHO NOSAN ENGEI-KYOKU (AGRICULTURE PRODUCTION BUREAU, MINISTRY OF AGRICULTURE AND FORESTRY)
2-1, 1-CHOME, KASUMIGASEKI, CHIYODA-KU, TOKYO 100
JAPANESE

HOJO TO DOJO 3317
(FARM AND SOIL)
NIHON DOJO KYOKAI (JAPANESE SOIL ASSOCIATION)
C/O SANSHI KAIKAN, 1-7, YURAKU-CHO, CHIYODA-KU, TOKYO 100
JAPANESE

HOKKAIDO DAIGAKU NOGAKUBU FUZOKU NOJO HOKOKU 3318
(RESEARCH BULLETIN OF THE UNIVERSITY FARM, FACULTY OF AGRICULTURE, HOKKAIDO UNIVERSITY)
HOKKAIDO DAIGAKU NOGAKUBU FUZOKU NOJO (UNIVERSITY FARM, FACULTY OF AGRICULTURE, HOKKAIDO UNIVERSITY)
NISHI 10-CHOME, KITA 11-JO, CHUO-KU, SAPPORO-SHI 060
JAPANESE AND ENGLISH

HOKKAIDO DAIGAKU NOGAKUBU HOBUN KIYO 3319
(MEMOIRS OF THE FACULTY OF AGRICULTURE, HOKKAIDO UNIVERSITY)
HOKKAIDO DAIGAKU NOGAKUBU (FACULTY OF AGRICULTURE, HOKKAIDO UNIVERSITY)
NISHI 9-CHOME, KITA 9-JO, CHUO-KU, SAPPORO-SHI 060
JAPANESE, WITH ENGLISH SUMMARY
INDEXED: CHE, EXC

HOKKAIDO NOCHIKUSANBUTSU SEISANHI 3320
(SURVEY REPORT OF AGRICULTURAL AND LIVE-STOCK PRODUCTS COST, HOKKAIDO)
NORIN-SHO HOKKAIDO TOKEI JOHO JIMUSHO (HOKKAIDO STATISTICS AND INFORMATION OFFICE, MINISTRY OF AGRICULTURE AND FORESTRY)
NISHI 4-CHOME, KITA 3-JO, CHUO-KU, SAPPORO-SHI 060
JAPANESE

HOKKAIDO NOGYO KANSOKU 3321
(OUTLOOK OF AGRICULTURE, HOKKAIDO)
HOKKAIDO NOMU-BU (AGRICULTURAL ADMINISTRATION DIVISION, HOKKAIDO PREFECTURAL GOVERNMENT)
NISHI 6-CHOME, KITA 3-JO, CHUO-KU, SAPPORO-SHI 063
JAPANESE

HOKKAIDO NOGYO KIHON CHOSA KEKKA HOKOKUSHO 3322
(FUNDAMENTAL RESEARCH REPORT OF AGRICULTURE, HOKKAIDO)
HOKKAIDO KAIHATSU CHOSEI-BU (DEVELOPING AND COORDINATE DIVISION, HOKKAIDO PREFECTURAL GOVERNMENT)
NISHI 6-CHOME, KITA 3-JO, CHUO-KU, SAPPORO-SHI 060
JAPANESE

HOKKAIDO NOGYO KYOIKU KENKYU 3323
(HOKKAIDO JOURNAL OF VOCATIONAL AGRICULTURE)
HOKKAIDO CHIKU SOGO NOGAKU KENKYUKAI (HOKKAIDO RESEARCH ASSOCIATION OF VOCATIONAL AGRICULTURE)
C/O OBIHIRO CHIKUSAN DAIGAKU, INADA-CHO, OBIHIRO-SHI 089-11
JAPANESE

HOKKAIDO NOGYO SHIKENJO DOSEI CHOSA HOKOKU 3324
(SOIL SURVEY REPORT, HOKKAIDO NATIONAL AGRICULTURAL EXPERIMENT STATION))
NORIN-SHO HOKKAIDO NOGYO SHIKENJO (HOKKAIDO NATIONAL AGRICULTURAL EXPERIMENT STATION, MINISTRY OF AGRICULTURE AND FORESTRY)
1, HITSUJIGAOKA, TOYOHIRA-KU, SAPPORO-SHI 061-01
JAPANESE AND ENGLISH
INDEXED: BIB, BIO, CHE

HOKKAIDO NOGYO SHIKENJO KENKYU HOKOKU 3325
(RESEARCH BULLETIN OF THE HOKKAIDO NATIONAL AGRICULTURAL EXPERIMENT STATION)
NORIN-SHO HOKKAIDO NOGYO SHIKENJO (HOKKAIDO NATIONAL AGRICULTURAL EXPERIMENT STATION, MINISTRY OF AGRICULTURE AND FORESTRY)
1, HITSUJIGAOKA, TOYOHIRA-KU, SAPPORO-SHI 061-01
JAPANESE AND ENGLISH
INDEXED: BIB, BIO, CHE, EXC, JICST

HOKKAIDO NOGYO SHIKENJO KENKYU SHIRYO 3326
(MISCELLANEOUS PUBLICATION OF THE HOKKAIDO NATIONAL AGRICULTURAL EXPERIMENT STATION)
NORIN-SHO HOKKAIDO NOGYO SHIKENJO (HOKKAIDO NATIONAL AGRICULTURAL EXPERIMENT STATION, MINISTRY OF AGRICULTURE AND FORESTRY)
1, HITSUJIGAOKA, TOYOHIRA-KU, SAPPORO-SHI 061-01
JAPANESE AND ENGLISH

HOKKAIDO NOGYO SHIKENJO NENPO 3327
(ANNUAL REPORT OF HOKKAIDO NATIONAL AGRICULTURAL EXPERIMENT STATION)
NORIN-SHO HOKKAIDO NOGYO SHIKENJO (HOKKAIDO NATIONAL AGRICULTURAL EXPERIMENT STATION, MINISTRY OF AGRICULTURE AND FORESTRY)
1, HITSUJIGAOKA, TOYOHIRA-KU, SAPPORO-SHI 061-01
JAPANESE
INDEXED: JICST

HOKKAIDO NOGYO TOKEIHYO 3328
(AGRICULTURAL STATISTICS, HOKKAIDO)
HOKKAIDO NOMU-BU (DEPARTMENT OF AGRICULTURE, HOKKAIDO PREFECTURAL GOVERNMENT)
NISHI 6-CHOME, KITA 3-JO, CHUO-KU, SAPPORO-SHI 060
JAPANESE

HOKKAIDO NORIN KENKYU 3329
(STUDIES ON AGRICULTURE AND FORESTRY IN HOKKAIDO)
HOKKAIDORITSU SOGO KEIZAI KENKYUSHO (HOKKAIDO PREFECTURAL INSTITUTE OF SYNTHETIC ECONOMIC RESEARCH)
NISHI 6-CHOME, KITA 3-JO, CHUO-KU, SAPPORO-SHI 060
JAPANESE

HOKKAIDO NORIN SUISAN TOKEI NENPO 3330
(STATISTICAL YEARBOOK OF AGRICULTURE, FORESTRY AND FISHERY, HOKKAIDO)
NORIN-SHO HOKKAIDO TOKEI JOHO JIMUSHO (HOKKAIDO STATISTICS AND INFORMATION OFFICE, MINISTRY OF AGRICULTURE AND FORESTRY)
NISHI 4-CHOME, KITA 3-JO, CHUO-KU, SAPPORO-SHI 060
JAPANESE

HOKKAIDO NORIN SUISANGYO DOKO SOKUHO 3331
(QUICK ANNOUNCEMENT OF TENDENCY OF AGRICULTURE, FORESTRY AND FISHERY, HOKKAIDO)
NORIN-SHO HOKKAIDO CHIHO RENRAKUKAIGI JIMUKYOKU
C/O HOKKAIDO NOGYO SHIKENJO, HITSUJIGAOKA, TOYOHIRA-KU, SAPPORO-SHI 061-01
JAPANESE

HOKKAIDO NOSAKUMOTSU SHICHOSONBETSU TOKEI 3332
(STATISTICS OF AGRICULTURAL PRODUCTS BY TOWNS, HOKKAIDO PREFECTURE)
NORIN-SHO HOKKAIDO TOKEI JOHO JIMUSHO (HOKKAIDO STATISTICS AND INFORMATION OFFICE, MINISTRY OF AGRICULTURE AND FORESTRY)
NISHI 4-CHOME, KITA 3-JO, CHUO-KU, SAPPORO-SHI 060
JAPANESE

HOKKAIDO RINGYO SHIDOJO GEPPO 3333
(HOKKAIDO FOREST PRODUCTS RESEARCH INSTITUTE, MONTHLY REPORTS)
HOKKAIDO RINGYO SHIDOJO
ASAHIGAWA
JAPANESE
INDEXED: CHE

HOKKAIDORITSU CHUO NOGYO SHIKENJO NENPO 3334
(ANNUAL REPORT OF HOKKAIDO CENTRAL AGRICULTURAL EXPERIMENT STATION)
HOKKAIDORITSU CHUO NOGYO SHIKENJO
HIGASHI 6-SEN KITA 15, NAGANUMA-CHO, YUBARI-GUN, HOKKAIDO 069-13
JAPANESE

HOKKAIDORITSU CHUO NOGYO SHIKENJO SHIRYO 3335
(REPORT OF HOKKAIDO CENTRAL AGRICULTURAL EXPERIMENT STATION)
HOKKAIDORITSU CHUO NOGYO SHIKENJO (HOKKAIDO CENTRAL AGRICULTURAL EXPERIMENT STATION)
HIGASHI 6-SEN KITA 15, NAGANUMA-CHO, YUBARI-GUN, HOKKAIDO 069-13
JAPANESE

HOKKAIDORITSU DONAN NOGYO SHIKENJO NENPO 3336
(ANNUAL REPORT OF DONAN AGRICULTURAL EXPERIMENT STATION)
HOKKAIDORITSU DONAN NOGYO SHIKENJO (DONAN AGRICULTURAL EXPERIMENT STATION)
680, HON-CHO, ONO-CHO, KAMEDA-GUN, HOKKAIDO 041-12
JAPANESE

HOKKAIDORITSU KAMIKAWA NOGYO SHIKENJO GYOMU NENPO 3337
(OPERATION YEARBOOK OF KAMIKAWA AGRICULTURAL EXPERIMENT STATION)
HOKKAIDORITSU KAMIKAWA NOGYO SHIKENJO
NAGAYAMA-CHO, ASAHIKAWA-SHI 078-02
JAPANESE

HOKKAIDORITSU KAMIKAWA NOGYO SHIKENJO KENKYU NYUSU 3338
(HOKKAIDO-PREFECTURE KAMIKAWA AGRICULTURAL EXPERIMENT STATION RESEARCH NEWS)
JAPANESE

HOKKAIDORITSU KAMIKAWA NOGYO SHIKENJO NENPO 3339
(ANNUAL REPORT OF KAMIKAWA AGRICULTURAL EXPERIMENT STATION)
HOKKAIDORITSU KAMIKAWA NOGYO SHIKENJO
NAGAYAMA-CHO, ASAHIKAWA-SHI 078-02
JAPANESE

HOKKAIDORITSU KITAMI NOGYO SHIKENJO NENPO 3340
(ANNUAL REPORT OF KITAMI AGRICULTURAL EXPERIMENT STATION)
HOKKAIDORITSU KITAMI NOGYO SHIKENJO
52, YAYOI, KUNNEPPU-MACHI, TOKORO-GUN, HOKKAIDO 099-14
JAPANESE

HOKKAIDORITSU KONSEN NOGYO SHIKENJO NENPO 3341
(ANNUAL REPORT OF KONSEN AGRICULTURAL EXPERIMENT STATION)
HOKKAIDORITSU KONSEN NOGYO SHIKENJO
HIGASHI 1-JO MINAMI 6, NAKASHIBETSU, SHIBETSU-GUN, HOKKAIDO 086-11
JAPANESE

HOKKAIDORITSU NOGYO SHIKENJO HOKOKU 3342
(REPORT OF HOKKAIDO PREFECTURAL AGRICULTURAL EXPERIMENTAL STATION)
HOKKAIDORITSU CHUO NOGYO SHIKENJO (HOKKAIDO CENTRAL AGRICULTURAL EXPERIMENT STATION)
HIGASHI 6-SEN KITA 15, NAGANUMA-CHO, YUBARI-GUN, HOKKAIDO 069-13
JAPANESE, WITH ENGLISH SUMMARY
INDEXED: BIB, BIO, CHE, JICST

HOKKAIDORITSU NOGYO SHIKENJO SHUHO 3343
(BULLETIN OF THE HOKKAIDO PREFECTURAL AGRICULTURAL EXPERIMENT STATION)
HOKKAIDORITSU CHUO NOGYO SHIKENJO (HOKKAIDO CENTRAL AGRICULTURAL EXPERIMENT STATION)
HIGASHI 6-SEN KITA 15, NAGANUMA-CHO, YUBARI-GUN, HOKKAIDO 069-13
JAPANESE AND ENGLISH
INDEXED: BIB, CHE, JICST

HOKKAIDORITSU TENPOKU NOGYO SHIKENJO NENPO 3344
(ANNUAL REPORT OF TENPOKU AGRICULTURAL EXPERIMENT STATION)
HOKKAIDORITSU TENPOKU NOGYO SHIKENJO
MIDORIGAOKA, HAMATONBETSU-CHO, ESASHI-GUN, HOKKAIDO 098-57
JAPANESE

HOKKAIDORITSU TOKACHI NOGYO SHIKENJO SHIRYO 3345
(MATERIALS FROM HOKKAIDO PREFECTURAL AGRICULTURAL EXPERIMENT STATION)
HOKKAIDORITSU TOKACHI NOGYO SHIKENJO (HOKKAIDO PREFECTURAL AGRICULTURAL EXPERIMENT STATION)
SHINSEI, MEMURO-MACHI, KASAI-GUN, HOKKAIDO 082
JAPANESE

HOKUNO 3346
(AGRICULTURE IN HOKKAIDO)
HOKUNO-KAI
C/O HOKKAIDO JICHI KAIKAN, NISHI 6-CHOME, KITA 6-JO, CHUO-KU, SAPPORO-SHI 060
JAPANESE
INDEXED: BIB, JICST

HOKUREN KAIHATSU KENKYUSHO HOKOKU 3347
(REPORT OF THE FEDERATION OF AGRICULTURAL COOPERATION SOCIETY OF HOKKAIDO)
HOKUREN (NOGYO KYODO KUMIAI RENGOKAI) KAIHATSU KENKYUSHO (THE FEDERATION OF AGRICULTURAL COOPERATIVE SOCIETIES OF HOKKAIDO)
375, HIGASHI 7-CHOME, KITA 6-JO, HIGASHI-KU, SAPPORO-SHI 065
JAPANESE

HOKURIKU CHIIKI SHIKEN KENKYU UCHIAWASE KAIGI HOKOKU 3348
(REPORT OF AGRICULTURAL SURVEY AND RESEARCH CONFERENCE IN HOKURIKU)
NORIN-SHO HOKURIKU NOGYO SHIKENJO (HOKURIKU NATIONAL AGRICULTURAL EXPERIMENT STATION, MINISTRY OF AGRICULTURE AND FORESTRY)
INADA, JOETSU-SHI 943-01
JAPANESE

HOKURIKU NOGYO JOSEI HOKOKU 3349
(ANNUAL REVIEW OF AGRICULTURE, HOKURIKU DISTRICT)
NORIN-SHO HOKURIKU NOSEI-KYOKU (HOKURIKU AGRICULTURAL ADMINISTRATION BUREAU, MINISTRY OF AGRICULTURE AND FORESTRY)
2-60, 2-CHOME, HIROSAKA, KANAZAWA-SHI 920
JAPANESE

HOKURIKU NOGYO KENKYU SHIRYO 3350
(RESEARCH REPORT OF AGRICULTURE IN HOKURIKU AREA)
NORIN-SHO HOKURIKU NOGYO SHIKENJO (HOKURIKU NATIONAL AGRICULTURAL EXPERIMENT STATION, MINISTRY OF AGRICULTURE AND FORESTRY)
INADA, JOETSU-SHI 943-01
JAPANESE

HOKURIKU NOGYO KIHO 3351
(QUARTERLY REPORT OF AGRICULTURAL BUSINESS IN HOKURIKU AREA)
NORIN-SHO HOKURIKU NOSEI-KYOKU (HOKURIKU AGRICULTURAL ADMINISTRATION BUREAU, MINISTRY OF AGRICULTURE AND FORESTRY)
2-60, 2-CHOME, HIROSAKA, KANAZAWA-SHI 920
JAPANESE

HOKURIKU NOGYO KIHON SHIHYO 3352
(FUNDAMENTAL DATA OF AGRICULTURE IN HOKURIKU)
NORIN-SHO HOKURIKU NOSEI-KYOKU (HOKURIKU AGRICULTURAL ADMINISTRATION BUREAU, MINISTRY OF AGRICULTURE AND FORESTRY)
2-60, 2-CHOME, HIROSAKA, KANAZAWA-SHI 920
JAPANESE

HOKURIKU NOGYO SHIKENJO HOKOKU 3353
(BULLETIN OF THE HOKURIKU AGRICULTURAL EXPERIMENT STATION)
NORIN-SHO HOKURIKU NOGYO SHIKENJO (HOKURIKU NATIONAL AGRICULTURAL EXPERIMENT STATION, MINISTRY OF AGRICULTURE AND FORESTRY)
KAMIINADA, JOETSU-SHI 943-01
JAPANESE AND ENGLISH
INDEXED: BIB, CHE

HOKURIKU NORIN SUISAN TOKEI 3354
(STATISTICS OF AGRICULTURE, FORESTRY AND FISHERY, HOKURIKU)
NORIN-SHO HOKURIKU NOSEI-KYOKU (HOKURIKU AGRICULTURAL ADMINISTRATION BUREAU, MINISTRY OF AGRICULTURE AND FORESTRY)
2-60, 2-CHOME, HIROSAKA, KANAZAWA-SHI 920
JAPANESE

HOKURIKU NOSEIKYOKU KYOKUHO 3355
(HOKURIKU AGRICULTURAL ADMINISTRATION BUREAU REPORT)
JAPANESE

HOPPO NOGYO 3356
(NORTHERN AGRICULTURE)
HOKKAIDO NOGYO KAIGI (HOKKAIDO AGRICULTURAL EXTENSION COUNCIL
NISHI 6-CHOME, KITA 3-JO, CHUO-KU, SAPPORO-SHI 060
JAPANESE

HOSHASEN IKUSHUJO KENKYU HOKOKU 3357
(ACTA RADIOBOTANICA ET GENETICA (BULLETIN OF THE INSTITUTE OF RADIATION BREEDING))
NORIN-SHO NOGYO GIJUTSU KENKYUSHO HOSHASEN IKUSHUJO (INSTITUTE OF RADIATION BREEDING)
OMIYA-CHO, NAKA-GUN, IBARAKI 319-22
JAPANESE AND ENGLISH
INDEXED: BIB, JICST

HOSHASEN IKUSHUJO NENPO 3358
(ANNUAL REPORT OF THE INSTITUTE OF RADIATION BREEDING)
NORIN-SHO NOGYO GIJUTSU KENKYUSHO HOSHASEN IKUSHUJO (INSTITUTE OF RADIATION BREEDING)
OMIYA-CHO, NAKA-GUN, IBARAKI 319-22
JAPANESE

HYOGO NORIN SUISAN TOKEI NENPO 3359
(STATISTICAL YEARBOOK OF AGRICULTURE, FORESTRY AND FISHERY, HYOGO)
NORIN-SHO KINKI NOSEI-KYOKU HYOGO TOKEI JOHO JIMUSHO (HYOGO PREFECTURE STATISTICS AND INFORMATION OFFICE, KINKI AGRICULTURAL ADMINISTRATION BUREAU, MINISTRY OF AGRICULTURE AND FORESTRY)
C/O KOBE CHIHO GODOCHOSHA, KAIGAN-DORI, IKUTA-KU, KOBE-SHI 650
JAPANESE

HYOGO NORIN TOKEI 3360
(HYOGO AGRICULTURE AND FORESTRY STATISTICS)
HYOGO NORIN TOKEI KYOKAI (HYOGO AGRICULTURE AND FORESTRY STATISTICS ASSOCIATION)
JAPANESE

HYOGO-KEN RINGYO SHIKENJO GYOMUSEISEKI HOKOKU 3361
(REPORT OF THE HYOGO PREFECTURAL FOREST EXPERIMENT STATION)
HYOGO-KEN RINGYO SHIKENJO
YAMASAKI
JAPANESE
INDEXED: CHE

HYOGO-KENRITSU NOGYO SHIKENJO GYOMU NENPO 3362
(ANNUAL REPORT OF HYOGO PREFECTURE AGRICULTURAL EXPERIMENT STATION)
HYOGO-KENRITSU NOGYO SHIKENJO (HYOGO PREFECTURE AGRICULTURAL EXPERIMENT STATION)
365,KITA OJI-MACHI, AKASHI-SHI 673
JAPANESE

HYOGO-KENRITSU NOGYO SHIKENJO KENKYU HOKOKU 3363
(BULLETIN OF THE HYOGO PREFECTURAL AGRICULTURAL EXPERIMENT STATION)
HYOGO-KENRITSU NOGYO SHIKENJO (HYOGO PREFECTURAL AGRICULTURAL EXPERIMENT STATION)
365, KITA OJI-MACHI, AKASHI-SHI 673
JAPANESE
INDEXED: CHE, JICST

IBARAKI DAIGAKU NOGAKUBU GAKUJUTSU HOKOKU 3364
(SCIENTIFIC REPORTS OF THE FACULTY OF AGRICULTURE IBARAKI UNIVERSITY)
IBARAKI DAIGAKU NOGAKUBU (FACULTY OF AGRICULTURE, IBARAKI UNIVERSITY)
AMI-MACHI, INASHIKI-GUN, IBARAKI 300-03
JAPANESE AND ENGLISH
INDEXED: BIB, CHE, EXC, JICST

IBARAKI NORIN SUISAN TOKEI NENPO 3365
(STATISTICAL YEARBOOK OF AGRICULTURE, FORESTRY AND FISHERY, IBARAKI)
NORIN-SHO KANTO NOSEI-KYOKU IBARAKI TOKEI JOHO JIMUSHO (IBARAKI STATISTICS AND INFORMATION OFFICE, KANTO AGRICULTURAL ADMINISTRATION BUREAU, MINISTRY OF AGRICULTURE AND FORESTRY)
1-22, KITAMI-CHO, MITO-SHI 310
JAPANESE
INDEXED: BIB

IBARAKI-KEN HISHIRYO KENSASHO HIRYO RYUTSU CHOSA HOKOKU 3366
(REPORT ON FERTILIZERS CIRCULATION)
IBARAKI-KEN HISHIRYO KENSASHO (IBARAKI PREFECTURAL FEEDS AND FERTILIZERS INSPECTION OFFICE)
4-1, ATAGO-CHO, MITO-SHI 310
JAPANESE

IBARAKI-KEN HISHIRYO KENSASHO JIGYO HOKOKU 3367
(OPERATION REPORT OF IBARAKI PREFECTURAL FEEDS AND FERTILIZERS INSPECTION OFFICE)
IBARAKI-KEN HISHIRYO KENSASHO (IBARAKI PREFECTURAL FEEDS AND FERTILIZERS INSPECTION OFFICE)
4-1, ATAGO-CHO, MITO-SHI 310
JAPANESE

IBARAKI-KEN NOGYO KIHON CHOSA KEKKA HOKOKUSHO 3368
(CENSUS OF FUNDAMENTAL AGRICULTURE IN IBARAKI PREFECTURE)
IBARAKI-KEN
5-38, 1-CHOME, SANNOMARU, MITO-SHI 310
JAPANESE

IBARAKI-KEN NOGYO SHIKENJO GYOMU NENPO 3369
(ANNUAL REPORT OF THE IBARAKI PREFECTURE AGRICULTURAL EXPERIMENT STATION)
IBARAKI-KEN NOGYO SHIKENJO (IBARAKI PREFECTURE AGRICULTURAL EXPERIMENT STATION)
KAMIKUNII-CHO, MITO-SHI 311-42
JAPANESE

IBARAKI-KEN NOGYO SHIKENJO KENKYU HOKOKU 3370
(BULLETIN OF THE IBARAKI AGRICULTURAL EXPERIMENT STATION)
IBARAKI-KEN NOGYO SHIKENJO (IBARAKI AGRICULTURAL EXPERIMENT STATION)
KAMIKUNII-CHO, MITO-SHI 311-42
JAPANESE

IBARAKI-KEN NOGYO SHIKENJO TOKUBETSU KENKYU HOKOKU 3371
(SPECIAL BULLETIN OF THE IBARAKI AGRICULTURAL EXPERIMENT STATION)
IBARAKI-KEN NOGYO SHIKENJO (IBARAKI PREFECTURE AGRICULTURAL EXPERIMENT STATION)
KAMIKUNII-CHO, MITO-SHI 311-42
JAPANESE

IBARAKI-KEN NORIN SUISANGYO SEISAN SHISU 3372
(PRODUCTS INDEX OF AGRICULTURE, FORESTRY AND FISHERY, IBARAKI PREFECTURE)
IBARAKI-KEN (IBARAKI PREFECTURAL GOVERNMENT)
5-38, 1-CHOME, SANNOMARU, MITO-SHI 310
JAPANESE

IKUSHUGAKU ZASSHI 3373
(JAPANESE JOURNAL OF BREEDING)
NIHON IKUSHU GAKKAI (JAPANESE SOCIETY OF BREEDING)
C/O TOKYO DAIGAKU NOGAKUBU, 1-1, 1-CHOME, YAYOI, BUNKYO-KU, TOKYO 113
JAPANESE AND ENGLISH
INDEXED: BIB, BIO, CHE

INSTITUTE FOR BREEDING RESEARCH TOKYO UNIVERSITY OF AGRICULTURE BULLETIN 3374
TOKYO NOGYO DAIGAKU IKUSHUGAKU KENKYUSHO (INSTITUTE FOR BREEDING RESEARCH, TOKYO UNIVERSITY OF AGRICULTURE)
4-28, 2-CHOME, KAMIYOGA, SETAGAYA-KU, TOKYO 158
ENGLISH

ISHIKAWA NOGYO NO KENKYU 3375
(STUDIES ON AGRICULTURE IN ISHIKAWA PREFECTURE)
ISHIKAWA-KEN NOGYO TANKI DAIGAKU (ISHIKAWA PREFECTURAL COLLEGE OF AGRICULTURE)
NONOICHI-MACHI, ISHIKAWA-GUN, ISHIKAWA 921
JAPANESE

ISHIKAWA NORIN SUISAN TOKEI NENPO 3376
(STATISTICAL YEARBOOK OF AGRICULTURE, FORESTRY AND FISHERY, ISHIKAWA)
NORIN-SHO HOKURIKU NOSEI-KYOKU ISHIKAWA TOKEI JOHO-BU (ISHIKAWA STATISTICS AND INFORMATION OFFICE, HOKURIKU AGRICULTURAL ADMINISTRATION BUREAU, MINISTRY OF AGRICULTURE AND FORESTRY)
C/O KANAZAWA GODOCHOSHA, 2-60, 2-CHOME, HIROSAKA, KANAZAWA-SHI 920
JAPANESE

ISHIKAWA-KEN CHIKU SHICHOSONBETSU NORIN SUISAN TOKEISHO 3377
(STATISTICS OF AGRICULTURE, FORESTRY AND FISHERY BY TOWNS, ISHIKAWA PREFECTURE)
NORIN-SHO HOKURIKU NOSEI-KYOKU TOKEI CHOSA-BU (HOKURIKU AGRICULTURAL ADMINISTRATION BUREAU, MINISTRY OF AGRICULTURE AND FORESTRY)
2-60, 2-CHOME, HIROSAKA, KANAZAWA-SHI 920
JAPANESE

ISHIKAWA-KEN NO NORINGYO 3378
(AGRICULTURE AND FORESTRY IN ISHIKAWA PREFECTURE)
ISHIKAWA-KEN (ISHIKAWA PREFECTURAL GOVERNMENT)
1-1, 3-CHOME, HIROSAKA, KANAZAWA-SHI 920
JAPANESE

ISHIKAWA-KEN NOGYO SHIKENJO KENKYU GYOMU HOKOKU 3379
(RESEARCH AND OPERATION REPORT OF ISHIKAWA-KEN AGRICULTURAL EXPERIMENT STATION)
ISHIKAWA-KEN NOGYO SHIKENJO
NAKABAYASHI, NONOICHI-MACHI, ISHIKAWA-GUN, ISHIKAWA 921
JAPANESE

ISHIKAWA-KEN NOGYO SHIKENJO KENKYU HOKOKU 3380
(BULLETIN OF THE ISHIKAWA-KEN AGRICULTURAL EXPERIMENT STATION)
ISHIKAWA-KEN NOGYO SHIKENJO (ISHIKAWA-KEN AGRICULTURAL EXPERIMENT STATION)
NAKABAYASHI, NONOICHI-MACHI, ISHIKAWA-GUN, ISHIKAWA 921
JAPANESE, WITH ENGLISH SUMMARY

ISHIKAWA-KEN NOGYO SHIKENJO TOKUBETSU KENKYU HOKOKU 3381
(SPECIAL RESEARCH BULLETIN OF THE ISHIKAWA-KEN AGRICULTURAL EXPERIMENT STATION)
ISHIKOW-KEN NOGYO SHIKENJO (ISHIKAWA PREFECTURAL AGRICULTURAL EXPERIMENT STATION)
JAPANESE

ISHIKAWA-KEN NOGYO TANKI DAIGAKU KENKYU HOKOKU 3382
(BULLETIN OF ISHIKAWA PREFECTURAL COLLEGE OF AGRICULTURE)
ISHIKAWA NOGYO TANKI DAIGAKU (ISHIKAWA PREFECTURAL COLLEGE OF AGRICULTURE)
SUEMATSU, NONOICHI-MACHI, ISHIKAWA-GUN, ISHIKAWA 921
JAPANESE, WITH ENGLISH SUMMARY

ISHIKAWA-KEN NOGYO TANKI DAIGAKU TOKUBETSU KENKYU HOKOKU 3383
(SPECIAL BULLETIN OF ISHIKAWA PREFECTURAL COLLEGE OF AGRICULTURE)
ISHIKAWA-KEN NOGYO TANKI DAIGAKU
SUEMATSU, NONOICHI-MACHI, ISHIKAWA-GUN, ISHIKAWA 921
JAPANESE, WITH ENGLISH SUMMARY

IWATE DAIGAKU NOGAKUBU HOKOKU 3384
(JOURNAL OF THE FACULTY OF AGRICULTURE, IWATE UNIVERSITY)
IWATE DAIGAKU NOGAKUBU (FACULTY OF AGRICULTURE, IWATE UNIVERSITY)
3-18, UEDA, MORIOKA-SHI 020
JAPANESE AND EUROPEAN
INDEXED: BIB, BIO, CHE, JICST

IWATE NORIN SUISAN TOKEI NENPO 3385
(STATISTICAL YEARBOOK OF AGRICULTURE, FORESTRY AND FISHERY, IWATE)
NORIN-SHO TOHOKU NOSEI-KYOKU IWATE TOKEI JOHO JIMUSHO (IWATE STATISTICS AND INFORMATION OFFICE, TOHOKU AGRICULTURAL ADMINISTRATION, BUREAU, MINISTRY OF AGRICULTURE & FORESTRY)
7-25, UCHIMARU, MORIOKA-SHI 020
JAPANESE

IWATE-KEN NO NOGYO 3386
(CENSUS OF AGRICULTURE, IWATE PREFECTURE)
IWATE-KEN SOMU-BU (GENERAL AFFAIRS DIVISION, IWATE PREFECTURAL GOVERNMENT)
10-1, UCHIMARU, MORIOKA-SHI 020
JAPANESE

IWATE-KEN NOGYO DOKO NENPO 3387
(ANNUAL REPORT OF TENDENCY OF AGRICULTURE, IWATE PREFECTURE)
IWATE-KEN NOSEI-BU (AGRICULTURAL ADMINISTRATION DEPARTMENT, IWATE PREFECTURAL GOVERNMENT)
10-1, UCHIMARU, MORIOKA-SHI 020
JAPANESE

IWATE-KENRITSU NOGYO SHIKENJO KENKYU HOKOKU 3388
(BULLETIN OF THE IWATE-KEN AGRICULTURAL EXPERIMENT STATION)
IWATE-KENRITSU NOGYO SHIKENJO (IWATE-KEN AGRICULTURAL EXPERIMENT STATION)
TAKIZAWA-MURA, IWATE-GUN, IWATE 020-01
JAPANESE, WITH EUROPEAN SUMMARY

JAPAN AGRICULTURAL COOP NEWS 3389
ENGLISH
INDEXED: BIB

JARQ; JAPAN AGRICULTURAL RESEARCH QUARTERLY 3390
NORIN-SHO NETTAI NOGYO KENKYU SENTA (TROPICAL AGRICULTURE RESEARCH CENTER)
2-1, 2-CHOME, NISHIGAHARA, KITA-KU, TOKYO 114
ENGLISH
INDEXED: BIB, CHE, EXC

JAS NYUSU 3391
(JAS NEWS)
NIHON NORIN KIKAKU KYOKAI (JAPAN AGRICULTURAL STANDARDS ASSOCIATION)
25, SHIBA NISHIKUBO AKEFUNE-CHO, MINATO-KU, TOKYO 105
JAPANESE

JETRO NOSUISAN GEPPO 3392
(JETRO AGRICULTURAL AND FISHERY MONTHLY)
NIHON BOEKI SHINKOKAI NOSUISANBU (AGRICULTURAL AND FISHERY DIVISION , JAPAN EXTERNAL TRADE ORGANIZATION)
JAPANESE AND ENGLISH

JITSUYOKA GIJUSTU REPOTO 3393
(TECHNICAL REPORT FOR PRACTICAL USE IN AGRICULTURE)
NORIN-SHO SUISAN GIJUTSU KAIGI (AGRICULTURE, FORESTRY AND FISHERIES RESEARCH COUNCIL)
2-1, 1-CHOME, KASUMIGASEKI, CHIYODA-KU, TOKYO 100
JAPANESE

JOURNAL OF THE FACULTY OF AGRICULTURE, HOKKAIDO UNIVERSITY 3394
HOKKAIDO DAIGAKU NOGAKUBU (FACULTY OF AGRICULTURE, HOKKAIDO UNIVERSITY)
NISHI 9-CHOME, KITA 9-JO, CHUO-KU, SAPPORO-SHI 060
ENGLISH
INDEXED: BIB, JICST

JOURNAL OF THE FACULTY OF AGRICULTURE, KYUSHU UNIVERSITY 3395
KYUSHU DAIGAKU NOGAKUBU (FACULTY OF AGRICULTURE, KYUSHU UNIVERSITY)
3575-1, HAKOZAKI, HIGASHI-KU, FUKUOKA-SHI 812
ENGLISH
INDEXED: BIB, BIO, CHE

JOURNAL OF THE FACULTY OF AGRICULTURE, TOTTORI UNIVERSITY 3396
TOTTORI DAIGAKU NOGAKUBU (FACULTY OF AGRICULTURE, TOTTORI UNIVERSITY)
1-1, KOYAMA-CHO, TOTTORI-SHI 680
EUROPEAN
INDEXED: BIB, BIO, CHE, JICST

KAGAWA DAIGAKU NOGAKUBU GAKUJUTSU HOKOKU 3397
(TECHNICAL BULLETIN OF FACULTY OF AGRICULTURE, KAGAWA UNIVERSITY)
KAGAWA DAIGAKU NOGAKUBU (FACULTY OF AGRICULTURE, KAGAWA UNIVERSITY)
MIKI-CHO, KIDA-GUN, KAGAWA 761-07
JAPANESE AND ENGLISH
INDEXED: BIB, BIO, CHE, JICST

KAGAWA DAIGAKU NOGAKUBU KIYO 3398
(MEMOIRS OF FACULTY OF AGRICULTURE, KAGAWA UNIVERSITY)
KAGAWA DAIGAKU NOGAKUBU (FACULTY OF AGRICULTURE, KAGAWA UNIVERSITY)
MIKI-CHO, KIDA-GUN, KAGAWA 761-07
JAPANESE AND ENGLISH
INDEXED: BIB, CHE, JICST

KAGAWA NORIN SUISAN TOKEI NENPO 3399
(STATISTICAL YEARBOOK OF AGRICULTURE, FORESTRY AND FISHERY, KAGAWA)
NORIN-SHO CHUGOKU SHIKOKU NOSEI-KYOKU KAGAWA TOKEI JOHO JIMUSHO (KAGAWA STATISTICS AND INFORMATION OFFICE, CHUGOKU SHIKOKU AGRICULTURAL ADMINISTRATION BUREAU, MINISTRY OF AGRICULTURE AND FORESTRY)
JAPANESE

KAGAWA-KEN NOGYO SHIKENJO GYOMU NENPO 3400
(ANNUAL REPORT OF KAGAWA AGRICULTURAL EXPERIMENT STATION)
KAGAWA-KEN NOGYO SHIKENJO (KAGAWA AGRICULTURAL EXPERIMENT STATION)
220, KO, BUSSHOZAN-CHO, TAKAMATSU-SHI 761-11
JAPANESE

KAGAWA-KEN NOGYO SHIKENJO KENKYU HOKOKU 3401
(BULLETIN OF KAGAWA AGRICULTURAL EXPERIMENT STATION)
KAGAWA-KEN NOGYO SHIKENJO (KAGAWA PREFECTURE AGRICULTURAL EXPERIMENT STATION)
220, KO, BUSSHOZAN-CHO, TAKAMATSU-SHI 761-11
JAPANESE AND ENGLISH
INDEXED: BIO, CHE

KAGOSHIMA DAIGAKU NOGAKUBU GAKUJUTSU HOKOKU 3402
(BULLETIN OF THE FACULTY OF AGRICULTURE, KAGOSHIMA UNIVERSITY)
KAGOSHIMA DAIGAKU NOGAKUBU (FACULTY OF AGRICULTURE, KAGOSHIMA UNIVERSITY)
1946, KAMI ARADA-MACHI, KAGOSHIMA-SHI 890
JAPANESE
INDEXED: BIB, CHE, JICST

KAGOSHIMA DAIGAKU NOGAKUBU NOJO KENKYU HOKOKU 3403
(BULLETIN OF THE EXPERIMENTAL FARM FACULTY OF AGRICULTURE, KAGOSHIMA UNIVERSITY)
JAPANESE

KAGOSHIMA NOGYO NO UGOKI 3404
(CENSUS OF AGRICULTURE, KAGOSHIMA PREFECTURE)
KAGOSHIMA-KEN
14-50, YAMASHITA-CHO, KAGOSHIMA-SHI 892
JAPANESE

KAGOSHIMA-KEN NOGYO SHIKENJO GYOMU HOKOKU 3405
(OPERATION REPORT OF KAGOSHIMA AGRICULTURAL EXPERIMENT STATION)
KAGOSHIMA-KEN NOGYO SHIKENJO (KAGOSHIMA AGRICULTURAL EXPERIMENT STATION)
5500, KAMIFUKUMOTO-CHO, KAGOSHIMA-SHI 891-01
JAPANESE, WITH ENGLISH SUMMARY

KAGOSHIMA-KEN NOGYO SHIKENJO GYOSEKI 3406
(ANNUAL REPORT OF KAGOSHIMA PREFECTURE AGRICULTURAL EXPERIMENT STATION)
KAGOSHIMA-KEN NOGYO SHIKENJO (KAGOSHIMA AGRICULTURAL EXPERIMENT STATION)
5500, KAMIFUKUMOTO-CHO, KAGOSHIMA-SHI 891-01
JAPANESE

KAGOSHIMA-KEN NOGYO SHIKENJO KENKYU HOKOKU 3407
(RESEARCH REPORT OF KAGOSHIMA AGRICULTURAL EXPERIMENT STATION)
KAGOSHIMA-KEN NOGYO SHIKENJO (KAGOSHIMA AGRICULTURAL EXPERIMENT STATION)
5500, KAMIFUKUMOTO-CHO, KAGOSHIMA-SHI 891-01
JAPANESE, WITH ENGLISH SUMMARY
INDEXED: BIB

KAGOSHIMA-KEN NOGYO SHIKENJO NYUSU 3408
(KAGOSHIMA AGRICULTURAL EXPERIMENT STATION NEWS)
KAGOSHIMA-KEN NOGYO SHIKENJO (KAGOSHIMA AGRICULTURAL EXPERIMENT STATION)
5500, KAMIFUKUMOTO-CHO, KAGOSHIMA-SHI 891-01
JAPANESE

KAGOSHIMA-KEN NOGYO SHIKENJO OSHIMA SHIJO GYOMU HOKOKU 3409
(OPERATION REPORT OF OSHIMA BRANCH, KAGOSHIMA AGRICULTURAL EXPERIMENT STATION)
KAGOSHIMA-KEN NOGYO SHIKENJO OSHIMA SHIJO
503, URAKAMI, NASE-SHI 894
JAPANESE

KAIGAI NOGYO KAIHATSU 3410
(OVERSEAS AGRICULTURE DEVELOPMENT)
KAIGAI NOGYO KAIHATSU KYOKAI
(OVERSEAS AGRICULTURE DEVELOPMENT ASSOCIATION)
JAPANESE

KAIGAI NOGYO KYORYOKU JIHO 3411
(INTERNATIONAL AGRICULTURAL COOPERATION TIMES)
NORIN-SHO NORIN KEIZAI-KYOKU (ECONOMIC AFFAIRS BUREAU, MINISTRY OF AGRICULTURE AND FORESTRY)
2-1, 1-CHOME, KASUMIGASEKI, CHIYODA-KU, TOKYO 100
JAPANESE

KAIGAI NOGYO SEMINA 3412
(SEMINAR OF FOREIGN AGRICULTURE)
KAIGAI NOGYO KAIHATSU ZAIDAN
3-32, 8-CHOME, AKASAKA, MINATO-KU, TOKYO 107
JAPANESE

KAIGAI NORINGYO KAIHATSU GIJUTSU JOHO 3413
(REPORT OF DEVELOPMENT TECHNOLOGY OF THE OVERSEAS AGRICULTURE AND FARMING INDUSTRY)
KOKUSAI KYORYOKU JIGYODAN (INTERNATIONAL COOPERATION ASSOCIATION)
JAPANESE

KAKUSHU SHIKEN SEISEKI OYOBI CHOSA KEKKA HOKOKUSHO 3414
(EXPERIMENTAL REPORT OF KYOTO PREFECTURAL INSTITUTE OF AGRICULTURE)
KYOTO-FURITSU NOGYO KENKYUSHO (KYOTO PREFECTURAL INSTITUTE OF AGRICULTURE)
9, WAKUNARI, AMARUBE-CHO, KAMEOKA-SHI 621
JAPANESE

KANAGAWA NORIN SUISAN TOKEI NENPO 3415
(STATISTICAL YEARBOOK OF AGRICULTURE, FORESTRY AND FISHERY, KANAGAWA)
NORIN-SHO KANTO NOSEI-KYOKU KANAGAWA TOKEI JOHO JIMUSHO (KANAGAWA STATISTICS AND INFORMATION OFFICE, KANTO AGRICULTURAL ADMINISTRATION BUREAU, MINISTRY OF AGRICULTURE AND FORESTRY)
5-57, KITA NAKA-DORI, NAKA-KU, YOKOHAMA-SHI 231
JAPANESE

KANAGAWA-KEN NOGYO SHIKEN KENKYU KIKAN KYODO KENKYU HOKOKU 3416
(BULLETIN OF THE KANAGAWA PREFECTURAL AGRICULTURAL CORPORATED EXPERIMENTAL RESEARCH ORGANIZATION)
KANAGAWA-KEN NOGYO SOGO KENKYUSHO (AGRICULTURAL RESEARCH INSTITUTE OF KANAGAWA PREFECTURE)
496, TERADANAWA, HIRATSUKA-SHI 259-12
JAPANESE, WITH ENGLISH SUMMARY

KANAGAWA-KEN NOGYO SOGO KENKYUSHO GYOMU NENPO 3417
(ANNUAL REPORT ON THE OPERATION OF AGRICULTURAL RESEARCH INSTITUTE OF KANAGAWA PREFECTURE)
KANAGAWA-KEN NOGYO SOGO KENKYUSHO (AGRICULTURAL RESEARCH INSTITUTE OF KANAGAWA PREFECTURE)
496, TERADANAWA, HIRATSUKA-SHI 259-12
JAPANESE

KANAGAWA-KEN NOGYO SOGO KENKYUSHO KENKYU HOKOKU 3418
(BULLETIN OF THE AGRICULTURAL RESEARCH INSTITUTE OF KANAGAWA PREFECTURE)
KANAGAWA-KEN NOGYO SOGO KENKYUSHO (AGRICULTURAL RESEARCH INSTITUTE OF KANAGAWA PREFECTURE)
496, TERADANAWA, HIRATSUKA-SHI 259-12
JAPANESE, WITH ENGLISH SUMMARY
INDEXED: BIO

KANAGAWA-KEN NORINGYOGYO DOKO NENPO 3419
(ANNUAL REPORT OF AGRICULTURE, FORESTRY AND FISHERY IN KANAGAWA)
KANAGAWA-KEN NOSEI-BU (AGRICULTURAL ADMINISTRATION DIVISION, KANAGAWA PREFECTURAL GOVERNMENT)
NIHON-ODORI, NAKA-KU, YOKOHAMA-SHI 231
JAPANESE

KANTO NO SAKUMOTSU 3420
(FARM PRODUCTS IN KANTO)
NORIN-SHO KANTO NOSEI-KYOKU (KANTO AGRICULTURAL ADMINISTRATION BUREAU, MINISTRY OF AGRICULTURE AND FORESTRY)
3-3, 1-CHOME, OTE-MACHI, CHIYODA-KU, TOKYO 100
JAPANESE

KANTO NOGYO JOSEI HOKOKU 3421
(ANNUAL REVIEW OF AGRICULTURE, KANTO DISTRICT)
NORIN-SHO KANTO NOSEI-KYOKU (KANTO AGRICULTURAL ADMINISTRATION BUREAU, MINISTRY OF AGRICULTURE AND FORESTRY)
C/O OTE-MACHI GODOCHOSHA 3-GOKAN, 3-3, 1-CHOME, OTE-MACHI, CHIYODA-KU, TOKYO 100
JAPANESE

KANTO NOGYO NO UGOKI 3422
(AGRICULTURE IN KANTO DISTRICT)
NORIN-SHO KANTO NOSEI-KYOKU (KANTO AGRICULTURAL ADMINISTRATION BUREAU, MINISTRY OF AGRICULTURE AND FORESTRY) AND NORIN TOKEI KYOKAI
11-14, 2-CHOME, MEGURO, MEGURO-KU, TOKYO 153
JAPANESE

KANTO NORIN SUISAN RUIKEI NENPO 3423
(TOTAL YEARBOOK OF AGRICULTURE, FORESTRY AND FISHERIES IN KANTO)
NORIN-SHO KANTO NOSEI-KYOKU (KANTO AGRICULTURAL ADMINISTRATION BUREAU, MINISTRY OF AGRICULTURE AND FORESTRY)
3-3, 1-CHOME, OTE-MACHI, CHIYODA-KU, TOKYO 100
JAPANESE

KANTO TOSAN NOGYO SHIKENJO KENKYU HOKOKU 3424
(JOURNAL OF THE KANTO TOSAN AGRICULTURAL EXPERIMENT STATION)
JAPANESE

KEMIKARU NOGYO 3425
(AGRICULTURAL CHEMICALS)
KUMIAI KAGAKU KOGYO K.K. (KUMIAI CHEMICAL INDUSTRY COMPANY, LIMITED)
6-2, 2-CHOME, OTE-MACHI, CHIYODA-KU, TOKYO 100
JAPANESE

KENKYU NO SHIRYO TO KIROKU 3426
(RECORDS OF LAND RECLAMATION RESEARCHES)
TOKYO DAIGAKU NOGAKUBU NOCHI KOGAKU KENKYU-SHITSU (LABORATORY OF LAND RECLAMATION, FACULTY OF AGRICULTURE, UNIVERSITY OF TOKYO)
1-1, 1-CHOME, YAYOI, BUNKYO-KU, TOKYO 113
JAPANESE

KENKYU SEISEKI (NOGYO KIKAIKA KENKYUSHO 3427
(REVIEWS OF INSTITUTE OF AGRICULTURAL MACHINERY)
NOGYO KIKAIKA KENKYUSHO (INSTITUTE OF AGRICULTURAL MACHINERY)
40-2, 1-CHOME, NISSHIN-CHO, OMIYA-SHI 330
JAPANESE

KENKYU SOSHO 3428
(RESEARCH BULLETIN)
NOGYO SOGO KENKYU-SHO
JAPANESE
INDEXED: BIB

KIHO NORIN TOKEI 3429
(AGRICULTURAL STATISTICS QUARTERLY)
ZENKOKU NORIN TOKEI KYOKAI RENGOKAI
20-3, 2-CHOME, SHINKAWA, CHUO-KU, TOKYO 104
JAPANESE

KIKAIKA NOGYO 3430
(FARMING MECHANIZATION)
SHIN NORINSHA
2-7, KANDA NISHIKI-CHO, CHIYODA-KU, TOKYO 101
JAPANESE
INDEXED: BIB, JICST

KIKAKUKEN TSUHO 3431
(BULLETIN OF JAPAN AGRICULTURAL STANDARDS INSPECTION INSTITUTE)
NORIN-SHO NORIN KIKAKU KENSAJO (JAPAN AGRICULTURAL STANDARDS INSPECTION INSTITUTE, MINISTRY OF AGRICULTURE AND FORESTRY)
4-7, 4-CHOME, KONAN, MINATO-KU, TOKYO 108
JAPANESE

KINDAI NOGAKUBU SHUHO (KINKI DAIGAKU) 3432
(REPORTS OF THE AGRICULTURE SCHOOL, KINKI UNIVERSITY)
JAPANESE
INDEXED: JICST

KINKI DAIGAKU NOGAKUBU KIYO 3433
(MEMOIRS OF THE FACULTY OF AGRICULTURE OF KINKI UNIVERSITY)
KINKI DAIGAKU NOGAKUBU (FACULTY OF AGRICULTURE, KINKI UNIVERSITY)
321, KOWAKAE, HIGASHI OSAKA-SHI 577
JAPANESE AND ENGLISH
INDEXED: CHE, JICST

KINKI NO NOSEI 3434
(AGRICULTURE IN KINKI)
NORIN-SHO KINKI NOSEI-KYOKU (KINKI AGRICULTURAL ADMINISTRATION BUREAU, MINISTRY OF AGRICULTURE AND FORESTRY)
102, SHIMOCHOJA-MACHI SAGARU, NISHINOTOIN-DORI, KAMIGYO-KU, KYOTO-SHI 602
JAPANESE

KINKI NOGYO JOSEI HOKOKU 3435
(ANNUAL REVIEW OF AGRICULTURE, KINKI DISTRICT)
NORIN-SHO KINKI NOSEI-KYOKU (KINKI AGRICULTURAL ADMINISTRATION BUREAU, MINISTRY OF AGRICULTURE AND FORESTRY)
102, SHIMOCHOJA-MACHI SAGARU, NISHINOTOIN-DORI, KAMIGYO-KU, KYOTO-SHI 602
JAPANESE

KINKI NOKYO KENKYU 3436
(KINKI AGRICULTURAL COOPERATIVE RESEARCH)
KINKI NOGYO KYODO KUMIAI KENKYUKAI, NOGYO KAIHATSU KENKYU SENTA (KINKI AGRICULTURAL COOPERATIVE RESEARCH GROUP: AGRICULTURAL DEVELOPMENT RESEARCH CENTER)
JAPANESE

KINKI SAKUMOTSU, IKUSHU DANWAKAIHO 3437
(JOURNAL OF KINKI FARM PRODUCTS AND BREEDING RESEARCH ASSOCIATION)
NIHON SAKUMOTSU GAKKAI KINKI SHIBU, NIHON IKUSHU GAKKAI KINKI CHIIKI DANWAKAI
C/O KOBE DAIGAKU NOGAKUBU, 1, ROKKODAI-MACHI, NADA-KU, KOBE-SHI 657
JAPANESE AND ENGLISH

KOBE DAIGAKU NOGAKUBU KENKYU HOKOKU 3438
(SCIENCE REPORTS OF FACULTY OF AGRICULTURE, KOBE UNIVERSITY)
KOBE DAIGAKU NOGAKUBU (FACULTY OF AGRICULTURE, KOBE UNIVERSITY)
1, ROKKODAI-MACHI, NADA-KU, KOBE-SHI 657
JAPANESE AND ENGLISH
INDEXED: BIB, BIO, CHE, JICST

KOBE DAIGAKU NOGYO KEIZAI 3439
(KOBE UNIVERSITY AGRICULTURAL ECONOMICS)
KOBE DAIGAKU NOGAKUBU NOGYO KEIZAI KEIEI KENKYU SHITSU
JAPANESE

KOCHI DAIGAKU GAKUJUTSU KENKYU HOKOKU, NOGAKU 3440
(RESEARCH REPORTS OF THE KOCHI UNIVERSITY, AGRICULTURAL SCIENCE)
KOCHI DAIGAKU (KOCHI UNIVERSITY)
1000, ASAKURA, KOCHI-SHI 780
JAPANESE AND ENGLISH
INDEXED: BIB, CHE, JICST

KOCHI DAIGAKU NOGAKUBU KIYO 3441
(MEMOIRS OF THE FACULTY OF AGRICULTURE, KOCHI UNIVERSITY)
KOCHI DAIGAKU NOGAKUBU (FACULTY OF AGRICULTURE, KOCHI UNIVERSITY)
2200, MONONOBE, NANKOKU-SHI 783
JAPANESE, WITH EUROPEAN SUMMARY
INDEXED: JICST

KOCHI NORIN SUISAN TOKEI NENPO 3442
(STATISTICAL YEARBOOK OF AGRICULTURE, FORESTRY AND FISHERY, KOCHI)
NORIN-SHO CHUGOKU SHIKOKU NOSEI-KYOKU KOCHI TOKEI JOHO JIMUSHO (KOCHI STATISTICS AND INFORMATION OFFICE, CHUGOKU SHIKOKU AGRICULTURAL ADMINISTRATION BUREAU, MINISTRY OF AGRICULTURE AND FORESTRY)
6-6, MASUGATA, KOCHI-SHI 780
JAPANESE

KOCHI OYOBI SAKUTSUKE MENSEKI TOKEI 3443
(STATISTICS OF ACREAGE OF ARABLE LAND AND PLANTED FIELD)
NORIN-SHO NORIN KEIZAI-KYOKU TOKEI JOHO-BU (STATISTICS AND INFORMATION DIVISION, ECONOMIC AFFAIRS BUREAU, MINISTRY OF AGRICULTURE AND FORESTRY)
1-2, KASUMIGASEKI, CHIYODA-KU, TOKYO 100
JAPANESE

KOCHI-KEN NOJI SHIKENJO GYOMU HOKOKU 3444
(ANNUAL REPORT OF KOCHI PREFECTURAL AGRICULTURAL EXPERIMENT STATION)
KOCHI-KEN NOJI SHIKENJO
1110, HATAEDA, NANKOKU-SHI 783
JAPANESE

KOCHI-KEN NOJI SHIKENJO KENKYU HOKOKU 3445
(RESEARCH REPORT OF THE KOCHI AGRICULTURAL EXPERIMENT STATION)
KOCHI-KEN NOJI SHIKENJO (KOCHI AGRICULTURAL EXPERIMENT STATION)
1110, HATAEDA, NANKOKU-SHI 783
JAPANESE

KOCHI-KEN NOJI SHIKENJO SHIKEN KENKYU JISSEKI HOKOKU 3446
(RESEARCH RESULTS OF THE KOCHI AGRICULTURAL EXPERIMENT STATION)
KOCHI-KEN NOJI SHIKENJO (KOCHI AGRICULTURAL EXPERIMENT STATION)
1110, HATAEDA, NANKOKU-SHI 783
JAPANESE

KOCHI-KEN NORIN GIJUTSU KENKYU HOKOKU 3447
(BULLETIN OF THE KOCHI PREFECTURAL INSTITUTE OF AGRICULTURAL AND FOREST SCIENCE)
KOCHI-KEN NORIN GIJUTSU KENKYUSHO (KOCHI PREFECTURAL INSTITUTE OF AGRICULTURAL AND FOREST SCIENCE)
287-4, HAKAWA, INO-MACHI, AGAWA-GUN KOCHI 781-21
JAPANESE AND ENGLISH

KOKUSAI NOGYO KIKAIKA KENKYUKAI NEWS LETTER 3448
(INTERNATIONAL AGRICULTURAL MECHANIZATION RESEARCH ASSOCIATION NEWS LETTER)
JAPANESE

KOKUSAI NOSEI DOKO 3449
(TENDENCY OF INTERNATIONAL AGRICULTURAL ADMINISTRATION)
NORIN-SHO NORIN KEIZAI-KYOKU (ECONOMIC AFFAIRS BUREAU, MINISTRY OF AGRICULTURE AND FORESTRY)
2-1, 1-CHOME, KASUMIGASEKI, CHIYODA-KU, TOKYO 100
JAPANESE

KOSO NO SEKAI 3450
(WORLD OF ENZYME)
KOSO NO SEKAISHA
MIZUKUCHI-CHO KOGA-GUN, SHIGA 528
JAPANESE

KUMAMOTO NORIN SUISAN TOKEI NENPO 3451
(STATISTICAL YEARBOOK OF AGRICULTURE, FORESTRY AND FISHERY, KUMAMOTO)
NORIN-SHO KYUSHU NOSEI-KYOKU TOKEI JOHO-BU (STATISTICS AND INFORMATION DIVISION, KYUSHU AGRICULTURAL ADMINISTRATION BUREAU, MINISTRY OF AGRICULTURE AND FORESTRY)
C/O KUMAMOTO GODOCHOSHA, 1-2, NINOMARU, KUMAMOTO-SHI 860
JAPANESE

KUMAMOTO-KEN AMAKUSA NOGYO KENKYU SHIDOSHO GYOMU HOKOKUSHO 3452
(REPORT OF AMAKUSA AGRICULTURAL RESEARCH TRAINING CENTER OF JUMAMOTO)
KUMAMOTO-KEN AMAKUSA NOGYO KENKYU SHIDOSHO (AMAKUSA AGRICULTURAL RESEARCH TRAINING CENTER OF KUMAMOTO)
642, HONTO BABA, HONDO-CHO, HONDO-SHI 863
JAPANESE

KUMAMOTO-KEN KUMA NOGYO KENKYU SHIDOSHO HOKOKU 3453
(REPORT OF KUMA DISTRICT AGRICULTURAL RESEARCH AND SERVICE STATION, KUMAMOTO PREFECTURE)
KUMAMOTO-KEN KUMA NOGYO KENKYU SHIDOSHO (KUMA DISTRICT AGRICULTURAL RESEARCH AND SERVICE STATION, KUMAMOTO PREFECTURE)
UE-MURA, KUMA-GUN, KUMAMOTO 868-04
JAPANESE

KUMAMOTO-KEN NOGYO SHIKENJO HOKOKU 3454
(REPORT OF KUMAMOTO PREFECTURAL AGRICULTURAL EXPERIMENT STATION)
KUMAMOTO-KEN NOGYO SHIKENJO (KUMAMOTO PREFECTURAL AGRICULTURAL EXPERIMENT STATION)
501, SAIKI, KAMINOGO-MACHI, KUMAMOTO-SHI 861-41
JAPANESE

KUMAMOTO-KEN NOGYO SHIKENJO SEISEKISKO 3455
(RESULT OF KUMAMOTO PREFECTURAL AGRICULTURAL EXPERIMENT STATION)
KUMAMOTO-KEN NOGYO SHIKENJO YASHIRO SHIJO
KAGAMI-CHO, YASHIRO--GUN, KUMAMOTO 869-42
JAPANESE

KYOTO NORIN SUISAN TOKHI NENPO 3456
(STATISTICAL YEARBOOK OF AGRICULTURE, FORESTRY AND FISHERY, KYOTO)
NORIN-SHO KINKI NOSEI-KYOKU TOKEI JOHO-BU (STATISTICS AND INFORMATION DIVISION, KINKI AGRI-CULTURAL ADMINISTRATION BUREAU, MINISTRY OF AGRICULTURE AND FORESTRY)
SHIMOCHOJA-MACHI SAGARU, NISHINOTOIN-DORI, KAMIGYO-KU, TYOTO-SHI 602
JAPANESE

KYOTO-FU NOGYO SHIDOSHO GYOMU NENPO 3457
(ANNUAL REPORT OF AGRICULTURE DEVELOPMENTAL STATION, KYOTO PREFECTURE)
KYOTO-FU NOGYO SHIDOSHO
IDEN-CHO, AYABE-SHI 623-02
JAPANESE

KYOTO-FURITSU DAIGAKU GAKUJUTSU HOKOKU, NOGAKU 3458
(SCIENTIFIC REPORTS OF KYOTO PREFECTURAL UNIVERSITY AGRICULTURE)
KYOTO-FURITSU DAIGAKU GAKUJUTSU HOKOKU HENSHU IINKAI
KYOTO-FURITSU DAIGAKU (KYOTO PREFECTURAL UNIVERSITY)
SHIMOGAMO HANMOKU-CHO, SAKYO-KU, KYOTO-SHI 606
JAPANESE AND ENGLISH
INDEXED: BIB, CHE, JICST

KYOTO-FURITSU NOGYO KENKYUSHO GYOMU NENPO 3459
(REPORT OF THE KYOTO PREFECTURAL INSTITUTE OF AGRICULTURE)
KYOTO-FURITSU NOGYO KENKYUSHO (KYOTO PREFECTURAL INSTITUTE OF AGRICULTURE)
9, WAKUNARI, AMARUBE-CHO, KAMEOKA-SHU 621
JAPANESE

KYOTO-FURITSU NOGYO KENKYUSHO KENKYU HOKOKU 3460
(BULLETIN OF THE KYOTO PREFECTURAL INSTITUTE OF AGRICULTURE)
KYOTO-FURITSU NOGYO KENKYUSHO (KYOTO PREFECTURAL INSTITUTE OF AGRICULTURE)
9, WAKUNARI, AMARUBE-CHO, KAMEOKA-SHI 621
JAPANESE

KYOTO-FURITSU NOGYO KENKYUSHO NENPO 3461
(ANNUAL REPORT OF THE KYOTO PREFECTURAL INSTITUTE OF AGRICULTURE)
KYOTO-FURITSU NOGYO KENKYUSHO (KYOTO PREFECTURAL INSTITUTE OF AGRICULTURE)
9, WAKUNARI, AMARUBE-CHO, KAMEOKA-SHI 621
JAPANESE
INDEXED: BIB

KYUSHU DAIGAKU NOGAKU-BU NOJO HOKOKU 3462
(BULLETIN OF THE KYUSHU UNIVERSITY FARM)
KYUSHU DAIGAKU NOGAKU-BU (KYUSHU UNIVERSITY, DEPARTMENT OF AGRICULTURE)
JAPANESE

KYUSHU DAIGAKU NOGAKUBU GAKUGEI ZASSHI 3463
(SCIENCE BULLETIN OF THE FACULTY OF AGRICULTURE, KYUSHU UNIVERSITY)
KYUSHU DAIGAKU NOGAKUBU (FACULTY OF AGRICULTURE, KYUSHU UNIVERSITY)
3575-1, HAKOZAKI-CHO, HIGASHI-KU, FUKUOKA-SHI 812
JAPANESE, WITH ENGLISH SUMMARY
INDEXED: CHE, JICST

KYUSHU NOGYO JOSEI HOKOKU 3464
(STATE OF AGRICULTURE IN THE KYUSHU REGION)
NORIN-SHO KYUSHU NOSEI-KYOKU KIKAKU CHOSEISHITSU (KYUSHU AGRICULTURAL ADMINISTRATION BUREAU, MINISTRY OF AGRICULTURE AND FORESTRY)
1-2, NINOMARU-CHO, KUMAMOTO-SHI 860
JAPANESE

KYUSHU NOGYO KENKYU 3465
(KYUSHU AGRICULTURAL RESEARCH)
KYUSHU NOGYO SHIKEN KENKYU KIKAN KYOGIKAI (ASSOCI-ATION OF THE KYUSHU AGRICULTURAL RESEARCH INSTITUTE)
IZUMI,CHIKUGO-SHI 833
JAPANESE

KYUSHU NOGYO SHIKENJO HOKOKU 3466
(BULLETIN OF THE KYUSHU AGRICULTURAL EXPERIMENT STATION)
NORIN-SHO KYUSHU NOGYO SHIKENJO (KYUSHU NATIONAL AGRICULTURAL EXPERIMENT STATION, MINISTRY OF AGRICULTURE AND FORESTRY)
IZUMI, CHIKUGO-SHI 833
JAPANESE AND EUROPEAN
INDEXED: BIB, BIO, CHE

KYUSHU NOGYO SHIKENJO KENKYU SHIRYO 3467
(MEMOIRS OF KYUSHU NATIONAL AGRICULTURAL EXPERIMENT STATION)
NORIN-SHO KYUSHU NOGYO SHIKENENJO (KYUSHU NATIONAL AGRICULTURAL EXPERIMENT STATION, MINISTRY OF AGRICULTURE AND FORESTRY)
IZUMI, CHIKUGO-SHI 833
JAPANESE AND ENGLISH

KYUSHU NOGYO SHIKENJO NYUSU 3468
(NEWS OF KYUSHU NATIONAL AGRICULTURAL EXPERIMENT STATION)
NORIN-SHO KYUSHU NOGYO SHIKENJO (KYUSHU NATIONAL AGRICULTURAL EXPERIMENT STATION, MINISTRY OF AGRICULTURE AND FORESTRY)
IZUMI, CHIKUGO-SHI 833
JAPANESE

KYUSHU NOSHI NENPO 3469
(ANNUAL REPORT OF THE KYUSHU AGRICULTURAL EXPERIMENT STATION)
NORIN-SHO KYUSHU NOGYO SHIKENJO (KYUSHU NATIONAL AGRICULTURAL EXPERIMENT STATION, MINISTRY OF AGRICULTURE AND FORESTRY)
IZUMI, CHIKUGO-SHI 833
JAPANESE
INDEXED: BIB

MEIJI DAIGAKU NOGAKUBU FUZOKU NOJO HOKOKU 3470
JAPANESE

MEIJI DAIGAKU NOGAKUBU KENKYU HOKOKU 3471
(BULLETIN OF THE FACULTY OF AGRICULTURE, MEIJI UNIVERSITY)
MEIJI DAIGAKU NOGAKUBU (FACULTY OF AGRICULTURE, MEIJI UNIVERSITY)
IKUTA, TAMA-KU, KAWASAKI-SHI 214
JAPANESE AND ENGLISH
INDEXED: BIB, BIO, CHE, JICST

MEIJO DAIGAKU NOGAKUBU GAKUJUTSU HOKOKU 3472
(SCIENTIFIC REPORTS OF THE FACULTY OF AGRICULTURE, MEIJO UNIVERSITY)
MEIJO DAIGAKU NOGAKUBU (FACULTY OF AGRICULTURE, MEIJO UNIVERSITY)
69-49, URAYAMA, YAGOTO, TENPAKU-CHO, SHOWA-KU, NAGOYA-SHI 468
JAPANESE AND EUROPEAN
INDEXED: BIB

MEMOIRS OF THE COLLEGE OF AGRICULTURE, KYOTO UNIVERSITY, AGRICULTURAL ECONOMY SERIES 3473
KYOTO DAIGAKU NOGAKUBU (COLLEGE OF AGRICULTURE, KYOTO UNIVERSITY)
KITASHIRAKAWA OIWAKE-CHO, SAKYO-KU, KYOTO-SHI 606
ENGLISH
INDEXED: BIB, CHE, JICST

MEMOIRS OF THE COLLEGE OF AGRICULTURE, KYOTO UNIVERSITY, ANIMAL SCIENCE SERIES 3474
KYOTO DAIGAKU NOGAKUBU (COLLEGE OF AGRICULTURE, KYOTO UNIVERSITY)
KITASHIRAKAWA OIWAKE-CHO, SAKYO-KU, KYOTO-SHI 606
ENGLISH
INDEXED: BIB, CHE, JICST

MEMOIRS OF THE COLLEGE OF AGRICULTURE, KYOTO UNIVERSITY, BOTANICAL SERIES 3475
KYOTO DAIGAKU NOGAKUBU (COLLEGE OF AGRICULTURE, KYOTO UNIVERSITY)
KITASHIRAKAWA OIWAKE-CHO, SAKYO-KU, KYOTO-SHI 606
ENGLISH
INDEXED: BIB, CHE, JICST

MEMOIRS OF THE COLLEGE OF AGRICULTURE, KYOTO UNIVERSITY, CHEMICAL SERIES 3476
KYOTO DAIGAKU NOGAKUBU (COLLEGE OF AGRICULTURE, KYOTO UNIVERSITY)
KITASHIRAKAWA OIWAKE-CHO, SAKYO-KU, KYOTO-SHI 606
ENGLISH
INDEXED: BIB, CHE, JICST

MEMOIRS OF THE COLLEGE OF AGRICULTURE, KYOTO UNIVERSITY, CROP SCIENCE SERIES 3477
KYOTO DAIGAKU NOGAKUBU (COLLEGE OF AGRICULTURE, KYOTO UNIVERSITY)
KITASHIRAKAWA OIWAKE-CHO, SAKYO-KU, KYOTO-SHI 606
ENGLISH
INDEXED: BIB, CHE, JICST

MEMOIRS OF THE COLLEGE OF AGRICULTURE, KYOTO UNIVERSITY, ENTOMOLOGICAL SERIES 3478
KYOTO DAIGAKU NOGAKUBU (COLLEGE OF AGRICULTURE, KYOTO UNIVERSITY)
KITASHIRAKAWA OIWAKE-CHO, SAKYO-KU, KYOTO-SHI 606
ENGLISH
INDEXED: BIB, CHE, JICST

MEMOIRS OF THE COLLEGE OF AGRICULTURE, KYOTO UNIVERSITY, FISHERIES SERIES 3479
KYOTO DAIGAKU NOGAKUBU (COLLEGE OF AGRICULTURE, KYOTO UNIVERSITY)
KITASHIRAKAWA OIWAKE-CHO, SAKYO-KU, KYOTO-SHI 606
ENGLISH
INDEXED: BIB, CHE, JICST

MEMOIRS OF THE COLLEGE OF AGRICULTURE, KYOTO UNIVERSITY, FOOD SCIENCE AND TECHNOLOGY SERIES 3480
KYOTO DAIGAKU NOGAKUBU (COLLEGE OF AGRICULTURE, KYOTO UNIVERSITY)
KITASHIRAKAWA OIWAKE-CHO, SAKYO-KU, KYOTO-SHI 606
ENGLISH
INDEXED: BIB, CHE, JICST

MEMOIRS OF THE COLLEGE OF AGRICULTURE, KYOTO UNIVERSITY, GENETICAL SERIES 3481
KYOTO DAIGAKU NOGAKUBU (COLLEGE OF AGRICULTURE, KYOTO UNIVERSITY)
KITASHIRAKAWA OIWAKE-CHO, SAKYO-KU, KYOTO-SHI 606
ENGLISH
INDEXED: BIB, CHE, JICST

MEMOIRS OF THE COLLEGE OF AGRICULTURE, KYOTO UNIVERSITY, HORTICULTURAL SERIES 3482
KYOTO DAIGAKU NOGAKUBU (COLLEGE OF AGRICULTURE, KYOTO UNIVERSITY)
KITASHIRAKAWA OIWAKE-CHO, SAKYO-KU, KYOTO-SHI 606
ENGLISH
INDEXED: BIB, CHE, JICST

MEMOIRS OF THE COLLEGE OF AGRICULTURE, KYOTO UNIVERSITY, PHYTOPATHOLOGICAL SERIES 3483
KYOTO DAIGAKU NOGAKUBU (COLLEGE OF AGRICULTURE, KYOTO UNIVERSITY)
KITASHIRAKAWA OIWAKE-CHO, SAKYO-KU, KYOTO-SHI 606
ENGLISH
INDEXED: BIB, CHE, JICST

MEMOIRS OF THE COLLEGE OF AGRICULTURE, KYOTO UNIVERSITY, PLANT BREEDING SERIES 3484
KYOTO DAIGAKU NOGAKUBU (COLLEGE OF AGRICULTURE, KYOTO UNIVERSITY)
KITASHIRAKAWA OIWAKE-CHO, SAKYO-KU, KYOTO SHI 606
ENGLISH
INDEXED: BIB, CHE, JICST

MEMOIRS OF THE COLLEGE OF AGRICULTURE, KYOTO UNIVERSITY, WOOD SCIENCE AND TECHNOLOGY SERIES 3485
KYOTO DAIGAKU NOGAKUBU (COLLEGE OF AGRICULTURE, KYOTO UNIVERSITY)
KITASHIRAKAWA OIWAKE-CHO, SAKYO-KU, KYOTO-SHI 606
ENGLISH
INDEXED: BIB, CHE, JICST

MEMOIRS OF THE FACULTY OF AGRICULTURE, KAGOSHIMA UNIVERSITY 3486
KAGOSHIMA DAIGAKU NOGAKUBU (FACULTY OF AGRICULTURE KAGOSHIMA UNIVERSITY)
1946, KAMIARATA-MACHI, KAGOSHIMA-SHI 892
ENGLISH
INDEXED: BIB, CHE, JICST

MEMOIRS OF THE FACULTY OF AGRICULTURE, NIIGATA UNIVERSITY 3487
NIIGATA DAIGAKU NOGAKUBU (FACULTY OF AGRICULTURE NIIGATA UNIVERSITY)
106, KOGANE-CHO, NIIGATA-SHI 950
JAPANESE AND ENGLISH
INDEXED: BIB, CHE, JICST

MEMOIRS OF THE FACULTY OF AGRICULTURE, UNIVERSITY OF MIYAZAKI 3488
JAPANESE
INDEXED: JICST

MEMOIRS OF THE TOKYO UNIVERSITY OF AGRICULTURE 3489
TOKYO NOGYO DAIGAKU GAKUJUTSU ZASSHI HENSHU IINKAI (EDITORIAL BOARD FOR ACADEMIC JOURNAL, THE TOKYO UNIVERSITY OF AGRICULTURE) AND TOKYO NOGYO DAIGAKU (TOKYO UNIVERSITY OF AGRICULTURE)
1-1, 1-CHOME, SAKURAGAOKA, SETAGAYA-KU, TOKYO 156
ENGLISH
INDEXED: BIO, CHE, JICST

MICHURIN NOGYO 3490
(MICHURIN AGRICULTURE)
NIHON MICHURINKIA (JAPAN MICHURIN SOCIETY)
5-5, 2-CHOME, YOYOGI, SHIBUYA-KU, TOKYO 151
JAPANESE

MIE DAIGAKU NOGAKUBU GAKUJUTSU HOKOKU 3491
(BULLETIN OF THE FACULTY OF AGRICULTURE, MIE UNIVERSITY)
MIE DAIGAKU NOGAKUBU (FACULTY OF AGRICULTURE, MIE UNIVERSITY)
UEHAMA-CHO, TSU-SHI 514
JAPANESE AND ENGLISH
INDEXED: BIB, CHE, JICST

MIE NORIN SUISAN TOKEI NENPO 3492
(STATISTICAL YEARBOOK OF AGRICULTURE, FORESTRY AND FISHERY, MIE)
NORIN-SHO TOKAI NOSEI-KYOKU MIE TOKEI JOHO JIMUSHO (MIE STATISTICS AND INFORMATION OFFICE, TOKAI AGRICULTURAL ADMINISTRATION BUREAU, MINISTRY OF AGRICULTURE AND FORESTRY)
26-8, MARUNOUCHI, TSU-SHI 514
JAPANESE

MIE-KEN NOGYO GIJUTSU SENTA GYOMU NENPO 3493
(OPERATION YEARBOOK OF MIE AGRICULTURAL TECHNICAL CENTER)
MIE-KEN NOGYO GIJUTSU SENTA (MIE AGRICULTURAL TECHNICAL CENTER)
530, KAWAKITA, URESHINO-CHO, ISSHIGUN, MIE 515-22
JAPANESE

MIE-KEN NOGYO GIJUTSU SENTA KENKYU HOKOKU 3494
(BULLETIN OF THE MIE AGRICULTURAL TECHNICAL CENTER)
MIE-KEN NOGYO GIJUTSU SENTA (MIE AGRICULTURAL TECHNICAL CENTER)
530, KAWAKITA, URESHIMO-CHO, ISSHIGUN, MIE 515-22
JAPANESE
INDEXED: JICST

MIE-KEN NORIN GYOGYO NO UGOKI 3495
(TENDENCY OF AGRICULTURE, FORESTRY AND FISHERY, MIE PREFECTURE)
MIE-KEN (MIE PREFECTURAL GOVERNMENT)
13, KOMEI-CHO, TSU-SHI 514
JAPANESE

MISCELLANEOUS INFORMATION, THE TOKYO UNIVERSITY FORESTS 3496
TOKYO UNIVERSITY EXPERIMENTAL FOREST
JAPANESE AND ENGLISH
INDEXED: BIB

MIYAGI NORIN SUISAN TOKEI NENPO 3497
(STATISTICAL YEARBOOK OF AGRICULTURE, FORESTRY AND FISHERY, MIYAGI)
NORIN-SHO TOHOKU NOSEI-KYOKU TOKEI JOHO-BU (STATISTICS AND INFORMATION DIVISION, TOHOKU AGRICULTURAL ADMINISTRATION BUREAU, MINISTRY OF AGRICULTURE AND FORESTRY)
3-1, 3-CHOME, HON-CHO, SENDAI-SHI 980
JAPANESE

MIYAGI-KEN NOGYO SENTA GYOMU NENPO 3498
(ANNUAL REPORT OF THE MIYAGI PREFECTURAL AGRICULTURAL CENTER)
MIYAGI-KEN NOGYO SENTA (MIYAGI PREFECTURAL AGRICULTURAL RESEARCH CENTER)
TAKADATE, NATORI-SHI 981-12
JAPANESE

MIYAGI-KEN NOGYO SENTA KENKYU HOKOKU 3499
(BULLETIN OF THE MIYAGI PREFECTURAL AGRICULTURAL RESEARCH CENTER)
MIYAGI-KEN NOGYO SENTA (MIYAGI PREFECTURAL AGRICULTURAL RESEARCH CENTER)
TAKADATE, NATORI-SHI 981-12
JAPANESE, WITH EUROPEAN SUMMARY

MIYAGI-KEN NOGYO TANKI DAIGAKU GAKUJUTSU HOKOKU 3500
(SCIENTIFIC REPORTS OF THE MIYAGI AGRICULTURAL COLLEGE)
MIYAGI-KEN NOGYO TANKI DAIGAKU TOSHOKAN UN'EI IINKAI AND MIYAGI-KEN NOGYO TANKI DAIGAKU (MIYAGI AGRICULTURAL COLLEGE)
1-16, HATATATE, YAMADA, SENDAI-SHI 982
JAPANESE, WITH ENGLISH SUMMARY
INDEXED: CHE, JICST

MIYAGI-KEN NORIN SUISANGYO CHOSA NOGYO KIHON CHOSA KEKKA HOKOKUSHO 3501
(REPORT OF AGRICULTURAL, FORESTRY AND FISHERY SURVEY, MIYAGI PREFECTURE)
MIYAGI-KEN KIKAKU-BU (PLANNING DIVISION, MIYAGI PREFECTURAL GOVERNMENT)
8-1, 3-CHOME, HON-CHO, SENDAI-SHI 980
JAPANESE

MIYAZAKI DAIGAKU NOGAKUBU KENKYU HOKOKU 3502
(BULLETIN OF THE FACULTY OF AGRICULTURE, UNIVERSITY OF MIYAZAKI)
MIYAZAKI DAIGAKU NOGAKUBU (FACULTY OF AGRICULTURE, UNIVERSITY OF MIYAZAKI)
100, FUNAZUKA-CHO, MIYAZAKI-SHI 880
JAPANESE AND ENGLISH
INDEXED: BIB, BIO, CHE, JICST

MIYAZAKI NORIN SUISAN TOKEI NENPO 3503
(STATISTICAL YEARBOOK OF MIYAZAKI AGRICULTURE, FORESTRY AND FISHERY)
NORIN-SHO KYUSHU NOSEI-KYOKU MIYAZAKI TOKEI JOHO JIMUSHO (MIYAZAKI STATISTICS AND INFORMATION OFFICE, KYUSHU AGRICULTURAL ADMINISTRATION BUREAU, MINISTRY OF AGRICULTURE AND FORESTRY)
JAPANESE

MIYAZAKI-KEN NO NORINGYO 3504
(AGRICULTURE AND FORESTRY IN MIYAZAKI PREFECTURE)
MIYAZAKI-KEN (MIYAZAKI PREFECTURAL GOVERNMENT)
10-1, 2-CHOME, TACHIBANA-DORI HIGASHI, MIYAZAKI-SHI 880
JAPANESE

MIYAZAKI-KEN SOGO NOGYO SHIKENJO GYOMU NENPO 3505
(ANNUAL REPORT OF MIYAZAKI PREFECTURE AGRICULTURAL EXPERIMENT STATION)
MIYAZAKI-KEN SOGO NOGYO SHIKENJO (MIYAZAKI PREFECTURE AGRICULTURAL EXPERIMENT STATION)
SHIMONAKA, SADOHARA-CHO, MIYAZAKI-GUN, MIYAZAKI 880-02
JAPANESE

MIYAZAKI-KEN SOGO NOGYO SHIKENJO KENKYU HOKOKU 3506
(BULLETIN OF THE MIYAZAKI AGRICULTURAL EXPERIMENT STATION)
MIYAZAKI-KEN SOGO NOGYO SHIKENJO (MIYAZAKI PREFECTURE AGRICULTURAL EXPERIMENT STATION)
SHIMONAKA, SADOHARA-CHO, MIYAZAKI-GUN, MIYAZAKI 880-02
JAPANESE, WITH ENGLISH SUMMARY

MIZU TO TSUCHI 3507
(IRRIGATION, DRAINAGE AND RECLAMATION)
NOGYO DOBOKU GIJUTSU KENKYUKAI (RESEARCH ASSOCIATION OF IRRIGATION, DRAINAGE AND RECLAMATION ENGINEERING)
C/O NOGYO DOBOKU KAIKAN, 34-4, 5-CHOME, SHINBASHI, MINATO-KU, TOKYO 105
JAPANESE
INDEXED: JICST

MONBU-SHO KAGAKU KENKYUHI NI YORU KENKYU HOKOKU SHUROKU, NOGAKU-HEN 1 3508
(REPORTS ON RESEARCHES BY GRANTEES MINISTRY OF EDUCATION, AGRICULTURE, PART 1)
NIHON GAKUJUTSU SHINKOKAI (JAPAN SOCIETY FOR THE PROMOTION OF SCIENCE)
3-1, 5-CHOME, KOJI-MACHI, CHIYODA-KU, TOKYO 102
JAPANESE

MONBU-SHO KAGAKU KENKYUHI NI YORU KENKYU HOKOKU SHUROKU, NOGAKU-HEN 2 3509
(REPORTS ON RESEARCHES BY GRANTEES MINISTRY OF EDUCATION, AGRICULTURE, PART 2)
NIHON GAKUJUTSU SHINKOKAI (JAPAN SOCIETY FOR THE PROMOTION OF SCIENCE)
3-1, 5-CHOME, KOJI-MACHI, CHIYODA-KU, TOKYO 102
JAPANESE

MONBU-SHO KAGAKU KENKYUHI NI YORU KENKYU HOKOKU SHUROKU, NOGAKU-HEN 3 3510
(REPORTS ON RESEARCHES BY GRANTEES MINISTRY OF EDUCATION, AGRICULTURE, PART 3)
NIHON GAKUJUTSU SHINKOKAI (JAPAN SOCIETY FOR THE PROMOTION OF SCIENCE)
3-1, 5-CHOME, KOJI-MACHI, CHIYODA-KU, TOKYO 102
JAPANESE

MOUNTAINS LAND USE RESEARCH LABORATORY, BULLETIN 3511
IWATE DAIGAKU NOGAKUBU FUZOKU SANCHI RIYO KENKYU SHISETSU (MOUNTAINS LAND USE RESEARCH LABORATORY, IWATE UNIVERSITY)
18-8, 3-CHOME, UEDA, MORIOKA-SHI 020
JAPANESE AND ENGLISH

NAGANO NORIN SUISAN TOKEI NENPO 3512
(STATISTICAL YEARBOOK OF AGRICULTURE, FORESTRY AND FISHERY, NAGANO)
NORIN-SHO KANTO NOSEI-KYOKU NAGANO TOKEI JOHO JIMUSHO (NAGANO STATISTICS AND INFORMATION OFFICE, KANTO AGRICULTURAL ADMINISTRATION BUREAU, MINISTRY OF AGRICULTURE AND FORESTRY)
C/O NAGANO GODOCHOSHA, 1108, ASAHIMACHI, NAGANO-SHI 380
JAPANESE

NAGANO-KEN NO NORINGYO 3513
(AGRICULTURE AND FORESTRY IN NAGANO PREFECTURE)
NAGANO-KEN (NAGANO PREFECTURAL GOVERNMENT)
692, HABASHITA, MINAMINAGANO, NAGANO-SHI 380
JAPANESE

NAGANO-KEN NOGYO SHIKENJO HOKOKU 3514
(BULLETIN OF THE NAGANO PREFECTURAL AGRICULTURAL EXPERIMENT STATION)
NAGANO-KEN NOGYO SHIKENJO (THE NAGANO PREFECTURAL AGRICULTURAL EXPERIMENT STATION
363, NAKAGOSHO, NAGANO-SHI 380
JAPANESE

NAGANO-KEN NORINGYO SHICHOSON-BETSU TOKEISHO 3515
(STATISTICS OF AGRICULTURE AND FORESTRY BY TOWNS, NAGANO PREFECTURE)
NORIN-SHO KANTO NOSEI-KYOKU (KANTO AGRICULTURAL ADMINISTRATION BUREAU, MINISTRY OF AGRICULTURE AND FORESTRY)
3-3, 1-CHOME, OTE-MACHI, CHIYODA-KU, TOKYO 100
JAPANESE

NAGASAKI NORIN SUISAN TOKEI NENPO 3516
(STATISTICAL YEARBOOK OF AGRICULTURE, FORESTRY AND FISHERY, NAGASAKI)
NORIN-SHO KYUSHU NOSEI-KYOKU NAGASAKI TOKEI JOHO JIMUSHO (NAGASAKI STATISTICS AND INFORMATION OFFICE, KYUSHU AGRICULTURAL ADMINISTRATION BUREAU, MINISTRY OF AGRICULTURE AND FORESTRY)
16-16, IWAKAWA-CHO, NAGASAKI-SHI 852
JAPANESE

NAGASAKI-KEN SOGO NORIN SHIKENJO GYOMU HOKOKU 3517
(ANNUAL REPORT OF THE NAGASAKI AGRICULTURAL AND FORESTRY EXPERIMENT STATION)
NAGASAKI-KEN SOGO NORIN SHIKENJO (NAGASAKI AGRICULTURAL AND FORESTRY EXPERIMENT STATION)
3118, KAIZU-CHO, ISAHAYA-SHI 854
JAPANESE

NAGASAKI-KEN SOGO NORIN SHIKENJO IHO 3518
(BULLETIN OF THE NAGASAKI AGRICULTURAL AND FORESTRY EXPERIMENT STATION)
NAGASAKI-KEN SOGO NORIN SHIKENJO (NAGASAKI AGRICULTURAL AND FORESTRY EXPERIMENT STATION)
3118, KAIZU-CHO, ISAHAYA-SHI 854
JAPANESE

NAGASAKI-KEN SOGO NORIN SHIKENJO KENKYU HOKOKU 3519
(REPORT OF THE NAGASAKI AGRICULTURAL AND FORESTRY EXPERIMENT STATION)
NAGASAKI-KEN SOGO NORIN SHIKENJO (NAGASAKI AGRI-CULTURAL AND FORESTRY EXPERIMENT STATION)
3118, KAIZU-CHO, ISAHAYA-SHI 854
JAPANESE AND ENGLISH

NAGOYA HISHIRYO KENSAJO JIGYO HOKOKU HIRYO NO BU 3520
(ANNUAL REPORT OF THE NAGOYA FERTILIZER AND FEED INSPECTION STATION, FERTILIZER SECTION)
NORIN-SHO NAGOYA HISHIRYO KENSAJO (NAGOYA FERTILIZER AND FEED INSPECTION STATION, MINISTRY OF AGRICULTURE AND FORESTRY)
C/O NORIN SOGOCHOSHA, 2-2, 1-CHOME, SAN'NOMARU, NAKA-KU, NAGOYA-SHI 450
JAPANESE

NARA NORIN SUISAN TOKEI NENPO 3521
(STATISTICAL YEARBOOK OF AGRICULTURE, FORESTRY AND FISHERY, NARA)
NORIN-SHO KINKI NOSEI-KYOKU NARA TOKEI JOHO JIMUSHO (NARA STATISTICS AND INFORMATION OFFICE, KINKI AGRICULTURAL ADMINISTRATION BUREAU, MINISTRY OF AGRICULTURE AND FORESTRY)
832, KIDERA-CHO, NARA-SHI 630
JAPANESE

NARA-KEN NOGYO SHIKENJO KENKYU HOKOKU 3522
(BULLETIN OF THE NARA AGRICULTURAL EXPERIMENT STATION)
NARA-KEN NOGYO SHIKENJO (NARA PREFECTURAL AGRICULTURAL EXPERIMENT STATION)
SHIJO, KASHIHARA-SHI 634
JAPANESE AND ENGLISH

NETTAI NOGYO 3523
(JAPANESE JOURNAL OF TROPICAL AGRICULTURE)
NETTAI NOGYO GAKKAI (TROPICAL AGRICULTURE RESEARCH ASSOCIATION, JAPAN)
C/O TOKYO KYOIKU DAIGAKU NOGAKUBU, 19-1, 2-CHOME, KOMABA, MEGURO-KU, TOKYO 153
JAPANESE AND ENGLISH
INDEXED: BIB, JICST

NETTAI NOKEN SHUHO 3524
(COLLECTED PAPERS OF TROPICAL AGRICULTURE RESEARCH CENTER)
NORIN-SHO NETTAI NOGYO KENKYU SENTA (TROPICAL AGRICULTURE RESEARCH CENTER, MINISTRY OF AGRICULTURE AND FORESTRY)
2-1, 2-CHOME, NISHIGAHARA, KITA-KU, TOKYO 114
JAPANESE

NETTAI RINGYO 3525
(TROPICAL FORESTRY)
NETTAI RINGYO KYOKAI (TROPICAL FORESTRY ASSOCIATION)
JAPANESE

NIHON DOJO HIRYO GAKKAI KOEN YOSHISHU 3526
(SUMMARY OF THE LECTURES OF THE JAPAN SOIL FERTILIZER SOCIETY)
NIHON DOJO HIRYO GAKKAI (JAPAN SOIL FERTILIZER SOCIETY)
JAPANESE

NIHON HIRYO SHINBUN 3527
(JAPAN FERTILIZER NEWS)
HIRYO KYOKAI SHINBUN-BU (JAPAN FERTILIZER ASSOCIATION)
2-3, KYOBASHI, CHUO-KU, TOKYO 104
JAPANESE

NIHON NO NOGYO (ASU E NO AYUMI) 3528
(BULLETIN OF JAPANESE AGRICULTURE)
NOSEI CHOSA IINKAI (AGRICULTURAL POLICY RESEARCH COMMITTEE)
26-3, 1-CHOME, NISHIGAHARA, KITA-KU, TOKYO 115
JAPANESE
INDEXED: BIB

NIHON NOGAKU BUNKEN KIJI SAKUIN 3529
(JAPANESE AGRICULTURAL SCIENCES INDEX)
NIHON NOGAKU TOSHOKAN KYOGIKAI (JAPAN ASSOCIATION OF AGRICULTURAL LIBRARIES AND DOCUMENTALISTS)
AND NORIN TOKEI KYOKAI
11-14, 2-CHOME, MEGURO, MEGURO-KU, TOKYO 153
JAPANESE
INDEXED: BIB

NIHON NOGAKU SHINPO NENPO 3530
(ANNUAL REPORT OF THE DEVELOPMENT OF AGRICULTURE IN JAPAN)
NIHON GAKUJUTSU KAIGI, NIHON NOGAKKAI (SCIENCE COUNCIL OF JAPAN, JAPANESE ASSOCIATION OF AGRICULTURAL SCIENCE SOCIETY)
22-34, 7-CHOME, ROPPONGI, MINATO-KU, TOKYO 106
JAPANESE

NIHON NOGEI KAGAKKAI KANTO SHIBU TAIKAI KOEN YOSHI 3531
(ABSTRACTS OF THE KANTO BRANCH CONGRESS OF THE AGRICULTURAL CHEMICAL SOCIETY OF JAPAN)
NIHON NOGEI KAGAKKAI KANTO SHIBU (AGRICULTURAL CHEMICAL SOCIETY OF JAPAN)
C/O NIIGATA DAIGAKU NOGAKUBU, NOGEI KAGAKKA, 106, KOGANE-CHO, NIIGATA-SHI 950
JAPANESE

NIHON NOGEI KAGAKKAI TAIKAI KOEN YOSHISHU 3532
(ABSTRACTS OF SYMPOSIUM OF AGRICULTURAL CHEMICAL SOCIETY OF JAPAN)
NIHON NOGEI KAGAKKAI (AGRICULTURAL CHEMICAL SOCIETY OF JAPAN)
C/O GAKKAI SENTA BUILDING, 4-16, 2-CHOME, YAYOI, BUNKYO-KU, TOKYO 113
JAPANESE

NIHON NOGEI KAGAKKAISHI 3533
(JOURNAL OF THE AGRICULTURAL CHEMICAL SOCIETY OF JAPAN)
NIPPON NOGEI KAGAKKAI (AGRICULTURAL CHEMICAL SOCIETY OF JAPAN)
C/O GAKKAI SENTA BUILDING, 4-16, 2-CHOME, YAYOI, BUNKYO-KU, TOKYO 113
JAPANESE, WITH ENGLISH SUMMARY
INDEXED: BIB, CHE, JICST

NIHON NOGYO GIJUTSU KONDANKAI NENPO 3534
(ANNUAL BULLETIN OF JAPAN AGRICULTURAL TECHNOLOGICAL SOCIETY)
NIHON NOGYO GIJUTSU KONDANKAI
KANESHO K.K. (KANESHO COMPANY, LIMITED)
4-1, 2-CHOME, MARUNOUCHI, CHIYODA-KU, TOKYO 100
JAPANESE

NIHON NOGYO KENKYUSHO NENPO 3535
(ANNUAL REPORT OF JAPAN AGRICULTURAL RESEARCH INSTITUTE)
NIHON NOGYO KENKYUSHO
3, KIOI-CHO, CHIYODA-KU, TOKYO 102
JAPANESE
INDEXED: BIB, CHE, JICST

NIHON NOGYO KYOIKU GAKKAISHI 3536
(JOURNAL OF JAPAN AGRICULTURAL EDUCATION SOCIETY)
NIHON NOGYO KYOIKU GAKKAI
C/O TOKYO NOGYO DAIGAKU NOSON SHAKAI KENKYUSHITSU, 1-1, 1-CHOME, SAKURAGAOKA, SETAGAYA-KU, TOKYO 156
JAPANESE

NIHON NOGYO NO CHIIKI KOZO 3537
(REGIONAL AGRICULTURE CONSTRUCTION)
NORIN-SHO TOKEI JOHO-BU (STATISTICS AND INFORMATION DIVISION, MINISTRY OF AGRICULTURE AND FORESTRY)
AND NORIN TOKEI KYOKAI
11-14, 2-CHOME, MEGURO, MEGURO-KU, TOKYO 153
JAPANESE

NIHON NOGYO NO UGOKI 3538
(TRENDS IN JAPANESE AGRICULTURE)
NORIN TOKEI KYOKAI (AGRICULTURAL AND FORESTRY STATISTICS ASSOCIATION)
JAPANESE

NIHON NOGYO SHINBUN 3539
(JAPAN AGRICULTURAL PRESS)
ZENKOKU SHINBUN JOHO NOGYO KYODO KUMIAI RENGOKAI (NATIONAL PRESS AND INFORMATION OF AGRICULTURAL CORPORATIVES)
2-3, AKIHABARA, TAITO-KU, TOKYO 110
JAPANESE

NIHON NOMIN SHINBUN 3540
(JAPAN FARMERS NEWS)
NIHON NOMIN SHINBUNSHA
13-14, 2-CHOME, NISHI SHINBASHI, MINATO-KU, TOKYO 105
JAPANESE

NIHON NOSAN KOGYO K.K. CHUO KENKYUSHO NENPO 3541
(ANNUAL REPORT OF THE RESEARCH CENTER, NIHON NOSAN KOGYO COMPANY)
NIHON NOSAN KOGYO K.K. CHUO KENKYUSHO (RESEARCH CENTER, NIHON NOSAN KOGYO COMPANY)
1, DAIMURA-CHO, MIDORI-KU, YOKOHAMA-SHI 226
JAPANESE

NIHON SAKUMOTSU GAKKAI KIJI 3542
(PROCEEDINGS OF THE CROP SCIENCE SOCIETY OF JAPAN)
NIHON SAKUMOTSU GAKKAI (CROP SCIENCE SOCIETY OF JAPAN)
C/O TOKYO DAIGAKU NOGAKUBU, 1-1, YAYOI, BUNKYO-KU, TOKYO 113
JAPANESE AND ENGLISH
INDEXED: BIB, BIO, CHE, JICST

NIHON SAKUMOTSU GAKKAI KYUSHU SHIBU KAIHO 3543
(REPORT OF THE KYUSHU BRANCH OF THE CROP SCIENCE SOCIETY OF JAPAN)
NIHON SAKUMOTSU GAKKAI KYUSHU SHIBU
C/O KYUSHU DAIGAKU NOGAKUBU SAKUMOTSUGAKU KYOSHITSU, HAKOZAKI, HIGASHI-KU, FUKUOKA-SHI 812
JAPANESE
INDEXED: JICST

NIIGATA NORIN KENKYU 3544
(NIIGATA AGRICULTURAL SCIENCE)
NIIGATA DAIGAKU NOGAKUBU (FACULTY OF AGRICULTURE, NIIGATA UNIVERSITY)
106, KOGANE-CHO, NIIGATA-SHI 950
JAPANESE AND EUROPEAN
INDEXED: BIB, CHE, JICST

NIIGATA NORIN SUISAN TOKEI NENPO 3545
(STATISTICAL YEARBOOK OF AGRICULTURE, FORESTRY AND FISHERY, NIIGATA)
NORIN-SHO HOKURIKU NOSEI-KYOKU NIIGATA TOKEI JOHO JIMUSHO (NIIGATA STATISTICS AND INFORMATION OFFICE, HOKURIKU AGRICULTURAL ADMINISTRATION BUREAU, MINISTRY OF AGRICULTURE AND FORESTRY)
601, NISHI OHATA-CHO, NIIGATA-SHI 951
JAPANESE

NIIGATA-KEN NOGYO KYOIKU KENKYU KAISHI 3546
(BULLETIN OF NIIGATA (PREFECTURE) AGRICULTURE EDUCATION RESEARCH SOCIETY)
NIIGATA-KEN KOTO GAKKO KYOIKU KENKYU KAI NOGYO BUKAI
JAPANESE

NIIGATA-KEN NOGYO NO DOKO 3547
(TENDENCY OF AGRICULTURE IN NIIGATA PREFECTURE)
NIIGATA-KEN (NIIGATA PREFECTURAL GOVERNMENT)
602, ICHIBAN-CHO, GAKKODORI, NIIGATA-SHI 951
JAPANESE

NIIGATA-KEN NOGYO SHIKENJO KENKYU HOKOKU 3548
(JOURNAL OF NIIGATA AGRICULTURAL EXPERIMENT STATION)
NIIGATA-KEN NOGYO SHIKENJO (NIIGATA AGRICULTURAL EXPERIMENT STATION)
857, NAGAKURA-MACHI, NAGAOKA-SHI 940
JAPANESE, WITH ENGLISH SUMMARY
INDEXED: CHE, JICST

NIIGATA-KEN NOGYO SHIKENJO NENPO 3549
(ANNUAL REPORT OF NIIGATA AGRICULTURAL EXPERIMENT STATION)
NIIGATA-KEN NOGYO SHIKENJO (NIIGATA AGRICULTURAL EXPERIMENT STATION)
857, NAGAKURA-MACHI, NAGAOKA-SHI 940
JAPANESE
INDEXED: JICST

NIIGATA-KEN NORIN SUISAN KANKEI SHIKEN KENKYU SEISEKI NENPO 3550
(ANNUAL REPORT OF INSTITUTIONS AND EXPERIMENT STATIONS OF AGRICULTURE AND FORESTRY, NIIGATA PREFECTURE)
NIIGATA-KEN NORIN-BU (AGRICULTURE AND FORESTRY DIVISION, NIIGATA PREFECTURE)
602, ICHIBAN-CHO, GAKKODORI, NIIGATA-SHI 951
JAPANESE

NIKKAN NOGYO SHINBUN 3551
(DAILY AGRICULTURAL NEWS)
NIKKAN NOGYO SHINBUNSHA
5-4, 1-CHOME, NAGATA-CHO, CHIYODA-KU, TOKYO 100
JAPANESE

NIPPON DOJO HIRYOGAKU ZASSHI 3552
(JOURNAL OF THE SCIENCE OF SOIL AND MANURE, JAPAN)
NIPPON DOJO HIRYO GAKKAI (SOCIETY OF THE SCIENCE OF SOIL AND MANURE, JAPAN)
C/O NORIN-SHO NOGYO GIJUTSU KENKYUSHO, 1-7, 2-CHOME, NISHIGAHARA, KITA-KU, TOKYO 153
JAPANESE
INDEXED: BIB, CHE, JICST

NO EICHI SHINPO 3553
(AGRICULTURE: KNOWLEDGE AND PROGRESS)
NOSEI CHOSA IINKAI
JAPANESE

NOCHI 3554
(FARMLAND)
NORINSHO NOCHIKYOKU (MINISTRY OF AGRICULTURE, AGRICULTURAL LAND BUREAU)
JAPANESE

NOCHI KAIHATSU 3555
(AGRICULTURAL LAND DEVELOPMENT)
HOKKAIDO NOCHI KAIHATSUBU (HOKKAIDO AGRICULTURAL LAND DEVELOPMENT OFFICE)
JAPANESE

NOCHI KAIHATSU KIKAI KODAN GYOMU NENPO 3556
(ANNUAL REPORT ON OPERATION OF AGRICULTURAL LAND DEVELOPMENT MACHINERY PUBLIC CORPORATION)
NOCHI KAIHATSU KIKAI KODAN
2-2, 1-CHOME, UCHISAIWAI-CHO, CHIYODA-KU, TOKYO 100
JAPANESE

NOCHI NENPO, HOKURIKU NOSEI-KYOKU 3557
(ANNUAL REPORT OF AGRICULTURAL LAND, HOKURIKU REGIONAL AGRICULTURAL ADMINISTRATION OFFICE, MINISTRY OF AGRICULTURE AND FORESTRY)
NORIN-SHO HOKURIKU NOSEI-KYOKU (HOKURIKU AGRICULTURAL ADMINISTRATION BUREAU, MINISTRY OF AGRICULTURE AND FORESTRY)
2-60, 2-CHOME, HIROSAKA, KANAZAWA-SHI 920
JAPANESE

NOCHI NO IDO TO TEN'YO 3558
(CHANGE AND DIVERSION OF AGRICULTURAL LAND)
NORIN-SHO KOZO KAIZEN-KYOKU (AGRICULTURAL STRUCTURE IMPROVEMENT BUREAU, MINISTRY OF AGRICULTURE AND FORESTRY)
2-1, 1-CHOME, KASUMIGASEKI, CHIYODA-KU, TOKYO 100
JAPANESE

NOCHI NOGYOYO SHISETSU SAIGAI TOKEI 3559
(STATISTICS OF DAMAGES FROM AGRICULTURAL LAND AND INSTALLATIONS)
NORIN-SHO KOZO KAIZEN-KYOKU (AGRICULTURAL STRUCTURE IMPROVEMENT BUREAU, MINISTRY OF AGRICULTURE AND FORESTRY)
2-1, 1-CHOME, KASUMIGASEKI, CHIYODA-KU, TOKYO 100
JAPANESE

NOCHIKUSANBUTSU RYUTSU TOKEI NENPO 3560
(CIRCULATION STATISTICS OF AGRICULTURAL AND ANIMAL PRODUCTS)
NORIN-SHO TOHOKU NOSEI-KYOKU FUKUSHIMA TOKEI CHOSA JIMUSHO (FUKUSHIMA STATISTICS AND INFORMATION OFFICE, HOKURIKU AGRICULTURAL ADMINISTRATION BUREAU, MINISTRY OF AGRICULTURE AND FORESTRY)
4-15, YAMASHITA-CHO, FUKUSHIMA-SHI 960
JAPANESE

NODEN DAYORI 3561
(NEWS OF AGRICULTURAL ELECTRIFICATION ASSOCIATION)
NOGYO DENKA KYOKAI CHUBU SHIBU
10-1, HIGASHI SHINMACHI, HIGASHI-KU, NAGOYA-SHI 461
JAPANESE

NOGAKU KENKYU 3562
(REPORT OF THE OHARA INSTITUTE FOR AGRICULTURAL BIOLOGY)
OKAYAMA DAIGAKU NOGYO SEIBUTSU KENKYUSHO (OHARA INSTITUTE FOR AGRICULTURAL BIOLOGY, OKAYAMA UNIVERSITY)
20-1, 2-CHOME, CHUO, KURASHIKI-SHI 710
JAPANESE
INDEXED: BIB, BIO, CHE, JICST

NOGIKEN NYUSU 3563
(NEWS OF NATIONAL INSTITUTE OF AGRICULTURAL SCIENCES)
NORIN-SHO NOGYO GIJUTSU KENKYUSHO
1-7, 2-CHOME, NISHIGAHARA, KITA-KU, TOKYO 114
JAPANESE

NOGYO 3564
(JOURNAL OF THE AGRICULTURAL SOCIETY OF JAPAN)
DAINIHON NOKAI (AGRICULTURAL SOCIETY OF JAPAN)
SANKAIDO BUILDING, 9-13, 1-CHOME, AKASAKA, MINATO-KU, TOKYO 107
JAPANESE
INDEXED: CHE

NOGYO AICHI 3565
(AGRICULTURE IN AICHI)
AICHI NOGYO FUKYU KYOKAI (PROPAGATION ASSOCIATION FOR AGRICULTURE IN AICHI)
C/O AICHIKENCHO NOGYO GIJUTSU-KA, 1-2, 3-CHOME SANNOMARU, NAKA-KU, NAGOYA-SHI 460
JAPANESE

NOGYO CHIBA 3566
(AGRICULTURE IN CHIBA PREFECTURE)
NOGYO CHIBA KANKO-BU
CHIBAKENCHO NOGYO KAIRYO-KA, 1-1, ICHIBA-MACHI, CHIBA-SHI 280
JAPANESE

NOGYO CHIIKI TOKEI 3567
(AGRICULTURE STATISTICS BY REGION)
KANAGAWA-KEN NOSEI-BU (AGRICULTURAL ADMINISTRATION DEPARTMENT, KANAGAWA PREFECTURAL GOVERNMENT)
1, NIHON'-ODORI, NAKA-KU, YOKOHAMA-SHI 231
JAPANESE

NOGYO CHOSA KEKKA HOKOKUSHO (KINKI NOSEI-KYOKU) 3568
(CENSUS OF AGRICULTURE, KINKI AGRICULTURAL ADMINISTRATION BUREAU)
NORIN-SHO KINKI NOSEI-KYOKU (KINKI AGRICULTURAL ADMINISTRATION BUREAU, MINISTRY OF AGRICULTURE AND FORESTRY)
102, SHIMOCHOJA-MACHI SAGARU, NISHINOTOIN-DORI, KAMIGYO-KU, KYOTO-SHI 602
JAPANESE

NOGYO CHOSA KEKKA HOKOKUSHO (NORIN-SHO) 3569
(CENSUS OF AGRICULTURE, MINISTRY OF AGRICULTURE AND FORESTRY))
NORIN-SHO NORIN KEIZAI-KYOKU TOKEI JOHO-BU (STATISTICS AND INFORMATION DEPARTMENT, ECONOMIC AFFAIRS BUREAU, MINISTRY OF AGRICULTURE AND FORESTRY)
11-14, 2-CHOME, MEGURO, MEGURO-KU, TOKYO 153
JAPANESE

NOGYO CHOSA KEKKA HOKOKUSHO SOKATSU-HEN 3570
(CENSUS OF AGRICULTURE, GENERAL SERIES)
NORIN-SHO NORIN KEIZAI-KYOKU TOKEI JOHO-BU (STATISTICS AND INFORMATION DEPARTMENT, ECONOMIC AFFAIRS BUREAU, MINISTRY OF AGRICULTURE AND FORESTRY)
2-1, 1-CHOME, KASUMIGASEKI, CHIYODA-KU, TOKYO 100
JAPANESE

NOGYO DENKA 3571
(AGRICULTURAL ELECTRIFICATION)
NOGYO DENKA KYOKAI (AGRICULTURAL ELECTRIFICATION ASSOCIATION)
1-13, 1-CHOME, SHINBASHI, MINATO-KU, TOKYO 105
JAPANESE

NOGYO DOBOKU 3572
(AGRICULTURAL ENGINEERING)
ZENKOKU NOGYO DOBOKU GIJUTSU RENMEI (NATIONAL AGRICULTURAL ENGINEERS' ASSOCIATION IN JAPAN)
34-4, 5-CHOME, SHINBASHI, MINATO-KU, TOKYO 105
JAPANESE

NOGYO DOBOKU GAKKAI RONBUNSHU 3573
(TRANSACTIONS OF THE JAPANESE SOCIETY OF IRRIGATION, DRAINAGE AND RECLAMATION ENGINEERING)
NOGYO DOBOKU GAKKAI (JAPANESE SOCIETY OF IRRIGATION DRAINAGE AND RECLAMATION ENGINEERING)
C/O NOGYO DOBOKU KAIKAN, 34-4, 5-CHOME, SHINBASHI, MINATO-KU, TOKYO 105
JAPANESE, WITH EUROPEAN SUMMARY
INDEXED: BIB, EXC, JICST

NOGYO DOBOKU GAKKAISHI 3574
(JOURNAL OF THE JAPANESE SOCIETY OF IRRIGATION, DRAINAGE AND RECLAMATION ENGINEERING)
NOGYO DOBOKU GAKKAI (JAPANESE SOCIETY OF IRRIGATION DRAINAGE AND RECLAMATION ENGINEERING)
C/O NOGYO DOBOKU KAIKAN, 34-4, 5-CHOME, SHINBASHI, MINATO-KU, TOKYO 105
JAPANESE
INDEXED: BIB, CHE, JICST

NOGYO DOBOKU KENKYU RONBUNSHU 3575
(PAPERS ON AGRICULTURAL ENGINEERING RESEARCH)
JAPANESE

NOGYO DOBOKU NO KIKAIKA 3576
(MECHANIZATION OF AGRICULTURAL ENGINEERING)
NOGYO DOBOKU KIKAIKA KYOKAI
34-4, 5-CHOME, SHINBASHI, MINATO-KU, TOKYO 105
JAPANESE
INDEXED: JICST

NOGYO DOBOKU SHIKENJO GIHO, A, TOCHI KAIRYO 3577
(TECHNICAL REPORT OF THE NATIONAL RESEARCH INSTITUTE OF AGRICULTURAL ENGINEERING, SERIES A, LAND IMPROVEMENT)
NORIN-SHO NOGYO DOBOKU SHIKENJO (NATIONAL RESEARCH INSTITUTE OF AGRICULTURAL ENGINEERING, MINISTRY OF AGRICULTURE AND FORESTRY)
1943, YAHATA, HIRATSUKA-SHI 254
JAPANESE
INDEXED: JICST

NOGYO DOBOKU SHIKENJO GIHO, B, SUIRI 3578
(TECHNICAL REPORT OF THE NATIONAL RESEARCH INSTITUTE OF AGRICULTURAL ENGINEERING, SERIES B, HYDRAULICS)
NORIN-SHO NOGYO DOBOKU SHIKENJO (NATIONAL RESEARCH INSTITUTE OF AGRICULTURAL ENGINEERING, MINISTRY OF AGRICULTURE AND FORESTRY)
1943, YAHATA, HIRATSUKA-SHI 254
JAPANESE
INDEXED: JICST

NOGYO DOBOKU SHIKENJO GIHO, C, ZOKO 3579
(TECHNICAL REPORT OF THE NATIONAL RESEARCH INSTITUTE OF AGRICULTURAL ENGINEERING, SERIES C, CONSTRUCTION)
NORIN-SHO NOGYO DOBOKU SHIKENJO (NATIONAL RESEARCH INSTITUTE OF AGRICULTURAL ENGINEERING, MINISTRY OF AGRICULTURE AND FORESTRY)
1943, YAHATA, HIRATSUKA-SHI 254
JAPANESE
INDEXED: JICST

NOGYO DOBOKU SHIKENJO GIHO, E, SAGA SHIJO 3580
(TECHNICAL REPORT OF THE NATIONAL RESEARCH INSTITUTE OF AGRICULTURAL ENGINEERING, SERIES E, SAGA)
NORIN-SHO NOGYO DOBOKU SHIKENJO (NATIONAL RESEARCH INSTITUTE OF AGRICULTURAL ENGINEERING, MINISTRY OF AGRICULTURE AND FORESTRY)
1943, YAHATA, HIRATSUKA-SHI 254
JAPANESE
INDEXED: JICST

NOGYO DOBOKU SHIKENJO GIHO, F, SOGO 3581
(TECHNICAL REPORT OF THE NATIONAL RESEARCH INSTITUTE OF AGRICULTURAL ENGINEERING, SERIES F, GENERAL)
NORIN-SHO NOGYO DOBOKU SHIKENJO (NATIONAL RESEARCH INSTITUTE OF AGRICULTURAL ENGINEERING, MINISTRY OF AGRICULTURE AND FORESTRY)
1943, YAHATA, HIRATSUKA-SHI 254
JAPANESE
INDEXED: JICST

NOGYO DOBOKU SHIKENJO HOKOKU 3582
(BULLETIN OF THE NATIONAL RESEARCH INSTITUTE OF AGRICULTURAL ENGINEERING)
NORIN-SHO NOGYO DOBOKU SHIKENJO (NATIONAL RESEARCH INSTITUTE OF AGRICULTURAL ENGINEERING, MINISTRY OF AGRICULTURE AND FORESTRY)
1943, YAHATA, HIRATSUKA-SHI 254
JAPANESE AND ENGLISH
INDEXED: BIB, JICST

NOGYO DOKO NENPO (KUMAMOTO-KEN) 3583
(ANNUAL REPORT OF AGRICULTURAL TENDENCY, KUMAMOTO PREFECTURE)
KUMAMOTO-KEN (KUMAMOTO PREFECTURAL GOVERNMENT)
18-1, 6-CHOME, SUIZENJI, KUMAMOTO-SHI 862
JAPANESE
INDEXED: JICST

NOGYO FUKYU 3584
(AGRICULTURAL PROPAGATION)
IWATE-KEN NOGYO KAIRYO FUKYUKAI (IWATE PREFECTURE AGRICULTURAL PROPAGATION ASSOCIATION)
C/O IWATE-KEN NOMU-BU, UCHIMARU, MORIOKA-SHI 020
JAPANESE

NOGYO FUMIN 3585
(AGRICULTURE AND BETTER FARMING)
FUMIN KYOKAI SHUPPAN-BU (BETTER FARMING ASSOCIATION)
C/O MAINICHI SHINBUNSHA, DOJIMA, KITA-KU, OSAKA-SHI 530
JAPANESE

NOGYO GIFU 3586
(AGRICULTURE OF GIFU PREFECTURE)
GIFU-KEN NOGYO SHIDO KENKYUKAI (GIFU PREFECTURAL SOCIETY OF THE GUIDE FOR AGRICULTURE)
C/O GIFU-KENCHO NOGYO KAIGI, YABUTA, GIFU-SHI 500
JAPANESE

NOGYO GIJUTSU 3587
(JOURNAL OF AGRICULTURAL SCIENCE)
NOGYO GIJUTSU KYOKAI (ASSOCIATION FOR ADVANCEMENT OF AGRICULTURAL SCIENCES)
26-3, 1-CHOME, NISHIGAHARA, KITA-KU, TOKYO 114
JAPANESE
INDEXED: BIB

NOGYO GIJUTSU KENKYU 3588
(TECHNICAL STUDY OF AGRICULTURE)
SHIZUOKA-KEN NOGYO KYODO KUMIAI CHUOKAI (CENTRAL COMMITTEE, SHIZUOKA PREFECTURE AGRICULTURAL CO-OPERATIVE UNION)
8-1, 3-CHOME, MAGARIGANE, SHIZUOKA-SHI 420
JAPANESE

NOGYO GIJUTSU KENKYUJO HOSHASEN IKUSHUJO NENPO 3589
(ANNUAL REPORT OF RADIATION NURSERY, AGRICULTURAL TECHNOLOGY RESEARCH INSTITUTE)
JAPANESE
INDEXED: JICST

NOGYO GIJUTSU KENKYUSHO HOKOKU, A, (BUTSURI, TOKEI) 3590
(BULLETIN OF THE NATIONAL INSTITUTE OF AGRICULTURAL SCIENCES, SERIES A (PHYSICS AND STATISTICS))
NORIN-SHO NOGYO GIJUTSU KENKYUSHO (NATIONAL INSTITUTE OF AGRICULTURAL SCIENCES)
1-7, 2-CHOME, NISHIGAHARA,KITA-KU, TOKYO 114
JAPANESE AND ENGLISH
INDEXED: BIB, CHE, JICST

NOGYO GIJUTSU KENKYUSHO HOKOKU, D, (SEIRI, IDEN, SAKUMOTSU IPPAN) 3591
(BULLETIN OF THE NATIONAL INSTITUTE OF AGRICULTURAL SCIENCES, SERIES D (PLANT PHYSIOLOGY, GENETICS AND CROPS IN GENERAL))
NORIN-SHO NOGYO GIJUTSU KENKYUSHO (NATIONAL INSTITUTE OF AGRICULTURAL SCIENCES)
1-7, 2-CHOME, NISHIGAHARA, KITA-KU, TOKYO 114
JAPANESE AND ENGLISH
INDEXED: BIB, BIO, CHE, JICST

NOGYO GIJUTSU KENKYUSHO HOKOKU, H, (KEIEI, TOCHI RIYO) 3592
(BULLETIN OF THE NATIONAL INSTITUTE OF AGRICULTURAL SCIENCES, SERIES H (FARM MANAGEMENT AND LAND UTILIZATION))
NORIN-SHO NOGYO GIJUTSU KENKYUSHO (NATIONAL INSTITUTE OF AGRICULTURAL SCIENCES)
1-7, 2-CHOME, NISHIGAHARA, KITA-KU, TOKYO 114
JAPANESE
INDEXED: BIB, CHE, JICST

NOGYO GIJUTSU KENKYUSHO NENPO 3593
(ANNUAL REPORT OF THE NATIONAL INSTITUTE OF AGRICULTURAL SCIENCES)
NORIN-SHO NOGYO GIJUTSU KENKYUSHO (NATIONAL INSTITUTE OF AGRICULTURAL SCIENCES, MINISTRY OF AGRICULTURE AND FORESTRY)
1-7, 2-CHOME, NISHIGAHARA, KITA-KU, TOKYO 114
JAPANESE
INDEXED: JICST

NOGYO GIJUTSU KENKYUSHO SHIRYO, B, (DOJO, HIRYO) 3594
(MISCELLANEOUS PUBLICATION OF THE NATIONAL INSTITUTE OF AGRICULTURAL SCIENCES, SERIES B (SOILS AND FERTILIZERS))
NORIN-SHO NOGYO GIJUTSU KENKYUSHO (NATIONAL INSTITUTE OF AGRICULTURAL SCIENCES, MINISTRY OF AGRICULTURE AND FORESTRY)
1-7, 2-CHOME, NISHIGAHARA, KITA-KU, TOKYO 114
JAPANESE
INDEXED: BIB, CHE

NOGYO GIJUTSU KENKYUSHO SHIRYO, D, (SEIRI IDEN) 3595
(DATA OF THE AGRICULTURAL TECHNOLOGY RESEARCH INSTITUTE, SERIES D, PHYSIOLOGICAL GENETICS)
JAPANESE

NOGYO GIJUTSU KENKYUSHO SHIRYO, H, (KEIEI, TOCHI RIYO) 3596
(MISCELLANEOUS PUBLICATION OF THE NATIONAL INSTITUTE OF AGRICULTURAL SCIENCES, SERIES H (FARM MANAGEMENT AND LAND UTILIZATION))
1-7, 2-CHOME, NISHIGAHARA, KITA-KU, TOKYO 114
JAPANESE

NOGYO IBARAKI 3597
(AGRICULTURE IN IBARAKI)
IBARAKI-KEN NOGYO GIJUTSU KENKYUKAI (IBARAKI PREFECTURE AGRICULTURE TECHNIQUE RESEARCH ASSOCIATION)
C/O IBARAKI-KEN KYOIKU FUKYU-KA, SANNOMARU, MITO-SHI 310
JAPANESE

NOGYO JOHO (FUKUOKA-KEN) 3598
(INFORMATION ON SUPPLY AND DEMAND OF AGRICULTURAL PRODUCTS, FUKUOKA PREFECTURE))
FUKUOKA-KEN NOSEI-BU (AGRICULTURAL ADMINISTRATION DIVISION, FUKUOKA PREFECTURAL GOVERNMENT)
1-1, 1-CHOME, TENJIN, CHUO-KU, FUKUOKA-SHI 810
JAPANESE

NOGYO JOHO (ISHIKAWA-KEN) 3599
(AGRICULTURAL NEWS)
ISHIKAWA-KEN NORINBU (ISHIKAWA PREFECTURE AGRICULTURE AND FORESTRY DEPARTMENT)
JAPANESE

NOGYO JOHO (KAGOSHIMA-KEN) 3600
(INFORMATION ON SUPPLY AND DEMAND OF AGRICULTURAL PRODUCTS, KAGOSHIMA PREFECTURE))
KAGOSHIMA-KEN NOSEI-BU (AGRICULTURAL ADMINISTRATION DIVISION, KAGOSHIMA PREFECTURAL GOVERNMENT)
14-50, YAMASHITA-CHO, KAGOSHIMA-SHI 892
JAPANESE

NOGYO JOHO (TOCHIGI-KEN) 3601
(INFORMATION ON SUPPLY AND DEMAND OF AGRICULTURAL PRODUCTS, TOCHIGI PREFECTURE)
TOCHIGI-KEN NOMU-BU (AGRICULTURAL ADMINISTRATION DIVISION, TOCHIGI PREFECTURAL GOVERNMENT)
504, HANAWADA-CHO, UTSUNOMIYA-SHI 320
JAPANESE

NOGYO JOHO (TOKUSHIMA-KEN) 3602
(INFORMATION ON SUPPLY AND DEMAND OF AGRICULTURAL PRODUCTS, TOKUSHIMA PREFECTURE)
TOKUSHIMA-KEN NORIN-BU (AGRICULTURE AND FORESTRY DIVISION, TOKUSHIMA PREFECTURAL GOVERNMENT)
TOKUSHIMA-KEN NORIN TOKEI KYOKAI
1-1, MANDAI-CHO, TOKUSHIMA-SHI 770
JAPANESE

NOGYO KAGAWA 3603
(AGRICULTURE IN KAGAWA)
KAGAWA-KEN NOGYO KAIRYO FUKYUKAI (KAGAWA ASSOCIATION FOR AGRICULTURAL IMPROVEMENT)
C/O KAGAWA-KENCHO, BAN-CHO, TAKAMATSU-SHI 760
JAPANESE

NOGYO KAGOSHIMA 3604
(AGRICULTURE IN KAGOSHIMA)
KAGOSHIMA-KEN NOGYO KAIRYO FUKYUKAI (KAGOSHIMA ASSOCIATION FOR AGRICULTURAL IMPROVEMENT)
3-1, NAN'EI, KAGOSHIMA-SHI 891-01
JAPANESE

NOGYO KAIRYO SHIKIN GIJUTSU DONYU SHIKIN TOKUNIN JIGYO KEIKAKU 3605
(SPECIAL APPROVAL PROJECT BY FUND FOR AGRICULTURAL IMPROVEMENT AND HIGHER TECHNIQUE INTRODUCTION)
NORIN-SHO NOSAN ENGEI-KYOKU (AGRICULTURAL PRODUCTION BUREAU, MINISTRY OF AGRICULTURE AND FORESTRY)
2-1, 1-CHOME, KASUMIGASEKI, CHIYODA-KU, TOKYO 100
JAPANESE

NOGYO KANKEI NO SHIKEN KENKYU SEIKA (FUKUOKA-KEN) 3606
(ANNUAL STATISTICS OF TESTING FOR AGRICULTURE, FUKUOKA PREFECTURE)
FUKUOKA-KEN NOSEI-BU (AGRICULTURAL ADMINISTRATION DEPARTMENT, FUKUOKA PREFECTURE)
1-1, 1-CHOME, TENJIN, CHUO-KU, FUKUOKA-SHI 810
JAPANESE

NOGYO KANKEI SHIKEN KENKYU NENPO 3607
(ANNUAL REPORT OF TESTING FOR AGRICULTURE)
IWATE-KEN NOSEI-BU (AGRICULTURAL ADMINISTRATION DEPARTMENT, IWATE PREFECTURAL GOVERNMENT)
10-1, UCHIMARU, MORIOKA-SHI 020
JAPANESE

NOGYO KANSOKU (FUKUOKA-KEN) 3608
(OUTLOOK OF AGRICULTURE, FUKUOKA PREFECTURE)
FUKUOKA-KEN NOSEI-BU (AGRICULTURAL ADMINISTRATION DIVISION, FUKUOKA PREFECTURAL GOVERNMENT)
1-1, 1-CHOME, TENJIN, CHUO-KU, FUKUOKA-SHI 810
JAPANESE

NOGYO KANSOKU (KAGOSHIMA-KEN) 3609
(OUTLOOK OF AGRICULTURE, KAGOSHIMA PREFECTURE)
KAGOSHIMA-KEN (KAGOSHIMA PREFECTURAL GOVERNMENT)
14-50, YAMASHITA-CHO, KAGOSHIMA-SHI 892
JAPANESE

NOGYO KANSOKU (KANAGAWA-KEN) 3610
(OUTLOOK OF AGRICULTURE, KANAGAWA PREFECTURE)
KANAGAWA-KEN NOGYO SOGO KENKYUSHO (AGRICULTURAL RESEARCH INSTITUTE OF KANAGAWA PREFECTURE)
496, TERADANAWA, HIRATSUKA-SHI 259-12
JAPANESE

NOGYO KANSOKU (NORIN-SHO) 3611
(OUTLOOK OF AGRICULTURE, MINISTRY OF AGRICULTURE AND FORESTRY)
NORIN-SHO DAIJIN KANBO (MINISTER'S SECRETARIAT, MINISTRY OF AGRICULTURE AND FORESTRY)
NORIN TOKEI KYOKAI
11-4, 2-CHOME, MEGURO, MEGURO-KU, TOKYO 153
JAPANESE

NOGYO KANSOKU (SHIZUOKA-KEN) 3612
(OUTLOOK OF AGRICULTURE, SHIZUOKA PREFECTURE)
SHIZUOKA-KEN KEIZAI-BU (ECONOMY DIVISION, SHIZUOKA PREFECTURAL GOVERNMENT)
251, OUTE-MACHI, SHIZUOKA-SHI 420
JAPANESE

NOGYO KANSOKU JOHO 3613
(INFORMATION OF AGRICULTURAL OUTLOOK)
NORIN-SHO DAIJIN KANBO (MINISTER'S SECRETARIAT, MINISTRY OF AGRICULTURE AND FORESTRY)
NORIN TOKEI KYOKAI
11-14, 2-CHOME, MEGURO, MEGURO-KU, TOKYO 153
JAPANESE

NOGYO KANSOKU TOKEIHYO IPPAN KEIZAI, NOGYO KEIZAI, NOGYO SHIZAI, TOKUYO SAKUMOTSU 3614
(STATISTICS OF AGRICULTURAL OUTLOOK, GENERAL ECONOMIC, AGRICULTURAL ECONOMIC, AGRICULTURAL MATERIALS, INDUSTRIAL CROPS)
FUKUOKA-KEN NOSEI-BU (AGRICULTURAL ADMINISTRATION DEPARTMENT, FUKUOKA PREFECTURE)
1-1, 1-CHOME, TENJIN, CHUO-KU, FUKUOKA-SHI 810
JAPANESE

NOGYO KEIEI TSUSHIN 3615
(AGRICULTURAL MANAGEMENT COMMUNICATION)
NORINSHO NOGYO BIJUTSU KENKYUSHO NUMBER 113
JAPANESE

NOGYO KEISANGAKU KENKYU 3616
(FARM ACCOUNTING STUDIES)
KYOTO DAIGAKU NOGAKU-BU NOGYO BOKI KENKYU SHISETSU
(FARM ACCOUNTING INSTITUTE, FACULTY OF AGRICULTURE, KYOTO UNIVERSITY)
KITASHIRAKAWA OIWAKE-CHO, SAKYO-KU, KYOTO-SHI 606
JAPANESE
INDEXED: BIB

NOGYO KEIZAI KENKYU 3617
(JOURNAL OF RURAL ECONOMICS)
JAPANESE

NOGYO KENKYUJO DAYORI 3618
(AGRICULTURAL RESEARCH INSTITUTE LETTER)
KYOTO FURITSU NOGYO KENKYUJO (KYOTO PREFECTURAL AGRICULTURAL RESEARCH INSTITUTE)
JAPANESE

NOGYO KIHON CHOSA KEKKA HOKOKU (IBARAKI-KEN) 3619
(REPORT ON THE RESULT OF AGRICULTURAL BASIC SURVEY, IBARAKI PREFECTURE)
IBARAKI-KEN SOGO KAIHATSU-BU (GENERAL DEVELOPMENT DIVISION, IBARAKI PREFECTURAL GOVERNMENT)
5-38, 1-CHOME, SANNOMARU, MITO-SHI 310
JAPANESE

NOGYO KIHON CHOSA KEKKA HOKOKUSHO (TOYAMA-KEN) 3620
(CENSUS OF AGRICULTURE, TOYAMA PREFECTURE)
TOYAMA-KEN SOMU-BU (GENERAL AFFAIRS DIVISION, TOYAMA PREFECTURAL GOVERNMENT)
1-7, SHINSOGAWA, TOYAMA-SHI 930
JAPANESE

NOGYO KIHON CHOSA NO KEKKA GAIYO, CHIBA-KEN NOGYO TOKEI CHOSA 3621
(CENSUS OF AGRICULTURE, OUTLINE, CHIBA PREFECTURE)
CHIBA-KEN KIKAKU-BU (PLANNING DEPARTMENT, CHIBA PREFECTURAL GOVERNMENT)
1-1, ICHIBA-MACHI, CHIBA-SHI 280-91
JAPANESE

NOGYO KIKAI GAKKAISHI 3622
(JOURNAL OF THE SOCIETY OF AGRICULTURAL MACHINERY, JAPAN)
NOGYO KIKAI GAKKAI (SOCIETY OF AGRICULTURAL MACHINERY)
1-1, 1-CHOME, YAYOI, BUNKYO-KU, TOKYO 113
JAPANESE AND ENGLISH
INDEXED: BIB, JICST

NOGYO KIKAI JOHO 3623
(FARM MACHINERY INFORMATION)
NIHON NOGYO KIKAI KOGYOKAI (JAPAN FARM MACHINERY MANUFACTURERS' ASSOCIATION)
C/O KIKAI SHINKO KAIKAN, 5-8, 3-CHOME, SHIBAKOEN, MINATO-KU, TOKYO 105
JAPANESE
INDEXED: JICST

NOGYO KIKAI NENKAN 3624
(FARM MACHINERY YEARBOOK)
SHIN-NORINSHO COMPANY, LIMITED
JAPANESE
INDEXED: BIB

NOGYO KIKAIKA KENKYUSHO HOKOKU 3625
(INSTITUTE OF AGRICULTURAL MACHINERY TECHNICAL REPORT)
NOGYO KIKAIKA KENKYUSHO (INSTITUTE OF AGRICULTURAL MACHINERY)
40-2, 1-CHOME, NISSHIN-CHO, OMIYA-SHI 330
JAPANESE

NOGYO KIKAIKA KENKYUSHO JIGYO HOKOKU 3626
(OPERATION REPORT OF AGRICULTURAL MACHINERY)
NOGYO KIKAIKA KENKYUSHO (INSTITUTE OF AGRICULTURAL MACHINERY)
40-2, 1-CHOME, NISSHIN-CHO, OMIYA-SHI 330
JAPANESE

NOGYO KIKAIKA KENKYUSHO NENPO 3627
(ANNUAL REPORT OF THE INSTITUTE OF AGRICULTURAL MACHINING)
NOGYO KIKAIKA KENKYUSHO (INSTITUTE OF AGRICULTURAL MACHINERY)
40-2, 1-CHOME, NISSHIN-CHO, OMIYA-SHI 330
JAPANESE

NOGYO KOZO KAIZEN 3628
(AGRICULTURAL IMPROVEMENT)
ZENKOKU NOGYO KOZO KAIZEN KYOKAI (NATIONAL ASSOCIATION OF AGRICULTURAL IMPROVEMENT)
C/O ZENKOKU CHOSON KAIKAN, 11-35, 1-CHOME, NAGATA-CHO, CHIYODA-KU, TOKYO 100
JAPANESE

NOGYO KOZO MONDAI KENKYU 3629
(RESEARCH ON PROBLEMS IN AGRICULTURAL STRUCTURE)
NOGYO KOZO MONDAI KENKYUKAI (RESEARCH SOCIETY FOR PROBLEMS IN AGRICULTURAL STRUCTURE)
JAPANESE

NOGYO KYODO KUMIAI 3630
(AGRICULTURAL COOPERATIVES)
JAPANESE
INDEXED: BIB

NOGYO MAINICHI SHINBUN 3631
(DAILY AGRICULTURE)
NOGYO MAINICHI SHINBUNSHA
5-3, SUMINOE-CHO, SUMIYOSHI-KU, OSAKA-SHI 558
JAPANESE

NOGYO MONDAI 3632
(PROBLEM OF AGRICULTURE)
NOGYO MONDAI KENKYU KAIGI
C/O YUGETA BUILDING, 1-5, 5-CHOME, HONGO, BUNKYO-KU, TOKYO 113
JAPANESE

NOGYO NO DOKO NI KANSURU NENJI HOKOKU 3633
(ANNUAL REPORT OF THE AGRICULTURE)
NORIN-SHO (MINISTRY OF AGRICULTURE AND FORESTRY)
2-1, 1-CHOME, KASUMIGASEKI, CHIYODA-KU, TOKYO 100
JAPANESE

NOGYO NO MITOSHI (SAITAMA-KEN) 3634
(OUTLOOK OF AGRICULTURE, SAITAMA PREFECTURE)
SAITAMA-KEN NORIN-BU (AGRICULTURE AND FORESTRY DIVISION, SAITAMA PREFECTURAL GOVERNMENT)
15-1, 3-CHOME, TAKASAGO-CHO, URAWA-SHI, 336
JAPANESE

NOGYO OYOBI ENGEI 3635
(AGRICULTURE AND HORTICULTURE)
YOKENDO
30-15, 5-CHOME, HONGO, BUNKYO-KU, TOKYO 113
JAPANESE
INDEXED: BIB, CHE, JICST

NOGYO SEISAN JISSEKI 3636
(STATISTICAL DATA OF THE AGRICULTURAL PRODUCTION)
KAGOSHIMA-KEN NOSEI-BU (AGRICULTURAL ADMINISTRATION DIVISION, KAGOSHIMA PREFECTURAL GOVERNMENT)
14-50, YAMASHITA-CHO, KAGOSHIMA-SHI 892
JAPANESE

NOGYO SHIDOJO DAYORI 3637
(LETTERS OF THE AGRICULTURAL DEVELOPMENT INSTITUTE)
KYOTO-FU NOGYO SHIDOJO (KYOTO MUNICIPAL AGRICULTURAL DEVELOPMENT INSTITUTE)
JAPANESE

NOGYO SHIKENJO JOHO 3638
(BULLETIN OF THE AGRICULTURAL RESEARCH INSTITUTE)
NORINSHO (MINISTRY OF AGRICULTURE AND FORESTRY)
JAPANESE

NOGYO SHIKENJO KENKYU NENJI HOKOKUSHO 3639
(ANNUAL REPORT OF AGRICULTURAL EXPERIMENTAL STATION RESEARCH)
JAPANESE
INDEXED: JICST

NOGYO SHIKENJO NENPO 3640
(ANNUAL REPORT OF THE TOKYO METROPOLITAN AGRICULTURAL EXPERIMENT STATION)
TOKYO-TO NOGYO SHIKENJO (TOKYO METROPOLITAN AGRICULTURAL EXPERIMENT STATION)
10, 3-CHOME, FUJIMI-CHO, TACHIKAWA-SHI, TOKYO 190
JAPANESE

NOGYO SHISETSU 3641
(JOURNAL OF THE SOCIETY OF AGRICULTURAL STRUCTURES, JAPAN.)
NOGYO SHISETSU KENKYUKAI (SOCIETY OF AGRICULTURAL STRUCTURES)
JAPANESE

NOGYO SOGO KENKYU 3642
(QUARTERLY JOURNAL OF AGRICULTURAL ECONOMY)
NORIN-SHO NOGYO SOGO KENKYUSHO (NATIONAL RESEARCH INSTITUTE OF AGRICULTURE, MINISTRY OF AGRICULTURE AND FORESTRY)
2-1, 2-CHOME, NISHIGAHARA, KITA-KU, TOKYO 114
JAPANESE
INDEXED: BIB

NOGYO SOGO KENKYUSHO NENPO 3643
(ANNUAL REPORT OF RESEARCHES)
NORIN-SHO NOGYO SOGO KENKYUSHO (NATIONAL RESEARCH INSTITUTE OF AGRICULTURE, MINISTRY OF AGRICULTURE AND FORESTRY)
2-1, 2-CHOME, NISHIGAHARA, KITA-KU, TOKYO 114
JAPANESE
INDEXED: BIB

NOGYO SOGO KENKYUSHO SEKISETSU CHIHO SHISHO KENKYU NOTO 3644
(MEMOIRS OF SNOWY AREA BRANCH, NATIONAL RESEARCH INSTITUTE OF AGRICULTURE)
NOGYO SOGO KENKYUSHO SEKISETSU CHIHO SHISHO (SNOWY AREA BRANCH, NATIONAL RESEARCH INSTITUTE OF AGRICULTURE, MINISTRY OF AGRICULTURE AND FORESTRY)
4-13, ISHIKAWA-MACHI, SHINJO-SHI 996
JAPANESE

NOGYO SOGO KENKYUSHO SEKISETSU CHIHO SHISHO KENKYU SHIRYO 3645
(RESEARCH REPORT OF THE SNOWY AREA BRANCH, NATIONAL RESEARCH INSTITUTE OF AGRICULTURE, MINISTRY OF AGRICULTURE AND FORESTRY)
NORIN-SHO NOGYO SOGO KENKYUSHO SEKISETSU CHIHO SHISHO (SNOWY AREA BRANCH, NATIONAL RESEARCH INSTITUTE OF AGRICULTURE, MINISTRY OF AGRICUL-TURE AND FORESTRY)
4-13, ISHIKAWA-MACHI, SHINJO-SHI 990
JAPANESE

NOGYO TO KEIZAI 3646
(AGRICULTURE AND ECONOMY)
FUMIN KYOKAI
C/O MAINICHI SHINBUNSHA, DOJIMA, KITA-KU, OSAKA-SHI 530
JAPANESE

NOGYO YOSUI SUISHITSU CHOSA HOKOKUSHO 3647
(REPORT OF AGRICULTURAL WATER QUALITY)
HOKKAIDO NOCHI KAIHATSU-BU (AGRICULTURAL LAND DEVELOPMENT DEPARTMENT, HOKKAIDO PREFECTURAL GOVERNMENT)
NISHI 6-CHOME, KITA 3-JO, CHUO-KU, SAPPORO-SHI 060
JAPANESE

NOGYOYO DENKI KYOKYU TOKEI 3648
(STATISTICS OF AGRICULTURAL ELECTRICITY SUPPLY)
NOGYO DENKA KYOKAI (AGRICULTURAL ELECTRIFICATION ASSOCIATION)
1-13, 1-CHOME, SHINBASHI, MINATO-KU, TOKYO 105
JAPANESE

NOJI SHIKENJO JOHO 3649
(BULLETIN OF CENTRAL AGRICULTURAL EXPERIMENT STATION)
NORIN-SHO NOJI SHIKENJO (CENTRAL AGRICULTURAL EXPERIMENT STATION, MINISTRY OF AGRICULTURE AND FORESTRY)
1227, KONOSU, KONOSU-SHI 365
JAPANESE

NOJI SHIKENJO KENKYU HOKOKU 3650
(JOURNAL OF THE CENTRAL AGRICULTURAL EXPERIMENT STATION)
NORIN-SHO NOJI SHIKENJO (CENTRAL AGRICULTURAL EXPERIMENT STATION. MINISTRY OF AGRICULTURAL AND FORESTRY)
1227, KONOSU, KONOSU-SHI 365
JAPANESE, WITH ENGLISH SUMMARY
INDEXED: BIB, BIO, CHE, JICST

NOJI SHIKENJO NENPO 3651
(ANNUAL REPORT OF THE CENTRAL AGRICULTURAL EXPERIMENT STATION)
NORIN-SHO NOJI SHIKENJO (CENTRAL AGRICULTURAL EX-PERIMENT STATION, MINISTRY OF AGRICULTURE AND FORESTRY)
1227, KONOSU, KONOSU-SHI 365
JAPANESE
INDEXED: JICST

NOJO KENSHUSHI 3652
(TRAINING REPORT OF THE UNIVERSITY FARM ATTACHED TO THE FACULTY OF AGRICULTURE, KYUSHU UNIVERSITY)
KYUSHU DAIGAKU NOGAKUBU FUZOKU NOJO (UNIVERSITY FARM ATTACHED TO THE FACULTY OF AGRICULTURE, KYUSHU UNIVERSITY)
11, HARA-MACHI, KASUYA-CHO, KASUYA-GUN, FUKUOKA 811-23
JAPANESE

NOKEI SHINPO 3653
(AGRICULTURAL ECONOMIC TIMES)
NOKEI SHINPOSHA
12-10, 1-CHOME, SHINBASHI, MINATO-KU, TOKYO 105
JAPANESE

NOKI SHINBUN 3654
(AGRICULTURAL MACHINERY WEEKLY)
SHIN NORINSHA
2-7, KANDA NISHIKI-CHO, CHIYODA-KU, TOKYO 101
JAPANESE

NOKI TAIMUSU 3655
(AGRICULTURAL MACHINERY TIMES)
KINDAI NOGYOSHA
6-3, 3-CHOME, SHINBASHI, MINATO-KU, TOKYO 105
JAPANESE

NOKO JOHO 3656
(AGRICULTURE AND INDUSTRY INFORMATION)
NOSON CHIIKI KOGYO DONYU SOKUSHIN SENTA
11-35, 1-CHOME, NAGATA-CHO, CHIYODA-KU, TOKYO 100
JAPANESE

NOKO KENKYUSHO KENKYU HOKOKU 3657
(SCIENTIFIC REPORTS OF THE RESEARCH INSTITUTE OF AGRICULTURAL TECHNOLOGY)
TOKYO KYOIKU DAIGAKU NOGAKUBU FUZOKU NOKO KENKYUSHO (RESEARCH INSTITUTE OF AGRICULTURAL TECHNOLOGY, FACULTY OF AGRICULTURE, TOKYO UNIVERSITY OF EDUCATION)
19-1, 2-CHOME, KOMABA-CHO, MEGURO-KU, TOKYO 153
JAPANESE AND ENGLISH

NOKO TO ENGEI 3658
(AGRICULTURIST AND HORTICULTURIST)
SEIBUNDO SHINKOSHA
1-5, KANDA NISHIKI-CHO, CHIYODA-KU, TOKYO 101
JAPANESE

NORIN GYOGYO DOKO NENPO 3659
(ANNUAL REPORT ON TENDENCY OF AGRICULTURE, FORESTRY AND FISHERY)
KANAGAWA-KEN NOSEI-BU (AGRICULTURAL ADMINISTRATION DIVISION, KANAGAWA PREFECTURE)
1, NOHON-ODORI, NAKA-KU, YOKOHAMA-SHI 231
JAPANESE

NORIN KENKYU KEISAN SENTA HOKOKU, A 3660
(BULLETIN OF THE COMPUTING CENTRE FOR RESEARCH IN AGRICULTURE, FORESTRY AND FISHERY)
NORIN-SHO NORIN SUISAN GIJUTSU KAIGI (AGRICULTURE FORESTRY AND FISHERIES RESEARCH COUNCIL, MINISTRY OF AGRICULTURE AND FORESTRY)
2-1, 1-CHOME, KASUMIGASEKI, CHIYODA-KU, TOKYO 100
JAPANESE

NORIN KOKU 3661
(AGRICULTURAL AVIATION)
NORIN SUISAN KOKU KYOKAI (AGRICULTURAL AVIATION ASSOCIATION)
11-35, 1-CHOME, NAGATA-CHO, CHIYODA-KU, TOKYO 100
JAPANESE

NORIN SAITAMA 3662
(AGRICULTURE AND FORESTRY IN SAITAMA)
SAITAMA-KEN NORIN-BU (AGRICULTURE AND FORESTRY DEPARTMENT, SAITAMA PREFECTURAL GOVERNMENT)
15-1, 3-CHOME, TAKASAGO-CHO, URAWA-SHI 336
JAPANESE

NORIN SUISAN GIJUTSU KAIGI GIJUTSU KAIGI CHOSA SHIRYO 3663
(TECHNOLOGY CONFERENCE SURVEY DATA, CONFERENCE ON AGRICULTURE, FORESTRY, FISHERIES TECHNOLOGY)
JAPANESE
INDEXED: JICST

NORIN SUISAN GIJUTSU KAIGI JIMUKYOKU KENKYU SEIKA 3664
(SCIENCE REPORT OF SECRETARIAT, AGRICULTURE, FORESTRY AND FISHERIES RESEARCH COUNCIL)
NORIN-SHO NORIN SUISAN GIJUTSU KAIGI
2-1, 1-CHOME, KASUMIGASEKI, CHIYODA-KU, TOKYO 100
JAPANESE
INDEXED: JICST

NORIN SUISAN KENKYU JOHO 3665
(INFORMATION ON STUDIES OF AGRICULTURE, FORESTRY AND FISHERY)
NORIN-SHO NORIN SUISAN GIJUTSU KAIGI (AGRICULTURE, FORESTRY AND FISHERIES COUNCIL, MINISTRY OF AGRICULTURE AND FORESTRY) AND HATACHI NOGYO SHINKOKAI
C/O KIYOSU BUILDING, 8-8, 2-CHOME, NISHI SHINBASHI, MINATO-KU, TOKYO 105
JAPANESE
INDEXED: CHE

NORIN SUISAN KOKU JIGYO KAIHATSU SHIKEN SEISEKISHO 3666
(SURVEY REPORT OF AGRICULTURAL NAVIGATION DEVELOPMENT)
NORIN SUISAN KOKU KYOKAI (AGRICULTURAL NAVIGATION ASSOCIATION)
11-35, 1-CHOME, NAGATA-CHO, CHIYODA-KU, TOKYO 100
JAPANESE

NORIN SUISAN KOKU NENPO 3667
(ANNUAL REPORT OF AGRICULTURAL AVIATION)
NORIN SUISAN KOKU KYOKAI (AGRICULTURAL AVIATION ASSOCIATION)
11-35, 1-CHOME, NAGATA-CHO, CHIYODA-KU, TOKYO 100
JAPANESE

NORIN SUISAN SHIKEN KENKYU NENPO, NOGYO-HEN, KOKURITSU 3668
(ANNUAL REPORT ON EXPERIMENTS IN AGRICULTURE, FORESTRY AND FISHERIES, SERIES ON AGRICULTURE, NATIONAL ORGANIZATIONS)
NORIN-SHO NORIN SUISAN GIJUTSU KAIGI JIMUKYOKU (AGRICULTURE, FORESTRY AND FISHERIES RESEARCH COUNCIL, MINISTRY OF AGRICULTURE AND FORESTRY)
11-14, 2-CHOME, MEGURO, MEGURO-KU, TOKYO 153
JAPANESE
INDEXED: JICST

NORIN SUISAN SHIKEN KENKYU NENPO, NOGYO-HEN, KORITSU 3669
(ANNUAL REPORT OF AGRICULTURE, FORESTRY, FISHERIES EXPERIMENT RESEARCH. AGRICULTURE, NATIONAL.)
JAPANESE
INDEXED: JICST

NORIN SUISAN SHIKEN KENKYU NENPO, SHIKEN KENKYU KADAI ICHIRAN 3670
(ANNUAL REPORT ON EXPERIMENTS IN AGRICULTURE, FORESTRY AND FISHERIES, SERIES RESEARCH PROJECTS)
NORIN-SHO NORIN SUISAN GIJUTSU KAIGI JIMUKYOKU (AGRICULTURE, FORESTRY AND FISHERIES RESEARCH COUNCIL, MINISTRY OF AGRICULTURE AND FORESTRY)
11-14, 2-CHOME, MEGURO, MEGURO-KU, TOKYO 153
JAPANESE

NORIN SUISAN TOKEI GEPPO 3671
(MONTHLY STATISTICS OF AGRICULTURE, FORESTRY AND FISHERIES)
NORIN-SHO NORIN KEIZAI-KYOKU TOKEI JOHO-BU (STATISTICS AND INFORMATION DIVISION, AGRICULTURE AND FORESTRY ECONOMIC BUREAU, MINISTRY OF AGRICULTURE AND FORESTRY)
11-14, 2-CHOME, MEGURO, MEGURO-KU, TOKYO 153
JAPANESE

NORIN SUISAN TOKEI HOKOKU 3672
(AGRICULTURE, FORESTRY AND FISHERIES STATISTICAL REPORT)
JAPANESE
INDEXED: BIB

NORIN SUISAN TOKEI NO GAIYO 3673
(INTRODUCTION TO AGRICULTURAL, FORESTRY AND FISHERY STATISTICS))
NORIN-SHO NORIN KEIZAI-KYOKU TOKEI JOHO-BU (STATISTICS AND INFORMATION DEPARTMENT, AGRICULTURE AND FORESTRY ECONOMIC BUREAU, MINISTRY OF AGRICULTURE AND FORESTRY)
11-14, 2-CHOME, MEGURO, MEGURO-KU, TOKYO 153
JAPANESE

NORIN SUISAN TOKEI SHIHYO 3674
(STATISTICAL INDEX ON AGRICULTURE, FORESTRY AND FISHERIES)
NORIN-SHO NORIN KEIZAI-KYOKU TOKEI JOHO-BU (STATISTICS AND INFORMATION DIVISION, AGRICULTURE AND FORESTRY ECONOMIC BUREAU, MINISTRY OF AGRICULTURE AND FORESTRY)
11-14, 2-CHOME, MEGURO, MEGURO-KU, TOKYO 153
JAPANESE

NORIN SUISANGYO SEISAN SHISU 3675
(INDEX NUMBERS OF PRODUCTION OF AGRICULTURE, FORESTRY AND FISHERIES)
NORIN-SHO NORIN KEIZAI-KYOKU TOKEI JOHO-BU (STATISTICS AND INFORMATION DIVISION, AGRICULTURE AND FORESTRY ECONOMIC BUREAU, MINISTRY OF AGRICULTURE AND FORESTRY)
2-1, 1-CHOME, KASUMIGASEKI, CHIYODA-KU, TOKYO 100
JAPANESE

NORIN SUISANGYO SEISANGAKU (IWATE-KEN) 3676
(PRODUCTION COST OF AGRICULTURAL, FORESTRY AND FISHERY PRODUCTS)
IWATE-KEN (IWATE PREFECTURAL GOVERNMENT)
10-1, UCHIMARU, MORIOKA-SHI 020
JAPANESE

NORIN SUISANGYO TOKUBETSU SHIKEN KENKYUHI HOJOKIN NI YORU SHIKEN KENKYU KEKKA GAIYOSHO 3677
NORIN-SHO NORIN SUISAN GIJUTSU KAIGI JIMUKYOKU (AGRICULTURE, FORESTRY AND FISHERIES RESEARCH COUNCIL, MINISTRY OF AGRICULTURE AND FORESTRY)
2-1, 1-CHOME, KASUMIGASEKI, CHIYODA-KU, TOKYO 100
JAPANESE

NORIN TOKEI CHOSA 3678
(STATISTICAL SURVEY OF AGRICULTURE AND FORESTRY)
NORIN-SHO (MINISTRY OF AGRICULTURE AND FORESTRY) AND NORIN TOKEI KYOKAI
11-14, 2-CHOME, MEGURO, MEGURO-KU, TOKYO 153
JAPANESE
INDEXED: BIB

NORIN TOSHO SHIRYO GEPPO 3679
(MONTHLY OF MINISTRY OF AGRICULTURE AND FORESTRY DATA)
JAPANESE
INDEXED: JICST

NORIN-SHO IRAI KENKYUIN KENKYU HOKOKUSHO 3680
(REPORT OF STUDIES AT REQUESTS OF MINISTRY OF AGRICULTURE AND FORESTRY)
YAMANASHI-KEN (YAMANASHI PREFECTURAL GOVERNMENT)
6-1, 1-CHOME, MARUNOUCHI, KOFU-SHI 400
JAPANESE AND ENGLISH

NORIN-SHO KANKEI HOSHANO CHOSA KENKYU NENPO 3681
(ANNUAL REPORT ON RADIOACTIVITY RESEARCH, MINISTRY OF AGRICULTURE AND FORESTRY)
NORIN-SHO NORIN SUISAN GIJUTSU KAIGI JIMUKYOKU (AGRICULTURE, FORESTRY AND FISHERIES RESEARCH COUNCIL SECRETARIAT, MINISTRY OF AGRICULTURE AND FORESTRY)
2-1, 1-CHOME, KASUMIGASEKI, CHIYODA-KU, TOKYO 100
JAPANESE

NORIN-SHO NENPO 3682
(ANNUAL REPORT OF MINISTRY OF AGRICULTURE AND FORESTRY)
NORIN-SHO DAIJIN KANBO (MINISTER'S SECRETARIAT, MINISTRY OF AGRICULTURE AND FORESTRY) AND NIHON NOSON CHOSAKAI
C/O NORIN-SHO, 2-1, 1-CHOME, KASUMIGASEKI, CHIYODA-KU, TOKYO 100
JAPANESE
INDEXED: BIB

NORIN-SHO RINGYO SHIKENJO NENPO 3683
(ANNUAL OF FORESTRY EXPERIMENT STATION, MINISTRY OF AGRICULTURE AND FORESTRY)
JAPANESE
INDEXED: JICST

NORIN-SHO TOKEIHYO 3684
(STATISTICAL YEARBOOK OF THE MINISTRY OF AGRICULTURE AND FORESTRY, JAPAN)
NORIN-SHO TOKEI JOHO-BU (STATISTICS AND INFORMATION DIVISION, MINISTRY OF AGRICULTURE AND FORESTRY) AND NORIN TOKEI KYOKAI
11-14, 2-CHOME, MEGURO, MEGURO-KU, TOKYO 153
JAPANESE
INDEXED: BIB

NORINGYO MONDAI KENKYU 3685
(JOURNAL OF RURAL PROBLEMS)
KANSAI NOGYO KEIZAI GAKKAI (KANSAI AGRICULTURAL ECONOMIC SOCIETY) AND FUMIN KYOKAI
2-36, DOJIMAKAMI, KITA-KU, OSAKA-SHI 530
JAPANESE
INDEXED: BIB

NORINGYO NO UGOKI (SHIMANE-KEN) 3686
(TENDENCY OF AGRICULTURE AND FORESTRY, SHIMANE PREFECTURE)
SHIMANE-KEN (SHIMANE PREFECTURAL GOVERNMENT)
1, TONO-MACHI, MATSUE-SHI 690
JAPANESE

NOSAN NENPO 3687
(ANNUAL REPORT OF CEREAL CROP PRODUCTION)
NORIN-SHO NOSEI-KYOKU (AGRICULTURAL ADMINISTRATION BUREAU, MINISTRY OF AGRICULTURE AND FORESTRY)
2-1, 1-CHOME, KASUMIGASEKI, CHIYODA-KU, TOKYO 100
JAPANESE

NOSANBUTSU KENSA TOKUHON 3688
(GUIDE TO INSPECTION OF AGRICULTURAL PRODUCTS)
NIHON NOMIN SHINBUNSHA
13-14, 2-CHOME, NISHISHINBASHI, MINATO-KU, TOKYO 105
JAPANESE

NOSANBUTSU NO MITOSHI (GUNMA-KEN) 3689
(OUTLOOK OF AGRICULTURE, GUNMA PREFECTURE)
GUNMA-KEN NOSEI-BU (AGRICULTURAL ADMINISTRATION DIVISION, GUNMA PREFECTURAL GOVERNMENT)
1-1, 1-CHOME, OTE-MACHI, MAEBASHI-SHI 371
JAPANESE

NOSANBUTSU NO SEISAN RYUTSU TOKEI 3690
(STATISTICS OF AGRICULTURAL PRODUCTS CIRCULATION)
HOKKAIDO NOMU-BU (AGRICULTURAL ADMINISTRATION DEPARTMENT, HOKKAIDO PREFECTURAL GOVERNMENT)
NISHI-6-CHOME, KITA-3-JO, CHUO-KU, SAPPORO-SHI 060
JAPANESE

NOSANBUTSU SEISANHI CHOSA HOKOKU, YASAI SEISANHI 3691
(COST OF PRODUCTION OF IMPORTANT AGRICULTURAL PRODUCTS)
NORIN-SHO TOKEI JOHO-BU (STATISTICS AND INFORMATION DIVISION, MINISTRY OF AGRICULTURE AND FORESTRY) AND NORIN TOKEI KYOKAI
11-14, 2-CHOME, MEGURO, MEGURO-KU, TOKYO 153
JAPANESE

NOSANBUTSU SHIJO NO UGOKI (HYOGO-KEN) 3692
(OUTLOOK OF AGRICULTURE, HYOGO PREFECTURE)
HYOGO-KEN NORIN-BU (AGRICULTURE AND FORESTRY DIVISION, HYOGO PREFECTURAL GOVERNMENT)
5-1, SHIMOYAMATE-DORI, IKUTA-KU, KOBE-SHI 650
JAPANESE

NOSEI CHOSA JIHO 3693
(AGRICULTURAL ADMINISTRATION INQUIRY REVIEW)
ZENKOKU NOGYO KAIGIJO (NATIONAL AGRICULTURAL ASSEMBLY HALL)
JAPANESE

NOSEI JOHO 3694
(JOURNAL OF AGRICULTURAL ADMINISTRATION)
ZENKOKU NOGYO KAIGISHO (JAPAN CHAMBER OF AGRICULTURE)
1-7, YURAKU-CHO, CHIYODA-KU, TOKYO 100
JAPANESE

NOSEI KYUSHU 3695
(AGRICULTURAL ADMINISTRATION IN KYUSHU)
NORIN-SHO KYUSHU NOSEI-KYOKU KIKAKU CHOSEISHITSU (KYUSHU AGRICULTURAL ADMINISTRATION BUREAU, MINISTRY OF AGRICULTURE AND FORESTRY)
23-18, 4-CHOME, KASUGA, KUMAMOTO-SHI 860
JAPANESE

NOSEI SHIRYO 3696
(AGRICULTURAL ADMINISTRATION DATA)
NOSEI CHOSAKAI (AGRICULTURAL ADMINISTRATIVE BOARD OF INQUIRY)
JAPANESE

NOSEIKEN NYUSU 3697
(AGRICULTURAL BIOLOGY RESEARCH NEWS)
NIHON NOGYO SEIBUTSUGAKU KENKYUKAI (JAPAN AGRICULTURAL BIOLOGY RESEARCH SOCIETY)
JAPANESE

NOSHI REPOTO 3698
(REPORT OF ISHIKAWA AGRICULTURAL EXPERIMENT STATION)
ISHIKAWA-KEN NOGYO SHIKENJO (ISHIKAWA AGRICULTURAL EXPERIMENT STATION)
86, NAKABAYASHI, NONOICHI-MACHI, ISHIKAWA-GUN, ISHIKAWA 921
JAPANESE

NOSON KAIHATSU 3699
(DEVELOPMENT OF RURAL COMMUNITY)
SHIMANE DAIGAKU NOGAKUBU NOSANSON CHIIKI KAIHATSU KENKYU CHOSASHITSU
1060, NISHIKAWAZU-CHO, MATSUE-SHI 690
JAPANESE

NOSON KENKYU 3700
(JOURNAL OF RURAL COMMUNITY STUDIES)
TOKYO NOGYO DAIGAKU NOGYO KEIZAI GAKKAI (AGRICULTURAL ECONOMIC SOCIETY, TOKYO AGRICULTURAL UNIVERSITY)
1-1, 1-CHOME, SAKURAGAOKA, SETAGAYA-KU, TOKYO 156
JAPANESE
INDEXED: JICST

NOSON KIKI NYUSU 3701
(NEWS OF AGRICULTURAL MACHINERY NEWS)
KOKUSAI NOGYOSHA
20-8, 3-CHOME, UENO, TAITO-KU, TOKYO 110
JAPANESE

NOSON SEIKATSU KENKYU 3702
(JOURNAL OF THE RURAL LIFE SOCIETY OF JAPAN)
NIHON NOSON SEIKATSU KENKYUKAI (RURAL LIFE SOCIETY OF JAPAN)
C/O NOGYO GIJUTSU KENKYUSHO, 1-7, 2-CHOME, NISHIGAHARA-CHO, KITA-KU, TOKYO 114
JAPANESE AND ENGLISH

NOSON TO TOSHI O MUSUBU 3703
(JOURNAL OF COMMUNICATION BETWEEN RURAL COMMUNITIES AND TOWNS)
NOSON TO TOSHI O MUSUBU HENSHU-BU
C/O ZEN'NORIN, 2-1, 1-CHOME, KASUMIGASEKI, CHIYODA-KU, TOKYO 100
JAPANESE

NOSUISAN GEPPO 3704
(MONTHLY BULLETIN OF AGRICULTURE AND FISHERIES)
NIHON BOEKI SHINKOKAI NOSUISAN-BU (JAPAN EXTERNAL TRADE ORGANIZATION)
2, AKASAKA AOI-CHO, MINATO-KU, TOKYO 107
JAPANESE

NOYU 3705
(AGRICULTURE IN FUKUSHIMA)
FUKUSHIMA-KEN NOYUKAI (FUKUSHIMA PREFECTURE AGRICULTURAL ASSOCIATION)
C/O FUKUSHIMA-KEN NOGYO SHIKENJO, 20, WAKAMIYAMAE, KORIYAMA-SHI 963
JAPANESE

OITA NORIN SUISAN TOKEI NENPO 3706
(STATISTICAL YEARBOOK OF AGRICULTURE, FORESTRY AND FISHERY, OITA)
NORIN-SHO KYUSHU NOSEI-KYOKU OITA TOKEI JOHO JIMUSHO (OTIS STATISTICS AND INFORMATION OFFICE, KYUSHU AGRICULTURAL ADMINISTRATION BUREAU, MINISTRY OF AGRICULTURE AND FORESTRY)
1-36, 2-CHOME, SHINKAWA-CHO, OITA-SHI 870
JAPANESE

OITA-KEN NOGYO GIJUTSU SENTA GIJUTSU HOKOKU 3707
(OITA PREFECTURE AGRICULTURAL TECHNOLOGY CENTER TECHNICAL BULLETIN)
JAPANESE

OITA-KEN NOGYO GIJUTSU SENTA GYOMU HOKOKU 3708
(OPERATION REPORT OF OITA PREFECTURAL AGRICULTURAL RESEARCH CENTER)
OITA-KEN NOGYO GIJUTSU CENTER (OITA PREFECTURAL AGRICULTURAL RESEARCH CENTER)
65, KITAUSA, USA-SHI 872-01
JAPANESE

OITA-KEN NOGYO GIJUTSU SENTA KENKYU HOKOKU 3709
(BULLETIN OF THE OITA PREFECTURAL AGRICULTURAL RESEARCH CENTER)
OITA-KEN NOGYO GIJUTSU SENTA (OITA PREFECTURAL AGRICULTURAL RESEARCH CENTER)
65, KITAUSA, USA-SHI 872-01
JAPANESE

OITA-KEN NOGYO GIJUTSU SENTA KNEKYU SOKUHO 3710
(RESEARCH REPORT OF OITA PREFECTURAL AGRICULTURAL RESEARCH CENTER)
OITA-KEN NOGYO GIJUTSU SENTA (OITA PREFECTURAL AGRICULTURAL RESEARCH CENTER)
65, KITAUSA, USA-SHI 872-01
JAPANESE

OITA-KEN NOGYO GIJUTSU SENTA SHIKEN NYUSU 3711
(OITA PREFECTURE AGRICULTURAL TECHNOLOGY CENTER TESTING NEWS)
JAPANESE

OITA-KEN NOGYO NO UGOKI 3712
(CENSUS OF AGRICULTURE IN OITA PREFECTURE)
NORIN-SHO KYUSHU NOSEI-KYOKU OITA TOKEI JOHO JIMUSHO (OITA STATISTICS AND INFORMATION OFFICE, KYUSHU AGRICULTURAL ADMINISTRATION BUREAU, MINISTRY OF AGRICULTURE AND FORESTRY)
JAPANESE

OITA-KEN ONSEN NETSU RIYO NOGYO KENKYUSHO SHIKEN SEISEKISHO 3713
(REPORT OF OITA RESEARCH STATION FOR AGRICULTURAL UTILIZATION OF HOTSPRING)
OITA-KEN ONSEN NETSU RIYO NOGYO KENKYUSHO
TSURUMI, BEPPU-SHI 874
JAPANESE

OKAYAMA DAIGAKU NOGAKUBU DOJO HIRYOGAKU KYOSHITSU HOKOKU 3714
(REPORTS OF THE LABORATORY OF SOILS AND FERTILIZERS, THE FACULTY OF AGRICULTURE, OKAYAMA UNIVERSITY)
OKAYAMA DAIGAKU NOGAKUBU DOJO HIRYOGAKU KYOSHITSU (LABORATORY OF SOILS AND FERTILIZERS, THE FACULTY OF AGRICULTURE, OKAYAMA UNIVERSITY)
TUSHIMA, OKAYAMA-SHI 700
JAPANESE, WITH ENGLISH SUMMARY
INDEXED: BIO, CHE

OKAYAMA DAIGAKU NOGAKUBU GAKUJUTSU HOKOKU 3715
(SCIENTIFIC REPORTS OF THE FACULTY OF AGRICULTURE, OKAYAMA UNIVERSITY)
OKAYAMA DAIGAKU NOGAKUBU (FACULTY OF AGRICULTURE, OKAYAMA UNIVERSITY)
TSUSHIMA, OKAYAMA-SHI 700
JAPANESE AND ENGLISH
INDEXED: BIB, JICST

OKAYAMA NORIN SUISAN TOKEI NENPO 3716
(STATISTICAL YEARBOOK OF AGRICULTURE, FORESTRY AND FISHERY, OKAYAMA)
NORIN-SHO CHUGOKU SHIKOKU NOSEI-KYOKU TOKEI JOHO-BU (CHUGOKU SHIKOKU AGRICULTURAL ADMINISTRATION BUREAU, MINISTRY OF AGRICULTURE AND FORESTRY)
9-24, TENJIN-CHO, OKAYAMA-SHI 700
JAPANESE

OKAYAMA-KENRITSU NOGYO SHIKENJO KENKYU HOKOKU 3717
(BULLETIN OF THE OKAYAMA PREFECTURAL AGRICULTURAL EXPERIMENT STATION)
JAPANESE

OKAYAMA-KENRITSU NOGYO SHIKENJO KENKYU NENPO 3718
(ANNUAL REPORT OF OKAYAMA PREFECTURAL AGRICULTURAL EXPERIMENT STATION)
OKAYAMA-KENRITSU NOGYO SHIKENJO (OKAYAMA PREFECTURAL AGRICULTURAL EXPERIMENT STATION)
KODA,SANYO-MACHI, AKAIWA-GUN, OKAYAMA 709-08
JAPANESE

OKAYAMA-KENRITSU NOGYO SHIKENJO RINJI HOKOKU 3719
(SPECIAL BULLETIN OF THE OKAYAMA PREFECTURAL AGRICULTURAL EXPERIMENT STATION)
OKAYAMA-KENRITSU NOGYO SHIKENJO (OKAYAMA PREFECTURAL AGRICULTURAL EXPERIMENT STATION)
KODA, SANYO-MACHI, AKAIWA-GUN, OKAYAMA 709-08
JAPANESE
INDEXED: CHE

OKINAWA NOGYO SENSASU HOKOKUSHO 3720
(CENSUS OF AGRICULTURE, OKINAWA)
OKINAWA-KEN (OKINAWA PREFECTURAL GOVERNMENT)
2-14, 1-CHOME, SENZAKI, NAHA-SHI 900
JAPANESE

OKINAWA NORIN SUISAN TOKEI NENPO 3721
(ANNUAL STATISTICS OF AGRICULTURE, FORESTRY AND FISHERY, OKINAWA))
OKINAWA-KEN NORIN SUISAN-BU (AGRICULTURAL, FORESTRY AND FISHERY DEPARTMENT, OKINAWA PREFECTURAL GOVERNMENT)
C/O MATSUYA SANGYO BUILDING, 21-5, 2-CHOME, MAEJIMA, NAHA-SHI 900
JAPANESE

OKINAWA-KEN NOGYO SHIKENJO GYOMU NENPO 3722
(REPORT ON ACTIVITIES OF OKINAWA AGRICULTURAL EXPERIMENTAL STATION)
OKINAWA-KEN NOGYO SHIKENJO (OKINAWA AGRICULTURAL EXPERIMENTAL STATION)
4-222, SHURI SAKIYAMA-CHO, NAHA-SHI 900
JAPANESE

OKINAWA-KEN NOGYO SHIKENJO KENKYU HOKOKU 3723
(BULLETIN, THE OKINAWA AGRICULTURAL EXPERIMENTAL STATION)
OKINAWA-KEN NOGYO SHIKENJO (OKINAWA AGRICULTURAL EXPERIMENTAL STATION)
4-222, SHURI SAKIYAMA-CHO, NAHA-SHI 900
JAPANESE

OKUNOTO KAIHATSU PAIROTTO NOJO RYUSHUTSU KIKO NI KANSURU CHOSA SEISEKISHO 3724
(LIST OF THE RESEARCH ON THE MECHANISM OF EFFLUXION IN OKUNOTO DEVELOPMENT PILOT FARM)
ISHIKAWA-KEN NORIN-BU (AGRICULTURE AND FORESTRY DEPARTMENT, ISHIKAWA PREFECTURAL GOVERNMENT)
1-1, 2-CHOME, HIROSAKA, KANAZAWA-SHI 920
JAPANESE

OSAKA HISHIRYO KENSAJO JIGYO HOKOKU, HIRYO NO BU 3725
(ANNUAL REPORT OF THE OSAKA FERTILIZER AND FEED INSPECTION STATION, FERTILIZER SECTION)
NORIN-SHO OSAKA HISHIRYO KENSAJO (OSAKA FERTILIZER AND FEED INSPECTION STATION, MINISTRY OF AGRICULTURE AND FORESTRY)
C/O OSAKA GODOCHOSHA, 6-25, HOENZAKA-CHO, HIGASHI-KU, OSAKA-SHI 540
JAPANESE

OSAKA NO NOGYO (OSAKA-FU NOGYO CHOSA KEKKA HOKOKU) 3726
(REPORT ON AGRICULTURAL CENSUS, OSAKA PREFECTURE)
OSAKA-FU KIKAKU-BU (PLANNING DEPARTMENT, OSAKA PREFECTURAL GOVERNMENT)
2, OTEMAENO-CHO, HIGASHI-KU, OSAKA-SHI 540
JAPANESE

OSAKA NO NORIN GYOGYO 3727
(AGRICULTURE, FORESTRY AND FISHERIES OF OSAKA)
OSAKA-FU NORIN-BU (AGRICULTURE AND FORESTRY DIVISION, OSAKA PREFECTURAL GOVERNMENT)
2, OTEMAENO-CHO, HIGASHI-KU, OSAKA-SHI 540
JAPANESE

OSAKA NO SAKUMOTSU TOKEI 3728
(STATISTICS OF AGRICULTURAL PRODUCTS IN OSAKA PREFECTURE)
NORIN-SHO KINKI NOSEI-KYOKU OSAKA TOKEI JOHO KIMUSHO (OSAKA STATISTICS AND INFORMATION OFFICE, KINKI AGRICULTURAL ADMINISTRATION BUREAU, MINISTRY OF AGRICULTURE AND FORESTRY)
OTEMAENO-CHO, HIGASHI-KU, OSAKA-SHI 540
JAPANESE

OSAKA NOGYO 3729
(AGRICULTURE OF OSAKA PREFECTURE)
OSAKA-FU NORIN GIJUTSU SENTA (OSAKA AGRICULTURAL RESEARCH CENTER)
442, SHAKUDO, HABIKINO-SHI 583
JAPANESE

OSAKA NORIN SUISAN TOKEI NENPO 3730
(STATISTICAL YEARBOOK OF AGRICULTURE, FORESTRY AND FISHERY, OSAKA)
NORIN-SHO KINKI NOSEI-KYOKU OSAKA TOKEI JOHO KIMUSHO (OSAKA STATISTICS AND INFORMATION OFFICE, KINKI AGRICULTURAL ADMINISTRATION BUREAU, MINISTRY OF AGRICULTURE AND FORESTRY)
C/O OSAKA GODOCHOSHA, OTEMAENO-CHO, HIGASHI-KU, OSAKA-SHI 532
JAPANESE

OSAKA-FU NOGYO NO UGOKI 3731
(TENDENCY OF AGRICULTURE IN OSAKA PREFECTURE)
NORIN-SHO KINKI NOSEI-KYOKU (KINKI AGRICULTURAL ADMINISTRATION BUREAU, MINISTRY OF AGRICULTURE AND FORESTRY)
102 SHIMOCHOJA-MACHISAGARU, NISHINOTOIN-DORI, KAMIGYO-KU, KYOTO-SHI 602
JAPANESE

OSAKA-FU NORIN GIJUTSU SENTA GYOMU NENPO 3732
(ANNUAL REPORT ON ACTIVITIES OF OSAKA AGRICULTURAL RESEARCH CENTER)
OSAKA-FU NORIN GIJUTSU SENTA (OSAKA AGRICULTURAL RESEARCH CENTER)
442, SHAKUDO, HABIKINO-SHI 583
JAPANESE

OSAKA-FU NORIN GIJUTSU SENTA KENKYU HOKOKU 3733
(BULLETIN OF THE OSAKA AGRICULTURAL RESEARCH CENTER)
OSAKA-FU NORIN GIJUTSU SENTA (OSAKA AGRICULTURAL RESEARCH CENTER)
442, SHAKUDO, HABIKINO-SHI 583
JAPANESE, WITH ENGLISH SUMMARY

OSAKA-FU NORIN GIJUTSU SENTA NYUSU 3734
(OSAKA AGRICULTURAL RESEARCH CENTER NEWS)
OSAKA-FU NORIN GIJUTSU SENTA (OSAKA AGRICULTURAL RESEARCH CENTER)
442, SHAKUDO, HABIKINO-SHI 583
JAPANESE

OSAKA-FU NORIN SUISAN SHICHOSONBETSU TOKEISHO 3735
(STATISTICS OF AGRICULTURE, FORESTRY AND FISHERY BY TOWNS, OSAKA PREFECTURE)
NORIN-SHO KINKI NOSEI-KYOKU OSAKA TOKEI JOHO JIMUSHO (OSAKA STATISTICS AND INFORMATION OFFICE, KINKI AGRICULTURAL ADMINISTRATION BUREAU, MINISTRY OF AGRICULTURE AND FORESTRY)
JAPANESE

POKETTO KANAGAWA NORIN SUISAN TOKEI 3736
(POCKET STATISTICAL YEARBOOK OF AGRICULTURE, FORESTRY AND FISHERY, KANAGAWA)
NORIN-SHO KANTO NOSEI-KYOKU (KANTO AGRICULTURAL ADMINISTRATION BUREAU, MINISTRY OF AGRICULTURE AND FORESTRY)
3-3, 1-CHOME, OTE-MACHI, CHIYODA-KU, TOKYO 100
JAPANESE

POKETTO NORIN SUISAN TOKEI 3737
(POCKET STATISTICAL BOOK OF AGRICULTURE, FORESTRY AND FISHERIES)
NORIN-SHO TOKEI JOHO-BU (STATISTICS AND INFORMATION DEPARTMENT, MINISTRY OF AGRICULTURE AND FORESTRY) AND NORIN TOKEI KYOKAI (ASSOCIATION OF AGRI-CULTURE-FORESTRY STATISTICS)
C/O OTORI BUILDING, 11-14, 2-CHOME, MEGURO, MEGURO-KU, TOKYO 153
JAPANESE
INDEXED: BIB

RECORDS OF RESEARCHES IN THE FACULTY OF AGRI-CULTURE, UNIVERSITY OF TOKYO 3738
JAPANESE
INDEXED: CHE, JICST

REPORTS OF THE INSTITUTE FOR AGRICULTURAL RESEARCH, TOHOKU UNIVERSITY 3739
TOHOKU DAIGAKU NOGAKU KENKYUSHO (INSTITUTE FOR AGRICULTURAL RESEARCH, TOHOKU UNIVERSITY)
1-34, 1-CHOME, KOMEGFUKURO, SENDAI-SHI 980
ENGLISH
INDEXED: BIB, BIO, CHE, JICST

RINGYO TOKEI YORAN 3740
(OUTLINE OF FORESTRY STATISTICS)
NORINSHO SANRINKYOKU
JAPANESE
INDEXED: BIB

RINO GIJUTSU 3741
(AGRICULTURAL TECHNOLOGY)
NIHON RINO KYOKAI (JAPANESE ASSOCIATION FOR AGRICULTURAL TECHNOLOGY)
14-1, 3-CHOME, NISHIHARA, SHIBUYA-KU, TOKYO 151
JAPANESE

RYUAN KYOKAIHO 3742
(BULLETIN OF JAPAN AMMONIUM SULPHATE INDUSTRY ASSOCIATION)
NIHON RYUAN KOGYO KYOKAI (JAPAN AMMONIUM SULPHATE INDUSTRY ASSOCIATION)
C/O HOKKAI BUILDING, 3-13, 1-CHOME, NIHONBASHI-DORI, CHOI-KU, TOKYO 103
JAPANESE

RYUKYU DAIGAKU NOGAKUBU GAKUJUTSU HOKOKU 3743
(SCIENCE BULLETIN OF THE COLLEGE OF AGRICULTURE, UNIVERSITY OF THE RYUKYUS)
RYUKYU DAIGAKU NOGAKUBU (COLLEGE OF AGRICULTURE, UNIVERSITY OF THE RYUKYUS)
TONOKURA-CHO, SHURI, NAHA-SHI 903
JAPANESE AND ENGLISH
INDEXED: BIB, BIO, CHE, JICST

RYUTSU RIYO SHIKEN SEISEKISHO 3744
(LIST OF THE EXAMINATION IN CIRCULATING UTILIZATION)
AICHI-KEN NOGYO SOGO SHIKENJO KISO KENKYU-BU (BASIC RESEARCH DEPARTMENT, AICHI-KEN AGRICULTURE RESEARCH CENTER)
NAGAKUTE-MURA, AICHI-GUN, AICHI 480-11
JAPANESE

SABRAO JOURNAL 3745
(SOCIETY FOR THE ADVANCEMENT OF BREEDING RESEARCH IN ASIA AND OCEANIA)
ENGLISH
INDEXED: BIB

SAGA DAIGAKU NOGAKU IHO 3746
(AGRICULTURAL BULLETIN OF THE SAGA UNIVERSITY)
SAGA DAIGAKU NOGAKUBU (FACULTY OF AGRICULTURE, SAGA UNIVERSITY)
1, HONJO-CHO, SAGA-SHI 840
JAPANESE AND ENGLISH
INDEXED: BIB, CHE, JICST

SAGA NORIN SUISAN TOKEI NENPO 3747
(STATISTICAL YEARBOOK OF AGRICULTURE, FORESTRY AND FISHERY, SAGA)
NORIN-SHO KYUSHU NOSEI-KYOKU SAGA TOKEI JOHO JIMUSHO (SAGA STATISTICS AND INFORMATION OFFICE, KYUSHU AGRICULTURAL ADMINISTRATION BUREAU, MINISTRY OF AGRICULTURE AND FORESTRY)
10-20, 2-CHOME, JONAI, SAGA-SHI 840
JAPANESE

SAGA-KEN HATACHI EINO SHIDOSHO GYOMU HOKOKU 3748
(BULLETIN OF THE SAGA-KEN GENERAL CROP FARMER GUIDANCE CENTER)
SAGA-KEN HATACHI EINO SHIDOSHO
EZARUGI, KARATSU-SHI 847-01
JAPANESE

SAGA-KEN NOGYO SHIKENJO GYOMU NENPO 3749
(ANNUAL REPORT OF THE SAGA AGRICULTURAL EXPERIMENT STATION)
SAGA-KEN NOGYO SHIKENJO (SAGA PREFECTURE AGRICUL-TURAL EXPERIMENT STATION)
NANRI, KAWAZOE-CHO, SAGA-GUN, SAGA 840-23
JAPANESE

SAGA-KEN NOGYO SHIKENJO KENKYU HOKOKU 3750
(BULLETIN OF THE SAGA AGRICULTURAL EXPERIMENT STATION)
SAGA-KEN NOGYO SHIKENJO (SAGA PREFECTURE AGRICULTURAL EXPERIMENT STATION)
NANRI, KAWAZOE-CHO, SAGA-GUN, SAGA 840-23
JAPANESE

SAITAMA NORIN TOKEI NENPO 3751
(STATISTICAL YEARBOOK OF AGRICULTURE AND FORESTRY, SAITAMA)
NORINSHO SAITAMA KANTO NOSEI-KYOKU TOKEI JOHO JIMUSHO (SAITAMA STATISTICS AND INFORMATION OFFICE, KANTO AGRICULTURAL ADMINISTRATION BUREAU, MINISTRY OF AGRICULTURE AND FORESTRY)
8-11, 5-CHOME, KISHI-MACHI, URAWA-SHI 336
JAPANESE

SAITAMA-KEN NOGYO SHIKENJO KENKYU HOKOKU 3752
(BULLETIN OF THE SAITAMA PREFECTURAL AGRICULTURAL EXPERIMENT STATION)
SAITAMA-KEN NOGYO SHIKENJO (SAITAMA PREFECTURAL AGRICULTURAL EXPERIMENT STATION)
1372, JUBOJIMA, JUMAGAYA-SHI 360
JAPANESE
INDEXED: JICST

SAITAMA-KEN NOGYO SHIKENJO NENPO 3753
(ANNUAL REPORT OF THE SAITAMA PREFECTURAL AGRICULTURAL EXPERIMENT STATION)
SAITAMA-KEN NOGYO SHIKENJO (SAITAMA PREFECTURAL AGRICULTURAL EXPERIMENT STATION)
1372, KUBOJIMA, JUMAGAYA-SHI 360
JAPANESE

SAKKYO KISO SHIKEN KENKYU SEISEKI 3754
(REPORT ON STUDIES OF CLIMATIC EFFECTS ON CROP GROWTH AND YIELDS)
NORIN-SHO NORIN KEIZAI-KYOKU TOKEI JOHO-BU (STATISTICS AND INFORMATION DIVISION, AGRICULTURE AND FORESTRY ECONOMIC BUREAU, MINISTRY OF AGRICULTURE AND FORESTRY)
JAPANESE

SAKUMOTSU TOKEI 3755
(CROP PRODUCTION STATISTICS)
NORIN-SHO NORIN KEIZAI-KYOKU TOKEI JOHO-BU (STATISTICS AND INFORMATION DIVISION, AGRICULTURE AND FORESTRY ECONOMIC BUREAU, MINISTRY OF AGRICULTURE AND FORESTRY)
11-14, 2-CHOME, MEGURO, MEGURO-KU, TOKYO 153
JAPANESE

SAKYU KENKYU 3756
(SAND DUNE RESEARCH)
NIHON SAKYU KENKYUKAI (RESEARCH ASSOCIATION OF SAND DUNE)
C/O TOTTORI DAIGAKU NOGAKUBU SAKYU KENKYUSHO, HAMASAKA, TOTTORI-SHI 680
JAPANESE

SAPPORO HISHIRYO KENSAJO JIGYO HOKOKU, HIRYO NO BU 3757
(ANNUAL REPORT OF THE SAPPORO FERTILIZER AND FEED INSPECTION STATION, FERTILIZER SECTION)
NORIN-SHO SAPPORO HISHIRYO KENSAJO (SAPPORO FERTILIZER AND FEED INSPECTION STATION, MINISTRY OF AGRICULTURE AND FORESTRY)
NISHI-19-CHOME, KITA-2-JO, CHUO-KU, SAPPORO-SHI 060
JAPANESE

SAROBETSU JIKKEN NOJO CHOSA HOKOKUSHO 3758
(REPORT OF SAROBETSU AGRICULTURAL EXPERIMENT FARM)
HOKKAIDO KAIHATSU-CHO HOKKAIDO KAIHATSU-KYOKU NOGYO SUISAN-BU (HOKKAIDO DEVELOPMENT BUREAU)
NISHI-4-CHOME, KITA-3-JO, CHUO-KU, SAPPORO-SHI 060
JAPANESE

SCIENCE REPORT OF SHIMA MARINELAND 3759
SHIMA MARINELAND
KASHIKOJIMA, SHIMA, MIE 309
JAPANESE AND ENGLISH
INDEXED: BIO

SEKAI NO NORIN SUISAN 3760
(AGRICULTURE, FORESTRY AND FISHERY OF THE WORLD)
KOKUSAI SHOKURYO NOGYO KYOKAI (JAPAN FAO ASSOCIATION)
C/O BAJI CHIKUSAN KAIKAN, 1-2, KANDA SURUGADAI, CHIYODA-KU, TOKYO 101
JAPANESE

SENDAI HISHIRYO KENSAJO JIGYO HOKOKU, HIRYO NO BU 3761
(ANNUAL REPORT OF THE SENDAI FERTILIZER AND FEED INSPECTION STATION, FERTILIZER SECTION)
NORIN-SHO SENDAI HISHIRYO KENSAJO (SENDAI FERTILIZER AND FEED INSPECTION STATION, MINISTRY OF AGRICULTURE AND FORESTRY)
2-23, 3-CHOME, HON-CHO, SENDAI-SHI 980
JAPANESE

SENSHU DAIGAKU BIBAI NOKO TANKI DAIGAKU NENPO 3762
(JOURNAL OF HOKKAIDO JUNIOR COLLEGE SENSHU UNIVERSITY)
SENSHU DAIGAKU BIBAI NOKO TANKI DAIGAKU
1610-1, BIBAI, BIBAI-SHI 072
JAPANESE

SHIGA NO NOGYO 3763
(AGRICULTURE IN SHIGA PREFECTURE)
SHIGA-KEN NOGYO SHIKENJO (SHIGA PREFECTURE AGRICULTURAL EXPERIMENT STATION)
8-4, 2-CHOME, NISHISHIBUKAWA, KUSATSU-SHI 525
JAPANESE

SHIGA NORIN SUISAN TOKEI NENPO 3764
(STATISTICAL YEARBOOK OF AGRICULTURE, FORESTRY AND FISHERY, SHIGA)
NORIN-SHO KINKI NOSEI-KYOKU SHIGA TOKEI JOHO KIMUSHO (SHIGA STATISTICS AND INFORMATION OFFICE, KINKI AGRICULTURAL ADMINISTRATION BUREAU, MINISTRY OF AGRICULTURE AND FORESTRY)
6-5, MIYUKI-CHO, OTSU-SHI 520
JAPANESE

SHIGA-KEN NOGYO SHIKENJO GYOMU HOKOKU 3765
(ANNUAL REPORTS OF THE SHIGA PREFECTURE AGRICULTURAL EXPERIMENT STATION)
SHIGA-KEN NOGYO SHIKENJO (SHIGA PREFECTURE AGRICULTURAL EXPERIMENT STATION)
8-4, 2-CHOME, NISHSHIBUKAWA, KUSATSU-SHI 525
JAPANESE

SHIGA-KEN NOGYO SHIKENJO KENKYU HOKOKU 3766
(SCIENTIFIC REPORTS OF SHIGA PREFECTURE AGRICULTURE EXPERIMENT STATION)
SHIGA-KEN NOGYO SHIKENJO (SHIGA PREFECTURE AGRICULTURAL EXPERIMENT STATION)
8-4, 2-CHOME, NISHISHIBUKAWA, KUSATSU-SHI 525
JAPANESE, WITH ENGLISH SUMMARY

SHIGA-KEN NOGYO SHIKENJO TOKUBETSU HOKOKU 3767
(SPECIAL REPORT OF SHIGA PREFECTURE AGRICULTURAL EXPERIMENT STATION)
SHIGA-KEN NOGYO SHIKENJO (SHIGA PREFECTURE AGRICULTURAL EXPERIMENT STATION)
8-4, 2-CHOME, NISHISHIBUKAWA, KUSATSU-SHI 525
JAPANESE, WITH ENGLISH SUMMARY

SHIGA-KEN SHICHOSONBETSU NORIN SUISAN TOKEISHO 3768
(STATISTICS OF AGRICULTURE, FORESTRY AND FISHERY BY TOWNS, SHIGA PREFECTURE)
NORIN-SHO KINKI NOSEI-KYOKU SHIGA TOKEI JOHO JIMUSHO (SHIGA STATISTICS AND INFORMATION OFFICE, KINKI AGRICULTURAL ADMINISTRATION BUREAU, MINISTRY OF AGRICULTURE AND FORESTRY)
JAPANESE

SHIKEN KENKYU NO UCHI JITSUYOKA NO IKI NI TASSHITA SEISEKI GAIYO 3769
(SUMMARY REPORTS ON THE EXPERIMENTAL RESEARCH AVAILABLE FOR PRACTICAL USE)
IBARAKI-KEN NOGYO SHIKENJO (IBARAKI PREFECTURAL EXPERIMENT STATION)
KAMIKUNII-CHO, MITO-SHI 311-42
JAPANESE

SHIKOKU NOGYO NO SHINGIJUTSU 3770
(ZENTSUJI, SHIKOKU AGRICULTURAL EXPERIMENT STATION)
JAPANESE
INDEXED: BIB

SHIKOKU NOGYO SHIKENJO HOKOKU 3771
(BULLETIN OF THE SHIKOKU AGRICULTURAL EXPERIMENT STATION)
NORIN-SHO SHIKOKU NOGYO SHIKENJO (SHIKOKU AGRICULTURAL EXPERIMENT STATION, MINISTRY OF AGRICULTURE AND FORESTRY)
3, ZENTSUJI-CHO, ZENTSUJI-SHI 760
JAPANESE AND ENGLISH
INDEXED: BIB, BIO, CHE

SHIKOKU NOSHI NYUSU 3772
(SHIKOKU AGRICULTURAL EXPERIMENT STATION NEWS)
NORIN-SHO SHIKOKU NOGYO SHIKENJO (SHIKOKU AGRICULTURAL EXPERIMENT STATION, MINISTRY OF AGRICULTURE AND FORESTRY)
3, ZENTSUJI-CHO, ZENTSUJI-SHI 765
JAPANESE

SHIMANE DAIGAKU NOGAKUBU KENKYU HOKOKU 3773
(BULLETIN OF THE FACULTY OF AGRICULTURE, SHIMANE UNIVERSITY)
SHIMANE DAIGAKU NOGAKUBU (FACULTY OF AGRICULTURE, SHIMANE UNIVERSITY)
1060, NISHIKAWATSU-CHO, MATSUE-SHI 690
JAPANESE AND ENGLISH
INDEXED: BIB, BIO, CHE, EXC

SHIMANE NORIN SUISAN TOKEI NENPO 3774
(STATISTICAL YEARBOOK OF AGRICULTURE, FORESTRY AND FISHERY, SHIMANE)
NORIN-SHO CHUGOKU SHIKOKU NOSEI-KYOKU SHIMANE TOKEI JOHO JIMUSHO (SHIMANE STATISTICS AND INFORMATION OFFICE, CHUGOKU SHIKOKU AGRICULTURAL ADMINISTRATION BUREAU)
552, KASUGA-CHO, MATSUE-SHI 690
JAPANESE

SHIMANE-KEN NOGYO SHIKENJO DAYORI 3775
(LETTERS OF THE SHIMANE PREFECTURE AGRICULTURAL EXPERIMENT STATION)
JAPANESE

SHIMANE-KEN NOJI SHIKENJO GYOMU HOKOKU 3776
(ANNUAL REPORT OF SHIMANE AGRICULTURAL EXPERIMENT STATION)
SHIMANE-KEN NOJI SHIKENJO (SHIMANE AGRICULTURAL EXPERIMENT STATION)
1940, EN'YA-CHO, IZUMO-SHI 693
JAPANESE

SHIMANE-KEN NOJI SHIKENJO KENKYU HOKOKU 3777
(BULLETIN OF THE SHIMANE AGRICULTURAL EXPERIMENT STATION)
SHIMANE-KEN NOJI SHIKENJO (SHIMANE AGRICULTURAL EXPERIMENT STATION)
1940, EN'YA-CHO, IZUMO-SHI 693
JAPANESE, WITH ENGLISH SUMMARY
INDEXED: CHE

SHIMANE-KEN NOKA CHOSA KEKKA HOKOKUSHO 3778
(CENSUS OF FARMERS, SHIMANE PREFECTURE)
SHIMANE-KEN (SHIMANE PREFECTURAL GOVERNMENT)
1, TONO-MACHI, MATSUE-SHI 690
JAPANESE

SHINRIN KONSABESHON 3779
(FOREST CONSERVATION)
NIHON CHISAN CHISUI KYOKAI (JAPAN FLOOD-CONTROL ASSOCIATION)
JAPANESE

SHINRIN REKURIESHON KENKYU 3780
(FOREST RECREATION RESEARCH)
TOKYO NOKO DAIGAKU NOGAKUBU RINGAKKA (TOKYO AGRICULTURE AND INDUSTRY UNIVERSITY, AGRICULTURE DEPARTMENT, FORESTRY SECTION)
JAPANESE

SHINSHU DAIGAKU NOGAKUBU KIYO 3781
(JOURNAL OF THE FACULTY OF AGRICULTURE, SHINSHU UNIVERSITY)
SHINSHU DAIGAKU NOGAKUBU (FACULTY OF AGRICULTURE, SHINSHU UNIVERSITY)
8304, MINAMIMINOWA-MURA, KAMIINA-GUN, NAGANO 399-45
JAPANESE AND ENGLISH
INDEXED: BIB, CHE, JICST

SHIZEN NOHO KENKYU SHIRIZU 3782
(RESEARCH SERIES ON LAW OF NATURE IN FARMING)
KANKYO KAGAKU SOGO KENKYUJO
JAPANESE
INDEXED: BIB

SHIZUOKA DAIGAKU NOGAKUBU KENKYU HOKOKU 3783
(BULLETIN OF THE FACULTY OF AGRICULTURE, SHIZUOKA UNIVERSITY)
SHIZUOKA DAIGAKU NOGAKUBU (FACULTY OF AGRICULTURE SHIZUOKA UNIVERSITY)
836, OYA, SHIZUOKA-SHI 420
JAPANESE AND ENGLISH
INDEXED: BIB, CHE, JICST

SHIZUOKA NORIN SUISAN TOKEI NENPO 3784
(STATISTICAL YEARBOOK OF AGRICULTURE, FORESTRY AND FISHERY, SHIZUOKA)
NORIN-SHO KANTO NOSEI-KYOKU SHIZUOKA TOKEI JOHO JIMUSHO (SHIZUOKA STATISTICS AND INFORMATION OFFICE, KANTO AGRICULTURAL ADMINISTRATION BUREAU, MINISTRY OF AGRICULTURE AND FORESTRY)
15-12, 2-CHOME, TAKASHO, SHIZUOKA-SHI 420
JAPANESE

SHIZUOKA-KEN NOGYO SHIKENJO GYOMU SEISEKI 3785
(ANNUAL REPORT OF SHIZUOKA AGRICULTURAL EXPERIMENT STATION)
SHIZUOKA-KEN NOGYO SHIKENJO (SHIZUOKA AGRICULTURAL EXPERIMENT STATION)
27-1, 4-CHOME, KITAANDO, SHIZUOKA-SHI 420
JAPANESE

SHIZUOKA-KEN NOGYO SHIKENJO KENKYU HOKOKU 3786
(BULLETIN OF SHIZUOKA AGRICULTURAL EXPERIMENT STATION)
SHIZUOKA-KEN NOGYO SHIKENJO (SHIZUOKA AGRICULTURAL EXPERIMENT STATION)
27-1, 4-CHOME, KITAANDO SHIZUOKA-SHI 420
JAPANESE AND ENGLISH
INDEXED: CHE, JICST

SHIZUOKA-KEN NOGYO SHIKENJO SEISEKI MOKUROKUSHU 3787
(ABSTRACT FROM STUDIES OF SHIZUOKA AGRICULTURAL EXPERIMENT STATION)
SHIZUOKA-KEN NOGYO SHIKENJO (SHIZUOKA AGRICULTURAL EXPERIMENT STATION)
27-1, 4-CHOME, KITAANDO, SHIZUOKA-SHI 420
JAPANESE

SHIZUOKA-KEN NOGYO SHIKENJO SHIRYO 3788
(RESEARCH MATERIALS OF SHIZUOKA AGRICULTURAL EXPERIMENT STATION)
SHIZUOKA-KEN NOGYO SHIKENJO (SHIZUOKA AGRICULTURAL EXPERIMENT STATION)
27-1, 4-CHOME, KITAANDO, SHIZUOKA-SHI 420
JAPANESE

SHUBYO KENSA NENPO 3789
(ANNUAL REPORT OF THE SEED TESTING LABORATORY)
NIHON SANSHI ENGEI-KYOKU
JAPANESE
INDEXED: BIB

SHUKAN NORIN 3790
(AGRICULTURE AND FORESTRY WEEKLY)
NORIN SHUPPANSHA
C/O IZUMI BUILDING, 1-9, KANDA AWAJI-CHO, CHIYODA-KU, TOKYO 101
JAPANESE

SHUYO SHIKEN SEISEKI NO SOKUHO 3791
(CURRENT REPORT OF PRINCIPAL EXPERIMENT RESULTS)
KYOTO-FURITSU NOGYO KENKYUSHO (KYOTO PREFECTURAL RESEARCH INSTITUTE OF AGRICULTURE)
JAPANESE

SOGO NOGAKU 3792
(VOCATIONAL AGRICULTURE)
SOGO NOGAKU GAKKAI (RESEARCH ASSOCIATION OF VOCATIONAL AGRICULTURE)
C/O TOKYO KYOIKU DAIGAKU NOGAKUBU, 19-1, 2-CHOME, KOMABA, MEGURO-KU, TOKYO 153
JAPANESE, WITH ENGLISH SUMMARY

SOGO NOSHI DAYORI 3793
(MIYAZAKI AGRICULTURAL EXPERIMENT STATION NEWS)
MIYAZAKI-KEN SOGO NOGYO SHIKENJO (MIYAZAKI AGRICULTURAL EXPERIMENT STATION)
SHIMONAKA, SADOHARA-CHO, MIYAZAKI-GUN, MIYAZAKI 880-02
JAPANESE

SOIL SCIENCE AND PLANT NUTRITION 3794
NIHON DOJO HIRYO GAKKAI (SOCIETY OF THE SCIENCE OF SOIL AND MANURE, JAPAN)
C/O NORIN-SHO NOGYO GIJUTSU KENKYUSHO, 1-7, 2-CHOME, NISHIGAHARA, KITA-KU, TOKYO 114
ENGLISH
INDEXED: BIB, BIO, CHE, EXC, JICST

SP DAYORI 3795
(SP NEWS)
AOMORI-KEN NORIN-BU (AGRICULTURE AND FORESTRY DEPARTMENT, AOMORI PREFECTURAL GOVERNMENT)
1-1, 1-CHOME, NAGASHIMA, AOMORI-SHI 030
JAPANESE

STATE OF AGRICULTURE 3796
JAPAN NORIN DAIJIN KAMBO CHOSA-KA
ENGLISH
INDEXED: BIB

SUIRIKUTO SAKKYO SHIKEN, SHIKEN SEKKEI NARABINI SHIKEN SEISEKI 3797
(INVESTIGATION IN GROWING CONDITIONS OF RICE PLANT, THE PLAN AND THE RESULTS)
NORIN-SHO NORIN KEIZAI-KYOKU TOKEI JOHO-BU (STATISTICS AND INFORMATION DEPARTMENT, ECONOMIC AFFAIRS BUREAU, MINISTRY OF AGRICULTURE AND FORESTRY)
2-1, 1-CHOME, KASUMIGASEKI, CHIYODA-KU, TOKYO 100
JAPANESE

TAMAGAWA DAIGAKU NOGAKUBU KENKYU HOKOKU 3798
(BULLETIN OF THE FACULTY OF AGRICULTURE, TAMAGAWA UNIVERSITY)
TAMAGAWA DAIGAKU NOGAKUBU (FACULTY OF AGRICULTURE TAMAGAWA UNIVERSITY)
1-1, 6-CHOME, TAMAGAWAGAKUEN, MACHIDA-SHI, TOKYO 194
JAPANESE AND ENGLISH
INDEXED: BIB, BIO, CHE

TECHNICAL BULLETIN OF TROPICAL AGRICULTURE RESEARCH CENTER 3799
NORIN-SHO NETTAI NOGYO KENKYU SENTA (TROPICAL AGRICULTURE RESEARCH CENTER, MINISTRY OF AGRICULTURE AND FORESTRY)
2-1, 2-CHOME, NISHIGAHARA, KITA-KU, TOKYO 114
ENGLISH
INDEXED: BIB

TEKUNIKARU NYUSU (HOSHASEN IKUSHUJO) 3800
(TECHNICAL NEWS (INSTITUTE OF RADIATION BREEDING))
NORIN-SHO NOGYO GIJUTSU KENKYUSHO HOSHASEN IKUSHUJO (INSTITUTE OF RADIATION BREEDING)
OMIYA-MACHI, NAKA-GUN, IBARAKI 319-22
JAPANESE AND ENGLISH

TOCHI KAIRYO SHINBUN 3801
(JAPAN LAND IMPROVEMENT NEWS)
TOCHI KAIRYO SHINBUNSHA
9-12, 1-CHOME, NISHISHINBASHI, MINATO-KU, TOKYO 105
JAPANESE

TOCHI SEIDO SHIGAKU 3802
(STUDY OF LAND SYSTEM HISTORY)
TOCHI SEIDO SHI GAKKAI
JAPANESE

TOCHI TO NOGYO 3803
(LAND AND AGRICULTURE)
ZENKOKU NOCHI HOYU GORIKA KYOKAI
11-35, 1-CHOME, NAGATA-CHO, CHIYODA-KU, TOKYO 100
JAPANESE

TOCHIGI NORIN TOKEI NENPO 3804
(STATISTICAL YEARBOOK OF AGRICULTURE, FORESTRY AND FISHERY, TOCHIGI)
NORIN-SHO KANTO NOSEI-KYOKU TOCHIGI TOKEI JOHO JIMUSHO (TOCHIGI STATISTICS AND INFORMATION OFFICE, KANTO AGRICULTURAL ADMINISTRATION BUREAU, MINISTRY OF AGRICULTURE AND FORESTRY)
1-13, 5-CHOME, SAKURA, UTSUNOMIYA-SHI 320
JAPANESE

TOCHIGI-KEN NOGYO SHIKENJO GYOMU NENPO 3805
(REPORT ON ACTIVITIES OF TOCHIGI AGRICULTURAL EXPERIMENT STATION)
TOCHIGI-KEN NOGYO SHIKENJO (TOCHIGI AGRICULTURAL EXPERIMENT)
1080, KAWARAYA-CHO, UTSUNOMIYA-SHI 320
JAPANESE

TOCHIGI-KEN NOGYO SHIKENJO GYOSEKI HOKOKU 3806
(REPORT OF THE TOCHIGI AGRICULTURAL EXPERIMENT STATION)
TOCHIGI-KEN NOGYO SHIKENJO (TOCHIGI AGRICULTURAL STATION)
1080, KAWARAYA-CHO, UTSUNOMIYA-SHI 320
JAPANESE

TOCHIGI-KEN NOGYO SHIKENJC KENKYU HOKOKU 3807
(BULLETIN OF THE TOCHIGI AGRICULTURAL EXPERIMENT STATION)
TOCHIGI-KEN NOGYO SHIKENJO (TOCHIGI AGRICULTURAL EXPERIMENT STATION)
1080, KAWARAYA-CHO, UTSUNOMIYA-SHI 320
JAPANESE

TOCHIGI-KEN NOGYO SHIKENJO MINAMI KAWACHI TOKUBETSU KENKYU HOKOKU 3808
(SPECIAL RESEARCH REPORT OF THE TOCHIGI PREFECTURE AGRICULTURAL EXPERIMENT STATION MINAMI KAWACHI SUBSTATION)
JAPANESE

TOHOKU DAIGAKU NOGAKU KENKYUSHO HOKOKU 3809
(BULLETIN OF THE INSTITUTE FOR AGRICULTURAL RESEARCH, TOHOKU UNIVERSITY)
TOHOKU DAIGAKU NOGAKU KENKYUSHO (INSTITUTE FOR AGRICULTURAL RESEARCH, TOHOKU UNIVERSITY)
1-34, 1-CHOME, KOMEGAFUKURO, SENDAI-SHI 980
JAPANESE, WITH ENGLISH SUMMARY
INDEXED: BIB, CHE, JICST

TOHOKU JOURNAL OF AGRICULTURAL RESEARCH 3810
TOHOKU DAIGAKU NOGAKUBU (FACULTY OF AGRICULTURE, TOHOKU UNIVERSITY)
1-1, TSUTSUMI-TORI, AMAMIYA-CHO, SENDA-SHI 980
ENGLISH
INDEXED: BIB, BIO, CHE, JICST

TOHOKU NOGYO JOSEI HOKOKU 3811
(SECOND ANNUAL REPORT ON AGRICULTURE AND ITS RELATED POLICIES IN THE TOHOKU DISTRICT)
NORIN-SHO TOHOKU NOSEI-KYOKU (TOHOKU AGRICULTURAL ADMINISTRATION BUREAU, MINISTRY OF AGRICULTURE AND FORESTRY)
C/O SENDAI GODOCHOSHA, 3-1, 3-CHOME, HON-CHO, SENDAI-SHI 980
JAPANESE

TOHOKU NOGYO KENKYU 3812
(TOHOKU AGRICULTURAL RESEARCH)
TOHOKU NOGYO SHIKEN KENKYU KYOGIKAI (ASSOCIATION OF THE TOHOKU AGRICULTURAL RESEARCH INSTITUTE)
4, SHIMOKURIYAGAWA AKAHIRA, MORIOKA-SHI 020-01
JAPANESE

TOHOKU NOGYO SHIKENJO KENKYU HOKOKU 3813
(BULLETIN OF THE TOHOKU NATIONAL AGRICULTURAL EXPERIMENT STATION)
NORIN-SHO TOHOKU NOGYO SHIKENJO (TOHOKU NATIONAL AGRICULTURAL EXPERIMENT STATION, MINISTRY OF AGRICULTURE AND FORESTRY)
4, SHIMOKURIYAGAWA AKAHIRA, MORIOKA-SHI 020-01
JAPANESE AND EUROPEAN
INDEXED: BIB, BIO, CHE, EXC, JICST

TOHOKU NOGYO SHIKENJO KENKYU SOKUHO 3814
(REPORT OF THE TOHOKU NATIONAL AGRICULTURAL EXPERIMENT STATION)
NORIN-SHO TOHOKU NOGYO SHIKENJO (TOHOKU NATIONAL AGRICULTURAL EXPERIMENT STATION, MINISTRY OF AGRICULTURE AND FORESTRY)
4, SHIMOKURIYAGAWA AKAHIRA, MORIOKA-SHI 020-01
JAPANESE
INDEXED: CHE, JICST

TOHOKU NOSEI 3815
(QUARTERLY JOURNAL ON AGRICULTURE OF THE TOHOKU DISTRICT)
NORIN-SHO TOHOKU NOSEI-KYOKU (TOHOKU AGRICULTURAL ADMINISTRATION BUREAU, MINISTRY OF AGRICULTURE AND FORESTRY)
C/O SENDAI GODOCHOSHA, 3-1, 3-CHOME, HON-CHO, SENDAI-SHI 980
JAPANESE

TOHOKU NOSEI-KYOKU DAYORI 3816
(MONTHLY BULLETIN OF THE TOHOKU AGRICULTURAL ADMINISTRATION BUREAU)
NORIN-SHO TOHOKU NOSEI-KYOKU (TOHOKU AGRICULTURAL ADMINISTRATION BUREAU, MINISTRY OF AGRICULTURE AND FORESTRY)
C/O SENDAI GODOCHOSHA, 3-1, 3-CHOME, HON-CHO, SENDAI-SHI 980
JAPANESE

TOKAI KINKI NOGYO SHIKENJO KENKYU HOKOKU 3817
(TOKAI-KINKI NATIONAL AGRICULTURAL EXPERIMENT STATION, TSU, JAPAN BULLETIN)
TOKAI-KINKI NATIONAL AGRICULTURAL EXPERIMENT STATION
ISHINDEN-OGOSO, TSU-CITY
ENGLISH
INDEXED: BIO, CHE, EXC

TOKAI NO NOSEI 3818
(AGRICULTURAL ADMINISTRATION IN TOKAI REGION)
NORIN-SHO TOKAI NOSEI-KYOKU (TOKAI AGRICULTURAL ADMINISTRATION BUREAU, MINISTRY OF AGRICULTURE AND FORESTRY)
2-2, 1-CHOME, SANNOMARU, NAKA-KU, NAGOYA-SHI 460
JAPANESE

TOKUBETSU KENKYU SEIKA NO GAIYO 3819
(OUTLINE OF SPECIAL RESEARCH RESULTS)
NORIN-SHO NORIN SUISAN GIJUTSU KAIGI JIMU-KYOKU (AGRICULTURE, FORESTRY AND FISHERIES RESEARCH COUNCIL SECRETARIAT, MINISTRY OF AGRICULTURE AND FORESTRY)
2-1, 1-CHOME, KASUMIGASEKI, CHIYODA-KU, TOKYO 100
JAPANESE

TOKUSAN NOSAKUMOTSU SEISAN JISSEKI 3820
(STATISTICS OF SPECIAL AGRICULTURAL PRODUCTS)
NORIN-SHO NOSAN ENGEI-KYOKU (AGRICULTURAL PRODUCTION BUREAU, MINISTRY OF AGRICULTURE AND FORESTRY)
2-1, 1-CHOME, KASUMIGASEKI, CHIYODA-KU, TOKYO 100
JAPANESE

TOKUSHIMA NORIN SUISAN TOKEI NENPO 3821
(STATISTICAL YEARBOOK OF AGRICULTURE, FORESTRY AND FISHERY, TOKUSHIMA)
NORIN-SHO CHUGOKU SHIKOKU NOSEI-KYOKU TOKUSHIMA TOKEI JOHO JIMUSHO (TOKUSHIMA STATISTICS AND INFORMATION OFFICE, CHUGOKU SHIKOKU AGRICULTURAL ADMINISTRATION BUREAU)
1-6, 5-CHOME, SHIMOSUKETO-CHO, TOKUSHIMA-SHI 770
JAPANESE

TOKUSHIMA-KEN NOGYO SHIKENJO GYOMU KOTEI 3822
(ANNUAL BULLETIN OF THE TOKUSHIMA AGRICULTURAL EXPERIMENT STATION)
TOKUSHIMA-KEN NOGYO SHIKENJO (TOKUSHIMA AGRICULTURAL EXPERIMENT STATION)
1660, ISHII, ISHII-CHO, MYOZAI-GUN, TOKUSHIMA 779-32
JAPANESE

TOKUSHIMA-KEN NOGYO SHIKENJO SHIKEN KENKYU HOKOKU 3823
(TOKUSHIMA AGRICULTURAL EXPERIMENT STATION REPORTS)
TOKUSHIMA-KEN NOGYO SHIKENJO (TOKUSHIMA AGRICULTURAL EXPERIMENT STATION)
1660, ISHII, ISHII-CHO, MYOZAI-GUN, TOKUSHIMA 779-32
JAPANESE AND ENGLISH

TOKUSHIMA-KEN NOGYO SHIKENJO TOKUBETSU HOKOKU 3824
(SPECIAL BULLETIN OF THE TOKUSHIMA AGRICULTURAL EXPERIMENT STATION)
TOKUSHIMA-KEN NOGYO SHIKENJO (TOKUSHIMA PREFECTURAL AGRICULTURAL EXPERIMENT STATION)
JAPANESE

TOKUYO RINSANBUTSU SHIKYO GEPPO 3825
(MARKET CONDITIONS FOR SPECIAL FOREST PRODUCTS MONTHLY)
RINYA-CHO RINSAN-KA (FORESTRY AGENCY, FOREST PRODUCTS DIVISION)
JAPANESE

TOKYO DAIGAKU NOGAKUBU ENSHURIN HOKOKU 3826
(BULLETIN OF THE TOKYO UNIVERSITY FORESTS)
TOKYO DAIGAKU NOGAKUBU FUZOKU ENSHURIN
1-1-1, YOYOI, BUNKYO-KU TOKYO
JAPANESE
INDEXED: BIB, CHE, JICST

TOKYO EIRIN-KYOKU KOHO 3827
(OFFICIAL REPORT OF THE TOKYO FORESTRY MANAGEMENT BUREAU)
JAPANESE

TOKYO HISHIRYO KENSAJO JIGYO HOKOKU HIRYO NO BU 3828
(ANNUAL REPORT OF THE TOKYO FERTILIZER AND FEED INSPECTION STATION, FERTILIZER SECTION)
NORIN-SHO TOKYO HISHIRYO KENSAJO (TOKYO FERTILIZER AND FEED INSPECTION STATION, MINISTRY OF AGRICULTURE AND FORESTRY)
3-3, 1-CHOME, OTE-MACHI, CHIYODA-KU, TOKYO 100
JAPANESE

TOKYO KYOIKU DAIGAKU NOGAKUBU KIYO 3829
(MEMOIRS OF THE FACULTY OF AGRICULTURE, TOKYO UNIVERSITY OF EDUCATION)
TOKYO KYOIKU DAIGAKU NOGAKUBU (FACULTY OF AGRICULTURE, TOKYO UNIVERSITY OF EDUCATION)
862, KOMABA-CHO, MEGURO-KU, TOKYO 153
JAPANESE, WITH EUROPEAN SUMMARY
INDEXED: BIB, JICST

TOKYO NOGYO DAIGAKU NOGAKU SHUHO 3830
(JOURNAL OF AGRICULTURAL SCIENCE, TOKYO NOGYO DAIGAKU)
TOKYO NOGYO DAIGAKU GAKUJUTSU ZASSHI HENSHU IINKAI (EDITORIAL BOARD FOR ACADEMIC JOURNALS, TOKYO UNIVERSITY OF AGRICULTURE) AND TOKYO NOGYO DAIGAKU (TOKYO UNIVERSITY OF AGRICULTURE)
1-1, 1-CHOME, SAKURAGAOKA, SETAGAYA-KU, TOKYO 156
JAPANESE AND ENGLISH
INDEXED: CHE, JICST

TOKYO NOKO DAIGAKU NOGAKUBU GAKUJUTSU HOKOKU 3831
(BULLETIN OF THE FACULTY OF AGRICULTURE, TOKYO UNIVERSITY OF AGRICULTURE AND TECHNOLOGY)
TOKYO NOKO DAIGAKU NOGAKUBU (FACULTY OF AGRICULTURE, TOKYO UNIVERSITY OF AGRICULTURE AND TECHNOLOGY)
5-8, 3-CHOME, SAIWAI-CHO, FUCHU-SHI, TOKYO 183
JAPANESE, WITH EUROPEAN SUMMARY
INDEXED: CHE, JICST

TOKYO NOKO DAIGAKU NOGAKUBU NOJO KENKYU HOKOKU 3832
(BULLETIN OF THE FARMS, TOKYO UNIVERSITY OF AGRICULTURE AND TECHNOLOGY)
TOKYO NOKO DAIGAKU NOGAKUBU FUZOKU NOJO
5-8, 3-CHOME, SAIWAI-CHO, FUCHU-SHI, TOKYO 183
JAPANESE, WITH ENGLISH SUMMARY

TOKYO NORIN SUISAN TOKEI NENPO 3833
(STATISTICAL YEARBOOK OF AGRICULTURE, FORESTRY AND FISHERY, TOKYO)
NORIN-SHO KANTO NOSEI-KYOKU TOKEI JOHO-BU (STATISTICS AND INFORMATION DIVISION, KANTO AGRICULTURAL ADMINISTRATION BUREAU, MINISTRY OF AGRICULTURE AND FORESTRY)
3-3, 1-CHOME, OTE-MACHI, CHIYODA-KU, TOKYO 100
JAPANESE

TOKYO-TO NOGYO SHIKENJO KENKYU HOKOKU 3834
(BULLETIN OF THE TOKYO-TO AGRICULTURAL EXPERIMENT STATION)
TOKYO-TO NOGYO SHIKENJO (TOKYO METROPOLITAN AGRICULTURAL EXPERIMENT STATION)
3-10, FUJIMI-CHO, TACHIKAWA-SHI, TOKYO 190
JAPANESE, WITH ENGLISH SUMMARY

TOKYO-TO NOGYO SHIKENJO KENKYU NENPO 3835
(ANNUAL REPORT OF TOKYO METROPOLITAN AGRICULTURAL EXPERIMENT STATION)
TOKYO-TO NOGYO SHIKENJO (TOKYO METROPOLITAN AGRICULTURAL EXPERIMENT STATION)
3-10, FUJIMI-CHO, TACHIKAWA-SHI, TOKYO 190
JAPANESE

TOKYO-TO NOGYO SHIKENJO TOKUBETSU HOKOKU 3836
(SPECIAL REPORT OF THE TOKYO METROPOLITAN AGRICULTURAL EXPERIMENT STATION)
JAPANESE

TOTTORI DAIGAKU NOGAKUBU KENKYU HOKOKU 3837
(BULLETIN OF THE FACULTY OF AGRICULTURE, TOTTORI UNIVERSITY)
TOTTORI DAIGAKU NOGAKUBU (FACULTY OF AGRICULTURE, TOTTORI UNIVERSITY)
1-1, KOYAMA-CHO, TOTTORI-SHI 680
JAPANESE AND EUROPEAN
INDEXED: BIB, BIO, CHE, JICST

TOTTORI DAIGAKU NOGAKUBU SAKYU KENKYUSHO HOKOKU 3838
(BULLETIN OF THE SAND DUNE RESEARCH INSTITUTE, TOTTORI UNIVERSITY)
TOTTORI DAIGAKU NOGAKUBU FUZOKU SAKYU RIYO KENKYU SHISETSU (SAND DUNE RESEARCH INSTITUTE, THE FACULTY OF AGRICULTURE, TOTTORI UNIVERSITY)
HAMASAKA, TOTTORI-SHI 680-11
JAPANESE AND ENGLISH

TOTTORI NORIN SUISAN TOKEI NENPO 3839
(STATISTICAL YEARBOOK OF AGRICULTURE, FORESTRY AND FISHERY, TOTTORI)
NORIN-SHO CHUGOKU SHIKOKU NOSEI-KYOKU TOTTORI TOKEI JOHO JIMUSHO (TOTTORI STATISTICS AND INFORMATION OFFICE, CHUGOKU SHIKOKU AGRICULTURAL ADMINISTRATION BUREAU, MINISTRY OF AGRICULTURE AND FORESTRY)
1-353, YUTOKORO-CHO, TOTTORI-SHI 680
JAPANESE

TOTTORI-KEN NOGYO NO UGOKI 3840
(CENSUS OF AGRICULTURE IN TOTTORI PREFECTURE)
NORIN-SHO CHUGOKU SHIKOKU NOSEI-KYOKU TOTTORI TOKEI JOHO JIMUSHO (TOTTORI STATISTICS AND INFORMATION OFFICE, CHUGOKU SHIKOKU AGRICULTURAL ADMINISTRATION BUREAU)
1-353, YUTOKORO-CHO, TOTTORI-SHI 680
JAPANESE

TOTTORI-KEN NOGYO SHIKENJO KENKYU HOKOKU 3841
(BULLETIN OF THE TOTTORI AGRICULTURAL EXPERIMENT STATION)
TOTTORI-KEN NOGYO SHIKENJO (TOTTORI AGRICULTURAL EXPERIMENTAL STATION)
260, HASHIMOTO, TOTTORI-SHI 680-11
JAPANESE AND ENGLISH
INDEXED: CHE

TOTTORI-KEN NOGYO SHIKENJO NENPO 3842
(ANNUAL REPORT OF TOTTORI AGRICULTURAL EXPERIMENT STATION)
TOTTORI-KEN NOGYO SHIKENJO (TOTTORI AGRICULTURAL EXPERIMENT STATION)
260, HASHIMOTO, TOTTORI-SHI 680-11
JAPANESE

TOTTORI-KEN NOGYO SHIKENJO TOKUBETSU KENKYU HOKOKU 3843
(SPECIAL BULLETIN OF THE TOTTORI AGRICULTURAL EXPERIMENT STATION)
TOTTORI-KEN NOGYO SHIKENJO (TOTTORI AGRICULTURAL EXPERIMENT STATION)
260, HASHIMOTO, TOTTORI 680-11
JAPANESE

TOYAMA NORIN SUISAN TOKEI NENPO 3844
(STATISTICAL YEARBOOK OF AGRICULTURE, FORESTRY AND FISHERY, TOYAMA)
NORIN-SHO HOKURIKU NOSEI-KYOKU TOYAMA TOKEI JOHO JIMUSHO (TOYAMA STATISTICS AND INFORMATION OFFICE, HOKURIKU AGRICULTURAL ADMINISTRATION BUREAU, MINISTRY OF AGRICULTURE AND FORESTRY)
11-7, USHIJIMASHIN-MACHI, TOYAMA-SHI 930
JAPANESE

TOYAMA-KEN NOGYO NO UGOKI 3845
(REPORT OF TOYAMA PREFECTURE FARMING TREND)
TOYAMA-KEN NORIN SUISAN-BU (AGRICULTURE AND FISHERIES DIVISION, TOYAMA PREFECTURAL GOVERNMENT)
1-7, SHINSOGAWA, TOYAMA-SHI 930
JAPANESE

TOYAMA-KEN NOGYO SHIKENJO KENKYU HOKOKU 3846
(BULLETIN OF THE TOYAMA AGRICULTURAL EXPERIMENT STATION)
TOYAMA-KEN NOGYO SHIKENJO (TOYAMA AGRICULTURAL EXPERIMENT STATION)
1124-1, YOSHIOKA, TOYAMA-SHI 930-11
JAPANESE

TOYAMA-KEN NOGYO SHIKENJO NENPO 3847
(ANNUAL REPORT OF TOYAMA AGRICULTURAL EXPERIMENT STATION)
TOYAMA-KEN NOGYO SHIKENJO (TOYAMA AGRICULTURAL EXPERIMENT STATION)
1124-1, YOSHIOKA, TOYAMA-SHI 930-11
JAPANESE

TOYAMA-KEN NOGYO SHIKENJO TOKUBETSU HOKOKU 3848
(SPECIAL REPORT OF THE TOYAMA AGRICULTURAL EXPERIMENT STATION)
TOYAMA-KEN NOGYO SHIKENJO (TOYAMA AGRICULTURAL EXPERIMENT STATION)
194, TAROMARU, TOYAMA-SHI 930
JAPANESE, WITH ENGLISH SUMMARY

TRANSACTIONS OF THE TOTTORI SOCIETY OF AGRICULTURAL SCIENCE 3849
ENGLISH
INDEXED: BIB

TROPICAL AGRICULTURE RESEARCH SERIES 3850
NORIN-SHO NETTAI NOGYO KENKYU SENTA (TROPICAL AGRICULTURE RESEARCH CENTER)
2-1, 2-CHOME, NISHIGAHARA, KITA-KU, TOKYO 114
ENGLISH

UTSUNOMIYA DAIGAKU NOGAKUBU GAKUJUTSU HOKOKU 3851
(BULLETIN OF THE COLLEGE OF AGRICULTURE, UTSUMONIYA UNIVERSITY)
UTSUNOMIYA DAIGAKU NOGAKUBU (COLLEGE OF AGRICULTURE, UTSUNOMIYA UNIVERSITY)
350, MINE-MACHI, UTSUNOMIYA-SHI 320
JAPANESE AND ENGLISH
INDEXED: BIB, BIO, CHE, JICST

UTSUNOMIYA DAIGAKU NOGAKUBU GAKUJUTSU HOKOKU TOKUSHU 3852
(SPECIAL BULLETIN OF THE COLLEGE OF AGRICULTURE, UTSUNOMIYA UNIVERSITY)
UTSUNOMIYA DAIGAKU NOGAKUBU
350, MINE-MACHI, UTSUNOMIYA-SHI 320
JAPANESE, WITH ENGLISH SUMMARY
INDEXED: BIB, BIO, CHE, JICST

WAKAYAMA NORIN SUISAN TOKEI NENPO 3853
(STATISTICAL YEARBOOK OF AGRICULTURE, FORESTRY AND FISHERY, WAKAYAMA)
NORIN-SHO KINKI NOSEI-KYOKU WAKAYAMA TOKEI JOHO JIMUSHO (WAKAYAMA STATISTICS AND INFORMATION OFFICE, KINKI AGRICULTURAL ADMINISTRATION BUREAU, MINISTRY OF AGRICULTURE AND FORESTRY)
2-6, KATAOKA-CHO, WAKAYAMA-SHI 640
JAPANESE

WAKAYAMA-KEN NOGYO SHIKENJO KENKYU HOKOKU 3854
(BULLETIN OF THE WAKAYAMA AGRICULTURAL EXPERIMENT STATION)
WAKAYAMA-KEN NOGYO SHIKENJO (WAKAYAMA PREFECTURE AGRICULTURAL EXPERIMENT STATION)
TAKAO, KISHIGAWA-CHO, NAGA-GUN, WAKAYAMA 640-04
JAPANESE

WAKAYAMA-KEN NOGYO SHIKENJO NENPO 3855
(ANNUAL REPORT OF WAKAYAMA AGRICULTURAL EXPERIMENT STATION)
WAKAYAMA-KEN NOGYO SHIKENJO (WAKAYAMA PREFECTURE AGRICULTURAL EXPERIMENT STATION)
TAKAO, KISHIGAWA-CHO, NAGA-GUN, WAKAYAMA 640-04
JAPANESE

YAMAGATA DAIGAKU KIYO, NOGAKU 3856
(BULLETIN OF THE YAMAGATA UNIVERSITY, AGRICULTURAL SCIENCE)
YAMAGATA DAIGAKU (YAMAGATA UNIVERSITY)
KOSHIROKAWA-MACHI, YAMAGATA-SHI 990
JAPANESE AND EUROPEAN
INDEXED: BIB, CHE, JICST

YAMAGATA NORIN GAKKAIHO 3857
(JOURNAL OF THE YAMAGATA AGRICULTURE AND FORESTRY SOCIETY)
YAMAGATA NORIN GAKKAI (YAMAGATA AGRICULTURE AND FORESTRY SOCIETY)
C/O YAMAGATA DAIGAKU NOGAKUBU, 1-23, WAKABA-CHO, TSURUOKA-SHI 997
JAPANESE, WITH EUROPEAN SUMMARY
INDEXED: BIB, BIO, CHE, JICST

YAMAGATA NORIN SUISAN TOKEI NENPO 3858
(STATISTICAL YEARBOOK OF AGRICULTURE, FORESTRY AND FISHERY, YAMAGATA)
NORIN-SHO TOHOKU NOSEI-KYOKU YAMAGATA TOKEI JOHO JIMUSHO (YAMAGATA STATISTICS AND INFORMATION OFFICE, TOHOKU AGRICULTURAL ADMINISTRATION BUREAU, MINISTRY OF AGRICULTURE AND FORESTRY)
5-27, KASUGA-MACHI, YAMAGATA-SHI 990
JAPANESE

YAMAGATA-KEN NO NOGYO 3859
(CENSUS OF AGRICULTURE, YAMAGATA PREFECTURE)
YAMAGATA-KEN KIKAKU-BU (PLANNING DIVISION, YAMAGATA PREFECTURAL GOVERNMENT)
4-51, 3-CHOME, HATAGO-MACHI, YAMAGATA-SHI 990
JAPANESE

YAMAGATA-KENRITSU NOGYO SHIKENJO KENKYU HOKOKU 3860
(BULLETIN OF THE YAMAGATA PREFECTURAL AGRICULTURAL EXPERIMENT STATION)
YAMAGATA-KENRITSU NOGYO SHIKENJO (YAMAGATA PREFECTURAL AGRICULTURAL EXPERIMENT STATION)
10-75, 2-CHOME, TEPPO-CHO, YAMAGATA-SHI 990
JAPANESE

YAMAGATA-KENRITSU NOGYO SHIKENJO TOKUBETSU HOKOKU 3861
(SPECIAL BULLETIN OF THE YAMAGATA PREFECTURAL AGRICULTURAL EXPERIMENT STATION)
YAMAGATA-KENRITSU NOGYO SHIKENJO (YAMAGATA PREFECTURAL AGRICULTURAL EXPERIMENT STATION)
10-75, 2-CHOME, TEPPO-CHO, YAMAGATA-SHI 990
JAPANESE

YAMAGUCHI DAIGAKU NOGAKUBU GAKUJUTSU HOKOKU 3862
(BULLETIN OF THE FACULTY OF AGRICULTURE, YAMAGUCHI UNIVERSITY)
YAMAGUCHI DAIGAKU NOGAKUBU (FACULTY OF AGRICULTURE, YAMAGUCHI UNIVERSITY)
YOSHIDA, YAMAGUCHI-SHI 753
JAPANESE AND ENGLISH
INDEXED: BIB, BIO, CHE, EXC, JICST

YAMAGUCHI NORIN SUISAN TOKEI NENPO 3863
(STATISTICAL YEARBOOK OF AGRICULTURE, FORESTRY AND FISHERY, YAMAGUCHI)
NORIN-SHO CHUGOKU SHIKOKU NOSEI-KYOKU YAMAGUCHI TOKEI JOHO JIMUSHO (YAMAGUCHI STATISTICS AND INFORMATION OFFICE, CHUGOKU SHIKOKU AGRICULTURAL ADMINISTRATION BUREAU, MINISTRY OF AGRICULTURE
142, ODONOOJI, YAMAGUCHI-SHI 753
JAPANESE

YAMAGUCHI-KEN NOGYO SHIKENJO BUNKEN MOKUROKU 3864
(BIBLIOGRAPHICAL REPORT, YAMAGUCHI PREFECTURAL AGRICULTURAL EXPERIMENT STATION)
YAMAGUCHI-KEN NOGYO SHIKENJO (YAMAGUCHI PREFECTURAL AGRICULTURAL EXPERIMENT STATION)
OUCHI MIHORI, YAMAGUCHI-SHI 747-13
JAPANESE

YAMAGUCHI-KEN NOGYO SHIKENJO GYOMU NENPO 3865
(ANNUAL REPORT OF THE YAMAGUCHI PREFECTURAL AGRICULTURAL EXPERIMENT STATION)
YAMAGUCHI-KEN NOGYO SHIKENJO (YAMAGUCHI PREFECTURAL AGRICULTURAL EXPERIMENT STATION)
OUCHI MIHORI, YAMAGUCHI-SHI 747-13
JAPANESE

YAMAGUCHI-KEN NOGYO SHIKENJO KENKYU HOKOKU 3866
(BULLETIN OF THE YAMAGUCHI PREFECTURAL AGRICULTURAL EXPERIMENT STATION)
YAMAGUCHI-KEN NOGYO SHIKENJO (YAMAGUCHI PREFECTURAL AGRICULTURAL EXPERIMENT STATION)
OUCHI MIHORI, YAMAGUCHI-SHI 747-13
JAPANESE

YAMAGUCHI-KEN NOGYO SHIKENJO TOKUBETSU KENKYU HOKOKU 3867
(SPECIAL BULLETIN OF THE YAMAGUCHI PREFECTURAL AGRICULTURAL EXPERIMENT STATION)
YAMAGUCHI-KEN NOGYO SHIKENJO (YAMAGUCHI PREFECTURAL AGRICULTURAL EXPERIMENT STATION)
OUCHI MIHORI, YAMAGUCHI-SHI 747-13
JAPANESE, WITH ENGLISH SUMMARY

YAMANASHI NORIN SUISAN TOKEI NENPO 3868
(STATISTICAL YEARBOOK OF AGRICULTURE, FORESTRY AND FISHERY, YAMANASHI)
NORIN-SHO KANTO NOSEI-KYOKU YAMANASHI TOKEI JOHO JIMUSHO (YAMANASHI STATISTICS AND INFORMATION OFFICE, KANTO AGRICULTURAL ADMINISTRATION BUREAU, MINISTRY OF AGRICULTURE AND FORESTRY)
8-10, 1-CHOME, AIOI, KOFU-SHI 400
JAPANESE

YAMANASHI-KEN NOGYO NO UGOKI 3869
(TENDENCY OF AGRICULTURE, YAMANASHI PREFECTURE)
YAMANASHI-KEN NOMU-BU (AGRICULTURAL ADMINISTRATION DEPARTMENT, YAMANASHI PREFECTURAL GOVERNMENT)
C/O YAMANASHI-KEN NOGYO GIJUTSU-SHITSU, 6-1, 1-CHOME, MARUNOUCHI, KOFU-SHI 400
JAPANESE

YAMANASHI-KEN NOGYO SHIKEN KENKYU KIKAN GAIYO 3870
(OUTLINE OF AGRICULTURAL EXPERIMENT STATION IN YAMANASHI PREFECTURE)
YAMANASHI-KEN NOMU-BU (AGRICULTURAL ADMINISTRATION DEPARTMENT, YAMANASHI PREFECTURAL GOVERNMENT)
C/O YAMANASHI-KEN NOGYO GIJUTSU-SHITSU, 6-1, 1-CHOME, MARUNOUCHI, KOFU-SHI 400
JAPANESE

YAMANASHI-KEN NOGYO SHIKENJO HOKOKU 3871
(BULLETIN OF THE YAMANASHI PREFECTURAL AGRICULTURAL EXPERIMENT STATION)
YAMANASHI-KEN NOGYO SHIKENJO (YAMANASHI PREFECTURAL AGRICULTURAL EXPERIMENTAL STATION)
1100, SHIMOIMAI, FUTABA-CHO, KITAKOMA-GUN, YAMANASHI 407-01
JAPANESE AND ENGLISH

YAMANASHI-KEN NORINGYO SHICHOSON-BETSU TOKEISHO 3872
(AGRICULTURE AND FORESTRY STATISTICS BY TOWNS, YAMANASHI PREFECTURE)
NORIN-SHO KANTO NOSEI-KYOKU (KANTO AGRICULTURAL ADMINISTRATION BUREAU, MINISTRY OF AGRICULTURE AND FORESTRY)
3-3, 1-CHOME, OTE-MACHI, CHIYODA-KU, TOKYO 100
JAPANESE

YAMAZAKI NOGYO KENKYUJO JOHO 3873
(REPORT OF YAMAZAKI AGRICULTURE RESEARCH INSTITUTE)
JAPANESE

YOKOHAMA-SHI NO NOGYO 3874
(AGRICULTURE IN YOKOHAMA)
YOKOHAMA-SHI SOMU-KYOKU GYOSEI-BU (ADMINISTRATION DEPARTMENT, GENERAL AFFAIRS BUREAU, YOKOHAMA MUNICIPAL OFFICE)
1-1, MINATO-CHO, NAKA-KU, YOKOHAMA-SHI 231
JAPANESE

YOSUI TO EINO 3875
(IRRIGATION AND AGRICULTURE)
NIHON IRIGESHON KURABU (JAPAN IRRIGATION CLUB)
3, SHIBA KOTOHIRA-CHO, MINATO-KU, TOKYO 105
JAPANESE

ZENKOKU NOGYO KAIGISHO CHOSA KENKYU SHIRYO 3876
(RESEARCH DATA OF NATIONAL CHAMBER OF AGRICULTURE)
ZENKOKU NOGYO KAIGISHO (NATIONAL CHAMBER OF AGRICULTURE)
1-7, YURAKU-CHO, CHIYODA-KU, TOKYO 100
JAPANESE

ZENKOKU NOGYO SHIKENJOCHOKAI KAIHO 3877
(BULLETIN OF DIRECTERS' ASSOCIATION OF AGRICULTURAL EXPERIMENT STATIONS)
ZENKOKU NOGYO SHIKENJOCHOKAI
C/O NOGYO GIJUTSU KYOKAI, 26-3, 1-CHOME, NISHIGAHARA, KITA-KU, TOKYO 114
JAPANESE

ZENKOKU NOGYO SHINBUN 3878
(JAPAN AGRICULTURAL NEWS)
ZENKOKU NOGYO KAIGISHO (NATIONAL CHAMBER OF AGRICULTURE)
1-7, YURAKU-CHO, CHIYODA-KU, TOKYO 100
JAPANESE

ZENKOKU SHIITAKE FUKYUKAI KINSHIN KENKYUJO KENKYU HOKOKU 3879
(MUSHROOM INSTITUTE REPORT OF THE ALL-JAPAN SHIITAKU MUSHROOM POPULARIZATION SOCIETY)
JAPANESE
INDEXED: JICST

ZENKOKU TOCHI KAIRYO JOHO 3880
(INFORMATION NEWS OF NATIONAL FEDERATION OF LAND IMPROVEMENT ASSOCIATION)
ZENKOKU TOCHI KAIRYO JIGYO DANTAI RENGOKAI (NATIONAL FEDERATION OF LAND IMPROVEMENT ASSOCIATION)
C/O ZENKOKU CHOSON KAIKAN, 1, NAGATA-CHO, CHIYODA-KU, TOKYO 100
JAPANESE

ZUSETSU NORINGYO NO DOKO (AOMORI-KEN) 3881
(AGRICULTURE AND FORESTRY TENDENCY GRAPH, AOMORI PREFECTURE)
AOMORI-KEN (AOMORI PREFECTURAL GOVERNMENT)
1-1, 1-CHOME, NAGASHIMA, AOMORI-SHI 030
JAPANESE

Plant Injuries, Diseases, Pests, Crop Damage and Protection

BOCHU KAGAKU 3882
(SCIENTIFIC PEST CONTROL)
BOCHU KAGAKU KENKYUSHO (INSTITUTE OF INSECT CONTROL)
C/O KYOTO DAIGAKU NOGAKUBU, KITA SHIRAKAWA, SAKYO-KU, KYOTO-SHI 606
JAPANESE AND ENGLISH
INDEXED: BIB, CHE, JICST

BYOGAICHU HASSEI YOSATSU JIGYO NENPO (FUTSU SAKUMOTSU BYOGAICHU) 3883
(ANNUAL REPORT OF FORECASTING BUSINESS ON CROPS' DISEASES AND NOXIOUS INSECTS, SERIES GENERAL CROPS)
AOMORI-KEN NOGYO SHIKENJO (AOMORI PREFECTURE AGRICULTURAL EXPERIMENT STATION)
1, SUNAMORI, KUROISHI, KUROISHI-SHI 036-03
JAPANESE

BYOGAICHU HASSEI YOSATSU JIGYO NENPO (TOTTORI-KEN) 3884
(ANNUAL REPORT OF FORECASTING BUSINESS ON CROPS' DISEASES AND NOXIOUS INSECTS, TOTTORI PREFECTURE)
TOTTORI-KEN NOGYO SHIKENJO (TOTTORI AGRICULTURAL EXPERIMENT STATION)
260, HASHIMOTO, TOTTORI-SHI 860-11
JAPANESE

BYOGAICHU HASSEI YOSATSU JIGYO SEISEKI NENPO 3885
(ANNUAL REPORT OF THE RESULT OF FORECASTING BUSINESS ON CROPS' DISEASES AND NOXIOUS INSECTS)
MIYAGI-KENRITSU NOGYO SHIKENJO (MIYAGI PREFECTURE AGRICULTURAL EXPERIMENT STATION)
TAKADATE, NATORI-SHI 981-12
JAPANESE

FORSCHUNGEN AUF DEM GEBIET PFLANZENKRANKHEITEN 3886
SHOKUBUTSU BYOGAKU KENKYUKAI
KYOTO
JAPANESE AND ENGLISH
INDEXED: BIB

FUTSU SAKUMOTSU BYOGAICHU HASSEI YOSATSU JIGYO NENPO 3887
(ANNUAL REPORT OF FORECASTING BUSINESS ON GENERAL CROPS' DISEASES AND NOXIOUS INSECTS)
YAMAGUCHI-KEN NOGYO SHIKENJO
OUCHI MIHORI, YAMAGUCHI-SHI 747-13
JAPANESE

HOKURIKU BYOGAICHU KENKYUKAIHO 3888
(PROCEEDINGS OF THE ASSOCIATION FOR THE PLANT PROTECTION OF HOKURIKU)
HOKURIKU BYOGAICHU KENKYUKAI (ASSOCIATION FOR PLANT PROTECTION OF HOKURIKU)
C/O NORIN-SHO HOKURIKU NOGYO SHIKENJO, KAMIINADA, JOETSU-SHI 943-01
JAPANESE

JAPAN PESTICIDE INFORMATION 3889
NIHON SHOKUBUTSU BOEKI KYOKAI (JAPAN PLANT PROTECTION ASSOCIATION)
43-11, 1-CHOME, KOMAGOME, TOSHIMA-KU, TOKYO 170
ENGLISH
INDEXED: BIB, CHE, EXC

KAJU-TO (CHA) BYOGAICHU HASSEI YOSATSU HOKOKUSHO 3890
(REPORT OF FORECASTING BUSINESS ON FRUIT TREES' DISEASES AND NOXIOUS INSECTS)
SHIZUOKA-KEN CHAGYO SHIKENJO (SHIZUOKA PREFECTURAL TEA EXPERIMENT STATION)
1706, KURASAWA, KIKUGAWA-CHO, OGASA-GUN, SHIZUOKA 439
JAPANESE AND ENGLISH

KAJU-TO SAKUMOTSU BYOGAICHU HASSEI YOSATSU JIGYO NENPO 3891
(ANNUAL REPORT OF FORECASTING BUSINESS ON FRUIT TREES' DISEASES AND NOXIOUS INSECTS)
YAMAGUCHI-KEN NOGYO SHIKENJO (YAMAGUCHI AGRICULTURAL EXPERIMENT STATION)
OUCHI MIHORI, YAMAGUCHI-SHI 747-13
JAPANESE

KANSAI BYOCHUGAI KENKYUKAIHO 3892
(PROCEEDINGS OF THE KANSAI PLANT PROTECTION SOCIETY)
KANSAI BYOCHUGAI KENKYUKAI (KANSAI PLANT PROTECTION SOCIETY)
C/O NORIN-SHO YASAI SHIKENJO, OGOSO, ISHINDEN-MACHI, TSU-SHI 541-01
JAPANESE AND ENGLISH
INDEXED: JICST

KANTO TOSAN BYOGAICHU KENKYUKAI NENPO 3893
(PROCEEDINGS OF THE KANTO TOSAN PLANT PROTECTION SOCIETY)
KANTO TOSAN BYOGAICHU KENKYUKAI (KANTO TOSAN PLANT PROTECTION SOCIETY)
C/O NOJI SHIKENJO, KONOSU, KONOSU-SHI 365
JAPANESE
INDEXED: JICST

KITANIHON BYOGAICHU KENKYUKAI KAIHO 3894
(ANNUAL REPORT OF SOCIETY OF PLANT PROTECTION OF NORTH JAPAN)
KITA NIHON BYOGAICHU KENKYUKAI (SOCIETY OF PLANT PROTECTION OF NORTH JAPAN)
C/O NORIN-SHO TOHOKU NOGYO SHIKENJO, 3, SHIMOFURMICHI, YOTSUYA, OMAGARI-SHI 014-01
JAPANESE AND ENGLISH
INDEXED: CHE, JICST

KOBE SHOKUBUTSU BOEKI JOHO 3895
(KOBE PLANT PROTECTION AND PLANT QUARANTINE INFORMATION)
KOBE-SHI SHOKUBUTSU BOEKISHO (KOBE PLANT PROTECTION STATION)
3, SHINKO-CHO, IKUTA-KU, KOBE-SHI 650
JAPANESE
INDEXED: BIB

KYUSHU BYOGAICHU KENKYU KAIHO 3896
(PROCEEDINGS OF THE ASSOCIATION FOR PLANT PROTECTION OF KYUSHU)
KYUSHU BYOGAICHU KENKYUKAI (ASSOCIATION FOR PLANT PROTECTION OF KYUSHU)
C/O KYUSHU NOGYO SHIKENJO, IZUMI, CHIKUGO-SHI 833
JAPANESE, WITH ENGLISH SUMMARY
INDEXED: BIB, JICST

KYUSHU BYOGAICHU KENKYU-KAI TOKUBETSU HOKOKU 3897
(MEMOIRS OF THE ASSOCIATION FOR PLANT PROTECTION OF KYUSHU)
KYUSHU BYOGAICHU KENKYU-KAI (ASSOCIATION FOR PLANT PROTECTION OF KYUSHU)
JAPANESE

KYUSHU SHOKUBUTSU BOEKI 3898
(KYUSHU PLANT PROTECTION)
NORIN-SHO MOJI SHOKUBUTSU BOEKISHO (MOJI PLANT PROTECTION STATION, MINISTRY OF AGRICULTURE AND FORESTRY)
3-10, 1-CHOME, NISHI KAIGAN, MOJI-KU, KITA KYUSHU-SHI 801
JAPANESE

MIYAZAKI-KEN NOGYO KISHO SAIGAI SOKUHO 3899
(CURRENT REPORT ON WEATHER-CAUSED DAMAGE TO AGRICULTURE IN MIYAZAKI PREFECTURE)
MIYAZAKI CHIHO KISHODAI (MIYAZAKI REGION METEOROLOGICAL OBSERVATORY)
JAPANESE

NAHA SHOKUBUTSU BOEKI JOHO 3900
(NAHA, PLANT PROTECTION AND PLANT QUARANTINE INFORMATION)
NORIN-SHO NAHA SHOKUBUTSU BOEKI JIMUSHO (NAHA PLANT QUARANTINE OFFICE, MINISTRY OF AGRICULTURE AND FORESTRY)
112, TONDO-CHO, NAHA-SHI 900
JAPANESE

NIHON NOYAKUGAKU KAISHI 3901
(JOURNAL OF PESTICIDE SCIENCE)
PESTICIDE SCIENCE SOCIETY OF JAPAN
C/O NATIONAL INSTITUTE OF AGRICULTURAL SCIENCES, NISHIGAHARA, KITA-KU, TOKYO 309
JAPANESE AND ENGLISH
INDEXED: BIO

NIHON SHOKUBUTSU BYORI GAKKAIHO 3902
(ANNALS OF THE PHYTOPATHOLOGICAL SOCIETY OF JAPAN)
HOKKAIDO DAIGAKU NOGAKUBU (COLLEGE OF AGRICULTURE, HOKKAIDO UNIVERSITY) AND NIHON SHOKUBUTSU BYORI GAKKAI (PHYTOPATHOLOGICAL SOCIETY OF JAPAN)
C/O NORIN-SHO NOGYO GIJUTSU KENKYUSHO, NISHIGAHARA, KITA-KU, TOKYO 114
JAPANESE AND ENGLISH
INDEXED: JICST

NOGYO GIJUTSU KENKYUSHO HOKOKU, C, (BYORI KONCHU) 3903
(BULLETIN OF THE NATIONAL INSTITUTE OF AGRICULTURAL SCIENCES, SERIES C (PLANT PATHOLOGY AND ENTOMOLOGY))
NORIN-SHO NOGYO GIJUTSU KENKYUSHO (NATIONAL INSTITUTE OF AGRICULTURAL SCIENCES)
1-7, 2-CHOME, NISHIGAHARA, KITA-KU, TOKYO 114
JAPANESE AND ENGLISH
INDEXED: BIB, BIO, CHE, JICST

NORIN-SHO SHOKUBUTSU UIRUSU KENKYUJO NYUSU 3904
(NATIONAL INSTITUTE OF PLANT VIRUS RESEARCH NEWS)
NATIONAL INSTITUTE OF PLANT VIRUS RESEARCH
JAPANESE

NOSAKUMOTSU YUGAI DOSHOKUBUTSU HASSEI YOSATSU JIGYO NENPO 3905
(ANNUAL REPORT OF FORECASTING BUSINESS ON CROPS' DISEASES AND NOXIOUS INSECTS)
YAMAGUCHI-KEN (YAMAGUCHI PREFECTURAL GOVERNMENT)
1-1, TAKI-MACHI, YAMAGUCHI-SHI 753
JAPANESE

NOSAKUMOTSU YUGAI DOSHOKUBUTSU HASSEI YOSATSU TOKUBETSU HOKOKU 3906
(SPECIAL REPORT OF FORECASTING BUSINESS ON CROPS' DISEASES AND NOXIOUS INSECTS)
NORIN-SHO NOSAN ENGEI-KYOKU (AGRICULTURAL PRODUCTION BUREAU, MINISTRY OF AGRICULTURE AND FORESTRY)
2-1, 1-CHOME, KASUMIGASEKI, CHIYODA-KU, TOKYO 100
JAPANESE, WITH ENGLISH SUMMARY

NOYAKU 3907
(PESTICIDES)
NOYAKU KANKOKAI
C/O NIHON NOYAKU K.K., 2-5, 1-CHOME, NIHONBASHI, CHUO-KU, TOKYO 103
JAPANESE

NOYAKU GURAFU 3908
(AGRICULTURAL CHEMICALS MAGAZINE)
NIHON TOKUSHU NOYAKU SEIZO K.K.
2-8, NIHONBASHI MURO-MACHI, CHUO-KU, TOKYO 103
JAPANESE

NOYAKU JIDAI 3909
(AGCHEM AGE)
NIHON SODA K.K.
C/O SHIN OTE-MACHI BUILDING, 2-1, 2-CHOME, OTE-MACHI, CHIYODA-KU, TOKYO 100
JAPANESE
INDEXED: CHE, JICST

NOYAKU JIKKEN TENJIHO SEISEKISHO 3910
(RESULT OF AGRICULTURAL CHEMICALS EXPERIMENTATION IN FIELD)
ISHIKAWA-KEN NOGYO SEISAN SHIZAITO JIKKEN TENJIHO SETCHI JIMUKYOKU
NAKABAYASHI, NONOICHI-MACHI, ISHIKAWA-GUN, ISHIKAWA 921
JAPANESE

NOYAKU KAGAKU 3911
(JOURNAL OF PESTICIDE SCIENCE)
NOYAKU KAGAKU KENKYUKAI (SOCIETY OF PESTICIDE SCIENCE)
C/O NIHONBASHI KURABU KAIKAN, 1-8, NIHONBASHI MURO-MACHI, CHUO-KU, TOKYO 103
JAPANESE AND ENGLISH
INDEXED: BIB, BIO JICST

NOYAKU KENKYU 3912
(AGRICULTURAL CHEMICALS RESEARCH)
NIHON TOKUSHU NOYAKU SEIZO K.K.
2-8, NIHONBASHI MURO-MACHI, CHUO-KU, TOKYO 103
JAPANESE
INDEXED: CHE, JICST

NOYAKU KENSASHO HOKOKU 3913
(BULLETIN OF THE AGRICULTURAL CHEMICALS INSPECTION STATION)
NORIN-SHO NOYAKU KENSASHO (AGRICULTURAL CHEMICALS INSPECTION STATION)
2-772, SUZUKI-CHO, KODAIRA-SHI, TOKYO 187
JAPANESE, WITH ENGLISH SUMMARY
INDEXED: BIB, BIO, CHE

NOYAKU SHUNJU 3914
(FERTILIZER, SPRING AND AUTUMN)
JAPANESE
INDEXED: JICST

NOYAKU TSUSHIN 3915
(AGRICULTURAL CHEMICALS NEWS)
IHARA NOYAKU (IHARA AGRICULTURAL CHEMICALS)
JAPANESE

NOYAKU TSUSHIN 3916
(KUMIAI CHEMICAL NEWS)
KUMIAI KAGAKU KOGYO K.K. (KUMIAI CHEMICAL INDUSTRY COMPANY, LIMITED)
C/O NIHON BUILDING, 6-2, 2-CHOME, OTE-MACHI, CHIYODA-KU, TOKYO 100
JAPANESE

NOYAKU YORON 3917
(COMPREHENSIVE LIST OF AGRICULTURAL CHEMICALS)
JAPAN NOSEIKYOKU SHOKUBUTSU BOEKIKA: NIHON SHOKUBUTSU BOEKI KYOKAI
JAPANESE
INDEXED: BIB

OSAKA SHOKUBUTSU BOEKI 3918
(OSAKA PLANT PROTECTION)
JAPANESE
INDEXED: JICST

REVIEW OF PLANT PROTECTION RESEARCH 3919
NIHON SHOKUBUTSU BYORI GAKKAI (PHYTOPATHOLOGICAL SOCIETY OF JAPAN) AND NIHON OYO DOBUTSU KONCHU GAKKAI (JAPANESE SOCIETY OF APPLIED ENTOMOLOGY AND ZOOLOGY)
C/O NORIN-SHO NOGYO GIJUTSU KENKYUSHO, 1-7, 2-CHOME, NISHIGAHARA, KITA-KU, TOKYO 114
ENGLISH
INDEXED: BIB, BIO, CHE

SHIKOKU SHOKUBUTSU BOEKI KENKYU 3920
(SHIKOKU PLANT PROTECTION RESEARCH)
SHIKOKU SHOKUBUTSU BOEKI KENKYU KYOGIKAI (SHIKOKU COUNCIL ON THE PLANT PROTECTION RESEARCH)
C/O NORIN-SHO SHIKOKU NOGYO SHIKENO, ZENTSUJI-MACHI, ZENTSUJI-SHI 765
JAPANESE
INDEXED: JICST

SHIMANE BYOGAICHU KENKYUKAI KAIHO 3921
(BULLETIN OF THE ASSOCIATION FOR PLANT PROTECTION OF SHIMANE PREFECTURE)
SHIMANE BYOGAICHU KENKYUKAI (ASSOCIATION FOR PLANT PROTECTION OF SHIMANE PREFECTURE)
C/O SHIMANE DAIGAKU NOGAKUBU, NISHIKAWAZU-MACHI, MATSUE-SHI 690
JAPANESE

SHIN NOYAKU 3922
(NEW AGRICULTURAL CHEMICALS)
SANKYO K.K.
10-17, 3-CHOME, GINZA, CHUO-KU, TOKYO 104
JAPANESE
INDEXED: CHE, JICST

SHOKUBUTSU BOEKI 3923
(PLANT PROTECTION)
NIHON SHOKUBUTSU BOEKI KYOKAI (JAPAN PLANT PROTECTION ASSOCIATION)
43-11, 1-CHOME, KOMAGOME, TOSHIMA-KU, TOKYO 170
JAPANESE
INDEXED: BIB, CHE, JICST

SHOKUBUTSU BOEKISHO CHOSA KENKYU HOKOKU 3924
(RESEARCH BULLETIN OF PLANT PROTECTION SERVICE, JAPAN)
NORIN-SHO YOKOHAMA SHOKUBUTSU BOEKISHO (YOKOHAMA PLANT PROTECTION STATION, MINISTRY OF AGRICULTURE OF FORESTRY)
5-57, KITA-NAKA-DORI, NAKA-KU, YOKOHAMA-SHI 231
JAPANESE AND ENGLISH
INDEXED: BIB, CHE, JICST

SHOKUBUTSU BYOGAI KENKYU 3925
(PLANT DISEASES RESEARCH)
DR. SHIGEYASU AKAI
KYOTO DAIGAKU, KYOTO
JAPANESE
INDEXED: CHE

TABAKO BYOGAICHU SOGO BOJO TENJI NOJO SEISEKI 3926
(RESULT OF TOBACCO PLANT PROTECTION RESEARCH)
NIHON SENBAI KOSHA SEISANSEI HONBU
2-1, AKASAKA AOI-CHO, MINATO-KU, TOKYO 107
JAPANESE

TAIKI OSEN TO SHOKUBUTSU HIGAI, CHIBA-KEN ICHIHARA CHIKU NO NASHI HIGAI O CHUSHIN TO SHITE 3927
(AIR POLLUTION AND ITS INFLUENCE TO PLANTS)
CHIBA-KEN ICHIHARACHIKU NASHITO HIGAI CHOSA IINKAI
1-1, ICHIBA-MACHI, CHIBA-SHI 280-91
JAPANESE

YASAI BYOGAICHU HASSEI YOSATSU JIKKEN JIGYO NENPO 3928
(ANNUAL REPORT OF FORECASTING BUSINESS ON VEGETABLES DISEASES AND NOXIOUS INSECTS)
YAMAGUCHI-KEN NOGYO SHIKENJO (YAMAGUCHI AGRICULTURAL EXPERIMENT STATION)
OUCHI MIHORI, YAMAGUCHI-SHI 747-13
JAPANESE

YOKOHAMA SHOKUBUTSU BOEKI NYUSU 3929
(YOKOHAMA PLANT PROTECTION NEWS)
NORIN-SHO YOKOHAMA SHOKUBUTSU BOEKISHO (YOKOHAMA PLANT PROTECTION STATION, MINISTRY OF AGRICULTURE AND FORESTRY)
5-57, KITA-NAKA-DORI, NAKA-KU, YOKOHAMA-SHI 231
JAPANESE

YUSHUTSUNYU SHOKUBUTSU KEN'EKI TOKEI 3930
(IMPORTS AND EXPORTS PLANT QUARANTINE STATISTICS)
NORIN-SHO YOKOHAMA SHOKUBUTSU BOEKISHO (YOKOHAMA PLANT PROTECTION STATIONS, MINISTRY OF AGRICULTURE AND FORESTRY)
5-57, KITA-NAKA-DORI, NAKA-KU, YOKOHAMA-SHI 231
JAPANESE

ZASSO BOJO KENKYU SHISETSU HOKOKU (UTSUNOMIYA DAIGAKU) 3931
(REPORT OF INSTITUTE OF WEED CONTROL SCIENCE, UTSUNOMIYA UNIVERSITY)
UTSUNOMIYA DAIGAKU NOGAKUBU ZASSO BOJO KENKYU SHISETSU (INSTITUTE OF WEED CONTROL SCIENCE, UTSUNOMIYA UNIVERSITY)
350, MINE-MACHI, UTSUNOMIYA-SHI 320
JAPANESE

ZASSO KENKYU 3932
(WEED RESEARCH, JOURNAL OF THE WEED SCIENCE SOCIETY OF JAPAN)
NIHON ZASSO GAKKAI
C/O NOGYO GIJUTSU KYOKAI, 26-3, 1-CHOME, NISHIGAHARA, KITA-KU, TOKYO 114
JAPANESE, WITH ENGLISH SUMMARY
INDEXED: CHE

Field Crops, Industrial Crops, Fruit Growing

AKITA-KEN KAJU SHIKENJO GYOMU HOKOKU 3933
(OPERATION REPORT OF AKITA FRUIT TREE EXPERIMENT STATION)
AKITA-KEN KAJU SHIKENJO (AKITA FRUIT TREE EXPERIMENT STATION)
DAIGO, HIRAKA-CHO, HIRAKA-GUN, AKITA 019-05
JAPANESE
INDEXED: BIB

AKITA-KEN KAJU SHIKENJO KENKYU HOKOKU 3934
(RESEARCH REPORT OF AKITA FRUIT TREE EXPERIMENT STATION)
AKITA-KEN KAJU SHIKENJO (AKITA FRUIT TREE EXPERIMENT STATION)
DAIGO, HIRAKA-CHO, HIRAKA-GUN, AKITA 019-05
JAPANESE AND ENGLISH
INDEXED: BIB, CHE

AOMORI-KEN HATASAKU ENGEI SHINKENJO KENKYU HOKOKU 3935
(BULLETIN OF THE AOMORI FIELD CROPS AND HORTICULTURAL EXPERIMENT STATION)
AOMORI-KEN HATAKEMONO ENGEI SHINKENJO (AOMORI FIELD CROPS AND HORTICULTURAL EXPERIMENT STATION)
JAPANESE
INDEXED: BIB

AOMORI-KEN RINGO SHIKENJO GYOMU NENPO 3936
(ANNUAL REPORT OF THE AOMORI APPLE EXPERIMENT STATION)
AOMORI-KEN RINGO SHIKENJO
24, FUKUTAMI, BOTANDAIRA, KUROISHI-SHI 036-03
JAPANESE

AOMORI-KEN RINGO SHIKENJO HOKOKU 3937
(BULLETIN OF THE AOMORI APPLE EXPERIMENT STATION)
AOMORI-KEN RINGO SHIKENJO (AOMORI APPLE EXPERIMENT STATION)
24, FUKUTAMI, BOTANDAIRA, KUROISHI-SHI 036-03
JAPANESE, WITH ENGLISH SUMMARY
INDEXED: BIB

AOMORI-KEN RINGO TOKEIHYO 3938
(APPLE STATISTICS, AOMORI PREFECTURE)
AOMORI-KEN NORIN-BU (AGRICULTURAL AND FORESTRY DIVISION, AOMORI PREFECTURAL GOVERNMENT)
1-1, 1-CHOME, NAGASHIMA, AOMORI-SHI 030
JAPANESE

BEIBAKU NIPPO 3939
(DAILY RICE AND WHEAT)
SHOKURYO SANGYO SHINBUNSHA
8-1, 1-CHOME, HAMAMATSU-CHO, MINATO-KU, TOKYO 105
JAPANESE

BEIBAKU SHUKA NENPO 3940
(ANNUAL REPORT OF GOVERNMENT PURCHASING OF CEREALS)
SHOKURYO-CHO (FOOD AGENCY)
2-1, 1-CHOME, KASUMIGASEKI, CHIYODA-KU, TOKYO 100
JAPANESE

BEIBAKU-TO NO TORIHIKI KAKAKU CHOSA GEPPO 3941
(MONTHLY SURVEY OF THE TRANSACTION PRICES OF RICE, BARLEY, ETC.)
SHOKURYO-CHO SOMUBU CHOSAKA (FOOD AGENCY GENERAL OPERATIONS DIVISION, SURVEY DEPARTMENT)
JAPANESE

BEIBAKU-TO NO TORIHIKI KAKAKU NENPO 3942
(ANNUAL REPORT OF FOOD MARKETING PRICE)
SHOKURYO-CHO SOMU-BU (GENERAL AFFAIRS DIVISION, FOOD AGENCY)
2-1, 1-CHOME, KASUMIGASEKI, CHIYODA-KU, TOKYO 100
JAPANESE

BEIKOKU NO HINSHUBETS DEMAWARI JOKYO 3943
(STATISTICS OF RICE SUPPLY, BY KINDS)
SHOKURYO-CHO (FOOD AGENCY)
2-1, 1-CHOME, KASUMIGASEKI, CHIYODA-KU, TOKYO 100
JAPANESE

BEIKOKU NO HINSHUBETSU SAKUTSUKE JOKYO 3944
(REPORT ON CROP CONDITIONS OF RICE BY KINDS)
SHOKURYO-CHO (RESEARCH AND ANALYSIS DIVISION, FOOD AGENCY)
2-1, 1-CHOME, KASUMIGASEKI, CHIYODA-KU, TOKYO 100
JAPANESE

BEIKOKU SHINBUN 3945
(RICE NEWS)
BEIKOKU SHINBUNSHA
1-9, NIHONBASHI KAKIGARA-CHO, CHUO-KU, TOKYO 103
JAPANESE

CHA 3946
(TEA)
SHIZUOKA CHAGYOKAI GISHO
(SHIZUOKA TEA INDUSTRY CONSULTATION CENTER)
JAPANESE

CHA TOKEI NENPO 3947
(ANNUAL STATISTICS OF TEA)
NORIN-SHO NORIN KEIZAI-KYOKU TOKEI JOHO-BU (STATISTICS AND INFORMATION DIVISION, AGRICULTURAL AND FORESTRY ECONOMIC BUREAU, MINISTRY OF AGRICULTURE AND FORESTRY)
2-1, 1-CHOME, KASUMIGASEKI, CHIYODA-KU, TOKYO 100
JAPANESE

CHAGYO GIJUTSU 3948
(TEA TECHNICAL JOURNAL)
SAITAMA-KEN CHAGYO GIJUTSU KYOKAI (SAITAMA PREFECTURE TEA TECHNICIANS' ASSOCIATION)
244-2, SHIMOYAGANUKI, IRUMA-SHI 358
JAPANESE

CHAGYO GIJUTSU KENKYU 3949
(STUDY OF TEA)
NORIN-SHO CHAGYO SHIKENJO (NATIONAL RESEARCH INSTITUTE OF TEA, MINISTRY OF AGRICULTURE AND FORESTRY)
2769, KANAYA, KANAYA-CHO, HAIBARA-GUN, SHIZUOKA 428
JAPANESE AND ENGLISH
INDEXED: CHE, JICST

CHAGYO KENKYU HOKOKU 3950
(RESEARCH REPORT OF TEA)
NIHON CHAGYO GIJUTSU KYOKAI (JAPAN TEA TECHNICAL ASSOCIATION)
2769, KANAYA, KANAYA-CHO, HAIBARA-GUN, SHIZUOKA 428
JAPANESE, WITH ENGLISH SUMMARY
INDEXED: CHE, JICST

CHAGYO SHIKENJO KENKYU HOKOKU 3951
(BULLETIN OF THE NATIONAL RESEARCH INSTITUTE OF TEA)
NORIN-SHO CHAGYO SHIKENJO (NATIONAL RESEARCH INSTITUTE OF TEA, MINISTRY OF AGRICULTURE AND FORESTRY)
2769, KANAYA, KANAYA-CHO, HAIBARA-GUN, SHIZUOKA-428
JAPANESE AND ENGLISH
INDEXED: BIB, CHE, JICST

CHAGYOKAI 3952
(TEA MANAGEMENT)
CHAGOKAISHA
4-5, NISHI CHIYODA-MACHI, SHIZUOKA-SHI 420
JAPANESE

EHIME-KEN KAJU SHIKENJO GYOMU HOKOKU 3953
(ANNUAL REPORT OF THE EHIME PREFECTURAL FRUIT TREE EXPERIMENT STATION)
EHIME-KENRITSU KAJU SHIKENJO (EHIME PREFECTURAL FRUIT TREE EXPERIMENT STATION)
227, HIGASHINO-MACHI, MATSUYAMA-SHI 790
JAPANESE

EHIME-KENRITSU KAJU SHIKENJO KENKYU HOKOKU 3954
(RESEARCH REPORT OF EHIME FRUIT TREE EXPERIMENT STATION)
EHIME-KENRITSU KAJU SHIKENJO (EHIME FRUIT TREE EXPERIMENT STATION)
4, HIGASHINO-MACHI, MATSUYAMA-SHI 790
JAPANESE, WITH ENGLISH SUMMARY

FUKUOKA-KENRITSU NOGYO SHIKENJO CHAGYO SHIDOSHO GYOMU NENPO 3955
(ANNUAL REPORT OF THE TEA INDUSTRY RESEARCH INSTITUTE FUKUOKA PREFECTURAL AGRICULTURAL EXPERIMENT STATION)
FUKUOKA-KENRITSU NOGYO SHIKENJO CHAGYO SHIDOSHO (TEA INDUSTRY RESEARCH INSTITUTE, FUKUOKA PREFECTURAL AGRICULTURAL EXPERIMENT STATION)
3266-1, HONBUN, KUROGI-MACHI, YAME-GUN, FUKUOKA 834-12
JAPANESE

HATABAKO KENKYU 3956
(RESEARCH OF TOBACCO LEAF)
NIHON SENBAI KOSHA SEISAN-HONBU KOSAKU SHIDO-KA (CULTIVATION GUIDANCE DIVISION, TOBACCO LEAF PRODUCTION DEPARTMENT, JAPAN MONOPOLY CORPORATION)
11, SHIBA NISHIKUBU SAKURAGAWA-CHO, MINATO-KU, TOKYO 105
JAPANESE

HATANO TABAKO SHIKENJO HOKOKU 3957
(REPORT OF THE HATANO TOBACCO EXPERIMENTAL STATION)
JAPANESE
INDEXED: BIB

HIROSHIMA-KEN KAJU SHIKENJO KENKYU HOKOKU 3958
(RESEARCH REPORT OF THE FRUIT TREE EXPERIMENT STATION OF HIROSHIMA PREFECTURE)
HIROSHIMA-KEN KAJU SHIKENJO (FRUIT TREE EXPERIMENT STATION OF HIROSHIMA PREFECTURE)
AKITSU-MACHI, TOYOTA-GUN, HIROSHIMA 729-24
JAPANESE AND ENGLISH

HOKURIKU SAKUMOTSU GAKKAIHO 3959
(HOKURIKU CROP SCIENCE)
NIHON SAKUMOTSU GAKKAI HOKURIKU SHIBU (JAPAN CROP SCIENCE SOCIETY HOKURIKU BRANCH)
JAPANESE

IWATA TABAKO SHIKENJO HOKOKU 3960
(BULLETIN OF THE IWATA TOBACCO EXPERIMENT STATION)
JAPANESE

KAGOSHIMA-KEN CHAGYO SHIKENJO GYOMU NENPO 3961
(ANNUAL REPORT OF KAGOSHIMA PREFECTURE TEA EXPERIMENT STATION)
KAGOSHIMA-KEN CHAGYO SHIKENJO
3964, NAGASATO, CHIRAN-CHO, KAWANABE-GUN, KAGOSHIMA 897-03
JAPANESE
INDEXED: JICST

KAGOSHIMA-KEN CHAGYO SHIKENJO KENKYU HOKOKU 3962
(REPORT OF TEA EXPERIMENT STATION OF KAGOSHIMA PREFECTURE)
KAGOSHIMA-KEN CHAGYO SHIKENJO (KAGOSHIMA PREFECTURE TEA EXPERIMENT STATION)
3964, NAGASATO, CHIRAN-CHO, KAWANABE-GUN, KAGOSHIMA 897-03
JAPANESE

KAGOSHIMA-KEN KAJU SHIKENJO GYOMU HOKOKU 3963
(ANNUAL REPORT OF KAGOSHIMA FRUIT TREE EXPERIMENT STATION)
KAGOSHIMA-KEN KAJU SHIKENJO (KAGOSHIMA FRUIT TREE EXPERIMENT STATION)
1452, HONJO, TARUMIZU-SHI 891-21
JAPANESE

KAGOSHIMA-KEN KAJU SHIKENJO SHIKEN SEISEKI-SHO 3964
(EXPERIMENT REPORT OF THE KAGOSHIMA PREFECTURE FRUIT TREE EXPERIMENT STATION)
KAGOSHIMA-KEN KAJU SHIKENJO
1452, HONJO, TARUMIZU, KAGOSHIMA-KEN 891-21
JAPANESE
INDEXED: CHE

KAJITSU NIHON 3965
(JAPANESE FRUITS)
NIHON ENGEI NOGYO KYODO KUMIAI RENGOKAI (JAPAN FRUIT GROWER COOPERATIVE ASSOCIATION)
14-1, 4-CHOME, SOTO KANDA, CHIYODA-KU, TOKYO 101
JAPANESE
INDEXED: BIB

KAJU 3966
(FRUIT TREE)
OKAYAMA-KEN KEIZAI NOGYO KYODO KUMIAI RENGOKAI
C/O OKAYAMA-KEN NOGYO KAIKAN, 18-201, 9-CHOME, TOGIYA-MACHI, OKAYAMA-SHI 700
JAPANESE

KAJU EIYO SHINDAN JIGYO SEISEKISHO 3967
(NUTRITION STUDY OF FRUIT TREE. HIROSHIMA PREFECTURE)
HIROSHIMA-KEN (HIROSHIMA PREFECTURE)
HIROSHIMA-KEN KAJITSU NOGYO KYODO KUMIAI RENGOKAI (HIROSHIMA PREFECTURAL FRUIT FEDERATION OF AGRICULTURAL CO-OP ASSOCIATION)
4395-41, TADANOUMI-CHO, TAKEHARA-SHI 729-33
JAPANESE

KAJU KAKI NENPO 3968
(ANNUAL REPORT OF FRUIT TREE AND FLOWERING PLANT)
NORIN-SHO NOSAN ENGEI-KYOKU (AGRICULTURAL PRODUCTION BUREAU, MINISTRY OF AGRICULTURE AND FORESTRY)
2-1, 1-CHOME, KASUMIGASEKI, CHIYODA-KU, TOKYO 100
JAPANESE

KAJU NO SHUKAKURYO RUINEN TOKEI 3969
(CUMULATIVE STATISTICS OF FRUIT HARVEST)
NORIN-SHO NORIN KEIZAI-KYOKU TOKEI-JOHO-BU (STATISTICS AND INFORMATION DEPARTMENT, ECONOMIC AFFAIRS BUREAU, MINISTRY OF AGRICULTURE AND FORESTRY)
2-1, 1-CHOME, KASUMIGASEKI, CHIYODA-KU, TOKYO 100
JAPANESE

KAJU SHIKEN SEISEKISHO (ISHIKAWA-KEN SAKYUCHI NOGYO SHIKENJO) 3970
(TEST REPORT ON FRUIT TREES, ISHIKAWA PREFECTURE SAND DUNE AGRICULTURE EXPERIMENT STATION))
ISHIKAWA-KEN SAKYUCHI NOGYO SHIKENJO
6-2, UCHIHISUMI, UNOKEMACHI KAHOKU-GUN ISHIKAWA 929-11
JAPANESE

KAJU SHIKENJO HOKOKU, A (HIRATSUKA) 3971
(BULLETIN OF THE FRUIT TREE RESEARCH STATION, MINISTRY OF AGRICULTURE AND FORESTRY, SERIES A (HIRATSUKA))
NORIN-SHO KAJU SHIKENJO (FRUIT TREE RESEARCH STATION, MINISTRY OF AGRICULTURE AND FORESTRY) (HIRATSUKA)
1-24, OHARA, HIRATSUKA-SHI 254
JAPANESE, WITH ENGLISH SUMMARY
INDEXED: BIB, CHE

KAJU SHIKENJO HOKOKU, B (OKITSU) 3972
(BULLETIN OF THE FRUIT TREE RESEARCH STATION, MINISTRY OF AGRICULTURE AND FORESTRY, SERIES B (OKITSU))
NORIN-SHO KAJU SHIKENJO OKITSU SHIJO (FRUIT TREE RESEARCH STATION, MINISTRY OF AGRICULTURE AND FORESTRY) (OKITSU)
OKITSU NAKA-MACHI, SHIMIZU-SHI 424-02
JAPANESE, WITH ENGLISH SUMMARY
INDEXED: BIB, CHE

KAJU SHIKENJO HOKOKU, C (MORIOKA) 3973
(BULLETIN OF THE FRUIT TREE RESEARCH STATION, MINISTRY OF AGRICULTURE AND FORESTRY, SERIES C (MORIOKA))
NORIN-SHO KAJU SHIKENJO MORIOKA SHIJO (FRUIT TREE RESEARCH STATION, MINISTRY OF AGRICULTURE AND FORESTRY) (MORIOKA)
92, NABEYASHIKI, SHIMO KURIYAGAWA, MORIOKA-SHI 020-01
JAPANESE, WITH ENGLISH SUMMARY
INDEXED: BIB

KAJU SHIKENJO HOKOKU, SERIES E (AKITSU) 3974
(BULLETIN OF THE FRUIT TREE RESEARCH STATION)
AKITSU BRANCH, FRUIT TREE RESEARCH STATION, MINISTRY OF AGRICULTURE AND FORESTRY
AKITSU, HIROSHIMA, 309
JAPANESE
INDEXED: BIB, BIO

KAJU SHIKENJO NYUSU 3975
(FRUIT TREE EXPERIMENT STATION NYUSU)
NORINSHO KAJU SHIKENJO (MINISTRY OF AGRICULTURE AND FORESTRY, FRUIT TREE EXPERIMENT STATION)
JAPANESE

KANKITSU 3976
(CITRUS)
SHIZUOKA-KEN KANKITSU NOGYO KYODO KUMIAI RENGOKAI (SHIZUOKA PREFECTURE CITRUS AGRICULTURAL CORPORATION)
4-12, 1-CHOME, TSUJI, SHIMIZU-SHI 424
JAPANESE

KINJIN 3977
(FUNGI)
NIHON KINOKO SENTA
1-84, TOMIYASU, TOTTORI-SHI 680
JAPANESE

KINJIN KENKYUSHO KENKYU HOKOKU 3978
(REPORTS OF THE TOTTORI MYCOLOGICAL INSTITUTE)
NIPPON KINOKO SENTA KINJIN KENKYUSHO (TOTTORI MYCOLOGICAL INSTITUTE, (A FOUNDATION) THE JAPAN KINOKO RESEARCH CENTER)
211, HIROHATA, FURUKOGE, TOTTORI-SHI 680
JAPANESE AND ENGLISH
INDEXED: BIB, BIO

KINOKO 3979
(MUSHROOMS)
BISEIBUTSU KEIZAI KENKYUSHO
C/O KYODO BUILDING, 9-5, 3-CHOME, NIHONBASHI HON-CHO, CHUO-KU, TOKYO 103
JAPANESE

KOCHI-KEN KAJU SHIKENJO GYOMU HOKOKU 3980
(REPORT OF KOCHI FRUIT TREE EXPERIMENT STATION)
KOCHI-KEN KAJU SHIKENJO (KOCHI FRUIT TREE EXPERIMENT STATION)
271, ASAKURA-MACHI, KOCHI-SHI 780
JAPANESE

KOGEI NOSAKUMOTSU TOKEI NENPO 3981
(ANNUAL STATISTICS OF INDUSTRIAL CROPS)
NORIN-SHO NORIN KEIZAI-KYOKU (ECONOMIC AFFAIRS BUREAU, MINISTRY OF AGRICULTURE AND FORESTRY)
2-1, 1-CHOME, KASUMIGASEKI, CHIYODA-KU, TOKYO 100
JAPANESE

KUMA NOKEN SHIKEN SEISEKISHO (CHA, KURI) 3982
(SURVEY RESULT OF KUMA AGRICULTURAL EXPERIMENT STATION, TEA AND CHESTNUT)
KUMAMOTO-KEN KUMA NOGYO KENKYU SHIDOSHO
2248-16, UEMURA, KUMA-GUN, KUMAMOTO 858-04
JAPANESE

KUMAMOTO-KEN CHAGYO SHIKENJO GYOMU HOKOKU 3983
(REPORT ON ACTIVITIES OF KUMAMOTO TEA EXPERIMENT STATION)
KUMAMOTO-KEN CHAGYO SHIKENJO
2614, KOMINE, KENGUN-MACHI, KUMAMOTO-SHI 862
JAPANESE

KUMAMOTO-KEN CHAGYO SHIKENJO HOKOKU 3984
(REPORT OF KUMAMOTO TEA EXPERIMENT STATION)
KUMAMOTO-KEN CHAGYO SHIKENJO (KUMAMOTO TEA EXPERIMENT STATION)
2614, KOMINE, KENGUN-MACHI, KUMAMOTO-SHI 862
JAPANESE

KUMAMOTO-KEN KAJU SHIKENJO GYOMU HOKOKU 3985
(ANNUAL REPORT OF KUMAMOTO FRUIT TREE EXPERIMENT STATION)
KUMAMOTO-KEN KAJU SHIKENJO
2566, TOYOBUKU, MATSUBASE-MACHI, SHIMOMASHIKI-GUN, KUMAMOTO 869-05
JAPANESE

KUMAMOTO-KEN KAJU SHIKENJO KENKYU HOKOKU 3986
(REPORT OF KUMAMOTO FRUIT TREE EXPERIMENT STATION)
KUMAMOTO-KEN KAJU SHIKENJO
2566, TOYOBUKU, MATSUBASE-MACHI, SHIMOMASHIKI-GUN, KUMAMOTO 869-05
JAPANESE

KYOTO-FURITSU CHAGYO KENKYUSHO GYOMU HOKOKUSHO 3987
(TEA RESEARCH JOURNAL, KYOTO)
KYOTO-FURITSU CHAGYO KENKYUSHO (RESEARCH INSTITUTE OF TEA INDUSTRY, KYOTO)
1, SHIRAKAWA NAKANOEN, UJI-SHI 611
JAPANESE

KYOTO-FURITSU CHAGYO KENKYUSHO SHIKEN SEISEKI 3988
(REPORT OF KYOTO PREFECTURAL RESEARCH INSTITUTE OF TEA INDUSTRY)
KYOTO-FURITSU CHAGYO KENKYUSHO (KYOTO PREFECTURAL RESEARCH INSTITUTE OF TEA INDUSTRY)
1, SHIRAKAWA NAKONOEN, UJI-SHI 611
JAPANESE

MIZUHO NIHON 3989
(RICE CROPS JAPAN)
AIZEN MIZUHOKAI
79, KITA KOSE, KOSE-CHO, KAMEOKA-SHI 621
JAPANESE

MUGIRUI RYUTSU CHOSA KEKKAHYO 3990
(STATISTICAL REPORTS OF THE CIRCULATION OF WHEAT)
SHOKURYO-CHO SOMU-BU (GENERAL AFFAIRS DIVISION, FOOD AGENCY)
2-1, 1-CHOME, KASUMIGASEKI, CHIYODA-KU, TOKYO 100
JAPANESE

NIHON SENBAI KOSHA CHUBU SHISHA JIGYO TOKEI 3991
(OPERATION STATISTICS, CHUBU BRANCH, JAPAN MONOPOLY CORPORATION)
NIHON SENBAI KOSHA CHUBU SHISHA (CHUBU BRANCH, JAPAN MONOPOLY CORPORATION)
5-2, 1-CHOME, SANNOMARU, NAKA-KU, NAGOYA-SHI 460
JAPANESE

NIHON SENBAI KOSHA CHUGOKU SHISHA JIGYO TOKEI 3992
(OPERATION STATISTICS, CHUGOKU BRANCH, JAPAN MONOPOLY CORPORATION)
NIHON SENBAI KOSHA CHUGOKU SHISHA (CHUGOKU BRANCH, JAPAN MONOPOLY CORPORATION)
5-36, TAKARA-MACHI, HIROSHIMA-SHI 730
JAPANESE

NIHON SENBAI KOSHA CHUO KENKYUSHO KENKYU HOKOKU 3993
(SCIENTIFIC PAPERS OF THE CENTRAL RESEARCH INSTITUTE, JAPAN MONOPOLY CORPORATION)
NIHON SENBAI KOSHA CHUO KENKYUSHO (CENTRAL RESEARCH INSTITUTE, JAPAN MONOPOLY CORPORATION)
6-2, UMEGAOKA, MIDORI-KU, YOKOHAMA-SHI 227
JAPANESE, WITH ENGLISH SUMMARY
INDEXED: BIB, CHE, JICST

NIHON SENBAI KOSHA HAKODATE KOJO KOJO GAIYO 3994
(REPORT OF HAKODATE FACTORY, JAPAN MONOPOLY CORPORATION)
NIHON SENBAI KOSHA HAKODATE KOJO (HAKODATE FACTORY OF JAPAN MONOPOLY CORPORATION)
403-2, KIKYO-MACHI, KAMEDA-SHI 040
JAPANESE

NIHON SENBAI KOSHA HIRATSUKA SEIZO SHIKENJO SHIKEN HOKOKU 3995
(BULLETIN OF HIRATSUKA TOBACCO MANUFACTURING EXPERIMENT STATION)
NIHON SENBAI KOSHA HIRATSUKA SEIZO SHIKENJO (HIRATSUKA TOBACCO MANUFACTURING EXPERIMENT STATION OF JAPAN MONOPOLY CORPORATION)
1-31, KUROBEOKA, HIRATSUKA-SHI 254
JAPANESE, WITH ENGLISH SUMMARY
INDEXED: JICST

NIHON SENBAI KOSHA IWATA TABAKO SHIKENJO HOKOKU 3996
(BULLETIN OF THE IWATA TOBACCO EXPERIMENT STATION)
NIHON SENBAI KOSHA IWATA TABAKO SHIKENJO (IWATA TOBACCO EXPERIMENT STATION OF THE JAPAN MONOPOLY CORPORATION)
700, HIGASHIBARA, TOYOTA-MURA, IWATA-GUN, SHIZUOKA 438
JAPANESE
INDEXED: BIB

NIHON SENBAI KOSHA KAGOSHIMA CHIHO-KYOKU JIGYO TOKEI 3997
(OPERATION STATISTICS, KAGOSHIMA REGIONAL BUREAU, JAPAN MONOPOLY CORPORATION)
NIHON SENBAI KOSHA KAGOSHIMA CHIHO-KYOKU (KAGOSHIMA REGIONAL BUREAU, JAPAN MONOPOLY CORPORATION)
352, KAMIARATA-MACHI, KAGOSHIMA-SHI 890
JAPANESE

NIHON SENBAI KOSHA KAGOSHIMA TABAKO SHIKENJO GYOTEI HOKOKU 3998
(REPORT OF THE KAGOSHIMA TOBACCO EXPERIMENT STATION)
NIHON SENBAI KOSHA KAGOSHIMA TABAKO SHIKENJO (KAGOSHIMA TOBACCO EXPERIMENT STATION, JAPAN MONOPOLY CORPORATION)
5648-2, KAMIFUKUMOTO-CHO, KAGOSHIMA-SHI 891-01
JAPANESE

NIHON SENBAI KOSHA KAGOSHIMA TABAKO SHIKENJO HOKOKU 3999
(BULLETIN OF THE KAGOSHIMA TOBACCO EXPERIMENT STATION)
NIHON SENBAI KOSHA KAGOSHIMA TABAKO SHIKENJO (KAGOSHIMA TOBACCO EXPERIMENT STATION, JAPAN MONOPOLY CORPORATION)
5648-2, KAMIFUKUMOTO-CHO, KAGOSHIMA-SHI 891-01
JAPANESE, WITH ENGLISH SUMMARY
INDEXED: BIB, CHE

NIHON SENBAI KOSHA KAGOSHIMA TABAKO SHIKENJO SOKUHO 4000
(PROMPT REPORT OF THE KAGOSHIMA TOBACCO EXPERIMENT STATION)
NIHON SENBAI KOSHA KAGOSHIMA TABAKO SHIKENJO (KAGOSHIMA TOBACCO EXPERIMENT STATION, JAPAN MONOPOLY CORPORATION)
5648-2, KAMIFUKUMOTO-CHO, KAGOSHIMA-SHI 891-01
JAPANESE

NIHON SENBAI KOSHA KANAZAWA CHIHO-KYOKU JIGYO TOKEI 4001
(OPERATION STATISTICS, KANAZAWA REGIONAL BUREAU, JAPAN MONOPOLY CORPORATION))
NIHON SENBAI KOSHA KANAZAWA CHIHO-KYOKU (KANAZAWA REGIONAL BUREAU, JAPAN MONOPOLY CORPORATION)
94, KAWAGISHI, NAGA-MACHI, KANAZAWA-SHI 920
JAPANESE

NIHON SENBAI KOSHA KORIYAMA CHIHO-KYOKU JIGYO TOKEI 4002
(OPERATION STATISTICS, KORIYAMA REGIONAL BUREAU, JAPAN MONOPOLY CORPORATION)
NIHON SENBAI KOSHA KORIYAMA CHIHO-KYOKU (KORIYAMA REGIONAL BUREAU, JAPAN MONOPOLY CORPORATION)
4-2, OZUTSUMI, KORIYAMA-SHI 963
JAPANESE

NIHON SENBAI KOSHA KUMAMOTO CHIHO-KYOKU JIGYO TOKEI 4003
(OPERATION STATISTICS, KUMAMOTO REGIONAL BUREAU, JAPAN MONOPOLY CORPORATION))
NIHON SENBAI KOSHA KUMAMOTO CHIHO-KYOKU (KUMAMOTO REGIONAL BUREAU, JAPAN MONOPOLY CORPORATION)
1-21, CHIBAJO-MACHI, KUMAMOTO-SHI 860
JAPANESE

NIHON SENBAI KOSHA MORIOKA TABAKO SHIKENJO GYOTEI HOKOKU 4004
(BUSINESS REPORT OF MORIOKA TOBACCO EXPERIMENT STATION, JAPAN MONOPOLY CORPORATION)
NIHON SENBAI KOSHA MORIOKA TABAKO SHIKENJO (MORIOKA TOBACCO EXPERIMENT STATION, JAPAN MONOPOLY CORPORATION)
13-1, 2-CHOME, MIDORIGAOKA, MORIOKA-SHI 020-01
JAPANESE
INDEXED: JICST

NIHON SENBAI KOSHA MORIOKA TABAKO SHIKENJO HOKOKU 4005
(BULLETIN OF THE MORIOKA TOBACCO EXPERIMENT STATION)
NIHON SENBAI KOSHA MORIOKA TABAKO SHIKENJO (MORIOKA TOBACCO EXPERIMENT STATION, JAPAN MONOPOLY CORPORATION)
13-1, 2-CHOME, MIDORIGAOKA, MORIOKA-SHI 020-01
JAPANESE, WITH ENGLISH SUMMARY
INDEXED: BIB, CHE

NIHON SENBAI KOSHA MORIOKA TABAKO SHIKENJO KENKYU SOKUHO 4006
(TECHNICAL NOTES OF THE MORIOKA TOBACCO EXPERIMENT STATION, JAPAN MONOPOLY CORPORATION))
NIHON SENBAI KOSHA MORIOKA TABAKO SHIKENJO (MORIOKA TOBACCO EXPERIMENT STATION, JAPAN MONOPOLY CORPORATION)
13-1, 2-CHOME, MIDORIGAOKA, MORIOKA-SHI 020-01
JAPANESE
INDEXED: BIB, BIO, CHE

NIHON SENBAI KOSHA OKAYAMA CHIHO-KYOKU JIGYO TOKEI 4007
(OPERATION STATISTICS, OKAYAMA REGIONAL BUREAU, JAPAN MONOPOLY CORPORATION)
NIHON SENBAI KOSHA OKAYAMA CHIHO-KYOKU (OKAYAMA REGIONAL BUREAU, JAPAN MONOPOLY CORPORATION)
10-2, 2-CHOME, SHIMOISHII, OKAYAMA-SHI 700
JAPANESE

NIHON SENBAI KOSHA OKAYAMA TABAKO SHIKENJO GYOTEI HOKOKU 4008
(REPORT OF OKAYAMA TOBACCO EXPERIMENT STATION)
NIHON SENBAI KOSHA OKAYAMA TABAKO SHIKENJO (OKAYAMA TABACCO EXPERIMENT STATION, JAPAN MONOPOLY CORPORATION)
5250, KASHIWAJIMA, TAMASHIMA, KURASHIKI-SHI 713
JAPANESE
INDEXED: JICST

NIHON SENBAI KOSHA OKAYAMA TABAKO SHIKENJO HOKOKU 4009
(BULLETIN OF THE OKAYAMA TOBACCO EXPERIMENT STATION)
NIHON SENBAI KOSHA OKAYAMA TABAKO SHIKENJO (OKAYAMA TOBACCO EXPERIMENT STATION, JAPAN MONOPOLY CORPORATION)
5250, KASHIWAJIMA, TAMASHIMA, KURASHIKI-SHI 713
JAPANESE AND ENGLISH
INDEXED: BIB, CHE

NIHON SENBAI KOSHA SHIKOKU SHISHA JIGYO TOKEI 4010
(OPERATION STATISTICS, SHIKOKU BRANCH, JAPAN MONOPOLY CORPORATION)
NIHON SENBAI KOSHA SHIKOKU SHISHA (SHIKOKU BRANCH, JAPAN MONOPOLY CORPORATION)
12-16, 1-CHOME, NISHIKI-CHO, TAKAMATSU-SHI 760
JAPANESE

NIHON SENBAI KOSHA SHISHA CHIHO-KYOKU TABAKO KOSAKU SHIKEN SEISEKI 4011
(EXPERIMENT RESULTS OF TOBACCO CULTURE IN LOCAL BUREAU)
NIHON SENBAI KOSHA SEISAN HONBU (PRODUCTION DEPARTMENT OF JAPAN MONOPOLY CORPORATION)
2-1, AKASAKA AOI-CHO, MINATO-KU, TOKYO 107
JAPANESE

NIHON SENBAI KOSHA TAKASAKI CHIHO-KYOKU JIGYO TOKEI 4012
(OPERATION STATISTICS, TAKASAKI REGIONAL BUREAU, JAPAN MONOPOLY CORPORATION)
NIHON SENBAI KOSHA TAKASAKI CHIHO-KYOKU (TAKASAKI REGIONAL BUREAU, JAPAN MONOPOLY CORPORATION)
5, TAKAMATSU-CHO, TAKASAKI-SHI 370
JAPANESE

NIHON SENBAI KOSHA TOHOKU SHISHA JIGYO TOKEI 4013
(OPERATION STATISTICS, TOHOKU BRANCH, JAPAN MONOPOLY CORPORATION)
NIHON SENBAI KOSHA TOHOKU SHISHA (TOHOKU BRANCH, JAPAN MONOPOLY CORPORATION)
10, SHIMIZU-KOJI, SENDAI-SHI 980
JAPANESE

NIHON SENBAI KOSHA TOKYO KOJO JIGYO GAIYO 4014
(OPERATION STATISTICS, TOKYO FACTORY OF JAPAN MONOPOLY CORPORATION)
NIHON SENBAI KOSHA TOKYO KOJO (TOKYO FACTORY OF JAPAN MONOPOLY CORPORATION)
17-7, 1-CHOME, YOKOKAWA, SUMIDA-KU, TOKYO 130
JAPANESE

NIHON SENBAI KOSHA UTSUNOMIYA CHIHO-KYOKU JIGYO TOKEI 4015
(OPERATION STATISTICS, UTSUNOMIYA REGIONAL BUREAU, JAPAN MONOPOLY CORPORATION)
NIHON SENBAI KOSHA UTSUNOMIYA CHIHO-KYOKU (UTSUNOMIYA REGIONAL BUREAU, JAPAN MONOPOLY CORPORATION)
2314, MUTSU-MACHI, UTSUNOMIYA-SHI 325
JAPANESE

NIHON SENBAI KOSHA UTSUNOMIYA TABAKO SHIKENJO GYOTEI HOKOKU 4016
(REPORT OF THE UTSUNOMIYA TOBACCO EXPERIMENT STATION)
NIHON SENBAI KOSHA UTSUNOMIYA TABAKO SHIKENJO (UTSUNOMIYA TOBACCO EXPERIMENT STATION, JAPAN MONOPOLY CORPORATION)
1900, IDEI, OYAMA-SHI 323
JAPANESE

NIHON SENBAI KOSHA UTSUNOMIYA TABAKO SHIKENJO HOKOKU 4017
(BULLETIN OF UTSUNOMIYA TOBACCO EXPERIMENT STATION)
NIHON SENBAI KOSHA UTSUNOMIYA TABAKO SHIKENJO (UTSUNOMIYA TOBACCO EXPERIMENT STATION, JAPAN MONOPOLY CORPORATION)
1900, IDEI, OYAMA-SHI 323
JAPANESE, WITH ENGLISH SUMMARY

NIHON SENBAI KOSHA UTSUNOMIYA TABAKO SHIKENJO KENKYU SOKUHO 4018
(CURRENT REPORT OF THE UTSUNOMIYA TOBACCO EXPERIMENT STATION)
NIHON SENBAI KOSHA UTSUNOMIYA TABAKO SHIKENJO (UTSUNOMIYA TOBACCO EXPERIMENT STATION, JAPAN MONOPOLY CORPORATION)
1900, IDEI, OYAMA-SHI 323
JAPANESE

NIHON SOCHI GAKKAISHI 4019
(JOURNAL OF JAPANESE SOCIETY OF GRASSLAND SCIENCE)
NIHON SOCHI GAKKAI (JAPANESE SOCIETY OF GRASSLAND SCIENCE)
C/O SOCHI SHIKENJO, NISHINASUNO-CHO, NASU-GUN, TOCHIGI 329-27
JAPANESE AND ENGLISH
INDEXED: BIB, BIO, CHE

NORIN-SHO CHAGYO SHIKENJO NYUSU 4020
(NEWS OF TEA RESEARCH STATION)
NORIN-SHO CHAGYO SHIKENJO (TEA RESEARCH STATION, MINISTRY OF AGRICULTURE AND FORESTRY)
2769, KANAYA, KANAYA-CHO, HAIBARA-GUN, SHIZUOKA 428
JAPANESE

NOSANBUTSU SEISANHI CHOSA HOKOKU, KOME SEISANHI 4021
(COST OF PRODUCTION OF RICE)
NORIN-SHO TOKEI JOHO-BU (STATISTICS AND INFORMATION DIVISION, MINISTRY OF AGRICULTURE AND FORESTRY) AND NORIN TOKEI KYOKAI
11-14, 2-CHOME, MEGURO, MEGURO-KU, TOKYO 153
JAPANESE

NOSANBUTSU SEISANHI CHOSA HOKOKU, MUGI, KOGEI SAKUMOTSUTO NO SEISANHI 4022
(COST OF PRODUCTION OF WHEAT AND BARLEYS)
NORIN-SHO TOKEI JOHO-BU (STATISTICS AND INFORMATION DIVISION, MINISTRY OF AGRICULTURE AND FORESTRY) AND NORIN TOKEI KYOKAI
11-14, 2-CHOME, MEGURO, MEGURO-KU, TOKYO 153
JAPANESE

OGATA JIKKEN NOJO JISSHI GAIYO 4023
(REPORT OF TOBACCO LARGE EXPERIMENTAL FIELD)
NIHON SENBAI KOSHA (JAPAN MONOPOLY CORPORATION)
2-1, AKASAKA AOI-CHO, MINATO-KU, TOKYO 107
JAPANESE

OITA-KEN CHIKUSAN SHIKENJO SOCHI-BU HOKOKU 4024
(REPORT OF OITA LIVESTOCK EXPERIMENT STATION)
OITA-KEN CHIKUSAN SHIKENJO (OITA LIVESTOCK EXPERIMENT STATION)
3989-1, KUSUMI, KUSUMI-MACHI, NAOIRI-GUN, OITA 878-02
JAPANESE

OITA-KEN IGYO SHIDOSHO NENPO 4025
(ANNUAL REPORT OF THE OITA PREFECTURAL RUSH RESEARCH INSTITUTE)
OITA-KEN IGYO SHIDOSHO (OITA PREFECTURAL RUSH RESEARCH INSTITUTE)
510, MINAMIKITSUKI, KITSUKI-SHI 873
JAPANESE

RINGO NYUSU 4026
(APPLE NEWS)
AOMORI-KEN RINGO KYOKAI (AOMORI PREFECTURE APPLE ASSOCIATION)
16, KITAKAWARAGA-CHO, HIROSAKI-SHI 036
JAPANESE

SAGA-KEN KAJU SHIKENJO GYOMU NENPO 4027
(OPERATION REPORT OF SAGA FRUIT TREE EXPERIMENT STATION)
SAGA-KEN KAJU SHIKENJO
TERAURA, OGI-MACHI, OGI-GUN, SAGA 845
JAPANESE

SAGA-KEN KAJU SHIKENJO KENKYU HOKOKU 4028
(BULLETIN OF THE SAGA FRUIT TREE EXPERIMENT STATION)
SAGA-KEN KAJU SHIKENJO (SAGA PREFECTURAL FRUIT TREE EXPERIMENT STATION)
TERAURA, OGI-MACHI, OGI-GUN, SAGA 845
JAPANESE

SAITAMA-KEN CHAGYO SHIKENJO GYOMU HOKOKU 4029
(ANNUAL REPORT OF THE SAITAMA PREFECTURE TEA EXPERIMENT STATION)
SAITAMA-KEN CHAGYO SHIKENJO (SAITAMA PREFECTURE TEA EXPERIMENT STATION)
244-2, SHIMOYAGANUKI, IRUMA-SHI 358
JAPANESE

SAITAMA-KEN CHAGYO SHIKENJO KENKYU HOKOKU 4030
(REPORT OF SAITAMA PREFECTURE TEA EXPERIMENT STATION)
SAITAMA-KEN CHAGYO SHIKENJO (SAITAMA PREFECTURE TEA EXPERIMENT STATION)
244-2, SHIMOYAGANUKI, IRUMA-SHI 358
JAPANESE

SENBAI TOKEI NENPO 4031
(STATISTICAL YEARBOOK OF JAPAN MONOPOLY CORPORATION)
NIHON SENBAI KOSHA (JAPAN MONOPOLY CORPORATION)
2-1, AKASAKA AOI-CHO, MINATO-KU, TOKYO 107
JAPANESE

SHIGA NO CHAGYO 4032
(TEA INDUSTRY IN SHIGA)
SHIGA-KEN CHA SEISANSHA DANTAI RENGOKAI
3-38, 4-CHOME, KYO-MACHI, OTSU-SHI 520
JAPANESE

SHIGA NO KAJU 4033
(FRUIT TREE IN SHIGA)
SHIGA-KEN KAJU KUMIAI RENGOKAI
4, KYO-MACHI, OTSU-SHI 520
JAPANESE

SHIGA-KEN CHAGYO SHIDOSHO GYOMU HOKOKU 4034
(REPORT OF THE TEA RESEARCH INSTITUTE OF SHIGA PREFECTURE)
SHIGA-KEN CHAGYO SHIDOSHO (TEA RESEARCH INSTITUTE OF SHIGA PREFECTURE)
HATOGADAIRA, MINAKUCHI-CHO, KOGA-GUN, SHIGA 528
JAPANESE

SHIITAKE SEISANHI CHOSA HOKOKU 4035
(STATISTICS OF 'SHIITAKE' PRODUCTION COST)
NORIN-SHO NORIN KEIZAI-KYOKU TOKEI JOHO-BU (STATISTICS AND INFORMATION DEPARTMENT, ECONOMIC AFFAIRS BUREAU, MINISTRY OF AGRICULTURE AND FORESTRY)
2-1, 1-CHOME, KASUMIGASEKI, CHIYODA-KU, TOKYO 100
JAPANESE

SHIRYO SAKUMOTSU SHUKAKURYO SHICHOSON-BETSU TOKEIHYO 4036
(STATISTICS OF HARVEST OF FEED PLANT BY TOWNS)
NORIN-SHO CHUGOKU SHIKOKU NOSEI-KYOKU (CHUGOKU SHIKOKU AGRICULTURAL ADMINISTRATION BUREAU, MINISTRY OF AGRICULTURE AND FORESTRY)
9-24, TENJIN-CHO, OKAYAMA-SHI 700
JAPANESE

SHIZUOKA-KEN CHAGYO SHIKENJO KENKYU HOKOKU 4037
(BULLETIN OF THE SHIZUOKA PREFECTURE TEA EXPERIMENT STATION)
SHIZUOKA-KEN CHAGYO SHIKENJO (SHIZUOKA PREFECTURE TEA EXPERIMENT STATION)
1706, KURASAWA, KIKUKAWA-CHO, OGASA-GUN, SHIZUOKA 439
JAPANESE

SHIZUOKA-KEN KANKITSU SHIKENJO KENKYU HOKOKU 4038
(BULLETIN OF THE SHIZUOKA PREFECTURAL CITRUS EXPERIMENT STATION)
SHIZUOKA-KEN KANKITSU SHIKENJO (SHIZUOKA PREFECTURAL CITRUS EXPERIMENT STATION)
2712, KOMAGOE, SHIMIZU-SHI 424
JAPANESE, WITH ENGLISH SUMMARY
INDEXED: BIB, JICST

SHIZUOKA-KEN KANKITSU SHIKENJO TOKUBETSU HOKOKU 4039
(SPECIAL BULLETIN (SHIZUOKA PREFECTURAL CITRUS EXPERIMENT STATION))
JAPANESE

SHIZUOKA-KEN NOGYO SHIKENJO, MIKATAHARA TAHATA, SHIKENJO TOKUBETSU HOKOKU 4040
(SHIZUOKA PREFECTURE AGRICULTURAL EXPERIMENT STATION, MIKATAHARA (P.N.) FIELDS, SPECIAL REPORT OF THE EXPERIMENT STATION)
JAPANESE

SOCHI SHIKENJO KENKYU HOKOKU 4041
(BULLETIN OF THE NATIONAL GRASSLAND RESEARCH INSTITUTE)
NORIN-SHO SOCHI SHIKENJO (NATIONAL GRASSLAND RESEARCH INSTITUTE, MINISTRY OF AGRICULTURE AND FORESTRY)
768, NISHINASUMO-CHO NASU-GUN, TOCHIGI 329-27
JAPANESE AND ENGLISH
INDEXED: BIB, BIO, CHE

SOCHI SHIKENJO NENPO 4042
(ANNUAL REPORT OF THE NATIONAL GRASSLAND RESEARCH INSTITUTE)
NORIN-SHO SOCHI SHIKENJO (NATIONAL GRASSLAND RESEARCH INSTITUTE, MINISTRY OF AGRICULTURE AND FORESTRY)
768, NISHINASUNO-CHO, NASU-GUN, TOCHIGI 329-27
JAPANESE

SUIRIKUTO SHUKAKURYO, ZENKOKU, TODOFUKEN-BETSU, SHICHOSON-BETSU 4043
(STATISTICS OF RICE PLANT HARVEST BY TOWNS, PREFECTURES AND NATIONAL WIDE)
NORIN-SHO NORIN KEIZAI-KYOKU TOKEI JOHO-BU (STATISTICS AND INFORMATION DEPARTMENT, ECONOMIC AFFAIRS BUREAU, MINISTRY OF AGRICULTURE AND FORESTRY)
2-1, 1-CHOME, KASUMIGASEKI, CHIYODA-KU, TOKYO 100
JAPANESE

TABAKO KOSAKU GIJUTSU KAIHATSU SHIKEN SEISEKISHO 4044
(RESULTS OF TECHNICAL DEVELOPMENT IN TOBACCO CULTIVATION)
NIHON SENBAI KOSHA SEISAN HON-BU (PRODUCTION DEPARTMENT OF JAPAN CORPORATION)
2-1, AKASAKA AOI-CHO, MINATO-KU, TOKYO 107
JAPANESE

TABAKO KOSAKU KAIZEN KEKKA CHUO KENTOKAI KENTO KEKKA HOKOKUSHO 4045
(REPORT ON THE RESULTS OF THE EXAMINATION, CENTRAL CONFERENCE FOR THE RESULTS OF TOBACCO CROP IMPROVEMENT)
ZENKOKU TABAKO KOSAKU KUMIAI CHUOKAI (CENTRAL ORGAN, NATIONAL TOBACCO GROWERS' ASSOCIATION)
10-1, 1-CHOME, SHIBA DAIMON, MINATO-KU, TOKYO 105
JAPANESE

TABAKO KOSAKU SHIRIZU 4046
(SERIES OF TOBACCO CULTURE)
NIHON SENBAI KOSHA SEISAN HON-BU (TOBACCO TECHNICAL SECTION OF JAPAN MONOPOLY CORPORATION) AND SENBAI JIGYO KYOKAI
11, SHIBA NISHIKUBO SAKURAGAWA-CHO, MINATO-KU, TOKYO 105
JAPANESE

TENSAI GIJUTSU RENRAKU KENKYUKAI HAPPYO RONBUNSHU 4047
(PROCEEDINGS OF SUGAR BEET TECHNICAL CONFERENCE)
KANMI SHIGEN SHINKOKAI
4-1, 2-CHOME, AKASAKA, MINATO-KU, TOKYO 107
JAPANESE

TENSAI NENPO 4048
(ANNUAL REPORT OF THE SUGAR BEET INDUSTRY)
KANMI SHIGEN SHINKOKAI (SUGAR CROP DEVELOPMENT FUND)
4-1, 2-CHOME, AKASAKA, MINATO-KU, TOKYO 107
JAPANESE AND ENGLISH
INDEXED: JICST

TOKAI KINKI INASAKU TENKO NO TAME NO GIJUTSU SHIRYO 4049
(TECHNICAL DATA ON RICE CROP ROTATION IN THE TOKAI KINKI REGION)
TOKAI KINKI CHI'IKU GIJUTSU RENRAKU KAIGI JIMU KYOKU (TOKAI KINKI REGION, TECHNICAL LIAISON CONFERENCE, SECRETARIAT)
JAPANESE

TOKUSHIMA-KEN KAJU SHIKENJO GYOMU HOKOKU 4050
(OPERATION REPORT OF TOKUSHIMA HORTICULTURAL EXPERIMENTAL STATION)
TOKUSHIMA-KEN KAJU SHIKENJO (TOKUSHIMA HORTICULTURAL EXPERIMENTAL STATION)
NUMAE, KATSUURA-MACHI, KATSUURA-GUN, TOKUSHIMA 771-43
JAPANESE

TOTTORI-KEN KAJU SHIKENJO GYOMU NENPO 4051
(ANNUAL REPORT OF TOTTORI FRUIT TREE EXPERIMENT STATION)
TOTTORI-KEN KAJU SHIKENJO (TOTTORI FRUIT TREE EXPERIMENT STATION)
MATSUTANI, AKASAKI-CHO, TOHAKU-GUN, TOTTORI 689-25
JAPANESE

TOTTORI-KEN KAJU SHIKENJO KENKYU HOKOKU 4052
(BULLETIN OF THE TOTTORI FRUIT TREE EXPERIMENT STATION)
TOTTORI-KEN KAJU SHIKENJO (TOTTORI FRUIT TREE EXPERIMENT STATION)
MATSUTANI, AKASAKI-CHO, TOHAKU-GUN, TOTTORI 689-25
JAPANESE AND ENGLISH

TSUCHI TO KENKO 4053
(SOIL AND HEALTH)
YUKI NOGYO KENKYUKAI (ORGANIC AGRICULTURE RESEARCH ASSOCIATION)
JAPANESE

WAKAYAMA NO KAJU 4054
(FRUIT PRODUCTION IN WAKAYAMA PREFECTURE)
WAKAYAMA-KEN KAJITSU NOGYO KYODO KUMIAI RENGOKAI (WAKAYAMA PREFECTURE FRUIT GROWERS CORPORATION ASSOCIATION)
1-1, 5-CHOME, MISONO-CHO, WAKAYAMA-SHI 640
JAPANESE

WAKAYAMA-KEN KAJU ENGEI SHIKENJO KENKYU HOKOKU 4055
(BULLETIN OF THE WAKAYAMA FRUIT TREE EXPERIMENT STATION)
WAKAYAMA-KEN KAJU ENGEI SHIKENJO (WAKAYAMA FRUIT TREE EXPERIMENT STATION)
753, OKI, KIBI-MACHI, ARIDA-GUN, WAKAYAMA 643
JAPANESE

WAKAYAMA-KEN KAJU ENGEI SHIKENJO SHIKEN KENKYU NENPO 4056
(ANNUAL REPORT OF WAKAYAMA FRUIT TREE EXPERIMENT STATION)
WAKAYAMA-KEN KAJU ENGEI SHIKENJO (WAKAYAMA FRUIT TREE EXPERIMENT STATION)
753, OKI, KIBI-MACHI, ARIDA-GUN, WAKAYAMA 643
JAPANESE

YAMANASHI-KEN KAJU SHIKENJO KENKYU HOKOKU 4057
(BULLETIN OF THE YAMANASHI FRUIT TREE EXPERIMENT STATION)
YAMANASHI-KEN KAJU SHIKENJO (YAMANASHI FRUIT TREE EXPERIMENT STATION)
1530, MANRIKI, YAMANASHI-SHI 405
JAPANESE AND ENGLISH

YASAI SHIKENJO HOKOKU. B. (MORIOKA) 4058
(BULLETIN OF THE VEGETABLE AND ORNAMENTAL CROPS RESEARCH STATION. (SERIES B - MORIOKA))
NORINSHO, YASAI SHIKENJO MORIOKA-SHIJO (MINISTRY OF AGRICULTURE AND FORESTRY, VEGETABLE AND ORNAMENTAL CROPS RESEARCH STATION, MORIOKA DIVISION)
JAPANESE

YASAI SHIKENJO HOKOKU. C. (KURUME) 4059
(BULLETIN OF THE VEGETABLE AND ORNAMENTAL CROPS RESEARCH STATION. (SERIES C))
NORINSHO YASAI SHIKENJO KURUME-SHIJO (MINISTRY OF AGRICULTURE AND FORESTRY, VEGETABLE AND ORNAMENTAL CROPS RESEARCH STATION, KURUME DIVISION)
JAPANESE

YASAI SHIKENJO HOKOKU. A. 4060
(BULLETIN OF THE VEGETABLE AND ORNAMENTAL CROPS RESEARCH STATION. (SERIES A))
NORINSHO YASAI SHIKENJO
ISHINDEN-OGOSO, TSU, 514-01 MIE
JAPANESE, WITH ENGLISH SUMMARY
INDEXED: BIB, BIO, CHE

Arboriculture, Trees, Shrubs

ABSTRACTS OF JAPANESE LITERATURE IN FOREST GENETICS AND RELATED FIELDS, PART A 4061
RINGYO SHIKENJO (FORESTRY EXPERIMENT STATION)
37-21, 5-CHOME, SHIMOMEGURO, MEGURO-KU, TOKYO 153
ENGLISH

ABSTRACTS OF JAPANESE LITERATURE IN FOREST GENETICS AND RELATED FIELDS, PART B 4062
RINGYO SHIKENJO (FORESTRY EXPERIMENT STATION)
37-21, 5-CHOME, SHIMOMEGURO, MEGURO-KU, TOKYO 153
ENGLISH

AICHI-KEN RINGYO SHIKENJO HOKOKU 4063
(BULLETIN OF THE AICHI PREFECTURE FOREST EXPERIMENT STATION)
AICHI-KEN RINGYO SHIKENJO (AICHI PREFECTURE FOREST EXPERIMENT STATION)
OTSU-ARATA, KAMIYOSHIDA, HORAI-CHO, MINAMISHIDARA-GUN, AICHI 441-17
JAPANESE

AKITA EIRIN-KYOKU JIGYO TOKEISHO 4064
(ANNUAL STATISTICS OF AKITA REGIONAL FOREST OFFICE)
RINYA-CHO AKITA EIRIN-KYOKU (AKITA REGIONAL FOREST OFFICE, FORESTRY AGENCY)
9-16, 5-CHOME, NAKA-DORI, AKITA-SHI 010
JAPANESE

AKITA MOKUZAI TSUSHIN 4065
(AKITA LUMBER PRESS)
AKITA MOKUZAI TSUSHINSHA
3-25, OTE-MACHI, NOSHIRO-SHI 016
JAPANESE

AKITA-KEN RINGYO SHIKENJO GAIYO 4066
(OUTLINE OF AKITA PREFECTURE FOREST EXPERIMENT STATION)
AKITA-KEN RINGYO SHIKENJO
1-1, SHISHIGAMORI, SHAKANAI, ODATE-SHI 017
JAPANESE

AKITA-KEN RINGYO TOKEI 4067
(FORESTRY STATISTICS OF AKITA PREFECTURE)
AKITA-KEN RINMU-BU (FOREST ADMINISTRATION DEPARTMENT, AKITA PREFECTURAL GOVERNMENT)
1-1, 4-CHOME, SANNO, AKITA-SHI 010
JAPANESE

ANETTAI RINGYO KENKYU ITAKU JIGYO HOKOKUSHO 4068
(REPORT OF CONTRACT RESEARCH ON SUBTROPICAL FORESTRY)
KAGOSHIMA-KEN RINGYO SHIKENJO OSHIMA BUNJO (FOREST EXPERIMENT STATION OF KAGOSHIMA PREFECTURE OSHIMA BRANCH)
1032, OGACHI, TATSUGO-MURA, OSHIMA-GUN, KAGOSHIMA 894-01
JAPANESE

AOMORI EIRIN-KYOKU JIGYO TOKEISHO 4069
(ANNUAL STATISTICS OF AOMORI REGIONAL FOREST OFFICE)
RIN'YA-CHO AOMORI EIRIN-KYOKU (AOMORI REGIONAL FOREST OFFICE, FORESTRY AGENCY)
4-37, 2-CHOME, YANAGAWA, AOMORI-SHI 030
JAPANESE
INDEXED: BIB

AOMORI RIN'YU 4070
(FRIENDS OF AOMORI FORESTS)
RIN'YA KOSAIKAI AOMORI-SHIBU (SOCIETY FOR PRESERVATION BENEFIT OF FORESTS, AOMORI CITY BRANCH)
JAPANESE

AOMORI-KEN NO RINGYO 4071
(FOREST IN AOMORI PREFECTURE)
AOMORI-KEN NORIN-BU (AGRICULTURE AND FORESTRY DEPARTMENT, AOMORI PREFECTURAL GOVERNMENT)
1-1, 1-CHOME, NAGASHIMA, AOMORI-SHI 030
JAPANESE

AOMORI-KEN RINGYO SHIKENJO HOKOKU 4072
(BULLETIN OF THE AOMORI PREFECTURAL FOREST EXPERIMENT STATION)
AOMORI-KEN RINGYO SHIKENJO (AOMORI PREFECTURAL FOREST EXPERIMENTAL STATION)
KOMINATO SHINDO, HIRANAI-MACHI, HIGASHITSUGARU-GUN, AOMORI 039-33
JAPANESE

ASAHIKAWA EIRIN-KYOKU DOJO CHOSA HOKOKU 4073
(AGRONOMICAL SURVEY REPORT, ASAHIKAWA REGIONAL FOREST OFFICE)
RIN'YA-CHO ASAHIKAWA EIRIN-KYOKU (ASAHIKAWA REGIONAL FOREST OFFICE, FORESTRY AGENCY)
5-419, KAGURA 4-JO, ASAHIKAWA-SHI 070
JAPANESE

ASAHIKAWA EIRIN-KYOKU JIGYO TOKEISHO 4074
(ANNUAL STATISTICS OF ASAHIKAWA REGIONAL FOREST OFFICE)
RIN'YA-CHO ASAHIKAWA EIRIN-KYOKU (ASAHIKAWA REGIONAL FOREST OFFICE, FORESTRY AGENCY)
5-419, KAGURA 4-JO, ASASHIKAWA-SHI 070
JAPANESE

CHIBA-KEN RINGYO NO DOKO, CHIBA-KEN NORINGYO NENJI HOKOKUSHO 4075
(ANNUAL REPORT OF FORESTRY IN CHIBA PREFECTURE)
CHIBA-KEN NORIN-BU (AGRICULTURE AND FORESTRY DEPARTMENT, CHIBA PREFECTURAL GOVERNMENT)
1-1, ICHIBA-CHO, CHIBA-SHI 280-91
JAPANESE

CHIBA-KEN RINGYO SHIKENJO HOKOKU 4076
(REPORT OF CHIBA PREFECTURAL FOREST EXPERIMENT STATION)
CHIBA-KEN RINGYO SHIKENJO (CHIBA PREFECTURAL FOREST EXPERIMENT STATION)
1887-1, HANIYA, SANPU-MACHI, SANPU-GUN, CHIBA 289-12
JAPANESE
INDEXED: JICST

CHISAN RINDO KOHO 4077
(JOURNAL OF JAPAN ASSOCIATION OF FORESTRY CONSERVATION)
NIHON CHISAN CHISUI KYOKAI (JAPAN ASSOCIATION OF FORESTRY CONSERVATION) AND NIHON RINDO KYOKAI
C/O NAGATA-CHO BUILDING, 4-3, 2-CHOME, NAGATA-CHO, CHIYODA-KU, TOKYO 100
JAPANESE

DANTAIRIN 4078
(SUBTROPICAL FOREST)
RIN'YA-CHO KUMAMOTO EIRIN-KYOKU (KUMAMOTO REGIONAL FOREST OFFICE, FORESTRY AGENCY)
RIN'YA-KOSAI-KAI KUMAMOTO SHIBU
2-7, KYO-MACHI HON-CHO, KUMAMOTO-SHI 860
JAPANESE

DOYU RIN'YA JIGYO TOKEISHO 4079
(STATISTICS OF HOKKAIDO PREFECTURAL FOREST)
HOKKAIDO RINMU-BU (FORESTRY AFFAIRS DIVISION, HOKKAIDO PREFECTURAL GOVERNMENT)
NISHI-6-CHOME, KITA-3-JO, CHUO-KU, SAPPORO-SHI 060
JAPANESE

EHIME DAIGAKU NOGAKUBU ENSHURIN HOKOKU 4080
(BULLETIN OF THE EHIME UNIVERSITY FOREST)
EHIME DAIGAKU NOGAKUBU FUZOKU ENSHURIN (FOREST, COLLEGE OF AGRICULTURE, EHIME UNIVERSITY)
5-7, 3-CHOME, TARUMI-CHO, MATSUYAMA-SHI 790
JAPANESE, WITH EUROPEAN SUMMARY
INDEXED: BIB, BIO

EHIME-KEN RINGYO SHIKENJO NENJIBETSU GYOMU HOKOKUSHO 4081
(ANNUAL REPORT OF THE EHIME PREFECTURE FOREST PRODUCTS RESEARCH INSTITUTE)
EHIME-KEN RINGYO SHIKENJO (EHIME PREFECTURE FOREST PRODUCTS RESEARCH INSTITUTE)
KO-2782, SUNOUCHI, KAWAUCHI-MACHI, ONSEN-GUN, EHIME 791-03
JAPANESE

FUKUI-KEN RINGYO SHIKENJO HOKOKU 4082
(REPORT OF FUKUI PREFECTURAL FORESTRY EXPERIMENT STATION)
FUKUI-KEN RINGYO SHIKENJO
2-10, 66-CHOME, WAKISANGA-CHO, FUKUI-SHI 910-21
JAPANESE

FUKUI-KEN RINGYO SHIKENJO KENKYU HOKOKU 4083
(BULLETIN OF THE FUKUI PREFECTURE FOREST EXPERIMENT STATION)
FUKUI-KEN RINGYO SHIKENJO (FUKUI PREFECTURE FOREST EXPERIMENT STATION)
2-10, 66-CHOME, WAKISANGA-CHO, FUKUI-SHI 910-21
JAPANESE

FUKUOKA-KEN RINGYO SHIKENJO JIHO 4084
(BULLETIN OF FUKUOKA-KEN FOREST EXPERIMENT STATION)
FUKUOKA-KEN RINGYO SHIKENJO (FUKUOKA PREFECTURE FOREST EXPERIMENT STATION)
KUROKI-CHO, YAME-GUN, FUKUOKA 834-12
JAPANESE, WITH ENGLISH SUMMARY
INDEXED: BIB, JICST

FUKUOKA-KEN RINGYO SHIKENJO KENKYU SHIRYO 4085
(MEMOIRS OF FUKUOKA PREFECTURE FOREST EXPERIMENT STATION)
FUKUOKA-KEN RINGYO SHIKENJO (FUKUOKA PREFECTURE FOREST EXPERIMENT STATION)
KUROKI-CHO, YAME-GUN, FUKUOKA 834-12
JAPANESE

FUKUSHIMA-KEN RINGYO SHIKENJO KENKYU HOKOKU 4086
(REPORT OF FUKUSHIMA PREFECTURE FOREST EXPERIMENT STATION)
FUKUSHIMA-KEN RINGYO SHIKENJO
1, NISHISHIMASAKA, NARITA, ASAKA-MACHI, KORIYAMA-SHI 963-01
JAPANESE

GIFU-KEN KANREICHI RINGYO SHIKENJO GYOMU HOKOKU 4087
(REPORT OF THE GIFU PREFECTURAL COLD DISTRICT FOREST EXPERIMENT STATION)
GIFU-KEN KANREICHI RINGYO SHIKENJO
1554, YAMADA-CHO, TAKAYAMA-SHI 506
JAPANESE

GIFU-KEN KANREICHI RINGYO SHIKENJO SHIKEN HOKOKU 4088
(BULLETIN OF THE GIFU PREFECTURAL COLD DISTRICT FOREST EXPERIMENT STATION)
GIFU-KEN KANREICHI RINGYO SHIKENJO (GIFU PREFECTURAL COLD DISTRICT FOREST EXPERIMENT STATION)
1554, YAMADA-CHO, TAKAYAMA-SHI 506
JAPANESE

GIFU-KEN RINGYO SENTA KENKYU HOKOKU 4089
(RESEARCH REPORT OF THE GIFU PREFECTURE FORESTRY CENTER)
GIFU-KEN RINGYO SENTA
128-1, KOJOSHITA, SODAI, MINO-SHI 501-37
JAPANESE

GIJUTSU KAIHATSU HOKOKUSHO 4090
(REPORT OF FORESTRY ENGINEERING DEVELOPMENT)
RIN'YA-CHO AOMORI EIRIN-KYOKU (AOMORI NATIONAL FOREST REGIONAL OFFICE, FORESTRY AGENCY)
4-37, 2-CHOME, YANAGAWA, AOMORI-SHI 030
JAPANESE

GIN'YO 4091
(SILVERY LEAVES)
RIN'YA-CHO HAKODATE EIRIN-KYOKU (HAKODATE REGIONAL FOREST OFFICE, FORESTRY AGENCY)
RIN'YA KOSAIKAI HAKODATE SHIBU
5-3, KOMABA-CHO, HAKODATE-SHI 042
JAPANESE
INDEXED: JICST

GUNMA-KEN RINGYO SHIKENJO HOKOKU 4092
(REPORT OF GUNMA PREFECTURAL FOREST EXPERIMENT STATION)
GUNMA-KEN RINGYO SHIKENJO (GUNMA PREFECTURAL FOREST EXPERIMENT STATION)
2935, ARAI, SHINTO-MURA, KITAGUNMA-GUN, GUNMA 370-35
JAPANESE

GURIN EIJI 4093
(MONTHLY JOURNAL OF FOREST INDUSTRIES (GREEN AGE))
SHINRIN SHIGEN SOGO TAISAKU KYOGIKAI (COUNCIL FOR FOREST RESOURCES)
C/O SHIN OTE-MACHI BUILDING, 2-1, 2-CHOME, OTE-MACHI, CHIYODA-KU, TOKYO 100
JAPANESE

HAKODATE EIRIN-KYOKU GYOMU KENKYU RONBUNSHU 4094
(REPORT OF HAKODATE REGIONAL FOREST OFFICE)
RIN'YA-CHO HAKODATE EIRIN-KYOKU (HAKODATE REGIONAL FOREST OFFICE)
4-9, KOMABA-CHO, HAKODATE-SHI 042
JAPANESE

HAKODATE EIRIN-KYOKU JIGYO TOKEISHO 4095
(ANNUAL STATISTICS OF HAKODATE REGIONAL FOREST OFFICE)
RIN'YA-CHO HAKODATE EIRIN-KYOKU (HAKODATE REGIONAL FOREST OFFICE, FORESTRY AGENCY)
4-9, KOMABA-CHO, HAKODATE-SHI 042
JAPANESE

HAYASHI 4096
(FOREST)
HOKKAIDO ZORIN SHIKO KYOKAI
NISHI-5-CHOME, KITA-4-JO, CHUO-KU, SAPPORO-SHI 060
JAPANESE

HIROSHIMA-KENRITSU RINGYO SHIKENJO GYOMU HOKOKU 4097
(REPORT ON ACTIVITIES OF HIROSHIMA PREFECTURAL FOREST EXPERIMENT STATION)
HIROSHIMA-KENRITSU RINGYO SHIKENJO (HIROSHIMA PREFECTURAL FOREST EXPERIMENT STATION)
KOBIRA, TOKAICHI-MACHI, MIYOSHI-SHI 728
JAPANESE

HIROSHIMA-KENRITSU RINGYO SHIKENJO KENKYU HOKOKU 4098
(BULLETIN OF HIROSHIMA PREFECTURAL FOREST EXPERIMENT STATION)
HIROSHIMA-KENRITSU RINGYO SHIKENJO (HIROSHIMA PREFECTURAL FOREST EXPERIMENT STATION)
KOBIRA, TOKAICHI-MACHI, MIYOSHI-SHI 728
JAPANESE

HIROSHIMA-KENRITSU RINGYO SHIKENJO KENKYU HOKOKU, DAIJESUTO 4099
(RESEARCH REPORT OF HIROSHIMA PREFECTURAL FORESTRY EXPERIMENT STATION)
JAPANESE

HOANRIN 4100
(JOURNAL OF THE JAPANESE INSTITUTE OF PROTECTION FOREST)
NIHON HOANRIN KENKYUSHO (JAPANESE INSTITUTE OF PROTECTION FOREST)
22-14, 5-CHOME, MORINO, MACHIDA-SHI, TOKYO 194
JAPANESE

HOKKAIDO DAIGAKU NOGAKUBU ENSHURIN KENKYU HOKOKU 4101
(RESEARCH BULLETINS OF THE COLLEGE EXPERIMENT FORESTS, HOKKAIDO UNIVERSITY)
HOKKAIDO DAIGAKU NOGAKUBU ENSHURIN (COLLEGE EXPERIMENT FORESTS, FACULTY OF AGRICULTURE, HOKKAIDO UNIVERSITY)
NISHI-9-CHOME, KITA-9-JO, KITA-KU, SAPPORO-SHI 060
JAPANESE AND EUROPEAN
INDEXED: BIB, JICST

HOKKAIDO NO RINGYO 4102
(FORESTRY IN HOKKAIDO)
HOKKAIDO RINMU-BU (FOREST ADMINISTRATION DEPARTMENT, HOKKAIDO PREFECTURAL GOVERNMENT)
C/O RINGYO KAIKAN, NISHI-5-CHOME, KITA-4-JO, CHUO-KU, SAPPORO-SHI 060
JAPANESE

HOKKAIDO RINGYO SHIKENJO HOKOKU 4103
(BULLETIN OF THE HOKKAIDO FOREST EXPERIMENT STATION)
HOKKAIDORITSU RINGYO SHIKENJO (HOKKAIDO FOREST EXPERIMENT STATION)
KOSHUNAI-MACHI, BIBAI-SHI 072
JAPANESE, WITH EUROPEAN SUMMARY
INDEXED: CHE, JICST

HOKKAIDO RINGYO TOKEI 4104
(FORESTRY STATISTICS OF HOKKAIDO)
HOKKAIDO RIN'MU-BU (FORESTRY AFFAIRS DIVISION, HOKKAIDO PREFECTURAL GOVERNMENT)
NISHI-6-CHOME, KITA-3-JO, CHUO-KU, SAPPORO-SHI 060
JAPANESE

HOKKAIDORITSU RINSAN SHIKENJO GEPPO 4105
(JOURNAL OF HOKKAIDO FOREST PRODUCTS RESEARCH INSTITUTE)
HOKKAIDORITSU RINSAN SHIKENJO (HOKKAIDO FOREST PRODUCTS RESEARCH INSTITUTE)
12, MIDORI-CHO, ASAHIKAWA-SHI 070
JAPANESE
INDEXED: BIB, CHE, JICST

HOKKAIDORITSU RINSAN SHIKENJO KENKYU HOKOKU 4106
(REPORT, HOKKAIDO FOREST PRODUCTS RESEARCH INSTITUTE)
HOKKAIDORITSU RINSAN SHIKENJO (HOKKAIDO FOREST PRODUCTS RESEARCH INSTITUTE)
12, MIDORI-CHO, ASAHIKAWA-SHI 070
JAPANESE, WITH ENGLISH SUMMARY
INDEXED: BIB, CHE, JICST

HOKKAIDORITSU RINSAN SHIKENJO NENPO 4107
(ANNUAL REPORT OF THE HOKKAIDO FOREST PRODUCTS RESEARCH INSTITUTE)
HOKKAIDORITSU RINSAN SHIKENJO (HOKKAIDO FOREST PRODUCTS RESEARCH INSTITUTE)
12, MIDORI-CHO, ASAHIKAWA-SHI 070
JAPANESE
INDEXED: BIB, JICST

HOPPO RINGYO 4108
(NORTHERN FORESTRY)
HOPPO RINGYO HENSHU IINKAI
13-CHOME, TOYODE-5-JO, TOYOHIRA-KU, SAPPORO-SHI 062
JAPANESE AND ENGLISH

HYOGO-KENRITSU RINGYO SHIKENJO GYOMU NENPO 4109
(ANNUAL REPORT OF HYOGO PREFECTURAL FOREST EXPERIMENT STATION)
HYOGO-KENRITSU RINGYO SHIKENJO (HYOGO PREFECTURAL FOREST EXPERIMENT STATION)
YAMASAKI-CHO, SHISO-GUN, HYOGO 671-25
JAPANESE
INDEXED: JICST

HYOGO-KENRITSU RINGYO SHIKENJO KENKYU HOKOKU 4110
(RESEARCH REPORT OF HYOGO PREFECTURAL FOREST EXPERIMENT STATION)
HYOGO-KENRITSU RINGYO SHIKENJO (HYOGO PREFECTURAL FOREST EXPERIMENT STATION)
95, SHIKASAWA, YAMASAKI-CHO, SHISO-GUN, HYOGO 671-25
JAPANESE

IBARAKI-KEN RINGYO SHIKENJO GYOMU HOKOKU 4111
(REPORT ON ACTIVITIES OF IBARAKI PREFECTURAL FOREST EXPERIMENT STATION)
IBARAKI-KEN RINGYO SHIKENJO
4692, TO, NAKA-MACHI, NAKA-GUN, IBARAKI 319-21
JAPANESE

IBARAKI-KEN RINGYO SHIKENJO KENKYU HOKOKU 4112
(REPORTS OF THE IBARAKI PREFECTURAL FORESTRY EXPERIMENT STATION)
IBARAKI-KEN RINGYO SHIKENJO (IBARAKI PREFECTURAL FOREST EXPERIMENT STATION)
4692, TO, NAKA-MACHI, NAGA-GUN, IBARAKI 319-21
JAPANESE, WITH ENGLISH SUMMARY
INDEXED: CHE

IKURINHI CHOSA HOKOKU 4113
(REPORT OF ARBORICULTURAL COST)
NORIN-SHO NORIN KEIZAI-KYOKU TOKEI JOHO-BU (STATISTICS AND INFORMATION DEPARTMENT, ECONOMIC AFFAIRS BUREAU, MINISTRY OF AGRICULTURE AND FORESTRY)
11-14, 2 CHOME, MEGURO, MEGURO-KU, TOKYO 153
JAPANESE

ISHIKAWA NO RINGYO 4114
(FORESTRY IN ISHIKAWA)
ISHIKAWA-KEN SANRIN KYOKAI (ISHIKAWA PREFECTURE FORESTRY ASSOCIATION)
RI-38-1, MASUIZUMI-CHO, KANAZAWA-SHI 921
JAPANESE

ISHIKAWA-KEN RINGYO SHIKENJO GYOMU HOKOKU 4115
(REPORT ON ACTIVITIES OF ISHIKAWA-KEN FOREST EXPERIMENT STATION)
ISHIKAWA-KEN RINGYO SHIKENJO
SANNOMIYA, TSURUKI-CHO, ISHIKAWA-GUN, ISHIKAWA 920-21
JAPANESE

ISHIKAWA-KEN RINGYO SHIKENJO KENKYU HOKOKU 4116
(BULLETIN OF THE ISHIKAWA-KEN FOREST EXPERIMENT STATION)
ISHIKAWA-KEN RINGYO SHIKENJO
SANNOMIYA, TSURUKI-CHO, ISHIKAWA-GUN, ISHIKAWA 920-21
JAPANESE

IWATE DAIGAKU NOGAKUBU ENSHURIN HOKOKU 4117
(BULLETIN OF THE IWATE UNIVERSITY FORESTS)
IWATE DAIGAKU NOGAKUBU FUZOKU ENSHURIN HON-BU (IWATE UNIVERSITY FORESTS)
18-8, 3-CHOME, UEDA, MORIOKA-SHI 020
JAPANESE, WITH EUROPEAN SUMMARY
INDEXED: BIO

IWATE-KEN RINGYO DOKO NENPO 4118
(ANNUAL REPORT ON FORESTRY IN IWATE PREFECTURE)
IWATE-KEN RINGYO SUISAN-BU (FORESTRY AND FISHERY DEPARTMENT, IWATE PREFECTURAL GOVERNMENT)
10-1, UCHIMARU, MORIOKA-SHI 020
JAPANESE

IWATE-KEN RINGYO SHIKENJO GYOMU HOKOKUSHO 4119
(REPORT ON ACTIVITIES OF IWATE PREFECTURAL FOREST EXPERIMENT STATION)
IWATE-KEN RINGYO SHIKENJO (IWATE PREFECTURAL FOREST EXPERIMENT STATION)
TAKIZAWA-MURA, IWATE-GUN, IWATE 020-01
JAPANESE

IWATE-KEN RINGYO SHIKENJO SEIKA HOKOKU 4120
(TEST REPORT OF IWATE PREFECTURAL FOREST EXPERIMENT STATION)
IWATE-KEN RINGYO SHIKENJO (IWATE PREFECTURAL FOREST EXPERIMENT STATION)
TAKIZAWA-MURA, IWATE-GUN, IWATE 020-10
JAPANESE

JAPAN LUMBER JOURNAL 4121
ENGLISH
INDEXED: BIB

JIGYO JOHO 4122
(INFORMATION OF FORESTRY OPERATION AND ENGINEERING)
RIN'YA-CHO MAEBASHI EIRIN-KYOKU (MAEBASHI REGIONAL FOREST OFFICE, FORESTRY AGENCY)
16-25, 4-CHOME, IWAGAMI-CHO, MAEBASHI-SHI 371
JAPANESE

JUHYO 4123
(SILVER THAW)
RIN'YA-CHO OBIHIRO EIRIN-KYOKU (OBIHIRO REGIONAL FOREST OFFICE, FORESTRY AGENCY)
RIN'YA KOSAIKAI OBIHIRO SHIBU
MINAMI-13-CHOME, HIGASHI-8-JO, OBIHIRO-SHI 080
JAPANESE

KAGAWA-KEN RINGYO SHIDOSHO GYOMU SEISEKI HOKOKU 4124
(REPORT OF KAGAWA PREFECTURAL FOREST EXPERIMENT STATION)
KAGAWA-KEN RINGYO SHIDOSHO (KAGAWA PREFECTURAL FOREST EXPERIMENT STATION)
823-1, SHINMOKU MATSUODANI, SOGO, CHUNAN-MACHI, NAKATADO-GUN, KAGAWA 769
JAPANESE

KAGOSHIMA DAIGAKU NOGAKUBU FUZOKU ENSHURIN HOKOKU 4125
(BULLETIN OF THE KAGOSHIMA UNIVERSITY FORESTS)
KAGOSHIMA DAIGAKU NOGAKUBU FUZOKU ENSHURIN (RESEARCH INSTITUTE OF UNIVERSITY FORESTS, KAGOSHIMA UNIVERSITY)
1946, UERATA-CHO, KAGOSHIMA-SHI 890
JAPANESE, WITH ENGLISH SUMMARY
INDEXED: CHE

KAGOSHIMA-KEN RINGYO SHIKENJO GYOMU HOKOKU 4126
(REPORT ON ACTIVITIES OF KAGOSHIMA PREFECTURAL FOREST EXPERIMENT STATION)
KAGOSHIMA-KEN RINGYO SHINKENJO (KAGOSHIMA PREFECTURAL FOREST EXPERIMENT STATION)
182-1, KAMIGYOTOKU, GAMO-CHO, AIRA-GUN, KAGOSHIMA 899-53
JAPANESE

KAIGAI RINGYO SHIRYO 4127
(OVERSEAS FORESTRY INFORMATION)
RIN'YA-CHO (FORESTRY AGENCY)
2-1, 1-CHOME, KASUMIGASEKI, CHIYODA-KU, TOKYO 100
JAPANESE

KANAGAWA-KEN RINGYO SHIKENJO GYOMU HOKOKU 4128
(REPORT OF KANAGAWA PREFECTURAL FOREST EXPERIMENT STATION)
KANAGAWA-KEN RINGYO SHIKENJO (KANAGAWA PREFECTURAL FOREST EXPERIMENT STATION)
657, HYUGAHARA, NANASAWA, ATSUGI-SHI 243-1
JAPANESE

KANAGAWA-KEN RINGYO SHIKENJO KENKYU HOKOKU 4129
(BULLETIN OF THE KANAGAWA PREFECTURE FOREST EXPERIMENT STATION)
JAPANESE

KANKYO RYOKKA SUISHIN JIGYO CHOSA KENKYU HOKOKU 4130
(RESEARCH REPORT ON TREE PLANTING PROMOTION)
HYOGO-KENRITSU RINGYO SHIKENJO (HYOGO PREFECTURAL FOREST EXPERIMENT STATION)
HYOGO-KEN SEIKATSU-BU (COMMUNITY LIFE DEPARTMENT, HYOGO PREFECTURAL GOVERNMENT)
5, SHIMOYAMATE-DORI, IKUTA-KU, KOBE-SHI 650
JAPANESE

KANSAI RINBOKU IKUSHUJO NENPO 4131
(ANNUAL REPORT OF THE KANSAI FOREST TREE BREEDING STATION, MINISTRY OF AGRICULTURE AND FORESTRY)
RIN'YA-CHO KANSAI RINBOKU IKUSHUJO (KANSAI FOREST TREE BREEDING STATION, MINISTRY OF AGRICULTURE AND FORESTRY)
SHOO-CHO, KATSUTA-GUN, OKAYAMA 709-43
JAPANESE

KANTO RINBOKU IKUSHUJO NENPO 4132
(ANNUAL REPORT OF KANTO FOREST TREE BREEDING STATION)
NORIN-SHO KANTO RINBOKU IKUSHUJO (KANTO FOREST TREE BREEDING STATION, MINISTRY OF AGRICULTURE AND FORESTRY)
978, KASAHARA-CHO, MITO-SHI 310
JAPANESE AND ENGLISH
INDEXED: BIB

KI 4133
(WOOD)
SHINODA MEIBOKUTEN
C/O KANEMASA BUILDING, 2-11, TAKARA-CHO, CHUO-KU, TOKYO 104
JAPANESE

KIKAIKA RINGYO 4134
(JOURNAL OF THE FORESTRY MECHANIZATION SOCIETY)
RINGYO KIKAIKA KYOKAI (FORESTRY MECHANIZATION SOCIETY)
9-13, 1-CHOME, AKASAKA, MINATO-KU, TOKYO 107
JAPANESE

KIKAN MORI 4135
(FOREST)
KOHO-SHITSU (RINNO-CHO)
JAPANESE

KIRYU-SHI NI OKERU RINGYO NO JOKYO 4136
(STATE OF FORESTRY IN KIRYU CITY)
KIRYU-SHI (KIRYU MUNICIPAL OFFICE)
1-1, ORIHIME-CHO, KIRYU-SHI 376
JAPANESE

KITAMI EIRIN-KYOKU JIGYO TOKEISHO 4137
(ANNUAL STATISTICS OF KITAMI REGIONAL FOREST OFFICE)
RIN'YA-CHO KITAMI EIRIN-KYOKU (KITAMI REGIONAL FOREST OFFICE, FORESTRY AGENCY)
KIYOMI-CHO, KITAMI-SHI 090
JAPANESE

KITAMI RIN'YU 4138
(FORESTRY IN KITAMI)
RIN'YA-CHO KITAMI EIRIN-KYOKU (KITAMI REGIONAL FOREST OFFICE, FORESTRY AGENCY)
RIN'YA KOSAIKAI KITAMI SHIBU
C/O KITAMI EIRIN-KYOKU, 70, KIYOMI-CHO, KITAMI-SHI 090
JAPANESE

KITO RINGYO CHIIKI NO RINGYO SHINKO NI KANSURU CHIIKI SHINDAN 4139
(SURVEY OF FORESTRY PROMOTION IN KITO FOREST AREA)
TOKUSHIMA-KEN (TOKUSHIMA PREFECTURAL GOVERNMENT)
1-1, BANDAI-CHO, TOKUSHIMA-SHI 770
JAPANESE

KOCHI EIRIN-KYOKU JIGYO TOKEISHO 4140
(STATISTICAL YEARBOOK OF KOCHI REGIONAL FOREST OFFICE)
RIN'YA-CHO KOCHI EIRIN-KYOKU (KOCHI REGIONAL FOREST OFFICE, FORESTRY AGENCY)
3-30, 1-CHOME, MARUNOUCHI, KOCHI-SHI 780
JAPANESE

KOCHI RIN'YU 4141
(FORESTRY IN KOCHI)
RIN'YA-CHO KOCHI EIRIN-KYOKU (KOCHI REGIONAL FOREST OFFICE, FORESTRY AGENCY) AND RINYA KYOSAIKAI KOCHI SHIBU
3-30, 1-CHOME, MARUNOUCHI, KOCHI-SHI 780
JAPANESE

KOCHI-KEN RINGYO SHIKENJO GYOMU HOKOKU 4142
(REPORT OF THE KOCHI PREFECTURE FOREST EXPERIMENTAL OFFICE)
KOCHI-KEN RINGYO SHIKENJO (KOCHI PREFECTURE FOREST EXPERIMENTAL OFFICE)
MIYANOKUCHI, TOSAYAMADA-CHO, KAMI-GUN, KOCHI 782
JAPANESE

KOCHI-KEN RINGYO SHIKENJO KENKYU HOKOKU 4143
(RESEARCH REPORT OF KOCHI PREFECTURE FOREST EXPERIMENTAL OFFICE)
KOCHI-KEN RINGYO SHIKENJO (KOCHI PREFECTURE FOREST EXPERIMENTAL OFFICE)
MIYANOKUCHI, TOSAYAMADA-CHO, KAMI-GUN, KOCHI 782
JAPANESE

KOKUYU RIN'YA JIGYO TOKEISHO 4144
(NATIONAL FOREST STATISTICS)
RIN'YA-CHO (FORESTRY AGENCY)
2-1, 1-CHOME, KASUMIGASEKI, CHIYODA-KU, TOKYO 100
JAPANESE

KOKUYU RIN'YA NO KATSUYO NI KANSURU CHOSA KEKKA 4145
(REPORT OF PRACTICAL USE OF NATIONAL FOREST)
ZENKOKU NOGYO KAIGISHO (JAPAN AGRICULTURE COUNCIL)
C/O SANSHI KAIKAN, 1-7, YURAKU-CHO, CHIYODA-KU, TOKYO 100
JAPANESE

KOKUYURIN GIJUTSU KENKYU HAPPYOSHU 4146
(PROCEEDINGS OF TECHNICAL RESEARCH ON NATIONAL FOREST)
SURI EMU KENKYUKAI
28, ICHIGAYA HONMURA-CHO, SHINJUKU-KU, TOKYO 162
JAPANESE

KOSHUNAI KIHO 4147
(QUARTERLY REPORT OF HOKKAIDO FOREST EXPERIMENT STATION)
HOKKAIDORITSU RINGYO SHIKENJO (HOKKAIDO FOREST EXPERIMENT STATION)
HIGASHIYAMA, KOSHUNAI-MACHI, BIBAI-SHI 072
JAPANESE

KUMAMOTO EIRIN-KYOKU DOJO CHOSA HOKOKU 4148
(REPORT OF AGRONOMICAL SURVEY, KUMAMOTO REGIONAL FOREST OFFICE)
RIN'YA-CHO KUMAMOTO EIRIN-KYOKU (KUMAMOTO REGIONAL FOREST OFFICE, FORESTRY AGENCY)
2-7, KYO-MACHI HONCHO, KUMAMOTO-SHI 860
JAPANESE

KUMAMOTO EIRIN-KYOKU JIGYO TOKEISHO 4149
(OPERATION STATISTICS OF KUMAMOTO REGIONAL FOREST OFFICE)
RIN'YA-CHO KUMAMOTO EIRIN-KYOKU (KUMAMOTO NATIONAL FOREST REGIONAL OFFICE, FORESTRY AGENCY)
2-7, KYO-MACHI HONCHO, KUMAMOTO-SHI 860
JAPANESE

KUMAMOTO NO RINGYO 4150
(FORESTRY IN KUMAMOTO)
KOKUDO RYOKKA KUMAMOTO-KEN SUISHIN IINKAI
C/O KUMAMOTO-KEN RINMU KANKO-BU, 18-1, 6-CHOME, SUIZENJI, KUMAMOTO-SHI 862
JAPANESE

KUMAMOTO-KEN RINGYO KENKYU SHIDOSHO GYOMU SEISEKI HOKOKUSHO 4151
(REPORT OF KUMAMOTO-KEN FORESTRY RESEARCH STATION)
KUMAMOTO-KEN RINGYO KENKYU SHIDOSHO
163, SHIMIZU-MACHI, KUMAMOTO-SHI 860
JAPANESE

KYOTO DAIGAKU NOGAKUBU ENSHURIN HOKOKU 4152
(BULLETIN OF THE KYOTO UNIVERSITY FORESTS)
KYOTO DAIGAKU NOGAKUBU FUZOKU ENSHURIN (KYOTO UNIVERSITY FORESTS)
KITASHIRAKAWA OIWAKE-CHO, SAKYO-KU, KYOTO-SHI 606
JAPANESE AND ENGLISH
INDEXED: CHE

KYOTO DAIGAKU NOGAKUBU ENSHURIN KISHO HOKOKU 4153
(WEATHER REPORT, KYOTO UNIVERSITY FORESTS)
KYOTO DAIGAKU NOGAKUBU FUZOKU ENSHURIN (KYOTO UNIVERSITY FORESTS)
KITASHIRAKAWA OIWAKE-CHO, SAKYO-KU, KYOTO-SHI 606
JAPANESE

KYOTO DAIGAKU NOGAKUBU ENSHURIN SHUHO 4154
(REPORTS OF THE KYOTO UNIVERSITY FORESTS)
KYOTO DAIGAKU NOGAKUBU FUZOKU ENSHURIN (KYOTO UNIVERSITY FORESTS)
KITASHIRAKAWA OIWAKE-CHO, SAKYO-KU, KYOTO-SHI 606
JAPANESE
INDEXED: BIB

KYOTO-FU RINGYO SHIKENJO CHOSA SHIKEN KENKYU HOKOKU 4155
(RESEARCH REPORT OF FOREST EXPERIMENT STATION OF KYOTO PREFECTURE)
KYOTO-FU RINGYO SHIKENJO
HONJO, WACHI-MACHI, FUNAI-GUN, KYOTO 629-11
JAPANESE

KYOTO-FU RINGYO SHIKENJO SHIKEN KENKYU NENPO 4156
(ANNUAL REPORT OF FOREST EXPERIMENT STATION OF KYOTO PREFECTURE)
KYOTO-FU RINGYO SHIKENJO
HONJO, WACHI-MACHI, FUNAI-GUN, KYOTO 629-11
JAPANESE

KYOTO-FU RINGYO TOKEI 4157
(FORESTRY STATISTICS, KYOTO PREFECTURE)
KYOTO-FU NORIN-BU (AGRICULTURE AND FORESTRY DEPARTMENT, KYOTO PREFECTURAL GOVERNMENT)
YABUNOUCHI-MACHI, SHIMODACHIYURI-DORI SHIN-MACHI NISHIIRU KAMIGYO-KU, KYOTO-SHI 602
JAPANESE

KYOTO-FURITSU DAIGAKU NOGAKUBU ENSHURIN HOKOKU 4158
(BULLETIN OF THE KYOTO PREFECTURAL UNIVERSITY FORESTS)
KYOTO-FURITSU DAIGAKU NOGAKUBU FUZOKU ENSHURIN (KYOTO PREFECTURAL UNIVERSITY FORESTS)
1, SHIMOKAMO HANGI-CHO, SAKYO-KU, KYOTO-SHI 606
JAPANESE AND EUROPEAN
INDEXED: JICST

KYUSHU DAIGAKU NOGAKUBU ENSHURIN HOKOKU 4159
(BULLETIN OF THE KYUSHU UNIVERSITY FORESTS)
KYUSHU DAIGAKU NOGAKUBU FUZOKU ENSHURIN (RESEARCH INSTITUTE OF UNIVERSITY FORESTS, FACULTY OF AGRICULTURE, KYUSHU UNIVERSITY)
HAKOZAKI, HIGASHI-KU, FUKUOKA-SHI 812
JAPANESE, WITH EUROPEAN SUMMARY
INDEXED: BIB, CHE, JICST

KYUSHU DAIGAKU NOGAKUBU ENSHURIN SHUHO 4160
(REPORT OF THE KYUSHU UNIVERSITY FORESTS)
KYUSHU DAIGAKU NOGAKUBU
3575-1, HAKOZAKI, HIGASHI-KU, FUKUOKA-SHI 812
JAPANESE AND ENGLISH
INDEXED: CHE, JICST

KYUSHU RINZAI SHINBUN 4161
(KYUSHU LUMBER NEWS)
KYUSHU RINZAI SHINBUNSHA
9-15, YAMASHITA-CHO, KAGOSHIMA-SHI 892
JAPANESE

MAEBASHI EIRIN-KYOKU DOJO CHOSA HOKOKU 4162
(AGRONOMICAL SURVEY OF MAEBASHI REGIONAL FOREST OFFICE)
RIN'YA-CHO MAEBASHI EIRIN-KYOKU (MAEBASHI REGIONAL FOREST OFFICE, FORESTRY AGENCY)
16-25, 4-CHOME, IWAGAMI-CHO, MAEBASHI-SHI 371
JAPANESE, WITH ENGLISH SUMMARY

MAEBASHI EIRIN-KYOKU JIGYO TOKEISHO 4163
(ANNUAL STATISTICS OF MAEBASHI REGIONAL FOREST OFFICE)
RIN'YA-CHO MAEBASHI EIRIN-KYOKU (MAEBASHI REGIONAL FOREST OFFICE, FORESTRY AGENCY)
16-25, 4-CHOME, IWAGAMI-CHO, MAEBASHI-SHI 371
JAPANESE

MIDORI 4164
(GREEN)
RIN'YA-CHO NAGOYA EIRIN-KYOKU (NAGOYA REGIONAL FOREST OFFICE, FORESTRY AGENCY) AND RIN'YA KOSAIKAI
6-1, 2-CHOME, SANNO-MARU, NAKA-KU, NAGOYA-SHI 460
JAPANESE

MIE DAIGAKU NOGAKUBU ENSHURIN HOKOKU 4165
(BULLETIN OF THE MIE UNIVERSITY FORESTS)
MIE DAIGAKU NOGAKUBU FUZOKU ENSHURIN (FACULTY OF AGRICULTURE, MIE UNIVERSITY)
UEHAMA-CHO, TSU-SHI 514
JAPANESE AND ENGLISH

MIN'EI NAEHATA SEISAN JITTAI CHOSASHO 4166
(SURVEY OF SAPLING PRODUCT IN CIVIL FIELD)
HOKKAIDO RINMU-BU (FORESTRY ADMINISTRATION DEPARTMENT, HOKKAIDO PREFECTURAL GOVERNMENT)
NISHI-6-CHOME, KITA-3-JO, CHUO-KU, SAPPORO-SHI 060
JAPANESE

MIYAGI-KEN RINGYO SHIKENJO GYOMU SEISEKISHO 4167
(REPORT OF MIYAGI PREFECTURAL FOREST EXPERIMENT STATION)
MIYAGI-KEN RINGYO SHIKENJO (MIYAGI PREFECTURE FOREST EXPERIMENT STATION)
OHIRA-MURA, KUROKAWA-GUN, MIYAGI 981-36
JAPANESE

MIYAMA 4168
(OSAKA REGIONAL FOREST OFFICE NEWS)
RIN'YA-CHO OSAKA EIRIN-KYOKU (OSAKA REGIONAL FORESTRY OFFICE, FORESTRY AGENCY) AND RIN'YA KOSAIKAI OSAKA SHIBU
6, HOENZAKA-CHO, HIGASHI-KU, OSAKA-SHI 540
JAPANESE

MIYAZAKI-KEN RINGYO SHIKENJO GYOMU HOKOKU 4169
(ANNUAL REPORT OF MIYAZAKI PREFECTURAL FORESTRY EXPERIMENTAL STATION)
MIYAZAKI-KEN RINGYO SHIKENJO (MIYAZAKI PREFECTURAL FORESTRY EXPERIMENTAL STATION)
KASHIWABARA, MIYAZAKI-SHI 880-21
JAPANESE

MOKUZAI GAKKAISHI 4170
(JOURNAL OF THE JAPAN WOOD RESEARCH SOCIETY)
NIHON MOKUZAI GAKKAI (JAPAN WOOD RESEARCH SOCIETY)
C/O NORIN-SHO RINGYO SHIKENJO, 37-21, 5-CHOME, SHIMOMEGURO, MEGURO-KU, TOKYO 153
JAPANESE AND EUROPEAN
INDEXED: BIB, CHE, JICST

MOKUZAI JUKYU HOKOKUSHO 4171
(REPORT OF DEMAND AND SUPPLY OF LUMBER)
NORIN-SHO NORIN KEIZAI-KYOKU TOKEI JOHO-BU (STATISTICS AND INFORMATION DIVISION, MINISTRY OF AGRICULTURE AND FORESTRY) AND NORIN TOKEI KYOKAI
11-14, 2-CHOME, MEGURO, MEGURO-KU, TOKYO 153
JAPANESE

MOKUZAI TO GIJUTSU 4172
(WOOD PRODUCTS AND TECHNIQUE)
TOYAMA-KEN MOKUZAI SHIKENJO (WOODS PRODUCTS RESEARCH INSTITUTE, TOYAMA PREFECTURE)
KUROKAWASHIN, KOSUGI-CHO, IZUMI-GUN, TOYAMA 939-03
JAPANESE
INDEXED: CHE

NAGANO EIRIN-KYOKU GYOMU KENKYU HAPPYOSHU 4173
(FOREST TECHNICAL REPORT OF THE NAGANO REGIONAL FOREST OFFICE)
RIN'YA-CHO NAGANO EIRIN-KYOKU (NAGANO REGIONAL FOREST OFFICE, FORESTRY AGENCY)
715-5, KURIDA, NAGANO-SHI 380
JAPANESE

NAGANO EIRIN-KYOKU JIGYO TOKEISHO 4174
(OPERATION STATISTICS OF NAGANO REGIONAL FOREST OFFICE)
RIN'YA-CHO NAGANO EIRIN-KYOKU (NAGANO REGIONAL FOREST OFFICE, FORESTRY AGENCY)
715-5, KURIDA, NAGANO-SHI 380
JAPANESE

NAGANO KOHO 4175
(NEWS OF NAGANO REGIONAL FOREST OFFICE)
RIN'YA-CHO NAGANO EIRIN-KYOKU (NAGANO REGIONAL FOREST OFFICE, FORESTRY AGENCY)
715-5, KURIDA, NAGANO-SHI 380
JAPANESE

NAGANO RIN'YU 4176
(FORESTRY IN NAGANO)
RIN'YA-CHO NAGANO EIRIN-KYOKU (NAGANO REGIONAL FOREST OFFICE, FORESTRY AGENCY) AND RIN'YA KOSAIKAI NAGANO SHIBU
715-5, KURIDA, NAGANO-SHI 380
JAPANESE

NAGANO-KEN RINGYO SHIDOSHO GYOMU HOKOKU 4177
(REPORT ON ACTIVITY OF NAGANO PREFECTURE FORESTRY RESEARCH INSTITUTE)
NAGANO-KEN RINGYO SHIDOSHO
73, KIKYOGAHARA, SOGA, SHIOJIRI-SHI 399-64
JAPANESE

NAGASAKI-KEN NO RINGYO TOKEI 4178
(FORESTRY STATISTICS IN NAGANO PREFECTURE)
NAGASAKI-KEN (NAGANO PREFECTURAL GOVERNMENT)
2-13, EDO-MACHI, NAGASAKI-SHI 850
JAPANESE

NAGOYA DAIGAKU NOGAKUBU ENSHURIN HOKOKU 4179
(BULLETIN OF THE NAGOYA UNIVERSITY FORESTS)
NAGOYA DAIGAKU NOGAKUBU FUZOKU ENSHURIN (NAGOYA UNIVERSITY FORESTS)
INABU-CHO, KITASHITARA-GUN, AICHI 441-25
JAPANESE, WITH ENGLISH SUMMARY
INDEXED: BIB

NAGOYA EIRIN-KYOKU GYOMU KENKYU HAPPYO RONBUNSHU 4180
(REPORT OF OPERATION RESEARCH CONFERENCE, NAGOYA REGIONAL FOREST OFFICE)
RIN'YA-CHO NAGOYA EIRIN-KYOKU (NAGOYA REGIONAL FOREST OFFICE, FORESTRY AGENCY)
6-1, 2-CHOME, SANNOMARU, NAKA-KU, NAGOYA-SHI 460
JAPANESE

NAGOYA EIRIN-KYOKU JIGYO TOKEISHO 4181
(OPERATION STATISTICS OF NAGOYA REGIONAL FOREST OFFICE)
RIN'YA-CHO NAGOYA EIRIN-KYOKU (NAGOYA REGIONAL FOREST OFFICE, FORESTRY AGENCY)
6-1, 2-CHOME, SANNOMARU, NAKA-KU, NAGOYA-SHI 460
JAPANESE

NARA-KEN RINGYO SHIDOSHO GYOMU NENPO 4182
(ANNUAL REPORT OF THE NARA PREFECTURAL FORESTRY RESEARCH INSTITUTE)
NARA-KEN RINGYO SHIDOSHO
KIBI TAKATORI-MACHI, TAKAICHI-GUN, NARA 635-01
JAPANESE

NARA-KEN RINGYO SHIDOSHO SHIKEN KENKYU HOKOKU 4183
(RESEARCH REPORT OF THE NARA PREFECTURAL FORESTRY RESEARCH INSTITUTE)
NARA-KEN RINGYO SHIDOSHO
KIBI TAKATORI-MACHI, TAKAICHI-GUN, NARA 635-01
JAPANESE

NIHON MOKUZAI GAKKAI TAIKAI KENKYU HAPPYO YOSHI 4184
(ABSTRACT PAPERS OF THE SYMPOSIUM OF JAPAN WOOD RESEARCH SOCIETY)
NIHON MOKUZAI GAKKAI
37-21, 5-CHOME, SHIMOMEGURO, MEGURO-KU, TOKYO 105
JAPANESE
INDEXED: JICST

NIHON RINGAKKAI TAIKAI KOENSHU 4185
(PROCEEDINGS OF THE SYMPOSIUM OF JAPANESE FORESTRY SOCIETY)
NIHON RINGAKKAI (JAPANESE FORESTRY SOCIETY)
37-21, 5-CHOME, SHIMOMEGURO, MEGURO-KU, TOKYO 105
JAPANESE

NIHON RINGAKKAISHI 4186
(JOURNAL OF THE JAPANESE FORESTRY SOCIETY)
NIHON RIN GAKKAI (JAPANESE FORESTRY SOCIETY)
C/O RINGYO SHIKENJO, 37-21, 5-CHOME, SHIMOMEGURO, MEGURO-KU, TOKYO 153
JAPANESE AND EUROPEAN
INDEXED: BIB, CHE, JICST

NIHON RINGYO 4187
(JAPANESE FORESTRY)
NIHON RINGYO KYOKAI (JAPAN FORESTRY ASSOCIATION
4-3, 2-CHOME, NAGATA-CHO, CHIYODA-KU, TOKYO 100
JAPANESE

NIHON RINZAI SHINBUN 4188
(JAPAN LUMBER NEWS)
NIHON RINZAI SHINBUNSHA
10-10, 7-CHOME, NISHISHINJUKU, SHINJUKU-KU, TOKYO 160
JAPANESE

NIIGATA DAIGAKU NOGAKUBU ENSHURIN HOKOKU 4189
(BULLETIN OF THE NIIGATA UNIVERSITY FORESTS)
NIIGATA DAIGAKU NOGAKUBU FUZOKU ENSHURIN (FACULTY OF AGRICULTURE, NIIGATA UNIVERSITY)
106, KOGANE-MACHI, NIIGATA-SHI 950
JAPANESE, WITH EUROPEAN SUMMARY
INDEXED: BIO

NIIGATA-KEN RINGYO SHIKENJO KENKYU HOKOKU 4190
(REPORT OF NIIGATA PREFECTURE FORESTRY EXPERIMENT STATION)
NIIGATA-KEN RINGYO SHIKENJO
UNOTORO, ASAHI-MURA, IWAFUNE-GUN, NIIGATA 958
JAPANESE
INDEXED: BIB

NIKKAN MOKUZAI SHINBUN 4191
(DAILY LUMBER NEWS)
NIKKAN MOKUZAI SHINBUNSHA
10-12, 2-CHOME, TSUKIJI, CHUO-KU, TOKYO 104
JAPANESE

NONEZUMI 4192
(FIELD MOUSE)
HOKKAIDO SHINRIN BOEKI KYOKAI
C/O HOKKAIDO RINGYO KAIKAN, NISHI-5-CHOME, KITA-4-JO, CHUO-KU, SAPPORO-SHI 060
JAPANESE

NORIN SUISAN SHIKEN KENKYU NENPO (RINGYO-HEN) 4193
(ANNUAL REPORT ON EXPERIMENTS IN AGRICULTURE, FORESTRY AND FISHERIES, SERIES FORESTRY))
NORIN_SHO NORIN SUISAN GIJUTSU KAIGI (AGRICULTURE, FORESTRY AND FISHERIES RESEARCH COUNCIL, MINISTRY OF AGRICULTURE AND FORESTRY) AND NORIN TOKEI KYOKAI
11-4, 2-CHOME, MEGURO, MEGURO-KU, TOKYO 153
JAPANESE
INDEXED: JICST

NORINSHO RINGYO SHIKENJO JOHO 4194
(BULLETIN OF THE MINISTRY OF AGRICULTURE AND FORESTRY, FORESTRY EXPERIMENT STATION)
NORINSHO RINGYO SHIKENJO CHOSASHITSU KANKOSHITSU (MINISTRY OF AGRICULTURE AND FORESTRY, FORESTRY EXPERIMENT STATION, INVESTIGATIVE SECTION, PUBLICATION ROOM)
JAPANESE

OBIHIRO 4195
(NEWS OF OBIHIRO REGIONAL FOREST OFFICE)
RIN'YA-CHO OBIHIRO EIRIN-KYOKU (OBIHIRO REGIONAL FOREST OFFICE)
MINAMI-13-CHO, HIGASHI-8-JO, OBIHIRO-SHI 080
JAPANESE

OBIHIRO EIRIN-KYOKU GYOMU KENKYU HAPPYO RONEUNSHU 4196
(REPORT OF THE OBIHIRO REGIONAL FOREST OFFICE)
RIN'YA-CHO OBIHIRO EIRIN-KYOKU (OBIHIRO REGIONAL FOREST OFFICE, FORESTRY AGENCY)
MINAMI-13-CHOME, HIGASHI-8-JO, OBIHIRO-SHI 080
JAPANESE

OBIHIRO EIRIN-KYOKU JIGYO TOKEI YORAN 4197
(STATISTICS OUTLINE OF OBIHIRO REGIONAL FOREST OFFICE)
RIN'YA-CHO OBIHIRO EIRIN-KYOKU (OBIHIRO REGIONAL FOREST OFFICE, FORESTRY AGENCY)
MINAMI-13-CHOME, HIGASHI-8-JO, OBIHIRO-SHI 080
JAPANESE

OBIHIRO EIRIN-KYOKU JIGYO TOKEISHO 4198
(STATISTICS OF WORKS OF OBIHIRO REGIONAL FOREST OFFICE)
RIN'YA-CHO OBIHIRO EIRIN-KYOKU (OBIHIRO REGIONAL FOREST OFFICE, FORESTRY AGENCY)
MINAMI-13-CHOME, HIGASHI-8-JO, OBIHIRO-SHI 080
JAPANESE

OBIHIRO EIRIN-KYOKU KANNAI GAIYO 4199
(OUTLINE OF FOREST WITHIN JURISDICTION OF OBIHIRO REGIONAL FORESTRY OFFICE)
RIN'YA-CHO OBIHIRO EIRIN-KYOKU (OBIHIRO REGIONAL FORESTRY OFFICE, FORESTRY AGENCY)
MINAMI-13-CHOME, HIGASHI-8-JO, OBIHIRO-SHI 080
JAPANESE

OHOTSUKU 4200
(OKHOTSK)
KITAMI EIRIN KYOKU (KITAMI FOREST MANAGEMENT BUREAU)
JAPANESE

OITA-KEN RINGYO SHIKENJO HOKOKU 4201
(REPORT OF FOREST EXPERIMENT STATION OF OITA)
OITA-KEN RINGYO SHIKENJO (FOREST EXPERIMENT STATION OF OITA)
SAJIHARA, ARITA, HITA-SHI 877-13
JAPANESE
INDEXED: CHE

OKAYAMA-KEN RINGYO SHIKENJO HOKOKU 4202
(BULLETIN OF OKAYAMA PREFECTURE FOREST EXPERIMENT STATION)
OKAYAMA-KEN RINGYO SHIKENJO (OKAYAMA PREFECTURE FOREST EXPERIMENT STATION)
SHOO-CHO, KATSUTA-GUN, OKAYAMA 709-43
JAPANESE

OKAYAMA-KEN RINGYO SHIKENJO KENKYU HOKOKU 4203
(RESEARCH REPORT OF THE OKAYAMA PREFECTURE FORESTRY EXPERIMENT STATION)
JAPANESE

OKINAWA-KEN RINGYO SHIKENJO KENKYU HOKOKU 4204
(RESEARCH REPORT OF FORESTRY EXPERIMENTAL STATION, OKINAWA PREFECTURE)
OKINAWA-KEN RINGYO SHIKENJO
3626, NAGO, NAGO-SHI 905
JAPANESE

OSAKA EIRIN-KYOKU JIGYO TOKEISHO 4205
(OPERATION STATISTICS OF OSAKA REGIONAL FOREST OFFICE)
RIN'YA-CHO OSAKA EIRIN-KYOKU (OSAKA REGIONAL FOREST OFFICE, FORESTRY AGENCY)
1-6, HOENZAKA-CHO, HIGASHI-KU, OSAKA-SHI 540
JAPANESE

OSAKA MOKUZAI SHINBUN 4206
(OSAKA LUMBER NEWS)
OSAKA MOKUZAI SHINBUNSHA
1-12, 3-CHOME, SAKURAGAWA, NANIWA-KU, OSAKA-SHI 556
JAPANESE

POKETTO RINGYO TOKEI 4207
(POCKET EDITION OF FORESTRY STATISTICS)
NIHON RINGYO GIJUTSU KYOKAI (JAPAN FORESTRY TECHNICS ASSOCIATION)
7, ROKUBAN-CHO, CHIYODA-KU, TOKYO 102
JAPANESE

RINBOKU IKUSHU JOHO 4208
(FOREST TREE BREEDING NEWS)
RINBOKU IKUSHU KYOKAI (FOREST TREE BREEDING ASSOCIATION)
C/O NIHON RINGYO GIJUTSU KYOKAI BUILDING, 7, ROKUBAN-CHO, CHIYODA-KU, TOKYO 102
JAPANESE

RINBOKU NO IKUSHU 4209
(FOREST TREE BREEDING)
RINBOKU IKUSHU KYOKAI (FOREST TREE BREEDING ASSOCIATION)
37-21, 5-CHOME, SHIMOMEGURO, MEGURO-KU, TOKYO 153
JAPANESE

RINGYO DOTAI CHOSA KEKKA HOKOKUSHO 4210
(DYNAMIC STATISTICS OF FORESTRY)
NORIN-SHO NORIN KEIZAI-KYOKU TOKEI JOHO-BU (STATISTICS AND INFORMATION DEPARTMENT, ECONOMIC AFFAIRS BUREAU, MINISTRY OF AGRICULTURE AND FORESTRY)
2-1, 1-CHOME, KASUMIGASEKI, CHIYODA-KU, TOKYO 100
JAPANESE

RINGYO GIJUTSU 4211
(FORESTRY TECHNIQUE)
NIHON RINGYO GIJUTSU KYOKAI (JAPAN FORESTRY TECHNICS ASSOCIATION)
7, ROKUBAN-CHO, CHIYODA-KU, TOKYO 102
JAPANESE
INDEXED: BIB

RINGYO GIJUTSU JOHO 4212
(FORESTRY ENGINEERING INFORMATION)
NAGANO-KEN RINGYO SHIDOSHO (NAGANO PREFECTURE FORESTRY RESEARCH INSTITUTE)
73, KIKYOGAHARA, SOGA, SHIOJIRI-SHI 399-64
JAPANESE

RINGYO GIJUTSU KENKYU SHUROKU, AOMORI EIRIN-KYOKU 4213
(RESEARCH REPORT OF THE FORESTRY TECHNOLOGY)
RIN'YA-CHO AOMORI EIRIN-KYOKU (AOMORI REGIONAL FOREST OFFICE, FORESTRY AGENCY)
4-37, 2-CHOME, YANAGAWA, AOMORI-SHI 030
JAPANESE

RINGYO GIJUTSU KENKYU SHUROKU, MAEBASHI EIRIN-KYOKU 4214
(COMPILATION OF RESEARCH PUBLICATION ON FORESTRY TECHNOLOGY)
RIN'YA-CHO MAEBASHI EIRIN-KYOKU (MAEBASHI REGIONAL FOREST OFFICE, FORESTRY AGENCY)
16-25, 4-CHOME, IWAGAMI-CHO, MAEBASHI-SHI 371
JAPANESE

RINGYO IBARAKI 4215
(FORESTRY IN IBARAKI)
IBARAKI-KEN RINGYO KAIRYO FUKYU KYOKAI (IBARAKI PREFECTURE FORESTRY IMPROVEMENT AND PROPAGATION ASSOCIATION)
5-88, 1-CHOME, KITASANNOMARU, MITO-SHI 310
JAPANESE

RINGYO KANKEI KENSETSU GYOMU TOKEI 4216
(STATISTICS OF CONSTRUCTION WORKS FOR FORESTRY)
RIN'YA-CHO SHIDO-BU (FORESTRY DEVELOPMENT DEPARTMENT, FORESTRY AGENCY)
2-1, 1-CHOME, KASUMIGASEKI, CHIYODA-KU, TOKYO 100
JAPANESE

RINGYO KEIZAI 4217
(FORESTRY ECONOMICS)
JAPANESE

RINGYO KENKYU HAPPYO TAIKAI RONBUNSHU 4218
(PRECEEDINGS OF CONFERENCE ON FORESTRY)
HOKKAIDO RINGYO KAIRYO FUKYU KYOKAI
KITA-2-JO, NISHI-19-CHOME, CHUO-KU, SAPPORO-SHI 060
JAPANESE

RINGYO NIIGATA 4219
(FORESTRY IN NIIGATA)
NIIGATA-KEN RINGYO KAIRYO KYOKAI (NIIGATA PREFECTURE FORESTRY IMPROVEMENT ASSOCIATION)
C/O NIIGATA-KEN RINSEI-KA, 5290, ICHIBAN-CHO, GAKKO-CHO, NIIGATA-SHI 950
JAPANESE

RINGYO NO DOKO NI KANSURU NENJI HOKOKU 4220
(ANNUAL REPORT OF FORESTRY TREND)
RIN'YA-CHO (FORESTRY AGENCY)
2-1, 1-CHOME, KASUMIGASEKI, CHIYODA-KU, TOKYO 100
JAPANESE

RINGYO SEISAN TOKEI NENPO 4221
(ANNUAL REPORT OF THE FORESTRY PRODUCTS)
NORIN-SHO TOKEI JOHO-BU (STATISTICS AND INFORMATION SERVICE DIVISION, MINISTRY OF AGRICULTURE AND FORESTRY) AND NORIN TOKEI KYOKAI
11-14, 2-CHOME, MEGURO, MEGURO-KU, TOKYO 153
JAPANESE

RINGYO SHIKEN HOKOKU (AKITA-KEN RINGYO SHIKENJO) 4222
(REPORT OF FORESTRY EXPERIMENT STATION OF AKITA)
AKITA-KEN RINGYO SHIKENJO
SHISHIGAMORI, SHAKANAI, ODATE-SHI 017
JAPANESE

RINGYO SHIKEN KENKYU HOKOKU 4223
(FORESTRY EXPERIMENTAL BULLETIN OF THE ITSUKAICHI OFFICE, TOKYO METROPOLITAN AGRICULTURAL EXPERIMENT STATION)
TOKYO-TO NOGYO SHIKENJO ITSUKAICHI BUNJO (TOKYO METROPOLITAN AGRICULTURAL EXPERIMENT STATION, ITSUKAICHI OFFICE)
853, TOKURA, ITSUKAICHI-MACHI, NISHITAMA-GUN, TOKYO 190-01
JAPANESE
INDEXED: JICST

RINGYO SHIKENJO GEPPO 4224
(MONTHLY REPORT OF THE FORESTRY EXPERIMENT STATION)
HOKKAIDO RITSU RINGYO SHIKENJO
(HOKKAIDO PREFECTURAL FORESTRY EXPERIMENT STATION)
JAPANESE

RINGYO SHIKENJO HOKKAIDO SHIJO NENPO 4225
(ANNUAL REPORT OF THE HOKKAIDO BRANCH, GOVERNMENT FOREST EXPERIMENT STATION)
RIN'YA-CHO RINGYO SHIKENJO HOKKAIDO SHIJO (HOKKAIDO BRANCH FOREST EXPERIMENT STATION, FORESTRY AGENCY)
5-JO, 13-CHOME, TOYOHIRA-KU, SAPPORO-SHI 062
JAPANESE, WITH EUROPEAN SUMMARY
INDEXED: BIB

RINGYO SHIKENJO HOKOKU 4226
(REPORT OF THE FOREST INDUSTRY EXPERIMENT STATION)
CHIBA-KEN RINGYO SHIKENJO (CHIBA PREFECTURE FOREST INDUSTRY EXPERIMENT STATION)
JAPANESE

RINGYO SHIKENJO JOHO 4227
(STATION NEWS (FOREST EXPERIMENT STATION, MINISTRY OF AGRICULTURE AND FORESTRY))
RIN'YA-CHO RINGYO SHIKENJO (FOREST EXPERIMENT STATION, FORESTRY AGENCY)
37-21, 5-CHOME, SHIMOMEGURO, MEGURO-KU, TOKYO 153
JAPANESE

RINGYO SHIKENJO KANSAI SHIJO NENPO 4228
(ANNUAL REPORT, GOVERNMENT FOREST EXPERIMENT STATION, KANSAI BRANCH)
RIN'YA-CHO RINGYO SHIKENJO KANSAI SHIJO (GOVERNMENT EXPERIMENT STATION, KANSAI BRANCH, FORESTRY AGENCY)
MOMOYAMA-CHO, FUSHIMI-KU, KYOTO-SHI 612
JAPANESE

RINGYO SHIKENJO KENKYU HOKOKU 4229
(RESEARCH REPORT OF THE FORESTRY EXPERIMENT STATION)
TAKUSHIMA-KEN RINGYO SHIKENJO
(TAKUSHIMA PREFECTURE FORESTRY EXPERIMENT STATION)
JAPANESE

RINGYO SHIKENJO KENKYU HOKOKU 4230
(BULLETIN OF THE GOVERNMENT FOREST EXPERIMENT STATION)
RIN'YA-CHO RINGYO SHIKENJO (GOVERNMENT FOREST EXPERIMENT STATION, FORESTRY AGENCY) AND RIN'YA KOSAIKAI
7-8, 1-CHOME, KORAKU, BUNKYO-KU, TOKYO 112
JAPANESE AND EUROPEAN
INDEXED: BIB, BIO, CHE

RINGYO SHIKENJO KENKYU HOKOKU; KAISETSU 4231
(RESEARCH REPORT OF THE FOREST INDUSTRY EXPERIMENT STATION; COMMENTARY)
NORINSHO RINGYO SHIKENJO (MINISTRY OF AGRICULTURE AND FORESTRY, FOREST INDUSTRY EXPERIMENT STATION)
JAPANESE

RINGYO SHIKENJO KYUSHU SHIJO NENPO 4232
(ANNUAL REPORT OF THE KYUSHU BRANCH, GOVERNMENT FOREST EXPERIMENT STATION)
RIN'YA-CHO RINGYO SHIKENJO KYUSHU SHIJO (KYUSHU BRANCH, GOVERNMENT FOREST EXPERIMENT STATION, FORESTRY AGENCY)
11-16, 4-CHOME, KUROKAMI, KUMAMOTO-SHI 860
JAPANESE

RINGYO SHIKENJO NENPO 4233
(ANNUAL REPORT OF GOVERNMENT FOREST EXPERIMENT STATION)
RIN'YA-CHO RINGYO SHIKENJO (GOVERNMENT FOREST EXPERIMENT STATION, FOREST AGENCY)
37-21, 5-CHOME, SHIMOMEGURO, MEGURO-KU, TOKYO 153
JAPANESE

RINGYO SHIKENJO SHIKEN KENKYU HOKOKU (TOTTORI) 4234
(RESEARCH REPORT OF TOTTORI PREFECTURE FORESTRY EXPERIMENT STATION))
TOTTORI-KEN RINGYO SHIKENJO (TOTTORI PREFECTURE FORESTRY EXPERIMENT STATION)
5, TACHIKAWA-MACHI, TOTTORI-SHI 680
JAPANESE

RINGYO SHIKENJO TOHOKU SHIJO NENPO 4235
(ANNUAL REPORT OF THE TOHOKU BRANCH, GOVERNMENT FOREST EXPERIMENT STATION)
RIN'YA-CHO RINGYO SHIKEN-JO TOHOKU SHIJO (TOHOKU BRANCH, GOVERNMENT FOREST EXPERIMENT STATION, FORESTRY AGENCY)
72, NABEYASHIKI, SHIMOKURIYAGAWA, MORIOKA-SHI 020-01
JAPANESE

RINGYO SHIN CHISHIKI 4236
(NEW KNOWLEDGE OF FORESTRY)
RIN'YA-CHO (FORESTRY AGENCY)
ZENKOKU RINGYO KAIRYO FUKYU KYOKAI
9-13, 1-CHOME, AKASAKA, MINATO-KU, TOKYO 107
JAPANESE

RINGYO TO YAKUZAI 4237
(FORESTRY AND FOREST CHEMICALS)
RINGYO YAKUZAI KYOKAI (FOREST CHEMICAL ASSOCIATION OF JAPAN)
SHIN OTE-MACHI BUILDING, 2-1, 2-CHOME, OTE-MACHI, CHIYODA-KU, TOKYO 100
JAPANESE

RINMU KANKEI JIGYO NO GAIYO 4238
(OUTLINE OF FORESTRY OPERATIONS)
KYOTO-FU RINMU-BU (FORESTRY ADMINISTRATION DEPARTMENT, KYOTO PREFECTURAL GOVERNMENT)
YABUNOUCHI-MACHI, SHIMODACHIYURI-DORI SHIN-MACHI NISHIIRU, KAMIGYO-KU, KYOTO-SHI 602
JAPANESE

RINSHI JOHO 4239
(JOURNAL OF GOVERNMENT FOREST EXPERIMENT STATION, FORESTRY AGENCY)
RIN'YA-CHO RINGYO SHIKENJO (GOVERNMENT FOREST EXPERIMENT STATION, FORESTRY AGENCY)
37-21, 5-CHOME, SHIMOMEGURO, MEGURO-KU, TOKYO 153
JAPANESE

RIN'YA JIHO 4240
(CURRENT REPORT OF JAPAN FORESTRY)
RIN'YA-CHO (FORESTRY AGENCY)
AND RIN'YA KOSAIKAI
7-8, 1-CHOME, KORAKU, BUNKYO-KU, TOKYO 112
JAPANESE
INDEXED: BIB

RIN'YA SHIRYO GEPPO 4241
(MONTHLY NEWS OF FORESTRY)
RIN'YA-CHO RIN'YA SHIRYOKAN (FORESTRY MATERIALS LIBRARY, FORESTRY AGENCY)
37-21, 5-CHOME, SHIMOMEGURO, MEGURO-KU, TOKYO 153
JAPANESE

RIN'YAKUKYO NYUSU 4242
(NEWS OF FOREST CHEMICALS ASSOCIATION)
RINGYO YAKUZAI KYOKAI (FOREST CHEMICALS ASSOCIATION OF JAPAN)
2-1, 2-CHOME, OTE-MACHI, CHIYODA-KU, TOKYO 100
JAPANESE

RINYU TSUSHIN 4243
(FRIENDS OF THE FOREST NEWS)
NIHON HOAN RIN KENKYUJO (JAPAN FOREST CONSERVATION RESEARCH INSTITUTE)
JAPANESE

RINZAI ANZEN 4244
(PREVENTION OF FORESTRY LABOR ACCIDENT)
RINGYO RODO SAIGAI BOSHI KYOKAI (ASSOCIATION FOR THE PREVENTION OF FORESTRY LABOR ACCIDENT)
35-1, 5-CHOME, SHIBA, MINATO-KU, TOKYO 108
JAPANESE

RINZAI JOHO 4245
(JAPAN LUMBER GAZETTE)
NIHON RINZAI JOHOSHA (JAPAN LUMBER INTELLIGENCE COMPANY)
256, KITABESSHO, KUWANA-SHI 511
JAPANESE

RINZAI SHINBUN 4246
(LUMBER NEWS)
RINZAI SHINBUNSHA
7-4, 2-CHOME, FUKAGAWA, KOTO-KU, TOKYO 135
JAPANESE

ROKU ENSHURIN KYODO SHIKEN SHIRYO 4247
(JOINT EXPERIMENTAL DATA OF THE SIX EXPERIMENTAL FORESTS)
KYUSHU DAIGAKU ENSHURIN KENKYUJO (KYUSHU UNIVERSITY EXPERIMENTAL FOREST RESEARCH DIVISION)
JAPANESE

SAGA-KEN RINGYO SHIKENJO GYOMU HOKOKUSHO 4248
(REPORT ON ACTIVITY OF SAGA PREFECTURE FOREST EXPERIMENT STATION)
SAGA-KEN RINGYO SHIKENJO (SAGA PREFECTURE FOREST EXPERIMENT STATION)
3408, IKENOUE, YAMATO-CHO, SAGA-GUN, SAGA 840-02
JAPANESE

SAITAMA-KEN RINGYO SHIKENJO GYOMU SEISEKI HOKOKU 4249
(REPORT OF SAITAMA-KEN FOREST EXPERIMENT STATION)
SAITAMA-KEN RINGYO SHIKENJO (SAITAMA-KEN FOREST EXPERIMENT STATION)
2609, HACHIGATA, YORII-MACHI, OSATO-GUN, SAITAMA 369-12
JAPANESE

SANRIN 4250
(FORESTRY)
DAINIHON SANRINKAI (JAPAN FORESTRY ASSOCIATION)
9-13, 1-CHOME, AKASAKA, MINATO-KU, TOKYO 107
JAPANESE

SAPPORO EIRIN-KYOKU JIGYO TOKEISHO 4251
(OPERATION STATISTICS OF SAPPORO REGIONAL FOREST OFFICE)
RIN'YA-CHO SAPPORO EIRIN-KYOKU (SAPPORO REGIONAL FOREST OFFICE, FORESTRY AGENCY)
NISHI-1-CHOME, KITA-2-JO, CHUO-KU, SAPPORO-SHI 060
JAPANESE

SAPPORO RIN'YU 4252
(FORESTRY IN HOKKAIDO)
RIN'YA-CHO SAPPORO EIRIN-KYOKU (SAPPORO REGIONAL FOREST OFFICE, FORESTRY AGENCY)
NISHI-1-CHOME, KITA-2-JO, CHUO-KU, SAPPORO-SHI 060
JAPANESE

SHIMANE DAIGAKU NOGAKUBU ENSHURIN HOKOKU 4253
(BULLETIN OF THE SHIMANE UNIVERSITY FORESTS)
SHIMANE DAIGAKU NOGAKUBU FUZOKU ENSHURIN (SHIMANE UNIVERSITY FORESTS)
1060, NISHIKAWAZU-CHO, MATSUE-SHI 690
JAPANESE
INDEXED: CHE, JICST

SHIMANE-KEN RINGYO SHIKENJO KENKYU HOKOKU 4254
(BULLETIN OF SHIMANE PREFECTURE FOREST EXPERIMENT STATION)
SHIMANE-KEN RINGYO SHIKENJO
1586, SHINJI, SHINJI-CHO, YATSUKA-GUN, SHIMANE 699-04
JAPANESE
INDEXED: CHE

SHINANO NO MIDORI 4255
(FORESTRY IN SHINANO DISTRICT)
NAGANO-KEN RYOKKA RENMEI (TREE-PLANTING FEDERATION IN NAGANO PREFECTURE)
C/O NAGANO KENCHO RINMU-BU CHISANKA, MINAMI NAGANO, NAGANO-SHI 380
JAPANESE

SHINRIN 4256
(FOREST)
RYOKUCHI KENKYUKAI HEN; DOI RINGAKU SHINKOKAI (GREENERY RESEARCH ASSOCIATION; DOI FORESTRY PROMOTION ASSOCIATION)
JAPANESE

SHINRIN BOEKI 4257
(FOREST PROTECTION)
RIN'YA-CHO (FORESTRY AGENCY)
AND ZENKOKU SHINRIN BYOCHU JUGAI BOJO KYOKAI
C/O ZENKOKU CHOSON KAIKAN, 1-17, NAGATA-CHO, CHIYODA-KU, TOKYO 100
JAPANESE
INDEXED: BIB

SHINRIN BYOGAICHUTO HIGAI HOKOKU 4258
(ANNUAL REPORT ON THE DAMAGE CAUSED BY FOREST PESTS IN JAPAN)
RIN'YA-CHO (FORESTRY AGENCY)
2-1, 1-CHOME, KASUMIGASEKI, CHIYODA-KU, TOKYO 100
JAPANESE

SHINRIN KOSOKU 4259
(FOREST AERIAL SURVEY)
NIHON RINGYO GIJUTSU KYOKAI (JAPAN FOREST TECHNICAL ASSOCIATION)
7, ROKUBAN-CHO, CHIYODA-KU, TOKYO 102
JAPANESE
INDEXED: JICST

SHINRIN KUMIAI 4260
(NEWS OF FORESTRY ASSOCIATION)
ZENKOKU SHINRIN KUMIAI RENGOKAI
1-12, 1-CHOME, UCHIKANDA, CHIYODA-KU, TOKYO 101
JAPANESE

SHINRIN RITCHI 4261
(FOREST AND ENVIRONMENT)
SHINRIN RICCHI KONWAKAI (LOCATION OF FORESTRY MEETING
C/O RINGYO SHIKENJO DOJO-BU, 37-21, 5-CHOME, SHIMOMEGURO, MEGURO-KU, TOKYO 153
JAPANESE
INDEXED: JICST

SHINRIN TO HIBAI 4262
(FOREST AND FERTILIZATION)
NIHON RINCHI HIBAI KYOKAI (FOREST FERTILIZATION ASSOCIATION OF JAPAN)
C/O SHINRIN SHIGEN SOGO TAISAKU KYOGIKAI, SHIN OTE-MACHI BUILDING, 2-1, 2-CHOME, OTE-MACHI, CHIYODA-KU, TOKYO 100
JAPANESE

SHINSHU DAIGAKU NOGAKUBU ENSHURIN HOKOKU 4263
(BULLETIN OF THE SHINSHU UNIVERSITY FORESTS)
SHINSHU DAIGAKU NOGAKUBU FUZOKU ENSHURIN (FACULTY OF AGRICULTURE, SHINSHU UNIVERSITY FORESTS)
INA P.O. BOX 1, INA-SHI 396
JAPANESE, WITH EUROPEAN SUMMARY
INDEXED: CHE

SHIZUOKA DAIGAKU NOGAKUBU ENSHURIN HOKOKU 4264
(BULLETIN OF THE SHIZUOKA UNIVERSITY FORESTS)
SHIZUOKA DAIGAKU NOGAKUBU FUZOKU ENSHURIN (SHIZUOKA UNIVERSITY FORESTS)
4300, MITSUKE, IWATA-SHI 438
JAPANESE, WITH ENGLISH SUMMARY

SHIZUOKA-KEN RINGYO SHIKENJO GYOMU SEISEKI HOKOKUSHO 4265
(ANNUAL REPORT OF THE SHIZUOKA PREFECTURE FOREST EXPERIMENT STATION)
SHIZUOKA-KEN RINGYO SHIKENJO (SHIZUOKA PREFECTURE FOREST EXPERIMENT STATION)
2542-8, NEGATA, HAMAKITA-SHI 434
JAPANESE

SHIZUOKA-KEN RINGYO SHIKENJO KENKYU CHOSA SHIRYO 4266
(RESEARCH AND INVESTIGATION INFORMATION OF SHIZUOKA PREFECTURE FORESTRY EXPERIMENT STATION)
SHIZUOKA-KEN RINGYO SHIKENJO
2542-8, NEGATA, HAMAKITA-SHI 434
JAPANESE

SHIZUOKA-KEN RINGYO SHIKENJO KENKYU HOKOKU 4267
(RESEARCH REPORT OF THE SHIZUOKA PREFECTURE FORESTRY EXPERIMENT STATION)
SHIZUOKA-KEN RINGYO SHIKENJO (SHIZUOKA PREFECTURE FORESTRY EXPERIMENT STATION)
2542-8, NEGATA, HAMAKITA-SHI 434
JAPANESE AND ENGLISH
INDEXED: CHE

SHUKAKU SHIKENCHI CHOSA HOKOKU 4268
(REPORT OF INVESTIGATION FOR TESTED SOIL)
RIN'YA-CHO RINGYO SHIKENJO (GOVERNMENT FOREST EXPERIMENT STATION, FORESTRY AGENCY)
37-21, 5-CHOME, SHIMOMEGURO, MEGURO-KU, TOKYO 153
JAPANESE

SORIN 4269
(GREEN FOREST)
RIN'YA-CHO AKITA EIRIN-KYOKU (AKITA REGIONAL FOREST OFFICE, FORESTRY AGENCY)
RIN'YA KOSAI-KAI AKITA-SHIBU
9-16, 5-CHOME, NAKA-DORI, AKITA-SHI 010
JAPANESE

SOZAI SEISANHI DOKO CHOSA KEKKA HOKOKU 4270
(COST STATISTICS OF LUMBER PRODUCTION)
RIN'YA-CHO RINSEI-BU (FORESTRY AGENCY)
2-1, 1-CHOME, KASUMIGASEKI, CHIYODA-KU, TOKYO 100
JAPANESE

TEKUNIKARU NOTO 4271
(TECHNICAL NOTE (OJI PAPER COMPANY, LIMITED))
OJI SEISHI K.K. RINBOKU IKUSHU KENKYUSHO (OJI INSTITUTE FOR FOREST TREE IMPROVEMENT, OJI PAPER COMPANY, LIMITED)
90, YUJI, KURIYAMA-CHO, YUBARI-GUN, HOKKAIDO 069-15
JAPANESE

TOCHIGI-KEN RINGYO SENTA GYOMU HOKOKU 4272
(REPORT OF TOCHIGI PREFECTURE FORESTRY CENTER)
TOCHIGI-KEN RINGYO SENTA (TOCHIGI PREFECTURE FORESTRY CENTER)
280, SHIMOKOIKE-CHO, UTSUNOMIYA-SHI 321-22
JAPANESE

TOCHIGI-KEN RINGYO SENTA KENKYU HOKOKU 4273
(BULLETIN OF THE TOCHIGI PREFECTURE FORESTRY CENTER)
TOCHIGI-KEN RINGYO SENTA (TOCHIGI PREFECTURE FORESTRY CENTER)
280, SHIMOKOIKE-CHO, UTSUNOMIYA-SHI 321-11
JAPANESE

TOCHIGI-KEN RINGYO SENTA NENPO 4274
(ANNUAL REPORT OF TOCHIGI PREFECTURE FORESTRY CENTER)
TOCHIGI-KEN RINGYO SENTA (TOCHIGI PREFECTURE FORESTRY CENTER)
280, SHIMOKOIKE-CHO, UTSUNOMIYA-SHI 321-22
JAPANESE

TOHOKU NO RINBOKU IKUSHU 4275
(FOREST TREE IMPROVEMENT IN TOHOKU DISTRICT)
RIN'YA-CHO TOHOKU RINBOKU IKUSHUJO (TOHOKU FOREST TREE IMPROVEMENT STATION, FORESTRY AGENCY)
TAKIZAWA, TAKIZAWA-MURA, IWATE-GUN, IWATE 020-01
JAPANESE

TOHOKU RINBOKU IKUSHUJO NENPO 4276
(ANNUAL REPORT OF THE TOHOKU FOREST TREE IMPROVEMENT STATION, FORESTRY AGENCY AND FORESTRY))
RINYA-CHO TOHOKU RINBOKU IKUSHUJO (TOBOKU FOREST TREE IMPROVEMENT STATION, MINISTRY OF AGRICULTURE AND FORESTRY)
TAKIZAWA, TAKIZAWA-MURA, IWATE-GUN, IWATE 020-01
JAPANESE

TOKUSHIMA-KEN RINGYO SHIKENJO KENKYU HOKOKU 4277
(REPORT OF TOKUSHIMA PREFECTURE FOREST EXPERIMENT STATION)
TOKUSHIMA-KEN RINGYO SHIKENJO
5-69, MINAMISHO-MACHI, TOKUSHIMA-SHI 770
JAPANESE AND ENGLISH

TOKYO (EIRIN-KYOKU KOHO) 4278
(INFORMATION OF TOKYO REGIONAL FOREST OFFICE)
RIN'YA-CHO TOKYO EIRIN-KYOKU KOHO-SHITSU (PUBLIC INFORMATION OFFICE, TOKYO REGIONAL FOREST OFFICE, FORESTRY AGENCY)
24-6, 2-CHOME, KAMIOSAKI, SHINAGAWA-KU, TOKYO 141
JAPANESE

TOKYO EIRIN-KYOKU JIGYO TOKEISHO 4279
(ANNUAL STATISTICS OF TOKYO REGIONAL FOREST OFFICE)
RIN'YA-CHO TOKYO EIRIN-KYOKU (TOKYO REGIONAL FOREST OFFICE, FORESTRY AGENCY)
24-6, 2-CHOME, KAMIOSAKAI, SHINAGAWA-KU, TOKYO 141
JAPANESE

TOKYO NOKO DAIGAKU NOGAKUBU ENSHURIN HOKOKU 4280
(BULLETIN OF THE EXPERIMENT FORESTS, TOKYO UNIVERSITY OF AGRICULTURE AND TECHNOLOGY)
TOKYO NOKO DAIGAKU NOGAKUBU FUZOKU ENSHURIN (THE EXPERIMENT FORESTS, TOKYO UNIVERSITY OF AGRICULTURE AND TECHNOLOGY)
SAIWAI-CHO, FUCHU-SHI, TOKYO 183
JAPANESE AND ENGLISH
INDEXED: BIB, CHE, JICST

TOKYO RIN'YU 4281
(FORESTRY IN TOKYO)
RIN'YA-CHO TOKYO EIRIN-KYOKU (TOKYO REGIONAL FOREST OFFICE, FORESTRY AGENCY)
TOKYO RIN'YA KYOSAIKAI
24-6, 2-CHOME, KAMIOSAKI, SHINAGAWA-KU, TOKYO 141
JAPANESE

TOKYO-TO NO RINGYO 4282
(FORESTRY STATISTICS IN TOKYO)
TOKYO-TO KEIZAI-KYOKU NORIN RYOKUSEI-BU (AGRICULTURE, FORESTRY AND TREE-PLANTING ADMINISTRATION DEPARTMENT, ECONOMIC BUREAU, TOKYO METROPOLITAN GOVERNMENT)
8-1, 3-CHOME, MARUNOUCHI, CHIYODA-KU, TOKYO 100
JAPANESE

TOTTORI DAIGAKU NOGAKUBU ENSHURIN HOKOKU 4283
(BULLETIN OF THE TOTTORI UNIVERSITY FORESTS)
TOTTORI DAIGAKU NOGAKUBU ENSHURIN (THE TOTTORI UNIVERSITY FORESTS)
1-1, KOYAMA-CHO, TOTTORI-SHI 680
JAPANESE AND EUROPEAN
INDEXED: BIO, CHE

TOYAMA-KEN RINGYO SHIKENJO GYOMU HOKOKUSHO 4284
(REPORT OF TOYAMA PREFECTURE FOREST EXPERIMENT STATION)
TOYAMA-KEN RINGYO SHIKENJO (TOYAMA PREFECTURE FOREST EXPERIMENT STATION)
YOSHIMINE, TATEYAMA-MACHI, NAKANIIKAWA-GUN, TYOAMA 930-13
JAPANESE

UTSUNOMIYA DAIGAKU NOGAKUBU ENSHURIN HOKOKU 4285
(BULLETIN OF THE UTSUNOMIYA UNIVERSITY FORESTS)
UTSUNOMIYA DAIGAKU NOGAKUBU (FACULTY OF AGRICULTURE, UTSUNOMIYA UNIVERSITY)
7556, FUNYU, SHIOYA, SHIOYA-MACHI, SHIOYA-GUN, TOCHIGI 329-29
JAPANESE AND ENGLISH

WAKAYAMA-KEN RINGYO SHIKENJO GYOMU SEISEKI HOKOKUSHO 4286
(REPORT OF WAKAYAMA PREFECTURE FORESTRY EXPERIMENT STATION)
WAKAYAMA-KEN RINGYO SHIKENJO
TAKAIKE, KOZAGAWA-CHO, HIGASHIMURO-GUN, WAKAYAMA 649-41
JAPANESE

WOOD RESEARCH 4287
KYOTO DAIGAKU MOKUZAI KENKYUSHO (WOOD RESEARCH INSTITUTE, KYOTO UNIVERSITY)
GOKANOSHO, UJI 611
ENGLISH
INDEXED: BIB, CHE, JICST

YAMAGATA-KEN RINGYO SHIKENJO KENKYU HOKOKU 4288
(RESEARCH REPORT OF YAMAGATA-KEN FOREST EXPERIMENT STATION)
YAMAGATA-KEN RINGYO SHIKENJO (YAMAGATA-KEN FOREST EXPERIMENT STATION)
2707, SAGAE, SAGAE-SHI 991
JAPANESE

YAMAGATA-KEN RINGYO TOKEI 4289
(FORESTRY STATISTICS, YAMAGATA PREFECTURE)
YAMAGATA-KEN NORIN-BU (AGRICULTURE AND FORESTRY DEPARTMENT, YAMAGATA PREFECTURAL GOVERNMENT)
4-51, 3-CHOME, HATAGO-CHO, YAMAGATA-SHI 990
JAPANESE

YAMAGUCHI-KEN RINGYO SHIKENJO GYOMU NENPO 4290
(REPORT ON ACTIVITIES OF YAMAGUCHI PREFECTURAL FOREST EXPERIMENT STATION)
YAMAGUCHI-KEN RINGYO SHIKENJO (YAMAGUCHI PREFECTURAL FOREST EXPERIMENT STATION)
2538, MIYANOKAMI, YAMAGUCHI-SHI 753
JAPANESE

YAMAGUCHI-KEN RINGYO SHIKENJO HOKOKU 4291
(REPORT OF YAMAGUCHI PREFECTURE FORESTRY EXPERIMENT STATION)
JAPANESE

YAMAGUCHI-KEN RINGYO SHIKENJO SHIKEN HOKOKU 4292
(RESEARCH REPORT OF YAMAGUCHI PREFECTURAL FOREST EXPERIMENT STATION)
YAMAGUCHI-KEN RINGYO SHIKENJO (YAMAGUCHI PREFECTURAL FOREST EXPERIMENT STATION)
2538, MIYANOKAMI, YAMAGUCHI-SHI 753
JAPANESE

YAMANAMI 4293
(MOUNTAIN CHAIN)
RIN'YA-CHO MAEBASHI EIRIN-KYOKU (MAEBASHI NATIONAL FOREST REGIONAL OFFICE, FORESTRY AGENCY)
16-25, 4-CHOME, IWAGAMI-MACHI, MAEBASHI-SHI 371
JAPANESE

YAMANASHI-KEN RINGYO SHIKENJO HOKOKU 4294
(BULLETIN OF THE YAMANASHI PREFECTURAL FOREST EXPERIMENT STATION)
YAMANASHI-KEN RINGYO SHIKENJO (YAMANASHI PREFECTURAL FOREST EXPERIMENT STATION)
688, TSUTSUJIGASAKI, IWAKUBO-CHO, KOFU-SHI 400
JAPANESE AND ENGLISH
INDEXED: BIB

YAMANASHI-KEN RINGYO SHIKENJO JIGYO HOKOKU 4295
(REPORT OF YAMANASHI PREFECTURAL FOREST EXPERIMENT STATION)
YAMANASHI-KEN RINGYO SHIKENJO
688, TSUTSUJIGASAKI, IWAKUBO-CHO, KOFU-SHI 400
JAPANESE

YAMANASHI-KEN RINGYO TOKEISHO 4296
(FORESTRY STATISTICS OF YAMANASHI PREFECTURE)
YAMANASHI-KEN RIN'YA-BU (FORESTRY DEPARTMENT, YAMANASHI PREFECTURAL GOVERNMENT)
1-CHOME, MARUNOUCHI, KOFU-SHI 400
JAPANESE

ZEN SHIN REN JIHO 4297
(ALL FORESTRY COOPERATIVES REVIEW)
JAPANESE
INDEXED: BIB

ZENKOKU SHU-UNZAI KASEN GISHI KYOKAIHO 4298
(BULLETIN OF THE JAPAN WOOD TRANSPORTATION AND WIRING ASSOCIATION)
ZENKOKU SHUUNZAI KASEN GISHI KYOKAI (JAPAN WOOD TRANSPORTATION AND WIRING ASSOCIATION)
C/O RINGYO RODO SAIGAI BOSHI KYOKAI, 35-1, 5-CHOME, SHIBA, MINATO-KU, TOKYO 108
JAPANESE

ZENMOKUREN JIHO 4299
(NEWS OF ALL JAPAN FEDERATION OF LUMBER CORPORATION)
ZENKOKU MOKUZAI KUMIAI RENGOKAI (ALL JAPAN FEDERATION OF LUMBER CORPORATION)
5-4, 2-CHOME, FUKAGAWA, KOTO-KU, TOKYO 135
JAPANESE

ZORIN JIGYO JISSEKI 4300
(OPERATION STATISTICS OF AFFORESTATION)
HOKKAIDO RINMU-BU (FORESTRY ADMINISTRATION DEPARTMENT, HOKKAIDO PREFECTURAL GOVERNMENT)
NISHI-6-CHOME, KITA-3-JO, CHUO-KU, SAPPORO-SHI 060
JAPANESE

ZORIN JIKKEN EIRINSHO KENKYU HOKOKU 4301
(RESEARCH REPORT OF AFFORESTATION EXPERIMENT)
RIN'YA-CHO KOCHI EIRIN-KYOKU (KOCHI REGIONAL FOREST OFFICE, FORESTRY AGENCY)
9, MARUNOUCHI, KOCHI-SHI 780
JAPANESE

ZORIN TSUSHIN 4302
(REFORESTATION AND REPRODUCTION)
RIN'YA-CHO MAEBASHI EIRIN-KYOKU (MAEBASHI REGIONAL FOREST OFFICE, FORESTRY AGENCY)
16-25, 4-CHOME, IWAGAMI-CHO, MAEBASHI-SHI 371
JAPANESE

Horticulture, Gardens, Gardening

AICHI-KEN ENGEI SHIKENJO KENKYU HOKOKU 4303
(RESEARCH REPORT OF AICHI PREFECTURE HORTICULTURAL EXPERIMENT STATION)
JAPANESE
INDEXED: JICST

AICHI-KEN NOGYO SOGO SHIKENJO KENKYU HOKOKU, B, ENGEI 4304
(RESEARCH BULLETIN OF THE AICHI-KEN AGRICULTURAL RESEARCH CENTER, SERIES B)
AICHI-KEN NOGYO SOGO SHIKENJO (AICHI-KEN AGRICULTURAL RESEARCH CENTER)
NAGAKUTE-CHO, AICHI-GUN, AICHI 480-11
JAPANESE, WITH ENGLISH SUMMARY
INDEXED: CHE

AICHI-KEN NOGYO SOGO SHIKENJO TOKUBETSU KENKYU HOKOKU, B, ENGEI 4305
(SPECIAL RESEARCH BULLETIN OF THE AICHI-KEN AGRICULTURAL RESEARCH CENTER, B)
AICHI-KEN NOGYO SOGO SHIKENJO (AICHI-KEN AGRICULTURAL RESEARCH CENTER)
NAGAKUTE-CHO, AICHI-GUN, AICHI 480-11
JAPANESE, WITH ENGLISH SUMMARY

BAREISHO GENGENSHU NOJO CHOSA KENKYU HOKOKU 4306
(RESEARCH BULLETIN OF THE NATIONAL POTATO FOUNDATION STOCK SEED FARM)
NORIN-SHO TSUMAGOI BAREISHO GENGENSHU NOJO (NATIONAL POTATO FOUNDATION STOCK SEED FARM)
1017-1, TASHIRO, TSUMAGOI-MURA, AZUMA-GUN, GUNMA 377-16
JAPANESE

BULLETIN OF THE NAGANO HORTICULTURAL EXPERIMENT STATION 4307
JAPANESE AND ENGLISH
INDEXED: BIB

CHIBA DAIGAKU ENGEIGAKUBU GAKUJUTSU HOKOKU 4308
(TECHNICAL BULLETIN OF THE FACULTY OF HORTICULTURE, CHIBA UNIVERSITY)
CHIBA DAIGAKU ENGEIGAKUBU (FACULTY OF HORTICULTURE, CHIBA UNIVERSITY)
648, TOJO, MATSUDO-SHI 271
JAPANESE AND ENGLISH
INDEXED: BIB, CHE, EXC

CHIBA DAIGAKU ENGEIGAKUBU TOKUBETSU HOKOKU 4309
(TRANSACTIONS OF FACULTY OF HORTICULTURE, CHIBA UNIVERSITY)
CHIBA DAIGAKU ENGEIGAKUBU (FACULTY OF HORTICULTURE, CHIBA UNIVERSITY)
648, TOJO, MATSUDO-SHI 271
JAPANESE AND ENGLISH
INDEXED: BIB, BIO, CHE

CHIBA-KEN DANCHI ENGEI SHIKENJO KENKYU HOKOKU 4310
(REPORT OF CHIBA HORTICULTURAL EXPERIMENT STATION)
CHIBA-KEN DANCHI ENGEI SHIKENJO (CHIBA HORTICULTURAL EXPERIMENT STATION)
1762, YAMAMOTO, TATEYAMA-SHI 294
JAPANESE

CHOSA SEISEKISHO (NORIN-SHO SHIRIBESHI BAREISHO GENGENSHU NOJO 4311
(INVESTIGATION REPORT OF THE SHIRIBESHI NATIONAL POTATO FOUNDATION STOCK SEED FARM)
NORIN-SHO SHIRIBESHI BAREISHO GENGENSHU NOJO
MAKARI-MURA, ABUTA-GUN, HOKKAIDO 048-16
JAPANESE

ENGEI GAIDO 4312
(GARDENING GUIDE)
SHUFU NO TOMOSHA
1-6, KANDA SURUGADAI, CHIYODA-KU, TOKYO 101
JAPANESE

ENGEI GAKKAI ZASSHI 4313
(JOURNAL OF THE JAPANESE SOCIETY FOR HORTICULTURAL SCIENCE)
ENGEI GAKKAI (JAPANESE SOCIETY FOR HORTICULTURAL SCIENCE)
C/O TOKYO DAIGAKU NOGAKUBU ENGEI DAI-1, KYOSHITSU, 1-1, 1-CHOME, YAYOI, BUNKYO-KU, TOKYO 113
JAPANESE AND ENGLISH
INDEXED: BIB, CHE, EXC, JICST

ENGEI SHIKENJO HOKOKU A - HIRATSUKA 4314
(BULLETIN OF THE HORTICULTURAL RESEARCH STATION (MINISTRY OF AGRICULTURE AND FORESTRY) SERIES A (HIRATSUKA))
HORTICULTURAL RESEARCH STATION, MINISTRY OF AGRICULTURE AND FORESTRY
HIRATSUKA, KANAGAWA 309
JAPANESE
INDEXED: BIB, BIO

ENGEI SHIKENJO HOKOKU B - OKITSU 4315
(BULLETIN OF THE HORTICULTURAL RESEARCH STATION (MINISTRY OF AGRICULTURE AND FORESTRY) SERIES B (OKITSU))
HORTICULTURAL RESEARCH STATION, MINISTRY OF AGRICULTURE AND FORESTRY
SHIMIZU, OKITSU, SHIZUOKA 309
JAPANESE
INDEXED: BIO

ENGEI SHIKENJO HOKOKU C - MORIOKA 4316
(BULLETIN OF THE HORTICULTURAL RESEARCH STATION (MINISTRY OF AGRICULTURE AND FORESTRY) SERIES C (MORIOKA))
MORIOKA BRANCH, FRUIT TREE RESEARCH STATION
SHIMOKURIYAGAWA, MORIOKA 309
JAPANESE
INDEXED: BIO

ENGEI SHIN CHISHIKI 4317
(NEW INFORMATION ON HORTICULTURE)
TAKII SHUBYO K.K. SHUPPAN-BU (TAKII SHUBYO COMPANY, LIMITED, PUBLISHING DIVISION)
180, INOKUMA HIGASHIIRU, UMEKOJI, SHIMOGYO-KU, KYOTO-SHI 600
JAPANESE
INDEXED: BIB

ENGEI SHIN CHISHIKI (HANA NO GO) 4318
(NEW INFORMATION ON HORTICULTURE, THE PART OF FLOWER))
TAKII SHUBYO K.K. SHUPPAN-BU (TAKII SHUBYO COMPANY, LIMITED, PUBLISHING DIVISION)
180, INOKUMA HIGASHIIRU, UMEKOJI, SHIMOGYO-KU, KYOTO-SHI 600
JAPANESE
INDEXED: BIB

ENGEI SHUNJU 4319
(JOURNAL OF GARDENING)
KYOTO ENGEI KURABU (KYOTO GARDENING CLUB)
C/O KYOTO-FURITSU SHOKUBUTSUEN, SHIMOKAMO HANGI-CHO, SAKYO-KU, KYOTO-SHI 606
JAPANESE

ENGEI TSUSHIN 4320
(HORTICULTURAL TIMES)
SAKATA SHUBYO K.K. SHUPPAN-BU
2, KIRIHATA, KANAGAWA-KU, YOKOHAMA-SHI 221
JAPANESE

FUKUOKA-KENRITSU ENGEI SHIKENJO KENKYU HOKOKU 4321
(BULLETIN OF THE FUKUOKA HORTICULTURAL EXPERIMENT STATION)
FUKUOKA-KENRITSU ENGEI SHIKENJO (FUKUOKA HORTICULTURAL EXPERIMENT STATION)
571, KASHIWAHARA, MINAMI-KU, FUKUOKA-SHI 814
JAPANESE, WITH ENGLISH SUMMARY

FUKUSHIMA-KEN ENGEI SHIKENJO KENKYU HOKOKU 4322
(REPORT OF FUKUSHIMA HORTICULTURAL EXPERIMENT STATION)
FUKUSHIMA-KEN ENGEI SHIKENJO (FUKUSHIMA HORTICULTURAL EXPERIMENT STATION)
1, DANNOHIGASHI, HIRANO, IZAKA-MACHI, FUKUSHIMA-SHI 960-02
JAPANESE
INDEXED: JICST

GADEN RAIFU 4323
(GARDEN LIFE)
SEIBUNDO SHINKOSHA
1-5, KANDA NISHIKI-CHO, CHIYODA-KU, TOKYO 101
JAPANESE

GEKKAN SATSUKI KENKYU 4324
(MONTHLY STUDY OF AZALEA)
NIHON SATSUKI BUNKA SHINKOKAI AND GEKKAN SATSUKI KENKYUSHA
1724, TENJIN-CHO, KANUMA-SHI 322
JAPANESE

IBARAKI-KEN ENGEI SHIKENJO KENKYU HOKOKU 4325
(BULLETIN OF THE IBARAKI HORTICULTURAL EXPERIMENT STATION)
IBARAKI-KEN ENGEI SHIKENJO (IBARAKI HORTICULTURAL EXPERIMENT STATION)
AMI-MACHI, INASHIKI-GUN, IBARAKI 300-03
JAPANESE, WITH ENGLISH SUMMARY

IWATE-KEN ENGEI SHIKENJO GYOMU NENPO 4326
(ANNUAL REPORT OF THE IWATE PREFECTURE HORTICULTURAL EXPERIMENT STATION)
IWATE-KEN ENGEI SHIKENJO
NARITA, IITOYO-CHO, KITAKAMI-SHI 024
JAPANESE

KAKIJOHO 4327
(INFORMATION OF FLOWERING PLANT)
NORIN-SHO NOSAN ENGEI-KYOKU (MINISTRY OF AGRICULTURE AND FORESTRY)
NIHON KAKI SEISAN KYOKAI
2-1, 1-CHOME, KASUMIGASEKI, CHIYODA-KU, TOKYO 100
JAPANESE

KANAGAWA-KEN ENGEI SHIKENJO GYOMU NENPO 4328
(ANNUAL REPORT OF THE KANAGAWA HORTICULTURAL EXPERIMENT STATION)
KANAGAWA-KEN ENGEI SHIKENJO (KANAGAWA HORTICULTURAL EXPERIMENT STATION)
1217, NINOMIYA, NINOMIYA-MACHI, NAKA-GUN, KANAGAWA 259-01
JAPANESE

KANAGAWA-KEN ENGEI SHIKENJO KENKYU HOKOKU 4329
(BULLETIN OF KANAGAWA HORTICULTURAL EXPERIMENT STATION)
KANAGAWA-KEN ENGEI SHIKENJO (KANAGAWA HORTICULTURAL EXPERIMENT STATION)
1217, NINOMIYA, NINOMIYA-MACHI, NAKA-GUN, KANAGAWA 259-01
JAPANESE, WITH ENGLISH SUMMARY
INDEXED: BIB, BIO, JICST

KUMA NOKEN SHIKEN SEISEKISHO, YASAI 4330
(SURVEY RESULT OF KUMA AGRICULTURAL EXPERIMENT STATION, VEGETABLE)
KUMAMOTO-KEN KUMA NOGYO KENKYU SHIDOSHO (KUMA AGRICULTURAL EXPERIMENT STATION)
2248-16, UE-MURA, KUMA-GUN, KUMAMOTO 868-04
JAPANESE

KYOTO ENGEI 4331
(KYOTO HORTICULTURE)
KYOTO ENGEI KURABU (KYOTO HORTICULTURAL CLUB)
JAPANESE

NAGANO-KEN ENGEI SHIKENJO HOKOKU 4332
(REPORT OF THE NAGANO HORTICULTURAL EXPERIMENT STATION)
NAGANO-KEN ENGEI SHIKENJO (NAGANO HORTICULTURAL EXPERIMENT STATION)
492, OGAWARA, SUZAKA-SHI 382
JAPANESE AND ENGLISH

NIIGATA-KEN ENGEI SHIKENJO KENKYU HOKOKU 4333
(BULLETIN OF THE NIIGATA HORTICULTURAL EXPERIMENT STATION)
NIIGATA-KEN ENGEI SHIKENJO (NIIGATA HORTICULTURAL EXPERIMENT STATION)
IKEDA, NANOKA-MACHI, NIITSU-SHI 956
JAPANESE, WITH ENGLISH SUMMARY

NIIGATA-KEN ENGEI SHIKENJO TOKUBETSU HOKOKU 4334
(TRANSACTIONS OF THE NIIGATA HORTICULTURAL EXPERIMENT STATION)
NIIGATA-KEN ENGEI SHIKENJO (NIIGATA HORTICULTURAL EXPERIMENT STATION)
IKEDA, NANOKA-MACHI, NIITSU-SHI 956
JAPANESE

NIPPON MAMERUI SHIRYO 4335
(BEANS NEWS)
NIPPON MAMERUI KIKIN KYOKAI
8-3, 1-CHOME, OTE-MACHI, CHIYODA-KU, TOKYO 100
JAPANESE

NIWA (BESSATSU) 4336
(GARDEN (SPECIAL ISSUE))
KENCHIKU SHIRYO KENKYUSHA (ARCHITECTURAL DATA RESEARCH PUBLISHER)
JAPANESE

NOGYO KANSOKU TOKEIHYO ENGEI OYOBI KOGEI SAKUMOTSU 4337
(STATISTICAL TABLE OF THE OUTLOOK OF AGRICULTURE, HORTICULTURAL CROPS AND SPECIAL CROPS)
FUKUOKA-KEN NOSEI-BU (FUKUOKA PREFECTURE AGRICULTURAL ADMINISTRATION DIVISION)
1-1, 1-CHOME, TENJIN, FUKUOKA-SHI 810
JAPANESE

RAN 4338
(JAPAN ORCHID SOCIETY BULLETIN)
NIHON RAN KYOKAI (JAPAN ORCHID SOCIETY)
C/O TAKEDA YAKUHIN KOGYO K.K. HISHOSHITSU,
2-27, DOSHO-MACHI, HIGASHI-KU, OSAKA-SHI 541
JAPANESE

RYOKUCHIGAKU KENKYU 4339
(BULLETIN OF THE INSTITUTE OF LANDSCAPE ARCHITECTURE)
TOKYO DAIGAKU NOGAKUBU ENGEIGAKU DAI-2 (RYOKUCHIGAKU) KENKYUSHITSU (LABORATORY OF LANDSCAPE ARCHITECTURE, THE UNIVERSITY OF TOKYO)
1-1, 1-CHOME, YAYOI, BUNKYO-KU, TOKYO 113
JAPANESE AND ENGLISH

SABOTEN NIHON 4340
(SUCCULENTARUM JAPONICA)
JAPANESE

SAITAMA-KEN ENGEI SHIKENJO KENKYU HOKOKU 4341
(BULLETIN OF THE SAITAMA HORTICULURAL EXPERIMENT STATION)
SAITAMA-KEN ENGEI SHIKENJO (SAITAMA HORTICULTURAL EXPERIMENT STATION)
ROKUMANBU, KUKI-SHI 346
JAPANESE, WITH ENGLISH SUMMARY
INDEXED: CHE, JICST

SAITAMA-KEN ENGEI SHIKENJO NENPO 4342
(ANNALS OF THE SAITAMA HORTICULTURAL EXPERIMENTAL STATION)
SAITAMA-KEN ENGEI SHIKENJO (SAITAMA HORTICULTURAL EXPERIMENTAL STATION)
ROKUMANBU, KUKI-SHI 346
JAPANESE

SHIBAKUSA KENKYU 4343
(JOURNAL OF JAPAN TURF GRASS RESEARCH ASSOCIATION)
NIHON SHIBAKUSA KENKYUKAI (JAPAN TURF GRASS RESEARCH ASSOCIATION)
C/O KARI KENKYUKAI, 1-1, YURAKU-CHO, CHIYODA-KU, TOKYO 100
JAPANESE

SHIN KAKI 4344
(NEW FLORAE)
NIHON KAKI ENGEI KYOKAI (JAPAN FLORICULTURE ASSOCATION)
TAKII SHUBYO K.K. SHUPPAN-BU
180, UMEKOJI INOKUMA HIGASHIIRU, SHIMOGYO-KU, KYOTO-SHI 600
JAPANESE

SHISETSU ENGEI 4345
(HOTHOUSE GARDENING)
ONSHITSU KENKYUSHA
ISHIWARA-CHO, IWATA-SHI 438
JAPANESE

SHUMI NO ENGEI 4346
(GARDENING FOR HOBBY)
NHK SABISU SENTA (NHK SERVICE CENTER)
41-1, UDAGAWA-CHO, SHIBUYA-KU, TOKYO 150
JAPANESE

SOSAI 4347
(VEGETABLES)
OKAYAMA-KEN KEIZAI NOGYO KYODO KUMIAI RENGOKAI
18-201, 9-CHOME, TOGIYA-CHO, OKAYAMA-SHI 700-91
JAPANESE

SOSAI SHIKEN SEISEKISHO (ISHIKAWA-KEN SAKYUCHI NOGYO SHIKENJO) 4348
(TEST REPORT ON VEGETABLES, ISHIKAWA PREFECTURE SAND DUNE AGRICULTURE EXPERIMENT STATION)
ISHIKAWA-KEN SAKYUCHI NOGYO SHIKENJO
6-2, UCHIHISUMI, UNOKE-MACHI, KAHOKU-GUN, ISHIKAWA 929-11
JAPANESE

TOCHIGI NO ENGEI TOKUSAN 4349
(HORTICULTURAL SPECIAL PRODUCT IN TOCHIGI)
TOCHIGI-KEN (TOCHIGI PREFECTURAL GOVERNMENT)
504, HANAWADA-CHO, UTSUNOMIYA-SHI 320
JAPANESE

TOKUSHU NOSANBUTSU JOHO 4350
(REPORT ON SPECIAL AGRICULTURAL PRODUCTS)
NORINSHO ENGEI TOKUSANKA (MINISTRY OF AGRICULTURE AND FORESTRY, HORTICULTURE AND SPECIAL PRODUCTS SECTION)
JAPANESE

TOKYO DAIGAKU NOGAKUBU ENGEI JIKKENSHO KENKYU HOKOKU 4351
(BULLETIN OF KEMIGAWA ARBORETUM, FACULTY OF
AGRICULTURE, UNIVERSITY OF TOKYO)
TOKYO DAIGAKU NOGAKUBU FUZOKU ENGEI JIKKENSHO
(KEMIGAWA ARBORETUM, UNIVERSITY OF TOKYO)
1051, HATA-MACHI, CHIBA-SHI 281
JAPANESE, WITH ENGLISH SUMMARY

TOYAMA-KEN NOGYO SHIKENJO TONAMI ENGEI BUNJO KENKYU 4352
HOKO-KU
(RESEARCH REPORT OF TONAMI GARDEN PLOT, TOYAMA
PREFECTURE AGRICULTURE EXPERIMENT STATION)
JAPANESE
INDEXED: JICST

YASAI SHIKENJO NYUSU 4353
(NEWS OF VEGETABLE AND ORNAMENTAL CROPS RESEARCH
STATION)
NORIN-SHO YASAI SHIKENJO
670, OGOSO, ISSHINDEN, TSU-SHI 514-01
JAPANESE

ZATTO NI KANSURU SHIRYO 4354
(INFORMATION OF MISCELLANEOUS BEANS)
NIHON MAMERUI KIKIN KYOKAI
8-3, 1-CHOME, OTE-MACHI, CHIYODA-KU, TOKYO 100
JAPANESE

Animal Husbandry, Stockbreeding, Animal Produce, Dairy and Other Animal Produce, Comparative Pathology, Veterinary Medicine

ACTA VETERINARIA JAPONICA 4355
NIHON DAIGAKU JUIGAKU KENKYUSHO (RESEARCH INSTITUTE FOR VETERINARY SCIENCE, NIHON UNIVERSITY)
SHIMOUMA, SETAGAYA-KU, TOKYO 154
ENGLISH

AICHI-KEN NOGYO SOGO SHIKENJO KENKYU HOKOKU, C, YOKEI 4356
(RESEARCH BULLETIN OF THE AICHI-KEN AGRICULTURAL RESEARCH CENTER, SERIES C)
AICHI-KEN NOGYO SOGO SHIKENJO (AICHI-KEN AGRICULTURAL RESEARCH CENTER)
NAGAKUTE-CHO, AICHI-GUN, AICHI 480-11
JAPANESE, WITH ENGLISH SUMMARY

AICHI-KEN NOGYO SOGO SHIKENJO KENKYU HOKOKU, E, CHIKUSAN 4357
(RESEARCH BULLETIN OF THE AICHI-KEN AGRICULTURE RESEARCH CENTER, SERIES E)
AICHI-KEN NOGYO SOGO SHIKENJO (AICHI-KEN AGRICULTURE RESEARCH CENTER)
NAGAKUTE-CHO, AICHI-GUN, AICHI 480-11
JAPANESE, WITH ENGLISH SUMMARY

AICHI-KEN SHUKEI SENTA JIGYO GAIYO 4358
(OUTLINE OF ACTIVITIES OF AICHI PREFECTURAL POULTRY BREEDING CENTER)
AICHI-KEN SHUKEI SENTA
30-6, KOBAYASHI, SASAME-CHO, ANJO-SHI 446
JAPANESE

AKITA-KEN CHIKUSAN SHIKENJO SHIKEN KENKYU SEISEKISHO 4359
(RESEARCH REPORT OF THE AKITA PREFECTURAL LIVESTOCK EXPERIMENT STATION)
AKITA-KEN CHIKUSAN SHIKENJO (AKITA PREFECTURAL LIVESTOCK EXPERIMENT STATION)
2, JINGUJI TAKANO, KAMIOKA-CHO, SENBOKU-GUN, AKITA 019-17
JAPANESE

AOMORI-KEN CHIKUSAN SHIKENJO HOKOKU 4360
(BULLETIN OF THE AOMORI ZOOTECHNICAL EXPERIMENT STATION)
AOMORI-KEN CHIKUSAN SHIKENJO (AOMORI ZOOTECHNICAL EXPERIMENT STATION)
51, BIWANO, NOHEJI, NOHEJI-MACHI, KAMIKITA-GUN, AOMORI 039-31
JAPANESE
INDEXED: JICST

AOMORI-KEN YOKEI SHIKENJO GYOMU NENPO 4361
(ANNUAL REPORT OF AOMORI PREFECTURAL POULTRY EXPERIMENT STATION)
AOMORI-KEN YOKEI SHIKENJO (AOMORI PREFECTURAL POULTRY EXPERIMENTAL STATION)
NAESHIROZAWA, GONOHE-MACHI, SANNOHE-GUN, AOMORI 039-15
JAPANESE

AOMORI-KEN YOKEI SHIKENJO SHIKEN KENKYU HOKOKU 4362
(BULLETIN OF AOMORI PREFECTURAL POULTRY EXPERIMENT STATION)
AOMORI-KEN YOKEI SHIKENJO (AOMORI PREFECTURAL POULTRY EXPERIMENT STATION)
NAESHIROZAWA, GONOHE-MACHI, SANNOHE-GUN, AOMORI 309-15
JAPANESE

AZABU JUIKA DAIGAKU KENKYU HOKOKU 4363
(BULLETIN OF THE AZABU VETERINARY COLLEGE)
AZABU JUIKA DAIGAKU (THE AZABU VETERINARY COLLEGE)
17-71, 1-CHOME, FUCHINOBE, SAGAMIHARA-SHI 229
JAPANESE AND ENGLISH
INDEXED: BIB, BIO, CHE, JICST

AZABU JUIKA DAIGAKU KYOYO KATEI KENKYU HOKOKU 4364
(RESEARCH REPORTS OF THE AZABU VETERINARY COLLEGE, GENERAL EDUCATION CURRICULUM)
JAPANESE

BULLETIN OF MEAT AND MEAT PRODUCTS 4365
ENGLISH

BUTA NO SANNIKU NORYOKU KENTEI SEISEKISHO 4366
(RESEARCH REPORT OF PIG PROGENY TESTING)
AICHI-KEN NOGYO SOGO SHIKENJO CHIKUSAN KENKYUSHO (ANIMAL HUSBANDRY RESEARCH DIVISION OF AICHI-KEN AGRICULTURAL RESEARCH CENTER)
1-1, SANGAMINE, YAZAKO NAGAKUTE-CHO, AICHI-GUN, AICHI 480-11
JAPANESE

BUTA SANNIKU NORYOKU KENTEI SEISEKI OYOBI CHOSA KENKYU HOKOKU 4367
(ANNUAL REPORT OF RESEARCH AND PIG PROGENY TESTING)
NORIN-SHO SHIRAKAWA SHUCHIKU BOKUJO IBARAKI SHIJO (IBARAKI BRANCH, SHIRAKAWA LIVESTOCK BREEDING STATION, MINISTRY OF AGRICULTURE AND FORESTRY)
FUJIGAYA, SEKIJO-MACHI, MAKABE-GUN, IBARAKI 308-01
JAPANESE

BYOSEI KANTEI GYOSEKI SHUROKU 4368
(REPORTS OF DIAGNOSIS)
AOMORI KACHIKU HOKEN EISEISHO (AOMORI LIVESTOCK HYGIENE SERVICE CENTER)
26-3, TSUKURIMICHI SAWADA, AOMORI-SHI 030
JAPANESE

CHIBA-KEN MINEOKA NYUGYU SHIKENJO GYOMU HOKOKU 4369
(REPORT OF THE CHIBA PREFECTURE MINEOKA MILK-CATTLE EXPERIMENT STATION)
CHIBA-KEN MINEOKA NYUGYU SHIKENJO (CHIBA PREFECTURE MINEOKA MILK-CATTLE EXPERIMENT STATION)
OI, MARUYAMA-MACHI, AWA-GUN, CHIBA 299-26
JAPANESE

CHIBA-KEN RAKUNO SHIKENJO KENKYUSHITSU SHIKEN SEISENKISHO 4370
(TEST REPORT OF CHIBA PREFECTURAL DAIRY EXPERIMENT STATION)
CHIBA-KEN RAKUNO SHIKENJO (CHIBA PREFECTURAL DAIRY EXPERIMENT STATION)
TAKIDAI, YACHIMATA-MACHI, INBA-GUN, CHIBA 289-11
JAPANESE

CHIBA-KEN YOKEI SHIKENJO SHIKEN KENKYU HOKOKU 4371
(RESEARCH REPORT OF THE POUTLRY EXPERIMENT STATION, CHIBA PREFECTURE)
CHIBA-KEN YOKEI SHIKENJO (CHIBA PREFECTURAL POULTRY EXPERIMENT STATION)
TAKIDAI, YACHIMATA-MACHI, INBA-GUN, CHIBA 289-11
JAPANESE

CHIBA-KEN YOTON SHIKENJO KENKYU HOKOKU 4372
(RESEARCH REPORT OF THE CHIBA PREFECTURAL SWINE EXPERIMENT STATION)
CHIBA-KEN YOTON SHIKENJO (CHIBA PREFECTURAL SWINE EXPERIMENT STATION)
TAKIDAI, YACHIMATA-MACHI, INBA-GUN, CHIBA 289-11
JAPANESE

CHIKUSAN CHIIKI KOZO KAIZEN CHOSA SHIDO JIGYO HOKOKUSHO 4373
(SURVEY OF STRUCTURE IMPROVEMENT IN LIVESTOCK FARMING AREA)
ZENKOKU NOGYO KOZO KAIZEN KYOKAI (JAPAN AGRICULTURAL STRUCTURE IMPROVEMENT ASSOCIATION)
11-35, 1-CHOME, NAGATA-CHO, CHIYODA-KU, TOKYO 100
JAPANESE

CHIKUSAN GIJUTSU 4374
(ANIMAL INDUSTRY)
CHIKUSAN GIJUTSU RENMEI (UNION OF ANIMAL INDUSTRY)
20-9, 3-CHOME, YUSHIMA, BUNKYO-KU, TOKYO 113
JAPANESE

CHIKUSAN HYOGO 4375
(ANIMAL INDUSTRY IN HYOGO PREFECTURE)
HYOGO-KEN CHIKUSAN NOGYO KYODO KUMIAI RENGOKAI (HYOGO PREFECTURE LIVESTOCK FARMING COOPERATIVE ASSOCIATION)
2-8, SHIMOYAMATE-DORI, IKUTA-KU, KOBE-SHI 650
JAPANESE

CHIKUSAN IWATE 4376
(LIVESTOCK FARMING IN IWATE)
IWATE-KEN (IWATE PREFECTURAL GOVERNMENT)
10-1, UCHIMARU, MORIOKA-SHI 020
JAPANESE

CHIKUSAN KANKEI KEIZAI TOKEI GEPPO 4377
(MONTHLY ECONOMIC STATUS ON ANIMAL HUSBANDRY)
CHIKUSAN NYOKU
JAPANESE

CHIKUSAN KANKEI TOKEI SHIRYO 4378
(STATISTICS ON ANIMAL INDUSTRY, HOKKAIDO PREFECTURE)
HOKKAIDO NOSEI-BU (AGRICULTURAL ADMINISTRATION DEPARTMENT, HOKKAIDO PREFECTURAL GOVERNMENT)
NISHI-6-CHOME, KITA-3-JO, CHUO-KU, SAPPORO-SHI 060
JAPANESE

CHIKUSAN KONSARUTANTO 4379
(LIVESTOCK CONSULTANT)
CHUO CHIKUSANKAI (CENTRAL ASSOCIATION OF LIVESTOCK INDUSTRY)
11-35, 1-CHOME, NAGATA-CHO, CHIYODA-KU, TOKYO 100
JAPANESE

CHIKUSAN NO KENKYU 4380
(ANIMAL HUSBANDRY)
YOKENDO
30-15, 5-CHOME, HONGO, BUNKYO-KU, TOKYO 113
JAPANESE
INDEXED: BIB, CHE, JICST

CHIKUSAN SHIKENJO KENKYU HOKOKU 4381
(BULLETIN OF THE NATIONAL INSTITUTE OF ANIMAL INDUSTRY)
NORIN-SHO CHIKUSAN SHIKENJO (NATIONAL INSTITUTE OF ANIMAL INDUSTRY, MINISTRY OF AGRICULTURE AND FORESTRY)
959, AOBA-CHO, CHIBA-SHI 280
JAPANESE, WITH ENGLISH SUMMARY
INDEXED: BIB, BIO, CHE, JICST

CHIKUSAN SHIKENJO NENPO 4382
(ANNUAL REPORT OF NATIONAL INSTITUTE OF ANIMAL INDUSTRY)
NORIN-SHO CHIKUSAN SHIKENJO (NATIONAL INSTITUTE OF ANIMAL INDUSTRY, MINISTRY OF AGRICULTURE AND FORESTRY)
959, AOBA-CHO, CHIBA-SHI 280
JAPANESE
INDEXED: JICST

CHIKUSAN SHINKO JIGYODAN GYOMU GAIKYO 4383
(JOURNAL OF THE LIVESTOCK INDUSTRY PROMOTION CORPORATION)
CHIKUSAN SHINKO JIGYODAN (LIVESTOCK INDUSTRY PROMOTION CORPORATION)
C/O SHIRYO KAIKAN, 2-1, 2-CHOME, AZABUDAI, MINATO-KU, TOKYO 106
JAPANESE

CHIKUSAN SHINKO JIGYODAN NENPO 4384
(ANNUAL REPORT OF LIVESTOCK INDUSTRY PROMOTION CORPORATION)
CHIKUSAN SHINKO JIGYODAN (LIVESTOCK INDUSTRY PROMOTION CORPORATION)
C/O SHIRYO KAIKAN, 2-1, 2-CHOME, AZABUDAI, MINATO-KU, TOKYO 106
JAPANESE

CHIKUSAN TOKEI 4385
(STATISTICS OF ANIMAL INDUSTRY)
NORIN-SHO NORIN KEIZAI-KYOKU TOKEI JOHO-BU (STATISTIC AND INFORMATION DEPARTMENT, ECONOMIC AFFAIRS BUREAU, MINISTRY OF AGRICULTURE AND FORESTRY) AND NORIN TOKEI KYOKAI
11-14, 2-CHOME, MEGURO, MEGURO-KU, TOKYO 153
JAPANESE

CHIKUSANBUTSU SEISANHI CHOSA SEISEKI 4386
(SURVEY OF THE COST OF LIVESTOCK PRODUCTS)
NORIN-SHO NORIN KEIZAI-KYOKU TOKEI JOHO-BU (STATISTICS AND INFORMATION DEPARTMENT, ECONOMIC AFFAIRS BUREAU, MINISTRY OF AGRICULTURE AND FORESTRY) AND NORIN TOKEI KYOKAI
11-14, 2-CHOME, MEGURO, MEGURO-KU, TOKYO 153
JAPANESE

CHOSA SEISEKI (NAGANO SHUCHIKU BOKUJO) 4387
(EXAMINATION RESULT)
NORIN-SHO NAGANO SHUCHIKU BOKUJO (NAGANO LIVESTOCK BREEDING STATION, MINISTRY OF AGRICULTURE AND FORESTRY)
SHINKODEN, SAKU-SHI 385
JAPANESE AND ENGLISH

CHUGOKU NOGYO SHIKENJO HOKOKU, B, CHIKUSAN-BU 4388
(BULLETIN OF THE CHUGOKU NATIONAL AGRICULTURAL EXPERIMENT STATION, SERIES B)
CHUGOKU NOGYO SHIKENJO (CHUGOKU NATIONAL AGRICULTURAL EXPERIMENT STATION)
HIGASHIFUKATSU-MACHI, FUKUYAMA-SHI 720
JAPANESE AND ENGLISH
INDEXED: BIB, CHE, JICST

COLLIE FANCY 4389
NIHON KORI KURABU (JAPAN COLLIE CLUB)
4-1, 2-CHOME, GINZA, CHUO-KU, TOKYO 104
JAPANESE

CONTRIBUTION FROM THE SCHOOL OF VETERINARY SCIENCE, OBIHIRO ZOOTECHNICAL UNIVERSITY 4390
OBIHIRO CHIKUSAN DAIGAKU (OBIHIRO ZOOTECHNICAL UNIVERSITY)
INADA-CHO, OBIHIRO-SHI 080
JAPANESE AND ENGLISH

DAIRY YEAR BOOK 4391
SHIN NORIN SHA
TOKYO
JAPANESE
INDEXED: BIB

DERI JAPAN 4392
(DAIRY JAPAN)
DERI JAPANSHA (DAIRY JAPAN COMPANY, LIMITED)
C/O ENPAIA KOPU, 28, DAIKYO-CHO, SHINJUKU-KU, TOKYO 160
JAPANESE

DERIMAN 4393
(DAIRYMAN)
DERIMANSHA (DAIRYMAN COMPANY)
HIGASHI-2-CHOME, KITA-3-JO, CHUO-KU, SAPPORO-SHI 062
JAPANESE
INDEXED: BIB

DOBUTSU IYAKUHIN KENSAJO NENPO 4394
(ANNUAL REPORT OF THE NATIONAL VETERINARY ASSAY LABORATORY)
NORIN-SHO DOBUTSU IYAKUHIN KENSAJO (THE NATIONAL VETERINARY ASSAY LABORATORY, MINISTRY OF AGRICULTURE AND FORESTRY)
15-1, 1-CHOME, TOKURA, KOKUBUNJI-SHI, TOKYO 185
JAPANESE AND ENGLISH
INDEXED: BIB

DOBUTSU KEN'EKI NENPO 4395
(ANNUAL REPORT OF ANIMAL QUARANTINE)
NORIN-SHO DOBUTSU KEN'EKISHO (ANIMAL QUARANTINE STATION, MINISTRY OF AGRICULTURE AND FORESTRY)
11-1, HARA-MACHI, ISOGO-KU, YOKOHAMA-SHI 235
JAPANESE

DOBUTSU YAKUJI 4396
(VETERINARY CHEMICALS)
NIHON DOBUTSU YAKUJI KYOKAI (JAPAN VETERINARY PHARMACEUTICAL ASSOCIATION)
1-2, KANDA SURUGADAI, CHIYODA-KU, TOKYO 101
JAPANESE

DOYAKU KENKYU 4397
(RESEARCH IN VETERINARY MEDICINE)
NIPPON TOKUSHU NOYAKU SEIZO K.K.
2-8, NIHONTASHI SHITSU-CHO, CHUO-KU, TOKYO 103
JAPANESE
INDEXED: CHE

EGG 4398
YOKEIKAI
3-18, EKIMAE-ODORI, TOYOHASHI-SHI 440
JAPANESE

EHIME-KEN CHIKUSAN SHIKENJO KENKYU HOKOKU 4399
(REPORT OF EHIME PREFECTURAL LIVESTOCK EXPERIMENT STATION)
EHIME-KEN CHIKUSAN SHIKENJO
AGE, NOMURA-CHO, HIGASHIUWA-GUN, EHIME 797-12
JAPANESE, WITH ENGLISH SUMMARY

EHIME-KENRITSU SHUCHIKUJO GYOMU KOTEI HOKOKUSHO 4400
(REPORT OF THE EHIME PREFECTURE ANIMAL HUSBANDRY STATION)
EHIME-KENRITSU SHUCHIKUJO (EHIME PREFECTURE LIVESTOCK BREEDING STATION)
AGE, NOMURA-CHO, HIGASHIUWA-GUN, EHIME 797-12
JAPANESE

EHIME-KENRITSU SHUCHIKUJO SHIKEN SEISEKI HOKOKUSHO 4401
(REPORT OF THE EHIME PREFECTURE LIVESTOCK BREEDING STATION)
EHIME-KENRITSU SHUCHIKUJO (EHIME PREFECTURE LIVESTOCK BREEDING STATION)
AGE, NOMURA-CHO, HIGASHIUWA-GUN, EHIME 797-12
JAPANESE

EHIME-KENRITSU YOKEI SHIKENJO HOKOKU 4402
(REPORT OF EHIME PREFECTURAL POULTRY EXPERIMENT STATION)
EHIME-KENRITSU YOKEI SHIKENJO
OTSU-159, FUKUJOJI, TOYO-SHI 799-14
JAPANESE

EHIME-KENRITSU YOKEI SHIKENJO KENKYU HOKOKU 4403
(RESEARCH REPORT OF EHIME PREFECTURAL POULTRY EXPERIMENT STATION)
EHIME-KENRITSU YOKEI SHIKENJO (EHIME PREFECTURAL POULTRY EXPERIMENT STATION)
OTSU-159, FUKUJOJI, TOYO-SHI 799-14
JAPANESE

FEED TRADE 4404
SHIRYO YUSHUTSUNYU (JAPAN FEED TRADE ASSOCIATION)
C/O KOIZUMI BUILDING, 3-13, 4-CHOME, GINZA, CHUO-KU, TOKYO 104
JAPANESE
INDEXED: BIB

FUKUI-KEN CHIKUSAN SHIKENJO HOKOKU 4405
(BULLETIN OF THE FUKUI EXPERIMENT STATION OF DOMESTIC ANIMAL INDUSTRY)
FUKUI-KEN CHIKUSAN SHIKENJO (FUKUI EXPERIMENT STATION OF DOMESTIC ANIMAL INDUSTRY)
HIRAYAMA, MIKUNI-CHO, SAKAI-GUN, FUKUI 913
JAPANESE

FUKUI-KEN CHIKUSAN SHIKENJO NENPO 4406
(ANNUAL REPORT OF THE FUKUI EXPERIMENT STATION OF DOMESTIC ANIMAL INDUSTRY)
FUKUI-KEN CHIKUSAN SHIKENJO (FUKUI EXPERIMENT STATION OF DOMESTIC ANIMAL INDUSTRY)
HIRAYAMA, MIKUNI-CHO, SAKAI-GUN, FUKUI 913
JAPANESE

FUKUOKA HISHIRYO KENSAJO JIGYO HOKOKU, SHIRYO NO BU 4407
(ANNUAL REPORT OF THE FUKUOKA FERTILIZER AND FEED INSPECTION STATION, FEED SECTION)
NORIN-SHO FUKUOKA HISHIRYO KENSAJO (FUKUOKA FERTILIZER AND FEED INSPECTION STATION, MINISTRY OF AGRICULTURE AND FORESTRY)
1-1, CHIHAYA, HIGASHI-KU, FUKUOKA-SHI 813
JAPANESE

FUKUOKA-KEN NO CHIKUSAN 4408
(ANIMAL INDUSTRY IN FUKUOKA PREFECTURE)
FUKUOKA-KEN (FUKUOKA PREFECTURAL GOVERNMENT)
1-1, 1-CHOME, TENJIN, CHUO-KU, FUKUOKA-SHI 810
JAPANESE

FUKUOKA-KEN NO YOKEI 4409
(POULTRY IN FUKUOKA PREFECTURE)
FUKUOKA-KEN (FUKUOKA PREFECTURAL GOVERNMENT)
1-1, 1-CHOME, TENJIN, CHUO-KU, FUKUOKA-SHI 810
JAPANESE

FUKUOKA-KEN SHUCHIKUJO GYOMU HOKOKU 4410
(BUSINESS REPORT OF THE FUKUOKA PREFECTURE LIVESTOCK BREEDING STATION)
FUKUOKA-KEN SHUCHIKUJO (FUKUOKA PREFECTURE LIVESTOCK BREEDING STATION)
4839-7, KITANAKAO, OGORI-SHI 838-01
JAPANESE

FUKUOKA-KEN SHUCHIKUJO SHIKEN KENKYU HOKOKU 4411
(RESEARCH REPORT OF THE FUKUOKA PREFECTURE LIVESTOCK BREEDING STATION)
FUKUOKA-KEN SHUCHIKUJO (FUKUOKA PREFECTURE LIVESTOCK BREEDING STATION)
4839-7, KITANAKAO, OGORI-SHI 838-01
JAPANESE

FUKUOKA-KEN SHUKEIJO KENKYU HOKOKU 4412
(BULLETIN OF THE FUKUOKA PREFECTURAL POULTRY BREEDING STATION)
FUKUOKA-KEN SHUKEIJO (FUKUOKA PREFECTURAL POULTRY BREEDING STATION)
DAZAIFU, DAZAIFU-MACHI, CHIKUSHI-GUN, FUKUOKA 818-01
JAPANESE

FUKUSHIMA CHIKUSHI KIHO 4413
(QUARTERLY REPORT OF FUKUSHIMA PREFECTURE ANIMAL HUSBANDRY EXPERIMENT STATION)
FUKUSHIMA-KEN CHIKUSAN SHIKENJO (FUKUSHIMA PREFECTURE ANIMAL HUSBANDRY EXPERIMENT STATION)
KO-18, JIZOHARA, ARAI, FUKUSHIMA-SHI 960-21
JAPANESE

FUKUSHIMA-KEN CHIKUSAN SHIKENJO GYOMU SEISEKI OYOBI SHIKEN SEISEKI HOKOKU 4414
(REPORT OF FUKUSHIMA PREFECTURAL ANIMAL HUSBANDRY EXPERIMENT STATION)
FUKUSHIMA-KEN CHIKUSAN SHIKENJO (FUKUSHIMA PREFECTURAL ANIMAL HUSBANDRY EXPERIMENT STATION)
KO-18, JIZOHARA, ARAI, FUKUSHIMA-SHI 960-21
JAPANESE

FUKUSHIMA-KEN YOKEI SHIKENJO SHIKEN SEISEKI HOKOKU 4415
(REPORT OF FUKUSHIMA PREFECTURAL POULTRY BREEDING EXPERIMENT STATION)
FUKUSHIMA-KEN YOKEI SHIKENJO
2, MANSUIDA, TOMIDA-MACHI, KORIYAMA-SHI 963
JAPANESE

GEINZU SHINPOJUMU REPOTO 4416
(GAINES SYMPOSIUM REPORT)
GEINZU PUROFESSHONARU SABISU (GAINES PROFESSIONAL SERVICE)
8, HONSHIO-CHO, SHINJUKU-KU, TOKYO 160
JAPANESE

GEKKAN SHOKUNIKUKAI 4417
(MEAT JOURNAL)
SHOKUNIKU TSUSHINSHA
IKEDA BUILDING, 10-11, 2-CHOME, UCHIKANDA, CHIYODA-KU, TOKYO 101
JAPANESE

GIFU-KEN CHIKUSAN SHIKENJO GYOMU HOKOKU 4418
(REPORT OF THE GIFU PREFECTURAL ANIMAL HUSBANDRY EXPERIMENT STATION)
GIFU-KEN CHIKUSAN SHIKENJO
1152-3, MAEHIRA, OTA-CHO, MINOKAMO-SHI 505
JAPANESE

GIFU-KEN SHUCHIKUJO SHIKEN SEISEKI 4419
(REPORT OF GIFU PREFECTURAL BREEDING-STOCK STATION)
GIFU-KEN SHUCHIKUJO
4393-1, MAKIGAHORA, KIYOMI-MURA, ONO-GUN, GIFU 506-01
JAPANESE

GIFU-KEN SHUKEIJO KENKYU HOKOKU 4420
(REPORT OF GIFU PREFECTURAL POULTRY BREEDING STATION)
GIFU-KEN SHUKEIJO (GIFU PREFECTURAL POULTRY BREEDING STATION)
2726-1, HASAMA, SEKI-SHI 501-32
JAPANESE

GIFU-KEN SHUKEIJO SHIKEN HOKOKU 4421
(RESEARCH BULLETIN OF THE GIFU PREFECTURAL POULTRY BREEDING STATION)
GIFU-KEN SHUKEIJO (GIFU PREFECTURAL POULTRY BREEDING STATION)
2726-1, HASAMA, SEKI-SHI 501-32
JAPANESE

GUNMA-KEN CHIKUSAN SHIKENJO KENKYU SEISEKI 4422
(RESEARCH REPORT OF THE GUNMA ANIMAL HUSBANDRY EXPERIMENT STATION)
GUNMA-KEN CHIKUSAN SHIKENJO (GUNMA ANIMAL HUSBANDRY EXPERIMENT STATION)
2425, KOGURE, FUJIMI-MURA, SETA-GUN, GUNMA 371-01
JAPANESE

GUNMA-KEN CHIKUSAN SHIKENJO NENPO 4423
(ANNUAL REPORT OF GUNMA ANIMAL HUSBANDRY EXPERIMENT STATION)
GUNMA-KEN CHIKUSAN SHIKENJO (GUNMA ANIMAL HUSBANDRY EXPERIMENT STATION)
2425, KOGURE, FUJIMI-MURA, SETA-GUN, GUNMA 371-01
JAPANESE

GYUNYU NYUSEIHIN NI KANSURU TOKEI 4424
(STATISTICS OF COW MILK AND PRODUCTS)
NORIN-SHO NORIN KEIZAI-KYOKU TOKEI JOHO-BU (STATISTICS AND INFORMATION DEPARTMENT, ECONOMIC AFFAIRS BUREAU, MINISTRY OF AGRICULTURE AND FORESTRY) AND NORIN TOKEI KYOKAI
11-14, 2-CHOME, MEGURO, MEGURO-KU, TOKYO 153
JAPANESE

HANSHOKU GIJUTSU 4425
(TECHNICAL FOR ANIMAL REPRODUCTION)
HOKKAIDO KACHIKU JINKO JUSEISHI KYOKAI (HOKKAIDO ARTIFICIAL INSEMINATION TECHNICIAN ASSOCIATION)
C/O HOKUNOREN, NISHI-1-CHOME, KITA-4-JO, SAPPORO-SHI 060
JAPANESE

HIROSHIMA NO CHIKUSAN 4426
(ANIMAL HUSBANDRY IN HIROSHIMA PREFECTURE)
HIROSHIMA-KEN CHIKUSANKAI
7-2, 4-CHOME, OTE-MACHI, HIROSHIMA-SHI 730
JAPANESE

HIROSHIMA-KEN CHIKUSAN KANKEI GYOSEKI HAPPYOKAI SHUROKU 4427
(PROCEEDINGS OF THE ANNUAL MEETING OF THE LIVESTOCK, HIROSHIMA PREFECTURE)
HIROSHIMA-KEN (HIROSHIMA PREFECTURAL GOVERNMENT)
10-52, MOTO-MACHI, HIROSHIMA-SHI 730
JAPANESE

HIROSHIMA-KENRITSU CHIKUSAN SHIKENJO MIYOSHI SHIJO SHIKEN KENKYU HOKOKUSHO 4428
(RESEARCH REPORT OF HIROSHIMA PREFECTURAL ANIMAL HUSBANDRY EXPERIMENT STATION, MIYOSHI BRANCH)
HIROSHIMA-KENRITSU CHIKUSAN SHIKENJO MIYOSHI SHIJO (HIROSHIMA PREFECTURAL ANIMAL HUSBANDRY EXPERIMENT STATION, MIYOSHI BRANCH)
SHIJIKKAN-MACHI, MIYOSHI-SHI 728
JAPANESE

HIROSHIMA-KENRITSU CHIKUSAN SHIKENJO SHIKEN KENKYU HOKOKUSHO 4429
(RESEARCH REPORT OF HIROSHIMA PREFECTURAL ANIMAL HUSBANDRY EXPERIMENT STATION)
HIROSHIMA-KENRITSU CHIKUSAN SHIKENJO (HIROSHIMA PREFECTURAL ANIMAL HUSBANDRY EXPERIMENT STATION)
584, NANATSUKA-CHO, SHOBARA-SHI 727
JAPANESE

HIROSHIMA-KENRITSU CHIKUSAN SHIKENJO YUKI SHIJO GYOMU KOTEI HOKOKUSHO 4430
(OPERATION REPORT OF YUKI BRANCH, HIROSHIMA PREFECTURAL ANIMAL HUSBANDRY EXPERIMENT STATION)
HIROSHIMA-KENRITSU CHIKUSAN SHIKENJO YUKI SHIJO
YUKI-CHO, JINSEKI-GUN, HIROSHIMA 720-18
JAPANESE

HIROSHIMA-KENRITSU CHIKUSAN SHIKENJO YUKI SHIJO SHIKEN SEISEKI 4431
(BULLETIN OF YUKI BRANCH, HIROSHIMA PREFECTURAL ANIMAL HUSBANDRY EXPERIMENT STATION)
HIROSHIMA-KENRITSU CHIKUSAN SHIKENJO YUKI SHIJO
YUKI-CHO, JINSEKI-GUN, HIROSHIMA 720-18
JAPANESE

HOKKAIDO CHIKUSAN TOKEI 4432
(LIVESTOCK STATISTICS IN HOKKAIDO PREFECTURE)
NORIN-SHO HOKKAIDO TOKEI JOHO JIMUSHO (HOKKAIDO STATISTICS AND INFORMATION OFFICE, MINISTRY OF AGRICULTURE AND FORESTRY)
NISHI-4-CHOME, KITA-3-JO, CHUO-KU, SAPPORO-SHI 060
JAPANESE

HOKKAIDO DAIGAKU NOGAKUBU FUZOKU BOKUJO KENKYU HOKOKU 4433
(RESEARCH BULLETINS OF THE LIVESTOCK FARM, FACULTY OF AGRICULTURE, HOKKAIDO UNIVERSITY)
LIVESTOCK FARM, FACULTY OF AGRICULTURE, HOKKAIDO UNIVERSITY
JAPANESE

HOKKAIDO JUISHIKAI ZASSHI 4434
(JOURNAL OF THE HOKKAIDO VETERINARY MEDICAL ASSOCIATION)
HOKKAIDO JUISHIKAI (HOKKAIDO VETERINARY MEDICAL ASSOCIATION)
NISHI-1-CHOME, KITA-4-JO, CHUO-KU, SAPPORO-SHI 060
JAPANESE

HOKKAIDO RAKUNO KENSASHO JIGYO SEISEKI HOKOKUSHO 4435
(TEST REPORT OF THE HOKKAIDO DAIRY INSPECTION STATION)
HOKKAIDO RAKUNO KENSASHO
C/O RAKUNO SENTA, NISHI-7-CHOME, KITA-3-JO, CHUO-KU, SAPPORO-SHI 060
JAPANESE

HOKKAIDORITSU SHINTOKU CHIKUSAN SHIKENJO KENKYU HOKOKU 4436
(BULLETIN OF THE HOKKAIDO PREFECTURAL SHINTOKU ANIMAL HUSBANDRY EXPERIMENT STATION)
HOKKAIDORITSU SHINTOKU CHIKUSAN SHIKENJO (HOKKAIDO PREFECTURAL SHINTOKU ANIMAL HUSBANDRY EXPERIMENT STATION)
SHINTOKU-MACHI, KAMIKAWA-GUN, HOKKAIDO 081
JAPANESE, WITH ENGLISH SUMMARY
INDEXED: JICST

HOKKAIDORITSU SHINTOKU CHIKUSAN SHIKENJO NENPO 4437
(ANNUAL REPORT OF THE HOKKAIDO PREFECTURAL SHINTOKU ANIMAL HUSBANDRY EXPERIMENT STATION)
HOKKAIDORITSU SHINTOKU CHIKUSAN SHIKENJO (HOKKAIDO PREFECTURAL SHINTOKU ANIMAL HUSBANDRY EXPERIMENT STATION)
SHINTOKU-MACHI, KAMIKAWA-GUN, HOKKAIDO 081
JAPANESE

HORUSUTAIN KAIHO 4438
(HOLSTEIN SOCIETY REPORT)
JAPANESE
INDEXED: JICST

HYOGO-KENRITSU CHIKUSAN SHIKENJO KENKYU HOKOKU 4439
(RESEARCH REPORT OF HYOGO PREFECTURAL EXPERIMENT STATION OF ANIMAL HUSBANDRY)
HYOGO-KENRITSU CHIKUSAN SHIKENJO (HYOGO PREFECTURAL EXPERIMENT STATION OF ANIMAL HUSBANDRY)
KAMIONO, HIMEJI-SHI 670
JAPANESE AND ENGLISH

IBARAKI NO CHIKUSAN TOKEI 4440
(LIVESTOCK STATISTICS IN IBARAKI PREFECTURE)
NORIN-SHO KANTO NOSEI-KYOKU IBARAKI TOKEI JOHO JIMUSHO (IBARAKI STATISTICS & INFORMATION OFFICE, KANTO REGIONAL OFFICE OF AGRICULTURAL ADMINISTRATION, MINISTRY OF AGRICULTURE AND FORESTRY)
1-11, KITAMI-CHO, MITO-SHI 310
JAPANESE

IBARAKI-KEN CHIKUSAN SHIKENJO NENPO 4441
(ANNUAL REPORT OF IBARAKI PREFECTURAL EXPERIMENTAL STATION OF ANIMAL INDUSTRY)
IBARAKI-KEN CHIKUSAN SHIKENJO
1718, TAIRA, TOMOBE-MACHI, NISHIIBARAKI-GUN, IBARAKI 309-17
JAPANESE

IBARAKI-KEN CHIKUSON SHIKENJO KENKYU HOKOKU 4442
(RESEARCH BULLETIN OF THE IBARAKI PREFECTURAL LIVESTOCK EXPERIMENT STATION)
IBARAKI-KEN CHIKUSON SHIKENJO (IBARAKI PREFECTURAL LIVESTOCK EXPERIMENT STATION)
JAPANESE

IBARAKI-KEN YOKEI SHIKENJO KENKYU HOKOKU 4443
(REPORT OF IBARAKI PREFECTURAL POULTRY EXPERIMENT STATION)
IBARAKI-KEN YOKEI SHIKENJO
OBATA, IBARAKI-MACHI, HIGASHIIBARAKI-GUN, IBARAKI 311-36
JAPANESE

IBARAKI-KEN YOTON SHIKENJO GYOMU NENPO 4444
(ANNUAL REPORT ON OPERATION OF IBARAKI PREFECTURAL INSTITUTE OF SWINE INDUSTRY)
IBARAKI-KEN YOTON SHIKENJO
3240, SAKURA, EDOSAKI-MACHI, INASHIKI-GUN, IBARAKI 300-05
JAPANESE

IBARAKI-KEN YOTON SHIKENJO SHIKEN KENKYU HOKOKU 4445
(RESEARCH REPORT OF IBARAKI PREFECTURAL INSTITUTE OF SWINE INDUSTRY)
IBARAKI-KEN YOTON SHIKENJO
3240, SAKURA, EDOSAKI-MACHI, INASHIKI-GUN, IBARAKI 300-05
JAPANESE

IKUSHUGAKU SAIKIN NO SHINPO 4446
(LATEST ADVANCEMENT IN BREEDING)
JAPANESE
INDEXED: BIB

ISHIKAWA NO CHIKUSAN 4447
(ANIMAL INDUSTRY IN ISHIKAWA PREFECTURE)
ISHIKAWA-KEN NORIN SUISAN-BU (AGRICULTURE, FORESTRY AND FISHERY DEPARTMENT, ISHIKAWA PREFECTURAL GOVERNMENT)
1-1, 2-CHOME, HIROSAKA, KANAZAWA-SHI 920
JAPANESE

ISHIKAWA-KEN CHIKUSAN SHIKENJO SHIKEN KENKYU SEISEKI HOKOKU 4448
(REPORT OF ISHIKAWA PREFECTURAL LIVESTOCK EXPERIMENT STATION)
ISHIKAWA-KEN CHIKUSAN SHIKENJO
TSUBOYAMA, OSHIMIZU-CHO, HAKUI-GUN, ISHIKAWA 929-13
JAPANESE

ISHIKAWA-KEN NIKUUSHI SEISAN SHIDOJO GYOMU JISSEKI HOKOKUSHO 4449
(OPERATION REPORT OF ISHIKAWA PREFECTURAL BEEF CATTLE BREEDING EXPERIMENT STATION)
ISHIKAWA-KEN NIKUUSHI SEISAN SHIDOJO
UCHIURA-MACHI, SUZU-GUN, ISHIKAWA 927-06
JAPANESE

ISHIKAWA-KEN NIKUUSHI SEISAN SHIDOJO SHIKEN KENKYU HOKOKUSHO 4450
(RESEARCH REPORT OF ISHIKAWA PREFECTURAL BEEF CATTLE BREEDING EXPERIMENT STATION)
ISHIKAWA-KEN NIKUUSHI SEISAN SHIDOJO
UCHIURA-MACHI, SUZU-GUN, ISHIKAWA 927-06
JAPANESE

IWATE-KEN CHIKUSAN SHIKENJO KENKYU HOKOKU 4451
(BULLETIN OF IWATE PREFECTURAL ANIMAL HUSBANDRY)
IWATE-KEN CHIKUSAN SHIKENJO (IWATE PREFECTURAL ANIMAL HUSBANDRY EXPERIMENT STATION)
TAKISAWA, TAKISAWA-MURA, IWATE-GUN, IWATE 020-01
JAPANESE

JAJI 4452
(JERSEY)
NIHON JAJI TOROKU KYOKAI (JERSEY CATTLE ASSOCIATION OF JAPAN)
38-13, 4-CHOME, HON-CHO, NAKANO-KU, TOKYO 164
JAPANESE
INDEXED: JICST

JAPANESE JOURNAL OF VETERINARY RESEARCH 4453
HOKKAIDO DAIGAKU JUIGAKUBU (FACULTY OF VETERINARY MEDICINE, HOKKAIDO UNIVERSITY)
NISHI-9-CHOME, KITA-18-JO, KITA-KU, SAPPORO-SHI 060
ENGLISH
INDEXED: BIB, BIO, CHE, EXC, IND, JICST

JITCHUKEN, ZENRINSHO KENKYUHO 4454
(CENTRAL INSTITUTE FOR EXPERIMENTAL ANIMALS, PRECLINICAL REPORTS)
CENTRAL INSTITUTE FOR EXPERIMENTAL ANIMALS
1433 NOGAWA, TAKATSU-KU, KAWASAKI 211
JAPANESE AND ENGLISH
INDEXED: BIO, CHE

JUI CHIKUSAN SHINPO 4455
(JOURNAL OF THE VETERINARY MEDICINE)
BUN'EIDO
27-18, 2-CHOME, HONGO, BUNKYO-KU, TOKYO 113
JAPANESE
INDEXED: CHE

JUI GIJUTSU 4456
(VETERINARY TECHNIQUE)
NIHON CHUO KEIBAKAI KYOSOBA HOKEN KENKYUSHO (EQUINE HEALTH LABORATORY, THE JAPAN RACING ASSOCIATION)
27-7, 5-CHOME, TSURUMAKI, SETAGAYA-KU, TOKYO 154
JAPANESE

JUIKAI 4457
(VETERINARY WORLD)
TAKEDA YAKUHIN KOGYO K.K. CHIKUSAN JIGYO-BU (TAKEDA CHEMICAL INDUSTRIES, LIMITED, ANIMAL HEALTH PRODUCTS DIVISION)
12-10, 2-CHOME, NIHONBASHI, CHUO-KU, TOKYO 103
JAPANESE

KACHIKU EISEI NO SHINPO 4458
(PROGRESS OF ANIMAL HEALTH SCIENCE IN JAPAN)
NORIN-SHO CHIKUSAN-KYOKU (BUREAU OF ANIMAL INDUSTRY, MINISTRY OF AGRICULTURE AND FORESTRY)
2-1, 1-CHOME, KASUMIGASEKI, CHIYODA-KU, TOKYO 100
JAPANESE

KACHIKU EISEI SHIRYO 4459
(INFORMATION OF DOMESTIC ANIMAL HEALTH)
ISHIKAWA-KEN NORIN-BU (AGRICULTURE AND FORESTRY DEPARTMENT, ISHIKAWA PREFECTURAL GOVERNMENT)
1-1, 2-CHOME, HIROSAKA, KANAZAWA-SHI 920
JAPANESE

KACHIKU EISEI SHUHO 4460
(ANIMAL HEALTH WEEKLY)
NORIN-SHO CHIKUSAN-KYOKU (ANIMAL INDUSTRY BUREAU, MINISTRY OF AGRICULTURE AND FORESTRY) AND NORIN KOSAIKAI
2-1, 1-CHOME, KASUMIGASEKI, CHIYODA-KU, TOKYO 100
JAPANESE

KACHIKU EISEI TOKEI 4461
(STATISTICS OF ANIMAL HYGIENE)
NORIN-SHO CHIKUSAN-KYOKU (ANIMAL INDUSTRY BUREAU, MINISTRY OF AGRICULTURE AND FORESTRY) AND NORIN KYOKAI
2-1, 1-CHOME, KASUMIGASEKI, CHIYODA-KU, TOKYO 100
JAPANESE AND ENGLISH
INDEXED: BIB

KACHIKU HANSHOKU KENKYUKAISHI 4462
(JAPANESE JOURNAL OF ANIMAL REPRODUCTION)
TOKYO NOKO DAIGAKU NOGAKUBU KACHIKU SEIRIGAKU KYOSHITSU (TOKYO UNIVERSITY OF AGRICULTURE AND TECHNOLOGY) AND KACHIKU HANSHOKU KENKYUKAI
C/O NORIN-SHO KACHIKU EISEI SHIKENJO KENKYU DAI-3-BU, 1500, JOSUIHON-CHO, KODAIRA-SHI, TOKYO 187
JAPANESE, WITH ENGLISH SUMMARY
INDEXED: BIB, JICST

KACHIKU HOKEN EISEI GIJUTSU KENKYUKAIHO 4463
(BULLETIN OF THE ASSOCIATION OF THE TECHNIC FOR ANIMAL HYGIENE)
KYUSHU KACHIKU HOKEN EISEI GIJUTSU KENKYUKAI (ASSOCIATION OF THE TECHNIC FOR ANIMAL HYGIENE)
C/O KACHIKU EISEI SHIKENJO KYUSHU SHIJO, 2702, CHUZAN-CHO, KAGOSHIMA-SHI 891-01
JAPANESE
INDEXED: CHE

KACHIKU JINKO JUSEI 4464
(ARTIFICIAL INSEMINATION OF ANIMALS)
NIHON KACHIKU JINKO JUSEISHI KYOKAI (JAPAN ARTIFICIAL INSEMINATORS ASSOCIATION)
11-35, 1-CHOME, NAGATA-CHO, CHIYODA-KU, TOKYO 100
JAPANESE

KACHIKU JINKO JUSEI SEISEKI CHOSASHO 4465
(REPORT ON RESULT OF ARTIFICIAL INSEMINATION OF ANIMALS)
YAMAGATA-KENRITSU CHIKUSAN SHIKENJO (YAMAGATA PREFECTURAL ANIMAL HUSBANDRY EXPERIMENT STATION)
KO-7057, HIGASHINE, HIGASHINE-SHI 999-37
JAPANESE

KACHIKU NO KANRI 4466
(JAPANESE JOURNAL OF LIVESTOCK MANAGEMENT)
NIHON KACHIKU KANRI KENKYUKAI (SOCIETY OF SCIENTIFIC LIVESTOCK MANAGEMENT JAPAN)
5-8, 3-CHOME, SAIWAI-CHO, FUCHU-SHI, TOKYO 183
JAPANESE AND ENGLISH

KACHIKU SHIYO TOHASU 4467
(NUMERICAL DATA OF BREEDING LIVESTOCK, FUKUOKA PREFECTURE)
FUKUOKA-KEN NOSEI-BU (AGRICULTURAL ADMINISTRATION BUREAU, FUKUOKA PREFECTURAL GOVERNMENT)
1-1, 1-CHOME, TENJIN, CHUO-KU, FUKUOKA-SHI 810
JAPANESE

KACHIKU TOHASU CHOSA SHUKEIHYO 4468
(NUMERICAL DATA OF BREEDING LIVESTOCK, NIIGATA PREFECTURE)
NIIGATA-KEN NORIN-BU (AGRICULTURE AND FORESTRY DEPARTMENT, NIIGATA PREFECTURAL GOVERNMENT)
602, GAKKODORI ICHIBAN-CHO, NIIGATA-SHI 951
JAPANESE

KAGAKU SHIRYO 4469
(SYNTHETIC FEEDS)
NIHON KAGAKU SHIRYO KYOKAI (SCIENTIFIC FEEDS ASSOCIATION, JAPAN)
1-2, KANDA SURUGADAI, CHIYODA-KU, TOKYO 101
JAPANESE
INDEXED: JICST

KAGAWA-KEN CHIKUSAN SHIKENJO KENKYU HOKOKU 4470
(REPORT OF KAGAWA PREFECTURE LIVESTOCK EXPERIMENT STATION)
KAGAWA-KEN CHIKUSAN SHIKENJO (KAGAWA PREFECTURE LIVESTOCK EXPERIMENT STATION)
2706, SHIMOTAKAOKA, MIKI-CHO, KITA-GUN, KAGAWA 761-07
JAPANESE

KAGAWA-KEN CHIKUSAN SHIKENJO SHIKEN SEISEKI HOKOKU 4471
(EXPERIMENTAL REPORT OF KAGAWA PREFECTURE LIVESTOCK EXPERIMENT STATION)
KAGAWA-KEN CHIKUSAN SHIKENJO (KAGAWA PREFECTURE LIVESTOCK EXPERIMENT STATION)
2706, SHIMOTAKAOKA, MIKI-CHO, KITA-GUN, KAGAWA 761-07
JAPANESE

KAGOSHIMA-KEN CHIKUSAN SHIKENJO SHIKEN KENKYU HOKOKUSHO 4472
(REPORT OF KAGOSHIMA PREFECTURAL LIVESTOCK EXPERIMENT STATION)
KAGOSHIMA-KEN CHIKUSAN SHIKENJO (KAGOSHIMA PREFECTURAL LIVESTOCK EXPERIMENT STATION)
2440, UENODAN, KOKUBU-SHI 899-44
JAPANESE

KAGOSHIMA-KEN YOKEI SHIKENJO JIGYO SEISEKISHO 4473
(ANNUAL REPORT OF KAGOSHIMA PREFECTURE POULTRY EXPERIMENT STATION)
KAGOSHIMA-KEN YOKEI SHIKENJO (KAGOSHIMA PREFECTURE POULTRY EXPERIMENT STATION)
217, KUMANOJO-CHO, SENDAI-SHI 895
JAPANESE

KAGOSHIMA-KEN YOKEI SHIKENJO SHIKEN SEISEKI HOKOKUSHO 4474
(REPORT ON RESULT OF RESEARCH OF KAGOSHIMA PREFECTURE POULTRY EXPERIMENT STATION)
KAGOSHIMA-KEN YOKEI SHIKENJO
217, KUMANOJO-CHO, SENDAI-SHI 895
JAPANESE

KANAGAWA-KEN CHIKUSAN SHIKENJO KENKYU HOKOKU 4475
(BULLETIN OF THE KANAGAWA LIVESTOCK EXPERIMENT STATION)
KANAGAWA-KEN CHIKUSAN SHIKENJO (KANAGAWA LIVESTOCK EXPERIMENT STATION)
3910, HONGO, EBINA-SHI 243-04
JAPANESE

KEIBYO KENKYUKAIHO 4476
(JOURNAL OF THE JAPANESE SOCIETY ON POULTRY DISEASES)
KEIBYO KENKYUKAI
C/O RAKUNO KAIKAN, 1-37, YOYOGI, SHIBUYA-KU, TOKYO 151
JAPANESE, WITH ENGLISH SUMMARY

KEIKAI JOHO 4477
(POULTRY BREEDING INFORMATION)
NIWATORI NO KENKYUSHA
4-1, 2-CHOME, MARUNOUCHI, CHIYODA-KU, TOKYO 100
JAPANESE
INDEXED: BIB

KEIRAN SEISANHI CHOSA SEISEKI 4478
(SURVEY OF THE COST OF EGG PRODUCTS)
AOMORI-KEN YOKEI SHIDOSHO
ASHINO, KANAKI-MACHI, KITA-TSUGARU-GUN, AOMORI 037-02
JAPANESE

KEIRAN TOKEI 4479
(EGG PRODUCTS STATISTICS)
ZENKOKU KEIRAN KAKAKU ANTEI KIKIN
8-3, 1-CHOME, OTE-MACHI, CHIYODA-KU, TOKYO 100
JAPANESE

KEIYO 4480
(POULTRY FRIEND)
KEIYUSHA
364, NAKASUNA-CHO, SHOWA-KU, NAGOYA-SHI 468
JAPANESE

KEMONO 4481
(BEASTS)
KONAGAWA DAIGAKU YAMAMOTO ZEMINARU
JAPANESE

KOCHI-KEN CHIKUSAN SHIKENJO KENKYU CHOSA SEISEKISHO 4482
(BULLETIN OF KOCHI PREFECTURAL LIVESTOCK EXPERIMENT STATION)
KOCHI-KEN CHIKUSAN SHIKENJO (KOCHI PREFECTURAL LIVESTOCK EXPERIMENT STATION)
1247, NAKAGUMI, SAGAWA-CHO, TAKAOKA-GUN, KOCHI 789-12
JAPANESE

KUMAMOTO SHUCHIKU BOKUJO ASO SHIJO CHOSA HOKOKU 4483
(REPORT OF ASO BRANCH STATION KUMAMOTO NATIONAL LIVESTOCK BREEDING STATION, NORIN-SHO)
KUMAMOTO SHUCHIKU BOKUJO ASO SHIJO (THE ASO BRANCH STATION. KUMAMOTO NATIONAL LIVESTOCK BREEDING STATION, MINISTRY OF AGRICULTURE AND FORESTRY)
KAWAYO, CHOYO-MURA, ASO-GUN, KUMAMOTO 869-21
JAPANESE

KUMAMOTO-KEN CHIKUSAN SHIKENJO JIGYO SEISEKISHO 4484
(ANNUAL REPORT OF THE KUMAMOTO PREFECTURE LIVESTOCK EXPERIMENT STATION)
KUMAMOTO-KEN CHIKUSAN SHIKENJO (KUMAMOTO PREFECTURE LIVESTOCK EXPERIMENT STATION)
3802, SAKAI, KOSHI-MACHI, KIKUCHI-GUN, KUMAMOTO
JAPANESE

KUMAMOTO-KEN CHIKUSAN SHIKENJO SHIKEN CHOSA SEISEKISHO 4485
(ANNUAL REPORT OF THE KUMAMOTO LIVESTOCK EXPERIMENT STATION)
KUMAMOTO-KEN CHIKUSAN SHIKENJO (KUMAMOTO PREFECTURE LIVESTOCK EXPERIMENT STATION)
3802, SAKAI, KOSHI-MACHI, KIKUCHI-GUN, KUMAMOTO 861-11
JAPANESE

KUMAMOTO-KEN YOKEI SHIKENJO KENKYU HOKOKU 4486
(BULLETIN OF KUMAMOTO PREFECTURAL POULTRY EXPERIMENT STATION)
KUMAMOTO-KEN YOKEI SHIKENJO (KUMAMOTO PREFECTURAL POULTRY EXPERIMENT STATION)
2250, IKURAKITAKATA, TAMANA-SHI 865
JAPANESE

KUMAMOTO-KEN YOKEI SHIKENJO SHIKEN SEISEKI HOKOKUSHO 4487
(REPORT OF KUMAMOTO PREFECTURAL POULTRY EXPERIMENT STATION)
KUMAMOTO-KEN YOKEI SHIKENJO
2250, IKURAKITAKATA, TAMANA-SHI 865
JAPANESE

KYOSO BARIKAGAKU KENKYUJO HOKOKU 4488
(BULLETIN OF LABORATORY OF RACING CHEMISTRY)
JAPANESE

KYOSOBA HOKEN KENKYUSHO HOKOKU 4489
(EXPERIMENTAL REPORTS OF EQUINE HEALTH LABORATORY)
NIHON CHUO KEIBAKAI KYOSOBA HOKEN KENKYUSHO (EQUINE HEALTH LABORATORY, JAPAN RACING ASSOCIATION)
27-7, 5-CHOME, TSURUMAKI, SETAGAYA-KU, TOKYO 154
JAPANESE AND ENGLISH
INDEXED: CHE

KYOTO-FU NOGYO SHIDOSHO CHIKUSAN SHIKEN-BU SHIKEN KENKYU HOKOKU 4490
(REPORT OF THE DEPARTMENT OF ANIMAL BREEDING, AGRICULTURE DEVELOPMENTAL STATION, KYOTO PREFECTURE)
KYOTO-FU NOGYO SHIDOSHO CHIKUSAN SHIKEN-BU
IDEN-CHO, AYABE-SHI 623-02
JAPANESE
INDEXED: CHE

KYOTO-FU NOGYO SHIDOSHO TANGO CHIKUSAN SHIKEN BUNJO GYOMU HOKOKU 4491
(OPERATION REPORT OF TANGO BRANCH LIVESTOCK EXPERIMENT STATION, AGRICULTURE DEVELOPMENTAL STATION, PREFECTURE OF KYOTO)
KYOTO-FU NOGYO SHIDOSHO TANGO CHIKUSAN SHIKEN BUNJO
NOMURA, INE-CHO, YOSA-GUN, KYOTO 626-03
JAPANESE

KYOTO-FU NOGYO SHIDOSHO TANGO CHIKUSAN SHIKEN BUNJO SHIKEN HOKOKU 4492
(SURVEY REPORT OF TANGO BRANCH LIVESTOCK EXPERIMENT STATION, AGRICULTURAL DEVELOPMENTAL STATION, PREFECTURE OF KYOTO)
KYOTO-FU NOGYO SHIDOSHO TANGO CHIKUSAN SHIKEN BUNJO
NOMURA, INE-CHO, YOSA-GUN, KYOTO 626-03
JAPANESE

KYOTO-FU NOGYO SHIDOSHO YOKEI SHIKEN-BU SHIKEN KENKYU HOKOKU 4493
(RESEARCH REPORT OF THE DEPARTMENT OF POULTRY, AGRICULTURE DEVELOPMENTAL STATION, PREFECTURE OF KYOTO)
KYOTO-FU NOGYO SHIDOSHO YOKEI SHIKEN-BU
IDEN-CHO, AYABE-SHI 623-02
JAPANESE

LACTOLOGY 4494
MORINAGA NYUGYO K.K. CHUO KENKYUSHO (CENTRAL RESEARCH LABORATORY, MORINAGA MILK INDUSTRY COMPANY, LIMITED)
4-22, 4-CHOME, MEGURO, MEGURO-KU, TOKYO 153
JAPANESE

MAKIBA 4495
(MAKIBA: QUARTERLY REPORT OF THE WAKAYAMA PREFECTURAL LIVESTOCK EXPERIMENT STATION)
WAKAYAMA-KEN CHIKUSAN SHIKENJO (WAKAYAMA PREFECTURAL LIVESTOCK EXPERIMENT STATION)
1, MIROTSU, SUSAMI-CHO, NISHIMURO-GUN, WAKAYAMA 649-31
JAPANESE

MIYAZAKI SHUCHIKU BOKUJO KAGOSHIMA SHIJO CHOSA SHIKEN SEISEKI HOKOKU 4496
(RESULT OF THE EXAMINATION, KAGOSHIMA BRANCH, MIYAZAKI LIVESTOCK BREEDING STATION)
NORIN-SHO MIYAZAKI SHUCHIKU BOKUJO KAGOSHIMA SHIJO
TSUKINO, OSUMI-CHO, SOO-GUN, KAGOSHIMA 899-82
JAPANESE

MIYAZAKI-KEN SOGO NOGYO SHIKENJO YOKEI SHIJO SHIKEN SEISEKI HOKOKUSHO 4497
(REPORT OF POULTRY BRANCH STATION, MIYAZAKI AGRICULTURAL EXPERIMENT STATION)
MIYAZAKI-KEN SOGO NOGYO SHIKENJO YOKEI SHIJO
SHIMONAKA, SADOHARA-CHO, MIYAZAKI-GUN, MIYAZAKI 880
JAPANESE

NAGANO-KEN CHIKUSAN GIJUTSU KENKYU HAPPYOSHU 4498
(COLLECTED RESEARCH FINDINGS FROM NAGANO PREFECTURE ON TECHNOLOGY FOR ANIMAL HUSBANDRY)
NAGANO-KEN NOSEIBU CHIKUSAN-KA, NIHON CHIKUSAN-GAKKAI HOKURIKU SHIBU NAGANO-KEN BUNKAI
(NAGANO PREFECTURE, AGRICULTURE ADMINISTRATION DIV., LIVESTOCK SECTION & JAPAN LIVESTOCK STUDY SOCIETY)
JAPANESE

NAGANO-KEN CHIKUSAN SHIKENJO KENKYU HOKOKU 4499
(RESEARCH REPORT OF ANIMAL HUSBANDRY EXPERIMENTAL STATION OF NAGANO PREFECTURE)
NAGANO-KEN CHIKUSAN SHIKENJO (ANIMAL HUSBANDRY EXPERIMENTAL STATION OF NAGANO PREFECTURE)
10847, KATAOKA, SHIOJIRI-SHI 399-07
JAPANESE

NAGOYA HISHIRYO KENSAJO JIGYO HOKOKU, SHIRYO NO BU 4500
(ANNUAL REPORT OF THE NAGOYA FERTILIZER AND FEED INSPECTION STATION, FEED SECTION)
NORIN-SHO NAGOYA HISHIRYO KENSAJO (NAGOYA FERTILIZER AND FEED INSPECTION STATION, MINISTRY OF AGRICULTURE AND FORESTRY)
C/O NORIN GODO CHOSHA, 2-2, 1-CHOME, SANNOMARU, NAKA-KU, NAGOYA-SHI 460
JAPANESE

NARA-KEN CHIKUSAN SHIKENJO GYOMU NARABINI SHIKEN KENKYU HOKOKUSHO 4501
(REPORT OF NARA PREFECTURAL LIVESTOCK EXPERIMENT STATION)
NARA-KEN CHIKUSAN SHIKENJO
SHIMOTAKE, OUDA-CHO, UDA-GUN, NARA 633-21
JAPANESE

NATIONAL INSTITUTE OF ANIMAL HEALTH QUARTERLY 4502
NORIN-SHO KACHIKU EISEI SHIKENJO (NATIONAL INSTITUTE OF ANIMAL HEALTH, MINISTRY OF AGRICULTURE AND FORESTRY)
1500, JOSUIHON-CHO, KODAIRA-SHI, TOKYO 187
EUROPEAN
INDEXED: BIB, BIO, EXC, JICST

NIHON CHIKUSAN GAKKAI HOKURIKU SHIBU TOYAMA-KEN BUNKAIHO 4503
(TOYAMA PREFECTURE CHAPTER REPORT OF THE HOKURIKU BRANCH OF THE JAPANESE SOCIETY OF ZOOTECHNICAL SCIENCE)
NIHON CHIKUSAN GAKKAI HOKURIKU SHIBU TOYAMA-KEN BUNKAI (TOYAMA PREFECTURE CHAPTER OF THE HOKURIKU BRANCH OF THE JAPANESE SOCIETY OF ZOOTECHNICAL SCIENCE)
C/O TOYAMA DAIGAKU KYOIKUGAKUBU, GOHUKU, TOYAMA-SHI 930
JAPANESE

NIHON CHIKUSAN GAKKAI KANTOBU KAIHO 4504
(REPORT OF THE JAPAN ANIMAL HUSBANDRY SOCIETY, KANTO BRANCH)
JAPAN ANIMAL HUSBANDRY SOCIETY, KANTO BRANCH
JAPANESE

NIHON CHIKUSAN GAKKAI TOHOKU SHIBU KAIHO 4505
(REPORT OF THE TOHOKU BRANCH OF THE JAPANESE SOCIETY OF ZOOTECHNICAL SCIENCE)
NIHON CHIKUSAN GAKKAI TOHOKU SHIBU (TOHOKU BRANCH, THE JAPANESE SOCIETY OF ZOOTECHNICAL SCIENCE)
C/O TOHOKU DAIGAKU NOGAKU-BU, 1-1, TSUTSUMI-TORI AMAMIYA-CHO, SENDAI-SHI 980
JAPANESE

NIHON CHIKUSAN GAKKAIHO 4506
(JAPANESE JOURNAL OF ZOOTECHNICAL SCIENCE)
NIHON CHIKUSAN GAKKAI (JAPANESE SOCIETY OF ZOOTECHNICAL SCIENCE)
C/O TOKYO DAIGAKU NOGAKUBU CHIKUSAN JUIGAKKA, 1-1, 1-CHOME, YAYOI, BUNKYO-KU, TOKYO 113
JAPANESE, WITH EUROPEAN SUMMARY
INDEXED: BIB, CHE, JICST

NIHON CHOGAKKAI NYUSU 4507
(JAPAN POULTRY STUDIES ASSOCIATION NEWS)
NIHON CHOGAKKAI, KOKURITSU KAGAKU HAKUBUTSUKANNAI (JAPAN POULTRY STUDIES ASSOCIATION WITHIN THE NATIONAL SCIENCE MUSEUM)
JAPANESE

NIHON DAIGAKU NOJUI GAKKAI KAISHI 4508
(JOURNAL OF THE AGRICULTURE AND VETERINARY MEDICINE NIHON UNIVERSITY)
NIHON DAIGAKU NOJUIGAKKAI (SOCIETY OF AGRICULTURE AND VETERINARY MEDICINE, NIHON UNIVERSITY)
34-1, 3-CHOME, SHIMOUMA, SETAGAYA-KU, TOKYO 154
JAPANESE

NIHON DAIGAKU NOJUIGAKUBU GAKUJUTSU KENKYU HOKOKU 4509
(BULLETIN OF THE COLLEGE OF AGRICULTURE AND VETERINARY MEDICINE, NIHON UNIVERSITY)
NIHON DAIGAKU NOJUIGAKKAI (SOCIETY OF AGRICULTURE AND VETERINARY MEDICINE, NIHON UNIVERSITY)
34-1, 3-CHOME, SHIMOUMA, SETAGAYA-KU, TOKYO 154
JAPANESE AND EUROPEAN
INDEXED: BIB, BIO, CHE, EXC, JICST

NIHON DAIGAKU NOJUIGAKUBU IPPAN KYOYO KENKYU KIYO 4510
(GENERAL EDUCATION REVIEW, COLLEGE OF AGRICULTURE AND VETERINARY MEDICINE, NIHON UNIVERSITY)
NIHON DAIGAKU NOJUIGAKUBU (COLLEGE OF AGRICULTURE AND VETERINARY MEDICINE, NIHON UNIVERSITY)
34-1, 3-CHOME, SHIMOUMA, SETAGAYA-KU, TOKYO 154
JAPANESE AND ENGLISH

NIHON JUI SHIGAKU ZASSHI 4511
(JAPAN VETERINARY SCIENCE HISTORICAL MAGAZINE)
NIHON JUI SHIGAKU KENKYUKAI (JAPAN VETERINARY SCIENCE HISTORICAL RESEARCH SOCIETY)
JAPANESE

NIHON JUIGAKU ZASSHI 4512
(JAPANESE JOURNAL OF VETERINARY SCIENCE)
NIHON JUIGAKKAI (JAPANESE SOCIETY OF VETERINARY SCIENCE)
C/O TOKYO DAIGAKU NOGAKUBU, 1-1, 1-CHOME, YAYOI, BUNKYO-KU, TOKYO 113
JAPANESE AND ENGLISH
INDEXED: BIB, CHE, JICST

NIHON JUISHIKAI ZASSHI 4513
(JOURNAL OF THE JAPAN VETERINARY MEDICAL ASSOCIATION)
NIHON JUISHIKAI (JAPAN VETERINARY MEDICAL ASSOCIATION)
5-18, 1-CHOME, MINAMIAOYAMA, MINATO-KU, TOKYO 107
JAPANESE AND ENGLISH
INDEXED: BIB, BIO, CHE, JICST

NIHON KAKIN GAKKAISHI 4514
(JAPANESE POULTRY SCIENCE)
NIHON KAKIN GAKKAI (JAPANESE POULTRY SCIENCE ASSOCIATION)
C/O NORIN-SHO CHIKUSAN SHIKENJO, AOBA-CHO, CHIBA-SHI 280
JAPANESE AND ENGLISH
INDEXED: BIB, CHE

NIHON MEN'YO KENKYUKAISHI 4515
(JAPANESE JOURNAL OF SHEEP SCIENCE)
NIHON MEN'YO KENKYUKAI
20-9, 3-CHOME, YUSHIMA, BUNKYO-KU, TOKYO 113
JAPANESE

NIHON NO NIKUGYU 4516
(JAPAN'S BEEF CATTLE)
ZENKOKU NIKUYUGYU KYOKAI (NATIONAL BEEF CATTLE ASSOCIATION)
JAPANESE

NIHON SHOKUNIKU KAKO JOHO 4517
(JAPAN MEAT PROCESSING JOURNAL)
NIHON SHOKUNIKU KAKO KYOKAI (JAPAN MEAT PROCESSORS ASSOCIATION)
5-6, 1-CHOME, EBISU, SHIBUYA-KU, TOKYO 150
JAPANESE
INDEXED: JICST

NIHON YOTON KENKYUKAISHI 4518
(JOURNAL OF JAPAN PIG BREEDING RESEARCH ASSOCIATION)
NIHON YOTON KENKYUKAI
C/O RAKUNO KAIKAN, 37-20, 1-CHOME, YOYOGI, SHIBUYA-KU, TOKYO 151
JAPANESE

NIHONKEI NO KENKYU 4519
(JAPANESE BREEDS OF FOWLS)
ZENNIHON KAKIN KYOKAI
462, NOBE, AKIGAWA-SHI, TOKYO 197
JAPANESE

NIIGATA-KEN CHIKUSAN SHIKENJO KENKYU HOKOKU 4520
(BULLETIN OF THE NIIGATA ANIMAL HUSBANDRY EXPERIMENT STATION)
NIIGATA ANIMAL HUSBANDRY EXPERIMENT STATION
JAPANESE

NIIGATA-KEN CHIKUSAN SHIKENJO NENPO 4521
(ANNUAL REPORT OF THE NIIGATA ANIMAL HUSBANDRY EXPERIMENT STATION)
NIIGATA-KEN CHIKUSAN SHIKENJO (NIIGATA ANIMAL HUSBANDRY EXPERIMENT STATION)
178, TANAHIRE, SHITADA-MURA, MINAMIKANBARA-GUN, NIIGATA 955-02
JAPANESE

NIIGATA-KEN KACHIKU HOKEN EISEI GYOSEKI HAPPYOKAI SHUROKU 4522
(PROCEEDINGS OF THE ANNUAL MEETING OF THE LIVESTOCK HYGIENE SERVICE, NIIGATA PREFECTURE)
NIIGATA-KEN NORIN-BU (NIIGATA PREFECTURAL GOVERNMENT)
602, GAKKO-DORI ICHIBAN-CHO, NIIGATA-SHI 951
JAPANESE

NIIGATA-KEN YOKEI SHIKENJO GYOMU NENPO 4523
(ANNUAL REPORT ON ACTIVITIES OF NIIGATA PREFECTURAL POULTRY EXPERIMENT STATION)
NIIGATA-KEN YOKEI SHIKENJO
NIKA, MAKI-MACHI, NISHIKANBARA-GUN, NIIGATA 953
JAPANESE

NIIGATA-KEN YOKEI SHIKENJO SHIKEN KENKYU HOKOKU 4524
(RESEARCH REPORT OF NIIGATA PREFECTURAL POULTRY EXPERIMENT STATION)
NIIGATA-KEN YOKEI SHIKENJO
NIKA, MAKI-MACHI, NISHIKANBARA-GUN, NIIGATA 953
JAPANESE

NIKU NO KAGAKU 4525
(MEAT SCIENCE)
NIHON SHOKUNIKU KENKYUKAI (JAPAN MEAT RESEARCH ASSOCIATION)
C/O HAMU SOSEIJI KAIKAN, 5-6, 1-CHOME, EBISU, SHIBUYA-KU, TOKYO 150
JAPANESE

NIPPON JUI CHIKUSAN DAIGAKU KIYO 4526
(BULLETIN OF THE NIPPON VETERINARY AND ZOOTECHNICAL COLLEGE)
NIPPON JUI CHIKUSAN DAIGAKU (NIPPON VETERINARY AND ZOOTECHNICAL COLLEGE)
7-1, 1-CHOME, SAKAIMINAMI-CHO, MUSASHINO-SHI, TOKYO 180
JAPANESE AND EUROPEAN
INDEXED: BIB

NIPPON KAGAKU SHIRYO KYOKAI KENKYU SHIRYO 4527
(RESEARCH MATERIALS, JAPAN SCIENCE MATERIALS ASSOCIATION)
JAPANESE
INDEXED: JICST

NIWATORI NO KENKYU 4528
(POULTRY RESEARCHES)
NIWATORI NO KENKYUSHA (POULTRY RESEARCHES PUBLISHING COMPANY)
C/O NO. 623, MARU BUILDING, 4-1, 2-CHOME, MARUNOUCHI, CHIYODA-KU, TOKYO 100
JAPANESE
INDEXED: BIB, CHE

NOGYO KANSOKU TOKEIHYO, CHIKUSAN-BU 4529
(AGRICULTURAL PROSPECT STATISTICS, SERIES ANIMAL INDUSTRY)
FUKUOKA-KEN NOSEI-BU (AGRICULTURE ADMINISTRATION DEPARTMENT, FUKUOKA PREFECTURAL GOVERNMENT)
1-1, 1-CHOME, TENJIN, FUKUOKA-SHI 810
JAPANESE

NORIN-SHO KACHIKU EISEI SHIKENJO KENKYU HOKOKU 4530
(BULLETIN OF THE NATIONAL INSTITUTE OF ANIMAL HEALTH)
NORIN-SHO KACHIKU EISEI SHIKENJO (NATIONAL INSTITUTE OF ANIMAL HEALTH, MINISTRY OF AGRICULTURE AND FORESTRY)
1500, JOSUIHON-CHO, KODAIRA-SHI, TOKYO 180
JAPANESE AND EUROPEAN
INDEXED: BIB, CHE, JICST

NORIN-SHO KACHIKU EISEI SHIKENJO NENPO 4531
(ANNUAL REPORT OF NATIONAL INSTITUTE OF ANIMAL HEALTH)
NORIN-SHO KACHIKU EISEI SHIKENJO (NATIONAL INSTITUTE OF ANIMAL HEALTH, MINISTRY OF AGRICULTURE AND FORESTRY)
1500, JOSUIHON-CHO, KODAIRA-SHI, TOKYO 187
JAPANESE
INDEXED: JICST

NYUGIKYO SHIRYO 4532
(BULLETIN OF JAPAN DAIRY TECHNICAL ASSOCIATION)
NIHON NYUGYO GIJUTSU KYOKAI (JAPAN DAIRY TECHNICAL ASSOCIATION)
3, KIOI-CHO, CHIYODA-KU, TOKYO 102
JAPANESE
INDEXED: JICST

NYUGYO JANARU 4533
(MILK JOURNAL)
NYUGYO JANARU (MILK JOURNAL)
C/O MITSUBISHI NAKA 28-GOKAN, 4-1, 1-CHOME, MARUNOUCHI, CHIYODA-KU, TOKYO 100
JAPANESE
INDEXED: JICST

NYUGYO JIHO 4534
(MILK TIMES)
NIHON NYUSEIHIN KYOKAI (JAPAN DAIRY PRODUCTS ASSOCIATION)
3, KIOI-CHO, CHIYODA-KU, TOKYO 102
JAPANESE

OBIHIRO CHIKUSAN DAIGAKU GAKUJUTSU KENKYU HOKOKU, DAI-1-BU 4535
(RESEARCH BULLETIN OF OBIHIRO ZOOTECHNICAL UNIVERSITY, SERIES 1)
OBIHIRO CHIKUSAN DAIGAKU (OBIHIRO ZOOTECHNICAL UNIVERSITY)
INADA-CHO, OBIHIRO-SHI 080
JAPANESE AND ENGLISH
INDEXED: BIB, BIO, CHE, JICST

OBIHIRO CHIKUSAN DAIGAKU GAKUJUTSU KENKYU HOKOKU, DAI-2-BU 4536
(RESEARCH BULLETIN OF OBIHIRO ZOOTECHNICAL UNIVERSITY, SERIES 2)
OBIHIRO CHIKUSAN DAIGAKU (OBIHIRO ZOOTECHNICAL UNIVERSITY)
INADA-CHO, OBIHIRO-SHI 080
JAPANESE AND ENGLISH

OITA-KEN CHIKUSAN SHIKENJO SHIKEN SEISEKI HOKOKUSHO 4537
(TEST REPORT OF THE OITA LIVESTOCK EXPERIMENT STATION)
OITA-KEN CHIKUSAN SHIKENJO (OITA LIVESTOCK EXPERIMENT STATION)
3989-1, KUSUMI, KUSUMI-MACHI, NAOIRI-GUN, OITA 878-02
JAPANESE
INDEXED: CHE

OKAYAMA-KEN RAKUNO SHIKENJO HOKOKU 4538
(BULLETIN OF OKAYAMA PREFECTURAL DAIRY EXPERIMENT STATION)
OKAYAMA-KEN RAKUNO SHIKENJO (OKAYAMA PREFECTURAL DAIRY EXPERIMENT STATION)
904, OTA, TSUYAMA-SHI 708
JAPANESE

OKAYAMA-KEN WAGYU SHIKENJO GYOMU HOKOKUSHO 4539
(REPORT OF OKAYAMA PREFECTURAL EXPERIMENT STATION ON JAPANESE BLACK BREED OF CATTLE)
OKAYAMA-KEN WAGYU SHIKENJO
2240-1, KOSAKABE, OSA-CHO, ATETSU-GUN, OKAYAMA 719-35
JAPANESE

OKAYAMA-KEN WAGYU SHIKENJO NENPO 4540
(ANNUAL REPORT OF OKAYAMA PREFECTURAL EXPERIMENT STATION ON JAPANESE BLACK BREED OF CATTLE)
OKAYAMA-KEN WAGYU SHIKENJO (OKAYAMA PREFECTURAL EXPERIMENT STATION ON JAPANESE BLACK BREED OF CATTLE)
2240-1, KOSAKABE, OSA-CHO, ATETSU-GUN, OKAYAMA 719-35
JAPANESE

OKAYAMA-KEN YOKEI SHIKENJO GYOMU GAIYO 4541
(REPORT OF OKAYAMA PREFECTURAL POULTRY EXPERIMENT STATION)
OKAYAMA-KEN YOKEI SHIKENJO (OKAYAMA PREFECTURAL POULTRY EXPERIMENT STATION)
111, ITA, MITSU-MACHI, MITSU-GUN, OKAYAMA 709-22
JAPANESE

OKAYAMA-KEN YOKEI SHIKENJO SHIKEN KENKYU HOKOKU 4542
(REPORT OF OKAYAMA POULTRY EXPERIMENT STATION)
OKAYAMA-KEN YOKEI SHIKENJO
111, ITA, MITSU-MACHI, MITSU-GUN, OKAYAMA 709-22
JAPANESE

OKINAWA-KEN CHIKUSAN SHIKENJO KENKYU HOKOKU 4543
(RESEARCH REPORT OF OKINAWA PREFECTURAL EXPERIMENTAL STATION OF ANIMAL HUSBANDRY)
OKINAWA-KEN CHIKUSAN SHIKENJO (OKINAWA PREFECTURAL EXPERIMENTAL STATION OF ANIMAL HUSBANDRY)
260, ARAKAWA KAMIBARU, HAEBARU-SON, SHIMAJIRI-GUN, OKINAWA 901
JAPANESE

OKINAWA-KEN CHIKUSAN SHIKENJO SHIKEN KENKYU HOKOKU 4544
(TEST AND RESEARCH BULLETIN OF THE OKINAWA PREFECTURE ANIMAL HUSBANDRY EXPERIMENT STATION)
JAPANESE

OKINAWA-KEN KACHIKU EISEI SHIKENJO NENPO 4545
(ANNUAL REPORT OF OKINAWA PREFECTURAL INSTITUTE OF ANIMAL HEALTH)
OKINAWA-KEN KACHIKU EISEI SHIKENJO (OKINAWA PREFECTURAL INSTITUTE OF ANIMAL HEALTH)
112, KOHAGURA, NAHA-SHI 902
JAPANESE

OSAKA HISHIRYO KENSAJO JIGYO HOKOKU SHIRYO NO BU 4546
(ANNUAL REPORT OF THE OSAKA FERTILIZER AND FEED INSPECTION STATION, FEED SECTION)
NORIN-SHO OSAKA HISHIRYO KENSAJO (OSAKA FERTILIZER AND FEED INSPECTION STATION, MINISTRY OF AGRICULTURE AND FORESTRY)
C/O 2-GOKAN BEKKAN, OSAKA GODO CHOSHA, 6-25, HOENZAKA-CHO, HIGASHI-KU, OSAKA-SHI 540
JAPANESE

PURAZUMA 4547
(PLASMA)
RAKUNO DAIGAKU ZASSHI KANKO KAI
(DAIRY FARMING COLLEGE JOURNAL PUBLICATION ASSOCIATION)
JAPANESE

RAKUNO GAKUEN DAIGAKU KIYO 4548
(JOURNAL OF THE COLLEGE OF DAIRYING)
RAKUNO GAKUEN DAIGAKU (COLLEGE OF DAIRYING)
582, NISHINOPPORO, EBETSU-SHI 069-01
JAPANESE AND EUROPEAN
INDEXED: BIB, BIO, CHE

RAKUNO JIJO 4549
(DAIRY FARMING)
RAKUNO JIJOSHA
C/O KINMON BUILDING, 1-42, KANDA JINBO-CHO, CHIYODA-KU, TOKYO 102
JAPANESE

RAKUNO KAGAKU NO KENKYU 4550
(JAPANESE JOURNAL OF DAIRY SCIENCE)
NIHON RAKUNO KAGAKU KENKYUKAI (JAPANESE DAIRY SCIENCE ASSOCIATION)
C/O TOHOKU DAIGAKU NOGAKUBU, 1-1, TSUTSUMI-DORI AMAMIYA-CHO, SENDAI-SHI 980
JAPANESE, WITH ENGLISH SUMMARY
INDEXED: BIB, BIO, CHE, JICST

RAKUNO KAGAKU SHOKUHIN NO KENKYU 4551
(JAPANESE JOURNAL OF DAIRY SCIENCE)
JAPANESE

SAGA-KEN CHIKUSAN SHIKENJO SHIKEN KENKYU HOKOKU 4552
(REPORT OF SAGA INSTITUTE OF ANIMAL INDUSTRY)
SAGA-KEN CHIKUSAN SHIKENJO
HIRAO, OKAWACHI-CHO, IMARI-SHI 848
JAPANESE

SAGA-KEN CHIKUSAN SHIKENJO SHIKEN KENKYU SEISEKISHO 4553
(JOURNAL OF SAGA INSTITUTE OF ANIMAL INDUSTRY)
SAGA-KEN CHIKUSAN SHIKENJO
HIRAO, OKAWACHI-CHO, IMARI-SHI 848
JAPANESE

SAGA-KEN YOKEI SHIKENJO GYOMU HOKOKU 4554
(OPERATION REPORT OF SAGA PREFECTURAL STATION OF POULTRY INDUSTRY)
SAGA-KEN YOKEI SHIKENJO
SHIMOTOGAWA, USHIZU-CHO, OGI-GUN, SAGA 849-03
JAPANESE

SAGA-KEN YOKEI SHIKENJO SHIKEN KENKYU HOKOKU 4555
(RESEARCH REPORT OF SAGA PREFECTURAL STATION OF POULTRY INDUSTRY)
SAGA-KEN YOKEI SHIKENJO
SHIMOTOGAWA, USHIZU-CHO, OGI-GUN, SAGA 849-03
JAPANESE

SAITAMA-KEN CHIKUSAN SHIKENJO GYOMU KOTEI HOKOKUSHO 4556
(OPERATION REPORT OF SAITAMA PREFECTURE ZOOTECHNICAL EXPERIMENT STATION)
SAITAMA-KEN CHIKUSAN SHIKENJO
784, SUGAHIRO, KONAN-MURA, OSATOGUN, SAITAMA 360-01
JAPANESE

SAITAMA-KEN CHIKUSAN SHIKENJO SHIKEN CHOSA SEISEKISHO 4557
(RESEARCH REPORT OF SAITAMA PREFECTURE ZOOTECHNICAL EXPERIMENT STATION)
SAITAMA-KEN CHIKUSAN SHIKENJO
784, SUGAHIRO, KONAN-MURA, OSATO-GUN, SAITAMA 360-01
JAPANESE

SAITAMA-KEN YOKEI SHIKENJO GYOMU HOKOKU 4558
(OPERATION REPORT OF SAITAMA PREFECTURAL POULTRY EXPERIMENT STATION)
SAITAMA-KEN YOKEI SHIKENJO
105, MORITOSHINDEN, HIDAKA-CHO, IRUMA-GUN, SAITAMA 350-12
JAPANESE

SAITAMA-KEN YOKEI SHIKENJO KENKYU HOKOKU 4559
(BULLETIN OF SAITAMA POULTRY EXPERIMENT STATION)
SAITAMA-KEN YOKEI SHIKENJO
105, MORITOSHINDEN, HIDAKA-CHO, IRUMA-GUN, SAITAMA 350-12
JAPANESE

SANNIKU NORYOKU KENTEI SEISEKI HOKOKUSHO 4560
(IBARAKI-KEN)
(MEAT PRODUCTIVITY TEST REPORT, IBARAKI PREFECTURE)
IBARAKI-KEN YOTON SHIKENJO (IBARAKI PREFECTURAL INSTITUTE OF SWINE INDUSTRY)
3240, SAKURA, EDOSAKI-MACHI, INASHIKI-GUN, IBARAKI 300-05
JAPANESE

SAPPORO HISHIRYO KENSAJO JIGYO HOKOKU SHIRYO NO BU 4561
(ANNUAL REPORT OF THE SAPPORO FERTILIZER AND FEED INSPECTION STATION, FEED SECTION)
NORIN-SHO SAPPORO HISHIRYO KENSAJO (SAPPORO FERTILIZER AND FEED INSPECTION STATION, MINISTRY OF AGRICULTURE AND FORESTRY)
NISHI-19-CHOME, KITA-2-JO, CHUO-KU, SAPPORO-SHI 060
JAPANESE

SENDAI HISHIRYO KENSAJO JIGYO HOKOKU, SHIRYO NO BU 4562
(ANNUAL REPORT OF THE SENDAI FERTILIZER AND FEED INSPECTION STATION, FEED SECTION)
NORIN-SHO SENDAI HISHIRYO KENSAJO (SENDAI FERTILIZER AND FEED INSPECTION STATION, MINISTRY OF AGRICULTURE AND FORESTRY)
SENDAI DAI-2 GODO CHOSHA, 2-23, 3-CHOME, HON-CHO, SENDAI-SHI 980
JAPANESE

SETSUIN NYUGYO GIJUTSU KENKYUSHO HOKOKU 4563
JAPANESE
INDEXED: BIB

SHIGA-KEN SHUCHIKUJO GYOMU KOTEI HOKOKU 4564
(OPERATION REPORT OF SHIGA PREFECTURAL ANIMAL BREEDING FARM)
SHIGA-KEN SHUCHIKUJO
YAMAMOTO, HINO-MACHI, GAMOU-GUN, SHIGA 529-16
JAPANESE

SHIGA-KEN SHUCHIKUJO HOKOKU 4565
(BULLETIN OF THE SHIGA PREFECTURAL ANIMAL BREEDING FARM)
SHIGA-KEN SHUCHIKUJO
YAMAMOTO, HINO-MACHI, GAMOU-GUN, SHIGA 529-16
JAPANESE

SHIGA-KEN SHUKEIJO GYOMU HOKOKU 4566
(BUSINESS REPORT OF THE SHIGA PREFECTURAL POULTRY BREEDING STATION)
SHIGA-KEN SHUKEIJO (SHIGA PREFECTURAL POULTRY BREEDING STATION)
YAGURA-CHO, KUSATSU-SHI 525
JAPANESE

SHIGA-KEN SHUKEIJO KENKYU HOKOKU 4567
(RESEARCH REPORT OF THE SHIGA PREFECTURAL POULTRY BREEDING STATION)
SHIGA-KEN SHUKEIJO (SHIGA PREFECTURAL POULTRY BREEDING STATION)
YAGURA-CHO, KUSATSU-SHI 525
JAPANESE

SHIMANE-KEN CHIKUSAN SHIKENJO GYOMU KOTEI HOKOKU 4568
(OPERATION REPORT OF SHIMANE INSTITUTE OF ANIMAL INDUSTRY)
SHIMANE-KENRITSU CHIKUSAN SHIKENJO
3775, KOSHI-CHO, IZUMO-SHI 693
JAPANESE

SHIMANE-KEN CHIKUSAN SHIKENJO SHIKEN SEISEKI HOKOKU 4569
(RESEARCH REPORT OF SHIMANE INSTITUTE OF ANIMAL INDUSTRY)
SHIMANE-KENRITSU CHIKUSAN SHIKENJO
3775, KOSHI-CHO, IZUMO-SHI 693
JAPANESE

SHIRYO GEPPO 4570
(MONTHLY FEED STATISTICS)
NORIN-SHO CHIKUSAN-KYOKU (ANIMAL INDUSTRY BUREAU, MINISTRY OF AGRICULTURE AND FORESTRY) AND NIHON SHIRYO KYOKAI (MAF JAPAN FEED COUNCIL)
C/O SHIRYO KAIKAN, 2-1, 2-CHOME, AZABUDAI, MINATO-KU, TOKYO 106
JAPANESE

SHIRYO TO CHIKUSAN 4571
(FEEDS AND ANIMAL INDUSTRY)
SHIRYO HIFPOSHA
5-8, 2-CHOME, UCHIKANDA, CHIYODA-KU, TOKYO 101
JAPANESE

SHIRYO TO SHIRYO KOGYO 4572
(FEEDS AND FEED INDUSTRY)
CHIKUSAN SHUPPANSHA
2-140, TOTSUKA-CHO, SHINJUKU-KU, TOKYO 162
JAPANESE

SHIZUOKA-KEN CHIKUSAN SHIKENJO KENKYU HOKOKU 4573
(RESEARCH REPORT OF THE SHIZUOKA PREFECTURE LIVE-STOCK BREEDING FARM)
SHIZUOKA-KEN CHIKUSAN SHIKENJO
1622, UCHINO, FUJINOMIYA-SHI 418-02
JAPANESE

SHIZUOKA-KEN KACHIKU EISEI KENKYUSHO KENKYU HOKOKU 4574
(REPORT OF THE LIVESTOCK SANITARY LABORATORY PRE-FECTURE OF SHIZUOKA)
SHIZUOKA-KEN KACHIKU EISEI KENKYUSHO (LIVESTOCK SANITARY LABORATORY, PREFECTURE OF SHIZUOKA)
333, SHIMOAOJIMA FUJIEDA-SHI 426
JAPANESE

SHIZUOKA-KEN YOKEI SHIKENJO SHIKEN KENKYU HOKOKU 4575
(RESEARCH REPORT OF SHIZUOKA PREFECTURAL POULTRY EXPERIMENT STATION)
SHIZUOKA-KEN YOKEI SHIKENJO
2770-1, SHIMOTARUKI, KAKEGAWA-SHI 436-02
JAPANESE

SHIZUOKA-KEN YOTON SHIKENJO KENKYU HOKOKU 4576
(RESEARCH REPORT OF THE SHIZUOKA PIG'S EXPERIMENT STATION)
SHIZUOKA-KEN YOTON SHIKENJO (SHIZUOKA PIG'S EXPERI-MENT STATION
3981, KOMATSU, HAMAKITA-SHI 434
JAPANESE

SHOKUNIKU KANKEI SHIRYO 4577
(MEAT STATISTICS IN JAPAN)
NORIN-SHO CHIKUSAN-KYOKU (ANIMAL INDUSTRY BUREAU, MINISTRY OF AGRICULTURE AND FORESTRY)
NIHON SHOKUNIKU KYOGIKAI (JAPAN MEAT COUNCIL)
1-2, KANDA SURUGADAI, CHIYODA-KU, TOKYO 101
JAPANESE

SHOKUNIKU RYUTSU TOKEI 4578
(STATISTICS ON MEAT MARKETING)
NORIN-SHO NORIN KEIZAI-KYOKU TOKEI JOHO-BU (STATISTICS AND INFORMATION DEPARTMENT, ECONOMIC AFFAIRS BUREAU, MINISTRY OF AGRICULTURE AND FORESTRY)
2-1, 1-CHOME, KASUMIGASEKI, CHIYODA-KU, TOKYO 100
JAPANESE

SHUCHIKU BOKUJO JIGYO HOKOKUSHO, SHIRYO NO BU 4579
(ANNUAL REPORT OF THE LIVESTOCK BREEDING STATION, SERIES FEED STUFF)
NORIN-SHO CHIKUSAN-KYOKU (ANIMAL INDUSTRY BUREAU, MINISTRY OF AGRICULTURE AND FORESTRY)
2-1, 1-CHOME, KASUMIGASEKI, CHIYODA-KU, TOKYO 100
JAPANESE

SHUCHIKU BOKUJO JIGYO SEISEKI HOKOKUSHO, SHUCHIKU NO BU 4580
(ANNUAL REPORT OF THE LIVESTOCK BREEDING STATION SERIES BREEDING STOCK)
NORIN-SHO CHIKUSAN-KYOKU (ANIMAL INDUSTRY BUREAU, MINISTRY OF AGRICULTURE AND FORESTRY)
2-1, 1-CHOME, KASUMIGASEKI, CHIYODA-KU, TOKYO 100
JAPANESE

SOTEI 4581
(SHOEING A HORSE)
NIHON SOTEISHIKAI (JAPAN FARRIERS' ASSOCIATION)
23-8, 1-CHOME, DAIZAWA, SETAGAYA-KU, TOKYO 155
JAPANESE

TAKIKAWA CHIKUSAN SHIKENJO KENKYU HOKOKU 4582
(BULLETIN OF THE HOKKAIDO PREFECTURAL TAKIKAWA ANIMAL HUSBANDRY EXPERIMENT STATION)
HOKKAIDORITSU TAKIKAWA CHIKUSAN SHIKENJO (TAKIKAWA ANIMAL HUSBANDRY EXPERIMENT STATION OF HOKKAIDO)
735, HIGASHITAKIKAWA, TAKIKAWA-SHI 073
JAPANESE
INDEXED: JICST

TAKIKAWA CHIKUSAN SHIKENJO NENPO 4583
(ANNUAL REPORT OF TAKIKAWA ANIMAL HUSBANDRY EXPERIMENT STATION OF HOKKAIDO)
HOKKAIDORITSU TAKIKAWA CHIKUSAN SHIKENJO (HOKKAIDO TAKIKAWA ANIMAL HUSBANDRY EXPERIMENT STATION OF HOKKAIDO)
735, HIGASHITAKIKAWA, TAKIKAWA-SHI 073
JAPANESE

TOCHIGI-KEN CHIKUSAN SHIKENJO SHIKEN SEISEKI HOKOKU 4584
(RESULT OF THE EXAMINATION, TOCHIGI PREFECTURE ANIMAL INDUSTRY EXPERIMENT STATION)
TOCHIGI-KEN CHIKUSAN SHIKENJO (TOCHIGI PREFECTURE ANIMAL INDUSTRY EXPERIMENT STATION)
HAGA-CHO, HAGA-GUN, TOCHIGI 321-33
JAPANESE

TOCHIGI-KEN CHIKUSAN SHIKENJO SHIRYO 4585
(REPORT OF TOCHIGI PREFECTURE EXPERIMENT STATION OF ANIMAL INDUSTRY)
TOCHIGI-KEN CHIKUSAN SHIKENJO (TOCHIGI PRE-FECTURE EXPERIMENTAL STATION OF ANIMAL INDUSTRY)
HAGA-CHO, HAGA-GUN, TOCHIGI 321-33
JAPANESE

TOCHIGI-KEN KACHIKU EISEI KENKYUSHO NENPO 4586
(ANNUAL REPORT OF ANIMAL HEALTH LABORATORY, PREFECTURE OF TOCHIGI)
TOCHIGI-KEN KACHIKU EISEI KENKYUSHO (ANIMAL HEALTH LABORATORY, PREFECTURE OF TOCHIGI)
2634-2, TOMATSURI-CHO, UTSUNOMIYA-SHI 320
JAPANESE

TOCHIGI-KEN RAKUNO SHIKENJO GYOMU HOKOKU 4587
(OPERATION REPORT OF TOCHIGI DAIRY EXPERIMENTAL STATION)
TOCHIGI-KEN RAKUNO SHIKENJO (TOCHIGI DAIRY EXPERI-MENTAL STATION)
298, MISHIMA, NISHINASUNO-MACHI, NASU-GUN, TOCHIGI 329-27
JAPANESE

TOCHIGI-KEN RAKUNO SHIKENJO KENKYU HOKOKU 4588
(BULLETIN OF THE TOCHIGI DAIRY EXPERIMENTAL STATION)
TOCHIGI-KEN RAKUNO SHIKENJO (TOCHIGI DAIRY EX-PERIMENTAL STATION)
298, MISHIMA, NISHINASUNO-MACHI, NASU-GUN, TOCHIGI 329-27
JAPANESE

TOJO JIGYO HOKOKU 4589
(SLAUGHTERHOUSE REPORT)
TOKYO-TO CHUO OROSHIURI SHIJO SHOKUNIKU SHIJO (MEAT MARKET, TOKYO METROPOLITAN GOVERNMENT)
2-1, 5-CHOME, TSUKIJI, CHUO-KU, TOKYO 104
JAPANESE

TOKUSHIMA-KEN CHIKUSAN SHIKENJO KENKYU HOKOKU 4590
(STUDIES OF TOKUSHIMA PREFECTURAL LIVESTOCK EXPERIMENT STATION)
TOKUSHIMA-KEN CHIKUSAN SHIKENJO (TOKUSHIMA PRE-FECTURAL LIVESTOCK EXPERIMENT STATION)
IZUMIDANI, KAMIITA-MACHI, ITANO-GUN, TOKUSHIMA 771-13
JAPANESE

TOKUSHIMA-KEN YOKEI SHIKENJO KENKYU HOKOKU 4591
(RESEARCH REPORT OF TOKUSHIMA PREFECTURAL POULTRY EXPERIMENT STATION)
TOKUSHIMA-KEN YOKEI SHIKENJO (TOKUSHIMA PREFECTURAL POULTRY EXPERIMENT STATION)
5-77, MINAMISHO-MACHI, TOKUSHIMA-SHI 770
JAPANESE

TOKYO HISHIRYO KENSAJO JIGYO HOKOKU, SHIRYO NO BU 4592
(ANNUAL REPORT OF THE NATIONAL FEEDS AND FERTILIZERS INSPECTION OFFICE, FEED SECTION)
NORIN-SHO TOKYO HISHIRYO KENSAJO (NATIONAL FEEDS AND FERTILIZER INSPECTION OFFICE, MINISTRY OF AGRICULTURE AND FORESTRY)
3-3, 1-CHOME, OTE-MACHI, CHIYODA-KU, TOKYO 100
JAPANESE

TOKYO HISHIRYO KENSAJO SHIRYO KENKYU HOKOKU 4593
(BULLETIN OF NATIONAL FEEDS AND FERTILIZERS INSPECTION OFFICE)
NORIN-SHO TOKYO HISHIRYO KENSAJO (NATIONAL FEEDS AND FERTILIZERS INSPECTION OFFICE, MINISTRY OF AGRICULTURE AND FORESTRY)
3-3, 1-CHOME, OTE-MACHI, CHIYODA-KU, TOKYO 100
JAPANESE AND ENGLISH

TOKYO JUIGAKU CHIKUSANGAKU ZASSHI 4594
(JOURNAL OF TOKYO SOCIETY OF VETERINARY AND ZOO TECHNICAL SCIENCE)
TOKYO JUI CHIKUSAN GAKKAI (TOKYO SOCIETY OF VETERINARY AND ZOOTECHNICAL SCIENCE)
C/O NIHON DAIGAKU NOJUIGAKUBU, 34-1, 3-CHOME, SHIMOUMA, SETAGAYA-KU, TOKYO 154
JAPANESE
INDEXED: BIB, CHE

TOKYO-TO CHIKUSAN SHIKENJO GYOMU NENPO 4595
(ANNUAL REPORT OF THE TOKYO METROPOLITAN LIVESTOCK EXPERIMENT STATION)
TOKYO-TO CHIKUSAN SHIKENJO (TOKYO METROPOLITAN LIVESTOCK EXPERIMENT STATION)
715, SHIN-MACHI, OME-SHI, TOKYO 198
JAPANESE

TOKYO-TO CHIKUSAN SHIKENJO SHIKEN KENKYU CHOSA HOKOKU 4596
(BULLETIN OF TOKYO METROPOLITAN LIVESTOCK EXPERIMENT STATION)
TOKYO-TO CHIKUSAN SHIKENJO (TOKYO METROPOLITAN LIVESTOCK EXPERIMENT STATION)
715, SHIN-MACHI, OME-SHI, TOKYO 198
JAPANESE

TOKYO-TO KACHIKU HOKEN EISEIJO JIGYO GAIYO 4597
(REPORT OF LIVESTOCK HYGIENE CENTER, TOKYO)
TOKYO-TO KACHIKU HOKEN EISEIJO
3-137, FUJIMI-CHO, TACHIKAWA-SHI, TOKYO 190
JAPANESE

TOKYO-TO NO CHIKUSAN 4598
(ANIMAL INDUSTRY IN TOKYO)
TOKYO-TO KEIZAI-KYOKU (ECONOMIC BUREAU, TOKYO METROPOLITAN GOVERNMENT)
8-1, 3-CHOME, MARUNOUCHI, CHIYODA-KU, TOKYO 100
JAPANESE

TOKYO-TO TAMA SHOKUNIKU EISEI KENSASHO JIGYO GAIYO 4599
(ANNUAL REPORT OF TAMA INSPECTION STATION FOR MEAT HYGIENE, TOKYO)
TOKYO-TO TAMA SHOKUNIKU EISEI KENSASHO
7-238, FUJIMI-CHO, TACHIKAWA-SHI, TOKYO 190
JAPANESE

TOTTORI-KEN CHIKUSAN SHIKENJO GYOMU HOKOKU 4600
(OPERATION REPORT OF TOTTORI ANIMAL HUSBANDRY EXPERIMENTAL STATION)
TOTTORI-KEN CHIKUSAN SHIKENJO
606, MATSUTANI, AKASAKI-CHO, TOHAKU-GUN, TOTTORI 689-25
JAPANESE

TOTTORI-KEN CHIKUSAN SHIKENJO SHIKEN KENKYU SEISEKI HOKOKU 4601
(RESEARCH REPORT OF TOTTORI ANIMAL HUSBANDRY EXPERIMENTAL STATION)
TOTTORI-KEN CHIKUSAN SHIKENJO
606, MATSUTANI, AKASAKI-CHO, TOHAKU-GUN, TOTTORI 689-25
JAPANESE

TOTTORI-KEN CHUSHO KACHIKU SHIKENJO GYOMU HOKOKU 4602
(OPERATION REPORT OF TOTTORI PREFECTURAL SWINE AND POULTRY EXPERIMENT STATION)
TOTTORI-KEN CHUSHO KACHIKU SHIKENJO (TOTTORI PREFECTURAL SWINE AND POULTRY EXPERIMENT STATION)
3210, RYOMITSUYANGI, YONAGO-SHI 683
JAPANESE

TOTTORI-KEN CHUSHO KACHIKU SHIKENJO KENKYU HOKOKU 4603
(RESEARCH REPORT OF TOTTORI PREFECTURAL SWINE AND POULTRY EXPERIMENT STATION)
TOTTORI-KEN CHUSHO KACHIKU SHIKENJO (TOTTORI PREFECTURAL SWINE AND POULTRY EXPERIMENT STATION)
3210, RYOMITSUYANAGI, YONAGO-SHI 683
JAPANESE

TOYAMA NO CHIKUSAN 4604
(ANIMAL INDUSTRY IN TOYAMA PREFECTURE)
TOYAMA-KEN CHIKUSAN-KA (ANIMAL INDUSTRY SECTION, TOYAMA PREFECTURAL GOVERNMENT)
1-7, SHINSOGAWA, TOYAMA-SHI 930
JAPANESE

TOYAMA-KEN CHIKUSAN SHIKENJO GYOMU SEISEKI HOKOKU 4605
(INVESTIGATION REPORT OF TOYAMA PREFECTURAL LIVESTOCK EXPERIMENT STATION)
TOYAMA-KEN CHIKUSAN SHIKENJO (TOYAMA PREFECTURAL LIVESTOCK EXPERIMENT STATION)
205, KASUGA, OSAWANO-MACHI, KAMINIIKAWA-GUN, TOYAMA 939-22
JAPANESE

TOYAMA-KEN CHIKUSAN SHIKENJO KENKYU HOKOKU 4606
(BULLETIN OF THE TOYAMA PREFECTURAL LIVESTOCK EXPERIMENT STATION)
TOYAMA-KEN CHIKUSAN SHIKENJO (TOYAMA PREFECTURAL LIVESTOCK EXPERIMENT STATION)
205, KASUGA, OSAWANO-MACHI, KAMINIIKAWA-GUN, TOYAMA 939-22
JAPANESE AND ENGLISH

TOYAMA-KEN CHIKUSAN SHIKENJO NENPO 4607
(ANNUAL REPORT OF TOYAMA PREFECTURAL LIVESTOCK EXPERIMENT STATION)
TOYAMA-KEN CHIKUSAN SHIKENJO (TOYAMA PREFECTURAL LIVESTOCK EXPERIMENT STATION)
205, KASUGA, OSAWANO-MACHI, KAMINIIKAWA-GUN, TOYAMA 939-22
JAPANESE

WAKAYAMA-KEN YOKEI SHIKENJO GYOMU NARABINI SHIKEN KENKYU HOKOKU 4608
(REPORT OF BUSINESS AND RESEARCH OF WAKAYAMA-KEN POULTRY EXPERIMENT STATION)
WAKAYAMA-KEN YOKEI SHIKENJO (WAKAYAMA-KEN POULTRY EXPERIMENT STATION)
1796, SHIGA, HIDAKA-CHO, HIDAKA-GUN, WAKAYAMA 644-01
JAPANESE

YAMAGATA-KENRITSU CHIKUSAN SHIKENJO SHIKEN KENKYU HOKOKU 4609
(REPORT OF YAMAGATA PREFECTURAL ANIMAL HUSBANDRY EXPERIMENT STATION)
YAMAGATA-KENRITSU CHIKUSAN SHIKENJO
KO-7057, HIGASHINE, HIGASHINE-SHI 999-37
JAPANESE

YAMAGUCHI JUIGAKU ZASSHI 4610
(YAMAGUCHI JOURNAL OF VETERINARY MEDICINE)
YAMAGUCHI-KEN JUIGAKKAI (YAMAGUCHI PREFECTURE VETERINARY SCIENCE ASSOCIATION)
JAPANESE

YAMAGUCHI-KEN SHUCHIKUJO GYOMU HOKOKUSHO 4611
(ANNUAL REPORT OF ANIMAL BREEDING FARM, YAMAGUCHI PREFECTURAL GOVERNMENT)
YAMAGUCHI-KEN SHUCHIKUJO (ANIMAL BREEDING FARM, YAMAGUCHI PREFECTURAL GOVERNMENT)
1200, KAWARA, ISA-CHO, MINE-SHI 759-22
JAPANESE

YAMAGUCHI-KEN SHUCHIKUJO SHIKEN SEISEKI 4612
(RESULT OF EXPERIMENTATION OF YAMAGUCHI PREFECTURAL POULTRY BREEDING STATION)
YAMAGUCHI-KEN SHUCHIKUJO
1200, KAWARA, ISA-CHO, MINE-SHI 759-22
JAPANESE

YAMAGUCHI-KEN SHUKEIJO KENKYU HOKOKU 4613
(REPORT OF YAMAGUCHI PREFECTURAL POULTRY BREEDING STATION)
YAMAGUCHI-KEN SHUKEIJO
261, MURE, HOFU-SHI 747
JAPANESE

YAMANASHI-KEN CHIKUSAN SHIKENJO SHIKEN KENKYU HOKOKU 4614
(RESEARCH REPORT OF THE YAMANASHI PREFECTURAL LIVESTOCK EXPERIMENT STATION)
YAMANASHI-KEN CHIKUSAN SHIKENJO (YAMANASHI PREFECTURAL LIVESTOCK EXPERIMENT STATION)
1821, TAKEHARADA, ICHINOMIYA-CHO, HIGASHIYATSUSHIRO-GUN, YAMANASHI 409-14
JAPANESE

YAMANASHI-KEN RAKUNO SHIKENJO KENKYU HOKOKU 4615
(BULLETIN OF YAMANASHI PREFECTURAL DAIRY EXPERIMENT STATION)
YAMANASHI-KEN RAKUNO SHIKENJO (YAMANASHI PREFECTURAL DAIRY EXPERIMENT STATION)
621, NAGASAKAKAMIJO, NAGASAKA-CHO, KITAKOMA-GUN, YAMANASHI 408
JAPANESE

YOKEI NO TOMO 4616
(POULTRY FRIEND)
NIHON CHIKUSAN SHINKOKAI (JAPANESE LIVESTOCK INDUSTRY PROMOTION CORPORATION)
C/O RAKUNO KAIKAN, 37-20, 1-CHOME, YOYOGI, SHIBUYA-KU, TOKYO 151
JAPANESE

YOKEI SEKAI 4617
(WORLD POULTRY JOURNAL)
YOKEI SEKAISHA
229, NISHIKI-MACHI, TOYOHASHI-SHI 440
JAPANESE

YOKEI SHIKENBU KENKYU HOKOKU 4618
(RESEARCH REPORT OF THE POULTRY EXPERIMENT DIVISION)
KYOTO-FU NOGYO SHIDOSHO (KYOTO MUNICIPAL AGRICULTURE DEVELOPMENT OFFICE)
JAPANESE

YOTON DAYORI 4619
(HOG RAISING LETTER)
NIHON SHUTON TOROKU KYOKAI (JAPAN SWINE REGISTRATION ASSOCIATION)
JAPANESE

YOTON DAYORI 4620
(PIG BREEDING NEWS)
ZENKOKU YOTON KYOKAI (JAPAN PIG BREEDING ASSOCIATION)
C/O RAKUNO KAIKAN, 37-20, 1-CHOME, YOYOGI, SHIBUYA-KU, TOKYO 151
JAPANESE

YOTON KAI 4621
(HOG WORLD)
CHIKUSAN SHUPPANSHA
JAPANESE

YOTON NO TOMO 4622
(PIG FRIENDS)
NIHON CHIKUSAN SHINKOKAI (JAPAN LIVESTOCK INDUSTRY PROMOTION CORPORATION)
C/O RAKUNO KAIKAN, 37-20, 1-CHOME, YOYOGI, SHIBUYA-KU, TOKYO 151
JAPANESE

YUKIJIRUSHI NYUGYO GIJUTSU KENKYUSHO HOKOKU 4623
(REPORT OF RESEARCH LABORATORY, SNOW BRAND MILK PRODUCTS COMPANY, LIMITED)
YUKIJIRUSHI NYUGYO K.K. GIJUTSU KENKYUSHO (RESEARCH LABORATORY, SNOW BRAND MILK PRODUCTS COMPANY, LIMITED)
1-2, 1-CHOME, MINAMIDAI, KAWAGOE-SHI 350
JAPANESE, WITH ENGLISH SUMMARY
INDEXED: CHE, JICST

ZAIRAI KACHIKU KENKYUKAI HOKOKU 4624
(REPORT OF THE SOCIETY FOR RESEARCHES ON NATIVE LIVESTOCK)
ZAIRAI KACHIKU KENKYUKAI AND NAGOYA CHIKUSANGAKU KENKYUSHO
20, OKITA, IZUMI-CHO, ANJO-SHI 444-12
JAPANESE

ZENKOKU KACHIKU HOKEN EISEISHO GYOSEKI HAPPYOKAI SHUROKU 4625
(PROCEEDINGS OF THE ANNUAL MEETING OF THE LIVESTOCK HYGIENE SERVICE CENTERS)
NORIN-SHO CHIKUSAN-KYOKU (ANIMAL INDUSTRY BUREAU, MINISTRY OF KANDA MIKURA-CHO, CHIYODA-KU, TOKYO 101
2-1, 1-CHOME, KASUMIGASEKI, CHIYODA-KU, TOKYO 100
JAPANESE
INDEXED: BIB

ZENKOKU KACHIKU HOKEN EISEISHO GYOSEKI SHOROKU 4626
(BULLETIN OF LIVESTOCK HYGIENE SERVICE CENTERS)
DOBUTSUYO SEIBUTSUGAKUTEKI SEIZAI KYOKAI
C/O SANKEI BUILDING, 12, CHIYODA-KU, TOKYO 101
JAPANESE

ZENRAKU SHINPO 4627
(DAIRY FARMING NEWS)
ZENKOKU RAKUNO KYOKAI (JAPAN DAIRY FARMING ASSOCIATION)
37-2, 1-CHOME, YOYOGI, SHIBUYA-KU, TOKYO 151
JAPANESE

Insect and Reptile Management, Breeding Industries, Sericulture

ACTA ODONATOLOGIA 4628
JAPANESE
INDEXED: BIB

AICHI-KEN NOGYO SOGO SHIKENJO KENKYU HOKOKU, D, SANGYO 4629
(RESEARCH BULLETIN OF THE AICHI-KEN AGRICULTURAL RESEARCH CENTER, SERIES D)
AICHI-KEN NOGYO SOGO SHIKENJO (AICHI-KEN AGRICULTURAL RESEARCH CENTER)
NAGAKUTE-CHO, AICHI-GUN, AICHI 480-11
JAPANESE, WITH ENGLISH SUMMARY
INDEXED: CHE

AKITA-KEN SANGYO SHIDO SENTA GYOMU NENPO 4630
(ANNUAL REPORT OF SERICULTURAL GUIDANCE CENTER, AKITA PREFECTURE)
AKITA-KEN SANGYO SHIDO SENTA
26, DOCHI, SEKIGUCHI, YUZAWA-SHI 012
JAPANESE

BI SAIENSU 4631
(BEE SCIENCE)
YOHO KENKYUSHO (BEE CULTURE LABORATORY)
SUISHOEN, OBATA, MORIYAMA-KU, NAGOYA-SHI 463
JAPANESE
INDEXED: BIB

BULLETIN OF THE JAPAN ENTOMOLOGICAL ACADEMY 4632
JAPAN ENTOMOLOGICAL ACADEMY
C/O BIOLOGICAL LABORATORY, NANZAN UNIVERSITY
18 YAMAZATOCHO, SHOWAU, NAGOYA 309
ENGLISH
INDEXED: BIO

CHIBA-KEN SANGYO SHIKENJO HOKOKU 4633
(REPORT OF CHIBA-KEN SERICULTURAL EXPERIMENT STATION)
CHIBA-KEN SANGYO SHIKENJO (CHIBA-KEN SERICULTURAL EXPERIMENT STATION)
1055-1, YUI, TOGANE-SHI, 238
JAPANESE

CHIBA-KEN SANGYO SHIKENJO NENPO 4634
(ANNUAL REPORT OF CHIBA-KEN SERICULTURAL EXPERIMENT STATION)
CHIBA-KEN SANGYO SHIKENJO
1055-1, YUI, TOGANE-SHI 283
JAPANESE

CHIBA-KEN SANGYO SHIKENJO SHIKEN SEISEKI GAIYO 4635
(BULLETIN OF CHIBA-KEN SERICULTURAL EXPERIMENT STATION)
CHIBA-KEN SANGYO SHIKENJO
1055-1, YUI, TOGANE-SHI 283
JAPANESE

FUKUOKA-KENRITSU NOGYO SHIKENJO SANGYO-BU SHIKEN SEISEKI NENPO 4636
(ANNUAL REPORT OF SERICULTURAL DEPARTMENT, FUKUOKA AGRICULTURAL EXPERIMENT STATION)
FUKUOKA-KENRITSU NOGYO SHIKENJO SANGYO-BU
232, KAMI-KOGA, CHIKUSHINO-SHI 818
JAPANESE

FUKUSHIMA-KEN SANGYO SHIKENJO YOHO 4637
(BULLETIN OF THE FUKUSHIMA PREFECTURE SERICULTURAL EXPERIMENT STATION)
FUKUSHIMA-KEN SANGYO SHIKENJO (FUKUSHIMA PREFECTURE SERICULTURAL EXPERIMENT STATION)
21, KITAMOTO-CHO, YANAGAWA-CHO, DATE-GUN, FUKUSHIMA 960-07
JAPANESE
INDEXED: CHE

FUKUSHIMA-KEN YOSAN TOKEI NENPO 4638
(ANNUAL REPORT OF SERICULTURAL STATISTICS, FUKUSHIMA)
NORIN-SHO TOHOKU NOSEI-KYOKU FUKUSHIMA TOKEI JOHO JIMUSHO (FUKUSHIMA STATISTICS AND TOHOKU NOSEI-KYOKU INFORMATION OFFICE, TOHOKU AGRICULTURAL ADMINISTRATION BUREAU)
4-15, YAMASHITA-CHO, FUKUSHIMA-SHI 960
JAPANESE

GIFU-KEN SANGYO SHIKENJO HOKOKU 4639
(BULLETIN OF THE GIFU PREFECTURE SERICULTURAL EXPERIMENT STATION)
GIFU-KEN SANGYO SHIKENJO (GIFU PREFECTURE SERICULTURAL EXPERIMENT STATION)
HACHIYA-CHO, MINOKAMO-SHI 505
JAPANESE

GIFU-KEN SANGYO SHIKENJO YOHO 4640
(REPORT OF GIFU SERICULTURAL EXPERIMENT STATION)
GIFU-KEN SANGYO SHIKENJO (GIFU SERICULTURAL EXPERIMENT STATION)
3481-2, KAMIHACHIYA, HACHIYA, MINO KAMOSHI 505
JAPANESE

GITO: KAGAKU TO ZUISO 4641
(ANT TOWER: SCIENCE AND STRAY THOUGHTS)
JAPANESE
INDEXED: BIB

GUNMA-KEN SANGYO SHIKENJO HOKOKU 4642
(REPORT OF GUNMA PREFECTURAL SERICULTURAL EXPERIMENT STATION)
GUNMA-KEN SANGYO SHIKENJO (GUNMA-KEN SERICULTURAL EXPERIMENT STATION)
2326-2, SOJA, SOJA-MACHI, MAEBASHI-SHI 371
JAPANESE
INDEXED: JICST

GUNMA-KEN SANGYO SHIKENJO NENPO 4643
(ANNUAL REPORT OF GUNMA PREFECTURAL SERICULTURAL EXPERIMENT STATION)
GUNMA-KEN SANGYO SHIKENJO
2326-2, SOJA, SOJA-MACHI, MAEBASHI-SHI 371
JAPANESE
INDEXED: JICST

GUNSAN YOHO 4644
(REPORT OF GUNMA PREFECTURAL SERICULTURAL EXPERIMENT STATION)
JAPANESE
INDEXED: JICST

HIROSHIMA-KEN SANGYO SHIDOSHO SHIKEN HOKOKU 4645
(BULLETIN OF THE HIROSHIMA PREFECTURE SERICULTURAL LEAD OFFICE)
HIROSHIMA-KEN SANGYO SHIDOSHO (HIROSHIMA PREFECTURE SERICULTURAL LEAD OFFICE)
HON-MACHI, SHOBARA-SHI 727
JAPANESE
INDEXED: JICST

HYOGO-KENRITSU SANGYO SHIKENJO GIJUTSU SHIRYO 4646
(TECHNICAL BULLETIN OF HYOGO PREFECTURE SERICULTURAL EXPERIMENT STATION)
HYOGO-KENRITSU SANGYO SHIKENJO
HIDAKA-CHO, KINOSAKI-GUN, HYOGO 669-53
JAPANESE

HYOGO-KENRITSU SANGYO SHIKENJO HOKOKU 4647
(BULLETIN OF THE HYOGO PREFECTURAL SERICULTURAL EXPERIMENT STATION)
HYOGO-KENRITSU SANGYO SHIKENJO (HYOGO PREFECTURE SERICULTURAL EXPERIMENT STATION)
HIDAKA-CHO, KINOSAKI-GUN, HYOGO 669-53
JAPANESE

HYOGO-KENRITSU SANGYO SHIKENJO KENKYU HAPPYOKAI SHIRYO 4648
(REPORT OF THE SYMPOSIUM OF HYOGO PREFECTURE SERICULTURAL EXPERIMENT STATION)
HYOGO-KENRITSU SANGYO SHIKENJO
HIDAKA-CHO, KINOSAKI-GUN, HYOGO 669-53
JAPANESE

IBARAKI-KEN BYOGAICHU KENKYU KAIHO 4649
(REPORT OF THE IBARAKI PREFECTURE SOCIETY FOR RESEARCH ON HARMFUL INSECTS)
IBARAKI-KEN BYOGAICHU KENKYUKAI (IBARAKI PREFECTURAL HARMFUL INSECTS RESEARCH GROUP)
JAPANESE

IBARAKI-KEN SANGYO SHIKENJO HOKOKU 4650
(BULLETIN OF THE IBARAKI PREFECTURE SERICULTURAL EXPERIMENT STATION)
IBARAKI-KEN SANGYO SHIKENJO (IBARAKI PREFECTURE SERICULTURAL EXPERIMENT STATION)
SEKIMOTO SEKIJO-MACHI, MAKABE-GUN, IBARAKI 308-01
JAPANESE

IBARAKI-KEN SANGYO SHIKENJO YOHO 4651
(REPORT OF IBARAKI PREFECTURE SERICULTURAL EXPERIMENT STATION)
IBARAKI-KEN SANGYO SHIKENJO (IBARAKI PREFECTURE SERICULTURAL EXPERIMENT STATION)
SEKIMOTO, SEKIJO-MACHI, MAKABE-GUN, IBARAKI 308-01
JAPANESE

IBARAKI-KEN SANSHIGYO TOKEI 4652
(ANNUAL STATISTICS OF SERICULTURE, IBARAKI PREFECTURE)
IBARAKI-KEN NORIN SUISAN-BU
5-38, 1-CHOME, SANNOMARU, MITO-SHI 310
JAPANESE

IWATE-KEN SANGYO SHIKENJO HOKOKU 4653
(BULLETIN OF THE IWATE PREFECTURE SERICULTURAL EXPERIMENT STATION)
IWATE-KEN SANGYO SHIKENJO (IWATE PREFECTURE SERICULTURAL EXPERIMENT STATION)
61, RYUGABABA, MIZUSAWA-SHI 023
JAPANESE

IWATE-KEN SANGYO SHIKENJO NENPO 4654
(ANNUAL REPORT OF IWATE PREFECTURE SERICULTURAL EXPERIMENT STATION)
IWATE-KEN SANGYO SHIKENJO (IWATE PREFECTURE SERICULTURAL EXPERIMENT STATION)
61, RYUGABABA, MIZUSAWA-SHI 023
JAPANESE

IWATE-KEN SANGYO SHIKENJO YOHO 4655
(MANUAL OF THE IWATE PREFECTURE SERICULTURE EXPERIMENT STATION)
JAPANESE

IWATE-KEN SANSHI TOKEI 4656
(ANNUAL STATISTICS OF SERICULTURE, IWATE PREFECTURE)
IWATE-KEN NOSEIBU (IWATE PREFECTURAL GOVERNMENT)
10-1, UCHIMARU, MORIOKA-SHI 020
JAPANESE

JAPANESE BEE JOURNAL 4657
YOHO GIJUTSU KYOKAI
JAPANESE
INDEXED: BIB

JOURNAL OF THE FACULTY OF TEXTILE SCIENCE AND TECHNOLOGY, SHINSHU UNIVERSITY, SERIES E. AGRICULTURE AND SERICULTURE 4658
SHINSHU DAIGAKU SEN'I GAKUBU (FACULTY OF TEXTILE SCIENCE AND TECHNOLOGY, SHINSHU UNIVERSITY)
15-1, 3-CHOME, TOKIDA, UEDA-SHI 386
JAPANESE AND ENGLISH
INDEXED: BIB, BIO, CHE

KAGOSHIMA-KEN SANGYO SHIKENJO HOKOKU 4659
(ANNUAL REPORT OF KAGOSHIMA SERICULTURAL EXPERIMENT STATION)
KAGOSHIMA-KEN SANGYO SHIKENJO (KAGOSHIMA PREFECTURE SERICULTURAL EXPERIMENT STATION)
YUDA, HIGASHI ICHIKI-MACHI, HICKI-GUN, KAGOSHIMA 899-22
JAPANESE

KANAGAWA NO SANSHIGYO 4660
(ANNUAL STATISTICS OF SERICULTURE, KANAGAWA PREFECTURE)
KANAGAWA-KEN NOSEI-BU (KANAGAWA PREFECTURAL GOVERNMENT)
1, NIHON ODORI, NAKA-KU, YOKOHAMA-SHI 231
JAPANESE

KANAGAWA-KEN SANGYO SENTA GYOMU HOKOKU 4661
(BUSINESS REPORT OF THE SERICULTURAL CENTER OF KANAGAWA PREFECTURE)
KANAGAWA-KEN SANGYO SENTA (SERICULTURAL CENTER OF KANAGAWA PREFECTURE)
2010, NAKASHINDEN, EBINA-SHI 243
JAPANESE

KANAGAWA-KEN SANGYO SENTA SHIKEN KENKYU HOKOKU 4662
(RESEARCH REPORT OF KANAGAWA PREFECTURE SERICULTURAL CENTER)
KANAGAWA-KEN SANGYO SENTA (KANAGAWA PREFECTURAL SERICULTURAL CENTER)
2010, NAKASHINDEN, EBINA-SHI 243
JAPANESE

KANAGAWA-KEN SANGYO SENTA SHIKENJO CHOSA SEISEKI GAIYO 4663
(BULLETIN OF KANAGAWA PREFECTURE SERICULTURAL CENTER)
KANAGAWA-KEN SANGYO SENTA
2010, NAKASHINDEN, EBINA-SHI 243
JAPANESE

KANAGAWA-KEN YOSANGYO JITTAI CHOSA SEISEKI 4664
(STATISTICAL SURVEY OF SERICULTURE IN KANAGAWA PREFECTURE)
KANAGAWA-KEN NOSEI-BU (KANAGAWA PREFECTURAL GOVERNMENT)
1, NIHON-ODORI, NAKA-KU YOKAMA SHI 231
JAPANESE

KOCHI-KEN SANGYO SHIKENJO KENKYU YOHO 4665
(BULLETIN OF THE KOCHI PREFECTURE SERICULTURAL EXPERIMENT STATION)
KOCHI-KEN SANGYO SHIKENJO (KOCHI PREFECTURE SERICULTURAL EXPERIMENT STATION)
1619-2, KUSUME, TOSAYAMA-MACHI, KAMI-GUN, KOCHI 782
JAPANESE

KUMAMOTO-KEN SANGYO SHIKENJO HOKOKU 4666
(BULLETIN OF THE KUMAMOTO SERICULTURAL EXPERIMENT STATION)
KUMAMOTO-KEN SANGYO SHIKENJO (KUMAMOTO PREFECTURE SERICULTURAL EXPERIMENT STATION)
238-1, IWANO, UEKI-MACHI, KAMOTO-GUN, KUMAMOTO 861-01
JAPANESE
INDEXED: CHE

KUWA TO KAIKO 4667
(MULBERRY AND SILKWORM)
KUMAMOTO-KEN SANGYO SHIKENJO (KUMAMOTO PREFECTURE SERICULTURAL EXPERIMENT STATION)
238-1, IWANO, UEKI-MACHI, KAMOTO-GUN, KUMAMOTO 861-01
JAPANESE

KYOTO-FU NOGYO SHIDOSHO SANGYO SHIKENBU KENKYU HOKOKU 4668
(REPORT OF SERICULTURAL DEPARTMENT, KYOTO PREFECTURAL AGRICULTURAL GUIDANCE OFFICE)
KYOTO-FU NOGYO SHIDOSHO
HINOKIMAE, IDEN-CHO, AYABE-SHI 623-02
JAPANESE
INDEXED: CHE

MAYU KENTEI GIJUTSU KENKYUKAISHI 4669
(ANNUAL REPORT OF THE SOCIETY FOR THE STUDY OF COCOON CERTIFICATION)
MAYU KENTEI GIJUTSU KENKYUKAI (SOCIETY FOR THE COCCON CERTIFICATION)
C/O YAMANASHI MAYU KENTEISHO, 11-13, 3-CHOME, ASAHI, KOFU-SHI 400
JAPANESE
INDEXED: JICST

MAYU KENTEI SEISEKI (EHIME-KEN) 4670
(SCIENTIFIC PAPERS OF COCOON GRADING, EHIME PREFECTURE)
EHIME-KEN MAYU KENTEISHO
7-1, TENJIN-CHO, UWAJIMA-SHI 798
JAPANESE

MAYU KENTEI SEISEKI (IWATE-KEN) 4671
(SCIENTIFIC PAPERS OF COCOON GRADING, IWATE PREFECTURE)
IWATE-KEN MAYU KENTEISHO
31-3, SHIMONO, SENHOKU-CHO, MORIOKA-SHI 020
JAPANESE

MAYU KENTEI SEISEKI (NORIN-SHO) 4672
(SCIENTIFIC PAPERS OF THE COCOON GRADING)
NORIN-SHO NOSANENGEI-KYOKU
2-1, 1-CHOME, KASUMIGASEKI, CHIYODA-KU, TOKYO 100
JAPANESE

MAYU SEISANHI CHOSA HOKOKU 4673
(SURVEY OF COST OF COCOON PRODUCTION)
NORIN-SHO NORIN KEIZAI-KYOKU TOKEI JOHO-BU AND NORIN TOKEI KYOKAI
11-14, 2-CHOME, MEGURO, MEGURO-KU, TOKYO 153
JAPANESE

MIE-KEN SANGYO SHIKENJO NENPO 4674
(ANNUAL REPORT OF MIE PREFECTURE SERICULTURAL EXPERIMENT STATION)
MIE-KEN SANGYO SHIKENJO (MIE PREFECTURE SERICULTURAL EXPERIMENT STATION)
FUJIKATA, TSU-SHI 514
JAPANESE

MIYAGI-KEN SANGYO SHIKENJO GIJUTSU SHIRYO 4675
(SERICULTURAL MATERIAL OF MIYAGI-KEN SERICULTURAL EXPERIMENTAL STATION)
MIYAGI-KEN SANGYO SHIKENJO (MIYAGI-KEN SERICULTURAL EXPERIMENTAL STATION)
70, TATEMINAMI, WATARI-CHO, WATARI-GUN, MIYAGI 989-23
JAPANESE

MIYAGI-KEN SANGYO SHIKENJO HOKOKU 4676
(REPORT OF MIYAGI-KEN SERICULTURAL EXPERIMENT STATION)
MIYAGI-KEN SANGYO SHIKENJO (MIYAGI-KEN SERICULTURAL EXPERIMENT STATION
70, TATEMINAMI, WATARI-CHO, WATARI-GUN, MIYAGI 989-23
JAPANESE

MIYAGI-KEN SANGYO SHIKENJO SANSO YOHO 4677
(SURVEY OF MIYAGI-KEN SERICULTURAL EXPERIMENT STATION)
MIYAGI-KEN SANGYO SHIKENJO (MIYAGI-KEN SERICULTURAL EXPERIMENT STATION)
70, TATEMINAMI, WATARI-CHO, WATARI-GUN, MIYAGI 989-23
JAPANESE

NAGANO SANSHI YOHO 4678
(REPORT OF THE NAGANO SERICULTURAL EXPERIMENT STATION)
NAGANO-KEN SANGYO SHIKENJO (NAGANO PREFECTURE SERICULTURAL EXPERIMENT STATION)
OHANA, SHIMOSHIOJIRI, UEDA-SHI 386
JAPANESE

NAGANO-KEN MAYU KENTEISHO GYOMU HOKOKU 4679
(REPORT OF COCOON TESTING STATION OF NAGANO PREFECTURE)
NAGANO-KEN MAYU KENTEISHO
7-23, 2-CHOME, SHONAI, MATSUMOTO-SHI
JAPANESE

NAGANO-KEN MAYU KENTEISHO KENKYU HAPPYO SHUROKU 4680
(PAPERS FROM THE COCOON TESTING STATION OF NAGANO PREFECTURE)
NAGANO-KEN MAYU KENTEISHO (COCOON TESTING STATION OF NAGANO PREFECTURE)
7-23, 2-CHOME, SHONAI, MATSUMOTO-SHI 390
JAPANESE

NAGANO-KEN SANGYO SHIKENJO HOKOKU 4681
(BULLETIN OF THE NAGANO SERICULTURAL EXPERIMENT STATION)
NAGANO-KEN SANGYO SHIKENJO (NAGANO PREFECTURE SERICULTURAL EXPERIMENT STATION)
OHANA, SHIMOSHIOJIRI, UEDA-SHI 386
JAPANESE, WITH ENGLISH SUMMARY

NAGANO-KEN SANSHIGYO TOKEI 4682
(SILK STATISTICS OF NAGANO PREFECTURE)
NAGANO-KEN NOSEI-BU (AGRICULTURAL ADMINISTRATION DIVISION, NAGANO PREFECTURAL GOVERNMENT)
HABASHITA, MINAMI NAGANO, NAGANO-SHI 380
JAPANESE

NAPI NEWS 4683
NAGOYA KONCHU DOKOKAI (NAGOYA INSECT SOCIETY)
JAPANESE AND ENGLISH

NIHON SANSHI SHINBUN 4684
(JAPAN SERICULTURAL NEWS)
NIHON SANSHI SHINBUNSHA
1-7, YURAKU-CHO, CHIYODA-KU, TOKYO 100
JAPANESE

NIHON SANSHIGAKU ZASSHI 4685
(JOURNAL OF SERICULTURAL SCIENCE OF JAPAN)
NIHON SANSHI GAKKAI (SERICULTURE SOCIETY OF JAPAN)
C/O NORIN-SHO SANSHI SHIKENJO, 55-30, 3-CHOME, WADA, SUGINAMI-KU, TOKYO 166
JAPANESE AND EUROPEAN
INDEXED: BIB, CHE, JICST

NIIGATA-KEN SANGYO SHIKENJO YOHO 4686
(REPORT OF NIIGATA-KEN SERICULTURAL LABORATORY)
NIIGATA-KEN SANGYO SHIKENJO (NIIGATA PREFECTURE SERICULTURAL LABORATORY)
USHIGAJIMA, KAWAGUCHI-MACHI, KITAUONUMA-GUN NIIGATA 949-75
JAPANESE

NIPPON NOYAKU GAKKAISHI 4687
(JOURNAL OF PESTICIDE SCIENCE)
PESTICIDE SCIENCE SOCIETY OF JAPAN
C/O JAPAN PLANT PROTECTION ASSOCIATION, 43-11, 1-CHOME, KOMAGOME, TOSHIMA-KU, TOKYO 170
JAPANESE
INDEXED: CHE, JICST

NOGYO CHOSA YOSAN BUMON KEKKA HOKOKUSHO 4688
(STATISTICS ON SERICULTURE)
NORIN-SHO NORIN KEIZAI-KYOKU TOKEI JOHO-BU (STATISTICS AND INFORMATION DIVISION, AGRICULTURE AND FORESTRY ECONOMIC BUREAU, MINISTRY OF AGRICULTURE AND FORESTRY)
2-1, 1-CHOME KASUMIGASEKI, CHIYODA-KU, TOKYO 100
JAPANESE

NOSANBUTSU SEISANHI CHOSA HOKOKU, MAYU SEISANHI 4689
(COST OF PRODUCTION OF COCOON)
NORIN-SHO NORIN KEIZAI-KYOKU TOKEI JOHO-BU (STATISTICS AND INFORMATION DIVISION, AGRICULTURE AND FORESTRY ECONOMIC BUREAU, MINISTRY OF AGRICULTURE AND FORESTRY)
11-14, 2-CHOME, MEGURO, MEGURO-KU, TOKYO 153
JAPANESE

ROYARUZERI NO KENKYU 4690
(ROYAL JELLY RESEARCH)
NIHON ROYARUZERI KENKYUKAI (JAPAN ROYAL JELLY RESEARCH SOCIETY) AND YOHO KENKYUSHO SHUPPAN-KYOKU (BEE CULTURE LABORATORY PRESS)
SUISHOEN, OBATA, MORIYAMA-KU, NAGOYA-SHI 463
JAPANESE

SAITAMA NO YOSAN 4691
(SERICULTURE IN SAITAMA PREFECTURE)
NORIN-SHO KANTO NOSEI-KYOKU SAITAMA TOKEI JOHO JIMUSHO (SAITAMA STATISTICS AND INFORMATION OFFICE, KANTO REGIONAL AGRICULTURAL ADMINISTRATION BUREAU, MINISTRY OF AGRICULTURE & FORESTRY)
14-19, 7-CHOME, KISHI-MACHI, URAWA-SHI 336
JAPANESE

SAITAMA-KEN SANGYO SHIKENJO GYOMU KOTEI 4692
(ANNUAL REPORT OF SAITAMA-KEN SERICULTURAL EXPERIMENT STATION)
SAITAMA-KEN SANGYO SHIKENJO (SAITAMA-KEN SERICULTURAL EXPERIMENT STATION)
1201, KUMAGAYA, KUMAGAYA-SHI 360
JAPANESE

SAITAMA-KEN SANGYO SHIKENJO HOKOKU 4693
(BULLETIN OF THE SAITAMA-KEN SERICULTURAL EXPERIMENT STATION)
SAITAMA-KEN SANGYO SHIKENJO (SAITAMA PREFECTURE SERICULTURAL EXPERIMENT STATION)
ISHIWARA, KUMAGAYA-SHI 360
JAPANESE

SAITAMA-KEN SANGYO SHIKENJO KENKYU YOHO 4694
(ESSENTIAL BULLETIN OF THE SAITAMA-KEN SERICULTURAL EXPERIMENT STATION)
SAITAMA-KEN SANGYO SHIKENJO (SAITAMA PREFECTURE SERICULTURAL EXPERIMENT STATION)
ISHIWARA, KUMAGAYA-SHI 360
JAPANESE

SAITAMA-KEN YOSAN TOKEI NENPO 4695
(ANNUAL REPORT OF SERICULTURAL STATISTICS, SAITAMA)
NORIN-SHO KANTO NOSEI-KYOKU SAITAMA TOKEI JOHO JIMUSHO (SAITAMA STATISTICS AND INFORMATION OFFICE, KANTO REGIONAL AGRICULTURAL ADMINISTRATION BUREAU, MINISTRY OF AGRICULTURE AND FORESTRY)
8-11, 5-CHOME, KISHI-MACHI, URAWA-SHI 335
JAPANESE

SANGYO FUKYU DAYORI 4696
(LETTERS ON SERICULTURE PROMOTION)
NORINSHO SANSHIKYOKU GIJUTSU KAIRYOKA (MINISTRY OF AGRICULTURE AND FORESTRY, SILKTHREAD DIVISION, TECHNICAL SKILL IMPROVEMENT SECTION)
JAPANESE

SANGYO SHIKEN SEISEKISHO 4697
(SERICULTURAL EXPERIMENT REPORT, MIYAZAKI PREFECTURE)
MIYAZAKI-KEN SOGO NOGYO SHIKENJO SANGYO-BU
4000, SHIMONAKA, SADOHARA-CHO, MIYAZAKI-GUN, MIYAZAKI 880-02
JAPANESE

SANSHI DAYORI 4698
(SERICULTURAL NEWS)
TOKYO-TO SANSHI SHIDOSHO
107, AKINO, HIKITA, AKIKAWA-SHI 197
JAPANESE

SANSHI KAGAKU KENKYUSHO IHO 4699
(REPORTS OF THE SILK SCIENCE RESEARCH INSTITUTE)
DAINIHON SANSHI-KAI, SANSHI KAGAKU KENKYU-SHO (SILK SCIENCE RESEARCH INSTITUTE OF DAINIHON RAW SILK FOUNDATION)
25-2, 3-CHOME, HYAKUNIN-CHO, SHINJUKU-KU, TOKYO 160
JAPANESE
INDEXED: JICST

SANSHI KAGAKU TO GIJUTSU 4700
(RAW SILK SCIENCE AND TECHNOLOGY)
ZENKOKU YOSAN NOGYO KYODO KUMIAI RENGOKAI
C/O SANSHI KAIKAN, 1-7, YURAKU-CHO, CHIYODA-KU, TOKYO 100
JAPANESE
INDEXED: CHE, JICST

SANSHI KANKEI SHIKEN SEISEKISHO (SHIZUOKA-KEN) 4701
(EXPERIMENTAL REPORT OF SERICULTURE, SHIZUOKA PREFECTURE)
SHIZUOKA-KEN SANGYO SHIKENJO (SHIZUOKA PREFECTURE SERICULTURAL EXPERIMENT STATION)
1-70, TA-MACHI, SHIZUOKA-SHI 420
JAPANESE

SANSHI KENKYU 4702
(ACTA SERICOLOGICA)
NORIN-SHO SANSHI SHIKENJO (SERICULTURAL EXPERIMENT STATION, MINISTRY OF AGRICULTURE AND FORESTRY)
55-30, 3-CHOME, WADA, SUGINAMI-KU, TOKYO 166
JAPANESE, WITH ENGLISH SUMMARY
INDEXED: BIB, CHE, JICST

SANSHI KINU JOHO 4703
(INFORMATION OF SERICULTURE AND RAW SILK PRODUCTION)
SANSHI KINU JOHOSHA
19-20, 1-CHOME, KINGUMAE, SHIBUYA-KU, TOKYO 150
JAPANESE

SANSHI NENKAN 4704
(YEARBOOK OF SERICULTURE)
NIHON SANSHI KOHO KYOKAI
JAPANESE
INDEXED: BIB

SANSHI NO HIKARI 4705
(GLORY OF RAW SILK)
ZENKOKU YOSAN NOGYO KYODO KUMIAI RENGOKAI (NATIONAL SILK RAISING AGRICULTURAL COOPERATIVE UNION)
C/O SANSHI KAIKAN, 1-7, YURAKUCHO, CHIYODA-KU, TOKYO 100
JAPANESE

SANSHI SHIKENJO HOKOKU 4706
(BULLETIN OF THE SERICULTURAL EXPERIMENT STATION)
NORIN-SHO SANSHI SHIKENJO (SERICULTURAL EXPERIMENT STATION, MINISTRY OF AGRICULTURE AND FORESTRY)
55-30, 3-CHOME, WADA, SUGINAMI-KU, TOKYO 166
JAPANESE, WITH ENGLISH SUMMARY
INDEXED: BIB, BIO, CHE, JICST

SANSHI SHIKENJO IHO 4707
(TECHNICAL BULLETIN OF THE SERICULTURAL EXPERIMENT STATION)
NORIN-SHO SANSHI SHIKENJO (SERICULTURAL EXPERIMENT STATION, MINISTRY OF AGRICULTURE & FORESTRY)
55-30, 3-CHOME, WADA, SUGINAMI-KU, TOKYO 166
JAPANESE
INDEXED: JICST

SANSHI SHIKENJO NENPO 4708
(ANNUAL REPORT OF THE SERICULTURAL EXPERIMENT STATION)
NORIN-SHO SANSHI SHIKENJO (SERICULTURAL EXPERIMENT STATION, MINISTRY OF AGRICULTURE AND FORESTRY)
55-30, 3-CHOME, WADA, SUGINAMI-KU, TOKYO 166
JAPANESE
INDEXED: JICST

SANSHI SHIKENJO SHIRYO 4709
(REPORT OF SERICULTURAL EXPERIMENT STATION, MINISTRY OF FORESTRY AND AGRICULTURE)
NORIN-SHO SANSHI SHIKENJO
55-30, 3-CHOME, WADA, SUGINAMI-KU, TOKYO 166
JAPANESE
INDEXED: JICST

SANSHI TOKEI GEPPO 4710
(RAW SILK STATISTICAL MONTHLY)
NORIN-SHO NOSAN ENGEI-KYOKU (RAW SILK AND HORTICULTURE BUREAU, MINISTRY OF AGRICULTURE AND FORESTRY) AND CHUO SANSHI KYOKAI
C/O SANSHI KAIKAN, 1-7, YURAKU-CHO, CHIYODA-KU, TOKYO 100
JAPANESE
INDEXED: JICST

SANSHI TOKEI OYOBI CHOSAHYO 4711
(STATISTICAL REPORT OF SERICULTURE, MIYAGI PREFECTURE)
MIYAGI-KEN NOSEI-BU
8-1, 3-CHOME, HON-CHO, SENDAI-SHI 980
JAPANESE

SANSHIGYO YORAN 4712
(HANDBOOK OF SERICULTURAL INDUSTRY)
NORIN-SHO NOSAN ENGEI-KYOKU (RAW SILK AND HORTICULTURE BUREAU, MINISTRY OF AGRICULTURE AND FORESTRY) AND NIHON SANSHI SHINBUNSHA
1-7, YURAKU-CHO, CHIYODA-KU, TOKYO 100
JAPANESE

SANSHU NI KANSURU TOKEI TO SHIRYO 4713
(STATISTICS ON SILKWORM SPECIES)
NORIN-SHO NOSAN ENGEI-KYOKU (RAW SILK AND HORTICULTURE BUREAU, MINISTRY OF AGRICULTURE AND FORESTRY)
2-1, 1-CHOME KASUMIGASEKI, CHIYODA-KU, TOKYO 100
JAPANESE

SANSO GIJUTSU SODAN 4714
(SERICULTURE TECHNIQUE CONSULTATION)
NORIN-SHO NOSAN ENGEI-KYOKU (RAW SILK AND HORTICULTURE BUREAU, MINISTRY OF AGRICULTURE AND FORESTRY) AND NIHON SANSHI SHINBUNSHA
1-7, YURAKU-CHO, CHIYODA-KU, TOKYO 100
JAPANESE

SANSO KENKYUSHO IHO 4715
(JOURNAL OF SERICULTURE RESEARCH LABORATORY, SERICULTURAL DIVISION, GUNZE COMPANY, LIMITED)
GUNZE K.K. SANSHI JIGYOBU SANSO KENKYUSHO
AONO-CHO, AYABE-SHI 623
JAPANESE

SHICHOSON-BETSU YOSAN TOKEI 4716
(STATISTICAL REPORT OF SERICULTURE BY TOWN)
NORIN-SHO NORIN KEIZAI KYOKU TOKEI CHOSA-BU
2-1, 1-CHOME, KASUMIGASEKI, CHIYODA-KU, TOKYO 100
JAPANESE

SHIGA-KEN SANGYO SHIKENJO HOKOKU 4717
(BULLETIN OF THE SHIGA PREFECTURE SERICULTURAL EXPERIMENT STATION)
SHIGA-KEN SANGYO SHIKENJO (SHIGA PREFECTURE SERICULTURAL EXPERIMENT STATION)
4-38, JIFUKUJI-CHO, NAGAHAMA-SHI 526
JAPANESE

SHIGA-KEN SANSHI GIJUTSU KENKYUKAI DAYORI 4718
(NEWS OF THE RESEARCH SOCIETY OF SERICULTURE, SHIGA PREFECTURE)
SHIGA-KEN SANGYO SHIKENJO
4-38, JIFUKUJI-CHO, NAGAHAMA-SHI 526
JAPANESE

SHIGA-KEN SANSHI SANSO GIJUTSU SOKUHO 4719
(TECHNICAL REPORT OF SERICULTURAL EXPERIMENT STATION, SHIGA PREFECTURE)
SHIGA-KEN SANGYO SHIKENJO
4-38, JIFUKUJI-CHO, NAGAHAMA-SHI 526
JAPANESE

SHIGA-KEN SANSHIGYO TOKEI 4720
(STATISTICAL REPORT OF SERICULTURE, SHIGA PREFECTURE)
SHIGA-KEN
1-1, 4-CHOME, KYO-MACHI, OTSU-SHI 520
JAPANESE

SHIMANE NO SANSHI 4721
(STATISTICAL REPORT OF SERICULTURE, SHIMANE PREFECTURE)
SHIMANE-KEN
1, TONO-MACHI, MATSUE-SHI 690
JAPANESE

SOEN OYOBI SOBYO NI KANSURU CHOSA 4722
(REPORT OF MULBERRY GARDEN)
NORIN-SHO NOSAN ENGEI-KYOKU
2-1, 1-CHOME, KASUMIGASEKI, CHIYODA-KU, TOKYO 100
JAPANESE

TOCHIGI-KEN SANGYO SHIKENJO KENKYU YOHO 4723
(SUMMARY OF THE TOCHIGI PREFECTURE SERICULTURAL EXPERIMENT STATION)
TOCHIGI-KEN SANGYO SHIKENJO (TOCHIGI PREFECTURE SERICULTURAL EXPERIMENT STATION)
3311-122, YAKUSHIJI, MINAMI KAWACHI-MACHI, KAWACHI-GUN, TOCHIGI 329-04
JAPANESE

TOKUSHIMA-KEN MAYU KENTEISHO KENKYU CHOSA HOKOKU 4724
(INVESTIGATION REPORT OF THE COCOON TESTING STATION OF TOKUSHIMA PREFECTURE)
TOKUSHIMA-KEN MAYU KENTEISHO (COCOON TESTING STATION OF TOKUSHIMA PREFECTURE)
87-1, KAMOJIMA-MACHI, OE-GUN, TOKUSHIMA 776
JAPANESE

TOKUSHIMA-KEN SANGYO SHIKENJO HOKOKU 4725
(BULLETIN OF THE TOKUSHIMA PREFECTURE SERICULTURAL EXPERIMENT STATION)
TOKUSHIMA-KEN SANGYO SHIKENJO (TOKUSHIMA PREFECTURE SERICULTURAL EXPERIMENT STATION)
KAMOJIMA-CHO, OE-GUN, TOKUSHIMA 776
JAPANESE
INDEXED: JICST

TOKUSHIMA-KEN SANGYO SHIKENJO NENPO 4726
(ANNUAL REPORT OF TOKUSHIMA PREFECTURE SERICULTURAL EXPERIMENT STATION)
TOKUSHIMA-KEN SANGYO SHIKENJO (TOKUSHIMA PREFECTURE SERICULTURAL EXPERIMENT STATION)
KAMOJIMA-CHO, OE-GUN, TOKUSHIMA 776
JAPANESE

TOKYO-TO SANGYO SHIKENJO NENPO 4727
(ANNUAL OF TOKYO METROPOLITAN SERICULTURAL EXPERIMENTAL STATION)
JAPANESE
INDEXED: JICST

TOKYO-TO SANSHI SHIDOSHO NENPO 4728
(ANNUAL REPORT OF TOKYO METROPOLITAN SERICULTURE GUIDANCE OFFICE)
TOKYO-TO SANSHI SHIDOSHO
107, AKINO, HIKITA, AKIKAWA-SHI, TOKYO 197
JAPANESE

TOKYO-TO SANSHI SHIDOSHO SHIKEN YOHO 4729
(BULLETIN OF TOKYO METROPOLITAN SERICULTURE GUIDANCE OFFICE)
TOKYO-TO SANSHI SHIDOSHO (TOKYO METROPOLITAN SERICULTURE GUIDANCE OFFICE)
107, AKINO, HIKIDA, AKIGAWA-SHI, TOKYO 197
JAPANESE
INDEXED: JICST

TORISAN YOHO 4730
(SUMMARY OF SERICULTURE IN TOTTORI PREFECTURE)
SANGYO GIJUTSU RENRAKU KYOGIKAI (ASSOCIATION FOR SERICULTURAL TECHNOLOGY)
C/O TOTTORI-KEN SANGYO SHIKENJO, 883-85, OTANI, KURAYOSHI-SHI 682
JAPANESE

TOTTORI-KEN SANGYO SHIKENJO HOKOKU 4731
(BULLETIN OF THE TOTTORI PREFECTURE SERICULTURAL EXPERIMENT STATION)
TOTTORI-KEN SANGYO SHIKENJO (TOTTORI PREFECTURE SERICULTURAL EXPERIMENT STATION)
883-85, OTANI, KURAYOSHI-SHI 682
JAPANESE
INDEXED: JICST

TOYAMA-KEN SANGYO SHIKENJO KENKYU HOKOKU 4732
(RESEARCH REPORT OF TOYAMA SERICULTURAL EXPERIMENT STATION)
TOYAMA-KEN SANGYO SHIKENJO (TOYAMA SERICULTURAL EXPERIMENT STATION)
2780, KAMISHIN-MACHI, YATSUO-MACHI, NEI-GUN, TOYAMA 939-23
JAPANESE

TOYAMA-KEN SANGYO SHIKENJO SEISEKI 4733
(ANNUAL REPORT OF TOYAMA PREFECTURAL SERICULTURAL EXPERIMENT STATION)
TOYAMA-KEN SANGYO SHIKENJO
2870, KAMISHIN-MACHI, YATSUO-MACHI, NEI-GUN, TOYAMA 939-23
JAPANESE

YAMAGATA NO YOSAN 4734
(ANNUAL REPORT OF SERICULTURAL STATISTICS, YAMAGATA)
NORIN-SHO TOHOKU NOSEI-KYOKU YAMAGATA TOKEI JOHO JIMUSHO (YAMAGATA STATISTICS AND SURVEY OFFICE, TOHOKU REGIONAL AGRICULTURAL ADMINISTRATION BUREAU, MINISTRY OF AGRICULTURE AND FORESTRY)
5-27, KASUGA-CHO, YAMAGATA-SHI 990
JAPANESE

YAMAGATA-KEN MAYU KENTEISHO KENKYU HOKOKU 4735
(RESEARCH REPORT OF THE COCOON-TESTING STATION OF YAMAGATA)
YAMAGATA-KEN MAYU KENTEISHO (COCOON TESTING STATION OF YAMAGATA)
5275, TATEOKA, MURAYAMA-SHI 995
JAPANESE

YAMAGATA-KEN SANGYO SHIKENJO YOHO 4736
(SUMMARY REPORT OF YAMAGATA PREFECTURE SERICULTURAL EXPERIMENT STATION)
YAMAGATA-KEN SANGYO SHIKENJO (YAMAGATA PREFECTURE SERICULTURAL EXPERIMENT STATION)
4, HIGASHI HARA-MACHI, YAMAGATA-SHI 990
JAPANESE
INDEXED: JICST

YAMANASHI-KENRITSU SANGYO SHIKENJO HOKOKU 4737
(BULLETIN OF THE YAMANASHI PREFECTURAL SERICULTURAL EXPERIMENT STATION)
YAMANASHI-KENRITSU SANGYO SHIKENJO (YAMANASHI PREFECTURAL SERICULTURAL EXPERIMENT STATION)
1700, SHOBUZAWA, FUTABA-CHO, KITAKOMA-GUN, YAMANASHI 407-01
JAPANESE

YAMANASHI-KENRITSU SANGYO SHIKENJO KENKYU YOHO 4738
(REPORT OF THE YAMANASHI PREFECTURAL SERICULTURAL EXPERIMENT STATION)
YAMANASHI-KENRITSU SANGYO SHIKENJO
1700, SHOBUSAWA, FUTABA-CHO, KITAKOMA-GUN YAMANSHI 407-01
JAPANESE

YOSAN NI KANSURU SANKO TOKEI 4739
(STATISTICAL REPORT OF SERICULTURE, MINISTRY OF FORESTRY AND AGRICULTURE)
NORIN-SHO NOSAN ENGEI-KYOKU
2-1, 1-CHOME, KASUMIGASEKI, CHIYODA-KU, TOKYO 100
JAPANESE

YOSAN TOKEI NENPO 4740
(ANNUAL STATISTICS OF SERICULTURE)
NORIN-SHO, NORIN KEIZAI-KYOKU, TOKEI JOHO-BU (STATISTICS AND INFORMATION DIVISION, AGRICULTURE AND FORESTRY ECONOMIC BUREAU, MINISTRY OF AGRICULTURE AND FORESTRY)
2-1, 1-CHOME, KASUMIGASEKI, CHIYODA-KU, TOKYO 100
JAPANESE

Fisheries

ABASHIRI SUISHI DAYORI 4741
(NEWSLETTER OF THE HOKKAIDO ABASHIRI FISHERIES EXPERIMENTAL STATION)
HOKKAIDORITSU ABASHIRI SUISAN SHIKENJO (HOKKAIDO ABASHIRI FISHERIES EXPERIMENTAL STATION)
35, MASUURA, ABASHIRI-SHI 099-31
JAPANESE

ADVANCES IN FISHERIES OCEANOGRAPHY 4742
SUISAN KAIYO KENKYU-KAI (JAPANESE SOCIETY OF FISHERIES OCEANOGRAPHY)
C/O TOKYO SUISAN DAIGAKU GYOJO KANKYOGAKU KYOSHITSU, 5-7, 4-CHOME, KONAN, MINATO-KU, TOKYO 108
ENGLISH

AICHI-KEN SUISAN SHIKENJO GYOMU HOKOKU 4743
(REPORT OF AICHI PREFECTURE FISHERY EXPERIMENTAL STATION)
AICHI-KEN SUISAN SHIKENJO (AICHI PREFECTURE FISHERY EXPERIMENTAL STATION)
97, WAKAMIYO, MIYA-CHO, GAMAGORI-SHI 443
JAPANESE

AKARUI GYOSON 4744
(BETTER LIVING FROM THE SEA)
SUISAN-CHO KENKYU KAIHATSU-BU (DEVELOPMENT AND EXTENSION DIVISION, FISHERIES AGENCY)
2-1, 1-CHOME, KASUMIGASEKI, CHIYODA-KU, TOKYO 100
JAPANESE

AKITA-KEN SUISAN SHIKENJO JIGYO HOKOKUSHO 4745
(REPORT OF AKITA PREFECTURE FISHERIES EXPERIMENT STATION)
AKITA-KEN SUISAN SHIKENJO
FUNAGAWA MINATO-MACHI, OGA-SHI 010-05
JAPANESE

AOMORI NORIN SUISAN TOKEI NENPO SUISAN-HEN 4746
(ANNUAL STATISTICS OF AGRICULTURE, FORESTRY AND FISHERY IN AOMORI PREFECTURE, SERIES FISHERY))
NORIN-SHO TOHOKU NOSEI-KYOKU AOMORI TOKEI JOHO JIMUSHO (AOMORI STATISTICS AND INFORMATION OFFICE, TOHOKU REGIONAL AGRICULTURAL ADMINISTRATION BUREAU, MINISTRY OF AGRICULTURE AND FORESTRY)
20-35, 3-CHOME, CHUO, AOMORI-SHI 030
JAPANESE

AOMORI-KEN SUISAN SHIKENJO JIGYO GAIYO 4747
(BULLETIN OF AOMORI PREFECTURAL FISHERIES RESEARCH LABORATORY)
AOMORI-KEN SUISAN SHIKENJO (AOMORI PREFECTURAL FISHERIES RESEARCH LABORATORY)
196, HON-CHO, AJIGASAWA-MACHI, NISHI TSUGARU-GUN, AOMORI 038-27
JAPANESE

AOMORI-KEN SUISAN ZOSHOKU SENTA JIGYO GAIYO 4748
(SURVEY OF OPERATION OF AQUACULTURE CENTER, AOMORI PREFECTURE)
AOMORI-KEN SUISAN ZOSHOKU SENTA (AQUACULTURE CENTER, AOMORI PREFECTURE)
HIRANAI-MACHI, HIGASHITSUGARU-GUN, AOMORI 039-34
JAPANESE

AOMORI-KEN SUISANBUTSU KAKO KENKYUSHO SHIKEN KENKYU HOKOKUSHO 4749
(REPORT OF AOMORI PREFECTURAL FISHERIES MANUFACTURING RESEARCH INSTITUTE)
AOMORI-KEN SUISANBUTSU KAKO KENKYUSHO
2-10, CHIKKOGAI, HACHINOHE-SHI 031
JAPANESE

ARIAKEKAI NO GYOGYO 4750
(FISHERIES OF ARIAKE SEA)
SUISAN-CHO ARIAKEKAI GYOGYO CHOSEI JIMU-KYOKU (REGULATION OFFICE OF ARIAKE SEA FISHERIES, FISHERIES AGENCY)
CHIKU-MACHI, OMUTA-SHI 836
JAPANESE

CHIBA-KEN NAIWAN SUISAN SHIKENJO SHIKEN CHOSA HOKOKUSHO 4751
(REPORT OF THE CHIBA PREFECTURE INNER-BAY FISHERIES EXPERIMENT STATION)
CHIBA-KEN NAIWAN SUISAN SHIKENJO (CHIBA PREFECTURE INNER-BAY FISHERIES EXPERIMENT STATION)
5-1, 3-CHOME, INAGE KAIGAN, CHIBA-SHI 281
JAPANESE

CHIBA-KEN SUISAN SHIKENJO JIGYO HOKOKU 4752
(OPERATION REPORT OF CHIBA PREFECTURAL FISHERIES EXPERIMENTAL STATION)
CHIBA-KEN SUISAN SHIKENJO (CHIBA PREFECTURAL FISHERIES EXPERIMENTAL STATION)
2492, HIRAISO, CHIKURA-MACHI, AWA-GUN CHIBA 295
JAPANESE

CHIBA-KEN SUISAN SHIKENJO KENKYU HOKOKU 4753
(SCIENCE REPORT OF CHIBA PREFECTURAL FISHERIES EXPERIMENTAL STATION)
CHIBA-KEN SUISAN SHIKENJO (CHIBA PREFECTURAL FISHERIES EXPERIMENTAL STATION)
1510, TATEYAMA, TATEYAMA-SHI 294
JAPANESE

CHIBAMARU MAGURO GYOGYO CHOSA HOKOKUSHO 4754
(TUNA FISHERY REPORT THROUGH CHIBA-MARU)
CHIBA-KEN GYOGYO SHIDOSEN UN'EI JIGYOSHO
2, ICHIBA-MACHI, CHIBA-SHI 280
JAPANESE

CHOKI GYOKYO YOHO, TOKAIKU 4755
(LONG-TERM FISHING FORECAST, TOKAI DISTRICT)
TOKAIKU SUISAN KENKYUJO (EASTERN SEA REGION FISHERIES RESEARCH INSTITUTE)
JAPANESE

CHOSASEN SHOYOMARU HOKOKUSHO 4756
(REPORT OF RESEARCH SHIP, SHOYO-MARU)
SUISAN-CHO CHOSA KENKYU-BU (RESEARCH AND DEVELOPMENT DEPARTMENT, FISHERY AGENCY)
2-1, 1-CHOME, KASUMIGASEKI, CHIYODA-KU TOKYO 100
JAPANESE

COLLECTED REPRINTS OF THE HOKKAIDO DAIGAKU RESEARCH INSTITUTE OF NORTH PACIFIC FISHERIES 4757
HOKKAIDO DAIGAKU RESEARCH INSTITUTE OF NORTH PACIFIC FISHERIES
ENGLISH
INDEXED: BIB

CONTRIBUTIONS FROM THE DEPARTMENT OF FISHERIES AND THE FISHERY RESEARCH LABORATORY, KYUSHU UNIVERSITY 4758
KYUSHU DAIGAKU NOGAKUBU SUISAN GAKKA (DEPARTMENT OF FISHERIES, FACULTY OF AGRICULTURE, KYUSHU UNIVERSITY)
3575-1, HAKOZAKI, HIGASHI-KU, FUKUOKA-SHI 812
JAPANESE AND ENGLISH

DATA RECORD OF OCEANOGRAPHIC OBSERVATIONS AND EXPLORATORY FISHING 4759
HOKKAIDO DAIGAKU SUISANGAKUBU (FACULTY OF FISHERIES, HOKKAIDO UNIVERSITY)
1-1, 3-CHOME, MINATO-MACHI, HAKODATE-SHI 040
JAPANESE AND EUROPEAN
INDEXED: BIO

EHIME-KEN SUISAN SHIKENJO JIGYO HOKOKUSHO 4760
(REPORT OF FISHERIES EXPERIMENT STATION OF EHIME-KEN)
EHIME-KEN SUISAN SHIKENJO (EXPERIMENT FISHERIES STATION OF EHIME-KEN)
SAKASHIZU, UWAJIMA-SHI 798
JAPANESE

ENGAN JUYO SHIGEN ITAKU CHOSA KEKKA HOKKAKUSHO 4761
(REPORT ON SURVEY ON CONSIGNMENT OF COASTAL RESOURCES)
AOMORI-KEN SUISAN SHIKENJO (AOMORI PREFECTURAL FISHERIES EXPERIMENT STATION)
196, HON-CHO, AJIGASAWA-MACHI, NISHI TSUGARU-GUN AOMORI 038-27
JAPANESE

EN'YO SUISAN KENKYUSHO KENKYU HOKOKU 4762
(BULLETIN FAR SEAS FISHERIES RESEARCH LABORATORY)
SUISAN-CHO ENYO SUISAN KENKYUSHO (FAR SEAS FISHERIES RESEARCH LABORATORY, FISHERY AGENCY)
1000, ORIDO, SHIMIZU-SHI 424
JAPANESE AND ENGLISH
INDEXED: JICST

FAR SEAS FISHERIES RESEARCH LABORATORY, S SERIES 4763
SUISAN-CHO EN'YO SUISAN KENKYUSHO (FAR SEAS FISHERIES RESEARCH LABORATORY, FISHERY AGENCY)
ORIDO, SHIMIZU-SHI 424
JAPANESE, WITH ENGLISH SUMMARY
INDEXED: BIO, JICST

FISHERIES STATISTICS OF JAPAN 4764
NORIN-SHO NORIN KEIZAI-KYOKU TOKEI JOHO-BU (STATISTICS AND INFORMATION DIVISION, AGRICULTURE AND FORESTRY ECONOMIC BUREAU, MINISTRY OF AGRICULTURE AND FORESTRY)
2-1, 1-CHOME, KASUMIGASEKI, CHIYODA-KU, TOKYO 100
EUROPEAN

FUKUI-KEN SUISAN SHIKENJO KENKYU HOKOKU 4765
(REPORT OF FUKUI PREFECTURAL FISHERIES EXPERIMENT STATION)
FUKUI-KEN SUISAN SHIKENJO
URAZOKO, TSURUGA-SHI 914
JAPANESE AND ENGLISH

FUKUOKA-KEN ARIAKE SUISAN SHIKENJO JIGYO HOKOKUSHO 4766
(OPERATION REPORT OF ARIAKE FISHERIES EXPERIMENT STATION, FUKUOKA PREFECTURE)
FUKUOKA-KEN ARIAKE SUISAN SHIKENJO (ARIAKE FISHERIES EXPERIMENT STATION, FUKUOKA PREFECTURE)
612-1, TSUKUSHI-CHO, YANAGAWA-SHI 832
JAPANESE

FUKUOKA-KEN ARIAKE SUISAN SHIKENJO KENKYU HOKOKU 4767
(REPORT OF ARIAKE FISHERIES EXPERIMENT STATION, FUKUOKA PREFECTURE)
FUKUOKA-KEN ARIAKE SUISAN SHIKENJO (ARIAKE FISHERIES EXPERIMENT STATION, FUKUOKA PREFECTURE)
612-1, TSUKUSHI-CHO, YANAGAWA-SHI 832
JAPANESE

FUKUOKA-KEN BUZEN SUISAN SHIKENJO KENKYU GYOMU HOKOKU 4768
(REPORT OF BUZEN FISHERIES EXPERIMENT STATION, FUKUOKA PREFECTURE)
FUKUOKA-KEN BUZEN SUISAN SHIKENJO
UJIMA, BUZEN-SHI 828
JAPANESE

FUKUOKA-KEN FUKUOKA SUISAN SHIKENJO KENKYU GYOMU HOKOKU 4769
(REPORT OF FUKUOKA FISHERIES EXPERIMENT STATION, FUKUOKA PREFECTURE)
FUKUOKA-KEN FUKUOKA SUISAN SHIKENJO
1141-1, IMATSU, NISHI-KU, FUKUOKA-SHI 819-01
JAPANESE

FUKUOKA-KEN NO SUISAN 4770
(FISHERIES IN FUKUOKA PREFECTURE)
FUKUOKA-KEN SHOKO SUISAN-BU (COMMERCE, INDUSTRY AND FISHERY DEPARTMENT, FUKUOKA PREFECTURAL GOVERNMENT)
1-1, 1-CHOME, TENJIN, FUKUOKA-SHI 810
JAPANESE

FUKUSHIMA-KEN GYOGYO NO DOKO 4771
(TENDENCY OF FISHERY IN FUKUSHIMA PREFECTURE)
NORIN-SHO TOHOKU NOSEI KYOKU FUKUSHIMA TOKEI JOHO JIMUSHO (FUKUSHIMA STATISTICS AND INFORMATION OFFICE, TOHOKU REGIONAL AGRICULTURAL ADMINISTRATION BUREAU, MINISTRY OF AGRICULTURE AND FORESTRY)
4-15, YAMASHITA-CHO, FUKUSHIMA-SHI 960
JAPANESE

FUKUSHIMA-KEN NAISUIMEN SUISAN SHIKENJO JIGYO HOKOKUSHO 4772
(REPORT OF FUKUSHIMA PREFECTURAL FRESH WATER FISHERY EXPERIMENT STATION)
FUKUSHIMA-KEN NAISUIMEN SUISAN SHIKENJO
5-33, MINAMI SENGOKU-MACHI, AIZUWAKAMATSU-SHI 965
JAPANESE

FUKUSHIMA-KEN SUISAN SHIKENJO JIGYO HOKOKUSHO 4773
(REPORT OF FUKUSHIMA PREFECTURAL FISHERIES EXPERIMENTAL STATION)
FUKUSHIMA-KEN SUISAN SHIKENJO (FUKUSHIMA PREFECTURAL FISHERIES EXPERIMENTAL STATION)
13-2, MATSUSHITA, ONAHAMA-SHIMOKAJIRO, IWAKI-SHI 970-03
JAPANESE

FUKUSHIMA-KEN SUISAN SHIKENJO KENKYU HOKOKU 4774
(RESEARCH REPORT OF FUKUSHIMA PREFECTURAL FISHERIES EXPERIMENTAL STATION)
FUKUSHIMA-KEN SUISAN SHIKENJO (FUKUSHIMA PREFECTURAL FISHERIES EXPERIMENTAL STATION)
13-2, MATSUSHITA, ONAHAMA-SHIMOKAJIRO, IWAKI-SHI 970-03
JAPANESE

GIFU-KEN SUISAN SHIKENJO KENKYU HOKOKU 4775
(REPORT OF GIFU PREFECTURAL FISHERIES EXPERIMENTAL STATION)
GIFU-KEN SUISAN SHIKENJO
2066-2, HANE, OGIWARA-MACHI, MASUDA-GUN, GIFU 509-25
JAPANESE

GIJUTSU KENKYU NENPO 4776
(ANNUAL REPORT OF FOOD TECHNOLOGY)
NICHIRO GYOGYO K.K. GIJUTSUBU (FOOD TECHNOLOGY AND ENGINEERING DEPARTMENT, NICHIRO GYOGYO COMPANY, LIMITED)
1-11, YURAKU-CHO, CHIYODA-KU, TOKYO 100
JAPANESE

GYOBYO KENKYU 4777
(FISH PATHOLOGY)
GYOBYO KENKYU DANWAKAI (JAPAN RESEARCH GROUP OF FISH PATHOLOGY)
C/O TOKYO DAIGAKU NOGAKUBU SUISANGAKKA, 1-1, 1-CHOME YAYOI, BUNKYO-KU, TOKYO 113
JAPANESE

GYOBYO TAISAKU NI KANSURU KENKYU HOKOKU 4778
(RESEARCH REPORT ON FISH PATHOLOGY)
SHIZUOKA-KEN SUISAN SHIKENJO (SHIZUOKA PREFECTURAL FISHERIES EXPERIMENT STATION)
3690, OGAWA SHIOIRI, YAIZU-SHI 425
JAPANESE

GYOGYO DOTAI CHOSA 4779
(REPORT ON INTERCENSAL FISHERIES MANAGEMENT SURVEY)
NORIN-SHO NORIN KEIZAI-KYOKU TOKEI JOHO-BU (STATISTICS AND INFORMATION DIVISION, AGRICULTURE AND FORESTRY ECONOMIC BUREAU, MINISTRY OF AGRICULTURE AND FORESTRY)
2-1, 1-CHOME, KASUMIGASEKI, CHIYODA-KU, TOKYO 100
JAPANESE

GYOGYO MUSEN 4780
(WIRELESS TELECOMMUNICATION FOR FISHERY)
ZENKOKU GYOGYO MUSEN KYOKAI (ALL JAPAN ASSOCIATION OF WIRELESS TELECOMMUNICATION FOR FISHERIES)
7-14, 2-CHOME, KITA AOYAMA, MINATO-KU, TOKYO 107
JAPANESE

GYOGYO MUSEN GEPPO 4781
(MONTHLY REPORT OF WIRELESS TELECOMMUNICATION FOR FISHERIES)
ZENKOKU GYOGYO MUSEN KYOKAI (ALL JAPAN ASSOCIATION OF WIRELESS TELECOMMUNICATION FOR FISHERIES)
7-14, 2-CHOME, KITA AOYAMA, MINATO-KU, TOKYO 107
JAPANESE

GYOGYO NO DOKOTO NI KANSURU NENJI HOKOKU 4782
(ANNUAL REPORT OF FISHERY)
NORIN-SHO (MINISTRY OF AGRICULTURE AND FORESTRY)
2-1, 1-CHOME, KASUMIGASEKI, CHIYODA-KU, TOKYO 100
JAPANESE
INDEXED: BIB

GYOGYO SENSASU 4783
(FISHERIES CENSUS)
NORIN-SHO NORIN KEIZAI-KYOKU TOKEI JOHO-BU (STATISTICS AND INFORMATION DIVISION, AGRICULTURE AND FORESTRY ECONOMIC BUREAU, MINISTRY OF AGRICULTURE AND FORESTRY)
2-1, 1-CHOME, KASUMIGASEKI, CHIYODA-KU, TOKYO 100
JAPANESE AND ENGLISH

GYOGYO SHIGEN KENKYU KAIGI-HO 4784
(REPORT OF THE CONFERENCE OF THE FISHERIES AGENCY, JAPANESE GOVERNMENT FOR FISHERIES RESOURCES INVESTIGATION)
SUISAN-CHO SUISAN KENKYUSHO GYOGYO SHIGEN KENKYU-KAIGI (CONFERENCE OF THE FISHERIES AGENCY, JAPANESE GOVERNMENT FOR FISHERIES RESOURCES INVESTIGATIONS, FISHERIES RESEARCH LABORATORY)
5-1, 5-CHOME, KACHIDOKI, CHUO-KU, TOKYO 104
JAPANESE AND ENGLISH

GYOGYO YOSHOKUGYO SEISAN TOKEI NENPO 4785
(ANNUAL STATISTICS OF FISHERY AND FISH CULTURE)
NORIN-SHO NORIN KEIZAI-KYOKU TOKEI JOHO-BU (STATISTICS AND INFORMATION DIVISION, AGRICULTURE AND FORESTRY ECONOMIC BUREAU, MINISTRY OF AGRICULTURE AND FORESTRY) AND NORIN TOKEI KYOKAI
11-14, 2-CHOME, MEGURO, MEGURO-KU, TOKYO 153
JAPANESE

GYOJO KAIKYO GAIHO (SEIKAI-KU) 4786
(REPORT OF SURVEY ON STATE OF SEA AND FISHING AREA, SEIKAI REGION)
SUISAN-CHO SEIKAI-KU SUISAN KENKYUSHO (SEIKAI REGIONAL FISHERIES RESEARCH LABORATORY)
49, KOKUBU-CHO, NAGASAKI-SHI 850
JAPANESE

GYOJO OSEN TAISAKU KENKYU KANREN NO KIO NO BUNKEN SHIRYO 4787
(RECORDS OF THE COUNTERPLANNING FOR WATER POLLUTION IN THE FISHING PLACE)
GYOJO OSEN TAISAKU KENKYU KYOGIKAI (RESEARCH ASSOCIATION OF THE COUNTERPLANNING FOR THE WATER POLLUTION IN THE FISHING PLACE) AND NANSEIKAIKU SUISAN KENKYUSHO
7782-9, MARUISHI, ONO-MACHI, SAIKI-GUN, HIROSHIMA 739-04
JAPANESE

GYOKAIKYO CHOSA GEPPO (AOMORI-KEN) 4788
(MONTHLY REPORT ON SURVEY OF STATE OF SEA AND FISHING AREA, AOMORI PREFECTURE)
AOMORI-KEN SUISAN SHIKENJO (AOMORI PREFECTURAL FISHERIES EXPERIMENT STATION)
196, HON-CHO, AJIGASAWA-CHO, NISHI TSUGARU-GUN, AOMORI 038-27
JAPANESE

GYOKAIKYO SOKUHO (CHIBA-KEN) 4789
(REPORT ON SURVEY OF STATE OF FISHING AREA AND SEA, CHIBA PREFECTURE)
CHIBA-KEN SUISAN SHIKENJO (CHIBA PREFECTURE FISHERIES EXPERIMENTAL STATION)
2492, HIRAISO, CHIKURA-MACHI, AWA-GUN, CHIBA 295
JAPANESE

GYOKO 4790
(FISHING PORTS)
JAPANESE
INDEXED: JICST

GYOKO GEPPO 4791
(FISHING PORTS MONTHLY)
ZENKOKU GYOKO KYOKAI (ALL JAPAN FISHING PORTS ASSOCIATION)
JAPANESE

GYOKO NO KOSEISHU 4792
(PAPER OF FISHING PORT CONDITION)
SUISAN-CHO GYOKO-BU (FISHING PORT DEPARTMENT, FISHERY AGENCY)
2-1, 1-CHOME, KASUMIGASEKI, CHIYODA-KU, TOKYO 100
JAPANESE

GYOKYO KAIKYO CHOSA GEPPO (AICHI-KEN) 4793
(MONTHLY REPORT ON SURVEY OF STATE OF SEA AND FISHING AREA, AICHI PREFECTURE)
AICHI-KEN SUISAN SHIKENJO (AICHI PREFECTURAL FISHERIES EXPERIMENT STATION)
97, WAKAMIYA, MITANI-CHO, GAMAGORI-SHI 443
JAPANESE

GYOKYO KAIKYO GAIHO (FUKUI-KEN) 4794
(REPORT ON SURVEY OF STATE OF SEA AND FISHING AREA, FUKUI PREFECTURE)
KUKUI-KEN SUISAN SHIKENJO (FUKUI PREFECTURE FISHERIES EXPERIMENTAL STATION)
URASOKO, TSURUGA-SHI 914
JAPANESE

GYOKYO KAIKYO GEPPO (SHIZUOKA-KEN) 4795
(MONTHLY REPORT ON SURVEY OF STATE OF SEA AND FISHING AREA, SHIZUOKA PREFECTURE)
SHIZUOKA-KEN SUISAN SHIKENJO (SHIZUOKA PREFECTURAL FISHERIES EXPERIMENT STATION)
3690, OGAWA SHIOIRI, YAIZU-SHI 425
JAPANESE

GYOKYO KAIKYO SOKUHO (SHIZUOKA-KEN) 4796
(REPORT ON SURVEY OF STATE OF SEA AND FISHING AREA, QUICK EDITION, SHIZUOKA PREFECTURE)
SHIZUOKA-KEN SUISAN SHIKENJO (SHIZUOKA PREFECTURAL FISHERIES EXPERIMENT STATION)
3690, OGAWA SHIOIRI, YAIZU-SHI 425
JAPANESE

GYOKYO KAIKYO YOHO JIGYO KEKKA HOKOKUSHO (ISHIKAWA-KEN) 4797
(REPORT ON SURVEY OF STATE OF SEA AND FISHING AREA, ISHIKAWA PREFECTURE)
ISHIKAWA-KEN SUISAN SHIKENJO (ISHIKAWA PREFECTURAL FISHERIES EXPERIMENT STATION)
USHITSU, NOTO-MACHI, FUGESHI-GUN, ISHIKAWA 927-04
JAPANESE

GYOKYO KAIKYO YOHO JIGYO KEKKA HOKOKUSHO (KANAGAWA-KEN) 4798
(REPORT ON SURVEY OF STATE OF SEA AND FISHING AREA, KANAGAWA PREFECTURE)
KANAGAWA-KEN SUISAN SHIKENJO (KANAGAWA PREFECTURAL FISHERIES EXPERIMENTAL STATION)
YOROSHI, JOGASHIMA, MISAKI-MACHI, MIURA-SHI 238
JAPANESE

GYOKYO KAIKYO YOHO JIGYO KEKKA HOKOKUSHO (MIE-KEN) 4799
(REPORT ON SURVEY OF STATE OF SEA AND FISHING AREA, MIE PREFECTURE)
MIE-KEN HAMAJIMA SUISAN SHIKENJO (MIE PREFECTURAL HAMAJIMA FISHERIES EXPERIMENT STATION)
HAMAJIMA-CHO, SHIMA-GUN, MIE 517-04
JAPANESE

GYOKYO KAIKYO YOHO JIGYO KEKKA HOKOKUSHO (SAGA-KEN) 4800
(REPORT ON SURVEY OF STATE OF SEA AND FISHING AREA, SAGA PREFECTURE)
SAGA-KEN SUISAN SHIKENJO (SAGA PREFECTURAL FISHERIES EXPERIMENTAL STATION)
6, TOBO, KARATSU-SHI 847-01
JAPANESE

GYOKYO KAIKYO YOHO JIGYO KEKKA HOKOKUSHO (TOKYO-TO) 4801
(REPORT ON SURVEY OF STATE OF SEA AND FISHING AREA, TOKYO)
TOKYO-TO SUISAN SHIKENJO (TOKYO METROPOLITAN FISHERIES EXPERIMENT STATION)
3-1, 6-CHOME, HIGASHI KOJIYA, OTA-KU, TOKYO 144
JAPANESE

GYOSON 4802
(FISHING VILLAGES)
GYOSON BUNKA KYOKAI (FISHING VILLAGES CULTURAL SOCIETY)
9-13, 1-CHOME, AKASAKA MINATO-KU, TOKYO 107
JAPANESE

HAMANA 4803
(LAKE HAMANA)
SHIZUOKA-KEN SUISAN SHIKENJO HAMANAKO BUNJO (SHIZUOKA PREFECTURAL FISHERIES EXPERIMENT STATION, LAKE HAMANA BRANCH)
3550, BENTENJIMA, MAISAKA-MACHI, HAMANA-GUN, SHIZUOKA 431-02
JAPANESE

HIROSHIMA DAIGAKU SUI-CHIKUSAN GAKUBU KIYO 4804
(JOURNAL OF THE FACULTY OF FISHERIES AND ANIMAL HUSBANDRY, HIROSHIMA UNIVERSITY)
HIROSHIMA DAIGAKU SUI-CHIKUSAN GAKUBU (FACULTY OF FISHERIES AND ANIMAL HUSBANDRY, HIROSHIMA UNIVERSITY)
2-17, MIDORI-CHO, FUKUYAMA-SHI 720
JAPANESE AND EUROPEAN
INDEXED: BIB, BIO, CHE, EXC, JICST

HIROSHIMA-KEN SUISAN SHIKENJO JIGYO HOKOKU 4805
(ANNUAL REPORT OF THE HIROSHIMA FISHERIES EXPERIMENTAL STATION)
HIROSHIMA-KEN SUISAN SHIKENJO (HIROSHIMA FISHERIES EXPERIMENTAL STATION)
5233-2, ONDO-CHO, AKI-GUN, HIROSHIMA 737-12
JAPANESE
INDEXED: CHE

HIROSHIMA-KEN SUISAN SHIKENJO KENKYU HOKOKU 4806
(BULLETIN OF THE HIROSHIMA FISHERIES EXPERIMENTAL STATION)
HIROSHIMA-KEN SUISAN SHIKENJO (HIROSHIMA FISHERIES EXPERIMENTAL STATION)
5233-2, ONDO-CHO, AKI-GUN, HIROSHIMA 737-12
JAPANESE AND ENGLISH
INDEXED: CHE

HIROSHIMA-KEN TANSUIGYO SHIDOSHO CHOSA KENKYU HOKOKU 4807
(RESEARCH REPORT OF HIROSHIMA PREFECTURAL FRESH WATER FISH EXPERIMENT STATION)
HIROSHIMA-KEN TANSUIGYO SHIDOSHO (HIROSHIMA PREFECTURAL FRESH WATER FISH EXPERIMENT STATION)
23-1, KAWATE-MACHI, SHOBARA-SHI 727
JAPANESE

HIROSHIMA-KEN TANSUIGYO SHIDOSHO JIGYO JISSEKI 4808
(ANNUAL REPORT OF THE HIROSHIMA PREFECTURAL FRESH WATER FISH EXPERIMENT STATION)
HIROSHIMA-KEN TANSUIGYO SHIDOSHO (HIROSHIMA PREFECTURAL FRESH WATER FISH EXPERIMENT STATION)
23-1, KAWATE-MACHI, SHOBARA-SHI 727
JAPANESE

HOKKAIDO DAIGAKU SUISAN GAKUBU HOKUYO SUISAN KENKYU SHISETSU GYOSEKISHU 4809
(RESEARCH INSTITUTE OF NORTH PACIFIC FISHERIES, FACULTY OF FISHERIES HOKKAIDO UNIVERSITY COLLECTED REPRINTS)
HOKKAIDO DAIGAKU SUISAN GAKUBU HOKUYO SUISAN KENKYU SHISETSU (RESEARCH INSTITUTE OF NORTH PACIFIC FISHERIES, FACULTY OF FISHERIES, HOKKAIDO UNIVERSITY)
1-1, 3-CHOME, MINATO-CHO, HAKODATE-SHI 040
JAPANESE AND ENGLISH

HOKKAIDO DAIGAKU SUISAN GAKUBU KENKYU HOKOKU 4810
(MEMOIRS OF THE FACULTY OF FISHERIES, HOKKAIDO UNIVERSITY)
HOKKAIDO DAIGAKU SUISAN GAKUBU (FACULTY OF FISHERIES, HOKKAIDO UNIVERSITY)
1-1, 3-CHOME, MINATO-CHO, HAKODATE-SHI 040
ENGLISH
INDEXED: BIB, BIO, CHE

HOKKAIDO DAIGAKU SUISAN GAKUBU KENKYU IHO 4811
(BULLETIN OF THE FACULTY OF FISHERIES, HOKKAIDO UNIVERSITY)
HOKKAIDO DAIGAKU SUISAN GAKUBU (FACULTY OF FISHERIES, HOKKAIDO UNIVERSITY)
1-1, 3-CHOME, MINATO-CHO, HAKODATE-SHI 040
JAPANESE AND EUROPEAN
INDEXED: BIB, BIO, CHE, JICST

HOKKAIDO ENGAN GYOGYOTO NO UGOKI 4812
(TENDENCY OF COASTAL FISHERY IN HOKKAIDO PREFECTURE)
NORIN-SHO HOKKAIDO TOKEI JOHO JIMUSHO (HOKKAIDO STATISTICS AND INFORMATION OFFICE, MINISTRY OF AGRICULTURE AND FORESTRY)
NISHI-4-CHOME, KITA-3-JO, CHUO-KU, SAPPORO-SHI 060
JAPANESE

HOKKAIDO GYOGYO KENKYU 4813
(JOURNAL OF FISHERIES ECONOMY OF HOKKAIDO)
HOKKAIDORITSU SOGO KEIZAI KENKYUSHO (HOKKAIDO PREFECTURAL INSTITUTE OF SYNTHETIC ECONOMIC RESEARCH)
NISHI-6-CHOME, KITA-3-JO, CHUO-KU, SAPPORO-SHI 060
JAPANESE

HOKKAIDO NO TSURI 4814
(HOKKAIDO FISHING)
SUIKOSHA
JAPANESE

HOKKAIDO OKIAI SOKOBIKIAMI GYOGYO GYOJO-BETSU GYOKAKU TOKEI NENPO 4815
(ANNUAL STATISTICS OF HAUL OF FISH BY DRAGNET FISHERY IN HOKKAIDO)
SUISAN-CHO HOKKAIDO GYOGYO CHOSEI JIMUSHO AND HOKKAIDO KISEN GYOGYO KYODO KUMIAI RENGOKAI
1, NISHI-6-CHOME, KITA-4-JO, CHUO-KU, SAPPORO-SHI 060
JAPANESE

HOKKAIDO SAKE, MASU, FUKAJO JIGYO SEISEKISHO 4816
(RESULT PAPER OF HOKKAIDO SALMON HATCHERY)
SUISAN-CHO HOKKAIDO SAKE MASU FUKAJO (HOKKAIDO SALMON HATCHERY, FISHERIES AGENCY)
2-CHOME, NAKANOSHIMA-2-JO, TOYOHIRA-KU, SAPPORO-SHI 062
JAPANESE

HOKKAIDO SAKE, MASU, FUKAJO KENKYU HOKOKU 4817
(SCIENTIFIC REPORTS OF THE HOKKAIDO SALMON HATCHERY)
SUISAN-CHO HOKKAIDO SAKE, MASU FUKAJO (HOKKAIDO SALMON HATCHERY, FISHERIES AGENCY)
2-CHOME, NAKANOSHIMA-2-JO, TOYOHIRA-KU, SAPPORO-SHI 062
JAPANESE, WITH EUROPEAN SUMMARY
INDEXED: BIO, JICST

HOKKAIDO SAKE, MASU, FUKAJO SHIRYO 4818
(DATA OF THE HOKKAIDO SALMON HATCHERY)
SUISAN-CHO HOKKAIDO SAKE MASU FUKAJO (HOKKAIDO SALMON HATCHERY, FISHERIES AGENCY)
2-CHOME, NAKANOSHIMA-2-JO, TOYOHIRA-KU, SAPPORO-SHI 062
JAPANESE

HOKKAIDO SUISAN GENSEI 4819
(PRESENT FISHERIES SITUATION IN HOKKAIDO)
HOKKAIDO SUISAN-BU (FISHERIES DIVISION, HOKKAIDO PREFECTURAL GOVERNMENT)
NISHI-5-CHOME, KITA-3-JO, CHUO-KU, SAPPORO-SHI 060
JAPANESE

HOKKAIDO-KU SUISAN KENKYUSHO KENKYU HOKOKU 4820
(BULLETIN OF THE HOKKAIDO REGIONAL FISHERIES RESEARCH LABORATORY)
SUISAN-CHO HOKKAIDO-KU SUISAN KENKYUSHO (HOKKAIDO FISHERIES RESEARCH LABORATORY, FISHERIES AGENCY)
REGIONAL FISHERIES RESEARCH LABORATORY, FISHERIES AGENCY, HOKKAIDO 046
JAPANESE AND ENGLISH
INDEXED: BIB, BIO,, CHE, JICST

HOKKAIDORITSU ABASHIRI SUISAN SHIKENJO JIGYO HOKOKUSHO 4821
(REPORT OF HOKKAIDO ABASHIRI FISHERIES EXPERIMENTAL STATION)
HOKKAIDORITSU ABASHIRI SUISAN SHIKENJO
31, MASUURA, ABASHIRI-SHI 099-31
JAPANESE

HOKKAIDORITSU HAKODATE SUISAN SHIKENJO JIGYO HOKOKUSHO 4822
(REPORT OF THE HOKKAIDO HAKODATE FISHERIES EXPERIMENTAL STATION)
HOKKAIDORITSU HAKODATE SUISAN SHIKENJO
12, MIDORI-CHO, ASAHIKAWA-SHI 070
JAPANESE

HOKKAIDORITSU KUSHIRO SUISAN SHIKENJO JIGYO HOKOKUSHO 4823
(REPORT OF HOKKAIDO KUSHIRO FISHERIES EXPERIMENTAL STATION)
HOKKAIDORITSU KUSHIRO SUISAN SHIKENJO
2-6, HAMA-CHO, KUSHIRO-SHI 085
JAPANESE

HOKKAIDORITSU SUISAN FUKAJO JIGYO SEISEKISHO 4824
(OPERATION REPORT OF HOKKAIDO FISH HATCHERY)
HOKKAIDORITSU SUISAN FUKAJO (HOKKAIDO FISH HATCHERY)
2-CHOME, NAKANOSHIMA-2-JO, TOYOHIRA-KU, SAPPORO-SHI 062
JAPANESE

HOKKAIDORITSU SUISAN FUKAJO KENKYU HOKOKU 4825
(SCIENTIFIC REPORTS OF THE HOKKAIDO FISH HATCHERY)
HOKKAIDORITSU SUISAN FUKAJO (HOKKAIDO FISH HATCHERY)
2-CHOME, NAKANOSHIMA-2-JO, TOYOHIRA-KU, SAPPORO-SHI 062
JAPANESE AND ENGLISH

HOKKAIDORITSU SUISAN SHIKENJO HOKOKU 4826
(SCIENTIFIC REPORTS OF THE HOKKAIDO FISHERIES EXPERIMENTAL STATION)
HOKKAIDORITSU CHUO SUISAN SHIKENJO (HOKKAIDO CENTRAL FISHERIES EXPERIMENT STATION)
HAMANAKA-CHO, YOICHI-MACHI, YOICHI-GUN, HOKKAIDO 046
JAPANESE AND ENGLISH
INDEXED: BIB, BIO, CHE, JICST

HOKKAIDORITSU WAKKANAI SUISAN SHIKENJO GYOSEKISHU 4827
(COLLECTED REPRINTS FROM THE HOKKAIDO WAKKANAI FISHERIES EXPERIMENTAL STATION)
HOKKAIDO WAKKANAI FISHERIES EXPERIMENTAL STATION
JAPANESE

HOKKAIDORITSU WAKKANAI SUISAN SHIKENJO JIGYO SEISEKISHO 4828
(REPORT OF HOKKAIDO WAKKANAI FISHERIES EXPERIMENTAL STATION)
HOKKAIDORITSU WAKKANAI SUISAN SHIKENJO
5-4, 4-CHOME, HORAI, WAKKANAI-SHI 097
JAPANESE

HOKUSUIKEN NYUSU 4829
(HOKKAIDO REGIONAL FISHERIES RESEARCH LABORATORY NEWS)
SUISAN-CHO HOKKAIDO-KU SUISAN KENKYUSHO
238, HAMANAKA-CHO, YOICHI-MACHI, YOICHI-GUN HOKKAIDO 046
JAPANESE

HOKUSUISHI GEPPO 4830
(MONTHLY REPORT FROM THE HOKKAIDO FISHERIES EXPERIMENT STATION)
HOKKAIDORITSU CHUO SUISAN SHIKENJO (HOKKAIDO CENTRAL FISHERIES EXPERIMENT STATION)
238, HAMANAKA-CHO, YOICHI-MACHI, YOICHI-GUN, HOKKAIDO 046
JAPANESE
INDEXED: CHE, JICST

HYOGO-KENRITSU SUISAN SHIKENJO JIGYO HOKOKU 4831
(OPERATION REPORT OF HYOGO PREFECTURAL FISHERIES EXPERIMENT STATION)
HYOGO-KENRITSU SUISAN SHIKENJO
6-8, 1-CHOME, NAKASAKI, AKASHI-SHI 673
JAPANESE

HYOGO-KENRITSU SUISAN SHIKENJO SHIKEN HOKOKU 4832
(RESEARCH REPORT OF HYOGO PREFECTURAL FISHERIES EXPERIMENT STATION)
HYOGO-KENRITSU SUISAN SHIKENJO
6-8, 1-CHOME, NAKASAKI, AKASHI-SHI 673
JAPANESE

IBARAKI-KEN NAISUIMEN SUISAN SHIKENJO CHOSA KENKYU HOKOKU 4833
(REPORT OF FRESH WATER FISHERIES EXPERIMENT STATION, IBARAKI PREFECTURE)
IBARAKI-KEN NAISUIMEN SUISAN SHIKENJO (FRESH WATER FISHERIES EXPERIMENT STATION, IBARAKI PREFECTURE)
TAMATSUKURI-MACHI, NAMEKATA-GUN, IBARAKI 311-35
JAPANESE, WITH EUROPEAN SUMMARY

IBARAKI-KEN SUISAN KAKO KENKYUSHO HOKOKU 4834
(IBARAKI PREFECTURAL FISHERY PRODUCT EXPERIMENT STATION)
IBARAKI-KEN SUISAN KAKO KENKYUSHO
3551-8, MITSUZUKA, HIRAISO-MACHI, NAKAMINATO-SHI 311-12
JAPANESE

IBARAKI-KEN SUISAN SHIKENJO JIGYO HOKOKU 4835
(OPERATION REPORT OF THE FISHERIES EXPERIMENTAL STATION OF IBARAKI-KEN)
IBARAKI-KEN SUISAN SHIKENJO
3551-8, MITSUZUKA, HIRAISO-MACHI, NAKAMINATO-SHI 311-12
JAPANESE

IBARAKI-KEN SUISAN SHIKENJO SHIKEN HOKOKU 4836
(BULLETIN OF THE FISHERIES EXPERIMENTAL STATION OF IBARAKI-KEN)
IBARAKI-KEN SUISAN SHIKENJO (FISHERIES EXPERIMENTAL STATION OF IBARAKI-KEN)
3551-8, MITSUZUKA, HIRAISO-MACHI, NAKAMINATO-SHI 311-12
JAPANESE

ISHIKAWA-KEN SUISAN SHIKENJO JIGYO HOKOKUSHO 4837
(ANNUAL REPORT OF ISHIKAWA PREFECTURE FISHERIES EXPERIMENT STATION)
ISHIKAWA-KEN SUISAN SHIKENJO
USHITSU, NOTO-MACHI, FUGESHI-GUN, ISHIKAWA 927-04
JAPANESE

ISHIKAWA-KEN SUISAN SHIKENJO KENKYU HOKOKU 4838
(RESEARCH REPORT OF THE ISHIKAWA PREFECTURAL FISHERIES EXPERIMENT STATION)
ISHIKAWA-KEN SUISAN SHIKENJO (ISHIKAWA PREFECTUAL FISHERIES EXPERIMENT STATION)
JAPANESE

ISHIKAWA-KEN ZOSHOKU SHIKENJO JIGYO HOKOKUSHO 4839
(OPERATION REPORT OF ISHIKAWA PREFECTURE MARINE CULTURE)
ISHIKAWA-KEN ZOSHOKU SHIKENJO (ISHIKAWA PREFECTURE MARINE CULTURE STATION)
MAGARI, NOTOJIMA-MACHI, KASHIMA-GUN, ISHIKAWA 926-02
JAPANESE

ISHIKAWA-KEN ZOSHOKU SHIKENJO KENKYU HOKOKU 4840
(BULLETIN OF THE ISHIKAWA PREFECTURE MARINE CULTURE STATION)
ISHIKAWA-KEN ZOSHOKU SHIKENJO (ISHIKAWA PREFECTURE MARINE CULTURE STATION)
MAGARI, NOTOJIMA-MACHI, KASHIMA-GUN, ISHIKAWA 926-02
JAPANESE AND ENGLISH

IWATE NO SUISAN 4841
(FISHERIES IN IWATE PREFECTURE)
IWATE-KEN KEIZAI-BU (ECONOMIC BUREAU, IWATE PRE-FECTURAL GOVERNMENT)
10-1, UCHIMARU, MORIOKA-SHI 020
JAPANESE

IWATE-KEN SUISAN SHIKENJO KENKYU HOKOKU 4842
(BULLETIN OF IWATE PREFECTURAL FISHERIES EXPERIMENTAL STATION)
IWATE-KEN SUISAN SHIKENJO (IWATE PREFECTURAL FISHERIES EXPERIMENTAL STATION)
4-21, 1-CHOME, SHINHAMA-CHO, KAMAISHI-SHI 026
JAPANESE AND ENGLISH

IWATE-KEN SUISAN SHIKENJO NENPO 4843
(ANNUAL REPORT OF THE IWATE PREFECTURAL FISHERIES EXPERIMENTAL STATION)
IWATE-KEN SUISAN SHIKENJO
20, TAKIZAWA-MURA, IWATE-GUN, IWATE 020-01
JAPANESE

IZU BUNJO DAYORI 4844
(IZU BRANCH NEWS, SHIZUOKA PREFECTURAL FISHERIES EXPERIMENT STATION)
SHIZUOKA-KEN SUISAN SHIKENJO IZU BUNJO
251, SHIRAHAMA, SHIMODA-SHI 415
JAPANESE

JISSEN GYOJO SETTEI CHOSA HOKOKU 4845
(REPORT OF SURVEY ON PRACTICAL ESTABLISHMENT OF FISHING AREA)
KOCHI-KEN (KOCHI PREFECTURAL GOVERNMENT)
2-20, 1-CHOME, MARUNOUCHI, KOCHI-SHI 780
JAPANESE

JOURNAL OF THE TOKYO UNIVERSITY OF FISHERIES 4846
TOKYO SUISAN DAIGAKU (TOKYO UNIVERSITY OF FISHERIES)
5-7, 4-CHOME, KONAN, MINATO-KU, TOKYO 108
JAPANESE AND EUROPEAN
INDEXED: BIB, JICST

KAGAWA-KEN SUISAN SHIKENJO JIGYO HOKOKU 4847
(REPORT OF KAGAWA PREFECTURAL FISHERIES EXPERIMENT STATION)
KAGAWA-KEN SUISAN SHIKENJO
YASHIMAHIGASHI-CHO, TAKAMATSU-SHI 761-01
JAPANESE

KAGAWA-KEN SUISAN TOKEI NENPO 4848
(ANNUAL STATISTICS OF FISHERIES IN KAGAWA PREFECTURE)
NORIN-SHO CHUGOKU SHIKOKU NOSEI-KYOKU KAGAWA TOKEI JOHO JIMUSHO (KAGAWA STATISTICS & INFORMATION OFFICE CHUGOKU SHIKOKU REGIONAL AGRICULTURAL ADMINISTRATION BUREAU, MINISTRY OF AGRICULTURE AND FORESTRY)
17-33, 1-CHOME, MATSUSHIMA-CHO, TAKAMATSU-SHI 760
JAPANESE

KAGOSHIMA DAIGAKU SUISANGAKUBU KIYO 4849
(MEMOIRS OF THE FACULTY OF FISHERIES, KAGOSHIMA UNIVERSITY)
KAGOSHIMA DAIGAKU SUISANGAKUBU (FACULTY OF FISHERIES KAGOSHIMA UNIVERSITY)
470, SHIMOARADA-MACHI, KAGOSHIMA-SHI 890
JAPANESE AND EUROPEAN
INDEXED: CHE, JICST

KAGOSHIMA-KEN SUISAN SHIKENJO JIGYO HOKOKUSHO 4850
(REPORT OF KAGOSHIMA PREFECTURAL FISHERIES EXPERIMENT STATION)
KAGOSHIMA-KEN SUISAN SHIKENJO
21-1, KINKO-CHO, KAGOSHIMA-SHI 892
JAPANESE

KAGOSHIMA-KEN SUISAN SHIKENJO KIYO 4851
(BULLETIN OF KAGOSHIMA PREFECTURAL FISHERIES EXPERIMENT STATION)
KAGOSHIMA-KEN SUISAN SHIKENJO
21-1, KINKO-CHO, KAGOSHIMA-SHI 892
JAPANESE

KAIGAI SUISAN NYUSU 4852
(OVERSEAS FISHERY NEWS)
SUISAN-CHO. SUISAN SHIRYOKAN (FISHERY AGENCY, FISHERY DOCUMENTS HALL)
JAPANESE

KAIYO CHOSA HOKOKU 4853
(MARINE RESEARCH REPORT)
CHIBA-KEN SUISAN SHIKENJO (CHIBA PREFECTURE FISHERIES EXPERIMENTAL STATION)
2492, HIRAISO, CHIKURA-MACHI, AWA-GUN, CHIBA 295
JAPANESE

KAIYOMARU CHOSA KOKAI HOKOKUSHO 4854
(REPORT OF NAVIGATION FOR SURVEY OF KAIYO-MARU)
SUISAN-CHO (FISHERY AGENCY)
2-1, 1-CHOME, KASUMIGASEKI, CHIYODA-KU, TOKYO 100
JAPANESE

KANAGAWA-KEN NO SUISAN 4855
(FISHERIES IN KANAGAWA PREFECTURE)
KANAGAWA-KEN NOSEI-BU (KANAGAWA PREFECTURAL GOVERNMENT)
1, NIHON-ODORI, NAKA-KU, YOKOHAMA-SHI 231
JAPANESE

KANAGAWA-KEN SUISAN SHIKENJO GYOMU GAIYO 4856
(REPORT OF KANAGAWA PREFECTURE FISHERIES EXPERIMENT STATION)
KANAGAWA-KEN SUISAN SHIKENJO (KANAGAWA PREFECTURE FISHERIES EXPERIMENT STATION)
YOROSHI, JOGASHIMA, MISAKI-CHO, MIURA-SHI 238
JAPANESE

KANAGAWA-KEN SUISAN SHIKENJO SAGAMIWAN SHISHO HOKOKU 4857
(REPORT OF KANAGAWA PREFECTURAL FISHERIES EXPERIMENTAL STATION, SAGAMI BRANCH)
KANAGAWA-KEN SUISAN SHIKENJO SAGAMIWAN SHISHO (KANAGAWA PREFECTURAL FISHERIES EXPERIMENTAL STATION, SAGAMI BRANCH)
8-1, 1-CHOME, HAYAKAWA, ODAWARA-SHI 250
JAPANESE

KANAGAWA-KEN SUISAN SHIKENJO SHIDO FUKYU SHIRYO 4858
(GUIDING INFORMATION OF KANAGAWA PREFECTURAL FISHERIES EXPERIMENT STATION)
KANAGAWA-KEN SUISAN SHIKENJO
YOROSHI, JOGASHIMA, MISAKI-CHO, MIURA-SHI 238
JAPANESE

KANAGAWA-KEN SUISAN SHIKENJO SHIRYO 4859
(SPECIAL BULLETIN OF THE KANAGAWA PREFECTURAL FISHERIES EXPERIMENT STATION)
KANAGAWA-KEN SUISAN SHIKENJO (KANAGAWA PREFECTURAL FISHERIES EXPERIMENT STATION)
YOROSHI, JOGASHIMA, MISAKI-CHO, MIURA-SHI 238
JAPANESE

KANAGAWA-KEN TANSUIGYO ZOSHOKUJO HOKOKU 4860
(REPORT OF KANAGAWA PREFECTURAL FRESHWATER FISHERIES STATION)
KANAGAWA TANSUIGYO ZOSHOKUJO (KANAGAWA PREFECTURAL FRESHWATER FISHERIES STATION)
1903-2, SHIMOMIZO, SAGAMIHARA-SHI 229
JAPANESE

KASUMIGAURA KITAURA SUISAN KAKOGYO NO GENKYO 4861
(ACTUAL STATE OF FISHERY PRODUCTS INDUSTRY IN KASUMIGAURA AND KITAURA)
IBARAKI-KEN KASUMIGAURA KITAURA SUISAN JIMUSHO (KASUMIGAURA KITAURA FISHERIES OFFICE)
603, TONOSATO-MACHI, TSUCHIURA-SHI 300
JAPANESE

KATSUO MAGURO TSUSHIN 4862
(BONITO AND TUNA NEWS)
SUISAN SHINCHOSHA
10-10, 1-CHOME, KUDANKITA, CHIYODA-KU, TOKYO 102
JAPANESE

KATSUO SAOZURI GYOGYO GYOJOBETSU TOKEI CHOSA KEKKA HOKOKU 4863
(ANNUAL REPORT OF EFFORT AND CATCH STATISTICS BY AREA JAPANESE SKIPJACK BUTBOAT FISHERY)
SUISAN-CHO KENKYU KAIHATSU-BU (RESEARCH AND DEVELOPMENT DEPARTMENT, FISHERY AGENCY)
2-1, 1-CHOME, KASUMIGASEKI, CHIYODA-KU, TOKYO 100
JAPANESE AND ENGLISH

KATSUO TO MAGURO 4864
(BONITO AND TUNA FISH)
NIHON KATSUO MAGURO GYOGYO KUMIAI RENGOKAI (FEDERATION OF JAPAN FISHERMEN'S COOPERATIVE ASSOCIATION)
3-22, 2-CHOME, KUDANKITA, CHIYODA-KU, TOKYO 102
JAPANESE

KENKYU HOKOKU (SUISAN-CHO GYOSEN KENKYUSHITSU) 4865
(SCIENTIFIC REPORT (FISHING BOAT LABORATORY, FISHERIES AGENCY))
SUISAN-CHO GYOSEN KENKYUSHITSU
5-5, KACHIDOKI, CHUO-KU, TOKYO 104
JAPANESE AND ENGLISH

KINKI DAIGAKU SUISAN KENKYUSHO HOKOKU 4866
(REPORT OF THE FISHERIES LABORATORY OF KINKI UNIVERSITY)
KINKI DAIGAKU SUISAN KENKYUSHO (FISHERIES LABORATORY OF KINKI UNIVERSITY)
3153, SHIRAHAMA-MACHI, NISHIMURO-GUN, WAKAYAMA 649-22
JAPANESE

KITA-TAIHEIYO GYOGYO KOKUSAI IINKAI KENKYU HOKOKU 4867
(RESEARCH REPORT OF THE INTERNATIONAL FISHERIES COMMISSION FOR THE NORTHERN PACIFIC)
JAPANESE

KOCHI-KEN SUISAN SHIKENJO JIGYO HOKOKUSHO 4868
(REPORT ON ACTIVITIES OF KOCHI PREFECTURAL FISHERIES EXPERIMENTAL STATION)
KOCHI-KEN SUISAN SHIKENJO (KOCHI PREFECTURAL FISHERIES EXPERIMENTAL STATION)
URANOUCHI, SUZAKI-SHI 785-01
JAPANESE

KOKU KANSOKU GYOKAIKYO CHOSA HOKOKU 4869
(FISHING AREA SURVEY BY AERONAUTICAL OBSERVATORY)
MIYAZAKI-KEN SUISAN SHIKENJO NICHINAN BUNJO (NICHINAN BRANCH STATION, MIYAZAKI PREFECTURAL FISHERIES EXPERIMENT STATION)
588-6, NISHI-MACHI, NICHINAN-SHI 887
JAPANESE

KOKURITSU SHINJU KENKYUSHO HOKOKU 4870
(BULLETIN OF THE NATIONAL PEARL RESEARCH LABORATORY)
SUISAN-CHO SHINJU KENKYUSHO (NATIONAL PEARL RESEARCH LABORATORY, FISHERIES AGENCY)
KASHIKOJIMA, AGO-MACHI, SHIMA-GUN, MIE 517-05
JAPANESE AND ENGLISH
INDEXED: BIO, JICST

KOKURITSU SHINJU KENKYUSHO NYUSU 4871
(NATIONAL PEARL RESEARCH LABORATORY NEWS)
SUISAN-CHO SHINJU KENKYUSHO
KASHIKOJIMA, AGO-MACHI, SHIMA-GUN, MIE 517-05
JAPANESE

KOKURITSU SHINJU KENKYUSHO SHIRYO 4872
(MATERIAL OF THE NATIONAL PEARL RESEARCH LABORATORY)
SUISAN-CHO SHINJU KENKYUSHO (FISHERIES AGENCY, NATIONAL PEARL RESEARCH LABORATORY)
KASHIKOJIMA, AGO-MACHI, SHIMA-GUN, MIE 517-05
JAPANESE
INDEXED: JICST

KUMAMOTO-KEN NORI KENKYUSHO GYOMU HOKOKU 4873
(ANNUAL REPORT OF KUMAMOTO PREFECTURE LAVER RESEARCH INSTITUTE)
KUMAMOTO-KEN NORI KENKYUSHO (KUMAMOTO PREFECTURE LAVER RESEARCH INSTITUTE)
MISUMI-MACHI, UTO-GUN, KUMAMOTO 869-32
JAPANESE

KUMAMOTO-KEN NORI KENKYUSHO SHIRYO 4874
(RESEARCH REPORT OF KUMAMOTO PREFECTURE LAVER RESEARCH INSTITUTE)
KUMAMOTO-KEN NORI KENKYUSHO (KUMAMOTO PREFECTURE LAVER RESEARCH INSTITUTE)
MISUMI-MACHI, UTO-GUN, KUMAMOTO 869-32
JAPANESE

KUMAMOTO-KEN NORI YOSHOKU JITTAI CHOSA SHIRYO 4875
(REPORT OF ACTUAL STATE OF LAVER CULTURE IN KUMAMOTO PREFECTURE)
KUMAMOTO-KEN NORI KENKYUSHO (KUMAMOTO PREFECTURE LAVER RESEARCH INSTITUTE)
MISUMI-MACHI, UTO-GUN, KUMAMOTO 869-32
JAPANESE

KUMAMOTO-KEN SUISAN SHIKENJO HOKOKUSHO 4876
(SCIENCE REPORT OF KUMAMOTO PREFECTURAL FISHERIES EXPERIMENTAL STATION)
KUMAMOTO-KEN SUISAN SHIKENJO (KUMAMOTO PREFECTURAL FISHERIES EXPERIMENTAL STATION)
OMON, KUSUURA-CHO, HONDO-SHI 863
JAPANESE

KUMAMOTO-KEN SUISAN SHIKENJO JIGYO HOKOKUSHO 4877
(ANNUAL REPORT OF THE KUMAMOTO PREFECTURAL FISHERIES EXPERIMENTAL STATION)
KUMAMOTO-KEN SUISAN SHIKENJO (KUMAMOTO PREFECTURAL FISHERIES EXPERIMENTAL STATION)
OMON, KUSUURA-CHO, HONDO-SHI 863
JAPANESE

KUSHIRO SUISHI DAYORI 4878
(KUSHIRO FISHERIES EXPERIMENT STATION LETTER)
HOKKAIDORITSU KUSHIRO SUISAN SHIKENJO (HOKKAIDO PREFECTURAL KUSHIRO FISHERIES EXPERIMENT STATION)
JAPANESE

KYOTO-FU SUISAN SHIKENJO GYOSEKI HOKOKUSHO 4879
(REPORT OF FISHERIES RESEARCH STATION OF KYOTO PREFECTURE)
KYOTO-FU SUISAN SHIKENJO (FISHERIES RESEARCH STATION OF KYOTO PREFECTURE)
TSURUGA, MIYAZU-SHI 626
JAPANESE

KYUSHU DAIGAKU NOGAKUBU FUZOKU SUISAN JIKKENSHO GYOSEKISHU 4880
(COLLECTED REPRINTS OF FISHERIES RESEARCH LABORATORY, KYUSHU UNIVERSITY)
KYUSHU DAIGAKU NOGAKUBU FUZOKU SUISAN JIKKENSHO (FISHERIES RESEARCH LABORATORY, KYUSHU UNIVERSITY FACULTY OF AGRICULTURE)
TSUYAZAKI-MACHI, MUNAKATA-GUN, FUKUOKA 811-33
JAPANESE AND ENGLISH

KYUSHU DAIGAKU NOGAKUBU FUZOKU SUISAN JIKKENSHO HOKOKU 4881
(REPORT OF FISHERIES RESEARCH LABORATORY, KYUSHU UNIVERSITY)
KYUSHU DAIGAKU NOGAKUBU FUZOKU SUISAN JIKKENSHO (FISHERIES RESEARCH LABORATORY, FACULTY OF AGRICULTURE, KYUSHU UNIVERSITY)
TSUYASAKI-MACHI, MUNAKATA-GUN, FUKUOKA 811-33
JAPANESE AND ENGLISH

MAGURO HAENAWA GYOGYO GYOJOBETSU TOKEI CHOSA KEKKA HOKOKU 4882
(ANNUAL REPORT OF EFFORT AND CATCH STATISTICS BY AREA ON JAPANESE TUNA LONG LINE FISHERY)
SUISAN-CHO KENKYU KAIHATSU-BU (RESEARCH AND DEVELOPMENT DEPARTMENT, FISHERY AGENCY)
2-1, 1-CHOME, KASUMIGASEKI, CHIYODA-KU, TOKYO 100
JAPANESE AND ENGLISH

MAGUROKAN NYUSU 4883
(CANNED TUNA NEWS)
MAGURO KANZUME SHINKOKAI
C/O IDA BUILDING, 1-3, YAESU, CHUO-KU, TOKYO 103
JAPANESE

MATOYA-WAN YOREI KENKYUSHO HOKOKU 4884
(RESEARCH REPORT OF THE MATOYA OYSTER RESEARCH LABORATORY)
MATOYA-WAN YOREI KENKYUSHO (MATOYA OYSTER RESEARCH LABORATORY)
889, MATOYA, ISOBE-CHO, SHIMA-GUN, MIE 517-03
JAPANESE

MEMOIRS OF THE FACULTY OF FISHERIES, HOKKAIDO UNIVERSITY 4885
(HOKKAIDO DAIGAKU SUISANGAKUBU KIYO)
HOKKAIDO DAIGAKU SUISANGAKUBU (FACULTY OF FISHERIES, HOKKAIDO UNIVERSITY)
1-1, 3-CHOME, MINATO-MACHI, HAKODATE-SHI 040
EUROPEAN
INDEXED: JICST

MIE DAIGAKU SUISANGAKUBU KENKYU HOKOKU 4886
(BULLETIN OF THE FACULTY OF FISHERIES, MIE UNIVERSITY)
MIE UNIVERSITY, FACULTY OF FISHERIES
2-80 EDOTASHI, TSU
JAPANESE AND ENGLISH
INDEXED: BIB, CHE

MIE-KEN HAMAJIMA SUISAN SHIKENJO NENPO 4887
(ANNUAL REPORT OF THE FISHERIES EXPERIMENT STATION OF MIE PREFECTURE)
MIE-KEN HAMAJIMA SUISAN SHIKENJO (HAMAJIMA EXPERIMENTAL STATION OF FISHERIES, MIE PREFECTURE)
HAMAJIMA-CHO, SHIMA-GUN, MIE 517-04
JAPANESE

MIE-KEN ISEWAN SUISAN SHIKENJO JIGYO HOKOKU 4888
(REPORT OF ISE BAY FISHERIES EXPERIMENT STATION OF MIE PREFECTURE)
MIE-KEN ISEWAN SUISAN SHIKENJO
6270-5, BORADAME, SHIROKO-CHO, SUZUKA-SHI 510-02
JAPANESE

MIE-KEN NAISUIMEN SUISAN SHIKENJO JIGYO HOKOKU 4889
(REPORT OF INLAND WATER FISHERIES EXPERIMENTAL STATION OF MIE PREFECTURE)
MIE-KEN NAISUIMEN SUISAN SHIKENJO (INLAND WATER FISHERIES EXPERIMENTAL STATION OF MIE PREFECTURE)
1028, KAMORI, KOMONO-MACHI, MIE-GUN, MIE 510-12
JAPANESE

MIE-KENRITSU SUISAN SHIKENJO KENKYU HOKOKU 4890
(BULLETIN OF THE MIE PREFECTURAL FISHERIES EXPERIMENTAL STATION)
MIE-KEN HAMAJIMA SUISAN SHIKENJO (HAMAJIMA EXPERIMENT STATION OF FISHERIES, MIE PREFECTURE)
HAMAJIMA-MACHI, SHIMA-GUN, MIE 517-04
JAPANESE

MIYAGI-KEN GYOGYO NO UGOKI 4891
(TENDENCY OF FISHERIES IN MIYAGI PREFECTURE)
NORIN-SHO TOHOKU NOSEI-KYOKU TOKEI JOHO-BU (STATISTICS AND INFORMATION OFFICE, TOHOKU REGIONAL AGRICULTURAL ADMINISTRATION BUREAU, MINISTRY OF AGRICULTURE AND FORESTRY)
3-1, 3-CHOME, HON-CHO, SENDAI-SHI 980
JAPANESE

MIYAGI-KEN KESENNUMA SUISAN SHIKENJO JIGYO HOKOKU 4892
(REPORT OF THE MIYAGI PREFECTURAL FISHERIES EXPERIMENTAL STATION AT KESENNUMA)
MIYAGI-KEN KESENNUMA SUISAN SHIKENJO (MIYAGI PREFECTURAL FISHERIES EXPERIMENTAL STATION AT KESENNUMA)
120, HAJIKAMI UCHINUMA, KESENNUMA-SHI 988-02
JAPANESE
INDEXED: CHE

MIYAGI-KEN SUISAN KAKO KENKYUSHO JIGYO HOKOKU 4893
(REPORT OF MIYAGI PREFECTURAL FISHERIES MANUFACTURING RESEARCH INSTITUTE)
MIYAGI-KEN SUISAN KAKO KENKYUSHO (MIYAGI PREFECTURAL FISHERIES MANUFACTURING RESEARCH INSTITUTE)
11-6, NAGAHAMA-CHO, ISHINOMAKI-SHI 986-21
JAPANESE

MIYAGI-KEN SUISAN KAKO KENKYUSHO KENKYU HOKOKU 4894
(RESEARCH REPORT OF MIYAGI PREFECTURAL FISHERIES MANUFACTURING RESEARCH INSTITUTE)
MIYAGI-KEN SUISAN KAKO KENKYUSHO (MIYAGI PREFECTURAL FISHERIES MANUFACTURING RESEARCH INSTITUTE)
11-6, NAGAHAMA-CHO, ISHINOMAKI-SHI 986-21
JAPANESE

MIYAGI-KEN SUISAN SHIKENJO KENKYU HOKOKU 4895
(BULLETIN OF MIYAGI PREFECTURAL FISHERIES EXPERIMENTAL STATION)
MIYAGI-KEN SUISAN SHIKENJO (MIYAGI PREFECTURAL FISHERIES EXPERIMENTAL STATION)
11-6, NAGAHAMA-CHO, ISHINOMAKI-SHI 986-21
JAPANESE

MIYAZAKI DAIGAKU SUISAN JIKKENJO HOKOKU 4896
(REPORTS OF THE FISHERIES LABORATORY, MIYAZAKI UNIVERSITY)
MIYAZAKI DAIGAKU NOGAKUBU FUZOKU SUISAN JIKKENJO (MIYAZAKI UNIVERSITY, DEPARTMENT OF AGRICULTURE, FISHERIES LABORATORY)
JAPANESE

MIYAZAKI-KEN SUISAN SHIKENJO JIGYO HOKOKU 4897
(REPORT OF MIYAZAKI PREFECTURAL FISHERIES EXPERIMENT STATION)
MIYAZAKI-KEN SUISAN SHIKENJO (MIYAZAKI PREFECTURAL FISHERIES EXPERIMENT STATION)
KAMISHIRAHAMA, ORIUZAKO, MIYAZAKI-SHI 889-22
JAPANESE

NAGASAKI DAIGAKU SUISAN GAKUBU KENKYU HOKOKU 4898
(BULLETIN OF THE FACULTY OF FISHERIES, NAGASAKI UNIVERSITY)
NAGASAKI DAIGAKU SUISAN GAKUBU (FACULTY OF FISHERIES NAGASAKI UNIVERSITY)
1-14, BUNKYO-MACHI, NAGASAKI-SHI 852
JAPANESE
INDEXED: CHE, JICST

NAGASAKI-KEN SUISAN SHIKENJO KENKYU HOKOKU 4899
(RESEARCH REPORT OF THE NAGASAKI PREFECTURE FISHERIES EXPERIMENT STATION)
NAGASAKI-KEN SUISAN SHINKENJO (NAGASAKI PREFECTURE FISHERIES EXPERIMENT STATION)
JAPANESE

NAGASAKI-KEN SUISAN SHIKENJO RONBUN SHU 4900
(COLLECTED PAPERS OF THE NAGASAKI PREFECTURAL FISHERIES EXPERIMENT STATION)
JAPANESE

NANSEI SUIKEN NYUSU 4901
(SOUTHWEST SEA (REGION) FISHERIES RESEARCH NEWS)
NANSEIKAIKU SUISAN KENKYUJO (SOUTHWEST SEA REGION, FISHERIES RESEARCH INSTITUTE)
JAPANESE

NANSEIKAI-KU GYOJO KAIKYO GAIHO 4902
(SURVEY OF FISHING AREA, NANSEI REGION)
SUISAN-CHO NANSEIKAI-KU SUISAN KENKYUSHO (NANSEI REGIONAL FISHERIES RESEARCH LABORATORY, FISHERY AGENCY)
1-21, 6-CHOME, SANBASHI-DORI, KOCHI-SHI 780
JAPANESE

NANSEIKAI-KU SUISAN KENKYUSHO KENKYU HOKOKU 4903
(BULLETIN OF THE NANSEI REGIONAL FISHERIES RESEARCH LABORATORY)
SUISAN-CHO NANSEIKAI-KU SUISAN KENKYUSHO (NANSEI REGIONAL FISHERIES RESEARCH LABORATORY, FISHERIES AGENCY)
7782-9, MARUISHI, ONO-MACHI, SAIKI-GUN, HIROSHIMA 739-04
JAPANESE AND ENGLISH
INDEXED: BIB, CHE, JICST

NANSEIKAI-KU SUISAN KENKYUSHO NENJI HOKOKU 4904
(ANNUAL REPORT OF NANSEI REGIONAL FISHERIES RESEARCH LABORATORY)
SUISAN-CHO NANSEIKAI-KU SUISAN KENKYUSHO (NANSEI REGIONAL FISHERIES RESEARCH LABORATORY, FISHERIES AGENCY)
7782-9, MARUISHI, ONO-MACHI SAIKI-GUN, HIROSHIMA 739-04
JAPANESE
INDEXED: CHE

NANSEIKAIKU SUISAN KENKYUJO NYUSU 4905
(SOUTHWEST SEA DISTRICT FISHERIES RESEARCH INSTITUTE)
JAPANESE

NIHON SUISAN GAKKAI TOHOKU SHIBU KAIHO 4906
(BULLETIN OF TOHOKU BRANCH OF THE JAPANESE SOCIETY OF SCIENTIFIC FISHERIES)
NIHON SUISAN GAKKAI TOHOKU SHIBU (TOHOKU BRANCH OF THE JAPANESE SOCIETY OF SCIENTIFIC FISHERIES)
C/O TOHOKU DAIGAKU NOGAKUBU SUISANGAKKA, 1-1, AMAMIYA-CHO, TSUTSUMI-DORI, SENDAI-SHI 980
JAPANESE

NIHON SUISAN GAKKAISHI 4907
(BULLETIN OF THE JAPANESE SOCIETY OF SCIENTIFIC FISHERIES)
NIHON SUISAN GAKKAI (JAPANESE SOCIETY OF SCIENTIFIC FISHERIES)
C/O TOKYO SUISAN DAIGAKU, 5-7, 4-CHOME, KONAN, MINATO-KU, TOKYO 108
JAPANESE AND EUROPEAN
INDEXED: BIO, CHE, JICST

NIHON SUISAN SHIGEN HOGO KYOKAI GEPPO 4908
(MONTHLY BULLETIN OF THE JAPAN FISHERIES RESOURCES CONSERVATION ASSOCIATION)
NIHON SUISAN SHIGEN HOGO KYOKAI (JAPAN FISHERIES RESOURCES CONSERVATION ASSOCIATION)
11-35, 1-CHOME, NAGATA-CHO, CHIYODA-KU, TOKYO 100
JAPANESE

NIHON SUISAN SHIGEN HOGO KYOKAI NENPO 4909
(JAPAN FISHERIES RESOURCE CONSERVATION ASSOCIATION, ANNUAL REPORT)
NIHON SUISAN SHIGEN HOGO KYOKAI (JAPAN FISHERIES RESOURCE CONSERVATION ASSOCIATION)
11-35, 1-CHOME, NAGATA-CHO, CHIYODA-KU, TOKYO 100
JAPANESE

NIHON SUISAN SHINBUN 4910
(JAPAN FISHERIES NEWS)
DAINIHON SUISANKAI (JAPAN FISHERIES ASSOCIATION)
9-13, 1-CHOME, AKASAKA, MINATO-KU, TOKYO 107
JAPANESE

NIHONKAI BUROKKU GYOKYO KAIKYO RENGOKAIGI KENKYU HAPPYO HOKOKUSHU 4911
(RESEARCH FINDINGS, REPORT OF THE JAPAN SEA BLOCK FISHING AND SEA CONDITIONS LIAISON COUNCIL)
NIHONKAIKU SUISAN KENKYUJO (JAPAN SEA REGION FISHERIES RESEARCH INSTITUTE)
JAPANESE

NIHONKAI SAIBAI GYOGYO SHIGEN SEITAI CHOSA HOKOKUSHO (ISHIKAWA-KEN) 4912
(ECOLOGICAL SURVEY OF CULTURAL FISHERY RESOURCES IN THE JAPAN SEA)
ISHIKAWA-KEN SUISAN SHIKENJO (ISHIKAWA PREFECTURAL FISHERIES EXPERIMENT STATION)
NOTO-MACHI, FUGESHI-GUN, ISHIKAWA 927-04
JAPANESE

NIHONKAI SURUMEIKA CHOKAI GYOKAIKYO YOHO 4913
(FISHERY FORECAST OF CUTTLEFISH IN THE JAPAN SEA)
SUISAN-CHO NIHONKAI-KU SUISAN KENKYUSHO (JAPAN SEA REGIONAL FISHERIES RESEARCH LABORATORY, FISHERIES AGENCY)
5939-22, HAMAURA, NISHI FUNAMI-CHO, NIIGATA-SHI 951
JAPANESE

NIHONKAI-KU SUISAN KENKYUSHO KENKYU HOKOKU 4914
(BULLETIN OF THE JAPAN SEA REGIONAL FISHERIES RESEARCH LABORATORY)
SUISAN-CHO NIHONKAI-KU SUISAN KENKYUSHO (JAPAN SEA REGIONAL FISHERIES RESEARCH LABORATORY, FISHERIES AGENCY)
5939-22, HAMAURA, NISHIFUNAMI-CHO, NIIGATA-SHI 951
JAPANESE AND ENGLISH
INDEXED: BIO, JICST

NIHONKAI-KU SUISAN SHIKEN KENKYU RENRAKU NYUSU 4915
(JAPAN SEA FISHERIES RESEARCH NEWS)
SUISAN-CHO NIHONKAI-KU SUISAN KENKYUSHO (JAPAN SEA REGIONAL FISHERIES RESEARCH LABORATORY, FISHERIES AGENCY)
5939-22, HAMAURA, NISHI FUNAMI-CHO, NIIGATA-SHI 951
JAPANESE

NIIGATA-KEN SUISAN SHIKENJO JIGYO HOKOKUSHO 4916
(REPORT OF NIIGATA PREFECTURAL FISHERY EXPERIMENT STATION)
NIIGATA-KEN SUISAN SHIKENJO
2-4, BANDAISHIMA, NIIGATA-SHI 950
JAPANESE

NIIGATA-KEN SUISAN SHIKENJO KENKYU HOKOKU 4917
(RESEARCH REPORT OF THE NIIGATA PREFECTURE FISHERIES EXPERIMENT STATION)
JAPANESE

NIIGATA-KEN SUISAN TOKEI 4918
(FISHERY STATISTICS, NIIGATA PREFECTURE)
NORIN-SHO HOKURIKU NOSEI-KYOKU NIIGATA TOKEI JOHO JIMUSHO (NIIGATA STATISTIC AND INFORMATION OFFICE, HOKURIKU REGIONAL AGRICULTURAL ADMINISTRATIONAL BUREAU, MINISTRY OF AGRICULTURE AND FORESTRY)
601, NISHIOHATA-MACHI, NIIGATA-SHI 951
JAPANESE

NIIGATA-KENNAI SUIMEN SUISON SHIKENJO CHOSA KENKYU HOKOKU 4919
(RESEARCH REPORT OF THE NIGATA PREFECTURE SURFACE WATER FISHERIES EXPERIMENT STATION INVESTIGATION)
NIGATA PREFECTURE SURFACE WATER FISHERIES EXPERIMENT STATION
JAPANESE

NIJIMASU 4920
(RAINBOW TROUT)
NAGANO-KEN SUISAN SHIDOSHO (NAGANO FISHERIES STATION)
AKASHINA-MACHI, HIGASHICHIKUMA-GUN, NAGANO 399-71
JAPANESE

NIPPON SUISAN K.K. CHUO KENKYUJO HOKOKU 4921
(REPORTS OF THE CENTRAL RESEARCH LABORATORY, NIPPON SUISAN COMPANY, LIMITED)
JAPANESE

NISSUI DAYORI 4922
(NEWS OF NIHON SUISAN COMPANY)
NIHON SUISAN K.K.
6-2, 2-CHOME, OTE-MACHI, CHIYODA-KU, TOKYO 100
JAPANESE

NOGYO DOBOKU SHIKENJO GIHO, D, SUISAN DOBOKU 4923
(TECHNICAL REPORT OF THE AGRICULTURAL ENGINEERING RESEARCH STATION, SERIES D, FISHERY ENGINEERING)
NORIN-SHO NOGYO DOBOKU SHIKENJO (AGRICULTURAL ENGINEERING RESEARCH STATION, MINISTRY OF AGRICULTURE AND FORESTRY)
1943, YAHATA, HIRATSUKA-SHI 254
JAPANESE
INDEXED: JICST

NORIKEN DAYORI 4924
(NEWS OF KUMAMOTO PREFECTURE LAVER RESEARCH INSTITUTE)
KUMAMOTO-KEN NORI KENKYUSHO (KUMAMOTO PREFECTURE LAVER RESEARCH INSTITUTE)
MISUMIURA, MISUMI-MACHI, UTO-GUN, KUMAMOTO 869-32
JAPANESE

NORIN GYOGYO GENCHI JOHO 4925
(AGRICULTURE AND FORESTRY ON-THE-SPOT REPORTS ON THE FISHING INDUSTRY)
NORINSHO NORIN KEIZAI KYOKU TOKEI JOHOBU (MINISTRY OF AGRICULTURE AND FORESTRY, AGRICULTURE AND FORESTRY ECONOMICS BUREAU, STATISTICS DEPARTMENT)
JAPANESE

NORIN GYOGYO JOHO; KANTO NOSEIKYOKU TOKEI JOHOBU-HEN 4926
(AGRICULTURE AND FORESTRY FISHING INDUSTRY INFORMATION)
TOKYO NORIN TOKEI KYOKAI (TOKYO AGRICULTURE AND FORESTRY STATISTICS ASSOCIATION)
JAPANESE

NORIN SUISAN KANKEI SHIKENJO DAYORI 4927
(AGRICULTURE AND FORESTRY DIVISION EXPERIMENTAL STATION LETTER ON FISHERIES)
NIIGATA-KEN NORINBU NOSEIKIKAKUKA (NIIGATA PREFECTURAL AGRICULTURAL AND FORESTRY DEPARTMENT, AGRICULTURAL ADMINISTRATIVE PLANNING SECTION)
JAPANESE

NORIN SUISAN SHIKEN KENKYU NENPO SUISAN-HEN 4928
(ANNUAL REPORT ON EXPERIMENTS IN AGRICULTURE, FORESTRY AND FISHERIES, SERIES FISHERIES)
NORIN SUISAN GIJUTSU KAIGI JIMU-KYOKU (AGRICULTURE FORESTRY AND FISHERIES RESEARCH COUNCIL, MINISTRY OF AGRICULTURE AND FORESTRY) AND NORIN TOKEI KYOKAI
11-4, 2-CHOME, MEGURO, MEGURO-KU, TOKYO 153
JAPANESE

NUMAZU BUNSHITSU DAYORI 4929
(NUMAZU BRANCH NEWS)
SHIZUOKA-KEN SUISAN SHIKENJO NUMAZU BUNSHITSU (NUMAZU BRANCH, SHIZUOKA PREFECTURAL FISHERIES EXPERIMENT STATION)
KUCHINO, NUMAZU-SHI 410-01
JAPANESE

OITA-KEN NAISUIMEN GYOGYO SHIKENJO JIGYO HOKOKU 4930
(BUSINESS REPORTS OF THE OITA PREFECTURAL INTERIOR WATER FISHERIES EXPERIMENT STATION)
OITA-KEN NAISUIMEN GYOGYO SHIKENJO (OITA PREFECTURAL INTERIOR WATER FISHERIES EXPERIMENT STATION)
SHO, AJIMU-MACHI, USA-GUN, OITA 872-05
JAPANESE

OITA-KEN SENKAI GYOGYO SHIKENJO KENKYU HOKOKU 4931
(REPORT OF OITA PREFECTURAL SHALLOW SEA FISHERIES EXPERIMENTAL STATION)
OITA-KEN SENKAI GYOGYO SHIKENJO (OITA PREFECTURAL SHALLOW SEA FISHERIES EXPERIMENTAL STATION)
3008, TAKADA, BUNGO TAKADA-SHI 879-06
JAPANESE

OITA-KEN SUISAN SHIKENJO CHOSA KENKYU HOKOKU 4932
(BULLETIN OF OITA PREFECTURAL FISHERIES EXPERIMENTAL STATION)
OITA-KEN SUISAN SHIKENJO
KAMIURA-MACHI, MINAMI AMABE-GUN, OITA 879-26
JAPANESE

OITA-KEN SUISAN SHIKENJO JIGYO HOKOKU 4933
(OPERATION REPORT OF OITA PREFECTURAL FISHERIES EXPERIMENTAL STATION)
OITA-KEN SUISAN SHIKENJO
KAMIURA-MACHI, MINAMI AMABE-GUN, OITA 879-26
JAPANESE

OITA-KEN SUISAN SHIKENJO SHIKEN KENKYU HOKOKU 4934
(SCIENCE REPORT OF OITA PREFECTURAL FISHERIES EXPERIMENTAL STATION)
OITA-KEN SUISAN SHIKENJO
KAMIURA-MACHI, MINAMI AMABE-GUN, OITA 879-26
JAPANESE

OKAYAMA-KEN SUISAN SHIKENJO JIGYO HOKOKUSHO 4935
(REPORT OF FISHERIES EXPERIMENT STATION, OKAYAMA PREFECTURE)
OKAYAMA-KEN SUISAN SHIKENJO (FISHERIES EXPERIMENT STATION, OKAYAMA PREFECTURE)
KASHINO, USHIMADO-CHO, OKU-GUN, OKAYAMA 701-43
JAPANESE

OKINAWA-KEN SUISAN SHIKENJO JIGYO HOKOKUSHO 4936
(REPORT OF THE OKINAWA PREFECTURE FISHERIES EXPERIMENTAL STATION)
OKINAWA-KEN SUISAN SHIKENJO (OKINAWA PREFECTURE FISHERIES EXPERIMENTAL STATION)
KAIGAN UMETATECHI, ITOMAN NISHI-KU, ITOMAN-SHI 901-03
JAPANESE

OSAKA-FU SUISAN SHIKENJO JIGYO HOKOKU 4937
(REPORT OF OSAKA PREFECTURAL FISHERIES EXPERIMENT STATION)
OSAKA-FU SUISAN SHIKENJO (OSAKA PREFECTURAL FISHERIES EXPERIMENT STATION)
2926-1, TANAGAWA TANIGAWA, MISAKI-CHO, SENNAN-GUN, OSAKA 599-03
JAPANESE

OSAKA-FU SUISAN SHIKENJO KENKYU HOKOKU 4938
(BULLETIN OF THE OSAKA PREFECTURAL FISHERIES EXPERIMENT STATION)
OSAKA-FU SUISAN SHIKENJO (OSAKA PREFECTURAL FISHERIES EXPERIMENT STATION)
2926-1, TANAGAWA TANIGAWA, MISAKI-CHO, SENNAN-GUN, OSAKA 599-03
JAPANESE

OSAKA-FU TANSUIGYO SHIKENJO GYOMU HOKOKU 4939
(OPERATION REPORT OF OSAKA PREFECTURAL FRESHWATER FISH EXPERIMENT STATION)
OSAKA-FU TANSUIGYO SHIKENJO (OSAKA PREFECTURAL FRESHWATER FISH EXPERIMENT STATION)
10-4 KIYAMOTO-MACHI, NEYAGAWA-SHI 572
JAPANESE

OSAKA-FU TANSUIGYO SHIKENJO KENKYU HOKOKU 4940
(BULLETIN OF THE OSAKA PREFECTURAL FRESHWATER FISH EXPERIMENT STATION)
OSAKA-FU TANSUIGYO SHIKENJO (OSAKA PREFECTURAL FRESHWATER FISH EXPERIMENT STATION)
10-4, KIYA MOTO-MACHI, NEYAGAWA-SHI 572
JAPANESE

SABA CHOKI RYOKYO YOHO; TOHOKUKAIKU 4941
(OUTLOOK FOR THE LONG-TERM MACKERAL CATCH NORTHEASTERN SEA REGION)
TOHOKUKU SUISANKENKYUJO, TOKAIKU SUISAN KENKYUJO (NORTHEASTERN REGION FISHERIES RESEARCH INSTITUTE, EASTERN SEA REGION RESEARCH INSTITUTE)
JAPANESE

SAGA-KEN ARIAKE SUISAN SHIKENJO HOKOKU 4942
(REPORT OF THE SAGA PREFECTURE ARIAKE FISHERIES EXPERIMENT STATION)
SAGA PREFECTURE ARIAKE FISHERIES EXPERIMENT STATION
JAPANESE

SAGA-KEN SUISAN SHIKENJO GYOMU HOKOKUSHO 4943
(ANNUAL REPORT OF SAGA PREFECTURAL FISHERIES EXPERIMENT STATION)
SAGA-KEN SUISAN SHIKENJO (SAGA PREFECTURAL FISHERIES EXPERIMENT STATION)
4948-9, 6-CHOME, TOBO, KARATSU-SHI 847-01
JAPANESE

SAGA-KEN YOSHOKU SHIKENJO HOKOKU 4944
(REPORT OF SAGA PREFECTURAL CULTURE FISHERIES EXPERIMENT STATION)
SAGA-KEN YOSHOKU SHIKENJO
KITAFUNATSU, HAMA-CHO, KASHIMA-SHI 849-13
JAPANESE

SAIBAI GYOGYO JISSEN GYOJO SETTEI CHOSA JIGYO HOKOKU 4945
(REPORT OF SURVEY ON PRACTICAL ESTABLISHMENT OF FISHING AREA FOR FISH FARMING)
HYOGO-KEN NORIN-BU, HYOGO-KENRITSU SUISAN SHIKENJO (AGRICULTURE AND FORESTRY DIVISION, HYOGO PREFECTURE GOVERNMENT, HYOGO PREFECTURAL FISHERIES EXPERIMENT STATION)
5-1, SHIMOYAMATE-DORI, IKUTA-KU, KOBE-SHI 650
JAPANESE

SAITAMA-KEN SUISAN SHIKENJO KENKYU HOKOKU 4946
(REPORT OF SAITAMA PREFECTURAL FISHERIES EXPERIMENT STATION)
SAITAMA-KEN SUISAN SHIKENJO (SAITAMA PREFECTURAL FISHERIES EXPERIMENTAL STATION)
1060, KITAKOHAMA, KAZO-SHI 347
JAPANESE

SAKANA 4947
(FISH)
SUISAN-CHO TOKAI-KU SUISAN KENKYUSHO (TOKAI REGIONAL FISHERY RESEARCH LABORATORY, FISHERY AGENCY)
5-1, 5-CHOME, KACHIDOKI, CHUO-KU, TOKYO 104
JAPANESE

SAKE TO MASU 4948
(SALMON AND TROUT)
NIHON SAKE MASU SHIGEN HOGO KYOKAI (JAPAN SALMON AND TROUT CONSERVATION ASSOCIATION)
9-13, 1-CHOME, AKASAKA, MINATO-KU, TOKYO 107
JAPANESE

SAN'IN TOTTORIOKI KINCHAKU GYOKYO JUNPO 4949
(NEWSLETTER OF CIRCULAR MOORED NET FISHERY IN TOTTORI COAST)
TOTTORI-KEN SUISAN SHIKENJO (TOTTORI PREFECTURAL FISHERY EXPERIMENT STATION)
OTANI, IWAMI-CHO, IWAMI-GUN, TOTTORI 681
JAPANESE

SATSUNAN KAIIKI NI OKERU KATSUO GYOKYO TO GYOJO 4950
(STATE OF BONITO FISHING AND FISHING AREA IN SATSUNAN)
MIYAZAKI-KEN SUISAN SHIKENJO NICHINAN BUNJO (NICHINAN BRANCH, MIYAZAKI PREFECTURAL FISHERIES EXPERIMENT STATION)
588-6, NISHI-MACHI, NICHINAN-SHI 887
JAPANESE

SEIKAI-KU SUISAN KENKYUSHO KENKYU HOKOKU 4951
(BULLETIN OF THE SEIKAI REGIONAL FISHERIES RESEARCH LABORATORY)
SUISAN-CHO SEIKAI-KU SUISAN KENKYUSHO (SEIKAI REGIONAL FISHERIES RESEARCH LABORATORY, FISHERIES AGENCY)
49, KOKUBU-CHO, NAGASAKI-SHI 850
JAPANESE, WITH ENGLISH SUMMARY
INDEXED: JICST

SEIKAI-KU SUISAN KENKYUSHO NYUSU 4952
(NEWS OF SEIKAI REGIONAL FISHERIES RESEARCH LABORATORY)
SUISAN-CHO SEIKAI-KU SUISAN KENKYUSHO (SEIKAI REGIONAL FISHERIES RESEARCH LABORATORY, FISHERY AGENCY)
49, KOKUBU-CHO, NAGASAKI-SHI 850
JAPANESE

SHIGA-KEN SAMEGAI YOSON SHIKENJO JIGYO HOKOKUSHO 4953
(REPORT ON THE WORKS OF SHIGA PREFECTURAL SAMEGAI TROUT EXPERIMENTAL STATION)
SHIGA-KEN SAMEGAI YOSON SHIKENJO (SHIGA PREFECTURAL SAMEGAI TROUT EXPERIMENTAL STATION)
KAMINYU, MAIHARA-MACHI, SAKATA-GUN, SHIGA 521-01
JAPANESE

SHIGA-KEN SUISAN SHIKENJO JIGYO SEISEKI 4954
(OPERATION REPORT OF SHIGA PREFECTURAL FISHERIES EXPERIMENT STATION)
SHIGA-KEN SUISAN SHIKENJO
2183-3, HASSAKA-CHO, HIKONE-SHI 522
JAPANESE

SHIGA-KEN SUISAN SHIKENJO KENKYU HOKOKU 4955
(BULLETIN OF THE SHIGA PREFECTURAL FISHERIES EXPERIMENTAL STATION)
SHIGA-KEN SUISAN SHIKENJO (SHIGA PREFECTURAL FISHERIES EXPERIMENTAL STATION)
2183-3, HASSAKA-CHO, HIKONE-SHI 522
JAPANESE

SHIMANE SUISHI KENKYU HOKOKUSHO 4956
(RESEARCH REPORT OF SHIMANE PREFECTURAL FISHERIES EXPERIMENT STATION)
SHIMANE-KEN SUISAN SHIKENJO (SHIMANE PREFECTURAL FISHERIES EXPERIMENT STATION)
24-1, SETOGASHIMA-MACHI, HAMADA-SHI 697
JAPANESE

SHIMANE-KEN KAIMEN GYOGYO KIHON CHOSA KEKKA HOKOKUSHO 4957
(REPORT OF FUNDAMENTAL SURVEY OF SEA FISHERIES IN SHIMANE PREFECTURE)
SHIMANE-KEN (SHIMANE PREFECTURAL GOVERNMENT)
1, TONO-MACHI, MATSUE-SHI 690
JAPANESE

SHIMANE-KEN NAISUIMEN GYOGYO KIHON CHOSA KEKKA HOKOKU 4958
(REPORT OF FUNDAMENTAL SURVEY OF INTERIOR WATER FISHERIES IN SHIMANE PREFECTURE)
SHIMANE-KEN SOMU-BU (GENERAL AFFAIRS DEPARTMENT, SHIMANE PREFECTURAL GOVERNMENT)
1, TONO-MACHI, MATSUE-SHI 690
JAPANESE

SHIMANE-KEN SUISAN SHIKENJO JIGYO HOKOKU 4959
(REPORT OF SHIMANE PREFECTURAL FISHERIES EXPERIMENT STATION)
SHIMANE-KEN SUISAN SHIKENJO
SETOGASHIMA-CHO, HAMADA-SHI 697
JAPANESE

SHINJU KENSASHO NENPO 4960
(ANNUAL REPORT OF PEARL INSPECTION STATION)
SUISAN-CHO SHINJU KENSASHO (PEARL INSPECTION STATION, FISHERY AGENCY)
9-35,3-CHOME, KONAN, MINATO-KU, TOKYO 108
JAPANESE

SHINJU SHINBUN 4961
(PEARL NEWS)
SHINJU SHINBUNSHA
10-10, 1-CHOME, KUDANKITA, CHIYODA-KU, TOKYO 102
JAPANESE

SHIZUOKA GYOREN SUISAN JOHO 4962
(REPORT OF THE SHIZUOKA FISHING INDUSTRIES FEDERATION)
SHIZUOKA-KEN GYOGYO KYODO KUMIAI GOKAI (SHIZUOKA PREFECTURE FISHING INDUSTRY COOPERATIVE)
JAPANESE

SHIZUOKA-KEN FUJI YOSONJO JIGYO HOKOKU 4963
(OPERATION REPORT OF SHIZUOKA PREFECTURAL FUJI TROUT CULTURAL LABORATORY)
SHIZUOKA-KEN FUJI YOSONJO (SHIZUOKA PREFECTURAL FUJI TROUT CULTURAL LABORATORY)
579-2, INOKASHIRA, FUJINOMIYA-SHI 418-02
JAPANESE

SHIZUOKA-KEN FUJI YOSONJO KENKYU HOKOKU 4964
(BULLETIN OF THE SHIZUOKA PREFECTURAL FUJI TROUT CULTURAL LABORATORY)
SHIZUOKA-KEN FUJI YOSONJO (SHIZUOKA PREFECTURAL FUJI TROUT CULTURAL LABORATORY)
579-2, INOGASHIRA, FUJINOMIYA-SHI 418-02
JAPANESE

SHIZUOKA-KEN SUISAN SHIKENJO JIGYO HOKOKU 4965
(ANNUAL REPORT OF SHIZUOKA PREFECTURE FISHERIES EXPERIMENT STATION)
SHIZUOKA-KEN SUISAN SHIKENJO (SHIZUOKA PREFECTURE FISHERIES EXPERIMENT STATION)
3690, SHIOIRI, KOGAWA, YAIZU-SHI 425
JAPANESE
INDEXED: CHE, JICST

SHIZUOKA-KEN SUISAN SHIKENJO KENKYU HOKOKU 4966
(SCIENCE REPORT OF SHIZUOKA PREFECTURE FISHERIES EXPERIMENT STATION)
SHIZUOKA-KEN SUISAN SHIKENJO (SHIZUOKA PREFECTURE FISHERIES EXPERIMENT STATION)
3690, SHIOIRI, KOGAWA, YAIZU-SHI 425
JAPANESE AND ENGLISH
INDEXED: CHE, JICST

SHUKAN HOKKAI SUISAN 4967
(HOKKAIDO FISHERIES WEEK)
HOKKAI SUISAN SHINBUNSHA
NISHI-14-CHOME, MINAMI-13-JO, CHUO-KU, SAPPORO-SHI 064
JAPANESE

SOREN HOKUYO GYOGYO KANKEI BUNKENSHU 4968
(INFORMATION OF NORTHERN SEAS FISHERIES)
DAI NIHON SUISANKAI (JAPAN FISHERIES ASSOCIATION)
9-13, 1-CHOME, AKASAKA, MINATO-KU, TOKYO 107
JAPANESE

SUISAN DAIGAKKO KENKYU HOKOKU 4969
(JOURNAL OF THE SHIMONOSEKI UNIVERSITY OF FISHERIES)
SUISAN-CHO SUISAN DAIGAKKO (SHIMONOSEKI UNIVERSITY OF FISHERIES, FISHERIES AGENCY)
1944, NAGATA HON-CHO, YOSHIMI, SHIMONOSEKI-SHI 750
JAPANESE AND ENGLISH
INDEXED: BIO, CHE, JICST

SUISAN DOBOKU 4970
(FISHERIES ENGINEERING)
JAPANESE
INDEXED: JICST

SUISAN FUKAJO KENKYU HOKOKU 4971
(FISH HATCHERY RESEARCH REPORT)
JAPANESE
INDEXED: JICST

SUISAN GIJUTSU TO KEIEI 4972
(FISHERIES TECHNIQUE AND MANAGEMENT)
SUISAN GIJUTSU KEIEI KENKYUKAI (RESEARCH ASSOCIATION FOR FISHERIES TECHNIQUE AND MANAGEMENT)
C/O TOKYO-TO CHUO ICHIBA, 2-1, 5-CHOME, TSUKIJI, CHUO-KU, TOKYO 104
JAPANESE

SUISAN JOHO 4973
(SHIZUOKA FISHERIES JOURNAL)
SHIZUOKA-KEN GYOGYO KYODO KUMIAI RENGOKAI (SHIZUOKA-KEN FISHERIES COOPERATIVE UNION)
9-18, OTE-MACHI, SHIZUOKA-SHI 420
JAPANESE

SUISAN KAIYO KENKYU KAIHO 4974
(BULLETIN OF THE JAPANESE SOCIETY OF FISHERIES OCEANOGRAPHY)
SUISAN KAIYO KENKYUKAI (JAPANESE SOCIETY OF FISHERIES OCEANOGRAPHY)
C/O TOKYO SUISAN DAIGAKU GYOJO KANKYOGAKU KENKYUSHITSU, 5-7, 4-CHOME, KONAN, MINATO-KU, TOKYO 108
JAPANESE

SUISAN SEKAI 4975
(FISHERY MAGAZINE)
NORIN KEIZAI KENKYU-SHO (INSTITUTE OF AGRICULTURAL AND FORESTRY ECONOMICS)
10-14, 5-CHOME, MINAMI AOYAMA, MINATO-KU, TOKYO 107
JAPANESE

SUISAN SHINCHO 4976
(FISHERIES NEWS)
SUISAN SHINCHOSHA
10-10, 1-CHOME, KUDAN-KITA, CHIYODA-KU, TOKYO 102
JAPANESE

SUISAN SHINKO 4977
(FISHERIES PROMOTION)
TOKYO SUISAN SHINKOKAI (TOKYO ASSOCIATION OF FISHERIES PROMOTION)
14-17, 6-CHOME, GINZA, CHUO-KU, TOKYO 104
JAPANESE

SUISAN SHIRYO SHIKIHO 4978
(QUARTERLY REPORT OF FISHERIES DATA)
SUISAN SHIRYO KAN
JAPANESE

SUISAN SHUHO 4979
(FISHING AND FOOD INDUSTRY WEEKLY)
SUISANSHA
8, SAN'EI-CHO, SHINJUKU-KU, TOKYO 160
JAPANESE

SUISAN TOKEI SHIHYO 4980
(INDEX OF FISHERIES STATISTICS)
SUISAN-CHO GYOSEI-BU (FISHERIES ADMINISTRATION DEPARTMENT, FISHERY AGENCY)
2-1, 1-CHOME, KASUMIGASEKI, CHIYODA-KU, TOKYO 100
JAPANESE

SUISAN ZOSHOKU 4981
(FISHERY CULTURE)
SUISAN ZOSHOKU DANWAKAI
C/O HOKKAIDO DAIGAKU SUISAN GAKUBU, 1-1, 3-CHOME, MINATO-MACHI, HAKODATE-SHI 040
JAPANESE
INDEXED: CHE

SUISAN ZOSHOKU: RINJIGO 4982
(INCREASING FISHERIES: SPECIAL NUMBER)
SUISAN ZOSHOKU DANWAKAI (INCREASING FISHERIES ROUNDTABLE)
JAPANESE

SUISAN-CHO SHUHO 4983
(FISHERY AGENCY'S WEEKLY)
SUISAN-CHO (FISHERY AGENCY)
2-1, 1-CHOME, KASUMIGASEKI, CHIYODA-KU, TOKYO 100
JAPANESE

SUISANBUTSU RYUTSU TOKEI NENPO 4984
(ANNUAL REPORT OF DISTRIBUTION STATISTICS ON FISHERIES PRODUCTS)
NORIN-SHO NORIN KEIZAI-KYOKU TOKEI JOHO-BU (STATISTICS AND INFORMATION DIVISION, AGRICULTURE AND FORESTRY ECONOMIC BUREAU, MINISTRY OF AGRICULTURE AND FORESTRY)
11-14, 2-CHOME, MEGURO, MEGURO-KU, TOKYO 153
JAPANESE

SUISANGYO KIHON CHOSA KEKKA HOKOKUSHO 4985
(REPORT OF FUNDAMENTAL SURVEY OF FISHERIES)
TOYAMA-KEN (TOYAMA PREFECTURAL GOVERNMENT)
1-7, SHINSOGAWA, TOYAMA-SHI 930
JAPANESE

SUISANGYO NO DOKO 4986
(TENDENCY OF FISHERIES)
CHIBA-KEN SUISAN-BU (FISHERY DEPARTMENT, CHIBA PREFECTURAL GOVERNMENT)
1-1, ICHIBA-CHO, CHIBA-SHI 280-91
JAPANESE

SUISANKAI 4987
(FISHERIES WORLD)
DAINIHON SUISANKAI (JAPAN FISHERIES ASSOCIATION)
9-13, 1-CHOME, AKASAKA, MINATO-KU, TOKYO 107
JAPANESE
INDEXED: JICST

SUISHI GEPPO 4988
(MONTHLY INVESTIGATIONAL REPORT OF THE FISHERIES EXPERIMENTAL STATION OF SHIMANE PREFECTURE)
SHIMANE-KEN SUISAN SHIKENJO (SHIMANE PREFECTURE FISHERIES EXPERIMENTAL STATION)
SETOGASHIMA-CHO, HAMADA-SHI 697
JAPANESE

SUISHI ITO 4989
(REPORT OF ITO BRANCH, SHIZUOKA PREFECTURE FISHERIES EXPERIMENT STATION)
SHIZUOKA-KEN SUISAN SHIKENJO ITO BUNJO
11-12, SHIZUMI-CHO, ITO-SHI 414
JAPANESE

TANSUI-KU SUISAN KENKYUSHO KENKYU HOKOKU 4990
(BULLETIN OF THE FRESHWATER FISHERIES RESEARCH LABORATORY)
SUISAN-CHO TANSUI-KU SUISAN KENKYUSHO (FRESHWATER FISHERIES RESEARCH LABORATORY, FISHERIES AGENCY)
399, MIYA , HINO-SHI, TOKYO 191
JAPANESE AND ENGLISH
INDEXED: BIO, CHE, EXC, JICST

TANSUI-KU SUISAN KENKYUSHO SHIRYO A SHIRIZU 4991
(GAIKOKU BUNKEN BASSUI)
(INFORMATION OF FRESHWATER FISHERIES RESEARCH LABORATORY, SERIES A, ABSTRACT FROM FOREIGN LITERATURE)
SUISAN-CHO TANSUI-KU SUISAN KENKYUSHO (FRESHWATER FISHERIES RESEARCH LABORATORY, FISHERIES AGENCY)
399, MIYA, HINO-SHI TOKYO 191
JAPANESE
INDEXED: CHE, JICST

TANSUIGYO 4992
(FRESHWATER FISH)
TANSUIGYO HOGO KYOKAI (SOCIETY FOR THE PROTECTION OF FRESHWATER FISH)
JAPANESE

TANSUIKEN NYUSU 4993
(FRESH WATER RESEARCH NEWS)
TANSUIKU SUISAN KENKYUJO TOSHO JOHO IINKAI (FRESHWATER DIVISION, FISHERIES RESEARCH PUBLICATIONS REPORT COMMITTEE)
JAPANESE

TOCHIGI-KEN SUISAN SHIKENJO GYOMU HOKOKUSHO 4994
(REPORT OF TOCHIGI PREFECTURAL FISHERIES EXPERIMENT STATION)
TOCHIGI-KEN SUISAN SHIKENJO
436, OZO-MACHI, UTSUNOMIYA-SHI 320
JAPANESE

TODAI SUISAN JIKKENSHO GYOSEKI 4995
(CONTRIBUTIONS OF THE FISHERIES LABORATORY, FACULTY OF AGRICULTURE, UNIVERSITY OF TOKYO)
TOKYO DAIGAKU NOGAKUBU FUZOKU SUISAN JIKKENSHO (FISHERIES LABORATORY, FACULTY OF AGRICULTURE, UNIVERSITY OF TOKYO)
BENTENJIMA, MAISAKA-MACHI, HAMANA-GUN, SHIZUOKA 431-02
JAPANESE AND ENGLISH

TOHOKU-KU SUISAN KENKYUSHO KENKYU HOKOKU 4996
(BULLETIN OF TOHOKU REGIONAL FISHERIES RESEARCH LABORATORY)
SUISAN-CHO TOHOKU-KU SUISAN KENKYUSHO (TOHOKU REGIONAL FISHERIES RESEARCH LABORATORY, FISHERIES AGENCY)
119-1, 3-CHOME, NIIHAMA-CHO, SHIOGAMA-SHI 985
JAPANESE AND ENGLISH
INDEXED: CHE, JICST

TOKAI KOKAI NI OKERU TEIGYO SHIGEN NO KENKYU 4997
(INVESTIGATIONS OF DEMERSAL FISH RESOURCES IN THE EAST CHINA AND YELLOW SEAS)
SUISAN-CHO SEIKAI-KU SUISAN KENKYUSHO (SEIKAI REGIONAL FISHERIES RESEARCH LABORATORY, FISHERIES AGENCY)
49, KOKUBU-CHO, NAGASAKI-SHI 850
JAPANESE AND ENGLISH

TOKAI-KU CHOKI GYOKYO YOHO 4998
(LONG-TERM FORECASTING ON THE DISTRIBUTION AND ABUNDANCE OF THE IMPORTANT FISHERY RESOURCES AND THE RELATED OCEANOGRAPHICAL CONDITIONS IN TOKAI REGION)
SUISAN-CHO TOKAI-KU SUISAN KENKYUSHO (TOKAI REGIONAL FISHERIES RESEARCH LABORATORY, FISHERIES AGENCY)
5-1, 5-CHOME, KACHIDOKI, CHUO-KU, TOKYO 104
JAPANESE

TOKAI-KU SUISAN KENKYUSHO GYOSEKI C-SHU SAKANA 4999
(RESULTS OF THE TOKAI DISTRICT REGIONAL FISHERIES RESEARCH LABORATORY, COLLECTION-C, FISH)
SUISAN-CHO, TOKAI SUISAN KENKYUSHO (FISHERY AGENCY, TOKAI REGIONAL FISHERY RESEARCH LABORATORY)
JAPANESE

TOKAI-KU SUISAN KENKYUSHO GYOSEKI MOKUROKU 5000
(ABSTRACTS FROM RESEARCHES OF TOKAI REGIONAL FISHERIES RESEARCH LABORATORY)
SUISAN-CHO TOKAI-KU SUISAN KENKYUSHO (TOKAI REGIONAL FISHERIES RESEARCH LABORATORY, FISHERY AGENCY)
5-1, 5-CHOME, KACHIDOKI, CHUO-KU, TOKYO 104
JAPANESE

TOKAI-KU SUISAN KENKYUSHO GYOSEKISHU 5001
(COLLECTED REPRINTS FROM THE TOKAI REGIONAL FISHERIES RESEARCH LABORATORY)
SUISAN-CHO TOKAI-KU SUISAN KENKYUSHO (TOKAI REGIONAL FISHERIES RESEARCH LABORATORY, FISHERIES AGENCY)
5-1, 5-CHOME, KACHIDOKI, CHUO-KU, TOKYO 104
JAPANESE, WITH EUROPEAN SUMMARY
INDEXED: BIB, JICST

TOKAI-KU SUISAN KENKYUSHO KENKYU HOKOKU 5002
(BULLETIN OF THE TOKAI REGIONAL FISHERIES RESEARCH LABORATORY)
SUISAN-CHO TOKAI-KU SUISAN KENKYUSHO (TOKAI REGIONAL FISHERIES RESEARCH LABORATORY, FISHERIES AGENCY)
5-1, 5-CHOME, KACHIDOKI, CHUO-KU, TOKYO 104
JAPANESE AND EUROPEAN
INDEXED: CHE, JICST

TOKAI-KU SUISAN KENKYUSHO KENKYU SHIRYOSHU 5003
(DATA FOR RESEARCH, TOKAI REGIONAL FISHERIES RESEARCH LABORATORY)
SUISAN-CHO TOKAI-KU SUISAN KENKYUSHO (TOKAI REGIONAL FISHERIES RESEARCH LABORATORY, FISHERIES AGENCY)
5-1, 5-CHOME, KACHIDOKI, CHUO-KU, TOKYO 104
JAPANESE
INDEXED: BIO

TOKUSHIMA-KEN SHUYO GYOGYO CHIKUBETSU SOGO TOKEI 5004
(FISHERY STATISTICS OF MAIN FISHERIES AREA IN TOKUSHIMA PREFECTURE)
NORIN-SHO CHUGOKU SHIKOKU NOSEI-KYOKU TOKUSHIMA TOKEI JOHO JIMUSHO (TOKUSHIMA STATISTICS AND INFORMATION OFFICE, CHUGOKU SHIKOKU REGIONAL AGRICULTURAL ADMINISTRATION BUREAU)
1-6, 5-CHOME, SHIMOSUKETO-MACHI, TOKUSHIMA-SHI 770
JAPANESE

TOKUSHIMA-KEN SUISAN SHIKENJO CHOSA KENKYU GYOSEKI 5005
(RESEARCH REPORT OF TOKUSHIMA PREFECTURAL FISHERIES EXPERIMENT STATION)
TOKUSHIMA-KEN SUISAN SHIKENJO
HIWASAURA, HIWASA-MACHI, KAIFU-GUN, TOKUSHIMA 779-23
JAPANESE

TOKUSHIMA-KEN SUISAN SHIKENJO JIGYO HOKOKU 5006
(ANNUAL REPORT OF TOKUSHIMA PREFECTURAL FISHERIES EXPERIMENT STATION)
TOKUSHIMA-KEN SUISAN SHIKENJO
HIWASAURA, HIWASA-MACHI, KAIFU-GUN, TOKUSHIMA 779-23
JAPANESE

TOKYO SUISAN DAIGAKU GAKUHO 5007
(REPORT OF THE TOKYO UNIVERSITY OF FISHERIES)
TOKYO SUISAN DAIGAKU, SHOMUKA (TOKYO UNIVERSITY OF FISHERIES, GENERAL AFFAIRS SECTION)
JAPANESE

TOKYO SUISAN DAIGAKU KENKYU HOKOKU 5008
(JOURNAL OF THE TOKYO UNIVERSITY OF FISHERIES)
JAPANESE

TOKYO-TO NO SUISAN 5009
(FISHERY OF TOKYO)
TOKYO-TO KEIZAI-KYOKU (BUREAU OF ECONOMIC AFFAIRS, TOKYO METROPOLITAN GOVERNMENT)
5-1, 3-CHOME, MARUNOUCHI, CHIYODA-KU, TOKYO 100
JAPANESE

TOKYO-TO SUISAN SHIKENJO CHOSA KENKYU YOHO 5010
(JOURNAL OF THE INVESTIGATION OF THE TOKYO METROPOLITAN FISHERIES EXPERIMENT STATION)
TOKYO-TO SUISAN SHIKENJO (TOKYO METROPOLITAN FISHERIES EXPERIMENT STATION)
3374, MIZUMOTO KOAI-CHO, KATSUSHIKA-KU, TOKYO 125
JAPANESE

TOKYO-TO SUISAN SHIKENJO JIGYO HOKOKU 5011
(ANNUAL REPORT OF THE TOKYO METROPOLITAN FISHERIES EXPERIMENT STATION)
TOKYO-TO SUISAN SHIKENJO (TOKYO METROPOLITAN FISHERIES EXPERIMENT STATION)
3374, MIZUMOTO KOAI-CHO, KATSUSHIKA-KU, TOKYO 125
JAPANESE

TOTTORI-KEN GYOGYO NO DOKO 5012
(FISHERIES STATISTICS IN TOTTORI PREFECTURE)
NORIN-SHO CHUGOKU SHIKOKU NOSEI-KYOKU TOTTORI TOKEI JOHO JIMUSHO (TOTTORI STATISTICS AND INFORMATION OFFICE, CHUGOKU SHIKOKU REGIONAL AGRICULTURAL ADMINISTRATION BUREAU, MINISTRY OF AGRICULTURE AND FORESTRY)
1-353, YUDOKORO-CHO, TOTTORI-SHI 680
JAPANESE

TOTTORI-KEN SUISAN SHIKENJO HOKOKU 5013
(REPORT OF TOTTORI PREFECTURE FISHERIES EXPERIMENT STATION)
TOTTORI-KEN SUISAN SHIKENJO (TOTTORI PREFECTURE FISHERIES EXPERIMENT STATION)
OTANI, IWAMI-MACHI, IWAMI-GUN, TOTTORI 681
JAPANESE

TOTTORI-KEN SUISAN SHIKENJO NENPO 5014
(ANNUAL REPORT OF THE TOTTORI PREFECTURAL FISHERY EXPERIMENTAL STATION)
TOTTORI-KEN SUISAN SHIKENJO (TOTTORI PREFECTURAL FISHERY EXPERIMENTAL STATION)
OTANI, IWAMI-MACHI, IWAMI-GUN, TOTTORI 681
JAPANESE

TOTTORI-KEN SUISAN SHIKENJO SHIRYO A 5015
(INFORMATION OF TOTTORI PREFECTURAL FISHERY EXPERIMENT STATION, SERIES A)
TOTTORI-KEN SUISAN SHIKENJO (TOTTORI PREFECTURAL FISHERY EXPERIMENT STATION)
OTANI, IWAMI-CHO, IWAMI-GUN, TOTTORI 681
JAPANESE

TOTTORI-KEN SUISAN SHIKENJO SHIRYO B 5016
(INFORMATION OF TOTTORI PREFECTURAL FISHERY EXPERIMENT STATION, SERIES B)
TOTTORI-KEN SUISAN SHIKENJO (TOTTORI PREFECTURAL FISHERY EXPERIMENT STATION)
OTANI-IWAMI-CHO, IWAMI-GUN, TOTTORI 681
JAPANESE

TOYAMA-KEN SUISAN SHIKENJO JIGYO HOKOKUSHO 5017
(REPORT OF TOYAMA PREFECTURAL FISHERIES EXPERIMENT STATION)
TOYAMA-KEN SUISAN SHIKENJO
1, TAKATSUKI, NAMERIKAWA-SHI 936
JAPANESE

TOYAMA-KEN SUISANGYO KIHON CHOSA KEKKA HOKOKUSHO 5018
(REPORT OF FUNDAMENTAL SURVEY OF FISHERIES IN TOYAMA PREFECTURE)
TOYAMA-KEN NOGYO SUISAN-BU (AGRICULTURE AND FISHERY DEPARTMENT, TOYAMA PREFECTURAL GOVERNMENT)
1-7, SHINSOGAWA, TOYAMA-SHI 930
JAPANESE

TRANSACTIONS OF THE TOKYO UNIVERSITY OF FISHERIES 5019
TOKYO UNIVERSITY OF FISHERIES LIBRARY
5-7 KONAN 4, MINATO-KU, TOKYO
ENGLISH
INDEXED: CHE

UO TO MIZU 5020
(FISH AND WATER)
HOKKAIDORITSU SUISAN FUKAJO (HOKKAIDO FISH HATCHERY)
2-CHOME, NAKANOSHIMA-2-JO, TOYOHIRA-KU, SAPPORO-SHI 062
JAPANESE

UO TO TAMAGO 5021
(FISH AND EGGS)
SUISAN-CHO HOKKAIDO SAKE MASU FUKAJO (HOKKAIDO SALMON HATCHERY, FISHERIES AGENCY)
2-CHOME, NAKANOSHIMA-2-JO, TOYOHIRA-KU, SAPPORO-SHI 062
JAPANESE

USHIO 5022
(TIDE, REPORT OF KAGOSHIMA PREFECTURAL FISHERIES EXPERIMENT STATION)
KAGOSHIMA-KEN SUISAN SHIKENJO
21-1, KINKO-CHO, KAGOSHIMA-SHI 892
JAPANESE

WAKAYAMA-KEN SUISAN SHIKENJO CHOSA KENKYU HOKOKU 5023
(REPORT OF WAKAYAMA PREFECTURAL FISHERIES EXPERIMENT STATION)
WAKAYAMA-KEN SUISAN SHIKENJO (WAKAYAMA PREFECTURAL FISHERIES EXPERIMENT STATION)
KUSHIMOTO-MACHI, NISHIMURO-GUN, WAKAYAMA 649-35
JAPANESE

WAKAYAMA-KEN SUISAN SHIKENJO JIGYO HOKOKU 5024
(OPERATION REPORT OF WAKAYAMA PREFECTURAL FISHERIES EXPERIMENT STATION)
WAKAYAMA-KEN SUISAN SHIKENJO
KUSHIMOTO-MACHI, NISHIMURO-GUN, WAKAYAMA 649-35
JAPANESE

WATAKUSHITACHI NO NORI KENKYU 5025
(OUR STUDIES OF PORPHYRA INDUSTRY)
SENKAI ZOSHOKU KENKYU CHUO KYOGIKAI, ZENKOKU NORI KAIRUI GYOGYO KYODO KUMIAI RENGOKAI (CENTRAL COMMITTEE, FOR STUDY OF PROMOTING PORPHYRA, CENTRAL PORPHYRA CULTIVATION ASSOCIATION)
16-5, 2-CHOME, TAKANAWA, MINATO-KU, TOKYO 108
JAPANESE

YAMAGATA SUISHI SHIRYO 5026
(BULLETIN OF THE YAMAGATA PREFECTURAL FISHERIES EXPERIMENTAL STATION)
YAMAGATA-KEN SUISAN SHIKENJO (YAMAGATA PREFECTURAL FISHERIES EXPERIMENTAL STATION)
KAMO, TSURUOKA-SHI 997-12
JAPANESE

YAMAGATA-KEN SUISAN SHIKENJO JIGYO HOKOKU 5027
(REPORT OF YAMAGATA PREFECTURAL FISHERIES EXPERIMENTAL STATION)
YAMAGATA-KEN SUISAN SHIKENJO (YAMAGATA PREFECTURAL FISHERIES EXPERIMENTAL STATION)
KAMO, TSURUOKA-SHI 997-12
JAPANESE

YAMAGATA-KEN TANSUIGYO SHIDOSHO JIGYO HOKOKU 5028
(REPORT OF THE YAMAGATA PREFECTURAL FRESHWATER FISHERIES EXPERIMENTAL STATION)
YAMAGATA-KEN TANSUIGYO SHIDOSHO (YAMAGATA PREFECTURAL FRESHWATER FISHERIES EXPERIMENTAL STATION)
BANGAICHI, IZUMI-CHO, YONEZAWA-SHI 992
JAPANESE

YAMAGUCHI-KEN GAIKAI SUISAN SHIKENJO CHOSA KENKYU HOKOKU 5029
(SCIENCE REPORT OF YAMAGUCHI PREFECTURE OPEN SEA FISHERIES EXPERIMENT STATION)
YAMAGUCHI-KEN GAIKAI SUISAN SHIKENJO
SENZAKI-OTOMARI, NAGATO-SHI 759-41
JAPANESE

YAMAGUCHI-KEN GAIKAI SUISAN SHIKENJO JIGYO HOKOKU 5030
(OPERATION REPORT OF YAMAGUCHI PREFECTURE OPEN SEA FISHERIES EXPERIMENT STATION)
YAMAGUCHI-KEN GAIKAI SUISAN SHIKENJO
SENZAKI-OTOMARI, NAGATO-SHI 759-41
JAPANESE

YAMAGUCHI-KEN NAIKAI SUISAN SHIKENJO HOKOKU 5031
(REPORT OF YAMAGUCHI PREFECTURAL INLAND SEA FISHERIES EXPERIMENT STATION)
YAMAGUCHI-KEN NAIKAI SUISAN SHIKENJO (YAMAGUCHI PREFECTURAL NAIKAI FISHERIES EXPERIMENT STATION)
NAGAHAMA, AIHOFUTAJIMA, YAMAGUCHI-SHI 754
JAPANESE
INDEXED: CHE

YAMAGUCHI-KEN NAIKAI SUISAN SHIKENJO JIGYO HOKOKU 5032
(ANNUAL REPORT OF YAMAGUCHI PREFECTURAL INLAND FISHERIES EXPERIMENTAL STATION)
YAMAGUCHI-KEN NAIKAI SUISAN SHIKENJO
NAGAHAMA, AIHOFUTAJIMA, YAMAGUCHI-SHI 754
JAPANESE

YAMAGUCHI-KEN NAIKAI SUISAN SHIKENJO KENKYU GYOSEKI 5033
(BULLETIN OF YAMAGUCHI PREFECTURAL INLAND SEA FISHERIES EXPERIMENTAL STATION)
YAMAGUCHI-KEN NAIKAI SUISAN SHIKENJO
NAGAHAMA, AIHOFUTAJIMA, YAMAGUCHI-SHI 754
JAPANESE

YAMANASHI-KEN GYOBYO SENTA JIGYO HOKOKUSHO 5034
(REPORT OF YAMANASHI PREFECTURAL FISHERY CULTURE CENTER)
YAMANASHI-KEN GYOBYO SENTA
497, USHIKU, SHIKISHIMA-CHO, NAKAKOMA-GUN, YAMANASHI 400-01
JAPANESE

YOSHOKU 5035
(FISH CULTURE)
JAPANESE

ZENKOKU GYOJO KANKYO HOZEN KISO CHOSA HOKOKUSHO 5036
(REPORT OF FUNDAMENTAL SURVEY OF ANTI-POLLUTION MEASUREMENTS IN FISHING AREA)
SHIZUOKA-KEN SUISAN SHIKENJO HAMANAKO BUNJO (HAMANAKO BRANCH, SHIZUOKA PREFECTURE FISHERIES EXPERIMENT STATION)
3550, BENTENJIMA, MAISAKA-CHO, HAMANA-GUN, SHIZUOKA 431-02
JAPANESE

APPLIED SCIENCES

Applied Sciences in General

AICHI IKA DAIGAKU KISO KAGAKUKA KIYO 5037
(BULLETIN OF THE LIBERAL ARTS AND SCIENCE COURSE, AICHI MEDICAL UNIVERSITY)
AICHI MEDICAL UNIVERSITY
JAPANESE

AICHI KYOIKU DAIGAKU KENKYU HOKOKU, GEIJUTSU, HOKEN-TAIIKU, KASEI GIJUTSU, NOGYO KAGAKU 5038
(SCIENCE REPORTS OF AICHI UNIVERSITY OF EDUCATION, ARTS, HEALTH AND PHYSICAL EDUCATION, HOME ECONOMICS, TECHNOLOGY, AGRICULTURAL SCIENCE)
AICHI KYOIKU DAIGAKU
1, HIROSAWA, IGAYA-CHO KARIYA-SHI 448
JAPANESE AND EUROPEAN

AICHI TOSHOKAN SANGYO SHIRYO JOHO 5039
(REPORT ON INDUSTRIAL DOCUMENTS OF AICHI LIBRARY (ALTERNATE NAME FOR NO.43-56: AICHI LIBRARY INDUSTRIAL DOCUMENTS))
AICHI-KEN BUNKAKAKAN AICHI TOSHOKAN SANGYO SHIRYO SOKUHO (CURRENT REPORT OF THE AICHI LIBRARY INDUSTRIAL DOCUMENTS, CULTURAL CENTER OF AICHI PREFECTURE)
JAPANESE

AKASHI TANKI DAIGAKU KENKYU KIYO 5040
(RESEARCH BULLETIN OF THE AKASHI JUNIOR COLLEGE)
AKASHI JUNIOR COLLEGE
JAPANESE

ANJO GAKUEN DAIGAKU KENKYU RONSHU 5041
(RESEARCH PAPERS OF ANJO GAKUEN UNIVERSITY)
ANJO GAKUEN DAIGAKU, ANJO TANKI DAIGAKUBU, ANJO GAKUEN JOSHI TANKI DAIGAKU (ANJO GAKUEN UNIVERSITY, JUNIOR COLLEGE & WOMEN'S JUNIOR COLLEGE)
JAPANESE

AOBA GAKUEN TANKI DAIGAKU KIYO 5042
(BULLETIN OF AOBA GAKUEN JUNIOR COLLEGE)
AOBA GAKUEN TANKI DAIGAKU (AOBA GAKEN JUNIOR COLLEGE)
JAPANESE

AOMORI TANKI DAIGAKU KIYO 5043
(BULLETIN OF THE AOMORI JUNIOR COLLEGE)
JAPANESE

ARUTESU RIBERARESU, IWATE DAIGAKU KYOYO BU HOKOKU 5044
(REPORT OF THE IWATE UNIVERSITY, DEPARTMENT OF GENERAL EDUCATION)
JAPANESE
INDEXED: JICST

ATOMI GAKUEN TANKI DAIGAKU KIYO 5045
(BULLETIN OF THE ATOMI GAKUEN JUNIOR COLLEGE)
ATOMI GAKUEN JUNIOR COLLEGE
JAPANESE

BAIKA TANKI DAIGAKU KENKYU KIYO 5046
(RESEARCH BULLETIN OF BAIKA JUNIOR COLLEGE)
JAPANESE
INDEXED: JICST

BAIKO JOGAKUIN DAIGAKU RONSHU 5047
(PAPERS OF BAIKO GAKUIN UNIVERSITY)
JAPANESE

CHIBA TANKI DAIGAKU KIYO 5048
(BULLETIN OF CHIBA JUNIOR COLLEGE)
CHIBA TANKI DAIGAKU KENKYU-SHITSU (CHIBA JUNIOR COLLEGE RESEARCH LABORATORIES)
JAPANESE

CHUKYO TANKI DAIGAKU RONSO 5049
(REVIEW OF CHUKYO JUNIOR COLLEGE)
CHUKYO JUNIOR COLLEGE
JAPANESE

DOTO TANKI DAIGAKU KIYO 5050
(BULLETIN OF DOTO JUNIOR COLLEGE)
JAPANESE

FUJI JOSHI DAIGAKU. FUJI JOSHI TANKI DAIGAKU KIYO: DAI 1 BU 5051
(BULLETIN OF FUJI WOMEN'S UNIVERSITY AND FUJI WOMEN'S JUNIOR COLLEGE: PART 1)
JAPANESE

FUJI JOSHU DAIGAKU. FUJI JOSHI TANKI DAIGAKU KIYO: DAI 2 BU 5052
(BULLETIN OF FUJI WOMEN'S COLLEGE: SERIES 2)
JAPANESE

FUKUI DAIGAKU KYOIKUGAKUBU KIYO, DAI-5-BU, OYO KAGAKU, GIJUTSU-HEN 5053
(MEMOIR OF FACULTY OF EDUCATION, FUKUI UNIVERSITY, SERIES 5, APPLIED SCIENCE AND TECHNOLOGY)
FUKUI DAIGAKU KYOIKUGAKUBU (FACULTY OF EDUCATION, FUKUI UNIVERSITY)
9-1, 3-CHOME, BUNKYO, FUKUI-SHI 910
JAPANESE, WITH ENGLISH SUMMARY

FUKUI DAIGAKU KYOIKUGAKUBU KIYO, DAI-5-BU, OYO KAGAKU, KASEIGAKU-HEN 5054
(MEMOIR OF THE FACULTY OF EDUCATION, FUKUI UNIVERSITY SERIES 5, APPLIED SCIENCE AND HOME ECONOMICS)
FUKUI DAIGAKU KYOIKUGAKUBU (FACULTY OF EDUCATION, FUKUI UNIVERSITY)
9-1, 3-CHOME, BUNKYO, FUKUI-SHI 910
JAPANESE, WITH ENGLISH SUMMARY

FUKUI DAIGAKU KYOIKUGAKUBU KIYO, DAI-5-BU, OYO KAGAKU, NOGAKU-HEN 5055
(MEMOIR OF THE FACULTY OF EDUCATION, FUKUI UNIVERSITY, SERIES 5, APPLIED SCIENCE, AGRICULTURAL SCIENCE)
FUKUI DAIGAKU KYOIKUGAKUBU (FACULTY OF EDUCATION, FUKUI UNIVERSITY)
9-1, 3-CHOME, BUNKYO, FUKUI-SHI 910
JAPANESE, WITH ENGLISH SUMMARY

FUKUOKA KYOIKU DAIGAKU KIYO, DAI-5-BU, GEIJUTSU, HOKEN TAIIKU, KASEI, GIKUTSUKA-HEN 5056
(BULLETIN OF FUKUOKA UNIVERSITY OF EDUCATION, PART 5, ART, HEALTH AND PHYSICAL EDUCATION, HOUSE KEEPING AND TECHNICS)
FUKUOKA KYOIKU DAIGAKU (FUKUOKA UNIVERSITY OF EDUCATION)
729, AKAMA, MUNAKATA-MACHI, MUNAKATA-GUN, FUKUOKA 811-41
JAPANESE, WITH ENGLISH SUMMARY

FUKUOKA-KEN JOSHI SENMON GAKKO KIYO 5057
(BULLETIN OF FUKUOKA PREFECTURE WOMEN'S VOCATIONAL SCHOOL)
JAPANESE

GEKKAN BUNKEN JANARU 5058
(MONTHLY BIBLIOGRAPHICAL JOURNAL)
JAPANESE
INDEXED: JICST

GUNMA DAIGAKU KYOIKUGAKUBU KIYO, GEIJUTSU, GIJUTSU-HEN 5059
(ANNUAL REPORT OF THE FACULTY OF EDUCATION, GUNMA UNIVERSITY, ART AND TECHNOLOGY SERIES)
GUNMA DAIGAKU KYOIKUGAKUBU (FACULTY OF EDUCATION, GUNMA UNIVERSITY)
1375, ARAMAKI-MACHI, MAEBASHI-SHI 371
JAPANESE AND ENGLISH
INDEXED: CHE, JICST

HAGOROMO GAKUEN TANKI DAIGAKU KENKYU KIYO 5060
(RESEARCH BULLETIN OF HAGOROMO GAKUEN JUNIOR COLLEGE)
JAPANESE

HIGASHI-OSAKA TANKI DAIGAKU KENKYU KIYO 5061
(STUDY REPORTS OF HIGASHIOSAKA JUNIOR COLLEGE)
JAPANESE

HIROSAKI GAKUIN DAIGAKU. HIROSAKI GAKUIN TANKI DAIGAKU KIYO 5062
(BULLETIN OF HIROSAKI GAKUIN COLLEGE)
JAPANESE

HIROSHIMA BUNKYO JOSHI DAIGAKU KENKYU KIYO 5063
(RESEARCH REPORT OF HIROSHIMA BUNKYO WOMEN'S COLLEGE)
HIROSHIMA BUNKYO JOSHI DAIGAKU BUNGAKUBU TANKI DAIGAKU BU (FACULTY OF LITERATURE, JUNIOR COLLEGE)
JAPANESE

HOKKAIDO KYOIKU DAIGAKU KIYO, DAI-2-BU, C, KATEI, TAIIKU-HEN 5064
(JOURNAL OF HOKKAIDO UNIVERSITY OF EDUCATION, SECTION 2, C, DOMESTIC SCIENCE, PHYSICAL EDUCATION)
HOKKAIDO KYOIKU DAIGAKU (HOKKAIDO UNIVERSITY OF EDUCATION)
NISHI-13-CHOME, MINAMI-24-JO, CHUO-KU, SAPPORO-SHI 064
JAPANESE, WITH EUROPEAN SUMMARY
INDEXED: JICST

HOKKAIDO SANGYO TANKI DAIGAKU KIYO 5065
(BULLETIN OF HOKKAIDO INDUSTRIAL JUNIOR COLLEGE)
JAPANESE

HOKKAIDO-RITSU CHIKA SHIGEN CHOSAJO HOKOKU 5066
(REPORT OF THE HOKKAIDO PREFECTURAL UNDERGROUND RESOURCES SURVEY CENTER)
JAPANESE
INDEXED: JICST

HOSEN GAKUEN TANKI DAIGAKU KENKYU KIYO 5067
(RESEARCH BULLETIN OF HOSEN GAKUEN JUNIOR COLLEGE)
JAPANESE

IBARAKI KIRISUTO KYO TANKI DAIGAKU KENKYU KIYO 5068
(RESEARCH BULLETIN OF THE IBARAKI CHRISTIAN JUNIOR COLLEGE)
JAPANESE

IKEBO TANKI DAIGAKU KIYO 5069
(BULLETIN OF IKEBO JUNIOR COLLEGE)
JAPANESE

IZUMI TANKI DAIGAKU KENKYU KIYO 5070
(RESEARCH BULLETIN OF THE IZUMI JUNIOR COLLEGE)
JAPANESE

KACHO TANKI DAIGAKU KENKYU KIYO 5071
(KACHO JUNIOR COLLEGE RESEARCH BULLETIN)
JAPANESE

KAGAKU BUNKEN. DENSHIRYOKU GENSHIRYOKU, KAGAKU KINZOKU NO GOHENSHI 5072
(SCIENCE BIBLIOGRAPHY: COMBINED ISSUE; ELECTRONIC ENGINEERING, ATOMIC ENERGY, CHEMISTRY, METALS)
SANGYO KAGAKU SENTA (INDUSTRIAL CHEMISTRY CENTER)
JAPANESE

KAGAKU SARON 5073
(SCIENCE SALON)
TOKAI DAIGAKU SHUPPANKAI (TOKAI UNIVERSITY PRESS)
JAPANESE

KAGOSHIMA TANKI DAIGAKU KENKYU KIYO 5074
(KAGOSHIMA JUNIOR COLLEGE RESEARCH BULLETIN)
KAGOSHIMA TANKI DAIGAKU (KAGOSHIMA JUNIOR COLLEGE)
JAPANESE

KAGOSHIMA-KENRITSU TANKI DAIGAKU KENKYU NENPO 5075
(KAGOSHIMA PREFECTURAL JUNIOR COLLEGE RESEARCH ANNUAL)
KAGOSHIMA KENRITSU TANKI DAIGAKU CHIIKI KENKYUJO (KAGOSHIMA PREFECTURAL JUNIOR COLLEGE REGIONAL RESEARCH INSTITUTE)
JAPANESE

KAIGAI SHINSEIHIN BURETEN 5076
(OVERSEAS NEW PRODUCTS BULLETIN)
DJK INTERNATIONAL
JAPANESE

KAIHATSU ORAI 5077
(DEVELOPMENT ONGOINGS)
JAPANESE

KAIHATSU RONSHU 5078
(DEVELOPMENT PAPERS)
HOKKAIDO GAKUEN DAIGAKU KAIHATSU KENKYUJO (HOKKAIDO GAKUEN UNIVERSITY DEVELOPMENT RESEARCH INSTITUTE)
JAPANESE

KANSAI GAKUIN DAIGAKU SANGYO KENKYUJO SHIRYO GEPPO 5079
(KANSAI GAKUIN UNIVERSITY INDUSTRIAL RESEARCH INSTITUTE MATERIALS MONTHLY)
JAPANESE

KANTO TANKI DAIGAKU KIYO 5080
(KANTO JUNIOR COLLEGE BULLETIN)
JAPANESE

KIHO NIRA 5081
(NIRA QUARTERLY REPORT)
SOGO KENKYU KAIHATSU KIKO (GENERAL RESEARCH DEVELOPMENT ORGANIZATION)
JAPANESE

KINKI DAIGAKU TANKI DAIGAKU RONSHU 5082
(PAPERS OF THE JUNIOR COLLEGE, KINKI UNIVERSITY)
KINKI DAIGAKU TANKIDAIGAKU-BU (JUNIOR COLLEGE DIVISION, KINKI UNIVERSITY)
JAPANESE

KINRAN TANKI DAIGAKU KENKYUSHI 5083
(RESEARCH JOURNAL OF KINRAN JUNIOR COLLEGE)
KINRAN TANKI DAIGAKU (KINRAN JUNIOR COLLEGE)
JAPANESE

KIRIGAOKA TANKI DAIGAKU KIYO 5084
(BULLETIN OF KIRIGAOKA JUNIOR COLLEGE)
KIRIGAOKA TANKI DAIGAKU (KIRIGAOKA JUNIOR COLLEGE)
JAPANESE

KITAKYUSHU SANGYO SHAKAI KENKYUJO KIYO 5085
(BULLETIN OF KITAKYUSHU INDUSTRIAL SOCIETY RESEARCH INSTITUTE)
JAPANESE
INDEXED: JICST

KITAKYUSHU SANGYO SHAKAI KENKYUJO SHOHO 5086
(REPORT OF THE KITAKYUSHU INDUSTRIAL SOCIETY RESEARCH INSTITUTE)
KITAKYUSHU DAIGAKU KITAKYUSHU SANGYO SHAKAI KENKYUJO (KITAKYUSHU INDUSTRIAL SOCIETY RESEARCH INSTITUTE KITAKYUSHU UNIVERSITY)
JAPANESE

KITAKYUSHU TANKI DAIGAKU KENKYU RONSHU 5087
(COLLECTED RESEARCH PAPERS OF KITAKYUSHU JUNIOR COLLEGE)
JAPANESE

KOGYO KYOIKU 5088
(INDUSTRIAL EDUCATION)
TOA TSUSHINSHA (TOA COMMUNICATIONS COMPANY)
JAPANESE

KOKUSAI HYOJUNKA SHIRYO 5089
(INTERNATIONAL STANDARDIZATIONS DATA)
JAPANESE
INDEXED: JICST

KOKYO DAYORI 5090
(INDUSTRIAL EDUCATION LETTER)
NIHON KOGYO KYOIKU KYOKAI (JAPAN INDUSTRIAL EDUCATION ASSOCIATION)
JAPANESE

KOMAZAWA TANKI DAIGAKU KENKYU KIYO 5091
(BULLETIN OF THE KOMAZAWA JUNIOR COLLEGE)
JAPANESE

KORIYAMA JOSHI DAIGAKU KIYO 5092
(BULLETIN OF THE KORIYAMA WOMEN'S COLLEGE)
JAPANESE

KOSHIEN TANKI DAIGAKU KIYO 5093
(BULLETIN OF THE KOSHIEN JUNIOR COLLEGE)
KOSHIEN JUNIOR COLLEGE
JAPANESE

KUMAMOTO TANDAI RONSHU 5094
(PAPERS OF THE KUMAMOTO JUNIOR COLLEGE)
KUMAMOTO TANKI DAIGAKU (KUMAMOTO JUNIOR COLLEGE)
JAPANESE

KURASHIKI-SHIRITSU TANKI DAIGAKU KENKYU KIYO 5095
(BULLETIN OF KURASHIKI JUNIOR COLLEGE)
KURASHIKI JUNIOR COLLEGE
JAPANESE

KUSHIRO TANKI DAIGAKU KIYO 5096
(BULLETIN OF KUSHIRO JUNIOR COLLEGE)
KUSHIRO JUNIOR COLLEGE
JAPANESE

KYOTO NORITSU 5097
(KYOTO EFFICIENCY)
JAPANESE
INDEXED: JICST

KYUSHU JOSHI GAKUIN TANKI DAIGAKU GAKUJUTSU KIYO 5098
(ARTS AND SCIENCES BULLETIN OF THE JUNIOR COLLEGE, KYUSHU WOMEN'S ACADEMY)
KYUSHU JOSHI GAKUIN TANKI DAIGAKU (KYUSHU WOMEN'S COLLEGE, JUNIOR COLLEGE)
JAPANESE

MATSUSHO TANDAI RONSHU 5099
(PAPERS OF THE MATSUSHO JUNIOR COLLEGE)
MATSUSHO GAKUEN TANKI DAIGAKU (MATSUSHO GAKUEN JUNIOR COLLEGE)
JAPANESE

MATSUYAMA HIGASHIGUMO TANKI DAIGAKU KENKYU RONSHU 5100
(RESEARCH PAPERS OF THE MATSUYAMA HIGASHIGUMO JUNIOR COLLEGE)
MATSUYAMA HIGASHIGUMO JUNIOR COLLEGE
JAPANESE

MEIJI DAIGAKU DAIGAKUIN KIYO, 5 KO, NOGAKU-HEN 5101
(MEMOIRS OF THE GRADUATE SCHOOL, MEIJI UNIVERSITY 5: ENGINEERING AND AGRICULTURE)
JAPANESE

MEIJI DAIGAKU TANKI DAIGAKU KIYO 5102
(BULLETIN OF THE MEIJI UNIVERSITY JUNIOR COLLEGE)
MEIJI UNIVERSITY JUNIOR COLLEGE
JAPANESE

NAGAOKA TANKI DAIGAKU KENKYU KIYO 5103
(BULLETIN OF NAGAOKA JUNIOR COLLEGE)
JAPANESE

NAGOYA TANKI DAIGAKU KENKYU KIYO 5104
(RESEARCH BULLETIN OF NAGOYA JUNIOR COLLEGE)
JAPANESE

NANZAN TANKI DAIGAKU KIYO 5105
(JOURNAL OF NANZAN JUNIOR COLLEGE)
JAPANESE

NOMURA SOGO KENKYUJO N R I KENKYU SHIRIZU 5106
(NOMURA GENERAL RESEARCH INSTITUTE NRI RESEARCH SERIES)
JAPANESE
INDEXED: JICST

NOMURA SOGO KENKYUJO SOGO KENKYU 5107
(NOMURA GENERAL RESEARCH INSTITUTE GENERAL RESEARCH)
JAPANESE
INDEXED: JICST

NORITSU OKAYAMA 5108
(EFFICIENCY OKAYAMA)
JAPANESE
INDEXED: JICST

NRC SEMINA (NIPPON RESEARCH CENTRE) 5109
(NRC SEMINAR)
JAPANESE
INDEXED: JICST

O & M JOHO 5110
(O & M INFORMATION)
JAPANESE
INDEXED: JICST

OKINAWA KIRISUTOKYO TANKI DAIGAKU KIYO 5111
(BULLETIN OF OKINAWA CHRISTIAN JUNIOR COLLEGE)
OKINAWA CHRISTIAN JUNIOR COLLEGE
JAPANESE

ONOMICHI TANKI DAIGAKU KENKYU KIYO 5112
(RESEARCH BULLETIN OF ONOMICHI JUNIOR COLLEGE)
JAPANESE

OSAKA CHIYODA TANKI DAIGAKU KIYO 5113
(BULLETIN OF OSAKA CHIYODA JUNIOR COLLEGE)
JAPANESE

OSAKA FURITSU SANGYO NORITSU KENKYUJO NENPO 5114
(ANNUAL OF OSAKA PREFECTORAL INDUSTRIAL EFFICIENCY RESEARCH INSTITUTE)
JAPANESE
INDEXED: JICST

OTANI JOSHI DAIGAKU KIYO 5115
(BULLETIN OF OTANI WOMEN'S UNIVERSITY)
JAPANESE

OTEMAE JOSHI DAIGAKU RONSHU 5116
(PAPERS OF OTEMAE WOMEN'S UNIVERSITY)
JAPANESE

OTSUKI TAUDAI RONSHU 5117
(PAPERS OF THE OTSUKI JUNIOR UNIVERSITY)
JAPANESE

PROCEEDINGS OF THE CONFERENCE ON SOLID STATE DEVICES 5118
JAPANESE
INDEXED: CHE, JICST

PURU GAKUIN TANKI DAIGAKU KENKYU KIYO 5119
(RESEARCH BULLETIN OF PURU GAKUIN JUNIOR COLLEGE)
JAPANESE

SAKATA TANKI DAIGAKU KENKYU KIYO 5120
(BULLETIN OF THE SAKATA JUNIOR COLLEGE)
SAKATA JUNIOR COLLEGE
JAPANESE

SAN'IKU GAKUIN TANKI DAIGAKU KIYO 5121
(BULLETIN OF THE SAN'IKU GAKUIN JUNIOR COLLEGE)
SAN'IKU GAKUIN JUNIOR COLLEGE
JAPANESE

SAPPORO OTANI TANKI DAIGAKU KIYO 5122
(BULLETIN OF THE SAPPORO OTANI JUNIOR COLLEGE)
SAPPORO OTANI JUNIOR COLLEGE
JAPANESE

SEIBI GAKUEN TANKI DAIGAKU KENKYU RONSO 5123
(COLLECTED RESEARCH PAPERS OF SEIBI GAKUEN JUNIOR COLLEGE)
JAPANESE

SEIJO DAIGAKU TANKI DAIGAKUBU KIYO 5124
(RESEARCH REPORTS OF THE JUNIOR COLLEGE, SEIJO UNIVERSITY)
JUNIOR COLLEGE, SEIJO UNIVERSITY
JAPANESE

SEISEN JOSHI DAIGAKU KIYO 5125
(BULLETIN OF SEISEN WOMEN'S UNIVERSITY)
JAPANESE

SHIKOKU GAKUIN DAIGAKU RONSHU 5126
(PAPERS OF THE SHIKOKU GAKUIN UNIVERSITY)
JAPANESE

SHIMANE JOSHI TANKI DAIGAKU KIYO 5127
(BULLETIN OF THE SHIMANE WOMEN'S JUNIOR COLLEGE)
JAPANESE

SHINWA JOSHI DAIGAKU KENKYU RONSO 5128
(RESEARCH REVIEW OF THE SHINWA WOMEN'S COLLEGE)
JAPANESE

SHIRAUME GAKUEN TANKI DAIGAKU KIYO 5129
(BULLETIN OF THE SHIRAUME GAKUEN JUNIOR COLLEGE)
JAPANESE

SHITEN OJI JOSHI DAIGAKU KIYO 5130
(BULLETIN OF THE SHITEN OJI WOMEN'S COLLEGE)
JAPANESE

SHIZUOKA JOSHI DAIGAKU KENKYU KIYO 5131
(BULLETIN OF SHIZUOKA WOMEN'S COLLEGE)
JAPANESE

SHOHI KENKYU 5132
(CONSUMER'S RESEARCH)
JAPANESE
INDEXED: JICST

SHOHOKU KIYO 5133
(BULLETIN OF THE SHOHOKU JUNIOR COLLEGE)
SHOHOKU TANKI DAIGAKU IPPAN KYOIKU KIYO HENSHU IINKAI (SHOHOKU JUNIOR COLLEGE, GENERAL EDUCATION BULLETIN, EDITORIAL BOARD)
JAPANESE

SHOWA GAKUIN TANKI DAIGAKU KIYO 5134
(BULLETIN OF THE SHOWA GAKUIN JUNIOR COLLEGE)
JAPANESE

SHUKO TANKI DAIGAKU KIYO 5135
(SHUKO JUNIOR COLLEGE BULLETIN)
JAPANESE

SHUKUTOKU DAIGAKU KENKYU KIYO 5136
(RESEARCH BULLETIN OF THE SHUKUTOKU UNIVERSITY)
JAPANESE

SONI GAKUEN KENKYU KIYO 5137
(RESEARCH BULLETIN OF THE SONY ACADEMY)
SONI GAKUEN (SONY ACADEMY)
JAPANESE

SUGINO JOSHI DAIGAKU KIYO 5138
(BULLETIN OF SUGINO WOMEN'S UNIVERSITY)
JAPANESE

TACHIBANA JOSHI DAIGAKU KENKYU KIYO 5139
(MEMOIRS OF TACHIBANA WOMEN'S COLLEGE)
JAPANESE

TACHIKAWA TANDAI KIYO 5140
(BULLETIN OF THE TACHIKAWA JUNIOR COLLEGE)
TOKYO-TORITSU TACHIKAWA TANKI DAIGAKU (TOKYO METROPOLITAN TACHIKAWA JUNIOR COLLEGE)
JAPANESE

TAMAKI JOSHI TANKI DAIGAKU KENKYU KIYO 5141
(RESEARCH REPORT OF THE TAMAKI WOMEN'S JUNIOR COLLEGE)
JAPANESE

TOHO GAKUHO 5142
(TOHO REPORT)
TOHO GAKUEN (TOHO ACADEMY)
JAPANESE

TOHO GAKUSHI 5143
(JOURNAL OF THE TOHO ACADEMY)
TOHO GAKUEN TANKI DAIGAKU (TOHO ACADEMY JUNIOR COLLEGE)
JAPANESE

TOKYO BUNKA TANKI DAIGAKU KIYO 5144
(TOKYO BUNKA JUNIOR COLLEGE MEMOIRS)
JAPANESE

TOKYO GAKUGEI DAIGAKU KIYO, GEIJUTSU, TAIIKU 5145
(BULLETIN OF TOKYO GAKUGEI UNIVERSITY, ARTS, PHYSICAL EDUCATION)
JAPANESE

TOKYO KIRISUTOKYO TANKI DAIGAKU RONSHU 5146
(PAPERS OF THE TOKYO CHRISTIAN JUNIOR COLLEGE)
JAPANESE

TOMAKOMAI KOMAZAWA TANKI DAIGAKU KENKYU KIYO 5147
(RESEARCH BULLETIN OF TOMAKOMAI KOMAZAWA JUNIOR COLLEGE)
JAPANESE

TOYAMA-KEN TAKAOKA SANGYO KYOIKUKAN KENKYU KIYO 5148
(RESEARCH REPORT OF THE TAKAOKA INDUSTRY EDUCATION HALL, TOYAMA PREFECTURE)
JAPANESE

TOYO DAIGAKU TANKI DAIGAKU KIYO 5149
(BULLETIN OF THE TOYO UNIVERSITY JUNIOR COLLEGE)
JAPANESE

UBE TANKI DAIGAKU GAKUJUTSU HOKOKU 5150
(ACADEMIC REPORT OF UBE JUNIOR COLLEGE)
JAPANESE

YAMANASHI EIWA TANKI DAIGAKU KIYO 5151
(BULLETIN OF YAMANASHI EIWA JUNIOR COLLEGE)
YAMANASH EIWA JUNIOR COLLEGE
JAPANESE

YAMAWAKI GAKUEN TANKI DAIGAKU KIYO 5152
(BULLETIN OF YAMAWAKI GAKUEN JUNIOR COLLEGE)
JAPANESE

Inventions, Patents

AIPPI (KOKUSAI KOGYO SHOYUKEN HOGO KYOKAI NIHON BUKAI GEPPO) 5153
(JOURNAL OF THE JAPANESE GROUP OF INTERNATIONAL ASSOCIATION FOR THE PROTECTION INDUSTRIAL PROPERTY)
KOKUSAI KOGYO SHOYUKEN HOGO KYOKAI NIHONBUKAI (JAPANESE GROUP OF INTERNATIONAL ASSOCIATION FOR THE PROTECTION OF INDUSTRIAL PROPERTY)
JAPANESE AND ENGLISH

BEIKOKU TOKKYO SHOROKU 5154
(US PATENT ABSTRACTS)
HATSUMEI KYOKAI (INVENTION ASSOCIATION)
17, SHIBA NISHIKUBO AKEFUNE-CHO, MINATO-KU, TOKYO 105)
JAPANESE

GAIKOKU TOKKYO SOKUHO, KAGAKU-HEN 5155
(FOREIGN PATENT NEWS)
NIHON KAGAKU GIJUTSU JOHO SENTA (THE JAPAN INFORMATION CENTER OF SCIENCE AND TECHNOLOGY)
1-1, 2-CHOME, NAGATA-CHO, CHIYODA-KU, TOKYO 100
ENGLISH
INDEXED: JICST

GEKKAN TOKKYO TO KIGYO 5156
(PATENT AND ENTERPRISE)
NIHON KAGAKU SHINKO ZAIDAN NIHON PATENTO AKADEMI
3-2, 3-CHOME, KASUMIGASEKI, CHIYODA-KU, TOKYO 100
JAPANESE
INDEXED: JICST

HATSUMEI 5157
(INVENTION)
HATSUMEI KYOKAI (JAPAN INSTITUTE OF INVENTION AND INNOVATION)
17, SHIBA NISHIKUBO AKEFUNE-CHO, MINATO-KU, TOKYO 105
JAPANESE
INDEXED: JICST

HATSUMEI HOKKAIDO 5158
(HOKKAIDO INVENTION ASSOCIATION NEWS)
HOKKAIDO HATSUMEI KYOKAI RENGOKAI
19, CHUO-KU ODORI NISHI, SAPPORO-SHI 060
JAPANESE

HATSUMEI KOAN NO SHOKAI 5159
(INVENTION AND CONTRIVANCE INFORMATION)
TOKKYO-CHO
4-3, 3-CHOME, KASUMIGASEKI, CHIYODA-KU, TOKYO 100
JAPANESE
INDEXED: JICST

HATSUMEI KYOKAI KOKAI GIHO 5160
(TECHNICAL REPORT FOR PUBLIC ANNOUNCEMENT OF THE INVENTIONS ASSOCIATION)
INVENTIONS ASSOCIATION
JAPANESE

HATSUMEI TO SEIKATSU 5161
(INVENTION & LIFE)
NIHON HATSUMEI SHINKO KYOKAI (THE JAPAN SOCIETY FOR THE ADVANCEMENT OF INVENTIONS)
22, 4-CHOME, SAKURAGAOKA-MACHI, SHIBUYA-KU TOKYO 150
JAPANESE

HATSUMEI TOKKYO SHINBUN 5162
(PATENTS NEWS)
HATSUMEI TOKKYO SHINBUN-SHA (PATENT NEWS PUBLISHING COMPANY)
10, 4-CHOME, SHIBA, MINATO-KU, TOKYO 108
JAPANESE

HOKKAIDO HATSUMEI KOAN NENPO 5163
(ANNUAL REPORT OF INVENTION IN HOKKAIDO)
ASAHIKAWA-SHI SANGYO GIJUTSU JOHO SENTA JIMKYOKU (SECRETARIAT OF ASAHIKAWA INFORMATION CENTER FOR INDUSTRY AND TECHNOLOGY)
TOKIWA-KOEN, ASAHIKAWA-SHI 070
JAPANESE

ISHO KOHO 5164
(DESIGN GAZETTE)
TOKKYO-CHO (PATENT OFFICE)
4-3, 3-CHOME, KASUMIGASEKI, CHIYODA-KU, TOKYO 100
JAPANESE

JAPATIC NEWS 5165
NIHON TOKKYO JOHO SENTA (JAPAN PATENT INFORMATION CENTER)
C/O DAI-15 MORI BUILDING, 16, SHIBA NISHIKUBO AKEFUNE-CHO, MINATO-KU, TOKYO 105
JAPANESE

JITSUYO SHIN'AN KOHO 5166
(UTILITY MODEL GAZETTE)
TOKKYO-CHO (PATENT OFFICE)
4-3, 3-CHOME, KASUMIGASEKI, CHIYODA-KU, TOKYO 100
JAPANESE

JITSUYO SHIN'AN SHUTSUGAN KOKOKU SAIBUNRUIBETSU SOMOKUROKU 5167
(GENERAL INDEX TO DETAILED CLASSIFICATION OF PUBLICATION OF UTILITY MODEL APPLICATION)
HATSUMEI KYOKAI (JAPAN INSTITUTE OF INVENTION AND INNOVATION)
17, SHIBA NISHIKUBO AKEFUNE-CHO, MINATO-KU, TOKYO 105
JAPANESE
INDEXED: JICST

JITSUYO SHIN'AN SHUTSUGAN SHOBUN KENSAKUHYO 5168
(REFERENCE TABLE OF DISPOSAL OF UTILITY MODEL APPLICATION)
HATSUMEI KYOKAI (JAPAN INSTITUTE OF INVENTION AND INNOVATION)
17, SHIBA NISHIKUBO AKEFUNE-CHO, MINATO-KU, TOKYO 105
JAPANESE
INDEXED: JICST

JITSUYO SHIN'AN TAISHOHYO 5169
(CROSS REFERENCE TABLES OF UTILITY MODEL)
HATSUMEI KYOKAI (JAPAN INSTITUTE OF INVENTION AND INNOVATION)
17, SHIBA NISHIKUBO AKEFUNE-CHO, MINATO-KU, TOKYO 105
JAPANESE
INDEXED: JICST

KOGYO GIJUTSU-IN TOKKYO SHOROKUSHU 5170
(PATENT ABSTRACTS, AGENCY OF INDUSTRIAL SCIENCE AND TECHNOLOGY)
TSUSHO SANGYO-SHO KOGYO GIJUTSU-IN (AGENCY OF INDUSTRIAL SCIENCE AND TECHNOLOGY MINISTRY OF INTERNATIONAL TRADE AND INDUSTRY)
DAI-20 MORI BUILDING, 7-3, 2-CHOME, NISHI-SHINBASHI, MINATO-KU, TOKYO 105
JAPANESE

KOGYO SHOYUKEN-HO KENKYU 5171
(STUDIES OF THE INDUSTRIAL PROPERTY LAWS)
HAYABUSA KOGYO SHOYUKEN KENKYUSHO (THE HAYABUSA INSTITUTE FOR THE PRODUCTION OF INDUSTRIAL PROPERTY)
SHUFUNOTOMO BUILDING, 6, 1-CHOME, KANDA SURUGADAI, CHIYODA-KU, TOKYO 101
JAPANESE

KOKAI JITSUYO SHIN'AN KOHO 5172
(PUBLIC UTILITY MODEL GAZETTE)
TOKKYO-CHO (PATENT OFFICE)
4-3, 3-CHOME, KASUMIGASEKI, CHIYODA-KU, TOKYO 100
JAPANESE

KOKAI JITSUYO SHIN'AN. BUNRUI SAKUIN 5173
(NEW IDEAS FOR PUBLIC DISSEMINATION AND UTILITY CLASSIFICATION INDEX)
JAPANESE
INDEXED: JICST

KOKAI JITSUYO SHIN'AN. SHUTSUGANNIN SAKUIN 5174
(NEW IDEAS FOR PUBLIC DISSEMINATION AND UTILITY. APPLICANT INDEX)
JAPANESE
INDEXED: JICST

KOKAI TOKKYO BUNRUI SAKUIN 5175
(PUBLIC PATENTS. CLASSIFICATION INDEX)
JAPANESE
INDEXED: JICST

KOKAI TOKKYO JITSUYO SHIN'AN SAKUIN 5176
(INDEX OF PUBLIC PATENT AND UTILITY MODEL)
NIHON TOKKYO JOHO SENTA
C/O MORI BUILDING, NO. 22, 1, SHIBA NIHSIKUBO SHIROYAMA-CHO, MINATO-KU, TOKYO 105
JAPANESE

KOKAI TOKKYO KOHO 5177
(PUBLIC PATENT GAZETTE)
TOKKYO-CHO (PATENT OFFICE)
4-3, 3-CHOME, KASUMIGASEKI, CHIYODA-KU
TOKYO 100
JAPANESE

KOKAI TOKKYO SHUTSUGAN SHOROKU 5178
(ABSTRACTS OF APPLICATION FOR PUBLIC PATENT)
NIHON TOKKYO JOHO SENTA (JAPAN PATENT INFORMATION CENTER)
3-BANCHI, SHIBAKOTOHIRA-CHO, TOKYO 105
JAPANESE

KOKAI TOKKYO SHUTSUGANNIN SAKUIN 5179
(INDEX OF APPLICANT FOR PUBLIC PATENT)
NIHON TOKKYO JOHO SENTA
C/O MORI BUILDING, NO. 22, 1, SHIBA NISHIKUBO SHIROYAMA-CHO, MINATO-KU, TOKYO 105
JAPANESE
INDEXED: JICST

KOKOKU JITSUYO SHIN'AN. SHUTSUGANNIN SAKUIN, BUNRUI SAKUIN 5180
(NEW IDEAS FOR PUBLIC ANNOUNCEMENT AND UTILITY. APPLICANT INDEX, CLASSIFICATION INDEX)
JAPANESE
INDEXED: JICST

KOKOKU TOKKYO. SHUTSUGANNIN SAKUIN, BUNRUI SAKUIN 5181
(PUBLIC PATENTS. APPLICANT INDEX, CLASSIFICATION INDEX)
JAPANESE
INDEXED: JICST

MOJI SHOHYOSHU 5182
(TRADE MARKS COMPOSED ONLY OF WRITTEN WORDS OR LETTERS)
BENRISHIKAI (THE PATENT ATTORNEYS ASSOCIATION OF JAPAN)
4-4, 3-CHOME, KASUMIGASEKI, CHIYODA-KU, TOKYO 100
JAPANESE

NAIGAIKOKU KOGYO SHOYUKEN KOHORUI KAIDAI MOKUROKU 5183
(ANNOTATED CATALOGUE OF THE OFFICIAL GAZETTES ON THE INDUSTRIAL PROPERTY RIGHT, JAPAN AND OVERSEAS)
TOKKYO-CHO BANKOKU KOGYO SHOYUKEN SHIRYOKAN (INDUSTRIAL PROPERTY LABORATORY, PATENT OFFICE)
17, SHIBA NISHIKUBO AKEFUNE-CHO, MINATO-KU, TOKYO 105
JAPANESE

NIPPON TOKKYO JISTUYO SHIN'AN HYOJUN SAKUIN. SOKUHO-BAN 5184
(STANDARD INDEX TO NIPPON TOKKYO JITSUYO SHIN'AN (NEW IDEAS FOR JAPAN PATENT UTILIZATION) CURRENT REPORT EDITIONS)
JAPANESE
INDEXED: JICST

NIPPON TOKKYO SAKUIN: SHUTSUGANNIN SAKUIN, BUNRUI SAKUIN 5185
(JAPAN PATENT INDEX: APPLICANT INDEX, CLASSIFICATION INDEX)
JAPANESE
INDEXED: JICST

NIPPON YUNIBAKKU SOKEN KIYO 5186
(GENERAL RESEARCH BULLETIN OF JAPAN UNIVAC)
JAPANESE
INDEXED: JICST

OSUPIKKU JOHO 5187
(OSPIC NEWS)
KAIGAI TOKKYO JOHO SENTA (OVERSEAS PATENT INFORMATION CENTER)
JAPANESE

PATENTO 5188
(PATENT)
BENRISHIKAI PATENTO HENSHU IINKAI BENRISHIKAI (THE PATENT ATTORNEYS ASSOCIATION OF JAPAN)
4-4, 3-CHOME, KASUMIGASEKI, CHIYODA-KU, TOKYO 100
JAPANESE
INDEXED: JICST

SHINKETSU KOHO 5189
(TRIAL DECISION JOURNAL)
TOKKYO-CHO (PATENT OFFICE)
4-3, 3-CHOME, KASUMIGASEKI, CHIYODA-KU, TOKYO 100
JAPANESE
INDEXED: JICST

SHOHYO KOHO 5190
(TRADE MARKS GAZETTE)
TOKKYO-CHO (PATENT OFFICE)
4-3, 3-CHOME, KASUMIGASEKI, CHIYODA-KU, TOKYO 100
JAPANESE

SHOHYO TAISHOHYO 5191
(CROSS REFERENCE TABLES OF TRADE MARKS)
HATSUMEI KYOKAI (JAPAN INSTITUTE OF INVENTION AND INNOVATION)
17, SHIBA NISHIKUBO AKEFUNE-CHO, MINATO-KU, TOKYO 105
JAPANESE
INDEXED: JICST

TAKEMURA KENKYUJO KIYO 5192
(BULLETIN OF THE TAKEMURA RESEARCH INSTITUTE)
TAKEMURA KENKYU JIMUSHO (TAKEMURA RESEARCH OFFICE)
JAPANESE

TOKKYO KANRI 5193
(PATENT MANAGEMENT)
NIHON TOKKYO KYOKAI (JAPAN PATENT ASSOCIATION)
3-1, 3-CHOME, MARUNOUCHI, CHIYODA-KU, TOKYO 100
JAPANESE
INDEXED: JICST

TOKKYO KANRI SHIRYO 5194
(PATENT ADMINISTRATION DOCUMENTS)
JIGYOSHA KOGYO SHOYUKEN KYOKAI (ASSOCIATION FOR ENTREPRENEUR PROPERTY RIGHTS)
JAPANESE

TOKKYO KOHO 5195
(PATENT GAZETTE)
TOKKYO-CHO (PATENT OFFICE)
4-3, 3-CHOME, KASUMIGASEKI, CHIYODA-KU, TOKYO 100
JAPANESE
INDEXED: JICST

TOKKYO KOHO 5196
(PATENT GAZETTE)
HATSUMEI KAI
JAPANESE
INDEXED: JICST

TOKKYO KOHO; DAI-1 SANGYO BUMON; NO SUISAN KANKEI 5197
(OFFICAL REPORT ON PATENTS; PART 1, INDUSTRIAL BRANCH; AGRICULTURE & FISHERIES AND RELATED FIELDS)
TOKKYO CHO (PATENT OFFICE)
JAPANESE

TOKKYO KOHO; DAI-2 SANGYO BUMON; DAI-1 KUBUN SAIKO, KINZOKU, BUTSURI KAGAKU, MUKIKAGAKU KANKEI 5198
(OFFICIAL REPORT ON PATENTS; PART 2, INDUSTRY BRANCH; DIVISION ONE; MINING, METALS, PHYSICAL CHEMISTRY, INORGANIC CHEMISTRY & RELATED SUBJECTS)
TOKKYO CHO (PATENT OFFICE)
JAPANESE

TOKKYO KOHO; DAI-2 SANGYO BUMON; DAI-2 KUBUN YUKIKAGOBUTSU KANKEI 5199
(OFFICIAL REPORT ON PATENTS; PART 2, INDUSTRY BRANCH; DIVISION TWO; ORGANIC COMPOUNDS AND RELATED SUBJECTS)
TOKKYO CHO (PATENT OFFICE)
JAPANESE

TOKKYO KOHO; DAI-2 SANGYO BUMON; DAI-3 KUBUN; SEIZOKAGAKU KOGYO, KOBUNSHI KAGOBUTSU KANKEI 5200
(OFFICIAL REPORT ON PATENTS; PART 2, INDUSTRY BRANCH DIVISION TWO; CHEMICAL INDUSTRY; HIGH MOLECULAR COMPOUNDS & RELATED SUBJECTS)
TOKKYO CHO (PATENT OFFICE)
JAPANESE

TOKKYO KOHO; DAI-2 SANGYO BUMON; DAI-4 KUBUN YAKUHIN, SHOKUHIN, BISEIBUTSU KOGYO KANKEI 5201
(OFFICIAL REPORT ON PATENTS; PART 2, INDUSTRY BRANCH ; DIVISION FOUR; PHARMACEUTICALS, FOODSTUFFS, INDUSTRIAL MICROBES & RELATED FIELDS)
TOKKYO CHO (PATENT OFFICE)
JAPANESE

TOKKYO KOHO; DAI-3 SANGYO BUMON; SEN'I KANKEI 5202
(OFFICIAL REPORT ON PATENTS; PART 3, INDUSTRY BRANCH; TEXTILES AND RELATED FIELDS)
TOKKYO CHO (PATENT OFFICE)
JAPANESE

TOKKYO KOHO; DAI-4 SANGYO BUMON; DAI-1 KUBUN GENDOKI, KIKAI YOSO, GENSHIRYOKU KANKEI 5203
(OFFICIAL REPORT ON PATENTS; PART 4, INDUSTRY BRANCH DIVISION ONE; MOTORS, MECHANICAL COMPONENTS, NUCLEAR POWER AND RELATED FIELDS)
TOKKYO CHO (PATENT OFFICE)
JAPANESE

TOKKYO KOHO; DAI-4 SANGYO BUMON; DAI-2 KUBUN; DENKI KANKEI 5204
(OFFICIAL REPORT ON PATENTS; PART 4 INDUSTRY BRANCH, DIVISION 2; ELECTRICAL AND RELATED SUBJECTS)
TOKKYO CHO (PATENT OFFICE)
JAPANESE

TOKKYO KOHO; DAI-4 SANGYO BUMON; DAI-3 KUBUN KOSAKU KIKAI, SANGYO KIKAI KANKEI 5205
(OFFICIAL REPORT ON PATENTS; PART 4 INDUSTRY BRANCH DIVISION 2; CONSTRUCTION MACHINERY, INDUSTRIAL MACHINERY AND RELATED FIELDS)
TOKKYO CHO (PATENT OFFICE)
JAPANESE

TOKKYO KOHO; DAI-5 SANGYO BUMON; DAI-1 KUBUN; UN'YU KANKEI 5206
(OFFICIAL REPORT ON PATENTS; PART 5 INDUSTRY BRANCH DIVISION ONE; TRANSPORTATION AND RELATED FIELDS)
TOKKYO CHO (PATENT OFFICE)
JAPANESE

TOKKYO KOHO; DAI-5 SANGYO BUMON; DAI-2 KUBUN KENCHIKU, EISEI KANKEI 5207
(OFFICIAL REPORT ON PATENTS; PART 5 INDUSTRY BRANCH DIVISION ONE; ARCHITECTURE, HYGIENE AND RELATED FIELDS)
TOKKYO CHO (PATENT OFFICE)
JAPANESE

TOKKYO KOHO; DAI-6 SANGYO BUMON; DAI-1 KUBUN DENSHI, TSUSHIN KANKEI 5208
(OFFICIAL REPORT ON PATENTS; PART 6 INDUSTRY BRANCH; DIVISION ONE; ELECTRONICS, COMMUNICATIONS AND RELATED FIELDS)
TOKKYO CHO (PATENT OFFICE)
JAPANESE

TOKKYO KOHO; DAI-6 SANGYO BUMON; DAI-2 KUBUN KOGAKU, KEISOKU KANKEI 5209
(OFFICIAL REPORT ON PATENTS; PART 6 INDUSTRY BRANCH DIVISION 2; OPTICS, MEASURING INSTRUMENTS AND RELATED FIELDS)
TOKKYO CHO (PATENT OFFICE)
JAPANESE

TOKKYO KOHO; DAI-7 SANGYO BUMON; JIMU, KATEI YOHIN KANKEI 5210
(OFFICIAL REPORT ON PATENTS; PART 7 INDUSTRY BRANCH OFFICE SUPPLIES, HOUSEHOLD SUPPLIES AND RELATED FIELDS)
TOKKYO KOHO (PATENT OFFICE)
JAPANESE

TOKKYO KOHO JITSUYO SHINAN KOHO SHUTSUGANSHAMEI SAKUIN 5211
(APPLICANT INDEX OF PATENT GAZETTE AND UTILITY MODEL GAZETTE)
OSAKA FURITSU YUHIGAOKA TOSHOKAN (OSAKA PREFECTURAL YUHIGAOKA LIBRARY)
KANSAI BUNKEN SENTA KYOGIKAI
C/O OSAKA-FURITSU YUHIGAOKA TOSHOKAN, REIJIN-CHO, TENNOJI-KU, OSAKASHI 543
JAPANESE

TOKKYO KOHORUI NO GO ANNAI 5212
(GUIDE TO THE VARIETIES OF PATENT PUBLIC REPORTS)
HATSUMEI KYOKAI (HATSUMEI ASSOCIATION)
JAPANESE

TOKKYO MOKUROKU TOKKYO BANGO SHUTSUGAN KOKOKU BANGO SAKUIN 5213
(CATALOGUE OF PATENTS; INDEX OF PATENT NUMBERS, APPLICANT ANNOUNCEMENT NUMBERS)
JAPANESE
INDEXED: JICST

TOKKYO NYUSU 5214
(PATENT NEWS)
JAPANESE
INDEXED: JICST

TOKKYO RYUTSU JOHO 5215
(PATENT CLEARING INFORMATION)
HATSUMEI KYOKAI (HATSUMEI ASSOCIATION)
JAPANESE

TOKKYO SHIN'AN SHUHO 5216
(ABSTRACTS OF PATENT AND UTILITY MODEL)
GIHODO
3-6, 1-CHOME, AKASAKA TAMEIKE-CHO, MINATO-KU, TOKYO 107
JAPANESE
INDEXED: JICST

TOKKYO SHIRYO GIFU 5217
(PATENT INFORMATION, GIFU)
GIFU-KENRITSU TOSHOKAN (GIFU PREFECTURAL LIBRARY)
1, OMIYA-MACHI, GIFU-SHI 500
JAPANESE

TOKKYO SHUTSUGAN KOKOKU SAIBUNRUIBETSU SOMOKUROKU 5218
(GENERAL INDEX TO DETAILED CLASSIFICATION OF PUBLICATION OF PATENT APPLICATION)
HATSUMEI KYOKAI (JAPAN INSTITUTE OF INVENTION AND INNOVATION
17, SHIBA NISHIKUBO AKEFUNE-CHO, MINATO-KU, TOKYO 105
JAPANESE
INDEXED: JICST

TOKKYO SHUTSUGAN SHOBUN KENSAKUHYO 5219
(REFERENCE TABLES OF DISPOSED PATENT APPLICATION)
HATSUMEI KYOKAI (JAPAN INSTITUTE OF INVENTION AND INNOVATION)
17, SHIBA NISHIKUBO AKEFUNE-CHO, MINATO-KU, TOKYO 105
JAPANESE
INDEXED: JICST

TOKKYO TAISHOHYO 5220
(CROSS REFERENCE TABLES OF PATENT)
HATSUMEI KYOKAI (JAPAN INSTITUTE OF INVENTION AND INNOVATION)
17, SHIBA NISHIKUBO AKEFUNE-CHO, MINATO-KU, TOKYO 105
JAPANESE
INDEXED: JICST

TOKKYO-CHO BANKOKU KOGYO JOYUKEN SHIRYOKUN GEPPO 5221
(MONTHLY REPORT OF THE PATENT OFFICE OF THE DOCUMENTS ARCHIVES RELATING TO UNIVERSAL INDUSTRIAL PROPERTY RIGHTS)
TOKKYO-CHO, BANKOKU KOGYO JOYUKEN SHIRYO KAN, KOKURITSU KOKKAI TOSHOKAN SHIBU TOKKYO-CHO TOSHOKAN (PATENT AGENCY, UNIVERSAL INDUSTRY PROPERTY RIGHTS DOCUMENTS ARCHIVES, NATIONAL DIET LIBRARY BRANCH)
JAPANESE
INDEXED: JICST

TOKKYO-CHO KOHO 5222
(PATENT REPORT)
TOKKYO-CHO (PATENT OFFICE)
4-3, 3-CHOME KASUMIGASEKI, CHIYODA-KU, TOKYO 100
JAPANESE
INDEXED: JICST

TOKKYO-CHO KOHO KOJI-GO 5223
(PATENT OFFICE OFFICIAL REPORT. PUBLIC ANNOUNCEMENT NUMBER)
JAPANESE
INDEXED: JICST

TOKKYO-CHO KOHO TOROKU. SHOHYO MOKUROKU 5224
(PATENT OFFICE REPORT. REGISTERED TRADEMARK CATALOGUE)
JAPANESE
INDEXED: JICST

TOKKYO-CHO KOHO. TOKKYO-CHO NENPO 5225
(PATENT OFFICE OFFICIAL REPORT. PATENT OFFICE ANNUAL)
JAPANESE
INDEXED: JICST

TOKKYO-CHO KOHO. TOROKU JITSUYO SHIN'AN MOKUROKU 5226
(PATENT OFFICE REPORT. CATALOGUE OF REGISTERED
NEW IDEAS FOR UTILIZATION)
JAPANESE
INDEXED: JICST

TOKKYO-CHO NENPO 5227
(PATENT OFFICE, ANNUAL GAZETTES)
TOKKYO-CHO, SOMU-BU, SOMU-KA (GENERAL
ADMINISTRATION DIVISION, PATENT OFFICE)
3-1, 1-CHOME, KASUMIGASEKI, CHIYODA-KU,
TOKYO 100
JAPANESE
INDEXED: JICST

YUASA AND HARA PATENT NEWS 5228
ENGLISH

YUASA. HARA TOKKYO NYUSU 5229
(YUASA. HARA PATENT NEWS)
YUASA. HARA HORITSU TOKKYO JIMUSHO
(YUASA. HARA PATENT LAW OFFICE)
JAPANESE

Engineering and Technology (Including Information Science)

AI UEA 5230
(EYE WEAR)
AI UEA (EYE WEAR)
JAPANESE

AICHI KOGYO DAIGAKU KENKYU HOKOKU 5231
(BULLETIN OF AICHI INSTITUTE OF TECHNOLOGY)
AICHI KOGYO DAIGAKU (AICHI INSTITUTE OF TECHNOLOGY)
1-18, WAKAMIZU-CHO, CHIKUSA-KU, NAGOYA-SHI 464
JAPANESE
INDEXED: JICST

AICHI KYOIKU DAIGAKU KYOIKU KOGAKU SENTA KIYO 5232
(MEMOIRS OF CENTER FOR EDUCATIONAL TECHNOLOGY FACULTY OF EDUCATION, AICHI UNIVERSITY OF EDUCATION)
AICHI KYOIKU DAIGAKU KYOIKU KOGAKU SENTA (CENTER FOR EDUCATIONAL TECHNOLOGY; FACULTY OF EDUCATION, AICHI UNIVERSITY OF EDUCATION)
1, HIROSAWA, IGAYA-CHO, KARIYA-SHI 448
JAPANESE AND ENGLISH

AICHI-KEN KOGYO SHIDOSHO HOKOKU 5233
(REPORTS OF THE INDUSTRIAL RESEARCH INSTITUTE, AICHI PREFECTURE)
AICHI-KEN KOGYO SHIDOSHO (INDUSTRIAL RESEARCH INSTITUTE, AICHI PREFECTURE)
3-58, HANADA-CHO, CHIKUSA-KU, NAGOYA-SHI 464
JAPANESE
INDEXED: CHE, JICST

AICHI-KEN KOGYO SHIDOSHO JIGYO HOKOKUSHO 5234
(OPERATION REPORT OF INDUSTRIAL RESEARCH INSTITUTE, AICHI PREFECTURE)
AICHI-KEN KOGYO SHIDOSHO (INDUSTRIAL RESEARCH INSTITUTE, AICHI PREFECTURE)
3-58, HANADA-CHO, CHIKUSA-KU, NAGOYA-SHI 464
JAPANESE
INDEXED: JICST

AICHI-KEN KOGYO SHIDOSHO JIGYO KEIKAKUSHO 5235
(OPERATION PROGRAMMES FOR INDUSTRIAL RESEARCH INSTITUTE, AICHI PREFECTURE)
AICHI-KEN KOGYO SHIDOSHO (INDUSTRIAL RESEARCH INSTITUTE, AICHI PREFECTURE)
3-58, HANADA-CHO, CHIKUSA-KU, NAGOYA-SHI 464
JAPANESE
INDEXED: JICST

AICHI-KEN KOGYO SHIDOSHO NENPO 5236
(ANNUAL REPORT OF THE AICHI PREFECTURE INDUSTRIAL DEVELOPMENT INSTITUTE)
JAPANESE
INDEXED: JICST

AICHI-KEN SEISAN DOTAI TOKEI NENPO 5237
(ANNUAL REPORT OF ACTUAL PRODUCTION STATISTICS, AICHI PREFECTURE)
AICHI-KEN KIKAKU-BU (PLANNING DIVISION, AICHI PREFECTURAL GOVERNMENT)
1-2, 3-CHOME, SANNOMARU, NAKA-KU, NAGOYA-SHI 460
JAPANESE

AIKOSHI NYUSU 5238
(AICHI PREFECTURAL INDUSTRIAL RESEARCH INSTITUTE NEWS)
AICHI-KEN KOGYO SHIDOSHO (INDUSTRIAL RESEARCH INSTITUTE, AICHI PREFECTURE)
3-58, HANADA-CHO, CHIKUSA-KU, NAGOYA-SHI 464
JAPANESE
INDEXED: CHE, JICST

AKASHI KOGYO KOTO SENMON GAKKO KENKYU KIYO 5239
(MEMOIRS OF THE AKASHI TECHNICAL COLLEGE)
AKASHI KOGYO KOTO SENMON GAKKO TOSHOKAN IINKAI (LIBRARY COMMITTEE OF THE AKASHI TECHNICAL COLLEGE)
NISHIOKA, UOZUMI-CHO, AKASHI-SHI 674
JAPANESE

AKITA KOGYO KOTO SENMON GAKKO KENKYU KIYO 5240
(RESEARCH REPORT OF AKITA TECHNICAL COLLEGE)
AKITA KOGYO KOTO SENMON GAKKO (AKITA TECHNICAL COLLEGE)
70, NAGAYAMASHITA, IIJIMA, AKITA-SHI 011
JAPANESE AND ENGLISH

AKITA-KEN KOGYO SHIKENJO GYOMU NENPO 5241
(ANNUAL REPORT OF AKITA INDUSTRIAL RESEARCH INSTITUTE)
AKITA-KEN KOGYO SHIKENJO (AKITA INDUSTRIAL RESEARCH INSTITUTE)
4-82, 1-CHOME, BARAJIMA, AKITA-SHI 010
JAPANESE

AKITA-KEN NO KOGYO, KOGYO TOKEI CHOSA KEKKA 5242
(MANUFACTURES IN AKITA PREFECTURE, CENSUS OF MANUFACTURE)
AKITA-KEN SOMU-BU (AKITA PREFECTURAL GOVERNMENT)
1-1, 4-CHOME, SANNO, AKITA-SHI 010
JAPANESE

ANAN KOGYO KOTO SENMON GAKKO KENKYU KIYO 5243
(RESEARCH REPORT OF THE ANAN TECHNICAL COLLEGE)
ANAN KOGYO KOTO SENMON GAKKO (THE ANAN TECHNICAL COLLEGE)
265, AOKI, MINOHAYASHI-CHO, ANAN-SHI 774
JAPANESE AND ENGLISH
INDEXED: MAT

ANNUAL REPORT OF THE TECHNICAL COOPERATION OF THE JAPANESE GOVERNMENT 5244
KAIGAI GIJUTSU KYORYOKU JIGYODAN (OVERSEAS TECHNICAL COOPERATION AGENCY)
42, ICHIGAYA HONMURA-CHO, SHINJUKU-KU, TOKYO 162
ENGLISH

AOMORI-KEN KOGYO SHIKENJO GYOMU HOKOKUSHO 5245
(OPERATION REPORT OF INDUSTRY RESEARCH INSTITUTE, AOMORI PREFECTURE)
AOMORI-KEN KOGYO SHIKENJO
75, FUKURO-MACHI, HIROSAKI-SHI 036
JAPANESE

ARIAKE KOGYO KOTO SENMON GAKKO KIYO 5246
(RESEARCH REPORT OF THE ARIAKE TECHNICAL COLLEGE)
ARIAKE KOGYO KOTO SENMON GAKKO (ARIAKE TECHNICAL COLLEGE)
150, HIGASHIHAGIO-CHO, OMUTA-SHI 836
JAPANESE, WITH EUROPEAN SUMMARY
INDEXED: JICST

ASAHIKAWA KOGYO KOTO SENMON GAKKO KENKYU HOBUN 5247
(JOURNAL OF THE ASAHIKAWA TECHNICAL COLLEGE)
ASAHIKAWA KOGYO KOTO SENMON GAKKO (ASAHIKAWA TECHNICAL COLLEGE)
2-JO, 2-CHOME, SHUNKODAI, ASAHIKAWA-SHI 070
JAPANESE AND ENGLISH

ASG ZASSHI 5248
(ASG MAGAZINE)
ASG HONBU (ASG HOME OFFICE)
JAPANESE AND ENGLISH

ASHIKAGA KOGYO DAIGAKU KENKYU SHUROKU 5249
(COLLECTED RESEARCH OF THE ASHIKAGA UNIVERSITY)
JAPANESE

ATSURYOKU GIJUTSU 5250
(PRESSURE ENGINEERING)
NIHON KOATSURYOKU GIJUTSU KYOKAI (JAPAN HIGH PRESSURE INSTITUTE)
1-11, KANDA SAKUMA-CHO, CHIYODA-KU, TOKYO 101
JAPANESE

BIAC JOHO 5251
(BIAC REPORT)
BIAC NIHON IINKAI (BIAC JAPAN COMMITTEE)
JAPANESE AND ENGLISH

BIBLIOGRAPHY OF LIBRARY SCIENCE AND DOCUMENTATION 5252
ENGLISH

BIBUROSU 5253
(BIBLOS)
JAPANESE
INDEXED: JICST

BOEI-CHO GIJUTSU KENKYU HONBU GIHO 5254
(TECHNICAL REPORT OF TECHNICAL RESEARCH AND DEVELOPMENT INSTITUTE, JAPAN DEFENSE AGENCY)
BOEI-CHO GIJUTSU KENKYU HONBU (TECHNICAL RESEARCH AND DEVELOPMENT INSTITUTE, JAPAN DEFENSE AGENCY)
2-24, 1-CHOME, IKEJIRI, SETAGAYA-KU, TOKYO 154
JAPANESE
INDEXED: JICST

BULLETIN OF FACULTY OF ENGINEERING, TOKUSHIMA UNIVERSITY 5255
TOKUSHIMA DAIGAKU KOGAKUBU (FACULTY OF ENGINEERING, TOKUSHIMA UNIVERSITY)
2-1, NANJO MISHIMA-MACHI, TOKUSHIMA-SHI 770
EUROPEAN
INDEXED: CHE, INS, JICST

BULLETIN OF THE FACULTY OF ENGINEERING, YOKOHAMA NATIONAL UNIVERSITY 5256
YOKOHAMA KOKURITSU DAIGAKU KOGAKUBU (FACULTY OF ENGINEERING, YOKOHAMA NATIONAL UNIVERSITY)
31-1, 2-CHOME, OOKA, MINAMI-KU, YOKOHAMASHI 233
ENGLISH
INDEXED: APP, CER, INS, WAA, JICST

BULLETIN OF THE NATIONAL RESEARCH LABORATORY OF METROLOGY 5257
TSUSHO SANGYO-SHO KOGYO GIJUTSU-IN KEIRYO KENKYUSHO (NATIONAL RESEARCH LABORATORY OF METROLOGY, AGENCY OF INDUSTRIAL SCIENCE AND TECHNOLOGY, MINISTRY OF INTERNATIONAL TRADE AND INDUSTRY)
10-4, 1-CHOME, KAGA, ITABASHI-KU, TOKYO 173
ENGLISH
INDEXED: CHE, JICST

BULLETIN OF THE RESOURCES EXPLOITATION INSTITUTE 5258
SHIGEN SOGO KAIHATSU KENKYUSHO (RESOURCES EXPLOITATION INSTITUTE)
SHIBUYA PARU HOMU, 29-25, SAKURAGAOKA-CHO, SHIBUYA-KU, TOKYO 150
ENGLISH

BULLETIN OF THE TOKYO INSTITUTE OF TECHNOLOGY 5259
TOKYO KOGYO DAIGAKU (TOKYO INSTITUTE OF TECHNOLOGY)
12-1, 2-CHOME, OOKAYAMA, MEGURO-KU, TOKYO 152
ENGLISH
INDEXED: CHE, INS, IAA, MET, WAA, JICST

BULLETIN OF THE UNIVERSITY OF OSAKA PREFECTURE, SERIES A, ENGINEERING AND NATURAL SCIENCES 5260
OSAKA-FURITSU DAIGAKU (UNIVERSITY OF OSAKA PREFECTURE)
142, MOZU HIGASHI-CHO, SAKAI-SHI 591
EUROPEAN
INDEXED: APP, ENG, INS, IAA, JICST

BURIJISUTON KOGYO YOHIN NYUSU 5261
(BRIDGESTONE INDUSTRIAL PRODUCTS NEWS)
JAPANESE

CHIBA DAIGAKU KOGAKUBU KENKYU HOKOKU 5262
(JOURNAL OF THE FACULTY OF ENGINEERING, CHIBA UNIVERSITY))
CHIBA DAIGAKU KOGAKUBU (FACULTY OF ENGINEERING, CHIBA UNIVERSITY
1-33, YAYOI-CHO, CHIBA-SHI 280
JAPANESE
INDEXED: CER, INS, MAT, JICST

CHIBA DAIGAKU KOGYO TANKI DAIGAKUBU KENKYU HOKOKU 5263
(RESEARCH REPORTS OF JUNIOR ENGINEERING COLLEGE, CHIBA UNIVERSITY)
CHIBA DAIGAKU KOGYO TANKI DAIGAKUBU (JUNIOR ENGINEERING COLLEGE, CHIBA UNIVERSITY)
1-33, YAYOI-CHO, CHIBA-SHI 280
JAPANESE AND ENGLISH
INDEXED: JICST

CHIBA KOGYO DAIGAKU KENKYU HOKOKU 5264
(REPORT OF CHIBA INSTITUTE OF TECHNOLOGY)
CHIBA KOGYO DAIGAKU (CHIBA INSTITUTE OF TECHNOLOGY)
7-1916, YATSU-MACHI, NARASHINO-SHI 275
JAPANESE AND EUROPEAN
INDEXED: JICST

CHIBA-KEN KOGYO SHIKENJO NENPO 5265
(ANNUAL REPORT OF INDUSTRIAL RESEARCH INSTITUTE OF CHIBA PREFECTURE)
CHIBA-KEN KOGYO SHIKENJO (INDUSTRIAL RESEARCH INSTITUTE OF CHIBA PREFECTURE)
889, KASORI-CHO, CHIBA-SHI 280
JAPANESE

CHIKUJI KANKOBUTSU BUNKAKAI HOKOKU 5266
(REPORT OF THE SUBCOMMITTEE ON SERIAL PUBLICATIONS)
SHIRITSU DAIGAKU TOSHOKAN KYOKAI TOCHIKU KENKYU BUKAI CHIKUJI KANKOBUTSU BUNKAKAI
JAPANESE

CHOSA TO GIJUTSU 5267
(SURVEY AND TECHNOLOGY)
JAPANESE
INDEXED: JICST

CHUBU KOGYO DAIGAKU KIYO 5268
(MEMOIRS OF CHUBU INSTITUTE OF TECHNOLOGY)
CHUBU KOGYO DAIGAKU (CHUBU INSTITUTE OF TECHNOLOGY)
1200, MATSUMOTO-CHO, KASUGAI-SHI 487
JAPANESE, WITH EUROPEAN SUMMARY
INDEXED: APP, CHE, ENG, INS, JICST

CHUGOKU DENRYOKU SHAHO 5269
(REPORT OF THE CHUGOKU ELECTRICAL POWER COMPANY)
JAPANESE
INDEXED: JICST

CHUGOKU KOGYO GIJUTSU 5270
(ENGINEERING TECHNIQUE, CHUGOKU)
TSUSHO SANGYO-SHO KOGYO GIJUTSU-IN CHUGOKU KOGYO GIJUTSU SHIKENJO (GOVERNMENT INDUSTRIAL TECHNOLOGY RESEARCH INSTITUTE, CHUGOKU AGENCY OF INDUSTRY SCIENCE AND TECHNOLOGY)
KOTSUBO-DANCHI, HIRO-MACHI, KURE-SHI 737-01
JAPANESE

CHUGOKU KOGYO GIJUTSU SHIKENJO HOKOKU 5271
(REPORTS OF THE GOVERNMENT INDUSTRIAL RESEARCH INSTITUTE, CHUGOKU)
KOGYO GIJUTSUIN CHUGOKU KOGYO GIJUTSU SHIKENJO
JAPANESE

CHUGOKU KOGYO GIJUTSU SHIKENJO NENPO 5272
(ANNUAL REPORT OF GOVERNMENT INDUSTRIAL RESEARCH INSTITUTE, CHUGOKU)
TSUSHO SANGYO-SHO KOGYO GIJUTSU-IN CHUGOKU KOGYO GIJUTSU SHIKENJO (GOVERNMENT INDUSTRIAL TECHNOLOGY RESEARCH INSTITUTE, CHUGOKU, AGENCY OF INDUSTRY SCIENCE AND TECHNOLOGY)
KOTSUBO-DANCHI, HIRO-MACHI, KURE-SHI 737-01
JAPANESE

CHUGOKU KOGYO TSUJIN 5273
(CHINESE INDUSTRY TIMES)
ODOSHA
20-14, 1-CHOME, SHINJUKU, SHINJUKU-KU, TOKYO 160
JAPANESE

CHUKOSHI NYUSU 5274
(NATIONAL INDUSTRIAL RESEARCH INSTITUTE OF CHUGOKU NEWS)
TSUSHO SANGYO-SHO KOGYO GIJUTSU-IN CHUGOKU KOGYO GIJUTSU SHIKENSHO (GOVERNMENT INDUSTRIAL TECHNOLOGY RESEARCH INSTITUTE, CHUGOKU, AGENCY OF INDUSTRIAL SCIENCE AND TECHNOLOGY)
KOTSUBO-DANCHI, HIRO-MACHI, KURE-SHI 737-01
JAPANESE

CHUSHO KIGYO KAIGAI GIJUTSU JOHO 5275
(TECHNICAL INFORMATION OF OVERSEAS SMALLER ENTERPRISES)
NIHON KAGAKU GIJUTSU JOHO SENTA (JAPAN INFORMATION CENTER OF SCIENCE AND TECHNOLOGY)
1-1, 2-CHOME, NAGATA-CHO, CHIYODA-KU, TOKYO 100
JAPANESE
INDEXED: JICST

CHUSHO KIGYO KOKUNAI GIJUTSU JOHO 5276
(REPORT ON DOMESTIC TECHNOLOGY FOR SMALL- AND MEDIUM-SIZED ENTERPRISES)
CHUSHYO KIGYO JOHO SENTA (SMALL-MIDDLE SIZED INDUSTRY INFORMATION CENTER)
JAPANESE

COLOR 5277
NIHON SHIKISAI KENKYUSHO (JAPAN COLOR RESEARCH INSTITUTE)
1-19, 3-CHOME, NISHIAZABU, MINATO-KU, TOKYO 106
JAPANESE

COLOR SPACE 5278
MURAKAMI SHIKISAI GIJUTSU KENKYUSHO (MURAKAMI COLOR RESEARCH LABORATORY)
11-1, 3-CHOME, KACHIDOKI, CHUO-KU, TOKYO 104
JAPANESE

DAIDO KOGYO DAIGAKU KIYO 5279
(BULLETIN OF DAIDO TECHNICAL COLLEGE)
DAIDO KOGYO DAIGAKU (DAIDO TECHNICAL COLLEGE)
DAIDO-CHO, MINAMI-KU, NAGOYA-SHI 457
JAPANESE AND EUROPEAN

DAIGAKU TOSHOKAN KENKYU 5280
(UNIVERSITY LIBRARIES RESEARCH)
GAKUJUTSU BUNKEN FUKYUKAI
JAPANESE

DAIKOSHI NYUSU 5281
(NEWS OF THE GOVERNMENT INDUSTRIAL RESEARCH INSTITUTE, OSAKA)
TSUSHO SANGYO-SHO KOGYO GIJITSU-IN OSAKA KOGYO GIJITSU SHIKENJO (GOVERNMENT INDUSTRIAL RESEARCH INSTITUTE, OSAKA, AGENCY OF INDUSTRIAL SCIENCE)
8-31, 1-CHOME, MIDORIGAOKA, IKEDA-SHI 563
JAPANESE
INDEXED: JICST

DENNETSU KENKYU 5282
(RESEARCH ON HEAT CONDUCTION)
NIHON DENNETSU KENKYUKAI
C/O TOKYO DAIGAKU KOGAKUBU, 3-1, 7-CHOME, HONGO, BUNKYO-KU, TOKYO 113
JAPANESE

DETA TSUSHIN 5283
(DATA COMMUNICATION AND PROCESSING)
JOHO KENKYU SHUPPANKAI
1-7, 9-CHOME, AKASAKA, MINATO-KU, TOKYO 107
JAPANESE
INDEXED: JICST

DOKUMENTESHON KENKYU 5284
(DOCUMENTATION STUDY)
NIHON DOKUMENTESHON KYOKAI (THE JAPAN DOCUMENTATION SOCIETY)
C/O SASAKI BUILDING, 5-7, 2-CHOME, KOISHIKAWA, BUNKYO-KU, TOKYO 113
JAPANESE, WITH ENGLISH SUMMARY
INDEXED: JICST

DOKYUMENTESHON KONDANKAI KAIHO 5285
(DOCUMENTATION KONDANKAI KAIHO)
DOKYUMENTESHON KONDANKAI
C/O SHIRYO KENKYUSHO, 15-6, 6-CHOME, SOTOKANDA, CHIYODA-KU, TOKYO 101
JAPANESE
INDEXED: JICST

DORYOKU 5286
(POWER)
NIHON DORYOKU KYOKAI (JAPAN POWER ASSOCIATES)
C/O DALDO BUILDING, 7-13, 1-CHOME, NISHISHINBASHI, MINATO-KU, TOKYO 105
JAPANESE
INDEXED: JICST

DOSHISHA DAIGAKU RIKOGAKU KENKYU HOKOKU 5287
(SCIENCE AND ENGINEERING REVIEW OF DOSHISHA UNIVERSITY)
DOSHISHA DAIGAKU RIKOGAKUBU (SCIENCE AND ENGINEERING RESEARCH INSTITUTE OF DOSHISHA UNIVERSITY)
IMADEGAWA HIGASHIIRU, KARASUMARU, KAMIGYO-KU, KYOTO-SHI 602
JAPANESE AND ENGLISH
INDEXED: CER, INS, JICST

EHIME DAIGAKU KIYO, DAI-3-BU, KOGAKU 5288
(MEMOIRS OF THE EHIME UNIVERSITY, SECTION 3, ENGINEERING)
EHIME DAIGAKU KOGAKUBU (THE FACULTY OF ENGINEERING, EHIME UNIVERSITY)
3, BUNKYO-CHO, MATSUYAMA-SHI 790
JAPANESE AND EUROPEAN
INDEXED: CHE, MAT, JICST

EHIME-KEN KOGYO SHIKENJO KENKYU HOKOKU 5289
(BULLETIN OF THE INDUSTRIAL RESEARCH INSTITUTE OF EHIME PREFECTURE)
EHIME-KEN KOGYO SHIKENJO (INDUSTRIAL RESEARCH INSTITUTE OF EHIME PREFECTURE)
5-11, 1-CHOME, MIYANISHI, MATSUYAMA-SHI 790
JAPANESE AND ENGLISH
INDEXED: CHE

EIKOKU KOGYO NYUSU 5290
(BRITISH INDUSTRIAL NEWS)
EIKOKU TAISHIKAN KOHO-BU
ICHIBAN-CHO, CHIYODA-KU, TOKYO 102
JAPANESE
INDEXED: JICST

ENERGY 5291
ESSO SUTANDADO SEKIYU K.K. (ESSO STANDARD OIL COMPANY)
3-3, 5-CHOME, AKASAKA, MINATO-KU, TOKYO 107
JAPANESE

ENERUGEIA 5292
(ENERGY)
ENERUGEIA KANKOKAI (ENERGY PUBLICATION ASSOCIATION)
JAPANESE

ENERUGI 5293
(ENERGY)
NIHON KOGYO SHINBUNSHA
7-2, 1-CHOME, OTE-MACHI, CHIYODA-KU, TOKYO 100
JAPANESE
INDEXED: JICST

ENERUGI KEIZAI 5294
(ENERGY ECONOMY)
NIHON ENERUGII KEIZAI KENKYUJO (JAPAN ENERGY ECONOMY RESEARCH INSTITUTE)
JAPANESE

ENERUGI-SHI KENKYU NOTO 5295
(NOTES ON ENERGY HISTORY RESEARCH)
ENERUGII-SHI KENKYUKAI (KYUSHU DAIGAKU KEIZAI GAKUBU NIHON KEIZAISHI KENKYUSHITSUNAI) NISHI NIHON BUNKA KYOKAI (ENERGY HISTORY RESEARCH SOCIETY KYUSHU UNIVERSITY, FACULTY OF ECONOMICS)
JAPANESE

ENGINEERING INDUSTRIES OF JAPAN 5296
KIKAI SHINKOKYOKAI KEIZAI KENKYUSHO
C/O KIKAI SHINKO KAIKAN, 1-5, 21-GOCHI, SHIBA KOEN, MINATO-KU, TOKYO 105
ENGLISH

ENGINEERS 5297
NIHON KAGAKU GIJUTSU RENMEI (JAPAN SCIENCE TECHNOLOGICAL FEDERATION)
JAPANESE AND ENGLISH

ENJINIA 5298
(ENGINEER)
SHIN GIJUTSU KAIHATSU SENTA (TECHNICAL CONSULTANTS INCORPORATED)
C/O HANAZONO BUILDING, 1, SANKO-CHO, SHINJUKU-KU, TOKYO 160
JAPANESE

ENJINIASU 5299
(ENGINEERS)
NIHON KAGAKU GIJUTSU RENMEI (UNION OF JAPANESE SCIENTISTS AND ENGINEERS)
10-11, 5-CHOME, SENDAGAYA, SHIBUYA-KU, TOKYO 151
JAPANESE
INDEXED: JICST

EUPHORIA ET CACOPHORIA 5300
PROFESSOR KASUKE ITO
PO BOX 120, GIFU
ENGLISH
INDEXED: CHE, EXC

FILLER BOOK 5301
SHIRAISHI KOGYO K.K. (SHIRAISHI KOGYO KAISHA LTD.)
4-78, MOTOHAMA-CHO, AMAGASAKI-SHI 660
JAPANESE AND ENGLISH

FQC (GENBA TO QC) 5302
(QUALITY CONTROL FOR THE FOREMEN)
NIHON KAGAKU GIJUTSU RENMEI (THE UNION OF JAPANESE SCIENTIST AND ENGINEERS)
10-11, 5-CHOME, SENDAGAYA, SHIBUYA-KU, TOKYO 151
JAPANESE
INDEXED: JICST

FUKUGO ZAIRYO 5303
(COMPOSITE MATERIALS AND STRUCTURES)
KOGYO CHOSAKAI
14-7, 2-CHOME, HONGO, BUNKYO-KU, TOKYO 113
ENGLISH

FUKUGO ZAIRYO SEKKEI 5304
(COMPOSITE MATERIALS AND DESIGN)
KOGYO SHUPPAN (INDUSTRIAL PUBLISHING)
JAPANESE

FUKUI DAIGAKU KOGAKUBU KENKYU HOKOKU 5305
(MEMOIRS OF THE FACULTY OF ENGINEERING, FUKUI UNIVERSITY)
FUKUI DAIGAKU KOGAKUBU (FACULTY OF ENGINEERING, FUKUI UNIVERSITY)
9-1, 3-CHOME, BUNKYO, FUKUI-SHI 910
JAPANESE AND ENGLISH
INDEXED: CHE, JICST

FUKUI KOGYO KOTO SENMON GAKKO KENKYU KIYO, SHIZEN KAGAKU, KOGAKU 5306
(RESEARCH REPORTS OF FUKUI TECHNICAL COLLEGE, NATURAL SCIENCE AND ENGINEERING)
FUKUI KOGYO KOTO SENMON GAKKO
GESU-MACHI, SABAE-SHI 916
JAPANESE, WITH ENGLISH SUMMARY
INDEXED: JICST

FUKUI-KEN KOGYO GIJUTSU JOHO 5307
(TECHNICAL REPORT OF THE FUKUI PREFECTURE INDUSTRIES)
FUKUI-KEN KOGYO GIJUTSU JOHO SHITSU (FUKUI PREFECTURE TECHNICAL ENGINEERING INFORMATION OFFICE)
JAPANESE

FUKUI-KEN KOGYO SHIKENJO KENKYU SEIKA FUKYU KOSHUKAI SHIRYO 5308
(RESEARCH RESULTS AND SEMINAR INFORMATION DATA OF THE FUKUI PREFECTURE INDUSTRIAL RESEARCH INSTITUTE)
FUKUI-KEN KOGYO SHIKENJO
920, WAKAI-CHO, FUKUI 910
JAPANESE
INDEXED: CHE

FUKUI-KEN KOGYO SHIKENJO NENPO 5309
(ANNUAL REPORT OF FUKUI PREFECTURE INDUSTRIAL RESEARCH INSTITUTE)
FUKUI-KEN KOGYO SHIKENJO
920, WAKAE-CHO, FUKUI-SHI 910
JAPANESE
INDEXED: CHE

FUKUI-KEN KOGYO TOKEIHYO 5310
(CENSUS OF MANUFACTURES, FUKUI PREFECTURE)
FUKUI-KEN SOMU-BU (FUKUI PREFECTURAL GOVERNMENT)
17-1, 3-CHOME, OTE, FUKUI-SHI 910
JAPANESE

FUKUOKA DAIGAKU KOGAKU SHUHO 5311
(FUKUOKA UNIVERSITY REVIEW OF TECHNICAL SCIENCES)
FUKUOKA DAIGAKU KENKYUSHO (THE INSTITUTE FOR ADVANCED RESEARCH, FUKUOKA UNIVERSITY)
11, NANAKUMA, NISHI-KU, FUKUOKA-SHI 814
JAPANESE, WITH ENGLISH SUMMARY

FUKUOKA KOGYO DAIGAKU KENKYU RONSHU 5312
(RESEARCH PAPERS OF FUKUOKA INDUSTRIAL UNIVERSITY)
JAPANESE

FUKUOKA KYOIKU DAIGAKU KYOIKU KOGAKU SENTA RETA 5313
(NEWS LETTER OF THE CENTER FOR EDUCATIONAL TECHNOLOGY FUKUOKA UNIVERSITY OF EDUCATION)
FUKUOKA KYOIKU DAIGAKU KYOIKU KOGAKU SENTA (CENTER FOR EDUCATIONAL TECHNOLOGY, FUKUOKA UNIVERSITY OF EDUCATION)
JAPANESE

FUKUOKA-KEN FUKUOKA KOGYO SHIKENJO KENKYU HOKOKU 5314
(RESEARCH REPORT OF THE FUKUOKA INDUSTRIAL RESEARCH INSTITUTE, FUKUOKA PREFECTURE)
FUKUOKA-KEN FUKUOKA KOGYO SHIKENJO (FUKUOKA INDUSTRIAL RESEARCH INSTITUTE, FUKUOKA PREFECTURE)
28-45, 1-CHOME, KATAKASU, HAKATA-KU, FUKUOKA-SHI 812
JAPANESE

FUKUOKA-KEN FUKUSHIMA KOGYO SHIKENJO KENKYU HOKOKU 5315
(RESEARCH REPORT OF THE FUKUSHIMA INDUSTRIAL RESEARCH INSTITUTE, FUKUOKA PREFECTURE)
FUKUOKA-KEN FUKUSHIMA KOGYO SHIKENJO (FUKUSHIMA INDUSTRIAL RESEARCH INSTITUTE, FUKUOKA PREFECTURE)
499, INADOMI, YAME-SHI 834
JAPANESE

FUKUOKA-KEN NO KOGYO 5316
(CENSUS OF MANUFACTURES, FUKUOKA PREFECTURE)
FUKUOKA-KEN KIKAKU KAIHATSU-BU (PLANNING AND DEVELOPMENT DIVISION, FUKUOKA PREFECTURAL GOVERNMENT)
1-1, 1-CHOME, TENJIN, CHUO-KU, FUKUOKA-SHI 810
JAPANESE

FUKUOKA-KEN SHOKO JIHO 5317
(FUKUOKA PREFECTURE COMMERCIAL AND INDUSTRIAL REVIEW)
FUKUOKA-KEN SHOKO SUISANBU (FUKUOKA PREFECTURE (CHAMBER OF) COMMERCE AND INDUSTRY, FISHERIES BUREAU)
JAPANESE

FUKUSHIMA KOGYO KOTO SENMON GAKKO KIYO 5318
(RESEARCH REPORTS, FUKUSHIMA TECHNICAL COLLEGE)
FUKUSHIMA KOGYO KOTO SENMON GAKKO (FUKUSHIMA TECHNICAL COLLEGE)
1-CHOME, TAIRA, IWAKI-SHI 970
JAPANESE AND ENGLISH
INDEXED: JICST

FUKUSHIMA KOGYO SHIKENJO GIJUTSU JOHO 5319
(TECHNICAL INFORMATION OF INDUSTRIAL RESEARCH INSTITUTE, FUKUSHIMA)
FUKUSHIMA KOGYO SHIKENJO
FUNOKAWA, SAKURASHITA, FUKUSHIMA-SHI 960-21
JAPANESE

FUKUSHIMA KOGYO SHIKENJO HOKOKU 5320
(REPORT OF INDUSTRIAL RESEARCH INSTITUTE, FUKUSHIMA)
FUKUSHIMA KOGYO SHIKENJO
FUNOKAWA, SAKURA-SHITA, FUKUSHIMA-SHI 960-21
JAPANESE

FUKUSHIMA-KEN AIZU WAKAMATSU KOGYO SHIKENJO SHIKEN KENKYU HOKOKU 5321
(REPORT OF FUKUSHIMA PREFECTURAL AIZUWAKAMATSU LABORATORY OF INDUSTRY)
FUKUSHIMA-KEN AIZU WAKAMATSU KOGYO SHIKENJO
IDERA, MONDEN-CHO, AIZU WAKAMATSU-SHI 965
JAPANESE

GAIKOKU GIJUTSU DONYU NENJI HOKOKU 5322
(ANNUAL REPORT ON INTRODUCTION OF FOREIGN TECHNIQUE)
KAGAKU GIJUTSU-CHO SHINKO-KYOKU (THE SCIENCE AND TECHNOLOGY AGENCY)
OKURA-SHO INSATSU-KYOKU
2, AKASAKA AOI-CHO, MINATO-KU, TOKYO 107
JAPANESE

GEKKAN ENERUGI DAIJESUTO 5323
(ENERGY DIGEST)
TOSHI ENERUGI KYORYOKUKAI
C/O ARAKAWA BUILDING, 35-4, 2-CHOME, NISHI-SHINBASHI, MINATO-KU, TOKYO 105
JAPANESE

GEKKAN GIJUTSU HYORON 5324
(MONTHLY TECHNICAL REVIEW)
GIJUTSU HYORONSHA (TECHNICAL REVIEW COMPANY)
JAPANESE

GEKKAN MEIKOKEN 5325
(MONTHLY JOURNAL OF THE NAGOYA MUNICIPAL INDUSTRIAL RESEARCH INSTITUTE)
NAGOYA-SHI KOGYO KENKYUSHO (INDUSTRIAL RESEARCH INSTITUTE, NAGOYA MUNICIPAL GOVERNMENT)
3-24, ROKUBAN-CHO, ATSUTA-KU, NAGOYA-SHI 456
JAPANESE
INDEXED: JICST

GENDAI GIJUTSU HYORON 5326
(REVIEW OF MODERN TECHNOLOGY)
GENDAI GIJUTSU HYORONSHA
10-7, 2-CHOME, DOGENZAKA, SHIBUYA-KU, TOKYO 150
JAPANESE
INDEXED: JICST

GENDAI NO TOSHOKAN 5327
(CONTEMPORARY LIBRARIES)
JAPANESE
INDEXED: JICST

GIEN 5328
(TECHNOLOGY)
KANSAI DAIGAKU KOGYO GIJUTSU KENKYUSHO (INDUSTRIAL TECHNOLOGY INSTITUTE, KANSAI UNIVERSITY)
3-35, 3-CHOME, YAMATE-CHO, SUITA-SHI 564
JAPANESE
INDEXED: JICST

GIFU DAIGAKU KOGAKUBU KENKYU HOKOKU 5329
(RESEARCH REPORTS, FACULTY OF ENGINEERING, GIFU UNIVERSITY)
GIFU DAIGAKU KOGAKUBU (THE FACULTY OF ENGINEERING, GIFU UNIVERSITY)
3-1, NAKA MONZEN-MACHI, KAGAMIHARA-SHI 504
JAPANESE AND ENGLISH
INDEXED: CHE, JICST

GIFU KOGYO KOTO SENMON GAKKO KIYO 5330
(MEMOIRS OF GIFU TECHNICAL COLLEGE)
GIFU KOGYO SENMON GAKKO (GIFU TECHNICAL COLLEGE)
SHINSEI-CHO, MOTOSU-GUN, GIFU 501-04
JAPANESE AND ENGLISH
INDEXED: JICST

GIFU-KEN GIJUTSU JOHO 5331
(GIFU PREFECTURE TECHNOLOGICAL REPORT)
GIFU-KEN KOGYO GIJUTSU SENTA (GIFU PREFECTURE INDUSTRIAL TECHNOLOGY CENTER)
JAPANESE

GIFU-KEN KOGYO GIJUTSU SENTA JIGYO NENPO 5332
(OPERATION YEARBOOK OF INDUSTRIAL RESEARCH INSTITUTE OF GIFU PREFECTURE)
GIFU-KEN KOGYO GIJUTSU SENTA (INDUSTRIAL RESEARCH INSTITUTE OF GIFU PREFECTURE)
KITAOYOBI, KASAMATSU-CHO, HASIMA-GUN, GIFU 501-61
JAPANESE

GIFU-KEN KOGYO GIJUTSU SENTA KENKYU HOKOKU 5333
(RESEARCH REPORT OF INDUSTRIAL RESEARCH INSTITUTE OF GIFU PREFECTURE)
GIFU-KEN KOGYO GIJUTSU SENTA (INDUSTRIAL RESEARCH INSTITUTE OF GIFU PREFECTURE)
KITAOYOBI, KASAMATSU-CHO, HASHIMA-GUN, GIFU 501-61
JAPANESE
INDEXED: CHE

GIFU-KENRITSU GIFU KOGYO KOTOGAKKO KENKYU HOKOKU 5334
(RESEARCH REPORT OF THE GIFU PREFECTURAL GIFU INDUSTRIAL HIGH SCHOOL)
JAPANESE

GIJUTSU JOHO 5335
(TECHNICAL REPORT)
OSAKA FURITSU KOGYO GIJUTSU KENKYUJO GIJUTSU JOHOSHITSU (TECHNICAL INFORMATION OFFICE, OSAKA PREFECTURAL INDUSTRIAL TECHNOLOGY RESEARCH)
JAPANESE

GIJUTSU JOHO 5336
(TECHNICAL REPORT)
WAKAYAMA-KEN KOGYO SHIKENJO (WAKAYAMA PREFECTURE INDUSTRIAL EXPERIMENT STATION)
JAPANESE

GIJUTSU JOHO 5337
(TECHNICAL REPORT)
IWATE-KEN KOGYO SHIKENJO. IWATE-KEN JOZO SHOKUHIN SHIKENJO (IWATE PREFECTURE EXPERIMENT STATION AND IWATE PREFECTURE BREWING & FOOD EXPERIMENT STATION)
JAPANESE

GIJUTSU JOHO (KOCHI-KEN KOGYO SHIKENJO) 5338
(TECHNICAL INFORMATION, INDUSTRIAL RESEARCH INSTITUTE, KOCHI PREFECTURE)
KOCHI-KEN KOGYO SHIKENJO
11-15, 2-CHOME, SANBASHI-DORI, KOCHI-SHI 780
JAPANESE

GIJUTSU KYOIKU 5339
(TECHNICAL EDUCATION)
SANGYO KYOIKU KENKYU RENMEI (RESEARCH INSTITUTE FOR TECHNICAL EDUCATION) AND KOKUDOSHA
17-6, 1-CHOME, MEJIRODAI, BUNKYO-KU, TOKYO 112
JAPANESE
INDEXED: JICST

GIJUTSU KYOIKU KENKYU 5340
(TECHNICAL EDUCATION RESEARCH)
GIJUTSU KYOIKU KENKYUKAI (TECHNICAL EDUCATION RESEARCH SOCIETY)
JAPANESE

GIJUTSU KYORYOKU NENPO 5341
(ANNUAL REPORT OF OVERSEAS TECHNICAL COOPERATION AGENCY)
KAIGAI GIJUTSU KYORYOKU JIGYODAN (OVERSEAS TECHNICAL COOPERATION AGENCY)
42, ICHIGAYA HONMURA-CHO, SHINJUKU-KU, TOKYO 162
JAPANESE
INDEXED: JICST

GIJUTSU SHINBUN 5342
(TECHNOLOGICAL NEWS)
GIJUTSU SHINBUNSHA
6-6, 2-CHOME, TAITO, TAITO-KU, TOKYO 110
JAPANESE

GIJUTSU TO KEIZAI 5343
(MONTHLY REPORT 'JAPAN TECHNO-ECONOMICS SOCIETY')
KAGAKU GIJUTSU TO KEIZAI NO KAI (JAPAN TECHNO-ECONOMICS SOCIETY)
4-5, 2-CHOME, IIDABASHI, CHIYODA-KU, TOKYO 102
JAPANESE
INDEXED: JICST

GIJUTSU TO KIGYO 5344
(TECHNOLOGY AND INDUSTRY)
OSAKA KAGAKU GIJUTSU SENTA (OSAKA SCIENCE AND TECHNOLOGY CENTER)
1-118, UTSUBO, NISHI-KU, OSAKA-SHI 550
JAPANESE
INDEXED: JICST

GIJUTSU TO NINGEN 5345
(TECHNOLOGY AND HUMAN BEING)
AGUNE
3-1, 3-CHOME, SHIBUYA, SHIBUYA-KU, TOKYO 150
JAPANESE
INDEXED: JICST

GIJUTSU YOSOKU SHINPOJIUMU RONBUNSHU 5346
(PAPERS OF THE SYMPOSIUM ON TECHNOLOGY FORECASTING)
JAPANESE
INDEXED: JICST

GIJUTSUSHI KENKYU (GENDAI GIJUTSUSHI KENKYUKAI KAISHI) 5347
(STUDY IN TECHNICAL HISTORY (REPORTS OF STUDY GROUP FOR MODERN TECHNICAL HISTORY))
GENDAI GIJUTSUSHI KENKYUKAI (STUDY GROUP FOR MODERN TECHNICAL HISTORY)
C/O REX BUILDING, 502, 49-2, 5-CHOME, HIGASHI-NIPPORI, ARAKAWA-KU, TOKYO 115
JAPANESE

GIKEN SHOHO 5348
(JOURNAL OF TECHNICAL RESEARCH INSTITUTE)
KIKAI SHINKO KYOKAI GIJUTSU KENKYUSHO (TECHNICAL RESEARCH INSTITUTE, JAPAN SOCIETY FOR THE PROMOTION OF MACHINE INDUSTRY)
JAPANESE
INDEXED: JICST

GINO YANGU 5349
(LEADER OF THE YOUNG TECHNICIANS)
GINO KENSHUSHA
C/O MORITA BUILDING, 3F., 1-12, 1-CHOME, MISAKI-CHO, CHIYODA-KU, TOKYO 101
JAPANESE

GINOSHI NO TOMO 5350
(MECHANICS' FRIEND)
TAIGA SHUPPAN (TAIGA PUBLISHING COMPANY, LTD.)
1-13, KANDA AWAJI-CHO, CHIYODA-KU, TOKYO 101
JAPANESE
INDEXED: JICST

GIREN 5351
(NEWS OF TECHNICAL ASSOCIATION OF HOSEI UNIVERSITY)
HOSEI DAIGAKU GIJUTSU RENMEI (TECHNICAL ASSOCIATION OF HOSEI UNIVERSITY)
7-2, 3-CHOME, KAJINO-MACHI, KOGANEI-SHI, TOKYO 184
JAPANESE

GKS NEWS 5352
GIFU-KEN KOGYO GIJUTSU SENTA (INDUSTRIAL RESEARCH INSTITUTE OF GIFU PREFECTURE)
KITAOYOBI, KASAMATSU-CHO, HASHIMA-GUN, GIFU 501-61
JAPANESE

GUNMA KOGYO KOTO SENMON GAKKO KENKYU HOKOKU 5353
(RESEARCH REPORTS OF GUNMA TECHNICAL COLLEGE)
GUNMA KOGYO KOTO SENMON GAKKO (GUNMA TECHNICAL COLLEGE)
580, TORIBA-MACHI, MAEBASHI-SHI 371
JAPANESE, WITH ENGLISH SUMMARY
INDEXED: JICST

GUNMA-KEN KOGYO SHIKENJO NENPO 5354
(ANNUAL REPORT OF GUNMA-KEN INDUSTRIAL RESEARCH LABORATORY)
GUNMA-KEN KOGYO SHIKENJO
190, TORIBA-MACHI, MAEBASHI-SHI 371
JAPANESE

HACHINOHE KOGYO KOTO SENMON GAKKO KIYO 5355
(RESEARCH REPORTS OF HACHINOHE TECHNICAL COLLEGE)
HACHINOHE KOGYO KOTO SENMON GAKKO (HACHINOHE TECHNICAL COLLEGE)
UENO-TAIRA, TAMONOGI, HACHINOHE-SHI 031
JAPANESE AND ENGLISH
INDEXED: JICST

HAKODATE DAIGAKU HOKKAIDO SANGYO KAIHATSU KENKYUJO KIYO 5356
(BULLETIN OF HAKODATE UNIVERSITY HOKKAIDO INDUSTRIAL DEVELOPMENT RESEARCH INSTITUTE)
JAPANESE

HAKODATE KOGYO KOTO SENMON GAKKO KIYO 5357
(RESEARCH REPORTS, HAKODATE TECHNICAL COLLEGE)
HAKODATE KOGYO KOTO SENMON GAKKO
226, TOKURA-CHO, HAKODATE-SHI 042
JAPANESE, WITH ENGLISH SUMMARY

HI-HAKAI KENSA 5358
NIHON HI-HAKAI KENSA KYOKAI (THE JAPANESE SOCIETY FOR NONDESTRUCTIVE INSPECTION)
1-11, KANDA SAKUMA-CHO, CHIYODA-KU, TOKYO 101
JAPANESE AND ENGLISH
INDEXED: CHE, JICST

HIGASHI OSAKA-SHI NO KOGYO 5359
(INDUSTRY IN HIGASHI-OSAKA CITY)
HIGASHI-OSAKA-SHI KIKAKU-BU (PLANNING DIVISION, HIGASHI-OSAKA MUNICIPAL GOVERNMENT)
1-1, 1-CHOME, INABA, HIGASHI OSAKA-SHI 578
JAPANESE

HIJIRIBASHI KOGYO KOTO SENMON GAKKO KENKYU KIYO 5360
(RESEARCH BULLETIN OF THE HIJIRIBASHI HIGHER TECHNICAL SCHOOL)
HIJIRIBASHI HIGHER TECHNICAL SCHOOL
JAPANESE

HIMEJI KOGYO DAIGAKU KENKYU HOKOKU 5361
(REPORTS OF HIMEJI INSTITUTE OF TECHNOLOGY)
HIMEJI KOGYO DAIGAKU (HIMEJI INSTITUTE OF TECHNOLOGY)
2167, SHOSHA, HIMEJI-SHI 617-22
JAPANESE AND ENGLISH
INDEXED: CHE, INS, JICST

HIMEJI KOGYO DAIGAKU KENKYU HOKOKU, IPPAN KYOIKU KANKEI 5362
(RESEARCH REPORT OF HIMEJI INSTITUTE OF TECHNOLOGY, GENERAL ARTS)
HIMEJI KOGYO DAIGAKU (HIMEJI INSTITUTE OF TECHNOLOGY)
2167, SHOSHA, HIMEJI-SHI 671-22
JAPANESE AND ENGLISH

HINSHITSU 5363
(QUALITY)
JAPANESE
INDEXED: JICST

HIROSHIMA DAIGAKU KOGAKUBU KENKYU HOKOKU 5364
(BULLETIN OF THE FACULTY OF ENGINEERING, HIROSHIMA UNIVERSITY)
HIROSHIMA DAIGAKU KOGAKUBU (FACULTY OF ENGINEERING, HIROSHIMA UNIVERSITY)
SENDA-MACHI, HIROSHIMA-SHI 730
JAPANESE, WITH EUROPEAN SUMMARY
INDEXED: CHE, INS, MAT, JICST

HIROSHIMA DAIGAKU KYOIKUGAKUBU KIYO, DAI-4-BU 5365
(BULLETIN OF THE FACULTY OF EDUCATION, HIROSHIMA UNIVERSITY, PART 4 ARTS & TECHNICS)
HIROSHIMA DAIGAKU KYOIKU GAKUBU (FACULTY OF EDUCATION, HIROSHIMA UNIVERSITY)
HIGASHISENDA-MACHI, HIROSHIMA-SHI 730
JAPANESE, WITH ENGLISH SUMMARY

HIROSHIMA DENKI DAIGAKU, HIROSHIMA JIDOSHA KOGYO TANKI DAIGAKU KENKYU HOKOKU 5366
(RESEARCH REPORTS OF THE HIROSHIMA COLLEGE OF ELECTRIC MACHINERY AND THE HIROSHIMA JUNIOR COLLEGE OF AUTOMOTIVE ENGINEERING)
595, NAKANO, SENOKAWA-MACHI, AKI-GUN, HIROSHIMA 739-03
JAPANESE, WITH ENGLISH SUMMARY

HIROSHIMA KOGYO DAIGAKU KENKYU KIYO 5367
(RESEARCH BULLETIN OF THE HIROSHIMA INSTITUTE OF TECHNOLOGY)
HIROSHIMA KOGYO DAIGAKU (HIROSHIMA INSTITUTE OF TECHNOLOGY)
MIYAKE, ITSUKAICHI-MACHI, SAEKI-GUN, HIROSHIMA 738
JAPANESE, WITH ENGLISH SUMMARY
INDEXED: JICST

HIROSHIMA-KEN CHUSHO KIGYO JOHO; KIKAI, KINZOKU, KAGAKU-HEN 5368
(BULLETIN OF THE HIROSHIMA PREFECTURE SMALL AND MEDIUM SIZED INDUSTRY; MACHINERY, METALS, CHEMISTRY EDITION)
HIROSHIMA-KEN GIJUTSU JOHO SHITSU (HIROSHIMA PREFECTURE TECHNICAL INFORMATION OFFICE)
JAPANESE

HIROSHIMA-KEN KOGYO GIJUTSU JOHO: KIKAI, KINZOKU, KAGAKU-HEN 5369
(HIROSHIMA PREFECTURE INDUSTRIAL TECHNOLOGY REPORT: MACHINERY, METALS, CHEMISTRY)
HIROSHIMA-KEN GIJUTSU JOHO SHITSU (HIROSHIMA PREFECTURE TECHNOLOGY INFORMATION DEPARTMENT)
JAPANESE

HIROSHIMA-KEN KOGYO GIJUTSU JOHO: KOGEI-HEN 5370
(HIROSHIMA PREFECTURE INDUSTRIAL TECHNOLOGY REPORT)
HIROSHIMA-KEN GIJUTSU JOHO SHITSU
JAPANESE

HIROSHIMA-KENRITSU KURE KOGYO SHIKENJO HOKOKU 5371
(REPORT OF HIROSHIMA PREFECTURAL INDUSTRIAL RESEARCH INSTITUTE, KURE)
HIROSHIMA-KENRITSU KURE KOGYO SHIKENJO (HIROSHIMA PREFECTURAL INDUSTRIAL RESEARCH INSTITUTE, KURE)
6-21, 3-CHOME, NISHICHUO, KURE-SHI 737
JAPANESE

HITACHI 5372
(HITACHI)
HITACHI HYORONSHA
5-1, 1-CHOME, MARUNOUCHI, CHIYODA-KU, TOKYO 100
JAPANESE
INDEXED: JICST

HITACHI HYORON 5373
(HITACHI REVIEW, JAPANESE EDITION)
HITACHI HYORONSHA
5-1, 1-CHOME, MARUNOUCHI, CHIYODA-KU, TOKYO 100
JAPANESE AND ENGLISH
INDEXED: COP, JICST

HITACHI REVIEW 5374
HITACHI SEISAKUSHO SENDEN-BU (ADVERTISING DEPARTMENT, HITACHI LIMITED)
6-2, 2-CHOME, OTE-MACHI, CHIYODA-KU, TOKYO 100
ENGLISH
INDEXED: CHE, ENG, INS, WAA, JICST

HIYOSHI KIYO 5375
(MEMOIRS OF FACULTY OF ENGINEERING, KEIO UNIVERSITY)
KEIO GIJUKU DAIGAKU KOGAKUBU
832, HIYOSHI-MACHI, KOHOKU-KU, YOKOHAMA-SHI 223
JAPANESE AND ENGLISH

HOKKAIDO DAIGAKU KOGAKUBU KENKYU HOKOKU 5376
(BULLETIN OF THE FACULTY OF ENGINEERING, HOKKAIDO UNIVERSITY)
HOKKAIDO DAIGAKU KOGAKUBU (THE FACULTY OF ENGINEERING, HOKKAIDO UNIVERSITY)
NISHI-8-CHOME, KITA-13-JO, KITA-KU, SAPPOROSHI 060
JAPANESE AND EUROPEAN
INDEXED: CHE, JICST

HOKKAIDO KAIHATSU-KYOKU GIJUTSU KENKYU HAPPYOKAI RONBUNSHU 5377
(REPORT OF THE MEETING, HOKKAIDO DEVELOPMENT BUREAU)
HOKKAIDO KAIHATSU-CHO HOKKAIDO KAIHATSU-KYOKU (HOKKAIDO DEVELOPMENT AGENCY, HOKKAIDO DEVELOPMENT BUREAU)
NISHI-4-CHOME, KITA-3-JO, CHUO-KU, SAPPORO-SHI 060
JAPANESE
INDEXED: JICST

HOKKAIDO KOGYO DAIGAKU KENKYU KIYO 5378
(MEMOIRS OF THE HOKKAIDO INSTITUTE OF TECHNOLOGY)
HOKKAIDO INSTITUTE OF TECHNOLOGY
JAPANESE

HOKKAIDO KOGYO KAIHATSU SHIKENJO HOKOKU 5379
(REPORT OF THE GOVERNMENT INDUSTRIAL DEVELOPMENT LABORATORY OF HOKKAIDO)
TSUSHO SANGYO-SHO KOGYO GIJUTSU-IN HOKKAIDO KOGYO KAIHATSU SHIKENJO (GOVERNMENT INDUSTRIAL DEVELOPMENT LABORATORY, HOKKAIDO, AGENCY OF INDUSTRIAL SCIENCE AND TECHNOLOGY)
41-2, HIGASHITSUKISAMU, TOYOHIRA-KU, SAPPORO-SHI 061-01
JAPANESE, WITH ENGLISH SUMMARY
INDEXED: CHE, WAA, JICST

HOKKAIDO KOGYO KAIHATSU SHIKENJO NENPO 5380
(ANNUAL REPORT OF GOVERNMENT INDUSTRIAL DEVELOPMENT LABORATORY, HOKKAIDO)
TSUSHO SANGYO-SHO KOGYO GIJUTSU-IN HOKKAIDO KOGYO KAIHATSU SHIKENJO (GOVERNMENT INDUSTRIAL DEVELOPMENT LABORATORY, HOKKAIDO, AGENCY OF INDUSTRIAL SCIENCE AND TECHNOLOGY)
41-2, HIGASHITSUKISAMU, TOYOHIRA-KU, SAPPORO-SHI 061-01
JAPANESE

HOKKAIDO KOGYO KENKYU SHIKENJO GIJUTSU SHIRYO 5381
(TECHNICAL DATA OF THE HOKKAIDO INDUSTRIAL RESEARCH LABORATORY)
KOGYO GIJUTSUIN HOKKAIDO KOGYO KAIHATSU KENKYUJO (INSTITUTE OF INDUSTRIAL TECHNOLOGY, HOKKAIDO RESEARCH INSTITUTE OF INDUSTRIAL DEVELOPMENT)
JAPANESE

HOKKAIDO SANGYO GEPPO 5382
(MONTHLY REPORT OF HOKKAIDO INDUSTRY)
HOKKAIDO SANGYO CHOSA KYOKAI
C/O KITA ICHIJO BUILDING, NISHI-5-CHOME, KITA-1-JO, CHUO-KU, SAPPORO-SHI, HOKKAIDO 060
JAPANESE
INDEXED: JICST

HOKKAIDORITSU KOGYO SHIKENJO HOKOKU 5383
(REPORTS OF THE HOKKAIDO INDUSTRIAL RESEARCH INSTITUTE)
HOKKAIDORITSU KOGYO SHIKENJO (HOKKAIDO INDUSTRIAL RESEARCH INSTITUTE)
95, NISHI-3-CHOME, HAKKEN-1-JO, NISHI-KU, SAPPORO-SHI 063
JAPANESE AND ENGLISH
INDEXED: CHE, JICST

HOKKAIDORITSU KOGYO SHIKENJO JIGYO HOKOKU 5384
(OPERATION REPORT OF THE HOKKAIDO INDUSTRIAL RESEARCH INSTITUTE)
HOKKAIDORITSU KOGYO SHIKENJO (HOKKAIDO INDUSTRIAL RESEARCH INSTITUTE)
95, NISHI-3-CHOME, HAKKEN-1-JO, NISHI-KU, SAPPORO-SHI 063
JAPANESE

HOKKAISHI NYUSU 5385
(NEWS OF THE GOVERNMENT INDUSTRIAL DEVELOPMENT LABORATORY, HOKKAIDO)
TSUSHO SANGYO-SHO KOGYO GIJUTSU-IN HOKKAIDO KOGYO KAIHATSU SHIKENJO (GOVERNMENT INDUSTRIAL DEVELOPMENT LABORATORY, HOKKAIDO)
41-2, HIGASHI TSUKISAMU, TOYOHIRA-KU, SAPPORO-SHI 061-01
JAPANESE
INDEXED: JICST

HOKKOSHI SHIRYO 5386
(LITERATURE OF DATA IN HOKKAIDO INDUSTRIAL RESEARCH INSTITUTE)
HOKKAIDORITSU KOGYO SHIKENJO
95, NISHI-3-CHOME,HAKKEN-1-JO, NISHI-KU, SAPPORO-SHI 063
JAPANESE AND ENGLISH

HOKUTO KOGYO TANKI DAIGAKU KENKYU KIYO 5387
(BULLETIN OF THE HOKUTO TECHNICAL JUNIOR COLLEGE)
HOKUTO TECHNICAL JUNIOR COLLEGE
JAPANESE

HONDA MEMORIAL SERIES ON MATERIALS SCIENCE 5388
HONDA MEMORIAL FOUNDATION
C/O AIZAWA BUILDING, 1-20-3, NIHONBASHI, CHUO-KU TOKYO
ENGLISH
INDEXED: CHE

HOSEI DAIGAKU KOGAKUBU KENKYU HOKOKU 5389
(REPORT OF THE COLLEGE OF ENGINEERING OF HOSEI UNIVERSITY)
HOSEI DAIGAKU KOGAKUBU KENKYU IINKAI (COMMITTEE OF RESEARCH REPORTS, COLLEGE OF ENGINEERING, HOSEI UNIVERSITY)
7-2, 3-CHOME, KAJINO-CHO, KOGANEI-SHI, TOKYO 184
ENGLISH
INDEXED: CHE, JICST

HOSEI DAIGAKU KOGAKUBU KENKYU SHUHO 5390
(BULLETIN OF THE COLLEGE OF ENGINEERING, HOSEI UNIVERSITY)
HOSEI DAIGAKU KOGAKUBU KENKYU IINKAI (COMMITTEE OF RESEARCH REPORTS, COLLEGE OF ENGINEERING, HOSEI UNIVERSITY)
7-2, 3-CHOME, KAJINO-CHO, KOGANEI-SHI, TOKYO 184
JAPANESE AND ENGLISH
INDEXED: ENG, JICST

HOSEI DAIGAKU KOGYO UOTO GAKKO KIYO 5391
(BULLETIN OF HOSEI UNIVERSITY INDUSTRIAL HIGH SCHOOL)
JAPANESE

HOSEI DAIGAKU TOSHOKANHO 5392
(REPORT OF HOSEI UNIVERSITY LIBRARY)
JAPANESE
INDEXED: JICST

HYOGO-KEN KOGYO GIJUTSU JOHO: SOGO-HEN 5393
(HYOGO PREFECTURE INDUSTRIAL TECHNOLOGY REPORT: GENERAL)
HYOGO-KENRITSU KOGYO SHIKENJO (HYOGO-KENRITSU PREFECTURAL INDUSTRIAL EXPERIMENT STATION)
JAPANESE

HYOGO-KEN KOGYO GIJUTSU JOHO: ZAKKA-HEN 5394
(HYOGO PREFECTURE INDUSTRIAL TECHNOLOGY REPORT: MISCELLANEOUS GOODS)
HYOGO KENRITSU NOGYO SHIKENJO (HYOGO PREFECTURAL INDUSTRIAL EXPERIMENT STATION)
JAPANESE

HYOGO-KENRITSU KOGYO SHIKENJO GYOMU HOKOKUSHO NENPO 5395
(OPERATION REPORT OF INDUSTRIAL RESEARCH INSTITUTE, HYOGO PREFECTURE)
HYOGO-KENRITSU KOGYO SHIKENJO (INDUSTRIAL RESEARCH INSTITUTE, HYOGO PREFECTURE)
3-1, YUKIHIRA-CHO, SUMA-KU, KOBE-SHI 654
JAPANESE

HYOGO-KENRITSU KOGYO SHIKENJO KENKYU HOKOKU 5396
(REPORTS OF INDUSTRIAL RESEARCH INSTITUTE, HYOGO PREFECTURE)
HYOGO-KENRITSU KOGYO SHIKENJO (INDUSTRIAL RESEARCH INSTITUTE, HYOGO PREFECTURE)
3-1, YUKIHIRA-CHO, SUMA-KU, KOBE-SHI 654
JAPANESE, WITH ENGLISH SUMMARY

HYOJUNKA JANARU 5397
(JOURNAL OF STANDARDIZATION)
NIHON KIKAKU KYOKAI (JAPANESE STANDARDS ASSOCIATION)
1-24, 4-CHOME, AKASAKA, MINATO-KU, TOKYO 107
JAPANESE

HYOJUNKA TO HINSHITSU KANRI 5398
(STANDARDIZATION AND QUALITY CONTROL)
NIHON KIKAKU KYOKAI (JAPANESE STANDARDS ASSOCIATION)
1-24, 4-CHOME, AKASAKA, MINATO-KU, TOKYO 107
JAPANESE
INDEXED: JICST

IBARAKI DAIGAKU KOGAKUBU KENKYU SHUHO 5399
(JOURNAL OF THE FACULTY OF ENGINEERING, IBARAKI UNIVERSITY)
IBARAKI DAIGAKU KOGAKUBU (FACULTY OF ENGINEERING, IBARAKI UNIVERSITY)
12-1, 4-CHOME, NAKANARUSAWA-CHO, HITACHI-SHI 316
JAPANESE AND ENGLISH
INDEXED: CHE, INS, JICST

IBARAKI DAIGAKU KOGAKUBU KIYO 5400
(BULLETIN OF THE FACULTY OF ENGINEERING, IBARAKI UNIVERSITY)
IBARAKI DAIGAKU KOGAKUBU (FACULTY OF ENGINEERING, IBARAKI UNIVERSITY)
12-1, 4-CHOME, NAKANARUSAWA-CHO, HITACHI-SHI 316
JAPANESE, WITH EUROPEAN SUMMARY
INDEXED: CHE, INS, JICST

IBARAKI KOGYO KOTO SENMON GAKKO KENKYU IHO 5401
(RESEARCH REPORTS OF IBARAKI TECHNICAL COLLEGE)
IBARAKI KOGYO KOTO SENMON GAKKO (IBARAKI TECHNICAL COLLEGE)
866, FUKAYATSU, NAKANE, KATSUTA-SHI 312
JAPANESE

IBARAKI NO KOGYO (KOGYO TOKEI CHOSA KEKKA HOKOKUSHO 5402
(CENSUS OF MANUFACTURES, IBARAKI PREFECTURE)
IBARAKI-KEN SOGO KAIHATSU-BU (GENERAL DEVELOPMENT DIVISION, IBARAKI PREFECTURAL GOVERNMENT)
5-38, 1-CHOME, SANNOMARU, MITO-SHI 310
JAPANESE

IBARAKI-KEN KOGAI GIJUTSU SENTA HOCHO SHIRYO 5403
(IBARAKI PREFECTURE ENVIRONMENTAL POLLUTION RESEARCH CENTER RADIATION SURVEY DATA)
IBARAKI-KEN KOGAI GIJUTSU SENTA
4040-36, 1-CHOME, ISHIKAWA, MITO 310
JAPANESE
INDEXED: CHE

IBARAKI-KEN KOGYO SHIKENJO NENPO 5404
(REPORT OF INDUSTRIAL RESEARCH INSTITUTE OF IBARAKI PREFECTURE)
IBARAKI-KEN KOGYO SHIKENJO (INDUSTRIAL RESEARCH INSTITUTE OF IBARAKI PREFECTURE)
NAGAOKA, IBARAKI-MACHI, HIGASHIBARAKI-GUN, IBARAKI 311-31
JAPANESE

ICHINOSEKI KOGYO KOTO SENMON GAKKO KENKYU KIYO 5405
(RESEARCH REPORTS OF ICHINOSEKI TECHNICAL COLLEGE)
ICHINOSEKI KOGYO KOTO SENMON GAKKO (ICHINOSEKI TECHNICAL COLLEGE)
TAKANASHI, HAGISHO, ICHINOSEKI-SHI 021
JAPANESE AND ENGLISH
INDEXED: JICST

IDEA CENTER NEWS 5406
AIDEIA SENTA (IDEA CENTER)
JAPANESE AND ENGLISH

IE (THE MAGAZINE OF INDUSTRIAL ENGINEERING) 5407
NIHON NORITSU KYOKAI (JAPAN MANAGEMENT ASSOCIATION)
1-22, 3-CHOME, SHIBA-KOEN, MINATO-KU, TOKYO 105
JAPANESE
INDEXED: JICST

IE REVIEW 5408
NIHON INDASUTORIARU ENJINIARINGU KYOKAI (JAPAN INSTITUTE OF INDUSTRIAL ENGINEERING)
C/O SAN'YO KOKUSAKU BUILDING, 7-8, 1-CHOME, SHIBUYA, SHIBUYA-KU, TOKYO 150
JAPANESE
INDEXED: JICST

IHI ENGINEERING REVIEW 5409
ISHIKAWAJIMA HARIMA JUKOGYO K. K. (ISHIKAWAJIMA HARIMA HEAVY INDUSTRIES COMPANY, LIMITED)
2-16, 3-CHOME, TOYOSU, KOTO-KU, TOKYO 135
ENGLISH
INDEXED: CER, CHE, IAA, WAA, JICST

IKUEI KOGYO KOTO SENMON GAKKO KENKYU KIYO 5410
(BULLETIN OF IKUEI TECHNICAL COLLEGE)
IKUEI KOGYO KOTO SENMON GAKKO (IKUEI TECHNICAL COLLEGE)
35-11, 2-CHOME, IGUSA, SUGINAMI-KU, TOKYO 167
JAPANESE
INDEXED: CHE, JICST

IKUTOKU KOGYO DAIGAKU KENKYU HOKOKU B ERIKOGAKU-HEN 5411
(RESEARCH REPORT OF IKUTOKU INDUSTRIAL COLLEGE B SCIENCE AND ENGINEERING EDITION)
IKUTOKU KOGYO DAIGAKU (IKUTOKU INDUSTRIAL COLLEGE)
JAPANESE

IKUTOKU KOGYO KOTO SENMON GAKKO KIYO 5412
(MEMOIRS OF IKUTOKU TECHNICAL COLLEGE)
IKUTOKU KOGYO KOTO SENMON GAKKO (IKUTOKU TECHNICAL COLLEGE)
SHIMO OGINO, ATSUGI-SHI 243-02
JAPANESE, WITH ENGLISH SUMMARY

INDUSTRIA 5413
DAIAMONDO RIDOSHA (DIAMOND LEAD COMPANY, LTD.)
4-2, 1-CHOME KASUMIGASEKI, CHIYODA-KU, TOKYO 100
ENGLISH

INDUSTRIAL NEWS REPORT 5414
KOGYO JIJI TSUSHINSHA (THE INDUSTRIAL NEWS)
3-10, KANDA OGAWA-CHO, CHIYODA-KU, TOKYO 101
ENGLISH

INDUSTRIAL REVIEW OF JAPAN 5415
NIHON KEIZAI SHINBUNSHA (THE JAPAN ECONOMIC JOURNAL)
9-5, 1-CHOME, OTE-MACHI, CHIYODA-KU, TOKYO 100
ENGLISH
INDEXED: BIO

INDUSTRIES OF JAPAN 5416
MAINICHI SHINBUNSHA EIBUN MAINICHI SHUPPAN HENSHU-BU (MAINICHI DAILY NEWS PUBLICATIONS DEPARTMENT, MAINICHI NEWSPAPERS)
1-1, 1-CHOME, HITOTSUBASHI, CHIYODA-KU, TOKYO 100
ENGLISH

INFOMESYON PURAZA 5417
(INFORMATION PLAZA)
KINDAI JOHO BUNKEN KENKYUKAI
JAPANESE

IRYO JOHO SHISUTEMU KOKUSAI SHINPOJIUMU 5418
(INTERNATIONAL SYMPOSIUM ON MEDICAL INFORMATION SYSTEM)
JAPANESE
INDEXED: JICST

ISHIKAWA KOGYO KOTO SENMON GAKKO KIYO 5419
(MEMOIRS OF ISHIKAWA TECHNICAL COLLEGE)
ISHIKAWA KOGYO KOTO SENMON GAKKO (ISHIKAWA TECHNICAL COLLEGE)
KITA NAKAJO, TSUBATA-MACHI, KOHOKU-GUN, ISHIKAWA 929-03
JAPANESE AND ENGLISH
INDEXED: JICST

ISHIKAWA KOSHI GIJUTSU JOHO 5420
(ISHIKAWA INDUSTRIAL EXPERIMENT STATION INFORMATION)
ISHIKAWA-KEN KOGYO SHIKENJO (ISHIKAWA INDUSTRIAL EXPERIMENT STATION)
JAPANESE

ISHIKAWA-KEN KOGYO KYOIKU 5421
(INDUSTRIAL EDUCATION IN ISHIKAWA PREFECTURE)
ISHIKAWA-KEN KOGYO KYOIKU KENKYUKAI (INDUSTRIAL EDUCATION RESEARCH ASSOCIATION IN ISHIKAWA PREFECTURE)
C/O ISHIKAWA-KENRITSU KANAZAWA SHORYO KOGYO KOKO, YOSHIWARA-CHO, KANAZAWA-SHI 920-01
JAPANESE

ISHIKAWA-KEN KOGYO SHIKENJO HOKOKU 5422
(REPORT OF INDUSTRIAL RESEARCH INSTITUTE, ISHIKAWA)
ISHIKAWA-KEN KOGYO SHIKENJO (INDUSTRIAL RESEARCH INSTITUTE, ISHIKAWA)
4-133, YONEIZUMI-CHO, KANAZAWA-SHI 921
JAPANESE
INDEXED: JICST

ISHIKAWA-KEN KOGYO SHIKENJO NENPO 5423
(ANNUAL REPORT OF THE INDUSTRIAL RESEARCH INSTITUTE, ISHIKAWA PREFECTURE)
ISHIKAWA-KEN KOGYO SHIKENJO (INDUSTRIAL RESEARCH INSTITUTE, ISHIKAWA)
4-133, YONEIZUMI-CHO, KANAZAWA-SHI 921
JAPANESE

ISHIKAWAJIMA HARIMA GIHO 5424
(ISHIKAWAJIMA-HARIMA ENGINEERING REVIEW)
ISHIKAWAJIMA-HARIMA JUKOGYO K. K. GIJUTSU-BU (TECHNICAL MANAGEMENT DEPARTMENT, ISHIKAWAJIMA-HARIMA HEAVY INDUSTRIES COMPANY, LTD.)
2-16, 3-CHOME, TOYOSU, KOTO-KU, TOKYO 135
JAPANESE, WITH ENGLISH SUMMARY
INDEXED: JICST

ISHIKAWAJIMA HARIMA GIHO BESSATSU 5425
(ISHIKAWAJIMA-HARIMA ENGINEERING REVIEW, SPECIAL ISSUE)
ISHIKAWAJIMA-HARIMA JUKOGYO K.K. GIJUTSU-BU (TECHNICAL MANAGEMENT DEPARTMENT, ISHIKAWAJIMA-HARIMA HEAVY INDUSTRIES COMPANY, LTD.)
2-16, 3-CHOME, TOYOSU, KOTO-KU, TOKYO 135
JAPANESE, WITH ENGLISH SUMMARY

IWATE DAIGAKU KOGAKUBU KENKYU HOKOKU 5426
(REPORT ON TECHNOLOGY OF IWATE UNIVERSITY)
IWATE DAIGAKU KOGAKUBU (FACULTY OF ENGINEERING, IWATE UNIVERSITY)
3-5, 4-CHOME, UEDA, MORIOKA-SHI 020-01
JAPANESE, WITH ENGLISH SUMMARY
INDEXED: CHE, JICST

IWATE-KEN KOGYO SHIKENJO GYOMU HOKOKU 5427
(OPERATION REPORT OF INDUSTRIAL RESEARCH INSTITUTE, IWATE PREFECTURE)
IWATE-KEN KOGYO SHIKENJO
TSUSHIDA, TONAN-MURA, SHIBA-GUN, IWATE 020
JAPANESE

IWATE-KEN KOGYO SHIKENJO KENKYU HOKOKU 5428
(REPORT OF INDUSTRIAL RESEARCH INSTITUTE, IWATE PREFECTURE)
IWATE-KEN KOGYO SHIKENJO
TSUSHIDA, TONAN-MURA, SHIBA-GUN, IWATE 020
JAPANESE

IWATE-KEN NO KOGYO 5429
(CENSUS OF MANUFACTURES, IWATE PREFECTURE)
IWATE-KEN SHOKO RODO-BU (IWATE PREFECTURAL GOVERNMENT)
10-1, UCHIMARU, MORIOKA-SHI 020
JAPANESE

J.E.T.R.O. JAPAN'S INDUSTRIAL AND TECHNICAL INFORMATION 5430
J.E.T.R.O.
ENGLISH

J.S.W. TECHNICAL REVIEW 5431
ENGLISH

JAPAN SEWING MACHINE TRADE NEWS 5432
ENGLISH

JIDO KENSA 5433
(LABORATORY AUTOMATION)
KYOWA KIKAKU TSUSHIN (KYOWA PLANNING NEWS)
JAPANESE

JIDOKA GIJUTSU 5434
(AUTOMATION TECHNIQUE)
KOGYO CHOSAKAI (JAPAN INDUSTRIAL RESEARCH ASSOCIATION)
14-7, 2-CHOME, HONGO, BUNKYO-KU, TOKYO 113
JAPANESE
INDEXED: JICST

JITA NYUSU 5435
(JITA NEWS)
NIHON SANGYO GIJUTSU SHINKO KYOKAI (JAPAN INDUSTRIAL TECHNOLOGY ASSOCIATION)
C/O MORI BUILDING, 2-7, NISHI SHINBASHI, MINATO-KU, TOKYO 105
JAPANESE
INDEXED: CHE

JOHO KANRI 5436
(INFORMATION AND DOCUMENTATION)
NIHON KAGAKU GIJUTSU JOHO SENTA (THE JAPAN INFORMATION CENTER OF SCIENCE AND TECHNOLOGY)
5-2, 2-CHOME, NAGATA-CHO, CHIYODA-KU, TOKYO 100
JAPANESE
INDEXED: CHE, JICST

JOURNAL OF THE FACULTY OF ENGINEERING, UNIVERSITY OF TOKYO, SERIES B 5437
TOKYO DAIGAKU KOGAKUBU (THE FACULTY OF ENGINEERING, UNIVERSITY OF TOKYO)
3-1, 7-CHOME, HONGO, BUNKYO-KU, TOKYO 113
EUROPEAN
INDEXED: CHE, COP, ENG, MAT, MET, JICST

JOURNAL OF THE FACULTY OF TEXTILE SCIENCE AND TECHNOLOGY, SHINSHU UNIVERSITY, SERIES B, ENGINEERING 5438
SHINSHU DAIGAKU SEN'I GAKUBU (THE FACULTY OF TEXTILE SCIENCE AND TECHNOLOGY, SHINSHU UNIVERSITY)
15-1, 3-CHOME, TOKIDA, UEDA-SHI 386
EUROPEAN
INDEXED: CHE, JICST

JOURNAL OF HUMAN ERGOLOGY 5439
JINRUI DOTAIGAKU KENKYUKAI (THE HUMAN ERGOLOGY RESEARCH ASSOCIATION)
4-16, 2-CHOME, YAYOI, BUNKYO-KU, TOKYO 113
ENGLISH
INDEXED: BIO, EXC

JRI DETA 5440
(JRI DATA)
NIHON SOGO KENKYUSHO (GENERAL RESEARCH ASSOCIATION)
JAPANESE AND ENGLISH

JSQC NYUZU 5441
(JSQC NEWS)
JAPANESE
INDEXED: JICST

JUKOGYO KOHO 5442
(JAPAN HEAVY INDUSTRY NEWS)
NIHON JUKOGYO KENKYUKAI (JAPAN HEAVY INDUSTRY RESEARCH ASSOCIATION)
2-4, KYOBASHI, CHUO-KU, TOKYO 104
JAPANESE

JUKOGYO SHINBUN 5443
(HEAVY INDUSTRIAL NEWS)
JUKOGYO SHINBUNSHA
4-5, NIHONBASHI HON-CHO, CHUO-KU, TOKYO 103
JAPANESE

KAGAKU GIJUTSU-CHO SHIGEN CHOSAJO KIHO 5444
(QUARTERLY REPORT OF NATIONAL INSTITUTE OF RESOURCES)
JAPANESE
INDEXED: JICST

KAGAWA-KEN NO KOGYO 5445
(CENSUS OF MANUFACTURES, KAGAWA PREFECTURE)
KAGA-KEN KIKAKU-BU (PLANNING DIVISION, KAGAWA PREFECTURAL GOVERNMENT)
1-10, 4-CHOME, BAN-CHO, TAKAMATSU-SHI 760
JAPANESE

KAGOSHIMA DAIGAKU KOGAKUBU KENKYU HOKOKU 5446
(RESEARCH REPORTS, FACULTY OF ENGINEERING, KAGOSHIMA UNIVERSITY)
KAGOSHIMA DAIGAKU KOGAKUBU (THE FACULTY OF ENGINEERING, KAGOSHIMA UNIVERSITY)
52, KAMOIKE-CHO, KAGOSHIMA-SHI 890
JAPANESE, WITH EUROPEAN SUMMARY
INDEXED: CHE, JICST

KAGOSHIMA DAIGAKU KOGAKUBU KIYO 5447
(MEMOIRS OF THE FACULTY OF ENGINEERING, KAGOSHIMA UNIVERSITY)
KAGOSHIMA DAIGAKU KOGAKUBU (FACULTY OF ENGINEERING KAGOSHIMA UNIVERSITY)
52, KAMOIKE-CHO, KAGOSHIMA-SHI 890
JAPANESE, WITH EUROPEAN SUMMARY
INDEXED: CHE, JICST

KAGOSHIMA KOGYO KOTO SENMON GAKKO KENKYU HOKOKU 5448
(RESEARCH REPORT OF KAGOSHIMA TECHNICAL COLLEGE)
KAGOSHIMA KOGYO KOTO SENMON GAKKO (KAGOSHIMA TECHNICAL COLLEGE)
SHINKO, HAYATO-CHO, AIRA-GUN, KAGOSHIMA 899-51
JAPANESE AND ENGLISH

KAGOSHIMA-KEN KOGYO SHIKENJO GYOMU HOKOKUSHO 5449
(REPORT OF THE TECHNICAL RESEARCH LABORATORY OF KAGOSHIMA PREFECTURE)
KAGOSHIMA-KEN KOGYO SHIKENJO (TECHNICAL RESEARCH LABORATORY OF KAGOSHIMA PREFECTURE)
100, TAKE-MACHI, KAGOSHIMA-SHI 890
JAPANESE
INDEXED: JICST

KAGOSHIMA-KEN KOGYO SHIKENJO NENPO 5450
(ANNUAL REPORT OF THE INDUSTRIAL RESEARCH INSTITUTE, KAGOSHIMA PREFECTURE)
KAGOSHIMA-KEN KOGYO SHIKENJO
100, TAKE-MACHI, KAGOSHIME-SHI 890
JAPANESE

KAGOSHIMA-KEN NO KOGYO 5451
(CENSUS OF MANUFACTURES, KAGOSHIMA PREFECTURE)
KAGOSHIMA-KEN KIKAKU-BU (PLANNING DIVISION, KAGOSHIMA PREFECTURAL GOVERNMENT)
14-50, YAMASHITA-CHO KAGOSHIMA-SHI 890-91
JAPANESE

KAIGAI GIJUTSU HAIRAITO 5452
(TECHNICAL HIGHLIGHTS)
NIHON KAGAKU GIJUTSU JOHO SENTA (JAPAN INFORMATION CENTER OF SCIENCE AND TECHNOLOGY)
5-2, 2-CHOME, NAGATA-CHO, CHIYODA-KU, TOKYO 100
JAPANESE
INDEXED: JICST

KAIGAI GIJUTSU JOHO 5453
(OVERSEAS TECHNICAL INFORMATION)
SHIGENKA GIJUTSA KYOKAI
C/O DAI-9 MORI BUILDING, 1-3, SHIBA ATAGO-CHO, MINATO-KU, TOKYO 105
JAPANESE

KAIGAI GIJUTSU KYORYOKU 5454
(OVERSEAS TECHNICAL COOPERATION)
KAIGAI GIJUTSU KYOROKU JIGYODAN (OVERSEAS TECHNICAL COOPERATION AGENCY)
C/O KEIZAI KYORYOKU SENTA BUILDING, 42, ICHIGAYA HONMURA-CHO, SHINJUKU-KU, TOKYO 160
JAPANESE
INDEXED: JICST

KAIGAI JOHO INDEKKUSU 5455
(OVERSEAS TECHNICAL INFORMATION INDEX)
NIHON JOHO SHORI KAIHATSU SENTA (INFORMATION PROCESSING DEVELOPMENT CENTER)
C/O KIKAI SHINKO KAIKAN, 5-8, 3-CHOME, SHIBA KOEN, MINATO-KU, TOKYO 105
JAPANESE

KAIGAI KIGYO TO GIJUTSU 5456
(OVERSEAS ENTERPRISES AND TECHNOLOGY)
NIHON SHOKO KAIGISHO KAIGAI KIGYO GIJUTSU KYORYOKU ASSEN HONBU (THE JAPAN ECONOMIC CORPORATION CENTER OF THE JAPAN CHAMBER OF COMMERCE AND INDUSTRY)
C/O SEKAI BOEKI SENTA BUILDING, 505, 4-1, 2-CHOME, HAMAMATSU-CHO, MINATO-KU, TOKYO 105
JAPANESE

KAIGAI SANGYO NENPO 5457
(ANNUAL REPORT OF OVERSEA INDUSTRIES)
KOKUMIN KEIZAI KENKYU KYOKAI (RESEARCH INSTITUTE OF NATIONAL ECONOMY)
C/O GINZA CHUO BUILDING, 10-2, 5-CHOME, GINZA, CHUO-KU, TOKYO 104
JAPANESE

KAIGAI SHIGEN JOHO 5458
(OVERSEAS RESOURCES INFORMATION)
NIHON BOEKI SHINKOKAI (JAPAN EXTERNAL TRADE ORGANIZATION)
2, AKASAKA AOI-CHO, MINATO-KU, TOKYO 107
JAPANESE

KAIJO HOAN DAIGAKKO KENKYU HOKOKU, DAI-2-BU 5459
(REPORTS JAPAN MARITIME SAFETY ACADEMY, SECTION 2)
KAIJO HOAN-CHO KAIJO HOAN DAIGAKKO (MARITIME SAFETY ACADEMY, MARITIME SAFETY AGENCY)
1, WAKABA-CHO, KURE-SHI 737
JAPANESE AND ENGLISH
INDEXED: CHE, JICST

KANAGAWA DAIGAKU KOGAKUBU KENKYU HOKOKU 5460
(REPORTS OF THE FACULTY OF TECHNOLOGY, KANAGAWA UNIVERSITY)
KANAGAWA DAIGAKU (KANAGAWA UNIVERSITY)
3, ROKKAKUBASHI, KANAGAWA-KU, YOKOHAMA-SHI 221
JAPANESE AND EUROPEAN
INDEXED: CHE, JICST

KANAGAWA-KEN KOGYO SHIKENJO GYOMU HOKOKU 5461
(BUSINESS REPORTS, INDUSTRIAL RESEARCH INSTITUTE OF KANAGAWA PREFECTURE)
KANAGAWA-KEN KOGYO SHIKENJO (INDUSTRIAL RESEARCH INSTITUTE OF KANAGAWA PREFECTURE)
3173, SHOWA-MACHI, KANAZAWA-KU, YOKOHAMA-SHI 236
JAPANESE

KANAGAWA-KEN KOGYO SHIKENJO KENKYU HOKOKU 5462
(BULLETIN OF THE INDUSTRIAL RESEARCH INSTITUTE OF KANAGAWA PREFECTURE)
KANAGAWA-KEN KOGYO SHIKENJO (THE INDUSTRIAL RESEARCH INSTITUTE OF KANAGAWA PREFECTURE)
3173, SHOWA-MACHI, KANAZAWA-KU, YOKOHAMASHI 236
JAPANESE
INDEXED: CHE, JICST

KANAGAWA-KEN KOGYO SHIKENJO NEWS 5463
(KANAGAWA PREFECTURE INDUSTRIAL EXPERIMENT STATION NEWS)
JAPANESE

KANAGAWA-KEN KOSHI KYOKAI GIJUTSU NYUSU 5464
(KANAGAWA PREFECTURAL INDUSTRIAL EXPERIMENT ASSOCIATION TECHNICAL NEWS)
KANAGAWA-KEN KOSHI KYOKAI (KANAGAWA PREFECTURAL INDUSTRIAL EXPERIMENT ASSOCIATION)
C/O KANAGAWA-KEN KOGYO SHIKENJO, 3173, SHOWA-MACHI, KANAZAWA-KU, YOKOHAMA-SHI 236
JAPANESE

KANAZAWA DAIGAKU KOGAKUBU KIYO 5465
(MEMOIRS OF THE FACULTY OF TECHNOLOGY, KANAZAWA UNIVERSITY)
KANAZAWA DAIGAKU KOGAKUBU (FACULTY OF TECHNOLOGY, KANAZAWA UNIVERSITY)
40-20, 2-CHOME, KODATSUNO, KANAZAWA-SHI 920
JAPANESE AND EUROPEAN
INDEXED: CHE, ENG, INS, JICST

KANAZAWA KOGYO DAIGAKU KENKYU KIYO 5466
(MEMOIRS OF KANAZAWA INSTITUTE OF TECHNOLOGY)
KANAZAWA KOGYO DAIGAKU (KANAZAWA INSTITUTE OF TECHNOLOGY)
7-1, OGIGAOKA, NONOICHI-MACHI, NANKO, KANAZAWA-SHI 921
JAPANESE, WITH ENGLISH SUMMARY
INDEXED: CHE

KANAZAWA KOGYO DAIGAKU KENKYU KIYO B 5467
(RESEARCH BULLETIN OF KANAZAWA INDUSTRIAL COLLEGE)
KANAZAWA INDUSTRIAL COLLEGE
JAPANESE

KANAZAWA-SHI NO KOGYO, KOGYO TOKEI CHOSA KEKKA 5468
(CENSUS OF MANUFACTURES, KANAZAWA CITY)
KANAZAWA-SHI KIKAKU-SHITSU (PLANNING AGENCY, KANAZAWA MUNICIPAL GOVERNMENT)
1-1, 1-CHOME, HIROSAKA, KANAZAWA-SHI 920
JAPANESE

KANKYO BUNKA KENKYUSHO SHIRYO JOHO 5469
(INFORMATION OF THE RESEARCH INSTITUTE FOR ENVIRONMENTAL DEVELOPMENT)
KANKYO BUNKA KENKYUSHO (RESEARCH INSTITUTE FOR ENVIRONMENTAL DEVELOPMENT)
23-5, 3-CHOME, NISHI SHINBASHI, MINATO-KU, TOKYO 105
JAPANESE

KANKYO KAGAKU RAIBURARI 5470
(ENVIRONMENTAL SCIENCE LIBRARY)
DAINIPPON TOSHO K. K.
1-9-10, GINZA, CHUO-KU TOKYO
JAPANESE
INDEXED: CHE

KANSAI BUNKEN SENTA NYUSU 5471
(KANSAI LITERATURE CENTER NEWS)
KANSAI BUNKEN SENTA KYOGIKAI
C/O OSAKA-FURITSU YUHIGAOKA TOSHOKAN, 21-1, REIJIN-CHO, TENNOJI-KU, OSAKA-SHI 543
JAPANESE

KANSAI DAIGAKU KOGAKUBU KENKYU HOKOKU 5472
(MEMOIRS OF FACULTY OF TECHNOLOGY, KANSAI UNIVERSITY)
KANSAI DAIGAKU KOGAKUBU
3-35, 3-CHOME, YAMATE-CHO, SUITA-SHI 564
ENGLISH

KANSEI GAKUIN DAIGAKU SANKEN RONSHU 5473
(REVIEW OF ECONOMICS AND BUSINESS MANAGEMENT)
KANSEI GAKUIN DAIGAKU SANGYO KENKYUSHO (INSTITUTE OF RESEARCH FOR INDUSTRY KANSEI GAKUIN UNIVERSITY)
UEGAHARA, NISHINOMIYA-SHI 662
JAPANESE

KANTO GAKUIN DAIGAKU KOGAKKAI KAIHO 5474
BULLETIN OF ENGINEERING ASSOCIATION, KANTO GAKUIN UNIVERSITY
KANTO GAKUIN DAIGAKU KOGAKUBU (THE FACULTY OF ENGINEERING, KANTO GAKUIN UNIVERSITY)
4834, MUTSUURA-MACHI, KANAZAWA-KU, YOKOHAMA-SHI 236
JAPANESE

KANTO GAKUIN DAIGAKU KOGAKUBU KENKYU HOKOKU 5475
(JOURNAL OF TECHNOLOGICAL RESEARCHES, THE COLLEGE OF ENGINEERING, KANTO GAKUIN UNIVERSITY)
KANTO GAKUIN DAIGAKU KOGAKUBU (THE FACULTY OF ENGINEERING, KANTO GAKUIN UNIVERSITY)
4834, MUTSUURA-CHO, KANAZAWA-KU, YOKOHAMASHI 236
JAPANESE AND ENGLISH
INDEXED: CHE, JICST

KANTO KOGYO KYOIKU KYOKAI SHIRYO 5476
(BULLETIN OF THE KANTO SOCIETY FOR ENGINEERING EDUCATION)
KANTO KOGYO KYOIKU KYOKAI (KANTO SOCIETY FOR ENGINEERING EDUCATION)
19-10, 2-CHOME, SHINBASHI, MINATO-KU, TOKYO 105
JAPANESE AND ENGLISH

KAWASAKI GIHO 5477
(KAWASAKI TECHNICAL REVIEW)
KAWASAKI JUKOGYO K. K. (KAWASAKI HEAVY INDUSTRIES, LIMITED)
1-1, KAWASAKI-CHO, AKASHI-SHI 673
JAPANESE, WITH ENGLISH SUMMARY
INDEXED: CHE, COP, JICST

KAWASAKI-SHI NO KOGYO 5478
(CENSUS OF MANUFACTURE, KAWASAKI CITY)
KAWASAKI-SHI SOMU-KYOKU (ADMINISTRATIVE BUREAUS, KAWASAKI MUNICIPAL GOVERNMENT)
1, MIYAMOTO-MACHI, KAWASAKI-KU, KAWASAKI-SHI 210
JAPANESE

KEIO ENGINEERING REPORTS 5479
KEIO GIJUKU DAIGAKU KOGAKUBU (FACULTY OF ENGINEERING, KEIO UNIVERSITY)
832, HIYOSHI-CHO, KOHOKU-KU, YOKOHAMA-SHI 223
ENGLISH
INDEXED: CER, ENG, MAT, JICST

KEIO GIKUKU DAIGAKU KOGAKUBU KENKYU HOKOKU BESSATSU 5480
(RESEARCH REPORT SUPPLEMENT, KEIO UNIVERSITY, SCHOOL OF ENGINEERING)
JAPANESE
INDEXED: JICST

KENKYU GYOSEKI ICHIRAN (TOHOKU KOGYO DAIGAKU) 5481
(LIST OF ACADEMIC ACHIEVEMENTS, TOHOKU INSTITUTE OF TECHNOLOGY)
TOHOKU KOGYO DAIGAKU TOSHOKAN (THE LIBRARY, TOHOKU INSTITUTE OF TECHNOLOGY)
19, NAGAMACHI KOEJI, SENDAI-SHI 982
JAPANESE

KENKYU HOKOKU, MISHIMAGAKUEN JOSHI TANKI DAIGAKU TO JOSHI DAIGAKU 5482
(REPORTS ON THE RESEARCH AT MISHIMAGAKUEN WOMEN'S JUNIOR COLLEGE AND SENIOR COLLEGE)
MISHIMAGAKUEN JOSHI TANKI DAIGAKU TO JOSHI DAIGAKU
3 SHIMIZU-KOJI, SENDAI
JAPANESE
INDEXED: CHE

KENKYU KAIHATSU OYOBI GIJUTSU KORYU NI KANSURU CHOSA HOKOKUSHO 5483
(RESEARCH REPORT ON TECHNICAL INTERCHANGE AND RESEARCH DEVELOPMENT)
TSUSHO SANGYO-SHO KOGYO GIJUTSU-IN (AGENCY OF INDUSTRIAL SCIENCE AND TECHNOLOGY, MINISTRY OF INTERNATIONAL TRADE & INDUSTRY)
3-1, 1-CHOME, KASUMIGASEKI, CHIYODA-KU, TOKYO 100
JAPANESE

KENKYU SEIKA (KOGYO GIJUTSU-IN) 5484
(ABSTRACTS OF RESEARCH REPORT, AGENCY OF INDUSTRIAL SCIENCE AND TECHNOLOGY)
TSUSHO SANGYO-SHO KOGYO GIJUTSU-IN (AGENCY OF INDUSTRIAL SCIENCE AND TECHNOLOGY, MINISTRY OF INTERNATIONAL TRADE & INDUSTRY)
3-1, 1-CHOME, KASUMIGASEKI, CHIYODA-KU, TOKYO 100
JAPANESE

KENSHU 5485
(NEWS OF THE ASSOCIATION FOR OVERSEAS TECHNICAL SCHOLARSHIP)
KAIGAI GIJUTSUSHA KENSHU KYOKAI (ASSOCIATION FOR OVERSEAS TECHNICAL SCHOLARSHIP
12-13, 2-CHOME, HONKONMAGOME, BUNKYO-KU, TOKYO 113
JAPANESE

KENSHU: TRAINEES QUARTERLY (ASSOCIATION FOR OVERSEAS TECHNICAL SCHOLARSHIP 5486
KAIGAI GIJUTSUSHA KENSHU KYOKAI (ASSOCIATION FOR OVERSEAS TECHNICAL SCHOLARSHIP)
12-13, 2-CHOME, HONKOMAGOME, BUNKYO-KU, TOKYO 113
JAPANESE AND ENGLISH

KIGYO SEIHIN DOKO TOKEI 5487
(STATISTICS OF ENTERPRISE PRODUCTS)
NIHON NORITSU KYOKAI SANGYO KENKYUSHO
NIHON NORITSU KYOKAI (JAPAN MANAGEMENT ASSOCIATION)
1-22, 3-CHOME, SHIBAKOEN, MINATO-KU, TOKYO 105
JAPANESE

KINKI DAIGAKU DAI-2 (KYUSHU) KOGAKUBU KYOYO RONSHU 5488
(KINKI REVIEW OF ARTS AND SCIENCES)
KINKI DAIGAKU DAI-2 KYUSHU KOGAKUBU (THE 2ND FACULTY OF TECHNOLOGY, KINKI UNIVERSITY)
11-6, KASHINOMORI, IIZUKA-SHI 820
JAPANESE AND ENGLISH

KINKI DAIGAKU KOGAKU-BU KENKYU HOKOKU 5489
(RESEARCH REPORTS OF THE FACULTY OF ENGINEERING, KINKI UNIVERSITY)
KINKI DAIGAKU KOGAKUBU (THE FACULTY OF ENGINEERING, KINKI UNIVERSITY)
YASUNAGA SHINKAI, HIRO-MACHI, KURE-SHI 737-01
JAPANESE
INDEXED: JICST

KIRYU-SHI NI OKERU KOGYO NO JOKYO 5490
(CENSUS OF MANUFACTURE, KIRYU CITY)
KIRYU-SHI JOHO KANRI-KA (INFORMATION CONTROL SECTION, KIRYU MUNICIPAL GOVERNMENT)
1-1, ORIHIME-CHO, KIRYU-SHI 376
JAPANESE

KISARAZU KOGYO KOTO SENMON GAKKO KIYO 5491
(BULLETIN OF KISARAZU TECHNICAL COLLEGE)
KISARAZU KOGYO KOTO SENMON GAKKO (KISARAZU TECHNICAL COLLEGE)
834, GION, KISARAZU-SHI 292
JAPANESE AND ENGLISH
INDEXED: CHE, JICST

KITAKYUSHU KOGYO KOTO SENMON GAKKO KENKYU HOKOKU 5492
(RESEARCH REPORT OF KITAKYUSHU TECHNICAL COLLEGE)
KITAKYUSHU KOGYO KOTO SENMON GAKKO (KITAKYUSHI TECHNICAL COLLEGE)
140, SHII, KOKURA-KU, KITAKYUSHU-SHI 803
JAPANESE
INDEXED: CHE, JICST

KITAMI KOGYO DAIGAKU GAKUHO 5493
(REPORT OF THE KITAMI INSTITUTE OF TECHNOLOGY)
JAPANESE

KITAMI KOGYO DAIGAKU KENKYU HOKOKU 5494
(MEMOIRS OF THE KITAMI INSTITUTE OF TECHNOLOGY)
KITAMI KOGYO DAIGAKU (KITAMI INSTITUTE OF TECHNOLOGY)
165, KOEN-MACHI, KITAMI-SHI 090
JAPANESE
INDEXED: CHE, MAT, JICST

KOBE DAIGAKU KOGAKU SHUHO 5495
(COLLECTED PAPERS OF THE FACULTY OF ENGINEERING, KOBE UNIVERSITY)
KOBE DAIGAKU KOGAKUBU (THE FACULTY OF ENGINEERING, KOBE UNIVERSITY)
ROKKODAI-MACHI, NADA-KU, KOBE-SHI 657
JAPANESE
INDEXED: JICST

KOBE JOSHI DAIGAKU KIYO 5496
(BULLETIN OF KOTE WOMEN'S COLLEGE)
SOCIETY OF KOBE WOMEN'S COLLEGE
2-1, AOYAMA, HIGASHI SUMA, SUMA-KU, KOBE
JAPANESE
INDEXED: CHE

KOCHI KOGYO KOTO SENMON GAKKO GAKUJUTSU KIYO 5497
(BULLETIN OF KOCHI TECHNICAL COLLEGE)
KOCHI KOGYO KOTO SENMON GAKKO (KOCHI TECHNICAL COLLEGE)
MONOBE, NANGOKU-SHI 783
JAPANESE AND EUROPEAN

KOCHI-KEN KOGYO SHIKENJO GYOMU NENPO 5498
(OPERATION REPORT OF INDUSTRIAL RESEARCH INSTITUTE, KOCHI PREFECTURE)
KOCHI-KEN KOGYO SHIKENJO
11-15, 2-CHOME, SANBASHI-DORI, KOCHI-SHI 78[illegible]
JAPANESE
INDEXED: JICST

KOCHI-KEN KOGYO SHIKENJO HOKOKU 5499
(REPORT OF INDUSTRIAL RESEARCH INSTITUTE, KOCHI PREFECTURE)
KOCHI-KEN KOGYO SHIKENJO
11-15, 2-CHOME, SANBASHI-DORI, KOCHI-SHI 780
JAPANESE
INDEXED: CHE

KOGAKU JOHO 5500
(ENGINEERING INFORMATION)
TOKYO DENKI DAIGAKKO YUKAI
(FRIENDS OF THE TOKYO ELECTRICAL EQUIPMENT COLLEGE)
JAPANESE

KOGAKU KENKYU 5501
(ENGINEERING RESEARCH)
JAPANESE
INDEXED: JICST

KOGAKU TO GIJUTSU 5502
(ENGINEERING AND TECHNOLOGY)
KANSAI DAIGAKU KOGAKKAI
10-1, 3-CHOME, SENRIYAMA HIGASHI, SUITA-SHI 565
JAPANESE, WITH ENGLISH SUMMARY
INDEXED: INS, JICST

KOGAKUIN DAIGAKU KENKYU HAPPYO KOENKAI KOEN YOSHI 5503
(SUMMARY OF THE KOGAKUIN UNIVERSITY RESEARCH ANNOUNCEMENT)
KOGAKUIN DAIGAKU
24-2, 1-CHOME, NISHISHINJUKU, SHINJUKU-KU, TOKYO 160
JAPANESE

KOGAKUIN DAIGAKU KENKYU HOKOKU 5504
(RESEARCH REPORTS OF THE KOGAKUIN UNIVERSITY)
KOGAKUIN DAIGAKU (KOGAKUIN UNIVERSITY)
24-2, 1-CHOME, NISHISHINJUKU, SHINJUKU-KU, TOKYO 160
JAPANESE
INDEXED: CHE, JICST

KOGEI NYUSU 5505
(INDUSTRIAL ARTS NEWS)
KOGYO GIJUTSUIN SANGYO KOGEI SHIKENJO (AGENCY OF INDUSTRIAL SCIENCE AND TECHNOLOGY INDUSTRIAL ARTS EXPERIMENT STATION)
JAPANESE

KOGEISHI NEWS 5506
(INDUSTRIAL ARTS EXPERIMENT STATION NEWS)
GIFUKEN KOGEI SHIKENJO (GIFU PREFECTURE INDUSTRIAL ARTS EXPERIMENT STATION)
JAPANESE

KOGIKEN NYUSU 5507
(INDUSTRIAL TECHNOLOGY RESEARCH NEWS)
KANSAI DAIGAKU KOGYO GIJUTSU KENKYUJO (INDUSTRIAL TECHNOLOGY RESEARCH INSTITUTE, KANSAI UNIVERSITY)
JAPANESE

KOGYO 5508
(INDUSTRY)
OSAKA KOGYOKAI (THE OSAKA INDUSTRIAL ASSOCIATION)
C/O OSAKA SHOKO KAIGISHO BUILDING, 58-7, HASHIZUME-CHO, UCHIHON-MACHI, HIGASHI-KU, OSAKA-SHI 540
JAPANESE
INDEXED: JICST

KOGYO GIJUTSU 5509
(INDUSTRIAL SCIENCE AND TECHNOLOGY)
TSUSHO SANGYO-SHO KOGYO GIJUTSU-IN (AGENCY OF INDUSTRIAL SCIENCE AND TECHNOLOGY, MINISTRY OF INTERNATIONAL TRADE AND INDUSTRY)
20-8, 2-CHOME, NISHISHINBASHI, MINATO-KU, TOKYO 105
JAPANESE
INDEXED: JICST

KOGYO GIJUTSU NENPO 5510
(INDUSTRY AND TECHNOLOGY YEARBOOK)
NIIGATA-KEN KOGYO GIJUTSU SENTA (INDUSTRY AND TECHNOLOGY CENTER, NIIGATA PREFECTURE)
504-1, ABUMI, NIIGATA-SHI 950
JAPANESE

KOGYO GIJUTSU-IN NENPO 5511
(ANNUAL REPORT OF THE AGENCY OF INDUSTRIAL SCIENCE AND TECHNOLOGY)
TSUSHO SANGYO-SHO KOGYO GIJUTSU-IN (AGENCY OF INDUSTRIAL SCIENCE AND TECHNOLOGY, MINISTRY OF INTERNATIONAL TRADE AND INDUSTRY)
JAPANESE

KOGYO GIJUTSU-IN SHIKEN KENKYUSHO KENKYU KEIKAKU 5512
(RESEARCH PROGRAMMES FOR INSTITUTE OF AGENCY OF INDUSTRIAL SCIENCE AND TECHNOLOGY)
TSUSHO SANGYO-SHO KOGYO GIJUTSU-IN (AGENCY OF INDUSTRIAL SCIENCE AND TECHNOLOGY, MINISTRY OF INTERNATIONAL TRADE & INDUSTRY)
7-3, 2-CHOME, NISHISHINBASHI, MINATO-KU, TOKYO 105
JAPANESE

KOGYO JIJI TSUSHIN, SHIGEN ENERUGI-BAN 5513
(INDUSTRY NEWS, EDITION OF RESOURCE AND ENERGY)
KOGYO JIJI TSUSHINSHA (THE INDUSTRIAL NEWS)
3-10, KANDA OGAWA-CHO, CHIYODA-KU, TOKYO 101
JAPANESE

KOGYO KYOIKU 5514
(ENGINEERING EDUCATION)
NIHON KOGYO KYOIKU KYOKAI (THE JAPAN SOCIETY FOR ENGINEERING EDUCATION)
C/O KURAMAE KOGYO KAIKAN, 10-19, 2-CHOME, SHINBASHI, MINATO-KU, TOKYO 105
JAPANESE AND ENGLISH

KOGYO NO SHINPO 5515
(ADVANCES IN ENGINEERING)
OSAKA-FU GIJUTSU KYOKAI (OSAKA PREFECTURE ENGINEERING ASSOCIATION)
C/O OSAKA-FURITSU KOGYO GIJUTSU KENKYUSHO, ENOKOKIMA KAMINO-CHO, NISHI-KU, OSAKA-SHI 550
JAPANESE

KOGYO NYUSU 5516
(INDUSTRIAL NEWS)
KANAGAWA-KEN KOGEI SHIDOJO (KANAGAWA PREFECTURE INDUSTRIAL ARTS DEVELOPMENT INSTITUTE)
JAPANESE

KOGYO RITCHI 5517
(INDUSTRIAL LOCATION)
NIHON KOGYO RITCHI SENTA (JAPAN INDUSTRIAL LOCATION CENTER
C/O TORANOMON TOYO BUILDING, 2-1, SHIBA KOTOHIRA-CHO, MINATO-KU, TOKYO 105
JAPANESE

KOGYO SEISAN TOKEI CHOSA KEKKA HOKOKU 5518
(REPORT ON RESULT OF SURVEY AND STATISTICS OF INDUSTRIAL OUTPUT)
KANAGAWA-KEN (KANAGAWA PREFECTURAL GOVERNMENT)
1-1, NIHON-ODORI, NAKA-KU, YOKOHAMA-SHI 231
JAPANESE

KOGYO SEISAN TOKEI CHOSA SOKUHO (KANAGAWA-KEN) 5519
(PROMPT REPORT OF CENSUS OF MANUFACTURES, KANAGAWA PREFECTURE)
KANAGAWA-KEN KIKAKU CHOSA-BU (PLANNING AND SURVEY DIVISION, KANAGAWA PREFECTURAL GOVERNMENT)
1-1, NIHON-ODORI, NAKA-KU, YOKOHAMA-SHI 231
JAPANESE

KOGYO TAIMUSU 5520
(INDUSTRIAL TIMES)
SOGO KEIZAI KENKYUSHO (JAPAN ECONOMIC RESEARCH INSTITUTE)
25-3, 5-CHOME, SHINBASHI, MINATO-KU, TOKYO 105
JAPANESE

KOGYO TO SEIHIN 5521
(INDUSTRY AND INDUSTRIAL PRODUCTS)
GAKKAI GYOMU KYOGIKAI (JAPANESE TECHNICAL REPORT PUBLICATION BUREAU)
5-3, KOJI-MACHI, CHIYODA-KU, TOKYO 102
JAPANESE
INDEXED: JICST

KOGYO TOKEI (ISHIKAWA-KEN) 5522
(CENSUS OF MANUFACTURES, ISHIKAWA PREFECTURE)
ISHIKAWA-KEN KIKAKU KAIHATSU-BU (PLANNING AND DEVELOPMENT DIVISION, ISHIKAWA PREFECTURAL GOVERNMENT)
2, HIROSAKA, KANAZAWA-SHI 920
JAPANESE

KOGYO TOKEI CHOSA HOKOKU (TOKYO-TO) 5523
(CENSUS OF MANUFACTURES, TOKYO METROPOLIS)
TOKYO-TO SOMU-KYOKU TOKEI-BU (STATISTICS DIVISION, ADMINISTRATIVE DEPARTMENT, TOKYO METROPOLITAN GOVERNMENT)
1-5, 3-CHOME, MARUNOUCHI, CHIYODA-KU, TOKYO 100
JAPANESE

KOGYO TOKEI CHOSA KARA MITA HACHINOHE NO KOGYO 5524
(CENSUS OF MANUFACTURES, HACHINOHE CITY)
HACHINOHE-SHI KIKAKU-SHITSU (PLANNING BUREAU, HACHINOHE MUNICIPAL GOVERNMENT)
2-1, HORIBATA-CHO, HACHINOHE-SHI 031
JAPANESE

KOGYO TOKEI CHOSA KEKKA GAISU SOKUHO (MIYAGI-KEN) 5525
(CENSUS OF MANUFACTURES, QUICK ANNOUNCEMENT, MIYAGI PREFECTURE)
MIYAGI-KEN KIKAKU-BU (PLANNING DIVISION, MIYAGI PREFECTURAL GOVERNMENT)
8-1, 3-CHOME, HON-CHO, SENDAI-SHI 980
JAPANESE

KOGYO TOKEI CHOSA KEKKA HOKOKU (HIROSHIMA-KEN) 5526
(CENSUS OF MANUFACTURES, HIROSHIMA PREFECTURE)
HIROSHIMA-KEN KIKAKU-BU (PLANNING DIVISION, HIROSHIMA PREFECTURAL GOVERNMENT)
10-52, MOTO-MACHI, HIROSHIMA-SHI 730
JAPANESE

KOGYO TOKEI CHOSA KEKKA HOKOKU (HYOGO-KEN) 5527
(CENSUS OF MANUFACTURES, HYOGO PREFECTURE)
HYOGO-KEN KIKAKU-BU (PLANNING DIVISION, HYOGO PREFECTURAL GOVERNMENT)
5-1, SHIMO YAMATE-DORI, IKUTA-KU, KOBE-SHI 650
JAPANESE

KOGYO TOKEI CHOSA KEKKA HOKOKU (KANAGAWA-KEN) 5528
(CENSUS OF MANUFACTURES, KANAGAWA PREFECTURE)
KANAGAWA-KEN KIKAKU CHOSA-BU (PLANNING RESEARCH DIVISION, KANAGAWA PREFECTURAL GOVERNMENT)
1-1, NIHON-ODORI, NAKA-KU, YOKOHAMA-SHI 231
JAPANESE

KOGYO TOKEI CHOSA KEKKA HOKOKUSHO (AICHI-KEN) 5529
(CENSUS OF MANUFACTURES, AICHI PREFECTURE)
AICHI-KEN KIKAKU-BU (PLANNING DIVISION, AICHI PREFECTURAL GOVERNMENT)
1-2, 3-CHOME, SANNOMARU, NAKA-KU, NAGOYA-SHI 460
JAPANESE

KOGYO TOKEI CHOSA KEKKA HOKOKUSHO (EHIME-KEN) 5530
(CENSUS OF MANUFACTURES, EHIME PREFECTURE)
EHIME-KEN SOMU-BU (ADMINISTRATIVE DIVISION, EHIME PREFECTURAL GOVERNMENT)
4-4, ICHIBAN-CHO, MATSUYAMA-SHI 791
JAPANESE

KOGYO TOKEI CHOSA KEKKA HOKOKUSHO (FUKUSHIMA-KEN) 5531
(CENSUS OF MANUFACTURES, FUKUSHIMA PREFECTURE)
FUKUSHIMA-KEN SOMU-BU TOKEI-KA (STATISTICS SECTION, GENERAL AFFAIRS DIVISION, FUKUSHIMA PREFECTURAL GOVERNMENT)
16, SUGIZUMA-CHO, FUKUSHIMA-SHI 960
JAPANESE

KOGYO TOKEI CHOSA KEKKA HOKOKUSHO (SHIGA-KEN) 5532
(CENSUS OF MANUFACTURES, SHIGA PREFECTURE)
SHIGA-KEN SOMU-BU (ADMINISTRATIVE DIVISION, SHIGA PREFECTURE)
1-1, 4-CHOME, KYO-MACHI OTSU-SHI 520
JAPANESE

KOGYO TOKEI CHOSA KEKKA HOKOKUSHO (SHIMANE-KEN) 5533
(CENSUS OF MANUFACTURES, SHIMANE PREFECTURE)
SHIMANE-KEN SOMU-BU (ADMINISTRATIVE DIVISION, SHIMANE PREFECTURAL GOVERNMENT)
1-TONO-MACHI, MATSUE-SHI 690
JAPANESE

KOGYO TOKEI CHOSA KEKKA HOKOKUSHO (SHIZUOKA-KEN) 5534
(CENSUS OF MANUFACTURES, SHIZUOKA PREFECTURE)
SHIZUOKA-KEN KIKAKU CHOSEI-BU (PLANNING DIVISION SHIZUOKA PREFECTURAL GOVERNMENT)
251, OITE-MACHI, SHIZUOKA-SHI 420
JAPANESE

KOGYO TOKEI CHOSA KEKKA HOKOKUSHO (YAMAGATA-KEN) 5535
(CENSUS OF MANUFACTURES, YAMAGATA PREFECTURE)
YAMAGATA-KEN KIKAKU-BU (PLANNING DIVISION, YAMAGATA PREFECTURAL GOVERNMENT)
4-51, 3-CHOME, HATAGO-MACHI, YAMAGATA-SHI 990
JAPANESE

KOGYO TOKEI CHOSA KEKKA SOKUHO (HIROSHIMA-KEN) 5536
(CENSUS OF MANUFACTURES, QUICK ANNOUNCEMENT, HIROSHIMA PREFECTURE)
HIROSHIMA-KEN KIKAKU-BU (PLANNING DIVISION, HIROSHIMA PREFECTURAL GOVERNMENT)
10-52, MOTO-MACHI, HIROSHIMA-SHI 730
JAPANESE

KOGYO TOKEI CHOSA KEKKA SOKUHO (IBARAKI-KEN) 5537
(CENSUS OF MANUFACTURES, QUICK ANNOUNCEMENT, IBARAKI PREFECTURE)
IBARAKI-KEN SOGO KAIHATSU-BU (GENERAL DEVELOPING DIVISION, IBARAKI PREFECTURAL GOVERNMENT)
5-38, 1-CHOME, SANNOMARU, MITO-SHI 310
JAPANESE

KOGYO TOKEI CHOSA KEKKA SOKUHO (ISHIKAWA-KEN) 5538
(CENSUS OF MANUFACTURES, QUICK ANNOUNCEMENT, ISHIKAWA PREFECTURE)
ISHIKAWA-KEN KIKAKU KAIHATSU-BU (PLANNING AND DEVELOPING DIVISION, ISHIKAWA PREFECTURAL GOVERNMENT)
1-1, 2-CHOME, HIROSAKA, KANAZAWA-SHI 920
JAPANESE

KOGYO TOKEI CHOSA KEKKA SOKUHO (TOYAMA-KEN) 5539
(MONTHLY REPORT OF THE CENSUS OF MANUFACTURES, TOYAMA PREFECTURE)
TOYAMA-KEN SOMU-BU (ADMINISTRATIVE DIVISION, TOYAMA PREFECTURAL GOVERNMENT)
1-7, SHINSOGAWA, TOYAMA-SHI 930
JAPANESE

KOGYO TOKEI CHOSA KEKKAHYO (KOBE-SHI) 5540
(CENSUS OF MANUFACTURES, KOBE CITY)
KOBE-SHI KIKAKU-KYOKU (PLANNING BUREAU, KOBE MUNICIPAL GOVERNMENT)
6-7, KANO-MACHI, IKUTA-KU, KOBE-SHI 650
JAPANESE

KOGYO TOKEI CHOSA KEKKAHYO (OSAKA-FU) 5541
(CENSUS OF MANUFACTURES, OSAKA PREFECTURE)
OSAKA-FU KIKAKU-BU (PLANNING DIVISION, OSAKA PREFECTURAL GOVERNMENT)
OTEMAENO-CHO, HIGASHI-KU, OSAKA-SHI 540
JAPANESE

KOGYO TOKEI CHOSA KEKKAHYO (TOYAMA-KEN) 5542
(CENSUS OF MANUFACTURES, TOYAMA PREFECTURE)
TOYAMA-KEN SOMU-BU (ADMINISTRATIVE DIVISION, TOYAMA PREFECTURAL GOVERNMENT)
1-7, SHINSOGAWA, TOYAMA-SHI 930
JAPANESE

KOGYO TOKEI CHOSA NO KEKKA (KURE-SHI) 5543
(CENSUS OF MANUFACTURES, KURE CITY)
KURE-SHI KIKAKU KANRI-BU (PLANNING AND ADMINISTRATIVE DIVISION, KURE MUNICIPAL GOVERNMENT)
1-6, 4-CHOME, CHUO, KURE-SHI 737
JAPANESE

KOGYO TOKEI CHOSA SHI-SHUKEI KEKKAHYO (AMAGASAKI-SHI) 5544
(CENSUS OF MANUFACTURES, AMAGASAKI CITY)
AMAGASAKI-SHI KIKAKU CHOSEI-KYOKU (PLANNING AND COORDINATION BUREAU, AMAGASAKI MUNICIPAL GOVERNMENT)
23-1, 1-CHOME, HIGASHINANAMATSU-CHO, AMAGASAKI-SHI 660
JAPANESE

KOGYO TOKEI KEKKAHYO (OKAYAMA-KEN) 5545
(CENSUS OF MANUFACTURES, OKAYAMA PREFECTURE)
OKAYAMA-KEN KIKAKU-BU (PLANNING DIVISION, OKAYAMA PREFECTURAL GOVERNMENT)
4-6, 2-CHOME, UCHISANGE, OKAYAMA-SHI 700
JAPANESE

KOGYO TOKEI SOKUHO (TSUSHO SANGYO-SHO) 5546
(QUICK ANNOUNCEMENT ON CENSUS OF MANUFACTURES, MINISTRY OF INTERNATIONAL TRADE & INDUSTRY)
TSUSHO SANGYO-SHO DAIJIN KANBO CHOSA TOKEI-BU (RESEARCH AND STATISTICS DEPARTMENT, MINISTER'S SECRETARIAT, MINISTRY OF INTERNATIONAL TRADE)
3-1, 1-CHOME, KASUMIGASEKI, CHIYODA-KU, TOKYO 100
JAPANESE

KOGYO TOKEIHYO, BUNPU SOKAN-HEN 5547
(CENSUS OF MANUFACTURES, REPORT BY MUTUAL DISTRIBUTION)
TSUSHO SANGYO-SHO DAIKIN KANBO CHOSA TOKEI-BU (RESEARCH AND STATISTICS DIVISION, MINISTER'S SECRETARIAT, MINISTRY OF INTERNATIONAL TRADE)
3-1, 1-CHOME, KASUMIGASEKI, CHIYODA-KU, TOKYO 100
JAPANESE

KOGYO TOKEIHYO, GENZAIRYO, NENRYO-HEN 5548
(CENSUS OF MANUFACTURES, MATERIALS, FUELS AND ELECTRIC ENERGY)
TSUSYO SANGYO-SHO DAIKIN KANBO CHOSA TOKEI-BU (STATISTICS DIVISION, MINISTER'S SECRETARIAT, MINISTRY OF INTERNATIONAL TRADE AND INDUSTRY)
3-1, 1-CHOME, KASUMIGASEKI, CHIYODA-KU
TOKYO 100
JAPANESE

KOGYO TOKEIHYO, HINMOKU-HEN 5549
(CENSUS OF MANUFACTURES, REPORT BY COMMODITIES)
TSUSHO SANGYO-SHO DAIJIN KANBO CHOSA TOKEI-BU (RESEARCH AND STATISTICS DIVISION, MINISTER'S SECRETARIAT, MINISTRY OF INTERNATIONAL TRADE)
3-1, 1-CHOME, KASUMIGASEKI, CHIYODA-KU, TOKYO 100
JAPANESE

KOGYO TOKEIHYO, KIGYO-HEN 5550
(CENSUS OF MANUFACTURES, REPORT BY ENTERPRISES)
TSUSHO SANGYO-SHO DAIJIN KANBO CHOSA TOKEI-BU (RESEARCH AND STATISTICS DIVISION, MINISTER'S SECRETARIAT, MINISTRY OF INTERNATIONAL TRADE)
3-1, 1-CHOME, KASUMIGASEKI, CHIYODAKU
TOKYO 100
JAPANESE

KOGYO TOKEIHYO, SAITAMA-KEN 5551
(CENSUS OF MANUFACTURES, SAITAMA PREFECTURE)
SAITAMA-KEN KIKAKU ZAISEI-BU (PLANNING & FINANCE DIVISION, SAITAMA PREFECTURAL GOVERNMENT)
15-1, 3-CHOME, TAKASAGO, URAWA-SHI 336
JAPANESE

KOGYO TOKEIHYO, SANGYO-HEN 5552
(CENSUS OF MANUFACTURES, REPORT BY INDUSTRIES)
TSUSHO SANGYO-SHO DAIJIN KANBO CHOSA TOKEI-BU (RESEARCH AND STATISTICS DIVISION, MINISTER'S SECRETARIAT, MINISTRY OF INTERNATIONAL TRADE AND INDUSTRY)
3-1, 1-CHOME, KASUMIGASEKI, CHIYODA-KU TOKYO 100
JAPANESE AND ENGLISH

KOGYO TOKEIHYO, SHICHOSON-HEN 5553
(CENSUS OF MANUFACTURES, REPORT BY CITIES, TOWNS AND VILLAGES)
TSUSHO SANGYO-SHO DAIJIN KANBO CHOSA TOKEI-BU (RESEARCH AND STATISTICS DIVISION, MINISTER'S SECRETARIAT, MINISTRY OF INTERNATIONAL TRADE AND INDUSTRY)
3-1, 1-CHOME, KASUMIGASEKI, CHIYODA-KU TOKYO 100
JAPANESE

KOGYO TOKEIHYO, TSUSHO SANGYO-SHO 5554
(CENSUS OF MANUFACTURES, MINISTRY OF INTERNATIONAL TRADE AND INDUSTRY)
TSUSHO SANGYO-SHO DAIJIN KANBO CHOSA TOKEI-BU (RESEARCH AND STATISTICS DEPARTMENT, MINISTER'S SECRETARIAT, MINISTRY OF INTERNATIONAL TRADE)
3-1, 1-CHOME, KASUMIGASEKI, CHIYODA-KU, TOKYO 100
JAPANESE

KOGYO TOKEIHYO, YOCHI, YOSUI-HEN 5555
(CENSUS OF MANUFACTURES, REPORT BY LAND AND WATER)
TSUSHO SANGYO-SHO DAIJIN KANBO CHOSA TOKEIBU (RESEARCH AND STATISTICS DIVISION, MINISTER'S SECRETARIAT, MINISTRY OF INTERNATIONAL TRADE AND INDUSTRY)
3-1, 1-CHOME, KASUMIGASEKI, CHIYODA-KU, TOKYO 100
JAPANESE

KOGYO YOHIN NYUSU 5556
(INDUSTRIAL MATERIALS NEWS)
BURIJISUTON TAIYA K.K. (BRIDGESTONE TIRE COMPANY)
C/O KANSEN BUILDING, 12-4, 4-CHOME, HACCHOBORI, CHUO-KU, TOKYO 104
JAPANESE

KOGYO ZAIRYO 5557
(ENGINEERING MATERIALS)
NIKKAN KOGYO SHINBUNSHA (INDUSTRIAL DAILY NEWS)
8-10, 1-CHOME, KUDANKITA, CHIYODA-KU, TOKYO 102
JAPANESE
INDEXED: CHE, JICST

KOJO KOSAKUSHA 5558
(FACTORY ENGINEERS)
'KOJO KOSAKUSHA' NENSHU GURUPU ('FACTORY ENGINEERS' EDITING GROUP)
JAPANESE

KOKEN DAYORI 5559
(OSAKA MUNICIPAL TECHNICAL RESEARCH INSTITUTE NEWS)
OSAKA-SHIRITSU KOGYO KENKYUSHO (OSAKA MUNICIPAL TECHNICAL RESEARCH INSTITUTE)
38, KITAOGI-MACHI, KITA-KU, OSAKA-SHI 530
JAPANESE
INDEXED: JICST

KOKURITSU KOKKAI TOSHOKAN GEPPO 5560
(DIET LIBRARY MONTHLY)
JAPANESE
INDEXED: JICST

KOKURITSU KOKKAI TOSHOKAN. NEPO 5561
(DIET LIBRARY ANNUAL)
JAPANESE
INDEXED: JICST

KOKUSAI KOGYO SHOYUKEN HOGO KYOKAI NIPPON BUKAI GEPPO 5562
(INTERNATIONAL ASSOCIATION FOR THE PRESERVATION OF INDUSTRIAL PROPRIETORY RIGHTS, JAPAN BRANCH MONTHLY)
JAPANESE
INDEXED: JICST

KOKUSAN SHIN GIJUTSU CHOSAHYO 5563
(SURVEY ON TECHNOLOGICAL DEVELOPMENT IN JAPAN)
SHIN GIJUTSU KAIHATSU ZAIDAN (NEW TECHNOLOGY FOUNDATION)
13-7, 6-CHOME, GINZA, CHUO-KU, TOKYO 104
JAPANESE
INDEXED: JICST

KOKUSHIKAN DAIGAKU KOGAKUBU KIYO 5564
(TRANSACTIONS OF THE KOKUSHIKAN UNIVERSITY DEPARTMENT OF ENGINEERING)
KOKUSHIKAN DAIGAKU KOGAKUBU (DEPARTMENT OF ENGINEERING, KOKUSHIKAN UNIVERSITY)
28-1, 4-CHOME, SETAGAYA, SETAGAYA-KU, TOKYO 154
JAPANESE, WITH ENGLISH SUMMARY
INDEXED: CHE, JICST

KUMAMOTO DAIGAKU KOGAKU KENKYU KIKI SENTA HOKOKU 5565
(REPORT OF EXPERIMENTAL MACHINERY AND TOOLS CENTER, FACULTY OF ENGINEERING, KUMAMOTO UNIVERSITY)
KUMAMOTO DAIGAKU KOGAKU KENKYU KIKI SENTA
39-1, 2-CHOME, KUROKAMI, KUMAMOTO-SHI 860
JAPANESE AND ENGLISH
INDEXED: JICST

KUMAMOTO DAIGAKU KOGAKUBU KENKYU HOKOKU 5566
(TECHNICAL REPORTS OF THE KUMAMOTO UNIVERSITY)
KUMAMOTO DAIGAKU KOGAKUBU (THE FACULTY OF ENGINEERING, KUMAMOTO UNIVERSITY)
39-1, 2-CHOME, KUROKAMI, KUMAMOTO-SHI 860
JAPANESE
INDEXED: CHE, JICST

KUMAMOTO KOGYO DAIGAKU KENKYU HOKOKU 5567
(BULLETIN OF THE KUMAMOTO INSTITUTE OF TECHNOLGY)
KUMAMOTO INSTITUTE OF TECHNOLGY
JAPANESE

KUMAMOTO-KEN KOGYO SHIKENJO KENKYU HOKOKU 5568
(RESEARCH REPORTS OF THE KUMAMOTO INDUSTRIAL RESEARCH INSTITUTE)
KUMAMOTO-KEN KOGYO SHIKENJO (THE KUMAMOTO INDUSTRIAL RESEARCH INSTITUTE)
3-11, HIGASHI-MACHI, KUMAMOTO-SHI 862
JAPANESE
INDEXED: JICST

KUMAMOTO-KEN NO KOGYO 5569
(MANUFACTURES IN KUMAMOTO PREFECTURE)
KUMAMOTO-KEN KIKAKU-BU (PLANNING DIVISION, KUMAMOTO PREFECTURAL GOVERNMENT)
18-1, 6-CHOME, SUIZENJI, KUMAMOTO-SHI 862
JAPANESE

KURAMAE KOGYOKAISHI 5570
(JOURNAL OF THE KURAMAE INDUSTRIAL ASSOCIATION)
KURAMAE KOGYOKAI (KURAMAE INDUSTRIAL ASSOCIATION)
19-10, 2-CHOME, SHINBASHI, MINATO-KU, TOKYO 105
JAPANESE

KURE KOGYO KOTO SENMON GAKKO KENKYU HOKOKU 5571
(MEMOIRS OF THE KURE TECHNICAL COLLEGE)
KURE KOGYO KOTO SENMON GAKKO (THE KURE TECHNICAL COLLEGE)
2-11, 2-CHOME, AGAMINAMI, KURE-SHI 737
JAPANESE, WITH ENGLISH SUMMARY
INDEXED: JICST

KURE KOGYO SHIKENJO HOKOKU 5572
(REPORT OF THE KURE INDUSTRIAL LABORATORY)
JAPANESE
INDEXED: JICST

KURUME KOGYO KOTO SENMON GAKKO KENKYU HOKOKU 5573
(RESEARCH REPORTS OF KURUME TECHNICAL COLLEGE)
KURUME KOGYO KOTO SENMON GAKKO (THE KURUME TECHNICAL COLLEGE)
1232, KOMORINO-CHO, KURUME-SHI 830
JAPANESE AND ENGLISH
INDEXED: CHE, JICST

KUSHIRO KOGYO KOTO SENMON GAKKO KIYO 5574
(RESEARCH REPORTS, KUSHIRO TECHNICAL COLLEGE)
KUSHIRO KOGYO KOTO SENMON GAKKO (KUSHIRO TECHNICAL COLLEGE)
128, OTANOSHIKE, KUSHIRYO-SHI 084
JAPANESE AND ENGLISH

KYOKOSHI NYUSU 5575
(KYOTO INDUSTRIAL RESEARCH INSTITUTE NEWS)
KYOTO-SHI KOGYO SHIKENJO
1-2, MINAMIDA-MACHI, NISHI-9-JO, MINAMI-KU, KYOTO-SHI 601
JAPANESE

KYOTO DAIGAKU KOGAKU KENKYU 5576
(ENGINEERING RESEARCHES, KYOTO UNIVERSITY)
KYOTO DAIGAKU KOGAKUBU (FACULTY OF ENGINEERING
KYOTO UNIVERSITY)
YOSHIDA HON-MACHI, SAKYO-KU, KYOTO-SHI 606
JAPANESE
INDEXED: JICST

KYOTO DAIGAKU KOGAKUBU FUZOKU OTOMESHON KENKYU 5577
SHISETSU KENKYU HOKOKU
(TECHNICAL REPORT OF THE AUTOMATION RESEARCH
LABORATORY, FACULTY OF ENGINEERING, KYOTO UNIVERSITY)
KYOTO DAIGAKU KOGAKUBU FUZOKU OTOMESHON KENKYU
SHISETSU (AUTOMATION RESEARCH LABORATORY, FACULTY
OF ENGINEERING, KYOTO UNIVERSITY)
GOKASHO, UJI-SHI, KYOTO 611
JAPANESE AND ENGLISH

KYOTO SANGYO DAIGAKU RONSHU 5578
(KYOTO SANGYO UNIVERSITY ESSAYS)
KYOTO SANGYO DAIGAKU
KAMIKAMO MOTOYAMA, KITA-KU, KYOTO-SHI 603
JAPANESE AND ENGLISH
INDEXED: CHE

KYOTO TECHNICAL UNIVERSITY. FACULTY OF INDUSTRIAL 5579
ARTS. MEMOIRS: SCIENCE AND TECHNOLOGY
KYOTO TECHNICAL UNIVERSITY, FACULTY OF INDUSTRIAL
ARTS
MATSUGASAKI, SAKYO-KU, KYOTO-SHI 606
ENGLISH
INDEXED: CHE, MAT

KYOTO-FU NO KOGYO 5580
(MANUFACTURES IN KYOTO PREFECTURE)
KYOTO-FU (KYOTO PREFECTURAL GOVERNMENT)
YABUNOUCHI-MACHI, SHIMOTACHIYURIDORI SHIN-MACHI,
NISHIIRU, KAMIGYO-KU, KYOTO-SHI 602
JAPANESE

KYOTO-FU SANGYO NO TENBO 5581
(PROSPECT OF MANUFACTURES IN KYOTO PREFECTURE)
KYOTO-FURITSU CHUSHO KIGYO SOGO SHIDOSHO
(GENERAL GUIDANCE INSTITUTE FOR SMALL ENTER-
PRISES, KYOTO PREFECTURE)
31, YAHATA-CHO, NISHI-7-JO, SHIMOGYO-KU,
KYOTO-SHI 600
JAPANESE

KYOTO-SHI KOGYO SHIKENJO GYOMU HOKOKU 5582
(ANNUAL REPORT OF THE INDUSTRIAL RESEARCH INSTITUTE,
KYOTO CITY)
KYOTO-SHI KOGYO SHIKENJO (INDUSTRIAL RESEARCH
INSTITUTE, KYOTO CITY)
1-2, MINAMIDA-MACHI, NISHI-9-JO, MINAMI-KU,
KYOTO-SHI 601
JAPANESE

KYOTO-SHI KOGYO SHIKENJO KENKYU HOKOKU 5583
(REPORT OF INDUSTRIAL RESEARCH INSTITUTE, KYOTO CITY)
KYOTO-SHI KOGYO SHIKENJO
1-2, MINAMIDA-CHO, NISHI-9-JO, MINAMI-KU
KYOTO-SHI 601
JAPANESE
INDEXED: JICST

KYOTO-SHI NO KOGYO (KOGYO TOKEI CHOSA KEKKA 5584
HOKOKUSHO
(INDUSTRY IN KYOTO (CENSUS OF MANUFACTURES))
KYOTO-SHI TOKEI SENTA (KYOTO MUNICIPAL STATISTICAL
CENTER)
C/O KYOTO SHIYAKUSHO, OIKE, TERAMACHI-DORI,
NAKAGYO-KU, KYOTO-SHI 604
JAPANESE

KYUKOSHI NYUSU 5585
(NEWS OF NATIONAL INDUSTRIAL RESEARCH INSTITUTE OF
KYUSHU)
TSUSHO SANGYO-SHO KOGYO GIJUTSU-IN KYUSHU KOGYO
GIJUTSU SHIKENSHO (GOVERNMENT INDUSTRIAL RESEARCH
INSTITUTE OF KYUSHU, SCIENCE & TECHNOLOGY AGENCY)
SHUKU-MACHI, TOSU-SHI 841
JAPANESE
INDEXED: JICST

KYUSAN GIREN (KYUSHU SANGYO GIJUTSU RENMEI) 5586
(FEDERATION OF KYUSHU INDUSTRIAL TECHNOLOGY)
JAPANESE
INDEXED: JICST

KYUSHU DAIGAKU KOGAKU SHUHO 5587
(TECHNOLOGY REPORTS OF THE KYUSHU UNIVERSITY)
KYUSHU DAIGAKU KOGAKUBU (FACULTY OF ENGINEERING,
KYUSHU UNIVERSITY)
3576, HAKOZAKI, HIGASHI-KU, FUKUOKA-SHI 812
JAPANESE
INDEXED: CHE, INS, JICST

KYUSHU DAIGAKU SEISAN KAGAKU KENKYUSHO HOKOKU 5588
(REPORTS OF RESEARCH INSTITUTE OF INDUSTRIAL SCIENCE,
KYUSHU UNIVERSITY)
KYUSHU DAIGAKU SEISAN KAGAKU KENKYUSHO
(THE RESEARCH INSTITUTE OF INDUSTRIAL SCIENCE,
KYUSHU UNIVERSITY)
HAKOZAKI, FUKUOKA-SHI 812
JAPANESE AND ENGLISH
INDEXED: CHE, INS, JICST

KYUSHU GEIJUTSU KOKA DAIGAKU IPPAN. KISO KYOKU 5589
KEIRETSU KEKYU RONSHU
(KYUSHU ARTS & ENGINEERING UNIVERSITY. GENERAL
AND BASIC EDUCATION)
HENSHU IIN KAI
JAPANESE

KYUSHU KOGYO DAIGAKU KENKYU HOKOKU, KOGAKU 5590
(BULLETIN OF THE KYUSHU INSTITUTE OF TECHNOLOGY,
ENGINEERING)
KYUSHU KOGYO DAIGAKU (THE KYUSHU INSTITUTE OF
TECHNOLOGY)
SENSUI-CHO, TOBATA-KU, KITAKYUSHU-SHI 804
JAPANESE, WITH EUROPEAN SUMMARY
INDEXED: CHE, INS, JICST

KYUSHU KOGYO GIJUTSU SHIKENSHO HOKOKU 5591
(REPORT OF GOVERNMENT INDUSTRIAL RESEARCH
INSTITUTE, KYUSHU)
TSUSHO SANGYO-SHO KOGYO GIJUTSU-IN KYUSHU
KOGYO GIJUTSU SHIKENJO (GOVERNMENT INDUSTRIAL
RESEARCH INSTITUTE OF KYUSHU, SCIENCE AND
TECHNOLOGY AGENCY)
SHUKU-MACHI, TOSU-SHI 841
JAPANESE AND ENGLISH
INDEXED: CHE, JICST

KYUSHU KOGYO GIJUTSU SHIKENSHO NENPO 5592
(ANNUAL REPORT OF GOVERNMENT INDUSTRIAL RESEARCH
INSTITUTE, KYUSHU)
TSUSHO SANGYO-SHO KOGYO GIJUTSU-IN KYUSHU KOGYO
GIJUTSU SHIKENSHO (GOVERNMENT INDUSTRIAL RESEARCH
INSTITUTE OF KYUSHU, SCIENCE AND TECHNOLOGY AGENCY.
MINISTRY OF INTERNATIONAL TRADE AND INDUSTRY)
SHUKU-MACHI, TOSU-SHI 841
JAPANESE

KYUSHU SANGYO DAIGAKU KOGAKUBU KENKYU HOKOKU 5593
(BULLETIN OF THE FACULTY OF ENGINEERING, KYUSHU
SANGYO UNIVERSITY)
KYUSHU SANGYO DAIGAKU KOGAKUBU
2-327, MATSUGADAI, HIGASHI-KU, FUKUOKA-SHI 812
JAPANESE AND ENGLISH
INDEXED: JICST

KYUSHU TSUSHO SANGYO NENPO 5594
(ANNUAL REPORT ON INTERNATIONAL TRADE AND INDUSTRY
IN KYUSHU)
TSUSHO SANGYO-SHO FUKUOKA TSUSHO SANGYO-KYOKU
(THE FUKUOKA BUREAU OF INTERNATIONAL TRADE AND
INDUSTRY)
13-24, 2-CHOME, HAKATAEKI HIGASHI, FUKUOKA-SHI
812
JAPANESE

LIBRARIAN 5595
SENMON TOSHOKAN KANSAI CHIKU KYOGIKA
JAPANESE

LIBRARY ABSTRACT BULLETIN 5596
JAPANESE
INDEXED: JICST

LIBRARY AND INFORMATION SCIENCE 5597
MITA TOSHOKAN JOHO GAKKAI (MITA SOCIETY FOR LIBRARY
AND INFORMATION SCIENCE)
15-15, 2-CHOME, MITA, MINATO-KU, TOKYO 108
JAPANESE AND ENGLISH
INDEXED: INS, JICST

LIBRARY AND INFORMATION SCIENCE NEWS 5598
KIHARA SHOZO DO LISN HENSHUSHITSU (LISN
EDITING OFFICE, KIHARA SHOZO PUBLISHING COMPANY)
JAPANESE AND ENGLISH

M. C. J. NEWS 5599
ENGLISH

M.H.I. TECHNICAL REVIEW 5600
MITSUBISHI JUKOGYO K. K. (MITSUBISHI HEAVY
INDUSTRIES LIMITED)
5-1, 2-CHOME, MARUNOUCHI, CHIYODA-KU, TOKYO 100
ENGLISH
INDEXED: AUT. CHE, COP, IAA, POL, JICST

MAIZURU KOGYO KOTO SENMON GAKKO KIYO 5601
(RESEARCH REPORTS OF MAIZURU TECHNICAL COLLEGE)
MAIZURU KOGYO KOTO SENMON GAKKO (THE MAIZURU TECHNICAL COLLEGE)
234, SHIRAYA, MAIZURU-SHI 625
JAPANESE

MATSUE KOGYO KOTO SENMON GAKKO KENKYU KIYO 5602
(RESEARCH REPORTS OF MATSUE TECHNICAL COLLEGE)
MATSUE KOGYO KOTO SENMON GAKKO (MATSUE TECHNICAL COLLEGE)
14-4, NISHIKUMA-CHO, MATSUE-SHI 690
JAPANESE, WITH ENGLISH SUMMARY

MCC NEWS 5603
MITSUBISHI SOGO KENKYUSHO (MITSUBISHI GENERAL RESEARCH INSTITUTE)
JAPANESE AND ENGLISH

MEIJI DAIGAKU KOGAKUBU KENKYU HOKOKU 5604
(RESEARCH REPORTS OF THE FACULTY OF ENGINEERING, MEIJI UNIVERSITY)
MEIJI DAIGAKU KOGAKUBU (FACULTY OF ENGINEERING, MEIJI UNIVERSITY)
5158, IKUTA, TAMA-KU, KAWASAKI-SHI 214
JAPANESE AND EUROPEAN
INDEXED: CHE, INS, JICST

MEIKOKEN. NAGOYA MUNICIPAL INDUSTRIAL RESEARCH 5605
(NAGOYA MUNICIPAL INDUSTRIAL RESEARCH)
NAGOYASHI KOGYO KENKYUJO (NAGOYA MUNICIPAL INDUSTRIAL RESEARCH INSTITUTE)
JAPANESE

MEIKOSHI GIJUTSU NYUSU 5606
(TECHNICAL NEWS OF GIRI, NAGOYA)
TSUSHO SANGYO-SHO KOGYO GIJUTSU-IN NAGOYA KOGYO GIJUTSU SHIKENSHO (GOVERNMENT INDUSTRIAL RESEARCH INSTITUTE, NAGOYA, AGENCY OF INDUSTRIAL SCIENCE AND TECHNOLOGY, MINISTRY OF INTERNATIONAL TRADE/INDUSTRY
1, HIRATE-CHO, KITA-KU, NAGOYA-SHI 462
JAPANESE
INDEXED: JICST

MEMOIRS OF THE FACULTY OF ENGINEERING, HIROSHIMA UNIVERSITY 5607
HIROSHIMA DAIGAKU KOGAKUBU (FACULTY OF ENGINEERING HIROSHIMA UNIVERSITY)
SENDA-MACHI, HIROSHIMA-SHI 730
EUROPEAN
INDEXED: CHE, INS, MAT, JICST

MEMOIRS OF THE FACULTY OF ENGINEERING, HOKKAIDO UNIVERSITY 5608
HOKKAIDO DAIGAKU KOGAKUBU (FACULTY OF ENGINEERING HOKKAIDO UNIVERSITY)
NISHI-8-CHOME, KITA-13-JO, KITA-KU, SAPPOROSHI 065
EUROPEAN
INDEXED: CHE, ENG, WAA, JICST

MEMOIRS OF THE FACULTY OF ENGINEERING, KOBE UNIVERSITY 5609
KOBE DAIGAKU KOGAKUBU (FACULTY OF ENGINEERING, KOBE UNIVERSITY)
ROKKODAI-CHO, NADA-KU, KOBE-SHI 657
ENGLISH
INDEXED: CHE, ENG, INS, MAT, JICST

MEMOIRS OF THE FACULTY OF ENGINEERING, KUMAMOTO UNIVERSITY 5610
KUMAMOTO DAIGAKU KOGAKUBU (FACULTY OF ENGINEERING KUMAMOTO UNIVERSITY)
39-1, 2-CHOME, KUROKAMI, KUMAMOTO-SHI 860
ENGLISH
INDEXED: CHE, ENG, INS, WAA, JICST

MEMOIRS OF THE FACULTY OF ENGINEERING, KYOTO UNIVERSITY 5611
KYOTO DAIGAKU KOGAKUBU (FACULTY OF ENGINEERING, KYOTO UNIVERSITY)
YOSHIDA HON-MACHI, SAKYO-KU, KYOTO-SHI 606
EUROPEAN
INDEXED: APP, CHE, COP, ENG, INS, IAA, MAT, MET, JICST

MEMOIRS OF THE FACULTY OF ENGINEERING, KYUSHU UNIVERSITY 5612
KYUSHU DAIGAKU KOGAKUBU (FACULTY OF ENGINEERING, KYUSHU UNIVERSITY)
3576, HAKOZAKI, HIGASHI-KU, FUKUOKA-SHI 812
EUROPEAN
INDEXED: APP, CHE, ENG, INS, MET, JICST

MEMOIRS OF THE FACULTY OF ENGINEERING, MIYAZAKI UNIVERSITY 5613
MIYAZAKI DAIGAKU KOGAKUBU (FACULTY OF ENGINEERING, MIYAZAKI UNIVERSITY)
118, NISHIMARUYAMA-CHO, MIYAZAKI-SHI 880
ENGLISH
INDEXED: CHE, MAT, JICST

MEMOIRS OF THE FACULTY OF ENGINEERING, NAGOYA UNIVERSITY 5614
NAGOYA DAIGAKU KOGAKUBU (FACULTY OF ENGINEERING, NAGOYA UNIVERSITY)
FURO-CHO, CHIKUSA-KU, NAGOYA-SHI 464
EUROPEAN
INDEXED: APP, CHE, ENG, INS, IAA, WAA, JICST

MEMOIRS OF THE FACULTY OF ENGINEERING, OSAKA CITY UNIVERSITY 5615
OSAKA-SHIRITSU DAIGAKU KOGAKUBU (FACULTY OF ENGINEERING, OSAKA CITY UNIVERSITY)
459, SUGIMOTO-CHO, SUMIYOSHI-KU, OSAKA-SHI 558
ENGLISH
INDEXED: CHE, INS, WAA. JICST

MEMOIRS OF THE FACULTY OF INDUSTRIAL ARTS, KYOTO TECHNICAL UNIVERSITY, SCIENCE AND TECHNOLOGY 5616
KYOTO KOGEI SEN'I DAIGAKU, KOGEIGAKUBU (THE FACULTY OF INDUSTRIAL ARTS, KYOTO TECHNICAL UNIVERSITY)
MATSUGASAKI, SAKYO-KU, KYOTO-SHI 606
ENGLISH
INDEXED: JICST

MEMOIRS OF THE FACULTY OF TECHNOLOGY, THE TOKYO METROPOLITAN UNIVERSITY 5617
TOKYO-TORITSU DAIGAKU KOGAKUBU (FACULTY OF TECHNOLOGY, THE TOKYO METROPOLITAN UNIVERSITY)
1-1, 2-CHOME, FUKAZAWA, SETAGAYA-KU, TOKYO 158
ENGLISH
INDEXED: CHE, INS, ENG, JICST

MEMOIRS OF THE INSTITUTE OF SCIENTIFIC AND INDUSTRIAL RESEARCH, OSAKA UNIVERSITY 5618
OSAKA DAIGAKU SANGYO KAGAKU KENKYUSHO (INSTITUTE OF SCIENTIFIC AND INDUSTRIAL RESEARCH, OSAKA UNIVERSITY)
YAMADAKAMI, SUITA-SHI 565
EUROPEAN
INDEXED: CHE, JICST

MEMOIRS OF THE KYUSHU INSTITUTE OF TECHNOLOGY, ENGINEERING 5619
KYUSHU KOGYO DAIGAKU (KYUSHU INSTITUTE OF TECHNOLOGY)
1, SENSUI-CHO, TOBATA-KU, KITAKYUSHU-SHI 804
ENGLISH
INDEXED: JICST

MEMOIRS OF THE SCHOOL OF ENGINEERING, OKAYAMA UNIVERSITY 5620
OKAYAMA DAIGAKU KOGAKUBU (SCHOOL OF ENGINEERING, OKAYAMA UNIVERSITY)
TSUSHIMA, OKAYAMA-SHI 700
EUROPEAN
INDEXED: CHE, MAT, JICST

MIE-KEN KOGYO GIJUTSU SENTA NENPO 5621
(ANNUAL REPORT OF THE INDUSTRIAL TECHNOLOGY CENTER, MIE PREFECTURE)
MIE-KEN KOGYO GIJUTSU SENTA
3485, KOMORI-MACHI, TAKAJAYA, TSU-SHI 514
JAPANESE

MIE-KEN KOGYO GIJUTSU SENTA NYUSU 5622
(MIE PREFECTURE INDUSTRIAL TECHNOLOGY CENTER NEWS)
MIE PREFECTURE INDUSTRIAL TECHNOLOGY CENTER
JAPANESE

MITSUBISHI JUKO 5623
(MITSUBISHI HEAVY INDUSTRIES' BULLETIN)
MITSUBISHI JUKOGYO K.K. (MITSUBISHI HEAVY INDUSTRIES, LTD.)
5-1, 2-CHOME, MARUNOUCHI, CHIYODA-KU, TOKYO 100
JAPANESE
INDEXED: JICST

MITSUBISHI JUKO GIHO 5624
(MITSUBISHI HEAVY INDUSTRIES TECHNICAL REVIEW)
MITSUBISHI JUKOGYO K.K. (MITSUBISHI HEAVY INDUSTRIES LTD.)
5-1, 2-CHOME, MARUNOUCHI, CHIYODA-KU, TOKYO 100
JAPANESE, WITH ENGLISH SUMMARY
INDEXED: JICST

MITSUBISHI TECHNICAL BULLETIN 5625
MITSUBISHI JUKOGYO K.K. (MITSUBISHI HEAVY INDUSTRIES LTD.)
5-1, 2-CHOME, MARUNOUCHI, CHIYODA-KU, TOKYO 100
JAPANESE, WITH ENGLISH SUMMARY
INDEXED: JICST

MIYAGI KOGYO KOTO SENMON GAKKO KENKYU KIYO 5626
(RESEARCH REPORTS, MIYAGI TECHNICAL COLLEGE)
MIYAGI KOGYO KOTO SENMON GAKKO (MIYAGI TECHNICAL COLLEGE)
48, NODAYAMA, SHIOTE, MEDESHIMA, NATORI-SHI 981-12
JAPANESE AND ENGLISH
INDEXED: MAT

MIYAGI-KEN NO KOGYO 5627
(CENSUS OF MANUFACTURES, MIYAGI PREFECTURE)
MIYAGI-KEN SOMU-BU (GENERAL AFFAIRS DIVISION, MIYAGI PREFECTURAL GOVERNMENT)
27, KOTODAI-DORI, SENDAI-SHI 980
JAPANESE

MIYAKONOJO KOGYO KOTO SENMON GAKKO KENKYU HOKOKU 5628
(MEMOIRS OF MIYAKONOJO TECHNICAL COLLEGE)
MIYAKONOJO KOGYO KOTO SENMON GAKKO (MIYAKONOJO TECHNICAL COLLEGE)
473-1, YOSHIO-CHO, MIYAKONOJO-SHI 885
JAPANESE AND EUROPEAN
INDEXED: CHE, JICST

MIYAZAKI DAIGAKU KOGAKUBU KENKYU HOKOKU 5629
(BULLETIN OF THE FACULTY OF ENGINEERING, MIYAZAKI UNIVERSITY)
MIYAZAKI DAIGAKU KOGAKUBU (FACULTY OF ENGINEERING, MIYAZAKI UNIVERSITY)
118, NISHIMARUYAMA-CHO, MIYAZAKI-SHI 880
JAPANESE AND ENGLISH
INDEXED: MAT, JICST

MIYAZAKI-KEN KOGYO SHIKENJO KENKYU HOKOKU 5630
(REPORT OF THE INDUSTRIAL RESEARCH INSTITUTE, MIYAZAKI PREFECTURE)
MIYAZAKI-KEN KOGYO SHIKENJO (INDUSTRIAL RESEARCH INSTITUTE OF MIYAZAKI PREFECTURE)
3515-1, KITABARA, TSUNEHISA, MIYAZAKI-SHI 880
JAPANESE AND ENGLISH

MIYAZAKI-KEN NO KOGYO 5631
(CENSUS OF MANUFACTURES, MIYAZAKI PREFECTURE)
MIYAZAKI-KEN KIKAKU CHOSEI-BU (PLANNING AND REGULATION DIVISION, MIYAZAKI PREFECTURAL GOVERNMENT)
10-1, 2-CHOME, TACHIBANA-DORI HIGASHI, MIYAZAKI-SHI 880
JAPANESE

MOL 5632
JAPANESE
INDEXED: JICST

MONBU-SHO KAGAKU KENKYUHI NI YORU KENKYU HOKOKU SHUROKU, KOGAKU-HEN 1 5633
(REPORTS ON RESEARCHES BY GRANTEES MINISTRY OF EDUCATION, ENGINEERING, PART 1)
NIHON GAKUJUTSU SHINKOKAI (JAPAN SOCIETY FOR THE PROMOTION OF SCIENCE)
3-1, 5-CHOME, KOJI-MACHI, CHIYODA-KU, TOKYO 102
JAPANESE

MONBU-SHO KAGAKU KENKYUHI NI YORU KENKYU HOKOKU SHUROKU, KOGAKU-HEN 2 5634
(REPORTS OF RESEARCHES BY GRANTEES MINISTRY OF EDUCATION, ENGINEERING, PART 2)
NIHON GAKUJUTSU SHINKOKAI (JAPAN SOCIETY FOR THE PROMOTION OF SCIENCE)
3-1, 5-CHOME, KOJI-MACHI, CHIYODA-KU, TOKYO 102
JAPANESE

MONBU-SHO KAGAKU KENKYUHI NI YORU KENKYU HOKOKU SHUROKU, KOGAKU-HEN 3 5635
(REPORTS ON RESEARCHES BY GRANTEES MINISTRY OF EDUCATION, ENGINEERING, PART 3)
NIHON GAKUJUTSU SHINKOKAI (JAPAN SOCIETY FOR THE PROMOTION OF SCIENCE)
3-1, 5-CHOME, KOJI-MACHI, CHIYODA-KU, TOKYO 102
JAPANESE

MUKI ZAISHITSU KENKYUSHO KENKYU HOKOKUSHO 5636
(SCIENCE REPORT OF THE NATIONAL INSTITUTE FOR RESEARCHES IN INORGANIC MATERIALS)
KAGAKU GIJUTSU-CHO MUKI ZAISHITSU KENKYUSHO (NATIONAL INSTITUTE FOR RESEARCHES IN INORGANIC MATERIALS, SCIENCE AND TECHNOLOGY AGENCY)
KURAGAKE, SAKURA-MURA, NIIHARI-GUN IBARAKI 300-31
JAPANESE
INDEXED: CHE

MUKI ZAISHITSU KENKYUSHO KENKYU RONBUNSHU 5637
(COLLECTED PAPERS OF NATIONAL INSTITUTE FOR RESEARCHES IN INORGANIC MATERIALS)
KAGAKU GIJUTSU-CHO MUKI ZAISHITSU KENKYUSHO (NATIONAL INSTITUTE FOR RESEARCHES IN INORGANIC MATERIALS, SCIENCE AND TECHNOLOGY AGENCY)
KURAGAKE, SAKURA-MURA, NIIHARI-GUN, IBARAKI 300-31
JAPANESE, WITH ENGLISH SUMMARY

MUKI ZAISHITSU KENKYUSHO KENKYU SHIRYO 5638
(RESEARCH REPORT OF THE NATIONAL INSTITUTE FOR RESEARCHES IN INORGANIC MATERIALS)
KAGAKU GIJUTSU-CHO MUKI ZAISHITSU KENKYUSHO (NATIONAL INSTITUTE FOR RESEARCHES IN INORGANIC MATERIALS, SCIENCE AND TECHNOLOGY AGENCY)
KURAGAKE, SAKURA-MURA, NIIHARI-GUN, IBARAKI 300-31
JAPANESE

MUKI ZAISHITSU KENKYUSHO NENPO 5639
(ANNUAL REPORT OF THE NATIONAL INSTITUTE FOR RESEARCHES IN INORGANIC MATERIALS)
KAGAKU GIJUTSU-CHO MUKI ZAISHITSU KENKYUSHO (NATIONAL INSTITUTE FOR RESEARCHES IN INORGANIC MATERIALS, SCIENCE AND TECHNOLOGY AGENCY)
KURAGAKE, SAKURA-MURA, NIIHARI-GUN, IBARAKI 300-31
JAPANESE
INDEXED: CHE

MUKIZAIKEN NYUSU 5640
(NATIONAL INSTITUTE FOR RESEARCHES IN INORGANIC MATERIALS NEWS)
KAGAKU GIJUTSU-CHO MUKI ZAISHITSU KENKYUSHO (NATIONAL INSTITUTE FOR RESEARCHES IN INORGANIC MATERIALS, SCIENCE AND TECHNOLOGY AGENCY)
KURAGAKE, SAKURA-MURA, NIIHARI-GUN IBARAKI 300-31
JAPANESE
INDEXED: CHE

MURORAN KOGYO DAIGAKU KENKYU HOKOKU 5641
(MEMOIRS OF THE MURORAN INSTITUTE OF TECHNOLOGY)
MURORAN KOGYO DAIGAKU (MURORAN INSTITUTE OF TECHNOLOGY)
27-1, MIZUMOTO-CHO, MURORAN-SHI 050
JAPANESE AND ENGLISH
INDEXED: CHE, JICST

NAGANO KOGYO KOTO SENMON GAKKO HOKOKU 5642
(REPORT OF THE NAGANO TECHNICAL COLLEGE)
JAPANESE

NAGANO KOGYO KOTO SENMON GAKKO KIYO 5643
(MEMOIRS OF NAGANO TECHNICAL COLLEGE)
NAGANO KOGYO KOTO SENMON GAKKO (NAGANO TECHNICAL COLLEGE)
716, TOKUMA, NAGANO-SHI 380
JAPANESE, WITH ENGLISH SUMMARY
INDEXED: JICST

NAGANO-KEN KOGYO SHIKENJO GYOMU-HOKOKU 5644
(ANNUAL REPORT OF NAGANO INDUSTRIAL RESEARCH INSTITUTE)
NAGANO-KEN KOGYO SHIKENJO (NAGANO INDUSTRIAL RESEARCH INSTITUTE)
188, WAKASATO, NAGANO-SHI 380
JAPANESE

NAGANO-KEN NO KOGYO, KOGYO TOKEI CHOSA KEKKA 5645
(MANUFACTURES IN NAGANO PREFECTURE, CENSUS OF MANUFACTURES)
NAGANO-KEN SOMU-BU (ADMINISTRATIVE DIVISION, NAGANO PREFECTURAL GOVERNMENT)
692-2, HABASHITA, MINAMINAGANO, NAGANO-SHI 380
JAPANESE

NAGAOKA KOGYO KOTO SENMON GAKKO KENKYU KIYO 5646
(RESEARCH REPORTS OF THE NAGAOKA TECHNICAL COLLEGE)
NAGAOKA KOGYO KOTO SENMON GAKKO (NAGAOKA TECHNICAL COLLEGE)
888, NISHIKATAKAI-MACHI, NAGAOKA-SHI 940
JAPANESE AND ENGLISH
INDEXED: CHE, MAT, INS, JICST

NAGAOKA NO KOGYO 5647
(MANUFACTURES IN NAGAOKA CITY)
NAGAOKA-SHI (NAGAOKA MUNICIPAL GOVERNMENT)
2-1, YANAGIHARA-MACHI, NAGAOKA-SHI 940
JAPANESE

NAGASAKI DAIGAKU KOGAKUBU KENKYU HOKOKU 5648
(REPORTS OF THE FACULTY OF ENGINEERING, NAGASAKI UNIVERSITY)
NAGASAKI DAIGAKU KOGAKUBU (FACULTY OF ENGINEERING, NAGASAKI UNIVERSITY)
1-14, BUNKYO-MACHI, NAGASAKI-SHI 852
JAPANESE, WITH ENGLISH SUMMARY
INDEXED: CHE, JICST

NAGASAKI-KEN NO KOGYO, KOGYO TOKEI CHOSA KEKKAHYO 5649
(MANUFACTURES IN NAGASAKI PREFECTURE, CENSUS OF MANUFACTURE)
NAGASAKI-KEN KIKAKU-BU (PLANNING DIVISION, NAGASAKI PREFECTURAL GOVERNMENT)
2-13, EDO-MACHI, NAGASAKI-SHI 850
JAPANESE

NAGASAKI-SHI NO KOGYO 5650
(MANUFACTURES IN NAGASAKI CITY)
NAGASAKI-SHI (NAGASAKI MUNICIPAL GOVERNMENT)
2-22, SAKURA-MACHI, NAGASAKI-SHI 850
JAPANESE

NAGOYA DAIGAKU KOGAKUBU DOATSU KENKYU SHISETSU SHISETSU RONBUNSHU 5651
(COLLECTION OF REPORTS OF EARTH PRESSURE RESEARCH LABORATORY, FACULTY OF ENGINEERING, NAGOYA UNIVERSITY)
NAGOYA DAIGAKU KOGAKUBU FUZOKU DOATSU KENKYU SHISETSU (EARTH PRESSURE LABORATORY, FACULTY OF ENGINEERING, NAGOYA UNIVERSITY)
FURO-CHO, CHIKUSA-KU, NAGOYA-SHI 464
JAPANESE AND ENGLISH

NAGOYA DAIGAKU KOGAKUBU JIDO SEIGYO KENKYU SHISETSU HOKOKU 5652
(RESEARCH REPORTS OF AUTOMATIC CONTROL LABORATORY, FACULTY OF ENGINEERING, NAGOYA UNIVERSITY)
NAGOYA DAIGAKU KOGAKUBU FUZOKU JIDO SEIGYO KENKYU SHISETSU (AUTOMATIC CONTROL LABORATORY, FACULTY OF ENGINEERING, NAGOYA UNIVERSITY)
FURO-CHO, CHIKUSA-KU, NAGOYA-SHI 464
JAPANESE AND ENGLISH
INDEXED: INS, CHE

NAGOYA KOGYO DAIGAKU GAKUHO 5653
(BULLETIN OF NAGOYA INSTITUTE OF TECHNOLOGY)
NAGOYA KOGYO DAIGAKU (NAGOYA INSTITUTE OF TECHNOLOGY)
GOKISHO-CHO, SHOWA-KU, NAGOYA-SHI 466
JAPANESE AND EUROPEAN
INDEXED: INS, MAT, JICST

NAGOYA KOGYO DAIGAKU ZAIRYO KAIHATSU KENKYU SHISETSU KENKYU HOKOKUSHU 5654
(COLLECTED RESEARCH REPORTS OF THE NAGOYA INSTITUTE OF TECHNOLOGY MATERIALS DEVELOPMENT RESEARCH FACILITIES)
NAGOYA INSTITUTE OF TECHNOLOGY
JAPANESE

NAGOYA KOGYO GIJUTSU SHIKENJO CHOSA HOKOKU 5655
(SURVEY REPORT OF THE NAGOYA INDUSTRIAL TECHNOLOGY LABORATORY)
JAPANESE
INDEXED: JICST

NAGOYA KOGYO GIJUTSU SHIKENSHO HOKOKU 5656
(REPORTS OF THE GOVERNMENT INDUSTRIAL RESEARCH INSTITUTE, NAGOYA)
TSUSHO SANGYO-SHO KOGYO GIJUTSU-IN NOGOYA KOGYO GIJUTSU SHIKENSHO (GOVERNMENT INDUSTRIAL RESEARCH INSTITUTE, NAGOYA, AGENCY OF INDUSTRIAL SCIENCE AND TECHNOLOGY)
1, HIRATE-CHO, KITA-KU, NAGOYA-SHI 462
JAPANESE AND ENGLISH
INDEXED: CHE, INS, MET, WAA, JICST

NAGOYA KOGYO GIJUTSU SHIKENSHO KENKYU HOKOKU 5657
(MEMOIRS OF THE GOVERNMENT INDUSTRIAL RESEARCH INSTITUTE, NAGOYA)
TSUSHO SANGYO-SHO KOGYO GIJUTSU-IN NAGOYA KOGYO GIJUTSU SHIKENSHO (GOVERNMENT INDUSTRIAL RESEARCH INSTITUTE, NAGOYA, AGENCY OF INDUSTRIAL SCIENCE AND TECHNOLOGY)
1, HIRATE-CHO, KITA-KU, NAGOYA-SHI 462
JAPANESE AND ENGLISH
INDEXED: CHE, JICST

NAGOYA-SHI KOGYO KENKYUJO GIJUTSU JOHO 5658
(TECHNICAL INFORMATION FROM THE NAGOYA MUNICIPAL INDUSTRIAL RESEARCH INSTITUTE)
NAGOYA MUNICIPAL INDUSTRIAL RESEARCH INSTITUTE
JAPANESE

NAGOYA-SHI KOGYO KENKYUSHO KENKYU HOKOKU 5659
(RESEARCH REPORTS ON THE NAGOYA MUNICIPAL INDUSTRIAL RESEARCH INSTITUTE)
NAGOYA-SHI KOGYO KENKYUSHO (NAGOYA MUNICIPAL INDUSTRIAL RESEARCH INSTITUTE)
3-24, ROKUBAN-CHO, ATSUTA-KU, NAGOYA-SHI 456
JAPANESE, WITH EUROPEAN SUMMARY
INDEXED: CHE, JICST

NAHA-SHI NO KOGYO, KOGYO TOKEI CHOSA KEKKA HOKOKU 5660
(MANUFACTURES IN NAHA CITY, CENSUS OF MANUFACTURES)
NAHA-SHI KIKAKU-BU (PLANNING DIVISION, NAHA MUNICIPAL OFFICE)
1-1, 1-CHOME, IZUMISAKI, NAHA-SHI 900
JAPANESE

NAKOSHI NYUSU 5661
(PREFECTURAL INDUSTRIAL RESEARCH INSTITUTE NEWS, NARA)
NARA-KEN KOGYO SHIKENJO (PREFECTURAL INDUSTRIAL RESEARCH INSTITUTE, NARA)
KASHIWAGI-CHO, NARA-SHI 630
JAPANESE

NANIWA KOGYO KOTO GAKKO KENKYU KIYO 5662
(RESEARCH BULLETIN OF NANIWA INDUSTRIAL HIGH SCHOOL)
NANIWA KOGYO KOTO GAKKO (NANIWA INDUSTRIAL HIGH SCHOOL)
JAPANESE

NARA KOGYO KOTO SENMON GAKKO KENKYU KIYO 5663
(RESEARCH REPORTS OF NARA TECHNICAL COLLEGE)
NARA KOGYO KOTO SENMON GAKKO (NARA TECHNICAL COLLEGE)
22, YATA-CHO, YAMATOKORIYAMA-SHI 639-11
JAPANESE, WITH ENGLISH SUMMARY
INDEXED: JICST

NARA-KEN KOGYO SHIKENJO GYOMU HOKOKU 5664
(OPERATION REPORT OF PREFECTURAL INDUSTRIAL RESEARCH INSTITUTE, NARA)
NARA-KEN KOGYO SHIKENJO (PREFECTURAL INDUSTRIAL RESEARCH INSTITUTE, NARA)
KASHIWAGI-CHO, NARA-SHI 630
JAPANESE

NATIONAL DIET LIBRARY NEWSLETTER 5665
JAPANESE AND ENGLISH
INDEXED: JICST

NETSU KANRI TO KOGAI 5666
(HEAT MANAGEMENT AND POLLUTION CONTROL)
NIHON NETSU ENERUGI GIJUTSU KYOKAI (JAPAN MANAGEMENT ASSOCIATION)
1-22, 3-CHOME, SHIBAKOEN, MINATO-KU, TOKYO 105
JAPANESE
INDEXED: CHE, JICST

NETSU SHORI 5667
(HEAT TREATMENT)
NIHON NETSU SHORI GIJUTSU KYOKAI (THE JAPAN SOCIETY FOR HEAT TREATMENT)
C/O SHINSEN BUILDING, 8-2, SHINSEN-CHO, SHIBUYA-KU, TOKYO 150
JAPANESE
INDEXED: JICST

NETSU SOKUTEI 5668
(CALORIMETRY AND THERMAL ANALYSIS)
SOCIETY OF CALORIMETRY AND THERMAL ANALYSIS,
JAPAN DAIICHI KANAMORI BUILDING, 1-5-31, YUSHIMA, BUNKYO-KU TOKYO 113
JAPANESE AND ENGLISH
INDEXED: CHE

NEW JAPAN SEWING MACHINE NEWS 5669
ENGLISH

NICHI-FUTSU KOGYO GIJUTSU 5670
NICHI-FUTSU KOGYO GIJUTSUKAI (SOCIETY FRANCO-JAPONAISE DE TECHNIQUES INDUSTRIELLES)
2-3, KANDA SURUGADAI, CHIYODA-KU, TOKYO 101
JAPANESE
INDEXED: JICST

NIHON DAIGAKU KOGAKUBU KIYO 5671
(JOURNAL OF THE COLLEGE OF ENGINEERING OF NIHON UNIVERSITY)
NIHON DAIGAKU KOGAKUBU (COLLEGE OF ENGINEERING, NIHON UNIVERSITY)
NAKAGAWARA, TOKUSADA, TAMURA-MACHI, KORIYAMA-SHI 963
JAPANESE AND ENGLISH
INDEXED: WAA, JICST

NIHON DAIGAKU KOGAKUBU KIYO A 5672
(JOURNAL OF THE COLLEGE OF ENGINEERING OF NIHON UNIVERSITY (SERIES A))
JAPANESE

NIHON DAIGAKU KOGAKUBU KIYO B 5673
(JOURNAL OF THE COLLEGE OF ENGINEERING OF NIHON UNIVERSITY (SERIES B))
JAPANESE

NIHON DAIGAKU RIKOGAKU KENKYUSHO SHOHO 5674
(JOURNAL OF THE RESEARCH INSTITUTE OF SCIENCE AND TECHNOLOGY, NIHON UNIVERSITY)
NIHON DAIGAKU RIKOGAKU KENKYUSHO (RESEARCH INSTITUTE OF SCIENCE AND TECHNOLOGY, NIHON UNIVERSITY)
1-8, KANDA SURUGADAI, CHIYODA-KU, TOKYO 101
JAPANESE AND ENGLISH
INDEXED: APP, CHE, INS, JICST

NIHON DAIGAKU RIKOGAKUBU IPPAN KYOIKU KYOSHITSU IHO 5675
(BULLETIN OF THE DEPARTMENT OF GENERAL EDUCATION, COLLEGE OF SCIENCE AND TECHNOLOGY, NIHON UNIVERSITY)
NIHON DAIGAKU RIGOGAKUBU IPPAN KYOIKU KYOSHITSU (DEPARTMENT OF GENERAL EDUCATION, COLLEGE OF SCIENCE AND TECHNOLOGY, NIHON UNIVERSITY)
1-8, KANDA SURUGADAI, CHIYODA-KU, TOKYO 101
JAPANESE AND ENGLISH
INDEXED: MAT

NIHON DAIGAKU SEISAN KOGAKUBU HOKOKU 5676
(JOURNAL OF THE COLLEGE INDUSTRIAL TECHNOLOGY, NIHON UNIVERSITY)
NIHON DAIGAKU SEISAN KOGAKUBU (COLLEGE OF INDUSTRIAL TECHNOLOGY, NIHON UNIVERSITY)
2-1, 1-CHOME, IZUMI-CHO, NARASHINO-SHI 275
JAPANESE AND ENGLISH
INDEXED: CHE, JICST

NIHON FUKUGO ZAIRYO GAKKAISHI 5677
(JOURNAL OF THE JAPAN SOCIETY FOR COMPOSITE MATERIALS)
JAPAN SOCIETY FOR COMPOSITE MATERIALS
JAPANESE

NIHON GAKUJUTSU KAIGI DAI-5-BU SHUHO 5678
(REPORT OF THE 5TH DIVISION, SCIENCE COUNCIL OF JAPAN)
NIHON GAKUJUTSU KAIGI (SCIENCE COUNCIL OF JAPAN)
22-34, 7-CHOME, ROPPONGI, MINATO-KU, TOKYO 106
JAPANESE
INDEXED: JICST

NIHON KOGAKKAI NENPO 5679
(ANNUAL REPORT OF THE JAPAN FEDERATION OF ENGINEERING SOCIETY)
NIHON KOGAKKAI (THE JAPAN FEDERATION OF ENGINEERING SOCIETY)
5-4, 8-CHOME, GINZA, CHUO-KU, TOKYO 104
JAPANESE

NIHON KOGAKKAI NYUSUS 5680
(JFES NEWS)
KYOEI TSUSHINSHA
2-1, 8-CHOME, GINZA, CHUO-KU, TOKYO 104
JAPANESE AND ENGLISH

NIHON KOGYO KURABU KAIHO 5681
(BULLETIN OF THE JAPAN INDUSTRIAL CLUB)
JAPANESE

NIHON KOGYO RITCHI SENTA KENKYU NENPO 5682
(ANNALS OF THE JAPAN INDUSTRIAL LOCATION CENTER)
NIHON KOGYO RITCHI SENTA (JAPAN INDUSTRIAL LOCATION CENTER)
2-1, SHIBA KOTOHIRA-CHO, MINATO-KU, TOKYO 105
JAPANESE

NIHON KOGYO SHINBUN 5683
(JAPAN INDUSTRIAL JOURNAL)
NIHON KOGYO SHINBUNSHA
7-2, 1-CHOME, OTE-MACHI, CHIYODA-KU, TOKYO 100
JAPANESE

NIHON SANGYO GIJUTSU KYOIKU GAKKAISHI 5684
(JOURNAL OF THE JAPAN INDUSTRIAL TECHNOLOGY EDUCATION SOCIETY)
JAPANESE

NIHON SANGYO GIJUTSU SHINKO KYOKAI GIJUTSU SHIRYO 5685
(TECHNICAL BULLETIN OF JAPAN INDUSTRIAL TECHNOLOGY ASSOCIATION)
NIHON SANGYO GIJUTSU SHINKO KYOKAI (JAPAN INDUSTRIAL TECHNOLOGY ASSOCIATION)
2-7, NISHISHINBASHI, MINATO-KU, TOKYO 105
JAPANESE

NIHON TOKEI GEPPO 5686
(MONTHLY STATISTICS OF JAPAN)
SORI-FU TOKEI-KYOKU (BUREAU OF STATISTICS, PRIME MINISTER'S OFFICE) AND NIHON TOKEI KYOKAI (JAPAN STATISTICAL ASSOCIATION)
95, WAKAMATSU-CHO, SHINJUKU-KU, TOKYO 162
JAPANESE AND ENGLISH
INDEXED: JICST

NIIGATA DAIGAKU KOGAKUBU KENKYU HOKOKU 5687
(RESEARCH REPORTS OF THE FACULTY OF ENGINEERING, NIIGATA UNIVERSITY)
NIIGATA DAIGAKU KOGAKUBU (FACULTY OF ENGINEERING, NIIGATA UNIVERSITY)
2-1, 1-CHOME, GAKKO-CHO, NAGAOKA-SHI 940
JAPANESE AND ENGLISH
INDEXED: CHE, INS, MAT, JICST

NIIGATA-KEN KOGYO GIJUTSU SENTA KENKYU HOKOKUSHO 5688
(REPORT OF NIIGATA PREFECTURAL CENTER OF INDUSTRIAL SCIENCE AND TECHNOLOGY)
NIIGATA-KEN KOGYO GIJUTSU SENTA (NIIGATA PREFECTURAL CENTER OF INDUSTRIAL SCIENCE AND TECHNOLOGY)
504-1, ABUMI, NIIGATA-SHI 950
JAPANESE

NIIGATA-KEN KOGYO GIJUTSU SENTA NENPO 5689
(ANNUAL REPORT OF NIIGATA PREFECTURAL CENTER OF INDUSTRIAL SCIENCE AND TECHNOLOGY)
NIIGATA-KEN KOGYO GIJUTSU SENTA (NIIGATA PREFECTURAL CENTER OF INDUSTRIAL SCIENCE AND TECHNOLOGY)
504-1, ABUMI, NIIGATA-SHI 950
JAPANESE

NIIGATA-KEN NO KOGYO, KOGYO TOKEI CHOSA KARA 5690
(MANUFACTURES IN NIIGATA PREFECTURE, CENSUS OF MANUFACTURES)
NIIGATA-KEN KIKAKU KAIHATSU-BU (PLANNING AND DEVELOPMENT DIVISION, NIIGATA PREFECTURAL GOVERNMENT)
602, ICHIBAN-CHO GAKKO-DORI, NIIGATA-SHI 951
JAPANESE

NIIHAMA KOGYO KOTO SENMON GAKKO KIYO, RIKOGAKU-HEN 5691
(MEMOIRS OF THE NIIHAMA TECHNICAL COLLEGE)
NIIHAMA KOGYO KOTO SENMON GAKKO (NIIHAMA TECHNICAL COLLEGE)
7-1, YAGUMO-CHO, NIIHAMA-SHI 792
JAPANESE, WITH ENGLISH SUMMARY
INDEXED: JICST

NIKKAN KOGYO SHINBUN 5692
(INDUSTRIAL DAILY NEWS)
NIKKAN KOGYO SHINBUNSHA (INDUSTRIAL DAILY NEWS COMPANY, LTD.)
8-10, 1-CHOME, KUDANKITA, CHIYODA-KU, TOKYO 102
JAPANESE
INDEXED: JICST

NINGEN KOGAKU 5693
(JAPANESE JOURNAL OF ERGONOMICS)
NIHON NINGEN KOGAKKAI (JAPAN ERGONOMICS RESEARCH SOCIETY)
KYOWA GINKO IKEBUKURO SHITEN KOTSU IGAKU KENKYU ZAIDAN, 9-3, 1-CHOME, HIGASHIIKEBUKURO TOSHIMA-KU TOKYO 170
JAPANESE, WITH ENGLISH SUMMARY
INDEXED: JICST

NIPPON GAKUJUTSU KAIGI, KYABITESHON NI KANSURU SHINPOJIUMU 5694
(JAPAN SCIENCE ASSOCIATION. SYMPOSIUM ON CAVITATION)
JAPANESE
INDEXED: JICST

NIPPON HINSHITSU KANRI GAKKAI NENJI TAIKAI KENKYU HAPPYO YOSHISHU 5695
(SUMMARIES OF PAPERS PRESENTED AT THE ANNUAL CONFERENCE OF THE JAPAN QUALITY CONTROL SOCIETY)
JAPANESE
INDEXED: JICST

NIPPON KOGAKUKAI SHINPOJIUMU TEKISUTO 5696
(TEXT OF THE SYMPOSIUM OF JAPAN ENGINEERING SOCIETY)
JAPANESE
INDEXED: JICST

NIPPON KOGYO DAIGAKU KENKYU HOKOKU 5697
(REPORT OF RESEARCHES, NIPPON INSTITUTE OF TECHNOLOGY)
NIPPON KOGYO DAIGAKU (NIPPON INSTITUTE OF TECHNOLOGY)
900 SUGA, MIYASHIRO-MACHI, MINAMISAITAMA-GUN, SAITAMA 345
JAPANESE AND ENGLISH
INDEXED: JICST

NIPPON NINGEN KOGAKUKAI TAIKAI RONBUNSHU 5698
(PAPERS OF THE BIG CONFERENCE OF THE JAPAN HUMAN ENGINEERING SOCIETY)
JAPANESE
INDEXED: JICST

NIPPON NO TOSHOKAN 5699
(JAPANESE LIBRARIES)
JAPANESE
INDEXED: JICST

NIPPON ZAIRYO KYODO GAKKAISHI 5700
(JOURNAL OF THE JAPANESE SOCIETY FOR STRENGTH AND FRACTURE OF MATERIALS)
NIPPON ZAIRYO KYODO GAKKAI (JAPANESE SOCIETY FOR STRENGTH AND FRACTURE OF MATERIALS)
C/O TOHOKU DAIGAKU KOGAKUBU ZAIRYO KYODO KENKYU SHISETSU, AOBA, ARAMAKI, SENDAI-SHI 980
JAPANESE AND ENGLISH
INDEXED: JICST

NITTSUKO GIHO (NIPPON TSUSHIN KOGGO K.K.) 5701
(REPORT OF NIPPON TSUSHIN NOGYO K.K.)
JAPANESE
INDEXED: JICST

NODA SANGYO KAGAKU KENKYUSHO KENKYU HOKOKU 5702
(REPORT OF THE NODA INSTITUTE FOR SCIENTIFIC RESEARCH)
NODA SANGYO KAGAKU KENKYUSHO (NODA INSTITUTE FOR SCIENTIFIC RESEARCH)
399, NODA, NODA-SHI 278
ENGLISH

NOHON SHUHO 5703
(WEEKLY REPORT OF ACQUISITION BOOKS)
DIET LIBRARY
JAPANESE
INDEXED: JICST

NORYOKU KAIHATSU KOGAKU SENTA KENKYU SHIRYO 5704
(RESEARCH MATERIALS FROM CENTER FOR ABILITY DEVELOPMENT ENGINEERING)
NORYOKU KAIHATSU KOGAKU SENTA (CENTER FOR ABILITY DEVELOPMENT ENGINEERING)
18-2, 4-CHOME, IGUSA, SUGINAMI-KU, TOKYO 167
JAPANESE AND ENGLISH

NUMAZU KOGYO KOTO SENMON GAKKO KENKYU HOKOKU 5705
(RESEARCH REPORTS OF NUMAZU TECHNICAL COLLEGE)
NUMAZU KOGYO KOTO SENMON GAKKO (NUMAZU TECHNICAL COLLEGE)
3600, OOKA, NUMAZU-SHI 410
JAPANESE AND ENGLISH
INDEXED: JICST

NYU ENERUGI 5706
(NEW ENERGY)
TOSHI ENERUGI KYORYOKUKAI
C/O KOJI-MACHI AKIYAMA BUILDING, 5-3, KOJI-MACHI, CHIYODA-KU, TOKYO 102
JAPANESE

OGATA PUROJEKUTO NYUSU 5707
(BIG PROJECT NEWS)
TSUSHO SANGYO-SHO KOGYO GIJUTSU-IN SOMU-BU (AGENCY INDUSTRIAL SCIENCE AND TECHNOLOGY, MINISTRY OF INTERNATIONAL TRADE AND INDUSTRY)
3-1, 1-CHOME, KASUMIGASEKI, CHIYODA-KU, TOKYO 100
JAPANESE

OITA KOGYO DAIGAKU KIYO 5708
(BULLETIN OF OITA INSTITUTE OF TECHNOLOGY)
JAPANESE

OITA KOGYO KOTO SENMON GAKKO KENKYU HOKOKU 5709
(RESEARCH REPORT OF OITA TECHNICAL COLLEGE)
OITA KOGYO KOTO SENMON GAKKO (OITA TECHNICAL COLLEGE)
1666, MAKI, OITA-SHI 870-01
JAPANESE, WITH ENGLISH SUMMARY
INDEXED: JICST

OITA-KEN BEPPU SANGYO KOGEI SHIKENSHO KENKYU HOKOKUSHO 5710
(RESEARCH REPORT OF BEPPU INDUSTRIAL ARTS INSTITUTE, OITA PREFECTURAL LABORATORY)
OITA-KEN BEPPU SANGYO KOGEI SHIKENSHO (BEPPU INDUSTRIAL ARTS INSTITUTE, OITA PREFECTURAL LABORATORY)
3-3, HIGASHISOEN-MACHI, BEPPU-SHI 874
JAPANESE

OITA-KEN KOGYO SHIKENJO NENPO 5711
(ANNUAL REPORT OF THE OITA-KEN INDUSTRIAL RESEARCH INSTITUTE)
OITA-KEN KOGYO SHIKENJO (OITA-KEN INDUSTRIAL RESEARCH INSTITUTE)
3239-4, SHIMOGORI, OITA-SHI 870
JAPANESE

OITA-KEN NO KOGYO (KOGYO TOKEI CHOSA KEKKA) 5712
(CENSUS OF MANUFACTURES, OITA PREFECTURE)
OITA-KEN KIKAKU-BU (PLANNING DIVISION, OITA PREFECTURAL GOVERNMENT)
1-1, 3-CHOME, OTE-MACHI, OITA-SHI 870
JAPANESE

OKAYAMA-KEN KOGAKU GIJUTSU SENTA HOKOKU 5713
(REPORT OF THE OKAYAMA PREFECTURE INDUSTRIAL TECHNOLOGY CENTER)
JAPANESE

OKAYAMA-KEN KOGYO SHIKENJO GYOMU HOKOKU 5714
(OPERATION REPORT OF OKAYAMA INDUSTRIAL RESEARCH INSTITUTE)
OKAYAMA-KEN KOGYO SHIKENJO
3-18, 4-CHOME, IFUKU-CHO, OKAYAMA-SHI 700
JAPANESE

OKAYAMA-KEN KOGYO SHIKENJO HOKOKU 5715
(REPORT OF THE OKAYAMA INDUSTRIAL RESEARCH INSTITUTE)
OKAYAMA-KEN KOGYO SHIKENJO
3-18, 4-CHOME, IFUKU-CHO, OKAYAMA-SHI 700
JAPANESE

OKINAWA-KEN KOGYO SHIKENJO GYOMU HOKOKU 5716
(REPORT OF OKINAWA INDUSTRIAL EXPERIMENT STATION)
OKINAWA-KEN KOGYO SHIKENJO
314, YORIMIYA, NAHA-SHI 902
JAPANESE

OKINAWA-KEN NO KOGYO 5717
(MANUFACTURES IN OKINAWA)
OKINAWA-KEN (OKINAWA PREFECTURAL GOVERNMENT)
2-1, 1-CHOME, IZUMISAKI, NAHA-SHI 900
JAPANESE

OMUTA-SHI KOGEI SHIDOSHO GYOMU HOKOKU 5718
(REPORT OF THE OMUTA INDUSTRIAL ART TRAINING INSTITUTE)
OMUTA-SHI KOGEI SHIDOSHO
48, ARIAKE-MACHI, OMUTA-SHI 836
JAPANESE

OSAKA KOGYO DAIGAKU CHUO KENKYUSHOHO 5719
(REPORT OF CENTRAL RESEARCH LABORATORY, OSAKA INSTITUTE OF TECHNOLOGY))
OSAKA KOGYO DAIGAKU CHUO KENKYUSHO (CENTRAL RESEARCH LABORATORIES, OSAKA INSTITUTE OF TECHNOLOGY)
16-1, 5-CHOME, NAKAMIYA, ASAHI-KU, OSAKASHI 535
JAPANESE
INDEXED: MAT, JICST

OSAKA KOGYO GIJUTSU SHIKENJO HOKOKU 5720
(REPORT OF THE GOVERNMENT INDUSTRIAL RESEARCH INSTITUTE, OSAKA)
JAPANESE

OSAKA KOGYO GIJUTSU SHIKENJO KIHO 5721
(BULLETIN OF THE GOVERNMENT INDUSTRIAL RESEARCH INSTITUTE, OSAKA)
JAPANESE

OSAKA KOGYO GIJUTSU SHIKENSHO HOKOKU 5722
(REPORT OF THE GOVERNMENT INDUSTRIAL RESEARCH INSTITUTE, OSAKA)
TSUSHO SANGYO-SHO KOGYO GIJUTSU-IN OSAKA KOGYO GIJUTSU SHIKENSHO (GOVERNMENT INDUSTRIAL RESEARCH INSTITUTE, OSAKA, AGENCY OF INDUSTRIAL SCIENCE AND TECHNOLOGY)
8-31, 1-CHOME, MIDORIGAOKA, IKEDA-SHI 563
JAPANESE AND ENGLISH
INDEXED: CHE, JICST

OSAKA KOGYO GIJUTSU SHIKENSHO KIHO 5723
(BULLETIN OF THE GOVERNMENT INDUSTRIAL RESEARCH INSTITUTE, OSAKA)
TSUSHO SANGYO-SHO KOGYO GIJUTSU-IN OSAKA KOGYO GIJUTSU SHIKENSHO (GOVERNMENT INDUSTRIAL RESEARCH INSTITUTE, OSAKA, AGENCY OF INDUSTRIAL SCIENCE AND TECHNOLOGY)
8-31, 1-CHOME, MIDORIGAOKA IKEDA-SHI 563
JAPANESE AND ENGLISH
INDEXED: CHE, JICST

OSAKA KOGYO GIJUTSU SHIKENSHO NENPO 5724
(ANNUAL REPORT OF THE GOVERNMENT INDUSTRIAL RESEARCH INSTITUTE, OSAKA)
TSUSHO SANGYO-SHO KOGYO GIJUTSU-IN OSAKA KOGYO GIJUTSU SHIKENSHO (GOVERNMENT INDUSTRIAL RESEARCH INSTITUTE, OSAKA AGENCY OF INDUSTRIAL SCIENCE AND TECHNOLOGY)
8-31, 1-CHOME, MIDORIGAOKA, IKEDA-SHI 563
JAPANESE
INDEXED: JICST

OSAKA NO KOGYO TOKEI SOKUHO 5725
(QUICK ANNOUNCEMENT OF INDUSTRIAL STATISTICS IN OSAKA PREFECTURE)
OSAKA-FU KIKAKU-BU (PLANNING DIVISION, OSAKA PREFECTURAL GOVERNMENT)
2, OTEMAENO-CHO, HIGASHI-KU, OSAKA-SHI 540
JAPANESE

OSAKA-FURITSU KOGYO GIJUTSU KENKYUSHO GYOMU NENPO 5726
(OPERATION YEARBOOK OF OSAKA PREFECTURAL INDUSTRIAL RESEARCH INSTITUTE)
OSAKA-FURITSU KOGYO GIJUTSU KENKYUSHO
KAMINO-CHO, ENOKOJIMA, NISHI-KU, OSAKA-SHI 550
JAPANESE

OSAKA-FURITSU KOGYO GIJUTSU KENKYUSHO HOKOKU 5727
(REPORTS OF OSAKA PREFECTURAL INDUSTRIAL RESEARCH INSTITUTE)
OSAKA-FURITSU KOGYO GIJUTSU KENKYUSHO (OSAKA PREFECTURAL INDUSTRIAL RESEARCH INSTITUTE)
KAMINO-CHO, ENOKOJIMA, NISHI-KU, OSAKA-SHI 550
JAPANESE AND ENGLISH
INDEXED: CHE, JICST

OSAKA-FURITSU KOGYO GIJUTSU KENKYUSHO NYUSU 5728
(NEWS OF OSAKA PREFECTURAL INDUSTRIAL RESEARCH INSTITUTE)
OSAKA-FURITSU KOGYO GIJUTSU KENKYUSHO (OSAKA PREFECTURAL INDUSTRIAL RESEARCH INSTITUTE)
KAMINO-CHO, ENOKOJIMA, NISHI-KU, OSAKA-SHI 550
JAPANESE

OSAKA-FURITSU KOGYO KOTO SENMON GAKKO KENKYU KIYO 5729
(BULLETIN OF OSAKA PREFECTURAL TECHNICAL COLLEGE)
OSAKA-FURITSU KOGYO KOTO SENMON GAKKO
26-12, SAIWAI-CHO, NEYAGAWA-SHI 572
JAPANESE AND ENGLISH
INDEXED: JICST

OSAKA-SHIRITSU KOGYO KENKYUJO HOKOKU 5730
(BULLETIN OF THE OSAKA MUNICIPAL TECHNICAL RESEARCH INSTITUTE)
JAPANESE

OSAKA-SHIRITSU KOGYO KENKYUSHO GYOMU NENPO 5731
(ANNUAL REPORT OF ACTIVITIES OF THE OSAKA MUNICIPAL TECHNICAL RESEARCH INSTITUTE)
OSAKA-SHIRITSU KOGYO KENKYUSHO (THE OSAKA MUNICIPAL TECHNICAL RESEARCH INSTITUTE)
38, KITAOGI-MACHI, KITA-KU, OSAKA-SHI 530
JAPANESE
INDEXED: JICST

OSAKA-SHIRITSU KOGYO KENKYUSHO HOKOKU 5732
(BULLETIN OF THE OSAKA MUNICIPAL TECHNICAL RESEARCH INSTITUTE)
OSAKA-SHIRITSU KOGYO KENKYUSHO (OSAKA MUNICIPAL TECHNICAL RESEARCH INSTITUTE)
38, KITAOGI-MACHI, KITA-KU, OSAKA-SHI 530
JAPANESE
INDEXED: CHE, JICST

OTOMESHON 5733
(AUTOMATION)
NIKKAN KOGYO SHINBUNSHA (INDUSTRIAL DAILY NEWS COMPANY, LTD.)
8-10, 1-CHOME, KUDANKITA, CHIYODA-KU, TOKYO 102
JAPANESE
INDEXED: JICST

OYAMA KOGYO KOTO SENMON GAKKO KENKYU KIYO 5734
(RESEARCH REPORTS OF OYAMA TECHNICAL COLLEGE)
OYAMA KOGYO KOTO SENMON GAKKO (OYAMA TECHNICAL COLLEGE)
771, NAKAKUKI, OYAMA-SHI 323
JAPANESE AND ENGLISH
INDEXED: JICST

PARUSU 5735
(PULSE)
NIHON KAGAKU GIJUTSU KENSHUSHO
10-11, 5-CHOME, SENDAGAYA, SHIBUYA-KU, TOKYO 151
JAPANESE

PIS: PULANNINGU, INFOMESHON, SHISUTEMU 5736
(PIS: PLANNING, INFORMATION, SYSTEM)
JAPANESE
INDEXED: JICST

PROCEEDINGS OF THE JAPAN CONGRESS ON MATERIALS RESEARCH 5737
NIHON ZAIRYO GAKKAI (SOCIETY OF MATERIALS SCIENCE, JAPAN)
1-101, YOSHIDA IZUMIDONO-CHO, SAKYO-KU, KYOTO-SHI 606
ENGLISH
INDEXED: APP, CHE, ENG, JICST

PURANTO ENJINIA 5738
(PLANT ENGINEER)
NIHON NORITSU KYOKAI (JAPAN MANAGEMENT ASSOCIATION)
C/O KYORITSU GUILDING, 1-22, 3-CHOME, SHIBAKOEN, MINATO-KU, TOKYO 105
JAPANESE
INDEXED: JICST

PUROFIRU 5739
(PROFILE)
SUMITOMO SURI EMU K.K. (SUMITOMO 3M LTD.)
33-1, 2-CHOME, TAMAGAWADAI, SETAGAYA-KU, TOKYO 158
JAPANESE

R.I.E.E.C. RESEARCH BULLETIN 5740
ENGLISH

RASA KOGYO K.K. GIJUTSU KENKYUBU KENKYU HOKOKU 5741
(RESEARCH REPORT OF MIYAKO LABORATORY, TECHNICAL RESEARCH DEPARTMENT, RASA INDUSTRIES LTD.))
RASA KOGYO K.K. GIJUTSU KENKYU-BU MIYAKO KENKYUSHO (MIYAKO LABORATORY, TECHNICAL RESEARCH DEPARTMENT, RASA INDUSTRIES LTD.)
7, DAI-6-CHIWARI, KOYAMADA, MIYAKO-SHI 027
JAPANESE

REPORT OF THE RESEARCH INSTITUTE OF INDUSTRIAL TECHNOLOGY - NIHON UNIVERSITY 5742
ENGLISH

REPORT OF THE RESEARCH LABORATORY OF ENGINEERING MATERIALS, TOKYO INSTITUTE OF TECHNOLOGY 5743
RESEARCH LABORATORY OF ENGINEERING MATERIALS, TOKYO INSTITUTE OF TECHNOLOGY
O-OKAYAMA, MEGURO-KU, TOKYO 152
ENGLISH
INDEXED: CHE

REPORT OF THE RESEARCH LABORATORY OF RESOURCES UTILIZATION 5744
TOKYO KOGYO DAIGAKU SHIGEN KAGAKU KENKYUSHO (RESEARCH LABORATORY OF RESOURCES UTILIZATION, TOKYO INSTITUTE OF TECHNOLOGY)
1-45, 1-CHOME, ISHIKAWA-CHO, OTA-KU, TOKYO 145
ENGLISH
INDEXED: JICST

REPORT OF THE SCIENCE AND ENGINEERING RESEARCH LABORATORY, WASEDA UNIVERSITY 5745
WASEDA DIAGAKU RIKOGAKU KENTYUSHO (WASEDA UNIVERSITY, SCIENCE AND ENGINEERING RESEARCH LABORATORY)
17, KIKUI-CHO, SHINJUKU-KU, TOKYO 162
ENGLISH

REPORTS OF THE RESEARCH INSTITUTE FOR STRENGTH AND FRACTURE OF MATERIALS, TOHOKU UNIVERSITY 5746
TOHOKU DAIGAKU KOGAKUBU ZAIRYO KYODO KENKYU SHISETSU (RESEARCH INSTITUTE FOR STRENGTH AND FRACTURE OF MATERIALS, TOHOKU UNIVERSITY)
AOBA, ARAMAKI, SENDAI-SHI 980
ENGLISH
INDEXED: APP, JICST

RESEARCH ACTIVITIES OF THE ENGINEERING RESEARCH INSTITUTE, KYOTO UNIVERSITY 5747
KYOTO UNIVERSITY, ENGINEERING RESEARCH INSTITUTE
ENGLISH

RESEARCH REPORT OF THE RESEARCH INSTITUTE OF FUNDAMENTAL INFORMATION SCIENCE, KYUSHU UNIVERSITY 5748
JAPANESE
INDEXED: MAT

RESEARCH REPORTS OF THE FACULTY OF ENGINEERING, MIE UNIVERSITY 5749
ENGLISH

RITE REVIEW 5750
DENKI TSUSHIN SOGO KENKYUSHO (TELECOMMUNICATIONS GENERAL RESEARCH INSTITUTE)
JAPANESE

ROBOTTO 5751
(ROBOT)
NIHON SANGYO YO ROBOTTO KOGYOKAI
(JAPAN INDUSTRIAL SOCIETY OF ROBOTS FOR PRODUCTION USE)
JAPANESE

ROKU SHA 5752
(DEER CART)
SHIBAURA KOGYO DAIGAKU KOTO GAKKO (SHIBAURA INDUSTRIAL UNIVERSITY HIGH SCHOOL)
JAPANESE

RYUKYU DAIGAKU RIKO GAKUBU KIYO: KOGAKU-HEN 5753
(BULLETIN OF THE RYUKYU UNIVERSITY FACULTY OF SCIENCE AND ENGINEERING, ENGINEERING EDITION)
JAPANESE

SAGA RYUTANI TANKI DAIGAKU KIYO 5754
(BULLETIN OF THE SAGA RYUTANI JUNIOR COLLEGE)
SAGA RYUTANI GAKKAI (SAGA RYUTANI LEARNED SOCIETY)
JAPANESE

SAGA-KEN KOGYO SHIKENJO GYOMU HOKOKUSHO 5755
(OPERATION REPORT OF INDUSTRIAL RESEARCH INSTITUTE, SAGA PREFECTURE)
SAGA-KEN KOGYO SHIKENJO
IPPONMATSU, KONO-MACHI, SAGA-SHI 840
JAPANESE

SAGA-KEN KOGYO SHIKENJO KENKYU HOKOKUSHO 5756
(RESEARCH REPORT OF THE INDUSTRIAL RESEARCH INSTITUTE, SAGA PREFECTURE)
SAGA-KEN KOGYO SHIKENJO (INDUSTRIAL RESEARCH INSTITUTE, SAGA PREFECTURE)
IPPONMATSU, KONO-MACHI, SAGA-SHI 840
JAPANESE

SAGA-KEN NO KOGYO, KOGYO TOKEI CHOSA KEKKA HOKOKUSHO 5757
(MANUFACTURES IN SAGA PREFECTURE, CENSUS OF MANUFACTURES)
SAGA-KEN KIKAKU-BU (PLANNING DIVISION, SAGA PREFECTURAL GOVERNMENT)
1-59, 1-CHOME, JONAI, SAGA-SHI 840
JAPANESE

SAGAMI KOGYO DAIGAKU KIYO 5758
(BULLETIN OF SAGAMI INSTITUTE OF TECHNOLOGY)
SAGAMI KOGYO DAIGAKU
1-25, 1-CHOME, NISHIKAIGAN, TSUJIDO, FUJISAWA-SHI 251
JAPANESE, WITH EUROPEAN SUMMARY
INDEXED: CHE

SAITAMA DAIGAKU KIYO, RIKOGAKUBU, KOGAKUKEI 5759
(SCIENCE AND ENGINEERING REPORTS OF SAITAMA UNIVERSITY, SERIES C)
SAITAMA DAIGAKU RIKOGAKUBU
255, SHIMOOKUBO, URAWA-SHI 338
JAPANESE AND ENGLISH
INDEXED: INS, JICST

SAITAMA-KEN KOGEI SHIKENJO GYOMU HOKOKU 5760
(REPORT OF INDUSTRIAL ARTS INSTITUTE OF SAITAMA)
SAITAMA-KEN KOGEI SHIKENJO (INDUSTRIAL ARTS INSTITUTE OF SAITAMA)
SHIMOKIZAKI, URAWA-SHI 338
JAPANESE

SAITAMA-KENRITSU URAWA KOGYO KOTO GAKKO KIYO 5761
(BULLETIN OF THE SAITAMA PREFECTURAL URAWA TECHNICAL HIGH SCHOOL)
SAITAMA PREFECTURAL URAWA TECHNICAL HIGH SCHOOL
JAPANESE

SANGYO GURAFU 5762
(GRAPHIC MAGAZINE OF INDUSTRY)
NIHON KEIZAI KYOIKU SENTA
C/O DAI-11 MORI BUILDING 11, SHIBA NISHIKUBO AKEFUNE-CHO, MINATO-KU, TOKYO 105
JAPANESE

SANGYO KAGAKU 5763
(INDUSTRIAL SCIENCES)
SANGYO KAGAKU KYOKAI (INDUSTRIAL SCIENCES ASSOCIATION)
8, ICHIGAYA KAWADA-CHO, SHINJUKU-KU, TOKYO 160
JAPANESE

SANGYO KANKYO KOGAKU 5764
(INDUSTRIAL ENVIRONMENTAL ENGINEERING)
JAPANESE
INDEXED: JICST

SANGYO KANKYO NENPO 5765
(INDUSTRIAL ENVIRONMENT ANNUAL)
JAPANESE
INDEXED: JICST

SANGYO NO NAGOYA 5766
(INDUSTRIAL NAGOYA)
NAGOYA-SHI KEIZAI-KYOKU (ECONOMIC BUREAU, NAGOYA MUNICIPAL GOVERNMENT)
1-1, 3-CHOME, SANNOMARU, NAKA-KU, NAGOYA-SHI 460
JAPANESE

SANGYO NORITSU RONSHU 5767
(PAPERS ON INDUSTRIAL EFFICIENCY)
OSAKA FURITSU SANGYO NORITSU KENKYUJO (OSAKA PREFECTURAL INDUSTRIAL EFFICIENCY RESEARCH INSTITUTE)
JAPANESE
INDEXED: JICST

SANGYO NORITSU TANKI DAIGAKU KIYO 5768
(BULLETIN OF THE INDUSTRIAL EFFICIENCY JUNIOR COLLEGE)
INDUSTRIAL EFFICIENCY JUNIOR COLLEGE
JAPANESE

SANGYO SHIKENJO IHO 5769
(BULLETIN OF THE INDUSTRIAL EXPERIMENT STATION)
JAPANESE
INDEXED: BIB

SANGYO TO ROPU 5770
(ROPES FOR INDUSTRY)
TOKYO SEIKO K.K. (TOKYO ROPE COMPANY, LTD.)
2-8, NIHONBASHI MURO-MACHI, CHUO-KU, TOKYO 103
JAPANESE

SANKEN 5771
(INDUSTRIAL RESEARCH)
JAPANESE
INDEXED: JICST

SANKEN RONSHU 5772
(PAPERS ON INDUSTRIAL RESEARCH)
KANSAI GAKUIN DAIGAKU SANGYO KENKYUJO (KANSAI GAKUIN UNIVERSITY, INDUSTRIAL RESEARCH INSTITUTE)
JAPANESE

SAN'YO GIJUTSU ZASSHI 5773
(JOURNAL OF THE SAN'YO ASSOCIATION FOR ADVANCEMENT OF SCIENCE & TECHNOLOGY)
SAN'YO GIJUTSU SHINKOKAI (SAN'YO ASSOCIATION FOR ADVANCEMENT OF SCIENCE & TECHNOLOGY)
1-1, 16-CHOME, HON-MACHI, KURASHIKI-SHI 710
JAPANESE
INDEXED: BIB, JICST

SASEBO KOGYO KOTO SENMON GAKKO KENKYU HOKOKU 5774
(RESEARCH REPORTS OF SASEBO TECHNICAL COLLEGE)
SASEBO KOGYO KOTO SENMON GAKKO (SASEBO TECHNICAL COLLEGE)
1-1, OKISHIN-MACHI, SASEBO-SHI 857-11
JAPANESE AND ENGLISH
INDEXED: JICST

SEIHIN KAGAKU KENKYUSHO KENKYU HOKOKU 5775
(BULLETIN OF INDUSTRIAL PRODUCTS RESEARCH INSTITUTE)
TSUSHO SANGYO-SHO KOGYO GIJUTSU-IN SEIHIN KAGAKU KENKYUSHO (INDUSTRIAL PRODUCTS RESEARCH INSTITUTE, AGENCY OF INDUSTRIAL SCIENCE AND TECHNOLOGY, MINISTRY OF INTERNATIONAL TRADE AND INDUSTRY)
21-2, 4-CHOME, SHIMOMARUKO, OTA-KU, TOKYO 144
JAPANESE, WITH ENGLISH SUMMARY
INDEXED: JICST

SEIHIN KAGAKU KENKYUSHO NENPO 5776
(ANNUAL REPORT OF THE INDUSTRIAL PRODUCTS RESEARCH INSTITUTE)
TSUSHO SANGYO-SHO KOGYO GIJUTSU-IN SEIHIN KAGAKU KENKYUSHO (INDUSTRIAL PRODUCTS RESEARCH INSTITUTE, AGENCY OF INDUSTRIAL SCIENCE AND TECHNOLOGY, MINISTRY OF INTERNATIONAL TRADE AND INDUSTRY)
JAPANESE
INDEXED: JICST

SEIKAKEN NYUSU 5777
(I P R I NEWS)
TSUSHO SANGYO-SHO KOGYO GIJUTSU-IN SEIHIN KAGAKU KENKYUSHO (INDUSTRIAL PRODUCTS RESEARCH INSTITUTE, AGENCY OF INDUSTRIAL SCIENCE AND TECHNOLOGY, MINISTRY OF INTERNATIONAL TRADE & INDUSTRY)
21-2, 4-CHOME, SHIMOMARUKO, OTA-KU, TOKYO 144
JAPANESE

SEIKEI DAIGAKU KOGAKUBU KOGAKU HOKOKU 5778
(TECHNOLOGY REPORTS OF THE SEIKEI UNIVERSITY)
SEIKEI DAIGAKU KOGAKUBU (COLLEGE OF TECHNOLOGY, SEIKEI UNIVERSITY)
3-1, 3-CHOME, KICHIJOJI-KITA-MACHI, MUSASHINO-SHI TOKYO 180
JAPANESE, WITH ENGLISH SUMMARY
INDEXED: CHE, INS, JICST

SEIKEI DAIGAKU KOGAKUBU KOGYO TOKUBETSU HOKOKU 5779
(SPECIAL ENGINEERING REPORT OF THE SEIKEI UNIVERSITY ENGINEERING FACULTY)
JAPANESE

SEIKOSHI NYUSU (SHIZUOKA-KEN KOGYO SHIKENJO) 5780
(SHIZUOKA PREFECTURAL INDUSTRIAL LABORATORY NEWS)
JAPANESE
INDEXED: JICST

SEISAN KENKYU 5781
(MONTHLY JOURNAL OF THE INSTITUTE OF INDUSTRIAL SCIENCE, UNIVERSITY OF TOKYO)
TOKYO DAIGAKU SEISAN GIJUTSU KENKYUSHO (INSTITUTE OF INDUSTRIAL SCIENCE, UNIVERSITY OF TOKYO)
22-1, 7-CHOME, ROPPONGIN, MINATO-KU, TOKYO 106
JAPANESE AND ENGLISH
INDEXED: APP, CHE, COP, JICST

SEISAN KENKYU 5782
(MONTHLY JOURNAL OF THE INSTITUTE OF INDUSTRIAL SCIENCE, UNIVERSITY OF TOKYO)
JAPANESE

SEISAN SHUKKA ZAIKO TOKEI SOKUHO 5783
(STATISTICS OF PRODUCTION, FORWARDING AND STOCK)
TSUSHO SANGYO-SHO DAIJIN KANBO CHOSA TOKEI-BU (RESEARCH AND STATISTICS DIVISION, MINISTER'S SECRETARIAT, MINISTRY OF INTERNATIONAL TRADE AND INDUSTRY) YOSHITOMI SHOKAI
4-24, 1-CHOME, ROPPONGI, MINATO-KU, TOKYO 106
JAPANESE

SEISAN TO GIJUTSU 5784
(PRODUCTION AND TECHNIQUE)
SEISAN GIJUTSU SHINKO KYOKAI
C/O OSAKA DAIGAKU KOGYO KAIKAN, 125-18, 5-CHOME, FUJISHIRODAI, SUITA-SHI 565
JAPANESE
INDEXED: JICST

SEISAN ZAIKO SHISU (AICHI-KEN) 5785
(INDEX OF MINING AND MANUFACTURING, AICHI PREFECTURE)
AICHI-KEN KIKAKU-BU (PLANNING DIVISION, AICHI PREFECTURAL GOVERNMENT)
1-2, 3-CHOME, SANNOMARU, NAKA-KU, NAGOYA-SHI 460
JAPANESE

SEISANSEI TOKEI GEPPO 5786
(MONTHLY JOURNAL OF PRODUCTIVITY STATISTICS)
NIHON SEISANSEI HONBU SEISANSEI KENKYUSHO (PRODUCTIVITY RESEARCH INSTITUTE)
1-1, 3-CHOME, SHIBUYA, SHIBUYA-KU, TOKYO 150
JAPANESE

SENDAI-SHI NO KOGYO 5787
(MANUFACTURES IN SENDAI CITY)
SENDAI-SHI SOMU-KYOKU (SENDAI MUNICIPAL GOVERNMENT)
7-1, 3-CHOME, KOKUBUN-CHO, SENDAI-SHI 980
JAPANESE

SENMON TOSHOKAN 5788
(SPECIAL LIBRARIES)
JAPANESE
INDEXED: JICST

SENSUI GIJUTSU SHINPOJIUMU GIJIROKU 5789
(JAPAN WORKING DIVERS SYMPOSIUM)
JAPANESE
INDEXED: JICST

SENTAN GIJUTSU SOSETSU 5790
(PAPERS ON VANGUARD TECHNOLOGY)
JAPANESE
INDEXED: JICST

SETOUCHI SANGYO BUNKA KENKYU 5791
(STUDIES OF SETOUCHI INDUSTRY AND CULTURE)
MOMOYAMA GAKUIN TANKI DAIGAKU SETOUCHI SANGYO BUNKA KENKYUJO (SETOUCHI INDUSTRIAL AND CULTURAL RESEARCH INSTITUTE, MOMOYAMA JUNIOR COLLEGE)
JAPANESE

SHIBAURA KOGYO DAIGAKU DAIGAKUIN KENKYU HOKOKU 5792
(RESEARCH REPORT OF GRADUATE SCHOOL, SHIBAURA INSTITUTE OF TECHNOLOGY)
SHIBAURA KOGYO DAIGAKU DAIGAKUIN
9-14, 3-CHOME, SHIBAURA, MINATO-KU, TOKYO 108
JAPANESE

SHIBAURA KOGYO DAIGAKU KENKYU HOKOKU 5793
(RESEARCH REPORTS OF SHIBAURA INSTITUTE OF TECHNOLOGY)
SHIBAURA KOGYO DAIGAKU (SHIBAURA INSTITUTE OF TECHNOLOGY)
9-14, 3-CHOME, SHIBAURA, MINATO-KU, TOKYO 108
JAPANESE, WITH ENGLISH SUMMARY
INDEXED: JICST

SHIBAURA KOGYO DAIGAKU KIYO 5794
(REVIEW OF ARTS AND SCIENCES)
SHIBAURA KOGYO DAIGAKU (SHIBAURA INSTITUTE OF TECHNOLOGY)
9-14, 3-CHOME, SHIBAURA, MINATO-KU, TOKYO 108
JAPANESE

SHIBAURA KOGYO DAIGAKU KOGAKUBU KIYO 5795
(BULLETIN OF THE SHIBAURA INSTITUTE OF TECHNOLOGY, DEPARTMENT OF ENGINEERING)
SHIBAURA INSTITUTE OF TECHNOLOGY
JAPANESE

SHIBAURA KOGYO TANKI DAIGAKU KIYO 5796
(REVIEW OF TECHNOLOGY AND LIBERAL ARTS)
SHIBAURA KOGYO TANKI DAIGAKU (SHIBAURA JUNIOR COLLEGE OF TECHNOLOGY)
9-14, 3-CHOME, SHIBAURA, MINATO-KU, TOKYO 108
JAPANESE AND EUROPEAN

SHIGEN 5797
(RESOURCES)
KAGAKU GIJUTSU-CHO SHIGEN CHOSASHO (NATIONAL INSTITUTE OF RESOURCES, SCIENCE AND TECHNOLOGY AGENCY)
6-3, 3-CHOME, NISHISHINBASHI, MINATO-KU, TOKYO 105
JAPANESE
INDEXED: CHE, JICST

SHIGEN CHOSASHO SHIRYO 5798
(INFORMATION NEWS OF THE NATIONAL INSTITUTE OF RESOURCES)
KAGAKU GIJUTSU-CHO SHIGEN CHOSASHO (NATIONAL INSTITUTE OF RESOURCES, SCIENCE AND TECHNOLOGY AGENCY)
2-1, 2-CHOME, KASUMIGASEKI, CHIYODA-KU, TOKYO 100
JAPANESE

SHIGEN GIJUTSU SHIKENJO HOKOKU 5799
(NATURAL RESOURCES TECHNOLOGY EXPERIMENT STATION REPORT)
KOGYO GIJUTSUIN SHIGEN GIJUTSU SHIKENJO (AGENCY OF INDUSTRIAL SCIENCE & TECHNOLOGY, NATURAL RESOURCES TECHNOLOGY EXPERIMENT STATION)
JAPANESE

SHIGEN SOGO KAIHATSU KENKYUSHO KENKYU HOKOKU 5800
(REPORT OF THE RESOURCES EXPLOITATION INSTITUTE)
SHIGEN SOGO KAIHATSU KENKYUSHO (RESOURCES EXPLOITATION INSTITUTE)
C/O SHIBUYA PARU HOMU, 29-25, SAKURAGAOKA-CHO, SHIBUYA-KU, TOKYO 150
JAPANESE

SHIGUMA M.H.I. 5801
(SIGMA M.H.I.)
MITSUBISHI JUKOGYO K.K. (MITSUBISHI HEAVY INDUSTRIES, LTD.)
5-1, 2-CHOME, MARUNOUCHI, CHIYODA-KU, TOKYO 100
ENGLISH

SHIKISAI JOHO 5802
(COLOR REPORT)
KARA PURANINGU SENTA (COLOR PLANNING CENTER)
JAPANESE

SHIKISAI KENKYU 5803
(STUDIES OF COLOR)
NIHON SHIKISAI KENKYUSHO (JAPAN COLOR RESEARCH INSTITUTE)
C/O ROPPONGI KOYAMA BUILDING, 1-19, 3-CHOME, NISHIAZABU, MINATO-KU, TOKYO 106
JAPANESE, WITH ENGLISH SUMMARY
INDEXED: JICST

SHIKIZAI KYOKAISHI 5804
(JOURNAL OF THE JAPAN SOCIETY OF COLOR MATERIALS)
SHIKIZAI KYOKAI (JAPAN SOCIETY OF COLOR MATERIAL)
C/O SAITO BUILDING, 1-9, NISHIKANDA, CHIYODA-KU, TOKYO 101
JAPANESE, WITH ENGLISH SUMMARY
INDEXED: CHE, JICST

SHIKOKU KOGYO GIJUTSU SHIKENSHO HOKOKU 5805
(REPORTS OF THE GOVERNMENT INDUSTRIAL RESEARCH INSTITUTE, SHIKOKU)
TSUSHO SANGYO-SHO KOGYO GIJUTSU-IN SHIKOKU KOGYO GIJUTSU SHIKENSHO (GOVERNMENT INDUSTRIAL RESEARCH INSTITUTE, SHIKOKU, SCIENCE AND TECHNOLOGY AGENCY, MINISTRY OF INTERNATIONAL TRADE AND INDUSTRY)
3-3, 2-CHOME, HANANOMIYA-CHO, TAKAMATSU-SHI 760
JAPANESE, WITH ENGLISH SUMMARY
INDEXED: JICST

SHIKOKU KOGYO GIJUTSU SHIKENSHO NENPO 5806
(ANNUAL REPORT OF THE GOVERNMENT INDUSTRIAL RESEARCH INSTITUTE, SHIKOKU)
TSUSHO SANGYO-SHO KOGYO GIJUTSU-IN SHIKOKU KOGYO GIJUTSU SHIKENSHO (GOVERNMENT INDUSTRIAL RESEARCH INSTITUTE, SHIKOKU, SCIENCE AND TECHNOLOGY AGENCY, MINISTRY OF INTERNATIONAL TRADE & INDUSTRY)
3-3, 2-CHOME, HANANOMIYA-CHO, TAKAMATSU-SHI 760
JAPANESE

SHIKOKU KOKEN KAIHO 5807
(REPORT OF THE SHIKOKU ENGINEERING ASSOCIATION)
SHIKOKU KOGYO KENKYUKAI (SHIKOKU ENGINEERING ASSOCIATION)
C/O SHIKOKU KOGYO GIJUTSU SHIKENSHO, 3-3, 2-CHOME HANANOMIYA-CHO, TAKAMATSU-SHI 760
JAPANESE
INDEXED: JICST

SHIKOKU TSUSAN TOKEI NENPO 5808
(ANNUAL STATISTICS OF INTERNATIONAL TRADE AND INDUSTRY IN SHIKOKU)
TSUSHO SANGYO-SHO SHIKOKU TSUSHO SANGYO-KYOKU (INTERNATIONAL TRADE AND INDUSTRY BUREAU, SHIKOKU, MINISTRY OF INTERNATIONAL TRADE AND INDUSTRY)
10-6, 1-CHOME, BAN-CHO, TAKAMATSU-SHI 760
JAPANESE

SHIKOSHI NYUSU 5809
(SHIKOKU INDUSTRIAL TECHNOLOGY EXPERIMENT STATION NEWS)
KOGYO GIJUTSUIN, SHIKOKU KOGYO GIJUTSU SHIKENJO (AGENCY OF INDUSTRIAL SCIENCE AND TECHNOLOGY, SHIKOKU INDUSTRIAL TECHNOLOGY EXPERIMENT STATION)
JAPANESE

SHIMANE DAIGAKU BUNRI GAKUBU KIYO. RIGAKUKA-HEN 5810
(BULLETIN OF THE SHIMANE UNIVERSITY ARTS AND SCIENCES FACULTY. SCIENCE DEPARTMENT EDITION)
JAPANESE
INDEXED: JICST

SHIMANE-KEN KOGYO SHIKENJO GYOMU HOKOKU 5811
(OPERATION REPORT OF INDUSTRIAL RESEARCH INSTITUTE, SHIMANE PREFECTURE)
SHIMANE-KEN KOGYO SHIKENJO
501, KOSHIHARA-CHO, MATSUE-SHI 690
JAPANESE

SHIMANE-KEN KOGYO SHIKENJO HOKOKUSHO 5812
(REPORT OF THE INDUSTRIAL RESEARCH INSTITUTE, SHIMANE PREFECTURE)
SHIMANE-KEN KOGYO SHIKENJO
501, KOSHIHARA-CHO, MATSUE-SHI 690
JAPANESE

SHIN GIJUTSU JOHO FAIRU 5813
(NEW TECHNOLOGY INFORMATION FILE)
TOYO SHO (TOYO COMPANY)
JAPANESE

SHIN GIJUTSU KAIHATSU JIGYODAN NENPO 5814
(ANNUAL REPORT OF RESEARCH DEVELOPMENT CORPORATION OF JAPAN)
SHIN GIJUTSU KAIHATSU JIGYODAN (RESEARCH DEVELOPMENT CORPORATION OF JAPAN)
5-2, 2-CHOME, NAGATA-CHO, CHIYODA-KU, TOKYO 100
JAPANESE

SHIN-SEIHIN 5815
(NEW PRODUCTS)
JAPANESE
INDEXED: JICST

SHINKOSHI NYUSU 5816
(NEWS OF THE INDUSTRIAL RESEARCH INSTITUTE, KANAGAWA PREFECTURE)
KANAGAWA-KEN KOGYO SHIKENJO (INDUSTRIAL RESEARCH INSTITUTE OF KANAGAWA PREFECTURE)
3173, SHOWA-MACHI, KANAZAWA-KU, YOKOHAMA-SHI 236
JAPANESE

SHINSHIKEN (KANAGAWA-KEN SHIRYO-SHITSU KENKYUKAI) 5817
(KANAGAWA PREFECTURE MATERIALS DEPARTMENT RESEARCH SOCIETY)
JAPANESE
INDEXED: JICST

SHINSHU DAIGAKU KOGAKUBU KIYO 5818
(JOURNAL OF THE FACULTY OF ENGINEERING, SHINSHU UNIVERSITY)
SHINSHU DAIGAKU KOGAKUBU (FACULTY OF ENGINEERING, SHINSHU UNIVERSITY)
500, WAKASATO, NAGANO-SHI 380
JAPANESE AND ENGLISH
INDEXED: CHE, INS, JICST

SHINSHU DAIGAKU KYOIKU GAKUBU KENKYU RONSHU 5819
(RESEARCH PAPERS OF SHINSHU UNIVERSITY, DEPARTMENT OF EDUCATION)
JAPANESE
INDEXED: JICST

SHINTO GIHO 5820
(SHINTO TECHNICAL REPORT)
SHINTO KOGYO K.K.
C/O TOYOTA GUILDING, 1, SASAJIMA-CHO, NAKAMURA-KU, NAGOYA-SHI 450
JAPANESE

SHISUTEMU TO SEIGYO 5821
(SYSTEMS AND CONTROL)
NIHON JIDO SEIGYO KYOKAI (JAPAN ASSOCIATION OF AUTOMATIC CONTROL)
1, YAMABANA, ITCHODA-CHO, SAKYO-KU, KYOTO-SHI 606
JAPANESE AND ENGLISH
INDEXED: INS, MAT, JICST

SHIZUKOSHI NYUSU 5822
(NEWS (PREFECTURAL INDUSTRIAL INSTITUTE, SHIZUOKA)
SHIZUOKA-KEN KOGYO SHIKENJO
9-3, 5-CHOME, KOMAGATA-DORI, SHIZUOKA-SHI 420
JAPANESE

SHIZUOKA DAIGAKU KOGAKUBU KENKYU HOKOKU 5823
(REPORTS OF THE FACULTY OF ENGINEERING, SHIZUOKA UNIVERSITY)
SHIZUOKA DAIGAKU KOGAKUBU (FACULTY OF ENGINEERING SHIZUOKA UNIVERSITY)
5-1, 3-CHOME, JOHOKU, HAMAMATSU-SHI 432
JAPANESE AND ENGLISH
INDEXED: CHE, INS, JICST

SHIZUOKA-KEN KOGYO GIJUTSA JOHO, KOGEI-HEN 5824
(SHIZUOKA PREFECTURE INDUSTRIAL TECHNOLOGY REPORT, INDUSTRIAL ARTS EDITION)
SHIZUOKA-KEN KOGYO SHIKENJO
(SHIZUOKA PREFECTURE INDUSTRIAL EXPERIMENT STATION)
JAPANESE

SHIZUOKA-KEN KOGYO SHIKENJO HOKOKU 5825
(REPORTS OF THE PREFECTURAL INDUSTRIAL RESEARCH INSTITUTE, SHIZUOKA)
SHIZUOKA-KEN KOGYO SHIKENJO (PREFECTURAL INDUSTRIAL RESEARCH INSTITUTE, SHIZUOKA)
9-3, 5-CHOME, KOMAGATA-DORI, SHIZUOKA-SHI 420
JAPANESE AND ENGLISH
INDEXED: CHE, JICST

SHIZUOKA-KEN KOGYO SHIKENJO NENPO 5826
(ANNUAL REPORT OF PREFECTURAL INDUSTRIAL RESEARCH INSTITUTE, SHIZUOKA)
SHIZUOKA-KEN KOGYO SHIKENJO (PREFECTURAL INDUSTRIAL RESEARCH INSTITUTE, SHIZUOKA)
9-3, 5-CHOME, KOMAGATA-DORI, SHIZUOKA-SHI 420
JAPANESE

SHOHIN KENKYU 5827
(COMMODITIES RESEARCH)
JAPANESE
INDEXED: JICST

SHOKUGYO KUNREN DAIGAKKO KIYO 5828
(BULLETIN OF THE INSTITUTE OF VOCATIONAL TRAINING)
INSTITUTE OF VOCATIONAL TRAINING
JAPANESE

SHOKUGYO KUNREN KENKYU 5829
(VOCATIONAL TRAINING RESEARCH)
SHOKUGYO KUNREN DAIGAKKO CHOSA KENKYUBU (INSTITUTE OF VOCATIONAL TRAINING, INVESTIGATION AND RESEARCH DEPARTMENT)
JAPANESE

SHORYOKU TO JIDOKA 5830
(JOURNAL OF LABOR SAVING & AUTOMATION)
OMUSHA
3-1, KANDA-NISHIKI-CHO, CHIYODA-KU, TOKYO 101
JAPANESE
INDEXED: JICST

SHUKAN CHUGOKU KOGYO NYUSU 5831
(CHINESE INDUSTRY NEWS)
KOYOSHA
C/O USUFUKU BUILDING, 1-2, 1-CHOME HAKUSAN, BUNKYO-KU, TOKYO 112
JAPANESE

SHUPPAN NYUSU 5832
(PUBLISHING NEWS)
JAPANESE
INDEXED: JICST

SOGO ENERUGI TOKEI 5833
(GENERAL ENERGY STATISTICS)
TSUSHO SANGYO-SHO DAIJIN KANBO (MINISTER'S SECRETARIAT, MINISTRY OF INTERNATIONAL TRADE AND INDUSTRY) AND TSUSHO SANGYO KENKYUSHO
12-4, 2-CHOME, AOYAMAKITA, MINATO-KU TOKYO 107
JAPANESE

SOGO KAIHATSU 5834
(GENERAL DEVELOPMENT OF CENTRAL JAPAN)
CHUBU NIHON SOGO KAIHATSU KENKYUSHO (INSTITUTE FOR THE GENERAL DEVELOPMENT OF CENTRAL JAPAN)
18-13, 3-CHOME, MARUNOUCHI, NAKA-KU, NAGOYA-SHI 460
JAPANESE

SOGO SHIKENJO NENPO 5835
(ANNUAL REPORT OF THE ENGINEERING RESEARCH INSTITUTE, FACULTY OF ENGINEERING, UNIVERSITY OF TOKYO)
TOKYO DAIGAKU KOGAKUBU FUZOKU SOGO SHIKENJO (THE ENGINEERING RESEARCH INSTITUTE, FACULTY OF ENGINEERING, UNIVERSITY OF TOKYO)
11-16, 2-CHOME, YAYOI, BUNKYO-KU, TOKYO 113
JAPANESE AND EUROPEAN

SUZUKA KOGYO KOTO SENMON GAKKO KIYO 5836
(MEMOIRS OF SUZUKA COLLEGE OF TECHNOLOGY)
SUZUKA KOGYO KOTO SENMON GAKKO
SHIRAKO-MACHI, SUZUKA-SHI 510-02
JAPANESE AND ENGLISH
INDEXED: JICST

TAKAMATSU KOGYO KOTO SENMON GAKKO KENKYU KIYO 5837
(ANNUAL REPORTS OF TAKAMATSU TECHNICAL COLLEGE)
TAKAMATSU KOGYO KOTO SENMON GAKKO
355, CHOKUSHI-MACHI, TAKAMATSU-SHI 760
JAPANESE AND ENGLISH

TAKEDA RIKEN GIJUTSU REPOTO 5838
(REPORT OF TAKEDA SCIENCE RESEARCH AND TECHNOLOGY)
JAPANESE
INDEXED: JICST

TAMAGAWA DAIGAKU KOGAKUBU KIYO 5839
(MEMOIRS OF THE FACULTY OF ENGINEERING, TAMAGAWA UNIVERSITY)
TAMAGAWA DAIGAKU KOGAKUBU
1-1, 6-CHOME, TAMAGAWAGAKUEN, MACHIDA-SHI 194
JAPANESE, WITH ENGLISH SUMMARY
INDEXED: APP, CHE, JICST

TECHNIQUES INDUSTRIALLES DU JAPON 5840
NICHI-FUTSU KOGYO GIJUTSUKAI (SOCIETE FRANCO-JAPONAISE DES TECHNIQUES INDUSTRIELLES)
2-3, KANDA SURUGADAI, CHIYODA-KU, TOKYO 101
EUROPEAN

TECHNOCRAT 5841
FUJI GAIKOKU SHIJO CHOSA K.K. (FUJI MARKETING RESEARCH COMPANY, LIMITED)
11-17, 6-CHOME, ROPPONGI, MINATO-KU, TOKYO 106
ENGLISH
INDEXED: BIB

TECHNOLOGY REPORTS OF THE IWATE UNIVERSITY 5842
IWATE DAIGAKU KOGAKUBU (FACULTY OF ENGINEERING, IWATE UNIVERSITY)
3-5, 4-CHOME, UEDA, MORIOKA-SHI 020
ENGLISH
INDEXED: APP, CHE, INS, JICST

TECHNOLOGY REPORTS OF THE KANSAI UNIVERSITY 5843
KANSAI DAIGAKU KOGAKUBU (FACULTY OF ENGINEERING, KANSAI UNIVERSITY)
3-35, 3-CHOME, YAMATE-CHO, SUITA-SHI 564
ENGLISH
INDEXED: CHE, ENG, INS, MAT, JICST

TECHNOLOGY REPORTS OF THE OSAKA UNIVERSITY 5844
OSAKA DAIGAKU KOGAKUBU (FACULTY OF ENGINEERING, OSAKA UNIVERSITY)
YAMADAKAMI, SUITA-SHI 565
ENGLISH
INDEXED: APP, CHE, INS, MAT, MET, JICST

TECHNOLOGY REPORTS OF THE TOHOKU UNIVERSITY 5845
TOHOKU DAIGAKU KOGAKUBU (FACULTY OF ENGINEERING, TOHOKU UNIVERSITY)
AOBA, ARAMAKI, SENDAI-SHI 980
ENGLISH
INDEXED: APP, BIB, CHE, ENG, INS, JICST

TECHNOLOGY REPORTS OF THE YAMAGUCHI UNIVERSITY 5846
YAMAGUCHI DAIGAKU KOGAKUBU (FACULTY OF ENGINEERING, YAMAGUCHI UNIVERSITY)
TOKIWADAI, UBE-SHI 755
ENGLISH

TEION KOGAKU 5847
(CRYOGENIC ENGINEERING)
TEION KOGAKU KYOKAI
C/O TOKYO DAIGAKU BUSSEI KENKYUSHO, 22-1, 7-CHOME, ROPPONGI, MINATO-KU, TOKYO 106
JAPANESE, WITH ENGLISH SUMMARY
INDEXED: JICST

TIME AND SPACE 5848
FUKUSHIMA DAIYONKI KENKYU GURUPU (FUKUSHIMA FOURTH PERIOD RESEARCH GROUP)
JAPANESE AND ENGLISH

TOCHIGI-KEN KENNAN KOGYO SHIDOSHO GYOMU HOKOKUSHO NARABINI KENKYU HOKOKUSHO 5849
(REPORT OF BUSINESS AND RESEARCH OF KENNAN INDUSTRIAL RESEARCH INSTITUTE OF TOCHIGI PREFECTURE)
TOCHIGI-KEN KENNAN KOGYO SHIDOSHO (KENNAN INDUSTRIAL RESEARCH INSTITUTE OF TOCHIGI PREFECTURE)
950, TENJIN-CHO, SANO-SHI 327
JAPANESE

TOCHIGI-KEN NO KOGYO, KOGYO TOKEI CHOSA KEKKA HOKOKU 5850
(MANUFACTURES IN TOCHIGI PREFECTURE, CENSUS OF MANUFACTURES)
TOCHIGI-KEN KIKAKU-BU (PLANNING DIVISION, TOCHIGI PREFECTURAL GOVERNMENT)
504, HANAWADA-CHO, UTSUNOMIYA-SHI 320
JAPANESE

TODOFUKEN-BETSU SEISAN SHISU 5851
(INDUSTRY PRODUCTS INDEX BY PREFECTURE)
TSUSHO SANGYO-SHO DAIJIN KANBO TOKEI-BU (STATISTICS DIVISION, MINISTER'S SECRETARIAT, MINISTRY OF INTERNATIONAL TRADE AND INDUSTRY)
3-1, 1-CHOME, KASUMIGASEKI, CHIYODA-KU, TOKYO 100
JAPANESE

TOHOKU GAKUIN DAIGAKU KOGAKUBU KENKYU HOKOKU 5852
(SCIENCE AND ENGINEERING REPORTS OF TOHOKU GAKUIN UNIVERSITY)
TOHOKU GAKUIN DAIGAKU KOGAKKAI
TOMEGAYA, TAGAJO-SHI 985
JAPANESE, WITH ENGLISH SUMMARY

TOHOKU KOGYO GIJUTSU SHIKENSHO HOKOKU 5853
(REPORT OF GOVERNMENT INDUSTRIAL TECHNOLOGY RESEARCH INSTITUTE, TOHOKU)
TSUSHO SANGYO-SHO KOGYO GIJUTSU-IN TOHOKU KOGYO GIJUTSU SHIKENSHO (GOVERNMENT INDUSTRIAL TECHNOLOGY RESEARCH INSTITUTE, TOHOKU, AGENCY OF INDUSTRIAL SCIENCE & TECHNOLOGY)
JAPANESE

TOHOKU KOGYO GIJUTSU SHIKENSHO HOKOKU 5854
(REPORT OF GOVERNMENT INDUSTRIAL TECHNOLOGY RESEARCH INSTITUTE, TOHOKU)
TSUSHO SANGYO-SHO KOGYO GIJUTSU-IN TOHOKU KOGYO GIJUTSU SHIKENSHO (GOVERNMENT INDUSTRIAL TECHNOLOGY RESEARCH INSTITUTE, TOHOKU, AGENCY OF INDUSTRIAL SCIENCE & TECHNOLOGY)
HARANO-MACHI NIGATAKE, SENDAI-SHI 983
JAPANESE, WITH ENGLISH SUMMARY

TOHOKU KOGYO GIJUTSU SHIKENSHO NENPO 5855
(ANNUAL REPORT OF GOVERNMENT INDUSTRIAL TECHNOLOGY RESEARCH INSTITUTE, TOHOKU)
TSUSHO SANGYO-SHO KOGYO GIJUTSU-IN TOHOKU KOGYO GIJUTSU SHIKENSHO (GOVERMENT INDUSTRIAL TECHNOLOGY RESEARCH INSTITUTE, TOHOKU, AGENCY OF INDUSTRIAL SCIENCE AND TECHNOLOGY)
HARANO-MACHI NIGATAKE-SENDAI-SHI 983
JAPANESE

TOHOKU KOSHI NYUSU 5856
(NEWS OF GOVERNMENT INDUSTRIAL TECHNOLOGY RESEARCH INSTITUTE, TOHOKU)
TSUSHO SANGYO-SHO KOGYO GIJUTSU-IN TOHOKU KOGYO GIJUTSU SHIKENJO (GOVERNMENT INDUSTRIAL TECHNOLOGY RESEARCH INSTITUTE, TOHOKU. AGENCY OF INDUSTRIAL SCIENCE AND TECHNOLOGY)
HARANO-MACHI NIGATAKE, SENDAI-SHI 983
JAPANESE

TOHOKU TSUSAN TOKEI NENPO 5857
(ANNUAL STATISTICS OF INTERNATIONAL TRADE AND INDUSTRY IN TOHOKU)
TSUSHO SANGYO-SHO SENDAI TSUSHO SANGYO-KYOKU INTERNATIONAL TRADE AND INDUSTRY BUREAU, SENDAI, MINISTRY OF INTERNATIONAL TRADE AND INDUSTRY)
C/O MIYAGI SHOKURYO KAIKAN, 8-3, 4-CHOME, CHUO, SENDAI-SHI 980
JAPANESE

TOJO KOGYO KOTO SENMON GAKKO KENKYU HOKOKU 5858
(TOJO INDUSTRIAL COLLEGE RESEARCH REPORT)
TOJO INDUSTRIAL COLLEGE
JAPANESE

TOKAI DAIGAKU KIYO, KOGAKUBU 5859
(PROCEEDINGS OF THE FACULTY OF ENGINEERING OF TOKAI UNIVERSITY)
TOKAI DAIGAKU KOGAKUBU (FACULTY OF ENGINEERING, TOKAI UNIVERSITY)
TOKAI DAIGAKU SHUPPANKAI (TOKAI UNIVERSITY PRESS)
C/O TOKAI BUILDING, 27-4, 3-CHOME, SHINJUKU, SHINJUKU-KU, TOKYO 160
JAPANESE, WITH EUROPEAN SUMMARY
INDEXED: JICST

TOKAI HOKURIKU TSUSAN TOKEI NENPO 5860
(ANNUAL STATISTICS OF INTERNATIONAL TRADE AND INDUSTRY IN TOKAI AND HOKURIKU)
TSUSHO SANGYO-SHO NAGOYA TSUSHO SANGYO-KYOKU SOMU-BU (ADMINISTRATIVE DIVISION, INTERNATIONAL TRADE AND INDUSTRY BUREAU, NAGOYA, MINISTRY OF INTERNATIONAL TRADE AND INDUSTRY)
5-2, 2-CHOME, SANNOMARU, NAKA-KU, NAGOYA-SHI 460
JAPANESE

TOKUSHIMA DAIGAKU KOGAKUBU KENKYU HOKOKU 5861
(SCIENTIFIC PAPERS OF FACULTY OF ENGINEERING, TOKUSHIMA UNIVERSITY)
TOKUSHIMA DAIGAKU KOGAKUBU (FACULTY OF ENGINEERING, TOKUSHIMA UNIVERSITY)
2-1, MINAMIJOSANJIMA-CHO, TOKUSHIMA-SHI 770
JAPANESE, WITH ENGLISH SUMMARY
INDEXED: CHE, INS, JICST

TOKUSHIMA-KEN KOGYO SHIKENJO, GYOMU HOKOKUSHO 5862
(OPERATION REPORT OF INDUSTRIAL RESEARCH INSTITUTE, TOKUSHIMA PREFECTURE)
TOKUSHIMA-KEN KOGYO SHIKENJO
2-6, KITAMAEKAWA-CHO, TOKUSHIMA-SHI 770
JAPANESE

TOKUSHIMA-KEN KOGYO SHIKENJO, KENKYU HOKOKUSHO 5863
(REPORT OF INDUSTRIAL RESEARCH INSTITUTE, TOKUSHIMA PREFECTURE)
TOKUSHIMA-KEN KOGYO SHIKENJO
2-6, KITAMAEKAWA-CHO, TOKUSHIMA-SHI 770
JAPANESE

TOKUSHIMA-KEN KYOIKU KENSHU SENTA KENKYU HOKOKUSHO 5864
(TOKUSHIMA PREFECTURAL EDUCATIONAL RESEARCH CENTER RESEARCH REPORT)
TOKUSHIMA-KEN KYOIKU KENSHU SENTA
5-22-9, BANDAI-CHO TOKUSHIMA 770
JAPANESE
INDEXED: CHE

TOKUSHIMA-KEN KYOIKU KENSHU SENTA KENKYU KIYO 5865
(TOKUSHIMA PREFECTURAL EDUCATIONAL RESEARCH CENTER RESEARCH BULLETIN)
TOKUSHIMA-KEN KYOIKU KENSHU SENTA
5-22-9, BANDAI-CHO, TOKUSHIMA 770
JAPANESE
INDEXED: CHE

TOKUSHIMA-KEN NO KOGYO 5866
(CENSUS OF MANUFACTURES, TOKUSHIMA PREFECTURE)
TOKUSHIMA-KEN KIKAKU KAIHATSU-BU (PLANNING AND DEVELOPMENT DIVISION, TOKUSHIMA PREFECTURAL GOVERNMENT)
1-1, BANDAI-CHO, TOKUSHIMA-SHI 770
JAPANESE

TOKUYAMA KOGYO KOTO SENMON GAKKO KENKYU KIYO 5867
(RESEARCH REPORTS OF TOKUYAMA HIGHER TECHNICAL SCHOOL)
TOKUYAMA KOGYO KOTO SENMON GAKKO (TOKUYAMA HIGHER TECHNICAL SCHOOL)
JAPANESE

TOKYO DAIGAKU KISO KAGAKU 5868
(UNIVERSITY OF TOKYO BASIC READINGS IN ENGINEERING)
TOKYO DAIGAKU SHUPPANKAI
7-3-1, HONGO, BUNKYO-KU, TOKYO 113
JAPANESE
INDEXED: CHE

TOKYO DAIGAKU KOGAKUBU FUZOKU SOGO SHIKENJO NENPO 5869
(ANNUAL REPORT OF THE ENGINEERING RESEARCH INSTITUTE, FACULTY OF ENGINEERING, UNIVERSITY OF TOKYO)
TOKYO DAIGAKU KOGAKUBU FUZOKU SOGO SHIKENJO (ENGINEERING RESEARCH INSTITUTE, FACULTY OF ENGINEERING, UNIVERSITY OF TOKYO)
11-16, 2-CHOME, YAYOI, BUNKYO-KU, TOKYO 113
JAPANESE AND ENGLISH
INDEXED: CHE, INS, JICST

TOKYO DAIGAKU KOGAKUBU KIYO, A 5870
(JOURNAL OF THE FACULTY OF ENGINEERING, UNIVERSITY OF TOKYO, SERIES A)
TOKYO DAIGAKU KOGAKUBU (FACULTY OF ENGINEERING UNIVERSITY OF TOKYO)
3-1, 7-CHOME, HONGO, BUNKYO-KU, TOKYO 113
JAPANESE, WITH ENGLISH SUMMARY
INDEXED: CHE, INS, JICST

TOKYO DAIGAKU SEISAN GIJUTSU KENKYUSHO HOKOKU 5871
(REPORTS OF THE INSTITUTE OF INDUSTRIAL SCIENCE, UNIVERSITY OF TOKYO)
TOKYO DAIGAKU SEISAN GIJUTSU KENKYUSHO (INSTITUTE OF INDUSTRIAL SCIENCE, UNIVERSITY OF TOKYO)
22-1, 7-CHOME, ROPPONGI, MINATO-KU, TOKYO 106
JAPANESE AND ENGLISH
INDEXED: APP, CHE, ENG, INS, IAA, JICST

TOKYO KOGYO DAIGAKU GAKUHO 5872
(BULLETIN OF THE TOKYO INSTITUTE OF TECHNOLOGY)
TOKYO KOGYO DAIGAKU (TOKYO INSTITUTE OF TECHNOLOGY)
12-1, 2-CHOME, OOKAYAMA, MEGURO-KU, TOKYO 152
JAPANESE AND EUROPEAN
INDEXED: INS, JICST

TOKYO KOGYO DAIGAKU KOGABU FUZOKU KOGYO KOTO GAKKO KENKYU HOKOKU 5873
(RESEARCH REPORT OF THE INDUSTRIAL HIGH SCHOOL OF THE TOKYO INDUSTRIAL COLLEGE, ENGINEERING FACULTY)
JAPANESE

TOKYO KOGYO DAIGAKU SOGO JOHO SHORI SENTA KOHO 5874
(OFFICIAL REPORT OF THE TOKYO INSTITUTE OF TECHNOLOGY INFORMATION PROCESSING CENTER)
JAPANESE

TOKYO KOGYO KOTO SENMON GAKKO KENKYU HOKOKUSHO 5875
(RESEARCH REPORT OF TOKYO TECHNICAL COLLEGE)
KOKURITSU TOKYO KOGYO KOTO SENMON GAKKO
1220-2, KUNUGIDA-CHO, HACHIOJI-SHI, TOKYO 193
JAPANESE AND ENGLISH
INDEXED: JICST

TOKYO KOGYO SHIKENSHO HOKOKU 5876
(JOURNAL OF THE NATIONAL CHEMICAL LABORATORY FOR INDUSTRY)
TSUSHO SANGYO-SHO KOGYO GIJUTSU-IN TOKYO KOGYO SHIKENSHO (NATIONAL CHEMICAL LABORATORY FOR INDUSTRY, AGENCY OF INDUSTRIAL SCIENCE AND TECHNOLOGY)
1-5, 1-CHOME, HON-MACHI, SHIBUYA-KU, TOKYO 150
JAPANESE AND ENGLISH
INDEXED: CHE, JICST

TOKYO KOGYO SHIKENSHO NENPO 5877
(ANNUAL REPORT OF THE NATIONAL CHEMICAL LABORATORY FOR INDUSTRY)
TSUSHO SANGYO-SHO KOGYO GIJUTSU-IN TOKYO KOGYO SHIKENSHO (NATIONAL CHEMICAL LABORATORY FOR INDUSTRY, AGENCY OF INDUSTRIAL SCIENCE AND TECHNOLOGY, MINISTRY OF INTERNATIONAL TRADE)
1-5, 1-CHOME, HON-MACHI, SHIBUYA-KU, TOKYO 150
JAPANESE
INDEXED: JICST

TOKYO NO SANGYO 5878
(INDUSTRIES IN TOKYO)
TOKYO-TO KEIZAI-KYOKU (ECONOMIC BUREAU, TOKYO METROPOLITAN GOVERNMENT)
8-1, 3-CHOME, MARUNOUCHI, CHIYODA-KU, TOKYO 100
JAPANESE

TOKYO NOKO DAIGAKU FUZOKU TOSHOKAN KOGAKUBU BUNKAN TOSHOKANHO 5879
(REPORT OF THE ENGINEERING FACULTY BRANCH LIBRARY OF THE TOKYO AGRICULTURAL AND INDUSTRIAL COLLEGE LIBRARY)
JAPANESE

TOKYO TSUSAN GEPPO 5880
(MONTHLY REPORT ON INTERNATIONAL TRADE AND INDUSTRY IN TOKYO)
TSUSHO SANGYO CHOSAKAI (INTERNATIONAL TRADE AND INDUSTRY RESEARCH ASSOCIATION)
15-1, 6-CHOME, GINZA, CHUO-KU, TOKYO 104
JAPANESE
INDEXED: JICST

TOKYO TSUSAN TOKEI NENPO 5881
(ANNUAL STATISTICS OF INTERNATIONAL TRADE AND INDUSTRY IN TOKYO)
TSUSHO SANGYO-SHO TOKYO TSUSHO SANGYO-KYOKU (INTERNATIONAL TRADE AND INDUSTRY BUREAU, MINISTRY OF INTERNATIONAL TRADE AND INDUSTRY)
3-3, 1-CHOME, OTE-MACHI, CHIYODA-KU, TOKYO 100
JAPANESE

TOKYO-TORITSU DAIGAKU KOGAKUBU KENKYU ICHIRAN 5882
(MEMOIRS OF FACULTY OF TECHNOLOGY, TOKYO METROPOLITAN UNIVERSITY)
TOKYO-TORITSU DAIGAKU KOGAKUBU (FACULTY OF TECHNOLOGY, TOKYO METROPOLITAN UNIVERSITY)
1-1, 2-CHOME, FUKAZAWA, SETAGAYA-KU, TOKYO 158
JAPANESE AND ENGLISH

TOKYO-TORITSU KOGYO GIJUTSU SENTA KENKYU HOKOKU 5883
(REPORT OF THE TOKYO METROPOLITAN INDUSTRIAL TECHNIC INSTITUTE)
TOKYO-TORITSU KOGYO GIJUTSU SENTA (TOKYO METROPOLITAN INDUSTRIAL TECHNOLOGY CENTER)
13-10, 3-CHOME, NISHIGAOKA, KITA-KU, TOKYO 115
JAPANESE AND ENGLISH
INDEXED: CHE, JICST

TOKYO-TORITSU KOGYO GIJUTSU SENTA NYUSU 5884
(NEWS OF THE TOKYO METROPOLITAN INDUSTRIAL TECHNOLOGY CENTER)
TOKYO-TORITSU KOGYO GIJUTSU SENTA (TOKYO METROPOLITAN INDUSTRIAL TECHNOLOGY CENTER)
13-10, 3-CHOME, NISHIGAOKA, KITA-KU, TOKYO 115
JAPANESE
INDEXED: JICST

TOKYO-TORITSU KOGYO KOTO SENMON GAKKO KENKYU HOKOKU 5885
(REPORTS OF THE TOKYO METROPOLITAN TECHNICAL COLLEGE)
TOKYO-TORITSU KOGYO KOTO SENMON GAKKO (TOKYO METROPOLITAN TECHNICAL COLLEGE)
10-40, 1-CHOME, HIGASHIOI, SHINAGAWA-KU, TOKYO 140
JAPANESE, WITH ENGLISH SUMMARY
INDEXED: JICST

TOKYO-TORITSU KOGYO SHOREIKAN HOKOKU 5886
(REPORTS OF THE TOKYO METROPOLITAN INDUSTRIAL RESEARCH INSTITUTE)
JAPANESE

TOKYO-TORITSU KOKA TANKI DAIGAKU KENKYU HOKOKU 5887
(MEMOIRS OF THE METROPOLITAN COLLEGE OF TECHNOLOGY)
TOKYO-TORITSU KOKA TANKI DAIGAKU GAKUJUTSU KENKYU UN'EIKAI (THE METROPOLITAN TECHNICAL COLLEGE)
6-6, ASAHIGAOKA, HINO-SHI, TOKYO 191
JAPANESE AND EUROPEAN
INDEXED: JICST

TOMAKOMAI KOGYO KOTO SENMON GAKKO KIYO 5888
MEMOIRS OF THE TOMAKOMAI TECHNICAL COLLEGE
(MEMOIRS OF THE TOMAKOMAI TECHNICAL COLLEGE)
TOMAKOMAI KOGYO KOTO SENMON GAKKO (TOMAKOMAI TECHNICAL COLLEGE)
443, NISHIKIOKA, TOMAKOMAI-SHI 059-12
JAPANESE AND ENGLISH
INDEXED: CHE, JICST

TOSHOKAN GAKKAI NEMPO 5889
(JAPAN SOCIETY OF LIBRARY SCIENCE ANNALS)
NIPPON TOSHOKAN GAKKAI (JAPAN SOCIETY OF LIBRARY SCIENCE)
C/O NATIONAL COLLEGE OF LIBRARY SCIENCE, 1-4-1 SHIMOUMA, SETAGAYA-KU, TOKYO
JAPANESE, WITH ENGLISH SUMMARY
INDEXED: JICST

TOSHOKAN GIJUTSU KENKYUKAI KENKYU JOHO 5890
(RESEARCH REPORT OF THE LIBRARY TECHNOLOGY RESEARCH ASSOCIATION)
JAPANESE

TOSHOKAN HYORON 5891
(LIBRARY REVIEW)
TOSHOKAN MONDAI KENKYUKAI
JAPANESE

TOSHOKAN KENKYU SHIRIZU 5892
(LIBRARY RESEARCH SERIES)
KOKURITSU KOKKAI TOSHOKAN (NATIONAL DIET LIBRARY)
JAPANESE

TOSHOKAN TANKI DAIGAKU KIYO 5893
(BULLETIN OF THE LIBRARY JUNIOR COLLEGE)
JAPANESE
INDEXED: JICST

TOSHOKAN ZASSHI 5894
(LIBRARY JOURNAL)
NIHON TOSHOKAN KYOKAI (JAPAN LIBRARY ASSOCIATION)
1-1-10 TAISHIDO, SETAGAYA-KU
JAPANESE
INDEXED: JICST

TOSHOKAN-KAI 5895
(LIBRARY WORLD)
NIPPON TOSHOKAN KENKYUKAI (JAPAN INSTITUTION FOR LIBRARY SCIENCE)
TOSHOKANGAKU KENKYUSHITSU, TENRI UNIVERSITY, 1050 SOMANOUCHI-CHO, TENRI, NARA
JAPANESE
INDEXED: JICST

TOSHOKANGAKU 5896
(LIBRARY SCIENCE)
NISHI NIHON TOSHOKAN GAKKAI (WESTERN JAPAN LIBRARY ASSOCIATION)
JAPANESE

TOSHOKANGAKU KENKYU HOKOKU 5897
(RESEARCH REPORT ON LIBRARY SCIENCE)
TOHOKU DAIGAKU FUZOKU TOSHOKAN (LIBRARY OF TOHOKU UNIVERSITY)
JAPANESE

TOTTORI DAIGAKU KOGAKUBU KENKYU HOKOKU 5898
(REPORTS OF THE FACULTY OF ENGINEERING, TOTTORI UNIVERSITY)
TOTTORI DAIGAKU KOGAKUBU (FACULTY OF ENGINEERING, TOTTORI UNIVERSITY)
1-1, KOYAMA-CHO, TOTTORI-SHI 680
JAPANESE AND ENGLISH
INDEXED: CHE, JICST

TOTTORI-KEN KOGYO SHIKENJO NENPO 5899
(ANNUAL REPORT OF INDUSTRIAL RESEARCH INSTITUTE TOTTORI PREFECTURE)
TOTTORI-KEN KOGYO SHIKENJO
GYOTOKU, TOTTORI-SHI 680
JAPANESE

TOYAMA DAIGAKU KOGAKUBU KIYO 5900
(BULLETIN OF FACULTY OF ENGINEERING, TOYAMA UNIVERSITY)
TOYAMA DAIGAKU KOGAKUBU (FACULTY OF ENGINEERING, TOYAMA UNIVERSITY)
1-1, NAKAGAWASONO-MACHI, TAKAOKA-SHI 933
JAPANESE, WITH ENGLISH SUMMARY
INDEXED: CHE, WAA, JICST

TOYAMA KOGYO KOTO SENMON GAKKO KIYO 5901
(BULLETIN OF TOYAMA TECHNICAL COLLEGE)
TOYAMA KOGYO KOTO SENMON GAKKO
13, HONGO-CHO, TOYAMA-SHI 930
JAPANESE AND ENGLISH
INDEXED: JICST

TOYAMA-KEN KOGYO SHIKENJO NENPO 5902
(REPORT OF INDUSTRIAL RESEARCH INSTITUTE OF TOYAMA PREFECTURE)
TOYAMA-KEN KOGYO SHIKENJO (INDUSTRIAL RESEARCH INSTITUTE OF TOYAMA PREFECTURE)
1-10, 1-CHOME, NAKAGAWA, TAKAOKA-SHI 933
JAPANESE
INDEXED: CHE, JICST

TOYAMA-KEN NO SANGYO 5903
(MANUFACTURES IN TOYAMA PREFECTURE)
HOKURIKU GINKO CHOSA-BU (THE ECONOMIC RESEARCH DEPARTMENT OF HOKURIKU BANK)
2-26, 1-CHOME, TSUTSUMICHO-DORI, TOYAMA-SHI 930
JAPANESE

TOYAMA-KENRITSU GIJUTSU TANKI DAIGAKU KENKYU HOKOKU 5904
(BULLETIN OF COLLEGE OF TECHNOLOGY OF TOYAMA)
TOYAMA-KENRITSU GIJUTSU TANKI DAIGAKU
KUROKAWA, KOSUGI-MACHI, IMIZU-GUN, TOYAMA 939-03
JAPANESE AND ENGLISH

TOYO DAIGAKU KOGAKUBU KENKYU HOKOKU 5905
(RESEARCH REPORTS OF THE FACULTY OF ENGINEERING, TOYO UNIVERSITY)
TOYO DAIGAKU KOGAKUBU KENKYU HOKOKU IINKAI
TOYO DAIGAKU KOGAKUBU (FACULTY OF ENGINEERING, TOYO UNIVERSITY)
2100, NAKANODAI, KUJIRAI, KAWAGOE-SHI 350
JAPANESE AND ENGLISH
INDEXED: JICST

TOYO DAIGAKU KOGAKUBU KYOYO KATEI KENKYU HOKOKU 5906
(RESEARCH REPORTS OF GENERAL EDUCATION COURSE, FACULTY OF ENGINEERING, TOYO UNIVERSITY)
TOYO DAIGAKU KOGAKUBU (FACULTY OF ENGINEERING, TOYO UNIVERSITY)
2100, NAKANODAI, KUJIRAI, KAWAGOE-SHI 350
JAPANESE AND ENGLISH

TOYOTA KOGYO KOTO SENMON GAKKO KENKYU KIYO 5907
(JOURNAL OF TOYOTA TECHNICAL COLLEGE)
TOYOTA KOGYO KOTO SENMON GAKKO (TOYOTA TECHNICAL COLLEGE)
2-1, EISEI-CHO, TOYOTA-SHI 471
JAPANESE AND ENGLISH
INDEXED: JICST

TRADE AND INDUSTRY OF JAPAN 5908
NIHON BOEKI SHINKOKAI (JAPAN EXTERNAL TRADE ORGANIZATION)
2, AKASAKA AOI-CHO, MINATO-KU, TOKYO 107
ENGLISH

TRANSACTIONS OF THE JAPANESE SOCIETY FOR COMPOSITE MATERIALS 5909
ENGLISH

TSUDOI 5910
(MEETING)
KYOTO-SHI KOGYO SHIKENJO NAISHO KENKAI (KYOTO MUNICIPAL INDUSTRIAL LABORATORY, NAISHO RESEARCH ACADEMY)
JAPANESE

TSURUOKA KOGYO KOTO SENMON GAKKO KENKYU KIYO 5911
(RESEARCH REPORTS OF THE TSURUOKA TECHNICAL COLLEGE)
TSURUOKA KOGYO KOTO SENMON GAKKO (THE TSURUOKA TECHNICAL COLLEGE)
104, SAWADA, IOKA, TSURUOKA-SHI 997
JAPANESE AND ENGLISH
INDEXED: JICST

TSUSAN JANARU 5912
(TRADE AND INDUSTRY JOURNAL)
MINISTRY OF TRADE & INDUSTRY
JAPANESE
INDEXED: JICST

TSUSAN TOKEI 5913
(STATISTICAL REPORT OF TRADE AND INDUSTRY)
TSUSHO SANGYO-SHO DAIJIN KANBO CHOSA TOKEI-BU (RESEARCH AND SURVEY DIVISION, MINISTER'S SECRETARIAT, MINISTRY OF INTERNATIONAL TRADE AND INDUSTRY) AND TSUSHO SANGYO CHOSAKAI
C/O TSUSHO SANGYO-SHO, 3-1, 1-CHOME, KASUMIGASEKI, CHIYODA-KU, TOKYO 100
JAPANESE
INDEXED: JICST

TSUSHO SANGYO CHOSAKAI KAIHO 5914
(REPORT OF THE TRADE AND INDUSTRY SURVEY COMMITTEE)
JAPANESE
INDEXED: JICST

TSUSHO SANGYO-SHO NENPO 5915
(ANNUAL REPORT OF THE MINISTRY OF THE INTERNATIONAL TRADE AND INDUSTRY)
TSUSHO SANGYO-SHO DAIJIN KANBO CHOSA-KA (RESEARCH SECTION, MINISTER'S SECRETARIAT, MINISTRY OF INTERNATIONAL TRADE AND INDUSTRY)
3-1, 1-CHOME, KASUMIGASEKI, CHIYODA-KU, TOKYO 100
JAPANESE

TSUYAMA KOGYO KOTO SENMON GAKKO KIYO 5916
(RESEARCH REPORTS OF TSUYAMA TECHNICAL COLLEGE)
TSUYAMA KOGYO KOTO SENMON GAKKO (TSUYAMA TECHNICAL COLLEGE)
624-1, NUMA, TSUYAMA-SHI 708
JAPANESE AND ENGLISH

UBE KOGYO KOTO SENMON GAKKO KENKYU HOKOKU 5917
(RESEARCH REPORTS OF UBE TECHNICAL COLLEGE)
UBE KOGYO KOTO SENMON GAKKO (UBE TECHNICAL COLLEGE)
TOKIWADAI, UBE-SHI 755
JAPANESE AND ENGLISH
INDEXED: JICST

UNMANNED AGE 5918
JAPANESE
INDEXED: JICST

UNTEI 5919
TENRI DAIGAKU TOSHO KANGAKU KENKYUSHATSU (TENRI UNIVERSITY, SCHOOL OF LIBRARY SCIENCE)
JAPANESE

WAKAYAMA KOGYO KOTO SENMON GAKKO KENKYU KIYO 5920
(MEMOIRS OF THE WAKAYAMA TECHNICAL COLLEGE)
WAKAYAMA KOGYO KOTO SENMON GAKKO (WAKAYAMA TECHNICAL COLLEGE)
77, NOSHIMA, NADA-CHO, GOBO-SHI 649-15
JAPANESE AND EUROPEAN

WAKAYAMA-KEN KOGYO SHIKENJO NENPO 5921
(ANNUAL REPORT OF WAKAYAMA PREFECTURAL INDUSTRIAL RESEARCH INSTITUTE)
WAKAYAMA-KEN KOGYO SHIKENJO (WAKAYAMA PREFECTURAL INDUSTRIAL RESEARCH INSTITUTE)
60, OGURA, WAKAYAMA-SHI 649-62
JAPANESE

WAKAYAMA-KEN NO KOGYO, KOGYO TOKEI CHOSA KEKKA HOKOKUSHO 5922
(MANUFACTURES IN WAKAYAMA PREFECTURE, CENSUS OF MANUFACTURES)
WAKAYAMA-KEN KIKAKU-BU (PLANNING DIVISION, WAKAYAMA PREFECTURAL GOVERNMENT)
1-1, KOMATSUBARA-DORI, WAKAYAMA-SHI 640
JAPANESE

WARERA NO KAGAKU TO GIJUTSU 5923
(OUR SCIENCE AND TECHNOLOGY)
JAPANESE
INDEXED: JICST

WASEDA DAIGAKU SEISAN KENKYUSHO KIYO 5924
(BULLETIN OF THE INSTITUTE FOR RESEARCH IN PRODUCTIVITY, WASEDA UNIVERSITY)
WASEDA DAIGAKU SEISAN KENKYUSHO (INSTITUTE FOR RESEARCH IN PRODUCTIVITY, WASEDA UNIVERSITY)
4-170, NISHIOKUBO, SHINJUKU-KU, TOKYO 160
JAPANESE AND ENGLISH

WASEDA DAIGAKU SEISAN KENKYUSHOHO 5925
(JOURNAL OF IRP, WASEDA UNIVERSITY)
WASEDA DAIGAKU SEISAN KENKYUSHO (INSTITUTE FOR RESEARCH IN PRODUCTIVITY, WASEDA UNIVERSITY)
4-170, NISHIOKUBO, SHINJUKU-KU, TOKYO 160
JAPANESE AND ENGLISH
INDEXED: JICST

YAKUSHIN 5926
(DANCING AHEAD (OR RUSHING ONWARD))
NIHON KIKAI DO BOKU (JAPAN MECHANICAL AND CIVIL ENGINEERING)
JAPANESE

YAMAGATA DAIGAKU KIYO, KOGAKU 5927
(BULLETIN OF THE YAMAGATA UNIVERSITY ENGINEERING)
YAMAGATA DAIGAKU (YAMAGATA UNIVERSITY)
KOSHIRAKAWA-MACHI, YAMAGATA-SHI 990
JAPANESE AND ENGLISH
INDEXED: CHE, INS, IAA, JICST

YAMAGATA KOGYO SHIKENJO GYOMU HOKOKUSHO 5928
(OPERATION REPORT OF INDUSTRIAL RESEARCH INSTITUTE, YAMAGATA)
YAMAGATA-KENRITSU YAMAGATA KOGYO SHIKENJO
2-30, 1-CHOME, DO-MACHI, YAMAGATA-SHI 990
JAPANESE

YAMAGATA KOGYO SHIKENJO HOKOKU 5929
(REPORT OF INDUSTRIAL RESEARCH INSTITUTE, YAMAGATA)
YAMAGATA-KENRITSU YAMAGATA KOGYO SHIKENJO
2-30, 1-CHOME, DO-MACHI, YAMAGATA-SHI 990
JAPANESE

YAMAGATA-KEN KOGYO SHIKENJO HOKOKU 5930
(REPORT OF THE YAMAGATA PREFECTURAL INDUSTRIAL LABORATORY)
JAPANESE
INDEXED: JICST

YAMAGUCHI DAIGAKU KOGAKUBU KENKYU HOKOKU 5931
(MEMOIRS OF THE FACULTY OF ENGINEERING, YAMAGUCHI UNIVERSITY)
YAMAGUCHI DAIGAKU KOGAKUBU (FACULTY OF ENGINEERING, YAMAGUCHI UNIVERSITY)
TOKIWADAI, UBE-SHI 755
JAPANESE AND EUROPEAN
INDEXED: CHE, JICST

YAMAGUCHI-KEN NO KOGYO, KOGYO TOKEI CHOSA KEKKA 5932
(CENSUS OF MANUFACTURES, YAMAGUCHI PREFECTURE)
YAMAGUCHI-KEN KIKAKU-BU (PLANNING DIVISION, YAMAGUCHI PREFECTURAL GOVERNMENT)
YAMAGUCHI-KEN TOKEI KYOKAI (STATISTICS ASSOCIATION YAMAGUCHI PREFECTURAL GOVERNMENT)
1-1, TAKI-MACHI, YAMAGUCHI-SHI 753
JAPANESE

YAMAGUCHI-KEN SHOKO SHIDO SENTA KENKYU HOKOKU 5933
(BULLETIN OF COMMERCE AND INDUSTRY CENTER FOR GUIDANCE, YAMAGUCHI PREFECTURE)
YAMAGUCHI-KEN SHOKO SHIDO SENTA
585-1, YUGAKI, ASADA, YAMAGUCHI-SHI 753
JAPANESE

YAMANASHI DAIGAKU KOGAKUBU KENKYU HOKOKU 5934
(REPORTS OF THE FACULTY OF ENGINEERING, YAMANASHI UNIVERSITY)
YAMANASHI DAIGAKU KOGAKUBU (FACULTY OF ENGINEERING YAMANASHI UNIVERSITY)
3-11, 4-CHOME, TAKEDA, KOFU-SHI 400
JAPANESE AND EUROPEAN
INDEXED: CHE, JICST

YAMANASHI NO SHOKOGYO, KOGYO-HEN 5935
(COMMERCE AND INDUSTRY IN YAMANASHI PREFECTURE, SECTION INDUSTRY)
YAMANASHI-KEN SHOKO RODO-BU (COMMERCE, INDUSTRY AND LABOUR DIVISION, YAMANASHI PREFECTURAL GOVERNMENT)
6-1, 1-CHOME, MARUNOUCHI, KOFU-SHI 400
JAPANESE

YOKKAICHI NO KOGYO 5936
(MANUFACTURES IN YOKKAICHI CITY)
YOKKAICHI SHIYAKUSHO (YOKKAICHI MUNICIPAL GOVERNMENT)
1-5, SUWA-CHO, YOKKAICHI-SHI 510
JAPANESE

YOKOHAMA KOKURITSU DAIGAKU KOGAKUBU ZAIRYO KISO KOGAKU KENKYU 5937
(JOURNAL OF INSTITUTE OF MATERIALS SCIENCE AND ENGINEERING, YOKOHAMA NATIONAL UNIVERSITY)
YOKOHAMA KOKURITSU DAIGAKU KOGAKUBU ZAIRYO KISO KOGAKU KENKYU SHISETSU (INSTITUTE OF MATERIALS SCIENCE AND ENGINEERING, YOKOHAMA NATIONAL UNIVERSITY)
31-1, 2-CHOME, OOKA, MINAMI-KU, YOKOHAMA-SHI 233
JAPANESE, WITH ENGLISH SUMMARY

YOKOHAMA-SHI NO KOGYO 5938
(MANUFACTURES IN YOKOHAMA)
YOKOHAMA-SHI SOMU-KYOKU GYOSEI-BU (ADMINISTRATIVE DIVISION, GENERAL BUREAU, YOKOHAMA MUNICIPAL GOVERNMENT)
1-1, MINATO-CHO, NAKA-KU, YOKOHAMA-SHI 231
JAPANESE

YOMO KAGAKU KENKYU KAISHI 5939
(JOURNAL OF THE WOOL SCIENCE RESEARCH SOCIETY)
JAPANESE
INDEXED: JICST

YONAGO KOGYO KOTO SENMON GAKKO KENKYU HOKOKU 5940
(REPORT OF YONAGO TECHNICAL COLLEGE)
YONAGO KOGYO KOTO SENMON GAKKO (YONAGO TECHNICAL COLLEGE)
4448, OTAKASUNA, HIKONA-CHO, YONAGO-SHI 683
JAPANESE AND ENGLISH

YONKOSHI NYUSU 5941
(NEWS OF GOVERNMENT INDUSTRIAL RESEARCH INSTITUTE, SHIKOKU)
TSUSHO SANGYO-SHO KOGYO GIJUTSU-IN SHIKOKU KOGYO GIJUTSU SHIKENSHO (GOVERNMENT INDUSTRIAL RESEARCH INSTITUTE SHIKOKU, SCIENCE AND TECHNOLOGY AGENCY, MINISTRY OF INTERNATIONAL TRADE AND INDUSTRY)
3-3, 2-CHOME, HANANOMIYA-CHO, TAKAMATSU-SHI 760
JAPANESE

YUSHUTSU ZAKKA NYUSU 5942
(EXPORT (GOODS) NEWS)
JAPANESE
INDEXED: JICST

ZAIRYO 5943
(JOURNAL OF SOCIETY OF MATERIALS SCIENCE, JAPAN)
NIHON ZAIRYO GAKKAI (SOCIETY OF MATERIAL SCIENCE, JAPAN)
1-101, YOSHIDA IZUMIDONO-CHO, SAKYO-KU, KYOTO-SHI 606
JAPANESE AND ENGLISH
INDEXED: APP, INS, MET, WAA, JICST

ZAIRYO KAGAKU 5944
(JOURNAL OF THE MATERIALS SCIENCE SOCIETY OF JAPAN)
NIHON ZAIRYO KAGAKUKAI
C/O TOKYO KOGYO DAIGAKU, OOKAYAMA, MEGURO-KU, TOKYO 152
JAPANESE AND ENGLISH
INDEXED: JICST

ZAIRYO SHIKEN HOKOKU 5945
(REPORT OF THE PUBLIC WORKS MATERIAL LABORATORY)
KENSETSU-SHO CHUGOKU CHIHO KENSETSU-KYOKU CHUGOKU GIJUTSU JIMUSHO (CHUGOKU TECHNICAL OFFICE, CHUGOKU DISTRICT CONSTRUCTION BUREAU, MINISTRY OF CONSTRUCTION)
1920-1, FUNAKOSHI-CHO, AKI-GUN, HIROSHIMA 736
JAPANESE
INDEXED: JICST

Mechanical Engineering and Machinery

BANE RONBUNSHU 5946
(RESEARCH PAPERS ON SPRING)
NIHON BANE KOGYOKAI
C/O SHOYU KAIKAN, 3-3, 3-CHOME, KASUMIGASEKI, CHIYODA-KU, TOKYO 100
JAPANESE
INDEXED: JICST

BARKUKA REBYU 5947
(VALQUA REVIEW)
NIPPON BARUKA KOGYO K.K. (NIPPON VALQUA INDUSTRIES, LIMITED)
C/O SHIN TOKYO BUILDING, 3-1, 3-CHOME, MARUNOUCHI, CHIYODA-KU, TOKYO 100
JAPANESE
INDEXED: CHE, JICST

BEARINGU 5948
(BEARING)
NIHON BEARINGU KOGYOKAI (JAPAN BEARING ASSOCIATION)
5-8, 3-CHOME, SHIBA KOEN, MINATO-KU, TOKYO 105
JAPANESE
INDEXED: JICST

BEARINGU ENJINIA 5949
(BEARING ENGINEER)
TOYO BEARINGU SEIZO K.K. (THE TOYO BEARING MANUFACTURING COMPANY, LTD.)
1-25, KYOMACHIBORI, NISHI-KU, OSAKA-SHI 550
JAPANESE, WITH ENGLISH SUMMARY
INDEXED: JICST

BOIRA GISHI 5950
(BOILERMEN)
NENRYO OYOBI NENSHOSHA
21, KITAYAMA-CHO, TENNOJI-KU, OSAKA-SHI 543
JAPANESE
INDEXED: CHE

BOIRA KENKYU 5951
(JOURNAL OF THE JAPAN BOILER ASSOCIATION)
NIHON BOIRA KYOKAI (JAPAN BOILER ASSOCIATION)
C/O ANZEN KAIKAN, 35-4, 5-CHOME, SHIBA, MINATO-KU, TOKYO 108
JAPANESE
INDEXED: CHE, JICST

BOIRA NYUSU 5952
(BOILER NEWS)
NIHON BOIRA KYOKAI (JAPAN BOILER ASSOCIATION)
35-4, 5-CHOME, SHIBA, MINATO-KU, TOKYO 108
JAPANESE

BULLETIN OF JSME 5953
NIHON KIKAI GAKKAI (JAPAN SOCIETY OF MECHANICAL ENGINEERS)
C/O SANSHIN HOKUSEI BUILDING, 4-9, 2-CHOME, YOYOGI, SHIBUYA-KU, TOKYO 151
EUROPEAN
INDEXED: APP, AUT, CHE, ENG, IAA, INS, MAT, MET, SCI, WAA, JICST

BULLETIN OF MECHANICAL ENGINEERING LABORATORY 5954
TSUSHO SANGYO-SHO KOGYO GIJUTSU-IN KIKAI GIJUTSU KENKYUSHO (MECHANICAL ENGINEERING LABORATORY, AGENCY OF INDUSTRIAL SCIENCE AND TECHNOLOGY, MINISTRY OF INTERNATIONAL TRADE AND INDUSTRY)
12-1, 4-CHOME, IGUSA, SUGINAMI-KU, TOKYO 167
ENGLISH
INDEXED: APP, INS, JICST

CHIBA-KEN KIKAI KINZOKU SHIKENJO JIGYO HOKOKU 5955
(REPORT OF THE MACHINERY AND METALLURGY RESEARCH INSTITUTE OF CHIBA PREFECTURE)
CHIBA-KEN KIKAI KINZOKU SHIKENJO
206, TENDAI-CHO, CHIBA-SHI 280
JAPANESE

CHIBA-KEN KIKAI KINZOKU SHIKENJO KENKYU HOKOKU 5956
(RESEARCH REPORTS OF THE MACHINERY AND METALLURGY RESEARCH INSTITUTE OF CHIBA PREFECTURE)
CHIBA-KEN KIKAI KINZOKU SHIKENJO
206, TENDAI-CHO, CHIBA-SHI 280
JAPANESE

CHINO NYUSU 5957
(CHINO TECHNICAL NEWS)
CHINO SEISAKUJO (CHINO WORKS, LTD.)
DENSHI KEISOKU SYUPANSHA
C/O YUZANKAKU BUILDING, 6-9, 2-CHOME, FUJIMI, CHIYODA-KU, TOKYO 102
JAPANESE

CHUTETSUKAN 5958
(CAST IRON PIPE)
NIHON CHUTESUKAN KYOKAI (JAPAN CAST IRON PIPE MANUFACTURERS ASSOCIATION)
8-9, 4-CHOME, KUDAN MINAMI, CHIYODA-KU, TOKYO 101
ENGLISH

DANBO 5959
(HEATERS)
JAPANESE
INDEXED: JICST

DESIGN AGE 5960
NIHON SANGYO DEZAIN SHINKO KAI
JAPANESE AND ENGLISH

DESIGN NEWS 5961
NIHON SANGYO DEZAIN SHINKO KAI (ASSOCIATION FOR PROMOTION OF JAPANESE INDUSTRIAL DESIGN)
JAPANESE

EHARA JIHO 5962
(EBARA TIMES)
EBARA SEISAKUSHO (EBARA MANUFACTURING COMPANY, LIMITED)
11-1, HANEDA ASAHI-CHO, OTA-KU, TOKYO 144
JAPANESE, WITH ENGLISH SUMMARY
INDEXED: CHE, JICST

EREBETAKAI 5963
(ELEVATOR INDUSTRY)
NIHON EREBETA KYOKAI
C/O KYODO BUILDING, 11-2, 5-CHOME, MINAMI-AOYAMA, MINATO-KU, TOKYO 107
JAPANESE

FUJIKOSHI GIHO 5964
(FUIJIKOSHI ENGINEERING REVIEW)
FUIJIKOSHI EIGYO KANRI-BU (FUJIKOSHI LTD.)
C/O BOEKI SENTA BUILDING, 4-1, 2-CHOME, HAMAMATSU-CHO, MINATO-KU, TOKYO 105
JAPANESE
INDEXED: JICST

GEKKAN MASHINARI 5965
(MONTHLY MACHINERY)
JAPANESE
INDEXED: JICST

GIKEN NYUSU 5966
(TECHNICAL RESEARCH NEWS)
KIKAI SHINKO KYOKAI GIJUTSU KENKYU CHOSA FUKYU-SHITSU (MACHINERY PROMOTION ASSOCIATION, TECHNOLOGICAL RESEARCH INSTITUTE, INFORMATION SECTION)
JAPANESE

HAGURUMA 5967
(GEAR)
OSAKA KOGYO DAIGAKU KIKAI KOGAKU-BU KENKYUSHO (MECHANICAL ENGINEERING CIRCLE, OSAKA INSTITUTE OF TECHNOLOGY)
16-1, 5-CHOME, OMIYA, ASAHI-KU, OSAKA-SHI 535
JAPANESE

HAGURUMA 5968
(GEARS)
NIKKAN KOGYO SHINBUNSHA (DAILY INDUSTRIAL NEWSPAPER COMPANY)
JAPANESE

HAIKAN GIJUTSU 5969
(PIPING ENGINEERING)
NIHON KOGYO SHUPPAN K.K.
C/O ISHIZAKA BUILDING, 3-18, 2-CHOME, NISHI-KANDA, CHIYODA-KU, TOKYO 101
JAPANESE
INDEXED: CHE, JICST

HAIKAN TO SOCHI 5970
(PIPING AND PROCESS MACHINERY)
HAIKAN GIJUTSU KENKYU KYOKAI (THE SOCIETY FOR SCIENTIFIC RESEARCH OF PIPING ENGINEERING)
C/O NICHIASU BUILDING, 1-26, 1-CHOME, HIBADAIMON, MINATO-KU, TOKYO 105
JAPANESE
INDEXED: JICST

HAKUNAIKO 5971
(SHIPBOARD INDUSTRY)
JAPANESE
INDEXED: JICST

HOSO GIJUTSU 5972
(PACKAGING TECHNIQUE)
NIHON HOSO GIJUTSU KYOKAI (JAPAN PACKAGING TECHNIQUE ASSOCIATION)
12-8, 5-CHOME, GINZA, CHUO-KU, TOKYO 104
JAPANESE
INDEXED: JICST

HOSO KIKAI SHINBUN 5973
(PACKAGING MACHINERY NEWS)
HOSO KIKAI KOGYOKAI (ASSOCIATION OF PACKAGING MACHINERY INDUSTRY)
C/O DAI 2 NAN'O BUILDING, 20-1, 2-CHOME, NISHI-SHINBASHI, MINATO-KU, TOKYO 105
JAPANESE

HOSO TAIMUSU 5974
(PACKAGING TIMES)
NIPPO
4-5, 4-CHOME, IIDABASHI, CHIYODA-KU, TOKYO 102
JAPANESE

HYOGO-KEN KIKAI KINZOKU KOGYO SHIDOSHO KENKYU HOKOKU 5975
(RESEARCH PAPERS OF THE MECHANICAL AND METAL INDUSTRY RESEARCH AND GUIDANCE INSTITUTE OF HYOGO PREFECTURE))
HYOGO-KEN KIKAI KINZOKU KOGYO SHIDOSHO
240-1, FUKE, HIRATA, MIKI-SHI 673-04
JAPANESE

HYOGO-KEN KIKAI KINZOKU KOGYO SHIDOSHO NENPO 5976
(ANNUAL REPORT OF THE MECHANICAL AND METAL INDUSTRY RESEARCH AND GUIDANCE INSTITUTE OF HYOGO PREFECTURE)
HYOGO-KEN KIKAI KINZOKU KOGYO SHIDOSHO
240-1, FUKE, HIRATA, MIKI-SHI 673-04
JAPANESE

HYOGO-KEN KOGYO GIJUTSU JOHO: KIKAI-HEN 5977
(HYOGO PREFECTURE INDUSTRIAL TECHNOLOGY REPORT: MACHINERY)
HYOGO KENRITSU KOGYO SHIKENJO (HYOGO PREFECTURAL INDUSTRIAL LABORATORY)
JAPANESE

JANOME MISHIN GIHO 5978
(TECHNICAL REPORTS OF JANOME SEWING MACHINE COMPANY)
JANOME MISHIN KOGYO K.K.
1463, HAZAMA-MACHI, HACHIOJI-SHI, TOKYO 193
JAPANESE
INDEXED: JICST

JAPAN MASHINISUTO 5979
(JAPAN MACHINIST)
JAPAN MASHINISUTOSHA (JAPAN MACHINIST PUBLISHING COMPANY, LTD.)
5-2, 2-CHOME, IIDABASHI, CHIYODA-KU, TOKYO 102
JAPANESE
INDEXED: JICST

JAPAN SCIENCE REVIEW, MECHANICAL & ELECTRICAL ENGINEERING 5980
NIHON DOKUMENTESHON KYOKAI (THE JAPAN DOCUMENTATION SOCIETY)
5-7, 2-CHOME, KOISHIKAWA, BUNKYO-KU, TOKYO 112
JAPANESE
INDEXED: JICST

JETRO KIKAI NYUSU 5981
(JETRO MACHINERY NEWS)
NIHON BOEKI SHINKOKAI KIKAI-BU (JAPAN EXTERNAL TRADE)
2, AKASAKA AOI-CHO, MINATO-KU, TOKYO 107
JAPANESE

JIMU KIKI GAIDO 5982
(GUIDE TO OFFICE/BUSINESS MACHINERY)
JAPANESE
INDEXED: JICST

JOURNAL OF MECHANICAL LABORATORY OF JAPAN 5983
JAPANESE
INDEXED: JICST

JUKI MAGAJIN 5984
(JUKI MAGAZINE)
TOKYO JUKI KOGYO K.K. KOGYO SEIHIN HANBAI-BU
23, KABUKI-CHO, SHINJUKU-KU, TOKYO 160
JAPANESE

JUNKATSU 5985
(JOURNAL OF JAPAN SOCIETY OF LUBRICATION ENGINEERS)
NIHON JUNKATSU GAKKAI (JAPAN SOCIETY OF LUBRICATION ENGINEERS)
C/O NO. 407-2 KIKAI SHINKO KAIKAN, 5-8, 3-CHOME, SHIBA KOEN, MINATO-KU, TOKYO 105
JAPANESE AND ENGLISH
INDEXED: CHE, ENG, JICST

KAGAKU GIJUTSU BUNKEN SOKUHO, KIKAI KOGAKU-HEN 5986
(CURRENT BIBLIOGRAPHY ON SCIENCE AND TECHNOLOGY, MECHANICAL ENGINEERING)
NIHON KAGAKU GIJUTSU JOHO SENTA (THE JAPAN INFORMATION CENTER OF SCIENCE AND TECHNOLOGY)
1-1, 2-CHOME, NAGATA-CHO, CHIYODA-KU, TOKYO 100
JAPANESE
INDEXED: JICST

KAGOSHIMA-KEN KIKAI KINZOKU GIJUTSU SHIDO SENTA GIJUTSU NYUSU 5987
(TECHNICAL NEWS FROM THE MACHINE AND METALS ENGINEERING RESEARCH CENTER, KAGOSHIMA PREFECTURE)
KAGOSHIMA-KEN KIKAI KINZOKU GIJUTSU SHIDO SENTA
3799-1, USHUKU-CHO, KAGOSHIMA-SHI 890
JAPANESE

KAGOSHIMA-KEN KIKAI KINZOKU GIJUTSU SHIDO SENTA GYOMU HOKOKU 5988
(REPORT OF THE MACHINE AND METALS ENGINEERING RESEARCH CENTER, KAGOSHIMA PREFECTURE)
KAGOSHIMA-KEN KIKAI KINZOKU GIJUTSU SHIDO SENTA
3799-1, USHUKU-CHO, KAGOSHIMA-SHI 890
JAPANESE

KAIGAI KIKAI KOGAKU BUNKENSHU 5989
(LITERATURE ON FOREIGN MECHANICAL ENGINEERING)
JAPANESE
INDEXED: JICST

KANAGATA 5990
(DIE AND MOLD)
KANAGATA TSUSHINSHA
C/O MEISEI BUILDING, 6, SHIMOMIYABI-CHO, SHINJUKU-KU, TOKYO 162
JAPANESE

KANKOJI KOGYO 5991
(PLUMBING AND HEATING CONTRACTOR)
NIHON KANKOJI KOGYO KYOKAI (NIHON PLUMBING AND HEATING CONTRACTORS ASSOCIATION)
3-6, KYOBASHI, CHUO-KU, TOKYO 140
JAPANESE
INDEXED: JICST

KENKOSHI NYUSU 5992
(NEWS BULLETIN OF THE YAMANASHI PREFECTURAL LAPIDARY RESEARCH INSTITUTE)
YAMANASHI-KENRITSU KENMA KOGYO SHIDOSHO
838-2, SADOYOSHI-CHO, KOFU-SHI 400
JAPANESE

KIKAI 5993
(MACHINE)
NIHON SENBAI KOSHA KIKAI SEISAKUSHO (MACHINE FACULTY JAPAN MONOPOLY CORPORATION)
20-46, 2-CHOME, HORIFUNE, KITA-KU, TOKYO 114
JAPANESE

KIKAI GIJUTSU 5994
(MECHANICAL ENGINEERING)
NIKKAN KOGYO SHINBUNSHA (INDUSTRIAL DAILY NEWS)
8-10, 1-CHOME, KUDAN KITA, CHIYODA-KU, TOKYO 102
JAPANESE
INDEXED: CHE, JICST

KIKAI GIJUTSU KENKYUSHO HOKOKU 5995
(TECHNICAL REPORT OF MECHANICAL ENGINEERING LABORATORY)
TSUSHO SANGYO-SHO KOGYO GIJUTSU-IN KIKAI GIJUTSU KENKYUSHO (MECHANICAL ENGINEERING LABORATORY, AGENCY OF INDUSTRIAL SCIENCE AND TECHNOLOGY, MINISTRY OF INTERNATIONAL TRADE & INDUSTRY)
12-1, 4-CHOME, IGUSA, SUGINAMI-KU, TOKYO 167
JAPANESE AND ENGLISH
INDEXED: APP, CHE, ENG, JICST

KIKAI GIJUTSU KENKYUSHO NENPO 5996
(ANNUAL REPORT OF THE GOVERNMENT MECHANICAL ENGINEERING LABORATORY)
TSUSHO SANGYO-SHO KOGYO GIJUTSU-IN KIKAI GIJUTSU KENKYUSHO (MECHANICAL ENGINEERING LABORATORY, AGENCY OF INDUSTRIAL SCIENCE AND TECHNOLOGY, MINISTRY OF INTERNATIONAL TRADE & INDUSTRY)
12-1, 4-CHOME, IGUSA, SUGINAMI-KU, TOKYO 167
JAPANESE
INDEXED: JICST

KIKAI GIJUTSU KENKYUSHO SHIRYO 5997
(BULLETIN OF THE GOVERNMENT MECHANICAL ENGINEERING LABORATORY)
TSUSHO SANGYO-SHO KOGYO GIJUTSU-IN KIKAI GIJUTSU KENKYUSHO (MECHANICAL ENGINEERING LABORATORY, AGENCY OF INDUSTRIAL SCIENCE AND TECHNOLOGY, MINISTRY OF INTERNATIONAL TRADE & INDUSTRY)
12-1, 4-CHOME, IGUSA, SUGINAMI-KU, TOKYO 167
JAPANESE, WITH ENGLISH SUMMARY
INDEXED: JICST

KIKAI GIJUTSU KENKYUSHO SHOHO 5998
(JOURNAL OF MECHANICAL ENGINEERING LABORATORY)
TSUSHO SANGYO-SHO KOGYO GIJUTSU-IN KIKAI GIJUTSU KENKYUSHO (MECHANICAL ENGINEERING LABORATORY, AGENCY OF INDUSTRIAL SCIENCE AND TECHNOLOGY, MINISTRY OF INTERNATIONAL TRADE & INDUSTRY)
12-1, 4-CHOME, IGUSA, SUGINAMI-KU, TOKYO 167
JAPANESE, WITH EUROPEAN SUMMARY
INDEXED: APP, CHE, INS, MET, WAA, JICST

KIKAI GIJUTSU KYOKAI KAIHO 5999
(TRANSACTION OF ASSOCIATION OF MECHANICAL TECHNOLOGY)
KIKAI GIJUTSU KYOKAI (ASSOCIATION OF MECHANICAL TECHNOLOGY)
12-1, 4-CHOME, IGUSA, SUGINAMI-KU, TOKYO 167
JAPANESE

KIKAI JUCHU JISSEKI CHOSA TOKEIHYO 6000
(STATISTICAL SURVEY TABLES OF MACHINE ORDERS GIVEN AND RECEIVED)
KEIZAI KIKAKUCHO CHOSAKYOKU TOKEIKA (STATISTICS SECTION, SURVEY BUREAU, ECONOMIC PLANNING AGENCY)
JAPANESE

KIKAI JUCHU JISSEKI TOKEI CHOSA HOKOKU 6001
(STATISTICAL SURVEY REPORT OF MACHINE ORDERS GIVEN AND RECEIVED)
KEIZAI KIKAKUCHO CHOSAKYOKU TOKEIKA (STATISTICS SECTION, SURVEY BUREAU, ECONOMIC PLANNING AGENCY)
JAPANESE

KIKAI JUCHU TOKEI CHOSA NENPO 6002
(YEARLY SURVEY OF ORDERS RECEIVED FOR MACHINERY)
KEIZAI KIKAKU-CHO CHOSA-KYOKU (ECONOMIC PLANNING AGENCY)
3-2, KASUMIGASEKI, CHIYODA-KU, TOKYO 100
JAPANESE

KIKAI KEIZAI KENKYU 6003
(RESEARCH ON THE ECONOMICS OF MECHANIZATION)
KIKAI SHINKOKAI KEIZAI KENKYUSHO (ECONOMIC RESEARCH INSTITUTE FOR THE PROMOTION OF MECHANIZATION)
JAPANESE

KIKAI KIGU RYUTSU TOKEI GEPPO 6004
(MONTHLY REPORT ON THE SURVEY ON MACHINERY DISTRIBUTION)
TSUSHO SANGO-SHO DAIJIN KANBO CHOSA TOKEI-BU (RESEARCH AND STATISTICS DIVISION, MINISTER'S SECRETARIAT, MINISTRY OF INTERNATIONAL TRADE AND INDUSTRY)
C/O TSUSAN-SHO, GINZACHOSHA, 15-1, 6-CHOME, GINZA, CHUO-KU, TOKYO 104
JAPANESE

KIKAI KIGU RYUTSU TOKEI SOKUHO 6005
(NEWS OF CIRCULATION OF MACHINE AND INSTRUMENTS)
TSUSHO SANGYO-SHO DAIJIN KANBO CHOSA TOKEI-BU (RESEARCH AND STATISTICS DIVISION, MINISTER'S SECRETARIAT, MINISTRY OF INTERNATIONAL TRADE AND INDUSTRY)
3-1, 1-CHOME, KASUMIGASEKI, CHIYODA-KU TOKYO 100
JAPANESE

KIKAI KOGAKU NENKAN 6006
(MECHANICAL ENGINEERING ANNUAL)
JAPANESE
INDEXED: JICST

KIKAI KOGYO KAIGAI JOHO 6007
(OVERSEAS REPORTS ON ENGINEERING INDUSTRY)
KIKAI SHINKO KYOKAI KEIZAI KENKYUSHO (ECONOMIC RESEARCH INSTITUTE JAPAN SOCIETY FOR THE PROMOTION OF MACHINE INDUSTRY)
5-8, 3-CHOME, SHIBA KOEN, MINATO-KU, TOKYO 105
JAPANESE
INDEXED: JICST

KIKAI KOGYO KEIZAI KENKYU SHIRYO 6008
(ECONOMIC RESEARCH DATA ON THE MACHINE INDUSTRY)
JAPANESE
INDEXED: JICST

KIKAI KOGYO SETSUBI CHOSA HOKOKU 6009
(REPORT OF A SURVEY OF MACHINE INDUSTRY FACILITIES)
JAPANESE
INDEXED: JICST

KIKAI KOGYO SHINBUN 6010
(MACHINERY INDUSTRY NEWS)
KIKAI KOGYO SHINBUNSHA (THE MACHINERY INDUSTRY NEWS, LTD.)
1-7, 1-CHOME, SHIBA KOEN, MINATO-KU, TOKYO 105
JAPANESE

KIKAI NO KENKYU 6011
(SCIENCE OF MACHINERY)
YOKENDO
30-15, 5-CHOME, HONGO, BUNKYO-KU, TOKYO 113
JAPANESE
INDEXED: CHE, JICST

KIKAI NO SONGAI 6012
(MACHINERY DAMAGE)
NIHON KIKAI HOKEN RENMEI (UNION OF MACHINERY INSURERERS OF JAPAN)
C/O IDA BUILDING, 3-8, 1-CHOME, YAESU, CHUO-KU, TOKYO 103
JAPANESE
INDEXED: CHE, JICST

KIKAI SEKKEI 6013
(MECHANICAL DESIGN)
NIKKAN KOGYO SHINBUNSHA (INDUSTRIAL DAILY NEWS)
8-10, 1-CHOME, KUDANKITA, CHIYODA-KU, TOKYO 102
JAPANESE
INDEXED: JICST

KIKAI SEKKEI HANDOBUKKU 6014
(HANDBOOK OF MACHINE DESIGNS)
OKAWA SHUPPAM
(OKAWA PUBLISHING COMPANY)
JAPANESE

KIKAI SHINKO 6015
(PROMOTING MACHINE INDUSTRY IN JAPAN)
KIKAI SHINKO KYOKAI (JAPAN SOCIETY FOR THE PROMOTION OF MACHINE INDUSTRY)
C/O KIKAI SHINKO KAIKAN, 5-8, 3-CHOME, SHIBA KOEN, MINATO-KU, TOKYO 105
JAPANESE
INDEXED: JICST

KIKAI TO KOGU 6016
(TOOL ENGINEER)
KOGYO CHOSAKAI (KOGYO CHOSAKAI PUBLISHING COMPANY, LTD.)
14-7, 2-CHOME, HONGO, BUNKYO-KU, TOKYO 113
JAPANESE
INDEXED: JICST

KIKAI TOKEI GEPPO 6017
(MONTHLY REPORT OF MACHINERY STATISTICS)
TSUSHO SANGYO-SHO DAIJIN KANBO CHOSA TOKEI-BU (RESEARCH AND STATISTICS DIVISION, MINISTER'S SECRETARIAT, MINISTRY OF INTERNATIONAL TRADE AND INDUSTRY)
15-1, 6-CHOME, GINZA, CHUO-KU, TOKYO 104
JAPANESE
INDEXED: JICST

KIKAI TOKEI NENPO 6018
(YEARBOOK OF MACHINERY STATISTICS)
TSUSHO SANGYO-SHO DAIJIN KANBO CHOSA TOKEI-BU (RESEARCH AND STATISTICS DIVISION, MINISTER'S SECRETARIAT, MINISTRY OF INTERNATIONAL TRADE AND INDUSTRY)
15-1, 6-CHOME, GINZA, CHUO-KU, TOKYO 104
JAPANESE

KIKAI TOKEI SOKUHO 6019
(STATISTICAL BULLETIN ON MACHINERY)
TSUSHO SANGYO-SHO DAIJIN-KANBO CHOSA TOKEI-BU (RESEARCH AND STATISTICS DIVISION, MINISTER'S SECRETARIAT, MINISTRY OF INTERNATIONAL TRADE AND INDUSTRY)
JAPANESE

KIKAI YUSHUTSU 6020
(MACHINE EXPORTS)
NIHON KIKAI YUSHUTSU KUMIAI (JAPAN UNION OF MACHINE EXPORTERS)
JAPANESE

KIKAIKEN NYUSU 6021
(NEWS OF MECHANICAL ENGINEERING LABORATORY)
TSUSHO SANGYO-SHO KOGYO GIJUTSU-IN KIKAI GIJUTSU KENKYUSHO (MECHANICAL ENGINEERING LABORATORY, AGENCY OF INDUSTRIAL SCIENCE & TECHNOLOGY, MINISTRY OF INTERNATIONAL TRADE & INDUSTRY)
12-1, 4-CHOME, IGUSA, SUGIMAMI-KU, TOKYO 167
JAPANESE
INDEXED: JICST

KIKAIRUI NO YUNYU TOKEI CHOSAHYO: GEPPO 6022
(STATISTICAL SURVEY TABLES FOR MACHINERY IMPORTS: MONTHLY)
NIHON KIKAI YUNYU KYOKAI (JAPAN MACHINERY IMPORTS ASSOCIATION)
JAPANESE

KIKOSHI NYUSU 6023
(NEWS FROM THE MACHINE AND METALS RESEARCH INSTITUTE, YAMANASHI PREFECTURE)
YAMANASHI-KEN KIKAI KINZOKU KOGYO SHIDOSHO
838-2, SATOYOSHI-CHO, KOFU-SHI 400
JAPANESE

KOMATSU GIHO 6024
(KOMATSU TECHNICAL REVIEW)
KOMATSU SEISAKU-SHO (KOMATSU, LTD.)
3-6, 2-CHOME, AKASAKA, MINATO-KU, TOKYO 107
JAPANESE, WITH ENGLISH SUMMARY
INDEXED: JICST

KOSAKU KIKAI NYUSU 6025
(MACHINE TOOL NEWS)
NIHON KOSAKU KIKAI KOGYOKAI (JAPAN MACHINE TOOL BUILDERS ASSOCIATION)
5-8, 3-CHOME, SHIBA KOEN, MINATO-KU, TOKYO 105
JAPANESE

KOYO ENJINIARINGU JANARU 6026
(KOYO ENGINEERING JOURNAL)
KOYO SEIKO K.K. (KOYO SEIKO COMPANY, LTD.)
2, UNAGIDANI NISHINO-CHO, MINAMI-KU, OSAKA-SHI 542
JAPANESE
INDEXED: JICST

KYOKI NYUSU 6027
(KYOKI NEWS)
MITSUBISHI JUKOGYO K.K., KYOTO SEIKI SEISAKUJO
1, UZUMASA TATSUMI-CHO, UKYO-KU, KYOTO-SHI 616
JAPANESE

LOOK JAPAN 6028
ENGLISH

MASHIN DEZAIN 6029
(MACHINE DESIGN)
KOGYO CHOSAKAI (INDUSTRIAL SURVEY ASSOCIATION)
JAPANESE

MASHINISUTO 6030
(MACHINIST (WORLD PROGRESS OF PRODUCTION ENGINEERING))
MASHINISUTO K.K.
C/O TAKAKUBO BUILDING, 28-2, 3-CHOME, KANDA OGAWA-MACHI, CHIYODA-KU, TOKYO 101
JAPANESE
INDEXED: JICST

MIE-KEN KOGYO GIJUTSU JOHO, KIKAI KINZOKU-HEN 6031
(MIE PREFECTURE INDUSTRIAL TECHNOLOGY INFORMATION: MACHINE AND METALS)
MIE-KEN KOGYO GIJUTSU SENTA (THE INDUSTRIAL TECHNOLOGY CENTER OF MIE PREFECTURE)
3485, TAKAJAYA KOMORI-CHO, TSU-SHI 514
JAPANESE

MISHIN 6032
(SEWING MACHINE)
NIHON MISHIN YUSHUTSU KUMIAI (JAPAN SEWING MACHINE EXPORT ASSOCIATION)
JAPANESE

NAGATA SEISAKUJOHO 6033
(REPORT OF NAGATA MANUFACTURING PLANT)
JAPANESE
INDEXED: JICST

NAINEN KIKAN 6034
(INTERNAL COMBUSTION ENGINE)
SANKAIDO
15, SAIKU-CHO, SHINJUKU-KU, TOKYO 162
JAPANESE
INDEXED: JICST

NEJI 6035
(FASTENERS)
NIHON NEJI KOGYO KYOKAI (THE FASTNERS INSTITUTE OF JAPAN)
C/O KIKAI SHINKO KAIKAN, 5-8, 3-CHOME, SHIBA KOEN, MINATO-KU, TOKYO 105
JAPANESE

NEJI NO SEKAI 6036
(INDUSTRIAL FASTENERS REVIEW)
NEJI NO SEKAISHA
1-9, KANDA SARUGAKU-CHO, CHIYODA-KU, TOKYO 101
JAPANESE
INDEXED: JICST

NIHON BARUBU KOGYOKAI KAIHO 6037
(NEWS OF THE JAPAN VALVE MANUFACTURER'S ASSOCIATION)
NIHON BARUBU KOGYOKAI (THE JAPAN VALVE MANUFACTURER'S ASSOCIATION)
C/O KIKAI SHINKO KAIKAN, 5-8, 3-CHOME, SHIBA KOEN, MINATO-KU, TOKYO 105
JAPANESE

NIHON DAIKASUTO KYOKAI KAIHO 6038
(BULLETIN OF THE JAPAN DIE CASTING ASSOCIATION)
NIHON DAIKASUTO KYOKAI
C/O KIKAI SHINKO KAIKAN, 5-8, 3-CHOME, SHIBA KOEN, MINATO-KU, TOKYO 105
JAPANESE
INDEXED: JICST

NIHON KIKAI GAKKAI KOEN RONBUNSHU (KOENKAI) 6039
(PROCEEDINGS OF THE SYMPOSIUM OF THE JAPAN SOCIETY OF MECHANICAL ENGINEERS, LECTURE MEETING)
NIHON KIKAI GAKKAI (THE JAPAN SOCIETY OF MECHANICAL ENGINEERS)
4-9, 2-CHOME, YOYOGI, SHIBUYA-KU, TOKYO 151
JAPANESE
INDEXED: JICST

NIHON KIKAI GAKKAI KOEN RONBUNSHU (ZENKOKU TAIKAI, TSUJO SOKAI) 6040
(PROCEEDINGS OF THE SYMPOSIUM OF THE JAPAN SOCIETY OF MECHANICAL ENGINEERS)
NIHON KIKAI GAKKAI (THE JAPAN SOCIETY OF MECHANICAL ENGINEERS)
4-9, 2-CHOME, YOYOGI, SHIBUYA-KU, TOKYO 151
JAPANESE
INDEXED: JICST

NIHON KIKAI GAKKAI RONBUNSHU 6041
(TRANSACTIONS OF THE JAPAN SOCIETY OF MECHANICAL ENGINEERS)
NIHON KIKAI GAKKAI (JAPAN SOCIETY OF MECHANICAL ENGINEERS)
4-9, 2-CHOME, YOYOGI, SHIBUYA-KU, TOKYO 151
JAPANESE
INDEXED: CHE, COP, ISM, JICST

NIHON KIKAI GAKKAI ZADANKAI SHIRYO 6042
(MEMOIRS OF DISCUSSION MEETING OF THE JAPAN SOCIETY OF MECHANICAL ENGINEERS)
NIHON KIKAI GAKKAI (THE JAPAN SOCIETY OF MECHANICAL ENGINEERS)
4-9, 2-CHOME, YOYOGI, SHIBUYA-KU, TOKYO 151
JAPANESE

NIHON KIKAI GAKKAISHI 6043
(JOURNAL OF THE JAPAN SOCIETY OF MECHANICAL ENGINEERS)
NIHON KIKAI GAKKAI (JAPAN SOCIETY OF MECHANICAL ENGINEERS)
4-9, 2-CHOME, YOYOGI, SHIBUYA-KU, TOKYO 151
JAPANESE
INDEXED: APP, CHE, COP, JICST

NIHON KIKAI KOGYO RENGOKAI JIGYO HOKOKU 6044
(REPORT OF THE JAPAN MACHINERY FEDERATION)
NIHON KIKAI KOGYO RENGOKAI
5-8, 3-CHOME, SHIBA KOEN, MINATO-KU, TOKYO 105
JAPANESE

NIHON KIKAI YUSHUTSU KUMIAI 6045
(JAPAN MACHINERY EXPORTER'S ASSOCIATION)
NIHON KIKAI YUSHUTSU KUMIAI (JAPAN MACHINERY EXPORTER'S ASSOCIATION)
C/O KIKAI SHINKO KAIKAN, 5-8, 3-CHOME, SHIBA KOEN, MINATO-KU, TOKYO 105
JAPANESE

NIHON KIKO SHINBUN 6046
(JAPAN MACHINERY JOURNAL)
NIHON KIKO SHINBUNSHA
14-15, 5-CHOME, SHIBA, MINATO-KU, TOKYO 108
JAPANESE

NIHON NEJI KENKYU KYOKAISHI 6047
(JOURNAL OF THE JAPAN INSTITUTE FOR SCREW THREADS AND FASTENERS)
NIHON NEJI KENKYU KYOKAI
C/O KIKAI SHINKO KAIKAN, 5-8, 3-CHOME, SHIBA KOEN, MINATO-KU, TOKYO 105
JAPANESE
INDEXED: JICST

NIHON REITO KYOKAI GAKUJUTSU KOENKAI KOEN RONBUNSHU 6048
(PROCEEDINGS OF THE SYMPOSIUM OF THE JAPANESE ASSOCIATION OF REFRIGERATION))
NIHON REITO KYOKAI (JAPANESE ASSOCIATION OF REFRIGERATION)
8, SAN'EI-CHO, SHINJUKU-KU, TOKYO 160
JAPANESE
INDEXED: CHE

NIHON SUPINDORU GIHO 6049
(NIHON SPINDLE TECHNICAL REPORT)
NIHON SUPINDORU SEIZO K.K. (NIHON SPINDLE MANUFACTURING COMPANY, LTD.)
2-1, SHIOE NISHI SOKE, AMAGASAKI-SHI 660
JAPANESE
INDEXED: JICST

NIKKAN JUKOGYO 6050
(DAILY HEAVY INDUSTRY NEWS)
JUKOGYO SHINBUNSHA (THE HEAVY INDUSTRY PRESS)
4-5, NIHONBASHI HON-CHO, CHUO-KU, TOKYO 103
JAPANESE

NIKKIREN GEPPO 6051
(JOURNAL OF THE JAPAN MACHINERY FEDERATION)
NIHON KIKAI KOGYO RENGOKAI KOHO-GRUPU (THE JAPAN MACHINERY FEDERATION)
C/O KIKAI SHINKO KAIKAN 5-8, 3-CHOME, SHIBA KOEN, MINATO-KU, TOKYO 105
JAPANESE
INDEXED: JICST

NIPPON KIKAI YUNYU KYOKAI KAIHO 6052
(BULLETIN OF JAPAN MACHINERY IMPORTERS ASSOCIATION)
NIPPON KIKAI YUNYU KYOKAI (JAPAN MACHINERY IMPORTERS ASSOCIATION)
C/O TAIYO BUILDING, 14, SHIBA KOTOHIRA-CHO, MINATO-KU, TOKYO 105
JAPANESE

NSK BEARING JOURNAL 6053
NIHON SEIKO K.K.
C/O YUSEN BUILDING, 3-2, 2-CHOME, MARUNOUCHI, CHIYODA-KU, TOKYO 100
JAPANESE
INDEXED: JICST

NYU MASHIN TSURU 6054
(NEW MACHINE TOOLS)
NYU MASHIN TSURU
JAPANESE

OBARU NYUSU 6055
(OVAL NEWS)
OBARU KIKI KOGYO K.K.
10-8, 3-CHOME, KAMI OCHIAI, SHINJUKU-KU, TOKYO 161
JAPANESE
INDEXED: JICST

OYO KIKAI KOGAKU 6056
(MECHANICAL ENGINEERING APPLICATIONS)
TAIGA SUPPAN
1-13, KANDA AWAJI-CHO, CHIYODA-KU, TOKYO 101
JAPANESE
INDEXED: JICST

PACKS 6057
NIPPO K.K. (NIPPO COMPANY, LTD.)
4-5, 4-CHOME, IIDABASHI, CHIYODA-KU, TOKYO 102
JAPANESE

PAKKEIJINGU 6058
(PACKAGING)
PAKKEIJINGUSHA
1-53, KITA UENO, TAITO-KU, TOKYO 110
JAPANESE
INDEXED: JICST

PAKKUSU 6059
(PACKS)
NIHON HOSO TAIMUSU SHA
JAPANESE
INDEXED: JICST

PURESU DAIJESUTO 6060
(PRESS DIGEST)
NIHON KINZOKU PURESU KOGYOKAI (JAPAN METAL STAMPING ASSOCIATION)
5-8, 3-CHOME, SHIBA KOEN, MINATO-KU, TOKYO 105
JAPANESE

RIFUTO GAKKAI KAISHI 6061
(JOURNAL OF LIFT RESEARCH ASSOCIATION)
RIFUTO KENKYUKAI
C/O NIHON HATSUMEI GAKKAI, 2-30, NAKAMURA, NERIMA-KU, TOKYO 176
JAPANESE

RIKEN GIJUTSU NYUSU 6062
(RIKEN TECHNICAL NEWS)
RIKEN PISUTON RINGU KOGYO K.K. (RIKEN PISTON INDUSTRY COMPANY, LTD.)
7-13, 1-CHOME, NISHI SHINBASHI, MINATO-KU, TOKYO 105
JAPANESE

RIKUYO NAINEN KIKAN 6063
(LAND INTERNAL COMBUSTION ENGINE)
RIKUYO NAINEN KIKAN KYOKAI (LAND INTERNAL COMBUSTION ENGINE MANUFACTURER'S ASSOCIATION)
1-2, ICHIGAYA SADOHARA-CHO, SHINJUKU-KU, TOKYO 162
JAPANESE
INDEXED: JICST

RYUTAI KOGAKU 6064
(FLUID ENGINEERING)
SANGYO KAIHATSUSHA
11-10, 1-CHOME, UCHIKANDA, CHIYODA-KU, TOKYO 101
JAPANESE
INDEXED: JICST

RYUTSU: HOSO GIJUTSU GAKKO KAIHO 6065
(REPORT OF THE PACKAGING TECHNOLOGY SCHOOL; DISTRIBUTION)
JAPANESE
INDEXED: JICST

SANGYO KIKAI 6066
(INDUSTRIAL MACHINERY)
NIHON SANGYO KIKAI KOGYOKAI (THE JAPAN SOCIETY OF INDUSTRIAL MACHINERY MANUFACTURERS)
C/O KIKAI SHINKO KAIKAN, 5-8, 3-CHOME, SHIBA KOEN, MINATO-KU, TOKYO 105
JAPANESE
INDEXED: JICST

SANKI MANSURI 6067
(SANKI MONTHLY)
SANKI KOGYO K.K. (SANKI ENGINEERING COMPANY, LTD.)
1-10, YURAKU-CHO, CHIYODA-KU, TOKYO 100
JAPANESE

SHARINGU KAIHO 6068
(SHEARING WORLD)
ZENKOKU SHEARINGU KUMIAI
3-16, NIHONBASHI KAYABA-CHO, CHUO-KU, TOKYO 103
JAPANESE

SHIGA-KENRITSU KIKAI KINZOKU KOGYO SHIDOSHO GYOMU HOKOKU 6069
(SHIGA PREFECTURAL INSTITUTE FOR MACHINE & METALS INDUSTRY REPORT)
SHIGA-KENRITSU KIKAI KINZOKU KOGYO SHIDOSHO (SHIGA PREFECTURAL INSTITUTE FOR MACHINE & METALS INDUSTRY)
52, OKA-MACHI, HIKONE-SHI 522
JAPANESE

SHIMAZU HYORON 6070
(SHIMAZU REVIEW)
SHIMAZU HYORON HENSYU-BU
C/O SHIMAZU SEISAKUSHO, 1, NISHINOKYO KUWABARA-MACHI, NAKAGYO-KU, KYOTO-SHI 604
JAPANESE, WITH ENGLISH SUMMARY
INDEXED: CHE, INS, JICST

SHIN NIHON KOKI TEKUNIKARU REPOTO 6071
(TECHNICAL REPORT OF ENGINEERING DEPARTMENT, SHIDAYAMA WORKS, SHIN NIHON MACHINERY COMPANY, LTD.)
SHIN NIHON KOKI K.K. SHIDAYAMA KOJO GIJUTSU-BU
500, HISHIKI, SAKAI-SHI 593
JAPANESE

SHIZUOKA-KEN KIKAI GIJUTSU SHIDOSHO KENKYU HOKOKU 6072
(REPORT OF THE MECHANICAL RESEARCH INSTITUTE OF SHIZUOKA PREFECTURE)
SHIZUOKA-KEN KIKAI GIJUTSU SHIDOSHO (THE MECHANICAL RESEARCH INSTITUTE OF SHIZUOKA PREFECTURE)
2444, KOIKE-CHO, HAMAMATSU-SHI 435
JAPANESE

SHIZUOKA-KEN KOGYO GIJUTSU JOHO, KIKAI-HEN 6073
(SHIZUOKA PREFECTURE INDUSTRIAL TECHNOLOGY REPORT, MECHANICAL EDITION)
SHIZUOKA-KEN KOGYO SHIKENJO (SHIZUOKA PREFECTURE INDUSTRIAL EXPERIMENT STATION)
JAPANESE

SHOKUHIN NO HOSO 6074
(PACKAGING OF FOOD STUFF)
HOSO SHOKUHIN GIJUTSU KYOKAI
C/O AICHI-KEN SHOKUHIN KOGYO SHIKENJO, 2-1, SHINPUKUJI-MACHI, NISHI-KU, NAGOYA-SHI 451
JAPANESE
INDEXED: JICST

SUBARU GIHO 6075
(SUBARU TECHNICAL REPORT)
FUJI JUKOGYO K.K. SUBARU GIJUTSU HONBU HATSUDOKI GIJUTSU-BU (ENGINE TECHNOLOGY DIVISION, SUBARU TECHNOLOGY DEPARTMENT, FUJI HEAVY INDUSTRY COMPANY, LTD.)
9-6, 3-CHOME, OSAWA, MITAKA-SHI, TOKYO 181
JAPANESE

SUMITOMO JUKIKAI GIHO 6076
(SUMITOMO SHIPBUILDING & MACHINERY)
SUMITOMO JUKIKAI KOGYO K.K. (SUMITOMO SHIPBUILDING & MACHINERY COMPANY, LTD.)
C/O SHIN OTEMACHI BUILDING, 2-1, 2-CHOME, OTE-MACHI, CHIYODA-KU, TOKYO 100
JAPANESE, WITH ENGLISH SUMMARY
INDEXED: JICST

TABO KIKAI 6077
(TURBOMACHINERY)
JAPANESE
INDEXED: JICST

TANGAROI 6078
(TUNGALOY)
TOSHIBA TANGAROI K.K.
1-7, TSUKAKOSHI, SAIWAI-KU, KAWASAKI-SHI 210
JAPANESE

TEIKETSU TO SETSUGO 6079
(FASTENING & JOINING)
NEJI NO SEKAISHA
2-8, KANDA JINBO-CHO, CHIYODA-KU, TOKYO 101
JAPANESE
INDEXED: JICST

TOKEITEN 6080
(WATCHMAKER)
JAPANESE
INDEXED: JICST

TOKICO REVIEW 6081
TOKICO K.K.
6-3, 1-CHOME, FUJIMI, KAWASAKI-KU, KAWASAKI-SHI 210
JAPANESE
INDEXED: CHE, JICST

TOSHIBA KIKAI NYUSU 6082
(TOSHIBA MACHINE NEWS)
TOSHIBA KIKAI K.K. (TOSHIBA MACHINE COMPANY, LTD.
2-11, 4-CHOME, GINZA, CHUO-KU, TOKYO 104
JAPANESE

TOSHO 6083
(TOSHIBA ELEVATOR)
TOSHIBA SHOKOKI SABISU TOSHO HENSHU IINKAI (TOSHIBA ELEVATOR SERVICE 'TOSHO' EDITING COMMITTEE)
JAPANESE

TOYOTA GIHO 6084
(TOYOTA TECHNICAL REVIEW)
TOYOTA KOKI K.K.
1-1, ASAHI-CHO, KARIYA-SHI 448
JAPANESE
INDEXED: JICST

TSUGAMI NEWS 6085
(TSUGAMI NYUSU)
TSUGAMI SHOJI K.K. (TSUGAMI TRADING COMPANY)
23-1, 2-CHOME, NISHISHINBASHI, MINATO-KU, TOKYO 105
JAPANESE

YAKUSHIN SURU KIKI 6086
(MACHINERY FOR REMARKABLE DEVELOPMENT)
JAPANESE
INDEXED: JICST

YAMANASHI-KENRITSU KENMA KOGYO SHIDOSHO KENKYU HOKOKU 6087
(RESEARCH REPORTS OF THE YAMANASHI PREFECTURAL LAPIDARY RESEARCH INSTITUTE)
838-2, SATOYOSHI-CHO, KOFU-SHI 400
JAPANESE

YUATSU GAIKOKU BUNKEN SHOROKUSHU 6088
(ABSTRACTS OF FOREIGN OIL HYDARULIC PAPERS)
KANSAI YUATSU KON'WA-KAI (KANSAI OIL HYDRAULIC ASSOCIATION)
C/O OSAKA-FURITSU KOGYO GIJUTSU KENKYUSHO, ENOKOJIMA KAMINO-CHO, NISHI-KU, OSAKA-SHI 550
JAPANESE
INDEXED: JICST

YUATSU GIJUTSU 6089
(HYDRAULICS AND PNEUMATICS)
NIHON KOGYO SUPPAN K.K.
C/O ISHIZAKA BUILDING, 3-18, 2-CHOME, NISHIKANDA, CHIYODA-KU, TOKYO 102
JAPANESE
INDEXED: JICST

YUATSU TO KUKIATSU 6090
(JOURNAL OF THE JAPAN HYDRAULICS & PNEUMATICS SOCIETY)
NIHON YU KUKIATSU KYOKAI (JAPAN HYDRAULICS & PNEUMATICS SOCIETY)
C/O KIKAI SHINKO KAIKAN, 5-8, 3-CHOME, SHIBA KOEN, MINATO-KU, TOKYO 105
JAPANESE, WITH ENGLISH SUMMARY

YUKUATSUKA SEKKEI 6091
(HYDRAULICS AND PNEUMATICS)
JAPANESE
INDEXED: JICST

Nuclear Technology

AISOTOPU RIYO KENKYU SEISEKI NENPO 6092
(ANNUAL REPORT OF RESULTS OF RESEARCH ON ISOTOPE USE)
JAPANESE
INDEXED: JICST

AMERIKA GENSHIRYOKU IINKAI KIZO BUNKEN MOKUROKU 6093
(U.S. ATOMIC ENERGY COMMISSION'S BIBLIOGRAPHY)
JAPANESE
INDEXED: JICST

ANNUAL REPORT OF THE ATOMIC ENERGY COMMISSION, JAPAN 6094
GENSHIRYOKU IINKAI (ATOMIC ENERGY COMMISSION, JAPAN)
2-1, 2-CHOME, KASUMIGASEKI, CHIYODA-KU, TOKYO 100
ENGLISH

ANNUAL REPORT OF THE ATOMIC ENERGY RESEARCH INSTITUTE, NIHON UNIVERSITY 6095
NIHON DAIGAKU GENSHIRYOKU KENKYUSHO (ATOMIC ENERGY RESEARCH INSTITUTE, NIHON UNIVERSITY)
1-8, KANDA SURUGADAI, CHIYODA-KU, TOKYO 101
ENGLISH

ANNUAL REPORT OF THE INSTITUTE FOR NUCLEAR STUDY, UNIVERSITY OF TOKYO 6096
TOKYO DAIGAKU GENSHIKAKU KENKYUSHO TOSHOSHITSU (LIBRARY, INSTITUTE FOR NUCLEAR STUDY, UNIVERSITY OF TOKYO)
2-1, 3-CHOME, MIDORI-CHO, TANASHI-SHI, TOKYO 188
ENGLISH
INDEXED: CHE, JICST

ANNUAL REPORT OF O U L N S 6097
OSAKA DAIGAKU RIGAKUBU FUZOKU GENSHIKAKU KENKYU SHISETSU (LABORATORY OF NUCLEAR STUDIES, FACULTY OF SCIENCE, OSAKA UNIVERSITY)
1-1, MACHIKANEYAMA-CHO, TOYONAKA-SHI 560
ENGLISH

ANNUAL REPORT OF THE RADIATION CENTER OF OSAKA PREFECTURE 6098
OSAKA-FURITSU HOSHASEN CHUO KENKYUSHO (RADIATION CENTER OF OSAKA PREFECTURE)
704, SHINKE-CHO, SAKAI-SHI 593
ENGLISH
INDEXED: BIB, JICST

ANNUAL REPORT OF THE TOKYO METROPOLITAN ISOTOPE RESEARCH 6099
JAPANESE
INDEXED: JICST

ANNUAL REPORTS OF THE RESEARCH REACTOR INSTITUTE, KYOTO UNIVERSITY 6100
KYOTO DAIGAKU GENSHIRO JIKKENSHO (RESEARCH REACTOR INSTITUTE, KYOTO UNIVERSITY)
KUMATORI-CHO, SENNAN-GUN, OSAKA 590-04
ENGLISH
INDEXED: BIO, CHE, INS, JICST

ATOMS IN JAPAN 6101
NIHON GENSHIRYOKU SANGYO KAIGI (JAPAN ATOMIC INDUSTRIAL FORUM, INCORPORATED)
1-13, 1-CHOME, SHINBASHI, MINATO-KU, TOKYO 105
ENGLISH

ATOMU 6102
(ECONOMIC JOURNAL 'ATOM')
GENSHIRYOKU KEIZAI KENKYUSHO
5, ICHIBAN-CHO, CHIYODA-KU, TOKYO 102
JAPANESE
INDEXED: JICST

BULLETIN OF THE RESEARCH LABORATORY FOR NUCLEAR REACTORS (TOKYO INSTITUTE OF TECHNOLOGY) 6103
RESEARCH LABORATORY FOR NUCLEAR REACTORS, TOKYO INSTITUTE OF TECHNOLOGY
OHOKAYAMA, MEGURO-KU, TOKYO
ENGLISH
INDEXED: CHE, INS

CONTENTS WEEKLY 6104
NIHON GENSHIRYOKU KENKYUSHO GIJUTSU JOHO-BU
TOKAI-MURA, NAKA-GUN, IBARAKI 319-11
ENGLISH

DONEN 6105
(NEWS LETTER OF POWER REACTOR AND NUCLEAR FUEL DEVELOPMENT CORPORATION)
DORYOKURO KAKU NENRYO KAIHATSU JIGYODAN (POWER REACTOR AND NUCLEAR FUEL DEVELOPMENT CORPORATION)
9-13, 1-CHOME, AKASAKA, MINATO-KU, TOKYO 107
JAPANESE

DORYOKURO GIHO 6106
(POWER REACTOR TECHNOLOGY)
DORYOKURO KAKU NENRYO KAIHATSU JIGYODAN (POWER REACTOR AND NUCLEAR FUEL DEVELOPMENT CORPORATION)
9-13, 1-CHOME, AKASAKA, MINATO-KU, TOKYO 107
JAPANESE
INDEXED: JICST

DORYOKURO KAKU NENRYO KAIHATSU JIGYODAN NENPO 6107
(ANNUAL REPORT OF THE POWER REACTOR AND NUCLEAR FUEL DEVELOPMENT CORPORATION)
DORYOKURO KAKU NENRYO KAIHATSU JIGYODAN (POWER REACTOR AND NUCLEAR FUEL DEVELOPMENT CORPORATION)
9-13, 1-CHOME, AKASAKA, MINATO-KU, TOKYO 107
JAPANESE
INDEXED: JICST

DORYOKURO KAKUNENRYO KAIHATSU JIGYODAN TOKAI JIGYOSHO GIJUTSU REPOTO 6108
(TECHNICAL REPORT OF TOKAI WORKS, POWER REACTOR AND NUCLEAR FUEL DEVELOPMENT CORPORATION)
JAPANESE
INDEXED: JICST

DORYOKURO KAKUNENRYO KAIHATSU JIGYODAN TOKAI JIGYOSHO SEIKA YOSHISHU 6109
(SUMMARY OF ACHIEVEMENTS OF TOKAI WORKS, POWER REACTOR AND NUCLEAR FUEL DEVELOPMENT CORPORATION)
DORYOKURO KAKU NENRYO KAIHATSU JIGYODAN
MURAMATSU, TOKAI-MURA, NAKA-GUN, IBARAKI 319-11
JAPANESE
INDEXED: JICST

FAPIG 6110
DAI-ICHI GENSHIRYOKU SANGYO GURUPU JIMUKYOKU (THE FIRST ATOMIC POWER INDUSTRY GROUP)
C/O NISSHO IWAI BUILDING, 4-5, 2-CHOME, AKASAKA, MINATO-KU, TOKYO 107
JAPANESE
INDEXED: JICST

FURANSU GENSHIRYOKU 6111
(ATOME-FRANCE DOCUMENTATION NUCLEAIRE FRANCAISE)
FURANSU GENSHIRYOKU SHIRYO KENKYUSHO (CENTRE DE DOCUMENTATION NUCLEAIRE FRANCAISE)
C/O D.F. BUILDING, 2-8, 2-CHOME, MINAMI-AOYAMA, MINATO-KU, TOKYO 107
JAPANESE
INDEXED: JICST

GAIKOKU GENSHIRYOKU KAAKEI KIKAN KANKO SHIRYO MOKUROKU 6112
(CATALOGUE OF PUBLISHED MATERIALS BY ORGANIZATION, CONNECTED WITH FOREIGN NUCLEAR POWER)
JAPANESE
INDEXED: JICST

GAIKOKU GENSHIRYOKU KIKAN KANKO SHIRYO GEPPO 6113
(MONTHLY LIST OF SELECTED ATOMIC ENERGY PUBLICATIONS)
KOKURITSU KOKKAI TOSHOKAN SANKOSHOSHI-BU (REFERENCE AND BIBLIOGRAPHY DIVISION, NATIONAL DIET LIBRARY)
10-1, 1-CHOME, NAGATA-CHO, CHIYODA-KU, TOKYO 100
ENGLISH
INDEXED: JICST

GEN'AN KYO BUNKEN SOKUHO 6114
(NUCLEAR SAFETY ASSOCIATION BIBLIOGRAPHY, CURRENT REPORT)
GENSHIRYOKU ANZEN KENKYU KYOKAI (NUCLEAR POWER SAFETY RESEARCH ASSOCIATION)
JAPANESE

GEN'AN KYO DAYORI 6115
(NUCLEAR SAFETY ASSOCIATION LETTER)
GENSHIRYOKU ANZEN KENKYU KYOKAI (NUCLEAR POWER SAFETY RESEARCH ASSOCIATION)
JAPANESE

GEN'AN KYO SHIRYO SHITSU JOHO 6116
(NUCLEAR POWER SAFETY ASSOCIATION, DOCUMENTS OFFICE)
GENSHIRYOKU ANZEN KENKYU KYOKAI (NUCLEAR POWER SAFETY RESEARCH ASSOCIATION)
JAPANESE

GEN'ANKYO HOKOKU 6117
(NSRA REPORT)
GENSHIRYOKU ANZEN KENKYU KYOKAI (NUCLEAR SAFETY RESEARCH ASSOCIATION)
2-2, 1-CHOME, UCHISAIWAI-CHO, CHIYODA-KU, TOKYO 100
JAPANESE

GENDEN DAYORI 6118
(ATOMIC-POWERED ELECTRICAL GENERATION LETTER)
NIHON GENSHIRYOKU HATSUDEN (JAPAN ATOMIC-POWERED ELECTRICAL GENERATION)
JAPANESE

GENKEN KENKYU SEIKA SHOROKUSHU 6119
(JAERI REPORTS ABSTRACTS)
NIHON GENSHIRYOKU KENKYUSHO (JAPAN ATOMIC ENERGY RESEARCH INSTITUTE)
TOKAI-MURA, NAKA-GUN, IBARAKI 319-11
JAPANESE

GENSAN NENJI TAIKAI GIJIROKU 6120
(PROCEEDINGS OF JAPAN ATOMIC INDUSTRIAL FORUM, INCORPORATED)
NIHON GENSHIRYOKU SANGYO KAIGI (JAPAN ATOMIC INDUSTRIAL FORUM, INCORPORATED)
1-13, 1-CHOME, SHINBASHI, MINATO-KU, TOKYO 105
JAPANESE
INDEXED: JICST

GENSHIKAKU KENKYU 6121
(NUCLEAR STUDY)
GENSHIKAKU KENKYU HENSHU-GAKARI (EDITORIAL OFFICE OF NUCLEAR STUDY)
TOKYO DAIGAKU GENSHIKAKU KENKYUSHO TOSHOSHITSU, 2-1, 3-CHOME, MIDORI-CHO, TANASHI-SHI, TOKYO 188
JAPANESE
INDEXED: CHE, JICST

GENSHIRYOKU ANZEN KENKYU SOGO HAPPOKAI YOKOSHU 6122
(PREPRINTS OF A GENERAL CONFERENCE ON NUCLEAR POWER SAFETY)
NUCLEAR SAFETY RESEARCH ASSOCIATION
JAPANESE
INDEXED: JICST

GENSHIRYOKU ANZEN KYOKAI JUTAKU KENKYU (SJ) 6123
(CONSIGNED STUDIES, NUCLEAR SAFETY RESEARCH ASSOCIATION)
GENSHIRYOKU ANZEN KYOKAI (NUCLEAR SAFETY RESEARCH ASSOCIATION)
C/O OSAKA BUILDING, NO. 2, UCHISAIWAI-CHO, CHIYODA-KU, TOKYO 100
JAPANESE, WITH ENGLISH SUMMARY

GENSHIRYOKU CHOSA JIHO 6124
(REVIEW OF ATOMIC INDUSTRIES)
NIHON GENSHIRYOKU SANGYO KAIGI (JAPAN ATOMIC INDUSTRIES FORUM, INCORPORATED)
1-13, 1-CHOME, SHINBASHI, MINATO-KU, TOKYO 105
JAPANESE
INDEXED: JICST

GENSHIRYOKU CHOSA SHIRYO 6125
(SURVEY DATA ON ATOMIC ENERGY)
JAPANESE
INDEXED: JICST

GENSHIRYOKU HEIWA RIYO KENKYU SEIKA HOKOKUSHO 6126
(RESEARCH CONTRACTS ON PEACEFUL USES OF NUCLEAR ENERGY)
KAGAKU GIJUTSU-CHO GENSHIRYOKU-KYOKU (ATOMIC ENERGY BUREAU, SCIENCE & TECHNOLOGY AGENCY)
2-1, 2-CHOME, KASUMIGASEKI, CHIYODA-KU, TOKYO 100
JAPANESE
INDEXED: JICST

GENSHIRYOKU IINKAI GEPPO 6127
(JOURNAL OF ATOMIC ENERGY COMMISSION)
KAGAKU GIJUTSU-CHO GENSHIRYOKU-KYOKU (ATOMIC ENERGY BUREAU, SCIENCE AND TECHNOLOGY AGENCY)
OKURA-SHO INSATSU-KYOKU
2, AKASAKA AOI-CHO, MINATO-KU, TOKYO 107
JAPANESE
INDEXED: JICST

GENSHIRYOKU KAIGAI JOHO 6128
(NUCLEAR INFORMATION FROM ABROAD)
KAGAKU GIJUTSU-CHO GENSHIRYOKU-KYOKU (ATOMIC ENERGY BUREAU, SCIENCE & TECHNOLOGY AGENCY)
2-1, 2-CHOME, KASUMIGASEKI, CHIYODA-KU, TOKYO 100
JAPANESE

GENSHIRYOKU KOGYO 6129
(NUCLEAR ENGINEERING)
NIKKAN KOGYO SHINBUNSHA (INDUSTRIAL DAILY NEWS)
8-10, 1-CHOME, KUDANKITA, CHIYODA-KU, TOKYO 102
JAPANESE
INDEXED: CHE, JICST

GENSHIRYOKU NAIGAI JOHO 6130
(FOREIGN AND DOMESTIC INFORMATION ON ATOMIC ENERGY)
JAPANESE
INDEXED: JICST

GENSHIRYOKU NENPO 6131
(ANNUAL REPORT OF THE ATOMIC ENERGY COMMISION, JAPAN)
GENSHIRYOKU IINKAI (ATOMIC ENERGY COMMISION, JAPAN)
2-1, 2-CHOME, KASUMIGASEKI, CHIYODA-KU, TOKYO 100
JAPANESE

GENSHIRYOKU SANGYO SHINBUN 6132
(ATOMIC INDUSTRY NEWS)
NIHON GENSHIRYOKU SANGYO KAIGI (JAPAN ATOMIC INDUSTRIAL FORUM, INCORPORATED)
1-13, 1-CHOME, SHINBASHI, MINATO-KU, TOKYO 105
JAPANESE

GENSHIRYOKU SHIRYO 6133
(NUCLEAR POWER INFORMATION)
NIHON GENSHIRYOKU SANGYO KAIGI (JAPAN ATOMIC INDUSTRIAL FORUM, INCORPORATED)
1-13, 1-CHOME, SHINBASHI, MINATO-KU, TOKYO 105
JAPANESE
INDEXED: JICST

GENSHIRYOKU SOGO SHINPOJUMU YOKOSHU 6134
(PREPRINT OF THE SYMPOSIUM OF ATOMIC ENERGY SOCIETY OF JAPAN)
NIHON GENSHIRYOKU GAKKAI (ATOMIC ENERGY SOCIETY OF JAPAN)
C/O NIHON GENSHIRYOKU KENKYUSHO, 1-13, 1-CHOME SHINBASHI, MINATO-KU, TOKYO 105
JAPANESE
INDEXED: JICST

GENSHIRYOKUSEN 6135
(NUCLEAR SHIPS)
NIHON GENSHIRYOKUSEN KAIHATSU JIGODAN (JAPAN NUCLEAR SHIP DEVELOPMENT AGENCY)
35, SHIBA KOTOHIRA-CHO, MINATO-KU, TOKYO 105
JAPANESE
INDEXED: JICST

GENTSU 6136
(ATOMIC PRESS)
GENTSU
4-12, 1-CHOME, SARUGAKU-CHO, CHIYODA-KU, TOKYO 101
JAPANESE

HOKEN BUTSURI TO ANZEN KANRI 6137
(HEALTH PHYSICS AND SAFETY IN JAERI)
NIHON GENSHIRYOKU KENKYUSHO HOKEN BUTSURI-BU (HEALTH PHYSICS DIVISION, JAERI)
TOKAI-MURA, NAKA-GUN, IBARAKI 319-11
JAPANESE

HOSHA KAGAKU TORONKAI KOEN YOKOSHU 6138
(PREPRINTS FROM A SYMPOSIUM ON RADIATION CHEMISTRY)
JAPANESE
INDEXED: JICST

HOSHANO CHOSA KENKYU SEIKA HAPPYOKAI RONBUN SHOROKUSKU 6139
(ABSTRACTS OF PAPERS PRESENTED AT A CONFERENCE ON THE SURVEY AND RESEARCH RESULTS WITH REGARD TO RADIOACTIVITY)
JAPANESE
INDEXED: JICST

HOSHASEN BUTSURI KENKYU 6140
(RADIATION PHYSICS RESEARCH)
JAPANESE
INDEXED: JICST

HOSHASEN KAGAKU 6141
(RADIATION CHEMISTRY)
NIHON HOSHASEN KAGAKKAI (JAPANESE SOCIETY OF RADIATION CHEMISTRY)
C/O NIHON GENSHIRYOKU KENKYUSHO OSAKA KENKYUSHO, 508, MITSUI, NEYAGAWA-SHI 572
JAPANESE

HOSHASEN KAGAKU KOGYO NYUSU 6142
(RADIATION CHEMICAL INDUSTRY NEWS)
GUNMA-KEN HOSHASEN KAGAKU KOGYO KENKYUKAI
C/O GUNMA-KEN SEN'I KOKOGYO-KA, 1-1, 1-CHOME, OTE-MACHI, MAEBASHI-SHI 371
JAPANESE

HOSHASEN RIYO TOKEI 6143
(STATISTICS ON THE USE OF RADIATION IN JAPAN)
KAGAKU GIJUTSU-CHO, GENSHIRYOKU-KYOKU (ATOMIC ENERGY BUREAU, SCIENCE AND TECHNOLOGY AGENCY) AND NIHON AISOTOPU KYOKAI (JAPAN RADIOISTOPE ASSOCIATION)
28-45, 2-CHOME, HONKOMAGOME, BUNKYO-KU, TOKYO 113
JAPANESE AND ENGLISH
INDEXED: JICST

IAERU REPORT 6144
RIKKYO DAIGAKU GENSHIRYOKU KENKYUSHO (INSTITUTE FOR ATOMIC ENERGY, RIKKYO UNIVERSITY)
10, MATSUKOSHI, SAJIMA, YOKOSUKA-SHI 240-01
JAPANESE AND ENGLISH

IBARAKI-KEN NI OKERU HOSHANO CHOSA 6145
(RADIATION MONITORING IN IBARAKI PREFECTURE)
IBARAKI-KEN KOGAI GIJUTSU SENTA
4040-36, 1-CHOME, ISHIKAWA, MITO-SHI 310
JAPANESE

INS REPORT 6146
TOKYO DAIGAKU GENSHIKAKU KENKYUSHO (INSTITUTE FOR NUCLEAR STUDY, UNIVERSITY OF TOKYO)
2-1, 3-CHOME, MIDORI-CHO, TANASHI-SHI, TOKYO 188
ENGLISH

INS-J 6147
TOKYO DAIGAKU GENSHIKAKU KENKYUSHO (INSTITUTE FOR NUCLEAR STUDY, UNIVERSITY OF TOKYO)
2-1, 3-CHOME, MIDORI-CHO, TANASHI-SHI, TOKYO 188
EUROPEAN
INDEXED: CHE

INS-PH 6148
TOKYO DAIGAKU GENSHIKAKU KENKYUSHO (INSTITUTE FOR NUCLEAR STUDY, UNIVERSITY OF TOKYO)
2-1, 3-CHOME, MIDORI-CHO, TANASHI-SHI, TOKYO 188
JAPANESE

INS-PT 6149
TOKYO DAIGAKU GENSHIKAKU KENKYUSHO (INSTITUTE FOR NUCLEAR STUDY, UNIVERSITY OF TOKYO)
2-1, 3-CHOME, MIDORI-CHO, TANASHI-SHI, TOKYO 188
JAPANESE

INS-TCA 6150
TOKYO DAIGAKU GENSHIKAKU KENKYUSHO TOSHOSHITSU (LIBRARY, INSTITUTE FOR NUCLEAR STUDY, UNIVERSITY OF TOKYO)
2-1, 3-CHOME, MIDORI-CHO, TANASHI-SHI, TOKYO 188
JAPANESE, WITH ENGLISH SUMMARY
INDEXED: CHE

INS-TCB 6151
TOKYO DAIGAKU GENSHIKAKU KENKYUSHO (INSTITUTE FOR NUCLEAR STUDY, UNIVERSITY OF TOKYO)
2-1, 3-CHOME, MIDORI-CHO, TANASHI-SHI, TOKYO 188
JAPANESE, WITH ENGLISH SUMMARY

INS-TCH 6152
TOKYO DAIGAKU GENSHIKAKU KENKYUSHO (INSTITUTE FOR NUCLEAR STUDY, UNIVERSITY OF TOKYO)
2-1, 3-CHOME, MIDORI-CHO, TANASHI-SHI, TOKYO 188
JAPANESE, WITH ENGLISH SUMMARY
INDEXED: CHE

INS-TEC 6153
TOKYO DAIGAKU GENSHIKAKU KENKYUSHO (INSTITUTE FOR NUCLEAR STUDY, UNIVERSITY OF TOKYO)
2-1, 3-CHOME, MIDORI-CHO, TANASHI-SHI, TOKYO 188
JAPANESE, WITH ENGLISH SUMMARY

INS-TH 6154
TOKYO DAIGAKU GENSHIKAKU KENKYUSHO (INSTITUTE FOR NUCLEAR STUDY, UNIVERSITY OF TOKYO)
2-1, 3-CHOME, MIDORI-CHO, TANASHI-SHI, TOKYO 188
JAPANESE, WITH ENGLISH SUMMARY
INDEXED: CHE

INS-TL 6155
TOKYO DAIGAKU GENSHIKAKU KENKYUSHO (INSTITUTE FOR NUCELAR STUDY, UNIVERSITY OF TOKYO)
2-1, 3-CHOME, MIDORI-CHO, TANASHI-SHI, TOKYO 188
JAPANESE, WITH ENGLISH SUMMARY

INS-TS 6156
TOKYO DAIGAKU GENSHIKAKU KENKYUSHO (INSTITUTE FOR NUCLEAR STUDY, UNIVERSITY OF TOKYO)
2-1, 3-CHOME, MIDORI-CHO, TANASHI-SHI, TOKYO 188
JAPANESE, WITH ENGLISH SUMMARY

INSTITUTE FOR NUCLEAR STUDY, UNIVERSITY OF TOKYO 6157
JAPANESE
INDEXED: JICST

IPCR CYCLOTRON PROGRESS REPORT 6158
RIKAGAKU KENKYUSHO SAIKUROTORON KENKYUSHITSU (INSTITUTE OF PHYSICAL AND CHEMICAL RESEARCH)
2-1, HIROSAWA, WAKO-SHI 351
ENGLISH
INDEXED: CHE

IPCR CYCLOTRON REPORT 6159
RIKAGAKU KENKYUSHO SAIKUROTORON KENKYUSHITSU (INSTITUTE OF PHYSICAL AND CHEMICAL RESEARCH)
2-1, HIROSAWA, WAKO-SHI 351
EUROPEAN

IPCR CYCLOTRON TECHNICAL REPORT 6160
RIKAGAKU KENKYUSHO SAIKUROTORON KENKYUSHITSU (INSTITUTE OF PHYSICAL AND CHEMICAL RESEARCH)
2-1, HIROSAWA, WAKO-SHI 351
ENGLISH

ISOTOPE NEWS 6161
NIHON AISOTOPU KYOKAI (JAPAN RADIOISTOPE ASSOCIATION)
28-45, 2-CHOME, HONKOMAGOME, BUNKYO-KU, TOKYO 113
JAPANESE
INDEXED: JICST

JAERI-M REPOTO 6162
(JAERI-M REPORT)
NIHON GENSHIRYOKU KENKYUSHO GIJUTSU JOHO-BU (DIVISION OF TECHNICAL INFORMATION SERVICE, JAPAN ATOMIC ENERGY RESEARCH INSTITUTE)
TOKAI-MURA, NAKA-GUN, IBARAKI 319-11
JAPANESE AND ENGLISH

JNDC NYUSU 6163
(JNDC NEWS)
NIHON GENSHIRYOKU GAKKAI SHIGUMA IINKAI (JAPAN NUCLEAR DATA COMMITTEE) AND NIHON GENSHIRYOKU KENKYUSHO (JAPAN ATOMIC ENERGY RESEARCH INSTITUTE)
TOKAI-MURA, NAKA-GUN, IBARAKI 319-11
JAPANESE AND ENGLISH
INDEXED: JICST

JOURNAL OF NUCLEAR SCIENCE AND TECHNOLOGY 6164
NIHON GENSHIRYOKU GAKKAI (ATOMIC ENERGY SOCIETY OF JAPAN)
C/O NIHON GENSHIRYOKU KENKYUSHO, 1-13, 1-CHOME, SHINBASHI, MINATO-KU, TOKYO 105
ENGLISH
INDEXED: CHE, ENG, EXC, MAT, SCI, JICST

KAGAKU GIJUTSU BUNKEN SOKUHO, GENSHIRYOKU KOGAKU-HEN 6165
(CURRENT BIBLIOGRAPHY ON SCIENCE AND TECHNOLOGY, SERIES OF NUCLEAR ENGINEERING)
NIHON KAGAKU GIJUTSU JOHO SENTA (JAPAN INFORMATION CENTER OF SCIENCE AND TECHNOLOGY)
5-2, 2-CHOME, NAGATA-CHO, CHIYODA-KU, TOKYO 100
JAPANESE
INDEXED: JICST

KAIGAI ATR.FBR KAIHATSU JOKYO GAIYO 6166
(SUMMARY OF OVERSEAS ATR AND FBR DEVELOPMENTS)
DORYOKURO.KAKUNENRYO KAIHATSU JIGYODAN (POWER REACTOR AND NUCLEAR FUEL DEVELOPMENT CORPORATION)
JAPANESE

KAIGAI URAN SHIGEN CHOSA HOKOKUSHO 6167
(REPORT OF OVERSEAS URANIUM RESOURCES SURVEY)
DORYOKURO, KAKU NENRYO KAIHATSU JIGYODAN (POWER REACTOR AND NUCLEAR FUEL DEVELOPMENT CORPORATION)
C/O SANKAIDAO BUILDING, 9-13, 1-CHOME, AKASAKA, MINATO-KU, TOKYO 107
JAPANESE

KAKU YUGO KENKYU 6168
(NUCLEAR FUSION)
KAKU YUGO KONDANKAI BUNKEN SENTA (RESEARCH INFORMATION CENTER, NUCLEAR RESEARCH GROUP)
C/O NAGOYA DAIGAKU PURAZUMA KENKYUSHO, FURO-CHO, CHIKUSA-KU, NAGOYA-SHI 464
JAPANESE
INDEXED: CHE, JICST

KAKURIKEN KENKYU HOKOKU 6169
(RESEARCH REPORT OF LABORATORY OF NUCLEAR SCIENCE)
TOHOKU DAIGAKU RIGAKUBU FUZOKU GENSHIKAKU RIGAKU KENKYU SHISETSU (LABORATORY OF NUCLEAR SCIENCE, TOHOKU UNIVERSITY)
1, TOMIZAWA KANAYAMA, SENDAI-SHI 982
JAPANESE

KINKI DAIGAKU GENSHIRYOKU KENKYUSHO NENPO 6170
(ANNUAL REPORTS OF KINKI UNIVERSITY, ATOMIC ENERGY RESEARCH INSTITUTE)
KINKI DAIGAKU GENSHIRYOKU KENKYUSHO (ATOMIC ENERGY RESEARCH INSTITUTE, KINKI UNIVERSITY)
321, KOWAKAE, HIGASHIOSAKA-SHI 577
JAPANESE AND ENGLISH
INDEXED: BIO, JICST

KYOTO DAIGAKU GENSHI ENERUGI KENKYUSHO IHO 6171
(BULLETIN OF THE INSTITUTE OF ATOMIC ENERGY, KYOTO UNIVERSITY)
KYOTO DAIGAKU GENSHI ENERUGI KENKYUSHO (INSTITUTE OF ATOMIC ENERGY, KYOTO UNIVERSITY)
GOKASHO, UJI-SHI 611
JAPANESE
INDEXED: CHE, JICST

KYOTO DAIGAKU GENSHIRO JIKKENSHO GAKUJUTSU KOENKAI KOEN YOSHISHU 6172
(ABSTRACT OF REPORT AT SCIENTIFIC MEETING OF THE RESEARCH REACTOR INSTITUTE, KYOTO UNIVERSITY)
KYOTO DAIGAKU GENSHIRO JIKKENSHO
NODA, KUMATORI-CHO, SENNAN-GUN, OSAKA 590-04
JAPANESE
INDEXED: CHE, JICST

KYOTO DAIGAKU GENSHIRYO JIKKENJO TECHNICAL REPORT 6173
(KYOTO UNIVERSITY ATOMIC REACTOR (EXPERIMENT STATION) TECHNICAL REPORT)
JAPANESE
INDEXED: JICST

MITSUBISHI GENSHIRYOKU GIHO 6174
(TECHNICAL REPORT OF ENGINEERING DEVELOPMENT DIVISION, MITSUBISHI ATOMIC POWER COMPANY, LTD.)
MITSUBISHI GENSHIRYOKU KOGYO K.K. KENKYUSHO (ENGINEERING DEVELOPMENT DIVISION, MITSUBISHI ATOMIC POWER COMPANY, LTD.)
1-297, KITABUKURO-CHO, OMIYA-SHI 330
JAPANESE

NICHIDAI GENSHIRYOKU KENKYUSHO HOKOKU 6175
(REPORT OF THE ATOMIC ENERGY RESEARCH INSTITUTE, NIHON UNIVERSITY)
NIHON DAIGAKU GENSHIRYOKU KENKYUSHO (ATOMIC ENERGY RESEARCH INSTITUTE, NIHON UNIVERSITY)
1-8, KANDA SURUGADAI, CHIYODA-KU, TOKYO 101
JAPANESE

NIHON AISOTOPU KAIGI HOBUNSHU 6176
(PROCEEDINGS OF THE JAPAN CONFERENCE ON RADIOSOTOPES)
NIHON GENSHIRYOKU SANGYO KAIGI (JAPAN ATOMIC INDUSTRIAL FORUM, INCORPORATED)
1-13, 1-CHOME, SHINBASHI, MINATO-KU, TOKYO 105
JAPANESE AND ENGLISH
INDEXED: CHE, JICST

NIHON GENSHIRYOKU GAKKAISHI 6177
(JOURNAL OF THE ATOMIC ENERGY SOCIETY OF JAPAN)
NIHON GENSHIRYOKU GAKKAI (ATOMIC ENERGY SOCIETY OF JAPAN)
C/O NIHON GENSHIRYOKU KENKYUSHO, 1-13, 1-CHOME, SHINBASHI, MINATO-KU, TOKYO 105
JAPANESE, WITH ENGLISH SUMMARY
INDEXED: CHE, INS, WAA, JICST

NIHON GENSHIRYOKU KENKYUSHO NENPO 6178
(ANNUAL REPORT OF THE JAPAN ATOMIC ENERGY RESEARCH INSTITUTE)
NIHON GENSHIRYOKU KENKYUSHO (JAPAN ATOMIC ENERGY RESEARCH INSTITUTE)
1-13, 1-CHOME, SHINBASHI, MINATO-KU, TOKYO 105
JAPANESE
INDEXED: CHE, JICST

NIHON GENSHIRYOKU SHINBUN 6179
(JAPAN ATOMIC POWER PRESS)
NIHON GENSHIRYOKU SHINBUNSHA (JAPAN ATOMIC POWER PRESS)
674, KAMIODANAKA, NAKAHARA-KU, KAWASAKI-SHI 211
JAPANESE AND ENGLISH

NIHON GENSHIRYOKUSEN KAIHATSU JIGYODAN HOKOKU (JNS) 6180
(REPORT OF THE JAPAN NUCLEAR SHIPS DEVELOPMENT AGENCY, JNS)
NIHON GENSHIRYOKUSEN KAIHATSU JIGYODAN (JAPAN NUCLEAR SHIPS DEVELOPMENT AGENCY)
35, SHIBA KOTOHIRA-CHO, MINATO-KU, TOKYO 105
JAPANESE
INDEXED: JICST

NIHON GENSHIRYOKUSEN KAIHATSU JIGYODAN NENPO 6181
(ANNUAL REPORT OF THE JAPAN NUCLEAR SHIP DEVELOPMENT AGENCY)
NIHON GENSHIRYOKUSEN KAIHATSU JIGYODAN
C/O SENPAKU SHINKO BUILDING, 35, SHIBA KOTOHIRA-CHO, MINATO-KU, TOKYO 105
JAPANESE
INDEXED: JICST

NIHON NO GENSHIRYOKU SANGYO 6182
(ATOMIC INDUSTRY IN JAPAN)
TSUSHO SANGYO-SHO (MINISTRY OF INTERNATIONAL TRADE AND INDUSTRY) AND DENKI TAIMUSU
3-7, 2-CHOME, SHINBASHI, MINATO-KU, TOKYO 105
JAPANESE

NIPPON AISOTOPU KAIGI RONBUN SHOROKUSHU 6183
(ABSTRACTS OF PAPERS PRESENTED AT THE JAPAN ISOTOPE CONFERENCE)
JAPANESE
INDEXED: JICST

NIPPON GENSHIRYOKU GAKKAI AKI NO BUNKAKAI YOKOSHU 6184
(PREPRINTS OF THE AUTUMN SECTION OF THE JAPAN ATOMIC ENERGY SOCIETY)
ATOMIC ENERGY SOCIETY
JAPANESE
INDEXED: JICST

NIPPON GENSHIRYOKU GAKKAI KAGAKU, KAGAKU KOGAKU BUNKAKAI YOKOSHU 6185
(PREPRINTS OF THE CHEMICAL ENGINEERING SECTION OF THE JAPAN ATOMIC ENERGY SOCIETY)
JAPANESE
INDEXED: JICST

NIPPON GENSHIRYOKU GAKKAI NENKAI YOSHISBU 6186
(ABSTRACTS OF THE ANNUAL MEETING OF JAPAN ATOMIC ENERGY SOCIETY)
JAPANESE
INDEXED: JICST

NIPPON GENSHIRYOKU KENKYUJO GIJUTSU JOHOBU SHIRYO JOHO 6187
(DATA INFORMATION, THE JAPAN ATOMIC ENERGY RESEARCH INSTITUTE, TECHNICAL INFORMATION SECTION)
JAPANESE
INDEXED: JICST

NIPPON GENSHIRYOKU KENKYUJO MEMO 6188
(JAPAN ATOMIC ENERGY RESEARCH INSTITUTE MEMO)
JAPANESE
INDEXED: JICST

NIPPON GENSHIRYOKU KENKYUJO NENJI HOKOKU 6189
(ANNUAL REPORT OF THE JAPAN ATOMIC ENERGY INSTITUTE)
JAPANESE
INDEXED: JICST

NIPPON GENSHIRYOKU KENKYUJO SHIRYO 6190
(JAPAN ATOMIC ENERGY INSTITUTE DATA)
JAPANESE
INDEXED: JICST

NIPPON GENSHIRYOKU KENKYUSHO PANFURETTO-RUI 6191
(PAMPHLET, JAPAN ATOMIC ENERGY RESEARCH INSTITUTE)
NIPPON GENSHIRYOKU KENKYUSHO
TOKAI, NAKA-GUN, IBARAKI-KEN
JAPANESE
INDEXED: CHE

NIPPON HOSHASEI DOI GENSO KYOKAI RIGAKUBU-KAI BUNKEN SHORO KUSHU 6192
(ABSTRACTS OF PAPERS AT JAPAN ASSOCIATION OF RADIOISOTOPES. SCIENCE AND ENGINEERING SECTION.)
JAPANESE
INDEXED: JICST

NIPPON HOSHASEI DOI GENSO KYOKAI. NOGAKU. SEIBUTSUGAKU BUKAI BUNKEN SHOROKUSHU 6193
(ABSTRACTS OF PAPERS AT JAPAN ASSOCIATION OF RADIOISOTOPES. AGRICULTURAL AND BIOLOGICAL SECTIONS)
JAPANESE
INDEXED: JICST

NIPPON HOSHASEN KOBUNSHI KENKYU KYOKAI NENPO 6194
(ANNUAL REPORT OF THE JAPAN RADIATION PARTICLE RESEARCH SOCIETY)
JAPANESE
INDEXED: JICST

NIPPON KAGAKUKAI. X-SEN BUNSEKI TORONKAI KOEN YOKOSHU 6195
(PREPRINTS OF PAPERS FROM THE SYMPOSIUM ON X-RAY ANALYSIS. JAPAN CHEMICAL SOCIETY.)
JAPANESE
INDEXED: JICST

NORIN-SHO KANKEI HOSHANO CHOSA KENKYU HOKOKU 6196
(ANNUAL REPORT OF RADIOACTIVE CONTAMINATION, MINISTRY OF AGRICULTURE AND FORESTRY)
NORIN-SHO NORIN SUISAN KAIGI JIMU-KYOKU (AGRICULTURE, FORESTRY AND FISHERIES COUNCIL, MINISTRY OF AGRICULTURE AND FORESTRY)
2-1, 1-CHOME, KASUMIGASEKI, CHIYODA-KU, TOKYO 100
JAPANESE

NSRA MEMO 6197
GENSHIRYOKU ANZEN KENKYU KYOKAI (NUCLEAR SAFETY RESEARCH ASSOCIATION)
2-2, 1-CHOME, UCHISAIWAI-CHO, CHIYODA-KU, TOKYO 100
JAPANESE

NUCLEAR SCIENCE INFORMATION OF JAPAN 6198
NIHON GENSHIRYOKU KENKYUSHO (JAPAN ATOMIC ENERGY RESEARCH INSTITUTE)
TAKAI-MURA, NAKA-GUN, IBARAKI 319-11
ENGLISH
INDEXED: CHE, JICST

OSAKA-FURITSU HOSHASEN CHUO KENKYUSHO DAYORI 6199
(INFORMATION BULLETIN OF THE RADIATION CENTER OF OSAKA PREFECTURE)
OSAKA-FURITSU HOSHASEN CHUO KENKYUSHO
704, SHINKE-CHO, SAKAI-SHI 593
JAPANESE
INDEXED: BIO, CHE

OULNS REPORT 6200
OSAKA DAIGAKU GENSHIKAKU KENKYU SHISETSU (OSAKA UNIVERSITY, LABORATORY OF NUCLEAR STUDY)
1-1, MACHIKANEYAMA-CHO, TOYONAKA-SHI 560
ENGLISH

PNCT 6201
DORYOKURO, KAKU NENRYO KAIHATSU JIGYODAN TOKAI EIGYOSHO (POWER REACTOR AND NUCLEAR FUEL DEVELOPMENT COMPANY, TOKAI WORKS)
MURAMATSU, TOKAI-MURA, NAKA-GUN, IBARAKI 319-11
JAPANESE

PROGRESS REPORT 6202
NIHON GENSHIRYOKU KENKYUSHO (JAPAN ATOMIC ENERGY RESEARCH INSTITUTE)
TOKAI-MURA, NAKA-GUN, IBARAKI 319-11
ENGLISH

RADIOACTIVITY SURVEY DATA IN JAPAN 6203
HOSHASEN IGAKU SOGO KENKYUSHO (NATIONAL INSTITUTE OF RADIOLOGICAL SCIENCES)
9-1, 4-CHOME, ANAGAWA, CHIBA-SHI 280
ENGLISH
INDEXED: JICST

RADIOISOTOPES 6204
NIHON HOSHASEI DOIGENSO KYOKAI (JAPAN RADIOISOTOPE ASSOCIATION)
28-45, 2-CHOME, HONKOMAGOME, BUNKYO-KU, TOKYO 113
JAPANESE AND ENGLISH
INDEXED: CHE, EXC, JICST

RAJIO AISOTOPU BUNKEN, DAIMOKUSHU 6205
(BIBLIOGRAPHY ON RADIOISOTOPES)
NIHON HOSHASEI DOIGENSO KYOKAI (THE JAPAN RADIO-ISOTOPE ASSOCIATION)
28-45, 2-CHOME, HONKOMAGOME, BUNKYO-KU, TOKYO 113
JAPANESE AND ENGLISH

REPORT OF THE INSTITUTE FOR ATOMIC ENERGY, RIKKYO UNIVERSITY 6206
RIKKYO UNIVERSITY, INSTITUTE FOR ATOMIC ENERGY
10, MATSUKOSHI, SAJIMA, YOKOSUKA-SHI 240-01,
JAPANESE AND ENGLISH

REPORT OF THE INSTITUTE FOR NUCLEAR STUDY 6207
TOKYO DAIGAKU GENSHIKAKU KENKYUSHO TOSHOSHITSU (LIBRARY INSTITUTE FOR NUCLEAR STUDY, UNIVERSITY OF TOKYO) AND TOKYO DAIGAKU GENSHIKAKU KENKYUSHO (INSTITUTE FOR NUCLEAR STUDY, UNIVERSITY OF TOKYO)
2-1, 3-CHOME, MIDORI-CHO, TANASHI-SHI, TOKYO 188
ENGLISH

REPORT OF THE LABORATORY OF NUCLEAR STUDY, OSAKA UNIVERSITY 6208
OSAKA UNIVERSITY, LABORATORY OF NUCLEAR STUDY
1-1, MACHIKANEYAMA-CHO, TOYONAKA-SHI 560
ENGLISH

RESEARCH ACTIVITIES OF THE INSTITUTE OF ATOMIC ENERGY, KYOTO UNIVERSITY 6209
KYOTO DAIGAKU GENSHI ENERUGI KENKYUSHO (THE INSTITUTE OF ATOMIC ENERGY, KYOTO UNIVERSITY)
GOKASHO, UJI-SHI 611
ENGLISH
INDEXED: JICST

RIKKYO DAIGAKU GENSHIRYOKU KENKYUJO GENSHIRO RIYO JISSEKI H OKOKU 6210
(REPORT OF THE ACHIEVEMENTS OF UTILIZING NUCLEAR REACTORS, ATOMIC ENERGY INSTITUTE, RIKKYO UNIVERSITY)
JAPANESE
INDEXED: JICST

RIKKYO DAIGAKU GENSHIRYOKU KENKYUJO PUREPURINTO 6211
(PREPRINT OF RIKKYU UNIVERSITY ATOMIC ENERGY INSTITUTE)
JAPANESE
INDEXED: JICST

SHIRYO JOHO (NIHON GENSHIRYOKU KENKYUSHO) 6212
(INFORMATION SERVICE OF JAPAN ATOMIC ENERGY RESEARCH INSTITUTE))
NIHON GENSHIRYOKU KENKYUSHO, TOKAI KENKYUSHO GIJUTSU JOHO-BU (JAPAN ATOMIC ENERGY RESEARCH INSTITUTE)
TOKAI-MURA, NAKA-GUN, IBARAKI 319-11
JAPANESE

SOKUHO GENSHI ENERUGI 6213
(ATOMIC ENERGY)
ATOMIKKU JANARUSHA (ATOMIC JOURNAL COMPANY)
18-27, 2-CHOME, MITA, MEGURO-KU, TOKYO 153
JAPANESE

SUMMARY OF THE OBSERVATION RESULT FOR NUCLEAR RADIATION IN THE WORLD DURING IGY-IGC 6214
JAPANESE
INDEXED: JICST

TAIKI HOSHANO KANSOKU SEISEKI 6215
(BULLETIN OF THE ATOMOSPHERIC RADIOACTIVITY)
KISHO-CHO (THE JAPAN METEOROLOGICAL AGENCY)
3-4, 1-CHOME, OTE-MACHI, CHIYODA-KU, TOKYO 100
JAPANESE

TECHNICAL REPORTS OF THE INSTITUTE OF ATOMIC ENERGY, KYOTO UNIVERSITY 6216
KYOTO DAIGAKU GENSHI ENERUGI KENKYUSHO (INSTITUTE OF ATOMIC ENERGY, KYOTO UNIVERSITY)
GOKASHO, UJI-SHI 611
ENGLISH
INDEXED: CHE, ENG, INS, MAT, JICST

TECHNICAL REPORTS OF THE RESEARCH REACTOR INSTITUTE, KYOTO UNIVERSITY (KURRI-TR) 6217
KYOTO DAIGAKU GENSHIRO JIKKENSHO (THE RESEARCH REACTOR INSTITUTE, KYOTO UNIVERSITY)
KUMATORI-CHO, SENNAN-GUN, OSAKA 590-04
JAPANESE, WITH ENGLISH SUMMARY

TOA KENKYU NYUSU 6218
(TOKYO METROPOLITAN ISOTOPE RESEARCH NEWS)
TOKYO TORITSU AISOTOPU SOGO KENKYUSHO (TOKYO METROPOLITAN ISOTOPE GENERAL RESEARCH INSTITUTE)
JAPANESE

TOKAI WORKS SEMI ANNUAL PROGRESS REPORT 6219
DORYOKURO, KAKU NENRYO KAIHATSU JIGYODAN (POWER REACTOR AND NUCLEAR FUEL DEVELOPMENT CORPORATION)
MURAMATSU, TOKAI-MURA, NAKA-GUN, IBARAKI 319-11
ENGLISH
INDEXED: CHI

TOKYO-TORITSU AISOTOPU SOGO KENKYUSHO NENPO 6220
(ANNUAL REPORT OF THE TOKYO METROPOLITAN ISOTOPE RESEARCH CENTER)
TOKYO-TORITSU AISOTOPU SOGO KENKYUSHO (TOKYO METROPOLITAN ISOTOPE RESEARCH CENTER)
11-1, 2-CHOME, FUKAZAWA, SETAGAYA-KU, TOKYO 158
JAPANESE
INDEXED: JICST

Electrical Engineering (Including Communication)

ABU TECHNICAL REVIEW 6221
ASIAN BROADCASTING UNION,BROADCASTING CENTRE
2-2-1 JINNAN, SHIBUYA-KU, TOKYO 150
JAPANESE AND ENGLISH
INDEXED: INS

AEU; JOURNAL OF ASIA ELECTRONICS UNION 6222
DENPA SHINBUNSHA
11-15, 1-CHOME, HIGASHIGOTANDA, SHINAGAWA-KU, TOKYO 141
ENGLISH
INDEXED: CHE, INS

ANDO GIHO 6223
(ANDO TECHNICAL NEWS)
ANDO DENKI K.K.
19-7, 4-CHOME, KAMATA, OTA-KU, TOKYO 144
JAPANESE
INDEXED: JICST

ANNUAL REPORT OF THE CENTRAL RESEARCH INSTITUTE OF ELECTRIC POWER INDUSTRY, TECHNICAL LABORATORY NUMBER 1 6224
DENRYOKU CHUO KENKYUSHO GIJUTSU DAI-1 KENKYUSHO (CENTRAL RESEARCH INSTITUTE OF ELECTRIC POWER INDUSTRY, TECHNICAL LABORATORY NUMBER 1)
1229, IWADO, KOMAE-SHI, TOKYO 182
ENGLISH

ANNUAL REPORT OF THE NIPPON TELEGRAPH & TELEPHONE PUBLIC COMPANY 6225
NIPPON DENSHIN DENWA KOSHA (NIPPON TELEGRAPH AND TELEPHONE PUBLIC COMPANY)
1-6, 1-CHOME, UCHISAIWAI-CHO, CHIYODA-KU, TOKYO 100
ENGLISH

ANRITSU TEKUNIKARU 6226
(ANRITSU TECHNICAL BULLETIN)
ANRITSU DENKI K.K. (ANRITSU ELECTRIC COMPANY)
12-20, 4-CHOME, MINAMIAZABU, MINATO-KU, TOKYO 106
JAPANESE, WITH ENGLISH SUMMARY
INDEXED: JICST

AV SCIENCE 6227
TOSHIBA SHOJI K.K.
2-1, 5-CHOME, GINZA, CHUO-KU, TOKYO 104
JAPANESE

BATTERI NYUSU 6228
(BATTERY NEWS)
SHIN KOBE DENKI K.K. (SHIN KOBE ELECTRIC MACHINERY COMPANY, LTD.
4-5, 1-CHOME, MARUNOUCHI, CHIYODA-KU, TOKYO 100
JAPANESE

BIJINESU KOMYUNIKESHON 6229
(BUSINESS COMMUNICATION)
KIKAKU SENTA (PLANNING CENTER COMPANY)
C/O KIMURA BUILDING, KANDA OGAWA-CHO, CHIYODA-KU, TOKYO 101
JAPANESE
INDEXED: JICST

BIKUTA AV TSUSHIN 6230
(VICTOR AV COMMUNICATION)
NIHON VICTOR K.K. (VICTOR CORPORATION OF JAPAN, LTD.)
4-1, NIHONBASHI HON-CHO, CHUO-KU, TOKYO 103
JAPANESE

CATALOGUE OF DATA IN WDC C2 CENTER OF IONOSPHERE 6231
YUSEI-SHO DENPA KENKYUSHO (RADIO RESEARCH LABORARORY, MINISTRY OF POST & TELECOMMUNICATIONS)
2-1, 4-CHOME, NUKUIKITA-MACHI, KOGANEI-SHI, TOKYO 184
JAPANESE

CATV JANARU 6232
(CATV JOURNAL)
HOSO JANARUSHA
3-13, 7-CHOME, GINZA, CHUO-KU, TOKYO 104
JAPANESE

CBC GIJUTSU HOKOKUKAI 6233
(TECHNICAL CONFERENCE OF CBC)
CHUBU NIPPON HOSO K.K. GIJUTSU-KYOKU HOSO SHISETSU-BU
4-15, SHINSAKAE-MACHI, NAKA-KU, NAGOYA-SHI 460
JAPANESE

CEC JOHO 6234
(INFORMATION FROM CHUBU ELECTRONICS CENTER)
CHUBU DENSHI KOGYO GIJUTSU SENTA
9959, OKAYA-SHI 394
JAPANESE

CHIIKI KAIHATSU NYUSU 6235
(REGIONAL DEVELOPMENT NEWS)
TOKYO DENRYOKU K.K. EIGYO-BU (TOKYO ELECTRIC POWER COMPANY, INCORPORATED)
1-3, 1-CHOME, UCHISAIWAI-CHO, CHIYODA-KU, TOKYO 100
JAPANESE

CHIKUDENCHI 6236
(STORAGE BATTERY)
NIHON CHIKUDENCHI KOGYOKAI (JAPAN STORAGE BATTERY ASSOCIATION)
C/O KIKAI SHINKO KAIKAN, 5-8, 3-CHOME, SHIBAKOEN, MINATO-KU, TOKYO 105
JAPANESE

CHOKOATSU DENRYOKU KENKYUSHO KENKYU HOKOKUSHO 6237
(RESEARCH REPORT OF HIGH VOLTAGE POWER LABORATORY)
CHOKOATSU DENRYOKU KENKYUSHO
6-1, 1-CHOME, OTE-MACHI, CHIYODA-KU, TOKYO 100
JAPANESE
INDEXED: JICST

CHOKOGATA DENSHI KAIRO KANKEI TOKKYO KOKOKU ICHIRANHYO 6238
(SURVEY OF PATENT REPORTS RELATING TO SUB-MINATURE ELECTRONIC CIRCUITRY)
NIHON DENSHI KOGYO SHINKO KYOKAI (JAPAN ELECTRONICS INDUSTRY PROMOTION ASSOCIATION)
JAPANESE

CHUBU DENRYOKU K.K. KENKYU SHIRYO 6239
(MEMOIRS OF CHUBU ELECTRIC POWER COMPANY, INCORPORATED)
CHUBU DENRYOKU K.K. (CHUBU ELECTRIC POWER COMPANY, INCORPORATED)
10-1, TOSHIN-CHO, HIGASHI-KU, NAGOYASHI 461
JAPANESE
INDEXED: CHE, JICST

CHUBU DENRYOKU SHINBUN 6240
(CHUBU ELECTRIC POWER NEWS)
CHUBU DENRYOKU K.K. (CHUBU ELECTRIC POWER COMPANY LTD.)
10-1, TOSHIN-CHO, HIGASHI-KU, NAGOYA-SHI 461
JAPANESE

CHUGOKU DENRYOKU K.K. GIKEN JIHO 6241
(REPORT OF THE TECHNICAL LABORATORY, CHUGOKU ELECTRIC POWER COMPANY, INCORPORATED)
CHUGOKU DENRYOKU K.K. GIJUTSU KENKYUSHO
4-32, 4-CHOME, OSU, HIROSHIMA-SHI 730
JAPANESE
INDEXED: JICST

CQ HAM RADIO 6242
CQ SHUPPAN K.K. (CQ PUBLISHING COMPANY, LTD.)
14-2, 1-CHOME, SUGAMO, TOSHIMA-KU, TOKYO 170
JAPANESE
INDEXED: JICST

DAINICHI-NIPPON DENSEN JIHO 6243
(DAINICHI-NIPPON CABLES REVIEW)
DAINICHI-NIPPON DENSEN K.K. GIJUTSU KANRISHITSU (TECHNICAL ADMINISTRATION DIVISION, DAINICHI-NIPPON CABLES COMPANY, LIMITED)
7-3, UMEDA, KITA-KU, OSAKA-SHI 530
JAPANESE, WITH ENGLISH SUMMARY
INDEXED: ENG, JICST

DEN-DEN JANARU 6244
(DEN-DEN JOURNAL)
TOKYO SHUPPAN SENTA
47-9, 1-CHOME, HIGASHIKEBUKURO, TOSHIMA-KU, TOKYO 170
JAPANESE

DEN-DEN KENSETSU 6245
(CONSTRUCTION OF TELECOMMUNICATIONS SYSTEM)
DENSHIN DENWA KOJI KYOKAI (TELEPHONE AND TELEGRAPH CONSTRUCTION ASSOCIATION)
3-3, SARUGAKU-CHO, SHIBUYA-KU, TOKYO 150
JAPANESE

DEN-DEN KENSETSU NYUSU 6246
(TELEPHONE AND TELEGRAPH CONSTRUCTION NEWS)
NIHON SOGO TSUSHINSHA
5-8, SHINBASHI, MINATO-KU, TOKYO 105
JAPANESE

DENCHI TORONKAI KOEN YOSHISHU 6247
(SUMMARY OF THE ELECTRIC CELL SYMPOSIUM)
NIHON KAGAKKAI (THE CHEMICAL SOCIETY OF JAPAN)
1-5, KANDA SURUGADAI, CHIYODA-KU, TOKYO 101
JAPANESE

DENDEN JIDAI 6248
(ELECTRICAL AGE)
TOKYO SHUPPAN SENTA (TOKYO PUBLISHING CENTER)
JAPANESE

DENDEN KENCHIKU 6249
(NTT ARCHITECTURE)
JAPANESE
INDEXED: JICST

DENGEN 6250
(POWER RESOURCES)
DENGEN KAIHATSU K.K. (ELECTRIC POWER DEVELOPMENT COMPANY, LTD.)
8-2, 1-CHOME, MARUNOUCHI, CHIYODA-KU, TOKYO 100
JAPANESE

DENGEN KAIHATSU K.K. CHOSA SHIRYO 6251
(RESEARCH BULLETIN OF ELECTRIC POWER DEVELOPMENT COMPANY, LTD.)
DENGEN KAIHATSU K.K. (ELECTRIC POWER DEVELOPMENT COMPANY, LTD.)
8-2, 1-CHOME, MARUNOUCHI, CHIYODA-KU, TOKYO 100
JAPANESE
INDEXED: JICST

DENGEN KAIHATSU K.K. SHIKEN HOKOKU 6252
(RESEARCH REPORTS OF ELECTRIC POWER DEVELOPMENT COMPANY, LTD.)
DENGEN KAIHATSU K.K. DOBOKU SHIKENSHO
2000, CHIGASAKI, CHIGASAKI-SHI 253
JAPANESE

DENGEN NYUSU 6253
(DENGEN NEWS)
DENGENSHA SEISAKUSHO
1730, IKUTA, TAMA-KU, KAWASAKI-SHI 214
JAPANESE

DENKAI CHIKUDENKI HYORON 6254
(REVIEW OF THE ELECTROLYTIC STORAGE BATTERY)
JAPANESE
INDEXED: JICST

DENKAI SHINGO 6255
(TRUE ELECTRONICS WORLD)
DENKAI SHIASO SHA
(TRUE ELECTRONICS WORLD COMPANY)
JAPANESE

DENKENPO 6256
(NEWS OF THE CENTRAL RESEARCH INSTITUTE OF ELECTRIC POWER INDUSTRY)
DENRYOKU CHUO KENKYUSHO
6-1, 1-CHOME, OTE-MACHI, CHIYODA-KU, TOKYO 100
JAPANESE
INDEXED: JICST

DENKI 6257
(ELECTRIC MACHINE)
NIHON DENKI KOGYOKAI (THE JAPAN ELECTRIC MACHINE INDUSTRY ASSOCIATION)
4-15, 2-CHOME, NAGATA-CHO, CHIYODA-KU, TOKYO 100
JAPANESE
INDEXED: JICST

DENKI DOHO 6258
(MONTHLY REPORT ON ELECTRIC APPLIANCES)
KAGAKU SHINBUNSHA
8-1, 1-CHOME, HAMAMATSU-CHO, MINATO-KU, TOKYO 105
JAPANESE

DENKI GAKKAI DENJIKAI RIRON KENKYUKAI SHIRYO 6259
(PAPERS OF TECHNICAL MEETING ON ELECTRO-MAGNETIC THEORY, IEEJ)
DENKI GAKKAI (THE INSTITUTE OF ELECTRICAL ENGINEERS OF JAPAN)
1-11, YURAKU-CHO, CHIYODA-KU, TOKYO 100
JAPANESE

DENKI GAKKAI DENKI KAGAKU, DENNETSU KENKYUKAI SHIRYO 6260
(PAPERS OF TECHNICAL MEETING ON ELECTRO-CHEMISTRY AND HEATING, IEEJ)
DENKI GAKKAI (THE INSTITUTE OF ELECTRICAL ENGINEERS OF JAPAN)
1-11, YURAKU-CHO, CHIYODA-KU, TOKYO 100
JAPANESE

DENKI GAKKAI DENKI ONKYO KENKYUKAI SHIRYO 6261
(PAPERS OF TECHNICAL MEETING ON ELECTROACOUSTICS, IEEJ)
DENKI GAKKAI (THE INSTITUTE OF ELECTRICAL ENGINEERS OF JAPAN)
1-11, YURAKU-CHO, CHIYODA-KU, TOKYO 100
JAPANESE

DENKI GAKKAI DENKI TETSUDO KENKYUKAI SHIRYO 6262
(PAPERS OF TECHNICAL MEETING ON RAILWAY AND TRANSPORTATION, IEEJ)
DENKI GAKKAI (THE INSTITUTE OF ELECTRICAL ENGINEERS OF JAPAN)
1-11, YURAKU-CHO, CHIYODA-KU, TOKYO 100
JAPANESE

DENKI GAKKAI DENRYOKU GIJUTSU KENKYUKAI SHIRYO 6263
(PAPERS OF TECHNICAL MEETING ON POWER ENGINEERING, IEEJ)
DENKI GAKKAI (THE INSTITUTE OF ELECTRICAL ENGINEERS OF JAPAN)
1-11, YURAKU-CHO, CHIYODA-KU, TOKYO 100
JAPANESE

DENKI GAKKAI DENRYOKU OYO KENKYUKAI SHIRYO 6264
(PAPERS OF TECHNICAL MEETING ON INDUSTRY APPLICATIONS, IEEJ)
DENKI GAKKAI (THE INSTITUTE OF ELECTRICAL ENGINEERS OF JAPAN)
1-11, YURAKU-CHO, CHIYODA-KU, TOKYO 100
JAPANESE

DENKI GAKKAI DENSEN KENKYUKAI SHIRYO 6265
(PAPERS OF TECHNICAL MEETING ON ELECTRIC CABLE AND WIRE, IEEJ)
DENKI GAKKAI (THE INSTITUTE OF ELECTRICAL ENGINEERS OF JAPAN)
1-11, YURAKU-CHO, CHIYODA-KU, TOKYO 100
JAPANESE

DENKI GAKKAI DENSHI KAIRO KENKYUKAI SHIRYO 6266
(PAPERS OF TECHNICAL MEETING ON ELECTRONIC CIRCUITS, IEEJ)
DENKI GAKKAI (THE INSTITUTE OF ELECTRICAL ENGINEERS OF JAPAN)
1-11, YURAKU-CHO, CHIYODA-KU, TOKYO 100
JAPANESE

DENKI GAKKAI DENSHI KEISOKU KENKYUKAI SHIRYO 6267
(PAPERS OF TECHNICAL MEETING ON ELECTRONIC MEASUREMENTS, IEEJ)
DENKI GAKKI (THE INSTITUTE OF ELECTRICAL ENGINEERS OF JAPAN)
1-11, YURAKU-CHO, CHIYODA-KU, TOKYO 100
JAPANESE

DENKI GAKKAI DENSHI SOCHI KENKYUKAI SHIRYO 6268
(PAPERS OF TECHNICAL MEETING OF ELECTRONIC DEVICES, IEEJ)
DENKI GAKKAI (THE INSTITUTE OF ELECTRICAL ENGINEERS OF JAPAN)
1-11, YURAKU-CHO, CHIYODA-KU, TOKYO 100
JAPANESE

DENKI GAKKAI DENSHI ZAIRYO KENKYUKAI SHIRYO 6269
(PAPERS OF TECHNICAL MEETING OF ELECTRONIC MATERIALS, IEEJ)
DENKI GAKKAI (THE INSTITUTE OF ELECTRICAL ENGINEERS OF JAPAN)
1-11, YURAKU-CHO, CHIYODA-KU, TOKYO 100
JAPANESE

DENKI GAKKAI GIJUTSU HOKOKU 1 -BU 6270
(TECHNICAL REPORT OF THE INSTITUTE OF ELECTRICAL ENGINEERS OF JAPAN, PART 1)
DENKI GAKKAI (THE INSTITUTE OF ELECTRICAL ENGINEERS OF JAPAN)
1-11, YURAKU-CHO, CHIYODA-KU, TOKYO 100
JAPANESE
INDEXED: JICST

DENKI GAKKAI GIJUTSU HOKOKU 2-BU 6271
(TECHNICAL REPORT OF THE INSTITUTE OF ELECTRICAL
ENGINEERS OF JAPAN, PART 2)
DENKI GAKKAI (INSTITUTE OF ELECTRICAL ENGINEERS
OF JAPAN)
1-11, YURAKU-CHO, CHIYODA-KU, TOKYO 100
JAPANESE
INDEXED: JICST

DENKI GAKKAI HODEN KENKYUKAI SHIRYO 6272
(PAPERS OF TECHNICAL MEETING OF ELECTRIC
DISCHARGE, IEEJ)
DENKI GAKKAI (INSTITUTE OF ELECTRICAL ENGINEERS
OF JAPAN)
1-11, YURAKU-CHO, CHIYODA-KU, TOKYO 100
JAPANESE

DENKI GAKKAI JIDO SEIGYO KENKYUKAI SHIRYO 6273
(PAPERS OF TECHNICAL MEETING ON AUTOMATIC CONTROL,
IEEJ)
DENKI GAKKAI (INSTITUTE OF ELECTRICAL ENGINEERS OF
JAPAN)
1-11, YURAKU-CHO, CHIYODA-KU, TOKYO 100
JAPANESE

DENKI GAKKAI JIKI OYO KENKYUKAI SHIRYO 6274
(PAPERS OF TECHNICAL MEETING OF APPLIED MAGNETICS,
IEEJ)
DENKI GAKKAI (INSTITUTE OF ELECTRICAL ENGINEERS OF
JAPAN)
1-11, YURAKU-CHO, CHIYODA-KU, TOKYO 100
JAPANESE

DENKI GAKKAI JISEI ZAIRYO KENKYUKAI SHIRYO 6275
(PAPERS OF TECHNICAL MEETING ON MAGNETIC MATERIALS,
IEEJ)
DENKI GAKKAI (INSTITUTE OF ELECTRICAL ENGINEERS OF
JAPAN)
1-11, YURAKU-CHO, CHIYODA-KU, TOKYO 100
JAPANESE

DENKI GAKKAI JOHO SHORI KENKYUKAI SHIRYO 6276
(PAPERS OF TECHNICAL MEETING ON INFORMATION
PROCESSING, IEEJ)
DENKI GAKKAI (INSTITUTE OF ELECTRICAL ENGINEERS OF
JAPAN)
1-11, YURAKU-CHO, CHIYODA-KU, TOKYO 100
JAPANESE

DENKI GAKKAI KAIHEI HOGOSOCHI KENKYUKAI SHIRYO 6277
(PAPERS OF TECHNICAL MEETING ON SWITCHING AND
PROTECTIVE DEVICES, IEEJ)
DENKI GAKKAI (INSTITUTE OF ELECTRICAL ENGINEERS OF
JAPAN)
1-11. YURAKU-CHO, CHIYODA-KU, TOKYO 100
JAPANESE

DENKI GAKKAI KAITENKI KENKYUKAI SHIRYO 6278
(PAPERS OF TECHNICAL MEETING ON ROTATING MACHINES,
IEEJ)
DENKI GAKKAI (INSTITUTE OF ELECTRICAL ENGINEERS OF
JAPAN)
1-11, YURAKU-CHO, CHIYODA-KU, TOKYO 100
JAPANESE

DENKI GAKKAI KINZOKU ZAIRYO KENKYUKAI SHIRYO 6279
(PAPERS OF TECHNICAL MEETING ON METALLIC MATERIALS,
IEEJ)
DENKI GAKKAI (INSTITUTE OF ELECTRICAL ENGINEERS OF
JAPAN)
1-11, YURAKU-CHO, CHIYODA-KU, TOKYO 100
JAPANESE

DENKI GAKKAI KOGEN KANREN SOCHI KENKYUKAI 6280
SHIRYO
(PAPERS OF TECHNICAL MEETING ON LIGHT SOURCES AND
SYSTEMS, IEEJ)
DENKI GAKKAI (INSTITUTE OF ELECTRICAL ENGINEERS OF
JAPAN)
1-11, YURAKU-CHO, CHIYODA-KU, TOKYO 100
JAPANESE

DENKI GAKKAI KYOIKU KENKYU GIJUTSU KENKYUKAI 6281
SHIRYO
(PAPERS OF TECHNICAL MEETING ON RESEARCH AND
EDUCATION, IEEJ)
DENKI GAKKAI (INSTITUTE OF ELECTRICAL ENGINEERS OF
JAPAN)
1-11, YURAKU-CHO, CHIYODA-KU, TOKYO 100
JAPANESE

DENKI GAKKAI PURAZUMA KENKYUKAI SHIRYO 6282
(PAPERS OF TECHNICAL MEETING ON PLASMA, IEEJ)
DENKI GAKKAI (INSTITUTE OF ELECTRICAL ENGINEERS OF
JAPAN)
1-11, YURAKU-CHO, CHIYODA-KU, TOKYO 100
JAPANESE

DENKI GAKKAI RONBUNSHI A 6283
(TRANSACTIONS OF THE INSTITUTE OF ELECTRICAL
ENGINEERS OF JAPAN, PART A)
DENKI GAKKAI (INSTITUTE OF ELECTRICAL ENGINEERS
OF JAPAN)
1-11, YURAKU-CHO, CHIYODA-KU, TOKYO 100
JAPANESE, WITH ENGLISH SUMMARY
INDEXED: INS, JICST

DENKI GAKKAI RONBUNSHI B 6284
(TRANSACTIONS OF THE INSTITUTE OF ELECTRICAL
ENGINEERS OF JAPAN, PART B)
DENKI GAKKAI (INSTITUTE OF ELECTRICAL ENGINEERS
OF JAPAN)
1-11, YURAKU-CHO, CHIYODA-KU, TOKYO 100
JAPANESE, WITH ENGLISH SUMMARY
INDEXED: INS, JICST

DENKI GAKKAI RONBUNSHI C 6285
(TRANSACTIONS OF THE INSTITUTE OF ELECTRICAL
ENGINEERS OF JAPAN, PART C)
DENKI GAKKAI (INSTITUTE OF ELECTRICAL ENGINEERS
OF JAPAN)
1-11, YURAKU-CHO, CHIYODA-KU, TOKYO 100
JAPANESE, WITH ENGLISH SUMMARY
INDEXED: INS, JICST

DENKI GAKKAI SEIGYO HENKAN SOCHI KENKYUKAI SHIRYO 6286
(PAPERS OF TECHNICAL MEETING ON POWER CONVERTORS
AND CONTROL EQUIPMENTS, IEEJ)
DENKI GAKKAI (INSTITUTE OF ELECTRICAL ENGINEERS OF
JAPAN)
1-11, YURAKU-CHO, CHIYODA,KU, TOKYO 100
JAPANESE

DENKI GAKKAI SEISHIKI KENKYUKAI SHIRYO 6287
(PAPERS OF TECHNICAL MEETING ON TRANSFORMERS AND
CAPACITORS, IEEJ)
DENKI GAKKAI (INSTITUTE OF ELECTRICAL ENGINEERS OF
JAPAN)
1-11, YURAKU-CHO, CHIYODA-KU, TOKYO 100
JAPANESE

DENKI GAKKAI SEITETSU KOGYO KENKYUKAI SHIRYO 6288
(PAPERS OF TECHNICAL MEETING ON IRON AND STEEL
INDUSTRY, IEEJ)
DENKI GAKKAI (INSTITUTE OF ELECTRICAL ENGINEERS
OF JAPAN)
1-11, YURAKU-CHO, CHIYODA-KU, TOKYO 100
JAPANESE

DENKI GAKKAI SHIKAKU JOHO KENKYUKAI SHIRYO 6289
(PAPERS OF TECHNICAL MEETING ON VISUAL INFORMATION,
IEEJ)
DENKI GAKKAI (INSTITUTE OF ELECTRICAL ENGINEERS OF
JAPAN)
1-11, YURAKU-CHO, CHIYODA-KU, TOKYO 100
JAPANESE

DENKI GAKKAI TOKYO-SHIBU TAIKAI RONBUNSHU 6290
(PAPERS FROM TOKYO CHAPTER MEETING OF IEEJ)
DENKI GAKKAI TOKYO-SHIBU
C/O SHIN YURAKU BUILDING, 1-11, YURAKU-CHO,
CHIYODA-KU, TOKYO 100
JAPANESE

DENKI GAKKAI ZASSHI 6291
(JOURNAL OF THE INSTITUTE OF ELECTRICAL
ENGINEERS OF JAPAN)
DENKI GAKKAI (INSTITUTE OF ELECTRICAL ENGINEERS
OF JAPAN)
1-11, YURAKU-CHO, CHIYODA-KU, TOKYO 100
JAPANESE
INDEXED: CHE, ENG, JICST

DENKI GAKKAI ZENKOKU TAIKAI KOEN RONBUNSHU 6292
(PAPERS FROM THE IEEJ NATIONAL CONVENTION)
DENKI GAKKAI (INSTITUTE OF ELECTRICAL ENGINEERS OF
JAPAN)
C/O SHIN YURAKU-CHO BUILDING, 1-11, YURAKU-CHO,
CHIYODA-KU, TOKYO 100
JAPANESE
INDEXED: JICST

DENKI GAKKAI ZETSUEN ZAIRYO KENKYUKAI SHIRYO 6293
(PAPERS OF TECHNICAL MEETING ON ELECTRICAL
INSULATING MATERIALS, IEEJ)
DENKI GAKKAI (INSTITUTE OF ELECTRICAL ENGINEERS OF
JAPAN)
1-11, YURAKU-CHO, CHIYODA-KU, TOKYO 100
JAPANESE

DENKI GENBA GIJUTSU 6294
(ON SITE ELECTROTECHNICS)
DENKI JOHOSHA
C/O SANKYO BUILDING, 1-3, KOJI-MACHI, CHIYODA-KU,
TOKYO 102
JAPANESE

DENKI GIJUTSU 6295
(ELECTRICAL TECHNOLOGY)
JAPANESE

DENKI GIJUTSUSHA 6296
(ELECTROTECHNICIAN)
DENKI GIJUTSUSHA KYOKAI
5-3, 1-CHOME, KORAKU, BUNKYO-KU, TOKYO 112
JAPANESE
INDEXED: CHE

DENKI HYORON 6297
(ELECTRICAL REVIEW)
DENKI HYORON HENSHU-BY (ELECTRICAL REVIEW EDITORIAL COMMITTEE)
49, TANAKAOI-CHO, SAKYO-KU, KYOTO-SHI 606
JAPANESE
INDEXED: JICST

DENKI JIGYO NENPO (CHUGOKU DENRYOKU) 6298
(ANNUAL REPORT OF THE CHUGOKU ELECTRIC POWER COMPANY, INCORPORATED)
CHUGOKU DENRYOKU K.K.
4-33, KOMACHI, HIROSHIMA-SHI 730
JAPANESE

DENKI JIGYO NENPO (HOKKAIDO DENRYOKU) 6299
(ANNUAL REPORT OF THE HOKKAIDO ELECTRIC POWER COMPANY, INCORPORATED)
HOKKAIDO DENRYOKU K.K.
1-CHOME, ODORI HIGASHI, CHUO-KU, SAPPORO-SHI 060
JAPANESE

DENKI JIGYO NENPO (HOKURIKU DENRYOKU) 6300
(ANNUAL REPORT OF THE HOKURIKU ELECTRIC POWER COMPANY, INCORPORATED)
HOKURIKU DENRYOKU K.K.
3-1, SAKURABASHI-DORI, TOYAMA-SHI 930
JAPANESE

DENKI JIGYO NENPO (KANSAI DENRYOKU) 6301
(ANNUAL REPORT OF THE KANSAI ELECTRIC POWER COMPANY, INCORPORATED)
KANSAI DENRYOKU K.K.
3-5, NAKANOSHIMA, KITA-KU, OSAKA-SHI 530
JAPANESE

DENKI JIGYO NENPO (KYUSHU DENRYOKU) 6302
(ANNUAL REPORT OF THE KYUSHU ELECTRIC POWER COMPANY, INCORPORATED)
KYUSHU DENRYOKU K.K.
1-82, 2-CHOME, WATANABE-DORI, CHUO-KU, FUKUOKA-SHI 810-91
JAPANESE

DENKI JIGYO NENPO (SHIKOKU DENRYOKU) 6303
(ANNUAL REPORT OF THE SHIKOKU ELECTRIC POWER COMPANY, INCORPORATED)
SHIKOKU DENRYOKU K.K.
2-5, MARUNOUCHI, TAKAMATSU-SHI 760
JAPANESE

DENKI JIGYO NENPO (TOHOKU DENRYOKU) 6304
(ANNUAL REPORT OF TOHOKU ELECTRIC POWER COMPANY, LTD.)
HOKURIKU DENRYOKU K.K.
7-2, 3-CHOME, ICHIBAN-CHO, SENDAI-SHI 980
JAPANESE

DENKI JIGYO NENPO (TOKYO DENRYOKU) 6305
(ANNUAL REPORT OF THE TOKYO ELECTRIC POWER COMPANY, INCORPORATED)
TOKYO DENRYOKU K.K.
1-3, 1-CHOME, UCHISAIWAI-CHO, CHIYODA-KU, TOKYO 100
JAPANESE

DENKI JIGYO NO GENJO 6306
(SURVEY OF ELECTRICAL INDUSTRY)
TSUSHO SANGYO-CHO KOEKI JIGYO-KYOKU (PUBLIC UTILITIES BUREAU, MINISTRY OF INTERNATIONAL TRADE & INDUSTRY) AND NIHON DENKI KYOKAI
1-3, YURAKU-CHO, CHIYODA-KU, TOKYO 100
JAPANESE

DENKI JIGYO TOKEI 6307
(STATISTICS ON ELECTRICAL INDUSTRY)
DENKI JIGYO RENGOKAI TOKEI IINKAI (STATISTICS COMMITTEE, THE FEDERATION OF ELECTRIC POWER COMPANY)
C/O DENKI KYOKAI BUILDING, 1-3, YURAKU-CHO, CHIYODA-KU, TOKYO 100
JAPANESE

DENKI JIHO 6308
(JOURNAL OF JAPAN ELECTRIC CONSTRUCTION WORKS)
DENKI JIHOSHA
1-14, NINGYO-CHO, NIHONBASHI-CHUO-KU, TOKYO 103
JAPANESE
INDEXED: JICST

DENKI JIKI ZAIRYO KENKYU HOKOKU 6309
(SCIENCE REPORT OF THE RESEARCH INSTITUTE OF ELECTRIC AND MAGNETIC ALLOYS)
DENKI JIKI ZAIRYO KENKYUSHO (RESEARCH INSTITUTE OF ELECTRIC AND MAGNETIC ALLOYS)
107-62, HIGASHIHACHIBANCHO, SENDAI-SHI 980
JAPANESE AND ENGLISH

DENKI JOHO 6310
(ELECTRICAL INFORMATION)
DENKI JOHOSHA
C/O SANKYO BUILDING, 1-3, KOJI-MACHI, CHIYODA-KU, TOKYO 102
JAPANESE

DENKI KAKO GAKKAISHI 6311
(JOURNAL OF THE JAPAN SOCIETY OF ELECTRICAL MACHINING ENGINEERS)
JAPANESE
INDEXED: JICST

DENKI KEISAN 6312
(ELECTRICAL CALCULATION)
DENKI SHOIN
1-55, KANDA JINBO-CHO, CHIYODA-KU, TOKYO 101
JAPANESE
INDEXED: JICST

DENKI KENTEISHO GIHO 6313
(JEMIC TECHNICAL REPORT)
NIHON DENKI KEIKI KENTEISHO (JAPANESE ELECTRIC METERS INSPECTION COMPANY) AND OMUSHA
3-1, KANDA NISHIKI-CHO, CHIYODA-KU, TOKYO 101
JAPANESE, WITH ENGLISH SUMMARY
INDEXED: JICST

DENKI KOGAKU NENPO 6314
(ANNUAL REPORT OF ELECTRICAL ENGINEERING)
DENKI GAKKAI (INSTITUTE OF ELECTRICAL ENGINEERS OF JAPAN)
1-11, YURAKU-CHO, CHIYODA-KU, TOKYO 100
JAPANESE

DENKI KOJI NO TOMO 6315
(ELECTRICAL WORKER'S COMPANION)
KANTO DENKI KYOKAI (KANTO ELECTRIC ASSOCIATION)
1-3, YURAKU-CHO, CHIYODA-KU, TOKYO 100
JAPANESE
INDEXED: JICST

DENKI KOJIGYO 6316
(MONTHLY: THE ELECTRICAL CONTRACTING INDUSTRY)
TOKYO-TO DENKI KOJI KOGYO KUMIAI
16-7, 1-CHOME, OJI HON-CHO, KITA-KU, TOKYO 114
JAPANESE

DENKI KOJISHI 6317
(ELECTRICAL CONSTRUCTION ENGINEER)
TOKYO-TO DENKI KOJISHI KYOKAI
7-8, 1-CHOME, MOTOAKASAKA, MINATO-KU, TOKYO 107
JAPANESE

DENKI KYODO KENKYU 6318
(ELECTRICAL COOPERATIVE RESEARCH)
DENKI KYODO KENKYUKAI (SOCIETY OF ELECTRICAL COOPERATIVE RESEARCH)
11-35, 1-CHOME, NAGATA-CHO, CHIYODA-KU, TOKYO 100
JAPANESE
INDEXED: JICST

DENKI KYOKAI ZASSHI 6319
(JOURNAL OF JAPAN ELECTRIC ASSOCIATION)
NIHON DENKI KYOKAI (JAPAN ELECTRIC ASSOCIATION)
1-3, YURAKU-CHO, CHIYODA-KU, TOKYO 100
JAPANESE
INDEXED: JICST

DENKI NIPPO 6320
(ELECTRICAL DAILY NEWS)
DENKI NIPPOSHA
11-4, 6-CHOME, SHINBASHI, MINATO-KU, TOKYO 105
JAPANESE

DENKI OYOBI SETSUBI KOJI SEKKEI KIHON NINKU KIJUN 6321
(INTEGRATING STANDARD LABOR OF ELECTRIC EQUIPMENT WORKS)
DENKI JIHOSHA
1-4, NINGYO-CHO, NIHONBASHI, CHUO-KU, TOKYO 103
JAPANESE

DENKI SANGYO SHINBUN 6322
(ELECTRICAL INDUSTRY NEWS)
DENKI SANGYO SHINBUNSHA
5-11, 2-CHOME, SHIBA DAIMON, MINATO-KU, TOKYO 105
JAPANESE

DENKI SHIGAKKAI RENGO TAIKAI KOEN RONBUNSHU 6323
(PAPERS FROM THE JOINT CONFERENCE OF THE INSTITUTE OF ELECTRICAL ENGINEERS OF JAPAN AND OTHER ELECTRICAL SOCIETIES)
DENKI GAKKAI, DENSHI TSUSHIN GAKKAI, SHOMEI GAKKAI, TEREBIJON GAKKAI
C/O SHIN YURAKU-CHO BUILDING, 1-11, YURAKU-CHO, CHIYODA-KU, TOKYO 100
JAPANESE

DENKI SHINBUN 6324
(ELECTRICAL NEWS)
NIHON DENKI KYOKAI (JAPAN ELECTRIC ASSOCIATION)
1-3, YURAKU-CHO, CHIYODA-KU, TOKYO 100
JAPANESE

DENKI SHOHIN 6325
(ELECTRICAL MERCHANDISE)
DENKI SHOHIN RENMEI (FEDERATION FOR ELECTRICAL MERCHANDISE)
C/O SHIN YURAKU-CHO BUILDING, 1-11, YURAKU-CHO, CHIYODA-KU, TOKYO 100
JAPANESE
INDEXED: JICST

DENKI TAIMUSU 6326
(ELECTRICAL TIMES)
DENKI TAIMUSU (ELECTRICAL TIMES)
3-7,. 2-CHOME, SHINBASHI, MINATO-KU, TOKYO 105
JAPANESE

DENKI TO GASU 6327
(ELECTRICITY AND GAS)
TSUSHO SANGYO-SHO KOEKI JIGYO-KYOKU (PUBLIC UTILITIES BUREAU, MINISTRY OF INTERNATIONAL TRADE AND INDUSTRY) AND TSUSHO SANGYO CHOSAKAI
15-1, 6-CHOME, GINZA, CHUO-KU, TOKYO 104
JAPANESE

DENKI TO KOJI 6328
(ELECTRICITY AND CONSTRUCTION)
OMUSHA
3-1, KANDA NISHIKI-CHO, CHIYODA-KU, TOKYO 101
JAPANESE
INDEXED: JICST

DENKI TSUSHIN 6329
(TELECOMMUNICATIONS)
DENKI TSUSHIN KYOKAI (TELECOMMUNICATIONS ASSOCIATION)
1-11, YURAKU-CHO, CHIYODA-KU, TOKYO 100
JAPANESE
INDEXED: JICST

DENKI TSUSHIN DAIGAKU DENKI TSUSHIN KENKYU SHISETSU NENPO 6330
(ANNUAL REPORT OF THE RESEARCH LABORATORY OF COMMUNICATION SCIENCES, UNIVERSITY OF ELECTRO-COMMUNICATIONS)
DENKI TSUSHIN DAIGAKU DENKI TSUSHIN KENKYU SHISETSU
5-1, 1-CHOME, CHOFUGAOKA, CHOFU-SHI, TOKYO 181
JAPANESE
INDEXED: CHE

DENKI TSUSHIN DAIGAKU GAKUHO 6331
(REPORTS OF THE UNIVERSITY OF ELECTRO-COMMUNICATIONS)
DENKI TSUSHIN DAIGAKU (UNIVERSITY OF ELECTRO-COMMUNICATIONS)
5-1, 1-CHOME, CHOGUGAOKA, CHOFU-SHI, TOKYO 182
JAPANESE, WITH ENGLISH SUMMARY
INDEXED: CHE, MAT, INS, JICST

DENKI TSUSHIN KENKYU SHISETSU NENPO 6332
(TELECOMMUNICATIONS RESEARCH FACILITIES ANNUAL)
JAPANESE
INDEXED: JICST

DENKI TSUSHIN KENKYUJO SEIKA HOKOKU 6333
(TELECOMMUNICATIONS RESEARCH INSTITUTE, FINDINGS REPORT)
JAPANESE
INDEXED: JICST

DENKI TSUSHIN KENKYUSHO KENKYU HAPPYOKAI RONBUNSHU 6334
(ELECTRICAL COMMUNICATION LABORATORY, TECHNICAL MEETING DOCUMENTS)
NIHON DENSHIN DENWA KOSHA MUSASHINO DENKI TSUSHIN KENKYUSHO (MUSASHINO ELECTRICAL COMMUNICATION LABORATORY, NIHON TELEGRAPH AND TELEPHONE PUBLIC CORPORATION)
9-11, 3-CHOME, MIDORI-CHO, MUSASHINO-SHI, TOKYO 180
JAPANESE

DENKI TSUSHIN KENKYUSHO KENKYU JITSUYOKA HOKOKU 6335
(ELECTRICAL COMMUNICATION LABORATORY TECHNICAL JOURNAL)
NIHON DENSHIN DENWA KOSHA MUSASHINO DENKI TSUSHIN KENKYUSHO (MUSASHINO ELECTRICAL COMMUNICATION LABORATORY, NIHON TELEGRAPH AND TELEPHONE PUBLIC CORPORATION)
9-11, 3-CHOME, MIDORI-CHO, MUSASHINO-SHI, TOKYO 180
JAPANESE
INDEXED: CHE, JICST

DENKI TSUSHIN KENKYUSHO KENKYU JITSUYOKA HOKOKU, BESSATSU 6336
(REPORT ON RESEARCH UTILIZATION, TELECOMMUNICATIONS RESEARCH INSTITUTE, SPECIAL ISSUE)
NIHON DENSHIN DENWA KOSHA DENKI TSUSHIN KENKYUSHO (JAPAN TELEGRAPH AND TELEPHONE PUBLIC CORPORATION, TELECOMMUNICATIONS RESEARCH INSTITUTE)
JAPANESE

DENKI TSUSHIN KENKYUSHO NENPO 6337
(ANNUAL REPORT OF THE ELECTRICAL COMMUNICATION LABORATORY)
NIHON DENSHIN DENWA KOSHA MUSASHINO DENKI TSUSHIN KENKYUSHO
9-11, 3-CHOME, MIDORI-CHO, MUSASHINO-SHI, TOKYO 180
JAPANESE
INDEXED: JICST

DENKI TSUSHIN SHISETSU 6338
(TELECOMMUNICATIONS ENGINEERING)
NIHON DENSHIN DENWA KOSHA SHISETSU-KYOKU (NIHON TELEGRAPH AND TELEPHONE PUBLIC CORPORATION) AND DENKI TSUSHIN KYOKAI
1-11, YURAKU-CHO, CHIYODA-KU, TOKYO 100
JAPANESE
INDEXED: JICST

DENKI TSUSHIN SHISETSU SOMOKUJI 6339
(CUMULATIVE INDEX TO TELECOMMUNICATIONS ENGINEERING)
NIHON DENSHIN DENWA KOSHA (NIHON TELEGRAPH AND TELEPHONE PUBLIC CORPORATION) AND DENKI TSUSHIN KYOKAI (TELECOMMUNICATION ASSOCIATION)
1-11, YURAKU-CHO, CHIYODA-KU, TOKYO 100
JAPANESE

DENKI TSUSHIN SOGO KENKYUSHO ZASSHI KIJI SAKUIN 6340
(INDEX TO PERIODICALS OF THE RESEARCH INSTITUTE OF TELECOMMUNICATIONS AND ECONOMICS)
DENKI TSUSHIN SOGO KENKYUSHO (RESEARCH INSTITUTE OF TELECOMMUNICATIONS AND ECONOMICS)
6-19, 1-CHOME, AZABUDAI, MINATO-KU, TOKYO 106
JAPANESE

DENKI YON-GAKKAI RENGO TAIKAI KOEN RONBUNSHU 6341
(PAPERS OF THE FOURTH ELECTRON SOCIETY MEETING IN CONJUNCTION WITH THE GENERAL MEETING)
JAPANESE
INDEXED: JICST

DENKITEN 6342
(ELECTRIC RETAILER)
DENPA SHINBUNSHA (DENPA PUBLICATIONS INCORPORATED)
11-15, 1-CHOME, HIGASHIGOTANDA, SHINAGAWA-KU, TOKYO 141
JAPANESE

DENKYO KOGYO KAIHO 6343
(BULLETIN OF ELECTRIC LAMP MANUFACTURES ASSOCIATION)
NIHON DENKYO KOGYOKAI (JAPAN ELECTRIC LAMP MANUFACTURES ASSOCIATION)
C/O TOKO BUILDING, 1-3, YURAKU-CHO, CHIYODA-KU, TOKYO 100
JAPANESE
INDEXED: JICST

DENPA GIJUTSU 6344
(RADIO TV HIFI & ELECTRONICS)
KINDAI KAGAKUSHA (KINDAI KAGAKU LTD.)
2-10, ICHIGAYA TA-MACHI, SHINJUKU, TOKYO 162
JAPANESE

DENPA GIJUTSU KYOKAIHO 6345
(JOURNAL OF THE RADIO ENGINEERING ASSOCIATION)
DENPA GIJUTSU KYOKAI (RADIO ENGINEERING ASSOCIATION)
1-11, YURAKU-CHO, CHIYODA-KU, TOKYO 100
JAPANESE

DENPA JIHO 6346
(JOURNAL OF RADIO WAVE CONTROL BUREAU)
YUSEI-SHO DENPA KANRI-KYOKU (RADIO WAVE CONTROL BUREAU, MINISTRY OF POSTS AND TELECOMMUNICATION)
6-13, AZABU IIKURA, MINATO-KU, TOKYO 106
JAPANESE
INDEXED: JICST

DENPA JIKKEN 6347
(RADIO EXPERIMENTS)
RAJIKON GIJUTSU SHA (RAJIKON TECHNOLOGY COMPANY)
JAPANESE

DENPA KAGAKU 6348
(RADIO TV TECHNIQUE AND SCIENCE)
NIHON HOSO SHUPPAN KYOKAI
41-1, UDAGAWA-CHO, SHIBUYA-KU, TOKYO 150
JAPANESE
INDEXED: JICST

DENPA KENKYUSHO KIHO 6349
(REVIEW OF THE RADIO RESEARCH LABORATORY)
YUSEI-SHO DENPA KENKYUSHO (RADIO RESEARCH LABORATORY, MINISTRY OF POSTS AND TELECOMMUNICATIONS) AND DENPA SHINKOKAI
6-13, AZABU IIKURA, MINATO-KU, TOKYO 106
JAPANESE, WITH ENGLISH SUMMARY
INDEXED: INS, JICST

DENPA SHINBUN 6350
(RADIO WAVE NEWS)
DENPA SHINBUNSHA (DENPA PUBLICATIONS INCORPORATED)
11-15, 1-CHOME, HIGASHIGOTANDA, SHINAGAWA-KU, TOKYO 141
JAPANESE

DENPA TAIMUZU 6351
(RADIO WAVE TIMES)
DENPA TAIMUZUSHA
3-23, NISHISHINBASHI, MINATO-KU, TOKYO 105
JAPANESE

DENPA YOHO 6352
(PROPAGATION FORECASTS)
YUSEI-SHO DENPA KENKYUSHO (RADIO RESEARCH LABORATORY, MINISTRY OF POSTS AND TELECOMMUNICATIONS)
6-13, AZABU IIKURA, MINATO-KU, TOKYO 106
JAPANESE
INDEXED: JICST

DENPATSU 6353
(INFORMATION NEWS OF THE ELECTRIC POWER DEVELOPMENT COMPANY, LIMITED)
DENGEN KAIHATSU K.K. SOMU-BU, KOHOSHITSU
8-2, 1-CHOME, MARUNOUCHI, CHIYODA-KU, TOKYO 100
JAPANESE

DENRYOKU 6354
(ELECTRIC POWER)
JAPANESE

DENRYOKU CHOSA TOKEI GEPPO 6355
(MONTHLY REPORT ON ELECTRIC POWER STATISTICS)
TSUSHO SANGYO-SHO KOEKI JIGYO-KYOKU (PUBLIC UTILITIES BUREAU, MINISTRY OF INTERNATIONAL TRADE AND INDUSTRY) AND NIHON DENKI KYOKAI
1-3, YURAKU-CHO, CHIYODA-KU, TOKYO 100
JAPANESE

DENRYOKU CHUO KENKYUJO DENRYOKU GIJUTSU KENKYUJO HOKOKU 6356
(REPORT OF ELECTRICAL POWER CENTRAL RESEARCH INSTITUTE, ELECTRICAL POWER TECHNICAL RESEARCH INSTITUTE)
JAPANESE
INDEXED: JICST

DENRYOKU CHUO KENKYUJO ENERUGI, KANKYO GIJUTSU KENKYUJO HOKOKU 6357
(REPORT OF ELECTRICAL POWER CENTRAL RESEARCH INSTITUTE, ENERGY, ENVIRONMENT TECHNICAL RESEARCH INSTITUTE)
JAPANESE
INDEXED: JICST

DENRYOKU CHUO KENKYUJO GIJUTSU KENKYUJO SHOHO 6358
(REPORT OF DENKYOKU CENTRAL RESEARCH INSTITUTE, AGRICULTURAL RESEARCH INSTITUTE)
JAPANESE
INDEXED: JICST

DENRYOKU CHUO KENKYUSHO GIJUTSU DAI-1, DAI-2 KENKYUSHO KIHO 6359
(BULLETIN OF THE FIRST AND THE SECOND INSTITUTE OF TECHNOLOGY, CENTRAL RESEARCH INSTITUTE OF ELECTRIC POWER INDUSTRY)
6-1, 1-CHOME, OTE-MACHI, CHIYODA-KU, TOKYO 100
JAPANESE
INDEXED: JICST

DENRYOKU CHUO KENKYUSHO GIJUTSU DAI-1 KENKYUSHO HOKOKU 6360
(TECHNICAL LABORATORY NUMBER 1 CENTRAL RESEARCH INSTITUTE OF ELECTRIC POWER INDUSTRY, REPORT)
DENRYOKU CHUO KENKYUSHO GIJUTSU DAI-1 KENKYUSHO (TECHNICAL LABORATORY NUMBER 1 CENTRAL RESEARCH INSTITUTE OF ELECTRIC POWER INDUSTRY)
1229, IWATO, KOMAE-SHI, TOKYO 182
JAPANESE, WITH ENGLISH SUMMARY
INDEXED: CHE

DENRYOKU CHUO KENKYUSHO GIJUTSU DAI-1 KENKYUSHO SOGO HOKOKU 6361
(COLLECTIVE REPORT OF THE TECHNICAL LABORATORY NUMBER 1, CENTRAL RESEARCH INSTITUTE OF ELECTRIC POWER INDUSTRY)
1229, IWARO, KOMAE-SHI, TOKYO 182
JAPANESE, WITH ENGLISH SUMMARY
INDEXED: CHE, JICST

DENRYOKU CHUO KENKYUSHO GIJUTSU DAI-2 KENKYUSHO HOKOKU 6362
(RESEARCH REPORTS OF THE TECHNICAL LABORATORY NUMBER 2, CENTRAL RESEARCH INSTITUTE OF ELECTRIC POWER INDUSTRY)
DENRYOKU CHUO KENKYUSHO GIJUTSU DAI-2 KENKYUSHO
1646, ABIKO, ABIKO-SHI 270-11
JAPANESE AND EUROPEAN
INDEXED: JICST

DENRYOKU CHUO KENKYUSHO GIJUTSU DAI-2 KENKYUSHO SOGO HOKOKU 6363
(COLLECTIVE REPORT OF THE TECHNICAL LABORATORY NUMBER 2, CENTRAL RESEARCH INSTITUTE OF ELECTRIC POWER INDUSTRY)
1646, ABIKO, ABIKO-SHI 270-11
JAPANESE

DENRYOKU CHUO KENKYUSHO HOKOKU 6364
(TECHNICAL REPORT OF THE CENTRAL RESEARCH INSTITUTE OF ELECTRIC POWER INDUSTRY)
CENTRAL RESEARCH INSTITUTE OF ELECTRIC POWER INDUSTRY
1229 IWATO, KOMAE-SHI, TOKYO
JAPANESE
INDEXED: CHE

DENRYOKU CHUO KENKYUSHO JOHO SHORI KENKYU SENTA HOKOKU 6365
(REPORT OF THE INFORMATION PROCESSING RESEARCH CENTER, CENTRAL RESEARCH INSTITUTE OF ELECTRIC POWER INDUSTRY)
DENRYOKU CHUO KENKYUSHO JOHO SHORI KENKYU SENTA (INFORMATION PROCESSING RESEARCH CENTER, CENTRAL RESEARCH INSTITUTE OF ELECTRIC POWER INDUSTRY)
6-1, 1-CHOME, OTE-MACHI, CHIYODA-KU, TOKYO 100
JAPANESE, WITH ENGLISH SUMMARY

DENRYOKU CHUO KENKYUSHO NODEN KENKYUSHO KENKYU HOKOKU 6366
(RESEARCH REPORTS OF THE AGRICULTURAL LABORATORY CENTER RESEARCH INSTITUTE OF ELECTRIC POWER INSTITUTE))
DENRYOKU CHUO KENKYUSHO NODEN KENKYUSHO
1646 ABIKO, ABIKO-SHI 270-11
JAPANESE AND ENGLISH
INDEXED: BIO

DENRYOKU DOBOKU 6367
(POWER CIVIL ENGINEERING)
JAPANESE

DENRYOKU GIKEN NYUSU 6368
(ELECTRIC POWER AND TECHNOLOGY RESEARCH NEWS)
JAPANESE
INDEXED: JICST

DENRYOKU JUYO SHIHYO 6369
(MONTHLY INDICATORS OF ELECTRICITY CONSUMPTION)
DENRYOKU CHUO KENKYUSHO (CENTRAL RESEARCH INSTITUTE OF ELECTRIC POWER INDUSTRY)
C/O OTEMACHI BUILDING, 6-1, 1-CHOME, OTE-MACHI, CHIYODA-KU, TOKYO 100
JAPANESE AND ENGLISH

DENRYOKU KEIZAI KENKYU 6370
(ECONOMICS & PUBLIC UTILITIES)
DENRYOKU CHUO KENKYUSHO KEIZAI KENKYUSHO (CENTRAL RESEARCH INSTITUTE OF ELECTRIC POWER INDUSTRY, ECONOMIC RESEARCH INSTITUTE)
C/O OTE-MACHI BUILDING, 6-1, 1-CHOME, OTE-MACHI, CHIYODA-KU, TOKYO 100
JAPANESE

DENRYOKU MANSURI 6371
(ELECTRIC POWER MONTHLY)
KANTO DENKI KYOKAI (KANTO ELECTRIC ASSOCIATION)
1-3, YURAKU-CHO, CHIYODA-KU, TOKYO 100
JAPANESE
INDEXED: JICST

DENRYOKU SHINPO 6372
(ELECTRIC POWER NEWS)
DENRYOKU SHINPOSHA
10-13, 5-CHOME, GINZA, CHUO-KU, TOKYO 104
JAPANESE
INDEXED: JICST

DENSEI 6373
(DENSEI TECHNICAL JOURNAL)
NIHON DENKI SEIKI K.K.
19-18, 1-CHOME TSUTSUMI-DORI, SUMIDA-KU, TOKYO 131
JAPANESE
INDEXED: JICST

DENSEN JIHO 6374
(ELECTRIC WIRE AND CABLE TIMES)
NIHON DENSEN KOGYOKAI (JAPANESE ELECTRIC WIRE & CABLE MAKER'S ASSOCIATION)
12-22, 1-CHOME, TSUKJI, CHUO-KU, TOKYO 104
JAPANESE
INDEXED: JICST

DENSEN TOKEI NENPO 6375
(ELECTRIC WIRE AND CABLE REVIEW)
NIHON DENSEN KOGYOKAI (JAPANESE ELECTRIC WIRE AND CABLE MAKER'S ASSOCIATION)
3-10, TSUKIJI, CHUO-KU, TOKYO 104
JAPANESE

DENSETSU GAIDO 6376
(GUIDE TO ELECTRIC POWER EQUIPMENTS)
TOSHIBA SHOJI K.K. TOKUHAN EIGYO HONBU
2-1, 5-CHOME, GINZA, CHUO-KU, TOKYO 104
JAPANESE

DENSETSU KOGYO 6377
(ELECTRICAL CONSTRUCTION ENGINEERING)
NIHON DENSETSU KOGYO KYOKAI (JAPAN ELECTRICAL CONTRACTOR'S ASSOCIATION)
7-8, 1-CHOME, MOTOKASAKA, MINATO-KU, TOKYO 107
JAPANESE
INDEXED: JICST

DENSETSU SHIZAI 6378
(ELECTRICAL INSTALLATION MATERIALS)
DENSETSU SHUPPAN (ELECTRICAL INSTALLATION PUBLISHING)
JAPANESE

DENSHI 6379
(ELECTRONICS)
DENSHI KIKAI KOGYOKAI (ELECTRONIC INDUSTRIES ASSOCIATION OF JAPAN)
C/O TOSHO BUILDING, 2-2, 3-CHOME, MARUNOUCHI, CHIYODA-KU, TOKYO 100
JAPANESE
INDEXED: JICST

DENSHI GIJUTSU 6380
(ELECTRONIC TECHNOLOGY)
NIKKAN KOGYO SHINBUNSHA (INDUSTRIAL DAILY NEWS)
8-10, 1-CHOME, KUDAN KITA, CHIYODA-KU, TOKYO 102
JAPANESE
INDEXED: CHE, JICST

DENSHI GIJUTSU CHOSA IINKAI SHIRYO. A. 6381
(DATA OF THE ELECTRONIC TECHNOLOGY SURVEY COMMITTEE. A.)
JAPANESE
INDEXED: JICST

DENSHI GIJUTSU IINKAI SHIRYO. B. 6382
(DATA OF THE ELECTRONIC TECHNOLOGY COMMITTEE. B.)
JAPANESE
INDEXED: JICST

DENSHI GIJUTSU SOGO KENKYUJO IHO 6383
(BULLETIN OF ELECTRONIC TECHNOLOGY CENTRAL RESEARCH INSTITUTE)
JAPANESE AND ENGLISH

DENSHI GIJUTSU SOGO KENKYUSHO CHOSA HOKOKU 6384
(CIRCULARS OF THE ELECTROTECHNICAL LABORATORY)
TSUSHO SANGYO-SHO KOGYO GIJUTSU-IN DENSHI GIJUTSU SOGO KENKYUSHO (ELECTROTECHNICAL LABORATORY, AGENCY OF INDUSTRIAL SCIENCE AND TECHNOLOGY, MINISTRY OF INTERNATIONAL TRADE AND INDUSTRY)
6-1, 2-CHOME, NAGATA-CHO, CHIYODA-KU, TOKYO 100
JAPANESE, WITH ENGLISH SUMMARY
INDEXED: CHE, ENG, INS, IAA, JICST

DENSHI GIJUTSU SOGO KENKYUSHO IHO 6385
(BULLETIN OF THE ELECTROTECHNICAL LABORATORY)
TSUSHO SANGYO-SHO KOGYO GIJUTSU-IN DENSI GIJUTSU SOGO KENKYUSHO (ELECTROTECHNICAL LABORATORY, AGENCY OF INDUSTRIAL SCIENCE AND TECHNOLOGY, MINISTRY OF INTERNATIONAL TRADE AND INDUSTRY)
JAPANESE AND ENGLISH
INDEXED: CHE, ENG, INS, JICST

DENSHI GIJUTSU SOGO KENKYUSHO KENKYU HOKOKU 6386
(RESEARCHES OF THE ELECTROTECHNICAL LABORATORY)
TSUSHO SANGYO-SHO KOGYO GIJUTSU-IN DENSHI GIJUTSU SOGO KENKYUSHO (ELECTROTECHNICAL LABORATORY, AGENCY OF INDUSTRIAL SCIENCE AND TECHNOLOGY, MINISTRY OF INTERNATIONAL TRADE AND INDUSTRY)
6-1, 2-CHOME, NAGATA-CHO, CHIYODA-KU, TOKYO 100
JAPANESE AND ENGLISH
INDEXED: CHE, INS, JICST

DENSHI GIJUTSU SOGO KENKYUSHO NENPO 6387
(ANNUAL REPORT OF THE ELECTROTECHNICAL LABORATORY, AGENCY OF INDUSTRIAL SCIENCE AND TECHNOLOGY)
TSUSHO SANGYO-SHO KOGYO GIJUTSU-IN DENSHI GIJUTSU SOGO KENKYUSHO
6-1, 2-CHOME, NAGATA-CHO, CHIYODA-KU, TOKYO 100
JAPANESE
INDEXED: JICST

DENSHI GIJUTSU SOGO KENKYUSHO TANASHI KEISANKISHITSU REPOTO 6388
(REPORT OF COMPUTER LABORATORY, TANASHI BRANCH OF ELECTROTECHNICAL LABORATORY)
TSUSHO SANGYO-SHO KOGYO GIJUTSUIN DENSHI GIJUTSU SOGO KENKYUSHO TANASHI KEISANKISHITSU
4-1, 5-CHOME, MUKODAI-MACHI, TANASHI-SHI, TOKYO 188
JAPANESE

DENSHI KAGAKU 6389
(ELECTRONIC SCIENCE)
SANPO
10-17, 1-CHOME, HAMAMATSU-CHO, MINATO-KU, TOKYO 105
JAPANESE
INDEXED: JICST

DENSHI KOGYO 6390
(ELECTRONICS INDUSTRY)
JAPANESE
INDEXED: JICST

DENSHI KOGYO GEPPO 6391
(JOURNAL OF ELECTRONIC INDUSTRY)
NIHON DENSHI KOGYO SHINKO KYOKAI (THE JAPAN ELECTRONIC INDUSTRY DEVELOPMENT ASSOCIATION)
C/O KIKAI SHINKO KAIKAN, 5-8, 3-CHOME, SHIBAKOEN, MINATO-KU, TOKYO 105
JAPANESE
INDEXED: JICST

DENSHI TENBO 6392
(ELECTRONICS REVIEW)
SEIBUNDO SHINKOSHA
1-5, KANDA, NISHIKI-CHO, CHIYODA-KU, TOKYO 101
JAPANESE

DENSHI TSUSHIN GAKKAI ANTENA DENPA KENKYUKAI SHIRYO 6393
(PAPER OF TECHNICAL GROUP ON ANTENNAS AND PROPAGATION)
DENSHI TSUSHIN GAKKAI (INSTITUTE OF ELECTRONICS AND COMMUNICATION ENGINEERS OF JAPAN)
5-8, 3-CHOME, SHIBA KOEN, MINATO-KU, TOKYO 105
JAPANESE
INDEXED: JICST

DENSHI TSUSHIN GAKKAI CHOONPA KENKYUKAI SHIRYO 6394
(PAPER OF TECHNICAL GROUP ON ULTRASONICS)
DENSHI TSUSHIN GAKKAI (INSTITUTE OF ELECTRONICS AND COMMUNICATION ENGINEERS OF JAPAN)
5-8, 3-CHOME, SHIBA KOEN, MINATO-KU, TOKYO 105
JAPANESE
INDEXED: JICST

DENSHI TSUSHIN GAKKAI DENKI ONKYO KENKYUKAI SHIRYO 6395
(PAPER OF TECHNICAL GROUP ON ELECTROACOUSTICS)
DENSHI TSUSHIN GAKKAI (INSTITUTE OF ELECTRONICS AND COMMUNICATION ENGINEERS OF JAPAN)
5-8, 3-CHOME, SHIBA KOEN, MINATO-KU, TOKYO 105
JAPANESE
INDEXED: JICST

DENSHI TSUSHIN GAKKAI DENSHI BUHIN, ZAIRYO KENKYUKAI SHIRYO 6396
(PAPER OF TECHNICAL GROUP ON COMPONENT PARTS AND MATERIALS))
DENSHI TSUSHIN GAKKAI (INSTITUTE OF ELECTRONICS AND COMMUNICATION ENGINEERS OF JAPAN)
5-8, 3-CHOME, SHIBA KOEN, MINATO-KU, TOKYO 105
JAPANESE

DENSHI TSUSHIN GAKKAI DENSHI KEISANKI KENKYUKAI SHIRYO 6397
(PAPER OF TECHNICAL GROUP ON ELECTRONIC COMPUTERS)
DENSHI TSUSHIN GAKKAI (INSTITUTE OF ELECTRONICS AND COMMUNICATION ENGINEERS OF JAPAN)
5-8, 3-CHOME, SHIBA KOEN, MINATO-KU, TOKYO 105
JAPANESE
INDEXED: JICST

DENSHI TSUSHIN GAKKAI DENSHI SOCHI KENKYUKAI SHIRYO 6398
(PAPER OF TECHNICAL GROUP ON ELECTRON DEVICES)
DENSHI TSUSHIN GAKKAI (INSTITUTE OF ELECTRONICS AND COMMUNICATION ENGINEERS OF JAPAN)
5-8, 3-CHOME, SHIBA KOEN, MINATO-KU, TOKYO 105
JAPANESE
INDEXED: JICST

DENSHI TSUSHIN GAKKAI GAZO KOGAKU KENKYUKAI SHIRYO 6399
(PAPER OF TECHNICAL GROUP ON IMAGE ENGINEERING)
DENSHI TSUSHIN GAKKAI (INSTITUTE OF ELECTRONICS AND COMMUNICATION ENGINEERS OF JAPAN)
5-8, 3-CHOME, SHIBA KOEN, MINATO-KU, TOKYO 105
JAPANESE

DENSHI TSUSHIN GAKKAI GIJUTSU KENKYU HOKOKU 6400
(TECHNOLOGY RESEARCH REPORT OF THE INSTITUTE OF ELECTRONICS AND COMMUNICATION ENGINEERS OF JAPAN)
DENSHI TSUSHIN GAKKAI (INSTITUTE OF ELECTRONICS AND COMMUNICATION ENGINEERS OF JAPAN)
5-8, 3-CHOME, SHIBA KOEN, MINATO-KU, TOKYO
JAPANESE
INDEXED: CHE, JICST

DENSHI TSUSHIN GAKKAI HANDOTAI TORANJISUTA KENKYUKAI SHIRYO 6401
(PAPER OF TECHNICAL GROUP ON SEMICONDUCTORS AND SEMICONDUCTOR DEVICES)
DENSHI TSUSHIN GAKKAI (INSTITUTE OF ELECTRONICS AND COMMUNICATION ENGINEERS OF JAPAN)
5-8, 3-CHOME, SHIBA KOEN, MINATO-KU, TOKYO 105
JAPANESE
INDEXED: JICST

DENSHI TSUSHIN GAKKAI HIKARI. RYOSHI EREKUTORONIKUSU KENKYUKAI SHIRYO 6402
(PAPER OF TECHNICAL GROUP ON OPTICAL AND QUANTUM ELECTRONICS)
DENSHI TSUSHIN GAKKAI (INSTITUTE OF ELECTRONICS AND COMMUNICATION ENGINEERS OF JAPAN)
5-8, 3-CHOME, SHIBA KOEN, MINATO-KU, TOKYO 105
JAPANESE
INDEXED: JICST

DENSHI TSUSHIN GAKKAI HISENKEI MONDAI KENKYUKAI SHIRYO 6403
(PAPER OF TECHNICAL GROUP ON NONLINEAR PROBLEMS)
DENSHI TSUSHIN GAKKAI (INSTITUTE OF ELECTRONICS AND COMMUNICATION ENGINEERS OF JAPAN)
5-8, 3-CHOME, SHIBA KOEN, MINATO-KU, TOKYO 105
JAPANESE
INDEXED: JICST

DENSHI TSUSHIN GAKKAI IYO DENSHI SEITAI KOGAKU KENKYUKAI SHIRYO 6404
(PAPER OF TECHNICAL GROUP ON MEDICAL ELECTRONICS AND BIOLOGICAL ENGINEERING)
DENSHI TSUSHIN GAKKAI (INSTITUTE OF ELECTRONICS AND COMMUNICATION ENGINEERS OF JAPAN)
5-8, 3-CHOME, SHIBA KOEN, MINATO-KU, TOKYO 105
JAPANESE
INDEXED: JICST

DENSHI TSUSHIN GAKKAI JIKI KIROKU KENKYUKAI SHIRYO 6405
(PAPER OF TECHNICAL GROUP ON MAGNETIC RECORDING)
DENSHI TSUSHIN GAKKAI (INSTITUTE OF ELECTRONICS AND COMMUNICATION ENGINEERS OF JAPAN)
5-8, 3-CHOME, SHIBA KOEN, MINATO-KU, TOKYO 105
JAPANESE
INDEXED: JICST

DENSHI TSUSHIN GAKKAI KAIRO TO SHISUTEMU RIRON KENKYUKAI SHIRYO 6406
(PAPER OF TECHNICAL GROUP ON CIRCUIT AND SYSTEM THEORY)
DENSHI TSUSHIN GAKKAI (INSTITUTE OF ELECTRONICS AND COMMUNICATION ENGINEERS OF JAPAN)
5-8, 3-CHOME, SHIBA KOEN, MINATO-KU, TOKYO 105
JAPANESE
INDEXED: JICST

DENSHI TSUSHIN GAKKAI KIKO BUHIN KENKYUKAI SHIRYO 6407
(PAPER OF TECHNICAL GROUP ON ELECTROMECHANICAL COMPONENTS)
DENSHI TSUSHIN GAKKAI (INSTITUTE OF ELECTRONICS AND COMMUNICATION ENGINEERS OF JAPAN
5-8, 3-CHOME, SHIBA KOEN, MINATO-KU, TOKYO 105
JAPANESE
INDEXED: JICST

DENSHI TSUSHIN GAKKAI KOKAN KENKYUKAI SHIRYO 6408
(PAPER OF TECHNICAL GROUP ON SWITCHING ENGINEERING)
DENSHI TSUSHIN GAKKAI (INSTITUTE OF ELECTRONICS AND COMMUNICATION ENGINEERS OF JAPAN)
5-8, 3-CHOME, SHIBA KOEN, MINATO-KU, TOKYO 105
JAPANESE
INDEXED: JICST

DENSHI TSUSHIN GAKKAI KYOIKU GIJUTSU KENKYUKAI SHIRYO 6409
(PAPER OF TECHNICAL EDUCATIONAL TECHNOLOGY)
DENSHI TSUSHIN GAKKAI (INSTITUTE OF ELECTRONICS AND COMMUNICATION ENGINEERS OF JAPAN)
5-8, 3-CHOME, SHIBA KOEN, MINATO-KU, TOKYO 105
JAPANESE
INDEXED: JICST

DENSHI TSUSHIN GAKKAI MAIKUROHA KENKYUKAI SHIRYO 6410
(PAPER OF TECHNICAL GROUP ON MICROWAVES)
DENSHI TSUSHIN GAKKAI (INSTITUTE OF ELECTRONICS AND COMMUNICATION ENGINEERS OF JAPAN
5-8, 3-CHOME, SHIBA KOEN, MINATO-KU, TOKYO 105
JAPANESE
INDEXED: JICST

DENSHI TSUSHIN GAKKAI OTOMATON TO GENGO KENKYUKAI SHIRYO 6411
(PAPER OF TECHNICAL GROUP ON AUTOMATA AND LANGUAGES)
DENSHI TSUSHIN GAKKAI (INSTITUTE OF ELECTRONICS AND COMMUNICATION ENGINEERS OF JAPAN)
5-8, 3-CHOME, SHIBA KOEN, MINATO-KU, TOKYO 105
JAPANESE
INDEXED: JICST

DENSHI TSUSHIN GAKKAI PATAN NINSHIKI TO GAKUSHU KENKYUKAI SHIRYO 6412
(PAPER OF TECHNICAL GROUP ON PATTERN RECOGNITION AND LEARNING)
DENSHI TSUSHIN GAKKAI (INSTITUTE OF ELECTRONICS AND COMMUNICATION ENGINEERS OF JAPAN)
5-8, 3-CHOME, SHIBA KOEN, MINATO-KU, TOKYO 105
JAPANESE
INDEXED: JICST

DENSHI TSUSHIN GAKKAI RONBUNSHI, A 6413
(TRANSACTION OF THE INSTITUTE OF ELECTRONICS AND COMMUNICATION ENGINEERS OF JAPAN, A)
DENSHI TSUSHIN GAKKAI (INSTITUTE OF ELECTRONICS AND COMMUNICATION ENGINEERS OF JAPAN)
C/O KIKAI SHINKO KAIKAN, 5-8, 3-CHOME, SHIBA KOEN, MINATO-KU, TOKYO 105
JAPANESE
INDEXED: ENG, INS, IAA, SCI, JICST

DENSHI TSUSHIN GAKKAI RONBUNSHI, B 6414
(TRANSACTIONS OF THE INSTITUTE OF ELECTRONICS AND COMMUNICATION ENGINEERS OF JAPAN, B)
DENSI TSUSHIN GAKKAI (INSTITUTE OF ELECTRONICS AND COMMUNICATION ENGINEERS OF JAPAN)
C/O KIKAI SHINKO KAIKAN, 5-8, 3-CHOME, SHIBA KOEN, MINATO-KU, TOKYO 105
JAPANESE
INDEXED: ENG, INS, IAA, SCI, JICST

DENSHI TSUSHIN GAKKAI RONBUNSHI, C 6415
(TRANSACTIONS OF THE INSTITUTE OF ELECTRONICS AND COMMUNICATION ENGINEERS OF JAPAN, C)
DENSHI TSUSHIN GAKKAI (INSTITUTE OF ELECTRONICS AND COMMUNICATION ENGINEERS OF JAPAN)
C/O KIKAI SHINKO KAIKAN, 5-8, 3-CHOME, SHIBA KOEN, MINATO-KU, TOKYO 105
JAPANESE
INDEXED: ENG, INS, IAA, SCI, JICST

DENSHI TSUSHIN GAKKAI RONBUNSHI, D 6416
(TRANSACTIONS OF THE INSTITUTE OF ELECTRONICS AND COMMUNICATION ENGINEERS OF JAPAN, D)
DENSHI TSUSHIN GAKKAI (INSTITUTE OF ELECTRONICS AND COMMUNICATION ENGINEERS OF JAPAN)
C/O KIKAI SHINKO KAIKAN, 5-8, 3-CHOME, SHIBA KOEN, MINATO-KU, TOKYO 105
JAPANESE
INDEXED: ENG, INS, IAA, SCI, JICST

DENSHI TSUSHIN GAKKAI SHIHRAISEI KENKYUKAI SHIRYO 6417
(PAPER OF TECHNICAL GROUP ON RELIABILITY)
DENSHI TSUSHIN GAKKAI (INSTITUTE OF ELECTRONICS AND COMMUNICATION ENGINEERS OF JAPAN)
5-8, 3-CHOME, SHIBA KOEN, MINATO-KU, TOKYO 105
JAPANESE
INDEXED: JICST

DENSHI TSUSHIN GAKKAI SOGO ZENKOKU TAIKAI RONBUNSHU 6418
(PAPER OF THE TELE-COMMUNICATIONS SOCIETY GENERAL ALL-JAPAN MEETING)
JAPANESE
INDEXED: JICST

DENSHI TSUSHIN GAKKAI TSUSHIN HOSHIKI KENKYUKAI SHIRYO 6419
(PAPER OF TECHNICAL GROUP ON COMMUNICATION SYSTEMS)
DENSHI TSUSHIN GAKKAI (INSTITUTE OF ELECTRONICS AND COMMUNICATION ENGINEERS OF JAPAN)
5-8, 3-CHOME, SHIBA KOEN, MINATO-KU, TOKYO 105
JAPANESE
INDEXED: JICST

DENSHI TSUSHIN GAKKAI UCHU KOKO EREKUTORONIKUSU KENKYUKAI SHIRYO 6420
(PAPER OF TECHNICAL GROUP ON SPACE, AERONAUTICAL AND NAVIGATION ELECTRONICS)
DENSHI TSUSHIN GAKKAI (INSTITUTE OF ELECTRONICS AND COMMUNICATION ENGINEERS OF JAPAN)
5-8, 3-CHOME, SHIBA KOEN, MINATO-KU, TOKYO 105
JAPANESE
INDEXED: JICST

DENSHI TSUSHIN GAKKAI ZENKOKU TAIKAI KOEN RONBUNSHU 6421
(PAPERS FROM THE ANNUAL MEETING OF THE INSTITUTE OF ELECTRONICS AND COMMUNICATION ENGINEERS OF JAPAN)
DENSHI TSUSHIN GAKKAI
C/O KIKAI SHINKO KAIKAN, 5-8, 3-CHOME, SHIBA KOEN, MINATO-KU, TOKYO 105
JAPANESE

DENSHI TSUSHIN GAKKAISHI 6422
(JOURNAL OF THE INSTITUTE OF ELECTRONICS COMMUNICATION ENGINEERS OF JAPAN)
DENSHI TSUSHIN GAKKAI (INSTITUTE OF ELECTRONICS AND COMMUNICATION ENGINEERS OF JAPAN)
C/O KIKAI SHINKO KAIKAN, 5-8, 3-CHOME, SHIBA KOEN, MINATO-KU, TOKYO 105
JAPANESE
INDEXED: APT, CHE, INS, JICST

DENSHI TSUSHIN NO GENJO TO TENBO 6423
(REVIEW OF ELECTRONICS AND COMMUNICATIONS)
KAGAKU GIJUTSU-CHO (SCIENCE AND TECHNOLOGY AGENCY)
2-1, 2-CHOME, KASUMIGASEKI, CHIYODA-KU, TOKYO 100
JAPANESE

DENSHI ZAIRYO 6424
(ELECTRONIC PARTS AND MATERIALS)
KOGYO CHOSAKAI (INDUSTRY INFORMATION CENTER)
14-7, 2-CHOME, HONGO, BUNKYO-KU, TOKYO 113
JAPANESE
INDEXED: JICST

DENSHIN DENWA 6425
(TELEGRAPH AND TELEPHONE)
NIHON DENSHIN DENWA KOSHA (NIHON TELEGRAPH & TELEPHONE PUBLIC CORPORATION)
126, YARAI-CHO, SHINJUKU-KU, TOKYO 162
JAPANESE

DENSHIN DENWA GYOMU KENKYU 6426
(TELEGRAPH AND TELEPHONE BUSINESS STUDY)
NIHON DENSHIN DENWA KOSHA GYOMUKANRI-KYOKJ
TSUSHIN BUNKA SHINKOKAI
C/O SANKEI BUILDING, 7-2, 1-CHOME, OTE-MACHI, CHIYODA-KU, TOKYO 100
JAPANESE
INDEXED: JICST

DENSHIN DENWA KEIEI GEPPO 6427
(TELEGRAPH AND TELEPHONE MANAGEMENT MONTHLY)
JAPANESE
INDEXED: JICST

DENSOKEN NYUSU 6428
(ETL NEWS)
TSUSHO SANGYO-SHO KOGYO GIJUTSU-IN DENSHI GIJUTSU SOGO KENKYUSHO (ELECTROTECHNICAL LABORATORY, AGENCY OF INDUSTRIAL SCIENCE AND TECHNOLOGY, MINISTRY OF INTERNATIONAL TRADE AND INDUSTRY)
6-1, 2-CHOME, NAGATA-CHO, CHIYODA-KU, TOKYO 100
JAPANESE
INDEXED: JICST

DEVELOPMENTS IN MULTIWIRE NETWORKS 6429
MONOGRAPHY SERIES OF THE RESEARCH INSTITUTE OF APPLIED ELECTRICITY, HOKKAIDO UNIVERSITY
JAPANESE
INDEXED: JICST

ECL TECHNICAL PUBLICATION 6430
NIHON DENSHIN DENWA KOSHA MUSASHINO DENKI TSUSHIN KENKYUSHO (MUSASHINO ELECTRICAL COMMUNICATION LABORATORY, N.T.T.)
9-11, 3-CHOME, MIDORI-CHO, MUSASHINO-SHI, TOKYO 180
ENGLISH

EISEI TSUSHIN NENPO 6431
(ANNUAL REPORT ON TELECOMMUNICATION SATELLITE)
YUSEI-SHO DAIJIN KANBO AND KOKUSAI DENSHIN DENWA K.K.
2-5, 3-CHOME, KASUMIGASEKI, CHIYODA-KU, TOKYO 100
JAPANESE

ELECTRIC POWER INDUSTRY IN JAPAN 6432
KAIGAI DENRYOKU CHOSAKAI (OVERSEAS ELECTRICAL INDUSTRY SURVEY INSTITUTE, INCORPORATED)
1-13, 1-CHOME, SHINBASHI, MINATO-KU, TOKYO 105
ENGLISH
INDEXED: JICST

ELECTRIC POWER SURVEY 6433
NIHON DENRYOKU CHOSA IINKAI (JAPAN ELECTRIC POWER SURVEY COMMITTEE)
C/O TOUDEN KYUKAN, 1-13, 1-CHOME, SHINBASHI, MINATO-KU, TOKYO 105
ENGLISH

ELECTRICAL TRADE NEWS 6434
ENGLISH

ELECTRONICS IN JAPAN 6435
JAPANESE
INDEXED: JICST

ELECTROPHOTOGRAPHY 6436
JAPANESE

ELECTROTECHNICAL JOURNAL OF JAPAN: ETJ OF JAPAN 6437
JAPANESE
INDEXED: JICST

EREKUTORONIKU SERAMIKUSU 6438
(ELECTRONIC CERAMICS)
GAKKENSHA (GAKKENSHA COMPANY, LTD.)
1-3, 3-CHOME, KOISHIKAWA, BUNKYO-KU, TOKYO 112
JAPANESE
INDEXED: JICST

EREKUTORONIKUSU 6439
(ELECTRONICS MAGAZINE)
NIHON DENSHI KIKAI KOGYOKAI GIJUTSU HON'IINKAI
OMUSHA
3-1, KANDA NISHIKI-CHO, CHIYODA-KU, TOKYO 101
JAPANESE
INDEXED: JICST

EREKUTORONIKUSU DAIJESUTO 6440
(ELECTRONICS DIGEST)
EREKUTORONIKUSU DAIJESUTO (ELECTRONICS DIGEST COMPANY)
C/O YUZANKAKU BUILDING, 6-9, 2-CHOME, FUJIMI-CHO, CHIYODA-KU, TOKYO 102
JAPANESE
INDEXED: JICST

EREKUTORONIKUSU NYUSU 6441
(ELECTRONICS NEWS)
EREKUTORONIKUSU KYOGIKAI (THE ELECTRONICS ASSOCIATION)
1-13, 1-CHOME, SHINBASHI, MINATO-KU, TOKYO 105
JAPANESE AND ENGLISH
INDEXED: JICST

FUJI ELECTRIC REVIEW 6442
FUJI DENKI SEIZO K.K. (FUJI ELECTRIC MANUFACTURING COMPANY, LIMITED)
C/O SHIN YURAKU BUILDING, 1-11, YURAKU-CHO, CHIYODA-KU, TOKYO 100
EUROPEAN
INDEXED: CHE, ENG, INS, JICST

FUJI JIHO 6443
(FUJI ELECTRIC JOURNAL)
FUJI DENKI SEIZO K.K.
1-11, YURAKU-CHO, CHIYODA-KU, TOKYO 100
JAPANESE
INDEXED: CHE, JICST

FUJIKURA DENSEN GIHO 6444
(FUJIKURA TECHNICAL REVIEW)
FUJIKURA DENSEN K.K. (THE FUJIKURA CABLE WORKS COMPANY, LTD.)
5-1, 1-CHOME, KIBA, KOTO-KU, TOKYO 135
JAPANESE, WITH ENGLISH SUMMARY
INDEXED: JICST

FUJIKURA TECHNICAL REVIEW 6445
FUJIKURA DENSEN K.K.
(FUJIKURA CABLE WORKS COMPANY, LIMITED)
2-5, 3-CHOME, KASUMIGASEKI, CHIYODA-KU, TOKYO 100
ENGLISH
INDEXED: CHE, INS

FUJITSU 6446
FUJITSU K.K. FUJITSU KENKYUSHO (FUJITSU LIMITED FUJITSU LABORATORY LIMITED)
1015, KAMIKODANAKA, NAKAHARA-KU, KAWASAKISHI 211
JAPANESE AND ENGLISH
INDEXED: ENG, MET, WAA, JICST

FUJITSU SCIENTIFIC & TECHNICAL JOURNAL 6447
FUJITSU K.K. FUJITSU KENKYUSHO (FUJITSU LIMITED, FUJITSU LABORATORY LIMITED)
1015, KAMIKODANAKA, NAKAHARA-KU, KAWASAKI-SHI 211
ENGLISH
INDEXED: CHE, INS, IAA, JICST

FURUKAWA DENKO JIHO 6448
(FURUKAWA ELECTRIC REVIEW)
FURUKAWA DENKI KOGYO K.K. (FURUKAWA ELECTRIC COMPANY, LIMITED)
6-1, 2-CHOME, MARUNOUCHI, CHIYODA-KU, TOKYO 100
JAPANESE, WITH ENGLISH SUMMARY
INDEXED: COP, INS, WAA, JICST

GAISHI REBYU 6449
(NGK REVIEW)
NIHON GAISHI K.K. (NGK INSULATOR LTD.)
2-56, SUDA-CHO, MIZUHO-KU, NAGOYA-SHI 467
JAPANESE AND ENGLISH

GAZO DENSHI GAKKAI KENKYUKAI KOEN YOKO 6450
(PREPRINTS OF PAPERS PRESENTED AT INSTITUTE OF IMAGE ELECTRONICS, ENGINEERS RESEARCH SOCIETY SYMPOSIUM)
JAPANESE
INDEXED: JICST

GAZO DENSHI GAKKAI ZENKOKU TAIKAI YOKOSHU 6451
(PREPRINTS OF THE INSTITUTE OF IMAGE ELECTRONIC ENGINEERS, ALL-JAPAN CONFERENCE)
JAPANESE
INDEXED: JICST

GAZO DENSHI GAKKAISHI 6452
(JOURNAL OF THE INSTITUTE OF IMAGE ELECTRONICS ENGINEERS OF JAPAN)
GAZO DENSHI GAKKAI
C/O SEIKEI DAIGAKU KOGAKUBU DENKI KOGAKUKA, 3-1 3-CHOME, KICHIJOJI KITA-MACHI, MUSASHINO-SHI, TOKYO 180
JAPANESE AND ENGLISH
INDEXED: JICST

GAZO GIJUTSU 6453
(IMAGE TECHNOLOGY)
EIZO HYORONSHA
3-21, KANDA NISHIKI-CHO, CHIYODA-KU, TOKYO 101
JAPANESE
INDEXED: JICST

GEKKAN ODIO 6454
(MONTHLY AUDIO)
DENPA SHINBUNSHA (RADIO NEWSPAPER COMPANY)
JAPANESE

GENKYO TO TENBO 6455
(PRESENT SITUATION AND THE VIEW)
DENGEN KAIHATSU K.K. (DEVELOPMENT OF POWER RESOURCES COMPANY, LTD.)
MARUNOUCHI, CYIYODA-KU, TOKYO 100
JAPANESE

GIJUTSU NYUSU 6456
(TECHNOLOGY NEWS)
HOKURIKU DENRYOKU KABUSHIKI KAISHA GIJUTSU KENKYUJO (RESEARCH INSTITUTE OF THE HOKURIKU POWER COMPANY, LTD.)
JAPANESE

GIJUTSU NYUSU (NIPPON SINGO K.K.) 6457
(TECHNICAL NEWS, THE NIPPON SIGNAL COMPANY, LIMITED)
NIPPON SHINGO K.K. GIJUTSU HONBU (THE NIPPON SIGNAL COMPANY, LTD.)
13-8, 1-CHOME, KIZAKI, URAWA-SHI 338
JAPANESE

GIKEN JIHO 6458
(JOURNAL OF THE TECHNICAL LABORATORY OF CHUGOKU ELECTRIC POWER COMPANY, INCORPORATED)
CHUGOKU DENRYOKU K.K. GIJUTSU KENKYUSHO
4-32, 4-CHOME, OSU, HIROSHIMA-SHI 730
JAPANESE

GIKEN NYUSU (SHIKOKU DENRYOKU GIJUTSU KENKYUSHO 6459
(NEWS OF THE TECHNICAL LABORATORY)
SHIKOKU DENRYOKU K.K. GIJUTSU KENKYUSHO
1-7, GOTO-CHO, TAKAMATSU-SHI 760
JAPANESE

GS NEWS 6460
NIHON DENCHI (JAPAN BATTERY)
JAPANESE AND ENGLISH

GS NEWS TECHNICAL REPORT 6461
NIHON DENCHI K.K. (JAPAN STORAGE BATTERY COMPANY, LIMITED)
NISHINOSHO, KISSHOIN, MINAMI-KU, KYOTO-SHI 601
JAPANESE, WITH EUROPEAN SUMMARY
INDEXED: CHE, JICST

HAM RAIFU 6462
(HAM LIFE)
DENPA SHINBUNSHA
11-15, 1-CHOME, HIGASHIGOTANDA, SHINAGAWA-KU, TOKYO 141
JAPANESE

HATSUDEN SUIRYOKU 6463
(HYDRO ELECTRIC POWER)
HATSUDEN SUIRYOKU KYOKAI (JAPANESE ASSOCIATION OF HYDRO ELECTRIC POWER)
C/O CHUGIN TORANOMON BUILDING, 8-8, 1-CHOME, NISHI SHINBASHI, MINATO-KU, TOKYO 105
JAPANESE
INDEXED: JICST

HEN'ATSUKI GIHO 6464
(TRANSFORMER ENGINEERING REVIEW)
OSAKA HEN'ATSUKI K.K. HEN'ATSUKI JIGYO-BU
1-11, 2-CHOME, TAGAWA, YODOGAWA-KU, OSAKA-SHI 532
JAPANESE AND ENGLISH

HODEN KAKO GIJUTSU 6465
(JOURNAL OF THE JAPAN SOCIETY OF ELECTRIC MACHINING)
HODEN KAKO GIJUTSU KENKYUKAI
4-12, 1-CHOME, SARUGAKU-CHO, CHIYODAKU, TOKYO 101
JAPANESE
INDEXED: JICST

HOKKAI GAKUEN DAIGAKU KOGAKUBU KENKYU HOKOKU 6466
(RESEARCH REPORT OF HOKKAI GAKUEN UNIVERSITY, ENGINEERING FACULTY)
JAPANESE
INDEXED: JICST

HOKKAIDO DAIGAKU OYO DENKI KENKYUSHO HOKOKU 6467
(BULLETIN OF THE RESEARCH INSTITUTE OF APPLIED ELECTRICITY, HOKKAIDO UNIVERSITY)
HOKKAIDO DAIGAKU OYO DENKI KENKYUSHO (RESEARCH INSTITUTE OF APPLIED ELECTRICITY, HOKKAIDO UNIVERSITY)
NISHI-6-CHOME, KITA-12-JO, KITA-KU, SAPPOROSHI 060
JAPANESE AND ENGLISH
INDEXED: CHE, INS, JICST

HOKKAIDO DENRYOKU CHOSA GEPPO 6468
(MONTHLY REPORT OF THE HOKKAIDO ELECTRIC POWER COMPANY, LTD.)
HOKKAIDO DENRYOKU K.K.
1, ODORI HIGASHI, CHUO-KU, SAPPORO-SHI 060
JAPANESE
INDEXED: JICST

HOKKAIDO DENRYOKU JUKYU GAIKYO 6469
(SURVEY OF ELECTRIC DEMAND AND SUPPLY IN HOKKAIDO)
TSUSHO SANGYO-SHO TSUSHO SANGYO-KYOKU KOEKI JIGYO-BU
C/O SAPPORO DAIICHI GODO CHOSHA NISHI-4-CHOME, KITA-3-JO, CHUO-KU, SAPPORO-SHI 060
JAPANESE

HOKKAIDO DENRYOKU K.K. GIJUTSU KENKYUSHO KENKYU HOKOKU 6470
(RESEARCH REPORT OF THE TECHNICAL LABORATORY, THE HOKKAIDO ELECTRIC POWER COMPANY, INCORPORATED)
HOKKAIDO DENRYOKU K.K. GIJUTSU KENKYUSHO (TECHNICAL LABORATORY, THE HOKKAIDO ELECTRIC POWER COMPANY, INCORPORATED)
1-2, 1-CHOME, NAEBO-CHO, HIGASHI-KU, SAPPORO-SHI 065
JAPANESE

HOKKAIDO DENRYOKU K.K. GIJUTSU KENKYUSHO NENPO 6471
(ANNUAL REPORT OF THE TECHNICAL LABORATORY, THE HOKKAIDO ELECTRIC POWER COMPANY, INCORPORATED))
HOKKAIDO DENRYOKU K.K. GIJUTSU KENKYUSHO (TECHNICAL LABORATORY, THE HOKKAIDO ELECTRIC POWER COMPANY, INCORPORATED)
1-2, 1-CHOME, NAEBO-CHO, HIGASHI-KU, SAPPOROSHI 065
JAPANESE
INDEXED: JICST

HOKKAIDO DENRYOKU KABUSHIKI KAISHA CHOSA GEPPO 6472
(MONTHLY REPORT OF THE HOKKAIDO ELECTRIC COMPANY (K.K.))
HOKKAIDO DENRYOKU K.K. KIKAKU SHITSU (HOKKAIDO ELECTRIC K.K., PLANNING OFFICE
JAPANESE

HOKUDEN GIHO (HOKURIKU DENRYOKU K.K.) 6473
(TECHNICAL REPORT OF HOKURIKU ELECTRIC K.K.)
JAPANESE
INDEXED: JICST

HOKUDEN NYUSU 6474
(HOKUDEN NEWS)
HOKURIKU DENRYOKU K.K. GIJUTSU KENKYUSHO
2-54, HISAKATA-MACHI, TOYAMA-SHI 930
JAPANESE

HOKURIKU DENRYOKU K.K. GIJUTSU KENKYUSHO GIJUTSU KENKYU HOKOKU 6475
(REPORT OF TECHNICAL RESEARCH, HOKURIKU ELECTRIC POWER COMPANY, INCORPORATED))
HOKURIKU DENRYOKU K.K. GIJUTSU KENKYUSHO (HOKURIKU ELECTRIC POWER COMPANY, INCORPORATED. TECHNICAL RESEARCH INSTITUTE)
2-54, HISAKATA-MACHI, TOYAMA-SHI 930
JAPANESE
INDEXED: JICST

HOKURIKU DENRYOKU NO GENJO 6476
(CURRENT ACTIVITIES OF THE HOKURIKU ELECTRIC POWER COMPANY, INCORPORATED)
HOKURIKU DENRYOKU K.K. (HOKURIKU ELECTRIC POWER COMPANY, INCORPORATED)
3-1, SAKURABASHI-DORI, TOYAMA-SHI 930
JAPANESE

HOKURIKU DENRYOKUSHAHO 6477
(HOKURIKU ELECTRIC POWER COMPANY REPORT)
JAPANESE

HOKURIKU NO DENKI TO KOGYO 6478
(ELECTRICITY AND INDUSTRY IN HOKURIKU DISTRICT)
NAGOYA TSUSHO SANGYO-KYOKU KOEKI JIGYO TOYAMASHIKYOKU
5-2, 2-CHOME, SANNOMARU, NAKA-KU, NAGOYASHI 460
JAPANESE

HOKUSHIN NYUSU 6479
(HOKUSHIN NEWS)
HOKUSHIN DENKI SEISAKUSHO (HOKUSHIN ELECTRIC WORKS, LTD.)
30-1, 3-CHOME, SHIMOMARUKO, OTA-KU, TOKYO 144
JAPANESE
INDEXED: JICST

HOSO BUNKA 6480
(BROADCASTING CULTURE)
JAPANESE

HOSO GIJUTSU 6481
(BROADCAST ENGINEERING)
KENROKUKAN SHUPPAN K.K. (KENROKUKAN PUBLISHING COMPANY, LTD.)
8-7, 2-CHOME, IIDABASHI, CHIYODA-KU, TOKYO 102
JAPANESE
INDEXED: JICST

HOSOGAKU KENKYU 6482
(BROADCASTING RESEARCH)
NIHON HOSO SHUPPAN KYOKAI (NHK PUBLISHERS)
JAPANESE

IBARAKI DAIGAKU KOGAKUBU DENKI KOGAKKA SOZOSEI KAIHATSU SHIRIZU 6483
(DEVELOPMENT OF CREATIVE ABILITY SERIES, SCHOOL OF ELECTRICAL ENGINEERING, DEPARTMENT OF ENGINEERING IBARAKI UNIVERSITY)
IBARAKI DAIGAKU KOGAKU-BU DENKI KOGAKKA
12-1, 4-CHOME, NAKA NARISAWA-CHO, HITACHI-SHI 316
JAPANESE, WITH ENGLISH SUMMARY
INDEXED: JICST

ITU KENKYU 6484
(INTERNATIONAL TELECOMMUNICATIONS UNION RESEARCH)
NIHON ITU KENKYU (JAPAN INTERNATIONAL TELECOMMUNICATIONS UNION)
JAPANESE AND ENGLISH

IWATSU GIHO 6485
(IWATSU TECHNICAL REPORTS)
IWASAKI TUSHINKI K.K. (IWATSU COMPANY, LTD.)
7-47, 1-CHOME, KUGAYAMA, SUGINAJI-KU, TOKYO 168
JAPANESE
INDEXED: JICST

JAPAN TELECOMMUNICATIONS REVIEW 6486
NIHON DENSHIN DENWA KOSHA (JAPAN TELEGRAPH AND TELEPHONE PUBLIC CORPORATION)
AND DENKI TSUSHIN KYOKAI
1-11, YURAKU-CHO, CHIYODA-KU, TOKYO 100
ENGLISH
INDEXED: ENG, EXC, INS, JICST

JEE. JAPAN ELECTRONIC ENGINEERING 6487
DENPA SHINBUNSHA (DENPA PUBLICATIONS INCORPORATED)
11-15, 1-CHOME, HIGASHI GOTANDA, SHINAGAWA-KU, TOKYO 141
ENGLISH
INDEXED: CHE, INS, JICST

JEI. JAPAN ELECTRONIC INDUSTRY 6488
DENPA SHINBUNSHA (DENPA PUBLICATIONS, INCORPORATED)
11-15, 1-CHOME, HIGASHIGOTANDA, SHINAGAWA-KU, TOKYO 141
ENGLISH
INDEXED: CHE, INS, JICST

JEOL R & D 6489
NIPPON DENSHI K.K. BUNSEKI KIKI JIGYO-BU (ANALYTICAL INSTRUMENT DIVISION, JEOL LTD.)
1418, NAKAGAMI-CHO, AKISHIMA-SHI, TOKYO 196
JAPANESE AND ENGLISH

JOURNAL OF INFORMATION PROCESSING 6490
ENGLISH

JOURNAL OF THE RADIO RESEARCH LABORATORY 6491
YUSEI-SHO DENPA KENKYUSHO (RADIO RESEARCH LABORATORY MINISTRY OF POSTS AND TELECOMMUNICATIONS)
2-1, 4-CHOME, NUKUI KITA-MACHI, KOGANEI-SHI, TOKYO 184
ENGLISH
INDEXED: INS, JICST

JTIF 6492
DENKI TSUSHIN SANGYO RENMEI (TELECOMMUNICATIONS PRODUCTION FEDERATION)
JAPANESE AND ENGLISH

KADEN SHINBUN 6493
(HOME ELECTRONIC APPLIANCE NEWS)
KADEN SHINBUNSHA
7-25, 2-CHOME, HAKATAEKI HIGASHI, HAKATA-KU, FUKUOKA-SHI 812
JAPANESE

KAGAKU GIJUTSU BUNKEN SOKUHO, DENKI KOGAKU-HEN 6494
(CURRENT BIBLIOGRAPHY ON SCIENCE AND TECHNOLOGY, SERIES ELECTRICAL ENGINEERING)
NIHON KAGAKU GIJUTSU JOHO SENTA (JAPAN INFORMATION CENTER OF SCIENCE AND TECHNOLOGY)
1-1, 2-CHOME, NAGATA-CHO, CHIYODA-KU, TOKYO 100
JAPANESE
INDEXED: JICST

KAIGAI DENKI JIGYO TOKEI 6495
(OVERSEAS ELECTRIC POWER INDUSTRY STATISTICS)
KAIGAI DENRYOKU CHOSAKAI (OVERSEAS ELECTRICAL INDUSTRY SURVEY INSTITUTE, INCORPORATED)
1-13, 1-CHOME, SHINBASHI, MINATO-KU, TOKYO 105
JAPANESE

KAIGAI DENKI TSUSHIN 6496
(OVERSEAS TELECOMMUNICATIONS JOURNAL)
DENKI TSUSHIN SOGO KENKYUSHO (RESEARCH INSTITUTE OF TELECOMMUNICATIONS AND ECONOMICS)
6-19, 1-CHOME, AZABUDAI, MINATO-KU, TOKYO 106
JAPANESE
INDEXED: JICST

KAIGAI DENRYOKU JOHO 6497
(OVERSEAS ELECTRIC POWER INFORMATION)
KAIGAI DENRKYOKU CHOSAKAI (OVERSEAS ELECTRICAL INDUSTRY SURVEY INSTITUTE, INCORPORATED)
C/O TODEN KYUKAN, 1-13, 1-CHOME, SHINBASHI, MINATO-KU, TOKYO 105
JAPANESE
INDEXED: JICST

KANDENCHI 6498
(DRY BATTERY)
NIHON KANDENCHI KOGYOKAI (JAPAN DRY BATTERY INDUSTRY ASSOCIATION)
C/O DAI-9, MORI-BUILDING, 1-3, SHIBA ATAGO-CHO, MINATO-KU, TOKYO 105
JAPANESE AND ENGLISH
INDEXED: JICST

KANSAI DENRYOKU K.K. KENSETSU-BU GIJUTSU KENKYU SHIRYO 6499
(TECHNICAL DOCUMENT OF CONSTRUCTION DEPARTMENT, THE KANSAI ELECTRIC POWER COMPANY, INCORPORATED)
KANSAI DENRYOKU K.K. KENSETSU-BU (CONSTRUCTION DEPARTMENT, THE KANSAI ELECTRIC POWER COMPANY, INCORPORATED)
3-5, NAKANOSHIMA, KITA-KU, OSAKA-SHI 530
JAPANESE

KANSAI DENRYOKU K.K. SOKEN HOKOKU 6500
(TECHNICAL RESEARCH CENTER OF THE KANSAI ELECTRIC POWER COMPANY, INCORPORATED, REPORT)
KANSAI DENRYOKU K.K. SOGO GIJUTSU KENKYUSHO (TECHNICAL RESEARCH CENTER OF THE KANSAI ELECTRIC POWER COMPANY, INCORPORATED)
1, ICHINOTSUBO, NYAKOJI, AMAGASAKI-SHI 661
JAPANESE

KANSAI DENRYOKU K.K. SOKEN NYUSU 6501
(NEWS OF THE TECHNICAL RESEARCH CENTER)
KANSAI DENRYOKU K.K. SOGO GIJUTSU KENKYUSHO (TECHNICAL RESEARCH CENTER OF THE KANSAI ELECTRIC POWER COMPANY, INCORPORATED)
1, ICHINOTSUBO, NYAKOJI, AMAGASAKI-SHI 661
JAPANESE

KARYOKU GENSIRYOKU HATSUDEN 6502
(THERMAL AND NUCLEAR POWER)
KARYOKU HATSUDEN GIJUTSU KYOKAI (SOCIETY OF THERMAL POWER ENGINEERS OF JAPAN)
4-16, AKASAKA AOYAMA KITA-MACHI, MINATOKU, TOKYO 105
JAPANESE
INDEXED: JICST

KAWASAKI DENKI GIHO 6503
(TECHNICAL REPORT OF KAWASAKI ELECTRIC)
JAPANESE
INDEXED: JICST

KIKAN TEPU SAUNDO 6504
(TAPE SOUND QUARTERLY)
STEREO SOUND
2-5, 5-CHOME, ROPPONGI, MINATO-KU TOKYO 106
JAPANESE

KINKI 6505
(KINKI)
DENKI TSUSHIN KYOSOIKAI, KINKI SHIBU (ELECTRONIC COMMUNICATION MUTUAL AID SOCIETY: KINKI CHAPTER)
JAPANESE

KINKI SHISETSU 6506
(PLANT ENGINEERING OF KINKI TELECOMMUNICATIONS)
NIHON DENSHIN DENWA KOSHA KINKI DENKI TSUSHIN-KYOKU GIJUTSU CHOSA-BU (ENGINEERING DIVISION OF KINKI TELECOMMUNICATIONS BUREAU, NIHON TELEGRAPH & TELEPHONE PUBLIC CORPORATION)
557-5, TAMABORI-CHO, HIGASHI-KU, OSAKA-SHI 540
JAPANESE
INDEXED: JICST

KOJI TO JUKEN 6507
(CONSTRUCTION AND TESTING)
DENKI SHOIN (ELECTRICITY PUBLISHING COMPANY)
JAPANESE

KOJO NO DENKA 6508
(ELECTRIFICATION)
NIKKAN DENKI TSUSHINSHA
4-12, 3-CHOME, SHINBASHI, MINATO-KU, TOKYO 105
JAPANESE

KOKASEN 6509
(OVERHEAD WIRES)
SOBOSHA
JAPANESE

KOKUSAI DENKI TSUSHIN RENGO TO NIHON 6510
(ITU AND JAPAN)
NIHON ITU KYOKAI (JAPAN-INTERNATIONAL TELECOMMUNICATION UNION ASSOCIATION)
6-19, 1-CHOME, AZABUDAI, MINATO-KU, TOKYO 106
JAPANESE

KOKUSAI DENSHIN DENWA 6511
(INTERNATIONAL TELECOMMUNICATION)
KOKUSAI DENSHIN DENWA K.K. KOHOSHITSU (KOKUSAI DENSHIN DENWA COMPANY, LTD. PUBLIC RELATIONS) AND KOKUSAI TSUSHIN BUNKA KYOKAI
C/O NIHON GASUKYOKAI BUILDING 38, SHIBA KOTOHIRA-CHO MINATO-KU, TOKYO 105
JAPANESE AND ENGLISH
INDEXED: JICST

KOKUSAI DENSHIN DENWA NENPO 6512
(ANNUAL REPORT ON INTERNATIONAL TELEGRAPH AND TELEPHONE))
KOKUSAI DENSHIN DENWA K.K.
1-5, OTE-MACHI, CHIYODA-KU, TOKYO 100
JAPANESE AND ENGLISH
INDEXED: JICST

KOKUSAI TSUSHIN NI KANSURU SHOMONDAI 6513
(VARIOUS PROBLEMS RELATED TO INTERNATIONAL COMMUNICATIONS)
KOKUSAI DENSHIN DENWA KABUSHIKI KAISHA (JAPAN INTERNATIONAL TELECOMMUNICATION UNION ASSOCIATION) (ITU)
JAPANESE

KOKUSAI TSUSHIN NO KENKYU 6514
(KDD TECHNICAL JOURNAL)
KOKUSAI DENSHIN DENWA K.K. (KOKUSAI DENSHIN DENWA COMPANY, LIMITED)
2-5, 3-CHOME, KASUMIGASEKI, CHIYODA-KU, TOKYO 100
JAPANESE, WITH ENGLISH SUMMARY
INDEXED: INS, JICST

KOKUSAI TSUSHIN TOKEI GEPPO 6515
(INTERNATIONAL TELECOMMUNICATION STATISTICS MONTHLY)
KOKUSAI DENSHIN DENWA K.K. KEIEI CHOSASHITSU (MANAGEMENT RESEARCH OFFICE KOKUSAI DENSHIN DENWA COMPANY, LTD. (KDD)
8-1, 3-CHOME, KASUMIGASEKI, CHIYODA-KU, TOKYO 100
JAPANESE

KOMYUNIKESHON KENKYU 6516
(COMMUNICATIONS RESEARCH)
JOCHI DAIGAKU KOMYUNIKESHON GAKKAI (SOPHIA UNIVERSITY COMMUNICATIONS ASSOCIATION)
JAPANESE

KONDENSA HYORON 6517
(CONDENSER REVIEW)
KONDENSA KENKYUKAI
C/O DENSHI KIKAI KOGYOKAI, 2-2, 3-CHOME, MARUNOUCHI, CHIYODA-KU, TOKYO 100
JAPANESE
INDEXED: JICST

KONPUTOPIA 6518
(COMPUTOPIA)
JAPANESE
INDEXED: JICST

KUMAMOTO DENPA KOGYO KOTO SENMON GAKKO KENKYU KIYO 6519
(KUMAMOTO ELECTRICAL INDUSTRY TECHNICAL SCHOOL, RESEARCH BULLETIN)
JAPANESE

KUNREN KENKYU 6520
(TRAINING STUDY)
CHUO DENKI TSUSHIN GAKUEN (CENTRAL TELECOMMUNICATION SCHOOL)
1-44, IRIMA-CHO, CHOFU-SHI, TOKYO 182
JAPANESE
INDEXED: JICST

KUOTARI TEKUNIKARU REPORTO 6521
(QUARTERLY TECHNICAL REPORT OF DEVELOPMENT AND ENGINEERING DEPARTMENT, CLARION COMPANY, LIMITED)
KURARION K.K. KAIHATSU SEKKEI-BU (DEVELOPMENT & ENGINEERING DEPARTMENT, CLARION COMPANY, LTD.)
50, KAMITODA, TODA-SHI 335
JAPANESE

KURASHI NO DENKA 6522
(ELECTRIFIED LIFE)
KURASHI NO DENKA SHUPPANSHA
38-6, 5-CHOME, CHUO, NAKANO-KU, TOKYO 164
JAPANESE

KYOSAN SAKYURA 6523
(KYOSAN CIRCULAR)
KYOSAN SEISAKUJO
2-29, HEIAN-CHO, TSURUMI-KU, YOKOHAMA-SHI 230
JAPANESE

KYOWA GIHO 6524
(KYOWA ENGINEERING NEWS)
KYOWA DENGYO (KYOWA ELECTRONIC INSTRUMENTS COMPANY, LTD.)
5-1, 3-CHOME, CHOFUGAOKA, CHOFU-SHI, TOKYO 182
JAPANESE
INDEXED: JICST

KYU-DENRYOKU KAISHA DENKI JIGYO NENPO 6525
(ANNUAL REPORT OF ACTIVITIES OF THE NINE ELECTRIC POWER COMPANIES)
DENKI JIGYO RENGOKAI
9-4, 1-CHOME, OTE-MACHI, CHIYODA-KU, TOKYO 100
JAPANESE

KYU-DENRYOKU KAISHA HATSU-JUDEN JISSEKI, NENRYO JISSEKI SOKUHO 6526
(PRELIMINARY REPORT OF GENERATION AND PURCHASE OF ELECTRICITY, AND FUEL CONSUMPTION BY THE NINE ELECTRICAL POWER COMPANIES IN JAPAN)
C/O KEIDANREN BUILDING, 9-4, 1-CHOME, OTE-MACHI, CHIYODA-KU, TOKYO 100
JAPANESE

KYU-DENRYOKU KAISHA JUYO JISSEKI 6527
(ENERGY SALES BY THE NINE ELECTRIC POWER COMPANIES IN JAPAN)
DENKI JIGYO RENGOKAI
C/O KEIDANREN BUILDING, 9-4, 1-CHOME, OTE-MACHI, CHIYODA-KU, TOKYO 100
JAPANESE

KYUSHU DEN-DEN 6528
(NEWS OF THE KYUSHU TELECOMMUNICATION BUREAU)
NIHON DENSHIN DENWA KOSHA KYUSHU DENKI DENSHINKYOKU (DIRECTORS OFFICE, KYUSHU TELECOMMUNICATION BUREAU, NIHON TELEGRAPH AND TELEPHONE CORPORATION)
AND DENKI TSUSHIN KYOKAI KYUSHU SHIBU
3-1, SAKURA-MACHI, KUMAMOTO-SHI 860
JAPANESE

KYUSHU DENRYOKU K.K. SOGO KENKYUSHO KENKYU HOKOKUSHO 6529
(RESEARCH REPORTS (RESEARCH DEPARTMENT, THE KYUSHU ELECTRIC POWER COMPANY, INCORPORATRED))
KYUSHU DENRYOKU K.K. SOGO KENKYUSHO (RESEARCH DEPARTMENT, THE KYUSHU ELECTRIC POWER COMPANY, INCORPORATED)
497-1, AIZO, SHIOBARU, MINAMI-KU, FUKUOKA-SHI 815
JAPANESE

KYUSHU DENRYOKU K.K. SOGO KENKYUSHO KENKYU KIHO 6530
(REPORT OF THE RESEARCH DEPARTMENT, KYUSHU ELECTRIC POWER COMPANY INCORPORATED)
KYUSHU DENRYOKU K.K. SOGO KENKYUSHO (KYUSHU ELECTRIC POWER COMPANY, INCORPORATED, RESEARCH DEPARTMENT)
497-1, AIZO, SHIOBARU, MINAMI-KU, FUKUOKA-SHI 815
JAPANESE
INDEXED: JICST

KYUSHU DENRYOKU K.K. ZENSHA GIJUTSU KENKYU HAPPYOKAI YOKOSHU 6531
(KYUSHU ELECTRIC POWER K.K. ALL-COMPANY TECHNOLOGICAL RESEARCH REPORT PREPRINTS)
JAPANESE
INDEXED: JICST

MATSUSHITA DENKO GIHO 6532
(MATSUSHITA ELECTRIC WORKS TECHNICAL REPORT)
MATSUSHITA DENKO K.K. GIJUTSUSOMU-BU (MATSUSHITA ELECTRIC WORKS LIMITED TECHNICAL SERVICES & INTERNATIONAL)
1048, KADOMA, KADOMA-SHI 571
JAPANESE, WITH ENGLISH SUMMARY
INDEXED: CHE, JICST

MATSUSHITA DENSHI 6533
(MATSUSHITA ELECTRONICS)
MATSUSHITA DENSHI KOGYO K.K. (MATSUSHITA ELECTRONICS CORPORATION)
1-1, SAIWAI-CHO, TAKATSUKI-SHI 569
JAPANESE

MEIDEN JIHO 6534
(MEIDENSHA REVIEW)
MEIDENSHA (MEIDENSHA ELECTRIC MANUFACTURING COMPANY, LTD.)
C/O SHIN OTEMACHI BUILDING, 2-1, 2-CHOME, OTE-MACHI, CHIYODA-KU, TOKYO 100
JAPANESE
INDEXED: JICST

MEIDENSHA REVIEW 6535
MEIDENSHA (MEIDENSHA ELECTRIC MANUFACTURING COMPANY, LIMITED)
2-1, 2-CHOME, OTE-MACHI, CHIYODA-KU, TOKYO 100
ENGLISH
INDEXED: CHE, INS

MITSUBISHI DENKI CHUKEN SEIKEN NYUSU 6536
(MITSUBISHI DENKI LABORATORY NEWS)
MITSUBISHI DENKI K.K. CHUO KENKYUSHO (CENTRAL RESEARCH LABORATORY, MITSUBISHI ELECTRIC CORPORATION)
80, NAKANO, MINAMI SHIMIZU, AMAGASAKI-SHI 661
JAPANESE
INDEXED: JICST

MITSUBISHI DENKI GIHO 6537
(MITSUBISHI DENKI TECHNICAL REVIEW)
MITSUBISHI DENKI GIHOSHA
C/O MITSUBISHI DENKI BUILDING, 2-3, 2-CHOME, MARUNOUCHI, CHIYODA-KU, TOKYO 100
JAPANESE, WITH EUROPEAN SUMMARY
INDEXED: CHE, JICST

MITSUBISHI DENKI LABORATORY REPORTS 6538
MITSUBISHI DENKI K.K. CHUO KENKYUSHO (CENTRAL RESEARCH LABORATORY, MITSUBISHI ELECTRIC CORPORATION)
80, NAKANO, MINAMI SHIMIZU, AMAGASAKI-SHI 661
ENGLISH
INDEXED: CHE, COP, INS, JICST

MITSUBISHI ELECTRIC ENGINEER 6539
MITSUBISHI DENKI K.K. (MITSUBISHI ELECTRIC CORPORATION)
2-3, 2-CHOME, MARUNOUCHI, CHIYODA-KU, TOKYO 100
ENGLISH
INDEXED: INS

MOBIRU HAMU 6540
(MOBILE HAM)
DENPA JIKKENSHA
C/O KINOKUNIYA BUILDING, 4-2, 1-CHOME, EBISU-MINAMI, SHIBUYA-KU, TOKYO 150
JAPANESE

MOKEI TO RAJIA 6541
(MODELS AND RADIO)
JAPANESE

MONOGRAPH SERIES OF THE RESEARCH INSTITUTE OF APPLIED ELECTRICITY, HOKKAIDO UNIVERSITY 6542
HOKKAIDO DAIGAKU OYO DENKI KENKYUSHO (RESEARCH INSTITUTE OF APPLIED ELECTRICITY, HOKKAIDO UNIVERSITY)
NISHI-6-CHOME, KITA-12-JO, KITA-KU, SAPPOROSHI 060
ENGLISH
INDEXED: CHE, ENG, INS, JICST

MUSEN TO JIKKEN 6543
(RADIO EXPERIMENTER'S MAGAZINE)
SEIBUNDO SHINKOSHA
1-5, KANDA NISHIKI-CHO, CHIYODA-KU, TOKYO 101
JAPANESE
INDEXED: JICST

NASHONARU RINEI 6544
(NATIONAL RINEI)
MATSUSHITA DENKI SANGYO (MATSUSHITA ELECTRIC INDUSTRY, LIMITED)
JAPANESE

NATIONAL TECHNICAL REPORT 6545
MATSUSHITA DENKI SANGYO K.K. (MATSUSHITA ELECTRIC INDUSTRY COMPANY, LIMITED)
1006, KADOMA, KADOMA-SHI 571
JAPANESE, WITH ENGLISH SUMMARY
INDEXED: CHE, ENG, INS, JICST

NEC MAGAJIN 6546
(NEC MAGAZINE)
NIHON DENKI NEC OKAYKUSAMA SODAN SHITSU (JAPAN ELECTRIC CUSTOMER CONSULTATION OFFICE)
JAPANESE AND ENGLISH

NEC NEWS 6547
NIPPON DENKI K.K. (NIPPON ELECTRIC COMPANY, LTD.)
33-1, 5-CHOME, SHIBA, MINATO-KU, TOKYO 108
ENGLISH

NEC RESEARCH & DEVELOPMENT 6548
NIPPON DENKI K.K. (NIPPON ELECTRIC COMPANY, LIMITED)
C/O NEC BUILDING, 33-1, 5-CHOME, SHIBA, MINATO-KU, TOKYO 108
ENGLISH
INDEXED: MAT, JICST

NHK GIJUTSU KENKYU 6549
(NHK TECHNICAL JOURNAL)
NIPPON HOSO KYOKAI SOGO GIJUTSU KENKYUSHO (TECHNICAL RESEARCH LABORATORY, JAPAN BROADCASTING CORPORATION) AND NIPPON HOSO SHUPPAN KYOKAI (JAPAN BROADCAST PUBLISHING COMPANY, LIMITED)
41-1, UDAGAWA-CHO, SHIBUYA-KU, TOKYO 150
JAPANESE, WITH ENGLISH SUMMARY
INDEXED: CHE, INS, JICST

NHK GIKEN GEPPO 6550
(NHK TECHNICAL REPORT)
NIPPON HOSO KYOKAI SOGO GIJUTSU KENKYUSHO (TECHNICAL RESEARCH LABORATORY, JAPAN BROADCASTING CORPORATION) AND NIPPON HOSO SHUPPAN KYOKAI (JAPAN BROADCASTING PUBLISHING COMPANY, LTD.)
41-1, UDAGAWA-CHO, SHIBUYA-KU, TOKYO 150
JAPANESE
INDEXED: JICST

NHK HOSO HAKUBUTSUKAN DAYORI 6551
(NHK BROADCAST MUSEUM NEWS)
NHK HOSO HAKUBUTSUKAN (NHK BROADCAST MUSEUM)
1-10, SHIBA ATAGO-CHO, MINATO-KU, TOKYO 105
JAPANESE
INDEXED: JICST

NHK HOSO KAGAKU KISO KENKYUSHO HOKOKU 6552
(BULLETIN OF THE NHK BROADCASTING SCIENCE RESEARCH LABORATORY)
NIPPON HOSO KYOKAI HOSO KAGAKU KISO KENKYUSHO (BROADCASTING SCIENCE RESEARCH LABORATORY, JAPAN BROADCASTING CORPORATION)
10-11, 1-CHOME, KINUTA, SETAGAYA-KU, TOKYO 157
ENGLISH
INDEXED: JICST

NHK LABORATORIES NOTE 6553
NIPPON HOSO KYOKAI SOGO GIJUTSU KENKYUSHO (TECHNICAL RESEARCH LABORATORY, JAPAN BROADCASTING CORPORATION)
10-11, 1-CHOME, KINUTA, SETAGAYA-KU, TOKYO 157
ENGLISH
INDEXED: INS, JICST

NHK TECHNICAL MONOGRAPH 6554
NIPPON HOSO KYOKAI SOGO GIJUTSU KENKYUSHO (TECHNICAL RESEARCH LABORATORY, JAPAN BROADCASTING CORPORATION)
10-11, 1-CHOME, KINUTA, SETAGAYA-KU, TOKYO 157
ENGLISH
INDEXED: CHE, INS, JICST

NIHON DENKI KYOKAI KENKYUKAI KOENKAI RONBUNSHU 6555
(COLLECTION OF THESE ON TECHNOLOGIES IN ELECTRIC UTILITIES)
NIHON DENKI KYOKAI
1-3, YURAKU-CHO, CHIYODA-KU, TOKYO 100
JAPANESE
INDEXED: JICST

NIHON DENRYOKU CHOSA HOKOKUSHO 6556
(ELECTRIC POWER SURVEY, JAPANESE EDITION)
NIHON DENRYOKU CHOSA IINKAI
1-13, 1-CHOME, SHINBASHI, MINATO-KU, TOKYO 105
JAPANESE

NIHON DENSHI NYUSU 6557
(NIHON DENSHI NEWS)
NIHON DENSHI K.K. (JEOL LIMITED)
1418, NAKAGAMI-CHO, AKISHIMA-SHI, TOKYO 196
JAPANESE
INDEXED: CHE

NIHON DENSHI ZAIRYO GIJUTSU KYOKAI KAIHO 6558
(BULLETIN OF THE JAPAN ELECTRIC MATERIALS SOCIETY)
NIHON DENSHI ZAIRYO GIJUTSU KYOKAI (JAPAN ELECTRIC MATERIALS SOCIETY)
C/O MIYATA BUILDING, 1-38, YOYOGI, SHIBUYA-KU, TOKYO 151
JAPANESE

NIHON MUSEN GIHO 6559
(JRC REVIEW)
NIHON MUSEN K.K. (JAPAN RADIO COMPANY, LTD.)
C/O DAI-5 MORI BUILDING, 25, SHIBA NISHIKUBO SAKURAGAWA-CHO, MINATO-KU, TOKYO 105
JAPANESE, WITH ENGLISH SUMMARY
INDEXED: JICST

NIHON NO DENSHI KOGYO 6560
(ELECTRONICS INDUSTRY IN JAPAN)
DENSHI KIKAI KOGYOKAI (ELECTRONIC INDUSTRIES ASSOCIATION OF JAPAN)
2-2, 3-CHOME, MARUNOUCHI, CHIYODA-KU, TOKYO 100
JAPANESE

NIHON ODIO KYOKAISHI 6561
(JOURNAL OF JAPAN AUDIO SOCIETY)
NIHON ODIO KYOKAI (JAPAN AUDIO SOCIETY)
C/O MORI BUILDING, 14-34, 1-CHOME, JINGUMAE, SHIBUYA-KU, TOKYO 150
JAPANESE

NIHON RAJIO SHINBUN 6562
(JAPAN RADIO NEWS)
NIHON RAJIO SHINBUNSHA
14-15, 2-CHOME, KOMAGOME, TOSHIMA-KU, TOKYO 170
JAPANESE

NIHON SHOMEIKA KYOKAI ZASSHI 6563
(JOURNAL OF THE JAPANESE ASSOCIATION OF ILLUMINATION ENGINEERS)
NIHON SHOMEIKA KYOKAI
23-26, 1-CHOME, HYAKUNIN-CHO, SHINJUKUKU, TOKYO 160
JAPANESE

NIKKAN DENKI TSUSHIN 6564
(DAILY TELECOMMUNICATION)
NIKKAN DENKI TSUSHINSHA
-12, 3-CHOME, SHINBASHI, MINATO-KU, TOKYO 105
JAPANESE

NIKKEI EREKUTORONIKUSU 6565
(NIKKEI ELECTRONICS)
NIKKEI MAGUROHIRUSHA (NIKKEI MC GRAW-HILL COMPANY, LTD.)
C/O NIKKEI BEKKAN, 1-2, 2-CHOME, UCHI-KANDA, CHIYODA-KU, TOKYO 101
JAPANESE

NIPPON DENKI GIHO 6566
(TECHNICAL BULLETIN OF NIPPON ELECTRIC COMPANY, LIMITED)
NIPPON DENKI K.K. (NIPPON ELECTRIC COMPANY, LTD.)
C/O NEC BUILDING, 33-1, 5-CHOME, SHIBA, MINATO-KU TOKYO 108
JAPANESE, WITH ENGLISH SUMMARY
INDEXED: JICST

NIPPON DENKI GIHO TOKUSHU SHIRIZU 6567
(ELECTRIC TECHNICAL REPORT. SPECIAL ISSUE SERIES)
NIPPON DENKI (NIPPON ELECTRIC COMPANY, LTD.)
JAPANESE

NIPPON DENSHI KENBIKYO GAKKAI KAIHO 6568
(TRANSACTIONS OF THE JAPAN ELECTRON MICROSCOPE SOCIETY)
JAPANESE

NIPPON DENSHIN DENWA KOSHA DENKI TSUSHIN KENKYUJO KEIKA SHIRYO 6569
(NIPPON TELEGRAPH AND TELEPHONE PUBLIC CORPORATION TELECOMMUNICATIONS RESEARCH INSTITUTE, DEVELOPMENT MATERIALS)
JAPANESE
INDEXED: JICST

NIPPON DENSHIN DENWA KOSHAHO 6570
(JAPAN TELEGRAPH AND TELEPHONE PUBLIC CORPORATION REPORT)
NIPPON DENSHIN DENWA KOSHA (NIPPON TELEGRAPH AND TELEPHONE PUBLIC CORPORATION)
JAPANESE

NIPPON HOSO KYOKAI GIJUTSU KENKYUJO KENKYU NENPO 6571
(RESEARCH ANNUAL OF JAPAN BROADCASTING ASSOCIATION, TECHNICAL RESEARCH INSTITUTE)
JAPANESE
INDEXED: JICST

NIPPON HOSO KYOKAI SOGO GIJUTSU KENKYUSHO HOSO KAGAKU KISO KENKYUSHO KENKYU NENPO 6572
(ANNUAL REPORT OF TECHNICAL RESEARCH LABORATROY OF NHK, BROADCASTING SCIENCE RESEARCH LABORATORY OF NHK)
10-11, 1-CHOME, KINUTA SETAGAYA-KU, TOKYO 157
JAPANESE
INDEXED: JICST

NIPPON KOKUYU TETSUDO DENSHI GIJUTSU CHOSA IINKAI SHIRYO 6573
(JAPANESE NATIONAL RAILWAYS. ELECTRONIC TECHNOLOGY SURVEY COMMITTEE. MATERIALS)
JAPANESE
INDEXED: JICST

NISSHIN DENKI GIHO 6574
(NISSHIN ELECTRIC REVIEW)
NISSHIN DENKI K.K. (NISSHIN ELECTRIC COMPANY, LTD.)
47, UMEZU TAKASE-CHO, UKYO-KU, KYOTO-SHI 615
JAPANESE AND ENGLISH
INDEXED: JICST

NITTO GIHO 6575
(NITTO TECHNICAL REPORT)
NITTO DENKI KOGYO K.K. GIJUTSU KENKYUSHO (TECHNICAL RESEARCH INSTITUTE, NITTO ELECTRIC INDUSTRIAL COMPANY, LTD.)
1-2, 1-CHOME, SHIMOHOZUMI, IBARAKI-SHI 567
JAPANESE

NITTSUKO GIHO 6576
(TECHNICAL REPORT OF NITTSUKO LIMITED)
NIHON TSUSHIN KOGYO K.K.
260, KITAMIKATA, TAKATSU-KU, KAWASAKI-SHI 213
JAPANESE AND ENGLISH

NODEN KENKYUJO NENPO 6577
(ANNUAL OF AGRICULTURAL & ELECTRICAL RESEARCH INSTITUTE)
JAPANESE
INDEXED: JICST

NYU DENKA 6578
(WEEKLY NEWS OF ELECTRIC GOODS)
DENTSU KEIZAI K.K.
2-1, 2-CHOME, NISHISHINBASHI, MINATO-KU TOKYO 105
JAPANESE

ODIO SENKA 6579
(AUDIO MAGAZINE)
ONGEN SHUPPAN
15-6, 3-CHOME, YUSHIMA, BUNKYO-KU, TOKYO 113
JAPANESE

OEP; OFFICE EQUIPMENT & PRODUCTS 6580
DENPA SHINBUNSHA DENPA PUBLICATIONS
11-15, 1-CHOME HIGASHIGOTANDA, SHINAGAWA-KU, TOKYO 141
JAPANESE AND ENGLISH

OFFICE EQUIPMENT & PRODUCTS 6581
DENPA SHINBUNSHA (DENPA PUBLICATIONS INCORPORATED)
11-15, 1-CHOME HIGASHIGOTANDA, SHINAGAWA-KU, TOKYO 141
ENGLISH

OHM 6582
OMUSHA
3-1, KANDA NISHIKI-CHO, CHIYODA-KU, TOKYO 101
JAPANESE
INDEXED: JICST

OHM BULLETIN 6583
OHN SHA (OHM COMPANY)
JAPANESE AND ENGLISH

OKI DENKI JIHO 6584
(OKI REVIEW)
OKI DENKI KOGYO K.K. (OKI ELECTRIC INDUSTRY COMPANY, LTD.)
C/O HONTEN 3-GO ANNEX, 10-3, 4-CHOME, SHIBAURA, MINATO-KU, TOKYO 108
JAPANESE
INDEXED: JICST

OMRON TECHNICS 6585
TATEISHI DENKI K.K. NAGAOKA JIGYOSHO GIJUTSU JOHO SABISU-BU (R&D SERVICE GROUP TECHNICAL INFORMATION SERVICE DEPARTMENT, TATEISHI ELECTRONICS COMPANY AND TATEISHI DENKI K.K. HANBAI PR SENTA)
10, HANAZONO TSUCHIDO-CHO, UKYO-KU, KYOTO SHI 616
JAPANESE, WITH ENGLISH SUMMARY
INDEXED: CHE, INS, JICST

ONKYO KENKYUSHO HOKOKU 6586
(REPORT OF ACOUSTIC RESEARCH LABORATROY, MATSUSHITA ELECTRIC INDUSTRY COMPANY, LIMITED)
MATSUSHITA DENKI SANGYO K.K. ONKYO KENKYUSHO (ACOUSTIC RESEARCH LABORATORY, MATSUSHITA ELECTRIC INDUSTRY COMPANY, LTD.)
1006, KADOMA, KADOMA-SHI 571
JAPANESE

ORIJIN 6587
(ORIGIN TECHNICAL JOURNAL)
ORIJIN DENKI K.K.
18-1, 1-CHOME, TAKADA, TOSHIMA-KU, TOKYO 171
JAPANESE, WITH ENGLISH SUMMARY
INDEXED: CHE, JICST

OSAKA HEN'ATSUKI 6588
(OSAKA TRANSFORMER REVIEW)
OSAKA HEN'ATSUKI K.K. (OSAKA TRANSFORMER COMPANY, LTD.)
1-11, 2-CHOME, TAGAWA, HIGASHIYODOGAWAKU, OSAKA-SHI 532
JAPANESE
INDEXED: JICST

OTOSURAIDO 6589
(JYANARU JOURNAL)
OTOSURAIDO JYANARU
JAPANESE

PREPRINTS ON SYMPOSIUM. RESEARCH INSTITUTE OF ELECTRICAL COMMUNICATION, TOHOKU UNIVERSITY 6590
JAPANESE
INDEXED: JICST

PRESENTATION 6591
JAPANESE
INDEXED: JICST

RADIO AND SPACE DATA 6592
JAPANESE

RAJIKON GIJUTSU 6593
(RADIO CONTROL TECHNIQUE)
DENPA JIKKENSHA
C/O KINOKUNIYA BUILDING, 4-2, 1-CHOME, EBISU 1-CHOME, EBISU MINAMI, SHIBUYA-KU, TOKYO 150
JAPANESE

RAJIO DENKA SHINBUN 6594
(ELECTRICAL AND RADIO NEWS)
RAJIO DENKA SHINBUNSHA
4-3, 3-CHOME, NIHON BASHI, CHUO-KU, TOKYO 103
JAPANESE

RAJIO GIJUTSU 6595
(RADIO TECHNIQUES)
RAJIO GIJUTSUSHA
1-9, KANDA AWAJI-CHO, CHIYODA-KU, TOKYO 101
JAPANESE
INDEXED: JICST

RAJIO NO SEISAKU 6596
(AMATEUR RADIO BUILDING)
DENPA SHINBUNSHA
11-15, 1-CHOME, HIGASHIGOTANDA, SHINGAWA-KU, TOKYO 141
JAPANESE

REPORTS OF THE RESEARCH INSTITUTE OF ELECTRICAL COMMUNICATION, TOHOKU UNIVERSITY 6597
TOHOKU DAIGAKU KENKYUSHO HOKOKU (KOGAKUDENKI TSUSHIN) AND TOHOKU DAIGAKU DENKI TSUSHIN KENKYUSHO (RESEARCH INSTITUTE OF ELECTRICAL COMMUNICATION TOHOKU UNIVERSITY)
1-1, 2-CHOME, KATAHIRA, SENDAI-SHI 980
ENGLISH
INDEXED: CHE, ENG, INS, JICST

REVIEW OF THE ELECTRICAL COMMUNICATION LABORATORY 6598
NIHON DENSHIN DENWA KOSHA, MUSASHINO DENKI TSUSHIN KENKYUSHO (MUSASHINO ELECTRICAL COMMUNICATION LABORATORY, NIHON TELEGRAPH AND TELEPHONE PUBLIC CORPORATION)
9-11, 3-CHOME, MIDORI-MACHI, MUSASHINOSHI, TOKYO 180
ENGLISH
INDEXED: CHE, ENG, INS, IAA, JICST

RIEC TECHNICAL REPORT 6599
TOHOKU DAIGAKU DENKI TSUSHIN KENKYUSHO (RESEARCH INSTITUTE OF ELECTRICAL COMMUNICATION, TOHOKU UNIVERSITY)
1-1, 2-CHOME, KATAHIRA, SENDAI-SHI 980
ENGLISH

SANGYO TO DENKI 6600
(INDUSTRY AND ELECTRICITY)
KANSAI DENKI KYOKAI (KANSAI ELECTRIC ASSOCIATION)
2-9, DOJIMA NAKA, KITA-KU, OSAKA-SHI 530
JAPANESE
INDEXED: JICST

SANKEN GIHO 6601
(SANKEN TECHNICAL REPORT)
SANKEN DENKI K.K. (SANKEN ELECTRICAL COMPANY)
133, KITANO- NIIZA-SHI 352
JAPANESE, WITH ENGLISH SUMMARY
INDEXED: JICST

SAN'YO DENKI GIHO 6602
(SAN'YO TECHNICAL REVIEW)
SAN'YO DENKI K.K.
18-13, 1-CHOME, HASHIRIDANI, HIRAKATA-SHI 573
JAPANESE, WITH ENGLISH SUMMARY
INDEXED: JICST

SEIDENKI GAKKAISHI 6603
(PROCEEDINGS OF THE INSTITUTE OF ELECTROSTATICS JAPAN)
JAPANESE

SEISAN TO DENKI 6604
(PRODUCTION AND ELECTRICITY)
NIHON DENKI KYOKAI (JAPAN ELECTRIC ASSOCIATION)
1-3, YURAKU-CHO, CHIYODA-KU, TOKYO 100
JAPANESE
INDEXED: JICST

SHAPU GIHO 6605
(SHARP TECHNICAL JOURNAL)
SHAPU K.K. (SHARP ELECTRIC COMPANY, LTD.)
2613-1, ICHII HON-CHO, TENRI-SHI 632
JAPANESE
INDEXED: JICST

SHIKOKU DENKI JIGYO GAIYO 6606
(SHIKOKU ELECTRIC WORKS BULLETIN)
TSUSHO SANGYO-SHO SHIKOKU TSUSHO SANGYO-KYOKU KOEKI JIGYO-BU (PUBLIC WORKS DEPARTMENT, SHIKOKU TRADE & INDUSTRY BUREAU, MINISTRY OF INTERNATIONAL TRADE & INDUSTRY)
7-9, KAMEI-CHO, TAKAMATSU-SHI 760
JAPANESE

SHIKOKU DENRYOKU K.K. KENKYU KIHO GIJUTSU KENKYU-HEN 6607
(SEMI-ANNUAL TECHNICAL REPORT, SHIKOKU ELECTRIC POWER COMPANY, INCORPORATED)
SHIKOKU DENRYOKU K.K. GIJUTSU KENKYUSHO (TECHNICAL LABORATROY, SHIKOKU ELECTRIC POWER COMPANY, INCORPORATED)
1-7, GOTO-CHO, TAKAMATSU-SHI 760
JAPANESE
INDEXED: JICST

SHIN DENKI 6608
(NEW ELECTRICITY)
OMUSHA
3-1, KANDA NISHIKI-CHO, CHIYODA-KU, TOKYO 101
JAPANESE

SHIN NIPPON DENKI GIHO 6609
(NEW NIPPON ELECTRIC TECHNICAL REVIEW)
SHIN NIPPON DENKI K.K. (NEW NIPPON ELECTRIC COMPANY, LIMITED)
9-1, 2-CHOME, SEIRAN, OTSU-SHI 520
JAPANESE, WITH ENGLISH SUMMARY
INDEXED: CER, JICST

SHINKO DENKI 6610
(SHINKO ELECTRIC TECHNICAL REVIEW)
SHINKO DENKI K.K. (SHINKO ELECTRIC COMPANY, LTD.)
12-2, 3-CHOME, NIHONBASHI, CHUO-KU, TOKYO 103
JAPANESE AND ENGLISH
INDEXED: JICST

SHISETSU. NIPPON DENSHIN DENWA KOSHA SHISETSU-HEN 6611
(FACILITIES. NIPPON TELEGRAPH AND TELEPHONE PUBLIC CORPORATION FACILITIES EDITION)
DENKI TSUSHIN KYOKAI (ELECTRONIC COMMUNICATION ASSOCIATION)
JAPANESE

SHISUTEMA 6612
(SYSTEMA)
DAIYAMONDO SABISUSHA AND SHINKO DENKI K.K. (SHINKO ELECTRIC COMPANY, LTD.)
12-2, 3-CHOME, NIHONBASHI, CHUO-KU, TOKYO 103
JAPANESE

SHIZUOKA DAIGAKU DENSHI KOGAKU KENKYUSHO KENKYU HOKOKU 6613
(BULLETIN OF RESEARCH INSTITUTE OF ELECTRONCIS, SHIZUOKA UNIVERSITY))
SHUZUOKA DAIGAKU DENSHI KOGAKU KENKYUSHO (RESEARCH INSTITUTE OF ELECTRONICS, SHIZUOKA UNIVERSITY)
5-1, 3-CHOME, JOHOKU, HAMAMATSU-SHI 432
JAPANESE AND ENGLISH
INDEXED: CHE, JICST

SHOHO NO RAJIO 6614
(BEGINNER'S RADIO)
SEIBUNDO SHINKOSHA
1-5, KANDA NISHIKI-CHO, CHIYODA-KU, TOKYO 101
JAPANESE

SHOMEI 6615
(LIGHTING FIXTURES AND EQUIPMENT)
NIHON SHOMEI KIGU KOGYOKAI (JAPAN LIGHTING FIXTURES AND EQUIPMENT INDUSTRY ASSOCIATION)
C/O MURAKI BUILDING, 14-17, 6-CHOME, GINZA, CHUO-KU, TOKYO 104
JAPANESE

SHOMEI GAKKAI ZASSHI 6616
(JOURNAL OF THE ILLUMINATING ENGINEERING INSTITUTE OF JAPAN)
SHOMEI GAKKAI (THE ILLUMINATING ENGINEERING INSTITUTE OF JAPAN)
1-3, YURAKU-CHO, CHIYODA-KU, TOKYO 100
JAPANESE AND ENGLISH
INDEXED: JICST

SHOMEI GAKKAI ZENKOKU TAIKAI KOEN RONBUNSHU 6617
(PROCEEDINGS OF THE SYMPOSIUM OF THE ILLUMINATING ENGINEERING INSTITUTE OF JAPAN)
SHOMEI GAKKAI (THE ILLUMINATING ENGINEERING INSTITUTE OF JAPAN)
1-3, YURAKU-CHO, CHIYODA-KU, TOKYO 100
JAPANESE

SHOWA DENSEN DENRAN REBYU 6618
(SHOWA WIRE AND CABLE REVIEW)
SHOWA DENSEN DENRAN K.K. (SHOWA ELECTRIC WIRE & CABLE COMPANY, LIMITED)
1-1, 2-CHOME, ODASAKAE, KAWASAKI-KU, KAWASAKI-SHI 210
JAPANESE, WITH ENGLISH SUMMARY
INDEXED: CHE, INS, JICST

SHUKAN DENTSU TSUSHIN 6619
(TELECOMMUNICATION WEEKLY)
DENTSU KEIZAI
2-1, 2-CHOME, NISHI SHINBASHI, MINATO-KU, TOKYO 105
JAPANESE

SONY CORPORATION RESEARCH CENTER REPORTS 6620
SONY K.K. CHUO KENKYUSHO (SONY CORPORATION RESEARCH CENTER)
174, FUJITSUKA-CHO, HODOGAYA-KU, YOKOHAMA-SHI 240
JAPANESE AND ENGLISH
INDEXED: JICST

STANDARD FREQUENCY AND TIME SERVICE BULLETIN 6621
YUSEI-SHO DENPA KENKYUSHO (RADIO RESEARCH LABORATORY, MINISTRY OF POST & TELECOMMUNICATIONS)
2-1, 4-CHOME, NUKUI KITA-MACHI, KOGANEI-SHI, TOKYO 184
JAPANESE

SUMITOMO DENKI 6622
(SUMITOMO ELECTRIC REVIEW)
SUMITOMO DENKI K.K. (SUMITOMO ELECTRIC INDUSTRIES LIMITED)
5-15, KITAHAMA, HIGASHI-KU, OSAKA-SHI 541
JAPANESE, WITH ENGLISH SUMMARY
INDEXED: CHE, JICST

SUMITOMO ELECTRIC TECHNICAL REVIEW 6623
SUMITOMO DENKI KOGYO (SUMITOMO ELECTRIC INDUSTRIES, LIMITED)
C/O SHIN SUMITOMO BUILDING, 5-15, KITAHAMA, HIGASHI-KU, OSAKA-SHI 541
ENGLISH
INDEXED: BIB, COP, ENG, INS, WAA, JICST

SUMMARIES OF REPORTS OF THE ELECTROTECHNICAL LABORATORY 6624
TSUSHO SANGYO-SHO KOGYO GIJUTSU-IN DENSHI GIJUTSU SOGO KENKYUSHO (ELECTROTECHNICAL LABORATORY, AGENCY OF INDUSTRIAL SCIENCE AND TECHNOLOGY, MINISTRY OF INTERNATIONAL TRADE AND INDUSTRY)
6-1, 2-CHOME, NAGATA-CHO, CHIYODA-KU, TOKYO 100
ENGLISH
INDEXED: JICST

SUTOREN NYUSU 6625
(STRAIN NEWS)
SHINKO TSUSHIN KOGYO K.K. (SHINKO COMMUNICATION INDUSTRY COMPANY, LTD.)
12-10, 1-CHOME, SAKURAYAMA, ZUSHI-SHI 249
JAPANESE

TAKAOKA REBYU 6626
(TAKAOKA REVIEW)
TAKAOKA SEISAKUSHO (TAKAOKA ELECTRIC MANUFACTURING COMPANY, LTD.)
JAPANESE

TATSUTA DENSEN GIHO 6627
(TATSUTA WIRE AND CABLE REPORTS)
TATSUTA DENSEN K.K. (TATSUTA ELECTRIC WIRE & CABLE COMPANY, LIMITED)
3-1, 2-CHOME, IWATA-CHO, HIGASHIOSAKA-SHI 578
JAPANESE, WITH ENGLISH SUMMARY
INDEXED: CHE, JICST

TATSUTA TECHNICAL REVIEW 6628
TATSUTA DENSEN K.K. GIJUTSU KENKYUSHO (TATSUTA ELECTRIC WIRE & CABLE COMPANY, LTD.)
3-1, 2-CHOME, IWATA-CHO, HIGASHIOSAKA-SHI 578
ENGLISH

TECHNICAL REPORT (CENTRAL RESEARCH INSTITUTE OF ELECTRIC POWER INDUSTRY AGRICULTURAL LABORATORY) 6629
DENRYOKU CHUO KENKYUSHO NODEN KENKYUSHO (CENTRAL RESEARCH INSTITUTE OF ELECTRIC POWER INDUSTRY, AGRICULTURAL LABORATORY)
1646, ABIKO, ABIKO-SHI 270-11
ENGLISH

TECHNICAL REPORT (CENTRAL RESEARCH INSTITUTE OF ELECTRIC POWER INDUSTRY TECHNICAL LABORATORY NUMBER 2) 6630
DENRYOKU CHUO KENKYUSHO GIJUTSU DAI-2 KENKYUSHO (CENTRAL RESEARCH INSTITUTE OF ELECTRIC POWER INDUSTRY, TECHNICAL LABORATORY NUMBER 2)
1646, ABIKO, ABIKO-SHI 270-11
ENGLISH

TEREBI GIJUTSU 6631
(TELEVISION TECHNICS & ELECTRONICS)
DENSHI GIJUTSU SHUPPAN K.K.
26-3, 2-CHOME, HONGO, BUNKYO-KU, TOKYO 113
JAPANESE

TEREBIJON 6632
(JOURNAL OF THE INSTITUTE OF TELEVISION ENGINEERS JAPAN)
TEREBIJON GAKKAI (INSTITUTE OF TELEVISION ENGINEERS OF JAPAN)
C/O KIKAI SHINKO BUILDING, 5-8, 3-CHOME, SHIBAKOEN, MINATO-KU, TOKYO 105
JAPANESE, WITH ENGLISH SUMMARY
INDEXED: CHE, ENG, INS, JICST

TEREBIJON EIJI 6633
(TELEVISION AGE)
SHIKI SHUPPAN SHINSHA (SHIKI NEW PUBLISHING)
JAPANESE

TEREBIJON GAKKAI GAZO DENSO KENKYU IINKAI SHIRYO 6634
(SCIENCE REPORT OF THE RESEARCH COMMITTEE ON PICTURE TRANSMISSION)
TEREBIJON GAKKAI (THE INSTITUTE OF TELEVISION ENGINEERS OF JAPAN)
5-8, 3-CHOME, SHIBAKOEN, MINATO-KU, TOKYO 105
JAPANESE
INDEXED: JICST

TEREBIJON GAKKAI GAZO HYOJI KENKYU IINKAI SHIRYO 6635
(SCIENCE REPORT OF THE RESEARCH COMMITTEE ON PICTURE DISPLAY)
TEREBIJON GAKKAI (THE INSTITUTE OF TELEVISION ENGINEERS OF JAPAN)
5-8, 3-CHOME, SHIBAKOEN, MINATO-KU, TOKYO 105
JAPANESE
INDEXED: JICST

TEREBIJON GAKKAI GIJUTSU OYO KENKYU IINKAI SHIRYO 6636
(SCIENCE REPORT OF THE RESEARCH COMMITTEE ON APPLICATION OF TECHNIQUE)
TEREBIJON GAKKAI (THE INSTITUTE OF TELEVISION ENGINEERS OF JAPAN)
5-8, 3-CHOME, SHIBAKOEN, MINATO-KU, TOKYO 105
JAPANESE

TEREBIJON GAKKAI HIKARI FUIRUMU KENKYU IINKAI SHIRYO 6637
(SCIENCE REPORT OF THE RESEARCH COMMITTEE ON LIGHT AND FILM)
TEREBIJON GAKKAI (THE INSTITUTE OF TELEVISION ENGINEERS OF JAPAN)
5-8, 3-CHOME, SHIBAKOEN, MINATO-KU, TOKYO 105
JAPANESE

TEREBIJON GAKKAI KAIRO KENKYU IINKAI SHIRYO 6638
(INSTITUTE OF TELEVISION ENGINEERS SCIENCE REPORT OF THE RESEARCH COMMITTEE ON CIRCUITRY)
JAPANESE
INDEXED: JICST

TEREBIJON GAKKAI ROKUGA KENKYU IINKAI SHIRYO 6639
(SCIENCE REPORT OF THE RESEARCH COMMITTEE ON TELERECORDING)
TEREBIJON GAKKAI (THE INSTITUTE OF TELEVISION ENGINEERS OF JAPAN)
5-8, 3-CHOME, SHIBAKOEN, MINATO-KU, TOKYO 105
JAPANESE
INDEXED: JICST

TEREBIJON GAKKAI SHIKAKU KENKYU IINKAI SHIRYO 6640
(SCIENCE REPORT OF THE RESEARCH COMMITTEE ON THE SENSE OF SIGHT)
TEREBIJON GAKKAI (THE INSTITUTE OF TELEVISION ENGINEERS OF JAPAN)
5-8, 3-CHOME, SHIBAKOEN, MINATO-KU, TOKYO 105
JAPANESE
INDEXED: JICST

TEREBIJON GAKKAI TEREBIJON DENSHI SOCHI KENKYU IINKAI SHIRYO 6641
(SCIENCE REPORT OF THE RESEARCH COMMITTEE ON ELECTRONIC EQUIPMENT FOR TELEVISION)
TEREBIJON GAKKAI (THE INSTITUTE OF TELEVISION ENGINEERS OF JAPAN)
5-8, 3-CHOME, SHIBAKOEN, MINATO-KU, TOKYO 105
JAPANESE
INDEXED: JICST

TEREBIJON GAKKAI TEREBIJON FUIRUMU KENKYU IINKAI SHIRYO 6642
(INSTITUTE OF TELEVISION ENGINEERS SCIENCE REPORT OF THE RESEARCH COMMITTEE ON FILM FOR TELEVISION)
JAPANESE
INDEXED: JICST

TEREBIJON GAKKAI TEREBIJON GIJUTSU OYO KENKYU IINKAI SHIRYO 6643
(INSTITUTE OF TELEVISION ENGINEERS SCIENCE REPORT OF THE RESEARCH COMMITTEE ON APPLIED TECHNOLOGY FOR TELEVISION)
JAPANESE
INDEXED: JICST

TEREBIJON GAKKAI TEREBIJON HOSHIKI KAIRO KENKYU IINKAI SHIRYO 6644
(SCIENCE REPORT OF THE RESEARCH COMMITTEE ON TELEVISION WAY CIRCUIT)
TEREBIJON GAKKAI (THE INSTITUTE OF TELEVISION ENGINEERS OF JAPAN)
5-8, 3-CHOME, SHIBAKOEN, MINATO-KU, TOKYO 105
JAPANESE

TEREBIJON GAKKAI TEREBIJON MUSEN GIJUTSU KENKYU IINKAI SHIRYO 6645
(SCIENCE REPORT OF THE RESEARCH COMMITTEE ON RADIO ENGINEERING FOR TELEVISION)
TEREBIJON GAKKAI (THE INSTITUTE OF TELEVISION ENGINEERS OF JAPAN)
5-8, 3-CHOME, SHIBAKOEN, MINATO-KU, TOKYO 105
JAPANESE
INDEXED: JICST

TEREBIJON GAKKAI UHF HOSO GIJUTSU KENKYU IINKAI SHIRYO 6646
(INSTITUTE OF TELEVISION ENGINEERS SCIENCE REPORT OF THE RESEARCH COMMITTEE ON UHF BROADCASTING TECHNOLOGY)
JAPANESE
INDEXED: JICST

TEREBIJON GAKKAI ZENKOKU TAIKAI KOEN YOKOSHU 6647
(PREPRINTS OF THE SYMPOSIUM OF THE INSTITUTE OF TELEVISION ENGINEERS OF JAPAN)
TEREBIJON GAKKAI (THE INSTITUTE OF TELEVISION ENGINEERS OF JAPAN)
5-8, 3-CHOME, SHIBAKOEN, MINATO-KU, TOKYO 105
JAPANESE
INDEXED: JICST

TOA GIHO 6648
(TOA TECHNICAL REPORT)
TOA IYO DENSHI KABUSHIKI KAISHA GAKUJUTSU SENTA (TOA MEDICAL ELECTRONICS K.K. ACADEMIC CENTER)
JAPANESE

TODEN GURAFU 6649
(TODEN GRAPHICS)
TOKYO DENRYOKU K.K.
3-1, 1-CHOME, UCHISAIWAI-CHO, CHIYODA-KU, TOKYO 100
JAPANESE

TOHOKU DAIGAKU DENKI TSUSHIN KENKYUSHO SHUSAI SHINPOJUMU RONBUNSHU 6650
(PAPERS FROM SYMPOSIA SPONSORED BY THE RESEARCH LABORATORY OF ELECTRICAL COMMUNICATION, TOHOKU UNIVERSITY))
TOHOKU DAIGAKU DENKI TSUSHIN KENKYUSHO
1-1, 2-CHOME, KATAHIRA, SENDAI-SHI 980
JAPANESE

TOHOKU DAIGAKU DENTSU DANWAKAI KIROKU 6651
(RECORD OF ELECTRICAL AND COMMUNICATION ENGINEERING CONVERSATION, TOHOKU UNIVERSITY)
TOHOKU DAIGAKU DENKI TSUSHIN KENKYUSHO (RESEARCH LABORATORY OF ELECTRICAL COMMUNICATION, TOHOKU UNIVERSITY)
1-1, 2-CHOME, KATAHIRA, SENDAI-SHI 980
JAPANESE
INDEXED: INS, JICST

TOHOKU DENRYOKU KABUSHIKI KAISHA SOGO KENKYUJO DAYORI 6652
(LETTERS OF THE GENERAL RESEARCH INSTITUTE, TOHOKU ELECTRIC POWER LTD.)
JAPANESE

TOHOKU KOGYO DAIGAKU KIYO C: DENSHI, TSUSHIN KOGAKU-HEN 6653
(BULLETIN OF THE TOHOKU INSTITUTE OF TECHNOLOGY SECTION C: ELECTRICAL COMMUNICATION AND ELECTRONICS)
TOHOKU KOGYO DAIGAKU (TOHOKU INSTITUTE OF TECHNOLOGY)
19, KOSHIJI, NAGA-MACHI, SENDAI-SHI 982
JAPANESE, WITH ENGLISH SUMMARY
INDEXED: JICST

TOHOKU SHISETSU 6654
(PLANT ENGINEERING OF TOHOKU TELECOMMUNICATIONS))
NIHON DENSHIN DENWA KOSHA TOHOKU DENKI TSUSHIN-KYOKU (TOHOKU TELECOMMUNICATIONS BUREAU, NIHON TELEGRAPH & TELEPHONE PUBLIC CORPORATION)
10, SHIMIZUKOJI, SENDAI-SHI 980
JAPANESE

TOKYO DAIGAKU KOGAKUBU DENKI KOGAKUBU DENSHI KOGAKU IHO 6655
(BULLETIN OF THE ELECTRICAL ENGINEERING AND ELECTRONIC ENGINEERING DEPARTMENT, UNIVERSITY OF TOKYO)
TOKYO DAIGAKU KOGAKUBU DENKI KOGAKKA, DENSHI KOGAKKA (THE DEPARTMENT OF ELECTRICAL ENGINEERING AND THE ELECTRONIC ENGINEERING, UNIVERSITY OF TOKYO)
3-1, 7-CHOME, HONGO, BUNKYO-KU, TOKYO 113
JAPANESE AND ENGLISH
INDEXED: JICST

TOKYO DENKI DAIGAKU JOHO KOGAKU KENKYUJO KENKYU HOKOKU 6656
(RESEARCH REPORTS (INSTITUTE OF INFORMATION SCIENCE AND TECHNOLOGY, TOKYO ELECTRICAL ENGINEERING COLLEGE))
JAPANESE

TOKYO DENKI DAIGAKU KENKYU HOKOKU 6657
(RESEARCH REPORTS OF TOKYO ELECTRICAL ENGINEERING COLLEGE)
TOKYO DENKI DAIGAKU KOGAKUBU (TOKYO ELECTRICAL ENGINEERING COLLEGE)
2-2, KANDA NISHIKI-CHO, CHIYODA-KU, TOKYO 101
JAPANESE AND ENGLISH
INDEXED: MAT, JICST

TOKYO DENRYOKU K.K. SHAHO TODEN 6658
(JOURNAL OF TOKYO ELECTRIC POWER COMPANY, INCORPORATED)
TOKYO DENRYOKU K.K. (TOKYO ELECTRIC POWER COMPANY, INCORPORATED)
3-1, 1-CHOME, UCHISAIWAI-CHO, CHIYODA-KU, TOKYO 100
JAPANESE
INDEXED: JICST

TORANJISUTA GIJUTSU 6659
(SEMICONDUCTOR TECHNIQUE)
CQ SHUPPAN K.K.
14-2, 1-CHOME, SUGAMO, TOSHIMA-KU, TOKYO 170
JAPANESE

TORIKA 6660
(JOURNAL OF THE TOKAI RIKA COMPANY, LIMITED)
TOKAI RIKA DENKI SEISAKUSHO (TOKAI RIKA COMPANY, LTD.)
1-1, SHIMOOTAI, NISHIBIWAJIMA-MACHI, NISHIKASUGAI-GUN, AICHI 452
JAPANESE

TOSHIBA DENZAI GAIDO 6661
(TOSHIBA ELECTRIC MATERIALS GUIDE)
TOSHIBA DENZAI K.K.
1-43, 1-CHOME, SHIBAURA, MINATO-KU, TOKYO 105
JAPANESE

TOSHIBA DEUSETSU GAIDO 6662
(TOSHIBA ELECTRICAL EQUIPMENT GUIDE)
TOSHIBA DENZAI (TOSHIBA ELECTRICAL MATERIALS)
JAPANESE

TOSHIBA NEWS LETTER 6663
TOKYO SHIBAURA DENKI K.K.
2-1, 5-CHOME, GINZA, CHUO-KU, TOKYO 104
ENGLISH

TOSHIBA REBYU 6664
(TOSHIBA REVIEW)
TOSHIBA SOGO KENKYUSHONAI TOSHIBA REBYU HENSHUSHITSU AND TOSHIBA REBYU HAKKOSHO
72, HORIKAWA-CHO, SAIWAI-KU, KAWASAKI-SHI 210
JAPANESE AND ENGLISH
INDEXED: CHE, ENG, INS, IAA, WAA, JICST

TOSHIBA REVIEW, INTERNATIONAL EDITION 6665
TOSHIBA REBYU HENSYUSHITSU
1, KOMUKAI TOSHIBA-CHO, SAIWAI-KU, KAWASAKI-SHI 210
ENGLISH
INDEXED: EXC, JICST

TOSHIBA TSUSHIN SHISUTEMU NYUSU 6666
(TOSHIBA COMMUNICATION SYSTEM NEWS)
TOSHIBA SHOJI K.K. TSUSHIN EIGYO-BU (TOSHIBA COMMERCIAL COMPANY, LTD.)
2-1, 5-CHOME, GINZA, CHUO-KU, TOKYO 104
JAPANESE

TOYO DENKI GIHO 6667
(TOYO DENKI TECHNICAL JOURNAL)
TOYO DENKI SEIZO K.K.
5-7, YAESU, CHUO-KU, TOKYO 104
JAPANESE

TOYO'S TECHNICAL BULLETIN 6668
TOYO TSUSHINKI K.K. (TOYO COMMUNICATION EQUIPMENT COMPANY, LTD.)
3-484, TSUKAGOSHI, SAIWAI-KU, KAWASAKI-SHI 210
JAPANESE
INDEXED: JICST

TRANSACTIONS OF THE INSTITUTE OF ELECTRONICS AND COMMUNICATION ENGINEERS OF JAPAN, ABSTRACTS 6669
DENSHI TSUSHIN GAKKAI (THE INSTITUTE OF ELECTRONICS AND COMMUNICATION ENGINEERS OF JAPAN)
C/O KIKAI SHINKO KAIKAN, 5-8, 3-CHOME, SHIBAKOEN, MINATO-KU, TOKYO 105
ENGLISH

TRANSACTIONS OF THE INSTITUTE OF ELECTRONICS AND COMMUNICATION ENGINEERS OF JAPAN, SECTION E 6670
ENGLISH
INDEXED: ENG, JICST

TSUKEN GEPPO 6671
(MONTHLY JOURNAL OF THE ELECTRICAL COMMUNICATION LABORATORY)
NIHON DENSHIN DENWA KOSHA DENKI TSUSHIN KENKYUSHO (THE ELECTRICAL COMMUNICATION LABORATORY, NIHON TELEGRAPH AND TELEPHONE PUBLIC CORPORATION)
9-11, 3-CHOME, MIDORI-CHO, MUSASHINO-SHI, TOKYO 180
JAPANESE
INDEXED: JICST

TSUSHIN KOGYO 6672
(COMMUNICATION INDUSTRIES)
TSUSHIN KIKAI KOGYOKAI (COMMUNICATION INDUSTRIES ASSOCIATION OF JAPAN)
C/O SANKEI BUILDING, 7-2, 1-CHOME, OTE-MACHI, CHIYODA-KU, TOKYO 100
JAPANESE
INDEXED: JICST

TSUSHIN NI KANSURU GENJO HOKOKU 6673
(CURRENT SURVEY ON COMMUNICATION)
YUSEI-SHO (MINISTRY OF POSTS AND TELECOMMUNICATIONS)
3-2, 1-CHOME, KASUMIGASEKI, CHIYODA-KU, TOKYO 100
JAPANESE

TSUSHIN TAIMUSU 6674
(TELECOMMUNICATION TIMES)
TSUSHIN TAIMUSSHA
11-5, 4-CHOME, SHIBA, MINATO-KU, TOKYO 108
JAPANESE

VOLT, ELECTRICAL TRADE MONTHLY 6675
NIHON BOEKI SHUPPANSHA (JAPAN TRADE PUBLISHING COMPANY, LTD.)
2-1, YOTSUYA, SHINJUKU-KU, TOKYO 160
ENGLISH
INDEXED: JICST

YASUKAWA DENKI 6676
(YASUKAWA REVIEW)
YASUKAWA DENKI SEISAKUSHO (YASUKAWA ELECTRIC MANUFACTURING COMPANY, LTD.)
2346, FUJITA, YAHATANISHI-KU, KITAKYUSHU-SHI 806
JAPANESE
INDEXED: JICST

YASUKAWA DENKI SEISAKUSHO KENKYUSHO KENKYU HOKOKU 6677
(REPORT OF THE RESEARCH LABORATORY, YASUKAWA ELECTRIC MANUFACTURING COMPANY, LTD.)
YASUKAWA DENKI SEISAKUSHO KENKYUSHO (RESEARCH LABORATORY, YASUKAWA ELECTRIC MANUFACTURING COMPANY, LTD.)
2346, FUJITA, YAHATANISHI-KU, KITAKYUSHU-SHI 806
JAPANESE

YASUKAWA NYUZU 6678
(YASUKAWA NEWS)
YASUKAWA DENKI SEISAKUSHO (YASUKAWA ELECTRIC MANUFACTURING COMPANY, LTD.)
2346, FUJITA, YAHATANISHI-KU, KITAKYUSHU-SHI 806
JAPANESE

YOKOGAWA GIHO 6679
(YOKOGAWA TECHNICAL REPORT)
YOKOGAWA DENKI SEISAKUSHO (YOKOGAWA ELECTRIC WORKS, LIMITED)
9-32, 2-CHOME, NAKA-MACHI, MUSASHINOSHI, TOKYO 180
JAPANESE
INDEXED: CHE, JICST

YUDENKAI 6680
OSAKA DENKI TSUSHIN DAIGAKU YUDENKAI (OSAKA ELECTRONIC COMMUNICATION UNIVERSITY, ALUMNI ASSOCIATION)
JAPANESE

YUKON GIJUTSU 6681
(U/O TECHNIQUE)
DENPA JIKKENSHA
4-2, 1-CHOME, EBISUMINAMI, SHIBUYA-KU, TOKYO 150
JAPANESE

ZUSETSU DENKI 6682
(ILLUSTRATED ELECTRIC MAGAZINE)
DENKI SHOIN
1-55, KANDA JINBO-CHO, CHIYODA-KU, TOKYO 101
JAPANESE
INDEXED: JICST

Mining and Mineral Dressing

AKITA DAIGAKU KOZANGAKUBU CHIKA SHIGEN KENKYU SHISETSU HOKOKU 6683
(REPORT OF THE RESEARCH INSTITUTE OF UNDERGROUND RESOURCES, MINING COLLEGE, AKITA UNIVERSITY)
AKITA DAIGAKU KOZANGAKUBU CHIKA SHIGEN KENKYU SHISETSU (RESEARCH INSTITUTE OF UNDERGROUND RESOURCES, MINING COLLEGE, AKITA UNIVERSITY)
1-1, TEGATA GAKUEN-CHO, AKITA-SHI 010
JAPANESE AND ENGLISH
INDEXED: CHE, JICST

AKITA-KEN KOKOGYO SEISAN SHISU 6684
(INDEX OF MINING AND MANUFACTURING, AKITA PREFECTURE)
AKITA-KEN (AKITA PREFECTURAL GOVERNMENT)
1-1, 4-CHOME, SANNO, AKITA-SHI 010
JAPANESE

CHIBA-KEN KOKOGYO SEISAN SHISU 6685
(INDEX OF MINING AND MANUFACTURING, CHIBA PREFECTURE)
CHIBA-KEN KIKAKU-BU (PLANNING DIVISION, CHIBA PREFECTURE)
1-1, ICHIBA-CHO, CHIBA-SHI 280-91
JAPANESE

CHIKA SHIGEN CHOSA HOKOKU 6686
(REPORT OF UNDERGROUND RESOURCES SURVEY)
NAGANO-KEN SHOKO-BU (COMMERCE AND INDUSTRY DIVISION, NAGANO PREFECTURAL GOVERNMENT)
692, HABASHITA, MINAMINAGANO, NAGANO-SHI 380
JAPANESE

CHIKA SHIGEN CHOSASHO HOKOKU 6687
(REPORT OF THE GEOLOGICAL SURVEY OF HOKKAIDO)
HOKKAIDORITSU CHIKA SHIGEN CHOSASHO (GEOLOGICAL SURVEY OF HOKKAIDO)
NISHI-3-CHOME, MINAMI-11-JO, CHUO-KU, SAPPORO-SHI 064
JAPANESE, WITH ENGLISH SUMMARY

CHOSA KAISEKI IINKAI SHIRYO 6688
(BULLETIN OF SURVEY AND ANALYSIS COMMITTEE)
KINZOKU KOGYO JIGYODAN SHIRYO SENTA (METAL MINING AGENCY OF JAPAN)
6-3, NISHIKUBO SAKURAGAWA-CHO, MINATO-KU, TOKYO 105
JAPANESE

DAINIHON KOGYO K.K. SEISAN KANRI GIJUTSU NYUSU 6689
(TECHNICAL PRODUCTION CONTROL NEWS, DAINIHON MINING COMPANY, LIMITED)
DAINIHON KOGYO K.K. (DAINIHON MINING COMPANY, LTD.)
24-8, 4-CHOME, SHINBASHI, MINATO-KU, TOKYO 105
JAPANESE

EHIME-KEN KOKOGYO SEISAN SHISU 6690
(INDEX OF MINING AND MANUFACTURING, EHIME PREFECTURE)
EHIME-KEN KIKAKU-BU (PLANNING DIVISION, EHIME PREFECTURAL GOVERNMENT)
15, ICHIBAN-CHO, MATSUYAMA-SHI 790
JAPANESE

EHIME-KEN SHOKO-KA CHOSA KENKYU HOKOKU (EHIME-KEN CHIKA SHIGEN SHIRYO) 6691
(GEOLOGICAL SURVEY, EHIME PREFECTURE)
EHIME-KEN (EHIME PREFECTURAL GOVERNMENT)
ICHIBAN-CHO, MATSUYAMA-SHI 790
JAPANESE

FUKUI-KEN KOKOGYO SEISAN SHISU 6692
(INDEX OF MINING AND MANUFACTURING, FUKUI PREFECTURE)
FUKUI-KEN KIKAKU KAIHATSU-BU (PLANNING & DEVELOPMENT DIVISION, FUKUI PREFECTURAL GOVERNMENT)
17-1, 3-CHOME, OOTE, FUKUI-SHI 910
JAPANESE

FUKUOKA-KEN KOKOGYO SEISAN SHISU 6693
(INDEX OF MINING AND MANUFACTURING, FUKUOKA PREFECTURE)
FUKUOKA-KEN KIKAKU KAIHATSU-BU (PLANNING & DEVELOPMENT DIVISION, FUKUOKA PREFECTURAL GOVERNMENT)
1-1, 1-CHOME, TENJIN, CHUO-KU, FUKUOKA-SHI 810
JAPANESE

FUKUOKA-KEN KOKOGYO SHIKENJO NENPO 6694
(ANNUAL REPORT OF MINING AND MANUFACTURING RESEARCH INSTITUTE)
FUKUOKA-KEN KOKOGYO SHIKENJO (MINING AND MANUFACTURING RESEARCH INSTITUTE, FUKUOKA PREFECTURAL GOVERNMENT)
1-1, 1-CHOME, TENJIN, CHUO-KU, FUKUOKA-SHI 812
JAPANESE

FUKUOKA-KEN NOGATA KOGYO SHIKENJO KENKYU HOKOKU 6695
(REPORT OF NOGATA MINING INDUSTRIAL LABORATORY, FUKUOKA PREFECTURE)
FUKUOKA-KEN NOGATA KOGYO SHIKENJO (FUKUOKA PREFECTURAL NOGATA MINING INDUSTRIAL LABORATORY)
9-29, 2-CHOME, CHIKO, NOGATA-SHI 822
JAPANESE

FUKUOKA-KEN NOGATA KOGYO SHIKENJO NENPO 6696
(ANNUAL REPORT OF NOGATA MINING INDUSTRIAL LABORATORY, FUKUOKA PREFECTURE)
FUKUOKA-KEN NOGATA KOGYO SHIKENJO (FUKUOKA PREFECTURAL NOGATA MINING INDUSTRIAL LABORATORY)
9-29, 2-CHOME, CHIKO, NOGATA-SHI 822-14
JAPANESE

GIJUTSU DAIJESUTO 6697
(TECHNICAL DIGEST)
MITSUI SEKITAN KOGYO K.K. (MITSUI COAL MINING COMPANY, LTD.)
1-1, 2-CHOME, NIHONBASHI MURO-MACHI, CHUO-KU, TOKYO 103
JAPANESE

GORIKA NYUSU 6698
(RATIONALIZATION NEWS)
SEKITAN KOGYO GORIKA JIGYODAN
(ASSOCIATION FOR THE RATIONALIZATION OF THE COAL MINING INDUSTRY)
JAPANESE
INDEXED: JICST

HIROSHIMA-KEN KOKOGYO SEISAN, SHUKKA, ZAIKO SHISU, 6699
HIROSHIMA-KEN KOKOGYO SEISAN DOTAI TOKEI CHOSA
(INDEX OF MINING AND MANUFACTURING PRODUCTION, FORWARDING AND STOCK, HIROSHIMA PREFECTURE)
HIROSHIMA-KEN KIKAKU-BU (PLANNING DIVISION, HIROSHIMA PREFECTURAL GOVERNMENT)
10-52, MOTO-MACHI, HIROSHIMA-SHI 730
JAPANESE

HOAN GIJUTSU JOHO 6700
(MINING SAFETY TECHNIQUE INFORMATION)
KOGYO RODO SAIGAI BOSHI KYOKAI
35-1, 5-CHOME, SHIBA, MINATO-KU, TOKYO 108
JAPANESE
INDEXED: JICST

HOKKAIDO KOKOGYO KAIHATSU KEIKAKU CHOSA 6701
(DEVELOPMENT PLAN OF MINING AND INDUSTRY IN HOKKAIDO)
HOKKAIDO KAIHATSU-CHO (HOKKAIDO DEVELOPMENT AGENCY)
1-1, 3-CHOME, KASUMIGASEKI, CHIYODA-KU, TOKYO 100
JAPANESE
INDEXED: JICST

HOKKAIDO KOKOGYO KAIHATSU SHIKEN CHOSA 6702
(DEVELOPMENT TEST OF MINING AND INDUSTRY IN HOKKAIDO)
HOKKAIDO KAIHATSU-CHO (HOKKAIDO DEVELOPMENT AGENCY)
1-1, 3-CHOME, KASUMIGASEKI, CHIYODA-KU, TOKYO 100
JAPANESE

HOKKAIDO KOZANGAKU KAISHI 6703
(JOURNAL OF THE MINING INSTITUTE OF HOKKAIDO)
JAPANESE
INDEXED: JICST

HOKKAIDO SEKITAN TOKEI NENPO 6704
(COAL STATISTIC YEARBOOK, HOKKAIDO)
TSUSHO SANGYO-SHO SAPPORO TSUSHO SANGYO-KYOKU (INTERNATIONAL TRADE AND INDUSTRY BUREAU, SAPPORO, MINISTRY OF INTERNATIONAL TRADE AND INDUSTRY)
NISHI-4-CHOME, KITA-3-JO, CHUO-KU, SAPPORO-SHI 060
JAPANESE

HONPO KOGYO NO SUSEI 6705
(MINING YEARBOOK OF JAPAN)
TSUSHO SANGYO-SHO DAIJIN KANBO CHOSA TOKEI-BU (RESEARCH AND STATISTICS DEPARTMENT, MINISTER'S SECRETARIAT, MINISTRY OF INTERNATIONAL TRADE AND INDUSTRY) AND TSUSHO SANGYO CHOSAKAI
JAPANESE

IBARAKI-KEN KOKOGYO SEISAN SHISU 6706
(INDEX OF MINING AND MANUFACTURING, IBARAKI PREFECTURE)
IBARAKI-KEN SOGO KAIHATSU-BU
5-38, 1-CHOME, SANNOMARU, MITO-SHI 310
JAPANESE

JOURNAL OF THE MINING COLLEGE, AKITA UNIVERSITY, SERIES A, MINING GEOLOGY 6707
AKITA DAIGAKU KOZAN GAKUBU KOZAN CHISITSUGAKU KYOSHITSU (INSTITUTE OF MINING GEOLOGY, MINING COLLEGE, AKITA UNIVERSITY) AND AKITA DAIGAKU KOZAN GAKUBU (MINING COLLEGE, AKITA UNIVERSITY)
TEGATA GAKUEN-CHO, AKITA-SHI 010
ENGLISH
INDEXED: CHE, JICST

KAGOSHIMA-KEN KOKOGYO SEISAN NO DOKO 6708
(ANNUAL STATISTICS OF MINING AND MANUFACTURING, KAGOSHIMA PREFECTURE)
KAGOSHIMA-KEN KIKAKU-BU (PLANNING DIVISION, KAGOSHIMA PREFECTURAL GOVERNMENT)
15-50, YAMASHITA-CHO, KAGOSHIMA-SHI 890-91
JAPANESE

KAIGAI KOGYO JIJO CHOSA HOKOKUSHO 6709
(REPORT OF OVERSEAS MINING INVESTIGATION)
KINZOKU KOGYO JIGYODAN SHIRYO SENTA (DATA CENTER, METAL MINING AGENCY OF JAPAN)
6-3, SHIBA NISHIKUBO SAKURAGAWA-CHO, MINATO-KU, TOKYO 105
JAPANESE

KAIGAI KOGYO JOHO 6710
(OVERSEAS MINING INFORMATION)
KINZOKU KOGYO JIGYODAN SHIRYO SENTA (DATA CENTER, METAL MINING AGENCY OF JAPAN)
6-3, NISHIKUBO SAKURAGAWA-CHO, SHIBA, MINATO-KU, TOKYO 105
JAPANESE

KAIGAI URAN KOGYO JOHO SHOROKU 6711
(OVERSEAS URANIUM MINING ABSTRACT)
DORYOKURO, KAKU NENRYO KAIHATSU JIGYODAN (POWER REACTOR AND NUCLEAR FUEL COOPERATIVE)
C/O SANKAIDO BUILDING, 9-13, 1-CHOME, AKASAKA, MINATO-KU, TOKYO 107
JAPANESE

KENTEI KOHO 6712
(BULLETIN OF INSPECTION OF UNDERGROUND ARTICLES USED IN MINES)
TSUSHO SANGYO-SHO KOGYO GIJUTSU-IN KOGAI SHIGEN KENKYUSHO (NATIONAL RESEARCH INSTITUTE FOR POLLUTION AND RESOURCES, AGENCY OF INDUSTRIAL SCIENCE AND TECHNOLOGY)
188, KOTOBUKI-CHO, KAWAGUCHI-SHI 332
JAPANESE

KINZOKU JIHO 6713
(METALS REVIEW)
JAPAN KOGYO KYOKAI CHOSA-BU (JAPAN MINING ASSOCIATION SURVEY BUREAU)
JAPANESE

KINZOKU KOGYO JIGYODAN CHOSA KAISEKI IINKAI SHIRYO 6714
(DATA OF THE ANALYTICAL RESEARCH COMMITTEE, THE METAL MINING TRADE ASSOCIATION)
KINZOKU KOGYO JIGYODAN SHIRYO SENTA (DATA CENTER, THE METAL MINING AGENCY OF JAPAN)
6-3, SHIBA NISHIKUBO SAKURAGAWA-CHO, MINATO-KU, TOKYO 105
JAPANESE

KINZOKU KOGYO JIGYODAN SHIRYO SENTA KAIGAI SHIRYO 6715
(OVERSEAS INFORMATION, METAL MINING AGENCY OF JAPAN)
KINZOKU KOGYO JIGYODAN SHIRYO SENTA (DATA CENTER, METAL MINING AGENCY OF JAPAN)
6-3, SHIBA NISHIKUBO SAKURAGAWA-CHO, MINATO-KU, TOKYO 105
JAPANESE

KINZOKU KOZAN KENKYUKAI KAIHO 6716
(REPORT OF THE METALS MINING RESEARCH GROUP)
KINZOKU KOZAN KENKYUKAI (METALS MINING RESEARCH GROUP)
JAPANESE

KOGYO HAKUBUTSUKAN HOKOKU 6717
(ANNUAL REPORT OF MINERAL INDUSTRY MUSEUM)
AKITA DAIGAKU KOZAN GAKUBU FUZOKU KOGYO HAKUBUTSUKAN (MINERAL INDUSTRY MUSEUM, AKITA UNIVERSITY)
28-2, OSAWA, TEGATA, AKITA-SHI 010
JAPANESE

KOGYO HOAN JOHO 6718
(MINING SAFETY INFORMATION)
KOGYO RODO SAIGAI BOSHI KYOKAI (LABOR ACCIDENTS PREVENTION ASSOCIATION OF MINING)
35-1, 5-CHOME, SHIBA, MINATO-KU, TOKYO 108
JAPANESE
INDEXED: JICST

KOGYO SHINBUN 6719
(MINING NEWS)
KOGYO SHINBUNSHA
C/O FUTABA BUILDING, 24, SHIBA NISHIKUBO SAKURAGAWA-CHO, MINATO-KU, TOKYO 105
JAPANESE

KOKOGYO SEISAN GAIKYO 6720
(STATISTICS OF MINING AND MANUFACTURING PRODUCTS)
IWATE-KEN SOMU-BU (ADMINISTRATIVE DIVISION, IWATE PREFECTURAL GOVERNMENT)
10-1, UCHIMARU, MORIOKA-SHI 020
JAPANESE

KOKOGYO SEISAN KATSUDO 6721
(INDUSTRIAL PRODUCTION ACTIVITY)
TSUSHO SANGYO-SHO DAIJIN KANBO CHOSA TOKEI-BU (RESEARCH AND STATISTICS DEPARTMENT, MINISTRY OF INTERNATIONAL TRADE AND INDUSTRY)
3-1, 1-CHOME, KASUMIGASEKI, CHIYODA-KU, TOKYO 100
JAPANESE

KOKOGYO SEISAN NO UGOKI (MIE-KEN) 6722
(STATISTICAL REPORT OF MINING AND MANUFACTURING, MIE PREFECTURE)
MIE-KEN KIKAKU-BU (PLANNING DIVISION, MIE PREFECTURAL GOVERNMENT)
13, KOMEI-CHO, TSU-SHI 514
JAPANESE

KOKOGYO SEISAN SHISU (SHIZUOKA-KEN) 6723
(INDEX OF MINING AND MANUFACTURING, SHIZUOKA PREFECTURE)
SHIZUOKA-KEN KIKAKU CHOSEI-BU (SHIZUOKA PREFECTURAL GOVERNMENT)
9-6, OTE-MACHI, SHIZUOKA-SHI 420
JAPANESE

KOKOGYO SHISU NENPO 6724
(YEARBOOK OF MINING AND MANUFACTURING INDEX)
TSUSHO SANGYO-SHO DAIJIN KANBO CHOSA TOKEI-BU (RESEARCH AND STATISTICS DIVISION, MINISTER'S SECRETARIAT, MINISTRY OF INTERNATIONAL TRADE AND INDUSTRY) AND TSUSHO SANGYO KENKYUSHA
JAPANESE

KOKUSU HANBAIRYO CHOSA 6725
(COKE SALES SURVEY)
TSUSHO SANGYO SHO, DAIJIN KANBO, CHOSA TOKEI BU (MINISTRY OF INTERNATIONAL TRADE AND INDUSTRY, MINISTER'S SECRETARIAT, STATISTICAL SURVEY DIVISION)
JAPANESE

KOSHO CHOSA (SHISUI CHOSA) HOKOKU 6726
(BULLETIN OF HOKKAIDO UNDERGROUND RESOURCES INVESTIGATION (BORING INVESTIGATION))
HOKKAIDO KAIHATSU-CHO (HOKKAIDO DEVELOPMENT AGENCY)
1-1, 3-CHOME, KASUMIGASEKI, CHIYODA-KU, TOKYO 100
JAPANESE

KOTSUZAI SHIGEN 6727
(AGGREGATE MATERIALS)
KOTSUZAI SHIGEN KOGAKKAI (AGGREGATE MATERIALS ENGINEERING ASSOCIATION)
C/O WASEDA DAIGAKU RIKO GAKUBU SHIGEN KOGAKUKA, 4-170, NISHIOKUBO, SHINJUKU-KU, TOKYO 160)
JAPANESE
INDEXED: JICST

KOZAN 6728
(BULLETIN OF JAPAN MINING INDUSTRY ASSOCIATION)
NIHON KOGYO KYOKAI (JAPAN MINING INDUSTRY ASSOCIATION)
C/O SHIN HIBIYA BUILDING, 3-6, 1-CHOME, USCHISAIWAI-CHO, CHIYODA-KU, TOKYO 100
JAPANESE AND ENGLISH
INDEXED: CHE, JICST

KOZAN CHISHITSU 6729
(MINING GEOLOGY)
NIHON KOZAN CHISHITSU GAKKAI (SOCIETY OF MINING GEOLOGISTS OF JAPAN)
C/O NIHON KOGYO KAIKAN BUILDING, 5-4, 8-CHOME, GINZA, CHUO-KU, TOKYO 104
JAPANESE AND ENGLISH
INDEXED: CHE, JICST

KOZAN HOAN NENPO 6730
(MINING SAFETY YEARBOOK)
TSUSHO SANGYO-SHO RITCHI KOGAI-KYOKU (INDUSTRIAL LOCATION AND ENVIROMENTAL PROTECTION BUREAU, MINISTRY OF INTERNATIONAL TRADE AND INDUSTRY)
3-1, 1-CHOME, KASUMIGASEKI, CHIYODA-KU, TOKYO 100
JAPANESE

KOZAN HOAN TOKEI GEPPO 6731
(MINING SAFETY STATISTICAL MONTHLY)
TSUSHO SANGYO-SHO RITCHI KOGAI-KYOKU (INDUSTRIAL LOCATION AND ENVIRONMENTAL PROTECTION BUREAU, MINISTRY OF INTERNATIONAL TRADE AND INDUSTRY)
3-1, 1-CHOME, KASUMIGASEKI, CHIYODA-KU, TOKYO 100
JAPANESE

KOZAN HOAN TOKEI NENPO 6732
(MINING SAFETY STATISTICAL YEARBOOK)
TSUSHO SANGYO-SHO RITCHI KOGAI-KYOKU (INDUSTRIAL LOCATION AND ENVIRONMENTAL PROTECTION BUREAU, MINISTRY OF INTERNATIONAL TRADE & INDUSTRY)
3-1, 1-CHOME, KASUMIGASEKI, CHIYODA-KU, TOKYO 100
JAPANESE

KOZAN HOAN TOKEI NENPO GAIHO 6733
(MINING SAFETY STATISTICAL YEARBOOK, GENERAL REPORT)
TSUSHO SANGYO-SHO RITCHI KOGAI-KYOKU (INDUSTRIAL LOCATION AND ENVIRONMENTAL PROTECTION BUREAU, MINISTRY OF INTERNATIONAL TRADE & INDUSTRY)
3-1, 1-CHOME, KASUMIGASEKI, CHIYODA-KU, TOKYO 100
JAPANESE

KOZAN KONAI YOHIN KENTEI KOHO 6734
(GAZZETA OF APPROVED MINE SAFETY APPLIANCES)
TSUSHO SANGYO-SHO KOGYO GIJUTSU-IN KOGAI SHIGEN KENKYUSHO (NATIONAL RESEARCH INSTITUTE FOR POLLUTION & RESOURCES, AGENCY OF INDUSTRIAL SCIENCE AND TECHNOLOGY)
C/O KYOIKU GUILDING, 6-16, 4-CHOME, KOHINA, BUNKYO-KU, TOKYO 112
JAPANESE

KYOTO-FU KOKOGYO SEISAN SHISU 6735
(INDEX OF MINING AND MANUFACTURING, KYOTO PREFECTURE)
KYOTO-FU SOMU-BU (ADMINISTRATIVE DIVISION, KYOTO PREFECTURAL GOVERNMENT)
YABUNOUCHI-MACHI, SHIMODACHIYURIDORI SHIN-MACHI NISHIIRU, KAMIGYO-KU, KYOTO-SHI 602
JAPANESE

KYUSHU SANTAN CHIIKI NO GENKYO 6736
(ACTUAL CONDITION OF KYUSHU COAL MINE)
TSUSHO SANGYO-SHO FUKUOKA TSUSHO SANGYO-KYOKU (INTERNATIONAL TRADE AND INDUSTRY BUREAU, FUKUOKA, MINISTRY OF INTERNATIONAL TRADE AND INDUSTRY)
2-1, 2-CHOME, HAKATA-EKI HIGASHI, HAKATA-KU, FUKUOKA-SHI 812
JAPANESE

KYUSHU SEKITAN TOKEI GEPPO 6737
(KYUSHU MONTHLY RETURN OF COAL STATISTICS)
NIHON SEKITAN KYOKAI KYUSHU SHIBU (JAPAN COAL ASSOCIATION, KYUSHU BRANCH)
10-2, 3-CHOME, TENJIN, CHUO-KU, FUKUOKA-SHI 810
JAPANESE
INDEXED: JICST

MESARAITO NYUSU 6738
(MESALITE NEWS)
MITSUI KINZOKU KOGYO K.K. (MITSUI MINING AND SMELTING COMPANY, LTD.)
1-1, 2-CHOME, NIHONBASHI MURO-MACHI, CHUO-KU, TOKYO 103
JAPANESE

MITSUBISHI TANKO GIJUTSU 6739
(MITSUBISHI COAL MINING TECHNOLOGY)
MITSUBISHI KOGYO KABUSHIKI KAISHA (MITSUBISHI MINING COMPANY, LTD.)
JAPANESE
INDEXED: JICST

MIYAGI-KEN KOKOGYO SEISAN SHISU 6740
(INDEX OF MINING AND MANUFACTURING, MIYAGI PREFECTURE)
MIYAGI-KEN KIKAKU-BU (PLANNING DIVISION, MIYAGI PREFECTURAL GOVERNMENT)
8-1, 3-CHOME, HON-CHO, SENDAI-SHI 980
JAPANESE

NIHON KOGYOKAISHI 6741
(JOURNAL OF THE MINING AND METALLURGICAL INSTITUTE OF JAPAN)
NIHON KOGYOKAI (MINING AND METALLURGICAL INSTITUTE OF JAPAN)
5-4, 8-CHOME, GINZA, CHUO-KU, TOKYO 103
JAPANESE, WITH ENGLISH SUMMARY
INDEXED: COP, ENG, CHE, JICST

NIPPON KOGYOKAI KENKYU GYOSEKI HAPPYO KOENKAI KOEN YOSHISHU 6742
(COLLECTED OUTLINES OF LECTURES AT THE RESEARCH FINDINGS PRESENTATION MEETING OF THE JAPAN MINING ASSOCIATION)
JAPANESE
INDEXED: JICST

NIPPON KOGYOKAI SHUKI TAIKAI BUNKA KENKYUKAI KOENSHU 6743
(COLLECTED LECTURES FROM THE SPECIALIZED RESEARCH COMMITTEES OF THE JAPAN MINING ASSOCIATION AUTUMN GENERAL MEETING)
NIHON KOGYO KAI (JAPAN MINING ASSOCIATION)
JAPANESE
INDEXED: JICST

OKAYAMA-KEN CHIKA SHIGEN CHOSA HOKOKUSHO 6744
(REPORT OF THE GEOLOGICAL SURVEY OF OKAYAMA PREFECTURE)
OKAYAMA-KEN SHOKO-BU (COMMERCE AND INDUSTRY DIVISION, OKAYAMA PREFECTURAL GOVERNMENT)
4-6, 2-CHOME, UCHISANGE, OKAYAMA-SHI 700
JAPANESE

OKAYAMA-KEN KOKOGYO SEISAN SHISU 6745
(INDEX OF MINING AND MANUFACTURING, OKAYAMA PREFECTURE)
OKAYAMA-KEN KIKAKU-BU (PLANNING DIVISION, OKAYAMA PREFECTURAL GOVERNMENT)
4-6, 2-CHOME, UCHISANGE, OKAYAMA-SHI 700
JAPANESE

OSAKA-FU KOKOGYO SEISAN SHISU 6746
(INDEX OF MINING AND MANUFACTURING, OSAKA PREFECTURE)
OSAKA-FU KIKAKUBU (PLANNING DIVISION, OSAKA PREFECTURAL GOVERNMENT)
OTEMAENO-CHO, HIGASHI-KU, OSAKA-SHI 540
JAPANESE

SAIKO TO HOAN 6747
(MINING AND SAFETY)
TSUSHO SANGYO-SHO KOGYO GIJUTSU-IN KOGAI SHIGEN KENKYUSHO (NATIONAL RESEARCH INSTITUTE FOR POLLUTION AND RESOURCES, AGENCY OF INDUSTRIAL SCIENCE AND TECHNOLOGY)
1-4, 2-CHOME, SOTOKANDA, CHIYODA-KU, TOKYO 101
JAPANESE AND ENGLISH
INDEXED: CHE, JICST

SAKUGANKI 6748
(ROCK DRILL)
TOYO KOGYO K.K. (TOYO KOGYO COMPANY, LTD.)
6047, FUCHU-MACHI, AKI-GUN, HIROSHIMA 730-91
JAPANESE
INDEXED: JICST

SANTANCHI SHINKO 6749
(PROMOTION OF COAL AND MINE AREA)
KOGYO SAIHAICHI, SANTANCHIIKI SHINKO KODAN
C/O TORANOMON MITSUI BUILDING, 8-1, 3-CHOME, KASUMIGASEKI, CHIYODA-KU, TOKYO 100
JAPANESE

SEIKA GIHO 6750
(SEIKA TECHNICAL REPORT)
SUMITOMO SEKITAN KOGYO K.K. (SUMITOMO COAL MINING COMPANY, LTD.)
JAPANESE
INDEXED: JICST

SEIMITSU CHOSA HOKOKUSHO 6751
(CLOSE EXAMINATION REPORT, METAL MINING AGENCY OF JAPAN))
KINZOKU KOGYO JIGYODAN (METAL MINING AGENCY OF JAPAN)
6-3, SHIBA NISHIKUBO SAKURAGAWA-CHO, MINATO-KU, TOKYO 105
JAPANESE

SEKITAN GIJUTSU KENKYUSHO HOKOKU 6752
(REPORT OF THE COAL MINING RESEARCH CENTER, JAPAN)
SEKITAN GIJUTSU KENKYUSHO (THE COAL MINING RESEARCH CENTER, JAPAN)
C/O KYOIKU SHUPPAN BUILDING, 2-10, KANDA JINBO-CHO, CHIYODA-KU, TOKYO 101
JAPANESE
INDEXED: JICST

SEKITAN GIJUTSU KENKYUSHO SHIRYO 6753
(INFORMATION OF THE COAL MINING RESEARCH CENTER, JAPAN)
SEKITAN GIJUTSU KENKYUSHO (THE COAL MINING RESEARCH CENTER, JAPAN)
C/O KYOIKU SHUPPAN BUILDING, 2-10, KANDA JINBO-CHO, CHIYODA-KU, TOKYO 101
JAPANESE
INDEXED: JICST

SEKITAN GIKEN 6754
(MONTHLY NEWS, THE COAL MINING RESEARCH CENTER, JAPAN)
SEKITAN GIJUTSU KENKYUSHO (THE COAL MINING RESEARCH CENTER, JAPAN)
C/O KYOIKU SHUPPAN BUILDING, 2-10, KANDA JINBO-CHO, CHIYODA-KU, TOKYO 101
JAPANESE
INDEXED: JICST

SEKITAN JIHO 6755
(BULLETIN OF JAPAN COAL ASSOCIATION)
NIHON SEKITAN KYOKAI (JAPAN COAL ASSOCIATION)
C/O HIBIYA PARK BUILDING, 1-1, YURAKU-CHO, CHIYODA-KU, TOKYO 100
JAPANESE
INDEXED: JICST

SEKITAN TOKEI GEPPO 6756
(COAL STATISTICAL MONTHLY)
TSUSHO SANGYO-SHO DAIJIN KANBO CHOSA TOKEI-BU (RESEARCH AND STATISTICS DIVISION, MINISTER'S SECRETARIAT, MINISTRY OF INTERNATIONAL TRADE AND INDUSTRY)
C/O NIKKATSU KOKUSAI KAIKAN, 1-1, YURAKU-CHO, CHIYODA-KU, TOKYO 100
JAPANESE

SEKKAISEKI 6757
(LIMESTONE)
SEKKAISEKI KOGYO KYOKAI KAISHI HENSYU IINKAI (THE INSTITUTE OF LIMESTONE QUARRY)
C/O GINRAKU BUILDING, 4-1, 2-CHOME, GINZA, CHUO-KU, TOKYO 104
JAPANESE
INDEXED: JICST

SENTAN 6758
(COAL PREPARATION)
FUJI KOHO-SHA
TOKYO
JAPANESE
INDEXED: CHE, JICST

SHIGA-KEN KOKOGYO SEISAN SHISU 6759
(INDEX OF MINING AND MANUFACTURING, SHIGA PREFECTURE)
SHIGA-KEN SOMU-BU (ADMINISTRATIVE DIVISION, SHIGA PREFECTURAL GOVERNMENT)
1-1, 4-CHOME, KYO-MACHI, OTSU-SHI 520
JAPANESE

TANKO 6760
(PROSPECTING)
KINZOKU KOGYO JIGYODAN (METAL MINING AGENCY OF JAPAN)
6-3, SHIBA NISHIKUBO SAKURAGAWA-CHO, MINATO-KU, TOKYO 105
JAPANESE

TANKO GIJUTSU 6761
(COLLIERY ENGINEERING)
HOKKAIDO TANKO GIJUTSUKAI (HOKKAIDO ASSOCIATION OF COAL MINING TECHNOLOGISTS)
C/O DAIGO BUILDING, 5, ODORI NISHI, CHUO-KU, SAPPORO-SHI 060
JAPANESE
INDEXED: CHE, JICST

TANKO SETSUBI KIRIHA CHOSA 6762
(REPORT OF MINING EQUIPMENT AND PIT FACE)
TSUSHO SANGYO-SHO DAIJIN KANBO CHOSA TOKEI-BU (RESEARCH AND STATISTIC DIVISION, MINISTER'S SECRETARIAT, MINISTRY OF INTERNATIONAL TRADE AND INDUSTRY) AND TSUSHO SANGYO CHOSAKAI
15-1, 6-CHOME, GINZA, CHUO-KU, TOKYO 104
JAPANESE

TANKO SHIN GIJUTSU 6763
(NEW TECHNOLOGY OF MINING)
ZENKOKU TANKO GIJUTSUKAI
C/O HIBIYA PARK BUILDING, 1-1, YURAKU-CHO, CHIYODA-KU, TOKYO 100
JAPANESE

TOCHIGI-KEN KOKOGYO SEISAN SHISU (TOCHIGI-KEN SEIZOGYO SEISANSHA SEIHIN ZAIKO SHISU) 6764
(INDEX OF MINING AND MANUFACTURING, TOCHIGI PREFECTURE)
TOCHIGI-KEN KIKAKU-BU (PLANNING DIVISION, TOCHIGI PREFECTURAL GOVERNMENT)
504, HANAWADA-CHO, UTSUNOMIYA-SHI 320
JAPANESE

TOHOKU DAIGAKU SENKO SEIREN KENKYUSHO IHO 6765
(BULLETIN OF THE RESEARCH INSTITUTE OF MINERAL DRESSING AND METALLURGY, TOHOKU UNIVERSITY)
TOHOKU DAIGAKU SENKO SEIREN KENKYUSHO (RESEARCH INSTITUTE OF MINERAL DRESSING AND METALLURGY, TOHOKU UNIVERSITY)
1-1, 2-CHOME, KATAHIRA, SENDAI-SHI 980
JAPANESE, WITH EUROPEAN SUMMARY
INDEXED: CHE, WAA, JICST

TOKUSHIMA-KEN KOKOGYO SEISAN SHISU 6766
(INDEX OF MINING AND MANUFACTURING, TOKUSHIMA PREFECTURE)
TOKUSHIMA-KEN KIKAKU KAIHATSU-BU (PLANNING AND DEVELOPMENT DIVISION, TOKUSHIMA PREFECTURAL GOVERNMENT)
1, BANDAI-MACHI, TOKUSHIMA-SHI 770
JAPANESE

TOKYO-TO KOKOGYO SHISU 6767
(INDEX OF MINING AND MANUFACTURING, TOKYO)
TOKYO-TO SOMU-KYOKU (ADMINISTRATIVE BUREAU, TOKYO METROPOLITAN GOVERNMENT)
5-1, 3-CHOME, MARUNOUCHI, CHIYODA-KU, TOKYO 100
JAPANESE

TOTTORI-KEN KOKOGYO SEISAN SHISU 6768
(INDEX OF MINING AND MANUFACTURING, TOTTORI PREFECTURE)
TOTTORI-KEN KIKAKU-BU (PLANNING DIVISION, TOTTORI PREFECTURAL GOVERNMENT)
1-220, HIGASHI-MACHI, TOTTORI-SHI 680
JAPANESE

TOYAMA-KEN NO KOKOGYO SEISAN NO UGOKI 6769
(CENSUS OF MINING AND MANUFACTURING, TOYAMA PREFECTURE)
TOYAMA-KEN SOMU-BU (ADMINISTRATIVE DIVISION, TOYAMA PREFECTURAL GOVERNMENT)
1-7, SINSOGAWA, TOYAMA-SHI 930
JAPANESE

WAGAKUNI NO KOKOGYO 6770
(JAPANESE MINING AND MANUFACTURING)
TSUSHO SANGYO-SHO DAIJIN KANBO CHOSA TOKEI-BU (RESEARCH AND STATISTICS DEPARTMENT, MINISTER'S SECRETARIAT, MINISTRY OF INTERNATIONAL TRADE AND INDUSTRY)
3-1, 1-CHOME, KASUMIGASEKI, CHIYODA-KU, TOKYO 100
JAPANESE

YAMAGATA-KEN KOGYO KENKYUSHO SHIKEN KENKYU HOKOKUSHU 6771
(RESEARCH REPORT OF MINING RESEARCH INSTITUTE OF YAMAGATA PREFECTURE)
YAMAGATA-KEN KOGYO KENKYUSHO (MINING RESEARCH INSTITUTE OF YAMAGATA PREFECTURE)
1129-6, OKI KANAZAWA, SHINJO-SHI 996
JAPANESE

YAMAGATA-KEN KOKOGYO SEISAN SHISU 6772
(INDEX OF MINING AND MANUFACTURING, YAMAGATA PREFECTURE)
YAMAGATA-KEN KIKAKU-BU (PLANNING DIVISION, YAMAGATA PREFECTURAL GOVERNMENT)
13-51, 4-CHOME, NANOKA-MACHI, YAMAGATA-SHI 990
JAPANESE

ZENKOKU TANKO GIJUTSUKAI KENKYUKAI HOKOKUSHO 6773
(ANNUAL RESEARCH REPORT OF ALL JAPAN ASSOCIATION FOR COAL MINING TECHNIQUES)
ZENKOKU TANKO GIJUTSUKAI
18-38, 2-CHOME, TENJIN, FUKUOKA-SHI 810
JAPANESE

Military and Naval Engineering and Technology

BOEI ANTENA 6774
(DEFENSE ANTENNA)
BOEI KOSAI KAI (DEFENSE BENEFIT ASSOCIATION)
JAPANESE

BOEI NIPPO 6775
(DAILY DEFENCE NEWS)
BOEI SANGYO KYOKAI (DEFENCE INDUSTRY ASSOCIATION)
11-10, 2-CHOME, SHIBA DAIMON, MINATO-KU, TOKYO 105
JAPANESE

BOEI SHISETSU KOHO 6776
(OFFICIAL REPORT OF THE DEFENSE FACILITIES AGENCY)
BOEI SHISETSU-CHO SOMUBU SUMUKA (DEFENSE FACILITIES AGENCY GENERAL AFFAIRS DEPARTMENT, GENERAL AFFAIRS SECTION)
JAPANESE

BOEI-CHO GIJUTSU KENKYU HONBU NENPO 6777
(ANNUAL REPORT OF THE DEFENSE AGENCY TECHNICAL RESEARCH ADMINISTRATION)
BOEI-CHO GIJUTSU KENKYU HONBU (DENFENSE AGENCY TECHNICAL RESEARCH ADMINISTRATION)
JAPANESE
INDEXED: JICST

BOEI-CHO KOHO 6778
(OFFICIAL REPORT OF THE DEFENSE AGENCY)
BOEI-CHO (HOKAN KANBO SOMUKA)
JAPANESE

GUNJI KENKYU 6779
(MILITARY AFFAIRS RESEARCH)
GUNJI KENKYUSHA (MILITARY AFFAIRS RESEARCH COMPANY)
JAPANESE

HEIKI TO GIJUTSU 6780
(ORDNANCE AND TECHNOLOGY)
NIHON HEIKI KOGYOKAI (JAPAN ORDNANCE ASSOCIATION)
C/O DAI-3 KANESAKA BUILDING, 11-8, 3-CHOME, SHINBASHI, MINATO-KU, TOKYO 105
JAPANESE
INDEXED: JICST

KANTEI TO GIJUTSU 6781
(WARSHIPS AND TECHNOLOGY)
BOEI-CHO GIJUTSU KENKYU HONBU (DEFENSE AGENCY TECHNICAL RESEARCH ADMINISTRATION)
JAPANESE
INDEXED: JICST

KOKUBO 6782
(NATIONAL DEFENSE)
ASAGUMO SHINBUNSHA (ASAGUMO NEWSPAPER COMPANY)
JAPANESE

SHIN BOEI RONSHO 6783
(JOURNAL OF NATIONAL DEFENSE)
ASAKUMO SHIMBUNSHA
9 SHIBA SAKAECHO, MINATO-KU, TOKYO 105
JAPANESE

Civil, Structural Engineering

AOKI KENSETSU GIJUTSU KENKYUSHO GIHO 6784
(TECHNICAL REPORT)
AOKI KENSETSU GIJUTSU KENKYUSHO (TECHNICAL RESEARCH LABORATORY, AOKI CONSTRUCTION CO., LTD.)
17-3, 2-CHOME, SHIBUYA, SHIBUYA-KU, TOKYO 150
JAPANESE

BOSUI JANARU 6785
(WATERPROOF JOURNAL)
JAPANESE
INDEXED: JICST

CEMENT CONCRETE PRODUCTS 6786
JAPANESE
INDEXED: JICST

CHIHO KENSETSU-KYOKU IRAI CHOSA SHIKEN HOKOKU 6787
(RESEARCH AND TESTING REPORTS AT REGIONAL CONSTRUCTION BUREAU'S REQUEST)
KENSETSU-SHO DOBOKU KENKYUSHO (PUBLIC WORKS RESEARCH INSTITUTE, MINISTRY OF CONSTRUCTION)
28-32, 2-CHOME, HONKOMAGOME, BUNKYO-KU, TOKYO 113
JAPANESE

CHUBU KENSETSU SHINBUN 6788
(CHUBU CONSTRUCTION NEWS)
CHUBU KENSETSU SHINBUNSHA
3-13, UCHISAIWAI-CHO, TOYAMA-SHI 930
JAPANESE

CIVIL ENGINEERING IN JAPAN 6789
DOBOKU GAKKAI (JAPAN SOCIETY OF CIVIL ENGINEERS)
1-CHOME, YOTSUYA, SHINJUKU-KU, TOKYO 160
ENGLISH
INDEXED: APP, ENG, WAA

COASTAL ENGINEERING IN JAPAN 6790
DOBOKU GAKKAI (JAPAN SOCIETY OF CIVIL ENGINEERS)
1-CHOME, YOTSUYA, SHINJUKU-KU, TOKYO 160
ENGLISH
INDEXED: ENG

DOBOKU GAKKAI HOKKAIDO-SHIBU KENKYU HAPPYOKAI RONBUNSHU 6791
(RESEARCH REPORT AT THE ANNUAL MEETING OF HOKKAIDO BRANCH, JAPAN SOCIETY OF CIVIL ENGINEERS)
DOBOKU GAKKAI HOKKAIDO-SHIBU
NISHI-2-CHOME, MINAMI-1-JO, CHUO-KU, SAPPORO-SHI 060
JAPANESE
INDEXED: JICST

DOBOKU GAKKAI KAIGAN KOGAKU KOENKAI RONBUNSHU 6792
(PROCEEDINGS OF THE CONFERENCE ON COASTAL ENGINEERING)
DOBOKU GAKKAI KAIGAN KOGAKU IINKAI (COMMITTEE ON COASTAL ENGINEERING, JAPAN SOCIETY OF CIVIL ENGINEERS)
1-CHOME, YOTSUYA, SHINJUKU-KU, TOKYO 160
JAPANESE
INDEXED: JICST

DOBOKU GAKKAI KANSAI SHIBU DAYORI 6793
(LETTERS, JAPAN SOCIETY OF CIVIL ENGINEERS, KANSAI BRANCH)
JAPANESE

DOBOKU GAKKAI KANSAI SHIBU KENSHUKAI SHIRYO 6794
(RESEARCH GROUP MATERIALS, JAPAN SOCIETY OF CIVIL ENGINEERS, KANSAI BRANCH)
JAPANESE
INDEXED: JICST

DOBOKU GAKKAI KANSAI SHIBU NENJI GAKUJUTSU KOENKAI KOEN GAIYO 6795
(PROCEEDINGS OF ANNUAL CONFERENCE ON CIVIL ENGINEERS, KANSAI BRANCH, JAPAN SOCIETY OF CIVIL ENGINEERS)
DOBOKU GAKKAI KANSAI SHIBU
2-2, SENBA BHUO, HIGASHI-KU, OSAKA-SHI 541
JAPANESE
INDEXED: JICST

DOBOKU GAKKAI NENJI GAKUJUTSU KOENKAI KOEN GAIYOSHU 6796
(PROCEEDINGS OF ANNUAL CONFERENCE ON CIVIL ENGINEERS)
DOBOKU GAKKAI (JAPAN SOCIETY OF CIVIL ENGINEERS)
1-CHOME, YOTSUYA, SHINJUKU-KU, TOKYO 160
JAPANESE
INDEXED: JICST

DOBOKU GAKKAI RONBUN HOKOKUSHU 6797
(COLLECTED PAPERS OF JAPAN SOCIETY OF CIVIL ENGINEERS)
DOBOKU GAKKAI (JAPAN SOCIETY OF CIVIL ENGINEERS)
1-CHOME, YOTSUYA, SHINJUKU-KU, TOKYO 160
JAPANESE AND ENGLISH
INDEXED: APT, JICST

DOBOKU GAKKAI SEIBU KENKYU HAPPYOKAI RONBUNSHU 6798
(COLLECTED PAPERS OF THE RESEARCH CONFERENCE, JAPAN SOCIETY OF CIVIL ENGINEERS, WESTERN BRANCH)
JAPANESE
INDEXED: JICST

DOBOKU GAKKAISHI 6799
(JOURNAL OF THE JAPAN SOCIETY OF CIVIL ENGINEERS)
DOBOKU GAKKAI (JAPAN SOCIETY OF CIVIL ENGINEERS)
1-CHOME, YOTSUYA, SHINJUKU-KU, TOKYO 160
JAPANESE, WITH ENGLISH SUMMARY
INDEXED: APP, ENG, JICST

DOBOKU GIJUTSU 6800
(CIVIL ENGINEERING)
DOBOKU GIJUTSUSHA (CIVIL ENGINEERING)
4-6, 1-CHOME, SOTO KANDA, CHIYODA-KU, TOKYO 101
JAPANESE
INDEXED: JICST

DOBOKU GIJUTSU NENPO (MITSUI KENSETSU K.K.) 6801
(ANNUAL REPORT ON CIVIL ENGINEERING, MITSUI CONSTRUCTION COMPANY, LIMITED)
MITSUI KENSETSU K.K. DOBOKU GIJUTSU-BU
1-1, 2-CHOME, NIHONBASHI MURO-MACHI, CHUO-KU, TOKYO 103
JAPANESE

DOBOKU GIJUTSU SHIRYO 6802
(PUBLIC WORKS REVIEW)
DOBOKU KENKYUKAI (PUBLIC WORKS RESEARCH ASSOCIATION)
C/O KENSETSU-SHO DOBOKU KENKYUSHO, 28-32, 2-CHOME, KOMAGOME, HONGO, BUNKYO-KU, TOKYO 113
JAPANESE
INDEXED: JICST

DOBOKU KEIKAKUGAKU SHINPOKUMU 6803
(SYMPOSIUM FOR CIVIL PLANNING)
DOBOKU GAKKAI DOBOKU KEIKAKUGAKU IINKAI (JAPAN SOCIETY OF CIVIL ENGINEERS)
1-CHOME, YOTSUYA, SHINJUKU-KU, TOKYO 160
JAPANESE
INDEXED: JICST

DOBOKU KENCHIKU SHINPO 6804
(CIVIL ENGINEERING AND CONSTRUCTION NEWS)
DOBOKU KENCHIKU SHINPOSHA
C/O ETO BUILDING, 6-13, 1-CHOME, MAIZURU, CHUO-KU, FUKUOKA-SHI 810
JAPANESE

DOBOKU KENKYUSHO HOKAKU 6805
(REPORT OF THE PUBLIC WORKS RESEARCH INSTITUTE, MINISTRY OF CONSTRUCTION)
KENSETSU-SHO DOBOKU KENKYUSHO (PUBLIC WORKS RESEARCH INSTITUTE, MINISTRY OF CONSTRUCTION)
28-32, 2-CHOME, HONKOMAGOME, BUNKYO-KU, TOKYO 113
JAPANESE
INDEXED: JICST

DOBOKU KENKYUSHO IHO 6806
(BULLETIN OF PUBLIC WORKS RESEARCH INSTITUTE)
KENSETSU-SHO DOBOKU KENKYUSHO (PUBLIC WORKS RESEARCH INSTITUTE, MINISTRY OF CONSTRUCTION)
28-32, 2-CHOME, HONKOMAGOME, BUNKYO-KU, TOKYO 113
JAPANESE
INDEXED: JICST

DOBOKU KENKYUSHO NENPO 6807
(ANNUAL REPORT OF THE PUBLIC WORKS RESEARCH INSTITUTE)
KENSETSU-SHO DOBOKU KENKYUSHO (PUBLIC WORKS RESEARCH INSTITUTE, MINISTRY OF CONSTRUCTION)
28-32, 2-CHOME, HONKOMAGOME, BUNKYO-KU, TOKYO 113
JAPANESE
INDEXED: JICST

DOBOKU KENKYUSHO SHIRYO 6808
(PUBLIC WORKS RESEARCH INSTITUTE REVIEW)
KENSETSU-SHO DOBOKU KENKYUSHO (PUBLIC WORKS RESEARCH INSTITUTE, MINISTRY OF CONSTRUCTION)
28-32, 2-CHOME, HONKOMAGOME, BUNKYO-KU, TOKYO 113
JAPANESE

DOBOKU KENSETSU 6809
(CIVIL ENGINEERING CONSTRUCTION)
DENRYOKU KENSETSUGYO KYOKAI AND
DOBOKU KOGYO KYOKAI
5-1, 2-CHOME, HATCHOBORI, CHUO-KU, TOKYO 104
JAPANESE
INDEXED: JICST

DOBOKU SEKO 6810
(PUBLIC WORKS)
SANKAIDO
15, SAIKU-CHO, SHINJUKU-KU, TOKYO 162
JAPANESE
INDEXED: JICST

DOBOKU SHIKENJO GAIYO 6811
(OPERATIONAL OUTLINE OF THE CIVIL ENGINEERING RESEARCH INSTITUTE)
HOKKAIDO KAIHATSU-CHO DOBOKU SHIKENJO (CIVIL ENGINEERING RESEARCH INSTITUTE OF HOKKAIDO DEVELOPMENT AGENCY)
3-CHOME, 1-JO, HIRAGISHI, TOYOHIRA-KU, SAPPORO-SHI 062
JAPANESE

DOBOKU SHIKENJO GEPPO 6812
(MONTHLY REPORT OF CIVIL ENGINEERING RESEARCH INSTITUTE, HOKKAIDO DEVELOPMENT AGENCY)
HOKKAIDO KAIHATSU-CHO DOBOKU SHIKENJO (CIVIL ENGINEERING RESEARCH INSTITUTE, HOKKAIDO DEVELOPMENT AGENCY)
3-CHOME, 1-JO, HIRAGISHI, TOYOHIRA-KU, SAPPORO-SHI 062
JAPANESE

DOBOKU SHIKENJO HOKOKU 6813
(REPORT OF THE CIVIL ENGINEERING RESEARCH INSTITUTE, HOKKAIDO DEVELOPMENT AGENCY)
HOKKAIDO KAIHATSU-CHO DOBOKU SHIKENJO (CIVIL ENGINEERING RESEARCH INSTITUTE, HOKKAIDO DEVELOPMENT AGENCY)
3-CHOME, 1-JO, HIRAGISHI, TOYOHIRA-KU, SAPPORO-SHI 062
JAPANESE

DOBOKU SHIKENJO NENPO 6814
(ANNUAL REPORT OF THE CIVIL ENGINEERING RESEARCH INSTITUTE OF HOKKAIDO DEVELOPMENT AGENCY)
HOKKAIDO KAIHATSU-CHO DOBOKU SHIKENJO (CIVIL ENGINEERING RESEARCH INSTITUTE OF HOKKAIDO DEVELOPMENT AGENCY)
3-CHOME, 1-JO, HIRAGISHI, TOYOHIRA-KU, SAPPORO-SHI 062
JAPANESE

DOSHITSU KOGAKKAI RONBUN HOKOKUSHU 6815
(JOURNAL OF THE JAPANESE SOCIETY OF SOIL MECHANICS AND FOUNDATION ENGINEERING)
DOSHITSU KOGAKKAI (JAPANESE SOCIETY OF SOIL MECHANICS AND FOUNDATION ENGINEERING)
13-5, 1-CHOME, NISHI SHINBASHI, MINATO-KU, TOKYO 105
JAPANESE AND ENGLISH
INDEXED: JICST

DOSHITSU KOGAKU KENKYU HAPPOKAI KOENSHU 6816
(PROCEEDINGS OF THE ANNUAL MEETING OF SOIL MECHANICS AND FOUNDATION ENGINEERING)
DOSHITSU KOGAKKAI (JAPANESE SOCIETY OF SOIL MECHANICS AND FOUNDATION ENGINEERING)
13-5, 1-CHOME, NISHI SHINBASHI, MINATO-KU, TOKYO 105
JAPANESE
INDEXED: JICST

DOSHITSU KOGAKU SHINPOJUMU 6817
(SYMPOSIUM ON SOIL MECHANICS AND FOUNDATION ENGINEERING)
DOSHITSU KOGAKKAI (JAPANESE SOCIETY OF SOIL MECHANICS AND FOUNDATION ENGINEERING)
13-5, 1-CHOME, NISHI SHINBASHI, MINATO-KU, TOKYO 105
JAPANESE
INDEXED: JICST

EHIME KENSETSU TAIMUSU 6818
(EHIME CONSTRUCTION TIMES)
EHIME KENSETSU TAIMUSUSHA (EHIME CONSTRUCTION TIMES COMPANY)
4-4, NIBAN-CHO, MATSUYAMA-SHI 790
JAPANESE

EHIME-KEN KENSETSU KENKYUSHOHO 6819
(REPORT OF THE EHIME-KEN CONSTRUCTION LABORATORY)
EHIME-KEN KENSETSU KENKYUSHO (EHIME-KEN CONSTRUCTION LABORATORY)
440, HIGASHIISHII-MACHI, MATSUYAMA-SHI 790
JAPANESE

FUJITA KOGYO GIJUTSU KENKYUSHO SHOHO 6820
(JOURNAL OF TECHNICAL LABORATORY OF FUJITA CORPORATION)
FUJITA KOGYO K.K. (FUJITA CORPORATION)
74, OTANA-MACHI, KOHOKU-KU, YOKOHAMA-SHI 223
JAPANESE
INDEXED: JICST

GANBAN RIKIGAKU NI KANSURU SHINPOJIUMU KOEN GAIYO 6821
(PROCEEDINGS OF THE SYMPOSIUM ON ROCK MECHANICS)
JAPANESE
INDEXED: JICST

GEKKAN KENSETSU 6822
(MONTHLY MAGAZINE 'KENSETSU' (CONSTRUCTION ENGINEER))
ZEN NIHON KENSETSU GIJUTSU KYOKAI (JAPAN CONSTRUCTION ENGINEERS ASSOCIATION)
C/O A I U AKASAKA BLDG., 1-2, 3-CHOME, AKASAKA, MINATO-KU, TOKYO 107
JAPANESE
INDEXED: JICST

GEKKAN SEKO KANRI 6823
(CONSTRUCTION MANAGEMENT)
KENSETSU KEIESHA
2-12, KANDA OGAWA-CHO, CHIYODA-KU, TOKYO 101
JAPANESE

GIJUTSU KAIHATSU SENTA GIJUTSU KENKYU HOKOKUSHU 6824
(COLLECTED TECHNICAL RESEARCH REPORTS OF THE TECHNICAL DEVELOPMENT CENTER)
TODA KENSETSU GIJUTSU KAIHATSU SENTA (TODA CONSTRUCTION TECHNICAL DEVELOPMENT CENTER)
JAPANESE

GIJUTSU KANRI GYOMU NO SEIKA GAIYO 6825
(RESEARCH ON ENGINEERING MANAGEMENT, TOHOKU REGION)
KENSETSU-SHO TOHOKU CHIHO KENSETSU-KYOKU TOHOKU GIJUTSU JIMUSHO (TOHOKU TECHNICAL OFFICE, TOHOKU REGIONAL CONSTRUCTION BUREAU, MINISTRY OF CONSTRUCTION)
6-1, 3-CHOME, SAKURAGI, TAGAJO-SHI 985
JAPANESE

GIJUTSU KANRI GYOMU SEIKA HOKOKU 6826
(RESEARCH ON ENGINEERING MANAGEMENT, CHUGOKU REGION)
KENSETSU-SHO CHUGOKU CHIHO KENSETSU-KYOKU CHUGOKU GIJITSU JIMUSHO (CHUGOKU TECHNICAL, OFFICE CHUGOKU REGIONAL CONSTRUCTION BUREAU, MINISTRY OF CONSTRUCTION)
1920-1, FUNAKOSHI-MACHI, AKI-GUN, HIROSHIMA 736
JAPANESE

GIJUTSU KENKYU HOKOKU (TODA KENSETSU K.K. KAIHATSU SENTA MATSUIDO KENKYUSHO) 6827
(REPORT OF TECHNICAL RESEARCH LABORATORY)
TODA KENSETSU K.K. KAIHATSU SENTA MATSUDO KENKYUSHO
209, KAMI HONGO, MATSUDO-SHI 271
JAPANESE

GIKENHO (HOKKAIDO KAIHATSU-KYOKU KIKAI KOSAKUJO) 6828
(ENGINEERING REPORT (HOKKAIDO DEVELOPMENT BUREAU, CONSTRUCTION MACHINERY WORKS))
HOKKAIDO KAIHATSU-KYOKU KENSETSU KIKAI KOSAKUJO (HOKKAIDO DEVELOPMENT BUREAU, CONSTRUCTION MACHINERY WORKS) AND HOKKAIDO KAIHATSU-CHO (HOKKAIDO DEVELOPMENT AGENCY)
8, HIGASHI-2-JO, TSUKISAMU, TOYOHIRA-KU, SAPPORO-SHI 062
JAPANESE

HASHI 6829
(BRIDGES)
DOBOKU GAKKAI (JAPAN SOCIETY OF CIVIL ENGINEERS)
1-CHOME, YOTSUYA, SHINJUKU-KU, TOKYO 160
JAPANESE AND ENGLISH

HAZAMAGUMI KENKYU NENPO 6830
(TECHNICAL RESEARCH REPORT OF HAZAMA-GUMI LIMITED)
HAZAMAGUMI GIJUTSU-KYOKU (TECHNICAL BUREAU, HAZAMAGUMI LIMITED)
5-8, 2-CHOME, KITA AOYAMA, MINATO-KU, TOKYO 107
JAPANESE
INDEXED: JICST

HOKKAIDO KAIHATSU-KYOKU DOBOKU SHIKENJO GEPPO 6831
(CIVIL ENGINEERING LABORATORY MONTHLY, HOKKAIDO DEVELOPMENT BUREAU)
JAPANESE
INDEXED: JICST

HOKKAIDO KAIHATSU-KYOKU DOBOKU SHIKENJO HOKOKU 6832
(CIVIL ENGINEERING LABORATORY MONTHLY, HOKKAIDO DEVELOPMENT BUREAU)
JAPANESE
INDEXED: JICST

HOKKAIDO KAIHATSU-KYOKU DOBOKU SHIKENJO NENPO 6833
(CIVIL ENGINEERING LABORATORY ANNUAL, HOKKAIDO DEVELOPMENT BUREAU)
JAPANESE
INDEXED: JICST

I E C (INTERNATIONAL ENGINEERING CONSULTANTS) 6834
KOKUSAI KENSETSU GIJUTSU KYOKAI (INTERNATIONAL ENGINEERING CONSULTANTS ASSOCIATION)
C/O NEW KOJIMACHI BLDG., 5-3, KOJI-MACHI, CHIYODA-KU, TOKYO 100
JAPANESE
INDEXED: JICST

IWATE-KEN DOBOKU GAIYO 6835
(OUTLINE OF PUBLIC WORKS, IWATE PREFECTURE)
IWATE-KEN DOBOKU-BU (PUBLIC WORKS DIVISION, IWATE PREFECTURAL GOVERNMENT)
10-1, UCHIMARU, MORIOKA-SHI 020
JAPANESE

JARI JIHO 6836
(GRAVEL TIMES)
NIHON JARI KYOKAI (NIHON GRAVEL SOCIETY)
10-20, 3-CHOME, IIDABASHI, CHIYODA-KU, TOKYO 102
JAPANESE

JOURNAL OF RESEARCH PUBLIC WORKS RESEARCH INSTITUTE 6837
KENSETSU-SHO DOBOKU KENKYUSHO (PUBLIC WORKS RESEARCH INSTITUTE, MINISTRY OF CONSTRUCTION)
28-32, 2-CHOME, HONKOMAGOME, BUNKYO-KU, TOKYO 113
ENGLISH

JSSC 6838
NIHON KO KOZO KYOKAI (SOCIETY OF STEEL CONSTRUCTION OF JAPAN)
C/O SHIN TOKYO BLDG., 3-1, 3-CHOME, MARUNOUCHI, CHIYODA-KU, TOKYO 100
JAPANESE
INDEXED: JICST

JUTAKU KENKYU HOKOKU (DOBOKU KENKYUSHO) 6839
(RESEARCH REPORT AT REQUEST)
KENSETSU-SHO DOBOKU KENKYUSHO (PUBLIC WORKS RESEARCH INSTITUTE, MINISTRY OF CONSTRUCTION)
28-32, 2-CHOME, HONKOMAGOME, BUNKYO-KU, TOKYO 113
JAPANESE

KAGAKU GIJUTSU BUNKEN SOKUHO, DOBOKU, KENCHIKU KOGAKU-HEN 6840
(CURRENT BIBLIOGRAPHY ON SCIENCE AND TECHNOLOGY, SERIES ON CIVIL ENGINEERING AND ARCHITECTURE)
NIHON KAGAKU GIJUTSU JOHO SENTA (JAPAN INFORMATION CENTER OF SCIENCE AND TECHNOLOGY)
1-1, 2-CHOME, NAGATA-CHO, CHIYODA-KU, TOKYO 100
JAPANESE
INDEXED: JICST

KAIGAI NO KENSETSU 6841
(CONSTRUCTION IN FOREIGN COUNTRIES)
ZEN NIHON KENSETSU GIJUTSU KYOKAI (JAPAN CONSTRUCTION ENGINEERS ASSOCIATION)
C/O AIU AKASAK BUILDING, 1-2, 3-CHOME, AKASAKA, MINATO-KU, TOKYO 107
JAPANESE

KAJIMA KENSETSU GEPPO 6842
(KAJIMA MONTHLY)
KAJIMA KENSETSU KOHOBU
2-7, 1-CHOME, MOTOAKASAKA, MINATO-KU, TOKYO 105
JAPANESE AND ENGLISH

KAJIMA KENSETSU GIJUTSU KENKYUSHO NENPO 6843
(ANNUAL REPORT OF KAJIMA INSTITUTE OF CONSTRUCTION TECHNOLOGY)
KAJIMA KENSETSU GIJUTS KENKYUSHO (KAJIMA INSTITUTE OF CONSTRUCTION TECHNOLOGY)
19-1, 2-CHOME, TOBITAKYU, CHOFU-SHI, TOKYO 182
JAPANESE
INDEXED: JICST

KENCHIKU DAIJESUTO 6844
(ARCHITECTURE DIGEST)
NIHON KENCHIKU BUNKEN SENTA (JAPAN ARCHITECTURE BIBLIOGRAPHY CENTER)
JAPANESE

KENCHIKU GIJUTSU (MITSUI KENSETSU K.K. GIJUTSU KENKYUSHITSU) 6845
(BUILDING TECHNOLOGY, TECHNICAL RESEARCH LABORATORY MITSUI CONSTRUCTION COMPANY, LIMITED)
MITSUI KENSETSU K.K. GIJUTSU KENKYUSHITSU
1-1, 2-CHOME NIHONBASHI MURO-MACHI, CHUO-KU, TOKYO 103
JAPANESE

KENSETSU BUKKA 6846
(CONSTRUCTION PRICE)
KENSETSU BUKKA CHOSAKAI
15-14, 2-CHOME, TSUKIJI, CHUO-KU, TOKYO 104
JAPANESE

KENSETSU GEPPO 6847
(MONTHLY BULLETIN OF CONSTRUCTION)
KENSETSU-SHO DAIJIN KANBO KOHOSHITSU (MINISTER'S SECRETARIAT, MINISTRY OF CONSTRUCTION)
C/O NISHISHINBASHI CHUO BUILDING, 15-8, 3-CHOME, NISHI SHINBASHI, MINATO-KU, TOKYO 105
JAPANESE
INDEXED: JICST

KENSETSU GEPPO TOHOKU 6848
(MONTHLY REPORT OF TOHOKU REGIONAL CONSTRUCTION BUREAU)
KENSETSU-SHO TOHOKU KENSETSU-KYOKU (TOHOKU REGIONAL CONSTRUCTION BUREAU, MINISTRY OF CONSTRUCTION) AND TOHOKU KENSETSU KYOKAI
10-15, FUTSUKA-MACHI, SENDAI-SHI 980
JAPANESE
INDEXED: JICST

KENSETSU GIJUTSU NARABINI KENSETSU KOGYO GIJUTSU KENKYU HOJOKIN NI YORU KENKYU HOKOKU SHOROKU 6849
(ABSTRACTS FROM RESEARCH REPORTS GRANTED TO CONSTRUCTION TECHNIC AND CONSTRUCTION INDUSTRY ENGINEERING STUDIES)
KENSETSU-SHO KEIKAKU-KYOKU (PLANNING BUREAU, MINISTRY OF CONSTRUCTION)
1-3, 2-CHOME, KASUMIGASEKI, CHIYODA-KU, TOKYO 100
JAPANESE

KENSETSU GIKEN 6850
(CONSTRUCTION ENGINEERING)
KENSETSU GIJITSU KENKYUSHA
30-9, 2-CHOME, HONGO, BUNKYO-KU, TOKYO 113
JAPANESE

KENSETSU GYOMU TOKEI NENPO 6851
(ANNUAL REPORT OF CONSTRUCTION WORKS STATISTICS)
KENSETSU-SHO KEIKAKU-KYOKU (PLANNING BUREAU, MINISTRY OF CONSTRUCTION)
1-3, 2-CHOME, KASUMIGASEKI, CHIYODA-KU, TOKYO 100
JAPANESE

KENSETSU KEIZAI GEPPO 6852
(MONTHLY OF CONSTRUCTION STATISTICS)
KENSETSU BUKKA CHOSAKAI (CONSTRUCTION PRICES INVESTIGATION INSTITUTE)
15-14, 2-CHOME, TSUKIJI, CHUO-KU, TOKYO 104
JAPANESE

KENSETSU KIKAI 6853
(CONSTRUCTION MACHINERY)
NIHON KOGYO SHUPPANSHA
C/O ISHIZAKA BUILDING, 3-18, 2-CHOME, NISHI KANDA, CHIYODA-KU, TOKYO 101
JAPANESE
INDEXED: CHE, JICST

KENSETSU KIKAIKA KENKYUSHO NENPO 6854
(ANNUAL REPORT OF CONSTRUCTION METHOD AND MACHINERY INSTITUTE)
KENSETSU KIKAIKA KENKYUSHO (CONSTRUCTION METHOD AND MACHINERY RESEARCH INSTITUTE)
3154, OBUCHI, FUJI-SHI 417
JAPANESE
INDEXED: JICST

KENSETSU KINKI 6855
(KINKI REGIONAL CONSTRUCTION BUREAU MONTHLY)
KENSETSU-SHO KINKI CHIHO KENSETSU-KYOKU (KINKI REGIONAL CONSTRUCTION BUREAU, MINISTRY OF CONSTRUCTION) AND KINKI KENSETSU KYOKAI (KINKI ASSOCIATION OF CONSTRUCTION)
C/O OSAKA ORUGAN BUILDING, 1-27, TANI-MACHI, HIGASHI-KU, OSAKA-SHI 540
JAPANESE
INDEXED: JICST

KENSETSU KOGAKU KENKYU SHINKOKAI NENPO 6856
(ANNUAL REPORT OF THE ASSOCIATION FOR THE PROMOTION OF CONSTRUCTION ENGINEERING)
KENSETSU KOGAKU KENKYU SHINKOKAI
4-6, KITAMI-CHO, SENDAI-SHI 980
JAPANESE
INDEXED: CHE, JICST

KENSETSU KOGAKU KENKYUSHO HOKOKU 6857
(REPORTS OF THE CONSTRUCTION ENGINEERING RESEARCH INSTITUTE FOUNDATION)
KENSETSU KOGAKU KENKYUSHO (CONSTRUCTION ENGINEERING RESEARCH INSTITUTE FOUNDATION)
3-10, 1-CHOME, TSURUKO, NADA-KU, KOBE-SHI 657
JAPANESE
INDEXED: JICST

KENSETSU KOGYO BUKKA CHINGIN GEPPO 6858
(MONTHLY REPORT OF PRICE AND WAGE IN CONSTRUCTION ENGINEERING)
SHIMIZU KENSETSU K.K. (SHIMIZU CONSTRUCTION COMPANY, LIMITED)
1-1, 2-CHOME, TAKARA-CHO, CHUO-KU, TOKYO 104
JAPANESE

KENSETSU KOJI NO UGOKI 6859
(TENDENCY OF CONSTRUCTION WORKS)
NIKKAN KENSETSU TSUSHIN SHINSHA
C/O TACHIKAWA BUILDING, 11-4, 2-CHOME, NISHISHINBASHI, MINATO-KU, TOKYO 105
JAPANESE

KENSETSU KOJI SEKO TOKEI CHOSA HOKOKU 6860
(REPORT OF COMMENCEMENT STATISTICS OF CONSTRUCTION REPORT)
KENSETSU-SHO KEIKAKU-KYOKU (PLANNING BUREAU, MINISTRY OF CONSTRUCTION)
1-3, 2-CHOME, KASUMIGASEKI, CHIYODA-KU, TOKYO 100
JAPANESE

KENSETSU NO KIKAIKA 6861
(CONSTRUCTION MECHANIZATION)
NIHON KENSETSU KIKAIKA KYOKAI (JAPAN CONSTRUCTION MECHANIZATION ASSOCIATION)
C/O KIKAI SHINKO KAIKAN, 5-8, 3-CHOME, SHIBAKOEN, MINATO-KU, TOKYO 105
JAPANESE
INDEXED: JICST

KENSETSU SHIRYO 6862
(CONSTRUCTION MATERIALS)
KENSETSU SOGO SHIRYOSHA
C/O SAKUMA BUILDING, 5-6, 2-CHOME, NISHI KANDA, CHIYODA-KU, TOKYO 101
JAPANESE

KENSETSU SOGO KENKYU 6863
(COMPREHENSIVE CONSTRUCTION RESEARCH)
KENSETSU CHOSA KAI (CONSTRUCTION RESEARCH ASSOCIATION)
JAPANESE
INDEXED: JICST

KENSETSU SOGO TOKEI KEKKAHYO 6864
(GENERAL STATISTICS OF PUBLIC WORKS)
KENSETSU-SHO KEIKAKU-KYOKU (PLANNING BUREAU, MINISTRY OF CONSTRUCTION)
1-3, 2-CHOME, KASUMIGASEKI, CHIYODA-KU, TOKYO 100
JAPANESE

KENSETSU SOGO TOKEI NENDOHO 6865
(GENERAL STATISTICAL YEARBOOK OF CONSTRUCTION)
KENSETSU-SHO KEIKAKU-KYOKU (PLANNING BUREAU, MINISTRY OF CONSTRUCTION)
1-3, 2-CHOME, KASUMIGASEKI, CHIYODA-KU, TOKYO 100
JAPANESE

KENSETSU-KYOKU JIGYO GAIYO 6866
(SUMMARY REPORT OF THE BUREAU OF CONSTRUCTION, TOKYO METROPOLITAN GOVERNMENT)
TOKYO TO KENSETSU-KYOKU (BUREAU OF CONSTRUCTION, TOKYO METROPOLITAN GOVERNMENT)
5-1, 3-CHOME, MARUNOUCHI, CHIYODA-KU, TOKYO 100
JAPANESE

KENSETSU-SHO GIJUTSU KENKYUKAI HOKOKU 6867
(REPORT OF THE TECHNICAL RESEARCH ASSOCIATION, MINISTRY OF CONSTRUCTION)
JAPANESE
INDEXED: JICST

KENSETSU-SHO TOHOKU CHIHO KENSETSU-KYOKU JIGYO NENPO 6868
(ANNUAL REPORT OF TOHOKU DISTRICT BUREAU OF CONSTRUCTION.)
TOHOKU CHIHO KENSETSU-KYOKU
9-15, FUTSUKA-MACHI, SENDAI-SHI 980
JAPANESE

KENZAI KENKYUSHO KENKYU HOKOKU 6869
(RESEARCH REPORT OF RESEARCH MATERIAL, NITTO BOSEKI COMPANY))
NITTO BOSEKI K.K. KENZAI KENKYUSHO (RESEARCH CENTER OF BUILDING MATERIAL, NITTO BOSEKI COMPANY)
210, ROPPO-CHO, CHIBA-SHI 284
JAPANESE

KENZAI SHIKEN JOHO 6870
(BUILDING MATERIAL TESTING NEWS)
KENZAI SHIKEN SENTA (TESTING CENTER OF BUILDING MATERIALS)
C/O TSUSHO SANGYO-SHO BUNSHITSU, 15-1, 6-CHOME, GINZA, CHUO-KU, TOKYO 104
JAPANESE
INDEXED: JICST

KISOKO 6871
(FUNDAMENTAL ENGINEERING)
SOGO DOBOKU KENKYUJO (PUBLIC WORKS GENERAL RESEARCH INSTITUTE)
JAPANESE
INDEXED: JICST

KOGYO JIJI TSUSHIN, KENSETSU KIKAIBAN 6872
(INDUSTRIAL NEWS, CONSTRUCTION MACHINERY EDITION)
KOGYO JIJI TSUSHINSHA (THE INDUSTRIAL NEWS)
3-10, KANDA OGAWA-CHO, CHIYODA-KU, TOKYO 101
JAPANESE

KOJIYO ZAIRYO SHIKEN HOKOKU 6873
(TESTING REPORT OF CONSTRUCTION MATERIALS)
TOKYO-TO DOBOKU GIJUTSU KENKYUSHO (TOKYO METROPOLITAN CIVIL ENGINEERING INSTITUTE)
1-18, 1-CHOME, KONAN, MINATO-KU, TOKYO 108
JAPANESE

KOKUDO KENSETSU NO GENKYO 6874
(PRESENT STATE OF THE LAND CONSTRUCTION)
KENSETSU-SHO (MINISTRY OF CONSTRUCTION)
1-3, 2-CHOME, KASUMIGASEKI, CHIYODA-KU, TOKYO 100
JAPANESE

KOKYO KOJI CHAKKO TOKEI GEPPO, MINKAN DOBOKU KOJI CHAKKO TOKEI GEPPO 6875
(MONTHLY REPORT OF STATISTICS OF PUBLIC WORKS COMMENCEMENT)
KENSETSU-SHO KEIKAKU-KYOKU (PLANNING BUREAU, MINISTRY OF CONSTRUCTION)
1-3, 2-CHOME, KASUMIGASEKI, CHIYODA-KU, TOKYO 100
JAPANESE

KOKYO KOJI CHAKKO TOKEI NENDOHO 6876
(ANNUAL REPORT OF STATISTICS OF PUBLIC WORKS COMMENCEMENT)
KENSETSU-SHO KEIKAKU-KYOKU (PLANNING BUREAU, MINISTRY OF CONSTRUCTION)
1-3, 2-CHOME, KASUMIGASEKI, CHIYODA-KU, TOKYO 100
JAPANESE

KONKURITO KAGAKU 6877
(CONCRETE JOURNAL)
NIHON KONKURITO KAIGI (JAPAN CONCRETE CONFERENCE)
JAPANESE
INDEXED: JICST

KOZO KOGAKU SHINPOJIUMU 6878
(STRUCTURAL ENGINEERING SYMPOSIUM)
JAPANESE
INDEXED: JICST

KOZOBUTSU NO TAIFUSEI NI KANSURU SHINPOJUMU RONBUNSHU 6879
(PROCEEDINGS OF THE SYMPOSIUM ON WIND EFFECTS ON STRUCTURES IN JAPAN)
NIHON KISHO GAKKAI (METEOROLOGICAL SOCIETY OF JAPAN)
C/O KISHO-CHO, 1-7, OTE-MACHI, CHIYODA-KU, TOKYO 100
JAPANESE AND ENGLISH
INDEXED: JICST

KUMAGAIGUMI GIHO 6880
(TECHNICAL REPORTS OF THE INSTITUTE OF CONSTRUCTION TECHNOLOGY, KUMAGAI GUMI COMPANY)
KUMAGAIGUMI GIJUTSU KENKYUSHO (INSTITUTE OF CONSTRUCTION TECHNOLOGY, KUMAGAIGUMI COMPANY, LTD.)
C/O TOKAI BUILDING, 1, SHIMOMIYABI-CHO, SHINJUKU-KY, TOKYO 162
JAPANESE

KYORYO 6881
(BRIDGE ENGINEERING)
KYORYO HENSANKAI
C/O NIKKEN JINBO-CHO BUILDING, 3-1, KANDA JINBO-CHO, CHIYODA-KU, TOKYO 101
JAPANESE

KYORYO GENKYO CHOSHO 6882
(DATA ON BRIDGES)
HOKKAIDO DOBOKU-BU (PUBLIC WORKS DIVISION, HOKKAIDO PREFECTURAL GOVERNMENT)
NISHI-6-CHOME, KITA-3-JO, CHUO-KU, SAPPORO-SHI 060
JAPANESE

KYORYO TO KISO 6883
(BRIDGE AND FOUNDATION)
KENSETSU TOSHO
C/O KYODO BLDG., 2-17, 2-CHOME, SOTO KANDA, CHIYODA-KU, TOKYO 101
JAPANESE
INDEXED: JICST

MAEDA KENSETSU KOGYO GIJUTSU KENKYUBUHO 6884
(REPORT OF THE TECHNOLOGY RESEARCH DIVISION, MAEDA CONSTRUCTION COMPANY, LIMITED))
MAEDA KENSETSU KOGYO K.K. GIJUTSU KENKYU-BU TECHNOLOGY RESEARCH DIVISION, MAEDA CONSTRUCTION COMPANY, LIMITED)
10-26, 2-CHOME, FUJIMI, CHIYODA-KU, TOKYO 102
JAPANESE

NIHON SAISEKI SHINBUN 6885
(JAPAN CRUSHED STONE NEWS)
NIHON SAISEKI SHINBUNSHA
C/O MIZUSHIMA KAIKAN, 2-11, 3-CHOME, UCHIKANDA, CHIYODA-KU, TOKYO 101
JAPANESE

NIKKAN KENSETSU KOGYO SHINBUN 6886
(DAILY CONSTRUCTION ENGINEERING NEWS)
NIKKAN KENSETSU KOGYO SHINBUNSHA
12-11, 8-CHOME, GINZA, CHUO-KU, TOKYO 104
JAPANESE

NIKKAN KENSETSU SANGYO SHINBUN 6887
(DAILY CONSTRUCTION INDUSTRY NEWS)
NIKKAN KENSETSU SANGYO SHINBUNSHA
48-8, 1-CHOME, ITABASHI, ITABASHI-KU, TOKYO 173
JAPANESE

NIKKAN KENSETSU TSUSHIN 6888
(DAILY CONSTRUCTION NEWS)
NIKKAN KENSETSU TSUSHIN SHINSHA
11-4, 2-CHOME, NISHI SHINBASHI, MINATO-KU, TOKYO 105
JAPANESE

NISSAN KENSETSU K.K. GIJUTSU KAIHATSUSHITSU GIJUTSU REPOTO 6889
(TECHNICAL REPORT OF TECHNOLOGY DEVELOPMENT OFFICE, NISSAN CONSTRUCTION COMPANY, LIMITED))
NISSAN KENSETSU K.K. GIJUTSU KAIHATSUSHITSU
2-8, 2-CHOME, MINAMIAOYAMA, MINATO-KU, TOKYO 107
JAPANESE

OBAYASHI-GUMI GIJUTSU KENKYUSHOHO 6890
(REPORT OF THE TECHNICAL RESEARCH INSTITUTE, OBAYASHI-GUMI, LIMITED)
OBAYASHI-GUMI GIJUTSU KENKYUSHO (TECHNICAL RESEARCH INSTITUTE, OBAYASHI-GUMI, LIMITED)
2-3, KANDA TSUKASA-CHO, CHIYODA-KU, TOKYO 101
JAPANESE AND ENGLISH
INDEXED: CHE, JICST

OKUMURA-GUMI GIJITSU KENKYU GIHO 6891
(TECHNICAL REPORT OF OKUMURA CONSTRUCTION COMPANY, LIMITED, TECHNICAL RESEARCH INSTITUTE)
OKUMURA-GUMI GIJUTSU KENKYUSHO (OKUMURA CONSTRUCTION COMPANY, LTD. TECHNICAL RESEARCH INSTITUTE)
5-8, 3-CHOME, HAMAGUCHI NISHI, SUMINOE-KU, OSAKA-SHI 559
JAPANESE

OKUMURA-GUMI GIJUTSU KENKYUSHO NENPO 6892
(ANNUAL REPORT OF OKUMURA CONSTRUCTION COMPANY, LIMITED, TECHNICAL RESEARCH INSTITUTE)
OKUMURA-GUMI GIJUTSU KENKYUSHO (OKUMURA CONSTRUCTION COMPANY, LIMITED, TECHNICAL RESEARCH INSTITUTE)
5-8, 3-CHOME, HAMAGUCHI NISHI, SUMINOE-KU, OSAKA-SHI 559
JAPANESE
INDEXED: JICST

P E T (PLANT ENGINEERING & TECHNOLOGY) 6893
CHEMICAL DAILY COMPANY, LTD.
19-16, SHIBARA, 3-CHOME, MINATO-KU, TOKYO
ENGLISH

PROCEEDINGS OF THE JAPAN SOCIETY OF CIVIL ENGINEERS. TOKYO 6894
JAPANESE AND ENGLISH
INDEXED: APP

RESEARCH ACTIVITIES IN CIVIL ENGINEERING AND RELEATED FIELDS, KYOTO UNIVERSITY 6895
DEPARTMENT OF CIVIL ENGINEERING, KYOTO UNIVERSITY
ENGLISH

SAITAMA DAIGAKU RIKO GAKUBU KENSETSU KISO KOGAKU KENKYU HOKOKU 6896
(REPORT OF DEPARTMENT OF FOUNDATION, ENGINEERING FACILITY OF SCIENCE AND ENGINEERING, SAITAMA UNIVERSITY)
SAITAMA DAIGAKU RIKO GAKUBU KENSETSU KISO KOGAKU KYOSHITSU (DEPARTMENT OF FOUNDATION, ENGINEERING, FACULTY OF SCIENCE AND ENGINEERING, SAITAMA UNIVERSITY)
255, SHIMOOKUBO, URAWA-SHI 338
JAPANESE AND ENGLISH
INDEXED: JICST

SEKO GIJUTSU 6897
(BUILDING EXECUTION)
NIKKAN KOGYO SHINBUNSHA (INDUSTRIAL DAILY NEWS)
8-10, 1-CHOME, KUDAN-KITA, CHIYODA-KU, TOKYO 102
JAPANESE
INDEXED: JICST

SEKO KANRI 6898
(CONSTRUCTION MANAGEMENT)
KENSETSU KEIEISHA
2-12, KANDA OGAWA-CHO, CHIYODA-KU, TOKYO 101
JAPANESE
INDEXED: JICST

SHIMIZU KENSETSU KENKYUSHO KENKYU HOKOKU 6899
(SPECIAL REPORT OF RESEARCH LABORATORY OF SHIMIZU CONSTRUCTION COMPANY, LIMITED)
SHIMIZU KENSETSU K.K. (SHIMIZU CONSTRUCTION COMPANY, LTD.)
1-1, 2-CHOME, TAKARA-CHO, CHUO-KU, TOKYO 104
JAPANESE, WITH ENGLISH SUMMARY
INDEXED: JICST

SHIMIZU KENSETSU KENKYUSHOHO 6900
(REPORTS OF THE RESEARCH LABORATORY OF SHIMIZU CONSTRUCTION COMPANY, LIMITED)
SHIMIZU KENSETSU K.K. (SHIMIZU CONSTRUCTION COMPANY, LIMITED)
1-1, 2-CHOME, TAKARA-CHO, CHUO-KU, TOKYO 104
JAPANESE, WITH EUROPEAN SUMMARY
INDEXED: JICST

SHIRUDO KOJI BUNKENSHU 6901
(JOURNAL OF SHIELD METHOD CONSTRUCTION)
NIHON SHIRUDO ENGINIARINGU (JAPAN SHIELD ENGINEERING)
C/O NIHON BUILDING, 6-2, 2-CHOME, OTE-MACHI, CHIYODAKU, TOKYO 100
JAPANESE

SHUKAN KENSETSU NYUSU 6902
(WEEKLY NEWS OF CONSTRUCTION)
KENSETSU JANARUSHA
2-7, KANDA TSUKASA-CHO, CHIYODA-KU, TOKYO 101
JAPANESE

SOILS AND FOUNDATION 6903
DOSHITSU KOGAKKAI (JAPANESE SOCIETY OF SOIL MECHANICS AND FOUNDATION ENGINEERING)
13-5, 1-CHOME, NISHISHINBASHI, MINATO-KU, TOKYO 105
EUROPEAN
INDEXED: APP, BIB, ENG

TAISEI KENSETSU GIJUTSU KENKYUSHOHO 6904
(REPORTS OF THE ENGINEERING RESEARCH DEPARTMENT, TAISEI CORPORATION)
TAISEI KENSETSU GIJUTSU KENKYUSHO
3-12, 2-CHOME, GINZA, CHUO-KU, TOKYO 104
JAPANESE, WITH ENGLISH SUMMARY
INDEXED: JICST

TAISEI KUOTARI 6905
(TAISEI QUARTERLY)
TAISEI KENSETSU K.K. (TAISEI CONSTRUCTION COMPANY, LIMITED)
5-11, 2-CHOME, GINZA, CHUO-KU, TOKYO 104
ENGLISH
INDEXED: JICST

TAKENAKA GIJUTSU KENKYU HOKOKU 6906
(TAKENAKA TECHNICAL RESEARCH REPORT)
TAKENAKA GIJUTSU KENKYUSHO (TAKENAKA TECHNICAL RESEARCH INSTITUTE)
5-14, 2-CHOME, MINAMISUNA, KOTO-KU, TOKYO 136
JAPANESE AND ENGLISH
INDEXED: INS, JICST

TEKKEN KOJI JOHO 6907
(NEWS OF TEKKEN CONSTRUCTION COMPANY)
TEKKEN KENSETSU K.K.
5-3, 2-CHOME, MISAKI-CHO, CHIYODA-KU, TOKYO 101
JAPANESE

TO-DOBOKU GIKEN SHIRYO 6908
(RESEARCH MATERIALS OF TOKYO INSTITUTE OF CIVIL ENGINEERING, TOKYO METROPOLITAN GOVERNMENT)
TOKYO-TO DOBOKU GIJUTSU KENKYUSHO
1-18, 1-CHOME, KONAN, MINATO-KU, TOKYO 108
JAPANESE
INDEXED: JICST

TOBISHIMA GIHO 6909
(TECHNICAL REPORT OF THE TOBISHIMA CONSTRUCTION COMPANY, LIMITED)
TOBISHIMA KENSETSU K.K. KOMU HONBU (TOBISHIMA CONSTRUCTION COMPANY, LIMITED)
3-28, 2-CHOME, KUDAN, CHIYODA-KU, TOKYO 102
JAPANESE

TOHOKU DAIGAKU KOGAKUBU DOBOKU KYOSHITSU KENKYU HOKOKU 6910
(TECHNOLOGY REPORTS OF DEPARTMENT OF CIVIL ENGINEERING IN FACULTY OF ENGINEERING, TOHOKU UNIVERSITY))
TOHOKU DAIGAKU KOGAKUBU DOBOKU KYOSHITSU (DEPARTMENT OF CIVIL ENGINEERING IN FACULTY OF ENGINEERING, TOHOKU UNIVERSITY)
AOBA, ARAMAKI, SENDAI-SHI 980
JAPANESE

TOHOKU KOGYO DAIGAKU KIYO, E. DOBOKU KOGAKU-HEN 6911
(BULLETIN OF THE TOHOKU INSTITUTE OF TECHNOLOGY, SECTION E; CIVIL ENGINEERING)
TOHOKU KOGYO DAIGAKU (TOHOKU INSTITUTE OF TECHNOLOGY)
19, KOSHIJI, SENDAI-SHI 982
JAPANESE, WITH ENGLISH SUMMARY
INDEXED: JICST

TOKAI KOGYO GIJUTSU KENKYUSHO HOKOKUSHU 6912
(REPORTS OF RESEARCH LABORATORY, TOKAI KOGYO COMPANY, LIMITED)
TOKAI KOGYO GIJUTSU KENKYUSHO
C/O KENSHIN GYOREN BUILDING, 1-15, OTA-CHO, NAKE-KU, YOKOHAMA-SHI 231
JAPANESE

TOKEN GEPPO 6913
(TOKEN MONTHLY)
TOKYO KENSETSUGYO KYOKAI (ASSOCIATION IN THE GENERAL CONTRACTORS OF TOKYO)
5-1, 2-CHOME, HATCHOBORI, CHUO-KU, TOKYO 104
JAPANESE
INDEXED: JICST

TOKUBETSU KENKYU SOKUSHIN CHOSAHI, KOKURITSU KIKAN KOGAI BOSHI TO SHIKEN KENKYUHI KENSETSU KIKAI KAIHATSU CHOSAHI KENSETSU GIJUTSU KENKYU KAIHATSU 6914
(REPORT ON COST OF RESEARCH FOR PROMOTION OF SPECIAL INVESTIGATION...)
KENSETSU-SHO DOBOKU KENKYUSHO (PUBLIC WORKS RESEARCH INSTITUTE, MINISTRY OF CONSTRUCTION)
28-32, 2-CHOME, HONKOMAGOME, BUNKYO-KU, TOKYO 113
JAPANESE

TOKYO DAIGAKU KOGAKUBU DOBOKU KOGAKKA RONBUN SHUROKU 6915
(TRANSACTION OF DEPARTMENT OF CIVIL ENGINEERING, FACULTY OF ENGINEERING, UNIVERSITY OF TOKYO)
TOKYO DAIGAKU KOGAKUBU (DEPARTMENT OF CIVIL ENGINEERING, UNIVERSITY OF TOKYO)
3-1, 7-CHOME, HONGO, BUNKYO-KU, TOKYO 113
JAPANESE

TOKYO KOGYO DAIGAKU DOBOKU KOGAKKA KENKYU HOKOKU 6916
(TECHNICAL REPORT, DEPARTMENT OF CIVIL ENGINEERING, TOKYO INSTITUTE OF TECHNOLOGY)
TOKYO KOGYO DAIGAKU DOBOKU KOGAKKA
12-1, 2-CHOME, OOKAYAMA, MEGURO-KU, TOKYO 152
JAPANESE

TOKYO-TO DAI-1 KENSETSU JIMUSHO JIGYO GAIYO 6917
(OPERATION REPORT OF 1ST CONSTRUCTION OFFICE, TOKYO)
TOKYO-TO DAI-1 KENSETSU JIMUSHO
2-4, AKASHI-CHO, CHUO-KU, TOKYO 104
JAPANESE

TOKYO-TO DAI-2 KENSETSU JIMUSHO JIGYO GAIYO 6918
(OPERATION REPORT OF 2ND CONSTRUCTION OFFICE, TOKYO)
TOKYO-TO DAI-2 KENSETSU JIMUSHO
1-36, 2-CHOME, HIRO-MACHI, SHINAGAWA-KU, TOKYO 140
JAPANESE

TOKYO-TO DAI-3 KENSETSU JIMUSHO JIGYO GAIYO 6919
(OPERATION REPORT OF 3RD CONSTRUCTION OFFICE, TOKYO)
TOKYO-TO DAI-3 KENSETSU JIMUSHO
8-1, 4-CHOME, NAKANO, NAKANO-KU, TOKYO 164
JAPANESE

TOKYO-TO DAI-3 KUKAKU SEIRI JIMUSHO JIGYO GAIYO 6920
(OPERATION REPORT OF 3RD CITY IMPROVEMENTS OFFICE, TOKYO)
TOKYO-TO DAI-3 KUKAKU SEIRI JIMUSHO
1-510, IKEBUKURO, TOHIMA-KU, TOKYO 170
JAPANESE

TOKYO-TO DAI-4 KENSETSU JIMUSHO JIGYO GAIYO 6921
(OPERATION REPORT OF 4TH CONSTRUCTION OFFICE, TOKYO)
TOKYO-TO DAI-4 KENSETSU JIMUSHO
3-9, 1-CHOME, NISHISUGAMO, TOSHIMA-KU, TOKYO 170
JAPANESE

TOKYO-TO DAI-5 KENSETSU JIMUSHO JIGYO GAIYO 6922
(OPERATION REPORT OF 5TH CONSTRUCTION OFFICE, TOKYO)
TOKYO-TO DAI-5 KENSETSU JIMUSHO
14-11, 1-CHOME, HIGASHISHINKOIWA, KATSUSHIKA-KU, TOKYO 124
JAPANESE

TOKYO-TO DAI-6 KENSETSU JIMUSHO JIGYO GAIYO 6923
(OPERATION REPORT OF 6TH CONSTRUCTION OFFICE, TOKYO)
TOKYO-TO DAI-6 KENSETSU JIMUSHO
13-1, 1-CHOME, ODAI, ADACHI-KU, TOKYO 120
JAPANESE

TOKYO-TO DOBOKU GIJUTSU KENKYUSHO NENPO 6924
(ANNUAL REPORT TO TOKYO INSTITUTE OF CIVIL ENGINEERING)
TOKYO-TO DOBOKU GIJUTSU KENKYUSHO (TOKYO INSTITUTE OF CIVIL ENGINEERING, TOKYO METROPOLITAN GOVERNMENT)
1-18, 1-CHOME, KONAN, MINATO-KU, TOKYO 108
JAPANESE
INDEXED: JICST

TOKYO-TO KENCHIKU ZAIRYO KENSAJO HOKOKU 6925
(REPORT OF TOKYO METROPOLITAN TESTING LABORATORY OF BUILDING MATERIALS))
TOKYO-TO KENCHIKU ZAIRYO KENSAJO (TOKYO METROPOLITAN TESTING LABORATORY OF BUILDING MATERIALS)
12-20, 1-CHOME, HIGASHIOI, SHINAGAWA-KU, TOKYO 140
JAPANESE
INDEXED: JICST

TOKYO-TO KITATAMA KENSETSU JIMUSHO JIGYO GAIYO 6926
(WORKS OF KITATAMA CONSTRUCTION OFFICE, TOKYO METROPOLITAN GOVERNMENT)
TOKYO-TO KITATAMA KENSETSU JIMUSHO
8220, FUCHU-SHI, TOKYO 183
JAPANESE

TOKYO-TO MINAMITAMA KENSETSU JIMUSHO JIGYO GAIYO 6927
(OPERATION REPORT OF MINAMITAMA CONSTRUCTION OFFICE, TOKYO)
TOKYO-TO MINAMITAMA KENSETSU JIMUSHO
19-2, 3-CHOME, MYOJIN-CHO, HACHIOJI-SHI, TOKYO 192
JAPANESE

TOKYO-TO NISHITAMA KENSETSU JIMUSHO JIGYO GAIYO 6928
(OPERATION REPORT OF NISHITAMA CONSTRUCTION OFFICE, TOKYO)
TOKYO-TO NISHITAMA KENSETSU JIMUSHO
4-1, 6-CHOME, KABE-CHO, OME-SHI, TOKYO 198
JAPANESE

TONNERU GIJUTSU KOENKAI TEKISUTO 6929
(TEXT OF THE SYMPOSIUM ON TUNNEL TECHNOLOGY)
JAPANESE
INDEXED: JICST

TONNERU KOJI KOENSHU 6930
(PROCEEDINGS OF TUNNEL CONSTRUCTION CONFERENCE)
NIHON DAMU KYOKAI
C/O DAIWA GINKO BUILDING, 2-2, NIHONBASHI HORIDOME-CHO, CHUO-KU, TOKYO 103
JAPANESE

TONNERU TO CHIKA 6931
(TUNNELS AND UNDERGROUND)
DOBOKU KOGAKUSHA
4-6, 1-CHOME, KOISHIKAWA, BUNKYO-KU, TOKYO 112
JAPANESE
INDEXED: JICST

TONNERU TO CHIKA GIJUTSU KOENKAI TEKISUTO 6932
(TEXT OF THE SYMPOSIUM ON TUNNELS AND UNDERGROUND
TECHNOLOGY)
JAPANESE
INDEXED: JICST

TRANSACTIONS OF THE JAPAN SOCIETY OF CIVIL 6933
ENGINEERS
DOBOKU GAKKAI
1-CHOME, YOTSUYA, SHINJUKU-KU, TOKYO 160
ENGLISH
INDEXED: ENG, JICST

TSUCHI TO KISO 6934
(SOIL MECHANICS AND FOUNDATION ENGINEERING)
DOSHITSU KOGAKUKAI (JAPANESE SOCIETY OF SOIL
MECHANICS AND FOUNDATION ENGINEERING)
13-5, 1-CHOME, NISHI SHINBASHI, MINATO-KU,
TOKYO 105
JAPANESE
INDEXED: JICST

YOKOGAWA KYORYO GIHO 6935
(YOKOGAWA BRIDGE WORKS TECHNICAL REPORT)
YOKOGAWA KYORYO SEISAKUSHO (YOKOGAWA BRIDGE
WORKS)
4-44, 4-CHOME, SHIBAURA, MINATO-KU, TOKYO 108
JAPANESE
INDEXED: JICST

ZENIDAKA-GUMI KENKYUBU KENKYU BUHO 6936
(ZENIDAKA-GUMI RESEARCH DEPARTMENT REPORT)
JAPANESE
INDEXED: JICST

ZENKEN JANARU 6937
(ZENKEN JOURNAL)
ZENKOKU KENSETSU GYOMUKAI (ZENKOKU KENSETSU
GYOMUKAI)
5-1, 2-CHOME, HATCHO-BORI, CHUO-KU, TOKYO 104
JAPANESE
INDEXED: JICST

Railway, Highway Engineering

ARU 6938
NIHON KOKUYU TETSUDO KOHO-BU (INFORMATION DIVISION, JAPANESE NATIONAL RAILWAYS)
4-1, 3-CHOME, MARUNOUCHI, CHIYODA-KU, TOKYO 100
JAPANESE

CHIKATETSU 6939
(SUBWAY)
TEITO KOSOKUDO KOTSU EIDAN SOMU-BU (TEITO RAPID TRANSIT AUTHORITY)
19-6, 3-CHOME, UENO, TAITO-KU, TOKYO 110
JAPANESE

DAIKO 6940
(BULLETIN OF OSAKA CONSTRUCTION BUREAU, JAPANESE NATIONAL RAILWAYS)
NIHON KOKUYU TETSUDO OSAKA KOJI-KYOKU
UMEDA-CHO, KITA-KU, OSAKA-SHI 530
JAPANESE
INDEXED: JICST

DENKI TETSUDO 6941
(ELECTRIC RAILWAYS)
TETSUDO DENKA KYOKAI (RAILWAY ELECTRIFICATION ASSOCIATION)
6-4, 1-CHOME, MARUNOUCHI, CHIYODA-KU, TOKYO 100
JAPANESE
INDEXED: JICST

DENKISHA NO KAGAKU 6942
(RAILWAY ELECTRIC ROLLING STOCKS)
DENKISHA KENKYUKAI (INSTITUTE OF ELECTRIC ROLLING STOCK)
C/O SHIN KOKUSAI BUILDING, 4-1, 3-CHOME, MARUNOUCHI, CHIYODA-KU, TOKYO 100
JAPANESE
INDEXED: JICST

DENRYOKU TO TETSUDO 6943
(ELECTRIC LIGHTING & FACILITIES IN RAILWAYS)
TETSUDO DENKA KYOKAI (RAILWAY ELECTRIFICATION ASSOCIATION)
C/O KOTSU KOSHA BUILDING, 6-4, 1-CHOME, MARUNOUCHI, CHIYODA-KU, TOKYO 100
JAPANESE

DORO 6944
(ROAD)
NIHON DORO KYOKAI (JAPAN ROAD ASSOCIATION)
C/O SHOYU KAIKAN, 3-3, 3-CHOME, KASUMIGASEKI, CHIYODA-KU, TOKYO 100
JAPANESE
INDEXED: JICST

DORO GENKYO CHOSA 6945
(REPORT ON PRESENT STATE OF ROADS)
TOKYO-TO (TOKYO METROPOLITAN GOVERNMENT)
5-1, 3-CHOME, MARUNOUCHI, CHIYODA-KU, TOKYO 100
JAPANESE

DORO JIGYO CHOSAHI HOKOKU 6946
(REPORT ON RESEARCH EXPENSE FOR ROAD CONSTRUCTION)
KENSETSU-SHO DOBOKU KENKYUSHO (PUBLIC WORKS RESEARCH INSTITUTE, MINISTRY OF CONSTRUCTION)
28-32, 2-CHOME, HONKOMAGOME, BUNKYO-KU, TOKYO 113
JAPANESE

DORO JIGYO GAIYO 6947
(OUTLINE OF ROAD CONSTRUCTION)
KENSETSU-SHO DORO-KYOKU (ROAD BUREAU, MINISTRY OF CONSTRUCTION)
1-3, 2-CHOME, KASUMIGASEKI, CHIYODA-KU, TOKYO 100
JAPANESE

DORO KENSETSU 6948
(ROAD CONSTRUCTION)
NIHON DORO KENSETSUGYO KYOKAI (JAPAN ROAD CONSTRUCTIONS' ASSOCIATION)
3-1, 3-CHOME, MARUNOUCHI, CHIYODA-KU, TOKYO 100
JAPANESE
INDEXED: JICST

DORO SEMINA 6949
(ROAD SEMINAR)
ZENKOKU KAJO HOREI SUPPAN K.K.
17, SAN'EI-CHO, SHINJUKU-KU, TOKYO 160
JAPANESE
INDEXED: JICST

DORO TO KONKURITO 6950
(ROAD AND CEMENT)
SEMENTO KYOKAI (CEMENT ASSOCIATION OF JAPAN)
C/O HATTORI BUILDING, 1-1, KYOBASHI, CHUO-KU, TOKYO 104
JAPANESE
INDEXED: JICST

DORO TO SHIZEN 6951
(ROAD AND NATURE)
DORO TO SHIZEN HENSHU IINKAI
DORO RYOKKA HOZEN KYOKAI (ASSOCIATION OF TREE-PLANTING IN ROAD)
C/O DAI-18 MORI BUILDING, 20, NISHIKUBO AKEFUNE-CHO, MINATO-KU, TOKYO 105
JAPANESE
INDEXED: JICST

DORO TOKEI NENPO 6952
(ANNUALS OF ROAD STATISTICS)
KENSETSU-SHO DORO-KYOKU (ROAD BUREAU, MINISTRY OF CONSTRUCTION)
1-3, 2-CHOME, KASUMIGASEKI, CHIYODA-KU, TOKYO
JAPANESE

FOAMAN 6953
(FOREMAN)
TETSUDO KENKYU SHU
(RAILWAY RESEARCH COMPANY)
JAPANESE

GIJUTSU HOKOKU-KAI SHIRYO. DORO-HEN. 6954
(DOCUMENTS OF THE TECHNICAL REPORT GROUP. ROADS EDITION.)
KENSETSU-SHO CHUBU CHIHO KENSETSU-KYOKU (MINISTRY OF CONSTRUCTION, CHUBU REGIONAL CONSTRUCTION BUREAU)
JAPANESE

GIJUTSU SHIRYO (NIPPON SHARYO SEIZO K.K.) 6955
(TECHNICAL INFORMATION, JAPAN RAILWAY CAR COMPANY, LIMITED)
NIPPON SHARYO SEIZO K.K.
1-1, SANBONMATSU-CHO, ATSUTA-KU, NAGOYA-SHI 456-91
JAPANESE
INDEXED: JICST

HAIUEI GAIDO 6956
(HIGHWAY GUIDE)
TAIKASHA
2-9, 8-CHOME, GINZA, CHUO-KU, TOKYO 104
JAPANESE

HANSHIN HAIUEI 6957
(HANSHIN HIGHWAY)
HANSHIN KOSOHU DORO KODAN (HANSHIN EXPRESSWAY PUBLIC CORPORATION)
JAPANESE

HANSHIN KOSOKU DORO KODAN NENPO 6958
(ANNUAL REPORT OF HANSHIN EXPRESS PUBLIC CORPORATION)
HANSHIN KOSOKU DORO KODAN (HANSHIN EXPRESS PUBLIC CORPORATION)
4-68, KITAKYUTARO-CHO, HIGASHI-KU, OSAKA-SHI 541
JAPANESE

HOSO 6959
(PAVEMENT)
KENSETSU TOSHO
2-17, 2-CHOME, SOTOKANDA, CHIYODA-KU, TOKYO 101
JAPANESE
INDEXED: JICST

JAPANESE RAILWAY ENGINEERING 6960
NIHON TETSUDO GIJUTSU KYOKAI (JAPAN RAILWAY ENGINEER'S ASSOCIATION)
5-18, 2-CHOME, OTE-MACHI, CHIYODA-KU, TOKYO 100
ENGLISH
INDEXED: ENG, JICST

JARTS 6961
(JARTS)
KAIGAI TETSUDO GIJUTSU KYORYOKU KYOKAI (OVERSEAS RAILROAD TECHNOLOGICAL COOPERATIVE SOCIETY)
JAPANESE AND ENGLISH

JARTS NYUSU 6962
(JARTS NEWS)
KAIGAI TETSUDO GIJUTSU KYORYOKU KYOKAI (JAPAN RAILWAY TECHNICAL SERVICE)
2-1, 2-CHOME, MARUNOUCHI, CHIYODA-KU, TOKYO 100
JAPANESE
INDEXED: JICST

JIGYO NO GAIYO TO DORO TOKEI 6963
(ROAD STATISTICS AND OPERATION OUTLINE OF JAPAN HIGHWAY CORPORATION)
NIHON DORO KODAN SOMU-BU (ADMINISTRATIVE DIVISION, JAPAN HIGHWAY CORPORATION)
1-13, 1-CHOME, SHINBASHI, MINATO-KU, TOKYO 105
JAPANESE

JOKIKIKANSHA 6964
(STEAM LOCOMOTIVE)
KINEMA JUNHO SHA
(KINEMA JUNHO COMPANY)
JAPANESE

JREA 6965
NIHON TETSUDO GIJUTSU KYOKAI (JAPAN RAILWAY ENGINEERING ASSOCIATION)
5-18, 2-CHOME, OTE-MACHI, CHIYODA-KU, TOKYO 100
JAPANESE
INDEXED: JICST

KANSAI DORO KENKYUKAI KAIHO 6966
(KANSAI ROADS RESEARCH SOCIETY REPORT)
KANSAI DORO KENKYUKAI (KANSAI ROADS RESEARCH SOCIETY)
JAPANESE

KENSETSU-SHO CHUBU CHIHO KENSETSU-KYOKU GIJUTSU HOKOKUKAI RONBUNSHU, DORO-HEN 6967
(PROCEEDINGS OF ANNUAL MEETING OF CHUBU REGIONAL CONSTRUCTION BUREAU, ROAD ENGINEERING EDITION)
KENSETSU-SHO CHUBU CHIHO KENSETSU-KYOKU KIKAKU-BU (PLANNING DIVISION, CHUBU REGIONAL CONSTRUCTION BUREAU, MINISTRY OF CONSTRUCTION)
5-1, 2-CHOME, SANNOMARU, NAKA-KU, NAGOYA-SHI 460
JAPANESE
INDEXED: JICST

KINKI NIHON TETSUDO GIJUTSU KENKYUSHO GIHO 6968
(TECHNICAL REPORT OF THE RAILWAY TECHNICAL RESEARCH INSTITUTE, KINKI NIHON RAILWAYS)
KINTETSU GIJUTSU KENKYUSHO
344, AMAGATSUJI-MACHI KO, NARA-SHI 630
JAPANESE
INDEXED: JICST

KINTETSU NYUSU 6969
(KINKI NIPPON RAILWAY NEWS)
KINKI NIPPON TETSUDO (KINKI NIPPON RAILWAY COMPANY, LTD.)
JAPANESE

KISO KOZO NI KANSURU CHOSA KENKYU 6970
(STUDIES ON FOUNDATION ENGINEERING)
KOSOKU DORO CHOSAKAI (EXPRESS HIGHWAY RESEARCH FOUNDATION)
6-17, 1-CHOME, KUDANMINAMI, CHIYODA-KU, TOKYO 102
JAPANESE

KOKUTETSU TABIJI 6971
(NATIONAL RAILWAY JOURNEYS)
NAGOYA TETSUDO KANRI KYOKU, RYOKYAKU KAHEN, KOKUTETSU TABIJI NOKAI (NAGOYA RAILROAD MANAGEMENT BUREAU, PASSENGER DIVISION EDITION, 'NATIONAL RAILWAY JOURNEYS' COMMITTEE)
JAPANESE

KOKUTETSU TSUSHIN 6972
(NATIONAL RAILWAYS NEWS)
NIHON KOKUYU TETSUDO KOHO-BU (INFORMATION DIVISION, JAPANESE NATIONAL RAILWAYS)
6-5, 1-CHOME, MARUNOUCHI, CHIYODA-KU, TOKYO 100
JAPANESE

KOKUTETSUSEN 6973
(NATIONAL RAILROAD)
NIHON KOKUYU TETSUDO RYOKYAKU-KYOKU (JAPANESE NATIONAL RAILWAYS)
KOTSU KYORYOKUKAI (TRANSPORTATION COOPERATIVE ASSOCIATION)
4-1, 3-CHOME, MARUNOUCHI, CHIYODA-KU, TOKYO 100
JAPANESE

KOKUYU TETSUDO 6974
(NATIONAL RAILWAYS)
KOTSU KYORYOKUKAI (TRANSPORTATION COOPERATIVE ASSOCIATION)
4-1, 3-CHOME, MARUNOUCHI, CHIYODA-KU, TOKYO 100
JAPANESE

KOSOKU DORO TO JIDOSHA 6975
(EXPRESSWAYS AND AUTOMOBILES)
KOSOKU DORO CHOSAKAI (EXPRESS HIGHWAY RESEARCH FOUNDATION OF JAPAN)
6-17, 1-CHOME, KUDANMINAMI, CHIYODA-KU, TOKYO 102
JAPANESE
INDEXED: JICST

KOZOBUTSU SEKKEI SHIRYO 6976
(STRUCTURE DESIGN DATA)
NIHON TETSUDO SHISETSU KYOKAI (JAPAN RAILWAY CIVIL ENGINEERING ASSOCIATION)
18-7, 2-CHOME, HIGASHIUENO, TAITO-KU, TOKYO 110
JAPANESE
INDEXED: JICST

MICHI 6977
(HIGHWAY)
NIHON DORO KODAN (JAPAN HIGHWAY PUBLIC CORPORATION)
1-13, 1-CHOME, SHINBASHI, MINATO-KU, TOKYO 105
JAPANESE

MONORERU 6978
(MONORAIL)
NIHON MONORERU KYOKAI (JAPAN MONORAIL ASSOCIATION)
C/O NICHIGINMAE BUILDING, 4-1, 3-CHOME, NIHONBASHI HONGOKU-CHO, CHUO-KU, TOKYO 103
JAPANESE
INDEXED: JICST

NAGOYA TETSUDO K.K. GIJUTSU KENKYUSHO KENKYU HOKOKU 6979
(REPORT OF THE TECHNICAL RESEARCH INSTITUTE, NAGOYA RAILWAYS COMPANY, LTD.)
NAGOYA TETSUDO GIJUTSU KENKYUSHO
1-223, SASAJIMA-CHO, NAKAMURA-KU, NAGOYA-SHI 450
JAPANESE

NIHON DORO KAIGI RONBUNSHU 6980
(REPORTS OF JAPAN ROAD CONFERENCE)
NIHON DORO KAIGI (JAPAN ROAD CONFERENCE)
NIHON DORO KYOKAI
3-3, 3-CHOME, KASUMIGASEKI, CHIYODA-KU, TOKYO 100
JAPANESE

NIHON DORO KODAN GYOMU KENKYU HAPPYOKAI RONBUNSHU 6981
(JOURNAL OF OPERATION RESEARCH CONFERENCE)
NIHON DORO KODAN KIKAKU CHOSA-BU (PLANNING AND RESEARCH DIVISION, JAPAN HIGHWAY PUBLIC CORPORATION)
1-13, 1-CHOME, SHINBASHI, MINATO-KU, TOKYO 105
JAPANESE

NIHON DORO KODAN NENPO 6982
(ANNUAL REPORT OF JAPAN HIGHWAY PUBLIC CORPORATION)
NIHON DORO KODAN (JAPAN HIGHWAY PUBLIC CORPORATION)
1-13, 1-CHOME, SHINBASHI, MINATO-KU, TOKYO 105
JAPANESE

NIHON DORO KODAN SHIKENJO HOKOKU 6983
(LABORATORY REPORT OF JAPAN HIGHWAY PUBLIC CORPORATION)
NIHON DORO KODAN SHIKENJO (JAPAN HIGHWAY PUBLIC CORPORATION)
1789, YAMAZAKI-CHO, MACHIDA-CHI, TOKYO 194-01
JAPANESE
INDEXED: JICST

NIHON HODO K.K. GIJUTSU KENKYUSHOHO 6984
(REPORT OF NIPPON HODO TECHNICAL RESEARCH LABORATORY)
NIHON HODO K.K. GIJUTSU KENKYUSHO (NIPPON HODO TECHNICAL RESEARCH LABORATORY)
32-34, 3-CHOME, HIGASHISHINAGAWA, SHINAGAWA-KU, TOKYO 140
JAPANESE, WITH ENGLISH SUMMARY

NIHON KOKUYU TETSUDO ZAIGAI JIMUSHO JOHO 6985
(BULLETIN OF THE JAPAN NATIONAL RAILWAYS OVERSEAS OFFICES)
NIHON KOKUYU TETSUDO GAIMUBU
(FOREIGN OFFICE, JAPAN NATIONAL RAILWAYS)
JAPANESE

PERMANENT WAY 6986
NIPPON TETSUDO SHISETSU KYOKAI (JAPAN RAILWAY CIVIL ENGINEERING ASSOCIATION)
18-7, 2-CHOME, HIGASHIUENO, TAITO-KU, TOKYO 110
JAPANESE
INDEXED: ENG, JICST

QUARTERLY REPORT OF THE RAILWAY TECHNICAL RESEARCH INSTITUTE 6987
NIHON KOKUYU TETSUDO TETSUDO GIJUTSU KENKYUSHO (RAILWAY TECHNICAL RESEARCH INSTITUTE, JAPANESE NATIONAL RAILWAY) AND KEN'YUSHA
45-6, 1-CHOME, HIKARI-CHO, KOKUBUNJI-SHI, TOKYO 185
ENGLISH
INDEXED: ENG

SANWA SHINBUN 6988
(SANWA TIMES)
SANWA TEKKI KOGYO K.K.
5-19, 6-CHOME, MINAMISHINAGAWA, SHINAGAWA-KU, TOKYO 140
JAPANESE

SEIKAN TONNERU GIJUTSU KENKYUKAI KIROKU 6989
(RECORD OF THE TECHNICAL COMMITTEE ON SEIKAN TUNNEL)
NIHON TETSUDO KENSETSU KODAN SEIKAN KENSETSU-KYOKU (SEIKAN CONSTRUCTION BUREAU, JAPAN RAILWAY CONSTRUCTION CORPORATION)
15-7, WAKAMATSU-CHO, HAKODATE-SHI 040
JAPANESE

SEKAI KAKKOKU TETSUDO TOKEI 6990
(WORLD RAILWAY STATISTICS BY COUNTRIES)
NIHON KOKUYU TETSUDO GAIMU-BU (FOREIGN AFFAIRS DIVISION, JAPANESE NATIONAL RAILWAYS)
6-5, 1-CHOME, MARUNOUCHI, CHIYODA-KU, TOKYO 100
JAPANESE

SHARYO GIJUTSU 6991
(ROLLING STOCK ENGINEERING)
NIHON TETSUDO SHARYO KOGYOKAI (JAPAN ASSOCIATION OF ROLLING STOCK INDUSTRIES)
C/O TEKKO BUILDING, 8-2, 1-CHOME, MARUNOUCHI, CHIYODA-KU, TOKYO 100
JAPANESE
INDEXED: JICST

SHARYO KOGAKU 6992
(RAILWAY CAR ENGINEERING)
TETSUDO NIHONSHA
1-4, 2-CHOME, NISHIKANDA, CHIYODA-KU, TOKYO 101
JAPANESE

SHARYO TO DENKI 6993
(RAILWAY CAR AND ELECTRIC EQUIPMENT)
SHARYO DENKI KYOKAI (ASSOCIATION OF RAILWAY AND ELECTRIC EQUIPMENT)
14-11, 5-CHOME, UENO, TAITO-KU, TOKYO 110
JAPANESE

SHIN SENRO 6994
(MAINTENANCE OF JAPANESE RAILWAY)
TETSUDO GENGYOSHA
4-3, 1-CHOME, HIGASHISHINBASHI, MINATO-KU, TOKYO 105
JAPANESE

SHISETSU NENPO 6995
(ANNUAL REPORT OF JAPAN RAILWAY CIVIL ENGINEERING ASSOCIATES)
NIHON TETSUDO SHISETSU KYOKAI
18-7, 2-CHOME, HIGASHIUENO, TAITO-KU, TOKYO 110
JAPANESE
INDEXED: JICST

SHITETSU TOKEI NENPO 6996
(ANNUAL STATISTICS OF PRIVATE RAILWAY)
SEIFU SHIRYOTO FUKYU CHOSAKAI
7-1, 1-CHOME, AKASAKA, MINATO-KU, TOKYO 107
JAPANESE

SHUTO KOSOKU 6997
(TOKYO EXPRESS WAY)
SHUTO KOSOKU DORO KODAN (TOKYO EXPRESS WAY PUBLIC CORPORATION)
5, SHIBA TORANOMON, MINATO-KU, TOKYO 105
JAPANESE
INDEXED: JICST

SHUTO KOSOKU DORO KODAN NENPO 6998
(ANNUAL REPORT OF TOKYO EXPRESS WAY PUBLIC CORPORATION)
SHUTO KOSOKU DORO KODAN
C/O NITTOCHI BUILDING, 4-1, 1-CHOME, KASUMIGASEKI, CHIYODA-KU, TOKYO 100
JAPANESE

SYMPOSIUM ON THE USE OF CYBERNETICS ON THE RAILWAY 6999
NIHON TETSUDO SAIBANETIKKUSU KYOGIKAI (JAPAN RAILWAY CYBERNETICS ASSOCIATION)
2-4, OTE-MACHI, CHIYODA-KU, TOKYO 100
ENGLISH

TAISHIN SEKKEI NI KANSURU CHOSA KENKYU 7000
(STUDIES ON EARTHQUAKE-PROOF DESIGN)
KOSOKU DORO CHOSAKAI (EXPRESS WAY RESEARCH ASSOCIATION)
C/O CHIYODA KAIKAN, 6-17, 1-CHOME, KUDANMINAMI, CHIYODA-KU, TOKYO 102
JAPANESE

TAKAMATSU TETSUDO YUBIN KYOKUHO 7001
(REPORT OF THE TAKAMATSU RAILWAY POSTAL BUREAU)
JAPANESE

TETSUDO DENKI 7002
(ELECTRICAL ENGINEERING OF JAPANESE RAILWAY)
TETSUDO GENGYOSHA
4-3, 1-CHOME, HIGASHISHINBASHI, MINATO-KU, TOKYO 105
JAPANESE

TETSUDO DOBOKU 7003
(RAILWAY CIVIL ENGINEERING)
NIPPON TETSUDO SHISETSU KYOKAI (JAPAN RAILWAY CIVIL ENGINEERING ASSOCIATES)
18-7, 2-CHOME, HIGASHIUENO, TAITO-KU, TOKYO 110
JAPANESE
INDEXED: JICST

TETSUDO FAN 7004
(JAPAN RAILFAN MAGAZINE)
KOYUSHA
3, MIYANISHI-CHO, CHIKUSA-KU, NAGOYA-SHI 464
JAPANESE

TETSUDO GIJUTSU BUNKEN SHOROKU 7005
(RAILWAY ENGINEERING ABSTRACTS)
NIHON KOKUYU TETSUDO TETSUDO GIJUTSU KENKYUSHO (THE RAILWAY TECHNICAL RESEARCH INSTITUTE, JAPANESE NATIONAL RAILWAYS) AND KEN'YUSHA
45-6, 1-CHOME, HIKARI-MACHI, KOKUBUNJI-SHI, TOKYO 185
JAPANESE
INDEXED: JICST

TETSUDO GIJUTSU KENKYU HOKOKU 7006
(RAILWAY TECHNICAL RESEARCH REPORT)
NIHON KOKUYU TETSUDO TETSUDO GIJUTSU KENKYUSHO (THE RAILWAY TECHNICAL RESEARCH INSTITUTE, JAPANESE NATIONAL RAILWAYS)
8-3, 2-CHOME, HIKARI-MACHI, KOKUBUNJI-SHI, TOKYO 185
JAPANESE
INDEXED: JICST

TETSUDO GIJUTSU KENKYU SHIRYO 7007
(JOURNAL OF RAILWAY ENGINEERING RESEARCH)
NIHON KOKUYU TETSUDO TETSUDO GIJUTSU KENKYUSHO (RAILWAY TECHNICAL RESEARCH INSTITUTE, JAPANESE NATIONAL RAILWAYS) AND KEN'YUSHA
45-6, 1-CHOME, HIKARI-MACHI, KOKUBUNJISHI, TOKYO 185
JAPANESE
INDEXED: CHE, JICST

TETSUDO GIJUTSU KENKYUJO KOENKAI KOEN MAEZURISHU 7008
(PREPRINTS OF THE SYMPOSIUM OF THE RAILROAD TECHNOLOGICAL INSTITUTE)
JAPANESE
INDEXED: JICST

TETSUDO GIJUTSU KENKYUJO NENPO 7009
(ANNUAL OF RAILROAD TECHNOLOGICAL INSTITUTE)
JAPANESE
INDEXED: JICST

TETSUDO JANARU 7010
(STEAM LOCOMOTIVE JOURNAL)
TETSUDO JANARUSHA
8-6, 4-CHOME, IIDABASHI, CHIYODA-KU, TOKYO 102
JAPANESE

TETSUDO KENCHIKU NYUSU 7011
(RAILWAY ARCHITECTS NEWS)
TETSUDO KENCHIKU KYOKAI (THE ASSOCIATION OF RAILWAY ARCHITECTS)
7-1, 1-CHOME, UCHISAIWAI-CHO, CHIYODA-KU, TOKYO 100
JAPANESE
INDEXED: JICST

TETSUDO KOJO 7012
(RAILROAD PLANT)
KOTSU SHIRYOSHA
1-121, OGIKUBO SUGINAMI-KU, TOKYO 162
JAPANESE

TETSUDO NI OKERU SAIBANETIKKUSU RIYO KOKUNAI SINPOJUMU RONBUNSHU 7013
(SYMPOSIUM ON THE USE OF CYBERNETICS ON THE RAILWAY)
NIHON TETSUDO SAIBANETIKKUSU KYOGIKAI (JAPAN RAILWAY CYBERNETICS ASSOCIATION)
C/O NIHON TETSUDO GIJUTSU KYOKAI, 2-4, OTE-MACHI, CHIYODA-KU, TOKYO 100
JAPANESE, WITH ENGLISH SUMMARY

TETSUDO PIKUTORIARU 7014
(RAILWAY PICTORIAL)
TETSUDO TOSHO KANKOKAI
C/O SHIN KOKUSAI BUILDING, 4-1, 3-CHOME, MARUNOUCHI, CHIYODA-KU, TOKYO 100
JAPANESE

TETSUDO SENRO 7015
(MAINTENANCE OF WAY)
NIPPON TETSUDO SHISETSU KYOKAI (THE PERMANENT WAY SOCIETY OF JAPAN
18-7, 2-CHOME, HIGASHIUENO, TAITO-KU, TOKYO 110
JAPANESE
INDEXED: JICST

TETSUDO SHARYO 7016
(ROLLING INDUSTRY NEWS)
NIHON TETSUDO SHARYO KOGYO KYOKAI (JAPAN ASSOCIATION OF ROLLING STOCK MANUFACTURERS)
8-2, 1-CHOME, MARUNOUCHI, CHIYODA-KU, TOKYO 100
JAPANESE

TETSUDO SHARYO KOGYO 7017
(ROLLING STOCK INDUSTRY)
NIHON TETSUDO SHARYO KOGYOKAI (JAPAN ASSOCIATION OF ROLLING STOCK INDUSTRY)
8-2, 1-CHOME, MARUNOUCHI, CHIYODA-KU, TOKYO 100
JAPANESE
INDEXED: JICST

TETSUDO SHARYO KOGYO NENPO 7018
(ANNUAL REPORT OF THE JAPAN ASSOCIATION OF ROLLING STOCK INDUSTRY)
NIHON TETSUDO SHARYO KOGYOKAI (JAPAN ASSOCIATION OF ROLLING STOCK INDUSTRY)
8-2, 1-CHOME, MARUNOUCHI, CHIYODA-KU, TOKYO 100
JAPANESE
INDEXED: JICST

TETSUDO SHARYO YUSHUTSU KUMIAIHO 7019
(JOURNAL OF THE JAPAN ROLLING STOCK EXPORTERS' ASSOCIATION)
NIHON TETSUDO SHARYO YUSHUTSU KUMIAI (JAPAN ROLLING STOCK EXPORTERS' ASSOCIATION)
C/O TEKKO BUILDING, 8-2, 1-CHOME, MARUNOUCHI, CHIYODA-KU, TOKYO 100
JAPANESE

TETSUDO SHARYOTO SEISAN DOTAI TOKEI GEPPO 7020
(MONTHLY STATISTICS OF ACTUAL PRODUCTION OF RAILWAY CARS)
UN'YU-SHO DAIJIN KANBO JOHO KANRI-BU (DATA PROCESSING DIVISION, MINISTER'S SECRETARIAT, MINISTRY OF TRANSPORT)
1-3, 2-CHOME, KASUMIGASEKI, CHIYODA-KU, TOKYO 100
JAPANESE

TETSUDO SHARYOTO SEISAN DOTAI TOKEI NENPO 7021
(ANNUAL STATISTICS OF ACTUAL PRODUCTION OF RAILWAY CARS)
UN'YU-SHO DAIJIN KANBO JOHO KANRI-BU (DATA PROCESSING DIVISION, MINISTER'S SECRETARIAT, MINISTRY OF TRANSPORT)
1-3, 2-CHOME, KASUMIGASEKI, CHIYODA-KU, TOKYO 100
JAPANESE

TETSUDO SHISETSU JIMU 7022
(RAILWAY BUSINESS)
NIHON TETSUDO SHISETSU KYOKAI (JAPAN RAILWAY CIVIL ENGINEERING ASSOCIATION)
18-7, 2-CHOME, HIGASHIUENO, TAITO-KU, TOKYO 110
JAPANESE
INDEXED: JICST

TETSUDO SHIZAI NIPPO 7023
(RAILWAY MATERIALS NEWS)
KOKUTETSU KANTO SHIZAI KYORYOKUKAI
2-6, TAKARA-CHO, CHUO-KU, TOKYO 103
JAPANESE

TETSUDO TOKEI GEPPO 7024
(MONTHLY RAILWAY STATISTICS)
NIHON KOKUYU TETSUDO KONPYUTA-BU (COMPUTER BUREAU, JAPANESE NATIONAL RAILWAYS)
6-5, 1-CHOME, MARUNOUCHI, CHIYODA-KU, TOKYO 100
JAPANESE

TETSUDO TOKEI NENPO 7025
(RAILWAY STATISTICS YEARBOOK)
NIHON KOKUYU TETSUDO KONPYUTA-BU (COMPUTER BUREAU, JAPANESE NATIONAL RAILWAY)
6-5, 1-CHOME, MARUNOUCHI, CHIYODA-KU, TOKYO 100
JAPANESE

TETSUDO TSUSHIN 7026
(RAILWAY TELECOMMUNICATIONS & ELECTRONICS)
TETSUDO TSUSHIN KYOKAI (JAPAN ASSOCIATION OF RAILWAY ENGINEERS OF TELECOMMUNICATION AND ELECTRONICS)
34-8, 2-CHOME, NISHISHINBASHI, MINATO-KJ, TOKYO 105
JAPANESE
INDEXED: JICST

TETSUDOKAI 7027
(RAILWAY WORLD)
TETSUDO HYORONSHA
12-1, 2-CHOME, MISAKI-CHO, CHIYODA-KU, TOKYO 101
JAPANESE

TOBU 7028
(TOBU)
TOBU DENSHA KENKYU-KAI (TOBU ELECTRIC TRAIN RESEARCH GROUP)
JAPANESE

TOKO 7029
(RECORD OF WORKS)
NIHON KOKUYU TETSUDO TOKYO DAI-1 KOJI-KYOKU (TOKYO 1ST CONSTRUCTION BUREAU, JAPANESE NATIONAL RAILWAYS)
2-6, 2-CHOME, YOYOGI, SHIBUYA-KU, TOKYO 150
JAPANESE
INDEXED: JICST

TOKYU SHARYO GIHO 7030
(TOKYU SHARYO TECHNICAL REVIEW)
TOKYU SHARYO SEIZO K.K. (TOKYU CAR CORPORATION COMPANY, LTD.)
1, KAMARIYA-CHO, KANAZAWA-KU, YOKOHAMA-SHI 236
JAPANESE
INDEXED: JICST

TOKYU SHARYO NYUSU 7031
(TOKYU CAR CORPORATION NEWS)
TOKYU SHARYO SEIZO K.K. (TOKYU CAR CORPORATION)
1, KAMARIYA-CHO, KANAZAWA-KU, YOKOHAMA-SHI 236
JAPANESE

TONNERU TO CHIKA 7032
(TUNNELS AND UNDERGROUND)
SEIKI TSUSHIN SHA
(PRECISION MACHINERY NEWS COMPANY)
JAPANESE

TOTETSU JANARU 7033
(TOTETSU JOURNAL)
TOKYO TETSUDO KANRI KYOKU
(TOKYO RAILROAD ADMINISTRATION BUREAU)
JAPANESE

USHIO 7034
(TIDE)
KOKUTETSU DORYOKUSHA KENKYUKAI (JAPAN NATIONAL RAILWAYS RESEARCH COMMITTEE ON POWER-DRIVEN CARS)
JAPANESE

Hydraulic Engineering, Natural Waters, and Channels

AICHI-KEN KOWAN TOKEI NENPO 7035
(ANNUAL STATISTICS OF PORTS AND HARBOURS, AICHI PREFECTURE)
AICHI-KEN DOBOKU-BU (CIVIL ENGINEERING DIVISION, AICHI PREFECTURAL GOVERNMENT)
1-2, 3-CHOME, SANNOMARU, NAKA-KU, NAGOYA-SHI 460
JAPANESE

CHIBAKO TOKEI NENPO 7036
(ANNUAL STATISTICS OF PORTS AND HARBOURS, CHIBA PREFECTURE)
CHIBA-KEN (CHIBA PREFECTURAL GOVERNMENT)
1-1, ICHIBA-CHO, CHIBA-SHI 280-91
JAPANESE

CHOKKATSU KOWAN KOJI NENPO 7037
(ANNUAL REPORT OF HARBOR WORKS UNDER THE DIRECT CONTROL OF MINISTRY OF TRANSPORTATION)
UN'YU-SHO KOWAN-KYOKU (THE PORT AND HARBOR BUREAU, MINISTRY OF TRANSPORT)
1-3, 2-CHOME, KASUMIGASEKI, CHIYODA-KU, TOKYO 100
JAPANESE

DAI DAMU 7038
(LARGE DAMS (JOURNAL OF THE JAPANESE NATIONAL COMMITTEE ON LARGE DAMS))
NIHON DAI DAMU KAIGI (JAPANESE NATIONAL COMMITTEE OF LARGE DAMS)
C/O TODEN KYUKAN, 1-13, 1-CHOME, SHIBA TAMURA-CHO, MINATO-KU, TOKYO 105
JAPANESE
INDEXED: JICST

DAMU NIPPON 7039
(DAM DIGEST)
NIPPON DAMU KYOKAI (JAPAN DAM ASSOCIATION)
14-5, 2-CHOME, SHINTOMI, CHUO-KU, TOKYO 104
JAPANESE
INDEXED: JICST

GEKKAN NAGOYAKO 7040
(PORT OF NAGOYA MONTHLY)
GEKKAN NAGOYAKO KANKOKAI (PUBLISHING ASSOCIATION OF PORT OF NAGOYA MONTHLY)
C/O NAGOYAKO KANRI KUMIAI, 6-7, MINATO HON-MACHI, MINATO-KU, NAGOYA-SHI 455
JAPANESE

GIJUTSU HOKOKU-KAI SHIRYO. KASEN-HEN 7041
(DOCUMENTS OF THE TECHNICAL REPORT GROUP. RIVERS EDITION)
KENSETSU-SHO CHUBU CHIHO KENSETSU-KYOKU (MINISTRY OF CONSTRUCTION, CHUBU REGIONAL CONSTRUCTION BUREAU)
JAPANESE

GYOKO KENSETSU GIJUTSU KENKYU HAPPYOKAI KOENSHU 7042
(PROCEEDINGS OF FISHING PORT ENGINEERING)
SUISAN-CHO GYOKO-BU (FISHING PORT DIVISION, FISHERIES AGENCY)
2-1, 1-CHOME, KASUMIGASEKI, CHIYODA-KU, TOKYO 100
JAPANESE

GYOKO KENSETSU GIJUTSU SHIRYO 7043
(TECHNICAL INFORMATION ON FISHING PORT CONSTRUCTION)
SUISAN-CHO GYOKO-BU (FISHING PORT DIVISION, FISHERIES AGENCY)
2-1, 1-CHOME, KASUMIGASEKI, CHIYODA-KU, TOKYO 100
JAPANESE

HYOGO-KEN KOWAN TOKEI NENPO 7044
(ANNUAL STATISTICS OF PORTS AND HARBOURS, HYOGO PREFECTURE)
HYOGO-KEN DOBOKU-BU (CIVIL ENGINEERING DIVISION, HYOGO PREFECTURAL GOVERNMENT)
5-1, SHIMOYAMATE-DORI, IKUTA-KU, KOBE-SHI 650
JAPANESE

JAPAN PORT INFORMATION 7045
JAPAN INBESTAZUSHA (THE JAPAN INVESTORS LTD.)
12-8, 2-CHOME, KITAAOYAMA, MINATO-KU, TOKYO 107
ENGLISH

KAIGAN 7046
(SEA COAST)
ZENKOKU KAIGAN KYOKAI (JAPAN SEA COAST ASSOCIATION)
1-1, 3-CHOME, ROPPONGI, MINATO-KU, TOKYO 106
JAPANESE
INDEXED: JICST

KAIGAN JIGYO CHOSAHI HOKOKU 7047
(REPORT ON RESEARCH EXPENSE FOR COAST CONSTRUCTION)
KENSETSU-SHO DOBOKU KENKYUSHO (PUBLIC WORKS RESEARCH INSTITUTE, MINISTRY OF CONSTRUCTION)
28-32, 2-CHOME, HONKOMAGOME, BUNKYO-KU, TOKYO 113
JAPANESE

KAIGAN KOGAKU KOENKAI RONBUNSHU 7048
(PROCEEDINGS OF COAST CONSTRUCTION ENGINEERING)
DOBOKU GAKKAI (JAPAN SOCIETY OF CIVIL ENGINEERS)
1-CHOME, YOTSUYA, SHINJUKU-KU, TOKYO 160
JAPANESE

KAIGAN TOKEI 7049
(COASTAL STATISTICS)
KENSETSU-SHO KASEN-KYOKU (RIVER BUREAU, MINISTRY OF CONSTRUCTION)
1-3, 2-CHOME, KASUMIGASEKI, CHIYODA-KU, TOKYO 100
JAPANESE

KANAZAWA KOWAN TOKEI NENPO 7050
(ANNUAL STATISTICS OF KANAZAWA PORT)
ISHIKAWA-KEN KANAZAWA KOWAN JIMUSHO (KANAZAWA PORT OFFICE, ISHIKAWA PREFECTURAL GOVERNMENT)
4-12, MINATO, KANAZAWA-SHI 920-02
JAPANESE

KANNAI KOWAN TOKEI NENPO 7051
(ANNUAL STATISTICS OF PORTS AND HARBOURS, MINISTRY OF TRANSPORT)
UN'YU-SHO DAI-1 KOWAN KENSETSU-KYOKU (1ST DISTRICT PORT CONSTRUCTION BUREAU, MINISTRY OF TRANSPORT)
1-332, SHIROYAURA, NIIGATA-SHI 951
JAPANESE

KASEN 7052
(RIVER)
NIHON KASEN KYOKAI (JAPAN RIVER ASSOCIATION)
11-35, 1-CHOME, NAGATA-CHO, CHIYODA-KU, TOKYO 100
JAPANESE
INDEXED: JICST

KASEN JIGYO CHOSAHI HOKOKU 7053
(REPORT ON RESEARCH EXPENSE FOR RIVER CONSTRUCTION)
KENSETSU-SHO DOBOKU KENKYUSHO (PUBLIC WORKS RESEARCH INSTITUTE, MINISTRY OF CONSTRUCTION)
28-32, 2-CHOME, HONKOMAGOME, BUNKYO-KU, TOKYO 113
JAPANESE

KASEN SOGO KAIHATSU JIGYO CHOSAHI HOKOKU 7054
(REPORT ON RESEARCH EXPENSE FOR GENERAL RIVER DEVELOPING CONSTRUCTION)
KENSETSU-SHO DOBOKU KENKYUSHO (PUBLIC WORKS RESEARCH INSTITUTE, MINISTRY OF CONSTRUCTION)
28-32, 2-CHOME, HONKOMAGOME, BUNKYO-KU, TOKYO 113
JAPANESE

KAWASAKI KOWAN TOKEI NENPO 7055
(ANNUAL STATISTICS OF PORTS AND HARBOURS, KAWASAKI)
KAWASAKI-SHI KOWAN-KYOKU (KAWASAKI PORT AUTHORITY)
1, MIYAMOTO-CHO, KAWASAKI-KU, KAWASAKI-SHI 210
JAPANESE

KENSETSU-SHO CHUBU CHIHO KENSETSU-KYOKU GIJUTSU HOKOKUKAI RONBUNSHU, KASEN-HEN 7056
(TECHNICAL REPORT OF CHUBU REGIONAL CONSTRUCTION BUREAU, RIVER SERIES)
KENSETSU-SHO CHUBU CHIHO KENSETSU-KYOHU KIKAKU-BU (CHUBU REGIONAL CONSTRUCTION BUREAU, MINISTRY OF CONSTRUCTION)
5-1, 2-CHOME, SANNOMARU, NAKA-KU, NAGOYA-SHI 460
JAPANESE
INDEXED: JICST

KINDAI KOWAN. 7057
(MODERN PORTS AND HARBORS)
TA MINARUREPOTO SHA (TERMINAL REPORT COMPANY)
JAPANESE

KISARAZUKO TOKEI NENPO 7058
(ANNUAL STATISTICS OF KISARAZU PORT)
CHIBA-KEN (CHIBA PREFECTURAL GOVERNMENT)
1-1, ICHIBA-CHO, CHIBA-SHI 280-91
JAPANESE

KOGYO YOSUI 7059
(INDUSTRIAL WATER)
NIHON KOGYO YOSUI KYOKAI (JAPAN INDUSTRIAL WATER ASSOCIATION)
4-3, 2-CHOME, HIRAKAWA-CHO, CHIYODA-KU, TOKYO 100
JAPANESE
INDEXED: CHE, JICST

KOWAN 7060
(PORTS AND HARBORS)
NIHON KOWAN KYOKAI (JAPAN PORT AND HARBOR ASSOCIATION)
C/O TORANOMON KOTOHIRAKAIKAN BUILDING, 1, SHIBA KOTOHIRA-CHO, MINATO-KU, TOKYO 105
JAPANESE
INDEXED: JICST

KOWAN GIJUTSU KENKYUSHO HOKOKU 7061
(REPORT OF PORT AND HARBOUR TECHNICAL RESEARCH INSTITUTE, MINISTRY OF TRANSPORT)
UN'YU-SHO KOWAN GIJUTSU KENKYUSHO (PORT AND HARBOUR TECHNICAL RESEARCH INSTITUTE, MINISTRY OF TRANSPORT)
1-1, 3-CHOME, NAGASE, YOKOSUKA-SHI 239
JAPANESE, WITH EUROPEAN SUMMARY
INDEXED: JICST

KOWAN GIJUTSU KENKYUSHO KENKYU SEIKA MOKUROKU 7062
(BIBLIOGRAPHY OF PORT AND HARBOUR TECHNICAL RESEARCH INSTITUTE)
UN'YU-SHO KOWAN GIJUTSU KENKYUSHO (PORT AND HARBOUR TECHNICAL RESEARCH INSTITUTE, MINISTRY OF TRANSPORT)
1-1, 3-CHOME, NAGASE, YOKOSUKA-SHI 239
JAPANESE

KOWAN GIJUTSU KENKYUSHO NENPO 7063
(ANNUAL REPORT OF THE PORT AND HARBOUR TECHNICAL RESEARCH INSTITUTE)
UN'YU-SHO KOWAN GIJUTSU KENKYUSHO (PORT AND HARBOUR TECHNICAL RESEARCH INSTITUTE, MINISTRY OF TRANSPORT)
1-1, 3-CHOME, NAGASE, YOKOSUKA-SHI 239
JAPANESE
INDEXED: JICST

KOWAN GIJUTSU YOHO 7064
(PROCEEDINGS OF HARBOUR ENGINEERING)
UN'YU-SHO KOWAN-KYOKU (PORT AND HARBOUR BUREAU, MINISTRY OF TRANSPORT)
1-3, 2-CHOME, KASUMIGASEKI, CHIYODA-KU, TOKYO 100
JAPANESE
INDEXED: JICST

KOWAN GIKEN SHIRYO 7065
(TECHNICAL NOTE OF PORT AND HARBOUR TECHNICAL RESEARCH INSTITUTE, MINISTRY OF TRANSPORT)
UN'YU-SHO KOWAN GIJUTSU KENKYUSHO (PORT AND HARBOUR TECHNICAL RESEARCH INSTITUTE, MINISTRY OF TRANSPORT)
1-1, 3-CHOME, NAGASE, YOKOSUKA-SHI 239
JAPANESE, WITH ENGLISH SUMMARY
INDEXED: JICST

KOWAN JIJO SOKUHO 7066
(PORT AND HARBOUR NEWS)
KAIJO HOAN-CHO SUIRO-BU (HYDROGRAPHIC DIVISION, MARITIME SAFETY BOARD) AND HEIWA NO UMI KYOKAI
23-7, 5-CHOME, SHINBASHI, MINATO-KU, TOKYO 105
JAPANESE

KOWAN KOENKAI KOENSHU 7067
(PROCEEDINGS OF THE SYMPOSIUM ON PORT SCIENCE)
NIHON KOWAN KYOKAI (JAPAN PORT AND HARBOR ASSOCIATION)
C/O TORANOMON KOTOHIRA KAIKAN, 1, SHIBA KOTOHIRA-CHO, MINATO-KU, TOKYO 105
JAPANESE

KOWAN TOKEI (NENPO) 7068
(PORTS AND HARBOURS STATISTICS, YEARBOOK)
UN'YU-SHO, DAIJIN KANBO JOHO KANRI-BU (RESEARCH AND DATA PROCESSING DEPARTMENT, MINISTRY OF TRANSPORT)
1-3, 2-CHOME, KASUMIGASEKI, CHIYODA-KU, TOKYO 100
JAPANESE

MARINE AND INDUSTRY NEWS 7069
BURIJISUTON TAIYA K. K. (BRIDGESTONE TIRE COMPANY)
C/O KANSEN BUILDING, 12-4, 4-CHOME, HATCHOBORI, CHUO-KU, TOKYO 104
ENGLISH

MIZU TO NINGEN 7070
(WATER AND MAN)
BIWAKO MONDAI KENKYU KIKO (BIWAKO PROBLEM RESEARCH ORGANIZATION)
JAPANESE

NAGOYAKO NO JIBAN 7071
(FOUNDATION OF NAGOYA PORT)
NAGOYAKO KANRI KUMIAI (NAGOYA PORT AUTHORITY)
8-21, 1-CHOME, IRIFUNE-CHO, MINATO-KU, NAGOYA-SHI 455
JAPANESE

NAGOYAKO TOKEI GEPPO 7072
(MONTHLY REPORT OF STATISTICS AT PORT OF NAGOYA)
NAGOYAKO KANRI KUMIAI GIJUTSU-BU (PLANNING SECRETARY, NAGOYA PORT AUTHORITY)
6-7, MINATOHON-MACHI, MINATO-KU, NAGOYA-SHI 455
JAPANESE

NAGOYAKO TOKEI NENPO 7073
(ANNUAL STATISTICS OF NAGOYA PORT)
NAGOYAKO KANRI KUMIAI GIJUTSU-BU AND NAGOYAKO KANRI KUMIAI (NAGOYA PORT AUTHORITY)
8-21, 1-CHOME, IRIFUNE-CHO, MINATO-KU, NAGOYA-SHI 455
JAPANESE

NAOETSUKO TOKEI NENPO 7074
(ANNUAL STATISTICS OF NAOETSU PORT)
NIIGATA-KEN NAOETSU KOWAN JIMUSHO (NAOETSU PORTS AND HARBOURS OFFICE, NIIGATA)
MINATO-CHO, JOETSU-SHI 942
JAPANESE

NIHON NO KAWA 7075
(JAPANESE RIVERS)
JAPANESE
INDEXED: JICST

NIIGATAKO TOKEI NENPO 7076
(ANNUAL STATISTICS OF NIIGATA PORT)
NIIGATA KOWAN JIMUSHO (NIIGATA PORTS AND HARBOURS OFFICE)
6-3, 1-CHOME, RYUGASHIMA, NIIGATA-SHI 950
JAPANESE

NOTICES TO MARINERS 7077
KAIJO HOAN-CHO SUIRO-BU (HYDROGRAPHIC DEPARTMENT, MARITIME SAFETY AGENCY)
ENGLISH

OSAKAKO 7078
(PORT OF OSAKA)
OSAKA-SHI KOWAN-KYOKU (PORT AND HARBOUR BUREAU, CITY OF OSAKA)
8-24, 2-CHOME, CHIKKO, MINATO-KU, OSAKA-SHI 552
JAPANESE

OSAKAKO KOSEI 7079
(STATE OF OSAKA PORT)
OSAKA-SHI KOWAN-KYOKU (PORT AND HARBOUR BUREAU, CITY OF OSAKA)
8-24, 2-CHOME, CHIKKO, MINATO-KU, OSAKA-SHI 552
JAPANESE

OSAKAKO KOSEI IPPAN 7080
(STATISTICS OF PORT AND HARBOURS, OSAKA PREFECTURE)
OSAKA-SHI KOWAN-KYOKU (PORT AND HARBOUR BUREAU, CITY OF OSAKA)
8-24, 2-CHOME, CHIKKO, MINATO-KU, OSAKA-SHI 552
JAPANESE

OSAKAKO TOKEI GEPPO 7081
(MONTHLY STATISTICS OF OSAKA PORT)
OSAKA-SHI KOWAN-KYOKU (PORT AND HARBOUR BUREAU, CITY OF OSAKA)
8-24, 2-CHOME, CHIKKO, MINATO-KU, OSAKA-SHI 552
JAPANESE

PORT OF OSAKA 7082
OSAKA-SHI KOWAN-KYOKU (PORT AND HARBOUR BUREAU, CITY OF OSAKA)
8-24, 2-CHOME, CHIKKO, MINATO-KU, OSAKA-SHI 552
JAPANESE AND ENGLISH

PORTS AND HARBORS 7083
KOKUSAI KOWAN KYOKAI
1,SHIBA KOTOCHIRA-CHO, MINATO-KU, TOKYO 105
ENGLISH
INDEXED: JICST

SABO JIGYO CHOSAHI HOKOKU 7084
(REPORT ON RESEARCH EXPENSE FOR EROSION-CONTROL CONSTRUCTION)
KENSETSU-SHO DOBOKU KENKYUSHO (PUBLIC WORKS RESEARCH INSTITUTE, MINISTRY OF CONSTRUCTION)
28-32, 2-CHOME, HONKOMAGOME, BUNKYO-KU, TOKYO 113
JAPANESE

SABO KISO CHOSA HOKOKUSHO 7085
(FUNDAMENTAL RESEARCH REPORT OF EROSION-CONTROL)
KANAGAWA-KEN DOBOKU-BU (PUBLIC WORKS DIVISION, KANAGAWA PREFECTURAL GOVERNMENT)
1,NIHON-ODORI, NAKA-KU, YOKOHAMA-SHI 231
JAPANESE

SABO TO CHISUI 7086
(EROSION-CONTROL AND RIVER WORK)
ZENKOKU CHISUI SABO KYOKAI (JAPAN EROSION-CONTROL AND RIVER WORK ASSOCIATION)
7-5, 2-CHOME, HIRAKAWA-CHO, CHIYODA-KU, TOKYO 102
JAPANESE
INDEXED: JICST

SHIGEN SETSUYAKUKA NO MIZA-MONDAI TO SONO TAISAKU SEMINA TEKISUTO 7087
(TEST OF THE SEMINAR ON THE WATER PROBLEM UNDER RESOURCE SAVING AND COUNTERMEASURES)
JAPANESE
INDEXED: JICST

SHIN SABO 7088
(JOURNAL OF THE EROSION-CONTROL ENGINEERING SOCIETY)
SABO GAKKAI (THE EROSION-CONTROL ENGINEERING SOCIETY)
C/O SABO KAIKAN, 7-5, 2-CHOME, HIRAKAWA-CHO, CHIYODA-KU, TOKYO 102
JAPANESE
INDEXED: JICST

SUIDO NYUSU 7089
(WATERWORKS NEWS)
TOKYO-TO SUIDO KYOKU (TOKYO METROPOLITAN WATER-WORKS BUREAU)
JAPANESE

SUIRO TSUHO 7090
(NOTICES TO MARINERS, JAPANESE EDITION)
KAIJO HOAN-CHO SUIRO-BU (HYDROGRAPHIC DEPARTMENT, MARITIME SAFETY AGENCY)
3-1, 5-CHOME, TSUKIJI, CHUO-KU, TOKYO 104
JAPANESE

TAMOKUTEKI DAMU KANRI NENPO 7091
(MULTIPURPOSE DAM CONTROL, ANNUAL REPORT)
KENSETSU-SHO KASEN-KYOKU (RIVER BUREAU, MINISTRY OF CONSTRUCTION)
1-3, 2-CHOME, KASUMIGASEKI, CHIYODA-KU, TOKYO 100
JAPANESE

TOA KOWAN KOGYO K.K. DOSHITSU KENKYU HOKOKU 7092
(REPORTS OF SOIL RESEARCH LABORATORY, TEA PORTS AND COAST ENGINEERING COMPANY, LIMITED)
TOA KOWAN KOGYO K.K. DOSHITSU KENKYUSHITSU
1-3, TSURUMI-KU, YOKOHAMA-SHI 254
JAPANESE

TOKYO KOUN KYOKAIHO 7093
(NEWS OF THE TOKYO MARITIME TRANSPORT ASSOCIATION)
TOKYO KOUN KYOKAI (TOKYO MARITIME TRANSPORT ASSOCIATION)
7-11, 2-CHOME, TSUKIJI, CHUO-KU, TOKYO 104
JAPANESE

TOKYO-TO KOWAN-KYOKU JIGYO GAIYO 7094
(OPERATION REPORT OF PORT AND HARBOUR BUREAU, TOKYO METROPOLITAN GOVERNMENT)
TOKYO-TO KOWAN-KYOKU KAIHATSU-BU (DEVELOPMENT DIVISION, PORT AND HARBOUR BUREAU, TOKYO METROPOLITAN GOVERNMENT)
8-1, 3-CHOME, MARUNOUCHI, CHIYODA-KU, TOKYO 100
JAPANESE

TOKYOKO KOSEI 7095
(STATE OF TOKYO PORT)
TOKYO-TO KOWAN-KYOKU (PORT AND HARBOUR BUREAU, TOKYO METROPOLITAN GOVERNMENT)
8-1, 3-CHOME, MARUNOUCHI, CHIYODA-KU, TOKYO 100
JAPANESE

UMETATE TO SHUNSETSU 7096
(RECLAMATION AND DREDGE)
NIHON UMETATE KYOKAI (THE RECLAMATION AND DREDGING ASSOCIATION OF JAPAN)
1, SHIBA KOTOHIRA-CHO, MINATO-KU, TOKYO 105
JAPANESE
INDEXED: JICST

UN'YUSHO KOWAN GIJUTSU KENKYUJO SHIRYO, BUNKEN SHUHO 7097
(MINISTRY OF TRANSPORT PORT AND HARBOR RESEARCH INSTITUTE DOCUMENTS, COLLECTED BIBLIOGRAPHICAL REPORTS)
UN'YUSHO KOWAN GIJUTSU KENKYUJO KENSHU SHIRYOKA (MINISTRY OF TRANSPORT, PORTLAND HARBOR RESEARCH INSTITUTE, TRAINING MATERIALS SECTION)
JAPANESE

YOKOHAMA KOWAN TOKEI SOKUHO 7098
(QUICK ANNOUNCEMENT OF PORTS AND HARBOURS STATISTICS, YOKOHAMA)
YOKOHAMA-SHI KOWAN-KYOKU (PORT AND HARBOURS BUREAU YOKOHAMA MUNICIPAL GOVERNMENT)
1-1, MINATO-MACHI, NAKA-KU, YOKOHAMA-SHI 231
JAPANESE

YOKOHAMAKO TOKEI GEPPO 7099
(STATISTICAL MONTHLY REPORT, PORT OF YOKOHAMA)
YOKOHAMA-SHI KOWAN-KYOKU (PORT AND HARBOUR BUREAU, YOKOHAMA MUNICIPAL GOVERNMENT)
1-1, MINATO-MACHI, NAKA-KU, YOKOHAMA-SHI 231
JAPANESE

YOKOHAMAKO TOKEI NENPO 7100
(ANNUAL STATISTICS OF YOKOHAMA PORT HARBOR)
YOKOHAMA-SHI KOWAN-KYOKU (PORT AND HARBOR BUREAU, YOKOHAMA MUNICIPAL GOVERNMENT)
1-1, MINATO-MACHI, NAKA-KU, YOKOHAMA-SHI 226
JAPANESE

ZENKOKU KOWAN KOJI HOKOKUKAI HOKOKU GAIYO 7101
(SUMMARY REPORT OF PORT WORKS CONFERENCE)
NIHON KOWAN KYOKAI (JAPAN PORT ASSOCIATION)
C/O TORANOMON KOTOHIRA KAIKAN, SHIBA KOTOHIRA-CHO, MINATO-KU, TOKYO 105
JAPANESE
INDEXED: JICST

ZOSUI GIJUTSU 7102
(JOURNAL OF WATER RE-USE TECHNOLOGY)
ZOSUI SOKUSHIN SENTA (WATER RE-USE PROMOTION CENTER)
JAPANESE
INDEXED: JICST

Public Health Engineering

BIWAKO SUISHITSU CHOSA HOKOKUSHO 7103
(REPORT OF WATER RESEARCH OF THE LAKE BIWA)
SHIGA-KENRITSU EISEI KOGAI KENKYUSHO (SHIGA PREFECTURAL LABORATORY OF HEALTH AND PUBLIC NUISANCE)
13-45, GOTENHAMA, OTSU-SHI 520
JAPANESE

CHIKASUI TO IDO TO PONPU 7104
(WATER JOURNAL)
CHIKASUI GIJUTSU KYOKAI (TECHNICAL ASSOCIATION OF UNDER GROUND WATER)
11, KANDA MITOSHIRO-CHO, CHIYODA-KU, TOKYO 101
JAPANESE
INDEXED: JICST

COLLECTED PAPERS ON SCIENCES OF ATMOSPHERE AND HYDROSPHERE 7105
NAGOYA DAIGAKU RIGAKUBU SUISHITSU KAGAKU KENKYU SHISETSU (WATER RESEARCH INSTITUTE, NAGOYA UNIVERSITY)
FURO-CHO, CHIKUSA-KU, NAGOYA-SHI 464
JAPANESE AND ENGLISH

EBARA INFIRUKO JIHO 7106
JAPANESE
INDEXED: JICST

EISEI KOGAKU KENKYU TORONKAI KOEN RONBUNSHU 7107
(PROCEEDINGS OF SANITARY ENGINEERING)
DOBOKU GAKKAI (JAPAN SOCIETY OF CIVIL ENGINEERS)
1, YOTSUYA, SHINJUKU-KU, TOKYO 160
JAPANESE

FUKUOKA-SHI SUIDO JIGYO TOKEI NENPO 7108
(ANNUAL STATISTICS OF WATER WORKS, FUKUOKA CITY)
FUKUOKA-SHI SUIDO-KYOKU (BUREAU OF WATER WORKS, FUKUOKA MUNICIPAL GOVERNMENT)
8-1, 1-CHOME, TENJIN, CHUO-KU, FUKUOKA-SHI 810
JAPANESE

GESUI SUISHITSU SHIKEN NENPO (KYOTO-SHI) 7109
(ANNUAL REPORT OF SEWAGE WATER, KYOTO CITY)
KYOTO-SHI (KYOTO MUNICIPAL GOVERNMENT)
OIKE, TERAMACHI-DORI, NAKAGYO-KU, KYOTO-SHI 604
JAPANESE

GESUIDO 7110
(SEWERAGE)
TOKYO-TO GESUIDO-KYOKU (BUREAU OF SEWERAGE, TOKYO METROPOLITAN GOVERNMENT)
5-1, 3-CHOME, MARUNOUCHI, CHIYODA-KU, TOKYO 100
JAPANESE

GESUIDO JIGYO CHOSAHI HOKOKU 7111
(REPORT ON EXPENSE FOR SEWERAGE WORKS)
KENSETSU-SHO DOBOKU KENKYUSHO (PUBLIC WORKS RESEARCH INSTITUTE, MINISTRY OF CONSTRUCTION)
28-32, 2-CHOME, HONKOMAGOME, BUNKYO-KU, TOKYO 113
JAPANESE

GESUIDO KENKYU HAPPOKAI KOENSHU 7112
(PROCEEDINGS OF SEWERAGE RESEARCH CONFERENCE)
NIHON GESUIDO KYOKAI (JAPAN SEWERAGE WORKS ASSOCIATION)
4-3, 2-CHOME, HIRAKAWA-CHO, CHIYODA-KU, TOKYO 102
JAPANESE
INDEXED: JICST

GESUIDO KYOKAISHI 7113
(JOURNAL OF JAPAN SEWAGE WORKS ASSOCIATION)
NIHON GESUIDO KYOKAI (JAPAN SEWAGE WORKS ASSOCIATION)
C/O KOJIMACHI KAIKAN, 4-3, 2-CHOME, HIRAKAWA-CHO, CHIYODA-KU, TOKYO 102
JAPANESE
INDEXED: APT, JICST

HAIKIBUTSU 7114
(WASTE MATTER)
JAPANESE
INDEXED: JICST

HEDORO 7115
(SLUDGE)
TEISHITSU JOKA KYOKAI
JAPANESE

KANAGAWA-KEN SUISHITSU CHOSA NENPYO 7116
(ANNUAL PAPERS OF WATER QUANTITY RESEARCH, KANAGAWA PREFECTURE)
KANAGAWA-KEN KOGAI TAISAKU JIMU-KYOKU (KAWAGAWA PREFECTURE GOVERNMENT)
1, NIHON-ODORI, NAKA-KU, YOKOHAMA-SHI 231
JAPANESE

KOKYO GESUIDO TOKEI 7117
(ANNUAL STATISTICS OF PUBLIC SEWAGE)
NIHON GESUIDO KYOKAI (JAPAN SEWAGE WORKS ASSOCIATION)
4-3, 2-CHOME, HIRAKAWA-CHO, CHIYODA-KU, TOKYO 102
JAPANESE

KOKYO GESUIDO TOKEI NENPO (KYOTO-SHI) 7118
(ANNUAL STATISTICS OF PUBLIC SEWAGE, KYOTO CITY)
KYOTO-SHI GESUIDO-KYOKU (SEWAGE BUREAU, KYOTO MUNICIPAL GOVERNMENT)
12, HIGASHISANNO-CHO, HIGASHI-9-JO, MINAMI-KU, KYOTO-SHI 601
JAPANESE

KUCHO GAIDO 7119
(GUIDE TO AIR CONDITIONING ENGINEERING)
TOKYO SHIBAURA DENKI K.K. (TOKYO SHIBAURA ELECTRIC COMPANY, LIMITED)
2-1, 5-CHOME, GINZA, CHUO-KU, TOKYO 104
JAPANESE

KUKI CHOWA EISEI KOGAKKAI RONBUNSHU 7120
(TRANSACTIONS OF THE SOCIETY OF HEATING, AIR-CONDITIONING AND SANITARY ENGINEERS OF JAPAN)
KUKI CHOWA EISEI KOGAKKAI (SOCIETY OF HEATING, AIR CONDITIONING AND SANITARY ENGINEERS OF JAPAN)
JAPANESE

KUKI CHOWA, EISEI KOGAKU 7121
(JOURNAL OF THE SOCIETY OF HEATING, AIR-CONDITIONING AND SANITARY ENGINEERS OF JAPAN)
KUKI CHOWA, EISEI KOGAKKAI (SOCIETY OF HEATING, AIR-CONDITIONING AND SANITARY ENGINEERS OF JAPAN)
2-9, KYOBASHI, CHUO-KU, TOKYO 104
JAPANESE, WITH ENGLISH SUMMARY
INDEXED: APT, JICST

KUKI CHOWA EISEI KOGAKUKAI SEMINA TEKISUTO 7122
(SEMINAR TEST OF THE SOCIETY OF HEATING, AIR-CONDITIONING AND SANITARY ENGINEERS OF JAPAN)
JAPANESE
INDEXED: JICST

KUKI CHOWA REITO RENGO KOENKAI KOEN RONBUNSHU 7123
(PROCEEDINGS OF THE SYMPOSIUM OF THE FEDERATION OF HEATING, AIR-CONDITIONING AND REFRIGERATION ENGINEERS)
JAPANESE
INDEXED: JICST

KUKI CHOWA TO REITO 7124
(AIR-CONDITIONING AND REFRIGERATION)
NIHON KUCHO GIJUTSU SHUPPANSHA
3-1, 1-CHOME, SARUGAKU-CHO, CHIYODA-KU, TOKYO 101
JAPANESE
INDEXED: JICST

KUKI SEIJO 7125
(AIR CLEANING)
NIHON KUKI SEIJO KYOKAI (JAPAN AIR CLEANING ASSOCIATION)
7-5, 6-CHOME, SOTO KANDA, CHIYODA-KU, TOKYO 101
JAPANESE
INDEXED: CHE, JICST

MIZU 7126
(WATER)
GEKKAN 'MIZU' HAKKOSHO (MONTHLY MAGAZINE 'WATER' PUBLICATION OFFICE)
22-6, 2-CHOME, NISHISHINBASHI, MINATO-KU, TOKYO 105
JAPANESE
INDEXED: JICST

MIZU KEIZAI NENPO 7127
(WATER ECONOMICS ANNUAL)
JAPANESE
INDEXED: JICST

MIZU SHORI GIJUTSU 7128
(WATER PURIFICATION AND LIQUID WASTES TREATMENT)
NIHON MIZU SHORI GIJUTSU KENKYUKAI (WATER PURIFICATION RESEARCH INSTITUTE)
C/O OGI-MACHI BUILDING, 17, NISHIOGI-MACHI, KITA-KU, OSAKA-SHI 530
JAPANESE
INDEXED: CHE, JICST

MIZU TO TOMONI 7129
(BULLETIN OF THE WATER RESOURCES DEVELOPMENT PUBLIC CORPORATION)
MIZU SHIGEN KAIHATSU KODAN (THE WATER RESOURCES DEVELOPMENT PUBLIC CORPORATION)
3-6, 5-CHOME, AKASAKA, MINATO-KU, TOKYO 107
JAPANESE

NAGASAKI SOGO KOSHU EISEIGAKU ZASSHI 7130
(NAGASAKI GENERAL PUBLIC HEALTH MAGAZINE)
JAPANESE
INDEXED: JICST

NIHON GESUIDO KYOKAI SHIRYO 7131
(REVIEWS OF JAPAN SEWAGE WORKS ASSOCIATION)
NIHON GESUIDO KYOKAI (JAPAN SEWAGE WORKS ASSOCIATION)
4-3, 2-CHOME, HIRAKAWA-CHO, CHIYODA-KU, TOKYO 102
JAPANESE

NIHON GESUIDO SHINBUN 7132
(JAPAN SEWAGE WORKS NEWS)
NIHON SUIDO SHINBUNSHA (THE NIHON WATER WORKS PRESS)
8-9, 4-CHOME, KUDANMINAMI, CHIYODA-KU, TOKYO 102
JAPANESE

NIHON KOGAI EISEI SENTA JOHO 7133
(BULLETIN OF JAPAN ENVIRONMENTAL SANITATION CENTER)
JAPAN ENVIRONMENTAL SANITATION CENTER
JAPANESE

NIHON MIZU SHORI SEIBUTSU GAKKAISHI 7134
(JAPANESE JOURNAL OF WATER TREATMENT BIOLOGY)
NIHON MIZUSHORI SEIBUTSU GAKKAI (WATER TREATMENT BIOLOGY SOCIETY OF JAPAN)
C/O OSAKA-SHI SUIDO-KYOKU SUISHITSU SHIKENJO, 1, HAMA-CHO, HIGASHIYODOGAWA-KU, OSAKA-SHI 533
JAPANESE

NIHON REITO REIBO SHINBUN 7135
(JAPANESE FREEZING AND AIR CONDITIONING NEWS)
NIHON REITO REIBO SHINBUNSHA
11-7, 2-CHOME, TSUKIJI, CHUO-KU, TOKYO 104
JAPANESE

NIHON SUIDO SHINBUN 7136
(NIHON WATER WORKS PRESS)
NIHON SUIDO SHINBUNSHA (THE NIHON WATER WORKS PRESS)
8-9, 4-CHOME, KUDANMINAMI, CHIYODA-KU, TOKYO 102
JAPANESE

OSAKA-FU SUIDO-BU TOKEI NENPO 7137
(ANNUAL STATISTICS OF WATER WORKS BUREAU, OSAKA PREFECTURAL GOVERNMENT)
OSAKA-FU SUIDO-BU (WATER WORKS BUREAU, OSAKA PREFECTURAL GOVERNMENT)
2, OTEMAENO-CHO, HIGASHI-KU, OSAKA-SHI 540
JAPANESE

OSAKA-SHI SUIDOKYOKU SUISHITSU SHIKENJO CHOSA HOKOKU NARABINI SHIKEN SEISEKI 7138
(ANNUAL REPORT OF THE WATER EXAMINATION LABORATORY OF OSAKA MUNICIPAL WATER WORKS BUREAU)
OSAKA-SHI SUIDO-KYOKU SUISHITSU SHIKENJO (WATER EXAMINATION LABORATORY OF OSAKA MUNICIPAL WATER WORKS BUREAU)
1, HAMA-CHO, HIGASHIYODOGAWA-KU, OSAKA-SHI 533
JAPANESE
INDEXED: CHE, JICST

OTAGAWA SUIKEI SUISHITSU CHOSA HOKOKUSHO 7139
(RESEARCH REPORT OF WATER QUALITY, OTA RIVER)
KENSETSU-SHO CHUGOKU CHIHO KENSETSU-KYOKU CHUGOKU GIJUTSU JIMUSHO (CHUGOKU TECHNICAL OFFICE, CHUGOKU DISTRICT CONSTRUCTION BUREAU, MINISTRY OF CONSTRUCTION)
1920-1, FUNAKOSHI-CHO, AKI-GUN, HIROSHIMA 736
JAPANESE

REITO 7140
(REFRIGERATION)
NIHON REITO KYOKAI (JAPANESE ASSOCIATION OF REFRIGERATION)
8, SAN'EI-CHO, SHINJUKU-KU, TOKYO 160
JAPANESE, WITH ENGLISH SUMMARY
INDEXED: CHE, JICST

REITO KUCHO GIJUTSU 7141
(TECHNIQUES OF REFRIGERATION AND AIR-CONDITIONING)
NIHON REITO KYOKAI (JAPANESE ASSOCIATION OF REFRIGERATION)
8, SAN'EI-CHO, SHINJUKU-KU, TOKYO 160
JAPANESE
INDEXED: JICST

REITO TO KUCHO 7142
(REFRIGERATION AND AIR-CONDITIONING)
NIHON REITO KUCHO KOGYOKAI (JAPAN REFRIGERATION AND AIR-CONDITIONING ENGINEERING ASSOCIATION)
15-2, 6-CHOME, GINZA, CHUO-KU, TOKYO 104
JAPANESE
INDEXED: JICST

SANTAMA CHIIKI CHIKASUI RIYO KANORYO CHOSA HOKOKUSHO 7143
(SURVEY REPORT OF THE GROUNDWATER UTILIZATION IN SAN-TAMA AREA)
TOKYO-TO SHUTO SEIBI-KYOKU (CAPITAL DEVELOPMENT DEPARTMENT, TOKYO METROPOLITAN GOVERNMENT)
5-1, 3-CHOME, MARUNOUCHI, CHIYODA-KU, TOKYO 100
JAPANESE

SAROBETSU SOGO CHOSA CHUKAN HOKOKUSHO. MIZU BUMON 7144
(SAROBETSU GENERAL SURVEY, INTERIM REPORT, WATER SECTION)
JAPANESE
INDEXED: JICST

SEISO GIHO 7145
(TECHNICAL REPORT, BUREAU OF PUBLIC CLEANSING, TOKYO METROPOLITAN GOVERNMENT)
SEISO-KYOKU GIJUTSU RENGOKAI SEISO GIHO HENSHU IINKAI-HEN, TOKYO-TO SEISO-KYOKU SOMU-BU FUKYU-KA (BUREAU OF CLEANSING TECHNICAL INFORMATION, EDITORIAL COMMITTEE, CLEANSING BUREAU TECHNICAL FEDERATION)
JAPANESE

SEISO-KYOKU NENPO 7146
(ANNUAL REPORT OF THE BUREAU OF PUBLIC CLEANSING)
TOKYO-TO SEISO-KYOKU KIKAKU-BU (DIVISION OF PLANNING, BUREAU OF PUBLIC CLEANSING, TOKYO METROPOLITAN GOVERNMENT)
5-1, 3-CHOME, MARUNOUCHI, CHIYODA-KU, TOKYO 100
JAPANESE

SHUTOKEN NO MIZU SHIGEN NI KANSURU KISO CHOSA HOKOKUSHO 7147
(ANNUAL SURVEY REPORT ON THE WATER RESOURCES IN THE METROPOLITAN AREA)
YAMANSHI-KEN KIKAKU KANRISHITSU (PLANNING AND MANAGING OFFICE, YAMANASHI PREFECTURE)
6-1, 1-CHOME, MARUNOUCHI, KOFU-SHI 400
JAPANESE

SORA TO MIZU 7148
(SKY AND WATER)
TOKYO KUKI CHOWA EISEI KOJIGYO KYOKAI (TOKYO AIR-CONDITIONING AND SANITARY INDUSTRY ASSOCIATION)
3-11, KYOBASHI, CHUO-KU, TOKYO 104
JAPANESE

SUIDO JIGYO TOKEI NENPO (KANAGAWA-KEN) 7149
(ANNUAL STATISTICS OF WATER WORKS, KANAGAWA PREFECTURE)
KANAGAWA-KEN KIGYO-CHO (ENTERPRISE AGENCY, KANAGAWA PREFECTURAL GOVERNMENT)
1, NIHON-ODORI, NAKA-KU, YOKOHAMA-SHI 231
JAPANESE

SUIDO KORON 7150
(WATER WORKS REVIEW)
NIHON SUIDO SHINBUNSHA
8-9, 4-CHOME, KUDANMINAMI, CHIYODA-KU, TOKYO 102
JAPANESE
INDEXED: CHE

SUIDO KYOKAI ZASSHI 7151
(JOURNAL OF WATER WORKS ASSOCIATION)
NIHON SUIDO KYOKAI (JAPAN WATER WORKS ASSOCIATION)
8-9, 4-CHOME, KUDANMINAMI, CHIYODA-KU, TOKYO 102
JAPANESE, WITH ENGLISH SUMMARY
INDEXED: CHE, JICST

SUIDO TOKEI 7152
(WATER WORKS STATISTICS)
KOSEI-SHO KANKYO EISEI-KYOKU (ENVIRONMENTAL SANITATION BUREAU, MINISTRY OF HEALTH AND WELFARE) AND NIHON SUIDO KYOKAI (JAPAN WATER WORKS ASSOCIATION)
8-9, 4-CHOME, KUDANMINAMI, CHIYODA-KU, TOKYO 102
JAPANESE

SUIDO TOKEI NENPO, KYOTO-SHI 7153
(ANNUAL REPORT OF KYOTO CITY WATER WORKS)
KYOTO-SHI SUIDO-KYOKU (BUREAU OF WATER WORKS, KYOTO MUNICIPAL GOVERNMENT)
12, HIGASHISANNO-MACHI, HIGASHI-9-JO, MINAMI-KU, KYOTO-SHI 601
JAPANESE

SUIDO-KYOKU JIGYO NENPO, OSAKA-SHI 7154
(ANNUAL REPORT OF WATER WORKS, OSAKA CITY)
OSAKA-SHI SUIDO-KYOKU (THE WATER WORKS BUREAU OF OSAKA CITY)
7, MINAMIOGI-MACHI, KITA-KU, OSAKA-SHI 530
JAPANESE

SUIDO-KYOKUHO 7155
(BUREAU OF WATER WORKS BULLETIN)
TOKYO-TO SUIDO-KYOKU (BUREAU OF WATER WORKS, TOKYO METROPOLITAN GOVERNMENT)
5-1, 3-CHOME, MARUNOUCHI, CHIYODA-KU, TOKYO 100
JAPANESE

SUIMON TEKKAN 7156
(WATER GATE & PENSTOCK)
SUIMON TEKKAN KYOKAI (WATER GATE AND PENSTOCK ASSOCIATION)
C/O TORANOMON JITSUGYO KAIKAN, 8, SHIBA TORANOMON, MINATO-KU, TOKYO 105
JAPANESE
INDEXED: JICST

SUIRI KAGAKU 7157
(WATER SCIENCE)
SUIRI KAGAKU KENKYUSHO (INSTITUTE OF WATER UTILIZATION)
7-22, 1-CHOME, KORAKU, BUNKYO-KU, TOKYO 112
JAPANESE
INDEXED: BIB, CHE, JICST

SUISHITSU CHOSA HOKOKUSHO (KUMAMOTO-KEN) 7158
(REPORT OF WATER QUALITY EXAMINATION, KUMAMOTO PREFECTURE)
KUMAMOTO-KEN (KUMAMOTO PREFECTURAL GOVERNMENT)
18-1, 6-CHOME, SUIZENJI, KUMAMOTO-SHI 862
JAPANESE

SUISHITSU CHOSA NARABINI SHIKEN NENPO 7159
(ANNUAL OF WATER QUALITY SURVEY AND TESTING)
JAPANESE
INDEXED: JICST

SUISHITSU KANRI NENPO (YOKOHAMA-SHI) 7160
(ANNUAL STATISTICS OF WATER QUALITY CONTROL, YOKOHAMA CITY)
YOKOHAMA-SHI GESUIDO-KYOKU (SEWAGE BUREAU, YOKOHAMA MUNICIPAL GOVERNMENT)
1-1, MINATO-MACHI, NAKA-KU, YOKOHAMA-SHI 231
JAPANESE

SUISHITSU ODAKU BOSHI TAISAKU NI KANSURU CHOSA KENKYU HOKOKUSHO 7161
(RESEARCH REPORT OF PREVENTION OF WATER POLLUTION)
HOKKAIDORITSU ABASHIRI SUISAN SHIKENJO (ABASHIRI FISHERY EXPERIMENT STATION, HOKKAIDO PREFECTURAL GOVERNMENT)
31, MASUURA, ABASHIRI-SHI 099-31
JAPANESE

SUISHITSU ODAKU NI KANSURU KENKYU SHUHO 7162
(COLLECTED REPORTS OF WATER POLLUTION)
HOKKAIDO SEIKATSU KANKYO-BU (LIFE ENVIROMENTAL DIVISION, HOKKAIDO PREFECTURAL GOVERNMENT)
NISHI-6-CHOME, KITA-3-JO, CHUO-KU, SAPPORO-SHI 060
JAPANESE

SUISHITSU ODAKU TAISAKU SHIKEN SEISEKISHO 7163
(REPORT OF EXAMINATION IN WATER POLLUTION PREVENTION)
AICHI-KEN NOGYO SOGO SHIKENJO (GENERAL AGRICULTURAL EXPERIMENT STATION, AICHI PREFECTURE)
NAGAKUDE-MURA, AICHI-GUN, AICHI 480-11
JAPANESE

SURUGAWAN KAIYO CHOSA HOKOKU 7164
(SURVEY REPORT OF THE SURUGAWAN SEAS)
JAPANESE
INDEXED: JICST

TOKYO-TO GESUIDO JIGYO GAIYO 7165
(OUTLINE OF SEWERAGE WORKS)
TOKYO-TO GESUIDO-KYOKU (BUREAU OF SEWERAGE WORKS, TOKYO METROPOLITAN GOVERNMENT)
5-1, 3-CHOME, MARUNOUCHI, CHIYODA-KU, TOKYO 100
JAPANESE

TOKYO-TO GESUIDO JIGYO NENPO 7166
(ANNUAL REPORT OF THE SEWAGE BUREAU, TOKYO METROPOLITAN GOVERNMENT)
TOKYO-TO GESUIDO-KYOKU (SEWAGE BUREAU, TOKYO METROPOLITAN GOVERNMENT)
5-1, 3-CHOME, MARUNOUCHI, CHIYODA-KU, TOKYO 100
JAPANESE
INDEXED: JICST

TOKYO-TO SEISO KENKYUJO KENKYU HOKOKU 7167
(BULLETIN OF THE TOKYO METROPOLITAN SANITATION RESEARCH INSTITUTE)
JAPANESE

TOKYO-TO SUIDO JIGYO NENPO 7168
(ANNUAL REPORT ON TOKYO METROPOLITAN WATER WORKS)
TOKYO-TO SUIDO-KYOKU (BUREAU OF WATER WORKS, TOKYO METROPOLITAN GOVERNMENT)
5-1, 3-CHOME, MARUNOUCHI, CHIYODA-KU, TOKYO 100
JAPANESE

TOKYO-TO SUIDO-KYOKU CHOSA SHIRYO 7169
(NEWS OF THE BUREAU OF WATER WORKS, TOKYO METROPOLITAN GOVERNMENT)
TOKYO-TO SUIDO-KYOKU (BUREAU OF WATER WORKS, TOKYO METROPOLITAN GOVERNMENT)
5-1, 3-CHOME, MARUNOUCHI, CHIYODA-KU, TOKYO 100
JAPANESE
INDEXED: JICST

TOKYO-TO SUIDO-KYOKU SHISETSU GAIYO 7170
(OUTLINE OF EQUIPMENT, BUREAU OF WATER WORKS, TOKYO METROPOLITAN GOVERNMENT)
TOKYO-TO SUIDO-KYOKU (BUREAU OF WATER WORKS, TOKYO METROPOLITAN GOVERNMENT)
5-1, 3-CHOME, MARUNOUCHI, CHIYODA-KU, TOKYO 100
JAPANESE

TOKYO-TORITSU EISEI KENKYUJO KENKYU HOKOKU 7171
(REPORT OF TOKYO METROPOLITAN RESEARCH LABORATORY OF PUBLIC HEALTH)
JAPANESE
INDEXED: JICST

TONAI KASEN, NAIWAN NO SUISHITSU 7172
(WATER QUALITY OF RIVERS AND INNER BAY, TOKYO)
TOKYO-TO KOGAI-KYOKU (PUBLIC NUISANCE BUREAU, TOKYO METROPOLITAN GOVERNMENT)
5-1, 3-CHOME, MARUNOUCHI, CHIYODA-KU, TOKYO 100
JAPANESE

TOSHI SEISO 7173
(CITY CLEANING)
ZENKOKU TOSHI SEISO KAIGI (ALL JAPAN CITY CLEANING INSTITUTE)
C/O TOKYO-TO HAMAMATSU-CHO CHOSHA, 4-29 1-CHOME, KAIGAN, MINATO-KU, TOKYO 105
JAPANESE
INDEXED: JICST

TRANSACTIONS OF SHASE JAPAN 7174
KUKI CHOWA EISEI KOGAKKAI (SOCIETY OF HEATING, AIR-CONDITIONING AND SANITARY ENGINEERS OF JAPAN)
2-9, KYOBASHI, CHUO-KU, TOKYO 104
ENGLISH
INDEXED: JICST

WIP MIZUKOGAI YOHAISUI JOHO 7175
(WATER INFORMATION PILOT)
SUIYOSHA
1-18, 2-CHOME, KORAKU, BUNKYO-KU, TOKYO 112
JAPANESE AND ENGLISH

YAMATOGAWA SUISHITSU ODAKU CHOSA HOKOKU 7176
(RESEARCH REPORT OF WATER POLLUTION, YAMATO RIVER)
YAMATOGAWA SUISHITSU ODAKU BOSHI RENRAKU KYOGIKAI
C/O KINKI CHIHO KENSETSU-KYOKU YAMATO KOJI JIMUSHO, 10-8, 2-CHOME, TAISHO, KASHIHARA-SHI 582
JAPANESE

YOSUI TO HAISUI 7177
(JOURNAL OF WATER AND WASTE)
SANGYO YOSUI CHOSAKAI (INDUSTRIAL WATER INSTITUTE)
20-12, 5-CHOME, SENDAGAYA, SHIBUYAKU, TOKYO 151
JAPANESE AND ENGLISH
INDEXED: CHE, JICST

ZENKOKU SUIDO KENKYU HAPPYOKAI KOENSHU 7178
(PROCEEDINGS OF WATER WORKS RESEARCH CONFERENCE)
NIHON SUIDO KYOKAI (JAPAN WATER WORKS ASSOCIATION)
8-9, 4-CHOME, KUDANMINAMI, CHIYODA-KU
TOKYO 102
JAPANESE

Land and Road Vehicles (Excluding Railway)

AUTO SPORT 7179
SAN'EI SHOBO
1, SUWA-CHO, SHINJUKU-KU, TOKYO 160
JAPANESE

BASU GEPPO 7180
(BUS MONTHLY)
NIHON BASU KYOKAI
(JAPAN BUS ASSOCIATION)
JAPANESE

BESSATSU HORIDE OTO 7181
(HOLIDAY AUTO, SEPARATE ISSUE)
MOTA MAGAJIN SHA (MOTOR MAGAZINE COMPANY)
JAPANESE

BULLETIN OF JSAE 7182
JIDOSHA GIJUTSUKAI (SOCIETY OF AUTOMOTIVE ENGINEERS OF JAPAN)
16-15, 1-CHOME, TAKANAWA, MINATO-KU, TOKYO 108
ENGLISH
INDEXED: APP, AUT, ENG

CAR GRAPHIC 7183
NIGENSHA
2-42, KANDA JINBO-CHO, CHIYODA-KU, TOKYO 101
JAPANESE

CAR TOPPU 7184
(CAR TOP)
KOTSU TAIMZUSHA (TRAFFIC TIMES COMPANY)
JAPANESE AND ENGLISH

DORAIBA 7185
(DRIVER)
MOTASAIKURU SHUPPAN SHA (MOTORCYCLE PUBLISHING COMPANY)
JAPANESE

DORAIBA 7186
(DRIVER)
YAESU SHUPPAN
5-15, 1-CHOME, YAESU, CHUO-KU, TOKYO 103
JAPANESE

GAIKOKU JIDOSHA GIJUTSU BUNKEN SHOROKUSHU 7187
(TECHNICAL ABSTRACTS OF THE FOREIGN-MADE AUTOMOBILE)
JIDOSHA GIJUTSUKAI (THE SOCIETY OF AUTOMOTIVE ENGINEERS OF JAPAN)
16-15, 1-CHOME, TAKANAWA, MINATO-KU, TOKYO 108
JAPANESE

GEKKAN BESUTO KA GAIDO 7188
(BEST CAR GUIDE MONTHLY)
KODANSHA (KODAN PUBLISHING COMPANY)
JAPANESE

GEKKAN JIDOSHA YOHIN NO JITSUMU 7189
(AUTOMOBILE ACCESSORIES BUSINESS MONTHLY)
JIDOSHA SANGYO TSUSHIN SHA
(AUTOMOBILE INDUSTRY NEWS COMPANY)
JAPANESE

GEKKAN JIKAYOSHA 7190
(OWNER DRIVER)
NAIGAI SHUPPANSHA (NAIGAI PUBLISHING COMPANY, LTD.)
C/O KOMORI BUILDING, 18-14, 2-CHOME, HIGASHIUENO, TAITO-KU, TOKYO 110
JAPANESE

GEKKAN TAIYA 7191
(TIRE MONTHLY)
GEKKAN TAIYASHA
6-1, YAESU, CHUO-KU, TOKYO 104
JAPANESE

GIJUTSU CHOSA SHIRYO 7192
(DATA ON TECHNICAL INVESTIGATIONS)
NIHON JIDOSHA KENKYUJO (JAPAN AUTOMOTIVE RESEARCH INSTITUTE)
JAPANESE

GIJUTSU JOHO 7193
(AUTOMOTIVE SERVICE NEWS)
NIHON JIDOSHA SEIBI SHINKOKAI RENGOKAI (JAPAN AUTOMOBILE SERVICE PROMOTION ASSOCIATION)
3-2, 6-CHOME, ROPPONGI, MINATO-KU, TOKYO 106
JAPANESE

GIKEN NYUSU (JITENSHA SANGYO SHINKOKAI GIJUTSU KENKYUSHO) 7194
(TECHNICAL LABORATORY NEWS, BICYCLE INDUSTRY PROMOTION INSTITUTE)
JITENSHA SANGYO SHINKOKAI GIJUTSU KENKYUSHO
1-6, SHINKAWA, INUYAMA-SHI 484
JAPANESE

HINO DAYORI 7195
(HINO LETTER)
HINO JIDOSHA HANBAI (HINO AUTOMOBILE SALES)
JAPANESE

HINO GIHO 7196
(TECHNICAL REVIEW, HINO MOTORS LIMITED)
HINO JIDOSHA KOGYO K.K. KENKYU KANRI-BU HINO GIHO HENSHU-BU (EDITORIAL OFFICE OF TECHNICAL REVIEW, HINO MOTORS, HINO MOTORS LTD.)
1-1, 3-CHOME, HINODAI, HINO-SHI, TOKYO 191
JAPANESE
INDEXED: AUT, JICST

HINODE 7197
(HINO AUTOMOBILE NEWS)
HINO JIDOSHA HANBAI K.K.
11-3, 4-CHOME, SHIBA, MINATO-KU, TOKYO 108
JAPANESE

HITO TO KURUMA 7198
(MAN AND AUTOMOBILE)
ZEN NIHON KOTSU ANZEN KYOKAI (ALL JAPAN TRANSPORTATION SAFETY SOCIETY)
JAPANESE

HOKKAIDO JIDOSHA TANKI DAIGAKU KENKYU KIYO 7199
(MEMOIRS OF THE AUTOMOTIVE COLLEGE OF HOKKAIDO)
HOKKAIDO JIDOSHA TANKI DAIGAKU (AUTOMOTIVE COLLEGE OF HOKKAIDO)
6-CHOME, NAKANOSHIMA-2-JO, TOYOHIRA-KU, SAPPORO-SHI 062
JAPANESE, WITH ENGLISH SUMMARY

ISUZU GIHO 7200
(ISUZU TECHNICAL JOURNAL)
ISUZU JIDOSHA K.K. (ISUZU MOTORS LTD.)
22-10, 6-CHOME, MINAMIOI, SHINAGAWA-KU TOKYO 140
JAPANESE
INDEXED: JICST

ISUZU SHINBUN 7201
(ISUZU AUTOMOBILE NEWS)
ISUZU JIDOSHA K.K. (ISUZU MOTORS LTD.)
22-10, 6-CHOME, MINAMIOI, SHINAGAWA-KU, TOKYO 140
JAPANESE

JAF NYUSU 7202
(JAF NEWS)
NIHON JIDOSHA RENMEI (JAPAN AUTOMOBILE FEDERATION)
5-8, 3-CHOME, SHIBAKOEN, MINATO-KU, TOKYO 105
JAPANESE

JAPAN AUTOMOBILE NEWS 7203
NIKKAN JIDOSHA SHINBUNSHA
1-25, 2-CHOME, KAIGAN, MINATO-KU, TOKYO 105
ENGLISH
INDEXED: AUT

JAPAN MOTOR INDUSTRY 7204
ENGLISH

JAPAN'S AUTOMOTIVE INDUSTRY 7205
JAPANESE AND ENGLISH
INDEXED: AUT

JAPAN'S BICYCLE GUIDE 7206
NIHON JITENSHA KOGYOKAI (JAPAN BICYCLE INDUSTRY ASSOCIATION)
9-15, 1-CHOME, AKASAKA, MINATO-KU, TOKYO 107
ENGLISH

JIDOSHA BUHIN 7207
(PARTS OF AUTOMOBILE)
NIHON JIDOSHA BUHIN KOGYOKAI HENSHU-BU (EDITORIAL BOARD, JAPAN AUTOMOBILE PARTS INDUSTRY ASSOCIATION)
8-12, 3-CHOME, KUDANMINAMI, CHIYODA-KU, TOKYO 102
JAPANESE

JIDOSHA BUHIN SEISAN DOKO CHOSA 7208
(SURVEY OF AUTOMOBILE PARTS PRODUCTION)
NIHON JIDOSHA BUHIN KOGYOKAI (JAPAN AUTOMOBILE PARTS INDUSTRY ASSOCIATION)
16-15, 1-CHOME, TAKANAWA, MINATO-KU, TOKYO 108
JAPANESE

JIDOSHA CHOSA NENPO 7209
(ANNUAL REPORT OF THE AUTOMOBILE SURVEY)
NIHON JIDOSHA KAIGISHO (JAPAN AUTOMOBILE COMMITTEE)
8-3, 1-CHOME, MARUNOUCHI, CHIYODA-KU, TOKYO 100
JAPANESE

JIDOSHA GIJUTSU 7210
(JOURNAL OF THE SOCIETY OF AUTOMOTIVE ENGINEERS OF JAPAN, INCORPORATED)
JIDOSHA GIJUTSUKAI (THE SOCIETY OF AUTOMOTIVE ENGINEERS OF JAPAN, INCORPORATED)
16-15, 1-CHOME, TAKANAWA, MINATO-KU, TOKYO 108
JAPANESE, WITH EUROPEAN SUMMARY
INDEXED: APT, AUT, JICST

JIDOSHA GIJUTSUKAI CHUBU SHIBUHO 7211
(REPORT OF THE CENTRAL AND BRANCH OFFICES OF THE AUTOMOBILE ENGINEERS OF JAPAN)
AUTOMOBILE ENGINEERS OF JAPAN
JAPANESE

JIDOSHA GIJUTSUKAI GAKUJUTSU KOENKAI MAEZURISHU 7212
(PREPRINTS OF THE SYMPOSIUM OF THE SOCIETY OF AUTOMOBILE ENGINEERS OF JAPAN)
JAPANESE
INDEXED: JICST

JIDOSHA GIJUTSUKAI RONBUNSHU 7213
(TRANSACTIONS OF THE SOCIETY OF AUTOMOBILE ENGINEERS OF JAPAN, INCORPORATED)
JIDOSHA GIJUTSUKAI (SOCIETY OF AUTOMOBILE ENGINEERS OF JAPAN, INCORPORATED)
16-15, 1-CHOME, TAKANAWA, MINATO-KU, TOKYO 108
JAPANESE, WITH ENGLISH SUMMARY
INDEXED: JICST

JIDOSHA HANBAI 7214
(AUTOMOBILE DEALER)
NIHON JIDOSHA HANBAI KYOKAI RENGOKAI (JAPAN AUTOMOBILE DEALER ASSOCIATION)
52-2, 5-CHOME, JINGUMAE, SHIBUYA-KU, TOKYO 150
JAPANESE

JIDOSHA HANBAI JISSEKI CHO, KOKASAN SHINSHA, KOKUNAIBUN 7215
(SURVEY OF PERFORMANCE OF AUTOMOBILE SALES, DOMESTICALLY PRODUCED NEW AUTOMOBILES, DOMESTIC DIVISION)
TSUSHO SANGYO SHO DAIJIN KANBO CHOSA TOKEIBU, JIDOSHA KOGYOKAI (MITI, MINISTER'S SECRETARIAT, STATISTICAL RESEARCH DIVISION; AUTOMOBILE INDUSTRY ASSOCIATION
JAPANESE

JIDOSHA JANARU 7216
(AUTOMOBILE JOURNAL)
JIDOSHA JANARU SHA (AUTOMOBILE JOURNAL COMPANY)
JAPANESE

JIDOSHA KAIGAI JOHO 7217
(OVERSEAS INFORMATION OF AUTOMOTIVE INDUSTRY)
NIHON JIDOSHA KOGYOKAI (JAPAN AUTOMOBILE MANUFACTURES ASSOCIATION)
C/O OTEMACHI BUILDING, OTE-MACHI, CHIYODA-KU, TOKYO 100
JAPANESE
INDEXED: JICST

JIDOSHA KOGAKU 7218
(AUTOMOBILE ENGINEERING)
TETSUDO NIHONSHA
1-4, 2-CHOME, NISHIKANDA, CHIYODA-KU, TOKYO 101
JAPANESE
INDEXED: JICST

JIDOSHA KOGYO 7219
(AUTOMOBILE INDUSTRY)
NIHON JIDOSHA KOGYOKAI (JAPAN AUTOMOBILE MANUFACTURES ASSOCIATION)
6-1, 1-CHOME, OTE-MACHI, CHIYODA-KU, TOKYO 100
JAPANESE
INDEXED: JICST

JIDOSHA KOGYO SHINBUN 7220
(AUTOMOBILE ENGINEERS' NEWS)
JIDOSHA KOGYO SHINBUNSHA
5-28, 4-CHOME, EBISU, SHIBUYA-KU, TOKYO 150
JAPANESE

JIDOSHA KOTSU SHINBUN 7221
(AUTOMOBILE TRAFFIC NEWS)
JIDOSHA KOTSU SHINBUNSHA
11-7, 1-CHOME, SHINBASHI, MINATO-KU, TOKYO 105
JAPANESE

JIDOSHA NO JITSUMU 7222
(AUTOMOBILE PRACTICES)
KOBUNSHA
251, WASEDA TSURUMAKI-CHO, SHINJUKU-KU, TOKYO 162
JAPANESE

JIDOSHA SANGYO 7223
(AUTOMOBILE INDUSTRY)
KEIZAI HYORONSHA
C/O CHUGIN SHINBASHI BUILDING, 9-5, 2-CHOME, SHINBASHI, MINATO-KU, TOKYO 105
JAPANESE

JIDOSHA SEMINA 7224
(AUTOMOBILE SEMINAR)
KOTSU SHA (KOTSU COMPANY)
JAPANESE

JIDOSHA SHUHO 7225
(AUTOMOBILE WEEKLY NEWS)
JIDOSHA SHUHOSHA
1-13, 1-CHOME, KITASHINAGAWA, SHINAGAWA-KU, TOKYO 140
JAPANESE

JIDOSHA TAIMUSU 7226
(AUTO TIMES)
JIDOSHA TAIMUSUSHA (AUTO TIMES COMPANY)
1, SHIBA KAMIYA-CHO, MINATO-KU, TOKYO 105
JAPANESE

JIDOSHA TO SEIBI 7227
(AUTOMOBILE AND MAINTENANCE)
NISSEIREN SHUPPANSHA
7-13, 7-CHOME, ROPPONGI, MINATO-KU, TOKYO 106
JAPANESE
INDEXED: JICST

JIDOSHA TO SONO SEKAI 7228
(AUTOMOBILES AND THEIR WORLD)
TOYOTA JIDOSHA HANBAI KABUSHIKI KAISHA, HANBAI KAKUCHOBU (TOYOTA AUTOMOBILE SALES, LTD., SALES DIVISION)
JAPANESE

JIDOSHA TOKEI GEPPO 7229
(MONTHLY STATISTICS OF MOTOR VEHICLES)
NIHON JIDOSHA KOGYOKAI (JAPAN AUTOMOBILE MANUFACTURERS ASSOCIATION, INCORPORATED)
6-1, 1-CHOME, OTE-MACHI, CHIYODA-KU, TOKYO 100
JAPANESE
INDEXED: JICST

JIDOSHA TOKEI NENPYO 7230
(ANNUAL STATISTICS OF MOTOR VEHICLES)
NIHON JIDOSHA KOGYOKAI (JAPAN AUTOMOBILE MANUFACTURES ASSOCIATION)
6-1, 1-CHOME, OTE-MACHI, CHIYODA-KU, TOKYO 100
JAPANESE

JIDOSHA UNSO JIGYOYO JIDOSHA JIKO TOKEI NENPO 7231
(ANNUAL STATISTICS OF MOTOR VEHICLES ACCIDENTS OWING TO BUSINESS USED CARS)
UN'YU-SHO JIDOSHA-KYOKU (ROAD TRANSPORT BUREAU, MINISTRY OF TRANSPORT)
1-3, 2-CHOME, KASUMIGASEKI, CHIYODA-KU, TOKYO 100
JAPANESE

JIDOSHASU TOKEIHYO 7232
(ANNUAL REPORT OF AUTOMOBILE STATISTICS)
UN'YU-SHO DAIJIN KANBO JOHO KANRI-BU (DEPARTMENT OF RESEARCH AND DATA PROCESSING, MINISTER'S SECRETARIAT, MINISTRY OF TRANSPORT)
1-3, 2-CHOME, KASUMIGASEKI, CHIYODA-KU, TOKYO 100
JAPANESE

JISHINKYO GIJUTSU KENKYUSHO HOKOKU 7233
(REPORT OF TECHNICAL RESEARCH LABORATORY, JAPAN BICYCLE PROMOTION INSTITUTE)
JITENSHA SANGYO SHINKOKYOKAI SEISAN GIJUTSU-BU (PRODUCTION TECHNOLOGY RESEARCH DIVISION, JAPAN BICYCLE PROMOTION INSTITUTE)
1-6, SHINKAWA, INUYAMA-SHI 484
JAPANESE
INDEXED: JICST

JISHINKYOKAIHO 7234
(BULLETIN OF THE JAPAN BICYCLE PROMOTION INSTITUTE)
JITENSHA SANGYO SHINKO KYOKAI (JAPAN BICYCLE PROMOTION INSTITUTE)
9-15, 1-CHOME, AKASAKA, MINATO-KU, TOKYO 107
JAPANESE
INDEXED: JICST

JITENSHA SEISAN GIJUTSU 7235
(AUTOMOBILE PRODUCTION TECHNOLOGY)
JAPANESE
INDEXED: JICST

KA BITO 7236
(CAR BEAT)
MOMOZONO SHOBO (MOMOZONO BOOKS)
JAPANESE

KA RODO 7237
(CAR ROAD)
KOTSU TAIMSUSHA (KOTSU TIMES COMPANY)
JAPANESE

KA SUTAIRINGU 7238
(CAR STYLING)
SAN'EI SHOBO
1, SUWA-CHO, SHINJUKU-KU, TOKYO 160
JAPANESE AND ENGLISH

KAIGAI ASSENJO NYUSU 7239
(OVERSEAS MEDIATION NEWS)
JAPANESE
INDEXED: JICST

KEIJIDOSHA JOHO 7240
(LIGHT AUTOMOBILE REPORT)
ZENKOKU JIDOSHA KYOKAI RENGOKAI, ZENKOKU NIRINSHA ANZEN UNTEN FUKYU KYOGIKAI (ALL-JAPAN FEDERATION OF AUTOMOBILE ASSOCIATIONS, ALL-JAPAN COUNCIL ON ENCOURAGEMENT OF TWO-WHEEL VEHICLES)
JAPANESE

KOGATAKAI 7241
(SMALL-SIZED MOTOR VEHICLES)
NIHON KOGATA JIDOSHA HANBAI KYOKAI (JAPAN SMALL-SIZED MOTOR VEHICLES DEALERS ASSOCIATION)
46-32, 2-CHOME, HON-MACHI, NAKANO-KU, TOKYO 164
JAPANESE

KURUMA NO TECHO 7242
(AUTOMOBILE NOTEBOOK)
NISSAN JIDOSHA KOHOBU (NISSAN AUTOMOBILE COMPANY, PUBLIC RELATIONS DIVISION)
JAPANESE

KYOSEI TSUSHIN 7243
(NEWS OF KYOTO FACTORY, MITSUBISHI MOTORS COMPANY, LIMITED)
MITSUBISHI JIDOSHA KOGYO K.K. KYOTO SEISAKUSHO (MITSUBISHI MOTORS COMPANY)
1, UZUMASA TATSUMI-CHO, UKYO-KU, KYOTO-SHI 616
JAPANESE

MISUTA BAIKU 7244
(MISTER BIKE)
MOTA MAGAJINSHA (MOTOR MAGAZINE PUBLISHING COMPANY)
JAPANESE

MIZUSHIMA NYUSU 7245
(MIZUSHIMA NEWS)
MITSUBISHI JIDOSHA KOGYO K.K. MIZUSHIMA JIDOSHA SEISAKUSHO (MITSUBISHI MOTORS COMPANY)
1-1, KAIGAN-DORI, MIZUSHIMA, KURASHIKI-SHI 712
JAPANESE

MOTA BIKURU 7246
(MOTOR VEHICLE)
KUDAN SHOBO
1-15, 4-CHOME, KUDANKITA, CHIYODA-KU, TOKYO 102
JAPANESE
INDEXED: JICST

MOTA EJI 7247
(MOTOR AGE)
JAPANESE
INDEXED: JICST

MOTA FAN 7248
(MOTOR FAN)
SAN'EI SHOBO K.K.
1, SUWA-CHO, SHINJUKU-KU, TOKYO 160
JAPANESE
INDEXED: AUT, JICST

MOTA FAN GURAFITI 7249
(MOTOR FAN GRAFFITI)
SAN'EI SHOBO (SAN'EI PUBLISHING COMPANY)
JAPANESE

MOTA MAGAJIN 7250
(MOTOR MAGAZINE)
MOTA MAGAJINSHA
3-5, 6-CHOME, SHINBASHI, MINATO-KU, TOKYO 105
JAPANESE AND ENGLISH
INDEXED: JICST

MOTA MAINICHI 7251
(JOURNAL OF MOTOR CARS)
MOTA MAINICHI SHUPPAN-KYOKU
6-19, 2-CHOME, MINAMIAOYAMA, MINATO-KU, TOKYO 107
JAPANESE

MOTA MATERIARU 7252
(MOTOR MATERIAL)
MOTA MATERIARUSHA (MOTOR MATERIAL COMPANY, LTD.)
1-64, KANDA JINBO-CHO, CHIYODA-KU, TOKYO 101
JAPANESE
INDEXED: JICST

MOTA NYUZU DAIJESUTO 7253
(MOTOR NEWS DIGEST)
NISSAN JIDOSHA KABUSHIKI KAISHA KOHOBU KOHOKA (NISSAN AUTOMOBILE K.K., PUBLICITY DEPARTMENT)
JAPANESE

MOTA RODO 7254
(MOTOR ROAD)
NAIGAI SHUPPAN SHA (NAIGAI PUBLISHING COMPANY)
JAPANESE AND ENGLISH

MOTA SAIKURISUTO 7255
(MOTOR CYCLIST)
YAESU SHUPPAN
5-15, 1-CHOME, YAESU, CHUO-KU, TOKYO 102
JAPANESE
INDEXED: JICST

MOTARIZESHON 7256
(MOTORIZATION)
TOYO KOGYO K.K. KOHO-BU
6047, FUCHU-MACHI, AKI-GUN, HIROSHIMA 735
JAPANESE

MOTO RIDER 7257
SAN'EI SHOBO (SAN'EI PUBLISHING HOUSE)
JAPANESE AND ENGLISH

MOTOR AGE 7258
JIDOSHA SHUHOSHA
1-13, 1-CHOME, KITASHINAGAWA, SHINAGAWA-KU, TOKYO 140
JAPANESE
INDEXED: JICST

MOTOR INDUSTRY OF JAPAN 7259
JAPANESE AND ENGLISH
INDEXED: AUT

NAKANIHON JIDOSHA TANKI DAIGAKU RONSO 7260
(JOURNAL OF THE NAKANIHON AUTOMOTIVE JUNIOR COLLEGE)
NAKANIHON JIDOSHA TANKI DAIGAKU RONSO HENSHU IINKAI AND NAKANIHON JIDOSHA TANKI DAIGAKU (NAKANIHON AUTOMOTIVE JUNIOR COLLEGE)
1301, FUKAGAYA, SAKAHOGI-CHO, KAMO-GUN GIFU 505
JAPANESE AND ENGLISH
INDEXED: JICST

NIHON JIDOSHA BUHIN KOGYOKAI JIGYO HOKOKUSHO 7261
(OPERATION REPORT OF JAPAN AUTOMOBILE PARTS INDUSTRY ASSOCIATION))
NIHON JIDOSHA BUHIN KOGYOKAI
16-15, 1-CHOME, TAKANAWA, MINATO-KU, TOKYO 108
JAPANESE

NIHON JIDOSHA BUHIN KUMIAI RENGOKAI KAIHO 7262
(BULLETIN OF THE JAPAN FEDERATION OF AUTOMOTIVE PARTS UNIONS)
JAPANESE

NIHON JIDOSHA KAIGISHO CHOSA GEPPO 7263
(MONTHLY STATISTICS OF JAPAN AUTOMOBILE CHAMBER OF COMMERCE)
NIHON JIDOSHA KAIGISHO (JAPAN AUTOMOBILE CHAMBER OF COMMERCE)
8-3, 1-CHOME, MARUNOUCHI, CHIYODA-KU, TOKYO 100
JAPANESE

NIHON JIDOSHA KENKYUSHO KAKUGETSUHO 7264
(JARI BI-MONTHLY)
NIHON JIDOSHA KENKYUSHO (JAPAN AUTOMOBILE RESEARCH INSTITUTE, INCORPORATED)
YATABE-MACHI, TSUKUBA-GUN, IBARAKI 300-21
JAPANESE
INDEXED: JICST

NIHON JIDOSHA KENKYUSHO KENKYU HOKOKU 7265
(JARI TECHNICAL REPORT)
NIHON JIDOSHA KENKYUSHO (JAPAN AUTOMOBILE RESEARCH INSTITUTE, INCORPORATED)
YATABE-MACHI, TSUKUBA-GUN, IBARAKI 300-21
JAPANESE
INDEXED: JICST

NIHON JIDOSHA KENKYUSHO KENKYU SOKUHO 7266
(JARI TECHNICAL MEMORANDUM)
NIHON JIDOSHA KENKYUSHO (JAPAN AUTOMOBILE RELEASE INSTITUTE, INCORPORATED)
YATABE-MACHI, TSUKUBA-GUN, IBARAKI 300-21
JAPANESE, WITH ENGLISH SUMMARY
INDEXED: AUT, JICST

NIHON JIDOSHA KENKYUSHO NENPO 7267
(JARI ANNUAL REPORT)
NIHON JIDOSHA KENKYUSHO (JAPAN AUTOMOBILE RESEARCH INSTITUTE, INCORPORATED)
YATABE-MACHI, TSUKUBA-GUN, IBARAKI 300-21
JAPANESE

NIPPON JITENSHA KOGYOKAI KAIHO 7268
(REPORT OF THE JAPANESE AUTOMOBILE INDUSTRIAL ASSOCIATION)
JAPANESE
INDEXED: JICST

NISSAN GIHO 7269
(NISSAN TECHNICAL REVIEW)
NISSAN JIDOSHA K.K. NISSAN GIHO HENSHU IINKAI (EDITORIAL COMMITTEE OF NISSAN TECHNICAL REVIEW, NISSAN MOTOR COMPANY, LTD.)
17-1, 6-CHOME, GINZA, CHUO-KU, TOKYO 104
JAPANESE, WITH ENGLISH SUMMARY
INDEXED: APT, AUT, JICST

NISSAN GURAFU 7270
(NISSAN GRAPHIC)
NISSAN JIDOSHA K.K. KOHO-BU (PUBLIC INFORMATION DIVISION, NISSAN MOTOR COMPANY, LTD.)
17-1, 6-CHOME, GINZA, CHUO-KU, TOKYO 104
JAPANESE

NISSAN JIZERU GIHO 7271
(NISSAN DIESEL REVIEW)
NISSAN JIZERU KOGYO K.K. (NISSAN DIESEL MOTOR COMPANY, LTD.)
1-1, AGEO-SHI 362
JAPANESE
INDEXED: AUT, JICST

NISSEIREN NYUSU 7272
(NEWS OF REVIEW OF JASPA)
NIHON JIDOSHA SEIBI SHINKOKAI RENGOKAI (JAPAN AUTOMOBILE SERVICE PROMOTION ASSOCIATION)
3-2, 6-CHOME, ROPPONGI, MINATO-KU, TOKYO 106
JAPANESE

OTO BAI 7273
(AUTO BYE)
MOTA MAGAZINSHA
3-5, 6-CHOME, SHINBASHI, MINATO-KU, TOKYO 105
JAPANESE
INDEXED: JICST

OTO PATSU MAKETTO: OTO PATSU MAKETTO KENKYU KAI-HEN 7274
(AUTO PARTS MARKET: SOCIETY FOR AUTO PARTS MARKET RESEARCH EDITION)
OTO TO REDO JYANARU (AUTO TRADE JOURNAL)
JAPANESE

OTO TEKUNIKKU 7275
(AUTO TECHNIC)
SANKAIDO
5-18, 5-CHOME, HONGO, BUNKYO-KU, TOKYO 113
JAPANESE

PRINCE 7276
NISSAN PURINSU JIDOSHA HANBAI K.K. (NISSAN PRINCE SALES COMPANY, LTD.)
5-28, 3-CHOME, MITA, MINATO-KU, TOKYO 108
JAPANESE

RAIDINGU 7277
(RIDING)
NIHON MOTA SAIKURU KYOKAI (THE MOTORCYCLING FEDERATION OF JAPAN)
9-12, 1-CHOME, GINZA, CHUO-KU, TOKYO 104
JAPANESE

SANGYO SHARYO 7278
(INDUSTRIAL VEHICLES)
NIHON SANGYO SHARYO KYOKAI (JAPAN INDUSTRIAL VEHICLES ASSOCIATION)
5-26, 1-CHOME, MOTOAKASAKA, MINATO-KU, TOKYO 107
JAPANESE

SHATAI GIJUTSU 7279
(TOYOTA BODY TECHNIQUE)
TOYOTA SHATAI GIJUTSUKAI (TOYOTA BODY TECHNICAL ASSOCIATION)
2-1, SHOWA-CHO, KARIYA-SHI 448
JAPANESE
INDEXED: JICST

SHUYOKOKU JIDOSHA TOKEI 7280
(MOTOR VEHICLE STATISTICS OF SELECTED COUNTRIES)
NIHON JIDOSHA KOGYOKAI (JAPAN AUTOMOBILE MANUFACTURES ASSOCIATION, INCORPORATED)
C/O OTE-MACHI BLDG., 6-1, 1-CHOME, OTE-MACHI, CHIYODA-KU, TOKYO 100
JAPANESE

SOSEI NYUSU 7281
(SAGAMIHARA FACTORY NEWS, MITSUBISHI HEAVY INDUSTRIES COMPANY, LIMITED)
MITSUBISHI JUKOGYO K.K. SAGAMIHARA SEISAKUJO
3000, TANA, SAGAMIHARA-SHI 229
JAPANESE

TAURA GIHO 7282
(TECHNICAL JOURNAL OF KANTO AUTO WORKS COMPANY, LIMITED)
KANTO JIDOSHA KOGYO K.K. GIJUTSU-BU (KANTO AUTO WORKS COMPANY, LIMITED)
MUBANCHI, TAURAMINATO-MACHI, YOKOSUKA-SHI 237
JAPANESE

TOJI NYUSU 7283
(TOJI NEWS)
MITSUBISHI JIDOSHA KOGYO K.K. TOKYO JIDOSHA SEISAKUJO (TOKYO MOTOR FACTORY, MITSUBISHI MOTOR COMPANY, LIMITED)
21-1, 4-CHOME, SHIMOMARUKO, OTA-KU, TOKYO 133
JAPANESE

TORAKKU 7284
(TRUCK)
MITSUBISHI JIDOSHA HANBAI (MITSUBISHI AUTOMOBILE SALES)
JAPANESE

TOYOTA CHUO KENKYUSHO KENKYU HOKOKU 7285
(TECHNICAL REPORT OF TOYOTA RD CENTER)
TOYOTA CHUO KENKYUSHO (TOYOTA CENTRAL RESEARCH AND DEVELOPMENT LABORATORY, INCORPORATED)
2-12, HISAKATA, SHOWA-KU, NAGOYA-SHI 468
JAPANESE
INDEXED: CHE

TOYOTA CHUO KENKYUSHO RD REBYU 7286
(REVIEW OF TOYOTA RD CENTER)
TOYOTA CHUO KENKYUSHO (TOYOTA CENTRAL RESEARCH AND DEVELOPMENT LABORATORY, INCORPORATED)
2-12, HISAKATA, SHOWA-KU, NAGOYA-SHI 468
JAPANESE
INDEXED: AUT, CHE

TOYOTA GIJUTSU 7287
(TOYOTA ENGINEERING)
TOYOTA JIDOSHA KOGYO K.K. (TOYOTA MOTOR COMPANY, LTD.)
1, TOYOTA-CHO, TOYOTA-SHI 471
JAPANESE
INDEXED: AUT, CHE, JICST

TOYOTA ROAD CENTER REVIEW 7288
JAPANESE AND ENGLISH
INDEXED: AUT

WHEEL EXTENDED 7289
GENDAI BUNK KENKYUSHO (GENDAI ADVANCED STUDIES RESEARCH ORGANIZATION)
TOYOTA JIDOSHA HANBAI K.K.
3-18, 2-CHOME, KUDAHMINAMI, CHIYODA-KU, TOKYO 102
ENGLISH
INDEXED: AUT

YURYO JIDOSHA BUHIN 7290
(GOOD AUTOMOBILE PARTS)
JIDOSHA TAIMUSUSHA
1, SHIBA KAMIYA-CHO, MINATO-KU, TOKYO 105
JAPANESE

Shipbuilding, Ships, Boats

FUNE NO KAGAKU 7291
(MARINE ENGINEERING)
SENPAKU GIJUTSU KYOKAI (MARINE ENGINEERING ASSOCIATION)
22-5, 2-CHOME, NISHIAZABU, MINATO-KU, TOKYO 106
JAPANESE
INDEXED: JICST

FUNE NO ZASSHI 7292
(SHIPS AND BOATS JOURNAL)
KAIYO KYOKAI
SHISEIDO
1-9, KANDA KAJI-CHO, CHIYODA-KU, TOKYO 101
JAPANESE

FUNE TO HYOJUNKA 7293
(MARINE STANDARDIZATION)
NIHON SENPAKU HYOJUN KYOKAI (JAPANESE ASSOCIATION OF MARINE STANDARDIZATION)
C/O SUMITOMO TORANOMON BUILDING, 7, SHIBA KOTOCHIRA-CHO, MINATO-KU, TOKYO 105
JAPANESE

GEKKAN KODANSEN 7294
(JOURNAL OF SHIP ADJUSTMENT COOPERATION)
KAIKOSHA
3, KAIGAN-DORI, IKUTA-KU, KOBE-SHI 650
JAPANESE

GYOSEN 7295
(JOURNAL OF THE FISHING BOAT ASSOCIATION OF JAPAN)
GYOSEN KYOKAI (FISHING BOAT ASSOCIATION OF JAPAN)
C/O SENPAKU SHINKO BUILDING, 35, SHIBA KOTOHIRA-CHO, MINATO-KU, TOKYO 105
JAPANESE
INDEXED: JICST

GYOSEN KENKYU GIHO 7296
(RESEARCH REPORT OF FISHING BOATS)
FISHING BOAT LABORATORY, FISHERIES AGENCY
5-1, KACHIDOKI 5 CHOME, CHUO-KU, TOKYO 104
JAPANESE
INDEXED: CHE, JICST

GYOSEN KIKAN 7297
(JOURNAL OF FISHING-BOAT ENGINE)
GYOSEN KIKANSHI KYOKAI (FISHING-BOAT ENGINEERS' ASSOCIATION)
16-5, 2-CHOME, TAKANAWA, MINATO-KU, TOKYO 108
JAPANESE
INDEXED: JICST

GYOSEN TOKEIHYO SOGO HOKOKU 7298
(STATISTIC TABLES OF FISHING VESSELS, GENERAL REPORT)
SUISAN-CHO (FISHERY AGENCY)
2-1, 1-CHOME, KASUMIGASEKI, CHIYODA-KU, TOKYO 100
JAPANESE AND ENGLISH

GYOSEN TOROKU NI YORU GYOSEN TOKEIHYO 7299
(ANNUAL STATISTICS OF FISHING BOAT BY FISHING BOAT REGISTRATION)
SUISAN-CHO (FISHERY AGENCY)
2-1,1-CHOME, KASUMIGASEKI, CHIYODA-KU, TOKYO 100
JAPANESE AND EUROPEAN

HITACHI ZOSEN GIHO 7300
(HITACHI ZOSEN TECHNICAL REVIEW)
HITACHI ZOSEN K.K. GIJUTSU KENKYUSHO (TECHNICAL RESEARCH LABORATORY, HITACHI SHIPBUILDING AND ENGINEERING COMPANY, LTD.)
60, SAKURAJIMA KITANO-CHO, KONOHANA-KU, OSAKA-SHI 554
JAPANESE, WITH EUROPEAN SUMMARY
INDEXED: CHE, ENG, JICST

JAPAN SHIPBUILDING INFORMATION NOTES 7301
JAPAN FOUNDATION FOR SHIPBUILDING ADVANCEMENT, LIBRARY
35 SHIBA-KOTOHIRACHO, MINATO-KU, TOKYO 105
ENGLISH

JAPAN SHIPBUILDING AND MARINE ENGINEERING 7302
NIHON GIJUTSU KOHO KYOKAI (JAPAN ASSOCIATION FOR TECHNICAL INFORMATION)
7-107, 4-CHOME, YAMAZAKI-CHO, MACHIDA-SHI, TOKYO 194-01
ENGLISH
INDEXED: EXC, INS, JICST

JAPANESE INTERNAL COMBUSTION ENGINES FOR MARINE USE 7303
ENGLISH

JOURNAL OF THE JAPANESE INSTITUTE NAVAL ARCHITECTURE 7304
JAPANESE AND ENGLISH
INDEXED: COP

KAIJI TO SHIZAI 7305
(MARITIME AFFAIRS AND ASSETS (OR MATERIALS))
MABUNEGU SHINBUNSHA NEWSPAPER COMPANY
JAPANESE

KAIYO KIKI KAIHATSU 7306
(MARINE MACHINERY DEVELOPMENT)
NIHON HAKUYO KIKI KAIHATSU KYOKAI (JAPAN SHIP'S MACHINERY DEVELOPMENT ASSOCIATION)
35, SHIBA KOTOCHIRA-CHO, MINATO-KU, TOKYO 105
JAPANESE
INDEXED: JICST

KAIYO KOGAKU SHINPOJIUMU 7307
(MARINE ENGINEERING SYMPOSIUM)
JAPANESE
INDEXED: JICST

KAJI 7308
(YACHT, BOAT AND MOTOR BOAT MAGAZINE)
SHUTEI KYOKAI SHUPPAN-BU
5-2, 3-CHOME, GINZA, CHUO-KU, TOKYO 104
JAPANESE
INDEXED: JICST

KANSAI ZOSEN KYOKAISHI 7309
(JOURNAL OF THE KANSAI SOCIETY OF NAVAL ARCHITECTS, JAPAN)
KANSAI ZOSEN KYOKAI (KANSAI SOCIETY OF NAVAL ARCHITECTS, JAPAN)
C/O OSAKA DAIGAKU KOGAKUBU
ZOSENGAKU KYOSHITSU, YAMADAUE, SUITA-SHI 565
JAPANESE AND ENGLISH
INDEXED: JICST

KOBE SHOSEN DAIGAKU KIYO, DAI-2-RUI, KOKAI, KIKAN, GENSHI DORYOKU, RIGAKU-HEN 7310
(REVIEW OF KOBE UNIVERSITY OF MERCANTILE MARINE, PART 2, NAVIGATION, MARINE ENGINEERING, NUCLEAR ENGINEERING AND SCIENTIFIC SECTION)
KOBE SHOSEN DAIGAKU (KOBE UNIVERSITY OF MERCANTILE MARINE)
1-1, 5-CHOME, FUKAEMINAMI-CHO, HIGASHINADA-KU, KOBE-SHI 658
JAPANESE AND ENGLISH
INDEXED: CHE, JICST

KYUMEIKI GIJUTSU KENKYUKAISHI 7311
(BULLETIN OF RESCUE APPARATUS RESEARCH CENTER)
KYUMEIKI GIJUTSU KENKYUKAI JIMU-KYOKU (RESCUE APPARATUS RESEARCH CENTER)
C/O HIBIYA PARK BUILDING, 1-1, YURAKU-CHO, CHIYODA-KU, TOKYO 100
JAPANESE

MARIN ENJINIA 7312
(MARINE ENGINEER)
NIHON SENPAKU KIKANSHI KYOKAI (JAPAN MARITIME ENGINEERS' ASSOCIATION)
11-11, 1-CHOME, HAMAMATSU-CHO, MINATO-KU, TOKYO 105
JAPANESE
INDEXED: JICST

MARINE STANDARDIZATION 7313
NIHON SENPAKU HYOJUN KYOKAI (JAPANESE ASSOCIATION OF MARINE STANDARDIZATION)
C/O SUMITOMO TORANOMON BUILDING, 7, SHIBA KOTOCHIRA-CHO, MINATO-KU, TOKYO 105
ENGLISH

MARU 7314
JAPANESE

MARU EXTRA 7315
JAPANESE

MARU GRAPHIC QUARTERLY 7316
JAPANESE

MARU SPECIAL 7317
JAPANESE

MITSUI ZOSEN GIHO 7318
(MITSUI TECHNICAL REVIEW)
MITSUI ZOSEN K.K. GIJUTSU HONBU (MITSUI SHIPBUILDING AND ENGINEERING COMPANY, LTD. TECHNICAL DIVISION)
6-4, 5-CHOME, TSUKIJI, CHUO-KU, TOKYO 104
JAPANESE, WITH ENGLISH SUMMARY
INDEXED: INS, JICST

NAGASAKI ZOSEN DAIGAKU KENKYU HOKOKU 7319
(BULLETIN OF THE COLLEGE OF NAVAL ARCHITECTURE OF NAGASAKI)
NAGASAKI ZOSEN DAIGAKU (COLLEGE OF NAVAL ARCHITECTURE OF NAGASAKI)
536, AMIBA-CHO, NAGASAKI-SHI 851-01
JAPANESE AND ENGLISH
INDEXED: JICST

NAIKO SENPAKU YUSO TOKEI NENPO 7320
(ANNUAL STATISTICS OF DOMESTIC MARINE TRANSPORT)
UN'YU-SHO DAIJIN KANBO JOHO KANRI-BU (RESEARCH AND DATA PROCESSING DEPARTMENT, MINISTER'S SECRETARIAT, MINISTRY OF TRANSPORT)
1-3, 2-CHOME, KASUMIGASEKI, CHIYODA-KU, TOKYO 100
JAPANESE

NIHON HAKUYO KIKAN GAKKAISHI 7321
(JOURNAL OF THE MARINE ENGINEERING SOCIETY IN JAPAN)
NIHON HAKUYO KIKAN GAKKAI (MARINE ENGINEERING SOCIETY IN JAPAN)
2-2, 1-CHOME, UCHISAIWAI-CHO, CHIYODA-KU, TOKYO 100
JAPANESE
INDEXED: JICST

NIHON ZOSEN GAKKAI RONBUNSHU 7322
(JOURNAL OF THE SOCIETY OF NAVAL ARCHITECTS OF JAPAN)
NIHON ZOSEN GAKKAI (THE SOCIETY OF NAVAL ARCHITECTS OF JAPAN)
C/O SENPAKU SHINKO BUILDING, 35, SHIBA KOTOHIRA-CHO, MINATO-KU, TOKYO 105
JAPANESE, WITH ENGLISH SUMMARY
INDEXED: CHE, JICST

NIHON ZOSEN GAKKAISHI 7323
(BULLETIN OF THE SOCIETY OF NAVAL ARCHITECTS OF JAPAN)
NIHON ZOSEN GAKKAI (SOCIETY OF NAVAL ARCHITECTS OF JAPAN)
C/O SENPAKU SHINKO BUILDING, 35, SHIBA KOTOCHIRA-CHO, MINATO-KU, TOKYO 105
JAPANESE
INDEXED: JICST

NIHON ZOSEN KENKYU KYOKAI HOKOKU 7324
(REPORT OF THE SHIPBUILDING RESEARCH ASSOCIATION OF JAPAN)
NIHON ZOSEN KENKYU KYOKAI (SHIPBUILDING RESEARCH ASSOCIATION OF JAPAN)
C/O SENPAKU SHINKO BUILDING, 35, SHIBA KOTOCHIRA-CHO, MINATO-KU, TOKYO 105
JAPANESE, WITH ENGLISH SUMMARY
INDEXED: JICST

NIHON ZOSEN KENKYU KYOKAI KENKYU SHIRYO 7325
(BULLETIN OF THE SHIPBUILDING RESEARCH ASSOCIATION OF JAPAN)
NIHON ZOSEN KENKYU KYOKAI (SHIPBUILDING RESEARCH ASSOCIATION OF JAPAN)
35, SHIBA KOTOHIRA-CHO, MINATO-KU, TOKYO 105
JAPANESE
INDEXED: JICST

NIHON ZOSEN KOGYOKAI JIGYO HOKOKU 7326
(OPERATION REPORT OF JAPAN SHIPBUILDERS' ASSOCIATION)
NIHON ZOSEN KOGYOKAI
35, SHIBA KOTOCHIRA-CHO, MINATO-KU, 105
JAPANESE

NIPPON HAKUYO KIKAN GAKKAI KOENKAI KOEN MAEZURI 7327
(PREPRINT OF THE SYMPOSIUM OF THE MARINE ENGINEERING SOCIETY OF JAPAN)
JAPANESE
INDEXED: JICST

NIPPON ZOSEN KENKYU KYOKAI NENPO 7328
(ANNUAL OF THE JAPAN SHIPBUILDING RESEARCH ASSOCIATION)
JAPANESE

OGATA SENPAKUYO KIKAN NO DOKO 7329
(STATISTICS OF ENGINES FOR LARGE SHIPS)
UN'YU-SHO SENPAKU-KYOKU (SHIP BUREAU, MINISTRY OF TRANSPORT)
1-3, 2-CHOME, KASUMIGASEKI, CHIYODA-KU TOKYO 100
JAPANESE

PAPERS OF SHIP RESEARCH INSTITUTE 7330
UN'YU-SHO SENPAKU GIJUTSU KENKYUSHO (SHIP RESEARCH INSTITUTE, MINISTRY OF TRANSPORT)
38-1, 6-CHOME, SHINKAWA, MITAKA-SHI, TOKYO 181
EUROPEAN
INDEXED: APP, CHE, ENG, JICST

PUBLICATION LIST OF TECHNICAL REPORTS (ABSTRACTS) 7331
HITACHI ZOSEN K. K. GIJUTSU KENKYUSHO (TECHNICAL RESEARCH LABORATORY, HITACHI SHIPBUILDING & ENGINEERING COMPANY, LIMITED)
60, SAKURAJIMA KITANO-CHO, KONOHANA-KU, OSAKA-SHI 554
ENGLISH

RIMOTO SENSHINGU NYUSU 7332
(REMOTE SENSING NEWS)
NIHON ZOSEN SHINKO ZAIDAN
(JAPAN SHIPBUILDING PROMOTION FOUNDATION)
JAPANESE

RYOKAKUSEN 7333
(PASSENGER BOAT)
NIHON RYOKAKUSEN KYOKAI (JAPAN PASSENGER BOAT ASSOCIATION)
C/O IINO BUILDING, 1-1, 2-CHOME, UCHISAIWAI-CHO, CHIYODA-KU, TOKYO 100
JAPANESE
INDEXED: JICST

SAGYOSEN 7334
(DREDGER)
NIHON SAGYOSEN KYOKAI (JAPAN DREDGER TECHNICAL SOCIETY)
6-3, YAESU, CHUO-KU, TOKYO 104
JAPANESE
INDEXED: JICST

SEIBU ZOSENKAI KAIHO 7335
(TRANSACTIONS OF THE WEST-JAPAN SOCIETY OF NAVAL ARCHITECTS)
SEIBU ZOSENKAI (WEST-JAPAN SOCIETY OF NAVAL ARCHITECTS)
C/O KYUSHU DAIGAKU KOGAKUBU ZOSENGAKU KYOSHITSU, HAKOZAKI, HIGASHI-KU, FUKUOKA-SHI 812
JAPANESE AND EUROPEAN
INDEXED: JICST

SEKAI NO KANSEN 7336
(SHIPS OF THE WORLD)
KAIJINSHA
C/O ZENRIN GAKUSEI KAIKAN, 5-3, 1-CHOME, KORAKU, BUNKYO-KU, TOKYO 112
JAPANESE
INDEXED: JICST

SENKEN KOHO 7337
(BULLETIN OF THE SHIP RESEARCH INSTITUTE)
UN'YU-SHO SENPAKU GIJUTSU KENKYUSHO (SHIP RESEARCH INSTITUTE, MINISTRY OF TRANSPORT)
38-1, 6-CHOME, SHINKAWA, MITAKA-SHI, TOKYO 181
JAPANESE

SENPAKU 7338
(SHIPPING (TENNENSHA))
TENNENSHA
50, AKAGISHITA-MACHI, SHINJUKU-KU, TOKYO 162
JAPANESE
INDEXED: JICST

SENPAKU GIJUTSU KENKYUJO KENKYU HAPPYOKAI KOEN GAIYO 7339
(PROCEEDINGS OF THE SYMPOSIUM OF THE SHIP RESEARCH INSTITUTE)
JAPANESE
INDEXED: JICST

SENPAKU GIJUTSU KENKYUSHO HAPPYO RONBUN HYODAISHU 7340
(TITLE INDEX OF PUBLISHED ARTICLES OF SHIP RESEARCH INSTITUTE)
UN'YU-SHO SENPAKU GIJUTSU KENKYUSHO (SHIP RESEARCH INSTITUTE, MINISTRY OF TRANSPORT)
38-1, 6-CHOME, SHINKAWA, MITAKA-SHI, TOKYO 181
JAPANESE

SENPAKU GIJUTSU KENKYUSHO HOKOKU 7341
(REPORT OF THE SHIP RESEARCH INSTITUTE)
UN'YU-SHO SENPAKU GIJUTSU KENKYUSHO (SHIP RESEARCH INSTITUTE, MINISTRY OF TRANSPORT)
38-1, 6-CHOME, SHINKAWA, MITAKA-SHI, TOKYO 181
JAPANESE, WITH EUROPEAN SUMMARY
INDEXED: APP, CHE, JICST

SENPAKU GIJUTSU KENKYUSHO KENKYU SEIKA HOKOKUSHO 7342
(CONTRIBUTIONS FROM THE SHIP RESEARCH INSTITUTE)
UN'YU-SHO SENPAKU GIJUTSU KENKYUSHO (SHIP RESEARCH INSTITUTE, MINISTRY OF TRANSPORT)
38-1, 6-CHOME, SHINKAWA, MITAKA-SHI, TOKYO 181
JAPANESE

SENPAKU GIJUTSU KENKYUSHO NENPO 7343
(ANNUAL REPORTS OF THE SHIP RESEARCH INSTITUTE)
UN'YU-SHO SENPAKU GIJUTSU KENKYUSHO (SHIP RESEARCH INSTITUTE, MINISTRY OF TRANSPORT)
38-1, 6-CHOME, SHINKAWA, MITAKA-SHI, TOKYO 181
JAPANESE
INDEXED: JICST

SENPAKU HYOJUNKA GIHO 7344
(TECHNICAL JOURNAL OF MARINE STANDARDIZATION)
NIHON SENPAKU HYOJUN KYOKAI (JAPANESE ASSOCIATION OF MARINE STANDARDIZATION)
7, SHIBA KOTOCHIRA-CHO, MINATO-KU, TOKYO 105
JAPANESE

SENPAKU KOGAKU, KAIYO KOGAKU GIJUTSU BUNKEN SHUHO 7345
(NSS LIBRARY NEWS)
NIHON SENPAKU SHINKOKAI TOSHOSHITSU (LIBRARY, THE JAPAN SHIPBUILDING INDUSTRY FOUNDATION)
C/O SENPAKU SHINKO BUILDING, 35, SHIBA KOTOCHIRA-CHO MINATO-KU, TOKYO 105
JAPANESE AND ENGLISH

SENPAKU TOKEI 7346
(SHIP STATISTICS)
UN'YU-SHO DAIJIN KANBO JOHO KANRI-BU (RESEARCH AND DATA PROCESSING DEPARTMENT, MINISTER'S SECRETARIAT, MINISTRY OF TRANSPORT)
1-3, 2-CHOME, KASUMIGASEKI, CHIYODA-KU, TOKYO 100
JAPANESE

SHIPPING GAZETTE (A WEEKLY DIGEST OF TRADE, TRAVEL AND TRANSPORTATION NEWS) 7347
JAPAN PURESU K. K. (JAPAN PRESS, LIMITED)
12-8, 2-CHOME, KITAAOYAMA, MINATO-KU, TOKYO 107)
JAPANESE AND ENGLISH

SHOSEN KOTO SENMON GAKKO KIYO 7348
(MEMOIRS OF MERCANTILE MARINE COLLEGES)
TOYAMA SHOSEN KOTO SENMON GAKKO (TOYAMA MERCANTILE MARINE COLLEGE)
1-2, EBIE NERIAI, SHINMINATO-SHI 933-02
JAPANESE AND ENGLISH

SOCIETY OF NAVAL ARCHITECTS OF JAPAN, 60TH ANNIVERSARY SERIES 7349
SOCIETY OF NAVAL ARCHITECTS OF JAPAN, 60TH ANNIVERSARY SERIES
JAPANESE
INDEXED: JICST

SUISAN-CHO GYOSEN KENKYUSHITSU KENKYU HOKOKU 7350
(FISHING BOAT LABORATORY SCIENTIFIC REPORT)
SUISAN-CHO GYOSEN KENKYUSHITSU (FISHING BOAT LABORATORY, FISHERY AGENCY)
5-1, 5-CHOME, KACHIDOKI, CHUO-KU, TOKYO 104
JAPANESE, WITH ENGLISH SUMMARY

TOYAMA SHOSEN KOTO SENMON GAKKO KENKYU SHUROKU 7351
(JOURNAL OF TOYAMA MERCANTILE MARINE COLLEGE)
TOYAMA SHOSEN KOTO SENMON GAKKO (TOYAMA MERCANTILE MARINE COLLEGE)
1-2, EBIE NERIAI, SHINMINATO-SHI 933-02
JAPANESE AND ENGLISH

UMI NO SEKAI 7352
(MARITIME WORLD)
NIHON KAIJI KOHO KYOKAI (JAPAN MARITIME RESEARCH INSTITUTE)
23-17, 1-CHOME, SHINKAWA, CHUO-KU, TOKYO 104
JAPANESE

ZOSEN 7353
(SHIPBULDING)
TOKYO NYUSU TSUSHINSHA
8-10, GINZANISHI, CHUO-KU, TOKYO 104
ENGLISH

ZOSEN GIJUTSU 7354
(SHIPBUILDING & ENGINEERING)
JAPAN INDASUTORIARU PABURISHINGU (JAPAN INDUSTRIAL PUBLISHING)
10-6, 3-CHOME, AZABU JUBAN, MINATO-KU, TOKYO 106
JAPANESE
INDEXED: JICST

ZOSEN GIJUTSU KENKYU KAIHATSU CHOSHO 7355
(REPORTS OF DEVELOPMENT OF SHIPBUILDING TECHNIQUE)
ZOSEN GIJUTSU KAIHATSU KYOGI KIKO (ORGANIZATION FOR DEVELOPMENT OF SHIPBUILDING TECHNIQUE)
C/O NIHON ZOSEN KENKYU KYOKAI, 35, SHIBA KOTOCHIRA-CHO, MINATO-KU, TOKYO 105
JAPANESE

ZOSEN GIJUTSU KENKYU KAIHATSU KADAISHU 7356
(RESEARCH SUBJECTS FOR DEVELOPMENT OF SHIPBUILDING TECHNIQUE)
ZOSEN GIJUTSU KAIHAUSU KYOGI KIKO (ORGANIZATION FOR DEVELOPMENT OF SHIPBUILDING TECHNIQUE)
C/O NIHON ZOSEN KENKYU KYOKAI, 35, SHIBA KOTOCHIRA-CHO, MINATO-KU, TOKYO 105
JAPANESE
INDEXED: JICST

ZOSEN GIJUTSU KENKYU KAIHATSU YOYAKUSHU 7357
(DEVELOPMENT OF SHIPBUILDING TECHNIQUE, DIGEST REVIEW)
ZOSEN GIJUTSU KAIHATSU KYOGI KIKO (ORGANIZATION FOR DEVELOPMENT OF SHIPBUILDING TECHNIQUE)
C/O NIHON ZOSEN KENKYU KYOKAI, 35, SHIBA KOTOHIRA-CHO, MINATO-KU, TOKYO 105
JAPANESE
INDEXED: JICST

ZOSEN KANKEI GIJUTSU SHIRYO SOKUHO 7358
(JAFSA LIBRARY NEWS)
JAPANESE
INDEXED: JICST

ZOSEN KANKEI TOKEI SHIRYO, KOKUNAI-HEN 7359
(STATISTICS OF SHIPBUILDING, DOMESTIC EDITION)
NIHON SENPAKU YUSHUTSU KUMIAI BOEKI SHIRYOSHITSU (TRADE INFORMATION ROOM, JAPAN SHIP EXPORTERS' ASSOCIATION)
35, SHIBA KOTOHIRA-CHO, MINATO-KU, TOKYO 105
JAPANESE

ZOSEN KANKEI TOKEI SHIRYO, SEKAI-HEN 7360
(STATISTICS OF SHIPBUILDING, WORLD EDITION)
NIHON SENPAKU YUSHUTSU KUMIAI BOEKI SHIRYOSHITSU (TRADE INFORMATION ROOM, JAPAN SHIP EXPORTERS' ASSOCIATION)
35, SHIBA KOTOHIRA-CHO, MINATO-KU, TOKYO 105
JAPANESE

ZOSEN ZOKI TOKEI GEPPO 7361
(MONTHLY REPORT OF SHIPBUILDING AND MACHINE MAKING)
UN'YU-SHO DAIJIN KANBO JOHO KANRI-BU (RESEARCH AND DATA PROCESSING DEPARTMENT, MINISTER'S SECRETARIAT, MINISTRY OF TRANSPORT)
1-3, 2-CHOME, KASUMIGASEKI, CHIYODA-KU, TOKYO 100
JAPANESE

ZOSENGYO DAYORI 7362
(SHIPBUILDING NEWS)
NIHON KOGATA SENPAKU KOGYOKAI (JAPAN UTILITY SHIPBUILDERS' ASSOCIATION)
1-8, NIHONBASHI KAYABA-CHO, CHUO-KU, TOKYO 103
JAPANESE

ZOSENKAI 7363
(SHIPBUILDERS' NEWS)
NIHON ZOSEN KOGYOKAI (SHIPBUILDERS' ASSOCIATION OF JAPAN)
C/O SENPAKU SHINKO BUILDING, 35, SHIBA KOTOHIRA-CHO, MINATO-KU, TOKYO 105
JAPANESE
INDEXED: JICST

ZOSHU RENRAKU 7364
(REPORT ON SHIPBUILDING AND REPAIR)
KAIJO HOAN-CHO SENPAKU GIJUTSU-BU (SHIP TECHNOLOGICAL DEPARTMENT, MARITIME SAFETY AGENCY)
1-3, 2-CHOME, KASUMIGASEKI, CHIYODA-KU, TOKYO 100
JAPANESE

Aeronautical Engineering

AIR SERVICE NEWS TOKYO 7365
RENGO PURESUSHA
C/O DAI-2 OKURA BUILDING, 1-2, UCHISAIWAI-CHO, CHIYODA-KU, TOKYO 100
ENGLISH

CAS KENKYU HOKOKU 7366
(REPORT OF THE COMMITTEE OF AEROSPACE STRUCTURE)
KOKUKI KOZO KENKYU IINKAI (COMMITTEE OF AEROSPACE STRUCTURE)
C/O TOKYO DAIGAKU KOGAKUBU KOKUGAKKA, 3-1, 7-CHOME, HONGO, BUNKYO-KU, TOKYO 113
JAPANESE AND ENGLISH

CURRENT PAPERS OF THE DEPARTMENT OF AERONAUTICAL ENGINEERING, KYOTO UNIVERSITY 7367
DEPARTMENT OF AERONAUTICAL ENGINEERING, KYOTO UNIVERSITY, KYOTO
ENGLISH
INDEXED: APP, CHE

DENSHI KOHO KENKYUSHO HOKOKU 7368
(ELECTRONIC NAVIGATION RESEARCH INSTITUTE PAPERS)
UN'YU-SHO DENSHI KOHO KENKYUSHO (ELECTRONIC NAVIGATION RESEARCH INSTITUTE, MINISTRY OF TRANSPORT)
38-1,, 6-CHOME, SHINKAWA, MITAKA-SHI, TOKYO 181
JAPANESE AND ENGLISH
INDEXED: JICST

DENSHI KOHO KENKYUSHO NENPO 7369
(ANNUAL REPORT OF ELECTRONIC NAVIGATION RESEARCH INSTITUTE)
UN'YU-SHO DENSHI KOHO KENKYUSHO (ELECTRONIC NAVIGATION RESEARCH INSTITUTE, MINISTRY OF TRANSPORT)
38-1, 6-CHOME, SHINKAWA, MITAKA-SHI, TOKYO 181
ENGLISH

GAIKOKU KOKU, UCHU BUNKEN MOKUROKU 7370
(FOREIGN AERO/SPACE LITERATURES)
KOKURITSU KOKKAI TOSHOKAN SANKOSHOSHI-BU (REFERENCE AND BIBLIOGRAPHY DIVISION, NATIONAL DIET LIBRARY)
10-1, 1-CHOME, NAGATA-CHO, CHIYODA-KU, TOKYO 100
JAPANESE AND ENGLISH
INDEXED: JICST

HOKODA 7371
(RUDDER)
NIHON GAKUSEI KOKU RENMEI
C/O ASAHI SHINBUNSHA, 2-3, YURAKU-CHO CHIYODA-KU, TOKYO 100
JAPANESE

ICAO JANARU 7372
(ICAO JOURNAL)
KOKU SHINKO ZAIDAN ICAO JIGYU KATSUDO KENKYUKAI (INVESTIGATION COMMITTEE ON INTERNATIONAL CIVIL AVIATION ORGANIZATION ACTIVITIES, FOUNDATION OF AERONAUTICS PROMOTION)
38, SHIBA KOTOHIRA-CHO, MINATO-KU, TOKYO 105
JAPANESE

JAPAN'S CURRENT AEROSPACE AND DEFENSE 7373
ENGLISH

JETTO NYUSU 7374
(JET NEWS)
RUFUTOHANZA (LUTF HANSA)
2-6, 3-CHOME, KASUMIGASEKI, CHIYODA-KU TOKYO 100
JAPANESE

KIKAN OZORA 7375
(SKY)
NIHON KOKU K. K. KOHOSHITSU (PUBLIC INFORMATION BUREAU, JAPAN AIR LINES)
7-3, 2-CHOME, MARUNOUCHI, CHIYODA-KU, TOKYO 100
JAPANESE

KOGIKEN NYUSU 7376
(NAL NEWS)
KAGAKU GIJUTSU-CHO KOKU UCHU GIJUTSU KENKYUSHO (NATIONAL AERONAUTICAL LABORATORY, SCIENCE AND TECHNOLOGY AGENCY)
1880, JINDAIJI-MACHI, CHOFU-SHI, TOKYO 182
JAPANESE
INDEXED: JICST

KOKU DAIGAKKO KENKYU HOKOKU 7377
(REPORT OF AERONAUTICS COLLEGE)
UN'YU-SHO KOKU DAIGAKKO (AERONAUTICS COLLEGE, MINISTRY OF TRANSPORT)
AKAE-CHO, MIYAZAKI-SHI 880
JAPANESE, WITH ENGLISH SUMMARY
INDEXED: JICST

KOKU DENSHI GIHO 7378
(AVIATION ELECTRONICS TECHNICAL BULLETIN)
JAPAN AVIATION ELECTRONICS INDUSTRY, LTD.
NANPEIDAI TOKYO BUILDING, 21-6, 1-CHOME, DOGENZAKA, SHIBUYA-KU, TOKYO
JAPANESE
INDEXED: CHE

KOKU FAN 7379
(AERO FAN MAGAZINE)
BUNRINDO
1-55, KANDA JINBO-CHO, CHIYODA-KU, TOKYO 101
JAPANESE AND ENGLISH
INDEXED: JICST

KOKU GIJUTSU 7380
(AIRCRAFT ENGINEERING)
NIHON KOKU SEIBI KYOKAI (JAPAN AERONAUTICAL ENGINEERS' ASSOCIATION)
C/O HIKOKAN BUILDING, 18-1, 1-CHOME, SHINBASHI, MINATO-KU, TOKYO 105
JAPANESE
INDEXED: JICST

KOKU JOHO 7381
(AEROVIEW)
KANTOSHA
C/O KOJUN BUILDING, 8-7, 6-CHOME, GINZA, CHUO-KU, TOKYO 104
JAPANESE
INDEXED: JICST

KOKU JOHO SHIRYO 7382
(AIRCRAFT INFORMATION)
KOKU JIEITAI KOKU SHIRYO SAGYOTAI (AIR SELF DEFENSE FORCE, DEFENSE AGENCY)
7-14, 9-CHOME, AKASAKA, MINATO-KU, TOKYO 107
JAPANESE

KOKU KOGYO KOKUSAI JOHO 7383
(INTERNATIONAL REPORT ON THE AERONAUTIC ENGINEERING INDUSTRY)
NIHON KOKU KOGYOKAI (JAPAN AERONAUTIC INDUSTRY ASSOCIATION)
JAPANESE

KOKU TENBO 7384
(AVIATION ACTIVITIES)
KOKU NYUSUSHA
C/O DAI-3 MORI BUILDING, 4-10, 1-CHOME, SHINBASHI, MINATO-KU, TOKYO 105
JAPANESE

KOKU UCHU GIJUTSU DOKO 7385
(TRENDS IN AERONAUTICS AND SPACE TECHNOLOGY)
NIHON KOKU UCHU KOGYOKAI (JAPAN AERONAUTICS AND SPACE INDUSTRY ASSOCIATION)
JAPANESE

KOKU UCHU GIJUTSU KENKYUJO KENKYU HAPPYOKAI MAEZURI 7386
(PREPRINT OF THE SYMPOSIUM OF THE NATIONAL AEROSPACE LAB)
JAPANESE
INDEXED: JICST

KOKU UCHU GIJUTSU KENKYUSHO HOKOKU 7387
(NAL TECHNICAL REPORT)
KAGAKU GIJUTSU-CHO KOKU UCHU GIJUTSU KENKYUSHO (NATIONAL AEROSPACE LABORATORY, SCIENCE AND TECHNOLOGY AGENCY)
1880, JINDAIJI-MACHI, CHOFU-SHI, TOKYO 182
JAPANESE, WITH ENGLISH SUMMARY
INDEXED: JICST

KOKU UCHU GIJUTSU KENKYUSHO NENPO 7388
(NAL ANNUAL REPORT)
KAGAKU GIJUTSU-CHO KOKU UCHU GIJUTSU KENKYUSHO (NATIONAL AEROSPACE LABORATORY, SCIENCE AND TECHNOLOGY AGENCY)
1880, JINDAIJI-MACHI, CHOFU-SHI, TOKYO 182
JAPANESE
INDEXED: JICST

KOKU UCHU GIJUTSU KENKYUSHO SHIRYO 7389
(NAL TECHNICAL MEMORANDUM)
KAGAKU GIJUTSU-CHO KOKU UCHU GIJUTSU KENKYUSHO (NATIONAL AEROSPACE LABORATORY, SCIENCE AND TECHNOLOGY AGENCY)
1880, JINDAIJI-MACHI, CHOFU-SHI, TOKYO 182
JAPANESE
INDEXED: JICST

KOKU UCHU JOHO 7390
(AEROSPACE INFORMATION)
KAGAKU GIJUTSU-CHO KENKYU CHOSEI-KYOKU (RESEARCH COORDINATION BUREAU, SCIENCE AND TECHNOLOGY AGENCY)
2-1, 2-CHOME, KASUMIGASEKI, CHIYODA-KU, TOKYO 100
JAPANESE
INDEXED: JICST

KOKU UCHU KOGYO KOKUSAI JOHO 7391
(INTERNATIONAL REPORT OF THE AERONAUTICS AND SPACE INDUSTRY)
NIHON KOKU UCHU KOGYOKAI (JAPAN AERONAUTICS AND SPACE INDUSTRY ASSOCIATION)
JAPANESE

KOKU UCHU KOGYO KOKUSAI JOHO. BESSATSU. 7392
(INTERNATIONAL REPORT OF THE AERONAUTICS AND SPACE INDUSTRY. SPECIAL ISSUE.)
NIHON KOKU UCHU KOGYOKAI (JAPAN AERONAUTICS AND SPACE INDUSTRY ASSOCIATION)
JAPANESE

KOKU UCHU SHUYO KIJI SAKUIN 7393
(CURRENT LITERATURE OF AERONAUTICS AND ASTRONAUTICS)
NIHON KOKU KYOKAI (JAPAN AERONAUTICS ASSOCIATION)
18-1, 1-CHOME, SHINBASHI, MINATO-KU, TOKYO 105
JAPANESE

KOZO KYODO NI KANSURU KOENKAI KOENSHU 7394
(REPORT OF THE MEETING ON AIRPLANE STRUCTURE)
NIHON KOKU UCHU GAKKAI (JAPAN SOCIETY FOR AERONAUTICAL AND SPACE SCIENCES) AND NIHON KIKAI GAKKAI (JAPAN SOCIETY OF MECHANICAL ENGINEERS)
JAPANESE
INDEXED: JICST

KUKU GIJUTSU HOBUNSHU 7395
(PROCEEDINGS OF AIRPORT ENGINEERING)
UN'YU-SHO TOKYO KOKU-KYOKU (TOKYO REGIONAL CIVIL AVIATION BUREAU, MINISTRY OF TRANSPORT)
3-1, 1-CHOME, OTE-MACHI, CHIYODA-KU, TOKYO 100
JAPANESE

NASDA NYUSU 7396
(NASDA NEWS)
UCHU KAIHATSU JIGYODAN (NATIONAL SPACE DEVELOPMENT AGENCY OF JAPAN)
C/O SEKAI BOEKI SENTA BUILDING, 4-1, 2-CHOME, HAMAMATSU-CHO, MINATO-KU, TOKYO 105
JAPANESE
INDEXED: JICST

NIHON KOKU KOGYOKAI KAIHO 7397
(JOURNAL OF THE SOCIETY OF JAPANESE AIRCRAFT CONSTRUCTORS)
NIHON KOKU KOGYOKAI (SOCIETY OF JAPANESE AIRCRAFT CONSTRUCTORS)
C/O CHIYODA BUILDING, 1-2, 2-CHOME, MARUNOUCHI, CHIYODA-KU, TOKYO 100
JAPANESE
INDEXED: JICST

NIHON KOKU KYOKAI KAIHO 7398
(BULLETIN OF THE JAPAN AERONAUTICS ASSOCIATION)
JAPANESE

NIHON KOKU UCHU GAKKAISHI 7399
(JOURNAL OF THE JAPAN SOCIETY FOR AERONAUTICAL AND SPACE SCIENCES)
NIHON KOKU UCHU GAKKAI (JAPAN SOCIETY FOR AERONAUTICAL AND SPACE SCIENCES)
C/O HIKOKAN BUILDING, 18-1, 1-CHOME, SHINBASHI, MINATO-KU, TOKYO 105
JAPANESE, WITH ENGLISH SUMMARY
INDEXED: JICST

NIHON KOKUROSHI 7400
(AIP-JAPAN)
UN'YU-SHO KOKU-KYOKU (AERONAUTICAL BUREAU, MINISTRY OF TRANSPORT)
1-3, 2-CHOME, KASUMIGASEKI, CHIYODA-KU, TOKYO 100
JAPANESE

PAIROTTO 7401
(PILOT)
NIHON PAIROTTO KYOKAI (JAPAN PILOT ASSOCIATION)
3-1, KYOBASHI, CHUO-KU, TOKYO 103
JAPANESE

REPORT OF THE INSTITUTE OF SPACE AND AERONAUTICAL SCIENCE, UNIVERSITY OF TOKYO 7402
TOKYO DAIGAKU UCHU KOKU KENKYUSHO (INSTITUTE OF SPACE AND AERONAUTICAL SCIENCE, UNIVERSITY OF TOKYO)
7-6, 3-CHOME, KOMABA, MEGURO-KU, TOKYO 153
ENGLISH
INDEXED: CHE, ENG, IAA, JICST

ROCKET NEWS 7403
NIHON ROKETTO KYOKAI (JAPANESE ROCKET SOCIETY)
C/O YOMIURI SHINBUNSHA, 7-1, 1-CHOME, OTE-MACHI, CHIYODA-KU, TOKYO 100
JAPANESE
INDEXED: JICST

SEKAI NO TSUBASA 7404
(WORLD AIRPLANES)
ASAHI SHINBUNSHA
YURAKU-CHO, CHIYODA-KU, TOKYO 100
JAPANESE

SPACE IN JAPAN 7405
JAPAN KAGAKU GIJUTSUCHO (SCIENCE AND TECHNOLOGY AGENCY, PRIME MINISTER'S OFFICE)
3-4, KASUMIGASOKI, CHIYODA-KU, TOKYO
JAPANESE

SUJI DE MIRU KOKU 7406
(AERONAUTIC STATISTICS)
KOKU SHINKO ZAIDAN (AERONAUTIC PROMOTION FOUNDATION)
C/O NIHON GASU KYOKAI BUILDING, 38, SHIBA KOTOHIRA-CHO, MINATO-KU, TOKYO 105
JAPANESE

TECHNICAL REPORT OF NATIONAL AEROSPACE LABORATORY, TOKYO 7407
JAPANESE AND ENGLISH
INDEXED: APP

TOKYO DAIGAKU UCHU KOKU KENKYUSHO HOKOKU 7408
(BULLETIN OF THE INSTITUTE OF SPACE AND AERONAUTICAL SCIENCE, UNIVERSITY OF TOKYO)
TOKYO DAIGAKU UCHU KOKU KENKYUSHO (INSTITUTE OF SPACE AND AERONAUTICAL SCIENCE, UNIVERSITY OF TOKYO)
4-6-1, KOMATA, MEGURO-KU, TOKYO
JAPANESE
INDEXED: CHE, ENG, INS, WAA, JICST

TOKYO-TORITSU KOKU KOGYO KOTO SENMON GAKKO KENKYU KIYO 7409
(RESEARCH REPORT OF THE TOKYO METROPOLITAN COLLEGE OF AERONAUTICAL ENGINEERING)
TOKYO-TORITSU KOKU KOGYO KOTO SENMON GAKKO (TOKYO METROPOLITAN COLLEGE OF AERONAUTICAL ENGINEERING)
53-1, 8-CHOME, MINAMISENJU, ARAKAWA-KU, TOKYO 116
JAPANESE AND EUROPEAN

TRANSACTIONS OF THE JAPAN SOCIETY FOR AERONAUTICAL AND SPACE SCIENCES 7410
NIHON KOKU UCHU GAKKAI (THE JAPAN SOCIETY FOR AERONAUTICAL AND SPACE SCIENCES)
18-1, 1-CHOME, SHINBASHI, MINATO-KU, TOKYO 105
ENGLISH
INDEXED: APP, CHE, IAA, ENG, JICST

UCHU KAGAKU GIJUTSU RENGO KOENKAI KOENSHU 7411
(PROCEEDINGS OF THE SPACE SCIENTIFIC TECHNOLOGY FEDERATION)
JAPANESE
INDEXED: JICST

UCHU KAIHATSU IINKAI GEPPO 7412
(MONTHLY BULLETIN OF THE SPACE ACTIVITIES COMMISSION)
KAGAKU GIJUTSU-CHO KENKYU CHOSEI-KYOKU (RESEARCH COORDINATION BUREAU, THE SCIENCE AND TECHNOLOGY AGENCY)
2-1, 2-CHOME, KASUMIGASEKI, CHIYODA-KU, TOKYO 100
JAPANESE

UCHU KAIHATSU JIGYODAN GIJUTSU HOKOKU (NASDA-TR) 7413
(TECHNICAL REPORT OF NATIONAL SPACE DEVELOPMENT
AGENCY OF JAPAN)
UCHU KAIHATSU JIGYODAN (NATIONAL SPACE DEVELOPMENT
AGENCY OF JAPAN)
4-1, 2-CHOME, HAMAMATSU-CHO, MINATO-KU, TOKYO 105
JAPANESE, WITH ENGLISH SUMMARY

UCHU KAIHATSU JIGYODAN GIJUTSU NOTO (NASDA-TN) 7414
(TECHNICAL NOTE OF NATIONAL SPACE DEVELOPMENT
AGENCY OF JAPAN)
UCHU KAIHATSU JIGYODAN (NATIONAL SPACE
DEVELOPMENT AGENCY OF JAPAN)
4-1, 2-CHOME, HAMAMATSU-CHO, MINATO-KU, TOKYO 105
JAPANESE
INDEXED: JICST

UCHU KAIHATSU SUISHIN HONBU GIJUTSU HOKOKU 7415
(TECHNICAL REPORT OF THE HEADQUARTERS FOR THE
PROMOTION OF SPACE DEVELOPMENT)
JAPANESE
INDEXED: JICST

UCHU KAIHATSU SUISHIN HONBU GIJUTSU NOTO 7416
(TECHNICAL NOTES OF THE HEADQUARTERS FOR THE
PROMOTION OF SPACE DEVELOPMENT)
JAPANESE
INDEXED: JICST

WING 7417
KOKU SHINBUNSHA
4-18, 5-CHOME, GINZA, CHUO-KU, TOKYO 104
JAPANESE

WING JAPAN'S CURRENT AEROSPACE & DEFENSE 7418
JAPANESE

Domestic Science, Household Economy, Housekeeping

AICHI SHUKUTOKU TANKI DAIGAKU KENKYU KIYO 7419
(RESEARCH BULLETIN OF THE AICHI WOMEN'S JUNIOR COLLEGE)
AICHI SHUKUTOKU TANKI DAIGAKU KENKYU KIYO HENSHU IIN (RESEARCH BULLETIN OF THE AICHI WOMEN'S JUNIOR COLLEGE, EDITORIAL COMMITTEE)
JAPANESE

AIZEN JOSHI TANKI DAIGAKU KIYO 7420
(BULLETIN OF THE AIZEN WOMEN'S JUNIOR COLLEGE)
JAPANESE

AKENOHOSHI JOSHI TANKI DAIGAKU KIYO 7421
(PUBLICATIONS OF AKENOHOSHI WOMEN'S JUNIOR COLLEGE)
AKENOHOSHI JOSHI TANKI DAIGAKU (AKENOHOSHI WOMEN'S JUNIOR COLLEGE)
JAPANESE

AOYAMA GAKUIN JOSHI TANKI DAIGAKU KIYO 7422
(BULLETIN OF AOYAMA GAKUIN WOMEN'S JUNIOR COLLEGE)
JAPANESE
INDEXED: JICST

ATOMI GAKUEN JOSHI DAIGAKU KIYO 7423
(BULLETIN OF THE ATOMI GAKUEN WOMEN'S UNIVERSITY)
ATOMI GAKUEN WOMEN'S UNIVERSITY
JAPANESE

BEACON 7424
KENMEI JOSHI GAKUIN TANKI DAIGAKU (KENMEI JUNIOR COLLEGE)
68, HON-MACHI, HIMEJI-SHI 670
JAPANESE AND EUROPEAN

BEPPU JOSHI TANKI DAIGAKU KENKYU KIYO 7425
(RESEARCH BULLETIN OF BEPPU WOMEN'S JUNIOR COLLEGE)
JAPANESE

BUNKYO JOSHI TANKI DAIGAKU KIYO 7426
(BULLETIN OF BUNKYO WOMEN'S JUNIOR COLLEGE)
JAPANESE

CHOFU GAKUEN JOSHI TANDAI KIYO 7427
(BULLETIN OF THE CHOFU GAKUEN WOMEN'S JUNIOR COLLEGE)
CHOFU GAKUEN JOSHI TANKI DAIGAKU SHAGAKU KENKYUKAI (RESEARCH COOPERATIVE OF THE CHOFU GAKUEN WOMEN'S JUNIOR COLLEGE)
JAPANESE

CHUBU JOSHI TANKI DAIGAKU KIYO 7428
(JOURNAL OF CHUBU WOMEN'S JUNIOR COLLEGE)
CHUBU JOSHI TANKI DAIGAKU
4909, KURACHI MUKOYAMA, SEKI-SHI 501-32
JAPANESE

DOSHISHA JOSHI DAIGAKU GAKUJUTSU KENKYU NENPO, II 7429
(ANNUAL REPORTS OF STUDIES, DOSHISHA WOMEN'S COLLEGE OF LIBERAL ARTS PART 2)
DOSHISHA JOSHI DAIGAKU (DOSHISHA WOMEN'S COLLEGE OF LIBERAL ARTS)
IMADEGAWA TERA-MACHI, NISHI-IRU, KAMIGYO-KU, KYOYO-SHI 602
JAPANESE AND ENGLISH
INDEXED: CHE

EHIME JOSHI TANKI DAIGAKU KIYO 7430
(EHIME WOMEN'S JUNIOR COLLEGE BULLETIN)
EHIME WOMEN'S JUNIOR COLLEGE
JAPANESE

FUKUOKA JOGAKUIN TANKI DAIGAKU KIYO 7431
(BULLETIN OF FUKUOKA WOMEN'S GAUKIN JUNIOR COLLEGE)
JAPANESE
INDEXED: JICST

FUKUOKA JOSHI TANDAI KIYO 7432
(FUKUOKA WOMEN'S JUNIOR COLLEGE STUDIES)
FUKUOKA JOSHI TANKI DAIGAKU
DAZAIFU, TSUKUSHI-GUN, FUKUOKA 818-01
JAPANESE, WITH ENGLISH SUMMARY

FUKUSHIMA JOSHI TANKI DAIGAKU KENKYU KIYO 7433
(SUMMARY STUDY REPORTS)
FUKUSHIMA JOSHI TANKI DAIGAKU (FUKUSHIMA WOMEN'S JUNIOR COLLEGE)
1-1, MIYASHIRO NYUJIIKE, FUKUSHIMA-SHI 960-01
JAPANESE AND ENGLISH

FUKUYAMA-SHIRITSU JOSHI TANKI DAIGAKU KIYO 7434
(BULLETIN OF FUKUYAMA CITY JUNIOR COLLEGE FOR WOMEN)
JAPANESE

GAKUEN 7435
(REVIEW JOURNAL OF SHOWA WOMEN'S UNIVERSITY)
SHOWA JOSHI DAIGAKU KOYOKAI (SHOWA WOMEN'S UNIVERSITY)
1-7, TAISHIDO, SETAGAYA-KU, TOKYO 154
JAPANESE AND ENGLISH

GAKUSHUIN JOSHI TANKI DAIGAKU KIYO 7436
(BULLETIN OF GAKUSHUIN WOMEN'S JUNIOR COLLEGE)
GAKUSHUIN WOMEN'S JUNIOR COLLEGE
JAPANESE

GENBU KURININGUHO 7437
(GENBU CLEANING REPORT)
GENBU K. K.
C/O ASAHI SEIMEIKAN, 5-50, KORAIBASHI, HIGASHI-KU, OSAKA-SHI 541
JAPANESE

GIFU JOSHI DAIGAKU KIYO 7438
(BULLETIN OF GIFU WOMEN'S COLLEGE)
GIFU JOSHI DAIGAKU (GIFU WOMEN'S COLLEGE)
300, TAROMARU, GIFU-SHI 501-25
JAPANESE

GIFU JOSHI TANKI DAIGAKU KENKYU KIYO 7439
(RESEARCH BULLETIN OF THE GIFU WOMEN'S JUNIOR COLLEGE)
GIFU JOSHI TANKI DAIGAKU (GIFU WOMEN'S JUNIOR COLLEGE
2693, NAGARA FUKUMITSU, GIFU-SHI 502
JAPANESE AND ENGLISH

GUNMA JOSHI TANKI DAIGAKU KIYO 7440
(MEMOIRS OF GUNMA WOMEN'S JUNIOR COLLEGE)
GUNMA WOMEN'S JUNIOR COLLEGE
JAPANESE

HAYASHI GAKUEN JOSHI TANKI DAIGAKU KIYO 7441
(MEMOIRS OF HAYASHI GAKUEN WOMEN'S JUNIOR COLLEGE)
HAYASHI GAKUEN JOSHI TANKI DAIGAKU (HAYASHI GAKUEN WOMEN'S JUNIOR COLLEGE)
RYOTAKAYA, KONAN-SHI 483
JAPANESE

HEIAN JOGAKUIN TANKI DAIGAKU KIYO 7442
(MEMOIRS OF HEIAN WOMEN'S JUNIOR COLLEGE)
HEIAN JOGAKUIN TANKI DAIGAKU
SHIMODACHIYURI, MUROMACHI-DORI, KAMIGYO-KU, KYOTO-SHI 602
JAPANESE

HIJIYAMA JOSHI TANKI DAIGAKU KIYO 7443
(BULLETIN OF HIJIYAMA WOMEN'S JUNIOR COLLEGE)
JAPANESE

HIMEJI TANKI DAIGAKU KENKYU HOKOKU 7444
(RESEARCH REPORT, HIMEJI JUNIOR COLLEGE)
HIMEJI TANKI DAIGAKU (HIMEJI JUNIOR COLLEGE)
SHINZAIKE, HIMEJI-SHI 670
JAPANESE, WITH EUROPEAN SUMMARY

HINOMOTO GAKUEN TANKI DAIGAKU KENKYU KIYO 7445
(RESEARCH BULLETIN OF HINOMOTO GAKUEN JUNIOR COLLEGE)
JAPANESE

HIROSHIMA JOSHI DAIGAKU KASEIGAKUBU KIYO 7446
(BULLETIN OF THE FACULTY OF HOME ECONOMICS, HIROSHIMA WOMEN'S UNIVERSITY)
HIROSHIMA JOSHI DAIGAKU (HIROSHIMA WOMEN'S UNIVERSITY)
1-71, 1-CHOME, UJINA-HIGASHI HIROSHIMA-SHI 734
JAPANESE AND ENGLISH
INDEXED: CHE, JICST

HOKKAIDO JOSHI TANKI DAIGAKU KENKYU KIYO 7447
(RESEARCH REPORT OF HOKKAIDO WOMEN'S JUNIOR COLLEGE)
JAPANESE

HOKKAIDO MUSAHINO JOSHI TANKI DAIGAKU KIYO 7448
(REPORT OF HOKKAIDO MUSASHINO WOMEN'S JUNIOR COLLEGE)
JAPANESE

HOKURIKU GAKUIN TANKI DAIGAKU KIYO 7449
(STUDIES BY THE FACULTY OF HOKURIKU GAKUIN JUNIOR COLLEGE)
HOKURIKU GAKUIN TANKI DAIGAKU (HOKURIKU GAKUIN JUNIOR COLLEGE)
MITSUKOJI-MACHI, KANAZAWA-SHI 920-13
JAPANESE

HOKUSEI GAKUEN JOSHI TANKI DAIGAKU KIYO 7450
(BULLETIN OF HOKUSEI GAKUEN JUNIOR COLLEGE)
HOKUSEI GAKUEN JOSHI TANKI DAIGAKU (HOKUSEI GAKUEN JUNIOR COLLEGE)
HIGASHI-4-CHOME, MINAMI-1-JO, CHUO-KU, SAPPORO-SHI 060
JAPANESE, WITH ENGLISH SUMMARY
INDEXED: JICST

HYOGO JOSHI TANKI DAIGAKU RONSHU 7451
(BULLETIN OF HYOGO WOMEN'S JUNIOR COLLEGE)
HYOGO JOSHI TANKI DAIGAKU (HYOGO WOMEN'S JUNIOR COLLEGE)
2301, SHINZAIKE, HIROAKA-CHO, KAGOGAWA-SHI 675-01
JAPANESE

IBARAKI JOSHI TANKI DAIGAKU KIYO 7452
(BULLETIN OF THE IBARAKI WOMEN'S JUNIOR COLLEGE)
JAPANESE

ICHINOMIYA JOSHI TANKI DAIGAKU KIYO 7453
(BULLETIN OF ICHINOMIYA WOMEN'S JUNIOR COLLEGE)
ICHINOMIYA JOSHI TANKI DAIGAKU (ICHINOMIYA WOMEN'S JUNIOR COLLEGE)
6, NIKKO-MACHI, ICHINOMIYA-SHI 491
JAPANESE

IDA JOSHI TANKI DAIGAKU KENKYU KIYO 7454
(RESEARCH BULLETIN OF IDA WOMEN'S JUNIOR COLLEGE)
JAPANESE

IFUKU GAKKAI ZASSHI 7455
(JAPANESE JOURNAL OF SCIENCE OF CLOTHING)
IFUKU GAKKAI (SOCIETY FOR SCIENCE OF CLOTHING)
C/O KYOTO-FURITSU IKA DAIGAKU EISEIGAKU KYOSHITSU, HIROKOJI, KAWARA-MACHI-DORI, KAMIGYO-KU, KYOTO-SHI 602
JAPANESE
INDEXED: JICST

ISEIKATSU 7456
(CLOTHING LIFE)
ISEIKATSU KENKYUKAI (SOCIETY FOR THE CLOTHING LIFE)
C/O OSAWA BUILDING, 4-1, 2-CHOME, FUJIMI, CHIYODA-KU, TOKYO 102
JAPANESE
INDEXED: JICST

JISSEN JOSHI DAIGAKU KASEIGAKUBU KIYO 7457
(PROCEEDINGS OF THE JISSEN WOMEN'S UNIVERSITY FACULTY OF HOME ECONOMICS)
JISSEN JOSHI DAIGAKU (JISSEN WOMEN'S UNIVERSITY)
1-11, 1-CHOME, HIGASHI, SHIBUYA-KU, TOKYO 150
JAPANESE
INDEXED: CHE

JOSHI SEIGAKUIN TANKI DAIGAKU KIYO 7458
(BULLETIN OF THE 'JOSHI SEIGAKUIN' (WOMEN'S) JUNIOR COLLEGE)
JAPANESE

JOTO GAKUEN JOSHI TANKI DAIGAKU KIYO 7459
(BULLETIN OF JOTO GAKUEN WOMEN'S JUNIOR COLLEGE)
JOTO GAKUEN JOSHI TANKI DAIGAKU
TAKASE-MACHI, MITOYO-GUN, KAGAWA 767
JAPANESE

JUMONJI GAKUEN JOSHI TANKI DAIGAKU KENKYU KIYO 7460
(MEMOIRS OF JUMONJI GAKUEN WOMEN'S JUNIOR COLLEGE)
JUMONJI GAKUEN JOSHI TANKI DAIGAKU (JUMONJI GAKUEN WOMEN'S JUNIOR COLLEGE)
NOBIDOME, NIIZA-SHI 352
JAPANESE, WITH ENGLISH SUMMARY

JUNSHIN JOSHI TANKI DAIGAKU KIYO 7461
(BULLETIN OF THE JUNSHIN WOMEN'S JUNIOR COLLEGE)
JAPANESE

JUNSHIN KIYO 7462
(REPORT OF THE JUNSHIN WOMEN'S JUNIOR COLLEGE)
JUNSHIN JOSHI TANKI DAIGAKU (JUNSHIN WOMEN'S JUNIOR COLLEGE)
JAPANESE

KAGAWA-KEN MEIZEN TANKI DAIGAKU KENKYU KIYO 7463
(REPORTS OF RESEARCH KAGAWA-KEN MEIZEN JUNIOR COLLEGE)
KAGAWA-KEN MEIZEN TANKI DAIGAKU
1-10, KAMEOKA-MACHI, TAKAMATSU-SHI 760
JAPANESE
INDEXED: JICST

KAGOSHIMA JOSHI TANKI DAIGAKU KIYO 7464
(KAGOSHIMA WOMEN'S JUNIOR COLLEGE BULLETIN)
JAPANESE

KAGOSHIMA JUNSHIN JOSHI TANKI DAIGAKU KENKYU KIYO 7465
(MEMOIRS OF KAGOSHIMA JUNSHIN JUNIOR COLLEGE)
KAGOSHIMA JUNSHIN JOSHI TANKI DAIGAKU KIYO
1847, KAMOIKE-MACHI, KAGOSHIMA-SHI 890
JAPANESE AND ENGLISH

KASEI GAKKAISHI 7466
(DOMESTIC SCIENCE SOCIETY MAGAZINE)
HIROSHIMA JOGAKUIN DAIGAKU TANKI DAIGAKUBU KASEIGAKKAI (HIROSHIMA WOMEN'S ACADEMY COLLEGE, JUNIO COLLEGE FACULTY, DOMESTIC SCIENCE SOCIETY)
JAPANESE

KASEI KEIZAIGAKU RONSO 7467
(COLLECTION OF PAPERS ON HOME ECONOMICS)
NIHON JOSHI DAIGAKU KASEI KEIZAIGAKUKA (NIHON WOMEN'S COLLEGE DEPARTMENT OF HOME ECONOMICS)
JAPANESE

KASEI KENKYU 7468
(STUDIES OF HOME ECONOMICS)
MIE TANKI DAIGAKU KASEIGAKU KENKYUKAI (ASSOCIATION OF HOME ECONOMICS, MIE JUNIOR COLLEGE)
517, KURATSUKI, NAKANO, ISSHINDEN, TSU-SHI 514-01
JAPANESE
INDEXED: JICST

KASEIGAKU KENKYU 7469
(RESEARCH JOURNAL OF LIVING SCIENCE)
NARA JOSHI DAIGAKU KASEIGAKUBU (FACULTY OF HOME ECONOMICS, NARA WOMEN'S UNIVERSITY)
KITAUOYA NISHI-MACHI, NARA-SHI 630
JAPANESE, WITH EUROPEAN SUMMARY
INDEXED: JICST

KASEIGAKU KENKYUKAISHI 7470
(DOMESTIC SCIENCE RESEARCH SOCIETY MAGAZINE)
JOSHI TANKI DAIGAKU KASEIGAKU KENKYUKAI (WOMEN'S JUNIOR COLLEGE DOMESTIC SCIENCE RESEARCH SOCIETY)
JAPANESE

KASEIGAKU ZASSHI 7471
(JOURNAL OF HOME ECONOMICS)
NIHON KASEI GAKKAI (THE JAPAN HOME ECONOMICS ASSOCIATION)
1-15, 2-CHOME, NO. 502, OTSUKA, BUNKYO-KU, TOKYO 112
JAPANESE, WITH ENGLISH SUMMARY
INDEXED: BIB, CHE, JICST

KASSUI RONBUNSHU 7472
(KASSUI BULLETIN)
KASSUI JOSHI TANKI DAIGAKU
1-50, HIGASHI YAMATE-MACHI, NAGASAKI-SHI 850
JAPANESE

KATEI KAGAKU 7473
(HOME SCIENCE)
DAI NIHON JOSHI SHAKAI KYOIKUKAI KATEI KAGAKU KENKYUSHO (HOME SCIENCE INSTITUTE OF JAPAN WOMAN'S SOCIAL EDUCATION ASSOCIATION)
6-2, 2-CHOME, SHIBA KOEN, MINATO-KU, TOKYO 105
JAPANESE

KATEIKA KYOIKU 7474
(DOMESTIC SCIENCE EDUCATION)
KASEI KYOIKUSHA
JAPANESE

KEISEN JOGAKUEN TANKI DAIGAKU KENKYU KIYO 7475
(RESEARCH REPORT OF KEISEN WOMEN'S GAKUEN JUNIOR COLLEGE)
JAPANESE

KENKYU HAPPYO YOSHISHU (NIPPON KASEI GAKKAI KYUSHU SHIBU TAIKAI) 7476
(SUMMARY REPORTS FROM THE CONFERENCE OF THE KYUSHU BRANCH, THE JAPAN HOME ECONOMICS ASSOCIATION)
NIPPON KASEI GAKKAI KYUSHU SHIBU
C/O OITA DAIGAKU KYOIKUGAKUBU, 700, TANNOHARA, OITA-SHI 870-11
JAPANESE

KENRITSU NIIGATA JOSHI TANKI DAIGAKU KENKYU KIYO 7477
(BULLETIN OF THE NIIGATA WOMEN'S COLLEGE)
KENRITSU NIIGATA JOSHI TANKI DAIGAKU
EBIGASE, NIIGATA-SHI 950
JAPANESE AND ENGLISH

KINJO GAKUIN DAIGAKU RONSHU, KASEIGAKUHEN 7478
(PAPERS OF THE DEPARTMENT OF HOME ECONOMICS, KINJO GAKUIN UNIVERSITY)
JAPANESE

KITA NIHON GAKUIN DAIGAKU JOSHI TANKI DAIGAKUBU KIYO 7479
(BULLETIN OF KITA NIHON GAKUIN UNIVERSITY WOMEN'S COLLEGE)
KITA NIHON GAKUIN UNIVERSITY
JAPANESE

KOBE KAISEI JOSHI GAKUIN DAIGAKU, TANKI DAIGAKU KENKYU KIYO 7480
(RESEARCH BULLETIN OF THE JUNIOR COLLEGE, KOBE KAISEI WOMEN'S COLLEGE)
JAPANESE

KOBE TOKIWA TANKI DAIGAKU KIYO 7481
(BULLETIN OF KOBE TOKIWA JUNIOR COLLEGE)
TOKIWA TANKI DAIGAKU
2-13, OTANI-MACHI, NAGATA-KU, KOBE-SHI 653
JAPANESE AND ENGLISH

KOKA JOSHI DAIGAKU KOKA JOSHI TANKI DAIGAKU KENKYU KIYO 7482
(RESEARCH BULLETIN OF KOKA WOMAN'S UNIVERSITY AND KOKA WOMEN'S JUNIOR COLLEGE)
JAPANESE

KOKUGAKUIN DAIGAKU TOCHIGI TANKI DAIGAKU KIYO 7483
(JOURNAL OF TOCHIGI JUNIOR COLLEGE, KOKUGAKUIN UNIVERSITY STUDIES IN THE ARTS AND SCIENCES)
KOKUGAKUIN DAIGAKU TOCHIGI TANKI DAIGAKU
608, HIRAI-MACHI, TOCHIGI-SHI 328
JAPANESE

KOMAZAWA JOSHI TANKI DAIGAKU KENKYA KIYO 7484
(BULLETIN OF THE KOMAZAWA WOMEN'S JUNIOR COLLEGE)
JAPANESE

KONAN KASEI 7485
(KONAN DOMESTIC SCIENCE)
KONAN JOSHI DAIGAKU KASEI KENKYUKAI (KONAN WOMEN'S JUNIOR COLLEGE)
2-23, 6-CHOME, MORI KITA, HIGASHI NADA-KU, KOBE-SHI 658
JAPANESE

KORAN JOSHI TANKI DAIGAKU KENKYU KIYO 7486
(BULLETIN OF KORAN WOMEN'S JUNIOR COLLEGE)
KORAN JOSHI TANKI DAIGAKU
IJIRIBASHI, 1-CHOME, YOKOTE, MINAMI-KU, FUKUOKA-SHI 816
JAPANESE, WITH ENGLISH SUMMARY

KOSHIEN DAIGAKU KIYO 7487
(BULLETIN OF KOSHIEN UNIVERSITY)
KOSHIEN DAIGAKU
10-1, MOMIJIGAOKA, TAKARAZUKA-SHI 665
JAPANESE AND ENGLISH
INDEXED: CHE

KUMAMOTO JOSHI DAIGAKU GAKUJUTSU KIYO 7488
(JOURNAL OF KUMAMOTO WOMEN'S UNIVERSITY)
KUMAMOTO JOSHI DAIGAKU (KUMAMOTO WOMEN'S UNIVERSITY)
7-1, 2-CHOME, OE-MACHI, KUMAMOTO-SHI 862
JAPANESE AND EUROPEAN
INDEXED: CHE, JICST

KURININGU RAIFU 7489
(CLEANING LIFE)
SEIRYUSHA
1-42, KANDA JINBO-CHO, CHIYODA-KU, TOKYO 101
JAPANESE

KYORITSU JOSHI DAIGAKU KASEIGAKUBU KIYO 7490
(BULLETIN OF THE DEPARTMENT OF HOME ECONOMICS, KYORITSU WOMEN'S COLLEGE)
KYORITSU JOSHI DAIGAKU (KYORITSU WOMEN'S COLLEGE)
JAPANESE

KYORITSU JOSHI DAIGAKU TANKI DAIGAKUBU KASEIKA KIYO 7491
(COLLECTED ESSAYS BY THE MEMBERS OF THE FACULTY, KYORITSU WOMEN'S JUNIOR COLLEGE)
KYORITSU JOSHI DAIGAKU TANKI DAIGAKUBU (KYORITSU WOMEN'S JUNIOR COLLEGE)
3-27, KANDA JINBO-CHO, CHIYODA-KU, TOKYO 101
JAPANESE AND EUROPEAN
INDEXED: JICST

KYOTO JOSHI DAIGAKU HIFUKU GAKKAI ZASSHI 7492
(JOURNAL OF CLOTHING SOCIETY, KYOTO WOMEN'S UNIVERSITY)
KYOTO JOSHI DAIGAKU HIFUKU GAKKAI (CLOTHING SOCIETY, KYOTO WOMEN'S UNIVERSITY)
17, IMAKUMANO KITAHIYOSHI-MACHI, HIGASHIYAMA-KU, KYOTO-SHI 605
JAPANESE

KYOTO JOSHI DAIGAKU SHOKUMOTSU GAKKAISHI 7493
(JOURNAL OF FOOD SCIENCE, KYOTO WOMEN'S UNIVERSITY)
KYOTO JOSHI DAIGAKU SHOKUMOTSU GAKKAI (FOOD SCIENCE SEMINAR, KYOTO WOMEN'S UNIVERSITY)
C/O KYOTO JOSHI DAIGAKU KASEIGAKUBU, IMAKUMANO KITA HIYOSHI-MACHI, HIGASHIYAMA-KU, KYOTO-SHI 605
JAPANESE
INDEXED: JICST

KYOTO KASEI TANKI DAIGAKU KENKYU KIYO 7494
(KENKYU KIYO OF THE KYOTO KASEI COLLEGE)
KYOTO KASEI TANKI DAIGAKU (KYOTO KASEI COLLEGE)
80, SENZOKU, MAKISHIMA-CHO, UJI-SHI 611
JAPANESE

KYOTO TANKI DAIGAKU RONSHU 7495
(KYOTO JUNIOR COLLEGE REVIEW)
KYOTO TANKI DAIGAKU
NISHI KOTANIGAOKA, FUKUCHIYAMA-SHI 620
JAPANESE

MATSUSAKA JOSHI TANKI DAIGAKU RONSO 7496
(JOURNAL OF MATSUSAKA WOMEN'S JUNIOR COLLEGE)
MATSUSAKA JOSHI TANKI DAIGAKU GAKUJUTSU KENKYUKAI (MATSUSAKA WOMEN'S JUNIOR COLLEGE)
UMEMURA, KUBO-CHO, MATSUZAKA-SHI 515
JAPANESE

MEIWA JOSHI TANKI DAIGAKU KIYO 7497
(BULLETIN OF MEIWA WOMEN'S JUNIOR COLLEGE)
MEIWA JOSHI TANKI DAIGAKU KYOMUBU
3, SHOWA-MACHI, MAEBASHI-SHI 371
JAPANESE

MEJIRO GAKUEN JOSHI TANKI DAIGAKU KENKYU KIYO 7498
(RESEARCH BULLETIN OF THE MEJIRO GAKUEN WOMEN'S JUNIOR COLLEGE)
MEJIRO GAKUEN WOMEN'S JUNIOR COLLEGE
JAPANESE

MIMASAKA JOSHI DAIGAKU MIMASAKA TANKI DAIGAKU KENKYU KIYO 7499
(BULLETIN OF MIMASAKA WOMEN'S COLLEGE MIMASAKA JUNIOR COLLEGE)
MIMASAKA JOSHI DAIGAKU MIMASAKA TANKI DAIGAKU
500, YAMAKITA, TSUYAMA-SHI 708
JAPANESE

MISHIMA GAKUEN JOSHI TANKI DAIGAKU, DOJOSHI DAIGAKU KENKYU HOKOKU 7500
(RESEARCH REPORT OF MISHIMA GAKUEN WOMEN'S JUNIOR COLLEGE)
JAPANESE

MIYAGIGAKUIN JOSHI DAIGAKU SEIKATSU KAGAKU KENKYUSHO KENKYO HOKOKU 7501
(ANNUAL REPORT OF THE INSTITUTE OF LIVING SCIENCES)
MIYAGI GAKUIN JOSHI DAIGAKU SEIKATSU KAGAKU KENKYUSHO (INSTITUTE OF LIVING SCIENCES MIYAGI GAKUIN WOMEN'S COLLEGE)
6-1, 4-CHOME, CHUO, SENDAI-SHI 980
JAPANESE, WITH ENGLISH SUMMARY
INDEXED: JICST

MORIOKA TANKI DAIGAKU KENKYU HOKOKU 7502
(BULLETIN OF THE MORIOKA COLLEGE)
IWATE-KENRITSU MORIOKA TANKI DAIGAKU KENKYU-BU (MORIOKA JUNIOR COLLEGE)
1-48, SUMIYOSHI-CHO, MORIOKA-SHI 020
JAPANESE

MUKOGAWA JOSHI DAIGAKU KIYO, HIFUKU-HEN 7503
(BULLETIN OF MUKOGAWA WOMEN'S UNIVERSITY, CLOTHING SCIENCE)
MUKOGAWA JOSHI DAIGAKU (MUKOGAWA WOMEN'S UNIVERSITY)
6-46, IKEBIRAKI-CHO, NISHINOMIYA-SHI 663
JAPANESE, WITH ENGLISH SUMMARY
INDEXED: JICST

MUKOGAWA JOSHI DAIGAKU KIYO, SHOKUMOTSU-HEN 7504
(BULLETIN OF MUKOGAWA WOMEN'S UNIVERSITY, FOOD SCIENCE)
MUKOGAWA JOSHI DAIGAKU (MUKOGAWA WOMEN'S UNIVERSITY)
6-46, IKEBIRAKI-CHO, NISHINOMIYA-SHI 663
JAPANESE, WITH ENGLISH SUMMARY
INDEXED: JICST

NAGASAKI JOSHI TANKI DAIGAKU KIYO 7505
(BULLETIN OF THE NAGASAKI WOMEN'S JUNIOR COLLEGE)
JAPANESE

NAGASAKI-KENRITSU JOSHI TANKI DAIGAKU KENKYU KIYO 7506
(RESEARCH BULLETIN OF NAGASAKI PREFECTURAL WOMEN'S JUNIOR COLLEGE)
JAPANESE

NAGOYA JOSHI DAIGAKU KIYO 7507
(JOURNAL OF NAGOYA WOMEN'S COLLEGE)
NAGOYA JOSHI DAIGAKU SEIKATSU KAGAKU KENKYUSHO
3-40, SHIOJI-CHO, MIZUHO-KU, NAGOYA-SHI 467
JAPANESE, WITH ENGLISH SUMMARY

NAGOYA-SHIRITSU JOSHI TANKI DAIGAKU KENKYU KIYO 7508
(JOURNAL OF NAGOYA MUNICIPAL WOMEN'S JUNIOR COLLEGE)
NAGOYA-SHIRITSU JOSHI TANKI DAIGAKU GAKUJUTSU KENKYUKAI (SOCIETY FOR SCIENTIFIC RESEARCHES, NAGOYA MUNICIPAL WOMEN'S JUNIOR COLLEGE)
ATAGO-MAE, KITACHIKUSA-CHO, CHIKUSA-KU, NAGOYA-SHI 464
JAPANESE

NAKAMURA GAKUEN KENKYU KIYO 7509
(BULLETIN OF NAKAMURA GAKUEN)
NAKAMURA GAKUEN DAIGAKU, TANKI DAIGAKU KASEI GAKKAI
7-1, 5-CHOME, BEPPU, NISHI-KU, FUKUOKA-SHI 814
JAPANESE AND ENGLISH

NIHON JOSHI DAIGAKU KIYO, KASEIGAKUBU 7510
(JOURNAL OF THE HOME ECONOMICS DEPARTMENT, JAPAN WOMEN'S UNIVERSITY)
NIHON JOSHI DAIGAKU (JAPAN WOMEN'S UNIVERSITY)
8-1, 2-CHOME, MEJIRODAI, BUNKYO-KU, TOKYO 112
JAPANESE, WITH EUROPEAN SUMMARY

NIHON KATEIKA KYOIKU GAKKAISHI 7511
(JOURNAL OF THE JAPAN ASSOCIATION OF HOME ECONOMICS EDUCATION)
JAPAN ASSOCIATION OF HOME ECONONMICS EDUCATION
JAPANESE

NOTORUDAMU SEISHIN JOSHI DAIGAKU KASEI-GAKUBU JIHO 7512
(BULLETIN OF FACULTY OF HOME ECONOMICS, NOTRE DAME SEISHIN WOMEN'S UNIVERSITY)
NOTORUDAMU SEISHIN JOSHI DAIGAKU KASEI GAKUBU
16-9, 2-CHOME, IFUKU-CHO, OKAYAMA-SHI 700
JAPANESE AND ENGLISH

NOTORUDAMU SEISHIN JOSHI DAIGAKU KIYO; SEIKATSU KEIEIGAKU, SHONIGAKU, SHOKUHIN, EIYOGAKU-HEN 7513
(BULLETIN OF THE NOTRE DAME SACRED HEART WOMEN'S JUNIOR COLLEGE, LIFE ADMINISTRATION STUDIES, CHILD WELFARE STUDIES, FOOD AND NUTRITION STUDIES)
JAPANESE

NOTORUDAMU SEISHIN JOSHI TANKI DAIGAKU RANSO 7514
(PAPER OF THE NOTRE DAME SACRED HEART WOMEN'S JUNIOR COLLEGE)
JAPANESE

OBIHIRO OTANI TANKI DAIGAKU KIYO 7515
(BULLETIN OF OBIHIRO OHTANI JUNIOR COLLEGE)
OBIHIRO OTANI TANKI DAIGAKU (OBIHO OHTANI JUNIOR COLLEGE)
MINAMI-20, NISHI-5-JO, OBIHIRO-SHI 080
JAPANESE

OBIRIN TANKI DAIGAKU KIYO: KASEI GAKU-HEN 7516
(JOURNAL OF OBIRIN JUNIOR COLLEGE: DOMESTIC SCIENCE EDITION)
OBIRIN TANKI DAIGAKU (OBIRIN JUNIOR COLLEGE)
JAPANESE

ODAWARA JOSHI TANKI DAIGAKU KENKYU KIYO 7517
(RESEARCH BULLETIN OF ODAWARA WOMEN'S JUNIOR COLLEGE)
JAPANESE

OGAKI JOSHI TANKI DAIGAKU KENKYU KIYO 7518
(RESEARCH BULLETIN OF OGAKI WOMEN'S JUNIOR COLLEGE)
JAPANESE

OKAYAMA-KENRITSU TANKI DAIGAKU KENKYU KIYO 7519
(OKAYAMA JUNIOR COLLEGE ANNUAL REPORT OF STUDIES)
OKAYAMA-KENRITSU TANKI DAIGAKU (OKAYAMA JUNIOR COLLEGE)
1370-2, KAMI IFUKU, OKAYAMA-SHI 700
JAPANESE AND EUROPEAN

OKAZAKI JOSHI TANKI DAIGAKU KENKYU HOKOKU 7520
(RESEARCH REPORT OF THE OKAZAKI WOMEN'S JUNIOR COLLEGE)
JAPANESE

OSAKA JOSHI GAKUEN TANKI DAIGAKU KIYO 7521
(ANNUAL REPORTS OF STUDIES, OSAKA JOSHIGAKUEN JUNIOR COLLEGE)
OSAKA JOSHIGAKUEN TANKI DAIGAKU (OSAKA JOSHIGAKUEN JUNIOR COLLEGE)
8-5, IKUTAMADERA, TENNOJI-KU, OSAKA-SHI 543
JAPANESE AND EUROPEAN

OSAKA JOSHI TANKI DAIGAKU KIYO 7522
(BULLETIN OF OSAKA WOMEN'S JUNIOR COLLEGE)
JAPANESE

OSAKA KUN'EI JOSHI TANKI DAIGAKU KENKYU HOKOKU 7523
(RESEARCH REPORT OF OSAKA WOMEN'S JUNIOR COLLEGE)
JAPANESE

OSAKA SEIKEI JOSHI TANKI DAIGAKU KENKYU KIYO 7524
(BULLETIN OF OSAKA SEIKEI WOMEN'S JUNIOR COLLEGE)
OSAKA SEIKEI JOSHI TANKI DAIGAKU
2-5, AIKAWA NAKA-DORI, HIGASHI YODOGAWA-KU OSAKA-SHI 532
JAPANESE

OSAKA SHIN'AI JOGAKUIN TANKI DAIGAKU KIYO 7525
(BULLETIN OF OSAKA SHIN'AI WOMEN'S JUNIOR COLLEGE)
JAPANESE

OSAKA SHOIN JOSHI DAIGAKU RONSHU 7526
(JOURNAL OF THE DOMESTIC SCIENCE OF SHOIN WOMEN'S COLLEGE)
OSAKA SHOIN JOSHI DAIGAKU (OSAKA SHOIN WOMEN'S COLLEGE)
258, HISHIYA NISHI, HIGASHIOSAKA-SHI 577
JAPANESE

OSAKA-SHIRITSU DAIGAKU KASEIGAKUBU KIYO 7527
(REPORTS OF THE SCIENCE OF LIVING)
OSAKA-SHIRITSU DAIGAKU KASEIGAKUBU (FACULTY OF THE SCIENCE OF LIVING, OSAKA CITY UNIVERSITY)
459, SUGIMOTO-CHO, SUMIYOSHI-KU, OSAKA-SHI 558
JAPANESE, WITH ENGLISH SUMMARY
INDEXED: BIB, CHE, JICST

OSHIMO GAKUEN JOSHI TANKI DAIGAKU KENKYU SHUHO 7528
(BULLETIN OF OSHIMO WOMEN'S COLLEGE)
OSHIMO GAKUEN JOSHI TANKI DAIGAKU (OSHIMO WOMEN'S COLLEGE)
MINAMI SHIMOYASU, GION-MACHI, HIROSHIMA-SHI 731-01
JAPANESE AND ENGLISH

OTANI JOSHI TANKI DAIGAKU KIYO 7529
(BULLETIN OF OTANI WOMEN'S JUNIOR COLLEGE)
JAPANESE

OTSUMA JOSHI DAIGAKU KASEI GAKUBU KIYO 7530
(BULLETIN OF THE FACULTY OF DOMESTIC SCIENCE, OTSUMA WOMAN'S UNIVERSITY)
OTSUMA JOSHI DAIGAKU (OTSUMA WOMAN'S UNIVERSITY)
12, SANBAN-CHO, CHIYODAO-KU, TOKYO 102
JAPANESE, WITH EUROPEAN SUMMARY

PL GAKUEN JOSHI TANKI DAIGAKU KIYO 7531
(BULLETIN OF PL GAKUEN WOMEN'S JUNIOR COLLEGE)
GAKUEN WOMEN'S JUNIOR COLLEGE
JAPANESE

RISSHO JOSHI DAIGAKU TANKI DAIGAKUBU KENKYU KIYO 7532
(RESEARCH BULLETIN OF THE RISSHO WOMEN'S UNIVERSITY JUNIOR COLLEGE DIVISION)
JAPANESE

SAN'YO GAKUEN TANKI DAIGAKU KENKYU RONSHU 7533
(BULLETIN OF SAN'YO GAKUEN WOMEN'S JUNIOR COLLEGE)
SAN'YO GAKUEN TANKI DAIGAKU
2180, HIRAI, OKAYAMA-SHI 703
JAPANESE

SAN'YO JOSHI TANKI DAIGAKU KENKYU KIYO 7534
(BULLETIN OF THE SANYO WOMEN'S COLLEGE)
JAPANESE
INDEXED: JICST

SAPPORO DAIGAKU KYOYOBU SAPPORO DAIGAKU JOSHI TANKI DAIGAKUBU KIYO 7535
(JOURNAL OF FACULTY OF GENERAL EDUCATION AND WOMEN'S JUNIOR COLLEGE)
SAPPORO DAIGAKU KYOYOBU SAPPORO DAIGAKU JOSHI TANKI DAIGAKUBU KIYO
243-2, NISHIOKA, TOYOHIRA-KU, SAPPORO-SHI 062
JAPANESE AND ENGLISH

SAPPORO KORAN JOSHI TANKI DAIGAKU KENKYU KIYO 7536
(BULLETIN OF THE SAPPORO KORAN WOMEN'S JUNIOR COLLEGE)
SAPPORO KORAN WOMEN'S JUNIOR COLLEGE
JAPANESE

SEI KATARINA JOSHI TANKI DAIGAKU KIYO 7537
(RESEARCH BULLETIN OF ST. CATALINA WOMEN'S JUNIOR COLLEGE)
SEI KATARINA JOSHI TANKI DAIGAKU
660, HOJO, HOJO-SHI 799-24
JAPANESE

SEIAN JOSHI TANKI DAIGAKU KIYO 7538
(BULLETIN OF SEIAN WOMEN'S JUNIOR COLLEGE)
JAPANESE

SEIBO JOGAKUIN TANKI DAIGAKU KENKYU KIYO 7539
(RESEARCH BULLETIN OF OUR LADY WOMEN'S GAKUIN JUNIOR COLLEGE)
JAPANESE
INDEXED: JICST

SEIBO JOSHI GAKUIN TANKI DAIGAKU KASEI GAKKA KENKYU KIYO 7540
(RESEARCH BULLETIN OF SACRED MOTHER WOMEN'S JUNIOR COLLEGE, DEPARTMENT OF DOMESTIC SCIENCE)
KASEIGAKU KENKYUJO
JAPANESE

SEIEN, OTSUMA JOSHI DAIGAKU KASEI GAKKAI KENKYU KIKANSHI 7541
(BULLETIN OF THE DOMESTIC SCIENCE RESEARCH INSTITUTE OF OTSUMA WOMEN'S UNIVERSITY)
OTSUMA JOSHI DAIGAKU KASEI GAKKAI (DOMESTIC SCIENCE RESEARCH INSTITUTE OF OTSUMA WOMEN'S UNIVERSITY)
12, SANBAN-CHO, CHIYODA-KU, TOKYO 102
JAPANESE

SEIKA JOSHI TANKI DAIGAKU KIYO 7542
(JOURNAL OF SEIKA WOMEN'S JUNIOR COLLEGE)
SEIKA JOSHI TANKI DAIGAKU
51-1, ONISHIDA, ITATSUKE, HATATA-KU, FUKUOKA 816
JAPANESE

SEIKATSU KAGAKU, FUKUOKA JOSHI DAIGAKU 7543
(SCIENCE OF HUMAN LIFE)
FUKUOKA JOSHI DAIGAKU KASEIGAKUBU (FUKUOKA WOMEN'S UNIVERSITY)
1-1, 1-CHOME, KASUMIGAOKA, HIGASHI-KU, FUKUOKA-SHI 813
JAPANESE, WITH ENGLISH SUMMARY
INDEXED: CHE, JICST

SEIKATSU KAGAKU KENKYU 7544
(SCIENTIFIC RESEARCH OF DAILY LIFE)
KAGOSHIMA JOSHI TANKI DAIGAKU SEIKATSU KAGAKU (KENKYUJO KAGOSHIMA WOMEN'S JUNIOR COLLEGE, INSTITUTE FOR SCIENTIFIC RESEARCH ON DAILY LIFE)
JAPANESE

SEIKATSU KAGAKU KENKYUJO KENKYU HOKOKU 7545
(RESEARCH REPORT OF THE DOMESTIC SCIENCE RESEARCH INSTITUTE)
MIYAGI GAKUIN JOSHI DAIGAKU (MIYAGI GAKUIN WOMEN'S UNIVERSITY)
JAPANESE

SEIKATSU KAGAKU RONSO 7546
(BULLETIN OF DOMESTIC SCIENCE)
SHOIN JOSHI GAKUIN DAIGAKU GAKUJUTSU KENKYUKAI, SHOIN TANKI DAIGAKU GAKUJUTSU KENKYUKAI (SHOIN WOMEN'S UNIVERSITY SCIENTIFIC RESEARCH SOCIETY, SHOIN JUNIOR COLLEGE SCIENTIFIC RESEARCH SOCIETY)
409, HINOKURA, TAMON-CHO, TARUMI-KU, KOBE-SHI
655, 1, NAKAJIMA-DORI, FUKIAI-KU, KOBE-SHI 651
JAPANESE AND ENGLISH
INDEXED: JICST

SEIKATSU KAGAKU, SEIKATSU KAGAKU KENKYUKAI 7547
(DOMESTIC SCIENCE)
SEIKATSU KAGAKU KENKYUKAI (SOCIETY FOR DOMESTIC SCIENCE)
C/O SAN'YO BUILDING, NO. 2, 2-39, IZUMI-CHO, HIGASHI-KU, OSAKA-SHI 540
JAPANESE

SEIKATSU KAGAKU SIRIZU 7548
(DOMESTIC SCIENCE SERIES)
SEIKATSU KAGAKU KENKYUKAI (SOCIETY FOR DOMESTIC SCIENCE)
2-39, IZUMI-CHO, HIGASHI-KU, OSAKA-SHI 540
JAPANESE

SEINAN JOGAKUIN TANKI DAIGAKU KENKYU KIYO 7549
(STUDIES OF SEINAN JOGAKUIN JUNIOR COLLEGE)
SEINAN JOGAKUIN TANKI DAIGAKU (SEINAN JOGAKUIN JUNIOR COLLEGE)
1, IHORI, KOKURA-KU, KITA-KYUSHU-SHI 803
JAPANESE

SEIREI JOSHI TANKI DAIGAKU KIYO 7550
(BULLETIN OF SEIREI WOMEN'S JUNIOR COLLEGE)
JAPANESE

SEITO JOSHI TANKI DAIGAKU KIYO 7551
(MEMOIRS OF SEITO WOMEN'S JUNIOR COLLEGE)
KINKI DAIGAKU SEITO JOSHI TANKI DAIGAKU (SEITO WOMEN'S JUNIOR COLLEGE, KINKI UNIVERSITY)
JAPANESE

SEITOKU GAKUEN TANKI DAIGAKU KIYO 7552
(BULLETIN SEITOKU GAKUEN JUNIOR COLLEGE FOR WOMEN)
SEITOKU GAKUEN TANKI DAIGAKU
531, SAGAMIDAI, MATSUDO-SHI 271
JAPANESE

SEIWA JOSHI DAIGAKU RONSHU 7553
(STUDIES SEIWA WOMEN'S COLLEGE)
SEIWA JOSHI DAIGAKU
7-54, OKADAYAMA, NISHINOMIYA-SHI 662
JAPANESE

SEIWA JOSHI TANKI DAIGAKU KIYO 7554
(BULLETIN OF SEIWA WOMEN'S JUNIOR COLLEGE)
SEIWA JOSHI TANKI DAIGAKU
865, OTA, KISARAZU-SHI 292
JAPANESE

SENDAI SHIRAYURI TANKI DAIGAKU KIYO 7555
(MEMOIRS OF SENDAI SHIRAYURI JUNIOR COLLEGE)
SENDAI SHIRAYURI TANKI DAIGAKU
46, MATSUMORI HONDA, IZUMI-SHI 981-31
JAPANESE AND ENGLISH

SENTAKU NO KAGAKU 7556
(LAUNDRY RESEARCH)
SENTAKU KAGAKU KYOKAI (LAUNDRY RESEARCH ASSOCIATION)
4-14, KAMIYAMA-CHO, SHIBUYA-KU, TOKYO 150
JAPANESE

SHIGA JOSHI TANKI DAIGAKU KENKYU KIYO 7557
(MEMOIRS OF SHIGA WOMEN'S JUNIOR COLLEGE)
SHIGA WOMEN'S JUNIOR COLLEGE
JAPANESE

SHIKOKU JOSHI DAIGAKU, SHIKOKU JOSHI TANKI DAIGAKU KENKYU KIYO 7558
(RESEARCH BULLETIN OF THE SHIKOKU WOMEN'S COLLEGE, SHIKOKU WOMEN'S JUNIOR COLLEGE)
JAPANESE

SHIMANE-KENRITSU SHIMANE JOSHI TANKI DAIGAKU KIYO 7559
(RESEARCH REPORTS OF SHIMANE WOMEN'S JUNIOR COLLEGE)
SHIMANE-KENRITSU SHIMANE JOSHI TANKI DAIGAKU (SHIMANE WOMEN'S JUNIOR COLLEGE)
282-2, NOGIFUKUTOMI-CHO, MATSUE-SHI 690
JAPANESE

SHIRAYURI JOSHI DAIGAKU KENKYU KIYO 7560
(WHITELILY WOMEN'S UNIVERSITY RESEARCH BULLETIN)
JAPANESE

SHITENOJI JOSHI TANKI DAIGAKU KENKYU KIYO 7561
(RESEARCH BULLETIN OF THE SHITENOJI WOMEN'S JUNIOR COLLEGE)
JAPANESE

SHIZUOKA EIWA JOSHI GAKUIN TANKI DAIGAKU KIYO 7562
(BULLETIN OF SHIZUOKA EIWA WOMEN'S GAKUIN JUNIOR COLLEGE)
SHIZUOKA EIWA WOMEN'S GAKUIN JUNIOR COLLEGE
JAPANESE

SHIZUOKA JOSHI TANKI DAIGAKU KENKYU KIYO 7563
(ANNUAL REPORTS OF STUDIES IN HUMANITIES, SOCIAL SCIENCES AND NATURAL SCIENCES, SHIZUOKA WOMEN'S COLLEGE)
SHIZUOKA JOSHI TANKI DAIGAKU (SHIZUOKA WOMEN'S JUNIOR COLLEGE)
3, NUNOHASHI, HAMAMATSU-SHI 432
JAPANESE AND EUROPEAN
INDEXED: JICST

SHODO JOSHI GAKUIN TANKI DAIGAKU KENKYU HOKOKU 7564
(RESEARCH REPORT OF SHODO WOMEN'S GAKUIN JUNIOR COLLEGE)
JAPANESE

SHOEI TANKI DAIGAKU KENKYU KIYO 7565
(ANNUAL REPORTS OF STUDIES, SHOEI JUNIOR COLLEGE)
SHOEI TANKI DAIGAKU
6-30, NAKAYAMATE-DORI, IKUTA-KU, KOBE-SHI 650
JAPANESE

SHOIN JOSHIGAKUIN DAIGAKU SHOIN TANKI DAIGAKU KENKYU KIYO 7566
(SHOIN REVIEW)
SHOIN JOSHI GAKUIN DAIGAKU, SHOIN TANKI DAIGAKU (SHOIN WOMEN'S UNIVERSITY, SHOIN JUNIOR COLLEGE)
409, HINOKURA, TAMON-CHO, TARUMI-KU, KOBE-SHI 655
JAPANESE AND ENGLISH

SHOKEI TANKI DAIGAKU KENKYU KIYO 7567
(RESEARCH BULLETIN OF SHOKEI WOMEN'S JUNIOR COLLEGE)
SHOKEI WOMEN'S JUNIOR COLLEGE
JAPANESE

SHUKUTOKU TANKI DAIGAKU KENKYU KIYO 7568
(RESEARCH BULLETIN OF THE SHUKUTOKU JUNIOR COLLEGE)
JAPANESE

SOAI JOSHI DAIGAKU, SOAI JOSHI TANKI DAIGAKU KENKYU RANSHU 7569
(RESEARCH PAPERS OF THE SOAI WOMEN'S UNIVERSITY AND SOAI WOMEN'S JUNIOR COLLEGE)
SOAI WOMEN' UNIVERSITY AND SOAI WOMEN'S JUNIOR COLLEGE
JAPANESE

SOAI JOSHI DAIGAKU, SOAI JOSHI TANKI DAIGAKU KENKYU RONSHU; KOKUBUN, KASEI GAKKA-HEN 7570
(COLLECTED RESEARCH OF SOAI WOMEN'S UNIVERSITY AND SOAI WOMEN'S JUNIOR COLLEGE; DEPARTMENTS OF LITERATURE AND DOMESTIC SCIENCE)
JAPANESE

SUMIRE JOSHI TANKI DAIGAKU KENKYU KIYO 7571
(RESEARCH BULLETIN OF SUMIRE WOMEN'S JUNIOR COLLEGE)
NAKARISHI FAKUEN SHUPPANBU (NAKARISHI ACADEMY PUBLISHING BUREAU)
JAPANESE

TAKAMATSU TANKI DAIGAKU KENKYU KIYO 7572
(BULLETIN OF TAKAMATSU JUNIOR COLLEGE)
TAKAMATSU TANKI DAIGAKU
960, KASUGA-MACHI, TAKAMATSU-SHI 761-01
JAPANESE

TAMAGAWA GAKUEN JOSHI TANKI DAIGAKU RONSO 7573
(BULLETIN OF TAMAGAWA GAKUEN WOMEN'S JUNIOR COLLEGE)
TAMAGAWA GAKUEN JOSHI TANKI DAIGAKU
TAMAGAWA GAKUEN, MACHIDA-SHI, TOKYO 194
JAPANESE

TOITA JOSHI TANKI DAIGAKU KENKYU NENPO 7574
(BULLETIN OF THE TOITA WOMEN'S JUNIOR COLLEGE)
TOITA GAKUEN SHUPPANKYOKU
38-14, 3-CHOME, SHIBA MINATO-KU, TOKYO 105
JAPANESE AND ENGLISH

TOKAI GAKUEN JOSHI TANKI DAIGAKU KIYO 7575
(ANNUAL REPORT OF THE WOMEN'S COLLEGE OF TOKAI GAKUEN, NAGOYA)
TOKAI GAKUEN JOSHI TANKI DAIGAKU SEIKATSU KAGAKU KENKYUSHO (WOMEN'S COLLEGE OF TOKAI GAKUEN, NAGOYA)
TENPAKU HEISHIN, SHOWA-KU, NAGOYA-SHI 468
JAPANESE

TOKIWA GAKUEN TANKI DAIGAKU KENKYU KIYO 7576
(BULLETIN OF TOKIWA GAKUEN JUNIOR COLLEGE)
TOKIWA GAKUEN TANKI DAIGAKU
430, IGO, MIWADA-CHO, MITO-SHI 310
JAPANESE

TOKIWA JOSHI TANKI DAIGAKU KIYO 7577
(BULLETIN OF TOKIWA WOMEN'S JUNIOR COLLEGE)
JAPANESE

TOKIWAKAI TANKI DAIGAKU KIYO 7578
(BULLETIN OF TOKIWAKAI JUNIOR COLLEGE)
TOKIWAKAI TANKI DAIGAKU
112, HIRANO NODO-CHO, HIGASHI SUMIYOSHI-KU, OSAKA-SHI 546
JAPANESE

TOKIWAMATSU GAKUEN JOSHI TANKI DAIGAKU KENKYU KIYO 7579
(BULLETIN OF TOKIWAMATSU GAKUEN WOMEN'S JUNIOR COLLEGE)
TOKIWAMATSU GAKUEN JOSHI TANKI DAIGAKU
1204, KAMOSHIDA-CHO, MIDORI-KU, YOKOHAMA-SHI 227
JAPANESE

TOKYO GAKUGEI DAIGAKU KIYO, DAI-6-BU, SANGYO, GIJUTSU, KASEI 7580
(BULLETIN OF TOKYO GAKUGEI UNIVERSITY, SERIES 6, TECHNOLOGY AND HOME ECONOMICS)
TOKYO GAKUGEI DAIGAKU (TOKYO GAKUGEI UNIVERSITY)
1-1, 4-CHOME, NUKUIKITA-MACHI, KOGANEI-SHI TOKYO 184
JAPANESE AND ENGLISH

TOKYO KASEI DAIGAKU KENKYU KIYO 7581
(BULLETIN OF THE TOKYO COLLEGE OF DOMESTIC SCIENCE)
TOKYO KASEI DAIGAKU (THE TOKYO COLLEGE OF DOMESTIC SCIENCE)
18-1, 1-CHOME, KAGA, ITABASHI-KU, TOKYO 173
JAPANESE AND ENGLISH
INDEXED: CHE

TOKYO KASEI GAKUIN DAIGAKU KIYO 7582
(JOURNAL OF TOKYO KASEI GAKUIN COLLEGE)
TOKYO KASEI GAKUIN DAIGAKU, TOKYO KASEI GAKUIN TANKI DAIGAKU (TOKYO KASEI GAKUIN COLLEGE, TOKYO KASEI GAKUIN JUNIOR COLLEGE)
22, SANBAN-CHO, CHIYODA-KU, TOKYO 102
JAPANESE, WITH ENGLISH SUMMARY
INDEXED: JICST

TOKYO RISSHO JOSHI TANKI DAIGAKU RONSO 7583
(PAPERS OF TOKYO RISSHO WOMEN'S JUNIOR COLLEGE)
JAPANESE

TOKYO SEITOKU TANKI DAIGAKU KIYO 7584
(BULLETIN OF THE TOKYO SEITOKU JUNIOR COLLEGE)
JAPANESE

TOYO EIWA JOSHI-GAKUIN TANKI DAIGAKU KENKYU KIYO 7585
(RESEARCH BULLETIN OF THE TOYO EIWA JOSHI-GAKUIN JUNIOR COLLEGE)
JAPANESE

TOYO JOSHI TANKI DAIGAKU KIYO 7586
(BULLETIN OF THE TOYO WOMEN'S JUNIOR COLLEGE)
JAPANESE

TOYOKO GAKUEN JOSHI TANKI DAIGAKU KIYO 7587
(BULLETIN OF THE TOYOKO GAKUEN WOMEN'S JUNIOR COLLEGE)
JAPANESE

TSUKUSHI JOGAKUEN TANKI DAIGAKU KIYO 7588
(BULLETIN OF THE TSUKUSHI WOMEN'S JUNIOR COLLEGE)
TSUKUSHI WOMEN'S JUNIOR COLLEGE
JAPANESE

UEDA JOSHI TANKI DAIGAKU KIYO 7589
(BULLETIN OF UEDA WOMEN'S JUNIOR COLLEGE)
JAPANESE

WAYO JOSHI DAIGAKU DAIGAKU KIYO 7590
(JOURNAL OF WAYO WOMEN'S UNIVERSITY)
WAYO JOSHI DAIGAKU (WAYO WOMEN'S UNIVERSITY)
3-1, 2-CHOME, KONODAI, ICHIKAWA-SHI 272
JAPANESE

YAMAGATA JOSHI TANKI DAIGAKU KIYO 7591
(BULLETIN OF YAMAGATA WOMEN'S JUNIOR COLLEGE)
JAPANESE

YAMAGATA-KENRITSU YONEZAWA JOSHI TANKI DAIGAKU KIYO 7592
(BULLETIN OF YAMAGATA PREFECTURAL YONEZAWA WOMEN'S JUNIOR COLLEGE)
YONEZAWA JOSHI TANKI DAIGAKU
JAPANESE

YAMANASHI-KENRITSU JOSHI TANKI DAIGAKU KIYO 7593
(BULLETIN OF YAMANASHI PREFECTURAL JUNIOR COLLEGE FOR WOMEN)
YAMANASHI-KENRITSU JOSHI TANKI DAIGAKU
5, IIDA-MACHI, KOFU-SHI 400
JAPANESE

YASUDA JOSHI DAIGAKU KIYO 7594
(BULLETIN OF YASUDA WOMEN'S COLLEGE)
YASUDA JOSHI DAIGAKU, YASUDA JOSHI TANKI DAIGAKU
(YASUDA WOMEN'S COLLEGE, YASUDA WOMEN'S JUNIOR COLLEGE)
JAPANESE

Management

ANNALS OF THE SCHOOL OF BUSINESS ADMINISTRATION, KOBE UNIVERSITY 7595
(KOBE DAIGAKU KEIEI GAKUBU (SCHOOL OF BUSINESS ADMINISTRATION, KOBE UNIVERSITY))
2, ROKKODAI-MACHI, NADA-KU, KOBE-SHI 657
JAPANESE

BIJINESU REBYU 7596
(BUSINESS REVIEW)
DAIYAMONDOSHA (DIAMOND PUBLISHING COMPANY)
JAPANESE
INDEXED: JICST

BULLETIN OF UNIVERSITY OF OSAKA PREFECTURE, SERIES D: SCIENCES OF ECONOMY, COMMERCE AND LAW 7597
JAPANESE
INDEXED: JICST

CHAMBER 7598
JAPANESE
INDEXED: JICST

DENTSU'S JAPAN MARKETING/ADVERTISING 7599
DENTSU KOKOKU K. K. (SENTSU ADVERTISING LIMITED)
1-11, TSUKIJI, CHUO-KU, TOKYO 104
ENGLISH

EDP NEWS 7600
NIHON KEIEI KAGAKU KENKYUSHO (JAPAN MANAGEMENT SCIENCE RESEARCH INSTITUTE)
28-26, 4-CHOME, MINAMIAOYAMA, MINATO-KU, TOKYO 1C7
JAPANESE

HINSHITSU KANRI 7601
(STATISTICAL QUALITY CONTROL)
NIHON KAGAKU GIJUTSU RENMEI (THE UNION JAPANESE SCIENTISTS AND ENGINEERS)
10-11, 5-CHOME, SENDAGAYA, SHIBUYA-KU, TOKYO 151
JAPANESE
INDEXED: JICST

HR SERIES 7602
JAPANESE
INDEXED: JICST

JIMA NYUZU 7603
(JIMA NEWS)
NIHON KEIEI KOGAKKAI
(JAPAN INDUSTRIAL MANAGEMENT SOCIETY)
JAPANESE

JIMU KANRI 7504
(BUSINESS MANAGEMENT)
NIKKAN KOGYO SHINBUNSHA
8-10, 1-CHOME, KUDANKITA, CHIYODA-KU, TOKYO 102
JAPANESE

JIMU KIKAI 7605
(BUSINESS MACHINE)
NIHON JIMU KIKAI KOGYOKAI (JAPANESE BUSINESS MACHINE ASSOCIATION)
C/O DAI-1-MORI BUILDING, 12-1, 1-CHOME, NISHI-SHINBASHI, MINATO-KU, TOKYO 105
JAPANESE

JIMUKIKI SHINBUN 7606
(PAPER OF BUSINESS INSTRUMENT)
JIMUKIKI SHINBUNSHA (PAPER OF BUSINESS INSTRUMENT COMPANY, LIMITED)
C/O MARUYAMA BUILDING, 2-19, KANDA TSUKASA-CHO, CHIYODA-KU, TOKYO 101
JAPANESE

JOURNAL OF THE OPERATIONS RESEARCH SOCIETY OF JAPAN 7607
NIHON OPERESHONZU RISACHI GAKKAI (OPERATIONS RESEARCH SOCIETY OF JAPAN)
C/O GAKKAI SENTA BUILDING, 4-16, 2-CHOME, YAYOI, BUNKYO-KU, TOKYO 113
ENGLISH
INDEXED: INS, MAT, SCI, JICST

KAGAKU GIJUTSU BUNKEN SOKUHO, KANRI SHISUTEMU-HEN 7608
(CURRENT BIBLIOGRAPHY ON SCIENCE AND TECHNOLOGY, SERIES ON MANAGEMENT SCIENCE AND SYSTEM)
NIHON KAGAKU GIJUTSU JOHO SENTA (JAPAN INFORMATION CENTER OF SCIENCE AND TECHNOLOGY)
5-2, 2-CHOME, NAGATA-CHO, CHIYODA-KU, TOKYO 100
JAPANESE
INDEXED: JICST

KAGAKU KEIEI 7609
(SCIENTIFIC MANAGEMENT)
JAPANESE
INDEXED: JICST

KEIEI KAGAKU 7610
(MANAGEMENT SCIENCE)
NIHON KEIEI KAGAKU (JAPAN MANAGEMENT SCIENCE)
JAPANESE

KEIEI KAIZEN SHIRYO 7611
(MANAGEMENT IMPROVEMENT MATERIALS)
JAPANESE
INDEXED: JICST

KEIEI KENKYU 7612
(MANAGEMENT RESEARCH)
TOKYO RIKA DAIGAKU KOGAKUBU KEIEI KOGAKKA YANO KENKYUSHITSU (TOKYO SCIENCE UNIVERSITY, ENGINEERING FACULTY, OPERATIONS MANAGEMENT DEPARTMENT, YANO RESEARCH ROOM)
JAPANESE

KEIEI KENKYUJO KENKYU HOKOKU 7613
(RESEARCH REPORT OF THE BUSINESS ADMINISTRATION MANAGEMENT RESEARCH INSTITUTE)
TOYO DAIGAKU KEIEI KENKYUJO (TOYO UNIVERSITY BUSINESS ADMINISTRATION MANAGEMENT RESEARCH INSTITUTE)
JAPANESE

KEIEI KIKAIKA SHINPOJIUMU 7614
(SYMPOSIUM ON MANAGEMENT MECHANIZATION)
JAPANESE
INDEXED: JICST

KEIEI KOGAKU 7615
(INDUSTRIAL MANAGEMENT)
HIROSHIMA DAIGAKU KOGAKUBU KEIEI KOGAKKA (HIROSHIMA UNIVERSITY, ENGINEERING FACULTY, OPERATIONS MANAGEMENT DEPARTMENT)
JAPANESE
INDEXED: JICST

KEIEI RONSHU 7616
(JOURNAL OF BUSINESS ADMINISTRATION)
TOYO DAIGAKU KEIEI KENKYUJO (TOYO UNIVERSITY BUSINESS ADMINISTRATION MANAGEMENT RESEARCH INSTITUTE)
JAPANESE

KEIO BIJINESU FORAM 7617
(KEIO BUSINESS FORUM)
JAPANESE
INDEXED: JICST

KENSA NENPO 7618
(ANNUAL REPORT OF INSPECTION)
TSUSHO SANGYO-SHO KOGYOHIN KENSAJO (INDUSTRIAL MANUFACTURES INSPECTION INSTITUTE, MINISTRY OF INTERNATIONAL TRADE AND INDUSTRY
15-1, 6-CHOME, GINZA, CHUO-KU, TOKYO 104
JAPANESE

KIKAN SEISANSEI TOKEI 7619
(QUARTERLY JOURNAL OF PRODUCTIVITY STATISTICS)
NIHON SEISANSEI HONBU SEISANSEI KENKYUSHO (PRODUCTIVITY RESEARCH INSTITUTE, JAPAN PRODUCTIVITY CENTER)
1-3, 3-CHOME, SHIBUYA, SHIBUYA-KU, TOKYO 150
JAPANESE, WITH ENGLISH SUMMARY

KOBE ECONOMIC AND BUSINESS REVIEW, KOBE UNIVERSITY 7620
JAPANESE
INDEXED: JICST

KOJO KANRI 7621
(FACTORY MANAGEMENT)
NIKKAN KOGYO SHINBUNSHA
8-10, 1-CHOME, KUDANKITA, CHIYODA-KU, TOKYO 102
JAPANESE
INDEXED: JICST

KOKUSAI KEIZAI 7622
(INTERNATIONAL ECONOMICS)
JAPANESE
INDEXED: JICST

KYOTO DAIGAKU DEIZAI GAKKAI KEIZZI RONSO 7623
(ECONOMIC PAPERS OF THE KYOTO UNIVERSITY ECONOMICS SOCIETY)
JAPANESE
INDEXED: JICST

KYUSHU DAIGAKU KEIZAI GAKKAI KEIZAI RONKYU 7624
(ECONOMIC PAPERS OF THE KYUSHU UNIVERSITY ECONOMICS SOCIETY)
JAPANESE
INDEXED: JICST

KYUSHU SANGYO DAIGAKU SHOKEI RONSO 7625
(PAPERS ON COMMERCIAL ECONOMICS OF KYUSHU INDUSTRIAL COLLEGE)
JAPANESE
INDEXED: JICST

MANEJIMENTO 7626
(MANAGEMENT)
NIHON NARITSU KYOKAI (JAPAN EFFICIENCY ASSOCIATION)
JAPANESE
INDEXED: JICST

MANEJIMENTO GAIDO 7627
(MANAGEMENT GUIDE)
JAPANESE
INDEXED: JICST

MANEJIMENTO JANARU 7628
(MANAGEMENT JOURNAL)
JAPANESE
INDEXED: JICST

MANEJIMENTO JITSUMU 7629
(MANAGEMENT PRACTICE)
JAPANESE
INDEXED: JICST

MANEJIMENTO NYUZU 7630
(MANAGEMENT NEWS)
JAPANESE
INDEXED: JICST

MRC KONSARUTANTO MANSURI 7631
(MRC CONSULTANT MONTHLY)
MARUNOUCHI RISACHI SENTA (MARUNOUCHI RESEARCH CENTER)
C/O OTEMACHI BUILDING, 1-4, OTEMACHI, CHIYODA-KU, TOKYO 100
JAPANESE

NIHON JIMUKI SHINBUN 7632
(JAPAN BUSINESS MACHINE NEWS)
NIHON JIMUKI SHINBUNSHA (JAPAN BUSINESS MACHINE NEWS, COMPANY, LIMITED)
1-33, EDOBORI KITADORI, NISHI-KU, OSAKA-SHI 550
JAPANESE
INDEXED: JICST

NIPPON KEIEI KOGAKU KAISHI 7633
(JOURNAL OF JAPAN INDUSTRIAL MANAGEMENT ASSOCIATION)
NIHON KOGYO KEIEI GAKKAI (JAPAN INDUSTRIAL MANAGEMENT ASSOCIATION)
C/O KINOKUNIYA BUILDING, KEIO KOGAKKAI, 17-7, 3-CHOME, SHINJUKU, SHINJUKU-KU, TOKYO 160
JAPANESE, WITH ENGLISH SUMMARY
INDEXED: JICST

NIPPON KEIEI KOGAKUKAI, SHUKI KENKYU HAPPYOKAI YOKOSHU 7634
(PREPRINT OF THE AUTUMN SYMPOSIUM OF JAPAN INDUSTRIAL MANAGEMENT ASSOCIATION)
JAPANESE
INDEXED: JICST

NIPPON KEIEI KOGAKUKAI, SHUNKI KENKYU HAPPYOKAI YOKOSHU 7635
(PREPRINT OF THE SPRING SYMPOSIUM OF THE JAPAN INDUSTRIAL MANAGEMENT ASSOCIATION)
JAPANESE
INDEXED: JICST

NIPPON OPERSHONZU, RISACHI GAKKAI, SHUNKI KENKYU HAPPYOKAI ABUSUTORAKUTO-SHU 7636
(ABSTRACTS OF THE SPRING SYMPOSIUM OF THE JAPAN OPERATIONS RESEARCH ASSOCIATION)
JAPANESE
INDEXED: JICST

OKAYAMA DAIGAKU SANGYO KEIEI KENKYUKAI KENKYU HOKOKUSHO 7637
(RESEARCH PAPERS OF OKAYAMA UNIVERSITY INDUSTRIAL MANAGEMENT RESEARCH SOCIETY)
JAPANESE

OKINAWA SEISANSEI 7638
(OKINAWA PRODUCTIVITY)
OKINAWA SEISANSEI HONBU (OKINAWA PRODUCTIVITY HEADQUARTERS)
JAPANESE

OPERESHIONZU RISACHI 7639
(OPERATIONS RESEARCH AS A MANAGEMENT SCIENCE)
NIHON KAGAKU GIJUTSU RENMEI (UNION OF JAPANESE SCIENTISTS AND ENGINEERS)
10-11, 5-CHOME, SENDAGAYA, SHIBUYA-KU, TOKYO 151
JAPANESE
INDEXED: JICST

PUREJIDENTO 7640
(PRESIDENT)
JAPANESE
INDEXED: JICST

RIDAI KEIEI KAGAKU KENKYU 7641
(RESEARCH ON MANAGEMENT AND INDUSTRY BY THE TOKYO COLLEGE OF SCIENCE)
TOKYO RIKA DAIGAKU KOGAKUBU KEIEI KOGAKUKA (TOKYO COLLEGE OF SCIENCE, DEPARTMENT OF ENGINEERING, MANAGEMENT AND INDUSTRY SECTION)
JAPANESE

SANGYO KEIEI 7642
(INDUSTRIAL MANAGEMENT)
WASEDA DAIGAKU SANGYO KEIEI KENKYUJO (WASEDA UNIVERSITY INDUSTRIAL MANAGEMENT RESEARCH INSTITUTE)
JAPANESE

SANGYO KEIEI KENKYU 7643
(INDUSTRIAL MANAGEMENT RESEARCH)
MEIJO DAIGAKU SHOGAKKAI (MEIJO UNIVERSITY SOCIETY FOR THE STUDY OF COMMERCE)
JAPANESE

SANGYO KEIEI KENKYU SHOHO 7644
(REPORT OF THE INDUSTRIAL MANAGEMENT RESEARCH INSTITUTE)
KYUSHU SANGYO DAIGAKU (KYUSHU INDUSTRIAL UNIVERSITY)
JAPANESE
INDEXED: JICST

SANGYO KEIRI 7645
(INDUSTRIAL ACCOUNTING)
JAPANESE
INDEXED: JICST

SANGYO KUNREN 7646
(INDUSTRIAL TRAINING)
NIHON SANGYO KUNREN KYOKAI (JAPAN INDUSTRIAL VOCATIONAL TRAINING ASSOCIATION)
17-3, 2-CHOME, SHIBUYA, SHIBUYA-KU, TOKYO 150
JAPANESE
INDEXED: JICST

SANGYO NORITSU 7647
(INDUSTRIAL EFFICIENCY)
OSAKA NORITSU KYOKAI (OSAKA MANAGEMENT ASSOCIATION)
19, BUNGO-CHO, HIGASHI-KU, OSAKA-SHI 540
JAPANESE
INDEXED: JICST

SENSHU KEIEI KENKYU NENPO 7648
(SENSHU MANAGEMENT RESEARCH ANNUAL)
SENSHU DAIGAKU KEIEI KENKYUJO (SENSHU UNIVERSITY MANAGEMENT RESEARCH INSTITUTE)
JAPANESE

SHIZAI KANRI 7649
(PURCHASING AND MATERIALS MANAGEMENT)
NIHON SHIZAI KANRISHI KYOKAI (JAPAN MATERIALS MANAGEMENT ASSOCIATION)
C/O TERASHIMA BUILDING, 2-1, NIHONBASHI HON-CHO, CHUO-KU, TOKYO 103
JAPANESE
INDEXED: JICST

SHOTEN TO KOJO 7650
(STORES AND FACTORIES)
KANAGAWA-KEN SHOKO SHIDO SENTA (COMMERCE AND INDUSTRY COUNSULTING INSTITUTE FOR THE KANAGAWA PREFECTURAL GOVERNMENT)
16-5, 2-CHOME, TSURUYA-CHO, KANAGAWA-KU, YOKOHAMA-SHI 221
JAPANESE

SOKA KEIEI RONSHU 7651
(REVIEW OF BUSINESS ADMINISTRATION)
SOKA DAIGAKU KEIEI GAKKAI (SOKA UNIVERSITY MANAGEMENT SOCIETY)
JAPANESE

SOKO TOKEI GEPPO 7652
(MONTHLY REPORT OF WAREHOUSE STATISTICS)
UN'YU-SHO KOWAN-KYOKU (PORT BUREAU, MINISTRY OF TRANSPORT)
1-3, 2-CHOME, KASUMIGASEKI, CHIYODA-KU, TOKYO 100
JAPANESE

SOSHIKI KAGAKU 7653
(SYSTEMS SCIENCE)
JAPANESE
INDEXED: JICST

SYSTEMS 7654
JAPANESE
INDEXED: JICST

Typography and Printing

ASIAN PRINTER 7655
INSAТSU GAKKAI SHUPPAN-BU (PUBLISHING COMPANY, LIMITED OF THE TECHNICAL ASSOCIATION OF GRAPHIC ARTS OF JAPAN)
6-2, 5-CHOME, GINZA, CHUO-KU, TOKYO 104
ENGLISH
INDEXED: JICST

BESSATSU PURINTINGU INFOMESHON 7656
(PRINTING INFORMATION (SPECIAL ISSUE))
NIHON INSATSU GIJUTSU KYOKAI (JAPANESE PRINTING TECHNOLOGY ASSOCIATION)
JAPANESE

BULLETIN OF THE TECHNICAL ASSOCIATION OF GRAPHIC ARTS OF JAPAN, OVERSEAS 7657
NIPPON INSATSU GAKKAI (TECHNICAL ASSOCIATION OF GRAPHIC ARTS OF JAPAN)
16-8, 1-CHOME, SHINTOMI, CHUO-KU, TOKYO 104
ENGLISH

HAMADA NYUSU 7658
(HAMADA NEWS)
HAMADA INSATSUKI SEIZOJO (PRINTING MACHINERY WORKS)
JAPANESE

INSATSU BUNKEN ABUSUTORAKUTO 7659
(GRAPHIC ARTS ABSTRACTS)
NIHON INSATSU GIJUTSU KYOKAI (JAPAN ASSOCIATION OF GRAPHIC ARTS TECHNOLOGY)
1-29, WADA, SUGINAMI-KU, TOKYO 166
JAPANESE AND ENGLISH

INSATSU JANARU 7660
(PRINTING JOURNAL)
INSATSU JIHOSHA
9-9, 1-CHOME, UENO, TAITO-KU, TOKYO 110
JAPANESE
INDEXED: CHE

INSATSU JIHO 7661
(MONTHLY NEWS OF PRINTING)
INSATSU JIHOSHA
14-14, TAMATSUKURI MOTO-MACHI, TENNOJI-KU, OSAKA-SHI 543
JAPANESE

INSATSU JOHO 7662
(GRAPHIC ARTS TECHNICAL EDUCATION AND INFORMATION)
INSATSU SHUPPAN KENKYUSHO (PRINTING AND PUBLISHING RESEARCH INSTITUTE)
3-21, KANDA NISHIKI-CHO, CHIYODA-KU, TOKYO 101
JAPANESE
INDEXED: CHE

INSATSU KURABU 7663
(PRINTING REVIEW)
ASAHI SUPPAN K.K.
13-17, 2-CHOME, GINZA, CHUO-KU, TOKYO 104
JAPANESE

INSATSU SHINPO 7664
(PRINTING NEWS)
INSATSU SHUPPAN KENKYUSHO
3-21, KANDA NISHIKI-CHO, CHIYODA-KU, TOKYO 101
JAPANESE

INSATSU ZASSHI 7665
(JAPAN PRINTER)
INSATSU GAKKAI SHUPPAN-BU (TECHNICAL ASSOCIATION OF GRAPHIC ARTS OF JAPAN)
C/O SHIPPO BUILDING, 6-2, 5-CHOME, GINZA, CHUO-KU, TOKYO 104
JAPANESE
INDEXED: JICST

INSATSU-KYOKU KENKYUSHO HOKOKU 7666
(RESEARCH BULLETIN OF THE GOVERNMENT PRINTING BUREAU)
OKURA-SHO INSATSU-KYOKU KENKYUSHO (THE RESEARCH INSTITUTE, GOVERNMENT PRINTING BUREAU, MINISTRY OF FINANCE)
6-1, 1-CHOME, OJI, KITA-KU, TOKYO 114
JAPANESE, WITH EUROPEAN SUMMARY
INDEXED: CHE, JICST

INSATSU-KYOKU KENKYUSHO JIHO 7667
(BULLETIN OF THE RESEARCH INSTITUTE, GOVERNMENT PRINTING BUREAU)
OKURA-SHO INSATSU-KYOKU KENKYUSHO (THE RESEARCH INSTITUTE, GOVERNMENT PRINTING BUREAU, MINISTRY OF FINANCE)
6-1, 1-CHOME, OJI, KITA-KU, TOKYO 114
JAPANESE

INSATSUKAI 7668
(PRINTING WORLD)
NIHON INSATSU SHINBUNSHA (JAPAN PRINTING NEWS PUBLISHING COMPANY)
16-16, 2-CHOME, GINZA, CHUO-KU, TOKYO 104
JAPANESE
INDEXED: CHE, JICST

KAMI TO INSATSU 7669
(PAPER AND PRESS)
KYUSHU INSATSU BUNKA SHUPPANSHA
2-25, 1-CHOME, MAIZURU, CHUO-KU, FUKUOKA-SHI 810
JAPANESE

KEI INSATSU 7670
(LIGHT PRINTING)
NIHON KEI INSATSU KOGYO KAI (JAPAN SOCIETY OF LIGHT PRINTING INDUSTRIES)
JAPANESE

KUKI 7671
(AIR)
KUKI INSATSU KYOKU (AIR PRINTING BUREAU)
JAPANESE

NIHON INSATSU GAKKAI RONBUNSHU 7672
(BULLETIN OF THE TECHNICAL ASSOCIATION OF GRAPHIC ARTS OF JAPAN)
NIHON INSATSU GAKKAI (TECHNICAL ASSOCIATION OF GRAPHIC ARTS OF JAPAN)
16-8, 1-CHOME, SHINTOMI-CHO, CHUO-KU, TOKYO 104
JAPANESE, WITH EUROPEAN SUMMARY
INDEXED: CHE, JICST

NIHON INSATSU SHINBUN 7673
(JAPAN PRINTING NEWS)
NIHON INSATSU SHINBUNSHA
16-16, 2-CHOME, GINZA, CHUO-KU, TOKYO 104
JAPANESE

NIHON INSATSU SHINPO 7674
(PRINTING PRESS)
INSATSU SHUPPAN KENKYUSHO (PRINTING AND PUBLISHING RESEARCH INSTITUTE)
3-21, KANDA NISHIKI-CHO, CHIYODA-KU, TOKYO 101
JAPANESE

NIHON NO INSATSU 7675
(JAPANESE PRINTING)
NIHON INSATSU KOGYO KUMIAI RENGOKAI (ASSOCIATION FOR JAPAN PRINTING INDUSTRY)
16-8, 1-CHOME, SHINTOMI-CHO, CHUO-KU, TOKYO 104
JAPANESE

OKURASHO INSATSU KYOKU JIHO 7676
(REVIEW OF THE PRINTING BUREAU, MINISTRY OF FINANCE)
JAPANESE

P & P TEKUNIAKARU DAIJESUTO 7677
(PHOTO PRINT TECHNICAL DIGEST)
INSATSU SHASHIN KENKYUSHO
10-27, 2-CHOME, HIMON'YA, MEGURO-KU, TOKYO 152
JAPANESE

PRINTING INFORMATION DIGEST 7678
INSATSU KOGYO KAI (PRINTING INDUSTRY ASSOCIATION)
JAPANESE AND ENGLISH

PURANNA 7679
(PLANNER)
INSATSU SHUPPAN KENKYUJO (PRINTING AND PUBLISHING INSTITUTE)
3-21, KANDA NISHIKI-CHO, CHIYODA-KU, TOKYO 101
JAPANESE

PURINTAZU SAKURU 7680
(PRINTER'S CIRCLE)
NIHON INSATSU GIJUTSU KYOKAI (JAPAN PRINTING TECHNOLOGY ASSOCIATION)
JAPANESE

PURINTINGU INFORMEISHON 7681
(PRINTING INFORMATION)
NIHON INSATSU GIJUTSU KYOKAI
29-11, 1-CHOME, WADA, SUGINAMI-KU, TOKYO 166
JAPANESE

RICOH TEKUNKARU REBYU 7682
(RICOH TECHNICAL REVIEW)
RIKO GIJUTSU JOHO SENTA
3-6, 1-CHOME, NAKAMAGOME, OTA-KU, TOKYO 143
JAPANESE
INDEXED: JICST

SHINBUN GIJUTSU JOHO 7683
(NEWSPAPER TECHNICAL INFORMATION)
NIHON SHINBUN KYOKAI (NEWSPAPER SOCIETY OF JAPAN)
1-3, HIBIYA KOEN, CHIYODA-KU, TOKYO 100
JAPANESE

SHINBUN INSATSU GIJUTSU 7684
(NEWSPAPER PRINTING TECHNIC)
NIHON SHINBUN KYOKAI (JAPAN NEWSPAPER PUBLISHER AND EDITOR'S ASSOCIATION)
1-3, HIBIYA KOEN, CHIYODA-KU, TOKYO 100
JAPANESE

SHUPPAN GEPPO 7685
(PUBLISHERS MONTHLY)
JAPANESE
INDEXED: JICST

TOKYO NO INSATSU 7686
(PRINTING IN TOKYO)
TOKYO-TO INSATSU KOGYO KUMAI (TOKYO PRINTER'S COORDINATION UNION)
16-8, 1-CHOME, SHINTOMI-CHO, CHUO-KU, TOKYO 104
JAPANESE

TOPPAN INSATSU K.K. GIJUTSU KENKYUSHO KENKYU HOKOKU 7687
(RESEARCH REPORT OF THE TECHNICAL INSTITUTE, TOPPAN PRINTING COMPANY, LIMITED)
TOPPAN INSATSU K.K. GIJUTSU KENKYUSHO (TECHNICAL INSTITUTE, TOPPAN PRINTING COMPANY, LIMITED)
2-15, 1-CHOME, SUIDO, BUNKYO-KU, TOKYO 112
JAPANESE

Transport

ANZEH UNTEN KANRI 7688
(SAFE DRIVING MANAGEMENT)
TOKYO KOTSU ANZEN KYOKAI
(TOKYO TRAFFIC SAFETY ASSOCIATION)
JAPANESE

BUTSURYU SHINBUN 7689
(PHYSICAL DISTRIBUTION NEWS)
NIPPO
4-5, 4-CHOME, IIDABASHI, CHIYODA-KU, TOKYO 102
JAPANESE

CURRENT: KAIUN CHOSA 7690
(CURRENT: SEA TRANSPORT SURVEY)
CHORYUSHA (THE CURRENT COMPANY)
JAPANESE

DAI- 4 KANKU KAIJO HOAN HONBUHO 7691
(REPORT OF THE HEADQUARTERS FOR MARITIME SAFETY, FOURTH DISTRICT)
JAPANESE

DAI-11 KANKU KAIJO HOAN HONBUHO 7692
(REPORT OF THE HEADQUARTERS OF THE 11TH MARITIME SAFETY DISTRICT)
JAPANESE

DENPA KOHO 7693
(ELECTRONIC NAVIGATION REVIEW)
KENPA KOHO KENKYUKAI (THE JAPANESE COMMITTEE FOR RADIO AIDS TO NAVIGATION)
C/O KAIJO HOAN-CHO TODAI-BU, 1-3, 2-CHOME, KASUMIGASEKI, CHIYODA-KU, TOKYO 100
JAPANESE, WITH ENGLISH SUMMARY

FRP GYOSEN 7694
(FRP FISHING BOAT)
FRP GYOSEN KENKYO KAI (FRP SOCIETY OF FISHING BOAT RESEARCH)
JAPANESE

FUKUOKA HANSO 7695
(FUKUOKA CONVEYANCE)
FUKUOKA HANSO TSUSHINBU (FUKUOKA CONVEYANCE INFORMATION DEPARTMENT)
JAPANESE

GAIKOKU KAIJI JOHO 7696
(FOREIGN MARITIME INFORMATION)
KAIJI SANGYO KENKYUSHO (JAPAN MARITIME RESEARCH INSTITUTE)
6-4, 2-CHOME, HIRAKAWA-CHO, CHIYODA-KU, TOKYO 102
JAPANESE

GAIKOKU KOTSU CHOSA SHIRYO 7697
(OVERSEAS INFORMATION OF TRAFFIC AND TRANSPORTATION)
NIHON KOKUYU TETSUDO GAIMU-BU (JAPANESE NATIONAL RAILWAYS)
6-5, 1-CHOME, MARUNOUCHI, CHIYODA-KU, TOKYO 100
JAPANESE

GEKKAN KOTSU KOGAI 7698
(MONTHLY JOURNAL OF TRAFFIC NUISANCE)
GEKKAN KOTSU KOGAI HENSHU-BU
14-4, 3-CHOME, KANDA OGAWA-CHO CHIYODA-KU, TOKYO 101
JAPANESE

GEKKAN KOTSU. KEISATSU CHO KOTSUKYOKU-HEN 7699
(TRAFFIC MONTHLY. NATIONAL POLICE AGENCY, TRAFFIC BUREAU EDITION)
TOKYO HOREI SHUPPAN
(TOKYO LEGAL PUBLISHING COMPANY)
JAPANESE

GEKKAN TEION RYUTSU 7700
(FIXED TEMPERATURE DISTRIBUTION MONTHLY)
RYUTSU SHISUTEMU KENKYU SENTA
(DISTRIBUTION SYSTEM RESEARCH CENTER)
JAPANESE

GEKKAN TORAKKU KAI 7701
(MONTHLY TRUCK WORLD)
UCHIYAMA RYUTSU KEIZAI KENKYUSHO
(UCHIYAMA DISTRIBUTION ECONOMICS RESEARCH INSTITUTE)
JAPANESE

GIJUTSU SHIRYO (KAIJO HOAN-CHO) 7702
(TECHNOLOGICAL MATERIALS, MARITIME SAFETY AGENCY)
KAIJO HOAN-CHO KEIBI KYUNAN-BU (THE GUARD AND RESCUE DEPARTMENT, MARITIME SAFETY AGENCY)
2-1, 1-CHOME, KASUMIGASEKI, CHIYODA-KU, TOKYO 100
JAPANESE

HAN KOTSU 7703
(PAN-TRANSPORTATION)
NIHON KOTSU KYOKAI (JAPAN TRAFFIC ASSOCIATION)
C/O SHIN KOKUSAI BUILDING, 9TH FLOOR, 3-4, MARUNOUCHI, CHIYODA-KU, TOKYO 100
JAPANESE
INDEXED: JICST

HANDORINGU EJI 7704
(HANDLING AGE)
JAPANESE
INDEXED: JICST

IATSS RESEARCH 7705
KOKUSAI KOTSU ANZEN GAKKAI (INTERNATIONAL TRAFFIC SAFETY ASSOCIATION)
ENGLISH

IATSS REVIEW 7706
KOKUSAI KOTSU ANZEN GAKKAI (INTERNATIONAL TRAFFIC SAFETY ASSOCIATION)
JAPANESE AND ENGLISH

IWATE NO KOTSU TOKEI 7707
(TRAFFIC STATISTICS IN IWATE PREFECTURE)
IWATE-KEN KEISATSU HONBU
10-1, UCHIMARU, MORIOKA-SHI 020
JAPANESE

JAPANESE NATIONAL RAILWAYS. FACTS AND FIGURES 7708
JAPANESE NATIONAL RAILWAYS
6-5, 1-CHOME, MARUNOUCHI, CHIYODA-KU, TOKYO,JAPAN
JAPANESE

JAPANESE NATIONAL RAILWAYS. QUARTERLY BULLETIN 7709
JAPANESE NATIONAL RAILWAYS, INTERNATIONAL DEPARTMENT
1-6-5 MARUNOUCHI, CHIYODA-KU, TOKYO 100
JAPANESE

JIDOSHA YUSO TOKEI HOKOKUSHIO 7710
(ANNUAL REPORT OF AUTOMOBILE TRANSPORTATION STATISTICS)
UN'YU-SHO DAIJIN KANBO JOHO KANRI-BU
(DEPARTMENT OF RESEARCH AND DATA PROCESSING, MINISTER'S SECRETARIAT, MINISTRY OF TRANSPORT)
C/O NO. 3, GODO CHOSHA, 1-3, 2-CHOME, KASUMIGASEKI, CHIYODA-KU, TOKYO 100
JAPANESE

JUNKAN KOTSU 7711
(TRANSPORTATION NEWS)
KOTSU SHUPPANSHA
1-1, KANDA IZUMI-CHO, CHIYODA-KU, TOKYO 101
JAPANESE

KAGAKU GIJUTSU SHIKEN KENKYU HOJOKIN NI YORU SHIKEN KENKYU SEIKASHU 7712
(REPORTS ON EXPERIMENTS MADE WITH GRANTS FOR SCIENTIFIC AND TECHNOLOGICAL STUDIES)
UN'YUSHO (MINISTRY OF TRANSPORT)
2-1, 1-CHOME, KASUMIGASEKI, CHIYODA-KU, TOKYO 100
JAPANESE

KAGAKU KEISATSU KENKYUSHO HOKOKU, KOTSU-HEN 7713
(REPORTS OF THE NATIONAL RESEARCH INSTITUTE OF POLICE SCIENCE, RESEARCH ON TRAFFIC SAFETY AND REGULATION)
KEISATSU-CHO KAGAKU KEISATSU KENKYUSHO (THE NATIONAL RESEARCH INSTITUTE OF POLICE SCIENCE, NATIONAL POLICE AGENCY)
6, SANBAN-CHO, CHIYODA-KU, TOKYO 102
JAPANESE, WITH EUROPEAN SUMMARY
INDEXED: CHE, JICST

KAGAWA NO KOTSU 7714
(TRAFFIC SAFETY AND REGULATION IN KAGAWA)
KAGAWA-KEN KEISATSU HONBU
1-10, 4-CHOME, BAN-CHO, TAKAMATSU-SHI 760
JAPANESE

KAIGAI DAIGAKKO KENKYU HOKOKU 7715
(REVIEW OF THE MARINE TECHNICAL COLLEGE)
UN'YU-SHO KAIGI KAIGAKKO (MARINE TECHNICAL COLLEGE, MINISTRY OF TRANSPORT)
12-24, NISHIKURA-CHO, ASHIYA-SHI 659
JAPANESE, WITH EUROPEAN SUMMARY
INDEXED: JICST

KAIHATSU GIHO 7716
(DEVELOPMENT TECHNICAL REPORT)
KAIJO HOAN-CHO SEMPAKU GIJUTSU-BU (MARITIME SAFETY AGENCY SHIPPING TECHNOLOGY DIVISION)
JAPANESE

KAIJI KOTSU KENKYU 7717
(MARINE TRANSPORTATION RESEARCH)
KAIJI KOTSU BUNKA KENKYUJO
JAPANESE

KAIJI SANGYO KENKYUSHOHO 7718
(JOURNAL OF THE JAPAN MARITIME RESEARCH INSTITUTE)
KAIJI SANGYO KENKYUSHO (JAPAN MARITIME RESEARCH INSTITUTE)
6-4, 2-CHOME, HIRAKAWA-CHO, CHIYODA-KU, TOKYO 102
JAPANESE

KAIJI SHINKO RENMEI KAIHO 7719
(BULLETIN OF MARINE AFFAIRS PROMOTION LEAGUE)
KAIJI SHINKO RENMEI
6-4, 2-CHOME, HIRAKAWA-CHO CHIYODA-KU, TOKYO 102
JAPANESE

KAIJISHI KENKYU 7720
(JOURNAL OF THE HISTORY OF MARITIME)
NIHON KAIJISHI GAKKAI (JAPANESE SOCIETY OF THE HISTORY OF MARITIME)
23-17, 1-CHOME, SHINKAWA, CHUO-KU, TOKYO 104
JAPANESE

KAIJO HOAN DAIGAKKUO KENKYU HOKOKU DAI-ICHIBU 7721
(RESEARCH REPORT OF MARITIME SAFETY ACADEMY PART I)
KAIJO HOAN DAIGAKKO MARITIME SAFETY ACADEMY
JAPANESE

KAIJO HOAN TOKEI NENPO 7722
(ANNUAL STATISTICS OF MARITIME SAFETY)
KAIJO HOAN-CHO SOMU-BU (ADMINISTRATIVE DEPARTMENT MARITIME SAFETY AGENCY)
2-1, 1-CHOME, KASUMIGASEKI, CHIYODA-KU, TOKYO 100
JAPANESE

KAIJO YUSHO NO GENKYO, GAIKO YUSO JISSEKI NENPO 7723
(ANNUAL CONDITION OF SHIPPING, STATISTICAL YEARBOOK OF TRANSPORTATION BY OCEAN NAVIGATION)
KAIJI SANGYO KENKYUSHO (JAPAN MARITIME RESEARCH INSTITUTE)
2-2, 1-CHOME, UCHISAIWAI-CHO, CHIYODA-KU, TOKYO 100
JAPANESE

KAIUN CHOSA GEPPO 7724
(MONTHLY REVIEW OF SHIPPING)
OSAKA SHOSEN MITSUI SENPAKU K. K.
(MITSUI OSK LINES LIMITED)
3-3, 5-CHOME, AKASAKA, MINATO-KU, TOKYO 107
JAPANESE

KAIUN SHIRYO 7725
(SHIPPING INFORMATION)
NIHON SENSHU KYOKAI KIKAKUSHITSU (THE JAPANESE SHIPOWNERS' ASSOCIATION)
6-4, 2-CHOME, HIRAKAWA-CHO, CHIYODA-KU, TOKYO 102
JAPANESE

KAIUN TAIMUSU 7726
(KAIUN TIMES)
KAIUN TAIMUSUSHA
C/O KYODO BUILDING, 1-3, NIHONBASHI YOSHI-CHO, CHUO-KU, TOKYO 103
JAPANESE

KAIUN TOKUHO 7727
(SHIPPING NEWS)
NIHON KAIUN SHUKAIJO (THE NIHON SHIPPING EXCHANGE)
2-1, NIHONBASHI MURO-MACHI, CHUO-KU, TOKYO 103
JAPANESE

KAMOTSU 7728
(RAILWAY FREIGHT)
TETSUDO KAMOTSU KYOKAI (RAILWAY FREIGHT ASSOCIATION)
8-3, 1-CHOME, MARUNOUCHI, CHIYODA-KU, TOKYO 100
JAPANESE

KAMOTSU FOTTO 7729
(FREIGHT PHOTO)
NIKKAN NUYU TAIMUSUSHA (DAILY TRANSPORTATION TIMES)
JAPANESE

KAMOTSU KYOSHITSU 7730
(GOODS CIRCULATION)
KOTSU NIHONSHA
5-5, 2-CHOME, UCHIKANDA, CHIYODA-KU, TOKYO 101
JAPANESE

KANTO KAIUN-KYOKU GYOMU GAIYO 7731
(OUTLINE OF OPERATION, KANTO SHIPPING BUREAU)
UN'YU-SHO KANTO KAIUN-KYOKU (KANTO SHIPPING BUREAU, MINISTRY OF TRANSPORT)
SHINMINATO-MACHI, NAKA-KU, YOKOHAMA-SHI 231
JAPANESE

KEIRYO KEIKAKU KENKYUSHO REPOTO ABUSUTOR AKUTO 7732
(REPORT ABSTRACTS, INSTITUTE OF BEHAVIORAL SCIENCES)
KEIRYO KEIKAKU KENKYUSHO (INSTITUTE OF BEHAVIORAL SCIENCES)
35-7, 1-CHOME, YOYOGI, SHIBUYA-KU, TOKYO 151
JAPANESE

KENKYU HOKOKU (UN'YU-SHO KAKAI KUNRENSHO) 7733
(RESEARCH REPORT OF THE INSTITUTE FOR SEA TRAINING, MINISTRY OF TRANSPORT)
UN'YU-SHO KOKAI KUNRENSHO (INSTITUTE FOR SEA TRAINING, MINISTRY OF TRANSPORT)
1-3, 2-CHOME, KASUMIGASEKI, CHIYODA-KU, TOKYO 100
JAPANESE

KIKAN UN'YU KEIZAI KENKYU SENTA 7734
(MONTHLY TRANSPORT ECONOMICS RESEARCH CENTER)
JAPANESE
INDEXED: JICST

KODOMO NO KOTSU JIKO 7735
(CHILDREN'S TRAFFIC ACCIDENTS)
OSAKA-FU KEISATSU HONBU (OSAKA PREFECTURAL POLICE HEADQUARTERS)
9, OTEMAENO-CHO, HIGASHI-KU, OSAKA-SHI 540
JAPANESE

KOKAI KUNRENJO CHOSA KENKYU JIHO 7736
(JOURNAL OF THE INSTITUTE FOR SEA-TRAINING)
UN'YU-SHO KOKAI KUNRENJO (THE INSTITUTE FOR SEA-TRAINING, MINISTRY OF TRANSPORT)
2-1, 1-CHOME, KASUMIGASEKI, CHIYODA-KU, TOKYO 100
JAPANESE

KOKAI KUNRENJO CHOSA KENKYU ZAPPO 7737
(BULLETIN OF THE INSTITUTE FOR SEA-TRAINING)
UN'YU-SHO KOKAI KUNRENJO KENKYU CHOSA-BU (INSTITUTE FOR SEA-TRAINING, MINISTRY OF TRANSPORT)
2-1, 1-CHOME, KASUMIGASEKI, CHIYODA-KU, TOKYO 100
JAPANESE

KOKAI NIHON KOKAI GAKKAISHI 7738
(NAVIGATION, JOURNAL OF NAUTICAL SOCIETY OF JAPAN)
NIHON KOKAI GAKKAI (NAUTICAL SOCIETY OF JAPAN)
C/O TOKYO SHOSEN DAIGAKU, 1-6, 2-CHOME, ETCHUJIMA, KOTO-KU, TOKYO 135
JAPANESE
INDEXED: JICST

KOKU JANARU 7739
(AVIATION JOURNAL)
KOKU JANANARUSHA (AVIATION JOURNAL PUBLICATION)
JAPANESE

KOKU UN'YU TOKEI GEPPO 7740
(MONTHLY STATISTICS OF AIR TRANSPORT)
JAPANESE
INDEXED: JICST

KOKU YUSO TOKEI NENPO 7741
(ANNUAL STATISTICS OF AIR TRANSPORT)
UN'YU-SHO DAIJIN KANBO JOHO KANRI-BU (DEPARTMENT OF RESEARCH AND DATA PROCESSING, MINISTER'S SECRETARIAT, MINISTRY OF TRANSPORT)
1-3, 2-CHOME, KASUMIGASEKI, CHIYODA-KU, TOKYO 100
JAPANESE

KOKU YUSO TOKEI SOKUHO 7742
(MONTHLY STATISTICS OF AIR TRANSPORTATION)
UN'YU-SHO DAIJIN KANBO JOHO KANRI-BU (DEPARTMENT OF RESEARCH AND DATA PROCESSING, MINISTER'S SECRETARIAT, MINISTRY OF TRANSPORTATION)
1-3, 2-CHOME, KASUMIGASEKI, CHIYODA-KU, TOKYO 100
JAPANESE

KONTENARIZESHON 7743
(CONTAINERIZATION)
NIHON KAIJO KONTENA KYOKAI
(JAPAN OVERSEAS CONTAINER ASSOCIATION)
JAPANESE

KORO HYOSHIKI GIJUTSU YOHO 7744
(LIGHTHOUSE ENGINEERING)
KAIJO HOAN-CHO TODAI-BU (NAVIGATION AIDS DEPARTMENT, MARITIME SAFETY AGENCY)
1-3, 2-CHOME, KASUMIGASEKI, CHIYODA-KU, TOKYO 100
JAPANESE

KOTSU ANZEN DAYORI 7745
(TRAFFIC SAFETY LETTER)
HOKKAIDO KOTSU ANZEN TAISAKU JIMUKYOKU (HOKKAIDO TRAFFIC SAFETY (COUNCIL) GENERAL AFFAIRS OFFICE)
JAPANESE

KOTSU ANZEN KOGAI KENKYU-SHO HOKOKU 7746
(REPORT OF TRAFFIC SAFETY AND NUISANCE RESEARCH INSTITUTE)
UN'YU-SHO KOTSU ANZEN KOGAI KENKYUSHO (TRAFFIC SAFETY AND NUISANCE RESEARCH INSTITUTE, MINISTRY OF TRANSPORT)
JAPANESE AND ENGLISH
INDEXED: JICST

KOTSU ANZEN KOGAI KENKYUSHO HOKOKU 7747
(ANNUAL REPORT OF TRAFFIC SAFETY AND NUISANCE RESEARCH INSTITUTE)
UN'YU-SHO KOTSU ANZEN KOGAI KENKYUSHO (TRAFFIC SAFETY AND NUISANCE RESEARCH INSTITUTE, MINISTRY OF TRANSPORT)
38-1, 6-CHOME, SHINKAWA, MITAKA-SHI, TOKYO 181
JAPANESE

KOTSU ANZEN KYOIKU 7748
(TRAFFIC SAFETY EDUCATION)
NIHON KOTSU ANZEN KYOIKU FUKYU KYOKAI (TRAFFIC SAFETY EDUCATION PROMOTION ASSOCIATION OF JAPAN)
JAPANESE

KOTSU BUNKEN JOHO 7749
(TRANSPORTATION ABSTRACT)
UN'YU CHOSA-KYOKU
C/O CHUO TETSUDO GAKUEN, 5-6, 2-CHOME, IZUMI-CHO, KOKUBUNJI-SHI, TOKYO 185
JAPANESE

KOTSU GIJUTSU 7750
(RAILING ENGINEERING MAGAZINE)
KOTSU KYORYOKUKAI (TRANSPORTATION COOPERATIVE ASSOCIATION)
4-1, 3-CHOME, MARUNOUCHI, CHIYODA-KU, TOKYO 100
JAPANESE
INDEXED: JICST

KOTSU JIKO NO JOKYO OYOBI KOTSU ANZEN SHISAKU NO GENKYO 7751
(STATE OF TRAFFIC ACCIDENT AND TRAFFIC SAFETY)
SORI-FU (PRIME MINISTER'S OFFICE)
6-1, 1-CHOME, NAGATA-CHO, CHIYODA-KU, TOKYO 100
JAPANESE

KOTSU JIKO TOKEI NENPO 7752
(ANNUAL STATISTICS ON TRAFFIC ACCIDENTS)
KEISATSU-CHO KOTSU-KYOKU
1-2, 2-CHOME, KASUMIGASEKI, CHIYODA-KU, TOKYO 100
JAPANESE

KOTSU KAGAKU 7753
(TRAFFIC SCIENCE)
OSAKA KOTSU KAGAKU KENKYUKAI (OSAKA TRAFFIC SCIENCE RESEARCH ASSOCIATION)
JAPANESE

KOTSU KAGAKU KENKYU SHIRYO 7754
(TRAFFIC RESEARCH REPORT)
NIHON KOTSU KAGAKU KYOGIKAI (THE JAPAN COUNCIL OF TRAFFIC SCIENCE)
16-14, 2-CHOME, HIRAKAWA-CHO, CHIYODA-KU, TOKYO 102
JAPANESE

KOTSU KEIKAKU 7755
(TRAFFIC PLANNING)
NIHON KOTSU KEIKAKU KYOKAI
JAPANESE

KOTSU KENKYU KIYO 7756
(TRAFFIC RESEARCH BULLETIN)
UN'YU CHOSA KYOKU (TRANSPORTATION RESEARCH BUREAU)
JAPANESE

KOTSU KENKYUKAI KAIHO 7757
(REPORT OF THE TRAFFIC RESEARCH INSTITUTE ASSOCIATION)
HOKKAIDO KOTSU KENKYUKAI (KOKKAIDO TRAFFIC RESEARCH ASSOCIATION)
JAPANESE

KOTSU KOGAKU 7758
(TRAFFIC ENGINEERING)
KOTSU KOGAKU KENKYUKAI
C/O SHOYU KAIKAN, 3-3, 3-CHOME, KASUMIGASEKI, CHIYODA-KU, TOKYO 100
JAPANESE
INDEXED: JICST

KOTSU KONSARUTANTO 7759
(TRAFFIC CONSULTANT)
ZENKOKU KOTSU JIKO BOSHI KYOKAI RENGOKAI (NATIONAL TRAFFIC ACCIDENT PREVENTION ASSOCIATION)
JAPANESE

KOTSU KYOIKU 7760
(TRAFFIC EDUCATION)
KOTSU KYOIKU SHA (TRAFFIC EDUCATION COMPANY)
JAPANESE

KOTSU RANSO 7761
(TRAFFIC PAPERS)
TOKYO KOTSU TANKI DAIGAKU
JAPANESE

KOTSU SHINBUN 7762
(PRESS TRANSPORT)
KOTSU KYORYOKUKAI (TRANSPORTATION COOPERATIVE ASSOCIATION)
4-1, 3-CHOME, MARUNOUCHI, CHIYODA-KU, TOKYO 100
JAPANESE

KOTSU SHINPO 7763
(TRAFFIC NEWS)
KOTSU SHINPOSHA
C/O NODA BUILDING, 10, ICHIBAN-CHO, CHIYODA-KU, TOKYO 102
JAPANESE

KOTSU TOKEI (SAGA-KEN) 7764
(TRAFFIC ACCIDENT DATA, SAGA PREFECTURE))
SAGA-KEN KEISATSU HONBU (POLICE HEADQUARTERS, SAGA PREFECTURE)
1-59, 1-CHOME, JONAI, SAGA-SHI 840
JAPANESE

KOTSUGAKI KENKYU 7765
(TRAFFIC SCIENCE RESEARCH)
UN'YU CHOSAKYOKU (TRANSPORTATION RESEARCH BUREAU)
JAPANESE

KOTSURYO CHOSAHYO 7766
(PAPERS ON TRAFFIC QUANTITY)
AICHI-KEN (AICHI PREFECTURAL GOVERNMENT)
1-2, 3-CHOME, SANNOMARU, NAKA-KU, NAGOYA-SHI 460
JAPANESE

KOWAN KEIZAI KENKYU 7767
(ANNUAL OF JAPAN PORT ECONOMICS ASSOCIATION)
NIPPON KOWAN KEIZAI GAKKAI
JAPANESE

KOWAN NIYAKU 7768
(CARGO HANDLING)
KOWAN NIYAKU KIKAIKA KYOKAI (JAPAN CARGO HANDLING MECHANIZATION ASSOCIATION)
C/O KOTOHIRA KAIKAN BUILDING, 1, SHIBA KOTOHIRA-CHO, MINATO-KU, TOKYO 105
JAPANESE
INDEXED: JICST

KOWAN TOKEI (GEPPO) 7769
(HARBOR STATISTICS (MONTHLY))
UN'YUSHO DAIJIN KANBO JOHO KANRIBU TOKEIKA (MINISTRY OF TRANSPORT, MINISTER'S SECRETARIAT, DEPARTMENT OF RESEARCH AND DATA PROCESSING)
JAPANESE

KYOTO-SHI KOTSUKYOKU. JIGYO GAIYO 7770
(SUMMARY OF ENTERPRISES, CITY OF KYOTO TRANSPORTATION BUREAU)
KYOTOSHI KOTSUKYAKU (CITY OF KYOTO: TRANSPORTATION BUREAU)
JAPANESE

METORO NYUSU 7771
(METRO NEWS)
TEITO KOSOKUDO KOTSU EIDAN
(TEITO RAPID TRANSIT AUTHORITY)
JAPANESE

NAIKO SENPAKU YUSO TOKEI GEPPO 7772
(MONTHLY COASTWISE TRANSPORTATION STATISTICS)
UN'YU-SHO DAIJIN KANBO JOHO KANRI-BU (DEPARTMENT OF RESEARCH AND DATA PROCESSING, MINISTER'S SECRETARIAT, MINISTRY OF TRANSPORT)
1-3, 2-CHOME, KASUMIGASEKI, CHIYODA-KU, TOKYO 100
JAPANESE

NIHON BUTSURYU SHINBUN 7773
(JAPAN PHYSICAL DISTRIBUTION NEWS)
NIHON BUTSURYU SHINBUNSHA
C/O TATORU BUILDING, 2-6, HIGASHIAZABU, MINATO-KU, TOKYO 106
JAPANESE

NIHON KAIJI KYOKAI KAISHI 7774
(JAPAN MARITIME AFFAIRS ASSOCIATION MAGAZINE)
JAPANESE

NIHON KAIJI SHINBUN 7775
(JAPAN MARITIME DAILY)
NIHON KAIJI SHINBUNSHA
20-1, 3-CHOME, NISHISHINBASHI, MINATO-KU, TOKYO 105
JAPANESE AND ENGLISH

NIHON KOKAI GAKKAI RONBUNSHU 7776
(JOURNAL OF THE NAUTICAL SOCIETY OF JAPAN)
NIHON KOKAI GAKKAI (THE NAUTICAL SOCIETY OF JAPAN)
C/O TOKYO SHOSEN DAIGAKU, 1-6, 2-CHOME, ETCHUJIMA, KOTO-KU, TOKYO 135
JAPANESE AND ENGLISH

NIHON KOTSU KAGAKU KYOGIKAI KAIHO 7777
(REPORT OF THE JAPAN TRAFFIC SCIENCE COUNCIL)
JAPAN TRAFFIC SCIENCE COUNCIL
JAPANESE

NIHON RYUTSU SHINBUN 7778
(JAPAN DISTRIBUTION NEWS)
NIHON RYUTSU SHINBUNSHA
C/O TAISHIN BUILDING 1-6, NIHONBASHI KOBUNE-CHO, CHUO-KU, TOKYO 103
JAPANESE

NIKKAN UN'YU SHINBUN 7779
(DAILY TRANSPORT NEWS)
NIKKAN UN'YU SHINBUNSHA
9-2, 1-CHOME, SHINBASHI, MINATO-KU, TOKYO 105
JAPANESE

NIPPON KAIJI KYOKAISHI 7780
(TRANSACTIONS OF NIPPON KAIJI KYOKAI)
NIPPON KAIJI KYOKAI (JAPAN MARITIME AFFAIRS ASSOCIATION)
17-26, 2-CHOME, AKASAKA, MINATO-KU, TOKYO 107
JAPANESE, WITH ENGLISH SUMMARY
INDEXED: JICST

NIPPON KOTSU KAGAKU 7781
(TRAFFIC SCIENCE OF JAPAN)
JAPANESE
INDEXED: JICST

NIYAKU TO KIKAI 7782
(MECHANICAL HANDLING)
NIYAKU KENKYUSHO (MATERIALS HANDLING RESEARCH INSTITUTE OF JAPAN)
14-12, 2-CHOME, GINZA, CHUO-KU, TOKYO 104
JAPANESE
INDEXED: JICST

OSAKA NO KOTSU JIKO NO ARAMASHI 7783
(OUTLINE OF TRAFFIC ACCIDENT IN OSAKA PREFECTURE)
OSAKA-FU KEISATSU HONBU KOTSU-BU (TRAFFIC DIVISION, OSAKA PREFECTURAL POLICE HEADQUARTERS) AND OSAKA-FU KEISATSU HONBU (OSAKA PREFECTURAL POLICE HEADQUARTERS)
9, OTEMAENO-CHO, HIGASHI-KU, OSAKA-SHI 540
JAPANESE

PAKINGU PURESU 7784
(PARKING PLACE)
KIGYO KAIHATSU K. K.
C/O KYODO BUILDING, 1-24, NIHONBASHI KAKIGARA-CHO CHUO-KU, TOKYO 103
JAPANESE

PHYSICAL DISTRIBUTION 7785
NIHON FAIRINGU K. K.
C/O SHUFU NO TOMO BUILDING, KANDA SURUGADAI, CHIYODA-KU, TOKYO 101
JAPANESE

PHYSICAL DISTRIBUTION MANAGEMENT 7786
NIPPO
4-5, 4-CHOME, IIDABASHI, CHIYODA-KU, TOKYO 102
JAPANESE

PORT OF TOKYO 7787
BUREAU OF PORT & HARBOR, 8, MARUNOUCHI 3, CHIYODA-KU, TOKYO
JAPANESE

REPORT OF TRANSPORTATION TECHNICAL RESEARCH INSTITUTION 7788
TRANSPORTATION TECHNICAL RESEARCH INSTITUTE
JAPANESE
INDEXED: JICST

RIKUJO KOTSU TO ANZEN TAISAKU 7789
(SAFETY MEASURES OF TRAFFIC)
MIYAZAKI-KEN (MIYAZAKI PREFECTURAL GOVERNMENT)
10-1, 2-CHOME, TACHIBANA-DORI HIGASHI, MIYAZAKI-SHI 880
JAPANESE

RIKUUN TOKEI GEPPO 7790
(MONTHLY STATISTICS OF LAND TRANSPORT)
UN'YU-SHO DAIJIN KANBO JOHO KANRI-BU (DEPARTMENT OF RESEARCH AND DATA PROCESSING, MINISTER'S SECRETARIAT, MINISTRY OF TRANSPORT)
1-3, 2-CHOME, KASUMIGASEKI, CHIYODA-KU, TOKYO 100
JAPANESE

ROSEN TORAKKU CHOSA HOKOKUSHO 7791
(RESEARCH REPORT OF THE TRUCK TRANSPORTATION)
UN'YU-SHO DAIJIN KANBO JOHO KANRI-BU (DEPARTMENT OF RESEARCH AND DATA PROCESSING, MINISTER'S SECRETARIAT, MINISTRY OF TRANSPORT)
1-3, 2-CHOME, KASUMIGASEKI, CHIYODA-KU, TOKYO 100
JAPANESE

RYUTSU KEIZAI DAIGAKU RONSHU 7792
(PAPERS OF THE RYUTSU KEIZAI (DISTRIBUTION ECONOMICS UNIVERSTIY))
RYUTSU KEIZAI DAIGAKU GAKUJUTSU KENKYUKAI (RYUTSU KEIZAI UNIVERSITY ACADEMIC RESEARCH SOCIETY)
JAPANESE

SANGYO SHARYO TO NIYAKU KIKI 7793
(INDUSTRIAL VEHICLE AND TRANSPORTATION MACHINERY)
NIHON YUSOKI K. K. GIJUTSU KAIHATSUSHITSU (TECHNOLOGY DEVELOPMENT OFFICE, NIPPON YUSOKI COMPANY, LIMITED)
2, KOTARI TORIUCHIBATA, NAGAOKAKYO-SHI 617
JAPANESE

SEISAN TO UNPAN 7794
(MODERN MATERIAL HANDLING)
KOGYO CHOSAKAI (INDUSTRIAL INFORMATION CENTER COMPANY, LIMITED)
2-14, HONGO, BUNKYO-KU, TOKYO 113
JAPANESE
INDEXED: JICST

SENDAI RIKUUN KOHO 7795
(SENDAI LAND TRANSPORT OFFICIAL REPORT)
SENDAI RIKUUN KYOKAI (SENDAI LAND TRANSPORT ASSOCIATION)
JAPANESE

SEN'IN SAIKYOIKU CHOSA KENKYU HOKOKU 7796
(REPORT ON SURVEY RESEARCH ON THE REEDUCATION OF SHIP CREWMEN)
KAIGI DAIGAKU SEN'IN SAIKYOIKU CHOSA KENKYU-SHITSU
JAPANESE

SENKYO GEPPO 7797
(MONTHLY REPORT OF JAPANESE SHIPOWNERS' ASSOCIATION)
NIHON SENSHU KYOKAI (JAPANESE SHIPOWNERS' ASSOCIATION)
C/O KAIUN BUILDING, 6-4, 2-CHOME, HIRAKAWA-CHO, CHIYODA-KU, TOKYO 102
JAPANESE

SENKYO KAIUN NENPO 7798
(ANNUAL REPORT OF MARITIME TRANSPORT)
NIHON SENSHU KYOKAI (JAPANESE SHIPOWNERS' ASSOCIATION)
C/O KAIUN BUILDING, 6-4, 2-CHOME, HIRAKAWA-CHO, CHIYODA-KU, TOKYO 102
JAPANESE

SHIKEN KENKYU GAIYO (KAIJO HOAN SHIKEN KENKYU SENTA) 7799
SUMMARY OF RESEARCH
KAIJO HOAN-CHO KAIJO HOAN SHIKEN KENKYU SENTA (MARITIME SAFETY AGENCY RESEARCH CENTER)
6-64, KITANAKA-DORI, NAKA-KU, YOKOHAMA-SHI 231
JAPANESE

SHINGO HOAN 7800
(SIGNALING)
SHINGO HOAN KYOKAI (THE SIGNAL ASSOCIATION OF JAPAN)
14-7, 4-CHOME, GINZA, CHUO-KU, TOKYO 104
JAPANESE
INDEXED: JICST

TEITO KOSOKUDO KOTSU EIDAN CHOSA TOKEI GEPPO 7801
(RESEARCH AND STATISTICS REPORT (TEITO RAPID TRANSIT AUTHORITY))
TEITO KOSOKUDO KOTSU EIDAN KIKAKU CHOSASHITSU (PLANNING AND RESEARCH OFFICE, TEITO RAPID TRANSIT AUTHORITY)
19-6, 3-CHOME, HIGASHIUENO, TAITO-KU, TOKYO 110
JAPANESE

TETSUDO KAMOTSU YUSO GAIKYO (KOTUTETSU EIGYO SEISEKI NENPO BESSATSU) 7802
(DATA OF THE RAILWAY FREIGHT TRANSPORT)
NIHON KOKUYU TETSUDO (JAPANESE NATIONAL RAILWAY)
6-5, 1-CHOME, MARUNOUCHI, CHIYODA-KU, TOKYO 100
JAPANESE

TETSUDO TOKEI NENPO, TETSUDO UN'YU-HEN 7803
(ANNUAL RAILWAY STATISTICS, SERIES ON RAILWAY TRANSPORTATION)
NIHON KOKUYU TETSUDO MOJI TETSUDO KANRI-KYOKU KEIRI-BU (ACCOUNTING DEPARTMENT, MOJI RAILWAY ADMINISTRATIVE BUREAU, JAPANESE NATIONAL RAILWAY)
1, NISHIKAIGAN, MOJI-KU, KITAKYUSHU-SHI 801
JAPANESE

TODAIHYO 7804
(LIST OF LIGHTS)
KAIJO HOAN-CHO (MARITIME SAFETY AGENCY)
1-3, 2-CHOME, KASUMIGASEKI, CHIYODA-KU, TOKYO 100
JAPANESE

TOKO 7805
(LIGHT)
TOKOKAI
C/O MARIN BUILDING, 1-CHOME, SHINKAWA, CHUO-KU, TOKYO 104
JAPANESE

TOKYO RIKU'UN KYOKA HO 7806
(REPORT OF THE TOKYO DISTRICT LAND TRANSPORT BUREAU)
JAPANESE

TORAKKU 7807
(TRUCK)
NIHON TORAKKU KYOKAI (JAPAN TRUCK ASSOCIATION)
3-2, YOTSUYA, SHINJUKU-KU, TOKYO 160
JAPANESE

TORANSUPOTO 7808
(TRANSPORT)
UN'YU DAIJIN KANBO JOHO KANRI-BU (DEPARTMENT OF RESEARCH AND DATA PROCESSING, MINISTRY OF TRANSPORT)
2, AKASAKA AOI-CHO, MINATO-KU, TOKYO 107
JAPANESE

TOSHI KOTSU NENPO 7809
(ANNUAL REPORT ON URBAN TRANSPORTATION)
UN'YU KEIZAI KENKYU SENTA
7, SHIBA KOTOHIRA-CHO, MINATO-KU, TOKYO 105
JAPANESE

TSUKO SENPAKU JITTAI CHOSA HOKOKUSHO 7810
(SURVEY OF ACTUAL STATE OF NAVIGATING SHIPS)
KAIJO HOAN-CHO KEIBI KYUNAN-BU (GUARD AND RESCUE DEPARTMENT, MARITIME SAFETY AGENCY)
1-3, 2-CHOME, KASUMIGASEKI, CHIYODA-KU, TOKYO 100
JAPANESE

UMI 7811
(OCEAN)
UMI HAKKOJO (UMI PUBLISHERS)
JAPANESE

UMI 7812
(OCEAN)
KAIJI SHINKO RENMEI (FEDERATION FOR THE PROMOTION OF MARITIME AFFAIRS)
JAPANESE

UNPAN KOGAKU 7813
(HANDLING MANAGEMENT)
JCC NIHON KONSARUTINGU SENTA
2-80, UTSUBOHON-MACHI, NISHI-KU, OSAKA-SHI 550
JAPANESE
INDEXED: JICST

UN'YU 7814
(TRANSPORTATION)
UN'YUSHA
9-2, 1-CHOME, SHINBASHI, MINATO-KU, TOKYO 105
JAPANESE

UN'YU KOHO 7815
(REPORT ON TRANSPORTATION)
UN'YU KOHO SHA (UN'YU DAIJIN KANBO BUNSHOKA KANSHU) (TRANSPORTATION REPORTS COMPANY (UNDER SUPERVISION OF TRANSPORT MINISTER'S SECRETARIAT, DOCUMENTS SECTION))
JAPANESE

UN'YU KOHO 7816
(OFFICIAL REPORT ON TRANSPORTATION)
UN'YUSHO DAIJIN KANBO BUNSHOKA (TRANSPORT MINISTER'S SECRETARIAT, DOCUMENTS SECTION)
JAPANESE

UN'YU TO KEIZAI 7817
(TRANSPORT AND ECONOMICS)
JAPANESE
INDEXED: JICST

UN'YU TOKEI GEPPO 7818
(STATISTICAL MONTHLY OF TRANSPORTATION)
TOKYO-TO KOTSU KYOKU KEIEI KANRIBU KIKAKUKA (CITY OF TOKYO TRAFFIC BUREAU, MANAGEMENT DIVISION, PLANNING SECTION)
JAPANESE

UN'YU TOKEI KIHO 7819
(QUARTERLY BULLETIN OF TRANSPORT STATISTICS)
UN'YU-SHO DAIJIN KANBO JOHO KANRI-BU (DEPARTMENT OF RESEARCH AND DATA PROCESSING, MINISTER'S SECRETARIAT, MINISTRY OF TRANSPORT)
1-3, 2-CHOME, KASUMIGASEKI, CHIYODA-KU, TOKYO 100
JAPANESE

UN'YU TOKEI NENPO 7820
(ANNUAL STATISTICS OF TRANSPORT IN TOKYO)
TOKYO-TO-KOTSU-KYOKU (TRANSPORTATION BUREAU, TOKYO METROPOLITAN GOVERNMENT)
2-13, YURAKU-CHO, CHIYODA-KU, TOKYO 100
JAPANESE

UN'YUKAI 7821
(TRANSPORTATION WORLD)
CHUO SHOIN
8-11, 2-CHOME, SARUGAKU-CHO, CHIYODA-KU, TOKYO 101
JAPANESE

YUSO SHINBUN 7822
(TRANSPORT NEWS)
YUSO SHINBUNSHA
6-11, 3-CHOME, SOTO KANDA, CHIYODA-KU, TOKYO 101
JAPANESE

YUSO TENBO 7823
(JOURNAL OF DISTRIBUTION)
NITTSU SOGO KENKYUSHO (NITTSU RESEARCH CENTER, INCORPORATED)
12-9, 3-CHOME, SOTO KANDA, CHIYODA-KU, TOKYO 101
JAPANESE
INDEXED: JICST

Chemical Industry, Chemical Technology (Includes Oils, Fats, and Petroleum)

ABSTRACTS OF RESEARCHES, TOBACCO, SALT, CAMPHOR 7824
NIHON SENBAI JOSHA
2, AKASAKA AOI-CHO, MINATO-KU, TOKYO
ENGLISH
INDEXED: CHE

ACTA CHROMATICA 7825
ENGLISH

AEROSOL 7826
NIHON EAZORU KYOKAI (AEROSOL INDUSTRY ASSOCIATION OF JAPAN)
C/O NO. 319, MARU BUILDING, 4-1, 2-CHOME, MARUNOUCHI, CHIYODA-KU, TOKYO 100
JAPANESE
INDEXED: JICST

AKO ENGYO SHIRYOKANHO 7827
(REPORT OF THE AKO DATA BUREAU ON SALT INDUSTRIES)
AKO DATA BUREAU
JAPANESE

AKURIRU NYUSU 7828
(ACRYL NEWS)
MITSUBISHI REIYON (MITSUBISHI RAYON)
JAPANESE

AMMONIA TO KOGYO 7829
(JOURNAL OF AMMONIUM INDUSTRY)
NIHON RYUAN KOGYO KYOKAI (JAPAN AMMONIUM SULPHATE INDUSTRY ASSOCIATION)
C/O HOKKAI BUILDING, 3-13, 1-CHOME, NIHONBASHI-DORI, CHUO-KU, TOKYO 103
JAPANESE
INDEXED: JICST

ANNUAL INDEX OF THE REPORTS ON PLANT CHEMISTRY 7830
ENGLISH

ANNUAL STATISTICS FOR OILSEEDS, OILS, AND OILCAKES 7831
OIL AND FAT IMPORTERS AND EXPORTERS ASSOCIATION, TOKYO
JAPANESE
INDEXED: BIB

ARAKAWA NYUSU 7832
(ARAKAWA FOREST CHEMICAL INDUSTRIES LIMITED NEWS)
ARAKAWA RINSAN KAGAKU KOGYO K.K. (ARAKAWA FOREST CHEMICAL INDUSTRIES, LIMITED)
1-21, HIRANO-CHO, HIGASHI-KU, OSAKA-SHI 541
JAPANESE
INDEXED: CHE, JICST

AROMATIKKUSU 7833
(AROMATICS)
NIHON HOKOZOKU KOGYOKAI (JAPAN AROMATIC INDUSTRY ASSOCIATION, INCORPORATED)
2-3 CHOME, NIHONBASHI-KAYABACHO, CHUO-KU TOKYO 103
JAPANESE
INDEXED: CHE, JICST

ASUFARUTO 7834
(ASPHALT)
NIHON ASUFARUTO KYOKAI (JAPAN ASPHALT ASSOCIATION)
12, SHIBA NISHIKUBO AKEFUNE-CHO, MINATO-KU, TOKYO 105
JAPANESE

BANDO DAYORI 7835
(BANDO LETTER)
BANDO KAGAKU (BANDO CHEMICAL INDUSTRIES)
JAPANESE

BANDO ELASTOMER REPORT 7836
BANDO KAGAKU K.K. (BANDO CHEMICAL INDUSTRIES, LTD.)
2-1, MEIWA-DORI, HYOGO-KU, KOBE-SHI 652
JAPANESE

BESSATSU ASUFARUTO 7837
(ASPHALT (SPECIAL ISSUE))
NIHON ASUFARUTO NYOKAI (NIHON ASPHALT ASSOCIATION)
JAPANESE

BESSATSU KAGAKU KOGYO 7838
(JOURNAL OF CHEMICAL ENGINEERING (SPECIAL ISSUE))
KAGAKU KOGYOSHA
5-9, 4-CHOME, SENDAGAYA, SHIBUYA-KU, TOKYO 151
JAPANESE
INDEXED: JICST

BESSATSU SEKIYU TSUSHIN 7839
(OIL NEWSLETTER (SPECIAL ISSUE))
SEKIYU TSUSHINSHA
(OIL NEWSLETTER COMPANY)
JAPANESE

BRAINS OF CHEMICAL ENGINEER 7840
NIHON KAGAKU GIJUTSU K.K. (JAPAN CHEMICAL ENGINEERING COMPANY, LIMITED)
C/O SHOWA BUILDING, 3-94, UTSUBOHONMACHI, NISHI-KU, OSAKA-SHI 550
JAPANESE

BULLETIN OF THE JAPAN PETROLEUM INSTITUTE 7841
SEKIYU GAKKAI (JAPAN PETROLEUM INSTITUTE)
C/O NISSEKI BUILDING, 4-2, 3-CHOME, MARUNOUCHI, CHIYODA-KU, TOKYO 100
ENGLISH
INDEXED: CHE, ENG, JICST

CEER, CHEMICAL ECONOMY ENGINEERING REVIEW 7842
KAGAKU KEIZAI KENKYUSHO (CHEMICAL ECONOMY RESEARCH INSTITUTE, JAPAN)
13-7, 1-CHOME, UCHIKANDA, CHIYODA-KU, TOKYO 101
ENGLISH
INDEXED: API, BIB, CHE, EXC, JICST

CHEMICAL TIMES 7843
KANTO KAGAKU K.K.
3-7, NIHONBASHI HON-CHO, CHUO-KU, TOKYO 103
JAPANESE AND ENGLISH
INDEXED: JICST

DAIICHI KOGYO SEIYAKU SHAHO 7844
(DAIICHI KOGYO SEIYAKU REVIEW)
DAIICHI KOGYO SEIYAKU K.K. EIGYO HONBU
55, HIGASHI KUBO-CHO, NISHI-7-JO, SHIMOGYO-KU, KYOTO-SHI 600
JAPANESE
INDEXED: JICST

ENAJI TAIWA 7845
(ENERGY DIALOGUE)
ESSO SUTANDADO SEKIYU KABUSHIKI KAISHA KOHOBU
(ESSO STANDARD OIL COMPANY, PUBLIC RELATIONS DIVISION)
JAPANESE

ENERUGII TOKEI GEPPO 7846
JAPANESE

ESSO JAPAN 7847
ESSO SUTANDADO SEKIYU K.K. KOHO-BU (ESSO STANDARD OIL COMPANY, LIMITED)
3-3, 5-CHOME, AKASAKA, MINATO-KU, TOKYO 107
JAPANESE

ESSO OIRUWEIZU INTANASHONARU 7848
(ESSO OILWAYS INTERNATIONAL)
ESSO SUTANDADO SEKIYU K.K. KOHO-BU (ESSO STANDARD OIL COMPANY, LIMITED)
3-3, 5-CHOME, AKASAKA, MINATO-KU, TOKYO 107
JAPANESE

FANDORI NYUSU 7849
(FOUNDRY NEWS)
MITSUBISHI CHEMICAL AND SYNTHETIC INDUSTRY, LTD.
JAPANESE
INDEXED: JICST

FUNSAI 7850
(MICROMERITICS)
HOSOKAWA FUNTAI KOGAKU KENKYUSHO (HOSOKAWA MICROMERITIC LABORATORY)
1-9, TAJIKA, SHODAI, HIRAKATA-SHI 573
JAPANESE, WITH ENGLISH SUMMARY
INDEXED: CHE, JICST

FUNTAI KOGAKU 7851
(POWDER ENGINEERING)
JAPANESE
INDEXED: JICST

FUNTAI KOGAKU KENKYUKAISHI 7852
(JOURNAL OF THE RESEARCH ASSOCIATION OF POWDER TECHNOLOGY)
FUNTAI KOGAKU KENKYUKAI (THE RESEARCH ASSOCIATION OF POWDER TECHNOLOGY, JAPAN)
C/O NAGOYA KOGYO GIJUTSU SHIKENJO, 1, HIRATE-CHO, KITA-KU, NAGOYA-SHI 462
JAPANESE, WITH ENGLISH SUMMARY
INDEXED: CHE, JICST

FUNTAI OYOBI FUNMATSU YAKIN 7853
(JOURNAL OF THE JAPAN SOCIETY OF POWDER AND POWDER METALLURGY)
FUNTAI FUNMATSU YAKIN KYOKAI (JAPAN SOCIETY OF POWDER AND POWDER METALLURGY)
C/O SEISAN KAIHATSU KAGAKU KENKYUSHO, 15, SHIMOGAMO MORIMOTO-CHO, SAKYO-KU, KYOTO-SHI 606
JAPANESE AND ENGLISH
INDEXED: CHE, ENG, JICST

FUNTAI TO KOGYO 7854
(POWDER PROCESS ENGINEERING)
FUNTAI TO KOGYOSHA
3-5, KANDA SURUGADAI, CHIYODA-KU, TOKYO 101
JAPANESE

GANRYO 7855
(PIGMENT)
NIHON GANRYO GIJUTSU KYOKAI (JAPAN PIGMENT TECHNICAL ASSOCIATION)
5-10, KAWARA-CHO, HIGASHI-KU, OSAKA-SHI 541
JAPANESE

GASU JIGYO TOKEI GEPPO 7856
(MONTHLY STATISTICS OF GAS INDUSTRY)
TSUSHO SANGYO-SHO KOEKI JIGYO-KYOKU (PUBLIC WORKS BUREAU, MINISTRY OF INTERNATIONAL TRADE AND INDUSTRY) AND NIHON GASU KYOKAI (JAPAN GAS ASSOCIATION)
38, SHIBA KOTOHIRA-CHO, MINATO-KU, TOKYO 105
JAPANESE

GASU JIGYO TOKEI NENPO 7857
(ANNUAL STATISTICS OF GAS INDUSTRY)
TSUSHO SANGYO-SHO KOEKI JIGYO-KYOKU (PUBLIC WORKS BUREAU, MINISTRY OF INTERNATIONAL TRADE AND INDUSTRY) AND NIHON GASU KYOKAI (JAPAN GAS ASSOCIATION)
38, SHIBA KOTOHIRA-CHO, MINATO-KU, TOKYO 105
JAPANESE

GASU SHIRYOKAN NENPO 7858
(ANNUAL REPORT OF GAS MUSEUM)
TOKYO GASU K.K. (TOKYO GAS COMPANY)
1-2, YAESU, CHUO-KU, TOKYO 103
JAPANESE

GASUTO 7859
(OSAKA GAS NEWS)
OSAKA GASU K.K. FUKURI-BU
5-1, HIRANO-MACHI, HIGASHI-KU, OSAKA-SHI 541
JAPANESE

GEKKAN GASORIN SUTANDO 7860
(MONTHLY GASOLINE STAND)
GEKKAN GASORIN SUTANDOSHA
2-3, 3-CHOME, SHINBASHI, MINATO-KU, TOKYO 105
JAPANESE

GEKKAN KYUYUSHO NIHON 7861
(MONTHLY JOURNAL OF GASOLINE SERVICE STATION, JAPAN)
YUGYO HOCHI SHINBUNSHA
15-19, 2-CHOME, SHINKAWA, CHUO-KU, TOKYO 104
JAPANESE

GEKKAN NENRYO KEIZAI 7862
(FUEL ECONOMICS MONTHLY)
JAPANESE
INDEXED: JICST

GEKKAN SEKIYU 7863
(PETROLEUM MONTHLY)
SEKIYU KOGYO JIHOSHA
16-5, 2-CHOME, MISAKI-CHO, CHIYODA-KU, TOKYO 101
JAPANESE
INDEXED: JICST

GIJUTSU PEN 7864
(TECHNICAL REVIEW (KAGAKU KOGYO GIJUTSU DOYUKAI))
KAGAKU KOGYO GIJUTSU DOYUKAI SHUPPAN-BU (THE ASSOCIATION OF PROMOTION FOR CHEMICAL INDUSTRY)
15-7, 1-CHOME, NIHONBASHI, CHUO-KU, TOKYO 103
JAPANESE
INDEXED: JICST

GIJUTSU SHIRYO (MITSUBISHI SEKIYU K.K.) 7865
(TECHNICAL REVIEW, MITSUBISHI OIL COMPANY, LIMITED)
MITSUBISHI SEKIYU KENKYUJO (MITSUBISHI OIL COMPANY, RESEARCH LABORATORY))
4-1, OGI-MACHI, KAWASAKI-KU, KAWASAKI-SHI 210
JAPANESE
INDEXED: ENG

HOSO TO ASUFARUTO NYUZAI 7866
(EMULSIFIED ASPHALT FOR ROAD WORK)
NIHON ASUFARUTO NYUZAI KYOKAI (NIHON EMULSIFIED ASPHALT ASSOCIATION)
C/O IDA BLDG., 3-8, 1-CHOME, YAESU, CHUO-KU, TOKYO 103
JAPANESE
INDEXED: JICST

HYOGO-KEN KOGYO GIJUTSU JOHO: KAGAKU-HEN 7867
(HYOGO PREFECTURE INDUSTRIAL TECHNOLOGY REPORT: CHEMISTRY)
HYOGO KENRITSU KOGYO SHIKENJO (HYOGO PREFECTURAL INDUSTRIAL LABORATORY)
JAPANESE

ICC BULLETIN 7868
KOGYO KEIZAI KENKYUSHO (INDUSTRIAL CHEMICAL CONSULTANTS INCORPORATED)
30, SHIBA KOTOHIRA-CHO, MINATO-KU, TOKYO 105
JAPANESE

IDEMITSU OIRU DAIZESUTO 7869
(IDEMITSU OIL DIGEST)
IDEMITSU KOSAN K.K.
1-1, 3-CHOME, MARUNOUCHI, CHIYODA-KU, TOKYO 100
JAPANESE
INDEXED: JICST

IDEMITSU SEKIYU GIJUTSU 7870
(IDEMITSU PETROLEUM JOURNAL)
IDEMITSU KOSAN K.K.
1-1, 3-CHOME, MARUNOUCHI, CHIYODA-KU, TOKYO 100
JAPANESE
INDEXED: JICST

JAPAN CHEMICAL REVIEW 7871
KAGAKU KOGYO NIPPOSHA (CHEMICAL DAILY COMPANY, LTD.)
19-16, 3-CHOME, SHIBAURA, MINATO-KU, TOKYO 108
ENGLISH
INDEXED: CHE

JAPAN CHEMICAL WEEK 7872
KAGAKU KOGYO NIPPOSHA (CHEMICAL DAILY COMPANY, LTD.)
19-16, 3-CHOME, SHIBAURA, MINATO-KU, TOKYO 108
ENGLISH
INDEXED: CHE, JICST

JAPAN PETROLEUM NEWS 7873
SENTORARU TSUSHINSHA
1-13, KANDA AWAJI-CHO, CHIYODA-KU, TOKYO 104
ENGLISH

JAPAN PETROLEUM WEEKLY 7874
NIHON SEKIYU KONSARUTANTO K.K. (JAPAN PETROLEUM CONSULTANTS LTD.)
C/O OKANA BUILDING, 13, SHIBA NISHIKUBO AKEFUNE-CHO, MINATO-KU, TOKYO 105
ENGLISH

JOURNAL OF CHEMICAL ENGINEERING OF JAPAN 7875
KAGAKU KOGAKU KYOKAI (THE SOCIETY OF CHEMICAL ENGINEERING, JAPAN)
C/O KYORITSU BUILDING, 6-19, 4-CHOME, KOHINATA, BUNKYO-KU, TOKYO 112
ENGLISH
INDEXED: API, CHE, ENG, JICST

JUNKAN SEKITSU 7876
(COAL AND PETROLEUM BULLETIN)
SEKITSU
7-2, 2-CHOME, NANOKAWA, CHUO-KU, FUKUOKA-SHI 810
JAPANESE
INDEXED: JICST

KAGAKU GIJUTSUSHI - MOL 7877
(CHEMICAL ENGINEERING - MOL)
OMUSHA
3-1, KANDA NISHIKI-CHO, CHIYODA-KU, TOKYO 101
JAPANESE

KAGAKU KEIZAI 7878
(CHEMICAL INDUSTRIAL ECONOMY)
KAGAKU KEIZAI KENKYUSHO (CHEMICAL ECONOMIC RESEARCH INSTITUTE)
13-7, 1-CHOME, UCHIKANDA, CHIYODA-KU, TOKYO 101
JAPANESE
INDEXED: JICST

KAGAKU KIKAI GIJUTSU 7879
(CHEMICAL MACHINE TECHNOLOGY)
KAGAKU KIOGAKU KYOKAI HEN
JAPANESE

KAGAKU KOGAKU 7880
(CHEMICAL ENGINEERING, JAPAN)
KAGAKU KOGAKU KYOKAI (THE SOCIETY OF CHEMICAL ENGINEERS, JAPAN)
C/O KYORITSU BUILDING, 6-19, 4-CHOME, KOHINATA, BUNKYO-KU, TOKYO 112
JAPANESE, WITH ENGLISH SUMMARY
INDEXED: CHE, JICST

KAGAKU KOGAKU RONBUNSHU 7881
(CHEMICAL ENGINEERING PAPERS)
KOGAKU KOGAKU KYOKAI (CHEMICAL ENGINEERING SOCIETY)
JAPANESE
INDEXED: JICST

KAGAKU KOGYO 7882
(CHEMICAL INDUSTRY)
KAGAKU KOGYOSHA
5-9, 4-CHOME, SENDAGAYA, SHIBUYA-KU, TOKYO 151
JAPANESE
INDEXED: BIB, CHE, JICST

KAGAKU KOGYO JIHO 7883
(CHEMICAL INDUSTRIAL TIMES)
KAGAKU KOGYO JIHOSHA (CHEMICAL INDUSTRIAL NEWS)
2-12, KANDA SAKUMA-CHO, CHIYODA-KU, TOKYO 101
JAPANESE AND ENGLISH

KAGAKU KOGYO KAIGAI NYUSU 7884
(CHEMICAL INDUSTRY FOREIGN NEWS)
HORIKOSHI KENKYUJO (HORIKOSHI RESEARCH INSTITUTE)
JAPANESE

KAGAKU KOGYO NIPPO 7885
(CHEMICAL DAILY NEWS)
KAGAKU KOGYO NIPPOSHA
19-16, 3-CHOME, SHIBAURA, MINATO-KU, TOKYO 108
JAPANESE

KAGAKU KOGYO TOKEI GEPPO 7886
(CHEMICAL INDUSTRY STATISTICAL MONTHLY)
TSUSHO SANGYO-SHO DAIJIN KANBO CHOSA TOKEI-BU (RESEARCH AND STATISTICS DIVISION, MINISTER'S SECRETARIAT, MINISTRY OF INTERNATIONAL TRADE AND INDUSTRY) AND TSUSHO SANGYO CHOSAKAI
15-1, 6-CHOME, GINZA, CHUO-KU, TOKYO 104
JAPANESE
INDEXED: JICST

KAGAKU KOGYO TOKEI NENPO 7887
(YEARBOOK OF CHEMICAL INDUSTRIES STATISTICS)
TSUSHO SANGYO-SHO DAIJIN KANBO CHOSA TOKEI-BU (RESEARCH AND STATISTICS DIVISION, MINISTER'S SECRETARIAT, MINISTRY OF INTERNATIONAL TRADE AND INDUSTRY) AND TSUSHO SANGYO CHOSAKAI
15-1, 6-CHOME, GINZA, CHUO-KU, TOKYO 104
JAPANESE

KAGAKU KOJO 7888
(CHEMICAL FACTORY)
NIKKAN KOGYO SHINBUNSHA
8-10, 1-CHOME, KUDAN-KITA, CHIYODA-KU, TOKYO 102
JAPANESE
INDEXED: CHE, JICST

KAGAKU SANGYO SHINBUN 7889
(CHEMICAL INDUSTRY NEWS)
KAGAKU SANGYO SHINBUNSHA
1-8, NIHONBASHI HON-CHO, CHUO-KU, TOKYO 103
JAPANESE

KAGAKU SOCHI 7890
(PLANT AND PROCESS)
KOGYO CHOSAKAI (INDUSTRY INFORMATION CENTER)
14-7, 2-CHOME, HONGO, BUNKYO-KU, TOKYO 113
JAPANESE
INDEXED: CHE, JICST

KAGAKU TO KOGYO (NIPPON KAGAKKAI) 7891
(CHEMISTRY AND CHEMICAL INDUSTRY)
NIPPON KAGAKUKAI (THE CHEMICAL SOCIETY OF JAPAN)
1-5, KANDA SURUGADAI, CHIYODA-KU, TOKYO 101
JAPANESE
INDEXED: BIB, CHE, JICST

KAIGAI KOGYO NYUSU TO GIJUTSU SHIRYO 7892
(OVERSEAS INDUSTRIAL NEWS AND TECHNICAL DATA)
HORIKOSHI KENKYUJO (HORIKOSHI RESEARCH INSTITUTE)
JAPANESE

KARIEN GEPPO 7893
(POTASSIUM FERTILIZER INDUSTRY)
KARIEN KONWAKAI
C/O HAYASHI BUILDING, 1-1, KANDA JINBO-CHO, CHIYODA-KU, TOKYO 101
JAPANESE

KASEI HIRYO KYOKAIHO 7894
(COMPOUND FERTILIZER, PHOSPHATE FERTILIZER)
NIHON KASEI HIRYO KYOKAI (JAPAN PHOSPHATIC AND COMPOUND FERTILIZERS MANUFACTURES' ASSOCIATION)
3-3, NIHONBASHI MURO-MACHI, CHUO-KU, TOKYO 103
JAPANESE

KASEIHIN NIPPO 7895
(CHEMICAL PRODUCTS NEWS)
KASEIHIN NIPPOSHA
8-8, 3-CHOME, UCHIKANDA, CHIYODA-KU, TOKYO 101
JAPANESE

KAWAMATA KAGAKU GEPPO 7896
(KAWAMATA CHEMICAL MONTHLY BULLETIN)
KAWAMATA KAGAKU KOGYO K.K. (THE KAWAMATA CHEMICAL INDUSTRY COMPANY, LTD.)
2-1, NIHONBASHI KAYABA-CHO, CHUO-KU, TOKYO 103
JAPANESE
INDEXED: CHE, JICST

KAYAKU TO HOAN 7897
(EXPLOSIVES AND SAFETY MAINTENANCE)
ZENKOKU KAYAKURUI HOAN KYOKAI
3-21, 2-CHOME, NIHONBASHI, CHUO-KU, TOKYO 103
JAPANESE

KAYAKURUI KOATSU GASU TORISHIMARI GEPPO 7898
(MONTHLY REPORT OF EXPLOSIVE AND HIGH PRESSURE GAS)
TSUSHO SANGYO-SHO RITCHI KOGAI-KYOKU
SEKIYU SANGYO SHINBUNSHA
13-9, 1-CHOME, NISHISHINBASHI, MINATO-KU, TOKYO 105
JAPANESE

KAYAKURUI TOKEI SHUROKU 7899
(ANNUAL STATISTICS OF INDUSTRIAL EXPLOSIVES)
NIHON SANGYO KAYAKUKAI (JAPAN INDUSTRIAL EXPLOSIVES ASSOCIATION)
C/O GUNMA BUILDING, 3-21, 2-CHOME, NIHONBASHI, CHUO-KU, TOKYO 103
JAPANESE

KEMIKARU ENJINIYARINGU 7900
(CHEMICAL ENGINEERING)
KAGAKU KOGYOSHA
5-9, 4-CHOME, SENDAGAYA, SHIBUYA-KU, TOKYO 151
JAPANESE
INDEXED: JICST

KINKI DAIGAKU OYO KAGAKU GAKKAISHI 7901
(JOURNAL OF THE KINKI UNIVERSITY ASSOCIATION OF APPLIED CHEMISTRY)
KINKI DAIGAKU RIKOGAKUBU (FACULTY OF SCIENCE AND TECHNOLOGY, KINKI UNIVERSITY)
JAPANESE

KINKI KAGAKU KOGYAKAI 7902
(KINKI CHEMICAL INDUSTRY WORLD)
KINKI KAGAKU KOGYOKAI (KINKI CHEMICAL INDUSTRY ASSOCIATION)
C/O OSAKA KAGAKU GIJUTSU SENTA, 1-118, UTSUBO, NISHI-KU, OSAKA-SHI 550
JAPANESE

KOATSU GASU 7903
(JOURNAL OF THE INSTITUTE FOR SAFETY OF HIGH PRESSURE GAS ENGINEERING)
KOATSU GASU HOAN KYOKAI (THE INSTITUTE FOR SAFETY OF HIGH PRESSURE GAS ENGINEERING)
C/O HIRAKAWA BUILDING, 11-28, 1-CHOME, NAGATA-CHO, CHIYODA-KU, TOKYO 100
JAPANESE
INDEXED: CHE, JICST

KOGYO KAYAKU KYOKAISHI 7904
(JOURNAL OF THE INDUSTRIAL EXPLOSIVE SOCIETY, JAPAN)
KOGYO KAYAKU KYOKAI (THE INDUSTRIAL EXPLOSIVE SOCIETY, JAPAN)
C/O NIHON SANGYO KAYAKUKAI, GUNMA BUILDING, 3-21, 2-CHOME, NIHONBASHI, CHUO-KU, TOKYO 103
JAPANESE AND ENGLISH
INDEXED: APP, CHE, JICST

KOKUSU SAKYURA 7905
(COKE CIRCULAR)
NENRYO KYOKAI KOKUSUBUKAI (COKE COMMITTEE, FUEL ASSOCIATION OF JAPAN)
5-4, 6-CHOME, SOTOKANDA, CHIYODA-KU, TOKYO 101
JAPANESE

KOKUSU TOKEI GEPPO 7906
(COKE STATISTICAL MONTHLY)
TSUSHO SANGYO-SHO DAIJIN KANBO CHOSA TOKEI-BU (RESEARCH AND STATISTICS DIVISION, MINISTER'S SECRETARIAT, MINISTRY OF INTERNATIONAL TRADE AND INDUSTRY)
C/O NIHON GASU KYOKAI BUILDING, 38, SHIBA KOTOHIRA-CHO, MINATO-KU, TOKYO 105
JAPANESE

KORYO 7907
(PERFUMERY AND FLAVOURING)
NIHON KORYO KYOKAI (JAPAN PERFUMERY AND FLAVOURING ASSOCIATION)
C/O NITTA BUILDING, 2-1, 8-CHOME, GINZA CHUO-KU, TOKYO 104
JAPANESE, WITH ENGLISH SUMMARY
INDEXED: CHE, JICST

KUORITI 7908
(QUALITY)
SHERU SEKIYU K.K. (SHELL OIL COMPANY)
C/O TOKYO BUILDING, 3-2, 2-CHOME, MARUNOUCHI, CHIYODA-KU, TOKYO 100
JAPANESE

KUREHA JIHO 7909
(KUREHA TIMES)
KUREHA KAGAKU KOGYO K.K. (KUREHA CHEMICAL INDUSTRY COMPANY, LIMITED)
1-8, NIHONBASHI HORIDOME, CHUO-KU, TOKYO 103
JAPANESE

KYOSEKI JIHO 7910
(KYODO SEKIYU OIL NEWS)
KYODO SEKIYU K.K.
11-2, 2-CHOME, NAGATA-CHO, CHIYODA-KU, TOKYO 100
JAPANESE

LP GASU 7911
(LIQUEFIED PETROLEUM GAS)
SANPO
10-17, 1-CHOME, HAMAMATSU-CHO, MINATO-KU, TOKYO 105
JAPANESE

LP GASU SANGYO NO GENJO 7912
(PRESENT STATE OF LIQUEFIED PETROLEUM GAS INDUSTRY)
TSUSHO SANGYO-SHO KOZAN SEKITAN-KYOKU (COAL MINING DEPARTMENT, MINISTRY OF INTERNATIONAL TRADE AND INDUSTRY) AND NIKKAN SEKIYU NYUSU
2-2, 2-CHOME, NIHONBASHI-DORI, CHUO-KU, TOKYO 103
JAPANESE

LP GASU SHIRYO NENPO 7913
(LIQUEFIED PETROLEUM GAS ANNUAL REPORT)
SEKIYU KAGAKU SHINBUNSHA
C/O KYODO BUILDING, 2, IWAMOTO-CHO, CHIYODA-KU, TOKYO 101
JAPANESE

LPG TSUSHIN 7914
(DAILY LPG INFORMATION)
SANPO
10-17, 1-CHOME, HAMAMATSU-CHO, MINATO-KU, TOKYO 105
JAPANESE
INDEXED: JICST

MARUZEN SEKIYU GIHO 7915
(MARUZEN OIL TECHNICAL REPORT)
MARUZEN SEKIYU K.K. CHUO KENKYUSHO, SHOHIN KENKYUSHO (MARUZEN OIL COMPANY, LTD., RESEARCH CENTER)
1134-2, GONGENDO, SATTE-CHO, KITAKATSUNSHIKA-GUN SAITAMA 340-01
JAPANESE AND ENGLISH
INDEXED: CHE, JICST

MIE-KEN KOGYO GIJUTSU JOHO, KAGAKU-HEN 7916
(INDUSTRIAL AND TECHNOLOGICAL INFORMATION, MIE PREFECTURE, SERIES ON CHEMISTRY)
MIE-KEN KOGYO GIJUTSU SENTA (THE INDUSTRIAL TECHNOLOGY CENTER OF MIE PREFECTURE)
3485, KOMORI-CHO, TAKAJAYA, TSU-SHI 514
JAPANESE

MOKUTAN, MAKI, SONOTA KATEI NENRYO TOKEI SHIRYO 7917
(STATISTICS OF CHARCOAL FIRE-WOOD AND OTHER DOMESTIC FUELS)
ZENKOKU NENRYO KYOKAI (JAPAN DOMESTIC FUEL DEALERS' ASSOCIATION)
12-15, 8-CHOME, GINZA, CHUO-KU, TOKYO 104
JAPANESE

MUKI YAKUHIN NO JISSEKI TO MITOSHI 7918
(RESULT AND PROSPECT OF INORGANIC CHEMICALS)
NIHON MUKI YAKUHIN KYOKAI
13-1, 1-CHOME, GINZA, CHUO-KU, TOKYO 104
JAPANESE

MUKIYAKU SHINPO 7919
(INORGANIC CHEMICALS INDUSTRY NEWS)
MUKIYAKU SHINPOSHA
C/O SANKO BUILDING, 13-1, 1-CHOME, GINZA, CHUO-KU, TOKYO 104
JAPANESE

NAIGAI SEKIYU SHIRYO 7920
(PETROLEUM INFORMATION OF THE WORLD)
SEKIYU RENMEI (PETROLEUM ASSOCIATION OF JAPAN)
9-4, 1-CHOME, OTE-MACHI, CHIYODA-KU, TOKYO 100
JAPANESE
INDEXED: JICST

NENRYO KYOKAISHI 7921
(JOURNAL OF THE FUEL SOCIETY OF JAPAN)
NENRYO KYOKAI (THE FUEL SOCIETY OF JAPAN)
5-4, 6-CHOME, SOTOKANDA, CHIYODA-KU, TOKYO 101
JAPANESE, WITH EUROPEAN SUMMARY
INDEXED: CHE, ENG, JICST

NENRYO OYOBI NENSHO 7922
(FUEL AND COMBUSTION)
NENRYO OYOBI NENSHOSHA
21, KITAYAMA-CHO, TENNOJI-KU, OSAKA-SHI 543
JAPANESE
INDEXED: CHE, JICST

NENSHO KENKYU 7923
(COMBUSTION RESEARCH)
NIHON NENSHO KENKYUKAI (COMBUSTION SOCIETY OF JAPAN)
C/O TOKYO DAIGAKU KOGAKU-BU HANNO KAGAKUKA, 3-1, 7-CHOME, HONGO, BUNKYO-KU, TOKYO 113
JAPANESE

NIHON GASU KYOKAISHI 7924
(JOURNAL OF THE JAPAN GAS ASSOCIATION)
NIHON GAS KYOKAI (THE JAPAN GAS ASSOCIATION)
38, SHIBA KOTOHIRA-CHO, MINATO-KU, TOKYO 105
JAPANESE
INDEXED: CHE, JICST

NIHON KESHOHIN GIJUTSUSHA RENGOKAI KAISHI 7925
(JOURNAL OF JCCA)
NIHON KESHOHIN GIJUTSUSHA RENGOKAI
C/O RAION HAMIGAKI K.K., 3-7, 1-CHOME, HONJO, SUMIDA-KU, TOKYO 130
JAPANESE, WITH ENGLISH SUMMARY
INDEXED: JICST

NIHON NO SEKIYU KAGAKU KOGYO 7926
(ANNUAL SURVEY OF PETROCHEMICAL INDUSTRIES IN JAPAN)
JUKAGAKU KOGYO TSUSHIN-SHA (THE HEAVY & CHEMICAL INDUSTRIES NEWS AGENCY)
C/O SANRITSU BUILDING, 11-12, 3-CHOME, HATCHOBORI, CHUO-KU, TOKYO 103
JAPANESE

NIHON PURANTO KYOKAI KAIHO 7927
(JOURNAL OF THE JAPAN CONSULTING INSTITUTE)
NIHON PURANTO KYOKAI (JAPAN CONSULTING INSTITUTE)
C/O HIGIYA PARK BUILDING, 1-1, YURAKU-CHO, CHIYODA-KU, TOKYO 100
JAPANESE

NIHON RYUTAN RYUSO KYOKAI KAIHO 7928
(BULLETIN OF JAPAN CARBON SULFIDE AND SODIUM SULFIDE INDUSTRY ASSOCIATION)
NIHON RYUTAN RYUSO KYOKAI
9-6, 2-CHOME, KANDA TA-CHO, CHIYODA-KU, TOKYO 101
JAPANESE

NIHON SEKKEN SENZAI KOGYOKAI, NIHON KAIMEN KASSEIZAI KOGYOKAI KAIHO 7929
(JOINT JOURNAL OF JAPAN SOAP AND DETERGENT ASSOCIATION AND JAPAN SURFACE-ACTIVE ASSOCIATION)
NIHON SEKKEN SENZAI KOGYOKAI, NIHON KAIMEN KASSEIZAI KOGYOKAI (JAPAN SOAP AND DETERGENT ASSOCIATION, THE JAPAN SURFACE-ACTIVE ASSOCIATION)
3-3, NIHONBASHI EDOBASHI, CHUO-KU, TOKYO 103
JAPANESE

NIHON SENBAI KOSHA ODAWARA SEIEN SHIKENJO SHIKEN HOKOKU 7930
(SCIENTIFIC PAPERS OF THE ODAWARA SALT EXPERIMENT STATION, JAPAN MONOPOLY CORPORATION)
NIHON SENBAI KOSHA ODAWARA SEIEN SHIKENJO (ODAWARA SALT EXPERIMENT STATION, JAPAN MONOPOLY CORPORATION)
2-3, SAKAWA, ODAWARA-SHI 256
JAPANESE, WITH ENGLISH SUMMARY
INDEXED: JICST

NIHON SENZAI SHINPO 7931
(JAPAN DETERGENT NEWS)
SENZAI SHINPOSHA
C/O OGASAWARA BUILDING, 6-1, NIHONBASHI HISAMATSU-CHO, CHUO-KU, TOKYO 103
JAPANESE

NIHON TORYO SHINBUN 7932
(PAINT INDUSTRY JOURNAL)
NIHON TORYO SHINBUNSHA
1-7, 3-CHOME, KOISHIKAWA, BUNKYO-KU, TOKYO 112
JAPANESE

NIHON YOSHA KYOKAISHI 7933
(JOURNAL OF THE SPRAY COATING SOCIETY OF JAPAN)
NIHON YOSHA KYOKAI
C/O KINKI DAIGAKU RIKOGAKUBU, 321, KOWAKAE, HIGASHIOSAKA-SHI 577
JAPANESE, WITH ENGLISH SUMMARY
INDEXED: JICST

NIKKAKYO GEPPO 7934
(JAPAN CHEMICAL INDUSTRY ASSOCIATION MONTHLY)
NIHON KAGAKU KOGYO KYOKAI KIKANSHI HENSYU IINKAI (EDITORIAL BOARD, JAPAN CHEMICAL INDUSTRY ASSOCIATION)
2-6, 3-CHOME, KASUMIGASEKI, CHIYODA-KU, TOKYO 100
JAPANESE
INDEXED: CHE, JICST

NIKKAN NENRYO YUSHI SHINBUN 7935
(DAILY FUELS AND FATS NEWS)
NENRYO YUSHI SHINBUNSHA
C/O GASORIN SUTANDO KAIKAN, 2-3, 3-CHOME, SHINBASHI, MINATO-KU, TOKYO 105
JAPANESE

NIKKAN YUGYO HOCHI SHINBUN 7936
(OILS AND FATS DAILY PRESS)
YUGYO HOCHI SHINBUNSHA
15-19, 2-CHOME, SHINKAWA, CHUO-KU, TOKYO 104
JAPANESE

NIKKAN YUSHI TOKUHO 7937
(DAILY OILS AND FATS BULLETIN)
SANGYO SHINBUNSHA
1-69, EDOBORI KITA-DORI, NISHI-KU, OSAKA-SHI 550
JAPANESE

NIPPON KAGAKUKAI SEKIYU KAGAKU TORONKAI KOEN YOKOSHU 7938
(PREPRINTS OF THE SYMPOSIUM ON PETROLEUM CHEMISTRY, JAPAN CHEMICAL SOCIETY)
JAPANESE
INDEXED: JICST

NIPPON KAGAKUKAI YUKAGAKU TORONKAI KOEN YOKOSHU 7939
(PREPRINT OF SYMPOSIUM PETROLEUM CHEMISTRY, JAPAN CHEMICAL SOCIETY)
JAPANESE
INDEXED: JICST

NIPPON KAGAKUKAI YUKI KAGOBUTSU NO MASUSUPEKUTOROMETORI TORONKAI KOEN YOKOSHU 7940
(PREPRINT OF SYMPOSIUM ON MASS SPECTROMETRY OF ORGANIC COMPOUNDS, JAPAN CHEMICAL SOCIETY)
JAPANESE
INDEXED: JICST

NIPPON TUNGSTEN REVIEW 7941
NIPPON TANGUSUTEN K.K. (NIPPON TUNGSTEN COMPANY, LIMITED)
460, SANNO, SHIOBARU, MINAMI-KU, FUKUOKA-SHI 815
ENGLISH
INDEXED: JICST

NISSEKI REBYU 7942
(NISSEKI TECHNICAL REVIEW)
NIHON SEKIYU K.K. KOHOSHITSU (PR OFFICE, NIHON OIL COMPANY, LIMITED)
3-12, 1-CHOME, NISHISHINBASHI, MINATO-KU, TOKYO 105
JAPANESE
INDEXED: JICST

NITTO KAGAKU KOGYO K.K. CHUO KENKYUSHO KENKYU GEPPO 7943
(MONTHLY BULLETIN OF THE CENTRAL LABORATORY, NITTO CHEMICAL INDUSTRY COMPANY, LIMITED)
NITTO KAGAKU KOGYO K.K. CHUO KENKYUSHO
10-1, DAIKOKU-MACHI, TSURUMI-KU, YOKOHAMA-SHI 230
JAPANESE

NITTOSO 7944
(JOURNAL OF JAPAN PAINTING CONTRACTORS ASSOCIATION)
NIHON TOSO KOGYOKAI (JAPAN PAINTING CONTRACTORS ASSOCIATION)
32-6, 2-CHOME, NISHISHINBASHI, MINATO-KU, TOKYO 105
JAPANESE
INDEXED: JICST

NITTOSO GEPPO 7945
(JAPAN PAINTING CONTRACTORS ASSOCIATES MONTHLY)
NIHON TOSO KOGYOKAI (JAPAN PAINTING CONTRACTORS ASSOCIATION)
32-6, 2-CHOME, NISHISHINBASHI, MINATO-KU, TOKYO 105
JAPANESE

NOC (OUCHI SHINKO CHEMICALS INDUSTRIES COMPANY, LIMITED) 7946
OUCHI SHINKO KAGAKU KOGYO K.K. (OUCHI SHINKO CHEMICAL INDUSTRIAL COMPANY, LIMITED)
3-7, 1-CHOME, NIHONBASHI, KOBUNA-CHO, CHUO-KU, TOKYO 103
JAPANESE

NOGUCHI KENKYUSHO JIHO 7947
(ANNUAL REPORT OF NOGUCHI INSTITUTE)
NOGUCHI KENKYUSHO (THE NOGUCHI INSTITUTE)
8-1, 1-CHOME, KAGA, ITABASHI-KU, TOKYO 173
JAPANESE, WITH EUROPEAN SUMMARY
INDEXED: CHE, JICST

OCP GIJUTSU NYUSU 7948
(OCP TECHNOLOGICAL NEWS)
JAPANESE
INDEXED: JICST

OIRU UEIZU INTANASHONARU 7949
(OIL WAYS INTERNATIONAL)
ESSO SUTANDADO SEKIYU (ESSO STANDARD OIL)
JAPANESE

OIRU WARUDO; ENERGY BESSATSU 7950
(OIL WORLD; ENERGY SPECIAL ISSUE)
ESSO SUTANDADO SEKIGU KUBUSHIKU KAISHA KOHOBU (ESSO STANDARD OIL COMPANY, PUBLIC RELATIONS DEPARTMENT)
JAPANESE
INDEXED: JICST

ORIENTAL ECONOMIST (TOKYO) 7951
ENGLISH
INDEXED: CHE

OSAKA GASU 'SOKEN' 7952
(TECHNICAL REPORTS OF OSAKA GAS COMPANY)
OSAKA GASU K.K. SOGO KENKYUSHO (OSAKA GAS COMPANY, LIMITED RESEARCH CENTER)
9-119, TORISHIMA-CHO, KONOHANA-KU, OSAKA-SHI 554
JAPANESE AND ENGLISH

OSAKA KAGAKU GIJUTSU SENTA KENKYU HOKOKU 7953
(RESEARCH BULLETIN OF THE OSAKA SCIENCE TECHNOLOGY CENTER)
JAPANESE
INDEXED: JICST

OSAKA KAGAKU SEN'I TORIHIKIJO GEPPO 7954
(MONTHLY REPORT OF THE OSAKA SYNTHETIC FIBERS EXCHANGE)
JAPANESE

OUTLINE OF THE OIL HYDRAULICS INDUSTRY IN JAPAN 7955
NIHON YUATSU KOGYOKAI
C/O KIKAI SHINKO KAIKAN, 5-8, 3-CHOME, SHIBA KOEN, MINATO-KU, TOKYO 105
ENGLISH

PARAFIN WAKKUSU 7956
(PARAFFIN WAX)
NIPPON SEIRO K.K. (NIPPON SEIRO COMPANY, LIMITED)
C/O CHUO KORON BUILDING, 2-1, KYOBASHI, CHUO-KU, TOKYO 104
JAPANESE

PARUHAMU 7957
(PERFUME)
CHUO KORYO K.K. (CHUO PERFUMERY INC.)
2-1, NIHONBASHI, KOBUNA-CHO, CHUO-KU, TOKYO 103
JAPANESE

PE 7958
PORIOREFIN KONWAKAI KOATSU PORIECHIRIN BUKAI (POLYOLEFIN ROUND TABLE, HIGH PRESSURE POLYETHYLENE DIVISION)
JAPANESE AND ENGLISH

PEINTO 7959
(PAINT)
JAPANESE
INDEXED: JICST

PETORO KEMIKARU ENJINIARINGU 7960
(PETRO CHEMICAL ENGINEERING)
JUKAGAKU KOGYO TSUSHINSHA
C/O SANRITSU BUILDING, 11-12, 3-CHOME, HATCHOBORI, CHUO-KU, TOKYO 104-91
JAPANESE

PETROLEUM ECONOMIST 7961
SEKIYU HYORONSHU (THE OIL REVIEW)
JAPANESE
INDEXED: JICST

PETROLEUM INDUSTRY IN JAPAN 7962
SEKAI SEKIYU KAIGI NIHON KOKUNAI IINKAI (THE JAPANESE NATIONAL COMMITTEE OF THE WORLD PETROLEUM CONGRESS)
C/O KASAHARA BUILDING, 6-10, 1-CHOME, UCHI KANDA, CHIYODA-KU, TOKYO 101
ENGLISH
INDEXED: JICST

PETROLEUM PRESS SERVICE (JAPANESE EDITION) 7963
SEKIYU HYORONSHA
C/O NISSEKI BUILDING, 4-2, 3-CHOME, MARUNOUCHI, CHIYODA-KU, TOKYO 100
JAPANESE

PETROTECH 7964
JAPANESE

PUROPAN BUTAN GURAFU 7965
(PROPANE AND BUTANE GRAPH)
SEKIYU KAGAKU SHINBUN DHA (PETROCHEMICAL NEWSPAPER COMPANY)
JAPANESE

PUROPAN SHINBUN 7966
(PROPANE NEWS)
SANPO
10-17, 1-CHOME, HAMANATSU-CHO, MINATO-KU, TOKYO 105
JAPANESE

PURUBU 7967
(PULP)
TOYO REYON (TOYO RAYON)
JAPANESE

RINSAN HIRYO KAIGAI BUNKEN SORAN 7968
(REVIEW OF FOREIGN LITERATURE ON PHOSPHATIC FERTILIZER)
NIHON KASEI HIRYO KYOKAI
JAPANESE
INDEXED: JICST

RYOEN 7969
(JOURNAL OF JAPAN GAS APPLICANCES INSPECTION ASSOCIATION)
NIHON GASUKIKI KENSA KYOKAI (JAPAN GAS APPLIANCES INSPECTION ASSOCIATION)
C/O DAI-18 MORI BUILDING, 20, SHIBA NISHIKUBO AKEFUNE-CHO, MINATO-KU, TOKYO 105
JAPANESE

RYUSAN JIHO 7970
(SULFURIC ACID TIMES)
RYUSAN KYOKAI (SULFURIC ACID ASSOCIATION OF JAPAN)
C/O SHINBASHI EKIMAE BUILDING, 2-GO KAN, 21-1, 2-CHOME, SHINBASHI, MINATO-KU, TOKYO 105
JAPANESE

RYUSAN TO KOGYO 7971
(SULPHURIC ACID AND INDUSTRY)
RYUSAN KYOKAI GIJUTSUBU (THE SULPHURIC ACID ASSOCIATION OF JAPAN, TECHNICAL DIVISION) AND RYUSAN KYOKAI (THE SULPHURIC ACID ASSOCIATION OF JAPAN)
C/O SHINBASHI EKIMAE BUILDING, 2-GO KAN, 21-1, 2-CHOME, SHINBASHI, MINATO-KU, TOKYO 105
JAPANESE, WITH ENGLISH SUMMARY
INDEXED: CHE, JICST

SAFAKUTANTO 7972
JAPANESE
INDEXED: JICST

SAIKIN NO KAGAKU KOGAKU 7973
(RECENT CHEMICAL ENGINEERING)
KAGAKU KOGAKU KYOKAI (THE SOCIETY OF CHEMICAL ENGINEERS, JAPAN) AND MARUZEN K.K.
2-6, NIHONBASHI-DORI, CHUO-KU, TOKYO 103
JAPANESE

SAN'YO KASEI NYUSU 7974
(SAN'YO KASEI NEWS)
SAN'YO KASEI KOGYO K.K.
11-1, IKKYO NOMOTO-CHO, HIGASHIYAMA-KU, KYOTO-SHI 605
JAPANESE
INDEXED: JICST

SEKAI NO KAGAKU KOGYO 7975
(WORLD CHEMICAL INDUSTRIES)
NOGUCHI KENKYUSHO (NOGUCHI INSTITUTE) AND CMC (CMC COMPANY LIMITED)
11-13, 1-CHOME, UCHI KANDA, CHIYODA-KU, TOKYO 101
JAPANESE

SEKAI NO SEKIYU KAGAKU; DOKO TO JOHO 7976
(WORLD PETROLEUM CHEMISTRY TRENDS AND INFORMATION)
JAPANESE
INDEXED: JICST

SEKAI SEKIYU KAIGI NIPPON KOKUNAI IINKAI KAIHO 7977
(REPORT OF THE JAPANESE DOMESTIC COMMITTEE OF THE WORLD PETROLEUM CONFERENCE)
JAPANESE
INDEXED: JICST

SEKITAN, KOKUSU TOKEI NENPO 7978
(YEARBOOK OF COAL AND COKE STATISTICS)
TSUSHO SANGYO-SHO DAIJIN KANBO TOKEI CHOSA-BU TSUSHO SANGYO CHOSAKAI (RESEARCH AND STATISTIC DEPARTMENT, MINISTER'S SECRETARIAT, MINISTRY OF INTERNATIONAL TRADE AND INDUSTRY)
15-2, 6-CHOME, GINZA, CHUO-KU, TOKYO 100
JAPANESE

SEKITSU SOKUHO 7979
(SEKITSU INFORMATION NEWS)
SEKITSU
7-2, 2-CHOME, NANOKAWA, CHUO-KU, FUKUOKA-SHI 810
JAPANESE

SEKIYU GAKKAISHI 7980
(JOURNAL OF THE JAPAN PETROLEUM INSTITUTE)
SEKIYU GAKKAI (THE JAPAN PETROLEUM INSTITUTE)
C/O NISSEKI BUILDING, 4-2, 3-CHOME, MARUNOUCHI, CHIYODA-KU, TOKYO 100
JAPANESE, WITH EUROPEAN SUMMARY
INDEXED: API, CHE, COP, JICST

SEKIYU GIJUTSU KYOKAISHI 7981
(JOURNAL OF THE JAPANESE ASSOCIATION OF PETROLEUM TECHNOLOGISTS)
SEKIYU GIJUTSU KYOKAI (JAPANESE ASSOCIATION OF PETROLEUM TECHNOLOGISTS)
C/O SEKIYU KOGYO RENMEI, 9-4, 1-CHOME, OTE-MACHI, CHIYODA-KU, TOKYO 100
JAPANESE AND ENGLISH
INDEXED: CHE, JICST

SEKIYU KAGAKU TSUSHIN 7982
(PETRO CHEMICAL NEWS)
JUKAGAKU KOGYO TSUSHINSHA
11-12, 3-CHOME, HATCHOBORI, CHUO-KU, TOKYO 104
JAPANESE

SEKIYU KAIHATSU JIHO 7983
(PETROLEUM RESOURCES DEVELOPMENT TIMES)
SEKIYU KOGYO RENMEI (PETROLEUM PRODUCERS ASSOCIATION)
C/O KEIDANREN KAIKAN, 9-4, 1-CHOME, OTE-MACHI, CHIYODA-KU, TOKYO 100
JAPANESE

SEKIYU KAIHATSU KANKEI SHIRYO 7984
(PETROLEUM DEVELOPING NEWS)
SEKIYU KOGYO RENMEI (PETROLEUM MINING LEAGUE OF JAPAN)
9-4, 1-CHOME, OTE-MACHI, CHIYODA-KU, TOKYO 100
JAPANESE

SEKIYU KAIHATSU KODAN SEKIYU KAIHATSU GIJUTSU SENTA KENKYU HOKOKU 7985
(REPORT OF THE TECHNOLOGY RESEARCH CENTER, JAPAN PETROLEUM DEVELOPMENT CORPORATION)
JAPANESE
INDEXED: JICST

SEKIYU NO KAIHATSU 7986
(PETROLEUM DEVELOPING)
SEKIYU KAIHATSU KODAN
C/O HIBIYA NITCHU BUILDING, 1-4, 2-CHOME, UCHISAIWAI-CHO, CHIYODA-KU, TOKYO 100
JAPANESE
INDEXED: JICST

SEKIYU SANGYO 7987
(PETROLEUM INDUSTRY)
SANGYO JIHOSHA
11-10, 1-CHOME, NISHISHINBASHI, MINATO-KU, TOKYO 105
JAPANESE

SEKIYU; SEKIYU KAGAKU PURANTO GIJUTSU 7988
(PETROLEUM; PETROLEUM CHEMISTRY PLANT TECHNOLOGY)
JAPANESE
INDEXED: JICST

SEKIYU SHIRYO 7989
(PETROLEUM NEWS)
TSUSHO SANGYO-SHO SHIGEN ENERUGI-CHO SEKIYU-BU (PETROLEUM DEPARTMENT, AGENCY OF NATURAL RESOURCES AND ENERGY, MINISTRY OF INTERNATIONAL TRADE AND INDUSTRY)
16-1-501, 2-CHOME, SHINBASHI, MINATO-KU, TOKYO 105
JAPANESE

SEKIYU SHIRYO GEPPO 7990
(MONTHLY REPORT ON PETROLEUM INDUSTRY)
SEKIYU RENMEI (PETROLEUM ASSOCIATION OF JAPAN) AND SEKIYU TSUSHINSHA (PETROLEUM PRESS LTD.)
16-1, 2-CHOME, SHINBASHI, MINATO-KU, TOKYO 105
JAPANESE

SEKIYU SHUNJU 7991
(PETROLEUM INDUSTRY INFORMATION)
SEKIYU SHUNJUSHA
11-4, 2-CHOME, NISHI SHINBASHI, MINATO-KU, TOKYO 105
JAPANESE

SEKIYU TO SEKIYU KAGAKU 7992
(PETROLEUM AND PETROCHEMICALS)
SAIWAI SHOBO
C/O SUZUKI BUILDING, 5-2, 1-CHOME, SOTO KANDA, CHIYODA-KU, TOKYO 101
JAPANESE
INDEXED: CHE, JICST

SEKIYU TOKEI GEPPO 7993
(PETROLEUM STATISTICAL MONTHLY)
TSUSHO SANGYO-SHO DAIJIN KANBO TOKEI CHOSA-BU (RESEARCH AND STATISTICS DIVISION, MINISTER'S SECRETARIAT, MINISTRY OF INTERNATIONAL TRADE AND INDUSTRY)
15-1, 6-CHOME, GINZA, CHUO-KU, TOKYO 104
JAPANESE
INDEXED: JICST

SEKIYU TOKEI NENPO 7994
(PETROLEUM STATISTICAL YEARBOOK)
TSUSHO SANGYO-SHO DAIJIN KANBO TOKEI CHOSA-BU (RESEARCH AND STATISTICAL DEPARTMENT, MINISTER'S SECRETARIAT, MINISTRY OF INTERNATIONAL TRADE AND INDUSTRY)
15-1, 6-CHOME, GINZA, CHUO-KU, TOKYO 104
JAPANESE

SEKIYU TOKEIHYO 7995
(PETROLEUM STATISTICS)
NIHON SEKIYU K.K.
3-12, 1-CHOME, NISHISHINBASHI, MINATO-KU, TOKYO 105
JAPANESE

SEMEDAIN REBYU 7996
(SEMEDAIN REVIEW)
JAPANESE
INDEXED: JICST

SENRYO TO YAKUHIN 7997
(DYESTUFFS AND CHEMICALS)
KASEIHIN KOGYO KYOKAI (JAPAN DYESTUFF INDUSTRY ASSOCIATION)
18-7, 5-CHOME, ROPPONGI, MINATO-KU, TOKYO 106
JAPANESE
INDEXED: CHE, JICST

SENSHOKU TO KAKO 7998
(DYING AND PROCESSING)
JAPANESE

SHERU BURIFINGU SABISU 7999
(SHELL BRIEFING SERVICE)
SHERU SEKIYU K.K. (SHELL OIL COMPANY)
C/O TOKYO BUILDING, 3-2, 2-CHOME, MARUNOUCHI, CHIYODA-KU, TOKYO 100
JAPANESE

SHERU KAGAKU GIJUTSU NYUSU 8000
(SHELL SCIENCE AND TECHNOLOGY NEWS LETTER)
SHERU KOSAN K.K. KOHO SOMU-BU
C/O KASUMIGASEKI BUILDING, 2-5, 3-CHOME, KASUMIGASEKI, CHIYODA-KU, TOKYO 100
JAPANESE

SHIGEN TO ENERUGI 8001
(NATURAL RESOURCES AND ENERGY)
SHIGEN ENERUGI MONDAI KENKYUJO (RESEARCH INSTITUTE FOR THE PROBLEM OF ENERGY FROM NATURAL RESOURCES)
JAPANESE

SHIN NIHON SEITETSU KAGAKU KOGYO K.K. TOBATA SEIZOSHO GIJUTSU KENKYUSHO KENKYU GEPPO 8002
(MONTHLY REPORT OF TECHNICAL LABORATORY, TOBATA WORKS, SHIN NIHON STEEL CHEMICAL INDUSTRY COMPANY, LIMITED)
SHIN NIHON SEITETSU KAGAKU KOGYO K.K. TOBATA SEIZOSHO GIJUTSU KENKYUSHO
46-51, SAKINOHAMA, NAKAHARA, TOBATA-KU, KITA KYUSHU-SHI 804
JAPANESE

SHINTO KENKYU GIJUTSU YOHO 8003
(BULLETIN OF THE TECHNICAL DIVISION, APPLIED TECHNIQUE DEPARTMENT, SHINTO PAINT COMPANY)
SHINTO TORYO K.K. GIJUTSU HONBU OYO GIJUTSU-BU (TECHNICAL DIVISION, APPLIED TECHNIQUE DEPARTMENT, SHINTO PAINT COMPANY)
10-73, 6-CHOME, MINAMI TSUKAGUCHI-CHO, AMAGASAKI-SHI 661
JAPANESE

SHIONO KORYO SHOHO 8004
(SHIONO PERFUMERY JOURNAL)
SHIONO KORYO K.K.
3-11, DOSHO-MACHI, HIGASHI-KU, OSAKA-SHI 541
JAPANESE

SHIZUOKA-KEN KOGYO GIJUTSU JOHO, KAGAKU-HEN 8005
(SHIZUOKA PREFECTURE INDUSTRIAL TECHNOLOGY REPORT, CHEMISTRY EDITION)
SHIZUOKA-KEN KOGYO SHIKENJO (SHIZUOKA PREFECTURE INDUSTRIAL EXPERIMENT STATION)
JAPANESE

SHUKAN KAGAKU KOGYO SHIRYO 8006
(CHEMICAL AND ENGINEERING DIGEST WEEKLY, INFORMATION SERVICE EDITION)
HORIKOSHI KENKYUSHO
C/O FUJI BUILDING, KANDA JINBO-CHO, CHIYODA-KU, TOKYO 101
JAPANESE
INDEXED: JICST

SHUKAN SEKAI NO KAGAKU KOGYO 8007
(WEEKLY WORLD CHEMICAL INDUSTRY)
NOGUCHI KENKYUSHO (NOGUCHI INSTITUTE) AND CMC K.K.
C/O KUSUMOTO BUILDING, 11-13, 1-CHOME, UCHI KANDA CHIYODA-KU, TOKYO 101
JAPANESE

SHUKAN SEKIYU GASU SHIRYO 8008
(WEEKLY INFORMATION OF PETROLEUM OIL AND GAS)
SEKIYU GASU CHOSAKAI
1-5, KANDA SURUGADAI, CHIYODA-KU, TOKYO 101
JAPANESE

SODA KOGYO KAIGAI JOHO 8009
(OVERSEAS INFORMATION OF SODA INDUSTRY)
NIHON SODA KOGYOKAI (JAPAN SODA INDUSTRY ASSOCIATION)
C/O HORIUCHI BUILDING, 2-10, KANDA KAJI-CHO, CHIYODA-KU, TOKYO 101
JAPANESE

SODA TO ENSO 8010
(SODA AND CHLORINE)
NIHON SODA KOGYOKAI (JAPAN SODA INDUSTRY ASSOCIATION)
C/O HORIUCHI BUILDING, 2-10, KANDA KAJI-CHO, CHIYODA-KU, TOKYO 101
JAPANESE
INDEXED: CHE, JICST

STATISTICS ON OIL AND FAT PRODUCTS AND RAW MATERIALS IN JAPAN 8011
JAPAN SOAP AND DETERGENT ASSOCIATION
ENGLISH
INDEXED: BIB

SUISAN YUSHI 8012
(AQUATIC OIL)
NIHON SUISAN YUSHI KYOKAI (JAPAN AQUATIC OIL ASSOCIATION)
52-1, MOTOYOYOGI-CHO, SHIBUYA-KU, TOKYO 151
JAPANESE
INDEXED: JICST

SUISAN YUSHI NO SHIORI 8013
(GUIDEBOOK TO FISH FATS AND OILS)
NIHON SUISAN YUSHI KYOKAI (JAPAN FISHERY FATS AND OILS ASSOCIATION)
JAPANESE

SUMITOMO KAGAKU 8014
(SUMITOMO CHEMICAL)
JAPANESE
INDEXED: JICST

SUMITOMO KAGAKU TOKUSHUGO 8015
(SUMITOMO CHEMICAL REVIEW, SPECIAL ISSUE)
SUMITOMO KAGAKU KOGYO K.K. (SUMITOMO CHEMICAL COMPANY, LIMITED)
5-15, KITAHAMA, HIGASHI-KU, OSAKA-SHI 541
JAPANESE

TAKASAGO KORYO JIHO 8016
(TAKASAGO TIMES)
TAKASAGO KORYO KOGYO K.K. (TAKASAGO PERFUMERY COMPANY, LIMITED)
36-31, 5-CHOME, KAMATA, OTA-KU, TOKYO 144
JAPANESE
INDEXED: JICST

TANSO 8017
(CARBONS)
TANSO ZAIRYO KENKYUKAI (CARBON SOCIETY OF JAPAN)
C/O TEIO KOGAKKAI, KINOKUNIYA BUILDING, 17-7, 3-CHOME, TSUNOHAZU, SHINJUKU-KU, TOKYO 160
JAPANESE AND EUROPEAN
INDEXED: CHE, JICST

TCI MERU 8018
(TCI MAIL)
TOKYO KASEI KOGYO K.K. (TOKYO CHEMICAL INDUSTRY COMPANY, LIMITED)
15-9, 6-CHOME, TOSHIMA, KITA-KU, TOKYO 114
JAPANESE

TENNEN GASU 8019
(NATURAL GAS)
TENNEN GASU KOGYOKAI (NATURAL GAS ASSOCIATION)
C/O DAI-5 MORI BUILDING, 25, SHIBA NISHIKUBO SAKURAGAWA-CHO, MINATO-KU, TOKYO 105
JAPANESE
INDEXED: JICST

TENNEN GASU NENPO 8020
(ANNUAL REPORT OF NATURAL GAS)
TENNEN GASU KOGYOKAI (JAPAN NATURAL GAS MINING ASSOCIATION)
C/O DAI-5 MORI BLDG., SHIBA NISHIKUBO SAKURAGAWA-CHO, MINATO-KU, TOKYO 105
JAPANESE

TIDINGS (TECHNICAL INFORMATION FOR DYEING INDUSTRY IN NIPPON KAYAKU GENERAL SERVICE) 8021
NIHON KAYAKU K.K.
C/O TOKYO KAIJO BUILDING, SHINKAN, 2-1, 1-CHOME, MARUNOUCHI, CHIYODA-KU, TOKYO 100
JAPANESE

TOA 8022
TOA KOSAN KABUSHIKI KAISHA KOHO KA
(TOA OIL COMPANY, LIMITED, PUBLIC RELATIONS SECTION)
JAPANESE

TOHOKU DAIGAKU HISUIYOEKI KAGAKU KENKYUSHO HOKOKU 8023
(BULLETIN OF THE CHEMICAL RESEARCH INSTITUTE OF NON-AQUEOUS SOLUTIONS, TOHOKU UNIVERSITY)
TOHOKU DAIGAKU HISUIYOEKI KAGAKU KENKYUSHO (THE CHEMICAL RESEARCH INSTITUTE OF NON-AQUEOUS SOLUTIONS, TOHOKU UNIVERSITY)
1-1, 2-CHOME, KATAHIRA, SENDAI-SHI 980
JAPANESE, WITH ENGLISH SUMMARY
INDEXED: CHE, JICST

TOKAKO KAIHO 8024
(TOKAKO BULLETIN)
TOKAI KAGAKU KOGYOKAI
C/O CHUBU KAGAKU GIJUTSU SENTA, 17-22, 2-CHOME, SAKAE, NAKA-KU, NAGOYA-SHI 460
JAPANESE

TOKOSHI NYUSU, KAGAKU KOGYO SHIRYO 8025
(NEWS OF NATIONAL CHEMICAL LABORATORY FOR INDUSTRY, CHEMICAL ENGINEERS DIGEST)
TSUSHO SANGYO-SHO KOGYO GIJUTSU-IN TOKYO KOGYO SHIKENSHO (NATIONAL CHEMICAL LABORATORY FOR INDUSTRY, AGENCY OF INDUSTRIAL SCIENCE & TECHNOLOGY)
1-5, 1-CHOME, HON-MACHI, SHIBUYA-KU, TOKYO 150
JAPANESE
INDEXED: JICST

TOKYO GASU K.K. SOGO KENKYUSHO HOKOKU 8026
(REPORT OF THE CENTRAL RESEARCH LABORATORY, TOKYO GAS COMPANY, LIMITED)
TOKYO GASU K.K. SOGO KENKYUSHO (CENTRAL RESEARCH LABORATORY, TOKYO GAS COMPANY, LIMITED)
16-25, 1-CHOME, SHIBAURA, MINATO-KU, TOKYO 105
JAPANESE

TOKYO GASU REPOTO 8027
(TOKYO GAS REPORT)
TOKYO GASU KABUSHIKI KAISHA KOHOSHITSU
(TOKYO GAS LIMITED, PUBLIC RELATIONS OFFICE)
JAPANESE

TOPICS IN NONBENZENOID AROMATIC CHEMISTRY 8028
HIROKAWA PUBLISHING COMPANY
HONGO-3, BUNKYO-KU, TOKYO
ENGLISH
INDEXED: CHE

TORE JIHO 8029
(TORAY REVIEW (TOYO RAYON) REVIEW)
TOYO REYON (TOYO RAYON)
JAPANESE

TORYO 8030
(PAINTS)
NIPPON PEINTO K.K. (NIPPON PAINT COMPANY, LIMITED)
1-1, 2-CHOME, OYODO-CHO KITA, OYODO-KU, OSAKA-SHI 531
JAPANESE
INDEXED: JICST

TORYO HOCHI 8031
(PAINT NEWS)
TORYO HOCHI SHINBUNSHA
3, YARAI-CHO, SHINJUKU-KU, TOKYO 162
JAPANESE

TORYO NO KENKYU 8032
(STUDIES IN PAINT)
KANSAI PEINTO K.K.
5-27, FUSHIMI-CHO, HIGASHI-KU, OSAKA-SHI 541
JAPANESE
INDEXED: CHE, JICST

TORYO SHINPO 8033
(PAINT NEWS)
TORYO ASAHI
C/O KITAKAMI BUILDING, 1-6, KANDA IZUMI-CHO,
CHIYODA-KU, TOKYO 101
JAPANESE AND ENGLISH

TORYO TO TOSO 8034
(PAINT AND PAINTING)
SHINTO TORYO K.K. (SHINTO PAINT COMPANY, LIMITED)
10-73, 6-CHOME, MINAMITSUKAGUCHI-CHO,
AMAGASAKI-SHI 661
JAPANESE

TOSO GIJUTSU 8035
(PAINTING TECHNOLOGY)
RIKO SHUPPANSHA
6-8, 3-CHOME, IIDABASHI, CHIYODA-KU, TOKYO 102
JAPANESE
INDEXED: CHE, JICST

TOSO NO GIJUTSU 8036
(PAINTING ENGINEERING)
NIHON TOSO GIJUTSU KYOKAI
3-20, KANDA NISHIKI-CHO, CHIYODA-KU,
TOKYO 101
JAPANESE
INDEXED: JICST

TOSO TO TORYO 8037
(FINISH AND PAINT)
TORYO SHUPPANSHA
18-4, 2-CHOME, TAKANAWA, MINATO-KU, TOKYO 108
JAPANESE
INDEXED: CHE, JICST

TOYO SODA KENKYU HOKOKU 8038
(SCIENTIFIC REPORT OF TOYO SODA MANUFACTURING
COMPANY, LIMITED)
TOYO SODA KOGYO K.K. (TOYO SODA MANUFACTURING
COMPANY, LTD.)
4560, TONDA, SHINNANYO-SHI 746
JAPANESE AND ENGLISH
INDEXED: CHE, JICST

YAKUHIN SHINBUN 8039
(INDUSTRIAL CHEMICALS NEWS)
YAKUHIN SHINBUNSHA
C/O YASUDA BUILDING, 1-155, KYO-MACHIBORI, NISHI-KU,
OSAKA-SHI 550
JAPANESE

YOYUEN 8040
(MOLTEN SALT)
DENKI KAGAKU KYOKAI YOYUEN IINKAI (MOLTEN SALT
COMMITTEE, ELECTRICAL CHEMISTRY ASSOCIATION)
C/O KYOTO DAIGAKU KOGAKU-BU KOGYO KAGAKU KYOSHITSU,
YOSHIDA, SAKYO-KU, KYOTO-SHI 606
JAPANESE
INDEXED: CHE, JICST

YUASA JIHO 8041
(YUASA TECHNICAL INFORMATION)
YUASA DENCHI K.K. (YUASA BATTERY COMPANY, LTD.)
6-6, JOSAI-CHO, TAKATSUKI-SHI 569
JAPANESE, WITH ENGLISH SUMMARY
INDEXED: CHE, JICST

YUATSU 8042
(OIL PRESSURE)
NIHON YUATSU KOGYOKAI (JAPAN OIL PRESSURE INDUSTRY
ASSOCIATION)
C/O KIKAI SHINKO KAIKAN, 5-8, 3-CHOME, SHIBA KOEN,
MINATO-KU, TOKYO 105
JAPANESE
INDEXED: JICST

YUATSU GEPPO 8043
(OIL PRESSURE MONTHLY)
NIHON YUATSU KOGYOKAI
C/O KIKAI SHINKO KAIKAN, 5-8, 3-CHOME, SHIBA KOEN,
MINATO-KU, TOKYO 105
JAPANESE

YUKAGAKU 8044
(JOURNAL OF THE JAPAN OIL CHEMISTS' SOCIETY)
NIHON YUKAGAKU KYOKAI (THE JAPAN OIL CHEMISTS'
SOCIETY)
C/O YUSHI KOGYO KAIKAN, 13-11, 3-CHOME, NIHONBASHI,
CHUO-KU, TOKYO 103
JAPANESE, WITH ENGLISH SUMMARY
INDEXED: CHE, JICST

YUKI GOSEI KAGAKU KYOKAISHI 8045
(JOURNAL OF SYNTHETIC ORGANIC CHEMISTRY, JAPAN)
YUKI GOSEI KAGAKU KYOKAI (THE SOCIETY OF
SYNTHETIC ORGANIC CHEMISTRY, JAPAN)
2-12, 2-CHOME, AZBUDAI, MINATO-KU, TOKYO 106
JAPANESE, WITH ENGLISH SUMMARY
INDEXED: CHE. JICST

YUSHI 8046
(OIL AND FAT)
SAIWAI SHOBO
5-2, 1-CHOME, SOTO KANDA, CHIYODA-KU, TOKYO 101
JAPANESE
INDEXED: JICST

ZENNEN KAIHO 8047
(BULLETIN OF THE ALL JAPAN FUEL ASSOCIATION)
ZENKOKU NENRYO KYOKAI (ALL JAPAN FUEL ASSOCIATION)
12-15, 8-CHOME, GINZA, CHUO-KU, TOKYO 104
JAPANESE

ZEON NYUSU 8048
(GEON NEWS)
NIHON ZEON K.K. (JAPAN GEON COMPANY, LIMITED)
6-1, 2-CHOME, MARUNOUCHI, CHIYODA-KU, TOKYO 100
JAPANESE

Beverages, Stimulants, Food Industries, Preservation

AICHI-KEN SHOKUHIN KOGYO SHIKENSHO NEMPO 8049
(ANNUAL REPORT OF THE FOOD RESEARCH INSTITUTE, AICHI PREFECTURE)
AICHI-KEN SHOKUHIN KOGYO SHIKENJO (FOOD RESEARCH INSTITUTE, AICHI PREFECTURE)
2-1, SHINPUKUJI-CHO, NISHI-KU, NAGOYA-SHI 451
JAPANESE
INDEXED: CHE, JICST

AMINOSAN KYOKAI GIJUTSUBU KAIHO 8050
(BULLETIN OF THE TECHNICAL DEPARTMENT, AMINO ACIDS ASSOCIATION)
AMINOSAN KYOKAI (AMINO ACID ASSOCIATION)
5-8, 3-CHOME, NIHONBASHI, CHUO-KU, TOKYO 103
JAPANESE
INDEXED: JICST

ASAHI BIRU K.K. CHUO KENKYUSHO HOKOKU 8051
(REPORT OF CENTRAL RESEARCH LABORATORY, ASAHI BEER COMPANY, LIMITED))
ASAHI BIRU K.K. CHUO KENKYUSHO
13-1, 2-CHOME, OMORIKITA, OTA-KU, TOKYO 143
JAPANESE

ASHITA HO SHOKUHIN SANGYO 8052
(FOOD INDUSTRY FOR TOMORROW)
SHOKUHIN SANGYO SENTA (FOOD INDUSTRY CENTER)
JAPANESE

BIBLIOGRAPHY OF FERMENTATION MICROORGANIZMS 8053
KEIGAKU PUBLISHING COMPANY
JAPANESE
INDEXED: BIB

BIKOKEN NYUSU 8054
(FERMENTATION RESEARCH INSTITUTE NEWS)
TSUSHO SANGYO-SHO KOGYO GIJUTSU-IN BISEIBUTSU KOGYO GIJUTSU KENKYUSHO (FERMENTATION RESEARCH INSTITUTE, AGENCY OF INDUSTRIAL SCIENCE AND TECHNOLOGY, MINISTRY OF INTERNATIONAL TRADE AND INDUSTRY)
8-1, 5-CHOME, INAGEHIGASHI, CHIBA-SHI 281
JAPANESE
INDEXED: JICST

BISCUIT 8055
ZENKOKU BISUKETTO KYOKAI (JAPAN BISCUIT ASSOCIATION)
9-5, 6-CHOME, SHINBASHI, MINATO-KU, TOKYO 105
JAPANESE

BISEIBUTSU KOGYO GIJUTSU KENKYUSHO KENKYU HOKOKU 8056
(REPORT OF THE FERMENTATION RESEARCH INSTITUTE)
TSUSHO SANGYO-SHO KOGYO GIJUTSU-IN BISEIBUTSU KOGYO GIJUTSU KENKYUSHO (FERMENTATION RESEARCH INSTITUTE, AGENCY OF INDUSTRIAL SCIENCE AND TECHNOLOGY, MINISTRY OF INTERNATIONAL TRADE & INDUSTRY)
8-1, 5-CHOME, INAGEHIGASHI, CHIBA-SHI 281
JAPANESE AND ENGLISH
INDEXED: BIB, BIO, CHE

BISEIBUTSU KOGYO GIJUTSU KENKYUSHO NENPO 8057
(ANNUAL REPORT OF FERMENTATION RESEARCH INSTITUTE)
TSUSHO SANGYO-SHO KOGYO GIJUTSU-IN BISEIBUTSU KOGYO GIJUTSU KENKYUSHO (FERMENTATION RESEARCH INSTITUTE, AGENCY OF INDUSTRIAL SCIENCE AND TECHNOLOGY, MINISTRY OF INTERNATIONAL TRADE AND INDUSTRY)
8-1, 5-CHOME, INAGEHIGASHI, CHIBA-SHI 281
JAPANESE

BULLETIN OF BREWING SCIENCE 8058
JOZO KAGAKU KENKYUSHO (THE BREWING SCIENCE RESEARCH INSTITUTE)
4-1, 1-CHOME, MITA, MEGURO-KU, TOKYO 153
EUROPEAN
INDEXED: BIB, CHE, JICST

CHA TO KOHI 8059
(COFFEE AND TEA)
CHA TO KOHISHA
C/O ROOM NUMBER 308, KANTO KOTAKIBASHI BUILDING, 23-14, 5-CHOME, HIGASHINAKANO, NAKANO-KU, TOKYO 164
JAPANESE

CHAGYO KENKYU HOKOKU, SHIRYO 8060
(TEA RESEARCH REPORT DATA)
JAPAN TEA TECHNICAL ASSOCIATION, 2769, KANAYA, KANAYA-CHO, HAIBARA-GUN SHIZUOKA
JAPANESE
INDEXED: CHE

CHOSHI SHOYU K.K. KENKYU HOKOKU 8061
(RESEARCH REPORTS OF CHOSHI SHOYU COMPANY, LIMITED)
CHOSHI SHOYU K.K.
1-210, KOYA, CHOSHI-SHI 288
JAPANESE

CHUGOKU JOKAI SHOKUHIN SHINBUN 8062
(CHUGOKU BREWING AND FOOD INDUSTRY NEWS)
CHUGOKU JOKAI SHOKUHIN SHINBUNSHA
35-4, 3-CHOME, KOINAKA, HIROSHIMA-SHI 733
JAPANESE

DAIZU TANPAKU SHINBUN 8063
(SOYBEAN FOOD NEWS)
DAIZU TANPAKU SHINBUNSHA (SOYBEAN FOOD NEWS)
7-1 3-CHOME, MINAMIDAI, NAKANO-KU, TOKYO 164
JAPANESE

DENPUN 8064
(STARCH)
NIHON DENPUN GAKKAI CHUBU SHIBU (CHUBU BRANCH OFFICE, JAPANESE SOCIETY OF STARCH SCIENCE)
C/O AICHI-KEN SHOKUHIN KOGYO SHIKENJO, 2-1, SHINPUKUJI-CHO, NISHI-KU, NAGOYA-SHI 451
JAPANESE
INDEXED: CHE, JICST

DENPUN KAGAKU 8065
(JOURNAL OF THE JAPANESE SOCIETY OF STARCH SCIENCE)
NIHON DENPUN GAKKAI (THE JAPANESE SOCIETY OF STARCH SCIENCE)
C/O NORIN-SHO SHOKUHIN SOGO KENKYUSHO, 4-12, 1-CHOME, SHIOHAMA, KOTO-KU, TOKYO 135
JAPANESE AND ENGLISH
INDEXED: CHE, JICST

EHIME SHOKURYO JIMUSHO YORAN 8066
(EHIME FOOD OFFICE STATISTICS)
NORIN-SHO EHIME SHOKURYO JIMUSHO (EHIME FOOD OFFICE, MINISTRY OF AGRICULTURE AND FORESTRY)
2-33, TSUJI-MACHI, MATSUYAMA-SHI 790
JAPANESE

FLAVORIST 8067
SHIRAISHI KENKYUSHITSU (SHIRAISHI LABORATORY)
301, SHIMOFUDA, CHOFU-SHI, TOKYO 182
JAPANESE, WITH ENGLISH SUMMARY

FUDO NYUSU 8068
(FOOD NEWS)
FUDO RYUTSU KEIZAI KENKYUJO
(FOOD DISTRIBUTION ECONOMICS RESEARCH INSTITUTE
JAPANESE

FUDO SABISU DEZAIN 8069
(FOOD SERVICE DESIGN)
SHIBATA SHOTEN (SHIBATA BOOKS)
JAPANESE

FUSHIMI JOYUKAI KAISHI 8070
(JOURNAL OF THE BREWERS' ASSOCIATION, FUSHIMI)
FUSHIMI JOYUKAI (BREWERS' ASSOCIATION FUSHIMI)
OTE-MACHI, FUSHIMI-KU, KYOTO-SHI 612
JAPANESE
INDEXED: JICST

GATEAUX 8071
NIHON YOGASHI KYOKAI RENGOKAI (FEDERATION OF JAPAN CONFECTIONERY ASSOCIATION)
19-26, 1-CHOME, EBISU, SHIBUYA-KU, TOKYO 150
JAPANESE

GEKKAN KORUDO CHEN GIJUTSU 8072
(COLD CHAIN AND TECHNOLOGY)
NIHON KUCHO GIJUTSU SHUPPANSHA
3-1, 1-CHOME, SARUGAKU-CHO, CHIYODA-KU, TOKYO 101
JAPANESE

GEKKAN SHOKUHIN 8073
(MONTHLY FOOD)
SHOKUHIN KENKYUSHA (INSTITUTE OF FOOD SCIENCE)
9-14, 1-CHOME, YUSHIMA, BUNKYO-KU, TOKYO 113
JAPANESE

GEKKAN SHOKURYO 8074
(MONTHLY 'RICE')
ZENKOKU SHOKURYO JIGYO KYODO KUMIAI RENGOKAI KOHO-BU (FEDERATION OF JAPAN RICE DEALERS COOPERATIVE ASSOCIATION)
3-6, 3-CHOME, KOJI-MACHI, CHIYODA-KU, TOKYO 102
JAPANESE
INDEXED: CHE

GEKKAN SHOKURYOHIN RYUTSU KAIZEN 8075
(MONTHLY JOURNAL OF IMPROVEMENT IN FOOD DISTRIBUTION)
SHOKURYOHIN RYUTSU KAIZEN KYOKAI
C/O DAITO BUILDING, 3-3, KANDA NERIBEI-CHO,
CHIYODA-KU, TOKYO 101
JAPANESE

GIFU-KEN KANTEN KENKYUSHO HOKOKU 8076
(REPORT OF THE GIFU RESEARCH INSTITUTE OF AGAR-AGAR)
GIFU-KEN KANTEN KENKYUSHO (GIFU RESEARCH INSTITUTE
OF AGAR-AGAR)
SHIMOTOGE, YAMOAOKA-MACHI, ENA-GUN, GIFU 509-76
JAPANESE

GODO SHUSEI K.K. KENKYU KAIHATSU-BU CHUO KENKYUSHO 8077
NENPO
(ANNUAL REPORTS OF CENTRAL RESEARCH LABORATORY,
DEPARTMENT OF RESEARCH AND DEVELOPMENT, GODO
BREWERY COMPANY, LIMITED)
GODO SHUSEI K.K. CHUO KENKYUSHO (CENTRAL RESEARCH
LABORATORY, GODO BREWERY COMPANY, LIMITED)
250, NAKAHARA, KAMIHONGO, MATSUDO-SHI 271
JAPANESE

GYONIKU SOSEJI 8078
(JOURNAL OF FISH SAUSAGE)
NIHON GYONIKU SOSEJI KYOKAI (JAPAN FISH SAUSAGE
ASSOCIATION)
3-9, 1-CHOME, HIGASHISHINBASHI, MINATO-KU, TOKYO
105
JAPANESE
INDEXED: CHE

HAKKO KOGAKU ZASSHI 8079
(JOURNAL OF FERMENTATION TECHNOLOGY)
NIHON HAKKO KOGAKKAI (THE SOCIETY OF FERMENTATION
TECHNOLOGY, JAPAN)
C/O OSAKA DAIGAKU KOGAKUBU, YAMADAUE, SUITA-SHI
565
JAPANESE AND ENGLISH
INDEXED: BIB, CHE, JICST

HAKKO KYOKAISHI 8080
(JOURNAL OF THE FERMENTATION ASSOCIATION, JAPAN)
HAKKO KYOKAI (FERMENTATION ASSOCIATION)
C/O TSUSHO SANGYO-SHO GINZA HIGASHI BUNSHITSU, 15-1,
6-CHOME, GINZA, CHUO-KU, TOKYO 104
JAPANESE

HAKKO TO KOGYO 8081
(FERMENTATION AND INDUSTRY)
HAKKO KYOKAI (THE FERMENTATION ASSOCIATION)
C/O TSUSHO SANGYO-SHO GINZA HIGASHI BUNSHITSU,
15-1, 6-CHOME, GINZA, CHUO-KU, TOKYO 104
JAPANESE
INDEXED: BIB, CHE, JICST

HAKKO TO TAISHA 8082
(AMINO ACID AND NUCLEIC ACID)
JAPANESE
INDEXED: JICST

HIROSHIMA JOZO GAKKAISHI 8083
(JOURNAL OF HIROSHIMA BREWING SOCIETY)
JAPANESE
INDEXED: JICST

HIROSHIMA-KEN KOGYO GIJUTSU JOHO: SHOKUHIN-HEN 8084
(HIROSHIMA PREFECTURE INDUSTRIAL TECHNOLOGY REPORT:
FOOD STUFFS)
HIROSHIMA-KEN GIJUTSU JOHO SHITSU
JAPANESE

HIROSHIMA-KEN SHOKUHIN KOGYO SHIKENJO GYOMU NENPO 8085
(ANNUAL REPORT OF THE FOOD RESEARCH INSTITUTE,
HIROSHIMA PREFECTURE)
HIROSHIMA-KEN SHOKUHIN KOGYO SHIKENJO
12-70, HIJIYAMAHON-MACHI, HIROSHIMA-SHI 730
JAPANESE

HIROSHIMA-KEN SHOKUHIN KOGYO SHIKENJO KENKYU HOKOKU 8086
(RESEARCH REPORT OF FOOD RESEARCH INSTITUTE,
HIROSHIMA PREFECTURE)
HIROSHIMA-KEN SHOKUHIN KOGYO SHIKENJO (FOOD
RESEARCH INSTITUTE, HIROSHIMA PREFECTURE)
12-70, HIJIYAMAHON-MACHI, HIROSHIMA-SHI 730
JAPANESE
INDEXED: CHE, JICST

HIROSHOKU KOSHI NYUSU 8087
(FOOD RESEARCH INSTITUTE NEWS, HIROSHIMA PREFECTURE)
HIROSHIMA-KEN SHOKUHIM KOGYO SHIKENJO
12-70, HIJIYAMAHON-MACHI, HIROSHIMA-SHI 730
JAPANESE

IBARAKI-KEN KOGYO GIJUTSU JOHO; SHOKUHIN-KEN 8088
(IBARAKI PREFECTURAL INDUSTRIAL TECHNICAL
INFORMATION OFFICE, FOOD PRODUCTS EDITION)
IBARAKI-KEN GIJUTSU JOHO-SHITSU (IBARAKI
PREFECTURAL TECHNICAL INFORMATION OFFICE)
JAPANESE

IBARAKI-KEN SHOKUHIN SHIKENJO HOKOKU 8089
(REPORT OF THE FOOD RESEARCH INSTITUTE, IBARAKI
PREFECTURE)
IBARAKI-KEN SHOKUHIN SHIKENJO (FOOD RESEARCH
INSTITUTE, IBARAKI PREFECTURE)
NAGAOKA, IBARAKI-MACHI, HIGASHIIBARAKI-GUN,
IBARAKI 311-31
JAPANESE
INDEXED: CHE, JICST

IFO RESEARCH COMMUNICATIONS 8090
HAKKO KENKYUSHO (INSTITUTE FOR FERMENTATION, OSAKA)
4-54, JUSONISHINO-CHO, HIGASHIYODOGAWAKU,
OSAKA-SHI 532
ENGLISH

INSTITUTE FOR FERMENTATION, OSAKA, RESEARCH 8091
COMMUNICATIONS
HAKKO KENKYUSHO (INSTITUTE FOR FERMENTATION, OSAKA)
4-54, JUSONISHINO-CHO, HIGASHIYODAGAWA-KU, OSAKA-
SHI 532
ENGLISH
INDEXED: BIB, BIO, CHE, ENG, JICST

JAPAN FUDO SAIENSU 8092
(JAPAN FOOD SCIENCE)
NIHON SHOKUHIN SHUPPAN K.K.
5-1, 6-CHOME, SHINBASHI, MINATO-KU, TOKYO 105
JAPANESE
INDEXED: CHE

JAPAN SUGAR YEARBOOK 8093
JAPAN SUGAR REFINERS ASSOCIATION
TOKYO
ENGLISH
INDEXED: BIB

JOKAI SHINPO 8094
(BREWING NEWS)
JOKAI SHINPOSHA
1-25, EDOBORI KITA-DORI, NISHI-KU, OSAKA-SHI 550
JAPANESE

JOKAI TAIMUSU 8095
(JAPAN BREWERY TIMES)
JOKAI TAIMUSUSHA
27, ISHIGATSUJI-MACHI, TENNOJI-KU, OSAKA-SHI 543
JAPANESE

JOYU 8096
(BREWING SCIENTISTS)
NODAI JOYU KAI (BREWING SCIENTISTS ASSOCIATION,
TOKYO UNIVERSITY OF AGRICULTURE)
C/O TOKYO NOGYO DAIGAKU JOZOKA, 1-1, 1-CHOME,
SAKURAGAOKA, SETAGAYA-KU, TOKYO 156
JAPANESE
INDEXED: JICST

JOZO HOCHI 8097
(BREWING INFORMATION)
JOZO HOCHI SHINBUNSHA
EBISUGAWAAGARU, TAKAKURA-DORI, NAKAGYO-KU,
KYOTO-SHI 604
JAPANESE

JOZO RONBUNSHU 8098
(ANNUAL OF THE BREWER'S ASSOCIATION, JAPAN)
NIHON JOYUKAI
C/O TODOROKI KANDA BUILDING, 1-21, KANDA NISHIKI-CHO
CHIYODA-KU, TOKYO 101
JAPANESE
INDEXED: JICST

JOZO SHIKENJO HOKOKU 8099
(REPORT OF THE RESEARCH INSTITUTE OF BREWING)
KOKUZEI-CHO JOZO SHIKENJO (THE RESEARCH INSTITUTE OF
BREWING)
6-30, 2-CHOME, TAKINOGAWA, KITA-KU, TOKYO 114
JAPANESE AND EUROPEAN
INDEXED: BIO, JICST

JUNKAN SEISEN SHOKUHIN JOHO 8100
(TEN-DAY REPORT ON FRESH FOOD PRODUCTS)
SHOKU SEIKATSU JOHO SABISU SENTA (DIET IMPROVEMENT
INFORMATION SERVICE CENTER)
JAPANESE

KAGAWA-KEN HAKKO SHOKUHIN SHIKENJO HOKOKU 8101
(ANNUAL REPORT OF KAGAWA PREFECTURAL FERMENTATION FOOD EXPERIMENTAL STATION)
KAGAWA-KEN HAKKO SHOKUHIN SHIKENJO (KAGAWA PREFECTURAL FERMENTATION FOOD EXPERIMENTAL STATION)
NOMA, UCHINOMI-CHO, SHOZU-GUN, KAGAWA 761-44
JAPANESE
INDEXED: CHE, JICST

KAGYO SHOKUHIN SHINBUN 8102
(JOURNAL OF CONFECTIONERY)
KAGYO SHOKUHIN SHINBUNSHA
2-199, KAMIFUKUSHIMAMINAMI, FUKUSHIMA-KU, OSAKA-SHI 553
JAPANESE

KAIGAI SATO JOHO 8103
(FOREIGN SUGAR INFORMATION)
NIHON SEITO KOGYOKAI (JAPAN SUGAR REFINER'S ASSOCIATION)
5-7, SANBAN-CHO, CHIYODA-KU, TOKYO 102
JAPANESE

KAIGAI SHOKURYO DOKO 8104
(FOREIGN FOODS INFORMATION)
SHOKURYO-CHO SOMU-BU (ADMINISTRATIVE SECTION, FOOD AGENCY)
2-1, 1-CHOME, KASUMIGASEKI, CHIYODA-KU, TOKYO 100
JAPANESE

KANZUME JIHO 8105
(CANNERS JOURNAL)
NIHON KANZUME KYOKAI (CANNERS ASSOCIATION OF JAPAN)
4-1, 2-CHOME, MARUNOUCHI, CHIYODA-KU, TOKYO 100
JAPANESE
INDEXED: BIB, CHE, JICST

KASHI AME SHINBUN 8106
(CONFECTIONERY AND CANDY NEWS)
KASHI AME SHINBUNSHA
7-5, KIKUI-DORI, NISHI-KU, NAGOYA-SHI 451
JAPANESE

KASHI KOGYO 8107
(CONFECTIONERY INDUSTRY)
ZENKOKU KASHI KYOKAI (JAPAN CONFECTIONERY)
C/O ZENKOKU BISUKETTO KAIKAN, 9-5, 6-CHOME, SHINBASHI, MINATO-KU, TOKYO 105
JAPANESE

KASHI SHOKUHIN SHINPO 8108
(CONFECTION AND FOOD NEWS)
KASHI SHOKUHIN SHINPOSHA
1-29, SUGAMO, TOSHIMA-KU, TOKYO 170
JAPANESE

KENKYU TO TENBO 8109
(ANNUAL REPORT OF THE SHINJO BRANCH OF THE NATIONAL FOOD RESEARCH INSTITUTE)
NORIN-SHO SHOKUHIN SOGO KENKYUSHO SHINJO SHISHO (SHINJO BRANCH, NATIONAL FOOD RESEARCH INSTITUTE, MINISTRY OF AGRICULTURE AND FORESTRY)
4-19, ISHIKAWA-CHO, SHINJO-SHI 996
JAPANESE AND ENGLISH

KIKAN TOGYO SHIHO 8110
(QUARTERLY INFORMATION OF SUGAR INDUSTRY)
NIHON SEITO KOGYOKAI (JAPAN SUGAR REFINER'S ASSOCIATION)
5-7, SANBAN-CHO, CHIYODA-KU, TOKYO 102
JAPANESE

KOKUSAI SHOKUHIN 8111
(MAGAZINE FOR FOOD INDUSTRY)
KOKUSAI SHOGYO KENKYUSHO (INTERNATIONAL COMMERCIAL RESEARCH INSTITUTE)
JAPANESE

KOMENUKAABURA GIJUTSU KENKYUSHO HOKOKU 8112
(REPORT OF RICE BRAN OIL TECHNICAL RESEARCH INSTITUTE)
KOMENUKAABURA GIJUTSU KENKYUSHO (RICE BRAN OIL TECHNICAL RESEARCH INSTITUTE)
290, ENNAMI, YONO-SHI 338
JAPANESE AND ENGLISH

KOMUGI NIJI KAKOGYO JITTAI CHOSA KEKKA 8113
(RESEARCH REPORT OF ACTUAL CONDITION OF WHEAT SECONDARY PRODUCTS INDUSTRY)
SHOKURYO-CHO GYOMU-BU (OPERATION DEPARTMENT, FOOD AGENCY)
NIHON MUGIRUI KENKYUKAI (JAPAN WHEAT RESEARCH INSTITUTE)
2-47, NIHONBASHI KABUTO-CHO, CHUO-KU, TOKYO 103
JAPANESE

KONNYAKU SHINBUN 8114
(KONNYAKU NEWS)
KONNYAKU SHINBUNSHA
12-9, 5-CHOME, IGUSA, SUGINAMI-KU, TOKYO 167
JAPANESE

KORUDO CHEN 8115
(COLD CHAIN)
KORUDO CHEN KYOKAI (MODERN FOODS DISTRIBUTION SYSTEM ASSOCIATION)
1-21, KANDA AWAJI-CHO, CHIYODA-KU, TOKYO 101
JAPANESE

KORUDO CHEN KENKYU 8116
(JOURNAL OF THE INSTITUTE OF COLD CHAIN)
NIHON KORUDO CHIEN KENKYUKAI (JAPAN COLD CHAIN RESEARCH ASSOCIATION)
JAPANESE

KYOTO DAIGAKU SHOKURYO KAGAKU KENKYUSHO HOKOKU 8117
(BULLETIN OF THE RESEARCH INSTITUTE FOR FOOD SCIENCE, KYOTO UNIVERSITY)
KYOTO DAIGAKU SHOKURYO KAGAKU KENKYUSHO (THE RESEARCH INSTITUTE FOR FOOD SCIENCE, KYOTO UNIVERSITY)
GOKASHO, UJI-SHI 611
JAPANESE AND ENGLISH
INDEXED: BIB, CHE, JICST

LIST OF CULTURES 8118
HAKKO KENKYUSHO (INSTITUTE FOR FERMENTATION, OSAKA)
4-54, JUSONISHINO-CHO, HIGASHIYODOGAWA-KU, OSAKA-SHI 532
ENGLISH

MEMOIRS OF THE RESEARCH INSTITUTE FOR FOOD SCIENCE, KYOTO UNIVERSITY 8119
KYOTO DAIGAKU SHOKURYO KAGAKU KENKYUSHO (THE RESEARCH INSTITUTE OF FOOD SCIENCE, KYOTO UNIVERSITY)
GOKASHO, UJI-SHI 611
EUROPEAN
INDEXED: BIB, BIO, CHE, JICST

MIE-KEN JOZO SHIKENJO HOKOKU 8120
(REPORT OF THE RESEARCH INSTITUTE OF BREWING, MIE PREFECTURE)
MIE-KEN JOZO SHIKENJO
141-2, OTANI-CHO, TSU-SHI 514
JAPANESE

MISO NO KAGAKU TO GIJUTSU 8121
(SCIENCE AND TECHNOLOGY FOR MISO)
ZENKOKU MISO GIJUTSUKAI (JAPAN MISO TECHNICAL ASSOCIATION)
26-19, 1-CHOME, SHINKAWA, CHUO-KU, TOKYO 104
JAPANESE
INDEXED: CHE, JICST

NAGAOKA KOJO JIHO 8122
(NAGAOKA FACTORY REVIEW)
NIHON SENBAI KOSHA NAGAOKA KOJO (JAPAN TOBACCO AND SALT PUBLIC CORPORATION, NAGAOKA FACTORY)
JAPANESE

NIHON JOKAI SHINBUN 8123
(JAPAN BREWING INDUSTRY NEWS)
NIHON JOKAI SHINBUNSHA
3-3, TAKARA-CHO, CHUO-KU, TOKYO 104
JAPANESE

NIHON JOZO KYOKAI ZASSHI 8124
(JOURNAL OF THE SOCIETY OF BREWING, JAPAN)
NIHON JOZO KYOKAI (THE SOCIETY OF BREWING, JAPAN)
6-30, 2-CHOME, TAKINOGAWA, KITA-KU, TOKYO 114
JAPANESE AND ENGLISH
INDEXED: BIO, CHE, JICST

NIHON KAKO SHOKUHIN SHINBUN 8125
(JAPAN PROCESSING FOOD NEWS)
NIHON SHOKUHIN KEIZAISHA
C/O KAZUMI BUILDING, 2-2, 3-CHOME, MINATO-KU, TOKYO 105
JAPANESE

NIHON REITO SHOKUHIN KYOKAI KAIHO 8126
(BULLETIN OF THE JAPAN FROZEN FOOD ASSOCIATION)
NIHON REITO SHOKUHIN KYOKAI (JAPAN FROZEN FOOD ASSOCIATION)
1-29, AKASHI-CHO, CHUO-KU, TOKYO 104
JAPANESE
INDEXED: JICST

NIHON RYOIN SHINBUN 8127
(JAPAN RESTAURANT JOURNAL)
NIHON RYOIN SHINBUNSHA (JAPAN RESTAURANT JOURNAL COMPANY, LTD.)
14-3, 2-CHOME, NAKANO-CHO, MIYAKOJIMA-KU, OSAKA-SHI 534
JAPANESE

NIHON SHOKUHIN KOGYO GAKKAISHI 8128
(JOURNAL OF THE JAPAN FOOD INDUSTRY ASSOCIATION)
JAPANESE

NIHON SHOKURYO SHINBUN 8129
(JAPAN FOODSTUFF NEWS)
NOHON SHOKURYO SHINBUNSHA
9-9, 1-CHOME, YAESU, CHUO-KU, TOKYO 103
JAPANESE

NIHON SHOYU KENKYUSHO ZASSHI 8130
(JOURNAL OF THE JAPAN SOY SAUCE INSTITUTE)
NIHON SHOYU KENKYU-SHO
JAPANESE

NIHON WAYOSHU KANZUME SHINBUN 8131
(CANNED PROVISIONS AND INTOXICATING BEVERAGES JOURNAL)
NIHON WAYOSHU KANZUME SHINBUNSHA
1-12, NIHONBASHI KAKIGARA-CHO, CHUO-KU, TOKYO 103
JAPANESE

NIIGATA-KEN SHOKUHIN KENKYUSHO KENKYU HOKOKU 8132
(REPORT OF THE NIIGATA FOOD RESEARCH INSTITUTE)
NIIGATA-KEN SHOKUHIN KENKYUSHO (NIIGATA FOOD RESEARCH INSTITUTE)
456, KAMIJO, KAMO-SHI, NIIGATA 959-13
JAPANESE
INDEXED: CHE

NIKKAN SHOKURYO SHINBUN 8133
(DAILY FOOD NEWS)
NIKKAN SHOKURYO SHINBUNSHA
C/O TOKYO-TO CHUO OROSHIURI ICHIBA, 2-1, 5-CHOME, TSUKIJI, CHUO-KU, TOKYO 104
JAPANESE

NIPPON SHOKUHIN KOGYO GAKKAISHI 8134
(JOURNAL OF FOOD SCIENCE AND TECHNOLOGY)
NIPPON SHOKUHIN KOGYO GAKKAI (JAPANESE SOCIETY OF FOOD SCIENCE AND TECHNOLOGY)
C/O NORIN-SHO SHOKUHIN SOGO KENKYUSHO, 4-12, 1-CHOME, SHIOHAMA, KOTO-KU, TOKYO 135
JAPANESE
INDEXED: BIB, CHE, JICST

NIPPON SHOYU KENKYUJO ZASSHI 8135
(JOURNAL OF JAPAN SOY SAUCE RESEARCH INSTITUTE)
JAPANESE
INDEXED: CHE, JICST

NISSHIN SEIFUN KENKYU HOKOKU 8136
(RESEARCH REPORT OF THE CENTRAL LABORATORY OF NISSHIN FLOUR MILLING COMPANY, LIMITED)
NISSHIN SEIFUN K.K. CHUO KENKYUSHO (CENTRAL LABORATORY OF NISSHIN FLOUR MILLING COMPANY, LTD.)
TSURGAOKA, OI-MACHI, IRUMA-GUN, SAITAMA 354
JAPANESE

NYU FUDO INDASUTORI 8137
(NEW FOOD INDUSTRY)
SHOKUHIN SHIZAI KENKYUKAI (THE SOCIETY OF STUDY OF FOOD MATERIALS)
C/O KYODO BUILDING, 1-3, NIHONBASHI HON-CHO, CHUO-KU, TOKYO 103
JAPANESE
INDEXED: CHE, JICST

PAIN 8138
NIHON PAN GIJUTSUSHA KYOKAI (JAPAN BAKERS' ASSOCIATION)
24-14, 4-CHOME, SENDAGAYA, SHIBUYA-KU, TOKYO 151
JAPANESE

PAN KAGAKKAISHI 8139
(REPORT OF THE JAPAN INSTITUTE OF BAKING)
NIHON PAN KAGAKKAI KENKYUSHO
24-14, 4-CHOME, SENDAGAYA, SHIBUYA-KU, TOKYO 151
JAPANESE

POKETTO SATO TOKEI 8140
(POCKET SIZE STATISTICS OF SUGAR PRODUCTS)
NIHON SEITO KOGYOKAI (JAPAN SUGAR REFINERS' ASSOCIATION)
5-7, SANBAN-CHO, CHIYODA-KU, TOKYO 102
JAPANESE
INDEXED: BIB

REIKA GIJUTSU SHIRYO 8141
(REFRIGERATED CONFECTIONARY TECHNOLOGY MATERIALS)
SHIRAISHI KENKYUSHITSU
JAPANESE

REISHOKU SANGYO 8142
(REFRIGERATED FOOD INDUSTRY)
NIHON SHOKURYO SHINBUNSHA
C/O TOKYO TATEMONO BUILDING, 9-9, 1-CHOME, YAESU, CHUO-KU, TOKYO 103
JAPANESE

REPORT OF THE NODA INSTITUTE FOR SCIENTIFIC RESEARCH 8143
NODA SANGYO KAGAKU KENKYUSHO (NODA INSTITUTE FOR SCIENTIFIC RESEARCH)
399, NODA, NODA-SHI 278
ENGLISH
INDEXED: JICST

REPORT OF THE RESEARCH LABORATORY OF KIRIN BREWERY COMPANY, LIMITED 8144
KIRIN BIRU K.K. SOGO KENKYUSHO (RESEARCH LABORATORY KIRIN BREWERY COMPANY, LTD.)
3, MIYAHARA-CHO, TAKASAKI-SHI 370-12
ENGLISH
INDEXED: BIO, CHE, JICST

SAITAMA-KEN SHOKUHIN KOGYO SHIKENJO GYOMU NENPO 8145
(OPERATION REPORT OF SAITAMA RESEARCH INSTITUTE OF FOOD TECHNOLOGY)
SAITAMA-KEN SHOKUHIN KOGYO SHIKENJO (SAITAMA RESEARCH INSTITUTE OF FOOD TECHNOLOGY)
2-133, SUEHIRO, KUMAGAYA-SHI 360
JAPANESE

SAKEURI SHOKUHIN TOKEI GEPPO 8146
(WINE SPIRITS AND PROVISIONS MONTHLY STATISTICS)
NIKKAN KEIZAI TSUSHINSHA
18-3, 3-CHOME, UCHIKANDA, CHIYODA-KU, TOKYO 101
JAPANESE
INDEXED: JICST

SAN'EI NYUSU 8147
(SAN'EI NEWS)
SAN'EI KAGAKU KOGYO K.K. (SAN'EI CHEMICAL INDUSTRIES, LTD.)
1-11, 1-CHOME, SANWA-CHO, TOYONAKA-SHI 561
JAPANESE
INDEXED: JICST

SATO TOKEI NENKAN 8148
(SUGAR STASTICS YEARBOOK)
JAPAN SUGAR REFINERS ASSOCIATION
TOKYO
JAPANESE
INDEXED: BIB

SEIKA SEIPAN 8149
(BAKING AND CONFECTION)
SEIKA JIKKENSHA
13-11, 2-CHOME, MINAMISHINAGAWA, SHINAGAWA-KU, TOKYO 140
JAPANESE

SEIMAI KOGYO 8150
(RICE MILLING INDUSTRY)
NIHON SEIMAI KOGYOKAI (JAPAN RICE MILLERS ASSOCIATION)
C/O SHOKURYO KAIKAN, 3-6, 3-CHOME, KOJI-MACHI, CHIYODA-KU, TOKYO 102
JAPANESE

SEITO GIJUTSU KENKYUKAISHI 8151
(PROCEEDINGS OF THE RESEARCH SOCIETY OF JAPAN SUGAR REFINERIES' TECHNOLOGIST)
NIHON SEITO KOGYOKAI GIJUTSU KENKYUSHO (RESEARCH LABORATORY, JAPAN SUGAR REFINER'S ASSOCIATION)
5-7, SANBAN-CHO, CHIYODA-KU, TOKYO 102
JAPANESE AND ENGLISH
INDEXED: BIB, CHE, JICST

SEKIYU TO BISEIBUTSU 8152
(PETROLEUM AND MICROORGANISMS)
SEKIYU HAKKO KANKYUKAI (RESEARCH ASSOCIATION OF PETROLEUM FERMETATION)
C/O TOKYO DAIGAKU NOGAKU-BU NOGEI KAGAKUKA, 1-1 1-CHOME, YAYOI, BUNKYO-KU, TOKYO 113
JAPANESE

SHINSHU MISO KENKYUSHO KENKYU HOKOKU 8153
(REPORT OF THE SHINSHU MISO RESEARCH INSTITUTE)
SHINSHU MISO KENKYUSHO (SHINSHU MISO RESEARCH INSTITUTE)
1014, MINAMIGATA-MACHI, NAGANO-SHI 380
JAPANESE
INDEXED: JICST

SHIZUOKA-KEN KOGYO GIJUTSU JOHO, SHOKUHIN-HEN 8154
(SHIZUOKA PREFECTURE INDUSTRIAL TECHNOLOGY REPORT, FOODS EDITION)
SHIZUOKA-KEN KOGYO SHIKENJO (SHIZUOKA PREFECTURE INDUSTRIAL EXPERIMENT STATION)
JAPANESE

SHOKKEN 8155
(FOOD RESEARCH)
SHOKUMOTSUGAKU KYOIKU KENKYUKAI
1-7, 5-CHOME, JINGUMAE, SHIBUYA-KU, TOKYO 150
JAPANESE
INDEXED: JICST

SHOKU NO KAGAKU 8156
(FOOD SCIENCE)
JAPANESE
INDEXED: JICST

SHOKUHIN 8157
(FOOD PRODUCTS)
SHOKUHIN KENKYUSHO (FOOD PRODUCTS RESEARCH COMPANY)
JAPANESE

SHOKUHIN GURAFU 8158
(JAPAN PROVISIONS GRAPHIC)
SHOKUHIN SHINBUNSHA
2-6, OIMATSU-CHO, KITA-KU, OSAKA 530
JAPANESE

SHOKUHIN HOSO 8159
(FOOD PACKAGING)
BIJINESU SENTASHA
(BUSINESS CENTER COMPANY)
JAPANESE

SHOKUHIN KAIHATSU 8160
(UP-TO-DATE FOOD PROCESSING)
SHOKUHIN KENKYUSHA
9-14, 1-CHOME, YUSHIMA, BUNKYO-KU, TOKYO 113
JAPANESE
INDEXED: BIB

SHOKUHIN KIKAI KAIHO 8161
(JOURNAL OF FOOD MACHINE ENGINEERING ASSOCIATION)
NIHON SHOKUHIN KIKAI KOGYOKAI (JAPAN FOOD MACHINE ENGINEERING ASSOCIATION)
16-2, 2-CHOME, NISHISHINBASHI, MINATO-KU, TOKYO 105
JAPANESE

SHOKUHIN KIKAI NYUSU 8162
(FOOD PRODUCTS MACHINERY NEWS)
NIHON SHOKUHIN KIKAI KOGYOKAI (JAPAN FOOD PRODUCTS MACHINERY INDUSTRY ASSOCIATION)
JAPANESE

SHOKUHIN KIKAI SOCHI 8163
(MACHINERY AND EQUIPMENT FOR FOOD INDUSTRY)
BIJINESU SENTASHA
2-6, KANDA TSUKASA-CHO, CHIYODA-KU, TOKYO 101
JAPANESE

SHOKUHIN KOGYO 8164
(FOOD INDUSTRY)
KORIN SHOIN
2-12, 6-CHOME, UENO, TAITO-KU, TOKYO 110
JAPANESE AND ENGLISH
INDEXED: BIB, CHE, JICST

SHOKUHIN KOGYO GIJUTSU JOHO 8165
(INFORMATION ON TECHNOLOGY FOR FOOD INDUSTRY)
NIHON KAGAKU GIJUTSU JOHO SENTA (THE JAPAN INFORMATION CENTER OF SCIENCE AND TECHNOLOGY)
1-1, 2-CHOME, NAGATA-CHO, CHIYODA-KU, TOKYO 100
JAPANESE
INDEXED: JICST

SHOKUHIN NIPPON 8166
(FOODSTUFFS JAPAN)
SHOKUHIN SHUPPANSHA (FOOD PRODUCTS PUBLISHING COMPANY)
JAPANESE

SHOKUHIN SANGYO SHINBUN 8167
(FOOD INDUSTRY NEWS)
SHOKUHIN SANGYO SHINBUNSHA
8-1, 1-CHOME, HAMAMATSU-CHO, MINATO-KU, TOKYO 105
JAPANESE

SHOKUHIN SHIKENJO SHOCHOKAI KAIHO 8168
(BULLETIN OF THE FOODSTUFFS EXPERIMENT STATION CHIEF'S ASSOCIATION)
ZENKOKU SHOKUHIN KANKEI SHIKENJO SHOCHOKAI (NATIONAL FOODSTUFFS - RELATED EXPERIMENT STATION CHIEF'S)
JAPANESE

SHOKUHIN SHINBUN 8169
(JAPAN PROVISIONS NEWS)
SHOKUHIN SHINBUNSHA (THE JAPAN PROVISIONS NEWS COMPANY, LTD.)
2-6, OIMATSU-CHO, KITA-KU, OSAKA-SHI 530
JAPANESE

SHOKUHIN SHOSHA 8170
(FOOD IRRADIATION, JAPAN)
NIHON SHOKUHIN SHOSHA KENKYU KYOGIKAI (JAPANESE RESEARCH ASSOCIATION FOR FOOD IRRADIATION)
4-12, 1-CHOME, SHIOHAMA, KOTO-KU, TOKYO 135
JAPANESE AND ENGLISH
INDEXED: JICST

SHOKUHIN SOGO KENKYAJO KENKYU HOKOKU 8171
(RESEARCH REPORT OF THE GENERAL RESEARCH INSTITUTE OF FOOD PRODUCTS)
NOSON SHO SOGO KENKYUSHO (MINISTRY OF AGRICULTURE AND FORESTRY, GENERAL RESEARCH INSTITUTE)
JAPANESE

SHOKUHIN SOGO KENKYUSHO HOKOKU 8172
(REPORT OF THE NATIONAL FOOD RESEARCH INSTITUTE)
NORIN-SHO SHOKUHIN SOGO KENKYUSHO (NATIONAL FOOD RESEARCH INSTITUTE, MINISTRY OF AGRICULTURE AND FORESTRY)
4-12, 1-CHOME, SHIOHAMA, KOTO-KU, TOKYO 135
JAPANESE, WITH ENGLISH SUMMARY
INDEXED: BIB, BIO, CHE, JICST

SHOKUHIN SOGO KENKYUSHO NYUSU 8173
(NATIONAL FOOD RESEARCH INSTITUTE NEWS)
NORIN-SHO SHOKUHIN SOGO KENKYUSHO (NATIONAL FOOD RESEARCH INSTITUTE, MINISTRY OF AGRICULTURE AND FORESTRY)
4-12, 1-CHOME, SHIOHAMA, KOTO-KU, TOKYO 135
JAPANESE

SHOKUHIN TENKA-BUTSU BENRAN 8174
(FOOD ADDITIVES HANDBOOK)
JAPANESE
INDEXED: BIB

SHOKUHIN TO KAGAKU 8175
(FOOD SCIENCE)
SHOKUHIN TO KAGAKUSHA
35, TOMITA-CHO, KITA-KU, OSAKA-SHI 530
JAPANESE
INDEXED: JICST

SHOKUHIN TO YOKI 8176
(FOOD AND PACKAGING)
KANZUME GIJUTSU KENKYUKAI (INSTITUTE OF CANNING TECHNOLOGY)
C/O DAI-7 TAKABATAKE BUILDING, 38-5, 4-CHOME, RYOGOKU, SUMIDA-KU, TOKYO 130
JAPANESE
INDEXED: JICST

SHOKURYO (SONO KAGAKU TO GIJUTSU) 8177
(FOOD, ITS SCIENCE AND TECHNOLOGY)
NORIN-SHO SHOKUHIN SOGO KENKYUSHO (NATIONAL FOOD RESEARCH INSTITUTE, MINISTRY OF AGRICULTURE AND FORESTRY)
4-12, 1-CHOME, SHIOHAMA, KOTO-KU, TOKYO 135
JAPANESE
INDEXED: BIB, JICST

SHOKURYO GIJUTSU 8178
(FOOD ENGINEERING, JAPAN)
SHIRAISHI KENKYUSHITSU (SHIRAISHI LABORATORY)
29-11, 4-CHOME, FUDA, CHOFU-SHI, TOKYO 182
JAPANESE, WITH EUROPEAN SUMMARY

SHOKURYO GIJUTSU FUKYU SHIRIZU 8179
(SERIES FOR FOOD PROCESSING TECHNOLOGY)
NORIN-SHO SHOKUHIN SOGO KENKYUSHO (NATIONAL FOOD RESEARCH INSTITUTE, MINISTRY OF AGRICULTURE AND FORESTRY)
4-12, 1-CHOME, SHIOHAMA, KOTO-KU, TOKYO 135
JAPANESE
INDEXED: BIB, CHE, JICST

SHOKURYO GYOMU TOKEI NENPO 8180
(STATISTICAL YEARBOOK OF FOOD OPERATION)
ISHIKAWA SHOKURYO JIMUSHO (ISHIKAWA FOOD OFFICE, MINISTRY OF AGRICULTURE & FORESTRY)
1-23, 3-CHOME, NO-MACHI, KANAZAWA-SHI 921
JAPANESE

SHOKURYO JOHO 8181
(FOOD INFORMATION)
SHOKURYO-CHO SOMU-BU (GENERAL AFFAIRS DIVISION, FOOD AGENCY)
2-1, 1-CHOME, KASUMIGASEKI, CHIYODA-KU, TOKYO 100
JAPANESE

SHOKURYO JOKAI TSUSHIN 8182
(FOOD AND BREWING TIMES)
SHOKURYO JOKAI TSUSHINSHA
C/O MEIJIYA BUILDING, 2-18, MINAMI-HON-CHO, HIGASHI-KU, OSAKA-SHI 541
JAPANESE

SHOKURYO JUKYUHYO 8183
(FOOD BALANCE SHEETS)
NORIN-SHO DAIJIN KANBO (MINISTER'S SECRETARIAT, MINISTRY OF AGRICULTURE AND FORESTRY) AND NORIN TOKEI KYOKAI
11-14, 2-CHOME, MEGURO, MEGURO-KU, TOKYO 153
JAPANESE

SHOKURYO JUYO BUNSEKI 8184
(ANALYSIS OF DEMAND FOR FOOD)
NORIN-SHO DAIJIN KANBO (MINISTER'S SECRETARIAT, MINISTRY OF AGRICULTURE AND FORESTRY)
2-1, 1-CHOME, KASUMIGASEKI, CHIYODA-KU, TOKYO 100
JAPANESE

SHOKURYO KANKEI SHINBUN RONCHO 8185
(FOODSTUFFS NEWSPAPER)
SHOKURYO-CHO (FOOD AGENCY)
JAPANESE

SHOKURYO KANRI GEPPO 8186
(MONTHLY FOOD BULLETIN)
SHOKURYO-CHO SHOKKAN GEPPO HENSHU IINKAI (MONTHLY FOOD BULLETIN EDITORIAL COMMITTEE, FOOD AGENCY)
1-31, 6-CHOME, NISHIOI, SHINAGAWA-KU, TOKYO 140
JAPANESE
INDEXED: BIB

SHOKURYO KANRI TOKEI NENPO 8187
(STATISTICAL YEARBOOK OF FOODS)
SHOKURYO-CHO (FOOD AGENCY)
2-1, 1-CHOME, KASUMIGASEKI, CHIYODA-KU, TOKYO 100
JAPANESE
INDEXED: BIB

SHOKURYO SHINBUN 8188
(FOOD INDUSTRY NEWS)
SHOKURYO SHINBUNSHA
12-5, 2-CHOME, SOTOKANDA, CHIYODA-KU, TOKYO 101
JAPANESE

SHOYU TO MISO 8189
(SOY SAUCE AND BEAN PASTE)
JAPANESE
INDEXED: JICST

SHUKAN JOZO 8190
(BREWING WEEKLY)
SHUKAN JOZO SHINBUNSHA
8-20, TAKATSU-CHO, MINAMI-KU, OSAKA-SHI 542
JAPANESE

SHUKENKAIHO 8191
(BULLETIN OF THE SAKE-BREWING SOCIETY, KYOTO)
KYOTO SHUZO KOGYO KENKYUKAI (THE SAKE-BREWING SOCIETY, KYOTO)
C/O KYOTO-SHI KOGYO SHIKENJO, MINAMIDA-CHO, NISHI-9-JO, MINAMI-KU, KYOTO-SHI 601
JAPANESE

TATSUNO JOZO KENKYUKAI KAIHO 8192
(REPORTS OF THE BREWING SOCIETY OF TATSUNO)
TATSUNO JOZO KENKYUKAI (THE BREWING SOCIETY OF TATSUNO)
TATSUNO SHOYU KYODO KUMIAI SHIKENJO
126, KAMIKAJO, TATSUNO-CHO, TATSUNO-SHI 679-41
JAPANESE
INDEXED: JICST

TOCHIGI-KEN SHOKUHIN KOGYO SHIDOSHO GYOMU NENPO 8193
(OPERATING YEARBOOK OF FOOD RESEARCH INSTITUTE, TOCHIGI PREFECTURE)
TOCHIGI-KEN SHOKUHIN KOGYO SHIDOSHO
508, ICHINOSAWA-MACHI, UTSUNOMIYA-SHI 320
JAPANESE

TOGYO NENPO 8194
(ANNUAL REPORT OF SUGAR INDUSTRY)
OKINAWA-KEN
2-1, 1-CHOME, IZUMISAKI, NAHA-SHI 900
JAPANESE

TOKUSHIMA-KEN SHOKUHIN KAKO SHIKENJO KENKYU HOKOKU 8195
(REPORT OF THE TOKUSHIMA FOOD RESEARCH INSTITUTE)
TOKUSHIMA-KEN SHOKUHIN KAKO SHIKENJO (TOKUSHIMA FOOD RESEARCH INSTITUTE)
2, AKUI-CHO, TOKUSHIMA-SHI 770
JAPANESE
INDEXED: JICST

TOTTORI-KEN SHOKUHIN KAKO KENKYUSHO KENKYU HOKOKU 8196
(RESEARCH REPORT OF FOOD, TECHNOLOGICAL INSTITUTE OF TOTTORI PREFECTURE)
TOTTORI-KEN SHOKUHIN KAKO KENKYUSHO (FOOD TECHNOLOGICAL INSTITUTE OF TOTTORI PREFECTURE)
OSAWA, WATARI-CHO, SAKAIMINATO-SHI 684
JAPANESE

TOYO SHOKUHIN KOGYO TANKI DAIGAKU, TOYO SHOKUHIN KENKYUSHO HOKOKUSHO 8197
(REPORT OF TOYO JUNIOR COLLEGE OF FOOD TECHNOLOGY AND TOYO INSTITUTE OF FOOD TECHNOLOGY)
TOYO SHOKUHIN KOGYO TANKI DAIGAKU, TOYO SHOKUHIN KENKYUSHO
23-2, 4-CHOME, MINAMI HANAYASHIKI, KAWANISHI-SHI 666
JAPANESE AND ENGLISH
INDEXED: BIB, CHE, JICST

YAKURUTO KENKYUSHO KENKYU HOKOKUSHU 8198
(ANNUAL REPORT OF THE YAKULT INSTITUTE)
YAKURUTO HONSHA KENKYUSHO (SECTION OF RESEARCH MANAGEMENT, YAKULT INSTITUTE)
1796, YABO, KUNITACHI-SHI, TOKYO 186
JAPANESE AND ENGLISH

YAMANASHI DAIGAKU HAKKO KENKYUSHO KENKYU HOKOKU 8199
(BULLETIN OF THE RESEARCH INSTITUTE OF FERMENTATION, YAMANASHI UNIVERSITY)
YAMANASHI DAIGAKU KOGAKUBU HAKKO KAGAKU KENKYU SHISETSU (THE RESEARCH INSTITUTE OF FERMENTATION, YAMANASHI UNIVERSITY)
13-1, 1-CHOME, KITASHIN-MACHI, KOFU-SHI 400
JAPANESE, WITH EUROPEAN SUMMARY
INDEXED: CHE, JICST

YAMANASHI-KEN SHOKUHIN KOGYO SHIDOSHO KENKYU HOKOKU 8200
(REPORT OF FOOD RESEARCH INSTITUTE, YAMANASHI PREFECTURE)
YAMANASHI-KEN SHOKUHIN KOGYO SHIDOSHO (INSTITUTE FOR WINE AND FOOD TECHNOLOGY YAMANASHI PREFECTURE)
312-4, SHINKAI-CHO, KUBONAKAJIMA, ISAWA-CHO, HIGASHI YATSUSHIRO-GUN, YAMANASHI 406
JAPANESE AND ENGLISH

YUNYU SHOKURYO KYOGIKAIHO 8201
(NEWS OF THE GRAIN IMPORTER'S ASSOCIATION)
YUNYU SHOKURYO KYOGIKAI (THE GRAIN IMPORTER'S ASSOCIATION)
1-16, 2-CHOME, NIHONBASHI, CHUO-KU, TOKYO 103
JAPANESE

YURYO TOKEI NENPO 8202
(ANNUAL STATISTICS OF OILS AND FATS)
NORIN-SHO SHOKUHIN RYUTSU-KYOKU (FOOD AND MARKETING BUREAU, MINISTRY OF AGRICULTURE AND FORESTRY)
2-1, 1-CHOME, KASUMIGASEKI, CHIYODA-KU, TOKYO 100
JAPANESE

ZENKOKU JOKAI SHINBUN 8203
(ALL JAPAN BREWING INDUSTRY NEWS)
ZENKOKU JOKAI SHINBUNSHA
P.O. BOX 35, NAKAGYO YUBINKYOKU, NAKAGYO-KU, KYOTO-SHI 604
JAPANESE

Glass and Ceramic Industries

AICHI-KEN SETO YOGYO GIJUTSU SENTA JIGYO KEIKAKUSHO 8204
(PLAN SHEET OF OPERATION OF TECHNICAL CERAMIC CENTER, AICHI PREFECTURE, SETO)
AICHI-KEN SETO YOGYO GIJUTSU SENTA (TECHNICAL CERAMIC CENTER, AICHI PREFECTURE, SETO)
318, MINAMIYAMA, YAMAGUCHI, SETO-SHI 489
JAPANESE

AICHI-KEN SETO YOGYO GIJUTSU SENTA KENKYU HOKOKUSHO 8205
(RESEARCH REPORT OF THE TECHNICAL CERAMIC CENTER, AICHI PREFECTURE, SETO)
AICHI-KEN SETO YOGYO GIJUTSU SENTA (THE TECHNICAL CERAMIC CENTER, AICHI PREFECTURE, SETO)
318, MINAMIYAMA, YAMAGUCHI, SETO-SHI 489
JAPANESE

AICHI-KEN SETO YOGYO GIJUTSU SENTA NYUSU 8206
(TECHNICAL CERAMIC CENTER NEWS, AICHI PREFECTURE, SETO)
AICHI-KEN SETO YOGYO GIJUTSU SENTA (TECHNICAL CERAMIC CENTER, AICHI PREFECTURE, SETO)
318, MINAMIYAMA, YAMAGUCHI, SETO-SHI 489
JAPANESE

AICHI-KEN TOKONAME YOGYO GIJUTSU SENTA HOKOKU 8207
(REPORTS OF CERAMIC RESEARCH INSTITUTE, AICHI PREFECTURE, TOKONAME)
AICHI-KEN TOKONAME YOGYO GIJUTSU SENTA (CERAMIC RESEARCH INSTITUTE, TOKONAME, AICHI PREFECTURE)
48-88, MITSUIKE, TOKONAME-SHI 479
JAPANESE
INDEXED: JICST

AICHI-KEN TOKONAME YOGYO GIJUTSU SENTA JIGYO HOKOKUSHO 8208
(OPERATION REPORT OF CERAMIC RESEARCH INSTITUTE, AICHI PREFECTURE, TOKONAME)
AICHI-KEN TOKONAME YOGYO GIJUTSU SENTA (CERAMIC RESEARCH INSTITUTE, TOKONAME, AICHI PREFECTURE)
48-88, MITSUIKE, TOKONAME-SHI 479
JAPANESE

AIYOGI NYUSU 8209
(AICHI PREFECTURE CERAMIC TECHNIQUE NEWS)
AICHI-KEN YOGYO GIJUTSU SENTA
(AICHI PREFECTURE CERAMIC PRODUCTION TECHNICAL CENTER)
JAPANESE

ASAHI GARASU KENKYU HOKOKU 8210
(REPORT OF THE RESEARCH LABORATORY OF THE ASAHI GLASS COMPANY, LIMITED)
ASAHI GARASU K.K. KENKYUSHO (THE RESEARCH LABORATORY, OF THE ASAHI GLASS COMPANY, LTD.)
1150, MATSUBARA, HAZAWA-CHO, KANAGAWA-KU, YOKOHAMA-SHI 221
JAPANESE AND EUROPEAN
INDEXED: CHE, JICST

ASAHI GARASU KOGYO GIJUTSU SHOREIKAI KENKYU HOKOKU 8211
(ANNUAL REPORT, THE ASAHI GLASS FOUNDATION FOR THE CONTRIBUTION TO INDUSTRIAL TECHNOLOGY)
ASAHI GARASU KOGYO GIJUTSU SHOREIKAI (THE ASAHI GLASS FOUNDATION FOR THE CONTRIBUTION TO INDUSTRIAL TECHNOLOGY)
1-2, 2-CHOME, MARUNOUCHI, CHIYODA-KU, TOKYO 100
JAPANESE AND EUROPEAN
INDEXED: CHE, JICST

ASANO SURETO K.K. CHUO KENKYUSHO HOKOKU 8212
(RESEARCH REPORT OF RESEARCH CENTER, ASANO SLATE COMPANY)
ASANO SURETO K.K. CHUO KENKYUSHO (RESEARCH CENTER, ASANO SLATE COMPANY)
13-10, 3-CHOME, MORISHITA, KOTO-KU, TOKYO 135
JAPANESE

ASBESTOS 8213
NIHON ASUBESUTO K.K.
1-26, 1-CHOME, SHIBA DAIMON, MINATO-KU, TOKYO 105
JAPANESE

CLAY SCIENCE 8214
NIHON NENDO GAKKAI (THE CLAY SCIENCE SOCIETY OF JAPAN)
C/O WASEDA DAIGAKU RIKOGAKUBU SHIGENKOGAKUKA, 4-170, NISHI OKUBO, SHINJUKU-KU, TOKYO 160
EUROPEAN
INDEXED: CHE, JICST

EHIME-KEN YOGYO SHIKENJO GYOMU HOKOKUSHO 8215
(REPORT OF THE EHIME PREFECTURE CERAMIC EXPERIMENT STATION)
EHIME-KEN YOGYO SHIKENJO (EHIME PREFECTURE CERAMIC EXPERIMENT STATION)
GOHONMATSU, TOBE-MACHI, IYO-GUN, EHIME 791-21
JAPANESE

FUKUI-KEN YOGYO SHIKENJO KENKYU HOKOKU 8216
(REPORT OF CERAMIC RESEARCH INSTITUTE, FUKUI PREFECTURE)
FUKUI-KEN YOGYO SHIKENJO
MIYAZAKI-MURA, NYU-GUN, FUKUI 916-02
JAPANESE

FUKUI-KEN YOGYO SHIKENJO NENPO 8217
(ANNUAL REPORT OF THE CERAMIC RESEARCH INSTITUTE, FUKUI PREFECTURE)
FUKUI-KEN YOGYO SHIKENJO
MIYAZAKI-MURA, NYU-GUN, FUKUI 916-02
JAPANESE

GARASU SANGYO SHINBUN 8218
(GLASS INDUSTRY TIMES)
GARASU SANGYO SHINBUNSHA
4-8, 4-CHOME, IIDABASHI, CHIYODA-KU, TOKYO 102
JAPANESE

GARASU SEN'I 8219
(GLASS FIBER)
GARASU SEN'I KYOKAI (GLASS FIBER ASSOCIATION)
5-8, 1-CHOME, NISHISHINBASHI, MINATO-KU, TOKYO 105
JAPANESE

GF 8220
NIHON GARASU SEN'I K.K. (NIPPON GLASS FIBER COMPANY, LTD.)
TAKACHAYA, TSU-SHI 514
JAPANESE

GIFU-KEN TOJIKI SHIKENJO GYOMU HOKOKU 8221
(REPORT OF THE GIFU PREFECTURAL POTTERY LABORATORY)
GIFU-KEN TOJIKI SHIKENJO (GIFU PREFECTURAL POTTERY LABORATORY)
3-11, HOSHIGADAI, TAJIMI-SHI 507
JAPANESE

GIJUTSU KENKYUKAIHO (ASO SEMENTO) 8222
(BULLETIN OF TECHNICAL RESEARCH GROUP OF THE TAGAWA FACULTY, DEPARTMENT OF RESEARCH, ASO CEMENT COMPANY)
ASO SEMENTO K.K. TAGAWA KOJO (TAGAWA FACULTY, DEPARTMENT OF RESEARCH, ASO CEMENT COMPANY)
2877, YUGETA, TAGAWA-SHI 826
JAPANESE

GLASS 8223
GARASU KOGEI KENKYU KAI (ASSOCIATION FOR RESEARCH OF GLASS INDUSTRIAL ARTS)
JAPANESE AND ENGLISH

GLASS BLOCK AND BRICK 8224
NIPPON DENKI GARASU (NIPPON ELECTRIC GLASS COMPANY, LTD.)
C/O HIGASHIHANKYU BUILDING, 16, TAKAGAKI-CHO, KITA-KU, OSAKA-SHI 530
JAPANESE

HARIMA TAIKA GIHO 8225
(TECHNICAL REPORT OF HARIMA REFRACTORIES COMPANY, LIMITED)
HARIMA TAIKA RENGA K.K. GIJUTSU KENKYUSHO
(HARIMA REFRACTORIES COMPANY, LTD.)
3-1, 1-CHOME, ARAI, ARAI-CHO, TAKASAGO-SHI 676
JAPANESE, WITH EUROPEAN SUMMARY
INDEXED: JICST

IBARAKI-KEN YOGYO SHIDOSHO GYOMU HOKOKU 8226
(OPERATION REPORT OF CERAMIC RESEARCH INSTITUTE, IBARAKI PREFECTURE)
IBARAKI-KEN YOGYO SHIDOSHO
50, SHIMOICHIGE, KASAMA-SHI 309-16
JAPANESE

INA DAYORI 8227
(INA TIMES)
INA SEITO
SENMEN SHINDEN, TOKONAME-SHI 479
JAPANESE

INA REPORT 8228
INA SEITO (INA PORCELAIN MANUFACTURING)
JAPANESE AND ENGLISH

INDIAQUA; INDUSTRIAL DIAMOND QUARTERLY 8229
ENGLISH

ISHIWATA 8230
(JOURNAL OF ASBESTOS)
NIHON ISHIWATA KYOKAI (JAPAN ASBESTOS ASSOCIATION)
12-4, 7-CHOME, GINZA, CHUO-KU, TOKYO 104
JAPANESE
INDEXED: JICST

ISHIWATA SURETO KYOKAI GIJUTSU BUKAI RONBUNSHU 8231
(TRANSACTIONS OF TECHNICAL COMMITTEE, ASBESTOS AND SLATE ASSOCIATION)
ISHIWATA SURETO KYOKAI
C/O TAKAHASHI BUILDING, 10-8, 7-CHOME, GINZA, CHUO-KU, TOKYO 104
JAPANESE

ITAGARASU TO SHINJIDAI 8232
(GLASS AND NEW AGE)
SENTORARU GARASU K.K. (CENTRAL GLASS COMPANY)
7-1, 3-CHOME, KANDA NISHIKI-CHO, CHIYODA-KU, TOKYO 101
JAPANESE

KAIHATSU SEMENTO GIHO 8233
(TECHNICAL REPORT OF THE CONCRETE CENTER, TOHOKU KAIHATSU COMPANY, LIMITED)
TOHOKU KAIHATSU K.K. KONKURITO SENTA (CONCRETE CENTER, TOHOKU KAIHATSU COMPANY, LTD.)
15-1, 2-CHOME, KORIYAMA, SENDAI-SHI 982
JAPANESE

KIKAN SEMENTO SANGYO APUROCHI 8234
(QUARTERLY JOURNAL OF THE CEMENT INDUSTRY)
SEMENTO SANGYO KENKYU KAIHEN SEMENTO SHINBUNSHA (CEMENT INDUSTRIAL RESEARCH EDITION, CEMENT NEWSPAPER COMPANY)
JAPANESE

KITA KARA MINAMI KARA 8235
(NEWS OF CENTRAL GLASS COMPANY, LIMITED)
SENTORARU GARASU K.K. (CENTRAL GLASS COMPANY)
3-7, KANDA NISHIKI-CHO, CHIYODA-KU, TOKYO 101
JAPANESE

KONKURITO BUKKUSU 8236
(CONCRETE BOOKS)
SEMENTO KYOKAI (THE CEMENT ASSOCIATION OF JAPAN)
C/O HATTORI BUILDING, 1-1, KYOBASHI, CHUO-KU, TOKYO 104
JAPANESE

KONKURITO BUROKKU 8237
(CONCRETE BLOCK)
NIHON KONKURITO BUROKKU KYOKAI (JAPAN CONCRETE BLOCK ASSOCIATION)
C/O FUJITETSU BUILDING, 2-16, KANDA KAJI-CHO, CHIYODA-KU, TOKYO 101
JAPANESE
INDEXED: JICST

KONKURITO JANARU 8238
(CONCRETE JOURNAL)
NIHON KONKURITO KAIGI (JAPAN NATIONAL COUNCIL ON CONCRETE)
5-10, 1-CHOME, SHIBA KOEN, MINATO-KU, TOKYO 105
JAPANESE

KONKURITO PANFURETTO 8239
(CONCRETE PAMPHLET)
SEMENTO KYOKAI (CEMENT ASSOCIATION OF JAPAN)
C/O HATTORI BUILDING, 1-1, KYOBASHI, CHUO-KU
JAPANESE

KONKURITO SEIHIN 8240
(PRECAST CONCRETE)
HENSYU SENMON IINKAI
ZENKOKU KONKURITO SEIHIN KYOKAI
12-4, 2-CHOME, GINZA, CHUO-KU, TOKYO 104
JAPANESE
INDEXED: JICST

MIE-KEN KOGYO GIJUTSU JOHO; SEMENTO, KONKURITO-HEN 8241
(REPORT OF THE MIE PREFECTURE INDUSTRIAL TECHNOLOGY DEMENT AND CONCRETE EDITION)
MIE-KEN KOGYO GIJUTSU SENTA (MIE PREFECTURAL INDUSTRIAL TECHNOLOGY CENTER)
JAPANESE

MIE-KEN KOGYO GIJUTSU JOHO, YOGYO-HEN 8242
(TECHNICAL INDUSTRY INFORMATION IN MIE PREFECTURE CERAMIC SERIES)
MIE-KEN KOGYO GIJUTSU SENTA (INDUSTRIAL TECHNOLOGY CENTER OF MIE PREFECTURE)
3485, KOMORI-CHO, TAKAJAYA, TSU-SHI 514
JAPANESE

MIE-KEN YOGYO KENKYUJO NENPO 8243
(ANNUAL REPORT OF THE MIE PREFECTURE CERAMICS INDUSTRY RESEARCH INSTITUTE)
MIE-KEN YOGYO KENKYUJO (MIE PREFECTURE CERAMICS INDUSTRY RESEARCH INSTITUTE)
JAPANESE

MIE-KEN YOGYO SHIKENJO GYOMU HOKOKU 8244
(ANNUAL REPORT OF THE CERAMIC RESEARCH INSTITUTE, MIE PREFECTURE)
MIE-KEN YOGYO SHIKENJO
788, HIGASHI AKURAGAWA-CHO, YOKKAICHI-SHI 510
JAPANESE

NAGASAKI-KEN YOGYO SHIKENJO GYOMU HOKOKUSHO 8245
(OPERATION REPORT OF NAGASAKI PREFECTURAL CERAMIC RESEARCH INSTITUTE)
NAGASAKI-KEN YOGYO SHIKENJO (NAGASAKI PREFECTURAL CERAMIC RESEARCH INSTITUTE)
ISEKIGO, HASAMI-CHO, HIGASHI SONOGI-GUN, NAGASAKI 859-37
JAPANESE

NAMA KONKURITO TOKEI NENPO 8246
(YEARBOOK OF READY MIXED CONCRETE STATISTICS)
TSUSHO SANGYO-SHO DAIJIN KANBO CHOSA TOKEI-BU (RESEARCH AND STATISTICS DEPARTMENT, MINISTER'S SECRETARIAT, MINISTRY OF INTERNATIONAL TRADE AND INDUSTRY)
6-1, 1-CHOME, HATCHOBORI, CHUO-KU, TOKYO 104
JAPANESE

NAMA KONKURITO TOKEI SHIHANKIHO 8247
(QUARTERLY REPORT OF READY MIXED CONCRETE)
TSUSHO SANGYO-SHO DAIJIN KANBO CHOSA TOKEI-BU (RESEARCH AND STATISTICS DEPARTMENT, MINISTER'S SECRETARIAT, MINISTRY OF INTERNATIONAL TRADE AND INDUSTRY)
6-1, 1-CHOME, HATCHOBORI, CHUO-KU, TOKYO 104
JAPANESE

NENDO KAGAKU 8248
(JOURNAL OF THE CLAY SCIENCE SOCIETY OF JAPAN)
NIHON NENDO GAKKAI (THE CLAY SCIENCE SOCIETY OF JAPAN)
C/O WASEDA DAIGAKU RIKOGAKUBU SHIGENKOGAKKA, 4-170, NISHI OKUBO, SHINJUKU-KU, TOKYO 160
JAPANESE, WITH EUROPEAN SUMMARY
INDEXED: JICST

NGK REBYU 8249
(NGK REVIEW)
NIHON GARASU (JAPAN GLASS)
JAPANESE AND ENGLISH

NICHI ASU HIROBA 8250
(JAPAN ASBESTOS PLAZA)
NIHON ASUBESUTO (JAPAN ASBESTOS)
JAPANESE

NICHIASU GIJUTSU GIHO 8251
(JAPAN ASBESTOS TECHNICAL REVIEW)
NIHON ASUBESUTO (JAPAN ASBESTOS)
JAPANESE

NIHON TOGYO SHINBUN 8252
(POTTERY JAPAN)
BOEKI NIPPONSHA
21, SHIRAKABE-MACHI, HIGASHI-KU, NAGOYA-SHI 461
JAPANESE

NOZAWA GIKENHO 8253
(REPORTS OF TECHNICAL RESEARCH, NOZAWA CORPORATION)
NOZAWA GIJUTSU KENKYUSHO (TECHNICAL RESEARCH LABORATORY, NOZAWA CORPORATION)
SHIMOSHINDEN, TSURUGASHIMA-MACHI, IRUMA-GUN, SAITAMA 350-02
JAPANESE

ONODA KENKYU HOKOKU 8254
(JOURNAL OF RESEARCH OF THE ONODA CEMENT COMPANY)
ONODA SEMENTO K.K. CHUO KENKYUSHO (CENTRAL RESEARCH LABORATORY, ONODA CEMENT COMPANY, LTD.)
1-7, 1-CHOME, TOYOSU, KOTO-KU, TOKYO 135
JAPANESE, WITH EUROPEAN SUMMARY
INDEXED: CHE, JICST

P & P 8255
NIHON DENKI GARASU (JAPAN ELECTRIC AND GLASS)
JAPANESE AND ENGLISH

PURESUTORESUTO KONKURITO 8256
(JOURNAL OF THE JAPAN PRESTRESSED CONCRETE ENGINEERING ASSOCIATION)
PURESUTORESUTO KONKURITO GIJUTSU KYOKAI (PRE-STRESSED CONCRETE ENGINEERING ASSOCIATION)
12-4, 2-CHOME, GINZA, CHUO-KU, TOKYO 104
JAPANESE, WITH EUROPEAN SUMMARY
INDEXED: JICST

REVIEW OF THE GENERAL MEETING, CEMENT ASSOCIATION OF JAPAN 8257
SEMENTO KYOKAI (THE CEMENT ASSOCIATION OF JAPAN)
C/O HATTORI BUILDING, 1-1, KYOBASHI, CHUO-KU, TOKYO 104
JAPANESE
INDEXED: JICST

REVIEW OF THE GENERAL MEETING, SESSION TECHNICAL 8258
SEMENTO KYOKAI (CEMENT ASSOCIATION OF JAPAN)
1-1, KYOBASHI, CHUO-KU, TOKYO 104
ENGLISH

SAGA-KEN YOGYO SHIKENJO KENKYU HOKOKU 8259
(REPORT OF CERAMIC RESEARCH INSTITUTE, SAGA PREFECTURE)
SAGA-KEN YOGYO SHIKENJO
TANOTAIRA, NAKABE, ARITA-CHO, NISHIMATSUURA-GUN, SAGA 844
JAPANESE

SAN'YOSHI KENKYU HOKOKU 8260
(RESEARCH REPORT OF THE MIE CERAMIC RESEARCH CENTER)
MIE-KEN YOGYO SHIKENJO
788, HIGASHI AKURAGAWA, YOKKAICHI-SHI 510
JAPANESE
INDEXED: JICST

SEKKAI 8261
(LIME)
NIHON SEKKAI KYOKAI (JAPAN LIME ASSOCIATION)
C/O TORANOMON JITSUGYO KAIKAN, 8, SHIBA TORANOMON, MINATO-KU, TOKYO 105
JAPANESE

SEKKAI CHISSO JUKYU NENPO 8262
(ANNUAL REPORT OF THE SUPPLY AND DEMAND OF CALCIUM CYANAMIDE)
NIHON SEKKAI CHISSO KOGYOKAI (JAPAN CALCIUM CYANAMIDE INDUSTRY ASSOCIATION)
C/O FUKUOKA BUILDING, 6-1, YAESU, CHUO-KU, TOKYO 104
JAPANESE

SEKKO BODO NYUSU 8263
(GYPSUM BOARD NEWS)
NIHON SEKKO BODO KOGYO KUMIAI
13-12, 2-CHOME, NISHISHINBASHI, MINATO-KU, TOKYO 105
JAPANESE
INDEXED: JICST

SEKKO TO SEKKAI 8264
(GYPSUM AND LIME)
SEKKO SEKKAI GAKKAI (THE INSTITUTE OF GYPSUM AND LIME RESEARCH)
10-10, 7-CHOME, NISHISHINJUKU, SHINJUKU-KU, TOKYO 160
JAPANESE, WITH EUROPEAN SUMMARY
INDEXED: CHE, INS, JICST

SEMENTO GIJUTSU NENPO 8265
(PROCEEDINGS OF THE CEMENT ASSOCIATION OF JAPAN)
SEMENTO KYOKAI (CEMENT ASSOCIATION OF JAPAN)
C/O HATTORI BUILDING, 1-1, KYOBASHI, CHUO-KU, TOKYO 104
JAPANESE, WITH ENGLISH SUMMARY
INDEXED: CER, CHE, JICST

SEMENTO KOGYO 8266
(CEMENT INDUSTRY)
NIHON SEMENTO K.K. KENKYUSHO (NIHON CEMENT COMPANY, LTD.)
6-1, 1-CHOME, OTE-MACHI, CHIYODA-KU, TOKYO 100
JAPANESE
INDEXED: JICST

SEMENTO KONKURITO 8267
(CEMENT AND CONCRETE)
SEMENTO KYOKAI (CEMENT ASSOCIATION OF JAPAN)
C/O HATTORI BUILDING, 1-1, KYOBASHI, CHUO-KU, TOKYO 104
JAPANESE
INDEXED: CHE, JICST

SEMENTO KYOKAI FUNMATSUDO SENMON IINKAI HOKOKU 8268
(REPORT OF SPECIAL COMMITTEE FOR POWDER, CEMENT ASSOCIATION OF JAPAN)
SEMENTO KYOKAI (CEMENT ASSOCIATION OF JAPAN)
C/O HATTORI BUILDING, 1-1, KYOBASHI, CHUO-KU, TOKYO 104
JAPANESE
INDEXED: JICST

SEMENTO KYOKAI FUNSAI SENMON IINKAI HOKOKU (S) 8269
(REPORT OF SPECIAL COMMITTEE FOR FRAGMENTATION, CEMENT ASSOCIATION OF JAPAN)
SEMENTO KYOKAI (CEMENT ASSOCIATION OF JAPAN)
C/O HATTORI BUILDING, 1-1, KYOBASHI, CHUO-KU, TOKYO 104
JAPANESE
INDEXED: JICST

SEMENTO KYOKAI KAGAKU BUNSEKI SENMON IINKAI HOKOKU (I) 8270
(REPORT OF SPECIAL COMMITTEE FOR CHEMICAL ANALYSIS, CEMENT ASSOCIATES OF JAPAN))
SEMENTO KYOKAI (CEMENT ASSOCIATION OF JAPAN)
C/O HATTORI BUILDING, 1-1, KYOBASHI, CHUO-KU, TOKYO 104
JAPANESE

SEMENTO KYOKAI KONKURITO SENMON IINKAI HOKOKU (F) 8271
(REPORT OF SPECIAL COMMITTEE FOR CONCRETE, CEMENT ASSOCIATION OF JAPAN)
SEMENTO KYOKAI (CEMENT ASSOCIATION OF JAPAN)
C/O HATTORI BUILDING, 1-1, KYOBASHI, CHUO-KU, TOKYO 104
JAPANESE
INDEXED: JICST

SEMENTO KYOKAI NENRYO SENMON IINKAI HOKOKU (T) 8272
(REPORT OF SPECIAL COMMITTEE FOR FUEL, CEMENT ASSOCIATION OF JAPAN))
SEMENTO KYOKAI (CEMENT ASSOCIATION OF JAPAN)
C/O HATTORI BUILDING, 1-1, KYOBASHI, CHUO-KU, TOKYO 104
JAPANESE
INDEXED: JICST

SEMENTO KYOKAI SHUSHUKU SENMON IINKAI HOKOKU (H) 8273
(REPORT OF SPECIAL COMMITTEE FOR SHRINKING, CEMENT ASSOCIATION OF JAPAN)
SEMENTO KYOKAI (CEMENT ASSOCIATION OF JAPAN)
C/O HATTORI BUILDING, 1-1, KYOBASHI, CHUO-KU, TOKYO 104
JAPANESE
INDEXED: JICST

SEMENTO KYOKAI SUIWANETSU SENMON IINKAI HOKOKU (J) 8274
(REPORT OF SPECIAL COMMITTEE FOR HEAT OF HYDRATION, CEMENT ASSOCIATION OF JAPAN)
SEMENTO KYOKAI (CEMENT ASSOCIATION OF JAPAN)
C/O HATTORI BUILDING, 1-1, KYOBASHI, CHUO-KU, TOKYO 104
JAPANESE
INDEXED: JICST

SERAMIKKUSU 8275
(CERAMICS)
YOGYO KYOKAI (THE CERAMIC SOCIETY OF JAPAN)
22-17, 2-CHOME, HYAKUMIN-CHO, SHINJUKU-KU, TOKYO 160
JAPANESE
INDEXED: APT, CER, CHE, JICST

SHIGA-KENRITSU SHIGARAKI YOGYO SHIKENJO GYOMU HOKOKU 8276
(OPERATION REPORT OF SHIGARAKI CERAMIC RESEARCH INSTITUTE, SHIGA PREFECTURE)
SHIGA-KENRITSU SHIGARAKI YOGYO SHIKENJO
NAGANO, SHIGARAKI-CHO, KOGA-GUN, SHIGA 529-18
JAPANESE

SHINAGAWA GIHO 8277
(SHINAGAWA TECHNICAL REPORT)
SHINAGAWA SHIRO RENGA K.K. GIJUTSU KENKYUSHO (SHINAGAWA FIRE BRICK COMPANY, LTD. SERVICE ENGINEER)
707, INBE, BIZEN-SHI 705
JAPANESE, WITH EUROPEAN SUMMARY
INDEXED: JICST

SOGO TSUSHIN, TOGYOBAN 8278
(GENERAL TIMES, SERIES ON CERAMIC WARE)
SOGO TSUSHINSHA
10-19, 2-CHOME, MARUNOUCHI, NAKA-KU, NAGOYA-SHI 460
JAPANESE

TAIKA RENGA KYOKAI KAIHO 8279
(BULLETIN OF THE FIREBRICK ASSOCIATION OF JAPAN)
TAIKA RENGA KYOKAI (FIREBRICK ASSOCIATION OF JAPAN)
3-13, 7-CHOME, GINZA, CHUO-KU, TOKYO 104
JAPANESE

TAIKA ZAIRYO 8280
(REFRACTORY MATERIAL)
KUROSAKI YOGYO K.K. (KUROSAKI FIRE-BRICK MANUFACTURING COMPANY, LTD.)
1-1, HIGASHI-HAMA-CHO, YAHATA-KU, KITAKYUSHU-SHI 806
JAPANESE
INDEXED: CHE, JICST

TAIKABUTSU 8281
(REFRACTORIES)
TAIKABUTSU GIJUTSU KYOKAI (TECHNICAL ASSOCIATION OF REFRACTORIES JAPAN)
3-13, 7-CHOME, GINZA, CHUO-KU, TOKYO 104
JAPANESE, WITH ENGLISH SUMMARY
INDEXED: CER, CHE, JICST

TAJIMI-SHI TOJIKI ISHO KENKYUSHO HOKOKU 8282
(RESEARCH REPORT OF THE TAJIMI POTTERY AND PORCELAIN DESIGN LABORATORY)
TAJIMI-SHI TOJIKI ISHO KENKYUSHO (TAJIMI POTTERY AND PORCELAIN DESIGN LABORATORY)
2-77, MISAKA-CHO, TAJIMI-SHI 507
JAPANESE

TOCHIGI-KEN YOGYO SHIDOSHO GYOMU HOKOKUSHO 8283
(OPERATION REPORT OF CERAMIC RESEARCH INSTITUTE, TOCHIGI PREFECTURE)
TOCHIGI-KEN YOGYO SHIDOSHO
4350, MASHIKO, MASHIKO-MACHI, HAGA-GUN, TOCHIGI 321-42
JAPANESE

TOJIKI ISHO NO KENKYU 8284
(RESEARCH ON CERAMICS DESIGN)
TSUSHO SANGYO-SHO KOGYO GIJUTSU-IN NAGOYA KOGYO GIJUTSU SHIKENJO SETO BUNSHITSU (SETO BRANCH, GOVERNMENT INDUSTRIAL RESEARCH INSTITUTE, NAGOYA, AGENCY OF INDUSTRIAL SCIENCE AND TECHNOLOGY)
110, NISHI IBARA-CHO, SETO-SHI 489
JAPANESE

TOKEN KAIHO 8285
(REPORT OF THE KYOTO CERAMIC RESEARCH ASSOCIATION)
KYOTO TOJIKI KENKYUKAI (KYOTO CERAMIC RESEARCH ASSOCIATION)
C/O KYOTO-SHI KOGYO SHIKENJO, 1-2, NANDEN-CHO, NISHI-9-JO, MINAMI-KU, KYOTO-SHI 601
JAPANESE
INDEXED: JICST

YOGYO KENZAI TOKEI GEPPO 8286
(MONTHLY STATISTICS OF CERAMICS AND ARCHITECTURAL MATERIALS)
TSUSHO SANGYO-SHO DAIJIN KANBO CHOSA TOKEI-BU (RESEARCH AND STATISTICS DIVISION, MINISTER'S SECRETARIAT, MINISTRY OF INTERNATIONAL TRADE & INDUSTRY)
22-17, 2-CHOME, HYAKUNIN-CHO, SHINJUKU-KU, TOKYO 160
JAPANESE
INDEXED: JICST

YOGYO KYOKAISHI 8287
(JOURNAL OF THE CERAMIC SOCIETY OF JAPAN)
YOGYO KYOKAI (THE CERAMIC SOCIETY OF JAPAN)
22-17, 2-CHOME, HYAKUNIN-CHO, SHINJUKU-KU, TOKYO 160
JAPANESE AND ENGLISH
INDEXED: CER, CHE, ENG, INS, JICST

YOGYO TOKEI NENPO 8288
(YEARBOOK OF CERAMICS STATISTICS)
TSUSHO SANGYO-SHO DAIJIN KANBO CHOSA TOKEI-BU (RESEARCH AND STATISTICS DIVISION, MINISTER'S SECRETARIAT, MINISTRY OF INTERNATIONAL TRADE AND INDUSTRY)
15-1, 6-CHOME, GINZA, CHUO-KU, TOKYO 104
JAPANESE

Metallurgy (Including Workshop Practice Generally)

ABAN KUBOTA 8289
(URBAN KUBOTA)
KUBOTA TEKKO K.K. (KUBOTA LIMITED)
2-22, FUNADE-MACHI, NANIWA-KU, OSAKA-SHI 556
JAPANESE

ABSTRACTS OF JAPANESE PAPERS ON METALS 8290
JAPANESE

AEN TEPPAN 8291
(ZINC IRON PLATE)
AEN TEPPANKAI AND TEKKO SANGYO KENKYUSHO
3-16, NIHONBASHI KAYABA-CHO, CHUO-KU, TOKYO 103
JAPANESE

ALUMINUM NEWS 8292
JAPANESE
INDEXED: JICST

ARU 8293
(ALUMINUM JOURNAL)
KEIKINZOKU TSUSHINSHA
C/O NAKAMOTO BUILDING, 27, DAIKU-MACHI, KITA-KU, OSAKA-SHI 530
JAPANESE

ARUMI ATSUEN 8294
(ALUMINUM ROLLING)
KEIKINZOKU ATSUEN KOGYOKAI (LIGHT METAL ROLLING INDUSTRY ASSOCIATION)
12-22, 1-CHOME, TSUKIJI, CHUO-KU, TOKYO 104
JAPANESE
INDEXED: JICST

ARUMI EJI 8295
(ALUMINUM AGE)
KEIKINZOKU ATSUEN KOGYOKAI (LIGHT METAL ROLLING INDUSTRY ASSOCIATION)
12-22, 1-CHOME, TSUKIJI, CHUO-KU, TOKYO 103
JAPANESE

ARUMI SHOWA 8296
(NEWS OF SHOWA ALUMINUM COMPANY)
SHOWA ARUMINYUMU K.K. (SHOWA ALUMINUM COMPANY, LTD.)
6-224, KAIZAN-CHO, SAKAI-SHI 590
JAPANESE

ARUMINIUM 8297
JAPANESE
INDEXED: JICST

ARUMINIUMU GIJUTSU BUNKENSHU 8298
(COLLECTED BIBLIOGRAPHY ON ALUMINUM TECHNOLOGY)
JAPANESE
INDEXED: JICST

ARUMINYUMU 8299
(ALUMINUM)
KEIKINZOKU KYOKAI (THE LIGHT METALS ASSOCIATION)
1-3, 2-CHOME, NIHONBASHI, CHUO-KU, TOKYO 103
JAPANESE

ARUMINYUMU HYOMEN SHORI KENKYU CHOSA HOKOKU 8300
(RESEARCH REPORT ON ALUMINUM FINISHING)
KINZOKU HYOMEN GIJUTSU KYOKAI, ARUMINYUMU HYOMEN SHORI KENKYUKAI
C/O KYODO BUILDING, 2, KANDA IWAMOTO-CHO, CHIYODA-KU, TOKYO 101
JAPANESE
INDEXED: JICST

ARUMINYUMU KENKYU KAISHI 8301
(JOURNAL OF THE ALUMINUM RESEARCH ASSOCIATION)
THE JAPANESE ANODIZING ASSOCIATION
C/O THE METAL FINISHING SOCIETY OF JAPAN KYODO BUILDING, NO. 2, KANDA IWAMOTO-CHO, CHIYODA-KU TOKYO 101
JAPANESE
INDEXED: CHE

ARUTOPIA 8302
(ALTOPIA)
KEIKINZOKU SHUPPAN K.K.
1-3, TAKARA-CHO, CHUO-KU, TOKYO 104
JAPANESE
INDEXED: JICST

BOSEI KANRI 8303
(RUST PREVENTION AND CONTROL)
NIPPON BOSEI GIJUTSU KYOKAI
3-5-8 SHIBA-KOEN, MINATO-KU, TOKYO
JAPANESE
INDEXED: CHE, JICST

BOSHO KANRI 8304
(RUST PREVENTION AND CONTROL)
NIHON BOSHO KYOKAI (JAPAN RUST PREVENTION TECHNIC ASSOCIATION)
5-8, 3-CHOME, SHIBAKOEN, MINATO-KU, TOKYO 105
JAPANESE

BOSHOKU GIJUTSU 8305
(CORROSION ENGINEERING)
NIHON GAKUJUTSU SHINKOKAI FUSHOKU BOSHI DAI-97-IINKAI (THE 97TH COMMITTEE FOR THE CORROSION PREVENTION, JAPAN SOCIETY FOR THE PROMOTION OF SCIENCE)
1-2, 2-CHOME, HITOTSUBASHI, CHIYODA-KU, TOKYO 101
JAPANESE, WITH EUROPEAN SUMMARY
INDEXED: CHE, ENG, WAA, JICST

BULLETIN OF MEIJI UNIVERSITY 8306
JAPANESE AND ENGLISH
INDEXED: COP

CHIBA SEITETSUSHO NYUSU 8307
(NEWS, CHIBA WORKS, KAWASAKI STEEL COMPANY)
KAWASAKI SEITETSU CHIBA SEITETSUSHO (CHIBA WORKS, KAWASAKI STEEL COMPANY)
1, KAWASAKI-CHO, CHIBA-SHI 280
JAPANESE

CHITANYUMU JIRUKONYUMU 8308
(TITANIUM AND ZIRCONIUM)
CHITANYUMU KYOKAI (JAPAN TITANIUM SOCIETY)
12-22, 1-CHOME, TSUKIJI, CHUO-KU, TOKYO 104
JAPANESE
INDEXED: CHE, JICST

CHUKEN HOKOKU 8309
(TRANSACTIONS OF THE CASTINGS RESEARCH LABORATORY)
WASEDA DAIGAKU IMONO KENKYUSHO (THE CASTINGS RESEARCH LABORATORY, WASEDA UNIVERSITY)
1-500, TOTSUKA-CHO, SHINJUKU-KU, TOKYO 160
JAPANESE
INDEXED: JICST

CHUKO, TANKO SHIRYO 8310
(INFORMATION ON CAST STEEL AND FORGED STEEL)
NIHON CHUTANKOKAI (THE JAPAN CAST STEEL AND FORGED STEEL SOCIETY)
KOGYO TOSHO SHUPPANSHA
8-2, 1-CHOME, MARUNOUCHI, CHIYODA-KU, TOKYO 100
JAPANESE

CHUKO TO TANKO 8311
(CASTING AND FORGING OF STEEL)
NIHON CHUTANKOKAI (THE JAPAN CAST STEEL AND FORGED STEEL SOCIETY) AND KOGYO TOSHO SHUPPANSHA
8-2, 1-CHOME, MARUNOUCHI, CHIYODA-KU, TOKYO 100
JAPANESE
INDEXED: JICST

CHUTANZO 8312
(CASTING AND FORGING)
SHIN NIHON CHUTANZO KYOKAI (THE NEW JAPAN CASTING AND FORGING SOCIETY)
31, KORAI HASHIZUME-CHO, HIGASHI-KU, OSAKA-SHI 541
JAPANESE
INDEXED: CHE, WAA, JICST

DAIDO KOHAN NO KENZAI 8313
(DAIDO NEWS)
DAIDO KOHAN K.K.
22, KUISE, UMASHINDEN, AMAGASAKI-SHI 660
JAPANESE

DENKI SEIKO 8314
(ELECTRIC FURNACE STEEL)
DENKI SEIKO KENKYUKAI (ELECTRIC FURNACE STEEL RESEARCH ASSOCIATION)
2-30, DAIDO-CHO, MINAMI-KU, NAGOYA-SHI 457
JAPANESE, WITH ENGLISH SUMMARY
INDEXED: CHE, JICST

DO 8315
(COPPER & BRASS)
NIHON DO SENTA (COPPER & BRASS CENTER, JAPAN)
12-22, 1-CHOME, TSUKIJI, CHUO-KU, TOKYO 104
JAPANESE
INDEXED: JICST

FERO AROI 8316
(FERRO ALLOY)
NIHON FERO AROI KYOKAI GIJUTSU IINKAI
(TECHNICAL COMMITTEE, JAPAN FERRO ALLOY ASSOCIATION)
C/O TOKYO KURABU BUILDING, 2-6, 3-CHOME,
KASUMIGASEKI, CHIYODA-KU, TOKYO 100
JAPANESE

FERO AROI SHINBUN 8317
(FERRO ALLOY NEWS)
NIHON FERO AROI KYOKAI (JAPAN FERRO ALLOY
ASSOCIATION)
2-6, 3-CHOME, KASUMIGASEKI, CHIYODA-KU, TOKYO 100
JAPANESE

FUSEN 8318
(FLOTATION)
FUSEN KENKYUKAI (THE FLOTATION RESEARCH ASSOCIA-
TION OF JAPAN)
C/O HANKYU BUILDING, TENROKU, OYODO-KU, OSAKA-SHI
531
JAPANESE
INDEXED: CHE, JICST

FUTSU KOTETSU ZAIHIN SHUBETSU SUNPOBETSU 8319
SEISAN JISSEKI
(ACCUMULATED PRODUCTION OF STEEL MATERIALS
DIVIDED BY TYPE AND SIZE)
TETSUZAI KURABU (STEEL MATERIALS CLUB)
JAPANESE

GIFU-GUN KINZOKU SHIKENJO GYOMU HOKOKU 8320
(OPERATION REPORT OF GIFU PREFECTURAL METAL RESEARCH
INSTITUTE)
GIFU-KEN KINZOKU SHIKENJO (GIFU PREFECTURAL
METAL RESEARCH INSTITUTE)
1288, OZE, SEKI-SHI 501-32
JAPANESE

GIFU-KEN KINZOKU SHIKENJO KENKYU HOKOKU 8321
(REPORT OF THE GIFU PREFECTURAL METALS EXPERIMENT
STATION)
GIFU-KEN KINZOKU SHIKENJO (GIFU PREFECTURAL METALS
EXPERIMENT STATION)
JAPANESE

GIJUTSU KAIHO (NIPPON KINZOKU K. K.) 8322
(BULLETIN OF THE TECHNICAL RESEARCH INSTITUTE,
NIPPON KINZOKU COMPANY, LIMITED)
NIPPON KINZOKU K.K. GIJUTSU KENKYUSHO (TECHNICAL
RESEARCH INSTITUTE, NIPPON KINZOKU COMPANY, LIMITED)
JAPANESE

HITETSU KINZOKU SEIHIN TOKEI GEPPO 8323
(MONTHLY REPORT OF NON-FERROUS METAL PRODUCTS
STATISTICS)
TSUSHO SANGYO-SHO DAIJIN KANBO CHOSA TOKEI-BU
(RESEARCH AND STATISTICS DIVISION, MINISTER'S
SECRETARIAT, MINISTRY OF INTERNATIONAL TRADE AND
INDUSTRY)
1-3, 2-CHOME, NIHONBASHI, CHUO-KU, TOKYO 103
JAPANESE
INDEXED: JICST

HITETSU KINZOKU SEIHIN TOKEI NENPO 8324
(YEARBOOK OF NON-FERROUS METAL PRODUCTS STATISTICS)
TSUSHO SANGYO-SHO DAIJIN KANBO CHOSA TOKEI-BU
(RESEARCH AND STATISTICS DIVISION, MINISTER'S
SECRETARIAT, MINISTRY OF INTERNATIONAL TRADE
AND INDUSTRY)
C/O NIHONBASHI ASAHI SEIMEI-KAN, 1-3, 2-CHOME,
NIHONBASHI, CHUO-KU, TOKYO 103
JAPANESE

HITETSU KINZOKUTO JUKYU TOKEI NENPO 8325
(STATISTICAL YEARBOOK OF SUPPLY AND DEMAND OF NON-
FERROUS METAL PRODUCTS))
TSUSHO SANGYO-SHO DAIJIN KANBO CHOSA TOKEI-BU
(RESEARCH AND STATISTICS DIVISION, MINISTER'S
SECRETARIAT, MINISTRY OF INTERNATIONAL TRADE
AND INDUSTRY)
3-1, 1-CHOME, KASUMIGASEKI, CHIYODA-KU,
TOKYO 100
JAPANESE

HYOGO-KEN KOGYO GIJUTSU JOHO: KINZOKU-HEN 8326
(HYOGO-PREFECTURE INDUSTRIAL TECHNOLOGY REPORT:
METALS)
HYOGO KENRITSU KOGYO SHIKENJO
(HYOGO PREFECTURAL INDUSTRIAL LABORATORY)
JAPANESE

IKEGAI GIHO 8327
(IKEGAI TECHNICAL REPORT)
JAPANESE
INDEXED: JICST

IKEGAI NYUSU 8328
(IKEGAI NEWS)
IKEGAI TEKKO K. K. (THE IKEGAI IRON WORKS
COMPANY)
1-21, 4-CHOME, SHIBA, MINATO-KU, TOKYO 108
JAPANESE

IMONO 8329
(JOURNAL OF THE JAPAN FOUNDRYMEN'S SOCIETY)
NIHON IMONO KYOKAI (JAPAN FOUNDRYMEN'S SOCIETY)
12-13, 8-CHOME, GINZA, CHUO-KU, TOKYO 104
JAPANESE AND ENGLISH
INDEXED: CHE, CDP, MET, JICST

IMONO DAIJESUTO 8330
(CASTING DIGEST)
NIHON IMONO KOGYOKAI (NIHON CAST IRON FOUNDRY
ASSOCIATION)
5-8, 3-CHOME, SHIBA KOEN, MINATO-KU, TOKYO 105
JAPANESE
INDEXED: WAA

JAKUTO NYUSU 8331
(JACT NEWS)
CHUZU GIJUTSU FUKYU KYOKAI (JAPANESE
ASSOCIATION OF CASTING TECHNOLOGY)
9-13, 5-CHOME, GINZA, CHUO-KU, TOKYO 104
JAPANESE
INDEXED: JICST

JAPAN IRON AND STEEL FEDERATION STATISTICAL 8332
YEARBOOK
ENGLISH

JAPAN METAL BULLETIN 8333
ENGLISH

JAPAN STEEL WORKS TECHNICAL NEWS 8334
NIHON SEIKOSHO EIBUN GIHO
NIHON SEIKOSHO (JAPAN STEEL WORKS, LIMITED)
1-12, YURAKU-CHO, CHIYODA-KU, TOKYO 100
EUROPEAN

JAPAN'S IRON AND STEEL INDUSTRY 8335
KAWATA PABURISHITI K.K. (KAWATA PUBLICITY,
INCORPORATED)
C/O HATORI BUILDING, 5-6, 5-CHOME, KOISHIKAWA,
BUNKYO-KU, TOKYO 112
ENGLISH

JITSUMU HYOMEN GIJUTSU 8336
(METAL FINISHING PRACTICE)
KINZOKU HYOMEN GIJUTSU KYOKAI (THE METAL FINISHING
SOCIETY OF JAPAN)
C/O KYODO BUILDING, 2, KANDA IWAMOTO-CHO,
CHIYODA-KU, TOKYO 101
JAPANESE
INDEXED: CHE, WAA, JICST

JMI NYUSU 8337
(JMI NEWS)
KIKAI DENSHIN KENSA KENTEI KYOKAI (JAPAN
MACHINERY AND METALS INSPECTION INSTITUTE)
9-15, 1-CHOME, AKASAKA, MINATO-KU, TOKYO 107
JAPANESE
INDEXED: JICST

KAGAKU GIJUTSU BUNKEN SOKUHO, KINZOKU KOGAKU, KOZAN 8338
KOGAKU, CHIKYU NO KAGAKU-HEN
(CURRENT BIBLIOGRAPHY ON SCIENCE AND TECHNOLOGY,
SERIES ON EARTH SCIENCES, MINING AND METALLURGY)
NIHON KAGAKU GIJUTSU JOHO SENTA (THE JAPAN
INFORMATION CENTER OF SCIENCE AND TECHNOLOGY)
1-1, 2-CHOME, NAGATA-CHO, CHIYODA-KU, TOKYO 100
JAPANESE
INDEXED: JICST

KAIGAI TOKUSHUKO JOHO 8339
(WORLD SPECIAL STEEL NEWS)
TOKUSHUKO KURABU (SPECIAL STEEL ASSOCIATION OF
JAPAN)
3-16, NIHONBASHI KAYABA-CHO, CHUO-KU, TOKYO 103
JAPANESE
INDEXED: JICST

KANKYO GIJUTSU KENKYUSHO HOKOKU (KUBOTA TEKKO K.K.) 8340
(REPORT OF ENVIRONMENT CONTROL ENGINEERING PLANNING DEPARTMENT, KUBOTA LIMITED)
KUBOTA TEKKO K.K. KANKYO GIJUTSU KENKYUSHO (ENVIRONMENT CONTROL ENGINEERING PLANNING DEPARTMENT, KUBOTA, LTD.)
2-22, FUNADE-CHO, NANIWA-KU, OSAKA-SHI 556-91
JAPANESE
INDEXED: CHE, JICST

KARAMU 8341
(COLUMN)
SHINNIHON SEITETSU K.K.
6-3, 2-CHOME, OTE-MACHI, CHIYODA-KU, TOKYO 100
JAPANESE
INDEXED: JICST

KAWASAKI SEITETSU GIHO 8342
(KAWASAKI STEEL TECHNICAL REPORT)
KAWASAKI SEITETSU K.K.
1-11, YURAKU-CHO, CHIYODA-KU, TOKYO 100
JAPANESE, WITH ENGLISH SUMMARY
INDEXED: CHE, JICST

KEIKINZOKU 8343
(LIGHT METALS)
KEIKINZOKU GAKKAI (JAPAN INSTITUTE OF LIGHT METALS)
1-3, 2-CHOME, NIHONBASHI, CHUO-KU, TOKYO 103
JAPANESE AND ENGLISH
INDEXED: CHE, IAA, JICST

KEIKINZOKU ATSUEN KOGYO TOKEI NENPO 8344
(ANNUAL STATISTICS OF THE LIGHT METAL ROLLING INDUSTRY)
KEIKINZOKU ATSUEN KOGYOKAI (LIGHT METAL ROLLING INDUSTRY ASSOCIATION)
12-22, 1-CHOME, TSUKIJI, CHUO-KU, TOKYO 104
JAPANESE

KEIKINZOKU ATSUENHIN TOKEI GEPPO 8345
(LIGHT METAL WROUGHT PRODUCTS MONTHLY REPORT)
KEIKINZOKU ATSUEN KOGYOKAI (LIGHT METAL ROLLING INDUSTRY ASSOCIATION)
12-22, 1-CHOME, TSUKIJI, CHUO-KU, TOKYO 104
JAPANESE

KEIKINZOKU KOGYO TOKEI NENPO 8346
(LIGHT METALS STATISTICS IN JAPAN)
KEIKINZOKU KYOKAI (JAPAN LIGHT METAL ASSOCIATION)
C/O NIHONBASHI ASAHI SEIMEIKAN, 1-3, 2-CHOME, NIHONBASHI, CHUO-KU, TOKYO 102
JAPANESE

KEIKINZOKU SEIHIN KYOKAIHO 8347
(REPORT OF THE ASSOCIATION FOR LIGHT METAL PRODUCTS)
ASSOCIATION FOR LIGHT METAL PRODUCTS
JAPANESE

KEIKINZOKU YOSETSU 8348
(JAPAN LIGHT METAL WELDING)
KEIKINZOKU YOSETSU GIJUTSUKAI (JAPAN INSTITUTE OF THE LIGHT METAL WELDING)
1-3, 2-CHOME, NIHONBASHI, CHUO-KU, TOKYO 103
JAPANESE
INDEXED: CHE, WAS, JICST

KENKYU CHOSA HOKOKUSHO (JIDO CHUZO SOCHI KAIHATSU KENKYU HOKOKUSHO) 8349
(REPORT OF JAPAN GENERAL FOUNDRY CENTER FOUNDATION, FOUNDRY RESEARCH DEVELOPMENT & TRAINING INSTITUTE)
SOGO IMONO SENTA KAIHATSU KENKYUSHO (JAPAN GENERAL FOUNDRY CENTER FOUNDATION, FOUNDRY RESEARCH DEVELOPMENT & TRAINING INSTITUTE)
1-67, 1-CHOME, HACHIMAN-CHO, HIGASHIKURUME-SHI, TOKYO 180-03
JAPANESE

KENSETSUYO KOZAI SHURYOSHU 8350
(DATA OF STEEL FOR ARCHITECTURE)
NIHON KOKOZO KYOKAI
KOZAI KURABU
3-1, 3-CHOME, MARUNOUCHI, CHIYODA-KU, TOKYO 100
JAPANESE

KIKAI KINZOKU SHINBUN 8351
(MACHINE AND METALS NEWS)
KIKAI KINZOKU SHINBUNSHA
11, KITA SUMIYA-CHO, MINAMI-KU, OSAKA-SHI 542
JAPANESE

KINKI ARUMINYUMU HYOMEN SHORI KENKYUKAI KAISHI 8352
(JOURNAL OF ALUMINUM FINISHING SOCIETY OF KINKI)
KINKI ARUMINYUMU HYOMEN SHORI KENKYUKAI (ALUMINUM FINISHING SOCIETY OF KINKI)
C/O KINKI DAIGAKU RIKOGAKUBU OYO KAGAKUKA YOSHIMURA KENKYUSHITSU, 321, KOWAKAE, HIGASHIOSAKA-SHI 577
JAPANESE
INDEXED: JICST

KINZAI GIKEN NYUSU 8353
(NEWS OF THE NATIONAL RESEARCH INSTITUTE FOR METALS)
KAGAKU GIJUTSU-CHO KINZOKU ZAIRYO GIJUTSU KENKYUSHO (NATIONAL RESEARCH INSTITUTE FOR METALS, SCIENCE AND TECHNOLOGY AGENCY)
3-12, 2-CHOME, NAKAMEGURO, MEGURO-KU, TOKYO 153
JAPANESE
INDEXED: MET, WAA, JICST

KINZOKU 8354
(METALS)
AGUNESHA
3-10, 3-CHOME, SHIBUYA, SHIBUYA-KU, TOKYO 150
JAPANESE
INDEXED: CHE, COP, JICST

KINZOKU HYOMEN GIJUTSU 8355
(JOURNAL OF THE METAL FINISHING SOCIETY OF JAPAN)
KINZOKU HYOMEN GIJUTSU KYOKAI (THE METAL FINISHING SOCIETY OF JAPAN)
C/O KYODO BUILDING, 2, KANDA IWAMOTO-CHO, CHIYODA-KU, TOKYO 101
JAPANESE, WITH EUROPEAN SUMMARY
INDEXED: CHE, COP, INS, MET, WAA, JICST

KINZOKU HYOMEN SHORI GIJUTSU BUNKENSHU 8356
(ALUMINUM FINISHING TECHNOLOGY ABSTRACTS)
NIHON KAGAKU GIJUTSU JOHO SENTA (THE JAPAN INFORMATION CENTER OF SCIENCE AND TECHNOLOGY)
1-1, 2-CHOME, NAGATA-CHO, CHIYODA-KU, TOKYO 100
JAPANESE
INDEXED: JICST

KINZOKU JANARU 8357
(METALS JOURNAL)
KINZOKU TSUSHINSHA
14-5, 3-CHOME, GINZA, CHUO-KU, TOKYO 104
JAPANESE
INDEXED: JICST

KINZOKU JIHYO 8358
(CURRENT REVIEW ON NEW METALS)
KINZOKU JIHYOSHA
3-4, KOJI-MACHI, CHIYODA-KU, TOKYO 102
JAPANESE
INDEXED: JICST

KINZOKU PURESU 8359
(METAL PRESS)
NIHON KINZOKU PURESU KOGYO KYOKAI (JAPANESE ASSOCIATION OF METAL PRESS INDUSTRY)
C/O NAMIKI BUILDING, 16-3, SAKURAGAOKA-CHO, SHIBUYA-KU, TOKYO 150
JAPANESE
INDEXED: JICST

KINZOKU ZAIRYO 8360
(METALS IN ENGINEERING)
NIKKAN KOGYO SHINBUNSHA
8-10, 1-CHOME, KUDAN KITA, CHIYODA-KU, TOKYO 102
JAPANESE
INDEXED: CHE, JICST

KINZOKU ZAIRYO GIJUTSU KENKYUSHO KENKYU HOKOKU 8361
(REPORT OF NATIONAL RESEARCH INSTITUTE FOR METALS)
KAGAKU GIJUTSU-CHO KINZOKU ZAIRYO GIJUTSU KENKYUSHO (NATIONAL RESEARCH INSTITUTE FOR METALS, SCIENCE AND TECHNOLOGY AGENCY)
3-12, 2-CHOME, NAKAMEGURO, MEGURO-KU, TOKYO 153
JAPANESE
INDEXED: CHE, JICST

KOGYO KANETSU 8362
(INDUSTRIAL HEATING)
NIHON KOGYORO KYOKAI (THE JAPAN INDUSTRIAL FURNACE MANUFACTURER'S ASSOCIATION)
21-10, 2-CHOME, YOYOGI, SHIBUYA-KU, TOKYO 151
JAPANESE
INDEXED: JICST

KOGYO REA METARU 8363
(INDUSTRIAL RARE METALS)
ARUMU SHUPPANSHA
5-6, 2-CHOME, HONGO, BUNKYO-KU, TOKYO 113
JAPANESE
INDEXED: CHE, JICST

KOZAI KURABU JOHO 8364
(STEEL MATERIALS CLUB REPORT)
KOZAI KURABU JIMUKYOKU (STEEL MATERIAL CLUB, GENERAL AFFAIRS OFFICE)
JAPANESE

KOZAI KURABU TOKEI GEPPO 8365
(MONTHLY STATISTICS OF STEEL MATERIALS)
KOZAI KURABU
3-16, NIHONBASHI KAYABA-CHO, CHUO-KU, TOKYO 103
JAPANESE

KOZAI SHINBUN 8366
(STEEL MATERIALS NEWS)
KOZAI SHINBUNSHA
3, NIHONBASHI KODENMA-CHO, CHUO-KU, TOKYO 103
JAPANESE

KUBOTA GIHO 8367
(KUBOTA TECHNICAL REPORT)
KUBOTA TEKKO (KUBOTA STEEL)
JAPANESE

KUBOTA GURAFU 8368
(KUBOTA GRAPH)
JAPANESE
INDEXED: JICST

KYOJIN CHUTETSU NYUSU 8369
(HIGH GRADE CAST IRON NEWS)
NIHON KYOJIN CHUTETSU KYOKAI (JAPAN HIGH GRADE CAST IRON FOUNDER'S ASSOCIATION)
C/O KIKAI SHINKO KAIKAN, 5-8, 3-CHOME, SHIBA KOEN, MINATO-KU, TOKYO 105
JAPANESE

KYOTSU SHISETSU GIJUTSU KENKYU HOKOKU 8370
(TECHNICAL REPORT OF COMMON LABORATORY IN RESEARCH INSTITUTE FOR METALS, TOHOKU UNIVERSITY)
TOHOKU DAIGAKU KINZOKU ZAIRYO KENKYUSHO (RESEARCH INSTITUTE FOR IRON, STEEL AND OTHER METALS, TOHOKU UNIVERSITY)
1-1, 2-CHOME, KATAHIRA, SENDAI-SHI 980
JAPANESE

LIST OF PAPERS PUBLISHED BY THE MEMBERS OF THE RESEARCH INSTITUTE FOR IRON, STEEL AND OTHER METALS, TOHOKU UNIVERSITY 8371
TOHOKU DAIGAKU KINZOKU ZAIRYO KENKYUSHO (THE RESEARCH INSTITUTE FOR IRON, STEEL AND OTHER METALS, TOHOKU UNIVERSITY)
1-1, 2-CHOME, KATAHIRA, SENDAI-SHI 980
ENGLISH

MEKKI NO SEKAI 8372
(GILDING WORLD)
TOKYO MEKKI ZAIRYO KYODO KUMIAI (TOKYO GILDING MATERIALS FEDERATION)
JAPANESE

MIE-KEN KINZOKU SHIKENJO GYOMU HOKOKU 8373
(REPORT OF MIE PREFECTURE METAL INDUSTRY INSTITUTE)
MIE-KEN KINZOKU SHIKENJO (MIE PREFECTURE METAL INDUSTRY RESEARCH INSTITUTE)
2-38, SHIN'YATA, KUWANA-SHI 511
JAPANESE

MITSUBISHI SEIKO GIHO 8374
(MITSUBISHI STEEL MANUFACTURING TECHNICAL REVIEW)
MITSUBISHI STEEL MANUFACTURING COMPANY, LIMITED
9-31, 1-CHOME, SHINONOME, KOTO-KU, TOKYO 135
JAPANESE AND ENGLISH
INDEXED: JICST

NAMARI TO AEN 8375
(LEAD AND ZINC)
NIHON NAMARI AEN JUYO KENKYUKAI (JAPAN LEAD ZINC DEVELOPMENT ASSOCIATION)
C/O NIHON KOGYO KYOKAI, 3-6, 1-CHOME, UCHISAIWAI-CHO, CHIYODA-KU, TOKYO 100
JAPANESE
INDEXED: CHE, JICST

NIHON DAIKASUTO SHINPO 8376
(JAPAN DIECASTING NEWS)
NIHON DAIKASUTO SHINPOSHA
1-51, TSUKAMOTO-CHO, HIGASHIYODOGAWA-KU, OSAKA-SHI 532
JAPANESE

NIHON KANAMONO SHINBUN 8377
(JAPAN HARDWARE NEWS)
NIHON KANAMONO SHINBUNSHA
33, KANDA HIGASHIMATSUSHITA-CHO, CHIYODA-KU, TOKYO 101
JAPANESE

NIHON KINZOKU GAKKAI KAIHO 8378
(BULLETIN OF THE JAPAN INSTITUTE OF METALS)
NIHON KINZOKU GAKKAI (THE JAPAN INSTITUTE OF METALS)
C/O TOKATSU BUILDING, 1-13, 1-CHOME, O-MACHI, SENDAI-SHI, 980
JAPANESE
INDEXED: CHE, COP, MET, WAA, JICST

NIHON KINZOKU GAKKAISHI 8379
(JOURNAL OF THE JAPAN INSTITUTE OF METALS)
NIHON KINZOKU GAKKAI (THE JAPAN INSTITUTE OF METALS)
C/O TOKATSU BUILDING, 1-13, 1-CHOME, O-MACHI, SENDAI-SHI 980
JAPANESE, WITH EUROPEAN SUMMARY
INDEXED: CHE, COP, ENG, IAA, INS, MET, WAA, JICST

NIHON SEIKOSHO GIHO 8380
(JAPAN STEEL WORKS TECHNICAL REVIEW)
NIHON SEIKOSHO (THE JAPAN STEEL WORKS, LTD.)
1-12, YURAKU-CHO, CHIYODA-KU, TOKYO 100
JAPANESE AND ENGLISH
INDEXED: CHE, MET, WAA, JICST

NIHON SUTENRESU GIHO 8381
(NIHON STAINLESS TECHNICAL REPORT)
NIHON SUTENRESU K.K. GIJUTSU-BU (TECHNOLOGICAL DEPARTMENT, NIHON STAINLESS STEEL COMPANY, LTD.)
8-2, MOTOSHIO-CHO, SHINJUKU-KU, TOKYO 160
JAPANESE
INDEXED: JICST

NIHON TEKKO YUSHUTSU KUMIAI JUNPO 8382
(BULLETIN OF THE JAPAN IRON AND STEEL EXPORTERS' ASSOCIATION))
NIHON TEKKO YUSHUTSU KUMIAL (JAPAN IRON & STEEL EXPORTERS' ASSOCIATION)
3-16, NIHONBASHI KAYABA-CHO, CHUO-KU, TOKYO 103
JAPANESE

NIKKAN KINZOKU TOKUHO 8383
(DAILY METAL BULLETIN)
SANGYO SHINBUNSHA (THE SANGYO PRESS)
1-69, EDOBORI KITA-DORI, NISHI-KU, OSAKA-SHI 550
JAPANESE

NIKKAN TEKKO SHINBUN 8384
(DAILY IRON AND STEEL NEWS)
TEKKO SHINBUNSHA
9-1, 2-CHOME, UCHIKANDA, CHIYODA-KU, TOKYO 101
JAPANESE

NIPPON KOKAN GIHO 8385
(NIPPON KOKAN TECHNICAL REPORT)
NIPPON KOKAN K.K. GIJUTSU-BU
1-3, 1-CHOME, OTE-MACHI, CHIYODA-KU, TOKYO 100
JAPANESE AND ENGLISH
INDEXED: MET, WAA, JICST

NIPPON KOKAN TECHNICAL REPORT, OVERSEAS 8386
NIPPON KOKAN K.K. GIJUTSU-BU
1-2, 1-CHOME, MARUNOUCHI, CHIYODA-KU, TOKYO 100
ENGLISH
INDEXED: CHE, ENG, MET

NIPPON KOKOZO KYOKAI TAIKAI KENKYU SHUKAI. MATORIKKUSU KOZO KAISEKIHO KENKYU HAPPYO RONBUNSHU 8387
(SYMPOSIUM OF THE JAPAN STEEL STRUCTURE ASSOCIATION. PAPERS ON THE ANALYSIS OF MATRIX STRUCTURE.)
JAPANESE
INDEXED: JICST

NIPPON NYUSU 8388
(NIHON TEPPAN COMPANY NEWS)
NIHON TEPPAN K.K.
C/O EITARO BUILDING, 2-5, 1-CHOME, NIHONBASHI-DORI, CHUO-KU, TOKYO 103
JAPANESE

NIPPON STEEL NEWS 8389
NIPPON STEEL CORPORATION
2-6-3 OTEMACHI, CHIYODA-KU, TOKYO 100, JAPAN
ENGLISH

NISHI NIHON FUSHOKU BOSHOKU 8390
(CORROSION AND PROTECTION)
NISHI NIHON FUSHOKU BOSHOKU KENKYUKAI (THE CORROSIVE AND ANTICORROSIVE SOCIETY OF WESTERN JAPAN)
C/O KYUSHU KOGYO DAIGAKU, SEN'SUI-CHO, TOBATA-KU, KITAKYUSHU-SHI 804
JAPANESE

NISSHIN SEIKO GIHO 8391
(NISSHIN STEEL TECHNICAL REPORT)
NISSHIN SEIKO K. K. NISSHIN STEEL COMPANY, LTD.
4-1, 3-CHOME, MARUNOUCHI, CHIYODA-KU TOKYO 100
JAPANESE
INDEXED: JICST

NITTOKU GIHO 8392
(JAPAN SPECIAL STEEL TECHNICAL REVIEW)
NIHON TOKUSHUKO K. K. (JAPAN SPECIAL STEEL COMPANY, LTD.)
29-1, 1-CHOME, OMORI-HIGASHI, OTA-KU, TOKYO 143
JAPANESE AND ENGLISH

NRIM CREEP DATA SHEET (NATIONAL RESEARCH INSTITUTE FOR METALS) 8393
JAPANESE
INDEXED: JICST

NTK NYUSU 8394
(NTK NEWS)
NITTOKU KINZOKU KOGYO K.K.
1-1, 2-CHOME, YATO-MACHI, TANASHI-SHI, TOKYO 188
JAPANESE

PURESU GIJUTSU 8395
(PRESS TECHNOLOGY)
NIKKAN KOGYO SHINBUNSHA
8-10, 1-CHOME, KUDAN-KITA, CHIYODA-KU, TOKYO 102
JAPANESE
INDEXED: JICST

PURESU SUKURU 8396
(PRESS SCHOOL)
NIHON KINZOKU PURESU KOGYOKAI (JAPAN METAL PRESS ENGINEERING ASSOCIATION)
C/O NAMIKI BUILDING, 16-3, SAKURAGAOKA-CHO, SHIBUYA-KU, TOKYO 150
JAPANESE

R & D KOBE SEIKO GIHO 8397
(KOBE STEEL ENGINEERING REPORTS)
KOBE SEIKOSHO (KOBE STEEL WORKS, LTD.)
3-18, 1-CHOME, WAKINOHAMA-CHO, FUKIAI-KU, KOBE-SHI 651
JAPANESE, WITH EUROPEAN SUMMARY
INDEXED: COP, JICST

REA METARU NYUSU 8398
(RARE METAL NEWS)
ARUMU SHUPPAN-SHA
5-6, 2-CHOME, HONGO, BUNKYO-KU, TOKYO 113
JAPANESE

REPORT OF THE CASTINGS RESEARCH LABORATORY 8399
WASEDA DAIGAKU IMONO KENKYUSHO TOSHOSHITSU (THE CASTINGS RESEARCH LABORATORY, WASEDA UNIVERSITY)
1-500, TOTSUKA-CHO, SHINJUKU-KU, TOKYO 160
ENGLISH
INDEXED: APP, CHE, JICST

REPORTS OF THE INDUSTRIAL RESEARCH INSTITUTE, JAPAN 8400
JAPANESE AND ENGLISH
INDEXED: COP

S + C STEEL AND CONSTRUCTION 8401
NIPPON KOKAN K. K. SHIJO CHOSA-BU (MARKET RESEARCH DEPARTMENT, NIPPON KOKAN K. K.)
1-2, 1-CHOME, MARUNOUCHI, CHIYODA-KU, TOKYO 100
JAPANESE

SABI 8402
(CORROSION)
NIHON BOSHOKU KOGYO K. K. (THE NIHON CORROSION ENGINEERING COMPANY)
C/O NIHON KOTSU KOSHA BUILDING, 6-4, 1-CHOME, MARUNOUCHI, CHIYODA-KU, TOKYO 100
JAPANESE

SAITAMA-KEN IMONO KIKAI KOGYO SHIKENJO NENPO 8403
(ANNUAL REPORT OF THE FOUNDRY AND MACHINERY RESEARCH INSTITUTE, SAITAMA)
SAITAMA-KEN IMONO KIKAI KOGYO SHIKENJO (THE FOUNDRY AND MACHINERY RESEARCH INSTITUTE, SAITAMA)
1473, SHIBA SUGIBASHI, KAWAGUCHI-SHI 332
JAPANESE

SEIHIN NYUSU 8404
(SEIHIN NEWS)
NIPPON KOKAN KABUSHIKI KAISHI (JAPAN STEEL AND TUBE CORPORATION)
1-1-2, MARUNOUCHI CHIYODA-KU, TOKYO 100
JAPANESE

SEITETSU GIJUTSU JOHO 8405
(METALLURGICAL ABSTRACT)
NIHON TEKKO RENMEI (THE JAPAN IRON & STEEL FEDERATION)
9-4, 1-CHOME, OTE-MACHI, CHIYODA-KU, TOKYO 100
JAPANESE
INDEXED: JICST

SEITETSU GIJUTSU SORAN BESSATSU 8406
(REVIEW OF STEEL PRODUCTION TECHNOLOGY SUPPLEMENT)
JAPANESE
INDEXED: JICST

SEITETSU KENKYU 8407
(NIPPON STEEL TECHNICAL REPORT, JAPANESE EDITION)
SHINNIPPON SEITETSU K.K. KENKYU KAIHATSU HONBU (R & D BUREAU, NIPPON STEEL COMPANY)
6-3, 2-CHOME, OTE-MACHI, CHIYODA-KU, TOKYO 100
JAPANESE
INDEXED: CHE, ENG, WAA, JICST

SENZAI TO SONO SEIHIN 8408
(WIRE ROD & ITS PRODUCTS)
SENZAI SEIHIN KYOKAI NIHON SENZAI SEIHIN YUSHUTSU KUMIAI (WIRE PRODUCTS ASSOCIATION, JAPAN WIRE PRODUCTS EXPORTER'S ASSOCIATION)
3-16, NIHONBASHI KAYABA-CHO, CHUO-KU, TOKYO 103
JAPANESE

SHIN KINZOKU KOGYO 8409
(NEWER METAL INDUSTRY)
SHIN KINZOKU KYOKAI (JAPAN ASSOCIATION OF NEWER METAL)
1-13, 1-CHOME, SHINBASHI, MINATO-KU, TOKYO 105
JAPANESE
INDEXED: JICST

SHIN KINZOKU SHIRYO 8410
(JOURNAL OF THE JAPAN SOCIETY OF NEWER METALS)
SHIN KINZOKU KYOKAI (JAPAN SOCIETY OF NEWER METALS)
1-13, 1-CHOME, SHINBASHI, MINATO-KU, TOKYO 105
JAPANESE
INDEXED: JICST

SHINDO GEPPO 8411
(MONTHLY JOURNAL OF JAPAN BRASS MAKERS ASSOCIATION)
NIHON SHINDO KYOKAI (JAPAN BRASS MAKERS ASSOCIATION)
12-22, 1-CHOME, TSUKIJI, CHUO-KU, TOKYO 104
JAPANESE

SHINDO GIJUTSU KENKYUKAISHI 8412
(JOURNAL OF THE JAPAN COPPER AND BRASS RESEARCH ASSOCIATION)
NIHON SHINDO KYOKAI (JAPAN BRASS MAKERS ASSOCIATION)
12-22, 1-CHOME, TSUKIJI, CHUO-KU, TOKYO 104
JAPANESE, WITH ENGLISH SUMMARY
INDEXED: CHE, COP, JICST

SHINDO KOGYO NO GAIKYO 8413
(OUTLINE OF BRASS INDUSTRY IN JAPAN)
NIHON SHINDO KYOKAI (JAPAN BRASS MAKERS ASSOCIATION)
12-22, 1-CHOME, TSUKIJI, CHUO-KU, TOKYO 104
JAPANESE
INDEXED: JICST

SHINKO TAIMUSO 8414
(SHINKO TIMES)
JAPANESE
INDEXED: JICST

SHUKAN KINZOKU 8415
(METAL WEEKLY)
KINZOKU SHUPPANSHA
C/O TOKYO YOFUKU KAIKAN BUILDING, 13, ICHIGAYA HACHIMAN-CHO, SHINJUKU-KU, TOKYO 162
JAPANESE
INDEXED: JICST

SHUKAN SUCHIRU RAIFU 8416
(STEEL LIFE WEEKLY)
SUCHIRU RAIFUSHA
C/O ASAHI BUILDING, 4-4,4-CHOME, IIDABASHI, CHIYODA-KU, TOKYO 102
JAPANESE

SOGO IMONO 8417
(GENERAL FOUNDRY)
NIHON SOGO IMONO SENTA (JAPAN GENERAL FOUNDRY CENTER)
C/O KIKAI SHINKO KAIKAN NO. 207, 5-8, 3-CHOME, SHIBA KOEN, MINATO-KU, TOKYO 105
JAPANESE
INDEXED: CHE, JICST

SOGO IMONO SENTA KENKYU CHOSA HOKOKU 8418
(RESEARCH AND SURVEY REPORT OF THE JAPAN GENERAL FOUNDRY CENTER)
JAPAN GENERAL FOUNDRY CENTER
ROOM 207 KIKAI SHINKO BUILDING, 3-5-8 SHIBA KOEN, MINATO-KU, TOKYO
JAPANESE
INDEXED: CHE

SOSEI TO KAKO 8419
(JOURNAL OF JAPAN SOCIETY FOR TECHNOLOGY OF PLASTICITY)
NIHON SOSEI KAKO GAKKAI (JAPAN SOCIETY FOR TECHNOLOGICAL OF PLASTICITY)
2-5, 5-CHOME, ROPPONGI, MINATO-KU, TOKYO 106
JAPANESE, WITH ENGLISH SUMMARY
INDEXED: APP, CHE, COP, ENG, WAA, JICST

STEEL TODAY AND TOMORROW 8420
NIHON TEKKO YUSHUTSU KUMIAI (JAPAN IRON AND STEEL EXPORTER'S ASSOCIATION)
C/O TEKKO KAIKAN, 3-16, NIHONBASHI KAYABA-CHO, CHUO-KU, TOKYO 103
ENGLISH
INDEXED: MET

SUCHIRU DEZAIN 8421
(STEEL DESIGN)
SHIN NIPPON SEITETSU K.K. (NIPPON STEEL CORPORATION)
6-3, 2-CHOME, OTE-MACHI, CHIYODA-KU, TOKYO 100
JAPANESE

SUIYOKAISHI 8422
(TRANSACTIONS OF THE MINING AND METALLURGICAL ALUMNI ASSOCIATION)
SUIYO-KAI (THE MINING AND METALLURGICAL ALUMNI ASSOCIATION)
C/O KYOTO DAIGAKU KOGAKUBU YAKINGAKU KYOSHITSU, YOSHIDA HON-CHO, SAKYO-KU, KYOTO-SHI 606
JAPANESE, WITH EUROPEAN SUMMARY
INDEXED: CHE, MET, JICST

SUMITOMO KEIKINZOKU GIHO 8423
(SUMITOMO LIGHT METAL TECHNICAL REPORTS)
SUMITOMO KEIKINZOKU KOGYO K.K. (SUMITOMO LIGHT METAL INDUSTRIES, LTD.)
500, RO-NO-WARI, CHITOSE, MINATO-KU, NAGOYA-SHI 455
JAPANESE, WITH ENGLISH SUMMARY
INDEXED: CHE, COP, ENG, MET, WAA, JICST

SUMITOMO KINZOKU 8424
(SUMITOMO METALS)
SUMITOMO KINZOKU KOGYO K.K. (SUMITOMO METAL INDUSTRIES LTD.)
C/O SHIN SUMITOMO BUILDING, 5-15, KITAHAMA, HIGASHI-KU, OSAKA-SHI 541
JAPANESE, WITH EUROPEAN SUMMARY
INDEXED: CHE, JICST

SUMITOMO KINZOKU KOZAN K.K. CHUO KENKYUSHO KENKYU HOKOKU 8425
(REPORT OF THE CENTRAL LABORATORY, SUMITOMO METAL AND MINING COMPANY, LTD.)
SUMITOMO KINZOKU KOZAN K.K. CHUO KENKYUSHO
18-5, 3-CHOME, NAKAKOKUBU, ICHIKAWA-SHI 272
JAPANESE

SUMITOMO KINZOKU SEIHIN NYUSU 8426
(SUMITOMO KINZOKU PRODUCTS NEWS)
SUMITOMO KINZOKU K.K. SHIJO-BU
3-2, 1-CHOME, MARUNOUCHI, CHIYODA-KU, TOKYO 100
JAPANESE

SUMITOMO SEARCH 8427
SUMITOMO KINZOKU KOGYO K.K. CHUO GIJUTSU KENKYUSHO
1-3, NAGASUHONDORI, AMAGASAKI-SHI 660
ENGLISH
INDEXED: CHE, ENG, INS

SUMITOMO TOKUSHU KINZOKU GIHO 8428
(TECHNICAL REPORT OF SUMITOMO SPECIAL METALS)
SUMITOMO TOKUSHU KINZOKU K.K. (SUMITOMO SPECIAL METALS COMPANY, LTD.)
5-22, KITAHAMA, HIGASHI-KU, OSAKA-SHI 541
JAPANESE, WITH ENGLISH SUMMARY

SUTENRESU 8429
(STAINLESS)
SUTENRESU KYOKAI (JAPAN STAINLESS STEEL ASSOCIATION)
3-16, NIHONBASHI KAYABA-CHO, CHUO-KU, TOKYO 103
JAPANESE
INDEXED: JICST

SUTENRESU KOHAN YOTO-BETSU JUCHU TOKEI NENPO 8430
(STATISTICAL YEARBOOK OF THE ORDER AWARDED BY USES FOR STAINLESS STEEL PLATE)
SUTENRESU KYOKAI
C/O TEKKO KAIKAN, 3-16, NIHONBASHI KAYABA-CHO, CHUO-KU, TOKYO 103
JAPANESE

TAINETSU KINZOKU ZAIRYO IINKAI KENKYU HOKOKU 8431
(REPORT OF THE 123RD COMMITTEE ON HEAT-RESISTING METALS AND ALLOYS, JAPAN SOCIETY FOR THE PROMOTION SCIENCE)
NIHON GAKUJUTSU SHINKOKAI (JAPAN SOCIETY FOR THE PROMOTION OF SCIENCE)
3-1, 5-CHOME, KOJI-MACHI, CHIYODA-KU, TOKYO 102
JAPANESE

TATARA KENKYU 8432
(STUDIES OF HISTORICAL IRON AND STEEL)
TATARA KENKYUKAI (THE SOCIETY FOR HISTORICAL IRON AND STEEL STUDIES)
C/O HIROSHIMA DAIGAKU BUNGAKUBU KOKOGAKU KENKYUSHITSU, 1-89, 1-CHOME, HIGASHISENDA-CHO, HIROSHIMA-SHI 730
JAPANESE

TEKKO DAIGAKUSHI 8433
(BULLETIN OF IRON AND STEEL TECHNICAL COLLEGE)
TEKKO TANKI DAIGAKU (IRON AND STEEL TECHNICAL COLLEGE)
1-1, ISHIDA, NISHIKON'YO, AMAGASAKI-SHI 661
JAPANESE AND ENGLISH
INDEXED: JICST

TEKKO JOHO 8434
(JOURNAL OF THE JAPAN IRON AND STEEL WHOLESALERS' ASSOCIATION)
ZENKOKU TEKKO TON'YA KUMIAI (JAPAN IRON AND STEEL WHOLESALERS' ASSOCIATION)
3-16, NIHONBASHI KAYABA-CHO, CHUO-KU, TOKYO 103
JAPANESE

TEKKO JUKYU KANREN SHIRYO KIJI SAKUIN 8435
(ABSTRACTS OF CURRENT LITERATURE ON SUPPLY AND DEMAND OF IRON AND STEEL)
KOZAI KURABU
C/O TEKKO KAIKAN, 3-16, NIHONBASHI KAYABA-CHO, CHUO-KU, TOKYO 103
JAPANESE

TEKKO KAIHO 8436
(BULLETIN OF THE JAPAN IRON AND STEEL FEDERATION)
NIHON TEKKO RENMEI (THE JAPAN IRON & STEEL FEDERATION)
9-4, 1-CHOME, OTE-MACHI, CHIYODA-KU, TOKYO 100
JAPANESE

TEKKO KANREN SHINBUN KIJI SAKUIN 8437
(NEWSPAPER INDEX ON IRON AND STEEL)
KOZAI KURABU
3-16, NIHONBASHI KAYABA-CHO, CHUO-KU, TOKYO 103
JAPANESE

TEKKO NO IE 8438
(INDUSTRIAL ENGINEERING FOR IRON & STEEL INDUSTRY)
NIHON TEKKO RENMEI IE IINKAI
9-4, 1-CHOME, OTE-MACHI, CHIYODA-KU, TOKYO 100
JAPANESE
INDEXED: JICST

TEKKO TOKEI GEPPO 8439
(IRON AND STEEL STATISTICAL MONTHLY)
TSUSHO SANGYO-SHO DAIJIN KANBO CHOSA TOKEI-BU (RESEARCH AND STATISTICS DIVISION, MINISTER'S SECRETARIAT, MINISTRY OF INTERNATIONAL TRADE AND INDUSTRY)
9-4, 1-CHOME, OTE-MACHI, CHIYODA-KU, TOKYO 100
JAPANESE
INDEXED: JICST

TEKKO TOKEI JIHO 8440
(IRON AND STEEL STATISTICS TIMES)
NIHON TEKKO RENMEI (THE JAPAN IRON & STEEL FEDERATION)
9-4, 1-CHOME, OTE-MACHI, CHIYODA-KU, TOKYO 100
JAPANESE

TEKKO TOKEI NENPO 8441
(YEARBOOK OF IRON AND STEEL STATISTICS)
TSUSHO SANGYO-SHO DAIJIN KANBO CHOSA TOKEI-BU (RESEARCH AND STATISTICS DIVISION, MINISTER'S SECRETARIAT, MINISTRY OF INTERNATIONAL TRADE AND INDUSTRY)
9-4, 1-CHOME, OTE-MACHI, CHIYODA-KU, TOKYO 100
JAPANESE

TEKKOKAI 8442
(IRON AND STEEL INDUSTRY)
NIHON TEKKO RENMEI (THE JAPAN IRON AND STEEL FEDERATION)
9-4, 1-CHOME, OTE-MACHI, CHIYODA-KU, TOKYO 100
JAPANESE
INDEXED: CHE, JICST

TETSU 8443
(IRON AND STEEL)
KAWASAKI SEITETSU (KAWASAKI IRON & STEEL MANUFACTURERS)
JAPANESE

TETSU TO HAGANE 8444
(JOURNAL OF THE IRON AND STEEL INSTITUTE OF JAPAN)
NIHON TEKKO KYOKAI (THE IRON AND STEEL INSTITUTE OF JAPAN)
9-4, 1-CHOME, OTE-MACHI, CHIYODA-KU, TOKYO 100
JAPANESE, WITH EUROPEAN SUMMARY
INDEXED: CHE, COP, ENG, MET, WAA, JICST

TOHOKU KINZOKU GIHO 8445
(TOHOKU METAL TECHNICAL REVIEW)
TOHOKU KINZOKU KOGYO K.K. (TOHOKU METAL INDUSTRIES, LTD.)
13-10, 7-CHOME, GINZA, CHUO-KU, TOKYO 104
JAPANESE, WITH ENGLISH SUMMARY

TOKUSHUKO 8446
(SPECIAL STEEL)
TOKUSHUKO KURABU
3-16, NIHONBASHI KAYABA-CHO, CHUO-KU, TOKYO 103
JAPANESE
INDEXED: CHE, JICST

TOYO KOHAN 8447
TOYO KOHAN K.K. KUDAMATSU KOJO (KUDAMATSU PLANT, TOYO KOHAN COMPANY, LTD.)
1302, HIGASHITOYOI, KUDAMATSU-SHI 744
JAPANESE, WITH ENGLISH SUMMARY
INDEXED: CHE, JICST

TRANSACTIONS OF THE IRON AND STEEL INSTITUTE OF JAPAN 8448
NIHON TEKKO KYOKAI (THE IRON AND STEEL INSTITUTE OF JAPAN)
9-4, 1-CHOME, OTE-MACHI, CHIYODA-KU, TOKYO 100
ENGLISH
INDEXED: CHE, ENG, SCI, WAA, JICST

TRANSACTIONS OF THE JAPAN INSTITUTE OF METALS 8449
NIHON KINZOKU GAKKAI (THE JAPAN INSTITUTE OF METALS)
1-13, 1-CHOME, OMACHI, SENDAI-SHI 980
ENGLISH
INDEXED: CHE, COP, ENG, IAA, MET, SCI, WAA, JICST

TRANSACTIONS OF THE JAPAN WELDING SOCIETY 8450
YOSETSU GAKKAI (JAPAN WELDING SOCIETY)
1-11, KANDA SAKUMA-CHO, CHIYODA-KU, TOKYO 101
ENGLISH
INDEXED: CHE, COP, ENG

TRANSACTIONS OF JWRI 8451
OSAKA DAIGAKU YOSETSU KOGAKU KENKYUSHO (WELDING RESEARCH INSTITUTE OF OSAKA UNIVERSITY)
YAMADA-KAMI, SUITA-SHI 565
ENGLISH
INDEXED: CHE, JICST

TRANSACTIONS OF THE NATIONAL RESEARCH INSTITUTE FOR METALS 8452
KINZOKU ZAIRYO GIJUTSU KENKYUSHO OBUN KENKYU HOKOKU
KAGAKU GIJUTSU-CHO KINZOKU ZAIRYO GIJUTSU KENKYUSHO (NATIONAL RESEARCH INSTITUTE FOR METALS, SCIENCE AND TECHNOLOGY AGENCY)
EUROPEAN
INDEXED: CHE, COP, ENG, INS, MET WAA, JICST

WERUDING ENJINIA 8453
(WELDING ENGINEER)
SANPO
10-17, 1-CHOME, HAMAMATSU-CHO, MINATO-KU, TOKYO 105
JAPANESE

YOSETSU GAKKAI ZENKOKU TAIKAI KOEN GAIYO 8454
(SUMMARY REPORT OF THE CONFERENCE, THE JAPAN WELDING SOCIETY)
YOSETSU GAKKAI
1-11, KANDA SAKUMA-CHO, CHIYODA-KU, TOKYO 101
JAPANESE
INDEXED: JICST

YOSETSU GAKKAISHI 8455
(JOURNAL OF THE JAPAN WELDING SOCIETY)
YOSETSU GAKKAI (THE JAPAN WELDING SOCIETY)
1-11, KANDA SAKUMA-CHO, CHIYODA-KU, TOKYO 101
JAPANESE AND ENGLISH
INDEXED: CHE, COP, WAA, JICST

YOSETSU GIJUTSU 8456
(WELDING TECHNIQUE)
SANPO
10-17, 1-CHOME, HAMAMATSU-CHO, MINATO-KU, TOKYO 105
JAPANESE
INDEXED: CHE, JICST

YOSETSU JIHO 8457
(REVIEWS OF WELDING)
JIHO-SHA
33, TOMIHISA-CHO, SHINJUKU-KU, TOKYO 160
JAPANESE

YOSETSU NIPPO 8458
(DAILY WELDING INFORMATION)
SANPO
10-17, 1-CHOME, HAMAMATSU-CHO, MINATO-KU, TOKYO 105
JAPANESE

YOSETSU NYUSU 8459
(WELDING NEWS)
SANPO
10-17, 1-CHOME, HAMAMATSU-CHO, MINATO-KU, TOKYO 105
JAPANESE

ZAIRYO KYODO TO HAKAI KOKUNAI SOGO SHINPOJIUMU RONBUNSHU 8460
(PROCEEDINGS OF THE JAPAN NATIONAL SYMPOSIUM ON STRENGTH, FRACTURE AND FATIGUE)
JAPANESE
INDEXED: JICST

Industries and Crafts Based on Processible Materials

AOMORI-KEN MOKKO SHIDOSHO GYOMU HOKOKU 8461
(OPERATIONAL REPORT OF THE AOMORI PREFECTURAL WOODCRAFT LABORATORY)
AMORI-KEN MOKKO SHIDOSHO
14-11, 1-CHOME, MATSUBARA, AOMORI-SHI 030
JAPANESE

BANDO BELTING REPORT 8462
BANDO KAGAKU K.K. (BANDO CHEMICAL INDUSTRIES, LTD.)
2-1, MEIWA-DORI, HYOGO-KU, KOBE-SHI 652
JAPANESE

BENIA TAIMUSU 8463
(VENEER TIMES)
BENIA TAIMUSUSHA (VENEER TIMES COMPANY, LTD.)
C/O BENIA SHOKAI BUILDING, 6-3, SHINBASHI, MINATO-KU, TOKYO 105
JAPANESE

BIJINESU FANICHUA 8464
(BUSINESS FURNITURE)
JAPANESE
INDEXED: JICST

BUNGU JIMUKI GAIDO 8465
(STATIONERY)
SANGYO SHINBUNSHA
6-19, 4-CHOME, HIGASHI NODA-CHO, MIYAKOJIMA-KU, OSAKA-SHI 534
JAPANESE

BUNGU TO JIMUKI 8466
(STATIONERY AND BUSINESS MACHINE)
NICHIMA (NIPPON MARKETING INC.)
6-21, KAGURAZAKA, SHINJUKU-KU, TOKYO 162
JAPANESE

FANICHA 8467
(FURNITURE SALES)
FANICHA
4-15, 5-CHOME, AKASAKA, MINATO-KU, TOKYO 107
JAPANESE

GAKKI SHOHO 8468
(MUSICAL INSTRUMENTS NEWS)
GAKKI SHOHOSHA
C/O GAKKI KAIKAN, 18-21, 2-CHOME, SOTO KANDA, CHIYODA-KU, TOKYO 101
JAPANESE

GIFU-KEN KOGEI SHIKENJO GYOMU HOKOKU 8469
(OPERATIONAL REPORT OF THE INDUSTRIAL ARTS INSTITUTE, GIFU PREFECTURAL GOVERNMENT)
GIFU-KEN KOGEI SHIKENJO (INDUSTRIAL ARTS INSTITUTE GIFU PREFECTURE)
1554, YAMADA-CHO, TAKAYAMA-SHI 506
JAPANESE

GOBAN TOKEI 8470
(STATISTICS OF PLYWOOD)
NORIN-SHO NORIN KEIZAI-KYOKU TOKEI JOHO-BU (STATISTICS AND INFORMATION DIVISION, AGRICULTURE AND FORESTRY ECONOMIC BUREAU, MINISTRY OF AGRICULTURE AND FORESTRY)
C/O MEISAN BUILDING, 18-17, 1-CHOME, NISHI-SHINBASHI, MINATO-KU, TOKYO 105
JAPANESE

HADO BODO, PATIKURU BODO, INSHURESHON BODO 8471
(HARD BOARD, PARTICLE BOARD, INSULATION BOARD)
NIHON KOSHITSU SEN'IBAN KOGYOKAI (JAPAN HARD FIBER BOARD INDUSTRY ASSOCIATION)
JAPANESE

HADO BODO PATIKURU BODO KAIHO 8472
(BULLETIN OF HARD BOARD AND PARTICLE BOARD MANUFACTURE'S ASSOCIATION)
NIHON KOSHITSU SENIBAN KOGYOKAI (JAPAN HARD BOARD AND PARTICLE BOARD MANUFACTURE'S ASSOCIATION)
5-15, 1-CHOME, YAESU, CHUO-KU, TOKYO 103
JAPANESE
INDEXED: JICST

HAKIMONO NENPO 8473
(YEARBOOK OF FOOTWARE)
GOMU HOCHI SHINBUNSHA
4-19, NIHONBASHI KAKIGARA-CHO, CHUO-KU
TOKYO 103
JAPANESE

HIKAKU GIJUTSU 8474
(LEATHER TECHNOLOGY)
NIHON HIKAKU GIJUTSU KYOKAI (JAPANESE ASSOCIATION OF LEATHER TECHNOLOGY)
C/O LEATHER RESEARCH INSTITUTE, TOKYO NOKO UNIVERSITY, 5-8, 3-CHOME, SAIWAI-CHO, FUCHU-SHI
TOKYO 183
JAPANESE
INDEXED: CHE, JICST

HIKAKU KAGAKU 8475
(LEATHER CHEMISTRY)
NIHON HIKAKU GIJUTSU KYOKAI (JAPANESE ASSOCIATION OF LEATHER TECHNOLOGY)
C/O LEATHER RESEARCH INSTITUTE, TOKYO NOKO UNIVERSITY, 5-8, 3-CHOME, SAIWAI-CHO, FUCHU-SHI
TOKYO 183
JAPANESE, WITH ENGLISH SUMMARY
INDEXED: CHE, JICST

HIROSHIMA-KENRITSU KOGEI SHIKENJO HOKOKU 8476
(REPORT OF INDUSTRIAL ARTS INSTITUTE, HIROSHIMA PREFECTURE)
HIROSHIMA-KENRITSU KOGEI SHIKENJO (INDUSTRIAL ARTS INSTITUTE, HIROSHIMA PREFECTURE)
2252-18, YAMAIZU-CHO, FUKUYAMA-SHI 729-01
JAPANESE

HIROSHIMA-KENRITSU KOGEI SHIKENJO NENPO 8477
(ANNUAL REPORT OF INDUSTRIAL ARTS INSTITUTE, HIROSHIMA PREFECTURE)
HIROSHIMA-KENRITSU KOGEI SHIKENJO (INDUSTRIAL ART INSTITUTE, HIROSHIMA PREFECTURE)
2252-18, YAMAIZU-CHO, FUKUYAMA-SHI 729-01
JAPANESE

HIROSHIMA-SHI KOGEI SHODISHO GYOMU HOKOKU 8478
(REPORT OF INDUSTRIAL ART INSTITUTE OF HIROSHIMA CITY)
HIROSHIMA-SHI KOGEI SHIDOSHO (INDUSTRIAL ART INSTITUTE OF HIROSHIMA CITY)
1-2, HIGASHIHONURA-CHO, HIROSHIMA-SHI 734
JAPANESE

HONO TAIMUSU 8479
(BAGS AND POUCHES TIMES)
CHUO TSUSHINSHA
1-30, ASAKUSABASHI, TAITO-KU, TOKYO 111
JAPANESE

HYOGO-KEN HIKAKU KENKYUKAISHI 8480
(JOURNAL OF THE LEATHER RESEARCH LABORATORY OF HYOGO PREFECTURE)
HYOGO-KEN HIKAKU KENKYUKAI
C/O HYOGO-KEN HIKAKU KOGYO SHIDOSHO, 3, HIGASHIGAWARA, NOZATO, HIMEJI-SHI 670
JAPANESE
INDEXED: JICST

HYOGO-KEN HIKAKU KOGYO SHIDOSHO NENPO 8481
(ANNUAL REPORT OF THE LEATHER RESEARCH LABORATORY OF HYOGO PREFECTURE)
HYOGO-KEN HIKAKU KOGYO SHIDOSHO (THE LEATHER RESEARCH LABORATORY OF HYOGO PREFECTURE)
3, HIGASHIGAWARA, NOZATO, HIMEJI-SHI 670
JAPANESE

JDCA KAIHO 8482
(BULLETIN OF THE JAPAN DESIGNER CRAFTMAN ASSOCIATION)
NIHON DEZAINA KURAFUTOMAN KYOKAI (JAPAN DESIGNER CRAFTMAN ASSOCIATION)
21-13, 1-CHOME, SENDAGAYA, SHIBUYA-KU, TOKYO 151
JAPANESE

KAGOSHIMA-KEN MOKUZAI KOGYO SHIKENJO GYOMU HOKOKU 8483
(OPERATIONAL REPORT OF WOOD INDUSTRY RESEARCH INSTITUTE OF KAGOSHIMA PREFECTURE)
KAGOSHIMA-KEN MOKUZAI KOGYO SHIKENJO
1, TOKAI-CHO, KAGOSHIMA-SHI 891-01
JAPANESE

KAGU SANGYO 8484
(FURNITURE INDUSTRY)
KAGU SANGYO SHUPPANSHA (THE FURNITURE INDUSTRY PUBLISHING COMPANY)
JAPANESE

KAGU SHIDO SENTA KENKYU HOKOKU 8485
(RESEARCH BULLETIN OF THE FUNITURE DEVELOPMENT CENTER)
KANAGAWA KEN KAGU SHIDO SENTA
(THE KANAGAWA PREFECTURE FUNITURE DEVELOPMENT CENTER)
JAPANESE

KANAGAWA-KEN KAGU SHIDO SENTA NYUSU 8486
(KANAGAWA PREFECTURAL FURNITURE DEVELOPMENT CENTER NEWS)
KANAGAWA-KEN KAGU SHIDO SENTA (KANAGAWA FURNITURE DEVELOPMENT CENTER)
JAPANESE

KANAGAWA-KEN KOGEI SHIDOSHO NENPO 8487
(ANNUAL REPORT OF KANAGAWA PREFECTURAL RESEARCH INSTITUTE FOR INDUSTRIAL ARTS)
KANAGAWA-KEN KOGEI SHIDOSHO (KANAGAWA PREFECTURAL RESEARCH INSTITUTE FOR INDUSTRIAL ARTS)
7-53, 1-CHOME, HON-CHO, ODAWARA-SHI 250
JAPANESE

MIE-KEN KOGYO GIJUTSU JOHO, MOKKO-HEN 8488
(TECHNICAL INFORMATION OF THE INDUSTRIAL TECHNOLOGY CENTER OF MIE PREFECTURE, WOOD INDUSTRIAL SECTION)
3485, KOMORI-CHO, TAKAJAYA, TSU-SHI 514
JAPANESE

MOKKO KIKAI 8489
(WOODWORKING MACHINERY)
ZENKOKU MOKKO KIKAI KOGYOKAI (JAPAN WOOD-WORKING MACHINERY ASSOCIATION)
C/O KIKAI SHINKO KAIKAN, 5-8, 3-CHOME, SHIBA KOEN, MINATO-KU, TOKYO 105
JAPANESE

MOKKO SEISAN 8490
(WOODWORKS PRODUCTION)
NIHON MOKUSEIHIN GIJUTSU KYOKAI (WOOD WORKS TECHNOLOGICAL ASSOCIATION OF JAPAN)
15-7, 3-CHOME, TAKADA, TOSHIMA-KU, TOKYO 172
JAPANESE
INDEXED: JICST

MOKUZAI HOZON 8491
(WOOD PRESERVATION)
NIHON MOKUZAI HOZON KYOKAI (JAPAN WOOD PRESER-VATION ASSOCIATION)
JAPANESE

MOKUZAI JOHO 8492
(LUMBER REPORT)
NAGOYA EIRIN KYOKU
(NAGOYA REGIONAL FORESTRY OFFICE)
JAPANESE

MOKUZAI KENKYU SHURYO 8493
(WOOD RESEARCH REVIEW)
KYOTO DAIGAKU MOKUZAI KENKYUSHO (WOOD RESEARCH INSTITUTE, KYOTO UNIVERSITY)
GOKASHO, UJI-SHI 611
JAPANESE
INDEXED: JICST

MOKUZAI KOGYO 8494
(WOOD INDUSTRY)
NIHON MOKUZAI KAKO GIJUTSU KYOKAI (WOOD TECHNOLOGICAL ASSOCIATION OF JAPAN)
1-23, 1-CHOME, SHIBA DAIMON, MINATO-KU, TOKYO 105
JAPANESE, WITH ENGLISH SUMMARY
INDEXED: BIB, CHE, JICST

MOKUZAI SHIHYO 8495
(LUMBER INDEX)
MOKUZAI SHIGENRIYO GORIKA SUISHIN HONBU
(HOME OFFICE FOR THE PROMOTION OF THE RATIONALIZA-TION OF THE USE OF LUMBER RESOURCES)
JAPANESE

MOKUZAI SHIKYO GEPPO 8496
(LUMBER MARKETING CONDITIONS MONTHLY)
JAPANESE

NIHON BIJUTSU KOGEI 8497
(JAPAN HANDICRAFTS)
JAPAN HANDICRAFTS
JAPANESE

NIHON MATCHI KOGYOKAI KAIHO 8498
(PROCEEDINGS OF THE JAPAN MATCH INDUSTRIAL SOCIETY)
JAPANESE

NIHON SHIKKO 8499
(JAPANESE LACQUER WARE)
NIHON SHIKKO KYOKAI (JAPANESE LACQUER WARE ASSOCIATION)
1-4, KANDA KAJI-CHO, CHIYODA-KU, TOKYO 101
JAPANESE
INDEXED: JICST

NIPPON MATCH KOGYOKAI KAIHO 8500
(REPORT OF JAPAN MATCH INDUSTRY ASSOCIATION)
JAPANESE
INDEXED: JICST

OITA-KEN BEPPU SANGYO KOGEI SHIKENJO GYOMU HOKOKUSHO 8501
(OPERATIONAL REPORT OF BEPPU INDUSTRIAL ARTS INSTITUTE, OITA PREFECTURAL LABORATORY)
OITA-KEN BEPPU SANGYO KOGEI SHIKENJO
3-3, TOSHOEN, BEPPU-SHI 874
JAPANESE

OITA-KEN HITA SANGYO KOGEI SHIKENJO GYOMU HOKOKU 8502
(OPERATIONAL REPORT OF OITA PREFECTURAL HITA INDUSTRIAL ARTS INSTITUTE)
OITA-KEN HITA SANGYO KOGEI SHIKENJO
TAMAGAWA-MACHI, HITA-SHI 877
JAPANESE

PAIROTTO TAIMUSU 8503
(PILOT TIMES)
PAIROTTO MAN'NENHITSU K.K. (PILOT PEN CO., LTD.,)
JAPANESE

PAIROTTO TAIMUSU 8504
(PILOT TIMES)
PAIROTTO MAN'NENHITSU K.K. (PILOT PEN COMPANY, LTD.)
2-7, KYOBASHI, CHUO-KU, TOKYO 104
JAPANESE AND ENGLISH

SEIHONKAI 8505
(BOOKBINDING INDUSTRY)
TOKYO-TO SEIHON KOGYO KUMIAI
2-5, KANDA NISHIKI-CHO, CHIYODA-KU, TOKYO 101
JAPANESE

SHOKKI 8506
(LIVING ACCESSORY)
SOGO TSUSHINSHA (SOGO TSUSHIN COMPANY, LTD.)
19-19, 2-CHOME, MARUNOUCHI, NAKA-KU, NAGOYA-SHI 460
JAPANESE

TATEGU KOGEI 8507
(FURNISHING INDUSTRY)
TATEGU KOGEI KENKYUSHO
4-3, 5-CHOME, SOTO KANDA, CHIYODA-KU, TOKYO 101
JAPANESE

TOKYO NOKO DAIGAKU HIKAKU KENKYU SHISETSU HOKOKU 8508
(REPORT OF THE LEATHER RESEARCH INSTITUTE, FACULTY OF AGRICULTURE, TOKYO UNIVERSITY OF AGRICULTURE AND TECHNOLOGY)
TOKYO NOKO DAIGAKU NOGAKUBU FUZOKU HIKAKU KENKYU SHISETSU (LEATHER RESEARCH INSTITUTE FACULTY OF AGRICULTURE, TOKYO UNIVERSITY OF AGRICULTURE AND TECHNOLOGY)
5-8, 3-CHOME, SAIWAI-CHO, FUCHU-SHI, TOKYO 183
JAPANESE AND ENGLISH

YAMAGATA-KENRITSU SHONAI MOKKO SHIDOSHO GYOMU HOKOKU 8509
(OPERATIONAL REPORT OF SHONAI WOODCRAFT LABORATORY, YAMAGATA PREFECTURE)
YAMAGATA-KENRITSU SHONAI MOKKO SHIDOSHO
1-21, RYOU-CHO, SAKATASHI 998
JAPANESE

YAMANASHI-KEN MOKKO SHIDOSHO HOKOKU 8510
(REPORT OF WOOD CRAFT RESEARCH LABORATORY, YAMANASHI PREFECTURE)
YAMANASHI-KEN MOKKO SHIDOSHO
836, KAGOUCHI, SATOYOSHI-CHO, KOFU-SHI 400
JAPANESE

YUNYU MOKUZAI KENKYU HOKOKU 8511
(RESEARCH REPORT OF FOREIGN WOOD)
JAPANESE
INDEXED: JICST

ZAKKA TOKEI GEPPO 8512
(MONTHLY STATISTICS OF MISCELLANEOUS GOODS)
TSUSHO SANGYO-SHO DAIJIN KANBO CHOSA TOKEI-BU
(RESEARCH AND STATISTICS DIVISION, MINISTER'S
SECRETARIAT, MINISTRY OF INTERNATIONAL TRADE AND
INDUSTRY)
C/O TSUSHO SANGYO-SHO, 3-1, 1-CHOME, KASUMIGASEKI,
CHIYODA-KU, TOKYO 100
JAPANESE

ZAKKA TOKEI NENPO, HIKAKU-HEN 8513
(YEARBOOK OF HIDE AND LEATHER STATISTICS)
TSUSHO SANGYO-SHO DAIJIN KANBO CHOSA TOKEI-BU
(RESEARCH AND STATISTICS DIVISION, MINISTER'S
SECRETARIAT, MINISTRY OF INTERNATIONAL TRADE
AND INDUSTRY)
15-1, 6-CHOME, GINZA, CHUO-KU, TOKYO 104
JAPANESE

ZAKKA TOKEI NENPO NICHIYOHIN, TOJIKI-TO-HEN 8514
(YEARBOOK OF DAILY NECESSARIES POTTERY AND
GLASSWARE STATISTICS)
TSUSHO SANGYO-SHO DAIJIN KANBO CHOSA TOKEI-BU
(RESEARCH AND STATISTICS DIVISION, MINISTER'S
SECRETARIAT, MINISTRY OF INTERNATIONAL TRADE
AND INDUSTRY)
15-1, 6-CHOME, GINZA, CHUO-KU, TOKYO 104
JAPANESE

Papers and Pulp Industry

DANBORU KOGYO SETSUBITO JITTAI CHOSA HOKOKUSHO 8515
(REPORT OF CORRUGATED AND CARDBOARD INDUSTRY)
TSUSHO SANGYO-SHO DAIJIN KANBO CHOSA TOKEI-BU (RESEARCH & STATISTICS DIVISION, MINISTER'S SECRETARIAT, MINISTRY OF INTERNATIONAL TRADE AND INDUSTRY)
9-11, 3-CHOME, GINZA, CHUO-KU, TOKYO 104
JAPANESE

EHIME-KEN SEISHI SHIKENJO HOKOKU 8516
(REPORT OF EHIME PREFECTURAL PAPER-MAKING EXPERIMENTAL STATION)
EHIME-KEN SEISHI SHIKENJO (EHIME PREFECTURE PAPER-MAKING EXPERIMENTAL STATION)
ICHI, KAWANOE-SHI 799-01
JAPANESE
INDEXED: CHE

GEKKAN KAMI PARUPU GIJUTSU NO SHOTEN 8517
(FOCUS OF PAPER AND PULP TECHNOLOGY)
CHUGAI SANGYO CHOSAKAI (CHUGAI INDUSTRIAL RESEARCH ASSOCIATION)
4-2, 2-CHOME, MUKOGAOKA, BUNKYO-KU, TOKYO 113
JAPANESE
INDEXED: JICST

GIFU-KEN SEISHI SHIKENJO GYOMU HOKOKU 8518
(OPERATIONAL REPORT OF GIFU PAPER LABORATORY)
GIFU-KEN SEISHI SHIKENJO
777, MAENO, MINO-SHI 501-37
JAPANESE

HOWAITO 8519
(WHITE)
NIHON PARUPU ROGYO (JAPAN PULP INDUSTRY)
JAPANESE

HYAKUMAN TO 8520
(HYAKUMAN TO, PUBLICATION OF THE PAPER MUSEUM)
KAMI HAKUBUTSUKAN
JAPANESE

ITAGAMI TOKEI NENPO 8521
(ANNUAL STATISTICS OF PAPER BOARD PRODUCTION)
NIHON SEISHI RENGO-KAI ITAGAMI-BU AND NIHON SEISHI RENGO-KAI
C/O KAMI PARUPU KAIKAN, 9-11, 3-CHOME, GINZA, CHUO-KU, TOKYO 108
JAPANESE

JAPAN PULP AND PAPER 8522
SHINGYO TIMES SHA. LTD.
BUNSEI BUILDING 9-4, 3-CHOME, GINZA, CHUOKU TOKYO
ENGLISH
INDEXED: BIB, CHE

JP RIPOTO 8523
(JP REPORT)
NIHONSHI PARUPU SHOJI KABUSHIKI KAISHA (JAPANESE PAPER PULP COMPANY, LTD.)
JAPANESE AND ENGLISH

KAKO GIJUTSU 8524
(PAPER CONVERTING)
PARUPU, KAMI, HOSO GIJUTSU KYODO KENKYUKAI
30-6, 2-CHOME, HONGO, BUNKYO-KU, TOKYO 113
JAPANESE AND ENGLISH

KAMI AYOBI PARUPU 8525
(PAPER AND PULP)
JAPANESE

KAMI PA GIKYOSHI 8526
(JOURNAL OF THE JAPANESE TECHNICAL ASSOCIATION OF PULP AND PAPER INDUSTRY)
KAMI PARUPU GIJUTSU KYOKAI (THE JAPANESE TECHNICAL ASSOCIATION OF PULP AND PAPER INDUSTRY)
C/O KAMI PARUPU KAIKAN, 9-11, 3-CHOME, GINZA, CHUO-KU, TOKYO 104
JAPANESE, WITH ENGLISH SUMMARY
INDEXED: CHE

KAMI PARUPU 8527
(JOURNAL OF THE JAPAN PAPER ASSOCIATION)
NIHON SEISHI RENGOKAI (EDITORIAL STAFF, JAPAN PAPER ASSOCIATION)
9-11, 3-CHOME, GINZA, CHUO-KU, TOKYO 104
JAPANESE
INDEXED: JICST

KAMI PARUPU GIJUTSU BUNKEN DAIMEISHU 8528
(PAPER AND PULP TECHNOLOGY LITERATURE)
CHUGAI SANGYO CHOSAKAI (CHUGAI INDUSTRIAL RESEARCH ASSOCIATION)
4-2, 2-CHOME, MUKOGAOKA, BUNKYO-KU, TOKYO 113
JAPANESE

KAMI PARUPU GIJUTSU TAIMUSU 8529
(PAPER AND PULP TECHNOLOGY TIMES)
SHIGYO TAIMUSUSHA
9-4, 3-CHOME, GINZA, CHUO-KU, TOKYO 104
1
JAPANESE

KAMI PARUPU KAIGAI GIJUTSU 8530
(PAPER AND PULP, OVERSEAS TECHNOLOGY)
CHUGAI SANGYO CHOSAKAI (CHUGAI INDUSTRIAL RESEARCH ASSOCIATION)
4-2, 2-CHOME, MUKOGAOKA, BUNKYO-KU, TOKYO 113
JAPANESE

KAMI PARUPU NO GIJUTSU 8531
(PULP AND PAPER TECHNOLOGY)
SHIZUOKA-KEN KAMI OYOBI PARUPU GIJUTSU KYOKAI (SHIZUOKA PREFECTURE TECHNICAL ASSOCIATION OF PULP AND PAPER INDUSTRY)
C/O SHIZUOKA-KEN SEISHI KOGYO SHIKENJO, 3363 DENPO, FUJI-SHI 417
JAPANESE

KAMI PARUPU SHINBUN 8532
(PAPER AND PULP NEWS)
KAMI PARUPU SHINBUNSHA
16-6, 2-CHOME, NIHONBASHI, CHUO-KU, TOKYO 103
JAPANESE

KAMI PARUPU TOKEI GEPPO 8533
(PAPER AND PULP STATISTICAL MONTHLY)
TSUSHO SANGYO-SHO DAIJIN KANBO CHOSA TOKEI-BU (RESEARCH AND STATISTICS DIVISION, MINISTER'S SECRETARIAT, MINISTRY OF INTERNATIONAL TRADE AND INDUSTRY)
C/O KAMI PARUPU KAIKAN, 9-11, 3-CHOME, GINZA, CHUO-KU, TOKYO 104
JAPANESE

KAMI PARUPU TOKEI NENPO 8534
TSUSHO SANGYO-SHO DAIJIN KANBO CHOSA TOKEI-BU (RESEARCH AND STATISTICS DIVISION, MINISTER'S SECRETARIAT, MINISTRY OF INTERNATIONAL TRADE AND INDUSTRY)
C/O NIHON SEISHI KAIKAN, 9-11, 3-CHOME, GINZA, CHUO-KU, TOKYO 104
JAPANESE

KAMI RYUTSU TOKEI GEPPO 8535
(MONTHLY STATISTICS OF PAPER DISTRIBUTION)
TSUSHO SANGYO-SHO DAIJIN KANBO CHOSA TOKEI-BU (RESEARCH AND STATISTICS DIVISION, MINISTER'S SECRETARIAT, MINISTRY OF INTERNATIONAL TRADE AND INDUSTRY)
C/O KAMI PARUPU KAIKAN, 9-11, 3-CHOME, GINZA, CHUO-KU, TOKYO 104
JAPANESE
INDEXED: JICST

KAMI RYUTSU TOKEI NENPO 8536
(YEARBOOK OF PAPER DISTRIBUTION STATISTICS)
TSUSHO SANGYO-SHO DAIJIN KANBO CHOSA TOKEI-BU (RESEARCH AND STATISTICS DIVISION, MINISTER'S SECRETARIAT, MINISTRY OF INTERNATIONAL TRADE AND INDUSTRY)
C/O KAMI PARUPU KAIKAN, 9-11, 3-CHOME, GINZA, CHUO-KU, TOKYO 104
JAPANESE

KAMI TO PURASUCHIKKU 8537
(PAPER AND PLASTICS)
KAKO GIJUTSU KENKYUKAI
C/O DAIE BUILDING, 3-37, KANDA SAKUMA-CHO, CHIYODA-KU, TOKYO 101
JAPANESE

KAMI-PARUPU RENGOKAI 8538
(PAPER AND PULP STATISTICS)
ENGLISH
INDEXED: BIB

KAMI-PARUPU SANGYO NENKAN 8539
(PAPER-PULP INDUSTRY YEARBOOK)
JAPANESE
INDEXED: BIB

KASENSHI KENKYUKAISHI 8540
(JOURNAL OF CHEMICAL PAPER STUDY-GROUP)
KASENSHI KENKYUKAI
C/O EHIME-KEN SEISHI SHIKENJO, KAWANOE-CHO, KAWANOE-SHI 799-01)
JAPANESE

KOCHI-KEN SHIGYO SHIKENJO GYOMU HOKOKU 8541
(REPORT OF PAPER INDUSTRY EXPERIMENT STATION, KOCHI PREFECTURE)
KOCHI-KEN SHIGYO SHIKENJO
3-115, ASAHI-MACHI, KOCHI-SHI 780
JAPANESE

KOSHI SAISEI SOKUSHIN SENTA KAIHO 8542
(REPORT OF THE CENTER TO PROMOTE PAPER RECYCLING)
JAPANESE

PAPER JAPAN 8543
JAPAN PAPER EXPORTERS ASSOCIATION
ENGLISH
INDEXED: BIB

SAITAMA-KEN SEISHI KOGYO SHIKENJO GYOMU HOKOKU 8544
(REPORTS OF PAPER INDUSTRIAL EXPERIMENT STATION, SAITAMA PREFECTURE)
SAITAMA-KEN SEISHI KOGYO SHIKENJO (PAPER INDUSTRIAL EXPERIMENT STATION, SAITAMA PREFECTURE)
226, OGAWA, OGAWA-MACHI, HIKI-GUN, SAITAMA 355-03
JAPANESE

SEISAN GIJUTSU KENKYU 8545
(INDUSTRIAL TECHNOLOGY RESEARCH)
HOKUETSU SEISHI K.K. KENKYUSHO (RESEARCH LABORATORY, HOKUETSU PAPER INDUSTRY COMPANY, LTD.)
5-1, 3-CHOME, NISHI ZAO, NAGAOKA-SHI 940
JAPANESE
INDEXED: JICST

SHIGYO NICHINICHI SHINBUN 8546
(DAILY NEWS OF THE PAPER INDUSTRY)
SHIGYO NICHINICHI SHINBUNSHA
1-13, KANDA AWAJI-CHO, CHIYODA-KU, TOKYO 101
JAPANESE

SHIGYO NYUSU 8547
(PAPER INDUSTRY NEWS)
SAITAMA-KEN SEISHIKOGYO SHIKENJO (PAPER INDUSTRIAL EXPERIMENT STATION, SAITAMA PREFECTURE)
226, OGAWA, OGAWA-MACHI, HIKI-GUN, SAITAMA 355-03
JAPANESE

SHIGYO SHINBUN 8548
(PAPER INDUSTRY INFORMATION)
SHIGYO SHINBUNSHA
4-1, 3-CHOME, MARUNOUCHI, CHIYODA-KU, TOKYO 100
JAPANESE

SHIGYO TAIMUSU 8549
(PAPER INDUSTRY TIMES)
SHIGYO TAIMUSUSHA
9-4, 3-CHOME, GINZA, CHUO-KU, TOKYO 104
JAPANESE

SHIGYO TO INSATSU 8550
(PAPER INDUSTRY AND PRINTING)
INSATSU SHIGYOSHA (PAPER PRINTING COMPANY)
JAPANESE

SHIGYOKAI 8551
(PAPER INDUSTRY)
SHIGYO SHINBUNSHA
4-1, 3-CHOME, MARUNOUCHI, CHIYODA-KU, TOKYO 100
JAPANESE
INDEXED: JICST

SHISHI NYUSU 8552
(PAPER EXPERIMENTS NEWS)
GIFU-KEN SEISHI SHIKENJO (GIFU PREFECTURE PAPER MANUFACTURING EXPERIMENT STATION)
JAPANESE

SHIZUOKA-KEN NO SHIGYO 8553
(PAPER INDUSTRY IN SHIZUOKA PREFECTURE)
SHIZUOKA-KEN SHOKO-BU
9-6, OTE-MACHI, SHIZUOKA-SHI 420
JAPANESE

SHIZUOKA-KEN SEISHI KOGYO SHIKENJO GYOMU GAIYO 8554
(OPERATIONAL REPORT OF THE INSTITUTE OF PULF AND PAPER INDUSTRY, SHIZUOKA PREFECTURE)
SHIZUOKA-KEN SEISHI KOGYO SHIKENJO
3363, DENPO, FUJI-SHI 417
JAPANESE

SHIZUOKA-KEN SEISHI KOGYO SHIKENJO HOKOKU 8555
(REPORT OF PAPER INDUSTRY EXPERIMENT STATION, SHIZUOKA PREFECTURE)
SHIZUOKA-KEN SEISHI KOGYO SHIKENJO (THE INSTITUTE OF PULPS AND PAPER INDUSTRY, SHIZUOKA PREFECTURE)
3363, DENPO, FUJU-SHI 417
JAPANESE
INDEXED: CHE, JICST

SHUKAN BYURO 8556
(WEEKLY BURO)
ZENKOKU SHIKO TSUSHINSHA
18-2, 1-CHOME, HIGASHI KOMAGATA, SUMIDA-KU, TOKYO 130
JAPANESE

TOMOEGAWA GIHO 8557
(JOURNAL OF THE TECHNICAL RESEARCH LABORATORY, TOMOEGAWA PAPER MANUFACTURING COMPANY))
TOMOEGAWA SEISHISHO GIJUTSU KENKYUSHO (TECHNICAL RESEARCH LABORATORY, TOMOEGAWA PAPER MANUFACTURING COMPANY)
110, MOCHIMUNE, SHIZUOKA-SHI 421-01
JAPANESE

TOYAMA-KEN SEISHI SHIDOSHO GYOMU HOKOKU 8558
(OPERATIONAL REPORT OF TOYAMA PREFECTURAL PAPER-MAKING GUIDANCE INSTITUTE)
TOYAMA-KEN SEISHI SHIDOSHO
3355, KAMISHIN-MACHI, KAWAO-MACHI, NEI-GUN, TOYAMA 939-23
JAPANESE

Textile and Cordage Industry

AI 8559
(INDIGO)
SHOTOSHA
JAPANESE

AICHI-KEN MIKAWA SEN'I GIJUTSU SENTA KENKYU SHIRYO 8560
(RESEARCH DATA OF THE MIKAWA TEXTILE RESEARCH INSTITUTE, AICHI PREFECTURE)
109, IGAKUBO, OTSUKA-CHO, GAMAGORI-SHI 443
JAPANESE

AICHI-KEN ORIMONO KENKYUKAISHI 8561
(JOURNAL OF THE AICHI-KEN TEXTILE TECHNICAL CENTER)
AICHI-KEN OWARI SEN'I GIJUTSU SENTA (AICHI-KEN TEXTILE TECHNICAL CENTER)
35, MIYAURA, YAMATO-MACHI UMABIKI, ICHINOMIYA-SHI 491
JAPANESE

ANNUAL STATISTICAL REVIEW: COTTON AND ALLIED TEXTILE INDUSTRY IN JAPAN 8562
NIHON BOSEKI KYOKAI (JAPAN COTTON SPINNERS ASSOCIATION)
JAPANESE
INDEXED: BIB

BOKEN REPOTO 8563
(JSIF REPORT)
NIHON BOSEKI KENSA KYOKAI (JAPAN SPINNERS' INSPECTION FOUNDATION)
1-25, UCHIKYUHOJI-MACHI, HIGASHI-KU, OSAKA-SHI 540
JAPANESE
INDEXED: BIB, JICST

BOSEKI JIJO SANKOSHIO 8564
(STATISTICS ON JAPANESE SPINNING INDUSTRY)
NIHON BOSEKI KYOKAI (JAPAN SPINNER'S ASSOCIATION)
C/O MENGYO KAIKAN, 3-8, BINGO-MACHI, HIGASHI-KU, OSAKA-SHI 541
JAPANESE

BOSEKI TOKEI GEPPO 8565
(MONTHLY REPORT OF SPINNING STATISTICS)
NIHON BOSEKI KYOKAI (JAPAN SPINNER'S ASSOCIATION)
3-8, BINGO-MACHI, HIGASHI-KU, OSAKA-SHI 541
JAPANESE

COTTON STATISTICS MONTHLY 8566
NIHON MENKA KYOKAI, MENKA KEIZAI KENKYUSHO (THE COTTON ECONOMICS RESEARCH INSTITUTE, THE JAPAN COTTON TRADERS' ASSOCIATION)
1-120, UTSUBO, NISHI-KU, OSAKA-SHI 550
ENGLISH
INDEXED: BIB

ECHO 8567
SHIZUOKA-KEN HAMAMATSU SEN'I KOGYO SHIKENJO
617, KITATERASHIMA-CHO, HAMAMATSU-SHI 430
JAPANESE

FRAGRANCE JOURNAL 8568
JAPANESE

FUKUI DAIGAKU KOGAKUBU SEN'I KOGYO KENKYU SHISETSU HOKOKU 8569
(BULLETIN OF THE TEXTILE RESEARCH INSTITUTE, FACULTY OF ENGINEERING, FUKUI UNIVERSITY)
FUKUI DAIGAKU KOGAKUBU SEN'I KOGYO KENKYU SHISETSU (THE TEXTILE RESEARCH INSTITUTE, FACULTY OF ENGINEERING, FUKUI UNIVERSITY)
9-1, 3-CHOME, BUNKYO, FUKUI-SHI 910
JAPANESE, WITH ENGLISH SUMMARY
INDEXED: CHE, JICST

FUKUI-KEN SEN'I KOGYO SHIKENJO GYOMU HOKOKU 8570
(OPERATIONAL REPORT OF FUKUI PREFECTURAL TEXTILE ENGINEERING RESEARCH INSTITUTE)
FUKUI-KEN SEN'I KOGYO SHIKENJO
10-1, 3-CHOME, BUNKYO, FUKUI-SHI 910
JAPANESE

FUKUI-KEN SEN'I KOGYO SHIKENJO KENKYU HOKOKU 8571
(BULLETIN OF FUKUI PREFECTURAL TEXTILE ENGINEERING RESEARCH INSTITUTE)
FUKUI-KEN SEN'I KOGYO SHIKENJO
10-1, 3-CHOME, BUNKYO, FUKUI-SHI 910
JAPANESE
INDEXED: JICST

FUKUYAMA SEN'I KOGYO SHIKENJO HOKOKU 8572
(REPORT OF THE FUKUYAMA TEXTILE INDUSTRY LABORATORY)
HIROSHIMA-KENRITSU FUKUYAMA SEN'I KOGYO SHIKENJO (HIROSHIMA PREFECTURAL FUKUYAMA TEXTILE INDUSTRY LABORATORY)
JAPANESE

GISEN SHI DAYORI 8573
(LETTER OF THE GIFU TEXTILE EXPERIMENT STATION)
GIFU-KEN SENI SHIKENJO (GIFU PREFECTURAL TEXTILE EXPERIMENT STATION)
JAPANESE

GOSEN DOKUHON 8574
(BOOK ON SYNTHETIC FIBER)
NIHON SEN'I SHINBUNSHA
3-9, NIHONBASHI HON-CHO, CHUO-KU, TOKYO 103
JAPANESE

GUNMA-KEN ISESAKI SEN'I KOGYO SHIKENJO GYOMU GAIYO HOKOKUSHO KENKYU HOKOKUSHO 8575
(REPORT OF GUNMA PREFECTURAL ISESAKI TEXTILE INDUSTRY LABORATORY)
GUNMA-KEN ISESAKI SEN'I KOGYO SHIKENJO (GUNMA PREFECTURAL ISESAKI TEXTILE INDUSTRIAL LABORATORY)
27-8, KURUWA-CHO, ISESAKI-SHI 372
JAPANESE

GUNMA-KEN KIRYU SEN'I KOGYO SHIKENJO GYOMU KOKOKU 8576
(OPERATIONAL REPORT OF THE GUNMA PREFECTURAL KIRYU TEXTILE INDUSTRIAL LABORATORY)
GUNMA-KEN KIRYU SEN'I KOGYO SHIKENJO
5-11, EIRAKU-CHO, KIRYU-SHI 376
JAPANESE

GYOMU NARABINI KENKYU HOKOKUSHO 8577
(YAMANASHI-KEN SEN'I KOGYO SHIKENJO)
(ANNUAL REPORT OF THE TEXTILE RESEARCH INSTITUTE OF YAMANSHI PREFECTURE)
YAMANASHI-KEN SEN'I KOGYO SHIKENJO
2095, SHIMOYOSHIDA, FUJIYOSHIDA-SHI 403
JAPANESE

HIROSHIMA-KEN KOGYO GIJUTSU JOHO: SEN'I-HEN 8578
(HIROSHIMA PREFECTURE INDUSTRIAL TECHNOLOGY REPORT TEXTILES)
HIROSHIMA-KIEN GIJUTSU JOHO SHITSU
JAPANESE

HIROSHIMA-KENRITSU FUKUYAMA SEN'I KOGYO SHIKENJO GYOMU HOKOKU 8579
(OPERATIONAL REPORT OF FUKUYAMA TEXTILE RESEARCH INSTITUTE OF HIROSHIMA PREFECTURE)
HIROSHIMA-KENRITSU FUKUYAMA SEN'I KOGYO SHIKENJO (FUKUYAMA TEXTILE RESEARCH INSTITUTE OF HIROSHIMA PREFECTURE)
312, YAMATE-CHO, FUKUYAMA-SHI 720
JAPANESE

HIROSHIMA-KENRITSU FUKUYAMA SEN'I KOGYO SHIKENJO NENPO 8580
(ANNUAL REPORT OF THE FUKUYAMA TEXTILE RESEARCH INSTITUTE OF HIROSHIMA PREFECTURE)
HIROSHIMA-KENRITSU FUKUYAMA SEN'I KOGYO SHIKENJO (FUKUYAMA TEXTILE RESEARCH INSTITUTE OF HIROSHIMA PREFECTURE)
312, YAMATE-CHO, FUKUYAMA-SHI 720
JAPANESE

HYOGO-KEN KOGYO GIJUTSU JOHO: SEN'I-HEN 8581
(HYOGO PREFECTURE INDUSTRIAL TECHNOLOGY REPORT)
HYOGO KENRITSU KOGYO SHIKENJO
(HYOGO PREFECTURAL INDUSTRIAL EXPERIMENT STATION)
JAPANESE

HYOGO-KEN SEN'I KOGYO SHIDOSHO KENKYU HOKOKU 8582
(BULLETIN OF THE TEXTILE RESEARCH INSTITUTE OF HYOGO PREFECTURE)
HYOGO-KEN SEN'I KOGYO SHIDOSHO (TEXTILE RESEARCH INSTITUTE OF HYOGO PREFECTURE)
1790-255, UENODAN, NOMURA-CHO, NISHIWAKI-SHI 677
JAPANESE

HYOGO-KEN SEN'I KOGYO SHIDOSHO NENPO 8583
(ANNUAL REPORT OF THE TEXTILE RESEARCH INSTITUTE OF HYOGO PREFECTURE)
HYOGO-KEN SEN'I KOGYO SHIDOSHO (TEXTILE RESEARCH INSTITUTE OF HYOGO PREFECTURE)
1790-255, UENODAN, NOMURA-CHO, NISHIWAKI-SHI 677
JAPANESE

IBARAKI-KEN SEN'I KOGYO SHIDOSHO GYOMU HOKOKUSHO 8584
(REPORT OF THE IBARAKI PREFECTURE TEXTILE INDUSTRY CONSULTANT OFFICE)
IBARAKI-KEN SEN'I KOGYO SHIDOSHO (IBARAKI PREFECTURE TEXTILE INDUSTRY CONSULTANT OFFICE)
607, YUKI, YUKI-SHI 307
JAPANESE

IMPORTS OF COTTON INTO JAPAN 8585
JAPAN COTTON TRADERS ASSOCIATION
ENGLISH
INDEXED: BIB

IRYOHIN SHOHI KOBAI DOKO 8586
(TRENDS IN BUYING AND CONSUMPTION OF CLOTHING ITEMS)
SENI KOGYO KOZO KAIZEN JIGYO KYOKAI SEHI JOHO SENTA
(ASSOCIATION FOR THE STRUCTURAL REFORMATION OF THE TEXTILE INDUSTRY, TEXTILE INFORMATION CENTER)
JAPANESE

JAPAN COTTON STATISTICS AND RELATED DATA 8587
OSAKA COTTON ECONOMICS RESEARCH INSTITUTE
ENGLISH
INDEXED: BIB

JOURNAL OF THE FACULTY OF TEXTILE SCIENCE AND TECHNOLOGY, SHINSHU UNIVERSITY, SERIES D, ARTS 8588
SHINSHU DAIGAKU SEN'I GAKUBU (THE FACULTY OF TEXTILE SCIENCE AND TECHNOLOGY, SHINSHU UNIVERSITY)
15-1, 3-CHOME, TOKIDA, UEDA-SHI 386
JAPANESE AND EUROPEAN
INDEXED: CHE, JICST

JOURNAL OF THE TEXTILE MACHINERY SOCIETY OF JAPAN 8589
NIHON SEN'I KIKAI GAKKAI (THE TEXTILE MACHINERY SOCIETY OF JAPAN)
C/O OSAKA KAGAKU GIJUTSU SENTA, 1-118, UTSUBO, NISHI-KU, OSAKA-SHI 550
ENGLISH
INDEXED: BIB, CHE, ENG

KAGOSHIMA-KEN OSHIMA SENSHOKU SHIDOSHO GYOMU HOKOKU 8590
(OPERATIONAL REPORT OF THE OSHIMA DYEING AND WEAVING RESEARCH INSTITUTE, KAGOSHIMA PREFECTURE)
KAGOSHIMA-KEN OSHIMA SENSHOKU SHIDOSHO
5-37, KUSATO-CHO, NAZE-SHI 894
JAPANESE

KAIGAI KASEN SOKUHO 8591
(CURRENT NEWS ON MAN-MADE FIBRES OVERSEAS)
NIHON KAGAKU SEN'I KYOKAI (JAPAN CHEMICAL FIBRES ASSOCIATION)
3-3, NIHONBASHI MURO-MACHI, CHUO-KU, TOKYO 103
JAPANESE

KAIGAI SEN'I BUNKEN SHOROKU 8592
(DIGEST OF FOREIGN LITERATURE ON TEXTILES)
NIHON SEN'I KIKAI GAKKAI (TEXTILE MACHINERY SOCIETY OF JAPAN)
C/O OSAKA KAGAKU GIJUTSU SENTA BUILDING, 1-118, UTSUBO, NISHI-KU, OSAKA-SHI 550
JAPANESE

KAIGAI SEN'I GIJUTSU BUNKENSHU 8593
(FOREIGN TREATISES ON TEXTILE ENGINEERING)
NIHON SEN'I KIKAI GAKKAI (TEXTILE MACHINERY SOCIETY OF JAPAN)
C/O OSAKA KAGAKU GIJUTSU SENTA BUILDING, 1-118, UTSUBO, NISHI-KU, OSAKA-SHI 550
JAPANESE

KANAGAWA-KEN SEN'I KOGYO SHIDOSHO GYOMU HOKOKU 8594
(OPERATIONAL REPORT OF THE TEXTILE INDUSTRIAL RESEARCH INSTITUTE OF KANAGAWA PREFECTURE)
KANAGAWA-KEN SEN'I KOGYO SHIDOSHO (TEXTILE INDUSTRIAL RESEARCH INSTITUTE OF KANAGAWA PREFECTURE)
4408, HANBARA, AIKAWA-CHO, AIKO-GUN, KANAGAWA 243-03
JAPANESE

KANAGAWA-KEN SEN'I KOGYO SHIDOSHO NYUSU 8595
(KANAGAWA PREFECTURE TEXTILE INDUSTRY DEVELOPMENT DIVISION NEWS)
JAPANESE

KANEMATSU GOSHO YOMO NENPO 8596
(KANEMATSU-GOSHO ANNUAL WOOL REVIEW)
KANEMATSU-GOSHO, K. K.
2-5, TAKARA-CHO, CHUO-KU, TOKYO 104
JAPANESE
INDEXED: JICST

KARA DEZAIN 8597
(COLOR DESIGN)
NIHON SEN'I ISHO SENTA (JAPAN TEXTILE COLOR DESIGN CENTER)
C/O MENGYO KAIKAN, 3-8, HON-MACHI, HIGASHI-KU, OSAKA 541
JAPANESE
INDEXED: CHE

KASEN GEPPO 8598
(JAPAN CHEMICAL FIBERS MONTHLY)
NIHON KAGAKU SEN'I KYOKAI (JAPAN CHEMICAL FIBERS ASSOCIATION)
C/O MITSUI BEKKAN, 3-3, NIHONBASHI MURO-MACHI, CHUO-KU, TOKYO 103
JAPANESE
INDEXED: CHE

KASEN KURABU 8599
(SYNTHETIC FIBERS)
JAPANESE
INDEXED: JICST

KATAKURA KOGYO K. K. KENKYU HOKOKU 8600
(RESEARCH REPORT OF THE HACHIOJI SERICULTURAL STATION, KATAKURA INDUSTRY COMPANY)
KATAKURA KOGYO K. K. HACHIOJI KENKYUSHO (HACHIOJI SERICULTURAL STATION, KATAKURA INDUSTRY COMPANY)
610, NARAHARA-MACHI, HACHIOJI-SHI, TOKYO 193
JAPANESE

KATAKURA KOGYO KABUSHIKI KAISHA KENKYU HOKOKU TOKUBETSUGO 8601
(RESEARCH REPORT OF KATAKURA INDUSTRIES COMPANY, LIMITED (SPECIAL ISSUE))
KATAKURA KOGYO KABUSHIKI KAISHA HACHIOJI KENKYUJO (KATAKURA INDUSTRIES COMPANY, LTD. HACHIOJI RESEARCH INSTITUTE)
JAPANESE

KENKYU HAPPYOKAI SANKO SHIRYO (KOGYO GIJUTSU-IN SEN'I KOBUNSHI ZAIRYO KENKYUSHO) 8602
(DATA FOR THE ANNUAL MEETING OF THE RESEARCH INSTITUTE FOR POLYMER AND TEXTILES, AGENCY OF INDUSTRIAL SCIENCE AND TECHNOLOGY)
TSUSHO SANGYO-SHO KOGYO GIJUTSU-IN SEN'I KOBANSHI ZAIRYO KENKYUSHO (RESEARCH INSTITUTE FOR POLYMER AND TEXTILES, AGENCY OF INDUSTRIAL SCIENCE AND TECHNOLOGY)
4-1, SAWATARI, KANAGAWA-KU, YOKOHAMA-SHI 221
JAPANESE

KENKYU SHIRYO, MIKAWA SEN'I GIJUTSU SENTA 8603
(RESEARCH DATA, MIKAWA TEXTILE TECHNICAL CENTER)
AICHI-KEN MIKAWA SEN'I GIJUTSU SENTA (AICHI-KEN MIKAWA TEXTILE TECHNICAL CENTER)
109, IGAKUBO, OTSUKA-CHO, GAMAGORI-SHI 443
JAPANESE

KENKYUSHO HOKOKU (ENSHU SEISAKU K. K. CHUO KENKYUSHO) 8604
(LABORATORY REPORT, CENTRAL RESEARCH LABORATORY, THE ENSHU SEISAKU K. K.)
ENSHU SEISAKU K. K. CHUO KENKYUSHO
4888, TAKATSUKA, KAMI-MURA, HAMANAGUN, SHIZUOKA 432
JAPANESE

KENSA JIHO 8605
(TEST REVIEW OF YOKOHAMA RAW SILK INSPECTION HOUSE)
YOKOHAMA KIITO KENSAJO (YOKOHAMA RAW SILK INSPECTION HOUSE) YOKOHAMA KIITO TON'YA KYOKAI
C/O TEISAN BUILDING, 5-57, KITANAKA-DORI, NAKA-KU, YOKOHAMA-SHI 231
JAPANESE

KIITO (GUNZE K. K. KENKYUSHO) 8606
(RAW SILK, GUNZE K. K. KENKYUSHO)
GUNZE K. K. KENKYUSHO
AONO-CHO, AYABE-SHI 623
JAPANESE
INDEXED: JICST

KIITO (NIHON SEISHI KYOKAI) 8607
(RAW SILK, NIHON SEISHI KYOKAI)
NIHON SEISHI KYOKAI (JAPAN RAW SILK REELERS' ASSOCIATION)
1-7, YURAKU-CHO, CHIYODA-KU, TOKYO 100
JAPANESE

KIITO KENSASHO GEPPO 8608
(MONTHLY REPORT OF SILK CONDITIONING HOUSE)
NORIN-SHO YOKOHAMA KIITO KENSASHO, KOBE KIITO KENSASHO (THE YOKOHAMA AND KOBE SILK CONDITIONING HOUSE, MINISTRY OF AGRICULTURE AND FORESTRY)
5-57, KITANAKA-DORI, NAKA-KU, YOKOHAMA-SHI 231
JAPANESE

KIITO KENSASHO JIGYO SEISEKI HOKOKU 8609
(REPORT ON TESTING RESULTS OF THE SILK CONDITIONING HOUSES)
NORIN-SHO YOKOHAMA KIITO KENSASHO, KOBE KIITO KENSASHO (THE YOKOHAMA AND KOBE SILK CONDITIONING HOUSES, MINISTRY OF AGRICULTURE AND FORESTRY)
5-57, KITANAKA-DORI, NAKA-KU, YOKOHAMA-SHI 231
JAPANESE

KIITO KENSASHO KENKYU HOKOKU 8610
(RESEARCH REPORTS OF THE SILK CONDITIONING HOUSES)
NORIN-SHO YOKOHAMA KIITO KENSASHO, KOBE KITTO KENSASHO (THE YOKOHAMA AND KOBE SILK CONDITIONING HOUSES, MINISTRY OF AGRICULTURE AND FORESTRY)
5-57, KITANAKA-DORI, NAKA-KU, YOKOHAMA-SHI 231
JAPANESE, WITH ENGLISH SUMMARY
INDEXED: BIB, CHE, JICST

KINU 8611
(SILK)
OIWA SHUPPAN
C/O TOSEN BUILDING, 1-9, NIHONBASHI HORIDOME-CHO, CHUO-KU, TOKYO 103
JAPANESE

KINU JINSEN ORIMONO TOKEI GEPPO 8612
(STATISTICAL MONTHLY OF SILK AND RAYON WEAVERS)
NIHON KINU JINSEN ORIMONO KOGYOKAI (JAPAN SILK AND RAYON WEAVERS ASSOCIATION)
15-12, 1-CHOME, KUDANKITA, CHIYODA-KU, TOKYO 102
JAPANESE

KNIT DESIGN 8613
NIHON NITTO UEA, DEZAIN KYOKAI (JAPAN KNITWEAR DESIGNER'S ASSOCIATION)
C/O TOKO BUILDING, 3-3, 1-CHOME, YANAGIBASHI, TAITO-KU, TOKYO 111
JAPANESE

KOBE KIITO TORIHIKISHO GYOMU HOKOKUSHO 8614
(OPERATIONAL REPORT OF KOBE RAW SILK EXCHANGE)
KOBE KIITO TORIHIKISHO (KOBE RAW SILK EXCHANGE)
126, HIGASHI-MACHI, IKUTA-KU, KOBE-SHI 650
JAPANESE

KOKUSAI KENGYO KYOKAI KAIKO YAKUBUN 8615
(TRANSLATION OF THE PROCEEDINGS OF THE INTERNATIONAL SILK INDUSTRY ASSOCIATION)
JAPANESE
INDEXED: JICST

KOTTON PUROMOSHON 8616
(COTTON PROMOTION)
NIHON MENGYO SHINKOKAI (JAPAN COTTON PROMOTION INSTITUTE)
3-8, BINGO-CHO, HIGASHI-KU, OSAKA-SHI 541
JAPANESE

KURARE TAIMUSU 8617
(KURASHIKI RAYON TIMES)
KURASHIKI REIYON K.K.
8, UMEDA, KITA-KU, OSAKA-SHI 530
JAPANESE

KYOTO DAIGAKU NIPPON KAGAKUSENI KENKYUSHO KOENSHU 8618
(BULLETIN OF THE RESEARCH INSTITUTE FOR SYNTHETIC FIBERS, KYOTO UNIVERSITY)
KYOTO DAIGAKU, NIPPON KAGAKUSENI KENKYUSHO YOSHIDAHON-CHO, SAKYO-KU
KYOTO
JAPANESE
INDEXED: CHE, JICST

KYOTO KOGEI SEN'I DAIGAKU SEN'IGAKUBU GAKUJUTSU HOKOKU 8619
(BULLETIN OF THE FACULTY OF TEXTILE SCIENCE, KYOTO UNIVERSITY OF INDUSTRIAL ARTS AND TEXTILE FIBERS)
KYOTO KOGEI SEN'I DAIGAKU SEN'IGAKUBU (FACULTY OF TEXTILE SCIENCE, KYOTO UNIVERSITY OF INDUSTRIAL ARTS AND TEXTILE FIBERS)
MATSUGASAKI GOSHOKAIDO-CHO, SAKYO-KU, KYOTO-SHI 606
JAPANESE AND ENGLISH
INDEXED: BIB, CHE, JICST

KYOTO-FU ORIMONO SHIDOSHO GYOMU HOKOKUSHO 8620
(OPERATIONAL REPORT OF THE INSTITUTE OF KYOTO PREFECTURAL TEXTILE GUIDANCE)
KYOTO-FU ORIMONO SHIDOSHO (THE INSTITUTE OF KYOTO PREFECTURAL TEXTILE GUIDANCE)
TANBA, MINEYAMA-CHO, NAKA-GUN KYOTO 627
JAPANESE

KYOTO-FU ORIMONO SHIDOSHO KENKYU HOKOKUSHO 8621
(REPORT OF THE INSTITUTE OF KYOTO PREFECTURAL TEXTILE GUIDANCE)
KYOTO-FU ORIMONO SHIDOSHO (THE INSTITUTE OF KYOTO PREFECTURAL TEXTILE GUIDANCE)
TANBA, MINEYAMA-CHO, NAKA-GUN, KYOTO 627
JAPANESE

KYOTO-SHI SENSHOKU SHIKENJO GYOMU HOKOKUSHO 8622
(OPERATIONAL REPORT OF THE DYEING AND WEAVING LABORATORY, KYOTO CITY)
KYOTO-SHI SENSHOKU SHIKENJO
KARASUMA-DORI KAMIDACHIYURIAGARU, KAMIGYO-KU, KYOTO-SHI 602
JAPANESE

KYOZOME TO SEIREN SENSHOKU 8623
(KYOZOME-DYEING AND REFINE DYEING)
KYOZOME KENKYUKAI
C/O KYOTO-SHI SENSHOKU SHIKENJO, KARASUMA-DORI KAMIDACHIYURIAGARU, KAMIGYO-KU, KYOTO-SHI 602
JAPANESE

MEKOSHI NYUSU 8624
(KNITTING INDUSTRY ADVISORY NEWS)
YAMANASHI KENRITSU MERIYASU KOGYO SHIDOJO (YAMANASHI PREFECTURAL KNITTING INDUSTRY DEVELOPMENT BOARD)
JAPANESE

MEN SUFU ORIMONO JOHO 8625
ASSOCIATION
(JOURNAL OF JAPAN COTTON AND STAPLE FIBER WEAVER'S ASSOCIATION)
NIHON MEN SUFU ORIMONO KOGYO RENGOKAI (JAPAN COTTON & STAPLE FIBER WEAVER'S ASSOCIATION)
8-7, 1-CHOME, NISHIAZABU, MINATO-KU, TOKYO 105
JAPANESE

MERIYASU 8626
(HOSIERY)
NAIGAI AMIMONO K. K. GIJUTSU-BU (TECHNICAL DEPARTMENT NAIGAI AMIMONO K. K.)
4-5, 5-CHOME, TSUNASHIMANISHI, KOHOKU-KU, YOKOHAMA-SHI 223
JAPANESE
INDEXED: JICST

MIE-KEN KOGYO GIJUTSU JOHO, SEN'I-HEN 8627
(TECHNICAL INFORMATION OF THE INDUSTRIAL TECHNOLOGY CENTER OF MIE PREFECTURE, TEXTILE SECTION)
MIE-KEN KOGYO GIJUTSU SENTA (THE INDUSTRIAL TECHNOLOGY CENTER OF MIE PREFECTURE)
3485, KOMORI-MACHI, TAKAJAYA, TSU-SHI 514
JAPANESE

MIKAWA SEN'I KENKYU SHIRYO 8628
(MIKAWA TEXTILES RESEARCH MATERIALS)
JAPANESE
INDEXED: JICST

MOMEN 8629
(COTTON)
UCHIDARI
JAPANESE

NAGANO-KEN SEN'I KOGYO SHIKENJO GYOMU HOKOKU 8630
(REPORT OF THE TEXTILE INSTITUTE OF NAGANO PREFECTURE)
NAGANO-KEN SEN'I KOGYO SHIKENJO (TEXTILE INSTITUTE OF NAGANO PREFECTURE)
1234-5, YOSHIKAWA NOMIZO, MATSUMOTO-SHI 399-65
JAPANESE

NIHON BOSEKI GEPPO 8631
(MONTHLY REPORT OF JAPAN SPINNERS' ASSOCIATION)
NIHON BOSEKI KYOKAI (JAPAN SPINNERS' ASSOCIATION)
C/O MENGYO KAIKAN, 3-8, BINGO-MACHI, HIGASHI-KU, OSAKA-SHI 541
JAPANESE

NIHON KAGAKU SEN'I KENKYUSHO KOENSHU 8632
(ANNUAL REPORT OF THE RESEARCH INSTITUTE FOR CHEMICAL FIBERS, JAPAN))
C/O KYOTO DAIGAKU, YOSHIDA HON-CHO, SAKYO-KU, KYOTO-SHI 606
JAPANESE

NIHON KAGAKU SEN'I KYOKAI CHOSA SHIRYO 8633
(RESEARCH JOURNAL FOR MAN-MADE FIBRE)
NIHON KAGAKU SEN'I KYOKAI (JAPAN CHEMICAL FIBRES ASSOCIATION)
3-3, NIHONBASHI MURO-MACHI, CHUO-KU, TOKYO 103
JAPANESE
INDEXED: JICST

NIHON SEN'I SHINBUN 8634
(JAPAN TEXTILE NEWS)
NIHON SEN'I SHINBUNSHA
3-9, NIHONBASHI HON-CHO, CHUO-KU, TOKYO 103
JAPANESE
INDEXED: CHE, EXC

NIHON SENSHOKU SHINBUN 8635
(JAPAN DYEING AND DYESTUFF NEWS)
NIHON SENSHOKU SHINBUNSHA
1-20, AZUCHI-MACHI, HIGASHI-KU, OSAKA-SHI 541
JAPANESE

NIKKAN GOSEI SEN'I REPOTO 8636
(DAILY SYNTHETIC TEXTILE REPORTS)
K. K. GOKEN
2-15, UTSUBO HON-MACHI, NISHI-KU, OSAKA-SHI 55[illegible]
JAPANESE

NIKKEN SHUHO 8637
(JAPAN SILK INDUSTRY WEEKLY)
NIHON KINU JINSEN ORIMONO KOGYOKAI (JAPAN SILK AND RAYON WEAVER'S ASSOCIATION)
15-12, 1-CHOME, KUDANKITA, CHIYODA-KU, TOKYO 102
JAPANESE

NIPPON MENGYO GIJUTSU KENKYUJO KENKYU HOKOKU 8638
(REPORT OF THE JAPAN COTTON INDUSTRY TECHNICAL RESEARCH INSTITUTE)
JAPANESE
INDEXED: JICST

NORIN-SHO KIITO KENSASHO GEPPO 8639
(MONTHLY REPORTS OF THE SILK CONDITIONING HOUSES)
NORIN-SHO YOKOHAMA KIITO KENSASHO AND KOBE KIITO KENSASHO (THE YOKOHAMA AND KOBE SILK CONDITIONING HOUSE, MINISTRY OF AGRICULTURE AND FORESTRY
5-57, KITANAKA-DORI, NAKA-KU, YOKOHAMA-SHI 231
JAPANESE

ORIMONO 8640
(FABRICS)
NISHIJIN ORIMONO KENKYUKAI
C/O KYOTO-SHI SENSHOKU SHIKENJO, KARASUMA-DORI KAMIDACHIYURIAGARU, KAMIGYO-KU, KYOTO-SHI 602
JAPANESE

ORIMONO ISIKAWA 8641
(FABRIC OF ISHIKAWA)
ISHIKAWA-KEN SEN'I KYOKAI (ISHIKAWA PREFECTURE TEXTILE ASSOCIATION)
3-16, NISHI-MACHI, KANAZAWA-SHI 920
JAPANESE

OSAKA-FURITSU SEN'I GIJUTSU KENKYUSHO KENKYU HOKOKU 8642
(BULLETIN OF THE TEXTILE TECHNOLOGY RESEARCH INSTITUTE OF OSAKA PREFECTURE)
OSAKA-FURITSU SEN'I GIJUTSU KENKYUSHO
22-28, ASAHI-CHO, IZUMIOTSU-SHI 595
JAPANESE
INDEXED: JICST

OSAKA-FURITSU SEN'I GIJUTSU KENKYUSHO NENPO 8643
(ANNUAL REPORT OF THE TEXTILE TECHNOLOGY RESEARCH INSTITUTE OF OSAKA PREFECTURE)
OSAKA-FURITSU SEN'I GIJUTSU KENKYUSHO
22-28, ASAHI-CHO, IZUMIOTSU-SHI 595
JAPANESE

SAITAMA-KEN CHICHIBU SEN'I KOGYO SHIKENJO GYOMU HOKOKU 8644
(OPERATIONAL REPORT OF THE CHICHIBU TEXTILE INDUSTRIAL LABORATORY OF SAITAMA PREFECTURE)
SAITAMA-KEN CHICHIBU SEN'I KOGYO SHIKENJO (THE CHICHIBU TEXTILE INDUSTRIAL RESEARCH INSTITUTE OF SAITAMA PREFECTURE)
28-1, KUMAGI-MACHI, CHICHIBU-SHI 368
JAPANESE

SAITAMA-KEN GYODA SEN'I KOGYO SHIKENJO GYOMU JISSEKI GAIYO 8645
(OPERATIONAL REPORT OF GYODA TEXTILE RESEARCH INSTITUTE OF SAITAMA PREFECTURE)
GYODA-SHI SEN'I KOGYO SHIKENJO (GYODA TEXTILE RESEARCH INSTITUTE OF SAITAMA PREFECTURE)
7-29, MUKAI-MACHI, GYODA-SHI 361
JAPANESE

SAITAMA-KEN SEN'I KOGYO SHIKENJO GYOMU HOKOKU 8646
(REPORT OF TEXTILE RESEARCH INSTITUTE OF SAITAMA PREFECTURE)
SAITAMA-KEN SEN'I KOGYO SHIKENJO (TEXTILE RESEARCH INSTITUTE OF SAITAMA PREFECTURE)
766, BUSSHI, IRUMA-SHI 358
JAPANESE

SAITAMA-KEN SEN'I KOGYO SHIKENJO GYOMU JISSEKI GAIYO 8647
(OPERATIONAL REPORT OF TEXTILE RESEARCH INSTITUTE OF SAITAMA PREFECTURE)
SAITAMA-KEN SEN'I KOGYO SHIKENJO (TEXTILE RESEARCH INSTITUTE OF SAITAMA PREFECTURE)
766, BUSSHI, IRUMA-SHI 358
JAPANESE

SAKAMOTO SEN'I KIKAI KENKYUSHO HOKOKU 8648
(TECHNICAL REPORT OF SAKAMOTO TEXTILE MACHINERY RESEARCH LABORATORY, ENSHU MANUFACTURING COMPANY, LIMITED)
ENSHU SEISAKU K. K. SAKAMOTO SEN'I KIKAI KENKYUSHO
242, KATSUNE, TAIMA-MACHI, KITAKATSURAGI-GUN, NARA 639-02
JAPANESE

SANSHI SHIKENJO NYUSU 8649
(SILK THREAD EXPERIMENT STATION NEWS)
NORINSHO SANSHI SHIKENJO
(MINISTRY OF AGRICULTURE & FORESTRY, SILK THREAD RESEARCH ISNTITUTE)
JAPANESE

SEISHI TOKEI GEPPO 8650
(MONTHLY REPORT OF SILK-REELING STATISTICS)
NORINSHO SANSHI KYOKU SHISEIKA
(MINISTRY OF AGRICULTURE AND FORESTRY, SERICULTURE BUREAU, SILK THREAD ADMINISTRATION SECTION)
JAPANESE

SEKAI SEN'I NYUSU 8651
(WORLD TEXTILE NEWS)
NIHON BOSEKI KYOKAI CHOSA-BU (RESEARCH DIVISION, ALL JAPAN COTTON SPINNERS' ASSOCIATION)
C/O MENGYO KAIKAN, 3-8, BINGO-CHO, HIGASHI-KU, OSAKA-SHI 541
JAPANESE

SEN'I 8652
(TEXTILE FIBERS)
SEN'I GIJUTSU KENKYUSHA
4-3, 1-CHOME, SHINBASHI, MINATO-KU, TOKYO 105
JAPANESE
INDEXED: CHE

SEN'I GAKKAISHI 8653
(JOURNAL OF THE SOCIETY OF FIBER SCIENCE AND TECHNOLOGY, JAPAN)
SEN'I GAKKAI (SOCIETY OF FIBER SCIENCE AND TECHNOLOGY, JAPAN)
3-9-208, 3-CHOME, KAMINOSAKI, SHINAGAWA-KU, TOKYO 141
JAPANESE, WITH EUROPEAN SUMMARY
INDEXED: CHE, JICST

SEN'I GIJUTSU JOHO (GUNMA-KEN) 8654
(INFORMATION OF TEXTILE ENGINEERING, GUNMA PREFECTURE)
GUNMA-KEN ISESAKI SEN'I KOGYO SHIKENJO (GUNMA PREFECTURAL ISESAKI TEXTILE INDUSTRIAL LABORATORY)
27-8, KURUWA-CHO, ISESAKI-SHI 372
JAPANESE

SEN'I GIJUTSU NYUSU 8655
(TEXTILE TECHNICAL NEWS)
NIHON BOSEKI KYOKAI (ALL JAPAN COTTON SPINNERS' ASSOCIATION)
C/O MENGYO KAIKAN, 3-8, BINGO-CHO, HIGASHI-KU, OSAKA-SHI 541
JAPANESE

SEN'I JANARU 8656
(TEXTILE JOURNAL)
SEN'I JANARU (TEXTILE JOURNAL COMPANY, LTD.)
4-2, AWAZA KAMI-DORI, NISHI-KU, OSAKA-SHI 550
JAPANESE

SEN'I KAGAKU 8657
(TEXTILE SCIENCE)
NIHON SEN'I SENTA (JAPAN TEXTILE CENTER)
4-14, NIHONBASHI KAKIGARA-CHO, CHUO-KU, TOKYO 103
JAPANESE

SEN'I KAI 8658
(TEXTILE MAGAZINE)
JAPANESE

SEN'I KAKO 8659
(DYEING AND FINISHING)
SEN'I KENKYUSHA
6, KANDA KON'YA-CHO, CHIYODA-KU, TOKYO 101
JAPANESE
INDEXED: CHE, JICST

SEN'I KIKAI GAKKAISHI 8660
(JOURNAL OF THE TEXTILE MACHINERY SOCIETY OF JAPAN)
NIHON SEN'I KIKAI GAKKAI (THE TEXTILE MACHINERY SOCIETY OF JAPAN)
C/O OSAKA KAGAKU GIJUTSU SENTA BUILDING, 1-118, UTSUBO, NISHI-KU, OSAKA-SHI 550
JAPANESE, WITH ENGLISH SUMMARY
INDEXED: CHE, JICST

SEN'I KIKAI TOPIKKUSU 8661
(TEXTILE MACHINE TOPICS)
ISHIKAWA-KEN KOGYO SHIKENJO (INDUSTRIAL RESEARCH INSTITUTE, ISHIKAWA)
4-133, YONEIZUMI-MACHI, KANAZAWA-SHI 921
JAPANESE

SEN'I KOBUNSHI ZAIRYO KENKYUSHO KENKYU HOKOKU 8662
(BULLETIN OF THE RESEARCH INSTITUTE FOR POLYMERS AND TEXTILES)
TSUSHO SANGYO-SHO KOGYO GIJUTSU-IN SEN'I KOBUNSHI ZAIRYO KENKYUSHO (RESEARCH INSTITUTE FOR POLYMERS AND TEXTILES, AGENCY OF INDUSTRIAL SCIENCE AND TECHNOLOGY)
4, SAWATARI, KANAGAWA-KU, YOKOHAMA-SHI 221
JAPANESE, WITH ENGLISH SUMMARY
INDEXED: BIB, CHE

SEN'I KOBUNSHI ZAIRYO KENKYUSHO NENPO 8663
(ANNUAL REPORT OF THE RESEARCH INSTITUTE FOR POLYMERS AND TEXTILES)
TSUSHO SANGYO-SHO KOGYO GIJUTSU-IN SEN'I KOBUNSHI ZAIRYO KENKYUSHO (RESEARCH INSTITUTE FOR POLYMERS AND TEXTILES, AGENCY OF INDUSTRIAL SCIENCE AND TECHNOLOGY)
4, SAWATARI, KANAGAWA-KU, YOKOHAMA-SHI 221
JAPANESE
INDEXED: JICST

SEN'I KOBUNSHI ZAIRYO KENKYUSHO NYUSU 8664
(NEWS OF THE RESEARCH INSTITUTE FOR POLYMERS AND TEXTILES)
TSUSHO SANGYO-SHO KOGYO GIJUTSU-IN SEN'I KOBUNSHI ZAIRYO KENKYUSHO (RESEARCH INSTITUTE FOR POLYMERS AND TEXTILES, AGENCY OF INDUSTRIAL SCIENCE AND TECHNOLOGY)
4, SAWATARI, KANAGAWA-KU, YOKOHAMA-SHI 221
JAPANESE
INDEXED: JICST

SEN'I KOGYO ZASSHI (BOSHOKUKAI) 8665
(TEXTILE REVIEW)
BOSHOKU ZASSHISHA (TEXTILE JOURNAL AND BOOK PUBLISHING COMPANY)
C/O ASHIIKE BUILDING, 3-14, SHIOMACHI-DORI, MINAMI-KU, OSAKA-SHI 542
JAPANESE
INDEXED: CHE, JICST

SEN'I NEWS 8666
(TEXTILE NEWS)
OSAKU-FURITSU SEN'I GIJUTSU KENKYUJO
JAPANESE

SEN'I SEIHIN SHOHI KAGAKU 8667
(JOURNAL OF THE JAPAN RESEARCH ASSOCIATION FOR TEXTILE END-USES)
NIHON SEN'I SEIHIN SHOHI KAGAKKAI (THE JAPAN RESEARCH ASSOCIATION FOR TEXTILE END-USES)
C/O SHIN-OGIBASHI BUILDING, 19, SUEHIRO-CHO, KITA-KU, OSAKA-SHI 530
JAPANESE AND ENGLISH
INDEXED: JICST

SEN'I SENSHOKU JOZAI 8668
(TEXTILES DYEING. FOSTERING AGENTS)
JAPANESE
INDEXED: JICST

SEN'I TOKEI GEPPO 8669
(TEXTILE STATISTICS MONTHLY)
TSUSHO SANGYO-SHO DAIJIN KANBO CHOSA TOKEI-BU (RESEARCH AND STATISTICS DIVISION, MINISTER'S SECRETARIAT, MINISTRY OF INTERNATIONAL TRADE AND INDUSTRY)
C/O SEN'I KAIKAN, 3-9, NIHONBASHI HON-CHO, CHUO-KU, TOKYO 103
JAPANESE
INDEXED: JICST

SEN'I TOKEI NENPO 8670
(YEARBOOK OF TEXTILE STATISTICS)
TSUSHO SANGYO-SHO DAIJIN KANBO CHOSA TOKEI-BU (RESEARCH AND STATISTICS DIVISION, MINISTER'S SECRETARIAT, MINISTRY OF INTERNATIONAL TRADE AND INDUSTRY) AND NIHON SEN'I KYOGIKAI
C/O SEN'I KAIKAN, 3-9, NIHONBASHI HON-CHO, CHUO-KU, TOKYO 103
JAPANESE
INDEXED: BIB

SEN'IKAI 8671
(TEXTILE MAGAZINE)
NIPPON SEN'I KENKYUKAI (JAPAN TEXTILES AND FIBERS RESEARCH INSTITUTE)
C/O TOKO BUILDING, 1, EDOBORI, NISHI-KU, OSAKA-SHI 550
JAPANESE

SENKEN SHINBUN 8672
(TEXTILE RESEARCH NEWS)
SENKEN SHINBUNSHA
1-6, NIHONBASHI YOSHI-CHO, CHUO-KU, TOKYO 103
JAPANESE

SENKOSHI NYUSU 8673
(SENKOSHI NEWS, FUKUI PREFECTURAL TEXTILE RESEARCH INSTITUTE)
FUKUI-KEN SEN'I KOGYO SHIKENJO (FUKUI PREFECTURAL TEXTILE RESEARCH INSTITUTE)
10-1, 3-CHOME, BUNKYO, FUKUI-SHI 910
JAPANESE

SENSHI NYUSHU (YAMANASHI-KEN) 8674
(SENSHI NEWS, YAMANASHI PREFECTURE)
YAMANASHI-KEN SEN'I KOGYO SHIKENJO (TEXTILE INDUSTRY INSTITUTE OF PREFECTURAL YAMANASHI)
2095, SHIMOYOSHIDA, FUJIYOSHIDA-SHI 403
JAPANESE

SENSHOKU GIJUTSU 8675
(DYEING TECHNOLOGY)
JAPANESE
INDEXED: JICST

SENSHOKU JOHO 8676
(DYEING INFORMATION)
NIHON SENSHOKU SHINBUNSHA (JAPAN DYEING NEWSPAPER COMPANY)
JAPANESE

SENSHOKU JOHO 8677
(DYEING INFORMATION)
NIHON SENSHOKU KYOKAI (JAPAN DYEING ASSOCIATION)
JAPANESE

SENSHOKU KENKYU 8678
(DYEING STUDY)
KYOTO SENSHOKU KENKYUKAI
C/O KYOTO-SHI SENSHOKU SHIKENJO, KARASUMA-DORI KAMIDACHIYURIGARU, KAMIGYO-KU, KYOTO-SHI 602
JAPANESE
INDEXED: JICST

SENSHOKU KOGYO 8679
(DYEING INDUSTRY)
NIHON SENSHOKU KOGYO KENKYUKAI (SOCIETY OF JAPAN DYEING INDUSTRY RESEARCH)
C/O HANKYU BUILDING, 6, TENJINBASHI-SUJI, OYODO-KU, OSAKA-SHI 531
JAPANESE
INDEXED: CHE, JICST

SENSHOKU SHIKENJO NYUSU 8680
(NEWS OF THE DYEING AND WEAVING LABORATORY, KYOTO CITY)
KYOTO-SHI SENSHOKU SHIKENJO
KARASUMA-DORI KAMIDACHIYURIAGARU, KAMIGYO-KU, KYOTO-SHI 602
JAPANESE

SENSHOKU SHINPO 8681
(DYEING AND WEAVING NEWS)
SENSHOKU SHINPOSHA
C/O MIYOI BUILDING, TAKOYAKUSHI-KADO, SAKAI-MACHI-DORI, NAKAGYO-KU, KYOTO-SHI 604
JAPANESE

SENSHOKU SHUNJU 8682
(DYEING AND WEAVING YEAR-ROUND)
HAPPODO
JAPANESE

SENSHOKU SOKUHO 8683
(CURRENT REPORT OF TEXTILE DYEING AND FINISHING)
NIHON SENSHOKU KYOKAI (JAPAN TEXTILE FINISHERS ASSOCIATION)
4, YONBAN-CHO, CHIYODA-KU, TOKYO 100
JAPANESE

SHIGA-KEN SEN'I KOGYO SHIDOSHO DAYORI 8684
(LETTERS OF THE INSTITUTE FOR THE DEVELOPMENT OF THE SHIGA PREFECTURE TEXTILE INDUSTRY)
JAPANESE

SHIGA-KEN SEN'I KOGYO SHIDOSHO GYOMU HOKOKU 8685
(OPERATIONAL REPORT OF THE GUIDANCE INSTITUTE FOR TEXTILE INDUSTRIES IN SHIGA PREFECTURE)
SHIGA-KEN SEN'I KOGYO SHIDOSHO
27-39, MITSUYA MOTO-MACHI, NAGAHAMA-SHI 526
JAPANESE

SHINKEN TAIMUSU 8686
(SHINKEN TIMES, KOBE SILK CONDITIONING HOUSE)
NORIN-SHO KOBE KIITO KENSASHO (KOBE SILK CONDITIONING HOUSES, MINISTRY OF AGRICULTURE AND FORESTRY) AND KOBE KIITO TON'YA KYOKAI
C/O TORIHIKISHO BUILDING, 49, HARIMA-CHO, IKUTA-KU, KOBE-SHI 650
JAPANESE

SHIZUOKA-KEN HAMAMATSU SEN'I KOGYO SHIKENJO KENKYU HOKOKUSHO 8687
(REPORT OF THE HAMAMATSU TEXTILE RESEARCH INSTITUTE, SHIZUOKA PREFECTURE)
SHIZUOKA-KEN HAMAMATSU SEN'I KOGYO SHIKENJO
617, KITATERAJIMA-CHO, HAMAMATSU-SHI 430
JAPANESE

SHIZUOKA-KEN HAMAMATSU SEN'I KOGYO SHIKENJO NENPO 8688
(ANNUAL REPORT OF THE HAMAMATSU TEXTILE RESEARCH INSTITUTE, SHIZUOKA PREFECTURE)
SHIZUOKA-KEN HAMAMATSU SEN'I KOGYO SHIKENJO
617, KITATERAJIMA-CHO, HAMAMATSU-SHI 430
JAPANESE

SHIZUOKA-KEN NO SEN'I KOGYO 8689
(TEXTILE INDUSTRIES IN SHIZUOKA PREFECTURE)
SHIZUOKA-KEN SHOKO-BU
9-6, OTE-MACHI, SHIZUOKA-SHI 420
JAPANESE

SOME TO ORI 8690
(DYEING AND WEAVING)
SENSHOKU SHINPOSHA
C/O MIYOI BUILDING, TAKOYAKUSHI-KADO, SAKAI-MACHI-DORI, NAKAGYO-KU, KYOTO-SHI 604
JAPANESE

TDS 8691
(JOURNAL OF TEXTILE CHEMISTS, DYERS AND SURFACE CHEMISTS SOCIETY)
SEN'I SENSHOKU KAIMEN KASSEIZAI KENKYUKAI (TEXTILE CHEMISTS, DYERS AND SURFACE CHEMIST SOCIETY)
C/O NAGOYA-SHI KOGYO KENKYUJO, 3-24, ROKUBAN-CHO, ATSUTA-KU, NAGOYA-SHI 456
JAPANESE
INDEXED: JICST

TEIJIN INFORMATION 8692
TEIJIN K. K.
1-1, 2-CHOME, UCHISAIWAI-CHO, CHIYODA-KU, TOKYO 100
JAPANESE

TEIJIN TAIMUSU 8693
(TEIJIN TIMES)
TEIJIN K. K. KOHO SENDENSHITSU
C/O IINO BUILDING, 1-1, 2-CHOME, UCHISAIWAI-CHO, CHIYODA-KU, TOKYO 100
JAPANESE
INDEXED: JICST

TEXTILE JAPAN 8694
NIHON SEN'I SHINBUNSHA
3-9, NIHONBASHI HON-CHO, CHUO-KU, TOKYO 103
ENGLISH

TIDINGS 8695
NIHON KAYAKU KABUSHIKIKAISHA SENRYOBU (DYESTUFFS DIVISION, JAPAN CHEMICAL AND DRUGS COMPANY, LTD.)
JAPANESE

TOCHIGI-KEN SEN'I KOGYO SHIKENJO GYOMU HOKOKU 8696
(OPERATIONAL REPORT OF THE TOCHIGI-KEN TEXTILE INDUSTRIAL LABORATORY)
TOCHIGI-KEN SEN'I KOGYO SHIKENJO
2870, NISHIMIYA-MACHI, ASHIKAGA-SHI 326
JAPANESE

TOCHIGI-KEN SEN'I KOGYO SHIKENJO SHIKEN KENKYU HOKOKU 8697
(REPORT OF THE TOCHIGI-KEN TEXTILE INDUSTRIAL LABORATORY)
TOCHIGI-KEN SEN'I KOGYO SHIKENJO
2870, NISHIMIYA-MACHI, ASHIKAGA-SHI 326
JAPANESE

TOKYO SENKOSHI NYUSU 8698
(NEWSLETTER OF THE TOKYO METROPOLITAN TEXTILE RESEARCH INSTITUTE)
TOKYO-TORITSU SEN'I KOGYO SHIKENJO (THE TOKYO METROPOLITAN TEXTILE RESEARCH INSTITUTE)
19-1, 3-CHOME, MYOJIN-CHO, HACHIOJI-SHI, TOKYO 192
JAPANESE

TOKYO-TORITSU SEN'I KOGYO SHIKENJO KENKYU HOKOKU 8699
(REPORT OF THE TOKYO METROPOLITAN TEXTILE RESEARCH INSTITUTE)
TOKYO-TORITSU SEN'I KOGYO SHIKENJO (THE TOKYO METROPOLITAN TEXTILE RESEARCH INSTITUTE)
19-1, 3-CHOME, MYOJIN-CHO HACHIOJI-SHI, TOKYO 192
JAPANESE
INDEXED: JICST

TOYAMA-KEN SEN'I KOGYO SHIKENJO GYOMU HOKOKU 8700
(OPERATIONAL REPORT OF THE TOYAMA-KEN TEXTILE INDUSTRIAL LABORATORY)
TOYAMA-KEN SEN'I KOGYO SHIKENJO (TOYAMA-KEN TEXTILE INDUSTRIAL LABORATORY)
4846-2, FUKUNO-MACHI, HIGASHITONAMI-GUN, TOYAMA 939-15
JAPANESE

WAKORU NYUSU 8701
(WACOAL NEWS)
WAKORU K.K. (WACOAL COMPANY, LTD.)
29, NAKAJIMA-MACHI, KISSHOIN, MINAMI-KU, KYOTO-SHI 601
JAPANESE

YAMAGATA DAIGAKU KOGAKUBU SEN'I SEIZO KENKYU SHISETSU HOKOKU 8702
(BULLETIN OF THE TEXTILE INSTITUTE, FACULTY OF ENGINEERING, YAMAGATA UNIVERSITY)
YAMAGATA DAIGAKU SEN'I SEIZO KENKYU SHISETSU (TEXTILE INSTITUTE, FACULTY OF ENGINEERING, YAMAGATA UNIVERSITY)
3-16, 4-CHOME, JONAN, YONEZAWA-SHI 992
JAPANESE AND ENGLISH
INDEXED: CHE, JICST

YAMAGATA-KENRITSU TSURUOKA SEN'I KOGYO SHIKENJO GYOMU NARABINI KENKYU HOKOKU 8703
(REPORT OF TSURUOKA TEXTILE RESEARCH INSTITUTE OF YAMAGATA PREFECTURE)
YAMAGATA-KENRITSU TSURUOKA SEN'I KOGYO SHIKENJO
14-8, KACHUSHIN-MACHI, TSURUOKA-SHI 997
JAPANESE

YAMANASHI-KEN SEN'I KOGYO SHIKENJO KENKYU HOKOKUSHO 8704
(REPORT OF TEXTILE INDUSTRIAL RESEARCH INSTITUTE OF PREFECTURAL YAMANASHI)
YAMANASHI-KEN SEN'I KOGYO SHIKENJO (TEXTILE INDUSTRIAL RESEARCH INSTITUTE OF PREFECTURAL YAMANASHI)
2095, SHIMOYOSHIDA, FUJIYOSHIDA-SHI 403
JAPANESE

YAMANASHI-KEN SEN'I KOGYO SHIKENJO NENPO 8705
(RESEARCH REPORT OF YAMANASHI TEXTILE INDUSTRY INSTITUTE)
YAMANASHI-KEN SEN'I KOGYO SHIKENJO (TEXTILE INDUSTRY INSTITUTE OF PREFECTURAL YAMANASHI)
2095, SHIMOYOSHIDA, FUJIYOSHIDA-SHI 403
JAPANESE

YAMANASHI-KENRITSU MERIYASU KOGYO SHIDOSHO GYOMU NARABINI KENKYU HOKOKU 8706
(REPORT OF THE YAMANASHI PREFECTURAL KNITTING TECHNOLOGY RESEARCH INSTITUTE)
YAMANASHI-KENRITSU MERIYASU KOGYO SHIDOSHO (YAMANASHI PREFECTURAL KNITTING TECHNOLOGY RESEARCH INSTITUTE)
867, SATOYOSHI-CHO, KOFU-SHI 400
JAPANESE

YOKOHAMA KIITO KENSAJO KENSA JIHO 8707
(YOKOHAMA RAW SILK THREAD INSPECTION STATION)
YOKOHAMA KIITO TOIYA KYOKAI (YOKOHAMA RAW SILK THREAD WHOLESALERS ASSOCIATION)
JAPANESE

YOMO JIHO 8708
(BULLETIN OF THE JAPAN WOOL SPINNERS' ASSOCIATION)
NIHON YOMO BOSEKIKAI (JAPAN WOOL SPINNERS' ASSOCIATION)
C/O SEN'I KAIKAN, 3-9, NIHONBASHI HON-CHO, CHUO-KU, TOKYO 103
JAPANESE

YOMO TOKEI 8709
(STATISTICAL DATA ON WOOL INDUSTRY)
NIHON YOMO BOSEKIKAI (JAPAN WOOL SPINNERS' ASSOCIATION)
C/O SEN'I KAIKAN, 3-9, NIHONBASHI HON-CHO, CHUO-KU, TOKYO 103
JAPANESE AND ENGLISH
INDEXED: BIB

YORIITO JOHO 8710
(YARN NEWS)
NIHON YORIITO KOGYO KUMIAI RENGOKAI
(THE FEDERATION OF JAPAN YARN INDUSTRIAL UNIONS)
JAPANESE

YUSHUTSU KE ASA 8711
(WOOL AND JUTE NEWS FOR EXPORT)
NIHON KE ASA YUSHUTSU KUMIAI
4-4, BINGO-MACHI, HIGASHI-KU, OSAKA-SHI 541
JAPANESE

YUSHUTSU KINU KASEN 8712
(SILK AND SYNTHETIC TEXTILES FOR EXPORT)
NIHON KINU KASEN YUSHUTSU KUMIAI
4-4, BINGO-MACHI, HIGASHI-KU, OSAKA-SHI, 541
JAPANESE

Macromolecular Materials, Rubber and Plastics

EMARUJON JIHO 8713
(EMULSION TIMES)
EMARUJON JIHOSHA (EMULSION TIMES INCORPORATED)
C/O KOBUNSHI KANKOKAI, CHIEKOIN MARUTAMACHISAGARU, KAMIGYO-KU, KYOTO-SHI 602
JAPANESE
INDEXED: CHE

ENBI TO PORIMA 8714
(VINYLS AND POLYMERS)
PORIMA KOGYO KENKYUSHO (INSTITUTE OF POLYMER INDUSTRY, INCORPORATED)
C/O MARUNOUCHI YAESU BUILDING, 6-2, 2-CHOME, MARUNOUCHI, CHIYODA-KU, TOKYO 100
JAPANESE AND ENGLISH
INDEXED: CHE, JICST

ESAFUOMU NYUSU 8715
(FOAM NEWS)
ASAHI DAU KABUSHIKU KAISHA, HOSOZAI HANBAIBU ESAFUOMU TANTO (ASAHI DOW LTD., PACKAGING MATERIALS SALES DIVISION, FOAM MANAGER)
JAPANESE

ESURON TAIMUSU 8716
(ESURON TIMES)
KENRYOSHA (KENRYO INCORPORATED)
43, DAIKU-MACHI, KITA-KU, OSAKA-SHI 530
JAPANESE

FOMU TAIMUSU 8717
(FOAM TIMES)
FOMU TAIMUSUSHA (FOAM TIMES COMPANY, LTD.)
C/O HISAMATSU BUILDING, 7-1, 5-CHOME, SHINBASHI, MINATO-KU, TOKYO 105
JAPANESE

GEKKAN REZA 8718
(MONTHLY LEATHER)
PURASUTIKKU SHINPOSHA (PLASTICS JOURNAL OF COMPANY, LTD.)
3-11, 1-CHOME, YANAGIBASHI, TAITO-KU, TOKYO 111
JAPANESE
INDEXED: JICST

GOMU HOCHI SHINBUN 8719
(INFORMATION PAPER ON RUBBER)
GOMU HOCHI SHINBUNSHA
5-10, 2-CHOME, KANDA IWAMOTO-CHO, CHIYODA-KU, TOKYO 101
JAPANESE

GOMU SEIHIN TOKEI NENPO 8720
(YEARBOOK OF RUBBERS STATISTICS)
TSUSHO SANGYO-SHO DAIJIN KANBO CHOSA TOKEI-BU (RESEARCH AND STATISTICS DEPARTMENT, MINISTER'S SECRETARIAT, MINISTRY OF INTERNATIONAL TRADE AND INDUSTRY)
C/O TSUSHO SANGYO-SHO GINZA CHOSHA, 15-1, 6-CHOME, GINZA, CHUO-KU, TOKYO 104
JAPANESE

GOMU TAIMUSU 8721
(RUBBER TIMES, WEEKLY)
GOMU TAIMUSUSHA
3-11, 1-CHOME, YANAGIBASHI, TAITO-KU, TOKYO 111
JAPANESE

GOMU TOKEI GEPPO 8722
JAPANESE
INDEXED: JICST

GOSEI GOMU 8723
(SYNTHETIC RUBBER)
NIHON GOSEI GOMU K.K.
1-1, KYOBASHI, CHUO-KU, TOKYO 104
JAPANESE
INDEXED: JICST

GOSEI JUSHI 8724
(PLASTICS)
NIHON GOSEI JUSHI GIJUTSU KYOKAI (THE JAPAN SOCIETY OF PLASTICS TECHNOLOGY)
C/O TOKYO-TO CHUSHO KIGYO KAIKAN, 10-8, 2-CHOME, GINZA, CHUO-KU, TOKYO 104
JAPANESE
INDEXED: JICST

GOSEI JUSHI KENKYU KAIHO 8725
(BULLETIN OF KYOTO PLASTICS ASSOCIATION)
KYOTO GOSEIJUSHI KENKYUKAI (KYOTO PLASTICS ASSOCIATION)
C/O KYOTO-SHI KOGYO SHIKENJO, 1-2, MINAMIDA-CHO, NISHI-9-JO, MINAMI-KU, KYOTO-SHI 601
JAPANESE
INDEXED: JICST

GUDDOIYA NYUSU 8726
(GOODYEAR NEWS)
NIHON GUIDDOIYA KABUSHIKI KAISHA KOHOBU (JAPAN GOODYEAR, LTD., PUBLIC RELATIONS DIVISION)
JAPANESE

JAPAN PLASTICS (ENGLISH EDITION) 8727
KOGYO CHOSAKAI (INDUSTRY INFORMATION CENTER)
14-7, 2-CHOME, HONGO, BUNKYO-KU, TOKYO 113
ENGLISH
INDEXED: EXC

JAPAN PLASTICS AGE 8728
PURASUCHIKKUSU EJI K.K. (PLASTICS AGE COMPANY, LTD.)
1-118, AWAZA MINAMI-DORI, NISHI-KU, OSAKA-SHI 550
ENGLISH
INDEXED: CHE, RAP, JICST

JSR NEWS 8729
NIHON GOSEI GOMU K.K. TOKYO KENKYUSHO
7569, IKUTA, TAMA-KU, KAWASAKI-SHI 214
ENGLISH

KAIGAI KOBUNSHI KENKYU 8730
KOBUNSHI GAKKAI (THE SOCIETY OF POLYMER SCIENCE, JAPAN)
C/O HONSHU BUILDING, 12-8, 5-CHOME, GINZA, CHUO-KU, TOKYO 104
JAPANESE
INDEXED: JICST

KOBE GOMU TORIHIKIJO GEPPO 8731
(MONTHLY REPORT OF THE KOBE RUBBER EXCHANGE)
JAPANESE

KOBUNSHI 8732
(HIGH POLYMERS, JAPAN)
KOBUNSHI GAKKAI (THE SOCIETY OF POLYMER SCIENCE, JAPAN)
C/O HONSHU BUILDING, 12-8, 5-CHOME, GINZA, CHUO-KU, TOKYO 104
JAPANESE
INDEXED: CHE, JICST

KOBUNSHI GENRYO GIJUTSU KENKYU KUMIAISHI 8733
(HIGH POLYMER RAW MATERIALS TECHNOLOGY RESEARCH ASSOCIATION MAGAZINE)
JAPANESE

KOBUNSHI KAGAKU 8734
KOBUNSHI GAKKAI (THE SOCIETY OF POLYMER SCIENCE, JAPAN)
C/O HONSHU BUILDING, 12-8, 5-CHOME, GINZA, CHUO-KU, TOKYO 104
JAPANESE AND ENGLISH
INDEXED: BIB, CHE, RAP, SCI, JICST

KOBUNSHI KAGAKU NO TENBO 8735
(ANNUAL REVIEW (SOCIETY OF POLYMER SCIENCE, JAPAN))
KOBUNSHI GAKKAI (SOCIETY OF POLYMER SCIENCE, JAPAN)
12-8, 5-CHOME, GINZA, CHUO-KU, TOKYO 104
JAPANESE

KOBUNSHI KAKO 8736
(POLYMER APPLICATION)
KOBUNSHI KANKOKAI (HIGH POLYMER PUBLISHING ASSOCIATION)
CHIEKOIN MARUTAMACHISAGARU, KAMIGYO-KU, KYOTO-SHI 602
JAPANESE
INDEXED: CHE, JICST

KOBUNSHI KENKYUSHO NENPO 8737
(ANNUAL REPORT OF THE INSTITUTE FOR POLYMER RESEARCH)
KOBUNSHI KENKYUSHO (THE INSTITUTE FOR POLYMER RESEARCH)
1-1, MACHIKANEYAMA, TOYONAKA-SHI 560
JAPANESE AND ENGLISH
INDEXED: JICST

KOGYO GIJUTSU-IN SEN'I KOBUNSHI ZAIRYO KENKYUJO KENKYU HOKOKU 8738
(INDUSTRIAL TECHNOLOGY CENTER. RESEARCH REPORTS OF THE TEXTILE POLYMER MATERIALS RESEARCH INSTITUTE)
JAPANESE
INDEXED: JICST

KOMITSUDO PORIECHIRIN 8739
(HIGH DENSITY POLYETHYLINE)
PORIOREFUIN KONWAKAI, CHU TEIATSU PORIECHIRIN BUKAI (POLYOLEFIN ROUNDTABLE, MEDIUM-LOW PRESSURE POLYETHYLINE DIVISION)
JAPANESE

KYOKA PURASUCHIKKUSU 8740
(REINFORCED PLASTICS)
KYOKA PURASUCHIKKUSU GIJUTSU KYOKAI (THE SOCIETY OF REINFORCED PLASTICS, JAPAN)
C/O HANABUSA BUILDING, 14-2, 3-CHOME, GINZA, CHUO-KU, TOKYO 104
JAPANESE
INDEXED: CHE, JICST

MOULDERS 8741
NILLON PURASUCHIKKU SHIMPO-SHA (JAPAN PLASTICS NEW INFORMATION COMPANY)
JAPANESE AND ENGLISH

NIHON GOMU KOGYOKAI GEPPO 8742
(STATISTICAL BULLETIN OF JAPAN RUBBER MANUFACTURERS ASSOCIATION)
NIHON GOMU KOGYOKAI (JAPAN RUBBER MANUFACTURERS ASSOCIATION)
5-2, 1-CHOME, MOTOAKASAKA, MINATO-KU, TOKYO 107
JAPANESE
INDEXED: JICST

NIHON GOMU KYOKAISHI 8743
(JOURNAL OF THE SOCIETY OF RUBBER INDUSTRY, JAPAN)
NIHON GOMU KYOKAI (THE SOCIETY OF RUBBER INDUSTRY, JAPAN)
5-26, 1-CHOME, MOTOAKASAKA, MINATO-KU, TOKYO 107
JAPANESE, WITH ENGLISH SUMMARY
INDEXED: CHE, RAP, JICST

NIHON PURASUCHIKKUSU SHINPO 8744
(JAPAN PLASTICS JOURNAL)
NIHON PURASUCHIKKUSU SHINPOSHA (JAPAN PLASTICS JOURNAL LTD.)
38, MINAMISUMIYA-CHO, MINAMI-KU, OSAKA-SHI 552
JAPANESE

NIHON SETCHAKU KYOKAISHI 8745
(JOURNAL OF THE ADHESION SOCIETY OF JAPAN)
NIHON SETCHAKU KYOKAI (THE ADHESION SOCIETY OF JAPAN)
C/O OSAKA-FURITSU KOGYO GIJUTSU KENKYUSHO, ENOKOJIMA, NISHI-KU, OSAKA-SHI 550
JAPANESE, WITH EUROPEAN SUMMARY
INDEXED: CHE, JICST

ORU PURASUCHIKKUSU 8746
(ALL PLASTICS)
GOSEI JUSHI KOGYO SHINBUNSHA
9-33, TENJINBASHISUJI, OYODO-KU, OSAKA-SHI 531
JAPANESE
INDEXED: JICST

ORUGANO HAIRAITO 8747
(ORGANO HI-LITES)
ORUGANO HAIRAITOSHA (JAPAN ORGANO COMPANY)
5-16, 5-CHOME, HONGO, BUNKYO-KU, TOKYO 113
JAPANESE
INDEXED: JICST

PLASTICS AGE 8748
(PURASUCHIKKUSU EJI)
PURASUCHIKKUSE EJI K.K. (PLASTICS AGE COMPANY, LTD.)
1-118, AWAZA MINAMI-DORI, NISHI-KU, OSAKA-SHI 550
JAPANESE, WITH ENGLISH SUMMARY
INDEXED: CHE, JICST

PLASTICS INDUSTRY NEWS 8749
PORIMA KOGYO KENKYUSHO (INSTITUTE OF POLYMER INDUSTRY, INCORPORATED)
C/O MARUNOUCHI YAESU BUILDING, 6-2, 2-CHOME, MARUNOUCHI, CHIYODA-KU, TOKYO 100
ENGLISH
INDEXED: CHE, RAP

POLYMER JOURNAL 8750
KOBUNSHI GAKKAI (THE SOCIETY OF POLYMER SCIENCE, JAPAN)
12-8, 5-CHOME, GINZA, CHUO-KU, TOKYO 104
ENGLISH
INDEXED: CHE, ENG, SCI, JICST

POLYMER REPORT 8751
PORIMA KOGYO KENKYUSHO (INSTITUTE OF POLYMER INDUSTRY, INCORPORATED)
C/O MARUNOUCHI YAESU BUILDING, 6-2, 2-CHOME, MARUNOUCHI, CHIYODA-KU, TOKYO 100
ENGLISH
INDEXED: CHE, RAP

PORIECHIREN 8752
(POLYOLEFIN AND PLASTICS)
YUSEISHA
18-6, 1-CHOME, GINZA, CHUO-KU, TOKYO 104
JAPANESE
INDEXED: JICST

PORIMA NO TOMO 8753
(POLYMER FRIENDS FOR RUBBER, PLASTICS AND FIBER)
TAISEISHA
1-5, KYOBASHI, CHUO-KU, TOKYO 104
JAPANESE
INDEXED: JICST

PORIOREFIN JIHO 8754
(POLYOLEFIN TIMES)
PORIOREFIN JIHOSHA
1-19, EDOBORI KITA-DORI, NISHI-KU, OSAKA-SHI 550
JAPANESE

PROGRESS IN POLYMER SCIENCE, JAPAN 8755
KODANSHA LTD.
12-21 OTOWA 2-CHOME, BUNKYO-KU, TOKYO 112
ENGLISH
INDEXED: BIB, CHE

PURASUCHIKKU 8756
(PLASTICS)
MITSUBISHI JUSHI K.K.
5-12, 2-CHOME, MARUNOUCHI, CHIYODA-KU, TOKYO 100
JAPANESE
INDEXED: JICST

PURASUCHIKKU JIHO 8757
(PLASTICS TIMES)
YUSEISHA
18-6, 1-CHOME, GINZA, CHUO-KU, TOKYO 104
JAPANESE

PURASUCHIKKU MATERIARU 8758
(PLASTICS MATERIALS)
ZA PURASUCHIKKUSU (THE PLASTICS COMPANY, LTD.)
C/O ENOMOTO BUILDING, 5-12, 1-CHOME, FUJIMI, CHIYODA-KU, TOKYO 102
JAPANESE
INDEXED: JICST

PURASUCHIKKU SEIHIN TOKEI GEPPO 8759
(MONTHLY STATISTICS OF PLASTICS PRODUCTS)
TSUSHO SANGYO-SHO DAIJIN KANBO CHOSA TOKEI-BU (STATISTICS AND SURVEY DIVISION, MINISTER'S SECRETARIAT, MINISTRY OF INTERNATIONAL TRADE AND INDUSTRY)
C/O TOKYO KURABU BUILDING, 2-6, 3-CHOME, KASUMIGASEKI, CHIYODA-KU, TOKYO 100
JAPANESE

PURASUCHIKKU SEIHIN TOKEI NENPO 8760
(YEARBOOK OF PLASTICS STATISTICS)
TSUSHO SANGYO-SHO CHOSA TOKEI-BU (RESEARCH AND STATISTICS DEPARTMENT, MINISTER'S SECRETARIAT, MINISTRY OF INTERNATIONAL TRADE AND INDUSTRY)
NIHON PURASUCHIKKU KOGYO RENMEI
C/O TOKYO KURABU BUILDING, 2-6, 3-CHOME, KASUMIGASEKI, CHIYODA-KU, TOKYO 100
JAPANESE

PURASUCHIKKU SEIKEI KOGYO 8761
(PLASTIC MOLDING IN JAPAN)
ZEN NIHON PURASUCHIKKU SEIKEI KOGYO RENGOKAI (ASSOCIATION OF PLASTICS MOLDERS, JAPAN)
3-2, 1-CHOME, EBISUNISHI, SHIBUYA-KU, TOKYO 150
JAPANESE

PURASUCHIKKU SHINBUN 8762
(PLASTIC NEWS)
KASEIHIN NIPPOSHA
8-8, 3-CHOME, UCHIKANDA, CHIYODA-KU, TOKYO 101
JAPANESE

PURASUCHIKKU SUKOPU 8763
(PLASTICS SCOPE)
KOGYO KAIHATSU KENKYUKAI
C/O TOKO BUILDING, 3-32, KANDA SAKUMA-CHO, CHIYODA-KU, TOKYO 101
JAPANESE
INDEXED: JICST

PURASUCHIKKU TAIMUSU 8764
(PLASTIC TIMES OF JAPAN)
NIHON GOSEI JUSHI GIJUTSU KYOKAI (THE JAPAN SOCIETY OF PLASTICS TECHNOLOGY)
10-18, 2-CHOME, GINZA. CHUO-KU, TOKYO 104
JAPANESE

PURASUCHIKKUSU 8765
(JAPAN PLASTICS)
KOGYO CHOSAKAI (INDUSTRY INFORMATION CENTER)
14-7, 2-CHOME, HONGO, BUNKYO-KU, TOKYO 113
JAPANESE
INDEXED: CHE, RAP

PURASUCHIKKUSU 8766
(PLASTICS)
KOGYO CHOSA KAI (INDUSTRIAL RESEARCH ASSOCIATION)
JAPANESE

RABA DAIJESUTO 8767
(RUBBER DIGEST (A MONTHLY MAGAZINE OF RUBBER AND PLASTICS))
RABA DAIJESUTOSHA (RUBBER DIGEST PUBLISHING COMPANY)
45-17, 5-CHOME, MATSUBARA, SETAGAYA-KU, TOKYO 156
JAPANESE
INDEXED: CHE, JICST

RABA INDASUTORI 8768
(RUBBER INDUSTRY)
GOMU HOCHI SHINBUNSHA
4-19, NIHONBASHI KAKIGARA-CHO, CHUO-KU, TOKYO 103
JAPANESE

SEROFAN, CHOSA TO KENKYU 8769
(CELLOPHANE)
SEROHAN KOGYOKAI (JAPAN TRANSPARENT PAPER INDUSTRY ASSOCIATION)
C/O KAMI PARUPU KAIKAN, 5-11, 3-CHOME, GINZA, CHUO-KU, TOKYO 104
JAPANESE
INDEXED: JICST

SETCHAKU 8770
(ADHESION AND ADHESIVES)
KOBUNSHI KANKOKAI (HIGH POLYMER PUBLISHING COMPANY)
CHIEKOIN MARUTA-MACHISAGARU, KAMIGYO-KU, KYOTO-SHI 602
JAPANESE
INDEXED: CHE, JICST

SHIN'ETSU SHIRIKON REBYU 8771
(SHIN'ETSU SILICONE REVIEW)
SHIN'ETSU KAGAKU KOGYO K.K. SHIRIKON JIGYO-HONBU (SILICONE DEPARTMENT, SHIN'ETSU CHEMICAL INDUSTRY COMPANY, LTD.)
C/O ASAHI TOKAI BUILDING, 6-1, 2-CHOME, OTE-MACHI, CHIYODA-KU, TOKYO 100
JAPANESE
INDEXED: JICST

TAIYA NYUSU 8772
(TIRE NEWS)
BURIJISUTON TAIYA
(BRIDGESTONE TIRE)
JAPANESE
INDEXED: JICST

TENNEN GOMU 8773
(NATURAL RUBBER)
MAREISHIA GOMU BYURO (BUREAU OF MALAYSIAN RUBBER)
C/O 12 F. SEKAI BOEKI SENTA BUILDING, 4-1, 2-CHOME, HAMAMATSU-CHO, MINATO-KU, TOKYO 105
JAPANESE
INDEXED: JICST

TOKYO GOMU NIPPO 8774
(TOKYO RUBBER TIMES)
GOMU TAIMUSUSHA
1-3, YANAGIBASHI, TAITO-KU, TOKYO 111
JAPANESE

TOYOTA GOSEI GIHO 8775
(RUBBER, PLASTICS AND CORK)
TOYOTA GOSEI GIHO HENSHU IINKAI AND TOYOTA GOSEI K.K.
1, NAGAHATA, OCHIAI, KASUGA-MURA, NISHI-KASUGAI-GUN, AICHI 452
JAPANESE

YOKOHAMA GOMU K.K. KENKYUSHO GEPPO 8776
(MONTHLY REPORT OF THE RESEARCH LABORATORY, YOKOHAMA RUBBER COMPANY, LTD.)
YOKOHAMA GOMU K.K. KENKYUSHO (RESEARCH LABORATORY, YOKOHAMA RUBBER COMPANY, LTD.)
150, SHINJUKU, HIRATSUKA-SHI 254
JAPANESE

YOKOHAMA GOMU K.K. KENKYUSHO KENKYU HOKOKU 8777
(PAPERS FROM THE RESEARCH LABORATORY, YOKOHAMA RUBBER COMPANY, LTD.)
YOKOHAMA GOMU K.K. KENKYUSHO
150, SHINJUKU, HIRATSUKA-SHI 254
JAPANESE

YOKOHAMA GOMU NYUSU 8778
(YOKOHAMA RUBBER NEWS)
YOKOHAMA GOMU K.K. (YOKOHAMA RUBBER COMPANY, LTD.)
36-11, 5-CHOME, SHINBASHI, MINATO-KU, TOKYO 105
JAPANESE
INDEXED: JICST

Precision Mechanisms, Apparatus and Machines, Instruments, Computer

AOYAMA KONPYUTA SAIENSU 8779
(AOYAMA COMPUTER SCIENCE)
AOYAMA GAKUIN DAIGAKU FUZOKU JOHO KAGAKU KENKYU SENTA (INFORMATION SCIENCE RESEARCH CENTER, AOYAMA GAKUIN UNIVERSITY)
JAPANESE

BIT 8780
KYORITSU SHUPPAN K.K.
6-19, 4-CHOME, KOHINATA, BUNKYO-KU, TOKYO 112
JAPANESE
INDEXED: JICST

BULLETIN OF THE JAPAN SOCIETY OF PRECISION ENGINEERING 8781
SEIKI GAKKAI (THE JAPAN SOCIETY OF PRECISION ENGINEERING)
22-17, 2-CHOME, HYAKUNIN-CHO, SHINJUKU-KU, TOKYO 160
EUROPEAN
INDEXED: CHE, ENG, INS, ISM, MET, WAA

BULLETIN OF THE RESEARCH LABORATORY OF PRECISION MACHINERY AND ELECTRONICS 8782
ENGLISH
INDEXED: APP

CDR BULLETIN SOKUHO 8783
(CDR BULLETIN, QUICK ANNOUNCEMENT)
NIHON DENSHI KEISANKI
C/O SHIN-KOKUSAI-BUILDING, 4-1, 3-CHOME, MARUNOUCHI, CHIYODA-KU, TOKYO 100
JAPANESE

CITIZEN CLOCK 8784
SHICHIZUNSHA (SHICHIZUN COMPANY)
JAPANESE AND ENGLISH

COMPUTER NEWS 8785
KEIEI KINDAIKA KENKYUSHO (MANAGEMENT SCIENCE RESEARCH INSTITUTE)
C/O RAKUEN, NO. 101, 13-1, SHIBUYA-KU, TOKYO 150
JAPANESE

COMPUTER REPORT 8786
NIHON KEIEI KAGAKU KENKYUSHO (JAPAN MANAGEMENT SCIENCE INSTITUTE)
28-26, 4-CHOME, MINAMI-AOYAMA, MINATO-KU, TOKYO 107
JAPANESE
INDEXED: JICST

COMPUTER SALON 8787
JAPANESE AND ENGLISH

DECISION 8788
NIHON UNIBAKKU (JAPAN UNIVAC)
ENGLISH

DENKI SHIKENJO KEISAN SENTA KOHO 8789
(OFFICIAL REPORT OF THE COMPUTING CENTER, ELECTRONICS EXPERIMENT STATION)
DENKI SHIKENJO SOMUBU GYOMUKA (ELECTRONICS EXPERIMENT STATION, GENERAL AFFAIRS DIVISION, OPERATIONS SECTION)
JAPANESE

DENSHI KEISANKISHITSU NYUSU 8790
(COMPUTER ROOM NEWS)
KUMAMOTO DAIGAKU DENSHI KEISANKISHITSU (COMPUTER ROOM, KUMAMOTO UNIVERSITY) AND KUMAMOTO DAIGAKU
39-1, 2-CHOME, KUROKAMI, KUMAMOTO-SHI 860
JAPANESE

DENSHI KEISOKU 8791
(INSTRUMENTATION)
DENSHI KIKAI KOGYOKAI (ELECTRONIC INDUSTRIES ASSOCIATION OF JAPAN) AND DENSHI KEISOKU SHUPPANSHA
6-9, 2-CHOME, FUJIMI, CHIYODA-KU, TOKYO 102
JAPANESE
INDEXED: JICST

DETA PUROSESSHINGU 8792
(DATA PROCESSING)
NIPPON DATA PUROSESSHING KYOKAI (NIPPON DATA PROCESSING ASSOCIATION)
C/O SEIKYO KAIKAN, 1-13, 4-CHOME, SENDAGAYA, SHIBUYA-KU, TOKYO 151
JAPANESE

FACOM EDP RONBUNSHU 8793
(FACOM EDP PAPERS)
JAPANESE
INDEXED: JICST

FACOM JANARU 8794
(FACOM JOURNAL)
FUJITSU
JAPANESE AND ENGLISH

FACOM NYUSU 8795
(FACOM NEWS)
FUJITSU K.K. (FUJITSU COMPANY, LTD.)
6-1, 2-CHOME, MARUNOUCHI, CHIYODA-KU, TOKYO 100
JAPANESE

FURUIDIKUSU SHINPOJUMU KOEN RONBUNSHU 8796
(PAPERS OF THE FLUIDICS SYMPOSIUM)
KEISOKU JIDO SEIGYO GAKKAI (THE SOCIETY OF INSTRUMENT AND CONTROL ENGINEERS)
20, SHIBA KOTOHIRA-CHO, MINATO-KU, TOKYO 105
JAPANESE, WITH ENGLISH SUMMARY
INDEXED: JICST

GAKUSHU KONPYUTA 8797
(STUDY-COMPUTER)
GAKUSHU KENKYUSHA
40-5, 4-CHOME, KAMIKEDAI, OTA-KU, TOKYO 145
JAPANESE

GANKYO 8798
(SPECTACLES MONTHLY)
TOKYO GANKYO KOGAKU JIHOSHA
C/O SANKO BUILDING, 2-22, UENO, TAITO-KU, TOKYO 110
JAPANESE

GEKKAN KONPYUTOPIA 8799
(COMPUTOPIA)
KONPYUTA EIJISHA (COMPUTER AGE COMPANY, LTD.)
C/O 30TH F., KASUMIGASEKI BUILDING, 2-5, 3-CHOME, KASUMIGASEKI, CHIYODA-KU, TOKYO 100
JAPANESE

GUNOMON 8800
GUNOMONSHA
2-14, KANDA OGAWA-CHO, CHIYODA-KU, TOKYO 101
JAPANESE

HAITAKKU 8801
(HITAC)
HITACHI SEISAKUSHO (HITACHI LTD.)
JAPANESE

HOKKAIDO DAIGAKU OGATA KEISANKI SENTA NYUSU 8802
(CENTER OF LARGE SCALE COMPUTER NEWS, HOKKAIDO UNIVERSITY)
HOKKAIDO DAIGAKU OGATA KEISANKI SENTA
NISHI-5-CHOME, KITA-11-JO, KITA-KU, SAPPORO-SHI 060
JAPANESE

IBM REVIEW 8803
NIHON I.B.M. K.K.
2-12, 3-CHOME, ROPPONGI, MINATO-KU, TOKYO 105
JAPANESE

INFORMATION PROCESSING IN JAPAN 8804
JOHO SHORI GAKKAI (INFORMATION PROCESSING SOCIETY OF JAPAN)
C/O KIKAI SHINKO KAIKAN, 5-8, 3-CHOME, SHIBAKOEN, MINATO-KU, TOKYO 105
ENGLISH
INDEXED: MAT

IR NEWS 8805
NIHON INTANOSHONARU SEIRYUKI (JAPAN INTERNATIONAL RECTIFIER)
JAPANESE AND ENGLISH

JAPAN CAMERA TRADE NEWS 8806
ENGLISH

JAPAN COMPUTER NEWS 8807
ENGLISH

JAS KAIHO 8808
(JAS BULLETIN)
NIHON ODEIO KYOKAI (JAPAN AUDIO ORGANIZATION)
JAPANESE AND ENGLISH

JECC KONPYUTA DAINAMIKKU REPOTO 8809
(JECC COMPUTER DYNAMIC REPORT)
NIHON DENSHI KEISANKI K.K. (JAPAN ELECTRONIC COMPUTER COMPANY, LTD.)
4-1, 3-CHOME, MARUNOUCHI, CHIYODA-KU, TOKYO 100
JAPANESE

JEOL NEWS: ANALYTICAL INSTRUMENTS/APPLICATION 8810
NIHON DENSHI K.K. (JEOL LTD., INTERNATIONAL MARKETING DIVISION)
1418 NAKAGAMI-CHO, AKISHIMA-SHI, TOKYO 196
ENGLISH
INDEXED: CHE

JEOL NEWS: ELECTRON OPTICS INSTRUMENTS/APPLICATION 8811
JEOL LTD., INTERNATIONAL MARKETING DIVISION - NIHON DENSHI K.K.
1418 NAKAGAMI-CHO, AKISHIMA-SHI, TOKYO 196, JAPAN
ENGLISH
INDEXED: CHE

JIPDEC JANARU 8812
(JIPDEC JOURNAL)
NIHON JOHO SHORI KAIHATSU SENTA (JAPAN INFORMATION PROCESSING DEVELOPMENT CENTER)
C/O KIKAI SHINKO KAIKAN, 5-8, 3-CHOME, SHIBAKOEN, MINATO-KU, TOKYO 105
JAPANESE
INDEXED: JICST

JIPDEC REPORT 8813
NIHON JOHO SHORI KAIHATSU SENTA (JAPAN INFORMATION PROCESSING DEVELOPMENT CENTER)
C/O KIKAI SHINKO KAIKAN, 5-8, 3-CHOME, SHIBAKOEN, MINATO-KU, TOKYO 105
ENGLISH

JOHO 8814
(INFORMATION)
TOHOKU KOGYO DAIGAKU KEISAN SENTA (TOHOKU INSTITUTE OF TECHNOLOGY COMPUTER CENTER)
19, KOSHIJI, NAGA-MACHI, SENDAI-SHI 982
JAPANESE
INDEXED: JICST

JOHO KAGAKU 8815
(INFORMATION, INFORMATION PROCESSING, INFORMATION RETRIEVAL)
JOHO KAGAKU KENKYUSHO (INSTITUTE FOR INFORMATION SCIENCE)
6-21, 1-CHOME, GOTOKUJI, SETAGAYA-KU, TOKYO 154
JAPANESE
INDEXED: CHE, JICST

JOHO KAGAKU GIJUTSU KENKYU SHUKAI HAPPYO RONBUNSHU 8816
(PROCEEDINGS OF THE NATIONAL CONVENTION FOR THE STUDY ON INFORMATION AND DOCUMENTATION)
NIHON KAGAKU GIJUTSU JOHO SENTA (THE JAPAN INFORMATION CENTER OF SCIENCE AND TECHNOLOGY)
1-1, 2-CHOME, NAGATA-CHO, CHIYODA-KU, TOKYO 100
JAPANESE, WITH ENGLISH SUMMARY
INDEXED: JICST

JOHO SHORI 8817
(JOURNAL OF THE INFORMATION PROCESSING SOCIETY OF JAPAN)
JOHO SHORI GAKKAI (INFORMATION PROCESSING SOCIETY OF JAPAN)
C/O KIKAI SHINKOKAIKAN, 5-8, 3-CHOME, SHIBAKOEN, MINATO-KU, TOKYO 105
JAPANESE, WITH ENGLISH SUMMARY
INDEXED: INS, JICST

JOHO SHORI GAKKAI GETSUREIKAI SHIRYO 8818
(MATERIALS OF THE MONTHLY MEETING OF INFORMATION PROCESSING SOCIETY OF JAPAN)
JAPANESE
INDEXED: JICST

JOHO SHORI GAKKAI PUROGURAMINGU SHINPOJIAMA IIANKAI HOKOKUSU 8819
(REPORT OF THE COMMITTEE FOR THE PROGRAMMING SYMPOSIUM OF INFORMATION PROCESSING SOCIETY OF JAPAN)
JAPANESE
INDEXED: JICST

JOHO SHORI GAKKAI TAIKAI KOEN YOKOSHU 8820
(PREPRINTS OF THE SYMPOSIUM OF INFORMATION PROCESSING SOCIETY OF JAPAN)
JOHO SHORI GAKKAI (INFORMATION PROCESSING SOCIETY OF JAPAN)
5-8, 3-CHOME, SHIBAKOEN, MINATO-KU, TOKYO 105
JAPANESE
INDEXED: JICST

JOHO SHORI KENKYU 8821
(INFORMATION PROCESSING RESEARCH JOURNAL)
DENRYOKU CHUO KENKYUJO, JOHO SHORI KENKYU SENTA (INFORMATION PROCESSING RESEARCH CENTER, GENERAL RESEARCH INSTITUTE OF ELECTRIC POWER INDUSTRY)
C/O OTE-MACHI-BUILDING, 6-1, 1-CHOME, OTE-MACHI, CHIYODA-KU, TOKYO 100
JAPANESE
INDEXED: JICST

JOHOKA SEKKEI 8822
(DESIGN FOR INFORMATION PROCESSING)
SYSTEMS COMPANY, LTD.
1-10, 1-CHOME, SHINJUKU, SHINJUKU-KU, TOKYO 160
JAPANESE
INDEXED: JICST

JOURNAL OF ELECTRON MICROSCOPY 8823
NIHON DENSHI KENBIKYO GAKKAI (JAPANESE SOCIETY OF ELECTRON MICROSCOPY)
C/O GAKKAI SENTA BUILDING, 4-16, 2-CHOME, YAYOI, BUNKYO-KU, TOKYO 113
ENGLISH
INDEXED: BIB, BIO, CHE, EXC, IND, INS, SCI, JICST

KEIRYO KANRI, KEISOKU KEISO 8824
(JOURNAL OF THE INSTRUMENTATION CONTROL ASSOCIATION, JAPAN)
KEIRYO KANRI KYOKAI (THE INSTRUMENTATION CONTROL ASSOCIATION, JAPAN)
C/O NIHON KEIRYO KAIKAN, 25-1, NANDO-CHO, SHINJUKU-KU, TOKYO 162
JAPANESE
INDEXED: JICST

KEIRYO KENKYUSHO HOKOKU 8825
(REPORT OF THE NATIONAL RESEARCH LABORATORY OF METROLOGY)
TSUSHO SANGYO-SHO KOGYO GIJUTSU-IN KEIRYO KENKYUSHO (NATIONAL RESEARCH LABORATORY OF METROLOGY, AGENCY OF INDUSTRIAL SCIENCE AND TECHNOLOGY, MINISTRY OF INTERNATIONAL TRADE AND INDUSTRY)
10-4, 1-CHOME, KAGA, ITABASHI-KU, TOKYO 173
JAPANESE, WITH EUROPEAN SUMMARY
INDEXED: CHE, JICST

KEIRYO KENKYUSHO NENPO 8826
(ANNUAL REPORT OF NATIONAL RESEARCH LABORATORY OF METROLOGY)
TSUSHO SANGYO-SHO KOGYO GIJUTSU-IN KEIRYO KENKYUSHO (NATIONAL RESEARCH LABORATORY OF METROLOGY, AGENCY OF INDUSTRIAL SCIENCE AND TECHNOLOGY, MINISTRY OF INTERNATIONAL TRADE AND INDUSTRY)
10-4, 1-CHOME, KAGA, ITABASHI-KU, TOKYO 173
JAPANESE
INDEXED: JICST

KEIRYOKEN NYUSU 8827
(NEWS OF THE NATIONAL RESEARCH LABORATORY OF METROLOGY)
TSUSHO SANGYO-SHO KOGYO GIJUTSU-IN KEIRYO KENKYUSHO (NATIONAL RESEARCH LABORATORY OF METROLOGY, AGENCY OF INDUSTRIAL SCIENCE AND TECHNOLOGY, MINISTRY OF INTERNATIONAL TRADE AND INDUSTRY)
10-4, 1-CHOME, KAGA, ITABASHI-KU, TOKYO 173
JAPANESE
INDEXED: JICST

KEISO 8828
(INSTRUMENTATION)
KOGYO GIJUTSUSHA
2-19, KANDA TSUKASA-CHO, CHIYODA-KU, TOKYO 101
JAPANESE
INDEXED: CHE, JICST

KEISOKU GIJUTSU 8829
(INSTRUMENTATION AND AUTOMATION)
NIHON KOGYO SHUPPAN K.K.
3-18, 2-CHOME, NISHIKANDA, CHIYODA-KU, TOKYO 101
JAPANESE
INDEXED: CHE, JICST

KEISOKU JIDO SEIGYO GAKKAI RONBUNSHU 8830
(TRANSACTIONS OF THE SOCIETY OF INSTRUMENT AND CONTROL ENGINEERS)
KEISOKU JIDO SEIGYO GAKKAI (THE SOCIETY OF INSTRUMENT AND CONTROL ENGINEERS)
C/O KOTOHIRA ANNEX, MORI BUILDING, 39, SHIBA KOTOHIRA-CHO, MINATO-KU, TOKYO 105
JAPANESE, WITH ENGLISH SUMMARY
INDEXED: JICST

KEISOKU TO SEIGYO 8831
(JOURNAL OF THE SOCIETY OF INSTRUMENT AND CONTROL ENGINEERS)
KEISOKU JIDO SEIGYO GAKKAI (THE SOCIETY OF INSTRUMENT AND CONTROL ENGINEERS)
C/O KOTOCHIRA ANNEX, MORI BUILDING, 39, SHIBA KOTOCHIRA-CHO, MINATO-KU, TOKYO 105
JAPANESE, WITH ENGLISH SUMMARY
INDEXED: INS, JICST

KINDAI MEGANE 8832
(GLASSES)
KINDAI KOGAKU SHUPPANSHA
C/O DAI-1 HIGASHI BUILDING, 14-13, 1-CHOME, HIGASHI UENO, TAITO-KU, TOKYO 110
JAPANESE

KOGAKU GIJUTSU KONTAKUTO 8833
(OPTICAL ENGINEERING CONTACT)
KOGAKU KOGYO GIJUTSU KENKYU KUMIAI (JAPAN OPTICAL ENGINEERING RESEARCH ASSOCIATION)
C/O MORI BUILDING 9TH, 1-3, SHIBA ATAGO-CHO, MINATO-KU, TOKYO 105
JAPANESE
INDEXED: JICST

KOGAKU SANGYO SHINBUN 8834
(JAPAN BINOCULARS AND SPECTACLES NEWS)
KOGAKU SANGYO SHINBUNSHA (JAPAN OPTICAL NEWS COMPANY)
38-5, OYAMA KANAI-CHO, ITABASHI-KU, TOKYO 173
JAPANESE

KOKUSAI TOKEI TSUSHIN 8835
(HOROLOGICAL INTERNATIONAL CORRESPONDENCE)
KOKUSAI TOKEI TSUSHINSHA
10-4, 2-CHOME, UCHI KANDA, CHIYODA-KU, TOKYO 101
JAPANESE

KONPYUTA 8836
(COMPUTER)
HOKKAIDO-CHO SOMUBU DENSHI KEISANKA (HOKKAIDO DEVELOPMENT AGENCY, GENERAL AFFAIRS OFFICE, ELECTRONIC COMPUTING SECTION)
JAPANESE

KONPYUTA JUYOGO SAKUNSHU 8837
(KWIC INDEX OF INFORMATION SCIENCE)
JOHO KAGAKU KENKYUSHO (INSTITUTE FOR INFORMATION SCIENCE)
6-21, 1-CHOME, GOTOKUJI, SETAGAYA-KU, TOKYO 154
JAPANESE

KONPYUTA SEMINA 8838
(COMPUTER SEMINAR)
SHIN NIHON TOSHO (NEW JAPAN LIBRARY)
JAPANESE

KONTAKUTO POINTO 8839
(CONTACT POINT)
YAMATAKE HANEUERU K.K. (YAMATAKE HONEYWELL COMPANY, LTD.)
C/O MARUNOUCHI YAESU BUILDING, 2-6, MARUNOUCHI, CHIYODA-KU, TOKYO 100
JAPANESE

KOSEIDO 8840
(HIGH PRECISION)
JAPANESE
INDEXED: JICST

MISHIN 8841
(HOUSEHOLD SEWING MACHINE)
NIHON KATEIYO MISHIN KOGYOKAI (JAPAN HOUSEHOLD SEWING MACHINE MANUFACTURERS ASSOCIATION)
C/O NIHON MISHIN SENTA, 13, SAKAMACHI, SHINJUKU-KU, TOKYO 160
JAPANESE

MPC REPOTO 8842
(MPC REPORT)
MITSUBISHI PURESHIJON K.K. (MITSUBISHI PRECISION COMPANY, LTD.)
345, KAMIMACHIYA, KAMAKURA-SHI 247
JAPANESE

MUGENDAI 8843
(INFINITY)
NIHON AI-BI-EMU (JAPAN IBM)
JAPANESE

NAGANO-KEN SEIMITSU KOGYO SHIKENJO JIGYO GAIYO 8844
(REPORT OF THE NAGANO-KEN PRECISION INDUSTRY EXPERIMENTAL STATION)
NAGANO-KEN SEIMITSU KOGYO SHIKENJO (NAGANO-KEN PRECISION INDUSTRY EXPERIMENTAL STATION)
9959, OKAYA-SHI 394
JAPANESE

NAGOYA DAIGAKU KEISANKI SENTA NYUSU 8845
(COMPUTATION CENTER NEWS)
NAGOYA DAIGAKU OGATA KEISANKI SENTA (COMPUTER CENTER, NAGOYA UNIVERSITY)
FURO-CHO, CHIKUSA-KU, NAGOYA-SHI 464
JAPANESE

NIHON DENKI KEISOKUKI KOGYOKAI KAIHO 8846
(JOURNAL OF THE JAPAN ELECTRIC MEASURING INSTRUMENTS MANUFACTURER'S ASSOCIATION)
NIHON DENKI KEISOKUKI KOGYOKAI (JAPAN ELECTRIC MEASURING MANUFACTURER'S ASSOCIATION)
20, SHIBA KOTOHIRA-CHO, MINATO-KU, TOKYO 105
JAPANESE

NIHON DENSHI KEISANKI K.K. CHOSA KIHO 8847
(EDP MARKET RESEARCH QUARTERLY)
NIHON DENSHI KEISANKI K.K. (JAPAN ELECTRONIC COMPUTER COMPANY, LTD.)
4-1, 3-CHOME, MARUNOUCHI, CHIYODA-KU, TOKYO 100
JAPANESE
INDEXED: JICST

NIHON KEIRYO SHINPO 8848
(JAPAN MEASURING WEEKLY)
NIHON KEIRYO SHINPOSHA
2-7, 4-CHOME, KUDANKITA, CHIYODA-KU, TOKYO 102
JAPANESE

NIHON KIKINZOKU TOKEI SHINBUN 8849
(JAPAN PRECIOUS METALS AND WATCH NEWS)
NIHON KIKINZOKU TOKEI SHINBUNSHA
C/O TOKYO KIKO KAIKAN, 19-16, 2-CHOME, NEGISHI, TAITO-KU, TOKYO 110
JAPANESE

NIHON TOKEI GAKKAISHI 8850
(JOURNAL OF THE HOROLOGICAL INSTITUTE OF JAPAN)
NIHON TOKEI GAKKAI (THE HOROLOGICAL INSTITUTE OF JAPAN)
C/O NIHON GAKKAI JIMU CENTER, 4-16, 2-CHOME, YAYOI, BUNKYO-KU, TOKYO 113
JAPANESE, WITH EUROPEAN SUMMARY

NIKON TECHNICAL JOURNAL 8851
NIHON KOGAKU KOGYO
2-3, 3-CHOME, MARUNOUCHI, CHIYODA-KU, TOKYO 100
JAPANESE
INDEXED: JICST

OSAKA DAIGAKU OGATA KEISANKI SENTA NYUSU 8852
(NEWS BULLETIN OF COMPUTER CENTER, OSAKA UNIVERSITY)
OSAKA DAIGAKU OGATA KEISANKI SENTA (COMPUTATION CENTER, OSAKA UNIVERSITY)
YAMADAKAMI, SUITA-SHI 565
JAPANESE

OTEMESHON GIJUTSU 8853
(AUTOMATION TECHNOLOGY)
JAPANESE
INDEXED: JICST

PENTON'S MACHINE DESIGN 8854
INTUPRESU (INTER-PRESS)
JAPANESE AND ENGLISH

PICO 8855
(PACKAGED INFORMATION ON COMPUTERS)
PICO JOHO CHOSA KENKYUKAI AND KONPYUTA EIJISHA (COMPUTER AGE COMPANY, LTD.)
C/O KASUMIGASEKI BUILDING, 2-5, 3-CHOME, KASUMIGASEKI, CHIYODA-KU, TOKYO 100
JAPANESE

PUROGURAMINGU SHINPOJUMU HOKOKUSHU 8856
(REPORT OF THE PROGRAMMING SYMPOSIUM)
JOHO SHORI GAKKAI (INFORMATION PROCESSING SOCIETY OF JAPAN)
5-8, 3-CHOME, SHIBAKOEN, MINATO-KU, TOKYO 105
JAPANESE

REPORT OF THE COMPUTER CENTER, UNIVERSITY OF TOKYO 8857
TOKYO DAIGAKU OGATA KEISANKI SENTA (THE COMPUTER CENTER, UNIVERSITY OF TOKYO)
11-6, 2-CHOME, YAYOI, BUNKYO-KU, TOKYO 113
ENGLISH
INDEXED: CHE, MAT, JICST

REPORT OF DATA PROCESSING CENTER, KYOTO UNIVERSITY 8858
KYOTO DAIGAKU OGATA KEISANKI SENTA (DATA PROCESSING CENTER, KYOTO UNIVERSITY)
YOSHIDA HON-CHO, SAKYO-KU, KYOTO-SHI 606
ENGLISH

SAN'EI REPOTO, RIKO KANKEI 8859
(SAN'EI REPORT, SCIENCE AND TECHNOLOGY)
SAN'EI SOKKI K.K. (SAN'EI INSTRUMENT COMPANY, LTD.)
1-57, TENJIN-CHO, KODAIRA-SHI TOKYO 187
JAPANESE

SEIKI GIJUTSU NYUSU 8860
(SEIKI TECHNICAL NEWS)
HITACHI SEIKI K.K. (HITACHI MACHINERY COMPANY, LTD.)
C/O HIGASHIYAMA BUILDING, 4-9, NIHONBASHI HON-CHO, SHUO-KU, TOKYO 103
JAPANESE

SEIMITSU KAKO 8861
(PRECISION PROCESSING)
AGLNE
C/O SHUWA AOYAMA REJIDENSU, 3-10, 3-CHOME, SHIBUYA, SHIBUYA-KU, TOKYO 150
JAPANESE
INDEXED: JICST

SEIMITSU KIKAI 8862
(JOURNAL OF THE JAPAN SOCIETY OF PRECISION ENGINEERING)
SEIKI GAKKAI (THE JAPAN SOCIETY OF PRECISION ENGINEERING)
22-7, 2-CHOME, HYAKUNIN-CHO, SHINJUKU-KU, TOKYO 160
JAPANESE, WITH ENGLISH SUMMARY
INDEXED: CHE, ISM, INS, JICST

SHIMYURESHON GIJUTSU KENKYUKAI RONBUNSHU 8863
(RESEARCH PAPERS OF THE JSST)
SHIMYURESHON GIJUTSU KENKYUKAI (JAPAN SOCIETY FOR SIMULATION TECHNOLOGY)
C/O NIHON KAGAKU GIJUTSU RENMEI, 10-11, 5-CHOME, SENDAGAYA, SHIBUYA-KU, TOKYO 151
JAPANESE, WITH ENGLISH SUMMARY

SOFUTO UEA GIJUTSU 8864
(SOFT-WARE TECHNOLOGY)
NIHON KONPYUTA KYOKAI (JAPAN COMPUTER ASSOCIATION)
JAPANESE

SOKEN KIYO 8865
(REPORTS OF RESEARCH ACTIVITIES ON COMPUTER SCIENCE AND TECHNOLOGY)
NIPPON UNIVAC SOGO KENKYUSHO
20, SHIBA NISHIKUBO AKEFUNE-CHO, MINATO-KU, TOKYO 105
JAPANESE AND ENGLISH

SOUND MATE 8866
GIJUTSU SHINBUNSHA (TECHNICAL NEWSPAPER COMPANY)
JAPANESE AND ENGLISH

SYSTEMS MONTHLY 8867
AIOEMU
JAPANESE AND ENGLISH

TOHOKU DAIGAKU KAGAKU KEISOKU KENKYUSHO HOKOKU 8868
(BULLETIN OF THE RESEARCH INSTITUTE FOR SCIENTIFIC MEASUREMENTS, TOHOKU UNIVERSITY)
TOHOKU DAIGAKU KAGAKU KEISOKU KENKYUSHO (RESEARCH INSTITUTE FOR SCIENTIFIC MEASUREMENTS, TOHOKU UNIVERSITY)
19-1, SANJO-MACHI, SENDAI-SHI 980
JAPANESE, WITH EUROPEAN SUMMARY
INDEXED: CHE, INS, JICST

TOKYO DAIGAKU OGATA KEISANKI SENTA NENPO 8869
(ANNUAL REPORT OF THE UNIVERSITY OF TOKYO COMPUTER CENTER)
TOKYO DAIGAKU OGATA KEISANKI SENTA (UNIVERSITY OF TOKYO COMPUTER CENTER)
11-16, 2-CHOME, YAYOI, BUNKYO-KU, TOKYO 113
JAPANESE
INDEXED: JICST

TOKYO DAIGAKU OGATA KEISANKI SENTA NYUSU 8870
(NEWS OF COMPUTER CENTER, UNIVERSITY OF TOKYO)
TOKYO DAIGAKU OGATA KEISANKI SENTA
11-16, 2-CHOME, YAYOI, BUNKYO-KU, TOKYO 113
JAPANESE

TOKYO KEIKI NYUSU 8871
(TOKYO KEIKI NEWS)
TOKYO KEIKI SEIZOJO
1-31, NISHIGOTANDA, SHINAGAWA-KU, TOKYO 141
JAPANESE

TOKYO KOGYO DAIGAKU JOHO SHORI SENTA NENPO 8872
(ANNUAL REPORT OF THE INFORMATION PROCESSING CENTER, TOKYO INSTITUTE OF TECHNOLOGY)
TOKYO KOGYO DAIGAKU JOHO SHORI SENTA
12-1, 2-CHOME, OOKAYAMA, MEGURO-KU, TOKYO 152
JAPANESE

TOSBAC NEWS 8873
TOKYO SHIBAURA DENKI KASUTOMA RIRESHONZUBU (CUSTOMER RELATIONS DIVISION, TOKYO SHIBAURA ELECTRIC COMPANY, LTD)
2, NISHIKUBO SAKURAGAWA-MACHI, MINATO-KU TOKYO 105
JAPANESE

UNICON NEWS 8874
NIHON GAKUJUTSU SHINKOKAI DENSHI KEISANKI GAKUJUTSU RIYO IINKAI (JAPAN SOCIETY FOR THE PROMOTION OF SCIENCE)
1-2, 2-CHOME, HITOTSUBASHI, CHIYODA-KU, TOKYO 100
JAPANESE

YAMATAKE HANEUERU AUTOMATION 8875
(YAMATAKE HONEYWELL AUTOMATION)
JAPANESE

Building Industry, Physical Planning, Landscape, Architecture

AAIA 8876
KENCHIKU BIJUTSU KOGYO KAI (INDUSTRIAL SOCIETY FOR BEAUTIFUL ARCHITECTURE)
JAPANESE AND ENGLISH

ARCHITECTURE AND URBANISM 8877
JAPANESE

ARGUS, GASU TO KENCHIKU 8878
(GAS AND ARCHITECTURE)
TOKYO GASU K.K. (TOKYO GAS COMPANY, LTD.)
2-16, 1-CHOME, YAESU, CHUO-KU, TOKYO 103
JAPANESE
INDEXED: JICST

ATARASHII JUTAKU 8879
(NEW HOUSE)
JITSUGYO NO NIHONSHA
3-9, 1-CHOME, GINZA, CHUO-KU, TOKYO 104
JAPANESE

ATARASHII SUMAI NO SEKKEI 8880
(NEW HOUSE DESIGNING)
SANKEI SHINBUN SHUPPAN-KYOKU
3-15, KANDA NISHIKI-CHO, CHIYODA-KU, TOKYO 101
JAPANESE

AWCOM 8881
AWCOM NIHON HONSHITSU KOZO ZAIRYO KYOKAI (AWCOM JAPANESE BASIC STRUCTURAL MATERIALS ASSOCIATION)
JAPANESE
INDEXED: JICST

BIRU MENTENANSU 8882
(BUILDING MAINTENANCE)
ZENKOKU BIRU MENTENANSU KYOKAI
6-12, 3-CHOME, AKASAKA, MINATO-KU, TOKYO 107
JAPANESE

BIRUDINGU RETA 8883
(BUILDING LETTER)
NIHON KENCHIKU SENTA
14-16, 1-CHOME, HARUMI, CHUO-KU, TOKYO 104
JAPANESE
INDEXED: JICST

BIRUJINGU 8884
(BUILDING)
NIHON BIRUJINGU KYOKAI RENGOKAI
4-1, 2-CHOME, MARUNOUCHI, CHIYODA-KU, TOKYO 100
JAPANESE

BRI RESEARCH PAPER 8885
KENSETSU-SHO KENCHIKU KENKYUSHO (BUILDING RESEARCH INSTITUTE, MINISTRY OF CONSTRUCTION)
28-8, 3-CHOME, HYAKUNIN-CHO, SHINJUKU-KU, TOKYO 160
ENGLISH

BYOIN KENCHIKU 8886
(JOURNAL OF THE JAPAN INSTITUTE OF HOSPITAL ARCHITECTURE)
NIHON BYOIN KENCHIKU KYOKAI (JAPAN INSTITUTE OF HOSPITAL ARCHITECTURE)
C/O BYOIN KANRI KENKYUSHO, 1, TOYAMA-CHO, SHINJUKU-KU, TOKYO 162
JAPANESE

CHIIKI KAIHATSU 8887
(AREA DEVELOPMENT)
NIHON CHIIKI KAIHATSU SENTA (JAPAN CENTER FOR AREA DEVELOPMENT RESEARCH)
1-1, 2-CHOME, UCHISAIWAI-CHO, CHIYODA-KU, TOKYO 100
JAPANESE
INDEXED: JICST

CHIIKI KAIHATSU DETA SOKUHO 8888
(CURRENT REPORT ON REGIONAL DEVELOPMENT DATA)
DETA LABU (DATA LABORATORY)
JAPANESE

CHIIKI KENKYU 8889
(REGIONAL (STUDIES) RESEARCH)
RISSHO CHIRI GAKKAI (RISSHO ACADEMY OF GEOGRAPHY)
JAPANESE

CHOSA KENKYU KIHO 8890
(SURVEY RESEARCH REPORT)
NIHON JUTUKU KODAN KENCHIKU BU CHOSA KENKYU KA
JAPANESE

CONTENTS 8891
OSAKA-FU KENCHIKUSHIKAI JOHO SENTA (INFORMATION CENTER, OSAKA ARCHITECTURAL CONSULTANTS ASSOCIATION)
C/O NAKAHIRO BUILDING, 2-41, BAKURO-CHO, HIGASHI-KU, OSAKA-SHI 541
JAPANESE

DITERU 8892
(DETAIL (MAGAZINE FOR ARCHITECTS AND ENGINEERS))
SHOGOKUSHA
16-9, 2-CHOME, HIRAKAWA-CHO, CHIYODA-KU, TOKYO 102
JAPANESE

EIZEN NENPO (KOBE-SHI) 8893
(ANNUAL REPORT OF BUILDING AND REPAIRING, KOBE CITY)
KOBE-SHI JUTAKU-KYOKU EIZEN-BU (MAINTENANCE DIVISION, BUREAU OF BUILDINGS, KOBE CITY)
6-7, KANO-CHO, IKUTA-KU, KOBE-SHI 650
JAPANESE

EIZEN NENPO (OSAKA-FU) 8894
(ANNUAL REPORT OF BUILDING AND REPAIRING, OSAKA PREFECTURE)
OSAKA-FU KENCHIKU-BU (BUILDINGS DIVISION, OSAKA PREFECTURAL GOVERNMENT)
OTEMAEHO-CHO, HIGASHI-KU, OSAKA-SHI 540
JAPANESE

EIZEN NENPO (TOKYO-TO) 8895
(ANNUAL REPORT, BUILDING AND REPAIRING, TOKYO METROPOLITAN GOVERNMENT)
TOKYO-TO EIZEN KOJI-BU (BUILDING PROJECT SECTION, BUILDING DIVISION, BUREAU OF FINANCE, TOKYO METROPOLITAN GOVERNMENT)
5-1, 3-CHOME, MARUNOUCHI, CHIYODA-KU, TOKYO 100
JAPANESE

EIZEN NYUSU 8896
(BUILDING AND REPAIRS NEWS)
JAPANESE
INDEXED: JICST

GAKKO KENCHIKU NENPO, KOKURITSU GAKKO-HEN 8897
(ANNUAL REPORT ON SCHOOL BUILDINGS, NATIONAL SCHOOLS)
MONBU-SHO KANRI-KYOKU SHISETSU-BU (DEPARTMENT OF FACILITIES, BUREAU OF ADMINISTRATION, MINISTRY OF EDUCATION)
2-2, 3-CHOME, KASUMIGASEKI, CHIYODA-KU, TOKYO 100
JAPANESE

GAKKO KENCHIKU NENPO, KORITSU GAKKO-HEN 8898
(ANNUAL REPORT ON SCHOOL BUILDINGS, PREFECTURAL AND MUNICIPAL SCHOOLS)
MONBU-SHO KANRI-KYOKU SHISETSU-BU (DEPARTMENT OF FACILITIES, BUREAU OF ADMINISTRATION, MINISTRY OF EDUCATION)
2-2, 3-CHOME, KASUMIGASEKI, CHIYODA-KU, TOKYO 100
JAPANESE

GARASU 8899
(GLASS AND ARCHITECTURE)
SO KENCHIKU KENKYUJO (SO ASSOCIATION FOR BUILDING DEVELOPMENT)
10-18, 4-CHOME, KAMI KITAZAWA, SETAGAYA-KU, TOKYO 156
JAPANESE
INDEXED: JICST

GBRC: GENERAL BUILDING RESEARCH CORPORATION 8900
JAPANESE
INDEXED: JICST

GEKKAN DAIKENKYO 8901
(OSAKA CONSTRUCTION ASSOCIATION MONTHLY)
OSAKA KENSETSUGYO KYOKAI (OSAKA CONSTRUCTION INDUSTRY ASSOCIATION)
JAPANESE

GEKKAN JUSETSUTEN 8902
(HOUSING EQUIPMENT)
KUCHO REIDAN SHINBUNSHA
2, UMEGAE, KITA-KU, OSAKA-SHI 530
JAPANESE

GEKKAN KENCHIKU TOSO 8903
(BUILDING AND PAINTING)
KOKUCHOSHA
C/O HIRATA BUILDING, 8-16, 2-CHOME, SARUGAKU-CHO, CHIYODA-KU, TOKYO 101
JAPANESE

GEKKAN KOJI JOHO 8904
(CONSTRUCTION REPORT MONTHLY)
GEKKAN KENSETSU KOGYO SHINBUNSHA (CONSTRUCTION INDUSTRY MONTHLY NEWSPAPER COMPANY)
JAPANESE

GEPPO TOHOKU CHIKEN 8905
(MONTHLY TOHOKU CONSTRUCTION)
JAPANESE
INDEXED: JICST

GIJUTSU HOKOKU-KAI SHIRYO. KEIKAKU KOJI BUMON 8906
(DOCUMENTS OF THE TECHNICAL REPORT GROUP. PLANS AND CONSTRUCTION DEPARTMENT.)
KENSETSU-SHO CHUBU CHIHO KENSETSU-KYOKU (MINISTRY OF CONSTRUCTION, CHUBU REGIONAL CONSTRUCTION BUREAU)
JAPANESE

GUNMA NO TOCHI KAIRYO 8907
(GUNMA LAND IMPROVEMENT)
GUNMA TOCHI KAIRYO JIGYO DANTAI RENGOKAI (THE LEAGUE OF GUNMA PREFECTURE LAND IMPROVEMENT ENTERPRISES)
JAPANESE

GURAFU KENZAI 8908
(GRAPHIC KENZAI)
OSAKA MOKUZAI SHINBUNSHA
1-12, 3-CHOME, SAKURAGAWA, NANIWA-KU, OSAKA-SHI 556
JAPANESE

HIROBA (KINKI KENCHIKUSHIKAI KYOGIKAI) 8909
(JOURNAL OF THE KINKI ARCHITECTS SOCIETY)
KINKI KENCHIKUSHI KYOGIKAI (THE KINKI ARCHITECTS SOCIETY)
2-41, BAKURO-CHO, HIGASHI-KU, OSAKA-SHI 541
JAPANESE
INDEXED: JICST

HOKKAIDO TOSHI 8910
(HOKKAIDO CITIES)
JAPANESE
INDEXED: JICST

HOKKAIDORITSU KANCHI KENCHIKU KENKYUSHO CHOSA KENKYU HOKOKU 8911
(REPORT OF THE HOKKAIDO BUILDING RESEARCH INSTITUTE)
HOKKAIDORITSU KANCHI KENCHIKU KENKYUSHO (THE HOKKAIDO BUILDING RESEARCH INSTITUTE)
1-8, NIJUYONKEN 4-JO, NISHI-KU, SAPPORO-SHI 063
JAPANESE
INDEXED: JICST

HOKKAIDORITSU KANCHI KENCHIKU KENKYUSHO KENKYU SOKUHO 8912
(NEWS OF THE HOKKAIDO BUILDING RESEARCH INSTITUTE)
HOKKAIDORITSU KANCHI KENCHIKU KENKYUSHO (THE HOKKAIDO BUILDING RESEARCH INSTITUTE)
1-8, NIJUYONKEN 4-JO, NISHI-KU, SAPPORO-SHI 063
JAPANESE
INDEXED: JICST

HOKKAIDORITSU KANCHI KENCHIKU KENKYUSHO SHOHO 8913
(BULLETIN OF HOKKAIDO BUILDING RESEARCH INSTITUTE)
HOKKAIDORITSU KANCHI KENCHIKU KENKYUSHO (THE HOKKAIDO BUILDING RESEARCH INSTITUTE)
1-8, NIJUYONKEN 4-JO, NISHI-KU, SAPPORO-SHI 063
JAPANESE

HYOJUN KENCHIKUHI SHISU 8914
(INDEX OF STANDARD COST OF CONSTRUCTION)
KENSETSU KOGYO KEIEI KENKYUKAI (SOCIETY FOR CONTRACTORS MANAGEMENT RESEARCH)
C/O TOKYO KENSETSU KAIKAN, 5-1, 2-CHOME, HATCHOBORI, CHUO-KU, TOKYO 103
JAPANESE

IAS REPOTO 8915
(IAS REPORT)
CHIKI KAIHATSU KENKYUSHO (THE INSTITUTE FOR AREA STUDIES)
13-5, 1-CHOME, NISHISHINBASHI, MINATO-KU, TOKYO 105
JAPANESE

IE 8916
(HOUSE)
DAICHI SHUPPAN (DAICHI PUBLISHING)
JAPANESE

IE 8917
(HOUSE)
TOTTORI-KEN KENCHIKUSHIKAI (TOTTORI PREFECTURE ARCHITECTS ASSOCIATION)
C/O KENCHO KENCHIKU-KA, 1-220, AZUMA-MACHI, TOTTORI-SHI 680
JAPANESE

IE ZUKURI 8918
(HOUSE BUILDING)
KANGIN HAUJINGU SENTA
(KANGIN HOUSING CENTER)
JAPANESE

INTERIA 8919
(JAPAN INTERIOR DESIGN)
INTERIA SHUPPAN K.K. (JAPAN INTERIOR INCORPORATED)
1-22, YOTSUYA, SHINJUKU-KU, TOKYO 160
JAPANESE AND ENGLISH

ISHIKAWA KENCHIKUSHI 8920
(JOURNAL OF ISHIKAWA ARCHITECTS SOCIETY)
ISHIKAWA-KEN KENCHIKUSHIKAI (ISHIKAWA ARCHITECTS SOCIETY)
1-23, 2-CHOME, YAYOI, KANAZAWA-SHI 921
JAPANESE

ISUTORANDO 8921
(EASTLAND)
RANDO KIKAKU
(LAND PLANNING)
JAPANESE

JAPAN ARCHITECT (INTERNATIONAL EDITION OF SHIN KENCHIKU) 8922
SHIN KENCHIKUSHA
31-2, 2-CHOME, YUSHIMA, BUNKYO-KU, TOKYO 113
ENGLISH

JCCA KAIHO 8923
(JCCA BULLETIN)
KENSETSU KONSARUTANTSU KYOKAI
(CONSTRUCTION CONSULTANTS ASSOCIATION)
JAPANESE AND ENGLISH

JUKEN TAIMUSU 8924
(ARCHITECTURAL TIMES)
TOKEN TAIMUSU
8-16, 2-CHOME, SARUGAKU-CHO, CHIYODA-KU, TOKYO 101
JAPANESE

JUTAKU 8925
(MONTHLY OF THE HOUSING)
NIHON JUTAKU KYOKAI (JAPAN HOUSING ASSOCIATION)
C/O JUTAKU KAIKAN BUILDING, 19, SHIBA NISHIKUBO AKEFUNE-CHO, MINATO-KU, TOKYO 105
JAPANESE
INDEXED: JICST

JUTAKU GAHO 8926
(HOUSE GRAPHIC)
JUTAKU SHINPOSHA
10, SHIBA TORANOMON, MINATO-KU, TOKYO 105
JAPANESE

JUTAKU GURAFU 8927
(ARCHITECTURAL GRAPHIC)
SHIN KENZAI SHINBUNSHA
15-7, 6-CHOME, HONKOMAGOME, BUNKYO-KU, TOKYO 113
JAPANESE

JUTAKU KENCHIKU 8928
(HOUSING JOURNAL FOR BUILDERS AND DESIGNERS)
KENCHIKU SHIRYO-KENKYUSHA
(BUILDING MATERIALS RESEARCH ASSOCIATION)
JAPANESE

JUTAKU KENCHIKU KENKYU SHOHO 8929
(REPORT OF THE HOUSING CONSTRUCTION RESEARCH INSTITUTE)
JAPANESE
INDEXED: JICST

JUTAKU SANGYO JOHO 8930
(BUILDING INDUSTRY INFORMATION)
NIHON BIJINESU
C/O HARAJUKU TAUN HOMU BUILDING 203, 1-1, 1-CHOME, JINGUMAE, SHIBUYA-KU, TOKYO 150
JAPANESE

JUTAKU SANGYO TOKEI SHIRYO NENPO 8931
(ANNUAL STATISTICS OF HOUSING INDUSTRY IN JAPAN)
M. I. C.
2-37, ICHIGAYA TAMACHI, SHINJUKU-KU, TOKYO 162
JAPANESE

JUTAKU SEKKEI 8932
(HOUSE DESIGN)
NYU HAUSU SHUPPAN K.K. (NEW HOUSE PUBLISHING COMPANY, LTD.)
C/O TAISHOKAKU BUILDING, 4-3, 2-CHOME, FUJIMI, CHIYODA-KU, TOKYO 102
JAPANESE

KAGOSHIMA KENCHIKUSHI 8933
(KAGOSHIMA ARCHITECTS)
KAGOSHIMA-KEN KENCHIKUSHI KAI
(ASSOCIATION OF KAGOSHIMA PREFECTURE ARCHITECTS)
JAPANESE

KAIHATSU 8934
(DEVELOPMENT)
HOKKAIDO KAIHATSU-CHO HOKKAIDO KAIHATSU-KYOKU (HOKKAIDO DEVELOPMENT AGENCY, HOKKAIDO DEVELOPMENT BUREAU)
NISHI-4-CHOME, KITA-3-JO, CHUO-KU, SAPPORO-SHI 060
JAPANESE

KAIHATSU FOTO 8935
(DEVELOPMENT PHOTO)
HOKKAIDO KAIHATSU-CHO, HOKKAIDO KAIHATSU-KYOKU (HOKKAIDO DEVELOPMENT AGENCY, HOKKAIDO DEVELOPMENT BUREAU) AND HOKKAIDO DEVELOPMENT ASSOCIATION
NISHI-19-CHOME, KITA-2-JO, CHUO-KU, SAPPORO-SHI 060
JAPANESE

KAIHATSU KOHO 8936
(HOKKAIDO DEVELOPMENT NEWS)
HOKKAIDO KAIHATSU KYOKAI (HOKKAIDO DEVELOPMENT ASSOCIATION)
NISHI-19-CHOME, KITA-2-JO, CHUO-KU, SAPPORO-SHI 060
JAPANESE

KATEN WORU 8937
(CURTAIN-WALL)
NIHON KATEN WORU KOGYOKAI (JAPAN CURTAIN-WALL MANUFACTURES ASSOCIATION)
11-2, 5-CHOME, MINAMIAOYAMA, MINATO-KU, TOKYO 107
JAPANESE
INDEXED: JICST

KEIKAKU HYORON 8938
(PLANNING REVIEW)
TOSHI KEIKAKU KYOKAI (CITY PLANNING ASSOCIATION OF JAPAN)
16-14, 2-CHOME, HIRAKAWA-CHO, CHIYODA-KU, TOKYO 102
JAPANESE

KENCHIKU 8939
(ARCHITECTURE (A MONTHLY JOURNAL FOR ARCHITECTS AND DESIGNERS))
CHUGAI SHUPPAN K.K.
C/O FUKUROKU BUILDING, 2-7, KANDA TSUKASA-CHO CHIYODA-KU, TOKYO 101
JAPANESE
INDEXED: JICST

KENCHIKU 8940
(CONSTRUCTION)
SEIDOSHA
JAPANESE

KENCHIKU BOKA TO BOSAI 8941
(ARCHITECTURE FIRE AND DISASTER PREVENTION)
JAPANESE
INDEXED: JICST

KENCHIKU BUNKA 8942
(ARCHITECTURAL CULTURE)
SHOGOKUSHA (ARCHITECTURAL CULTURE)
16-9, 2-CHOME, HIRAKAWA-CHO, CHIYODA-KU, TOKYO 102
JAPANESE
INDEXED: JICST

KENCHIKU CHISHIKI 8943
(PRACTICAL KNOWLEDGE FOR BUILDING)
NIHON HOUSING CENTER (NIHON HOUSING CENTER COMPANY, LTD.)
C/O ODAKYU SHINANOMACHI MANSION, 5-3, MINAMIMOTO-MACHI, SHINJUKU-KU, TOKYO 160
JAPANESE
INDEXED: JICST

KENCHIKU DOTAI TOKEI GEPPO 8944
(MONTHLY REPORT OF STATISTICS OF BUILDING WORKS STARTED)
KENSETSU-SHO KEIKAKU-KYOKU (PLANNING BUREAU, MINISTRY OF CONSTRUCTION)
1-3, 2-CHOME, KASUMIGASEKI, CHIYODA-KU, TOKYO 100
JAPANESE

KENCHIKU GAHO 8945
(PHOTOGRAPHIC ARCHITECTURE (ARCHITECTURE ILLUSTRATED))
KENCHIKU KAKUHOSHA (ARCHITECTURAL PHOTOGRAPHIC REPORT COMPANY)
JAPANESE

KENCHIKU GIJUTSU 8946
(BUILDING ENGINEERING)
KENCHIKU GIJUTSUSHA
17-5, 1-CHOME, HYAKUNIN-CHO, SHINJUKU-KU, TOKYO 160
JAPANESE, WITH ENGLISH SUMMARY
INDEXED: JICST

KENCHIKU HYORON 8947
(BUILDING REVIEW)
KENCHIKU HYORONSHA
1-20, KITAHOREIDORI, NISHI-KU, OSAKA-SHI 550
JAPANESE

KENCHIKU JIDAI 8948
(ARCHITECTURAL AGE)
KOKUCHOSHA
C/O HIRATA BUILDING, 8-16, 2-CHOME, SARUGAKU-CHO, CHIYODA-KU, TOKYO 101
JAPANESE
INDEXED: JICST

KENCHIKU JOHO 8949
(CONSTRUCTION REPORT)
NIHON KENCHIKU DAIKU GINOSHIKAI (JAPAN CONSTRUCTION CARPENTER'S ASSOCIATION)
JAPANESE

KENCHIKU KEIZAI TOKEI SHIRYO 8950
(STATISTICS ON BUILDING ECONOMICS)
NIHON KENCHIKU GAKKAI KENCHIKU KEIZAI IINKAI (ARCHITECTURAL INSTITUTE OF JAPAN)
2-19, 3-CHOME, GINZA, CHUO-KU, TOKYO 104
JAPANESE

KENCHIKU KENKYU HOKOKU 8951
(REPORT OF THE BUILDING RESEARCH INSTITUTE)
KENSETSU-SHO KENCHIKU KENKYUSHO (BUILDING RESEARCH INSTITUTE, MINISTRY OF CONSTRUCTION)
28-8, 3-CHOME, HYAKUNIN-CHO, SHINJUKU-KU, TOKYO 160
JAPANESE, WITH ENGLISH SUMMARY
INDEXED: JICST

KENCHIKU KENKYU SHIRYO 8952
(REPORTS ON BUILDING RESEARCH)
KENSETSU-SHO KENCHIKU KENKYUSHO (BUILDING RESEARCH INSTITUTE, MINISTRY OF CONSTRUCTION)
28-8, 3-CHOME, HYAKUNIN-CHO, SHINJUKU-KU, TOKYO 160
JAPANESE

KENCHIKU KENKYUSHO CHOSA SHIKEN KENKYU GAIYO HOKOKU 8953
(RESEARCH REPORTS OF BUILDING RESEARCH INSTITUTE)
KENSETSU-SHO KENCHIKU KENKYUSHO (BUILDING RESEARCH INSTITUTE, MINISTRY OF CONSTRUCTION)
28-8, 3-CHOME, HYAKUNIN-CHO, SHINJUKU-BU, TOKYO 160
JAPANESE

KENCHIKU KENKYUSHO NENPO 8954
(RESEARCH WORK FOR THE PERIOD...., BUILDING RESEARCH INSTITUTE)
KENSETSU-SHO KENCHIKU KENKYUSHO (BUILDING RESEARCH INSTITUTE, MINISTRY OF CONSTRUCTION)
28-8, 3-CHOME, HYAKUNIN-CHO, SHINJUKU-KU, TOKYO 160
JAPANESE
INDEXED: JICST

KENCHIKU KENKYUSHO SHUKI KOENKAI KOGAISHU 8955
(ABSTRACT PAPERS AT THE AUTUMN MEETING OF THE BUILDING RESEARCH INSTITUTE)
KENSETSU-SHO KENCHIKU KENKYUSHO (BUILDING RE-SEARCH INSTITUTE, MINISTRY OF CONSTRUCTION)
28-8, 3-CHOME, HYAKUNIN-CHO, SHINJUKU-BU, TOKYO 160
JAPANESE

KENCHIKU KOJI SEKO TANKA SHIRYO 8956
(STATISTICAL REPORT OF THE COST OF CONSTRUCTION)
KENSETSU KOGYO KEIEI KENKYUKAI (SOCIETY FOR CONTRACTORS MANAGEMENT RESEARCH)
C/O TOKYO KENSETSU KAIKAN, 5-1, 2-CHOME, HATCHOBORI, CHUO-KU, TOKYO 104
JAPANESE

KENCHIKU NO GIJUTSU SEKO 8957
(ARCHITECTURE PRODUCTION ENGINEERING)
SHOGOKUSHA
16-9, 2-CHOME, HIRAKAWA-CHO, CHIYODA-KU, TOKYO 102
JAPANESE
INDEXED: JICST

KENCHIKU SEISAN 8958
(BUILDING MATERIALS)
KOGYO CHOSAKAI (INDUSTRY INFORMATION CENTER)
14-7, 2-CHOME, HONGO, BUNKYO-KU, TOKYO 113
JAPANESE
INDEXED: JICST

KENCHIKU SETSUBI 8959
(MAGAZINE OF BUILDING EQUIPMENT)
KENCHIKU SETSUBI SOGO KYOKAI (THE ENGINEERING ASSOCIATION OF BUILDING EQUIPMENT)
C/O MINAGAWA BUILDING, 1-6, GINZA, CHUO-KU, TOKYO 104
JAPANESE
INDEXED: JICST

KENCHIKU SETSUBI GIJUTSU KAIGI TEKISUTO 8960
(TEST OF THE CONFERENCE ON ARCHITECTURAL EQUIPMENT TECHNOLOGY)
JAPANESE
INDEXED: JICST

KENCHIKU SETSUBI TO HAIKAN KOJI 8961
(HEATING PIPING AND AIR CONDITIONING)
NIHON KOGYO SHUPPANSHA K.K.
C/O ISHIZAKA BUILDING, 3-18, 2-CHOME, NISHI KANDA, CHIYODA-KU, TOKYO 101
JAPANESE
INDEXED: JICST

KENCHIKU SETSUBISHI 8962
(ARCHITECTURAL EQUIPMENT INSTALLER)
JAPANESE
INDEXED: JICST

KENCHIKU TECHO 8963
(ARCHITECTURAL NOTEBOOK)
KINRYUDO
2-3, 5-CHOME, HIGASHI UENO, TAITO-KU, TOKYO 110
JAPANESE

KENCHIKU TO SEKISAN 8964
(BUILDING COSTS)
KEIZAI CHOSAKAI SHUPPAN-BU (THE ECONOMIC RESEARCH ASSOCIATION)
15-10, 3-CHOME, GINZA, CHUO-KU, TOKYO 104
JAPANESE

KENCHIKU TO SHAKAI 8965
(ARCHITECTURE AND SOCIETY)
NIHON KENCHIKU KYOKAI (THE ARCHITECTURAL ASSOCIATION OF JAPAN)
3-37, KITAHAMA, HIGASHI-KU, OSAKA-SHI 541
JAPANESE
INDEXED: JICST

KENCHIKU TO TOSHI 8966
(MONTHLY JOURNAL OF WORLD ARCHITECTURE AND URBANISM)
A & U
21-5, 1-CHOME, SENDAGAYA, SHIBUYA-KU, TOKYO 151
JAPANESE

KENCHIKU TOKEI NENPO (KENSETSU-SHO) 8967
(ANNUAL REPORT OF STATISTICS OF BUILDING WORKS AS STARTED)
KENSETSU-SHO KEIKAKU-KYOKU (PLANNING BUREAU, MINISTRY OF CONSTRUCTION)
1-3, 2-CHOME, KASUMIGASEKI, CHIYODA-KU, TOKYO 100
JAPANESE

KENCHIKU TOKEI NENPO (TOKYO-TO) 8968
(ANNUAL REPORT OF CONSTRUCTION STATISTICS, TOKYO METROPOLITAN GOVERNMENT))
TOKYO-TO SHUTO SEIBI-KYOKU KENCHIKU SHIDO-BU (BUREAU OF CAPITAL CITY DEVELOPMENT, TOKYO METROPOLITAN GOVERNMENT)
5-1, 3-CHOME, MARUNOUCHI, CHIYODA-KU, TOKYO 100
JAPANESE

KENCHIKU ZASSHI 8969
(JOURNAL OF ARCHITECTURE AND BUILDING SCIENCE)
NIHON KENCHIKU GAKKAI (ARCHITECTURAL INSTITUTE OF JAPAN)
2-19, 3-CHOME, GINZA, CHUO-KU, TOKYO 104
JAPANESE AND ENGLISH
INDEXED: JICST

KENCHIKUGYO KYOKAISHO SAKUHINSHU 8970
(BCS PRIZE WORKS)
KENCHIKUGYO KYOKAI (BUILDING CONTRACTORS SOCIETY)
5-1, 2-CHOME, HATCHOBORI, CHUO-KU, TOKYO 104
JAPANESE AND ENGLISH

KENCHIKUKA 8971
(ARCHITECT)
NIHON KENCHIKUKA KYOKAI (JAPAN ARCHITECTS ASSOCIATION)
3-16, 2-CHOME, JINGUMAE, SHIBUYA-KU, TOKYO 150
JAPANESE

KENCHIKUKAI 8972
(ARCHITECTURAL WORLD)
RIKO TOSHO K.K.
4-6, 1-CHOME, SOTO KANDA, CHIYODA-KU, TOKYO 101
JAPANESE
INDEXED: JICST

KENCHIKUSHI 8973
(ARCHITECTS)
NIHON KENCHIKUSHIKAI RENGOKAI (JAPAN FEDERATION OF ARCHITECTS AND BUILDING ENGINEERS' ASSOCIATION)
C/O KENCHIKU KAIKAN, 2-19, 3-CHOME, GINZA, CHUO-KU, TOKYO 104
JAPANESE

KENCHIKUSHI KENKYU 8974
(JOURNAL OF ARCHITECTURAL HISTORY)
KENCHIKUSHI KENKYUKAI (ARCHITECTURAL HISTORY SOCIETY OF JAPAN)
16-9, 2-CHOME, HIRAKAWA-CHO, CHIYODA-KU, TOKYO 102
JAPANESE

KENSETSU SETSUBI TO HAIKAN KOJI 8975
(BUILDING EQUIPMENT AND PIPING)
NIPPON KOGYO SHUPPAN
2-3-18, NISHI KANDA, CHIYODA-KU
TOKYO 101
JAPANESE
INDEXED: CHE

KENSETSU-SHO EIZEN GIJUTSU HOKOKUSHU 8976
(TECHNICAL REPORT OF BUILDING AND REPAIRING, MINISTRY OF CONSTRUCTION)
KENSETSU-SHO DAIJIN KANBO KANCHO EIZEN-BU (GOVERNMENT BUILDING DEVELOPMENT, MINISTRY OF CONSTRUCTION)
1-3, 2-CHOME, KASUMIGASEKI, CHIYODA-KU, TOKYO 100
JAPANESE

KENZAI RUTO 8977
(BUILDING MATERIALS NEWS)
KENZAI RUTO CHOSAKAI
1-6, MISAKI-CHO, CHIYODA-KU, TOKYO 101
JAPANESE

KENZAI TOKEI NENPO 8978
(YEARBOOK OF BUILDING MATERIALS STATISTICS)
TSUSHO SANGYO-SHO DAIJIN KANBO CHOSA TOKEI-BU (RESEARCH AND STATISTICS DIVISION, MINISTER'S SECRETARIAT, MINISTRY OF INTERNATIONAL TRADE AND INDUSTRY)
15-1, 6-CHOME, GINZA, CHOU-KU, TOKYO 104
JAPANESE

KIKAN KOKUDO 8979
(NATIONAL PLANNING)
KOKUDO KEIKAKU KYOKAI (NATIONAL PLANNING ASSOCIATION)
C/O DAI-5 MORI BUILDING, 21, SHIBA KOTOHIRA-CHO, MINATO-KU, TOKYO 105
JAPANESE
INDEXED: JICST

KIKAN KOZO KAIRYO 8980
(STRUCTURAL IMPROVEMENTS QUARTERLY)
KOZO KAIRYOSHA (STRUCTURAL IMPROVEMENT COMPANY)
JAPANESE

KINDAI KENCHIKU 8981
(MAGAZINE OF CONTEMPORARY ARCHITECTURE)
KINDAI KENCHIKUSHA
22-6, 6-CHOME, HONGO, BUNKYO-KU, TOKYO 113
JAPANESE

KOBE NO TOSHI KEIKAKU 8982
(CITY PLANNING OF KOBE)
KOBE-SHI TOSHI KEIKAKU-KYOKU (CITY PLANNING BUREAU, KOBE PREFECTURAL GOVERNMENT)
5-1, SHIMO YAMATE-DORI, IKUTA-KU, KOBE-SHI 650
JAPANESE

KODAN NO JUTAKU 8983
(HOUSE OF PUBLIC CORPORATION)
JUTAKU MODAI KENKYUKAI (RESEARCH ASSOCIATION FOR HOUSING QUESTION)
24, KANDA MATSUNAGA-CHO, CHIYODA-KU, TOKYO 101
JAPANESE

KOEN RYOKUCHI 8984
(PARKS AND RESERVATIONS)
NIHON KOEN RYOKUCHI KYOKAI (ASSOCIATION OF PARKS AND RESERVATIONS)
16-14, 2-CHOME, HIRAKAWA-CHO, CHIYODA-KU, TOKYO 102
JAPANESE
INDEXED: JICST

KOGYO JIJI TSUSHIN, JUTAKU SANGYO-BAN 8985
(INDUSTRIAL NEWS, HOUSING)
KOGYO JIJI TSUSHINSHA (THE INDUSTRIAL NEWS)
3-10, KANDA OGAWA-CHO, CHIYODA-KU, TOKYO 101
JAPANESE

KOGYO JIJI TSUSHIN, KOKUDO KAIHATSU-BAN 8986
(INDUSTRIAL NEWS, EDITION OF NATIONAL LAND CULTIVATION)
KOGYO JIJI TSUSHINSHA (THE INDUSTRIAL NEWS)
3-10, KANDA OGAWA-CHO, CHIYODA-KU, TOKYO 101
JAPANESE

KOKUDO JOHO 8987
(LAND PLANNING INFORMATION)
KOKUDO KEIKAKU KYOKAI (COUNTRY DEVELOPMENT SURVEY)
C/O DAI-5 MORI BUILDING BEKKAN, 21, SHIBA KOTOHIRA-CHO, MINATO-KU, TOKYO 105
JAPANESE
INDEXED: JICST

KOKUDO MONDAI 8988
(LAND PROBLEMS)
KOKUDO MONDAI KENKYUSHO (RESEARCH INSTITUTE FOR LAND PROBLEMS)
3-1, 1-CHOME, UCHISAIWAI-CHO, CHIYODA-KU, TOKYO 100
JAPANESE

KOKURITSU KOEN 8989
(NATIONAL PARKS)
KOKURITSU KOEN KYOKAI (NATIONAL PARKS ASSOCIATION)
C/O TORANOMON DENKI BUILDING, 15, SHIBA NISHIKUBO AKEFUNE-CHO, MINATO-KU, TOKYO 105
JAPANESE

KOKYO KENCHIKU 8990
(ARCHITECTURE OF PUBLIC)
EIZEN KYOKAI
16-4, 2-CHOME, HIRAKAWA-CHO, CHIYODA-KU, TOKYO 102
JAPANESE
INDEXED: JICST

KOSO JUTAKU 8991
(MODERN MANSION LIFE AND INTERIOR DESIGN)
YUHOSHA
3-6, KYOBASHI, CHUO-KU, TOKYO 104
JAPANESE
INDEXED: JICST

KOSUI SOKUHO 8992
(FLOOD REPORT)
JAPANESE
INDEXED: JICST

KUKAKU SEIRI 8993
(LAND ADJUSTMENT)
NIHON TOCHI KUKAKU SEIRI KYOKAI
16-14, 2-CHOME, HIRAKAWA-CHO, CHIYODA-KU, TOKYO 102
JAPANESE
INDEXED: JICST

KUME KENCHIKU JIMUSHO KENKYUHO 8994
(REPORTS OF KUME ARCHITECTS - ENGINEERS)
JAPANESE

MITSUMI HOMU 8995
(MITSUMI HOME)
MITSUMI DENKI (MITSUMI ELECTRIC COMPANY)
JAPANESE

NIHON JUTAKU KODAN CHOSA KENKYU HOKOKUSHU 8996
(COLLECTED BULLETINS OF THE FINDINGS OF THE JAPAN HOUSING CORPORATION)
JAPANESE

NIHON JUTAKU KODAN RYOSAN SHIKENJO GIJUTSU KENKYU HOKOKU 8997
(TECHNICAL RESEARCH REPORT OF JAPAN HOUSING CORPORATION, MASS PRODUCTION)
SHIKENJO
JAPANESE

NIHON KENCHIKU GAKKAI HOKURIKU SHIBU KENKYU HOKOKUSHU 8998
(RESEARCH REPORT OF THE JAPAN ARCHITECTURAL SOCIETY, HOKURIKU BRANCH)
JAPAN ARCHITECTURAL SOCIETY
JAPANESE

NIHON KENCHIKU GAKKAI RONBUN HOKOKUSHU 8999
(TRANSACTIONS OF THE ARCHITECTURAL INSTITUTE OF JAPAN)
NIHON KENCHIKU GAKKAI (ARCHITECTURAL INSTITUTE OF JAPAN)
2-19, 3-CHOME, GINZA, CHUO-KU, TOKYO 104
JAPANESE AND ENGLISH
INDEXED: JICST

NIHON KENCHIKU GAKKAI TAIKAI GAKUJUTSU KOEN KOGAISHU 9000
(ABSTRACT PAPERS AT THE ANNUAL CONFERENCE OF THE ARCHITECTURAL INSTITUTE OF JAPAN)
NIHON KENCHIKU GAKKAI
2-19, 3-CHOME, GINZA, CHUO-KU, TOKYO 104
JAPANESE
INDEXED: JICST

NIHON KENCHIKUKA KYOKAI NYUSU 9001
(JAPAN ARCHITECTS ASSOCIATION NEWS)
NIHON KENCHIKUKA KYOKAI (JAPAN ARCHITECTS ASSOCIATION)
C/O KENCHIKUKA KAIKAN, 3-16, 2-CHOME, JINGUMAE, SHIBUYA-KU, TOKYO 105
JAPANESE

NIHON TENPO SEKKEIKA KYOKAI KAIHO 9002
(JOURNAL OF JAPAN COMMERCIAL BUILDING DESIGNERS ASSOCIATION)
NIHON TENPO SEKKEIKA KYOKAI (JAPAN COMMERICAL BUILDING DESIGNERS ASSOCIATION)
4-17, 6-CHOME, KOJI-MACHI, CHIYODA-KU, TOKYO 102
JAPANESE

NIHON TOSHI GAKKAI NENPO 9003
(ANNUAL REPORT OF THE JAPAN SOCIETY FOR URBAN STUDIES)
NIHON TOSHI GAKKAI (JAPAN SOCIETY FOR URBAN STUDIES) AND CHIJIN SHOBO
1, SHIMO MINAMIDA-CHO, JODOJI, SAKYO-KU, KYOTO-SHI 606
JAPANESE

NIHON TOSHI KEIKAKU GAKKAI GAKUJUTSU KOENKAI RONBUNSHU 9004
(CITY PLANNING REVIEW, TRANSACTION OF SCIENCE LECTURE MEETING)
NIHON TOSHI KEIKAKU GAKKAI (THE CITY PLANNING INSTITUTE OF JAPAN)
C/O TOSHI KEIKAKU KAIKAN, 2-9, HIRAKAWA-CHO, CHIYODA-KU, TOKYO 102
JAPANESE
INDEXED: JICST

NIKKEI AKITEKUCHA 9005
(NIKKEI ARCHITECTURE)
NIKKEI MAGUROUHIRUSHA (NIKKEI MC GRAW-HILL PUBLISHING COMPANY)
JAPANESE

NIKKEN SEKKEI KOMU KABUSHIKI KAISHA GIHO 9006
(NIKKEN DESIGN AND ENGINEERING WORKS COMPANY, LTD. TECHNICAL BULLETIN)
JAPANESE

NIPPON JUTAKU KODAN CHOSA KENKYU KIHO 9007
(QUARTERLY REVIEW OF THE RESEARCH AND STUDYING SECTION, JAPAN HOUSING CORPORATION))
NIPPON JUTAKU KODAN KENCHIKU-BU (THE RESEARCH AND STUDYING SECTION, JAPAN HOUSING CORPORATION)
14-6, 1-CHOME, KUDANKITA, CHIYODA-KU, TOKYO 102
JAPANESE

NIPPON JUTAKU KODAN NENPO 9008
(ANNUAL REPORT OF THE JAPAN HOUSING CORPORATION)
NIPPON JUTAKU KODAN SOMU-BU (JAPAN HOUSING CORPORATION, GENERAL AFFAIRS DEPARTMENT)
14-6, 1-CHOME, KUDAN-KITA, CHIYODA-KU, TOKYO 102
JAPANESE AND ENGLISH

NIPPON JUTAKU KODAN RYOSAN SHIKENJO GIJUTSU KENKYU HOKOKU 9009
(TECHNICAL REPORTS, JAPAN HOUSING CORPORATION)
NIPPON JUTAKU KODAN RYOSAN SHIKENJO (MASS PRODUCTION RESEARCH LABORATORY, JAPAN HOUSING CORPORATION)
2683-3, ISHIKAWA-CHO, HACHIOJI-SHI, TOKYO 192
JAPANESE
INDEXED: JICST

NIPPON JUTAKU KODAN SHOKUIN RONBUNSHU 9010
(RESEARCH PAPERS BY THE STAFF OF THE JAPAN HOUSING CORPORATION)
NIPPON JUTAKU KODAN SOMU-BU (GENERAL AFFAIRS DEPARTMENT, JAPAN HOUSING CORPORATION)
14-6, 1-CHOME, KUDAN-KITA, CHIYODA-KU, TOKYO 102
JAPANESE

NIPPON KENCHIKU GAKKAI TAIKAI GAKUJUTSU KOEN KOGAISHU. KEIK AKU-KEI 9011
(SUMMARY OF THE PROCEEDING OF THE JAPAN ARCHITECTURE SOCIETY MEETING, PLANNING)
JAPANESE
INDEXED: JICST

NIPPON KENCHIKU GAKKAI TAIKAI GAKUJUTSU KOEN KOGAISHU. KOZ O-KEI 9012
(SUMMARY OF THE PROCEEDINGS OF THE JAPAN ARCHITECTURE SOCIETY MEETING, CONSTRUCTION)
JAPANESE
INDEXED: JICST

NIPPON SUWEDEN KENCHIKU SANGYO KAIGI HOKOKUSHO 9013
(REPORT OF THE CONFERENCE OF JAPAN, SWEDEN ARCHITEURAL INDUSTRY)
JAPANESE
INDEXED: JICST

NIWA 9014
(GARDEN)
NAGANO-KEN ZOCH KYOKAI (NAGANO PREFECTURE LAND-SCAPING ASSOCIATION)
JAPANESE

NOSON KENCHIKU 9015
(RURAL ARCHITECTURE)
NOSON KENCHIKU KENKYUKAI (RURAL ARCHITECTURAL INSTITUTE)
25-20, 3-CHOME, NOZAWA, SETAGAYA-KU, TOKYO 154
JAPANESE

NOSON KOGAKU KENKYU 9016
(FARMING VILLAGE ENGINEERING RESEARCH)
NOSEN KAIHATSU KAKAKU IINKAI (FARMING VILLAGE DEVELOPMENT PLANNING COUNCIL)
JAPANESE

NYU HAUSU 9017
(NEW HOUSE)
NYU HAUSU SHUPPAN K.K. (NEW HOUSE PUBLISHING COMPANY)
4-3, 2-CHOME, FUJIMI, CHIYODA-KU, TOKYO 102
JAPANESE
INDEXED: JICST

ONKYO GIJUTSU 9018
(ARCHITECTURAL ACOUSTICS AND NOISE CONTROL)
NIHON ONKYO ZAIRYO KYOKAI (ACOUSTICAL MATERIALS ASSOCIATION OF JAPAN)
13-12, 2-CHOME, NISHI SHINBASHI, MINATO-KU, TOKYO 105
JAPANESE
INDEXED: JICST

PROCESS: ARCHITECTURE 9019
PUROSESU AKITEKUCHUA
JAPANESE AND ENGLISH

PUREHABU JUTAKU. JUTAKU SETSUBI YUNITTO TOKEI SHI HANKIHO 9020
(QUARTERLY REPORT OF PREFAB HOUSING AND HOUSING INSTALLATION UNIT STATISTICS)
TSUSHO SANGYO DAIJIN KANBO CHOSA TOKEIBU (MITI, MINISTER'S SECRETARIAT, RESEARCH AND STATISTICS DEPARTMENT)
JAPANESE

PUREHABU KENCHIKU NYUSU 9021
(PREFABRICATED CONSTRUCTION NEWS)
PUREFABU KENCHIKU KYOKAI
10-5, 2-CHOME, SHINBASHI, MINATO-KU, TOKYO 105
JAPANESE
INDEXED: JICST

RANDO SUKEPU 9022
(JAPAN LANDSCAPE)
TOSHI KEIKAKU KENKYUSHO LANDSCAPE SHUPPANKAI
16-6, 2-CHOME, HIRAKAWA-CHO, CHIYODA-KU, TOKYO 102
JAPANESE
INDEXED: JICST

SAKAN KYOSHITSU 9023
(PLASTERERS' CLASSROOM)
KOKUCHOSHA
C/O HIRATA BUILDING, 8-16, 2-CHOME, SARUGAKU-CHO, CHIYODA-KU, TOKYO 101
JAPANESE

SALOON 9024
NITTETSU KENZAI (NITTETSU CONSTRUCTION MATERIALS)
JAPANESE

SAROBETSU SOGO CHOSA CHUKAN HOKOKUSHO. SHAKAI KEIZAI BUMON 9025
(INTERIM REPORT OF THE SAROBETSU GENERAL SURVEY. SOCIAL ECONOMICS SECTION)
JAPANESE
INDEXED: JICST

SAROBETSU SOGO CHOSA CHUKAN HOKOKUSHO. TSUCHI BUMON 9026
(INTERIM REPORT OF THE SAROBETSU GENERAL SURVEY. LAND SECTION)
JAPANESE
INDEXED: JICST

SENDAI-TOSHI KAGAKU KENKYUKAI KAIHO 9027
(CONFERENCE REPORT OF THE SENDAI URBAN SCIENCE RESEARCH SOCIETY)
JAPANESE

SETSUBI SEKKEI 9028
(ARCHITECTURAL EQUIPMENT PLAN)
NIHON SETSUBI SEKKEIKA KYOKAI (ARCHITECTURAL EQUIPMENT DESIGNER'S ASSOCIATION OF JAPAN)
C/O KORAKU BUILDING, 1-8, 1-CHOME, KORAKU, BUNKYO-KU, TOKYO 112
JAPANESE
INDEXED: JICST

SETSUBI TO KANRI 9029
(EQUIPMENTS AND MANAGEMENT)
OMUSHA
3-1, KANDA NISHIKI-CHO, CHIYODA-KU, TOKYO 101
JAPANESE
INDEXED: JICST

SHIMIZU BULLETIN 9030
SHINKENCHIKUSHA HENSHU KYORYOKU, SHIMIZU KENSETSU KOHO SHITSU (NEW ARCHITECTURAL FIRMS EDITORIAL CO-OPERATIVE, SHIMIZU CONSTRUCTION INFORMATION ROOM)
JAPANESE AND ENGLISH

SHIN JUTAKU 9031
(NEW DWELLING HOUSE)
SHIN JUTAKUSHA
17, JOAN-CHO, KITA-KU, OSAKA-SHI 530
JAPANESE

SHIN KENCHIKU 9032
(JAPAN ARCHITECT)
SHIN KENCHIKUSHA
31-2, 2-CHOME, YUSHIMA, BUNKYO-KU, TOKYO 113
JAPANESE
INDEXED: JICST

SHIN NIHON RETTO 9033
(REMODELING OF THE JAPANESE ARCHIPELAGO)
KOGYO SAIHAICHI, SANTANCHIIKI SHINKO KODAN (THE CORPORATION FOR THE RELOCATION OF INDUSTRY AND THE DEVELOPMENT OF COOL MINING AREAS)
8-1, 3-CHOME, KASUMIGASEKI, CHIYODA-KU, TOKYO 100
JAPANESE

SHIN TOSHI 9034
(NEW CITIES)
TOSHI KEIKAKU KYOKAI (CITY PLANNING ASSOCIATION OF JAPAN)
16-14, 2-CHOME, HIRAKAWA-CHO, CHIYODA-KU, TOKYO 102
JAPANESE

SHITSUNAI 9035
(INTERIOR)
KOSAKUSHA
29, SHIBA KOTOHIRA-CHO, MINATO-KU, TOKYO 105
JAPANESE

SHOTEN KENCHIKU 9036
(COMMERCIAL ARCHITECTURE)
SHOTEN KENCHIKUSHA
C/O BUNKA BUILDING, 1-456, NISHIOKUBO, SHINJUKU-KU 160
JAPANESE AND ENGLISH

SHUKAN JUTAKU JANARU 9037
(WEEKLY HOUSING JOURNAL)
SHINKENZAI SHINBUNSHA
C/O MOKKO KAIKAN, 15-7, 6-CHOME, HONKOMAGOME, BUNKYO-KU, TOKYO 113
JAPANESE

SOGO SETSUBI KONSARUTANTO GIJUTSU NEMPO 9038
(GENERAL FACILITIES CONSULTANT TECHNICAL ANNUAL)
SOGO SETSUBI KONSARUTANTO (GENERAL FACILITIES CONSULTANT)
JAPANESE

SOKEN 9039
(FOUNDATIONS)
NICHIDAI KOGAKUBU KENCHIKUGAKU KYOSHITSU (NIHON UNIVERSITY, FACULTY OF ENGINEERING, ARCHITECTURAL STUDIES DEPARTMENT)
JAPANESE

SONOIKE GIHO 9040
(SONOIKE TECHNICAL REPORT)
JAPANESE
INDEXED: JICST

SPACIOLOGY 9041
KYOTO DAIGAKU KOGAKUBU KENCHIKUGAKKA (DEPARTMENT OF ARCHITECTURE, KYOTO UNIVERSITY)
YOSHIDA HON-CHO, SAKYO-KU, KYOTO-SHI 606
ENGLISH

SUMAI 9042
(HOUSE AND LIVING)
SUMAI NO KENKYUSHA
15-8, 3-CHOME, NIHONBASHI, CHUO-KU, TOKYO 103
JAPANESE

SUPESU DEZAIN 9043
(SPACE DESIGN (S.D.))
KAJIMA KENKYUSHO SHUPPANKAI (KAJIMA INSTITUTE PUBLISHING COMPANY, LTD.)
5-13, 6-CHOME, AKASKA, MINATO-KU, TOKYO 107
JAPANESE, WITH ENGLISH SUMMARY

TAISEI KENSETSU K.K. GIJUTSU KENKYU SEIKA 9044
(TECHNICAL RESEARCH OF TAISEI CONSTRUCTION, INCORPORATED)
JAPANESE
INDEXED: JICST

TAKUZO JIHO 9045
(HOUSING TRACT DEVELOPMENT REVIEW)
ZENKOKU TAKUCHI ZOSEI RENGOKAI (NATIONAL HOUSING TRACT DEVELOPMENT FEDERATION)
JAPANESE

TEIKEN (NIHON TEIEN KENKYUKAI KAISHI) 9046
(GARDEN)
NIHON TEIEN KENKYUKAI (JAPANESE GARDEN RESEARCH ASSOCIATION)
30-4, 2-CHOME, AKAZUTSUMI, SETAGAYA-KU, TOKYO 156
JAPANESE

TEKKIN NO TSUGITE OYOBI TEICHAKU NO SEKKEI SEKO NI KANSURU SHINPOJIUMU HAPPYO HOBUNSHU 9047
(PROCEEDINGS OF THE SYMPOSIUM ON THE PLANNING AND OF SPLICING AND FASTENING REINFORCING STEEL RODS)
JAPANESE
INDEXED: JICST

TENPO DEZAIN 9048
(SHOP DESIGN)
TENPO DEZAIN
6-3, 7-CHOME, ROPPONGI, MINATO-KU, TOKYO 106
JAPANESE

TENPO TO KENCHIKU 9049
(SHOPS AND ARCHITECTURE)
KENCHIKU SHIRYO KENKYUSHA (ARCHITECTURAL MATERIALS RESEARCH COMPANY)
JAPANESE

TODA KENSETSU K.K. GIJUTSU KAIHATSU SENTA GIJUTSU KENKYU HOKOKUSHU 9050
(REPORTS OF TECHNICAL RESEARCH OF THE TECHNOLOGY DEVELOPMENT CENTER, TODA CONSTRUCTION COMPANY, LTD.)
JAPANESE
INDEXED: JICST

TOHOKU DAIGAKU KENCHIKU KEIKAKUGAKU KENKYU NENPO 9051
(ANNUAL REPORT OF THE RESEARCH IN ARCHITECTURAL PLANNING)
TOHOKU DAIGAKU KOGAKUBU KENCHIKU GAKKA KEIKAKU DAI 1, DAI 2 KOZA
(TOHOKU UNIVERSITY DEPARTMENT OF ENGINEERING, ARCHITECTURAL DIVISION, PLANNING SECTIONS ONE AND TWO)
JAPANESE

TOHOKU DAIGAKU KENCHIKUGAKUHO 9052
(ARCHITECTURAL REPORTS OF TOHOKU UNIVERSITY)
TOHOKU DAIGAKU KENCHIKUGAKKA (THE ARCHITECTURAL SECTION, FACULTY OF ENGINEERING, TOHOKU UNIVERSITY)
AOBA, ARAMAKI, SENDAI-SHI 980
JAPANESE
INDEXED: JICST

TOHOKU KOGYO DAIGAKU KIYO, D: KENCHIKUGAKUHEN 9053
(BULLETIN OF THE TOHOKU INSTITUTE OF TECHNOLOGY, SECTION D: ARCHITECTURE)
TOHOKU KOGYO DAIGAKU (TOHOKU INSTITUTE OF TECHNOLOGY)
19, KOSHIJI, NAGA-MACHI, SENDAI-SHI 982
JAPANESE AND ENGLISH
INDEXED: JICST

TOKYO-TO JUTAKU NENPO 9054
(ANNUAL REPORT ON HOUSING IN TOKYO)
TOKYO-TO JUTAKU-KYOKU
5-1, 3-CHOME, MARUNOUCHI, CHIYODA-KU, TOKYO 100
JAPANESE AND ENGLISH

TOKYO-TO TOSHI KEIKAKU GAIYO 9055
(CITY PLANNING OF TOKYO)
TOKYO-TO SHUTO SEIBI-KYOKU (THE BUREAU OF CAPITAL CITY DEPARTMENT TOKYO METROPOLITAN GOVERNMENT)
5-1, 3-CHOME, MARUNOUCHI, CHIYODA-KU, TOKYO 100
JAPANESE

TOKYU KENSETSU GIJUTSU KENKYU SHOHO 9056
(TECHNICAL RESEARCH REPORT. TOKYO CONSTRUCTION COMPANY, LTD.)
JAPANESE
INDEXED: JICST

TOSHI JUTAKU 9057
(URBAN HOUSING)
KAJIMA KENKYUSHO SHUPPANKAI (KAJIMA INSTITUTE PUBLISHING COMPANY, LTD.)
5-13, 6-CHOME, AKASAKA, MINATO-KU, TOKYO 107
JAPANESE

TOSHI KAGAKU 9058
(INSTITUTE OF GENERAL ARTS AND URBAN SCIENCE)
TOSHI KAGAKU KENKYUSHO (URBAN SCIENCE RESEARCH INSTITUTE)
JAPANESE

TOSHI KEIKAKU 9059
(CITY PLANNING REVIEW)
NIHON TOSHI KEIKAKU GAKKAI (THE CITY PLANNING INSTITUTE OF JAPAN)
16-14, 2-CHOME, HIRAKAWA-CHO, CHIYODA-KU, TOKYO 102
JAPANESE, WITH ENGLISH SUMMARY
INDEXED: JICST

TOSHI KEIKAKU BESSATSU 9060
(CITY PLANNING, SPECIAL EDITION)
NIHON TOSHI KEIKAKU GAKKAI
JAPANESE

TOSHI KEIKAKU NENPO 9061
(ANNUAL REPORT ON CITY PLANNING)
KENSETSU-SHO TOSHI-KYOKU (CITY PLANNING BUREAU, MINISTRY OF CONSTRUCTION)
1-3, 2-CHOME, KASUMIGASEKI, CHIYODA-KU, TOKYO 100
JAPANESE

TOSHI KENKYU BUNKEN MOKUROKU 9062
(BIBLIOGRAPHY ON URBAN RESEARCH)
TOKYO-TORITSU DAIGAKU TOSHI KENKYU IINKAI
1-1, 1-CHOME, YAKUMO, MEGURO-KU, TOKYO 152
JAPANESE

TOSHI KENKYU CHOSA HOKOKU 9063
(REPORT ON URBAN RESEARCH AND STUDIES)
TOKYO-TORITSU DAIGAKU TOSHI KENKYU IINKAI
1-1, 1-CHOME, YAKUMO, MEGURO-KU, TOKYO 152
JAPANESE

TOSHI KENKYU HOKOKU 9064
(REPORT ON URBAN RESEARCH)
TOKYO-TORITSU DAIGAKU TOSHI KENKYU IINKAI
1-1, 1-CHOME, YAKUMO, MEGURO-KU, TOKYO 152
JAPANESE

TOSHI KOEN 9065
(PUBLIC PARK MAGAZINE)
TOKYO-TO KOEN KYOKAI (THE PARK SOCIETY OF TOKYO)
C/O KOEN SHIRYOKAN, HIBIYA KOEN, CHIYODA-KU, TOKYO 100
JAPANESE
INDEXED: JICST

TOSHI MONDAI 9066
(MUNICIPAL PROBLEMS (THE JOURNAL OF THE TOKYO INSTITUTE FOR MUNICIPAL RESEARCH))
TOKYO SHISEI CHOSAKAI (THE TOKYO INSTITUTE FOR MUNICIPAL RESEARCH)
HIBIYA KOEN, CHIYODA-KU, TOKYO 100
JAPANESE
INDEXED: JICST

TOSHI MONDAI KENKYU 9067
(JOURNAL OF MUNICIPAL PROBLEMS)
TOSHI MONDAI KENKYUKAI (ASSOCIATION ON MUNICIPAL PROBLEMS)
C/O OSAKA-SHI SOMU-KYOKU, 1-4, NAKANOSHIMA, KITA-KU, OSAKA-SHI 530
JAPANESE

TOSHI SAIKAIHATSU 9068
(REDEVELOPMENT CITIES)
ZENKOKU TOSHI SAIKAIHATSU SOKUSHIN RENMEI (JAPAN FEDERATION OF URBAN REDEVELOPMENT PROMOTION CENTERS) AND TOKYO TOSHI SAIKAIHATSU SOKUSHINKAI (URBAN REDEVELOPMENT PROMOTION CENTER OF TOKYO)
C/O NIHON SHOKO KAIGISHO, 2-2, 3-CHOME, MARUNOUCHI, CHIYODA-KU, TOKYO 100
JAPANESE
INDEXED: JICST

TRANSACTIONS OF THE SOCIETY OF HEATING, AIR-CONDITIONING AND SANITARY ENGINEERS OF JAPAN. 9069
TOKYO
JAPANESE AND ENGLISH
INDEXED: APP

UTSUKUSHII NIHON NO KAGU TO INTERIA 9070
(GOOD DESIGN AND INTERIOR OF JAPAN)
ZENKAGUREN SHUPPANSHA
C/O ISHIDA BUILDING, 12-1, 3-CHOME, SHITAYA, TAITO-KU, TOKYO 110
JAPANESE AND ENGLISH

WASEDA KENCHIKU 9071
(WASEDA ARCHITECTURE)
TOMON KENCHIKUKAI
C/O WASEDA DAIGAKU RIKO KENKYUSHO, 17, KIKUI-CHO, SHINJUKU-KU, TOKYO 162
JAPANESE
INDEXED: JICST

WASEDA KENCHIKU NYUSU 9072
(WASEDA ARCHITECTURE NEWS)
SEMON KENCHIKUKAI (SEMON ARCHITECTURAL SOCIETY)
JAPANESE

WORKS OF ARCHITECTS 9073
NIHON KENCHIKUKA KYOKAI KANSAI SHIBU
C/O KENSETSU HOSHO BUILDING, 5 UEMACHI, HIGASHI-KU, (540) OSAKA, JAPAN
ENGLISH

YOCHI KOHO 9074
(OFFICIAL BULLETIN ON LAND FOR USE (SITES))
KENSETSU DAIJIN KANBO-KOKYO YOCHIKA
JAPANESE

ZOEN ZASSHI 9075
(JOURNAL OF THE JAPANESE INSTITUTE OF LANDSCAPE ARCHITECTS)
NIHON ZOEN GAKKAI (JAPANESE INSTITUTE OF LANDSCAPE ARCHITECTS)
C/O TOKYO DAIGAKU NOGAKUBU 1-GO-KAN, 1-1, 1-CHOME, YAYOI, BUNKYO-KU, TOKYO 113
JAPANESE AND ENGLISH
INDEXED: JICST

The Arts, Recreation, Entertainment, Sport, etc.

GEKIJO GIJUTSU 9076
(JOURNAL OF JAPANESE INSTITUTE FOR THEATRE TECHNOLOGY)
JAPANESE
INDEXED: JICST

GUN 9077
KOKUSAI SHUPPAN
C/O SHOWA BUILDING, 3-2, 5-CHOME, NAKANO, NAKANO-KU, TOKYO 164
JAPANESE

KURE NYUSU 9078
(CLAY SHOOTING NEWS)
ZENKOKU KURE KOGYO KUMIAI
3-5, KABUTO-CHO, NIHONBASHI, CHUO-KU, TOKYO 103
JAPANESE
INDEXED: JICST

MODERU ATO 9079
(MODEL ART)
MODERU ATOSHA
13-5, 2-CHOME, IDABASHI, CHIYODA-KU, TOKYO 102
JAPANESE

NICHIGAKU SHAHO 9080
(REPORT OF JAPAN MUSICAL INSTRUMENT MANUFACTURING COMPANY)
NIHON GAKKI (JAPAN MUSICAL INSTRUMENT MANUFACTURING COMPANY)
JAPANESE
INDEXED: JICST

SEKAI NO KESSAKUKI 9081
(FAMOUS AIRPLANES OF THE WORLD)
BUNRINDO
1-55, KANDA JINBO-CHO, CHIYODA-KU, TOKYO 101
JAPANESE

TETSUDO MOKEI SHUMI 9082
(HOBBY OF MODEL RAILROADING)
KIGEI SHUPPANSHA (KIGEI PUBLISHING COMPANY, LTD.)
15-11, 1-CHOME, SOSHIGAYA, SETAGAYA-KU, TOKYO 153
JAPANESE

Photography and Cinematography

ASAHI KAMERA 9083
(ASAHI CAMERA)
JAPANESE
INDEXED: JICST

BULLETIN OF THE SOCIETY OF PHOTOGRAPHIC SCIENCE AND TECHNOLOGY OF JAPAN 9084
NIHON SHASHIN GAKKAI (THE SOCIETY OF PHOTOGRAPHIC SCIENCE AND TECHNOLOGY OF JAPAN)
C/O TOKYO SHASHIN DAIGAKU, 9-5, 2-CHOME, HON-CHO, NAKANO-KU, TOKYO 164
EUROPEAN
INDEXED: JICST

DANSO SATSUEIHO KENKYUKAI ZASSHI 9085
(BULLETIN OF THE SOCIETY FOR THE STUDY OF SECTIONAL PHOTOGRAPHY)
DANSO SATSUEI HO KENKYU KAI
JAPANESE

DENSHI SHASHIN 9086
(ELECTRO PHOTOGRAPHY)
DENSHI SHASHIN GAKKAI (THE SOCIETY OF ELECTROPHOTOGRAPHY OF JAPAN)
C/O TOKYO SHASHIN DAIGAKU, 9-5, 2-CHOME, HON-CHO, NAKANO-KU, TOKYO 164
JAPANESE AND ENGLISH
INDEXED: CHE, INS, JICST

EIGA TEREBI GIJUTSU 9087
(MOTION PICTURE AND TV ENGINEERING)
NIHON EIGA TEREBI GIJUTSU KYOKAI (MOTION PICTURE AND TELEVISION ENGINEERING SOCIETY OF JAPAN, INCORPORATED)
C/O SANKEI BUILDING, 7-2, 1-CHOME, OTE-MACHI, CHIYODA-KU, TOKYO 100
JAPANESE

EISHA 9088
(PROJECTIONIST'S MAGAZINE)
ZEN NIHON EISHA GIJUTSUSHA RENMEI (ALL JAPAN PROJECTIONIST'S UNION)
C/O OTAKE BUILDING, 13-5, 6-CHOME, SHINBASHI, MINATO-KU, TOKYO 105
JAPANESE

FOTO FAMIRI 9089
(PHOTO FAMILY)
KONISHIRAKU SHASHIN KOGYO (KONISHIROKU PHOTOGRAPHIC INDUSTRY)
JAPANESE

FOTO RIPUROGURAFI 9090
(PHOTO REPROGRAPHY)
NAGASE SANGYO KODAKKU SEIHINBU: GURAFIKU RIPURODAKUSHON TEKUNIKARU SABISU SENTA (NAGASE INDUSTRIES KODAK PRODUCTS DIVISION GRAPHIC REPRODUCTION TECHNICAL SERVICE)
JAPANESE

FOTO TEKUNIKKU 9091
(PHOTOGRAPHIC TECHNIQUE)
GENKOSHA
1-5, 4-CHOME, IDABASHI, CHIYODA-KU, TOKYO 102
JAPANESE

FOTORISOGURAFI 9092
(PHOTOLITHOGRAPHY)
NAGASE SANGYO K.K. KODDAKU SEIHIN-BU
2-3, NIHONBASHI KOBUNE-CHO, CHUO-KU, TOKYO 104
JAPANESE

FUJI SHASHIN FIRUMU KENKYU HOKOKU 9093
(SCIENTIFIC PUBLICATIONS OF THE FUJI PHOTO FILM COMPANY, LTD.)
FUJI SHASHIN FIRUMU K.K. (FUJI PHOTO FILM COMPANY, LTD.)
NAKANUMA, MINAMIASHIGARA-SHI 250-01
JAPANESE AND ENGLISH
INDEXED: CHE, JICST

FUJI X-REI KENKYU 9094
(FUJI X-RAY REPORT)
FUJI SHASHIN FIRUMU K.K. (FUJI PHOTO FILM COMPANY, LTD.)
2-10, 3-CHOME, GINZA, CHUO-KU, TOKYO 104
JAPANESE

GREEN 9095
FUJI SHASHIN FIRUMU K.K. MAIKURO SHISUTEMU-BU (MICRO SYSTEM DEPARTMENT, FUJI PHOTO FILM COMPANY, LTD.)
26-30, 2-CHOME, NISHI-AZABU, MINATO-KU, TOKYO 106
JAPANESE

GURAFIKESHON 9096
(GRAPHICATION)
FUJI ZEROKKUS (FUJI XEROX)
JAPANESE

IMEJINGU DENSHI SHASHIN B SHIRIZU 9097
(IMAGING TODAY)
DENSHI SHASHIN GAKKAI (THE SOCIETY OF ELECTROPHOTOGRAPHY OF JAPAN)
9-5, 2-CHOME, HON-CHO, NAKANO-KU, TOKYO 164
JAPANESE AND ENGLISH

KAMERA ATO 9098
(CAMERART)
KAMERAATOSHA (CAMERART INCORPORATED)
C/O HINODE BUILDING, 2-11, KYOBASHI, CHUO-KU, TOKYO 104
ENGLISH

KAMERA JIDAI 9099
(CAMERA TIMES)
SHASHIN DOJINSHA
JAPANESE

KODAKKU INTANASHONARU FOTOGURAFI 9100
(KODAK INTERNATIONAL PHOTOGRAPHY)
NAGASE SANGYO (NAGASE AND COMPANY, LTD.)
JAPANESE

KOGATA EIGA 9101
(8MM HOME MOVIE)
GENKOSHA
1-5, 4-CHOME, IIDABASHI, CHIYODA-KU, TOKYO 102
JAPANESE

KOGATA EIGA BIGINA SHIRIZU 9102
(HOME MOVIE BEGINNERS SERIES)
GENKOSHA
1-5, 4-CHOME, IIDABASHI, CHIYODA-KU, TOKYO 102
JAPANESE

MAIKURO SHASHIN 9103
(JOURNAL OF MICROPHOTOGRAPHY)
NIHON MAIKURO SHASHIN KYOKAI (JAPAN MICRO-PHOTOGRAPHY ASSOCIATION)
3-5, KANDA KAJI-CHO, CHIYODA-KU, TOKYO 1[illegible]
JAPANESE
INDEXED: JICST

MINIKOPI 9104
(MINI-COPY)
JAPANESE
INDEXED: JICST

NIHON KAMERA 9105
(JAPAN CAMERA)
NIHON KAMERASHA (JAPAN CAMERA COMPANY)
JAPANESE

NIHON SHASHIN KOGYO TSUSHIN 9106
(JAPAN PHOTOGRAPHIC INDUSTRY NEWS)
NIHON SHASHIN KOGYO TSUSHISHA (JAPAN PHOTOGRAPHIC INDUSTRY NEWS COMPANY)
JAPANESE

NIHON SHASIN GAKKAI KAISHI 9107
(JOURNAL OF THE SOCIETY OF SCIENTIFIC PHOTOGRAPHY OF JAPAN)
NIHON SHASHIN GAKKAI (THE SOCIETY OF SCIENTIFIC PHOTOGRAPHY OF JAPAN)
C/O TOKYO SHASHIN DAIGAKU, 9-5, 2-CHOME, HON-CHO, NAKANO-KU, TOKYO 164
JAPANESE, WITH EUROPEAN SUMMARY
INDEXED: CHE, JICST

OLYMPUS PHOTOGRAPHY 9108
ZUIKONYUSU (ZUIKO NEWS)
JAPANESE AND ENGLISH

PHOTO ADVICES MONTHLY REPORT 9109
FOTO ADOBAISU K.K.
8, SAKA-MACHI, SHINJUKU-KU, TOKYO 160
JAPANESE

PHOTOGRAPHIC SENSITIVITY 9110
MARUZEN COMPANY LTD.
EXPORT DEPARTMENT, POB 5050, TOKYO INT 100-31
ENGLISH
INDEXED: CHE

PHOTOGRAPHY IN JAPAN 9111
NIHON SHASHIN KYOKAI (THE PHOTOGRAPHIC SOCIETY OF JAPAN)
C/O KYODO BUILDING, 2-2, KANDA NISHIKI-CHO, CHIYODA-KU, TOKYO 101
JAPANESE

SAKURA X-REI SHASHIN KENKYU 9112
(SAKURA X-RAY PHOTOGRAPHIC REVIEW)
KONISHIROKU SHASHIN KOGYO K.K. (KONISHIROKU PHOTO INDUSTRY COMPANY, LTD.)
3-1, NIHONBASHI MURO-MACHI, CHUO-KU, TOKYO 103
JAPANESE
INDEXED: JICST

SHASHIN KOGYO 9113
(PHOTOGRAPHIC INDUSTRIES)
SHASHIN KOGYO SHUPPANSHA
1-65, KANDA JINBO-CHO, CHIYODA-KU, TOKYO 101
JAPANESE
INDEXED: JICST

SHASHIN KOGYO 9114
(PHOTOGRAPHIC INDUSTRIES)
SHASHIN KOGYO SHUPPANSHA
1-65, KANDA JINBO-CHO, CHIYODA-KU, TOKYO 101
JAPANESE
INDEXED: CHE, JICST

TOKEN 9115
(TOKEN)
TOKYO SHASHIN KENKYUKAI (TOKYO PHOTOGRAPHY RESEARCH SOCIETY)
JAPANESE

TOKYO SHASHIN KENKYUKAI KAIHO 9116
(PROCEEDINGS OF THE TOKYO PHOTOGRAPHY RESEARCH SOCIETY)
JAPANESE

INDEX TO TITLES

www.ingramcontent.com/pod-product-compliance
Lightning Source LLC
LaVergne TN
LVHW062136200926
848484LV00007B/1817

Using Cross-Layer Techniques for Communication Systems

Habib F. Rashvand
University of Warwick, UK

Yousef S. Kavian
Shahid Chamran University, Iran

Information Science
REFERENCE

Managing Director:	Lindsay Johnston
Senior Editorial Director:	Heather A. Probst
Book Production Manager:	Sean Woznicki
Development Manager:	Joel Gamon
Development Editor:	Hannah Abelbeck
Acquisitions Editor:	Erika Gallagher
Typesetter:	Milan Vracarich, Jr.
Cover Design:	Nick Newcomer, Lisandro Gonzalez

Published in the United States of America by
Information Science Reference (an imprint of IGI Global)
701 E. Chocolate Avenue
Hershey PA 17033
Tel: 717-533-8845
Fax: 717-533-8661
E-mail: cust@igi-global.com
Web site: http://www.igi-global.com

Library of Congress Cataloging-in-Publication Data

Using cross-layer techniques for communication systems: techniques and applications / Habib F. Rashvand and Yousef S. Kavian, editors.
p. cm.
Includes bibliographical references and index.
ISBN 978-1-4666-0960-0 (hardcover) -- ISBN 978-1-4666-0961-7 (ebook) -- ISBN 978-1-4666-0962-4 (print & perpetual access) 1. Cross-layer designs (Telecommunications) I. Rashvand, Habib F. II. Kavian, Yousef S., 1978-
TK5105.66.U84 2012
621.382'1--dc23

2011051278

British Cataloguing in Publication Data
A Cataloguing in Publication record for this book is available from the British Library.

All work contributed to this book is new, previously-unpublished material. The views expressed in this book are those of the authors, but not necessarily of the publisher.

Editorial Advisory Board

Table of Contents

Detailed Table of Contents

A cognitive radio adapts to the environment in which it operates by sensing the spectrum and then opportunisticly exploiting unused and/or underused frequency bands in order to achieve certain performance goals. Due to the close coupling and interaction among protocol layers, the optimal design of opportunistic spectrum access and cognitive radio networks call for a cross-layer approach that integrates signal processing and networking with regulatory policy making. This chapter introduces basic concepts, design issues involved, and some recent development in this emerging technological field. Future research directions are also briefly examined.

Design and analysis of battery-less sensor networks from a cross layer point of view are studied in this chapter. The state of the art in fabrication and test of this new class of wireless sensor systems are also reviewed. Interactions between the lower layer with passive sensors and the upper layer with active sensors—a different perspective on cross layer—is exploited to achieve significant performance gains in terms of signal to noise and interference ratio, correlation peak to side-lobe ratio, operation range, and data rate.

This chapter describes the different aspects related to designing a suitable monitoring architecture for Cloud Computing, aiming to support cross-layer monitoring across all layers available in the Cloud stack. For this purpose, the importance of monitoring services in Cloud scenarios is outlined, followed by a comprehensible analysis of a wide set of distributed monitoring solutions. After that, the particular requirements related to cross-layer monitoring for Cloud Computing architectures are identified and explained. Then, diverse aspects, which may fit a monitoring architecture for fulfilling such requirements, are explained. Finally, some future research directions and conclusions are highlighted.

In this chapter, dependencies between the different protocol layers and across the hops through Wireless Mesh Networks (WMNs) will be exploited to deliver video streaming with high Quality of Experience (QoE). Specifically, a cross-layer optimization approach is applied to a replication strategy for video on-demand over WMNs. Additionally, a distributed implementation algorithm, namely Mod&Timer, to handle the placement of replicas for saving storage resources of the network is introduced. Simulation results are shown to demonstrate that the proposed cross-layer design outperforms many existing schemes in terms of QoE. More importantly, compared to other strategies without cross-layer interaction, the proposed cross-layer design satisfies the heterogeneous bandwidth constraints of end users.

This chapter presents some new results for the cross-layer design for packet data transmission applied to co-located Multiple-Input Multiple-Output (MIMO) with maximum ratio transmission. In this context, closed-form expressions for the average packet error rate, average achievable spectral efficiency, and outage probability are derived. This enables users to evaluate the performance of our cross-layer design under predetermined delay and packet loss constraints. Some important results in the high Signal-to-Noise Ratio (SNR) regime are also provided, revealing insights on how system and channel parameters affect system performance. More importantly, a mathematical analysis is proved to be a powerful tool to investigate the impact of practical issues such as channel estimation error, feedback delay, antenna correlation, and rank-deficient MIMO channel matrix on the performance of cross-layer design.

In this chapter, a cross-layer multidimensional discrete-time Markov chain-based queuing model is developed to jointly exploit the capabilities of (hybrid) automatic repeat request protocols and adaptive modulation and coding. Based on the stationary state probability distribution of this multidimensional discrete-time Markov chain, closed-form analytical expressions for performance metrics such as throughput, average packet delay, and packet loss rate are derived. Furthermore, the proposed analytical framework is used to formulate multidimensional and simplified two-dimensional constrained cross-layer optimization problems aiming at maximizing the system throughput under prescribed quality of service constraints.

Preface

With the recent technological developments towards making use of the ever increasing Information and Communication Technologies (ICT) it is not surprising to see a unified move towards removing this technological bottleneck from the forthcoming systems. To this effect, a brief search of one of the most popular research paper databases, IEEE Xplore, carried out on 7th December 2011, revealed 9,561 results for the term 'Cross Layer' being mentioned in the search and 2,074 results for the papers with the name used in the title, where over a quarter of them were published in this decade, indicating the amount of pressure on the energy and other scarce recourses of the new systems, such as smart sensors. This also indicates the need to move away from the traditional layering bottlenecks embedded in the infrastructure of our systems.

Considering that layering breaks up the holistic nature of any atomic action and its context apart from its valuable strategies, which derive their ability to manipulate their untouched properties, like any technological side effect, started to reveal some facts within the present decade. The traditional layered models of networks, while valuable conceptually and pedagogically, were never a realistic implementation of strategies, which is more obvious in autonomous systems.

Following the advent of the Open Systems Interconnected (OSI) and its further deployment for the Internet, stimulating the Internet as the main thrust for the communication infrastructure started some 30 years ago for an era where all ICT services use packets built upon layers and protocols to carry information and manipulate applications. Although these layers provide the timely desired precision and required reliability for effective services of the last five decades they all add significant overheads and associated processing-time and bandwidth overheads due to excessive moves of the packets up and down the layers for interactive processes required by the protocols making many new applications highly inefficient. This problem can become critically important and often represent a bottleneck for many multi-user applications using the wireless as their transmission media. The highly variable condition of the channel caused by distributed interference, multipath and continuously changing noise conditions can make most viable services undesirable. Therefore, due to the rapid rise of demand for wireless broadband and a ubiquitous access paradigm for future service developments the layering deficiency is important and should be resolved urgently. The Cross Layer Algorithms (CLA) in general provide new innovative ways to find suitable scheduling mechanisms for achieving optimized objectives such as minimized or bounded transmission power, delay and packet loss and maximized worst case throughput, fairness, and resource utilization.

Fundamentally, wireless comes with many potential and unique features over its wired counterparts in either forms of economical investment or ease of use paving new avenues providing a whole new range of services under a new wireless application paradigm in less than a quarter of a century under new flagships such as ubiquitous access applications and the Wireless Sensor Networks (WSN). There are, however, many serious problems hindering proper uses of this technology to deliver proper applications. With this effect and many other signs of true industrial impact from emerging wireless technological developments in various sectors of the economy, we can consider three main areas of concern: (a) extremely high consumption of processing power; (b) lack of visibility or knowledge of other layers' parameters and conditions; and (c) lack of control or flexibility to adaptively accommodate further intelligence in the communication networks.

In order to explain further details here we can look further into some details. A wireless channel is inherently a shared medium. In the past century, sharing has helped the telecommunications industry to grow extensively. However, efficiency of this valuable resource heavily depends on the physical-layer's statistical structure and the nature of the users' activity where the layering complexity can be regarded as nothing but a serious bottleneck for a proper use of the channel. For example, though knowing the state of channel has a potential to improve the system's performance significantly but due to the application's overheads in a fading channel the service in a poor channel is excessively more than one estimated upon the Signal-to-Noise Ratio (SNR) in a good channel state. This problem becomes more critical if the energy consumption at the node is of vital concern where for example without a cutting across the layers may result in much higher power consumption in the device due to considerable error rates at lower transmission power and higher interferences. The followings points may exemplify the need and the benefits gained from using cross layer concepts in various aspects of wireless communications.

- Most *adaptive* systems make use of automation supported by cross-layer decisions, which is normally embedded in the device. The adaptive loading of a wireless access environment such that channel condition can be used to decide on the optimum number of users.
- The *physical* channel of wireless links is subject to serious weaknesses in the networks, usually caused by fading and to combat this is regarded as one of the key challenges in wireless networks. If not controlled properly it fades naturally creating a highly disturbing fluctuating SNR variable in the system and as a result injects heavy uncontrollable bursts of error causing heavy packet losses and sometimes a long-term outage degrading many real time services very heavily.
- *Power* for transmission and processing signals in many new wireless networks used for long term continuous monitoring and surveillance systems is a critical factor for the nodes to stay active. Power consumption can be minimized by the use of some intelligence normally available in the higher layers and the application but due to the fact that the existing layer protocols can only communicate in a strict manner with the use of intelligence and excessive waste of power for going up and down the layers we often end up in a worse condition.
- *Scheduling* is a well-established and widely used method in network design. An intelligent scheduling method that can work better than a basic round robin requires an effective use of certain network information usually gathered from various resources and network parameters counted critical to the objectives of the design.

Although cross-layer is not fundamentally intended by choice to violate strict network layering infrastructure of the existing standards, many experts believe that it may happen one day so that we will have a more flexible stack of layering and protocols standards that gives more freedom to embed intelligent layer agents enabling the systems to accommodate better and smoother systems, networks, and applications. Let us classify the cross-layering approach into two main groups upon their adoption process. One approach is *emulation* of the existing standards for a better network design as incremental improvement, where new designs merely simplify each individual system internally. We call this group Cross-Layer Emulation (CLE) which can be divided into two subgroups of Passive Layer Emulation (PLE) where the cross-layer simplifies the system for a limited number of protocols and interfaces, and Active Layer Emulation (ALE) for those extending the use of the layers' capability with extra information made available to other layers without violating the infrastructure.

The *second* group's approach makes better use of intelligence throughout the network using layer decomposition where the network is optimized for the application design rather than conforming to the standards. As these approaches consider clear objectives and are normally adopted to solve serious layering problems within unstructured ad hoc networks, we can call them the Network Utility Maximization (NUM) group. The NUM can also be divided into two subgroups of Unstructured Network Design (UND) for those with a tendency to stay compatible with the OSI as long as none of the fundamental objectives are jeopardized with Unstructured Network Optimization (UNO) for those heading towards total freedom from the traditional infrastructure (see Figure 1).

Figure 1. Cross-layer classification

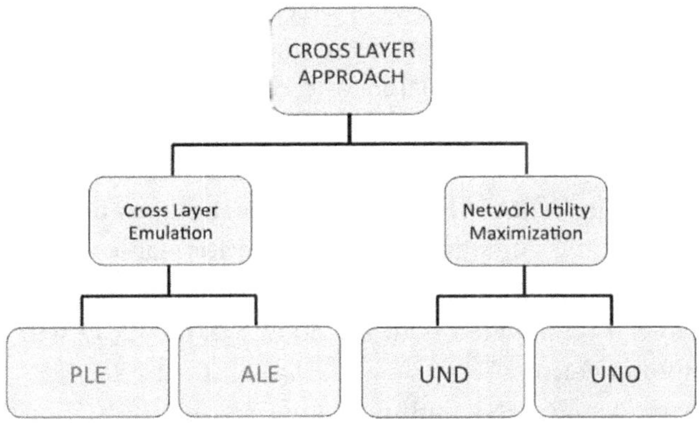

Cross-layer adaptive resource allocation techniques are found to be powerful techniques for achieving high throughput and high reliability over wireless fading channels. Recently, it has been revealed in the literature that cross-layer adaptation and optimization techniques can improve the overall system level Quality of Service (QoS) performance significantly over separate single layer adaptation and optimization techniques. **Chapter 1** discusses the novel cross-layer techniques that jointly consider the physical layer channel gain as well as the upper layer buffer occupancy and traffic information in order to find transmission rate and power policies that jointly optimize transmission power, buffer delay and packet overflow for an application specific bit error rate. A conceptual study of the cross-layer adaptation and optimization techniques, which fuels necessary motivation and direction on how to implement them in different wireless standards and devices, is provided. The associated system modeling, problem formulation and solution techniques are discussed and the benefits of cross-layer adaptation and optimization techniques as compared to single-layer counterpart with numerical results are demonstrated.

Wireless communications are a fast growing part of the telecommunication market. While new types of traffic and challenges related to the wireless medium are appearing, the methodologies for designing system architectures substantially remain the same. Under the increasing pressure of market, users, and physical medium issues, designers are in need of new approaches. Cross Layer then becomes a handy solution for coping with such problems. In fact, it allows both tighter optimization of the existing functionalities, and the introduction of new ones that do not fit within the traditional protocol stack design methodology. However, Cross Layer also carries a risk due to possibly unexpected and undesired effects. **Chapter 2** provides architecture designers with a set of guidelines synthesized from an analysis of the state of art, but enriched with the perspective of the development of the future generations of communication systems such as Cognitive Radio.

Quality of service has been an open issue from the beginning, due to the Internet paradigm of delivering the datagrams as simply as possible, on a hop-by-hop basis. As an answer to the inefficiency of the legacy solutions, rethinking of Internet principles brought among other ideas the self-management. Based on a clean-slate approach applied within the FP7-4WARD project, **Chapter 3** proposes a preliminary design and implementation of a key management capability of the near Future Internet, i.e. Cross-Layer QoS. Its applications are focused on congestion control, activation of network coding and QoS-aware routing based on a modified Dijkstra's algorithm.

Improvement of wireless and sensing technology enabled us to design a new network technology called Wireless Sensor Network (WSN). However, applications are greatly limited due to the resource constraint nature of wireless sensor nodes and the lack of flexibility in protocol design, until almost all the WSN protocols are developed for different layers independently based on traditional OSI layered architecture. Nowadays, researchers are envisioned to optimize resource utilization just breaking the barrier of layer abstraction. **Chapter 4** presents basic concepts of layered architecture design and its pros and cons for Wireless Sensor Network (WSN) applications followed by a survey of different applications of Cross-Layer Design (CLD) in WSN. All the different techniques used for CLD are also discussed in detail in this chapter. Then some guidelines are provided for stable and efficient cross-layer design. Finally, some open issues are indicated to show the future direction of research in cross-layer protocol design for WSN. The chapter is concluded with the hope that the CLD will have a widespread use in next generation WSN communication.

Body Area Sensor Networks (BASN) with miniature sensors providing wireless communications capabilities have become a promising tool for monitoring and logging vital parameters of patients suffering from chronic diseases such as diabetes, asthma and heart diseases. Particularly, with the technical development in BASN, a low-cost, high quality, convenient Electrocardiographic (ECG) diagnosis system becomes a future major tool for healthcare systems. However, numerous important research issues remain to be addressed in BASN for ECG transmissions. Among these, communication energy efficiency and security are the most important issues. **Chapter 5** introduces a survey of and analyzes effective cross-layer strategies for wireless ECG transmissions in BASN. The key idea of these cross-layer communication techniques is to take advantage of both source data properties and communication strategies for the optimization of the system energy efficiency while providing secure wireless ECG transmissions. The goal of improving communication energy efficiency is achieved by matching the source coding of ECG signals with the channel coding strategy. Thanks to the dependency property of the compressed ECG data, a selective encryption algorithm needs only to be applied on a very small portion of the transmitted data. At the same time, it provides a level of security equivalent to traditional full-scale encryption using block or stream ciphers without the burden of the associated energy and computational expense. In our example of an authentication scheme, sensors for the same body can authenticate each other by common traits such as inter-pulse intervals. Authors have analyzed the proposed cross-layer technique for ECG transmissions and validated the achieved energy efficiency improvements by both simulation and experimental results.

Cross-Layer Design is useful for wireless communication to improve the performance and efficiency. On the other hand, Cross-Layer Design is also helpful in enhancing network security. With the help of cross-layer information and interactions, the security overhead can be reduced and the security efficiency can be improved. **Chapter 6** reviews the existing literature and provides a classification that based on their methods and applications. According to the survey, these cross-layer security enhancements could be classified into four categories: 1) security protocol; 2) security policy; 3) key management; 4) intrusion detection.

The design of the mobility management is the key issue in the next-generation mobile networks. It is important to provide seamless service switching for a Mobile Station (MS) or a Mobile Node (MN) with the uninterrupted services during an IP-based session between different access networks, where IP convergence has led to the co-existence of several IP-based wireless access technologies and the emergence of next generation technologies. Many research results of cross-layer design in mobility recently had been able to minimize the data loss rate and delay time during switching so that users do not experience obvious and unacceptable interruptions during the handoff. The cross-layer design is the important approach for the mobility management. **Chapter 7** introduces and reviews existing protocols and then discusses the possible interest and useful applications of cross-layer in mobility management.

Chapter 8 presents joint interference suppression and power allocation algorithms for DS-CDMA and Multiple-Input Multiple-Output (MIMO) networks with multiple hops and amplify-and-forward and Decode-and-Forward (DF) protocols. A scheme for joint allocation of power levels across the relays and linear interference suppression is proposed. Furthermore, another strategy for joint interference suppression and relay selection that maximizes the diversity available in the system, is discussed. Simulations show that the proposed cross-layer optimization algorithms obtain significant gains in capacity and performance over existing schemes.

Because of the impact of noise, interference, fading and shadowing in a wireless network, there has been a realization that the strict layering of wireline networks may be unsuitable for wireless. It is the volatility over time that demands an adaptive solution and the basis of adaptation must arise by communication of the channel conditions along with the datalink settings. Video communication is particularly vulnerable because, except when reception is decoupled from distribution as in multimedia messaging, there are some real-time displays and decoding deadlines to be met. The predictive nature of video compression also makes it susceptible to temporal error propagation. In **Chapter 9**, case studies from the authors' experiences with broadband wireless access networks and personal area wireless networks serve to illustrate how information exchange across the layers can benefit received video quality. These schemes are adaptive and serve as a small sample of a much greater population of cross-layer techniques. Given the importance of multimedia communications as an engine of growth for networked communication, 'cross-layer' should be the first consideration in designing a video application.

Most of the wireless ad hoc networks have battery-powered nodes, so a lifetime of a network depends on the power of nodes. Power conservation is one of the main issues of wireless ad hoc networks. Power conservation can be achieved in two different ways: power control algorithms and power management algorithms. Transmission power control is one of the best examples of a cross-layer design problem. The purpose of **Chapter 10** is to discuss the transmission power control problem and one possible solution by using cross layer framework through a case study. This chapter describes two cross-layer power control protocols developed at two different layers and also presents the effect of applying the combination of these protocols through a simulation study using the GloMoSim system.

Wireless Sensor Networks are a promising way to interconnect smart objects sensing and acting on the environment, enabling the Internet of Things. However, this kind of network is particularly constrained: nodes have limited energy reserve, their CPU and memory resources are limited, and the radio bandwidth is very low. After having exposed the main approaches for the routing and MAC layers, **Chapter 11** explains why they must be jointly optimized, why a metric reflecting the radio link quality is necessary and how it should integrate different criteria in layers 2 and 3. Furthermore, a new architecture based on IEEE 802.15.4 and RPL is introduced so that both protocols work in symbiosis.

Radio spectrum has become a precious resource. Most frequency bands have been allocated for exclusive use in the US. However, studies have shown that a very large portion of the radio spectrum is unused or underused for long periods of time at a given geographic location. Therefore, allowing users without a license to operate in licensed bands while causing no interference to the license holder becomes a promising way to satisfy the fast growing need for spectrum resources. Dynamic spectrum access and cognitive radio are technologies for enabling opportunistic spectrum access and enhancing the efficiency and utilization of the spectrum. A cognitive radio adapts to the environment in which it operates by sensing the spectrum and then the opportunistic exploiting of unused and/or underused frequency bands in order to achieve certain performance goals. Due to the close coupling and interaction among protocol layers, the optimal design of opportunistic spectrum access and cognitive radio networks calls for a cross-layer approach that integrates signal processing and networking with regulatory policy making. **Chapter 12** introduces basic concepts, design issues involved, and some recent development in this emerging technological field. Future research directions are also briefly examined.

One of the major limiting factors preventing wide use of wireless sensor networks in practical scenarios is power consumption. Battery-less or passive sensors promise to alleviate this issue and enable a wide variety of embedded sensor applications such as structural health and vehicular monitoring, biomedical applications, smart homes, and smart grids. Embedding these sensors in structures without the need for changing batteries, their rugged design to withstand harsh environments and coded communication with multiple access features makes this technology a desirable candidate for a variety of applications. Design and analysis of these sensors from a cross layer point of view is studied in **Chapter 13**. State of the art in fabrication and test of this new class of wireless sensor systems is also reviewed. Interactions between lower layer with passive sensors and upper layer with active sensors—a different perspective on cross layer—is exploited to achieve significant performance gains in terms of signal to noise and interference ratio, correlation peak to side-lobe ratio, operation range, and data rate.

Chapter 14 describes the different aspects related to designing a suitable monitoring architecture for Cloud Computing, aiming to support cross-layer monitoring across all layers available in the Cloud stack. For this purpose, the importance of monitoring services in Cloud scenarios is outlined, followed by a comprehensible analysis of a wide set of distributed monitoring solutions. After that, the particular requirements related to cross-layer monitoring for Cloud Computing architectures are identified and explained in this chapter. Then, diverse aspects of which may fit a monitoring architecture for fulfilling such requirements are explained. Finally, some future research directions and conclusions are highlighted.

To pervasively and diversely support a wide range of multimedia services, several traditional replication techniques have been considered for improving data accessibility in Wireless Mesh Networks (WMN). The metrics used to evaluate data accessibility performance are often related to service integrity parameters of the lower layers such as the network, medium access control, and physical layers. However, these parameters cannot directly reflect the Quality of Experience (QoE) as perceived by end users. In contrast, QoE based parameters for video streaming applications are deduced from subjective tests and involve processing at the application layer including quality of playback in terms of spatial artifacts, smooth playback, and continuous playback. In **Chapter 15**, dependencies between the different protocol layers and across the hops through WMN will be exploited to deliver video streaming with high QoE. Specifically, a cross-layer optimization approach is applied to a replication strategy for video on-demand over WMN. Additionally, a distributed implementation algorithm, namely Mod&Timer, to handle the placement of replicas for saving storage resources of the network is introduced. Simulation results are shown to demonstrate that our proposed cross-layer design outperforms many existing schemes in terms of QoE. More importantly, compared to other strategies without cross-layer interaction, the proposed cross-layer design satisfies the heterogeneous bandwidth constraint of end users.

Multiple-Input Multiple-Output (MIMO) systems, where multiple antennas are deployed at both ends of the wireless links, can significantly increase the transmission rate and reliability of wireless networks. However, due to the increasing demand for high data rates and QoS, the improved performance of MIMO systems may reach its limit. In particular, deployment of many antennas at user devices is constrained by the desired compact designs of mobile equipment. On the other hand, by exploiting the time-varying nature of wireless channels, Adaptive Modulation and Coding (AMC) at the physical layer is utilized to enhance the throughput. In addition, Automatic Repeat Request (ARQ) at the link layer may be applied to circumvent the deleterious effects of fading channels where packets detected with errors can be retransmitted. In an attempt to maximize the spectral efficiency of co-located MIMO wireless systems under prescribed QoS constraints, a cross-layer design that jointly combines Adaptive Modulation (AM) at the physical layer and ARQ at the data link layer is considered. The main purpose of **Chapter 16** is to present a general framework for performance analysis of such a cross-layer design. The presented mathematical analysis can be extended to many transmission schemes with MIMO systems including spatial multiplexing schemes (e.g., the V-BLAST system), spatial diversity schemes (e.g., orthogonal space-time block codes), and opportunistic schemes (e.g., maximum transmission ratio) This chapter will therefore present some new results for the cross-layer design for packet data transmission applied to co-located MIMO with maximum ratio transmission. In this context, closed-form expressions for the average packet error rate, average achievable spectral efficiency, and outage probability are derived. Some important results in the high SNR regime are also provided, revealing insights on how system and channel parameters affect system performance. More importantly, the mathematical analysis is proved to be a powerful tool to investigate the impact of practical issues such as channel estimation error, feedback delay, antenna correlation, and rank-deficient MIMO channel matrix, on the performance of cross-layer design.

In order to provide a heterogeneous quality of service guarantees to applications, most wireless communications standards combine the error-correcting capability of the automatic repeat request protocols at the data link control layer with the adjustment ability of adaptive modulation and coding strategies at the physical layer. In **Chapter 17**, a cross-layer multidimensional discrete-time Markov chain based queuing model is developed to jointly exploit the capabilities of the automatic repeat request protocols and adaptive modulation and coding. Based on the stationary state probability distribution of this multidimensional discrete-time Markov chain, closed-form analytical expressions for performance metrics such as throughput, average packet delay and packet loss rate are derived. Furthermore, the proposed analytical framework is used to formulate multidimensional and simplified two-dimensional constrained cross-layer optimization problems aiming at maximizing the system throughput under prescribed quality of service constraints.

Designing a wireless video communication system is a challenging task due to high error rates of wireless channel, limited and dynamically varying bandwidth availability, and low energy and complexity requirements of portable multimedia devices. Scalable video coders having excellent rate-distortion performance are most suited to cope with the time varying bandwidth of wireless networks, but encoded bits are extremely sensitive to channel errors. **Chapter 18** presents a reliable video communication system exploring opportunities offered by various network layers for improved overall performance, while optimizing the resources. More specifically, cross-layer approach for Unequal Error Protection (UEP) of scalable video bit-stream is the main theme of this chapter. In UEP, the important bits are given a higher protection compared to the other bits. Conventionally, UEP is achieved by using Forward Error Correction (FEC) at the application layer. However, UEP can also be provided at the physical layer using a hierarchical modulation scheme. In this chapter, cross-layer design methodology for UEP that rely on interaction between the application layer and the physical layer to achieve reliable and high quality end-to-end performance in wireless environments is discussed. The discussion is mainly focused on wavelet-coded video, but it is applicable to other embedded bit-streams as well.

Multimedia applications are becoming the main driver for advancement in the telecommunications industry. Increased demand for anytime anywhere mobility has given rise to the development of ad hoc networks. Supporting multimedia applications over ad hoc networks is extremely challenging due to the stringent QoS requirements and the lack of infrastructure. In the recent years, cross-layer design techniques have emerged as a potential solution to address these challenges. **Chapter 19** takes a cross-layer design approach and presents the design and analysis of wireless ad hoc networks that support multimedia applications with scalable QoS. The chapter has contributed in two areas. Firstly, a novel, cross-layer, Hierarchical Clustering, Provisioning, and Routing (HCPR) scheme for ad hoc networks, is proposed, implemented and analyzed. Secondly, a novel methodology is developed for multimedia network analysis and its effectiveness is demonstrated by the analysis of the HCPR scheme and other well-known protocols. The HCPR scheme is implemented as an extension to the OPNET simulation software and is analyzed in detail for its QoS performance to deliver multimedia applications over ad hoc networks. It is compared with three well-known and widely used routing protocols, Ad Hoc On Demand Distance Vector (AODV), Optimized Link State Routing (OLSR), and Geographic Routing Protocol (GRP). Several networking scenarios have been carefully configured with variations in networks sizes, applications, codecs, and routing protocols to extensively analyze the proposed scheme. The HCPR enabled ad hoc network outperforms the well-known routing schemes, in particular for relatively large networks and high QoS network loads. These results are promising because many QoS schemes do work for small networks and low network loads but are unable to sustain performance for large networks and high QoS loads. Several directions to extend this research for future work are given.

As IEEE 802.11 wireless devices have become increasingly widespread, providing Quality of Service in the context of H.264/AVC, the video-coding standard for future multimedia networking has become an important issue in the fields of communication and networking. Cross-Layer Adaptive Video Prioritization (CAVP) is a cross-layer framework that prioritizes video frame transmission according to the application-layer information and the MAC layer transmission condition. In **Chapter 20**, a Peak Signal-to-Noise Ratio (PSNR) estimation method is proposed to sort out different priorities of H.264/AVC (Advanced Video Coding) video frames at the application layer to provide user-Centric media quality estimation. Compared to previous heuristic algorithms, a theoretic access delay estimator to monitor the wireless medium access delay at the MAC layer is investigated. In addition, an admission control is employed to serve the delay-sensitive video application and to give higher priority to those critical video frames. Video packets are dynamically classified into different 802.11e access categories according to the level of wireless medium access delay and the priority of the video frames. The myths of naïvely prioritizing video packets based on I/P/B types as well as naïvely assign packets to high priority access categories in 802.11e are resolved. Rather than creating a complex scheme that is unable to be implemented in practical scenarios, the proposed scheme is designed with practical implementability in mind. The proposed scheme is implemented with a Click kernel module and the MadWifi WLAN driver. The performance of the proposed CAVP design is evaluated by both NS-2 simulation and real test-bed experiments, and results show that it enhances receiving video quality in error-prone wireless networking environments.

New and diverse applications for Wireless Sensor Networks (WSN) have also led to new challenges. Cross-layer approaches have proven to be the most efficient optimization techniques for these problems, since they are able to take the behavior of the protocols at each layer into consideration to overcome the constraints for WSN. Thus, **Chapter 21** focuses on identifying the core problems of WSN and collects available cross layer solutions for them that have been proposed so far. The literatures on cross-layer protocols, protocol improvements, and design methodologies for WSN are reviewed and taxonomy is proposed in order to provide insights on the identification of cross layer design in WSN. The open issues are discussed in detail for future research and precautionary guidelines for cross layer design to WSN is indicated.

In general, we are very proud to be able to receive so many quality proposals from the experts in the field to enable us to compile such a valuable piece of technological advancement for such a challenging part of today's pressing technological development.

29 February 2012

Habib F. Rashvand
University of Warwick, Coventry, UK

Yousef S. Kavian
Shahid Chamran University of Ahvaz, Iran

Acknowledgment

We would like to thank the chapter authors for providing us with the fundamental constituents of this volume and the reviewers who provided detailed and constructive feedback. Also, thanks are due to members of staff at the publisher who have guided us through the publishing process.

Finally,
Habib dedicates this work to Liz, Leila, Cyrus, and Joan
Yousef thanks his wife, Mandana, for her patience, understanding, and support during the preparation of this manuscript.

Habib F. Rashvand
University of Warwick, UK

Yousef S. Kavian
Shahid Chamran University of Ahvaz, Iran

Chapter 1

Cross-Layer Adaptive Packet Scheduling over Fading Channel:
A Decision Theoretic Approach

Ashok K. Karmokar
Ryerson University, Canada

Alagan Anpalagan
Ryerson University, Canada

ABSTRACT

Cross-layer adaptive resource allocation techniques are found to be powerful techniques for achieving high throughput and high reliability over wireless fading channels. Recently, it has been revealed in the literature that cross-layer adaptation and optimization techniques can improve the overall system level Quality of Service (QoS) performance significantly over separate single layer adaptation and optimization techniques. In this chapter, the authors discuss the novel cross-layer techniques that jointly consider the physical layer channel gain as well as the upper layer buffer occupancy and traffic information in order to find transmission rate and power policies that jointly optimize transmission power, buffer delay, and packet overflow for an application specific bit error rate. They provide a conceptual study on the cross-layer adaptation and optimization techniques, which fuels necessary motivation and direction on how to implement them in different wireless standards and devices. The authors discuss the associated system modeling, problem formulation, and solution techniques as well as show the benefits of cross-layer adaptation and optimization techniques as compared to single-layer counterpart with numerical results.

DOI: 10.4018/978-1-4666-0960-0.ch001

INTRODUCTION

Adaptive resource adaptation schemes in different layers of Open System Interconnection (OSI) protocol stack have proven to be powerful techniques both in increasing high throughput and in achieving high reliability over time-varying fading channels. Adaptive resource allocation has already been used in several wireless standards. The key idea of these schemes is adaptation of some of the parameters, e.g., transmitter power level, symbol transmission rate, bandwidth, frame-length, packet retransmission rate, constellation size, Bit Error Rate (BER), coding rate/scheme, or any combination of these parameters with instantaneous channel conditions. Good performance of these schemes requires accurate channel estimation at the receiver and a reliable feedback path between that estimator and the transmitter. Adaptive modulation schemes using *M*-QAM and *M*-PSK for flat fading channels are examined in (Chuang & Goldsmith, 2001). The data rate, transmit power, and instantaneous BER are varied to maximize spectral efficiency, subject to average power and BER constraints.

Adaptive hybrid automatic repeat-request schemes are another important technique for increasing both the throughput and the reliability of packet transmission over time-varying fading channels. Hybrid Automatic Repeat-Request (HARQ) schemes include parity bits for both error detection and error correction. They combine the throughput efficiency of physical-layer Forward Error-Correction (FEC) coding and the reliability of data-link layer error-detection coding. The key idea of adaptive coding schemes is to use a lower-rate coding or more error protection when the channel condition is worse, and vice versa (Vucetic, 1991).

Over the last decade, various cross-layer techniques have been developed in the literature and proposed for different wireless networks and standards. We summarize few recent works as follows. In (Pantelidou, et al., 2011), the authors studied the problem of optimal scheduling of multicast traffic in time-varying wireless networks in the framework of utility maximization. An online gradient-based scheduling and rate control policy is introduced that identifies when to access the wireless medium and with waht rates. A cross-layer design of transmitting scalable video streams from a base station to multiple clients over a shared fading wireless network by jointly considering the application layer information and the wireless channel condition is considered in (Zhang, et al., 2010). In Argyriou (2010), a cross-layer framework for optimizing the performance of opportunistic network coding in wireless multihop ad hoc networks is presented considering backlogged nodes and multiple unicast packet flows. Considering the combinatorial complexity of providing an optimal policy, Pantelidou and Ephremides (2010) studied polynomial-time suboptimal alternative for solving the problem of joint scheduling and rate control for multicast traffic in wireless networks in order to jointly maximize both the sum throughput maximization and the proportional fairness.

On the assumption of independent users' states, Poggioni et al. (2010) analyzed the QoS of scheduling algorithms for heterogeneous users in multiuser wireless systems that take advantage from a cross-layer design with both *Adaptive Modulation and Coding* (AMC) and Automatic Repeat Request (ARQ). The average delay, the Packet-Loss Rate (PLR), and the throughput of a scheduling algorithm based on the channel condition, the buffer occupancy, and the number of retransmissions, of users belonging to different service classes are analytically evaluated using FSMC channel model. Utilizing machine learning techniques, an autonomous host-based intrusion detection system for detecting malicious sinking behavior that maximizes the detection accuracy by associating MAC layer feature with other cross-layer features to define a routing behavior is studied for ad hoc networks in Joseph et al. (2011). Kuran et al. (2010) discussed cross-layer routing-scheduling scheme to illustrate the application and realization of the cross-layer design for improving the performance of IEEE 802.16 WiMAX. In Pompili et al. (2010), the authors studied the cross-layer interactions of key underwater communication functionalities such as modulation, forward error correction, medium access control and routing, and developed a distributed cross-layer communication solution that allows multiple devices having delay and loss sensitivity to efficiently and fairly share the bandwidth limited high-delay underwater acoustic medium.

An adaptive cross-layer design for the downlink multiuser Orthogonal Frequency Division Multiplexing (OFDM) systems to maximize the weighted sum capacity of all users, where each user has multiple heterogeneous traffic queues simultaneously, is studied in Zhou et al. (2010). The weight design in packet dependent scheduling considers the delay, size and Quality of Service (QoS) priority level of packets. In Seo and Leung (2011), by exploiting cross-layer information on multiple access interference, contention resolution algorithms for immediate and deferred first-transmission protocols in a multi-packet reception slotted ALOHA system based on code division multiple access is studied. Melodia et al. (2010) discussed a cross-layer communication architecture based on the time-hopping impulse radio ultra wide band technology, where the objective is to reliably and flexibly deliver QoS to heterogeneous applications in WMSNs, by leveraging and controlling interactions among different layers of the protocol stack according to applications requirements. A cross-layer framework to accurately selecting modulation rates for time-varying channel conditions is discussed in Camp et al. (2010) for avoiding performance degradations due to rate overselection when channel conditions degrade or underselection when channel conditions improve.

The problem of downlink utility-maximization for wireless mesh networks with orthogonal frequency-division multiple access is formulated as a cross-layer design of transport-layer rate control and channel-aware and queue-aware OFDMA MAC/PHY scheduling to efficiently utilize the scarce radio resources in Fathi et al. (2010). Ramis et al. (2011) developed a cross-layer multidimensional discrete-time Markov chain based queuing model to combine the error correcting capability of hybrid automatic repeat request protocols at the data link control layer with the adaptation ability of adaptive modulation and coding strategies at the physical layer. jointly exploit the capabilities of HARQ and AMC. Closed-form analytical expressions for performance metrics such as throughput, average packet delay, and packet loss rate, which were caused either by buffer overflow or by exceeding the maximum number of allowed retransmissions, are derived. In Ding et al. (2011), cross-layer routing and path selection strategy are investigated for cooperative wireless vehicular ad hoc networks that achieve a better tradeoff between the transmission power consumption and end-to-end reliability.

Traditional resource adaptation, allocation and optimization schemes in the literature assume that the transmission buffer is of infinite length with enough packets in the buffer for transmission. There are also few cross-layer design techniques in the literature (Liu, et al., 2004) that consider only channel state to make transmission decision; however, in these techniques the buffer occupancy is looked at only when it has insufficient packets. These techniques do so only to calculate the throughput and to turn off the transmitter. They do not make transmission decision considering upper layer parameters, such as buffer occupancy, traffic type, etc. These traditional channel-dependent adaptive transmission techniques cannot guarantee application specific delay, packet overflow and so on requirements in the upper layers. The system level throughput suffers due to a large number of packet losses due to overflow. In addition, the transmission power is not controlled depending on the buffer information. Due to lack of intelligent control on the use of transmitter power levels, the transmitter may end up spending more than double power to transmit a given amount of data over a period. Generally, two sets of issues have been analyzed in virtual isolation from each other in the traditional resource adaptation literature: some authors have focused mainly on the physical layer channel modeling, coding, modulation and detection, while others have concentrated on data link layer throughput, delay, packet overflow, etc. Recently, it has been realized that the system performance, such as throughput, power, delay, overflow, BER can be improved significantly by adapting transmission parameters with the aid of the information from other layer parameters along with own layer parameter.

Intelligent decision making on transmission rate and power in wireless networks is crucial due to their limited resources, such as bandwidth, transmit power, etc. Future generation wireless networks would be included with a scheduler that considers green metric (such as energy efficiency) along with traditional throughput efficiency, etc., while making the transmission decision. Because of time-varying fading of the wireless channels, the decision on the transmission rate and power for wireless systems primarily depends on channel conditions. Modern and future wireless networks are envisioned to support numerous services with a wide range of QoS, such as delay, rate, BER, packet dropping, etc. requirements while minimizing carbon footprint and green house gas emission. Different type of traffic has different statistical properties. Also, different traffic demands different type of QoS requirements. Non-real time data traffic (e.g., file transfer) can tolerate delay, but has stringent BER requirement. On the other hand, real-time traffic (e.g., video streaming) can tolerate certain degree of errors, but has stringent delay requirement. The nature of the real network traffic is generally randomly varying, bursty and correlated. The randomly varying incoming traffic plays an important role in scheduling wireless resources, as the practical systems are equipped with finite storage buffer. Sometime, the buffer may be full and other time it may be empty. Therefore, optimal channel-adaptive schemes do not optimize the systems as a whole for all the time. Due to the randomness of incoming packets and channel gains, the exact scheduling policies cannot be determined using static channel-dependent optimization techniques.

Since the packet scheduling problem is inherently dynamic in nature, the use of stochastic dynamic programming algorithms seeking the optimal adaptation policies makes more sense. Dynamic programming methods are very popular techniques in the fields of artificial intelligence, control and finance to realize important decision-making task. In this chapter, we discuss dynamic programming based decision making techniques for transmission over fading channels considering both the physical layer and the data-link layer optimization goals. At the physical layer, our goal is to optimize the transmission power, BER and spectral efficiency. On the other hand, at the data-link layer, our goal is to optimize the effective throughput, the delay, the packet overflow, the packet error rate and the retransmission rate. Optimal policies can be computed off-line using Markov Decision Process (MDP) framework and these policies are stored in the memory of the wireless terminals. After determining the instantaneous state of the system, the optimal policy can be applied. With the advancement of modern low-power CMOS VLSI devices, the memory and computational speed of wireless devices are no longer a bottleneck like past.

In this chapter, we present a conceptual study on cross-layer adaptation techniques over fading channels based on MDP framework. The study provides the readers with very important insight on the benefits of cross-layer adaptation in wireless systems. We present a general framework for cross-layer adaptation scheme, and then discuss the related models with an example without loss of generality. We discuss the cross-layer adaptation of the data rate and transmission power when statistics of the traffic and the channel are fully known. We describe the system model along with

Figure 1. Conceptual illustration of cross-layer adaptation technique

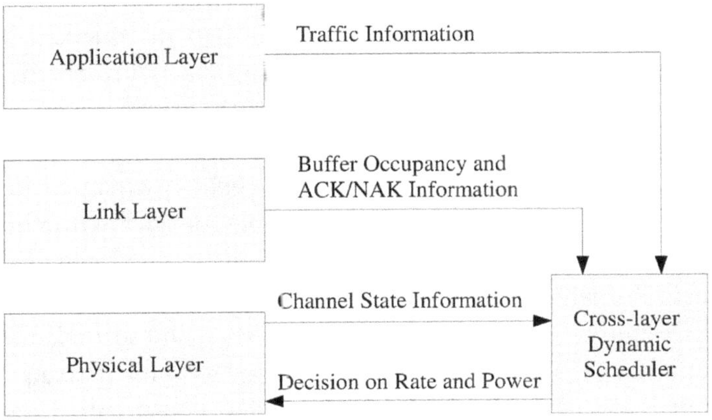

some discussion on the traffic, channel and buffer modeling. We then show how to formulate the adaptation problem as a MDP using different optimization criteria. We provide all the ingredients for the formulation considering three objectives, namely, power, delay and overflow, and define the related cost functions. We then discuss different possible optimal solution techniques for the problem and provide some results on the structure of the policy and the benefits of cross-layer policy as compared to single layer policy. We conclude the chapter with some future research directions.

BACKGROUND

Traditionally, transmission decision on rate and power at the physical layer is taken based only on the channel quality assuming that the buffer capacity is infinite and it always holds enough number of packets for transmission. The performance of wireless networks can be improved significantly by adapting system parameters with higher-layer parameters in addition to physical-layer time-varying channel gain. Therefore, the design of wireless protocols that rely on the interactions between various layers has received a significant attention in the wireless research community and become an active area of promising research. Cross-layer adaptation includes packet scheduling techniques, where the transmission decision is taken at the physical layer based on physical as well as upper layer information. The idea can be shown with a simple block diagram in Figure 1, where a packet scheduler is fed with the physical layer channel gain, link layer buffer occupancy and automatic repeat request feedback, and application layer nature of traffic. To the best knowledge of the authors, the study in this field is initiated in Collins and Cruz (1999). Due to its promise to increase data rate for delay-limited application and to decrease power for energy-limited systems, it is a growing potential area of research today.

Early work in Collins and Cruz (1999) developed a dynamic programming framework for transmission policies over a simple two-state Gilbert-Elliott channel model, with constraints on average delay and peak power. The authors consider critical backlog transmission policies and shows by numerical study that critical backlog policies are near optimal. Berry and Gallger (2002) addressed the issue of optimal trade-off between the average power and the average delay for a single user communicating over a memoryless block-fading channel from an information-theoretic standpoint and the behavior of this tradeoff is quantified in the regime of large delay. In Rajan et al. (2004), the utility of packet scheduling for independent and identically distributed (i.i.d.) traffic over both Additive White Gaussian Noise (AWGN) channel and fading channels is demonstrated. Convexity properties and characterization of the delay-power region of different schedulers are discussed. Goyal et al. (2008) have extended some structural results given in Berry and Gallager (2002) to develop optimal policies that minimize mean delay subject to an average power constraint for an i.i.d. channel. In Karmokar et al. (2006) and Djonin et al. (2008), the authors studied the optimal power and rate allocation problem for wireless systems, where the cross-layer adaptation policy is chosen to optimize the transmission power, the buffering delay, the BER and the packet overflow. The authors formulated the problem as average cost per stage MDP problem and sought optimal solutions based on both the unconstrained and constrained formulations. In Ho et al. (2009), a MDP-based user selection, power selection and rate selection technique is studied. A trade-off between the energy and the packet loss rate subject to a hard delay constraint is studied over correlated fading channels using finite horizon MDP framework in Ngo et al. (2009). It is shown that the optimal policy is threshold in nature under some constraints on costs. An optimization-based suboptimal scheduler is described in Wang and Mandayam (2004) for an i.i.d. block-fading channel. The authors optimized the average power subject to the average delay and the packet loss rate constraints.

An offline packet scheduling scheme for an AWGN channel has been analyzed in Biyikoglu et al. (2002), with the goal of minimizing energy, subject to a deadline constraint. An online lazy packet scheduling algorithm is also devised that varies transmission time according to backlog. It is shown to be more energy efficient than a deterministic scheduler with the same queue stability region and the similar delay. Study with similar objective is also carried out in Zafer and Modiano (2008), where the authors used cumulative curves methodology and a decomposition approach. A heuristic policy for the case of arbitrary packets arrivals to the queue with individual deadline constraints is presented. Optimal transmission scheduling with energy and deadline constraints for a satellite transmitter is given using a Shannon capacity equation in Fu et al. (2006). The authors consider two optimization problems: first, maximization of average data throughput under a given number of time-slots and a fixed amount of energy, and second, minimization of expected energy to send a fixed amount of data given deadline constraints. Djonin and Krishnamurthy (2007) studied Constrained MDP (CMDP) based techniques for optimal power and rate allocation in Multiple-Input Multiple-Output (MIMO) wireless systems. They used the concepts of stochastic dominance, submodularity and multimodularity to prove the monotonicity property of the optimal randomized policy. A systematic framework for cross-layer optimization, where each layer make autonomous decisions to maximize the individual discounted long-term utility by exchanging information among layers is studied in Fu and Schaar (2009). In this work, a layered dynamic programming operator is proposed that determines transmission strategies in each layer.

Because of the random variation of the channel, SNR errors occur in bursts. Hence, sometimes the perfect state of the current channel cannot be accurately predicted at the receiver and may not be available at the transmitter while transmitting. Rather, the packet error feedback signal is received from the receiver after the packet has been transmitted. In the following literature, no perfect CSI is assumed at the transmitter. In Choi et al. (1999), the authors formulate the decision process of determining transmission (when to attempt or suspend) over a Gilbert-Elliott fading channel as a Partially Observable Markov Decision Process (POMDP). The optimal policy for the throughput vs. energy efficiency tradeoff problem is derived for a time horizon less than or equal to thirteen, and shown to be a threshold rule that varies with the memory present in the channel error process. A suboptimal, limited-lookahead implementation of this policy is also simulated and its performance compared with the persistent retransmission and probing protocols. In Karmokar et al. (2006), policy heuristic based solution for packet scheduling is studied and shown to perform quite close to optimal policy.

CROSS-LAYER PACKET SCHEDULING

Let us consider a scenario where a pair of users is communicating with each other over a Multiple Input, Multiple Output (MIMO) wireless fading channel. The duration of the communication time is termed as horizon and denoted as H, which is divided into a finite number of discrete time-slots. Let us denote the duration of a time-slot by T_B seconds. The time-slot is the basic unit of the time that corresponds to a single block (also called physical layer frame) of N_f channel uses. That is, the transmitter transmits N_f modulated symbols over the channel in time T_B. In practice, the channel gain from one time-slot to next time-slot is correlated in time. The gain of the channel varies slowly and therefore it remains the same during a slot (Berry & Gallager, 2002). This type of channel can be described as a block-fading Finite State Markov Channel (FSMC). The transmitter is equipped with a buffer of limited storage capacity. Let B denote the maximum number of packets that the buffer can hold. The buffer receives time-varying traffic from the upper layer application. Since the incoming traffic is randomly varying and the transmitter takes a random number of packets for transmission in each time-slot, the buffer occupancy also varies randomly. Unless otherwise specified, in this chapter we use the superscript n to denote the value of certain variables at time-slot $n \in \mathcal{T} = \{1,2,\ldots H\}$. For example, b^n denotes the number of packets in the buffer in the beginning of time-slot $n \in \mathcal{T}$.

General System Description

The block diagram of the cross-layer adaptation system is shown in Figure 2. Let N_p denote the number of bits in a packet. At the beginning of each time-slot n, the transmitter takes w^n packets from the transmitter buffer and encodes the corresponding $k=w^nN_p$ bits using a high-rate error detection code (k',k). Usually, Cyclic Redundancy Check (CRC) is used for error detection purposes. We assume that it is capable of detecting all the errors perfectly in the received codeword. These encoded bits are subsequently encoded using a FEC code, such as, convolutional code, turbo code, block code, etc. The resulting codeword is modulated with an adaptive multi-level modulator and transmitted with controllable power level using multiple antennas over the fading channel in time-slot n.

An adaptive cross-layer power and rate scheduler at the transmitter is fed with various information from different layers. The scheduler has the information on the current channel state, the buffer state and the traffic state. Based on the information, the scheduler takes control action. The modulation and coding rate, and the transmission power are dependent on the action taken at the start of that particular time-slot. The control action chosen at the start of a time-slot is continued until the end of that time-slot. As mentioned before, it is assumed that all the packets transmitted in a time-slot experience the same channel gain because of slowly-varying block fading channel. Note that the processing unit at the data-link layer is a packet consisting of N_p bits and the processing unit at the physical layer is a frame consisting of N_f modulated symbols. We assume a Nyquist pulse shaping filter with bandwidth $W_f=1/T_s$, where $T_s=T_B/N_f$ is the symbol duration. Therefore, the number of symbols per frame can be expressed as:

Figure 2. Block diagram of the cross-layer adaptation system

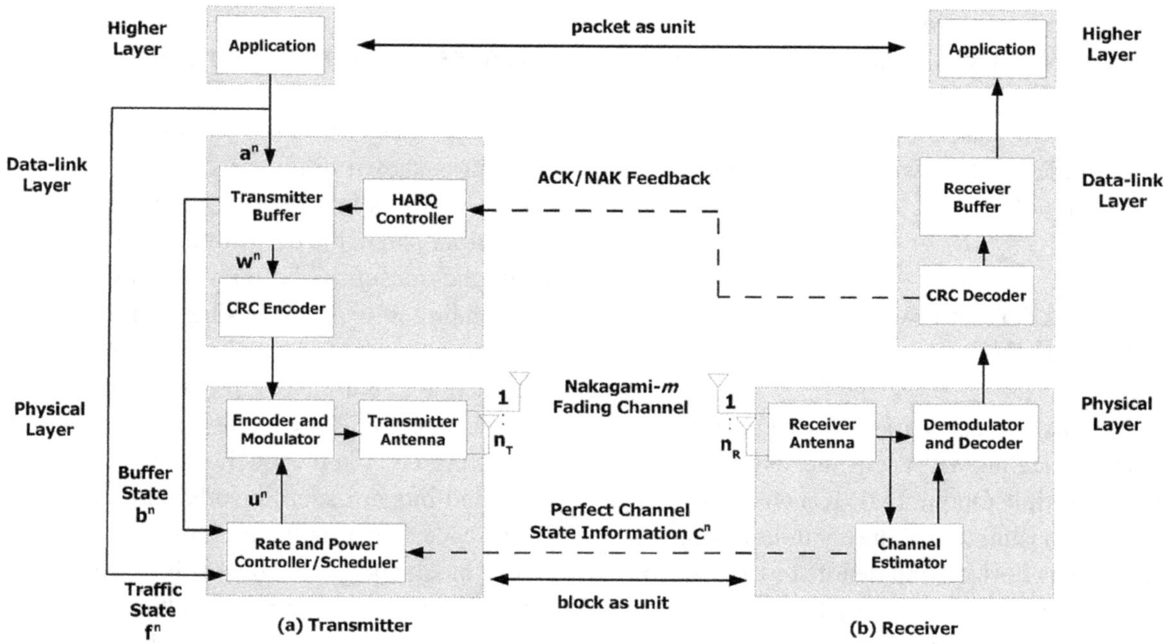

$$N_f = N_c + \frac{w^n N_p}{R_d R_c R_i} \qquad (1)$$

where N_c is the number of pilot and control symbols, and R_d=*k/k'* and R_c=*k'/n'* are the error detection and error correction coding rate, respectively. R_i=log$_2$(M) is the number of bits that modulates a single transmitted symbol in the frame corresponding to the constellation *M* of the *M*-ary modulation scheme and *n'* is the total number of bits to be transmitted after adding error detection and coding redundant bits with the information bits.

After being received at the receiver, the frame is demodulated. The error correction decoder first attempt to correct any errors in the received demodulated frame, then the decoded frame is checked for error detection. Observation feedback is sent to the transmitter over a noiseless and fast feedback channel. A Stop-and-Wait ARQ (SW-ARQ) or Selective-Repeat ARQ (SR-ARQ) protocol can be used in the data-link layer for handling retransmission of erroneous packets. If no errors are detected, the packets are delivered to the higher-layer application and an ACK observation feedback is sent to the transmitter. Accordingly, the transmitter removes those packets from the buffer. Otherwise, the receiver discards the packets and sends a NAK observation feedback requesting a retransmission of the same packets. The process is repeated until the packet is successfully received. We assume that the observation feedback is received without any delay. We denote the observation state space by $\mathcal{O}_\omega = \{\omega_1, \omega_2\} = \{ACK, NAK\}$.

Markov Traffic Modeling

Packet-based communications is the basis of the next-generation wireless system that is introduced to support wide range of QoS demands, such as throughput, delay, BER, etc. The multimedia traffic to be supported in the present and future wireless networks is broadly categorized into three classes: 1) voice, a relatively low bandwidth, Constant Bit Rate (CBR) real-time traffic, 2) a wideband, Variable Bit Rate (VBR) real-time traffic, for example, video streaming, tele-conferencing, games, etc., and 3) a VBR non real-time (best-effort) data traffic, for example, remote database access, client-server interaction, http access, etc. A mobile terminal may set up sessions for voice, data, as well as video and it may switch among them during a call. Real-time traffic (e.g., voice and video) can tolerate certain degree of bit errors, but have stringent delay requirements. On the other hand, non real-time traffic, such as data can tolerate substantial delay, but may require very low BER. The nature of the real network traffic is bursty and correlated, and therefore it can be modeled accurately with a tractable Markov Modulated Poisson Process (MMPP) (Babu, et al., 2001). While some traffic, such as interactive data and compressed video, is highly bursty, other traffic, such as large files, is continuous. Therefore, incorporation of traffic model into the analysis is important for adaptive packet transmission scheme in wireless networks.

In a Markov Modulated Poisson Process (MMPP) model, in any state the incoming traffic is Poisson distributed and the transitions between the states are governed by an underlying Markov chain. Let $\mathcal{F} = \{f_1, f_2, \ldots, f_F\}$ denotes the set of states (also called phase) of the incoming traffic and P_{f_i,f_j} denotes the probability of transition from state f_i to state f_j. In any state, the set of number of packet arrivals is given by $\mathcal{A} = \{a_0, a_1, \ldots, a_A\}$, where a_i represents i, i=0,1,…,A packet arrivals from the higher layer application. Also suppose λ_i is the average packet arrival rates in state f_j and $P(a_i|f_j)$ is the probability of a_i packets arrival in traffic state f_j; j=1,2,…,F. The row vector $\boldsymbol{\pi}^f = [\pi_1^f, \pi_2^f, \ldots, \pi_F^f]$ representing the stationary distribution of the incoming traffic state satisfies:

$$\boldsymbol{\pi}^f = \boldsymbol{\pi}^f \boldsymbol{P}^f \quad (2)$$

where $P^f := [P_{f_i,f_j}]_{F\times F}$ is the transition probability matrix for the underlying Markov chain governing transitions between traffic states. Arrival transition matrix $\boldsymbol{P}^f$ is a right stochastic matrix, so we can write $\sum_{j=1}^{F} P_{f_i,f_j} = 1$ for all i. Stationary distribution $\boldsymbol{\pi}^f$ corresponds to normalized left eigenvector of the transition matrix $\boldsymbol{P}^f$ associated with the eigenvalue 1 with normalizing condition $\sum_{j=1}^{F} \pi_j^f = 1$. The average arrival rate is $\bar{\lambda} = \sum_{i=1}^{F} \pi_i^f \lambda_i$.

Finite State Channel Modeling

Antenna diversity technique is a powerful technique to compensate the detrimental effect of channel fading and is often combined with Adaptive Modulation and Coding (AMC) to further improve data transmission rate. Due to inherent memory of the radio channel, the channel quality shows a significant degree of correlation among consecutive slots. The first order Finite-State Markov Channel (FSMC) model (Wang & Moayeri, 1995) is a very popular model to capture the memory of the fading channel, which is important to consider in order to accurately assessing the performance of a wireless system. Let $\mathcal{C} = \{c_1, c_2, \ldots, c_C\}$ denote the set of channel states, where C is the total number of states. The transition probability matrix of the channel can be written as $\boldsymbol{P}^c := [P_{c_i,c_j}]_{C\times C}$. When the fading of the considered channel is slow, the first order FSMC model can be approximated as a Birth-and-Death process. The crossover probabilities of the channel states can be approximately calculated using the following two equations,

$$P_{c_k,c_{k+1}} \approx \frac{N_{\gamma_k}}{R_{B_k}}, k = 1,2,\ldots,C-1 \text{ and } P_{c_k,c_{k-1}} \approx \frac{N_{\gamma_{k-1}}}{R_{B_k}}, k = 2,3,\ldots,C \tag{3}$$

In (3), N_{γ_k} is the expected number of times per second the received SNR passes from state c_k to c_{k+1}, called level crossing rate and R_{B_k} is the average number of time-slots per second during the channel remaining in state c_k.

The self-transition probabilities of the channel states can be found using the following relations of the associated Markov chain:

$$P_{c_k,c_k} = 1 - \sum_{j=k-1,\, j\neq k}^{k+1} P_{c_k,c_j}; k = 1,2,\ldots,C \tag{4}$$

where redundant probabilities $P_{c_1,c_0} = P_{c_C,c_{C+1}} = 0$ and other channel state probabilities in $\boldsymbol{P}^c$ are 0.

In both land-mobile and indoor-mobile multipath propagation environments, the received signal envelope has Nakagami-*m* distribution. The Nakagami-*m* distribution spans over a wide range of fading conditions and fading severity parameter *m* ranges from 1 to ∞. In a special case, when *m* = 1, it corresponds to Rayleigh distribution and the signal does not have a direct Line Of Sight (LOS) component. For *m*= 2, 3,…, it closely approximates the Rice distribution, which has a LOS component. As the value of *m* increases, the LOS component becomes gradually stronger. When $m \to \infty$, the Nakagami-*m* fading channel converges to a non-fading AWGN channel. In order to show the derivation of parameters for FSMC model in more details, we consider the FSMC model for Nakagami-*m* fading channel with selection diversity combining technique at the receiver (Iskander, et al., 2003). This model is a generalization of the classic Gilbert-Elliot channel model with only two states. Gilbert-Elliot model is then generalized for multi-states Rayleigh fading channel by Wang and Moayeri (1995).

When the wireless channel is slowly varying and has flat-fading, it can be modeled with an independent quasi-static Nakagami-*m* distribution. A necessary condition is that the spacing between antenna elements should be sufficient in order for the received SNR to remain constant for at least a time-slot. The time-slot over which the received SNR remains constant should satisfy the condition, $T_B \ll \frac{0.423}{f_m}$ (Rappaport, 2002), where f_m is the Doppler frequency. Suppose $\Gamma = \{\gamma_0, \gamma_1, \gamma_2, \ldots, \gamma_C\}$ is the set of normalized received SNR thresholds of the diversity combiner in increasing order with $\gamma_0 = 0$ and $\gamma_C = \infty$. The average received SNR of each independent branch is denoted by $\bar{\gamma} = \mathbb{E}\{\gamma\}$. The selection-combining (SC) diversity technique only processes one of the diversity branches, specifically the one with the largest instantaneous received SNR. The Probability Density Function (PDF) of the received output SNR of a n_R-branch diversity combiner is given by:

$$f_\gamma(\gamma) = n_R \left[\frac{\gamma\left(m, \frac{m\gamma}{\bar{\gamma}}\right)}{\Gamma(m)} \right]^{n_R-1} \frac{m^m \gamma^{m-1}}{\bar{\gamma}^m \Gamma(m)} e^{-\frac{m\gamma}{\bar{\gamma}}} \tag{5}$$

where $\Gamma(m) = \int_0^\infty t^{m-1} e^{-t} dt$ and $\gamma(m,x) = \int_0^x t^{m-1} e^{-t} dt$ are the complete and the lower incomplete Gamma functions, respectively.

As discussed earlier, the slowly varying Nakagami-*m* flat fading MIMO channel can be modeled as a FSMC by quantizing the received SNR at the output of the decoder into a finite number of states. The received instantaneous SNR can be

Figure 3. A typical variation of buffer occupancy in packets with respect to time-slots

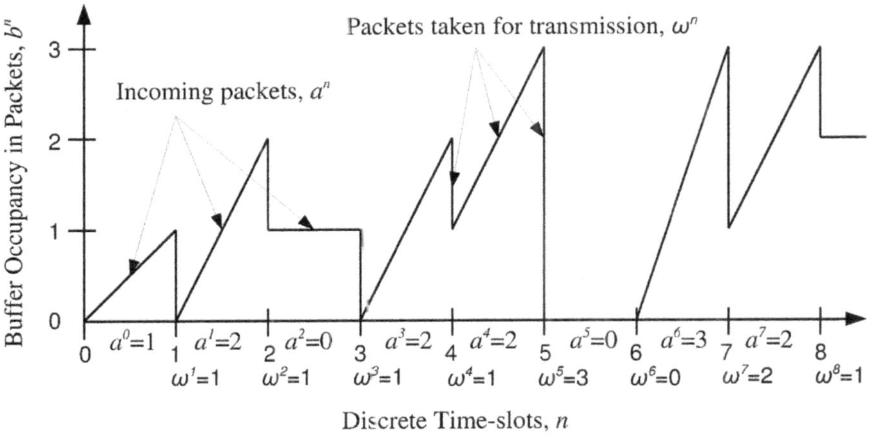

partitioned using several methods. Without loss of generality, we discuss equal probability method (EPM), where the instantaneous received SNR is partitioned so that the stationary probabilities of all channel states are the same. Thus, in EPM, the stationary probabilities are $\phi_k = \int_{\gamma_{k-1}}^{\gamma_k} f_\gamma(\gamma)\, d\gamma = \frac{1}{C}, \forall c_k \in \mathcal{C}$. Let γ_{k-1} and γ_k denote the lower and upper received SNR thresholds associated with channel state c_k. Using EPM, the received SNR thresholds can be obtained numerically from the following equation:

$$\left[\gamma\left(m, \frac{m\gamma_k}{\bar{\gamma}}\right)\right]^{n_R} - \left[\gamma\left(m, \frac{m\gamma_{k-1}}{\bar{\gamma}}\right)\right]^{n_R} = \frac{[\Gamma(m)]^{n_R}}{C} \tag{6}$$

The level crossing rate of state c_k can be given by:

$$N_{\gamma_k} = \frac{\sqrt{2\pi} f_m n_R}{[\Gamma(m)]^{n_R}} \left(\frac{m\gamma_k}{\bar{\gamma}}\right)^{m-\frac{1}{2}} e^{-\frac{m\gamma_k}{\bar{\gamma}}} \left[\gamma\left(m, \frac{m\gamma_k}{\bar{\gamma}}\right)\right]^{n_R - 1} \tag{7}$$

The derivation for MIMO system utilizing transmit-diversity is given in Karmokar and Bhargava (2006). The FSMC channel parameters for MIMO VBLAST spatial multiplexing system and other systems can be derived in the similar way.

Buffer Dynamics

Since in practical wireless transmitter, packets are stored in a buffer of finite capacity, unlike most of the adaptive transmission methods in the literature, we assume that the size of the transmission buffer is finite. Since the traffic is random in nature, the buffer may be empty sometimes and it may be full other times.

In Figure 3, a typical variation of the buffer occupancy is shown. It could be noted that the occupancy depends on the randomly varying number of packet arrivals as well as randomly varying number of packets taken in a particular time-slots. Let $\mathcal{B} = \{b_0, b_1, b_2, \ldots, b_B\}$ denote the state space of the buffer's packet occupancy, where b_1 corresponds to 1 packet in the buffer.

The dynamics of the buffer in terms of packet occupancy is then given by:

$$b^{n+1} = \min\{\max\left(b_0, b^n - \omega^n w^n + a^n\right), b_B\} \quad (8)$$

where the "max" operator ensures that the minimum buffer state is no less than b_0, whereas the "min" operator ensures that the maximum buffer state is no more than b_B. We assume that the transmitter can choose w^n based on the buffer state b^n, the channel state c^n, and the source state a^n. The value of observation ω^n equals 1 for ACK and 0 for NAK.

All packets that arrive in time-slot n can be transmitted only in time-slot $n+1$ or later. A natural constraint on w^n is that $0 \leq w^n \leq b^n$ for all n. That is, there is no transmission when the buffer is empty and the scheduler cannot transmit more packets than available in the queue. Since the buffer is of finite size, when it does not have enough space for all incoming packets, some packets will be dropped. The packet that comes first also is dropped first when FIFO service strategy is used. The queue is stable since it is finite in capacity and the excess packet arrivals in addition to the buffer capacity are dropped. However, the average overflow is bounded to an arbitrary small value. This is possible when the transmission power is variable. We assume that the average arrival rate is less than the average transmission rate so that the overflow rate can be controlled without incurring excessive overflow. Therefore, there is no possibility of growing the queue size infinitely and dropping a large number of packets from the finite buffer (Karmokar & Bhargava, 2009). We shall see later that the control action is chosen so that both the buffer delay and the overflow requirements can be maintained in cross-layer policy.

Problem Formulation and Solutions

In cross-layer adaptation techniques, the modulation and coding rate as well as the transmission power are judiciously chosen so that both the physical layer objectives, such as, BER, throughput, transmission power, and the upper layer objectives, such as, delay, packet dropping and overall throughput are jointly optimized. Due to stochastic nature of the traffic arrival, the buffer occupancy and the channel variation, the problem is inherently a dynamic optimization problem. Therefore, MDP based framework fits perfectly to formulate such problem. The cross-layer adaptation and optimization problem can be modeled as a MDP with the following ingredients:

System State Space: The system state space of the problem is composite and consists of traffic states, channel states and buffer states. Let $\mathcal{S} = \mathcal{F} \times \mathcal{C} \times \mathcal{B} = \{s_1, s_2, \ldots, s_S\}$ denotes the set of system states, where $S=F\times C\times(B+1)$ is the total number of states.

Action Space: At each time-slot n, the scheduler chooses an action u^n depending on the current composite state s^n. The control action of the cross-layer optimization problem corresponds to the rate and power chosen for a particular time-slot. A decision rule denoted with μ^n specifies the action at time-slot n. Decision rules for all time-slots constitute a policy of the problem. Policy (also called contingency plan) can be given by $\pi=\{\mu^1,\mu^2,\ldots,\mu^H\}$, where H is the horizon (total number of time-slots over which communications takes place) of the problem. Assume that there are total U actions available and therefore the set of all available actions can be written as $\mathcal{U} = \{u_1, u_2, \ldots, u_U\}$. Let $\mathcal{P}_t = \{P_{t_1}, P_{t_2}, \ldots, P_{t_P}\}$ and $\mathcal{W} = \{w_1, w_2, \ldots, w_W\}$ denote the set of all available transmission power levels and AMC rates, then we can write $P_t^n = \Phi(u^n)$ and $w^n = \Psi(u^n)$. Two mapping functions Φ and Ψ respectively map the action into the power level and the rate.

Transition Probability: After an action u^n is taken the system state moves from s^n to s^{n+1} with probability $P_{s^n,s^{n+1}}(u^n)$, called the transition probability.

The transition matrix for a particular action $u^n=u_i$ can be written by:

$$\boldsymbol{P}^s\left(u_i\right)=\begin{bmatrix} P_{s_1,s_1(u_i)} & \cdots & P_{s_1,s_S(u_i)} \\ \vdots & \ddots & \vdots \\ P_{s_S,s_1(u_i)} & \cdots & P_{s_S,s_S(u_i)} \end{bmatrix} \quad (9)$$

where the probability of system state transition from state $s^n=s_q=[f_j,c_k,b_l]$ to state $s^{n+1}=s_r=[f_x,c_y,b_z]$ for action $u^n=u_i$ can be given by:

$$P_{s_q,s_r}\left(u_i\right)=P_{f_jc_kb_l,f_xc_yb_z}= P\left(f_x,c_y,b_z|f_j,c_k,b_l\right)=P_{f_j,f_x}P_{c_k,c_y}P_{b_l,b_z} \quad (10)$$

where, for a particular action u_i, the probability of occupying buffer state $b^{n+1}=b_z$ from state $b^n=b_l$ is given by:

$$P_{b_l,b_z}=\sum_{\alpha=0}^{\alpha=A}\delta\left(b_z-b_l-a_\alpha+\omega^n\Psi\left(u_i\right)\right)P\left(a_\alpha|f_j\right),\forall f_j\in\mathcal{F} \quad (11)$$

where function $\delta(x)$ returns 1 when $x=0$ and returns 0 otherwise. The relations between system states, and traffic, channel and buffer states are as follows: $q=(j-1)(B+1)C+(k-1)(B+1)+1$ and $r=(x-1)(B+1)C+(y-1)(B+1)+z$.

- Rewards/Costs: The system also receives an immediate reward or incurs an immediate cost when a particular action u^n is chosen in a particular state s^n. Although cross-layer adaptation techniques can take many cross-layer objectives to optimize the performance of overall system. In this chapter, we discuss optimization of three conflicting goals: transmission power, buffer delay and packet overflow. Note that maximization of a reward is equivalent to minimization of negative reward (or cost). Therefore, in sequel we use the terms cost and reward interchangeably. Each objective is associated with a cost function $G_X(s^n,u^n)$.

Transmission Power Cost

Minimization of transmission power in wireless networks is important due to limited battery capacity of user equipment and to reduce interference. Transmission power cost in a particular time-slot is the actual transmitter power used for transmitting packets. Suppose, our target is to keep average BER of the transmission the same irrespective of modulation scheme and channel state. The BER requirement can be specified from the application being handled. For a certain channel state c_k and action u_i, and with a fixed specified average BER $\bar{P}_e$ for all channel states, the power cost $G_P=P_T$ can be found numerically using either following equation:

$$\bar{P}_e(c_k,u_j)=\frac{1}{\pi_k^c}\int_{\gamma_{k+1}}^{\gamma_k}P_e\left(\gamma\right)f_\gamma\left(\gamma\right)d\gamma$$

with appropriate BER expression or using (31) with instantaneous received SNR γ replaced by average received SNR $\bar{\gamma}=\frac{1}{\pi_k^c}\int_{\gamma_{k+1}}^{\gamma_k}f_\gamma\left(\gamma\right)d\gamma$.

Queueing Delay Cost

Delay is an important parameter to consider for communications systems involving transmission buffers. The maximum tolerable packet delay for a particular system depends on the QoS requirements of the application being handled. For example, real-time traffic must have very low delay. For this traffic, the received packets are useful only when the scheduler maintains the strict delay requirements. On the other hand, best-effort traffic is not real-time and is quite unsusceptible to delay. The delay experienced by a packet is composed of buffer-queuing, encoding, propagation and decoding delay. In this chapter, we consider only buffer delay, since encoding, propagation and decoding delay are usually fixed and are negligible compared to buffer delay. The average packet delay in the buffer is related to the average buffer occupancy via Little's theorem, as follows: $\bar{D} = \frac{1}{\lambda}\mathbb{E}[b^n]$. Therefore, we can write the immediate buffer delay cost by:

$$G_D\left(s^n, u^n\right) = \frac{b^n}{\lambda^n}, \forall c^n \in \mathcal{C}, f^n \in \mathcal{F}, u^n \in \mathcal{U} \tag{12}$$

Note that λ^n is the average packet arrival rate in traffic state f^n.

Packet Overflow Cost

When the buffer is nearly full and the empty space is smaller than the number of packets arrival, packet overflow happens. Suppose, current buffer state is b^n and scheduler takes $w^n = \Psi(u^n)$ packets from the buffer. Therefore, buffer can accommodate $r^n = b_B - b^n + w^n$ arriving packets in the current time-slot. Now, if arriving packets a^n in traffic state f^n is larger than r^n, $(a^n - r^n)$ packets will be dropped with probability one. Therefore, packer overflow rate, for buffer state b^n, traffic state f^n and action u^n, can be expressed by:

$$G_o\left(s^n, u^n\right) = \sum_{a^n \in \mathcal{A}} \varphi\left(a^n, \left(b_B - b^n + \omega^n w^n\right)\right) P\left(a^n | f^n\right), \forall f^n \in \mathcal{F} \tag{13}$$

where $\varphi(x, y)$ is a positive difference function, which returns the difference of x and y when $x > y$, and it returns 0 when $x \leq y$. Note in this chapter we consider actual packet dropping rate instead of packet dropping probability since packet dropping probability does not give an indication of how many packet has been dropped. It just tells us packet dropping events.

Unconstrained MDP Formulation

Sometimes in practice, in a multi-objective problem, the objectives are conflicting; hence we are interested in the trade-off between the objectives. The problem with two or more conflicting objectives can be formulated as an Unconstrained MDP (UMDP) by jointly optimizing a weighted combination of all the costs. Therefore, the combined immediate cost function can be written as:

$$G\left(s^n, \mu(s^n)\right) = \sum_{k=0}^{K-1} \frac{\beta_k}{\beta_0} G_{X_k}\left(s^n, \mu(s^n)\right) = G_P\left(s^n, \mu(s^n)\right) + \beta_1 G_D\left(s^n, \mu(s^n)\right) + \beta_2 G_O\left(s^n, \mu(s^n)\right) \tag{14}$$

where weighting factor β_k, $k=0,1,2,\ldots,K-1$ are determined by the relative importance of the objectives, normalizing factor $\beta_0=1$ and G_{X_k}, $k=0,1,2,\ldots,K-1$ are different costs for objectives to be optimized (power, delay and overflow are considered). Depending upon the combined cost, there are following three types of UMDP formulations in the literature.

Finite Horizon Total Costs

For a finite horizon total cost problem, the number of time-slots is finitely small. When there is a fixed deadline on the information data (such as in file transfer), the problem can be formulated as a finite horizon problem, where the deadline is less than or equal to the length of the horizon. The objective is to minimize transmission power or the energy of the wireless devices with limited energy under constraint on short-term delay, short-term power and deadline. Dynamic programming algorithm (also called backward induction algorithm) provides an efficient method for solving finite-horizon discrete-time MDP problem (Puterman, 1994). In Uysal-Biyikoglu et al. (2002), Zafer et al. (2008), and Ngo et al. (2009), the authors discussed cross-layer adaptation problem by formulating the problem as a finite horizon dynamic programming problem. The optimal policy for a finite horizon problem can be found using backward induction algorithm (also called dynamic programming algorithm) as follows:

Backward Induction Algorithm

Set n=H and $u^N\left(s^N\right) = G(s^N)$, for all $s^N \in \mathcal{S}$

Substitute n-1 for n and compute $u^n\left(s^n\right)$ for each $s^n \in \mathcal{S}$ by:

$$u^n\left(s^n\right) = \max_{u \in \mathcal{U}_{s^n}}\left[G\left(s^n, u\right) + \sum_{s_j \in \mathcal{S}} P_{s^n, s_j}\left(u\right) u^{n+1}\left(s_j\right)\right]$$

- If n=1, stop. Otherwise, return to second step.

where $G(s^N)$ is the terminal cost. It could be noted that the optimal policy for a finite horizon problem is non-stationary since the optimal action in a given state could change over time depending on the horizon size.

Infinite Horizon Discounted Total Costs

In an infinite horizon problem, the number of stages (or time-slots) is infinitely large. Infinite horizon problems are interesting because their analysis is elegant and insightful, and the implementation of optimal policies is often simple. In an infinite horizon discounted costs problem, the future costs are successively weighted by factor α. The expected long-term discounted total cost with stationary policy μ is given by:

$$G^{\mu} = \lim_{H \to \infty} \sum_{n=1}^{H} \mathbb{E}\{\alpha^n G\left(s^n, \mu(s^n)\right)\}, \tag{15}$$

where the discount factor satisfies 0≤α<1, which measures the value at time n of one unit cost incurred at time n+1. The significance of discount factor α is that a one-unit cost incurred n periods in the future has present value α.α… up to n times. In Fu et al. (2009), the authors presented cross-layer adaptation problem using infinite horizon discounted costs formulation. It could be noted that infinite horizon does not necessarily mean that all state sequences are infinite; rather it just means that there is no fixed deadline. The optimal policy for infinite horizon problem is stationary.

Infinite Horizon Average Costs per Stage

In many situations, discounting on future costs is inappropriate. In such situations, it is often meaningful to optimize the average cost per stage. The expected long-term average cost per stage with stationary policy μ is given by:

$$G^{\mu} = \lim_{H \to \infty} \frac{1}{H} \sum_{n=1}^{H} \mathbb{E}\{G\left(s^n, \mu(s^n)\right)\} \tag{16}$$

The cross-layer optimization problems considering infinite-horizon average cost per stage are discussed in Berry et al. (2002), Karmokar et al. (2006), Djonin et al. (2008), Karmokar et al. (2009), Rajan et al. (2004), and Goyal et al. (2008). Let Π denote the set of all the admissible policies π. For an ergodic system, the immediate cost and transition probability do not vary with respect to time-slot n. A policy is called stationary if it does not vary with time-slot n, i.e., μ^n=μ for all n. The

Figure 4. The state diagram and optimal action for each state of the cross-layer adaptive packet scheduling system

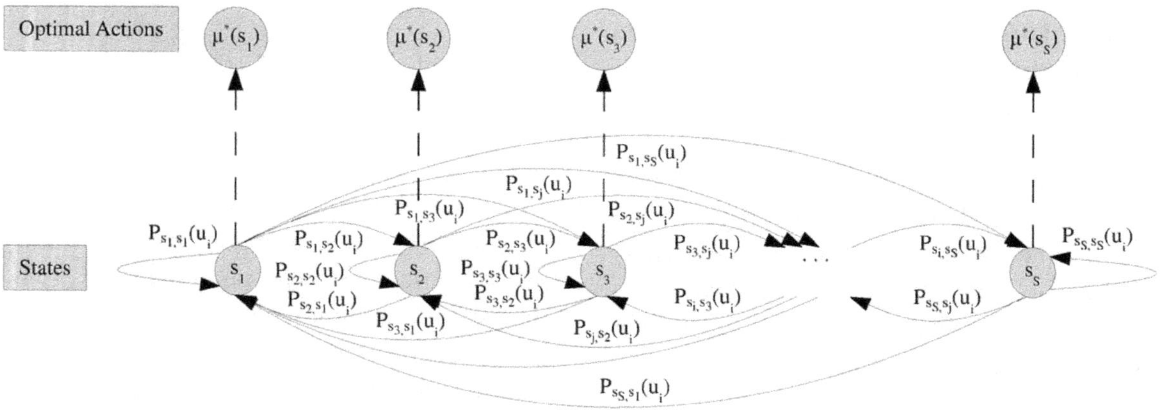

stationary policy has the form $\pi=\{\mu,\mu,\ldots,\mu\}$; for brevity it can be denoted by μ. The one-to-one relationship between the optimal stationary policy for infinite horizon formulation, $\mu^*(s_i)$ for state s_i is shown in Figure 4. The optimal policy $\mu^* \in \Pi$ can be found using any of the Dynamic Programming (DP) techniques. The detailed algorithms are discussed in the following subsections. In both infinite horizon cases, the policy μ^* over all stationary policies π that minimizes the total discounted cost (15) or the average cost per stage (16) is called the optimal policy.

The corresponding optimal cost per stage is given by:

$$G^{\mu^*} = \min_{\mu\in\Pi} G^{\mu} \qquad (17)$$

In general, the scenario being considered forms a weakly communicating MDP because some policies are multichain, yet there exist policies under which the transition probability matrix is unichain. Bellman's optimality equations of this model are given:

$$\min_{u\in\mathcal{U}_{s_i}}\left[\sum_{j=1}^{S} P_{s_i,s_j}(u)\lambda(s_j) - \lambda(s_i)\right] = 0 \qquad (18)$$

And:

$$\lambda(s_i) + h(s_i) = \min_{u\in\mathcal{U}_{s_i}}\left[G(s_i,u) + \sum_{j=1}^{S} P_{s_i,s_j}(u)h(s_j)\right] \qquad (19)$$

where $\lambda(s_i)$ is the gain or average cost for s_i, $h(s_i)$ is the relative cost or differential cost for each state s_i with respect to a reference state s_r, and $\mathcal{U}_{s_i}$ is the set of allowable actions in state s_i. For a unichain policy, the gain is constant, i.e., $\lambda(s_i)=\lambda$, for all s_i. Therefore, (18) vanishes for any stationary policy that is unichain. Also, for both the communicating and the weakly communicating models, the optimal gain is constant. Hence, the relative value iteration algorithm for the unichain model can be applied directly to both models to find a stationary ε-optimal policy and its corresponding gain (Puterman, 1994). The algorithm converges to a unique optimal deterministic policy finitely, and the optimal average cost is independent of the initial state. The problem can also be solved using a policy iteration algorithm. We have seen from our simulations that all of the improved policies are unichain, and hence the policy iteration algorithm terminates finitely with an optimal stationary deterministic policy.

Solution Techniques for UMDP Problem

In sequel, we discuss infinite horizon average cost per stage problem in detail. The solution techniques for infinite horizon discounted total cost problem discussed above is similar. There are two DP algorithms, namely Relative Value Iteration (RVI) and Policy Iteration (PI) that can be used to determine deterministic optimal policy for UMDP average cost per time-slot problem. The details of RVI algorithms are given below:

Relative Value Iteration Algorithm

In relative value iteration algorithm, the relative values of the expected cost for all the states with respect to a reference state are updated iteratively. The process is repeated until equilibrium is reached. The RVI algorithm for unichain discrete-time MDP can be described as follows:

- Set k=0, tolerance ε >0 and initial value of $h^{(0)}(s_i)$=0, for all s_i
- Select a state s_r as a reference state
- Find $h^{(k+1)}(s_i)$, for all s_i from the following equations:

$$h^{(k+1)}\left(s_i\right)=\left(Th^{(k)}\right)\left(s_i\right)-\left(Th^{(k)}\right)\left(s_r\right),\forall s_i\in\mathcal{S} \tag{20}$$

where the mapping function T has the form:

$$\left(Th^{(k)}\right)\left(s_i\right)=\min_{u\in\mathcal{U}_{s_i}}\left[G\left(s_i,u\right)+\sum_{j=1}^{S}P_{s_i,s_j}(u)h^{(k)}\left(s_j\right)\right] \tag{21}$$

Find $\underline{c}^{(k)}=\min_{s_i}\left[\left(Th^{(k)}\right)\left(s_i\right)-h^{(k)}\left(s_i\right)\right]$ and $\overline{c}^{(k)}=\max_{s_i}\left[\left(Th^{(k)}\right)\left(s_i\right)-h^{(k)}\left(s_i\right)\right]$. If $\overline{c}^{(k)}-\underline{c}^{(k)}$ < ε, go to next step. Otherwise, increment k by 1 and return to third step.

Find the optimal average cost using:

$$\lambda=\left(Th^{(k)}\right)\left(s_r\right) \tag{22}$$

and the optimal policy using:

$$\mu^{(k)}\left(s_i\right)=\min_{u\in\mathcal{U}_{s_i}}\left[G\left(s_i,u\right)+\sum_{j=1}^{S}P_{s_i,s_j}(u)h^{(k)}\left(s_j\right)\right],\forall s_i\in\mathcal{S} \tag{23}$$

Constrained MDP Formulation

Multi-objective dynamic problem can also be solved using CMDP formulation, where one type of cost (called objective cost) is minimized while keeping the other types of costs (called the constrained costs) below some given bounds (Altman, 1999). Therefore, our objective is to find the optimal stationary policy μ^* so that:

$$\min_{\mu} G^{\mu}_{X_0} \tag{26}$$

subject to:

$$G^{\mu}_{X_k}\leq\tilde{G}_{X_k},k=1,2,\ldots,K-1 \tag{27}$$

where $\tilde{G}_{X_k}$, k=1,2,...,K-1 denotes the bounds corresponding to the constrained costs.

Solution Technique for CMDP Problem

Now let us concentrate on the multi-objective problem where our goal is to optimize transmission power with constraints on delay and overflow. Let $x(s,u)$ represent the steady-state probability that the system is in state s and action u is applied.

The optimal policy μ^* can be determined by solving the following equivalent linear program (LP) for x:

$$\min_{x} \sum_{s\in\mathcal{S},u\in\mathcal{U}_s} G_P(s,u)\,x(s,u)$$

s.t. :

$$\sum_{s\in\mathcal{S},u\in\mathcal{U}_s} G_D(s,u)\,x(s,u) \le \tilde{G}_D$$

$$\sum_{s\in\mathcal{S},u\in\mathcal{U}_s} G_O(s,u)\,x(s,u) \le \tilde{G}_O$$

$$\sum_{u\in\mathcal{U}_s} x(t,u) = \sum_{s\in\mathcal{S},u\in\mathcal{U}_s} x(s,u)\,P_{s,t}(u), \forall t \in \mathcal{S}$$

$$\sum_{s\in\mathcal{S},u\in\mathcal{U}_s} x(t,u) = 1; x(t,u) \ge 0, \forall s \in \mathcal{S}, u \in \mathcal{U}_s \tag{28}$$

Note that not all actions are allowed in all states. Set $\mathcal{U}_s$ is the set of actions that are allowed in state s. It can be noted that the optimal control policy x^* is a probability distribution over $\mathcal{S}\times\mathcal{U}$. Suppose that there exists an optimal solution x^* to the above LP problems (28). Then, there exists a stationary policy μ^* that is optimal for the respective CMDP problem. The optimal policy μ^* for CMDP is randomized and is uniquely characterized with probability $\theta_{\mu^*(s)}(u)$ of applying policy u in state s, where:

$$\theta_{\mu^*(s)}(u) = \frac{x^*(s,u)}{\sum_{u'\in\mathcal{U}_s} x^*(s,u')} \; if \sum_{u'\in\mathcal{U}_s} x^*(s,u') > 0 \tag{29}$$

If $\sum_{u'\in\mathcal{U}_s} x^*(s,u') = 0$ for some $s \in \mathcal{S}$, an action that drives the system to the set of recurrent states $\mathcal{S}_{x^*} = \{s \in \mathcal{S} : \sum_{u'\in\mathcal{U}_s} x^*(s,u') > 0\}$ is chosen in each state. In general, LP can handle a problem with large number of variables and can give an optimal solution easily and quickly. Above linear program can be solved using modern linear programming techniques (such as, Interior point methods, etc.) and software tools (such as, MATLAB optimization toolbox, ILOG CPLEX, etc.) very easily and efficiently. There is a one-to-one correspondence between feasible (and hence optimal) solution of the LP and the feasible (and hence optimal) solution of CMDP. LP is feasible if and only if CMDP is feasible (Altman, 1999).

Numerical Analysis

In this section, we conduct numerical analysis to show how the cross-layer policy looks like structurally and to show the benefit that cross-layer policy offers as compared to single-layer policy. Like most packet adaptation schemes, we assume that perfect channel state is available at the beginning of a time-slot. Therefore, observation feedback is not needed in order to take transmission decision and it is assumed that the transmitted packets are received without any error (i.e., $\omega^n=1$ for all transmission). However, the scheme is equally valid for type-I HARQ when SW-ARQ or SR-ARQ is used at the link layer for added reliability of the successful reception of the packet. In this case, the HARQ controller updates the buffer according to the observation received. Without loss of generality, we show results for only rate adaptations across different layers. When power and coding rates are included, the number of actions is just extended. Rayleigh fading channel is considered with average transmission power $\bar{P} = 1mW$ and normalized average received SNR $\bar{\gamma} = 1$.

Analysis 1: Structure of the Optimal Policy

In this section, we show the monotonically non-decreasing trend of the optimal policy in Table 1. The optimal deterministic action obtained using UMDP formulation is shown as a function of buffer states and channel states. Either RVI or PI

Table 1. The optimal action (rate) for different channel and buffer states

Channel State → **Buffer State** ↓	c_1	c_2	c_3	c_4
b_0	u_1	u_1	u_1	u_1
b_1	u_1	u_2	u_2	u_2
b_2	u_1	u_2	u_3	u_3
b_3	u_1	u_3	u_3	u_4
b_4- b_6	u_1	u_3	u_4	u_5
b_7- b_9	u_2	u_3	u_4	u_5

dynamic programming algorithm can be used. However, PI algorithm for considered cross-layer adaptation problem converges 2-3 times faster than RVI. At time-slot n, when the transmit power is P_t and the average power is $\bar{P}$, the instantaneous received SNR is $\gamma \frac{P_t}{\bar{P}}$. We use information theoretic ergodic capacity (Goldsmith & Varaiya, 1997) over Rayleigh fading channel to obtain the transmission power for a discrete transmission rate. The results are shown for number of traffic states F=1, channel states C=4 and buffer states $B+1$=10. The weighting factors are β_1=1 and β_2=0, only transmission rate is adapted with number of possible rate U=5, fading rate f_mT_B=0.01, N_p/N_f=1 and average traffic arrival rate λ=1 packets/time-slot. Rate u_i corresponds to transmitting i-1 bits per symbol.

It is seen from Table 1 that the optimal actions are monotonically non-decreasing with both the buffer states and the channel states. Note that first buffer state b_0 corresponds to no packet in the buffer and first action u_1 corresponds to no transmission. Therefore, at buffer state b_0, first action is selected irrespective of channel state. However, in order to minimize power and delay, the scheduler opportunistically uses the channel gain. This is reflected in Table 1, for example, when buffer state is b_5, the scheduler takes action u_1 in channel state c_1, action u_3 in channel state c_2, action u_4 in channel state c_3, and action u_5 in channel state c_4. The scheduler also uses the buffer occupancy opportunistically. Therefore, when the buffer has lower number of packets, it chooses lower order action in order to minimize power. However, when the buffer has higher number of packets, it chooses higher order actions in order to meet the delay requirement. For example, in buffer state b_4, even if the buffer has 3 packets, remaining within average delay requirement, it is transmitting only 2 in order to save power. This tradeoff is more prominent when buffer size is larger. Because larger buffer (such as 50 packets) offer extended flexibility in manipulating transmission according to the requirement and present conditions.

In conclusion, the cross-layer adaptation methods take transmission based on both the channel and the buffer states judiciously so that both the power and the delay are both minimized jointly. However, single layer adaptation techniques based on only channel gain cannot take the opportunity of buffer occupancy to control transmission. This results in either loss of packets and/or loss of power.

Analysis 2: Comparison between Single-Layer and Cross-Layer Policy

In Figure 5, we show a comparison between the cross-layer policy discussed above and the channel dependent physical layer adaptation policy. The rate of the channel dependent policy is chosen as done in Liu et al. (2004), where the scheduler chooses adaptive modulation rate depending only on the channel state but it maintains a fixed specified BER or packet error rate. When the buffer has less number of packets than the rate determined by the channel, it takes only the number of packets available. The main limitation of this single-layer policy is that it cannot guarantee any delay or overflow requirements depending on the application being considered. In channel dependent adaptation scheme, we divide the channel into finite C=U number of states. The scheduler chooses action

Figure 5. Average power vs. average delay curves for different cross-layer scheduling policies with fading rates and traffic arrivals as parameters

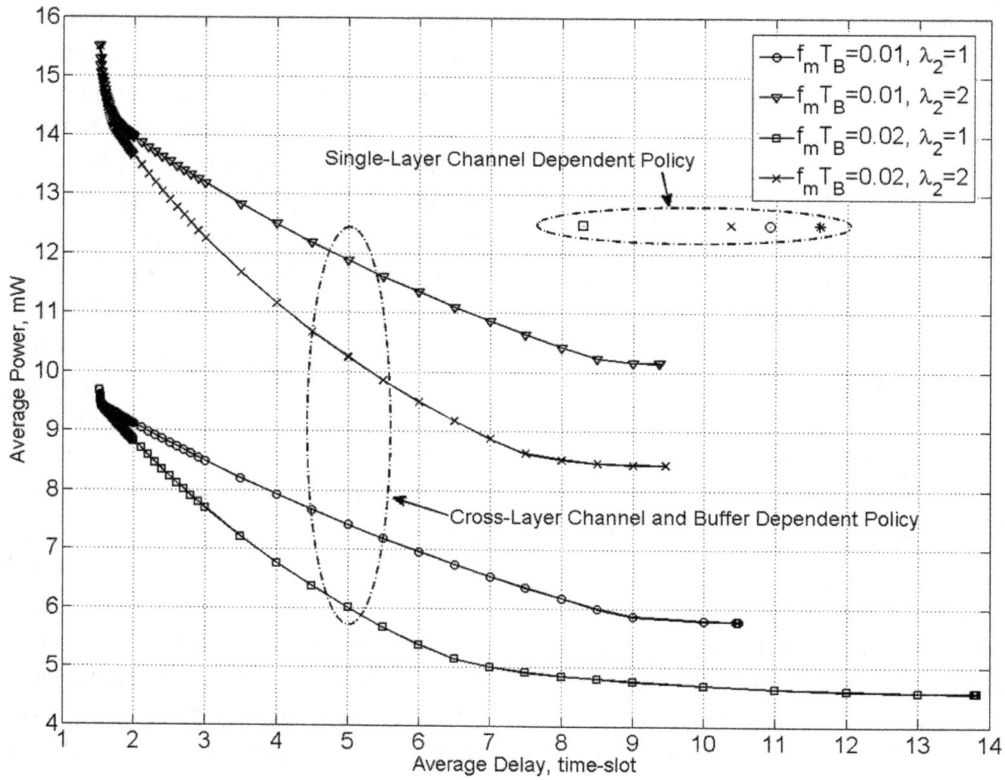

u_i when the gain of channel state c_i falls in the interval (γ_{i-1},γ_i). The cross-layer problem on the other hand is formulated as a CMDP and solved using LP technique.

The steps for determining the average power, delay and overflow are as follows:

1. for a particular values of data, the system transition matrix $\boldsymbol{P}_s$, and costs $G_P(s^n,u^n)$, $G_D(s^n,u^n)$ and $G_O(s^n,u^n)$ for all s^n and u^n are calculated,
2. then for cross-layer dynamic policy, the optimal randomized policy using LP method described above are calculated, and for single-layer policy, action u_i for corresponding channel state c_i irrespective of other states (buffer and traffic) are applied, and
3. finally the optimal transition matrix, and the optimal average power, delay and overflow for the optimal policy as determined by respective adaptation method are calculated.

Note that while the cross-layer policies are dynamic since they are different for different sets of bounds, the single-layer policies are static since they are the same. We consider receiver diversity technique using Selection Combining (SC) for this plot. The simulation is carried out for the number of traffic states F=2, the number of channel states C=4, the number of buffer states B+1=51, and the number of action U=4. We consider modulation scheme proposed for 3GPP standard, i.e., QPSK, *16*-QAM, and *64*-QAM. Coding rate is assumed to be one for the sake of simplicity. However, the techniques are equally valid for different coding rate. Suppose, if we have two coding rate, then total number of actions could be seven. Let $P_e(\gamma)$ denotes the instantaneous BER with received SNR γ. An approximate expression of BER for M-QAM is given in (Lu et al, 1999) and it can be expressed by:

$$P_e(\gamma) = \frac{2}{v}\left(1 - \frac{1}{\sqrt{M}}\right)\sum_{i=1}^{\frac{\sqrt{M}}{2}} \text{erfc}\left((2i-1)\sqrt{\frac{3v\gamma P_t}{2(M-1)\bar{P}}}\right) \tag{30}$$

where $v=\log_2(M)$ is the number of bits that modulates a 2^v-QAM symbol. Sum of the first two terms in (31) has been found to give best approximation to the system BER for all values of M and SNR.

Other parameters for the simulation are as follows: number of receiver diversity branches $n_R=1$, average channel BER $\bar{P}=10^{-4}$, average arrival rate in traffic state f_1 is $\lambda_1=1$ and in state f_2 it is varied between $\lambda_2=1$ and $\lambda_2=2$, the steady state probability of traffic states $\pi^f=[0.5\ 0.5]$ and maximum number of packet arrival $A=15$. Each point on a particular curve represents a particular optimal adaptation policy. We fixed overflow bound to $\tilde{G}_O = 10^{-4}$ packets per time-slot and varied delay bound $\tilde{G}_D$ to get different points for the curves. Since the adaptation law for single layer policy does not depend on the delay and the overflow bounds, we got only a point for each set of data.

It is seen from Figure 6 that the average power decreases as the average delay increases. This due to the fact that when the delay bound increases, the scheduler has extended facility to store more packets in the bad channel condition and to send them when channel condition improves. Therefore, for delay insensitive application (e.g., file transfer, web browsing), it is possible to save power significantly by setting larger value of delay bound as permitted by the QoS requirement of the application. On the other hand, for real-time application (e.g., video streaming), it is possible to guarantee exact delay and overflow by using variable power level at the transmitter. It is seen that for the same average delay and the same average arrival rates in traffic state f_2, the average transmission power requirement for the channel-dependent policy is significantly larger than the cross-layer policy. For $n_R=1$, $\lambda_2=1$ in Figure 5, approximately 38.37% power can be saved using cross-layer policy as compared to single-layer policy. Lowering transmission power intelligently also reduces interference it cause to other users/systems.

Now let us examine the curves for the same fading rates and the same delay, but for increased traffic arrivals. For cross-layer adaptation policy, it can be seen from the figure that the average transmission power increases as the arrival rate increases. Since the overflow is bounded to the same value for both traffic rates, in order to send the additional packets, the scheduler needs to raise the power level. For channel-dependent policy, the average transmission power remains the same for both arrival rates. However, it is also intuitively clear that both the delay and the overflow increase as the arrival rate increases.

The effect of receiver antennas on the power saving with delay is shown in Figure 5. It is seen from the figure that it is also possible to save power when using $n_R=2$ antennas at the receiver. In conclusion, the power decreases for any settings when delay bound is relaxed. The beauty of the cross-layer adaptation policy is that it is quite flexible and dynamic. The scheduler can find out optimal policy that suits well with application being handled. It is seen that for the same number of antennas, faster fading require less transmission power to maintain the same delay and overflow requirements. For faster fading channel, the channel moves between states more than for the slower fading channel, permitting less accumulation in the buffer. Since the channel moves to the higher channel states sooner, the buffer can send the accumulated packets with lower power.

Figure 6. Average power vs. average delay curves for different cross-layer scheduling policies with number of receiver diversity branches and fading rates as parameters

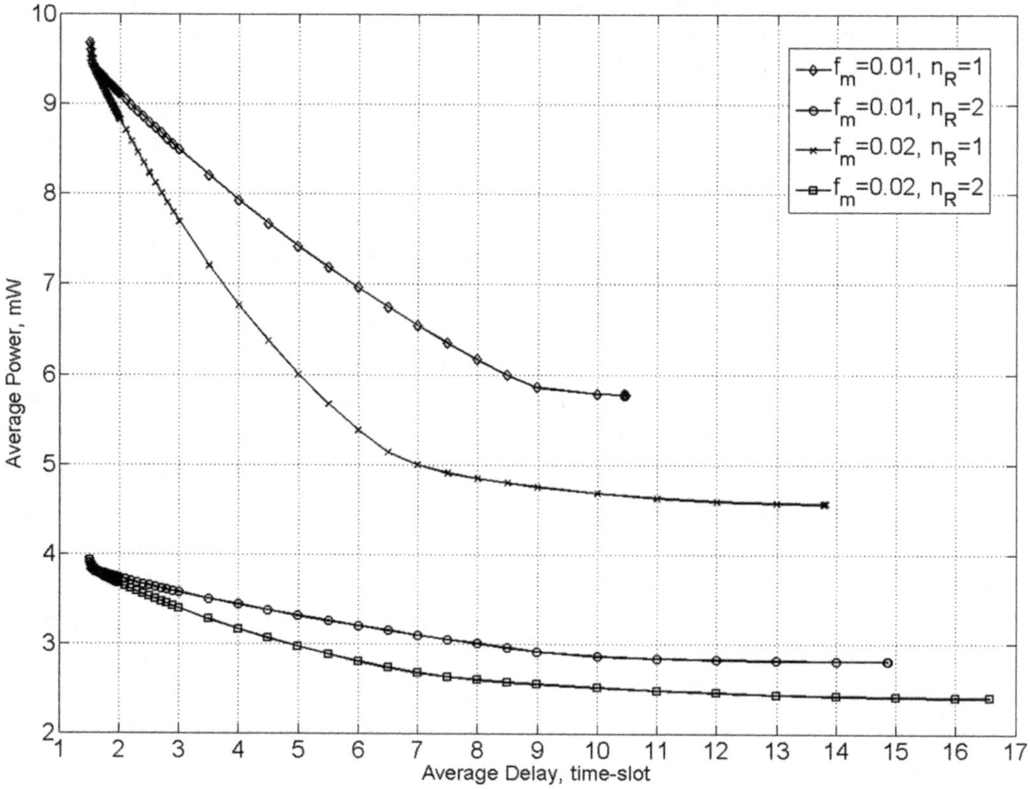

The policy for channel-dependent adaptation scheme is static and unique since it is only dependent on the current measured channel state. But, this channel-dependent policy causes unavoidable queuing delay and packet overflow because the controller does not have any control on them. On the contrary, cross-layer adaptation policy takes the buffer occupancy and traffic behavior into account to tackle delay and overflow in a goal-oriented way. For example, the transmitter needs different transmission powers in different channel states since channel gains are different there. However, the transmitter may also use higher power level to keep delay within limit and to avoid packet overflow when the buffer is nearly full and/or when the average arrival rate is higher and vice versa. With the fixed transmission power scheme, transmitter also cannot decrease power (whenever possible) to optimize transmission power in an optimal way (i.e., by maintaining specified delay and overflow).

FUTURE RESEARCH DIRECTIONS

The CO_2 emission from Information Technology (IT) industry sector is about 6% and increasing at a rapid pace due to dependence on more electrical and electronic systems in our daily lives. Energy-efficient communications systems are of prime importance for future sustainable world. Therefore, the energy-efficient dynamic scheduler design for multiuser systems considering cross-layer objectives and unknown system states and unknown environments could be a potential area of further research. The performance of the policies can be studied for more complex systems utilizing multiclass mix traffic, multiple antennas systems, multicarrier system, different diversity systems, etc. Performance results of the cross-layer policies for ad hoc, multi hop, cognitive radio, cooperative, and green radio communications networks are still not fully explored yet.

CONCLUSION

In this chapter, we described how to formulate the cross-layer rate and power scheduling problem with a Markov decision process formulation when the controller knows the states of the system perfectly. We showed how to model the system using first order Markov processes. We explained the ingredients of the MDP formulations and solutions techniques under different horizon length and different weighting of future rewards. We provided dynamic programming and linear programming based algorithms for finding the optimal policy. Through numerical study, we showed the general structure of the cross-layer adaptive optimal policy. We also showed the benefits of dynamic cross-layer adaptation techniques as compared to static single layer adaptation.

REFERENCES

Altman, E. (1999). *Constrained Markov decision processes: Stochastic modeling*. London, UK: Chapman and Hall/CRC.

Argyriou, A. (2010). Cross-layer and cooperative opportunistic network coding in wireless ad hoc networks. *IEEE Transactions on Vehicular Technology*, *59*(2), 803–812. doi:10.1109/TVT.2009.2035811

Berry, R. A., & Gallager, R. G. (2002). Communication over fading channels with delay constraints. *IEEE Transactions on Information Theory*, *48*(3), 1135–1149. doi:10.1109/18.995554

Camp, J., & Knightly, E. (2010). Modulation rate adaptation in urban and vehicular environments: Cross-layer implementation and experimental evaluation. *IEEE/ACM Transactions on Networking*, *18*(6), 1949–1962. doi:10.1109/TNET.2010.2051454

Choi, J. D., Wasserman, K. M., & Stark, W. E. (1999). Effect of channel memory on retransmission protocols for low energy wireless data communications. *Proceedings IEEE International Conference on Communications, 3*, 1552-1556.

Chung, S. T., & Goldsmith, A. J. (2001). Degrees of freedom in adaptive modulation: A unified view. *IEEE Transactions on Communications*, *49*, 1561–1571. doi:10.1109/26.950343

Collins, B., & Cruz, R. (1999). Transmission policy for time varying channel with average delay constraints. In *Proceedings of Allerton Conference on Communications, Control and Computing*, (pp. 709-717). Allerton.

Ding, Z., & Leung, K. K. (2011). Cross-layer routing using cooperative transmission in vehicular ad-hoc networks. *IEEE Journal on Selected Areas in Communications*, *29*(3), 571–581. doi:10.1109/JSAC.2011.110307

Djonin, D. V., Karmokar, A. K., & Bhargava, V. K. (2008). Joint rate and power adaptation for type-I hybrid ARQ systems over correlated fading channels under different buffer-cost constraints. *IEEE Transactions on Vehicular Technology*, *57*, 421–435. doi:10.1109/TVT.2007.901876

Djonin, D. V., & Krishnamurthy, V. (2007). Q-learning algorithms for constrained Markov decision processes with randomized monotone policies: Application to MIMO transmission control. *IEEE Transactions on Signal Processing*, *55*(5), 2170–2181. doi:10.1109/TSP.2007.893228

Fathi, M., Taheri, H., & Mehrjoo, M. (2010). Cross-layer joint rate control and scheduling for OFDMA wireless mesh networks. *IEEE Transactions on Vehicular Technology*, *59*(8), 3933–3941. doi:10.1109/TVT.2010.2064346

Fu, A., Modiano, E., & Tsitsiklis, J. N. (2006). Optimal transmission scheduling over a fading channel with energy and deadline constraints. *IEEE Transactions on Wireless Communications*, *5*, 630–641.

Fu, F., & vander Schaar, M. (2009). A new systematic framework for autonomous cross-layer optimization. *IEEE Transactions on Vehicular Technology*, *58*(4), 1887–1903. doi:10.1109/TVT.2008.2007418

Goldsmith, A. J., & Varaiya, P. P. (1997). Capacity of fading channels with channel side information. *IEEE Transactions on Information Theory*, *43*(6), 1986–1992. doi:10.1109/18.641562

Goyal, M., Kumar, A., & Sharma, V. (2008). Optimal cross-layer scheduling of transmissions over a fading multiaccess channel. *IEEE Transactions on Information Theory*, *54*(8), 3518–3537. doi:10.1109/TIT.2008.926335

Ho, Z. K. M., Lau, V. K. N., & Cheng, R. S.-K. (2009). Cross-layer design of FDD-OFDM systems based on ACK/NAK feedbacks. *IEEE Transactions on Information Theory*, *55*(10), 4568–4584. doi:10.1109/TIT.2009.2025565

Iskander, C.-D., & Mathiopoulos, P. T. (2003). Fast simulation of diversity Nakagami fading channels using finite-state Markov models. *IEEE Transactions on Broadcasting*, *49*, 269–277. doi:10.1109/TBC.2003.817096

Joseph, J. F. C., Lee, B., Das, A., & Seet, B. (2011). Cross-layer detection of sinking behavior in wireless ad hoc networks using SVM and FDA. *IEEE Transactions on Dependable and Secure Computing*, *8*(2), 233–245. doi:10.1109/TDSC.2009.48

Karmokar, A. K., & Bhargava, V. K. (2009). Performance of cross-layer optimal adaptive transmission techniques over diversity Nakagami-m fading channels. *IEEE Transactions on Communications*, *57*(12), 3640–3652. doi:10.1109/TCOMM.2009.12.080262

Karmokar, A. K., Djonin, D. V., & Bhargava, V. K. (2006). Optimal and suboptimal packet scheduling over time-varying flat fading channels. *IEEE Transactions on Wireless Communications*, *5*, 446–457. doi:10.1109/TWC.2006.1611068

Karmokar, A. K., Djonin, D. V., & Bhargava, V. K. (2006). POMDP-based coding rate adaptation for type-i hybrid ARQ systems over fading channels with memory. *IEEE Transactions on Wireless Communications*, *5*, 3512–3523. doi:10.1109/TWC.2006.256974

Krishnamurthy, V., & Djonin, D. V. (2007). Structured threshold policies for dynamic sensor scheduling—A partially observed Markov decision process approach. *IEEE Transactions on Signal Processing*, *55*(10), 4938–4957. doi:10.1109/TSP.2007.897908

Kuran, M. S., Gur, G., Tugcu, T., & Alagoz, F. (2010). Applications of the cross-layer paradigm for improving the performance of WiMax. *IEEE Wireless Communications*, *17*(3), 86–95. doi:10.1109/MWC.2010.5490983

Liu, Q., Zhou, S., & Giannakis, G. B. (2004). Cross-layer combining of adaptive modulation and coding with truncated ARQ over wireless links. *IEEE Transactions on Wireless Communications*, *3*, 1746–1755. doi:10.1109/TWC.2004.833474

Lu, J., Letaief, K. B., Chuang, J. C.-I., & Liou, M. L. (1999). M-PSK and M-QAM BER computation using signal-space concepts. *IEEE Transactions on Communications*, *47*(2), 181–184. doi:10.1109/26.752121

Melodia, T., & Akyildiz, I. F. (2010). Cross-layer QoS-aware communication for ultra wide band wireless multimedia sensor networks. *IEEE Journal on Selected Areas in Communications*, *28*(5), 653–663. doi:10.1109/JSAC.2010.100604

Ngo, M. H., & Krishnamurthy, V. (2009). Optimality of threshold policies for transmission scheduling in correlated fading channels. *IEEE Transactions on Communications*, *57*, 2474–2483. doi:10.1109/TCOMM.2009.08.070350

Pantelidou, A., & Ephremides, A. (2010). A cross-layer view of optimal scheduling. *IEEE Transactions on Information Theory*, *56*(11), 5568–5580. doi:10.1109/TIT.2010.2068970

Pantelidou, A., & Ephremides, A. (2011). Wireless multicast optimization: A cross-layer approach. *IEEE Transactions on Information Theory*, *57*(7), 4333–4343. doi:10.1109/TIT.2011.2145530

Poggioni, M., Rugini, L., & Banelli, P. (2010). QoS analysis of a scheduling policy for heterogeneous users employing AMC jointly with ARQ. *IEEE Transactions on Communications*, *58*(9), 2639–2652. doi:10.1109/TCOMM.2010.09.090276

Pompili, D., & Akyildiz, I. F. (2010). A multimedia cross-layer protocol for underwater acoustic sensor networks. *IEEE Transactions on Wireless Communications*, *9*(9), 2924–2933. doi:10.1109/TWC.2010.062910.100137

Puterman, M. L. (1994). *Markov decision processes: Discrete stochastic dynamic programming*. New York, NY: John Wiley.

Rajan, D., Sabharwal, A., & Aazhang, B. (2004). Delay-bounded packet scheduling of bursty traffic over wireless channels. *IEEE Transactions on Information Theory*, *50*, 125–144. doi:10.1109/TIT.2003.821989

Ramis, J., & Femenias, G. (2011). Cross-layer design of adaptive multirate wireless networks using truncated HARQ. *IEEE Transactions on Vehicular Technology*, *60*(3), 944–954. doi:10.1109/TVT.2011.2108324

Rappaport, T. S. (2002). *Wireless communications: Principles and practice*. Upper Saddle River, NJ: Prentice Hall.

Seo, J.-B., & Leung, V. C. M. (2011). Design and Analysis of cross-layer contention resolution algorithms for multi-packet reception slotted ALOHA systems. *IEEE Transactions on Wireless Communications*, *10*(3), 825–833. doi:10.1109/TWC.2011.122010.100032

Uysal-Biyikoglu, E., Gamal, A. E., & Prabhakar, B. (2002). Energy-eficient packet transmission over a wireless link. *IEEE/ACM Transactions on Networking*, *10*, 487–499. doi:10.1109/TNET.2002.801419

Vucetic, B. (1991). An adaptive coding scheme for time-varying channels. *IEEE Transactions on Communications*, *39*, 653–663. doi:10.1109/26.87156

Wang, H. S., & Moayeri, N. (1995). Finite-state Markov channel- A useful model for radio communication channels. *IEEE Transactions on Vehicular Technology*, *44*(1), 163–171. doi:10.1109/25.350282

Zafer, M., & Modiano, E. (2008). Optimal rate control for delay-constrained data transmission over a wireless channel. *IEEE Transactions on Information Theory*, *54*(9), 4020–4039. doi:10.1109/TIT.2008.928249

Zhang, H., Zheng, Y., Khojastepour, M., & Rangarajan, S. (2010). Cross-layer optimization for streaming scalable video over fading wireless networks. *IEEE Journal on Selected Areas in Communications*, *28*(3), 344–353. doi:10.1109/JSAC.2010.100406

Zhou, N., Zhu, X., Huang, Y., & Lin, H. (2010). Low complexity cross-layer design with packet dependent scheduling for heterogeneous traffic in multiuser OFDM systems. *IEEE Transactions on Wireless Communications*, *9*(6), 1912–1923. doi:10.1109/TWC.2010.06.081595

ADDITIONAL READING

Babu, T. V. J. G., Le-Ngoc, T., & Hayes, J. F. (2001). Performance of a priority-based dynamic capacity allocation scheme for wireless ATM systems. *IEEE Journal of Selected Areas in Commununications*, *19*(2), 355–369. doi:10.1109/49.914513

Bai, X., & Shami, A. (2008). Two dimensional cross-layer optimization for packet transmission over fading channel. *IEEE Transactions on Wireless Communications*, *7*, 3813–3822. doi:10.1109/T-WC.2008.070378

Berry, R. A., & Yeh, E. M. (2004). Fundamental performance limits for wireless fading channels cross-layer wireless resource allocation. *IEEE Signal Processing Magazine*, *21*, 59–68. doi:10.1109/MSP.2004.1328089

Bertsekas, D. P. (2001). *Dynamic programming and optimal control*. Belmont, MA: Athena Scientific.

Boyd, S., & Vandenberghe, L. (2004). *Convex optimization*. Cambridge, UK: Cambridge University Press.

Chen, W., Neely, M. J., & Mitra, U. (2008). Energy-efficient transmissions with individual packet delay constraints. *IEEE Transactions on Information Theory*, *54*(5), 2090–2109. doi:10.1109/TIT.2008.920344

Chockalingam, A., Zorzi, M., Milstein, L. B., & Venkataram, P. (1998). Performance of a wireless access protocol on correlated Rayleigh-fading channels with capture. *IEEE Transactions on Communications*, *46*, 644–655. doi:10.1109/26.668735

Cohen, R., Grebla, G., & Katzir, L. (2010). Cross-layer hybrid FEC/ARQ reliable multicast with adaptive modulation and coding in broadband wireless networks. *IEEE/ACM Transactions on Networking*, *18*(6), 1908–1920. doi:10.1109/TNET.2010.2050902

Djonin, D. V., & Krishnamurthy, V. (2007). MIMO transmission control in fading channels- A constrained Markov decision proces formulation with monotone randomized policies. *IEEE Transactions on Signal Processing*, *55*, 5069–5083. doi:10.1109/TSP.2007.897859

Duong, D. V., Øien, G. E., & Hole, K. J. (2006). Adaptive coded modulation with receive antenna diversity and imperfect channel knowledge at receiver and transmitter. *IEEE Transactions on Vehicular Technology*, *55*(3), 458–465. doi:10.1109/TVT.2005.863348

Ekbal, A., Song, K.-B., & Cioffi, J. M. (2004). QoS-constrained physical layer optimization for correlated flat-fading wireless channels. *Proceedings IEEE International Conference on Communications, 7*, 4211-4215.

Gilbert, E. N. (1960). Capacity of a burst-noise channels. *The Bell System Technical Journal*, *39*, 1253–1266.

Hoang, A. T., & Motani, M. (2008). Cross-layer adaptive transmission with incomplete system state information. *IEEE Transactions on Communications*, *56*(11), 1961–1971. doi:10.1109/TCOMM.2008.060618

Hoang, A. T., & Motani, M. (2008). Cross-layer adaptive transmission: Optimal strategies in fading channels. *IEEE Transactions on Communications*, *56*(5), 799–807. doi:10.1109/TCOMM.2008.060214

Johnston, L. A., & Krishnamurthy, V. (2006). Opportunistic file transfer over a fading channel: A POMDP search theory formulation with optimal threshold policies. *IEEE Transactions on Wireless Communications*, *5*, 394–405. doi:10.1109/TWC.2006.1611063

Karmokar, A. K., & Bhargava, V. K. (2006). Optimal packet scheduling using adaptive M-QAM and orthogonal STBC in MIMO Nakagami-m fading channels. In *Proceedings IEEE International Conference on Communications*, (pp. 5271-5276). IEEE Press.

Luccini, M., Shami, A., & Primak, S. (2010). Cross-layer optimization of network performance over multiple-input multiple-output wireless mobile channels. *IET Communications*, *4*(6), 683–696. doi:10.1049/iet-com.2009.0140

Luo, C., Yu, F. R., Ji, H., & Leung, V. C. M. (2010). Cross-layer design for TCP performance improvement in cognitive radio networks. *IEEE Transactions on Vehicular Technology*, *59*(5), 2485–2495. doi:10.1109/TVT.2010.2041802

Maltsev, A., Khoryaev, A., Lomayev, A., Maslennikov, R., Antonopoulos, C., & Avgeropoulos, K. (2010). MIMO and multihop cross-layer design for wireless backhaul: A testbed implementation. *IEEE Communications Magazine*, *48*(3), 172–179. doi:10.1109/MCOM.2010.5434390

Nanda, S., Balachandran, B., & Kumar, S. (2000). Adaptation techniques in wireless packet data services. *IEEE Communications Magazine*, *38*(1), 54–64. doi:10.1109/35.815453

Negi, R., & Cioffi, J. M. (2002). Delay-constrained capacity with causal feedback. *IEEE Transactions on Information Theory*, *48*, 2478–2494. doi:10.1109/TIT.2002.801471

Wang, H., & Mandayam, N. B. (2004). A simple packet-transmission scheme for wireless data over fading channels. *IEEE Transactions on Communications*, *52*, 1055–1059. doi:10.1109/TCOMM.2004.831354

Zhang, Y., Fu, F., & van der Schaar, M. (2010). On-line learning and optimization for wireless video transmission. *IEEE Transactions on Signal Processing*, *58*(6), 3108–3124. doi:10.1109/TSP.2010.2046040

Zheng, K., Wang, Y., Lei, L., & Wang, W. (2010). Cross-layer queueing analysis on multihop relaying networks with adaptive modulation and coding. *IET Communications*, *4*(3), 295–302. doi:10.1049/iet-com.2009.0380

KEY TERMS AND DEFINITIONS

Adaptive Modulation and Coding: The adaptation techniques in which the modulation and coding rates are varied with the channel conditions.

Cross-layer Adaptation and Optimization: The adaptation technique that chooses transmission coding and modulation rate, and power based on not only with physical layer channel gain, but also with upper layer parameters, such as, data link layer buffer occupancy, application layer traffic information. The joint optimization of different layers QoS goals with the different layer optimization variables is called the cross-layer optimization.

Markov Decision Process: It is also called sequential decision making process, where an agent makes decisions in a fully observable stochastic environment. It can be described by a set of states, a set of actions, markovian transition probability matrices and rewards. Dynamic programming and linear programming algorithms are used to solve the optimal policy, which dictates the optimal action for each states, of the Markov decision process problem. The optimal policy for infinite horizon problem does not depend on the initial state and is stationary. The optimal policy for finite horizon problem is however non-stationary.

Weakly Communicating Markov Chain: A Markov chain is a characterization of a system where the transition of the next state from the current state only depends on the current state and not on the past states. A stationary policy whose associated Markov chain has a single recurrent class and a possibly empty set of transient states is called a *unichain* policy. A stationary policy whose associated Markov chain has two or more closed irreducible classes is called a *multichain* policy. A MDP is referred to as *communicating* if, for every pair of states, there exists a deterministic stationary policy under which one state is accessible from another, even though some of the policies may be multichain. A MDP is said to be *weakly communicating* if there exists a closed set of states, with each state in that set accessible from every other state in that set under some deterministic stationary policy, plus a possibly empty set of states that is transient under every policy.

Chapter 2
Architecture Design Approaches and Issues in Cross Layer Systems

Andrea F. Cattoni
Aalborg University, Denmark

Troels B. Sørensen
Aalborg University, Denmark

Preben E. Mogensen
Aalborg University, Denmark

ABSTRACT

Wireless communications are a fast growing part of the telecommunication market. While new types of traffic and challenges related to the wireless medium are appearing, the methodologies for designing system architectures substantially remain the same. Under the increasing pressure of market, users, and physical medium issues, designers are in need for new approaches.

Cross Layer becomes, then, a handy solution for coping with such problems. In fact, it allows both tighter optimizations of the existing functionalities, and the introduction of new ones that do not fit within the traditional protocol stack design methodology.

However, Cross Layer also carries a risk due to possibly unexpected and undesired effects. In this chapter, the authors provide architecture designers with a set of guidelines synthesized from an analysis of the state of the art, but enriched with the perspective of the development of future generations of communication systems such as Cognitive Radio.

DOI: 10.4018/978-1-4666-0960-0.ch002

INTRODUCTION

In the last couple of decades, there has been a strong diffusion of wireless systems. Especially in the last five to ten years, the advent of high-throughput wireless technologies completely revolutionized the traditional concept of networking. Nowadays, users want to be always online, connected over the internet everywhere. The wireless traffic has grown exponentially within recent years, especially after the release of smartphones that allow a comfortable browsing and the usage of multimedia applications. Of course this dramatic increase in wireless traffic sets new needs for the next generation of wireless technologies. But this endless loop of traffic growth/new wireless possibilities is nowadays facing big challenges. This is mostly due to the aging of the traditional internet architecture based on the layered TCP/IP architecture (Stevens, 1994). Let us anyway considered the Open System Interconnect (OSI) protocol stack (Schwartz, 1988) as a generalization for layered architectures in the following of the discussion. Two of the main problems that have to be faced to overcome the OSI limitations are:

1. The extreme dynamism of the wireless medium: due to propagation effects such as fading, multipath, and phase rotations, the wireless medium rapidly varies and it can introduce errors and problems to higher layers of the protocol stack. These effects are emphasized when the users are in high mobility conditions, i.e. moving at high speed.
2. Sub-optimality of the stacked layers: the independent design of the stacked layers is suboptimal when compared to the overall stack. In order to properly face the traffic growth a tighter global optimization of the system is required.

Several years ago, pushed by the aforementioned limitations and problems, researchers started investigating the possibility of optimizing the OSI stack beyond the traditionally layered functionality and independence. This "Cross-Layer" (CL) process is the main core of this book and of this chapter.

Several techniques and approaches have been proposed in literature for improving the performances of specific layers or applications. The goal of the current chapter is to summarize all these techniques by identifying the design approach behind the method rather than listing them individually. The main principle is to provide researchers with a general overview helpful for understanding and designing CL Architectures (CLAs), based on the definition of information flows within the stacked architecture and beyond. This point of view will help the reader understand which are the benefits and the risks associated to each design methodology and how to properly use them to solve his or her own problem.

BEYOND THE STACKED ARCHITECTURE

In the current section, we will discuss the traditional reference architecture for data communications and compare it with the current problems and needs of modern wireless communications. It will be evident in the end of the section that new architectural design methodologies are needed in order to provide the market with more efficient technologies that keep into account the user satisfaction as well as the efficiency of the technology itself.

For these reasons, it is required to go beyond the stacked architecture, re-inventing the design approaches as well as the standardization ones in order to have the same flexibility and longevity of the traditional OSI model while having a tighter global optimization of the system.

Figure 1. Traditional OSI layered architectural approach

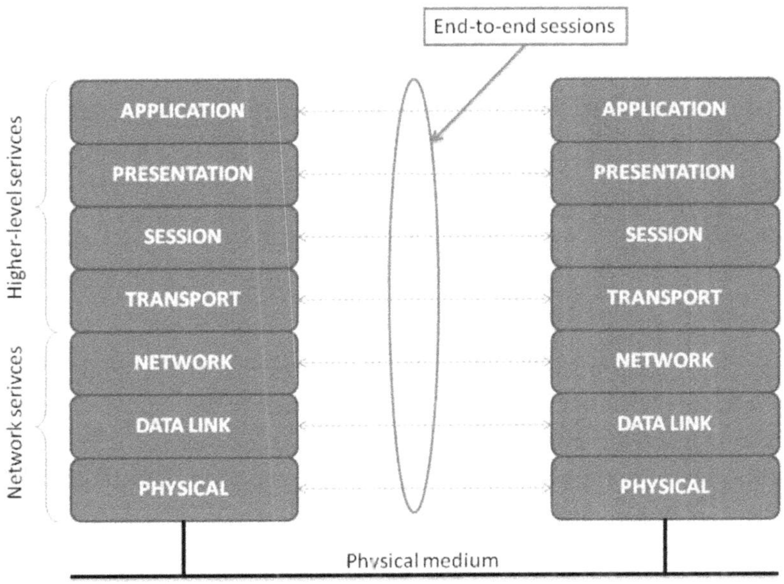

The Reference Model

As previously mentioned, let us take the traditional OSI layered model, shown in Figure 1, as reference for stacked architecture. It was standardized by ITU-T, as part of the X recommendations in 1988, subsequently updated in 1994, but was anyway already in use before 1988. As a matter of fact, it was designed to cope with the problems related to the wired communication systems.

Ethernet (Spurgeon, 2000) is a perfect example of the power of such an approach. The Medium Access Control (MAC) systems used for wired media quickly resolve the conflicts that arise, allowing the upper layers to have an almost transparent end-to-end connectivity. This is in fact the main idea behind such a model. In this way, each layer can be designed independently from the others. It can also easily be replaced by any other one that respects the interfaces with the upper and lower layers.

As shown in Figure 2, the definition of the encapsulation service guarantees the total independence of each layer, since the addition of headers and tails is limited to the end-to-end session between layers at the same level. All of this happens through a few simple primitives, and bi-directional calls and replies. Furthermore, another fundamental design rule used in such a model is the self-containment of each layer. Lower layers should never read the information contained in the header or tail added by an upper layer. This limits the possibilities of the lower layers, but it also avoids a specific coupling between layers (Schwartz, 1988).

One of the main impacts of the OSI model is the possibility of using heterogeneous networking technologies for connecting several entities in the network without caring about higher layers close to the application. One such example is illustrated in Figure 3 with the example of a router to interconnect two different physical media.

Figure 2. Definition of a service in the OSI model

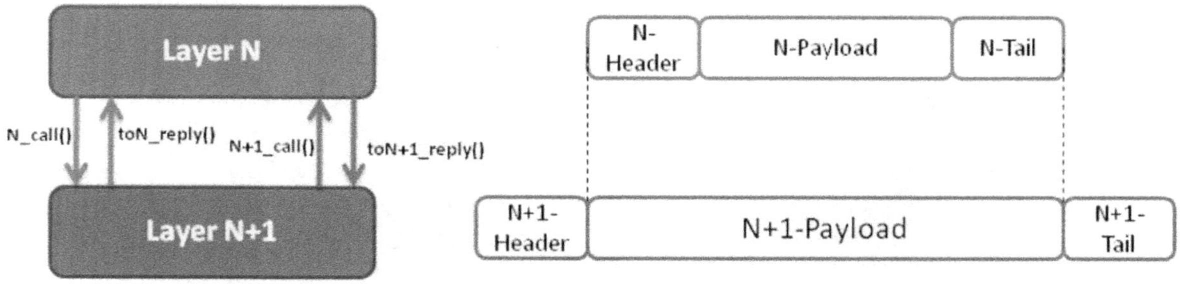

All these good properties have granted the OSI model longevity. Unfortunately, some signs of aging are becoming evident when the architecture is stressed by the presence of an extremely varying medium like the wireless one, and by requirements imposed by traffic with strict Quality of Service (QoS) needs.

Wireless Communication Needs

The new way of setting up Ethernet networks is substantially based on the usage of switch devices. This basically simplifies the role of the MAC layer since there are almost no collisions on the medium shared between the two end points of the connection. This allows increase of the speed of the physical layer (PHY) up to 10Gbps, but it still relies on a cable. The comfort of having a nomadic device is irreplaceable, and users can accept a performance degradation in terms of connection speed if this allows them to remove the wires. As a matter of fact, the wireless medium behaves more or less like a shared medium, and brings back the technology problems to the time when the Ethernet devices where connected through a hub. Of course all these problems and challenges are on the shoulders of designers and developers, since the users care only about the possibilities offered by the wireless medium and the perceived performances of the carried services, but let us analyze more in detail the problems related to the usage of the wireless medium and the new types of services that it has to carry.

Problems Related to the Lower Layers

The propagation of the electromagnetic waves is an extremely complex phenomenon. Luckily, from a telecommunication point of view, the macro effects that impact the transmission and reception of the signal can be modeled in an easier way by accepting imperfections in the final results. What is anyway important is the statistical accuracy, not the punctual one.

One of the most common propagation problems is the reception of multiple electromagnetic rays, or “paths” of the signal. This phenomenon, in fact called “multipath,” creates instantaneous variations of the received power, since the rays can be positively (constructive multipath) or negatively (destructive multipath, or multipath fading) added to the main signal, and dispersion of the signal in the time domain. In the case of destructive multipath, it can create deep fades and even completely cancel the received signal. It is easy to imagine that it is difficult to design receivers able to tolerate huge variations in the received signal and to reconstruct the information lost in the fades. Another propagation phenomenon that creates increasing difficulties by increasing the working frequency is the well-known Doppler effect (dispersion of the signal in the frequency domain). Especially in high speed conditions, for

Figure 3. Heterogeneous networking. The usage of a simple router grants the interfacing between different physical media.

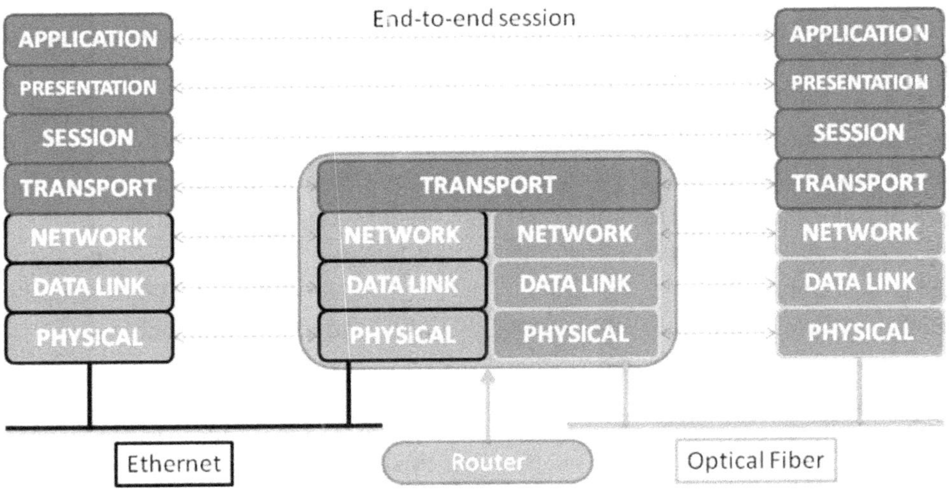

example mobile users on a highway, the deviations of the frequencies cause synchronization problems between each couple of communication peers. In order to have deeper insights on such problems, it is possible to refer to Vaughan and Bach Andersen (2003)(Vaughan & Bach Andersen, 2003) and Jeruchim, Balaban, and Shanmugan (2000) (Jeruchim, Balaban, & Shanmugan, 2000).

The impact of these effects on the system performance is emphasized when multiple users share the same medium. Properly accessing it, but also creating as less interference as possible within and outside the used channels are the main targets of the next generation mobile communication systems, and the number of users that share the same medium is increasing fast. The real-life example of the troubles within the presentation of the iPhone clearly explains that multiple-access in limited amount of spectrum has a significant impact on current commercial systems (Liu, 2011).

Over the past years there has been a huge interest in techniques at the physical and mac layer by which it is possible to overcome the detrimental effects of propagation and interference and at the same time allow multiple access from several users at a time (Gesbert, Hanly, Huang, Shitz, Simeone, & Yu, 2010).

Problems Related to the Higher Layers

The carried traffic is becoming substantially different from the past. Real time applications and heavy interactive websites are becoming a major threat for the experience of the user. From a general point of view it is possible to summarize these requirements in high transfer rates and reduced latencies. Of course, the trivial solution to the problem would be the availability of infinite bandwidth, but it is practically unfeasible. For this reason, it is necessary to introduce complex control mechanisms able to compensate for the limits in the physical resources.

The traditional methodologies used until now to provide a good user experience are unfortunately insufficient to cope with the dynamism of the wireless medium. As a matter of fact, the same methodologies used to provide robustness in the TCP protocol (Stevens, 1994) are sometimes counterproductive when used for the wireless medium. Congestion and Flow control mechanisms become serious limitations when they are used without awareness of the channel conditions.

Routing also becomes a problem when the new possibilities introduced by the wireless systems are in place. The most simple example is the handover (or sometimes called handoff) procedure. Moving the affiliation of a user from one access point (or base station, or cell) to another requires complex signaling mechanisms and routing, and at the same time, ensuring reduced latency becomes incompatible with mobility itself.

These are just a few examples of the issues that can arise when traditional wired-designed techniques are used in the new wireless scenarios. A huge number of methodologies have been proposed so far, going in the direction of Cross-Layering (Cross-layering), in order to tackle all these problems. Examples and deeper insights in the problem can for example be seen in Nokia (2002), Dixit and Prasad (2003), and Carneiro, Ruela, and Ricardo (2004).(Nokia, 2002)(Dixit & Prasad, 2003)(Carneiro, Ruela, & Ricardo, 2004)

Problems Related to the Spectrum Usage

One of the major problems in next generation communication systems worth mentioning is the so-called "Spectrum Management" problem. From a regulatory point of view, the spectrum allocated to different systems and services appears to be extremely overcrowded, and as a result there are huge difficulties in finding space for new systems.

Some studies and measurements performed by the Federal Communications Commission (FCC) (Spectrum Policy Task Force, 2002) evidenced that this overcrowding is only apparent. Real data show that in a specific location, at a specific time, the spectrum utilization is between 15% to 25% only. This would mean plenty of space for introducing new systems. Unfortunately, the available bandwidths are officially allocated either to public services or to licensed paying operators. The few freely available parts are therefore already crowded.

In this context, the new concept of "Spectrum Management" or "Spectrum Sharing" appeared (Akyildiz, Lee, Vuran, & Mohanty, 2008). It substantially consists in mechanisms for ensuring inter-system multiple access in specifically assigned bandwidths, like the well known "TV White Spaces" (Marcus, 2005).

Like for the two previous issues, this problem calls also for CL methodologies.

CROSS LAYER ARCHITECTURE DESIGN APPROACHES

In this section, an overview of the different approaches available for designing CLAs will be presented, together with their impact on structural stability and possible longevity perspectives.

In fact, it is possible to design CL components in different ways, ranging from the replacement of a traditional layer with a more interconnected one, to a complete re-design of the entire architecture.

Here, following an approach similar to the one presented in Srivastava and Motani (2007) (Srivastava & Motani, 2007), we will introduce such approaches, but first make a note about the possible impacts that CL design could have on the system.

The reader should be anyway aware that the classification approach chosen for this chapter represents one of the possible points of view. Other classification methods can be found in the literature. In fact, a very interesting additional reading is represented by Foukalas, Gazis, and Alonistioti (2008)(Foukalas, Gazis, & Alonistioti, 2008) and references within. Material in the Additional Reading Section tries also to cover possible specific applications, which are not mentioned along chapter.

Advantages and Drawbacks of Cross Layer Design

As previously mentioned, the traditional stacked architecture provides a reliable methodology for building telecommunication systems. This means that any modification to it can potentially bring both advantages and risks, associated to an untraditional new way of working. Nevertheless, these drawbacks did not stop researchers from trying more comprehensive and application-specific optimizations of the traditional protocol stack, even considering new out-of-the box ideas.

The advantages of a CL design are manifold. The main aspect is of course the possibility of having a tighter, stack-wise optimization. Involving multiple layers in the optimization process allows increase of the search-space of a global optimum, in this way improving the overall efficiency of the system. Such improvements can also be focused on a specific application or type of traffic, allowing the system to be polarized on use-cases of interest, like for example video streaming.

These are only the most visible aspects that anyone can easily identify when approaching the CL design. Other aspects, more hidden, but definitely more significant, can be identified with a more in-depth analysis of the peculiarities of the protocol stack. Let's take, as a working example, two layers at the opposite extremes of the traditional architecture: TCP and the Link Layer (LL). Both of them perform Automatic Retransmission Request (ARQ), which of course works on a different time granularity. Nevertheless, both the TCP and the LL layers are trying to perform the same task. This, of course, introduces a redundancy that affects the efficiency of the system. However, the worst-case scenario is when these redundancies conflict with each other (Carneiro, Ruela, & Ricardo, 2004). This happens, for example, when QoS requirements enter the game and force TCP to drop the packet while the ARQ is still trying to transmit it. In this case, it is quite obvious that a CL coordination between the two layers can lead to an overall benefit.

Unfortunately, nothing comes for free, and the CL approaches have dark sides that need to be highlighted in the design phase, in order to evaluate whether the associated benefits are worth the risk, or whether the required solution has, for example, a short expected life-span. This second aspect will be clarified in the following, but let us now try to identify which threats are preventing a number of researchers and developers to fully exploit the CL design potentialities.

Let's go back to the reliability of the traditionally stacked architecture. This reliability is intended not only in terms of easy replacement of each single layer, but also on the stability of the entire system. In fact, if each layer is self-consistent, self-contained, and intrinsically stable, also the entire architecture from top to bottom will be stable. By using a CL approach, all the previous assumptions do not hold anymore, thus making cross layering risky. Performance degradation is the most common drawback that one can face, if the design is not carefully planned with due account of all the impacts that a CL optimization can have on all the layers.

Even if a designer ensures the stability of his/her CLA, in most of the designs a coupling between two or more layers is introduced. It is then impossible to replace the involved layers in the same easy way allowed by the OSI model. This consideration impacts a range of aspects of system design:

- **Flexibility:** new layers either have to fit more stringent requirements dictated by the CL connections, or all the layers involved in cross layering have to be replaced.
- **Technology dependency:** when upper and lower layers are involved in cross layering, a technology dependency is introduced. This means that the designed architecture is not portable to multiple technologies. This is nowadays a severe drawback,

given the availability of multiple wireless communication methodologies in mobile terminals.

- **Implementation problems:** the implementation of CLAs is severely impacted. In fact the software code can be either non-resilient to changes or *"spaghetti-like"* (Kawadia & Kumar, 2005). While the latter is a tough problem that can only be partly managed by using advanced software development methodologies, e.g. design patterns (Freeman, Freeman, Bates, & Kathy, 2004), the former can be the result of a bad design or old code. In this case, a severe restructuring of all the code is required, increasing the needed implementation effort.

All the mentioned problems discourage a number of Research and Development (R&D) specialists and engineers from using CL as a daily work practice. As a matter of fact, especially when talking about products that have to be on the market, the longevity of an architecture is fundamental, since not even the biggest company can afford global re-working of an architecture in short-mid terms.

Another and more theoretical drawback is the increased communication overhead required to let multiple layers interact among each other in a CL way. The traditionally stacked model is optimal in this respect since it clearly defines and limits the type of services required from layer to layer. A non-careful CL design with *"hooks"* here and there for future expansions, could lead to the definition of broad information flows between the layers. What does then prevent the designer from using a totally *"white-box"* approach? This term is here intended as a complete unpacking of all the layers, making all the internal variables, messages, and procedures of each layer available to all the others. This approach is the extreme opposite of the traditional *"black-box"* foreseen by the OSI model, where each layer is self-contained and the internal structure is irrelevant to the other layers as long as the defined services are provided. In the following, some tools will be provided to the reader finding the proper CL trade-off, in order to paint each box available with the right shade of grey.

Design Principles and Guidelines

In literature, several approaches for tackling CL design and optimization are available. The choice of one or the other was mostly due to the specific objective of the proponents and their available knowledge. Nowadays the Cross-layering has reached enough maturity for having more structured design approaches. The goal of this section is to provide an overview of the most used methods, by categorizing them in terms of information flows.

Before entering into the details of each approach, some general design rules have to be set, in order to have a good design. It means that the CLA is designed in a way that tries already to minimize the aforementioned risks associated to Cross-layering. By approximately following the same exposition order of the issues, the rules should be the following:

- **Impact analysis:** when including CL functionalities or optimization, the designer should backtrace all the places where the added functionality can impact the architecture. This is one of the most critical steps, often overlooked by designers, in order to guarantee the self-stability of the impacted layers. Furthermore, it also ensures that there will be more gains than losses in terms of performances. Analyzing all the parameters impacted is not a trivial job, but it is a necessary effort when cross layering.
- **Technology decoupling:** when technology specific layers are involved, it is good practice not to directly expose technology specific variables, parameters, and services. An abstraction effort has to be done

in order to ensure the portability of the architecture over multiple physical media. A possible abstraction solution is to exchange with the other layers information, which is standardized or widely used/recognized. Such information could also be the result of processing by combining several technology specific pieces of information.

- **Minimization of shared info:** as previously mentioned, the *"white-box"* approach can have detrimental effects on the well functioning of the system. A *grey-shading* process is required in order to protect the stability of the modules and reduce the overheads. Good practice is to include only the information flows, which are really necessary for the design. In the implementation of the CLA it is of course possible to leave *"hooks"* that will be helpful when additional information is required.

Let's now enter into the details of the approaches in the following subsections.

Approaches to Cross Layer Design

As previously mentioned, there will be an attempt to categorize the approaches according to the kind of information flow foreseen. It is then possible to identify seven different approaches. The first two are based on implicit (or offline) information flows. The second chunk of three is conversely characterized by an explicit information flow among layers. Such a flow can be designed in addition to the traditional services in adjacent layers, or to connect non-adjacent ones. The last two approaches are then the most peculiar, since they rely on non-stacked functionalities. Let's proceed with the first two.

In both of the cases, the modifications to the stacked architecture are minimal, and the Cross-layering is performed through a carefully targeted design. Let's take for example the strategy adopted in Figure 4a: a single layer is designed and optimized in order to operate in conjunction with a *target* one. The positions of the target and of the optimized layers in the stack are irrelevant for the method. Figure 4a shows only an example. Of course, this kind of design introduces a strong coupling in the architecture, intrinsically reducing its flexibility. Nevertheless, it is a comfortable approach for product development, since this kind of specific tuning is easy to achieve in the implementation phase. This is also the way proprietary features can be introduced in the architecture. A possible example of such methodology can be found in Kliazovich, Gomes, Granelli, and da Fonseca (2009)(Kliazovich, Gomes, Granelli, & da Fonseca, 2009). In this case, the MAC layer is designed in order to provide good performances to the overlayed TCP.

The other approach relying on implicit information foresees the joint design of two (or more) adjacent layers (and sub-layers). It is exemplified in Figure 4b. The interfaces towards the other entities in the protocol stack remain intact, while the internal functioning can be optimized due to the higher amount of information available. This represents a good design strategy for the lower layers in applications like sensor networks, as foreseen in Rodrigues and Chen (2010)(P., Rodrigues, & Chen, 2010). As a matter of fact, the PHY and MAC layers of such systems are way more simple than for example in 4G standards. A joint design of the PHY and MAC procedures can allow improved communication capabilities and extended battery life.

Figure 5a shows an Upward information flow approach. The method is intended to provide a higher layer with operative information about a layer that is closer to the physical medium. The flow is mostly composed by measurement or notifications that would otherwise not be directly available. With such details the higher layer can properly tune the operational parameters. A possible example of such approach is the coordination between TCP-ARQ and LL-ARQ as presented by Kliazovich, Granelli, and Gerla (2007)(Kliazov-

Figure 4. Implicit information flow approaches: (a) target-based optimized design, (b) merging of adjacent layers

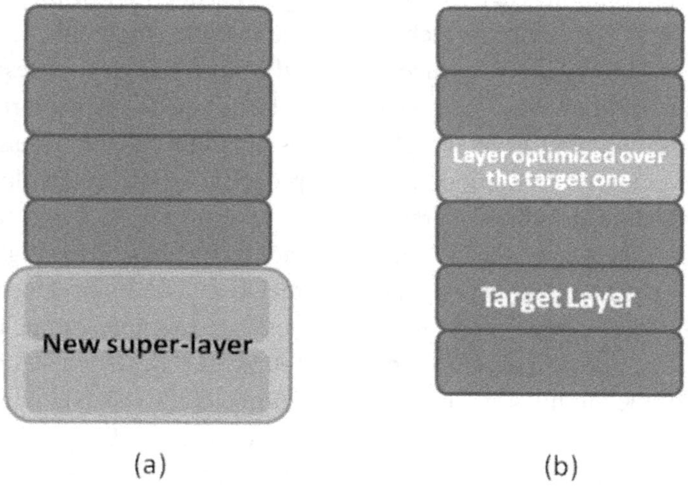

ich, Granelli, & Gerla, 2007). In the presented method, an additional CL agent is in charge of reducing the overhead due to the double ARQ job from the two layers. This is done through an explicit notification of LL-ACK reception from the LL to the CL agent, which occupies an intermediate slot between TCP and LL.

Figure 5b, instead, shows the Downward information flow approach. The method is intended to provide the higher layer with a substantial control of what happens in the lower one. In this case, the flow is mostly composed by either command or measurements/indications. In the former case, the higher layer is in charge of optimizing the lower one. In the latter, the lower layer self-optimizes through the indications provided by the higher layer, for example about the type of carried traffic. A possible application would rely on setting the maximum timers at the LL according to the QoS of the traffic, for example in delay-sensitive applications like video or voice streaming. Sometimes a new information flow is not even necessary, whenever the required information is already available in the encapsulated payload of the normal service.

The last approach that foresees an explicit information flow is of course characterized by a bi-directional information exchange, and it is shown in Figure 5c. This approach can be thought as a sum of the two previous ones, i.e. when the optimization is achieved in a distributed fashion between the layers, each one taking care of a subset of functionalities (Dimic, Sidiropoulos, & Zhang, 2004).

Anyway, it is possible to combine all the previously mentioned approaches. A simple example is the scheduling mechanism in modern communication systems like the 4G ones. The downlink scheduler gets information about the channel conditions and the traffic types from the MIP, TCP, and PHY layers in order to provide the users with the highest QoS possible. In this point of view it is possible to claim that, for example, the LTE protocol stack is somehow CL (Palat & Godin, 2009).

Anyway, there are cases where the addition of explicit information flows is not feasible or discouraged. In such cases, it is possible to use the two following approaches shown in Figure 5.

Figure 5. Explicit information flow approaches: (a) upward information flow, (b) downward information flow, (c) bi-directional information flow

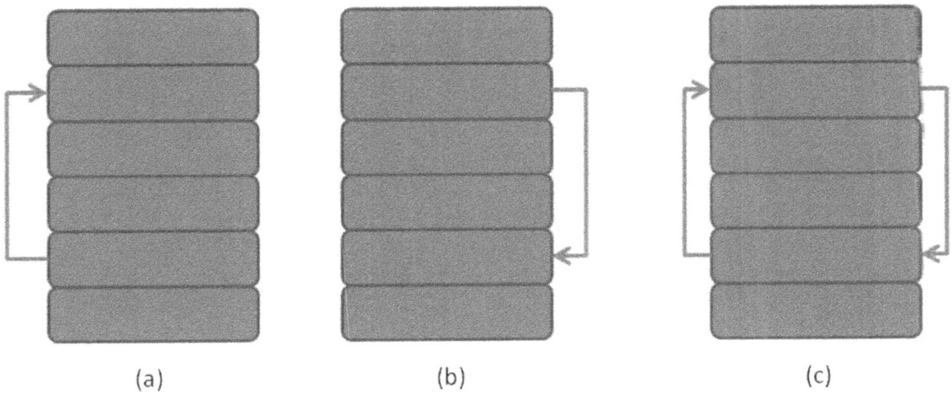

The last two approaches are the most interesting ones, since they break completely the traditional concept of the layered architecture. In Figure 6 there is a graphical exemplification.

These approaches are extremely interesting, because they are able to tackle the modern communication problems and be implemented with the latest state-of-art methodologies and tools. Between the two, the one depicted in Figure 6a is the less used in architecture design. The main example can be found in Braden, Faber, and Handley (2002). It anyway represents an interesting solution since it is very close to the software architectural design. This approach would allow developers to have a clearer, software engineering-oriented vision on how the entities should be connected, and which kind of services each of the nodes of the graph is able to provide to the others. Of course, by breaking the traditional layered structure there is the risk that the run-time functions could suffer from unexpected and undesired effects. Given the complex connectivity among the entities, one of the most common problems could be the amount of overhead that is introduced within the system and exchanged between wireless peers. Conversely, the stacked service-oriented architecture clearly defines where such overhead is introduced, and in which amount.

The last approach, shown in Figure 6b is instead widely used nowadays. Most of the CLAs and methodologies are today relying on such structure, with the help of a parallel CL entity. Given this widespread usage, more space will be spent on describing state-of-art techniques that apply it. Let us here introduce just the references where the reader can further investigate its usage: Raisinghani and Iyer (2004), Winter, Schiller, Nikaein, and Bonnet (2006), and Conti, Maselli, Turi, and Giordano (2004). More detailed analysis will be provided in the following.

Most of the aforementioned approaches rely on explicit information flows, but how are these flows defined? This is an important step of the design phase that not only simplifies, but also avoids misunderstandings in the implementation stage. Let us now clearly define a structured way for encapsulating the information flows: the concept of Application Protocol Interface (API) will be used as a reference in the following.

Let us define an API as a set of **Objects**, **Primitives**, and **Services**. The **Objects** define which data can be exchanged through the API and their format. Objects could for example be some proprietary parameters of a layer, a data packet, or a specific variable, like the SINR at the receiver. The **Primitives** are the functions that are in charge

Figure 6. Non-stacked approaches: (a) graph-based architecture representation, (b) parallel entity usage for non-stacked functionalities and optimizations

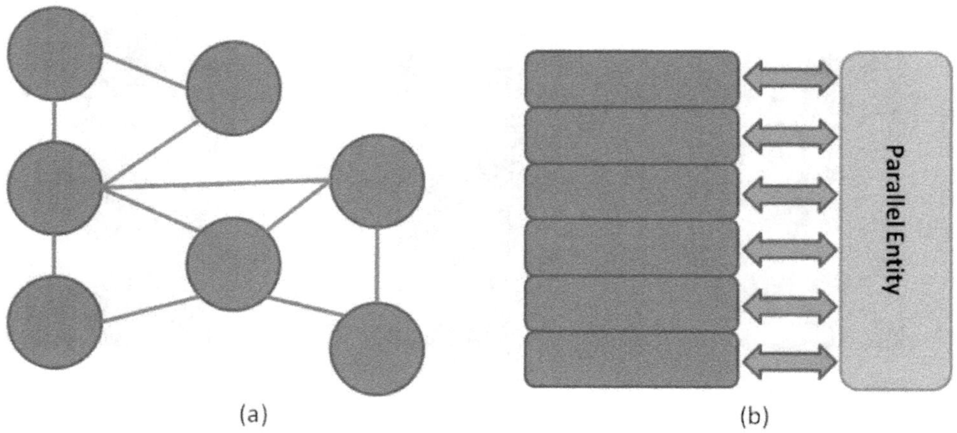

of dealing with the **Objects**. The **Primitives** can be called by any of the interacting layers. Finally, a **Service** is a sequence of **Primitives** that are called for in exchange of a specific **Object** or set of **Objects**.

In Figure 7, a simple example of one service for a hypothetic PHY Layer API is shown. The reader would probably have noticed that architectural design and the software design and implementation languages are now mixed. As previously mentioned, having a more implementation-aware design facilitates the communication between designers and developers, reducing also the time-to-market of the architecture. Of course the design of the API should not be considered as a replacement of the implementation, but conversely as a definition of the minimal needs and requirements of the designer.

Let's now analyze in more detail how to concretize the aforementioned approaches.

Vertical Cross Layer Interface Definition

A **Vertical Cross Layer Interface** is an API used for sharing CL information among layers, but still maintaining the stacked architecture. As a matter of fact, a number of designers are reluctant to abandon the stability and resiliency granted by such an approach. Anyway it should be possible to Cross-Layering and optimizing its performances by defining APIs between two or more vertically stacked layers. This interface suites the design approaches depicted in Figure 5 and previously described.

It is possible to design two kinds of vertical interfaces:

- Layer-to-Layer (L2L);
- Generalized API.

The former is suitable for coupling two layers together. The advantage of such an approach is the possibility of limiting to the minimum the type and number of **Objects** that are shared between the layers. This helps in respecting the minimization design rule previously set. On the other hand, especially if the API is defined for lower, technology-dependent layers, the same dependency is implicitly inherited by the higher coupled layer. This reduces the modularity and flexibility of the architecture itself, making it difficult to upgrade. Furthermore, if one layer is coupled to multiple others, there would be the need for defining multiple APIs afferent to

Figure 7. Simple pseudo-code example of entities involved in the definition of an API

PHY Layer API

- Object: double SINR
- Primitive: function Get_SINR_RESPONSE()

```
{
    return SINR;
}
```

- Service
  - Get_SINR_REQUEST()
  - Get_SINR_RESPONSE()

the same layer. Another undesired effect is the increased complexity of the impact analysis, due to this intricate connectivity. This would make the implementation complex, and slow down the processing since multiple parsers for interpreting the **Services** provided by each API would be needed.

In this case, it is better to use the other approach, a generalized API that contains all the **Services** and **Objects** that the layer can offer for Cross-layering. While this approach makes the impact of Cross-layering easier to analyze, the side effects could be more severe, due to the multiple access problem. As a matter of fact, multiple entities could compete for reading/writing the same parameters. Nevertheless, the architecture proves more resilient to changes, since each layer is still de-coupled from the others. A non-careful designer could be tempted to add more information than needed in the API. This would violate a design principle by exposing internal parameters that could be risky to make available for modifications.

Horizontal Cross Layer Interface Definition

A **Horizontal Cross Layer Interface** is an API used for sharing information with Parallel CL entities. The same API concept can also be applied to the graph-based approach. This API definition is nowadays extensively used, since the approach shown in Figure 6b is quite popular. This kind of approach helps in keeping the design within the aforementioned rules. As a matter of fact, this concept is similar to the generalized vertical interface in terms of **Services** and **Objects** that it provides. The main difference is that there is only one entity accessing the information, keeping the unintended interactions problem under control.

Even in this case, it is possible to adopt two different strategies:

- White box
- Gray box

These two terms were previously mentioned in this chapter. They represent the transparency of the layer borders in relation to the other “external” entities. White box means that it is possible to control (read/write) all the parameters of the layer from the outside. Gray box means that some more delicate parameters are hidden. The “shade” of Gray depends on how much the designer wants to trade between extreme, but risky, optimization and stability. Another advantage of the Gray box is the possibility of exchanging **Objects** which are not directly parameters of the layer, but also aggregated information that represents the behavior of the layer.

The vertical APIs are considered intrinsically Gray since for the sake of stability of the system it is better not to provide "sensitive" information about the layer functioning to other layers. In this case, the parallel entities represent a sort of omniscient layer that could need full control of all the parameters of the layer.

It should now be clear to the reader that these parallel entities are considered of high importance for CL design nowadays. Let us analyze in detail what these objects can be, and what their role is in Cross-layering.

Passive and Active Parallel Structures: From Databases to Resource Managers

The parallel entities can be defined in several ways, according to their role in the architecture, or also where they are located (locally, remotely). A good survey of these entities is provided in Foukalas, Gazis, and Alonistioti (2008). Such elements can, in the first place, be passive ones being used only as common repositories where the interacting layers read/write the **Objects** they need. The MobileMan (Conti, Maselli, Turi, & Giordano, 2004) CLA represents a good example of this type of design, having the *Network Status* component which is exactly a common repository for all the layers. Let us call this kind of solution as *Local Database*. It is worth saying that the API used in this case can be defined as a generalized horizontal one, where all the layers could potentially access all the available **Services**. A good example of this API definition is the *Cross Layer Bus* used by Baldo *et al.* (2008).

In order to solve the aforementioned problem of spectrum sharing, it is today considered a common solution to use *Remote Databases*. The difference between this type and the previous one is the need of a higher level API that is able to exchange messages through the network. This is the kind of solution that is currently investigated by the IEEE 1900.6 (Committee, 2011) standardization group and the Wireless Innovation Forum in order to tackle the requirements for dynamic spectrum sharing set by the FCC.

Shortly after the introduction of the passive parallel entities, researchers realized that by exploiting this parallelism in an active way it was possible to achieve tighter optimizations. Let us call this type of entity *Local Manager*. Its usage can of course be manifold. In the first place, it is extremely useful for introducing a small set of CL non-stacked functionalities within a traditional stacked architecture. A clear example is shown in Akyildiz, Lee, Vuran, and Mohanty (2008) for the implementation of the spectrum decision and mobility functionalities. Also, in the CL literature, the *Local Managers* are often used for placing the CL optimization algorithms. This is widely used in practice, as Raisinghani and Iyer (2004) and Carneiro, Ruela, and Ricardo (2004) clearly show. It is worth mentioning, that also the LTE Radio Resource Manager (Holma, Kolding, Laselva, Pedersen, Rosa, & Viering, 2009) could be considered an intrinsically CL entity, even though it is included in a standardized architecture. This is an extremely interesting and actual topic that could be deepened with additional reading like Toufik and Knopp (2009) or the original specification document, 3GPP (2011).

An interesting design approach used by some of these state-of-art solutions relies on the identification of a two-tiered *Local Manager*. Let us call the first tier, used as a technology de-coupler, *Abstractor* (*Tuning Layer* in Raisinghani and Iyer (2004), *Contextor* in Razzaque, Dobson, and Nixon (2007). This entity is usually coupled to one single layer through a White box horizontal API. It can either have an active role as a single-layer optimizer, or simply provide the second tier with a technology de-coupled abstraction. The communication with the second tier is based on a

Gray box API. This approach can help designers in better identifying which "shade of gray" they can use for the layer-to-manager communication. The second tier is then the effective CL *Local Manager* in charge to manage the CL interactions.

For the sake of completeness, the work carried out by the IEEE 1900 group also includes the definition of active decision makers, combined with the databases that will be able to enforce the spectrum regulations within the network. This approach can then be identified as a *Remote Manager*.

A combined usage of both remote and local parallel entities is more and more common. A very good example is the Cross-Talk CLA (Winter, Schiller, Nikaein, & Bonnet, 2006), where *Local Managers* are combined with a remote one in order to cooperatively optimize the performances of the entire network.

PERSPECTIVE TOWARDS COGNITIVE COMMUNICATIONS

In the last decade, a new paradigm for wireless communications emerged from the scientific community: Cognitive Radio (CR). First introduced by Mitola (1999) and then stressed by Haykin (2005), it was born as an intelligent extension of Software Defined Radios (SDRs). The underlying concept is substantially funded on the application of optimization, pattern recognition, and machine learning algorithms for exploiting the flexibility and reconfigurability of the SDRs in a totally automated way. This would create intelligent devices able to autonomously react to any environmental or context change, ranging from radio interface optimization, to improvements in users' everyday life. A very good example of the vision is presented by Mitola (1999) when he explains the "taxi scenario." A more detailed explanation of such services is presented in Campbell, Campbell, Lewandowski, and Weinstein (2006) instead.

Even if the status of research is far away from the realization of the concept, the architectural design of the CRs started assuming a more stable shape in the last few years. As a matter of fact, CR architectures are intrinsically CL, since the intelligence of the system should be able to adapt and optimize every single aspect of the CR device. This requires a global interaction among all the components, from the radio frequency front-end up to the graphical user interface, i.e. a CL approach. One of the most shared conceptual architectures is based on the same design principle explained in Figure 6b, and it foresees the usage of an abstract and parallel entity called *Cognitive Engine* (CE). It is of course specialized, from case to case, given the problem and objective that each researcher is tackling. In the following, some examples of CR architectures will be presented.

A good overview of the usage of Cross-layering in CR is presented in Arslan and Yarkan (2007). Their reference architecture foresees a CE and all the vertical information flows as well, for the sake of generality. Within such framework, the paper clearly explains which are the available state-of-the-art optimization techniques and how to apply them.

This approach is usually extended through the usage of multiple parallel tiers, involving also remote entities connected through the network. In Reddy and Bullmaster (2008), the proposed architecture follows exactly this approach, providing a basic and extremely synthetic framework. Their Cross-Layer manager is connected to the stack with a non-specified horizontal interface, possibly a bus. The network interface of the manager could also be used, in this case, for direct cooperation among *Local Managers*.

Another interesting framework is the one specified by Akyildiz, Lee, Vuran, and Mohanty (2008) for the specific problem of spectrum management. Despite the fact that it is not defined where each of the involved entities is located, locally or remotely, the paper clearly identifies which type of information needs to be exchanged at which level, helping designers and researchers in the definition of their own version and implementation.

Baldo and Zorzi (2008) specify from an information flow point of view an interesting architecture. Their first parallel tier is composed by a series of layer-specific *Cognitive Interfaces* used as *Abstractors* for obtaining a technology de-coupling between the protocol stack and the CE. The role of the interfaces is to create a set of technology-independent parameters that can assume fuzzy values. In the opposite direction of the information flow, such entities write the optimized parameter setting in the protocol stack.

Fuzzy reasoning is an interesting tool that proved its utility in the CR framework. Nevertheless, it is not the only one available. One of the most comprehensive and flexible CR architecture, based on a *Cognitive Resource Manager* and a possibly remote knowledge database, is the one proposed by Petrova and Mahonen (2007). A very detailed definition of the APIs used for the architecture and hints on the development phase are also presented in the chapter.

Moving from the design phase to the implementation one, among the most promising software platforms able to implement flexible CL functionalities is IRIS, developed at Trinity College Dublin. An overview of its implemented architecture and the offered functionalities is shown in Sutton et al. (2010). A point worth being stressed is the usage of XML as higher-level language for the definition of the architecture and structure of the node. Meta-languages have always been one of the key points of the CR design since its dawn. The same Mitola introduced this concept together with the CR paradigm, as well explained in the book by Mitola (2006).

Which is the perspective towards the design of optimized or cognitive radios? In the following, we try to identify a possible approach that takes into account the advantages and drawbacks of the different design approaches introduced until now.

Hierarchical Cross Layer Architecture for Cognitive Systems

One of the biggest challenges for designers is finding the right trade-off between the desired tightness of the optimization and the derived complexity and technology coupling of the algorithms used for the optimization process. This problem is even emphasized when parallel entities are involved, since:

1. They require additional management structures and computation resources;
2. There are information flows involved.

A possible solution is provided by the use of the *Abstractors* between the protocol stack and the *Manager*. In Figure 8 a design based on a localized optimizer, close to the layer is presented.

In this design solution, each layer is wrapped in a layer-specific *Optimizer/Abstractor*. The stacked vertical API provides the traditional services to the neighbor layers. Despite the graphical representation, this API is intended to be connected to both the lower and higher layers. In the figure a simplified depiction is presented for the sake of clarity.

The layer is also provided with a White Box API used to completely expose all the internal parameters of the layer to the *Optimizer/Abstractor*. While the usage of White Box APIs should generally be discouraged for Cross-layering, the approach could be considered in this case quite safe, since the parameters still remain localized within the layer-specific *Optimizer/Abstractor*. The other internal modules within the wrapper are the ones effectively carrying out the functionalities. While the *Layer-Specific Optimizer* has a clear role, the other two require further explanation.

It is possible to conceptually split a layer into two aspects: its "shape," defined by the variables that define the layer structure, and the information that runs through it, defined for example by a num-

Figure 8. Definition of a layer optimizer/abstractor for each layer. Several entities within the framework have specific roles. Two external APIs and an internal one are needed.

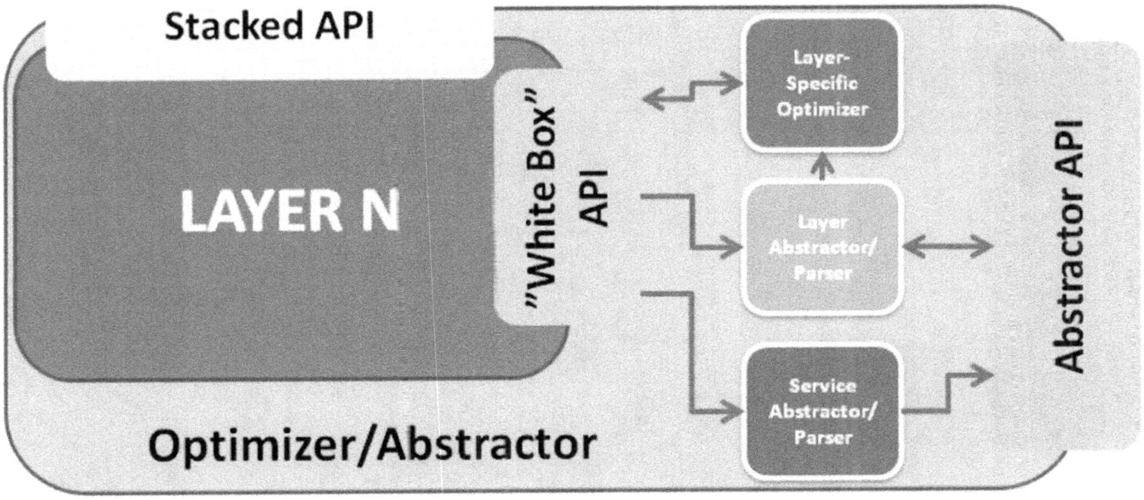

ber of performance measurements performed by the layer. The former aspect is taken into account by the *Layer Abstractor/Parser*, used to translate the internal variables and shape of the layer into a technology-de-coupled set of variables to be made available to the parallel entity. Of course it is valid also vice versa, meaning that whenever the parallel entity performs a global CL optimization and provides back the new optimized values of such generalized variables, the internal parser needs to translate those into requirements to be used by the local optimizer.

The latter aspect is instead covered by the *Service Abstractor/Parser* that performs the same abstraction operation. The only difference relies on the fact that this class of variables is used within the optimization metrics, but they are not subject of optimization themselves, since most likely they consist of measurements. This makes the parsing operation much more simplified.

Finally, the *Abstractor API* provides the horizontal CL interface towards the parallel entity that will be explained in detail in the following.

In Figure 9, the overall multi-tier architecture is presented. On the left side of the picture the traditionally stacked architecture is represented, with each layer wrapped in its Optimizer/Abstractor. They are then the first tier of the architecture, while the second tier consists of a Local Manager, i.e. the Architecture Manager shown in figure. This entity contains both an abstract representation of the layers and services available in the stack, and the non-layered (i.e. CL) functionalities and optimizers. As shown in the picture, this structure is defined through a graph-based representation. The motivations for such a design choice are fundamentally two. The first one is that in such a way it is possible to have complex architectures, but with a high modularity, allowing easy replacements and expansions of the functionalities. The second motivation is due to the desired development-oriented design previously mentioned. In fact, such representation can be easily achieved by means of usage of XML language, as proposed by Mitola (2006).

A toy example of XML architectural definition can be seen in Figure 10.

Figure 9. Definition of the general multi-tier architecture. The protocol stack is fully wrapped in single-layer optimizers/abstractors and they are all connected to the architecture manager that contains an abstracted representation of the stack and additional non-layered (CL) functionalities.

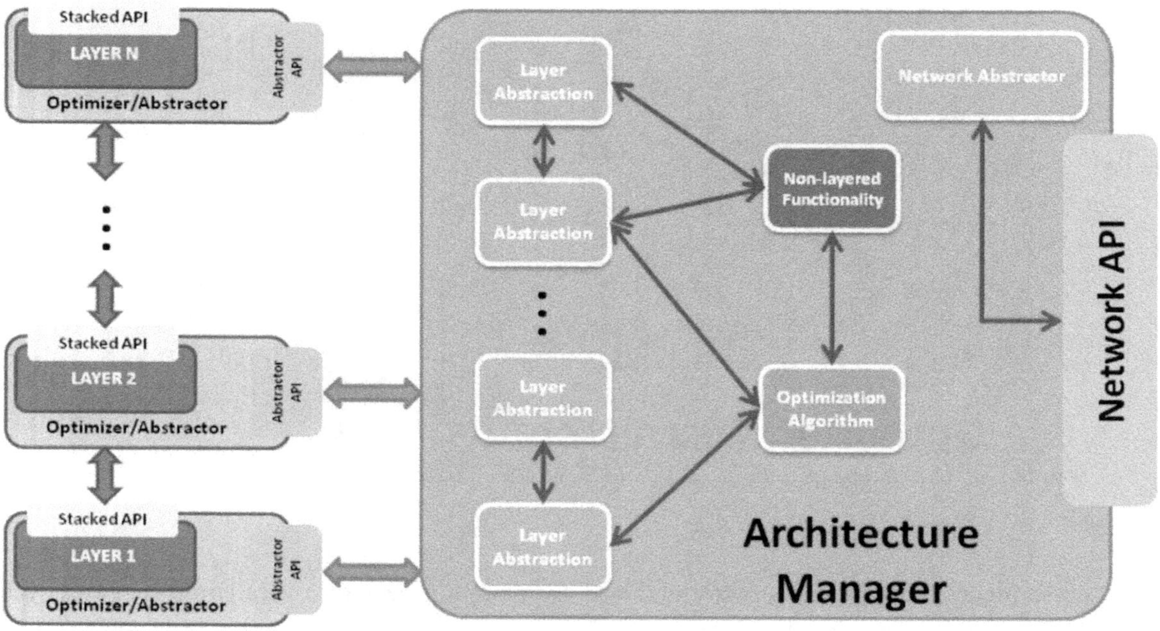

In the example of Figure 10 we assume the use of a Long Term Evolution (LTE) (Bogineni, et al., 2009) at the physical layer. It is possible to understand that in the abstracted representation, the LTE PHY is sending information to the Dynamic Spectrum Allocation function. The **<connection>** tag identifies that these two entities are connected within the *Architecture Manager*. The **<sender>** and **<receiver>** define the direction of the information flow, while within the **<link>** object, all the specific types of information exchanged in that specific direction between the two entities are defined. In this case, they are the Signal-to-Interference-plus-Noise-Ratio defined per carrier or per sub-channel. This is of course a toy explanatory example, designed for providing some insight on how to proceed. Every designer is anyway free to structure his/her own meta-structure as he or she prefers.

The *Architecture Manager* should also include a *Network Abstractor*, providing the necessary abstraction level to the parameters that are exposed to higher level parallel entities (such as network-based databases) or to the other *Managers* in the network through the *Network API.*

As pinpointed earlier, such graph-based representation could be complicated to analyze in terms of impact. Failing on this point would compromise the entire stability of the architecture, since one of the basic design principle, established in this chapter, will be violated. In order to cope with such weakness, each *Layer Abstraction* within the *Architecture Manager* is composed not only by the set of variables exposed by the *Layer Optimizer/Abstractor*, but also by their impact graph, as shown in Figure 11.

In this type of representation, all the variables that define the layer characteristics are interconnected based on their impact, i.e. if for example

Figure 10. Toy example of XML pseudo-code for an architectural definition in graph-based representations. In the example an LTE PHY layer is considered as technology specific use-case.

```
<architecture>
   <connection>
         <sender> LTE-PHY </sender>
         <receiver> Dynamic_Spectr_Alloc </receiver>
         <link>
                  <type> service </type>
                  <container> SINR_subchannel </container>
                  <container> SINR_carrier </container>
         </link>
   </connection>
   ...
   ...
   ...
</architecture>
```

Figure 11. Example of graph-based visualization of the impact connections among the variables that define the layer "shape"

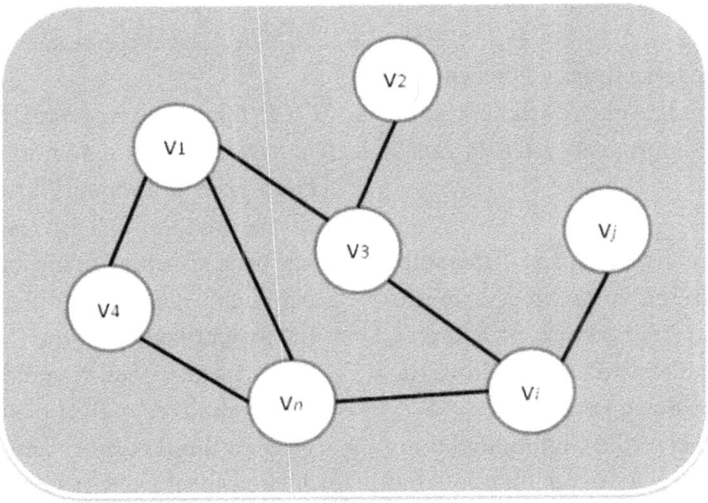

a change of value in a variable affects the proper value setting of any other. This kind of representation is in line with the structure of the Architecture Manager, since it can be also described through an XML scripting. In this way, there is a homogeneity in the development of the Manager, due to the usage of the same internal representations for all the levels of the abstractions.

This architecture wants to represent a summarization of the concepts presented in this chapter, and how to properly exploit the benefits of Cross-layering without compromising the stability and reliability of the traditional stacked model.

CONCLUSION

Wireless communications are becoming fundamental in everyday life. From the TV remote to the fanciest smartphone, wireless is the keyword of the future. Unfortunately, this increased attention on the wireless world also brings new and more stringent requirements for ensuring the satisfaction of the users. These new open challenges are gathering researchers and developers in order to overcome the limitations of the traditional layered model. *Cross-Layer* is then the passkey for the solution of the wireless problems. Over the years, there have been several ways of approaching the design of *Cross-Layer Architectures*. Each of them contributed to the accumulation of experiences that evidenced that *Cross-Layering* is a very powerful tool, but also that it contains risk factors whenever it is not carefully used.

It is hence possible to summarize this experience in a few guidelines that each designer should follow when he or she approaches a *Cross-Layer* design:

- Identify the expected lifespan of the solution: a proper design is required when the longevity of the architecture is a target. Short-term and short-living solutions do not require much design effort.
- Identify the impact of the CL modifications to the protocol stack: depending on the lifespan and the amount of layer impacted by the new CL functionality or optimization, the designer should choose the design approach that requires the minimal effort and modifications. For example, it would be useless to design an entire multi-tier architecture for exchanging only the ARQ parameters between TCP and PHY, when a simple vertical API is sufficient.
- Ensure the stability of the solution: the designer should use all the available tools for analyzing the impact of the cross layering on the services provided to the users, avoiding risks that could compromise the performance of the system. This can be achieved through a careful design, an accurate impact analysis, and the minimization of the information flows that run through the architecture. The motto of the designer should always be *"Only if it is needed"*[1].
- Use an implementation-oriented design: a good design, in general, should be easily understandable by developers to minimize the time-to-market of the solution. This is especially valid for CLAs, where the boundaries between the layers and functionalities become blurred. The use of terminologies and solutions that are similar to the ones used in software development eases the transfer process between designers and developers by maintaining the clarity of the solution.

In this chapter, several literature examples that motivate the aforementioned recommendations have been presented. Furthermore, the analysis of the state-of-art proves that Cross-Layering has been absorbed in the modern communication paradigms such as Cognitive Radio. A path towards the generalization of the Cognitive Communications has also been covered in the presentation of the generalized multi-tier architecture that should provide the flexibility and the abstraction levels desired for the CR systems.

REFERENCES

Akyildiz, I. F., Lee, W.-Y., Vuran, M. C., & Mohanty, S. (2008). A survey on spectrum management in cognitive radio networks. *IEEE Communications Magazine*, *46*(4), 40–48. doi:10.1109/MCOM.2008.4481339

Arslan, H., & Yarkan, S. (2007). Cross-layer adaptation and optimization for cognitive radio. In Arslan, H. (Ed.), *Cognitive Radio, Software Defined Radio, and Adaptive Wireless Systems* (pp. 421–452). Springer. doi:10.1007/978-1-4020-5542-3_14

Baldo, N., & Zorzi, M. (2008). Fuzzy logic for cross-layer optimization in cognitive radio networks.*IEEE Communications Magazine*, *46*(4), 64–71. doi:10.1109/MCOM.2008.4481342

Bogineni, K., Ludwig, R., Mogensen, P., Nandlall, V., Vucetic, V., & Yi, B. (2009). LTE part I: Core network.*IEEE Communications Magazine*, *47*(2). doi:10.1109/MCOM.2009.4785378

Bogineni, K., Ludwig, R., Mogensen, P., Nandlall, V., Vucetic, V., & Yi, B. (2009). LTE part II: Radio access.*IEEE Communications Magazine*, *47*(4). doi:10.1109/MCOM.2009.4907405

Braden, R., Faber, T., & Handley, M. (2002). From protocol stack to protocol heap - Role-based architecture. In *Hot Topics in Networking (HOTNETS I)*. Princeton, NJ: Princeton University Press.

Campbell, J. P., Campbell, W. M., Lewandowski, S. M., & Weinstein, C. J. (2006). Cognitive services for the user. In Fette, B. (Ed.), *Cognitive Radio Technology* (pp. 313–336). Burlington, MA: Newnes. doi:10.1016/B978-075067952-7/50011-8

Carneiro, G., Ruela, J., & Ricardo, M. (2004). Cross-layer design in 4G wireless terminals. *IEEE Wireless Communications*, *11*(2), 7–13. doi:10.1109/MWC.2004.1295732

Committee, I. D. (2011). *IEEE 1900.6 working group on spectrum sensing interfaces and data structures for dynamic spectrum access and other advanced radio communication systems*. Retrieved March 2011 from http://www.ieee.org.

Conti, M., Maselli, G., Turi, G., & Giordano, S. (2004). Cross-layering in mobile ad hoc network design.*IEEE Computer*, *37*(2), 48–51. doi:10.1109/MC.2004.1266295

Dimic, G., Sidiropoulos, N., & Zhang, R. (2004). Medium access control - Physical cross-layer design.*IEEE Signal Processing Magazine*, *21*(9), 40–50. doi:10.1109/MSP.2004.1328087

Dixit, S., & Prasad, R. (Eds.). (2003). *Wireless IP and building the mobile internet*. London, UK: Artech House.

Freeman, E., Freeman, E., Bates, B., & Kathy, S. (2004). *Head first design patterns*. New York, NY: O'Reilly Media.

Gesbert, D., Hanly, S., Huang, H., Shitz, S., Simeone, O., & Yu, W. (2010). Multi-cell MIMO cooperative networks: A new look at interference.*IEEE Journal on Selected Areas in Communications*, *28*(9), 1380–1408. doi:10.1109/JSAC.2010.101202

Haykin, S. (2005). Cognitive radio: Brain-empowered wireless communications.*IEEE Journal on Selected Areas in Communications*, *23*(2), 201–220. doi:10.1109/JSAC.2004.839380

Holma, H., Kolding, T., Laselva, D., Pedersen, K., Rosa, C., & Viering, I. (2009). Radio resource management. In Holma, H. T. (Ed.), *LTE for UMTS - OFDMA and SC-FDMA Based Radio Access* (pp. 181–212). Chichester, UK: John Wiley & Sons.

Jeruchim, M. C., Balaban, P., & Shanmugan, K. S. (2000). *Simulation of communication systems - Modeling, methodology, and techniques*. New York, NY: Kluver Academic/Plenum Publishers.

Kawadia, V., & Kumar, P. (2005). A cautionary perspective on cross-layer design.*IEEE Wireless Communications*, *12*(1), 3–11. doi:10.1109/MWC.2005.1404568

Kliazovich, D., Gomes, P. H., Granelli, F., & da Fonseca, N. L. (2009). Receiver-driven queue management for achieving RTT-fairness in Wi-Fi networks. In *Proceedings from Globecom*. Honolulu, HI: IEEE Communications Society. doi:10.1109/GLOCOM.2009.5426120

Kliazovich, D., Granelli, F., & Gerla, M. (2007). Performance improvement in wireless networks using cross-layer ARQ.*Computer Networks*, *51*, 4396–4411. doi:10.1016/j.comnet.2007.06.002

Liu, K. R. (2011). Cognitive-radio games. *IEEE Spectrum*. Retrieved from http://spectrum.ieee.org/telecom/wireless/cognitive-radio-and-game-theory.

Marcus, M. (2005). Unlicensed cognitive sharing of TV spectrum: The controversy at the federal communications commission.*IEEE Communications Magazine*, 24–25. doi:10.1109/MCOM.2005.1453413

Mitola, J. (1999). Cognitive radio: Making software radio more personal.*IEEE Journal on Selected Areas in Communications*, *17*(4), 514–538. doi:10.1109/49.761033

Mitola, J. I. (2006). *Cognitive radio architecture - The engineering foundation of radio XML*. Hoboken, NJ: John Wiley & Sons, Inc. doi:10.1002/0471773735

Nokia. (2002). *Mobile internet technical architecture,* (Vol. 1). New York, NY: IT Press, Edita Publishing Inc.

Palat, S., & Godin, P. (2009). Network architecture. In Sesia, S., Toufik, I., & Baker, M. (Eds.), *LTE - The UMTS Long Term Evolution - From Theory to Practice* (pp. 23–50). Chichester, UK: John Wiley & Sons. doi:10.1002/9780470742891.ch2

Petrova, M., & Mahonen, P. (2007). Cognitive resource manager. In Fitzek, F. H., & Katz, M. D. (Eds.), *Cognitive Wireless Networks* (pp. 397–422). Springer. doi:10.1007/978-1-4020-5979-7_20

Raisinghani, V., & Iyer, S. (2004). ECLAIR: An efficient cross layer architecture for wireless protocol stacks. In *Proceedings of the World Wireless Congress*. World Wireless Congress.

Razzaque, M., Dobson, S., & Nixon, P. (2007). Cross-layer architectures for autonomic communications.*Journal of Network and Systems Management*, *15*(1), 13–27. doi:10.1007/s10922-006-9051-8

Reddy, Y. B., & Bullmaster, C. (2008). Cross-layer design in wireless cognitive networks. In *Proceedings of the International Conference on Parallel and Distributed Computing, Applications and Technologies,* (pp. 462-467). IEEE Computer Society.

Rodrigues, J. J., & Chen, M. (2010). A cross-layer sleep and rate adaptation mechanism for slotted ALOHA wireless sensor networks. In *Proceedings of the International Conference on ICT Convergence*. Jeju Island, Korea: IEEE Communications Society.

Schwartz, M. (1988). *Telecommunication networks - Protocols, modeling and analysis*. Reading, MA: Addison-Wesley Publishing Company.

Spectrum Policy Task Force. (2002). *Spectrum policy task force report*. Washington, DC: Federal Communication Commission.

Spurgeon, C. (2000). *Ethernet - The definitive guide*. New York, NY: O'Reilly Media.

Srivastava, V., & Motani, M. (2007). Cross-layer design and optimization in wireless networks. In Mahmoud, Q. H. (Ed.), *Cognitive Networks* (pp. 121–146). Chichester, UK: John Wiley & Sons Ltd. doi:10.1002/9780470515143.ch6

Stevens, W. R. (1994). *TCP/IP illustrated* (*Vol. 1*). Reading, MA: Addison-Wesley Publishing Company.

Sutton, P. D., Jörg, L., Lahlou, H., Fahmy, S. A., Nolan, K. E., & Özgül, B. (2010). Iris: An architecture for cognitive radio networking testbeds. *IEEE Communications Magazine*, *48*(9), 114–122. doi:10.1109/MCOM.2010.5560595

Vaughan, R., & Bach Andersen, J. (2003). *Channels, propagation and antennas for mobile communications*. London, UK: The Institution of Electrical Engineers.

Winter, R., Schiller, J. H., Nikaein, N., & Bonnet, C. (2006). CrossTalk: Cross-layer decision support based on global knowledge. *IEEE Communications Magazine*, *44*(1), 93–99. doi:10.1109/MCOM.2006.1580938

ADDITIONAL READING

Akyildiz, I., & Wang, X. (2005). A survey on wireless mesh networks. *IEEE Communications Magazine*, *43*(9). doi:10.1109/MCOM.2005.1509968

Akyildiz, I., Xie, J., & Mohanty, S. (2004). A survey of mobility management in next-generation all-IP-based wireless systems. *IEEE Wireless Communications*, *11*(4), 16–28. doi:10.1109/MWC.2004.1325888

Akyildiz, I. F., Lee, W.-Y., Vuran, M. C., & Mohanty, S. (2006). NeXt generation/dynamic spectrum access/cognitive radio wireless networks: A survey. *Computer Networks*, *50*(13), 2127–2159. doi:10.1016/j.comnet.2006.05.001

Akyildiz, I. F., Wang, X., & Wang, W. (2005). Wireless mesh networks: A survey. *Computer Networks*, *47*(4), 445–487. doi:10.1016/j.comnet.2004.12.001

Chieochan, S., & Hossain, E. (2009). Adaptive radio resource allocation in OFDMA systems: A survey of the state-of-the-art approaches. *Wireless Communications and Mobile Computing*, *9*(4), 513–527. doi:10.1002/wcm.696

Conti, M., Gregori, E., & Turi, G. (2005). A cross-layer optimization of gnutella for mobile ad hoc networks. In *Proceedings of the 6th ACM International Symposium on Mobile Ad Hoc Networking and Computing*, (pp. 343-354). New York, NY: ACM.

Demirkol, I., Ersoy, C., & Alagoz, F. (2006). MAC protocols for wireless sensor networks: A survey. *IEEE Communications Magazine*, *44*(4), 115–121. doi:10.1109/MCOM.2006.1632658

Georgiadis, L., Neely, M., Neely, M., & Tassiulas, L. (2006). *Resource allocation and cross-layer control in wireless networks*. New York, NY: Now Pub.

Giambene, G., & Kota, S. (2006). Cross-layer protocol optimization for satellite communications networks: A survey. *International Journal of Satellite Communications and Networking*, *24*(5), 323–341. doi:10.1002/sat.853

Hanzo-II, L., & Tafazolli, R. (2007). A survey of QoS routing solutions for mobile ad hoc networks. *IEEE Communications Surveys & Tutorials*, *9*(2), 50-70.

Lin, X., & Shroff, N. (2005). The impact of imperfect scheduling on cross-layer rate control in wireless networks. In *Proceedings of INFOCOM 2005, 24th Annual Joint Conference of the IEEE Computer and Communications Societies*, (pp. 1804- 1814). IEEE Press.

Lin, X., Shroff, N., & Srikant, R. (2006). A tutorial on cross-layer optimization in wireless networks. *IEEE Journal on Selected Areas in Communications*, *24*(8), 1452–1463. doi:10.1109/JSAC.2006.879351

Madan, R., Cui, S., Lall, S., & Goldsmith, A. (2005). Cross-layer design for lifetime maximization in interference-limited wireless sensor networks. In *Proceedings of INFOCOM 2005. 24th Annual Joint Conference of the IEEE Computer and Communications Societies*, (pp. 1964- 1975). IEEE Press.

McNair, J., Tugcu, T., Wang, W., & Xie, J. (2005). A survey of cross-layer performance enhancements for mobile IP networks. *Computer Networks*, *49*(2), 119–146. doi:10.1016/j.comnet.2005.06.001

Miao, G., Himayat, N., Li, Y., & Swami, A. (2009). Cross-layer optimization for energy-efficient wireless communications: A survey. *Wireless Communications and Mobile Computing*, *9*(4), 529–542. doi:10.1002/wcm.698

Setton, E., Yoo, T., Zhu, X., Goldsmith, A., & Girod, B. (2005). Cross-layer design of ad hoc networks for real-time video streaming. *IEEE Wireless Communications*, *12*(4), 59–65. doi:10.1109/MWC.2005.1497859

van der Schaar, M., Krishnamachari, S., Choi, S., & Xu, X. (2003). Adaptive cross-layer protection strategies for robust scalable video transmission over 802.11 WLANs. *IEEE Journal on Selected Areas in Communications*, *21*(10), 1752–1763. doi:10.1109/JSAC.2003.815231

Wang, H., Qin, H., & Zhu, L. (2008). A survey on MAC protocols for opportunistic spectrum access in cognitive radio networks. In *Proceedings of the International Conference on Computer Science and Software Engineering,* (pp. 214-218). Los Alamitos, CA: IEEE Computer Society.

Xie, J., & Wang, X. (2008). A survey of mobility management in hybrid wireless mesh networks. *IEEE Network*, *22*(6), 34–40. doi:10.1109/MNET.2008.4694172

Yick, J., Mukherjee, B., & Ghosal, D. (2008). Wireless sensor network survey. *Computer Networks*, *52*(12), 2292–2330. doi:10.1016/j.comnet.2008.04.002

Zhu, H., Li, M., Chlamtac, I., & Prabhakaran, B. (2004). A survey of quality of service in IEEE 802.11 networks. *IEEE Wireless Communications*, *11*(4), 6–14. doi:10.1109/MWC.2004.1325887

KEY TERMS AND DEFINITIONS

Application Protocol Interface: Structured set of commands/function calls that define how two different layers interact with each other. This allows to replace any layer as long as the API is maintained.

Cognitive Radio: Futuristic vision for mobile radio communications. It foresees the usage of advanced and multidisciplinary techniques to allow network/nodes to be autonomous in the selection of the optimal parameter set.

Flexibility: Allowing changes in a particular place in the current architecture without affecting the rest of the system.

Hierarchical Cross Layer Approach: Usage of multiple Cross Layer techniques that define a hierarchy in the architecture. Each level of the hierarchy deals with a different abstraction level of information.

Information Flow: Exchange of information from one layer of the architecture to another one.

Stacked Architecture: Traditional telecommunication architecture as defined by the ISO-OSI model and the TCP/IP or IETF protocol stack.

Technology Dependency: Bonding of a particular layer to a specific underlying technology.

White/Gray/Black Box: Transparency of the layer; i.e. amount/type of information that can be sent out of the layer/interface.

ENDNOTE

1. This can be summarized with the old principle developed by William of Ockham, the famous *"Occam's Razor"*: *"Frustra fit per plura quod fieri potest per pauciora."* (it is useless to do with many things what can be done with few).

Chapter 3
Cross Layer QoS Implementations:
The Clean-Slate Approach

Virgil Dobrota
Technical University of Cluj-Napoca, Romania

Andrei Bogdan Rus
Technical University of Cluj-Napoca, Romania

ABSTRACT

Quality of service has been an open issue from the beginning, due to the Internet paradigm of delivering the datagrams as simple as possible, on a hop-by-hop basis. As an answer to the inefficiency of the legacy solutions, rethinking of Internet principles brought among other ideas the self-management. Based on a clean-slate approach applied within FP7-4WARD project, this chapter proposes preliminary design and implementation of a key management capability of the near Future Internet, i.e. Cross-Layer QoS. Its applications are focused on congestion control (triggering activation of network-coding techniques) and QoS-aware routing based on a modified Dijkstra's algorithm.

INTRODUCTION

Although quality of service in IP-based networks is still an important research topic in IT community, with a significant effort towards standardization for almost two decades, there is no widespread study of such techniques in commercial applications and infrastructures. As a result, the end user is still receiving best effort services. A major explanation of this situation is that the implementation of classical QoS mechanisms supposes an increased operational cost. Moreover it demands strong expertise, thus the operators rather prefer over-provisioning solutions. This approach might be suitable for wired networks, but it definitely wastes the resources when it is implemented in wireless environment. Nowadays, real time services are becoming very popular but the classical

DOI: 10.4018/978-1-4666-0960-0.ch003

approach for QoS guaranteeing is not suitable anymore. The traffic parameters cannot be always maintained in a strict range of values. Thus, the Quality of Experience (QoE) is very often not acceptable. The cross-layer paradigm was proposed to be a solution for this challenging issue. Thus the entities could exchange management information, obtaining a global view of the networks performances in real-time. Consequently, optimal decisions are expected to be taken. Thus, communication protocols from upper layers are able to adapt their behaviour according to the variation of the traffic parameters from lower ones.

Following the legacy solutions for guaranteeing the quality of service, the main focus of this chapter is to propose a clean-slate design and a preliminary implementation of cross-layer QoS architecture. The results are based on experience obtained within FP7-"4WARD—Architecture and Design for the Future Internet," during 2008-2010. According to Correia, Abramowicz, Johnsson, and Wünstel, 2011, the term "clean-slate approach stands for a coherent solution that breaks the current network's stagnation imposed by the need to support current technologies and solutions; it is the answer to the following question: with what we know today, if we were to start again from scratch, how would we design a global communications infrastructure?"

The original results presented herein reflect the authors view and not necessarily the position of the project. However, several publications, such as Rus et al. (2010), Rus, Dobrota, Vedinas, Boanea, and Barabas (2010), and Rus (2011), include parts of the proposed implementation, validating the ideas. The two major applications covered by this chapter are congestion control and QoS-aware routing. None of them would have been possible to be demonstrated with the existing solutions, because they involve the Cross-Layer QoS (CLQ) with self-management. First, a real-time implementation of a six-node testbed in a butterfly topology was deployed. Whenever congestion occurs on an existing link and over-provisioning and/or re-routing are not possible, the management capability could trigger the activation/de-activation of cooperative techniques, such as Network Coding. This application does not solve the congestion but it allows the network to face the situation and to improve dramatically the quality of experience. The second application is an OMNET++-based simulation of a 13-node topology. The Dijkstra's algorithm, used by the OSPF (Open Shortest Path First) routing protocol, can be modified due to CLQ self-management information exchanged permanently between the nodes. Note that the traffic parameters are obtained in real-time, and only the QoS-aware routing based on this modified algorithm has been simulated. The conclusions at the end of this chapter are followed by future work and references.

BACKGROUND

The classical approach where the communication protocols worked independently from each other was suitable for non real-time applications running over wired networks. In what concerns real-time services, the previously mentioned paradigm proved to be unsatisfying. A certain degree of adaptation is necessary to optimize the execution of high-level applications or protocols, using information received from the lower layers. In this section, we discuss first about the information exchanged between the layers through cross-layer techniques. Examples according to evolutionary approach are accompanying the survey, the material being organized with respect to the OSI Layer the information is acquired from. Later we commented the role of the clean-slate approach with respect to the step-by-step strategy of progress is this field.

Table 1. Information from physical layer through cross-layer techniques

From \ To	Physical Layer	Data Link Layer	Network Layer	Transport Layer	Application Layer
Physical Layer	N/A	BER influences the retransmission Channel state information (i.e. SNR, quality of each OFDMA sub-channel) could trigger the handover AMC scheme influences ARQ scheme	BER, energy consumption influence the route selection	Further work	BER can be used to choose the suitable audio/ video codec

Information from the Physical Layer through Cross-Layer Techniques

The typical parameters to be offered by the Physical Layer are the following: transmitting power, energy consumption, Bit Error Rate (BER), Signal to Noise Ratio (SNR), type of error correcting code, and type of modulation. However, not all of them are useful for the upper layers. As a general observation, the Adaptive Modulation and Coding (AMC) scheme has a great impact on Automatic Repeat Request (ARQ) protocol if the cross-layer techniques are activated. The channel status is relevant for the upper layers (except for the Transport Layer for whom we have not found yet an example, as in Table 1).

One major characteristic of the wireless channel is the fading phenomenon, being manifested through a variation of the received signal strength. An interesting proposal can be found in Kulkarni and Rosenberg (2003) and Gyasi-Agyei and Kim (2006). The presented approach uses the variation of the wireless links as a benefit. If multiple devices are communicating on the same radio channel, the base station will decide which one will be allowed to transmit data at a given moment. Due to the fact that the channel performance varies in time, there is a low probability that the same user will experience the best channel parameters during multiple consecutive time slots. Using these techniques, the users will transmit data only when the channel performs well. The simulated tests, demonstrated that the overall performance of the systems is superior to the classical approach. To implement this proposal, a cross-layer information exchange is necessary between the Physical Layer of the equipments and the Data Link Layer of the base station. The Physical Layer of each device will measure the SNR parameter on a pilot signal received. The results of the measurements will be transmitted to the base station that will decide what user will gain the control of the wireless channel, for the following time slot.

It seems that this exchange from Physical to Data Link Layer is among the most popular cross-layer techniques currently implemented. Thus, Carrasco, Ramis, and Femenias (2009) presents a framework for cross-layer design in wireless networks, combining adaptive modulation and coding with an automatic repeat request protocol. The AMC scheme uses multiple transmission modes, each of them corresponding to a particular combination of modulation and coding scheme. Note that the signallings are bi-directional so this should be considered also as an example where the beneficiary is the network infrastructure too.

The results described in Chellappa Doss and Pan (2006) demonstrate the benefits for MANET (Mobile Ad-hoc Network). The performances are superior whenever we consider the characteristics of the physical link to compute the metric of a route, compared to the classical approach that uses the number of hops or the transfer rate only. Changing the size of the data frames or the retransmission mechanism used at the Data Link Layer according to the level of the BER parameter or the signal to noise ratio can make the data transmission more efficient due to an improvement of the transfer rate and the consumed energy.

In Alonso and Agusti (2004), a cross-layer signalling mechanism, focused on exchanging information between the Physical Layer and the MAC Sub-Layer is being proposed too. Thus, the transfer rate can be adapted during run-time in order to improve the spectral efficiency of the transmission, while maintaining the packet delay at a low level.

A recent implementation of cross-layer techniques for multimedia transmission over cognitive radio networks discussed in Ali and Yu (2009) the sending of channel state information from Physical Layer to both Data Link and Application Layers. For the first one this could be useful to determine the best moment to access the channel, whilst the latter could use it to select the corresponding video parameters and to adjust the intra refreshing rate.

The real-time optimisation of the energy consumed by mobile devices connected to the network is an example of application that can benefit from implementing cross-layer techniques. Due to the limited battery power, any approach focused on decreasing the energy consumption while increasing the autonomy of the device is very important. For instance, at the Network Layer, the amount of energy consumed when forwarding a certain data packet can be used as a routing metric (Felice, 2008).

For transmitting multimedia or real-time data on the wireless channel, a strong communication between the protocol layers is requested so that any decisions that will be taken to be the optimal one. Zhang, Yang, and Zhu (2005) are proposing to change the coding rate at the Application Layer when processing audio/video streams, according to parameters related to the communication channel (i.e. delay, packets lost, available transfer rate). Furthermore, to adapt to the fast fluctuations of the network, the same authors are proposing to predict the channel status, allowing proactive measures to be taken.

Information from the Data-Link Layer through Cross-Layer Techniques

This layer is in charge with error-free communications. For example, ARQ mechanism is used to retransmit corrupt or lost frames. Examples of information available at this layer are the following: number of retransmitted frames, length of the frames, moment when the channel is available etc. In cellular radio networks, this layer has a major role in managing the handover process that is activated when a user is moving between two radio cells. The main information used by the handover mechanism is the SNR measured at the Physical Layer. A cross-layer information exchange with the Network Layer could help the Mobile IP-based radio communications to detect faster the moment when the device is leaving a network and is entering into another one. Whenever the network is changed, the access point or the base station the mobile device is connected to it also changed. The Data Link Layer could detect this situation and trigger a signaling procedure to the Network Layer. Hence, the Mobile IP protocol will be informed that the equipment is in a foreign network and specific decisions will be taken properly. With respect to cross-layer information exchanged between the Data Link Layer and the Transport Layer, the number of retransmitted frames is of great importance. This parameter could indicate the quality of the communication channel. Thus, instead of involving TCP to retransmit the lost data, the action performed at Data Link Layer could reduce the power consumed by the equipment. Cross-layer signaling with the Application Layer can be very important if we want to implement service differentiation. Hence, the data coming from multiple applications must be treated with different priorities. As an example, the data com-

Table 2. Information from data link layer through cross-layer techniques

From \ To	Physical Layer	Data Link Layer	Network Layer	Transport Layer	Application Layer
Data Link Layer	ARQ scheme influences AMC scheme	N/A	Access Point or Base Station MAC addresses changing could influence Mobile IP	Confirmation messages are relevant for LLE-TCP	Further work
	OFDMA sub-channel allocation to users requires transfer rate and reliability		Alternative Label Switch Paths (LSPs) influence IP routing	Number of voice frames with errors	

ing from a service that requires low delay can be stored into a higher priority queue. Thus, the Data Link Layer will be adapting its functionality according to the QoS parameters requested by each service or application that transmits information into the network. The needs of each flow will be sent through top-down cross-layer methods. There is one disadvantage though because all the extra processing is increasing the requirements of the nodes with respect to the processing power. See the summarized survey in Table 2.

An original approach is being proposed in Yuksel, Ramakrishnan, and Doverspike (2008), by designing a cross-layer mechanism for path restoration when implementing robust distribution of IPTV. The MPLS-based infrastructure uses the Protocol-Independent Multicast–Source Specific Multicast (PIM-SSM) routing protocol. Because both MPLS and PIM-SSM have their own path recovery mechanisms, when a link connecting two network nodes breaks down, two signalling flows transporting the same information, but at different level might use inefficient the resources of the network. Using cross-layer signalling one protocol can inform the other one about the situation in the network in order to find the alternative route, eliminating this way the redundant signalling flow. The performed tests illustrated two advantages of the mechanism: there was no duplicate signalling in the network (decreased management traffic) and the percentage of lost packets (when a link was failing) was reduced with 5-6% because the recovery mechanism was able to find faster an alternative path.

Bottom-up cross-layer techniques for improving Voice Transmissions over Wireless Local Area Networks (VoWLAN) uses a communication agent to eliminate congestion misinterpretation (Alshakhsi & Hasbullah, 2010). The agent receives an event from Data Link Layer and if it is a voice packet error then it will relay the event to Transport Layer which accordingly will adjust voice stream related parameters such as window size and compression ratio.

Another method for improving the performance of TCP over wireless channels is described in Granelli and Kliazovich (2005), under the name of LLE-TCP (Link Layer ARQ Exploitation TCP). The main idea consists of using the confirmation messages received at the Data Link Layer to inform the Transport layer that the transmitted segments were received at the destination. Thus, the overhead in the channel will be mitigated, eliminating the TCP acknowledgement messages. The simulations with ns-2 showed that depending on the number of hops, between the source and the destination, the performance was from 5% to 20% better than in the classical approach. As this mechanism was implemented only on the source and destination, the overall performance improvement is smaller when the number of nodes is getting higher.

Table 3. Information from network layer through cross-layer techniques

From \ To	Physical Layer	Data Link Layer	Network Layer	Transport Layer	Application Layer
Network Layer	Further work	Alternative IP routes influence LSP selection in MPLS	N/A	IP handoff prevents TCP not to activate the congestion avoidance mechanism	Further work
		Wireless Sensor Network cluster information influences scheduling for each cluster		Explicit Congestion Notification (RFC 3168)	

Information from the Network Layer through Cross-Layer Techniques

The main functions refer to routing, addressing, and selecting the output interface. The cross-layer interaction with the Application Layer could optimize the routing process. For instance, let us consider a device that has multiple network connections, each one utilizing a different transmission technology. A different output interface for each data flow should be chosen, according to the set of QoS parameters (delay, jitter, lost packets, or transfer rate) requested by each application. Another feature is to change a session from one interface to another without being sensed by the upper layers. Information exchanged between the Network and Transport Layers could be useful when IP handoff procedure is being activated. IP handoff is a mechanism specialized in helping a device to roam from one network to another. Because during the handoff operations, packets are being lost, a signaling message could inform TCP not to enable the congestion avoidance mechanism, keeping this way the output transfer rate unchanged and maintain the optimal level of channel usage. The synthesis of this survey is presented in Table 3.

We recall here the example (Yuksel, Ramakrishnan, & Doverspike, 2008) previously given. Due to cross-layer signalling it is the IP turn now to inform MPLS about the situation in the network in order to find the alternative route, eliminating the redundant signalling flows.

The paper (Ruiz, Gallardo, Villasenor-Gonzalez, Makrakis, & Mouftah, 2009) proposes a cross-layer mechanism in which the routing and medium access protocols collaborate together to achieve energy-efficient and QoS-aware medium-access mechanism for wireless sensor networks. The routing protocol at Network Layer deals with: path discovery, resource reservation, cluster formation, and gathering information regarding the interference among different clusters. The MAC Sub-Layer uses this information to create activity-window schedules for clusters. Moreover, it uses a collision-free protocol for communication within each cluster. The proposed mechanism saves energy by avoiding collisions and by allowing nodes to safely turn off their transceivers outside their activity windows without the risk of losing any relevant frame transmission.

Whenever we use TCP in radio networks, the available resources are not efficiently utilized. In case of packets lost due to high BER, the transport protocol misinterprets this event as being the effect of congestion and it reduces the transfer rate. A well-known mechanism to eliminate this disadvantage is called Explicit Congestion Notification (two least significant bits within Type of Service field from the IPv4 header) (Ramakrishnan, Floyd, & Black, 2001). This prevents TCP

Table 4. Information from transport layer through cross-layer techniques

From \ To	Physical Layer	Data Link Layer	Network Layer	Transport Layer	Application Layer
Transport Layer	Further work	Further work	Further work	N/A	Communication performances influences the codec selection for real-time data streaming

to activate congestion avoidance mechanism if it is not needed.

Information from the Transport Layer through Cross-Layer Techniques

This layer is in charge with an end-to-end transport between a source and a destination device on top of the network infrastructure. For a connection-oriented protocol like TCP, the set of parameters that might help includes Round-Trip Time (RTT), receiving window, congestion window, effective data transfer rate (transfer rate received and confirmed by the destination), etc. This information is a valuable input for a video streaming application for instance, by selecting properly the codec (see Table 4). Note that we did not find information from Transport Layer to be used by the lower layers.

An interesting example of cross-layer signalling can be found in Raisinghani and Iyer (2004) where the TCP protocol announces the Application Layer that the packets were lost, thus adaptation measures being taken. For instance, if a video flow is being streamed over the network, the application can adapt the codec (i.e. choosing one that uses a suitable compression level) according to the quality of the communication channel.

Information from the Application Layer through Cross-Layer Techniques

If the applications were initially designed to run in a wired network, their performances in hybrid/pure wireless networks are most of the time unsatisfying. A solution to this issue would be to adapt the applications behavior according to information received from lower layers. Examples of parameters are the following: maximum accepted delay, maximum accepted jitter, maximum number of packets lost, minimum transfer rate supported, etc. The Application Layer could inform TCP, using a cross-layer mechanism, about the needed QoS parameters for maintaining a certain quality of the transmission. For the lower layers in general per flow QoS requirements could be taken into consideration either directly, either through a cross-layer controller that intermediates the information exchange (see Table 5).

The authors in Du and Zhang (2009) propose a cross-layer framework for efficient multi-layer-video multicast with rate adaptation. A separate transmission queue is maintained to send data corresponding to each video layer. Due to the fact that different video layers require a different loss rate specified by Application Layer (e.g. base layer requires a higher reliability, while other ones can encounter transmissions lost). Based on the loss threshold received and on channel state information, at the Physical Layer the sending device regulates dynamically the transmission rate and allocates appropriate resources for the transmission of data existing in each queue.

More complex cross-layer techniques have been combined in Shah, Weifa, and Xiaojun (2010) to be applied in wireless multimedia sensor networks. The main target is to maximize the number of video streams sent while preserving the imposed QoS requirements (per hop transfer

Table 5. Information from application layer through cross-layer techniques

From \ To	Physical Layer	Data Link Layer	Network Layer	Transport Layer	Application Layer
Application Layer	Per flow QoS requirements influence directly or through a cross-layer controller the loss	Per flow QoS requirements influence directly or through a cross-layer controller the priority of the queues before transmission	Per flow QoS requirements influence directly or through a cross-layer controller the routing decisions	QoS requirements influence the transport decisions	N/A

rate, delay, and reliability). It is assumed that all the nodes, as well as the sink, are location-aware and the location of the sink is known to all the sensor nodes. The routing algorithm employed is stateless and the selection of the forwarding node is purely based on the decision made by the relaying node. This example illustrates signalling exchanges from Application to Data-Link, from Network to Physical and, respectively, from Physical to Network Layers.

The case of indirect exchange through a cross-layer controller could be found in Melodia and Akyildiz (2010), targeting reliable and flexible transmissions in Wireless Multimedia Sensor Network (WMSN). The main scope is to maximize the aggregate achievable data rate. Although the paper does not presents a practical solution but points out important design principles. Thus, the Application Layer imposes per each flow requirements (i.e. end-to-end delay, required transfer rate, maximum end-to-end packet error rate).

The solution controls all the lower layers (except Transport Layer) and gets local measurements from the Physical Layer.

A direct intervention of the Application Layer into the Data-Link Layer behaviour is illustrated by the paper, Ukil and Sen (2009). It presents a QoS aware cross-layer optimized resource allocation scheme for WiMAX networks. The main target is to optimize the system's performance while maintaining the end-to-end QoS of each individual user at a specific level. The main goal is to optimize the resource allocation in an OFDMA (Orthogonal Frequency Division Multiple Access) system. The idea is based on the fact that at any given time, in a large network, there is a high probability that at a given time a use experiences better performances than the other ones on a set of sub-channels. Thus, the overall system performance can be maximized by allocating the OFDMA subcarriers to the users with good channel conditions. This approach may lead to fairness issues but the authors are solving them implementing a proportional fair scheduler. Being very new, the proposal has not been yet optimized for real-time traffic. Note this example is again more complex, because it includes also bi-directional signalling between Physical and Data Link Layers.

Another cross-layer signaling solution towards the Data Link Layer was implemented in Lazar, Dobrota, and Blaga (2008), in order to account for the different QoS characteristics of video and elastic traffic. In addition, network support for efficient multi-queue transmission is enabled in the Linux wireless card driver. Measurements also illustrate the effects of changing several H.264 and IEEE 802.11e parameters in the player and driver software.

Evolutionary Approach vs. Clean-Slate Approach?

As the definition states, a clean-slate approach applied in Internet is an opportunity to start over and rethink its principles and architecture. GENI (Global Environment for Network Innovations) is one of the most visible research projects in this area. Its main target is to offer a dedicated environment for any research study, focused on designing of the Future Internet. When trying to find reasons for choosing between clean slate and evolutionary approaches, the authors in Rexford and Dovrolis (2010) are stating that both approaches are valid and they could complement each other. Thus, the radically new networking designs can guide the incremental ones towards enhancing the current Internet.

The reader should be aware that this chapter is focused mainly on Cross-Layer QoS implementations according to the clean-slate approach. However, we wanted to present in the previous paragraphs the evolutionary approach background. It seems that in 2011 the industry has a less optimistic vision (compared to 2008), meaning that the progress should be achieved more incremental rather than adopting revolutionary solutions. However, our survey is touching now other cross-layer QoS implementations that had the courage to start from scratch. For instance, the authors in ElRakabawy and Lindemann (2009) are proposing a new communications system that uses cross-layer techniques for reliable data delivery over IEEE 802.11 wireless mesh networks. After identifying the drawbacks of TCP, a new transport protocol that interacts directly with the routing mechanisms was designed. It managed to improve both goodput and fairness of the entire network. Another work trying to overpass the limitations of the current Transport Layer can be found in Miao et al. (2011) where the authors are proposing an experimental RTLA (Redesigned Transport Layer Architecture). The main idea is to split the transport services into three sub-layers and a set of cross-layer functional blocks. The result of the work is a scalable solution that can be easily improved with new functionalities. A study regarding existing clean-slate architectural proposals for new generation networks is being made in Teraoka (2009). A recursive layered model, designed to support network virtualization, combined with cross-layer techniques, by separating the data plane from the control one is the most suitable solution for an easy management.

We propose in the next Section to discuss a preliminary implementation of CLQ according to a clean-slate approach. We do not necessarily refer to layers and by the time we designed it we tried not to be influenced by the implementations quoted herein. Actually, several of them were proposed later than 2007 (when we started the work at our approach).

The former Physical Layer + MAC Sub-Layer are grouped in a Substrate Network, whilst the remaining layers are grouped in a Virtual Network. This does not mean the end of layers, and the legacy solutions may co-exist. The preliminary implementation of CLQ discussed is independent to TCP/IP suite (that eventually could be replaced in the next decade). Because an alternative has not been made available yet we are still using this protocol stack on top of MAC Sub-Layer. However, we could switch immediately to a new protocol suite governing the network + transport + application part.

DESIGN AND PRELIMINARY IMPLEMENTATION OF CROSS-LAYER QOS (CLQ)

Optimal Routing Decisions Issues

The IP-based networks are connectionless and without any confirmation of the successful delivery of packets. Although majority of Data-Link Layer technologies allow Internet Protocol on top of them, the routing decisions are sub-optimal. They are based on well-known traffic parameters (transfer rate, delay, and jitter) that are usually statically measured at Network Layer. Although a technology such as Fast Ethernet (IEEE 802.3u) considers a theoretical transfer rate (100 Mbps), in reality the link utilization (for data and overheads) is between 82.05% and 99.08%. This is due to additional 14 bytes (inter-frame gap IFG of 12 bytes plus 1 byte start-of-frame delimiter and 1 byte end-of-frame delimiter). For the minimum Fast Ethernet frame the link utilization is 64/(64+14)=82.05%, whilst for the maximum frame the value raises up to 1518/(1518+14)=99.08%. Do not confuse this parameter with the maximum throughput at Data Link Layer calculated for data only with respect to the total amount of bytes. Thus the throughput for the maximum efficiency is from 46/(46+32)x100=58.97 Mbps (minimum frame) up to 1500/(1500+32)x100= 97.91 Mbps (maximum frame). Also notice that for Ethernet and Gigabit Ethernet the link utilization has to be recalculated because the additional number of bytes is 20 (12 bytes IFG plus 7 bytes preamble plus 1 byte start-of-frame delimiter).

The existing link-state routing protocols (such as OSPF, IS-IS etc.) cannot seize the changes in the network (except failures and/or technology changes). To allow a better decision to be taken, the cross-layer QoS will bring real-time information measured on top of MAC Sub-Layer (so without depending on TCP/IP suite). This bottom-up approach is in a cycle with the top-down approach, being able to instruct the hardware (i.e. the Physical Layer and the MAC Sub-Layer) about a requested behaviour.

Preliminary Design and Implementation of Cross-Layer QoS Architecture

The Future Internet is expected to overpass the limitations of the existing networks, characterized mainly by suboptimal decisions, lack of security, low level of scalability and difficulties in supporting present and future innovations. It seems that despite the huge research effort, a new design of the whole Internet is needed, based on a clean-slate approach. Distributed management is the major trend of nowadays networks because the classical centralized approach is considered to be outdated due to multiple issues that are related with the single point of failure problem. In this light, the Cross-Layer QoS (CLQ) module, presented in this section, is being designed so that it can provide real time information in every node in the network, creating the possibility for management tasks to run on any machine. The main focus is to describe methods of inserting QoS capabilities into a network that implements In-Network Management (short: INM) (Nunzi, et al., 2009). A very important aspect that should be outlined is that the CLQ module is designed to have in the future a complex set of properties like:

- *self-configuration*: the module should be able to bootstrap itself and determine the suitable values for the configuration parameters;
- *composability*: this capability allows creating new composite services from a set of existing ones;
- *interoperability*: is the ability for multiple entities or equipments to collaborate, communicate and interact with one another, no mater of the manufacturer;
- *extendibility*: the module should allow future changes or improvements;

The CLQ module needs two types of protocol stacks (called Netlets) as shown in Figure 1. One type of stack has actually no protocols included and it is used for communicating directly with the hardware (i.e. Netlet Type 1). This is required to perform two major tasks:

a. bottom-up cross-layer data exchange while performing periodical measurements on top of the MAC Sub-Layer. This way customized probes can be sent on the communication link in order to perform active measurements between two connected devices;
b. top-down approach, can be used to impose a certain behaviour to the underlying technology and in case of emergencies, it can perform global resource allocation by replac-

Figure 1. Cross-layer QoS architecture in a clean slate approach

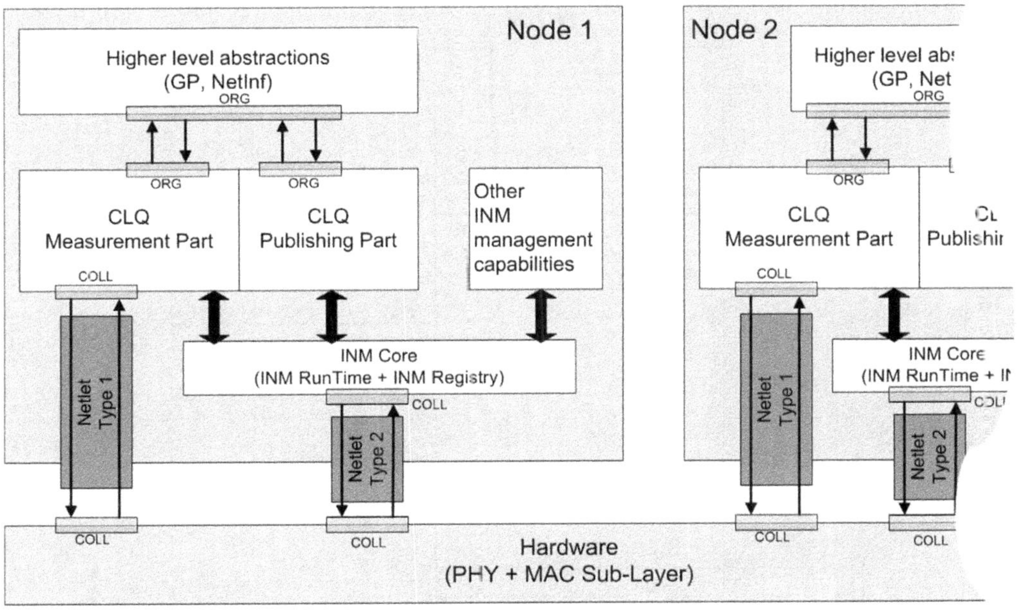

ing other entities dedicated for transporting user data. Hence, a hop-by-hop transport protocol can be employed to deliver data to the destination;

The Netlet Type 2 may include legacy or future protocols and it is needed to exchange information with other MCs located in different nodes (Gonzales, et al., 2010). The communication with other entities of the network is mandatory if using a distributed management framework. Hence, any device can receive or send information to other peers in the domain so that important data can be located in multiple places, avoiding this way the single point of failure issue. If for instance a specific entity (e.g. GP—Generic Path) needs to impose a set of traffic parameters (committed QoS parameters) to the network infrastructure, it will send a dedicated control message to the INM CLQ module, using the Organizational (ORG) interface. Further on, a similar XML management message will be sent to the underlying hardware through the Collaboration interface (COLL), imposing the requested behaviour. To verify the effect of the imposed changes, a monitoring entity that will supervise the channels performance, will be continuously running. Every time when the traffic information will be collected from the network, a QoS object will be created and it will be published to NetInf, i.e. Network of Information. This is a communication architecture based on an information-centric paradigm. It should make the access of information objects much easier, without the need of considering the underlying transport technologies. By the time we realised the implementation NetInf was not available, so we replaced it by a local database. See the interactions previously described represented in Figure 2

Traffic Parameters Measurements

Another action that can be performed with the help of the CLQ module is the composite metric calculation. This is a global performance indicator that includes the following traffic parameters: One-Way Delay (OWD), Available Transfer Rate (ATR), and Bit Error Rate (BER). These parameters are being measured on top of the MAC Sub-

Figure 2. QoS interactions

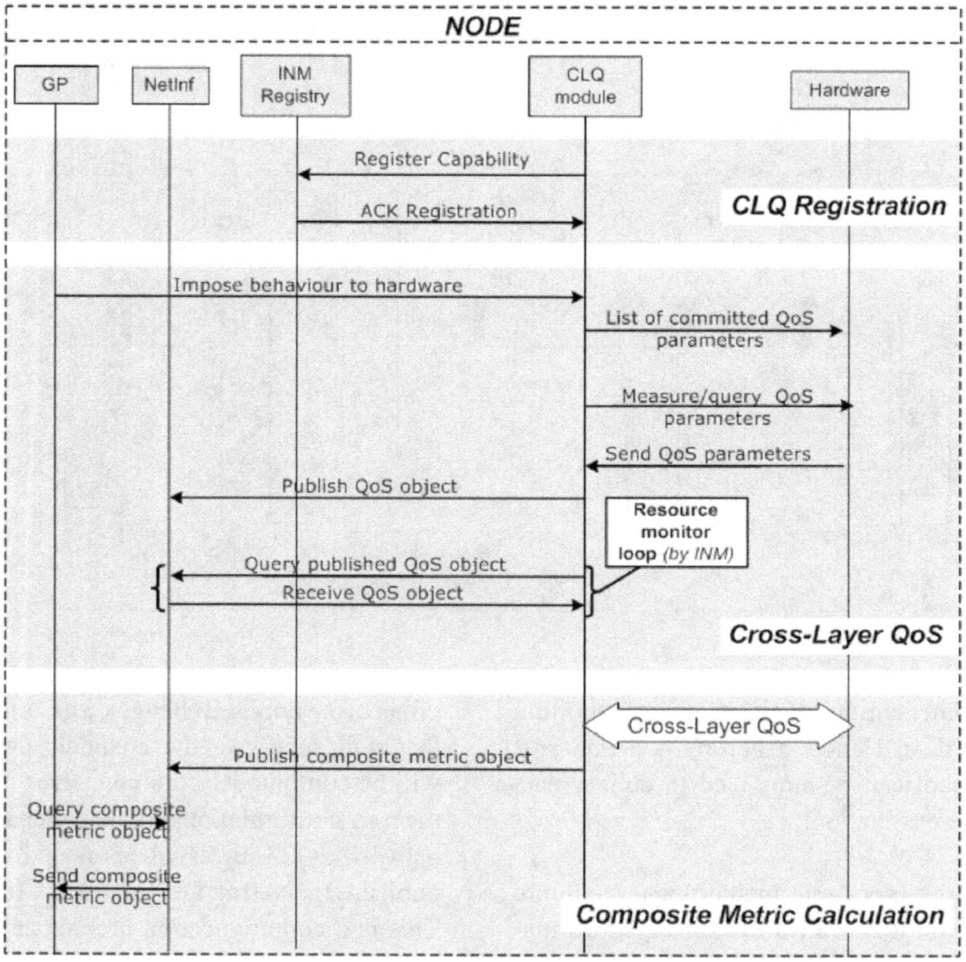

Layer (i.e. above the driver) using both active and passive measuring techniques. The advantage of collecting QoS parameters at a low level is that no further modifications will be necessary in the future, if the protocols running at higher layers will be changed.

One-Way Delay (OWD) Measurement

When performing delay measurements in networks with nodes that do not have their internal clocks synchronized, the only time related parameter that could be measured is the Round-Trip Time (RTT). Because we wanted to be as close as possible to the hardware when performing the measurements, the probes were sent directly over the Data Link Layer, actually on top of Physical Layer and MAC Sub-Layer. Moreover, the goal was to perform active measurements in the network so, a special kind of communication socket was used (i.e. the raw socket). The Linux operating system is the one that provides an API for working with these sockets, allowing customized measuring probes to be sent directly to the driver, avoiding the TCP/IP protocol stack. The measuring method used is similar with the one implemented by the ping tool when measuring round-trip time parameter at the Network Layer. A source node will send a

Figure 3. RTT measurement

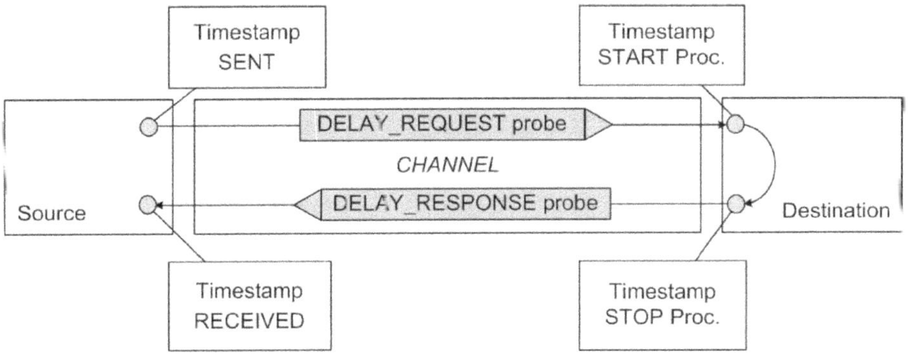

customized probe to a certain destination to which it is directly connected. In the next step, the probe will be sent back to the source device but with extra information stored in specific fields. The probes used for measuring the delay parameter are containing three timestamps (see Figure 3):

- The timestamp generated when the probe was sent from the source device (Timestamp SENT)
- The timestamp generated when the probe was received at the destination (Timestamp START Proc.)
- The timestamp generated when the probe was sent back from the destination to the source (Timestamp STOP Proc.)

The probe sent on the measured channel is a Data Link frame with a customized payload, illustrated in Figure 4.

Because the source and the destination device are not having their internal clocks synchronized, the only way to obtain a value for the one-way delay is to make an estimation of it. The procedure consists of taking into account the round trip time and subtracting from it the time needed for processing and finally dividing the result by two. Let us now consider the following notation:

- Ts_SENT = Time when the packet was sent from the source
- Ts_START_Proc = Time when the packet was received at the destination (timestamp when the processing started at the destination)
- Ts_STOP_Proc = Time when the packet was sent back to the source
- Ts_RECEIVED = Time when the packet was received back to the source node

The equation used to estimate the one way delay is the following:

$$Delay = \frac{(Ts\_RECEIVED - Ts\_SENT) - (Ts\_STOP\_Proc - Ts\_START\_Proc)}{2} [s] \quad (1)$$

Figure 4. Delay probe

6 bytes	6 bytes	2 bytes	8 bytes	8 bytes	8 bytes	22 bytes	4 bytes
Destination MAC Addr.	Source MAC Addr.	Ether. Type	Timestamp SENT	Timestamp START Proc.	Timestamp STOP Proc.	Unused Bytes	CRC32
			Customized Payload				

Figure 5. Measuring OWD

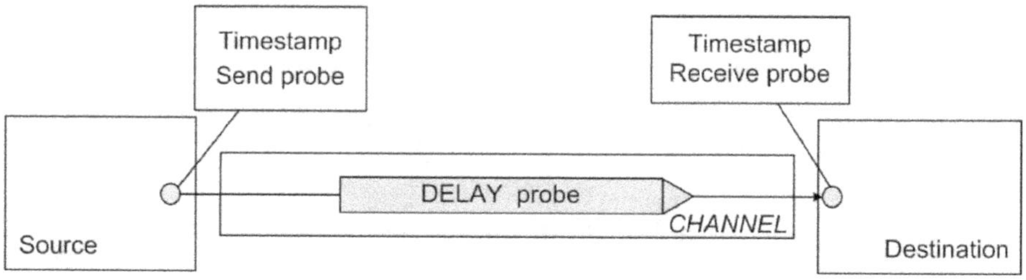

The inaccuracy is due to different occupancy of the transmission queues (source-to-destination and destination-to-source). If the nodes in the monitored network have their internal clocks synchronized (using for instance a GPS synchronization kit), the most suitable temporal parameter that could be measured is the One-Way Delay (OWD). For doing so, on both nodes connected to the monitored link, a specialised measuring program should be running (see Figure 5 and Figure 6).

Available Transfer Rate (ATR) Measurement

When passive measurements of the ATR parameter are being performed, the measuring tool tries to compute an estimation of the available traffic. Thus, the used transfer rate is being subtracted from the theoretical capacity of the underlying technology used. For a better accuracy of the results, it is envisaged that in the future to employ active measurements. This is needed mainly for wireless channels where the maximum transfer rate is not a fixed reference. It varies in time due to the adaptive modulation and coding scheme.

Bit Error Rate (BER) Estimation

When transmitting data on wireless communication channels, a very important parameter that influences directly the level of QoS is the Bit Error Rate. Thus, knowing the BER parameter of a communication channel would be an important indicator when optimizing the decisions taken at higher layers. Before publishing this information to other entities, the CLQ module must be able to retrieve the parameter in real-time. For doing so, two main approaches could be used: direct reading from the hardware (this option might not be available for all the transmission technologies) and measuring or estimating the parameter using a dedicated tool. Measuring directly the BER parameter consists of dividing the number of erroneous bits to the total number of bits received. If the requested information is available, the tool destined to perform this task seems to be quite simple. If the number of erroneous bits is unknown, the task is getting more complicated. However, information characterising the transmission could be also useful. For instance, if the signal to noise ratio together with the modulation used is being known, the BER parameter could be estimated

Figure 6. Customised OWD probe

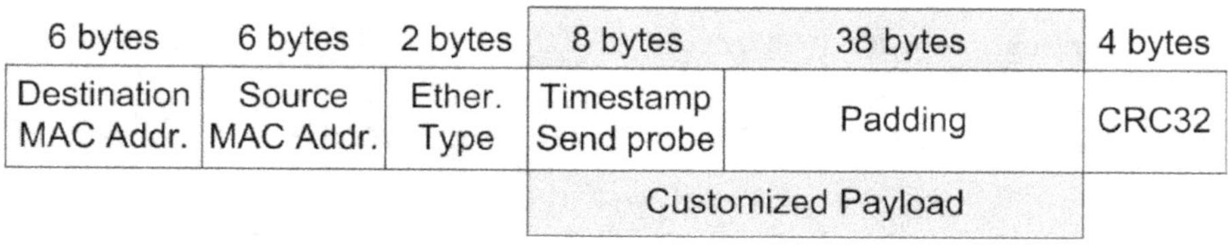

Table 6. Modulation used for various transfer rates in 802.11 a/b/g

Modulation	Data rates [Mbps]
BPSK	1.5, 6, 9
QPSK	3, 4.5, 11, 12, 18
16-QAM	24, 36
64-QAM	27, 48, 54

using the set of equations presented in this section. The technologies considered in this work are 802.11 a/b/g and xDSL. For other technologies used at the Physical and Data Link Layers, appropriate mathematical models that will not be presented herein should be used. When analysing the 802.11 a/b/g standard one can observe that multiple transmission speeds are available for data exchange between devices, each one with its own modulation type, being changed according to the actual performance of the radio channel. The modulation techniques used for various transmission rates can be found in the public standard and a selection of them are also presented in Table 6. Note that the data rates considered herein are only the ones that are being used together with the OFDM (Orthogonal Frequency-Division Multiplexing) transmission technique.

The mathematical model presented herein uses a couple of approximations. The first one consists of the presumption that the radio channel is disturbed only by Additive White Gaussian Noise (AWGN). The second approximation is that each OFDM sub-channel has the same noise power. Because all the OFDM sub-channels are using the same modulation technique at a given moment, we are considering that the SNR influencing the overall OFDM transmission is the same with the SNR parameter for one OFDM sub-channel:

$$SNR = 10\log_{10}\frac{P_S}{N_{Noise}} = 10\log_{10}\frac{n * E_b}{N_0 * B * T_S} \tag{2}$$

Where P_S = the power of the transmitted symbol, N_{Noise} = the noise power, n = the number of bits transmitted per symbol, E_b = the energy per bit, N_0 = the power spectral density of the noise, B = noise bandwidth, T_S = symbol period. Due to the fact that B can be approximated as being equal with the symbol frequency (F_S), equation (2) can be further processed as follows:

$$SNR = 10\log_{10}\frac{n * E_b}{N_0 * F_S * T_S} = 10\log_{10}\frac{n * E_b}{N_0 * \frac{1}{T_S} * T_S} = 10\log_{10}\frac{n * E_b}{N_0} \tag{3}$$

Because the available parameter for computing the BER value is the SNR variable we have to express the E_b/N_0 parameter as a function of SNR. For doing so, the equation (3) will be further processed as follows:

$$\frac{E_b}{N_0} = \frac{10^{SNR/10}}{n} \tag{4}$$

Due to the fact that we considered that the radio environment is disturbed only by AWGN, we have to use a noise margin (M) that will be subtracted from the SNR parameter received from the wireless interface ($SNR_{Interface}$), in order to approximate other interfering phenomenon that can appear on the wireless link. Because the data transmitted on the channel is usually coded for error correction, the actual SNR parameter can be considered higher than the one measured, due to the fact that there is a gain of the coding process (C). If considering the above presented aspects, the SNR parameter can be computed as follows:

$$SNR_{ef} = SNR_{Interface} - M + C \tag{5}$$

The M parameter is usually considered as being 6dB while the coding gain (C) has various values according to the coding scheme used. Some common values are in the range of 3.5 dB and 4 dB. When using the *BPSK* modulation the mathematical model used to estimate the BER parameter can be found in Rappaport (2002) and it is presented below:

$$BER_{BPSK} = P_b = Q\left(\sqrt{\frac{2E_b}{N_0}}\right) \tag{6}$$

The Q function can be further processed using the following expression:

$$Q(x) = \frac{1}{2} erfc\left(\frac{x}{\sqrt{2}}\right) \tag{7}$$

The reason for which we expressed the Q() function using the erfc() function is that most of the programming languages are providing specialized libraries that are implementing the erfc() expression. After analysing Equations (6) and (7), the bit error probability can be further developed as in the below equation:

$$BER_{BPSK} = \frac{1}{2} erfc\left(\sqrt{\frac{E_b}{N_0}}\right) \tag{8}$$

Because in the case of the BPSK modulation, the number of bits transmitted per symbol is n=1, if using the Equations (4) and (8), the BER parameter will be expressed as follows:

$$BER_{BPSK} = \frac{1}{2} erfc\left(\sqrt{10^{SNR_{ef}/10}}\right) \tag{9}$$

Thus, knowing the modulation type and the SNR parameter, one can observe that it is possible to estimate the bit error rate on one OFDM sub channel if applying the Equation (9). When the *QPSK* modulation is being used, the BER parameter is computed using the same equation as for BPSK (see Equation [6]). After replacing the Q function with Equation (7), the BER parameter will have the following equation:

$$BER_{QPSK} = \frac{1}{2} erfc\left(\sqrt{\frac{E_b}{N_0}}\right) \tag{10}$$

For the QPSK modulation, the number of bits per symbol is 2 (n=2), thus Equation (4) can be processed as below:

$$\frac{E_b}{N_0} = \frac{10^{SNR_{ef}/10}}{2} \tag{11}$$

Combining together Equations (10) and (11), the bit error rate probability when using the QPSK modulation will have the following equation:

$$BER_{QPSK} = \frac{1}{2} erfc\left(\sqrt{\frac{10^{SNR_{ef}/10}}{2}}\right) \tag{12}$$

The SNR_{ef} parameter will be computed using equation (12). If employing the *N-QAM* modulations, where N equals with 16 or 64 according to the above table, the formula used to compute the symbol error probability (P_e) was extracted from Rappaport (2002) and it is presented in Equation (13).

$$P_e = 4\left(1 - \frac{1}{\sqrt{N}}\right) \cdot Q\left(\sqrt{\frac{3E_{av}}{(N-1) \cdot N_0}}\right) \tag{13}$$

The E_{av} is the average energy per symbol. Using Equation (7) and expressing the Q function using the ercf() function, the bit error probability will have the following equation.

$$P_e = 2\left(1 - \frac{1}{\sqrt{N}}\right) \cdot erfc\left(\sqrt{\frac{3E_{av}}{2 \cdot (N-1) \cdot N_0}}\right) \quad (14)$$

The bit error probability can be further extracted from the symbol error probability using the following equation:

$$BER_{N-QAM} = P_b = \frac{P_e}{\log_2 N} \quad (15)$$

After processing Equations (14) and (15), the bit error probability can be expressed as follows:

$$BER_{N-QAM} = \frac{2\left(1 - \frac{1}{\sqrt{N}}\right) \cdot erfc\left(\sqrt{\frac{3E_{av}}{2 \cdot (N-1) \cdot N_0}}\right)}{\log_2 N} \quad (16)$$

The SNR parameter can be expressed as follows:

$$SNR = 10\log_{10} \frac{P_S}{N_{Noise}} = 10\log_{10} \frac{E_{av}}{N_0 * B * T_S} = 10\log_{10} \frac{E_{av}}{N_0} \quad (17)$$

If processing Equation (18), the E_{av}/N_0 parameter can be expressed as below:

$$\frac{E_{av}}{N_0} = 10^{SNR/10} \quad (18)$$

After combining (17) and (18) together, the bit error rate probability for any N-QAM modulation where N equals 16 or 64 can be expressed as in the below equation:

$$BER_{N-QAM} = \frac{2\left(1 - \frac{1}{\sqrt{N}}\right) \cdot erfc\left(\sqrt{\frac{3 \cdot 10^{SNR_{ef}/10}}{2 \cdot (N-1)}}\right)}{\log_2 N} \quad (19)$$

Where the SNR_{ef} parameter will be computed using Equation (5). After computing the BER parameter for each sub-channel of the OFDM symbol, the overall bit error rate can be computed using the following equation:

$$BER_{OFDM} = 1 - P_0 \quad (20)$$

In the above expression, P_0 is the probability of not having any erroneous bits on all the OFDM tones. In order to compute the P_0 parameter, the following expression could be used:

$$P_0 = \prod_{i=1}^{T} (1 - BER_{sub\_channel}) \quad (21)$$

Where $BER_{sub\_channel}$ is the bit error rate probability computed on each sub-channel of the OFDM transmission, using one of the Equations (9), (12), or (19), the SNR_{ef} and the modulation used.

Let us now consider that the node on which BER estimation should be performed is connected to the Internet using xDSL access technology (see Figure 7). Due to the fact that the optical fibre link between the DSLAM (Digital Subscriber Line Access Multiplexer) and BRAS (Broadband Remote Access Server), BER will not be considered. The interesting link for our analysis is the one between the xDSL modem and the DSLAM. The xDSL model presented is not trivial because it is a good practical example where this method of BER estimation could be applied. Effective SNR includes crosstalk, see Equation (22), where M is a noise margin related to it and BER is estimated based on SNR.

Figure 7. Generic topology of an xDSL connection

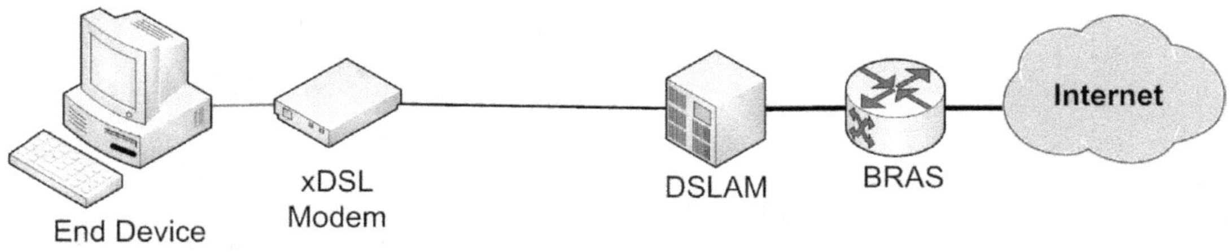

The data transmission on an xDSL connection is made using multiple sub-channels also known as tones, which are carrying bits of information. Depending on the physical channel characteristics, on each sub-channel a certain type of modulation is being used. The modulation usually employed is one of the following: BPSK, QPSK, 8-QAM, 16-QAM, 32-QAM, 64-QAM, 128-QAM, and 256-QAM. Because a different modulation could be used on each sub-channel, before estimating the overall BER probability, it is necessary to first compute the error probability on each tone. For doing so, we must obtain information about the modulation type used on the xDSL link. The mathematical models used to estimate the BER parameter for each modulation are the following: the BER_{BPSK} can be obtained using Equation (9), for estimating the BER_{QPSK} parameter Equation (12) could be used, while for computing an estimator for the $BER_{N\text{-}QAM}$ parameter, Equation (19) can be considered. The SNR_{ef} parameter used in this set of equations has the following expression:

$$SNR_{ef} = SNR_{Obtained\_from\_modem\_for\_each\_sub-channel} - M + C \quad (22)$$

The M parameter is a 6dB noise margin modelling the interfering phenomenon that can appear on the channel. Moreover, the C parameter is the coding gain, which is usually having values between 3.5 dB and 4 dB. For estimating the BER parameter for the whole transmission, we could use the previously computed BER indicators for each tone into the following mathematical Equations (20) and (21).

Alternative Solution to Over-Provisioning and Re-Routing Issues

Whenever congestion occurs on an existing link, the appropriate solution is to re-route the traffic and eventually perform updates on the infrastructure, by adding new resources (i.e. over-provisioning). However in rural wireless networks and other cases, none of the previously mentioned solution could be employed by the operator. Since cooperative techniques were launched, network coding would be an alternative treatment. The decision to activate/de-activate it is very important, because this mechanism needs some conditions to be fulfilled and it is not desirable to be activated before a pre-congestion is encountered. Thus, CLQ could trigger the situation.

Cross-Layer QoS Applied in Congestion Control

In this subsection, a preliminary design of cross-layer quality of service module, applied to congestion control in the Future Internet is being proposed. This is an alternative solution to QoS-aware routing, whenever the infrastructure operator cannot add new resources, and/or when re-routing is not possible. Dedicated software, running in each node, collects a list of local parameters such as available transfer rate and one-way delay between all neighbours. Then this real-time status information is being distributed to all the in-network management enabled nodes that are allowed to be reached. Due to the statistics regarding individual link traffic, a minimal network coding scheme, triggered by cross-layer quality of service, is temporarily activated. This system presents an enhanced distributed routing that preserves the performances of the running services, despite the congestion, which cannot be eliminated.

Other researches dealing with evaluation of Network-Coding (NC) techniques are currently under progress. For instance in Broustis et al. (2009), a wireless mesh testbed combines dedicated hardware (ORBIT platforms) with simulators (Click Router and/or NECO). The result is actually a packet-oriented peer-to-peer communication, simulating that throughput could be close to link capacity while maintaining low latency. Another approach refers to multi-point to multi-point flow-oriented transmission, where the synchronization aspects are delaying its deployment in practice. This is not a trivial task and theoretical solutions are proposed in Biermann et al. (2009). However, the target is to increase permanently the transmission quality for aggregated flows. Two clean-slate approaches, i.e. In-Network Management (INM) and Generic Path (GP), were introduced by FP7-4WARD project "Architecture and Design for the Future Internet." INM supposes inherent management (or at least integrated) into the monitored network, and not external as in the legacy solutions (Nunzi, et al., 2009). The generic path is the other key concept, being an abstraction facilitating the development of applications that use the data transport and enhancing the communication's reliability and quality (Guillemin, et al., 2009). Within this chapter, the target is to demonstrate that a combination between INM Cross-Layer QoS and Network Coding-based GP is feasible. By combining it with a simplified generic path, i.e. a multi-point-to-multi-point communication based on Network-Coding (NC), congestion control is one of its immediate applications. Note that our approach is totally different than in previously discussed (Broustis, et al., 2009; Biermann, et al., 2009), because we address the case of preserving the performances of the running services, despite the congestion which cannot be eliminated. Thus we offer an enhanced distributed routing on a real-time implementation on the existing hardware (no dedicated platforms are needed). This is valid for all types of networks and it involves dedicated software that could be automatically activated/deactivated by the Cross-Layer QoS module.

The new design principle for cross-layer QoS proposed herein is based on the assumption that the virtualization of the network is essential for the future Internet. Thus, the current Physical Layers and MAC Sub-Layers of the legacy or future network technologies are grouped in a substrate network, whilst the remaining ones are composing the virtual part of the network. It is realistic to consider TCP/IP as the dominant stack, although the clean-slate approaches consider that its replacement may happen in the next years (decade). Note that the CLQ is designed to have in the future a complex set of properties: self-configuration, composability, interoperability, and extendibility. This chapter does not cover the design principles referring to the complete list previously mentioned. We took a more realistic approach, i.e. we tried to identify the minimum architecture that demonstrates that the CLQ in the new paradigm is able to allow future Internet techniques to become available. Thus, we selected the congestion control as an immediate beneficiary. Having this vision, the CLQ proposed is using both bottom-up and top-down approaches between substrate and virtual networks, carrying out both status and control information, delivery method being in-band/out-of-band while the timing of the adaptation process is asynchronous. CLQ runs permanently to monitor the infrastructure, but it accepts also service requests (i.e. committed QoS parameters, network status, etc.) from GP. Moreover, the CLQ module computes in real time a global indicator (i.e. Composite Metric—CM) in order to detect the actual performance of the network. Our preliminary formula used in the testbed for an overall perspective of the links with neighbours was:

$$CM = \frac{k_0}{ATR[bps]} + \frac{OWD[s]}{k_1} + k_2 \times BER \quad (23)$$

where $k_0=10^9$[bps], $k_1=10^{-5}$[s] and $k_2=10^{12}$. This CM could be used by the management plane as criteria for triggering such mechanisms as network-coding, QoS-aware routing, etc. The formula should be interpreted in a similar way Cisco's EIGRP composite metric is used at the Network Layer. This means that the maximum ATR (Available Transfer Rate) envisaged was 1 Gbps, the minimum OWD (One-Way Delay) was 10 microseconds and the minimum BER is 10^{-12}. Thus in the most favourable case the contribution of each factor is 1. This means that the minimum CM is 1+1+1=3. Once the technologies evolve, the constants k_0, k_1 and could be updated. For instance if the 10 Gbps would be the reference, then $k_0=10^{10}$[bps]. Note that Layer 1 technology, the type of media plus the distance between neighbouring nodes have a great impact that cannot be seized by ATR and OWD. This is the reason we introduced BER. Unfortunately this traffic parameter cannot be measured directly, being calculated or estimated (based on SNR values). Additional work is needed to demonstrate completely that this formula seizes the dynamicity of the physical links. However, as a preliminary step, we appreciate the composite metric provided as useful. Let us discuss now in details this study case. Network coding techniques represents one possibility to deal with congestions that can appear in network nodes, by combining the flows transmitted on the congested links (Ahlswede, Cai, Li, & Yeung, 2000; Koetter & Medard, 2003). The simplest and most well known topology where NC techniques could be employed is represented by the butterfly topology, as in Figure 8 (Rus, et al., 2010). All the machines are running dedicated software under Fedora Core. The hardware requirements: minimum 1 GB RAM, up to 3 Fast Ethernet accesses provided by network interface cards and USB-FE adaptors. R1 needs to send data to R4 (dashed line) and respectively R2 to R3 (dotted line). If there is no congestion in the network, the data flows will use the paths R1-R5-R6-R4 and respectively R2-R5-R6-R3. We want to remind you that this is a real implementation, not a simulation.

It is possible to face congestion on the link R5–R6 by performing a simple XOR coding operation in the central node R5 (see the flow represented with the solid line in Figure 9) and by sending supplementary data on the links R1–R3 (dashed line) and R2–R4 (dotted line).

Even in the case of the simplest NC-aware topology, several problems need to be solved. These are related to flow synchronization, burstiness of the encoded flows and activation/deactivation of the encoding process. The later operation is mandatory to ensure an efficient use of the network resources, required by the transmission

Figure 8. Six-node asymmetric butterfly topology: no network coding

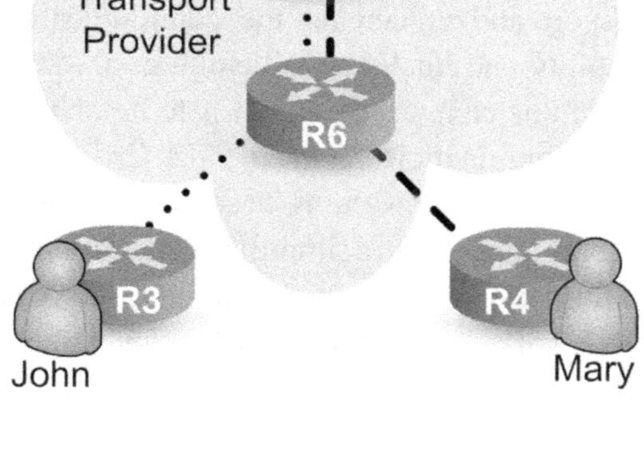

of data on the R1–R3 and R2–R4 links. To address these issues, a dynamic encoding process that is following a set of rules is being proposed:

1. Coding is performed only on the data packets that are generating congestion at a given moment. If no datagrams are transmitted in a flow, the packets of the other streams are sent not coded.
2. If the lengths of the packets to be coded are not the same, only the common parts (having the size equal with the shortest packet) are being combined.

When NC is activated, the operating mode of node R6 is modified in the following way: coded packets are multicast to both sink nodes while not coded ones are being routed towards their destination. Considering this modification, the conditions to activate the coding operations are the following:

$$ATR_{R1-R3} > D_{R1-R5-R6-R4} \quad (24)$$

$$ATR_{R2-R4} > D_{R2-R5-R6-R3} \quad (25)$$

The parameters from relations (24) and (25) are being provided by the CLQ mechanism. Moreover, the monitoring is made in real-time once a second so any change of the condition will be signalized to the NC mechanism as soon as possible. In the above expressions, $ATR_{R_i-R_j}$ represents the available transfer rate on the link Ri-Rj while the parameter $D_{R_i-R_k-R_l-R_m}$ is the used transfer rate on link Ri-Rk-Rl-Rm. Activation/deactivation of the NC coding process is performed based on medium term statistics applied to measurements on the R1-R3 and R2-R4 links. Moreover, the congestion state in node R5 is being considered. The main goal is to avoid the frequent activation/deactivation of the coding process, which besides the signalling problems, could create packet losses due to switching between coding and non-coding operation mode. The evaluation of the conditions necessary for activation/deactivation of the coding operations is realized independently by nodes R1, R2 (rate conditions) and R5 (congestion). The proposed NC algorithm does not require explicit resource allocation, coding operations being activated only

Figure 9. Six-node asymmetric butterfly topology: with network coding

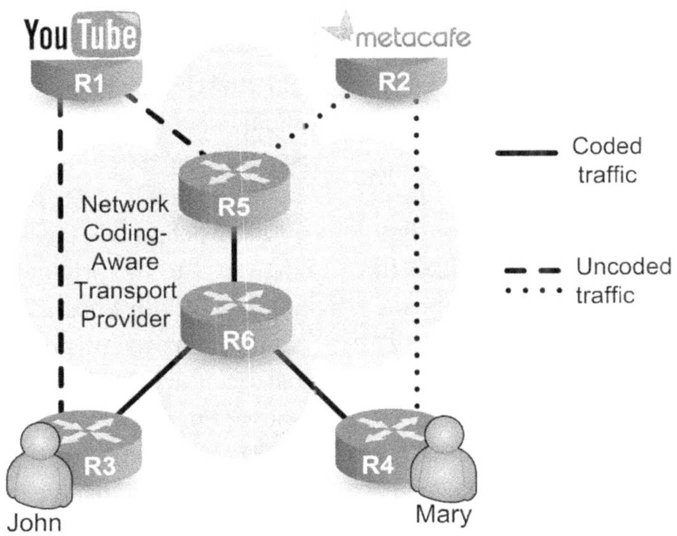

Figure 10. Shape of used rate for link R6-R3 for case 1

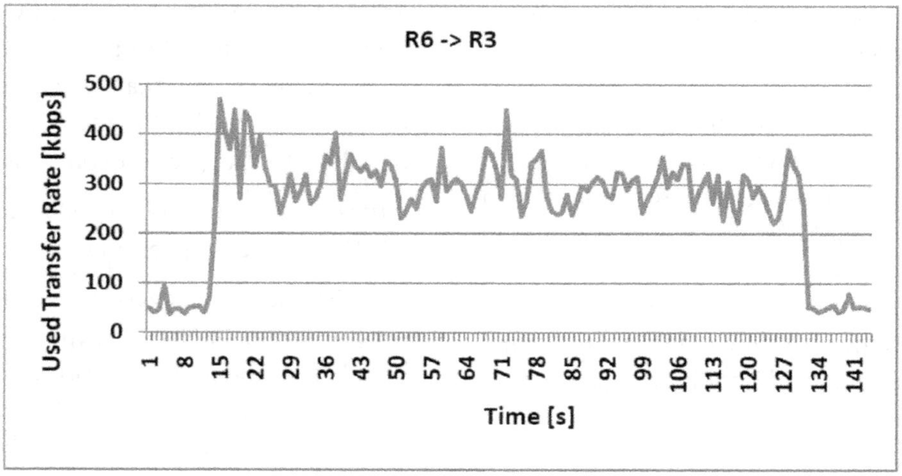

when the mentioned rate and congestion related conditions are fulfilled.

During the employed tests, two video streams, R1-R5-R6-R4 and R2-R5-R6-R3, were sent by VLC clients running in R1 and R2 to the nodes R4 and respectively R3. The performance evaluation was based on the following parameters measured per each video flow:

a. number of packets lost, calculated based on Sequence Number information, located into the RTP header;
b. time distribution of packets lost;

Case 1 (no congestion, no NC-based GP): due to enough available transfer rates on link R5-R6, the quality of experience at the destinations was very good. This was demonstrated for the flow R2-R5-R6-R3, as in Figure 10. The video flow on the link R6-R3 is similar to the one on link R2-R5 meaning that the quality of the received video was close to the original). The composite metric CM for link R5-R6, varied experimentally within the range 23…27.

Case 2 (congestion on link R5-R6, no NC-based GP): the shared link R5-R6 was congested due to the background traffic injected, the available transfer rate being less than the value (about 1 Mbps) required to transmit the video streams simultaneously. Both sink nodes, R3 and R4 experienced a bad quality of the movies because of the packets lost in the congested link. The degradation of the receiving flow could be observed in Figure 11. The composite metric CM for link R5-R6 varied between 300 and 14,000.

Case 3 (congestion on link R5-R6, with NC-based GP awareness): the link between R5 and R6 remained congested as in *Case 2*, but a NC-based GP was instantiated whenever the CLQ module triggered the situation. Note that a pre-congestion state was experimentally detected using the INM algorithm, when CM for the link R5-R6 is higher than 100, but less than 300. Thus, the mechanism was activated in advance, before severe congestion might occur. In Figure 12 the average used transfer rate of the flow, when employing adaptively the NC processing, is becoming close to the original one: around 300 kbps, instead of 200 kbps as in Case 2, although the congestion was still present. The shape of the flow has been improved, with respect to the original showed in Figure 10.

The measurement results presented in Figure 13 are showing that the number of packets lost is

Figure 11. Flow received by node R3 in case 2

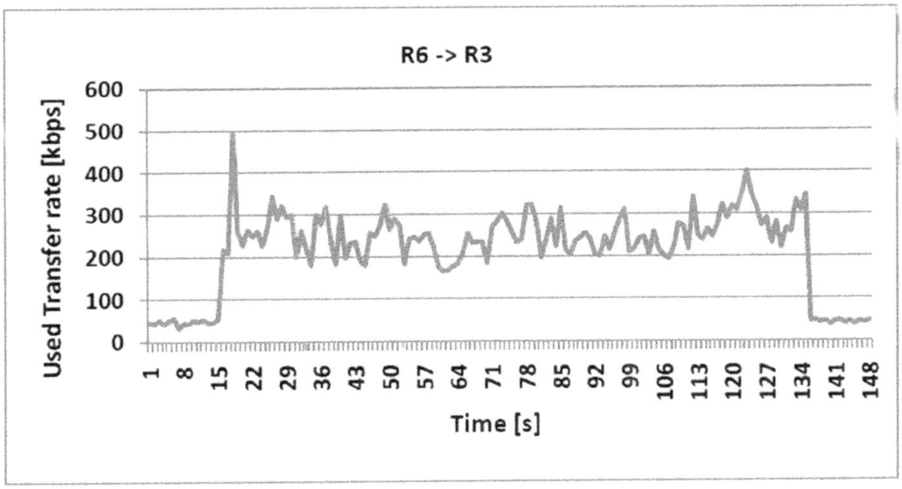

Figure 12. Used rates for link R6-R3 in case 3

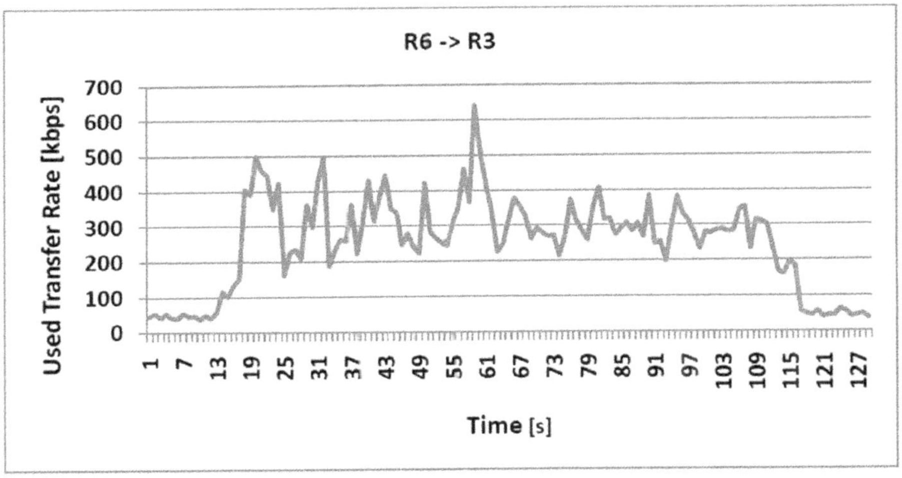

very low when activating the NC mechanism in the network in case o congestion (about 1%, in Case 3 compared to 20% in Case 2).

From quality of experience QoE point-of-view the video image of the stream received by the user when the data flow crossed a congested network and network coding operations were running, was relatively good. Due to software implementation of NC mechanism, the resolution of the video stream transmitted in the test-bed was limited to QCIF format (176x144). Another characteristic of this particular implementation of the Cross-Layer QoS module was that the measurements were performed constantly in the network, once every second, so that the system was able to react in time if any anomaly appeared and disappeared in a smaller time interval. Hence the whole mechanism (INM CLQ module collaborating with the NC implementation) managed to keep the service quality at a reasonably high level. The prototype presented within this chapter is suitable for the infrastructure operators that are not able to perform over-provisioning and/or re-routing in congested networks. Furthermore, the proposed

Figure 13. GP-aware network could face congestion (without eliminating it)

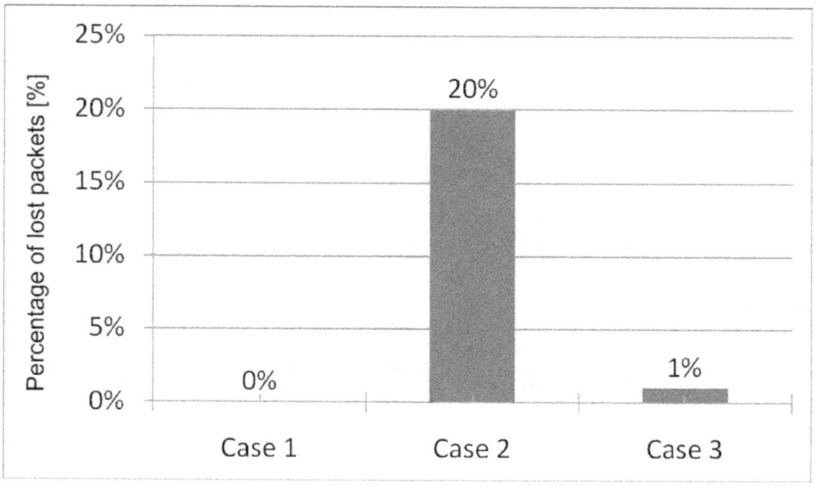

mechanism is useful immediately, whenever a more complex solution, such as QoS-aware routing (solving the congestion), is not available.

QoS-Aware Routing Issues

Because alternative solution such as network coding are not solving the congestion (it just allows the network to face the situation), any QoS-aware routing is welcomed. The idea is not new, but again the legacy solutions are involving complex signalling which still cannot guarantee the optimal decision. In case CLQ is involved, the available transfer rate, one way-delay and bit-error rate are available in real-time as a composite metric. Furthermore, by modifying the existing algorithm (Dijkstra's) while keeping the same routing protocol (OSPF), the routing table is dynamically changed and the decisions are more close to real status of the network.

Modified Dijkstra's Algorithm with Cross-Layer QoS

The routing protocols are an important factor influencing the quality of the services received by a user (Macfarlane, 2006; Medhi & Ramasamy, 2007). Unfortunately, the majority of them are using static information to choose the best path between multiple sources and a given destination node. The legacy routing protocols are aware about the theoretical capacity of the communication channel, given by the technology used at hardware level. The idea presented herein is to modify the Dijkstra's algorithm (the multiple sources, single destination version), from the metric point of view, as well as with respect to the method used to find the cost of the path between the source and the destination nodes. The proposed modifications are trying to correct the deficiencies of the current routing methods by taking into account the dynamicity of the QoS parameters at the Physical Layer and MAC Sub-layer. Moreover, in order to achieve a certain level of situation awareness in each device, the measurements performed by other peer nodes are published within the routing domain. Therefore a modified Dijkstra's Algorithm (DA) was implemented, in which the cost of a path is equal to a composite metric that is based on the following parameters: Available Transfer Rate, One Way Delay, and Bit Error Rate. This information is being received in real time from the Cross-Layer QoS software module. To validate the metric formula, experiments were performed within the OMNET++ simulator. Furthermore, a second issue addressed herein is related to real-time measurements of previously mentioned QoS parameters. Thus, if the values are not constantly updated according to the real situation in the underlying technology, a link with a nominal transfer rate theoretically higher than another one can perform worse if it encounters congestion. Thus, herein we propose to use a set of key performance indicators, monitored on top of MAC Sub-Layer, using both passive and active techniques. Based on these parameters, the spanning tree of the considered network will be constantly adapted to any change that may appear. Employing a set of cross-layer techniques and additionally involving a local database, information from lower layers is being offered to routing processes running at Network Layer (e.g. OSPF – Open Shortest Path First). The considered version of Dijkstra's algorithm is the one that solves the single destination problem, where the scope is to find the spanning tree that includes the shortest paths (with respect to the minimum cost) from each node in the topology to a given destination. In the following lines, a short description of the classical version of the algorithm will be made (see Figure 14). Notations:

- D_i: the cost of the path from the source node i to the destination node d
- P: the set of nodes i for which the shortest path was detected
- Q: the set of nodes for which the shortest path has not been detected yet
- d_{ij}: the distance of a link between nodes i and j

However, the OSPF protocol computes the cost function of a link between two directly connected nodes, (D_{ij}) where $D_{ij} \geq 0$, by taking into consideration a constant value for a given technology. The formula used for computing the cost is as follows:

Figure 14. Dijkstra's algorithm

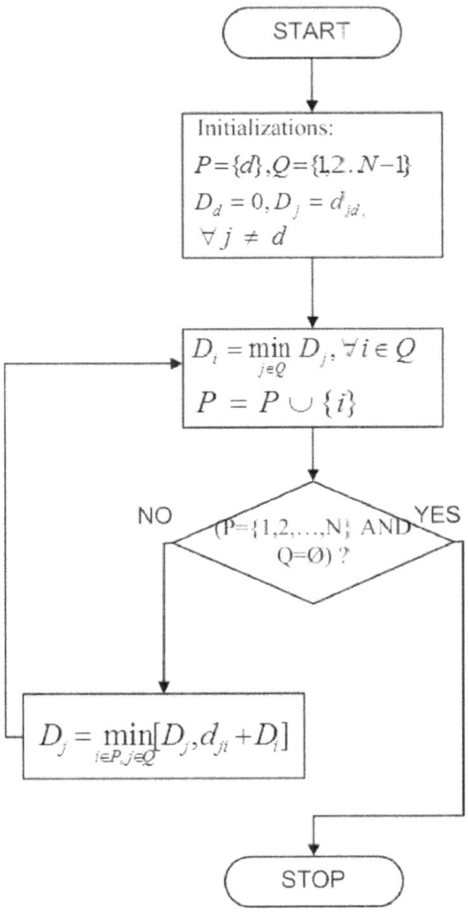

$$D_i = D_{ij} + D_j = \frac{10^9[bps]}{C_{ij}[bps]} + D_j \tag{26}$$

where: $C_{ij} \geq 0$ is the theoretical link capacity at Network Layer between node i and node j. Actually in the classical version of DA, used by the OSPF protocol, the total cost of the path starting at node i to a destination (D_i), is equal to the sum of all the costs characterising the links that are composing the route.

$$D_i = \sum_{j,k} D_{jk} = \sum_{j,k} \frac{10^9}{C_{jk}} \tag{27}$$

In the following subsection, the steps of the Dijkstra's algorithm will be presented as they are described in Dobrota (2011). Initially, the only node included within the set P is the destination one, while the rest are still in the set Q. Moreover, the cost of the path from the destination node to itself is always equal to zero. All the other ones have the initial value equal to the distance of the link between nodes *i* and destination node *d*.

$$P = \{d\}, Q = \{1, 2, ..N - 1\}, D_d = 0, D_j = d_{jd,} \forall j \neq d \quad (28)$$

In the first step of any iteration, the closest node to the destination is detected and included within the set P. Note that the verification is applied exclusively to the nodes that are within the set Q.

$$D_i = \min_{j \in Q} D_j, \forall i \in Q \qquad P = P \cup \{i\} \quad (29)$$

If all the nodes are in the set P, the algorithm will stop, otherwise it will continue with the following step.

The costs of the routes to the destination from all the nodes within the set Q will be updated. The algorithm chooses the path characterized by the minimum value of the cost from all available alternatives.

$$D_j = \min_{i \in P, j \in Q} [D_j, d_{ji} + D_i] \quad (30)$$

After updating the distances, the algorithm returns to the first step. As a major drawback, when calculating the cost, the classical Dijkstra's algorithm considers only the nominal transfer rate of the link. Thus, it is not capable of sensing the congestion that may occur. Note that the latency, i.e. one-way delay for the link between nodes *i* and *j*, could be significantly higher than the distance, because the average waiting time within the queue (service time not included) Tw_{ij} is not negligible.

$$OWD_{ij} = d_{ij} + T_{W_{ij}} \quad (31)$$

Further on it will be presented in detail how Dijkstra's algorithm was modified in order to take advantage of the services offered by the cross-layer QoS module (see Figure 15). When choosing the best available path between multiple sources to a destination, a set of key performance indicators KPIs (i.e. ATR, OWD, and BER) need to be monitored in real-time. If a method of evaluating globally the performance of a path with respect to the previously mentioned parameters is needed, a formula that aggregates all the values into one parameter called Composite Metric (CM) can be use. The CM is having the same role as the cost of a path has in the OSPF protocol that implements the classical Dijkstra's algorithm. The formula used to compute the CM parameter is presented in Equation (36). Different weights are given to the included parameters, according to their influence on the quality of the data services that are travelling through the network. The ATR_j, OWD_j, and BER_j parameters are calculated globally for all the concatenated links of a specific path having the starting point node *j* and the destination node *d*. However, their composability is different than in the classical approach.

The available transfer rate is computed globally for the whole path, between a node *j* included in the set Q to the destination node *d*, (ATR_j) as follows:

$$ATR_j = \min(ATR_{ji}, ATR_{ik}, ..., ATR_{ld}) \quad (32)$$

$$where \quad j \in Q \quad i, k, l, d \in P \quad i \neq k \neq .. \neq l \neq d$$

The one-way delay of a path (OWD_j) between nodes *j* and destination *d* is equal to the sum of the OWD of all the links that are composing it.

Figure 15. Modified Dijkstra's algorithm

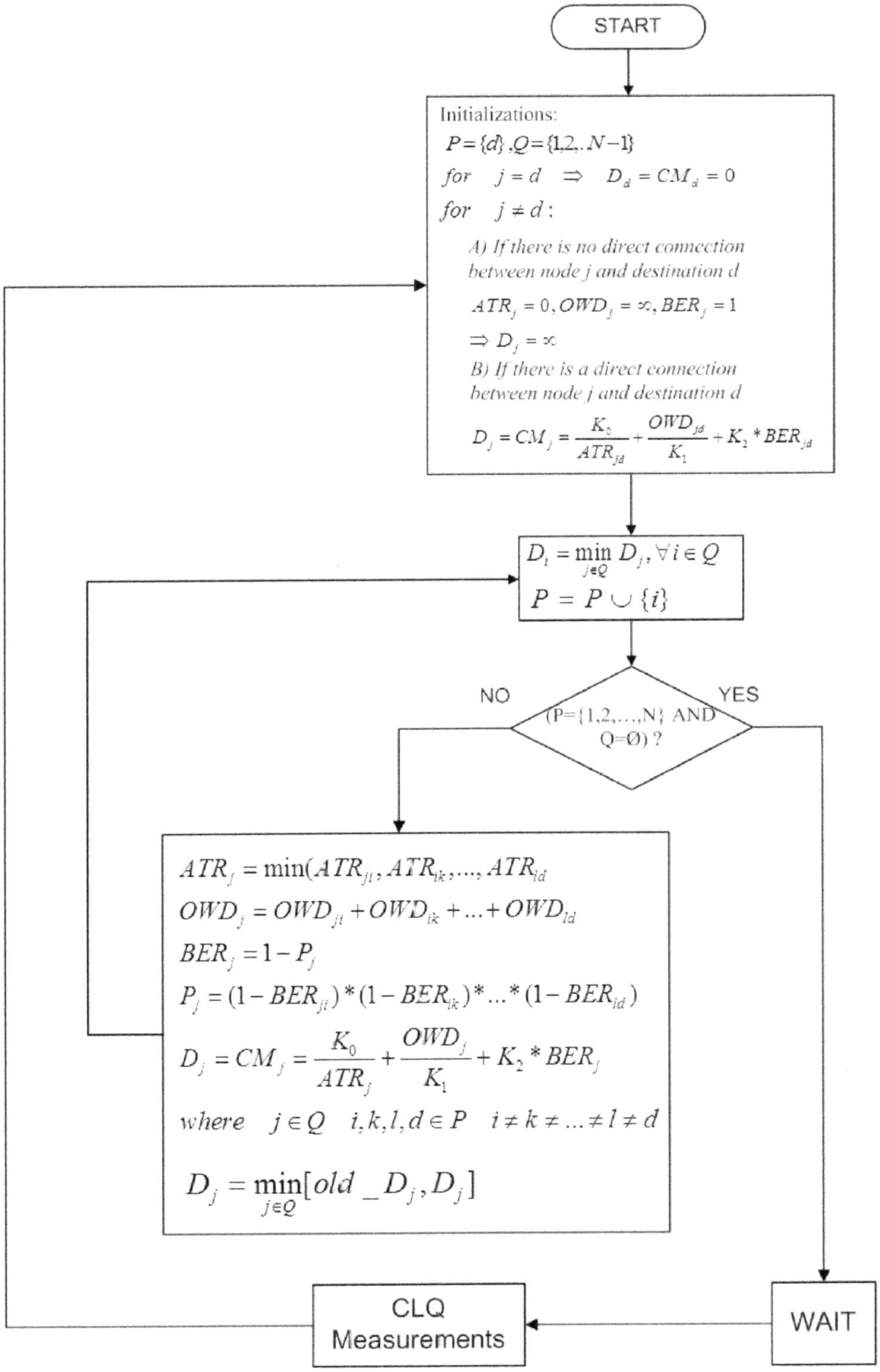

$$OWD_j = OWD_{ji} + OWD_{ik} + ... + OWD_{ld} \quad (33)$$

The bit error rate (BER_j) between nodes *j* and destination *d* is computed as below:

$$BER_j = 1 - P_j \quad (34)$$

$$P_j = (1 - BER_{ji}) * (1 - BER_{ik}) * ... * (1 - BER_{ld}) \quad (35)$$

$$where \quad j \in Q \quad i,k,l,d \in P \quad i \neq k \neq .. \neq l \neq d$$

P_j=probability of not having any erroneous bits on the whole path between nodes *j* and destination *d*

The cost (Dj) computed for the entire path, will aggregate into one composite metric (CMj) all the global traffic parameters. The formula used is the one below:

$$D_j = CM_j = \frac{K_0}{ATR_j} + \frac{OWD_j}{K_1} + K_2 * BER_j \quad (36)$$

Where: $K_0=10^9$[bps], $K_1=10^{-5}$[s] and $K_2=10^{12}$. The constants were chosen so that the metric to have a minimum value, for a link with the capacity of 1 Gbps, a one way delay of 10 μs and a level of bit error rate equal to 10^{-12}. For demonstrating the benefits of adding the Cross-Layer QoS mechanism to the routing protocols based on the Dijkstra's algorithm, a testbed was simulated in OMNET++ (see Figure 16). A number of 13 routers were included into the tested topology to ensure that enough multiple paths are available, between multiple sources and the single destination node. The network is abstracted by a graph with directed arcs, a full-duplex connection being represented by two oriented links. The traffic transmitted through the network will travel from the source node H0 to the sink one H12.

$$K_1 \to \infty, K_2 = 0$$

Note that there are several changes in the modified routing scheme proposed herein, besides the replacement of the simple metric given in Equation (27) with the composite metric from Equation (36). Thus the classical Dijkstra's algorithm stops when the spanning-tree of the shortest path is obtained (i.e. all the nodes are included into the set P). Unless major changes occur in the network (link failures, new links available or new technologies at Data Link Layer) the algorithm is not restarted. However, the modified Dijkstra's algorithm is running continuously, because at least the monitoring part of ATR, OWD, and BER is always activated due to Cross-Layer QoS. The classical algorithm is treated as a particular case of modified Dijkstra's if the following conditions are fulfilled: a) ; b) ATR is always equal to the link capacity; c) Cross-Layer QoS is not activated (Rus, Dobrota, Vedinas, Boanea, & Barabas, 2010).

Influence of Available Transfer Rate

The main purpose of the employed tests is to verify how the Modified Dijkstra's algorithm with CLQ mechanism seizes the dynamic changes of the available transfer rates, latencies and BER on different links within the network. The results will be compared with the ones obtained when employing the classical Dijkstra's algorithm, on the same topology and in the same conditions. Based on the spanning-tree computed separately for each algorithm, the routing protocol (i.e. OSPF) can detect the shortest path from all source nodes to the given destination. Moreover, the routes are being added into the routing table of each node in the network. Furthermore, the End-to-End Delay (EED) parameter of a video flow transmitted between the node H0 and H12 will be monitored, while varying the properties of a couple of links.

Figure 16. Testbed used to simulate Dijkstra's algorithm with cross-layer QoS

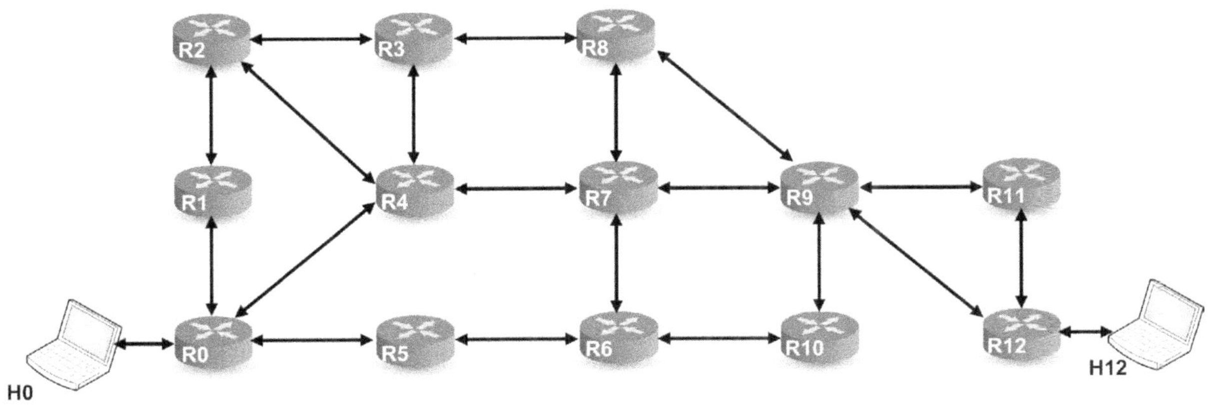

The scenario includes two tests (one for each algorithm) with three steps presented below:

- Step 1: ATR parameter of all the links in the simulated network are being set to a default value (≥1 Mbps)
- Step 2: ATR of the link R0-R4 changes from 5 Mbps to 500 kbps
- Step 3: ATR parameter of the link R4-R7 changes from 3 Mbps to 300 kbps

The influence on the routing protocol's decisions will be depicted in Figure 17-23. For each step, the spanning tree will be presented and immediately the path used to forward the test flows to the destination nodes. The dotted line will show the path used by data packets when running the modified Dijkstra's algorithm while the dashed one will show the path between H0 and H12 used when employing the classical version of the algorithm in the same topology and conditions.

Comparing Figure 18, Figure 20 and Figure 22, one can observe that the variation of the network's performance is transparent to the classical algorithm. This is due to the fact that the forwarding path between the source node H0 and the destination one H12 does not change at all during the three steps simulated. When activating the

Figure 17. Spanning tree computed by the classical and modified Dijkstra's algorithm in step 1

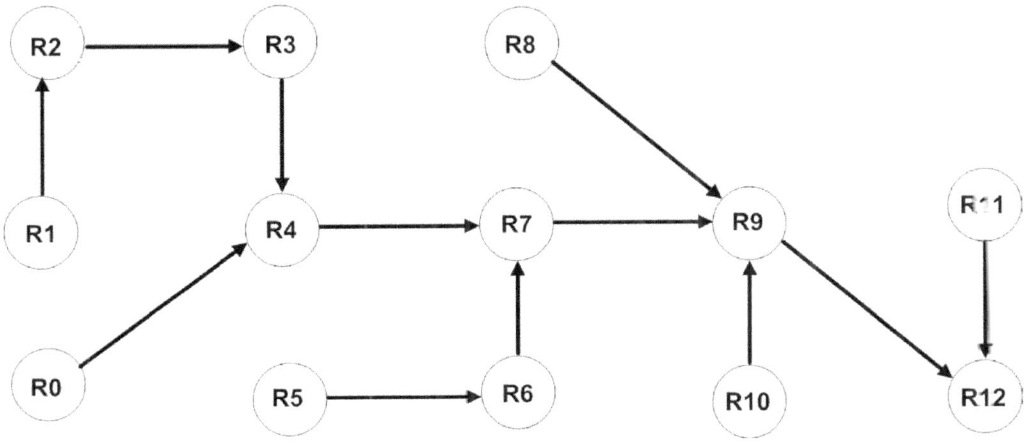

modified Dijkstra's algorithm, the spanning-tree is adapted whenever the performance of any link is being modified. Thus the best available routes are being used and the transmitted video stream is received at a better quality with respect to the End-to-End Delay (EED), comparing it to the classical approach (see Figure 23). Numerically, the average value of the EED was equal to 2.5 ms for the classical algorithm and 1.45 ms when using the modified version.

Influence of One-Way Delay

Similar set of tests could be performed, having the OWD parameter modified in a simulated scenario in three steps:

- Step 1: OWD parameter of all the links were set to a default value ($<10^{-3}$s)
- Step 2: OWD parameter of the link R0-R4 was changed from $2*10^{-4}$ s to $2*10^{-2}$ s
- Step 3: OWD parameter of the link R4-R7 was changed from $2*10^{-4}$ s to $2*10^{-2}$ s

Figure 18. Path between H0 and H12 in step 1

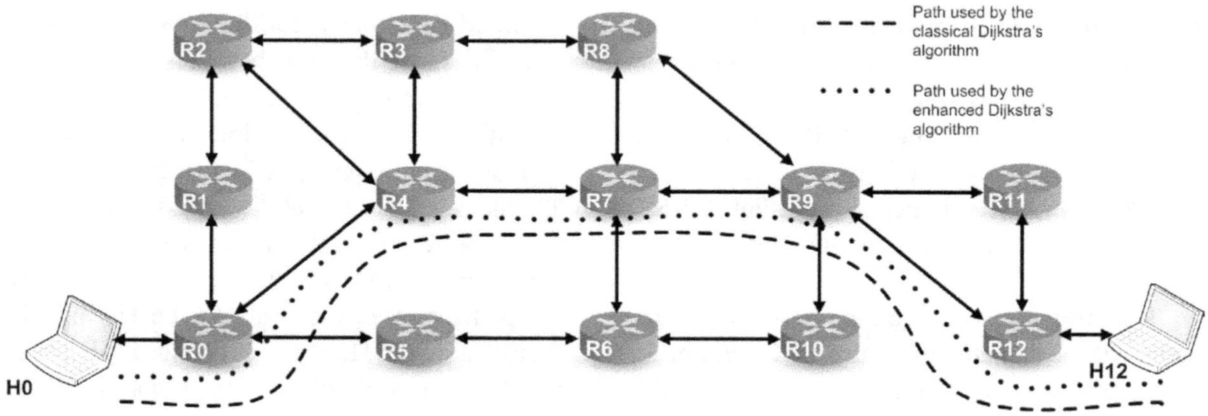

Figure 19. Spanning tree computed by the modified Dijkstra's algorithm in step 2

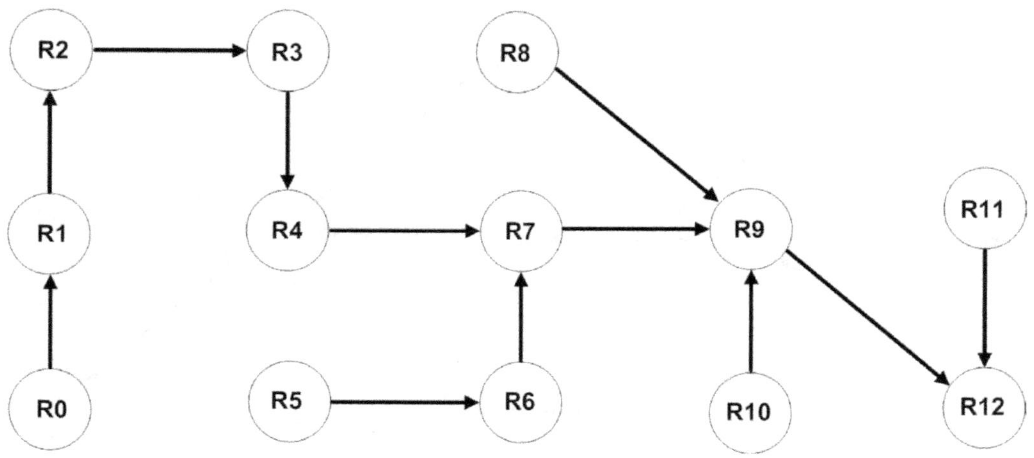

Figure 20. Path between H0 and H12 in step 2

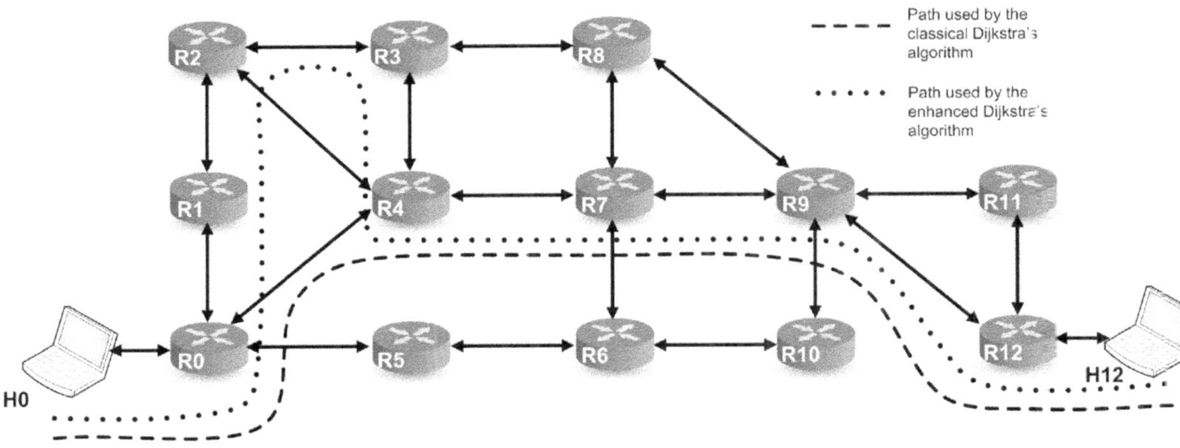

Figure 21. Spanning tree computed by the modified Dijkstra's algorithm in step 3

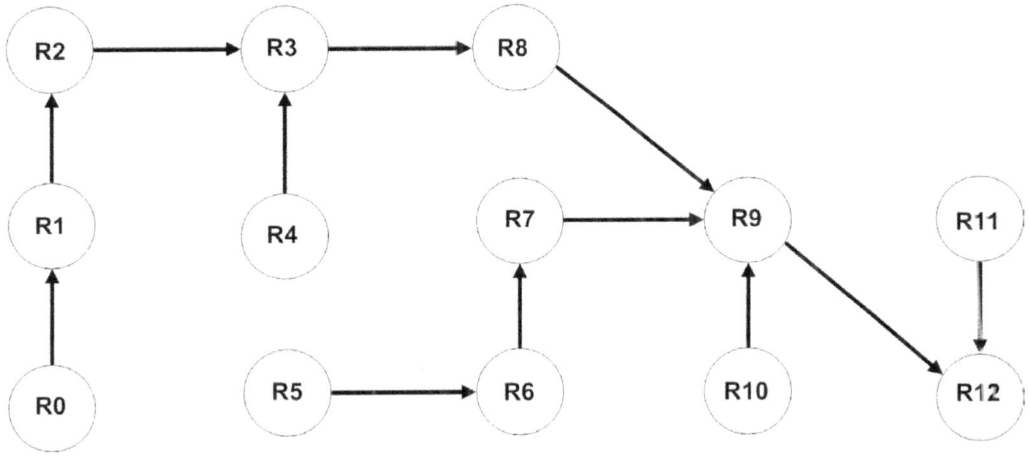

Figure 22. Path between H0 and H12 in step 3

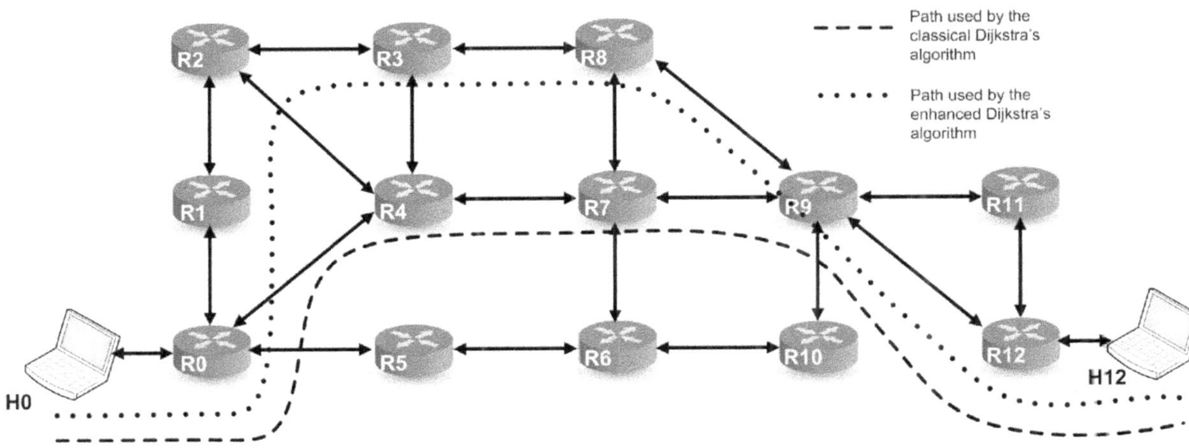

Figure 23. End-to-end delay at classical and modified Dijkstra's algorithms when varying ATR

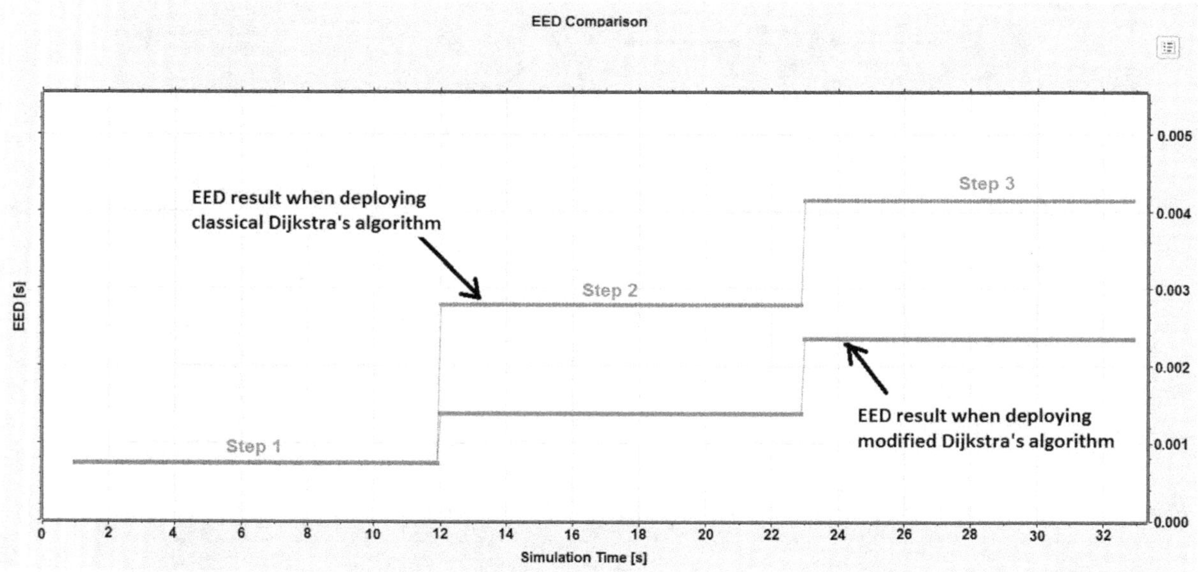

Comparing the average value of EED in both simulations, one can observe that the classical version of Dijkstra's algorithm performed worse (i.e. 2.1 ms) than the modified one (i.e. 0.5 ms). The evolution of the instant values of EED parameter can be seen in Figure 24.

Influence of Bit Error Rate

While maintaining ATR and OWD unchanged, we studied a 3-step scenario varying the BER as follows:

- Step 1: BER parameter of all the links is being set to a default value of around 10^{-11}
- Step 2: BER parameter of the link R0-R4 changes from $6*10^{-11}$ to $6*10^{-5}$
- Step 3: BER parameter of the link R4-R7 changes from 10^{-11} to 10^{-5}

The paths with higher error probability cannot be avoided when simulating the classical Dijkstra's algorithm. As a result, only 242 out of 320 packets sent were received at the destination node. For the new mechanism proposed herein, all the packets sent were received at the destination. However this had a price: because the shortest path was changed to another one characterized by a higher OWD, the EED parameter increased accordingly (see Figure 25). The average end-to-end delay measured for the video stream was 0.7 ms (Dijkstra) and 1.4 ms (modified version). We consider that the increased value of EED is an acceptable trade-off for keeping the percentage of lost video packets as low as possible.

Simultaneous Influence of ATR, OWD, and BER

The experiments ended with a set of tests where all three parameters (ATR, OWD, and BER) were modified as in previous scenarios, but simultaneously. The results can be observed in Figure 26.

One can see that the classical algorithm does not react to performance changes in the network even if the path's quality decreased considerably. Besides the fact that in the classical approach the EED parameter is higher than in the QoS-aware one, the video flow is affected additionally by lost packets due to an increased probability of errone-

Figure 24. End-to-end delay at classical and modified Dijkstra's algorithms when varying OWD

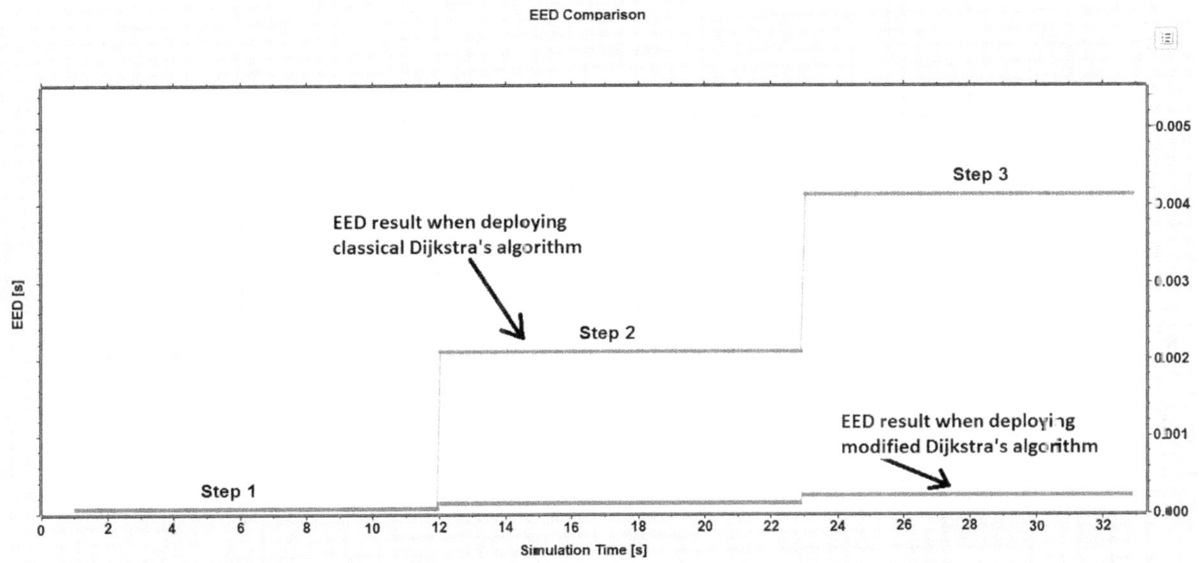

ous bits on the communication channel. The modified Dijkstra's algorithm provides an end-to-end delay about 10 times lower in the conditions of a link state deterioration (100 times OWD increases, ATR decreases of 10 times and a six orders of magnitude BER increases) on two network segments of the route.

Referring to the disadvantages, the mechanism presumes that each node is informed with respect to the performance indicators measured by other peers residing in the same domain. To accomplish

Figure 25. End-to-end delay at classical and modified Dijkstra's algorithms when varying BER

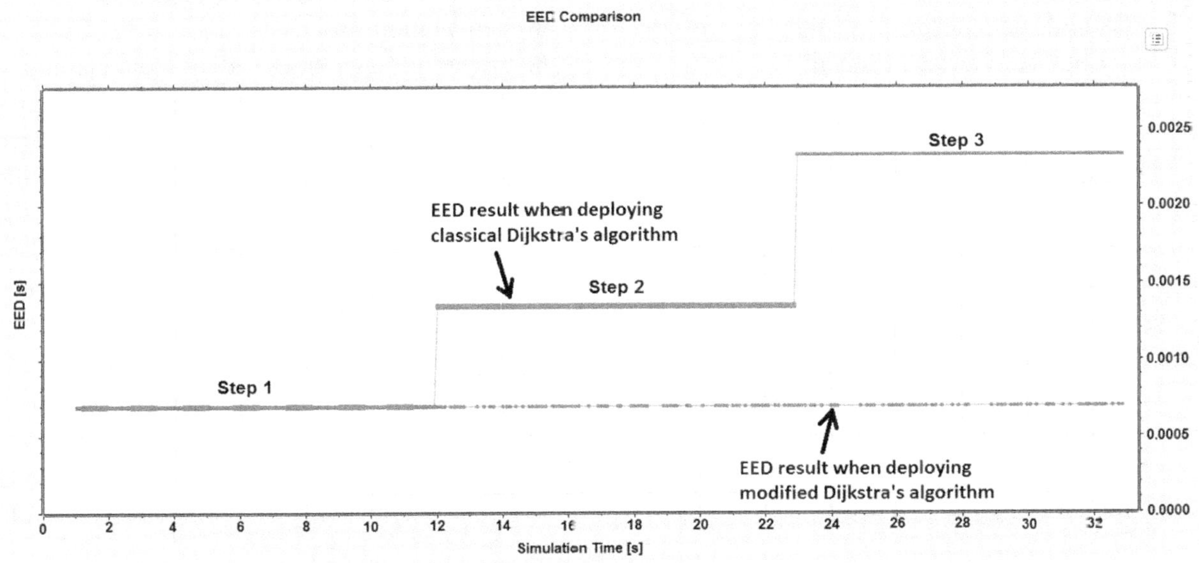

Figure 26. End-to-end delay at classical and modified Dijkstra's algorithms when varying ATR, OWD, and BER simultaneously

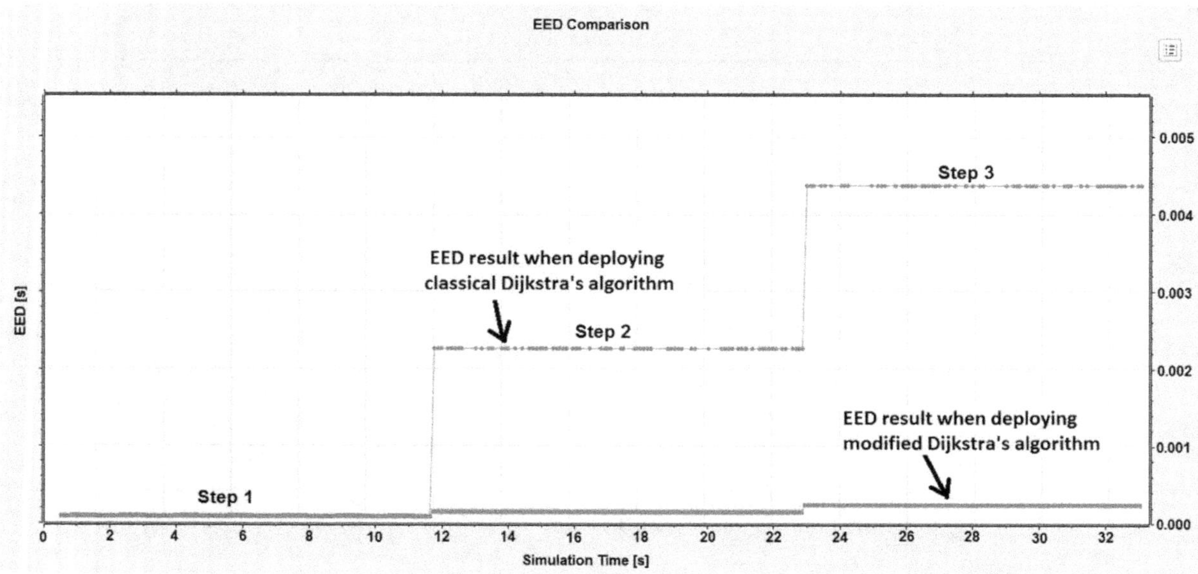

it, cross-layer management information should be exchanged between devices. Knowing the amount of time between consecutive messages transmitted on each interface (1 second) and the average length of the message, the total overhead was preliminary estimated at 600 bytes (including headers) (Rus, 2011). Sending information about ATR, BER and OWD once in a second, the additional traffic added when publishing QoS information was 4.8 kbps per node per interface, i.e. 40*4.8 kbps=192 kbps for our 40-interface testbed.

FUTURE RESEARCH DIRECTIONS

Internet was designed as an infrastructure specialized in connecting devices that needed to exchange data between them. This paradigm proved to be unsatisfying when new applications that used information stored on multiple devices appeared: a) cloud computing; b) peer-to-peer networks.

As future research directions in the clean-slate approach, we consider that the Internet of Objects and Resources (information centric network) is a key subject. Data is expected to be accessed by its name not by location. This would bring a major benefit with respect to mobility or portability issues.

From the evolutionary point of view, using cross-layer techniques for guarantying quality of service at the Application Layer remain a major task. Nowadays there are plenty cross-layer implementation proposals that can be found in the literature. Even though all of them claim that are improving the existing communication paradigms only a few deal with the stability problem that arises due to the increased complexity of the whole communication system. In order to convince the industry with respect to the real benefits of implementing cross-layer mechanism, there is a need to develop also various simulation models that can be used to further study the overheads vs. performance improvements achieved when deploying such a system.

Many cross-layer mechanisms do not deal with the amount of energy consumed (it is an important factor for mobile devices or sensors) for implementing the extra processing tasks. Thus, due to autonomy purposes it is mandatory to validate cross-layer implementation in order to by considering this aspect.

With respect to QoS-aware routing approaches (involving cross-layer techniques) there is a need to deal with the route-oscillation problem and the scalability. In the incremental approach, there is a need that any new proposal to support backwards compatibility, meaning that it should be able to work with classical devices. Many of the nowadays implementations/ simulations do not deal with this aspect. Nevertheless, this aspect should not be ignored because it may be one of the most important ones when trying to impose such an innovative solution.

Today there are several Cross Layer techniques available, but it is not clear which of these are most important. A thorough analysis must be done for all worth to be considered cross layer techniques in terms of performance vs. implementation complexity. In the future, the coexistence of different cross layer implementations should be analysed. If it is possible to have more than one cross layer technology working simultaneously, then the benefits of different cross layer can be combined. Moreover, designers should try to figure out which cross layers techniques have potential to work together.

Yet it is not clear if in the future the protocol layers division will remain the same as today and only the cross-layer techniques will evolve or we are talking also about an evolution regarding the protocol layers.

Most of the cross layer techniques involve enhancements at Physical Layer. When implementing the new paradigms, it should provide additional information regarding the transmission channel performances, so the enhanced role of Physical Layer should be revisited and standardized in the future.

CONCLUSION

This chapter presented the state-of-the-art in Cross-Layer QoS implementations, starting with the evolutionary perspective. The next section included an original contribution of the authors in a clean-slate approach. The design and preliminary implementation were realised within FP7-4WARD project, being a real-time software module including new concepts such as in-network management and generic path. Following this, we identified its benefits in optimal routing decisions (in general) and as alternative solutions to over-provisioning and re-routing, plus QoS-aware routing (in particular). The main idea is to consider a set of QoS parameters extracted in real-time on top of MAC Sub-Layer when computing the spanning tree. Thus, whenever the traffic parameters will change, the spanning tree will be modified too. The routing protocols aiming to find the best path between a source and a destination are the major beneficiaries of this approach. A specialized CLQ module performing both active and passive measurements is actually the entity that provides the real-time traffic parameters to the routing algorithm. Moreover, it publishes a set of key-performance indicator parameters using some bottom-up cross-layering techniques to the modified Dijkstra's algorithm. Even though we presented a preliminary version, the tests proved that the composite metric proposed herein is good enough to bring benefits, when compared it to the single metric. The second contribution would be the fact that by adding the CLQ mechanism, the composite metric will be constantly recomputed using real-time parameters measured or extracted directly from the hardware. Thus, the spanning tree will be able to adapt to any change that can appear in the network. The third contribution would be the way the global value of each parameter (i.e. ATR, OWD, and BER) is being computed for a given path. In the classical approach, the path cost is equal to the sum of all distances of its links. In our proposal, the latency remains cumulative, but the available transfer rate (which replaces the theoretical link capacity) is computed globally for the whole path. The BER has been added to seize Physical Layer behaviour, being calculated based on multiplicative probabilities of not having any erroneous bits on the links composing the path.

The Cross-Layer QoS mechanism presented in this chapter followed the clean-slate approach. However, it can be characterized as follows: it performs both *top-down* and *bottom-up* signalling. The implementation methods are *direct communication* between entities and *using a shared database.* The transported data was both *status* and *control,* exchanged *in band* and *out of band, inside the node* and *between entities residing in different nodes*. Finally, from the adaptation process point of view, the CLQ module is working *asynchronous* while running both *inside* and *outside* the communication process.

When focusing on the benefits of the CLQ mechanism, the following can be claimed: the percentage of lost packets was reduced from 20% to less than 1% if applied in congestion control and end-to-end delay of a video stream was 10 times smaller if alternative paths are available, when deploying the modified Dijkstra's algorithm.

REFERENCES

Ahlswede, R., Cai, N., Li, S.-Y. R., & Yeung, R. W. (2000). Network information flow. *IEEE Transactions on Information Theory, 46*(4), 1204–1216. doi:10.1109/18.850663

Ali, S., & Yu, F. R. (2009). Cross-layer QoS provisioning for multimedia transmissions in cognitive radio networks. In *Proceedings of the IEEE Wireless Communications and Networking Conference WCNC 2009,* (pp. 1-5). Budapest, Hungary: IEEE Press.

Alonso, L., & Agusti, R. (2004). Automatic rate adaptation and energy-saving mechanisms based on cross-layer information for packet-switched data networks. *IEEE Communications Magazine, 42*(3), S15–S20. doi:10.1109/MCOM.2004.1273769

Alshakhsi, S. A. A., & Hasbullah, H. (2010). Improving QoS of VoWLAN via cross-layer interaction approach. In *Proceedings of the 2010 International Symposium in Information Technology (ITSim),* (pp. 678–682). Kuala Lumpur, Malaysia: ITSim.

Biermann, T. (Ed.). (2009). D-5.2.0 description of generic path mechanism. *FP7-ICT-2007-1-216041-4WARD- "Architecture and Design for the Future Internet" Deliverables*. Retrieved January 2010, from http://www.4ward-project.eu/index.php?s=Deliverables.

Broustis, I. (Ed.). (2009). D4.2 first report on test-bed functionalities and implementation of network coding schemes. *FP7-ICT-215252 N-CRAVE Deliverables*. Retrieved December 2009, from http://www.n-crave.eu/deliverables.php.

Carrasco, L., Ramis, J., & Femenias, G. (2009). Multidimensional Markov models for the cross-layer design of multi-rate wireless systems using the effective capacity function. In *Proceedings of the IEEE 69th Vehicular Technology Conference VTC Spring 2009,* (pp.1-5). Barcelona, Spain: IEEE Press.

Chellappa Doss, R., & Pan, L. (2006). A case for cross layer design: The impact of physical layer properties on routing protocol performance in MANETs. In *Proceedings of the Australian Telecommunication Networks and Applications Conference,* (pp. 409-413). Melbourne, Australia: IEEE.

Correia, L. M., Abramowicz, H., Johnsson, M., & Wünstel, K. (2011). *Architecture and design for the future internet*. Berlin, Germany: Springer Science.

Dobrota, V. (2011). Switching and routing systems. *Technical University of Cluj-Napoca*. Retrieved March 2011, from http://el.el.obs.utcluj.ro/scr/.

Du, Q., & Zhang, X. (2009). A cross-layer framework for multi-layer-video multicast with QoS requirements in multirate wireless networks. *IEEE Communications Letters*, *13*(9), 658–660. doi:10.1109/LCOMM.2009.082211

ElRakabawy, S. M., & Lindemann, C. (2009). A clean-slate architecture for reliable data delivery in wireless mesh networks. In *Proceedings of the IEEE Wireless Communications and Networking Conference, WCNC 2009*, (pp. 1-6). Budapest, Hungary: IEEE Press.

Felice, M. (2008). *Cross-layer optimization in multi-hop ad-hoc networks*. Unpublished Doctoral Dissertation. Padova, Italy: Universita di Bologna.

Gonzales, A. (Ed.). (2010). D-4.3 in-network management design. *FP7-ICT-2007-1-216041-4WARD-"Architecture and Design for the Future Internet" Deliverables*. Retrieved February 2010 from http://www.4ward-project.eu/index.php?s=Deliverables.

Granelli, F., & Kliazovich, D. (2005). Cross-layering for performance improvement in multi-hop wireless networks. In *Proceedings of the 8th International Symposium on Parallel Architectures, Algorithms and Networks ISPAN 2005*, (pp. 536-541). Trento, Italy: ISPAN.

Guillemin, F. (Ed.). (2009). D-5.1 architecture of a generic path. *FP7-ICT-2007-1-216041-4WARD-"Architecture and Design for the Future Internet" Deliverables*. Retrieved November 2009 from http://www.4ward-project.eu/index.php?s=Deliverables.

Gyasi-Agyei, A., & Kim, S. L. (2006). Cross-layer multiservice opportunistic scheduling for wireless networks. *IEEE Communications Magazine*, *44*(6), 50–57. doi:10.1109/MCOM.2006.1668419

Koetter, R., & Medard, M. (2003). An algebraic approach to network coding. *IEEE/ACM Transactions on Networking*, *11*(5), 782–795. doi:10.1109/TNET.2003.818197

Kulkarni, S. S., & Rosenberg, C. (2003). Opportunistic scheduling policies for wireless systems with short term fairness constraints. In *Proceedings of IEEE GLOBECOM 2003*, (Vol. 1), (pp. 533–537). San Francisco, CA: IEEE Press.

Lazar, G., Dobrota, V., & Blaga, T. (2008). Practical evaluation of H.264 video streaming over IEEE 802.11e devices by cross-layer signaling. In *Proceedings of the 16th IEEE Workshop on Local and Metropolitan Area Networks LANMAN 2008*, (pp. 13-18). Cluj-Napoca, Romania: IEEE Press.

Macfarlane, J. (2006). *Understanding IP routing in Cisco systems*. New York, NY: John Wiley & Sons Inc.

Medhi, D., & Ramasamy, K. (2007). *Network routing algorithms, protocols, and architectures*. New York, NY: Morgan Kauffman Publishers.

Melodia, T., & Akyildiz, I. F. (2010). Cross-layer QoS-aware communication for ultra wide band wireless multimedia sensor networks. *IEEE Journal on Selected Areas in Communications*, *28*(5), 653–663. doi:10.1109/JSAC.2010.100604

Miao, X., Song, F., Qiu, F., Wan, M., Zhang, S., & Zhang, H. (2011). Redesigning transport layer architecture for future internet. In *Proceedings of the 7th International Conference on Wireless Communications, Networking and Mobile Computing (WiCOM 2011)*, (pp. 1-5). Wuhan, China: WiCOM.

Nunzi, G., & Dudkowski, D. (Eds.). (2009). D-4.2 in-network management concept. *FP7-ICT-2007-1-216041-4WARD – "Architecture and Design for the Future Internet" Deliverables*. Retrieved November 2009, from http://www.4ward-project.eu/index.php?s=Deliverables.

Raisinghani, V. T., & Iyer, S. (2004). Cross-layer design optimizations in wireless protocol stacks. *Computer Communications*, *27*(8), 720–724. doi:10.1016/j.comcom.2003.10.011

Ramakrishnan, K. K., Floyd, S., & Black, D. (2001). The addition of explicit congestion notification (ECN) to IP. *Internet Engineering Task Force RFC 3168*. Retrieved from http://www.ietf.org.

Rappaport, T. S. (2002). *Wireless communications: Principles and practice* (2nd ed.). New York, NY: Prentice Hall.

Rexford, J., & Dovrolis, C. (2010). Future Internet architecture: Clean-slate versus evolutionary research. *Communications of the ACM, 53*(9), 36–40. doi:10.1145/1810891.1810906

Ruiz, J., Gallardo, J. R., Villasenor-Gonzalez, L., Makrakis, D., & Mouftah, H. T. (2009). *QUATTRO: QoS-capable cross-layer MAC protocol for wireless sensor networks*. Paper presented at the IEEE Global Telecommunications Conference GLOBECOM 2009. Honolulu, HI.

Rus, A. B. (2011). *Quality of service through cross-layer techniques for future internet*. Unpublished Ph.D. Thesis. Cluj-Napoca, Romania: Technical University of Cluj-Napoca.

Rus, A. B., Barabas, M., Boanea, G., Kiss, Z., Polgar, Z., & Dobrota, V. (2010). Cross-layer QoS and its application in congestion control. In *Proceedings of the 17th IEEE Workshop on Local and Metropolitan Area Networks LANMAN 2010,* (pp. 1-6). Long Branch, NJ: IEEE Press.

Rus, A. B., Dobrota, V., Vedinas, A., Boanea, G., & Barabas, M. (2010). Modified Dijkstra's algorithm with cross-layer QoS. *ACTA TECHNICA NAPOCENSIS. Electronics and Telecommunications*, *51*(3), 75–80.

Shah, G. A., Weifa, L., & Xiaojun, S. (2010). Cross-layer design for QoS support in wireless multimedia sensor networks. In *Proceedings of the IEEE Global Telecommunications Conference GLOBECOM 2010,* (pp.1-5). Miami, FL: IEEE Press.

Teraoka, F. (2009). Redesigning layered network architecture for next generation networks. In *Proceedings of the IEEE GLOBECOM Workshops, 2009*, 1–6. doi:10.1109/GLOCOMW.2009.5360742

Ukil, A., & Sen, J. (2009). QoS aware cross-layer optimized resource allocation in WiMAX systems. In *Proceedings of the First International Conference on Wireless Communication, Vehicular Technology, Information Theory and Aerospace & Electronic Systems Technology, Wireless VITAE 2009,* (pp.818-822). Aalborg, Denmark: VITAE.

Yuksel, M., Ramakrishnan, K. K., & Doverspike, R. D. (2008). Cross-layer failure restoration techniques for a robust IPTV service. In *Proceedings of the 16th IEEE Workshop on Local and Metropolitan Area Networks LANMAN 2008,* (pp. 49-54). Cluj-Napoca, Romania: IEEE Press.

Zhang, Q., Yang, F., & Zhu, W. (2005). Cross-layer QoS support for multimedia delivery over wireless. *EURASIP Journal on Applied Signal Processing*, *2*, 207–219. doi:10.1155/ASP.2005.207

ADDITIONAL READING

Cao, B. Q., Li, B., & Xia, Q. M. (2009). In Jaatun, M. G. (Ed.), *A service-oriented QoS-assured and multi-agent cloud computing architecture* (pp. 644–649). Lecture Notes in Computer Science Berlin, Germany: Springer-Verlag.

Choi, S. H., Perry, D. E., & Nettles, S. M. (2008). A software architecture for cross-layer wireless network adaptations. In *Proceedings of the 7th Working IEEE/IFIP Conference on Software Architecture, WICSA 2008,* (pp. 281-284). Vancouver, Canada: IEEE Press.

Day, J. (2008). *Patterns in network architecture: A return to fundamentals*. Upper Saddle River, NJ: Pearson Education, Inc.

Doss, R., & Pan, L. (2006). *A case for cross layer design: The impact of physical layer properties on routing protocol performance in MANETs*. Paper presented at the Australian Telecommunication Networks and Applications Conference. Melbourne, Australia.

Ferreira, L. S., et al. (2010). D-5.4 in-network management for generic path. *FP7-ICT-2007-1-216041-4WARD– "Architecture and Design for the Future Internet" Deliverables*. Retrieved July 2010 from http://www.4ward-project.eu/index.php?s=Deliverables.

Ghosh, A., & Gigargis, T. (2009). QoS routing in wired sensor networks with partial updates. *World Academy of Science. Engineering and Technology, 55*(1), 389–393.

Gonzalez, A., et al. (2010). D-4.3 in-network management design. *FP7-ICT-2007-1-216041-4WARD– "Architecture and Design for the Future Internet" Deliverables*. Retrieved February 2010 from http://www.4ward-project.eu/index.php?s=Deliverables.

Kartsakli, E., Cateura, A., Alonso, L., Alonso-Zarate, J., & Verikoukis, C. (2008). Cross-layer enhancement for WLAN systems with heterogeneous traffic based on DQCA. *IEEE Communications Magazine, 46*(6), 60–66. doi:10.1109/MCOM.2008.4539467

Kawadia, V., & Kumar, P. R. (2005). A cautionary perspective on cross-layer design. *IEEE Wireless Communications, 12*(1), 3–11. doi:10.1109/MWC.2005.1404568

Khan, S., Peng, Y., Steinbach, E., Sgroi, M., & Keller, W. (2006). Application-driven cross-layer optimization for video streaming over wireless networks. *IEEE Communications Magazine, 44*(1), 122–130. doi:10.1109/MCOM.2006.1580942

Kiss, Z., Polgar, Z., Vinti, C., Varga, M., Rus, A. B., & Dobrota, V. (2010). Network-coding based congestion control at network layer: Protocol design and evaluation. *International Journal of Computer Networks & Communications, 3*(1), 119–138. doi:10.5121/ijcnc.2011.3108

Koucheryavy, Y. (Ed.). (2009). *Traffic and QoS management in wireless multimedia networks, COST290 final report*. Berlin, Germany: Springer-Verlag.

Mieghem, P. (2006). *Performance analysis of communications networks and systems*. Cambridge, UK: Cambridge University Press. doi:10.1017/CBO9780511616488

Nunzi, G., et al. (2010). D-4.4 in-network management system demonstrator. *FP7-ICT-2007-1-216041-4WARD– "Architecture and Design for the Future Internet" Deliverables*. Retrieved July 2010 from http://www.4ward-project.eu/index.php?s=Deliverables.

Polgar, Z., Kiss, Z., Rus, A. B., Boanea, G., Barabas, M., & Dobrota, V. (2009). Preliminary implementation of point-to-multi-point multicast transmission based on cross-layer QoS and network coding. In *Proceedings of the 17th International Conference on Software, Telecommunications & Computer Networks IEEE SOFTCOM 2009,* (pp. 131-135). Split-Hvar, Croatia: IEEE Press.

Roth, R., Wolff, F., & Zseby, T. (2009). D-4.1 definition of scenarios and use cases. *FP7-ICT-2007-1-216041-4WARD– "Architecture and Design for the Future Internet" Deliverables*. Retrieved April 2009 from http://www.4ward-project.eu/index.php?s=Deliverables.

Rus, A. B., Barabas, M., Boanea, G., & Dobrota, V. (2010). Implementation of QoS-aware virtual routers. In *Proceedings of the 9th International Symposium on Electronics and Telecommunications, ISETC 2010,* (pp.161–164). Timisoara, Romania: ISETC.

Rus, A. B., & Dobrota, V. (2009). Overview of the cross-layer paradigm evolving towards future internet. *ACTA TECHNICA NAPOCENSIS. Electronics and Telecommunications, 50*(2), 9–14.

Sadler, C. M., Kant, L., & Chen, W. (2005). *Cross-layer self-healing mechanisms in wireless networks*. Paper presented at the 6th World Wireless Congress. San Francisco, CA.

Setton, E., Yoo, T., Zhu, X., Goldsmith, A., & Girod, B. (2005). Cross-layer design of ad hoc networks for real-time video streaming. *IEEE Wireless Communications, 12*(4), 59–65. doi:10.1109/MWC.2005.1497859

Shakkottai, S., Rappaport, T. S., & Karlsson, P. C. (2003). Cross-layer design for wireless networks. *IEEE Communications Magazine, 41*(10), 74–80. doi:10.1109/MCOM.2003.1235598

Srivastava, V., & Motani, M. (2005). Cross-layer design: A survey and the road ahead. *IEEE Communications Magazine, 43*(12), 112–119. doi:10.1109/MCOM.2005.1561928

Triantafyllopoulou, D. K., Passas, N., & Kaloxylos, A. (2007). *A cross-layer optimization mechanism for multimedia traffic over 802.16 networks*. Paper presented at the 13th European Wireless Conference 2007. Paris, France.

Vinti, C., Varga, M., Polgar, Z., Rus, A. B., & Dobrota, V. (2010). Signaling protocol for network coding-based congestion control at network layer. In *Proceedings of the 9th RoEduNet International Conference Networking in Education and Research RoEDUNET 2010,* (pp. 304–309). Sibiu, Romania: RoEDUNET.

Wang, Q., & Abu-Rgheff, M. A. (2003). *Cross-layer signaling for next-generation wireless systems*. Paper presented at the Wireless Communications and Networking, 2003, WCNC 2003. New Orleans, LA.

Wolff, F., et al. (2010). D-4.5 evaluation of the in-network management approach. *FP7-ICT-2007-1-216041-4WARD– "Architecture and Design for the Future Internet" Deliverables*. Retrieved July 2010 from http://www.4ward-project.eu/index.php?s=Deliverables.

KEY TERMS AND DEFINITIONS

ATR: Available Transfer Rate is calculated as the difference between link capacity and current total transfer rate.

BER: Bit Error Rate is calculated as the number of bit errors over the total number of transmitted bits.

CLQ: Cross-Layer QoS is a dedicated module that facilitates cross-layer signalling between running entities, in order to exchange various sets of QoS parameters.

GP: Generic Path generalises a data transportation and/or transformation service across a physical or virtual network facility.

INM: In-Network Management approach involves embedding management intelligence in the network. As a consequence: the managed system now executes management functions on its own.

NETINF: Network of Information is a communication architecture based on an information-centric paradigm. It should make the access of information objects much easier, without the need of considering the underlying transport technologies.

NC: Network Coding is an in-network data processing mechanism that combines data packets together using cooperative techniques.

OWD: One-Way Delay is calculated as the latency between two synchronized devices.

Chapter 4
Cross-Layer Techniques and Applications in Wireless Sensor Networks

Sudip Misra
Indian Institute of Technology Kharagpur, India

Manas Khatua
Indian Institute of Technology Kharagpur, India

ABSTRACT

Improvement of wireless and sensing technology enabled the design of a new network technology called Wireless Sensor Network (WSN). However, its applications are greatly limited due to the resource constraint nature of wireless sensor nodes and the lack of flexibility in protocol design. Until recently, almost all the WSN protocols were developed for different layers independently based on traditional OSI layered architecture. Nowadays, researchers are envisioning optimized resource utilization by breaking the barrier of layer abstraction. In this chapter, the authors present basic concepts of layered architecture design and its pros and cons for WSN applications followed by a survey on different applications of Cross-Layer Design (CLD) in WSNs. All the different techniques used for CLD are also discussed in detail. They then highlight some guidelines provided for stable and efficient cross-layer design. Finally, the authors indicate some open issues to show the future direction of research in cross-layer protocol design for WSNs. They conclude the chapter with a great hope that the CLD will have widespread use in next generation WSNs communication.

DOI: 10.4018/978-1-4666-0960-0.ch004

INTRODUCTION

In recent years, wireless technologies have improved significantly in terms of cost, size, and efficiency. Microminiaturization has enabled the designing and fabrication of very small sized and low cost wireless sensor nodes (Akyildiz, et al., 2002a). These sensor nodes are designed to be very efficient, and the applications and protocols designed for them can function with limited amount of energy (Akyildiz, et al., 2002b). This advancement has spiraled into the development of a new network technology, called Wireless Sensor Network (WSN) (Akyildiz, et al., 2002b; Karl & Willig, 2003). In a WSN, the nodes sense the environment and communicate with one another through the wireless medium with certain objectives.

A WSN consists of very small-sized nodes (a.k.a "motes") having sensing, computation and wireless communication capabilities, and powered by small amount of energy unit. They have constraints on storage capacity, computation power and energy supply. In spite of these constraints, the WSNs have been envisioned to be used in a wide range of application domains such as surveillance, inventory system, environmental observation, event forecasting, health monitoring, habitat monitoring and military (Akyildiz, et al., 2002b). Despite the potential applicability of WSN, their actual use in solving real-life problems have been greatly limited due to different constraints on the network as a whole, the nodes that constitute it and the protocols and algorithms that are executed on them (Jambli & Tully, 2007). The adoption of conventional layered communication architectures that are akin to the conventional OSI model can help in handling the communication challenges in WSNs as well (Kawadia & Kumar, 2005).

In a layered architecture, there exist a series of layers endeavoring to reduce the complexity of the network by splitting it into smaller modules with different functionalities. Each layer receives certain services from the lower layer and offers certain others to the higher layers. This enables the shielding of other layers from the details of how the services are implemented in a specific layer. In order to meet these requirements, exhaustive research has been done to design protocols for supporting functionalities in different layers. Most of the communication protocols proposed so far have simple computation requirements and improved energy efficiency characteristics.

The layered design has many disadvantages like non-adaptive and non-optimized. In this regard, the network architecture and the protocol design based on cooperative approaches between layers, called *Cross-Layer Design* (*CLD*), shows a promising alternative to help solve some of the inefficiencies of the layered design.

In fact, a recent trend on research to develop protocol using multilayer integration and design techniques has shown a significant improvement in terms of energy conservation and efficiency. The issues are even more challenging in WSNs because of limited energy and computation power (Mendes & Rodrigues, 2011; Vuran, et al., 2005; Hoesel, et al., 2004; Safwat, et al., 2003). A considerable amount of work has been done using cross-layer design on different issues of WSNs such as *resource allocation and optimization, network lifetime and throughput maximization, providing real-time heterogeneous packet transmission, error control, adaptive routing, target tracking*, and *security*. However, a systematic classification of different applications based on cross-layer design is still missing. Keeping all these perspectives in mind, this chapter is written with two objectives. The first is to present a brief overview of the CLD approaches. In doing so, we make an overview of layered architecture used in WSN. Then, we present the advantages or stipulations of CLD over layered design in WSN. We discuss basic types of CLD techniques used in WSN with examples taken from literature. Finally, we present some concerns and precautionary guidelines regarding CLD in WSN. Our second objective is to present a brief survey on different applications of CLD in WSN. In doing so, we take stock of ongoing works and try to put in perspective to categorize them based on their application. We then draw attention to few potential open problems and new opportunities for CLD in WSN.

The remainder of this cross-layer chapter is organized as follows. In the second section, we provide an overview of the layered architecture design for network communication. There, we analyze the pros and cons of layered approaches from WSN point of view. Next, we introduce the CLD approaches to overcome the limitations of layered design, and we present different cross-layer techniques used in WSN. The different applications of CLD approach in WSN are shown in the fourth section. Then, we try to categorize them based on their applications. While designing protocols for WSN, different guidelines exist that are discussed in the fifth section. All these potential open issues, arise in the development of protocols using cross-layer techniques, are discussed in section six. Finally, the chapter is concluded with a summary.

LAYERED PROTOCOL STACK IN WSN

In the late 1970s, the International Standard Organization (ISO) started to design a framework to facilitate communication between different systems irrespective of the hardware (e.g. architecture) and software (e.g. protocol) used in the system. They introduced the Open System Interconnection (OSI) model, which suggests a layered architecture for network design along with a set of specific protocols. The main intention of the OSI model was to enable communication between different systems without any requirement of changing the logic and design of software and hardware. The OSI model consists of seven different but related layers. Each of the layers performs a specific task to fulfill the whole functionality. Each layer receives few services from the layer immediately beneath it and provides certain facilities for use by the higher layer, thereby shielding other layers from the details of how the services are implemented. Thus, the modular structure and abstraction model aids to reduce the complexity of the network. In brief, the key benefits of layered architecture are as follows:

- It reduces the complexity of understanding.
- Functional design complexity is broken down to different layers.
- It is possible to analyze and test a layer independently.
- It is possible to simplify, extend, or delete a layer from the stack at any time.
- Higher layer may get the benefit of services offered in lower layer.
- It increases the interoperability and compatibility of various network/internetwork components designed and manufactured by different vendor.

Over the years the OSI reference model, as shown in Figure 1, is being used as a building block of different types of network-based communication. Although, the TCP/IP protocol suite, as shown in Figure 2, has become the dominant architecture used in the Internet, it is loosely based on the OSI model. The hierarchy of layers provides modularization, abstraction, and tight synchronization between layers. The physical layer deals with signal and transmission medium, and provides services to carry stream of bits over communication medium. The data link layer provides access control, framing and physical addressing functionalities for hop-to-hop delivery. The network layer deals with the individual packet routing and forwarding from source to destination. The transport layer is responsible for either process-to-process reliable delivery or non-reliable delivery based on the protocol that has been used. The remaining three layers of the OSI model have been merged into the application layer in TCP/IP protocol suite. The main responsibility

Figure 1. The layered architecture of OSI model

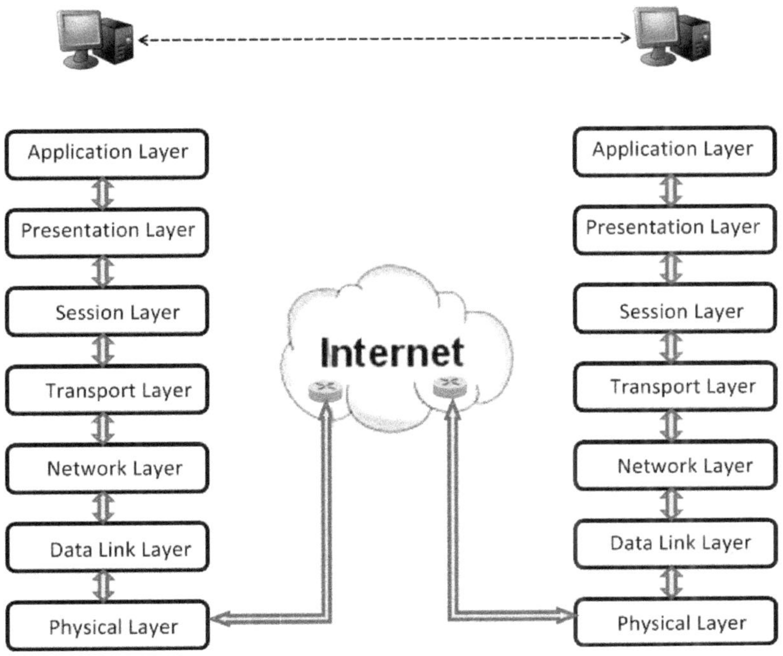

of the application layer is to provide the services to the user. This model provides a strict control on interaction between layers. The interaction can be performed by packet headers only which are attached with packet in every layer.

Many protocols have been designed for WSNs based on the layered design approaches even though several disadvantages and trade-offs exist in the layered architecture design. We write below some of these limitations in the context of WSN.

- *Number of layers and complexity*: In traditional layered architecture, a basic trade-off exists between the number of layers and design complexity. Decreasing the number of layers would increase the protocol design complexity, as many diverse functionalities need to be introduced in these layers. On the other hand, if all functionalities are placed in different layers, the number of layers may become unpredictable, thereby increasing the complexity of the architecture.

Figure 2. TCP/IP protocol stack

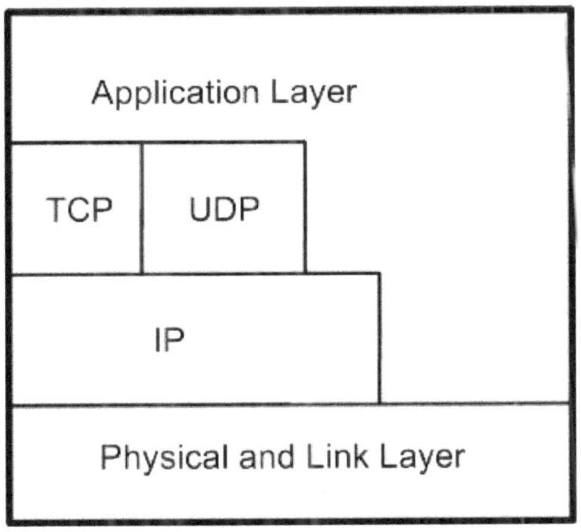

- *Resource allocation and optimization*: The resource allocation problem concerns with the distribution of resources among competing or demanding layers to achieve optimality in resource expenses. It has been observed that the adoption of layered architecture can help in successfully using optimum amount of resources for providing good quality of services (Pompili, et al., 2006). However, they are not suitable for optimizing the use of network resources while the network performance remains acceptable (Pompili, et al., 2006). In recent years, a number of pieces of literature (e.g., Chang & Tassiulas, 2000; Jain, et al., 2003; Melodia, et al., 2005) have been reported on resource allocation in wireless networks. On analyzing the resource optimal protocols, it has been observed that joint optimization of resource usage performs better resource conservation than individual optimization in different layers (Melodia, et al., 2006; Mendes & Rodrigues, 2011).

- *Non-adaptability*: Strict layered design is not adaptable enough to be acquainted with the dynamic environmental changes occurred in WSNs. Furthermore, due to the non-adaptability of protocols it sometimes happens that an erroneous decision is predicted while event changes. For example, in a wireless network, a packet loss due to feeble link may be treated as an indicator of network congestion by TCP sender (Srivastava & Motani, 2005; Smith, et al., 2009).
- *Application unawareness*: WSNs may be deployed in the same area for different purposes. In Wireless Multimedia Sensor Networks (WMSN), the applications are delay-sensitive; whereas, in Delay-Tolerant Networks (DTN), the applications are delay-insensitive. Therefore, an exciting challenge in sensor network is how to efficiently transmit real-time traffic such as voice, game, video and multimedia teleconferencing, and data traffic such as web browsing, messaging and file transfer over wireless links. Hence, the protocols should be application aware to identify heterogeneous traffic, thereby making it desirable to promote interaction between different layers.

CROSS-LAYER DESIGN TECHNIQUES

CLD advocates interaction between different layers for optimal network performance. According to Smith et al. (2009), CLD is "*a recent design principle that allows coordination, interaction, and joint design of protocols crossing different layers appropriately for wireless networks.*" So, the main objectives are to overcome the deficiencies present in layered design approach. Effective improvements in network performance and resource utilization are shown in many applications by breaking down the layer boundaries and explicitly passing information from one layer to others. There are few basic reasons which make CLD protocols attractive (Pompili, et al., 2006; Melodia, et al., 2006). Firstly, constraints present in WSN with respect to energy, computation and memory have created the necessity of more efficient and optimized utilization of resources. Secondly, recent studies reveal that designing a protocol considering channel condition and transceiver's status yield better network performance. Finally, application-aware communication protocol could provide proper support to event-centric requirements in WSNs.

There are several approaches that exist in CLD to perform coordination and interaction between layers. All these CLD approaches can be implemented using one of the following mechanisms:

- *Using interface between layers* (Smith, et al., 2009; Srivastava & Motani, 2005; Lachenmann, et al., 2005; Raisinghania & Iyer, 2004): Almost all cross-layer protocols require interfaces to communicate between layers, as illustrated in Figure 3a. The underlying functionality of these new interfaces is to be able to perform the sharing of required information at runtime. It can be further classified into three sub-categories based on the direction of information flow with respect to the OSI layer stack.

1. *Interface with upper layer*: Using this interface, a lower layer protocol can get information from an upper layer. For example, if the priorities of different applications are communicated from application layer to TCP, the receiving window can be adjusted accordingly to reduce the packet loss.
2. *Interface with lower layer*: Using this interface, an upper layer protocol can get information from a lower layer. For example, the network layer can select the less erroneous interface to route a packet, if the bit error rate for all interfaces are known.
3. *Interface to both directions*: Using this interface, information can move in both directions. At run-time, layers can collaboratively work to perform a specified task. For example, to reduce the power consumption, the network and MAC layers can jointly formulate the routing path and sleep/wakeup cycle.

- *Merging layers* (Smith, et al., 2009; Srivastava & Motani, 2005): Another approach of CLD is *layer merging* where two or more adjacent layers can be merged into a super layer, as shown in Figure 3b, to provide better Quality of Services (QoS) in WSNs. In this approach any new interface is not required to communicate between layers. Although no such protocol has been developed so far that follows this design process, the collaborative design still shows a tendency to blur the boundary between layers in WSNs to enhance the QoS.
- *Indirect Coupling* (Smith, et al., 2009; Srivastava & Motani, 2005): Sometimes coupling between layers is needed at design time to share information at runtime. In this case also any new interface is not required to perform the interaction between layers. This is referred to as *indirect coupling*. For example, if the physical layer is changed to be capable of receiving multiple packets concurrently, then the MAC layer needs to be redesigned to utilize this advancement.
- *Shared Database* (Smith, et al., 2009; Jambli & Tully, 2007; Lachenmann, et al., 2005): In this category of CLD approach, a *shared database*, as shown in Figure 3c, is used. In the *shared database* approach, all layers can communicate with only a common database, which contains global information. All layers can update the parameters and can also fetch the value of any parameter from the database. It is accomplished by the joint tuning of protocols to achieve better performance as the performance measure is seen as a function of different layer's parameters. Srivastava and Motani (2005) have referred to this approach as *vertical calibration across layers,* in which the parameters are set to optimize some metric at design time or run time for *static* or *dynamic vertical calibration* respectively.
- *Completely New Abstraction* (Jambli & Tully, 2007; Lachenmann, et al., 2005): The last category of CLD approach is to create completely new abstractions, as illustrated in Figure 3d, for better communication performance. Vuran and Akyildiz (2010) proposed a complete cross-layer communication solution (XLP) for WSN where the entire layered communication stack has been replaced by *cross-layer stack*. The main goal of this design approach is to develop a single communication module that can provide functionalities of all layers. It

Figure 3. Different implementation techniques of cross-layer approaches

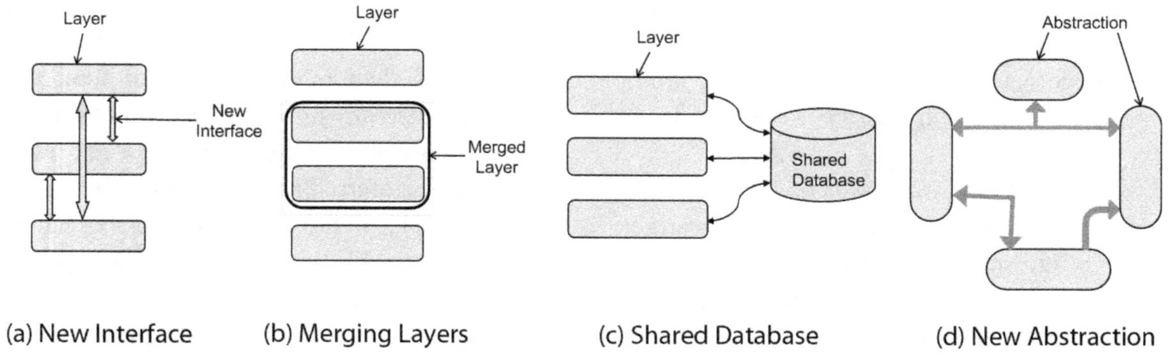

shows greater flexibility to both designer and developer. However, the advantages of modular design have been lost and create more complexity in maintenance and enhancement.

DIFFERENT CROSS-LAYER APPLICATIONS

Despite of the existence of many challenging issues, WSNs have been drawing increasing attention since the time they got attention. Along with the growth of WSN applications, the performance enhancement has come into picture. It is observed that the CLD approaches have better performances in terms of resource utilization, latency, throughput, physical awareness, and security compared to the layered approach. As mentioned in Section 3, there exist many cross-layer implementation techniques. In this Section, few applications of those techniques are discussed. Furthermore, it has also been discussed that why and how the CLD is providing better performance in different applications of WSNs.

Energy Conservation

One of the basic components of a sensor node is its *power supply* unit. Very often a small battery is attached with a node that supplies the required energy to perform the task. Of course, this type of battery has limited energy budget. In addition, it could be impossible to replace the battery in all cases as nodes may be deployed in a hostile environment. Recharging the batteries is not always possible, except in some specific cases in which the battery can be recharged from the external environment (Want, et al., 2005) (e.g. using solar cells as power supply). A large amount of energy is consumed to perform the network operation based on different sorts of applications. On the other hand, prolonging the network lifetime is desirable to provide the application support long time. Therefore, a fundamental question that arises in WSNs is: "*how to reduce the energy consumption to prolong the network lifetime?*"

It has been shown experimentally in Raghunathan et al. (2002) that the data transmission is much more expensive than data processing in terms of energy consumption. The energy is also consumed by the sensing subsystem although that is much less than the network subsystem. Therefore, all energy-saving techniques mainly focus on two aspects: saving energy in *networking subsystem* and in *sensing subsystem* (Anastasi, et al., 2009). The major factors that are responsible for power consumption in WSNs are route finding, packet receiving, data/control packet transmission/retransmission, packet collision, idle participation,

Figure 4. A drawback of adaptive S-MAC

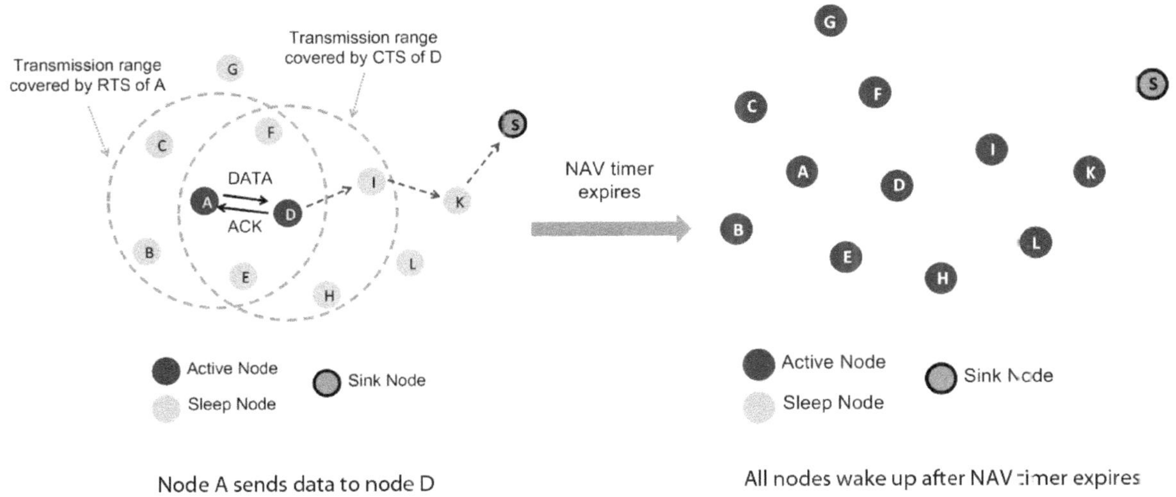

unbalanced utilization of available resources, and node's mobility. If anyhow it becomes possible to mitigate the impact of at least one of these problems, then that approach could conserve energy to some extent. There have been significant amount of research efforts aimed at reducing the energy consumption in WSNs. A detail survey on energy conservation techniques in WSNs is presented in Anastasi et al. (2009). In this section, we will only present few cross-layer techniques that can reduce the energy consumption, and thus increase the network lifetime.

All cross-layer approaches for energy conservation in WSN can be classified according to the message interaction and cooperation among OSI layers. Those protocols have used mainly the *new interface* approach to optimize energy consumption.

2-Layer-X

MAC+NET: In recent times, communication between two OSI layers is being explored by many researchers with an aim to minimize the energy consumption. In this regard, an energy-efficient cross-layer MAC protocol called MAC-CROSS was proposed by Suh et al. (2006), where the properties of MAC and network layers are explored. Their solution uses the Adaptive S-MAC (Ye, et al., 2004) protocol in the data link layer. However, the major disadvantage of Adaptive S-MAC is the *compulsory wake up* problem, which is shown in Figure 4.

Suppose a shortest path (A-D-I-K-S) between source A and sink S is already decided by the network layer protocol. At the outset, node A sends data to node D, while all other nodes are in sleep state for a period of Network Allocation Vector (NAV) time. After the NAV timer expires, all nodes wake up to participate in communication even though the node I only is required to continue communication according to the shortest path theory. Therefore, it creates a *compulsory wake up* problem for other nodes. If the MAC protocol is aware of the routing path information it can help to wake up only the required nodes, while the remaining nodes still remain in the sleep state for the next time cycle (as shown in Figure 5). This process continues until the data reaches to the destination. MAC-CROSS protocol can achieve this requirement by passing the routing information to MAC layer.

Figure 5. Solution of compulsory wake up problem in MAC-CROSS

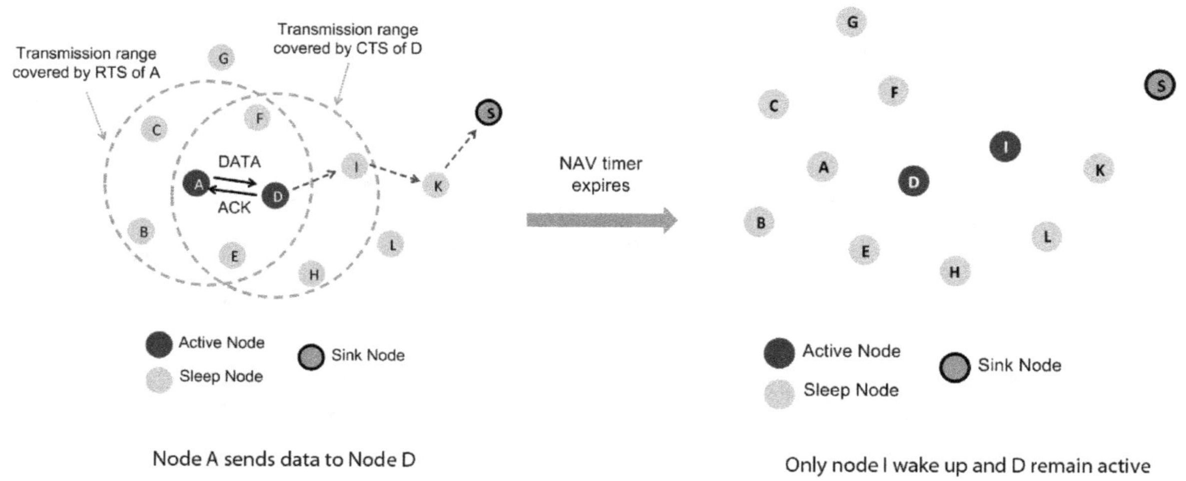

Another cross-layer based energy-efficient routing protocol has been proposed by Chilamkurti et al. (2009). Their solution uses the Dynamic Source Routing (DSR) algorithm interacting with the MAC layer in order to minimize the frequency of recomputation of routing path. The protocol recomputes the routing path only when a link failure occurs. In this manner, it can achieve up to 50% less path recomputation time compared to the original DSR, and hence can help to reduce the energy consumption.

3-Layer-X

PHY+MAC+NET: Many routing protocols follow the shortest path algorithm based on hop count, geographic position and transmission power to forward data from source to sink. The basic requirements of these protocols are minimizing delay and power consumption, and maximizing throughput. The main problem of this technique is that the nodes are quickly drained out of their energy along the shortest path, which may cause connection failure. Zhao et al. (2010b) named this problem as *Flow Equilibrium Assignment Problem* (FEAP) and proposed Path Cumulative Power Consumption (PCPC) routing to address this problem. The basic idea of PCPC is to assign all flows through "*lowest cost path*" at every time instant. This routing approach ensures that a packet will always be flown through the "*lowest cost path*" if there exist at least two available paths. It also ensures that any path will not exhaust their energy substantially prior to the other paths, which in turn helps to maximize the network lifetime. In addition, an optimal programming model based on cross-layer technique has been proposed to present a detail analysis of the proposed routing strategy.

An energy optimization protocol using cross-layer approach was proposed by Bai et al. (2008) for WSNs. It is a joint design among the physical, MAC and network layers. It presents a feedback transmission power control algorithm aiming at reducing the power consumption and interference between nodes. At the network layer, it utilizes the ISTH (Xing, et al., 2005) algorithm along with the nodes' neighbor table received from physical layer to design an energy-efficient routing protocol. This protocol considers the energy consumption for all states (e.g., transmit, receive, idle, and sleep) of a node. Finally, the MAC layer uses the S-MAC (Ye, et al., 2002) algorithm along with the nodes' routing tables received from the network layer to design the duty-cycle scheduling for each node, which again helps to reduce the energy consumption in the network.

Kwon et al. (2006) proposed a cross-layer strategy to maximize the network lifetime while a certain level of end-to-end reliability is maintained. It is shown that by optimizing the power allocation and applying a retry limit, this protocol can minimize the total energy consumption for each path and at the same time it can satisfy the given reliability constraint. The power consumption of a network will be increased if either the number of hops or the transmission power increases. On the other hand, the remaining energy of nodes, those are along the routing path, will be drained out quickly if all packets are forwarded through the minimum energy consuming path. Therefore, a heuristic cost-based algorithm is required for balanced energy consumption by the network, which implicitly increases the network lifetime.

Another energy-efficient cross-layer routing solution for lifetime maximization was proposed by Pudashine (2009), in which the physical layer properties, *Bit Error Rate (BER)* and *energy* were used in the MAC layer to distinguish between link failure and congestion. This conclusion was further used in routing to reduce the routing overhead as well as the energy consumption level when link failures occur.

NET+MAC+LLC: A joint design across network, MAC and logical link layer was proposed by Cui et al. (2005) to minimize the energy consumption in WSNs. It uses the Time Division Multiple Access (TDMA) scheme in MAC layer to eliminate the impact of interference. To reduce the transmission time in each routing path, this protocol considers variable transmission rate (i.e., link adaptation). Upon considering this, the energy minimization problem has been relaxed to a convex problem for what many efficient algorithms exist to provide near-optimal solutions. An extended version of this work was shown by Madan et al. (2006), where the transmission power is considered with transmission rate and link schedule to maximize the network lifetime.

TRN+MAC+NET: A cross-layer protocol, called XLP, was proposed by Vuran and Akyildiz (2010) to achieve the main functionalities of the transport, network and MAC layers in a common cross-layer module. The main objectives of their work were achieving the highly reliable communication with minimum energy consumption, making adaptive routing decision, and avoiding local congestion using the concept of *initiative determination*. The basic idea of *initiative determination* is the decision about the participation in communication, and it is taken by the receiver instead of the source node, based on parameters such as signal-to-noise ratio, traffic rate, buffer occupancy, and residual energy. Therefore, it can be achieved in a distributed manner by the interaction between the transport, network, and data link layers. Furthermore, the routing decision partly depends on the position of the receivers. Hence, a scenario may arise when no intended receiver exists to forward the packet. In this case it uses the *angle based routing* (Vuran & Akyildiz, 2010) approach to overcome the problem. The basic principle of *angle based routing* is the local source will forward the data to a neighbor whose angle formed with local source & final sink is minimum. Finally, the authors shown that XLP improves the network performance significantly in terms of energy, congestion, implementation complexity, reliability and throughput.

APP+PHY+NET: Gunasekaran and Qi (2008) proposed a routing protocol, called XLRP, for reducing the energy consumption in WSN using CLD. It dynamically changes the transmission power level based on the volume of data being transmitted at the time of routing; hence, conserves energy. Again, it uses a back-off mechanism by switching off the unintended receivers based on the power level of the received signal. In addition, this protocol uses piggybacking method to reduce the control message exchange by unintended receivers. Finally, the evaluation of this protocol shows that it performs better in terms of energy consumption, latency, scalability, and throughput.

Discussion

WSNs are broadcast in nature and have many resource constraints, of which energy is the prime concern. We could be able to increase the energy conservation of the network if anyhow we are able to reduce the protocol execution impact on energy consumption which is the primary resource used to perform any functionality. Many approaches have been proposed by researchers based on layered or cross-layer techniques. It reveals from the review that the cross-layer approaches seem to provide more energy conservation than the layered approaches. However, most of the cross-layer approaches proposed so far has considered only a few functionalities of the network to improve energy conservation. Therefore, there exists ample opportunity to improve it further by considering a variety of others.

Congestion Control

At present, there is an upsurge in the usage of a wide range of communication networks in every respect of peoples' lives. All these communication networks have few basic principles to work efficiently. Congestion is one of the major problems that prevent the network from providing full functionality. We can describe an event as network congestion if a node or link carries so much traffic that it deteriorates its quality-of-service (QoS). The main reasons of network congestion include high data rates with limited capacity, many-to-one data forwarding, packet collision and buffer overflow (Zhao, et al., 2010a; Chakravarthi, et al., 2010; Rahman, et al., 2008). Not only the packet loss increases but also the event reliability and system throughput decreases due to network congestion in WSN. It also increases the end-to-end delay and energy consumption. Therefore, congestion control is very important to improve the QoS and network lifetime.

WSN consists of thousands of sensor nodes and a limited number of sink nodes. Nodes are energy constrained and have limited link capacity. Thus, congestion control in WSN is one of the prime issues. Congestion control generally follows two steps—congestion detection and congestion mitigation or avoidance. Congestion detection is the process of detecting abnormality in the normal traffic flow. We can segregate it into two basic categories (Rahman, et al., 2008): *channel-sampling-based* and *queue-occupancy-based* congestion detection. A sensor node's buffer and channel load are monitored to detect congestion using *queue-occupancy-based* and *channel-sample-based* congestion detection methods respectively (Rahman, et al., 2008). On the other hand, congestion control mechanism can be classified into two categories: *resource management* and *traffic control*. In the *resource management* approach, attempts are made to increase the network resources to mitigate congestion when it occurs. On the other hand, in *traffic control* approach, the data rate at source or intermediate node is regulated to mitigate congestion.

Many protocols have been proposed for congestion detection and control based on layered network architecture. Still there exists significant requirement to jointly consider the layers for addressing the same problem and improve the overall network performance. The reasons of congestion in WSNs directly establish the necessity of cross-layer techniques in congestion control. Gunger et al. (2007) performed an analysis to understand the relation between local contention and network congestion. They concluded that adaptive cross-layer mechanism for efficient data delivery is required. In this article we present few cross-layer based congestion control mechanism in WSNs. Detailed survey of different congestion control techniques are available in Zhao et al. (2010a), Chakravarthi et al. (2010), Rahman et al. (2008), and Mendes and Rodrigues (2011).

2-Layer-X

In typical WSN applications, the *idle listening* state of a node wastes the network energy to a significant extent. To mitigate this unnecessary energy consumption, different MAC protocols such as S-MAC (Ye, et al., 2002; Ye, et al., 2004) and T-MAC (Dam & Langendoen, 2003), have been proposed based on the periodic listen-and-sleep (duty-cycle) scheme. However, the listen-and-sleep MAC protocol sacrifices the end-to-end latency and throughput to get the advantages over general MAC. Furthermore, when the traffic load and forwarding path length increases, it causes packet loss, which in turn causes network congestion. Therefore, Kim and Park (2009) have proposed an energy-efficient transport-controlled MAC (TC-MAC) to address the issues. TC-MAC uses the listen period of duty-cycle for multi-hop channel reservation and sleep period for data transmission. It reduces the end-to-end delay, as the data are forwarded as soon as they arrive at a node. To mitigate network congestion, TC-MAC implicitly uses the *backpressure* mechanism to reduce the data transmission rate of immediate neighbors and explicitly uses *Early Congestion Notification (ECN)* to reduce the data transmission rate of the original source. They experimentally showed that TC-MAC can achieve 6 times less energy consumption than other MAC protocols.

An Adaptive NAV-Assisted Routing (ANAR) protocol was proposed by Hsu and Feng (2008) for congestion mitigation in WSN. This work used the IEEE 802.11 protocol (IEEE802.11, 1997) in the MAC layer. So, the NAV value of RTS-CTS approach helps to calculate the congestion-free probability of each possible route. Based on the calculated congestion-free probability the best routing path is selected. The selection can also be switched adaptively to another path, while the congestion level will be changed. It is shown experimentally that the ANAR scheme can increase packet delivery ratio by 10% and reduce the end-to-end delay by 2 seconds over the AODV (Perkins, et al., 2003) scheme.

Wan et al. (2009) proposed a cross-layer active predictive congestion control (CL-APCC) scheme based on queuing theory. CL-APCC analyses the data flow rate and average memory occupancy to predict and solve the congestion in node as well as the grid levels. Then using the adjustment method of node level and system level, the data rate of sending nodes are controlled to mitigate system level congestion. Consequently, the IEEE 802.11 protocol ensures the mitigation of packet collision and unfairness depending on the packet service time and its original priority.

Chen and Zhang (2006) have proposed an *aggregate fairness* model to provide end-to-end fairness, and a distributed fairness algorithm, called *AFA*, to implement this model. AFA deals with two problems in data-aggregation application: *rate reduction* and *fairness control*. This method can mitigate congestion near to the base station for downstream data movement using dynamic forwarding rate adjustment and allocate bandwidth fairly among all data sources.

Akyildiz et al. (2006) suggested a replacement of the entire layered protocol architecture by a unified cross layer module, called XLM, to achieve resource efficiency and data reliability.

3-Layer-X

Wang et al. (2007) proposed a Priority-Based Congestion Control Protocol (PCCP) using cross-layer optimization in WSN. PCCP uses the ratio of *packet service time* over *packet inter-arrival time*, which is referred in the literature as "congestion degree," to detect congestion. Afther detection, the results are informed to the other nodes using an Implicit Congestion Notification (ICN) message. Finally, each node reduces its traffic rate based on the node's priority to control the congestion. To execute these operations this protocol includes two queues—*source traffic queue* and *transit traffic queue*—in between the network and the MAC layers. The former is used for locally generated traffic and the latter is used for the traffic routed through the node. This protocol also achieves the packet loss reduction, and in turn improves energy-efficiency and delay.

Discussions

There are many factors that cause network congestion in WSNs. Also, there are many protocols that have been proposed to address those issues. All protocols discussed so far have one or more assumptions and limitations that are discussed below.

- The upper and lower threshold values of a node's buffer have been assumed to indicate packet loss.
- Priority has been assumed for each node to discriminate packet.
- All protocols are designed for static network topology only.
- Countermeasures are designed for special kind of physical layer or MAC layer protocols only.
- Strict synchronization between nodes is required.
- Sometimes, the congestion detection and control are achieved with the sacrificing of end-to-end delay, network lifetime, and QoS.

Therefore, in spite of the existence of all the above cross-layer approaches, we should design an efficient protocol to optimize the resource utilization and improve network performance while the congestion control operation is performed.

Target Tracking

Tracking and localization of a moving target is considered as an important problem in WSN. It has wide range of applications (Zhao, et al., 2002) such as surveillance (Valera & Velastin, 2005), natural disaster relief (Wang, et al., 2003), habitat monitoring (Mainwaring, et al., 2002), traffic monitoring (Li, et al., 2009), and intruder tracking. In a target tracking application, the mobile target is sensed by the sensor at a particular space and time, and this activity is continued collaboratively by a group of sensors to whom the target is moved forward. Therefore, target tracking in WSN presents a number of challenges, some of which are as follows:

- Resource constraints which include limited energy supply, low bandwidth, short range of transmission, limited storage capacity and low computation speed.
- Distributed and centralized algorithm.
- Quality-of-Services (QoS).
- Traceable area or network size.

In the recent years, various architectures and protocols have been proposed for tracking a target. Detailed surveys of target tracking approaches in WSN are available in Li and Zhou (2010) and Bhatti and Xu (2009). In all target tracking algorithms, the basic tradeoff present is in between tracking accuracy and network resources such as energy, bandwidth, computation and communication overhead. Based on the network architecture, we can classify all target tracking protocols into two categories: *traditional layered approach* and *cross-layered approach*. A further classification and description of different protocols proposed in traditional layered-based approach is given in Li and Zhou (2010) and Bhatti and Xu (2009). In this article, a few cross-layered approaches for target tracking are discussed.

3-Layer-X

In Song and Hatzinakos (2007), a Low-Energy Self-Organizing Protocol (LESOP) is proposed for target tracking in dense WSN. Unlike the classical OSI layers used in communication networks, this protocol excludes the transport and network layers to simplify the algorithm. The basic idea of

Figure 6. Illustration of LESOP protocol

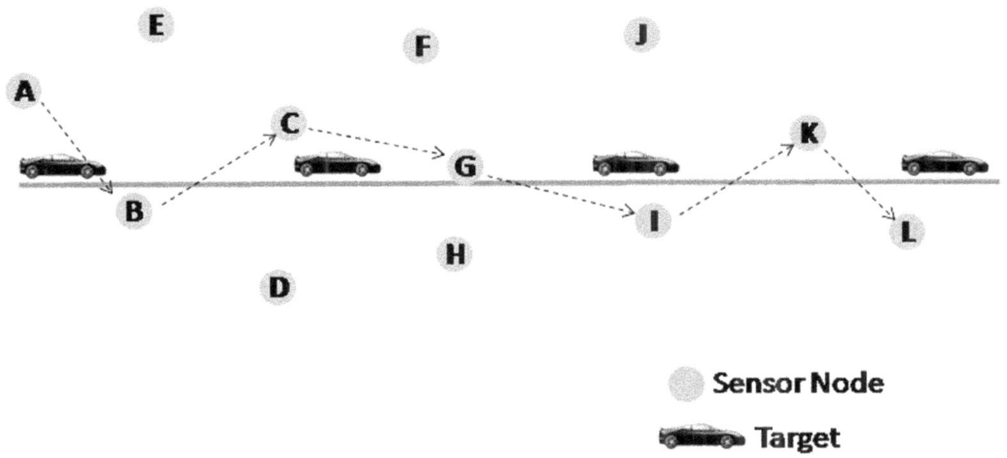

LESOP is at any instant of time a set of node first detect the target and then inform to their leader, say node A as shown in the Figure 6. Then target location estimation is performed based on the "detection information fusion" and the next leader, say B, is elected from the set of neighbors of A using a self-organizing approach. Node A passes the "track information" as well as the control to B after a calculated period of time. Same procedure is repeated for B and continued similarly until the target disappears from the surveillance region.

An energy-efficient and collaborative target tracking algorithm using cross-layer optimization of utility functions is presented in Arienzo (2010). It considers the node selection problem as a cross-layer optimization problem, and therefore, the optimal greedy-based algorithm is used to find an optimal solution. A new energy-based metric is proposed to evaluate the energy consumption by node selection algorithm. The proposed protocol uses both bootstrap and unscented particle filters (Arulampalam, et al., 2002) in both distributed and centralized manner to track the target.

Discussion

Protocols designed for target tracking using cross-layered approaches have many limitations like they considered the ideal scenario where intentional or unintentional node failure will not occur and no coverage-hole in the network. In addition, real time detection analysis for fixed and variable speed of target was not performed in the literature. Therefore, the special characteristics of WSN provoked researchers to work on the issues relating to target tracking. Few of them are as following:

- Channel-aware tracking: Wireless communication channels between source and sink are imperfect. So, the cross-layer techniques can be used to design more energy efficient and accurate tracking algorithm interacting with the channel statistics.
- Fault-tolerant tracking: The sensor nodes might be deployed in harsh environments in which the probability of node failure is high. Therefore, an efficient fault-tolerant target tracking algorithm is required for WSN, as the traditional fault-tolerant approaches developed for wired and wireless network are not suitable in this context (Ding & Cheng, 2009).
- Energy efficient target recovery.
- Tracking target with varying speed and direction.
- Multi-target tracking.

Finally, it should be emphasized that there exists many open issues for solving and enhancing the area of target tracking.

WSN Security

WSNs are deployed in the physical environment based on the application. The deployment of a network may be done in a remote place or in a harsh environment where continuous physical monitoring is nearly impossible. However, the WSN provide us the capability to monitor such real-world environment. WSNs are bare to all to be attacked by threats such as node capture, node tampering, denial-of-service, and traffic analysis. Therefore, a security mechanism should be equipped with WSNs to provide defense against different attacks. Building a security mechanism in WSNs is more challenging as it has some inherent constraints in node level and network level as discussed in section one and physical level too. A lot of work has been done on the development of security mechanism for different threats. Detail surveys on security threats and its counterattack or avoidance are described in Walters et al. (2006), Zia and Zomaya (2006), Wang et al. (2006), Sen (2009), Kalita and Kar (2009), and Padmavathi and Shanmugapriya (2009). Primarily four basic requirements have been considered in security mechansims Primarily four basic requirements have been considered in security mechanisms. They are as follows: Walters et al. (2006), Sen (2009), Xiao et al. (2006), and Zia and Zomaya (2006):

- *Availability*: It ensures that the node is entitled to use resources and the network is available to provide services.
- *Confidentiality*: The information exchanged should remain confidential. It should not be understood by active or passive attacker.
- *Integrity*: Ensuring that the message has not been tampered, modified, or replaced by attacker while the message was travelling through network.
- *Authentication*: It ensures the reliability of a message's origin.

Recently, many security techniques have been proposed to maintain the basic security requirements, and, in turn, protecting the WSNs from security attacks. Most of the security solutions have been developed with an aim to provide defense to a particular network layer. However, Xiao et al. (2006) claimed that cross-layer security solution perform better than single-layered security solution. They pointed out few limitations in the single-layered security approaches like redundant security provisioning, non-adaptive security solution, and power inefficient security solution. Therefore, in this Section, we present few cross-layer security solutions designed for WSNs.

Key Management

In a wireless network, data can be securely transmitted by encrypting it using a cryptographic key. The sender and receiver must hold the valid key to be competent enough to encrypt and decrypt the data. Therefore, the problem of distributing and updating the cryptographic key between the authenticated sender and receiver is known as *key management*. In resource-constrained networks like WSNs, key management adds extra overhead on resource, computation and communication. Lazos and Poovendran (2004a) proposed an energy-efficient multicast key distribution mechanism for ad-hoc networks. In their work, a sub-optimal cross-layer algorithm was considered as the energy, security and multicast routing are the properties of physical, application, and network layer, respectively. Another cross-layer algorithm for multicast key distribution is proposed in Lazos et al. (2004b) where Hamming code and group controls are used to find the common path from the sender to each node, and then routing energy is saved by sending data only once through the common path.

Denial-of-Services

Denial-of-Service (DoS) is an event that disrupts or destroys a network and diminishes the capacity of the network to provide services. In WSNs, several types of DoS attacks are possible such as jamming and tampering in physical layer; collision, exhaustion and unfairness in datalink layer; and flooding and spoofing in the network layer. Therefore, several approaches have been proposed for mitigating DoS attack in WSNs. In this section, we discuss few cross-layer approaches designed for detecting and avoiding the DoS attack.

JAMMING: Jamming is one type of DoS attack that happens in the physical layer of a communication network. In WSNs, it requires less effort by an adversary to launch jamming attack. In jamming attack the malicious node, called the jammer, continuously or intelligently transmits a radio signal with an aim to block the legitimate user from accessing the medium and/or interfere the data reception. Intelligent jamming may happens in upper layers also. Chiang and Hu (2011) proposed a cross-layer scheme that allows the receiver to detect and mitigate jamming in a wireless broadcast network. In this approach, a binary key-tree structure has been implemented along with an existing spread-spectrum technique to mitigate jamming in broadcast network. After that a *tree-remerging* technique has been proposed to optimize the spreading codes used in the network, which in turn reduces the power consumption and probability of jamming. A number of techniques have been proposed by researchers to detect, mitigate and avoid jamming in WSNs. Detailed surveys on jamming attacks are presented in Pelechrinis et al. (2011) and Mpitziopoulos et al. (2009). It is shown in this review that different jamming models exploit the parameters of different layers, which directly exposes the requirement of cross-layer techniques to create a unified anti-jamming system.

INTRUSION DETECTION: Another type of DoS attack present in the wireless network is malfunction intrusion. In this attack, a malicious entity can deny network services and connectivity of the network and causes unnecessary packet loss. Traditional approaches of intrusion prevention techniques such as secret key and message encryption cannot work successfully for insider attack. Therefore, an Intrusion Detection System (IDS) has been developed that resides in a gateway node and monitors the abnormalities in the network to indicate intrusion. In a wireless network, the dedicated gateway may not be always possible and the IDS should be intelligent enough to differentiate intrusion from packet loss due to valid reason such as network congestion, link failure, and topology change. Thamilarasu et al. (2005) proposed a Cross-Layer Based Intrusion Detection System (CIDS) for wireless ad-hoc network where a host based IDS is used to monitor all neighbour of that host. The host IDS exchanges the detection information across the network layers to provide more confirmed decision and trigger multiple levels of detection. This cross-layered IDS shows the increment on intrusion detection accuracy and reduces the false positive alarm compare to single-layered IDS.

Discussion

In addition to the above-mentioned attacks, there are many other types of attacks that are possible in WSNs. A list of different types of attacks and their countermeasure are discussed in Walters et al. (2006), Zia and Zomaya (2006), Wang et al. (2006), Sen (2009), Kalita and Kar (2009), and Padmavathi and Shanmugapriya (2009) surveys on security. Almost all security mechanisms have been developed based on the strict layered architecture design. However, a review on WSNs security done by Xiao et al. (2006) reveal that the security design based on the layered approach is often not suitable for WSNs. Furthermore, the security of a network is not determined by the individual layer security; rather, it is a combination of all security measures for all layers. Hence, it appears that developing a security solution based on the cross-layer design approach would produce better performance in terms of true detection ratio, power efficiency, and computational complexity in WSNs.

Error Control Scheme Selection

Wireless channels are more error prone because of multipath fading and poor channel quality. However, many applications demand error free communication in WSNs. Therefore, error control has become one of the fundamental requirements of data communication. Several techniques have been proposed by many researchers to detect and correct the error in data communication efficiently. All error control techniques can be broadly classified into three categories: Automatic Repeat Request (ARQ), Forward Error Correction (FEC), and Hybrid ARQ (HARQ).

- *Automatic Repeat Request*: In ARQ-based error control scheme, only error detection is considered, and instead of correcting the error, it is requested to retransmit the corrupted packet.
- *Forward Error Correction*: In FEC-based error control scheme, sender adds few systematically generated redundant bits to data packets before they are transmitted. These added bits are used by the receiver to detect and correct error at the receiving end.
- *Hybrid ARQ*: In HARQ-based error control scheme, the advantages of both FEC and ARQ have been used.

An overview of different error control schemes for networks is presented in Liu et al. (1997). In WSNs, the sensors are resource constrained and have limited computation power. Therefore, energy-efficient error control is a requirement for WSNs. On the other hand, end-to-end delay is more crucial for delay sensitive real-time applications. A detailed review of error control scheme selection for different state of affairs in WSN was performed by Vuran and Akyildiz (2006, 2009) using cross-layer techniques. This analysis shows that error correction capability is directly related to energy consumption, latency, and packet error rate.

PRECAUTIONARY GUIDELINES IN CROSS-LAYER DESIGN

In this section, we describe the design principles and precautionary guidelines that should be followed when a cross-layer approach is used for designing the architecture and/or protocol. It is explained in sections one and four that interaction between layers can lead to optimized utilization of node and network resources, and the system performs more efficiently than the layered design. Eventually, CLD has increased the attention of researchers to decouple dependencies across layers especially for resource constrained design. Based on these assumptions, many cross-layer protocols, as discussed in section four, have been proposed in the recent past. The objectives of those protocols were to improve the performance in terms of energy, throughput, delay, and security. However, these improvements are often obtained by sacrificing the architectural modularity or functional abstraction, and the logical separation between designers and developers (Pompili, et al., 2006). Consequently, the performance gained by a cross-layer solution needs to be evaluated against the possible downfall of its any characteristic.

Kawadia and Kumar (2005) examined holistically the issues centered on CLD and its architectural ramifications. Based on their study they presented some principles of good architecture for wireless networks. They exhibit some good architectural design examples, which include the Von Neumann (Burks, et al., 1963) architecture for computer system, the layered OSI architecture for networking, the Source-Channel Separation and Digital Communication System Architecture (Shannon & Weaver, 1949), and the feedback architecture for control system. All these good architectural design satisfies one basic principle that is long time survivability.

The various concerns and cautionary advices expressed in Aune (2004) and Kawadia and Kumar (2005) for CLD are as follows:

- *Degree of Modularity:* In a modular approach, the architecture or functionality of a system is broken down into small, separate, and interchangeable components, called modules, which are having strict logical boundaries between components. This design philosophy reduces the complexity of a problem by breaking down into easier sub-problem and the solution of each sub-problem can be independently tested and standardized. Therefore, it provides the abstractions and parallelization of effort needed for both design and implementation of a system yields the result of quick proliferation of the technology. In classical layered approach of networking the functionality of each layer and the interaction between layers are systematically specified. On the other hand, in CLD approach we lose coupling between layers as the degree of modularity is not specified strictly. This loose coupling may leads to impair design and implementation, and slow the innovation down.
- *System Enhancement:* Any system design will persist for the longer period of time if it is adaptable with the enhancement of its components. In traditional layered design, functionality of any layer can be easily upgraded as the strict logical boundary between layers is mentioned. Conversely, a CLD approach does not specify clearly the discrimination of different functionalities, which stand a barrier up in front of further design development and innovation. Furthermore, in some cases the entire system may need to be replaced instead of changing any component of the system. In addition to that, the maintenance cost and complexity increases in case of CLD.
- *Stability*: Another important consideration is the stability of a system. According to adaptive control theory (Kumar, 1985; Astrom & Wittenmark, 1995) stability is a paramount issue of system design. It is well known that any uncontrolled design leads to a "spaghetti" design (Kawadia & Kumar, 2005), which, in turn, increases the probability of instability. In addition, the probability of instability in CLD is much higher than the single-layered architectural design as it loses the strict logical boundary between modules and there may exists tow conflicting adaptation loops (Kawadia & Kumar, 2005). Hence, *dependency relation* may be checked between interchanging parameters to find any conflicting loops, and then *timescale separation* can be used to avoid this problem. Finally, a careful selection of a system design is needed to avoid any negative effect that can reduce the overall performance of the system.
- *Robustness*: Robustness is the ability of a system to continue its operation when unpredictable events occur such as node failure, loss of connectivity, security attack, reduction of channel capacity, or anomalies in communication (Liu, et al., 2005). In the CLD approach, the robustness of a system degrades along with the reduction of modularity.
- *Cost*: In single-layered architecture, the designers can easily develop a protocol for a particular layer without paying concern about the protocols of other layers, as there

is strict logical boundary between layers. Therefore, it is convenient to standardize a subsystem, which can be applied in many different applications. When the applicability of a subsystem increases, per unit cost of that subsystem goes down. In addition to that, parallelization can be applied to design and implement a system, which can reduce the development time, and in turn reduce the development cost. On the other hand, in CLD, the development cost may increase compared to layered design because of the attributes already mentioned above.

OPEN PROBLEMS

The unique characteristics of WSNs such as less computational ability, small memory storage, and limited energy supply necessitate the cross-layer implementation of data communication architecture and protocol. Recent works on cross-layer design shows a significant improvement in terms of energy conservation, efficiency and security in WSNs. However, as a consequence of the CLD techniques, few questions still remain unanswered. Therefore, in this Section, we present a brief overview of few open problems regarding cross-layer implementation in WSNs.

- *Comparison Metrics*: All existing metrics present in the literature do not specifically emphasize on CLD; rather, they can be used in any type of design. Furthermore, it is not verified whether they are satisfactory in CLD or not. Therefore, an exhaustive study is needed to identify the adequate metrics that will explicitly represent the design objectives of WSN using the cross-layer approaches.
- *Energy Conservation*: Some problems commonly appear in any type of wireless networks. Energy conservation is one of them, and it is still a prime concern in design and development of WSNs. Many cross-layer optimization protocols have been proposed with an objective to minimize energy consumption and/or to maximize network lifetime. It has been discussed in detail earlier. All these proposed approaches only consider few functionality of the network to improve energy conservation. Therefore, still we need to develop better cross-layer protocol for energy conservation in WSNs.
- *Mobility in CLD*: Almost all cross-layer protocol design did not consider the mobility of sensor nodes, which results in frequent topological changes. The list of neighbors of a sensor is continuously changing due to nodes' mobility that continuously creates and breaks links between nodes. Therefore, a new analytical model implemented with cross-layer design techniques is required to address the mobility of nodes.
- *Cross-Layer Network Simulator*: Presently all existing simulators such as OPNET (OPNET, 2000), NS-2 (NS-2, 1997), NS-3 (NS-3, 2006), J-Sim (J-Sim, 2001), GloMoSim (GloMoSim, 1998), and Aqua-Sim (Xie, et al., 2009) are discrete-event simulator and they have a layered architecture in core. Hence, these are not be suitable for implementation and testing cross-layer solutions. For this reason, a new software solution is needed to provide this new development paradigm so as to reduce the deployment complexity of cross-layer protocol and algorithm.
- *Co-existence of Cross-Layer Approaches*: The question of co-existence of the CLD approaches has come into picture as there are different types of application requirements that lead to interaction and communication between layers. For example, suppose in one requirement, the MAC layer

adjusts the data rate depending on channel condition. If this protocol is deployed first, then the question arises whether other scheme, that either having cross-layer coupling or not having any coupling, can be deployed too later time. Kawadia and Kumar (2005) stressed on finding cross-layer design proposals that may or may not be employed together.

- *Most Important Cross-Layer Coupling*: An exhaustive amount of work has been done on CLD techniques. Still it is not clear which design techniques have most significant impact on WSNs. To identify the most effective one, a thorough cost-benefit analysis of CLD techniques is needed. Cost-benefit analysis should consider the implementation complexity, performance enhancement, resource utilization, and data security.
- *Standardization of Interfaces*: In layered design, one advantage is having a well defined strict boundary between layers that helps to standardize the functionality of each layer. However, in CLD, this layered architecture has been violated in different ways. Therefore, challenges appear for the research community to resolve the boundary between layers (Srivastava & Motani, 2005). It also creates a question whether the CLD implies that we have to dispose of the OSI reference model or not. Furthermore, we need to find an answer, whether a completely new abstraction, as shown in Figure 3d, is necessary or still can have layered architecture along with new interfaces, as shown in Figure 3a, b, and c. If we need new interfaces, then what information should be exchanged between layers and whether we can standardize them or not. All these questions still remain open to researchers.
- *Heterogeneous Networks*: Another open issue regarding the CLD is to decide whether we should have different methodologies for different networks such as WSN, Ad-Hoc Network, Cellular Network, Wireless LAN, and MANET, or if a unified methodology is sufficient enough for any kind of network.
- *WSN Security*: Many wireless industries are trying to incorporate some level of CLD for power saving and solving the jamming problems (Smith, et al., 2009). Very little work has been done on designing security system to prevent different types of attack possible in WSNs. Hence, there is a scope to design better security system using cross-layer design.
- *Testbeds*: All the proposed cross-layer protocols have been tested in a network simulator and their advantages are analyzed theoretically. These protocols have generally not been implemented in practical scenario or at least on testbeds. Therefore, it is an open challenge to produce the same result in testbeds also.
- *Communication Model*: Last but not the least is what should be the right communication model for wireless communication in cross-layered networks, whereas, in wired network point-to-point communication model is followed in general (Srivastava & Motani, 2005).

SUMMARY

In this chapter, we have reviewed and classified the cross-layer protocols based on their implementation in WSN. We started with the explanation and limitations of layered design approach, and then it is shown that the CLD approaches have some advantages over layered approaches. We reviewed the techniques of CLD approaches that have been used in designing cross-layer protocols. Then, all cross-layer protocols were classified from their application point of view. Furthermore, we discussed about the precautionary guidelines of cross-layer design in WSNs. Finally, we conclude the chapter showing some open issues present in the development of cross-layer protocols in WSNs.

ACKNOWLEDGMENT

This work has been partially supported by a grant from the Department of Information Technology, Government of India, Grant No. 13(10)/2009-CC-BT, which the authors gratefully acknowledge.

REFERENCES

Akyildiz, I. F., Su, W., Sankarasubramaniam, Y., & Cayirci, E. (2002a). A survey on sensor networks. *IEEE Communications Magazine*, *40*(8), 102–114. doi:10.1109/MCOM.2002.1024422

Akyildiz, I. F., Su, W., Sankarasubramaniam, Y., & Cayirci, E. (2002b). Wireless sensor networks: A survey. *Computer Networks*, *38*(4), 393–422. doi:10.1016/S1389-1286(01)00302-4

Akyildiz, I. F., Vuran, M. C., & Akan, O. B. (2006). A cross-layer protocol for wireless sensor networks. In *Proceedings of the 40th Annual IEEE Conference on Information Sciences and Systems,* (pp. 1102 - 1107). Princeton, NJ: IEEE Press.

Anastasi, G., Conti, M., Francesco, M. D., & Passarella, A. (2009). Energy conservation in wireless sensor networks: A survey. *Ad Hoc Networks*, *7*(3), 537–568. doi:10.1016/j.adhoc.2008.06.003

Arienzo, L. (2010). An information-theoretic approach for energy-efficient collaborative tracking in wireless sensor networks. *EURASIP Journal on Wireless Communications and Networking.* Retrieved from http://downloads.hindawi.com/journals/wcn/2010/641632.pdf.

Arulampalam, M. S., Maskell, S., Gordon, N., & Clapp, T. (2002). A tutorial on particle filters for online nonlinear/non-gaussian bayesian tracking. *IEEE Transactions on Signal Processing*, *50*(2), 174–188. doi:10.1109/78.978374

Astrom, K. J., & Wittenmark, B. (1995). *Adaptive control*. Boston, MA: Addison-Wesley Publishing Company.

Aune, F. (2004). *Cross-layer design tutorial.* Technical Report. Retrieved from http://www.iet.ntnu.no/projects/cuban/archive/1812041.pdf.

Bai, Y., Liu, S., Sha, M., Lu, Y., & Xu, C. (2008). An energy optimization protocol based on cross-layer for wireless sensor networks. *The Journal of Communication*, *3*(6), 27–34.

Bhatti, S., & Xu, J. (2009). Survey of target tracking protocols using wireless sensor network. In *Proceedings of the 5th International Conference on Wireless and Mobile Communications,* (pp. 110-115). Cannes, France: IEEE.

Burks, A. W., Goldstine, H. H., & Neumann, J. V. (1963). *John von Neumann collected works* (Taub, A., Ed.). New York, NY: Pergamon Press.

Chakravarthi, R., Gomathy, C., Sebastian, S. K., Pushparaj, K., & Mon, V. B. (2010). A survey on congestion control in wireless sensor networks. *International Journal of Computer Science & Communication*, *1*(1), 161–164.

Chang, J. H., & Tassiulas, L. (2000). Energy conserving routing in wireless ad hoc networks. In *Proceedings of the IEEE Infocom,* (pp. 22-31). IEEE Press.

Chen, S., & Zhang, Z. (2006). Localized algorithm for aggregate fairness in wireless sensor networks. In *Proceedings of the 12th Annual International Conference on Mobile Computing and Networking,* (pp. 274-285). Los Angeles, CA: IEEE Press.

Chiang, J. T., & Hu, Y. C. (2011). Cross-layer jamming detection and mitigation in wireless broadcast networks. *IEEE/ACM Transactions on Networking, 19*(1), 286–298. doi:10.1109/TNET.2010.2068576

Chilamkurti, N., Zeadally, S., Vasilakos, A., & Sharma, V. (2009). Cross-layer support for energy efficient routing in wireless sensor networks. *Journal of Sensors*. Retrieved from downloads.hindawi.com/journals/js/2009/134165.pdf.

Cui, S., Madan, R., Goldsmith, A., & Lall, S. (2005). Joint routing, MAC, and link layer optimization in sensor networks with energy constraints. In *Proceedings of the IEEE International Conference on Communications,* (Vol. 2), (pp. 725-729). IEEE Press.

Dam, T. V., & Langendoen, K. (2003). An adaptive energy-efficient MAC protocol for wireless sensor networks. In *Proceedings of the International Conference on Embedded Networked Sensor Systems,* (pp. 171-180). Los Angeles, CA: IEEE Press.

Ding, M., & Cheng, X. (2009). Fault tolerant target tracking in sensor networks. In *Proceedings of the Tenth ACM International Symposium on Mobile Ad Hoc Networking and Computing,* (pp. 125-134). New Orleans, LA: ACM Press.

GloMoSim. (1998). *Glomosim: A scalable network simulation environment.* Retrieved from http://pcl.cs.ucla.edu/projects/glomosim/.

Gunasekaran, R., & Qi, H. (2008). Xlrp: Cross layer routing protocol for wireless sensor networks. In *Proceedings of the IEEE Wireless Communications and Networking Conference,* (pp. 2135-2140). Las Vegas, NV: IEEE Press.

Gungor, V. C., Vuran, M. C., & Akan, O. B. (2007). On the cross-layer interactions between congestion and contention in wireless sensor and actor networks. *Ad Hoc Networks, 5*(6), 897–909. doi:10.1016/j.adhoc.2007.02.007

Hoesel, L. V., Nieberg, T., Wu, J., & Havinga, P. J. M. (2004). Prolonging the lifetime of wireless sensor networks by cross-layer interaction. *IEEE Wireless Communications, 11*(6), 78–86. doi:10.1109/MWC.2004.1368900

Hsu, Y. P., & Feng, K. T. (2008). Cross-layer routing for congestion control in wireless sensor networks. In *Proceedings of the IEEE Radio and Wireless Symposium,* (pp. 783-786). Orlando, FL: IEEE Press.

IEEE802. 11. (1997). *IEEE standard 802.11: Wireless LAN medium access control (MAC) and physical layer (PHY) specifications* New York, NY: IEEE Computer Society.

J-Sim. (2001). *J-sim: An object-oriented library for discrete-time process-oriented simulation.* Retrieved from http://www.j-sim.zcu.cz/.

Jain, K., Padhye, J., Padmanabhan, V., & Qiu, L. (2003). Impact of interference on multi-hop wireless network performance. In *Proceedings of the ACM Mobicom.* San Diego, CA: ACM Press.

Jambli, M. N., & Tully, A. (2007). Cross-layer design for information dissemination in wireless sensor networks: State-of-the-art and research challenges. In *Proceedings of the 37th Annual IEEE/IFIP International Conference on Dependable Systems and Networks,* (pp. 13-18). Edinburgh, UK: IEEE Press.

Kalita, H. K., & Kar, A. (2009). Wireless sensor network security analysis. *International Journal of Next-Generation Networks, 1*(1), 1–10.

Karl, H., & Willig, A. (2003). *A short survey of wireless sensor networks. TKN Technical Report.* Berlin, Germany: Technical University Berlin.

Kawadia, V., & Kumar, P. R. (2005). A cautionary perspective on cross-layer design. *IEEE Wireless Communications, 12*(1), 3–11. doi:10.1109/MWC.2005.1404568

Kim, J., & Park, K. H. (2009). An energy-efficient, transport-controlled MAC protocol for wireless sensor networks. *Computer Networks*, *53*(11), 1879–1902. doi:10.1016/j.comnet.2009.03.002

Kumar, P. R. (1985). A survey of some results in stochastic adaptive control. *Society for Industrial and Applied Mathematics Journal on Control and Optimization*, *23*(3), 329–380.

Kwon, H., Kim, T. H., Choi, S., & Lee, B. G. (2006). A cross-layer strategy for energy-efficient reliable delivery in wireless sensor networks. *IEEE Transactions on Wireless Communications*, *5*(12), 3689–3699. doi:10.1109/TWC.2006.256992

Lachenmann, A., Marron, P. J., Minder, D., & Rothermel, K. (2005). An analysis of cross-layer interactions in sensor network applications. In *Proceedings of the International Conference on Intelligent Sensors, Sensor Networks and Information Processing,* (pp. 121-126). Melbourne, Australia: IEEE Press.

Lazos, L., & Poovendran, R. (2004a). Cross-layer design for energy-efficient secure multicast communications in ad hoc networks. In *Proceedings of the IEEE International Conference on Communications,* (Vol. 6), (pp. 3633-3639). Paris, France: IEEE Press.

Lazos, L., Salido, J., & Poovendran, R. (2004b). Vp3: Using vertex path and power proximity for energy efficient key distribution. In *Proceedings of the IEEE 60th Vehicular Technology Conference,* (Vol. 2), (pp. 1228-1232). Los Angeles, CA: IEEE Press.

Li, J., & Zhou, Y. (2010). *Wireless sensor networks: Application-centric design*. Retrieved from http://www.intechopen.com/articles/show/title/target-tracking-in-wireless-sensor-network.

Li, X., Shu, W., Li, M., Huang, H. Y., Luo, P. E., & Wu, M. Y. (2009). Performance evaluation of vehicle-based mobile sensor networks for traffic monitoring. *IEEE Transactions on Vehicular Technology*, *58*(4), 1647–1653. doi:10.1109/TVT.2008.2005775

Liu, H., Ma, H., Zarki, M. E., & Gupta, S. (1997). Error control schemes for networks: An overview. *Mobile Networks and Applications*, *2*(2), 167–182. doi:10.1023/A:1013676531988

Liu, Q., Zhou, S., & Giannakis, G. B. (2005). Cross-layer scheduling with prescribed QoS guarantees in adaptive wireless networks. *IEEE Journal on Selected Areas in Communications*, *23*(5), 1056–1066. doi:10.1109/JSAC.2005.845430

Madan, R., Cui, S., Lall, S., & Goldsmith, A. (2006). Layer design for lifetime maximization in interference-limited wireless sensor networks. *IEEE Transactions on Wireless Communications*, *5*(11), 3142–3152. doi:10.1109/TWC.2006.04770

Mainwaring, A., Polastre, J., Szewczyk, R., Culler, D., & Anderson, J. (2002). Wireless sensor networks for habitat monitoring. In *Proceedings of the 1st ACM International Workshop on Wireless Sensor Networks and Applications,* (pp. 88-97). New York, NY: ACM Press.

Melodia, T., Pompili, D., & Akyildiz, I. F. (2005). On the interdependence of distributed topology control and geographical routing in ad hoc and sensor networks. *Journal of Selected Areas in Communications*, *23*(3), 520–532. doi:10.1109/JSAC.2004.842557

Melodia, T., Vuran, M. C., & Pompili, D. (2006). *Wireless system/network architect*. Berlin, Germany: Springer-Verlag.

Mendes, L. D. P., & Rodrigues, J. J. P. C. (2011). A survey on cross-layer solutions for wireless sensor networks. *Journal of Network and Computer Applications*, *34*(2), 523–534. doi:10.1016/j.jnca.2010.11.009

Mpitziopoulos, A., Gavalas, D., Konstantopoulos, C., & Pantziou, G. (2009). A survey on jamming attacks and countermeasures in WSNS. *IEEE Communications Surveys & Tutorials*, *11*(4), 42–56. doi:10.1109/SURV.2009.090404

NS-2. (1997). *Ns-2: A discrete event network simulator.* Retrieved from http://www.isi.edu/nsnam/ns/.

NS-3. (2006). *Ns-3: A discrete-event network simulator for internet systems.* Retrieved from http://www.nsnam.org/.

OPNET. (2000). *OPNET modeler: A software tool for network modeling and simulation.* Retrieved from http://www.opnet.com/.

Padmavathi, G., & Shanmugapriya, D. (2009). A survey of attacks, security mechanisms, and challenges in wireless sensor networks. *International Journal of Computer Science and Information Security*, *4*(1-2), 117–125.

Pelechrinis, K., Iliofotou, M., & Krishnamurthy, S. V. (2011). Denial of service attacks in wireless networks: The case of jammers. *IEEE Communications Surveys & Tutorials*, *13*(2), 245–257. doi:10.1109/SURV.2011.041110.00022

Perkins, C., Belding-Royer, E., & Das, S. (2003). *Ad hoc on-demand distance vector routing.* Internet Engineering Task Force, RFC 3561. Retrieved from http://www.ietf.org.

Pompili, D., Vuran, M. C., & Melodia, T. (2006). *Sensor network and configuration: Fundamentals, standards, platforms, and applications*. Berlin, Germany: Springer-Verlag.

Pudashine, K. (2009). Energy efficient cross layer routing for life time maximization in wireless sensor networks. In *Proceedings of the first Asian Himalayas International Conference on Internet*, (pp. 1-3.). Kathmandu, Nepal: IEEE.

Raghunathan, V., Schurghers, C., Park, S., & Srivastava, M. (2002). Energy-aware wireless microsensor networks. *IEEE Signal Processing Magazine*, *19*(2), 40–50. doi:10.1109/79.985679

Rahman, M. A., Saddik, A. E., & Gueaieb, W. (2008). *Sensors*. Berlin, Germany: Springer-Verlag.

Raisinghania, V. T., & Iyer, S. (2004). Cross-layer design optimizations in wireless protocol stacks. *Computer Communications*, *27*(8), 720–724. doi:10.1016/j.comcom.2003.10.011

Safwat, A., Hassanein, H., & Mouftah, H. (2003). Optimal cross-layer designs for energy-efficient wireless ad hoc and sensor networks. In *Proceedings of the 22nd IEEE International Performance, Computing, and Communications Conference,* (pp. 123-128). Phoenix, AZ: IEEE Press.

Sen, J. (2009). A survey on wireless sensor network security. *International Journal of Communication Networks and Information Security*, *1*(2), 59–82.

Shannon, C., & Weaver, W. (1949). *The mathematical theory of communication*. Urbana, IL: University of Illinois Press.

Smith, N., Reddy, Y. B., & Gajendar, N. (2009). Cross-layer design approach for wireless networks to improve the performance. In *Proceedings of the Sixth International Conference on Information Technology: New Generations,* (pp. 552-557). Las Vegas, NV: IEEE.

Song, L., & Hatzinakos, D. (2007). A cross-layer architecture of wireless sensor networks for target tracking. *IEEE/ACM Transactions on Networking*, *15*(1), 145–158. doi:10.1109/TNET.2006.890084

Srivastava, V., & Motani, M. (2005). Cross-layer design: A survey and the road ahead. *IEEE Communications Magazine*, *43*(12), 112–119. doi:10.1109/MCOM.2005.1561928

Suh, C., Ko, Y. B., & Son, D. M. (2006). An energy efficient cross-layer MAC protocol for wireless sensor networks. In *Proceedings of the Eighth Asia Pacific Web Conference,* (pp. 410-419). Harbin, China: IEEE.

Thamilarasu, G., Balasubramanian, A., Mishra, S., & Sridhar, R. (2005). A cross-layer based intrusion detection approach for wireless ad hoc networks. In *Proceedings of the IEEE International Conference on Mobile Ad Hoc and Sensor Systems.* (pp. 855-861). Washington, DC: IEEE Press.

Valera, M., & Velastin, S. A. (2005). Intelligent distributed surveillance systems: A review. In *Proceedings of the IEEE Visual Image Signal Processing,* (Vol. 152), (pp. 192-204). IEEE Press.

Vuran, M. C., & Akyildiz, I. F. (2006). Cross-layer analysis of error control in wireless sensor networks. In *Proceedings of the 3rd Annual IEEE Communications Society on Sensor and Ad Hoc Communications and Networks,* (pp. 585-594). Reston, VA: IEEE Press.

Vuran, M. C., & Akyildiz, I. F. (2009). Error control in wireless sensor networks: A cross layer analysis. *IEEE/ACM Transactions on Networking, 17*(4), 1186–1199. doi:10.1109/TNET.2008.2009971

Vuran, M. C., & Akyildiz, I. F. (2010). XLP: A cross-layer protocol for efficient communication in wireless sensor networks. *IEEE Transactions on Mobile Computing, 9*(11), 1578–1591. doi:10.1109/TMC.2010.125

Vuran, M. C., Gungor, V. C., & Akan, O. B. (2005). On the interdependence of congestion and contention in wireless sensor networks. In *Proceedings of the Third International Workshop on Measurement, Modeling, and Performance Analysis of Wireless Sensor Networks.* San Diego, CA: IEEE.

Walters, J. P., Liang, Z., Shi, W., & Chaudhary, V. (2006). Security in distributed, grid, and pervasive computing. In Xiao, Y. (Ed.), *Wireless Sensor Network Security: A Survey.* Boca Raton, FL: CRC Press.

Wan, J., Xu, X., Feng, R., & Wu, Y. (2009). Cross-layer active predictive congestion control protocol for wireless sensor networks. *Sensors (Basel, Switzerland), 9*(10), 8278–8310. doi:10.3390/s91008278

Wang, C., Li, B., Sohraby, K., Daneshmand, M., & Hu, Y. (2007). Upstream congestion control in wireless sensor networks through cross-layer optimization. *IEEE Journal on Selected Areas in Communications, 25*(4), 786–795. doi:10.1109/JSAC.2007.070514

Wang, Q., Chen, W. P., Zheng, R., Lee, K., & Sha, L. (2003). Acoustic target tracking using tiny wireless sensor devices. In *Proceedings of the 2nd International Conference on Information Processing in Sensor Network,* (pp. 642-657). IEEE.

Wang, Y., Attebury, G., & Ramamurthy, B. (2006). A survey of security issues in wireless sensor networks. *IEEE Communications Surveys & Tutorials, 8*(2), 2–23. doi:10.1109/COMST.2006.315852

Want, R., Farkas, K. I., & Narayanaswami, C. (2005). Guest editors' introduction: Energy harvesting and conservation. *IEEE Pervasive Computing / IEEE Computer Society and IEEE Communications Society, 4*(1), 14–17. doi:10.1109/MPRV.2005.12

Xiao, M., Wang, X., & Yang, G. (2006). Cross-layer design for the security of wireless sensor networks. In *Proceedings of the Sixth World Congress on Intelligent Control and Automation,* (pp. 104-108). Dalian, China: IEEE.

Xie, P., Zhou, Z., Peng, Z., Yan, H., Hu, T., Cui, J.-H., et al. (2009). Aqua-sim: A ns-2 based simulator for underwater sensor networks. In *Proceedings of the IEEE/MTS Oceans*. IEEE Press.

Xing, G., Lu, C., Zhang, Y., Huang, Q., & Pless, R. (2005). Minimun power configration in wireless sensor networks. In *Proceeding of the 6th ACM International Symposium on Mobile Ad Hoc Networking and Computing*. Urbana-Champaign, IL: ACM Press.

Ye, W., Heidemann, J., & Estrin, D. (2002). An energy-efficient MAC protocol for wireless sensor networks. In *Proceeding of the Twenty-First Annual Joint Conference of the IEEE Computer and Communications Societies,* (Vol. 3), (pp. 1567-1576). IEEE Press.

Ye, W., Heidemann, J., & Estrin, D. (2004). Medium access control with coordinated adaptive sleeping for wireless sensor networks. *IEEE/ACM Transactions on Networking, 12*(3), 493–506. doi:10.1109/TNET.2004.828953

Zhao, F., Shin, J., & Reich, J. (2002). Information-driven dynamic sensor collaboration for tracking applications. *IEEE Signal Processing Magazine, 19*(2), 61–72. doi:10.1109/79.985685

Zhao, J., Wang, L., Li, S., Liu, X., Yuan, Z., & Gao, Z. (2010a). A survey of congestion control mechanisms in wireless sensor networks. In *Proceedings of the Sixth International Conference on Intelligent Information Hiding and Multimedia Signal Processing,* (pp. 719-722) Darmstadt, Germany: IEEE.

Zhao, T., Guo, T. D., Yang, W. G., & Chen, G. (2010b). An optimal cross-layer model for maximum lifetime routing in wireless sensor networks. In *Proceedings of the International Conference on Communications and Mobile Computing,* (Vol. 3), (pp. 244-248). Shenzhen, China: IEEE.

Zia, T., & Zomaya, A. (2006). Security issues in wireless sensor networks. In *Proceedings of the International Conference on Systems and Networks Communications,* (pp. 40-43). Tahiti, French Polynesia: IEEE.

Chapter 5
Using Cross-Layer Techniques for ECG Transmissions in Body Area Sensor Networks

Tao Ma
University of Nebraska – Lincoln, USA

Hamid Sharif
University of Nebraska – Lincoln, USA

Michael Hempel
University of Nebraska – Lincoln, USA

Fahimeh Rezaei
University of Nebraska – Lincoln, USA

Dongming Peng
University of Nebraska – Lincoln, USA

Pradhumna L Shrestha
University of Nebraska – Lincoln, USA

Hsiao-Hwa Chen
National Cheng Kung University, Taiwan

ABSTRACT

Body Area Sensor Networks (BASN) with miniature sensors providing wireless communications capabilities have become a promising tool for monitoring and logging vital parameters of patients suffering from chronic diseases such as diabetes, asthma, and heart diseases. Particularly, as the technical development in BASN, a low-cost, high-quality, convenient Electrocardiographic (ECG) diagnosis system becomes a future major tool for healthcare systems. However, numerous important research issues remain to be addressed in BASN for ECG transmissions. Among these, communication energy efficiency and security are the most concerning issues. In this chapter, the authors introduce, survey, and analyze effective cross-layer strategies for wireless ECG transmissions in BASN. The key idea of these cross-layer communication techniques is to take advantage of both source data properties and communication strategies for the optimization of the system energy efficiency while providing secure wireless ECG transmissions. The goal of improving communication energy efficiency is achieved by matching the source coding of ECG signals with the channel coding strategy. In addition one can leverage biometric ECG properties to implement an energy-efficient cross-layer security strategy. As an example the authors showcase two security methods in this chapter—selective encryption and self-authentication. Thanks to the dependency property of the compressed ECG data, a selective encryption algorithm needs only to be applied on a

DOI: 10.4018/978-1-4666-0960-0.ch005

very small portion of the transmitted data, and at the same time it provides a level of security equivalent to traditional full-scale encryption using block or stream ciphers without the burden of the associated energy and computational expense. In the example authentication scheme, sensors for the same body can authenticate each other by common traits such as inter-pulse intervals. The authors analyzed the proposed cross-layer techniques for ECG transmissions and validated the achieved energy efficiency improvements by both simulation and experimental results.

INTRODUCTION

Recent advances in wireless Body Area Sensor Networks (BASN) make feasible a wireless and mobile diagnosis system for delivering medical and healthcare services. In such systems, wearable or implantable medical sensors, around or inside the human body, form a wirelessly interconnected network. This system is then used to collect an individual's medical information ubiquitously and to perform critical and complicated tasks collaboratively. Communications among sensor nodes must be energy-efficient, and reliably protected against errors from wirelessly transmitted packets, as well as malicious or fraudulent actions.

Particularly, with the technical development in BASN, a low-cost, high-quality, convenient Electrocardiographic (ECG) diagnosis system becomes a future major tool for healthcare systems. In this chapter, an energy-efficient and secure scheme for ECG transmissions in BASN is proposed. Differently from previous research in the literature, in this technique, the diverse properties of compressed ECG data are extensively explored and exploited at the application layer before data transmission. The transmission schemes crossing the link and MAC layers are then adapted to diverse ECG data properties in a way that both data security and quality as well as communication energy efficiency are ensured. Our study's results have demonstrated a considerable degree of performance improvement in transmission quality, security and energy consumption.

In this chapter, we address the primary concern for energy efficiency in ECG transmissions for BASN utilizing cross-layer techniques. Because BASNs are constrained by resource limitations such as battery power, computing and processing power, communication link vulnerability and bandwidth availability, there is a strong need for employing cross-layer techniques to tackle diverse issues to achieve a low-cost, reliable and secure wireless collection and monitoring system for BASN applications.

To showcase the advantages of cross-layer designs we leverage our approach to implement a novel approach to secure transmissions for ECG-based BASN; using source authentication and data encryption. Based on this approach, we will investigate and summarize the common cross-layer methodologies being used in ECG systems using wireless BASNs.

BACKGROUND

Heart disease, stroke, cancer, chronic respiratory diseases, and diabetes are the major causes of mortality in the world, representing 60% of all deaths. Since chronic diseases often have asymptomatic or intermittent properties, long-term continuous health monitoring is essential in detecting and treating such diseases. For example, long-term monitoring of ECG data helps detecting these diseases at an early stage and improves the chance of successful treatments. For preventing strokes, real time ECG information monitoring is essential for heart rate control and prompt re-treatment. Long-term monitoring requires the transmission of a large amount of data in a reliable, convenient, energy-efficient and secure manner. Thus, it generates a huge challenge for today's technology.

Recently, with advances in wireless networks and sensor miniaturization, BASN becomes a promising solution to help meet this demand. A BASN consists of multiple wearable sensor nodes. These nodes are able to sample, process, and communicate vital signals of information such as heart rate, blood pressure, oxygen saturation. These sensors are usually so small in size that they can be placed on a human body as tiny patches or hidden in users' clothes. There have been many recent efforts focused on wearable systems for health monitoring and ECG data collection, which are briefly summarized in the following paragraphs.

MIThril, in coordination with MIT Media Lab researchers, has been developing and prototyping new techniques of human-computer interaction for body-worn applications (Rich DeVaul, 2003). Their wearable sensors include ECG, skin temperature, galvanic skin response sensors, tiny cameras, and microphones. These tiny sensors are able to sense and communicate with low power consumption. Zhang et al. (2010) developed an energy-efficient application-specific integrated circuit for related wearable applications, which occupies a die area of $5mm^2$, and consumes less than 10 nA of current in passive standby mode, and less than 10 μA in active standby mode. Lin, Sinclair, and Bittner (2006) proposed a BASN platform which is based on the Bluetooth protocol, providing services through Application-Programming Interfaces (APIs) running on the patient's phone with relatively low cost and high energy efficiency. They also introduced a wrist-worn device, called *Cache-watch,* into the BASN. As a result of all these efforts on miniature sensor design and hardware platform development, BASN is becoming a more realistic and promising future tool for wireless medical monitoring systems.

However, the current research still far lags behind the requirements for long-term medical applications. The primary concern is how to achieve energy efficiency when BASNs are required to be secure and robust.

An energy efficient and reliable transmission scheme is essential for ECG transmissions in BASN. Typically, medical data is vital information which requires high quality and reliability during transmission. However, packet loss is unavoidable in error-prone wireless channels. Moreover, for long-term monitoring applications, the volume of data is often accumulated to a very large size. For ECG transmissions in BASN with a sampling rate of 360 Hz and 11 bits/sample data resolution, a 24-hour record requires about 43Mbytes per ECG channel (Nave & Cohen, 1993). Therefore, an energy efficient transmission is very important.

Recently, some energy efficient communication schemes for BASN were proposed in the literature. Kim, Kim, and Yoo (2008) proposed a quadratic level ECG compression algorithm. They claimed that the ECG encoding delay and hardware cost can be reduced by their algorithm. The main idea of this algorithm is to assign different compression ratios to different types of data blocks. Blocks are strategically categorized according to different importance levels for ECG data reconstruction. Therefore, a high compression ratio can be guaranteed while the reconstructed signal quality is maintained. Another energy efficient approach for ECG transmission was proposed by Boskovic and Despotovic (2005). In their study, a predictive coding model is carefully designed. By using this model, a large amount of transmission data can be saved. Wang et al. (2010) presented a similar approach. They also divided the raw ECG data into different important and non-important portions, and assigned different communication resources to them in order to reduce energy consumption without compromising the transmission quality. In a recent study by our team (Ma, Hempel, Peng, & Sharif, 2010), the unequal importance levels of compressed ECG were explored, and a rate-switching algorithm was presented to adaptively assign different coding rates to different parts of the compressed ECG data in order to optimize the transmission quality in BASN. To sum up, the main idea of these proposed schemes was to adjust the channel coding strategy to match the source-coded ECG data. These schemes are good examples of how cross-layer techniques can be applied to BASNs to address a variety of problems.

Another area where cross-layer schemes are essential is BASN transmission security. Usually, medical signals in BASN carry private health information about the patients. They must be protected against unauthorized use for personal advantage or fraudulent actions that might be hazardous to a user's life. Security methods in BASN have also been addressed by many researchers. Sufi, Mahmoud, and Khalil (2008) presented an ECG encryption algorithm. They claim that encrypting only a segment (25%) of ECG data is enough for ensuring the security of the entire ECG data. In Poon, Zhang, and Bao's (2006) study, a biometric trait generated from a sequence of Inter-Pulse Intervals (IPI) has a high level of randomness which was used as a trust trait in their authentication method. Based on such trust traits, a unique authentication method was proposed. However, the research on BASN security is still in its infant stage.

In this chapter, we introduce and summarize a cross-layer technique to achieve improved energy efficiency even in the presence of our novel security mechanisms for transmissions in ECG BASNs. For this technique we investigate ECG data compression, and explore its unequal importance properties. Furthermore, we present our cross-layer techniques of Two-Rate Unequal Error Protection and Selective Encryption. By applying communication strategies to compressed medical data rather than directly to raw ECG data as in the aforementioned papers, significantly more energy can be saved, and more efficient encryption can be achieved. In addition, biometric authentication utilizing common trust traits is presented.

TWO-RATE UNEQUAL PROTECTION FOR ECG

In this technique, both source coding strength and channel coding strength are combined to provide energy efficient wireless ECG transmissions for BASN. Firstly, we give the background knowledge of the ECG set partition in hierarchy tree (SPIHT) source coding and the Rate-Compatible Punctured Convolutional Codes (RCPC) algorithms. These two codes are the basis of the Two-Rate Unequal Error Protection (UEP) scheme we present. Secondly, we explain the Two-Rate UEP technique itself.

ECG SPIHT Coding Algorithms

ECG coding algorithms have been studied for more than 30 years. Many coding methods have been proposed. Typically, these methods can be categorized into direct methods and transform methods.

Direct methods are time-domain techniques, where the samples of the signal are directly handled to provide the coding result. Coding by time-domain methods is based on the idea of extracting a subset of significant signal samples to represent the signal. Examples of the methods belonging to this group are FAN (Dipersio, 1985), Coordinate Reduction Time Encoding System (CORTES) (Tompkins, 1982), Amplitude Zone Time Epoch Coding (AZTEC) (Cox, Nolle, Fozzard, & Oliver, 1968), and Turning Point algorithms (Mueller, 1978).

The second group contains the transform-based methods, particularly utilizing wavelet transform. Wavelet-based methods have shown promising performance because of their good localization properties in the time and frequency domain. Therefore, wavelet transform methods have become the preferred choice for the study of ECG signal coding. Several wavelet-based coding algorithms have been proposed. Among these, the algorithm of Set Partitioning In Hierarchical

Table 1. ECG coding algorithm comparison for different compression ratios

Method	PRD (%)	CR	MIT-BIH Record
SPIHT (Lu, Kim & Pearlman, 2000)	5.0	21.6:1	119
ASEC (Zigel, Cohen, & Katz, Oct. 2000)	5.5	21.6:1	119

Method	PRD (%)	CR	MIT-BIH Record
SPIHT (Lu, Kim & Pearlman, 2000)	1.18	8:1	117
MSPIHT (Wang & Chen, 2008)	0.91	8:1	117
Hilton (Hilton, 1997)	2.6	8:1	117
Djohan (Djohan, Nguyen & Tompkins, 1995)	3.9	8:1	117

*CR: Compression Ratio, PRD: Percentage Root Mean Square Difference,
MIT-BIT Record: http://www.physionet.org/physiobank/database/mitdb/

Trees (SPIHT) has shown the best performance. Table 1 demonstrates a comparison among different ECG coding methods. In this chapter, we select SPIHT as the ECG coding method for the basis of our cross layer techniques.

The procedure of the SPIHT coding algorithm is illustrated in Figure 1. Firstly, ECG data is transformed from time domain into the wavelet domain. Here, the resulting coefficients have particular relationships. Because of inherent ECG data properties, the lower-frequency coefficients often have higher amplitude than the higher-frequency coefficients. In Figure 2, coefficient 2 is greater than coefficients 3 or 4. Similarly, coefficient 4 is typically greater than coefficients 7 or 8.

Based on this phenomenon, these wavelet coefficients can be encoded as follows. All coefficients are represented by their value and bit-map information in each bit plane. The threshold T is about half of the maximal value among all wavelet coefficients. In the first bit plane, all coefficients whose absolute values are greater than the threshold T are encoded. The outcome of the first bit-plane bit stream indicates all position information of these encoded coefficients and their sign bit information. Then in the second bit plane, the threshold is decreased to T/2. The coefficients whose values are greater than T/2 are encoded in this round. Their positions and sign bits are also recorded as in the first bit plane. In addition, the $\log_2(T / 2)$ bits of the coefficients in the first bit-plane are also recorded. The same process repeats until the expected compression ratio is reached.

Some terms used in the SPIHT algorithm are explained in the following.

Figure 1. SPIHT compression algorithm for ECG data

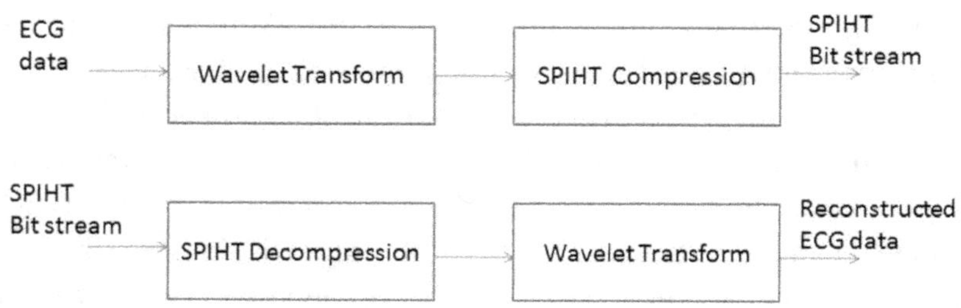

Figure 2. Wavelet coefficient relationship

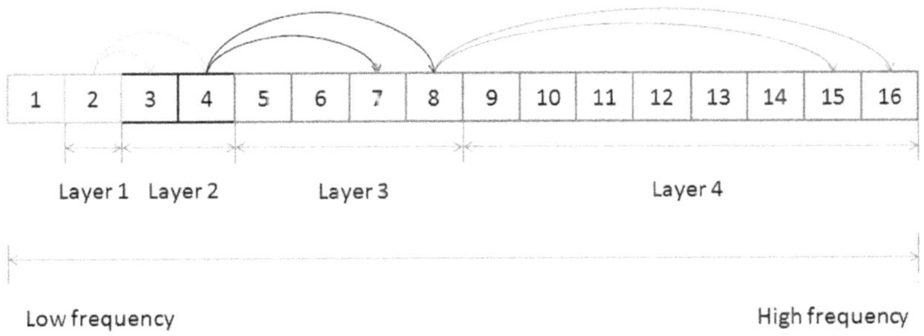

- LSP: List of Significant Pixel
- LIP: List of Insignificant Pixel
- LIS: List of Insignificant Set (two types)
  - Type 1. $D(i)$ set of coordinates of all descendants of node (i).
  - Type 2. $L(i)$ set of coordinates of all descendants of node (i), except offspring.

The detailed concept of SPIHT is as follows. For more details please refer to (Lu, Kim, & Pearlman, 2000).

INPUT: the ECG signal sample sequence $e(k)$.

OUTPUT: the compressed ECG binary stream $E(m)$.

1. *Initialization:* Perform 1-D 5-layer Discrete Wavelet Transform (DWT) on the input ECG sequence $e(k)$. Let the DWT result sequence be w. Set the list of significant points (LSP) as empty. Initialize the roots of similarity trees in the lists of insignificant points (LIP) and insignificant sets (LIS), i.e., copy the high-frequency wavelet coefficient domain at the first layer of w into LIP, and the high-frequency wavelet coefficient domains at other higher layers into LIS. Initialize n to be 10. Set the significance threshold to 2^n.
2. *Sorting Pass:* Search significant coefficients in LIP and LIS. If coefficients in LIP are found significant, they are queued in LSP. If coefficients in LIS are found significant, LIS is decomposed at lower layers until all new LISs are insignificant. Within LIS, if $D(i)$ has significant coefficients, its offspring is added to LSP or LIP depending on whether it is significant or not. At the same time, a new $L(i)$ is generated. If $L(i)$ has significant coefficients, it is decomposed into $D(2i)$, $D(2i\text{-}1)$. LIS decomposition goes on until no significance is found.
3. *Refinement Pass:* For each entry in the LSP, queue the nth most significant bit into the $E(m)$.
4. *Loop:* Decrement n by one and return to Step 2 until the specified bit rate is reached.

The process of the SPIHT algorithm is also visualized in the flowchart in Figure 3.

Rate Compatible Puncture Convolutional Code

We first introduce the puncture code by a simple example (Atkin & Tam, 1992). Consider two identical convolution codes with the same coding rate of 1/2 as shown in Figure 4. By puncturing one of the output bits, we can create 2/3 coding rate convolutional code. The mother code rate in this example is 1/2. On the receive side, dummy bits that do not affect the decoding metric are inserted in the appropriate places before decod-

Figure 3. Flowchart of the SPIHT algorithm

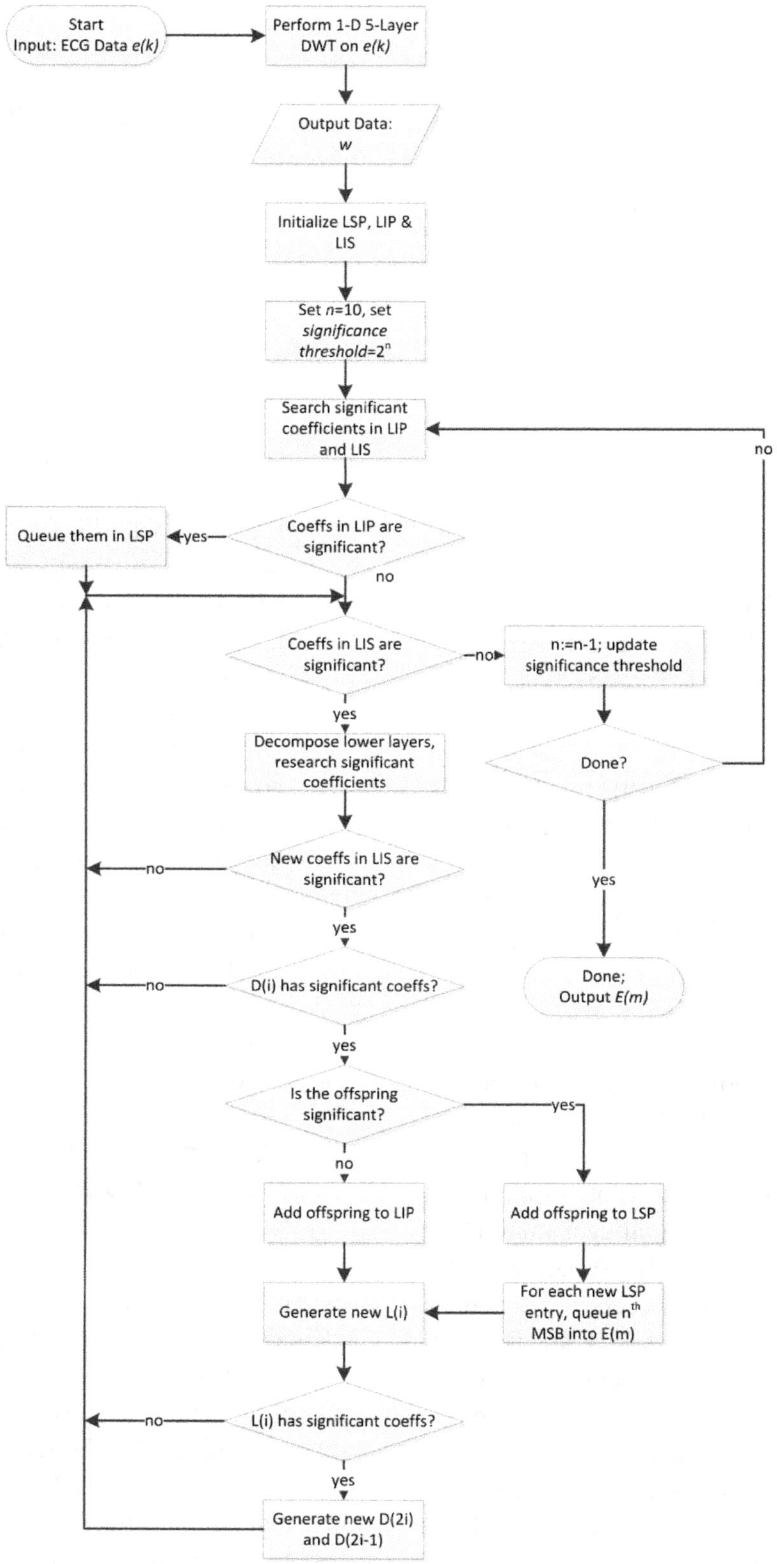

Figure 4. Puncture coding

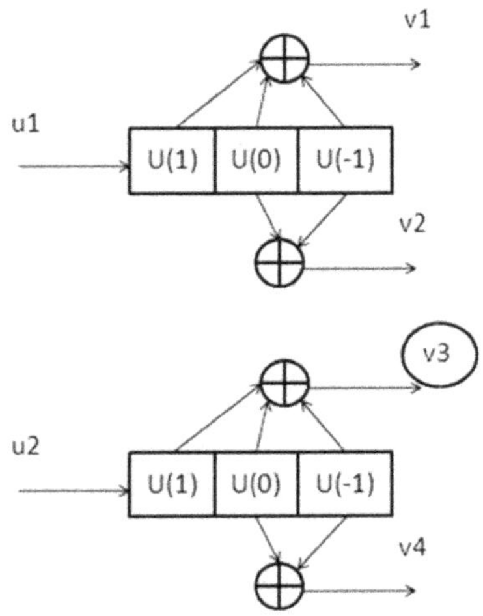

ing. In our cross layer technique, we use three convolutional codes with mother code rate of 1/3, with its specific configuration discussed in (Ma, Hempel, Peng, & Sharif, 2010).

Two-Rate Unequal Protection Technique

In ECG signals, there are several important features for cardiac disease diagnosis (Figure 5), one of which is that the QRS complex has sharp slopes. The center of its frequency is around 17Hz, and the frequency coefficients are distributed mostly between 1 and 40 Hz (Pan & Tompkins, 1985). The P wave normally appears before the QRS complex, and its frequency is usually below 10 Hz. The T wave appears after the QRS complex, and

Figure 5. ECG signal

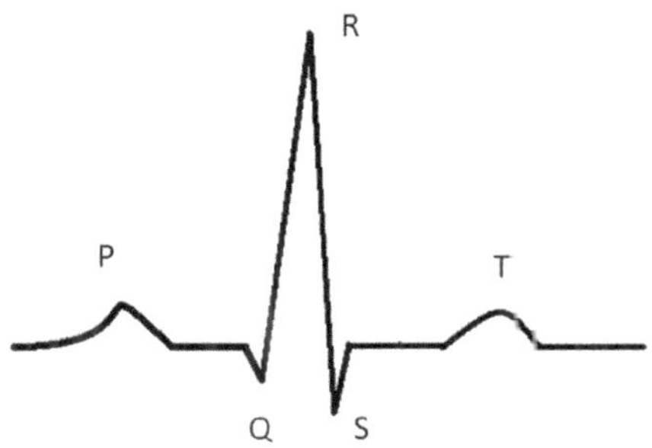

it can appear in various shapes (Khadra, 1997). Its frequency distribution is typically less than 6 Hz. S-T segments often occupy a low frequency range. Heuristically, the required wavelet decomposition level is dependent on the system's sample rate (Al-Fahoum & Howitt, 1999),

$$L = \log_2\left(f_s\right) - 2.96 \tag{1}$$

where f_s is the sample rate.

If the sample rate is 360Hz, the sufficient wavelet decomposition level is 5. The energy distribution of the wavelet coefficients in terms of frequency partitions is shown in Table 2 (Al-Fahoum, 2006). L5 is the approximation coefficients in level 5. H1, H2, H3, H4, H5 are the detail coefficients in Level 1, 2, 3, 4, and Level 5, respectively.

As a matter of fact, all feature points are below 40 Hz, which are mainly in L5, H5, H4, and H3. The number of coefficients in L5, H5, H4, and

Table 2. Wavelet coefficient distribution

Wavelet Frequency Bands	Ranges of Bands	Average Energy of Bands
L5	0 Hz- 5.625 Hz	6/27
H5	5.625 Hz- 11.25 Hz	9/27
H4	11.25 Hz- 22.5 Hz	7/27
H3	22.5 Hz- 45 Hz	3/27
H2	45 Hz- 90 Hz	1/27
H1	90 Hz- 180 Hz	1/27

H3 occupy only 1/4 of the total coefficients. This suggests that the wavelet-domain ECG data are unequally distributed. Another characteristic worth to note is the monotonically decreasing energy distribution of coefficients from H5 to H1. This phenomenon complies very well with the supposition of the SPIHT compression algorithm and explains why SPIHT compression works very well on ECG data.

In the following analysis, the Wavelet-Based Weighted Percent Root-Mean-Squared Difference (WWPRD) is utilized as the ECG quality measurement. It is defined as follows (Al-Fahoum, 2006):

$$WWPRD = \sum_{j=0}^{N_L} w_j \cdot WPRD_j \tag{2}$$

$N_L + 1$ is the number of subbands. For instance, five levels of wavelet decomposition result in six subbands L1, H1, H2, H3, H4, and H5. W_j is the normalized sum of wavelet coefficients in the jth subband, which is defined as (Al-Fahoum, 2006):

$$w_j = \frac{\sum_{i=1}^{n_j} \left|c_j(i)\right|}{\sum_{j=1}^{N_L} \sum_{i=1}^{n_j} \left|c_j(i)\right|} \tag{3}$$

where $c_j(i)$ is the coefficient i in subband j. The parameter n_j is the number of coefficients in subband j. $WPRD_j$ is the normalized wavelet percent root-mean-squared difference in subband j.

$$WPRD_j = \sqrt{\frac{\sum_{i=1}^{n_j} (c_j(i) - \widehat{c_j(i)})^2}{\sum_{i=1}^{n_j} c_j(i)^2}} \cdot 100\% \tag{4}$$

where $\widehat{c_j(i)}$ is the reconstructed wavelet coefficient.

In our simulation, the configuration is as follows: the ECG data sample frequency is 360Hz, the decomposition level is 5, and the wavelet basis is chosen as Daubechies 9, which has the best compression results. The tested ECG is Sample 105 in MIT-BIH Arrhythmia Database. As more bit planes are transmitted, the ECG transmission quality indicated by WWPRD is improved with a nonlinear slope. Figure 6 shows the reconstructed ECG data quality in terms of WWPRD with different numbers of received bit planes. As shown in the left graph of Figure 6 the first group of bit planes provides a higher contribution to the reconstruction quality than the rest of the bit planes. The right graph in Figure 6 shows the average bit quality in each bit plane, with the definition of average bit quality of bit plane as follows.

$$\lambda = \frac{\Delta\varepsilon}{L} \tag{5}$$

where $\Delta\varepsilon$ is the WWPRD decrement caused by decoding the current bit plane, and L is the length of this bit plane. Clearly, the average bit quality of the bit plane shows the unequal importance for each of the different bit planes.

By taking advantage of the unequal importance property of the compressed ECG data, we can therefore apply a Two-Rate unequal protection technique. In this scheme, the first m bit planes are channel-coded with a coding rate c_1, and the rest of the bit planes are channel-coded with a smaller code rate c_2, assigning fewer bits per coded value and thus less importance to c_2-coded bit-planes.

The channel coding is utilized as a concatenated FEC scheme. In this scheme, bit planes are first attached with an 8-bit Cyclic Redundancy Check (CRC) code. Secondly, a Rate-Compatible Punctured Convolutional code (RCPC) is applied.

The goal of the Two-Rate unequal protection is to maximize the average WWPRD quality within the total transmission bit budget (or the transmission energy budget).

Figure 6. Left figure shows ECG quality changes with increasing number of received bit-plane. Right figure shows average ECG quality per bit changes with increasing number of received bit-planes.

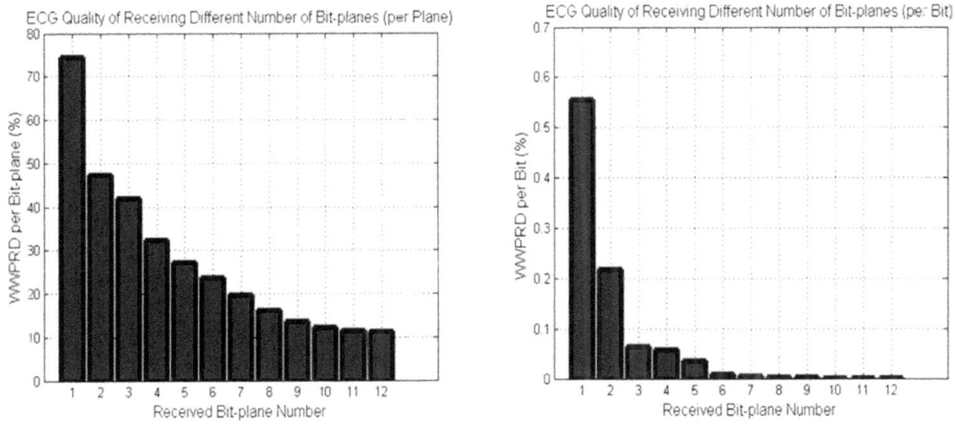

The average WWPRD quality ε is the sum of all possible WWPRD qualities weighted with their receiving rates. Let $WWPRD(k)$ be the WWPRD of the first *k* bit planes, and $P(k)$ denotes the possibility of the first *k* bit planes being received successfully. Then we can express the average WWPRD quality as follows.

$$\varepsilon = \sum_{k} WWPRD(k) \cdot P(k) \tag{6}$$

Furthermore, the communication energy consumption is dependent on the amount of bits to be sent with error protection. We can calculate the total number of bits to be sent as

$$n = \frac{1}{c_1}\sum_{k=1}^{m} B(k) + \frac{1}{c_2}\sum_{k=m}^{n} B(k) \tag{7}$$

where $B(k)$ is the number of bits in the kth bit plane. Therefore, the optimization problem can be formulated as finding optimal values of m, c_1 and c_2 in order to maximize the WWPRD quality with the constraints of communication energy and the transmission bit budget expressed in (7). To show how to solve this optimization problem, a typical ergodic method can be employed to search for the solution among a very limited discrete solution space.

Matlab simulation results are shown in Figure 7, where our proposed Two-Rate UEP is compared against the traditional Equal Error Protection (EEP) scheme. As demonstrated, the Two-Rate UEP achieves a significant improvement in ECG transmission quality with the same transmission energy cost in comparison with the EEP scheme. In other words, a large amount of communication energy is saved by the Two-Rate UEP while the same transmission quality is accomplished.

SELECTIVE ENCRYPTION IN BASN

The prospect of transferring and storing medical data and health information in electronic form raises concerns about patients' privacy. For instance, ECG signals contain sensitive physiological information about the patients (Muramatsu, 2006). Moreover, recent research indicated that even a single ECG signal without any patient's name can still release the person's identification information. Therefore, encrypting the ECG signal in order to protect the individual's privacy is indispensible. In contrast to traditional encryption algorithms, cross-layer encryption schemes in

Figure 7. Left figure shows comparison between WWPRD quality and overall coding rate. Right figure shows comparison between WWPRD quality and transmit energy consumption

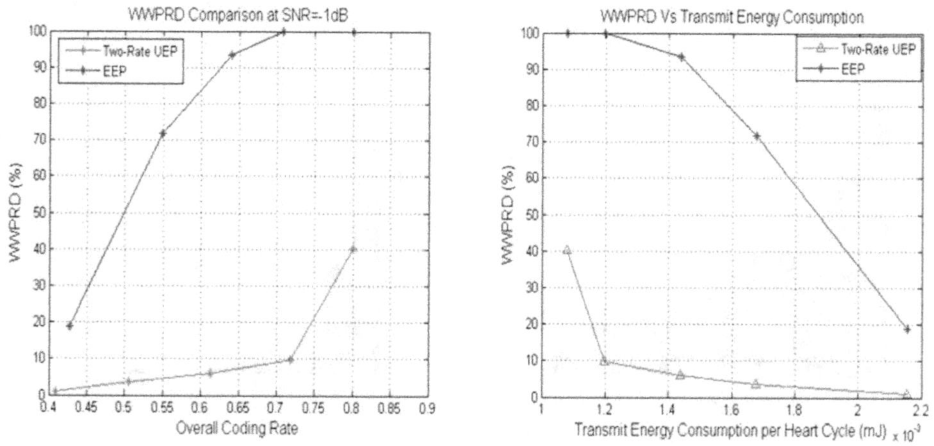

BASN can take into account the characteristics of the output source coding of ECG signals. By doing so, only a small portion of ECG data needs to be encrypted. Thus, the complexity and data overhead is greatly reduced without compromising the security level in handling and transmitting the ECG data. In the following section, selective encryption for ECG data is discussed as a representative case of cross-layer encryption in BASN.

We first analyze the dependency among the output bits of SPIHT-compressed ECG data. Secondly, we present the selective encryption scheme based on such a dependency analysis.

In the SPIHT-compression algorithm, the raw ECG signals are first transformed into the wavelet domain. Among the resulting wavelet coefficients, the low-frequency coefficients are statistically of higher magnitude than the high-frequency coefficients which are corresponding to the same time periods. This structure is represented by two trees, i.e., the List of Insignificant Points (LIP) and the List of Insignificant Sets (LIS). Bit-plane coding is then applied to these structured coefficients. In the bit-plane coding, the first threshold is set as $log_2(\text{maximal coefficients})$. A complete search is performed to find all coefficients whose values are greater than the threshold. These discovered coefficients belong to subset T_1. After the searching, all coefficients in T_1 are removed from the SPIHT structure which is represented by the deduction of LIP and LIS. In the first bit plane, the decoder compares the initial structure with the received bi-tree structure, consisting of the deducted LIS and LIP, to reconstruct the position information of the coefficients in subset T_1. Similarly in the second round, the threshold is reduced by half and another search is performed so that all coefficients which are greater than the current threshold are removed from the bi-tree structure. These coefficients belong to T_2. As a result, the decoder needs to combine the first bit-plane information $T-T_1$ and the second bit-plane information $(T-T_1-T_2)$ to get the positions of T_2 in the bi-tree structure. In the third bit-plane, the bi-tree structure $(T-T_1-T_2)$ continues to be searched. The coefficients T_3 are removed, and the bi-tree structure becomes $(T-T_1-T_2-T_3)$, etc. Similar procedures for further bit-plane coding continue until the quality requirements of ECG compression and transmission are satisfied.

As a result, this property of encoding ECG data leads to a strong dependency between the

Figure 8. WWPRD quality comparison VS overall coding rate

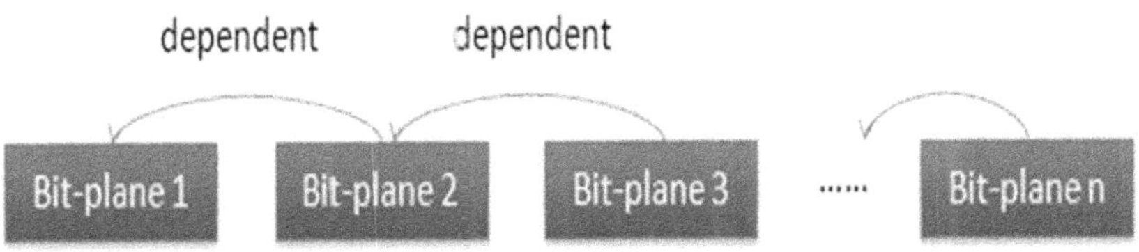

bit planes as shown in Figure 8. There is a chain effect among the bit planes. If one of the bit planes is missing during transmission or reconstruction, all other bit planes thereafter would be unreadable. So the first bit plane has the highest importance. If it contains any errors, an avalanche effect would cause all other segments to be incomprehensible to the decoder; and the entire message would become undecipherable.

By considering this property of SPIHT coding, we present an ECG selective encryption algorithm. Only the first bit plane is encrypted by using a standard encryption algorithm. The standard encryption algorithm can be either symmetric encryption (e.g., RC4), or public key encryption (e.g., RSA). In the selective encryption algorithm, a header is attached to the front of the transmitted data. It contains the size of the encoded ECG data, the bit plane number, the level of wavelet decomposition, and the length of the first encrypted bit plane. When the receiver receives this bit-stream of the encrypted ECG data, it first evaluates the length of the encrypted bit plane. This portion will be decrypted using the same cryptographic algorithm and the corresponding key. After the first bit plane is decrypted, a standard SPIHT-decompression process can be applied to reconstruct the original ECG data.

This selective encryption is very efficient and effective. Usually, the first bit plane only occupies a small part of the entire bit stream of the SPIHT-encoded ECG data. For instance, with respect to 1024-point ECG data with a 330Hz sampling rate and five decomposition levels, the first bit plane only occupies approximately 1% of the total compressed data among the total 11 bit planes. Also, it is very secure. Consider 2048 sample points of the SPIHT-encoded data for example. Given six wavelet decomposition levels, the positions of the first bit-plane coefficients could be anywhere among the last 2016 positions. Therefore, the brute-force attack needs to search $O(2^{2016})$ combinations before the correct position information can be deciphered. This number is considered to be safer than state-of-the-art encryption standards.

AUTHENTICATION IN BASN

Authentication is another important demand for ECG transmissions in BASN because fraudulent and malicious actions (for example, alteration of drug dosages and treatment procedures) could be extremely hazardous. In traditional networks, authentication includes several steps—key distribution, challenge request, and challenge response. The key distribution scheme assigns a common key to the trusted clients, which is used in both the challenge request and challenge response procedures. During challenge request, a challenge message is sent to the client to check whether it has the shared key or not. The client sends back a response to the request server to show its shared key authorship. However, since the key distribution needs to be performed frequently, and requires periodic updates, such an authentication procedure may not be realistic for resource-constrained BASNs.

In BASNs, different sensors are deployed around the human body. They form a self-organized body area wireless network. Biometric traits of heart rates can be sensed by different sensors with a high degree of correlation for sensors on the same patient. This property provides an opportunity for a flexible and innovative design of authentication for ECG transmissions in BASN.

First, we review a key distribution algorithm (Juels Ari, 1999) which can be utilized as an inspiration for authentication in BASN. In this algorithm, no key agreement and key exchanges are needed. Let the common trait be T_{com}, which can be derived from physiological signals in BASN. Session key k is committed by a fuzzy function with common trait T_{com} as follows (Juels Ari, 1999):

$$com = (h(k), k \oplus T_{com}) \tag{8}$$

In the above equation, two parts are concatenated, where the first part is the hash output of the session key, and the second part is the XOR of session key and common trait. When the receiver gets this concatenated message, it will de-commit the message in the following way. It uses its own witnessed trait T'_{com} to compute k by the following equation.

$$F\left(T'_{com} \oplus \left(k \oplus T_{com}\right)\right) = F(k \oplus \left(T_{com} \oplus T'_{com}\right)) \tag{9}$$

where F is an efficient decoding function with t bits of error correction capability (Togneri, 2002). As long as the different bits between T_{com} and T'_{com} are less than t bits, the output of the decoding function F would be k. Figure 9 illustrate such a procedure. In the key de-commit process, the same hash function is applied with the output of decoding function F. The result is then compared with the received $h(k)$ to check if they are equal.

A similar authentication method was proposed by Poon, Zhang, and Bao (2006), which also takes advantage of the common trait information. Figure 10 illustrates this process. However, it is only based on the Hamming distance between physiological signals, but has not shown how the common traits can be extracted under a context of statistic patterns.

Inspired from the schemes above, we can extract and employ physiological signals such as Inter-Pulse Interval (IPI) signals as a representative indicator for a common trait T_{com} (Wang, Hua, Hempel, Peng, Sharif, & Chen, 2010). The function of the physiological signal coder is to extract the distinct and representative indictor from a number of physiological samples. Such a coder can be developed based on the Gaussian Mixture Model (GMM) (Jang, 2007), which is a popular pattern recognition method and widely

Figure 9. Key distribution without key exchange in BASN

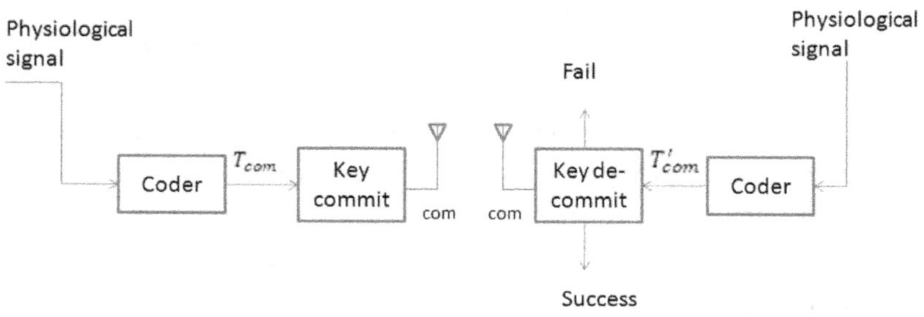

Figure 10. Authentication with common trait in BASN

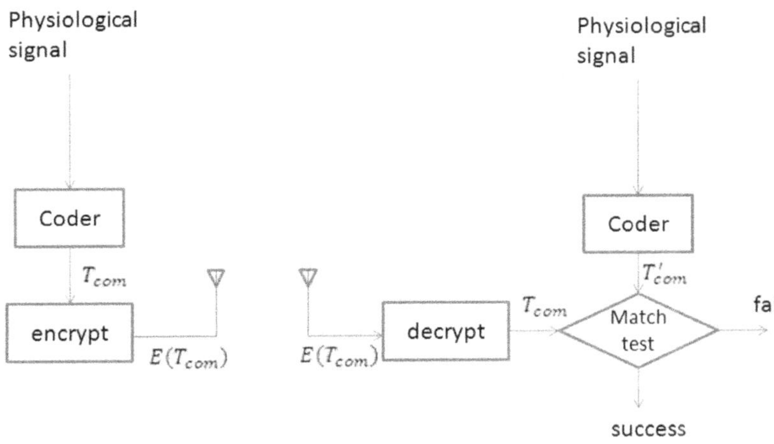

used in biometric systems (e.g., Memon, Lech, & Maddage, 2010).

As a matter of fact, any random sequence can be effectively represented by a linear combination of a number of Gaussian random variables with certain means and variances. Given a number of samples of ECG data $\{s_1, s_2, s_3, \cdots, s_n\}$, for example, we could express it by the following weighted Gaussian function (Wang, Hua, Hempel, Peng, Sharif, & Chen, 2010):

$$\sum_{i=1}^{M} w_i G_i(u_i, \Sigma_i) \quad (10)$$

where G_i is an n-variate Gaussian function whose means and covariances are u_i and Σ_i, respectively. The parameter w_i represents the weight of the Gaussian variable G_i. The parameters w_i, u_i, Σ_i, which maximize the following probability function, are the representative indictors for these samples of ECG data:

$$P(X = S) = \sum_{i=1}^{M} w_i g_i(x_1 = s_1, x_2 = s_2, x_3 = s_3, \cdots, x_n = s_n) \quad (11)$$

where g_i is the probability function of Gaussian variable G_i, which is expressed as follows

$$g_i = \frac{1}{(2\pi)^{n/2} \left|\Sigma_i\right|^{1/2}} \exp\{-\frac{1}{2}(X - u_i)' \Sigma_i^{-1} (X - u_i)\} \quad (12)$$

To be more concise, the collective indicative parameters can be denoted as follows (Wang, Hua, Hempel, Peng, Sharif, & Chen, 2010):

$$\lambda = \{w_i, u_i, \Sigma_i\} \qquad i = 1, \cdots, M \quad (13)$$

Finding an optimal λ given the samples of input $\{s_1, s_2, s_3, \cdots, s_n\}$ is a maximization problem. The estimated parameters can be obtained iteratively using a special case of the Expectation-Maximization (EM) algorithm.

Since common traits (e.g., IPI) are strongly correlated between each cycle among sensors, we can easily select a common covariance matrix for all Gaussian functions (Gauvain, 1994) to serve for the authentication process during the ECG data communication in BASN. To sum up, such GMM-based self-authentication mechanism in BASN involves cross-layer techniques to extract and embed the common trait of physiological signals in the data communications with a lower degree of complexity and higher degree of trustworthiness.

CONCLUDING REMARKS

Achieving communication energy efficiency is one of the most important issues for ECG transmissions in a Body Area Sensor Network (BASN). This is especially true when other requirements exist, such as robust communication and security. In this chapter, we presented a framework of cross-layer techniques to address this concern and showcase how it can be leveraged to achieve robust secure ECG transmissions with great energy efficiency.

To demonstrate the flexibility and effectiveness of cross-layer designs for BASNs, we first introduced Two-Rate Unequal Error Protection. Because there exists a great deal of unequal importance among the different bit planes of the compressed ECG data, the Two-Rate Unequal Error Protection can be utilized to adjust the available communication resources to adaptively assign different channel code rates to the different bit planes of the compressed ECG data. This way, significant savings in communication energy can be achieved without a reduction in the communication quality.

Secondly, a Selective Encryption cross-layer scheme was presented as an example of leveraging cross-layer designs for energy efficient security strategies. We analyzed the dependency among the bit planes of the SPIHT-compressed ECG data and, based on our findings, concluded that we only need to find the most important bit plane to encrypt. Other bit planes would be completely indecipherable without this most important bit plane. Therefore, the entire ECG data is secured, with a level of security equivalent to traditional encryption methods at greatly reduced energy and computational expense.

Finally, we also introduced a second example of cross-layer security techniques in ECG systems, demonstrating self-organized authentication in BASNs. Biometric traits of inter-pulse interval provide consistency in different parts of the body area, and are unique among different individuals. Hence, they can serve as a natural "authentication key" among sensors in the BASN. The challenge is how to utilize an efficient pattern recognition method in order to extract the biometric traits from physiological signals. In this chapter, we described a Gaussian Mixture Model based solution for such authentication key extraction and communication in BASN.

In summary, we presented several promising advances and developments utilizing cross-layer techniques to address communication energy efficiency and showcase how they can be utilized for simultaneously achieving other requirements such as link robustness and security for ECG transmissions in BASN. Our simulation results demonstrate that a higher performance in terms of the communication energy efficiency can be achieved with all of the introduced schemes compared to traditional approaches.

REFERENCES

Al-Fahoum, A., & Howitt, I. (1999). Combined wavelet transformation and radial basis neural networks for classifying life-threatening cardiac arrhythmias. *Journal of Medical and Biological Engineering and Computing, 37*(5), 566–573. doi:10.1007/BF02513350

Amjed, S. A. F. (2006). Quality assessment of ECG compression techniques using a wavelet-based diagnostic measure. *IEEE Transactions on Information Technology in Biomedicine, 10*(1), 182–191. doi:10.1109/TITB.2005.855554

Atkin, G., & Tam, C. (1992). Performance analysis of punctured convolutional codes for variable-bit-rate speech transmission. In *Proceedings of the IEEE 42nd Vehicular Technology Conference*, (pp. 183-186). IEEE Press.

Boskovic, A., & Despotovic, M. (2005). An efficient approach to ECG signal transmission via GPRS. In *Proceedings of the International Conference on Computer as a Tool*, (pp. 76-79). IEEE.

Cox, J. R., Nolle, F. M., Fozzard, H. A., & Oliver, G. C. (1968). AZTEC: A preprocessing program for real-time ECG rhythm analysis. *IEEE Transactions on Bio-Medical Engineering*, *15*(2), 128–129. doi:10.1109/TBME.1968.4502549

DeVaul, R., Sung, M., Gips, J., & Pentland, A. (2003). MIThril 2003: Applications and architecture. In *Proceedings of the 7th IEEE International Symposium on Wearable Computers*, (pp. 4-11) IEEE Press.

Dipersio, D. A. (1985). Evaluation of the fan method of adaptive sampling on human electrocardiograms. *Medical & Biological Engineering &. Computing*, *23*(5), 401–410.

Djohan, A., Nguyen, T., & Tompkins, W. (1995). ECG compression using discrete symmetric wavelet transform. In *Proceedings of the IEEE 17th Annual Conference on Engineering in Medicine and Biology Society*, (pp. 167-168). IEEE Press.

Gauvain, J. L. (1994). Maximum a posteriori estimation for multivariate Gaussian mixture observations of Markov chains. *IEEE Transactions on Speech and Audio Processing*, *2*(2), 291–298. doi:10.1109/89.279278

Hagenauer, J. (1988). Rate-compatible punctured convolutional codes (RCPC codes) and their applications. *IEEE Transactions on Communications*, *36*(4), 389–400. doi:10.1109/26.2763

Hilton, M. (1997). Wavelet and wavelet packet compression of electrocardiograms. *IEEE Transactions on Bio-Medical Engineering*, *44*(5), 394–402. doi:10.1109/10.568915

Jang, J. (2007). Gaussian mixture model. In *Data Clustering and Pattern Recognition.* Retrieved from http://mirlab.org/jang.

Juels, A., & Waternberg, M. (1999). A fuzzy commitment scheme. In *Proceedings of the 6th ACM Conference on Computer and Communications Security*, (pp. 28-36). ACM Press.

Khadra, L., Al-Fahoum, A., & Al-Nashash, H. (1997). Detection of lifethreatening cardiac arrhythmias using the wavelet transformation. *Medical & Biological Engineering & Computing*, *35*(6), 626–632. doi:10.1007/BF02510970

Kim, H., Kim, Y., & Yoo, H. J. (2008). A low cost quadratic level ECG compression algorithm and its hardware optimization for body sensor network system. In *Proceedings of the 30th Annual International Conference of the IEEE*, (pp. 5490-5493). IEEE Press.

Li, Z. S. (2007). Joint source-channel-authentication resource allocation and unequal authenticity protection for multimedia over wireless networks. *IEEE Transactions on Multimedia*, *9*(4), 837–850. doi:10.1109/TMM.2007.893338

Lin, Z., Sinclair, M., & Bittner, R. (2006). A phone-centered body sensor network platform cost, energy efficiency and user interface. In *Proceedings of the Wearable and Implantable Body Sensor Networks Workshop*, (pp. 179-182). IEEE.

Lu, Z., Kim, D. Y., & Pearlman, W. (2000). Wavelet compression of ECG signals by the set partitioning in hierarchical trees algorithm. *IEEE Transactions on Bio-Medical Engineering*, *47*(7), 849–856. doi:10.1109/10.846678

Ma, T., Hempel, M., Peng, D., & Sharif, H. (2010). Rate-switching unequal error protection for wireless electrocardiogram (ECG) transmission. In *Proceedings of the Military Communications Conference*, (pp. 1181-1186). IEEE.

Memon, S., Lech, M., & Maddage, N. (2010). Information theoretic expectation maximization based gaussian mixture modeling for speaker verification. In *Proceedings of the 20th International Conference on Pattern Recognition*, (pp. 4536-4540). IEEE.

Mueller, W. C. (1978). Arrhythmia detection program for an ambulatory ECG monitor. *Biomedical Sciences Instrumentation*, *14*, 1–85.

Muramatsu, D. K. (2006). A markov chain monte carlo algorithm for bayesian dynamic signature verification. *IEEE Transaction on Information Forensics and Security*, *1*(1), 22–34. doi:10.1109/TIFS.2005.863507

Nave, G., & Cohen, A. (1993). ECG compression using long-term prediction. *IEEE Transactions on Bio-Medical Engineering*, *40*(9), 877–885. doi:10.1109/10.245608

Pan, J., & Tompkins, W. J. (1985). A real-time QRS detection algorithm. *IEEE Transactions on Bio-Medical Engineering*, *32*(3), 230–236. doi:10.1109/TBME.1985.325532

Poon, C., Zhang, Y.-T., & Bao, S. D. (2006). A novel biometrics method to secure wireless body area sensor networks for telemedicine and m-health. *IEEE Communications Magazine*, *44*(4), 73–81. doi:10.1109/MCOM.2006.1632652

Sufi, F., Mahmoud, S., & Khalil, I. (2008). A wavelet based secured ECG distribution technique for patient centric approach. In *Proceedings of the 5th International Summer School and Symposium on Medical Devices and Biosensors*, (pp. 301-304). IEEE.

Togneri, J. R. (2002). *Fundamentals of information theory and coding design*. Boca Raton, FL: CRC Press.

Tompkins, J., & Abenstein, J. P. (1982). New data-reduction algorithm for real-time ECG analysis. *IEEE Transactions on Bio-Medical Engineering*, *29*(1), 43–48. doi:10.1109/TBME.1982.324962

Wang, H., Peng, D., Wang, W., Sharif, H., Chen, H. H., & Khoynezhad, A. (2010). Resource-aware secure ECG healthcare monitoring through body sensor networks. *IEEE Wireless Communications*, *17*(1), 12–19. doi:10.1109/MWC.2010.5416345

Wang, W., Hua, K., Hempel, M., Peng, D., Sharif, H., & Chen, H.-H. (2010). A stochastic biometric authentication scheme using uniformed GMM in wireless body area sensor networks. In *Proceedings of the 2010 IEEE 21st International Symposium on Personal Indoor and Mobile Radio Communications*, (pp. 1620-1624). IEEE Press.

Wang, Z., & Chen, Y. (2008). ECG signal compression based on MSPIHT algorithm. In *Proceedings of the International Conference on Information Technology and Applications in Biomedicine*, (pp. 415-419). IEEE.

Zhang, X., Jiang, H., Zhang, L., Zhang, C., Wang, Z., & Chen, X. (2010). An energy-efficient ASIC for wireless body sensor networks in medical applications. *IEEE Transactions on Biomedical Circuits and Systems*, *4*(1), 11–18. doi:10.1109/TBCAS.2009.2031627

Zigel, Y., Cohen, A., & Katz, A. (2000). ECG signal compression using analysis by synthesis coding. *IEEE Transactions on Bio-Medical Engineering*, *47*(10), 1308–1316. doi:10.1109/10.871403

KEY TERMS AND DEFINITIONS

Body Area Sensor Network (BASN): BASN is the centering wireless network aiming to provide convenient communications on, near and around the human body. It consists of several miniaturized body sensor units and a single body central unit worn at the human body.

Cyclic Redundancy Check (CRC): CRC is a cyclic error correcting code designed to detect errors of received data. Cyclic codes are not only simple to implement but have the benefit of being particularly well suited for the detection of burst errors, contiguous sequences of erroneous data symbols in messages.

Electrocardiography (ECG or EKG): ECG is a recording of the electrical activity of the heart over a period of time, as detected by electrodes attached to the outer surface of the skin.

Gaussian Mixture Model (GMM): GMM is a weighted sum of Gaussian density functions aiming to represent a target signal's probability function. GMMs are commonly used as features retrieval of biometric system.

Inter-Pulse Intervals (IPI): IPI is the time elapsed between successive nerve impulses. It is a important parameter in ECG for cardiac disease diagnosis.

Rate-Compatible Punctured Convolutional Codes (RCPC): RCRP is the extended channel coding of punctured convolutional code. By periodically puncturing the output of punctured code, a family of codes with different coding rate can be obtained. RCRC is famous of its and flexibility and easiness of generating different codes rates.

Set Partition in Hierarchy Tree (SPIHT): SPIHT is a wavelet coding algorithm which mainly explores the similarity of different subbands. This is very simple and fast, widely used in biometric signal compression.

Unequal Error Protection (UEP): UEP is a effective communication technique which allocate more communication resources to more important parts of data, less communication resource to less important parts of data in order to achieve overall optimal transmission with the given energy consumption budget.

Wavelet-based Weighted Percent Root-mean-squared Difference (WWPRD): WWPRD is a compressed ECG quality measurement. It is more accurate than traditional measurement percent root-mean-squared difference (PRD), and t correlate very well with subjective tests.

Chapter 6
Cross-Layer Design for Network Security Enhancement

Chi-Yuan Chen
National Dong Hwa University, Taiwan

Han-Chieh Chao
National Dong Hwa University, Taiwan & National I-Lan University, Taiwan

ABSTRACT

Cross-Layer Design is useful for wireless communication to improve the performance and efficiency. On the other hand, Cross-Layer Design is also helpful in enhancing network security. With the help of cross-layer information and interactions, the security overhead can be reduced and the security efficiency can be improved. In this chapter, the authors review the existing literature and provide a taxonomy based on their methods and applications. According to their survey, these cross-layer security enhancements could be classified into four categories: 1) security protocol; 2) security policy; 3) key management; 4) intrusion detection. The authors hope to attract more applications and research attention in this direction.

INTRODUCTION

The traditional Open System Interconnection (OSI) layered protocol stacks have been used for a long time. In OSI, the function of each layer is defined clearly that could reduce the complexity of network implementation and increase the flexibility. In the layered OSI architecture model, the protocol at each layer is designed independently for the different layers. The layered architecture does not allow direct communication between nonadjacent layers. Communication between adjacent layers must follow the pre-defined interfaces through procedure calls and responses. However, for the next generation wireless networks and mobile systems, the traditional layered network design cannot satisfy the user requirement on performance and efficiency (Chao, et al., 2009). Because of the nature of wireless communications such as time varying transmission of the wireless channel and the dynamic resource requirements of

DOI: 10.4018/978-1-4666-0960-0.ch006

different applications, the mobile communication is meeting the challenge of architectural design.

Cross-Layer Design (CLD) is still a hot research topic that actively exploits the dependence and interaction between different protocol layers to obtain performance gains. Srivastava and Motani have made a detailed survey on cross-layer design and depicted main reasons to motivate designers to violate the layered traditional architecture. Under the wireless environment, these problems include some unique problems created by wireless links such as the possibility of opportunistic communication on wireless links and the new modalities of communication offered by the wireless medium (Srivastava & Motani, 2005). For instance, a classic problem of wireless link is that TCP sender may mistake a packet error and believe that network congestion happened. Network element can cooperate with others or collect more information from different layers to improve the communication quality. Cross-Layer Design is a novel and practical mode for wireless communications in protocol or mechanism design (Zhou, et al., 2009; Zhou, et al., 2011; & Chang, et al., 2010).

In the aspect of network security, cross-layer design is helpful for improving the performance of existing protocols and enhancing the efficiency of existing mechanisms. For instance, due to the limitation of traditional layered architecture, application or protocol cannot communicate or cooperate with other nonadjacent ones. It may cause redundant processes to degrade the performance such as multiple encryptions in network security. We can utilize cross-layer design to deploy appropriate configuration for tradeoff between performance and security. Moreover, cross-layer information and interaction is also useful for designing security mechanisms such as key management and intrusion detection.

In this chapter, we will introduce the background of cross-layer design in next section. The background includes cross-layer design categories, methodology, signaling, and features. According to our survey of cross-layer design for network security, we classify these cross-layer security enhancements into four categories and describe them in order.

BACKGROUND

An important cross-layer design aspect is the management of cross-layer interaction that can guarantee the system operation. Such cross-layer entities (i.e., functions, programs, or protocol layers) may reside within the stack, in which case it is considered as an internal cross-layer entity or an external network node as shown in Figure 1 and Figure 2 (Foukalas, et al., 2008). Internal entity may be either an inter-layer entity that coordinates the operation of all protocol stack layers or an intra-layer entity that is located within a protocol layer. The external entities may be centralized by a specific network node or distributed over several network nodes.

Another important issue in cross-layer design is the signaling mechanism. Foukalas et al. (2008) performed a well-studied survey of different kinds of cross-layer signaling mechanisms and protocols. The ways of exchanging and indicating information are hints and notifications. A hint is the message sending from higher layer toward lower

Figure 1. Internal cross-layer entities

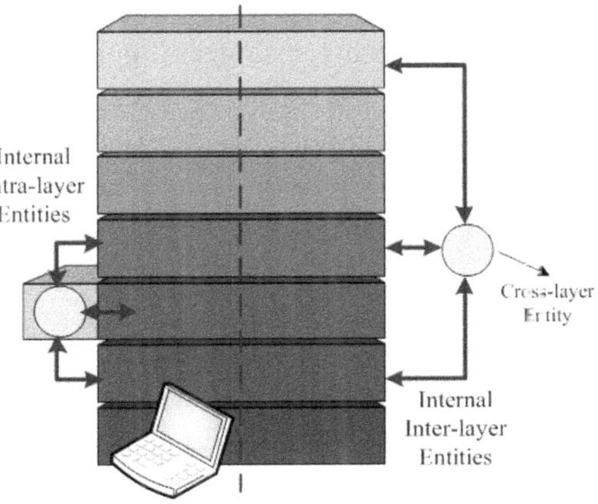

Figure 2. External cross-layer entities

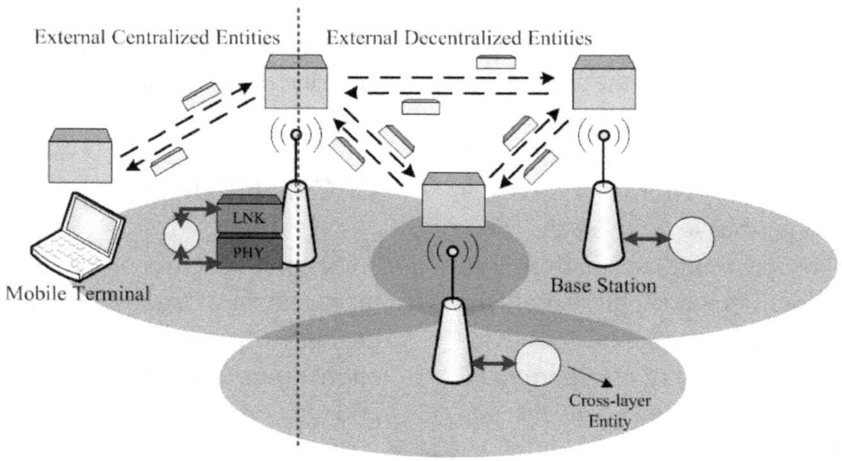

layer and a notification is the message triggering from the lower layer to notify the higher layer about the operation state (i.e., channel state or network status). Furthermore, the cross-layer signaling could be implemented by several ways: 1) header options; 2) additional header; 3) local profiles; 4) network service. In the header, cross-layer notification could be in-band protocol such as IP and TCP header options or out-band protocol such as ICMP or RTCP. In local profiles, Layer-Independent Descriptors (LIDs) that represent a collection of parameters is organized and stored. The network service could be realized through Wireless Channel Interface (WCI) and abstract format such as Extensible Markup Language (XML).

There are many studies discussing about the cross-layer solution for different topics. After a survey of several internal cross-layer design proposals from the literature, Srivastava and Motani (2005) noted that the layered architecture could be violated in the following basic ways.

- Creation of new interfaces;
- Merging of adjacent layers;
- Design coupling without new interfaces;
- Vertical calibration across layers.

While creating new interfaces, according to the interaction method between layers, there are three subtypes to create the new interfaces: upward, downward, and back and forth. The upward model is a bottom-up approach. Higher layer protocols may require the parameters from the lower layer at runtime. It could also be said that the lower layer notifies the higher layer about the network condition information. Most of the existing cross-layer literature are based on this approach. In the other hand, the downward model is a top-down approach. Higher layer protocols would set the parameters within the lower layer through the new interface at runtime. The lower layer could get the action hints of higher layer and take the more appropriate actions. The back and forth model which combines the bottom-up approach and the top-down method. Higher layer and lower layer can cooperate more closely through the new interfaces. In the second approach, merging of adjacent layers, a super layer composed of two or more adjacent layers is used to achieve the cross-layer function. The super layer could provide the same interfaces to the rest protocol layers or it could also create new interfaces for use. For the third approach, design coupling without new interfaces, the design of some protocol layers would keep in mind about the functionality of the other layers.

The coupled layers would be highly correlated with each other. Thus, the system architecture should be complex and those correlated layers need to be modified at the same time while some design changes need to be applied. Finally, the idea of the vertical calibration model is to find the optimal set of parameters used for different layers. Vertical calibration can be implemented using static or dynamic manners. For the static approach, the parameters across different layers are chosen at design time. In the dynamic manner, the parameters would be determined and adapted during run time. Unfortunately, the dynamic approach is very complex and incurs significant overhead and cost problems.

Most cross-layer design proposals in the current literature could be fitted into one of the above basic categories. Moreover, considering the implementation methods for the cross-layer interactions, these techniques can be summarized into three categories.

- Direct communication between layers;
- A shared database across the layers;
- Completely new abstractions.

For the direct communication between layers, the variables or parameters at one layer could be directly used by other layers through the new created interfaces or the vertical calibration model. Many methods could be used to make different layers directly communicate with each other. For example, the protocol headers could be used to contain the cross-layer information. The new packet format and information flow could be defined for cross-layer. Extra internal cross-layer entities could be designed to coordinate the cross-layer information. In the second category, a common shared database is used as the intermediate medium for the communication of different layers. New interfaces or vertical calibration model could be realized through the shared database approach. For the last category, the idea of layers is broken thoroughly and the new operation mechanism is re-designed. The flexibility of this architecture is great and can provide rich interactive communications. However, the system complexity is very high. Almost a complete new system should be designed for this approach. The kernel system level modifications would be touched for implementation of this new architecture.

Some of the cross-layer issues are described in more detail by Srivastava and Motani (2005). What is the role the physical layer would play is suggested to be an important question in cross-layer designs. It is suggested that new researchers who would like to study cross-layer design proposals can start to resolve the open challenges such as cross-layer coexisting problem, the selection consideration between the environmental condition and different cross-layer models, the impact of other layers and architecture, and the possibility of standardized model for cross-layer design, etc.

In evaluating the cross-layer design proposals, the trade-off between the performance and the architectural design needs to be fundamentally considered. Performance is considered to be more short-term and architecture is longer-term. Thus, a particular cross-layer suggestion may get an improvement in throughput or performance but lacks of a longer-term consideration. However, it is very difficult to evaluate the merit of proposals on the overall architecture. Kawadia and Kumar (2005) have performed two simulation studies to show that unintended cross-layer interactions can lead to undesirable consequences on overall system performance. It is emphasized that unbridled cross-layer design can lead to sophisticated design, stifle further innovation, and be difficult to maintain for longer-term operation. Because wireless networks maybe attract an increasing number of users, the importance of good architecture needs to be kept in mind. The cautionary information proposed by Kawadia and Kumar (2005) should be examined while engaging in cross-layer design.

Kawadia and Kumar (2005) have examined the cross-layer design issues and its architectural categories. They emphasized the importance of well-constructed architecture. Modularity in system could provide the abstractions of system and makes designers easy to understand the overall system. It can also accelerate development of both design and implementation. Designers can focus their effort on a particular subsystem and assure that the entire system will operate correctly. A good architectural design can make system easy to maintain and lead to longer life cycle. Based on the electronic transfer of information, Kawadia and Kumar also depicted that operating a wireless network based on a layered stack is a natural way to implement the multi-hop decode and forward strategy. A layered architecture can achieve optimal performance, within a constant, with regard to network capacity. They also claimed that all cross-layer design can only improve throughput by at most a constant factor, but cannot result in any unbounded improvements (see Kawadia and Kumar [2005] and Xie and Kumar [2004]).

CROSS-LAYER DESIGN FOR NETWORK SECURITY

Based on the existing literature, the cross-layer design for network security could be classified into cross-layer protocols and cross-layer applications. The cross-layer security applications include security policy, key management, and intrusion detection. We choose some typical proposals for introducing these approaches in the following.

Security Protocol

The performance and efficiency of security protocols also could be improved by adopting cross-layer design. Under the consideration of performance, the primary goal is to improve the tradeoff between network performance and security even under malicious activities. In other words, the efficiency of security protocol is the most challenge. Due to the broadcasting nature of wireless transmissions, wireless networks lack a physical boundary. For this reason, the security becomes a critical concern in the physical layer for wireless networks. Some related proposals for cross-layer based security protocol are listed in Table 1.

Ramachandran et al. (2007) designed a cross-layer architecture for multimedia application to protect protocol header and limit packet discarding. Due to IPsec is not suitable for UDP-Lite based application, they utilized cross layer signaling scheme with link layer to work with Authentication Header (AH) or Encapsulating Security Protocol (ESP) for authentication and encryption. Wu et al. (2007) proposed a cross-layer AAA (Authentication, Authorization, and Accounting) component for enhancing the key generation and data encryption mechanisms over heterogeneous networks. By adopting cross-layer AAA, the session key could be generated quickly during handoff and the traffic could only be decrypted and monitored by service provider. Silva et al. (2010) proposed a CRoss-layer and Adaptive scheme for Balancing performance and security (CRAB) over data routing in wireless mesh networks. This scheme is consisted of four cross-layer functions as evaluating path performance costs, calculating routing metric, assessing path security level, and exchanging information. Moreover, the cross-layer information is also useful for providing better privacy and anonymity. In order to improve the performance of source location privacy in sensor networks, Shao et al. (2009) utilizes cross layer beacons at the link layer instead of network-wide dummy messages.

For efficiency, Muraleedharan and Osadciw (2006b) proposed a cross-layer protocol in wireless sensor network to find an optimal route to avoid nodes under Denial of Service (DoS) attack. They utilize Swarm Intelligence by jointly considering information from the link layer and the network layer to improve packet delivery rate with the minimal energy consumption. Shao et al. (2008) studied a cross-layer dropping attack against video streaming over ad hoc networks.

Table 1. Proposals of security protocol

Proposals	Environment	Application	Design Goal	Cross layers
Ramachandran et al. (2007)	Unspecified	Multimedia	Performance	Link Layer, Transport Layer
Wu et al. (2007a) & Wu et al. (2007b)	Heterogeneous networks	Authentication, Authorization, and Accounting	Performance	Physical Layer Link Layer, Network Layer
Silva et al. (2010)	Wireless mesh networks	Routing	Performance	Link Layer, Network Layer
Shao et al. (2009)	Sensor networks	Broadcasting and Routing	Performance, Privacy	Link Layer, Network Layer
Muraleedharan & Osadciw (2006b)	Wireless sensor networks	Routing	Against DoS attack	Link Layer, Network Layer
Shao et al. (2008)	Ad-hoc Networks	Video Streaming	Against Dropping attack	Network Layer, Application Layer
Wen et al. (2010)	MIMO wireless networks	STBC	Security	Physical Layer, Upper Layer
Wang et al. (2010)	Cognitive radio networks	Trust Management	Against attacks	Physical Layer Link Layer.

They identified a network layer dropping attack could achieve destructive impact by leveraging the application layer information (video streaming). In other words, the attacker can selectively drop I frames by exploiting the application layer knowledge. In this work, the authors also discuss several possible diagnosis approaches and solutions. Wen et al. (2010) present a cross layer communication scheme based on Space Time Block Code (STBC) system. They employ a random flip-flops process controlled by upper-layer pseudo random sequence to confuse the attacker at the physical layer. Moreover, this scheme uses the upper-layer encryption techniques to control the transmission sequences in the physical layer. In cognitive radio networks, Wang et al. (2010) studied the report-false-sensing-data attack in physical layer and small-back-off-window attack in the link layer. In order to defend against attacks, they proposed a cross-layer trust defense scheme by adopting abnormal detection schemes in the physical layer and the link layer. Furthermore, they also develop a cross-layer trust manager for overall trust fusion.

Security Policy

Currently, each OSI layer has its own security mechanisms. More security mechanisms are enabled in the same time, the more secure computing environment is obtained. However, the computing power is limited and valuable in wireless communications. That will be a trade-off between the security strength and performance. It is better to enforce the security policy with cross-layer interactions according to the user requirement and device capability. The security policy is the chosen configuration of protocols (might consist of several mechanisms) between network nodes.

Multiple layers of encryption may be used to improve the security strength of communications. However, the additional security overhead will decrease system performance. Carneiro et al. (2004) proposed the concept of a coordination plane using cross-layer design. This coordination plane is designed to eliminate the multiple encryptions in different layer such as shown in the Figure 3. Several encryption protocols and security mechanisms based on different layers are currently in use. At the transport and the application layer,

Figure 3. Security plane

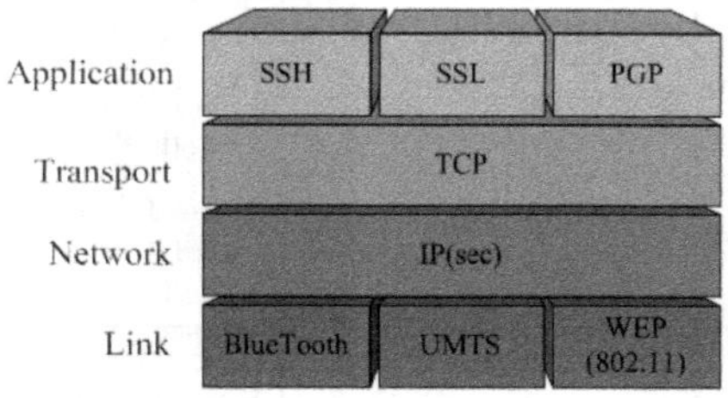

Security Shell (SSH), Secure Sockets Layer (SSL), and Pretty Good Privacy (PGP) provide end-to-end encryption mechanism. At the network layer, Internet Protocol Security (IPSec) is an encryption mechanism for end-to-end hosts or a secured tunnel between two networks. Unlike upper layers, which security mechanism should be applied is according to the different network technologies. Wired Equivalent Privacy (WEP) is defined by the 802.11 standard to provide a Wireless Local Area Network (WLAN) with a minimal level of security and privacy. Bluetooth is another wireless technology for Personal Area Networks (PANs) that provides encryption mechanism. The next generation mobile communication system, such as UMTS, also provides encryption mechanism. Under this mechanism, both data and signaling are encrypted between the user equipment and the radio network controller.

Enabling encryption mechanisms from multiple layers may enhance the security, but it would require more computing power and the delay would increase. Carneiro et al. also suggested disabling the encryptions on all layers except one. Encryption by upper-layer protocols is preferable to lower-layer ones. Intermediate layer flows such as signaling should use encryption at the highest layer. The suggested solution is that each datagram or data unit keeps a flag to indicate if it has already been encrypted by an upper layer protocol or not. Some related cross-layer based proposals are listed in the Table 2.

Some results demonstrate that the stronger the security it is, the more signaling and delay overhead it suffers. Agarwal et al. (2005) classified security protocols into individual and hybrid policies. They also introduced a new metric and relative security index to analyze security strength and overhead tradeoffs quantitatively. Finally, they concluded that there is a trade-off between security strength and performance. Their experimental results indicate that IPsec policy provides the best trade-off for authentication time, and 802.1x-EAP-TLS policy is suitable for those applications requiring low cryptographic overhead and better security strength. The implementation should be focused on the integration of different layers via security policies.

Roman et al. (2011) presented a transversal layer concept for containing all the different security mechanisms inside the software architecture and network stack in sensor networks. By adopting this integrated way, elements can interact with security mechanisms and the security mechanisms can have a holistic point of view. Urbano Fullana et al. (2011) proposed an algorithm for security inter-layer communication and a solution for Cross-Layer signaling between nodes in TCP/IP over IEEE 802.11 networks. In order to save

Table 2. Proposals of security policy

Proposals	Environment	Method	Interaction	Cross layers
Agarwal et al. (2005)	Public Access Wireless Networks	Experiment	Policy	All Layers
Roman et al. (2011)	Sensor networks	Transversal Layer	Interfaces	All Layers
Urbano Fullana et al. (2011)	TCP/IP. 802.11	Cross-Layer Algorithm	IP header	All Layers

Table 3. Proposals of key management

Proposals	Environment	Cryptography	Design Goal	Cross layers
Lazos & Poovendran (2004)	Ad-hoc Networks	Symmetric	Energy-efficient	Physical layer, Network layer
Li & Ephremides (2006)	Ad-hoc Networks	Symmetric	Computation Efficiency	Network layer, Application layer
Saglam & Dalkilic (2009)	Wireless sensor networks	Asymmetric	Energy-efficient	Link Layer, Network Layer
Zeng et al. (2010)	Ad-hoc Networks	Asymmetric	Computation Efficiency	All Layers

power, they create a new option at IP header to signaling for solving multiple encryptions.

Key Management

Key management is the most important basic mechanism for securing networks, especially in ad-hoc networks and wireless sensor networks (Chen & Chao, 2011). By joint considering the information from different layers, the energy consumption of re-keying and distribution could be reduced. Currently, several cross-layer based key management schemes have been proposed as listed and compared in the Table 3.

In symmetric key algorithm, Lazos and Poovendran (2004) proposed an energy-efficient key distribution scheme for ad-hoc networks called Routing-Aware Key distribution scheme (RAwKey). In order to minimize the energy consumption of re-keying, they consider the node transmission power in the physical layer and the multicast routing tree in the network layer. Li and Ephremides (2006) adopt a secret handshake scheme that allows two users to authenticate each other and set up a shared secret key. Moreover, they have taken joint consideration of the application and network layer for anonymous routing protocol in ad hoc networks. This scheme provides computationally and requires simple key management compared to public key management scheme.

On the other side, Saglam and Dalkilic (2009) have analyzed the cross-layer relations about the network connectivity, total energy cost of the key establishment and the achieved security level of wireless sensor networks. Simulation results indicate that key establishment cost can be reduced by controlling the number of secured links per sensor node according to the network configuration protocol. Based on key pre-distribution and Public Key Cryptography (PKC), they have proposed three different link key setup procedures for selection of links. Recently, Zeng et al. (2010) adopt cross-layer approach for improving the performance of threshold-based distributed CA (Certification Authority) key management scheme in ad hoc networks. Their key management scheme utilizes state information of every layer.

Intrusion Detection

Wireless ad hoc networks and sensor networks are vulnerable to various kinds of security threats. Most intrusion detection techniques of wired or wireless networks are not suitable for ad hoc or sensor networks. The major problem is due to the lack of a centralized infrastructure. Traditionally, most intrusion detection techniques are focused on extracting the misbehavior features from the network layer only. There are vulnerabilities of multiple layers in the wireless ad hoc network such as shown in the Table 4, and therefore we need an intrusion detection module that considers across layers. The cross-layer information is useful in modeling the malicious behaviors. For instance, a network node may drop packet due to malicious attacks or a poor wireless channel. Meanwhile, the cross-layer information (i.e, collision from link layer) could be used to identify the reason.

The more input information will increase the detection accuracy. Zhang and Lee (2000) regarded the better intrusion detection should be distributed, cooperative and take place in all networking layers in an integrated cross-layer manner. However, more input information will also increase the complex of detection mechanism. The related cross-layer based Intrusion Detection approaches are listed in Table 5.

There is a lot of research for improving the accuracy of intrusion detection by collecting information of collision from link layer and packet dropping, misdirection from network layer. For instance, Thamilarasu et al. (2005) proposed a Cross-Layer Based Intrusion Detection System (CIDS) to identify the malicious node in wireless ad-hoc networks. Bose and Kannan (2008) also proposed a Cross-Layer Based Intrusion Detection System (CIDS) to detect intruders on Ad Hoc On-Demand Distance Vector (AODV) protocol. Bansal and Sofat (2010) proposed an Intrusion Detection System (IDS) framework based on cross-layer interaction against Denial of Service (DoS) attacks in wireless mesh networks.

In the anomaly detection approach, Liu et al. (2005) defined a feature set from link layer and network layer to profile normal behavior for IDS in wireless ad-hoc networks. The collected feature sets includes flow direction (SEND, RECV, DROP), Send address, Destination address, MAC packet type (RTS, CTS, DATA, ACK), and Routing packet type (data packet, control packet). They also adapt a rule-based data mining technique for anomaly detection and utilized Maximal Frequent

Table 4. DoS attacks in wireless networks

Protocol layer	DoS Attacks
Physical Layer	Sinking
Link Layer	Collision
Network Layer	Packet Drop; Misdirection

Table 5. Proposals of intrusion detection

Proposals	Environment	Method	Goal	Cross layers
Thamilarasu et al. (2005)	Wireless ad-hoc networks	Detection Algorithm	Against DoS attack	Link Layer, Network Layer
Bose & Kannan (2008)	Wireless ad-hoc networks	Detection Algorithm	Against DoS attack	Link Layer, Network Layer
Bansal & Sofat (2010)	Wireless mesh networks	Detection Algorithm	Against DoS attack	Link Layer, Network Layer
Liu et al. (2005)	Wireless ad-hoc networks	Anomaly Detection	Accuracy	Link Layer, Network Layer
Muraleedharan & Osadciw (2006a)	Wireless sensor networks	Swarm Intelligence	Against DoS attack	Link Layer, Network Layer
Wang et al. (2009)	Wireless mesh networks	Machine Learning	Accuracy	Link Layer, Network Layer
Shrestha et al. (2010)	Mobile ad-hoc networks	Cluster data mining	Accuracy	Physical Layer, Link Layer, Network Layer
Joseph et al. (2008, 2011)	Mobile ad-hoc networks	SVM & FDA	Against Routing attack	Physical Layer, Link Layer, Network Layer
Chiang & Hu (2011)	Wireless Broadcast Networks	Detection Algorithm	Against Jamming attack	Physical Layer, Upper Layer

Item set (MFI) criteria to prune the rules. Muraleedharan and Osadciw (2006a) utilized Swarm Intelligence based on cross-layer approach (link layer and network layer) to detect DoS attack in wireless sensor networks. Wang et al. (2009) proposed a cross-layer based anomaly detection system for wireless mesh networks. The normal profile is trained from features collected from both link layer and network layer with different machine learning algorithms. Shrestha et al. (2010) presented a cooperative anomaly detection system utilizing cluster data mining technique in MANET. Joseph et al. presented a Cross-Layer Routing Attack Detection Systems (CRADS) (Joseph, et al., 2008) for detection routing attacks and an autonomous host-based intrusion detection system (Joseph, et al., 2011) for detecting malicious sinking behavior in wireless ad-hoc networks. They adopt nonlinear machine learning algorithm called Support Vector Machines (SVMs) and linear classification algorithm called Fisher Discriminant Analysis (FDA) to maximize the detection accuracy. Most importantly, they indicated that the cross-layer approach forms a bigger feature set made the nonlinear machine learning techniques infeasible. In order to reduce the feature set without reducing the information content and detection accuracy, they collect features from link layer are associated/correlated with features from other layers instead of using all cross-layer features.

Most existing works focused on typical wireless ad-hoc and sensor network. Recently, a different focus on wireless broadcast networks was proposed by Chiang and Hu (2011). They proposed a scheme for receivers to detect jamming by observing secondary message without the primary message. They also presented a code tree system that provides input to help the physical layer circumvent jammers.

Furthermore, the architectural design for intrusion detection also has been considered. Thamilarasu and Sridhar (2007a) discussed the functional advantages of the parameter interactions and provide probabilistic analysis based on Bayes rule for cross-layer based intrusion detection. They also presented the limitations and performance of cross-layer architectural modification for intrusion detection (Thamilarasu & Sridhar, 2007b). Furthermore, they considered the Vertical cross-layer optimizations and the Horizontal cross-layer optimizations for a Distributed Cross-Layer Framework (XLSEC) in Wireless Sensor Networks (Thamilarasu & Sridhar, 2009). Joseph et al. (2007) discuss Cross Layer versus Single Layer approaches for Intrusion Detection in MANETs. The cross-layer configurations are divided into Multiple data Collection and Multiple data Analysis (MCMA) and Multiple data Collection and a Single data Analysis (MCSA). They have shown that choosing the suitable combination of layers is critical and different combinations aid to detect different attacks.

FUTURE RESEARCH DIRECTIONS

In recent year, more research papers about network security aided with cross-layer design have been published. The previous works concerned utilizing cross-layer design to improve existing problems such as multiple encryptions, key management, and intrusion detection. However, in these approaches, more input information will increase the complexity of design. How to choose the minimum cross-layer information without loss of performance and efficiency is the most challenge.

Moreover, cross-layer design is also useful for designing new security protocols and mechanisms. It is worthy of note that cross-layer design is also be adopt for new type attacks such as Shao et al. (2008) and Wang et al. (2010). Hence, the more new security problems remain waiting for new solutions.

CONCLUSION

Due to the nature and limitations of wireless and layered networks, researchers have conducted different cross-layer techniques to improve the performance and efficiency of network security. In this chapter, we first introduce the cross-layer design approach and methodology. Then, the state-of-the-art techniques of existing solutions are categorized, described, and compared. We also highlight the future research directions in cross-layer design. Cross-layer design in network security is still an active research area. More studies are needed to address the challenges from traditional layered networks.

REFERENCES

Agarwal, A. K., Wang, W., & McNair, J. Y. (2005). An experimental study of cross-layer security protocols in public access wireless networks. In *Proceedings of the IEEE Global Telecommunications Conference (GLOBECOM 2005)*, (vol. 3), (pp. 1747–1751). IEEE Press.

Bansal, D., & Sofat, S. (2010). Use of cross layer interactions for detecting denial of service attacks in WMN. In *Proceedings of the 2010 14th International Telecommunications Network Strategy and Planning Symposium (NETWORKS)*, (pp. 1-6). NETWORKS.

Bose, S., & Kannan, A. (2008). Detecting denial of service attacks using cross layer based intrusion detection system in wireless ad hoc networks. In *Proceedings of the International Conference on Signal Processing, Communications and Networking (ICSCN 2008)*, (pp. 182-188). ICSCN.

Carneiro, G., Ruela, J., & Ricardo, M. (2004). Cross-layer design in 4G wireless terminals. *IEEE Wireless Communications*, *11*(2), 7–13. doi:10.1109/MWC.2004.1295732

Chang, R. I., Liu, C. T., Wang, T. C., & Wang, C. H. (2010). Improve QoS of mobile multimedia streaming via cross-layer predictive packet transmission over MIPv4. *Journal of Internet Technology*, *11*(4), 529–542.

Chao, H. C., Chang, C. Y., Chen, C. Y., & Chang, K. D. (2009). Survey of Cross-layer optimization techniques for wireless networks. In Adibi, S., Mobasher, A., & Tofigh, T. (Eds.), *The Fourth-Generation Wireless Networks: Applications and Innovations* (pp. 453–468). Hershey, PA: IGI Global. doi:10.4018/978-1-61520-674-2.ch021

Chen, C. Y., & Chao, H. C. (2011). A survey of key distribution in wireless sensor networks. *Security and Communication Networks*. Retrieved from http://onlinelibrary.wiley.com/doi/10.1002/sec.354/abstract.

Chiang, J. T., & Hu, Y. C. (2011). Cross-layer jamming detection and mitigation in wireless broadcast networks. *IEEE/ACM Transactions on Networking*, *19*(1), 286–298. doi:10.1109/TNET.2010.2068576

Foukalas, F., Gazis, V., & Alonistioti, N. (2008). Cross-layer design proposals for wireless mobile networks: A survey and taxonomy. *IEEE Communications Surveys & Tutorials*, *10*(1), 70–85. doi:10.1109/COMST.2008.4483671

Joseph, J. F. C., Das, A., Seet, B. C., & Lee, B. S. (2007). Cross layer versus single layer approaches for intrusion detection in MANETs. In *Proceedings of the 15th IEEE International Conference on Networks (ICON 2007)*, (pp. 194-200). IEEE Press.

Joseph, J. F. C., Das, A., Seet, B. C., & Lee, B. S. (2008). CRADS: Integrated cross layer approach for detecting routing attacks in MANETs. In *Proceedings of the IEEE Wireless Communications and Networking Conference (WCNC 2008)*, (pp. 1525-1531). IEEE Press.

Joseph, J. F. C., Lee, B. S., Das, A., & Seet, B. C. (2011). Cross-layer detection of sinking behavior in wireless ad hoc networks using SVM and FDA. *IEEE Transactions on Dependable and Secure Computing*, *8*(2), 233–245. doi:10.1109/TDSC.2009.48

Kawadia, V., & Kumar, P. R. (2005). A cautionary perspective on cross-layer design. *IEEE Wireless Communications*, *12*(1), 3–11. doi:10.1109/MWC.2005.1404568

Lazos, L., & Poovendran, R. (2004). Cross-layer design for energy-efficient secure multicast communications in ad hoc networks. In *Proceedings of the 2004 IEEE International Conference on Communications (ICC)*, (Vol. 6), (pp. 3633-3639). IEEE Press.

Li, S., & Ephremides, A. (2006). Anonymous routing: A cross-layer coupling between application and network layer. In *Proceedings of the 2006 40th Annual Conference on Information Sciences and Systems*, (pp. 783-788). IEEE.

Liu, Y., Li, Y., & Man, H. (2005). Short paper: A distributed cross-layer intrusion detection system for ad hoc networks. In *Proceedings of the First International Conference on Security and Privacy for Emerging Areas in Communications Networks (SecureComm 2005)*, (pp. 418-420). SecureComm.

Muraleedharan, R., & Osadciw, L. A. (2006a). Cross layer denial of service attacks in wireless sensor network using swarm intelligence. In *Proceedings of the 2006 40th Annual Conference on Information Sciences and Systems*, (pp. 1653-1659). IEEE.

Muraleedharan, R., & Osadciw, L. A. (2006b). Security: Cross layer protocol in wireless sensor network. In *Proceedings of the 25th IEEE International Conference on Computer Communications (INFOCOM 2006)*, (pp. 1-2). IEEE Press.

Ramachandran, S., Fairhurs, G., Luglio, M., Roseti, C., & Provenzano, S. (2007). Network layer security: Design for a cross layer architecture. In *Proceedings of the International Workshop on Satellite and Space Communications (IWSSC 2007)*, (pp. 271-275). IEEE.

Roman, R., Lopez, J., & Najera, P. (2011). A cross-layer approach for integrating security mechanisms in sensor networks architectures. *Wireless Communications and Mobile Computing*, *11*, 267–276. doi:10.1002/wcm.1006

Saglam, O., & Dalkilic, M. E. (2009). Cross layer implementation of key establishment protocols in wireless sensor networks. In *Proceedings of the 24th International Symposium on Computer and Information Sciences (ISCIS 2009)*, (pp. 346-351). ICIS.

Shao, M., Hu, W., Zhu, S., Cao, G., Krishnamurthy, S., & Porta, T. L. (2009). Cross-layer enhanced source location privacy in sensor networks. In *Proceedings of the 6th Annual IEEE Communications Society Conference on Sensor, Mesh and Ad Hoc Communications and Networks, (SECON 2009)*, (pp. 1-9). IEEE Press.

Shao, M., Zhu, S., Cao, G., Porta, T. L., & Mohapatra, P. (2008). A cross-layer dropping attack in video streaming over ad hoc networks. In *Proceedings of the 4th International Conference on Security and Privacy in Communication Networks, (SecureComm 2008)*. SecureComm.

Shrestha, R., Han, K. H., Choi, D. Y., & Han, S. J. (2010). A novel cross layer intrusion detection system in MANET. In *Proceedings of the 2010 24th IEEE International Conference on Advanced Information Networking and Applications (AINA)*, (pp. 647-654). IEEE Press.

Silva, H., Holanda, R., Nogueira, M., & Santos, A. (2010). A cross-layer and adaptive scheme for balancing performance and security on WMN data routing. In *Proceedings of the 2010 IEEE Global Telecommunications Conference (GLOBECOM 2010)*, (pp. 1-5). IEEE Press.

Srivastava, V., & Motani, M. (2005). Cross-layer design: A survey and the road ahead. *IEEE Communications Magazine*, *43*(12), 112–119. doi:10.1109/MCOM.2005.1561928

Thamilarasu, G., Balasubramanian, A., Mishra, S., & Sridhar, R. (2005). A cross-layer based intrusion detection approach for wireless ad hoc networks. In *Proceedings of the 2005 IEEE International Conference on Mobile Adhoc and Sensor Systems Conference (MASS) Workshop, WSNS 2005*, (pp. 861-868). IEEE Press.

Thamilarasu, G., & Sridhar, R. (2007a). Toward building a multi-level robust intrusion detection architecture for distributed mobile networks. In *Proceedings of the 27th International Conference on Distributed Computing Systems Workshops (ICDCSW 2007)*, (pp. 5-13). ICDCSW.

Thamilarasu, G., & Sridhar, R. (2007b). Exploring cross-layer techniques for security: Challenges and opportunities in wireless networks. In *Proceedings of the IEEE Military Communications Conference (MILCOM 2007)*, (pp. 1-6). MILCOM.

Thamilarasu, G., & Sridhar, R. (2009). XLSEC - A distributed cross-layer framework for security in wireless sensor networks. In *Proceedings of the 6th IEEE Consumer Communications and Networking Conference (CCNC 2009)*, (pp. 1-2). IEEE Press.

Urbano Fullana, A., Ferrer Gomila, J. L., Payeras Capellá, M., Hinarejos Campos, M., & Huguet Rotger, L. (2011). Paper. In *Proceedings of the 2011 4th IFIP International Conference on New Technologies, Mobility and Security (NTMS)*, (pp. 1-5). IFIP.

Wang, W., Sun, Y., Li, H., & Han, Z. (2010). Cross-layer attack and defense in cognitive radio networks. In *Proceedings of the 2010 IEEE Global Telecommunications Conference (GLOBECOM 2010)*, (pp. 1-6). IEEE Press.

Wang, X., Wong, J. S., Stanley, F., & Basu, S. (2009). Cross-layer based anomaly detection in wireless mesh networks. In *Proceedings of the Ninth Annual International Symposium on Applications and the Internet (SAINT 2009)*, (pp. 9-15). SAINT.

Wen, H., Gong, G., & Ho, P. H. (2010). MIMO cross-layer secure communication architecture based on STBC. In *Proceedings of the 2010 IEEE Global Telecommunications Conference (GLOBECOM 2010)*, (pp. 1-6). IEEE Press.

Wu, T. Y., Chen, C. Y., Chang, K. D., Chung, D., & Chao, H. C. (2007b). Pave the way to future smart living space - Cross-layer enhanced AAA for 4G core network. *International Journal of Smart Home*, *1*(2), 191–203.

Wu, T. Y., Chen, C. Y., Chen, C. H., Chang, K. D., Chung, J. H., & Chao, H. C. (2007a). Cross-layer AAA for various heterogeneous systems over the 4G core network environments. *Journal of Internet Technology*, *8*(2), 157–163.

Xie, L.-L., & Kumar, P. R. (2004). A network information theory for wireless communication: Scaling laws and optimal operation. *IEEE Transactions on Information Theory*, *50*(5), 748–767. doi:10.1109/TIT.2004.826631

Zeng, P., Hu, R., Yang, Y., & Song, J. (2010). A cross-layer key management scheme in ad hoc network. In *Proceedings of the 2010 2nd International Conference on Signal Processing Systems (ICSPS)*, (Vol. 1), (pp. 132-137). ICSPS.

Zhang, Y., & Lee, W. (2000). Intrusion detection in wireless ad-hoc networks. In *Proceedings of the 6th Annual International Conference on Mobile Computing and Networking (MobiCom)*, (pp. 275-283). MobiCom.

Zhou, L., & Chao, H. C. (2011). Multimedia traffic security architecture for internet of things. *IEEE Network*, *25*(3), 29–34. doi:10.1109/MNET.2011.5772059

Zhou, L., Geller, B., Wang, X., Wei, A., Zheng, B., & Chao, H. C. (2009). Multi-user video streaming over multiple heterogeneous wireless networks: A distributed, cross-layer design paradigm. *Journal of Internet Technology*, *10*(1), 1–11. doi:10.1145/1552291.1552295

ADDITIONAL READING

Carneiro, G., Ruela, J., & Ricardo, M. (2004). Cross-layer design in 4G wireless terminals. *IEEE Wireless Communications*, *11*(2), 7–13. doi:10.1109/MWC.2004.1295732

Chao, H. C., Chang, C. Y., Chen, C. Y., & Chang, K. D. (2009). Survey of cross-layer optimization techniques for wireless networks. In Adibi, S., Mobasher, A., & Tofigh, T. (Eds.), *The Fourth-Generation Wireless Networks: Applications and Innovations* (pp. 453–468). Hershey, PA: IGI Global. doi:10.4018/978-1-61520-674-2.ch021

Foukalas, F., Gazis, V., & Alonistioti, N. (2008). Cross-layer design proposals for wireless mobile networks: A survey and taxonomy. *IEEE Communications Surveys & Tutorials*, *10*(1), 70–85. doi:10.1109/COMST.2008.4483671

Kawadia, V., & Kumar, P. R. (2005). A cautionary perspective on cross-layer design. *IEEE Wireless Communications*, *12*(1), 3–11. doi:10.1109/MWC.2005.1404568

Srivastava, V., & Motani, M. (2005). Cross-layer design: A survey and the road ahead. *IEEE Communications Magazine*, *43*(12), 112–119. doi:10.1109/MCOM.2005.1561928

Chapter 7
Protocols and Applications of Cross-Layer in Mobility Management

Yuh-Shyan Chen
National Taipei University, Taiwan

Yun-Wei Lin
National Taipei University, Taiwan

ABSTRACT

The design of the mobility management is the key issue in the next-generation mobile networks. It is important to provide seamless service switching for a Mobile Station (MS) or a Mobile Node (MN) with uninterrupted services during an IP-based session between different access networks, where IP convergence has led to the co-existence of several IP-based wireless access technologies and the emergence of next generation technologies. Many recent research results of cross-layer design in mobility aim to minimize the data loss rate and delay time during switching so that users do not experience obvious and unacceptable interruptions during the handoff. The cross-layer design is the important approach for mobility management. This chapter mainly introduces and reviews existing protocols and then discusses the possible interest and useful applications of cross-layer in mobility management.

INTRODUCTION

With the rapid development of wireless technologies, mobile networks have become more popular to provide IP-based media applications. To provide the seamless handoff for a mobile station moving between different base stations of the mobile networks, a mobility management (Siddiqui, et al., 2006) with low handoff delay is very important for the next-generation mobile networks. Existing handoff results need to perform the layer 2 (link layer) handoff operation to establish new link to the new base station, and then perform the layer 3 handoff operation to maintain network connectiv-

DOI: 10.4018/978-1-4666-0960-0.ch007

ity. Existing handoff results thus offer the large handoff latency and the high packet loss rate. It is very useful to have the cross-layer design for the handoff mechanism to significantly improve the handoff latency and packet loss rate to provide a seamless handoff result.

Currently, various wireless technologies have been specified to provide wireless access to the internet, such as such as WLAN (IEEE 802.11) (Chen, et al., 2009), WiMAX (IEEE 802.16) (WiMAX Forum, 2011), and LTE (Long Term Evolution) (3G, 2011; 3GPP, 2009). The IEEE 802.11 specifications provide a wireless alternative to Ethernet LANs (Local Area Networks). First, the IEEE 802.11 standard has the small coverage area and contention based resource allocation, which results in frequent handoffs, ineffective resource allocation, and intolerable handoff latency. The IEEE 802.16-2004 standard (IEEE, 2004) initially defines physical and Medium Access Control (MAC) layer operations of Broadband Wireless Access (BWA) systems to support the last mile connectivity at high data rate and large transmission range in wireless metropolitan area networks. WiMAX (Worldwide Interoperability for Microwave Access) has been specified in IEEE 802.16 standards for metropolitan networks. The IEEE 802.16-2004 standards are only for the fixed wireless connections; a further amendment, the IEEE 802.16m/j standard (IEEE, 2005), is specified to support mobility for mobile devices. The Third Generation Partnership Project (3GPP) recently has proposed a 3G Long-Term Evolution (LTE) (3G, 2011; 3GPP, 2009) toward the fourth-generation (4G) cellular systems. LTE is based on the Universal Terrestrial Radio Access (UTRA) and High-Speed Down-Link Packet Access (HSDPA) and further strengthen its communications capacity to upload in order to enhance its quality of service. The LTE coexists with 2G/3G systems, WLAN, WiMAX, etc. Therefore, the 4G (Akyildiz, et al., 2004; Raivio, 2001; Raychaudhuri, 2004) heterogeneous networks (Bari, et al., 2009) aim to integrate many up-to-date wireless technologies, including IEEE 802.11, WiMAX, and LTE technologies, together to provide multimedia services. In 4G heterogeneous networks, a mobile multiple-mode device can move between wireless networks. The design of the cross-layer handover protocols is very important for the mobile networks.

Basically, the handover protocol can be divided into the Layer-2 (L2) handoff operation and the Layer-3 (L3) handoff operation. The L2 handoff operation (IEEE 802.16e, 2005) is that every BS exchanges channel information and Mobile Station's (MS) handoff parameters over the backbone network to speed up L2 handoff. The MSs are advertised by their serving BSs of neighbor BSs' information to support Mobile Station (MS) scanning and handoff procedures. The L2 handoff latency contains probe (channel scanning), authentication, and reassociation delay times (Chen, et al., 2008). With the support for the mobility in L2, a flawless mobility management still requires a feasible handoff operation in L3.

A Layer-3 (L3) handoff operation is required to deal with the problem of the IP mobility. Mobility support in internet protocol version 6 (MIPv6) (Johnson, et al., 2004) has been proposed by Internet Engineering Task Force (IETF) Request For Comments (RFC) 3775. In MIPv6 protocol, each MS is always identified by the home address. While situated away from the home network, a MS is also associated with a Care-of Address (CoA), which provides information about the MS's current location. The protocol enables IPv6 nodes to cache the binding of a MS's home address with its CoA, and then to send any packets destined for the MS directly to it by the CoA. MIPv6 protocol offers a solution to solve the IP mobility, but due to intolerable high data lost rate and long handoff latency. Many research works have been developed to improve MIPv6 protocol. An improved protocol, called hierarchical mobile IPv6 had been proposed by Internet Engineering Task Force (IETF) Request For Comments (RFC) 4140 (Soliman, et al., 2005), which extends MIPv6 protocol to allow for both micro mobility and macro mobility handling. Hierarchical mobility management for Mobile IPv6 is designed to reduce the amount of signaling between the MS, the Correspondent Node (CN), and the Home Agent (HA). The Mobility Anchor Point (MAP) is created to improve the performance of MIPv6 protocol in terms of handoff latency.

The SIP protocol is an IETF-defined signaling protocol (3GPP, 2006), widely used for controlling multimedia communication sessions such as voice and video calls over Internet Protocol (IP), which has been selected as the main signaling protocol in the UMTS/HSDPA IMS (IP Multimedia Subsystem), to support the multimedia service management and the session mobility scheme (Schulzrinne, et al., 2000). A main component in the IMS, Call Session Control Function (CSCF), is used to control an SIP session (Zeadally, et al., 2007). The SIP protocol cannot directly provide a seamless handover scheme to the mobile multiple-mode device for 4G heterogeneous networks. Therefore, mobile SCTP (mSCTP) (Stewart, et al., 2000; Koh, et al., 2004) had designed to support flexible association for wide application services in transport-layer (Layer 7 or L7) using multi-homing scheme. The multi-homing approach can supports soft and seamless handover scheme. To maintain connectivity and robustness of the connection, the mobile multiple-mode device requires a fast and seamless mobility management scheme for the heterogeneous networks. It is interest to develop a fast handover scheme with handover policy over the heterogeneous networks.

To reduce the handoff latency and enhance the reliability, many cross-layer mobility management protocols are proposed in the literature. The survey of existing cross-layer design in mobility is organized into five parts. The first part introduces a cross-layer relay-based mobility approach. The second part investigates a cross-layer session mobility approach. The third part discusses cross-layer MIH-based mobility approaches. The fourth part describes a cross-layer cognitive mobility approach. Finally, applications of cross-layer mobility are introduced.

In the first part, cross-layer relay-based mobility approaches are introduced, termed as P_HMIPv6 protocol, for WLAN (Chen, et al., 2009) and WiMAX (Chen, et al., 2011a) systems. The P_HMIPv6 protocol is a cross-layer, L2 + L3, approach, which is based on HMIPv6. The basic idea of the partner node is to perform the pre-handoff procedures. The Partner Station (PS) is a static MS which has the relay ability defined in IEEE 802.16j. With the assistance of PS, the MS detects the existence of neighbor BSs and then to request some specified PSs in the coverage of the next BS to perform L3 mobility before the MS initiates the L2 handoff. The PS can concurrently perform the Duplicate Address Detection (DAD) operation of the MS's new CoA before the MS initiates the handoff request. To achieve the goal, L3 control message is combined into L2 control message. It is observed that P_HMIPv6 protocol significantly reduces the handoff delay time and packet loss rate.

In the second part, a cross-layer session mobility approach for mobile multiple-mode devices in heterogeneous networks (WLAN and UMTS/HSDPA IMS framework) is discussed by taking advantage of cross-layer design (Chiu, et al., 2011a). This work concentrates on cross-layer design to utilize the SIP protocol to perform handover decision and maintain handover procedure in application-layer, and utilizes the mSCTP scheme to switch different heterogeneous networks in transport-layer. To achieve the seamless handover between the heterogeneous networks, a SIP-based mSCTP scheme, called as SmSCTP, is described (Chiu, et al., 2011a).

In the third part, cross-layer MIH-based mobility approaches are explained. IEEE 802.21 (Das, et al., 2011) is developing standards to enable handover and interoperability between heterogeneous network types including both 802 and non-802 networks. IEEE 802.21 standard working group has created framework that defines a Media

Independent Handover (MIH) Function at Layer 2.5 (L2.5), which will help mobile terminals to roam seamlessly across heterogeneous access networks. Some interested cross-layer results to combine SIP (L7) and MIH (L2.5) protocols are discussed.

In the fourth, a cross-layer cognitive mobility approach (Chen, et al., 2011b) is introduced. This is a cross-layer protocol of spectrum mobility (L2) and handover (L3) in cognitive LTE networks (Chen, et al., 2011b) is investigated. With the consideration of the Poisson distribution of spectrum resources, a cross-layer handoff protocol with the minimum expected transmission time is developed in cognitive LTE networks.

In the last part, some applications of cross-layer mobility are introduced. First, a cross-layer design vehicle-aided handover scheme (Chiu, et al., 2011b) is described in vehicular networks. Some other useful applications are also explained.

The remainder of this chapter is organized as follows. This chapter initially gives a basic concept of the cross-layer mobility managements. This chapter then provides the generic aspect of the cross-layer mobility managements. Applications of cross-layer mobility management are then discussed. Finally, we conclude this chapter and give some possible future works.

BASIC CONCEPT

This section describes the basic background of handover (L2/L3) and spectrum mobility (L2) protocols in WiMAX system, heterogeneous networks, and cognitive-based networks. This subsection provides the traditional mobile IPv6 based approach, introduces the existing mobility approaches in WiMax systems, and discusses the mobility approaches on heterogeneous networks. Furthermore, spectrum mobility approaches in cognitive-based networks are described.

Mobile IPv6 Based Approach

Mobile IPv6 (MIPv6) protocol (Johnson, et al., 2004) has originally been proposed to handle the IP mobility. This protocol allows a MS to maintain connections while moving between different subnets. Each MS is identified by its Home Address (HoA). When connecting to a foreign network, a MS receives the Router Advertisement (RA) to obtain the foreign network prefix and then autoconfigure a CoA. After configuring a CoA, the MS needs to perform the DAD (Duplicate Address Detection) to ensure that the CoA is unique. If the CoA is available, the MS sends the location information to a Home Agent (HA) to perform address binding. MIPv6 protocol (Johnson, et al., 2004) is capable of handling IP handoff between different subnets in a transparent way for the high level connections. The handoff latency resulting from MIPv6 protocol is often unacceptable for the real-time services; such as Voice Over IP (VoIP). Sharma *et al.* (2004) investigated the mobile IP mechanism in the network layers. This scheme must significantly modify the underlying networking infrastructure as well as mobile devices.

It is observed that Hierarchical Mobile IPv6 protocol (HMIPv6) (Soliman, et al., 2005) and Fast Handoff for Mobile IPv6 protocol (FMIPv6) (Koodli, et al., 2005) are recently proposed to reduce the handoff latency and packet loss. First, the basic idea of HMIPv6 is to use the hierarchical network architecture for the local mobility handling to reduce the number of control messages between the MS, Correspondent Nodes (CN), and Home Agent (HA). HMIPv6 protocol reduces the amount of messages. A useful component, Mobility Anchor Point (MAP), acts as a local HA for the micro mobility handling. A MS has two sub-parts of CoA, Local CoA (LCoA), and Regional CoA (RCoA). LCoA is used for binding LCoA with MAP and RCoA is used for binding with HA. When a MS is under the same MAP, a local binding update is only performed from MS to MAP. The main time cost of L3 handoff latency in

Table 1. Comparison of the traditional handover approaches

Properties	Methods			
	MIPv6	**HMIPv6**	**FMIPv6**	**SHMIPv6**
Handover type	Hard	Soft	Soft	Soft
Protocol layer	Layer 3	Layer 3	Layer 3	Layer 3
Handover latency	High	Medium	Low	Low
Header overhead	No	Yes	Yes	Yes

HMIPv6 is to perform DAD procedures (Thomson, et al., 2002). After finishing the DAD procedures for LCoA and RCoA, MS then performs binding update using the new generated LCoA and RCoA to HA and CN, respectively. The FMIPv6 (Koodli, et al., 2005) protocol focuses on the predicting and preparing the impending handoff in advance. By sending Fast Binding Update (FBU) before the L2 handoff operation, PAR is able to notify NAR to perform the DAD procedure and response to PAR for the DAD result. The tunnel between PAR and NAR is established, and the packets are buffered in the NAR until the MS has finished the L2 handoff. The purpose of the design of the fast handoff for MIPv6 is to eliminate the handoff latency. Unfortunately, anticipation of fast handoff does not guarantee that the MS can always have enough time to exchange messages for the fast handoff if those mobile nodes move with high speed. Most previous works improves the fast handoff for MIPv6 focused on how to have the results of a reliable anticipation.

Chao *et al.* (2003) proposed a micro-mobility mechanism in an integrated ad hoc and cellular IPv6 networks to provide a smooth handoff under a high-speed movement. This protocol utilizes dynamic access routers to pre-execute the sub-binding operation to CN for a MH. A multicast operation is used to send the same packets to a lot of access points to satisfy the purpose of the smooth handoff in a high speed moving. Unfortunately, the binding update time is still same as the MIPv6 protocol. Lai *et al.* (2005) proposed a Stealth-Time HMIPv6 (SHMPv6) protocol to further improve handoff latency. This approach reduces DAD delay time using the pre-handoff notification scheme and reduces the packet loss rate using the buffer technique. The main idea is to use buffer technique in the previous MAP (pMAP) to buffer the data packets from CN. Overall, SHMIPv6 protocol reduces the DAD time for LCoA, but SHMIPv6 still not significantly reduces the DAD time for RCoA. Lee *et al.* (2005) also proposed a new protocol, called as HMIPv6, by integrated IAPP (IEEE, 2002) and access router to reduce the handoff latency. This approach uses the IAPP multicast messages to notify the Access Router (AR) of new domain to send packets to new access point. This rendezvous time, the time to finding a new AR, is reduced by using the IAPP notification. However, this protocol is worked well under all ARs must support the IAPP function. Table 1 illustrates the comparison of traditional handover approaches. The major characteristics of each method are compared, and formal definitions are shown as follows. The "handover type" can point out that a node can set up more than one connection at the same time or only one. The "protocol type" shows the OSI layer which the protocol is designed. The "handover latency" can respond the amount of jitter. The "header overhead" denotes that the packet is transmitted using tunneling or not.

Mobility Approach in WiMAX System

To provide the session continuity during handoff in WiMAX (IEEE, 2005) network, some researches

Figure 1. The comparison of FMIPv6-based handoff schemes in WiMAX systems

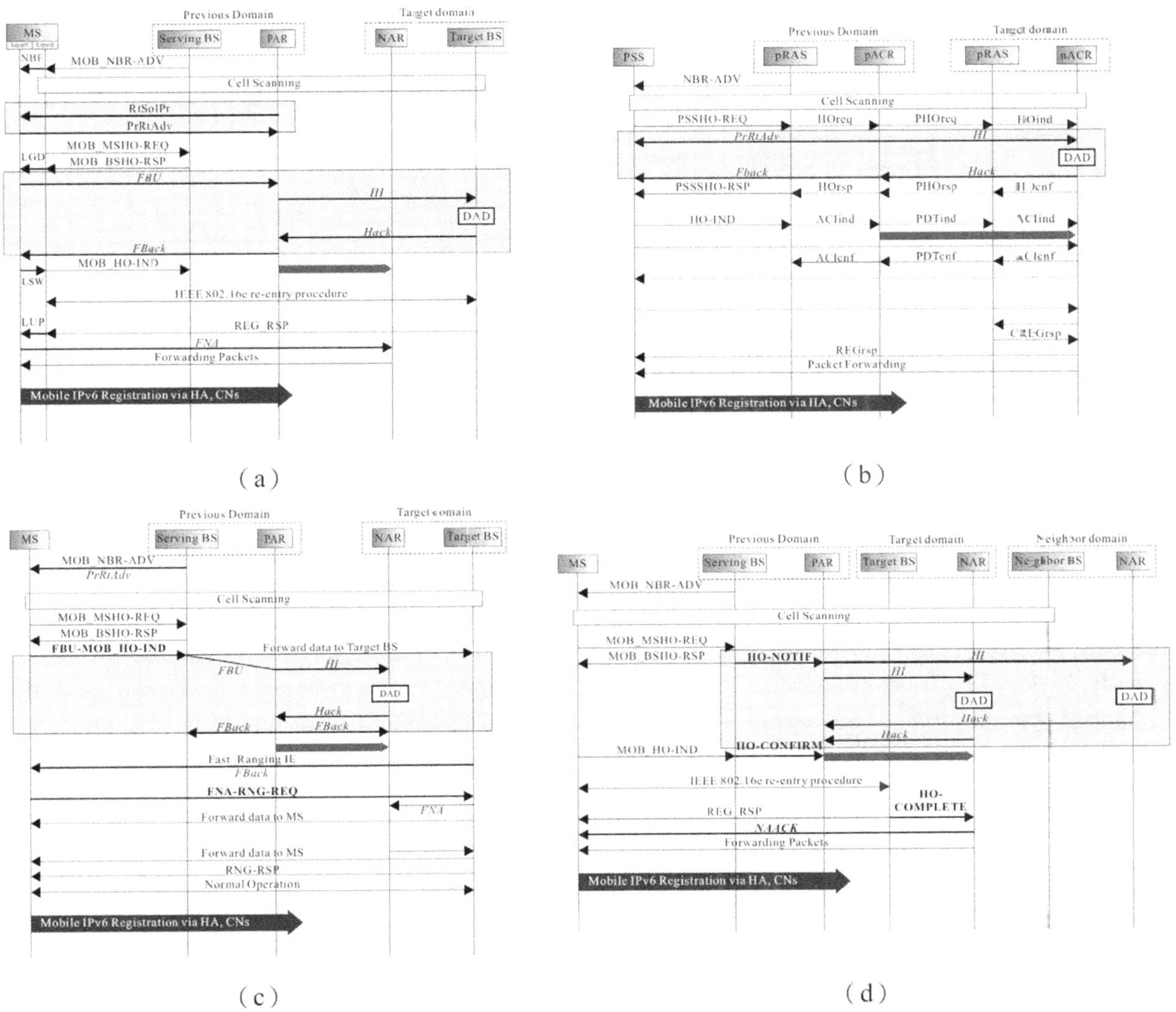

had been proposed to apply the L3 handoff protocol over the existing L2 handoff. Figure 1 gives the comparison of FMIPv6-based handoff schemes in WiMAX system.

Jang *et al.* (2007) first proposed FMIPv6 protocol in IEEE 802.16e systems. Figure 1a gives the example. The proposed scheme uses four triggers, named New BS Found (NBF), Link Going Down (LGD), Link Up (LUP), and Link Switch (LSW), defined in IEEE 802.21 (IEEE, 2006) to control the timing of mobility procedures in IP layer and link layer. In addition, Shim *et al.* (2006) proposed an FMIPv6-based scheme in WiBro system. Figure 1b shows that this scheme combines L3 messages with L2 message; the handoff latency is thus reduced. In WiBro system, the Radio Access Station (RAS) has the ability of the packet buffering. When Portable Subscriber Station (PSS) switches its access link, both the new streams and undelivered packets can be delivered at the same time. The out-of order problem has occurred; this problem is solved by using previous Access Control Router (pACR) buffering. Chen, *et al.* (2007) had proposed an FMIPv6-based

Table 2. Comparison of the handover approaches in WiMAX

Properties	Methods			
	FMIPv6	**NBF**	**FMIPv6-based scheme**	**FMIPv6-enhanced scheme**
Handover type	Soft	Soft	Soft	Soft
Protocol layer	Layer 3	Combination of L3 msg. with L2 msg.	Combination of L3 msg. with L2 msg.	Layer 3
Handover latency	High	Low	Low	Low
Header overhead	Yes	Yes	Yes	Yes

scheme in IEEE 802.16e systems. Figure 1c illustrates that the proposed scheme also combines L3 message with L2 message and uses PAR (Previous Access Router) for the data buffering. The main difference of this work is to change the timing to perform the DAD and tunnel between PAR and NAR (new access router). By overlapping the L2 handoff delay time and FMIPv6 procedure, the MS precisely predicts the next target BS. Finally, Park *et al.* (2006) proposed an FMIPv6-enhanced scheme in 802.16e system. Figure 1d shows that the serving BS notifies its Previous Access Router (PAR) of an impending handoff based on the MAC management messages. The proposed scheme not only reduces the handoff latency but also solves the problems related to the imprecise L2-trigger generation time. The proposed scheme minimizes the packet loss because the PAR buffers the packets destined for the MS until the tunnel is set up.

All of the mentioned protocols were developed based on FMIPv6 protocol in WiMAX system. It is observed that each access router must offer extra hardware functions to successfully support the FMIPv6 operation, such as tunnels between access routers and the buffer capability. In addition, the main problem of FMIPv6 is the timing of the L2 trigger. If the timing assumes to be lately, FMIPv6 cannot guarantee MSs have enough time to exchange messages for the fast handoff. If the timing assumes to be early, the tunnel between PAR and NAR has been established but the MS cannot perform L2 handoff or switch to a different BS in different subnet. This causes the wrong data forwarding. Table 2 shows the protocols comparison.

Mobility Approach on Heterogeneous Networks

The next generation mobile networks (4G) is expected to integrate a large number of heterogeneous wireless systems. The vertical handoff is a popular approach for the session mobility. The vertical handoff coordinates different wireless interfaces in a mobile device and can maintain a connection between two or more wireless heterogeneous networks. Schulzrinne *et al.* (2000) initially examined the concept of the application layer mobility using SIP mechanism. Inoiue *et al.* (2004) developed the Multimedia Integrated Network by Radio Access Innovation (MIRAI) server to provide the seamless interworking. It works both as a Basic Access Signaling (BAS) agent and mobile IP agent. But, the MIRAI server still suffers high handoff latency and high traffic overhead. Wu *et al.* (2005) presented an SIP-based hard handoff procedure by using re-INVITE mechanism to provide the session mobility on the heterogeneous networks. Azhari *et al.* (2005) had proposed the Vertical Handoff Support Node (VHSN) protocol on WLAN-to-Cellular networks to forward and redirect a connection to a mobile device via a cellular network. When an abrupt connection is lost on a WLAN, the VHSN intercepts and redirects the media flow through the local cellular base station. Sur *et al.* (2005) proposed a framework based on cross layer rules in the Wideband Code Division Multiple Access (WCDMA) and WLAN wireless networks, by introducing a core component, called the Rule Engine (RE), to collect the information from multiple layers of a protocol stack or from other databases. The RE uses the collected information to successfully predict the handoff. In addition, Smadi *et al.* (2006) had developed the Dynamically Anchored Conferencing Handoff (DACH) protocol for the soft handoff between WLAN and UMTS networks. The DACH protocol utilizes the concept of anchor to handle with the connection sessions. Salsano *et al.* (2006) introduced the Back-to-Back User Agent (B2BUA), a bottleneck component, for managing sessions for the purpose of seamless vertical handover based on SIP session border controllers. Kim *et al.* (2009) proposed a novel re-transmission scheme for mSCTP protocol to solve the packet re-ordering problem in heterogeneous wireless networks. While switching different SCTP connections, a mobile multiple-mode device can fast retransmit non-acknowledged packets through another high speed wireless interface. Finally, Oh *et al.* (2009) try to develop a seamless handover scheme in IPv6 mobile network.

Spectrum Mobility Approach in Cognitive-based Networks

Cognitive radio technique (Mitola, et al., 1999) is the next step toward efficient wireless bandwidth utilization. While some of the spectrum bands (unlicensed band) have been increasingly used, most of the other spectrum resources (licensed band) are underutilized. This drives the challenges of open spectrum and dynamic spectrum access concepts, which allows unlicensed users (or called Secondary Users, SUs) equipped with cognitive radios to opportunistically access the spectrum not used by licensed users (or called Primary Users, PUs). The Cognitive Radio (CR) concept is first presented by Mitola and Maguire (Mitola, et al., 1999). In CR networks, there are two types of users; one is licensed users (primary users or PUs), and another is unlicensed users (secondary user or SUs). PUs can access the wireless network resources anytime and anywhere. SUs equipped with cognitive radios to opportunistically access the spectrum not used by PUs. When a PU reclaims the radio spectrum resources, the SU detects the reclamation and immediately moves away the current radio spectrum resources and switch to idle radio spectrum resources.

To perform above operations of PUs and SUs in CR networks, the spectrum sensing, the spectrum management, the spectrum sharing, and the spectrum mobility (the L2 mobility problem) are four important functions. Spectrum sensing is the first important issue to find an idle radio spectrum and detect the appearance of PUs. The Dynamic Spectrum Allocation (DSA) is the main L2 task of CR system to exploit the temporal and spatial traffic statistics to allocate the underutilized spectrum to SUs. Jo and Cho (Jo, et al., 2008) proposed a spectrum matching algorithms to efficiently allocate spectrum holes to SUs in the CR architecture. To minimize the probability of spectrum mobility, the spectrum matching algorithm is executed based on the different sizes of the underutilized spectrum holes and the different holding times, while the holding time is the predicted holding time for a specific spectrum hole. The spectrum matching algorithm selects and assigns a spectrum hole which is the most closed to the real-spectrum holding time. Akbar and Tranter (Akbar, et al., 2007) had proposed a Markov-Based Channel Prediction Algorithm (MCPA), where the channel occupancy of PUs is assumed by Poisson distribution. This algorithm utilizes the Hidden Markov Model (HMM) to predict the appearance of spectrum holes in different spectrum bands. In addition, Hoyhtya *et al.* (2008) proposed a simple classification and learning of predictive channel selection method. To choose a channel with the largest idle time, this

Table 3. Comparison of the handover approaches in WiMAX

	Methods		
Properties	Spectrum matching algorithm	MCPA	Predictive channel selection method
Protocol layer	Layer 2	Layer 2	Layer 2
Spectrum mobility latency	Medium	Low	Low
Scalability	Medium	High	High
Flexibility	Medium	High	High

predictive channel selection method is performed based on the statistics of spectrum sensing results and the channel history. Table 3 gives the protocol comparison. The major characteristics of each method are compared, and formal definitions are shown as follows. The "protocol type" shows the OSI layer which the protocol is designed. The "spectrum mobility latency" can point out the time period while a node changes the communication spectrum. The "scalability" represents the ability to be enlarged to accommodate that network growth. The "flexibility" describes the ability to adapt a changeable environment.

GENERIC ASPECT

Some kinds of cross-layer mobility approaches are discussed to introduce the generic aspect of the design of cross-layer in mobility. This subsection explains the cross-layer relay-based mobility approach, introduces the cross-layer session mobility approach, and describes the cross-layer MIH-based mobility approach. Finally, the cross-layer cognitive mobility approach is presented.

Cross-Layer Relay-Based Mobility Approach

A cross-layer partner-assisted handoff, called as PHMIPv6, protocol which is L2+L3 protocol, developed by Chen *et al.* (2009) was introduced to utilize a Partner Node (PN), which is a *relay* node, with cross-layer design to improve the handoff latency during the handoff process. The Partner Node (PN) is defined below. Given a Mobile Node (MN), a neighboring node of the MN is denoted the partner Node (PN), where the MN and PN are located in different MAP (Mobility Anchor Point) domains. The PN can directly connect with an IP network through an AP (Access Point) and can directly communicate with the MN using an *ad hoc* network. The main task of the PN is to perform the pre-handoff procedure for the MN before the MN reaches a new MAP domain. Figure 2 illustrates the PHMIPv6 system architecture (Chen, et al., 2009) which is based on HMIPv6 system architecture. The PHMIPv6 protocol divides the network into two IPv6 subnet domains. MN sends the data packets from the AP and previous access router, AR (pAR), to the CN through the previous MAP (pMAP). That is, CN sends data packets to the RCoA of MN, and MAP then forward the packets to the LCoA of MN. While MN moves into a new MAP (nMAP) domain, MN performs the registration procedure to its nMAP. The macro-mobility is occurred if MN switches from a pMAP to a nMAP domain. Then, MN must acquire a new unique CoA to register the CoA to new access router (nAR) and nMAP. Observe that, in the PHMIPv6 protocol, MN performs the registration procedure with the assistance of PN if PN is existed during the macro-mobility.

Figure 3 gives the protocol stack of PHMIPv6. The bottom layer of the PHMIPv6 protocol stack

Figure 2. The system architecture of PHMIPv6

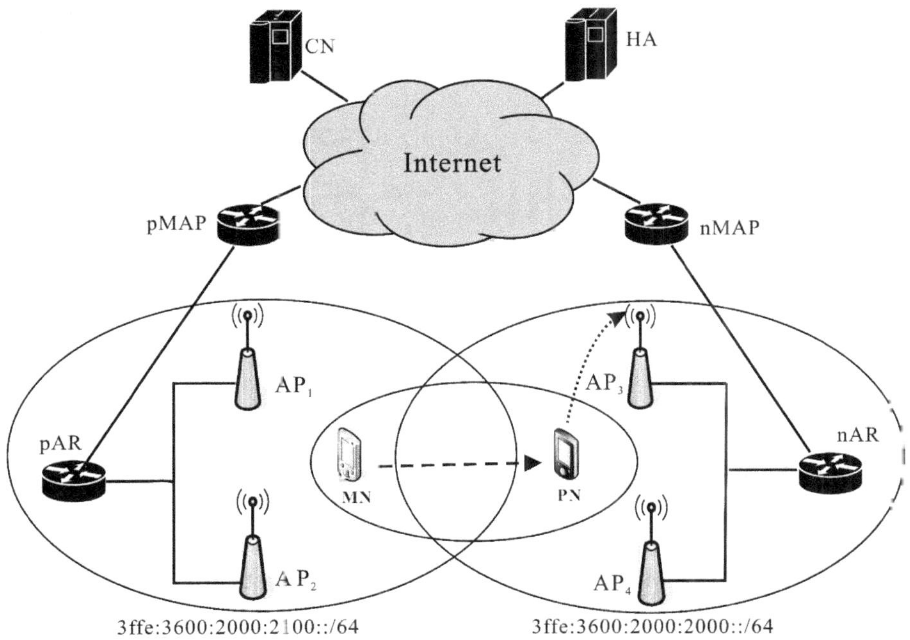

Figure 3. The protocol stack of the PHMIPv6

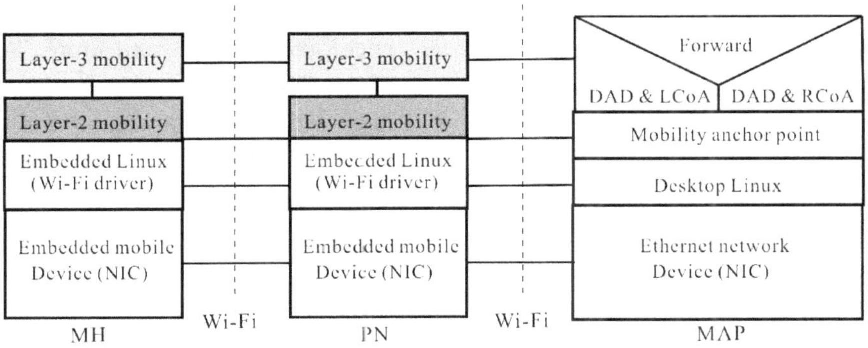

is the embedded mobile device (NIC). The second layer is the Wi-Fi card driver to control the operation of the Wi-Fi card, and the higher layer is L2 and L3 mobility management. In the PHMIPv6 protocol, the MH and PN both modify L2 and L3 mobility management parts. This work aims to investigate the cross-layer design between L2 and L3, and the upper layer is not changed. This work also does not modify any network entities of the HMIPv6 protocol. Therefore, the protocol stack of the MAP is the same as the original MAP defined in the HMIPv6 protocol. The L2 handoff procedure of the PHMIPv6 protocol adopts a DeuceScan scheme (Chen, et al., 2008), which aims to improve the accuracy of choosing a new AP depending on the quality of the surrounding APs. The L2 information is directly provided to L3. In addition, the L3 handoff procedure of the PHMIPv6 protocol uses a PN in the nMAP domain to assist the MH in pre-performing the handoff procedure.

Figure 4. The cross-layer (L2+L3) operation of the PHMIPv6

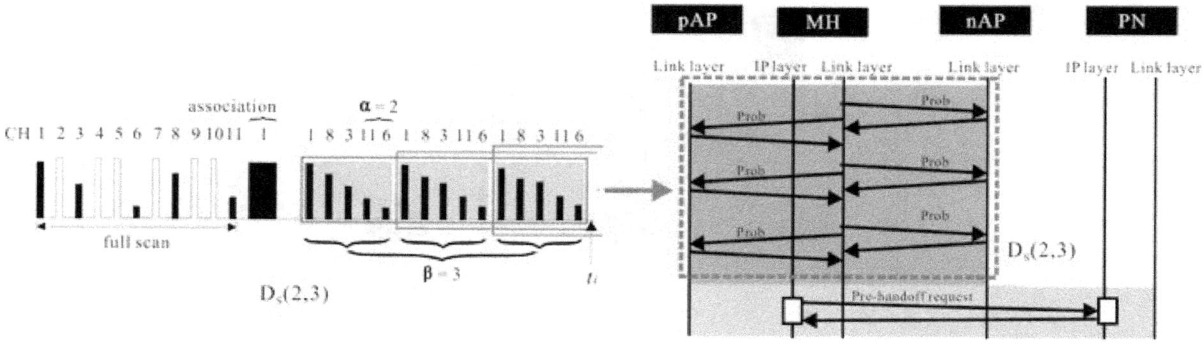

The L2 part in PHMIPv6 protocol utilizes DeuceScan protocol (Chen, et al., 2008). This protocol is a fast L2 handoff scheme to effectively reduce the probe delay for 802.11-based WLANs. The DeuceScan protocol is a pre-scan approach to significantly reduce the L2 layer handoff latency. Two factors of stable signal strength and variable of signal strength are considered in DeuceScan protocol. One key idea of the DeuceScan protocol is the deuce process, which is the partial pre-scanning operation. One more partial pre-scan operation can be executed in the same time period of one full pre-scan operation.

It is important that the designing of PHMIPv6 protocol in L3 reduces the DAD latency by using the PN to improve the handoff latency with the assistance of L2 information. Figure 4 shows the cross layer idea in PHMIPv6. The MH and the PN locate in different domains, while MH and PN communicate with each other in ad hoc mode. The DAD time for LCoA and RCoA can be reduced by the PN with the assistance of L2 information provided by DeuceScan protocol (Chen, et al., 2008). This is because that the pre-handoff procedure can be started and performed by the PN when the MH is still not switched to the new MAP domain, but when the MH is really switched to the MAP domain, the PN can immediately deliver the LCoA and RCoA to the MH, where the LCoA and RCoA are already checked by performing the DAD operation. One important result of this protocol is the cross-layer design for the fast handoff scheme. The cross-layer design is a merging of adjacent layers (L2 and L3) to improve the overall handoff latency.

The idea of partner-assisted handoff can be further applied into WiMAX system. Chen *et al.* had proposed P_HMIPv6 protocol in WiMAX (IEEE 802.16e) networks (Chen, et al., 2011a). The key difference is a new component has been specified in this architecture. When a Mobile Station (MS) wants to perform handoff due to the lack of the signal strength but the MS is near to the neighbor BSs' coverage area, this work tries to find the appropriate MSs in the neighbor BSs coverage area in nMAP domain. This component called as *partner station* or PS. The key idea of this scheme is that the PS assists MS to deal with partial impending mobility procedures in advance. A MS who wants to obtain the PSs' assistance needs to obtain PSs' DCD and UCD information to speed up the synchronization with these PSs. According to IEEE802.16e-2005 (IEEE, 2005), a BS broadcasts information about the neighboring BSs using a MOB NBR ADV message to facilitate the operation of L2 handoff procedures. To maintain the backward compatibility, this result assumes that the PS acts like a BS from the viewpoint of the attached MSs, so serving BS also broadcasts neighbor PSs' information using the MOB NBR ADV message. Because MS can

Figure 5. The PS registration operation in WiMAX systems

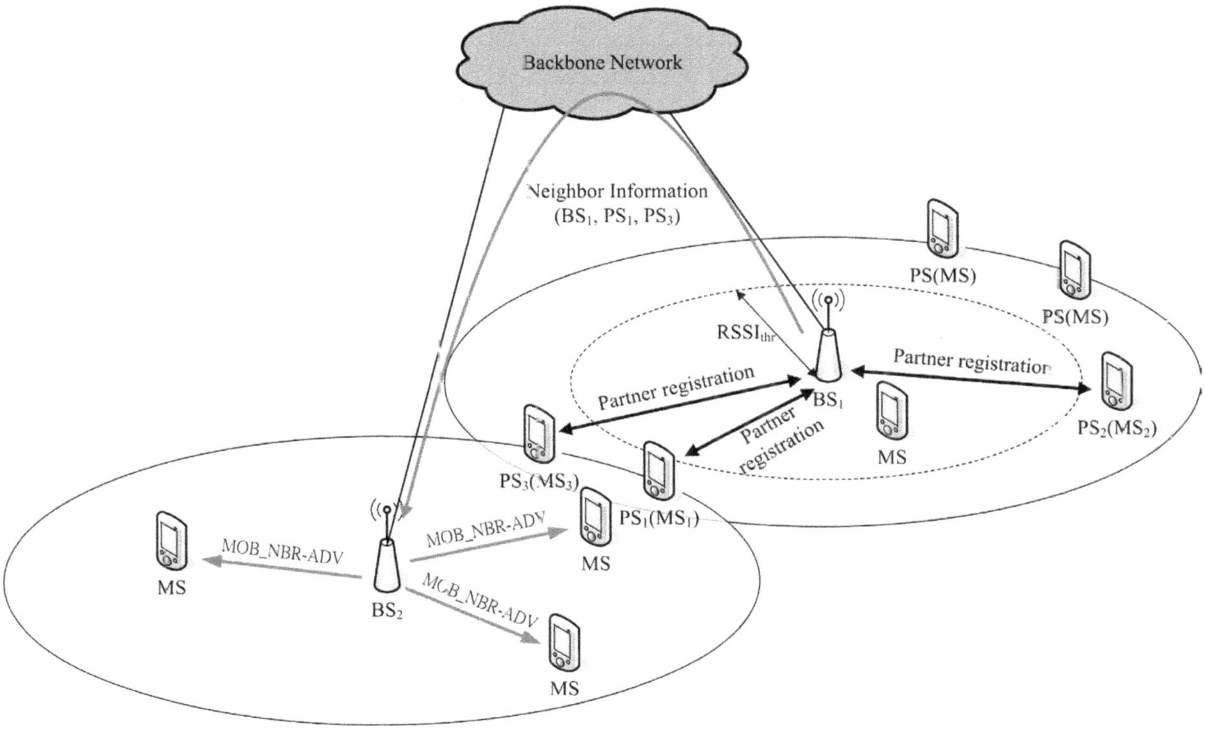

become a PS, the PS information is also included in the MOB NBR ADV message. The larger the overhead in distributing the neighbor BS and PS information and the longer the message is likely to take an MN to scan the neighbors. Figure 5 shows how to reduce the amount of neighbor PS information in MOB NBR ADV using a registration procedure. A MS who wants to become a PS needs to satisfy the following PS conditions: (1) The RSSI of the BS downlink to the MS must less than $RSSI_{thr}$ to ensure that the MS is nearly in the boundary of the neighbor BS's coverage. (2) The MS is static. (3) The MS must perform neighbor BS scanning to ensure its locality and report the scanning result to its serving BS. A MS needs to be in the boundary of neighbor BS's coverage because these PSs' main function is to extend neighbor BS's coverage to assist MS in performing impending L3 handoff. The MS assumes to be static to ensure that the MS to be PS can be stably stay in the boundary of the neighbor BS's coverage. The neighbor PSs' information is correct to all the MSs. The third condition is the main method to reduce the information of PSs in MOB NBR ADV. Because of the scanning results of PSs, the neighbor PSs' information can be shortened based on the locality. For example, Figure 5 shows that PS_1, PS_2, and PS_3 have registered the PSs to the neighbor BS_1, but with the scanning result of PSs, BS_1 only provides the information of PS_1 and PS_3, to BS_2. The information of PS_2 is not needed given by BS_2.

After satisfying the PS conditions, the MS registers with its serving BS. The purpose of the MS registration procedure is to inform the MS's serving BS if the MS satisfy the PS conditions and can to be a PS. The scanning results of PSs are reported to the serving BS in the registration procedure. Because the PSs need to connect to IP network through their serving BS and communicate with the MSs at the same time. It is assumed that a MS, who becomes PS, has the Relay Station

Table 4. The protocols comparison

Properties	Methods			
	MIPv6	HMIPv6	PHMIPv6	P_HMIPv6
Handover type	Hard	Soft	Soft	Soft
Protocol layer	Layer 3	Layer 3	Layer 2+3	Layer 2+3
Handover latency	High	Medium	Low	Low
Relay overhead	No	No	Yes	Yes
Header overhead	No	Yes	Yes	Yes

(RS) functions defined in IEEE 802.16j (IEEE, 2011). This handoff mechanism is performed as follows. The MS obtains the assistance of partner node (relay node) to perform the prehandoff procedure. The MS decides the target access point (AP) before MS finds out appropriate partner node. Therefore, the MS chooses one PS in the target AP to obtain the assistance of partner node using the ad-hoc mode communication. When the MS performs L2 handoff, the MS may not in the coverage of neighbor BSs if the overlap coverage area of two BSs is too small. The MS needs to find PSs before the target BS is selected. The unpredictable target BS results in unpredictable target PS. Due to no ad-hoc mode is defined in IEEE 802.16e, P_HMIPv6 assumes every MS has relay station functions defined in IEEE 802.16j (IEEE, 2011) to support the communication between the MS and the PS. The key idea of P_HMIPv6 is that MS requests the assistance from PS to pre-perform L3 handoff procedures before the MS actually reaches to a target BS coverage which is in a new MAP domain. As described in RFC 2461 (Narten, et al., 1998), each MS needs to wait for the RA messages for actively receiving network prefix or send Router Solicitation (RS) messages. The DAD procedure is described in RFC 2462, each MS sends Neighbor Solicitation (NS) messages whose target address is set to the address being checked to other MSs in the same domain to ensure that this address is unique, and wait for the Neighbor Advertisement (NA) messages at least one second. If no response of the NS message, the address can be assigned into the interface. The above statement shows that Neighbor Discovery and DAD procedures are the main bottleneck to reduce handoff latency of MIPv6-based scheme (Johnson, et al., 2004). The latency of macro mobility in HMIPv6 is more seriously because the handoff procedure includes the DAD procedures of the RCoA and LCoA. The basic idea of the L3 handoff approach is to preperform the DAD procedure and local binding update by the assistance of PS. Table 4 shows the protocols comparison. The major characteristics of each method are compared, and formal definitions are shown as follows. The "handover type" can point out that a node can set up more than one connection at the same time or only one. The "protocol type" shows the OSI layer which the protocol is designed. The "handover latency" can respond the amount of jitter. The "relay overhead" means that the packet is whether forwarded by a relay node or not. The "header overhead" denotes that the packet is transmitted using tunneling or not.

Cross-Layer Session Mobility Approach

Chiu *et al.* (2010a) proposed a SIP-Based mobile Stream Control Transmission Protocol (SmSCTP) over the heterogeneous (UMTS/HSDPA with WLAN) network. In heterogeneous networks, the UMTS/HSDPA network supports large transmission coverage to mobile multiple-mode devices. The WLAN network provides high transmission

Figure 6. The network architecture of SmSCTP protocol

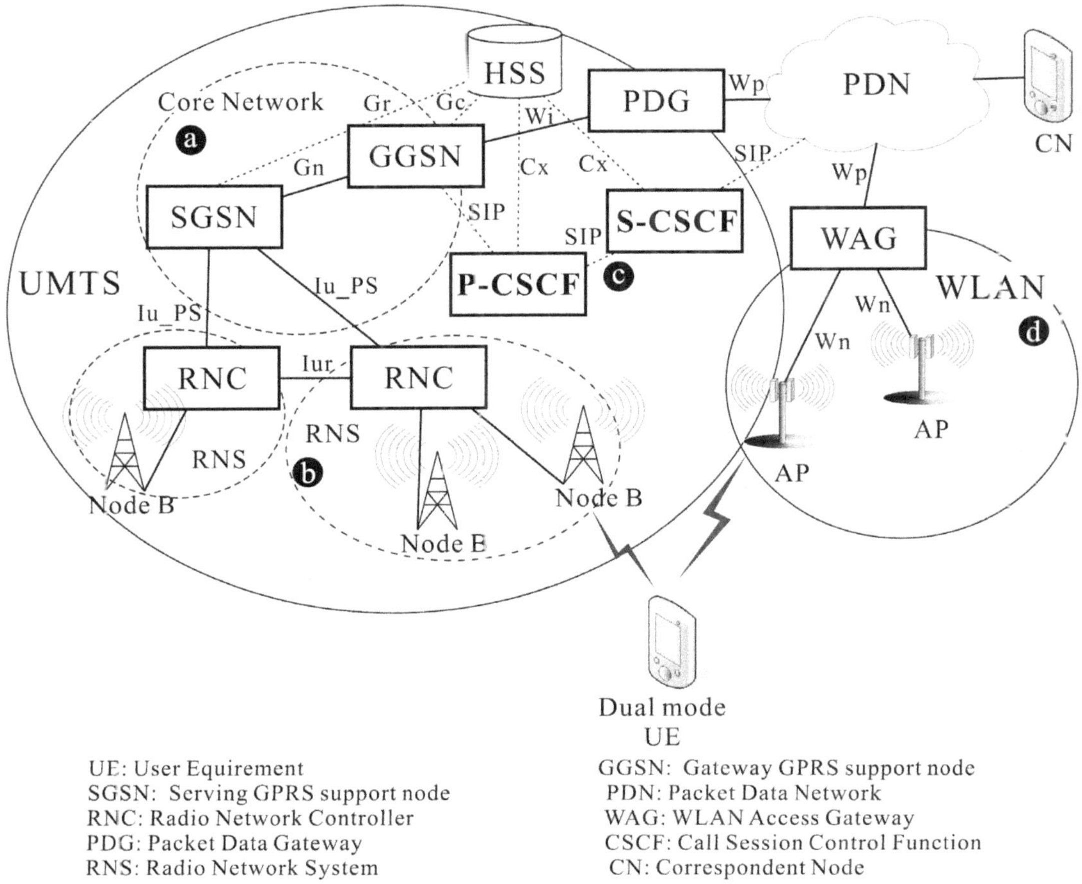

rate. The mobile multiple-mode device thus needs to switch different heterogeneous networks to acquire an acceptable communication. Figure 6 shows the proposed architecture. In this architecture, many mobile multiple-mode devices connect to the UMTS/HSDPA and WLAN networks, individually. The mobile multiple-mode device and User Equipment (UE) are interchangeable in this work. An UE can move arbitrarily between the WLAN and UMTS/HSDPA networks. Each part is shown separately as follows.

In the 3GPP R99 specification, the UMTS network can be divided into two parts-Circuit-Switched (CS) and Packet-Switched (PS) service domains. The UE connects to Public Switched Telephone Network (PSTN) through the CS service. The PS service allows the UE to access the Internet services. The PS domain connects to an external Packet Data Network (PDN) through a Serving GPRS Support Node (SGSN) and a Gateway GPRS Support Node (GGSN) (Figure 6a). The current position information of a user is stored in the SGSN, such that an incoming data packet is routed to the user immediately. The GGSN has a similar role as a gateway between the PDN and UMTS core network. The GGSN carries out tasks associated with network access control, packet routing and mobility management. In terms of mobility management, the GGSN tunnels to the SGSN using the GPRS Tunnel Protocol (GTP). When a user moves into another Radio Network System (RNS) (Figure 6b), which is controlled by the SGSN, the GGSN updates the SGSNID address and tunnels packets to the SGSN via the GTP.

The CSCF supports session control within the IMS (Figure 6c). The CSCF server is an SIP server that is responsible for interacting with Home Subscriber Server (HSS) for mobility and session control. The CSCF server can be divided into three principal components, Proxy-CSCF (P-CSCF), Serving-CSCF (S-CSCF), and Interrogating-CSCF (I-CSCF). The P-CSCF of the IMS is the first contact point for the UE. The main function of the P-CSCF is to forward SIP messages to the S-CSCF or I-CSCF depending on the message type or procedure. The S-CSCF is an important element for IMS. As the SIP proxy server, the S-CSCF forwards SIP messages to an appropriate node. The S-CSCF also supports the registration procedure. The I-CSCF function plays as the contact point of an operator network. Similar to a doorkeeper, the I-CSCF supports charging, routing, and forwarding SIP messages to an appropriate node. The I-CSCF of this work is ignored for the simple illustration (Figure 6c). Figure 6d displays the inter working between the UMTS/HSDPA and WLAN networks. The Wireless Access Gateway (WAG) (Raivio, et al., 2001) forwards data and messages from or to the PDN. The WAG can connect to more than one wireless access point. According to characteristics of the UE and proposed architecture, SmSCTP protocol (Chiu, et al., 2010a) makes the following assumptions: (1) a UE can always associate with the UMTS/HSDPA network prior to a WLAN network, (2) the MAC-layer of the UE provides physical and statistic information to upper layer, (3) the UE is authenticated by the UMTS/HSDPA authentication mechanism (Ouyang, et al., 2007).

The SmSCTP protocol (Chiu, et al., 2010a) highlights some of the major contribution to have handover policy and fast handover scheme in a UE. The handover policy is an *application*-layer algorithm, and the UE gathers and calculates the information of Signal-to-Noise Ratio (SNR), transmission rate, and transmission probability from both wireless networks periodically. The handover policy also collects the information of the battery. The UE connects to a better access point to acquire the maximum transmission good-put depending on the battery lifetime. This work also achieves the fast handover in the proposed infrastructure. The SmSCTP is a cross-layer scheme, which combines the SIP scheme at L7 and the mSCTP scheme at L4. In L7, with the flexibility of the SIP, the SmSCTP scheme develops two signaling messages to increase and switch access points. The SmSCTP scheme adopts the multi-homing mechanism using mSCTP in L4. The SmSCTP scheme performs the soft handover procedure when SIP sends a handover request to mSCTP protocol. The UE does not transmit control messages of the current mSCTP scheme to the Corresponding Node (CN). The control messages are replaced by the developed SIP messages. Two novel SIP signaling messages are designed for controlling session mobility. One is Annexing IP, called as ANNEXIP, and the other is Switching Session, teamed as SSESSION. The ANNEXIP message inserts location information of a UE into the databases of the CSCF server and the CN simultaneously. The ANNEXIP message can be continuously sent once the UE acquires a new IP address from different wireless networks. According to the cross-layer design, the SmSCTP scheme combines SIP (L7) with mSCTP (L4) to perform add IP address (ADD-IP) and the handoff procedures. The SSESSION message is able to update the location information of the UE in the CSCF server and CN. The SIP REGISTER message is cancelled. Table 5 shows the protocols comparison. The major characteristics of each method are compared, and formal definitions are shown as follows. The 'handover type' can point out that a UE can set up more than one connection at the same time or only one. The SIP re-INVITE scheme is a hard handover scheme. Thus, the connection of the SIP re-INVITE scheme should be broken before setting up a new one. The mSCTP

Table 5. The protocols comparison

Properties	Methods		
	SIP re-INVITE	mSCTP	SmSCTP
Handover type	Hard	Soft	Soft
Protocol layer	Layer 7	Layer 4	Layer 4 and Layer 7
Handover latency	High	Medium	Low
Multi-streaming	No	Yes	No
Multi-homing	No	Yes	Yes
Scalability	Medium	High	Medium
Flexibility	High	Low	Medium

and SmSCTP schemes are based on soft handover and enable to build up multi-connection before switching the wireless access point. In "protocol type," the SmSCTP scheme is a cross-layer method, and provides a handover policy and low signaling cost than the other two schemes. The "handover latency" can respond the comfort ability on call for a user. SmSCTP achieves better handover latency than other schemes. In "multi-homing," mSCTP and SmSCTP are able to set up more than one backup connection through different wireless networks. In addition, the SIP re-INVITE scheme needs to send another INVITE message to the CN to switch the wireless network. Before the INVITE message arrivals, the packet is lost. In "multi-streaming," the voice/video packets of SmSCTP are only transmitted using primary path. Therefore, SmSCTP does not support multi-streaming and the transmission efficiency of mSCTP is better than SmSCTP. In "scalability," the SIP re-INVITE and SmSCTP schemes use infrastructure to maintain all UEs. The SIP messages must be sent to the CSCF server. Therefore, both of these schemes cause a bottleneck in the CSCF server. The mSCTP scheme is an end-to-end protocol. The control messages of mSCTP are sent to another endpoint directly. The mSCTP can thus acquire a better scalability than the SIP re-INVITE and SmSCTP schemes. In "flexibility," the SIP re-INVITE and SmSCTP schemes belong to layer-7 protocol. The UE can easily add or modify the protocol using the software update mechanism. On the other hand, the user of mSCTP needs to modify the operation system to support a better performance or a new function.

Cross-Layer MIH-Based Mobility Approach

IEEE 802.21 (Das, et al., 2011) is developing standards to enable handover and interoperability between heterogeneous network types including both 802 and non 802 networks. IEEE 802.21 standard working group has created framework that defines a Media Independent Handover Function (MIHF), which will help mobile terminals to roam seamlessly across heterogeneous access networks. The MIHF consists of a Layer 2.5 (L2.5) that sits between the IP layer and different MACs layer which offers a standard interface to a number of access technologies. MIHF provides services to upper layers through a unified interface, which is known as Service Access Point (SAP). The lower layer protocols communicate with the MIHF via media dependent SAP. The MIHF provides three essential services to the higher layers:

- *Media Independent Event Service* (MIES) which provides triggered events corresponding to dynamic change in the link characteristics, link quality and link status.

As well as advance prediction (for example the link is going down).

- *Media Independent Command Service* (MICS) to provide the higher layers with a standard mechanism o control handover (such as handover initiation, preparation and completion).
- *Media Independent Information Service* (MIIS) that provides the capabilities for obtaining the necessary information about the serving and neighboring networks and their capabilities which allow network discovery and aid network selection.

The IEEE 802.21 MIH group (Das, et al, 2011) is now standardizing a L2.5 that gives access to additional information on the layer 2 link as well as standardizing handover commands. It also introduces a network discovery mechanism via an Information Service (IS). The MIH layer offers advanced services for handover such as event notification, L2 handover commands, and information on available networks.

Eastwood *et al.* (2008) initially show a system combining extensions of two radio access technologies, IEEE 802.11 and IEEE 802.16 networks. The focus of this work is to show how IEEE 802.21 (MIH services) supports "seamless" mobility between these two radio access technologies. Al Mosawi *et al.* (2009) then combined the MIH (L2.5) smart triggers with SIP (L7) mechanism to provide a fast handover solution. This work shows that the addition of MIH functions to a modified SIP handover mechanism can provide enhanced support for the vertical handover of real-time services in an all-IP network. Tang *et al.* (2010) had combined SIP Protocol and FMIP (Fast Mobile IPv6 Protocol), and MIH protocols by the cross-layer design to optimize those protocols' signaling flows and improve the performance of vertical handover. Huang *et al.* (2009) had proposed SIP-NEMO to achieve the route optimization. This work investigates how to extend the current SIP-NEMO to support multi-homing functionality. The proposed multi-homed SIPNEMO further integrated with the MIH standard by the cross-layer design. Liu *et al.* (2010) introduce a cross-layer design of the Inter-RAT handover between UMTS and WiMAX. The objective of this work is to propose a novel MIHF (Media-Independent Handover Function) variant, which is renamed Interworking (IW) sublayer, while the IW sublayer provides a seamless inter-RAT (Radio Access Technology) handover procedure between UMTS and WiMAX systems.

Recently, Proxy Mobile IPv6 (or PMIPv6, or PMIP) is a network-based mobility management protocol standardized by IETF (RFC 5213) (Gundavelli, et al., 2008). It is a protocol for building a common and access technology independent mobile core network, accommodating various access technologies such as WiMAX, 3GPP, 3GPP2, and WLAN based access architectures. Proxy Mobile IPv6 is the only network-based mobility management protocol standardized by IETF. Proxy Mobile IPv6 (PMIPv6) is an emerging network based localized mobility management scheme that enables IP mobility for a MN without requiring its participation in any mobility-related signaling. The network is responsible for managing IP mobility on behalf of the MN. PMIPv6 extends MIPv6 signaling and reuses many concepts such as the Home Agent (HA) functionality, although it supports an MN in its (PMIPv6) domain. The PMIPv6 domain introduces two new network functional entities called Local Mobility Agent (LMA) and Mobile Access Gateway (MAG) (Gundavelli, et al., 2008). The LMA is a local HA of the MN in the PMIPv6 domain. It has additional capabilities required to support PMIPv6. It is the topological anchor point for the MN's home network prefix and is the entity that manages the MN's binding state. That is, it maintains the reachability state to the MN's address while it moves within the PMIPv6 domain. The MAG is a functional entity that handles all mobility-related signaling on behalf of an MN attached to its access link. It typically runs in the default Access Router (AR) and tracks the movement of the MN, authenticates it and initiates the required mobility signaling with the LMA on behalf of the MN. Furthermore, the MAG establishes a bidirectional tunnel with the LMA for enabling the MN to use an address from its home network prefix. Consequently, it emulates the MN's home network on the access network for each MN. It supports a per-MN prefix model. When an MN is in the PMIPv6 domain, the serving network assigns a unique home network prefix, and conceptually this prefix always follows the MN wherever it moves within the PMIPv6 domain. Accordingly, it is needless to configure the Care-of-Address (CoA) at the MN. Implementation of a mobility stack in the MN is not necessary in a PMIPv6 domain, although a global mobility management protocol is necessary if the MN has to roam across domains.

Magagula *et al.* (2008) proposed a novel mobility management architectural framework based on cross-layer design interfacing with IEEE802.21 MIHF and utilizing a network-based localized mobility management protocol, PMIPv6, to reduce handover delay while making advanced handovers across heterogeneous wireless networks in NGWN. This framework's operation is confined to a PMIPv6 domain (e.g. an administrative domain with inter-networking heterogeneous networks). The model of the proposed architectural framework, which is implemented on the MN-side, while MIHF is implemented in the network infrastructure.

Cross-Layer Cognitive Mobility Approach

The LTE system (3G, 2011; 3GPP, 2009) is developed to integrate with various wireless systems; for instance, GERAN, UTRAN, non-3GPP system, and WLAN system. A LTE system can be called as a cognitive radio LTE system, CR-LTE system, if the LTE system adopts the Cognitive Radio (CR) technique. Existing heterogeneous system requires a wide range of agreements to integrate with different kinds of heterogeneous systems. The core of LTE system includes System Architecture Evolution (SAE) anchor, 3GPP anchor, and Mobility Management Entity (MME)/ User Plane Entity (UPE) to achieve the mobility management in the heterogeneous networks. The SAE anchor manages the mobility between 3PGG RATs (Radio Access Technology) and non-3GPP RATs. The 3GPP anchor manages the mobility between GERAN/UTRAN and the LTE system. The LTE system provides the frequency orthogonality for mobile users in the uplink and downlink communications. The performance of the LTE system, in terms of available data rates and spectrum efficiency, is limited by the inter-cell interference. The inter-cell interference coordination is a scheduling method to improve the date rate in the cell edge by taking the inter-cell interference into account. By restricting the transmission power of the spectrum, the interference appeared in the neighboring cells can be reduced to provide the higher data rates for user in the neighboring cell. One another interest characteristic of LTE systems is the spectrum agility. The LTE system supports more different bandwidth of channels. This characteristic of the LTE system is easily to adopt the Cognitive Radio (CR) concept to achieve the full frequency spectrum utilization.

Chen *et al.* (2011b) proposed a cross-layer protocol of spectrum mobility and handover in cognitive LTE networks. The system model used in Chen *et al.* (2011b) is given in Figure 7; the system model consists of the LTE core systems, licensed users (or called primary users, PUs), and unlicensed users (or called secondary users, SUs). For a CR network, it allows SUs equipped with cognitive radios to opportunistically access the spectrum not used by PUs. Chen *et al.* (2011b) considers the handover problem in a LTE-CR network. Figure 7 also shows a handover scenario in a CR-LTE network. A SU is moving from one cell into another cell. Two cases are considered. For the first case, the initial location of the SU is not in the overlapped area. If a PU is appeared and

Figure 7. A handover scenario in a CR-LTE network

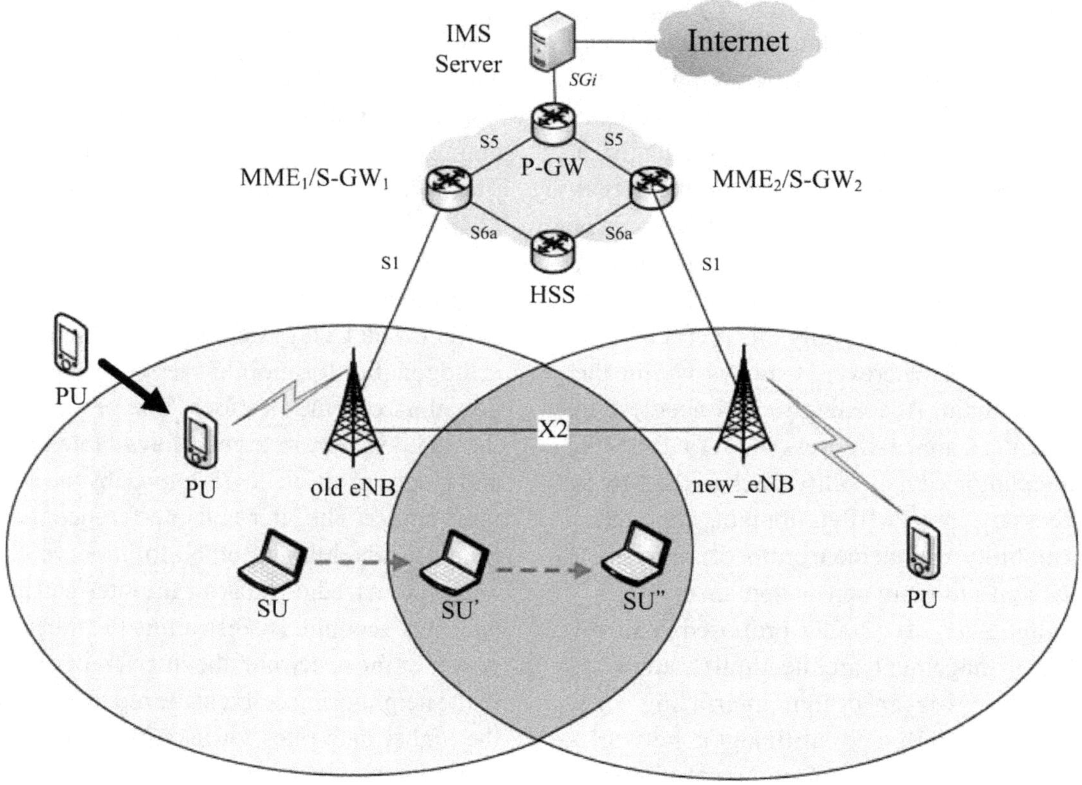

attempted to access and use the spectrum occupied by the SU, the SU performs the spectrum mobility processing to dynamically select a new spectrum hole, move to the new spectrum hole, and release the previously occupied spectrum for the PU. The second case is when the SU continually moves to a overlapped area between two adjacent eNBs. Figure 7 illustrates that the SU moves from location of *SU'* to that of *SU"*. Two possible events happen: (1) one is the spectrum mobility and (2) the another one is to handover to a new_eNB. A cross-layer protocol is developed of spectrum mobility and handover in cognitive LTE Networks with the consideration of the *minimum expected transmission time*.

To implement the above purpose, a SU is equipped with the SDR (Software Defined Radio) device to detect and use these spectrum holes. The PU and SU architectures are given in Figure 8. A PU device architecture is without any modification. In a SU device, components of L1 and L2 are modified to have the functions of spectrum sensing, spectrum analysis, spectrum decision, and spectrum mobility.

A SU with SDR device can be reconfigured for the different network protocols (802.11, 802.16, LTE, etc.) without addition devices and sense the different frequency band by one transceiver or two transceiver (Crow, et al., 1997) to achieve the spectrum management, including the functionalities of spectrum sensing, spectrum analysis, spectrum decision, and spectrum mobility. For example, a SU performs the spectrum sensing in the CR-LTE system and reports the sensing result of the LTE spectrum bands given in Figure 9. Obviously, the LTE spectrum is not fully utilized, and many underutilized spectrum holes can be used by SUs to improve the spectrum

Figure 8. The PU and SU devices

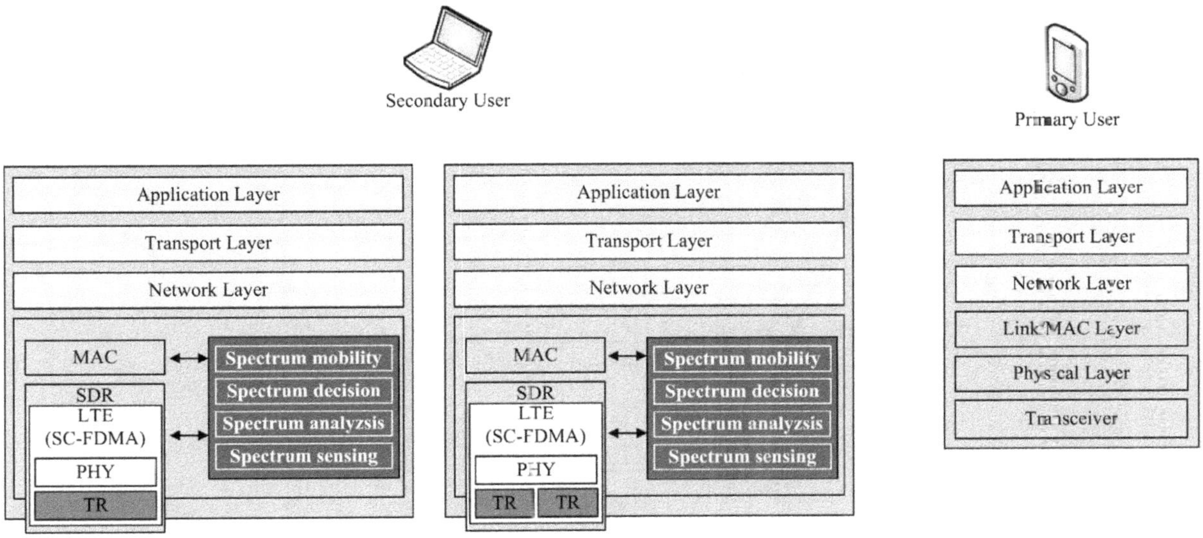

utilization. According to the LTE spectrum distribution, Resource Block (RB) is 180 kHz, where RB is the smallest unit of the LTE spectrum. The Spectrum Hole (SH) consists of one or more resource blocks (RB), this makes the SH with flexible size. For instance, PU_1 = {RB1, RB2}, PU_2 = {RB5}, PU_3 = {RB9, RB10}, and PU_4 = {RB15}.

The basic idea in Chen *et al.* (2011b) is to observe the spectrum occupied ratio to predict the probability of the resource reclaiming by PUs to perform the spectrum mobility on the serving eNB or the handover procedure to the new eNB. The goal is to calculate the minimum expected transmission time when selecting a new spectrum, where the spectrum is in serving eNB or the next eNB. Let $T_E(SH_i)$ denote as the expected transmission time of spectrum hole *i*, denoted as SH_i, for data service of a SU. The large value of $T_E(SH_i)$ is, the higher expected transmission time is. This indicates that the SU has the higher probability of resource reclaimed by a PU. This may frequently cause the operations of spectrum mobility and handover. Conversely, the smaller value of $T_E(SH_i)$ is, the higher successful data transmission is. This is because that there is no resource reclaimed by any PU. Therefore, the main contribution of this protocol is to propose a new scheme with the minimum $T_E(SH_i)$ in the CR-LTE systems. The function of $T_E(SH_i)$ is determined by the spectrum occupied ratio, the signal strength fluctuation, and the available bandwidth of the spectrum hole. It is observed that the high spectrum occupied ratio implies that the opportunity of the spectrum utilization is low. In addition, if the signal strength is strong and the bandwidth is large, then the higher transmission rate is obtained. The transmission rate is normally computed by the signal strength and the bandwidth. That is, a $T_E(SH_i)$ with strong signal strength and bandwidth can reduce the data transmission time. The main efforts in Chen *et al.* (2011b) is to develop a scheme with minimum expected transmission time, $T_E(SH_i)$. Chen *et al.* (2011b) follows the standard model of the cognitive cycle. Three phases of the cognitive cycle of Chen *et al.* (2011b) are described below:

- *Environment observation* phase: each SU periodically monitors the spectrum utilization, receives signal power, and detects the appearance of any PU using the SDR device.

Figure 9. The sensed spectrum distribution of a CR-LTE network

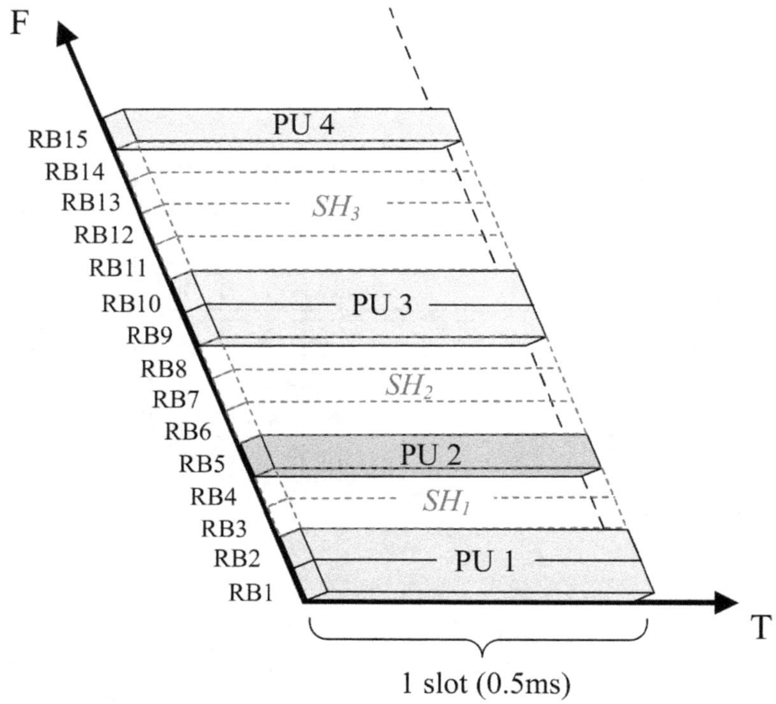

- *Computation and analysis* phase: each SU computes and analyzes the characteristics of spectrum hole, such as the transmission rate and the spectrum unoccupied rate, based on the observed information from environment observation phase.
- *Evaluation* and *transmission* phase: each SU evaluates all possible expected transmission times of the SU to determine to-move to new spectrum on the serving eNB or handover to new eNB.

This scheme (Chen, et al., 2011b) is a cross-layer protocol of spectrum mobility (L2) and handover (L3) in cognitive LTE networks. They consider two problems and provides a cross-layer protocol. Most existing protocols (Perkins, et al., 2002) only consider the spectrum mobility (L2) issue to switch a free spectrum when a PU reclaimed the resource occupied by a SU. Hoyhtya *et al.* (2008) recently proposed a predictive channel selection for cognitive radios. An example is shown in Figure 11a. Let T_{ON} denote a channel be occupied by PU, T_{OFF} denote a channel is idle, and T_{idle} denote a channel idle time. This predictive channel selection algorithm is to select the channel with the maximum idle time T_{idle}. For example as shown in Figure 9a, a SU performs the spectrum mobility from CH_4 to CH_1, where T_{idle} of CH_5, CH_3, and CH_1 are 8s, 10s, and 15s.

It is observed that considering only a channel with the maximum idle time cannot be optimal for the data transmission time. Because the channel size in LTE system may be different, a channel with the maximum idle time may be with the minimum channel size. If the selected channel has a maximum idle time and minimum size, the data transmission time is longer. It significantly reduces the data transmission time if the factor of the channel size is considered for designing a new predictive channel selection. In the proposed scheme, a channel with the maximum channel

Figure 10. The difference of (a) the maximum idle time scheme and (b) Chen et al.'s scheme

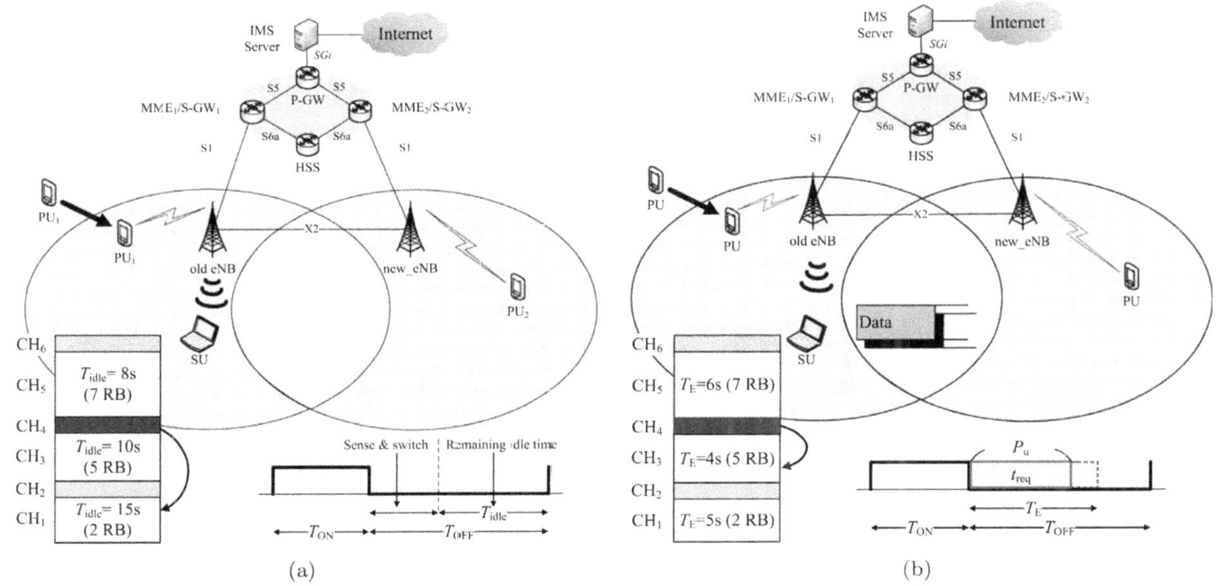

size and the maximum idle time is considered. Similar example of Figure 10a is given in Figure 10b. A SU performs the spectrum mobility from CH_4 to CH_3, where T_E of CH_5, CH_3, and CH_1 are 6s, 4s, and 5s. Let t_{req} denote the SU service required time, SH_i denote the i_{th} spectrum hole, and $P_u(SH_i,t_{req})$ denote the probability of spectrum hole unoccupied by PUs. The $P_u(SH_i,t_{req})$ is the unoccupied probability of SH_i within time period of t_{req}. The $T_E(SH_i)$ is calculated by $P_u(SH_i,t_{req})$. In addition, Chen *et al.* (2011b) further considers the handover (L3) procedure from the serving eNB to the next eNB with the consideration of the minimum T_E. Chen *et al.* (2011b) develops a cross-layer protocol of spectrum mobility (L2) and handover (L3) with the consideration of the minimum T_E in cognitive LTE networks. Figure 11 illustrates that when a SU in the overlapped area between old_eNB and new_eNB. Two possible results are generated. One is that SU still performs the spectrum mobility to move to SH_2 in original old_eNB, because that the minimum T_E in old_eNB is lower than the minimum T_E of new_eNB plus the handover delay time. Conversely, the SU is handover to SH_1 of new_eNB.

Recently, Software Defined Radio (SDR) is a revolutionary technology to implement the CR networks. To upgrade the inconvenience and inflexibility of hardware architecture, a SDR with the programmable and reconfigurable modules is integrated with different communication protocols. Finally, a mobile terminal (SU) with SDR device uses the sensing device in the physical layer to periodically sense the current signal strength and the idle spectrum resources. With the sensed information, the SU dynamically selects a suitable spectrum hole for the spectrum mobility. The Universal Software Radio Peripheral (USRP) can be used which can enable engineers to rapidly design and implement powerful, flexible software radio systems. A USRP-based cognitive testbed for implementing the cross-layer handover protocol built in Chen *et al.* (2011b) is illustrated in Figure 12. Table 6 shows the protocols comparison. The major characteristics of each method are compared, and formal definitions are shown as follows. The "handover type"

Figure 11. A mobility example in a CR-LTE network

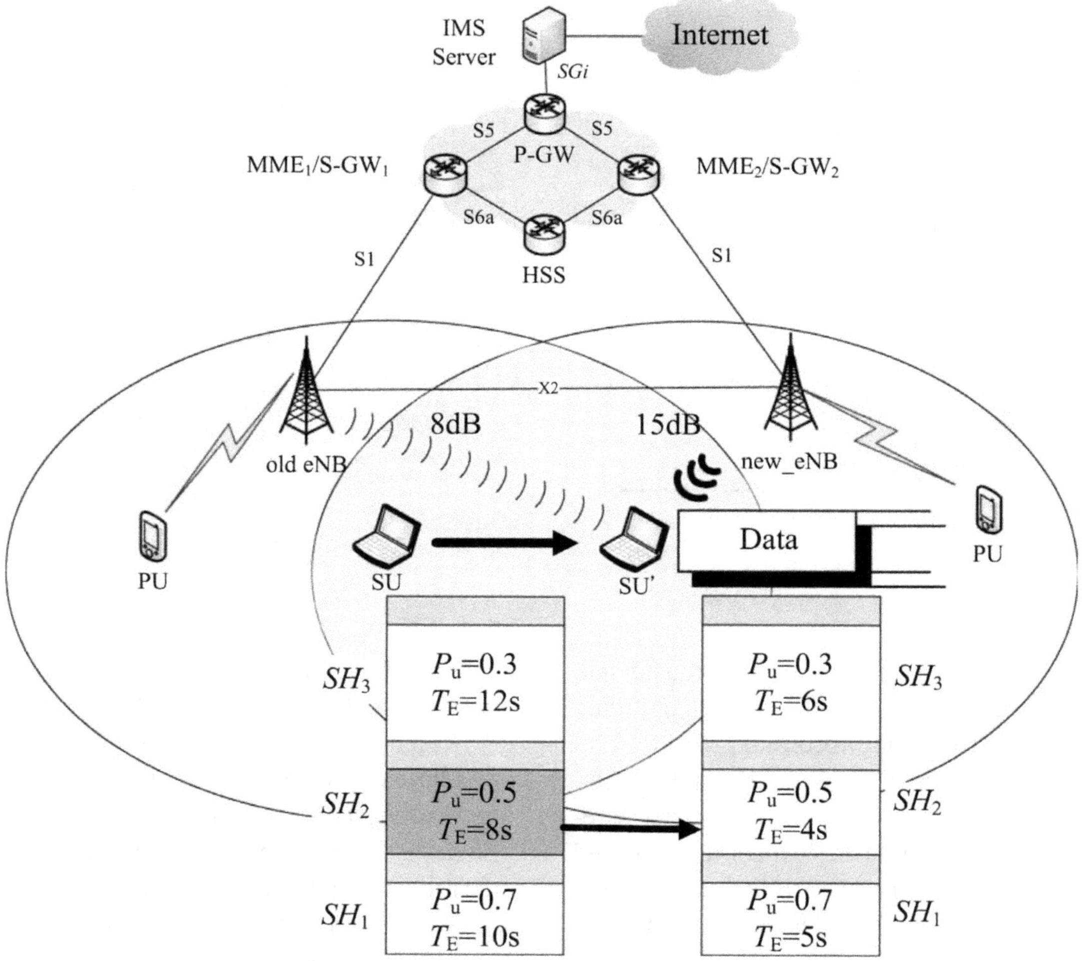

can point out that a node can set up more than one connection at the same time or only one. The "protocol type" shows the OSI layer which the protocol is designed. The "handover latency" can respond the amount of jitter. The "Spectrum interrupt rate" means the rate that the communication is interrupted by primary users. The "scalability" represents the ability to be enlarged to accommodate that network growth. The "flexibility" describes the ability to adapt a changeable environment.

APPLICATIONS

Some applications the design of cross-layer in mobility for vehicular networks are introduced. This subsection describes the cross-layer vehicle-aided mobility for vehicular networks, and then discusses other applications.

Cross-Layer Vehicle-Aided Mobility for Vehicular Networks

A cross-layer design Vehicle-Aided Handover Scheme (VFHS) developed by Chiu *et al.* (2011b) in VANETs. Figure 13 depicts a scenario of VFHS.

Figure 12. A USRP-based handover testbed for cognitive networks

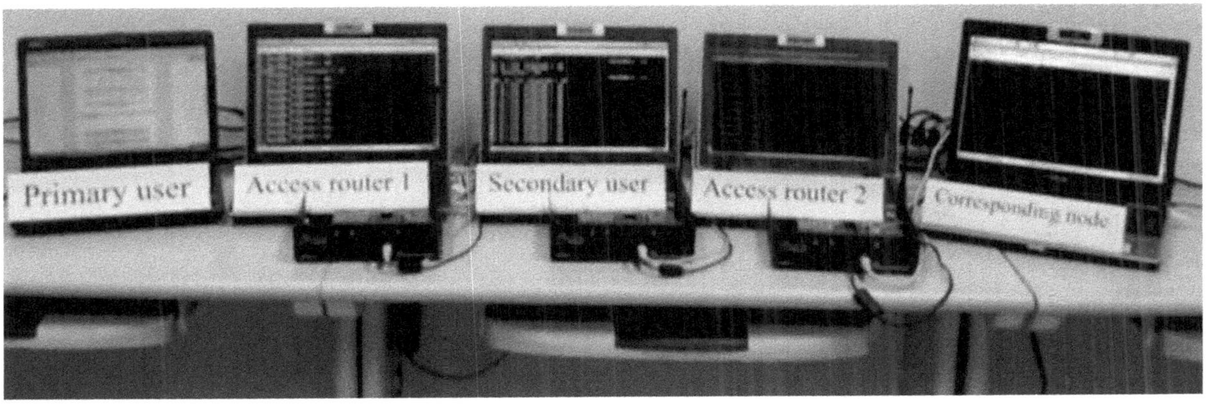

Table 6. The protocols comparison

	Methods	
Properties	**The maximum idle time scheme**	**Chen *et al.*'s scheme**
Handover type	Hard	Soft
Protocol layer	Layer 2	Layer 2 and Layer 3
Handover latency	Medium	Low
Spectrum interrupt rate	Medium	Low
Scalability	Medium	High
Flexibility	Medium	High

A number of vehicles of various sizes are traveling on the road. All vehicles are assumed to be equipped with a number of sensors to detect and gather information around the vehicle. A GPS adapter, an On-Board Diagnostic (OBD) and a WiMAX or communication interface are used for the inter-vehicle communication. The OBD is an on-broad unit, and able to control engine functions and perform self-diagnostic (OBD-II, 2010). All vehicle subsystems, e.g. GPS, wireless communication interface, and sensors, connect to the OBD to store and retrieve data. This work assumes that drivers or passengers can adopt a laptop or PDAs to access the Internet, and support navigation, traffic jam avoidance and alert report provided by the OBD. The model considers three kinds of vehicles according to their capability:

- *Relay Vehicle* (RV): An RV is a large vehicle which provides the capability of relay and mobile management of the MMR WiMAX network to its neighboring vehicles. A RV together with vehicles in its transmission range forms a cluster network.
- *Disconnected Vehicle* (DV): A DV is a small vehicle, which is outside of the coverage of any RV, and needs to transmit packets.
- *Oncoming Side Vehicle* (OSV): An OSV is a small vehicle driving on the oncoming direction, and has no packets to be transmitted. The OSV can accumulate the physical layer information of the RV located in the opposite direction (oncoming RV), and provide this information to DVs with the

Figure 13. System model of VFHS

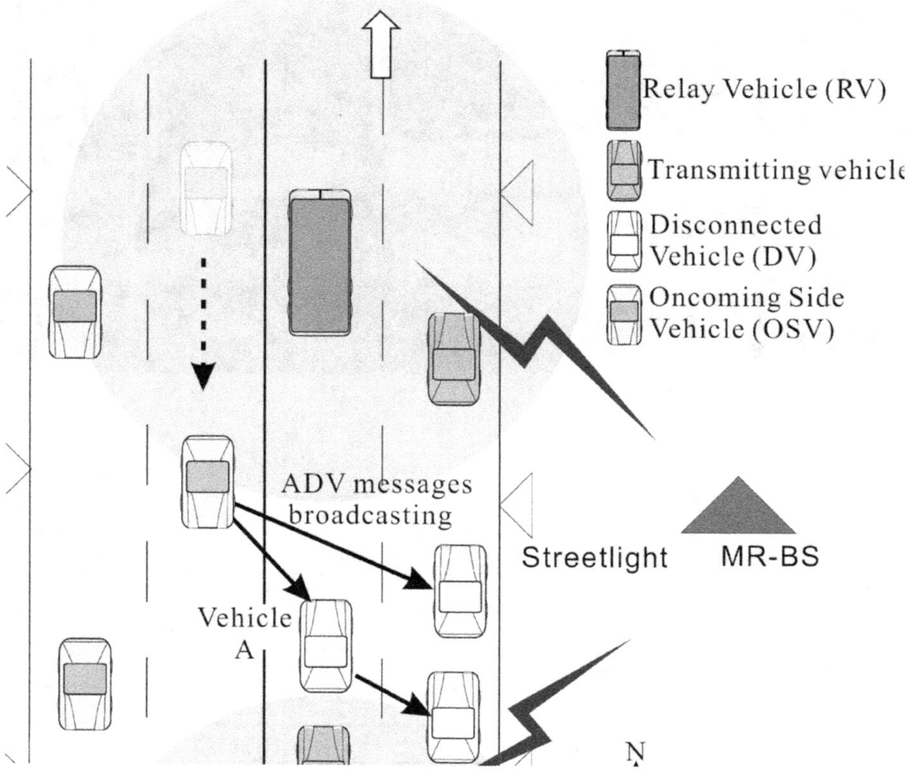

cross-layer Network Topology Message (NTM) for facilitating fast handover. The OSV uses a set of specific channels to broadcast NTM to DVs.

Besides vehicles on the road, a number of MR-BSs provide WiMAX multi-relay transmission environment are deployed at the roadside, as illustrated in Figure 13. The MR-BS and the RV are configured in Point-to-Multipoint (PMP) mode, and the MR-BS simultaneously manages a number of RVs. The MR-BS operates as a serving MR-BS when an RV connects to it for transmitting and receiving packets from the Internet. Figure 13 represents the relay links using black, bolt, and also displays gaps between RVs on the road. A vehicle located in the gap becomes a DV. The disconnection problem anticipated to be solved is explained below. Vehicle *A*, in the middle area of the Figure 13, travels from south to north. Vehicle *A* becomes a DV when it leaves the transmission coverage of the previous RV and losses its connectivity. Therefore, vehicle *A* attempts to handover to a nearby RV. According to the handover procedure of the standard WiMAX (IEEE, 2005), the previously RV may not receive or measure the signals of neighbor RVs located on the front. The MOB_NBR-ADV message comprises "empty" physical information about neighboring RVs. Vehicle *A* has to complete the entry procedure; because vehicle *A* does not have any information of neighbor RVs. The entry procedure consists of scanning, synchronization, ranging, and authorization. Scanning incurs the longest delay due to all channels must be scanned

To solve this problem, Chiu *et al.* (2011b) proposed a novel methodology for reducing the scanning latency. The proposed protocol allows

Figure 14. The protocol stack of cross-layer by integrating with the position information

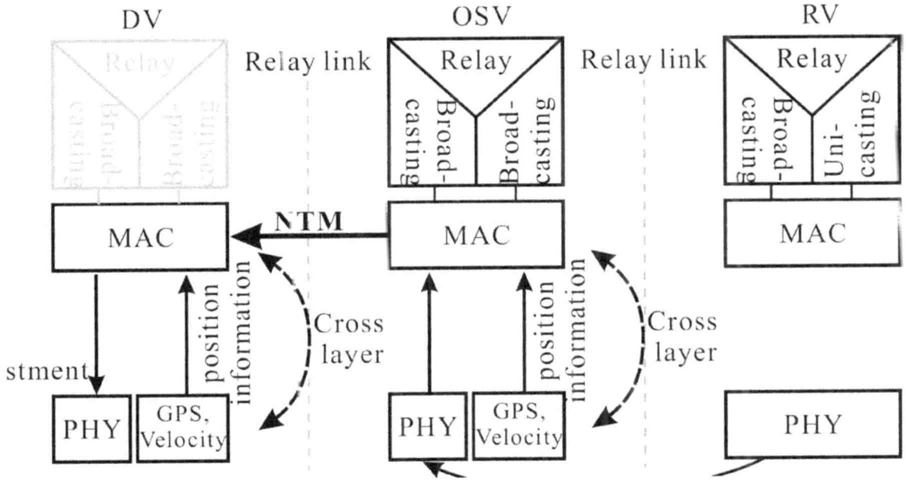

an OSV to accumulate the physical layer information of RVs. The OSV broadcasts NTM (Network Topology Message) to DVs using predefined broadcast channels upon acquiring the physical information of neighboring RVs. This NTM enables a DV to know which channel to listen when it enters the transmission range of the target RV on the front. According to NTM, a DV can adjust its physical setting, such as the channel, when it receives NTM, rather than waiting and receiving the neighbor advertisement message from the target RV. By comparing blocks of recorded time and velocities, the DV identifies the nearest RV on the front. The interaction of the handover procedure at MAC layer with position and channel information at physical layer delivers an interesting and efficient solution.

To support the dynamic adaptation, VFHS adopts the explicit cross-layer design (Kota, et al., 2007; Srivastava, et al., 2005) to provide the signaling message to cross MAC and physical layers. Figure 14 illustrates the cross-layer structure of the system under the consideration of cross-layer design. It is noted that the NTM is a cross-layer message, which comprises the information of the physical (position and channel) and MAC (WiMAX) layers. In the VFHS, the position and channel information of the oncoming RV are accumulated and managed by OSVs. An OSV inserts its position information into NTM, and broadcasts to DVs. Upon receiving a NTM, DVs adjust the setting of the channel frequency of the WiMAX adapter immediately by retrieving the channel and position information; instead of scanning all channel frequencies sequentially by the physical layer. According to the cross-layer design, VFHS protocol significantly reduces the scanning time. Table 7 shows the comparisons of existing frameworks in vehicular environments. The major characteristics of each investigation are compared, and formal definitions are shown as following. The "data transmission delay" is the amount of time required to transmit a packet from the source to the destination. The "push" is an actively broadcast or multicast method caused by a traffic condition. The "pull" is used to acquire specific information when a vehicle is willing to provide it. The "header overhead" denotes that the packet is transmitted using tunneling or not. Finally, the "relay overhead" means that the packet is whether forwarded by a relay node (vehicle) or not.

Table 7. The protocols comparison

Properties	Methods	
	FMIPv6	VFHS
Data transmission delay	Medium	Low
Push	No	Yes
Pull	No	Yes
Header overhead	Yes	No
Relay overhead	No	Yes

Other Applications

Some applications have recently been investigated for cross-layer design of mobility. Yang *et al.* (2007) proposed a cross-layer design WiMAX mesh network, called Coordinated External Peer Communication (CEPEC), to provide Internet-access services in the freeways environment. To support Internet-access services to a vehicle allocated outside of the coverage of the serving BS, CEPEC separates the road into multi-segment, and uses channel-sharing to relay data packets. Each segment has a segment head (SH) to perform local-packet collecting and aggregated packet relaying. Mussabbir *et al.* (2007) developed an enhanced FMIPv6 mechanism with IEEE 802.21 technology over VANETs. The authors proposed an improved FMIPv6 mechanism to support Network Mobility (NEMO) in vehicular environments and combine with IEEE 802.21 protocol to improve the handover performance by using cache mechanism.

FUTURE RESEARCH DIRECTIONS AND ADDITIONAL READING

One research direction is to develop cross-layer mobility managements in a two-tier macrocell/femtocell network. While the next-generation wireless technologies continue to evolve, mobile communications have also transformed from traditional voice services to IP-based services. The 4G networks, LTE-advanced and WiMax, have been developed to redefine the traditional physical-layer air interface to bring the high transmission rate. With the emerging technology; such as Orthogonal Frequency Division Multiple Access (OFDMA) (Cao, et al., 2011), both LTE-advanced and WiMAX are expected to improve end-user throughputs and network capacity with the full mobility. A femtocell (Chandrasekhar, et al., 2008) is a low power, low cost, and user-deployed cellular Base Stations (BSs) with a small coverage range (e.g. 30 ~ 40 meters in diameter). A large number of femtocells are deployed in the coverage area of macrocells. The main purpose is to increase the indoor coverage for voice and high-speed data service. A femtocell works on the licensed band and connects to the operator core network via DSL broadband backhaul. A hierarchical radio access network is called as a two-tier macrocell/femtocell network. It is observed that the deployment of femtocells (Chandrasekhar, et al., 2008) over LTE-advanced and WiMAX networks benefits both users and service providers. Subscribers accessing femtocells can achieve greater signal strength and better quality of service because of the short transmit-receive distance. Several issues, including resource, interference, and especially for mobility managements (including in-bound and out-bound mobility), needed to be carefully addressed and investigated. The in-bound mobility indicates that the mobility is from macrocell (LTE-advanced or WiMAX networks) to femtocell and the out-bound mobility implies that the mobility is from femtocell from the macrocell.

Existing results of the mobility management had been recently developed in OFDMA two-tier macrocell-femtocell networks. First, Moon *et al.* (2010) introduced a RSS-based handover decision protocol in hierarchical macro/femto-cell networks. The main idea is to combine the value of the received signal strength from a serving MBS and a target FBS with the consideration of the large asymmetry in their transmit powers. Not only considers the SNR, some other results further considers the UE speed for handover protocol designs (Wu, et al., 2009; Ulvan, et al., 2010). Wu *et al.* (2009) proposed a handover algorithm both considering signal strength and velocity in the two-hierarchy network. Ulvan *et al.* (2010) and Zhang *et al.* (2010) proposed new handover algorithms in LTE-based femtocell networks based on the speed of UE and Quality of Service (QoS). In addition, the real-time and non-real-time traffics have been considered as QoS parameters. The main purpose of above-mentioned algorithms is to achieve the seamless handover and reduce the handover latency. However, these mobility protocols are not considered the power-saving issues.

Another research direction is to investigate the green mobility management by the cross-layer designs for the next-generation mobile networks and the two-tier macrocell/femtocell networks. It is observed that over 90% of the power in mobile communications is consumed in the radio access network (Fettweis, et al., 2008). For the green communication, the power consumption of the radio access networks, including the base stations, must be reduced. According to the prediction of ABI Research (Chandrasekhar, et al., 2008), more than 36 million femtocells are expectedly deployed around the world by the end of 2012. Assumed that each femtocell requires a power cost of 12W, the total energy consumption of all deployed femtocells amount to 3.784×10^9 kWh/annum. Energy efficiency of femtocells becomes an urgent issue. In general, the femtocells in active mode usually produce in the order of 2.05 million tones of CO_2 per year. To reduce the impact of Information and Communication Technologies (ICTs) on environment, efforts for the energy efficiency recently have received lots of attention, and appearing a concept—"Green Radio," which means reducing the CO_2 emissions in ICTs. One and useful way is provide energy efficient protocol to reduce the unnecessary power consumption of femtocells.

The cell power consumption issues are recently raised more attentions (3GPP TS 36.927 V2.0.0, 2011; Ashraf, et al., 2010). The 3GPP TS 36.927 (release 10) (3GPP TS 36.927 V2.0.0, 2011) recently defined potential solutions for energy saving for E-UTRAN, which indicates that the cell can be totally switched off during the Energy Saving (ES) procedure. The ES procedure may be triggered in case of the light traffic or no traffic. Since there are thousands of femtocells within a macrocell area, the femtocell deployments may increase energy consumption except for the case of low femtocell/user densities and the use of femtocells with "idle" mode. Ashraf *et al.* (2010) recently proposed an improving energy efficiency of femtocell base statio via user activity detection. The proposed energy-saving procedure allows the femto BS to completely switch off its radio transmissions and associated processing when not involved in an active call. Consequently, it is important and interested to in-depth design more new cross-layer "green" handover results, protocols, and standards in the near future by significantly reduce the total power consumption to achieve the goal of the green communication.

CONCLUSION

This chapter had initially introduced the traditional mobile IPv6 based approach, existing mobility approaches in WiMax systems, mobility approaches on heterogeneous networks, and spectrum mobility approaches in cognitive-based networks to provide the general concept of mobility management. To introduce the cross-layer mobility management, some existing cross-layer protocols for relay-based mobility, session mobility, MIH-based mobility, and cognitive mobility approaches had been discussed in details. Applications of cross-layer mobility are additionally introduced. Finally, future works are discussed to investigate the cross-layer design of mobility management for supporting the green radio mobile networks, IOT (Internet of Things)-based cloud computing environment, and the next-generation mobile networks.

REFERENCES

Akbar, I. A., & Tranter, W. H. (2007). Dynamic Spectrum allocation in cognitive radio using hidden Markov models: Poisson distributed case. In *Proceedings of the IEEE Communications Society Conference on Sensor, Mesh and Ad Hoc Communications and Network*, (pp. 196–201). Richmond, VA: IEEE Press.

Akyildiz, I. F., Xie, J., & Mohanty, S. (2004). A survey of mobility management in next-generation all-IP-based wireless systems. *IEEE Wireless Communications*, *11*(4), 16–28. doi:10.1109/MWC.2004.1325888

Al Mosawi, T., Shuaib, H., & Aghvami, A. H. (2009). A fast handover scheme based on smart triggers and SIP. In *Proceedings of the IEEE 70th Vehicular Technology Conference Fall (VTC 2009-Fall)*, (pp. 1-5). Anchorage, AK: IEEE Press.

Ashraf, I., Ho, L. T. W., & Claussen, H. (2010). Improving energy efficiency of femtocell base stations via user activity detection. In *Proceedings of the IEEE Wireless Communications and Networking Conference (WCNC 2010)*, (pp. 1-5). Sydney, Australia: IEEE Press.

Azhari, V., & Todd, T. D. (2005). Fast connection recovery in WLAN-to-cellular vertical handoff. In *Proceedings of the IEEE International Conference on Mobile Adhoc and Sensor Systems*. Washington, DC: IEEE Press.

Bari, F., & Leung, V. C. (2009). Architectural aspects of automated network selection in heterogeneous wireless systems. *International Journal of Ad Hoc and Ubiquitous Computing*, *45*, 282–291. doi:10.1504/IJAHUC.2009.027478

Cao, G., Yang, D., An, R., Ye, X., Zheng, R., & Zhang, X. (2011). An adaptive sub-band allocation scheme for dense femtocell environment. In *Proceedings of the IEEE Wireless Communications and Networking Conference (WCNC 2011)*, (pp. 102-107). Quintana-roo, Mexico: IEEE Press.

Chandrasekhar, V., Andrews, J. G., & Gatherer, A. (2008). Femtocell networks: A survey. *IEEE Communications Magazine*, *46*(9), 59–67. doi:10.1109/MCOM.2008.4623708

Chao, H. C., & Huang, C. Y. (2003). Micro-mobility mechanism for smooth handoffs in an integrated ad-hoc and cellular IPv6 network under high-speed movement. *IEEE Transactions on Vehicular Technology*, *52*(6), 1576–1593. doi:10.1109/TVT.2003.819450

Chen, J., Wang, C. C., & Lee, J. D. (2007). Pre-coordination mechanism for fast handover in WiMAX networks. In *Proceedings of the International Conference on Wireless Broadband and Ultra Wideband Communications AusWireless*, (pp. 15-21). AusWireless.

Chen, L., Cai, X., Sofia, R., & Huang, Z. (2007). A cross-layer fast handover scheme for mobile WiMAX. In *Proceedings of the IEEE 66th VTC-2007 Fall Vehicular Technology Conference*, (pp. 1578-1582). IEEE Press.

Chen, Y. S., Chen, C. K., & Chuang, M. C. (2008). DeuceScan: Deuce-based fast handoff scheme in IEEE 802.11 wireless networks. *IEEE Transactions on Vehicular Technology*, *57*(2), 1126–1141. doi:10.1109/TVT.2007.907027

Chen, Y. S., Chiu, K. L., & Hwang, R. H. (2007). SmSCTP: SIP-based MSCTP scheme for session mobility over WLAN/3G heterogeneous networks. In *Proceedings of the IEEE Wireless Communications and Networking Conference*, (pp. 3307-3312). IEEE Press.

Chen, Y. S., Cho, C. H., Ilsun, Y., & Chao, H. C. (2011b). A cross-layer protocol of spectrum mobility and handover in cognitive LTE networks. *Simulation Modelling Practice and Theory, 19*(8), 916–928. doi:10.1016/j.simpat.2010.09.007

Chen, Y. S., Hsiao, W. H., & Chiu, K. L. (2009). A cross-layer partner-based fast handoff mechanism for IEEE 802.11 wireless networks. *International Journal of Communication Systems, 22*(12), 1515–1541. doi:10.1002/dac.1033

Chen, Y. S., & Wu, K. L. (2011a). A cross-layer partner-assisted handoff scheme for hierarchical mobile IPv6 in IEEE 802.16e systems. *Wireless Communications and Mobile Computing, 11*(4), 522–541. doi:10.1002/wcm.844

Chen, Y. W., & Hsieh, F. Y. (2007). A cross layer design for handoff in 802.16e network with IPv6 mobility. In *Proceedings of the IEEE Wireless Communications and Networking Conference*, (pp. 3844–3849). IEEE Press.

Chiu, K. L., Chen, Y. S., & Hwang, R. H. (2011a). Seamless session mobility scheme in heterogeneous wireless networks. *International Journal of Communication Systems, 24*(6), 789–809. doi:10.1002/dac.1189

Chiu, K. L., Hwang, R. H., & Chen, Y. S. (2011b) Cross-layer design vehicle-aided handover scheme in VANETs. *Wireless Communications and Mobile Computing, 11*(7), 916–928. doi:10.1002/wcm.861

Crow, B. P., Widjaja, I., Kim, J. G., & Sakai, P. T. (1997). IEEE 802.11 wireless local area networks. *IEEE Communications Magazine, 35*, 116–126. doi:10.1109/35.620533

Das, S., & Zuniga, J. C. (2011). *IEEE 802.21*. Retrieved from http://www.ieee802.org/21/.

Eastwood, L., Migaldi, S., Qiaobing, X., & Gupta, V. (2008). Mobility using IEEE 802.21 in a heterogeneous IEEE 802.16/802.11-based, IMT-advanced (4G) network. *IEEE Wireless Communications, 15*(2), 26–34. doi:10.1109/MWC.2008.4492975

Fettweis, G. P., & Zimmermann, E. (2008). ICT energy consumption – Trends and challenges. In *Proceedings of the International Symposium on Wireless Personal Multimedia Communications*. Lapland, Finland: IEEE.

3GPP. (2006). Technical specification group services and system aspects: IP multimedia subsystem (IMS): Stage 2. *3GPP Std*. Retrieved from http://www.3gpp.org.

3GPP. (2009). TS36.300: *Evolved universal terrestrial radio access (E-UTRA) and evolved universal terrestrial radio access network (E-UTRAN)*. Retrieved from http://www.3gpp.org.

3GPP. (2011). *Mobile group*. Retrieved from http://www.3gpp.org/.

3GPP TS 36.927 V2.0.0. (2011). *Potential solutions for energy saving for E-UTRAN (Release 10)*. Retrieved from http://www.3gpp.org.

Gundavelli, S., Leung, K., Devarapalli, V., Chowdhury, K., & Patil, B. (2008). Proxy mobile IPv6. *IETF RFC 5213*. Retrieved from http://www.ietf.org.

Hoyhtya, M., Pollin, S., & Mammela, A. (2008). Performance improvement with predictive channel selection for cognitive radios. In *Proceedings of the Cognitive Radio and Advanced Spectrum Management*, (pp. 1–5). Aalborg, Denmark: IEEE.

Huang, C. M., Chao-Hsien Lee, C. H., & Tseng, P. H. (2009). Multihomed SIP-based network mobility using IEEE 802.21 media independent handover. In *Proceedings of the IEEE International Conference on Communications (IEEE ICC 2009)*. IEEE Press.

IEEE. Computer Society. (2002). IEEE trial-use recommended practice for multi-vendor access point interoperability via an inter-access point protocol across distribution systems supporting IEEE 802.11 operation. *Internet Engineering Task Force (IETF) Internet-Draft*. Retrieved from http://www.ietf.org.

IEEE. (2004). Standard for local and metropolitan area networks—Part 16: Air interface for fixed broadband wireless access systems. *IEEE 802.16-REVd/D5-2004*. Retrieved from http://www.ieee.org.

IEEE. (2005). Standard for local and metropolitan area networks—Part 16: Air interface for fixed and mobile broadband wireless access systems—Amendment for physical and medium access control layers for combined fixed and mobile operation in licensed bands. *IEEE 802.16e/D8-2005*. Retrieved from http://www.ieee.org.

IEEE. (2006). Draft standard for local and metropolitan area networks: Media independent handover services. *IEEE P802.21/D03.00*. Retrieved from http://www.ieee.org.

IEEE. 802.16's Relay Task Group. (2011). *Website.* Retrieved from http://www.ieee802.org/16/relay/.

Inoiue, M., Mahmuhd, K., Murakaami, H., & Hasegawa, M. (2004). Novel out-of-band signaling for seamless interworking between heterogeneous networks. *IEEE Wireless Communications, 11*(2), 56–63. doi:10.1109/MWC.2004.1295740

Jang, H. J., Jee, J. H., Han, Y. H., & Cha, J. (2007). Mobile IPv6 fast handovers over IEEE 802.16e networks. *Internet Engineering Task Force (IETF), Internet-Draft*. Retrieved from http://www.ietf.org.

Jo, O., & Cho, D. H. (2008). Efficient spectrum matching based on spectrum characteristics in cognitive radio systems. In *Proceedings of the IEEE Wireless Telecommunications Symposium*, (pp. 230–235). Pomona, CA: IEEE Press.

Johnson, D., Perkins, C., & Arkko, J. (2004). Mobility support in IPv6. *Internet Engineering Task Force (IETF), RFC-3775*. Retrieved from http://www.ietf.org.

Kim, D. P., & Koh, S. J. (2009). Performance enhancement of mSCTP for vertical handover across heterogeneous wireless networks. *International Journal of Communication Systems, 22*(12), 1573–1591. doi:10.1002/dac.1038

Koh, S., Chang, M., & Lee, M. (2004). mSCTP for soft handover in transport layer. *IEEE Communications Letters, 8*(3), 189–191. doi:10.1109/LCOMM.2004.823432

Koodli, R. (2005). Fast handovers for mobile IPv6. *Internet Engineering Task Force (IETF), RFC-4068*. Retrieved from http://www.ietf.org.

Kota, S., Giambene, G., & Candio, N. L. (2007). Cross-layer approach for an air interface of GEO satellite communication networks. *International Journal of Satellite Communications and Networking, 25*(5), 481–499. doi:10.1002/sat.887

Lai, W. K., & Chiu, J. C. (2005). Improving handoff performance in wireless overlay networks by switching between two-layer IPv6 and one- layer IPv6 addressing. *IEEE Journal on Selected Areas in Communications, 23*(11), 621–628.

Lee, C. W., Chen, L. M., Chen, M. C., & Sun, Y. S. (2005). A framework of handoffs in wireless overlay networks based on mobile IPv6. *IEEE Selected Areas in Communications, 23*(11), 2118–2128. doi:10.1109/JSAC.2005.856833

Liu, B., Marins, P., & Bertin, P. (2010). Cross-layer design of the inter-RAT handover between UMTS and WiMAX. *EURASIP Journal on Wireless Communications and Networking*, 10.

Magagula, L. A., & Chan, H. A. (2008). IEEE 802.21-assisted cross-layer design and PMIPv6 mobility management framework for next generation wireless networks. In *Proceedings of the IEEE International Conference on Wireless & Mobile Computing, Networking & Communication (IEEE WiMob)*. Avignon, France: IEEE Press.

Mitola, J. III, & Maguire, G. Q. Jr. (1999). Cognitive radio: Making software radios more personal. *IEEE Personal Communications*, *6*(4), 13–18. doi:10.1109/98.788210

Moon, J. M., & Cho, D. H. (2010). Novel handoff decision algorithm in hierarchical macro/femtocell networks. In *Proceedings of the IEEE Wireless Communications and Networking Conference (WCNC 2010)*, (pp. 1-6). Sydney, Australia: IEEE Press.

Mussabbir, Q. B., Yao, W., Niu, Z., & Fu, X. (2007). Optimized FMIPv6 using IEEE 802.21 MIH services in vehicular networks. *IEEE Transactions on Vehicular Technology*, *56*(6), 3397–3407. doi:10.1109/TVT.2007.906987

Narten, T., & Nordmark, E. (1998). Neighbor discovery for IP version 6 (IPv6). *Internet Engineering Task Force (IETF), RFC-2461*. Retrieved from http://www.ietf.org.

OBD-II. (2010). *Website.* Retrieved from http://www.obdii.com/obdii.html.

Oh, H., Yoo, K., Na, J., & Kim, C. (2009). A seamless handover scheme in IPv6-based mobile networks. *International Journal of Ad Hoc and Ubiquitous Computing*, *4*(1), 54–60. doi:10.1504/IJAHUC.2009.021914

Ouyang, Y. C., Jang, C. B., & Chen, H. T. (2007). A secure authentication policy for UMTS and WLAN interworking. In *Proceedings of the IEEE International Conference on Communications*, (pp. 1552-1557) Glasgow, UK: IEEE Press.

Park, J., Kwon, D. H., & Suh, Y. J. (2006). An integrated handover scheme for fast mobile IPv6 over IEEE 802.16e systems. In *Proceedings of the IEEE Vehicular Technology Conference*, (pp. 1-5). IEEE Press.

Park, J. H., Han, K. Y., & Cho, D. H. (2007). Reducing inter-cell handover events based on cell ID information in multi-hop relay systems. In *Proceedings of the IEEE 65th VTC 2007-Spring Vehicular Technology Conference*, (pp.743–747). IEEE Press.

Perkins, C., et al. (2002). IP mobility support for IPv4. *RFC 3344*. Retrieved from http://www.ietf.org.

Raivio, Y. (2001). 4G-hype or reality. In *Proceedings of the International Conference on 3G Mobile Communication Technologies*, (Vol. 47), (pp. 346-350). London, UK: IEEE.

Raychaudhuri, D. (2004). *4G network architectures: WLAN hot-spots, infostations and beyond.* London, UK: International 4G Forum.

Salsano, S., Veltri, L., Martiniello, G., & Polidoro, A. (2006). Seamless vertical handover of VoIP calls based on SIP session border controllers. In *Proceedings of the IEEE International Conference on Communications*, (Vol. 5), (pp. 2040-2047). IEEE Press.

Schulzrinne, H., & Wedlund, E. (2000). Application layer mobility using SIP. *IEEE Service Portability and Virtual Customer Environments*, *4*(1), 29–36.

Sharma, S., Zhu, N., & Chiueh, T. C. (2004). Low-latency mobile IP handoff for infrastructure-mode wireless LANs. *IEEE Journal on Selected Areas in Communications*, *22*(4), 643–652. doi:10.1109/JSAC.2004.825988

Shim, M., Kim, H., & Lee, S. (2006). A fast handover mechanism for IPv6 based WiBro system. In *Proceedings of the IEEE International Conference on Advanced Communication Technology*. IEEE Press.

Siddiqui, F., & Zeadally, S. (2006). Mobility management across hybrid wireless networks: Trends and challenges. *Computer Communications*, *29*(9), 1363–1385. doi:10.1016/j.comcom.2005.09.003

Smadi, M., Todd, T. D., Zhao, D., & Kezys, V. (2006). Dynamically anchored conferencing handoff for dual-mode cellular/WLAN handsets. In *Proceedings of the IEEE International Conference on Communications*, (Vol. 5), (pp. 2028-2033). IEEE Press.

Soliman, H., Castelluccia, C., Malki, K., & Bellier, L. (2005). Hierarchical mobile IPv6 mobility management (HMIPv6). *Internet Engineering Task Force (IETF), RFC-4140*. Retrieved from http://www.ietf.org.

Srivastava, V., & Motani, M. (2005). Cross-layer design: A survey and the road ahead. *IEEE Communications Magazine*, *43*(12), 112–119. doi:10.1109/MCOM.2005.1561928

Stewart, R., et al. (2000). Stream control transmission protocol. *RFC 2960*. Retrieved from http://www.ietf.org.

Sur, A., & Sicker, D. C. (2005). Multi layer rules-based framework for vertical handoff. In *Proceedings of the International Conference on Broadband Networks*, (Vol. 1), (pp. 71-580). IEEE.

Thomson, S., & Narten, T. (2002). IPv6 stateless address autoconfiguration. *Internet Engineering Task Force (IETF) RFC-2462*. Retrieved from http://www.ietf.org.

Ulvan, A., Bestak, R., & Ulvan, M. (2010). The study of handover procedure in LTE-based femtocell network. In *Proceedings of the IEEE Wireless and Mobile Networking Conference (WMNC 2010)*, (pp. 1-6). Budapest, Hungary: IEEE Press.

WiMAX Forum. (2011). *Website*. Retrieved from http://www.wimaxforum.org/home/.

Wu, S., Zhang, X., Zheng, R., Yin, Z., Fang, Y., & Yang, D. (2009). Handover study concerning mobility in the two-hierarchy network. In *Proceedings of the IEEE Vehicular Technology Conference (VTC 2009-Spring)*, (pp. 1-5). Barcelona, Spain: IEEE Press.

Wu, W., Banerjee, N., Basu, K., & Dasu, S. K. (2005). SIP-based vertical handoff between WWANs and WLANs. *IEEE Wireless Communications*, *12*(3), 66–72. doi:10.1109/MWC.2005.1452856

Yang, K., Ou, S., Chen, H. H., & He, J. (2007). A multihop peer-communication protocol with fairness guarantee for IEEE 802.16-based vehicular networks. *IEEE Transactions on Vehicular Technology*, *56*(6), 3358–3370. doi:10.1109/TVT.2007.906875

Yuliang Tang, Y., Lin, C. C., Kou, G., & Deng, D. J. (2010). Cross-layer handover scheme for multimedia communications in next generation wireless networks. *EURASIP Journal on Wireless Communications and Networking*, 10.

Zeadally, S., & Siddiqui, F. (2007). An empirical analysis of handoff performance for SIP, mobile IP, and SCTP protocols. *Wireless Personal Communications*, *43*(2), 589–603. doi:10.1007/s11277-007-9251-x

Zhang, H., Wen, X., Wang, B., Zheng, W., & Sun, Y. (2010). A novel handover mechanism between femtocell and macrocell for LTE based networks. In *Proceedings of the International Conference on Communication Software and Networks (ICCSN 2010)*, (pp. 228-231). Singapore: ICCSN.

Chapter 8
Joint Resource Allocation and Interference Mitigation Techniques for Cooperative Wireless Networks

Rodrigo C. de Lamare
University of York, UK

Patrick Clarke
University of York, UK

ABSTRACT

This chapter presents joint interference suppression and power allocation algorithms for DS-CDMA and MIMO networks with multiple hops and amplify-and-forward and Decode-and-Forward (DF) protocols. A scheme for joint allocation of power levels across the relays and linear interference suppression is proposed. The authors also consider another strategy for joint interference suppression and relay selection that maximizes the diversity available in the system. Simulations show that the proposed cross-layer optimization algorithms obtain significant gains in capacity and performance over existing schemes.

INTRODUCTION

Multiple-antenna wireless communication systems can exploit the spatial diversity in wireless channels, mitigating the effects of fading and enhancing their performance and capacity. Due to the size and cost of mobile terminals, it is considered impractical to equip them with multiple antennas. However, spatial diversity gains can be obtained when single-antenna terminals establish a distributed antenna array via cooperation (Sendonaris, Erkip, & Aazhang, 2003; Laneman & Wornell, 2004). The use of cooperative strategies can lead to several types of gains (Dohler & Li, 2010; Pabst, et al., 2004), namely, pathloss, diversity, and multiplexing gains. Pathloss gains

DOI: 10.4018/978-1-4666-0960-0.ch008

allow a significant reduction in the transmitted power for an equivalent performance, can increase the coverage (Ettefagh, Kuhn, Hammerstrom, & Wittneben, 2006) and enhance the interference suppression capability (Dohler & Li, 2010; Pabst, et al., 2004). The diversity gains improve the performance of the wireless system with respect to the probability of error because the transmission of multiple copies of the signals reduce the probability that the message will not be received correctly. The multiplexing gains (So & Liang, 2005), which correspond to the additional number of bits that the system can transmit as compared to a single-antenna link, can be obtained when a designer can use relays to form independent channels and increase the rate of communication.

Despite the many advantages in terms of gains as previously outlined, cooperative communications also entail some disadvantages such as signalling overheads (Pabst, et al., 2004), more computationally complex scheduling algorithms (Dohler & Li, 2010) and increased latency (Lu, Steenkiste, & Chen, 2007). For this reason, it is important to weigh the pros and cons of cooperative techniques prior to their adoption and consider the practical scenarios of interest (Kyritsi, et al., 2007). Motivated by their performance and diversity gains, cooperative techniques are now being considered for the next generation of mobile networks (Scaglione, Goeckel, & Laneman, 2006; Fan & Thompson, 2007; Peters, Panah, Truong, & Heath, 2009). In cooperative systems, terminals or users relay signals to each other in order to propagate redundant copies of the same signals to the destination user or terminal. To this end, the designer must resort to a cooperation protocol such as Amplify-and-Forward (AF) (Laneman & Wornell, 2004), Decode-and-Forward (DF) (Laneman & Wornell, 2004; Huang, Hong, & Kuo, 2007), and Compress-and-Forward (CF) (Kramer, Gastpar, & Gupta, 2005).

In order to obtain the benefits of cooperative techniques, designers must address a number of problems that are encountered in cooperative wireless systems. These problems include physical-layer strategies such as synchronization, interference mitigation, and parameter estimation. However, designers also have to consider a number of associated problems that belong to higher protocol layers and include the allocation of resources such as power, relays and rate. These tasks present an opportunity to perform cross-layer design and to obtain very significant gains in performance and capacity for cooperative wireless networks. This chapter is concerned with cross-layer design techniques for cooperative wireless networks and investigates the benefits of approaches that jointly mitigate interference and perform resource allocation.

In this chapter, we will consider two types of schemes, namely, Direct-Sequence Code-Division Multiple Access (DS-CDMA) (Ziemer, Peterson, & Borth, 1995; Honig & Poor, 1998) and Multi-Input Multi-Output (MIMO) (Foschini & Gans, 1998; Telatar, 1999) systems. The former is of fundamental importance in wireless ad-hoc and sensor networks (Dohler & Li, 2010), whereas the latter is one of the main ingredients of future wireless cellular networks. When implementing cooperative techniques in wireless systems, designers often consider the transmission technologies available and their suitability to certain applications. Therefore, the concept of distributed antenna arrays can be easily extended to techniques such as MIMO (Foschini & Gans, 1998; Telatar, 1999) and DS-CDMA systems (Ziemer, Peterson, & Borth, 1995; Honig & Poor, 1998).

In the context of MIMO systems, one can obtain substantial multiplexing and diversity gains (Foschini & Gans, 1998; Telatar, 1999; Zheng & Tse, 2003) with the deployment of multiple antennas at both ends of the wireless system. MIMO technology is poised to equip most of the future wireless systems and can be incorporated in conjunction with other transmission systems. There are two basic configurations which exploit the nature of the wireless channel: spatial multiplexing (de Lamare & Sampaio-Neto, 2010b) and diversity (Xu & Tsatsanis, 2006). Spatial multiplexing relies on the concept of forming individual data stream between pais of transmit and receive antennas. The capacity gains of spatial multiplexing grow linearly with the minimum number of transmit and receive antennas (Foschini & Gans, 1998; Telatar, 1999) and allow a MIMO system to obtain a considerable increase in data rates. Diversity configurations adopt space-time codes (Alamout, 1998; Tarokh, 1999) to transmit data from the antennas at the transmitter and can obtain a lower probability of error.

DS-CDMA systems are a key multiple access technology for current and future wireless communication systems. Such systems rely on the idea of transmitting data with the aid of unique signatures, which are also known as spreading codes. These signatures are responsible for spreading the information in frequency, and allow the system to have multiple users on the same channel. The advantages of DS-CDMA include good performance in multi-path channels, flexibility in the allocation of channels, increased capacity in bursty and fading environments and the ability to share bandwidth with narrowband communication systems without deterioration of either's systems performance (Ziemer, Peterson, & Borth, 1995; Honig & Poor, 1998). Demodulating a desired user in a DS-CDMA network requires processing the received signal in order to mitigate different types of interference, namely, Narrowband Interference (NBI), Multi-Access Interference (MAI), Inter-Symbol Interference (ISI) and the noise at the receiver. The major source of interference in most CDMA systems is MAI, which arises due to the fact that users communicate through the same physical channel with non-orthogonal signals.

The similarities between MIMO and CDMA systems include their mathematically similar descriptions and their fundamental need for interference mitigation. Indeed, the data streams of MIMO systems operating in a spatial multiplexing configuration are equivalent to the users of a DS-CDMA system. In order to separate data streams or users, a designer must resort to detection techniques (Verdu, 1998), which are very similar when applied to either MIMO or DS-CDMA. The optimal Maximum Likelihood (ML) detector is often too complex to be implemented for systems with a large number of antennas. For this reason, designers often resort to suboptimal solutions that an attractive trade-off between performance and complexity. These include the Sphere Decoder (SD) algorithms (Hassibi & Vikalo, 2005), linear detectors (Verdu, 1998), the Successive Interference Cancellation (SIC) approach (de Lamare & Sampaio-Neto, 2010b), the Parallel Interference Cancellation (PIC) (Verdu, 1998), and the Decision Feedback (DF) detectors (de Lamare, Haardt, & Sampaio-Neto, 2008; Choi, Yu, & Lee, 2005) are techniques that can offer an attractive trade-off between performance and complexity. These detection algorithms can be combined with cross-layer design techniques for enhanced interference mitigation and improved overall performance. In this chapter, we are specifically interested in exploring the advantages of linear detection with power allocation, data stream and relay selection.

Prior and Related Work

Prior work on cross-layer design for cooperative and multihop communications has considered the problem of resource allocation (Luo, Blum, Cimini, Greenstein, & Haimovich, 2007; Long & Hossain, 2008) in generic networks. These include power and rate allocation strategies. Related work on cooperative multiuser DS-CDMA networks has focused on the assessment of the impact of Multiple Access Interference (MAI) and Intersymbol Interference (ISI), the problem of partner selection (Huang, Hong, & Kuo, 2007; Venturino, Wang, & Lops, 2006) the Bit Error Ratio (BER) and outage performance analysis (Vardhe, Reynolds, & Valenti, 2008), and training-based joint power allocation and interference mitigation strategies (Joung & Sayed, 2010; Souryal, Vojcic, & Pickholtz, 2006). Previous works have also considered the problem of antenna selection, Relay Selection (RS) and diversity maximization, which are central themes in the MIMO relaying literature (Ding, Liu, Luo, & Chen, 2010; Berenguer, Wang, & Krishnamurhty, 2005). However, current approaches are often limited to stationary, single relay systems and channels which assume the direct path from the source to the destination is negligible (Fang, Hua, & Koshy, 2006).

Most of these resource allocation and interference mitigation strategies require a higher computational cost to implement the power allocation and a significant amount of signalling, decreasing the spectral efficiency of cooperative networks. This problem is central to ad-hoc and sensor networks (Souryal, Vojcic, & Pickholtz, 2006) that employ spread spectrum systems and require multiple hops to communicate with nodes that are far from the source node. This is also of paramount importance in cooperative cellular networks.

Contributions

In this chapter, we present joint interference suppression and power allocation algorithms for DS-CDMA and MIMO networks with multiple hops and AF and DF protocols. A scheme that jointly considers the power allocation across the relays subject to group-based power constraints and the design of linear receivers for interference suppression is proposed. The idea of a group-based power allocation constraint is shown to yield close to optimal performance, while keeping the signalling and complexity requirements low. A constrained Minimum Mean-Squared Error (MMSE) design for the receive filters and the power allocation vectors is developed along with an MMSE channel estimator for the cooperative system under consideration. The linear MMSE receiver design is adopted due to its mathematical tractability and good performance. However, the incorporation of more sophisticated detection strategies including interference cancellation with iterative decoding (de Lamare & Sampaio-Neto, 2008a) and advanced parameter estimation methods (de Lamare & Sampaio-Neto, 2009a) is also possible. In order to solve the proposed optimization problem efficiently, a method to form an effective group of users and an alternating optimization strategy are presented with Recursive Alternating Least Squares (RALS) algorithms for estimating the parameters of the receiver, the power allocation and the channels. A joint relay selection and transmit diversity selection strategy for MIMO networks with linear receivers is also proposed which optimizes relay transmissions with minimal feedback requirements. Effectively a novel approach to 1-bit power allocation, two joint discrete optimization functions are formed which are solved using discrete stochastic algorithms.

Organisation of the Chapter

The chapter is organized as follows. Section 2 describes cooperative DS-CDMA and MIMO system models with multiple hops. Section 3 formulates the problem, details the constrained MMSE design of the receive filters and the power allocation vectors subject to a group-based power allocation constraint, and describes an MMSE channel estimator. An extension to cooperative MIMO systems is also presented and discrete optimization problems are formulated to jointly select the optimal relays and their transmit antennas. Section 4 presents an algorithm to form the group and the alternating optimization strategy along with RLS-type algorithms for estimating the parameters of the receiver, the power allocation and the channels. For the solution of the combinatorial problems posed by the relay selection strategy, a pair of discrete stochastic algorithms are introduced and their joint operation detailed. Section 5 presents and discusses the simulation results, and Section 6 draws the conclusions of this work.

Figure 1. (a) Uplink and (b) downlink of the cooperative DS-CDMA system

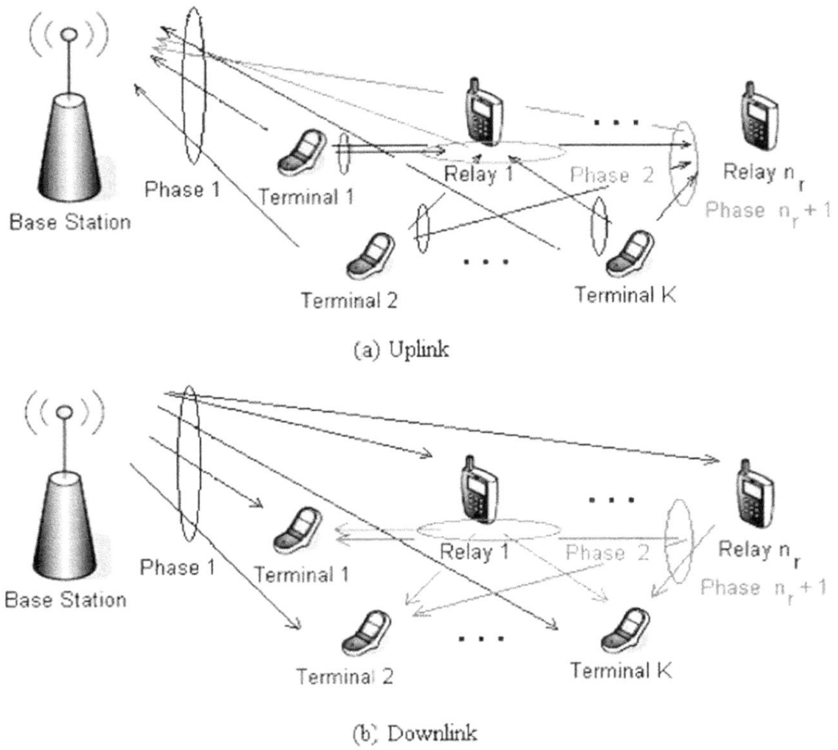

SYSTEM AND DATA MODELS OF COOPERATIVE WIRELESS SYSTEMS

In this section, we consider system and data models of cooperative wireless systems. The basic idea is to use a linear algebra approach to describe models of the cooperative systems of interest. In particular, we focus on DS-CDMA and MIMO systems and we present a unified approach to the description of these systems (see Figure 1).

Cooperative DS-CDMA System and Data Model

Let us first consider a synchronous DS-CDMA network with multipath channels. The DS-CDMA system operates with QPSK modulation, K users, N chips per symbol and L as the maximum number of propagation paths for each link. An outline of the system is depicted in (1). The system is equipped with AF and DF protocols that allow communication in multiple hops using n_r fixed relays in a repetitive fashion. We assume that the source node or terminal transmits data organized in packets with P symbols, there is enough control data to coordinate transmissions and cooperation, and the linear receivers at the relay and destination terminals are synchronized with their desired signals. The received signals are filtered by a matched filter, sampled at chip rate and organized into $M \times 1$ vectors $\mathbf{r}_{sd}$, $\mathbf{r}_{sr_j}$ and $\mathbf{r}_{r_j d}$, which describe the signal received from the source to the destination, the source to the relays, and the relays to the destination, respectively:

$$\begin{aligned}
\mathbf{r}_{sd} &= \sum_{k=1}^{K} a_{sd}^{k} \mathbf{C}_k \mathbf{h}_{sd,k} b_k + \eta_{sd} \\
&+ \mathbf{n}_{sd}, \\
\mathbf{r}_{sr_j} &= \sum_{k=1}^{K} a_{sr_j}^{k} \mathbf{C}_k \mathbf{h}_{sr_j,k} b_k + \eta_{sr_j} \\
&+ \mathbf{n}_{sr_1 j}, \\
\mathbf{r}_{r_j d} &= \sum_{k=1}^{K} a_{r_j d}^{k} \mathbf{C}_k \mathbf{h}_{r_j d,k} \tilde{b}_k + \eta_{r_j d} \\
&+ \mathbf{n}_{r_j d}, \\
j &= 1,\ldots,n_r, i = 1,\ldots,P
\end{aligned} \quad (1)$$

where $M = N + L - 1$, P is the number of packet symbols, $n_p = n_r + 1$ is the number of transmission phases or hops, and n_r is the number of relays. The vectors $\mathbf{n}_{sd}$, $\mathbf{n}_{sr_j}$ and $\mathbf{n}_{r_j d}$ are zero mean complex Gaussian vectors with variance σ^2 generated at the receivers of the destination and the relays from different links, and the vectors η_{sd}, η_{sr_j} and $\eta_{r_j d}$ represent the Intersymbol Interference (ISI). The amplitudes of the source to destination, source to relay and relay to destination links for user k are denoted by a_{sd}^k, $a_{sr_j}^k$ and $a_{r_j d}^k$, respectively. The quantities b_k and $\tilde{b}_k$ represent the original and reconstructed symbols by the AF or DF protocol at the relays, respectively. The $M \times L$ matrix $\mathbf{C}_k$ contains versions of the signature sequences of each user shifted down by one position at each column as described by:

$$\mathbf{C}_k = \begin{bmatrix} c_k(1) & & 0 \\ \vdots & \ddots & c_k(1) \\ c_k(N) & & \vdots \\ 0 & \ddots & c_k(N) \\ & & \end{bmatrix}, \quad (2)$$

where $\mathbf{c}_k = [c_k(1), c_k(2), \ldots, c_k(N)]$ stands for the signature sequence of user k, the $L \times 1$ channel vectors from source to destination, source to relay, and relay to destination are $\mathbf{h}_{sd,k}$, $\mathbf{h}_{sr_j,k}$, $\mathbf{h}_{r_j d,k}$, respectively. By collecting the data vectors in (5) (including the links from relays to the destination) into a $J \times 1$ received vector at the destination, where $J = (n_r + 1)M$, we obtain:

$$\begin{bmatrix} \mathbf{r}_{sd} \\ \mathbf{r}_{r_1 d} \\ \vdots \\ \mathbf{r}_{r_{n_r} d} \end{bmatrix} = \begin{bmatrix} \sum_{k=1}^{K} a_{sd}^{k} \mathbf{C}_k \mathbf{h}_{sd,k} b_k \\ \sum_{k=1}^{K} a_{r_1 d}^{k} \mathbf{C}_k \mathbf{h}_{r_1 d,k} \tilde{b}_k^{r_1 d} \\ \vdots \\ \sum_{k=1}^{K} a_{r_{n_r} d}^{k} \mathbf{C}_k \mathbf{h}_{r_{n_r} d,k} \tilde{b}_k^{r_{n_r} d} \end{bmatrix} + \eta + \mathbf{n} \quad (3)$$

Rewriting the above signals in a compact form yields:

$$\begin{aligned}
\mathbf{r}[i] &= \sum_{k=1}^{K} \tilde{\mathbf{B}}_k[i] \tilde{\mathbf{A}}_k[i] \underbrace{\tilde{\mathbf{C}}_k \mathbf{h}_k[i]}_{\mathbf{p}_k[i]} + \eta[i] + \mathbf{n}[i] \\
&= \sum_{k=1}^{K} \tilde{\mathbf{B}}_k[i] \tilde{\mathbf{A}}_k[i] \tilde{\mathbf{C}}_k \mathbf{h}_k[i] + \eta[i] + \mathbf{n}[i] \\
&= \sum_{k=1}^{K} \mathbf{P}_k[i] \mathbf{B}_k[i] \mathbf{a}_k[i] + \eta[i] + \mathbf{n}[i],
\end{aligned} \quad (4)$$

where the $J \times (n_r + 1)L$ matrix $\tilde{\mathbf{C}}_k = diag\{\mathbf{C}_k \ldots \mathbf{C}_k\}$ contains copies of $\mathbf{C}_k$ shifted down by M positions for each group of L columns and zeros elsewhere. The $Q \times 1$ vector $\mathbf{h}_k[i]$, where $Q = (n_r + 1)L$ contains the channel gains of the links between the source, the relays and the destination, and $\mathbf{p}_k[i]$ is the effec-

Figure 2. Block diagram of 2-phase cooperative MIMO system

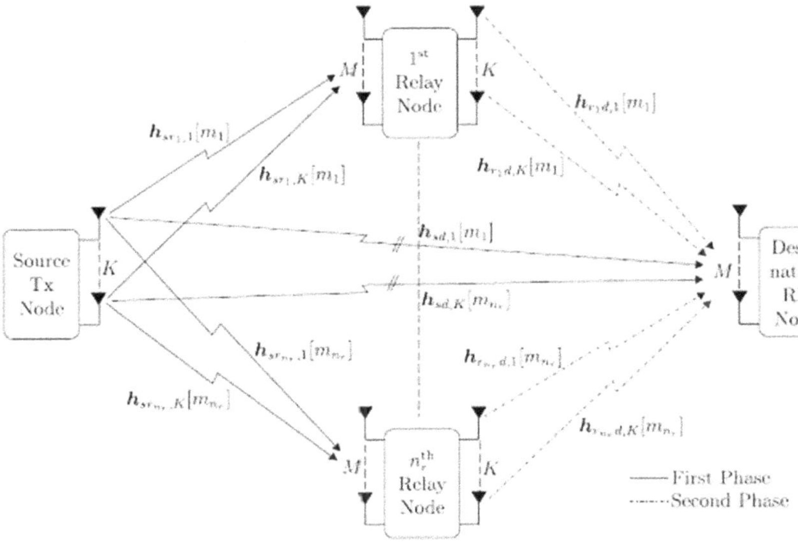

tive signature for user k. The $(n_r+1)\times(n_r+1)$ diagonal matrix

$$\mathbf{B}_k[i] = diag(b_k[i]\, \tilde{b}_k^{r_1 d}[i] \ldots \tilde{b}_k^{r_n d}[i])$$

contains the symbols transmitted from the source to the destination $(b_k[i])$ and the n_r symbols transmitted from the relays to the destination $(\tilde{b}_k^{r_1 d}[i] \ldots \tilde{b}_k^{r_n d}[i])$ on the main diagonal, and the $J\times J$ diagonal matrix

$$\tilde{\mathbf{B}}_k[i] = diag(b_k[i] \otimes \mathbf{I}_M\, \tilde{b}_k^{r_1 d}[i] \otimes \mathbf{I}_M \ldots \tilde{b}_k^{r_n d}[i] \otimes \mathbf{I}_M),$$

where $\otimes$ denotes the Kronecker product and $\mathbf{I}_M$ is an identity matrix with dimension M. The $(n_r+1)\times 1$ power allocation vector $\mathbf{a}_k[i] = [a_{sd}^k\; a_{r_1 d}^k \ldots a_{r_{n_r} d}^k]^T$ has the amplitudes of the links, the $(n_r+1)\times(n_r+1)$ diagonal matrix $\mathbf{A}_k[i]$ is given by $\mathbf{A}_k[i] = diag\{\mathbf{a}_k[i]\}$, and the $J\times J$ diagonal matrix

$$\tilde{\mathbf{A}}_k[i] = [a_{sd}^k \otimes \mathbf{I}_M\; a_{r_1 d}^k \otimes \mathbf{I}_M \ldots a_{r_{n_r} d}^k \otimes \mathbf{I}_M]^T.$$

The $J\times(n_r+1)$ matrix $\mathbf{P}_k$ has copies of the effective signature $\mathbf{p}_k[i]$ shifted down by M positions for each column and zeros elsewhere. The $J\times 1$ vector $\eta[i]$ represents the ISI terms and the $J\times 1$ vector $\mathbf{n}[i]$ has the noise components.

Cooperative MIMO System and Data Model

Let us now consider a synchronous MIMO system model, which has similarities with the DS-CDMA system model of the previous subsection. We consider a narrowband MIMO system with flat fading channels, QPSK modulation, K transmit antennas, and M receive antennas as illustrated in Figure 2. The cooperative MIMO network is equipped with DF protocol that allows communication in $n_p = 2$ hops using n_r fixed relays in a repetitive fashion where a non-negligible, direct source to destination link exists during the first phase. We assume that the source node or terminal transmits data organized in packets with P symbols, there is enough control data to coordinate transmissions and cooperation, and the linear receivers at the relay and destination terminals are synchronized with their desired signals. It should be noted that the MIMO and CDMA system and data models are mathematically equivalent and the main difference is that we employ for the MIMO version a spreading code matrix $\mathbf{C}_k = 1$.

The received signals are filtered by a matched filter, sampled at chip rate and organized into $M \times 1$ vectors $\mathbf{r}_{sd}$, $\mathbf{r}_{sr_j}$ and $\mathbf{r}_{rd}$, which describe the signal received from the source to the destination, the source to the relays, and the relays to the destination, respectively:

$$\begin{aligned}
\mathbf{r}_{sd} &= \sum_{k=1}^{K} a_{sd}^k \mathbf{h}_{sd,k} b_k[i] + \eta_{sd} \\
&\quad + \mathbf{n}_{sd}, \\
\mathbf{r}_{sr_j} &= \sum_{k=1}^{K} a_{sr_j}^k \mathbf{h}_{sr_j,k} b_k[i] + \eta_{sr_j} \\
&\quad + \mathbf{n}_{sr_j}, \\
\mathbf{r}_{r_j d} &= \sum_{k=1}^{K} a_{r_j d}^k \mathbf{h}_{r_j d,k} \tilde{b}_k[i] + \eta_{r_j d} \\
&\quad + \mathbf{n}_{r_j d}, \\
j &= 1, \ldots, n_r,\ i = 1, \ldots, P,\ p = 1,2
\end{aligned} \quad (5)$$

where P is the number of packet symbols, $n_p = 2$ is the number of transmission phases or hops, and n_r is the number of relays. The vectors $\mathbf{n}_{sd}$, $\mathbf{n}_{sr_j}$ and $\mathbf{n}_{r_j d}$ are zero mean complex Gaussian vectors with variance σ^2 generated at the receivers of the destination and the relays from different links, and the vectors η_{sd}, η_{sr_j} and $\eta_{r_j d}$ represent the Intersymbol Interference (ISI).

The amplitudes of the source to destination, source to relay and relay to destination links for user k are denoted by a_{sd}^k, $a_{sr_j}^k$ and $a_{r_j d}^k$, respectively. The quantities $b_k[i]$ and $\tilde{b}_k[i]$ represent the original and reconstructed symbols by the AF or DF protocol at the relays, respectively. The $M \times 1$ spatial channel vectors from source to destination, source to relay, and relay to destination are $\mathbf{h}_{sd,k}$, $\mathbf{h}_{sr_j,k}$, $\mathbf{h}_{r_j d,k}$, respectively. By collecting the data vectors in (5) (including the links from relays to the destination) into a $J \times 1$ received vector at the destination, where $J = n_p M$ for MIMO systems, we obtain:

$$\mathbf{r}[i] = \begin{bmatrix} \sum_{k=1}^{K} a_{sd}^k \mathbf{h}_{sd,k} b_k[i] \\ \sum_{j}^{n_r} \sum_{k=1}^{K} a_{r_{n_r} d}^k \mathbf{h}_{r_{n_r} d,k} \tilde{b}_k^{r_{n_r} d}[i] \end{bmatrix} + \mathbf{n}[i] \quad (6)$$

Rewriting the above signals in a compact form yields:

$$\begin{aligned}
\mathbf{r}[i] &= \sum_{k=1}^{Kn_r} \tilde{\mathbf{B}}_k[i] \tilde{\mathbf{A}}_k[i] \mathbf{h}_k[i] + \mathbf{n}[i] \\
&= \sum_{k=1}^{K} \mathbf{P}_k[i] \mathbf{B}_k[i] \mathbf{a}_k[i] + \mathbf{n}[i],
\end{aligned} \quad (7)$$

The $J \times 1$ vector $\mathbf{h}_k[i]$ contains the spatial channel gains of the links between the source, the relays and the destination. The $n_p \times n_p$ diagonal matrix $\mathbf{B}_k[i] = diag(b_k[i]\, \tilde{b}_k^{r_k d}[i])$, for $1 \leq k \leq K$, and $\mathbf{B}_k[i] = diag(0\, \tilde{b}_k^{r_k d}[i])$, for $K < k \leq Kn_r$, contains the symbols transmitted from the source to the destination ($b_k[i]$) and the n_r symbols transmitted from the relays to the destination ($\tilde{b}_k^{r_k d}[i]$) on the main diagonal, and the $J \times J$ diagonal matrix

$$\tilde{\mathbf{B}}_k[i] = diag(b_k[i] \otimes \mathbf{I}_M \; \tilde{b}_k^{r_k d}[i] \otimes \mathbf{I}_M),$$

for

$1 \le k \le K$,

and $\tilde{\mathbf{B}}_k[i] = diag(0 \otimes \mathbf{I}_M \; \tilde{b}_k^{r_k d}[i] \otimes \mathbf{I}_M)$, for $K < k \le Kn_r$. The $n_p \times 1$ power allocation vector $\mathbf{a}_k[i] = [a_{sd}^k \ldots a_{r_{n_k} d}^k]^T$ has the amplitudes of the links, the $n_p \times n_p$ diagonal matrix $\mathbf{A}_k[i]$ is given by $\mathbf{A}_k[i] = diag\{\mathbf{a}_k[i]\}$, and the $J \times J$ diagonalmatrix $\tilde{\mathbf{A}}_k[i] = [a_{sd}^k \otimes \mathbf{I}_M \ldots a_{r_{n_k} d}^k \otimes \mathbf{I}_M]^T$. The $J \times n_p$ matrix $\mathbf{P}_k$ has copies of the spatial signatures $\mathbf{h}_k[i]$ shifted down by M positions for each column and zeros elsewhere. The $J \times 1$ vector $\mathbf{n}[i]$ has the noise components.

JOINT MMSE RECEIVER DESIGN, POWER ALLOCATION, RELAY SELECTION, AND CHANNEL ESTIMATION

In this section, our aim is to describe techniques to mitigate interference and allocate the power and select the best relays according to the mean square error (MSE) criterion. We present a joint receiver design and power allocation strategy using constrained linear MMSE estimation and group-based power constraints along with a linear MMSE channel estimator. The interesting aspect of the group-based power constraints is that a designer can choose a subset of users or data streams for power adjustment. Another technique that is detailed here is a method called Transmit Diversity Selection (TDS) which operates with the linear MMSE receivers. With the TDS technique the relay selection is used to jointly optimize the selection of antennas in a strategy equivalent to a 1-bit transmit antenna power allocation.

In order to describe the techniques necessary for interference mitigation and resource allocation, we introduce an alternative way of expressing the $J \times 1$ received vector in (7). Our goal is to separate the subset of users or data streams that will be used for resource allocation with the group-based power constraints. The modified $J \times 1$ received vector can be expressed as:

$$\mathbf{r}[i] = \mathbf{P}_{\mathsf{S}}[i]\mathbf{B}_{\mathsf{S}}[i]\mathbf{a}_{\mathsf{S},k}[i] + \sum_{k \neq \mathsf{S}} \mathbf{P}_k[i]\mathbf{B}_k[i]\mathbf{a}_k[i] + \eta[i] + \mathbf{n}[i], \quad (8)$$

where $\mathsf{S} = \{\mathsf{S}_1, \mathsf{S}_2, \ldots, \mathsf{S}_G\}$ denotes the group of G users to consider in the design. The $J \times G(n_r+1)$ matrix $\mathbf{P}_{\mathsf{S}} = [\mathbf{P}_{\mathsf{S}_1} \mathbf{P}_{\mathsf{S}_2} \ldots \mathbf{P}_{\mathsf{S}_G}]$ contains the G effective signatures of the group of users. The $G(n_r+1) \times G(n_r+1)$ diagonal matrix

$\mathbf{B}_{\mathsf{S}}[i] = diag(b_{\mathsf{S}_1}[i]\, \tilde{b}_{\mathsf{S}_1}^{r_1 d}[i] \ldots \tilde{b}_{\mathsf{S}_1}^{r_n d}[i] \ldots b_{\mathsf{S}_G}[i]\, \tilde{b}_{\mathsf{S}_G}^{r_1 d}[i] \ldots \tilde{b}_{\mathsf{S}_G}^{r_n d}[i])$ contains the symbols transmitted from the sources to the destination and from the relays to the destination of the G users in the group on the main diagonal, the $G(n_r+1) \times 1$ power allocation vector

$\mathbf{a}_{\mathsf{S},k}[i] = [a_{sd}^{\mathsf{S}_1}[i]\, a_{r_1 d}^{\mathsf{S}_1}[i] \ldots a_{r_{n_r} d}^{\mathsf{S}_1}[i], \ldots, a_{sd}^{\mathsf{S}_G}[i]\, a_{r_1 d}^{\mathsf{S}_G}[i] \ldots a_{r_{n_r} d}^{\mathsf{S}_G}[i]]^T$ of the amplitudes of the links used by the G users or data streams in the group.

Linear MMSE Receiver Design and Power Allocation Scheme with Group-Based Constraints

The linear MMSE interference mitigation for user or data stream k is performed by the receive filter $\mathbf{w}_k[i] = [w_{k,1}[i], \ldots, w_{k,J}[i]]$ with J coefficients on the received data vector $\mathbf{r}[i]$ and yields:

$$z_k[i] = \mathbf{w}_k^H[i]\mathbf{r}[i], \tag{9}$$

where $z_k[i]$ is an estimate of the symbols, which are processed by a slicer $Q(\cdot)$ that performs detection and obtains the desired symbol as $\hat{b}_k[i] = Q(z_k[i])$.

Let us now detail the linear MMSE-based design of the receivers for user or data stream k represented by $\mathbf{w}_k[i]$ and for the computation of the $G(n_r+1)\times 1$ power allocation vector $\mathbf{a}_{\mathsf{S},k}[i]$. This problem can be cast as the following constrained optimization:

$$\begin{aligned}[w_k^{opt}, a_{\mathsf{S},k}^{opt}] &= \arg \min_{w_k[i], a_{\mathsf{S},k}[i]} E[(\mid b_k[i] - w_k^H[i]r[i] \mid^2] \\ & \textit{subject to } a_{\mathsf{S},k}^H[i]a_{\mathsf{S},k}[i] = P_G,\end{aligned} \tag{10}$$

In order to obtain expressions for the receive filter $\mathbf{w}_k[i]$ and the power allocation vector $\mathbf{a}_{\mathsf{S},k}[i]$ subject to the group-based power constraints, we need the help of the method of Lagrange multipliers (10) (Haykin, 2002) that transforms a constrained optimization into an unconstrained one. The MMSE expressions for $\mathbf{w}_k[i]$ and $\mathbf{a}_{\mathsf{S},k}[i]$ are given by:

$$\begin{aligned}\mathsf{L}_k &= E[\mid b_k[i] - \mathbf{w}_k^H[i](\mathbf{P}_\mathsf{S}[i]\mathbf{B}_\mathsf{S}[i]\mathbf{a}_{\mathsf{S},k}[i] \\ &\quad + \sum_{k\neq \mathsf{S}} \mathbf{P}_k[i]\mathbf{B}_k[i]\mathbf{a}_k[i] + \eta[i] + \mathbf{n}[i]) \mid^2] \\ &\quad + \lambda_k(\mathbf{a}_{\mathsf{S},k}[i] - P_G),\end{aligned} \tag{11}$$

where λ_k is a Lagrange multiplier.

Since the Lagrangian in (11) is a function of both $\mathbf{w}_k[i]$ and $\mathbf{a}_{\mathsf{S},k}[i]$, we need to employ a strategy for optimization the function with respect to both parameter vectors. The main idea is to fix one of the parameter vectors and compute the gradient terms with respect to the other parameter vector that minimizes the Lagrangian and obtain the expression of interest. In particular, an expression for $\mathbf{a}_{\mathsf{S},k}[i]$ is obtained by fixing $\mathbf{w}_k[i]$, taking the gradient terms of the Lagrangian and equating them to zero, which yields:

$$\mathbf{a}_{\mathsf{S},k}[i] = (\mathbf{R}_{\mathsf{S},k}[i] + \lambda_k \mathbf{I})^{-1}\mathbf{p}_{\mathsf{S},k}[i] \tag{12}$$

where the $G(n_r+1)\times G(n_r+1)$ covariance matrix $\mathbf{R}_{\mathsf{S},k}[i] = E[\mathbf{B}_\mathsf{S}^H[i]\mathbf{P}_\mathsf{S}^H[i]\mathbf{w}_k[i]\mathbf{w}_k^H[i]\mathbf{P}_\mathsf{S}[i]\mathbf{B}_\mathsf{S}[i]]$ and the vector $\mathbf{p}_{\mathsf{S},k}[i] = E[b_k[i]\mathbf{B}_\mathsf{S}^H[i]\mathbf{P}_\mathsf{S}^H[i]\mathbf{w}_k[i]]$ is a $G(n_r+1)\times 1$ cross-correlation vector. The Lagrange multiplier λ_k plays the role of a regularization term and has to be determined numerically due to the difficulty of evaluating its expression.

In order to compute the expression for $\mathbf{w}_k[i]$, we fix $\mathbf{a}_{\mathsf{S},k}[i]$, calculate the gradient terms of the Lagrangian and equate them to zero which leads to:

$$\mathbf{w}_k[i] = \mathbf{R}^{-1}[i]\mathbf{p}_k[i], \tag{13}$$

where the covariance matrix of the received vector is given by $\mathbf{R}[i] = E[\mathbf{r}[i]\mathbf{r}^H[i]]$ and $\mathbf{p}_k[i] = E[b_k^*[i]\mathbf{r}[i]]$ is a $J\times 1$ cross-correlation vector. The quantities $\mathbf{R}[i]$ and $\mathbf{p}_k[i]$ depend on the power allocation vector $\mathbf{a}_{\mathsf{S},k}[i]$. The expressions in (12) and (13) do not have a closed-form solution as they have a dependence on each other. Moreover, the expressions also require the estimation of the channel vector $\mathbf{h}_k[i]$. Thus, it is necessary to iterate (12) and (13) with initial values to obtain a solution and to estimate the channel. The network has to convey the information from the group of users which is necessary to compute the group-based power allocation including the filter $\mathbf{w}_k[i]$. The expressions in (12) and (13) require matrix inversions with cubic complexity $(O((J)^3)$ and $O((G(n_r+1))^3)$.

Transmit Diversity Selection and Relay Selection

In this subsection, we explore the idea of Transmit Diversity Selection (TDS) and Relay Selection (RS) and how they can be used to improve the performance of cooperative systems. In cooperative wireless systems with multiple relays, there are links that have very poor propagation conditions that can degrade the performance of the overall system. These links can be identified and removed from the operation of the system via TDS and RS. To this end, we formulate a TDS and RS strategy for a 2 DF MIMO network as a discrete combinatorial MSE problem which optimizes the use of the channels of the second phase via 1-bit power allocation (Clarke & de Lamare, 2011). It turns out that the problems of TDS and RS are combinatorial problems which require either an exhaustive search or some relaxation approach. We specify that a subset of K_{sub} antennas of the $n_r K$ relay antennas are active at each time instant in order to reduce the optimization complexity but also to ensure a minimum available level of diversity. The destination node's MSE TDS optimization function is given by:

$$\begin{aligned}\mathsf{T}_r^{opt} &= \arg\min_{\mathsf{T}_r \in \Omega_T} \mathsf{C}[i, \mathsf{T}_r, \mathbf{r}] \\ &= \operatorname*{argmin}_{\mathsf{T}_r \in \Omega_T} \sum_{k=1}^{K} E[\| b_k[i] - \mathbf{w}_k[i]\mathbf{r}[i] \|^2],\end{aligned} \tag{14}$$

where $\mathsf{T}_r = \operatorname{diag}(a^1_{r_1 d} a^K_{r_1 d},, a^1_{r_{n_r} d} a^K_{r_{n_r} d})$ and $a^k_{r_j d} = \{0,1\}$, and $\mathbf{w}_k$ is the linear MMSE filter for the kth symbol. Under the assumption of no inter-relay communication and that each data stream is allocated to its correspondingly numbered transmit antenna at each relay, the set Ω_T has a cardinality of $|\Omega_T| = \binom{n_r K}{K_{sub}}$ and contains all possible combinations of relay transmit antennas patterns. The performance and complexity of solutions to (14) depend on $|\Omega_T|$ and its elements. However, $|\Omega_T|$ is significant even for modest numbers of antennas and relays, e.g. $n_r \geq 4$ and $K \geq 2$. Further improvements can be achieved by a process we term RS which addresses the possibility of mismatching poor first phase channels with optimized second phase channels as well as reducing the cardinality of Ω_T.

By removing one or more relays based on their MSE performance from consideration by (14), Ω_T can be optimized and its cardinality improved without overly restricting the second-phase channels available to the TDS process. The selection of the single highest MSE relay can be expressed as a discrete maximization problem given by

$$\begin{aligned}j_{opt} &= \arg\max_{j \in \Omega_R} \mathsf{F}[i, \mathbf{r}_{sr_j}]. \\ &= \arg\max_{j \in \Omega_R} \sum_{k=1}^{K} E[\| b_k[i] - \mathbf{w}_{j,k}[i]\mathbf{r}_{sr_j}[i] \|^2],\end{aligned} \tag{15}$$

where Ω_R is the set of candidate relays and $\mathbf{w}_{j,k}[i]$ is the MMSE filter for the kth symbol at the jth relay. On the solution of (15), a refined subset, $\bar{\Omega}_T \in \Omega_T$, is generated by removing members of Ω_T which involve transmission from relay j_{opt}, i.e. members of Ω_T where $[a^1_{r_{j_{opt}} d} a^K_{r_{j_{opt}} d}] \neq 0$. TDS then operates with this subset, where $|\bar{\Omega}_T| = \binom{K(n_r-1)}{K_{sub}}$. Extension to the selection of multiple relays involves summing the MSE from candidate relays and populating Ω_R with sets of these relays. However, the selection of the number of relays to remove is vital, as too high a value will result in a overly restricting the second phase channels available to the TDS process therefore increasing the probability of a channel mismatch.

Cooperative MMSE Channel Estimation

The next task that is necessary for the interference mitigation and resource allocation is to compute the channel gains of the links of the cooperative system. In order to estimate the channel in the cooperative system under study, let us first consider the transmitted signal for user k, $\mathbf{x}_k[i] = \mathbf{B}_k[i]\,\tilde{\mathbf{A}}_k[i]\,\tilde{\mathbf{C}}_k\,\tilde{\mathbf{h}}_k[i] = \mathbf{Q}_k[i]\mathbf{h}_k[i]$, and the covariance matrix given by:

$$\begin{aligned}\mathbf{R} &= [\mathbf{r}[i]\mathbf{r}^H[i]] \\ &= \sum_{k=1}^{K}\mathbf{Q}_k[i]E[\mathbf{h}_k[i]\mathbf{h}_k^H[i]]\mathbf{Q}_k^H[i] + E[\eta[i]\eta^H[i]] + \sigma^2\mathbf{I} \\ &= \sum_{k=1}^{K}\mathbf{Q}_k[i]\mathbf{P}_{\mathbf{h}_k}\mathbf{Q}_k^H[i] + \mathbf{P}_\eta + \sigma^2\mathbf{I}\end{aligned} \quad (16)$$

A linear estimator of $\mathbf{h}_k[i]$ applied to $\mathbf{r}[i]$ can be represented as $\hat{\mathbf{h}}_k[i] = \mathbf{F}_k^H\mathbf{r}[i]$. The linear MMSE channel estimation problem for the cooperative system under consideration is formulated as:

$$\begin{aligned}\mathbf{F}_{k,opt} &= \arg\min_{\mathbf{F}_k} E[\|\mathbf{h}_k[i] - \hat{\mathbf{h}}_k[i]\|^2] \\ &= \arg\min_{\mathbf{T}_k} E[\|\mathbf{h}_k[i] - \mathbf{T}_k^H\mathbf{r}[i]\|^2].\end{aligned} \quad (17)$$

Computing the gradient terms of the argument and equating them to zero yields the MMSE solution:

$$\mathbf{F}_{k,opt} = \mathbf{R}^{-1}\mathbf{P}_k, \quad (18)$$

where $\mathbf{P}_k = E[\mathbf{r}[i]\mathbf{h}_k^H[i]] = \mathbf{Q}_k[i]E[\mathbf{h}_k[i]\mathbf{h}_k^H[i]] = \mathbf{Q}_k[i]\mathbf{P}_{\mathbf{h}_k}$. Using the relation $\hat{\mathbf{h}}_k[i] = \mathbf{F}_k^H\mathbf{r}[i]$, we obtain

$$\begin{aligned}\hat{\mathbf{h}}_k[i] &= \mathbf{F}_{k,opt}^H\mathbf{r}[i] = \mathbf{P}_k^H\mathbf{R}^{-1}\mathbf{r}[i] \\ &= \mathbf{P}_{\mathbf{h}_k}^H\mathbf{Q}_k^H[i](\sum_{k=1}^{K}\mathbf{Q}_k[i]\mathbf{P}_{\mathbf{h}_k}\mathbf{Q}_k^H[i] + \mathbf{P}_\eta + \sigma^2\mathbf{I})^{-1}\mathbf{r}[i],\end{aligned} \quad (19)$$

The expressions in (19) require matrix inversions with cubic complexity ($O(J^3)$), however, this matrix inversion is common to (13) and needs to be computed only once for both expressions. In what follows, computationally efficient algorithms with quadratic complexity ($O(J^2)$) based on an alternating optimization strategy will be detailed.

ADAPTIVE ALGORITHMS

In this section, we present algorithms to compute the parameters of interest and the expressions derived in the previous section with lower computational complexity. Specifically, we develop adaptive RALS algorithms using a method to build the group of G users based on the power levels, and then we employ an alternating optimization strategy for efficiently estimating the parameters of the receive filters, the power allocation vectors and the channels. Despite the joint optimization that is associated with a non-convex problem, the proposed RALS algorithms have been extensively tested and have not presented problems with local minima.

Group Allocation and Channel Estimation

The first step in the proposed strategy is to build the group of G users that will be used for the power allocation and receive filter design. A RAKE receiver (Ziemer, Peterson, & Borth, 1995), which is equivalent to a filter matched to the signature sequence of the desired user or the spatial signature of the desired data stream will be used for the group allocation. The RAKE receiver is employed to obtain

$$z_k^{RAKE}[i] = (\tilde{\mathbf{C}}_k \hat{\mathbf{h}}_k[i])^H \mathbf{r}[i] = \hat{\mathbf{p}}_k^H[i]\mathbf{r}[i].$$

The group is then formed according to:

$$\text{compute the } G \text{ largest } |z_k^{RAKE}[i]|, \; k = 1,2,\ldots,K. \tag{20}$$

The design of the RAKE and the other tasks require channel estimation. The power allocation, receive filter design and channel estimation expressions given in (12), (13), and (19), respectively, are solved by replacing the expected values with time averages, and RLS-type algorithms with an alternating optimization strategy. In order to solve (19) efficiently, we develop a variant of the RLS algorithm that is described by:

$$\hat{\mathbf{h}}_k[i] = \hat{\mathbf{P}}_{\mathbf{h}_k}^H[i]\mathbf{Q}_k^H[i]\hat{\mathbf{R}}^{-1}[i]\mathbf{r}[i], \tag{21}$$

where $\mathbf{Q}_k[i] = \tilde{\mathbf{B}}_k[i]\tilde{\mathbf{A}}_k[i]\tilde{\mathbf{C}}_k$, the estimate of the inverse of the covariance matrix $\hat{\mathbf{R}}^{-1}[i]$ is computed with the matrix inversion lemma (Haykin, 2002):

$$\mathbf{k}[i] = \frac{\alpha^{-1}\hat{\mathbf{R}}[i-1]\mathbf{r}[i]}{1+\alpha^{-1}\mathbf{r}^H[i]\hat{\mathbf{R}}[i-1]\mathbf{r}[i]}, \tag{22}$$

$$\hat{\mathbf{R}}[i] = \alpha^{-1}\hat{\mathbf{R}}[i-1] - \alpha^{-1}\mathbf{k}[i]\mathbf{r}^H[i]\hat{\mathbf{R}}[i-1], \tag{23}$$

and

$$\hat{\mathbf{P}}_{\mathbf{h}_k}[i] = \alpha\hat{\mathbf{P}}_{\mathbf{h}_k}[i-1] + \hat{\mathbf{h}}_k[i-1]\hat{\mathbf{h}}_k^H[i-1], \tag{24}$$

where α is a forgetting factor that should be close to but less than 1.

Joint Interference Suppression and Power Allocation

The approach for allocating the power within a group is to drop the constraint, estimate the quantities of interest and then impose the constraint via a subsequent normalization. The group-based power allocation algorithm is computed by:

$$\begin{aligned}\hat{\mathbf{a}}_{\mathrm{S},k}[i] &= \hat{\mathbf{R}}_{\mathrm{S},k}[i]\hat{\mathbf{p}}_{\mathrm{S},k}[i] \\ &= \hat{\mathbf{R}}_{\mathrm{S},k}[i](\alpha\hat{\mathbf{p}}_{\mathrm{S},k}[i-1] + b_k[i]\mathbf{v}_k[i]) \\ &= \hat{\mathbf{a}}_{\mathrm{S},k}[i-1] + \xi_a[i]\mathbf{k}_{\mathrm{S},k}[i],\end{aligned} \tag{25}$$

where $\xi_a[i] = b_k[i] - \hat{\mathbf{a}}_{\mathrm{S},k}^H[i-1]\mathbf{v}_k[i]$ is the a priori error, $\mathbf{v}_k[i] = \mathbf{B}_{\mathrm{S}}^H[i]\mathbf{P}_{\mathrm{S}}^H[i]\mathbf{w}_k[i]$ is the input signal to the recursion:

$$\mathbf{k}_{\mathrm{S},k}[i] = \frac{\alpha^{-1}\hat{\mathbf{R}}_{\mathrm{S},k}[i-1]\mathbf{v}_k[i]}{1+\alpha^{-1}\mathbf{v}_k^H[i]\hat{\mathbf{R}}_{\mathrm{S},k}[i-1]\mathbf{v}_k[i]} \tag{26}$$

$$\hat{\mathbf{R}}_{\mathrm{S},k}[i] = \alpha^{-1}\hat{\mathbf{R}}_{\mathrm{S},k}[i-1] - \alpha^{-1}\mathbf{k}_{\mathrm{S},k}[i]\mathbf{v}_k^H[i]\hat{\mathbf{R}}_{\mathrm{S},k}[i-1]. \tag{27}$$

The normalization

$$\hat{\mathbf{a}}_{\mathrm{S},k}[i] \leftarrow P_G\, \hat{\mathbf{a}}_{\mathrm{S},k}[i] / \|\hat{\mathbf{a}}_{\mathrm{S},k}[i]\|$$

is then performed to ensure the power constraint.

The linear receive filter is computed by:

$$\hat{\mathbf{w}}_k[i] = \hat{\mathbf{w}}_k[i-1] + \mathbf{k}[i]\xi^*[i], \quad (28)$$

where the a priori error is given by $\xi[i] = b_k[i] - \hat{\mathbf{w}}_k^H[i-1]\mathbf{r}[i]$ and $\mathbf{k}[i]$ is given by (22). The proposed scheme employs the algorithm in (20) to allocate the users in the group and the channel estimation approach of (21)-(24). The alternating optimization strategy uses the recursions (25) and (28) with $1\ or\ 2$ iterations per symbol i.

Transmit Diversity Selection and Relay Selection Based on Discrete Stochastic Gradient Algorithms

In this part, we describe a low-complexity solution to the joint TDS and RS problem based on a Discrete Stochastic Gradient Algorithm (DSA) that can compute the optimal combinatorial solution outline in (14) and (15) with a substantially reduced cost as compared with the exhaustive search. We present a pair of low-complexity DSA that was first reported in (Clarke & de Lamare, 2011), which jointly optimizes RS and TDS in accordance with (14) and (15), and converges to the optimal exhaustive solution.

The RS portion of the DSA is given by the algorithm of Table 1. At each iteration the MSE of a randomly chosen candidate relay (j^C) (step 2) and that of the worst performing relay currently known (j^W) are calculated (step 3). Via a comparison, the higher MSE relay is designated j^W for the next iteration (step 3). The current solution and the relay chosen for removal (j) is denoted as the current optimum and is the relay which has occupied j^W most frequently over the course of the packet up to the ith time instant; effectively an average of the occupiers of j^W. This averaging/selection process is performed by allocating each member of Ω_R a $|\Omega_R| \times 1$ unit vector, $\mathbf{v}_l$, which has a one in its corresponding position in Ω_R, i.e., $\mathbf{v}_{j^W}[i]$ is the label of the worst performing relay at the ith iteration. The current optimum is then chosen and tracked by means of a $|\Omega_R| \times 1$ state occupation probability (SOP) vector, π_R. This vector is updated at each iteration by adding $\mathbf{v}_{j^W}[i+1]$ and subtracting the previous value of π_R (step 4). The current optimum is then determined by selecting the largest element in π_R and its corresponding entry in Ω_R (step 5). Through this process, the current optimum converges towards and tracks the exhaustive solution (Andradottir, 1996). An alternative interpretation of the proposed algorithm is to view the transitions, $j^W[i] \rightarrow j^W[i+1]$, as a Markov chain and the members of Ω_R as the possible transition states. The current optimum can then be defined as the most visited state.

Once RS is complete at each time instant, set reduction ($\Omega_T \rightarrow \bar{\Omega}_T$, step 6) and TDS can take place. To perform TDS, modified versions of steps $1-5$ are used. The considered set is replaced, $\Omega_R \rightarrow \bar{\Omega}_T$; the structure of interest is replaced, $j \rightarrow \mathsf{T}_r$; the best performing matrix is sought $j^W \rightarrow \mathsf{T}_r^B$; the SOP vector is replaced $\pi_R \rightarrow \pi_{\mathsf{T}}$ and $\mathsf{C} \rightarrow \mathsf{F}$ from (14). Finally, the inequality of step 3 is reversed to enable convergence to the lowest MSE TDS matrix which is then feedback to the relays in the form of 1-bit per relay antenna.

Significant complexity savings result from the proposed algorithm; savings which increase with K, n_r and the number relays removed in the RS process. For example, when $n_r = 10$, $K = 2$ and 4 relays are removed, the number of complex multiplications for MMSE reception and exhaustive TDS, exhaustive TDS with RS, iterative TDS and iterative TDS with RS are 5.8×10^8, 1.7×10^8, 1.8×10^5, and 5.9×10^4, respectively, for each time instant.

Table 1. Proposed discrete stochastic joint TDS and RS algorithm

Step
1. Initialization
choose $j[1] \in \Omega_R$, $j^W[1] \in \Omega_R$, $\pi_R[1, j[1]] = 1$, $\pi_R[1, \bar{j}] = 0$
for $\bar{j} \neq j[1]$
2. For the time index $i = 1, 2, ..., N$
choose $j^C[i] \in \Omega_R$
3. Comparison and update of the worst performing relay
if $\mathsf{F}[i, \mathbf{r}_{sr_{j^C[i]}}] > \mathsf{F}[i, \mathbf{r}_{sr_{j^W[i]}}]$ then $j^W[i+1] = j^C[i]$
otherwise $j^W[i+1] = j^W[i]$t
4. State occupation probability (SOP) vector update
$\pi_R[i+1] = \pi_R[i] + \mu[i+1](\mathbf{v}_{j^W[i+1]} - \pi_R[i])$ where $\mu[i] = 1/i$
5. Determine largest SOP vector element and select the optimum relay
if $\pi_R[i+1, j^W[i+1]] > \pi_R[i+1, j[i]]$ then $j[i+1] = j^W[i+1]$
otherwise $j[i+1] = j[i]$
6. TDS Set Reduction
remove members of Ω_T which utilize re ay $j[i+1]$ ($\Omega_T \rightarrow \bar{\Omega}_T$)

ANALYSIS AND REQUIREMENTS OF THE ALGORITHMS

In this section, we assess the requirements of the proposed and existing algorithms for cross-layer design in terms of computational complexity and number of feedback bits. The basic idea is to show the computational cost of the algorithms presented and compare them with those of existing techniques for interference mitigation and/or resource allocation.

Computational Complexity Requirements

We discuss here the computational complexity of the proposed and existing algorithms. Specifically, we will detail the required number of complex additions and multiplications of the proposed JPAIS-GBC algorithms and compare them with interference suppression schemes without cooperation (NCIS) and with cooperation (CIS) using an equal power allocation across the relays. Both

Table 2. Computational complexity of algorithms with a global power constraint $G = K$

	Number of operations per symbol	
Parameter	**Additions**	**Multiplications**
	$2(J)^2$	$3(J)^2$
$\hat{\mathbf{W}}[i]$	$+2K(J)$	$+2K(J)$
	$-J+1$	$+3J+1$
	$3K(K(n_r+1))$	$K(K(n_r+1))^2$
	$+K(n_r+1)(L-1)$	$+4(K(n_r+1))^2$
$\hat{\mathbf{a}}_T[i]$	$+K(M(n_r+1))$	$+(K+L)(K(n_r+1))^2$
	$+K(K(n_r+1))$	$-(K(n_r+1))^2$
	$+6(K(n_r+1))^2$	$+K(MQ)$
	$+3K(n_r+1)+n_r+2$	$+n_r$
	$5(KQ)^2$	$+5(K(n_r+1)^2$
$\hat{\mathbf{h}}_k[i]$	$+5KQ$	$+6KQ$
	$+3$	$+1$

uplink and downlink scenarios are considered in the analysis. In Table 1 we show the computational complexity required by each recursion associated with a parameter vector/matrix for the JPAIS-GBC with $G = K$, which is more suitable for the uplink.

In Table 2 we show the computational complexity required by each recursion associated with a parameter vector for the JPAIS-GBC algorithm, which is suitable for both the uplink and the downlink. A noticeable difference between the JPAIS-GBC with $G = K$ and $G = 1$ is that the latter is employed for each user, whereas the former is used for all the K users in the system. Since the computation of the inverse of $\hat{\mathbf{R}}[i]$ is common to all users for the uplink in our system, the JPAIS-GBC with $G = K$ is more efficient than the JPAIS-GBC with $G = 1$ computed for all the K users.

The recursions employed for the proposed JPAIS-GBC with $G = K$ and the JPAIS-GBC with $G = 1$ are general and parts of them are used in the existing CIS and NIS algorithms. Therefore, we can use them to describe the required computational complexity of the existing algorithms. In Table 3 we show the required recursions for the proposed and existing algorithms, whose complexity is detailed in Tables 1 and 2.

In Figure 3, we illustrate the required computational complexity for the proposed and existing

Table 3. Computational complexity of algorithms with individual power constraints ($G=1$)

	Number of operations per symbol	
Parameter	**Additions**	**Multiplications**
	$2(J)^2$	$3(J)^2$
$\hat{\mathbf{w}}_k[i]$	$+J$	$+5J$
	$+1$	$+1$
	$2(n_r+1)^2$	$3(n_r+1)^2$
	$+3(n_r-1)$	$+7(n_r+1)$
$\hat{\mathbf{a}}_k[i]$	$+JL$	$+JL$
	$+Q$	$+Q$
	-3	$+3$
	$2(Q)^2$	$6(Q)^2$
$\hat{\mathbf{h}}_k[i]$	$-5MQ$	$+MQ$
	$-5(n_r+1)+3$	$+4(n_r+1)+1$

Figure 3. Computational complexity in terms of the number of complex multiplications of the proposed and existing schemes for the uplink

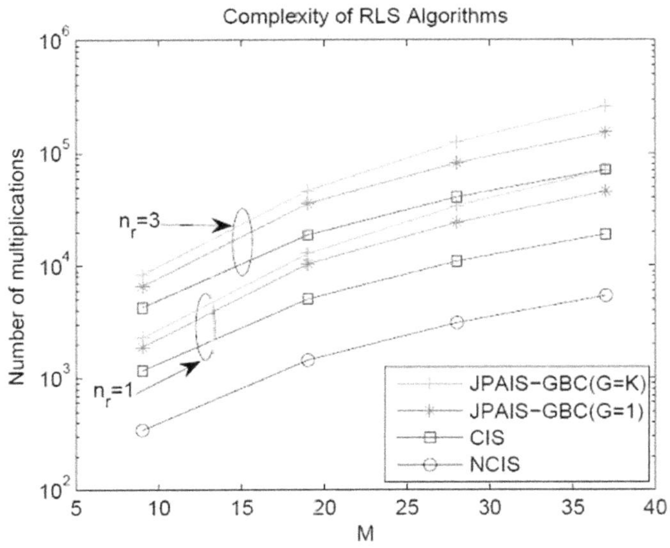

schemes for different number of relays (n_r). The curves show that the proposed JPAIS-GBC with $G = K$ and JPAIS-GBC with $G = 1$ are more complex than the CIS scheme and the NCIS. This is due to the fact that the power allocation and channel estimation recursions are employed. However, we will show in the next section that this additional required complexity (which is modest) can significantly improve the performance of the system (see Table 4).

Table 4. Computational complexity of the proposed JPAIS and existing algorithms

Algorithm	Recursions
JPAIS-GPC $(G = K)$ (Uplink)	$\hat{\mathbf{W}}[i]$, $\hat{\mathbf{a}}_T[i]$, $\hat{\mathbf{h}}_k[i]$
JPAIS-GBC $(G = 1)$ (Downlink)	$\hat{\mathbf{w}}_k[i]$, $\hat{\mathbf{a}}_k[i]$, $\hat{\mathbf{h}}_k[i]$
CIS (Uplink)	$\hat{\mathbf{W}}[i]$, $\hat{\mathbf{a}}_T[i]$ is fixed
CIS (Downlink)	$\hat{\mathbf{w}}_k[i]$, $\hat{\mathbf{a}}_k[i]$ is fixed
NCIS (Uplink)	$\hat{\mathbf{W}}[i]$ with $n_r = 0$
NCIS (Downlink)	$\hat{\mathbf{w}}_k[i]$ with $n_r = 0$

Feedback Channel Requirements

The JPAIS algorithms presented so far for cross-layer design require feedback signalling in order to allocate the power levels across the relays. In order to illustrate how these requirements are addressed, we can refer to Figure 4 which depicts the structure for both the data and feedback packets. The data packet comprises a number of allocated bits for training (N_{tr}), for synchronization and control (N_{sync}) and the transmitted data (N_{data}). The feedback packet requires the transmission of the power allocation vector $\mathbf{a}_T$ for the case of the JPAIS-GBC algorithm with $G = K$, whereas it requires the transmission of $\mathbf{a}_k$ for each user for JPAIS-GBC with $G = 1$. A typical number of bits n_b required to quantize each coefficient of the vectors $\mathbf{a}_T$ and $\mathbf{a}_k$ via scalar quantization is $n_b = 4$ bits. More efficient schemes employing vector quantization (Gersho & Gray, 1992; de Lamare & Alcaim, 2005c) and that take into account correlations between the coefficients are also possible.

For the uplink (or multiple-access channel), the base station (or access point) needs to feedback the power levels across the links to the K destination users in the system. With the JPAIS-GBC with $G = K$ algorithm, the parameter vector $\mathbf{a}_T$ with $(n_r + 1)Kn_b$ bits/packet must be broad-

Figure 4. Proposed structure of the data and feedback packets

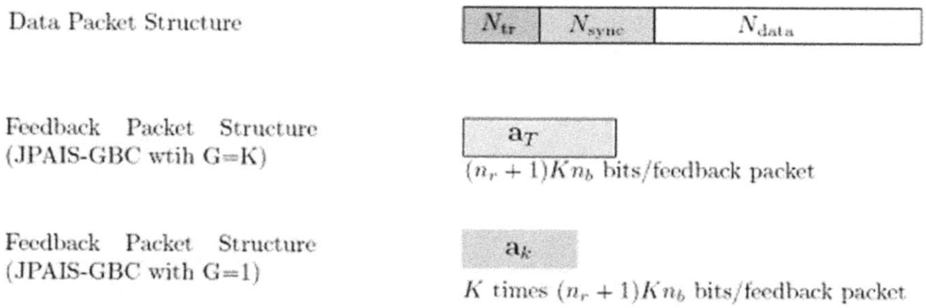

casted to the K users. For the JPAIS-GBC algorithm with $G=1$, a parameter vector $\mathbf{a}_k$ with $(n_r+1)n_b$ bits/packet must be broadcasted to each user in the systems. In terms of feedback, the JPAIS-GBC algorithm with $G=1$ is more flexible and may require less feedback bits if there is no need for a constant update of the power levels for all K users.

For the downlink (or broadcast channel), the K users must feedback the power levels across the links to the base station. With the JPAIS-GBC algorithm with $G=K$, the parameter vector $\mathbf{a}_T$ with $(n_r+1)Kn_b$ bits/packet must be computed by each user and transmitted to the base station, which uses the $\mathbf{a}_T$ vector coming from each user. An algorithm for data fusion or a simple averaging procedure can be used. For the JPAIS-GBC algorithm with $G=1$, a parameter vector $\mathbf{a}_k$ with $(n_r+1)n_b$ bits/packet must be transmitted from each user to the base station. In terms of feedback, the JPAIS-GBC algorithm with $G=1$ requires significantly less feedback bits than the JPAIS-GBC with $G=K$ in this scenario.

The MIMO TDS and RS scheme can be interpreted as a 1-bit power allocation scheme and therefore achieves performance improvements whilst utilizing the minimum number of feedback bits per antenna. Consequently, the feedback requirements per update of a cooperative MIMO system using TDS and RS is given by the total number of relay transmit antennas n_rK. This minimal feedback allows optimization of the system whilst maintaining the capacity of the system with regards to the transmission of useful data.

SIMULATIONS

In this section, we illustrate with Monte-Carlo simulations the performance of the cross-layer algorithms described in this chapter. Specifically, we assess the performance in terms of the Bit Error Ratio (BER) of the JPAIS scheme and adaptive algorithms with Group-Based Power Constraints (GBC). The JPAIS scheme and algorithms are compared with schemes without cooperation (NCIS) and with cooperation (CIS) (Venturino, Wang, & Lops, 2006) using an equal power allocation across the relays (the power allocation in the JPAIS scheme is disabled). We also assess the proposed algorithms for transmit diversity selection and relay selection (Iterative TDS with RS) are presented and comparisons drawn against the optimal exhaustive solutions (Exhaustive TDS with RS), the unmodified system (No TDS), and the direct transmission (Non-Cooperative).

DS-CDMA System

A DS-CDMA network with randomly generated spreading codes and a processing gain $N=16$ is considered. The fading channels are generated considering a random power delay profile with gains taken from a complex Gaussian variable with unit variance and mean zero, $L=5$ paths spaced by one chip, and are normalized for unit power. The power constraint parameter $P_{A,k}$ is set for each user so that the designer can control the SNR $(SNR = P_{A,k}/\sigma^2)$ and $P_T = P_G + (K-G)P_{A,k}$, whereas it follows a log-normal distribution for the users with associated standard deviation equal to 3 dB. The DF cooperative protocol is adopted and all the relays and the destination terminal use either linear MMSE, which have full channel and noise variance knowledge, or adaptive receivers. The receivers are adjusted with the proposed RALS with 2 iterations for the JPAIS scheme, and with RLS algorithms for the NCIS and CIS schemes. We employ packets with 1500 QPSK symbols and average the curves over 1000 runs. For the adaptive receivers, we provide training sequences with $N_{tr}=200$ symbols placed at the preamble of the

Figure 5. BER performance versus number of symbols. Parameters: AF protocol, $\lambda_T = \lambda_k = 0.025$ *(for MMSE schemes),* $\alpha = 0.998$, $\hat{\mathbf{R}}_{S,k}^{-1}[i] = 0.01\mathbf{I}$ *and* $\hat{\mathbf{R}}^{-1}[i] = 0.01\mathbf{I}$.

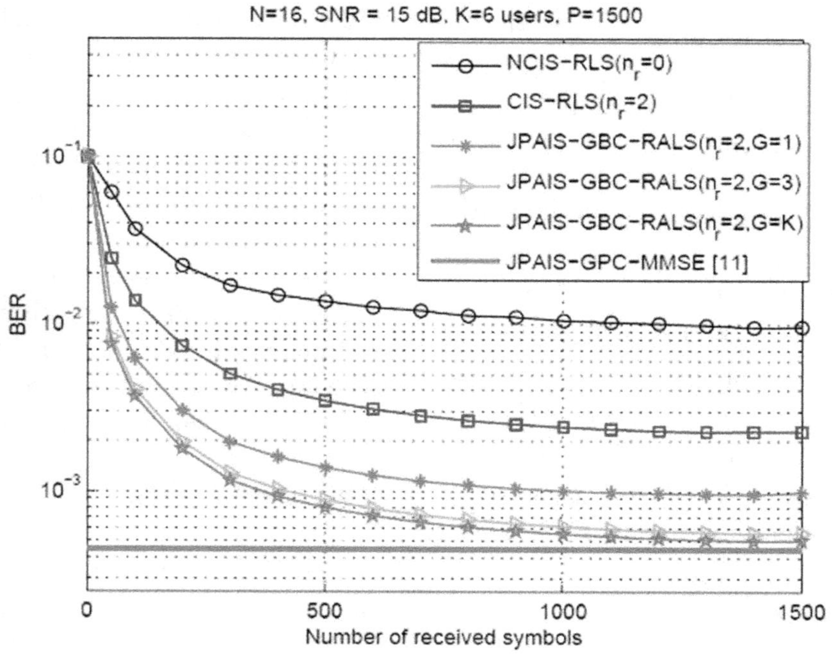

packets. After the training sequence, the adaptive receivers are switched to decision-directed mode.

The first experiment depicted in Figure 5 shows the BER performance of the proposed JPAIS scheme and algorithms against the NCIS and CIS schemes with $n_r = 2$ relays. The JPAIS scheme is considered with the group-based power constraints (JPAIS-GBC). All techniques employ MMSE or RLS-type algorithms for estimation of the channels, the receive filters and the power allocation for each user. The results show that as the group size G is increased the proposed JPAIS scheme and algorithms converge to approximately the same level of the cooperative JPAIS-MMSE scheme reported in (de Lamare, 2010a), which employs $G = K$ for power allocation, and has full knowledge of the channel and the noise variance.

The proposed JPAIS-GBC scheme is then compared with a non-cooperative approach (NCIS) and a cooperative scheme with equal power allocation (CIS) across the relays for $n_r = 1,2$ relays. The results shown in Figure 6 illustrate the performance improvement achieved by the JPAIS scheme and algorithms, which significantly outperform the CIS and the NCIS techniques. As the number of relays is increased so is the performance, reflecting the exploitation of the spatial diversity. In the scenario studied, the proposed JPAIS-GBC with $G = 3$ can accommodate up to 3 more users as compared to the CIS scheme and double the capacity as compared with the NCIS for the same BER performance. The curves indicate that the GBC for power allocation with only a few users is able to attain a performance close to the JPAIS-GBC with $G = K$ users, while requiring a lower complexity and less network signalling. A comprehensive study of the signalling requirements will be considered in a future work.

Figure 6. BER performance versus SNR and number of users for the optimal linear MMSE detectors. Parameters: AF protocol, $\alpha = 0.998$, $\hat{\mathbf{R}}_{S,k}^{-1}[i] = 0.01\mathbf{I}$ *and* $\hat{\mathbf{R}}^{-1}[i] = 0.01\mathbf{I}$.

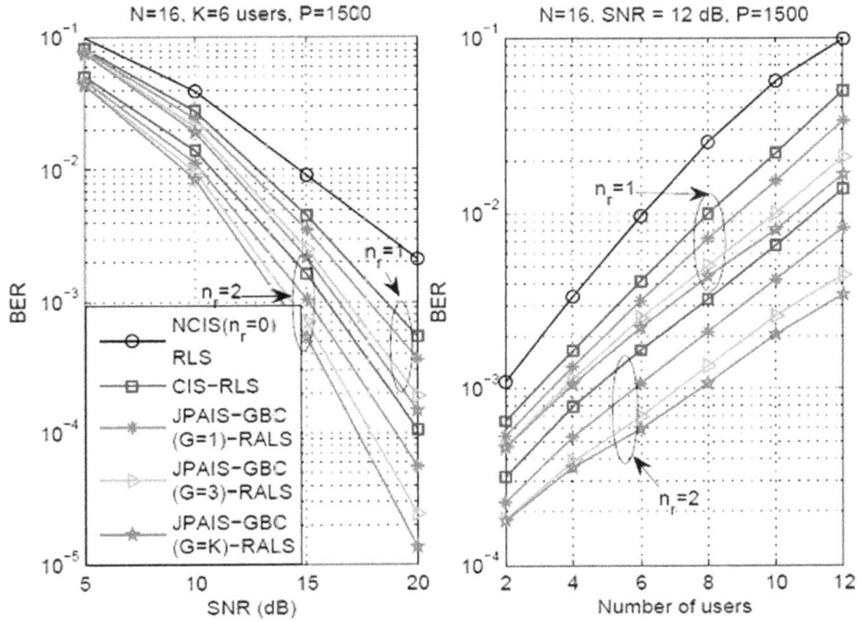

The next experiment considers the average BER performance against the normalized fading rate f_dT (cycles/symbol), as depicted in Figure 7. The idea is to illustrate a situation where the channel changes within a packet and the system transmits the power allocation vectors computed by the proposed JPAIS algorithms via a feedback channel. In this scenario, the JPAIS algorithms compute the parameters of the receiver and the power allocation vector, which is transmitted only once to the mobile users. This leads to a situation in which the power allocation becomes outdated. The results show that the gains of the proposed JPAIS algorithms decrease gradually as the f_dT is increased to the BER level of the existing CIS algorithms for both $n_r = 2$ and $n_r = 4$ relays, indicating that the power allocation is no longer able to provide performance advantages. This problem requires the deployment of a frequent update of the power allocation via feedback channels.

The last experiment, shown in Figure 8, illustrates the averaged BER performance versus the percentage of errors in the feedback channel for an uplink scenario. Specifically, the feedback packet structure is employed and each coefficient is quantized with 4 bits. Each feedback packet is constructed with a sequence of binary data (0s and 1s) and is transmitted over a Binary Symmetric Channel (BSC) with an associated probability of error P_e . We then evaluate the BER of the proposed JPAIS and the existing algorithms against several values of the P_e . The results show that the proposed JPAIS algorithms obtain significant gains over the existing CIS algorithm for values of $P_e < 0.1\%$. As we increase the rate of feedback errors, the performance of the proposed JPAIS becomes worse than the CIS algorithms. This suggests the use of error-control coding techniques to keep the level of errors in the feedback channel below a certain value.

Figure 7. BER performance versus $f_d T$ for the AF protocol. The parameters of the adaptive algorithms are optimized for each $f_d T$.

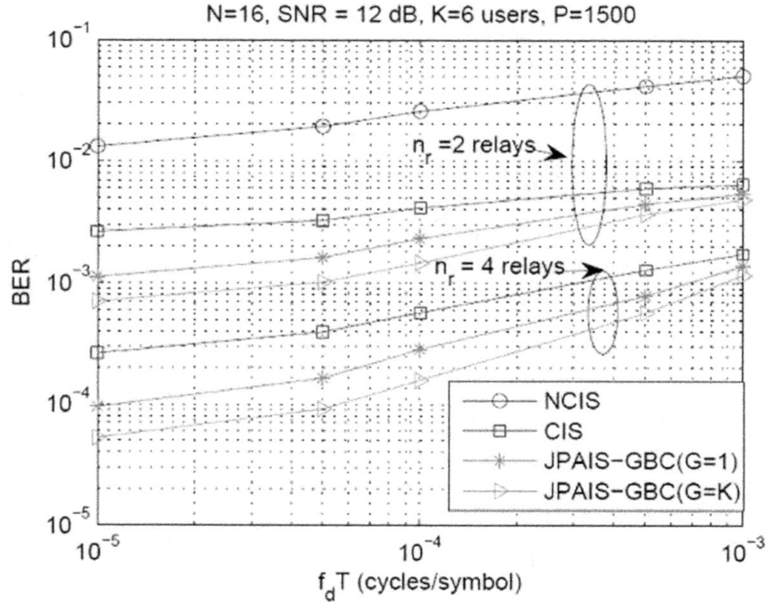

Figure 8. BER performance versus $f_d T$ for the AF protocol. The parameters of the adaptive algorithms are optimized for each $f_d T$.

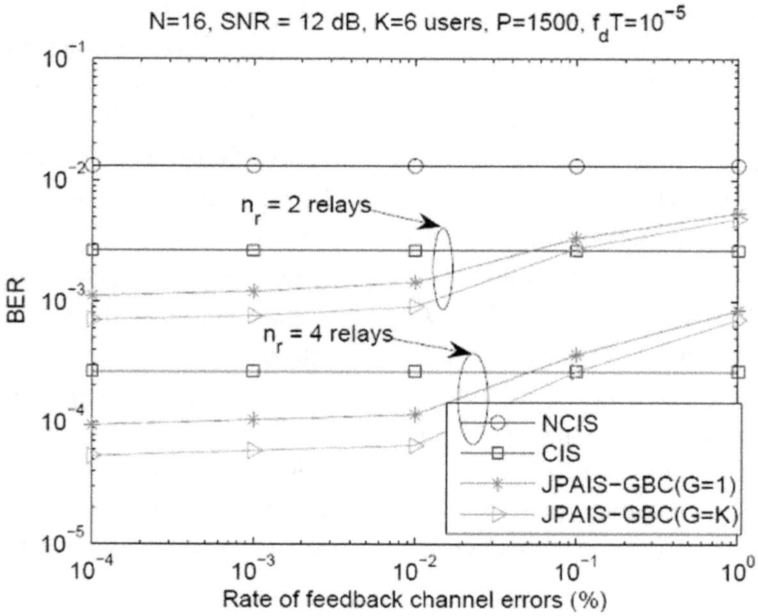

Figure 9. Cooperative DF MIMO BER performance versus the number of received symbols

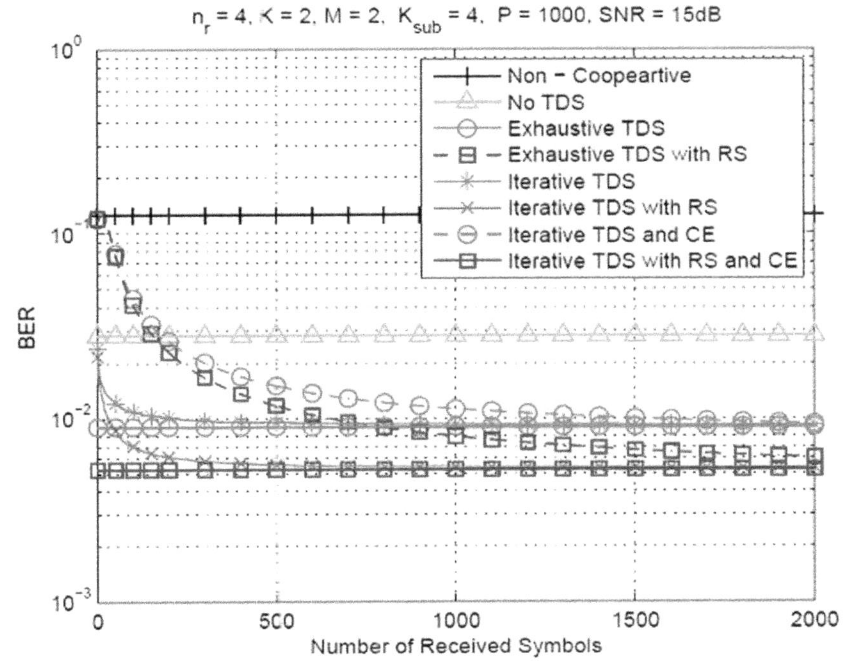

MIMO System

In this part, simulations of the proposed algorithms (Iterative TDS with RS) are presented and comparisons drawn against the optimal exhaustive solutions (Exhaustive TDS with RS), the unmodified system where all antennas are active (No TDS), and the direct transmission (Non-Cooperative). Plots of the schemes with TDS only (Exhaustive TDS, Iterative TDS) are also included to illustrate the performance improvement obtained by RS. Equal power allocation is maintained in each phase so that the total transmit bit power of the relays is unity. RLS Channel Estimation (CE) is used where all auxiliary matrices are initialized as identity matrices and estimation matrices are zero matrices, and the exponential forgetting factor is 0.9. Each simulation is averaged over 1000 packets (N_p), each with training sequences of 200 symbols.

Figure 9 gives the BER convergence performance of the proposed algorithms. The iterative TDS with RS algorithm converges to the optimal BER as does TDS with RS and CE, albeit in a delayed fashion due to the CE. The TDS with RS scheme exhibits quicker convergence and lower steady state BER. These results and the interdependence between elements of the algorithm confirm that both the RS and TDS portions of the algorithm converge to their exhaustive solutions.

Figure 10 shows the BER versus SNR performance of the proposed and conventional algorithms. Increased diversity has been achieved without sacrificing multiplexing gain and illustrates that although the maximum available diversity advantage decreases from $M(n_r+1)$ to $M(K_{sub}/K+1)$ with RS with TDS because fewer antennas are active, the actual diversity achieved has increased. These diversity effects can be attributed to the removal of poor paths and therefore a lower probability of first phase and second phase channel mismatch but also the increase in transmit power over the remaining paths. The largest gains in diversity are present in the

Figure 10. Cooperative FD MIMO BER performance versus SNR

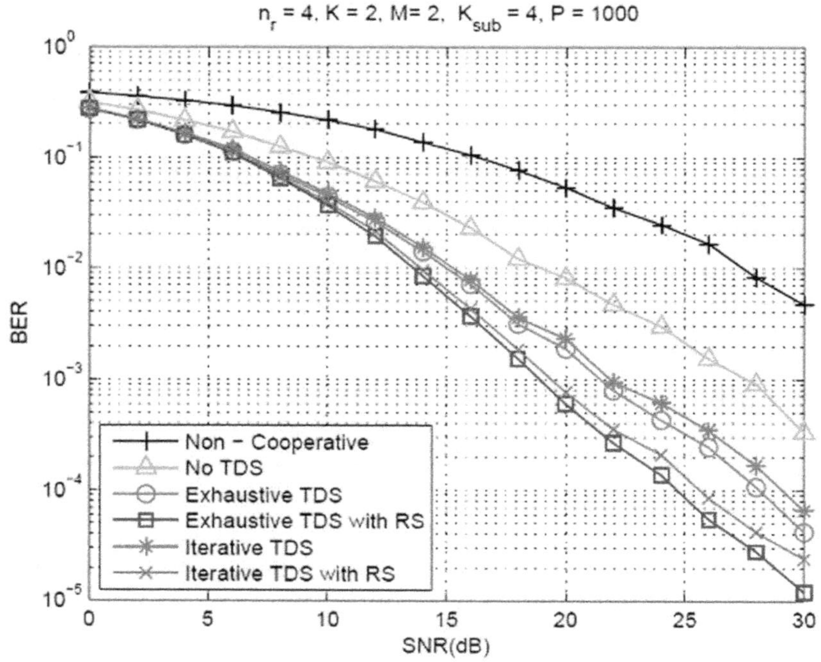

15 – 25 dB region and begin to diminish above this region because relay decoding becomes increasingly reliable and lower power paths become more viable for transmission.

EXTENSIONS AND SUGGESTIONS FOR FUTURE WORK

The algorithms for joint resource allocation and interference mitigation described in this chapter are quite general and can be employed in a variety of wireless communication systems that are equipped with cooperative techniques. These include Orthogonal-Frequency-Division-Multiplexing (OFDM) (Stuber, Barry, MacLaughlin, Li, Ingram, & Pratt, 2004), Single-Carrier systems with Frequency-Domain Equalisation (SC-FDE) (Falconer, Ariyavisitakul, Benyamin-Seeyar, & Edison, 2002) and Ultra-Wide Band (UWB) systems (Win & Scholtz, 2000).

Possible extensions include the incorporation of more advanced interference mitigation strategies than linear schemes. These include nonlinear detection techniques such as successive interference cancellation (de Lamare, Sampaio-Neto, & Hjorungnes, 2007; Wang & Poor, 1999), decision feedback strategies (de Lamare & Sampaio-Neto, 2003; de Lamare, Haardt, & Sampaio-Neto, 2008) and sphere decoders (Hassibi & Vikalo, 2005). The detection algorithms could also be considered with space-time coding schemes (Alamouti, 1998; Tarokh, 1999) channel coding and iterative processing approaches (de Lamare, Haardt, & Sampaio-Neto, 2008; Wang & Poor, 1999).

Another complementary set of techniques comprises algorithms for adaptive parameter estimation. These methods are fundamental to estimate key parameters such as channel gains, amplitudes and receive filters, whilst keeping the complexity low and being able to track variations of the environment. Amongst the adaptive parameter estimation techniques, a designer can choose between supervised and blind approaches (Haykin, 2002). Blind techniques (de Lamare & Sampaio-Neto, 2010b; Xu & Tsatsanis, 2006, de Lamare, Haardt, & Sampaio-Neto, 2008) are appealing as they can increase the spectral efficiency of wireless systems. This is especially relevant for cooperative systems as they require extra signalling for cross-layer design. Supervised adaptive algorithms usually rely on training sequences that are sent at the beginning of each data packet (Hassibi & Hochwald, 2003; Choi, Yu, & Lee, 2005). One fundamental issue in the choice of the adaptive parameter estimation algorithm is the speed of convergence and the tracking performance. The literature suggests that reduced-rank algorithms (de Lamare & Sampaio-Neto, 2005a; de Lamare & Sampaio-Neto, 2009a; Honig & Goldstein, 2002; Pados & Batalama, 1997) are very attractive choices when fast training and accurate tracking are important issues.

CONCLUDING REMARKS

We have presented in this work joint iterative power allocation and interference mitigation techniques for DS-CDMA and MIMO networks which employ multiple hops and the AF and DF cooperation protocols. A joint constrained optimization framework and algorithms that consider the allocation of power levels across the relays subject to group power constraints and the design of linear receivers for interference suppression were proposed. A scheme for joint transmit diversity optimisation and relay selection along with linear interference suppression has also been detailed and applied to MIMO systems. A study of the requirements of the proposed and existing algorithms in terms of computational complexity and feedback channels has also been conducted. The results of simulations have shown that the proposed algorithms obtain significant gains in performance and capacity over existing non-cooperative and cooperative schemes.

REFERENCES

Alamouti, S. (1998). A simple transmit diversity technique for wireless communications. *IEEE Journal on Selected Areas in Communications*, *16*, 1451–1458. doi:10.1109/49.730453

Andradottir, S. (1996). A global search method for discrete stochastic optimization. *SIAM Journal on Optimization*, *6*, 513–530. doi:10.1137/0806027

Berenguer, I., Wang, X., & Krishnamurhty, V. (2005). Adaptive MIMO antenna selection via discrete stochastic optimization. *IEEE Transactions on Signal Processing*, *53*, 4315–4329. doi:10.1109/TSP.2005.857056

Choi, J. H., Yu, H. J., & Lee, Y. H. (2005). Adaptive MIMO decision feedback equalization for receivers with time-varying channels. *IEEE Transactions on Signal Processing*, *53*, 4295–4303. doi:10.1109/TSP.2005.857053

Clarke, P., & de Lamare, R. C. (2011). Joint transmit diversity optimization and relay selection for multi-relay cooperative MIMO systems using discrete stochastic algorithms. *IEEE Communications Letters*, *15*, 1035–1037. doi:10.1109/LCOMM.2011.082611.102262

de Lamare, R. C. (2010a). Joint iterative power allocation and interference suppression algorithms for cooperative spread spectrum networks. In *Proceedings of the IEEE International Conference on Acoustics, Speech and Signal Processing*. Dallas, TX: IEEE Press.

de Lamare, R. C., & Alcaim, A. (2005c) Strategies to improve the performance of very low bit rate speech coders and application to a 1.2 kb/s codec. In *Proceedings of the IEEE Conference on Vision, Image and Signal Processing*, (p. 152). IEEE Press.

de Lamare, R. C., & Diniz, P. S. R. (2009b). Set-membership adaptive algorithms based on time-varying error bounds for CDMA interference suppression. *IEEE Transactions on Vehicular Technology*, *58*, 644–654. doi:10.1109/TVT.2008.926608

de Lamare, R. C., Haardt, M., & Sampaio-Neto, R. (2008). Blind adaptive constrained reduced-rank parameter estimation based on constant modulus design for CDMA interference suppression. *IEEE Transactions on Signal Processing*, 56.

de Lamare, R. C., & Sampaio-Neto, R. (2003). Adaptive MBER decision feedback multiuser receivers in frequency selective fading channels. *IEEE Communications Letters*, *7*(2), 73–75. doi:10.1109/LCOMM.2002.808373

de Lamare, R. C., & Sampaio-Neto, R. (2005a). Adaptive reduced-rank MMSE filtering with interpolated FIR filters and adaptive interpolators. *IEEE Signal Processing Letters*, 12.

de Lamare, R. C., & Sampaio-Neto, R. (2005b). Reduced-rank interference suppression for DS-CDMA based on interpolated FIR filters. *IEEE Communications Letters*, 9.

de Lamare, R. C., & Sampaio-Neto, R. (2005d). Blind adaptive code-constrained constant modulus algorithms for CDMA interference suppression in multipath channels. *IEEE Communications Letters*, *9*, 334–336.

de Lamare, R. C., & Sampaio-Neto, R. (2006). Low-complexity variable step-size mechanisms for stochastic gradient algorithms in minimum variance CDMA receivers. *IEEE Transactions on Signal Processing*, *54*, 2302–2317. doi:10.1109/TSP.2006.873651

de Lamare, R. C., & Sampaio-Neto, R. (2007b). Adaptive interference suppression for DS-CDMA systems based on interpolated FIR filters with adaptive interpolators in multipath channels. *IEEE Transactions on Vehicular Technology*, 56.

de Lamare, R. C., & Sampaio-Neto, R. (2007c). Reduced-rank adaptive filtering based on joint iterative optimization of adaptive filters. *IEEE Signal Processing Letters*, ▪▪▪, 14.

de Lamare, R. C., & Sampaio-Neto, R. (2008a). Minimum mean squared error iterative successive parallel arbitrated decision feedback detectors for DS-CDMA systems. *IEEE Transactions on Communications*, *56*, 778–789. doi:10.1109/TCOMM.2008.060209

de Lamare, R. C., & Sampaio-Neto, R. (2009a). Adaptive reduced-rank processing based on joint and iterative interpolation, decimation, and filtering. *IEEE Transactions on Signal Processing*, *57*, 2503–2514. doi:10.1109/TSP.2009.2018641

de Lamare, R. C., & Sampaio-Neto, R. (2010b). Reduced-rank space-time adaptive interference suppression with joint iterative least squares algorithms for spread-spectrum systems. *IEEE Transactions on Vehicular Technology*, *59*, 1217–1228. doi:10.1109/TVT.2009.2038391

de Lamare, R. C., & Sampaio-Neto, R. (2010c). Blind adaptive MIMO receivers for space-time block-coded DS-CDMA systems in multipath channels using the constant modulus criterion. *IEEE Transactions on Communications*, *58*, 21–27. doi:10.1109/TCOMM.2010.01.070549

de Lamare, R. C., Sampaio-Neto, R., & Haardt, M. (2011). Blind adaptive constrained constant-modulus reduced-rank interference suppression algorithms based on interpolation and switched decimation. *IEEE Transactions on Signal Processing*, *59*, 681–695. doi:10.1109/TSP.2010.2091274

de Lamare, R. C., Sampaio-Neto, R., & Hjørungnes, A. (2007a). Joint iterative interference cancellation and parameter estimation for CDMA systems. *IEEE Communications Letters*, *11*, 916–918. doi:10.1109/LCOMM.2007.070943

Ding, M., Liu, S., Luo, H., & Chen, W. (2010). MMSE based greedy antenna selection scheme for AF MIMO relay systems. *IEEE Signal Processing Letters*, *17*, 433–436. doi:10.1109/LSP.2010.2042531

Dohler, M., & Li, Y. (2010). *Cooperative communications: Hardware, Channel and PHY*. New York, NY: John Wiley.

Ettefagh, A., Kuhn, M., Hammerstrom, I., & Wittneben, A. (2006). On the range performance of decode-andforward relay: IEEE 802.11 WLANs. In *Proceedings of the IEEE 17th International Symposium Personal, Indoor and Mobile Radio Communications*. IEEE Press.

Falconer, D., Ariyavisitakul, L. S., Benyamin-Seeyar, A., & Eidson, B. (2002). Frequency domain equalization for single-carrier broadband wireless systems. *IEEE Communications Magazine*, *40*(4), 58–66. doi:10.1109/35.995852

Fan, Y., & Thompson, J. (2007). MIMO configurations for relay channels: Theory and practice. *IEEE Transactions on Wireless Communications*, *6*, 1774–1786. doi:10.1109/TWC.2007.360379

Fang, Z., Hua, Y., & Koshy, J. C. (2006). Joint source and relay optimzation for a non-regenerative MIMO relay. In *Proceedings of the Fourth IEEE Workshop on Sensor Array and Multichannel Processing*. Waltham, MA: IEEE Press.

Foschini, G. J., & Gans, M. J. (1998). On limits of wireless communications in a fading environment when using multiple antennas. *Wireless Personal Communications*, *6*, 311–335. doi:10.1023/A:1008889222784

Gersho, A., & Gray, R. M. (1992). *Vector quantization and signal compression*. Dordrecht, The Netherlands: Kluwer Academic Press/Springer.

Golden, G. D., Foschini, C. J., Valenzuela, R. A., & Wolniansky, P. W. (1999). Detection algorithm and initial laboratory results using V-BLAST space-time communication architecture. *Electronics Letters*, *35*(1). doi:10.1049/el:19990058

Hassibi, B., & Hochwald, B. M. (2003). How much training is needed in multiple-antenna wireless links? *IEEE Transactions on Information Theory*, *49*, 951–963. doi:10.1109/TIT.2003.809594

Hassibi, B., & Vikalo, H. (2005). On the sphere decoding algorithm: Part I, the expected complexity. *IEEE Transactions on Signal Processing*, *53*, 2806–2818. doi:10.1109/TSP.2005.850352

Haykin, S. (2002). *Adaptive filter theory* (4th ed.). Upper Saddle River, NJ: Prentice Hall.

Honig, M. L., & Goldstein, J. S. (2002). Adaptive reduced-rank interference suppression based on the multistage Wiener filter. *IEEE Transactions on Communications*, 50.

Honig, M. L., & Poor, H. V. (1998). Adaptive interference suppression. In Poor, H. V., & Wornell, G. W. (Eds.), *Wireless Communications: Signal Processing Perspectives*. Englewood Cliffs, NJ: Prentice-Hall.

Huang, W. J., Hong, Y. W., & Kuo, C. C. J. (2007). Decode-and-forward cooperative relay with multi-user detection in uplink CDMA networks. In *Proceedings of the IEEE GLOBECOM*, (pp. 4397-4401). IEEE Press.

Joung, J., & Sayed, A. H. (2010). Multiuser two-way amplify-and-forward relay processing and power control methods for beamforming systems. *IEEE Transactions on Signal Processing*, *58*, 1833–1846. doi:10.1109/TSP.2009.2038668

Kramer, G., Gastpar, M., & Gupta, P. (2005). Cooperative strategies and capacity theorems for relay networks. *IEEE Transactions on Information Theory*, *51*, 3037–3063. doi:10.1109/TIT.2005.853304

Kyritsi, P., Popovski, P., Eggers, P., Wang, Y., Khan, D. A., Bouaziz, A. E., et al. (2007). Cooperative transmission: A reality check using experimental data. In *Proceedings of the IEEE 65th Vehicular Technology Conference, 2007 (VTC 2007-Spring)*, (pp. 2281–2285). IEEE Press.

Laneman, J. N., & Wornell, G. W. (2003). Distributed space-time-coded protocols for exploiting cooperative diversity in wireless networks. *IEEE Transactions on Information Theory*, *49*, 2415–2425. doi:10.1109/TIT.2003.817829

Laneman, J. N., & Wornell, G. W. (2004). Cooperative diversity in wireless networks: Efficient protocols and outage behaviour. *IEEE Transactions on Information Theory*, *50*, 3062–3080. doi:10.1109/TIT.2004.838089

Long, L., & Hossain, E. (2008). Cross-layer optimization frameworks for multihop wireless networks using cooperative diversity. *IEEE Transactions on Wireless Communications*, *7*, 2592–2602. doi:10.1109/TWC.2008.060962

Lu, M.-H., Steenkiste, P., & Chen, T. (2007). Time-aware opportunistic relay for video streaming over WLANs. In *Proceedings of the IEEE International Conference in Multimedia and Expo*, (vol 7), (pp. 1782–1785). IEEE Press.

Luo, J., Blum, R. S., Cimini, L. J., Greenstein, L. J., & Haimovich, A. M. (2007). Decode-and-forward cooperative diversity with power allocation in wireless networks. *IEEE Transactions on Wireless Communications*, 6.

Pabst, R., Walke, B. H., Schultz, D. C., Herhold, P., Yanikomeroglu, H., & Mukherjee, S. (2004). Relay-based deployment concepts for wireless and mobile broadband radio. *IEEE Communications Magazine*, *42*, 80–89. doi:10.1109/MCOM.2004.1336724

Pados, D. A., & Batalama, S. N. (1997). Low complexity blind detection of DS/CDMA signals: Auxiliary-vector receivers. *IEEE Transactions on Communications*, *45*, 1586–1594. doi:10.1109/26.650237

Peters, S. W., Panah, A. Y., Truong, K. T., & Heath, R. W. (2009). Relay architextures for 3GPP LTE-advanced. *EURASIP Journal on Wireless Communications and Networking*. Retrieved from http://downloads.hindawi.com/journals/wcn/2009/618787.pdf.

Scaglione, A., Goeckel, D. L., & Laneman, J. N. (2006). Cooperative communications in mobile ad hoc networks. *IEEE Signal Processing Magazine*, *23*(5), 18–29. doi:10.1109/MSP.2006.1708409

Sendonaris, A., Erkip, E., & Aazhang, B. (2003). User cooperation diversity - Parts I and II. *IEEE Transactions on Communications*, *51*, 2386–2391.

So, A., & Liang, B. (2005). Effect of relaying on capacity improvement in wireless local area networks. In *Proceedings of the IEEE Wireless Communications and Networking Conference*, (vol 3), (pp. 1539–1544). IEEE Press.

Souryal, M. R., Vojcic, B. R., & Pickholtz, L. (2006). Adaptive modulation in ad hoc DS/CDMA packet radio networks. *IEEE Transactions on Communications*, *54*, 714–725. doi:10.1109/TCOMM.2006.873092

Stuber, G. L., Barry, J. R., MacLaughlin, S. W., Li, Y., Ingram, M. A., & Pratt, T. G. (2004). Broadband MIMO-OFDM wireless communications. *Proceedings of the IEEE, 92*.

Sun, Y., Tripathi, V., & Honig, M. L. (2005). Adaptive, iterative, reduced-rank (turbo) equalization. *IEEE Transactions on Wireless Communications*, *4*, 2789–2800.

Tarokh, V., Jafarkhani, H., & Calderbank, A. R. (1999). Space-time block codes from orthogonal designs. *IEEE Transactions on Information Theory*, *45*, 1456–1467. doi:10.1109/18.771146

Telatar, I. E. (1999). Capacity of multi-antenna gaussian channels. *European Transactions on Telecommunications*, *10*, 585–595. doi:10.1002/ett.4460100604

Vardhe, K., Reynolds, D., & Valenti, M. C. (2008). The performance of multi-user cooperative diversity in an asynchronous CDMA uplink. *IEEE Transactions on Wireless Communications*, *7*, 1930–1940. doi:10.1109/TWC.2008.070282

Venturino, L., Wang, X., & Lops, M. (2006). Multiuser detection for cooperative networks and performance analysis. *IEEE Transactions on Signal Processing*, 54.

Verdu, S. (1998). *Multiuser detection*. Cambridge, UK: Cambridge University Press.

Wang, X., & Poor, H. V. (1999). Iterative (turbo) soft interference cancellation and decoding for coded CDMA. *IEEE Transactions on Communications*, *47*, 1046–1061. doi:10.1109/26.774855

Win, M. Z., & Scholtz, R. A. (2000). Ultra-wide bandwidth time-hopping spread-spectrum impulse radio for wireless multiple-acess communications. *IEEE Transactions on Communications*, *48*, 679–689. doi:10.1109/26.843135

Xu, C., Feng, G., & Kwak, K. S. (2001). A modified constrained constant modulus approach to blind adaptive multiuser detection. *IEEE Transactions on Communications*, 49.

Xu, Z., & Liu, P. (2002). Code-constrained blind detection of CDMA signals in multipath channels. *IEEE Signal Processing Letters*, 9.

Xu, Z., & Tsatsanis, M. K. (2006). Blind adaptive algorithms for minimum variance CDMA receivers. *IEEE Transactions on Communications*, 49.

Zheng, L., & Tse, D. N. C. (2003). Diversity and multiplexing: A fundamental tradeoff in multiple-antenna channels. *IEEE Transactions on Information Theory*, *49*, 1073–1096. doi:10.1109/TIT.2003.810646

Ziemer, R. E., Peterson, R. L., & Borth, D. E. (1995). *Introduction to spread spectrum communications*. Upper Saddle River, NJ: Prentice-Hall.

Chapter 9
Applications of "Cross-Layer" in Video Communications over Wireless Networks

Martin Fleury
University of Essex, UK

Rouzbeh Razavi
University of Essex, UK

Laith Al-Jobouri
University of Essex, UK

Salah Al-Majeed
University of Essex, UK

Mohammed Ghanbari
University of Essex, UK

ABSTRACT

Because of the impact of noise, interference, fading, and shadowing in a wireless network, there has been a realization that the strict layering of wireline networks may be unsuitable for wireless. It is the volatility over time that demands an adaptive solution and the basis of adaptation must arise by communication of the channel conditions along with the datalink settings. Video communication is particularly vulnerable because, except when reception is decoupled from distribution as in multimedia messaging, there are real-time display and decode deadlines to be met. The predictive nature of video compression also makes it susceptible to temporal error propagation. In this chapter, case studies from the authors' experiences with broadband wireless access networks and personal area wireless networks serve to illustrate how information exchange across the layers can benefit received video quality. These schemes are all adaptive and serve as a small sample of a much greater population of cross-layer techniques. Given the importance of multimedia communications as an engine of growth for networked communication, "cross-layer" should be the first consideration in designing a video application.

DOI: 10.4018/978-1-4666-0960-0.ch009

INTRODUCTION

Traditional network design has largely taken place independently of the application. This form of design preserves protocol layering as a way of providing a well-defined interface between the layers and of isolating errors. Because of this, in network design it becomes easier to replace one protocol with another within the same layer. However, multimedia applications with real-time constraints and limited tolerance of data loss can significantly benefit from cross-layer design and optimization (Schaar, 2007). A cross-layer approach allows the exchange of information across protocol-layer boundaries and in some cases the approach allows system-wise optimization of performance. For example, sensor networks benefit from system-wise optimization (Jurdak, 2007), as they usually run dedicated applications such as video surveillance. The same applies to ad hoc networks in general, where video streaming can benefit (Setton, et al., 2005) from cross-layer design. Widening the remit, IEEE 802.11 wireless LANs (Schaar, et al., 2005) have also proved a fertile ground for multimedia applications of cross-layer design.

There are penalties in applying a cross-layer scheme (Kawadia & Kumar, 2003), namely it may result in a monolithic application that is hard to modify or evolve. However, for wireless communication (Srivastav & Motani, 2005) an adaptive scheme that leverages information across the layers can cope with the volatile state of the channel due to fading and shadowing and the constrained available bandwidth of the channel. It is not necessary to abandon layering altogether in a 'layerless' design but simply to communicate between the layers. Video applications break protocol boundaries with limited objectives in mind, though improvements in performance remain the goal. Performance may be defined variously in terms of reduction of delay, reduction of errors, throughput efficiency, and, in wireless networks, reduction of energy consumption. This list by no means exhausts the possible trade-offs that can be engineered through cross-layer exchange of information.

As an example consider cross-layer design for IEEE 802.16e (mobile WiMAX) (IEEE 802.16e-2005, 2005) broadband wireless access systems. IEEE 802.16e specifies the lower two layers of the protocol stack. In multicast mode, scheduling decisions for the real-time Polling Service queue in particular are suspended. This can cause excessive delay to multimedia applications. To avoid this, in Chang & Chou (2007) knowledge of the application types and their delay constraints is conveyed to the datalink layer, where the scheduling mode is decided upon. The network layer can also benefit from communication with the datalink layer in order to synchronize WiMAX and IP handoff management (Chen & Hsieh, 2007) and in that way reduce the number of control messages. For further general examples of cross-layer design in WiMAX, the reader should consult Kuhran et al. (2007). Specifically video applications using PHY layer information were targeted in Juan et al. (2009) and She et al. (2009). In Juan et al. (2009), layers of a scalable video stream are mapped onto different 802.16e connections. The base station (BS) periodically reports average available bandwidth to a collocated video server, which then dynamically allocates video packets to the connections. The base layer occupies one connection while the remaining enhancement layer(s) packets occupy the second connection. If base layer packets (and certain key pictures) are lost, then the BS only retransmits these if available bandwidth permits. In She et al. (2009), cross-layer design is applied to WiMAX IPTV multicast to guard against channel diversity between different receivers. The solution again utilizes scalable video layers but, instead of a mapping onto different connections, superposition coding is employed. In such coding, more important data are typically modulated at BPSK whereas enhancement layers are transmitted at higher order modulation such as 16QAM. A cross-layer unit performs the superposition at the BS, whereas at the subscriber stations layers are selected according to channel conditions. Both these schemes fall into the class of wireless medium-aware video streaming. However, neither of these papers explains how signaling between lower and higher level protocols can take place. In Neves et al. (2009) it is pointed out that IEEE 802.21 *Media Independent Handover (MIH)* services (IEEE 802.21 WG, 2008) already provides a framework for cross-layer signaling that could be enhanced for more general purposes. In fact, another WiMAX specific set of standardized communication primitives as IEEE 802.16g. However, it could be that legacy WiMAX systems will need to be provided with a different interface. In IEEE 802.21, a layer 2.5 is inserted between the level 2 link layer and the level 3 network layer. Upper-layer services, known as MIH users or MIHU communicate through this middleware to the lower-layer protocols.

The remainder of this chapter is organized principally as a series of short case studies from the authors' experiences. The first two mini-case studies are concerned with cross-layer techniques for video transmission over WiMAX. There follows a look at how cross-layer techniques can benefit Bluetooth (Haartsen, 2006) transmission. Thus the two sets of case studies are concerned with outdoor and indoor wireless systems respectively. However, firstly the immediately following section briefly outlines background to understanding networked video services.

VIDEO SERVICES

This Section introduces background to understanding video services in general and includes a primer on video compression. If the reader is sufficiently familiar with this field then it can be passed over and the cross-layer case studies can be directly accessed.

In networked video delivery, systems are classically divided (Chou, 2007) into streaming and broadcast systems. In the former, video is pre-encoded before storage and access by a server, while in the latter there is no storage before server access and multicast over a network. A further distinction in this model is that in streaming a control path exists, whereas the presence of many receivers in a broadcast system means that feedback implosion is a risk. Feedback can be used for congestion control but it can also return *Video Cassette Recorder (VCR)* commands, typically through the *Real Time Streaming Protocol (RTSP)* (Schulzrinne, et al., 1998). Nevertheless, it is possible to stream both pre-encoded and online or live video because, after feedback notification of congestion, the streaming rate can be changed through bitrate transcoding (Assunção & Ghanbari, 1997; Sun, et al., 2005). One problem that fast transcoding may face in the state-of-the-art H.264/AVC (Advanced Video Coding) codec (Wiegand, et al., 2007) is error drift when transcoding I-frames (Lefol, et al., 2006).

Scalable video also allows rate control as a response to network conditions or target device capability but a full discussion of the variety of multi-layer or scalable options such as Fine Grain Scalability (Radha, et al., 2001), Multiple Description Coding (Wang, 2005), Signal-to-Noise Ratio (SNR) scalability (Pesquet-Popescu, et al., 2006) would require another chapter. Rich though the scalable options are commercial Internet operators seem to prefer simple schemes such as simulcast as used by RealVideo. In simulcast, multiple streams are stored (or encoded online) at different rates and selected according to network conditions. In H264/AVC, stream switching frames allow a smoother transition between low and higher quality stream at lower cost in bandwidth than through switching at I-frames, though there is a cost in storage of stream switching frames.

At the target device, video is first buffered in a playout, decoder or client buffer (there are various alternative names) prior to access by the decoder. This buffer will vary in size depending on the capabilities of the device. Large buffers are not advisable for battery-powered devices because of both active and passive energy consumption. Nevertheless, some buffering is required to absorb variation of delay (jitter) over the network and this buffering can be adaptive to network conditions.

Because of motion-compensated prediction, coding it is always necessary to store packets prior to decode, especially if VBR is in use. An additional render buffer, able to store a few frames prior to display, is also generally present. Apart from buffer overflow in the intermediate buffers of routers through congestion, buffer overflow at the playout buffer is also possible. Packets arriving too late for their display or decode deadlines may also be dropped. It is also possible, because of jitter, for buffer underflow to occur. In fact, in the Windows Media system (Chou, 2007) the receiver monitors the buffer level to detect network congestion. Again like RealVideo, Windows Media uses simulcast, with the receiver signaling the server to swap to a lower rate stream when it detects congestion. However, the Windows Media receiver or client is not only reliant on buffer monitoring, because packet loss at the receiver is also taken into account.

In video streaming, the compressed video bitstream is transmitted across a network to the end user's decoder (prior to display) without the need for storage other than in temporary buffering. Its advantage over progressive download from a network point-of-view is that the throughput is only that required to render the video at the user's display. Download risks overloading the network by too high a throughput. If download is not progressive, then the user has to wait an intolerable time before (say) viewing a 2 hr movie. There are also issues of commercial confidentiality if the video is stored on the user's machine.

Downloading video does permit *Variable Bitrate (VBR)* to be transported. In VBR, the codec *Quantization Parameter (QP)* is fixed leading to a constant quality. The alternative is to set a target bit rate for *Constant Bitrate (CBR)* video and allow fluctuating quality but with a gain in controllability. The main problem with VBR is that due to a strong variation in the number of bits allocated to each of the frame types (Lakshman, et al., 1998) the rate is highly variable (Van der Auwera & Reisslein, 2009). Long video streams are also not statistically stationary in time, which causes a problem when attempting to model video input to a network. This variability is accentuated in the H.264/AVC and it is reported (Van der Auwera, et al., 2008) that the variability is accentuated the more so in the Scalable Video Coding (SVC) extension to H.264, with the result that prior smoothing of VBR streams is contemplated. (The reason for increased variability is attributable to the increased number of motion estimation modes in H.264/AVC and in H.264/SVC, the addition of hierarchical B-frames.)

Video is known as a delay-sensitive service but in fact there are varying levels of intolerance, and a limit of 200 ms has been mentioned. However, for one-way streaming the delay requirements are less stringent. For example, channel swapping or VCR-like control is restricted to 500 ms intervals, because anchor or key frames at which switching can occur are placed at these intervals within a stream. Another form of delay is start-up delay, with *Video-on-Demand (VoD)* services hoping to make this imperceptible (< 20 ms), which is perhaps possible on the Internet if the *Resource ReSerVation Protocol (RSVP)* (Zhang, et al., 1997) were to be widely deployed. Variation in delay (jitter) is also important in terms of media synchronization (between audio and video) (Blakowski & Steinmetz, 1996). However, there are also display deadlines to be met, implying that a jitter buffer should be dimensioned to absorb any variation in delivery (assuming Internet delivery). For reference frames (one used for predictive motion estimation), their data is still of value for decoding future frames even if they miss their display deadline. Too large a receiver buffer will lead to increased end-to-end latency and start-up delay, while too small a buffer may cause overflow. This

Figure 1. A general diagram of a GOP

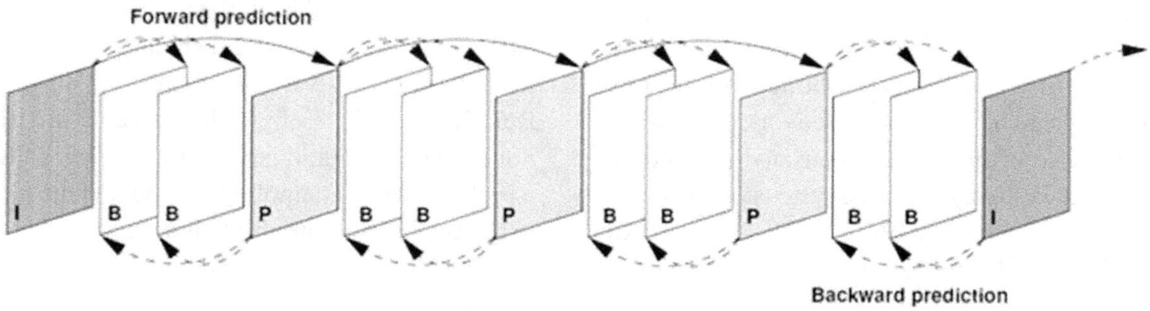

is why adaptive buffers have been contemplated in the research literature (Kalman, et al., 2002).

As background the following supplies essential video coding information based on a presentation in Fleury et al. (2011). Video compression or video coding refers to the reduction in the quantity of data used to represent video sequences. In the standard video method, compression is based on spatial and temporal redundancy reduction along with entropy coding[1] (static or dynamic Huffman, or Arithmetic) (MacKay, 2003). Spatial redundancy reduction results in Intra (I) frames, within which each block of the picture is predicted from its neighboring coded blocks without reference to any other frame. The difference or residue between the two blocks is typically transformed using the *Discrete Cosine Transform (DCT)*[2], which removes correlations. The wavelet transform has also found favor in the BBC's Dirac codec. The orthogonal frequency transform (or bi-orthogonal in the case of the wavelet) is followed by quantization of the transform coefficients and subsequent entropy coding. As the intra frames do not rely on other frames, they are used for a number of operations, typically for random access (video cassette recorder functionality) and confining temporal error propagation errors due to losses during transmission. Unfortunately, the BBC has withdrawn active support for Dirac, which will affect its commercial prospects.

Inter frames exploit the temporal redundancy between frames. In this process, the movements of objects in the neighboring frames are modeled by means of motion vectors. The resulting frame is subtracted from the original frame to be encoded, and the residue is again transformed before followed by entropy coding. Inter frames can be further categorized as Predictive (P) frames and Bi-predictive (B) frames. In P frames the prediction is made from earlier P- or I-frames, while in B-frames the prediction can be made from an earlier and/or later I- and/or P-frame, but not from a B-frame. In an H.264/AVC codec this restriction is removed and a B-frame can be used as reference for predicting other frames (Flierl & Girod, 2003) A sequence of frames grouped between two I frames is referred to as a *Group of Pictures (GOP)* as shown in Figure 1.

PHYSICAL TO APPLICATION LAYER INTERVENTION

In the following cross-layer scheme, information from the PHYsical layer is used to protect video streaming over a mobile WiMAX (IEEE 802.16e) link to a Mobile Subscriber Station (MS). Refer to the introduction section of the chapter for a discussion of cross-layer techniques for WiMAX. Protection is through an adaptive *Forward Error Correction (FEC)* scheme in which channel conditions as reported by channel estimation at the PHY layer serve to adjust the level of FEC. This flexibility is achieved by rateless channel coding

(MacKay, 2005) in the sense that the ratio of FEC to data is adjusted according to the information received from the PHY layer. The scheme also works in cooperation with PHY-layer FEC, which serves to filter out packet data in error, so that only correctly received data within a packet are passed up the layers to the video-streaming application.

The probability of channel Byte Loss (*BL*) serves to predict the amount of redundant data to be added to the payload. In an implementation, *BL* is found through estimation of channel conditions and then communicated from the PHY layer via the link layer to the application layer. The IEEE 802.16e standard IEEE 802.16e-2005, 2005) specifies that an MS should provide channel measurements, which can either be Received Signal Strength Indicators or may be Carrier-to-Noise-and-Interference Ratio measurements made over modulated carrier preambles. WiMAX is by no means alone among wireless technologies in specifying channel condition estimation, because of the widespread use of adaptive modulation and coding within *Orthogonal Frequency-Division Multiple Access (OFDMA)* at the physical layer.

If the original packet length is *L*, then the redundant data is given simply by:

$$\begin{aligned} R &= L \times BL + \left(L \times BL^2\right) + \left(L \times BL^3\right)\ldots \\ &= L / \left(1 - BL\right) \; - L, \end{aligned} \tag{1}$$

which adds successively smaller additions of redundant data, based on taking the previous amount of redundant data multiplied by *BL*. To achieve an incremental increase in redundant data, rateless channel coding is used.

Rateless code decoding in traditional form operates by a belief-propagation algorithm which is reliant upon the identification of clean symbols. This latter function is performed by PHY-layer FEC, which passes up correctly received blocks of data (through a cyclic redundancy check) but suppresses erroneous data. For example, in IEEE 802.16e, a binary, non-recursive convolutional encoder with a constraint length of 7 and a native rate of 1/2 operates at the PHY layer.

The following statistical model (Luby et al., 2007) applies:

$$\begin{aligned} P_f\left(m,k\right) &= 1, & \text{if } m < k, \\ &= 0.85 \times 0.567^{m-k} & \text{if } m \geq k \end{aligned} \tag{2}$$

where $P_f(m,k)$ is the decode failure probability of the code with k source symbols if m symbols have been successfully received (and 1 - P_f is naturally the success probability). Notice that Luby et al. (2007) remark and show that for $k > 200$ the model almost perfectly models the performance of the code, which implies that, if blocks are used, approximately 200 blocks should be received before reasonable behavior takes place. This observation motivated the choice of bytes within a packet as the symbols, to reduce latencies. Upon receipt of the correctly received data, decoding of the information symbols is attempted, which will fail with a probability given by (2) for $k > 200$.

If a packet cannot be decoded, despite the provision of redundant data then extra redundant data are added to the next packet. In Figure 2, packet X is corrupted to such an extent that it cannot be immediately decoded. Therefore, in packet X+1 some extra redundant data are included up to the level that decode failure is no longer certain. If according to (2), $m < k$ then a packet is corrupted to such an extent that it is effectively dropped. However, this was found to be a rare event. It is implied from (2) that if less than k symbols (bytes) in the payload are successfully received then a further k - m + e redundant bytes can be sent to reduce the risk of failure. This reduced risk arises because of the exponential decay of the risk that is evident from equation (2) and which gives rise to Raptor code's low error probability floor (Palanki & Yedidai, 2004). The exercise of re-transmission of extra redundant data is only repeated once to

Figure 2. Division of payload data in a packet between source data, original redundant data, and piggybacked data for a previous erroneous packet

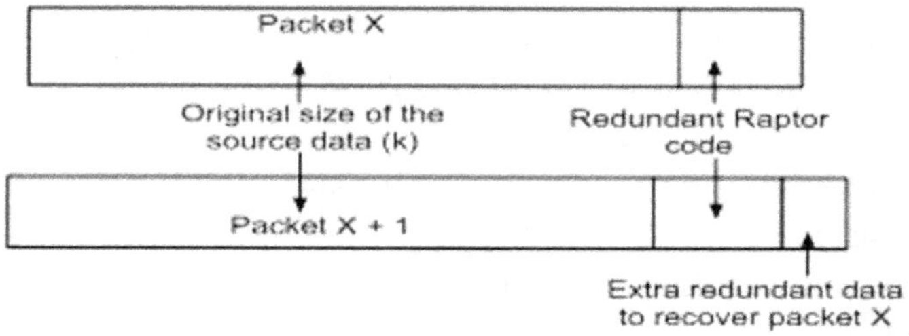

reduce latency. Requests for extra redundant data are easily implemented using IEEE 802.16e's *Time Division Duplex (TDD)* frame structure, with requests sent in the uplink sub-frame.

TRANSPORT LAYER INTERVENTION

In the following short case study, cross-layer intervention takes place at the transport layer to take advantage of the knowledge of video packet structure to make decisions about retransmission. In this cross-layer scheme, if an out-of-sequence packet arrives then a decision to send a *Negative Acknowledgment (NACK)* for a missing packet is made. However, there are two variants of the scheme: 1) in which all lost packets are requested for retransmission, and 2) in which only prioritized packets. In this short case study, priority is determined by the video frame type of the packet with priority given to Intra-coded I-pictures. However, other prioritizations are possible such as by data-partitioning partition type or favoring the base-layer of a scalable video stream. At the transport layer, application content is inspected to determine whether a packet should be retransmitted. At the receiver, a record is kept of packet sequence numbers available through the *Real-Time Transport Protocol (RTP)* header, and if an out-of-sequence packet arrives, a NACK is transmitted to the sender. The video source prevents transmission from its input buffer until a single retransmission of the missing packet in the sequence has taken place. Further retransmissions do not take place, because waiting packets could be delayed and because the failure of one retransmission may indicate continuing poor channel conditions at the broadband wireless link. Thus, the sender only transmits a NACK if this is the first time that particular packet has been lost. If it is the first time and the non-selective NACK version is in operation then a NACK is sent. However, if prioritized operation is in use then a decision is made according to the picture type of the video packet that has been lost, reflecting the importance to the reconstruction of the video of that packet. Figure 3 is a general representation of the processing involved in the scheme, showing the NACK response of the receiver.

The NACK selection scheme is but one example of cross-layer intervention. Görkemli et al. (2008) used the scalable video coding extension to the H.264/AVC codec in conjunction with *Data Congestion Control Protocol (DCCP)* (Kohler, et al., 2006) in a scenario in which there were wireless access links to the core network at the sender and receiver sites. Similarly to Fu et al. (1996), lower-layer information was employed to modify DCCP's estimate of packet loss rates. In this case, the estimate was based on the PHY layer *Automatic Repeat Request (ARQ)* information. A similar cross-layer scheme can be found in Yang et al. (2007).

Figure 3. Operation of NACK enhancement to UDP

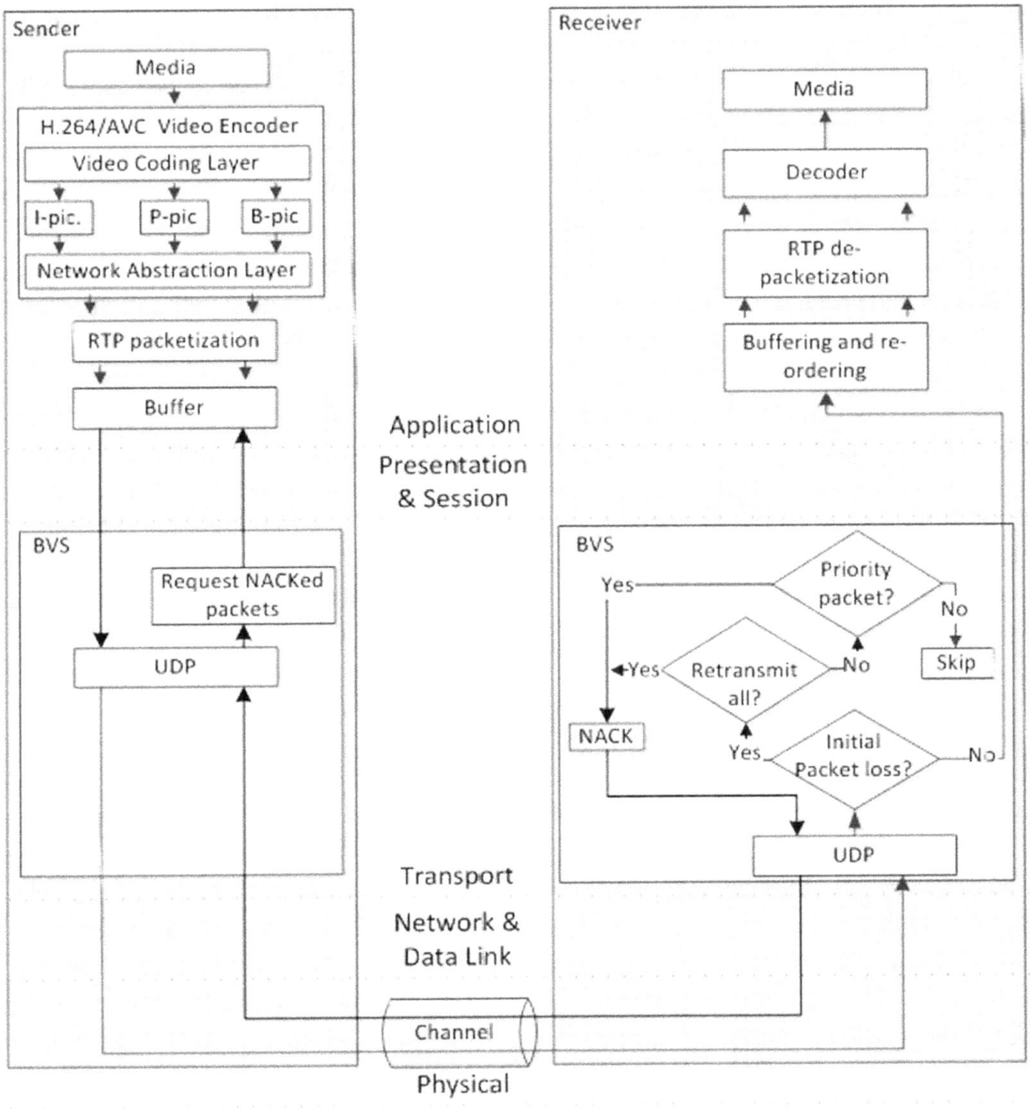

"CROSS-LAYER" FOR PERSONAL AREA NETWORK

Personal Area Networks (PANs) in the original conception allow short-range communication between desk-top or office devices. This includes cable replacement for single link communication. A PAN can be extended into something more like a sensor network but this is not considered herein. Instead, this section tackles the task of protecting video streaming over a volatile wireless channel when there is PAN congestion. This is performed in a cross layer manner, firstly through a form of adaptive modulation at the PHY layer that favors more important video data and secondly through datalink control of a transmit buffer that is partitioned according to video frame type. PANs originally came to prominence through the success of Bluetooth technology. The following summarizes the cross-layer scheme. For a much fuller presentation the reader is referred to work by the authors of this chapter in Razavi et al. (2008).

Bluetooth (Haartsen, 2000), standardized as IEEE 802.15.1, is a short-range radio-frequency interconnection, which can be expanded to form a piconet, with one master node and up to seven slaves. In Bluetooth, there is no direct slave-slave communication, as all cross-traffic must pass through a Bluetooth master node. This section investigates unequal protection of encoded video data transmitted from master to slave, in the face of cross-traffic passing from slave to slave via the Bluetooth piconet master. It is possible to apply modulation adaptation to achieve unequal protection, particularly in *Orthogonal Frequency Division Multiplexing (OFDM)* systems. As an example in Sajadiehet et al. (1998), adaptive modulation was traded against error coding. However, if data-link FEC is not available, it is still possible to apply adaptive modulation. In Bluetooth version 2.1, FEC is not implemented for *Enhanced Data Rate (EDR)* modes (EDR, 2007), possibly because low-cost devices could not cope with the computational requirements of coding at the higher data rates. On the other hand, Bluetooth version 2.1's EDR provides several forms of modulation, though not through OFDM. In fact, Bluetooth v. 2.1 under EDR supports gross air rates of both 3 and 2 Mbps (maximum gross user payload of 1.4485 Mbps), through, respectively, *π/4-Differential Quadrature Phase-Shift Keying (DQPSK)* or *Eight-Phase Differential Phase-Shift Keying (8DPSK)* modulation. This implies that, through adaptive modulation, a lower bit rate is available that can serve to give unequal protection to some of the packets of more important frame types, the *intra- (I-)* and *predictive- (P-)* anchor frames, as well as some *bi-predictive- (B-)* frame packets, depending on circumstances.

Protection by adaptive modulation can be combined with transmit buffer management to avoid packet loss from buffer congestion, through consideration of packet importance and wireless channel conditions. In this scheme, the buffer is partitioned into three zones that correspond to the three frame or picture types of video sequences. Tri-zone management of the transmit buffer for video stream packets can be based on the relative content importance (Ghanbari, 2003) of those differing video frame types. The combination of frame-packet-type and subsidiary macroblock-type frequency counts provides a clear means of regulating the zones. Notice that a macroblock is the unit of motion estimation in block-based predictive coding.

Because a lower bit rate is employed for priority packets, there is a risk of buffer overflow at the transmit buffer, compared to a situation in which all packets were sent at the higher bit rate. Therefore, a trizone buffer applies a different unequal protection policy for each zone. However, the fact that there are three zones does not mean that only I-frame packets occur in one zone, P-frame packets in another zone, and B-frame packets in the third zone. All packet types can occupy each zone, but the prioritization policy between each zone is different as a reflection of the greater fullness of the buffer as each successive zone is occupied. As the buffer fullness increases, packets of whatever type begin to fill the second and then the third zone, and the prioritization policy between frame-type packets changes accordingly. Ideally, the output at the lower bit rate should decrease linearly as buffer fullness increases. However, to achieve this, because of a varying number of packets between the frame types, a linear unequal protection policy based simply on frame type will not work. Therefore, in the second zone of the tri-zone buffer, the number of P-frame packets offered protection is controlled by the content importance and its predominance within the arriving P-frames.

The buffer zone boundaries are based on the frequency within a video stream of I-, P-, and B-frames, and they dynamically change according to the relative size ratio of the arriving frame-type packets. In other words, the ratio of data allocated to each frame type within an arriving video stream dynamically determines the zone sizes, while the frame type determines the UP policy applied within the zone. Zone 1 is first occupied by arriving packets. In this zone, not all B-frame packets are protected, and B-frame packets are not protected in zones 2 and 3. As zone 1 is the only zone in which B-frames receive some protection, it makes sense to allocate the size of zone 1 according to the relative amount of data arising from B-frame packets. Doing otherwise would bias the zone size against B-frame packets. It should be noted that, because of the GOP structure, B-frame packets occur with greater frequency than other frame-type packets. In zone 2, not all P-frame packets are protected and P-frame packets are not protected in zone 3. Therefore, in zone 2, when P-frame packets begin to lose the protection received in zone 1, the size of the zone determines, so to speak, how quickly they lose their protection. This rate is determined by the amount of P-frame type data within the stream to avoid unfairly biasing of the zone size against P-frame packets. Finally, a similar observation applies to zone 3. If there are packets occupying this zone, then the buffer would be at its fullest state and as a result, not all I-frame packets are protected in zone 3.

By monitoring transmitter buffer fullness, available through Bluetooth *Host Controller Interface (HCI)*, an adaptive unequal protection scheme is applied. It turns out that buffer fullness is an excellent indication of congestion within a Bluetooth piconet. Buffer fullness is responsive not only to buffer congestion from an arriving video stream but also to an increase in buffer service time when piconet cross-traffic is present. As buffer fullness reflects the congestion of the Bluetooth wireless channel, it can be used to regulate the unequal protection scheme. The channel condition should also be ascertained. As in the earlier discussion of the estimation of WiMAX channel condition, this can be achieved by *Received Signal Strength Indicator (RSSI)* (Krishnamachari, et al., 2003) or can rely on channel probing messages or channel condition feedback messages (Chen, et al., 2004). RSSI is an optional feature of Bluetooth implementations, though care should be taken as in Chen et al. (2004) it was found that the RSSI reported that Bluetooth channel quality oscillated rapidly.

A range of packet types exists in Bluetooth according to the number of timeslots occupied by a packet (1, 3, or 5) and the modulation type. The classical Bluetooth *Channel Quality-Driven Data Rate (CQDDR)* model assumes different packet types, and hence data rates are chosen depending on channel conditions. This model can be achieved by means of a *Look-Up Table (LUT)*, which effectively establishes the per-bit SNR boundaries between the differing packet types. Selecting the packet type by content type in addition to selection by channel quality overrides CQDDR. This is provided by offering up some video packets when traffic on the shared Bluetooth channel permits it. When channel conditions deteriorate and/or traffic congestion across the Bluetooth piconet increases, then the tri-zone policy effectively converges upon the CQDDR model.

The scheme has no implications such as changing the form of modulation for the Bluetooth EDR standard. Priority packet marking can take place above the HCI boundary within the host's software, which is available in open source form, such as the Bluez stack for the Linux operating system. However, firmware modification would be required at the data-link layer in order to recognize marked packets and apply adaptive modulation.

In another Bluetooth cross-layer scheme, input from lower layers can be used to determine the packetization strategy. In general, effective packetization is vital (Stockhammer, 2006) for video over wireless transmission. Bluetooth v. 2.1's EDR with multiple Bluetooth packet sizes and two data rates according to modulation scheme makes it difficult for a video codec source to anticipate the optimal slice sizes to suite channel conditions. Slices are defined for error resilience purposes to prevent temporal error propagation by means of decoder resynchronization markers.

However, a single slice/packet packetization may not efficiently use the predefined capacity of the Bluetooth packets. One solution is dynamic packetization (Razavi, et al., 2006), whereby slice boundaries are not preserved in outgoing Bluetooth packets. For higher input data rates, dynamic packetization maximizes throughput, regardless of content. However, radio frequency noise may result in an increased loss of synchronization at the decoder if packets bearing partial slices are lost. Alternatively, an integral number of slices, up to some maximum, may be placed in a Bluetooth packet, that is integral slice packetization. At lower input data rates, dynamic packetization may be unnecessary, while smaller integral slice packetizations, for example one slice per Bluetooth packet, maximize noise immunity.

By monitoring a Bluetooth master access point of both the incoming IP packet-size and the Carrier-to-Noise Ratio of the Bluetooth link, it becomes possible: firstly, to adapt to channel conditions by swapping between Bluetooth's EDR's two modulation schemes and, subsequently, to adapt the Bluetooth slice packetization method to the incoming data rate. The result of this scheme is a worthwhile improvement in received objective video quality.

FUTURE DEVELOPMENTS

This Chapter has concentrated on unicast delivery of video. Though value-added services such as Video-on-Demand (VoD) and catch-up TV employ unicast, along with real-time interactive services such as videophone and video conferencing, staple networked video may also use multicast. In Al-Suhail et al. (2011) the problem of adjusting differential streaming rates and FEC according to localized channel conditions is tackled in a cross-layer fashion. Because of the problem of feedback implosion, it is not possible to adjust the FEC rate for individual receivers in WLAN, and in fact a compromise between the set of groups of receivers may also be necessary. The video coding rate is jointly optimized depending on the intra-packet and inter-packet FEC rate, which in turn depends on feedback of channel conditions for each access point. In Villalón et al. (2007), information is passed upwards and downwards between the PHY, datalink, and application layers to achieve effective multicast from an access point within a IEEE 802.1 WLAN. The proposed scheme determines the worst channel conditions for any one receiver within the WLAN as a guide to setting the source-coding rate. A further method is to employ scalable coding so that those receivers with poor channel conditions can choose to only accept the video base layer packets. However, data partitioning was used in this scheme as a basis of layering. Partitions A and B containing parameters, motion vectors and macroblock addresses, along with intra-coded transform coefficients formed one layer, whereas the enhancement layer was formed from partition-C's inter-coded transform coefficients. At the *Medium Access Control (MAC)* sub-layer of the datalink layer prioritization is provided according to modulation type.

Another area that is suited to cross-layer techniques is providing greater robustness (Stockhammer, 2006) through the use of error resilience techniques. In the H.264/AVC codec a greater range of error resilience methods have been provided (Wenger, 2003) precisely because of increased video services over wireless. Stockhammer (2006) surveys available techniques from the perspective of packet-based 3G wireless systems. It is noted that if error rates of 10% can be tolerated then 10-25% more users can be accommodated compared to when there are just 1% errors. If packets are lost, as opposed to being received in error, retransmission can take place. However, if all packets are retransmitted then latency can accumulate, resulting in missed display deadlines. It is also worth remarking that it may still be worth retransmitting important packets, (say) containing anchor frame data, even if the display deadline is missed, because the data in these packets are needed to help reconstruct later frames. In Chen et al. (2010), information about the likely retransmission time from the MAC sub-layer is combined with content type information from the application layer to find an estimate of the number of retries. In Schaar et al. (2003), the retry limit is calculated based on MAC sub-layer estimations. In addition, the FEC level is adaptively selected and packet sizes also depend on reported channel conditions.

CONCLUSION

To achieve Quality-of-Service goals, the user should be presented with a stable video viewing environment in which the underlying instability of the wireless channel medium is suppressed. In the wireline Internet, this objective is solved by over provisioning so, for example, congestion is not a serious problem because of the excess capacity in the network core. However, in wireless networks there is always a competition for restricted bandwidth, as a common medium is shared by all communicators. Layered design provides a clean architecture but unfortunately does not readily allow adaptation. However, layer fusion is normally also not a solution, as it does not present a generic network architecture that is able concurrently to support a variety of applications. Cross-layer approaches take advantage of sharing of source coding, channel coding, physical layer and data-link parameters and settings. They also permit the exchange of video content and video content type, as well as channel condition estimation. Though this exchange contradicts Shannon's separation theory, it is now widely accepted that such a separation applies mostly in certain ideal situations. In this chapter, the potential of cross-layer design has been shown by showing through example how such breaches can take place. It is likely that many other ways exist in which adaptation can take advantage of such breaches and it is hoped that this chapter has shown why such an approach is worthwhile.

REFERENCES

Al-Suhail, G., Fleury, M., & Al-Majeed, S. S. (2011). Mobile video streaming over heterogeneous networks. In Khalil, I. (Ed.), *Mobile Multimedia Communications and Applications: New Technologies*. Hershey, PA: IGI Global. doi:10.4018/978-1-60960-563-6.ch013

Assunção, P. A. A., & Ghanbari, M. (1997). Congestion control of video traffic with transcoder. In *Proceedings of the International Conference on Image Processing*, (pp. 523-527). IEEE.

Blakowski, G., & Steinmetz, R. (1996). A media synchronization survey: Reference model, specification, and case studies. *IEEE Journal on Selected Areas in Communications*, *14*(1), 9–33. doi:10.1109/49.481691

Chang, B.-J., & Chou, C.-M. (2007). Cross-layer based delay constraint adaptive polling for high density subscribers in IEEE 802.16 WiMAX networks. *Wireless Personal Communications*, *46*(3), 285–304. doi:10.1007/s11277-007-9434-5

Chen, C.-M., Lin, C.-W., & Chen, Y.-C. (2010). Cross-layer packet retry limit adaptation for video transport over wireless LANs. *IEEE Transactions on Circuits and Systems for Video Technology*, *20*(11), 1448–1461. doi:10.1109/TCSVT.2010.2077475

Chen, L.-J., Kapoor, R., Sanadidi, M. Y., Lee, R., & Gerla, M. (2004). Audio streaming over Bluetooth: An adaptive ARQ timeout approach. In *Proceedings of the 24th International Conference on Distributed Computing Systems*, (pp. 196–201). IEEE.

Chen, Y.-W., & Hsieh, F.-Y. (2007). A cross-layer design for handoff in 802.16e network with IPv6 mobility. In *Proceedings of the IEEE Wireless Communications and Networking Conference*, (pp. 3844-3849). IEEE Press.

Chou, P. A. (2007). Streaming media on demand and live broadcast. In Chou, P. A., & van der Schaar, M. (Eds.), *Multimedia in IP and Wireless Networks* (pp. 453–501). Burlington, MA: Academic Press. doi:10.1016/B978-012088480-3/50015-1

EDR. (2007). *Core specification of the Bluetooth system, version 2.1 + EDR*. Retrieved from http://www.bluetooth.com.

Fleury, M., Altaf, M., Moiron, M., Qadri, M., & Ghanbari, M. (2011). Source coding methods for robust wireless video streaming. In Farrugia, R., & Debono, C. J. (Eds.), *Multimedia Networking and Coding: From Capture to Display*. Hershey, PA: IGI Global.

Flierl, M., & Girod, B. (2003). Generalized B pictures and the draft H.264/AVC video compression standard. *IEEE Transactions on Circuits and Systems for Video Technology*, *13*(7), 587–597. doi:10.1109/TCSVT.2003.814963

Fu, Y., Hu, R., Tian, G., & Wang, T. (1996). TCP-friendly rate control for streaming service over 3G network. In *Proceedings of the International Conference on Wireless Communications, Networking and Mobile Computing*, (pp. 1-4). IEEE.

Ghanbari, M. (2003). *Standard codecs: Image compression to advanced video coding*. Stevenage, UK: IET Press. doi:10.1049/PBTE049E

Görkemli, B., Sunay, M. O., & Tekalp, A. M. (2008). Video streaming over wireless DCCP. In *Proceedings of the IEEE International Conference on Image Processing,* (pp. 2028-2031). IEEE Press.

Haartsen, J. (2000). The Bluetooth radio system. *IEEE Personal Communications*, *7*(1), 28–36. doi:10.1109/98.824570

IEEE. 802.16e-2005. (2005). *IEEE standard for local and metropolitan area networks: Part 16: Air interface for fixed and mobile broadband wireless access systems*. Retrieved from http://www.ieee.org.

IEEE. 802.21 WG. (2008). *IEEE standard for local and metropolitan area networks, media independent handover services: IEEE 802.21-2008*. Retrieved from http://www.ieee.org.

Juan, H.-H., Huang, H.-C., Huang, C., & Chiang, T. (2009). Cross-layer mobile WiMAX MAC designs for the H.264/SVC scalable video coding. *Wireless Networks*, *16*(1), 113–123. doi:10.1007/s11276-008-0118-1

Jurdak, R. (2007). *Wireless ad hoc and sensor networks: A cross-layer design perspective*. Berlin, Germany: Springer Verlag.

Kalman, M., Steinbach, E., & Girod, B. (2002). Adaptive media playout for low delay video streaming over error-prone channels. In *Proceedings of the International Conference on Image Processing,* (pp. 841-851). IEEE.

Kawadia, V., & Kumar, P. R. (2003). A cautionary perspective on cross layer design. *IEEE Wireless Communications Magazine*, *12*(1), 3–11. doi:10.1109/MWC.2005.1404568

Kohler, E., Handley, M., & Floyd, S. (2006). *Datagram congestion control protocol (DCCP), RFC 4340*. Retrieved from http://www.ietf.org.

Krishnamachari, S., van der Schaar, M., Choi, S., & Xu, X. (2003). Video streaming over wireless LANs: A cross-layer approach. In *Proceedings of the 13th International Packet Video Workshop*. IEEE.

Kuhran, M. S., Gür, G., Togcu, T., & Alagöz, F. (2010). Applications of the cross-layer paradigm for improving the performance of WiMAX. *IEEE Wireless Communications*, *17*(3), 86–95. doi:10.1109/MWC.2010.5490983

Lakshman, T., Ortega, A., & Reibman, A. (1998). VBR video: Tradeoffs and potentials. *Proceedings of the IEEE*, *86*(5), 952–973. doi:10.1109/5.664282

Lefol, D., Bull, D., & Canagarajah, N. (2006). Performance evaluation of transcoding algorithms for H.264. *IEEE Transactions on Consumer Electronics*, *52*(1), 215–222. doi:10.1109/TCE.2006.1605050

Luby, M., Gasiba, T., Stockhammer, T., & Watson, M. (2007). Reliable multimedia download delivery in cellular broadcast networks. *IEEE Transactions on Broadcasting*, *53*(1), 235–246. doi:10.1109/TBC.2007.891703

MacKay, D. (2003). *Information theory, inference, and learning algorithms*. Cambridge, UK: Cambridge University Press.

MacKay, D. J. C. (2005). Fountain codes. *IEEE Proceedings on Communications*, *152*(6), 1062–1068. doi:10.1049/ip-com:20050237

Neves, P., Sargneto, S., Pentikousis, K., & Fontes, F. (2009). WiMAX cross-layer system for next generation heterogeneous environments. In Dalal, U. D. D., & Kosta, Y. P. (Eds.), *WiMAX: New Developments* (pp. 357–376). Vukovar, Croatia: In-Tech.

Palanki, R., & Yedidai, J. (2004). Rateless codes on noisy channels. In *Proceedings of the International Symposium on Information Theory*, (p. 37). IEEE.

Pesquet-Popescu, B., Li, S., & van der Schaar, M. (2007). Scalable video coding for adaptive streaming applications. In Chou, P. A., & van der Schaar, M. (Eds.), *Multimedia in IP and Wireless Networks* (pp. 117–158). Burlington, MA: Academic Press. doi:10.1016/B978-012088480-3/50006-0

Radha, H., van der Schaar, M., & Chen, Y. (2001). The MPEG-4 fine_grained scalable video coding method for multimedia streaming over IP. *IEEE Transactions on Multimedia*, *3*(1), 53–68. doi:10.1109/6046.909594

Razavi, R., Fleury, M., & Ghanbari, M. (2008). Unequal protection of video streaming through adaptive modulation with a tri-zone buffer over Bluetooth enhanced data rate. *EURASIP Journal on Wireless Communications and Networking*. Retrieved from http://jwcn.eurasipjournals.com/content/2008/1/658794.

Razavi, R., Fleury, M., Jammeh, E., & Ghanbari, M. (2006). An efficient packetization scheme for Bluetooth video transmission. *Electronics Letters*, *42*(20), 1143–1145. doi:10.1049/el:20061638

Sajadieh, M., Kschischung, F., & Leon-Garcia, A. (1998). Modulation-assisted unequal error protection over the fading channel. *IEEE Transactions on Vehicular Technology*, *47*(3), 900–908. doi:10.1109/25.704843

Schulzrinne, H., Rao, A., & Lampeter, R. (1998). *Real time streaming protocol (RTSP) proposed standard, RFC 2326*. Retrieved from http://www.ietf.org.

Setton, E., Yoo, T., Xhu, X., Goldsmith, A., & Girod, B. (2005). Cross-layer design of ad-hoc networks for realtime video streaming. *IEEE Wireless Communications Magazine*, *12*(4), 59–65. doi:10.1109/MWC.2005.1497859

She, J., Yu, X., Po, P.-H., & Yang, E. H. (2009). A cross-layer design framework for robust IPTV services over IEEE 802.16 network. *IEEE Journal on Selected Areas in Communications*, *29*(2), 235–245. doi:10.1109/JSAC.2009.090214

Srivastav, E., & Motani, M. (2005). Cross-layer design: A survey and the road ahead. *IEEE Communications Magazine*, *43*(12), 712–721.

Stockhammer, T. (2006). Robust system and cross-layer design for H.264/AVC-based wireless video applications. *EURASIP Journal on Applied Signal Processing*. Retrieved from http://asp.eurasipjournals.com/content/2006/1/089371.

Sun, H., Chen, X., & Chiang, T. (2005). *Digital video transcoding for transmission and storage*. Boca Raton, FL: CRC Press.

Van der Auwera, G., & Reisslein, M. (2009). Implications of smoothing on statistical multiplexing of H.264/AVC and SVC video streams. *IEEE Transactions on Broadcasting*, *55*(3), 541–558. doi:10.1109/TBC.2009.2027399

Van der Schaar, M. (2007). Cross-layer wireless multimedia. In van der Schaar, M., & Chou, P. A. (Eds.), *Multimedia over IP and Wireless Networks* (pp. 337–408). Burlington, MA: Academic Press. doi:10.1016/B978-012088480-3/50013-8

Van der Schaar, M., Krishnamachari, S., Choi, S., & Xu, X. (2003). Adaptive cross-layer protection strategies for robust scalable video transmission over 802.11 WLANs. *IEEE Journal on Selected Areas in Communications*, *21*(10), 1751–1763. doi:10.1109/JSAC.2003.815231

Villalón, J., Cuenca, P., Qrozco-Barbosa, L., Seok, Y., & Turletti, T. (2007). Cross-layer architecture for adaptive video multicast streaming over multirate wireless LANs. *IEEE Journal on Selected Areas in Communications*, *25*(4), 1–14. doi:10.1109/JSAC.2007.070507

Wang, Y., Reibman, A. R., & Lee, S. (2005). Multiple description coding for video delivery. *Proceedings of the IEEE*, *93*(1), 57–70. doi:10.1109/JPROC.2004.839618

Wenger, S. (2003). H.264/AVC over IP. *IEEE Transactions on Circuits and Systems for Video Technology*, *13*(7), 645–655. doi:10.1109/TCSVT.2003.814966

Wiegand, T., Sullivan, G. J., Bjontegaard, G., & Luthra, A. (2003). Overview of the H.264/AVC video coding standard. *IEEE Transactions on Circuits and Systems for Video Technology*, *13*(7), 560–576. doi:10.1109/TCSVT.2003.815165

Yang, Y., Zhang, Q., Zhu, W., & Zhang, Y.-Q. (2007). End-to-end TCP-friendly streaming protocol and bit allocation for scalable video over wireless Internet. *IEEE Journal on Selected Areas in Communications*, *22*(4), 777–790. doi:10.1109/JSAC.2004.826008

Zhang, L., et al. (1997). *Resource ReSerVation protocol (RSVP): RFC 2205*. Retrieved from http://www.ietf.org.

ADDITIONAL READING

Duhamel, P., & Kieffer, M. (2010). *Joint source-channel decoding: A cross-layer perspective with applications in video broadcasting*. Oxford, UK: Academic Press.

Ibnkahla, M. (Ed.). (2008). *Adaptation and cross-layer design in wireless networks*. Boca Raton, FL: CRC Press. doi:10.1201/9781420046106

Perez-Neira, A. I., & Campalans, M. R. (2009). *Cross-layer resource allocation in wireless communications*. Oxford, UK: Academic Press.

Richardson, I. E. G. (2003). *H.264 and MPEG-4 video compression: Video coding for next generation multimedia*. Chichester, UK: John Wiley & Sons. doi:10.1002/0470869615

Sadka, A. (2002). *Compressed video communications*. Chichester, UK: John Wiley & Sons. doi:10.1002/0470846712

Van der Schaar, S., & Chou, P. A. (Eds.). (2007). *Multimedia over wireless networks: Compression, networking, and systems*. Burlington, MA: Academic Press.

Wang, Y., Ostermann, J., & Zhang, Y.-Q. (2002). *Video processing and communications*. Upper Saddle River, NJ: Prentice Hall.

ENDNOTES

1 Entropy coding is a lossless form of data compression.

2 The DCT is now replaced by a reversible integer transform in an H.264/AVC codec to avoid arithmetical inaccuracies during the inverse transform.

Chapter 10
Cross-Layer Framework for Power Conservation in Wireless Ad Hoc Networks:
A Case Study

N. Kasi Viswanath
G. Pulla Reddy Engineering College, India

ABSTRACT

Most of the wireless ad hoc networks have battery-powered nodes, so the life of the network depends on the power of the nodes. Power conservation is one of the main issues of wireless ad hoc networks. Power conservation can be achieved in two different ways: power control algorithms and power management algorithms. Transmission power control is one example of a cross-layer design problem. The purpose of this chapter is to discuss transmission power control problems and one possible solution by using cross layer framework through a case study. This chapter describes two cross-layer power control protocols developed at two different layers and also presents the effect of applying the combination of these protocols through a simulation study using Glomosim.

INTRODUCTION

A critical design issue for future wireless ad hoc networks is to develop suitable communication architectures, protocols, and services that efficiently reduce power consumption of wireless devices and thus increasing their operational lifetime in the network. In wireless ad hoc networks, power conservation can be achieved in two different ways: power control algorithms and power management algorithms. Power management in

DOI: 10.4018/978-1-4666-0960-0.ch010

ad hoc networks means that a station can decide when to switch its radio transceiver off to save energy. However, turning its transceiver off means that other stations cannot send their packets to the station, if any. Therefore, it should be done with caution to avoid possible delays. The power management issue is challenging especially in a multihop ad hoc network because hosts may sometimes lose synchronization. On the other hand, power control refers to the technique of tuning hosts transmission powers to the proper range to conserve energy. In addition, power control can also reduce interference and improve spatial reuse of wireless channels. The scope of the chapter is on power control algorithms in wireless ad hoc networks.

A wireless node in an ad hoc network has limited battery power; therefore, it is important to reduce its energy consumption. A wireless node has four modes: transmit, receive, idle, and sleep. It consumes most power in the transmit mode. In the idle mode, it needs to sense the medium and consumes a little less than in the receive mode. The sleep mode consumes very little and can be ignored compared to the other modes (Nilson, 2001). To save energy, one direction is to force a node to enter the sleep mode when it is not necessary to be awake (Jamieson, Balakrishnan, & Morris, 2001; Jung & Vaidya, 2002). The other direction is to apply a Transmission Power Control (TPC) protocol (Gomez, Campbell, Naghshineh, & Bisdikian, 2001; Agarwal, Krishnamurthy, Katz, & Dao, 2001; Pursley, Russell, & Wysocarski, 2000; Jung & Vaidya, 2002). This chapter is focused on the second direction. TPC looks for the minimum required transmit power between the transmitter and the receiver.

TPC is a simple and fundamental technique which is used for communications and impacts the operational lifetime of devices in different ways. Reducing the transmission power to significantly improve operational life time for some devices may not be significant if the transmission power accounts only for a small percentage of the overall power consumed, (e.g., a Wireless Local Area Network [WLAN] radio attached to a notebook computer). In contrast, for small computing devices with built-in or attached radios (e.g., sensors) reducing the transmission power may significantly extend the operational lifetime of a device, thus, improving the overall user experience.

The power control problem in wireless ad hoc networks is that of choosing the transmit power for each packet in a distributed fashion at each node (Kawadia & Kumar, 2005). Selecting power level fundamentally affects the network at various layers as follows:

- Power control affects the **physical layer** because the transmit power level determines the quality of the signal received at the receiver.
- Transmit power level determines the range of a transmission. Since the contention for the medium depends on the number of other nodes within range, the power levels determine the performance of medium access control and thus power control affects the **mac layer**.
- The basic functionality of the network layer is to discover and maintain routes between sources and destinations. The transmit power determines the set of possible neighbors a node can reach and hence the potential set of next hops and the overall network connectivity. Power control affects the **network layer** since the transmission range affects routing.
- The transmit power level determines the magnitude of the interference it creates for the other receivers. Power control affects the **transport layer** because interference causes congestion.
- Also the transmit power control problem affects several key performance measures including throughput, delay and energy consumption.

We see now, the need of cross-layer design in wireless ad hoc networks. Traditionally, network protocols are divided into a series of several independent layers each one built on the one below it. Each layer is designed separately, and the interaction between layers is performed through a well-defined interface. The main advantage of this type of approach is architectural flexibility. Although this traditional layered approach simplifies the network design, the behavior of a protocol can vary significantly depending on the protocol above it or below it in the protocol stack. Understanding how, and to what extent, protocols interact with each other will ultimately yield improvements in network performance. Researchers have proposed to consider new protocol layering by decomposing the overall network-performance optimization (Chiang, Low, Calderbank, & Doyle, 2007). As long as the optimization decomposition is successfully done, each protocol in each layer works as an optimal module to achieve the best network performance; as a consequence, cross-layer optimization has been considered in protocol layering.

On the other hand, the protocol layering by optimization decomposition does not necessarily match the existing protocol stack that is widely adopted in wireless ad hoc networks. Depending on different representations of the optimization problem, different decomposition structures may be derived, which may result in different architectures for protocol layering. However, the protocol stack in wireless ad hoc networks follows the classical TCP/IP protocol stack. The mismatch between the classical TCP/IP protocol stack protocol and the new protocol layering based on optimization decomposition actually illustrate that cross-layer design is highly desired if the optimum network performance must be achieved.

Therefore, it is reasonable to believe that cross-layer optimization will continue to be one of the most important tasks in protocol design for wireless ad hoc networks. However, critical issues must be considered for cross-layer design (Kawadia & Kumar, 2005), because it has risks due to loss of protocol-layer abstraction, incompatibility with existing protocols, unexpected impact on the future design of the network, and difficulty in maintenance and management. Thus, certain guidelines need to be followed when carrying our cross-layer design.

Cross-layer design can significantly improve the network performance (Bhatia & Kodialam, 2001; Kozat, Koutsopoulos, & Tassiulas, 2004; Chiang, 2004). The design of cross-layer models can be done by using two main approaches: loosely coupled and tightly coupled cross-layer design. In the loosely coupled cross-layer design, optimization is carried out without crossing layers but focusing on one protocol layer. In order to improve the performance of this protocol layer, parameters in other protocol layers are taken into account. Thus, information in one layer must be passed to another layer. Typically, parameters in the lower protocol layers are reported to higher layers. For example, in Figure 1a, the packet-loss rate in the Medium Access Control (MAC) layer or channel condition in the physical layer can be reported to the transport layer so that a Transmission Control Protocol (TCP) protocol is able to differentiate congestion from packet loss. As another example, the physical layer can report the link quality to a routing protocol as an additional performance metric for the routing algorithms.

It should be noted that information from multiple layers can be used in another layer to perform cross-layer design. There are two different methods in utilizing such information. The first one is the simplest case of cross-layer design, in which the information in other layers works just as one of the parameters needed by the algorithm in a protocol layer. The performance of this algorithm can be improved because a better (more accurate or reliable) parameter is used, but the algorithm itself does not need a modification. For example, in Figure 1b, the physical layer and MAC layer can inform TCP layer of the channel quality and packet-loss rate respectively so that TCP can dif-

Figure 1. Cross-layer design approaches

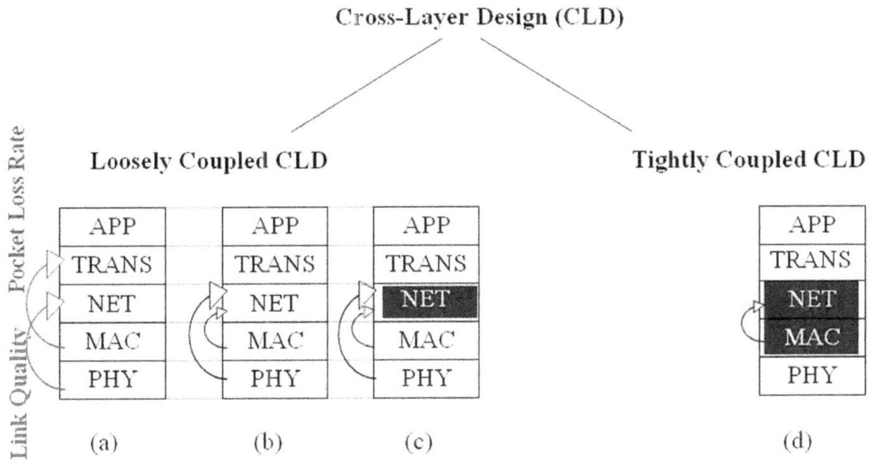

ferentiate real congestion from channel-quality degradation and thus, can carry out congestion control more intelligently. In the second method, based on the information from other layers, the algorithms of a protocol have to be modified. For example, in Figure 1c, if a MAC protocol can provide a routing protocol about its performance, the routing can perform multipath routing to utilize spatial diversity. However, the change from a single-path routing to multipath routing needs a significant modification in the routing protocol rather than just a parameter adaptation.

In the tightly coupled cross-layer design, only information sharing between layers is not enough. In this scheme, the algorithms in different layers are optimized altogether as one optimization problem. For example, in Figure 1d, for MAC and routing protocols in a multichannel Time Division Multiple Access (TDMA) Wireless Mesh Network (WMN), timeslots, channels, and routing path can be determined by one single algorithm.

Due to optimization across layers, it can be expected that better performance improvement can be achieved by the tightly coupled cross-layer design than the loosely coupled scheme. However, the advantage of the loosely coupled design is that it does not totally discard the transparency between protocol layers. An extreme case of tightly coupled cross-layer design is to merge different protocol layers into one layer. Merging multiple protocol layers into one layer keeps the advantage of tightly coupled cross-layer design. Furthermore, it can also eliminate the overhead in cross-layer information passing and is a trend in WMNs.

This chapter discusses about the cross-layer framework for the transmission power control problem and a possible solution in the context of wireless ad hoc networks. The proposed cross-layer design architecture takes the advantage of loosely coupled design paradigm where MAC and routing protocols exchanges information in packet level.

The remainder of this chapter is organized as follows: Section 2 describes the background knowledge required to understand the transmission power control problem. Section 3 explains related research work on TPC. Section 4 describes the design and analysis of a cross-layer framework proposed in this chapter for controlling transmission power in wireless ad hoc networks. Simulation results are analyzed in section 5, and finally section 6 concludes the chapter with future work to extend the proposed cross-layer framework.

BACKGROUND KNOWLEDGE

This section provides basic background knowledge that is needed to discuss the case study of TPC.

Radio Propagation Models

To understand how to control transmit power, we must first understand the basic knowledge about radio propagation. Radio propagation models can be broadly classified into two categories: large-scale propagation (or fading) models and small-scale fading models. Large-scale models characterize how the transmission power between two nodes changes over long distances and over a long time. Small-scale models account for the fact that small movements (in the order of the wavelength) may have large influence on the transmission quality. In this section, we provide a brief review for the propagation models that have been used in our studies.

In large-scale path loss models, the average received power level, P_r, is a function of the Transmitter-Receiver (T-R) distance, d. That is:

$$P_r = \beta \frac{P_t}{d^\alpha}$$

Where P_t is the transmit power, α is the path loss exponent, and β is a constant.

The free space propagation model and the two-ray propagation models are commonly used in wireless communication analysis. The free space propagation model is used when the transmitter and receiver have a clear line-of-sight path and there is no ground reflection between them. The path loss exponent α is 2, and the received power level is given by the Friis free space propagation model (Rappaport, 2002):

$$P_r = \frac{P_t G_t G_r \lambda^2}{(4\pi)^2 d^2 L} \tag{1}$$

Where G_t and G_r are the antenna gains of transmitter and receiver respectively, λ is the wavelength in meters, and L is the system loss factor not related to propagation ($L \geq 1$).

The two-ray propagation model considers both the direct path and a ground reflected propagation path between the transmitter and receiver (Rappaport, 2002). The path loss exponent α is 4, and the received power level is:

$$P_r = \frac{P_t G_t G_r h_t^2 h_r^2}{d^4} \tag{2}$$

Where h_t is the height of transmitter antenna and h_r is the height of receiver antenna. This model is an approximation that holds for long transmitter-receiver distance.

There are other large-scale propagation models corresponding to different environments and thus different path loss exponents. The case study in this chapter uses (1) to calculate data transmit power, P_t. Since our cross-layer framework is based on the distance between transmitter and receiver, we considered large-scale path loss model in our studies and ignored the small-scale fading model as it accounts for the shorter distances between sender and receiver.

Three Range Model

In order to facilitate theoretical analysis, a three-range model (Xu, Gerla, & Bae, 2002; Zhou & Nettles, 2005) is used to describe Carriers Sense Multiple Access (CSMA) based wireless transmissions. In this model, three thresholds are important: the carrier sensing threshold, T_{cs}, the receiving threshold, T_{rx}, and SINR threshold, $SINR_0$. All of these are characteristics of the wireless nodes physical layer implementation.

There are three ranges which are closely related to these thresholds. When a transmitter X transmits to a receiver Y, these three ranges are crucial: the Carrier Sense Range (CSR) around

X, the Transmit Range (TR) around X, and the Interference Range (IR) around Y. Each of these ranges is explained below.

Carrier Sense Range

The carrier sense range of the transmitter X, CSR_x, is the area around X, within which a node can detect the transmitted signal. That is, the signal power level P_r at the node is greater than T_{cs}. In a CSMA MAC, all nodes within the CSR of an ongoing transmission defer their own transmissions until the channel is idle. According to this definition and the large scale propagation models discussed earlier (Rappaport, 2002), the radius of the CSR is:

$$R_{cs} \propto \sqrt[\alpha]{\frac{P_t}{T_{cs}}}$$

Transmit Range

The transmit range of the transmitter X, TR_x, is the area around X, within which a receiver can detect and correctly decode the transmitted signal with high probability, assuming no interference. That is, the signal power level P_r at the receiver is greater than T_{rx}. According to this definition and the large scale propagation models (Rappaport, 2002), the radius of the TR is:

$$R_{tr} \propto \sqrt[\alpha]{\frac{P_t}{T_{rx}}}$$

Further, let γ denote the ratio of $\frac{R_{cs}}{R_{tr}}$, we have:

$$\gamma = \sqrt[\alpha]{\frac{T_{rx}}{T_{cs}}}$$

Because T_{rx} and T_{cs} are characteristics of the wireless nodes physical layer implementation, γ can be considered as a constant.

Interference Range

The interference range of the receiver Y, IR_y, is the area around Y, within which another node's simultaneous transmission would corrupt the received signal at Y. That is, within this region another transmission would cause the SINR at Y to fall below $SINR_0$. If all nodes transmit with the same power level, the radius of the IR is:

$$R_{ir} = d\sqrt[\alpha]{SINR_0}$$

Where d is the T-R distance (Xu, Gerla, & Bae, 2002). Depending on d, the IR may be wholly or partly covered by the CSR. Notice that it does not consider aggregate interference from multiple transmitters.

Signal Reception

It is also important to understand what happens when a radio attempts to receive a packet. In reality, a receiver simply attempts to decode the detected signals. As long as the checksum is valid, the signal is said to be correctly decoded. Signals are correctly decoded as long as the following two conditions are satisfied. First, the received power level must be greater than a certain threshold, T_{rx}. Second, the ratio of the received power level and the level of interference and noise, i.e., the Signal to Interference and Noise Ratio (SINR), at the receiver must be greater than a certain threshold, $SINR_0$. This is referred to as the capture effect. A collision occurs when another transmission drops the SINR below this threshold.

Carrier Sense Multiple Access

Our case study is based on MAC protocols that use Carrier Sense Multiple Access (CSMA) as part of their basic mechanism. It is a kind of Time Division Multiplexing (TDM). The basic idea is to listen for a carrier before accessing the medium, and then access the medium only if the channel is idle. CSMA is usually combined with a back-off scheme in the case of a busy medium so that some fairness can be achieved among competing nodes.

Based on what to do when the medium is busy, several versions of CSMA exist: non-persistent CSMA, p-persistent CSMA, and 1-persistent CSMA (Leon-Garcia & Widjaja, 2004). Additionally, combined with Collision Detection (CD) and Collision Avoidance (CA), CSMA results in other schemes: CSMA/CD and CSMA/CA (Leon-Garcia & Widjaja, 2004; Peterson & Davie, 2003). Both collision detection and collision avoidance are used to improve the performance of CSMA. Collision detection terminates transmission as soon as a collision is detected, while collision avoidance attempts to avoid collisions by reserving the network for individual transmitters. Collision avoidance is used in media such as radio, where reliable collision detection is not possible. Nowadays, CSMA/CA is widely used in WLANs.

In theoretical analysis, we use a model that accounts for when a carrier is likely to be detected. That is, a carrier can be detected as long as the received power level is greater than the carrier sense threshold, T_{cs}.

Hidden and Exposed Nodes

Hidden and exposed node problems are two essential challenges in wireless MAC protocol design.

The Hidden Node Problem

In general, a hidden node is one that the MAC allows to transmit despite the fact that such a transmission may cause a collision. For CSMA,

Figure 2. The hidden node problem

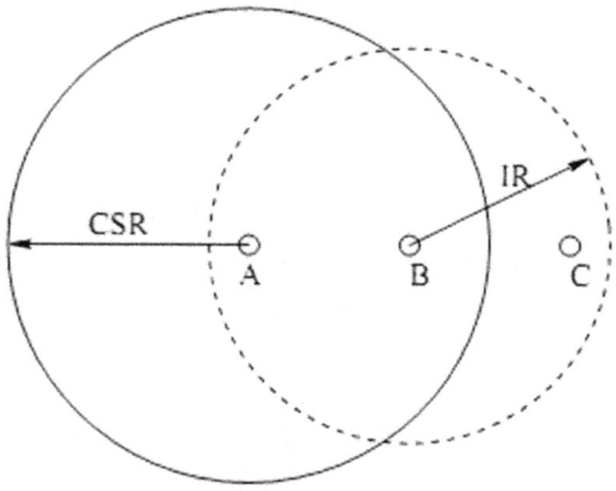

hidden nodes are the ones that are out of the CSR of a transmitter but within the IR of the intended receiver. Consider the case shown in Figure 2.

A is transmitting to B. Node C, which is far away from A but close to B, can not sense this transmission because it is out of CSR_A. Thus, C determines the channel to be idle, and might commence concurrent transmission corrupting the received signal at B. In this scenario, C is a hidden node. From the perspective of the three-range model, the receiver's IR determines the potential hidden nodes.

The Exposed Node Problem

In general, an exposed node is one which the MAC causes to be silent despite the fact that its transmission would not cause collision. For CSMA, exposed nodes are ones that are within the CSR of a transmitter but out of the IR of the intended receiver. Consider the case shown in Figure 3.

A is transmitting to B. Node C, which is far away from B but close to A, can sense this transmission because it is within CSR_A. Thus, C determines the channel to be busy and keeps silent, even though its concurrent transmission would not corrupt the received signal at B. In this scenario, C is an exposed node. From the perspective of the three-range model, the transmitter's CSR determines the potential exposed nodes.

The IEEE 802.11 MAC Protocol

Currently, the IEEE 802.11 MAC protocol (IEEE 802.11 standard, 1999) is the most widely deployed wireless LAN MAC protocol. It combines CSMA with a collision avoidance scheme based on Request To Send (RTS)/Clear To Send (CTS) exchange that attempts to solve the hidden and exposed node problems. Even though these problems are not completely eliminated and some other deficiencies exist, they have not prevented the 802.11 MAC from being extremely successful.

Figure 3. The exposed node problem

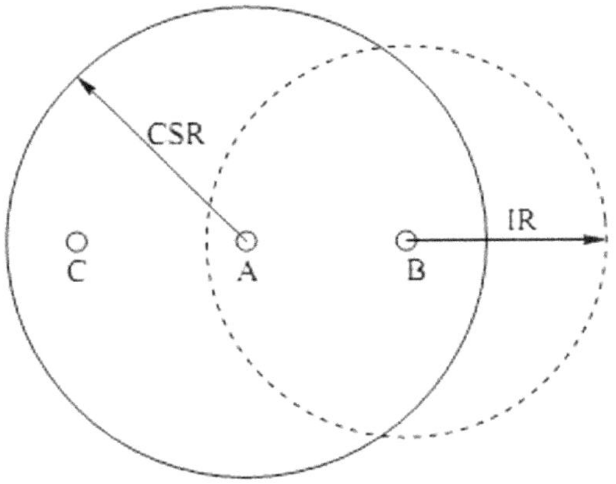

The 802.11 MAC works as follows. All transmissions use CSMA and only proceed when the channel is sensed to be idle. For broadcasts, the transmitter simply sends out broadcast messages. Nodes that receive this message do not reply with acknowledgements. For unicasts, reliable transmission is guaranteed by a four-way handshake between the transmitter and the desired receiver. Figure 4 illustrates this procedure.

Once transmitter A senses the channel to be idle, it sends an RTS to receiver B which, in turn, replies with a CTS. Meanwhile, the nodes in the vicinity that receive the RTS or CTS defer their own transmissions for a period that is long enough for the subsequent DATA/ACK exchange. When the RTS/CTS exchange completes, A sends data packet and B replies with an ACK if the data packet is correctly decoded. If A does not receive a corresponding CTS or ACK within a predefined time interval, it initiates retransmissions.

Two things determine which nodes remain silent in IEEE 802.11. First, CSMA, in which nodes are silent if the strength of some other transmissions, exceeds the carrier sense threshold. Second, nodes are silent if they have decoded either a RTS or CTS, each of which contain in-

Figure 4. A four-way handshake between the transmitter and the receiver

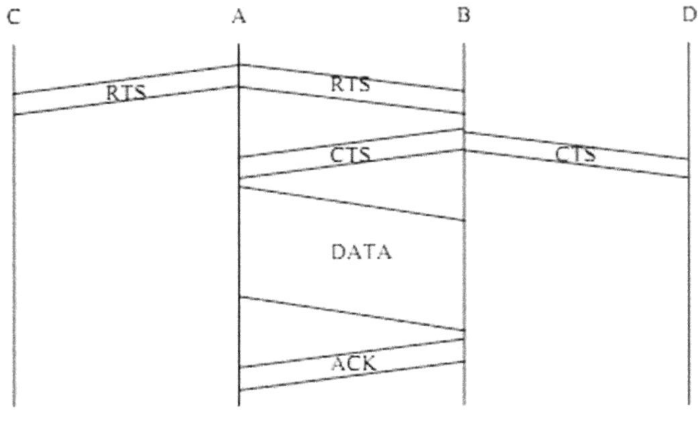

formation about the duration of another nodes' transmission. This duration information is used to create a "virtual" carrier.

802.11 MAC uses an exponential back-off scheme to avoid and resolve collisions. A node maintains a contention window size, w. After every transmission, it has to wait for m time slots before sensing carrier. Here, m is a number randomly chosen from 0 to w. w is doubled every time when the node sends an RTS but no CTS is received or when the node sends a data packet but no ACK is received. This exponential backoff is repeated until a retry limitation is reached. In this way, collisions could be reduced and every node has a fair chance of accessing the channel.

Routing Protocols for Multihop Wireless Networks

At the network level, routing protocols play a key role. Based on when and how routes are discovered and maintained (Royer & Keong Toh, 1999), routing protocols in multihop wireless networks can be divided into two categories: table-driven (proactive) and on-demand (reactive) routing. It is generally believed that on-demand routing is more appropriate for multihop wireless networks (Royer & Keong Toh, 1999).

Table-Driven Routing Protocols

For table-driven routing protocols, a node attempts to maintain one or more routing tables that contain up-to-date routing information to every other node in the network. The typical examples are Destination-Sequenced Distance-Vector Routing (DSDV) (Perkins & Bhagwat, 1994), Clusterhead Gateway Switch Routing (CGSR) (Chiang, Wu, Liu, & Gerla, 1997) and Wireless Routing Protocol (WRP) (Murthy & Garcia-Luna-Aceves, 1996).

To discover routes, each node periodically broadcasts to its neighbors its view of the distance to all the other nodes (distance vector) or to the whole network the status of its links (link status). Based on the advertised information, each node computes the path to all the others in the network. To maintain routes, each node periodically broadcasts routing updates to cope with network topology changes.

On-Demand Routing Protocols

For on-demand routing protocols, routes are created only when needed by the source node. Typical examples are Dynamic Source Routing (DSR) (Johnson & Maltz, 1996), Ad Hoc On-Demand Distance Vector Routing (AODV) (Perkins & Royer, 1999) and Temporally Ordered Routing Algorithm (TORA) (Park & Corson, 1997).

Most of the on-demand routing protocols employ a blind flooding technique to discover routes, which works as follows. The source broadcasts a Route Request (RREQ) packet which contains the source address, the destination address, and a unique identification number. The RREQ is flooded throughout the whole network until it reaches the destination or an intermediate node that has the routing information to the destination. During this process, a reverse path from the destination to the source is established. Then, the destination, or the intermediate node, generates a Route Reply (RREP) packet and sends it back to the source. To return the RREP, the responding node has three choices. First, if it has a route to the source, it can use that routing information. Second, if symmetric links are supported, the responding node can use the reverse route discovered by the RREQ. Third, the responding node can initiate its own route discovery and piggyback the RREP.

Dynamic Source Routing (DSR)

Because our proposed work is based on DSR (Johnson & Maltz, 1996), we describe it in brief. DSR is chosen because it is an on-demand routing protocol which is more appropriate for multihop wireless networks than table-driven routing protocols.

As with most on-demand routing protocols, DSR employs the blind flooding technique to discover routes. In DSR, the RREQ contains a route record field. Every time before a node forwards the RREQ, it adds its address into the record. Thus, the route record contains the sequence of hops the RREQ traversed. When the destination receives the RREQ, it copies the record into a RREP packet and replies with the RREP to the source. When the source sends a data packet, it copies the route record into the data packet header and the data packet is forwarded along the hops listed in the record.

RELATED WORK

This section is a survey on transmission power control protocols proposed so far in the literature for wireless ad hoc networks.

Intensive research has been conducted to improve the network performance in terms of network throughput, delay, and energy efficiency with transmission power control. Many protocols are proposed which try to integrate power control into the IEEE 802.11 MAC protocol. The main idea is to use different power levels to transmit Request To Send (RTS), Clear To Send (CTS), DATA, and ACK frames. In BASIC power control protocols (Agarwal, Krishnamurthy, Katz, & Dao, 2001; Gomez, Campbell, Naghshineh, & Bisdikian, 2001; Karn, 1990; Pursley, Russell, & Wysocarski, 2000) the stations use the maximum power to transmit RTS and CTS frames, and the minimum required power to transmit DATA and ACK frames. The RTS–CTS handshake is used to decide the minimum required power for the subsequent DATA–ACK transmissions. Ref. (Jung & Vaidya, 2002) proposes a new power control MAC protocol to prevent the potential collision problem in the BASIC protocol. The main difference is that the power level for transmitting DATA frames is periodically increased to the power level p_{max} from the power level $p_{desired}$ with a period of one EIFS.

Ref. (Wu, Tseng, & Sheu, 2000) also proposes a MAC protocol allowing nodes to use different transmission power levels to increase channel utilization. Busy Tones, are introduced to prevent other mobile hosts, unaware of the earlier RTS/CTS dialogues, from destroying the ongoing transmission. The channel is split into two subchannels: a data channel and a control channel. The control channel is for transmitting RTS/CTS dialogues. Also, two narrow-band busy tones, called transmit busy tone BT_t and receive busy tone BT_r, are placed on the spectrum at different frequencies with enough separation. Some other early works on power controlled MAC protocols are PCM (Jung & Vaidya, 2002; Jung & Vaidya, 2005), PCMA (Monks, Bhargavan, & Hwu, 2001), and so on. The POWMAC protocol (Muqattash & Krunz, 2005) is a comprehensive, throughput-oriented, and single-channel MAC protocol for wireless ad hoc networks. Instead of exchanging only one pair of control packets (RTS/CTS) during one four-way handshake, POWMAC uses an Access Window (AW) to allow a series of RTS/CTS exchanges to occur before the multiple, concurrent data packet transmissions. The length of an AW is dynamically adjusted based on local traffic load information such that the maximum number of concurrent transmissions in the vicinity can be achieved. Collision avoidance information is inserted into the CTS packet and is used to bound the transmit powers of potential interfering nodes, rather than to silence these nodes.

ElBatt and Ephremides (2004) propose a joint scheduling and power control algorithm to maximize the MAC layer throughput while conserving energy. The joint design is formulated as a constrained optimization problem, and the objective function is to minimize the total power of the scheduled links, and the performance metric concerned is the throughput of neighboring transmissions in the network. Kozat, Koutsopoulos, and Tassiulas (2004) formulate the joint problem of power control and scheduling similar to that of ElBatt and Ephremides (2004). The authors also try to minimize the total transmit power in each timeslot while guaranteeing the end-to-end quality of service for sessions in terms of bandwidth and bit error rate probabilities. They prove that the problem of finding feasible power assignment and schedule under given source data rate constraint and SINR requirements is NP-completeness, hence focus should be put on finding the near-optimal and efficient algorithms to perform the joint scheduling and power allocation.

Behzad, Rubin, and Mojibi-Yazdi (2003, 2004) propose a cross-layer design of power control and scheduling based on a Power-Based Interference Graph. Fang, and Rao (2003) formulate the resource allocation (power) problem of maximizing the sum of transmitter utilities subject to a minimum and maximum data rate constraint per link and peak power level constraint per node in a wireless multihop network. Neely, Modiano, and Rohrs (2005) considered dynamic routing and power allocation for a wireless network with time-varying channels. Authors proposed Dynamic Routing and Power Control (DRPC) algorithm to achieve the capacity region and offer delay guarantees with consideration of the full effects of queueing. Savas, Alanyali, and Yener (2004) considered the problem of joint routing and power control as a constrained optimization problem.

The COMPOW power control scheme (Narayanaswamy, Kawadia, Sreenivas, & Kumar (2002) is the simplest power control at the network layer. In COMPOW, each node chooses a common power level. The power level is chosen such that the network remains connected. This can be achieved by running multiple independent routing protocols, one for each feasible power level, and a COMPOW agent figures out the lowest possible power level that ensures the connectivity of the network. In the CLUSTERPOW scheme (Kawadia & Kumar, 2003), each node runs a routing protocol at each feasible power level pi, and it maintains a routing table RT_{pi} for each power level. In order to route a packet to destination d, a node looks up the routing table for d in increasing order of the corresponding transmit power level. So the network layer makes sure that a packet is transmitted with the lowest possible power at each hop. The LOADPOW scheme (Kawadia & Kumar, 2005) is similar to the COMPOW and CLUSTERPOW schemes, and it tries to adapt the transmit power according to the network load. It opportunistically uses a higher transmit power level whenever the network load is low, and vice versa.

Chiang (2004, 2005) proposes a Jointly Optimal Congestion Control and Power Control (JOCP) algorithm in wireless multi hop networks. CDMA-based medium access and fixed single-path routing are assumed in the algorithm, thus avoiding the problem of contention resolution in the MAC layer and routing in the network layer. The congestion control algorithm is based on TCP Vegas (Brakmo & Peterson, 1995), where the source adjusts the window size according to the delay between the source and destination. Author proposed joint algorithm which considers the interaction between power control and congestion control.

All the protocols proposed above are based on single layer approach i.e., they are implemented at a single layer (MAC, NETWORK, TRANSPORT) in the wireless protocol stack. They apply power tuning mechanism at a particular single layer only. This chapter proposes a new method, Dual Protocol for Power Conservation (DPPC), of applying two power controlled protocols, CLP-CMA (Kasiviswanath, Charles, & Reddy, 2007), and PDSR (Kasiviswanath, Charles, & Reddy, 2008), at two different layers (MAC, NETWORK) simultaneously in the network.

DESIGN AND ANALYSIS OF A CROSS-LAYER TRANSMISSION POWER CONTROL PROTOCOL

Our proposed case study applies at the same time two protocols of transmission power control at two different layers: mac layer and network layer.

Transmission Power Control at MAC Layer

Cross-Layer Power Controlled Multiple Access (CLPCMA) (Kasiviswanath, Charles, & Reddy, 2007) protocol is implemented at MAC layer. It is designed and developed based on the principles of CSMA and MACA protocols of IEEE 802.11. This is a distributed power controlled multiple access protocol for the ad hoc networks. This protocol will transfer the control packets prior to data transmission from one node (source) to another node (destination) in the network and follows CSMA/CA paradigm. The main objective of this protocol is to achieve minimum transmission power for a node(transmitter), that result in a successful packet reception at the receiver and to increase spectral reuse. This protocol improves the network spectral reuse between source and destination pairs.

CLPCMA Operation

Implementation of CLPCMA can be done first, by sending the control packets (RPTS/APTS) before data packets. Here the control packets are RPTS (Requested Power To Send) and APTS (Accepted Power To Send) which differ in size compared to RTS and CTS packets used in IEEE 802.11 MAC protocol standard, shown in Figures 5, 6, and 7. RTS packet header is modified to include power level and used as RPTS in CLPCMA. Similarly, CTS packet header is also modified and used as APTS. When a sending node is broadcasting RPTS packet, it includes its transmission power. On receiving the RPTS packet, the intended receiving node measures the signal power and calculates excess power transmitted using its signal to noise ratio threshold and the received signal power. Then while responding with the APTS packet, it piggybacks the excess power value. That helps the sender on receiving the APTS packet to tune its transmit power for data packet transmission to the required reduced level. RPTS and APTS packets are transmitted using maximum transmit power available at that time. Transmit power is reduced for the transmission of data packets only. CLPCMA operation is shown in Figure 8.

Assumptions and Dependencies

- Considered a wireless ad-hoc network consisting of 'n' nodes. There is no wireline infrastructure to connect the nodes, i.e, they can communicate only via the wireless medium.
- Each node is supported by an Omni-directional antenna.

Figure 5. General MAC frame header for RTS/CTS

Source Address	Destination Address	Payload size	Frame Type	Priority

Figure 6. CLPCMA MAC frame header for RPTS

Source Address	Destination Address	Payload size	Frame Type	Priority	Source power Tx

- Each node knows the geographical location of all neighboring nodes located one hop distance away from that node.
- The channel gains are same in both the directions.
- Each node generates information packets of fixed length, destined to all other nodes.
- Assumed that each generated packet is intended for a single neighbor only.
- Assumed a maximum power level that a node can use for transmission, path loss coefficient, and antenna gains of both source and destination.

CLPCMA Algorithm

The steps involved in the proposed algorithm to perform transmit power control for data packets are listed below:

Power level P_R of received RPTS control signal is measured

$$P_R = P_T * (\lambda/4*\Pi*d)^{n} *G_T *G_R \quad (3)$$

Where, λ is the wavelength of the carrier, d is the distance between the transmitter and the receiver, n is the path loss coefficient and G_T and

Figure 7. CLPCMA MAC frame header for APTS

Source Address	Destination Address	Payload size	Frame Type	Priority	Source power Tx

Figure 8. Flow diagram for power tuning at MAC Layer

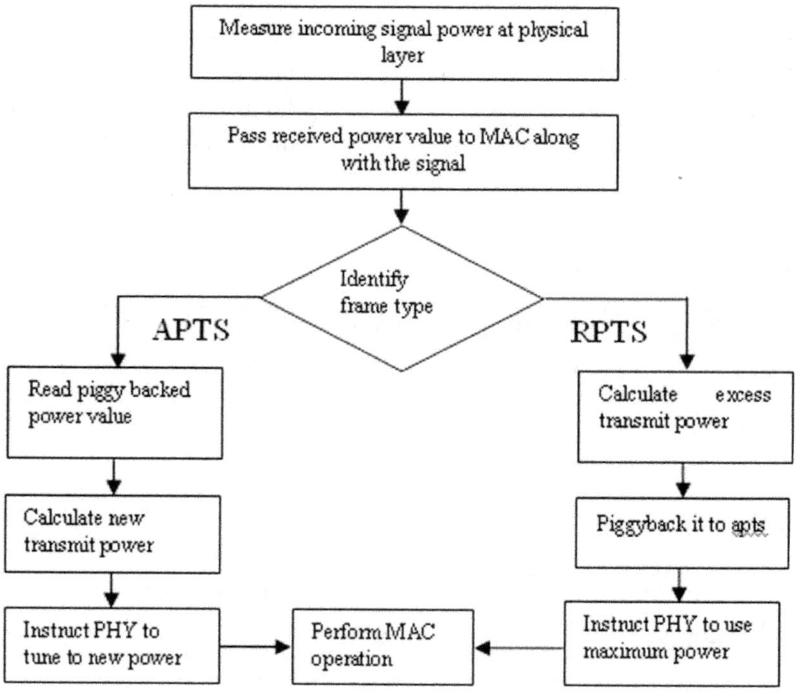

G_R are the transmitting and receiving antenna gains. Normally the λ, G_T and G_R are constants. The path loss coefficient 'n' may vary between 2 and 6, which depends on the environment. As the data packet transmission duration very short, n and d can also be considered as constants.

Excess received power P_{ER} is calculated as,

$$P_{ER} = P_R - (SNR_{th} * P_{noise}) \quad (4)$$

Where, SNR_{th} is the receiver SNR threshold and P_{noise} is the accumulated interference/noise power

Excess transmitted power P_{ET}, can be calculated from the general relationship between the received and transmitted powers as given below:

$$P_{ET} = P_{ER} * (P_T/P_R) \quad (5)$$

Where, P_T is the transmitted power value piggybacked in RPTS packet, P_R is the measured power level of the received RPTS packet calculated using equation(1) and P_{ER} is the excess received power value calculated using equation(4).

The power reduced for data packet transmissions is calculated by the sender when it receives the APTS packet piggybacked with the excess transmitted power using the following equation:

$$P_{reduced} = P_T - P_{ET} \quad (6)$$

As the power calculated is highly influenced by the distance between the sending and the receiving nodes, the performance of this transmit power control algorithm is studied by varying the speed of the nodes, varying the number of packets and varying the number of nodes and the results are presented in the next Section.

Transmission Power Control at Routing Layer

CLPCMA is a local level power conservation protocol for only data packets. Control packets RPTS and APTS are transmitted between sender and receiver with the maximum power level. To further achieve power conservation, control packets must also be transmitted with reduced power level. Power Dynamic Source Routing (PDSR) (Kasiviswanath, Charles, & Reddy, 2008) is the protocol designed for this purpose. It is a global level power conservation protocol which provides cross-layer interaction among Network, MAC and Physical layers. In the cross-layer design of PDSR, the received signal strength is measured at physical layer and made available to the access of top layers as shown in Figure 9. In our proposed cross-layer design implementations, we have modified the procedures at physical layer to pass the measured value of received signal strength to the MAC layer along the signal. This value is used in MAC layer computations if required or passed to routing layer along the routing control packets, where it is stored in routing/neighbor tables and used in some decision-making processes. Hence, the received signal strength concerned to physical layer is passed to the top layers as interlayer interaction parameter and used to improve the performance of wireless ad hoc networks by modifying medium access and routing protocols as per the required cross-layer design.

PDSR Operation

Dynamic Source Routing (DSR) (Johnson & Maltz, 1996) is a reactive protocol that creates

Figure 9. Cross-layer design of PDSR

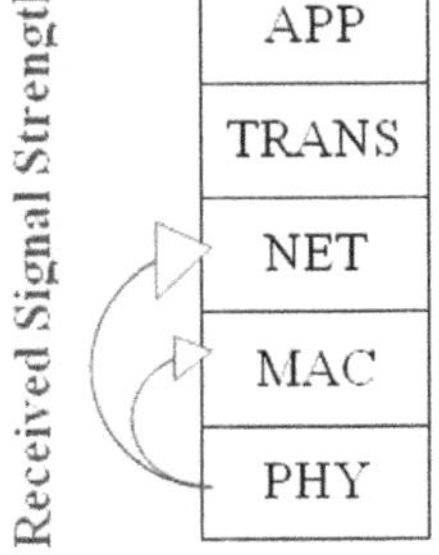

a route to the destination only when the source node has data to transmit by initiating a route discovery mechanism and maintains the route until it is required by the source. Since our cross-layer framework is based on control packets piggybacking information, we chosen DSR protocol in our studies, in fact any other reactive routing protocol using control packets is applicable like Ad Hoc On-demand Distance Vector (AODV) (Perkins, Royer, & Chakeres, 2003). PDSR is designed by modifying the DSR protocol. The changes are needed in the route discovery phase of DSR algorithm. The RREP packet is sent to the particular source with the transmission power in its header along with the route, shown in Figure 10. Upon receiving the RREP packet, the source node measures the received power of the RREP packet and collects the transmitted power of the RREP packet that is piggybacked in the same packet. Then the source node calculates the path loss of the RREP packet and calculates its required minimum transmission power using the receiver threshold. The receiver threshold along with the path loss gives the optimum power required for transmission at MAC layer. Controlling the transmit power at a given node increases not only the operating life of the battery but also the overall system capacity. The RREQ and RREP packets are transmitted with maximum (fixed) transmission power to main connectivity and bi-directionality of links. Where as all the MAC transmissions, the RTS-CTS-DATA-ACK sequences use the reduced minimum transmit power levels. The advantage of implementing the PDSR protocol in an ad hoc network is that the source-destination pairs can be more tightly packed into the network allowing a greater number of simultaneous transmissions (spectral reuse).

PDSR Algorithm

1. The RREP packet is sent to the particular source with the transmission power (P_T) in its header along with the route.
2. Upon receiving the RREP packet:
   a. If the node is intended source
      i. Received power (P_R) of the RREP packet is measured. $P_R = P_T * (\lambda/4*\Pi*d)^n * G_T * G_R$
      ii. The transmitted power of the RREP packet that is piggybacked (P_T) in the same packet is collected.
      iii. Then the node calculates the path loss of the packet using Path loss $= P_T - P_R$
      iv. Then the node calculates its required minimum transmission power using the receiver threshold. The receiver threshold along with the path loss gives the optimum power required for transmission at MAC layer. Required Min T_x Power = Path loss + Receiver Threshold
      v. This Min T_x Power calculated at routing layer is directly passed to MAC layer where the node can transmit control packets (RTS/CTS) and data packets to the destination with this reduced T_x Power.
   b. If the node is intermediate node

Figure 10. Changes are made in the header format of route reply packet

Source Address	Destination Address	Route	Power Level

i. Performs all the calculations same as the source node and relays this RREP packet to the intended source through some intermediate nodes.
ii. As the RREP is routed back along the reverse path, the nodes along the path will only setup forward route entries in their route table but also forced to enter the required power for transmission in the route table.

Dual Protocol for Power Conservation (DPPC)

CLPCMA saves the transmission power in the delivery of DATA and ACK packets from sender to receiver and uses maximum power to transmit RPTS and APTS control packets. PDSR achieves considerable amount of savings over CLPCMA by setting minimum required power level for RPTS and APTS packets along with DATA and ACK. PDSR determines this power level during router discovery process. Once the route is established between source and destination, MACA protocol uses this minimized power level for all four types of packets: RTS, CTS, DATA and ACK. Simulation results (Kasiviswanath, Charles, & Reddy, 2007; Kasiviswanath, Charles, & Reddy, 2008) show that CLPCMA and PDSR have significant performance in terms of transmission power conservation, network throughput, end to end delay and also they are scalable. These two protocols are best suitable for low mobility wireless ad hoc networks. This chapter studies the effect of applying the combination of CLPCMA and PDSR, under DPPC. So far in the literature, only one power tuning protocol is applied at a single layer in the protocol stack. The main idea behind DPPC is to study what happens in the wireless ad hoc network when two power controlled protocols are applied at two different layers at the same time. While simulating DPPC with Glomosim (UCLA, 2011; Bajaj, Takai, Ahuja, & Bagrodia, 1999), in the simulation setup, PDSR is selected as the routing protocol at network layer and CLPCMA as MAC protocol in the wireless network protocol stack. Simulation parameters set up for CLPCMA and PDSR can be found in Kasiviswanath, Charles, and Reddy (2007, 2008).

DPPC Operation

The main objective of DPPC protocol is to study what happens in a wireless ad hoc network when two transmission power control protocols developed at two different layers are applied simultaneously in the protocol stack. In DPPC, PDSR at the routing layer, sets the minimum required transmission power to be used by CLPCMA at the below layer (MAC) for the packets RPTS, APTS, DATA, and ACK. In DPPC, we modified the operation of CLPCMA so that a node sending RPTS refers to routing table to get the minimum required transmission power calculated by PDSR as shown in Figure 11a. CLPCMA calculates again for minimum required power for DATA and ACK packets when a node moves away from the transmission range determined by PDSR but the node present in the range reachable with maximum transmission power as shown in Figure 11b. In case of PDSR (Kasiviswanath, Charles, & Reddy, 2008), this situation is reported as link error by the MAC layer protocol to the routing layer protocol, PDSR which results in alternate route selection if any, or route discovery process initiation. But in DPPC, we modified CLPCMA to use maximum transmit power to send RPTS and APTS to maintain connectivity and computing required minimum transmit power for DATA and ACK packets shown in Figure 11. From the simulation results it is observed that DPPC can control link failures and performs similar to PDSR.

SIMULATION AND RESULTS

CLPCMA and PDSR are simulated with Glomosim which is a scalable simulation environment

Figure 11. (a) PDSR power calculation, (b) CLPCMA power calculation

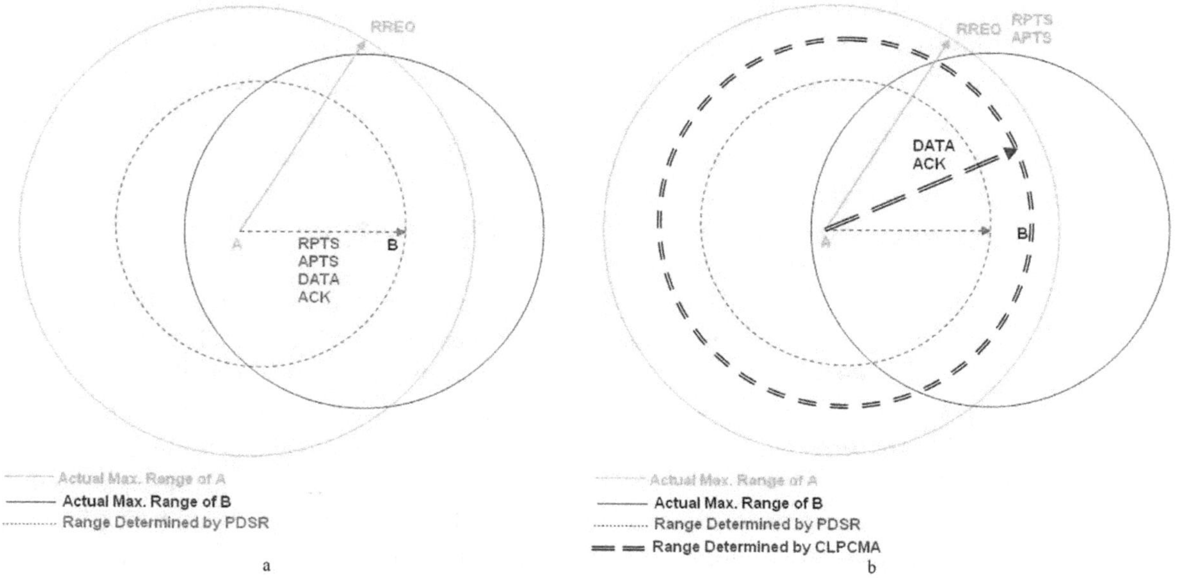

Figure 12. Simulation parameters

PARAMETER NAME	PARAMETER VALUE
SIMULATION-TIME	15M
SEED	1
NUMBER-OF-NODES	36
MOBILITY-POSITION-GRANULARITY	0.5
NODE-PLACEMENT	RANDOM
TERRAIN-DIMENSIONS	(500, 500)
PROPAGATION-LIMIT	-111.0
PROPAGATION-PATHLOSS	FREE SPACE
NOISE-FIGURE	10.0
RADIO-TYPE	RADIO-ACCNOISE
RADIO-FREQUENCY	2.4e9
RADIO-BANDWIDTH	2000000
RADIO-RX-TYPE	SNR-BOUNDED
RADIO-RX-SNR-THRESHOLD	10.0
RADIO-TX-POWER	15.0
RADIO-ANTENNA-GAIN	0.0
RADIO-RX-SENSITIVITY	-94.0
RADIO-RX-THRESHOLD	-81.0
MAC-PROTOCOL	CLPCMA
PROMISCUOUS-MODE	NO
ROUTING-PROTOCOL	PDSR
NETWORK-PROTOCOL	IP
NETWORK-OUTPUT-QUEUE-SIZE-PER-PRIORITY	100

for wireless network systems. It is designed using the parallel discrete-event simulation capability provided by Parsec (Meyer & Bagrodia, 1999). CLPCMA simulation analysis and results can be found in Kasiviswanath, Charles, & Reddy, 2007). Similarly, PDSR simulation results can be found in Kasiviswanath, Charles, & Reddy, 2008). This section presents simulation setup, results, and analysis of DPPC.

DPPC Simulation Setup

DPPC performance is analyzed against CLPCMA and PDSR with respect to network parameters—transmit power consumption, throughput, delay. Figures 13-17 give the simulated analysis of DPPC.

From the above simulation results, Figures 12, 13, 14, 15, 16, and 17, it is found that by using

Figure 13. Transmission power consumption of CLPCMA, PDSR vs DPPC for a network with different mobility

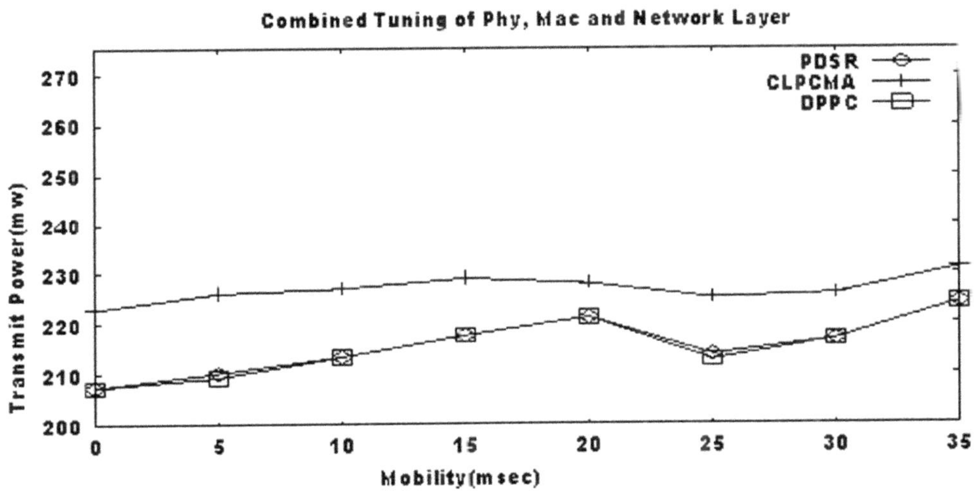

Figure 14. Throughput of CLPCMA, PDSR vs DPPC for a network with different mobility

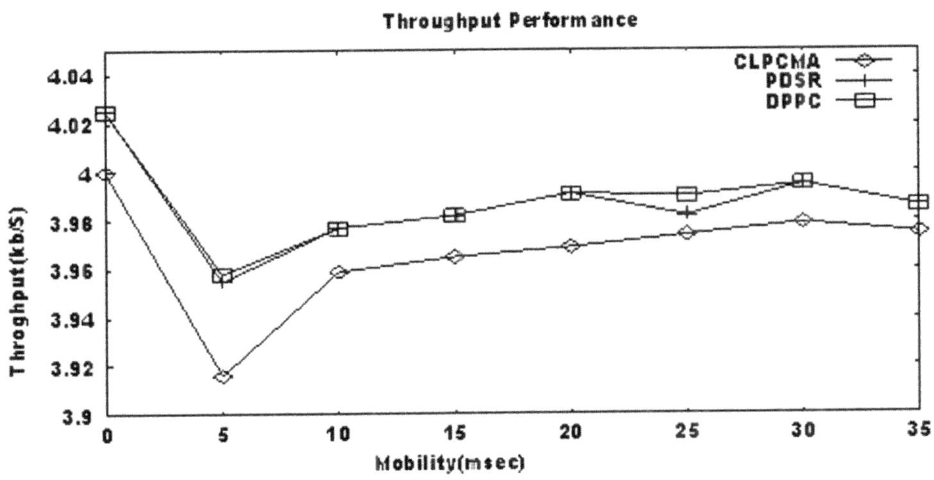

Figure 15. Delay of CLPCMA, PDSR vs DPPC for a network with different mobility

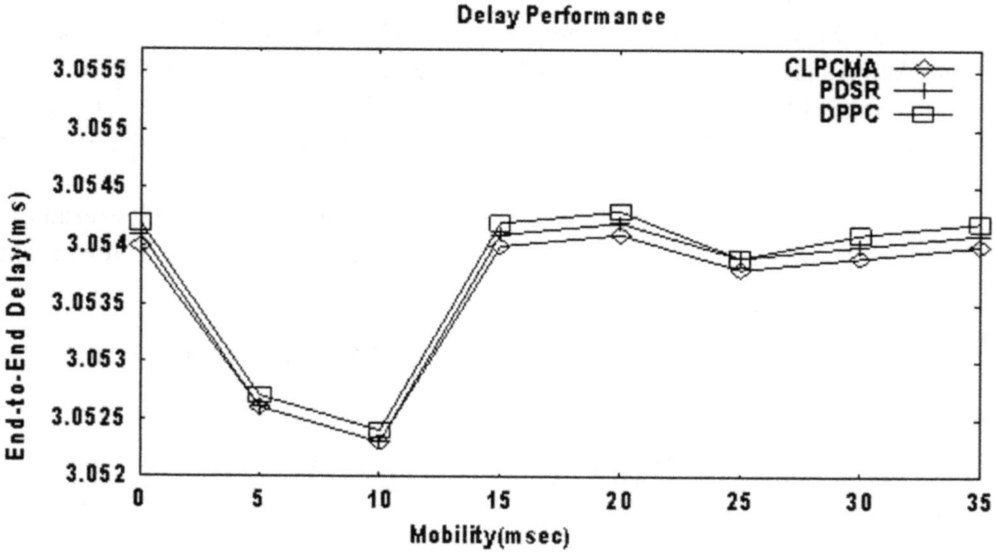

Figure 16. Transmission power consumption of CLPCMA, PDSR vs DPPC for a network with different number of nodes

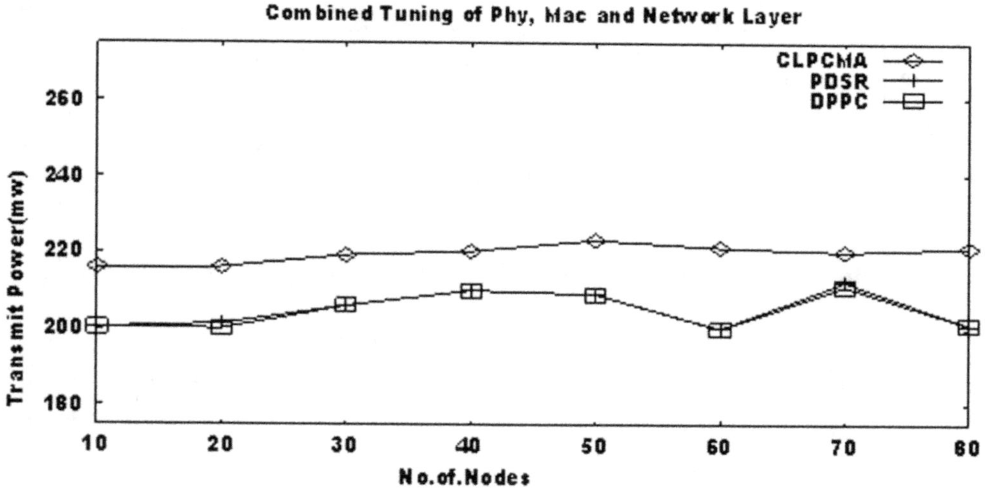

DPPC, the total transmission power consumption of the network is almost similar to PDSR. A little bit increase in network throughput is found when controlling link failures but average end-to-end delay has been increased slightly in this experiment. DPPC is a scalable and distributed protocol.

CONCLUSION AND FUTURE WORK

This chapter discussed about the problem of transmission power control in wireless ad hoc networks. The CLPCMA and PDSR protocols are explained with algorithms. DPPC is an experi-

Figure 17. Transmission power consumption of CLPCMA, PDSR vs DPPC for a network with different number of data packets

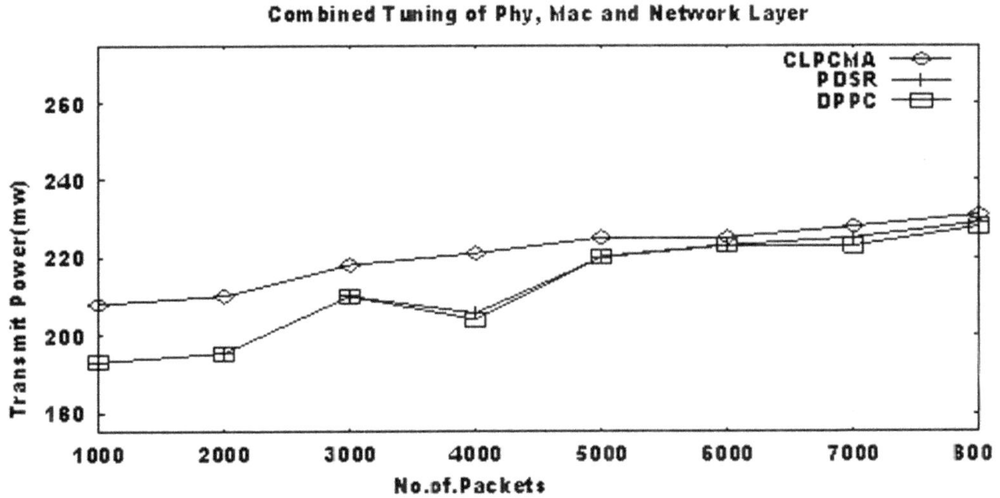

ment by using CLPCMA and PDSR protocols. All these protocols are illustrating the need of cross layer design in future wireless ad hoc networks to achieve better network resource utilization. This work can be extended if proper cross layer interaction is provided between transport layer and below layers.

REFERENCES

Agarwal, S., Krishnamurthy, S., Katz, R. H., & Dao, S. K. (2001). Distributed power control in ad-hoc wireless networks. In *Proceedings of PIMRC*. PIMRC.

Bajaj, M. G. L., Takai, M., Ahuja, R., & Bagrodia, R. (1999). *GloMoSim: A scalable network simulation environment*. Technical Report 990027. Los Angeles, CA: University of California.

Behzad, A., & Rubin, I. (2004). Multiple access protocol for power-controlled wireless access nets. *IEEE Transactions on Mobile Computing, 3*(4), 307–316. doi:10.1109/TMC.2004.44

Behzad, A., Rubin, I., & Mojibi-Yazdi, A. (2003). Distributed power controlled medium access control for ad-hoc wireless networks. In *Proceedings of IEEE 18th Annual Workshop on Computer Communications*, (pp. 47–53). IEEE Press.

Bhatia, R., & Kodialam, M. (2001). On power efficient communication over multi-hop wireless networks: Joint routing, scheduling, and power control. In *Proceedings of the IEEE Annual Conference INFOCOM*, (pp. 1457–1466). IEEE Press.

Brakmo, L. S., & Peterson, L. L. (1995). End to end congestion avoidance on a global internet. *IEEE Journal on Selected Areas in Communications, 13*(8), 1465–1480. doi:10.1109/49.464716

Chiang, C., Wu, H., Liu, W., & Gerla, M. (1997). Routing in clustered multihop, mobile wireless networks with fading channel. In *Proceedings of the Singapore International Conference on Networks, IEEE SICON 1997*, (pp. 197-211). IEEE Press.

Chiang, M. (2004). To layer or not to layer: Balancing transport and physical layers in wireless multihop networks. In *Proceedings of IEEE INFOCOM* (pp. 2525–2536). IEEE Press.

Chiang, M. (2005). Balancing transport and physical layers in wireless multihop networks: Jointly optimal congestion control and power control. *IEEE Journal on Selected Areas in Communications*, *23*(1), 104–116. doi:10.1109/JSAC.2004.837347

Chiang, M., Low, S. H., Calderbank, A. C., & Doyle, J. C. (2007). Layering as optimization decomposition: A mathematical theory of network architectures. *Proceedings of the IEEE*, *95*(1), 255–312. doi:10.1109/JPROC.2006.887322

ElBatt, T., & Ephremides, A. (2004). Joint scheduling and power control for wireless ad hoc networks. *IEEE Transactions on Wireless Communications*, *3*, 74–85. doi:10.1109/TWC.2003.819032

Fang, J. C., & Rao, R. R. (2003). An integrated and distributed scheduling and power control algorithm for maximizing network utility for wireless multihop networks. In *Proceedings of IEEE MILCOM,* (pp. 1011–1017). IEEE Press.

Gomez, J., Campbell, A. T., Naghshineh, M., & Bisdikian, C. (2001). Conserving transmission power in wireless ad hoc networks. In *Proceedings of IEEE ICNP*. IEEE Press.

IEEE. (1999). *802.11 standard for wireless LAN medium access control and physical layer specifications*. New York, NY: IEEE Computer Society LAN MAN Standards Committee.

Jamieson, B. K., Balakrishnan, H., & Morris, R. (2001). An energy-efficient coordination algorithm for topology maintenance in ad hoc wireless networks. In *Proceedings of MOBICOM*. MOBICOM.

Johnson, D. B., & Maltz, D. A. (1996). Dynamic source routing in ad hoc wireless networks. In *Mobile Computing* (pp. 153–181). Dordrecht, The Netherlands: Kluwer. doi:10.1007/978-0-585-29603-6_5

Jung, E.-S., & Vaidya, N. H. (2002a). An energy efficient mac protocol for wireless LANs. In *Proceedings of INFOCOM*. INFOCOM.

Jung, E.-S., & Vaidya, N. H. (2002b). A power control MAC protocol for ad hoc networks. In *Proceedings of ACM MobiCom,* (pp. 36–47). ACM Press.

Jung, E.-S., & Vaidya, N. H. (2005). A power control MAC protocol for ad hoc networks. *Wireless Networks*, *11*(1–2), 55–66. doi:10.1007/s11276-004-4746-9

Karn, P. (1990). MACA-A new channel access method for packet radio. In *Proceedings of 9th ARRL Computer Networking Conference*. ARRL.

Kasiviswanath, N., Charles, B. S., & Reddy, P. C. (2007). A combined tuning of power control in wireless ad hoc networks using cross layer architecture. *Indian Journal of Computing Technology*, *2*(2), 1–30.

Kasiviswanath, N., Charles, B. S., & Reddy, P. C. (2008). A new power conservative routing for wireless mobile ad hoc networks using cross layer design. *International Journal of Computer Science and Network Security*, *8*(3).

Kawadia, V., & Kumar, P. R. (2003). Power control and clustering in ad hoc networks. In *Proceedings of IEEE INFOCOM* (pp. 459–469). IEEE Press.

Kawadia, V., & Kumar, P. R. (2005). Principles and protocols for power control in wireless ad hoc networks. *IEEE Journal on Selected Areas in Communications*, *23*(1). doi:10.1109/JSAC.2004.837354

Kawadia, V., & Kumar, P. R. (2005). A cautionary perspective on cross-layer design. *IEEE Wireless Communications*, *12*(1), 3–11. doi:10.1109/MWC.2005.1404568

Kozat, U. C., Koutsopoulos, I., & Tassiulas, L. (2004). A framework for cross-layer design of energy-efficient communication with QoS provisioning in multi-hop wireless networks. In *Proceedings of IEEE INFOCOM* (pp. 1446–1456). IEEE Press. doi:10.1109/INFCOM.2004.1357029

Leon-Garcia, A., & Widjaja, I. (2004). *Communication networks: Fundamental concepts and key architectures* (2nd ed.). New Delhi, India: McGraw-Hill.

Meyer, R. A., & Bagrodia, R. (1999). *PARSEC user manual for PARSEC release 1.1*. Retrieved from http://www.parsec.org.

Monks, J. P., Bharghavan, V., & Hwu, W.-M. W. (2001). A power controlled multiple access protocol for wireless packet networks. In *Proceedings of IEEE INFOCOM* (*Vol. 1*, pp. 219–228). INFOCOM. doi:10.1109/INFCOM.2001.916704

Muqattash, A., & Krunz, M. (2005). POWMAC: A single-channel power-control protocol for throughput enhancement in wireless ad hoc networks. *IEEE Journal on Selected Areas in Communications*, *23*, 1067–1084. doi:10.1109/JSAC.2005.845422

Murthy, M., & Garcia-Luna-Aceves, J. J. (1996). An efficient routing protocol for wireless networks. *ACM Mobile Networks and Applications*, *1*(2), 183–197. doi:10.1007/BF01193336

Narayanaswamy, S., Kawadia, V., Sreenivas, R. S., & Kumar, P. R. (2002). Power control in ad hoc networks: Theory, architecture, algorithms and implementation of the COMPO protocol. In *Proceedings of European Wireless Conference*, (pp. 156–162). European Wireless Conference.

Neely, M. J., Modiano, E., & Rohrs, C. E. (2005). Dynamic power allocation and routing for time-varying wireless networks. *IEEE Journal on Selected Areas in Communications*, *23*(1), 89–103. doi:10.1109/JSAC.2004.837349

Nilsson, M. (2001). Investigating the energy consumption of a wireless network interface in an ad hoc networking. In *Proceedings of IEEE INFOCOM*. IEEE Press.

Park, V. D., & Corson, M. S. (1997). A highly adaptive distributed routing algorithm for mobile wireless networks. In *Proceedings of the 16th Annual Joint Conference of the IEEE Computer and Communications Societies, IEEE INFOCOM 1997*, (vol. 3), (pp. 1405-1413). IEEE Press.

Perkins, C., & Bhagwat, P. (1994). Highly dynamic destination-sequenced distance-vector routing (DSDV) for mobile computers. In *Proceedings of the Conference on Communications Architectures, Protocols and Applications, ACM SIG-COMM 1994*, (vol. 24), (pp. 234-244). ACM Press.

Perkins, C. E., & Royer, E. M. (1999). Ad-hoc on-demand distance vector routing. In *Proceedings of the 2nd IEEE Workshop on Mobile Computing Systems and Applications, IEEE WMCSA 1999*, (pp. 90-100). IEEE Press.

Perkins, C. E., Royer, E. M., & Chakeres, I. D. (2003). *Ad hoc on-demand distance vector (AODV) routing*. IETF Draft. Retrieved from http://www.ietf.org.

Peterson, L. L., & Davie, B. S. (2003). *Computer networks: A systems approach* (3rd ed.). San Francisco, CA: Morgan Kaufmann Publishers.

Pursley, M. B., Russell, H. B., & Wysocarski, J. S. (2000). Energy-efficient transmission and routing protocols for wireless multiple-hop networks and spread-spectrum radios. In *Proceedings of EUROCOMM*, (pp. 1–5). EUROCOMM.

Rappaport, T. S. (2002). *Wireless communications principles and practice* (2nd ed.). Upper Saddle River, NJ: Prentice Hall.

Royer, E. M., & Keong Toh, C. (1999). A review of current routing protocols for ad hoc mobile wireless networks. *IEEE Personal Communications*, *6*(2), 46–55. doi:10.1109/98.760423

Savas, O., Alanyali, M., & Yener, B. (2004). Joint route and power assignment in asynchronous multi-hop wireless networks. In *Proceedings MedHocNet.* MedHocNet.

UCLA. (2011). *Glomosim.* Retrieved from http://pcl.cs.ucla.edu.projects/glomosim.

Wu, S.-L., Tseng, Y.-C., & Sheu, J.-P. (2000). Intelligent medium access for mobile ad hoc networks with busy tones and power control. *IEEE Journal on Selected Areas in Communications*, *18*(9), 1647–1657. doi:10.1109/49.872953

Xu, K., Gerla, M., & Bae, S. (2002). How effective is the IEEE 802.11 RTS/CTS handshake in ad hoc networks? In *Proceedings of Global Telecommunications Conference, IEEE GLOBECOM 2002,* (vol. 1), (pp. 72 – 76). IEEE Press.

Zhou, Y., & Nettles, S. M. (2005). Balancing the hidden and exposed node problems with power control in CSMA/CA-based wireless networks. In *Proceedings of Wireless Communication and Networking Conference, IEEE WCNC 2005.* IEEE Press.

KEY TERMS AND DEFINITIONS

Carrier Sense Multiple Access / Collission Avoidance (CSMA/CA): It is the MAC protocol used to avoid collisions while accessing the common wireless medium.

Cross-Layer Design (CLD): A technique to optimize the network efficiency with proper interactions between protocol layers.

Dynamic Source Routing (DSR): It is an on-demand routing protocol initiated by the sender node.

On-Demand Routing Protocol: A routing protocol establishing a path between sender and receiver when it is needed.

SINR: Signal to Noise Ratio, It is a metric used to measure the signal quality.

Transmission Power Control (TPC): A mechanism used to control the transmission power of a sender node.

Transmission Range: An area covered by a sender node within which the successful transmission is possible.

Chapter 11
MAC and Routing Integration in Wireless Sensor Networks

Fabrice Theoleyre
CNRS, France

Bogdan Pavkovic
LIG, France

ABSTRACT

Wireless Sensor Networks are a promising way to interconnect smart objects sensing and acting on the environment, enabling the Internet of Things. However, this kind of network is particularly constrained: nodes have limited energy reserve, their CPU and memory resources are limited, and the radio bandwidth is very low. After exposing the main approaches for the routing and MAC layers, the authors explain why they must be jointly optimized. They explain why a metric reflecting the radio link quality is necessary and how it should integrate different criteria in layers 2 and 3. The authors also introduce a new architecture based on IEEE 802.15.4 and RPL so that both protocols work in symbiosis. Finally, they conclude detailing some open problems in this research area.

INTRODUCTION

Wireless Sensor Networks (WSN) have attracted much interest in the last few years. The IETF Routing Over Low-Power lossy Networks (ROLL) working group has already targeted several key applications:

- Industrial environments to monitor assembly lines or machines;
- Building automation to control the HVAC, lighting to detect fire, etc.
- Healthcare so that the patient can safely stay at home while being connected with the medical staff;
- Urban Sensor Networks to control urban lighting, noise and air pollution, monitor the containers filling.

DOI: 10.4018/978-1-4666-0960-0.ch011

Since the Machine2Machine market is expected to quickly grow in the next years, it is essential to design efficient protocols for this kind of environment. Unfortunately, existing solutions designed for wired networks cannot be applied as is in WSN, since we face to specific constraints that must be tackled in the networking stack:

1. Energy optimization: each node is battery-powered and has to save energy to prolong the network lifetime;
2. Scalability: a WSN may comprise hundreds of nodes. The topology is not hierarchized unlike wired networks;
3. Limited constraints: a node has very limited resources (CPU, memory), and protocols must maintain a very limited number of states;
4. Low-cost: for some applications (e.g. home automation), smart object have to be cheap to be integrated everywhere;
5. Fault-tolerance: a node may fail suddenly (e.g. cheap and faulty hardware, energy depletion). The network must react in an autonomous way;

A Wireless Sensor Network comprises several smart objects (also called sensors and/or actuators or PAN). One sink may collect the data, or as this is more and more frequent, a collection of gateways interconnects the WSN area to the Internet. Since the network may be deployed to cover a large area, transmissions may be multihop, i.e. smart objects have to forward packets after having constructed the routes dynamically.

In this chapter, we focus on the MAC and routing problems in WSN, and particularly on their cross-layer integration. Because of the specific constraints in WSN, we must implement a cross-layer approach: energy has to be optimized in all the layers. In the same way, information has to be mutualized in order to reduce the memory fingerprint. Since an exhaustive study is unpractical for such a vast topic, we illustrate this issue by examining first the metric definition problem, and then focus on the integration of the two emerging standards in WSN.

This chapter is organized as follows. We first introduce very shortly existing MAC and routing protocols in WSN, since we must understand how they work before putting the different pieces of the puzzle together. Then, in the next section, we discuss on the metrics we should use in the layers 2 and 3: could we use the same metric to measure the MAC and routing efficiency? Finally, we study the specific problem of the IEEE 802.15.4 and RPL integration (two major standards for WSN). Can they be executed without any close integration? How can we optimize their performances?

BACKGROUND

We present here the different MAC and routing protocols existing in Wireless Sensor Networks. Then, we expose why cross-layer is required in this kind of networks, and the current mainstream approaches.

MAC Protocols

Medium Access Control is in charge of distributing bandwidth to contending nodes: which node is allowed to transmit one frame, and when. Wireless Networks present several constraints:

- Low bitrate: because they consume little energy, radio components in wireless sensor networks offer a low bandwidth (typically 250kbps in IEEE 802.15.4). The MAC layer has to distribute efficiently this bandwidth: even with a small amount of data, congestion may appear quickly;
- Sleeping: to save energy, a smart sensor should switch-off its radio: as demonstrated by Feeney et al. (2001), this represents the only way to save energy efficiently;

Figure 1. S-MAC scheduling of the sleeping periods

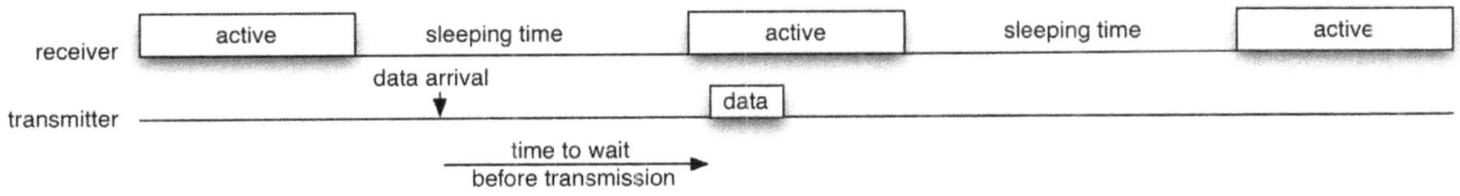

- Overhearing: radio transmissions are broadcast by nature. Thus, each neighbor (a node in the radio range of the transmitter) receives the packet, and consumes energy. We have to limit this impact.

We present here the existing mainstream approaches proposed in the last decade.

Low Duty-Cycle MAC

In the very beginning, researchers have focused on energy savings since they consider that WSN have little traffic to forward.

Ye et al. (2006) proposed S-MAC, one of the first low-duty cycle protocols. Each node chooses a sleeping schedule (Figure 1), piggybacked in the hellos broadcasted periodically. A transmitter has to find in its neighborhood table the schedule of the receiver. Then, it tries to send the data frame during the active period of the receiver. Obviously, the nodes must maintain a local synchronization; or else the transmitter would *miss* the active period of the receiver if its clock drift is too important.

Clearly, this method requires an adaptive routing layer: sleeping period impacts on the routing decision.

MAC with Preambles

When the WSN reports only alarms, the traffic may be very low. To save energy, Polastre et al. (2004) proposed to decrease the average energy spent for listening while increasing the energy required to transmit one packet. Each node listens to the medium periodically and sleeps most of the time, as in S-MAC, but neighbors are not aware of this schedule. The transmitter starts sending a preamble before the date frame. A preamble is a kind of jamming signal: when the receiver wakes-up, it senses the energy on the medium, meaning that it must stay awake to receive later incoming data frame. Thus, we do not need anymore to maintain synchronization, limiting the number of transmissions dedicated to control packets. On the other side, the preamble has to be sufficiently long to cover the interval of wake-up of the receiver.

However, transmitting the long preamble is expensive. El-Hoiydi et al. (2004) proposed a WiseMAC—a kind of adaptable preamble mechanism. Node learns and updates the wake-up periods of its neighbors from the Acknowledgement (ACK) packets. When a data packet has to be sent, the transmitter has to wait the receiver wakes-up, and to send a short preamble to cope with clock drifts. For high-traffic situations, the transmitter knows precisely when the receiver wakes-up: it can minimize the preamble length. Oppositely, in the case of infrequent traffic, wake-up information gets stale, extending the length of preamble duration in order to account for higher clock drifts.

This protocol was extensively modified to extend its capabilities. For instance, Bachir et al. (2006) proposed a micro-frames preamble sampling approach: the preamble is split into several micro-frames, containing the expected time of transmission and the destination id. The receiver, after having received one micro-frame,

Figure 2. Superframe in IEEE 802.15.4 (beacon-enabled mode)

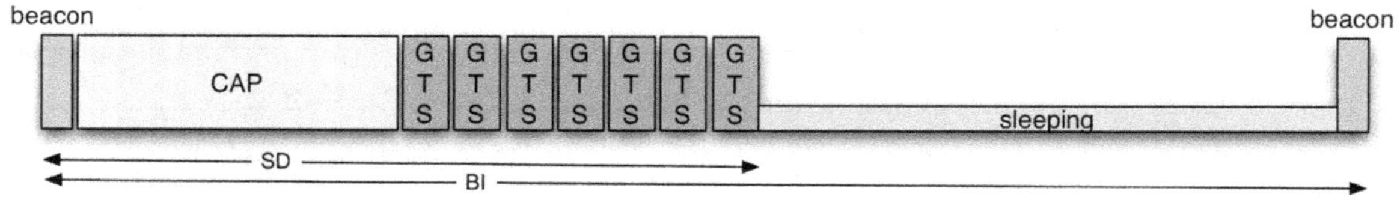

is able to sleep until the actual transmission, saving more energy.

Time-Division MAC

Time Division Multiple Access divides the signal into timeslots, allocated to one particular node for its transmissions. We may optimize frequency spatial re-use by allocating the same timeslot for two different transmitters that do not interfere with each other. Since interferences are also receiver-sensitive (i.e. two different receivers may not sense the same amount of interferences), the schedule must choose a timeslot for a pair of nodes, i.e. a radio link.

However, to find a proper TDMA scheduling in a distributed fashion represents a complicated task. Rhee et al. (2006) proposed each node chooses its timeslot and floods it to its 2-neighbors. Thereafter, an algorithm permits to solve the collisions between timeslots so that the TDMA schedule becomes globally valid, i.e. no pair of interfering nodes uses the same timeslot. Rhee et al. (2008) presented the associated MAC layer to achieve synchronization and to forward efficiently the packets, switching from the CSMA-CA mode under low contention to the TDMA mode.

Soldati et al. (2010) proposed to adapt the scheduling to deal with deadlines when transmissions are unreliable. The authors adopted a probabilistic approach, computing the optimal scheduling to upper bound the end-to-end delays.

IEEE 802.15.4

IEEE 802.15.4 has received a strong attention in the recent years since it may probably become the de facto standard in next generation WSN. It originally focused on the star topology: the PAN coordinator, at the center, is interconnected to the outside, collects data, and pushes the commands. In non-beacon mode, all the nodes use CSMA-CA to transmit their frames.

In the beacon-enabled mode, the PAN coordinator organizes its transmissions into a superframe structure (Figure 2). It first transmits a beacon transmitting several network parameters, including the BO and SO values. Beacon Order (BO) defines the Beacon Intervals (BI): the time between two beacons. Superframe Order (SO) gives the Superframe Duration (SD): the active period of the superframe. During the first part, children of the PAN coordinator use a CSMA-CA slotted mode to transmit their frames, while GTS slots are allocated on demand to one child for its periodic transmissions. This mode is more appropriated for energy saving since all the children can safely switch their radio off during the sleeping period.

The IEEE working group has extended this protocol to deal with multihop topologies. They propose to construct a *cluster-tree*: a tree rooted at the PAN coordinator. A node implements a *passive* discovery (it listens for a beacon) or an *active* discovery (it sends a solicitation) to find a parent to associate to. After having exchanged messages for the association, the new child is able to maintain its own superframe, sending a beacon,

Figure 3. Superframes organization in IEEE 802.15.4

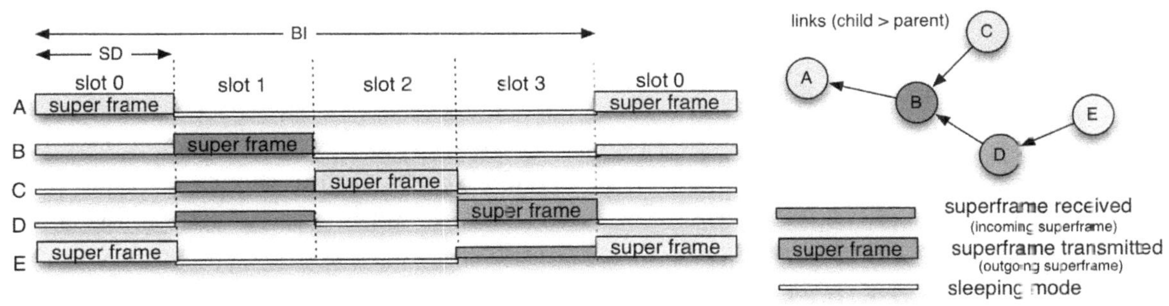

and maintaining the GTS slots. Thus, it becomes the *coordinator* of its own superframe.

The standard advocates the superframe has to follow directly the superframe of its parent. However, collisions may arise between e.g. siblings, wasting radio resources. The following solutions try to tackle this problem.

Beacon Only Period: Koubaa et al. (2008) proposed to schedule the beacons so that they do not collide with each other. The authors assume a low contention so that the data part of the superframes collide infrequently.

Superframe scheduling: Muthukumaran et al. (2009) proposed rather to schedule the superframes to forbid two interfering coordinators to maintain a superframe at the same instant. We may use a TDMA approach to schedule these superframes: two interfering coordinators are not authorized to maintain their superframe in the same timeslot. This scheduling does only work if all the nodes adopt the same superframe length: different superframe durations would mean the timeslots should have different lengths. Clearly, this TDMA would be complicated to obtain. Since the standard assumes a unique BO and SO value in the PAN, this does not constitute a strong assumption.

We can obtain the scheduling illustrated in Figure 3. We can notice that the nodes A and E can profit from the slot reuse since they are assumed sufficiently far from each other to avoid interferences and/or collisions. On the contrary, A and B must use a different timeslot since their beacons and data frames would else collide.

To construct this scheduling, we may adopt the DRAND approach, proposed by Rhee et al. (2006). In a general way, we can gather the slots chosen by interfering neighbors (we can assume that only 2-hop neighbors can be interfering nodes). Then, a node changes a slot it has already chosen if a 2-hop neighbor with a lower id has already chosen the same slot. We would converge quickly to a valid scheduling.

We can note that we must have at least as many timeslots as 2-hop neighbors (denoted N_{2hops}). Since the number of available timeslots depends directly on the superframe duration and on the beacon interval, the following inequality must consequently hold:

$$N_{2hops} \leq N_{slot} = 2^{BO-SO}$$

Villaverde et al. (2010) demonstrated the superframe scheduling is more efficient concerning the energy and capacity problems compared to the Beacon-Only Period (BOP) solution.

Zigbee defines an overset of IEEE 802.15.4, including the application and network layers. In particular, it defines how layer-3 addresses are allocated and presents a proactive routing scheme, using the hierarchized addresses. Pan et al. (2009) have proved this approach may present severe scalability limits. Zigbee is a cross-layer stack in Wireless Sensor Networks.

Routing Protocols for WSN

WSNs require specific routing protocols that minimize energy-consumption and signaling overhead. Moreover, a WSN constitutes a flat collection of nodes: classical routing protocols have to be adapted to deal with such anarchy. Routing protocols for WSN must conform to several specific concerns:

- Save energy by e.g. choosing the nodes with the maximum remaining energy, or the path that consumes on average less energy;
- Save bandwidth since radio transmissions present a low bitrate, particularly in WSN (e.g. 250 kbps);
- Be scalable to deal with a large flat collection of sensors.

We traditionally oppose the proactive and reactive approaches: while the former one creates routes a priori, the second approach constructs a route only when a packet has to be transmitted. Proactive approach, inherited from classical wired networks, may construct optimal routes by collecting information about all links. This approach is particularly suited for convergecast traffic: a single destination has to be announced in the network.

Reactive protocols do not maintain routes a priori: they construct a new one only when a packet has to be transmitted. They permit to reduce the overhead when the topology is highly dynamic, traffic is sporadic/bursty and destinations may change along the time. Route repair may be expensive in time and control packets since the nodes do not maintain any information about the topology to find alternative paths.

Energy Constrained Routing

Wireless Sensor Networks are highly constrained in energy. Thus, most propositions aim at constructing energy efficient routes.

Bachir et al. (2007) proposed to modify the routing mechanism to take into account energy. A node forwards a route-request after a delay inversely proportional to its energy reserve. Thus, routes are propagated quicker via a collection of nodes with a high level of energy.

Most of the routing protocols from this class have been guided by this energy constraint by:

- Reducing the overhead;
- Limiting unnecessary receptions (idle listening);
- Balancing the load to avoid a node runs quick out of energy;
- Permitting the network to operate in low-duty-cycle mode.

Data Centric Routing

Since Wireless Sensor Networks aim at collecting measures, using the semantic for routing could be efficient, saving useless transmissions. Intanagonwiwat et al. (2000) proposed in Directed Diffusion a data-centric approach. Consumers publish a list of their interests, propagated in the network. In this way, producers can direct their measures toward the interested consumers. Zabin et al. (2008) extended this work to deal with unreliable transmissions.

Madden et al. (2005) presented a very pioneering piece of work, proposing to consider a Wireless Sensor Network as a distributed database. The authors create the concept of *continuous queries*: measures are reported periodically, filtered in a distributed manner. Routing is implicit in this case.

However, such approach is very dependent on the semantic and changes the classical paradigm of networking.

Geographic Routing

If all the sensors know their geographic location, we may use a geographic routing approach: a node chooses to forward a packet to the neighbor closest to the destination. Ratsanamy et al. (2002) proposed to use a global Hash Table so that each node can retrieve distributively the geographic location of the destination.

However, the greedy rule often leads to voids: no neighbor is closer to the destination. Thus, Bose et al. (1999) proposed a *face routing* approach: after having planarized the graph, we may implement a localized strategy to surround the void. Recently, Sanchez et al (2009) proposed beaconless solutions.

Seada et al. (2004) proposed to modify geographic routing to integrate the energy constraint.

QoS Routing

As highlighted by Akyildiz et al. (2008), Multimedia Wireless Sensor Networks tend to recently emerge. In these conditions, guaranteeing a certain Quality of Service (QoS) for the transmissions is vital. Recently, Liu et al. (2008) adopted a biological inspired approach to reach this objective. By using several queues and properly adjusting probabilities, we may enforce differentiated classes of service.

Felemban et al. (2006) proposed a QoS enabled version of geographic routing: by properly choosing the next hop, a minimum *speed* is guaranteed (ratio of geographic improvement and transmission delay).

IPv6 Routing Framework of RPL

The IEFT ROLL working group focuses on the specification of an IP routing protocol suitable for embedded devices with limited power, memory and processing resources and intermittent operation for energy saving. Winter et al. (2010) described in details the RPL framework. A set of border routers is interconnected with the Internet and relays the packets outside the WSN area.

RPL adopts a distance vector approach that builds a DODAG (Destination Oriented Directed Acyclic Graph) anchored at the WSN border router: each node maintains a set of next hops toward one border router so that loops are forbidden. The RPL is built upon a *rank* representing a *virtual distance* to the border router and an objective function. The rank has to be monotonically increasing, but can be defined flexibly. Most implementations use the cumulative ETX metric (Expected Transmission Count): a node computes the number of transmissions required to send one data packet to the border router, taking into account the retransmissions.

A DODAG provides multiple routes to the border router, which fits convergecast traffic particularly well.

To maintain a certain consistency, each node has to advertise periodically its routing variables (e.g. its *rank*). However, this periodicity should adapt to the number of topology changes: stable topologies can safely increase the interval of control packet advertisements. Levis et al. (2010) proposed a trickle timer: when the node has stable information, the period is doubled, while it is reduced when several changes have occurred since the last advertisement. This self-stabilizing method is particularly accurate to cope with different environments.

Any2Any traffic (P2P) is also supported: any node may be a destination, even if it is not the sink. Such node may propagate a Destination Oriented Object (DAO) toward the sink, creating entries in the routing tables of intermediary nodes. The source generates a packet, forwarded along the DODAG until it reaches a relay owning a route toward the destination (a common ancestor in the DODAG). Then, the packets are forwarded down in the DODAG toward the destination. If nodes chose to store no routing information, the packet has to reach the sink before being forwarded to the destination (non-storing mode).

Figure 4. RPL traffic types

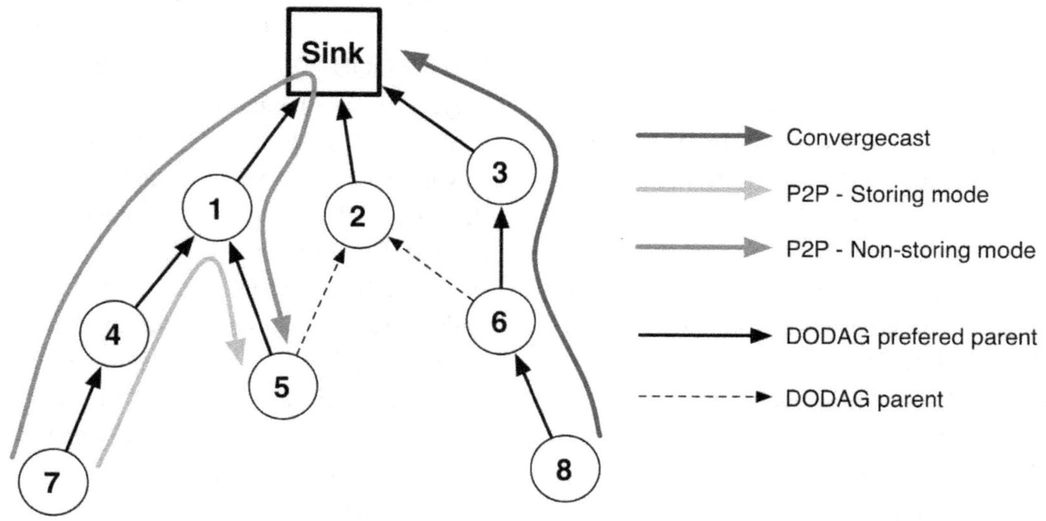

Even if almost all traffic patterns exist (cf. Figure 4), current RPL version is underspecified regarding to P2MP and P2P traffic types, not offering efficiency or robustness with respect to mobility.

Ko et al. (2011) exposed how RPL and 6LoWPAN can be used to enable the future Internet of Things, interconnecting the wireless multihop part to the Internet, re-implementing in particular the neighbor discovery and the fragmentation features of IPv6. Ko et al. (2011b) provided a TinyOS implementation.

Cross-Layer Approaches for MAC and Routing

Wireless Sensor Networks have specific requirements for which cross-layer design is particularly efficient:

- CPU and memory constraints make the integration of MAC and routing vital. For instance, to mutualize MAC and routing information permits to reduce memory fingerprint;
- WSN face often to a convergecast traffic pattern. Consequently, most radio links are not used to forward traffic: the MAC layer should not provision resource for them;
- Nodes should save energy to improve the network lifetime. A careful co-design permits to achieve more easily this feature;
- Routing depends heavily on the underlying MAC layer. For instance, a low duty-cycle MAC has an impact on the routing decisions: a node should choose the next hop according to the current scheduled reception time of each neighbor.

We may adopt two approaches: either we merge both layers to create a single protocol or we maintain two separated protocols, interacting with each other.

We now explain the existing approaches and their respective limits.

Design a Unified MAC/Routing Layer

When two different protocols are used, we can only optimize them separately. In other words, the local routing and MAC optima may be antagonist and would not lead to a global optimum. On the contrary, to execute a unified MAC and routing protocol permits to explore all the solutions, finding the optimal cross-layered one.

For instance, Kulkarni et al. (2006) proposed a simple opportunistic MAC and routing solution for convergecast networks. The protocol follows the classical RTS/CTS/DATA/ACK paradigm. A node asks for a candidate with a RTS: an overhearing node replies with a CTS and the data transmissions takes place. MAC and routing are mutually dependent. Zeng et al. (2010) aim at tackling the same problem with an optimization approach. They proposed to find the optimal MAC scheduling and routing schemes in a centralized way. Theoleyre et al. (2011) proposed to construct routes so that traffic is concentrated on a few nodes. Medium access is modified to optimize the access in this kind of topologies.

Removing the OSI layers architecture permits also to reduce the complexity. If the solutions are very simple, they may be implemented in a FPGA (Field-Programmable Gate Array) as Flynn et al. (2005) proposed. We can in this way reduce largely the energy consumption.

While such approach has clear benefits, it also presents severe drawbacks:

- **Non-flexibility:** since the solutions are monolithic, one modification may require to possibly redesign from scratch a new protocol;
- **Re-usability:** a new application requires designing a new *ad hoc* solution. On the contrary, a modular solution, designed with the classical layered structure, would have permitted to choose the most accurate pieces of protocols we need;
- **Interoperability:** profusion of proprietary solutions complicates interoperability. In particular, we may not be able to enable the new Internet of Things.

The OSI Layered (IETF) Philosophy for WSN

For the beginning of the Internet, researchers have promoted the layered architecture. It privileges inter-operability since some of the layers can be changed without rethinking the whole system.

We can remark that experiments in this research field often adopt implicitly this approach. In the Sensorscope project, Barrenetxea et al. (2008) have decided to design a very simple solution to deploy a 7 nodes weather-monitoring testbed. Although they face to a simple testbed, they chose to keep the layered architecture of the OSI model. In Zebranet, Juang et al. (2002) also implemented a classical stack of protocols for monitoring Zebras in their natural environment.

For a few years, the 6lowPAN [6lowpan] and ROLL [roll] working groups have been promoting the IP vision for the Wireless Sensor Networks. They consider the Internet of Things must execute IPv6 to enable new applications; this constitutes a *sine qua non condition* to its development. The ROLL working group advocates the M2M market has not yet known the expected growth mainly because it is currently a world of proprietary solutions.

Thus, we have to clearly make a distinction between:

- MAC layer in charge of radio bandwidth sharing: which node has the right to transmit a packet at a given instant. 6lowPAN assumes for instance that IEEE 802.15.4 is used for the transmissions;
- IP layer for the inter-operability layer: how a node should route packets, which protocol it uses for exchanging packets with the Internet.

We can mention here the packet switching approach of IEEE 802.15.5: multihop transmissions are handled in the MAC layer. Thus, IP layer just focuses on the inter-operability feature. However, we just move the *routing* part into the MAC layer while keeping the layered paradigm.

DEFINING ACCURATE METRICS FOR BOTH THE MAC AND ROUTING LAYERS

Problem Statement

In the beginning, researchers in Wireless Sensor Networks have tried to isolate some problems and to solve them separately. Although it provided very good insights of the specific challenges to tackle, putting all the pieces of the puzzle together have often lead to disappointing results.

Recent experimental results have demonstrated the network protocols often work inconsistently and lead to instabilities. For instance, routes and radio links cannot be considered independently because of collisions and interferences. A new flow may create more collisions in the MAC layer: unfairness arises, and routing may consider the radio link broken, rerouting the traffic to a new route. We could obtain a domino effect, and routing would never converge. To avoid these oscillations, we should guarantee self-stabilization: whatever the current state is, the network converges to a legal state in a finite time. In other words, we must have a close cooperation between MAC and routing layers to distinguish local and global congestion and to exhibit a self-stabilizing property.

Recently, Srinivasan et al. (2008) demonstrated experimentally a burstiness behavior of radio links: packet error rate is time variant. A radio link may be *good* for a small time (i.e. a few packets) and thereafter all packets begin to be dropped. Consequently, opportunistic routing is clearly advantageous: best radio links at a given instant should be used. We must implement a close interaction between the MAC and routing layers.

What Should be a Good Metric?

In wireless sensor networks, to define a good metric reflecting the radio link quality is essential for both the MAC and routing layers. It should in particular present the following properties:

- Dynamic: WSN environment is varying because of e.g. mobility, link's burstiness. The metric has to converge quickly to an accurate value. For instance, we should be able to capture quickly a bandwidth estimation even when traffic is varying;
- Stability: a protocol using a very volatile metric always re-adapts and changes its decisions. Worse, it may over-react to changes. A *good* metric has to avoid such oscillations;
- Minimized overhead/energy consumption: bandwidth and battery are scarce resources in WSN. Radio module represents the biggest energy consumer in WSN node. Thus, a node should try to use radio links exhibiting the best energy budget;
- Localized: the metric should use only local information to privilege convergence (a distributed version is acceptable), while limiting oscillations. However, a centralized solution is not scalable;
- Isolation: the metric should be able to differentiate the internal and external factors. For instance, the link quality depends on the radio propagation but also on transmissions of other nodes. While the node's decision has no impact on the former, the inter-dependencies of the later are apparent;
- Minimized computational complexity: a good metric should capture accurately the real nature of measured phenomena. However, it should be simple enough to be practically tractable, limiting CPU and memory consumption: these resources are very constrained in wireless sensor networks.

Figure 5. Box plot

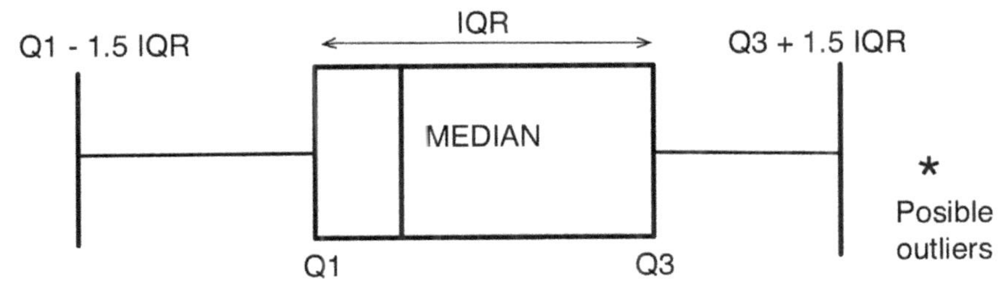

Statistical Tools to Filter Data

A metric has to filter the experimental measures it is based on: radio chips and sensors are often prone to errors. Algorithms have to detect errors with a limited complexity since smart objects have small computing capabilities.

Boxplots may help to characterize measures, since they represent their distribution. A quartile is the number larger than 25% of all data in data set sorted in the ascending order. The difference between the first (Q1, 25%) and the third (Q3, 75%) quartile is called Inter-Quartile Range (IQR). The second quartile (Q2, 50%) is also called the median.

The boxplots constitute a visual representation of the method to discard errors. (cf. Figure 5) An outlier is a value that conforms to the following condition:

$$x < Q_1 - 1.5 \cdot IQR \text{ or } x > Q_3 + 1.5 \cdot IQR$$

Single isolated outliers should be discarded since they are extreme, isolated values, corresponding surely to a transient behavior. On the contrary, multiple consecutive outliers could arise from drastic but pertinent transition in measured values and should be kept.

To compute a metric (e.g. radio link quality), we should reduce the memory fingerprint while estimating it accurately. To maintain exhaustively in memory all the measured values is impracticable. Pavkovic et al. (2010) have reported the number of outliers in a sliding window of 20 samples for the RSSI measured for a particular radio link in an outdoor testbed (Figure 6).

In more than 99% of the cases, we have 4 outliers or less. We can consequently use such sliding window to estimate a measure: the number of *correct* values is sufficient to estimate accurately the median and extreme values. A higher number of outliers per sliding window indicate that we should increase the buffer size to save more samples. In other words, we must calibrate the buffer size from the preliminary measures we may obtain from a testbed or during the pre-deployment.

We may consequently maintain a buffer of 20 samples. When a measure is estimated an outlier, it is tagged and placed on the right of the buffer (yellow locations in Figure 7). These tagged values should not be used to compute the median and Inter-Quartile Range.

Possibly, a new value could be detected as outlier although 4 values were already tagged as outliers. In this case, the outlier closest to the median value is untagged, and inter-quartile ranges and median values are re-computed. Finally, we verify that the newly tagged values remain considered outliers or not.

Using this approach, the quality metric is smoothed and inaccurate measures are discarded. Moreover, the approach is also reactive: changes in the observed variable are efficiently detected and its value is updated accordingly. We maintain

Figure 6. Number of consecutive outliers in an outdoor testbed

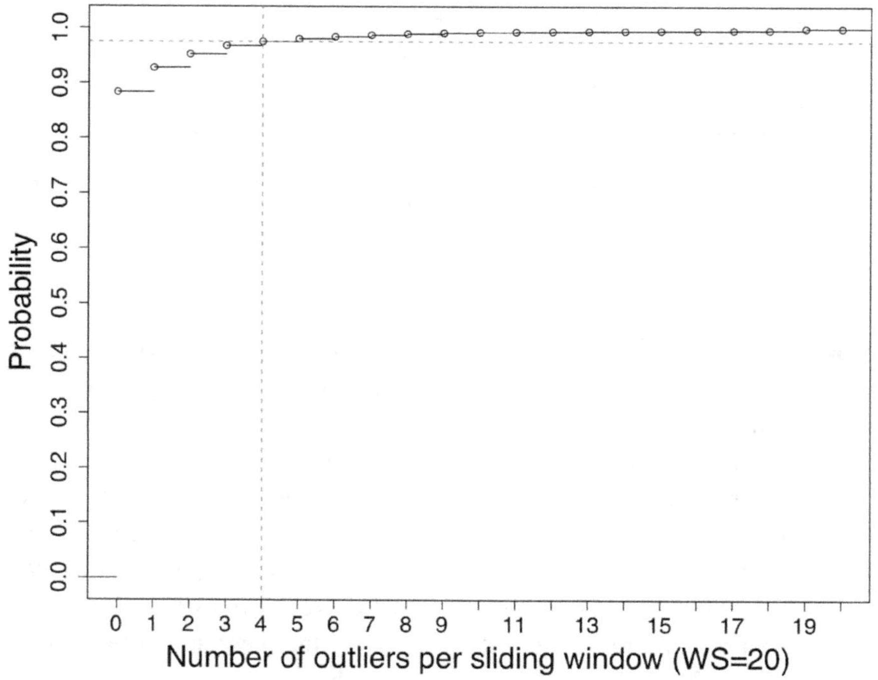

Figure 7. Sliding window method to dynamically detect and remove outliers

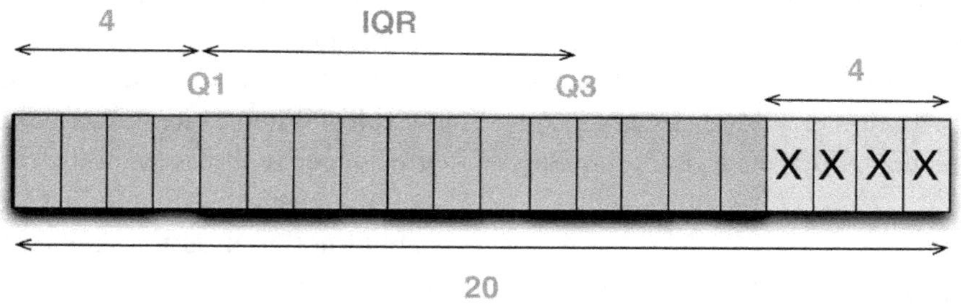

this property with a sliding window of only 20 samples.

Packet Delivery Ratio

Packet Delivery Ratio (PDR) measures the average number of packets correctly received by the receiver. It can be measured locally by the MAC layer (for a radio link) or by the routing protocol (for a route).

However, bursty links, which have a fluctuating quality, are frequent in Wireless Sensor Networks. In other words, their PDR is highly unstable, and we have difficulties to estimate it accurately.

Although PDR can be measured in the MAC, routing strategy has also an impact. For instance by re-routing traffic, it reduces the number of contenders. Thus, the Packet Delivery Ratio should be measured in isolation to estimate accurately the link quality: no interfering traffic should exist.

Figure 8. RSSI distribution for a particular radio link

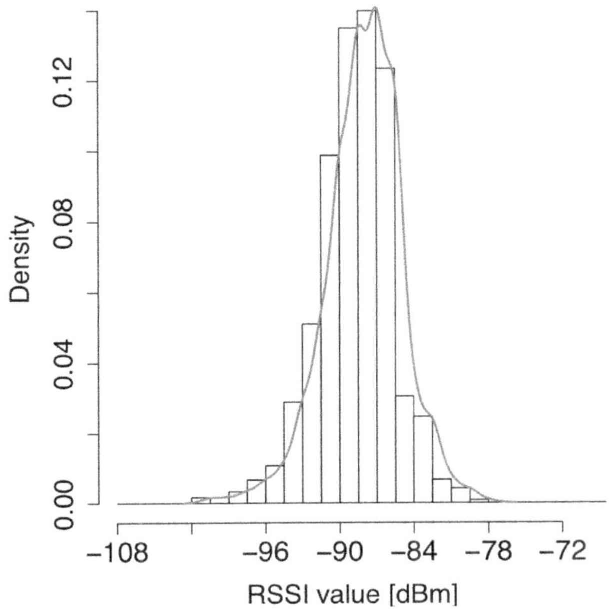

This metric created a complete family of routing metrics. In particular, ETX (defined by De Couto et al. [2003]) estimates the overhead for one transmission by using the packet delivery ratio to approximate the required number of retransmissions. Lukić et al. (2009) studies its practical impact in the real world WSN deployment.

The Packet Delivery Ratio, despite its imperfections, constitutes a good estimation of the radio link quality for both the MAC and routing layers. However, other metrics have also to be integrated in the routing decisions to avoid oscillations.

RSSI

RSSI presents another popular metric useful in WSN, as described by Srinivasan et al. (2006). The NIC offers to the higher layer an arbitrary value reflecting more or less the SINR (Signal to Interference and Noise Ratio). Although both measures are not directly proportional, many protocols try to exploit it as such.

We reported in Figure 8 the typical distribution of the RSSI values for one particular radio link in outdoor. For different pairs of links, the collections of values follow the same distribution according to the Kolmogorov-Smirnov test. However, one particular collection follows neither the Cauchy nor the Logistic distribution. Thus, no well-known distribution permits to have a generic model for such RSSI values.

We reported in Figure 9 the RSSI values for different classes of radio links. Each class corresponds to a particular link occurrence ratio, i.e. more or less its packet delivery ratio. Radio link quality increases from left to right. We can notice that the RSSI/quality relationship is not clearly apparent. We can only conclude small RSSI values mean that we face to a bad radio link. On the contrary, very stable RSSI means we have a larger probability to obtain a good link. This represents only a probability and not certitude.

For all these reasons, RSSI measure has a limited interest to reflect the link quality for both the MAC and routing layers.

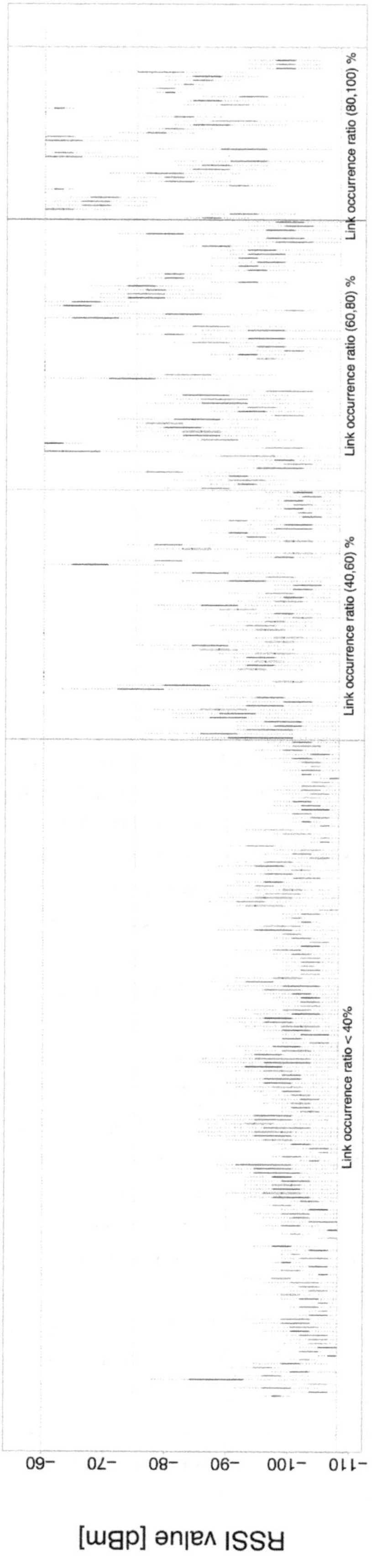

Figure 9. RSSI values for different classes of radio links

Bandwidth Estimation

Estimated available bandwidth would be one of the most interesting cross-layer metrics: the routing protocol can safely decide which route to use based on its requirements. If routing decisions also integrate energy consumption, we can have a smart cross-layer metric.

It must integrate in the MAC layer:

- The actual number of frames generated for each data packet (like the ETX metric);
- Number of interfering nodes and their load. For instance, siblings always interfere in IEEE 802.15.4: they share the same superframe. Thus, the capacity is shared among all the children of these sibling nodes;
- The link datarate since a better modulation may offer a larger bitrate;
- The amount of external interferences.

Besides, routing layer has also to take into account:

- **Intra-flow interferences:** packets that belong to the same data flow may interfere with each other. This would reduce the available bandwidth:
- **Inter-flow interferences:** re-routing some packets may distribute the load and reduce contention, offering a larger bandwidth.

While Zhao et al. (2009) and Chaudet et al. (2008) have already conducted such surveys for IEEE 802.11 based networks, to combine all these criteria in a single metric still constitutes an open-problem in WSN.

Open Challenges

Although we presented here the state-of-the-art thoughts about the metric problem in wireless sensor networks, several questions remain open.

Quantifying the Cross-Layer Impact

If we do not adopt a cross-layer approach, we could define one *good* metric per layer. If we consider the system as a whole, inter-dependencies would make the metrics not accurate. For instance, a good MAC metric would be impacted by the metric chosen for routing. In other words, a union of local optima in each layer may not constitute a global optimum in the whole stack.

Measuring the difference between the global optimum and the union of local optima would permit to quantify the asset of a cross-layer approach, definitively proving the accuracy of such design.

Energy Consumption

A metric reflecting the global energy consumption for a whole path toward a sink has still to be defined. It should take into account the link quality, the number of transmissions, the remaining energy, the queue lengths, interferences, overhearing, synchronization maintenance cost, etc. Quantifying exactly the impact on the energy of all these factors represents a challenging task.

Experimental Measurements

Recently, some researchers have tried to characterize and understand the multihop wireless topologies. Raman et al. (2009) focused on wireless mesh networks to verify the link abstraction paradigm can hold: the conclusions are transposable to WSN. Srinivasan et al. (2008) isolated the burstiness property of radio links: the quality may increase significantly for a short time. Thus, opportunistic approaches should be able to capture this effect, after having estimated correctly the quality increased. Liu et al. (2011) deployed a large-scale testbed to verify the current MAC protocols may cope with such applications. Surprisingly, bottlenecks far from the sink may appear in the network.

Although these results permit to understand better this kind of environments, a lot has still to be done to characterize better the link quality.

Combining Metrics

We presented here several candidate metrics for WSN to estimate the link quality. We have now to combine them linearly or not to constitute one single cross-layer metric. We should for instance combine bandwidth estimation and energy consumption while forbidding links with a bad RSSI.

We present in the next section a way to combine these metrics for routing in RPL.

Self-Stabilization

Current protocols do not focus on the self-stabilization property for Wireless Sensor Networks. Whatever the initial state is, the network must converge in a finite time (quickly) to a legal state. Since this property has not yet been much studied, we still have either to propose metrics or to find new protocol mechanisms to guarantee such behavior.

A CASE STUDY: INTEGRATION OF THE IEEE 802.15.4 AND RPL STANDARDS

Problem Statement

We focus in this part on the IEEE 802.15.4 and RPL standards: how can we specifically integrate these two protocols so that they work in symbiosis?

First of all, both protocols do not lead on the same structure. Worse, layer 3 leads on an overset of the layer 2 structure. Let consider Figure 10 illustrating the problem. IEEE 802.15.4 prunes some radio links to construct a cluster-tree, rooted at the PAN coordinator (i.e. the WSN gateway). RPL should construct a redundant Destination Oriented Directed Acyclic Graph (DODAG) to

Figure 10. Antagonist layer 2 and 3 structures

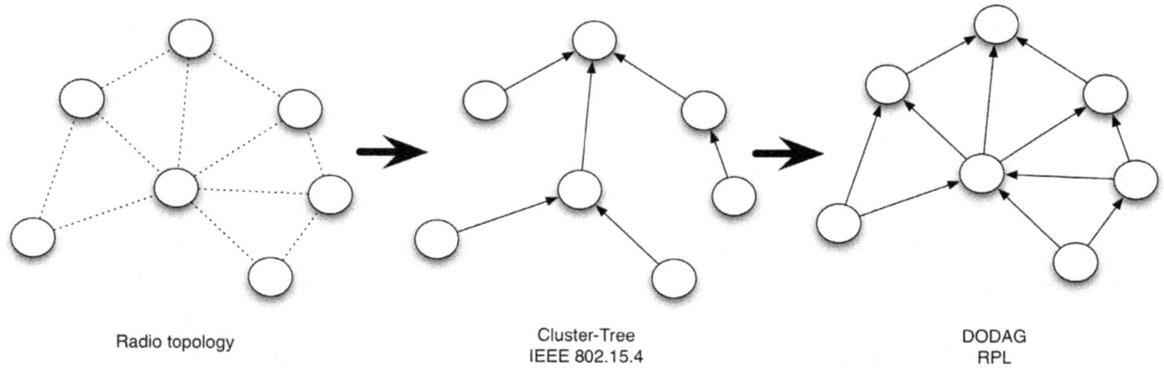

be efficient, as represented on the right of the figure. However, it cannot since the lower layer just offers one unique parent. In conclusion, both layers have to interact in a better manner.

Opportunistic routing and MAC have received recently much attention: they permit to be fault-tolerant. Indeed, they take a decision only just before usage to take into account transient faults. For instance, a node chooses a next hop just before a packet is transmitted. Clearly, the structure of IEEE 802.15.4 is inefficient for such consideration: the whole structure may become inefficient after only one radio link was broken.

Besides, the metric problem developed previously remains intact for the RPL/IEEE 802.15.4.

We now demonstrate how we can *adapt* these standards so that they can work efficiently together without putting back into question the core features of the protocols. This attests the interest of cross-layer in WSN.

Cross-Layer Structure for RPL and IEEE 802.15.4

As highlighted in the Background section page 5, IEEE 802.15.4 requires slight modifications to cope with multihop topologies. We adopt here the TDMA scheduling of superframes since it improves the network capacity while reducing the energy consumption.

We chose to modify slightly the IEEE 802.15.4 cluster-tree, so that we can use the same structure for both the MAC and routing layers. Instead of creating a tree, we authorize a node to *follow* several parents, i.e. to participate to several superframes as child.

Let take a look on the example illustrated in Figure 11. The DODAG is represented on the right of the figure. If the nodes A, B and C use a different timeslot, D can participate to each superframe.

If the superframe scheduling is properly organized, a node may have different parents, which maintain their superframe in a different timeslot. In particular, if each coordinator chooses its timeslot for its own superframe pseudo-randomly, we reduce the probability two potential parents use the same timeslot. In other words, we increase the probability redundant paths exist toward the sink in this DODAG structure. Possibly, these redundant paths may be exploited by the RPL protocol.

Participating to several superframes increases energy consumption very marginally: if a child has no packet to send, it just has to wake-up to receive the beacon. If it is not present in the pending destinations list in the beacon, the node sleeps again until the next superframe. The node may also

Figure 11. DODAG with superframe TDMA scheduling in IEEE 8021.5.4

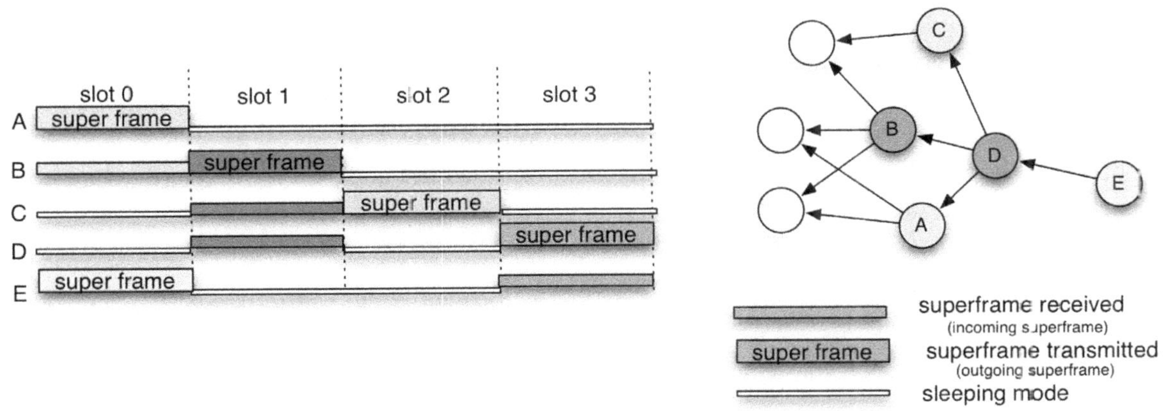

choose not to participate to all superframes: to be standard-compliant, it just has to receive one beacon every macTransactionPersistenceTime since the coordinator buffers the packet meanwhile. Besides, we may implement strategies to reduce energy consumption by adopting a power-saving method: the number of parents to follow, the period for listening their beacons depend on the traffic to forward and on the battery level of the child.

Clearly, this modified DODAG constitutes a cross-layer structure since it fits the requirements of both the MAC and routing protocols.

We now detail how this mutualized DODAG can be exploited to improve the performance of the union IEEE 802.15.4/RPL

Multipath with IEEE 802.15.4 and RPL

RPL constructs a DODAG to use a redundant, mesh structure, enabling multipath. Such routing strategy presents several assets, demonstrating the inter-dependency between the structures at the different layers:

- **Robustness:** a parent may die because of energy depletion, or may become inaccessible. Several parents offer an automatic recovery procedure;
- **Load balancing:** a node may choose to forward its packets according to a round-robin strategy, weighted by the link quality or the load of its parent. Hence, we would avoid overloading one of our parents;
- **Opportunistic routing:** we may have the choice of several parents, making the best choice at a given instant. Because of link burstiness property, one parent could become temporarily very good: we should use it in priority;
- **Delay optimization:** since we adopt a TDMA scheduling with long timeslots, delay may become problematic. Our approach permits to limit end-to-end delay;
- **Reliability:** to use several paths permits to introduce spatial and temporal diversity. Thus, we reduce the packet losses dependencies. If we design carefully the routing strategy, we would be able to guarantee certain reliability, as we demonstrate it further.

Lou et al (2006) presented a classification of multipath strategies. In particular, node disjoint multipath aims at guaranteeing two paths do not have a node in common. This represents a very complicated task to discover such paths in a distributed manner.

Thus, we adopt an opportunistic approach, particularly accurate in convergecast networks since we have a single destination: a node chooses the set of next hops only just before transmitting the packets. The routes are not constructed a priori and follow the DODAG structure.

We also reduce the overhead and energy consumption to construct and maintain independent routes. This opportunistic approach is clearly the most accurate in Wireless Sensor Networks with a convergecast traffic pattern.

Opportunistic Cross-Layer Routing for Delay Optimization

Smart objects report their measures either periodically or after having detected an event. While we could accept a long end-to-end delay for periodical measures, we would have to guarantee a small delivery delay for alarms.

Thus, we should implement a QoS version of RPL: we associate to each generated packet expiration time i.e. *deadline*. A packet with a short deadline would require more energy to be transmitted, but the network has to guarantee such end-to-end delay. Thus, a node maintains a buffer for all packets, sorted according to their deadline, following a *First Deadline, First Out* strategy.

When a node is authorized to transmit one packet by the MAC layer, it extracts the first packet of its buffer. If the deadline already elapsed, it drops the packets and looks for the next packet, until either one valid packet is found or the buffer is empty.

With opportunistic multipath, the node may have several parents in the DODAG: it should choose as next hop a parent that can respect the deadline constraint.

Thus, a node has to verify for each parent if it can respect the deadline. It first computes the time it has to transmit the packet to its parent *V*. It assumes the time until the deadline should be uniformly distributed among the hops in the path. Felemban et al. (2006) proceed in a similar way to guarantee a uniform *multipath speed* in MMSPEED.

The time budget allocated to this hop is consequently:

$$budget = \frac{deadline(p) - t}{depth(V)}$$

where *deadline(p)* represents the deadline of the packet *p*, *t* the current time and *depth(V)* the depth in hops announced by *V* (its distance to the sink).

The transmission time has to fit the budget. It depends on:

1. The IEEE 802.15.4 superframe scheduling delay: the time before the superframe of this parent begins.

If the superframe of this parent has already begun and the beacon was correctly received, this delay is zero. Otherwise, it includes the waiting time before a valid superframe. It depends on the slot used to transmit the superframe and on the packet delivery ratio of beacons for this parent. More formally, this superframe scheduling delay is in conclusion:

$$D_{sframe} = SD \cdot \left|slot(V) - slot(N)\right| + BI \cdot \max(0, \frac{1}{PDR_{beacon}(V)} - 1)$$

where the superframe duration is denoted SD, and the inter-beacon time BI (as in the background section). Furthermore, *slot(X)* denotes the slot chosen by the node X for its superframe, and $PDR_{beacon}(V)$ its packet delivery ratio for beacon messages.

Average delay before the packet is received correctly by the parent: it is estimated by the ETT (Expected Transmission Time) metric:

$$D_{tx} = \frac{size(p)}{PDR_{data}(V) \cdot datarate(V)}$$

where *size(p)* stands for packet *p* size, $PDR_{data}(V)$ for packet delivery ratio for data packets with neighbor V. Finally, *datarate(V)* is the datarate of the link toward *V*.

Finally, a parent represents a valid next hop if it satisfies the constraint:

$$budget \leq D_{sframe} + D_{tx}$$

If there are several candidates, a node chooses the one that minimizes the generated traffic since this also minimizes energy consumption. Thus, among the different candidates, a node chooses the next hop presenting the lowest ETX.

We see here how we can combine the metrics defined in the previous section to take a *good* routing decision.

Since nodes are very constrained, we must maintain a low memory and CPU consumption. A node must maintain for each of its neighbors the beacon used for its superframes, and estimate a PDR metric toward it. Since this information is also required in the original RPL version, we do not increase here the complexity. Finally, the routing algorithm iterates on each neighbor, to verify it respects the deadline constraint. In conclusion, the time complexity is in O(D) with D being the number of neighbors.

Simulation Results and Analysis

To illustrate the interest of this approach, we present here a few simulation results. We simulated the original RPL and its modified version implementing the opportunistic approach. We use topologies of 256 nodes, each node having on average 10 neighbors. We adopt the Rayleigh propagation model for the PHY layer. We mix best-effort periodical measures (70%) and delay-sensitive alarms (30%) with a deadline of either 360s or 180s.

We focus here only on the delay-sensitive data in Figure 12. If we focus on the packet delivery ratio (the ratio of packets, which are received by the sink), we can remark that the opportunistic version permits to deliver with a higher probability the alarms to the sink. This difference becomes significant when the deadline is shorter: the constraint is higher and the original RPL version does not succeed to deliver the packet before the deadline elapsed. Besides, a smaller deadline means also we drop more packets: the deadline has elapsed, and the network had to drop the corresponding packets. If we focus on the end-to-end delay (time between the packet generation and its reception by the sink), we can verify the opportunistic RPL achieves to exploit efficiently the DAG structure in the IEEE 802.15.4 layer. It chooses the best next hop, saving time for relaying the packets. The delay is reduced by 30%.

We verified that the energy consumption is marginally increased with opportunistic RPL (a few percent). Indeed, more packets are delivered to the sink increasing the global energy consumption. In the same way, alarms are forwarded across more expensive radio links to respect the deadline. However, this energy increase is the price to pay for QoS routing.

Opportunistic Multi-Path Routing with Respect to Reliability

Wireless Sensor Networks are unstable by nature. Radio links may be bursty: their quality fluctuates quickly. Thus, we have to guarantee the network operates efficiently in an unstable environment.

We have to integrate the reliability QoS criterion into the routing decision. To tackle this problem, we propose to use multipath to de-correlate

Figure 12. Comparison of the original and opportunistic versions of RPL

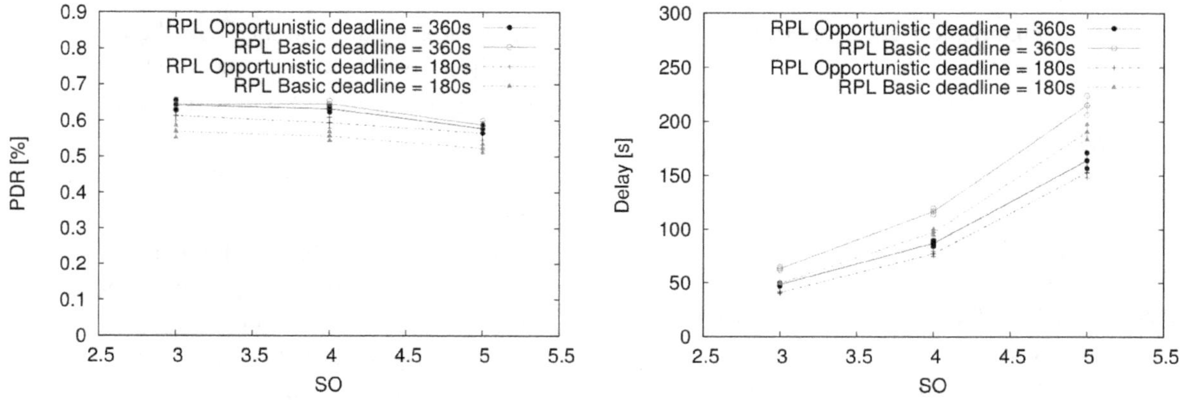

spatially and temporally the packet losses over independent paths.

To each packet, the source associates an expected deadline and minimum end-to-end reliability (p_{E2E}). As discussed previously, we should use localized metrics to accelerate the convergence. Thus, we can assume that the reliability for each radio link (p_{local}) is uniformly distributed along the path. When a node V is *depth(V)* hops far from the sink, we must guarantee a local reliability of:

$$p_{local} = \sqrt[depth(V)]{p_{E2E}}$$

We illustrate this with a simple example: if we have to guarantee an end-to-end reliability (p_{E2E}) of 90%, we have to guarantee that each transmission along a path of 4 hops presents a minimum reliability of $\sqrt[4]{0.9} \approx 97\%$.

Practically, a node monitors the packet delivery ratio toward its parent. The MAC layer may obtain this measure, reporting to the higher layer the number of failed and successful transmissions. After having discarded all the parents that cannot guarantee the deadline, the reliability constraint has also to be respected.

To increase the reliability, a node may transmit the same packet to *different parents*. Since transmissions are temporally separated, the packet losses are also uncorrelated. An alternative would consist in transmitting multiple copies to the *same parent*. By combining these methods, we can guarantee a local reliability:

1. A node verifies each parents respects the deadline constraint: they constitute the set of *candidates*;
2. While the local reliability is not respected (p_{local} <threshold):
   a. a node creates a new copy of the packet, it transmits one copy to all parents respecting the deadline and updates the local reliability according to the following recursive equation:

$$p_{local} = 1 - \left(\left(1 - p_{local}\right) * \prod_{V \in Canditates} \left(1 - PDR_{data}(V)\right) \right)$$

In short, the transmitter chooses to forward packets to all its parents respecting the deadline, and sends several copies if the reliability keeps on being insufficient.

p_{E2E} should be piggybacked in the packet so that all the chosen parents can globally guarantee the end-to-end reliability. Thus, each parent must guarantee after having received the packet the following reliability:

$$p_{E2E}^{next\_hop} = \sqrt[n_p]{1 - p_{E2E}}$$

where n_p represent the number of parents created and the end-to-end reliability that is later used by next hop forwarders.

High reliability demand produces in the beginning several branches (paths) and the reliability requirement reduction at each hop stops further explosion, calming it down when the packet arrives closer to the sink (i.e. the PAN gateway).

Opportunistic multi-path routing offers a way to maintain the reliability constant whatever the distance to the sink and the link qualities are. While we deal with good links, standard RPL routing with preferred parent is sufficient to respect expected reliability, but as soon as the link quality deteriorates we converge to a multi-path solution where reliability is provisioned by creating more paths.

Compared to the previous version, we do not increase the time complexity: we just have a final loop in the algorithm, generating several copies of the same packet to respect the reliability constraint. A naïve implementation would insert new copies directly in the buffers. We may choose rather to maintain a master queue storing real packets and one slave queue for each neighbor pointing to the master queue. The memory consumption is thus limited: one pointer and one counter is negligible compared to a packet.

MAC-Aware Buffering Strategies for Routing

IEEE 802.15.4 defines when a node is authorized to sleep: it does not participate to any superframe and can switch its radio off. However, this sleeping period has also a practical impact on the routing layer. A node should use an opportunistic decision to choose the next hop since it depends on the data arrival time, the scheduling of parents and their reliability.

In classical routing protocols, we have to look-up in the routing table the most accurate next hop and then buffer the packet in the corresponding NIC. However, in WSN, we should rather apply an opportunistic strategy: a buffer is maintained, sorted according to the QoS parameters of the packets. When a node is able to transmit one packet, it should choose the most accurate one in the buffer.

When we face to a convergecast network, all the packets have the same set of next hops. To maintain a buffer sorted according to their QoS parameters is sufficient. The most constrained packet has just to be forwarded first.

If we authorize Any2Any communications, different packets may not have the same set of next hops. Routing decisions would be in this case inter-dependent. Let consider the example illustrated in Figure 13. For sake of simplicity, we assume node A can only transmit one data packet in each superframe. If it extracts the first packet and transmits it in the superframe of B (a next hop candidate), node A has to drop the second packet in the buffer since the deadline is exceeded. However, a proper scheduling exists such that all packets can respect the deadline constraints: node A could transmit the second packet during the superframe of B and then the first packet during the superframe of C.

In conclusion, a node should find the correct scheduling by examining all the packets in its buffer. Although the buffer size is practically small, the problem is NP hard since it is clearly a scheduling problem. Thus, a greedy approach could be used. We extract the packets with a deadline elapsed in the next *K* Beacon Intervals. We sort them into ascending order according to the number of next hops: we should decide first for the most constrained packets. Finally, we apply a greedy rule: a packet is scheduled in the next available slot for this set of next hops.

A large *K* value would increase the complexity but would permit to foresee better the next transmissions to schedule. A node with low capacities chooses rather a small *K* value.

Figure 13. MAC-awareness for opportunistic routing in WSN

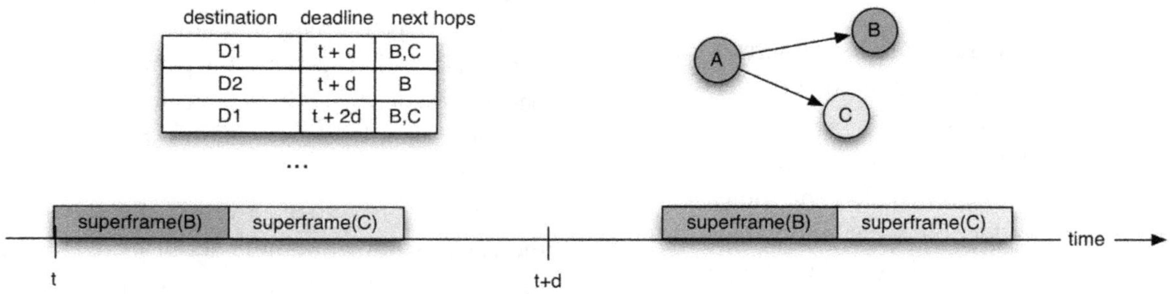

FUTURE RESEARCH DIRECTIONS

Wireless Sensor Networks represent a very active research area. Some protocols are currently standardized and we have a real opportunity to define in the next years what the Internet of Things will be.

Experimentally Driven Design

In the recent years, many researchers have set up a testbed to demonstrate the practical interest of Wireless Sensor Networks. For instance, Barrenetxea et al. (2008) presented the Sensorscope project, deploying a collection of weather monitoring nodes. This platform measured local variations, which would not have been detected with a single measurement node. In the same way, Ceriotti et al. (2009) deployed a testbed monitoring the deformation of the Torre Aquila in Italy to prove the building would be impacted by the construction of a tunnel. Recently, Liu et al. (2011) have evaluated the scalability of this kind of networks, isolating new problems. While these experiments provided guidelines to design a cross-layer set of protocols, they have practical limitations: networking requirements are simple, energy considerations are limited, and scalability is seldom evaluated.

The next step consist in designing efficient, scalable solutions, offering a certain QoS because the Internet of Things is heterogeneous, and may be mutualized by several applications.

We have to:

- Use bursty and asymmetrical links;
- Balance the load (at both the routing and MAC layers);
- Guarantee the network operates in a stable manner.

These typical problems are cross-layer by nature and have to be triggered by an experimental research.

Standardization

Several working groups are currently defining which standards may be used for the Internet of Things: ROLL studies how routing should be implemented; 6LowPAN defines how WSN can use IPv6; IEEE 802.15.4 works on defining a multihop MAC layer.

We have now to decide how these protocols cohabit and collaborate. The following points must be in particular studied:

- Cross layering complicates inter-operability: what is the best trade-off?
- What metrics should use each protocol? Can the different layers use the same metric?

- How QoS can be guaranteed in the whole networking stack?
- How the MAC and routing layers can self-adapt to the application layer: the criterion may be chosen by the application layer (e.g. energy versus delay).

We have now to define how we can put all these pieces together.

Data Centric Networking

Since Wireless Sensor Networks aim at collecting measures, data signification plays a major role. Thus, we should perhaps define new paradigms, where data has a central role. Alphand et al. (2009) advocate the usage of e.g. geocast (all the sensors in a particular location), subjectcast (addressing a topic), timecast (to introduce the time dimension), controlcast (to push the control actions to actuators).

We have now to explain how we can push these concepts in the MAC and routing layers. What modifications are required to enable this cross-layer data-centric semantic WSN?

Real-Time Traffic

As highlighted by Gungor et al. (2009), Wireless Sensor Networks become to be used for an industrial purpose or for real-time traffic. The mechanisms we present here to offer a certain service differentiation have to be extended to deal with more complicated scenarios, particularly for hard QoS constraints.

New mechanisms have to be proposed to guarantee probabilistically some properties, taking into account in particular the radio link time-variability.

CONCLUSION

Cross-layer techniques are a necessity in WSN due to their high integration constraints. They permit to save energy transversally in the networking stack, to mutualize information and thus to reduce overhead in these low-bandwidth networks. MAC and routing need particularly to be designed jointly: to decide when packets have to be transmitted has a direct impact on the routing decisions.

We have introduced the instability problem in WSN: antagonist metrics tend to create inconsistent control loops, leading to inconsistent decisions in the MAC and routing layers. Thus, we have to define metrics working in a symbiotic manner in the whole networking stack. However, these metrics depend heavily on the application: they require a careful design and measurement before deploying the solution.

Finally, we focused on the specific integration of the IEEE 802.15.4 and RPL standards. Because they were designed entirely independently, they lead on antagonist and incompatible structures. However, we explained how small modifications make a cross-layered efficient stack: the modified standards present a better reliability and offer easily a QoS-enabled version.

REFERENCES

Akyildiz, I., Melodia, T., & Chowdhury, K. (2008). Wireless multimedia sensor networks: Applications and testbeds. *Proceedings of the IEEE, 96*(10), 1588–1605. doi:10.1109/JPROC.2008.928756

Alphand, O., Duda, A., Heusse, M., Ponsard, B., Rousseau, F., & Theoleyre, F. (2010). Towards the future internet of sensors. In *The Internet of Things* (pp. 309–318). New York, NY: Springer. doi:10.1007/978-1-4419-1674-7_30

Bachir, A., Barthel, D., Heusse, M., & Duda, A. (2006). Micro-frame preamble mac for multihop wireless sensor networks. In *Proceedings of the International Conference on Communications (ICC)*, (vol 7), (pp. 3365–3370). IEEE Press.

Bachir, A., Barthel, D., Heusse, M., & Duda, A. (2007). O(1)-reception routing for sensor networks. *Computer Communications*, *30*(13), 2603–2614. doi:10.1016/j.comcom.2007.05.054

Barrenetxea, G., Ingelrest, F., Schaefer, G., & Vetterli, M. (2008). The hitchhiker's guide to successful wireless sensor network deployments. In *Proceedings of the Conference on Embedded Network Sensor Systems (Sensys)*, (pp. 43–56). Raleigh, NC: ACM Press.

Bose, P., Morin, P., Stojmenovic, I., & Urrutia, J. (1999). Routing with guaranteed delivery in ad hoc wireless networks. In *Proceedings of the International Workshop on Discrete Algorithms and Methods for Mobile Computing and Communications (DIALMPOMC), DIALM1999*, (pp. 48–55). New York, NY: ACM Press.

Ceriotti, M., Mottola, L., Picco, G. P., Murphy, A. L., Guna, S., & Corra, M. … Zanon, P. (2009). Monitoring heritage buildings with wireless sensor networks: The torre aquila deployment. In *Proceedings of the International Conference on Information Processing in Sensor Networks (IPSN)*, (pp. 277–288). Washington, DC: ACM Press.

El-Hoiydi, A., Decotignie, J.-D., & Hernandez, J. (2004). Low power mac protocols for infrastructure wireless sensor networks. In *Proceedings of the European Wireless Conference*, (pp. 563–569). Barcelona, Spain: European Wireless.

Feeney, L., & Nilson, M. (2001). Investigating the energy consumption of a wireless network interface in an ad hoc networking environment. In *Proceedings of INFOCOM*, (pp. 1548–1557). Anchorage, AK: IEEE Press.

Felemban, E., Lee, C.-G., & Ekici, E. (2006). Mmspeed: Multipath multi-speed protocol for qos guarantee of reliability and. timeliness in wireless sensor networks. *IEEE Transactions on Mobile Computing*, *5*(6), 738–754. doi:10.1109/TMC.2006.79

Gungor, V., & Hancke, G. (2009). Industrial wireless sensor networks: Challenges, design principles, and technical approaches. *IEEE Transactions on Industrial Electronics*, *56*(10), 4258–4265. doi:10.1109/TIE.2009.2015754

Intanagonwiwat, C., Govindan, R., & Estrin, D. (2000). Directed diffusion: A scalable and robust communication paradigm for sensor networks. In *Proceedings of the Annual International Conference on Mobile Computing and Networking (MOBICOM)*, (pp. 56–67). Boston, MA: ACM Press.

Juang, P., Oki, H., Wang, Y., Martonosi, M., Peh, L. S., & Rubenstein, D. (2002). Energy-efficient computing for wildlife tracking: Design tradeoffs and early experiences with zebranet. *SIGARCH Computer Architecture News*, *30*(5), 96–107. doi:10.1145/635506.605408

Ko, J., Dawson-Haggerty, S., Gnawali, O., Culler, D., & Terzis, A. (2011a). Evaluating the performance of rpl and 6lowpan in tinyos. In *Proceedings of the International Symposium on Information Processing in Sensor Networks (IPSN), Workshop*. IEEE Press.

Ko, J., Terzis, A., Dawson-Haggerty, S., Culler, D., Hui, J., & Levis, P. (2011b). Connecting low-power and lossy networks to the internet. *IEEE Communications Magazine*, *49*(4), 96–101. doi:10.1109/MCOM.2011.5741163

Koubâa, A., Cunha, A., Alves, M., & Tovar, E. (2008). Tdbs: A time division beacon scheduling mechanism for zigbee cluster-tree wireless sensor networks. *Real-Time Systems*, *40*(3), 321–354. doi:10.1007/s11241-008-9063-4

Kulkarni, S., Iyer, A., & Rosenberg, C. (2006). An address-light, integrated MAC and routing protocol for wireless sensor networks. *IEEE/ACM Transactions on Networking*, *14*(4), 793–806. doi:10.1109/TNET.2006.880163

Levis, P., et al. (2011). *The trickle algorithm.* draft 8, IETF. Retrieved from http://www.ietf.org.

Liu, Y., He, Y., Li, M., Wang, J., Liu, K., & Mo, L. … Li, X.-Y. (2011). Does wireless sensor network scale? A measurement study on greenorbs. In *Proceedings of INFOCOM,* (pp. 873–881). IEEE Press.

Liu, Z., Kwiatkowska, M., & Constantinou, C. (2005). A biologically inspired qos routing algorithm for mobile ad hoc networks. In *Proceedings of the International Conference on Advanced Information Networking and Applications (AINA)*, (vol 1), (pp. 426 – 431). IEEE Press.

Lukic, M., Pavkovic, B., Mitton, N., & Stojmenovic, I. (2009). Greedy geographic routing algorithms in real environment. In *Proceedings of the International Conference on Mobile Ad-Hoc and Sensor Networks (MSN)*, (pp. 86 –93). IEEE Press.

Madden, S. R., Franklin, M. J., Hellerstein, J. M., & Hong, W. (2005). Tinydb: An acquisitional query processing system for sensor networks. *ACM Transactions on Database Systems*, *30*(1), 122–173. doi:10.1145/1061318.1061322

Mulligan, G., et al. (2011). *IPv6 over IEEE 802.15.4 BOF (6lowplan)*. Technical report, IETF Working group. Retrieved from http://www.ietf.org.

Muthukumaran, P. S., de Paz, R., Spinar, R., & Pesch, D. (2009). Mesh-mac: Enabling mesh networking over IEEE 802.15.4 through distributed beacon scheduling. In *Proceedings of the International Conference on Ad Hoc Networks (AdHocNets)*. Niagara Falls, Canada: IEEE.

O'Flynn, B., Bellis, S., Delaney, K., Barton, J., O'Mathuna, S., & Barroso, A. … Sreenan, C. (2005). The development of a novel minaturized modular platform for wireless sensor networks. In *Proceedings of the International Conference on Information Processing in Sensor Networks (IPSN)*, (pp. 370 – 375). IPSN.

Pan, M.-S., Tsai, C.-H., & Tseng, Y.-C. (2009). The orphan problem in zigbee wireless networks. *IEEE Transactions on Mobile Computing*, *8*(11), 1573–1584. doi:10.1109/TMC.2009.60

Pavkovic, B., Theoleyre, F., Barthel, D., & Duda, A. (2010). Experimental analysis and characterization of a wireless sensor network environment. In *Proceedings of the International Symposium on Performance Evaluation of Wireless Ad Hoc, Sensor, and Ubiquitous Networks (PE-WASUN)*. Bodrum, Turkey: ACM Press.

Perkins, C., Belding-Royer, E., & Das, S. (2003). *Ad hoc on-demand distance vector (aodv) routing. rrfc 3561, IETF*. Retrieved from http://www.ietf.org.

Polastre, J., Hill, J., & Culler, D. (2004). Versatile low power media access for wireless sensor networks. In *Proceedings of the International Conference on Embedded Networked Sensor Systems (SENSYS)*, (pp. 95–107). Baltimore, MD: ACM Press.

Raman, B., Chebrolu, K., Gokhale, D., & Sen, S. (2009). On the feasibility of the link abstraction in wireless mesh networks. *IEEE Transactions on Networking*, *17*(2), 528–541. doi:10.1109/TNET.2009.2013706

Ratnasamy, S., Karp, B., Yin, L., Yu, F., Estrin, D., Govindan, R., & Shenker, S. (2002). GHT: A geographic hash table for data-centric storage. In *Proceedings of the International Workshop on Wireless Sensor Networks and Applications (WSNA)*, (pp. 78–87). ACM Press.

Rhee, I., Warrier, A., Aia, M., Min, J., & Sichitiu, M. L. (2008). Z-mac: Hybrid mac for wireless sensor networks. *IEEE/ACM Transactions on Networking*, *16*(3), 511–524. doi:10.1109/TNET.2007.900704

Rhee, I., Warrier, A., Min, J., & Xu, L. (2006). Drand: Distributed randomized tdma scheduling for wireless ad-hoc networks. In *Proceedings of the International Symposium on Mobile Ad Hoc Networking & Computing (MOBIHOC)*. Florence, Italy: ACM Press.

Sanchez, J., Ruiz, P., & Marin-Perez, R. (2009). Beacon-less geographic routing made practical: Challenges, design guidelines, and protocols. *IEEE Communications Magazine*, *47*(8), 85–91. doi:10.1109/MCOM.2009.5181897

Sarr, C., Chaudet, C., Chelius, G., & Guerin Lassous, I. (2008). Bandwidth estimation for IEEE 802.11-based ad hoc networks. *IEEE Transactions on Mobile Computing*, *7*(10), 1228–1241. doi:10.1109/TMC.2008.41

Seada, K., Zuniga, M., Helmy, A., & Krishnamachari, B. (2004). Energy-efficient forwarding strategies for geographic routing in lossy wireless sensor networks. In *Proceedings of the International Conference on Embedded Networked Sensor Systems (SenSys)*, (pp. 108–121). Baltimore, MD: ACM Press.

Srinivasan, K., Kazandjieva, M. A., Agarwal, S., & Levis, P. (2008). The β-factor: Measuring wireless link burstiness. In *Proceedings of the Conference on Embedded Network Sensor Systems (Sensys)*, (pp. 29–42). Raleigh, NC: ACM Press.

Srinivasan, K., & Levis, P. (2006). Rssi is under appreciated. In *EmNets*. Cambridge, MA: Harvard University Press.

Theoleyre, F. (2011). A route-aware mac for wireless multihop networks with a convergecast traffic pattern. *Computer Networks*, *55*(3), 822–837. doi:10.1016/j.comnet.2010.10.018

Vasseur, J., & Culler, D. (2011). Routing over low power and lossy networks (roll). *Working Group, IETF*. Retrieved from http://www.ietf.org.

Villaverde, B., De Paz Alberola, R., Rea, S., & Pesch, D. (2010). Experimental evaluation of beacon scheduling mechanisms for multihop IEEE 802.15.4 wireless sensor networks. In *Proceedings of the International Conference on Sensor Technologies and Applications (SENSORCOMM)*, (pp. 226–231). IARIA.

Ye, W., & Heidemann, J. (2006). Ultra-low duty cycle mac with scheduled channel polling. In *Proceedings of the Conference on Embedded Networked Sensor Systems (SenSys)*, (pp. 321–334). Boulder, CO: ACM Press.

Zabin, F., Misra, S., Woungang, I., Rashvand, H., Ma, N.-W., & Ahsan Ali, M. (2008). Reep: Data-centric, energy-efficient and reliable routing protocol for wireless sensor networks. *IET Communications*, *2*(8), 995–1008. doi:10.1049/iet-com:20070424

Zeng, W., Arora, A., & Shroff, N. (2010). Maximizing energy efficiency for convergecast via joint duty cycle and route optimization. In *Proceedings of INFOCOM*, (pp. 1 –5). IEEE Press.

Zhao, H., Garcia-Palacios, E., Wei, J., & Xi, Y. (2009). Accurate available bandwidth estimation in IEEE 802.11-based ad hoc networks. *Computer Communications*, *32*, 1050–1057. doi:10.1016/j.comcom.2008.12.031

ADDITIONAL READING

Al-Karaki, J. N., & Kamal, A. E. (2004). Routing techniques in wireless sensor networks: A survey. *IEEE Wireless Communications*, *11*(6), 6–28. doi:10.1109/MWC.2004.1368893

Bejerano, Y. (2004). Efficient integration of multihop wireless and wired networks with QoS constraints. *IEEE/ACM Transactions on Networking*, *12*(6), 1064–1078. doi:10.1109/TNET.2004.838599

Braginsky, D., & Estrin, D. (2002). Rumor routing algorithm for sensor networks. In *Proceedings of the International Workshop on Wireless Sensor Networks and Applications (WSNA)*, (pp. 22–31). Atlanta, GA: ACM Press.

Chehri, A., Fortier, P., & Tardif, P.-M. (2009). Cross-layer link adaptation design for UWB-based sensor networks. *Computer Communications*, *32*, 1568–1575. doi:10.1016/j.comcom.2009.05.015

Chilamkurti, N., Zeadally, S., Vasilakos, A., & Sharma, V. (2009). Cross-layer support for energy efficient routing in wireless sensor networks. *Journal of Sensors*. Retrieved from http://www.hindawi.com/journals/js/2009/134165/.

Cuomo, F., Cipollone, E., & Abbagnale, A. (2009). Performance analysis of IEEE 802.15.4 wireless sensor networks: An insight into the topology formation process. *Computer Networks*, *53*(18), 3057–3075. doi:10.1016/j.comnet.2009.07.016

Demirkol, I., Ersoy, C., & Alagoz, F. (2006). MAC protocols for wireless sensor networks: A survey. *IEEE Communications Magazine*, *44*(4), 115–121. doi:10.1109/MCOM.2006.1632658

Gungor, V. C., Vuran, M. C., & Akan, O. B. (2007). On the cross-layer interactions between congestion and contention in wireless sensor and actor networks. *Ad Hoc Networks*, *5*, 897–909. doi:10.1016/j.adhoc.2007.02.007

Han, B., Leblet, J., & Simon, G. (2009). Query range problem in wireless sensor networks. *IEEE Communications Letters*, *13*(1), 55–57. doi:10.1109/LCOMM.2009.081546

Huang, H.-J., Chen, L.-J., & Huang, P. (2008). Cross-layer performance evaluation of sensor networks: Routing over energy efficient mac. In *Proceedings of the Workshop on Performance Monitoring and Measurement of Heterogeneous Wireless and Wired Networks (PM2HW2N)*, (pp. 124–130). Vancouver, Canada: ACM Press.

Isik, S., Donmez, M. Y., & Ersoy, C. (2011). Cross layer load balanced forwarding schemes for video sensor networks. *Ad Hoc Networks*, *9*(3), 265–284. doi:10.1016/j.adhoc.2010.07.002

Karapistoli, E., Pavlidou, F.-N., Gragopoulos, I., & Tsetsinas, I. (2010). An overview of the IEEE 802.15.4a standard. *IEEE Communications Magazine*, *48*(1), 47–53. doi:10.1109/MCOM.2010.5394030

Kwon, H., Kim, T. H., Choi, S., & Lee, B. G. (2006). A cross-layer strategy for energy-efficient reliable delivery in wireless sensor networks. *IEEE Transactions on Wireless Communications*, *5*(12), 3689–3699. doi:10.1109/TWC.2006.256992

Labiod, H. (Ed.). (2007). *Wireless ad hoc and sensor networks*. New York, NY: ISTE.

Lee, M., Zhang, R., Zheng, J., Ahn, G.-S., Zhu, C., & Park, T. R. (2010). IEEE 802.15.5 WPAN mesh standard-low rate part: Meshing the wireless sensor networks. *IEEE Journal on Selected Areas in Communications*, *28*(7), 973–983. doi:10.1109/JSAC.2010.100902

Lin, X., Shroff, N. B., & Srikant, R. (2006). A tutorial on cross layer optimization in wireless networks. *IEEE Journal on Selected Areas in Communications*, *24*(8), 1452–1463. doi:10.1109/JSAC.2006.879351

Lou, W., Wei, L., & Zhang, Y. (2006). Performance optimization using multipath routing in mobile ad hoc and wireless sensor networks. *Combinatorial Optimization in Communication Networks*, *18*, 117–146. doi:10.1007/0-387-29026-5_5

Luo, H., Luo, J., & Liu, Y. (2005). Energy efficient routing with adaptive data fusion in sensor networks. In *Proceedings of the Workshop on Discrete Algothrithms and Methods for MOBILE Computing and Communications (DIALM POMC)*, (pp. 80–88). Cologne, Germany: DIALM POMC.

Mendes, L. D., & Rodrigues, J. J. (2011). A survey on cross-layer solutions for wireless sensor networks. *Journal of Network and Computer Applications*, *34*(2), 523–534. doi:10.1016/j.jnca.2010.11.009

Misic, J., Shafi, S., & Misic, V. (2006). Cross-layer activity management in an 802-15.4 sensor network. *IEEE Communications Magazine*, *44*(1), 131–136. doi:10.1109/MCOM.2006.1580943

Raman, B., & Chebrolu, K. (2008). Censor networks: A critique of "sensor networks" from a systems perspective. *SIGCOMM Computer Communication Review*, *38*, 75–78. doi:10.1145/1384609.1384618

Tang, X., & Xu, J. (2008). Optimizing lifetime for continuous data aggregation with precision guarantees in wireless sensor networks. *IEEE/ACM Transactions on Networking*, *16*, 904–917. doi:10.1109/TNET.2007.902699

van Hoesel, L., Nieberg, T., Wu, J., & Havinga, P. (2004). Prolonging the lifetime of wireless sensor networks by cross-layer interaction. *IEEE Wireless Communications*, *11*(6), 78–86. doi:10.1109/MWC.2004.1368900

Vuran, M., & Akyildiz, I. (2010). Xlp: A cross-layer protocol for efficient communication in wireless sensor networks. *IEEE Transactions on Mobile Computing*, *9*(11), 1578–1591. doi:10.1109/TMC.2010.125

Wang, H., Peng, D., Wang, W., Sharif, H., & Chen, H. (2008a). Cross-layer routing optimization in multirate wireless sensor networks for distributed source coding based applications. *IEEE Transactions on Wireless Communications*, *7*(10), 3999–4009. doi:10.1109/T-WC.2008.070516

Wang, H., Zhang, X., Abdesselam, F., & Khokhar, A. (2010). Cross-layer optimized mac to support multihop qos routing for wireless sensor networks. *IEEE Transactions on Vehicular Technology*, *59*(5), 2556–2563. doi:10.1109/TVT.2010.2042185

Wang, Y., Wu, H., & Tzeng, N.-F. (2008b). Cross-layer protocol design and optimization for delay/fault-tolerant mobile sensor networks (dft-msn's). *IEEE Journal on Selected Areas in Communications*, *26*(5), 809–819. doi:10.1109/JSAC.2008.080607

Willig, A., Karowski, N., & Hauer, J.-H. (2010). Passive discovery of IEEE 802.15.4-based body sensor networks. *Ad Hoc Networks*, *8*(7), 742–754. doi:10.1016/j.adhoc.2010.02.001

Zhang, H., Arora, A., Choi, Y.-R., & Gouda, M. G. (2007). Reliable bursty convergecast in wireless sensor networks. *Computer Communications*, *30*(13), 2560–2576. doi:10.1016/j.comcom.2007.05.046

KEY TERMS AND DEFINITIONS

Bursty Links: Fluctuant radio link, for which the quality can be temporarily good and then drop for a long time.

Coordinators: all the non-leaf nodes of a cluster-tree.

Destination Oriented Direct Acyclic Graph (DODAG): Structure rooted at the destination, comprising node→parent links and without loop. RPL uses a DODAG rooted at the PAN gateway/coordinator.

Multipath Routing: Protocol that discovers several path and may use a combination of them to route packets.

Routing over Low Power Lossy Networks (RPL): Routing protocol currently standardized for Wireless Sensor and Actuator Networks.

Received Signal Strength Indicator (RSSI): Measure reflecting the signal quality for a link. It is related to the Signal To Interference and Noise Ratio (SINR) value during the transmission.

IPv6 over Low power Wireless Personal Area Networks (6LowPAN): Working group proposing to enable IPv6 inside the Wireless Sensor Networks.

IEEE 8021.5.4: standardized MAC layer for low power networks.

Cluster-Tree: Structure used by IEEE 802.15.4, rooted at the PAN gateway/coordinator, forming a tree.

Chapter 12
Cross-Layer Design in Cognitive Radio Networks

Bin Wang
Wright State University, USA

Zhiqiang Wu
Wright State University, USA

Zhongmei Yao
University of Dayton, USA

ABSTRACT

Radio spectrum has become a precious resource. Most frequency bands have been allocated for exclusive use in the US. However, studies have shown that a very large portion of the radio spectrum is unused or underused for long periods of time at a given geographic location. Therefore, allowing users without a license to operate in licensed bands while causing no interference to the license holder becomes a promising way to satisfy the fast growing need for spectrum resources. Dynamic spectrum access and cognitive radio are technologies for enabling opportunistic spectrum access and enhancing the efficiency and utilization of the spectrum. A cognitive radio adapts to the environment in which it operates by sensing the spectrum and then opportunistically exploiting unused and/or underused frequency bands in order to achieve certain performance goals. Due to the close coupling and interaction among protocol layers, the optimal design of opportunistic spectrum access and cognitive radio networks calls for a cross-layer approach that integrates signal processing and networking with regulatory policy making. This chapter introduces basic concepts, design issues involved, and some recent development in this emerging technological field. Future research directions are also briefly examined.

DOI: 10.4018/978-1-4666-0960-0.ch012

INTRODUCTION

Today's spectrum usage follows a long-standing fixed allocation policy. The spectrum use is regulated by government agencies. Spectrum segments are licensed to users for exclusive use in specific geographic areas. In the United States, the authority for radio spectrum regulation is the Federal Communications Commission (FCC) for commercial applications (FCC, 2003). The past decades have witnessed several trends: wireless devices are getting smaller, data transfer rates are increasing, and wireless applications become more and more popular. With the booming wireless systems and applications, only a few small unlicensed bands are left open for anyone to use as long as certain power regulations are followed. As more devices such as laptops, cell phones, wireless sensors, and radio frequency ID go wireless, they must share a finite and overly crowded radio spectrum. As a result, these unlicensed bands have become crowded with everything from wireless networks to digital cordless phones. Radio spectrum over-crowding resulting in congestion is a problem concerning both commercial and military applications alike. Decreasing electromagnetic spectrum availability is a fact facing today's wireless communication as more wireless transmissions and applications are being deployed and envisioned, and the available contiguous spectrum becomes narrower.

On the other hand, current rigid spectrum control policies have resulted in severe spectrum under-utilization. As shown in Figure 1 (FCC Spectrum Policy Task Force, 2002), some frequency bands are largely unused most of the time and some other frequency bands are only seldom used. FCC measurements show that occupancy of approximately 700 MHz of spectrum below 1 GHz is less than 6-10% (FCC Spectrum Policy Task Force, 2002). A field spectrum measurement taken in New York City has also shown that the maximum total spectrum occupancy is only 13.1% from 30 MHz to 3 GHz frequency band (McHenry, 2005). Moreover, studies have shown that the spectrum usage varies significantly in time, frequency, and geographic locations. Meanwhile, a 20 MHz frequency band fetches multi-billion dollars in European 3G spectrum auction. In November 2002, the FCC released a report generated by the Spectrum Policy Task Force that reshaped the traditional models of spectrum allocation and control. As a result, allowing users without a license to operate in licensed bands while causing no interference to the license holder becomes a promising way to satisfy the fast growing need for spectrum resources.

Dynamic spectrum access and cognitive radio are technologies for enabling opportunistic spectrum access and enhancing the efficiency and utilization of the spectrum (Wyglinski, Nekovee, & Hou, 2010). A cognitive radio adapts to the environment in which it operates by sensing the spectrum and then opportunistic exploiting unused and/or underused frequency bands in order to achieve certain performance goals (Steenstrup, 2005; Akyildiz, et al., 2006). Using these technologies, the next generation communication system may adaptively detect the existence of unused and/or underused spectrum bands and intelligently exploit these potentially non-contiguous spectrum opportunities to improve the network performance in a complex and possibly hostile environment, e.g., military Mobile Ad-Hoc Networks (MANETs) such as Soldier Radio Waveform (SRW).

Due to close coupling and interaction among protocol layers, the optimal design of opportunistic spectrum access and cognitive radio networks calls for a cross-layer approach that integrates signal processing and networking with regulatory policy making. For instance, power control directly affects the receiving power at a destination node (signal power) and at other nodes (interference power), therefore profoundly influencing the interference relationship among the nodes and on spectrum sensing and access.

Figure 1. FCC measurements of spectrum use: occupancy of approximately 700 MHz of spectrum below 1 GHz

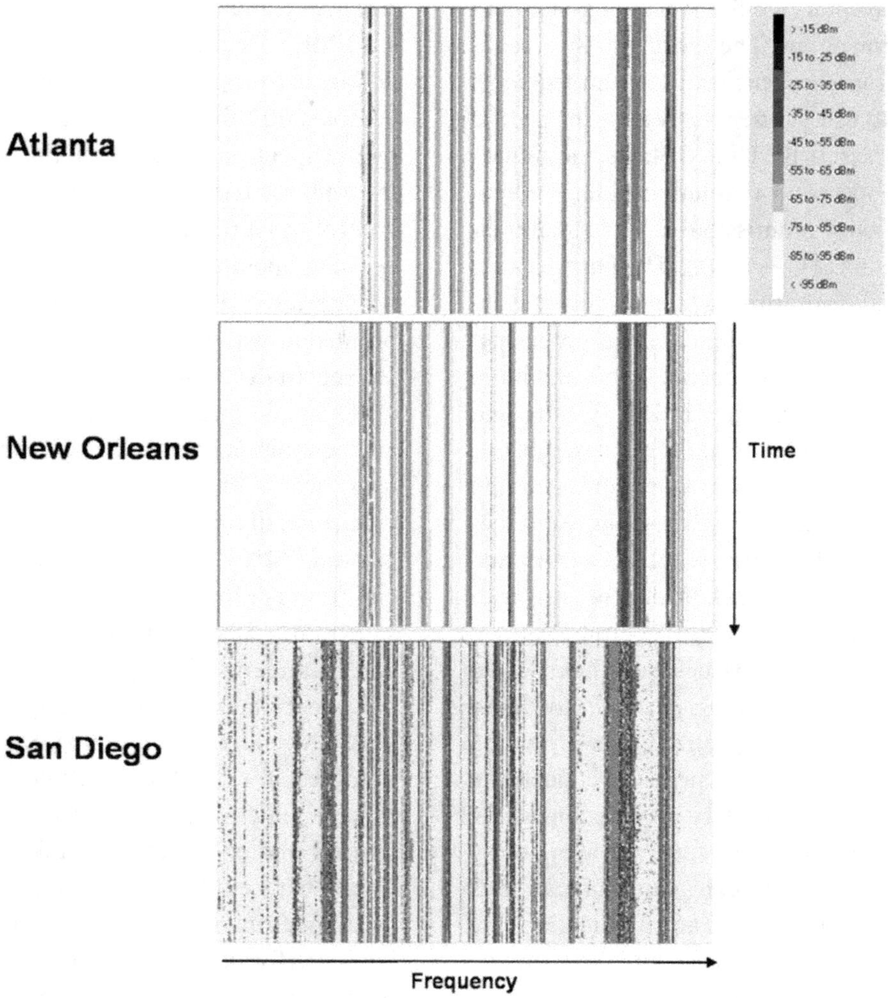

Moreover, power control, spectrum sensing, and spectrum access and scheduling determine link capacities, which, in turn, affect routing. Thus, a networking problem for cognitive radio networks is inherently cross layer in nature. This chapter introduces basic concepts, design issues involved, and some recent development in this emerging technological field. Future research directions are also briefly examined.

DYNAMIC SPECTRUM ACCESS AND COGNITIVE RADIO

As discussed in the introduction section, spectrum crowding will continue to increase as the demand for higher data rates grows and the number of wireless applications and users increase. Recent studies suggest that spectrum congestion is mainly due to inefficient spectrum allocation and usage rather than spectrum scarcity. Two techniques are

Figure 2. Dynamic spectrum access models

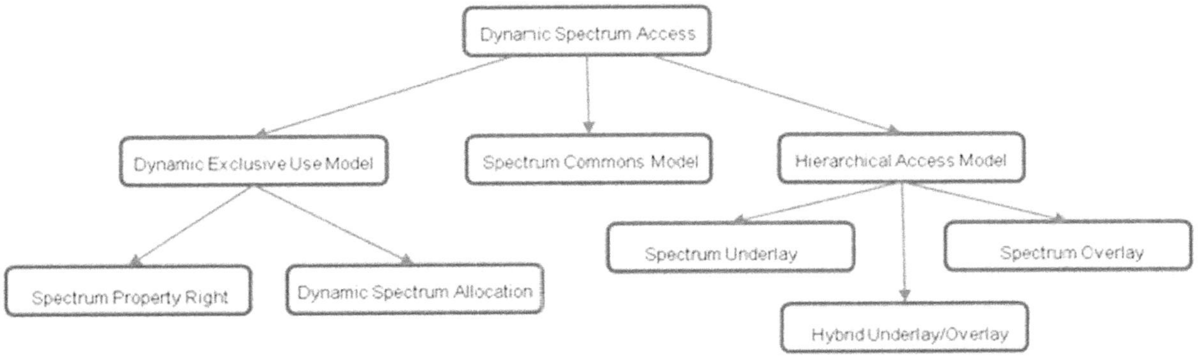

being considered by both the research community and the telecommunication industry to enhance the efficiency and utilization of the spectrum, namely *Dynamic Spectrum Access* (DSA) and *Cognitive Radio* (CR).

Dynamic Spectrum Access Models

Contrary to the current static spectrum management policy, Dynamic Spectrum Access (DSA) is a paradigm shift and comprises a host of approaches to spectrum management reform. The engineering, economics, and regulation communities have proposed diverse ideas and strategies for better spectrum management policies and techniques. As illustrated in Figure 2, dynamic spectrum access strategies can be broadly categorized under three DSA models (Zhao & Sadler, 2007).

Dynamic Exclusive Use Model

This model is the basic model in which the spectrum segments are licensed to services for exclusive use, yet it offers some flexibility in how and when spectrum segments are allocated and used. One approach named "Spectrum Property Rights" allows licensees to sell and trade spectrum and to freely choose technologies (Hatfield & Weiser, 2005). The other approach termed as "Dynamic Spectrum Allocation" (Xu, et al., 2000) tries to improve the spectrum efficiency by dynamically assigning the spectrum segments according to spatial and temporal traffic statistics of different services. This dynamic spectrum assignment occurs at a much faster time scale than the current fixed spectrum management policy allows. However, analogous to carrying bursty traffic using TDMA, this dynamic exclusive use model cannot fully take advantage of spectrum opportunities that are not actively used as a result of the bursty nature of wireless traffic.

Open Sharing/Spectrum Commons Model

Similar to a few open access frequency bands under the current spectrum management regulations such as the unlicensed Industrial, Scientific, and Medical (ISM) frequency band, open sharing/spectrum commons model allows peer users to freely share a spectrum region (Lehr & Crowcroft, 2005). The approaches of this model can be classified as "centralized spectrum sharing" where a central entity (such as a spectrum broker) controls the spectrum sharing (Ileri, Samardzija, & Mandayam, 2005)) and "distributed spectrum sharing" where various local policies govern the spectrum use (Etkin, Parekh, & Tse, 2005; Huang, Berry, & Honig, 2005; Peng, Zheng, & Zhao, 2006).

Figure 3. Underlay dynamic spectrum access

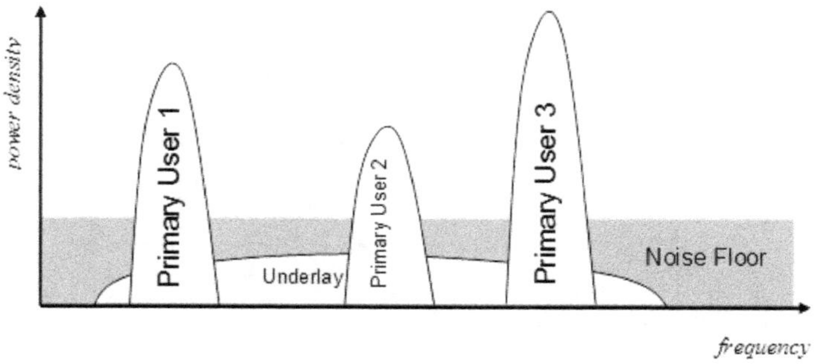

Hierarchical Access Model

Hierarchical access model is most compatible with the current spectrum management policy. It adopts a hierarchical access structure with primary users (PUs or the licensees of the spectrum) and secondary users (SUs or the unlicensed users). In this model, the licensed spectrum is opened to SUs while efforts are made to limit the interference by SUs to PUs. Three approaches have been proposed for spectrum sharing between PUs and SUs, namely spectrum underlay, spectrum overlay, and spectrum hybrid underlay/overlay.

Spectrum underlay is a spectrum management principle by which secondary signals with a very low spectral power density can coexist, as a secondary user, with the primary users of the frequency band(s) (Zhao & Sadler, 2007), e.g., by way of Ultra Wide Band transmission. Figure 3 illustrates the concept of spectrum underlay dynamic spectrum access. In spectrum underlay DSA, an Ultra Wideband (UWB) signal is employed by the radio to spread its signal over a very wide bandwidth (larger than 500 MHz) with very low power spectrum density to minimize interference to primary users. Hence, even though the radio transmission coexists with the primary users' transmission, the interference to primary users is negligible. The underlay approach imposes severe constraints on the transmission power of secondary signals so that they operate below the noise floor of primary signals. The *interference temperature* has been proposed as a measure of the RF power at an antenna generated by other transmitters and noise sources (Kolodzy, 2006). The spectral interference temperature helps a user to determine how much interference it would cause on other users in the locale to ensure the interference does not cross a certain threshold. This access model is suitable for sharing spectrum with legacy primary users without explicit signaling. However, the primary users are often not compensated in this model, and thus lack proper incentives to share the spectrum.

Spectrum overlay (Mitola, 2000) is a spectrum management principle whereby a secondary user uses the frequency band from a primary user only when it is not occupied. Spectrum overlay DSA techniques are based on a detect-and-avoid mechanism such as *dynamic frequency selection*. The secondary user senses the frequency spectrum. If a primary user is active, the frequency band will not be used as shown in Figure 4, e.g., the Radio Local Area Networks (RLANs) use a frequency band that is also used by radar systems. Most current work on DSA and CR assumes this mode of spectrum access and sharing.

Spectrum hybrid underlay/overlay is a spectrum management principle that combines underlay and overlay DSA techniques, which applies the overlay DSA approach on the unoccupied spectrum and applies the underlay DSA approach on the occupied spectrum. The spectrum can be

Figure 4. Overlay dynamic spectrum access

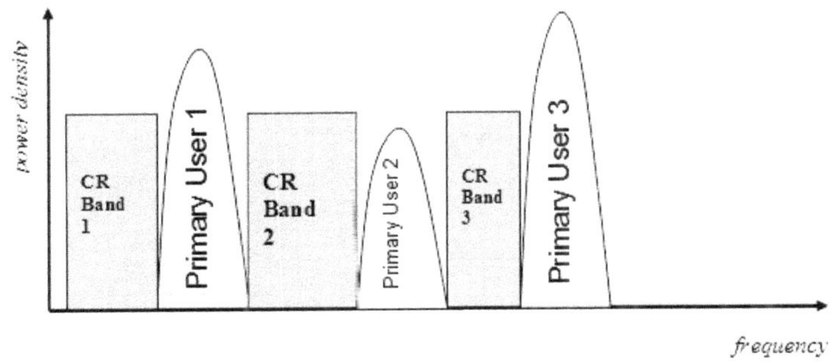

classified into three broadly defined types: (1) white space where the spectrum is free of RF interferers except for ambient noise; (2) grey space where the spectrum is partially occupied by low-power interferers; and (3) black space where the spectrum is occupied by high-power interferers. White spaces and grey spaces are obvious opportunities for use by SUs. The coexistence of underlay transmission and primary users indicates that all the primary users' transmissions can tolerate some level of interference. Hence, the channel capacity can be further increased by making a *soft decision* on the usability of each and every *used band*. If the interference tolerance level of each primary user can be determined, the cognitive radio can transmit over both the unused bands and the used bands to optimize the spectrum usage and maximize the channel capacity. Figure 5 shows such a *hybrid overlay/underlay dynamic spectrum access system*. The spectrum overlay DSA and the spectrum underlay DSA are just two special cases of this *hybrid* paradigm. By combining the benefits of both systems, hybrid overlay/underlay dynamic spectrum access offers better Bit-Error-Rate (BER) performance and larger channel capacity. Under *spectrum hybrid underlay/overlay*, the instantaneous interference tolerance level of all underused bands can be determined by the Signal-to-Noise Ratio (SNR) and the modulation schemes of the transmission of primary users. Based on the instantaneous interference tolerance level, a dynamic interference temperature floor can be used to distribute transmission power over the entire bandwidth including both unused bands and underused bands to maximize the available bandwidth and the channel capacity (Chakarvarthy, Li, Wu, & Temple, 2009; Chakarvarthy, Li, Zhou, & Wu, 2010).

Figure 5. Hybrid underlay/overlay dynamic spectrum access

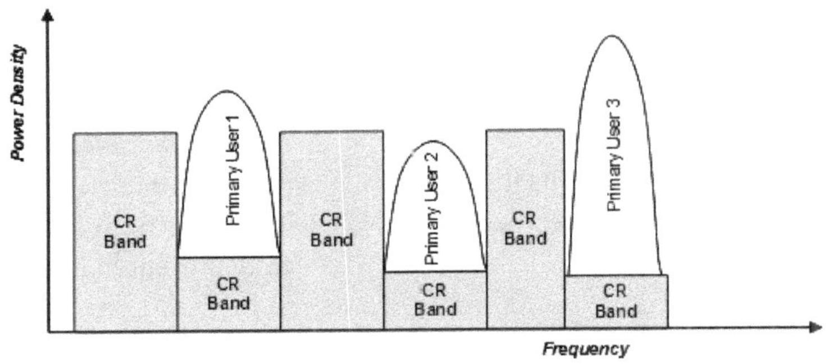

Figure 6. Cognition cycle of a cognitive radio

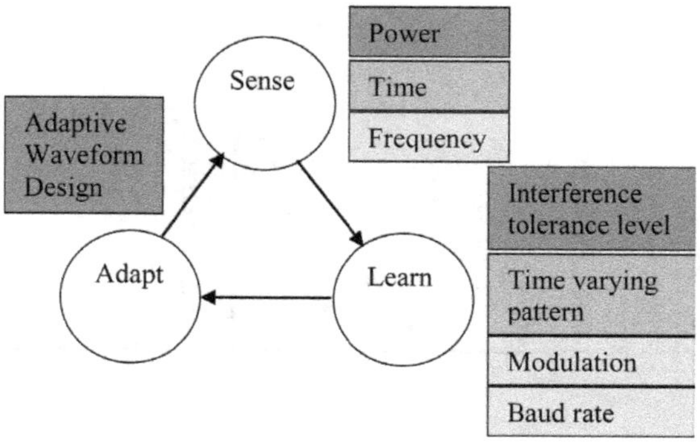

Cognitive Radio

Cognitive radio (Mitola, 2000) was proposed and promoted by Mitola and published in an article in 1999 (Mitola, 1999). Cognitive radio is a context-aware intelligent radio potentially capable of autonomous reconfiguration by learning from and adapting to the communication environment. Cognitive radio is a novel approach in wireless communications that Mitola later described as: "*The point in which wireless Personal Digital Assistants (PDAs) and the related networks are sufficiently computationally intelligent about radio resources and related computer-to-computer communications to detect user communications needs as a function of use context, and to provide radio resources and wireless services most appropriate to those needs.*"

Figure 6 shows the cognition cycle of a cognitive radio, which includes three components: *Sense, Learn,* and *Adapt*:

- *Sense*: the CR periodically senses the environment for spectrum availability;
- *Learn*: the CR learns knowledge from the radio environment;
- *Adapt*: the CR dynamically adapts its transmission based on the knowledge learned.

The functionalities of a cognitive radio include spectrum sensing, spectrum management, spectrum mobility, and spectrum sharing. Specifically,

- *Spectrum Sensing*: The cognitive radio needs to sense the radio environment to detect the presence of primary transmitters and/or receivers, in order to determine the availability of spectrum opportunities in the form of, for example, so called spectrum white space or holes if the spectrum bands are not used. Additionally, spectrum sensing can also determine other important radio frequency parameters such as modulation, Baud rate, power level, and so on. A secondary user can also make use of so called grey spaces (or spectrum opportunities that are underused) that refer to spectra occupied by low-power primary signals. Characterization and modeling of spectrum opportunities are of great importance for the planning and design of cognitive radio networks. Spectrum sensing techniques can be classified into three categories (Akyildiz, et al., 2006): (1) *Transmitter detection* (non-cooperative detection) where cognitive radios determine if a signal from a primary transmitter is locally present in a certain spectrum. Several approaches have been proposed including matched filter detection, energy detection, and cyclostationary feature detection. (2) *Cooperative detection* which refers to spectrum sensing methods where information from multiple cognitive radio users is incorporated for primary user detection. (3) *Interference-based detection* that measures interference at the receivers through the interference temperature model that imposes a maximum cap on aggregate RF energy from multiple transmissions. A SU can use the frequency band if the interference temperature is not exceeded.

- *Spectrum Management:* Cognitive radios should decide on the best spectrum band to meet the quality of service requirements over all available spectrum bands. Therefore, spectrum management functions are required for cognitive radios. These management functions can be classified as *spectrum analysis* and *spectrum decision.*
- *Spectrum Mobility*: Spectrum mobility (or spectrum handoff) is defined as the process in which a cognitive radio user exchanges its frequency of operation. Cognitive radio networks aim to use the spectrum in a dynamic manner by allowing the radio terminals to operate in the best available frequency band while maintaining seamless communication requirements during the transition to better spectrum.
- *Spectrum Sharing*: Cognitive radios and networks need to share spectrum opportunities, therefore requiring methods for spectrum scheduling based on the spectrum sensing results. This is one of the major challenges in dynamic spectrum access paradigm. In a sense, it is similar to generic media access control MAC problems in the existing wireless systems.

Software Defined Radio and Cognitive Radio Implementation

Software Defined Radio (SDR) has emerged in recent years as a powerful concept for current and future wireless communication system design and as a means for cognitive radio implementation. In SDR, key components of the radio are implemented in software. Figure 7 shows a typical diagram of SDR, in which the received RF signal goes through the RF frontend and feeds to an Analog to Digital Converter (ADC), then the output of ADC, the digital signal, goes into, and will be processed by, the software-based communication receiver. In SDR, no complicated and expensive analog circuitry is needed to perform the transmission. On the contrary, software defines the transmitted waveforms and demodulates the received waveforms. Software radio has led the trend in the wireless communication arena to design and build wireless communication systems using reconfigurable software rather than fixed hardware. One such SDR development platform is the off-the-shelf Universal Software Radio Peripheral (USRP) family of boards (Ettus Research LLC, 2011) and GNU software radio (GNU Radio, 2011). USRP boards are designed to allow general purpose computers to function as high bandwidth software radios. GNU radio provides a library of signal processing and communication blocks implemented in C++. By using these blocks, modifying these blocks, or implementing new blocks, users can easily develop SDR-based wireless communication systems. GNU software radio not only can provide the capability to develop new technologies, but also can improve the traditional wireless communication by utilizing the advanced algorithms available in GNU radio to enhance the signal quality.

Clearly, cognitive radio represents a broad paradigm where many aspects of communication systems can be improved through cognition. Dynamic spectrum access is a concrete application of the cognitive radio concept and often is imple-

Figure 7. Typical software radio block diagram

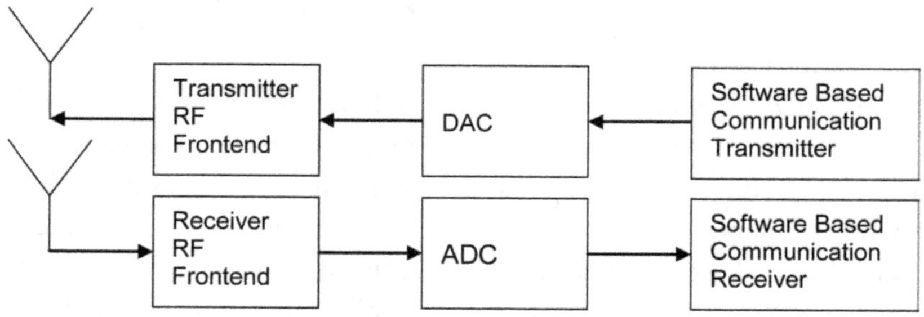

mented on a software defined radio platform where transceivers and protocols can be reconfigured through software to exploit opportunistic spectrum segments. Using the USRP and GNU software radio platform, we have demonstrated a SDR based non-contiguous *Spectrally Modulated Spectrally Encoded* (SMSE) cognitive radio in a *mobile* environment that won the Best Demo Award at the prestigious IEEE Globecom conference in December, 2010 (Zhou, et al., 2010). There are a few unique features of our cognitive radio implementation: (1) we have demonstrated real-time seamless video transmission without interference to primary users and from primary users; (2) our cognitive radio is capable of taking advantage of *multiple spectrum holes* and operating over *multiple noncontiguous* spectrum bands; (3) the cognitive radio dynamically adjusts which subcarriers to turn off according to the primary users' transmission; (4) the cognitive radio can also easily adjust other parameters such as the total number of subcarriers, center frequency, bandwidth of each subcarrier, etc, making it a very flexible and robust cognitive radio node; and (5) finally, the implementation and demonstration also incorporates state-of-the-art inter-carrier-interference cancelation techniques to improve the performance by combating the Doppler effect in a highly mobile environment.

COGNITIVE RADIO NETWORK ARCHITECTURE

Cognitive radios can form Cognitive Radio Networks (CRNs) by extending the radio link features and node functions to network layer functions and beyond. Figure 8 presents reference network architecture for CRNs that interwork with the primary network and other CRNs in licensed as well as unlicensed frequency bands. Depending on the dynamic spectrum access model under which cognitive radios operate, future CRNs can be organized as infrastructure-based networks or infrastructure-less or ad-hoc networks in environments that consist of both licensed and/or unlicensed frequency bands.

CRNs operating in unlicensed bands use open spectrum access where all network entities have the same right to access the spectrum. All CRNs then share the same spectrum, which requires intelligent (and fair) spectrum sharing algorithms in order to improve the efficiency of spectrum use as well as to provide quality of service assurance to users.

CRNs operating in licensed bands use DSA models other than the open spectrum access. In this case, CRNs need to detect the presence of primary transmitters and receivers, opportunistically exploit spectrum white spaces and/or grey spaces to form end-to-end paths, and vacate the spectrum band being used and move to an available spectrum band as needed (e.g., when the primary user becomes active).

Figure 8. Cognitive radio network architecture

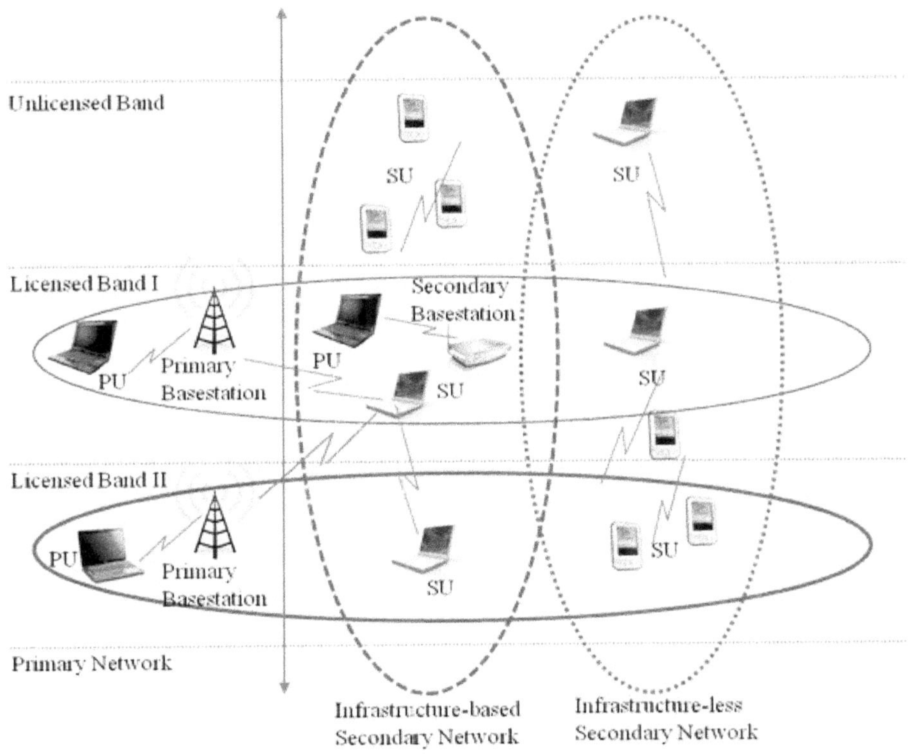

Users of CRNs obtain services by way of (1) *CRN access* where users can access their own CRN base-station on licensed and/or unlicensed spectrum bands; (2) *ad-hoc access* where users can communicate with other CRN users through ad-hoc multi-hop communication on licensed and/or unlicensed spectrum bands; (3) *primary network access* where users access base-stations of primary networks through the licensed band.

Most research on dynamic spectrum access assumes that the primary user is unaware of the presence of secondary users, and secondary users appear transparent to primary users and are only allowed to exploit spectrum opportunistically. The reference network architecture shown in Figure 8, however, also allows the primary users to use the secondary CRNs for their communication. This is indeed possible given the recent development in cooperative communication technology. The transmission rate of the primary users can be dramatically boosted by exploiting cooperative diversity if secondary users that have better channel conditions are exploited as cooperative relays for primary users' communication. As the transmission rate for the primary users improves, CRNs will be left with more spectrum opportunities because the time needed to serve the same amount of primary user traffic is reduced. This will have a positive impact on the performance of CRNs. The cooperation among the primary networks and the secondary CRNs has the potential to enable enhanced performance for both types of networks.

Based on the reference network architecture shown in Figure 8 and heterogeneity of the network environment, some communication functions unique to CRNs (i.e., functions for spectrum-aware communication) are required to support

Figure 9. Cognitive radio network communication: a layered perspective

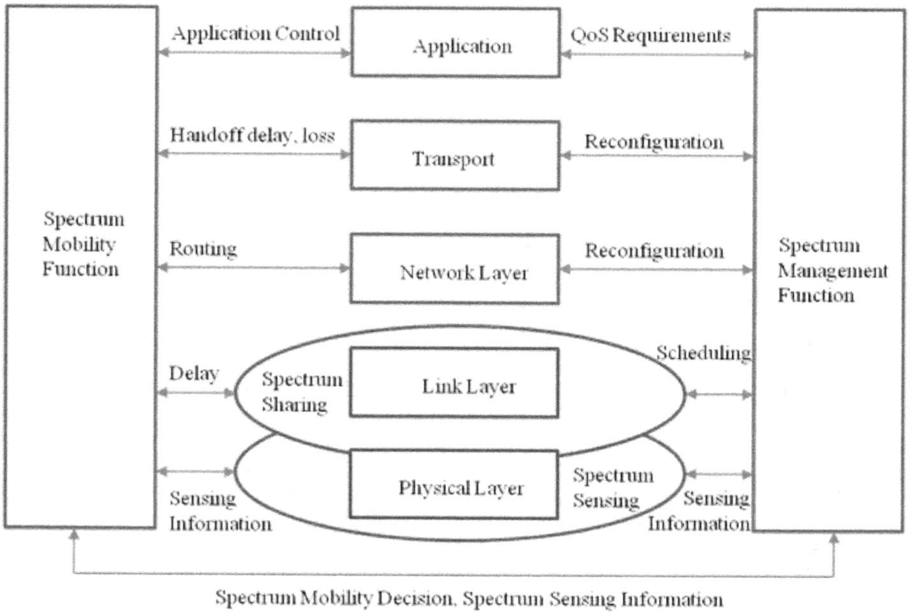

the network operation in a dynamic spectrum environment. Figure 9 presents a layered view of cognitive radio network communication functions and services. Clearly, the dynamics of radio environment have significant impacts on how protocols at various layers work as well as the performance thereof, and pose various challenges for networking in CRNs such as power control, design of scheduling/routing/transport protocols, and end-to-end quality of service assurance. For instance, power control directly affects the receiving power at a destination node (signal power) and at other nodes (interference power), therefore profoundly influencing the interference relationship among the nodes and on spectrum sensing and access. Moreover, power control, spectrum sensing, and spectrum access and scheduling determine link capacities, which, in turn, affect routing. Thus, a networking problem for cognitive radio networks is inherently cross layer in nature. Due to the close coupling and interaction among protocol layers, the optimal design of opportunistic spectrum access calls for a cross-layer approach that integrates signal processing and networking with regulatory policy making.

CROSS-LAYER DESIGN IN COGNITIVE RADIO NETWORKS

Cross-Layer Design Methodology

The dynamic nature of radio frequency environment poses significant challenges in designing new communication protocols for spectrum-aware communication at various layers. More importantly, the performance of CRNs fundamentally depends on the properties of appropriate spectrum opportunities selected. That is, the various spectrum management functionalities such as spectrum sensing, spectrum access and allocation, and spectrum handoff must work collaboratively with higher-layer protocols such as routing and transport protocols and beyond. It has been demonstrated that optimization of network system performance in many cases can be better achieved or can be achieved only through cross-layer design. For instance, a joint design of spectrum sensing at the physical layer and spectrum opportunity exploitation at the MAC layer is necessary to achieve the optimal performance.

There has been a large amount of published work on cross-layer approaches for wireless network design optimization (Shakkottai, Rappaport, & Karlsson, 2003; Akyildiz & Wang, 2008; Foukalas, Gazis, & Alonistioti, 2008). Protocols can be designed by violating the reference network protocol stack (e.g., the well-known 7-layer reference protocol stack) and by allowing direct communication among different protocol layers through sharing variables with nonadjacent layers to obtain performance gains. For example, in wireless networks, physical channel condition and state information can be exploited by the data link layer to choose the best modulation scheme to achieve better channel capacity.

The cross-layer design approaches can be classified based on the type of architectural violation exploited (Srivastava & Motani, 2005). Specifically, layered architectural violations include the following four types:

1. *Creation of new interfaces* where new interfaces not available in the layered architecture between the layers are created for information sharing between the layers at runtime. This category can be further divided into three subcategories depending on the direction of information flow along the new interfaces (a) upward: from lower layer(s) to a higher layer; (b) downward: from higher layer(s) to a lower layer; (c) back and forth: iterative flow between two layers.
2. *Merging of adjacent layers* where the functionalities of several adjacent layers are merged together to offer all the services of the constituent layers. Therefore, no new interface needs to be created in the stack.
3. *Design coupling without new interfaces* where the design of multiple layers is intertwined for information sharing at runtime though no new interface needs to be created. The result of this coupling is that if one layer changes, the design of other layers also need to be modified.
4. *Vertical calibration across layers* where essentially all protocol layers are intertwined.

Various approaches have been proposed to implement cross-layer interactions, including (1) *direct communication between layers* that makes the variables at one layer visible to other layers at runtime. Note however that the benefit of limiting direct communication only between adjacent layers without architectural violation is quite limited. (2) *a shared database across the layers* that can be accessed by all layers and provides the service of storage or retrieval of information to all layers; (3) *completely new abstractions* that present a new way to organize the protocols such as in heaps (Braden, Faber, & Handley, 2002) instead of in stacks in the case of layered architecture. In this case, the interactions of services and functionalities are among different building blocks, which offers greater flexibility both during design and at runtime, but requires complete new system-level implementation.

Figure 10 illustrates a cross-layer perspective of cognitive radio wireless networks. While there has been a great deal of work on various component services and functionalities of CRNs such as spectrum sensing, dynamic spectrum access, and so on, the research on cross-layer design in CRNs has been relatively recent. The need of constant spectrum-awareness in CRNs differentiates the cross-layer design approaches for CRNs from those proposed and studied for other wireless networks. We present a brief summary of some of these spectrum-aware cross-layer design issues in the next section.

Figure 10. A cross-layer perspective of the cognitive radio networks

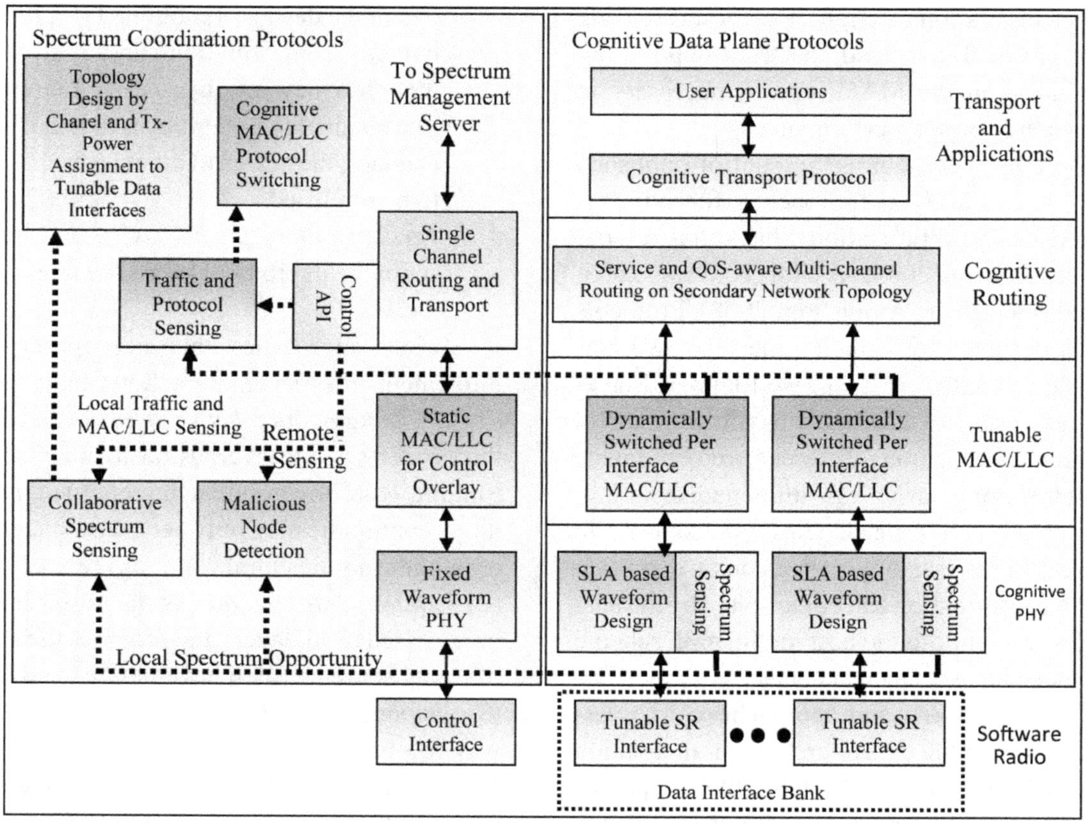

Cognitive Radio Network Cross-Layer Design Issues

Spectrum Sensing

One of the most important components and challenges of cognitive radio networks is the ability to sense the dynamic radio environment to identify the presence of primary users over a wide range of spectrum at a particular time and a specific geographic location with the objective of detecting spectrum opportunities. At the physical layer, many signal processing techniques have been proposed to sense the spectrum opportunities or "holes," including energy detection, matched filter detection, cyclostationary feature detection, wavelet detection, and hybrid detection. One of the most challenging issues of spectrum sensing is the hidden terminal problem which occurs when the cognitive radio is shadowed, in severe multipath fading or inside buildings with high penetration loss while a primary user is operating in the vicinity. As a result, spectrum sensing and spectrum opportunity detection are never perfect.

The merits of spectrum opportunity sensing schemes can be measured in terms of a few parameters such as *probability of detection* or alternatively probability of miss detection, *probability of false alarm*, and *channel detection time*. The detection of spectrum opportunities is, to some extent, equivalent to testing a hypothesis, i.e., whether or not a spectrum band is occupied by the primary user. A spectrum opportunity sensing scheme or a detector carries with it a certain level of uncertainty due to reasons such as measurement errors, modeling limits, and so on. When a false alarm occurs, a spectrum opportunity is not detected because the detection algorithm falsely believes that the primary user is present in the band. The rate of false alarm occurrence is measured by probability of false alarm. On the other hand, when the spectrum is used by the primary user but the detector does not realize that the spectrum is being used, a miss detection results, which potentially leads to a collision with the primary user when the spectrum band is used by a secondary user. The performance of a spectrum opportunity detector is therefore also measured by probability of miss detection. Unfortunately, the two performance metrics, probability of false alarm and probability of miss detection, cannot be minimized simultaneously. Instead, there is a tradeoff between probability of false alarm and probability of miss detection.

What operating characteristics in terms of false alarm versus miss detection probabilities of the spectrum opportunity detector are desirable should be addressed by taking into account the performance of other protocol layers such as the MAC protocol layer, e.g., the throughput of secondary users and the probability of colliding with primary users. On the other hand, spectrum access strategies at the MAC protocol layer should also consider the operating characteristics of the spectrum opportunity detector (Zhao, Tong, Swami, & Chen, 2007). In fact, a joint design of spectrum sensing at the physical layer and spectrum opportunity exploitation at the MAC layer is necessary to achieve the optimal performance.

Spectrum sensing is also hampered by the hidden-terminal problem that arises because of shadowing. Secondary users may be shadowed away from the primary's transmitter but there may be primary receivers close to the secondary users that are not shadowed from the primary transmitter. Hence, if the secondary transmits, it may interfere with the primary receiver's reception. To deal with the hidden terminal problem in CRNs for spectrum sensing, cooperative spectrum sensing has been proposed to improve the detection of the primary users (Unnikrishnan & Veeravalli, 2007). Specifically, multiple secondary users cooperate to conduct spectrum sensing. For example, neighboring secondary users exchange sensing information through a dedicated control channel and the collected sensing information is then fused together to achieve improved spectrum sensing outcomes. Cooperative sensing results in lower probability of false alarm and increases probability of detection. This approach also entails more overhead due to the coordination in spectrum sensing among secondary users. Hence, there is a tradeoff between the overhead and sensing performance.

However, the cooperative sensing approach is still not satisfactory because the existing spectrum sensing model is mainly based on the detection of primary transmitters, which does not give direct and accurate information of unused spectrum. Ultimately, spectrum opportunity detection needs to be performed jointly by the transmitter and receiver of secondary users, which involves both signal processing and networking aspects, calling for cross-layer design approaches.

Cooperative communications (Kramer, Maric, & Yates, 2006) have been proposed to exploit the broadcast nature of the wireless communication to form virtual antenna arrays for transmission, thereby achieving the spatial diversity gain inherent in multiuser wireless systems. This is made possible by allowing users to relay each other's messages, and thus forming multiple transmission paths to the destination. This mode of cooperation in wireless ad-hoc networks can be further extended by optimally allocating the energy and bandwidth resources among users based on the available channel state information at each node. Cooperative communications can be applied to cognitive radio networks to improve the overall network spectrum utilization, power efficiency, and end-to-end quality of service assurance. In this case, the radio parameter tuning is based on the updated information of the internal and external radio environment, such as radio frequency, user behavior, and network state. This again requires intelligent and accurate spectrum sensing and the use of cross-layer design approaches.

Spectrum Dynamics/Spectrum Opportunity Modeling

As discussed previously, in CRNs secondary users can access the spectrum bands not being used by the primary users. However, since the spectrum usage changes both temporally and spatially, secondary users need to keep track of the changes of the radio environment, such as primary users' spectrum activities (Zheng & Peng, 2005; Peng, Zheng, & Zhao, 2006). Because efficient utilization of spectrum by secondary users heavily depends on the spectrum availability, the prediction and modeling of spectrum opportunities are essential for effective spectrum allocation in CRNs and for evaluating new wireless communication and networking algorithms and techniques designed to perform dynamic spectrum access. Spectrum occupancy information can be used for both learning the usage of the spectrum and predicting the future occupancy status. Secondary users may predict the availability of a specific spectrum band based on the past spectrum sensing data. An effective spectrum availability prediction algorithm helps secondary users to make right decisions on spectrum allocation and to minimize the interference to primary users. Statistical methods are often used in spectrum occupancy and spectrum modeling and availability prediction research.

Numerous spectrum modeling and availability prediction approaches have been proposed and studied. They are mostly based on statistics of past spectrum usage of primary users. A time-varying statistical spectrum occupancy model is proposed in (Gosh, Pagadarai, Agrawal, & Wyglinski, 2010). The model is capable of emulating primary user occupancy for different sub-bands. Specifically, the utilization and idle periods are governed by two independent Poisson processes. Transmission power during a utilization period is modeled by a Gaussian distribution with mean and standard deviation computed from measurements. The primary user's selection of sub-bands in each of its utilization period follows a uniform distribution. The model is shown to accurately generate both the temporal and frequency behavior of various wireless transmissions by comparing with spectrum measurements obtained from frequency measurements in the paging band.

In Yarkan and Arslan (2007), the use of binary time series for spectrum occupancy characterization and prediction was proposed. At each time instant, CR decides if the bands observed are occupied or not by a threshold. At the decision instant, the occupancy status is converted into the binary form. Then, CR stamps this occupancy state with the time information. A binary time series model is obtained to represent the previous occupancy status and to allow one to establish near-field predictions. The performance of the predictor suffers from the non-deterministic nature of the binary series. In Wen et al. (2008), a second-order autoregressive model was used to predict the channel status transitions over fading channels with Gaussian noise. The parameters of the model are computed using the Yule-Walker algorithm. The spectrum occupancy is then predicted using a Kalman filter.

In Chittabrata et al. (2010), the authors used Beta distribution in modeling channel occupancy of primary users in a licensed spectrum band. It is validated that the Beta distribution is a good fit for the spectrum occupancy patterns in the 1500 MHz band centered at 770 MHz observed inside a modern office building at Aachen, Germany. Based on this occupancy model, the authors presented a Poisson-normal approximation method for spectrum availability modeling as well as a classification of available channels based on occupancy of its adjacent channels. Probabilistic modeling of different types of available channels was presented as well. The simulation results indicate that these analytical models of spectrum availability and channel type classification will facilitate efficient sensing over preferable sections of an operating spectrum.

Markov model and Hidden Markov Model (HMM) are also commonly used in spectrum availability prediction. In Chittabrata et al. (2009), the spectrum band occupancy at any time instant is considered as a state, which can be either free (un-occupied by a PU) or busy (occupied by a PU). The states of a spectrum band are monitored over *L* consecutive time periods. A Markov chain is used for modeling the utilization of each spectrum band by a PU over *L* time periods. The authors validated such a Markov model using real-time measurements collected in the paging band (928-948 MHz). Since the detection of idle spectrum bands by a cognitive radio is prone to errors, the sensing errors were probabilistically modeled using a Hidden Markov prediction model to predict the true states of the spectrum band. In Akbar and Tranter (2007), another Hidden Markov model based spectrum prediction was proposed. In Chen and Qiu (2010) the latency between spectrum sensing and data transmission was studied, and an approach for prediction of channel state based on spectrum sensing slots using higher-order HMM was proposed. However, to the best of our knowledge, not all frequency channels were validated to fit the property of Markov chain and Hidden Markov chain. In addition, different frequency channels have different transition matrix and initial probabilities. These parameters need to be stored for efficient Markov chain parameter estimation of channel availability, whereas most of the secondary users (devices) only have limited memory. By the same token, the Hidden Markov prediction model requires more memory to predict the channel availability. Even though the memory is sufficient enough, to guarantee the accuracy of the prediction of channel availability, a relatively long time is still needed to obtain the transition matrix and other parameters based on the statistical data from past sensing results. A neural network approach using the multi-layer perceptron network to predict the channel availability was proposed in Tumuluru, Wang, and Niyato (2010). The approach is essentially based on past sensing data set as well, which means a long sensing phase is needed to guarantee a decent channel availability prediction. If the past sensing data set is not large enough, the limited past sensing investigation may mislead the prediction to some extent.

Spectrum Allocation and Access

As an intelligent wireless communication technique, cognitive radio is made aware of the radio frequency environment, and can select communication parameters such as carrier frequency, bandwidth, and transmission power to optimize the spectrum usage and adapt its transmission and reception accordingly.

An important aspect of cognitive radio system design is to allow dynamic sharing, e g., dynamic allocation and access of spectrum opportunities.

There are two very different approaches to dynamic spectrum sharing between the PUs and the SUs: coexistence and cooperation (Akyildiz, et al., 2008). In the coexistence approach, the secondary users try to avoid interferences with the primary users without explicit signaling (e.g., the underlay DSA model introduced earlier). In the cooperation approach (e.g., the overlay DSA model introduced earlier), primary and secondary users may explicitly communicate with each other to reach a sharing arrangement. The cooperation approach makes flexible negotiation and economic compensation easier through explicit communication. The primary users can obtain performance improvement through the help of secondary users. The secondary users often have a better idea of the spectrum environment and therefore allow the system to achieve a better quality of service. However, the governmental spectrum sharing policy-making needs to support the cooperation approach.

The approaches for spectrum sharing among SUs can be divided into cooperative and non-cooperative approaches. Cooperative approaches consider the interference measurements of all secondary users (Brik, Rozner, & Banerjee, 2005; Huang, Berry, & Honig, 2005). In other words, the interference information of each secondary user is shared among the others. On the contrary, the non-cooperative approaches consider only the node at hand (Zhao, Tong, & Swami, 2005; Zheng & Cao, 2005). These approaches are commonly called selfish approaches. The advantage of non-cooperative approaches is that a secondary user only needs the minimal communication with other secondary users. A great deal of research work has been conducted in this area. For example, spectrum opportunity allocation among secondary users in a slowly varying RF environment is equivalent to graph coloring where secondary users and interference relationship between two secondary users are represented as vertices and edges of the graph, and spectrum bands are denoted by colors. Two vertices linked by an edge cannot be assigned the same color (i.e., interfering secondary users cannot be allocated the same spectrum band). The objective is to obtain a color assignment that optimizes a given utility function. The problem is *NP*-hard. Based on the architecture of cognitive radio networks, the spectrum allocation approaches can be divided into centralized approaches and distributed approaches. Centralized and distributed heuristic approaches have been devised and studied (Zhao, Zheng, & Yang, 2005; Peng, Zheng, & Zhao, 2006; Etkin, Parekh, & Tse, 2005; Huang, Berry, & Honig, 2005). A host of approaches based on game theory, auction, and bargaining have also been reported (Han, Pandana, Liu, 2007; Gandi, et al., 2007; Zhou, Gandi, Suri, & Zheng, 2008).

In terms of spectrum access after spectrum opportunity detection, cognitive radio needs to determine whether to transmit and how to transmit over a particular spectrum opportunity based on the detection result in the presence of detection errors. Cognitive radio also needs to decide what modulation and transmission power to use, and how to share opportunities among secondary users to achieve a network-wide optimization objective. The optimal access strategy should take into account the operating characteristics of the spectrum detector for reasons discussed previously. The strategy should achieve a proper tradeoff between minimizing overlooking spectrum opportunities and avoiding collisions with the primary users. Additionally, optimal spectrum allocation and access strategies also need to respond to, and can only be achieved through interaction with, higher protocol layers such as routing, transport, and application. This further demonstrates the benefits and the importance of cross-layer design approaches in CRNs.

Transmission Power and Modulation

Determining an appropriate level of transmission power of secondary users is also rather complicated and poses unique challenges in CRNs. The maximum transmission power of a secondary user depends on the detection range of its spectrum detector, the prescribed maximum interference level, and the path loss factor. Power control for secondary users should also take into account the transmission power of primary users, shadowing, fading, and aggregated interference from simultaneous transmission from multiple secondary users. Power control directly affects the receiving power at a destination node (signal power) and at other nodes (interference power), therefore profoundly influencing the interference relationship among the nodes and on spectrum sensing and access. Moreover, power control, spectrum sensing, and spectrum access and scheduling determine link capacities, which, in turn, affect routing. Thus, a networking problem for cognitive radio networks is inherently cross layer in nature.

Another closely coupled challenging issue is modulation. To make better use of spectrum, secondary users transceivers may transmit over multiple *non-contiguous* frequency bands as the available contiguous spectrum becomes narrower and available spectrum becomes fragmented. It is therefore natural to choose multi-carrier based technologies for the physical layer. In a multi-carrier based CR, the subcarriers located in the vicinity of a primary user need to be deactivated to avoid interference to the primary users. In the literature, these schemes are referred to as Non-Contiguous OFDM (NC-OFDM) or Non-Contiguous MC-CDMA (NCMC-CDMA).

We proposed overlay/underlay cognitive radio waveform design using the SD-SMSE framework (Chakravarthy, et al., 2009; Chakravarthy, et al., 2010), and Non-Contiguous Spread OFDM (NC-SOFDM) is an important application in the overlay cognitive radio waveform design. When M out of the total N subcarriers are deactivated, it has been shown in Rajbanshi et al. (2007) that in the NC-SOFDM case, deactivating the M subcarriers introduces a loss of orthogonality among different symbols' spreading codes, hence significantly decreasing the BER performance of NC-SOFDM. It is also pointed out in Rajbanshi et al. (2007) that the system performance can be improved by using adaptive spreading code selection to minimize (or even eliminate) the loss of orthogonality.

We investigated the BER performance of NC-SOFDM with Hadamard-Walsh spreading code and adaptive spreading code selection in a CRN. However, Hadamard-Walsh codes do not exist for every code length. Specifically, Hadamard-Walsh codes only exist when $N = 1$; 2 and 4m, where N is the code length and m is a natural number. To retain the orthogonality among spreading codes, the system may have to deactivate redundant subcarriers, which may also reduce the data rate. To maintain the data rate, traditional NC-SOFDM with Hadamard-Walsh spreading code employs a code length larger than the number of active subcarriers. The loss of orthogonality among spreading codes that cannot be eliminated is minimized by *adaptive code selection*. By treating the system as sub-systems, each sub-system has Hadamard-Walsh code. We can then generate binary orthogonal code set for all the sub-systems to maintain the data rate. On the other hand, if the spreading code is not limited to be binary, better spreading solutions exist for NCSOFDM.

Our work on carrier interferometry OFDM (CI/OFDM) (Li, et al., 2009) and CI/MC-CDMA (Wu, Ratazzi, Chakravarthy, & Hong, 2008), and other similar technologies such as single carrier OFDM (SCOFDM) (Kobayashi, et al., 2003) and single carrier transmission with frequency domain equalization (SCFDE) (Czylwik, 1997) are essentially the same technology that combines the benefit of multi-carrier transmission and single carrier transmission, by employing cyclic prefix to single carrier transmission and frequency domain processing. These technologies have received a lot of attention as the alternative transmission technique to the conventional OFDM because of their better performance and lower Peak to Average Power Ratio (PAPR) property as compared with the conventional OFDM. Besides the performance and PAPR benefit, a very nice feature of Carrier Interferometry (CI) code is that the code length can be of any integer value. Hence, by applying CI code as the spreading code for NC-SOFDM, we can *adaptively allocate orthogonal CI spreading codes*, eliminating the loss of orthogonality entirely among spreading codes, regardless the number of deactivated subcarriers. We have proposed two novel spreading code sets for NC-SOFDM for cognitive radio in a CRN. Both proposed code sets can maintain orthogonality regardless spreading code length. By employing any one of the proposed code sets, the NC-SOFDM outperforms the adaptive NCSOFDM with adaptive Hadamard-Walsh code set and the traditional NC-OFDM.

Recent Cross-Layer Design Approaches for CRNs

In this section, we present a few examples of cross-layer design approaches for some representative cross-layer interaction that address many of the design issues discussed in the previous section. Unless clearly pointed out, most current work assumes *underlay* dynamic spectrum access and sharing discussed in the Dynamic Spectrum Access Models section.

MAC/Physical Cross-Layer Design

As we discussed in the Cognitive Radio Network Cross-layer Design Issues section, in wireless ad hoc networks especially infrastructure-less cognitive radio networks, the MAC layer has to adapt to the changes in communication resources, network density, and mobility by reconfiguring the parameters of transceivers. In fact, there is always a close coupling between the MAC layer and the physical in order to achieve good performance and to be able to adapt to the environment and network dynamics. Advanced physical signal processing, modulation, coding, and antenna techniques such as MIMO, OFDM, UWB, provide a great potential for improving network performance in terms of delay, throughput, and packet loss rate. The improvement can be best harvested by cross-layer design that integrates the physical layer and the MAC layer.

Identify time-varying spectrum availability in CRNs poses a number of nontrivial design problems, especially in an ad hoc wireless networks where there is no centralized controllers such as a base-station.

The CR of secondary users must decide when and which band(s) they should choose to transmit/receive without causing undue inference to primary users. The work by Zhao, Tong, Swami, and Chen (2007) addressed the joint design of spectrum sensing at the physical layer and spectrum opportunity access and allocation at the MAC layer in the presence of sensing error while limiting the interference to the primary user. They developed an optimal spectrum opportunity access strategy based the Partially Observable Markov Decision Processes (POMDPs) framework. However, implementation-wise, the scheme is complicated. Su and Zhang (2008) proposed a cross-layer opportunistic multi-channel MAC protocol for wireless ad hoc networks. The proposed schemes integrate the spectrum sensing at the physical layer with packet scheduling at the MAC layer. Two transceivers are used: one is for the control channel and the other is for dynamic accessing spectrum opportunities detected. When multi-channel transceivers are not available, the secondary users can only transmit in a fixed frequency band. Normally, if only white space can be used, then the secondary users have to wait till the frequency band is vacant. This obviously may result in large packet delay and loss. To provide better quality of service, in Chen, Letaief, and Cao (2008), a joint coding and scheduling scheme for cognitive multiple access is proposed where a successive interference decoder (i.e., successive interference cancellation) is adopted to mitigate the secondary users' interference to the primary user. As a result, *the secondary users are allowed to share the frequency band with the primary user*. To minimize average packet delay under a given power constraint of the secondary users, a joint channel-aware and queue-aware scheduling is proposed to exploit the delay-power tradeoff of joint coding and scheduling by optimizing the scheduling parameters at the MAC layer. The joint coding and scheduling scheme significantly reduces the packet delay of the secondary users.

Network/Physical Cross-Layer Design

The dynamic nature of the underlying spectrum in cognitive radio networks also necessitates higher communication protocols to adapt to the wireless channel parameters. The behaviors of protocols at different layers affect the performance of other protocols. For example, in ad hoc cognitive radio networks, routing is affected by link failures that arise from spectrum mobility. The change in link error rate and the performance of the media access protocols also affect the link's usability, leading to potential re-routing. A major design choice for routing in CRNs is the collaboration between routing and spectrum sensing and access. This calls for novel spectrum-aware routing algorithms. Routing also needs to differentiate spectrum mobility from a node failure. Therefore, the interaction between routing and the spectrum management functionalities must be taken into account when designing a routing algorithm in this environment.

Based on the protocol interference model where two links either interfere destructively or do not interfere at all, in Wang and Zheng (2006), a source node finds candidate paths using dynamic source routing and collects information on link connectivity and quality. For each candidate route, the algorithm finds all feasible spectrum assignment combinations and estimates the end-to-end throughput performance for each combination. The route and spectrum assignment with the maximum throughput is then select and a conflict free channel is then scheduled for this route. In the work of Xin, Xie, and Shen (2005), a layered graph model is proposed where each layer corresponds to a channel. A route is determined to be the shortest path based on the layered graph. The work of Ding et al. (2010) considered a multi-hop cognitive ad hoc network where spectrum opportunities are location-dependent, and in a multi-hop path, the available spectrum bands may be different at each intermediate node. The work proposed a distributed algorithm that jointly addresses routing, dynamic spectrum allocation, scheduling, and power control functionalities under the physical interference model. This model computes the interference among secondary users using a SINSR-based model. The algorithm allows each cognitive radio to make real-time decisions on spectrum and power allocation based on locally available information to maximize the link capacity on the selected spectrum band. The authors showed that the proposed cross-layer solution that solves the routing and spectrum jointly at each hop outperforms approaches where routes are selected independently of the spectrum assignment.

Design across Multiple Protocol Layers

A fundamental problem for a wireless network is power control. Power control enables dynamic transmission power configuration within the permissible power limit. A lower transmitter power level allows more users to share the spectrum and to decrease the interference. Power control also enables dynamic network topology control and significantly affects upper protocol layer functionalities such as scheduling and routing. The effect is especially pronounced in CRNs. In fact, coupling relationship exists between power control and upper protocol layers such as scheduling and routing in CRNs. Power control can be done on a network-wide synchronized manner where transmission power at each node in the network is adjustable but is synchronized to be identical. This approach may not offer optimal network performance. The work in Shi and Hou (2009) investigated the challenging problem of per-node based power control for multi-hop CRNs. A mathematical framework is developed to optimize multiple layers jointly and to formulate a mixed integer non-linear program to support a set of user sessions with an objective of minimizing network resource usage which is measured by the bandwidth-footprint-product metric. A solution procedure based on branch-and-bound and convex full relaxation is devised to derive results for all three layers jointly. The solution guarantees *1-ε* optimal performance where ε is a small pre-specified error tolerance parameter. Shi and Hou (2008) designed a distributed algorithm for a multi-hop cognitive radio network with the objective of maximizing data rates for a set of user communication sessions. The work jointly considered power control, scheduling, and routing. The distributed optimization algorithm iteratively increases data rates for user communication sessions. In each iteration, two separate processes, a conservative iterative process and an aggressive iterative process, are employed that incorporate routing, minimalist scheduling, and power control/

scheduling. The distributed algorithm obtains results close to the upper bound of the objective function.

Baldo and Zorzi (2007) proposed a cognitive cross-layer architecture where fuzzy logic is used for the representation of cross-layer information and the implementation of optimization strategies. Cognitive radio relies on cross-layer information exchange and interactions to exploit different wireless interfaces and protocol stack configurations in order to achieve the best service quality. In practice, most information exported by the different layers is obtained from measurements, which are affected by errors in precision and accuracy. For example, SNR measurements are very inaccurate and BER performance has a very sharp transition from excellent to bad performance. Based on SNR alone, a bad channel might be misinterpreted as good and vice versa. The proposed fuzzy logic based architecture can naturally deal with imprecise information. For each layer in the protocol stack, a set of variables and parameters that are representative of the functionality provided by the layer are identified. These variables and parameters will be the fuzzy variables. For instance, a physical layer can exploit knowledge of the modulation and coding schemes being used to translate SNR measurements in dB into more abstract fuzzy attributes such as *bad* or *good* to allow the fuzzy logic to interpret even without explicit knowledge regarding modulation and coding. Similarly, measurements and settings obtained at other layers can be translated into fuzzy control variables. For example, at the transport layer, measurements could be congestion window size and retransmission timeout which then undergo a fuzzification process to become a fuzzy variable that can be used to interpret network congestion level as *high* or *low*. Fuzzy logic controllers can then be designed to implement the cross-layer joint optimization control schemes over multiple layers. A proof of concept for a cross-layer optimization scheme for TCP is presented. The architecture is shown to achieve benefits in terms of modularity, complexity, scalability, and usability. However, to attain the benefits, a significant standardization effort is required.

Cross-Layer Design for Transport Protocols

In cognitive radio networks, media access techniques affect the Round-Trip Time (RTT) for the transport layer protocols. Latency associated with spectrum mobility also increases the instantaneous RTT of the packet transmission, which lead to variations in packet transmission delay. Link failures may occur due to spectrum environment change, which results in packet loss apart from packet loss due to congestion.

Interdependences among functionalities of the communication stack, and their close coupling with the physical layer necessitates a cross-layer spectrum management function which ideally should consider physical (spectrum opportunity detection and identification), MAC, routing, transport, and application quality of service requirements. Therefore, transport protocols for CRNs should be spectrum-aware and exploit cooperative operation with the other communication layers.

Chowdhury, Felice, and Akyildiz (2009) proposed a window-based Tranpsort Protocol for CR Ad-Hoc Networks, or TP-CHAHN. The protocol is adapted from the classical TCP rate control algorithm and runs at the source to interact with the physical layer channel information, the link layer functions of spectrum sensing, buffer management, and a predictive mobility framework, apart from dealing with classical concerns of congestion, flow control and connection loss due to mobility. It distinguishes states that signify intermittent spectrum sensing, spectrum change due to activity of the licensed users of the spectrum, route failure as a result of spectrum availability, and predicted mobility. The protocol uses explicit congestion notification and combines explicit feedback from the intermediate nodes and the destination. Link-layer information, such as residual buffer space, observed link bandwidth, link latency, is piggybacked over the data packets to the destination, which is then sent back to the source through ACK packets. Based on the combined information, the source responds by appropriately adjusting its transmission rate.

The work by Luo, Yu, and Leung (2010) jointly considered the spectrum sensing, spectrum access decision, physical-layer modulation and coding scheme, and data-link layer frame size to maximize the TCP throughput in cognitive radio networks. A cross-layer optimization architecture is presented. Based on the historic observations and past action decisions, the secondary user will decide whether to sense the channel. If a sensing action is selected, the cognitive sensor will observe the channel and obtain the sensing outcomes. The sensing outcomes are then sent to the TCP layer. To maximize the TCP throughput, the secondary user will determine whether to access the channel and the corresponding modulation and coding scheme at the physical layer and the frame size at the data-link layer. Consequently, it will feed back the three decisions to the MAC layer, the physical layer, and the data-link layer, respectively. The wireless channel and the primary network and usage are modeled as a first-order, finite-state Markov process. The TCP throughput optimization problem is formulated as a Partially Observable Markov Decision Process (POMPDP) to obtain the optimal policy for channel sensing selection, sensor operating point, access decision, selection of the modulation and coding scheme, and the frame size to maximize the TCP throughput.

The above sample recent work has demonstrated the benefits and the importance of cross-layer design approaches in CRNs.

FUTURE RESEARCH DIRECTIONS

Future cognitive radio networks will include dynamic selection and reconfiguration of network protocols not only at the physical layer, but also at the MAC layer, the network/routing layer, and to some extent, at the transport layer. While the research community is currently developing spectral sensing and cognitive radio mechanisms primarily at the physical layer, it is appropriate to start developing models and protocol structures to extend the cognitive aspects of a network from the physical layer to the upper protocol layers. Cognitive features of a wireless communication system can be fully exploited only when the sensing, network cognition and protocol reconfiguration operations are all carried out at appropriate layers. A cross-layer architectural framework is therefore necessary for cognitive networks with emerging applications. Figure 10 illustrates a cross-layer perspective of cognitive radio wireless networks.

Available spectrum fragments in the form of white space or gray space appear as useable channels to a CR node with respect to the spectrum band in question. For a CR node, given its location, a set of available channels can be available at a given time. Such a set of available channels are referred to as the spectrum opportunities of the CR node. The spectrum opportunities can be highly dynamic depending on various factors including the environment spectrum usage pattern, channel conditions, and node mobility. Different CR nodes can also sense different spectrum opportunities at a given time. Therefore, low-cost more accurate and faster spectrum sensing techniques that can work over a wide spectrum band are sought for.

From an upper layer perspective, parameterized spectrum opportunities can be used as the cognitive information interface between the MAC, routing and transport layers, and the process of access dynamics at the physical layer. From this standpoint, significant amount of research and development will be needed in terms of measurement, characterization, modeling, and prediction of the spectrum opportunity dynamics for different spectrum segments, primary user community, and dominant applications. The success of any cross-layer network protocol design will depend on, and have to be measured and evaluated with such spectrum opportunity models.

A dynamic protocol sensing and protocol switching mechanism is needed which will not only dynamically reallocate communication resources at the physical layer to avoid jamming and interference, but will dynamically re-choose suitable higher layer protocols based on cognitive information gathered across the communication protocol stack.

Protocol sensing in a cognitive spectrum by a node can be accomplished by snooping and stochastic profiling of the control messages overheard from its radio vicinity. If the protocol data and control messages are encrypted then the only way the sensing node can detect the protocol is by statistical profiling the relative size and temporal sequence of the packets that are being exchanged by the group. The research in this area will explore protocol sensing in practical scenarios in which multiple groups may simultaneously use different protocols at different spectrum holes in the vicinity of the sensing node.

Mechanisms need to be developed for choosing and switching to optimal protocols based on the currently detected protocols in a secondary node's cognitive neighborhood. A simple example of such protocol switching for a group of wireless nodes within a spectrum hole is to switch from a random access MAC protocol (e.g. CSMA) to a TDMA based protocol when either the local traffic density increases or there is a nodal crowding. Models and distributed protocol-handoff mechanisms are necessary with minimal traffic disruption and optimal resource usage.

Finally, most research on dynamic spectrum access assumes that the primary user is unaware of the presence of secondary users, and secondary users appear transparent to primary users and are only allowed to exploit spectrum opportunistically. Given the recent development in cooperative communication technology, the transmission rate of the primary users can be dramatically boosted by exploiting cooperative diversity if secondary users that have better channel conditions are exploited as cooperative relays for primary communication. As the transmission rate for the primary users improves, CRNs will be left with more spectrum opportunities because the time needed to serve the same amount of primary traffic is reduced. This will have a positive impact on the performance of CRNs. The cooperation among the primary networks and the CRNs has the potential to enable enhanced performance for both networks.

CONCLUSION

Dynamic spectrum access and cognitive radio are emerging technologies for enabling opportunistic spectrum access and enhancing the efficiency and utilization of the spectrum. The dynamic nature of radio frequency environment poses significant challenges in designing new communication protocols for spectrum-aware communication in cognitive radio networks. Due to the close coupling and interaction among protocol layers in CRNs, the optimal design of opportunistic spectrum access and cognitive radio networks calls for a cross-layer approach that integrates signal processing and networking with regulatory policy making. In this chapter, we have introduced some basic concepts such as dynamic spectrum access and spectrum access models, cognitive radio, software defined radio, various cross-layer design issues, and some recent cross-layer design approaches in this emerging technological field. The recent work has clearly demonstrated the benefits and the importance of cross-layer design approaches in CRNs.

Finally, the chapter briefly examined future research directions.

REFERENCES

Akbar, I. A., & Tranter, W. H. (2007). Dynamic spectrum allocation in cognitive radio using hidden Markov models: Poisson distributed case. In *Proceedings of IEEE Southeast Conference,* (pp. 196-201). IEEE Press.

Akyildiz, I., Lee, W., Vuran, M., & Mohanty, S. (2006). Next generation/dynamic spectrum access/cognitive radio wireless networks: A survey. *Computer Networks, 50*(13), 2127–2159. doi:10.1016/j.comnet.2006.05.001

Akyildiz, I. F., Lee, W., Vuran, M. C., & Hohanty, S. (2008). A survey on spectrum management in cognitive radio networks. *IEEE Communications Magazine, 46*(4), 40–48. doi:10.1109/MCOM.2008.4481339

Akyildiz, I. F., & Wang, X. (2008). Cross-layer design in wireless mesh networks. *IEEE Transactions on Vehicular Technology, 57*(2), 1061–1076. doi:10.1109/TVT.2007.911615

Baldo, N., & Zorzi, M. (2007). Fuzzy logic for cross-layer optimization in cognitive radio networks. In *Proceedings of 4th IEEE Consumer Communications and Networking Conference,* (pp. 1128-1133). IEEE Press.

Braden, R., Faber, T., & Handley, M. (2002). From protocol stack to protocol heap-role-based architecture. In *Proceedings of Hot Topics in Networking (HOTNETS I)*. HOTNETS.

Brik, V., Rozner, E., & Banerjee, S. (2005). DSAP: A protocol for coordinated spectrum access. In *Proceedings of IEEE DySPAN,* (pp. 611-614). IEEE Press.

Chakarvarthy, V., Li, X., Wu, Z., Temple, M., Garber, F., Kannan, R., & Vasilakos, A. (2009). Novel overlay/underlay cognitive radio waveforms using SD-SMSE framework to enhance spectrum efficiency – Part I: Theoretical framework and analysis in AWGN channel. *IEEE Transactions on Communications, 57*(12), 3794–3804. doi:10.1109/TCOMM.2009.12.080400

Chakarvarthy, V., Li, X., Zhou, R., Wu, Z., & Temple, M. (2010). Novel overlay/underlay cognitive radio waveforms using SD-SMSE framework to enhance spectrum efficiency – Part II: Analysis in fading channel. *IEEE Transactions on Communications, 58*(6), 1868–1976. doi:10.1109/TCOMM.2010.06.090176

Chen, W., Letaief, K. B., & Cao, Z. (2008). A joint coding and scheduling method for delay optimal cognitive multiple access. *Proceedings of the IEEE, ICC*, 3558–3562. IEEE Press.

Chen, Z., & Qiu, R. C. (2010). Prediction of channel state for cognitive radio using higher-order hidden Markov model. In *Proceedings of IEEE Southeast Conference,* (pp. 276-282). IEEE Press.

Chittabrata, G., Cordeiro, C., Agrawal, D. P., & Rao, M. B. (2009). Markov chain existence and hidden Markov models in spectrum sensing. In *Proceedings of IEEE International Conference on Pervasive Computing and Communications,* (pp. 1-6). IEEE Press.

Chittabrata, G., Roy, S., Rao, M. B., & Agrawal, D. P. (2010). Spectrum occupancy validation and modeling using real-time measurements. In *Proceedings of ACM Workshop on Cognitive Radio Networks,* (pp. 25-30). ACM Press.

Chowdhury, K. R., Felice, M. D., & Akyildiz, I. F. (2009). TP-CRAHN: A transport protocol for cognitive radio ad-hoc networks. In *Proceedings of IEEE INFOCOM,* (pp. 2482-2490). IEEE Press.

Czylwik, A. (1997). Comparison between adaptive OFDM and single carrier modulation with frequency domain equalization. In *Proceedings of IEEE Vehicular Technology Conference,* (pp. 865-869). IEEE Press.

Ding, L., Melodia, T., Batalama, S. N., Matyjas, J. D., & Medley, M. J. (2010). Cross-layer routing and dynamic spectrum allocation in cognitive radio ad hoc networks. *IEEE Transactions on Vehicular Technology, 59*(4), 1969–1979. doi:10.1109/TVT.2010.2045403

Etkin, R., Parekh, A., & Tse, D. (2005). Spectrum sharing for unlicensed bands. In *Proceedings of IEEE DySPAN,* (pp. 251-258). IEEE Press.

Ettus Research LLC. (2011). *Website.* Retrieved from http://www.ettus.com/products.

FCC. (2003). Notice of proposed rule making and order. *ET Docket No. 03-222*. Washington, DC: FCC.

FCC Spectrum Policy Task Force. (2002). *Report of the spectrum efficiency working group.* Retrieved from http://www.fcc.gov/sptf/reports.html.

Foukalas, F., Gazis, V., & Alonistioti, N. (2008). Cross-layer design proposals for wireless mobile networks: A survey and taxonomy. *IEEE Communications Surveys & Tutorials, 10*(1), 70–85. doi:10.1109/COMST.2008.4483671

Gandhi, S., Buragohain, C., Cao, L., Zheng, H., & Suri, S. (2007). A general framework for wireless spectrum auction. In *Proceedings of IEEE DySPAN*. IEEE Press.

Ghosh, C., Pagadarai, S., Agrawal, D., & Wyglinski, A. (2010). A framework for statistical wireless spectrum occupancy modeling. *IEEE Transactions on Wireless Communications, 9*(1), 38–44. doi:10.1109/TWC.2010.01.081701

Han, Z., Pandana, C., & Liu, K. J. R. (2007). Distributive opportunistic spectrum access for cognitive radio using correlated equilibrium and no-regret learning. In *Proceedings of IEEE WCNC*. IEEE Press.

Hatfield, D., & Wiser, P. (2005). Property rights in spectrum: taking the next step. In *Proceedings of DySPAN* (pp. 43–55). IEEE Press. doi:10.1109/DYSPAN.2005.1542616

Huang, J., Berry, R., & Honig, M. (2005). Spectrum sharing with distributed interference compensation. In *Proceedings of IEEE DySPAN,* (pp.88-93). IEEE Press.

Ileri, O., Samardzija, D., & Manadayam, N. (2005). Demand responsive pricing and competitive spectrum allocation via a spectrum server. In *Proceedings of IEEE DySPAN,* (pp. 194-202). IEEE Press.

Kobayashi, H., Fukuhara, T., Yuan, H., & Takeuchi, Y. (2003). Proposal of single carrier OFDM technique with adaptive modulation method. In *Proceedings of IEEE Vehicular Technology Conference,* (pp. 1915-1919). IEEE Press.

Kolodzy, P. J. (2006). Interference temperature: A metric for dynamic spectrum utilization. *International Journal of Network Management, 16*(12).

Kramer, G., Maric, I., & Yates, R. D. (2006). Cooperative communications. *Foundations and Trends in Networking, 1*(3-4).

Lehr, W., & Crowcroft, J. (2005). Managing shared access to a spectrum commons. In *Proceedings of IEEE DySPAN,* (pp. 110-118). IEEE Press.

Li, X., Zhou, R., Chakravarthy, V., & Wu, Z. (2009). Intercarrier interference immune single carrier OFDM via magnitude shift keying modulation. In *Proceedings of IEEE Globecom,* (pp. 1-6). IEEE Press.

Luo, C., Yu, F. R., Ji, H., & Leung, V. C. M. (2010). Cross-layer design for TCP performance improvement in cognitive radio networks. *IEEE Transactions on Vehicular Technology*, *59*(5), 2485–2495. doi:10.1109/TVT.2010.2041802

McHenry, M. A. (2005). *NSF spectrum occupancy measurements project summary*. Shared Spectrum Company Report. Retrieved from http://www.sharedspectrum.com.

Mitola, J. (1999). Cognitive radio for flexible mobile multimedia communications. In *Proceedings of IEEE International Workshop on Mobile Multimedia Communications,* (pp. 3-10). IEEE Press.

Mitola, J. (2000). *Software radios: Wireless architecture for the 21st century*. New York, NY: Wiley. doi:10.1002/047121664X

Peng, C., Zheng, H., & Zhao, B. Y. (2006). Utilization and fairness in spectrum assignment for opportunistic spectrum access. *ACM/Springer. Mobile Networks and Applications*, *11*(4), 555–576. doi:10.1007/s11036-006-7322-y

Radio, G. N. U. (2011). *Website.* Retrieved from http://www.gnuradio.org.

Rajbanshi, R., Chen, Q., Wyglinski, A., Minden, G., & Evans, J. (2007). Quantitative comparison of agile modulation technique for cognitive radio transceivers. In *Proceedings of CCNC*. CCNC.

Shakkottai, S., Rappaport, T. S., & Karlsson, P. C. (2003). Cross-layer design for wireless networks. *IEEE Communications Magazine*, *41*(10), 74–80. doi:10.1109/MCOM.2003.1235598

Shi, Y., & Hou, Y. T. (2008). A distributed optimization algorithm for multi-hop cognitive radio networks. In *Proceedings of IEEE INFOCOM* (pp. 1966–1974). IEEE Press. doi:10.1109/INFOCOM.2008.186

Shi, Y., & Hou, Y. T. (2009). Per-node based optimal power control for multi-hop cognitive radio networks. *IEEE Transactions on Wireless Communications*, *8*(10), 5290–5299. doi:10.1109/TWC.2009.081702

Srivastava, V., & Motani, M. (2005). The road ahead for cross-layer design. *IEEE Communications Magazine*, *43*(12), 112–119. doi:10.1109/MCOM.2005.1561928

Steenstrup, M. E. (2005). Opportunistic use of radio-frequency spectrum: A network perspective. In *Proceedings of IEEE DySPAN,* (pp. 638-641). IEEE Press.

Su, H., & Zhang, X. (2008). Cross-layer based opportunistic MAC protocols for QoS provisioning over cognitive radio wireless networks. *IEEE Journal on Selected Areas in Communications*, *26*(1), 118–129. doi:10.1109/JSAC.2008.080111

Tumuluru, V. K., Wang, P., & Niyaoto, D. (2010). A neural network based spectrum prediction scheme for cognitive radio. *Proceedings of the IEEE, ICC*, 1–5.

Unnikrishnan, J., & Veeravalli, V. V. (2007). Cooperative spectrum sensing and detection for cognitive radio. In *Proceedings of IEEE Globecom*. IEEE Press.

Wang, Q., & Zheng, H. (2006). Route and spectrum selection in dynamic spectrum networks. In *Proceedings of CNNC*. CNNC.

Wen, Z., Luo, T., Xiang, W., Majhi, S., & Ma, Y. (2008). Autoregressive spectrum hole prediction model for cognitive radio systems. In *Proceedings of IEEE ICC Workshop on Communications*. (pp. 154-157). IEEE Press.

Wu, Z., Ratazzi, P., Chakravarthy, V., & Hong, L. (2008). Performance evaluation of adaptive non-contiguous MC-CDMA and non-contiguous CI/MC-CDMA for dynamic spectrum access. In *Proceedings of 3rd International Conference on Cognitive Radio Oriented Wireless Networks and Communcations*. IEEE.

Wyglinski, A. M., Nekovee, M., & Hou, Y. T. (2010). *Cognitive radio communications and networks: Principles and practice*. London, UK: Academic Press/Elsevier.

Xin, C., Xie, B., & Shen, C.-C. (2005). A novel layered graph model for topology formation and routing in dynamic spectrum access networks. In *Proceedings of IEEE DySPAN,* (pp. 308-317). IEEE Press.

Xu, L., Tonjes, R., Paila, T., Hansmann, W., Frank, M., & Albrecht, M. (2000). DRiVE-ing to the Internet: Dynamic radio for IP services in vehicular environments. In *Proceedings of 25th Annual IEEE Conference on Local Computer Networks,* (pp. 281-289). IEEE Press.

Yarkan, S., & Arslan, H. (2007). Binary time series approach to spectrum prediction for cognitive radio. In *Proceedings of IEEE Vehicular Technology Conference*. IEEE Press.

Zhao, J., Zheng, H., & Yang, G. (2005). Distributed coordination in dynamic spectrum allocation networks. In *Proceedings of IEEE DySPAN,* (pp. 259-268). IEEE Press.

Zhao, Q., & Sadler, M. (2007). A survey of dynamic spectrum access. *IEEE Signal Processing Magazine*, *24*(3), 79–89. doi:10.1109/MSP.2007.361604

Zhao, Q., Tong, L., Swami, A., & Chen, Y. (2007). Decentralized cognitive MAC for opportunistic spectrum access in ad hoc networks: A POMDP framework. *IEEE Journal on Selected Areas in Communications*, *25*(3), 589–600. doi:10.1109/JSAC.2007.070409

Zheng, H., & Cao, L. (2005). Device-centric spectrum management. In *Proceedings of IEEE DySPAN,* (pp. 56-65). IEEE Press.

Zheng, H., & Peng, C. (2005). Collaboration and fairness in opportunistic spectrum access. *Proceedings of the IEEE, ICC*, 3132–3136.

Zhou, R., Li, X., Wu, Z., Chakravarthy, V., & Li, H. (2010). The demonstration of SMSE based cognitive radio in mobile environment via software defined radio. In *Proceedings of the IEEE Globecom Demo Session.* IEEE Press.

Zhou, X., Gandi, S., Suri, S., & Zheng, H. (2008). eBay in the sky: Strategy-proof wireless spectrum auctions. In *Proceedings of ACM MobiCom*. ACM Press.

ADDITIONAL READING

Chakravarthy, V., Li, X., Zhou, R., Wu, Z., & Temple, M. (2009). A novel hybrid overlay/underlay cognitive radio waveform in frequency selective fading channel. In *Proceedings of IEEE Crowncom 2009.* IEEE Press.

Chakravarthy, V., Wu, Z., Shaw, A., & Garber, F. (2008). Cognitive radio centric overlay/underlay waveform. In *Proceedings of IEEE DySPAN 2008*. IEEE Press.

Chen, Y., Zhao, Q., & Swami, A. (2008). Joint design and separation principle for opportunistic spectrum access in the presence of sensing errors. *IEEE Transactions on Information Theory*, *54*(5), 2053–2071. doi:10.1109/TIT.2008.920248

Chiang, M. (2005). Balancing transport and physical layers in wireless multi-hop networks: Jointly optimal congestion control and power control. *IEEE Journal on Selected Areas in Communications*, *23*(1), 104–116. doi:10.1109/JSAC.2004.837347

Hong, S., Like, E., Wu, Z., & Tekin, C. (2010). Multi-user signal classification via spectral correlation. In *Proceedings of IEEE CCNC 2010*. IEEE Press.

Ji, Z., & Liu, K. J. R. (2007). Dynamic spectrum sharing: A game theoretical overview. *IEEE Communications Magazine*, *45*(5), 88–94. doi:10.1109/MCOM.2007.358854

Kim, H., & Shin, K. G. (2008). Efficient discovery of spectrum opportunities with MAC-layer sensing in cognitive radio networks. *IEEE Transactions on Mobile Computing*, *7*(5), 533–545. doi:10.1109/TMC.2007.70751

Lee, W., & Akyildiz, F. (2008). Optimal spectrum sensing framework for cognitive radio networks. *IEEE Transactions on Wireless Communications*, *7*(10), 3845–3857. doi:10.1109/TWC.2008.070391

Li, X., Chakravarthy, V., Wang, B., & Wu, Z. (2011). Spreading code design of adaptive non-contiguous SOFDM for cognitive radio and dynamic spectrum access. *IEEE Journal of Selected Topics in Signal Processing*, *5*(1), 190–196. doi:10.1109/JSTSP.2010.2055538

Li, X., Husnay, R., Chakravarthy, V., Wang, B., & Wu, Z. (2009). A novel highly accurate log skew normal approximation method to lognormal sum distributions. In *Proceedings of IEEE WCNC 2009*. IEEE Press.

Li, X., Zhou, R., Chakravarthy, V., & Wu, Z. (2009). Intercarrier interference immune single carrier OFDM via magnitude shift keying modulation. In *Proceedings of IEEE Globecom 2009*. IEEE Press.

Li, X., Zhou, R., Zhang, J., Wang, B., & Wu, Z. (2009). High performance frequency division MC-CDMA system via carrier interferometry codes. In *Proceedings of IEEE IWCMC 2009*. IEEE Press.

Like, E., Chakravarthy, V., & Wu, Z. (2009). Signal classification in fading channels using cyclic spectral analysis. *EURASIP Journal on Wireless Communications and Networking*. Retrieved from http://jwcn.eurasipjournals.com/content/2009/1/879812.

Like, E., Chakravarthy, V., & Wu, Z. (2010). Modulation recognition in multipath fading channels using cyclic spectral analysis. In *Proceedings of IEEE Globecom*. IEEE Press.

Like, E., Temple, M., & Wu, Z. (2009). SMSE waveform design using soft decision selection and dynamic assignment of subcarrier modulation order and power. *The Journal of Communication*, *4*(10), 766–780.

Like, E., Temple, M., & Wu, Z. (2010). Soft decision design of spectrally partitioned CI-SMSE waveforms for coexistent applications. *Proceedings of the IEEE, ICC*, 2010.

Mian, O., Zhou, R., Li, X., Hong, S., & Wu, Z. (2009). A software-defined radio based cognitive radio demonstration over FM band. In *Proceedings of IEEE IWCMC 2009*. IEEE Press.

Papadimitratos, P., Sankaranarayanan, S., & Mishra, A. (2005). A bandwidth sharing approach to improve licensed spectrum utilization. *IEEE Communications Magazine*, *43*(12), 10–14. doi:10.1109/MCOM.2005.1561918

Ugarte, D., & McDonald, A. B. (2005). On the capacity of dynamic spectrum access enabled networks. In *Proceedings of IEEE DySPAN*, (pp. 630-633). IEEE Press.

Unnikrishnan, J., & Veeravalli, V. V. (2008). Cooperative sensing for primary detection in cognitive radio networks. *IEEE Journal on Selected Topics on Signal Processing*, *2*(1), 18–27. doi:10.1109/JSTSP.2007.914880

Wang, F., Krunz, M., & Cui, S. (2008). Spectrum sharing in cognitive radio networks. In *Proceedings of IEEE INFOCOM* (pp. 1885–1893). IEEE Press.

Xie, B., Like, E., Temple, M., & Wu, Z. (2009). Performance of downlink MC-CDMA and CI/MC-CDMA systems in the presence of narrowband interference. In *Proceedings of IEEE IWCMC 2009*. IEEE Press.

Zhou, R., Li, X., Chakravarthy, V., Wang, B., Cooper, R., & Wu, Z. (2010). Software defined radio implementation of SMSE based overlay cognitive radio. In *Proceedings of IEEE DySPAN 2010*. IEEE Press.

Zhou, R., Li, X., Chakravarthy, V., Wang, B., & Wu, Z. (2009). Inter-carrier interference self-cancellation in synchronous downlink MC-CDMA system. In *Proceedings of IEEE IWCMC 2009*. IEEE Press.

Zhou, R., Mian, O., Li, X., Wang, B., & Wu, Z. (2010). A software-defined radio based cognitive radio demonstration over FM band. *Wireless Communication and Mobile Computing Journal*, *10*(1), 4–15. doi:10.1002/wcm.903

KEY TERMS AND DEFINITIONS

Cognitive Radio: Is a context-aware intelligent radio potentially capable of autonomous reconfiguration by learning from and adapting to the communication environment.

Cross-Layer Design: Refers to protocol design approach that actively exploits the interdependence between protocol layers to achieve performance gains.

Dynamic Spectrum Access: Is a set of techniques that improve efficiency and utilization of spectrum by enabling users to share spectrum in a more flexible manner, such as open sharing of spectrum, unlicensed users accessing spectrum without causing undue interference to the licensee.

Primary User: Refers to the licensees of a spectrum band.

Secondary User: Refers to the unlicensed user with respect to the licensees of a spectrum band that is being accessed the unlicensed user.

Software Defined Radio: Is a radio communication system where key components that have been typically implemented in hardware (e.g. amplifiers, modulators/demodulators, detectors, etc.) are instead implemented by means of software and the system can be easily reconfigured on demand.

Spectrum Opportunity: Refers to a spectrum band that is not currently used or is being underused by the primary user.

Chapter 13
Design and Analysis of Battery-Less Sensor Networks:
A Cross Layer Approach

Ali Abedi
University of Maine, USA

Kale Schrader
University of Maine, USA

ABSTRACT

One of the major limiting factors preventing wide use of wireless sensor networks in practical scenarios is power consumption. Battery-less or passive sensors promise to alleviate this issue and enable a wide variety of embedded sensor applications such as structural health and vehicular monitoring, biomedical applications, smart homes, and smart grids. Embedding these sensors in structures without the need for changing batteries, their rugged design to withstand harsh environments, and coded communication with multiple access features makes this technology a desirable candidate for a variety of applications. Design and analysis of these sensors from a cross layer point of view is studied in this book chapter. State of the art in fabrication and test of this new class of wireless sensor systems is also reviewed. Interactions between lower layer with passive sensors and upper layer with active sensors—a different perspective on cross layer—is exploited to achieve significant performance gains in terms of signal to noise and interference ratio, correlation peak to side-lobe ratio, operation range, and data rate.

DOI: 10.4018/978-1-4666-0960-0.ch013

INTRODUCTION

In this chapter, an overview of hybrid sensor systems with passive layer and active layer wireless sensors is presented. This is a different way of looking at cross layer design where interactions between sensors in passive layer and sensors in active layer provides a great advantage for practical applications. Particular application scenarios including distributed source estimation, impact localization, and shape monitoring are studied and relevant results are presented. After reading this chapter, you should be able to

- Understand operation of a battery-less wireless sensor
- Distinguish between various passive and active wireless sensor layers
- Design and analyze a battery-less wireless sensor system
- Use cross-layer approach to optimize battery-less wireless sensor network performance
- Apply knowledge gained in this chapter to practical scenarios

Required background to understand materials in this chapter include communication theory, introductory level information theory, and familiarity with electromagnetic and acoustic wave propagation. Most required concepts are introduced in appropriate locations in the text; however, the readers are encouraged to study the references cited for further information.

BACKGROUND

Battery-Less Wireless Sensors

One of the major problems in practical use of Wireless Sensor Networks (WSN) is power consumption. Numerous research and development works have been devoted to reduce power consumption through efficient power management strategies at physical, network, and upper layers. Considerable efforts have been invested toward lowering the power consumption by improving sensor network platforms, including selectable power states (off, sleep, standby), operating at low voltages, fine grained control of hardware, and efficient use of wireless broadcast medium (Rhee, Seetharam, & Liu, 2004). Modulation scaling is used for energy aware wireless packet scheduling to meet the low power requirement (Schurgers, Aberthorne, & Srivastava, 2001). Randomizing the choice of both measurement and transmit rate is proposed in Appadwedula, Veeravalli, and Jones (2005) to optimize detection performance with certain power constraint in time-varying noisy environments. Properties of sensor data is exploited in Mukhopadhyay, Panigrahi, and Dey (2004) to eliminate overhead on the sensor nodes and therefore save the energy consumed by WSN. Furthermore, numerous researches have been conducted on the optimization of protocol stack for the design of networks to achieve longer lifetimes for the systems (Heinzelman, Chandrakasan, & Balakrishnan, 2000). A protocol architecture is developed for WSN called Low-Energy Adaptive Clustering Hierarchy (LEACH), which includes a distributed cluster formation technique for self-organization. This protocol is energy efficient since it uses hierarchy to reduce the data collected before transmission to the cluster head. A Physical (PHY) layer driven approach is proposed in (Shih, et al., 2001) to design protocols and algorithms that minimize the energy consumption of the WSN and individual nodes, where energy per useful bit required to decode different rate convolutional codes with varying constraint lengths are measured.

These elegant solutions work fine in specific applications with short term deployment or access to wireless sensor nodes for battery replacement, while they fall short in long term deployments especially when sensors are embedded in structures or machines in hard to reach areas making the battery replacement impossible or very costly. In addition, usage of wireless sensors in ***harsh environments*** with extremely hot or cold temperatures is also limited due to battery/device limitations. Solutions proposed in the literature advanced the area via increasing temperature tolerance for both batteries and electronic circuits to about 400°C (Hunter, Neudeck, Okojie, Beheim, & Powell, 2003). These impediments reveal the important role of ***Battery-Less or Passive*** wireless sensors to enable practical applications in extreme conditions. Battery-less sensors can respond to several of these demanding applications by providing a number of important features: no battery, small size, low cost for high volume production, and capability to withstand harsh environments, such as on moving parts at high temperatures or pressures, or in the presence of hazardous chemicals (Scholl, et al., 2003).

A battery-less sensor consists of metal electrodes photo-lithographically printed onto a piezoelectric crystalline substrate. These electrodes can be used to implement matched filter transfer functions, which allow the identification of each individual sensor. This capability leads to new wireless and passive sensor applications in which multiple sensors in close proximity must be individually interrogated. These sensors can be designed to operate on crystals such as Lithium Niobate (LNO). Transducers need to be phase-coded based on design criteria established for matched filters used in correlation receivers (Morgan, 1991). The coded transducers are designed to respond to the coded sequences to which they are matched, allowing sensors to be addressed individually. The sensor output is the differential delay between the responses of the two coded transducers. The delay of each path is dependent on the wave propagation velocity on the substrate. The device can be used for physical sensing, because this velocity can be affected by variations in temperature, pressure, vibration, and mass loading. By depositing a thin film on one of the delay paths, sensitive and selective to a specific chemical or biological compound, the sensor can detect that particular measure.

One major practical challenge in designing battery-less sensors is orthogonal code design for sensor network applications. Quasi-orthogonal codes are required to be designed and implemented on the miniaturized wireless differential-delay sensor. The main objective of orthogonal code design is to add multiple access features to passive sensors. The proposed codes are selected based on their auto-correlation and cross-correlation characteristics. The following criteria are used to select the most appropriate codes for this application:

1. Largest peak-to-side lobe ratio of the auto-correlation response of a given code
2. Smallest maximum peak value of cross-correlation of that code with any other code in the set

Depending on the thresholds for the two criteria, one could generate larger or smaller final subsets of codes. The proposed scheme in this chapter is different from the methods based on finite projective planes (Woo, 2000), where only auto-correlation is minimized. The proposed code design method is developed based on the experimental results with battery-less sensor devices, where imperfection in the device fabrication process leads to undesired side lobes in the sensor response. As an example, a set of 31 codes has been designed in Dudzik, Abedi, Pereira da Cunha, and Hummels (2008). The process starts with a set of 1024 codes which is later reduced to a subset with a peak to side lobe ratio (of their auto-correlation) of 5 or greater. This drops the subset size down to 73 codes.

Hybrid Sensor Networks

Networking passive (battery-less) and active (battery operated) sensor nodes fully exploits the cross layer approach benefits by achieving goals of embedded sensors applications while, providing the added range and advantages of active sensors. This is a different way of looking at cross layer approach.

Estimation of a parameter in an inaccessible area using multiple imperfect observations, the so called Chief Executive Officer (CEO) problem, is still a challenging problem in the WSN (Wang & X, 2010). This problem can be addressed using distributed coding scheme implemented in a hybrid sensor network with double-sink configuration. Noting the high correlation among passive sensors' observations, the overall data flow efficiency of the system can be considerably improved using Distributed Source Coding (DSC) techniques based on the Slepian-Wolf theorem (Xiong, Fan, Liu, & Zhong, 2010). The sensor node can be simplified by combining distributed source and channel coding into a Distributed Joint Source Channel Coding (D-JSCC), without degradation in BER performance (Bajwa, Haupt, Sayeed, & Nowak, 2007).

Most reported D-JSCC schemes in the literature are still too complex for the computationally constrained sensors (Garcia-Frias & Xiong, 2005). A practical low complexity DJSCC solution is introduced in Razi, Yasami, and Abedi (2011) based on Recursive Systematic Convolutional Codes (RSC) equipped sensors that form a distributed version of Parallel Concatenated Convolutional Codes (PCCC). The proposed iterative decoding algorithm with self-correlation extraction property may be employed in a double-sink two-tiered WSN. In a traditional clustered two-tiered WSN structure, a sink node collects data from sensors inside a cluster and relays it to the data fusion center located at a base station using different relaying modes (Jiang, Yuan, & Zhao, 2009). A new system model is proposed in Razi, Afghah, and Abedi (2011), where two sink nodes are placed at each cluster. DeModulate and Forward (DMF) relaying mode is applied to sink nodes, since it enables the sink nodes to reformat the frames into a multi-frame and forward it to the base station. This it is not applicable to the widely used Amplify and Forward (AF) relaying mode.

The links from each sensor to the base station form a two-hop multi-relay communication system. It is shown that for a multi-relay system which employs Distributed Space-Time Block Codes (D-STBC) assisted AF relaying mode, the optimum power allocation between source and relay nodes, is equal power allocation (Jing & Hassibi, 2006). BER performance for these systems with equal power allocation is analyzed in Maham and Nader-Esfahani (2007).

MAIN FOCUS OF THE CHAPTER

The main focus of this chapter includes three different parts as case studies to demonstrate the advantages of using cross layer approach in the context of passive and active sensor layers. The first part studies the problem of source estimation using multiple imperfect measurements using passive sensors in a lower layer and coding and transmission using active sensors in the upper layer (cross layer communication).

A similar problem is studied in second part where impact localization is considered. The location of impact is most likely not on any specific sensor, but a cross layer approach aggregating passive sensors data and active sensors computation enables precise impact localization (cross layer computing).

Finally, the third part focuses on shape monitoring where passive sensors provide multiple independent measures for localization. These estimates are then combined using 3D image reconstruction techniques and mechanical properties of the material to determine shape (cross layer communication and computing).

Figure 1. System model for two-layer double-sink wireless sensor network

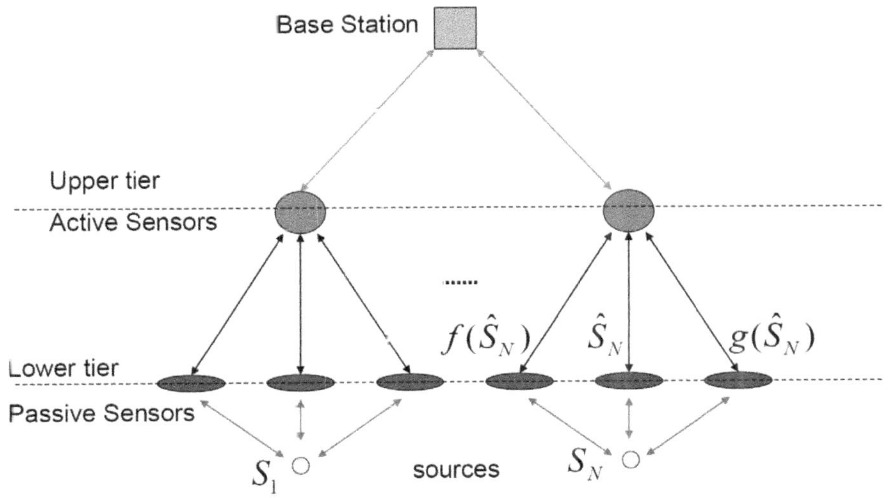

Distributed Source Estimation: Cross Layer Communications

Conventional design of wireless sensors includes separate sensor and radio design, while using cross layer approach both sensor and radio can be designed on the same platform. In other words, we can design a radio with varying parameters to act as sensor. In addition, we can augment this passive layer by adding an active layer to provide longer range communication and higher computational power for our application. The structure depicted in Figure 1 includes two layers, which require a cross layer approach for its design.

First layer of the system is composed of clusters of sensors. Each cluster includes a binary source observed by a group of sensors. The source data of different clusters assumed to be uncorrelated. There is no overlap between clusters, and each sensor belongs to a unique cluster. The virtual channels from sensors to the sink nodes are orthogonal in time and/or frequency.

The second layer is composed of two active nodes that forward the received signals from sensors to the base station. All channel coefficients are assumed to be independent zero-mean complex Gaussian random variables.

The optimum power allocation for DMF relaying mode is calculated in Razi, Afghah, and Abedi (2011) and an upper bound on the system BER performance as a function of sensor's observation accuracy, number of sensors, and end to end system SNR value is derived as follows.

$$p_e < \sum_{i=\frac{M}{2}+1}^{M} \binom{M}{i} p^i q^{M-i} + \frac{(-1)^M+1}{4} \binom{M}{\frac{M}{2}} (pq)^{\frac{M}{2}} \qquad (1)$$

where $p=q-1$ and the error probability between tier 1 and 2 for signal to noise ratios of γ_1 and γ_2 is

$$p = \frac{1}{4(\gamma_1+1)} + \frac{1}{2(\gamma_2+2)^2} - \frac{1}{4(\gamma_1+1)(\gamma_2+2)^2} \qquad (2)$$

Simulation results matches the analytical expression, specially for large SNR values, where the approximations are held better. One interesting result is that allocating more power to the sensor nodes, specially at large SNR values improves

Figure 2. Performance of two layer hybrid network (modified multi-turbo-decoder MTC) compared to conventional and modified turbo decoder (TC)

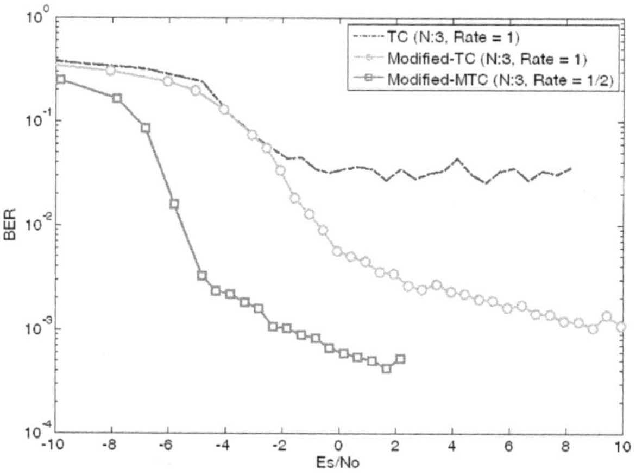

the performance. The optimum power allocation scenario can be chosen based on the average SNR value of the system. This result provides a new criterion to configure clustered WSN for CEO applications to achieve a desired performance with minimum number of sensors in a cross layer approach in hybrid wireless sensor networks (see Figure 2).

Impact Localization: Cross Layer Computing

A challenging aspect of impact detection is accurate localization of the impact over the structure's surface. The location of impact is most likely not on any specific sensor, but a cross layer approach aggregating passive sensors data and active sensors computation enables precise impact localization (passive sensors computing). Similar problems have been explored for years, ranging in application from radar detection of aircraft in an airport, to the position of a receiver using Global Positioning System (GPS). In this respect, the fundamentals of detecting an object's location is well researched and commonly known, however, its implementation requirements can change drastically from one environment to another. In this section, a hybrid time of arrival and amplitude based impact localization is presented as an example application using sensors in proximity of a source for precise source estimation.

The test-bed used for the experimental verification of this theory is illustrated in Figure 3. The cylinder is designed by ILC Dover company

Figure 3. Inflatable test-bed at UMaine used for impact localization

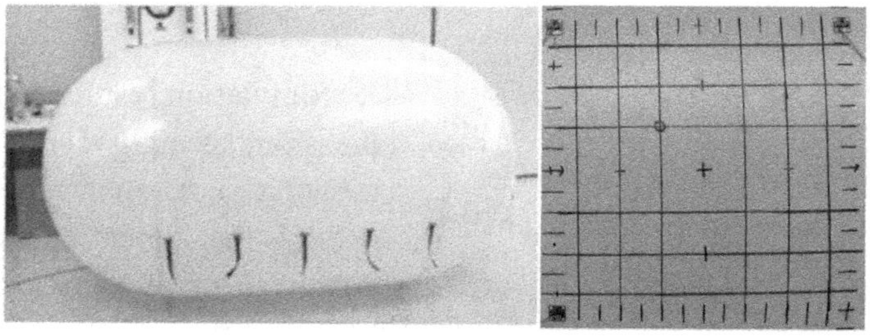

Figure 4. Sensor responses from impacts located in the x direction and y direction from sensors

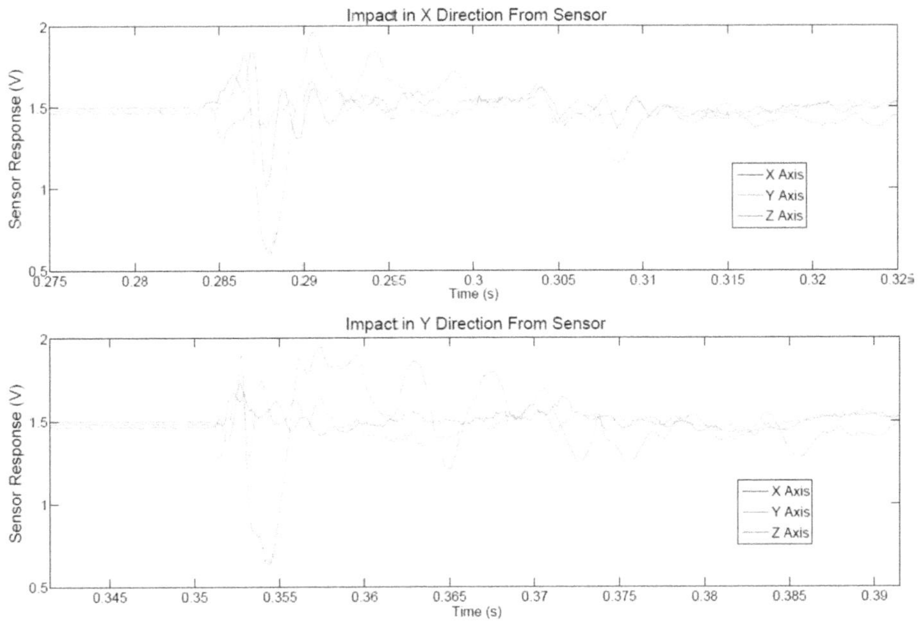

in collaboration with UMaine researchers and instrumented by UMaine's students with accelerometer sensors and data acquisition and analysis hardware and software. The grid on the right was used as reference coordinate for verification of sensor signal analysis results.

Once the structure is hit by a precise impact hammer, each sensor experiences different signal intensities at different times. Typical sensor responses in X-Y-Z directions are depicted in Figure 4. These responses are then analyzed using the algorithm detailed in Schrader, Abedi, and Caccese (2010) to produce precise location estimates as presented in Figure 5.

Perhaps the most intuitive way to localize an object is by amplitude analysis. The technique works in much the same way a person can roughly determine the distance between themselves and a running car, or a person speaking, solely by how much sound they are able to detect. More relevantly, this has been applied to WSNs using the Received Signal Strength (RSS) from different nodes to locate a transmitter. In this scenario, the signal strength from the transmitter attenuates in a known manner (Yang & Liu, 2008), often by a variation of the Friis Equation.

$$\frac{P_r}{P_t} = G_t G_R \left(\frac{\pi}{2\pi D} \right)^2 \tag{3}$$

By characterizing the environment to modify the Friis equation, one can make reasonable projections of the power received at an antenna at a given distance. Having this information, the transmitted power, and the received power, a distance from the transmitter can be calculated. By having more than one receiver, the system can determine the location of the transmitter via trilateration.

Similarly, this technique can be applied to the physical waves propagating across the surface of the structure. At the initial impact, a quantified amount of energy is deposited in the structure creating waves of a given amplitude. The structure naturally acts as a damper, attenuating the signal over time and propagation distance. This

Figure 5. Localization of impacts using hybrid time of arrival and amplitude based detection

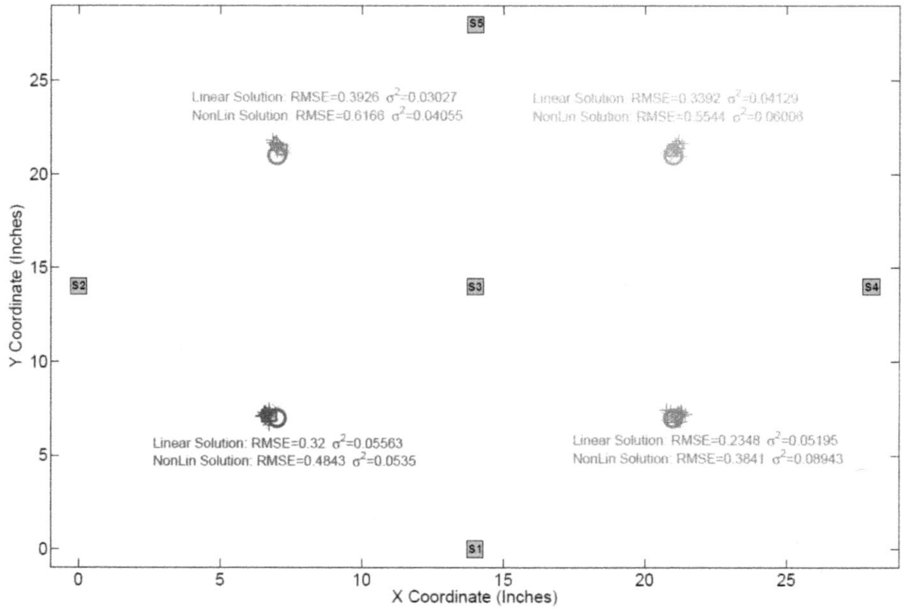

rate of attenuation can differ between wave types, yet assuming one type of wave can be detected reliably, it can be modeled in a similar manner as described for WSNs. This allows the possibility of amplitude based detection systems.

Time Difference of Arrival (TDOA) is a slightly more complex algorithm than the more commonly known Time of Arrival (TOA) (Yost, 1998) scheme. When applying the latter to a WSN, an event occurring at a known time instigates the transmission of a wireless signal. Upon receipt of this signal at the receiver, a Time of Flight (TOF) is calculated. Knowing the velocity of electromagnetic waves, the distance from the transmitter is easily deduced. If the location of the receiver is known, a sphere of location can be drawn around that point. Additional receivers and TOA values allow trilateration in two or three dimensions. This technique is proven on a larger scale as the method of operation behind the accurate and dependable GPS tracking system (Nardi & Pachter, 1998).

As TOA works using known transmission times and propagation velocity, TDOA is implemented in scenarios where the time of transmission is unavailable. Instead of knowing when a signal was instigated, only the time of receipt is detectable. A minimum of two localized receivers and a wave velocity are required to give a TDOA value. By solving a set of non-linear equations, a hyperbole of location can be traced with relation to the receivers as seen in Figure 6. Intuitively, additional receivers allow two and three dimensional localization.

Although the simple wave detection algorithms require little computation, and gave reasonable velocities, a more intensive alternative was considered to increase accuracy. Rayleigh waves travel at a distinct velocity, however they also travel with a unique frequency. Simple peak detection in the time domain proves tedious in isolating this wave, especially in the curved direction of the structure. However, the possibility exists that there could be artifacts in the frequency domain that would better extract TOA values. To get both frequency feedback, as well as time resolution, a wavelet transform was in-

Figure 6. Hyperbola of location using two receivers in TDOA system

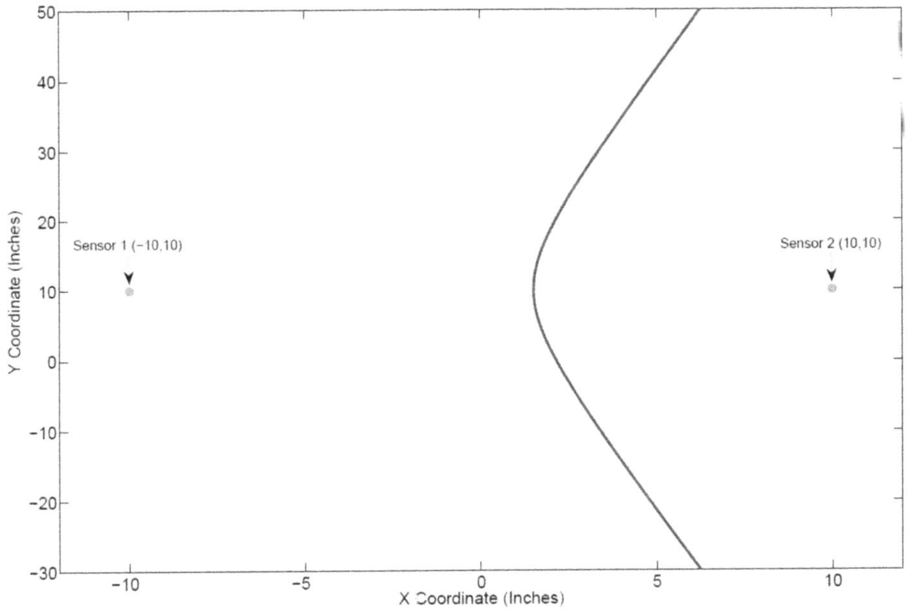

corporated into a complex detection algorithm. To instigate the complex detection algorithm, the original data was preprocessed to isolate the downward peaks of the signals. Then, the continuous wavelet transform was performed with a selected mother wavelets and scale. An example of the wavelet transform technique is shown in Figure 7.

After finding the TOA of the peak wavelet coefficient, the results were used to calculate velocities of the sections of the 1 dimensional scheme. To optimize the complex wave detection algorithm, the results from a variety of mother wavelets and scales were used to compute section velocities of the threshold testing scheme. Then, the parameters with the lowest variation in propagation velocity between different sections of the same dimension were selected. The results of the optimized complex wave detection algorithms versus the best simple detection algorithm illustrates the validity of previous claim of the extreme variations encountered when using the simple thresholding methods in the y direction as opposed to the x direction. However, one will notice the much more consistent results of the complex wave detection scheme.

This case study demonstrated the advantages of cross later computing in passive sensor networks, where electrical and mechanical properties of material, waves, and impacts are all taken into consideration. The next case study follows a similar approach, while using both cross layer communication and computing to achieve high resolution in shape monitoring.

Shape Monitoring: Cross Layer Communications and Computing

This part focuses on shape monitoring where passive sensors provide multiple independent measures for localization. These estimates are then combined using 3D image reconstruction techniques and mechanical properties of the material to determine shape. This is a perfect example for cross layer communication and computing.

Figure 7. Illustration of wavelet transform procedure and results

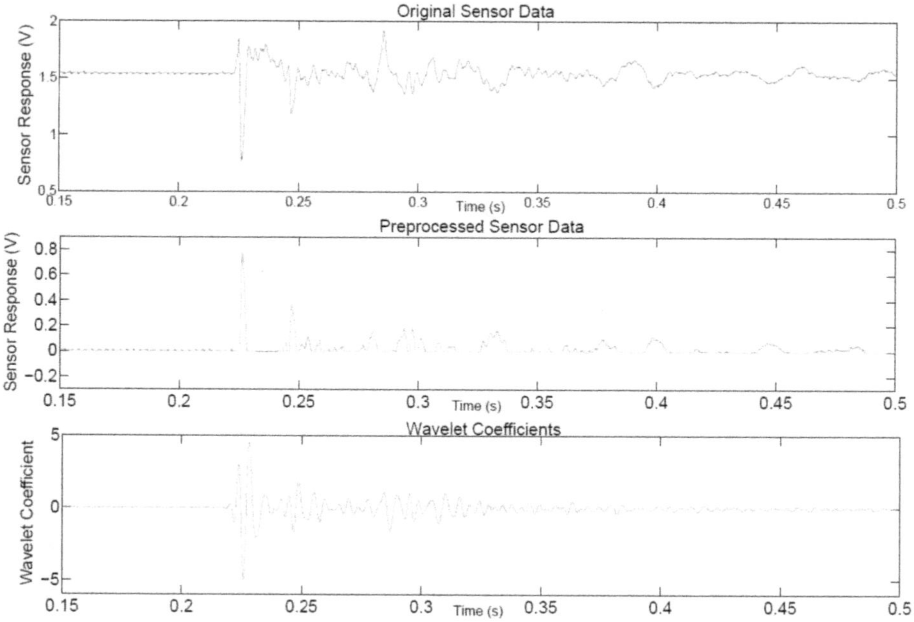

Localization of battery-less wireless sensors is emerging as an essential technology in continuous real-time shape monitoring systems. Inflatable space structures such as antennas, solar panels, radiation shields and decelerators will benefit from this technology to ensure their functionality and structural integrity. However, due to the lack of processing power on the sensor side, the localization problem in passive wireless sensor networks is more challenging than in cellular networks or in active wireless sensor networks. A parabolic dish type inflatable antenna to verify the hybrid localization algorithm with multiple embedded passive sensors is considered in this section as an example for cross layer approach to localization. Fundamental limits on various localization methods based on Received Signal Strength (RSS), signal Time Of Arrival (TOA) and Angle Of Arrival (AOA) to estimate the position of the sensor are studied in different layers. The orthogonal coding presented in the background section needs to be co-designed with localization goals in mind to avoid *multi-sensor interference*. Achieving high positioning accuracy within a few centimeters is often a requirement in passive wireless sensor networks.

RSS relies on path loss to find position. The signal strength attenuates at an exponential rate with the distance between the receiver and the sensor. With some amount of error due to noise and multipath interference, this technique can achieve a typical accuracy within a few feet. TOA measures the amount of time it takes for the signal to propagate to the sensor, the time the sensor delays the signal, and the amount of time it takes for the signal to return. AOA relies on multiple receiving antennas working together to extrapolate the position v based on the bearing angles of the incoming signals at the receiver.

RSS scheme is relatively inexpensive and simple to implement in wireless sensor networks where the strength of RF signals can be measured by each receiving antenna during normal data communication without using additional bandwidth or consuming extra energy. Environment-dependent errors caused by multipath fading and shadowing deteriorate the precision of RSS localization. TOA measures the time at which transmitted RF signal first arrives at the receiver and its measurement is more complicated than that of RSS method in terms of hardware implementations. AOA approach requires multiple antenna elements to measure bearing angles or directions of incoming signals; hence, it contributes much to receiver device cost, size, and complexity.

The hybrid RSS/TOA/AOA localization scheme exploits merits of both TOA and AOA methods offering high localization accuracy as well as taking advantage of simplicity and low-cost of RSS measurements to acquire a roughly initial location estimate. Several hybrid models including RSS/TOA, RSS/AOA, TOA/AOA and RSS/TOA/AOA are analyzed for obtaining the near optimum and accurate localization in different scenarios.

The equations of the location estimator based on multiple measurements are usually nonlinear and can be linearized using the Taylor-series estimation. However the Taylor series approach needs an initial location guess and suffers from the convergence problem of intensive matrix computation. To overcome the drawbacks of convergence, a two layer linear least-square estimator for the TOA/AOA location is proposed in Li and Weihua (2002), but it requires to find geometric relationships between TOA and AOA measurements. In our proposed hybrid model, without solving geometric constraints, wide-sense two-layer LS estimator for RSS/TOA, RSS/AOA, and TOA/AOA methods and three-layer LS estimator for the RSS/TOA/AOA method using the Taylor-series approximation are set up to solve equations of hybrid models. First, we use one of three available techniques (RSS, TOA, or AOA) to obtain an initial location estimate with estimation errors. Then the initial estimation is incorporated with additional measurements by others to estimate the final location. The RSS/TOA, RSS/AOA, TOA/AOA, and RSS/AOA/TOA models combine measurements from RSS and TOA, RSS and AOA, TOA and AOA, as well as RSS, TOA and AOA methods respectively into an estimation matrix to find positions under noisy environments.

In order to solve nonlinear RSS/TOA, RSS/AOA, TOA/AOA, RSS/TOA/AOA equations, we assume that: 1) the RSS, TOA, AOA measurement errors are independent Gaussian random variables with zero-mean and known variance which is often assumed in wireless communication systems and 2) there exists line of sight paths in AOA measurements. This hybrid cross layer approach combines the advantages of all these three methods to achieve a higher localization accuracy. Simulations are set up to test the effectiveness and accuracy of the AOA, TOA, and RSS techniques in passive wireless sensor networks and compare performances with hybrid model as seen in Figure 8.

The resolution of hybrid RSS/TOA, RSS/AOA, RSS/TOA/AOA localization schemes are very close to that of TOA, AOA and TOA/AOA respectively because RSS contributes much less than TOA, AOA, or TOA/AOA to the accuracy in the two-layer or three-layer LS estimators. However the RSS measurement is required to provide a rough initial location estimate for combined localization methods and ensure convergence of LS methods.

The combined TOA/AOA in Figure 8 offers more accurate positioning resolution than that of AOA only localization or TOA only localization. Moreover, TOA/AOA will eliminate the ambiguity of location by adding AOA measurements. As depicted in Figure 9, positioning error of TOA based schemes (TOA, RSS/TOA, TOA/AOA, or RSS/TOA/AOA) is shown to decrease as the signal bandwidth is increased in Figure 9, whereas the positioning error by RSS, AOA, RSS/AOA schemes does not change with the bandwidth of received signals. This is consistent with the fact that the arrival-time variance is inversely proportional to the effective bandwidth.

Ultra-wide bandwidth radios with more than 1 GHz bandwidth may provide high resolution

Figure 8. Positioning error as a function of the number of receivers

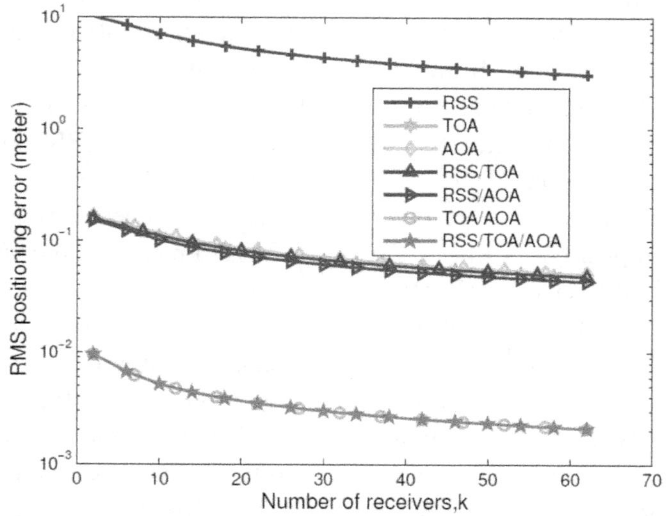

Figure 9. Positioning error as a function of bandwidth

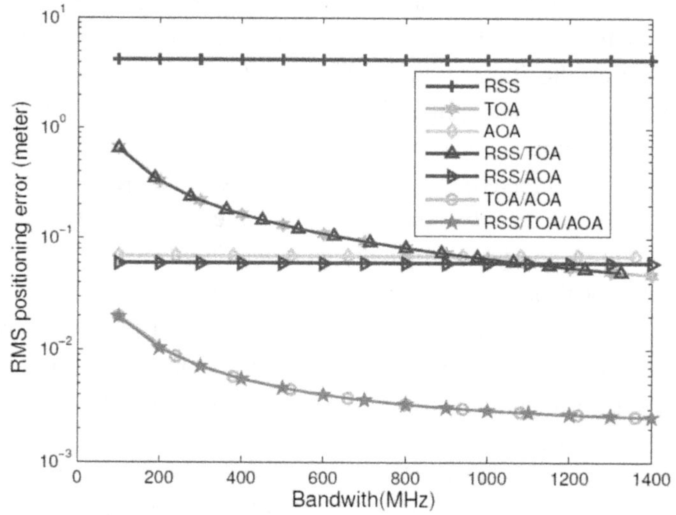

positioning with the accuracy close to one centimeter based on TOA technique. The combined TOA/AOA model offers better accuracy than TOA in terms of the effective bandwidth and it means the TOA/AOA based method needs less bandwidth of signals transmitted by the sensor. More information on this example and the analytical studies related to the limits discussed in this section may be found in Chen and Abedi (2010).

FUTURE RESEARCH DIRECTIONS

One of the specific scenarios considered in this chapter was impact detection. One of the most prominent difficulties in all aspects of this application scenario is overcoming hardware limitations during testing. The use of accelerometers with high sensing ranges could provide feedback in assessing large impact energies and possible non-linear effects impacts of that magnitude could cause. Also, given the scaling algorithms previously discussed, greater sensing range should improve the ability to predict the intensity of high energy impacts. It is the belief of the author that the techniques proposed in this chapter, both in localization and scaling, could be easily adapted to such a system. Optimization of the algorithms presented in this chapter should start with an approach to make the hardware more suitable for the system's intended purpose. To enhance the system applicability, low mass, small size, wireless nodes could be implemented. Although challenging, extracting the most reliable data from accelerometers on

the surface of the structure requires the minimization of the effects they have on wave propagation. The operation of the localization algorithm could also be affected by the orientation of the sensor array it operates. Although structure size is a limiting factor, larger habitats allow the possibility of greater node distances and an exceedingly large number of sensor locations, including non-symmetric orientations. While evaluating this system, errors were associated with both impacts located too close, as well as too far from sensors. Testing these combinations of orientations and spacing could provide optimal sensor layouts with respect to structure properties. To further the development of the TDOA localization algorithms for curved structures, one must realize a more refined approach would account for velocity shifts due to Rayleigh wave propagation variations. A strong evidence of this phenomena hint at the need to fully characterize the material for localization optimization.

Ultimately, a theoretical study of wave mechanics in an inflatable structure has the potential to increase performance of both the localization and scaling algorithms while possibly reducing computational requirements for the system. The wavelet transforms used in the complex wave detection were optimized for this system using all the possible mother wavelets MATLAB has by default, as well as an educated guess on the scales appropriate for precise wave detection. If one could isolate the signature of a Rayleigh wave propagating across a given sensor, a custom mother wavelet could be devised accordingly, providing even more exacting wave arrival results. Although this could prove a tedious task specialized to the hardware being used, the result has the potential to revolutionize the wave detection scheme.

The scaling algorithm in this thesis shows larger errors as impact severity increased due to nonlinear effects of the material on wave attenuation, as well as sensor inconsistencies. Meticulous characterization of both the sensor and structure material has the potential to alleviate this additional error. This could be obtained by simple repetitive testing, allowing a lookup table of characteristics, or as complex as a mathematical model to project these values. Regardless, optimal performance of an amplitude based scaling algorithm requires additional attenuation parameters for the environment, and perhaps a sensor technology more isotropic in sensing feedback.

One promising direction to continue this cross layer approach is redefining the whole concept of cooperative relaying in the context of batter-less sensor networks. Cooperative relaying can improve the overall throughput of a network in various communication scenarios in which a terminal acts as a traffic relay for other users. Examples of these scenarios can be found in home and consumer communication networks different applications of wireless sensor network to improve energy consumption and prolong network lifetime, such as environmental monitoring, target tracking, etc. This approach can take advantage of recent advancements in network coding.

Wireless network coding has received an increasing attention since network coding was introduced. This increasing interest is due to considerable gains that are foreseen in application of network coding in cooperative relaying scenarios with two-way or multi-way traffic. Effect of erroneous channel estimation on performance of physical layer network coding over fading channels has been investigated in (Yasami & Abedi, 2011), where the relay maps the superimposed noisy modulated data, received from the two end terminals, to network coded combination of the source packets. Channel estimation error is considered to be Gaussian distributed and formulate the network coding error by the distance between real and estimated points in the channel coefficients plane. Using this model, a statistical lower bound on variance of estimation error is presented that can be tolerated by the relay terminal without imposing a network coding error on the system. This study can determine the required channel estimation accuracy in design of practical relaying systems which use physical layer network coding. Extending these results to the case of battery-less sensor networks requires cross layer techniques as detailed in this chapter.

CONCLUSION

The main message in this chapter is that cross layer approach is not limited to the seven layers of a networked system, while passive and active sensors, multiple localization algorithms, and in multiple independent measurements can also be considered in a cross layer communication and computing framework.

In this chapter, a cross layer code design process for passive wireless sensors is developed to provide multiple access and high isolation among several sensor nodes. The selected codes from a set of orthogonal codes provide the largest peak-to-side lobe ratio in their auto-correlation, while they have minimal cross-correlation with other codes. The basic operation of the wireless sensor system concept has been validated with the prototype system. The system is capable of wirelessly interrogating the sensors and processing the received responses to the point where sensor measurement information could be extracted from them. Maintenance-free and battery-free features of this wireless sensor enable myriad of applications beyond those that exist today. Development is ongoing to characterize and improve system performance and range, and explore sensor applications.

Efficient use of these battery-less sensors is not possible without a hybrid network. The advantages of a hybrid network are described in the distributed source estimation section and mainly includes higher reliability and robustness. The CEO problem of estimating a binary source is considered in a cross layer context. System information capacity is derived in Razi, Yasami, and Abedi (2011) as a function of observation accuracy, channel quality and the number of sensors. Observation accuracy is mostly defined by the application and is relatively low for the applications that sensors cannot be placed at the exact data source location. Also, the channel SNR is limited by the power consumption constrained tiny sensors. The derived expression for the capacity defines the achievable capacity for different number of sensors. Inversely, if the coding rate is fixed which is the case in most practical applications, the minimum required number of sensors can be determined in order to achieve arbitrary low error rates for any given set of system parameters. A practical PCCC based distributed coding scheme is also proposed that provides acceptable performance using the intrinsic correlation among sensors data. The SNR values required to achieve low error rate using different number of sensors is very close to the limits derived from the system capacity curve. The fast convergence property and simple encoder structures of the proposed system makes it an appropriate choice for wireless sensor networks. Both analytical and simulation results show that the performance of the system is saturated by the SNR value. Consequently, the number of sensors are mainly imposed by the desired performance level and the observation accuracy.

Impact localization and shape monitoring are studied as two special application cases where cross layer approach plays a critical role. The focus of impact localization section was on a proof of concept for an accelerometer based impact detection system. A test-bed was constructed using relevant materials and monitoring hardware, providing a useful environment with characteristics similar to that of a final prototype. Algorithms were developed for accelerometer based impact localization, and impact scaling. Different techniques, researched in the field of structural health monitoring and WSNs, were considered. Through theoretical considerations, and experimental deduction, a series of final approaches were designed for each of these tasks. The algorithms were implemented, tested, and modified for enhanced performance on the inflatable structure testbed. These systems produce localization and scaling parameters in accordance with the project guidelines, and provide the framework for implementation on a final prototype. Finally, the shape monitoring using a combination of various localization schemes for battery-less sensors was presented. Theoretical models are set up to analyze fundamental limits of RSS, TOA, AOA, and hybrid localization techniques in passive wireless sensor networks. Numerical results based on proposed mathematical models show that the hybrid localization can achieve high resolution positioning with the accuracy within a centimeter using sensors having wide bandwidth of 1 GHz and high center frequency of 6 GHz and the receiver having a relatively large distance of 1 meter among a number of receiving antennas. Hybrid TOA/AOA and RSS/TOA/AOA schemes demonstrate up to 250 times improvements of the accuracy compared with RSS-only localization, TOA-only localization and AOA-only localization, respectively.

REFERENCES

Appadwedula, S., Veeravalli, V., & Jones, D. L. (2005). Energy efficient detection in sensor networks. *IEEE Selected Areas in Communications*, *23*(4), 693–702. doi:10.1109/JSAC.2005.843536

Bajwa, W., Haupt, J., Sayeed, A., & Nowak, R. (2007). Joint source-channel communication for distributed estimation in sensor networks. *IEEE Transactions on Information Theory*, *53*(10), 3629–3653. doi:10.1109/TIT.2007.904835

Chen, J., & Abedi, A. (2010). A hybrid framework for radio localization in broadband wireless systems. In *Proceedings of the IEEE Global Communications Conference,* (pp. 1-6). Miami, FL: IEEE Press.

Dudzik, E., Abedi, A., Pereira da Cunha, M., & Hummels, D. (2008). Orthogonal code design for passive wireless sensors. In *Proceedings of the 24th Queen's Biennial Symposium on Communications,* (pp. 316 - 319). Kingston, Canada: IEEE Press.

Garcia-Frias, J., & Xiong, Z. (2005). Distributed source and joint source channel coding: From theory to practice. In *Proceedings of the 30th IEEE Acoustics, Speech, and Signal Processing Conference,* (pp. 1093-1096). IEEE Press.

Heinzelman, W., Chandrakasan, A., & Balakrishnan, H. (2000). Energy-efficient communication protocol for wireless microsensor networks. In *Proceedings of the 33rd Hawaii International Conference on System Sciences*, (p. 10). IEEE.

Hunter, G. W., Neudeck, P. G., Okojie, R. S., Beheim, G. M., & Powell, J. A. (2003). An overview of high-temperature electronics and sensor development at NASA Glenn research center. *Turbomachinery*, *125*(4), 658–665. doi:10.1115/1.1579508

Jiang, C., Yuan, D., & Zhao, Y. (2009). Towards clustering algorithms in wireless sensor networks-A survey. In *Proceedings of the IEEE Wireless Communications and Networking Conference*, (pp. 1-6). IEEE Press.

Jing, Y., & Hassibi, B. (2006). Distributed space-time coding in wireless relay networks. *IEEE Transactions on Wireless Communications*, *5*(12), 3524–3536. doi:10.1109/TWC.2006.256975

Li, C., & Weihua, Z. (2002). Hybrid tdoa/aoa mobile user location for wideband CDMA cellular systems. *IEEE Transactions on Wireless Communications*, *1*(3), 439–447. doi:10.1109/TWC.2002.800542

Maham, B., & Nader-Esfahani, S. (2007). Performance analysis of distributed space-time codes in amplify-and-forward mode. In *Proceedings of the 8th IEEE Workshop on Signal Processing Advances in Wireless Communications*, (pp. 1-5). IEEE Press.

Morgan, D. P. (1991). *Surface-wave devices for signal processing*. London, UK: Elsevier Science Publishers.

Mukhopadhyay, S., Panigrahi, D., & Dey, S. (2004). Data aware, low cost error correction for wireless sensor networks. In *Proceedings of the IEEE Wireless Communications and Networking Conference,* (pp. 2492 - 2497). IEEE Press.

Nardi, S., & Pachter, M. (1998). GPS estimation algorithm using stochastic modeling. In *Proceedings of the 37th IEEE Conference on Decision and Control*, (pp. 4498–4502). IEEE Press.

Razi, A., Afghah, F., & Abedi, A. (2011). *Power optimized DSTBC assisted DMF relaying in wireless sensor networks with redundant super nodes*. Unpublished.

Razi, A., Yasami, K., & Abedi, A. (2011). On minimum number of wireless sensors required for reliable binary source estimation. In *Proceedings of the IEEE Wireless Communications and Networking Conference,* (pp. 1852 - 1857). Cancun, Mexico: IEEE Press.

Rhee, S., Seetharam, D., & Liu, S. (2004). Techniques for minimizing power consumption in low-data rate wireless sensor networks. In *Proceedings of the IEEE Wireless Communications and Networking Conference,* (pp. 1727 - 1731). IEEE Press.

Scholl, G., Korden, C., Riha, E., Ruppel, C. W., Wolff, U., & Riha, G. (2003). SAW-based radio sensor systems for short-range applications. *IEEE Microwave Magazine*, *4*(4), 68–76. doi:10.1109/MMW.2003.1266068

Schrader, K., Abedi, A., & Caccese, V. (2010). Impact localization and scaling for SHM of inflatable habitats. In *Proceedings of the 3rd IEEE/CANEUS Fly By Wireless Workshop*, (pp. 20-21). IEEE Press.

Schurgers, C., Aberthorne, O., & Srivastava, M. B. (2001). Modulation scaling for energy aware communication systems. In *Proceedings of the International Symposium on Low Power Electronics and Design*, (pp. 96 - 99). Huntington Beach, CA: IEEE.

Shih, E., Cho, S., Ickes, N., Min, R., Sinha, A., Wang, A., et al. (2001). Energy-efficient communication protocol for wireless microsensor networks physical layer driven protocol and algorithm design for energy-efficient wireless sensor networks. In *Proceedings of the 7th Annual International Conference on Mobile Computing and Networking*, (pp. 272--287). IEEE.

Wang, J. C., & X, W. (2010). On the sum rate of Gaussian multi-terminal source coding: New proofs and results. *IEEE Transactions on Information Theory*, *56*(8), 3946–3960. doi:10.1109/TIT.2010.2050960

Woo, T. K. (2000). Orthogonal code design for quasi-synchronous CDMA. *IET Electronics Letters*, *36*(19), 1632–1633. doi:10.1049/el:20001136

Xiong, Z. Y., Fan, X. P., Liu, S. Q., & Zhong, Z. (2010). Distributed image coding in wireless multimedia sensor networks: A survey. In *Proceedings of the Third International Workshop on Advanced Computational Intelligence*, (pp. 618 - 622). IEEE.

Yang, Z., & Liu, Y. (2008). Quality of trilateration: Confidence based iterative localization. In *Proceedings of the 28th International Conference on Distributed Computing Systems*, (pp. 446–453). IEEE.

Yasami, K., & Abedi, A. (2011). Effect of channel estimation error on physical layer network coding. In *Proceedings of the IEEE Consumer Communications and Networking Conference,* (pp. 973 - 974). Las Vegas, NV: IEEE Press.

Yost, G. A. (1998). Improvement in estimation of time of arrival (TOA) from timing advance (TA). In *Proceedings of the International Conference on Universal Personal Communications*, (pp. 1367–1372). IEEE.

Chapter 14
Cross-Layer Monitoring in Cloud Computing

Jose M. Alcaraz Calero
Hewlett-Packard Research Laboratories, UK

Benjamin König
Hewlett-Packard Research Laboratories, UK

Johannes Kirschnick
Hewlett-Packard Research Laboratories, UK

ABSTRACT

This book chapter describes the different aspects related to designing a suitable monitoring architecture for Cloud Computing, aiming to support cross-layer monitoring across all layers available in the Cloud stack. For this purpose, the importance of monitoring services in Cloud scenarios is outlined, followed by a comprehensible analysis of a wide set of distributed monitoring solutions. After that, the particular requirements related to cross-layer monitoring for Cloud Computing architectures are identified and explained. Then, diverse aspects which may fit a monitoring architecture for fulfilling such requirements are explained. Finally, some future research directions and conclusions are highlighted.

INTRODUCTION

Traditionally, businesses had to invest in the acquisition of their own dedicated IT infrastructure to satisfy their computational requirements. This led to a planning problem, as requirements for IT infrastructure in businesses are constantly changing according to current workload and business strategy. As a consequence, traditional businesses have to make a significant up-front investment in IT infrastructure to provide enough computational power and storage capacity to handle the highest peaks of workload and successfully face future challenges.

Cloud Computing (Armbrust, et al., 2010) is an upcoming computing technology with the potential to change how businesses implement their services. The idea behind Cloud Computing is to use virtualized resources which are provided by a third-party. These resources are acquired on-demand and typically paid for in a pay-per-usage model (Buyyaa, Shin Yeo, Venugopal, Broberg,

DOI: 10.4018/978-1-4666-0960-0.ch014

& Brandic, 2009). This paradigm enables Cloud customers to adapt the allocated IT infrastructure dynamically, based on changing business objectives and user demands. This leads to a significant reduction in investment needed to acquire IT infrastructure and creates a paradigm, which leads to a balance between own dedicated IT computation resources and IT resources rented from third parties for dealing with peaks of workload.

Cloud Computing provides services on a stack composed of three service layers (Hurwitz, Bloor, Kaufman, & Halper, 2009): *Infrastructure as a Service* (*IaaS*), *Platform as a Service* (*PaaS*), and *Software as a Service* (*SaaS*).

The *IaaS* layer exposes services for which the users can deploy virtual resources (e.g. virtual machines) on demand. This *IaaS* layer uses the underlying physical resources (e.g. physical machines, network, and storage components) to achieve the functionality. The *PaaS* layer relies on these virtualized resources to provide more complex middleware services. These middleware services are intended for being used as components and functionalities for other applications and services. Finally, the *SaaS* layer provides consumable high-level services to users (i.e. entire enterprise applications or web services). To provide these services, the *SaaS* layer uses the other layers of the cloud stack. Each layer available in the cloud stack is not restricted to end users but target different groups of users, i.e. developers and operations staff illustrates the types of services provided by the *IaaS*, *PaaS*, and *SaaS* layers and their primary user groups.

Each service layer may be provided by a different entity (*service provider*). Likewise, there may be different consumers for Cloud services (*service consumer*). In this context, a *private* Cloud is a usage scenario in which service provider and consumer are of the same party, whereas a *public* Cloud refers to a scenario in which provider and consumer are different parties and have to be from different administrative domains. Note that a private cloud can be considered as a software stack for internally governing the computational resources available in a company whereas a public cloud involves different providers of services

Currently, almost all research works are focused on public Cloud scenarios, which are the predominant option in Cloud Computing due to imminent advantages for the service consumer, such as, a reduction of acquisition cost in IT infrastructure and outsourcing of IT maintenance tasks to the service provider. This book chapter is focused on a public Cloud scenario in which a service provider offers Cloud services, generally by means of the Internet, to a variable number of service consumers.

The design of a stack for a public Cloud scenario is generally challenging as the architecture has to reflect the individual requirements of different service consumers and service providers. Many challenges associated to the design of the service layers are still open issues nowadays and they are regarded as drawbacks for which some businesses are reluctant to move to the Cloud. One of the main concerns reported by Cloud customers (Fujitsu, 2010) is how consumers get full control over virtualized resources rented on-demand.

Monitoring enable service providers a way to increase the control that customers have over computational resources by providing monitored information about the infrastructure and the services deployed therein.

Monitoring systems in cloud computing entail several problems, which are not correctly addressed in the current monitoring solutions. Firstly, as shown in Figure 1, monitoring services for the Cloud have to interact with all service layers available in the cloud stack. This requires an architecture for performing a monitoring across all the layers available in the cloud stack, i.e. a cross-layer monitoring rather than traditional monitoring solutions focused on monitoring homogeneous resources, i.e. a single layer. The reason for which cross-layer monitoring is critical is due to the high level of demanded control over the rented computational architecture. The final user demands

Figure 1. Cloud computing stack and monitoring services

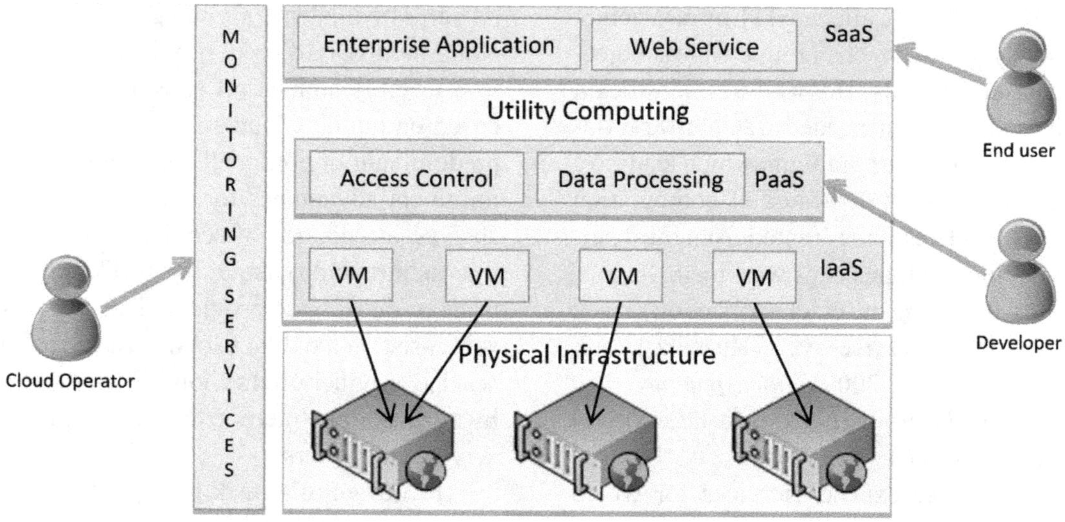

information and full control over all the involved layers rather than over a single one, being this a imperative feature to the successful of the cloud computing. Note that data has to be correlated and homogenized across different layers, which fuels the need for cross-layer support. For instance, to determine resource consumption for a given user, it is necessary to combine data held in the PaaS layer (i.e. user information) with data provided by the IaaS layer (i.e. processor utilization of virtual machines). This information may be exposed to the user and also be used internally by the service provider for accounting purposes and to enforce service level agreements.

Secondly, a cloud infrastructure is elastic by nature, i.e. the computation resources can be added and removed dynamically. This requires novel monitoring solutions flexible enough to be able to deal with this dynamic change in the network topology.

Another problem is the monitoring in multi-tenancy environments, directly related to public cloud scenarios, in which different administrative domains are using the same cloud infrastructure (being monitored). All these problems require the design of specific cross-layer monitoring solutions for cloud computing. Thus, this chapter describes diverse aspects, which may fit a monitoring architecture for fulfilling such requirements.

The following section presents an overview of existing monitoring systems and evaluates their suitability for monitoring a Cloud stack. This is followed by a list of requirements for cross-layer monitoring in Cloud Computing and an outline for our proposal of a suitable monitoring system for the Cloud.

BACKGROUND

There are some basic concepts associated to monitoring. The following provides an introduction to these concepts to enable the reader to fully understand the remainder of this chapter.

- Entity: An entity may be any networked resource, which is considered relevant for being monitored. Typical entities in a Cloud Computing scenario are: processors, memory, storage, processes, artifacts provided by applications, and network components.

Figure 2. Levels of monitoring architectures as proposed by Zanikolas and Sakellariou (2005)

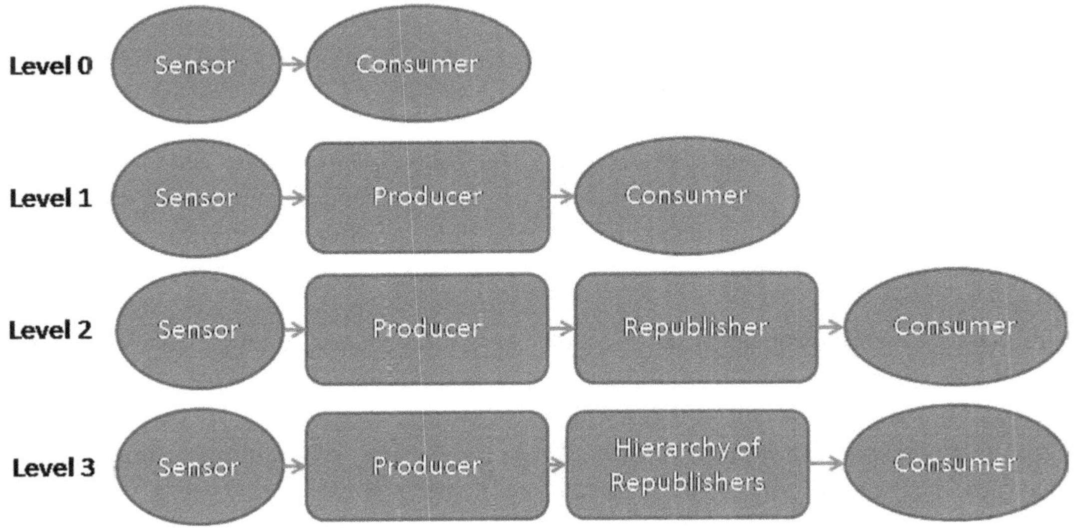

- Event*:* An event is defined as a collection of time-stamped information/data. Events can be associated to an originating entity and represented using a given structure.
- *Event Type:* The event type is defined as a unique identifier, which maps an *event* to a particular data structure.
- *Sensor:* A sensor may be any monitoring process that is in charge of monitoring an *entity*. Sensors produce new information in the form of *events*.
- *Actuator:* An actuator is an element, which directly influences a monitored entity. For instance, it may be a method which can be called on the entity to cause action (i.e. trigger a change in state or data).
- *Consumer:* A consumer is defined as an arbitrary software element, which receives events produced by *sensors*. In most monitoring systems, there is only a single consumer. This may be either a central repository, which collects monitored information, or the main data processor of the monitoring software.

Existing research has already provided a large body of work for monitoring IT infrastructures. Many different architectures, approaches, paradigms, and systems have been proposed to address the monitoring of information systems.

MONITORING ARCHITECTURES

Zanikolas and Sakellariou (2005) provide a complete taxonomy of monitoring architectures, which has been widely accepted in the scientific community, and evaluate their suitability for monitoring Grid architectures. This taxonomy is composed by four different levels of distributed monitoring architectures as illustrated in Figure 2.

In level 0 architectures events flow from *sensors* directly to *consumers*. Level 0 architectures are self-contained, monolithic, and their differentiating feature is the lack of a *producer API* which would enable the distribution of events to remotely located components.

Level 1 monitoring architectures enable the use of *sensors* which may be deployed in a remote location by using a *producer* component to communicate information to (a potentially variable) number of *consumers*. Basic network monitoring systems may be classified as level 1 monitoring solutions.

In addition to a producer, a level 2 monitoring architecture is distinguished by the presence of *republisher* components in the event flow between *sensors* and *consumers*. At this level, a *republisher* is a data processor unit with a fixed functionality, for instance, to aggregate or filter data provided by different sensors.

Finally, level 3 monitoring architectures are distinguished by the presence of an entire hierarchy of *republisher* elements which provide a wide range of data processing features over monitored data, such as aggregation, time-window based aggregation, data filtering, statistical evaluations, and trend analysis. In essence, level 3 architectures are the basis for highly flexible monitoring solutions in which all involved components can be deployed in a distributed environment.

MONITORING SOLUTIONS

Level-0/1 Centralized Monitoring Solutions

Nowadays, there are many different monitoring solutions available on the market. *ObserverNMS* (Frommel, 2010), *Cacti* (Lavlu, 2009), *Zenoss* (Bowling, 2008), and *IPHost Network Monitor* (ITelSib, Ltd., 2011) are some examples of free and open-source monitoring solutions. *TclMon* (Derevenetz, 2011), *AdRem, NetCrunch* (NetCrunch Adrem, 2011), and *FreeNATS* (PurplePixie Systems, 2011) are comparable commercial alternatives.

These level 0/1 solutions share the way in which they retrieve information from monitored hosts. They do not deploy any agent (sensor) on monitored hosts. Instead, they are entirely centralized. In order to retrieve information from involved hosts, these systemes rely on periodical polling. Typical polling methods include: SNMP, ping, TCP Connect, UDP Datagrams, WMI, FTP, and HTTP/HTTPS. This approach to monitoring is less intrusive than agent-based approaches but also may not fit well with Cloud Computing due to the heterogeneous nature of events occuring in all Cloud service layers and the lack of a uniform way to extend the monitoring capabilities of the software. Some of these monitoring solutions feature plugable monitoring extensions, but these only aim support new ways to retrieve information (e. g. new methods for polling) and do not explicity facilitate the inclusion of new types of monitored data.

As these monitoring solutions do not have distributed monitoring capability and only use a single node for data processing, they also introduce a bottleneck, which makes them less scalable. They are not suitable for monitoring the Cloud, as Cloud Computing is elastic by nature and requires a highly scalable monitoring solution which is able to cope with a varying number of users and a constant change in the number of virtualized resources available.

Level-2/3 Monitoring Solutions with Centralized Data Processing

Nagios (Barth, 2008), *Ganglia* (Massie, Chun, & Culler, 2004), *Zabbix* (Vladishev, 2010), *Pandora FMS* (Mateos, 2011), *OpenNMS* (Balog, 2005), and *Osmius* (Garcia, Llana, Malagon, & Pancorbo, 2010) are widely used open-source enterprise-class monitoring and alerting solutions for IT infrastructures. There are also commercial solutions such as *Accelops* (Accelops, 2011), *DNMS* (Dhyan Infotech, 2011), *DopplerVUE* (Kratos Network Solutions, 2011), *OpenKBM* (Rocket Software Inc., 2011), and *Zyrion Transverse* (Zyrion Inc., 2010), which are wide-spread monitoring systems for IT infrastructures.

All these solutions are designed based on level 2 and level 3 monitoring architectures in order to be flexible and powerful enough to satisfy all the demands of a competitive market. All of them provide support for distributed monitoring and are able to perform load-balancing between different hosts. Note that these features are especially relevant for cross-layer monitoring in Cloud Computing due to a constant change in the number of resources being provided and monitored in each service layer.

Nagios is a level 3 enterprise-class monitoring system which claims to have more than 1 million users nowadays. It provides advanced features, such as a complete hierarchy of *republishers*, using the *NDRP* (*Nagios Remote Data Processor*) protocol, and multi-tenancy support. Note that multi-tenancy is an inherit concept associated to Cloud Computing and a suitable monitoring solution for Cloud Computing has to support it. In this context, multi-tenancy support referres to the possibility to support different administrative domains (tenants) using the same monitoring software and control the visibility of data (data privacy) between tenants according to credentials provided to the monitoring system.

Ganglia is a scalable and distributed level 3 monitoring system for high-performance computing systems based on a hierarchical design targeted at groups of clusters. It makes use of daemons in the monitored hosts, which act as both *sensors* and *producers*. Moreover, this software enables the deployment of meta-daemons, which act as *republishers* to process the information received by the subscribed daemons.

Zabbix is level 3 monitoring solution that offers advanced monitoring, alerting and visualization features. Its authors claim that Zabbix has been successfully tested with 100.000 monitored nodes at the same time. This would clearly prove the scalability of the underlying architecture. Zabbix also enables to process data by means of *republishers*.

Pandora FMS, OpenNMS, and *Osmius* are level 3 open-source monitoring architectures which, similar to distributed monitoring solutions presented earlier, enable monitoring of the fulfillment of IP Service Level Agreements (SLAs) and provide insight into the state of the agreement. They also provide features for combining monitored information into groups and subgroups in order to process and visualize correlated data together. These advanced monitoring systems also provide auto-discovery services, which can (automatically) discover new entities in the monitored IT infrastructure, and additional alert features to provide notifications to administrators.

All previously described monitoring solutions have one feature in common: they have a central repository in which monitored information is being stored for queries and data processing over aggregated information. While this is almost mandatory and we foster this approach, we believe that some considerations have to be made in order to achieve a proposal suitable for cross-layer monitoring in Cloud Computing.

Most of these monitoring systems have a drawback: they simply gather and send monitored data to the central repository without any kind of data processing on the way. Data processing is only done in the central repository or in some other services used for load balancing in distributed monitoring. In these monitoring systems, data processing may be performed at different stages:

- periodically by daemons running in the background (periodical approach),
- when users acquire updated information using a graphical interface associated to the monitoring software (push approach),
- or just when new monitored data arrives at the repository (pull approach).

This approach may fit well in scenarios in which data processing is cheap. However, it does not fit once data volume or the complexity of data processing increases. This is especially the case in Cloud Computing. Here, a large user-base generates a significant data volume. Further, complex data processing and cross-layer correlation are essential features to compute desired information.

The analyzed monitoring systems differ significantly in their support of data processing features.

In general, all monitoring solutions can provide information gathered directly from monitored sensors. This information can later be evaluated to compute average, minimum, or maximum values, and detect peaks or trends over homogeneous monitored data. This allows to determine, for instance, CPU usage, service availably, network link status, and network usage. Moreover, all monitoring solutions additionally provide charts to visually represent monitored information, making it easy for the user to perform an analysis of the data. Almost all investigated monitoring solutions also provide statistics about monitored information for each sensor. These statistics represent a basic aggregation of the data provided by a given sensor. However, these solutions rarely allow users to combine information provided by different heterogeneous sensors, as required by cross-layer monitoring. Moreover, they only publish a couple of pre-defined reports and provide a way to customize your own reports, but do not provide any possibility to perform data filtering or data aggregation according to user-defined criteria and do not enable a correlation with other data.

Pandora FMS also provides statistics and charts over aggregated monitored data and enables a basic combination of data, called *Combined Report*, in which information provided by different and possibly heterogeneous sensors is shown in the same report and in the same chart. These artifacts provide the essential information to manually discover any correlation between data provided by different sensors. However, this is a mere combination of information rather than complex data correlation.

MonAMI claims to provide input for an arbitrary number of monitoring systems. Basically, it may be regarded as a universal sensor framework. Its idea is to gather information and send it to another monitoring framework. A direct usage of *MonAMI* is to deploy a sensor inside a computational resource to be able to gather information provided by other sensors. This smart sensor can be used for complex data processing of the monitored information and to spread the evaluation results to other monitoring systems.

Zyrion Transvere provides Root Cause Analysis (RCA) which is a feature for identifying the source of the problems by mining monitored information. It goes beyond traditional network level analysis. The root cause analysis engine is based on a service object model designed for analyzing end-to-end business impact instead of just stopping at the network layer. Obviously, this implies data correlation as there are complex relationships between IT elements, which have to be taken into account when a distributed application is delivered. However, only a closed set of different data correlations are supported with no ability to adapt the configuration for the correlations.

Monitoring solutions available on the market provide almost no support for user-defined queries over monitored information gathered from sensors, even though they provide a central repository, which could easily support data evaluation at run-time. Note that queries in this context do not refer to the possibility to export and extract monitored data, but instead refer to the use of a SQL-like language to process monitored data in order to compute high-level information and deliver useful results. The ability to post queries is an essential feature for a cross-layer monitoring solution as it supports complex data correlation and processing of information provided over the different service layers. As of now, such queries are rarely supported in distributed monitoring solutions.

These queries are often executed centrally. A viable alternative to centralized evaluation and data processing is the distributed execution of queries across the network. In general, a query may be defined using a query language and may be enforced by placing one or more *republishers* in the event flow from sensors to consumers. In essence, these republishers have to process the monitored information to achieve the semantics postulated by a given query. This approach avoids the necessity of having a central repository for processing data (as it is only required for data collecting purposes).

Level-3 Monitoring Solutions with Distributed Data Processing

R-GMA (Bonacorsi, et al., 2004) is a well-known distributed framework, which implements this approach to monitoring. The framework provides a homogeneous space for querying aggregated information using a *virtual table*. This table enables the insertion of complex SQL-like queries over data, which will be represented using *republishers* and distributed across the network to be processed and evaluated in a distributed fashion. Although R-GMA does not need a central data repository, it still has a central point of failure as it requires a dedicated central registry, which manages the location of data input for rendering virtual tables. This point of failure may not be adequate for cross-layer monitoring in Cloud Computing due to the critical nature of monitoring services in Cloud scenarios.

MDS4 (Schopf, et al., 2006) is another monitoring solution used in Globus Toolkit 4. It is a hybrid solution with a central repository for monitored information, but also with the capability to distribute aggregators across the network to provide basic data processing over information flowing between sensors and consumers.

These solutions all have a central point of failure, which is potentially susceptible of denial-of-service attacks. To address this issue, *MonALISA* (Newman, Legrand, Galvez, & Voicu, 2003) provides a multi-agent based approach with no central point of failure, powerful monitoring capabilities, as well as control and global optimization services for complex systems. However, this solution lacks in providing efficient query distribution services.

Regarding tasks achieved by the monitoring software, almost all the solutions provide a wide set of alert and notification methods in order to keep users and external software systems up to date with the current state of monitored entities. However, monitoring solutions very rarely provide a way to define policies, which have to be executed based on defined circumstances to achieve a policy-based management system. They also do not incorporate *actuators* into their architecture, which could invoke actions on managed entities in order to react to specific circumstances. Such a policy-based system is especially attractive in combination with cross-layer monitoring due to the capability to receive information from different layers and perform actions in consequence to correlated data.

MONITORING THE CLOUD INFRASTRUCTURE

This section illustrates the basic requirements for a suitable monitoring solution for Cloud Computing and presents an exemplary cross-layer monitoring framework, which meets these requirements. These requirements can be categorized as shown in Figure 3.

There are many unique features associated to Cloud Computing, which have to be taken into account to design an adequate monitoring system for such an environment. First of all, Cloud Computing is split into different conceptual layers which provide Cloud services of a varying degree of granularity and abstraction. Each layer is composed of a set of computational resources, which are totally heterogeneous across layers. For example, the *IaaS* layer may be composed of physical and virtual machines, whereas the *PaaS* layer may be composed of middleware services, which can be used by multiple tenants simultaneously. This fact makes data homogenization of monitored information an essential requirement for any Cloud monitoring system, as cross-layer data correlation is impossible without enforcing a uniform data representation.

Further, one of the main drawbacks associated to Cloud Computing is the lack of control over

Figure 3. Requirements for cross-layer monitoring in cloud computing

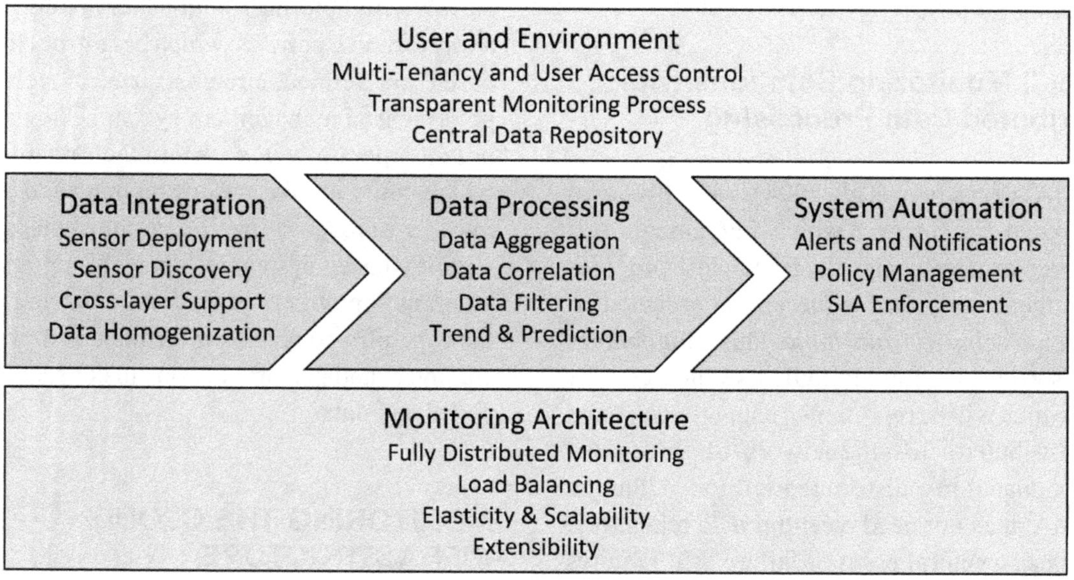

Cloud resources. An essential requirement to increase the level of control is to provide extended insight for users by providing a wide set of information across all service layers by means of a monitoring system. However, isolated information from different service layers may not be sufficient for Cloud users to retain confidence over the state of provided Cloud resources. Thus, an adequate monitoring solution has to be able to perform cross-layer data processing, data aggregation and data correlation to provide high-level information about Cloud resources. In this respect, the quality of information is essential. In fact, monitoring becomes a critical service in Cloud Computing due to the strong control requirements imposed by business users. As a consequence, dedicated points of failure are not welcomed and a fully distributed monitoring approach should be implemented.

Cloud infrastructure and its layers are elastic by design. Due to scaling infrastructure, the amount of monitored information may change rapidly. For this reason, it is essential to provide an elastic monitoring architecture that is able to scale in the number of computational resources being used for monitoring purposes. This ensures that the monitoring framework can be applied in different environments and enables load balancing insofar as the number of monitored entities is being augmented.

From an architecture perspective, a monitoring solution for Cloud Computing may be based on a level 3 monitoring architecture and rely on either agent, or agent-less approaches. Level 3 enables remote sensing features and distributed data processing across the entire event flow from sensors to consumers. An agent-based approach may be used as the preferred way to retrieve information since agents deployed remotely may act as a data homogenizer and also ensure that the monitoring architecture is extensible with respect to the kind of data being monitored. However, due to the heterogeneity of data provided by the different service layers, cross-layer monitoring may sometimes require a less intrusive agent-less approach for gathering information to make the monitoring process transparent for the customer. In this sense, monitoring information provided at the virtualization layer plays a critical role for this purpose and has to be exploited as much as possible in order to avoid user being aware of monitoring processes.

Another essential requirement is the auto-discovery of entities, which have to be monitored. Due to the highly dynamic nature of Cloud Computing environments, sensors have to be created and destroyed constantly and the monitoring system has to keep an inventory of the currently monitored entities. Moreover, logical grouping of monitored sensors is important since it enables a more accurate and precise processing of monitored information. Note that although entities are dynamically created on-demand, they still need a sensor in order to be monitored. Thus, auto-deployment of sensors for monitored entities may be required to efficiently manage the monitoring software and allow the system to scale. Obviously, the deployment of an adequate sensor has to be part of the monitoring life cycle.

Cloud Computing has to provide multi-tenancy support. Analogously, monitoring software has to provide support for multi-tenancy. This feature also allows enforcing access control for monitored entities, according to the administrative domain of a tenant that has been authenticated with the system, while still providing a framework with full control for the Cloud provider. Multi-tenancy may restrict the visibility of monitored information managed by the monitoring system to specific Cloud users. Multi-tenancy scenarios also enable federations in which monitored information may be extended beyond a single administrative domain by retrieving information from federated entities according to trust relationships established among tenants. This is much related to the requirement of a fine-grain access control over monitored information which may be necessary in this scenario to give Cloud user complete control.

Another issue is how to inform users. Almost all monitoring solutions offer a wide range of alert and notification methods in order to inform users about data monitored by the system. Obviously, alerts and notifications are also needed in Cloud Computing scenarios and these features have to be considered when designing the monitoring architecture. In fact, cross-layer monitoring enables to correlate data, which is essential to create smart alerts and notifications according to elaborated information achieved by combining data from different service layers. The way in which the alerts and notification functionality is inserted into the architecture may be pluggable to enable an easily extensible monitoring system.

Regarding data processing, a distributed and elastic way of performing data processing avoids any central point of failure. Moreover, any smart data processing would be welcome, especially processing features related to trend metering and trend prediction of monitored values. Note that this kind of information enables users to get more refined information about the monitored infrastructure and to predict future system states. Current analyzed monitoring solutions do not provide any way of fully customizing the data processing and correlation over monitored data, whereas this is an essential feature for enabling cross-layer monitoring. Thus, a query language may be exposed by the monitoring architecture. This language may be applied to specify the information to be retrieved and instruct complex data correlation and data processing. Generally speaking, this query language may be an SQL-like language which is able to identify sources of data, express information filters, data aggregations using a given criteria, filters over aggregated data, joins between data sources, and others. This language may be the first step to give a Cloud user enough control to define what information he wants to be monitored (selective monitoring) and how it has to be processed to achieve high-level information and knowledge about Cloud resources from monitored data. Probably, a query in this language may be enforced in a distributed fashion by placing customized republishers in the event flow between sender and consumer to process the information across the path. This is not in contradiction to the possible requirement to keep a centralized redundant repository of monitored data to be able to provide required information to customers at any time. A direct application for such a query language may be to provide aggregated information over monitored data (trending) and the correlation of heterogeneous inter-layer and intra-layer information. Note that data processing and correlation to get high-level elaborated information are important features to make the monitoring system flexible enough to be usable by any Cloud user and within any scenario.

Another application of such a query language may be to implement high-level monitoring features such as monitoring of Service Level Agreements (SLA). Basic SLAs, such as IP SLAs, which control the quality of service at IP network level may be directly implemented by mean of queries to monitor and correlate information provided by the *IaaS* layer. Moreover, note that Cloud Computing is very close to the establishment of complex SLAs between Cloud providers and users, which have to be fulfilled in all the layers of the Cloud stack. Thus, monitoring of complex SLAs may be established by using the query language to define and correlating monitored data.

A powerful query language is the first step to provide a way to define reactions of the Cloud infrastructure for certain events in the system. It has to be accompanied by a mechanism to express actions which closes the management loop and allows the monitored system to be self-managed. Policy-based systems are a suitable mechanism as they enable the definition of high-level policy rules, which are applied against the current state of the infrastructure, services, and monitored data. Thus, methods for implementing system reactions over the monitored information are becoming essential in order to manage efficiently a high number of machines used to host Cloud infrastructure. However, note that these systems have to be flexible enough to enable administrators to easily extend the set of actions to be executed within the system (at any layer) as system reactions.

Cross-Layer Monitoring Framework

In the previous section, we have identified the requirements for cross-layer monitoring systems in Cloud Computing. The reader may note that none of the monitoring solutions analyzed in this chapter supports all of these requirements. This section discusses different alternatives for designing an architecture, which satisfies these requirements.

First of all, let's analyze the topology of the monitoring architecture. Figure 4 shows four different topologies. They are evaluated for their suitability to implement a reliable monitoring service in Cloud Computing scenarios.

Master-slave topology. Applying this traditional topology to monitoring, we regard sensors as slave nodes and require a master node to host a central data repository. Sensors monitor entities and forward events to the central repository where data is collected and evaluated to provide high-level information to an administrator. This kind of topology has an implicit a point of failure, the master node, and thus is not adequate to implement a critical monitoring service.

Mesh topology. This topology requires every node in the network architecture to be connected with each other. While this improves fault tolerance, the overhead created by the architecture to establish, manage and update all the links between nodes is really high. Moreover, connections between sensors do not make much sense in monitoring systems as sensors are only interested in their assigned entity and do not have to exchange information with each other.

Multimaster-slave topology. This topology is an improved version of the tradition master-slave topology. In summary, there is a group or even hierarchy of several master nodes to provide fault tolerance as well as implement load balancing and distributed monitoring. This topology may be a viable alternative to implement a critical monitoring service for the Cloud as it is more flexible. However, this architecture does not exploit data

Figure 4. Topologies of monitoring systems

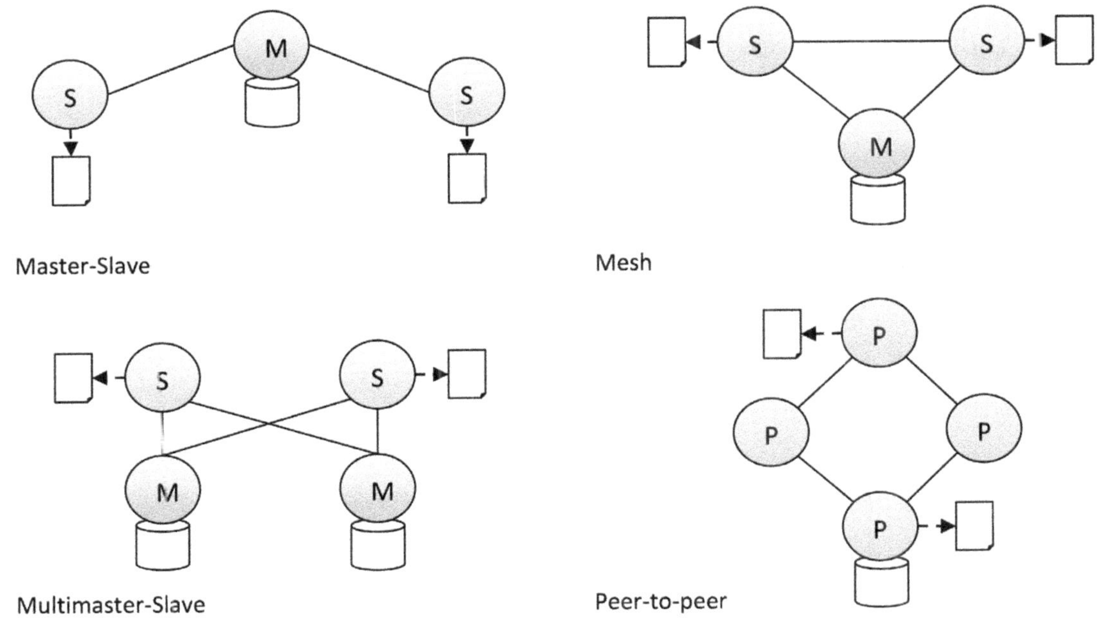

processing capabilities provided by slave nodes since data processing is carried out only in master nodes and this leads a waste of resources. For instance, slave nodes could take part in data processing by filtering unwanted information early on as this reduces network traffic. Moreover, master nodes may be targets of distributed denial-of-service attacks in this topology.

Peer-to-peer architecture. In this architecture every node has the same functionality. There is no distinction between master and slave nodes. In essence, there is no central point of failure and any node in the architecture may be in charge of data processing and sensing any data/resources in the monitored system. This architecture leads to a fully distributed and very flexible approach for monitoring Cloud infrastructure. This kind of architecture provides a good basis to implement critical services and to exploit distributed data processing capabilities.

Based on this preliminary analysis, we believe that both *multimaster-slave* and *peer-to-peer* are suitable topologies for implementing a cross-layer monitoring system. Note how both topologies enable to implement load balancing methods for carrying out distributed monitoring and data processing. However, a peer-to-peer based system is more flexible and can carry out distributed data processing more efficiently (as data processing is not restricted to master nodes) and thus has been selected in this book chapter.

Now that we have selected a suitable topology, let's analyze how cross-layer monitoring may be enabled. In general, the monitoring system needs a way to perform data homogenization of the heterogeneous information provided by different sensors (which are located in different layers). Different alternatives can be considered for pursuing this kind of monitoring: *1)* the design of a well-known interface for implementing sensors which requires providing monitored data in a common data structure. *2)* to define a domain schema such as *XML Schema, DTD, OWL* and to use this schema as the common language for

Figure 5. Interface for cross-layer data homogenization

```
public interface QueryOperator {
    // used for configuration
    void config(String configFile);

    // start data retrieval
    void start();

    // Result Publication
    void added(Reference sender, DataRow row) throws RemoteException;
    void removed(Reference sender, DataRow row) throws RemoteException;
    void done(Reference sender) throws RemoteException;

    // Call-backs for error and status messages
    void error(ErrorMsg msg) throws RemoteException;
    void status(String msg) throws RemoteException;

    // QueryObserver may be added to any query operator
    void addObserver(QueryObserver gui) throws RemoteException;

    // Data Structure monitored by this component
    List<DataColumns> getMonitoredDataStructure() throws RemoteException;
}
```

expressing monitored data by sensors. Both have advantages and disadvantages. The former is more imperative (method-oriented), and enables a detailed data life cycle control, but also is less declarative and lacks on providing a data schema. The latter is less imperative, since there is no data life-cycle control, and more declarative and thus provides a description of the data and data inter-relationships. We believe that monitoring in Cloud Computing should be a long-living service and thus demands a detailed life-cycle control of monitored data to keep track of data updates, aggregations, modifications, sampling, etc. For this reason, we have chosen the first approach based on the implementation of a common interface as a suitable solution to handle heterogeneous data in the monitoring system. This interface is depicted in *Figure 5*.

A simple table structure is an adequate common data structure since it has been demonstrated to be able to represent all kinds of information in traditional data bases. Then, all monitored information can be represented using the same data structure to provide a homogeneous way to access and further process monitored information (see method *getMonitoredDataStructure*). This data homogenization provides the foundation for carrying out a cross-layer data processing and complex data correlation. Moreover, this interface provides full life-cycle control over changes in the monitored data by means of the methods *start, added, removed, done, error, status* which matches with the proposed life cycle of the monitored information.

Several adaptors can be implemented to retrieve information from a variety of sources such as log files, JDBC sources, and even other complementary monitoring solutions. Moreover, this interface is also shared between the entities to be monitored and the different processing capabilities, which can be used in the query language.

The query language required to evaluate monitored information has to be designed with requirements related to Cloud Computing in mind:

1. *The query language has to be exposed to Cloud users.* It enables the definition of custom queries over information provided by sensors present in the user's Cloud resources. This contributes to the feeling of having complete control over Cloud infrastructure, which is one of the biggest issues in Cloud Computing nowadays.
2. *A query may involve diverse monitored resources.* The use of tables as a common data structure for monitored information enables the design of a SQL-like data manipulation language to define data processing over integrated information. In fact, SQL has been demonstrated to be useful for a vast number of applications during the last years. For this reason, its operators and semantics may be used as a basis for the design of an adequate data manipulation language. In fact, the language may include operators for determining the number of data sources used in the query (comparable to the *FROM clause* in SQL).
3. *The query language has to enable selection and filtering of monitored data.* The language should provide specific operators to select (*SELECT clause*) and filter (*WHERE clause*) monitored data according to some criteria.
4. *The query language has to provide features to perform complex data aggregation and data correlation.* The query language may group data according to an aggregator function (*GROUP BY clause*), such as average, minimum, maximum, count, and sum. This may be combined by a filter to only include data, which fulfills some given criteria (*HAVING clause*). The query language also has to correlate data by combining data from different entities (*JOIN clause*), even if the information comes from different layers.
5. *The query language has to provide high-level information.* By combining previously mentioned features in a single query, more complex queries can be defined which obtain high-level information from monitored data.
6. *The query language has to be executed in a distributed fashion.* Regardless of the syntax used for expressing queries, queries have to be processed and split into tasks which have to be distributed across resources and executed in a distributed way to achieve distributed data processing and even processing in parallel when possible. Basically, this avoids any processing bottleneck and at the same time ensures scalability of the monitoring system.

Features 1-5 are related to the expressiveness and semantics of the query language. Rather than selecting a given syntax (i.e. SQL), we believe that any syntax providing these features would be suitable for the query language.

A suitable approach is the usage of a dot notation to express queries, chaining references to data sources and performing evaluations along the way. A query is composed by two different kinds of operators: *sources* and *features*. As queries are evaluated from left to right, each query has to start with a *source*, i.e. a reference to a data source defined, i.e. instantiation of a given adaptor. A *feature* is a reference to an *operator*, which provides capabilities for evaluating and processing data. Different feature operators, e.g. selections, projections, filter for duplicated and empty data, aggregations, and aggregations over time may be necessary. Further, each "." within a query is translated into a *join operator*, which joins sets of data received from two child operators by evaluating a join predicate.

For example, the query customers.eq(level:gold).services retrieves the monitored information of all running Cloud services associated to customers with "gold" status in the running example. Note that the query does not specify where and how the query operators are evaluated. Note as well how cross layer monitoring is achieved since both PaaS and *SaaS* information is being monitored, homogenized and combined.

Feature 6 is related to the way in which the queries are parsed and executed in the monitoring system. A query is defined using the query language and may be enforced by placing *republishers* in event flows from sensors to consumers to process provided information in order to achieve the semantics stated by a given query. This leads to the fact that a suitable cross-layer monitoring architecture for Cloud Computing may be a level 3 architecture in which *republishers* are smartly deployed according to queries provided by Cloud users.

Deployment is an important aspect within this architecture as there are many different artifacts to be placed across resources provided by the Cloud. Firstly, sensors have to be deployed remotely within the monitored Cloud infrastructure. Secondly, *republisher* objects for data processing and data correlating have to be deployed remotely in computational resources for carrying out their respective tasks. Further, any artifacts required by the architecture may be deployed even more than once for fault tolerance purposes. In essence, the deployment of these software artifacts can be achieved by following one of two different approaches. On the one hand, a manual approach requires that administrators perform the deployment of these components either totally on their own or using a deployment strategy. For instance, a timesaving deployment strategy may be to provide all sensor implementations inside of a volume image booted into VMs and instantiate them as part of the VM boot-strapping process. This approach has serious limitations in Cloud scenarios:

1. It assumes that the location of sensors is already known before the run-time stage.
2. Any human involvement in the deployment process is error-prone and drastically reduces the scalability of the monitoring architecture.
3. Re-configuration of deployed artifacts at run-time is not covered.
4. Elasticity of the monitoring system is not ensured.

Due to these limitations, an automatic approach for carrying out the provisioning of software artifacts is needed, as it leads to a more complete and reliable solution for cross-layer monitoring in Cloud Computing. An automated deployment of components within the Cloud environment can be achieved using an automated deployment service such as *SmartFrog* (Goldsack, et al., 2009), *SLIM* (Kirschnick, Alcaraz Calero, Wilcock, & Edwards, 2010), or *Puppet* (Turnbull, 2007). These kinds of services provide the underlying functionality for carrying out an automated provisioning of services into interconnected nodes, i.e. physical or virtual machines. The process of service deployment may include several steps including: data transfer, installation, configuration and execution of an application. Usually, solutions for performing an automated deployment are composed of a daemon installed in each node, which provides the entry point to manage the node, and a master node, which is in charge of performing and orchestrating the deployment process. Although the monitoring architecture may not have a direct dependency to a particular deployment solution, the following features have to be taken into account when a distributed deployment solution has to be selected or designed. These features also match with those that should be fulfilled, in general, by a suitable monitoring architecture for cloud computing.

1. *Architecture*. Using a master-slave architecture leads to a potential point-of-failure. By contrast, pure peer-to-peer and multimaster-node architectures do not have a single point-of-failure.
2. *Life cycle*. The life-cycle software components follow has to be considered to exercise control over deployed components, manage the level of orchestration and determine the data to be monitored.

3. *Reliability.* The architecture has to support fail-over. In a sense that, if a virtual resource fails, a new resource is requested.
4. *Cloud-aware.* The architecture should be cloud-aware, that is, being able to request additional virtual resources on-demand from the Cloud provider as part of the deployment process. Note that cloud-awareness is essential to ensure elasticity of the cross-layer monitoring architecture.
5. *Complexity and Extensibility.* The range of software artifacts to be managed in the system is other selection criteria, as well as the amount of effort that is required to support extension with additional artifacts. Note that every sensor and *republisher* has to be deployed by the system.
6. Scalability. As Cloud infrastructure is elastic by nature, this entails that scalability plays a critical role and has to be considered when a solution is chosen.
7. Security. It has to be ensured that only trusted software components are deployed.

A solution fulfilling these requirements may provide a comprehensible support for deploying all software artifacts as required by the cross-layer monitoring architecture by using resources, which are provided on-demand by the Cloud infrastructure in order to support elasticity in the number of nodes used for monitoring purposes and for data processing and correlation. This elasticity is essential for a monitoring system in Cloud Computing scenarios and ensures that the monitoring system is able to adapt rapidly to new monitoring requirements according to the current size of the monitored infrastructure.

With regards to multi-tenancy support, the cross-layer monitoring architecture has to provide a monitoring service covering the whole Cloud infrastructure, performing the monitoring for every tenant. This requires a security model to be integrated into the monitoring architecture to cover the following aspects:

1. Access Control. Information has to be restricted to certain tenants and a security model is needed to enforce data privacy.
2. Information Exchange. Sensors, providers, *republishers,* and even consumer objects have to support authentication methods in order to ensure that only those objects belonging to the same tenant are able to share information.
3. Trust Model. Information may be shared between tenants, but this has to be controlled by a trust model, which decides what information can be shared, and between whom.

FUTURE RESEARCH DIRECTIONS

Currently, there are strong requirements regarding infrastructure control imposed by cloud users. There is a certain necessity for providing a suitable monitoring service for cloud computing. For this reason, current research work is addressing monitoring for cloud computing opening a wide range of future directions.

First of all, a necessary step is the integration of automated provisioning services into monitoring software in order to provide elasticity and cross-layer support. Note that this integration requires additional research work to design suitable lifecycle models for monitoring in cloud computing. Potential for further research is fueled by the need to design frameworks, which avoid the necessity to implement specific sensors for custom applications. Instead, alternative approaches are required, which are flexible enough to re-use existing sensors by means of configuring them to monitor a specific entity.

The design of new techniques for carrying out undetectable monitoring of entities provided by the Cloud has to be investigated. Monitoring virtual resources by using sophisticated techniques such as hypervisor-driven monitoring or transparent network monitoring are possible future directions.

New methods for data processing may also be investigated to provide novel ways of correlating and aggregating data in order to increase the level of information derived from monitored data.

More advanced and powerful data homogenization techniques may be developed in order to create a common framework for performing data management and data processing efficiently and also increasing performance of monitoring architecture. Work on smarter sensors is another required effort to provide additional metrics and information about monitored resources.

Cross-layer monitoring introduces new business models in Cloud Computing. For instance, *Sotec* (SOTEC Tachnical Solutions, Inc., 2011) provides monitoring-as-a-service to provide business value to its customers. This service provides multi-tenancy support to serve a wide range of users and helps them to monitor their infrastructure.

CONCLUSION

A comprehensible, comparative analysis of several distributed monitoring solutions has been done in this chapter concluding with the identification of the requirements for the design of a suitable cross-layer enabled monitoring framework for Cloud Computing.

Elastic and distributed data processing have been identified as critical features for any cloud-aware monitoring framework. These features enable the use of on-demand Cloud resources (e.g. virtual machines) for scaling the monitoring framework according to individual user requirements.

Cross-layer has been demonstrated to be essential. In order to enable cross-layer monitoring, data homogenization by sensors has been identified as a fundamental requirement for the monitoring framework. This enables the integration of heterogeneous data from different layers. As demonstrated in this chapter, a wide range of features (i.e. data aggregation, correlation, and filtering) provided by the monitoring framework is required to process and correlate integrated data in order to provide high-level information. This kind of information is the basis for later management tasks and required to enforce service level agreements. In the end, this may lead to a full automation of the Cloud infrastructure and provided services. Especially for implementing management tasks, the monitoring process has to be transparent to the user.

Regarding deployment, current monitoring solutions do not integrate any automated deployment service to instantiate sensors and *republishers*. However, this chapter has demonstrated that automated deployment is essential to facilitate monitoring in Cloud scenarios. Note how automated deployment enables any resource available in the Cloud infrastructure to be monitored simply by deploying an agent to retrieve information from an entity hosted in the Cloud.

ACKNOWLEDGMENT

Thanks the Fundación Seneca for sponsoring Jose M. Alcaraz Calero under the post-doctoral grant 15714/PD/10.

REFERENCES

Accelops. (2011). *Accelops*. Retrieved from http://www.accelops.net/.

Armbrust, M., Fox, A., Griffith, R., Joseph, A., Katz, R., & Konwinski, A. (2010). A view of cloud computing. *Communications of the ACM, 53*(4), 50–58. doi:10.1145/1721654.1721672

Balog, T. (2005). *Enterprise-wide network management with OpenNMS*. New York, NY: O'Really SysAdmin.

Barth, W. (2008). *Nagios: System and network monitoring*. New York, NY: No Starch Press.

Bonacorsi, D., Colling, D., Field, L., Fisher, S., Grandi, C., & Hobson, P. (2004). Scalability test of R-GMA based grid job monitoring system for CMS Monte Carlo data production. *IEEE Transactions on Nuclear Science*, *51*(6), 3026–3029. doi:10.1109/TNS.2004.839094

Bowling, J. (2008). Zenoss and the art of enterprise monitoring. *Linux Journal, 172*(6).

Buyyaa, R., Shin Yeo, C., Venugopal, S., Broberg, J., & Brandic, I. (2009). Cloud computing and emerging IT platforms: Vision, hype, and reality for delivering computing as the 5th utility. *Future Generation Computer Systems*, *25*(6), 599–616. doi:10.1016/j.future.2008.12.001

Derevenetz, O. (2011). *TclMon: Open source cross-platform network monitoring solution*. Retrieved from http://tclmon.sourceforge.net/.

Dhyan Infotech. (2011). *Dhyan network/element management system (DNMS)*. Retrieved from http://www.dhyanit.com/dnms_overview.html.

Frommel, O. (2010). Monitoring: Aus observerNMS wird observium. *Linux Magazine Online*. Retrieved from http://www.linux-mag.com.

Fujitsu. (2010). *Fujitsu research reveals global consumer attitudes to data privacy crucial to realizing benefits of cloud computing*. Retrieved from http://solutions.us.fujitsu.com/www/content/news/newsdetail.php?nf=10678411.nitf.

Garcia, R., Llana, L., Malagon, C., & Pancorbo, J. (2010). Event prediction in network monitoring systems: Performing sequential pattern mining in osmius monitoring tool. *LNCS Advances in Data Mining: Applications and Theoretical Aspects*, *6171*, 632–642. doi:10.1007/978-3-642-14400-4_49

Goldsack, P., Guijarro, J., Loughran, S., Coles, A., Farrell, A., & Lain, A. (2009). The SmartFrog configuration management framework. *ACM SIGOPS Operating Systems Review*, *43*, 16–25. doi:10.1145/1496909.1496915

Hurwitz, J., Bloor, R., Kaufman, M., & Halper, F. (2009). *Cloud computing for dummies*. New York, NY: For Dummies.

ITelSib, Ltd. (2011). *IPHost network monitor*. Retrieved from http://www.iphostmonitor.com/.

Kirschnick, J., Alcaraz Calero, J., Wilcock, L., & Edwards, N. (2010). Towards an architecture for the automated provisioning of cloud services. *IEEE Communications Magazine*, *48*(12). doi:10.1109/MCOM.2010.5673082

Kratos Network Solutions. (2011). *dopplerVUE*. Retrieved from http://www.kratosnetworks.com/products/dopplervue/.

Lavlu, S. M. (2009). *Cacti 0.8 network monitoring*. New York, NY: Packt Publishing.

Massie, M. L., Chun, B., & Culler, D. (2004). The ganglia distributed monitoring system: Design, implementation, and experience. *Parallel Computing*, *30*(7), 817–840. doi:10.1016/j.parco.2004.04.001

Mateos, R. (2011). *Pandora FMS: The free monitoring system*. Retrieved from http://pandora.sourceforge.net/en/index.php.

Millar, P. (2011). *MonAMI*. Retrieved from http://monami.sourceforge.net/.

NetCrunch Adrem. (2011). *NetCrunch 6*. Retrieved from http://www.adremsoft.com/netcrunch/.

Newman, H., Legrand, I., Galvez, P., & Voicu, C. (2003). MonALISA: A distributed monitoring service architecture. In *Proceedings of the Computing in High Energy and Nuclear Physics Conference*. La Jolla, CA: ACM Press.

Purple Pixie Systems. (2011). *FreeNATS network monitor*. Retrieved from http://www.purplepixie.org/freenats/.

Rocket Software Inc. (2011). *Rocket OpenKBM.* Retrieved from http://www.rocketsoftware.com/networks/enterprise-management/openkbm/overview.

Schopf, J. M., Pearlman, L., Miller, N., Kesselman, C., Foster, I., & D'Arcy, M. (2006). Monitoring the grid with the globus toolkit MDS4. *Journal of Physics: Conference Series, 46*(46), 521. doi:10.1088/1742-6596/46/1/072

SOTEC Technical Solutions, Inc. (2011). *SOTEC technical solutions.* Retrieved from http://sotec-technical.com/.

Turnbull, J. (2007). *Pulling strings with puppet.* New York, NY: FristPress. doi:10.1007/978-1-4302-0622-4

Vladishev, A. (2010). *Zabbix: An enterprise-class open source distributed monitoring solution.* Retrieved from http://www.zabbix.com/.

Zanikolas, S., & Sakellariou, R. (2005). A taxonomy of grid monitoring systems. *Future Generation Computer Systems, 21*(1), 163–188. doi:10.1016/j.future.2004.07.002

Zyrion Inc. (2010). *Zyrion transverse.* Retrieved from http://www.zyrion.com/.

ADDITIONAL READING

Abiteboul, S., & Marinoiu, B. (2007). Distributed monitoring of peer to peer systems. In *Proceedings of the 9th Annual ACM International Workshop on Web Information and Data Management,* (pp. 41-48). Lisbon, Portugal: ACM Press.

Bailey, M., Cooke, E., Jahanian, F., Nazario, J., & Watson, D. (2005). The internet motion sensor: A distributed blackhole monitoring system. In *Proceedings of Network and Distributed System Security Symposium,* (pp. 167-179). IEEE.

Calheiros, R. N., Ranjan, R., Beloglazov, A., Rose, C. A., & Buyya, R. (2011). CloudSim: A toolkit for modeling and simulation of cloud computing environments and evaluation of resource provisioning algorithms. *Software, Practice & Experience, 41*(1), 23–50. doi:10.1002/spe.995

Dejun, J., Pierre, G., & Chi, C.-H. (2009). EC2 performance analysis for resource provisioning of service-oriented applications. In *Proceedings of the 2009 International Conference on Service-Oriented Computing,* (pp. 197-207). Stockholm, Sweden: Springer-Verlag.

Eric Fabre, A. B. (2005). Distributed monitoring of concurrent and asynchronous systems. *Discrete Event Dynamic Systems, 15*(1), 33–84. doi:10.1007/s10626-005-5238-5

Gedik, B., & Liu, L. (2006). MobiEyes: A distributed location monitoring services using moving location queries. *IEEE Transactions on Mobile Computing, 5*(10), 1384–1402. doi:10.1109/TMC.2006.153

Graham, C., Kalampura, R., & Ramamirtham, J. (2006). Communication-efficient distributed monitoring of thresholded counts. In *Proceedings of the 2006 ACM SIGMOD International Conference on Management of Data,* (pp. 289-300). Chicago, IL: ACM Press.

IEEE. (2009). Interfaces for placement, migration, and monitoring of virtual machines in federated clouds. In *Proceedings of the Eighth International Conference on Grid and Cooperative Computing,* (pp. 253-260). IEEE Press.

Ivanović, D., Carro, M., & Hermenegildo, M. (2009). An initial proposal for data-aware resource analysis of orchestrations with applications to predictive monitoring. In *Proceedings of the 2009 International Conference on Service-Oriented Computing.* Stokholm, Sweden: Springer.

Jain, N., Mahajan, P., Kit, D., Yalagandula, P., Dahlin, M., & Zhang, Y. (2008). Network imprecision: A new consistency metric for scalable monitoring. In *Proceedings of the 8th USENIX Conference on Operating Systems Design and Implementation,* (pp. 87-102). San Diego, CA: USENIX Association.

Kazhamiakin, R., Pistore, M., & Zengin, A. (2009). Cross-layer adaptation and monitoring of service-based applications. In *Proceedings of the 2009 International Conference on Service-Oriented Computing,* (pp. 325-334). Stockholm, Sweden: Springer-Verlag.

Kutare, M., Eisenhauer, G., Wang, C., Schwan, K., Talwar, V., & Wolf, M. (2010). Monalytics: Online monitoring and analytics for managing large scale data centers. In *Proceeding of the 7th International Conference on Autonomic Computing,* (pp. 141-150). Washington, DC: ACM Press.

Mehedi Hassan, M., Song, B., & Huh, E.-N. (2009). A framework of sensor-cloud integration opportunities and challenges. In *Proceedings of the 3rd International Conference on Ubiquitous Information Management and Communication,* (pp. 618-626). Suwon: ACM Press.

Meier, A., Motani, M., Siquan, H., & Künzli, S. (2008). DiMo: Distributed node monitoring in wireless sensor networks. In *Proceedings of the 11th International Symposium on Modeling, Analysis and Simulation of Wireless and Mobile Systems,* (pp. 117-121). Vancouver, Canada: ACM Press.

Micalizio, R., & Torasso, P. (2007). On-line monitoring of plan execution: A distributed approach. *Knowledge-Based Systems*, *20*(2), 134–142. doi:10.1016/j.knosys.2006.11.010

Mietzner, R., & Leymann, F. (2008). Towards provisioning the cloud: On the usage of multi-granularity flows and services to realize a unified provisioning infrastructure for SaaS applications. In *Proceedings of the IEEE Congress on Services,* (pp. 3-10). IEEE Press.

Moreno-Vozmediano, R. S., Montero, R., & Llorente, I. (2009). Elastic management of cluster-based services in the cloud. In *Proceedings of the 1st Workshop on Automated Control for Datacenters and Clouds,* (pp. 19-24). Barcelona, Spain: ACM Press.

NASA. (2009). *NodeMon*. Retrieved from http://sourceforge.net/projects/nodemon/.

Newhall, T., Libeks, J., Greenwood, R., & Knerr, J. (2010). PeerMon: A peer-to-peer network monitoring system. In *Proceedings of the 24th International Conference on Large Installation System Administration,* (pp. 1-12). San Jose, CA: USENIX Association.

Paulo Salvador, R. V. (2007). *Network monitoring system with a peer-to-peer architecture*. Retrieved from http://www.google.es/url?sa=t&source=web&cd=1&ved=0CBwQFjAA&url=http%3A%2F%2Fciteseerx.ist.psu.edu%2Fviewdoc%2Fdownload%3Bjsessionid%3D3F5D7951625375CF98BB7C23EA701920%3Fdoi%3D10.1.1.135.9582%26rep%3Drep1%26type%3Dpdf&ei=v7CTTZmQO8vBswbyiu29CA&usg=AFQjCNF.

Payne, B., Carbone, M., Sharif, M., & Lee, W. (2008). Lares: An architecture for secure active monitoring using virtualization. In *Proceedings of the IEEE Symposium on Security and Privacy,* (pp. 233-247). IEEE Press.

Peer Hasselmeyer, N. D. (2010). Towards holistic multi-tenant monitoring. In *Proceedings of the Network Operations and Management Symposium Workshops,* (pp. 350 - 356). Osaka, Japan: IEEE Press.

Popescu, R., Staikopoulos, A., & Clarke, S. (2009). An extensible monitoring and adaptation framework. In *Proceedings of the 2009 International Conference on Service-Oriented Computing,* (pp. 314-324). Stockholm, Sweden: Springer.

Rajiv Ranjan, L. Z. (2010). Peer-to-peer cloud provisioning: Service discovery and load-balancing. *Computer Science Cloud Computing, 2,* 195–217. doi:10.1007/978-1-84996-241-4_12

Vidačković, K., Weiner, N., Kett, H., & Renner, T. (2009). Towards business-oriented monitoring and adaptation of distributed service-based applications from a process owner's viewpoint. In *Proceedings of the 2009 International Conference on Service-Oriented Computing,* (pp. 385-394). Stockholm, Sweden: Springer-Verlag.

Yalagandula, P., Sharma, P., Banerjee, S., Basu, S., & Lee, S.-J. (2006). S3: A scalable sensing service for monitoring large networked systems. In *Proceedings of the 2006 SIGCOMM Workshop on Internet Network Management,* (pp. 71-76). Pisa, Italy: ACM Press.

Zafeiropoulos, A., Liakopoulos, A., Davy, A., & Chaparadza, R. (2009). Monitoring within an autonomic network: A GANA based network monitoring framework. In *Proceedings of the 2009 International Conference on Service-Oriented Computing,* (pp. 303-313). Stockholm, Sweden: Springer.

Chapter 15
Replication Strategies for Video On-Demand over Wireless Mesh Networks:
A Cross-Layer Optimization Approach

Trung Q. Duong
Blekinge Institute of Technology, Sweden

Nguyen-Son Vo
Huazhong University of Science and Technology, China

Hans-Jürgen Zepernick
Blekinge Institute of Technology, Sweden

Lei Shu
Osaka University, Japan

Xu Du
Huazhong University of Science and Technology, China

Wenqing Cheng
Huazhong University of Science and Technology, China

ABSTRACT

To pervasively and diversely support a wide range of multimedia services, several traditional replication techniques have been considered for improving data accessibility in Wireless Mesh Networks (WMNs). The metrics used to evaluate data accessibility performance are often related to service integrity parameters of the lower layers such as the network, medium access control, and physical layers. However, these parameters cannot directly reflect the Quality of Experience (QoE) as perceived by end users. In contrast, QoE based parameters for video streaming applications are deduced from subjective tests and involve processing at the application layer including quality of playback in terms of spatial artifacts, smooth playback, and continuous playback.

In this chapter, dependencies between the different protocol layers and across the hops through WMNs will be exploited to deliver video streaming with high QoE. Specifically, a cross-layer optimization ap-

DOI: 10.4018/978-1-4666-0960-0.ch015

proach is applied to a replication strategy for video on-demand over WMNs. Additionally, the authors consider a distributed implementation algorithm, namely Mod&Timer, to handle the placement of replicas for saving storage resources of the network. Simulation results are shown to demonstrate that the proposed cross-layer design outperforms many existing schemes in terms of QoE. More importantly, compared to other strategies without cross-layer interaction, the proposed cross-layer design satisfies the heterogeneous bandwidth constraint of end users.

INTRODUCTION

Wireless mesh networks (WMNs) (Akyildiz, et al., 2005) have gained an important role in broadband wireless networks to serve a large number of users having high demands for multimedia services. WMNs rely on multihop wireless backbones consisting of relatively stationary mesh routers and gateways as illustrated in Figure 1. The salient advantages of WMNs, such as self-organization, self-configuration, self-healing, reliability, and scalability, provide an easy and economical network deployment and management. With the rapid growth of multimedia applications, many studies have focused on how to improve the efficiency and the performance of delivering multimedia streaming over WMNs by using feasible techniques such as scheduling, dropping, routing, caching, replication, and cross-layer design.

Multimedia streaming applications allow the users to initiate and commence playback simultaneously instead of waiting for downloading to complete. In general, multimedia streaming applications can be subcategorized into three scenarios: 1) On-demand streaming of stored multimedia, e.g., Video on-Demand (VoD), 2) live streaming, e.g., Internet live sport networks, Internet radio stations, and 3) real-time interactive streaming, e.g., on-line games, video conference, E-education. These three scenarios have several QoE objectives in common such as small distortions in the reconstructed application at the receiver, smooth playback, and continuous playback. However, live and real-time interactive streaming

Figure 1. Components of a typical wireless mesh network

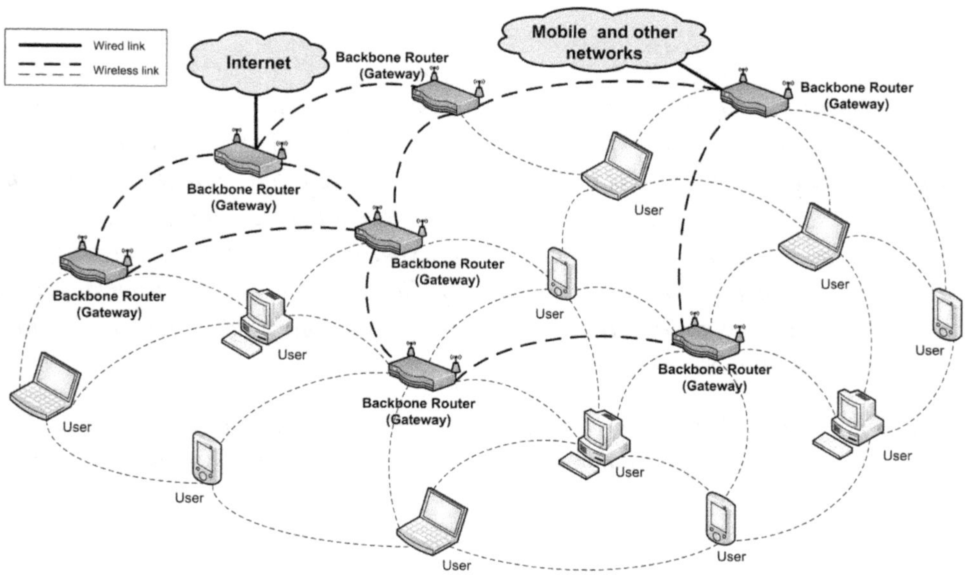

has their own characteristics and technical challenges including stringent startup delay constraint and different interactive requirements. In this chapter, we mainly focus on VoD over WMNs and evaluate its QoE performance as well as network resources for this scenario.

Aiming at improving the streaming performance of VoD, caching and replication strategies have been considered as efficient techniques to solve the problem of data accessibility for WMNs (Ghandeharizadeh, et al., 2004; Jin, 2004; Aazami, et al., 2004; Yu, et al., 2005; Hara & Madria, 2006; Jin & Wang, 2005; Ding, et al., 2008). The metrics to evaluate data accessibility performance such as delay, throughput, packet loss, and connectivity are related to service integrity parameters and statistics of the network, medium access control, and physical layers. However, service integrity parameters do not often correlate well with QoE obtained from mechanisms at the application layer such as 1) minimum average reconstructed distortion as quantified by Peak Signal-to-Noise Ratio (PSNR) or for high reconstructed quality playback, 2) minimum fluctuation of reconstructed distortion among received video frames for smooth playback, and 3) in time arrival of received video frames for continuous playback.

In this chapter, we will therefore consider the interrelationships between statistics of the network, medium access control, and physical layers and QoE metrics at the application layer including reconstructed distortion, smooth playback, and continuous playback. Specifically, a cross-layer design is proposed for the replication strategy of VoD streaming over WMNs. The cross-layer design exploits the mutual dependence among protocol layers and across hops to deliver video streaming with high QoE. Moreover, by utilizing the proposed cross-layer design, QoE performance can be optimized by formulating a constrained nonlinear optimization problem in terms of network resources, e.g., bandwidth and storage capacity.

It is noted that this chapter expands on our earlier research reported in (Vo, et al., 2011), where the cross-layer design has been initially applied for a replication strategy of videos streaming over general multihop wireless networks. However, the replication strategy is implemented at unknown source nodes of multihop networks leading to a high dynamically varying network topology and limited resources (Vo, et al., 2011). To circumvent this drawback, in this chapter, we have applied the cross-layer design framework for replication strategy in WMNs. In fact, the stationary and powerful gateways of WMNs are more applicable to deploy a replication strategy in a centralized fashion compared to such general multihop wireless networks (Vo, et al., 2011). More importantly, in this chapter, we will consider the effect of input parameters to an optimization problem on the proposed cross-layer approach and investigate the benefits and drawbacks of cross-layer optimization. Accordingly, the main content and contributions of this chapter may be summarized as follows:

1. We consider a Cross-Layer Replication Strategy (CRS) consisting of a Middleware Replication Strategy (MRS) layer with a Stack Profile (SP), and considered protocol layers. On this basis, protocol layers and hops can exchange their parameters; and network resources, e.g., resources of backbone mesh routers and users, can be efficiently managed.
2. We present tractable closed-form expressions associated with QoE criteria and network resources, which readily allow us to evaluate the performance of replication strategies.
3. Optimal replication density for each segment of a video stream is obtained to ensure that every segment arrives on time for continuous playback.
4. A constrained nonlinear optimization problem of optimal bit allocation for the video re-encoding process is formulated. As a result, QoE criteria and network resources are simultaneously optimized.

5. A distributed Implementation Service (IMS) with the help of Mod&Timer algorithm is presented. The Mod&Timer algorithm is installed at all mesh routers in order to control the placement of replicas of video segments according to their optimal replication densities. Consequently, unnecessary replicas are removed to conserve storage capacity.
6. Simulation results demonstrate that our proposed approach satisfies the heterogeneous bandwidth constraint of users and outperforms many existing schemes in terms of QoE performance.

The remainder of this chapter is organized as follows. We start with background on the replication strategies, cross-layer design approach, rate-distortion model, and bit allocation scheme in the next section. Section 3 proposes a cross-layer optimization approach for replication strategies consisting of a framework of the cross-layer replication strategy, parameter specifications, and optimization problems. In addition, to illustrate the significantly improved performance of our proposed cross-layer design compared to other existing schemes without cross-layer optimization, we also provide performance evaluations in this section. Subsequently, some future research directions are introduced in Section 4. Finally, we conclude the chapter in Section 5.

BACKGROUND

Replication Strategies

Existing Strategies

Existing replication strategies reported in the contemporary literature provide users with high data accessibility in multihop wireless networks (Ghandeharizadeh, et al., 2004; Jin, 2004; Aazami, et al., 2004; Yu, et al., 2005; Hara & Madria, 2006; Jin & Wang, 2005; Ding, et al., 2008). These strategies may only meet the continuous playback requirement. They neither satisfy the full aspects of delivering high QoE nor efficiently utilize network resources in comparison to the proposed cross-layer scheme. In the sequel, we will briefly summarize some representative existing replication strategies for video streaming applications.

In the field of wireless peer-to-peer networks, a replication technique for continuous media to enhance startup latency and investigate Storage capacity Requirement (STR) has been proposed in (Ghandeharizadeh, et al., 2004). It has been assumed that the beginning segments in a video are more urgent and important than the tail segments. However, in reality, users may randomly play forward or backward to access their preferred segments. Additionally, considering only startup latency and STR is not adequate for achieving high QoE and saving network resources (Ghandeharizadeh, et al., 2004). Specifically, considering startup latency is not enough to improve QoE since QoE consists of continuous playback, high reconstructed quality playback, and smooth playback. On the other hand, improving startup latency only provides continuous playback while the other two aspects of QoE (high reconstructed quality playback, and smooth playback) cannot be guaranteed. Also, STR is not adequate in terms of network resources which include both STR and bandwidth. Hence, this technique is not feasible to be applied to the real context of video streaming applications.

In Wireless Ad-Hoc Networks (WAHNs), Jin (2004) has proposed several topologies and replication strategies to mitigate the inherent characteristics of highly dynamic and frequent disconnection. By exploiting the access probabilities of video segments, the replicas of video segments can be discovered and retrieved with a minimum number of hops. Aazami et al. (2004) have introduced a method to obtain a near optimal proportional relationship between the number of video replicas and the product of its display bandwidth requirement and access rate to maximize the number of active users. However, the replication strategies in Jin (2004) and Aazami et al. (2004) do not consider QoE or take into account the STR and Bandwidth Requirement (BWR) of a network.

Other important replication solutions to enhance data accessibility for WAHNs are presented in Yu et al. (2005) and Hara and Madria (2006). A distributed hash table replication in which all mobile nodes are organized into non-overlapping clusters is proposed in Yu et al. (2005). At the same time, a two-level distributed replica information directory on the cluster heads is established in order to reduce the delay of query messages and update messages. To overcome the problem of frequent network partitioning, which causes the data accessibility of WAHNs to be lower than that of traditional fixed networks, three main replica allocation methods, i.e., Static Access Frequency (SAF), Dynamic Access Frequency and Neighborhood (DAFN), and Dynamic Connectivity-Based Grouping (DCG), and their extensions have been presented in (Hara & Madria, 2006). Although the continuous playback criterion can be satisfied, other QoE requirements could not be guaranteed since reconstructed distortion and smooth playback have not been examined.

In the area of WMNs, content and service replication strategies to access an object in two-dimensional (2D) mesh networks with minimum average cost have been investigated in Jin and Wang (2005). Their objective is to determine an optimal number of replicas based on the skewed inter popularity of the object. Although these replication strategies are applied for general data (Jin & Wang, 2005), they can ensure minimum average maximum transmission distance (continuous playback property) for video streaming applications (Vo, et al., 2011). As an extension of Ghandeharizadeh et al. (2004), Ding et al. (2008) considered the frequent online/offline behavior of users and the access probability of the videos.

Replication Strategy for WMNs

In general, replication aims at supporting users in retrieving video stream from the replicating mesh routers within the backbone of WMNs rather than from the Internet through gateways. When a gateway observes that a video stream from the Internet is frequently accessed, the video is first divided into a number of segments. The segments are proactively replicated in some indicated mesh routers in a centralized fashion for future use. In this way, the mesh routers only need to store a small number of segments instead of the entire video, to save storage resources. The gateway bottleneck is mitigated by sharing workload with mesh routers so that the use of bandwidth is equally spread over the backbone of WMNs. As a consequence, data accessibility performance can be improved. Additionally, the effect of dynamically varying properties of wireless environment is reduced for high QoE.

Cross-Layer Design Approach

In recent years, the number of cross-layer design approaches for video streaming applications in wireless networks has considerably increased. For example, a statistical analysis presented in Lindeberg et al. (2010) indicates that 65% of the proposed techniques for video streaming in mobile ad-hoc networks are associated with cross-layer design. The existing non cross-layer solutions adopt algorithms and protocols for "single layered" optimization. They cannot meet adequate requirements in video streaming applications over dense wireless networks due to the dynamically varying noise, induced interference, and associated mobility (Schaar & Shankar, 2005). For example, the resource management, protection strategies, and Quality of Service (QoS) in the lower layers cannot be optimized if we do not address the specific characteristics of video streaming applications such as huge file sizes, real-time delivery demands, and advanced QoE criteria. Moreover, video compression and streaming techniques in the application layer cannot be executed without mechanisms of the lower layers, e.g., error protection, scheduling, dropping, routing, and resource management.

Current cross-layer designs can be categorized into four groups as creation of new interfaces, merging of adjacent layers, coupling of interfaces without new interface creation, and vertical calibration across layers (Srivastava & Motani, 2005). These cross-layer designs are implemented based on three main ways of cross-layer interaction including direct communication between layers, a shared database across the layers, and completely new abstractions (Srivastava & Motani, 2005).

Although there have been many works considering either cross-layer design or the replication of video streaming over WMNs, none of them has attempted to investigate the cross-layer design of video streaming replication to improve the performance of VoD streaming over WMNs. In this chapter, we limit our discussion to physical layer, Medium Access Control (MAC) layer, network layer, and application layer. Furthermore, the adopted cross-layer design and cross-layer interaction are the creation of a new interface and a shared database across the layers, respectively. Thus, similarly (Conti, et al., 2004), our cross-layer framework is typically described in Figure 2.

Rate-Distortion Model

Rate-Distortion (RD) model represents the relationship between the encoding distortion and the number of bits, or sometimes the encoding bit rate of a video sequence. In other words, the RD model maps a given encoding bit rate (e.g., R_1, R_2) to the corresponding source distortion (e.g., D_1, D_2) as described in Figure 3. Many RD models have been proposed to optimize and evaluate the received video quality in streaming applications (Radha, et al., 2001; Hsu & Hefeeda, 2007; Chou & Miao, 2006; Chakareski, et al., 2002; Li, et al., 2004; Chakareski, et al., 2005; Chakareski & Chou, 2006), each of which is applied to different network conditions and has its own characteristics of accuracy and time complexity. Specifically, in Radha et al. (2001), a simple and flexible MPEG-4 Fine-Granularity-Scalability (FGS) video coding method was introduced to improve the RD performance applied to unicast and multicast streaming applications over IPs. By using the piece-wise linear rate-distortion model, Hsu and Hefeeda (2007) presented optimal bit allocation schemes so as to minimize the distortion of FGS video sequences at a single receiver supported by multiple senders in distributed streaming environments. Taking into account a given transmission opportunity over lossy packet networks, the authors in

Figure 2. A typical cross-layer framework

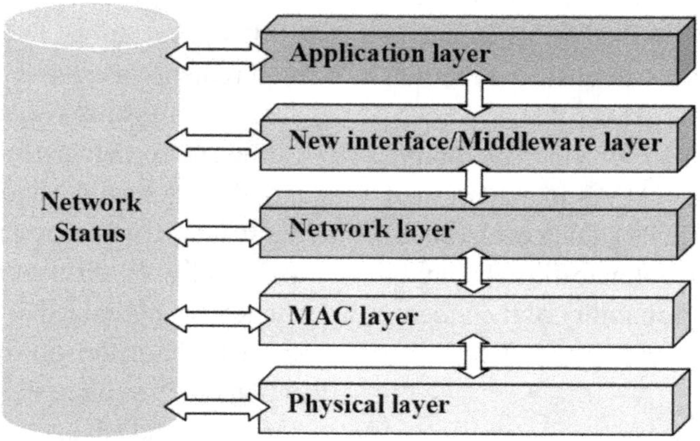

Chou and Miao (2006), Chakareski et al. (2002), Chakareski and Chou (2006) addressed the problem of streaming packetized media in which all considered packets will be optimally transmitted to minimize end-to-end distortion while satisfying a rate constraint. In a particular wireless network architecture, where a receiver accesses a video server via an access point in the infrastructure mode and simultaneously communicates with its peers in the ad-hoc model, a joint server/peer video streaming architecture was proposed to improve the performance of video streaming by using the application of asynchronous clocks and the RD framework (Li, et al., 2004). In Chakareski et al. (2005), the concept of RD hint track, which is suitable for online RD optimized streaming, was introduced to 1) summarize the RD attributes of a media streaming, 2) predict the distortion for a feasible packet schedule, and 3) determine the best packet schedule for streaming.

In this chapter, the FGS encoded video sequences (Li, 2011) following MPEG-4 standard is applied and the linear RD model reported in Hsu and Hefeeda (2007) is used to generate the source distortion model of a video stream at the application layer before replicating. Using this distortion model and taking lower layer parameters into account, the average reconstructed distortion is expressed as one of the objective functions for cross-layer optimization problem. The reconstructed distortion is then minimized by solving this optimization problem for optimal bit allocation.

Figure 3. A typical RD model

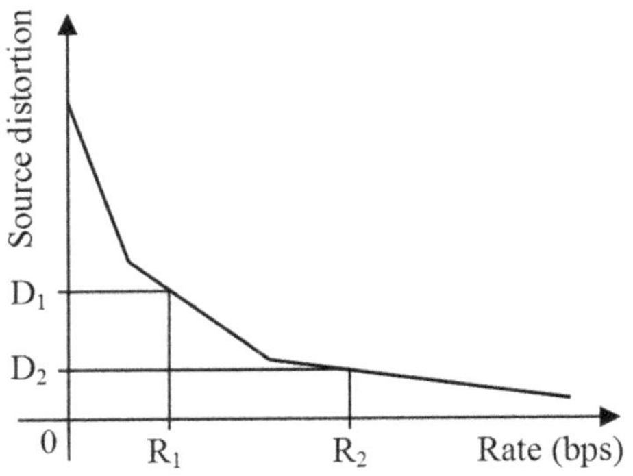

Bit Allocation Scheme

To efficiently meet the potential portfolio of resources available to heterogeneous users (e.g., PCs, laptops, PDAs, etc.), a replication strategy should take into account the encoding rate of a "hot" video stream. In the case of PDA users with relatively limited resources in terms of battery lifetime, computational processing speed, and screen resolution, a suitable replication strategy should avoid delivering a high definition video stream. If the arrival rate exceeds the users' capabilities, they cannot correctly decode and display the video content. In other words, before replicating, the video stream should be re-encoded at a proper rate to satisfy available resources. With this strategy, network bandwidth is efficiently used, i.e., consuming reasonable bandwidth.

Given an average encoding rate of an FGS encoded sequence in MPEG-4 standard, which is lower than the full rate of the highest reconstructed quality, every video frame should not be re-encoded at the same rate, i.e., the same number of bits. This yields much skewed distortion fluctuation among frames as displayed in Figure 4a. The reason is that according to the RD model, the interrelationship of RD for every frame is different and depends on the content of the frames. In other words, each frame with different content has its own RD relationship as illustrated in Figure 5, where the contents and the RD relationships of the *50*-th and *250*-th frames of the Common International Format (CIF) sample video 'Foreman' are shown. Thus, the distortion of each received frame is not constant even if the frame is encoded at the same rate. If the frames are encoded at different rates, the distortion fluctuation among them can be mitigated as shown in Figure 4b.

This skewed characteristic among video segments is even more pronounced if the bit allocation in the replication process is not optimized.

Figure 4. Encoding rate-distortion problem: (a) equal encoding rate for all frames and the related variation of distortions, (b) different encoding rates for each frame and the related reduced distortions

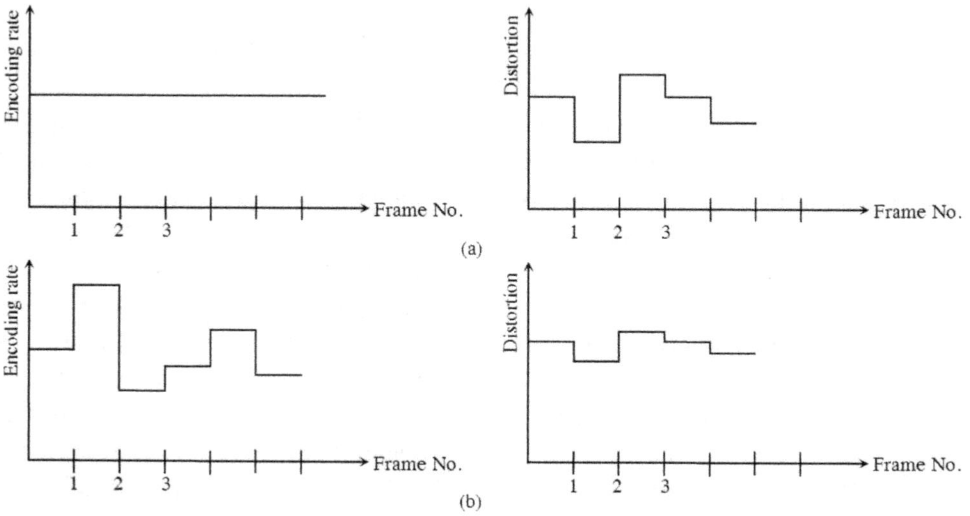

Figure 5. RD and content relationships of the 50-th frame (top) and 250-th frame (bottom) of sample video 'foreman'

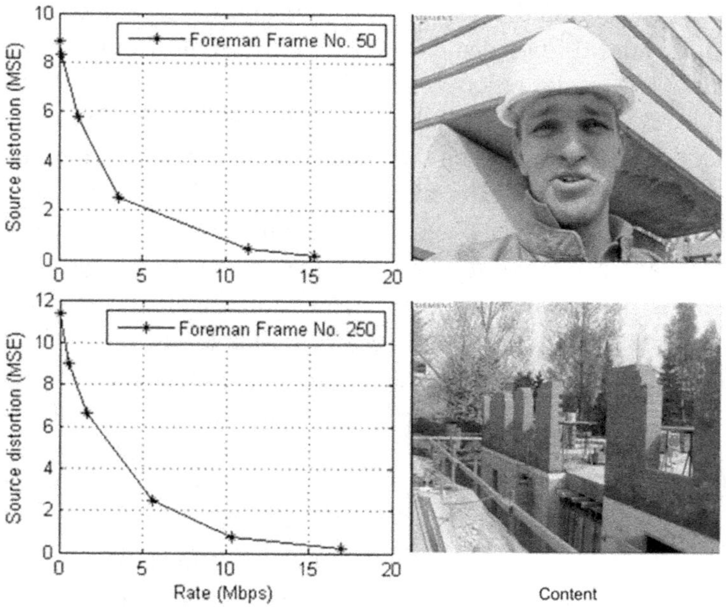

The segments with higher intra popularities (i.e., higher access rates or higher access probabilities), often replicated with higher densities, will be reconstructed with lower distortion. The highly skewed fluctuation in distortion of the reconstructed frames significantly disturbs user perception, i.e., causing unsmooth playback. Consequently, to prevent a large fluctuation in the reconstructed distortion, a replication strategy optimally allocates the number of bits to each frame of video segments before replicating. The objective of this so-called bit allocation scheme is to minimize the reconstructed distortion while satisfying a small level of fluctuation of distortion among reconstructed frames/segments.

CROSS-LAYER OPTIMIZATION APPROACH OF REPLICATION STRATEGY

Framework of Cross-Layer Replication Strategy

In the Cross-Layer Replication Strategy (CRS) proposed in Vo et al. (2011) and adopted here for WMNs, we simultaneously take into account four layers of the protocol stack, i.e. the application, network, medium access control, and physical layer. To ensure that the parameters of all considered layers are updated, stored and shared in the optimization of the replication strategy, a Stack Profile (SP) and a Middleware Replication Strategy (MRS) layer are introduced as illustrated in Figure 6. Specifically, the CRS utilizes five modules, namely, SP, Replication Management (RM), Pre-Replication Service (PRS), Implementation Service (IMS), and Post-Replication Service (POS).

First, at an arbitrary gateway, the application layer traces inter and intra popularities of every video stream flowing into the gateway from the Internet. It is important to note that unlike intra popularity, inter popularity represents the access rate or access probability of a video stream. If a video stream is popular enough, i.e., searched by a large number of users, the application layer sends some essential parameters such as video size, RD/source distortion model of each frame, current inter and intra popularities to the MRS for the replication processes.

Second, based on the current intra popularity, the video size, and the average storage capacity of the mesh routers, the MRS computes a suitable number of segments and the new intra popularity of each segment. Then, the optimal replication densities for all segments of the video stream are generated so that the Average Maximum Transmission Distance (AMTD) from the replicating mesh routers to users is minimized for continuous playback. Subsequently, the MRS reads the latest parameters of the lower layers such as the expected Bit Error Rate (BER), the guaranteed bandwidth, the transmission delay, and queuing delay for each hop from the SP. Here, we assume that these parameters can be periodically conveyed by an overlay network topology (Andreopoulos, et al., 2006). According to the optimal replica densities and the information of the video, an optimization process for QoE performance may be posed as to optimally allocate bits for each frame.

Figure 6. Framework of the considered cross-layer replication strategy (Vo, et al., 2011)

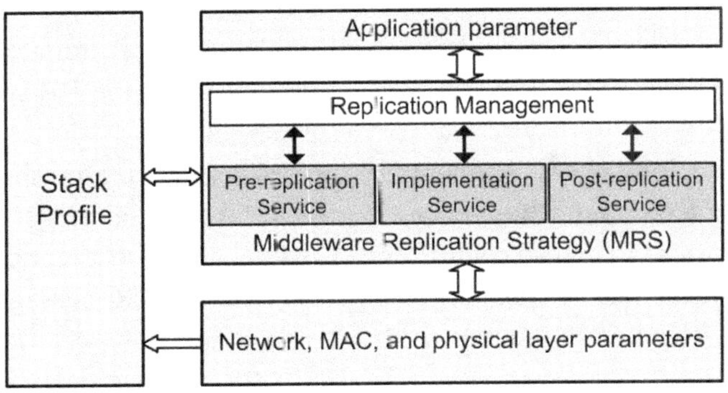

Third, following the optimal bit allocation for each frame, the MRS requires the application layer for re-encoding the video stream. Then, segments are replicated in some identified mesh routers corresponding to their optimal replica densities. In addition, the Mod&Timer algorithm as outlined below controls the placement of replicas to reduce unnecessary replicas for saving storage capacity in the network.

Finally, some necessary services will be deployed to maintain the replicas due to the random on/off of mesh routers, update the replicas due to changes in user behavior or/and network conditions, and release the replicas when the video stream becomes obsolete.

Replication Stack Profile

The replication SP extracts the latest information about resources at users and mesh routers and collects parameters from the lower layers such as BER per hop, guaranteed bandwidth per hop, and delay per hop. These details are conveyed to the SP and are also shared with the MRS. In addition, the SP stores current information about the video stream including inter and intra popularity, number of segments, Identities (IDs) of replicating mesh routers.

Middleware Replication Strategy

The MRS is responsible for the RM, which in the context of WMNs utilizes the following three related services:

1. ***Pre-replication service:*** The main functionalities and objectives of the PRS can be summarized as follows:
    a. Compute a suitable number of segments based on the video size and the average storage capacity of the mesh routers.
    b. Formulate unconstrained/constrained nonlinear optimization problems.
    c. Generate an optimal replication density for each segment and an optimal bit allocation to each frame of the video by solving the related optimization problems in order to optimize QoE while meeting the BWR constraint.
    d. Re-encode the video.
    e. Send the segments and their optimal replication densities to the IMS for replication.
2. ***Implementation service:*** The IMS uses the Mod&Timer algorithm to control the placement of replicas so that there exists only one replica for each segment within a replication density. The main functions of the Mod&Timer algorithm are described in the sequel:

Consider a gateway that stores the original video stream V consisting of v frames divided into n segments, s_i, $i=1,2,...,n$. Assuming that the optimal replication density/optimal radius of replication of the i-th segment is given as H_i^{max} hops, then there is only one replica of this segment in the radius range of H_i^{max} hops. With the Mod&Timer algorithm in the IMS, Mod is an operator used to place a replica of the i-th segment in a mesh router after jumping over every $2H_i^{max}$ hops and Timer is to ensure that there is one replica of the i-th segment in the radius range of H_i^{max} hops.

The gateway firstly broadcasts a Replica Requirement (RR) message into the network consisting of the 4-tuple (s_i, H_i^{max}, TTL, H_c). Here, the parameter TTL (Time To Live) is used to avoid the flooding loop problem in the network while hop count H_c is initiated by zero and increased by one when the RR message is received by a mesh router.

Afterwards, a mesh router, which is located $2H_i^{max}$ k (k is an integer) hops from the gateway, is likely to store a replica of the i-th segment if it receives the RR message. Before making a decision to store, it sets $Timer(i) = h\eta\ H_i^{max}$, where h is the average delay per hop, and timer coefficient $\eta \geq 1$. If *Timer(i)* has counted down to zero and the *i*-th segment has not been replicated by its neighbors in the radius range of $H_i^{\max}$ hops, it will decide to store this segment by requesting the gateway for a replica. At the same time, this router sends a confirm message (CM) to its neighbors in the radius range of $H_i^{\max}$ hops to declare that the *i*-th segment is already replicated. As a result, its neighbors do not need to store the replica of this segment anymore.

3. ***Post-replication service:*** After the operation of IMS, all replicas of the video stream are placed in indentified mesh routers. The POS is deployed to maintain the replicas due to the random on/off of mesh routers, update the replicas due to changes in user behavior and network conditions, and release the replicas when the video becomes obsolete.

Cross-Layer Replication Strategy Parameter Specification

Lower Layer Hop Parameters

In the replication strategy context, it is not necessary to consider an entire WMN with many gateways connected with their own mesh routers and users. A representative problem setting can be deduced from multiple copies of a typical mesh network with a limited number of mesh routers and one gateway. This typical model allows feedback parameters at the gateway not to be obsolete due to high variation of network conditions.

Assume that a WMN consists of *N* mesh routers and one gateway. An N_j mesh router consists of the (*j-1*)-th incoming hop and the *j*-th outgoing hop. As with general wireless multihop systems (Vo, et al., 2011), essential parameters of the *j*-th outgoing hop to be conveyed across the hops and exchanged between the protocol layers also includes the expected BER per hop BER_j, allocated guaranteed bandwidth per hop GBW_j, and delay per hop DEL_j (Vo, et al., 2011). Their definition shall be described for convenience in the sequel.

Expected BER per hop: At the physical layer, given a certain modulation and channel coding scheme, we can calculate the BER of an outgoing hop of mesh router N_j. Specifically, the BER of a Medium Access Control Service Data Unit (MSDU) comprising *L* bits is expressed as

$$BER_j(L) = 1 - (1 - BER_j)^L \quad (1)$$

Guaranteed bandwidth per hop: To avoid network contention, according to the Hybrid Control Channel Access (HCCA) protocol (IEEE, 2003), each multimedia service can reserve a predetermined transmission interval. Given the traffic specification parameters, the *j*-th outgoing hop can therefore provide a guaranteed bandwidth as (Ansel, et al., 2004):

$$GBW_j = \frac{t_j^{XTOP} L}{\left(L / R_j^{phy} + T_{oh}\right) t_j^{SI}} \quad (2)$$

where t_j^{XTOP} is the duration of transmission opportunity in HCCA admission control, t_j^{SI} is the duration of service interval, T_{oh} is the overhead delay such as polling and ACK policies, and R_j^{phy} represents the transmission rate at the physical layer.

Delay per hop: The delay per hop DEL_j of the *j*-th hop associated with mesh router N_j, consists of overhead delay, transmission delay, and queuing delay $T_j^{queuing}$. It can be expressed as (Vo, et al., 2011):

$$DEL_j\left(RTL_j^{\max}\right) = E\left(RTL_j^{\max}\right)\left[T_{oh} + \frac{L}{GBW_j}\left(1 + q_j\right)\right] \quad (3)$$

where q_j is the average queuing length when an MSDU arrives at mesh router N_j, parameter RTL_j^{max} denotes the retransmission limit and $E(RTL_j^{max})$ is the average number of transmissions for successfully receiving an MSDU. The latter retransmission limit is defined as:

$$E\left(RTL_j^{\max}\right) = \left[1 - BER_j(L)^{RTL_j^{\max}+1}\right]\left[1 - BER_j(L)\right]^{-1}$$

It is important to note that the term $q_i L/GBW_j$ on the right-hand side of (3) is obtained by computing the queuing delay $T_j^{queuing}$ at mesh router N_j. In particular, it is assumed that the average arrival rate of MSDUs is λ_j, the number of MSDUs at time t is $\varphi_j(t) = \lambda_j t$, the total waiting delay to be transmitted of all MSDUs at time t is $\xi_j(t)$, and the average queuing length is $q_j = \xi_j(t)/t$. Moreover, considering the queue in equilibrium, i.e., the arrival rate of MSDUs reaches the outgoing hop guaranteed bandwidth GBW_j, we have $\lambda_j = GBW_j/L$. The average queuing delay can now be calculated as $T_j^{queuing} = \xi_j(t)/\varphi_j(t) = q_i L/GBW_j$.

Application Layer Parameters

Similarly, a number of application layer parameters can be carried over from general wireless multihop networks reported in Vo et al. (2011) to the case of replication strategies for WMNs. The relevant parameters are provided for convenience in the sequel along with some additional background.

1. *Intra popularity of a video stream:* In video streaming applications, users may jump forward or backward to access their preferred video segments. This behavior can be characterized by the intra popularity of the video stream. Specifically, the access probability of the i-th most popular segment p_i follows a Zipf-like distribution as (Breslau, et al., 1999):

$$p_i = \frac{i^{-\alpha}}{\sum_{i=1}^{n} i^{-\alpha}} \tag{4}$$

where α denotes the intra popularity parameter.

2. *Radius of replication, average maximum transmission distance, and continuous playback*:

Definition 1: Given the aggregate capacity constraint $\sum_{i=1}^{n} r_i = 1$, where r_i is the replication density of the i-th segment, the maximum transmission distance between the replicating mesh routers and the users (i.e., the radius of replication of the i-th video stream segment) is given by (Vo, et al., 2011):

$$H_i^{\max} = \frac{\Omega}{\sqrt{r_i}} \tag{5}$$

where Ω is a predefined constant.

Definition 2: The Average Maximum Transmission Distance (AMTD) of a video stream from the replicating mesh routers to the users can be defined as:

$$\overline{H} = \sum_{i=1}^{n} p_i H_i^{\max} \tag{6}$$

Although the continuous playback criterion, i.e., every segment of a video stream shall arrive at the users on time, is a complex function of many parameters such as GBW_j, DEL_j, transmission distance, etc., it can basically be characterized by the average maximum transmission distance $\overline{H}$ for simplicity. To achieve continuous playback, we will minimize $\overline{H}$ by finding the optimal value of the maximum transmission distance $H_i^{\max}$ for the i-th segment with the access probability given as p_i.

3. *Source distortion and average reconstructed distortion:* The most important information conveyed from the application layer to the

MRS is the source distortion model of the video frame. This distortion can be computed at a particular rate by using RD models in practical video encoding. In fact, in order to control the rate, different RD models have been proposed and efficiently applied to corresponding video encoding standards, such as MPEG-2, H.263, MPEG-4, or H.264/AVC. Among the existing standards, although the RD model of H.264/AVC has been more widely adopted for its higher performance in terms of RD optimization than others, the rate control algorithm is very complicated (Ma, et al., 2005; Tian, et al., 2010). Thus, for simplicity, we utilize the RD model of Fine Granularity Scalability (FGS) encoded sequences following the MPEG-4 standard.

The FGS-encoded sequence (Li, 2011) consists of two streams including a non-scalable Base Layer (BL) and a scalable Enhancement Layer (EL). The BL offers a basic quality and the EL supports additional quality which is proportional to the number of bits encoded. Alternatively, the more bits the EL is encoded with, the lower distortion in the decoded video sequence. With the help of Microsoft MPEG-4 Video Reference Software (ISO, 2004), Hsu and Hefeeda (2007) proposed a method to build a linear RD model, which actually represents the relationship between the distortion and the number of bits (or sometimes the rate) of EL for every frame. Figure 7 shows the example of the RD model for the *200*-th frame of the sample video 'Foreman' when given in CIF size. In general, given an arbitrary rate f_m, the source distortion of the m-th video frame of an examined video is given by:

$$d_m(f_m) = d_m^{bl} - \Delta d_m^{el}(f_m) \quad (7)$$

where d_m^{bl} quantifies distortion when the m-th video frame is decoded without an EL frame. On the other hand, if the considered video frame is correctly decoded utilizing an EL frame at rate f_m, the source distortion $d_m(f_m)$ of this frame is reduced by an amount of $\Delta d_m^{el}(f_m)$.

Given a communications or broadcasting context where a video is transmitted from the source over an error-prone transmission channel to a destination, the average reconstructed distortion of a received video stream can be derived by taking the transmission errors into account. For this purpose, let us assume that BL and EL frames are both distributed in the WMN using the same replication strategy. Furthermore, the BL frames are assumed to be sent without loss

Figure 7. Linear RD model for the 200-th frame of sample video 'foreman' consisting of 300 frames

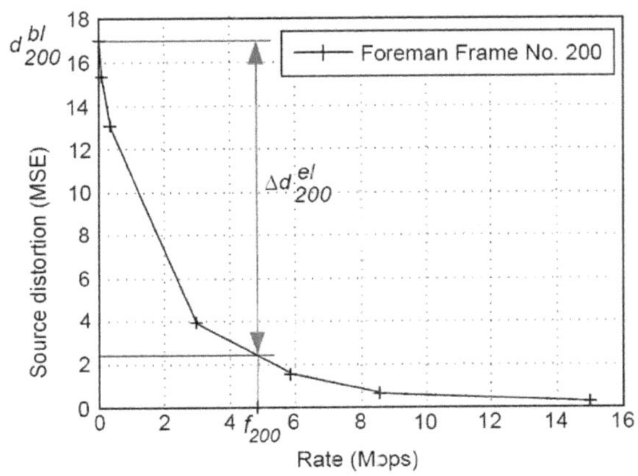

by using sufficiently strong forward error correction or incorporating backward error correction if the Round-Trip Time (RTT) is relatively small (Hsu & Hefeeda, 2007). This can be motivated by the fact that BL frames are of relatively small size and would also be subject to powerful error control coding due to their importance for the video reconstruction. As a result, we only need to consider the EL frames for analyzing the end-to-end distortion of the received video which enables us to derive the average reconstructed distortion as follows.

Let each video segment consist of S frames where the k-th frame in the i-th segment consists of $z_{i,k}$ MSDUs, for $k=1, 2, ..., S$. Thus, the number of MSDUs can be given by (Vo, et al., 2011):

$$z_{i,k} = \left\lceil \frac{f_{i,k}}{L} \right\rceil \tag{8}$$

where $\lceil A \rceil$ is the ceiling function applied here to the ratio of the number L of bits in a MSDU over the size $f_{i,k}$ of the k-th frame in the i-th segment.

Given that the i-th segment is replicated within a radius of $H_i^{\max}$ hops, all MSDUs of the k-th frame in the i-th segment experience an equivalent end-to-end error probability as:

$$B_i = 1 - \left[1 - \frac{1}{N}\sum_{j=1}^{N} BER_j(L)\right]^{H_i^{\max}} \tag{9}$$

However, in video streaming applications, a correctly decoded frame can still be considered as lost if it is outdated. As a result, the out-of-date probability of the k-th frame in the i-th segment $P(H_i^{\max})$, i.e., the probability that the arrival time of a frame is later than a predefined threshold τ_i, can be shown as (Vo, et al., 2011):

$$P\left(H_i^{\max}\right) = \int_{\tau_i}^{\infty} \frac{\mu}{\Gamma\left(H_i^{\max}+1\right)} \left[\mu\left(t-\kappa_i\right)\right]^{H_i^{\max}} \exp\left[-\mu\left(t-\kappa_i\right)\right] dt \tag{10}$$

where $t \geq \kappa_i$ with κ_i being the total delay over $H_i^{\max}+1$ mesh routers. The average waiting delay per node $1/\mu$ is modeled as a steady state M/M/1 queue (Ross, 1996) (Markov arrival process/Markov distribution of service time/Single server) and can be expressed as:

$$\frac{1}{\mu} = \frac{1}{N}\sum_{j=1}^{N} T_j^{queuing} E\left(RTL_j^{\max}\right) \tag{11}$$

The threshold τ_i, i.e. the lower limit of the integral in (10), can be expressed in the form of the startup delay coefficient β and the video frame rate f as:

$$\tau_i = \beta H_i^{\max} \frac{1}{N}\sum_{j=1}^{N} DEL_j\left(RTL_j^{\max}\right) + \frac{(i-1)v}{nf} \tag{12}$$

Finally, from (9), (10), (11), and (12), by taking into account both the impact of an error-prone transmission channel and the accumulated delay effects, we can obtain the end-to-end error probability as (Vo, et al., 2011):

$$P_i = B_i + (1-B_i)P\left(H_i^{\max}\right) \tag{13}$$

which readily enables us to obtain the reconstructed distortion of the k-th frame in the i-th segment $d_{i,k}^{r}(f_{i,k})$ as (Vo, et al., 2011):

$$d_{i,k}^{r}(f_{i,k}) = d_{i,k}^{bl} - \Delta d_{i,k}^{el}(f_{i,k})\left(1-P_i\right)^{z_{i,k}} \tag{14}$$

Finally, the average reconstructed distortion per frame of the whole video stream V can be given in the form of the reconstructed distortion $d^r_{i,k}(f_{i,k})$ as given in (14) by (Vo, et al., 2011):

$$\bar{d} = \sum_{i=1}^{n} \frac{p_i}{S} \sum_{k=1}^{S} d^r_{i,k}(f_{i,k}) = \sum_{i=1}^{n} \frac{p_i}{S} \sum_{k=1}^{S} \left[d^{bl}_{i,k} - \Delta d^{el}_{i,k}(f_{i,k}) \left(1 - P_i\right)^{z_{i,k}} \right] \quad (15)$$

Fluctuation of reconstructed distortion and smooth playback: To evaluate the smoothness of playback, we have to consider the fluctuation of reconstructed distortion that remains in the received video frames. This may be quantified by the variance of the reconstructed distortion in the received frames as:

$$\sigma_d^2 = \sum_{i=1}^{n} \frac{p_i}{S} \sum_{k=1}^{S} \left[d^r_{i,k}(f_{i,k}) - \bar{d} \right]^2 \quad (16)$$

Optimization Problems

In this section, using the cross-layer design described above, we will formulate a twofold optimization problem. One is to minimize the AMTD for continuous playback, the other is to minimize the average reconstructed distortion under the constraints of fluctuation of reconstructed distortion (i.e., smooth playback) and average received bandwidth of a huge number of heterogeneous users (e.g., PCs, laptops, and PDAs).

Minimum Average Maximum Transmission Distance

The optimal replica density of the *i*-th segment is strictly calculated by either the replication radius $H_i^{\max}$ or the replication density r_i. Recent research has attempted to optimize the replication density under specific objectives. Specifically, the authors in Ghandeharizadeh et al. (2004) and Ding et al. (2008) determined $H_i^{\max}$ to minimize startup latency, Jin and Wang (2005) proved that the replication density r_i must be proportional to $p_i^{2/3}$ to Minimize Access Cost (MAC), and the replication density r_i should be proportional to $p_i^{1/2}$ for a Minimum Query Cost (MQC) (Cohen & Shenker, 2002).

Unlike the objectives in Ghandeharizadeh et al. (2004), Ding et al. (2008), Jin and Wang (2005), and Cohen and Shenker (2002), our objective is to obtain the replication density r_i to minimize the AMTD, which can be used to represent the continuous playback metric relating to QoE. The key motivation for this choice is the fact that if the *i*-th segment is accessed at a higher rate, i.e., higher p_i, it should be replicated with higher density r_i, or with shorter radius of replication, $H_i^{\max}$, and thus shorter maximum transmission distance.

Theorem 1: The minimized AMTD can be achieved if and only if the replication density r_i of the *i*-th segment is proportional to $p_i^{2/3}$, i.e.:

$$r_i \propto p_i^{2/3} \quad (17)$$

Proof: From (5) and (6), the problem of minimizing AMTD is equivalent to minimizing the following:

$$\sum_{i=1}^{n} \frac{p_i}{\sqrt{r_i}} \quad (18)$$

This minimization problem can be solved by using the law of equi-marginal utility, which is usually used in economics. For this purpose, let us denote the utility function $U_i(r_i) = -p_i / \sqrt{r_i}$ and the marginal utility function then being written as:

$$U_i'(r_i) = \frac{p_i}{2} r_i^{-3/2} \quad (19)$$

We can observe that $U_i^{'}(r_i)$ is a monotonically decreasing function with respect to r_i, which allows us to adopt the law of equi-marginal utility. From (19), let $U_i^{'}(r_i) = U_j^{'}(r_j)$, we finally have $r_i \propto p_i^{2/3}$.

It should be noted that this result is similar to the work in Jin and Wang (2005), but our objective is to minimize the AMTD while (Jin & Wang, 2005) minimizes the average access cost.

Now, by substituting (17) into (4) and using the fact that $\sum_{i=1}^{n} r_i = 1$, the optimal replication density for the i-th segment is obtained as:

$$r_i = i^{-2\alpha/3}\left(\sum_{i=1}^{n} i^{-2\alpha/3}\right)^{-1} \tag{20}$$

From (20) and (5), we can obtain the optimal maximum transmission distance for the i-th segment and the minimum AMTD, respectively, as follows:

$$H_i^{\max} = \Omega\left(\sum_{i=1}^{n} i^{-2\alpha/3}\right)^{1/2} i^{\alpha/3} \tag{21}$$

$$\overline{H} = \Omega\left(\sum_{i=1}^{n} i^{-2\alpha/3}\right)^{3/2}\left(\sum_{i=1}^{n} i^{-\alpha}\right)^{-1} \tag{22}$$

which readily enable us to evaluate the continuous playback behavior of a video stream. By combining $H_i^{\max}$ given in (21) with (13), (14), and (15), we can formulate a constrained nonlinear optimization problem, which will be described in the next subsection.

Minimum Average Reconstructed Distortion

The second optimization problem is to aim at improving quality playback in terms of reconstructed distortion. To do so, we have to minimize the average reconstructed distortion given in (15) by optimally distributing the amount of reduction in distortion $\Delta d_{i,k}^{el}(f_{i,k})$ corresponding to the access probability of the i-th most popular segment p_i and the probability of the frame k in segment i is correctly received $(1-P_i)^{z_{i,k}}$. In other words, the average reconstructed distortion is minimized by allocating the number of bits, $f_{i,k}$ (or sometimes the bit rate), to frame k in segment i. In addition, we should consider the fluctuation of reconstructed distortion constraint for smooth playback and satisfy the heterogeneous received bandwidth constraint of users for efficiently using network bandwidth. Mathematically, this so-called constrained nonlinear optimization problem can be posed as:

$$\min_{f_{i,k}} \overline{d} \tag{23}$$

s.t.

$$\begin{cases} \overline{EER} \leq ERB \\ \left|d_{S(i-1)+l+1}^{r} - d_{S(i-1)+l}^{r}\right| \leq \delta \overline{d_s}, \quad l = 1,\ 2,\ldots,\ S-1 \\ f_{i,k}^{\min} \leq f_{i,k} \leq f_{i,k}^{\max} \end{cases} \tag{24}$$

where in the first constraint of (24), $\overline{EER}$ is the average encoding rate of video stream V, given by:

$$\overline{EER} = \frac{\mathrm{f}}{nS}\sum_{i=1}^{n}\sum_{k=1}^{S} f_{i,k} \tag{25}$$

and ERB is the encoding rate budget corresponding to the average received bandwidth of a user given by:

$$ERB = \frac{f \times EBB}{nS} \tag{26}$$

while EBB is the encoding bit budget of the video stream V.

The second constraint of (24) is used to limit the fluctuation of reconstructed distortion between two adjacent video frames for smooth playback by using the smoothness coefficient δ. The average encoding source distortion $\overline{d}_s$ is given by:

$$\overline{d}_s = \sum_{i=1}^{n} \frac{p_i}{S} \sum_{k=1}^{S} \left[d_{i,k}^{bl}(f_{i,k}) - \Delta d_{i,k}^{el}(f_{i,k}) \right] \tag{27}$$

In the third constraint, the lower bound is the minimum number of bits per frame required to achieve an acceptable video fidelity, e.g., peak signal-to-noise ratio (PSNR) > 27 dB (Khan, et al., 2009), while the upper bound accounts for the maximum number of bits, i.e., full encoding rate.

By substituting (25) and (26) into the first constraint of (24), the constrained nonlinear optimization problem is now rewritten as:

$$\min_{f_{i,k}} \overline{d} \tag{28}$$

s.t.

$$\begin{cases} \sum_{i=1}^{n} \sum_{k=1}^{S} f_{i,k} \le EBB \\ \left| d_{S(i-1)+l+1}^{r} - d_{S(i-1)+l}^{r} \right| \le \delta \overline{d}_s, \quad l = 1, 2, \ldots, S-1 \\ f_{i,k}^{\min} \le f_{i,k} \le f_{i,k}^{\max} \end{cases} \tag{29}$$

The above optimization problem can be solved by separable programming, i.e., a nonlinear programming methods, which approximately converts a nonlinear into a linear programming problem (Lieberman, 2005).

Since separable programming is an approximate closed-form method, it cannot provide an accurate solution. In contrast, nonlinear programming can support more accurate optimization although it may involve more complicated computation and inflict a longer start-up delay in video streaming applications. Fortunately, the gateway in WMNs is a powerful unit for implementing such a complicated optimization problem. Moreover, time delay incurred during the re-encoding and any other processing tasks in the replication strategy does not affect start-up delay. Thus, in this chapter, we use the 'fmincon' function of the Matlab Optimization Toolbox, i.e., a general nonlinear programming method, to solve the optimization problem in (28) and (29).

Performance Evaluations

In this section, we show that our Cross-Layer Replication Strategy (CRS) achieves higher QoE performance compared to other optimal replication strategies such as MAC (Jin & Wang, 2005), MQC (Cohen & Shenker, 2002), and Proportional Strategy (PRS), where we set $p_i \propto r_i$. Furthermore, we demonstrate that our IMS with the help of the Mod&Timer algorithm to control the placement of replicas significantly outperforms the Timer algorithm in Ghandeharizadeh et al. (2004) in terms of reducing unnecessary replicas.

Simulation Setup

We deploy a grid topology for the examined WMN with the system parameters used for the simulations listed in Table 1. In addition, the video coding standard follows the MPEG-4 Video Reference Software Version 2.5 (ISO, 2004) supported by Microsoft and is used to analyze the 19.64Mbytes EL stream of sample video 'Foreman' given in CIF (352×288 pixels) resolution,

Table 1. System parameter settings

Parameter	Symbol	Specification
Number of mesh routers	N	300 mesh routers
Number of segments	n	30 segments
Expected bit error rate	BER	10^{-6}
MSDU size	L	1000 Bytes
Average guaranteed bandwidth	$GBW = \frac{1}{N}\sum_{j=1}^{N} GBW_j$	36 Mbps
Maximum retransmission limit	RTL_j^{max}	1
Average queuing length	$q = \frac{1}{N}\sum_{j=1}^{N} q_j$	500
Total processing delay over $H_i^{max}+1$ hops	κ_i	0
Predefined constant	Ω	1
Startup delay coefficient	β	10

30 frames per second (fps), and 6 bit-planes. Due to limited and heterogeneous received bandwidth, it is assumed that the video is frequently required by the users whose received bandwidth is set to equal 70% of the full transmission rate of the video (e.g., 11Mbps). As a result, *EBB* in the first constraint of (29) is 13.75 MBytes. In this case, the PSNR of the source distortion of Foreman is 50 dB, which represents a very high quality before replication. To limit the fluctuation of reconstructed distortion, in the second constraint of (29), we select $\delta = 0.005$. The lower and upper threshold for the number of bits per frame, respectively, are set such that $f_{i,k}^{min}$ supports a PSNR of 27 dB while $f_{i,k}^{max}$ is set to the full encoding rate of each frame. Finally, the initial optimal value of $f_{i,k}$ is selected as 50% of $f_{i,k}^{max}$ as the requirement for the 'fmincon' function in Matlab and the system simulation block diagram is described in Figure 8.

Performance Metrics

The performance assessment of the considered replication strategy for WMNs shall be conducted in the following with respect to four characteristics. Specifically, these are the minimum average maximum transmission distance for continuous playback, minimum average reconstructed distortion, fluctuation of reconstructed distortion, and resources for storage of replicas. A comparison of the considered replication strategy to other contemporary replication approaches will also be provided.

Figure 8. System simulation block diagram

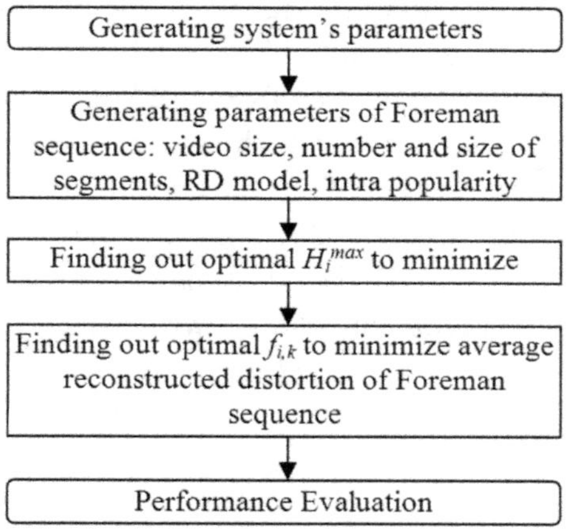

1. Minimum average maximum transmission distance for continuous playback: Figure 9 displays the continuous playback performance for different replication strategies by plotting AMTD versus intra popularity parameter α. We can observe that the AMTD of our CRS and MAC from (22) is lower than those of MQC and PRS. For all strategies, the performance is enhanced when α increases. It should be mentioned that keeping AMTD to a minimum ensures that the received video stream is played back with high continuity or a small effect of interruption.
2. Minimum average reconstructed distortion for high reconstructed quality playback: To evaluate the quality of playing back the received and reconstructed video stream, we use the average PSNR instead of the average reconstructed distortion measured by the Mean Square Error (MSE). The PSNR is calculated based on comparing the original video stream with the reconstructed video sequence in terms of distortion measured in MSE at the end users. Then, PSNR metric can be obtained from MSE by using the following relationship:

$$PSNR = 10\log_{10}\left(\frac{255^2}{MSE}\right) \quad (30)$$

Figure 10 shows the reconstructed quality playback performance of the received video stream by plotting PSNR versus the intra popularity parameter α. As can be seen from Figure 10, MAC and MQC surpass PRS since the latter cannot exploit the skewed intra popularity of each segment. However, both MAC and MQC cannot exploit the skewed property of each frame, i.e., $(1-P_i)^{z_{i,k}}$ as our CRS. In order to significantly enhance the PSNR, our CRS scheme takes full advantage of both skewed properties, i.e., in both segment and frame. Clearly, CRS outperforms all other replication strategies, i.e., MAC, MQC, and PRS. When α increases, the skewed intra popularity of segments also increases, and thus the improvement of our CRS scheme becomes more significant.

Figure 9. Continuous playback performance versus intra popularity parameter α for different replication strategies

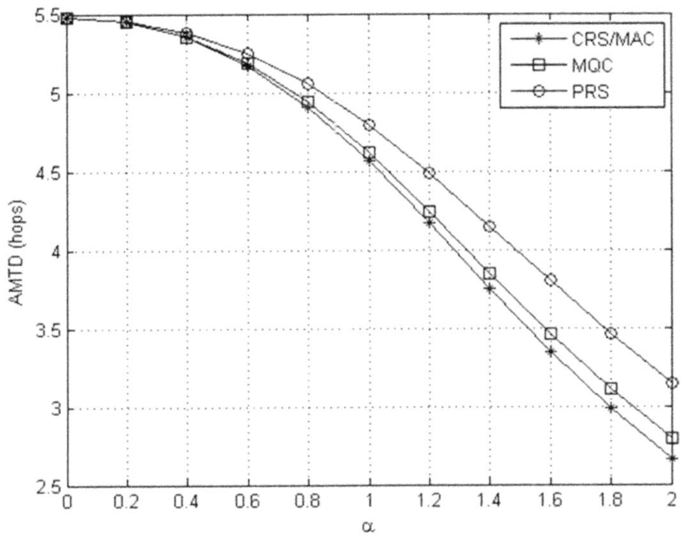

Figure 10. High quality playback performance versus intra popularity parameter α for different replication strategies

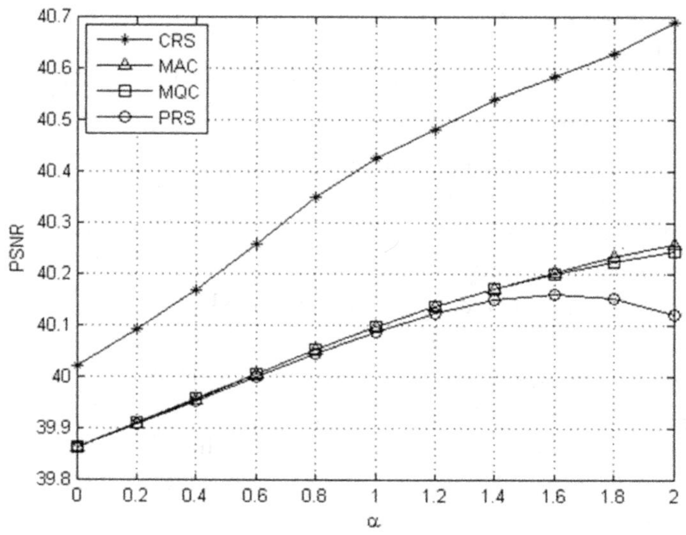

3. Small fluctuation of reconstructed distortion for smooth playback: Since a video stream is divided into a number of segments and replicated in some identified mesh routers with different densities, segments with higher intra popularity are often replicated with higher densities (shorter maximum transmission distances). We can see from Figure 11, i.e., generated by (21), that received segments with lower segment number have better performance on frame level in terms of a shorter radius of replication. As a consequence, extreme fluctuation of reconstructed distortion is observed among received frames, which significantly disturbs end users. In order to quantify the smoothness of playback of the received video stream, the variance of reconstructed PSNR of all frames is computed and shown in Figure 12. It can be seen from the figure that our CRS offers the smallest variance of reconstructed PSNR and as such supports smoothest playback compared to MAC, MQC, and PRS. It should be noted that the variance of reconstructed PSNR is similarly calculated as given in (16) by using (30) to convert from MSE to PSNR.

In addition, the value of smoothness coefficient δ must be carefully selected as it directly affects high quality playback and quality fluctuation performances. Assume that $f^*_{i,k}$ is the optimal result of the size of k-th frame in the i-th segment, if δ is too small then some optimal results will be removed and replaced by other close to optimal values. This process does not generate the highest quality playback but the smallest quality fluctuation. On the contrary, if δ is too large, the optimal algorithm can hardly converge to the optimal result from the initial values of $f_{i,k}$ and the quality fluctuation is high in this case. Figure 13 shows that for high reconstructed quality playback, δ should be in the range from 0.005 to 0.03. In Figure 14, however, the reconstructed quality fluctuation is relatively high under the range of δ from 0.01 to 0.03. Hence, to satisfy both high reconstructed quality playback and low reconstructed quality fluctuation, we set δ to 0.005.

Figure 11. The skewed replica density/radius of replication

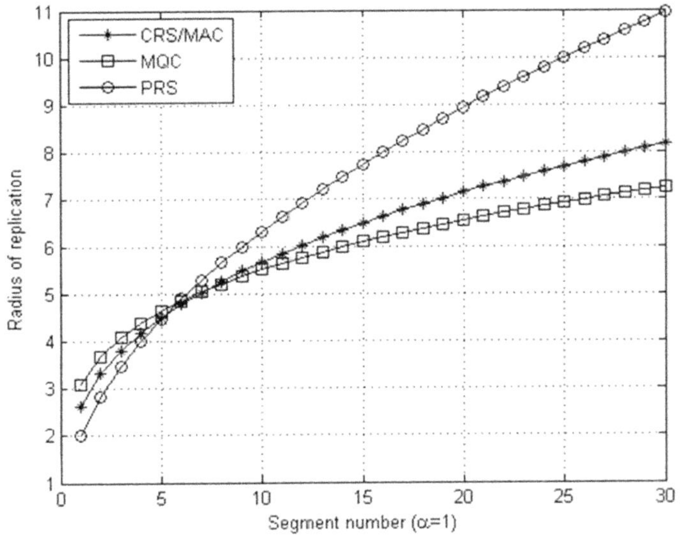

4. Small unnecessary replicas for saving STR in WMNs: As discussed in the previous section, IMS deployed with the Mod&Timer algorithm controlling the placement of replicas can conserve the storage capacity in the network. According to the Mod&Timer algorithm, the number of mesh routers which are likely to store replicas of the i-th segment is given as $2H_i^{\max} N\left[2\left(H_i^{\max}\right)^2 + 2H_i^{\max} + 1\right]^{-1}$. Note that in the Timer algorithm specified in Ghandeharizadeh et al. (2004), this number

Figure 12. Reconstructed quality fluctuation/smooth playback performance versus intra popularity parameter α for different replication strategies

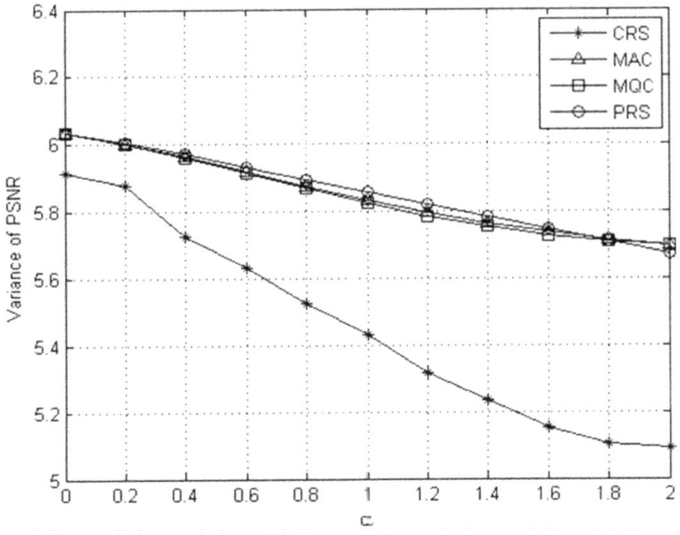

Figure 13. The effect of δ on the high reconstructed quality performance

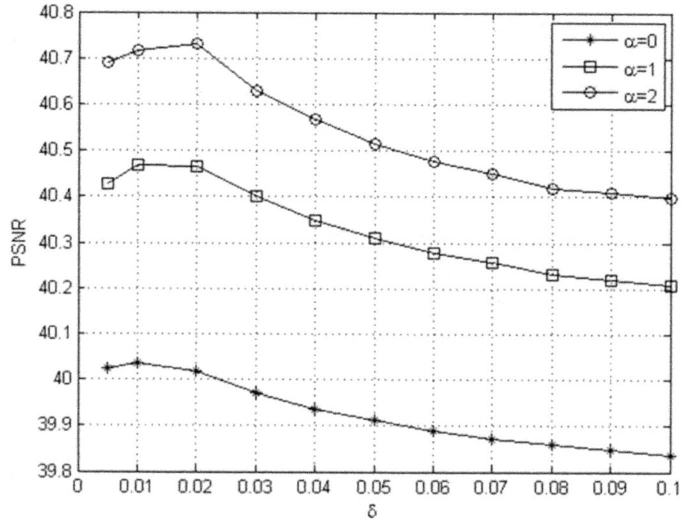

of mesh router equals to N for all segments.

Unnecessary replicas of a segment occur in the following case. Assume that a replicated mesh router stores a segment and sends a CM to its neighbors within the radius of H_i^{max} hops. If the neighboring mesh router does not receive a CM signal before the timer counts down to zero, it will decide to cache the replica. Since there is more than one replica within the radius of H_i^{max} from the replicating mesh router, this replication at the neighbor is unnecessary. The probability of a neighboring mesh router not receiving CM, i.e, the probability of occurring an unnecessary replica, can be similarly calculated as in (13). By

Figure 14. The effect of δ on the reconstructed quality fluctuation performance

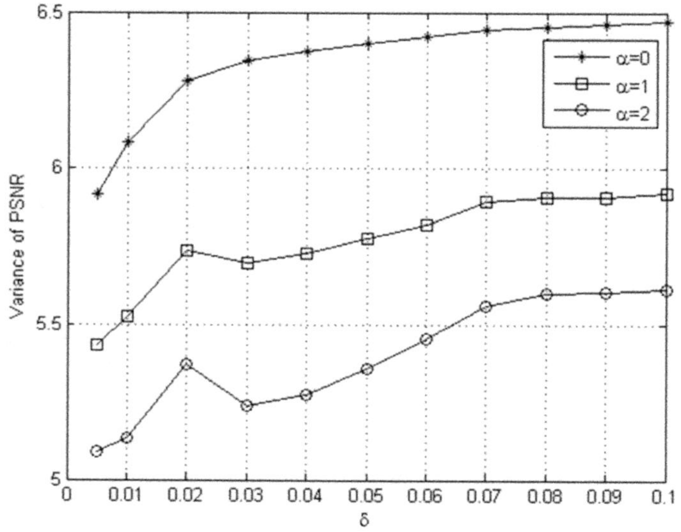

Figure 15. Unnecessary replica evaluation versus segment number

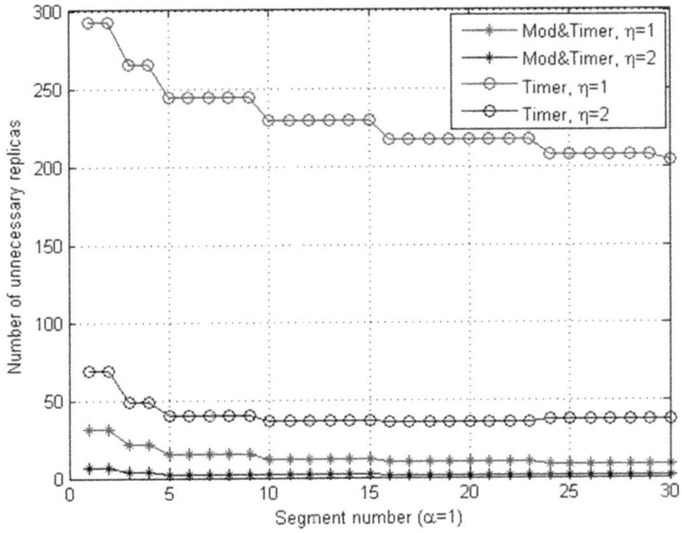

using this probability together with the number of potential replicated mesh routers, we can compare our proposed Mod&Timer algorithm with the Timer algorithm (Ghandeharizadeh, et al., 2004) in terms of the number of unnecessary replicas.

To evaluate the performance of the Mod&Timer algorithm, we assume that the probability of CM being lost due to BER is given as 0.01, the average delay per hop of CM is 10 ms, and intra popularity parameter α is set to 1. Figure 15 shows that our algorithm with Mod&Timer outperforms the Timer algorithm (Ghandeharizadeh, et al., 2004) in terms of reducing unnecessary replicas by a very large fraction. For the Timer algorithm in Ghandeharizadeh et al. (2004), it is likely that a considerable number of mesh routers is required to store the replicas of every segment. However, by using the operator "Mod" in our CRS, the number of mesh routers that are required for storage is substantially reduced.

The Trade-Off of the Proposed CRS

As can be seen from Figure 16, the video segments in CRS/MAC are replicated with highest densities leading to the fact that CRS/MAC certainly consumes a higher amount of STR compared to MQC and PRS. For a fair comparison, STR must be considered as a constraint in the minimized AMTD problem, i.e., finding $H_i^{\max}$. In this regard, STR consumption can be limited while satisfying a minimum AMTD. The amount of STR in our considered grid topology can be given by (Ghandeharizadeh, et al., 2004):

$$STR = N\sum_{i=1}^{n} \frac{S_i}{2\left(H_i^{\max}\right)^2 + 2H_i^{\max} + 1} \tag{31}$$

where S_i is the size of the i-th segment measured in bytes.

FUTURE RESEARCH DIRECTIONS

We have just gone through the proposed novel CRS. It can be observed that to gain a high performance of replication in VoD systems, the objective of CRS is associated with two main aspects. The first one is to enhance QoE and the other is

Figure 16. STR performance versus intra popularity parameter α for different replication strategies

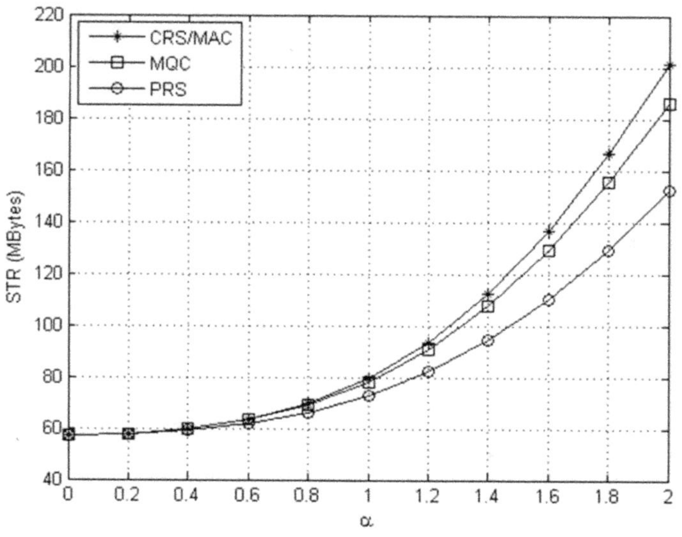

to conserve the resources of network (e.g., STR and bandwidth).

In consideration of QoE, we have been trying to present the three metrics that reflect the full feature of QoE perceived by users. These metrics include quality playback in terms of spatial artifacts, smooth playback, and continuous playback. Among them, the continuous playback, which is evaluated by AMTD, has not been considered well. Using AMTD leads to the problem that we ignore delay variant between two received adjacent segments which is an important property of continuous playback. To solve this problem, i.e., to limit this so-called segment jitter delay, we should take it into account as constraint for forming the optimal replication density of each segment to minimize AMTD.

With regard to the consumption of network resources, since CRS achieves a shorter AMTD (higher average replication density) as shown in Figure 9, CRS consumes more STR in the WMN than MQC and PRS. If a large amount of STR is used for high replication density, the QoE performance is significantly improved and vice versa. Based on the storage capacity, i.e., the STR budget of WMNs, we can deploy an optimal replication strategy. Because STR is also a function of replication density, to utilize STR efficiently, STR will be also considered as constraint in the optimization problem of minimizing AMTD for continuous playback by finding optimal replication density.

Based on the above discussions, to make the proposed CRS more efficient and robust, we will thus investigate the CRS performance by taking into account STR and segment jitter delay as constraints in the optimization problem in section 3.3.1. Consequently, we obtain a new constrained nonlinear optimization problem formulated as below:

$$\min_{H_i^{\max}} \overline{H} \tag{32}$$

$$\text{s.t.} \begin{cases} \left| D_{i+1} - D_i \right| \leq \gamma DEL \\ STR \leq STR_{ref} \\ 1 \leq H_i^{\max} \leq H_{\max} \end{cases} \tag{33}$$

where D_{i+1} and D_i are total delays of the $(i+1)$-th and i-th segments, respectively. DEL is the average delay per hop given by:

$$DEL = \frac{1}{N}\sum_{j=1}^{N} DEL_j \quad (34)$$

The coefficient γ is used to limit the jitter delay between two adjacent segments, namely - jitter delay coefficient. STR_{ref} is the budget of storage capacity of network, or the storage capacity constraint of network for limiting the consumption of storage capacity. In the last constraint of (33), the lower threshold makes sure that it is not permitted to replicate a segment in all mesh routers, i.e., $H_i^{max} = 0$. The upper threshold is given to avoid the worst case when a segment is replicated by only one replica in the entire network. In this case, the replication radius cannot be greater than the radius of WMN backbone, H_{max}. In grid topology with N mesh routers, the radius of WMN backbone is actually the positive root of the second order equation $2(H_{max})^2 + 2H_{max} + 1 = N$.

Solving the above constrained nonlinear optimization problem allows us to attain a more efficient and robust radius of replication, H_i^{max}. The new optimal result not only minimizes AMTD and satisfies the constraint of segment jitter delay for continuous playback but also meets a given budget of STR for reasonably using the storage capacity of network.

Finally, as in the considered CRS for VoD systems, we mainly focused on video streaming applications in which the FGS of MPEG-4 standard was used for analyzing RD model, which is the most important key of the optimization problem presented in section 3.3.2. Although this standard can improve rate scalability and error resilience, FGS involves a high overhead because of the lack of inter-layer data dependencies. And thus, the FGS of MPEG-4 has been superseded by H.264/AVC or/and H.264/SVC mainly due to the unpopularity of its model-based coding. In the case of the video stream encoded by H.264/AVC or/and H.264/SVC standards, new efficient RD models must be analyzed and applied to the optimization problem in section 3.3.2. In other words, the proposed CRS for VoD systems should be changed adaptively in terms of optimization problems. This will make the CRS for VoD systems become more adaptable and widely used in video communications. We believe that the proposed CRS is quite promising for practical over real-world networks, and can be implemented within the context of the popular H.264/AVC or/and H.264/SVC standard(s).

CONCLUSION

In this chapter, aiming at achieving high performance (e.g., continuous playback, high reconstructed quality playback, and smooth playback) and efficiently using network resources, we have considered a replication strategy based on a cross-layer design approach for video streaming in WMNs. Unlike other optimal replication strategies, our CRS scheme inherently considers many aspects of subjective human perception in video quality as well as bandwidth and storage capacity of the network. By way of example, simulation results demonstrate that CRS succeeds in providing users with high video playback performance compared to other optimal replication strategies while satisfying the heterogeneous received bandwidth constraint of users for efficiently using the network bandwidth. Furthermore, in CRS, the proposed MRS with the help of the Mod&Timer algorithm outperforms the Timer algorithm proposed in Ghandeharizadeh et al. (2004) in terms of reducing unnecessary replicas and hence conserves storage capacity.

REFERENCES

Aazami, A., Ghandeharizadeh, S., & Helmi, T. (2004). Near optimal number of replicas for continuous media in ad-hoc networks of wireless devices. In *Proceedings of the Multimedia Information Systems,* (pp. 40–49). College Park, MD: IEEE.

Akyildiz, I. F., Wang, X., & Wang, W. (2005). Wireless mesh networks: A survey. *Computer Networks*, *47*(4), 445–487. doi:10.1016/j.comnet.2004.12.001

Andreopoulos, Y., Mastronarde, N., & Schaar, M. V. D. (2006). Cross-layer optimized video streaming over wireless multihop mesh networks. *IEEE Journal on Selected Areas in Communications*, *24*(11), 2104–2115. doi:10.1109/JSAC.2006.881614

Ansel, P., Ni, Q., & Turletti, T. (2004). An efficient scheduling scheme for IEEE 802.11e. In *Proceedings of the IEEE Workshop on Modeling and Optimization in Mobile, Ad Hoc and Wireless Networks*. Cambridge, UK: IEEE Press.

Breslau, L., Cao, P., Fan, L., Phillips, G., & Shenker, S. (1999). Web caching and zipf-like distributions: Evidence and implications. In *Proceedings of IEEE Infocom* (pp. 126–134). New York, NY: IEEE Press. doi:10.1109/INFCOM.1999.749260

Chakareski, J., Apostolopoulos, J., Wee, S., Tan, W., & Girod, B. (2005). Rate-distortion hint tracks for adaptive video streaming. *IEEE Transactions on Circuits and Systems for Video Technology*, *15*(10), 1257–1269. doi:10.1109/TCSVT.2005.854227

Chakareski, J., Chou, P., & Girod, B. (2002). Rate-distortion optimized streaming from the edge of the network. In *Proceedings of IEEE Workshop on Multimedia Signal Processing,* (pp. 49–52). St. Thomas, US Virgin Islands: IEEE Press.

Chakareski, J., & Chou, P. A. (2006). Radio edge: Rate-distortion optimized proxy-driven streaming from the network edge. *IEEE/ACM Transactions on Networking*, *14*(6), 1302–1312. doi:10.1109/TNET.2006.886299

Chou, P. A., & Miao, Z. (2006). Rate-distortion optimized streaming of packetized media. *IEEE Transactions on Multimedia*, *8*(2), 390–404. doi:10.1109/TMM.2005.864313

Cohen, E., & Shenker, S. (2002). Replication strategies in unstructured peer-to-peer networks. In *Proceedings of ACM SIGCOMM,* (pp. 177–190). Pittsburgh, PA: ACM Press.

Conti, M., Maselli, G., Turi, G., & Giordano, S. (2004). Cross-layering in mobile ad hoc network design. *IEEE Computer Society*, *37*(2), 48–51. doi:10.1109/MC.2004.1266295

Ding, J.-W., Wang, W.-T., & Wang, C.-F. (2008). An efficient data replica scheme for peer-to-peer video streaming over wireless mesh community networks. In *Proceedings of the International Conference on Intelligent Information Hiding and Multimedia Signal Processing,* (pp. 767–770). Harbin Shi, China: IEEE.

Ghandeharizadeh, S., Krishnamachari, B., & Song, S. (2004). Placement of continuous media in wireless peer-to-peer networks. *IEEE Transactions on Multimedia*, *6*(2), 335–342. doi:10.1109/TMM.2003.822787

Hara, T., & Madria, S. K. (2006). Data replica for improving data accessibility in ad hoc networks. *IEEE Transactions on Mobile Computing*, *5*(11), 1515–1532. doi:10.1109/TMC.2006.165

Hsu, C., & Hefeeda, M. (2007). Optimal bit allocation for fine-grained scalable video sequences in distributed streaming environments. In *Proceedings of ACM/SPIE Multimedia Computing and Networking Conference,* (pp. 1–12). San Jose, CA: ACM Press.

IEEE. (2003). *Draft supplement to part 11: Wireless LAN medium access control (MAC) and physical layer (PHY) specifications: Medium access control (MAC) enhancements for quality of service (QoS), IEEE 802.11e/d5.0*. Retrieved from http://www.ieee.org.

ISO. (2004). MPEG-4 visual reference software. *ISO/IEC 14496*. Retrieved from http://www.iso.org.

Jin, S. (2004). Replication of partitioned media streams in wireless ad hoc networks. In *Proceedings of the ACM Multimedia,* (pp. 396–399). New York, NY: ACM Press.

Jin, S., & Wang, L. (2005). Content and service replica strategies in multi-hop wireless mesh networks. In *Proceedings of the International Symposium on Modeling Analysis and Simulation of Wireless and Mobile Systems,* (pp. 79–86). Montreal, Canada: IEEE.

Khan, A., Sun, L., & Ifeachor, E. (2009). Impact of video content on video quality for video over wireless networks. In *Proceedings of the International Conference on Autonomic and Autonomous Systems,* (pp. 277–282). Valencia, Spain: IEEE.

Li, D., Cheung, G., Chuah, C.-N., & Yoo, S. B. (2004). Joint server/peer receiver-driven rate-distortion optimized video streaming using asynchronous clocks. In *Proceedings of IEEE International Conference on Image Processing,* (pp. 3149–3152). Singapore: IEEE Press.

Li, L. (2011). Overview of fine granularity scalability in MPEG-4 video standard. *IEEE Transactions on Circuits and Systems for Video Technology, 11*(3), 301–317.

Lieberman, H. (2005). *Introduction to operations research.* New York, NY: McGraw Hill.

Lindeberg, M., Kristiansen, S., Plagemann, T., & Goebel, V. (2010). Challenges and techniques for video streaming over mobile ad-hoc networks. *Multimedia Systems, 17*(1), 51–82. doi:10.1007/s00530-010-0187-8

Ma, S., Gao, W., & Lu, Y., Y. (2005). Rate-distortion analysis for H.264/AVC video coding and its application to rate control. *IEEE Transactions on Circuits and Systems for Video Technology, 15*(12), 1533–1544. doi:10.1109/TCSVT.2005.857300

Radha, H., Schaar, M. V. D., & Chen, Y. (2001). The MPEG-4 fine-grained scalable video coding method for multimedia streaming over IP. *IEEE Transactions on Multimedia, 3*(1), 53–68. doi:10.1109/6046.909594

Ross, S. (1996). *Stochastic processes*. Chichester, UK: Wiley & Sons.

Schaar, M. V. D., & Shankar, N. S. (2005). Cross-layer wireless multimedia transmission: Challenges, principles, and new paradigms. *IEEE Transactions on Wireless Communications, 12*(4), 50–58. doi:10.1109/MWC.2005.1497858

Srivastava, V., & Motani, M. (2005). Cross-layer design: A survey and the road ahead. *IEEE Communications Magazine, 43*(12), 112–119. doi:10.1109/MCOM.2005.1561928

Tian, L., Zhou, Y., Sun, Y., & Sun, S. (2010). Rate distortion model analysis for H.264/AVC video coding. *Journal on Advanced Materials Research, 15*(2), 172–177. doi:10.4028/www.scientific.net/AMR.108-111.172

Vo, N.-S., Duong, T. Q., Shu, L., Du, X., Zepernick, H.-J., & Cheng, W. (2011). Cross-layer design for video replication strategy over multihop wireless networks. In *Proceedings of IEEE International Communications Conference.* Kyoto, Japan: IEEE Press.

Yu, H., Martin, P., & Hassanein, H. (2005). Cluster based replication for large-scale mobile ad-hoc networks. In *Proceedings of the International Conference on Wireless Networks, Communications and Mobile Computing,* (pp. 552–557). Hawaii, HI: IEEE.

Chapter 16
Cross-Layer Design for Packet Data Transmission in Maximum Ratio Transmission Systems with Imperfect CSI and Co-Channel Interference

Trung Q. Duong
Blekinge Institute of Technology, Sweden

Hans-Jürgen Zepernick
Blekinge Institute of Technology, Sweden

ABSTRACT

Multiple-Input Multiple-Output (MIMO) systems, where multiple antennas are deployed at both ends of the wireless links, can significantly increase the transmission rate and reliability of wireless networks. However, due to the increasing demand for high data rates and Quality of Service (QoS), the prosperous performance of MIMO systems may reach its limit. In particular, deployment of many antennas at user devices is constrained by the desired compact designs of mobile equipment. On the other hand, by exploiting the time-varying nature of wireless channels, Adaptive Modulation and Coding (AMC) at the physical layer is utilized to enhance the throughput. In addition, Automatic Repeat Request (ARQ) at the link layer may be applied to circumvent the deleterious effects of fading channels where packets detected with errors can be retransmitted.

In an attempt to maximize the spectral efficiency of co-located MIMO wireless systems under prescribed QoS constraints, the authors consider a cross-layer design that jointly combines Adaptive Modulation (AM) at the physical layer and ARQ at the data link layer. The main purpose of this chapter is to present a general framework for performance analysis of such a cross-layer design. The presented mathematical

DOI: 10.4018/978-1-4666-0960-0.ch016

analysis can be extended to many transmission schemes with MIMO systems including spatial multiplexing schemes (e.g., the V-BLAST system), spatial diversity schemes (e.g., orthogonal space-time block codes), and opportunistic schemes (e.g., maximum transmission ratio).

This chapter will therefore present some new results for the cross-layer design for packet data transmission applied to co-located MIMO with maximum ratio transmission. In this context, closed-form expressions for the average packet error rate, average achievable spectral efficiency, and outage probability are derived. This enables one to evaluate the performance of the cross-layer design under predetermined delay and packet loss constraints. Some important results in the high Signal-to-Noise Ratio (SNR) regime are also provided, revealing insights on how system and channel parameters affect system performance. More importantly, a mathematical analysis is proved to be a powerful tool to investigate the impact of practical issues such as channel estimation error, feedback delay, antenna correlation, and rank-deficient MIMO channel matrix, on the performance of cross-layer design.

INTRODUCTION

With the enormous growth in demand for multimedia services, wireless communications has become the bottleneck in end-to-end wired-wireless networks because radio resources are very scarce and expensive. In addition, the transmission of high data rate demanding services faces fundamental limitations due to impairments inflicted by multipath fading channels. Hence, achieving high data rates with reliable transmission over error-prone mobile radio channels is a major challenge for a mobile radio system design. Recently, being an efficient solution, the so-called cross-layer design paradigm has been proposed where a set of parameters across multiple layers of the protocol stack are exchanged. With this strategy, the overall system performance can be enhanced while keeping the Quality of Service (QoS) satisfactory.

In this chapter, we will investigate a cross-layer design that jointly considers Adaptive Modulation (AM) at the physical layer and Truncated-Automatic Repeat Request (T-ARQ) at the data link layer. As signal traversing from transmitter to receiver may experience reflection, diffraction, and scattering, it can be severely degraded. Instead of circumventing the time-varying effects of the wireless channel, we can exploit these rapid fluctuations to improve spectral efficiency. Specifically, adaptive modulation is deployed by changing to an appropriate modulation constellation, i.e., selecting suitable signal constellation size in accordance with the variation of the end-to-end received Signal-to-Noise Ratio (SNR) (Webb & Steele, 1995; Goldsmith & Chua, 1997, 1998; Alouini, et al., 1999; Alouini & Goldsmith, 2000). However, applying pure adaptive modulation cannot maximize spectral efficiency since the number of available signal constellations is limited. The ARQ scheme at the data link layer is often used to enhance transmission reliability by retransmitting packets detected being in error. Due to the finite buffer size and stringent delay constraints for multimedia services, the maximum number of retransmissions should be limited by using T-ARQ (Malkamaki & Leib, 2000). Under a predefined threshold, e.g., packet lost rate and delay, a cross-layer design combining AM and T-ARQ has gained great attention in the research community (see e.g., Liu, et al., 2004a, 2004b; Maaref & Aïssa, 2004a, 2004b, 2005; Qi, et al., 2010; Wu & Ci, 2006; and references therein).

Multiple-Input Multiple-Output (MIMO) systems, where multiple antennas are deployed at transceivers to significantly improve the system performance compared to its single antenna counterpart, can be subcategorized into spatial multiplexing gain, spatial diversity gain, and opportunistic beamforming transmission. The cross-layer design combining AM at the physical layer and ARQ at the data link layer has been well investigated for Orthogonal Space-Time Block Coding (OSTBC) to exploit the spatial diversity of MIMO systems. In particular, Maaref and Aïssa (2004a, 2004b) have presented the performance of cross-layer design for OSTBC transmission over Rayleigh and Nakagami-*m* fading channels, respectively. Taking into account the antenna correlation and rank-deficient channel effects, the cross-layer design has been investigated for Nakagami-*m* fading (Qi, et al., 2010). It has been shown that the packet error rate performance of the cross-layer design is significantly degraded by antenna correlation. More importantly, the keyhole fading makes the fluctuating behavior in the system performance of the cross-layer design less severe compared to the non-keyhole Nakagami-*m* case. Taking channel estimation errors into consideration, Maaref and Aïssa (2005) have shown that the average spectral efficiency of the cross-layer design can deteriorate by as much as 2.5 bits/s/Hz.

Together with the OSTBC transmission to exploit spatial diversity gain, the Maximum Ratio Transmission (MRT), also known as MIMO beamforming with Maximal Ratio Combining (MRC) (Dighe, et al., 2003a, 2003b; Lo, 1990), has been applied to maximize the end-to-end received SNR when the Channel State Information (CSI) is available at both the transmitter and receiver. Recently, the cross-layer design for MRT systems has been investigated (Qi & Aïssa, 2009; Duong & Zepernick, 2009). In particular, Qi and Aïssa (2009) have considered the effect of feedback delay and limited feedback on the multiuser MRT systems with cross-layer design. Different from Liu et al. (2004a, 2004b), Maaref and Aïssa (2004a, 2004b, 2005), Qi et al. (2010), and Wu and Ci (2006), where the conventional adaptive modulation scheme is combined with ARQ, Qi and Aïssa (2009) have applied an aggressive adaptive modulation for cross-layer design. For single user MRT systems, the effect of both feedback delay and channel estimation error has been investigated for the cross-layer design in the study of Duong and Zepernick (2009). Besides imperfect CSI, Co-Channel Interference (CCI) is also an important factor and has been well investigated for MRT systems in some research works (Rui, et al., 2007; Ahn, 2008; Kang & Alouini, 2004). However, these works only study the non cross-layer approach.

Motivated by all of the above, in this chapter, we therefore extend our previous work (Duong & Zepernick, 2009) by investigating the cross-layer design for MRT systems taking into account not only the imperfect CSI in terms of channel estimation errors at the receiver, feedback delay at the transmitter, but also include CCI from other co-existing systems. Our main target is to investigate how imperfect conditions influence the cross-layer design performance. To do so, we have mathematically derived the packet error rate, spectral efficiency, and outage probability for such systems. Our analytical closed-form expression has allowed us to assess the system performance, which reveals the fact that the cross-layer design performance is significantly decreased compared to its perfect counterpart. We also investigate the influence of several important system parameters such as the number of retransmission, the number of transceiver antennas, on the performance.

The remaining of this chapter is organized as follows. Section 2 contains the required background on cross-layer design jointly combining adaptive modulation at the physical layer and T-ARQ at the data link layer. Section 3 introduces the system and channel model for MRT systems and provides some statistical characteristics for the maximum eigenvalue of the related Wishart matrix. Section 4 presents the performance analysis of the cross-layer design for MRT systems. Section 5 contains and discusses numerical results for a variety of system scenarios. Sections 6 and 7

Figure 1. Example of an end-to-end communication link comprising wired and wireless components

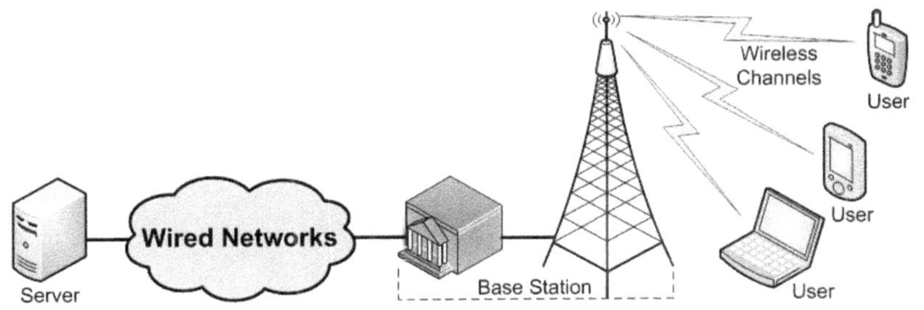

suggest future research directions and draw some conclusions, respectively.

CROSS-LAYER DESIGN COMBINING ADAPTIVE MODULATION AND ARQ

Let us briefly describe the end-to-end connection link including a wired network from a server to a base station and a wireless network where a base station and end users are equipped with multiple antennas as shown in Figure 1. For wireless channels, it is assumed that the CSI is available at the base station and the user so that the MRT technique can be deployed to maximize the received SNR. We consider a packet transmission scheme where arriving packets from the server are stored in the buffer at the base station and are processed in First-In First-Out (FIFO) order. At the physical layer, an adaptive modulation controller is implemented at the base station while the adaptive modulation mode is selected at the user which then sends feedback to the base station. The mode selection is decided by the instantaneous received SNR under the constraint that the average packet error rate $\overline{PER}$ is under a predefined threshold P_{loss}, i.e., $\overline{PER} \leq P_{loss}$. At the data link layer, a T-ARQ controller is implemented to retransmit erroneous packets. In practice, we cannot retransmit an infinite number of packets due to the finite-length buffer and the delay constraint imposed by application data. In particular, the number of retransmissions is decided on the condition that the total average delay for the end-to-end transmission must guarantee the required QoS of the underlying applications such as audio and video services (Liu, et al., 2004a).

In view of the above setting, we present a cross-layer design combining AM at the physical layer and T-ARQ at the data link layer in order to maximize spectral efficiency under predefined constraints such as packet loss rate and delay as shown in Figure 2. Following the approach of our earlier work (Duong & Zepernick, 2009), we here expand the related analysis by also including CCI into the formulation of the performance measures of the cross-layer design.

At the transmitter, AM with multiple signal constellation modes is implemented, each of which corresponds to a specific M-ary Quadrature Amplitude Modulation (M-QAM) scheme. Based on the channel condition, the receiver will estimate the received SNR and then compare it with the thresholds to select a suitable modulation mode. The selected mode will be fed back to the transmitter via a reverse channel which is assumed to be error-free.

To be more specific, let N_{mod} denote the total number of modulation modes. As a result, the received SNR range is divided into $N_{mod} + 1$ non-overlapping intervals. The n-th interval is specified by the two SNR values γ_n and γ_{n+1}. If the estimated SNR γ at the receiver falls into the interval $[\gamma_n, \gamma_{n+1})$, then the n-th mode is selected,

Figure 2. General framework of cross-layer design combining adaptive modulation and T-ARQ

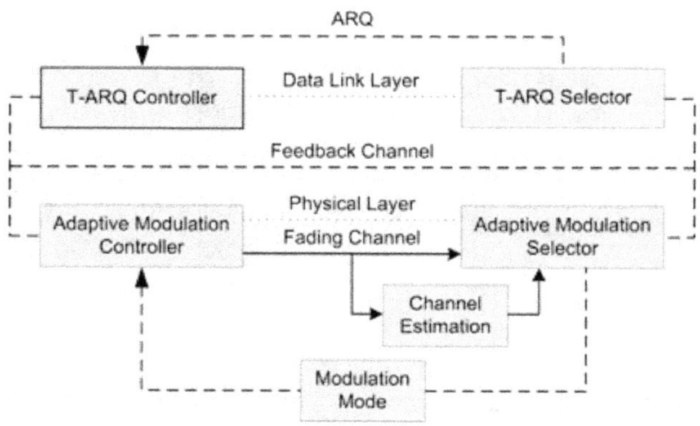

corresponding to an M_n-QAM constellation with transmission rate R_n=n bits/symbol, n=0,1,…,N_{mod}.

As for the overall interval, we define γ_0=0 and γ_{Nmod+1}=∞. To find the remaining switching thresholds γ_n, n=1,2,…, N_{mod}, we use the following approximation (Goldsmith & Chua, 1998; Liu, et al., 2004a):

$$PER_n = \begin{cases} 1 & if\ 0 < \gamma < \gamma_{pn} \\ a_n \exp\left(-g_n \gamma\right) & if \quad \gamma > \gamma_{pn} \end{cases} \tag{1}$$

where PER_n is the packet error rate of the M_n-QAM scheme over Additive White Gaussian Noise (AWGN) channels. The set of parameters $\{a_n, g_n, \gamma_n\}$ can be obtained by using least square curve fitting on the plot of the exact packet error rate as Liu et al. (2004a) described in their study. For example, the parameters in Table 1 are calculated for the packet length of N_p = 1080.

The threshold SNR for the n-th mode γ_n can be computed to be the minimum SNR required to achieve the target packet error rate P_t. As can be seen from (1), the tractable form of the approximated packet error rate allows us to obtain the switching thresholds for SNR by just simply inverting (1) as follows (Goldsmith & Chua, 1998; Liu, et al., 2004a):

$$\begin{cases} \gamma_0 = 0, \\ \gamma_n = \dfrac{1}{g_n} \ln\left(\dfrac{a_n}{P_t}\right), \text{ for } n=1, 2,\dots, N_{mod}, \\ \gamma_{N+1} = +\infty, \end{cases} \tag{2}$$

where P_t denotes the target packet error rate taking into account the effect of T-ARQ at the data link layer. We will describe how to calculate P_t from predefined QoS criterion of the average packet error rate at the physical layer P_{loss} in the sequel.

Table 1. Transmission mode with M-QAM modulation scheme (Liu, et al., 2004a)

	Mode 1	Mode 2	Mode 3	Mode 4	Mode 5	Mode 6	Mode 7
Modulation	BPSK	QPSK	8-QAM	16-QAM	32-QAM	64-QAM	128-QAM
R_n bits/sym	1	2	3	4	5	6	7
a_n	67.7328	73.8279	58.7332	55.9137	50.0552	42.5594	40.2559
g_n	0.9819	0.4945	0.1641	0.0989	0.0381	0.0235	0.0094
γ_{pn} (dB)	6.3281	9.3945	13.9470	16.0938	20.1103	22.0340	25.9677

It can be observed that the feedback information for the mode selection requires only a few bits and therefore can be assumed being without error and latency.

To protect transmitted packets against impairments induced by error-prone channels, a selective repeat T-ARQ is used at the data link layer to retransmit those packets being detected in error. If the packet is correctly received at the destination, the receiver will send an Acknowledgment (ACK); otherwise it will issue a Negative ACK (NACK) to the transmitter. At the transmitter, transmitted packets without ACK/NACK message will be stored in the buffer. When an ACK has been received, the corresponding packet will be removed from the buffer; otherwise, the transmitter will resend the packet. Since the buffer length is finite and the time delay must be limited, the transmitter cannot store many packets. In fact, the number of allowable retransmissions is set not to exceed N_{max}, which shall be related to the QoS requirements of underlying applications. For example, for an MPEG4 video requiring QoS in terms of delay being less than 400 ms, the maximum number of retransmissions N_{max} should not exceed four (Liu, et al., 2004a). Since the total number of transmissions is $N_{max}+1$, i.e., including the first transmission followed by N_{max} retransmissions, the overall packet error rate constraint becomes (Liu, et al., 2004a):

$$\overline{PER}^{N_{\max}+1} \leq P_{loss} \tag{3}$$

From (3), the constraint imposed on the PER P_{loss} of the physical layer can now be transformed to P_t of the cross-layer design as (Liu, et al., 2004a):

$$\overline{PER} \leq P_t = P_{loss}^{1/(N_{\max}+1)} \tag{4}$$

By substituting (4) and (1) into (2), we can obtain the SNR thresholds, which guarantee that the packet error rate of the cross-layer design is under the predefined constraint P_{loss}. This approach will lead to the target packet error rate given in (4) being restricted to the case of the AWGN channel which is too idealistic and induces a loss in spectral efficiency. Another approach following the threshold searching algorithm has been proposed by Liu et al. (2005) to ensure that the packet error rate constraint can be satisfied over fading channels. From this perspective, we aim at deriving the overall packet error rate by averaging the conditioned packet error rate for AWGN channels of the n-th mode given in (1) over the number of selected modes and the considered fading environment.

We first start to consider the probability that the n-th mode is selected. Let us denote γ_S as the instantaneous received SNR of the considered system. It is noticeable that depending on the specific model of the system under consideration (e.g., OSTBC, MRT) and the fading channel environment (e.g., Rayleigh, Nakagami-m), γ_S can be differently presented. The probability that γ_S falls within the interval $[\gamma_n, \gamma_{n+1})$ is given by:

$$\Pr_n = \int_{\gamma_n}^{\gamma_{n+1}} f_{\gamma_S}\left(\gamma\right) d\gamma \tag{5}$$

where $f_{\gamma_S}\left(\gamma\right)$ is the Probability Density Function (PDF) of γ_S. Given that the n-th mode is chosen, then the the n-th mode average packet error rate for the considered systems over fading channels is obtained by averaging PER_n given in (1) with respect to the distribution of γ_S as follows:

$$\overline{PER_n} = \int_{\gamma_n}^{\gamma_{n+1}} PER_n f_{\gamma_S}\left(\gamma\right) d\gamma \tag{6}$$

We now aim to find thresholds γ_n such that the prescribed constraint given in (4) can be achieved for all transmission modes, i.e.:

Figure 3. Block diagram for finding the switching thresholds of SNR

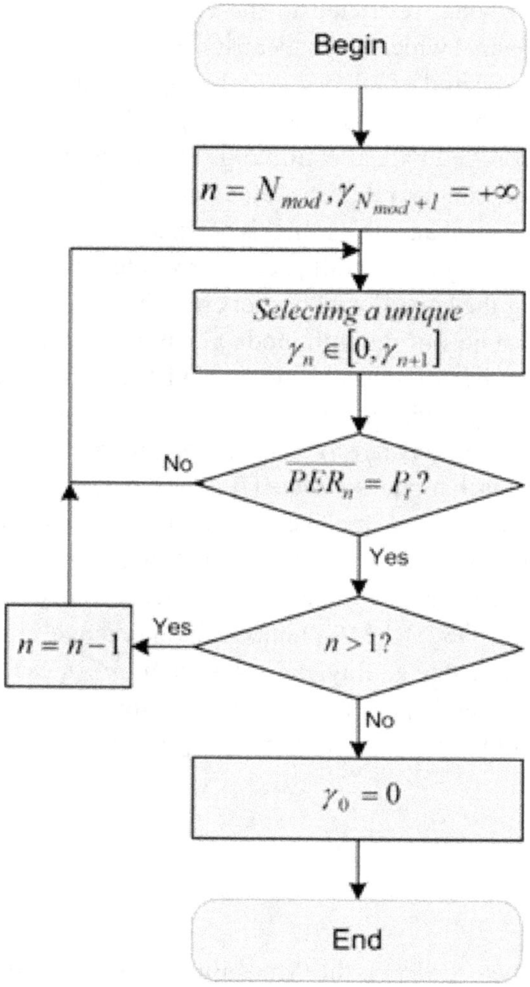

$$\overline{PER_n} = P_t \tag{7}$$

It is important to note that the constraint in (4) will immediately be satisfied when (7) holds. Using (6) and (7), the search algorithm determining the thresholds is described in the diagram shown in Figure 3 (Liu, et al., 2005).

MAXIMAL RATIO TRANSMISSION (MIMO-MRC) SYSTEMS

System and Channel Model

In this section, we consider an MRT system where the full CSI knowledge is exploited at both the transmitter and receiver. At the transmitter deploying n_T antennas, transmit beamforming is applied by multiplying the signal symbol s_t with the beam-stearing vector $\boldsymbol{v}_t$. Assuming that the transmitter consumes an average power per symbol P_s, the input-output relationship at discrete time t for an MRT system can be written as:

$$\boldsymbol{r}_t = \boldsymbol{H}_t \boldsymbol{v}_t s_t + \boldsymbol{n}_t \tag{8}$$

where $\boldsymbol{r}_t$ is an n_R x 1 received vector with n_R being the number of antennas at the receiver, $\boldsymbol{H}_t$ is an n_R x n_T channel matrix, and $\boldsymbol{n}_t$ is an AWGN vector whose elements are complex Gaussian random variables with zero mean and variance N_0. Here, the beam-stearing vector $\boldsymbol{v}_t$ is selected to be the eigenvector $\boldsymbol{u}_t$ corresponding to the largest eigenvalue of the Wishart matrix $\boldsymbol{H}_t^\dagger \boldsymbol{H}_t$, defined as $\lambda_{max}(\boldsymbol{H}_t^\dagger \boldsymbol{H}_t)$, where $(.)^\dagger$ is the conjugate transpose of a matrix.

At the receiver, the signals from n_R receiver antennas are combined using the MRC technique. As a result, the output of the MRC receiver can be shown as (Lo, 1990; Khatri, 1964): $\hat{s}_t = \boldsymbol{w}_t \left(\boldsymbol{H}_t \boldsymbol{u}_t s_t + \boldsymbol{n}_t \right)$. In order to maximize the SNR, the weighting vector $\boldsymbol{w}_t$ is selected to be $\boldsymbol{w}_t = \boldsymbol{u}_t^\dagger \boldsymbol{H}_t^\dagger$, which then yields the end-to-end instantaneous received SNR as:

$$\gamma_S = \gamma_0 \lambda_{max} \left(\boldsymbol{H}_t^\dagger \boldsymbol{H}_t \right) \tag{9}$$

where $\gamma_0 = P_s/N_0$ is the average SNR.

The above mentioned system and channel model has been investigated for the ideal case where we have assumed perfect feedback at the transmitter and perfect channel estimation at the receiver as well as absence of co-channel interference. In practice, the receiver estimates an approximated version of the channel matrix $\hat{\boldsymbol{H}}_t$ with a certain amount of error. Then it will send back $\hat{\boldsymbol{H}}_t$ to the transmitter. Due to the delay in the feedback path, the transmitter receives a delayed version of the channel matrix $\hat{\boldsymbol{H}}_t$ at the time t-τ, denoted as $\hat{\boldsymbol{H}}_{t-\tau}$. Therefore, the channel matrix for this non-ideal case can be characterized as (Yoo & Goldsmith, 2006; Au, et al., 2008):

$$\boldsymbol{H}_t = \rho\hat{\boldsymbol{H}}_{t-\tau} + \boldsymbol{E}_t + \boldsymbol{G}_t \tag{10}$$

where ρ is the correlation coefficient, $\boldsymbol{E}_t = \boldsymbol{H}_t - \hat{\boldsymbol{H}}_t$ is the channel estimation error whose elements are independent and identically distributed (i.i.d.) complex Gaussian random variables with zero mean and variance σ_e^2 denoted as $CN\left(0, \sigma_e^2\right)$ and uncorrelated with $\boldsymbol{H}_t$. Finally, $\boldsymbol{G}_t = \hat{\boldsymbol{H}}_t - \rho\hat{\boldsymbol{H}}_{t-\tau}$ is the channel estimation error induced by feedback delay whose elements are i.i.d. random variables of type $CN\left(0, \left(1-\sigma_e^2\right)\left(1-\rho^2\right)\right)$.

It is important to note that the entries of $\hat{\boldsymbol{H}}_{t-\tau}$ are i.i.d. random variables given by $CN\left(0, 1-\sigma_e^2\right)$ and are assumed to be perfectly known at the transmitter and receiver. As a result, from (10), the combining estimated signal can be obtained as follows (Duong & Zepernick, 2009):

$$\hat{s}_t = \rho\tilde{\boldsymbol{u}}_t^\dagger\hat{\boldsymbol{H}}_{t-\tau}^\dagger\hat{\boldsymbol{H}}_{t-\tau}\tilde{\boldsymbol{u}}_t s_t + \tilde{\boldsymbol{u}}_t^\dagger\hat{\boldsymbol{H}}_{t-\tau}^\dagger\boldsymbol{E}_t\tilde{\boldsymbol{u}}_t s_t + \tilde{\boldsymbol{u}}_t^\dagger\hat{\boldsymbol{H}}_{t-\tau}^\dagger\boldsymbol{G}_t\tilde{\boldsymbol{u}}_t s_t + \tilde{\boldsymbol{u}}_t^\dagger\hat{\boldsymbol{H}}_{t-\tau}^\dagger n_t \tag{11}$$

where $\tilde{\boldsymbol{u}}_t$ represents the normalized eigenvector corresponding the largest eigenvalue of the Wishart matrix $\hat{\boldsymbol{H}}_{t-\tau}^\dagger\hat{\boldsymbol{H}}_{t-\tau}$. As can be observed from (11), the first term is the signal part and the remaining terms are the noise parts. This observation comes from the fact that the MRC weighting vector at the receiver is generated by the delayed channel matrix $\hat{\boldsymbol{H}}_{t-\tau}$ but not the channel estimation errors $\boldsymbol{E}_t$ and $\boldsymbol{G}_t$. Although there exists the signal part $\boldsymbol{s}_t$ in the second and third terms of (11), we cannot treat these two terms as signal but noise. It has been described by Rui et al. (2007) that this kind of noise can be considered as self-interference. It is easy to see that the second term and the last term in (11) are complex Gaussian random variables with zero mean and variances $P_s\sigma_e^2\lambda_{\max}\left(\hat{\boldsymbol{H}}_{t-\tau}^\dagger\hat{\boldsymbol{H}}_{t-\tau}\right)$ and $N_0\lambda_{\max}\left(\hat{\boldsymbol{H}}_{t-\tau}^\dagger\hat{\boldsymbol{H}}_{t-\tau}\right)$, where $\lambda_{\max}\left(\hat{\boldsymbol{H}}_{t-\tau}^\dagger\hat{\boldsymbol{H}}_{t-\tau}\right)$ is the largest eigenvalue of the Wishart matrix $\hat{\boldsymbol{H}}_{t-\tau}^\dagger\hat{\boldsymbol{H}}_{t-\tau}$. However, for the third term in (11), the receiver knows $\boldsymbol{G}_t$ leading this term to be a complex Gaussian with zero mean and variance

$$P_s\mathbf{E}\left\{\tilde{\boldsymbol{u}}_t^\dagger\hat{\boldsymbol{H}}_{t-\tau}^\dagger\boldsymbol{G}_t\tilde{\boldsymbol{u}}_t\left(\tilde{\boldsymbol{u}}_t^\dagger\hat{\boldsymbol{H}}_{t-\tau}^\dagger\boldsymbol{G}_t\tilde{\boldsymbol{u}}_t\right)^\dagger\right\},$$

where $\mathbf{E}\{.\}$ denotes the expectation operator. This in turn yields the mathematical analysis intractable. To facilitate our derivation, we will approximate the variance of the third term by treating $\boldsymbol{G}_t$ as Gaussian noise which then results in the variance to be $P_s\left(1-\sigma_e^2\right)\left(1-\rho^2\right)\lambda_{\max}\left(\hat{\boldsymbol{H}}_{t-\tau}^\dagger\hat{\boldsymbol{H}}_{t-\tau}\right)$. Hence, the instantaneous end-to-end received SNR of the MRT system with channel estimation error and feedback delay can be now approximated as (Duong & Zepernick, 2009):

$$\gamma_S = \frac{\rho^2}{\sigma_e^2 + \left(1-\sigma_e^2\right)\left(1-\rho^2\right) + 1/\gamma_0}\lambda_{\max}\left(\hat{\boldsymbol{H}}_{t-\tau}^\dagger\hat{\boldsymbol{H}}_{t-\tau}\right) \tag{12}$$

The advantage of our approximation is that the previous results published in the literature will be particularized to its special cases. As expected, when $\rho^2 = 1$, (12) approaches a similar form as (Chen & Tellambura, 2005, eq. [6]) for the channel estimation error and no feedback delay. When $\rho^2 = 1$ and $\sigma_e^2 = 0$, (12) reduces to (9) for the ideal case of perfect channel estimation and no feedback delay.

Figure 4. CDF of the instantaneous SNR for a 3×3 MRT system with $\sigma_e^2 = 0.01$ and $\rho^2 = 0.98$

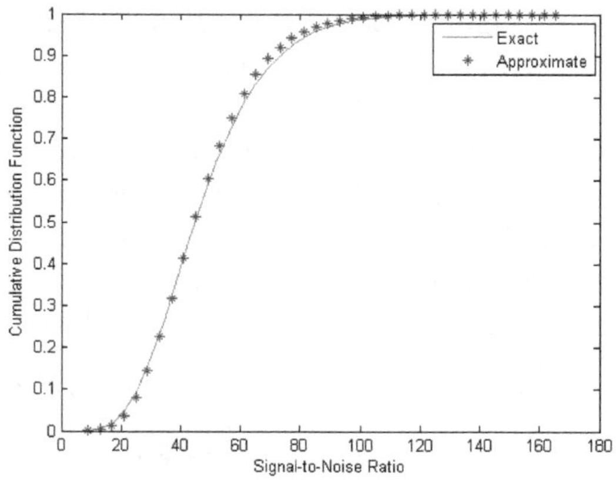

In order to illustrate the tightness of our approximation, we have plotted the Cumulative Distribution Function (CDF) and PDF in Figures 4 and 5, respectively. Results are shown for MRT systems equipped with $n_R = n_T = 3$ antennas, variance of the channel estimation error being $\sigma_e^2 = 0.01$ and correlation coefficient for the feedback delay set to $\rho^2 = 0.98$. As can be observed from these two figures, the approximated SNR given in (12) demonstrates good agreement with the exact results.

In a mobile cellular environment, there will always exist interference from other cells. As a result, the signal vector in (8) can be expressed as (Rui, et al., 2007; Ahn, 2008):

$$\boldsymbol{r}_t = \boldsymbol{H}_t \boldsymbol{v}_t s_t + \sqrt{P_I} \boldsymbol{H}_I \boldsymbol{s}_I + \boldsymbol{n}_t \qquad (13)$$

where P_I is the average received power of the *i*-th interferer, $\boldsymbol{H}_I$ is the $n_R \times L$ channel matrix of L interferers, and $\boldsymbol{s}_I$ represents the $L \times 1$ transmitted symbol vector of CCI. Here, we assume equal powers among CCIs which can be referred to as intra-cell interferers. Similarly as in (11), the

Figure 5. PDF of the instantaneous SNR for a 3×3 MRT system with $\sigma_e^2 = 0.01$ and $\rho^2 = 0.98$

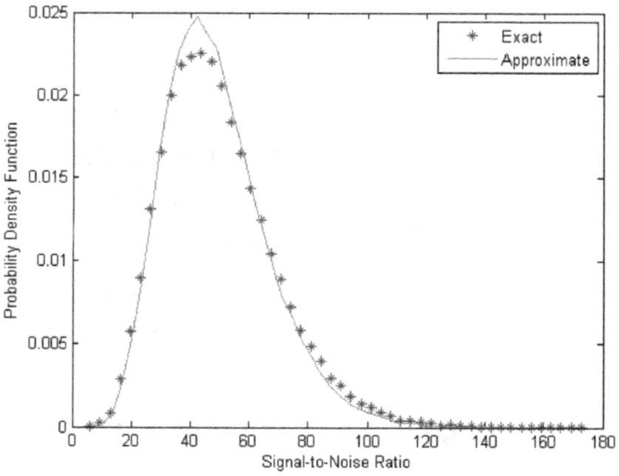

output of an MRC receiver for the case of CCIs can be expressed as:

$$\hat{s}_t = \rho \tilde{\boldsymbol{u}}_t^\dagger \hat{\boldsymbol{H}}_{t-\tau}^\dagger \hat{\boldsymbol{H}}_{t-\tau} \tilde{\boldsymbol{u}}_t s_t + \tilde{\boldsymbol{u}}_t^\dagger \hat{\boldsymbol{H}}_{t-\tau}^\dagger \boldsymbol{E}_t \tilde{\boldsymbol{u}}_t s_t + \tilde{\boldsymbol{u}}_t^\dagger \hat{\boldsymbol{H}}_{t-\tau}^\dagger \boldsymbol{G}_t \tilde{\boldsymbol{u}}_t s_t + \tilde{\boldsymbol{u}}_t^\dagger \hat{\boldsymbol{H}}_{t-\tau}^\dagger \sqrt{P_I} \boldsymbol{H}_I \boldsymbol{s}_I + \tilde{\boldsymbol{u}}_t^\dagger \hat{\boldsymbol{H}}_{t-\tau}^\dagger n_t \quad (14)$$

Compared to (11) and our earlier work (Duong & Zepernick, 2009), we have a new term in (14), namely, $\tilde{\boldsymbol{u}}_t^\dagger \hat{\boldsymbol{H}}_{t-\tau}^\dagger \sqrt{P_I} \boldsymbol{H}_I \boldsymbol{s}_I$. This term is a direct result from the CCIs and yields the instantaneous received SINR as follows:

$$SINR = \frac{\rho^2 \lambda_{\max}\left(\hat{\boldsymbol{H}}_{t-\tau}^\dagger \hat{\boldsymbol{H}}_{t-\tau}\right)}{\sigma_e^2 + \left(1-\sigma_e^2\right)\left(1-\rho^2\right) + P_I / P_S \sum_{i=1}^{L} \left\| \tilde{\boldsymbol{u}}_t^\dagger \hat{\boldsymbol{H}}_{t-\tau}^\dagger \boldsymbol{H}_{I,i} \right\|^2 / \lambda_{\max}\left(\hat{\boldsymbol{H}}_{t-\tau}^\dagger \hat{\boldsymbol{H}}_{t-\tau}\right) + 1/\gamma_0} \quad (15)$$

where $\boldsymbol{H}_{I,i}$ is the i-th column of $\boldsymbol{H}_I$. Conditioned on $\hat{\boldsymbol{H}}_{t-\tau}$, the term $\tilde{\boldsymbol{u}}_t^\dagger \hat{\boldsymbol{H}}_{t-\tau}^\dagger \boldsymbol{H}_{I,i}$ is a complex Gaussian random variable with zero mean while the variance is given as $\lambda_{\max}\left(\hat{\boldsymbol{H}}_{t-\tau}^\dagger \hat{\boldsymbol{H}}_{t-\tau}\right)$ (Kang & Alouini, 2004). Thus,

$$Y = \sum_{i=1}^{L} \left\| \tilde{\boldsymbol{u}}_t^\dagger \hat{\boldsymbol{H}}_{t-\tau}^\dagger \boldsymbol{H}_{I,i} \right\|^2 / \lambda_{\max}\left(\hat{\boldsymbol{H}}_{t-\tau}^\dagger \hat{\boldsymbol{H}}_{t-\tau}\right)$$

can be considered as a Gamma random variable with parameters 1 and L whose PDF can be formulated as:

$$f_Y\left(y\right) = \frac{y^{L-1} e^{-y}}{\left(L-1\right)!} \quad (16)$$

Finally, the SINR given in (15) can be simplified as:

$$SINR = \frac{\rho^2 \lambda_{\max}\left(\hat{\boldsymbol{H}}_{t-\tau}^\dagger \hat{\boldsymbol{H}}_{t-\tau}\right)}{\sigma_e^2 + \left(1-\sigma_e^2\right)\left(1-\rho^2\right) + 1/\gamma_0 + P_I / P_s Y} = \frac{\lambda_{\max}}{\alpha Y + \beta} \quad (17)$$

where $\lambda_{\max}$ is the maximum eigenvalue of a standard Wishart matrix; and α, β, respectively, are given by:

$$\alpha = \frac{\sigma_e^2 + \left(1-\sigma_e^2\right)\left(1-\rho^2\right) + 1/\gamma_0}{\rho^2\left(1-\sigma_e^2\right)}, \beta = \frac{P_I}{P_S \rho^2 \left(1-\sigma_e^2\right)}$$

Statistical Distribution of the Instantaneous Received SINR

The performance of the cross-layer design for MRT systems solely depends on the SINR as given in (17). For analyzing the performance of MRT systems, the statistical properties of the largest eigenvalue of the Wishart matrix have been introduced by Khatri (1964), where the CDF of $\lambda_{\max}$ can be written as:

$$F_{\lambda_{\max}}\left(\lambda\right) = \kappa \det\left(\Xi\left(\lambda\right)\right) \quad (18)$$

where $\kappa^{-1} = \prod_{i=1}^{M} \left(N-i\right)!\left(i-1\right)!$ with $M = \min(n_T, n_R), N = \max(n_T, n_R)$. Furthermore, the (i,j)-th entry of the $M \times M$ Hankel matrix $\Xi\left(\lambda\right)$ is given by:

$$\left\{\Xi\right\}_{i,j} = \left(N-M+i+j-2\right)! - \Gamma\left(N-M+i+j-1, \lambda\right) \quad (19)$$

where Γ(.) is the incomplete gamma function (Gradshteyn & Ryzhik, 2000, eq.[8.352.2]). The PDF of the largest eigenvalue $\lambda_{\max}$ can be easily obtained by taking the derivative of (18) with respect to λ, yielding (Maaref & Aïssa, 2005):

$$f_{\lambda_{\max}}\left(\lambda\right) = \kappa \sum_{k=1}^{M} \sum_{l=N-M}^{(N+M-2k)k} d_{k,l} \lambda^l e^{-k\lambda} \quad (20)$$

where the weighting coefficients $d_{k,l}$ can be computed by employing curve fitting on the plot of $\frac{d}{d\lambda}\det\left(\Xi\left(\lambda\right)\right)$ versus λ. Recently, a simple numerical algorithm to compute the exact coefficients $d_{k,l}$ has been reported by Maaref and Aïssa (2005).

From (17), the PDF of the SINR can be easily written as:

$$f_{SINR}\left(\gamma\right)=\int_{0}^{\infty}\left(\alpha y+\beta\right)f_{\lambda_{\max}}\left(\gamma\left(\alpha y+\beta\right)\right)f_{Y}\left(y\right)dy \tag{21}$$

By substituting (16) and (20) into (21), we get:

$$f_{SINR}\left(\gamma\right)=\kappa\sum_{k=1}^{M}\sum_{l=N-M}^{(N+M-2k)k}\sum_{j=0}^{l+1}\binom{l+1}{j}\frac{d_{k,l}\alpha^{j}\beta^{l+1-j}\gamma^{l}e^{-k\beta\gamma}}{\left(L-1\right)!}\int_{0}^{\infty}y^{L+j-1}e^{-\left(k\alpha\gamma+1\right)y}dy \tag{22}$$

where (22) is obtained by applying the binomial theorem, i.e., $\left(x+a\right)^{n}=\sum_{j=0}^{n}\binom{n}{j}x^{j}a^{n-j}$. With the help of (Gradshteyn & Ryzhik, 2000, eq. [3.351.3]), the PDF of the SINR given in (22) can be expressed in closed-form as follows:

$$f_{SINR}\left(\gamma\right)=\kappa\sum_{k=1}^{M}\sum_{l=N-M}^{(N+M-2k)k}\sum_{j=0}^{l+1}\binom{l+1}{j}\frac{\left(L+j-1\right)!\,d_{k,l}\alpha^{j}\beta^{l+1-j}}{\left(L-1\right)!}\frac{\gamma^{l}e^{-k\beta\gamma}}{\left(k\alpha\gamma+1\right)^{L+j}} \tag{23}$$

PERFORMANCE ANALYSIS OF CROSS-LAYER DESIGN FOR MRT SYSTEMS

The tractable expression of the PDF of SINR derived in the previous section will be utilized to obtain analytical results for the cross-layer design with MRT in terms of average packet error rate, average spectral efficiency, and outage probability. In order to evaluate the performance of the cross-layer design, we first calculate the probability in (5), which is the probability that the n-th transmission mode is chosen, as follows:

$$\Pr_{n}=\int_{\gamma_{n}}^{\gamma_{n+1}}f_{SINR}\left(\gamma\right)d\gamma=F_{SINR}\left(\gamma_{n+1}\right)-F_{SINR}\left(\gamma_{n}\right) \tag{24}$$

where $F_{SINR}\left(\gamma\right)$ is the CDF of the SINR and can be expressed as:

$$F_{SINR}\left(\gamma\right)=\int_{0}^{\infty}F_{\lambda_{\max}}\left(\gamma\left(\alpha y+\beta\right)\right)f_{Y}\left(y\right)dy \tag{25}$$

It is noted that (25) follows immediately from (17), and $F_{\lambda_{\max}}\left(\gamma\right)$ is the CDF of the largest eigenvalue $\lambda_{\max}$ which can be easily obtained from (20) as:

$$F_{\lambda_{\max}}\left(\lambda\right)=\int_{0}^{\lambda}f_{\lambda_{\max}}\left(x\right)dx=\kappa\sum_{k=1}^{M}\sum_{l=N-M}^{(N+M-2k)k}\frac{d_{k,l}}{k^{l+1}}\left(\Gamma\left(l+1\right)-\Gamma\left(l+1,k\lambda\right)\right) \tag{26}$$

where $\Gamma(n)$ is the gamma function (Gradshteyn & Ryzhik, 2000, eq.[8.310.1]) and $\Gamma(n,x)$ is the incomplete gamma function (Gradshteyn & Ryzhik, 2000, eq.[8.350.2]).

By substituting (16) and (26) into (25), the CDF of the SINR can be given by:

$$F_{SINR}\left(\gamma\right)=\kappa\sum_{k=1}^{M}\sum_{l=N-M}^{(N+M-2k)k}\frac{d_{k,l}}{k^{l+1}}\left(\Gamma\left(l+1\right)-I\left(\gamma\right)\right) \tag{27}$$

Box 1.

$$F_{SINR}\left(\gamma\right)=\kappa\sum_{k=1}^{M}\sum_{l=N-M}^{(N+M-2k)k}\frac{d_{k,l}}{k^{l+1}}\left[\Gamma\left(l+1\right)-\sum_{u=0}^{l}\frac{l!k^{u}}{u!\left(L-1\right)!}\sum_{v=0}^{u}\binom{u}{v}\alpha^{v}\beta^{u-v}\Gamma\left(L+v\right)\frac{\gamma^{u}e^{-k\beta\gamma}}{\left(k\alpha\gamma+1\right)^{L+v}}\right] \quad (30)$$

where:

$$I\left(\gamma\right)=\int_{0}^{\infty}\Gamma\left(l+1,k\gamma\left(\alpha y+\beta\right)\right)f_{Y}\left(y\right)dy \quad (28)$$

By substituting (16) into (28) and using (Gradshteyn & Ryzhik, 2000, eq. [8.352.7]) to expand the incomplete gamma function $\Gamma\left(.,.\right)$ into a finite sum, we get:

$$I\left(\gamma\right)=\sum_{u=0}^{l}\frac{l!k^{u}}{u!\left(L-1\right)!}\sum_{v=0}^{u}\binom{u}{v}\alpha^{n}\beta^{u-v}\Gamma\left(L+v\right)\frac{\gamma^{u}e^{-k\beta\gamma}}{\left(k\alpha\gamma+1\right)^{L+v}} \quad (29)$$

Finally, from (27) and (29), we can obtain the closed-form expression for the CDF of the SINR as shown in Box 1 which readily allows us to calculate the probability of the n-th mode being selected for transmission as given in (24).

Average Packet Error Rate

We aim at deriving the overall average packet error rate for the cross-layer design deployed with multimode transmission at the physical layer. To do so, we start with the average packet error rate for each transmission mode. As can be observed from (6), the average packet error rate for the n-th mode is obtained by taking the expectation of PER_n in AWGN channel given in (1) with respect to the SINR in (17) over the interval $[\gamma_n, \gamma_{n+1})$. From (1), (6), and (21), we have:

$$\overline{PER_n}= \int_{\gamma_n}^{\gamma_{n+1}} a_n e^{-g_n\gamma}\int_{0}^{\infty}\left(\alpha y+\beta\right)f_{\lambda_{\max}}\left(\gamma\left(\alpha y+\beta\right)\right)f_{Y}\left(y\right)dyd\gamma \quad (31)$$

By exchange the order of the variables y and γ in the double integral, we can rewrite (31) as:

$$\overline{PER_n}=\int_{0}^{\infty}a_n\left(\alpha y+\beta\right)f_{Y}\left(\gamma\right)Jdy \quad (32)$$

where:

$$J=\int_{\gamma_n}^{\gamma_{n+1}}f_{\lambda_{\max}}\left(\gamma\left(\alpha y+\beta\right)\right)e^{-g_n\gamma}d\gamma \quad (33)$$

To calculate the integral J, we substitute (20) into (33) and after several calculations obtain:

$$J=\kappa\sum_{k=1}^{M}\sum_{l=N-M}^{(N+M-2k)k}\frac{d_{k,l}\left(\alpha y+\beta\right)^{l}}{c^{l+1}}\left[\Gamma\left(l+1,c\gamma_n\right)-\Gamma\left(l+1,c\gamma_{n+1}\right)\right] \quad (34)$$

where $c=\left[k\left(\alpha y+\beta\right)+g_n\right]$. Then, substituting (16) and (34) into (32) yields Box 2.

The integration in (35) consists of powers, exponential and incomplete gamma functions. To solve this integral, we apply (Gradshteyn & Ryzhik, 2000, eq. [8.352.7]) and the binomial theorem (Gradshteyn & Ryzhik, 2000, eq. [1.111]). After several simple manipulations, the average packet error rate for the n-th mode can then be written as:

Box 2.

$$\overline{PER_n} = \kappa \sum_{k=1}^{M} \sum_{l=N-M}^{(N+M-2k)k} \frac{a_n d_{k,l}}{(L-1)!} \int_0^{\infty} y^{L-1} \left(\frac{\alpha y + \beta}{c} \right)^{l+1} e^{-y} \left[\Gamma\left(l+1, c\gamma_n\right) - \Gamma\left(l+1, c\gamma_{n+1}\right) \right] dy \tag{35}$$

$$\overline{PER_n} = \kappa \sum_{k=1}^{M} \sum_{l=N-M}^{(N+M-2k)k} \frac{a_n d_{k,l}}{(L-1)!} \left[\tilde{J}\left(\gamma_n\right) - \tilde{J}\left(\gamma_{n+1}\right) \right] \tag{36}$$

where Box 3 and $\Phi(.)$ is the confluent hyper geometric function (Gradshteyn & Ryzhik, 2000, eq. [9.211.4]). The overall average packet error rate $\overline{PER}$ over all transmission modes can be computed in closed-form as:

$$\overline{PER} = \frac{\sum_{n=1}^{N_{\text{mod}}} R_n \overline{PER_n}}{\sum_{n=1}^{N_{\text{mod}}} R_n \Pr_n} \tag{38}$$

where R_n is the transmission rate of the n-th mode, denoted as $R_n = \log_2\left(M_n\right) = n$. As can be observed from (38), the overall packet error rate at the physical layer is the ratio of the average number of packets in error over the total average number of transmitted packets. By pulling (24), (30), (36), and (38) together, an exact closed-form expression of the average packet error rate for the cross-layer design becomes:

$$\overline{PER} = \frac{\sum_{n=1}^{N_{\text{mod}}} \sum_{k=1}^{M} \sum_{l=N-M}^{(N+M-2k)k} \frac{n a_n d_{k,l}}{(L-1)!} \left[\tilde{J}\left(\gamma_n\right) - \tilde{J}\left(\gamma_{n+1}\right) \right]}{\sum_{n=1}^{N_{\text{mod}}} \sum_{k=1}^{M} \sum_{l=N-M}^{(N+M-2k)k} \frac{n d_{k,l}}{k^{l+1}} \left[I\left(\gamma_n\right) - I\left(\gamma_{n+1}\right) \right]} \tag{39}$$

Average Spectral Efficiency

The average spectral efficiency is an important performance metric to illustrate the advantage of the cross-layer design compared to the non cross-layer approach. In view of the T-ARQ implemented at the data link layer, a packet may be transmitted up to ($N_{\max}$ +1) times which include the first transmission and $N_{\max}$ retransmissions. As a result, the average spectral efficiency is computed by taking into account both physical and data link layer. In particular, the overage average spectral efficiency of the cross-layer design $\overline{SE}$ can be given by:

Box 3.

$$\tilde{J}\left(\gamma_n\right) = l! e^{-(k\beta + g_n)\gamma_n} \sum_{u=0}^{l+1} \alpha^u \beta^{l+1-u} \sum_{i=0}^{l} \frac{1}{i!} \sum_{j=0}^{i} \binom{i}{j} \left(k\alpha\gamma_n\right)^j \left[\left(k\beta + g_n\right)\gamma_n \right]^{i-j} \left(k\alpha\gamma_n + 1\right)^{l+1-L-j-u}$$
$$\times \left(k\alpha\right)^{-l-1} \Gamma\left(L + j + u\right) \Phi\left[l+1, l+2-L-j-u, \frac{\left(k\alpha\gamma_n + 1\right)\left(k\beta + g_n\right)}{k\alpha} \right] \tag{37}$$

$$\overline{SE} = \frac{\sum_{n=1}^{N_{\text{mod}}} R_n \Pr_n}{\bar{N}\left(\overline{PER}, N_{\max}\right)} \tag{40}$$

The numerator of (40) is the spectral efficiency for the physical layer which is equal to the transmission rate R_n of the n-th transmission mode weighted by the probability $\Pr_n$ that the n-th mode is selected. The denominator of (40), i.e., $\bar{N}\left(\overline{PER}, N_{\max}\right)$, corresponding to the effect of the data link layer, is defined as the average number of transmissions per packet and can be formulated as (Liu, et al., 2004a, eq. [10]):

$$\bar{N}\left(\overline{PER}, N_{\max}\right) = 1 + \overline{PER} + \overline{PER}^2 + \ldots + \overline{PER}^{N_{\max}} = \frac{1 - \overline{PER}^{N_{\max}+1}}{1 - \overline{PER}} \tag{41}$$

From (24), (30), (36), (38), and (41), the overall average spectral efficiency of the cross-layer design can be obtained in closed-form as shown in Box 4.

Outage Probability

When the received SINR falls below the threshold γ_1, the system is in outage stage, i.e., no data is transmitted. As a result, the cross-layer design suffers an outage probability P_{out}. From the view of the physical layer, it is easy to see that the outage probability is equal to:

$$P_{out} = \int_0^{\gamma_1} f_{SINR}\left(\gamma\right) d\gamma \tag{43}$$

However, the cross-layer design also employs T-ARQ for retransmitting the incorrectly received packets. As a result, the overall outage probability of the cross-layer design is obtained by powering the outage probability at the physical layer with the average number of packet transmissions $\bar{N}\left(\overline{PER}, N_{\max}\right)$ as shown in Box 5 where (44) follows immediately from (30).

Performance of Non Cross-Layer Design: Adaptive Modulation Only and T-ARQ Only

To highlight the advantage of the cross-layer design over non cross-layer schemes, we also provide the analytical results for T-ARQ with non-adaptive transmission and adaptive modulation only. Here, non cross-layer design approaches are investigated by taking into account the imperfect CSI as well as CCI for a fair comparison.

For adaptive modulation only, we can easily obtain its performance by exploiting the analytical results of the cross-layer design derived in the previous section. Specifically, by substituting $\bar{N}\left(\overline{PER}, N_{\max}\right) = 1$ into (40) and (44), we can obtain the average spectral efficiency and outage probability of adaptive modulation only, respectively, as follows and as shown in Box 6:

Box 4.

$$\overline{SE} = \frac{\kappa \sum_{n=1}^{N_{\text{mod}}} \sum_{k=1}^{M} \sum_{l=N-M}^{(N+M-2k)k} \left\{ \frac{nd_{k,l}}{k^{l+1}} \left[I\left(\gamma_n\right) - I\left(\gamma_{n+1}\right) \right] - \frac{na_n d_{k,l}}{(L-1)!} \left[\tilde{J}\left(\gamma_n\right) - \tilde{J}\left(\gamma_{n+1}\right) \right] \right\}}{1 - \overline{PER}^{N_{\max}+1}} \tag{42}$$

Box 5.

$$P_{out} = \left(\int_0^{\gamma_1} f_{SINR}(\gamma) d\gamma \right)^{\bar{N}(\overline{PER}, N_{\max})}$$

$$= \left\{ \kappa \sum_{k=1}^{M} \sum_{l=N-M}^{(N+M-2k)k} \frac{d_{k,l}}{k^{l+1}} \left[\Gamma(l+1) - \sum_{u=0}^{l} \frac{l! k^u}{u!(L-1)!} \sum_{v=0}^{u} \binom{u}{v} \alpha^v \beta^{u-v} \Gamma(L+v) \frac{\gamma_1^u e^{-k\beta\gamma_1}}{(k\alpha\gamma_1 + 1)^{L+v}} \right] \right\}^{\bar{N}(\overline{PER}, N_{\max})} \quad (44)$$

$$\overline{SE}_{AM} = \sum_{n=1}^{N_{\text{mod}}} \sum_{k=1}^{M} \sum_{l=N-M}^{(N+M-2k)k} \frac{n d_{k,l}}{k^{l+1}} \left[I(\gamma_n) - I(\gamma_{n+1}) \right] \quad (45)$$

The average packet error rate of adaptive modulation only is equivalent to the cross-layer design and can be expressed as in (39).

For T-ARQ only, since a fixed modulation scheme is applied, e.g., the n-th mode is selected, the average packet error rate is obtained by averaging the PER_n given in (1) with respect to the SINR and can be shown as follows:

$$\overline{PER}_{TARQ} = \int_0^{\infty} PER_n f_{SINR}(\gamma) d\gamma = \int_0^{\infty} a_n e^{-g_n \gamma} \int_0^{\infty} (\alpha y + \beta) f_{\lambda_{\max}}(\gamma(\alpha y + \beta)) f_Y(y) dy d\gamma \quad (47)$$

Similarly as in the derivation for the packet error rate of cross-layer design, we can compute (47) by exchanging the order of the two integrals. After some simple manipulations, we have:

$$\overline{PER}_{TARQ} = \kappa \sum_{k=1}^{M} \sum_{l=N-M}^{(N+M-2k)k} \frac{a_n d_{k,l} l!}{(L-1)!} \sum_{u=0}^{l+1} \alpha^u \beta^{l+1-u} (k\alpha)^{-L-u} (k\beta + g_n)^{L+u-l-1} \times \Gamma(L+u) \Phi\left[L+u, L+u-l, \frac{k\beta + g_n}{k\alpha}\right] \quad (48)$$

The average number of transmissions per packet for the T-ARQ only is therefore shown as:

$$\bar{N}(\overline{PER}_{TARQ}, N_{\max}) = \frac{1 - \overline{PER}_{TARQ}^{\,N_{\max}+1}}{1 - \overline{PER}_{TARQ}} \quad (49)$$

The average spectral efficiency for T-ARQ with non-adaptive modulation can be given by:

Box 6.

$$P_{out,AM} = \kappa \sum_{k=1}^{M} \sum_{l=N-M}^{(N+M-2k)k} \frac{d_{k,l}}{k^{l+1}} \left[\Gamma(l+1) - \sum_{u=0}^{l} \frac{l! k^u}{u!(L-1)!} \sum_{v=0}^{u} \binom{u}{v} \alpha^v \beta^{u-v} \Gamma(L+v) \frac{\gamma_1^u e^{-k\beta\gamma_1}}{(k\alpha\gamma_1 + 1)^{L+v}} \right] \quad (46)$$

$$\overline{SE}_{TARQ} = \begin{cases} \dfrac{R_n}{\bar{N}\left(\overline{PER}_{TARQ}, N_{\max}\right)} & if \quad \gamma_0 > \gamma_{n,th} \\ 0 & otherwise \end{cases} \tag{50}$$

where $\gamma_{n,th}$ is the SNR threshold for the n-th mode for T-ARQ without adaptive modulation.

Asymptotic Analysis: High SNR Behavior

The closed-form expressions of average packet error rate, average spectral efficiency, and outage probability derived in the previous subsections readily enable us to evaluate the effect of channel estimation error, feedback delay, and CCI on the performance of cross-layer design. Although these results are accurate and easily implemented by using some mathematical software packages such as MATHEMATICA, MAPLE, or MATLAB, they cannot reveal any insight into system performance. More simplified and tractable formulas are required to provide succinct performance assessment for cross-layer design. Asymptotic analysis is desired here as a powerful tool to reveal the impact of key parameters on system performance. In this section, we adopt the asymptotic analysis for MRT systems developed by (Zhou & Dai, 2006) to reveal the high SNR behavior of system performance. The PDF and CDF of $\lambda_{\max}$ can be shown, respectively, as (Zhou & Dai, 2006):

$$f_{\lambda_{\max}}\left(\lambda\right) \approx \frac{n_T n_R \prod_{k=0}^{M-1} k!}{\prod_{k=0}^{M-1}\left(N+k\right)!} \lambda^{n_T n_R - 1} \tag{51}$$

$$F_{\lambda_{\max}}\left(\lambda\right) \approx \frac{\prod_{k=0}^{M-1} k!}{\prod_{k=0}^{M-1}\left(N+k\right)!} \lambda^{n_T n_R} \tag{52}$$

Our aim now is to obtain the asymptotic expressions for the PDF of the SINR given in (21). To do so, we need to derive the first non-zero high order derivative of $f_{SINR}\left(\gamma\right)$ at zero value of γ, i.e., $\gamma = 0$. In particular, the t-th order derivative of $f_{SINR}\left(\gamma\right)$ can be given by:

$$\frac{\partial^t f_{SINR}\left(\gamma\right)}{\partial \gamma^t} = \int_0^\infty \left(\alpha y + \beta\right)^{t+1} \frac{\partial^t f_{\lambda_{\max}}\left(\gamma\right)}{\partial \gamma^t} f_Y\left(y\right) dy \tag{53}$$

From (51), we can select $t = n_T n_R - 1$ and after some calculations obtain:

$$\begin{aligned} \frac{\partial^{n_T n_R - 1} f_{SINR}\left(\gamma\right)}{\partial \gamma^{n_T n_R - 1}} &= \frac{\left(n_T n_R\right)! \prod_{k=0}^{M-1} k!}{\left(L-1\right)! \prod_{k=0}^{M-1}\left(N+k\right)!} \int_0^\infty \left(\alpha y + \beta\right)^{n_T n_R} y^{L-1} \epsilon^{-y} dy \\ &= \frac{\left(n_T n_R\right)! \left(\prod_{k=0}^{M-1} k!\right) \left[\sum_{k=0}^{n_T n_R} \binom{n_T n_R}{k} \alpha^k \beta^{n_T n_R - k} \Gamma\left(L+k\right)\right]}{\left(L-1\right)! \prod_{k=0}^{M-1}\left(N+k\right)!} \end{aligned} \tag{54}$$

The PDF of the SINR can be asymptotically approximated as:

$$f_{SINR}\left(\gamma\right) \overset{\gamma \to 0}{=} \Theta \gamma^{n_T n_R - 1} \tag{55}$$

where:

$$\Theta = \frac{1}{\left(n_T n_R - 1\right)!} \left. \frac{\partial^{n_T n_R - 1} f_{SINR}\left(\gamma\right)}{\partial \gamma^{n_T n_R - 1}} \right|_{\gamma = 0} \tag{56}$$

Figure 6. Outage probability of non cross-layer design for 3×2 and 2×2 MRT systems

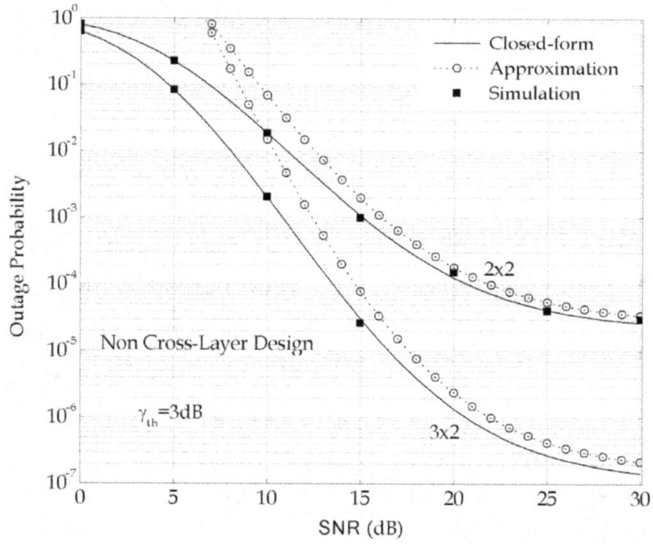

From (54) and (56), the asymptotic expression for the PDF of the SINR given in (55) can be shown as:

$$f_{SINR}(\gamma) \overset{\gamma\to 0}{=} \frac{n_T n_R\left(\prod_{k=0}^{M-1} k!\right)\left[\sum_{k=0}^{n_T n_R}\binom{n_T n_R}{k}\alpha^k \beta^{n_T n_R - k}\Gamma(L+k)\right]}{(L-1)!\prod_{k=0}^{M-1}(N+k)!}\gamma^{n_T n_R - 1} \quad (57)$$

The tractable form of $f_{SINR}(\gamma)$ given in (57) will be utilized to obtain the asymptotic performance by following a similar calculation as in the previous section. Specifically, the outage probability given in (44) can be asymptotically approximated as:

$$P_{out} \overset{\text{high SNR}}{\approx} \left(\Omega\gamma_1^{n_T n_R}\right)^{\overline{N}(\overline{PER}, N_{\max})} \quad (58)$$

where:

$$\Omega = \frac{\left(\prod_{k=0}^{M-1} k!\right)\left[\sum_{k=0}^{n_T n_R}\binom{n_T n_R}{k}\alpha^k \beta^{n_T n_R - k}\Gamma(L+k)\right]}{(L-1)!\prod_{k=0}^{M-1}(N+k)!} \quad (59)$$

To illustrate the tightness of the asymptotic approximation for the PDF and CDF of the SINR, we plot the outage probability for two MRT systems with minimum number of antennas M and maximum number of antennas N, respectively, being $M=3, N=2$ and $M=N=2$. For the purpose of verifying the tightness of our approximation, without loss of generality, we consider a non cross layer design, i.e., a conventional MRT system. Results are shown for $\sigma_e^2 = 0.01$, $\rho^2 = 0.98$, $L = 2$, and $P_I = N_0$. As can be seen from Figure 6, the results obtained from closed-form expression and simulation match very well. In addition, the approximated curves also asymptotically converge to the exact curves in the high SNR regime, which indicates the tightness of approximation.

Figure 7. Performance comparison for different channel estimation errors σ_e^2 and feedback delays ρ^2

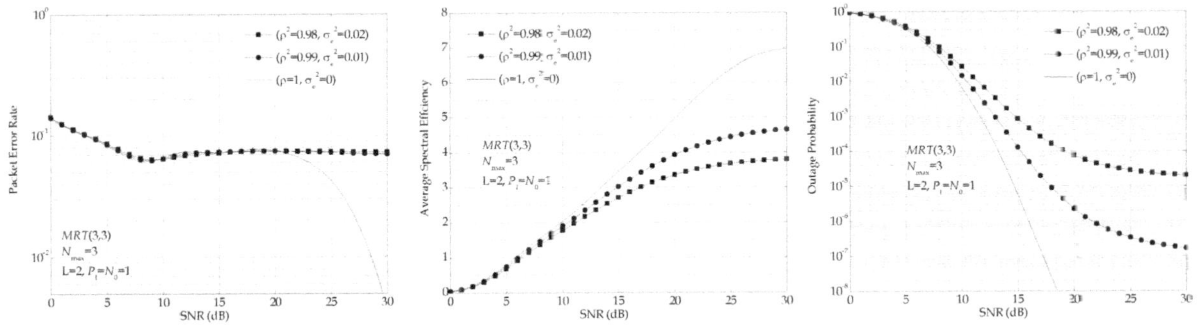

(a) packet error rate (b) average spectral efficiency (c) outage probability

NUMERICAL RESULTS

In this section, we present numerical results to illustrate the performance of cross-layer design for MRT systems in the presence of both imperfect CSI and CCI. The performance is investigated in terms of average packet error rate, average spectral efficiency, outage probability by using the closed-form expressions derived in the previous section. We set the packet length to N_p = 1080 bits and the number of transmission modes for adaptive modulation to N_{mod} = 7. The channel is assumed to experience quasi-static Rayleigh fading with unit channel mean power. Unless otherwise stated, we set the packet loss rate requirement to P_{loss}= 0.01 leading to the target packet error rate of P_t = 0.316. The CCI power is set to $P_I = N_0$. For shorthand notation, we denote the MRT system with minimum *M* antennas and maximum *N* antennas at the transceiver as *MRT*(3,3). We consider the impact of several important parameters, e.g., channel estimation error, feedback delay, the number of retransmissions, the number of transmitted and received antennas, and CCI, on system performance in the sequel.

Effect of Channel Estimation Error and Feedback Delay

We first consider the scenario with imperfect CSI, i.e., channel estimation error and feedback delay. Figure 7 shows the average packet error rate, average spectral efficiency, and outage probability for *MRT*(3,3) versus the average SNR with different parameters

$$\left(\rho^2 = 1, \sigma_e^2 = 0\right), \left(\rho^2 = 0.99, \sigma_e^2 = 0.01\right),$$

and $\left(\rho^2 = 0.98, \sigma_e^2 = 0.02\right)$.

Results are shown for two CCI interferers, i.e., *L* = 2. As can be observed from the figure, the performance of the cross-layer design is significantly degraded by the channel estimation error and feedback delay. The average packet error rate and outage probability curves yield an irreducible error floor in the high SNR regime for the imperfect cases. The spectral efficiency of the perfect CSI, i.e., $\left(\rho^2 = 1, \sigma_e^2 = 0\right)$, eventually reaches 7 bits/s/Hz when the SNR is high. However, the average spectral efficiencies for the two imperfect cases are only 3.5 and 4.5 bits/s/Hz.

Effect of Number of Retransmissions N_{max}

Next, we study the effect of cross-layer design on the performance of MRT systems with different number of retransmissions where N_{max} varies from 0 to 3. In this scenario, without loss of generality, we restrict our attention to the case $\left(\rho^2 = 0.99, \sigma_e^2 = 0.01\right)$ and L=2 co-channel interferers. Figure 8 depicts the performance of the $MRT(2,4)$ system versus average SNR γ_0. As N_{max} increases from 0 to 3, the packet error rate increases at the physical layer. However, to show the advantage of the cross-layer design employing T-ARQ scheme, besides the average packet error rate at the physical layer, we also display the average packet loss rate at the data link layer, i.e., $\overline{PER}^{N_{max}+1}$ where $\overline{PER}$ is the average packet error rate given in (38). As can be observed from this figure, the performance of packet loss rate significantly improves as N_{max} increases from 0 to 3. It clearly indicates that using the T-ARQ protocol increases the system performance as the outage probability diminishes with the increase of N_{max}. We also witness the effective role of T-ARQ to enhance the spectral efficiency from this figure. It is observed that the spectral efficiency gain of the cross-layer scheme using only one retransmission (N_{max}=1) exceeds that of adaptive modulation by approximately 0.3 bits/transmitted symbol. This is a significant gain in high-data rate wireless systems. At high average SNR, the average spectral efficiency curves for the cases of N_{max} =1, 2, 3 approach the highest transmission rate of 4.5 bits/s/Hz. It is obvious that the average spectral efficiency increases as N_{max} increases, however, this trend decreases when the maximum number of retransmission N_{max} increases. This observation implies that N_{max} needs not to be a large number; in some practical applications we need only one retransmission to increase the average spectral efficiency of the systems while satisfying delay criteria.

Effect of Number of Transceiver Antennas

It is worthwhile to investigate the dependence of our cross-layer design on the different antenna configurations of MRT systems. In this scenario, we set N_{max} =1, L=2, and assume perfect CSI, i.e., $\left(\rho^2 = 1, \sigma_e^2 = 0\right)$. Figure 9 displays the performance of a cross-layer design for four different number of transmitter/receiver antennas as $MRT(1,3)$, $MRT(2,2)$, $MRT(2,4)$, and $MRT(3,3)$. It can clearly be seen from these figures that the performance increases as the number of antennas increases. It should be mentioned that with an equal total number of antennas at the transmitter and receiver, higher performance can be achieved when M=N. In particular, for the two cases of $MRT(1,3)$ and $MRT(2,2)$ with the same total number of antennas of four, the latter yields higher performance. The same observation can be achieved by comparing the performance of the $MRT(2,4)$ and $MRT(3,3)$, i.e., $MRT(3,3)$ outperforms $MRT(2,4)$ although the two systems deploy an equal total of four antennas.

Effect of Co-Channel Interference

The performance degradation of a cross-layer design for an $MRT(3,3)$ system due to the influence of CCI is highlighted in Figure 10. In this figure, without loss of generality, we consider perfect CSI conditions by setting $\left(\rho^2 = 1, \sigma_e^2 = 0\right)$ and varying the number of interferes from L=1 to L=4. It can be observed that CCI has a deleterious impact on the performance. As for all the performance metrics, increasing the number of interferes reduces the performance. Moreover, the packet error rate and outage probability curves do not show an error floor in the high SNR regime as we assume perfect channel estimation and no feedback delay in this scenario. This assumption also affects the average spectral efficiency performance. Specifically, in the high SNR regime,

Figure 8. Performance comparison between different channel estimation error σ_e^2 and feedback delay ρ^2

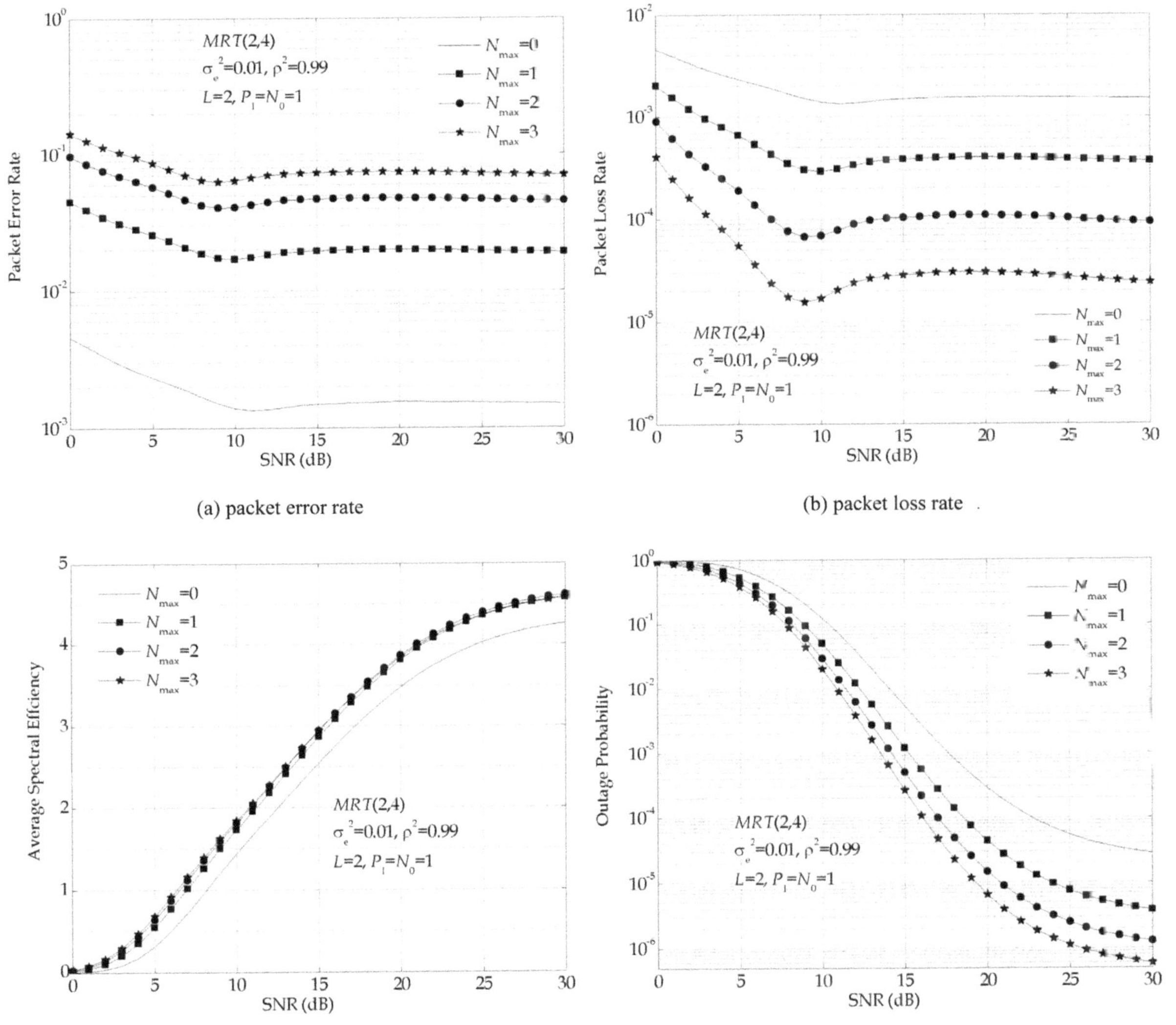

(a) packet error rate

(b) packet loss rate

(c) average spectral efficiency

(d) outage probability

all spectral efficiency curves when L varies from one to four converge to the highest transmission mode of 7 bits/s/Hz.

Comparison Cross-Layer Design with Adaptive Modulation and T-ARQ

We compare the average spectral efficiency of our cross-layer scheme for perfect/imperfect scenarios with adaptive modulation only and T-ARQ only under perfect conditions in Figure 11. In this figure, in particular, we consider the perfect cases of having no CCI, no feedback delay, and perfect channel estimation. As can be seen from this figure, the cross-layer design under such perfect conditions offers the best performance as expected. The non-adaptive T-ARQ fails to provide spectral efficiency in the low and medium SNR regime.

Figure 9. Performance comparison between different antenna configurations

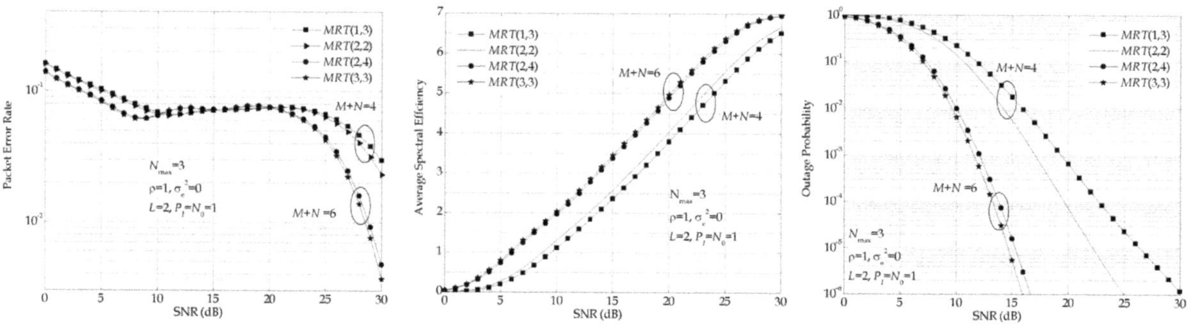

(a) packet error rate (b) average spectral efficiency (c) outage probability

Figure 10. Performance comparison between different numbers of CCI interferers, i.e. L=1,2,3,4

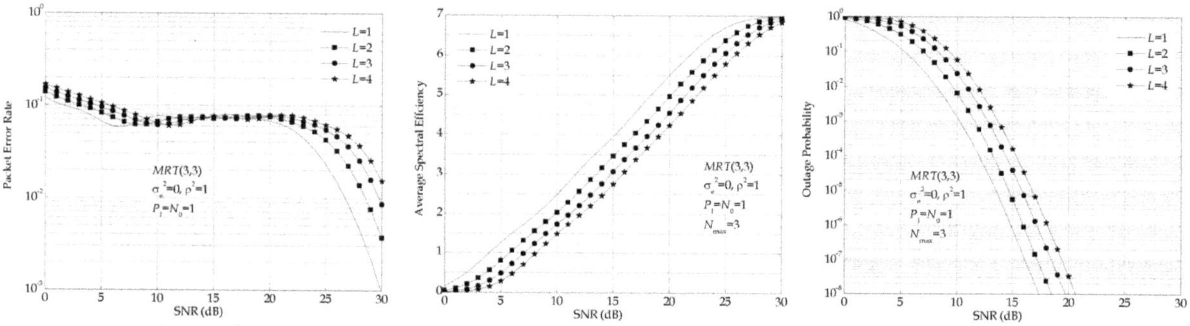

(a) packet error rate (b) average spectral efficiency (c) outage probability

Figure 11. Performance comparison among cross-layer design, adaptive modulation, and T-ARQ

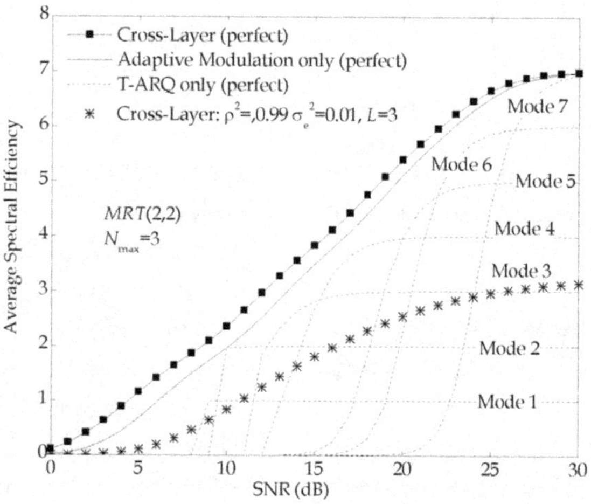

However, in the high SNR regime, it can obtain the same spectral efficiency as achieved by adaptive modulation only. The worst performance is the cross-layer scheme under imperfect condition. As displayed in Figure 11, the average spectral efficiency of a cross-layer scheme in an imperfect case is just slightly equal to that of T-ARQ mode 3, i.e., 3 bits/s/Hz, while the maximum achievable transmission rate is mode 7, i.e., 7 bits/s/Hz.

FUTURE RESEARCH DIRECTIONS

In future work, we will investigate the cross-layer design for multiuser MRT scheduling with channel estimation error and feedback delay.

In this chapter, we have approximated the self-interference induced by the feedback delay as a Gaussian random variable. Hence, an exact derivation is required for our future work.

Our MRT system has been assumed to enjoy rich scattering in which the channel matrix is of full-rank. However, in practice, due to the lack of scattering, the channel may be reduced to keyhole fading where its channel matrix rank is unit. This degeneration due to the keyhole effect is also very interesting to be considered in our future research.

In this work, we have assumed that the channels are independent. Deploying multiple antennas on small devices may incur noticeable antenna correlation due to their spatial proximity at portable devices.

Direct communication between a source and a destination may be inapplicable due to severe path loss and shadowing effects. In such circumstances, the multihop communication of relay networks leads to a remarkable coverage extension for cellular, wireless local area, and ad hoc networks. Recently, the extension of MIMO systems to multihop relay networks has received great attention since this combination can meet the desirable goal of current wireless systems with respect to providing better QoS for very high data rate communication. Hence, considering the cross-layer design for MRT dual-hop wireless relay networks is one of our future research topics.

CONCLUSION

This chapter has presented a cross-layer design for MRT systems by jointly combining adaptive modulation at the physical layer with T-ARQ at the data link layer. Practical effects have also been investigated in terms of channel estimation error, feedback delay, and CCI. We have mathematically obtained closed-form expressions for performance metrics of the cross-layer design by deriving the packet error rate, average spectral efficiency, and outage probability. Numerical results have shown that employing the cross-layer design can significantly improve spectral efficiency while satisfying predefined QoS levels in terms of packet loss rate and delay. However, under imperfect conditions, the cross-layer design performance has substantially degraded. Our mathematical analysis has been proven to be a powerful tool to assess the performance in such practical scenarios.

REFERENCES

Ahn, K. S. (2008). Performance analysis of MIMO-MRC systems with channel estimation error in the presence of co-channel interferences. *IEEE Signal Processing Letters*, *15*, 445–448. doi:10.1109/LSP.2008.923788

Alouini, M.-S., & Goldsmith, A. J. (2000). Adaptive modulation over Nakagami fading channels. *Wireless Personal Communications*, *13*(1-2), 119–143. doi:10.1023/A:1008979107539

Alouini, M.-S., Tang, X., & Goldsmith, A. J. (1999). An adaptive modulation scheme for simultaneous voice and data transmission over fading channels. *IEEE Journal on Selected Areas in Communications*, *17*(5), 337–350. doi:10.1109/49.768199

Au, E. K. S., Jin, S., McKay, M. R., Mow, W. H., Gao, X., & Collings, I. B. (2008). Analytical performance of MIMO-SVD systems in Ricean fading channels with channel estimation error and feedback delay. *IEEE Transactions on Wireless Communications*, *7*(4), 1315–1325. doi:10.1109/TWC.2008.060912

Chen, Y., & Tellambura, C. (2005). Performance analysis of maximum ratio transmission with imperfect channel estimation. *IEEE Communications Letters*, *9*(4), 322–324. doi:10.1109/LCOMM.2005.1413621

Dighe, P. A., Mallik, R. K., & Jamuar, S. S. (2003a). Analysis of transmit-receive diversity in Rayleigh fading. *IEEE Transactions on Communications*, *4*(51), 694–703. doi:10.1109/TCOMM.2003.810871

Dighe, P. A., Mallik, R. K., & Jamuar, S. S. (2003b). Analysis of K transmit dual-receive diversity with cochannel interferers over a Rayleigh fading channel. *Wireless Personal Communications*, *25*(2), 87–100. doi:10.1023/A:1023699628908

Duong, T. Q., & Zepernick, H.-J. (2009). Cross-layer design for MRT Systems with channel estimation error and feedback delay. *Wireless Personal Communications*, *58*(4), 681–694. doi:10.1007/s11277-009-9901-2

Goldsmith, A. J., & Chua, S.-G. (1997). Variable-rate variable-power MQAM for fading channels. *IEEE Transactions on Communications*, *45*(10), 218–1230. doi:10.1109/26.634685

Goldsmith, A. J., & Chua, S.-G. (1998). Adaptive coded modulation for fading channels. *IEEE Transactions on Communications*, *46*(5), 595–602. doi:10.1109/26.668727

Gradshteyn, I. S., & Ryzhik, I. M. (Eds.). (2000). *Table of integrals, series, and products* (6th ed.). San Diego, CA: Academic Press.

Kang, M., & Alouini, M.-S. (2004). A comparative study on the performance of MIMO MRC systems with and without cochannel interference. *IEEE Transactions on Communications*, *52*(18), 1417–1425. doi:10.1109/TCOMM.2004.833027

Khatri, C. G. (1964). Distribution of the largest or the smallest characteristic root under null hyperthesis concerning complex multivariate normal populations. *Annals of Mathematical Statistics*, *35*, 1807–1810. doi:10.1214/aoms/1177700403

Liu, Q., Zhou, S., & Giannakis, G. B. (2004a). Cross-layer combining of adaptive modulation and coding with truncated ARQ over wireless links. *IEEE Transactions on Wireless Communications*, *3*(5), 1746–1755. doi:10.1109/TWC.2004.833474

Liu, Q., Zhou, S., & Giannakis, G. B. (2004b). TCP performance in wireless access with adaptive modulation and coding. In *Proceedings of the IEEE International Communications Conference*, (pp. 3989-3993). Paris, France: IEEE Press.

Liu, Q., Zhou, S., & Giannakis, G. B. (2005, May). Queuing with adaptive modulation and coding over wireless links: Cross-layer analysis and design. *IEEE Transactions on Wireless Communications*, *4*(3), 1142–1153. doi:10.1109/TWC.2005.847005

Lo, T. K. Y. (1990). Maximum ratio transmission. *IEEE Transactions on Communications*, *47*(10), 1458–1461. doi:10.1109/26.795811

Maaref, A., & Aïssa, S. (2004a). A cross-layer design for MIMO Rayleigh fading channels. In *Proceedings of the Electrical and Computer Engineering Canadian Conference*, (pp. 2247-2250). Ontario, Canada: IEEE.

Maaref, A., & Aïssa, S. (2004b). Combined adaptive modulation and truncated ARQ for packet data transmission in MIMO systems. In *Proceedings of the IEEE Global Commununications Conference*, (pp. 3818-3822). Dallas, TX: IEEE Press.

Maaref, A., & Aïssa, S. (2005). Closed-form expressions for the outage and ergodic Shannon capacity of MIMO MRC systems. *IEEE Transactions on Communications*, *53*(7), 1092–1095. doi:10.1109/TCOMM.2005.851564

Maaref, A., & Aïssa, S. (2005, May). Cross-layer design for MIMO Nakagami fading channels in the presence of Gaussian channel estimation errors. In *Proceedings of the Electrical and Computer Engineering Canadian Conference*, (pp. 1380-1383). Saskatoon, Canada: IEEE.

Malkamaki, E., & Leib, H. (2000). Performance of truncated type-II hybrid ARQ schemes with noisy feedback over block fading channels. *IEEE Transactions on Communications*, *48*(9), 1477–1487. doi:10.1109/26.870015

Qi, J., & Aïssa, S. (2009). Cross-layer design for multiuser MIMO MRC systems with feedback constraints. *IEEE Transactions on Vehicular Technology*, *58*(7), 3347–3360. doi:10.1109/TVT.2009.2015668

Qi, J., Aïssa, S., & Maaref, A. (2010). Cross-layer design for MIMO over spatially correlated and keyhole Nakagami-*m* fading channels. *Wireless Communications and Mobile Computing*, *10*, 1055–1067.

Rui, X., Jin, R., & Geng, J. (2007). Performance analysis of MIMO-MRC systems in the presence of self-interference and co-channel interferences. *IEEE Signal Processing Letters*, *14*(11), 801–803. doi:10.1109/LSP.2007.903254

Webb, W. T., & Steele, R. (1995). Variable rate QAM for mobile radio. *IEEE Transactions on Communications*, *43*, 2223–2230. doi:10.1109/26.392965

Wu, D., & Ci, S. (2006). Cross-layer combination of hybrid ARQ and adaptive modulation and coding for QoS provisioning in wireless data networks. In *Proceedings of the 3rd International Conference on QoS in Heterogenoeous Wired/Wireless Networks*. Waterloo, Canada: IEEE.

Yoo, T., & Goldsmith, A. (2006). Capacity and power allocation for fading MIMO channels with channel estimation error. *IEEE Transactions on Information Theory*, *52*(5), 2203–2214. doi:10.1109/TIT.2006.872984

Zhou, Q., & Dai, H. (2006). Asymptotic analysis in MIMO MRT/MRC systems. *Eurasip Journal on Wireless Communications and Networking*. Retrieved from http://jwcn.eurasipjournals.com/content/2006/1/045831.

Chapter 17
Cross-Layer Design, Analysis, and Optimization of QoS-Constrained AMC/ARQ-Based Wireless Networks

Jaume Ramis
University of the Balearic Islands, Spain

Guillem Femenias
University of the Balearic Islands, Spain

ABSTRACT

In order to provide heterogeneous quality of service guarantees to applications, most wireless communications standards combine the error-correcting capability of (hybrid) automatic repeat request protocols at the data link control layer with the adjustment ability of adaptive modulation and coding strategies at the physical layer. In this chapter, a cross-layer multidimensional discrete-time Markov chain based queuing model is developed to jointly exploit the capabilities of (hybrid) automatic repeat request protocols and adaptive modulation and coding. Based on the stationary state probability distribution of this multidimensional discrete-time Markov chain, closed-form analytical expressions for performance metrics such as throughput, average packet delay, and packet loss rate are derived. Furthermore, the proposed analytical framework is used to formulate multidimensional and simplified two-dimensional constrained cross-layer optimization problems aiming at maximizing the system throughput under prescribed quality of service constraints.

DOI: 10.4018/978-1-4666-0960-0.ch017

INTRODUCTION

Context

In the last decade, an explosive development of wireless services and applications has taken place fuelled by the users' ever-increasing demands in terms of throughput and mobility. In such networks, the time and frequency selective nature of wireless fading channels poses a great challenge when designing solutions to support the Quality of Service (QoS) requirements, such as packet error rate and delay deadlines, for current and envisaged heterogeneous mobile applications.

In order to counteract the impact of wireless fading on the performance of the system, many transmission and reception strategies have been developed to improve the spectral and/or power efficiency of the Physical (PHY) layer. Among them, Adaptive Modulation and Coding (AMC) schemes have been intensely researched resulting in their adoption in most state-of-the-art wireless communications standards.

To guarantee the QoS constraints of new applications, powerful schemes to sustain or enhance link reliability are required. To that end, the use of error control strategies at the Data Link Control (DLC) layer, also called Medium Access Control (MAC) layer, has become an integral part in the design of modern communication systems. There are two categories of techniques for controlling transmission errors: the Forward Error Correction (FEC) scheme and the Automatic Repeat Request (ARQ) scheme. In an FEC system, an error-correction code is used. When a received codeword is detected in error and the number of errors is within the designed error-correcting capabilities of the code, the errors are corrected. In an ARQ system, a code with good error-detection capability is used. If a codeword is erroneously received, the transmitter is instructed, through the return channel, to retransmit the same codeword. In order to achieve the advantages of both strategies, most state-of-the-art wireless communications standards use combinations of ARQ and FEC.

Following the spirit of cross-layer design, in contrast to a local optimization of each of the layers, a joint optimization of different layers has been widely considered in order to improve the whole system performance. In this case, parameters and algorithms of different layers are jointly considered. For example, when the queuing model is taken into account, the throughput at the MAC layer is lower than that achieved at the PHY layer. The reasons for this loss are the buffer overflow due to finite-length queues, which may cause the drop of arriving packets, and the absence of packets waiting for transmission even if the channel is in good quality. Many recent cross-layer proposals coincide in combining AMC at the PHY layer with an ARQ protocol at the DLC layer, taking into consideration the queuing effects induced by both processes.

Adaptive Modulation and Coding

Adaptive modulation is a method to improve the spectral efficiency of a radio link for a given maximum required quality (error probability) by adapting the modulation and coding to the channel conditions (see, e.g., Svensson, 2007 and references therein). Since all practical wireless channels perturb the transmitted waveforms, the detection process is never free from errors thus making mandatory the use of error correction/detection techniques. Depending on the severity of the fading, different degrees of information protection are required. The basic premise is to estimate the channel at the receiver and feed this estimate back to the transmitter, so that the transmission scheme can be adapted relative to the channel characteristics. Modulation and coding techniques that do not adapt to fading conditions require a fixed link margin to maintain acceptable performance when the channel quality is poor. Thus, these systems are effectively designed for the worst-case channel conditions. Consequently, if the channel is not constant during the whole transmission, which is actually the case for most state-of-the-art wireless systems, adaptive schemes have a large performance advantage in comparison to fixed schemes. Adapting to the channel fading can increase average throughput, reduce required transmit power, or reduce average probability of bit error by taking advantage of favorable channel conditions to operate at higher data rates or lower power, and reducing the data rate or increasing power as the channel degrades (Goldsmith, 2005).

Automatic Repeat Request Strategies

ARQ error control strategies achieve high reliability by using an error detection scheme coupled with a block retransmission scheme (Lin & Costello, 2004). There are basically two types of ARQ: Stop and Wait (SW) ARQ and Continuous ARQ. With the first strategy, the transmitter sends a single packet and waits for a positive Acknowledgement (ACK) or a Negative Acknowledgement (NACK) before sending more packets. It is a very inefficient method due to the waste of time waiting for the receiver response. On the contrary, with the continuous ARQ strategy, packets are sent continuously until a NACK is received. Two continuous ARQ techniques can be distinguished: Go-Back-N (GBN) ARQ, in which the transmitter resends the erroneously received packet plus all the packets that followed until the NACK was received, and Selective Repeat-ARQ (SR-ARQ), in which only the erroneous packet is retransmitted. Both types of continuous ARQ offer improved throughput efficiency over SW ARQ at the cost of greater memory and higher processing requirements. Among all ARQ schemes, SR-ARQ is reported to show the best throughput performance, hence it will be the one adopted in our analysis.

In Type-I Hybrid FEC/ARQ schemes, both error detection and forward error correction bits are added to each packet prior to transmission (using, for instance, a concatenation of a Bit-Interleaved Coded Modulation [BICM] scheme and a Cyclic Redundancy Check [CRC] code). When a received codeword is detected in error, two situations may arise. If the number of errors is within the designed error-correcting capabilities of the code, the errors are corrected. Otherwise, when an uncorrectable error pattern is detected, the received coded data block is discarded and a retransmission is requested by the receiver, similar to standard ARQ. Retransmissions take place at either the same or a different code rate. We will consider both, an infinitely persistent ARQ protocol, in which packets will be removed from the buffer after being successfully received, and a truncated ARQ protocol with a maximum allowed number of ARQ retransmissions set to N_r. In this case, packets will leave the buffer either after being successfully received by the mobile host or after $N_r + 1$ failed attempts.

When using a more sophisticated form of hybrid FEC/ARQ, known as Hybrid Automatic Repeat Request (HARQ) (Cheng, 2006), if the receiver fails to decode a packet, any previously received signal is stored in a buffer and a retransmission request in the form of a NACK is fed back to the transmitter. In Type-I HARQ, upon reception of this NACK the transmitter sends the same coded packet again. At the receiver side, the optimal solution is to combine these multiple signals according to the Maximal Ratio Combining (MRC) principle (Chase, 1973, 1985), an approach usually known as Chase Combining (CC). Alternatively, in the so-called Type-II or Type-III (when each packet is self-decodable) HARQ, also known as the Incremental Redundancy (IR) scheme, when the receiver fails to decode a packet, the decoded bits (or log-likelihood ratios) are stored in a buffer and a retransmission is requested. The transmitter, upon reception of a NACK, instead of resending the same coded packet, transmits additional redundant information in each retransmission. Obviously, IR requires larger buffer size than

CC; nevertheless, although the CC scheme can improve link performance by coherently combining multiple copies of the received signal, the IR strategy can benefit from these and also certain coding gains, since the different coded versions jointly form a lower-rate code with stronger error protection capabilities (Cheng, 2006).

Related Work

The unique nature of AMC to improve the performance of upper-layer protocols has spurred the development of cross-layer designs with the goal of integrating QoS provisioning protocols at higher network layers with AMC at the PHY layer. Specifically, many of the most recent cross-layer design proposals between the PHY layer and the DLC layer coincide in combining AMC with an ARQ protocol, aiming at improving the spectral efficiency by jointly exploiting the adaptability of AMC to the wireless channel conditions and the error-correcting capability of ARQ.

Liu et al. (2004) proposed a scheme combining AMC at the PHY layer with a truncated ARQ protocol at the DLC layer in order to improve the overall system spectral efficiency. However, for analytical convenience, the fading channel coefficients corresponding to the original and the retransmitted packets were assumed independent and identically distributed random variables and, furthermore, queueing effects on the average packet delay were not taken into account. In Liu et al. (2005), the same authors also proposed a cross-layer design combining finite buffer queuing at the DLC layer with AMC at the PHY layer and applied finite-state Markov chain analysis to derive analytical expressions for the packet loss rate and throughput. However, that contribution did not take into account the possible performance improvement from the ARQ protocol (i.e., unsuccessfully transmitted packets were dropped) and ignored correlation in the traffic arrival process. Wang et al. (2007) tried to solve some of the previous weaknesses by generalizing the cross-layer combination of queuing with AMC in Liu et al. (2004) and the cross-layer combination of queuing with truncated ARQ and AMC in Liu et al. (2005). However, similarly to Liu et al. (2004) and Liu et al. (2005), a memory-less packet arrival process was assumed. Moreover, in order to facilitate mathematical tractability of the queuing process, they relied on the rather unrealistic assumption of a time slotted system where only one frame was transmitted per slot, and with each frame at the PHY layer containing at most one packet from the DLC layer. Traffic burstiness was considered by Le et al. (2006), where an analytical framework was provided for point-to-point wireless systems with infinite/finite buffer queuing and ARQ-based error control at the DLC layer, and AMC at the PHY layer. Nevertheless, infinitely persistent "pure" ARQ-based error control schemes were considered, while the generalization to more sophisticated truncated-HARQ protocols was not addressed at all.

One common shortcoming of all these cited works is that they rely on first-order Amplitude-Based Finite-State Markov Chains (AFSMC) to model the wireless fading channel. However, as was shown by Tan and Beaulieu (2000), first-order AFSMCs having an exponentially decaying Auto-Correlation Function (ACF) can not fit the hypergeometric ACF of the statistical Rayleigh fading process used to model wireless flat-fading channels (Clarke, 1968), thus compromising the design of higher layer protocols.

Based on the use of a first-order two-dimensional Finite State Markov Chain (FSMC) model, in Ramis et al. (2008), Femenias et al. (2009), and Ramis et al. (2009), a novel cross-layer analytical framework for AMC-based wireless systems using either infinitely persistent or truncated Type-I hybrid FEC/ARQ was proposed. Although this method does not require a large buffer at the receiver, it is a very inefficient way of exploiting ARQ. Hybrid ARQ (HARQ) schemes were analyzed in Ramis et al. (2010) and Ramis and

Femenias (2011) when used in combination with AMC-based PHY layers.

Chapter Aim and Objectives

Building on all previously referenced papers, the main objective of this chapter is to provide a unified view of the cross-layer design, analysis and optimization of AMC/(H)ARQ-based wireless systems, whose general architecture can be seen in Figure 1. The PHY layer model is based on the use of enhanced first-order two-dimensional FSMC models for the wireless flat-fading channel. In contrast to previous proposals in the literature, this model relies on the amplitude and the rate-of-change of the fading envelope. Moreover, it incorporates the implementation of the AMC scheme used in the TM selection subsystem, which is designed independently from the channel model. Additionally, based on this PHY layer model, a unified analytical link level queuing model of a point-to-point adaptive multi-rate wireless system using either infinitely persistent or truncated (H)ARQ is proposed. Using this approach, analytical expressions for performance metrics such as throughput, average packet delay and packet loss rate, both due to buffer overflow and due to exceeding the maximum number of allowed retransmissions, are derived. The importance of each metric depends on the required QoS, which is service-specific (for voice traffic, the delay has clearly to be minimized, whereas for file transfer services it is more interesting to maximize the throughput). The resulting analytical link level queuing model, which is based on the use of Discrete Time Markov Chains (DTMCs), provides then a roadmap towards effective cross-layer multidimensional design strategies that are thoroughly explored. A bi-dimensional simplification approach of this problem is also proposed, resulting in a simpler scheme at the cost of a performance loss. Particularly, it is shown that the analytical results not only allow a-priori design decisions to be made, but they also provide an insight that enables the derivation of dynamic reconfiguration strategies that take into account instantaneous Channel State Information (CSI) at the PHY layer and Queue State Information (QSI) at the DLC layer. Particularly, as illustrated in Figure 2, the cross-layer design will optimally determine specific design parameters at the physical layer, $\mathbf{X}_{\mathrm{PHY}}^{\mathrm{opt}}$ (e.g., AMC switching thresholds and set of AMC useful modes), and at the data-link layer, $\mathbf{Y}_{\mathrm{DLC}}^{\mathrm{opt}}$ (e.g., average packet arrival rate and buffer size), that maximize the average throughput η of the system, while satisfying prescribed QoS requirements. Analytical and Monte Carlo simulation results will be used to verify the validity of the proposed cross-layer framework, as well as to compare the performance of either infinitely persistent or truncated classical hybrid FEC/ARQ schemes with the truncated HARQ-CC and HARQ-IR strategies.

Figure 1. System model

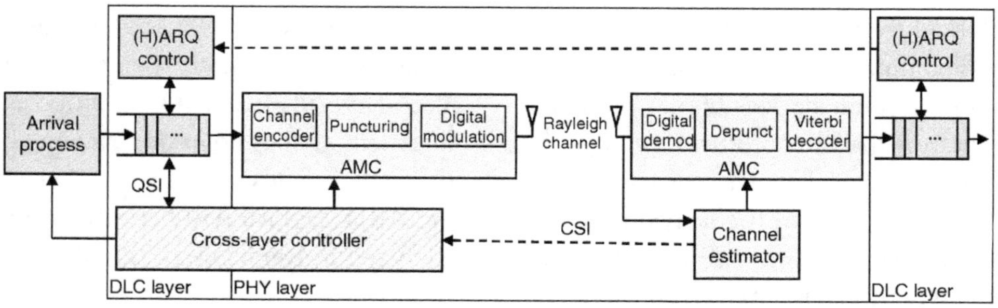

Figure 2. Cross-layer approach

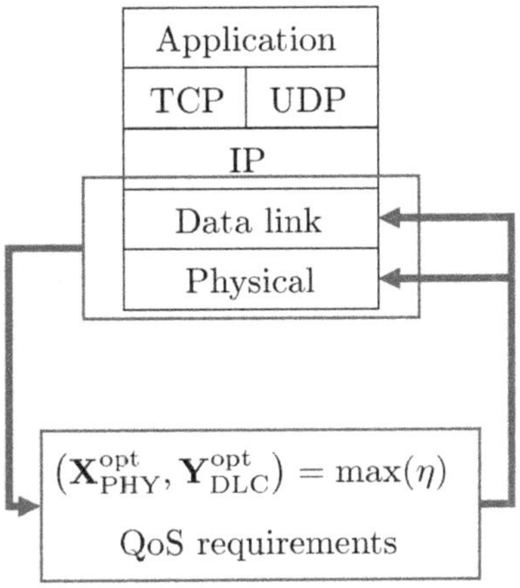

SYSTEM MODEL AND ASSUMPTIONS

The point-to-point wireless packet communication system under consideration is illustrated in Figure 1, showing the most important blocks involved in the transmission and reception processes.

Following the (H)ARQ controller, and based on the CSI feedback from the receiver on a frame-by-frame basis, the AMC scheme selects a transmission mode, which corresponds to a modulation and FEC convolutional code pair. Different code rates can be easily obtained through the combination of a fixed rate convolutional code followed by a puncturer. This technique, which will be further discussed in the next Section, consists in eliminating, according to a puncturing pattern, some of the bits at the output of the fixed rate convolutional encoder, obtaining in this way the set of allowed coding rates. At the receiver end, joint coherent demodulation and soft decoding are first employed to retrieve each coded packet. Then, after suitable depuncturing by conveniently padding dummy bits that do not affect the decoding metric, a Viterbi decoder is used to decode the received packet.

The system is assumed to support QoS-guaranteed traffic characterized by a maximum average packet delay $D_{l\max}$ and a target link layer packet loss rate $P_{l\max}$.

Transmitter

At the transmitter side, the (H)ARQ controller manages a buffer (queue) that operates in a First-In-First-Out (FIFO) mode and can store up to $\bar{Q}$ packets. As shown in Figure 3, the processing unit at the data link layer is a packet of fixed size equal to N_b bits, and the processing unit at the physical layer is a frame made of a variable number of packets that depends on the Transmission Mode (TM) selected by the AMC protocol. The AMC scheme is assumed to have a set $\mathcal{M} = \{0, \ldots, M-1\}$ of M TMs, each of which corresponding to a particular combination of modulation and coding strategies, including the case in which the transmitter does not transmit.

The instantaneous received Signal-to-Noise Ratio (SNR) during the ν th frame transmission is defined as $\gamma_\nu = E_s \mid h_\nu \mid^2 / N_0$, where E_s denotes the average power of the received signal, h_ν denotes the normalized (unit power) channel gain and the noise has been modeled as a zero-mean circularly symmetric complex Gaussian random variable with variance $N_0 / 2$ per dimension. The average received SNR is thus given by $\bar{\gamma} = E\{\gamma_\nu\} = E_s / N_0$. According to Liu et al. (2004), when implementing the AMC strategy, the entire SNR range is partitioned into a set of nonoverlapping intervals defined by the partition:

$$\mathbf{\Gamma}^m = \{[\gamma_0^m, \gamma_1^m), [\gamma_1^m, \gamma_2^m), \ldots, [\gamma_{M-1}^m, \gamma_M^m)\} \tag{1}$$

and mode n is selected when $\gamma_\nu \in \left[\gamma_n^m, \gamma_{n+1}^m\right)$, as illustrated in Figure 4.

Without loss of generality, convolutionally coded M-ary Quadrature Amplitude Modulation (M-QAM) schemes adopted from IEEE 802 16e standard (IEEE, 2004) will be used in the AMC pool. All considered TMs (except the non-transmission mode) are listed in Table 1.

Figure 3. Frame and packet structures

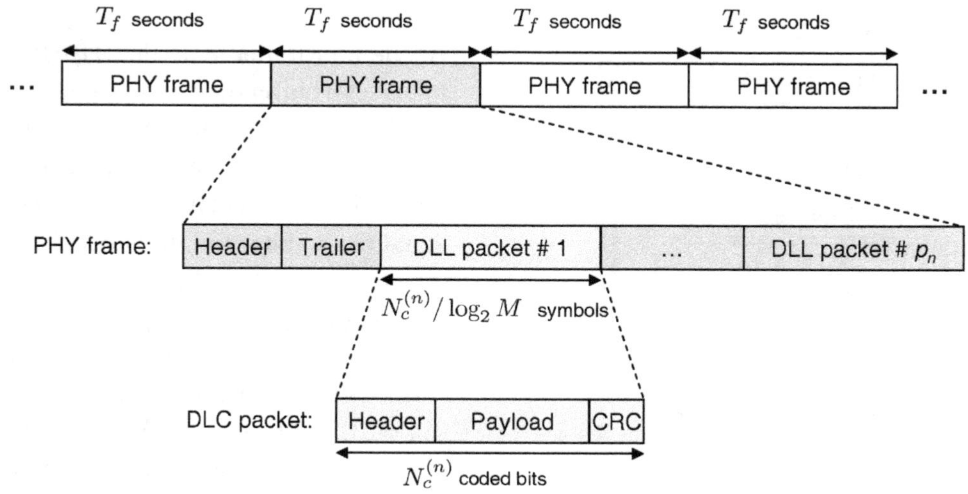

Figure 4. Transmission modes selection thresholds

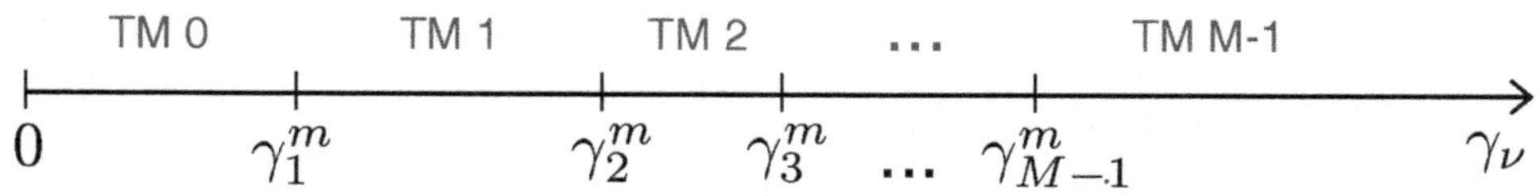

In this case, when using transmission mode $n \in \mathcal{M}$, a rate-1/2 convolutional encoder generates a sequence $\tilde{\mathbf{b}} = \{\tilde{b}_l\}_{l=1}^{2N_b}$ of encoded bits and, after puncturing, the system transmits $N_c^{(n)} = N_b / R_c^{(n)}$ coded bits per packet, where $R_c^{(n)}$ denotes the code rate obtained after puncturing when using TM n.

The puncturing patterns for the HARQ-IR scheme in the IEEE 802.16e standard (IEEE, 2004) are shown in Table 2, where the puncturing pattern corresponding to SPID (SubPacket IDentifier) $i+1$ is obtained as the left cyclic shift of the puncturing pattern corresponding to SPID i. A 0 in the puncturing pattern means that the corresponding output bit of the fixed rate convolutional encoder will not be transmitted. Figure 5 illustrates an example of the puncturing process corresponding to the first transmission (SPID=0) and a code rate $R_c^{(n)}$=3/4.

In Type-I hybrid FEC/ARQ and in HARQ schemes based on the CC strategy the same puncturing pattern is applied for each (re)transmission

Table 1. Transmission modes with convolutionally coded modulation

	Mode 1	Mode 2	Mode 3	Mode 4	Mode 5
Modulation	**BPSK**	**QPSK**	**QPSK**	**16QAM**	**16QAM**
Code rate $R_c^{(n)}$	1/2	2/3	5/6	2/3	5/6
R_n (bits/symbol)	1/2	4/3	5/3	8/3	10/3

Figure 5. An example of the puncturing process for the first transmission $i = 0$ and $R_c^{(n)}$ =3/4

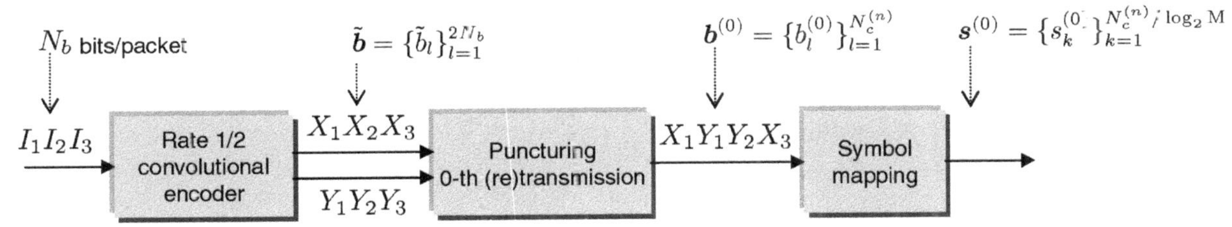

Table 2. Puncturing pattern definition for the HARQ-IR

		Code rate $R_c^{(n)}$			
		1/2	2/3	3/4	5/6
SPID=0	X	1	10	101	10101
	Y	1	11	110	11010
SPID=1	X	1	01	011	01011
	Y	1	11	101	10101
SPID=2	X	1	10	110	10110
	Y	1	11	011	01011
SPID=3	X	1	01	101	01101
	Y	1	11	110	10110

(without loss of generality, the puncturing pattern corresponding to SPID=0 is applied to all (re) transmissions). In contrast, when using the IR strategy, a different puncturing pattern can be applied to successive (re)transmissions of a packet (Cheng, 2006). As it can be observed in Figure 5, for a generic packet, the sequence of punctured coded bits corresponding to (re)transmission $i \in \{0,\ldots,N_r\}$ can be denoted as $\mathbf{b}^{(i)} = \{b_l^{(i)}\}_{l=1}^{N_c^{(n)}}$. These punctured coded bits are mapped onto a sequence of symbols $\mathbf{s}^{(i)} = \{s_k^{(i)}\}_{k=1}^{N_c^{(n)}/\log_2 \mathrm{M}}$ which are selected from the M-QAM constellation corresponding to TM n.

The number of transmitted packets per frame depends on the TM n and it is given by $p_n = bR_n$, where R_n denotes the number of information bits per symbol used by TM n (see Table 1) and b is a parameter to be set by the designer. For convenience, we will consider that $p_0 < \cdots < p_{M-1}$, with $p_0 = 0$ (i.e., transmission mode 0 corresponds to the absence of transmission) and $p_{M-1} \triangleq C$.

Wireless Channel

A Rayleigh block-fading channel model (Biglieri, et al., 2001) has been adopted in which the channel gain h_ν corresponding to the ν th frame transmission is characterized as a zero-mean circularly symmetric complex Gaussian random variable with unit power. It is assumed that the frame duration T_f is much shorter than the coherence time of the channel. In this case, although the channel is assumed to remain invariant over at least one time frame interval while it is allowed to vary across successive frame intervals, it is also highly probable that it will remain practically invariant over a large number of successive frame intervals (slow block-fading channel).

Receiver

The received signal corresponding to transmitted symbol $s_k^{(i)}$, $k = 1,\ldots,N_c^{(n)} / \log_2 \mathrm{M}$, can be expressed as:

$$r_k^{(i)} = s_k^{(i)} h_i + n_{i,k}, \qquad (2)$$

where, by convenient abuse of notation, h_i denotes the channel gain of the frame period corresponding to the ith (re)transmission and $n_{i,k}$ is a zero-mean circularly symmetric complex Gaussian noise with variance $N_0 / 2$ per dimension.

Chase combining: Assuming perfect CSI to be available at the receiver side, the HARQ-CC scheme combines multiple received signals according to the MRC principle. Thus, assuming without loss of generality that the 0th puncturing pattern has been used in all (re)transmissions, the combined signal after i (re)transmissions can be expressed as:

$$\hat{r}_k^{(i)} = \sum_{j=0}^{i} r_k^{(j)} h_j^* = s_k^{(0)} \rho_i + v_k^{(i)}, \tag{3}$$

where $\rho_i = \sum_{j=0}^{i} |h_j|^2$ and $v_k^{(i)}$ is a zero-mean circularly symmetric complex Gaussian noise with variance $\sigma_{v,i}^2 = \rho_i N_0$. The Logarithmic Likelihood Ratio (LLR) corresponding to bit $b_l^{(0)}$ mapped onto symbol $s_k^{(0)}$ on the ith (re)transmission can be expressed as:

$$\lambda_l^{(i)} = \log \frac{Pr\{b_l^{(0)} = 1 \mid \hat{r}_k^{(i)}, \rho_i\}}{Pr\{b_l^{(0)} = 0 \mid \hat{r}_k^{(i)}, \rho_i\}} = \log \frac{\sum_{\hat{s}_k \in S_{l,i}^{(1)}} \exp\left(- | \hat{r}_k^{(i)} - \hat{s}_k \rho_i |^2 / \sigma_{v,i}^2\right)}{\sum_{\hat{s}_k \in S_{l,i}^{(0)}} \exp\left(- | \hat{r}_k^{(i)} - \hat{s}_k \rho_i |^2 / \sigma_{v,i}^2\right)}, \tag{4}$$

where $S_{l,i}^{(0)}$ and $S_{l,i}^{(1)}$ are, respectively, the sets of symbols $\hat{s}_k$ with the bit indexed by l, corresponding to the ith (re)transmission, equal to zero or one, respectively. These LLRs are subsequently depunctured to obtain the sequence $\tilde{\lambda}^{(i)} = \{\tilde{\lambda}_l^{(i)}\}_{l=1}^{2N_b}$ that is then passed to the soft Viterbi decoder.

Incremental redundancy: For the IR-HARQ scheme, the LLR corresponding to bit $b_l^{(i)}$ mapped onto symbol $s_k^{(i)}$ on the ith (re)transmission is given by:

$$\lambda_l^{(i)} = \log \frac{Pr\{b_l^{(i)} = 1 \mid r_k^{(i)}, h_i\}}{Pr\{b_l^{(i)} = 0 \mid r_k^{(i)}, h_i\}} = \log \frac{\sum_{\hat{s}_k \in S_{l,i}^{(1)}} \exp\left(- | r_k^{(i)} - \hat{s}_k h_i |^2 / N_0\right)}{\sum_{\hat{s}_k \in S_{l,i}^{(0)}} \exp\left(- | r_k^{(i)} - \hat{s}_k h_i |^2 / N_0\right)}. \tag{5}$$

After suitable depuncturing, and assuming an add-LLR soft-output packet combining scheme (Aïssa & Aniba, 2008), the sequence of combined LLRs is obtained as:

$$\tilde{\lambda}^{(i)} = \left\{\sum_{j=0}^{i} \tilde{\lambda}_l^{(j)}\right\}_{l=1}^{2N_b}, \tag{6}$$

and it is then sent to the soft Viterbi decoder.

Type I hybrid FEC/ARQ: In the Type I hybrid FEC/ARQ scheme, when an uncorrectable pattern is detected, the received packet is discarded and a retransmission is requested. Therefore, this strategy is equivalent to the IR-HARQ scheme with no addition of LLRs after depuncturing and assuming that the 0th puncturing pattern is used in all (re)transmissions. Consequently, the LLRs at the input of the soft Viterbi decoder can be expressed as $\tilde{\lambda} = \{\tilde{\lambda}_l^{(0)}\}_{l=1}^{2N_b}$.

In all of the (H)ARQ strategies, after soft Viterbi decoding, an error detection process (e.g., CRC code) is performed and the corresponding ACK/NACK message is fed back to the ARQ controller. Given the short length of the ACK/NACK messages and the use of a high degree of FEC protection, an error-free and instantaneous ARQ feedback channel will be assumed in this chapter.

Characterizing the Packet Error Rate

When the Type I hybrid FEC/ARQ scheme is adopted, the PHY layer instantaneous Packet Error Rate (PER) at the output of the soft Viterbi decoder will depend on the selected TM and on the instantaneous received SNR γ. Thus, the instantaneous PER can be obtained as the ratio between the erroneously transmitted packets using TM n and the overall number of transmitted packets. When using an HARQ scheme, as all the (re)transmissions are utilized to decode the packet, the instantaneous PER will also depend on the (re)transmission number. Consequently, the instantaneous PER can be obtained as the ratio between the erroneously transmitted packets after (re)transmission number i using TM n and the overall number of transmitted packets (obviously, only those packets that have been erroneously transmitted in (re)transmission numbers $0,\ldots,i-1$ will also be transmitted in the ith retransmission). The instantaneous PER curves obtained through Monte-Carlo simulation with a packet length of $N_b = 1080$ bits are shown in Figure 6.

With the assumption of a slow block-fading channel model, the instantaneous packet error rate (PER) at the output of the soft Viterbi decoder when using TM n and Type I hybrid FEC/ARQ scheme, for all the (re)transmissions of a packet, can be approximated as:

$$\mathrm{PER}_n(\gamma) \approx \begin{cases} 1 & ,0 \le \gamma < \gamma_{pn,0} \\ a_{n,0} e^{-g_{n,0}\gamma} & ,\gamma \ge \gamma_{pn,0} \end{cases} \tag{7}$$

Figure 6. Instantaneous PER fitting curves

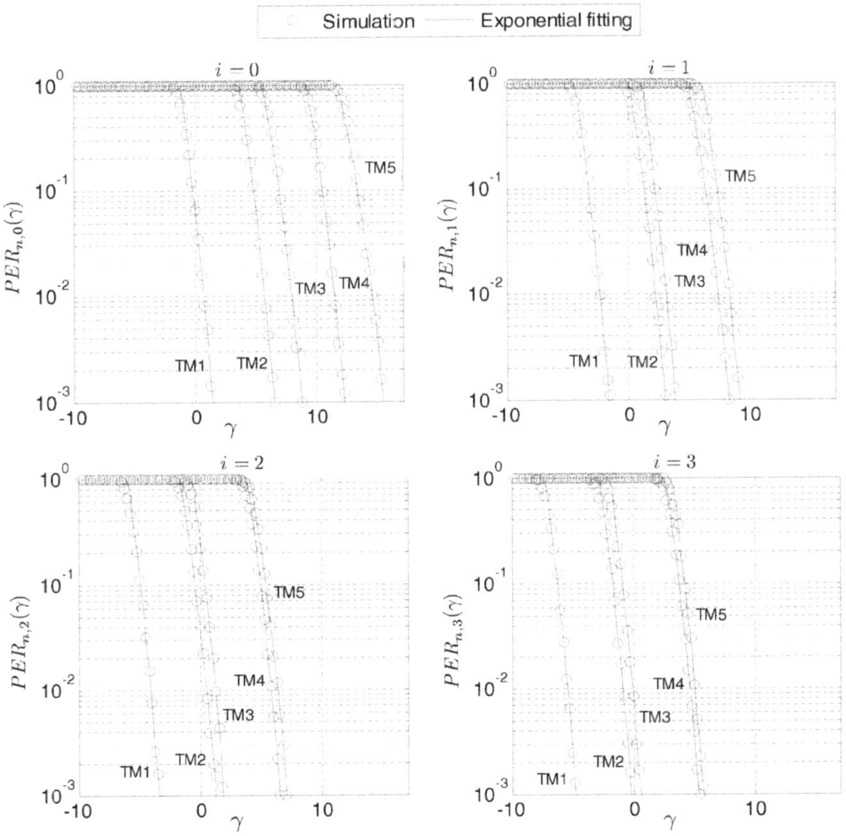

When using more sophisticated HARQ protocols, the instantaneous PER corresponding to the n th TM over the i th (re)transmission can be safely approximated as:

$$\mathrm{PER}_{n,i}(\gamma) \approx \begin{cases} 1 & ,0 \le \gamma < \dfrac{\gamma_{pn,0}}{i+1} \\ a_{n,0} e^{-g_{n,0}(i+1)\gamma} & ,\gamma \ge \dfrac{\gamma_{pn,0}}{i+1} \end{cases} \tag{8}$$

for the HARQ-CC strategy, and:

$$\mathrm{PER}_{n,i}(\gamma) \approx \begin{cases} 1 & ,0 \le \gamma < \gamma_{pn,i} \\ a_{n,i} e^{-g_{n,i}\gamma} & ,\gamma \ge \gamma_{pn,i} \end{cases} \tag{9}$$

for the HARQ-IR strategy. The parameters $a_{n,i}$, $g_{n,i}$ and $\gamma_{pn,i}$, listed in Table 3, are the fitting parameters for TM n and (re)transmission number i, which have been obtained by least-squares fitting the above approximate expressions for the PER to the curves obtained through Monte-Carlo simulation. It is clear from Figure 6 that both curves are close to each other, which justifies the utilization of the exponential expression as an approximation of the instantaneous PER.

Figure 7 depicts the PER curves corresponding to transmission modes TM1, TM3 and TM5 for (re)transmissions $i = \{0,1,2,3\}$ when both, HARQ-CC and HARQ-IR strategies are considered. We can observe that the curves are the same for both techniques in the first transmission, whereas a better performance is obtained with the HARQ-IR strategy for successive retransmissions. It can be clearly appreciated that the enhancement in the achieved coding gain when the HARQ-IR technique is selected increases with higher order TMs. The reason for this behaviour is that, traditionally, the effects of the HARQ-CC scheme are modeled as allowing the effective SNRs of the involved packets to be added together. The advantage of the HARQ-IR scheme can then be modeled as providing additional coding gains on top of the combined SNR (Cheng, 2006). Thus, as lower order TMs are more robust, they do not benefit from the additional coding gains provided by the HARQ-IR scheme.

Table 3. PER fitting parameters for (re)transmissions $i \in \{0,\ldots,3\}$

	Mode 1	Mode 2	Mode 3	Mode 4	Mode 5
$a_{n,0}$	4447.4	2068.5	514.7	850.9	142.9
$a_{n,1}$	2298.6	1344.3	3297.8	372.3	895.7
$a_{n,2}$	5944.9	1428.4	7247.9	3567.3	1057.1
$a_{n,3}$	5180.5	4374.3	8498.3	2335.5	5017.6
$g_{n,0}$	11.104	3.315	1.759	0.816	0.339
$g_{n,1}$	21.012	6.997	6.196	1.895	1.657
$g_{n,2}$	34.203	10.569	10.224	3.378	2.706
$g_{n,3}$	45.627	15.744	13.852	4.228	4.089
$\gamma_{pn,0}$ dB	-1.212	3.623	5.502	9.172	11.660
$\gamma_{pn,1}$ dB	-4.337	0.127	1.164	4.946	6.131
$\gamma_{pn,2}$ dB	-5.950	-1.629	-0.608	3.841	4.105
$\gamma_{pn,3}$ dB	-7.271	-2.737	-1.850	2.635	3.189

Figure 7. Instantaneous PER curves comparison for HARQ-CC and HARQ-IR

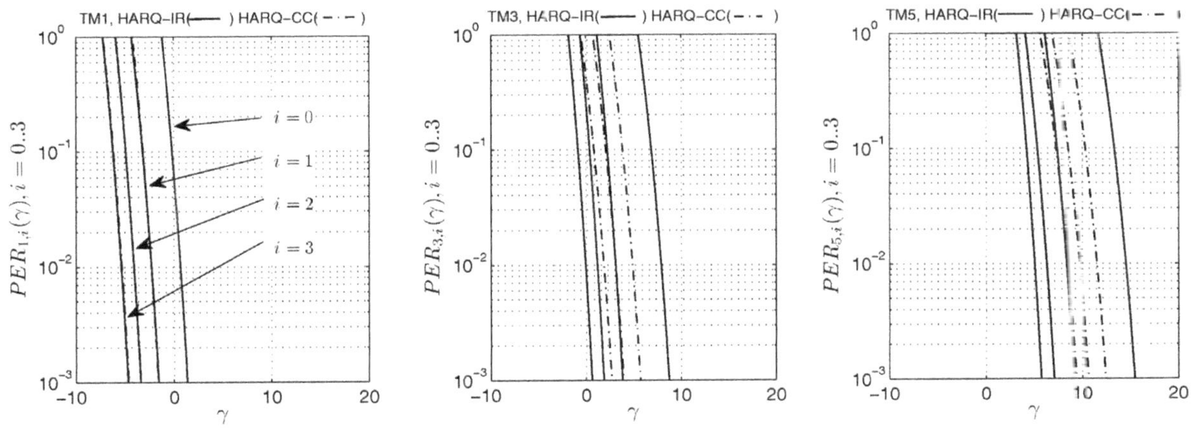

DISCRETE TIME MARKOV CHAIN MODEL AND ANALYSIS

Markov Chain-Based Modeling

Typically, PHY-layer channel processes are expressed as general correlated stochastic processes. Moreover, DLC-layer processes, such as queue arrival and departure, are described using Markovian models (Alfa, 2010). Hence, many cross-layer analysis techniques use Markov models to accurately describe the system dynamics and to simplify analysis. A key advantage of the Markov model-based system description is that it enables formulation and solution of the optimal cross-layer problem with standard optimization techniques.

In this section, we will build up an analytical link level queueing model for the previously described QoS-constrained AMC/ARQ-based wireless system. This model will be obtained based on the use of Discrete Time Markov Chains (DTMCs), through a comprehensive consideration of the following three pillars:

- the packet arrival process (blue rectangle in Figure 8);
- the PHY-layer, that will be modeled taking into account the AMC scheme and the wireless channel statistics (orange rectangle in Figure 8);

Figure 8. Markov chain-based system model approach

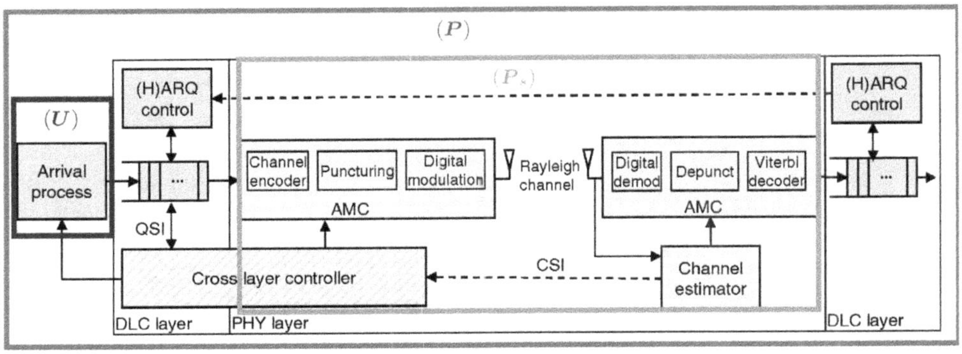

- the queueing process induced by both, the AMC scheme and the (H)ARQ protocol (green rectangle in Figure 8).

Arrival Process

The packet generation model adheres to a special case of the Discrete Batch Markovian Arrival Process (D-BMAP) (Blondia, 1993), which is characterized by a transition probability matrix:

$$\mathbf{U} = \sum_{a=0}^{\infty} \mathbf{U}_a = \begin{bmatrix} u(0,0) & \cdots & u(\mathcal{A}-1,0) \\ \vdots & & \vdots \\ u(0,\mathcal{A}-1) & \cdots & u(\mathcal{A}-1,\mathcal{A}-1) \end{bmatrix}, \quad (10)$$

where $u(a_\mu, a_{\mu'})$ denotes the probability of a transition from phase a_μ to phase $a_{\mu'}$ with a batch arrival of size a_μ packets, as illustrated in Figure 9. The sub-stochastic matrices $\mathbf{U}_a$, for all $a \in \{0,\ldots,\mathcal{A}-1\}$ are constructed by keeping only the $(a+1)$th row of $\mathbf{U}$ and setting all other rows to zero, and by definition of the D-BMAP, $\mathbf{U}_a = \mathbf{0}$ for all $a \geq \mathcal{A}$.

Owing to the Markovian property of the arrival process we have that $\boldsymbol{\omega} = \boldsymbol{\omega}\mathbf{U}$ and $\boldsymbol{\omega}\mathbf{1}_{\mathcal{A}} = 1$, where $\boldsymbol{\omega}$ denotes the D-BMAP steady-phase probability vector and $\mathbf{1}_{\mathcal{A}}$ is an all-ones column vector of length $\mathcal{A}$. Then the average arrival rate λ can be calculated as:

$$\lambda = \boldsymbol{\omega} \sum_{a=0}^{\mathcal{A}-1} a \, \mathbf{U}_a \mathbf{1}_{\mathcal{A}}. \quad (11)$$

Physical Layer Two-Dimensional Markov Model

The procedure for the development of the proposed Markov chain-based PHY layer model will take into consideration, on the one hand, the modeling of the channel, that will reproduce the amplitude and the rate-of-change of the fading envelope, and on the other hand, the implementation of the AMC scheme.

Two-Dimensional Markov Channel Modeling

Starting from early works of Gilbert (1960) and Elliot (1963), an effective representation of digital communication channels with memory is obtained by using finite state Markov chains (see, e.g., Sadeghi, et al., 2008 and references therein). An FSMC is a discrete valued time-varying channel, whose variation is determined by a finite-state Markov chain: each state corresponds to a known memory-less channel (Kanal & Sastry, 1978). For

Figure 9. Transition state diagram of the discrete batch Markovian arrival process

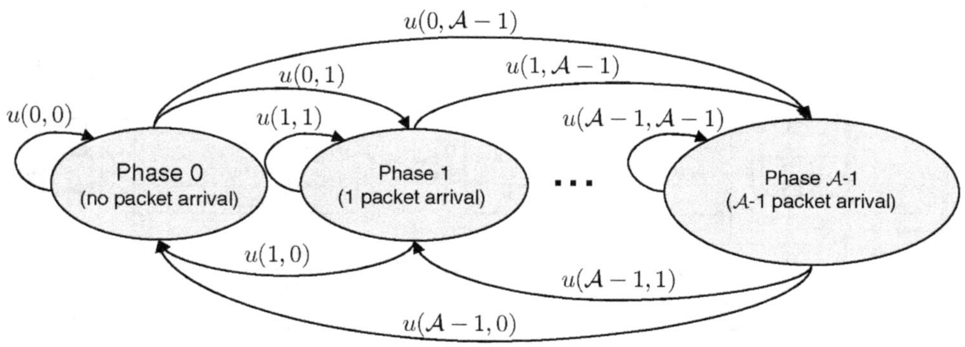

narrow band systems in flat fading with Additive White Gaussian Noise (AWGN), the SNR at the receiver is proportional to channel attenuation (fading). Fading can be discretized by sampling and quantization, wherein one fading state is associated to each interval of attenuation values (Babich, et al., 2000). Thus, each state corresponds to one of these discretized levels of the channel quality.

Let us consider the Rayleigh block-fading channel quantities γ_ν and $\delta_\nu = \gamma_{\nu-1} - \gamma_\nu$, representing the instantaneous SNR and rate of change of the SNR. Let us also partition the ranges of γ_ν and δ_ν into sets of nonoverlapping two-dimensional cells defined by the partitions:

$$\mathbf{\Gamma}^c = \left\{\left[\gamma_0^c, \gamma_1^c\right), \left[\gamma_1^c, \gamma_2^c\right), \ldots, \left[\gamma_{K-1}^c, \gamma_K^c\right)\right\}$$

with $\gamma_0^c = 0$ and $\gamma_K^c = \infty$,

and $\mathbf{\Delta} = \left\{(-\infty, 0), [0, \infty)\right\}$, respectively.

This pair of values represents the generic channel state. Thus, a first-order two-dimensional Markov channel model can be defined where each two-dimensional state is defined by the amplitude and the rate of change of the instantaneous SNR at the output of the fading channel. Each state of the channel corresponds to one of such cells, that is, the Markov chain state of the channel at time instant $t = \nu T_f$ can be denoted as $\zeta_\nu = \left(\chi_\nu, \Delta_\nu\right)$, $\nu = 0, 1, \ldots, \infty$, where $\chi_\nu = k$ if and only if $\gamma_k^c \leq \gamma_\nu < \gamma_{k+1}^c$ and $\Delta_\nu = 0$ (or $\Delta_\nu = 1$) if and only if $\delta_\nu < 0$ (or $\delta_\nu \geq 0$).

In our approach the partition $\mathbf{\Gamma}^c$ is designed assuming that the observable dummy output of our improved first-order two-dimensional Markov model at time instant $t = \nu T_f$ belongs to a codebook of nominal values of SNR $\mathbf{\Psi}^c = \left\{\Psi_1^c, \Psi_2^c, \ldots, \Psi_K^c\right\}$. The Max-Lloyd algorithm (Max, 1960; Lloyd, 1982), developed for the optimum design of non-uniform quantizers, is then used to determine the optimum partition and (dummy) codebook in the sense that minimize the mean square error between γ_ν and the (dummy) quantizer output.

Two-Dimensional Markov PHY-Layer Modeling

Keeping in mind both the TM selection process used by the AMC scheme, defined by the partition $\mathbf{\Gamma}^m$, and the first-order two dimensional Markov channel model, characterized by the partitions $\mathbf{\Gamma}^c$ and $\mathbf{\Delta}$, let us now partition the range of γ_ν into the set of non-overlapping intervals defined by the partition: $\mathbf{\Gamma}^{m,c} = \{[\gamma_0^{m,c}, \gamma_1^{m,c}), [\gamma_1^{m,c}, \gamma_2^{m,c}), \ldots, [\gamma_{N_{\text{PHY}}-1}^{m,c}, \gamma_{N_{\text{PHY}}}^{m,c})\}$, where $\gamma_0^{m,c} = 0$, $\gamma_{N_{\text{PHY}}}^{m,c} = \infty$, $\{\gamma_1^{m,c}, \ldots, \gamma_{N_{\text{PHY}}-1}^{m,c}\} = \text{sort}\left(\{\gamma_1^m, \ldots, \gamma_{M-1}^m\} \cup \{\gamma_1^c, \ldots, \gamma_{K-1}^c\}\right)$, and each partition interval $[\gamma_k^{m,c}, \gamma_{k+1}^{m,c})$ is characterized by a particular combination of TM and channel state. Figure 10 illustrates this process and shows how the partition $\mathbf{\Gamma}^{m,c}$ is obtained. As in Subsection 'Two-dimensional Markov channel modeling,' let us also consider the partition of δ_ν into the set of non-overlapping intervals $\mathbf{\Delta} = \left\{(-\infty, 0), [0, \infty)\right\}$. Using this two-dimensional partitioning, a first-order two-dimensional Markov model for the physical layer can be defined where each state corresponds to one of such two-dimensional rectangular-shaped cells. Furthermore, the physical layer Markov chain state at time instant $t = \nu T_f$ can be denoted as $\varsigma_\nu = \left(\varphi_\nu, \Delta_\nu\right)$, $\nu = 0, 1, \ldots, \infty$, where $\varphi_\nu \in \{0, \ldots, N_{\text{PHY}} - 1\}$ denotes the combination of TM and channel state in this frame interval and $\Delta_\nu \in \{0, 1\}$ is used to denote the *up* or *down* characteristic of the instantaneous SNR in time frame interval $t = (\nu - 1)T_f$ (if $\gamma_\nu < \gamma_{\nu-1}$ then the instantaneous SNR is descending and it can

Figure 10. PHY layer model

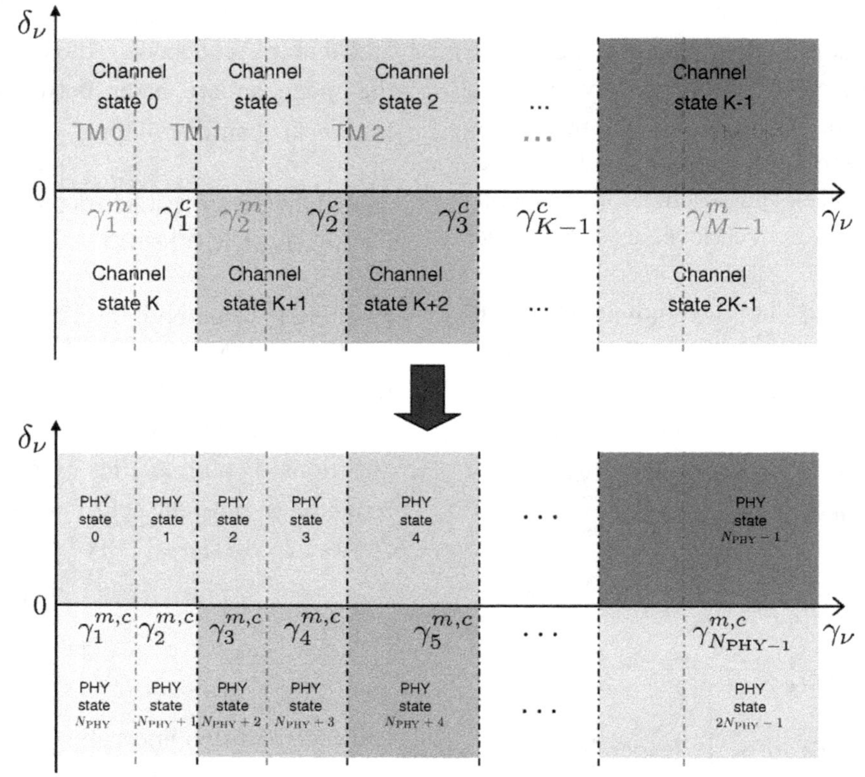

be tagged as *down*; on the contrary, if $\gamma_\nu \geq \gamma_{\nu-1}$ then the instantaneous SNR is ascending and it can be tagged as *up*). Equivalently, at any time instant $t = \nu T_f$ the physical-layer state can be univocally characterized by an integer number $n_\nu = 2\varphi_\nu + \Delta_\nu$ and obviously, $n_\nu \in \{0,\ldots,2N_{\text{PHY}}-1\}$. The PHY layer will be in a state $n \in \{0,\ldots,2N_{\text{PHY}}-1\}$ with a steady state probability $P^{\text{PHY}}(n)$ that can be calculated using (Femenias, et al., 2009, eqs. [8]-[9]).

Average PER Calculation

When the Type I hybrid FEC/ARQ scheme is used, each of these states will be characterized by an average PER that can be calculated as:

$$\overline{PER}_n^{\text{PHY}} = \begin{cases} \dfrac{\int_{\gamma_{n/2}^{m,c}}^{\gamma_{n/2+1}^{m,c}} \int_0^x PER_{\beta_n}(x) p_{\gamma_\nu,\gamma_{\nu-1}}\left(x,y\right) dy dx}{P^{\text{PHY}}(n)}, & \text{n even} \\ \dfrac{\int_{\gamma_{(n-1)/2}^{m,c}}^{\gamma_{(n+1)/2}^{m,c}} \int_x^{+\infty} PER_{\beta_n}(x) p_{\gamma_\nu,\gamma_{\nu-1}}\left(x,y\right) dy dx}{P^{\text{PHY}}(n)}, & \text{n odd} \end{cases} \quad (12)$$

that corresponds to the averaging of PER_{β_n}, which denotes the instantaneous PER of TM β_n corresponding to the nth PHY layer state, over the joint pdf of the random variables γ_ν and $\gamma_{\nu-1}$, denoted as $p_{\gamma_\nu,\gamma_{\nu-1}}\left(x,y\right)$. Note that the integration limits of γ_ν are the boundary thresholds of the nth PHY layer state. For an even value of n_ν, the integration interval of $\gamma_{\nu-1}$ matches all the values below γ_ν, which correspond to an ascending instantaneous SNR. On the contrary, for an odd value of n_ν, the integration interval of $\gamma_{\nu-1}$ matches all the values above γ_ν, which correspond to a descending instantaneous SNR.

When using more sophisticated HARQ protocols, these physical layer states will be characterized by a conditional average PER for the ith (re)transmission given by:

$$\overline{PER}_{n,i}^{\mathrm{PHY}} = \begin{cases} \dfrac{\int_{\gamma_{n/2}^{m,c}}^{\gamma_{n/2+1}^{m,c}} \int_0^x PER_{\beta_n,i|i-1}(x) p_{\gamma_\nu,\gamma_{\nu-1}}(x,y)\, dy dx}{P^{\mathrm{PHY}}(n)}, & \text{n even} \\ \dfrac{\int_{\gamma_{(n-1)/2}^{m,c}}^{\gamma_{(n+1)/2}^{m,c}} \int_x^{+\infty} PER_{\beta_n,i|i-1}(x) p_{\gamma_\nu,\gamma_{\nu-1}}(x,y)\, dy dx}{P^{\mathrm{PHY}}(n)}, & \text{n odd} \end{cases} \quad (13)$$

where $PER_{n,i|i-1}(\cdot)$ represents the probability that the ith (re)transmission attempt fails conditioned on that the previous transmissions have been erroneous, that is:

$$PER_{n,i|i-1}(\gamma) = \frac{PER_{n,i}(\gamma)}{PER_{n,i-1}(\gamma)}. \quad (14)$$

The Transition Probability Matrix

The physical-layer FSMC will be characterized by a transition probability matrix:

$$\mathbf{P}_s = \left[P_s(i,j)\right]_{i,j=0}^{2N_{\mathrm{PHY}}-1}, \quad (15)$$

whose elements $P_s(i,j)$ denote the probability of a transition from PHY layer state i to PHY layer state j, with $i,j \in \{0,\ldots,2N_{\mathrm{PHY}}-1\}$. These probabilities can be calculated following the procedure detailed in Femenias et al. (2009), eqs. (11)-(15).

In this paper, the steady-state probabilities, the conditional average packet error rates and the state-transition probabilities have all been computed either numerically or by simulation. Clarke's statistical Rayleigh fading process, characterized by a maximum normalized Doppler frequency $f_d T_f$, has been used to model the wireless flat-fading channel (Clarke, 1968).

Embedded Markov Chain

Once the PHY layer has been modeled, the DLC layer must also be introduced in the analytical model. The queueing process induced by both the AMC scheme and the (H)ARQ protocol can be formulated in discrete time with one time unit equal to one frame interval. The system states are observed at the beginning of each time unit.

Infinitely Persistent ARQ

Following the approach described in Le et al. (2006), let $\sigma_\nu = (q_\nu, a_\nu, n_\nu)$ denote the system state at time instant $t = \nu T_f$, where $q_\nu \in \{0,\ldots,\bar{Q}\}$ denotes the queue length, $a_\nu \in \{0,\ldots,\mathcal{A}-1\}$ represents the phase of the D-BMAP and $n_\nu \in \{0,\ldots,2N_{\mathrm{PHY}}-1\}$ represents the physical layer state. If we just look at the set of time instants $t = \nu T_f$, $\nu = 0,1,\ldots,\infty$, the transitions between states σ_ν are Markovian. Therefore, an embedded Markov chain can be used to describe the underlying queuing process. The state space of this embedded finite state Markov chain is $\mathcal{S} = \{\mathcal{S}_\mu\}_{\mu=1}^{N_s}$, where $N_s = 2N_{\mathrm{PHY}}\mathcal{A}(\bar{Q}+1)$ and $\mathcal{S}_\mu = (q_\mu, a_\mu, n_\mu)$.

Figure 11 illustrates the queue transition process between states (q_μ, a_μ, n_μ) and $(q_{\mu'}, a_{\mu'}, n_{\mu'})$, corresponding to two consecutive frame intervals. At the beginning of the first frame interval there are q_μ packets in the queue and, given that the number of packets that can be transmitted when the PHY-layer is in state n_μ is $\mathcal{C}_{n_\mu}$, the number of packets that will actually be transmitted is $\min\left\{\mathcal{C}_{n_\mu}, q_\mu\right\}$ (blue arrow). Among them, let us assume that h packets are correctly transmitted and leave the queue (green arrow), while $\min\left\{\mathcal{C}_{n_\mu}, q_\mu\right\} - h$ packets are received in error and, consequently, are kept in the queue (orange arrow). Obviously, the $q_\mu - \min\left\{\mathcal{C}_{n_\mu}, q_\mu\right\}$ packets that are not transmitted also remain in the queue (camel arrow). Thus, the number of packets in the queue at the end of this frame interval will correspond to $q_\mu - h$ packets plus the arriving packets (yellow arrow) that can be lodged in the queue. If there is no room for all the a_μ incoming packets, a buffer overflow will occur (red arrow). Consequently, the number of packets in the buffer at the end of this frame interval will be equal to $\min\left\{\overline{Q}, q_\mu + a_\mu - h\right\}$, and the number of lost packets due to buffer overflow will be given by $\max\left\{0, q_\mu + a_\mu - h - \overline{Q}\right\}$.

Taking into account that infinitely persistent Type I hybrid FEC/ARQ is used at the data-link layer, and assuming that the transmission outcomes (success or failure) of consecutive packets in a frame interval are independent, the probability that k packets are successfully transmitted (leave the queue) given $\mathcal{C}_{n_\mu}$ packets are transmitted when the physical layer is in state n_μ can be written as:

$$p_{k,\mathcal{C}_{n_\mu}}^{(n_\mu)} = \binom{\mathcal{C}_{n_\mu}}{k}\left(\overline{PER}_{n_\mu}^{\text{PHY}}\right)^{\mathcal{C}_{n_\mu}-k}\left(1-\overline{PER}_{n_\mu}^{\text{PHY}}\right)^{k}. \tag{16}$$

Considering that there are $q_\mu = q$ packets in the queue before transmission and that the PHY-layer is in state n_μ, then $\min\left\{\mathcal{C}_{n_\mu}, q\right\}$ packets will be transmitted (blue arrow). Thus, in this situation, the probability that h packets are successfully transmitted (green arrow) and the physical layer changes from state n_μ to state $n_{\mu'}$, can be expressed as:

Figure 11. Queue transition for an infinitely persistent ARQ scheme

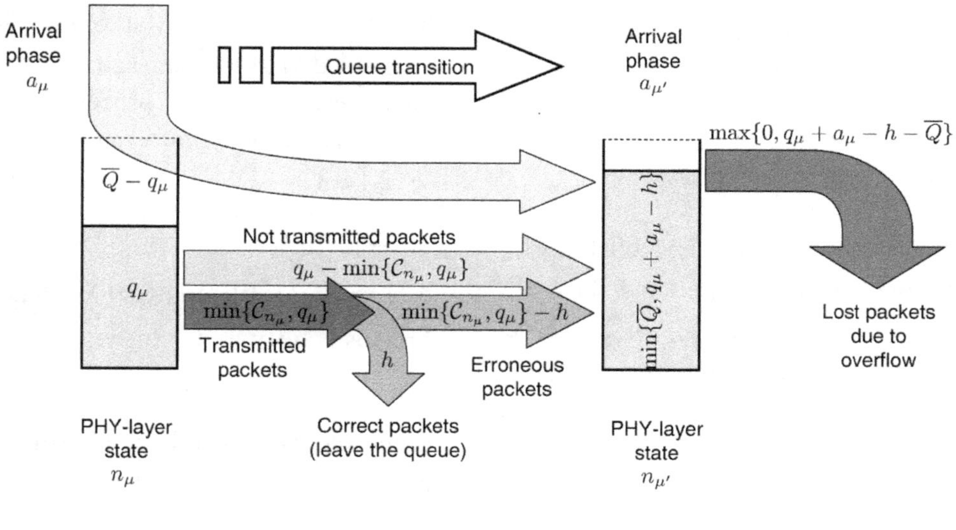

$$t_h^{(q)}\left(n_\mu, n_{\mu'}\right) = p_{h,\min\left\{\mathcal{C}_{n_\mu}, q\right\}}^{(n_\mu)} P_s(n_\mu, n_{\mu'}) \quad (17)$$

for $q \in \{0, \ldots, \bar{Q}\}$, $h \in \{0, \ldots, \min\{q, \mathcal{C}\}\}$ and $n_\mu, n_{\mu'} \in \left\{0, \ldots, 2N_{\text{PHY}} - 1\right\}$.

The $2N_{\text{PHY}} \times 2N_{\text{PHY}}$ terms in (17), capturing all the cases where h packets are successfully transmitted given that there are q packets in the queue before transmission, can be expressed in matrix form as:

$$\mathbf{T}_h^{(q)} = \mathcal{D}\left(\mathbf{p}_h^q\right) \mathbf{P}_s \quad (18)$$

for $q \in \{0, \ldots, \bar{Q}\}$ and $h \in \{0, \ldots, \min\{q, \mathcal{C}\}\}$, where $\mathbf{p}_h^q = \left(p_{h,\min\{\mathcal{C}_0, q\}}^{(0)} \cdots p_{h,\min\{\mathcal{C}_{2N_{\text{PHY}}-1}, q\}}^{(2N_{\text{PHY}}-1)}\right)$ and $\mathcal{D}(\mathbf{x})$ is used to denote a diagonal matrix with the elements of the vector $\mathbf{x}$ in its main diagonal. Notice that for $q \geq \mathcal{C}_{2N_{\text{PHY}}-1} = \mathcal{C}$, the number of transmitted packets is always $\mathcal{C}$ and, consequently, the probabilities in these matrices do not depend on q and $\mathbf{T}_h^{(q)} = \mathbf{T}_h^{(\mathcal{C})}$.

As shown in Figure 12, let $q_\mu = q$ and $q_{\mu'} = q + \mathcal{A} - l - 1$ be the number of packets in the queue in two consecutive frames. The maximum number of packet arrivals in one frame interval is equal to $\mathcal{A} - 1$, and the maximum number of successfully transmitted packets in one frame interval is equal to $\min\{q, \mathcal{C}\}$; thus, it is quite obvious that

$$q_\mu - \min\{q, \mathcal{C}\} \leq q_{\mu'} \leq q_\mu + \mathcal{A} - 1$$

or, equivalently,

$$0 \leq l \leq \mathcal{A} + \min\{q, \mathcal{C}\} - 1.$$

Furthermore, let $a_\mu = a$ and h be the number of arriving packets and the number of packets successfully transmitted in the first one of these

Figure 12. Arrival and departure of packets

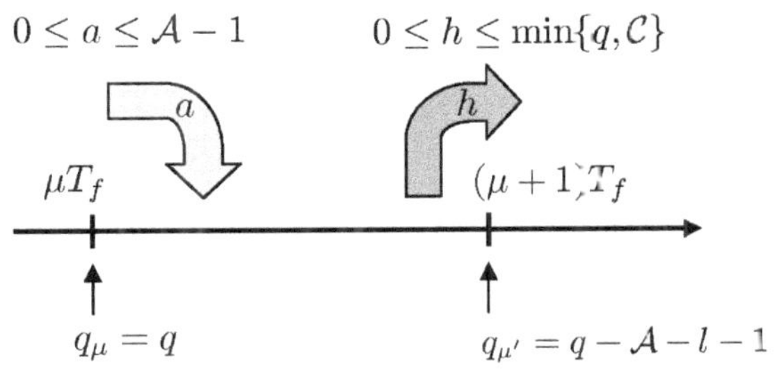

frame intervals, respectively (yellow and green arrows). In this case, $q_{\mu'} - q_\mu = \mathcal{A} - l - 1 = a - h$ or $h = l - \mathcal{A} + a + 1$. Thus, given that $0 \leq h = l - \mathcal{A} + a + 1 \leq \min\{q, \mathcal{C}\}$ or, equivalently, $\mathcal{A} - l - 1 \leq a \leq \mathcal{A} - l + \min\{q, \mathcal{C}\} - 1$, and $0 \leq a \leq \mathcal{A} - 1$, the probability that the queuing system changes from a generic state $\mathcal{S}_\mu = (q, a_\mu, n_\mu) \in \mathcal{S}$ to another generic state $\mathcal{S}_{\mu'} = (q + \mathcal{A} - l - 1, a_{\mu'}, n_{\mu'}) \in \mathcal{S}$ can be expressed as:

$$A_{q,l}\left(\mathcal{S}_\mu, \mathcal{S}_{\mu'}\right) = \sum_{a=a_{\min}}^{a_{\max}} u_a(a_\mu, a_{\mu'})\, t_{l-\mathcal{A}+a+1}^{(q)}\left(n_\mu, n_{\mu'}\right) \quad (19)$$

where, from the above mentioned inequalities, we obtain $a_{\min} = \max\{0, \mathcal{A} - l - 1\}$ and $a_{\max} = \min\{\mathcal{A} - 1, \mathcal{A} + \min\{q, \mathcal{C}\} - 1 - l\}$, for $q \in \{0, \ldots, \bar{Q}\}$, $l \in \{0, \ldots, \mathcal{A} + \min\{q, \mathcal{C}\} - 1\}$, $a_\mu, a_{\mu'} \in \{0, \ldots, \mathcal{A} - 1\}$

and $n_\mu, n_{\mu'} \in \left\{0, \ldots, 2N_{\text{PHY}} - 1\right\}$.

The $2\mathcal{A}N_{\text{PHY}} \times 2\mathcal{A}N_{\text{PHY}}$ terms capturing all the cases where the queue length changes from q packets in one frame interval to $q + \mathcal{A} - l - 1$ packets in the next frame interval can be expressed in matrix form as:

$$\mathbf{A}_{q,l} = \begin{cases} \sum_{a=a_{\min}}^{a_{\max}} \mathbf{U}_a \otimes \mathbf{T}_{l-\mathcal{A}+a+1}^{(q)} & , l \in \{0,\ldots,l_{\max}(q)\} \\ \mathbf{0} & , \text{otherwise} \end{cases} \tag{20}$$

for $q \in \{0,\ldots,\bar{Q}\}$, where $\otimes$ denotes the Kronecker product and $l_{\max}(q) = \mathcal{A} + \min\{q, \mathcal{C}\} - 1$. The resulting transition matrix of the Markov chain can then be written as:

$$\mathbf{P} = \begin{bmatrix} \mathbf{A}_{0,\mathcal{A}-1} & \cdots & \mathbf{A}_{0,\mathcal{A}-\bar{Q}} & \bar{\mathbf{A}}_{0,\mathcal{A}-\bar{Q}-1} \\ \vdots & & \vdots & \vdots \\ \mathbf{A}_{\bar{Q},\mathcal{A}+\bar{Q}-1} & \cdots & \mathbf{A}_{\bar{Q},\mathcal{A}} & \bar{\mathbf{A}}_{\bar{Q},\mathcal{A}-1} \end{bmatrix}, \tag{21}$$

where:

$$\bar{\mathbf{A}}_{q,i} = \sum_{a=0}^{i} \mathbf{A}_{q,a}. \tag{22}$$

$\pi\mathbf{P} = \pi$ Notice that, for $q \geq \mathcal{C}$, the transition probabilities in these matrix blocks do not depend on q and, therefore, for simplicity, this index can be omitted, that is, $\mathbf{A}_{q,l} = \mathbf{A}_l$ and $\bar{\mathbf{A}}_{q,l} = \bar{\mathbf{A}}_l$ for all $q \geq \mathcal{C}$.

To derive the system performance measures, we need to obtain the steady-state probability vectors corresponding to each level of the transition matrix, which can be calculated using the fact that the transition probability matrix $\mathbf{P}$ and steady-state probability vector

$$\boldsymbol{\pi} = \begin{bmatrix} \boldsymbol{\pi}_0 & \boldsymbol{\pi}_1 & \cdots & \boldsymbol{\pi}_{\bar{Q}} \end{bmatrix}$$

satisfy $\boldsymbol{\pi}\mathbf{P} = \boldsymbol{\pi}$ along with the normalization condition $\sum_{i=0}^{\bar{Q}} \boldsymbol{\pi}_i \mathbf{1} = 1$, where $\mathbf{1}$ is a column vector of all ones with the appropriate length.

Packet loss rate and throughput: In our finite buffering ARQ-based error control system with infinite persistence, the packet loss rate P_l (measured in packets per second) is simply equal to the buffer overflow probability $P_{l_{BO}}$. As in Le et al. (2006), we denote by $\mathbf{V}_k$ the stationary vector describing the probabilities that k packets are lost due to buffer overflow upon arrival of a burst of data packets. Assuming that a batch of a packets arrive at the link layer buffer (yellow arrow), if there are $q > \bar{Q} - a$ packets in the queue at the end of the previous frame interval and h packets are successfully transmitted (green arrow), then the number of packets that will be lost due to buffer overflow is $k = a - h - \bar{Q} + q$ (red arrow). Therefore, $\mathbf{V}_k$ can be written as:

$$\mathbf{V}_k = \sum_{a=1}^{\mathcal{A}-1} \sum_{q=\max\{0,\bar{Q}-a+1\}}^{\bar{Q}} \boldsymbol{\pi}_q \sum_{\substack{h=0 \\ a-h=\bar{Q}-q+k}}^{\mathcal{C}} \mathbf{U}_a \otimes \mathbf{T}_h^q. \tag{23}$$

Thus, the average number of lost packets due to buffer overflow N_l can be obtained as:

$$N_l = \sum_{k=1}^{\mathcal{A}-1} k\mathbf{V}_k \mathbf{1}. \tag{24}$$

Given the average number of arriving packets λ in one frame interval, the packet loss rate can then be calculated as:

$$P_l = N_l / \lambda = (1/\lambda)\sum_{k=1}^{\mathcal{A}-1} k\mathbf{V}_k \mathbf{1}, \tag{25}$$

and the average throughput (measured in packets per frame) can be obtained as:

$$\eta = \lambda(1 - P_l). \tag{26}$$

Average queue length and average packet delay: Due to the assumption of infinite persistence in the ARQ-based error control system, the packet blocking probability is equal to the packet loss rate. Thus, using the well known Little's formula (Kleinrock, 1975), the average delay for our embedded Markov chain can be calculated as:

$$D_l = \frac{L_q}{\lambda(1 - P_{l_{BO}})} = \frac{L_q}{\eta}, \tag{27}$$

where L_q denotes the average number of packets in the queue that can be obtained as:

$$L_q = \sum_{q=1}^{\bar{Q}} q\, \boldsymbol{\pi}_q \mathbf{1}. \tag{28}$$

Truncated (H)ARQ

Let $\sigma_\nu = (\mathbf{q}_\nu, a_\nu, n_\nu)$ denote the system state at time instant $t = \nu T_f$, where $\mathbf{q}_\nu = (q_{\nu,0}, \ldots, q_{\nu,N_r})$, with $q_{\nu,i}$ denoting the number of packets in the queue that have been already transmitted i times, and $Q_\nu \triangleq \sum_{i=0}^{N_r} q_{\nu,i} \in \{0, \ldots, \bar{Q}\}$ denoting the total number of packets in the queue. Like in the infinitely persistent scheme, the underlying queueing process can be characterized through an embedded Markov chain. The state space of this embedded finite state Markov chain is $\mathcal{S} = \{\mathcal{S}_\mu\}_{\mu=1}^{N_s}$ with size $N_s = 2N_{\text{PHY}}\mathcal{A}\sum_{k=0}^{\bar{Q}}\binom{k+N_r}{k}$.

The transition from state $\mathcal{S}_\mu = (\mathbf{q}_\mu, a_\mu, n_\mu) \in \mathcal{S}$ to state $\mathcal{S}_{\mu'} = (\mathbf{q}_{\mu'}, a_{\mu'}, n_{\mu'}) \in \mathcal{S}$ is illustrated in Figure 13. The queue state $\mathbf{q}_\mu = (q_{\mu,0}, \ldots, q_{\mu,N_r})$ contains the number of packets in the queue that have never been transmitted $q_{\mu,0}$, the number of packets in the queue that have been already transmitted one time $q_{\mu,1}$, an so on up to the number of packets in the queue that have been already transmitted N_r times q_{μ,N_r}, which is the maximum number of allowed retransmissions. Assuming that there are Q_μ packets in the queue at the beginning of the first frame interval and, given that the number of packets that can be transmitted when the PHY-layer is in state n_μ is C_{n_μ}, the number of packets that will be definitely transmitted is $\min\{C_{n_\mu}, Q_\mu\}$ (blue arrows). Transmission will start with packets that have been already transmitted N_r times. Obviously, these transmitted packets will leave the queue either due to a successful transmission (green arrow) or, in case of an erroneous transmission, due to exceeding the maximum number of allowed retransmissions (red arrow). In case that the transmission capacity, $\min\{C_{n_\mu}, Q_\mu\}$, allows the transmission of all q_{μ,N_r} packets and more packets can be also transmitted, the next selected packets for transmission will be chosen among the packets that have been already transmitted $N_r - 1$ times. Those among them that are correctly transmitted will leave the queue (green arrow), while those that are erroneously transmitted will contribute to form, in the next time interval, the set of packets that have been already transmitted N_r times (orange arrow). An analogous behaviour is observed for the packets that have been already transmitted $N_r - 2$ times, and so on until having transmitted the $\min\{C_{n_\mu}, Q_\mu\}$ packets. Obviously, the $Q_\mu - \min\{C_{n_\mu}, Q_\mu\}$ packets that are not transmitted will also remain in the queue (camel arrow). Moreover, the new arrived packets a_μ (yellow arrow) that can be stored in the buffer, will be included in the set of packets that have never been

Figure 13. Queue transition for a truncated ARQ scheme

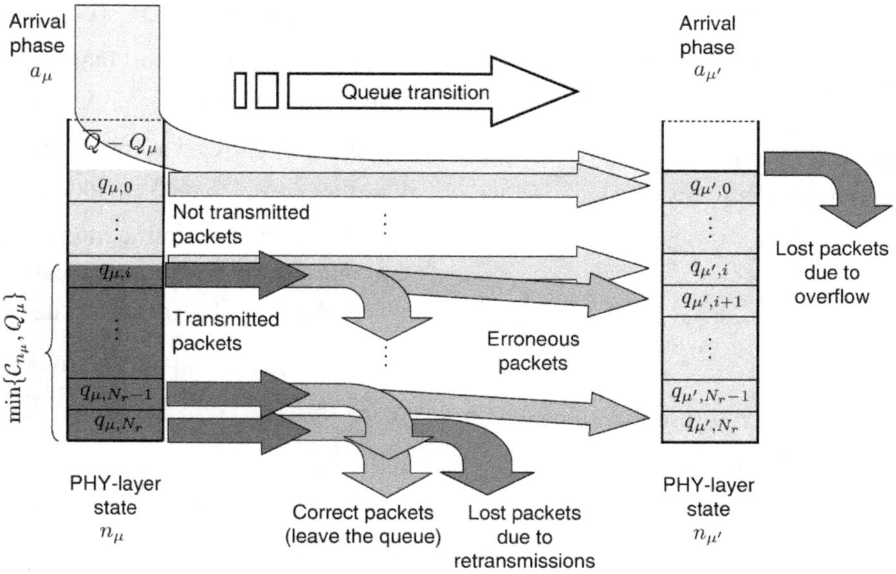

transmitted $q_{\mu',0}$. In case there is not enough room in the queue for all the new packets, those that will be dropped (pink arrow) will increase the packet loss rate due to buffer overflow.

The transition probability from state $\mathcal{S}_\mu$ to state $\mathcal{S}_{\mu'}$ can be written as:

$$P_{\mathcal{S}_\mu,\mathcal{S}_{\mu'}} = u\left(a_\mu,a_{\mu'}\right)P_s\left(n_\mu,n_{\mu'}\right)P_{\mathbf{q}_\mu,\mathbf{q}_{\mu'}|a_\mu,n_\mu}, \tag{29}$$

where $u\left(a_\mu,a_{\mu'}\right)$ denotes the transition probability between D-BMAP phases a_μ and $a_{\mu'}$, which can be obtained from matrix $\mathbf{U}$, $P_s\left(n_\mu,n_{\mu'}\right)$ is the physical layer transition probability between states n_μ and $n_{\mu'}$, which can be derived from matrix $\mathbf{P}_s$, and $P_{\mathbf{q}_\mu,\mathbf{q}_{\mu'}|a_\mu,n_\mu} = \prod_{i=0}^{N_r} P_{q_{\mu,i},q_{\mu',i}|a_\mu,n_\mu}$ corresponds to the queue transition probability from state $\mathbf{q}_\mu$ to state $\mathbf{q}_{\mu'}$ when the D-BMAP is in phase a_μ and the physical layer state is n_μ.

Let us consider that the system state is $\mathcal{S}_\mu$ and that $Q_{\mu,i} = \sum_{l=i}^{N_r} q_{\mu,l}$ represents the number of packets in the queue that have already been transmitted i or more times (obviously, $Q_{\mu,0} = Q_\mu$). Additionally, let $\tau_\mu = \min\left\{Q_\mu,\mathcal{C}_{n_\mu}\right\}$ denote the number of transmitted packets (blue arrows), $\tau_{\mu,i} = \min\left\{q_{\mu,i},\tau_\mu - Q_{\mu,i+1}\right\}$ the number of transmitted packets among those in the queue that have been already transmitted i times and $\epsilon_{\mu,i}$ the number of packets erroneously transmitted among those in the queue that have been already transmitted i times (orange and red arrows). Let us also define $\mathcal{P}_y^x(z) \triangleq \binom{x}{y} z^y(1-z)^{x-y}$ and assume a slow block-fading channel. Then, the queue transition probabilities, given the arrival phase is a_μ (yellow arrow) and the PHY-layer state is n_μ, can be safely approximated as:

$$P_{q_{\mu,N_r},q_{\mu',N_r}|a_\mu,n_\mu} = \begin{cases} 1 & , q_{\mu',N_r} = q_{\mu,N_r} - \tau_\mu, \ \tau_\mu \le q_{\mu,N_r} \\ \mathcal{P}_{q_{\mu',N_r}}^{\tau_{\mu,N_r-1}}\left(\overline{PER}_{n_\mu,N_r-1}^{\text{PHY}}\right) & , \tau_\mu > q_{\mu,N_r}, \ \tau_{\mu,N_r-1} \ge q_{\mu',N_r} \\ 0 & , \text{otherwise} \end{cases} \tag{30}$$

$$P_{q_{\mu,i},q_{\mu',i}|a_\mu,n_\mu} = \begin{cases} 1 & , q_{\mu,i} = q_{\mu',i}, \ \tau_\mu \le Q_{\mu,i+1} \\ 1 & , Q_{\mu,i+1} < \tau_\mu \le Q_{\mu,i}, \ q_{\mu',i} = Q_{\mu,i} - \tau_\mu \\ \mathcal{P}_{q_{\mu',i}}^{\tau_{\mu,i-1}}\left(\overline{PER}_{n_\mu,i-1}^{\text{PHY}}\right) & , \tau_\mu > Q_{\mu,i}, \ \tau_{\mu,i-1} \ge q_{\mu',i} \\ 0 & , \text{otherwise} \end{cases} \tag{31}$$

for $i \in \{1,\ldots,N_r - 1\}$, and:

$$P_{q_{\mu,0},q_{\mu',0}|a_\mu,n_\mu} = \begin{cases} 1 & , \tau_\mu \le Q_{\mu,1}, \ q_{\mu',0} = \min\left\{\bar{Q} - Q_{\mu',1}, q_{\mu,0} + a_\mu\right\} \\ 1 & , Q_{\mu,1} < \tau_\mu \le Q_\mu, \ q_{\mu',0} = \min\left\{\bar{Q} - Q_{\mu',1}, Q_\mu + a_\mu - \tau_\mu\right\} \\ 0 & , \text{otherwise.} \end{cases} \tag{32}$$

Notice that for the Type I hybrid FEC/ARQ strategy, as the conditional average PER only depends on the physical layer state n_μ, then $\overline{PER}_{n_\mu,i}^{\text{PHY}} = \overline{PER}_{n_\mu}^{\text{PHY}}$ for all $i \in \{0,\ldots,N_r\}$.

As in the infinitely persistent case, the transition probability matrix $\mathbf{P}$ and steady-state probability vector $\boldsymbol{\pi} = \left[\pi_{\mathcal{S}_1} \cdots \pi_{\mathcal{S}_{N_s}}\right]$ satisfy $\boldsymbol{\pi}\mathbf{P} = \boldsymbol{\pi}$ along with the normalization condition $\boldsymbol{\pi}\mathbf{1}_{N_s} = 1$. The complexity involved in calculating the steady-state probabilities increases exponentially with the number of phases in the arrival process, the number of states in the physical layer Markov model and the number of queue states in the system.

Packet loss rate and throughput: The number of lost packets due to buffer overflow when the system changes from state $\mathcal{S}_\mu$ to state $\mathcal{S}_{\mu'}$ (pink arrow) corresponds to the arriving packets that can not be stored in the queue due to lack of space. The available space in the queue at the end of this frame interval for the packets that have never been transmitted can be obtained as the difference between the buffer size, $\bar{Q}$, and the number of packets that will have been already transmitted 1 or more times at time instant $t = \mu' T_f$, $Q_{\mu',1}$, that is $\bar{Q} - Q_{\mu',1}$. This available space will be occupied by the packets that have never been transmitted at the end of this frame interval, which will correspond to $q_{\mu,0}$ if no packets among them are transmitted, and to $q_{\mu,0} - \tau_{\mu,0} = Q_\mu - \tau_\mu$ otherwise. Moreover, the incoming packets a_μ should be also stored in the buffer. Thus, the number of packets that overflow from the buffer can be obtained as:

$$N_{l_{BO}|\mathcal{S}_\mu,\mathcal{S}_{\mu'}} = \begin{cases} \max\{0, Q_{\mu',1} + q_{\mu,0} + a_\mu - \bar{Q}\} & , \tau_\mu \le Q_{\mu,1} \\ \max\{0, Q_{\mu',1} + Q_\mu - \tau_\mu + a_\mu - \bar{Q}\} & , \tau_\mu > Q_{\mu,1}. \end{cases} \tag{33}$$

Therefore, the average number of lost packets due to buffer overflow is given by:

$$\bar{N}_{l_{BO}} = \sum_{\mu=1}^{N_s}\sum_{\mu'=1}^{N_s} \pi_{\mathcal{S}_\mu} P_{\mathcal{S}_\mu,\mathcal{S}_{\mu'}} N_{l_{BO}|\mathcal{S}_\mu,\mathcal{S}_{\mu'}} \tag{34}$$

and the packet loss rate $P_{l_{BO}}$ (measured in packets per frame) can then be obtained as:

$$P_{l_{BO}} = \bar{N}_{l_{BO}} / \lambda. \quad (35)$$

The number of lost packets due to exceeding the maximum number of allowed retransmissions when the system state is $\mathcal{S}_\mu$ corresponds to those packets that are erroneously transmitted after having been already transmitted N_r, and can be calculated as:

$$N_{l_{\text{ARQ}}|\mathcal{S}_\mu} = \begin{cases} \sum_{l=0}^{\tau_{\mu,N_r}} l\, \mathcal{P}_l^{\tau_{\mu,N_r}} \left(\overline{PER}_{n_\mu}^{\text{PHY}}\right) & \text{(Type I hybrid FEC/ARQ)} \\ \sum_{l=0}^{\tau_{\mu,N_r}} l\, \mathcal{P}_l^{\tau_{\mu,N_r}} \left(\overline{PER}_{n_\mu,N_r}^{\text{PHY}}\right) & \text{(CC or IR)} \end{cases} \quad (36)$$

with $\tau_{\mu,N_r} = \min\left\{q_{\mu,N_r}, \mathcal{C}_{n_\mu}\right\}$, and the corresponding average number of lost packets can be obtained as:

$$\bar{N}_{l_{\text{ARQ}}} = \sum_{\mu=1}^{N_s} \pi_{\mathcal{S}_\mu} N_{l_{\text{ARQ}}|\mathcal{S}_\mu}. \quad (37)$$

Accordingly, the probability of packet loss due to exceeding N_r retransmissions can be expressed as:

$$P_{l_{\text{ARQ}}} = \bar{N}_{l_{\text{ARQ}}} / \lambda. \quad (38)$$

In our finite buffering truncated ARQ-based error control system, the packet loss rate P_l (measured in packets per frame) can be expressed as $P_l = P_{l_{\text{BO}}} + P_{l_{\text{ARQ}}}$, and given the packet loss rate P_l, the average throughput can be calculated as:

$$\eta = \lambda(1 - P_l). \quad (39)$$

Average queue length and average packet delay: Using the Little's formula (Kleinrock, 1975), and taking into consideration that the effective packet arrival rate is equal to $\lambda(1 - P_{l_{\text{BO}}})$, the average delay for our embedded Markov chain can be calculated as:

$$D_l = L_q / \lambda(1 - P_{l_{\text{BO}}}), \quad (40)$$

where L_q denotes the average number of packets in the queue that is given by:

$$L_q = \sum_{\mu=1}^{N_s} \pi_{\mathcal{S}_\mu} Q_\mu. \quad (41)$$

CROSS-LAYER OPTIMIZATION

Multidimensional Approach

As shown in previous sections, given a maximum afforded queue length $\bar{Q}$, an average SNR $\bar{\gamma}$ and a normalized maximum Doppler frequency $f_d T_f$, performance measures of the system like, for instance, throughput, average packet delay or packet loss rate, are a function of the AMC transmission mode switching levels $\mathbf{\Gamma}^m \in \mathbb{R}_+^{M+1}$, where $\mathbb{R}_+$ denotes the set of non-negative real numbers, and the measured or estimated arrival packet rate $\lambda \in \mathbf{\Theta}$, where $\mathbf{\Theta}$ is the range of feasible arrival rate values. Therefore, if the objective of the link adaptation scheme is to maximize the average throughput when supporting QoS-guaranteed traffic characterized by a maximum packet loss rate $P_{l\max}$ and a maximum average packet delay $D_{l\max}$, the system needs to jointly optimize the selected protocol parameters at the physical and data link control layers by solving the cross-layer optimization problem

$$\left(\mathbf{\Gamma}^m_{\text{opt}}, \lambda_{\text{opt}}\right) = \arg\max_{\mathbf{\Gamma}^m \in \mathbb{R}_+^{M+1}, \lambda \in \mathbf{\Theta}} \eta\left(\mathbf{\Gamma}^m, \lambda\right) \tag{42}$$

subject to the constraints:

$$P_l\left(\mathbf{\Gamma}^m_{\text{opt}}, \lambda_{\text{opt}}\right) \leq P_{l\max}, \quad D_l\left(\mathbf{\Gamma}^m_{\text{opt}}, \lambda_{\text{opt}}\right) \leq D_{l\max}. \tag{43}$$

The analytical expressions for η, P_l and D_l do not leave much room for developing efficient algorithms in solving our constrained optimization problem. However, considering that $\mathbf{\Gamma}^m$ and λ lie in a bounded space $\mathbb{R}_+^{M+1} \times \mathbf{\Theta}$, we could resort to a multidimensional exhaustive search to numerically solve the proposed cross-layer optimization problem.

Once the optimum values of $\mathbf{\Gamma}^m$ and λ have been obtained, these values ought to be accordingly adjusted, with the help of the cross-layer controller, at the PHY and DLC layers, respectively. There exist a lot of strategies to manage the packet arrival rate and to adapt it to the system conditions. In fact, to offer QoS in a network, traffic entering the service provider network needs to be policed on the network boundary routers to make sure the traffic rate stays within the service limit. A Token Bucket can be used to enforce a traffic contract since it provides a precise algorithmic definition for conforming packets and non-conforming packets and therefore, it provides a bound on the average rate. In addition, traffic shapers smooth traffic and send it out at a constant rate. When tokens are exhausted, they buffer packets and send them out later, when tokens are available (Vegesna, 2001).

Bidimensional Simplification Approach

In order to simplify this multidimensional optimization approach, let us define the AMC switching thresholds as the instantaneous SNR values for which the value of $PER_{n,N_r} = P_0$, that is:

$$\gamma_n^m = \begin{cases} \dfrac{1}{g_{n,0}} ln\left(\dfrac{a_{n,0}}{P_0}\right) & \text{(Type I hybrid FEC/ARQ)} \\[2ex] \dfrac{1}{(N_r+1)g_{n,0}} ln\left(\dfrac{a_{n,0}}{P_0}\right) & \text{(CC)} \\[2ex] \dfrac{1}{g_{n,N_r}} ln\left(\dfrac{a_{n,N_r}}{P_0}\right) & \text{(IR)} \end{cases} \tag{44}$$

for $n = 1, \ldots, M-1$, with $\gamma_0^m = 0$, and $\gamma_M^m = \infty$. In this case, the simplified constrained optimization problem can be formulated as:

$$\left(P_{0\text{opt}}, \lambda_{\text{opt}}\right) = \arg\max_{P_0 \in \mathbb{R}_+, \lambda \in \mathbf{\Theta}} \eta\left(P_0, \lambda\right) \tag{45}$$

subject to the constraints:

$$P_l\left(P_{0\text{opt}}, \lambda_{\text{opt}}\right) \leq P_{l\max}, \quad D_l\left(P_{0\text{opt}}, \lambda_{\text{opt}}\right) \leq D_{l\max}. \tag{46}$$

Obviously, because P_0 and λ lie in a bounded space $\mathbb{R}_+ \times \mathbf{\Theta}$, we can resort to a 2-D exhaustive search to numerically solve the proposed cross-layer optimization problem. The bidimensional optimization approach has less degrees of freedom than the multidimensional strategy, resulting in a simpler scheme at the cost of a reduction in performance, as it will be illustrated in the following section.

NUMERICAL RESULTS

In order to verify the validity of the proposed cross-layer framework, analytical results obtained with the Two-Dimensional FSMC (2D-FSMC) model are confronted with computer simulation results obtained using Clarke's statistical Rayleigh model. Unless otherwise specified, numerical results are obtained using the following default parameters: a normalized maximum Doppler frequency $f_d T_f = 0.02$, an average received SNR $\bar{\gamma} = 8\,\mathrm{dB}$, a buffer size $\bar{Q} = 8$, a number of channel states $K = 5$, and a parameter $b = 6$. Without loss of generality, a D-BMAP traffic model either characterized by the transition probability matrix:

$$\mathbf{U} = \begin{bmatrix} 0.8 & 0.1 & 0.05 & 0.05 \\ 0.05 & 0.8 & 0.1 & 0.05 \\ 0.05 & 0.05 & 0.8 & 0.1 \\ 0.05 & 0.05 & 0.1 & 0.8 \end{bmatrix}$$

or parameterized to obtain a truncated Poisson process with an average arrival rate λ. Recall that matrix $\mathbf{U}$ introduces memory in the packet arrival process. In fact, the probability of remaining on a given phase is larger than the transition probability, this leading to a high probability of generating packet batches of a given length.

Figure 14 depicts the dependence of the average throughput η on the target PER P_0 for the Type I hybrid FEC/ARQ, the HARQ-CC and the HARQ-IR schemes. Results have been obtained using both, the proposed analytical model and the real system simulation. It can be clearly observed that the greater the number of allowed retransmissions the higher the achieved throughput; however, the most significant gain is obtained when going from $N_r = 0$ to $N_r = 1$, and the advantage of using higher N_r values becomes marginal. Notice that the maximum achieved throughput with the classical Type I hybrid FEC/ARQ scheme, even with infinite persistence, is clearly lower than the optimum throughput corresponding to any of the HARQ strategies, even when only one retransmission is allowed, clearly demonstrating the superiority of HARQ over conventional Type I hybrid FEC/ARQ. Furthermore, these results show that larger P_0 values, which cause the selection of higher order TMs, lead to an increase of the queueing service rate and, consequently, to an improved throughput. However, after reaching the optimum, higher P_0 values imply the use of TMs that are not robust enough, which in turn leads to a reduction in η. In agreement with Frederiksen and Kolding (2002), it is worth noting that HARQ-IR and HARQ-CC strategies perform similarly when a link adaptation algorithm is used. Consequently, only results corresponding to the IR scheme are displayed from this point onwards.

The dependence of the average packet loss rate P_l, as well as its two components $P_{l_{BO}}$ and $P_{l_{ARQ}}$, on the target PER P_0, is shown in Figures 15 and 16. It is clear that the packet loss due to buffer overflow $P_{l_{BO}}$ is the dominant packet loss factor except for high P_0 values. Thus, as it is shown in Figure 15, the influence on the average packet loss rate of the packet losses due to exceeding the maximum number of allowed retransmissions is only noticeable for high values of the target PER P_0. The analysis of these graphs reveals that higher P_0 values bring along the utilization of less robust (e.g. higher order) TMs, thus causing an increase in the packet loss rate due to exceeding the maximum number of allowed retransmissions. Furthermore, higher P_0 values also lead to an increment of the queuing service rate and, consequently, to a decrease in the buffer overflow probability $P_{l_{BO}}$. Nevertheless, it can be observed that when the number of allowed retransmissions N_r grows, the increase of the service rate cannot cope with the huge number of required retransmissions, resulting in more packets remaining in the queue and causing $P_{l_{BO}}$ to increase accordingly. As expected, when more retransmissions are allowed, lower $P_{l_{ARQ}}$ values are obtained. In contrast with the behaviour of the HARQ schemes, where higher N_r values imply a decrease in $P_{l_{BO}}$, when

Figure 14. Average throughput vs. target PER

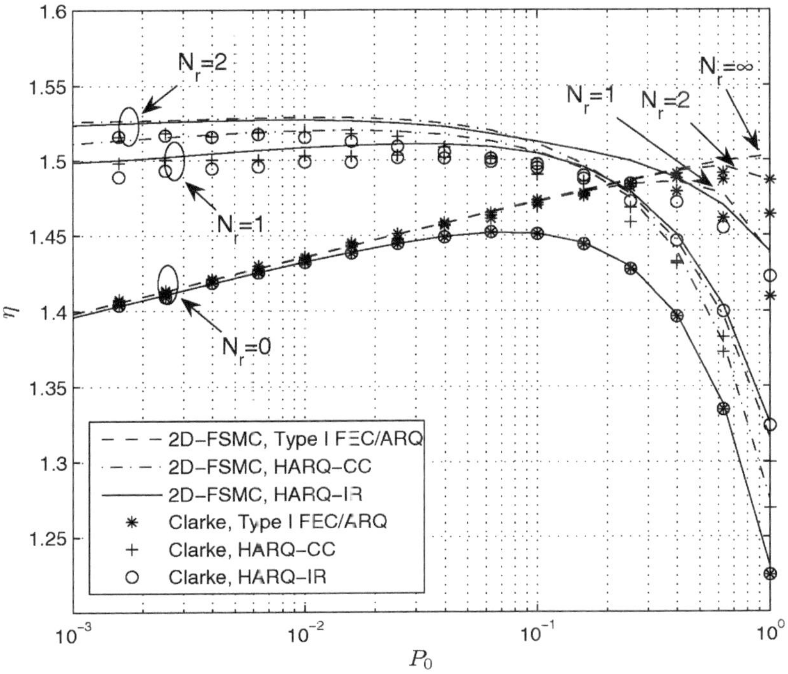

Figure 15. Total average packet loss rate vs. target PER

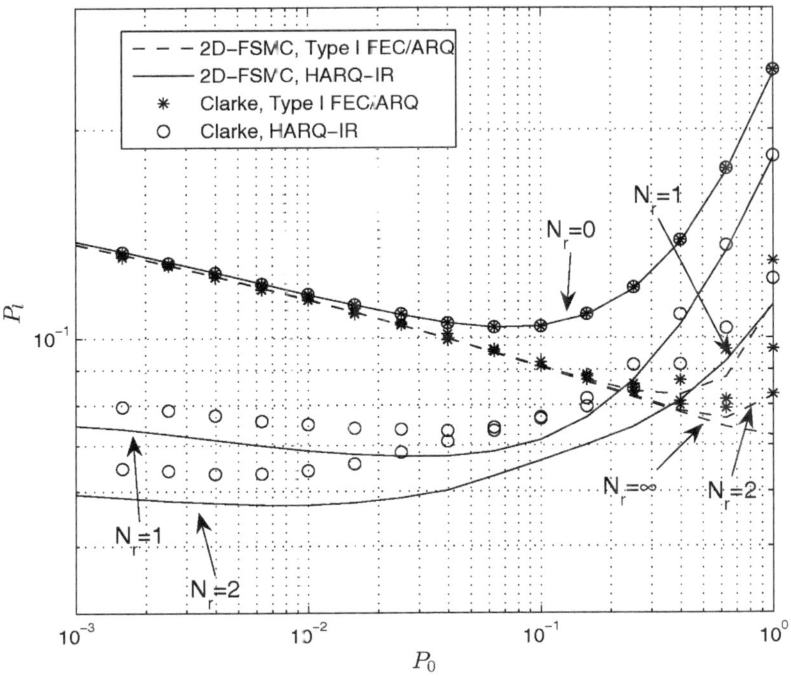

Figure 16. Average packet loss rate components, $P_{l_{BO}}$ and $P_{l_{ARQ}}$, vs. target PER

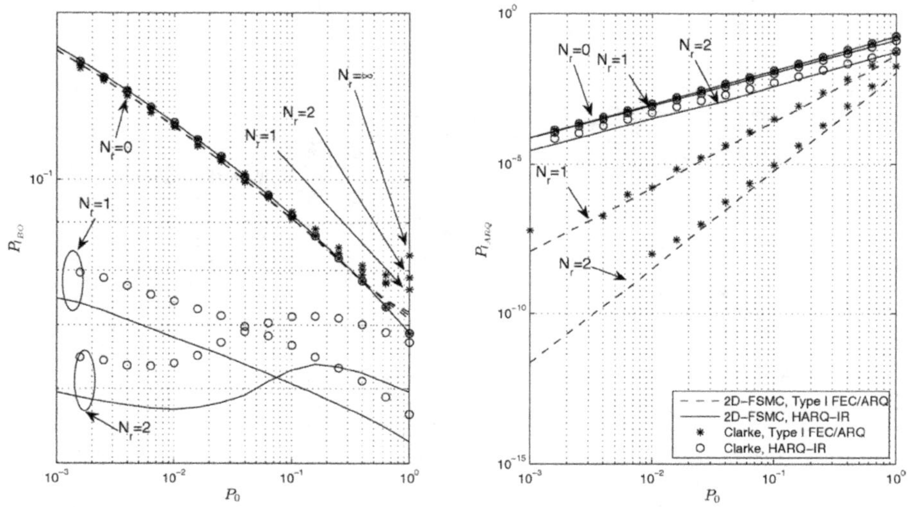

the Type I hybrid FEC/ARQ strategy is considered, if more retransmissions are allowed more packets remain in the queue and therefore $P_{l_{BO}}$ increases.

Results presented in Figures 14-16 reveal that, in all cases, the behaviour of a real FIFO queuing system using either an infinitely persistent or truncated (H)ARQ protocol is faithfully reproduced by the proposed analytical Markov model. The shape and location of the optimum minimum/maximum of the curves obtained by simulation (Clarke's model) coincide with those obtained using our analytical model, which is particularly important to ensure an optimal cross-layer design.

With the objective of analyzing the proposed cross-layer design, QoS-guaranteed traffic characterized by a maximum average packet loss rate $P_{l\max} = 0.1$ packets/frame and a maximum average packet delay $D_{l\max} = 4$ frames has been considered. Traffic has been generated using a truncated Poisson process with a truncation length equal to 4 packets, i.e. $\boldsymbol{\Theta} = (0, 4]$. Figure 17 plots the dependence of the obtained QoS parameters η and D_l on the maximum normalized Doppler frequency, for an average received SNR $\overline{\gamma} = 6\,\mathrm{dB}$. In Figure 18, the dependence of this cross-layer design on the average SNR with a normalized maximum Doppler frequency $f_d T_f = 0.02$ is depicted. These figures illustrate results obtained for both, the multidimensional and the simplified bidimensional optimization problems. Although not shown in the figures, the constraint on the maximum average packet loss rate has been satisfied with equality in all simulations. It can be inferred from the shape of the curves that for higher $f_d T_f$ values, corresponding to shorter fading durations, the optimum throughput η increases while ensuring the fulfillment of both QoS requirements. It can also be observed that for higher $\overline{\gamma}$ values, which lead to the use of higher order TMs and imply a decrease in P_l, a better throughput can be achieved.

Figures 17 and 18 reveal, in short, a significant improvement in the maximum achieved throughput when using the HARQ-IR error control protocol in comparison to the classical Type-I hybrid FEC/ARQ scheme. In fact, it can be noticed that the IR scheme with $N_r = 1$ reaches a much greater optimum throughput than the classical Type-I hybrid FEC/ARQ scheme, even with $N_r = 2$. As expected, when the AMC switching thresholds are found using the multidimensional optimization approach, a higher maximum throughput is obtained in comparison to the bidi-

Figure 17. Cross-layer optimization vs. maximum normalized Doppler frequency

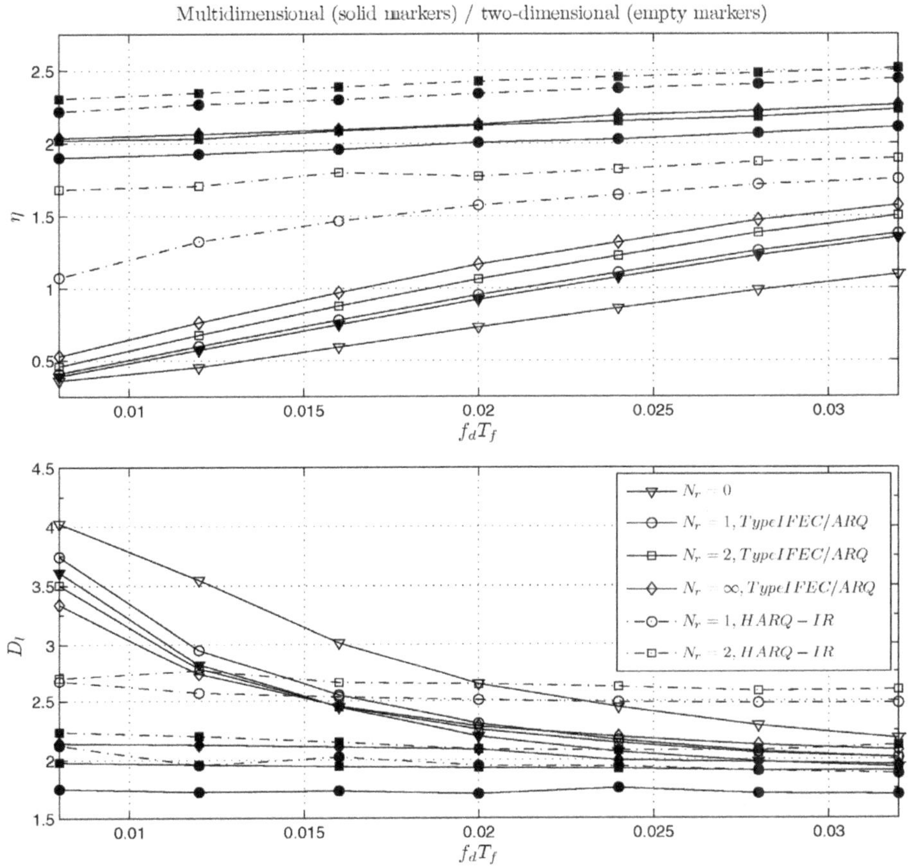

mensional optimization strategy, thus confirming the exchange of complexity for performance of the latter approach. As already mentioned, similar results have been obtained with the HARQ-CC due to the use of an AMC scheme at the physical layer.

OPEN ISSUES

The analytical framework proposed in this research work assumes the use of either infinitely persistent or truncated SR-ARQ protocols with instantaneous feedback of ACK/NACK at the data link layer. Furthermore, it has been assumed that perfect CSI is available at the receiver, which allows an ideal TM selection at the receiver AMC controller. The resultant TM is then fed back to the transmitter without error and latency. However, the CSI used to perform link adaptation in practical systems may be outdated due to the transmission delay and the processing imperfections both at the transmitter and the receiver. Hence, further considerations on system design with delayed or noisy CSI should be taken into account in future research work.

Moreover, one of the main drawbacks of cross-layer optimization approaches based on FSMCs is that the complexity involved in calculating the steady-state probabilities increases exponentially with the number of phases in the arrival process, the number of states in the physical layer Markov model, and the number of queue states in the system. Consequently, for large buffer sizes, efficient strategies should be devised to tackle this

Figure 18. Cross-layer optimization vs. average SNR

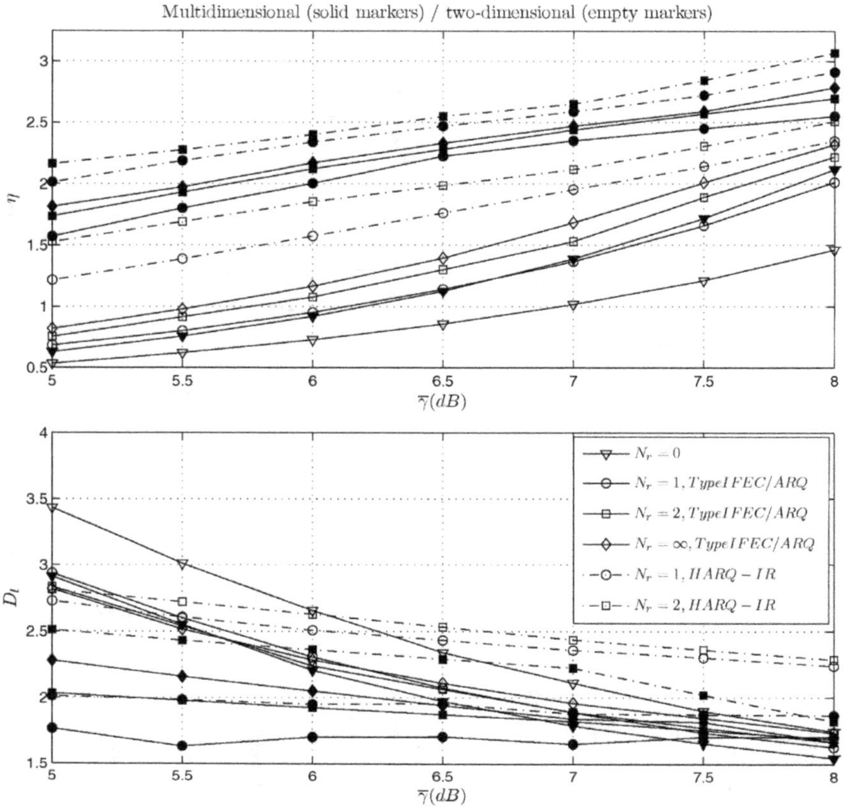

problem. This is an issue that has been left for further research.

The use of multiple antennas at transmitter and receiver, popularly known as Multiple Input Multiple Output (MIMO), is an emerging cost-effective technology that offers very high speed wireless links while providing QoS guarantees and range capability in Non-Line-Of-Sight (NLOS) environments. It constitutes a significant research and engineering challenge that has been taken into account in recent 3rd Generation Partnership Project (3GPP) standards, such as High-Speed Packet Access Plus (HSPA+), Long Term Evolution (LTE) and LTE-advanced, as well as in the IEEE 802.16e and IEEE 802.11n standards. MIMO physical layer techniques and their impact on the QoS provisioning of AMC/ARQ-based systems should also be considered using the analytical framework presented in this chapter.

Next generation wireless networks require not only the development of high-performance physical-layer technologies, but also powerful resource management strategies to provide high throughput and efficient use of resources. Scheduling algorithms are key components for QoS provisioning in wireless systems. Based on our theoretical framework, the analysis of different packet scheduling techniques for the resource management in wireless networks with heterogeneous QoS requirements remains to be done.

Ongoing work focuses on extending the proposed queuing model to include cooperative HARQ protocols at the DLC layer. Cooperative relaying has recently emerged as a powerful spatial diversity technique for improved performance. In cooperative HARQ protocols, relays can help the source to transmit its packets to the destination. Performance improvement brought by cooperation is due to a better average SNR of the relay-destination link compared with that of the source-destination link. It is also due to spatial diversity since the best relay is selected. This technique and its influence on the QoS provisioning performance at higher layers is being investigated by applying the proposed analytical framework.

CONCLUSION

A novel cross-layer DTMC-based queueing model jointly exploiting the error-correcting capability of either infinitely persistent or truncated HARQ protocols (DLC) and the adaptation ability of AMC strategies (PHY) has been developed. This DTMC model is based on a comprehensive consideration of packet multimedia traffic sources modeled as a D-BMAP, finite queue-length systems, HARQ protocols, AMC strategies, and a wireless channel first-order two-dimensional Markov model using both the amplitude and the rate of change of the fading envelope. The stationary state probability distribution of this multidimensional DTMC has been used to obtain closed-form analytical expressions for performance metrics such as throughput, average packet delay and packet loss rate, both due to buffer overflow and due to exceeding the maximum number of allowed retransmissions. Furthermore, this cross-layer analytical framework has been exploited to formulate multidimensional and simplified two-dimensional constrained optimization problems aiming at maximizing the system throughput under maximum packet loss rate and delay constraints.

To assess the validity of the proposed analytical cross-layer framework, computer simulation results have been presented showing that the derived closed-form analytical expressions can be used to faithfully reproduce results based on Clarke's statistical Rayleigh fading model. Numerical results have shown that both HARQ-CC and HARQ-IR schemes lead to a remarkable improvement on the maximum achieved throughput when compared to the classical Type-I hybrid FEC/ARQ scheme. The extent of this improvement largely depends on the channel conditions and the maximum number of allowed ARQ retransmissions. As in Frederiksen and Kolding (2002), it has also been shown that HARQ-IR protocols provide a small throughput gain when compared to HARQ-CC schemes due to the use of physical layer AMC algorithms. Numerical results have also confirmed the benefits of the cross-layer multidimensional optimization in attaining the maximum throughput in comparison to the simplified two-dimensional approach.

The obtained analytical results not only allow a-priori design decisions to be made, but they also provide an insight that enables the derivation of dynamic reconfiguration strategies that take into account instantaneous CSI at the PHY layer and QSI at the DLC layer.

ACKNOWLEDGMENT

This work has been supported in part by the MEC and FEDER under project COSMOS (TEC2008-02422/TEC).

REFERENCES

Aïssa, S., & Aniba, G. (2008). BER analysis of M-QAM with packet combining over space-time block coded MIMO fading channels. *IEEE Transactions on Wireless Communications*, *7*(3), 799–805. doi:10.1109/TWC.2008.060745

Alfa, A. S. (2010). *Queueing theory for telecommunications - Discrete time modeling of a single node system*. Berlin, Germany: Springer.

Babich, F., Kelly, O., & Lombardi, G. (2000). Generalized Markov modeling for flat fading. *IEEE Transactions on Communications*, *48*(4), 547–551. doi:10.1109/26.843121

Biglieri, E., Caire, G., & Taricco, G. (2001). Limiting performance of block fading channels with multiple antennas. *IEEE Transactions on Information Theory*, *47*(4), 1273–1289. doi:10.1109/18.923715

Blondia, C. (1993). A discrete time batch Markovian arrival process as b-isdn traffic model. *Belgian Journal of Operations Research. Statistics and Computer Science*, *32*(3/4), 3–23.

Chase, D. (1973). A combined coding and modulation approach for communications over dispersive channels. *IEEE Transactions on Communications*, *21*(3), 159–174. doi:10.1109/TCOM.1973.1091636

Chase, D. (1985). Code combining: A maximum-likelihood decoding approach for combining an arbitrary number of noisy packets. *IEEE Transactions on Communications*, *33*(5), 385–393. doi:10.1109/TCOM.1985.1096314

Cheng, J.-F. T. (2006). Coding performance of hybrid ARQ schemes. *IEEE Transactions on Communications*, *54*(6), 1017–1029. doi:10.1109/TCOMM.2006.876846

Clarke, R. H. (1968). A statistical theory of mobile radio reception. *The Bell System Technical Journal*, *47*(6), 957–1000.

Elliott, E. O. (1963). Estimates of error rates for codes on burst-noise channels. *The Bell System Technical Journal*, *42*, 1977–1997.

Femenias, G., Ramis, J., & Carrasco, L. (2009). Using two dimensional Markov models and the effective-capacity approach for cross-layer design in AMC/ARQ-based wireless networks. *IEEE Transactions on Vehicular Technology*, *58*(8), 4193–4203. doi:10.1109/TVT.2009.2020723

Frederiksen, F., & Kolding, T. (2002). Performance and modeling of WCDMA/HSDPA transmission/H-ARQ schemes. In *Proceedings of the VTC Fall*, (pp. 472-476). VTC.

Gilbert, E. N. (1960). Capacity of a burst-noise channel. *The Bell System Technical Journal*, *39*, 1253–1266.

Goldsmith, A. (2005). *Wireless communications*. Cambridge, UK: Cambridge University Press.

IEEE. (2004). *802.16 standard for local and metropolitan area networks, part 16: Air interface for fixed broadband wireless access systems*. New York, NY: IEEE Computer Society, IEEE Microwave Theory and Techniques Society.

Kanal, L., & Sastry, A. (1978). Models for channels with memory and their applications to error control. *Proceedings of the IEEE*, *66*(7), 724–744. doi:10.1109/PROC.1978.11013

Kleinrock, L. (1975). *Queueing systems* (*Vol. 1*). New York, NY: Wiley.

Le, L. B., Hossain, E., & Alfa, A. S. (2006). Radio link level performance evaluation in wireless networks using multi-rate transmission with ARQ-based error control. *IEEE Transactions on Wireless Communications*, *5*(10), 2647–2653. doi:10.1109/TWC.2006.04485

Lin, S., & Costello, D. J. (2004). *Error control coding: Fundamentals and applications* (2nd ed.). Englewood Cliffs, NJ: Prentice Hall.

Liu, Q., Zhou, S., & Giannakis, G. B. (2004). Cross-layer combining of adaptive modulation and coding with truncated ARQ over wireless links. *IEEE Transactions on Wireless Communications*, *3*(5), 1746–1755. doi:10.1109/TWC.2004.833474

Liu, Q., Zhou, S., & Giannakis, G. B. (2005). Queueing with adaptive modulation and coding over wireless links: Cross-layer analysis and design. *IEEE Transactions on Wireless Communications*, *4*(3), 1142–1153. doi:10.1109/TWC.2005.847005

Lloyd, S. P. (1982). Least squares quantization in PCM. *IEEE Transactions on Information Theory*, *28*, 129–137. doi:10.1109/TIT.1982.1056489

Max, J. (1960). Quantization for minimum distortion. *I.R.E. Transactions on Information Theory*, *6*, 7–12. doi:10.1109/TIT.1960.1057548

Ramis, J., Carrasco, L., & Femenias, G. (2008). A two-dimensional Markov model for cross-layer design in AMC/ARQ-based wireless networks. In *Proceedings of the IEEE GLOBECOM,* (pp. 4637-4642). IEEE Press.

Ramis, J., & Femenias, G. (2011). Cross-layer design of adaptive multirate wireless networks using truncated HARQ. *IEEE Transactions on Vehicular Technology*, *60*(3), 944–954. doi:10.1109/TVT.2011.2108324

Ramis, J., Femenias, G., & Carrasco, L. (2009). Cross-layer design of multi-rate wireless networks based on link layer truncated ARQ. In *Proceedings of IWCLD.* IWCLD.

Ramis, J., Femenias, G., Riera-Palou, F., & Carrasco, L. (2010). Cross-layer optimization of adaptive multirate wireless networks using truncated Chase combining HARQ. In *Proceedings of IEEE GLOBECOM.* IEEE Press.

Sadeghi, P., Kennedy, R., Rapajic, P., & Shams, R. (2008). Finite-state Markov modeling of fading channels - A survey of principles and applications. *IEEE Signal Processing Magazine*, *25*(5), 57–80. doi:10.1109/MSP.2008.926683

Svensson, A. (2007). An introduction to adaptive QAM modulation schemes for known and predicted channels. *Proceedings of the IEEE*, *95*(12), 2322–2336. doi:10.1109/JPROC.2007.904442

Tan, C. C., & Beaulieu, N. C. (1997). Infinite series representations of the bivariate Rayleigh and Nakagami-m distributions. *IEEE Transactions on Communications*, *45*(10), 1159–1161. doi:10.1109/26.634675

Tan, C. C., & Beaulieu, N. C. (2000). On first-order Markov modeling for the Rayleigh fading channel. *IEEE Transactions on Communications*, *48*(12), 2032–2040. doi:10.1109/26.891214

Vegesna, S. (2001). *IP quality of service*. New York, NY: Cisco Press.

Wang, X., Liu, Q., & Giannakis, G. B. (2007). Analyzing and optimizing adaptive modulation coding jointly with ARQ for QoS-guaranteed traffic. *IEEE Transactions on Vehicular Technology*, *56*(2), 710–720. doi:10.1109/TVT.2007.891465

Chapter 18
Cross-Layer Techniques for Reliable Wireless Video Communication

Athar Ali Moinuddin
Aligarh Muslim University, India

Mohd Ayyub Khan
Aligarh Muslim University, India

Ekram Khan
Aligarh Muslim University, India

Mohammed Ghanbari
University of Essex, UK

ABSTRACT

Designing wireless video communication *system is a challenging task due to high error rates of wireless channels, limited and dynamically varying bandwidth availability, and low energy and complexity requirements of portable multimedia devices. Scalable video coders having excellent rate-distortion performance are most suited to cope with time varying bandwidth of wireless networks, but encoded bits are extremely sensitive to channel errors. This chapter presents a reliable video communication system exploring opportunities offered by various network layers for improved overall performance, while optimizing the resources. More specifically, cross-layer approach for* Unequal Error Protection (UEP) of scalable video bitstream is the main theme of this chapter. In UEP, the important bits are given *a higher protection compared to the other bits. Conventionally, UEP is achieved by using* Forward Error Correction (FEC) *at the application layer. However, UEP can also be provided at the physical layer using* hierarchical modulation *scheme. In this chapter, the authors discuss* cross-layer design *methodology for UEP that rely on interaction between the application layer and the physical layer to achieve reliable and high quality end-to-end performance in wireless environments. The discussion is mainly focused for wavelet coded video, but it is applicable to other embedded bitstreams as well.*

DOI: 10.4018/978-1-4666-0960-0.ch018

INTRODUCTION

In recent years, wireless communications and networking have experienced unprecedented growth. Video communication over wireless networks is envisioned for a vast range of applications, such as videotelephony, videoconferencing, telemedicine, surveillance, security monitoring, disaster rescue, environment monitoring, gaming, and entertainment, wildlife activity monitoring, and multimedia systems in consumer electronics (Al-Mualla, 2002). However, the issues of limited and dynamically varying bandwidth and error prone nature of the wireless channels, limited processing power, and battery life of the portable multimedia devices, and network heterogeneity make the video delivery over wireless networks a difficult and challenging task (Etoh, 2005).

Compressed video bitstream is extremely sensitive to bit errors, and therefore, not suitable for error prone and dynamically varying wireless channels. A variety of solutions have been proposed to cope with these challenges (Wang, 1998; Girod, 2000). These include *Forward Error Correction (FEC)*, retransmission using Automatic Repeat Request (ARQ), error resilient *coding, hierarchical modulation, joint source-*channel coding, scalable coding with transport prioritization, multiple-description coding, and error concealment by post processing etc. Traditionally, these solutions are employed at various network layers (such as application, transport, Media Access Control [MAC], and physical layers) separately and independently. Although, the decoupled layered protocol reduces network design complexity, however, independently optimized protocol paradigm is not well suited for wireless networks due to dynamically varying bandwidth and error rates, user mobility, high interferences, limited power, and complexity (van der Schaar, 2007). *Cross-layer design* methodologies that rely on interaction among different protocol layers hold great promises for addressing these challenges and for providing reliable and high-quality end-to-end performance in wireless multimedia communications (Shakkottai, 2003; van der Schaar, 2005).

In recent years, *cross-layer designs* for *wireless video communication* have been widely reported in literature. Researchers have pursued *cross-layer design* from the perspectives of reliable communication (van der Schaar, 2003; Xu, 2004; Choi, 2004; Barmada, 2006; Pei, 2007; Maani, 2010), end-to-end Quality-of-Service (QoS) requirements (Kumwilaisak, 2003; Zhang, 2005; Melodia, 2010), rate-adaptation (Haratcherev, 2006), resource allocation (Katsaggelos, 2005), scheduling (Anton-Haro, 2006; Li, 2010), and optimization (Khan, 2006; Foh, 2007; Zhang, 2010), etc. However, the main focus of this chapter is on the opportunities offered by cross-layer for error protection to enable reliable *wireless video communication*. The idea is to jointly explore the error protection strategies at various network layers, in order to improve the transmission efficiency in terms of protection, bandwidth, and resource consumption.

VIDEO CODING: AN OVERVIEW

Digital video signals generally involve huge amount of data in raw or uncompressed form. For example, a relatively low resolution QCIF (Quarter Common Intermediate Format) video have uncompressed data rate of about 6 Mbps (megabits/second). On the other hand, HDTV (High Definition Television) resolution video sequences have uncompressed data rate of about 1.5 Gbps (gigabits/sec). Given the vast data volumes, and limited bandwidth of wireless channels, it is obvious that the most essential requirement for wireless video is high compression efficiency. Moreover, growing network heterogeneity and varying resolution of display devices coupled with time varying characteristics of the wireless channels requires video coder to be equipped with scalability feature. This section presents a general overview of *video coding*.

Video Basics

Video signals are generated at the output of a camera by scanning a Two-Dimensional (2-D) scene. To get a digital video, the output of the camera for each picture or frame is sampled at regular intervals to produce a set of two-dimensional samples (also called pixels) of a frame. Each pixel (picture element) is then quantized to eight-bit resolution. Sampling is repeated at intervals to produce a moving video. For colour videos, a camera generates three primary colour signals called Red, Green, and Blue (RGB). These three colour signals are sampled and quantized individually and are stored in three separate rectangular arrays for further processing, and thereby each colour pixel of a video requires twenty four-bits for its storage. However, for compatibility with monochrome system and because of the fact that the three colour signals are highly correlated, the RGB video are often transformed into *luminance-chrominance* colour space such as Y, U, V (or Y, C_b, C_r). The luminance component Y represents the brightness. The two chrominance components U, V represent the extent to which the colour deviates from gray toward blue and red, respectively. When full colour fidelity is required, then a sample of each colour component exists at every pixel position. This is referred as 4:4:4 sampling, indicating that for every four luminance samples there are four U and four V samples. However, since the Human Visual System (HVS) is more sensitive to luminance than the chrominance, often the chrominance components are down-sampled. For example, in 4:2:0 sampling, chrominance is represented with half the horizontal and vertical resolution of Y. This reduces the amount of bits required to represent the chrominance components without having an obvious effect on visual quality of video (Wang, 2002).

Video Compression

As stated earlier, a raw or uncompressed video generates huge amount of data, therefore, it would require prohibitively large amount of bandwidth for communication. Fortunately, however, video signals consist of lot of temporal, spatial, and statistical correlations. Temporal correlation in a video is due to similarity of the adjacent frames as changes are mainly due to camera or object motion, whereas spatial correlation is due to the similarity of neighbouring pixels within a frame. The statistical redundancy is related with probability distribution function of pixel intensities/colours in a frame. Efficient reduction of these correlations is utilized to achieve the video compression.

As adjacent video frames are highly correlated, independent coding of each frame generally generates a lot of redundant bits. Changes in adjacent video frames are typically due to the motion of objects. By performing Motion Estimation (ME) between current and reference (previously decoded) frames, mechanism for efficient exploitation of temporal correlation can be devised. A practical and widely-used method for ME involves dividing the current frame into square blocks (usually of size 16×16) and finding a best match of each block within a predetermined search region in the reference frame. The matching criterion may be any error measure such as Mean Square Error (MSE) or Sum of Absolute Difference (SAD). The best match of the current block in the reference frame is a block which results in minimum error. The displacement between the two blocks is the estimated Motion Vector (MV). Often, ME accuracy is improved by using fraction-pixel accuracy.

The MVs are used to reduce the temporal redundancies either by performing Motion-Compensated Prediction (MCP) (Sullivan, 2005) or Motion Compensated Temporal Filtering (MCTF) (Ohm, 2004). In MCP, a prediction of the current frame is generated from the reference frame by using motion-compensated prediction. By taking the difference between the current frame and its

MCP, a prediction error or residual is generated. If prediction is good, then residual will be having small energy compared to the original frame. On the other hand, MCTF is a temporal decomposition technique that performs wavelet decomposition along the motion trajectory. Using MCTF, a Group Of Frame (GOF) is temporally decomposed into a set of temporal frequency bands. The first-level decomposition results in a temporal low-pass and high-pass subbands. The temporal high-pass subband represents the residual signal with most of the coefficients having small values. Since, most of the energy is concentrated in the temporal low-pass subband, therefore it is recursively decomposed to the desired temporal level to produce a temporal low-pass subband capturing most of the energy of the GOF and number of temporal high-pass subbands representing the residual.

The residual frames obtained after temporal redundancy reduction has a lot of spatial redundancies as adjacent pixels are highly correlated. Often, this redundancy reduction is facilitated by 2-D signal transforms (Ohm, 2004). One of the beneficial properties of the transform is that they tends to compact most of the energy of the signal into a relatively small number of transform coefficients, with the majority of coefficients having little energy (Goyal, 2001). Efficient quantization of these low valued coefficients lead majority of them to the zero value, and with a small number of significant coefficients that provide a more compact representation of the residual frame. Over the years, a large number of signal transforms have been proposed. However, in video compression applications, only the Discrete Cosine Transform (DCT) and the Discrete Wavelet Transform (DWT) are widely used (Ohm, 2004). DCT is a block-based transform which is applied after partitioning a frame into non-overlapping blocks of pixels and then processing each block individually. On the contrary, the DWT is applied on entire frame rather than to separate blocks of pixels.

The quantized symbols occurred with varying probabilities. It is a well-known fact of information theory that symbols occurring with high probabilities have less information content compared to symbols having lower probabilities. This information is exploited to assign short codes to more probable symbols compared to less probable symbols, resulting in Variable Length Code (VLC). A near-optimal method of designing VLCs is the Huffman coding. A more efficient method which can also be more easily designed to adapt to varying symbol statistics is the arithmetic coding (Sullivan, 2005).

DCT-Based Compression

Almost all the standard video coders like MPEG-2, MPEG-4, H.261, and H.263 use a hybrid of MCP and DCT to exploit the temporal and spatial redundancies (Ghanbari, 2003). Figure 1 shows a generic hybrid video coder which works as follows. Each frame is split into blocks of typically 16×16 pixels called Macroblock (MB) and each MB is processed more or less independently. A MB may be coded in Inter- and/or Intra-mode. In Inter-mode, a MB may use a unidirectional prediction (P-picture) or a bidirectional prediction (B-picture). In P-picture, for each MB in the current frame, a block matching ME process is used to find the best match in the previous frame. Only the luminance component is used in ME process. The estimated MV is used to generate a prediction of the current MB using MCP. In B-picture, block matching finds two best matching blocks, one in a previous frame and another in a future frame, and uses a weighted average of the two matches as the prediction for the current MB. In this case, two MVs are associated with each block. A prediction error or residual signal is generated by taking the difference between the current MB with its prediction. The residual is transformed using 8×8 DCT and coefficients are then scaled, entropy coded, and transmitted together with MV bits. Since MVs of adjacent MBs are usually similar, they are differentially coded. In the block matching process if no suitable match of the current MB is found,

Figure 1. Block diagram of a generic hybrid video coder

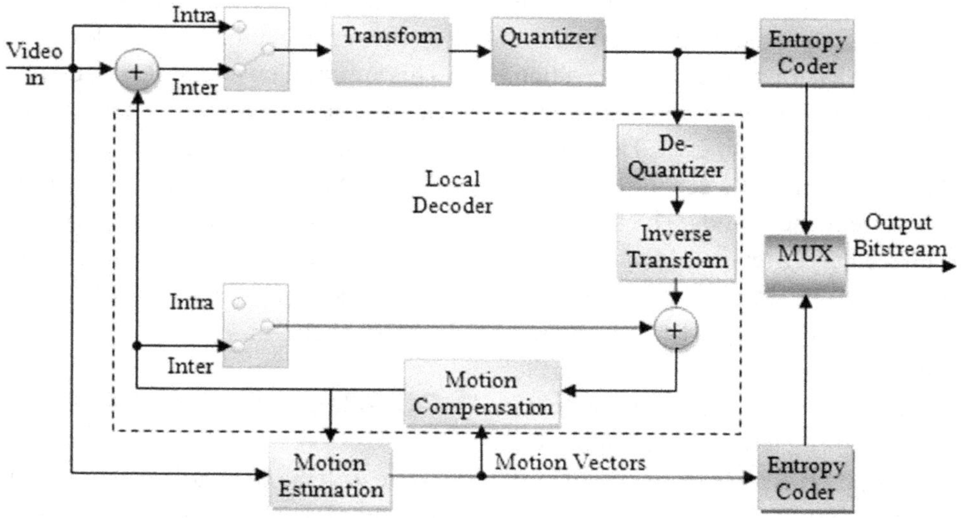

the block will be coded directly using transform coding. This is known as Intra-mode. In all the **video coding** standards, bit stream syntax follows hierarchical structure, in which a number of MBs form a Group Of Blocks (GOB) or a slice, and several GOBs or slices form a picture. The size and shape of GOBs, slices, and picture sizes are different in various coding standards, which can often be tailored to the application's needs.

A frame may be coded entirely in the intra-mode, and such a frame is called I-frame. The first frame of a sequence is always coded as an I-frame. A P-frame uses only a previous frame for prediction, and depending on the prediction accuracy, an MB can be coded in either intra- or P-mode. Finally, a B-frame uses a bidirectional prediction, and an MB in a B-frame can be coded in intra-, P-, or B-mode. Further, a B-frame can only be coded after the surrounding I- or P-frames are already coded (Wang, 2002).

The latest *video coding* standard H.264/AVC also uses a similar architecture but with the following notable differences. Instead of 8×8 DCT, it use a 4×4 integer transform. For more accurate MCP, quarter-pixel accuracy ME is used for luma and one eighth-pixel accuracy for chroma components. It also uses variable block size, bipredictive, multi-reference, and weighted MCP. In addition it uses a very powerful intra prediction, which was not available in previous standards. It also uses an in-loop adaptive de-blocking filter to overcome the problem of blocking artifacts. When used all together, these new tools provide approximately a 50% bit rate savings for equivalent perceptual quality relative to the performance of prior standards (Wiegand, 2003).

Wavelet-Based Compression

The outstanding performance of the wavelets for image compression has motivated researchers to extend them for *video coding* as well (Fowler, 2007). The straightforward approach is essentially an adaptation of the traditional hybrid architecture of Figure 1 replacing DCT with DWT (Dirac, 2011; Martucci, 1997; Moinuddin, 2006). A block-based ME with fractional-pixel accuracy is used to obtain the MVs. Overlapped Block Motion Compensation (OBMC) (Orchad, 1994) is then employed to remove temporal redundancy. Overlapping is necessary in order to remove the blocking artifacts in residual frames resulting from

Figure 2. Block diagram of a 3-D wavelet video codec

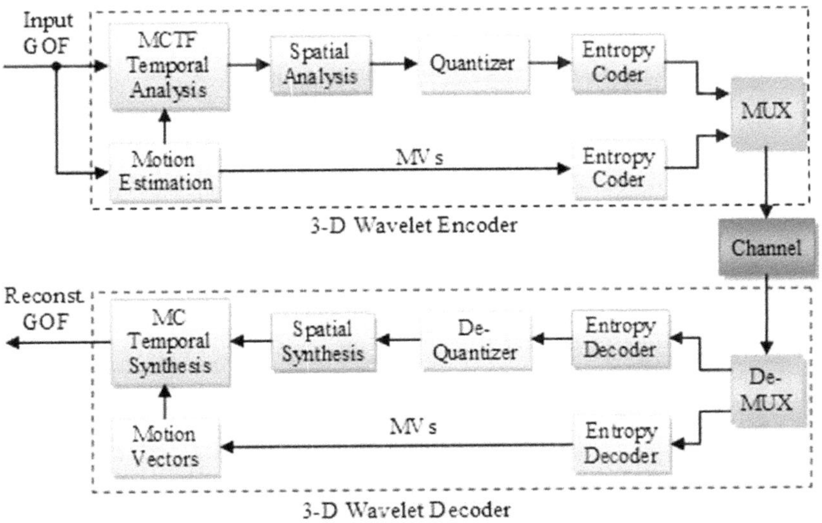

block matching. For block-based transform such as DCT, the blocking artifacts are not critical as long as the block boundaries of transformation and block matching are well aligned. However, wavelet-based coders that transform the entire residual frame will have to sacrifice much of their coding efficiency by coding the artificial high frequency information at the block boundaries of motion compensated frames.

A group of researchers have developed 3-D wavelet video coder as an alternative to hybrid video coder by extending wavelet transform in temporal direction. In contrast to the traditional hybrid video coding, 3-D video coding systems use 1-D temporal decomposition of a GOF using MCTF to exploit the temporal redundancy present in the video signal. The resulting temporal frames are then spatially decomposed using a 2-D wavelet transform followed by coefficients encoding as shown in Figure 2 (Ohm, 2004). MCTF is a temporal decomposition technique that adaptively performs the wavelet decomposition and reconstruction along the motion trajectories. It is conveniently implemented using a lifting scheme. Theoretically any filter could be used, however, research has shown that 5/3 filter has overall best performance for temporal decomposition (Golwelkar, 2007). For MCTF with 5/3 filter, the odd indexed frames are predicted from the adjacent even indexed frames to produce the highpass frames. The even indexed frames are then updated to generate lowpass frames using combination of the adjacent highpass frames. In both predict and update step, filtering is done along the motion direction. Temporal lowpass subband is recursively decomposed to give a lowpass subband and a number of highpass subbands. After temporal decomposition, each frame will undergo 2-D spatial wavelet decomposition, leading to 3-D spatio-temporal subbands as shown in Figure 3. For coding, any 3-D wavelet coding algorithms such as 3D-SPIHT, JPEG2000 multi-component, or MC-EZBC is applied to generate the final bitstream (Fowler 2007). MC-EZBC video coder (Chen, 2004) is currently considered to be the state-of-the-art in wavelet-based MCTF scalable coding.

Scalable Video Coding

Video delivery to diversified clients through heterogeneous networks with time varying char-

Figure 3. 3-D wavelet decomposition of a GOF of sixteen frames (a) input GOF and (b) the resulting spatio-temporal subbands for 4-levels of spatial and temporal decomposition

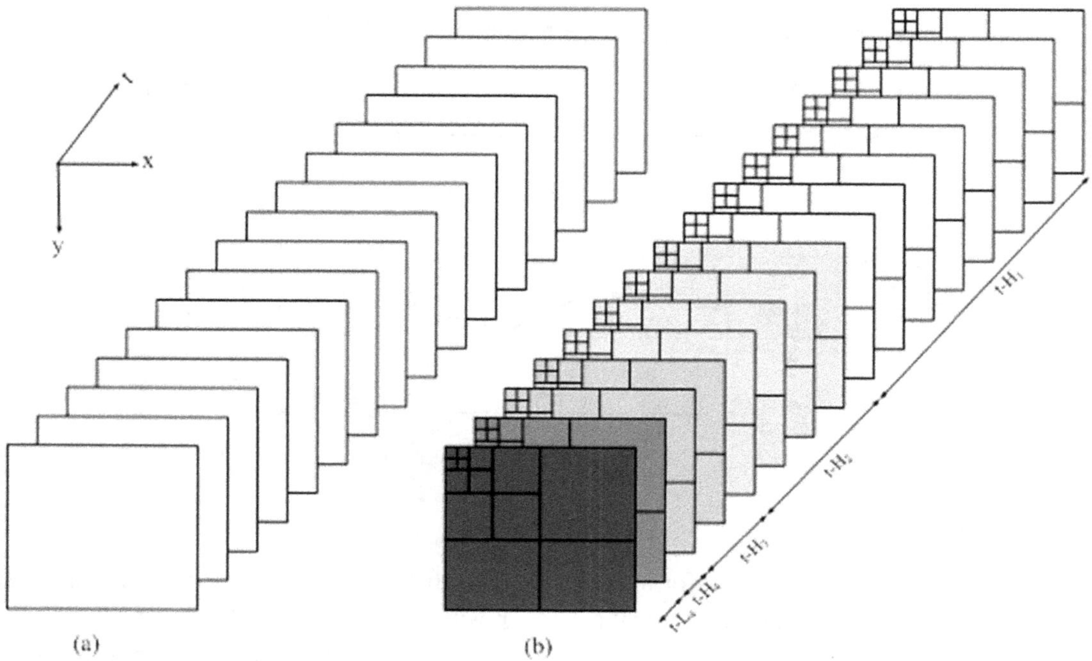

acteristics requires on-the-fly adaptation of the source bitstream. This is facilitated by Scalable Video Coding (SVC), whereby the coded bitstream is organized such that it has number of truncation points. Decoding only the partial streams will provide a range of scalability. The most common types of scalability are spatial, temporal, and quality (or SNR). In spatial scalability, complete decoding of the bitstream produces full resolution video, whereas partial decoding produces reduced spatial resolution. In temporal scalability, partial decoding leads to reduced decoded frame rates (temporal resolution), whereas in quality scalability, video quality (SNR) varies depending on how much of the bitstream is decoded (Ohm, 2005).

In most of the standard video coders using hybrid structure, scalability is achieved through a layered structure consisting of a base layer and a number of (usually two to three) enhancement layers as shown in Figure 4. The base layer is generally coded such that it has very low quality, resolution, and frame rate. The enhancement layers encodes the residual signal and when combined with base layer decoding, it provides a higher spatial resolution for spatial scalability, a higher frame rate for temporal scalability and a progressive quality improvement in case of SNR scalability.

Layered quality scalability is accomplished by using increasingly finer quantization step sizes. The base layer is obtained by applying a coarse quantizer to DCT coefficients giving low decoded quality. For any of the enhancement layer, the DCT coefficients of the original input frame is subtracted from the reconstructed DCT coefficients of the previous layer and the residual is quantized using smaller step size than that of the previous layer. This type of quality scalability provides only Coarse Granularity Scalability (CGS), in that the quality improvements are obtained with rate increases in large steps. On the other hand, in Fine Granularity Scalability

Figure 4. Principles of layered scalability

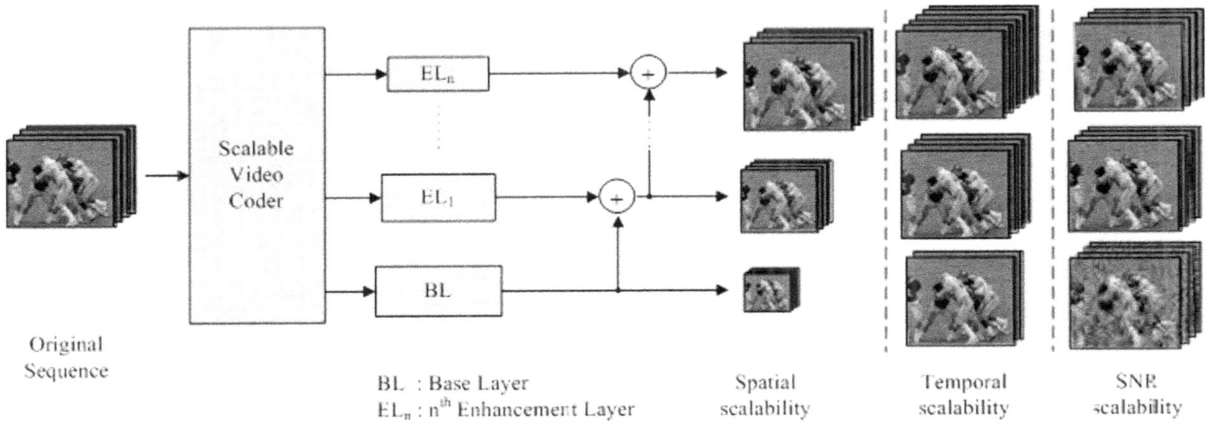

(FGS) a bitstream can provide continuously improving video quality with every additional bit. In FGS, a video is coded into a base layer and a fine granular enhancement layer. The base layer is coded with non-scalable technique, whereas the enhancement layer bitstream is generated by coding the difference between the original DCT coefficients and the reconstructed base layer coefficients using a bitplane coding technique (Pesquet-Popescu, 2007). FGS is used in MPEG-4 (Li, 2001) and SVC (Schwarz, 2007).

To achieve spatial scalability, the input video is decomposed into a spatial pyramid. Each spatial layer is encoded independently while the motion and temporal prediction are derived from the reference pictures at the same layer. To remove the redundancy among layers, the interlayer prediction comes from only the reconstructed frame of the most recent layer. Temporal scalability involves partitioning of the Group Of Pictures (GOP) into temporal layers. A simple way to achieve temporal scalability is to use the well-known IBBP prediction structure. Up to three frame rates are supported by decoding I-pictures only, both I- and P-pictures, or all of the I-, P-, and B-pictures, respectively. In the H.264/AVC and SVC, more levels of temporal scalability are possible with hierarchical B-pictures, which also improve the coding efficiency (Schwarz, 2007).

One of the problems with layered scalability is that every additional enhancement layer reduces the coding efficiency by about 15-20%, thereby restricting the number of enhancement layers to two-three only (Ghanbari, 2003). On the other hand, in 3-D wavelet video coders, quality scalability is achieved by embedded coding and bit-level granularity is possible. Temporal scalability of multiple levels is achieved using a combination of lowpass and highpass temporal subbands as shown in Figure 3. Resolution scalability is achieved by coding the 3-D coefficients in resolution pyramid (Pesquet-Popescu, 2007).

WIRELESS VIDEO COMMUNICATION

The past few years have seen unprecedented growth in wireless communication. Other than 3G cellular network, the use of Bluetooth and Wi-Fi enabled devices are very common in household and other section of the community. World wide deployment of the 4G networks would be the next ahead. On the other hand, wireless ad hoc network is getting popularity to connect multiple wireless devices and is very useful if somehow other network fails. Also the last mile in the internet connectivity is shifting from wired to wireless. This yields a huge demand of *wireless*

Figure 5. Multipath propagation

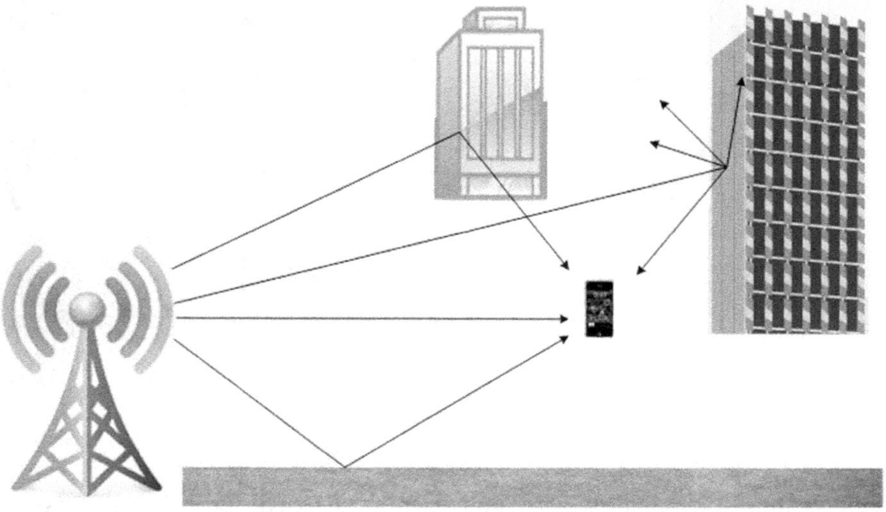

video communication for vast range of applications (Goldsmith, 2005). However, video communication over wireless channels faces challenges such as limited and dynamic channel bandwidth, signal fading, high-level of interferences, limited power, and low complexity requirements especially for portable devices. In this section, we discuss the characteristics of the wireless channel and challenges it poses for video communication. Finally, we discuss some of the proposed solutions to cope with these challenges.

Characteristics of Wireless Channel

Signal propagation in terrestrial mobile wireless environment is mainly due to reflection, scattering, transmission and diffraction. As a consequence, multiple replicas of the transmitted signal arrive at the receiver, each with random amplitude, path delay, phase shift and Doppler shift. This phenomenon is referred as multipath propagation (Figure 5) causing signal fading, time-varying delay spread, and spectral broadening (Goldsmith, 2005). These will be briefly discussed in the following.

Due to multipath propagation, the received signal consists of a series of attenuated, time delayed, phase shifted replicas of the transmitted signal. The resultant received signal is the vector sum of these individual signal components. Since the carrier frequency is very large, a small change in any of the path delay causes a large change in its phase. This causes individual multipath components to add constructively and destructively, thus giving random amplitude variation in the received signal. This small-scale signal variation is termed as fading. For a moving receiver, this space-variant signal strength is perceived as a time-variant channel, where the velocity of the mobile terminal determines the speed of fluctuation. In the absence of a dominant Line-Of-Sight (LOS) path, the envelope of the received signal is characterized by Rayleigh density function. On the other hand, if the channel has a fixed LOS component, the signal envelope is characterized by Rician density function. Rayleigh fading is a worst-case scenario. However, envelope is often characterized by a more general density function known as Nakagami density (Goldsmith, 2005).

Apart from envelope fading, multipath propagation also causes delay spread. For a single transmitted pulse, the time T_m between the first and last received components of significant amplitude represents the maximum excess delay, which is an important parameter to characterize the channel. If T_m is greater than the symbol duration T_S, neighboring symbols interfere with each other, causing Inter-Symbol-Interference (ISI). An equivalent parameter to characterize delay spread in frequency domain is the coherence bandwidth B_c, indicating the range of frequency over which two signals are correlated. For a given signal bandwidth B_S, if $B_c << B_S$, the signal will experience flat fading, otherwise frequency selective fading. The effect of ISI can be mitigated using equalization, Orthogonal Frequency Division Multiplexing (OFDM) and spread spectrum (Goldsmith, 2005).

Similar to the delay spread in the time-domain, the received signal can also be spread in the frequency-domain. This spectral broadening is caused by the Doppler shift of an electromagnetic wave when observed from a moving object. The amount of shift for each path depends on the incident direction relative to the velocity vector of the receiver. In frequency domain, spectral broadening is characterized by a Doppler power spectrum. The range of frequency over which Doppler power spectrum is non-zero is called the Doppler spread of the channel B_D. The Doppler power spectrum has an important influence on the time-varying behavior of the channel because it is directly related to the temporal correlation of the received signal amplitude via the Fourier transform. For a given carrier frequency, the correlation increases with decreasing mobile velocity, such that slowly moving receivers encounter longer fades (and error bursts). Therefore, B_D is often used to characterize how rapidly the fading amplitude changes.

Challenges in Wireless Video Delivery

Wireless video delivery faces several challenges, such as high error rate, channel capacity variation, delay (important for real-time applications), low power, and low complexity requirements (Etoh, 2005; Girod, 2000). Some of these issues are briefly discussed here.

As discussed previously, wireless channel introduces both random as well as burst errors. Use of spatial and temporal dependencies as well as entropy coding makes compressed video bitstream extremely sensitive to these errors. Also, different bits generated by the video coder have different degrees of vulnerability to the channel errors (Sadka, 2002). The error in some bits has a local effect as shown in Figure 6(b), for the original frame shown in Figure 6a. On the other hand, the effect of errors in some bits (such as header, MV bits, and critical bits) is more severe, damaging the video frame globally by disturbing the synchronization between the encoder and decoder (Figure 6c). Also, use of MCP or MCTF causes error propagation in temporal as well as spatial directions. Figures 6d-6f show how the effect of error in a frame propagates to the subsequent frames. Moreover, fading causes burst errors and for slow changing channel, burst duration is likely to be quite large. The burst error may lead to loss of whole MB or slice or even whole frame. This leads to severe degradation in the decoded video quality.

For real time video communication such as video conferencing, delay is also an important issue. Usually the real time video communication can tolerate an end-to-end delay of not more than 200 ms (van der Schaar, 2007). If delay increases beyond this limit, real time video communications will be annoying. Another problem in wireless channel is the capacity variation. The capacity of a communication channel depends on bandwidth and the Signal-to-Noise-Ratio (SNR). In wireless channel, fading causes time varying SNR, therefore, for a given bandwidth, wireless channel has a time varying capacity. If the Channel Side Information (CSI) is not known at the transmitter, then deep fading will severally degrade the capacity or it will even cause an outage. How-

Figure 6. Effect of error in compressed video: (a) original frame, (b) local effect, (c) global effect, and (d)-(f) error propagation from first frame to next two frames

ever, if CSI is available at the transmitter, then, transmitter can optimally adapt the rate to maximize the capacity (Goldsmith, 2005).

Moreover, there is a growing trend to access video services using hand-held portable multimedia devices such as smart-phone, PDA, laptop, etc. These devices have limited processing power and battery life. Contemporary video coders such as H.264/AVC (Wiegand, 2003) or MC-EZBC (Chen, 2004) give state-of-the-art compression performance, however, use of large number of optimized coding tools makes them highly complex. When used in hand-held multimedia devices, they drain up majority of the battery power. Thus, computational complexity is also a big issue in designing a video coder. Low complexity algorithms are essential for *wireless video communication* using handheld portable devices, having limited processing power and battery life.

In summary, *wireless video communication* has to cope with time-varying channel conditions such as high error rates, burst error, link outage in severe fading, and capacity variations as well as limited power and low complexity issues. However, since the main focus of this chapter is reliable video communication, we discuss some of the conventional solutions to achieve reliability.

Conventional Solutions

Compressed video bitstream is extremely sensitive to bit errors, and therefore, not suitable for error prone and dynamically varying wireless channels. A variety of solutions have been proposed to cope with these challenges (Wang, 1998; Girod, 2000). These include FEC, retransmission using ARQ, error resilient coding, *hierarchical modulation*, scalable coding with transport prioritization, multiple-description coding, and error concealment by post processing, etc. Traditionally, these solutions are employed at various network layers (such as application, transport, MAC, and physical layers) separately and independently. A comprehensive discussion of all of these approaches is beyond the scope of this chapter and reader may refer to (Wang, 1998; Girod, 2000). In the following, some of these approaches are briefly reviewed.

Forward Error Correction

FEC is a very powerful and commonly used technique for error detection as well as error correction for video communication. FEC can be applied using either the block or convolutional coding. In an (n, k) block coding, for a message block of k symbols, n symbols code word is generated by adding n-k parity symbols, resulting into coding rate of k/n. Usually, FEC codes are capable of correcting both random errors and erasures in a block of n symbols. Reed-Solomon (RS) and Low Density Parity Check (LDPC) codes are the most widely used block codes in video communication applications. RS code, a special class of linear non-binary block codes, is very popular and mostly used due to its high error correcting capability against channel errors and erasures. It achieves ideal error protection against packet loss since it is a Maximum Distance Separable (MDS) code. However, besides block size limitation, RS codes also suffer from computational complexity. In recent years, LDPC codes have attracted the researchers due to its low encoding/decoding complexity and capability to operate over a large block size. However, LDPC codes are not MDS code and suffer with poor coding efficiency. Recent studies have pointed out that LDPC codes are more suitable for large block transmission over unidirectional channels, while RS codes are more appropriate for small block size and real time video streams (Nafaa, 2008).

On the other hand, a convolutional code is generated by passing the information sequence to be transmitted through a linear finite-state shift register. It encodes k bit message sequence into n bit code word. The encoded bits not only depend on the current k input bits but also on past input bits. The main decoding strategy for convolutional codes is based on the widely used Viterbi algorithm. The Rate Compatible Punctured Convolution code (RCPC), turbo codes, and Trellis Coded Modulation (TCM) are popular convolutional codes and have been widely used in video communication. Moreover, FEC has been applied at application layer, transport layer, link layer, and physical layer. For slow fading channel, a deep fade will cause burst errors of long duration. To improve the performance under such conditions, interleaving is often used. Interleaving randomizes the bit errors which can be then corrected by FEC. However, use of interleaving introduces delay and therefore, care must be taken while designing an interleaving scheme. Various types of FEC coding has been studied and applied for video communications (Hagenauer, 1999; Youshi, 2002; Kang, 2007; Hamzaoui, 2007; Nafaa, 2008; Lin, 2008; Sung, 2008; Kang, 2009; Noh, 2009; Lin, 2010). Moreover, recently some new codes such as raptor codes (Noh, 2009; Cataldi, 2010) have emerged as a powerful alternative to classical channel codes.

Automatic Repeat Request

If feedback channel is available, then better QoS can be achieved using an ARQ scheme (Lin, 1984). Usually in data link layer, classical ARQ is incorporated for hop-to-hop communication. ARQ can also be utilized at the transport layer for end-to-end delivery so that encoder can be adopted according to channel conditions and decoder's performance (Wang, 1998). For real-time video transmission, however, classic ARQ method is often unacceptable due to inherent large delay. Schemes have been developed to use feedback channel for video communication by modifying the classical feedback techniques. For example, instead of retransmitting indefinitely to recover from packet loss errors, the number of retransmissions can be limited by the tolerable amount of delay. Another simple technique is that whenever the decoder detects an error, a request to the encoder is send so that instead of retransmitting the same frame it codes the next video frame in intra-mode so that the propagation of previous errors can be stopped (Wada, 1989). In, yet another scheme, when a video data unit is corrupted, a retransmission request is sent to the encoder. Instead of waiting for the arrival of retransmitted data, the corrupted video part is concealed by a chosen error-concealment method and normal decoding is continued. In the background, a trace of the affected pixels and their coding information is recorded. Upon the arrival of the retransmitted data, the affected pixels are corrected, so that they are reproduced as if no transmission loss had occurred.

Error Resilient Modulation

Another approach for error robustness could be the use of *hierarchical modulation* at physical layer. Due to high bit rate requirements of video applications, it is desirable to choose a bandwidth efficient modulation technique such as *M*-ary Phase Shift Keying (PSK) or Quadrature Amplitude Modulation (QAM). These higher order modulation schemes increase the transmission capacity by assigning more bits to each transmitted symbol. However, modulation schemes which allow a larger number of bits per symbol, have symbols closer to each other in the constellation diagram, and small errors can result in erroneous decoding. To overcome this problem, the uses of non-uniform signal space constellations are suggested to give different degrees of error protection (Mirabbasi, 2000). In *hierarchical modulation*, the signal constellation is partitioned so that specific blocks of the partition contain message points with the maximum possible Euclidean distance between them at the expense of message points in other blocks which are separated by a much smaller minimum distance. The coding of the message points is chosen so that the specific bits assigned to the widely separated points have lower error probabilities than the others. This feature could be used for *Unequal Error Protection (UEP)* by providing more protection to important information without any additional bandwidth requirement at the expense of low priority information (Ghandi, 2006; Li, 2010). This is in contrast to FEC-based approach where added parity bits either increase the channel bandwidth (bit rate) or for Constant Bit Rate (CBR) the quality is sacrificed.

Robust Source Coding

In contrast to the previous methods, video communication can also be made robust by keeping redundancy in the source encoding of the video. This redundant information can be utilized at the decoder to confine the channel error to very small region and control its propagation to other block/frames. Sometime it can provide better error concealment at the decoder. To minimize the effect of error propagation due to prediction, the prediction can be restricted within non-overlapping spatial and temporal regions. To reduce spatial prediction error, whole frame is divided into number of slices with each slice include group of macroblock. So prediction from inter-slice macroblock is avoided. For reducing temporal error propagation group of frames called thread is formed. Each thread is coded without using prediction from other threads (Wang, 1998). Another approach which helps the error concealment task at the decoder is by sending motion vectors for macroblocks that would not ordinarily use motion-compensated prediction for e.g. sending motion vectors for macroblock in I-frames. The corrupted macroblock in I-frames can be recovered with the help of surrounded error free macroblock and motion vectors.

Redundancies can also be added in entropy coding. One of the techniques is self-synchronized entropy coding. A code word called synchronization code is added in the entropy encoded streams. This codeword has the capability to resynchronize the entropy decoder if erroneous bits have corrupted the data (Maxted, 1985; Ferguson, 1984).

Thus, error propagation is kept confined between two synchronization code words. To avoid high bit rate overhead shorter synchronization coder words are used but it increases the probability that a bit error may make a false synchronization code words. On the other hand long codes increase bit overhead but are strong in synchronization and robustness.

In Error Resilient Entropy Coding (EREC) (Redmill, 1996) scheme, variable length bit-streams from individual blocks are distributed into slots of equal size. Initially, the coded data for each image block are placed into the designated slot for the block either fully or partially. Then, a predefined offset sequence is used to search for empty slots to place any remaining bits of blocks that are bigger than the slot size. This is done until all the bits are packed into one of the slots. With EREC, synchronization can be obtained at the start of the each block in the decoder operation. The redundancy introduced by using EREC is negligible. Reversible Variable-Length Code (RVLC) is also employed as error resilient entropy coding (Ghanbari, 2003). In RVLC the coded bitstream can be decoded backward from any synchronization code word. If error occurred in the bitstream, the subsequent coded bitstream can be recovered by decoding backward from the next synchronization code word detected. The RVLC codes give improved robustness to errors but at the cost of reduced coding efficiency.

CROSS-LAYER WIRELESS VIDEO COMMUNICATION

As discussed in the preceding section that most of the traditional solutions of *error resiliency* applied to the video at various network layers independently while sacrificing bandwidth efficiency, increased delay or complexity. However, the utilization of these resources can be optimized using cross-layer approach. The cross-layer approach takes into account the dependencies and interactions among layers and supports optimization of resources across layer boundaries (Shakkottai, 2003). It may be noted here that cross-layer is not an alternative to the traditional layered approach, but rather as a complement. For *error resiliency* purposes, strategies at application, MAC and physical layers can be utilized jointly in order to optimize the communication resources. In this section, first the need of *cross-layer design* is discussed, followed by the cross-layer solutions for *error resiliency*. At the end of this section, some results are presented to highlight the advantage of cross-layer approach for *error resiliency* over the traditional approaches.

Need for Cross-Layer Design

Network protocol currently follows the layered architecture in which each layer provides a separate solution to the challenges offered by the modern wireless networks. Although, the decoupled layered protocol reduces network design complexity, however, independently optimized protocol paradigm is not well suited for wireless networks. In order to optimize the overall system performance, interaction among the various protocol layers is needed. Therefore, a cross-layer strategy involving multiple protocol layers is required to improve the overall system performance (van der Schaar, 2005).

In recent years, *cross-layer designs* for *wireless video communication* have been widely reported (Foukalas, 2008). Researchers have pursued *cross-layer design* from the perspectives of reliable communications, end-to-end QoS requirements, scheduling, complexity minimization, resource allocation, application driven criteria and so on. However, the main focus of this chapter is on the opportunities offered by cross-layer for error protection to enable reliable *wireless video communication*. The idea is to jointly consider error protection strategies at various network layers, in order to improve the transmission efficiency in terms of protection, bandwidth and resource consumption.

Cross-Layer Solutions

Usually, the error control strategies are implemented at the following layers: physical layer (adaptive modulation and FEC), MAC layer (ARQ and FEC), application layer (prioritized packetization, FEC and ARQ) and transport layer (ARQ and FEC). An overview of cross-layer schemes for improving multimedia transmission over wireless LANs is provided in (van der Schaar, 2005). A number of researchers have used application layer FEC and MAC layer ARQ to provide cross-layer error protection for wireless LANs (van der Schaar, 2003; Xu, 2004; Choi, 2007). An adaptive application layer FEC using RS code, MAC layer ARQ and adaptive packetization for MPEG-4 FGS coded video over WLAN is used in (van der Schaar, 2003). It was found that when the channel is poor, using only retransmission is not very useful and a strong FEC is required to improve the performance. However, as the channel improves, ARQ becomes the best protection strategy as there is no overhead associated unlike FEC. The work in (Choi, 2007) combines cross-layer error protection schemes composed of error correction code in the MAC layer, erasure code in the application layer, and ARQ across the MAC and application layers. The proposed cross-layer error control scheme reduces ARQ overhead adaptive to channel Bit Error Rate (BER), maintaining similar end quality. For delay sensitive video applications, delay constraint cross-layer solutions have also been proposed. For scalable video delivery, a network adaptive application-level error control scheme using hybrid UEP and delay constrained ARQ was proposed in Wang (2001). Current and estimated round trip time is used at sender side to determine the maximum number of retransmission based on delay constraint. A cross-layer protection mechanism that employs layered coding for the compression of the video combined with adaptive protection strategies, like priority queueing, selective dropping, and unequal protection through real-time MAC-layer retry-limit adaptation for video transmission over WLAN is presented in Li (2004). Simulation results demonstrate the effectiveness of layered priority queuing and cross-layer adaptive error protection in improving the overall subjective video quality. The work proposed in Ksentini (2006) uses a cross-layer architecture combining Data Portioning (DP) mode of H.264 and QoS-based MAC protocol of 802.11e WLAN. Through employing DP, the H.264 encoder partitions the compressed data in separate units of different importance called partitions. Based on the QoS requirements of those different partitions, a marking algorithm at the MAC layer is specified that associates each partition with an access category (AC) provided by 802.11e enhanced distributed channel access (EDCA). Thus, it allows the application layer to pass its streams along with their requirements in order to protect the most important H.264 information, which guarantees low degradation of received H.264 stream.

However, for large networks such as cellular mobile, the delay introduced by MAC layer ARQ may not be suitable for delay sensitive video applications. Also, for applications such as multicast and broadcast, there is no provision of feedback channel to support ARQ protocol. Therefore, for such scenarios, cross-layer error protection employing different combinations of application layer FEC and adaptive modulation is a viable alternative (Barmada, 2006; Pei, 2007; Gross, 2004; Khan, 2010). For example, in (Barmada, 2006), a combination of turbo code and *Hierarchical QAM (HQAM)* is used to provide UEP for two-layer scalable H.264/AVC bitstream. Similarly, a combination of RCPC and non-uniform PSK modulation is suggested in Pei (2007) to achieve UEP for layered video coded bitstream. The authors in (Gross 2004) schedule packet transmission over Orthogonal Frequency-Division Multiplexing (OFDM) channels giving higher priority to the most important packets (I-frames). This approach takes the size of the queues into account in order to determine the number of OFDM subcarriers to allocate to each user. Also, the observed quality of the channels is explored to assign to each terminal a set of sub-carriers in a good transmission state. A cross-layer scheduling framework with adaptive modulation and coding is proposed in Liu (2005). In Bhatnagar (2009), *error resiliency* tools such as data partition etc at application layer are combined with asymmetric 8-ary PSK modulation to achieve UEP. The work in Pradas (2009) implements a *cross-layer design* using error correction code and *hierarchical modulation* in the physical layer, erasure code at the transport layer and FGS *video coding* at the application layer.

The authors in Wang (2010) have used application layer and IP layer FEC and the amount of FEC redundancy is determined by a Playable Frame Rate (PFR) model. Broadcast of scalable video stream in a CDMA2000 wireless system using Reed-Solomon coding and priority aware interleaving in MAC layer is presented in Kang (2011). The work in Bajic (2007) presents cross-layer schemes combining source rate selection, power management, and error control for wireless multicast of SNR scalable video. An analytical framework for a *cross-layer design* to control the transmission power and *forward error correction* on the basis of channel status and source information for embedded video delivery over a wireless link is given in Granelli (2007). A source adaptation-based cross-layer UEP is proposed in Wang (2001). Assuming no CSI at transmitter, it uses motion information for adaptation. Motion information inside each frame is used to assign different amount of redundancy bits to different parts of the frame, thus achieving UEP. This work also uses a novel interleaving scheme for bursty packet losses occurring on wireless networks without incurring any additional implementation complexity or delay.

Cross-Layer UEP for Wavelet Coded Scalable Video

As discussed earlier, the *wavelet-based video coder* equipped with embedded coding strategies such as Wavelet Block-Tree Coding (WBTC) generates scalable video bitstream (Moinuddin, 2006), in which the importance of bits decreases as coding progresses from most significant bit-plane to the least significant bitplane. Moreover, the bits generated by video coder have different degrees of sensitivity to channel errors. Some bits are highly sensitive to noise and a single bit error may cause the loss of synchronization between the encoder and decoder and may cause the failure of the reconstruction process after it occurs (Figure 6c). These bits are called critical bits and consist of headers, motion vector bits and significant test bits. The remaining bits, called non-critical bits, have less severe effects of errors. The error in non-critical bits has effects limited to a single coefficient or in its neighborhood and does not disturb the progression of the decoding process after it occurs (Figure 6b). The non-critical bits consist of the sign and refinement bits (Khan, 2010). Therefore, the network recourses should be utilized according to their importance in order to provide optimal UEP. Two commonly used approaches to provide UEP are the use of FEC and *hierarchical modulation*, which have their own limitations. The FEC-based UEP performed at the application layer is bandwidth inefficient. It offers a tradeoff between quality and bandwidth. On the other hand, *hierarchical modulation*-based UEP performed at the physical layer achieves protection of High Priority (HP) bits without requiring additional bandwidth, but at the cost of increased errors in the Low Priority (LP) bits. In order to improve the overall quality of the received video, lower priority bits also need to be protected, which can be achieved by only increased transmitter power. That is, the physical layer protection offers a tradeoff between quality and power. Video transmission over mobile devices is generally

Figure 7. Reliable video communication system combining FEC and HQAM

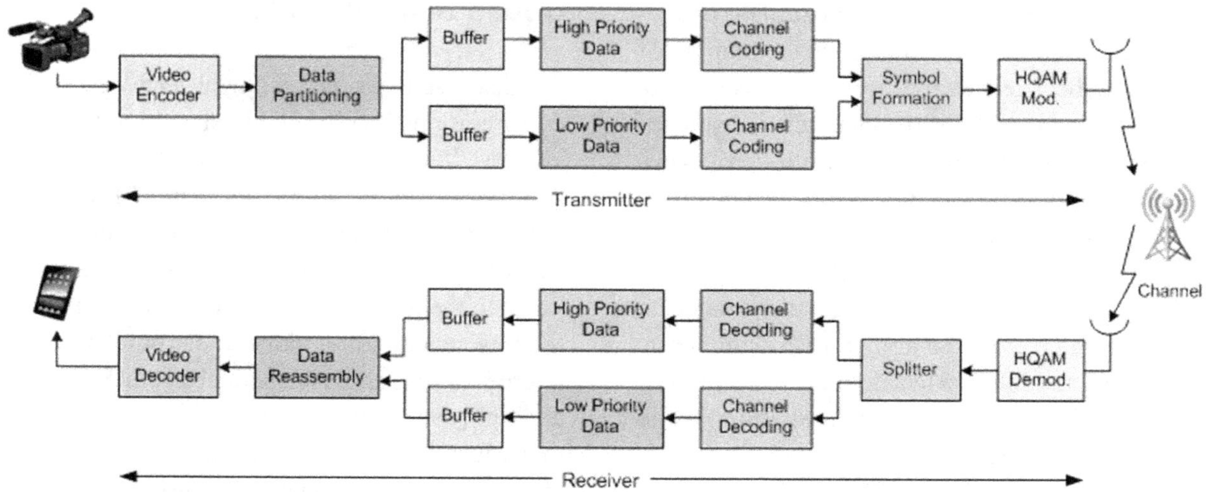

constrained by both powers (due to limited battery life) as well as bandwidth of wireless channels.

The cross-layer approach can be used to transmit scalable coded video over wireless network under power and bandwidth constrained environments to achieve the best overall quality. That is, to achieve better quality, the application layer's FEC need to be combined with physical layer's adaptive modulation such as *Hierarchical QAM (HQAM)* (Mirabbasi, 2000). The block diagram of such a video communication system to enhance reliability is shown in Figure 7 (Khan, 2010). At the application layer, the encoded video bitstream is split into two separate substreams: HP and LP. Then they are protected unequally using (n, k) RS codes. Let the FEC coding rate (k/n) of two substreams are *RH* and *RL* (*RH* < *RL*) respectively. The FEC coded HP and LP substreams are then multiplexed to form symbols to be mapped over asymmetrical constellations of 16-HQAM, shown in Figure 8. The value of modulation parameter $\alpha = a/b$ in HQAM and FEC rate for two substreams (*RH* and *RL*) can be adjusted appropriately to achieve the best overall quality.

Figure 8. Signal constellation diagram for 16-HQAM

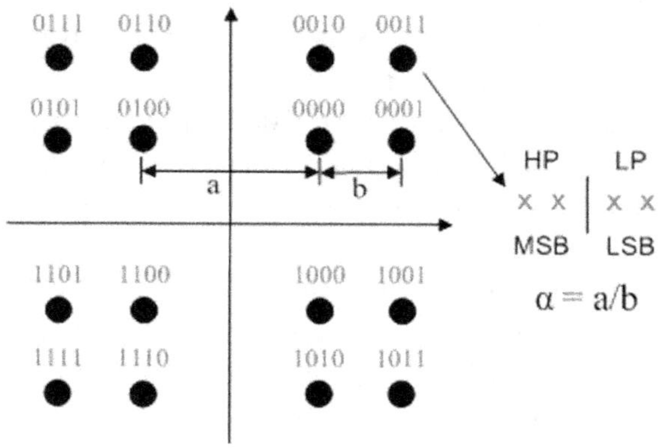

The layered architecture utilizing cross-layer approach for UEP in a video communication system is depicted in Figure 9. The function of the cross-layer allocator is to allocate the channel resources for video data with different priorities. It receives prioritized bitstreams from application layer, and based on the noise condition of the channel, allocates and sends FEC coding rates (*RH* and *RL*) to prioritized bitstreams of the application layer and modulation parameter α of HQAM to the physical layer.

In order to jointly utilize the properties of application layer FEC and physical layer HQAM, care should be taken that after FEC coding, the sizes of HP and LP substreams are such that they

Figure 9. A cross-layer approach for UEP in video communication system

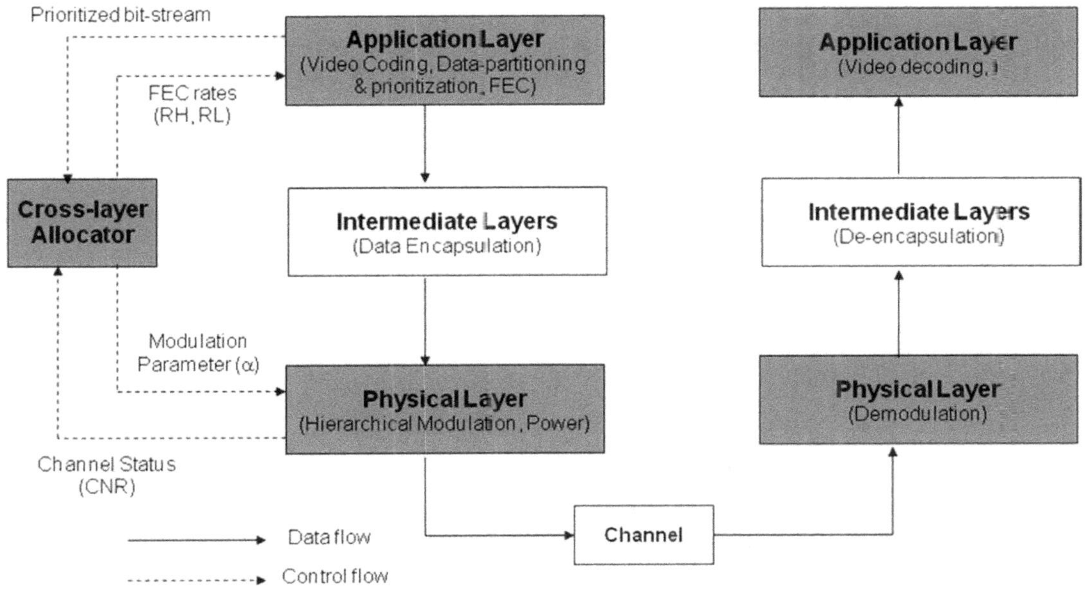

are easily mapped to 16-HQAM. For 16-HQAM, in which a symbol is formed by multiplexing two bits of HP and two bits of LP substream (see Figure 8), it is necessary that FEC protected HP and LP substreams must be of the same size. However, the motion vector bits (B_{mv}) and critical bits (B_C) of wavelet coefficient which have very severe effect in case of channel error and need to be protected strongly, consumes about 65-70% of bit budget, whereas non-critical bits (B_{NC}) that causes local degradation are only 30-35% of total bit budget. Under such condition, it is difficult to map symbols on the constellation diagram of 16-*HQAM*, without reorganization of bits. Since embedded coders have the property of generating most significant bits earlier than the least significant bits. Therefore, errors in earlier bits are more damaging than those which are generated later. This allows treating some of the critical bits (generated in later bitplanes) as part of LP substream as shown in Figure 10.

Therefore, in the scheme used here, within each video frame, HP substream consists of B_{mv} and B_C^{HP}, whereas LP substream consists of B_{NC} and B_C^{LP}, and thereby the FEC rates for HP and LP substreams, *RH* and *RL* should be selected such that the following constraint is satisfied (see Figure 10):

$$\frac{1}{RH}\left(B_{mv} + B_C^{HP}\right) = \frac{1}{RL}\left(B_C^{LP} + B_{NC}\right) \qquad (1)$$

Symbols are then formed by combining two bits (MSB) from FEC coded HP substream and remaining bits (LSB) from FEC coded LP substreams. These symbols are then modulated using *HQAM* with appropriate but fixed value of α to achieve UEP.

The performance of the cross-layer video communication system is evaluated through simulation. Video sequence (CIF resolution, 4:2:0 colour format *Football* sequence of 90 frames) is coded at the rate of 1000kbps using wavelet-based hybrid video coding structure (shown in Figure 1). The coder employs half-pixel accurate motion estimation, OBMC and WBTC algorithm for quantization and encoding of motion predicted

Figure 10. Bitstream partitioning for cross-layer UEP

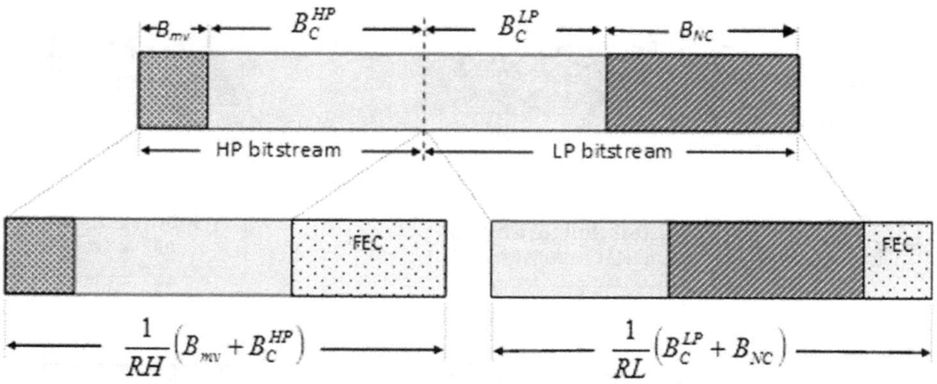

error frames. Motion vectors are arithmetic coded. At the application layer, FEC is applied to HP and LP substreams using (n, k) RS codes over a packet size of n = 255 bytes. The (n, k) may have any values from the set {(255, 255), (255, 215), (255, 195)} with the corresponding FEC rate k/n for these RS codes are1, 0.84 and 0.76 respectively. At the physical layer, 16-HQAM is used for adaptive protection of HP bitstream having value of modulation parameter from the set α = {1, 1.2, 1.4, 1.6, 1.8, and 2}. Owing to random nature of channel errors over the wireless networks and their impact on the compressed video, for each tested channel conditions (specified in terms of Carrier-to-Noise-Ratio [CNR]) and protection strategy, 20 different runs of the experiments were conducted. The objective video quality is measured in terms of average Peak-Signal-to-Noise-Ration (PSNR), which is determined by averaging the PSNR value of different runs of experiments. Figures 11 (a) and (b) shows performance of the cross-layer UEP for (RH = 0.76, RL = 1, $1 \leq \alpha \leq 2$) and (RH = 0.76, RL = 0.84, $1 \leq \alpha \leq 2$) respectively. Both these figures also include results for no protection (RL = 1, RH = 1 and α = 1) case for comparison purposes. Comparing results for (RH, RL, α) = (1, 1, 1) and (0.76, 1, 1) in Figure 11a, it can be seen that effect of FEC coding of HP bits is very dominant and improves the quality of decoded video drastically for CNR higher than 9 dB. Similar effect can also be observed for (RH, RL, α) = (1, 1, 1) and (0.76, 0.84, 1) in Figure 11b. Further comparing PSNR values for (RH, RL, α) = (0.76, 1, 1) and (0.76, 0.84, 1) Figures 11a and 11b, respectively, it can be observed that the best quality video can be obtained at lower CNR (equivalent to higher channel noise) for (RH, RL, α) = (0.76, 0.84, 1). This improved performance is due to the FEC coding of LP bits, which are otherwise not protected in results of Figure 11a.

Further, from these graphs, it can be seen that if RH and RL are fixed and α is increased, the decoded video quality improves at low CNR (high channel noise conditions), but at the same time, it deteriorates at CNR higher than about 11 dB. This is due to the fact that at low CNR (high noise power), the probability of errors in both HP and LP substream is high. There is room to improve HP data error performance for better visual quality. On the other hand, in high CNR condition, the error probability of HP substream is very low. The increased protection of HP bits cannot offer any further visual quality improvement. However, on the contrary, increasing α of HQAM increases the probability of error in LP bits. Since some of the significant test bits (of later bitplanes) are also parts of LP substream and errors in these bits cause degradation of visual quality due to error in more number of bits in LP substream.

Figure 11. Performance of cross-layer UEP with fixed modulation parameter for "football" sequence coded at 1000 kbps for (a) RH = 0.76, RL = 1, 1≤ α ≤ 2 and (b) RH = 0.76, RL = 0.84, 1≤ α ≤ 2

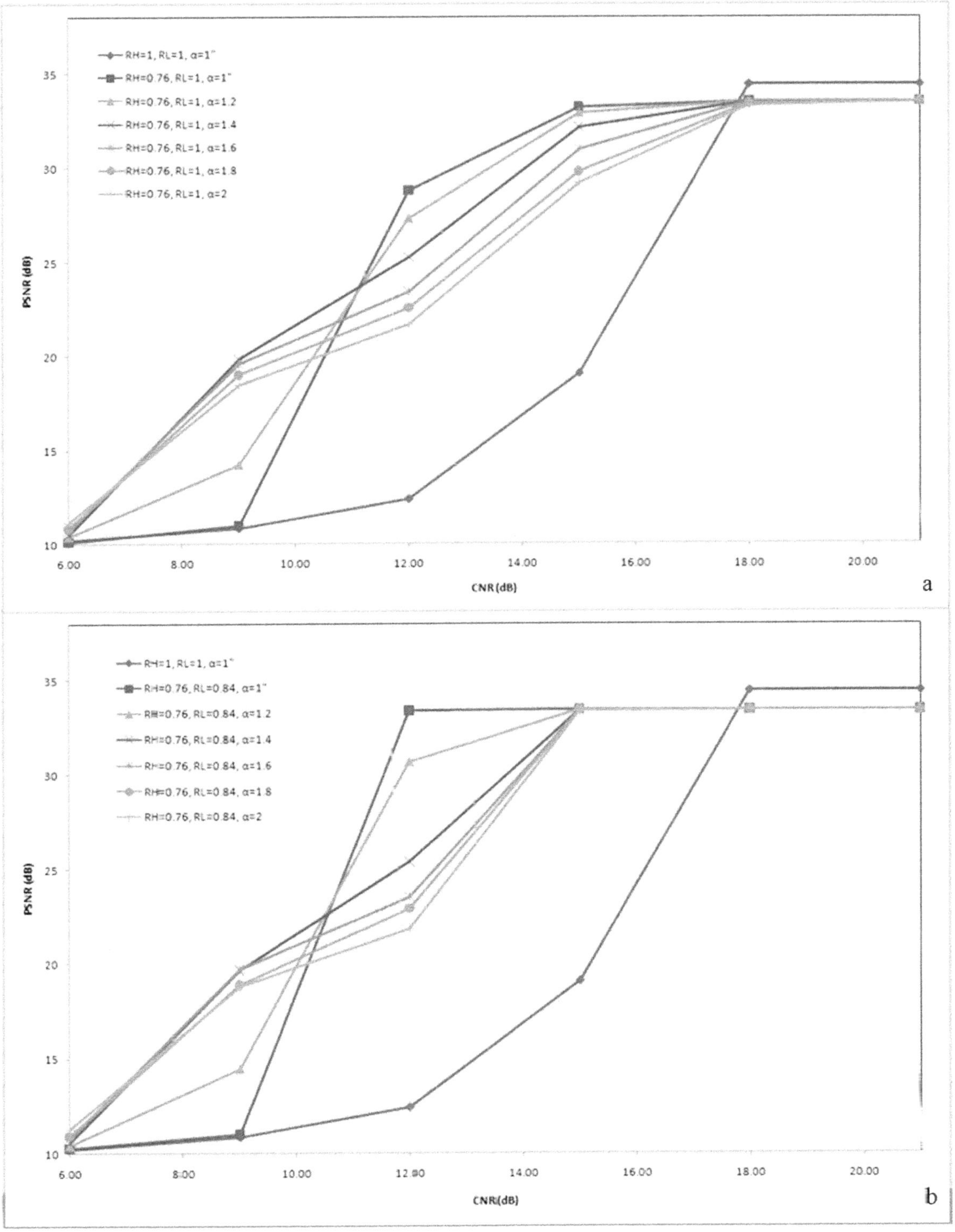

Figure 12. Partitioning and symbol formation by combining HP and LP bits in cross-layer UEP with adaptive α

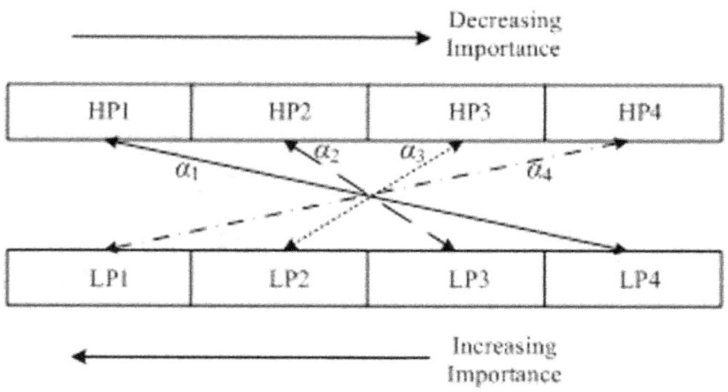

Cross-Layer UEP with Adaptive Modulation Parameter α

As discussed that video communication system with cross-layer UEP suffers from the problem that increasing modulation parameter α increases the protection to HP bits, but at the cost of LP bits, thereby reducing its performance at higher CNR, as evident from Figure 11. Further, the use of a constant value of α assigns the same level of protection to all HP bits relative to LP bits. However, in embedded bitstream, the importance of HP as well as LP bits decreases as coding proceeds. In fact, the initial bits have more importance than the later bits in either of HP or LP substreams.

In order to further improve the performance of the system, each of the two substreams (HP and LP) are further divided into four blocks each containing equal number of bits as shown in Figure 12. Due to the embedded nature of bitstream, the bits in HP1 block are of highest significance and should be protected strongly, whereas the bits of LP4 are of least significance and may be protected lightly. Therefore, for 16-HQAM, if symbols are formed by combining two MSB bits from HP1 block and two LSB bits from LP4 block, then these symbols can be modulated with largest value of α (say α_1). Similarly symbols formed by combining bits from HP2 and LP3 can be modulated using relatively smaller value of α (say $\alpha_2 < \alpha_1$), as bits of HP2 block are of relatively lesser importance than HP1 and LP3 bits are relatively higher importance than LP4. Similarly the values α_3 and α_4 are used while modulating symbols formed from bits of block pairs (HP3, LP2), and (HP4, LP1), respectively. We termed this scheme as adaptive HQAM (AHQAM). The scheme is expected to perform better than HQAM as the value of modulation parameter is reduced successively according to the importance of HP and LP blocks. That is $\alpha_1 > \alpha_2 > \alpha_3 > \alpha_4$ as the importance of bits has the order HP1 > HP2 > HP3 > HP4 > LP1 > LP2 > LP3 > LP4.

The results of cross-layer UEP with adaptive α are presented in Figure 13, for which most of the parameters, i. e. RH and RL are kept same as in Figure 11, except the modulation parameter of HQAM for bits in different blocks is varied as (α_1, α_2, α_3, α_4) = (1.6, 1.4, 1.2, 1.0). Figures 13 compares the performance of video communication system with adaptive α with that for fixed value of $\alpha = 1$, $\alpha = 1.3$, and $\alpha = 1.6$, for '*football*' sequence coded at 1000 kbps. It can be observed from the figure that cross-layer UEP with adaptive α has consistently better performance for whole range of CNR values as compared to that with fixed α.

Figure 13. Performance comparison of cross-layer UEP having fixed and adaptive α for "football" sequence coded at 1000 kbps for (a) RH = 0.76, RL = 1, 1 ≤ α ≤ 2 and (b) RH = 0.76, RL = 0.84, 1 ≤ α ≤ 2

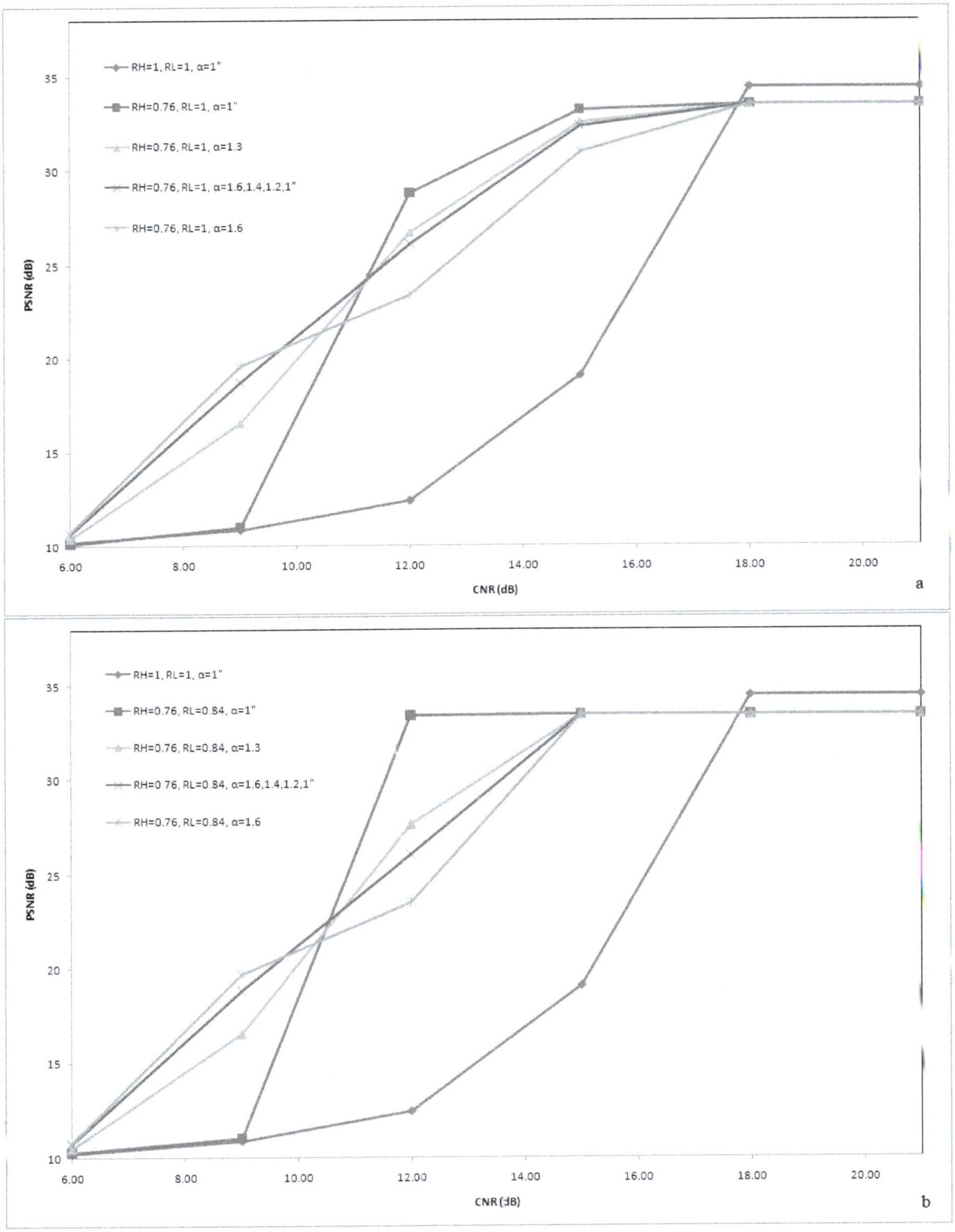

CONCLUSION

In this chapter, cross-layer approach for reliable video communication over wireless networks has been discussed. It has been shown that establishing interaction between application and physical layers to design an UEP system can lead to an important improvement in overall performance of the system. The emphasis has been given to the error resilient *hierarchical modulation* (fixed or adaptive) at physical layers and efficient partitioning (based on *error resiliency*) and use of FEC-based UEP at the application layer. It was emphasized in the chapter that appropriate selection of FEC coding rates at application layer and adaptively varying the modulation parameter value of *HQAM* at physical layer can result in overall performance improvement of the system for wide range of channel noise conditions.

REFERENCES

Al-Mualla, M. E., Canagrajah, C. N., & Bull, D. R. (2002). *Video coding for mobile communication*. San Diego, CA: Academic Press.

Anton-Haro, C. (2006). Cross-layer scheduling for multi-user MIMO systems. *IEEE Communications Magazine*, *44*(9), 39–45. doi:10.1109/MCOM.2006.1705977

Bajic, I. V. (2007). Efficient cross-layer error control for wireless video multicast. *IEEE Transactions on Broadcasting*, *53*(1), 276–285. doi:10.1109/TBC.2006.889211

Barmada, B., Ghandi, M. M., Jones, E. V., & Ghanbari, M. (2006). Combined turbo coding and hierarchical QAM for unequal protection of H.264 coded video. *Signal Processing Image Communication*, *21*(1), 390–395. doi:10.1016/j.image.2006.01.001

Bhatnagar, A., Kumar, S., Janarthanan, A., Annamalai, A., Jr., Medley, M., & John Matyjas, J. (2009). A multilevel unequal error protection scheme for robust H.264/AVC video transmission over wireless channels. In *Proceedings of Military Communication Conference*. Military Communication Conference.

Cataldi, P., Grangetto, M., Tillo, T., Magli, E., & Olmo, G. (2010). Sliding-window raptor codes for efficient scalable wireless video broadcasting with unequal loss protection. *IEEE Transactions on Image Processing*, *19*(6), 1491–1503. doi:10.1109/TIP.2010.2042985

Chen, P., & Woods, J. W. (2004). Bidirectional MC-EZBC with lifting implementation. *IEEE Transactions on Circuits and Systems for Video Technology*, *14*(10), 1183–1194. doi:10.1109/TCSVT.2004.833165

Choi, J.-Y., & Shin, J. (2007). Cross-layer error-control with low-overhead ARQ for H.264 video transmission over wireless LANs. *Computer Communications*, *30*(1), 1476–1486. doi:10.1016/j.comcom.2006.12.027

Dirac Video Project. (2011). *Website*. Retrieved from http://diracvideo.org.

Etoh, M., & Yoshimura, T. (2005). Advances in wireless video delivery. *Proceedings of the IEEE*, *93*(1), 111–122. doi:10.1109/JPROC.2004.839605

Ferguson, T. J., & Ranowitz, J. H. (1984). Self-synchronizing Huffman codes. *IEEE Transactions on Information Theory*, *30*, 687–693. doi:10.1109/TIT.1984.1056931

Foh, C. H., Zhang, Y., Ni, Z., Cai, J., & Ngan, K. N. (2007). Optimized cross-layer design for scalable transmission over IEEE 802.11e network. *IEEE Transactions on Circuits and Systems for Video Technology*, *17*(12), 1665–1678. doi:10.1109/TCSVT.2007.903808

Foukalas, F., Gazis, V., & Alonistioti, N. (2008). Cross-layer design proposals for wireless mobile networks: A survey and taxonomy. *IEEE Communications Surveys, 10*(1), 72–85. doi:10.1109/COMST.2008.4483671

Fowler, J. E., & Pesquet-Popescu, B. (2007). An overview on wavelets in source coding, communications, and networks. *EURASIP Journal on Image and Video Processing, 4*, 1–27. doi:10.1155/2007/60539

Ghanbari, M. (2003). *Standard codecs: Image compression to advanced video coding*. London, UK: IEE. doi:10.1049/PBTE049E

Ghandi, M. M., & Ghanbari, M. (2006). Layered H.264 video transmission with hierarchical QAM. *Journal of Visual Communication and Image Representation, 17*(8), 451–466. doi:10.1016/j.jvcir.2005.05.005

Girod, B., & Farber, N. (2000). Wireless video. In Sun, M.-T., & Reibman, A. R. (Eds.), *Compressed Video over Networks* (pp. 1–38). New York, NY: Marcel Dekker.

Goldsmith, A. (2005). *Wireless communications*. New Delhi, India: Cambridge University Press.

Golwelkar, A., & Woods, J. W. (2007). Motion compensated temporal filtering and motion vector coding using biorthogonal filters. *IEEE Transactions on Circuits and Systems for Video Technology, 17*(4), 417–428. doi:10.1109/TCSVT.2007.893834

Goyal, V. (2001). Theoretical foundations of transform coding. *IEEE Signal Processing Magazine, 18*(9), 9–21. doi:10.1109/79.952802

Granelli, F., Costa, C. E., & Katsaggelos, A. K. (2007). A study on the usage of cross-layer power control and forward error correction for embedded video transmission over wireless links. *Advances in Multimedia*. Retrieved from http://www.hindawi.com/journals/am/2007/095807/abs/.

Gross, J. (2004). Cross-layer optimization of OFDM transmission systems for MPEG-4 video streaming. *Computer Communications, 27*, 1044–1055. doi:10.1016/j.comcom.2004.01.010

Hagenauer, J., & Stockhammer, T. (1999). Channel coding and transmission aspects for wireless multimedia. *Proceedings of the IEEE, 87*(10), 1764–1777. doi:10.1109/5.790636

Hamzaoui, R., Stankovic, V., & Xiong, Z. (2007). Forward error control for packet loss and corruption. In van der Schaar, M., & Chou, P. (Eds.), *Multimedia over IP and Wireless Networks: Compression, Networking, and Systems* (pp. 271–291). California: Academic Press. doi:10.1016/B978-012088480-3/50010-2

Haratcherev, I., Taal, J., Langendoen, K., Lagendijk, R., & Sips, H. (2006). Optimized video streaming over 802.11 by cross-layer signaling. *IEEE Communications Magazine, 44*(1), 115–121. doi:10.1109/MCOM.2006.1580941

Kang, K., Cho, Y., & Shin, H. (2007). Energy-efficient MAC-layer error recovery for mobile multimedia applications in 3GPP2 BCMCS *IEEE Transactions on Broadcasting, 53*(1), 338–349. doi:10.1109/TBC.2007.891699

Kang, K., & Kim, T. (2009). Improved error control for real-time video broadcasting over CDMA2000 networks. *IEEE Transactions on Vehicular Technology, 58*(1), 188–197. doi:10.1109/TVT.2008.926077

Kang, K., Won, J., & Jeon, W. J. (2012). Differentiated protection of video layers to improve perceived quality. To appear in *IEEE Transactions on Mobile Computing, 11*(2), 292-304.

Katsaggelos, A. K., Eisenberg, Y., Zhai, F., Berry, R., & Pappas, T. N. (2005). Advances in efficient resource allocation for packet-based real-time video transmission. *Proceedings of the IEEE, 93*(1), 135–147. doi:10.1109/JPROC.2004.839621

Khan, M. A., Moinuddin, A. A., Khan, E., & Ghanbari, M. (2010). Reliable transmission of wavelet-based scalable video over wireless networks using cross-layer approach. *IET Communications*, *4*(11), 1325–1336. doi:10.1049/iet-com.2009.0412

Khan, S., Peng, Y., Steinbach, E., Sgroi, M., & Kellerer, W. (2006). Application driven cross-layer optimization for video streaming over wireless networks. *IEEE Communications Magazine*, *44*(1), 122–130. doi:10.1109/MCOM.2006.1580942

Kim, S. W., Kim, S.-Y., Kim, S., & Heo, J. (2008). Performance analysis of forward error correcting codes in IPTV. *IEEE Transactions on Consumer Electronics*, *54*(2), 376–380. doi:10.1109/TCE.2008.4560102

Ksentini, A., Naimi, M., & Gueroui, A. (2006). Toward an improvement of H.264 video transmission over IEEE 802.11e through a cross-layer architecture. *IEEE Communications Magazine*, *44*(1), 107–114. doi:10.1109/MCOM.2006.1580940

Kumwilaisak, W., Hou, Y. T., Zhang, Q., Zhu, W., Kuo, C.-C. J., & Zhang, Y.-Q. (2003). A cross-layer quality-of-service mapping architecture for video delivery in wireless network. *IEEE Journal on Selected Areas in Communications*, *21*(10), 1685–1698. doi:10.1109/JSAC.2003.816445

Li, F., Liu, G., & He, L. (2010). Cross-layer scheduling for multiuser H.264 video transmission over wireless networks. *IET Communications*, *4*(8), 1012–1025. doi:10.1049/iet-com.2009.0618

Li, P., Chang, Y., Feng, N., & Yang, F. (2010). A novel hierarchical QAM-based unequal error protection scheme for H.264/AVC video over frequency-selective fading channels. *IEEE Transactions on Consumer Electronics*, *56*(4), 2741–2746. doi:10.1109/TCE.2010.5681164

Li, Q., & van der Schaar, M. (2004). Providing adaptive QoS to layered video over wireless local area networks through real-time retry limit adaptation. *IEEE Transactions on Multimedia*, *6*(2), 278–290. doi:10.1109/TMM.2003.822792

Li, W. (2001). Overview of fine granularity scalability in MPEG-4 video standard. *IEEE Transactions on Circuits and Systems for Video Technology*, *11*(3), 301–317. doi:10.1109/76.911157

Lin, C.-H., Ce-Kuen Shieh, N., Chilamkurti, K., Ke, C.-H., & Hwang, W.-S. (2008). A RED-FEC mechanism for video transmission over WLANs. *IEEE Transactions on Broadcasting*, *54*, 517–524. doi:10.1109/TBC.2008.2001713

Lin, S., Costello, D. J., & Miller, M. J. (1984). Automatic repeat error control schemes. *IEEE Communications Magazine*, *22*(12), 5–17. doi:10.1109/MCOM.1984.1091865

Lin, T.-L., & Cosman, P. C. (2010). Efficient optimal RCPC code rate allocation with packet discarding for pre-encoded compressed video. *IEEE Signal Processing Letters*, *17*(5), 505–508. doi:10.1109/LSP.2010.2044938

Liu, Q., Zhou, S., & Giannakis, G. (2005). Cross-layer scheduling with prescribed QoS guarantees in adaptive wireless networks. *IEEE Journal on Selected Areas in Communications*, *23*(5), 1056–1066. doi:10.1109/JSAC.2005.845430

Maani, E., & Katsaggelos, A. K. (2010). Unequal error protection for robust streaming of scalable video over packet lossy networks. *IEEE Transactions on Circuits and Systems for Video Technology*, *20*(3), 407–416. doi:10.1109/TCSVT.2009.2035846

Martucci, S. A., Sodagar, I., Chiang, T., & Zhang, Y.-Q. (1997). A zerotree wavelet video coder. *IEEE Transactions on Circuits and Systems for Video Technology*, *7*(2), 109–118. doi:10.1109/76.554422

Maxted, J. C., & Robinson, J. P. (1985). Error recovery for variable length codes. *IEEE Transactions on Information Theory*, *31*, 794–801. doi:10.1109/TIT.1985.1057110

Melodia, T., & Akyildiz, I. F. (2010). Cross-layer QoS-aware communication for ultra wide band wireless multimedia sensor networks. *IEEE Journal on Selected Areas in Communications*, *28*(5), 653–663. doi:10.1109/JSAC.2010.100604

Mirabbasi, S., & Martin, K. (2000). Hierarchical QAM: A spectrally efficient DC-free modulation scheme. *IEEE Communications Magazine*, *38*, 140–146. doi:10.1109/35.883503

Moinuddin, A. A., Khan, E., & Ghanbari, M. (2006). Wavelet-based low bit rate video coding using block-tree approach. In *Proceedings of IET International Conference on Visual Information Engineering* (*VIE 2006*), (pp. 287-291). Bangalore, India: IET.

Nafaa, A., Taleb, T., & Murphy, L. (2008). Forward error correction strategies for media streaming over wireless networks. *IEEE Communications Magazine*, *46*(12), 72–79. doi:10.1109/MCOM.2008.4427233

Noh, K., Yoon, S., & Heo, J. (2009). Performance analysis of network coding with raptor codes for IPTV. *IEEE Transactions on Consumer Electronics*, *55*(1), 83–87. doi:10.1109/TCE.2009.4814418

Ohm, J.-R. (2004). *Multimedia communication technology*. Berlin, Germany: Springer-Verlag.

Ohm, J.-R. (2005). Advances in scalable video coding. *Proceedings of the IEEE*, *93*(1), 42–56. doi:10.1109/JPROC.2004.839611

Orchad, M. T., & Sullivan, G. J. (1994). Overlapped block motion compensation: an estimation-theoretic approach. *IEEE Transactions on Image Processing*, *3*, 693–699. doi:10.1109/83.334974

Pei, Y., & Modestino, J. W. (2007). Cross-layer design for video transmission over wireless Rician slow-fading channels using an adaptive multiresolution modulation and coding scheme. *EURASIP Journal on Advances in Signal Processing*. Retrieved from http://downloads.hindawi.com/journals/asp/2007/086915.pdf.

Pesquet-Popescu, B., Li, S., & van der Schaar, M. (2007). Scalable video coding for adaptive streaming applications. In van der Schaar, M. (Ed.), *Multimedia over IP and Wireless Networks: Compression, Networking, and Systems* (pp. 117–158). California: Academic Press. doi:10.1016/B978-012088480-3/50006-0

Pradas, D., Bouabdallah, A., Lacan, J., Vazquez Castro, M. A., & Michel Bousquet, M. (2009). Cross-layer optimization of unequal protected layered video over hierarchical Modulation. In *Proceedings of the IEEE Global Telecommunications Conference 2009* (*GLOBECOM 2009*). IEEE Press.

Redmill, D. W., & Kingsbury, N. G. (1996). The EREC: An error resilient technique for coding variable length blocks of data. *IEEE Transactions on Image Processing*, *5*(4), 565–574. doi:10.1109/83.491333

Sadka, A. H. (2002). *Compressed video communications*. London, UK: John Wiley & Sons, Ltd. doi:10.1002/0470846712

Schwarz, H., Marpe, D., & Wiegand, T. (2007). Overview of the scalable video coding extension of the H.264/AVC standard. *IEEE Transactions on Circuits and Systems for Video Technology*, *17*, 1103–1120. doi:10.1109/TCSVT.2007.905532

Shakkottai, S., Rappaport, T. S., & Karrisson, P. C. (2003). Cross-layer design for wireless networks. *IEEE Communications Magazine*, *41*(10), 74–80. doi:10.1109/MCOM.2003.1235598

Sullivan, G. J., & Wiegand, T. (2005). Video compression-from concepts to the H.264/AVC standard. *Proceedings of the IEEE, 93*(1), 18–31. doi:10.1109/JPROC.2004.839617

van der Schaar, M., & Chou, P. (2007). *Multimedia over IP and wireless networks: Compression, networking, and systems*. California: Academic Press.

van der Schaar, M., Krishnamachari, S., Choi, S., & Xu, X. (2003). Adaptive cross-layer protection strategies for robust scalable video transmission over 802.11 WLANs. *IEEE Journal on Selected Areas in Communications, 21*(10), 1752–1763. doi:10.1109/JSAC.2003.815231

van der Schaar, M., & Shankar, S. N. (2005). Cross-layer wireless multimedia transmission: Challenges, principles, and new paradigms. *IEEE Wireless Communication Magazine, 12*(4), 50–58. doi:10.1109/MWC.2005.1497858

Wada, W. (1989). Selective recovery of video packet loss using error concealment. *IEEE Journal on Selected Areas in Communications, 7*(6), 807–814. doi:10.1109/49.32344

Wang, G., Zhang, Q., Zhu, W., & Zhang, Y.-Q. (2001). Channel-adaptive unequal error protection for scalable video transmission over wireless channel. In *Proceedings of SPIE VCIP*, (pp. 648-655). San Jose, CA: SPIE.

Wang, Y., Ostermann, J., & Zhang, Y.-Q. (2002). *Video processing and communications*. Upper Saddle River, NJ: Prentice-Hall.

Wang, Y., & Zhu, Q.-F. (1998). Error control and concealment for video communication: A review. *Proceedings of the IEEE, 86*(5), 974–997. doi:10.1109/5.664283

Wang, Y.-C., Lin, C.-H., Tsai, M.-F., & Shieh, C.-K. (2010) Cross-layer unequal error protection mechanism with an interleaved strategy for video streaming over wireless networks. In *Proceedings of the IEEE WCNC 2010*. IEEE Press.

Wiegand, T., Sullivan, G. J., Bjontegaard, G., & Luthra, A. (2003). Overview of the H.264/AVC video coding standard. *IEEE Transactions on Circuits and Systems for Video Technology, 13*(7), 560–576. doi:10.1109/TCSVT.2003.815165

Xu, X., van der Schaar, M., Krishnamachari, S., Choi, S., & Wang, Y. (2004). Fine-granular-scalability video streaming over wireless LANs using cross-layer error control. In *Proceedings of the IEEE International Conference on Acoustic, Speech and Signal Processing (ICASSP 2004)*, (pp. 989-992). IEEE Press.

Xu, Y., & Zhang, T. (2002). Variable shortened-and-punctured Reed-Solomon codes for packet loss protection. *IEEE Transactions on Broadcasting, 48*(3), 237–245. doi:10.1109/TBC.2002.803706

Zhang, Q., Yang, F., & Zhu, W. (2005). Cross-layer QoS support for multimedia delivery over wireless Internet. *EURASIP Journal on Applied Signal Processing, 2*, 207–219. doi:10.1155/ASP.2005.207

Zhang, Y., Fu, F., & van der Schaar, M. (2010). On-line learning and optimization for wireless video transmission. *IEEE Transactions on Signal Processing, 58*(6), 3108–3124. doi:10.1109/TSP.2010.2046040

ADDITIONAL READING

Adami, N., Signoroni, A., & Leonardi, R. (2007). State-of-the-art and trends in scalable video compression with wavelet-based approaches. *IEEE Transactions on Circuits and Systems for Video Technology, 17*(9), 1238–1255. doi:10.1109/TCSVT.2007.906828

Akansu, A. N., & Haddad, P. R. (2001). *Multiresolution signal decomposition: Transforms, subbands, wavelets*. London, UK: Academic Press.

Hagenauer, J., & Stockhammer, T. (1988). Rate-compatible punctured convolutional (RCPC) codes and their applications. *IEEE Transactions on Communications*, *36*, 389–400. doi:10.1109/26.2763

Sikora, T. (2005). Trends and perspectives in image and video coding. *Proceedings of the IEEE*, *93*(1), 6–17. doi:10.1109/JPROC.2004.839601

Zhang, Y. Q., Yang, F., & Zhu, W. (2005). Cross-layer QoS support for multimedia delivery over wireless Internet. *EURASIP Journal on Applied Signal Processing*, *2*, 207–219. doi:10.1155/ASP.2005.207

Chapter 19
Cross-Layer Multimedia QoS Provisioning over Ad Hoc Networks

Raad Alturki
Imam Mohammed Bin Saud University, Saudi Arabia

Rashid Mehmood
Swansea University, UK

ABSTRACT

Multimedia applications are becoming the main driver for the advancements in the telecommunications industry. Increased demand for anytime anywhere mobility has given rise to the development of ad hoc networks. Supporting multimedia applications over ad hoc networks is extremely challenging due to the stringent Quality of Service (QoS) requirements and the lack of infrastructure. In recent years, cross-layer design techniques have emerged as a potential solution to address these challenges.

This chapter takes a cross-layer design approach and presents the design and analysis of wireless ad hoc networks that support multimedia applications with scalable Quality of Service (QoS). The chapter has contributed in two areas. Firstly, a novel, cross-layer, Hierarchical Clustering, Provisioning, and Routing (HCPR) scheme for ad hoc networks is proposed, implemented, and analysed. Secondly, a novel methodology is developed for multimedia network analysis and its effectiveness is demonstrated by the analysis of the HCPR scheme and other well-known protocols.

The HCPR scheme is implemented as an extension to the OPNET simulation software and is analysed in detail for its QoS performance to deliver multimedia applications over ad hoc networks. It is compared with three well-known and widely used routing protocols: Ad Hoc On Demand Distance Vector (AODV), Optimised Link State Routing (OLSR), and Geographic Routing Protocol (GRP). Several networking scenarios have been carefully configured with variations in networks sizes, applications, codecs, and

DOI: 10.4018/978-1-4666-0960-0.ch019

routing protocols to extensively analyse the proposed scheme. The HCPR enabled ad hoc network outperforms the well-known routing schemes, in particular for relatively large networks and high QoS network loads. These results are promising because many QoS schemes do work for small networks and low network loads but are unable to sustain performance for large networks and high QoS loads. Several directions to extend this research for future work are given.

INTRODUCTION

Multimedia applications have been the key driving force behind the convergence of fixed, mobile, and IP networks. The current trends suggest that, in the future, multimedia services will be delivered over multiple heterogeneous network platforms all employing IP-based technologies. The Internet is a loosely coupled distributed collection of autonomous networks, which delivers traffic of different types on a best effort basis. In contrast, multimedia applications usually have stringent QoS (Quality of Service) requirements. Various IP-based heterogeneous wired/wireless networking technologies are being connected to the ever-growing Internet infrastructure, which is now being referred to as 'internet of things.' However, it is still a significant challenge for such an infrastructure to support QoS requirements of various applications particularly those involving real time media (such as voice and video). Moreover, increased demands for mobility and anytime anywhere communications have given rise to the concept of ad hoc networks[1], and this has exacerbated the QoS support challenges due to certain unique features of ad hoc networks namely its autonomic and multi-hop nature and the lack or scarcity of infrastructure. It is, therefore, of paramount importance to manage networks for scalable end-to-end QoS provisioning, and understand the QoS requirements of multimedia applications and the performance of heterogeneous networks carrying such multimedia traffic.

Cross-layer design techniques have emerged in the recent years as a potential solution to address the multimedia QoS provision challenges plaguing the development of heterogeneous networks. The traditional approach for network design is based on the OSI (Open Systems Interconnection) model, which divides a communication system into multiple layers where each layer carries out a specific job, receives services from the layer below and provides services to its upper layers. For several reasons this layered architecture provides excellent modularity for networks, such as the Internet, to scale. However, a disadvantage of such layered architecture is its inability to provide service differentiation and QoS, as, for example, is required for multimedia delivery over ad hoc networks. The cross-layer ideology is to optimise the design and performance by increased and effective interaction between layers and optimisation at a global level (multiple layers) rather than at the local level (single layer).

Chapter Aims and Contributions

The aim of our research is the design and analysis of wireless ad hoc networks that support multimedia applications with scalable QoS. We are convinced that cross-layer approaches have the potential to address these QoS provisioning challenges. In this context, this chapter proposes HCPR (Hierarchical Clustering, Provisioning, and Routing), a cross-layer scheme to address QoS challenges to deliver multimedia over ad hoc networks. The HCPR scheme is based on an intelligent protocol that optimises multimedia QoS provisioning by enabling and exploiting interactions between Application, Transport and Network layers. The cross-layer aspects of the HCPR are explained further in Section 4 where the Scheme is described in detail. The motivation and novelty for this work have been established through a state of the art literature review of the relevant cross-layer design approaches for multimedia and ad hoc networks. Most of the cross-layer schemes for multimedia ad hoc networks have focused on interfacing, or merging, the Physical and MAC layers, MAC and Network layers, Application and Transport layers, or Application and Physical layers etc: no work has considered the delicate balance of layered interaction between Application, Transport and Network layers, as proposed in HCPR. This is discussed further in Section 3 where we present a review of the relevant approaches.

The HCPR scheme is implemented as an extension to the OPNET simulation software and is analysed in detail for its QoS performance to deliver multimedia applications over ad hoc networks. It is compared with three well-known and widely used routing protocols; Ad Hoc On Demand Distance Vector (AODV), Optimised Link State Routing (OLSR), and Geographic Routing Protocol (GRP). The HCPR enabled ad hoc network outperforms the well-known routing schemes, in particular for relatively large networks and high QoS network loads. These results are promising because many QoS schemes do work for small networks and low network loads but are unable to sustain performance for large networks and high QoS loads.

Another novel contribution of our work lies in the analysis methodology that we have developed for configuring multimedia applications and networking scenarios. The approach is to simulate and evaluate realistic multimedia applications scenarios for networks with a sufficiently large number and type of applications, both elastic and non-elastic. This will be made evident later when we present an extensive literature review on the design and analysis of multimedia networks.

The rest of the chapter is organised as follows. Section 2 describes the fundamental topics related to this Chapter including IEEE 802.11 WLAN ad hoc mode, network traffic classifications, QoS and routing algorithms. Section 3 gives the related literature review on cross-layer and layered approaches for design and analysis of multimedia networking. Section 4 describes the HCPR scheme and its implementation as an extension to the OPNET simulation software. Section 5 presents the results and analysis of the HCPR scheme in comparison to other well-known protocols. Finally, Section 6 concludes this chapter with directions for future work.

FUNDAMENTALS

This section provides the background material related to this book chapter. We begin with an introduction to the IEEE 802.11 WLAN (both infrastructure and ad hoc networking modes), the wireless networking technologies that we have used in this work. Subsequently, we discuss in detail multimedia applications, Quality of Service (QoS), and the challenges associated with wireless ad hoc networks.

IEEE 802.11 WLAN: Infrastructure and Ad Hoc Modes

The IEEE 802.11 WLAN, also known as Wi-Fi, is widely used these days because of its flexibility and the low cost of equipment and deployability. The IEEE 802.11 is a family of standards that specifies the MAC and Physical layers of a Local Area Wireless Network (WLAN) (O'Hara & Petrick, 2005). The main purpose of 802.11 standard, according to the published standard (IEEE, 2007), is to provide wireless connectivity for mobile and stationary devices within a certain range.

WLAN can be configured in two modes: infrastructure mode or ad hoc mode. The infrastructure mode requires nodes to be connected to centralised node called Access Point (AP) in order to initiate any connection with any other device. Most WLAN applications are based on the infrastructure mode. Nodes within the Basic Set Service (BSS) can exchange data easily within the WLAN or with the outside worlds (e.g. the internet). On the other hand, the ad hoc mode allows nodes to connect to each other directly. This mode allows peer to peer communication between nodes within their Independent Basic Service Set (IBSS). In this chapter, the WLAN performance we only consider networks configured in ad hoc mode. The proposed scheme will be analysed in comparison to the well-known protocols found in the literature.

Elastic and Inelastic Network Traffic

Rapid developments in computing and communication technologies over the last decade or so have led to an unprecedented growth in multimedia content generation and multimedia delivery technologies over the networks. Multimedia which includes text, images, voice, and video is being used in all possible ways for businesses, social interactions and professional communications. This is in turn has led to the development of a whole new range of applications and services based on multimedia content. Examples of emerging multimedia services include mobile TV, online gaming, video telephony, mobile video conferencing, video mails, and multimedia messaging services. These next generation multimedia services have paved the way for network convergence, and have placed additional stricter QoS demands on the IP-based best effort networks.

The term *multimedia* literally means more than one media (Tanenbaum, 2010) such as text, voice, images, and video. The combination of text and figures in a webpage or book, hence, would qualify as multimedia, however, usually multimedia refers to one, or a combination of more than one type of, *continuous* media (Tanenbaum, 2010). Network traffic based on its characteristics can be classified into two main categories—*elastic* and *inelastic* traffic. It is typically easier to deliver elastic traffic on a network because it is tolerant to variations in delay and bandwidth. TCP/IP networks were designed for this kind of traffic. Elastic traffic is usually delivered using TCP and UDP transport protocols. Application layer traffics that can be classified as elastic traffic include web (HTTP), File Transfer Protocol (FTP), and electronic mail service (Stallings, 2006). Inelastic traffic, on the other hand, has stringent requirements on the network resources such as bandwidth, jitter and delay (see Section 5 for definitions of these performance metrics). Any failure by the network in meeting the requirements of an inelastic flow would usually result in degradation of the quality experienced by the user. Real time applications are an example of inelastic traffic. Examples of such applications are video and voice streaming, video conferencing and Voice over IP (VoIP).

We have used both elastic (HTTP traffic comprising web pages containing text and images) and inelastic (interactive voice and streaming video) traffic for performance analysis in this study.

Quality of Service (QoS)

Rapid developments in computing and communication technologies over the last decade have led to an unprecedented growth in multimedia content generation and multimedia delivery technologies over the networks. Multimedia which includes text, images, voice, and video is being used in all possible ways for businesses, social interactions, and professional communications. This is in turn has led to the development of a whole new range of applications and services based on multimedia content. Examples of emerging multimedia services include mobile TV, online gaming, video telephony, mobile video conferencing, video mails, and multimedia messaging services. These next generation multimedia services have paved the way for network convergence, and have placed additional stricter QoS demands on the IP-based best effort networks.

Quality of Service (QoS) is the performance level of a service offered by the network to the application (Murthy & Manoj, 2004). Applications may vary in their requirements from the network; some of them need to have fast packet delivery but could be flexible in security or confidentiality of the packets. The QoS is the service that prioritises users' requirements and then aims to guarantee satisfying their requirements without affecting other users' requirements. Some common QoS parameters for applications include delay, jitter (variation in delay), throughput, and (tolerance in) packet losses. Networks create delay, jitter, and packet losses due to, for example, limited bandwidth availability, buffering and switching delays, lack of buffer space and transmission errors. An application can use its QoS parameters to negotiate QoS with the underlying network. The network designer or operator can consider these QoS parameters for network design purposes to negotiate service level agreements, or to develop and implement policies and procedures required to guarantee service level agreements. This has led to many developments and standardisation activities to provide end-to-end QoS over the Internet, including the IPv4 to IPv6 evolution.

Best effort networks are not enough for some applications where, for example, delay cannot be tolerated, and that is why there have been many methods and techniques used to overcome such problems. Each QoS technique has its own pros and cons and there is no technique that can fit all QoS requirements. We discuss some well-known QoS techniques in the following subsections.

Buffering

Buffering techniques are used to buffer packets received at a destination to overcome later packets being used by an application before the earlier ones. It increases the delay slightly but does not affect reliability or bandwidth. This method is important in applications like voice and video where jitter could lead to degrading the performance.

Shaping

Unlike buffering, this technique is used to arrange packet movements at source whereas buffering arranges packets at destination. This method is ideal in situations where network traffic is not regular, and implementing this method might help in reducing collisions and, therefore, improving performance. The leaky bucket algorithm and token bucket algorithm are examples of some implementation of traffic shaping.

Resource Reservation

This method is used to reserve network resources to the applications stream. The traffic shaping techniques mentioned earlier allow smooth transmission of packets and help in avoiding collision. However, packets could take different routes to destinations which could lead to unwanted results. That is why making a reservation for traffic streams from source to destination could be ideal in some applications. That is done by choosing the Bandwidth, buffer space and CPU cycle of all routers on the way to destination. This method is ideal in situations where a reliable connection is very important.

Admission Control

This technique will make the decision to give and grant resource to specific flow as requested. When a traffic flow needs to reserve some resources across the route, the admission control will check resources and grant that accordingly.

Multipath Routing

Sending packets in the same route could lead to congestion in some areas of the network, which could lead to delay and packet dropping. That is why having packets for the same traffic flow taking different routes to a destination could lead to spreading traffic across the network, which will lead to better packet delivery.

Packet Scheduling

This technique schedule packets processing at each point (Router) in the network from source to destination. This idea came about after the observation that sometimes flows are not processed at the router. This is because of the heavy packet flow coming from a specific source to destination, or because of the priorities that some QoS sensitive traffic flows are taking most of the bandwidth (Bhatti & Crowcroft, 2000).

Provisioning

This technique allows part of, or the whole, network resource to be reserved between source and destination. An example of this is the traditional telephone system where phones lines are available at all times. This method leads to excellent results but it has the drawback of being very expensive.

QoS Architectures

There have been some QoS architectures developed to support end-to-end QoS, including Integrated Services (Braden, Clark, & Shenker, 1994), Differentiated Services, (Blake, et al., 1998), and Muti-Protocol Label Switching (Rosen, Viswanathan, & Callon, 2001). The Integrated Services, or IntServ, architecture is based on fine grained, flow-based methods to provide end-to-end QoS using the Resource Reservation Protocol (RSVP). Differentiated Services, or DiffServ, on the other hand, provides a coarse grained, class-based approach to guarantee QoS using Differentiated Services Code Point (DSCP) byte. MPLS was first proposed by several router vendors to make a label in each packet and make decisions based on that instead of the destination address (Tanenbaum, 2010). Then it was standardised by the Internet Engineering Task Force (IETF) as in Rosen et al. (2001).

Routing

In the study of general networking, there are many routing algorithms, which differ in their way of finding the best path, network topology and the required performance. However, most of them fall into two major protocol categories: Distance Vector protocol and Link State protocol. Distance Vector protocol (DV) uses the Bellman-Ford algorithm to calculate paths (Peterson & Davie, 2011). It requires the host to inform its neighbour hosts about the topology changes in the case of any topology changes occurring. On the other hand, Link State protocol requires the host to inform all the network hosts of any changes in the network topology, and every node has a table of the connectivity of the network (Peterson & Davie, 2011). It uses a hop-by-hop method for forwarding the packets, which means every node makes an independent decision for any passing by packets that need to be forward. This classification is based on how the node will inform other nodes if anything has happened.

In addition, routing protocols can be classified in terms of how ready they are to send the packet into two classes: reactive algorithms and proactive algorithms. Reactive algorithms are the routing algorithms that determine the route of the packet when it is needed. These algorithms are called 'on demand' algorithms. On the other hand, proactive algorithms are continuously updating every node in the network of any change in the network. That means the route to any node is determined even if there is no need to send any packets. Every class of these has its advantage over the other. Sending packets in reactive algorithms may take more time than for proactive algorithms; this is because before sending any message, the node should send a route request message to the destination in order to know the best path to the destination. Then, after it receives the reply, it sends the message accordingly, but in proactive protocols, if a node decides to send a message, the message will be sent directly because the route to the destination

Table 1. List of ad hoc routing protocols

No	Protocol		No	Protocol		No	Protocol
1	DSDV		12	MMWN		23	PLBR
2	WRP		13	CGSR		24	FORP
3	DBF		14	HSR		25	LAR
4	GSR		15	AODV		26	CBRP
5	TBRPF		16	DSR		27	ZRP
6	FSLS		17	TORA		28	LANMAR
7	STAR		18	ABR		29	CEDAR
8	DREAM		19	SSA		30	SHARP
9	GPSR		20	RDMAR		31	ZHLS
10	FSR		21	LMR		32	SLURP
11	OLSR		22	ARA			

is already known. Proactive protocols on the other hand, have more overheads than the reactive protocols because proactive protocols update the routing information in all nodes periodically, and too much information is exchanged at all times, but reactive algorithms only send a routing request when it is required to send a data message.

There are some other classification methods mentioned in the literature, such as whether the routing protocol is flat or hierarchical, source routing or hop-by-hop, location based or not, uniform or non-uniform (Liu & Kaiser, 2003), and single or multiple channels. In fact, those classifications are interleaved with other classifications mentioned earlier, but each writer gives his or her own insight and opinion about their views on classifications.

In fact, there are more than 30 reported ad hoc routing protocols (see Table 1). Each routing protocol has been designed either to solve a specific problem or to improve on other routing protocols. In the rest of this section, a brief description will be given of four well-known routing protocols that are popular in the research community, and these will be used in Section 5 where we discuss the performance of our proposed scheme (HCPR) in comparison to these protocols.

Temporally-Ordered Routing Algorithm (TORA)

Temporally-Ordered Routing Algorithm (TORA) (Park & Corson, 2001; Park & Corson, 1997) is a distributed loop free multipath routing protocol. It is destination oriented routing, which supports a mix of reactive and proactive routing. It is based on link reversal protocol where link failure will result in reversing the direction of links. It has three main functions: create route, maintain route and erase route. When a node requires a connection to a destination, the create route function will flood a request for route to destination. Then a reply message will be returned back to the sender if a route is discovered. The main route function will use the link reversal algorithm to correct any link failure. Invalid routes sometimes exist, which will trigger the erase route function to flood a message to erase the route in a network. TORA reduces the routing overheads by localising the algorithm reaction to any topology change. The Route Optimality (Shortest path) is considered to be second priority in TORA design where it always considers having lower reactivity and higher bandwidth conservation as more important. It has been reported that TORA has high routing overheads which could lead to packet dropping in high traffic situations (Broch, Maltz, Johnson, Hu, & Jetcheva, 1998).

Dynamic Source Routing (DSR)

Dynamic Source Routing (DSR) (Johnson, Hu, & Maltz, 2007) is an on demand source routing protocol. Source routing algorithms make the decision about the route from source to destination at the source node. Then, the route to destination will be carried out in the packet. This idea has the advantage of not worrying about nodes in the middle to update their tables. DSR uses two mechanisms to maintain its table: firstly, Route Discovery, and secondly, Route Maintenance. Route Discovery allows a node wishing to send a data packet to a destination to obtain the route. That is done by broadcasting a message requesting a route to the destination. Reply messages will be sent back with the route from the destination, or any other intermediate nodes that know a route to the destination. In order to reduce the cost of Route Discovery, nodes overhearing the messages will keep a cache of source routes discovered. The Maintenance mechanism will allow the sender to check whether any previous route to a destination is still correct or not. Like other on-demand routing protocols, DSR has higher delays than some other routing protocols (Layuan, Chunlin, & Peiyan, 2007). Despite the fact that DSR carries the source route in the packet, which could increase the routing load, there have been some results which show that it has lower routing load than other on-demand routing protocols (Broch, et al., 1998). Also, it has been reported that it has good throughput results in comparison to other routing protocols (Layuan, et al., 2007).

Ad Hoc On-Demand Distance Vector (AODV)

Ad Hoc On-Demand Distance Vector (AODV) was first published as an internet draft by Charles Perkins in 1997, then published in 2003 as RFC. It is a reactive routing protocol that is based on a distance vector routing approach. Nodes are kept quiet until a connection is required. The node that requires the connection will broadcast a route request message asking for a route to destination. A reply to this message is expected from any node that has the route to destination, or from the destination itself. After that, it will choose the route to destination with the least number of hops. To avoid the problem of "counting to infinity" associated with distance vector algorithm, it uses a sequence number with routing messages (Perkins, Belding-Royer, & Das, 2003). It has the advantage of low complexity and conserves the bandwidth. Also, the on demand feature leads to no calculation being necessary and memory is not consumed unless needed. The problem with AODV is that it needs to establish a new connection each time there is no fresh connection to destination. That will cause a delay in sending and could lead to packet loss, especially in dynamic environments (Broch, et al., 1998).

Optimised Link State Routing (OLSR)

Optimised Link State Routing (OLSR) (Clausen & Jacquet, 2003) is a link state proactive routing protocol that maintains its routing and forwarding table with disregard of packet arrivals. It sends a periodic hello message to exchange information about network. In order to avoid extensive routing maintenance messages from consuming the limited wireless bandwidth, it has some chosen nodes called Multi Point Relays (MPRs) to reduce packets. Those MPR nodes are selected nodes that do two main jobs: a) generate topology control messages, b) act as data packet forwarders for other nodes. The advantage of using MPRs is clearly observed when a network is big and tense in comparison to pure link state routing or other MANET routing protocols. Also, when a network is small and nodes are sparse, OLSR becomes a pure link state routing protocol. Packets are routed in a hop-by-hop basis—that is nodes in the route will make the forwarding decision according the local routing table.

AD HOC NETWORK DESIGN: LITERATURE REVIEW

In the previous section, we have explained the fundamental topics related to this chapter. In this section, we give a review of the literature relevant to the core topic of this chapter, i.e. multimedia networking: in this context, we have reviewed both traditional (layered[2]) as well as cross-layer design approaches for the design and analysis of multimedia over ad hoc networks. We had initially thought to include, in this chapter, a literature review on the cross-layer design approaches alone. However, there are both proponents and opponents of cross-layer design approach and, therefore, for the sake of fairness, we felt inclined to also include a review of literature on traditional, layered, design approaches. Moreover, a comprehensive literature review of both cross-layer and layered approaches has allowed us to develop a novel analysis methodology for multimedia networking by identifying the shortcomings of the current performance analysis techniques.

Cross Layer Design in Multimedia Networking

This subsection reviews the literature on cross-layer design techniques within the scope of multimedia delivery over ad hoc networks. The traditional layered protocol network architecture is based on a series or stack of layers, each layer provides services to the higher layers, allowing an abstraction, and hiding implementation complexities, for the higher layer. This layered approach reduces complexity through modularisation of the network and allows manageable design, interoperability, extensibility, scalability, and standardisation. There are, however, well-known disadvantages of the layered design. Firstly, the layered design does not allow global optimisation and one can only optimise the functions, locally, within a single layer, and as a result, it has been unable to provide service differentiation and QoS, as, for example, is required for multimedia delivery over ad hoc networks. In particular, anytime anywhere communications, requiring wireless communications with mobility, and increasing QoS demands, have given rise to heterogeneous networks and technologies. These heterogeneous systems have their own challenges, demanding more than what the layered, isolated, approach could offer.

The so-called cross-layer design techniques have emerged in the recent years as a potential solution to address the multimedia QoS provision challenges plaguing the development of heterogeneous networks. The cross-layer ideology is to optimise the design and performance by increased and effective interaction between layers and optimisation at a global level (multiple layers) rather than at the local level (single layer). That is, the networks can be optimised to deliver differentiated QoS by sharing knowledge across different layers. However, the cross layer design approaches have also been criticised due to their increased complexity, reduced flexibility and interoperability, and reduced modularisation that hampers R&D and standardisation. We present, in the following, a literature review of the works on cross-layer design approaches in multimedia networking, and use this review to demonstrate in Section 3.1.1 that our proposed HCPR makes a novel contribution to this area.

Ramathen and Redi (2002) have described the cross-layer approach as a promising method which could solve problems related to QoS. By allowing layers to use other layers' information and accessing them freely, layers could adapt to the changes using other layers' information. Therefore, the QoS requirements can be satisfied accordingly. Conti, Maselli, Turi, and Giordano (2004) have described how the current challenges of MANET cannot be solved by the traditional structural layers design. The suggested approach is to allow layers to interact, which will result in significant performance improvements. However, having a cross-layering design for network protocols will result in having a "Spaghetti like code" that cannot be debugged or reused easily. In order to gain the benefit of a cross layering design without violating the benefits that structural layer design provides, the authors have presented a MobilMan architecture for ad hoc networks. The MobileMan architecture keeps the structural layering design with the ability of layer interactions. The layers of protocols are still separate despite having to co-operate with each other. The co-operation between layers is done by allowing data sharing across layers. For example, functions from the network layer could access data in the MAC layer.

Kitae, Ahmed, and Kuo (2005) have discussed the effects of congestion and MAC contention on the performance of the interaction of the transport layer and on-demand routing protocols. Their study shows that many problems related to the interaction are a result of the lack of the interaction between these layers and the MAC layer. They have observed that the TCP throughput can be improved by changing the window mechanism used because the current window mechanism acts aggressively toward the on demand routing protocols, which makes such protocols to over react. In order to solve this problem, they propose a Fraction Window increment window mechanism that reduces the TCP aggressiveness and, consequently, the on-demand routing protocol reaction to change will be acceptable. Lee and Song (2010) have proposed a cross layers algorithm for video streaming over ad hoc network. The algorithm chooses the efficient PHY mode and the re-transmission limit of WLAN for each node in a distributed way. The algorithm makes the choice depending on the information available at other layers Application, MAC, and PHY.

Pompili and Akyildiz (2010) have studied the multimedia underwater delivery in the acoustic sensor networks. They mentioned that the previous underwater communication works have followed a traditional stack layered approach, which was firstly designed for wired networks. Authors claim that the cross layer approach will improve the multimedia and delay sensitive applications in underwater communications environments. They studied the interactions of underwater communications functionalities like modulation, forward error correction, MAC and routing. In addition, they proposed a new distributed cross layer solution that allows devices to share the underwater acoustic medium fairly and efficiently which is characterised by having high-delay and bandwidth limitations. Gharavi and Ban (2003) have proposed a cross-layer feedback control scheme for video transmission over ad hoc networks. The proposed scheme was designed to overcome the current problems of variations of delay and bandwidth. The current schemes like RTP have limitations in mobile environments, which causes the network to be congested. The proposed scheme is zero-delay and traffic free channel assessment scheme that allow the sending node to adapt to the current network conditions. The scheme takes the critical information (such as number of hops) from the underlying routing layer. Those extracted information are also used in other proposed schemes like packet recovery and redundant packet transmission.

Zhiguo and Leung (2011) have studied the cross-layer routing using the cooperative transmission and proposed a new path selection strategy for vehicular ad hoc networks that have a better trade-off between the power consumptions and end-to-end reliability. The study of cooperative transmission and its influence to the wireless medium cost shows that the quality of wireless links could be improved and showed how that could affect the routing decision. Two types of the routing optimisations were investigated. Firstly is to optimize the routing protocols to have the maximum end-to-end reliability with a constraint in power consumption. Secondly is to have the minimum transmission power with the constraint in the end-to-end reliability. Setton, Taesang, Xiaoqing, Goldsmith, and Girod (2005) have investigated the potential that cross layer approach can offer to support real-video by allowing layers to exchange information, and showed how end-to-end performance was optimised by adapting to such information. The results show that the video-stream performance has improved substantially.

Conti, Maselli, and Turi (2006) have shown how cross layering concept is achieved using a vertical interface which preserves the architectural stacked modules and they showed the proof of its usability. The main contribution of this work is designing interface called (XL-interface) that allow layers to interact without violating the traditional stacked network model. To support such claim, two cases were presented in the paper. The first case shows how cross-layer approach could help in reducing the problem of TCP performance degradation. The second case involves improving the performance of a well know unstructured overlay platform called Gnutella. Mundarath, Ramanathan, and Veen (2009) have proposed a new cross-layer approach for multiple antenna ad hoc networks that consider QoS and reduce power consumption called QoS-Aware Smart Antenna Protocol (QSAP). The proposed approach adapts the degree of freedom present in the multi antenna in each node to achieve less energy consumption while keeping Application's QoS requirements assured. The results show that using cross-layer approach in the design saves considerable energy consumptions in comparison to other QoS schemes that use strict layering approach.

Melodia and Akyildiz (2010) have proposed new cross-layer communication architecture for multimedia delivery over Wireless Sensor Network (WSN) based on time hopping impulse radio Ultra Wide Band (UWB). The architecture aims to deliver heterogeneous application traffic over WSN reliably and with flexibility by controlling the interactions between the layers. The resulted simulations show that the architecture achieved the aimed target without affecting the modularity of the stacked design.

Contributions of HCPR to Cross Layer Design Approaches

As it has been mentioned, the aim of our research is the design and analysis of wireless ad hoc networks that support multimedia applications with scalable QoS. The adoption of cross-layer designs in multimedia networking has been on rise and we are convinced that cross-layer approaches have the potential to address these QoS provisioning challenges. Our proposed HCPR is a cross-layer scheme to address QoS challenges to deliver multimedia over ad hoc networks. The HCPR scheme is based on an intelligent protocol that optimises multimedia QoS provisioning by enabling and exploiting interactions between Application, Transport, and Network layers. The literature review presented earlier in this section shows that most of the cross-layer schemes for multimedia ad hoc networks have focused on interfacing, or merging, the Physical and MAC layers (e.g. [21]), MAC, Physical and Network layers (e.g. [23]), MAC, Application and Physical layers (e.g. [22]), etc. No work has identified the potential of, and considered, the delicate balance of layered interaction between Application, Transport, and Network layers, as we proposed in HCPR. The interfacing of Application, Transport, and Network layers in HCPR allows HCPR to scale well, and provide QoS, to large networks under heavy multimedia traffic. The cross-layer aspects of the HCPR will become clearer in Section 4 where the Scheme is described in detail. The analysis of the HCPR scheme is presented in Section 5.

Layered Approaches in Multimedia Ad Hoc Networks

A huge amount of literature is available on the design and performance analysis of ad hoc networks. The aim here is to study what has been done in this area in terms of scenarios and methodologies used with some focus on multimedia. In fact, there are many possible network scenarios where small changes could affect the overall network and individual node's performance. Such possibilities come from the varieties that can be set for scenario parameters, including application type, traffic intensity, number of connection, number of nodes, nodes' locations and so on. Varying those parameters will result in having different results and drawing different conclusions. Results can be interpreted by many evaluation metrics, such as throughput and delay, and depend on application and the details of the scenario used. Choosing the right scenario and metrics used is vital for the accuracy and reliability of the conclusions reached. In order to choose the right scenario's settings and metrics, literature has been studied that (1) proposes new ideas and includes performance evaluation of new work and (2) describes work that focuses on multimedia over ad hoc networks. Each of the coming sub-sections start by giving an overview of the literature under the category, followed by a brief description of the work included in that category.

Approaches to Improve Multimedia over Ad Hoc Networks

Studies that have proposed new methods to improve multimedia over ad hoc networks and evaluate it include Hongqi, Jiying, and Yang (2008), Li, Jannotti, Couto, Karger, and Morris (2000), Li and Cuthbert (2005), and Utsu, Chow, Ishii, Xue, and Ganz (2003). In Hongqi et al. (2008), the authors have proposed a new method to reduce packet loss, whereas the authors in Utsu et al. (2008) propose a new approach to make video transmission smoother. Li and Cuthbert (2005) claim that they have made improvements to the Ad hoc performance for multimedia. Xue and Ganz (2003) propose a new end-end QoS protocol, while the authors of Li et al. (2000) propose a scalable location service. Each of these works will now be discussed in more detail.

Hongqi et al (2008) have studied the performance of voice over wireless ad hoc networks, and proposes a new scheme to improve the performance by reducing packet loss. They have used MOS as the main measure for the performance, besides packet loss and jitter. Utsu et al (2008) have studied the problem of video over ad hoc networks and suggest new methods to make video transmission smoother. In order to evaluate their results, they used NS2 as a simulation tool with some experimental work. The number of nodes simulated in NS2 scenarios was 20 nodes placed in 1 x 0.6 km and a random way point mobility model was used. The bandwidth used was 2 Mbps with a 250m transmission range. They used DSR as routing protocol and a video traffic application of 128 kbps and frame size of 1024 byte, and bit sequence traffic to simulate that. Throughput was the main evaluating metric. Li and Cuthbert (2005) have proposed a new QoS multipath routing method that improves the performance of multimedia. They have shown some performance improvements by simulating 50 nodes in 1x1 km area and 250 transmission range using OPNET 8.1 modeler. Traffic was populated with 512 byte CBR data packets. Five to 20 sources were chosen randomly to send 80 kbps Expedited Forwarding Packets. Another 20 nodes sent eight kbps background packets in best effort manner. The source-destination pairs were chosen randomly with variety in the number of connections for EF packets. They used packet delivery ratio and packet end-end delay as metrics for performance evaluation. The simulation was run for 800 seconds.

Li and Cuthbert (2005) have introduced a new resource reservation-based routing and signalling protocol called AQOR, which provides end-end QoS support and has been implemented in OPNET modeler. They have used 802.11 DCF enabled nodes with 2 Mbps and 250 meters transmission range, and they have used small and large network sizes to see how the protocol behaves. Small networks are used to discover the QoS recovery time, which consists of five nodes- four statics and one mobile with one flow (400kbps) from source to destination. The large network is used to study the performance of the proposed protocol, and consists of 50 nodes placed in a 1x0.5 km area and moving randomly. To simulate multimedia streaming, they have used CBR traffic with 512 packet size and sent 10 packets a second. They have populated 10 and 15 traffic flows with random sources and destinations. In order to study performance, they used the following metrics: traffic admission ratio, end-to-end delivery ratio, average end-to-end delay, ratio of late packet and normalised control overhead. They defined end-to-end delivery ratio as "End-to-end delivery ratio, ratio between the number of data packets received at the destinations and the number of data packets sent from the sources. This metric indicates the reliability of the admitted flows." The average end-to-end delay defined as the latency incurred by the packets between their generation time and their arrival time at the destination. This metric indicates the performance of the admitted flows."

Li et al (2000) propose a distributed location service for mobile ad hoc network routing that scales well when the network gets bigger. They use a NS2 simulator with up to 600 nodes, and 2Mbps bandwidth with a 200 metre transmission range. The traffic is Constant Bit Rate (CBR) with number of connections half the number of nodes. For each connection, four 128-KB data packets per second are sent to a destination for 20 seconds. Connections are generated at random times in the simulation time. Packet Delivery ratio metric is used to view the traffic performance.

Multimedia Traffic over Ad Hoc Networks

The Second category of interest is work that focuses on multimedia over wireless and ad hoc networks, as in An, Loyola, and Aad (2007), Gottron et al. (2009), Jeong, Kakumanu, Tsao, and Sivakumar (2009), Santos (2009), and Sondi, Gantsou, and Lecomte (2010). In Gottron et al. (2009), the authors show the feasibility of voice communication in larger scale networks, whereas in An et al. (2007), the authors have made a performance analysis study of voice over static wireless. The authors in Jeong et al. (2009) have made the observation that only five simultaneous VoIP calls can be made in 802.11, whereas the authors in Sondi et al. (2010) have done a performance evaluation of voice communication using OLSR-QoS extension. This master thesis (Santos, 2009) studies the performance of VoIP over ad hoc network using OLSR routing protocol.

Gottron et al (2009) show the feasibility of voice communication in larger scale ad hoc networks if the right settings are used. They have used Jist/SWANS simulation tools to conduct their studies, and scenarios with 50,100,200 and 500 nodes. They have evaluated the results according to (1) Voice quality where MOS is used as the measure (2) Transmission delay and (3) Packet loss. They have varied the network load between five to 20 voice streams of G.711 voice codec. The transmission range used was 250 meters and nodes were surrounded by an average of eight nodes. There is no clear description of how the source and destination were chosen. An et al (2007) have carried out a performance analysis study of voice over a static wireless ad hoc network. They used throughput, delay and jitter as metrics for the evaluation using simulation (NS2) and experimental work. Eight Nodes have been placed to form a ring topology and a source is sending to the destination node, which is the last node in the ring and should be accessible through the other six nodes. They used 1 Mbps and G.711 codec.

Sondi et al. (2010) have done a performance evaluation of OLSR-QoS extension by using OPNET simulator. They populated 50 nodes in a 1 x 1 km area and nodes were equipped with an 802.11g network card. They have simulated scenarios with statics and mobile nodes, with the aim of evaluating voice communication with some background. Therefore, they have made one voice connection between two nodes using G.711 codec (64 kbps) with the two nodes having FTP traffic as background traffic. For evaluation, they have used IP number of hops, jitter, packet end-to-end delay, MOS, and traffic sent and received as evaluation metrics. It has been observed that voice-packet-end-end delay were around 0.06 sec in most scenarios. Santos in a master thesis (Santos, 2009) has studied the performance of VoIP over ad hoc networks using OLSR routing protocol. It varied nodes intensities, number of data streams and mobility in the simulations to obtain 18 different scenarios using OPNET simulator. It focuses on studying the impact of the number of nodes, number of streams and node mobility on the performance of ad hoc networks using end-to-end delay and packet loss. The author has simulated 10 to 50 nodes placed randomly in a 1 x 1 km area to send streams of VoIP with fixed packet lengths of 200 bytes using the G.711 codec. The author has defined the end-to-end delay but not packet loss. However, the author did not show how end-to-end delay was obtained clearly and whether voice application or IP statistics were used. Their definition of end-to-end delay is defined as delay measured from the instant a packet leaves the sender's Network Interface Card (NIC) to the instant it is received at the destination's NIC. Still, it is not clear how that has been obtained from the OPNET simulator, and from our experience there is nothing defined as such in voice application.

Jeong et al. (2009) have studied the observation that only five simultaneous VoIP calls can be made in 802.11 using NS2 and show the reason behind that. Then, they proposed a new algorithm at the MAC sub-layer to improve performance. They have used Jitter, loss rate, delay and call capacity as metrics for measurement. They used a network consisting of wireless and wired nodes. Two laptops communicated a single VoIP application over the wireless network and the rest of the communications were within the wired network. The detailed network setting and results of MOS for this study can be found in the paper. They report that theoretically 85 calls can be made in 802.11b but, in practice, only five can be supported with QoS.

Contributions to Analysis Methodology in Multimedia Networking

A selection from an extensive literature review (over one hundred sources) that we had carried out at the beginning of our research in multimedia networking is presented. Subsequent to this literature review, we were convinced that the existing methodologies to evaluate multimedia networking studies lack sufficient details and fail to provide assessment of realistic multimedia networking environments. To address this, we have developed a novel analysis methodology for multimedia networks (the methodology is used in our research, see Al Turki & Mehmood, 2008; Rashid Mehmood & Alturki, 2010; Rashid Mehmood, Alturki, & Zeadally, 2011), and in this chapter for the analysis of the HCPR and other schemes. The approach is to simulate and evaluate realistic multimedia applications scenarios for networks with a sufficiently large number and type of applications, both elastic and non-elastic. The network QoS is explored for a set of different applications (as may typically be found in realistic situations), so that the mutual effects of network applications can be taken into account. Specifically, the networks have been populated with VoIP, Video and HTTP applications. Networks typically have to share multiple types of applications including voice, video, and data. We have seen that most of the studies that exist in the literature have focused on studying capacity and performance of multimedia applications in isolation to other applications, i.e. the networks are populated by the traffic that only belongs to one application type. These studies are therefore unable to capture the dynamics of realistic networking environments, because they do not take into account the mutual effects that various multimedia applications will have on each other, while sharing a single network resource. While there are some studies that have reported on multimedia performance while sharing the network resources with other applications, those studies are limited in their approach to setting up the applications and networking scenarios (e.g. the number and types of applications, and the analyses are limited). Furthermore, the multimedia traffic models used in the literature are high-level such as Constant Bit Rate (CBR) while we have used specific models including voice codecs and HTTP traffic.

HCPR SCHEME: ARCHITECTURE AND IMPLEMENTATION

We describes here our proposed HCPR scheme in detail and briefly reports on its implementation. HCPR scheme has been implemented as an extension to the OPNET tool. Communication systems functions and protocols, in OPNET software, are implemented as models. The necessary background on OPNET is given, including an overview of the software, the common communications models implemented in OPNET, the architecture of MANET models implemented in OPNET, and, as an example, depiction of the architecture of WLAN workstation model in OPNET.

Overview of the OPNET Software

OPNET (http://www.opnet.com/) is a comprehensive tool to model and analyse the behaviour and performance of communication and distributed systems using discrete event simulations. Discrete event simulations represent the system behaviour as a set of discrete states making transition from one state to another caused by certain events. In OPNET, events are listed in chronological order in order to be executed. Events happen randomly and discretely where two events could happen independently from each other and at the same time. An example of that is to simulate the sending and receiving of packets from two nodes. The two nodes can send each other simultaneously and the simulator should consider that. That will be done by making the two sending event in the list of events and the time of their sending is the same. In OPNET, the software kernel takes care of organising and scheduling the events in an event list for the simulated scenario. The models and objects within a scenario generate certain events.

OPNET Models

OPNET is featured by the implementation of common network and communication system models. Those models implemented by OPNET and called Standard library models where they come with the modeler. Users can access and change the code of any standard model. When users develop their models and give it to the OPNET users' community for no charges, it is called contributed models and called third party products if there is a charge for it. OPNET Standard models includes common protocols and technologies in all network stack like HTTP, TCP, IPv4, BGP WLAN, and MANET, etc.

OPNET MANET Model

As has been mentioned in the last section about the standard OPNET models, it can be seen that MANET (Mobile Ad hoc Network) is one of the models supported by OPNET. OPNET support MANET by providing some known MANET routing protocols like AODV and OLSR. Also, they have designed some nodes with the MANET routing models embedded in their design model. That means nodes can connect to each other wirelessly in ad hoc manner. OPNET MANET models provides the ability for ad hoc network to be connected to infrastructure IP network. The MANET node models provided have the ability to create raw traffic flows. The list of MANET node models in OPNET are as following:

- **Wireless LAN workstations and servers**: Those nodes are used to generate standard traffics like HTTP, FTP, E-MAIL, Video and Voice. Those applications sent over IP over WLAN and can be used with any MANET routing protocols. See Figure 1 as an example of this kind of node design which will be taken as case study.
- **MANET Stations**: This kind of nodes is used to send and receive raw IP traffics. The user needs to set the source and destination and with any MANET routing protocol.
- **Wireless LAN routers and MANET gateway**: This kind of nodes is a standard Access Point (AP) with the functionality to act as gateway between MANET and IP network. There is limitation of the routing protocols that can be used in such nodes.

MANET Architecture

MANET nodes are normal WLAN nodes with some changes in the network layer to allow MANET routings process to control the routing operations if MANET is enabled. It can be said that MANET main task is to take care of the routing operations includes generating the forwarding table, handle routing packets and forward the required packets. In OPNET, there are five MANET routing protocols supported in the standard model library:

- Ad Hoc On Demand Distance Vector (AODV)
- Dynamic Source Routing (DSR)
- Geographic Routing Protocol (GRP)
- Optimised Link State Routing (OLSR)
- Temporally Ordered Routing Algorithm (TORA)

OPNET uses IP as the base for MANET routing protocols. As in Figure 1, all MANET routing operations have to go through IP process. In fact, MANET uses IP for addressing and forwarding and supports both versions of IP: IPv4 and IPv6. To make that clearer, we will discuss the architecture of WLAN workstation node model and where MANET process is residing. Figure 1 shows the architecture of WLAN workstation. The workstation architecture has been designed to follow the TCP/IP layering model starting by the physical layer from bottom going up to the application layer at the top. Every layer has one or more module. Module could be processor, queue, transmitter, or receiver modules. Process module is represented normally in blank squares and it is the main building block of the node. Queue is a superset of the process module with some extra functionality of buffering and managing data collections. Transmitter and receiver modules work as interfaces for communication link outside the nodes for outbound and inbound traffics respectively.

Hierarchical Clustering, Provisioning, and Routing (HCPR) Scheme

This section introduces the proposed Hierarchical Clustering, Provisioning, and Routing (HCPR) scheme by providing an algorithmic view of the scheme. HCPR is a cross-layer scheme that works on application, transport and network (routing) layers. The idea of cross-layer in HCPR was inspired by the insight gained from a literature review of the previous works in multimedia networking. Literature reviewed provides us a clear opportunity to propose and develop a scheme that addresses QoS at network, transport and application layers for ad hoc networks. This section explains the proposed HCPR architecture from a high-level perspective. We focus in this section on a static ad hoc networking environment where the nodes' location is fixed and we assume that

Figure 1. WLAN node architecture

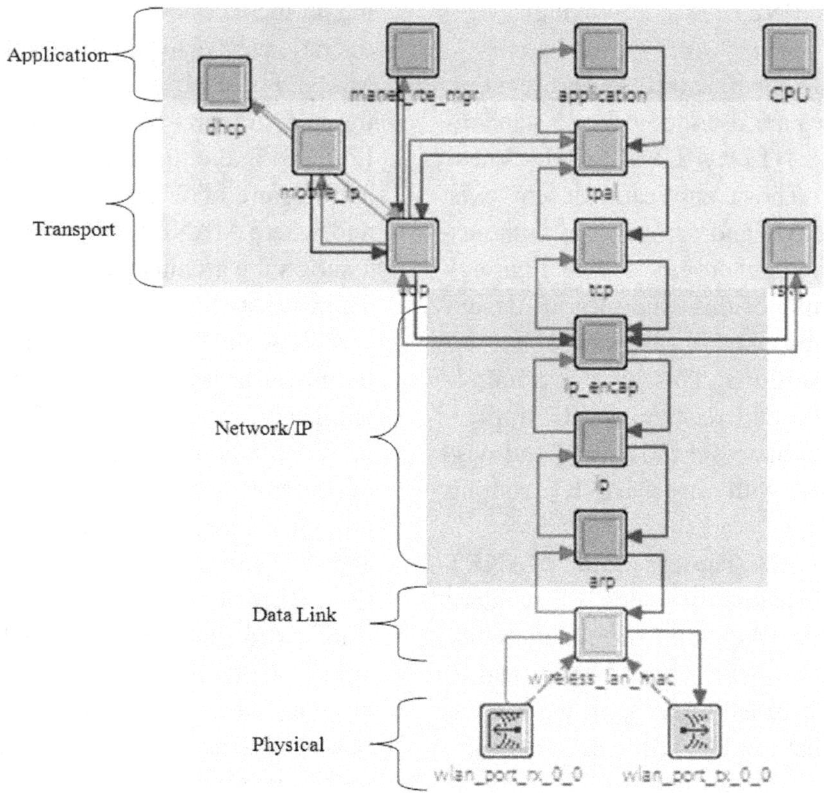

the clusterhead nodes do not fail. Later in this section, we give ideas to extend this architecture to a dynamic environment where network nodes move around and possibly fail.

Consider an initial scenario where N Wi-Fi enabled nodes are spread across a geographical area of $X x Y$ square meters. Each node is equipped with an appropriate positioning technology. The nodes in the region initially have no knowledge of other nodes, their positions, and the geographical boundaries of the network. For the purpose of explanation, we differentiate between two major phases of the network: the network formation phase and the network operational phase.

Network Formation Phase

The network formation phase prepares the network in order for the network to move to the operational phase. The network formation phase enables the nodes to find each other, their positions, to form node clusters and an overlay network connecting the clusterheads (by clusterhead we mean the Head or leader of a cluster).

The network formation phase is divided into the network discovery phase and the cluster overlay formation phase. The network discovery phase allows the nodes to know the number of nodes, their respective positions, and the geographic boundaries of the network. The cluster overlay formation phase is used to build multiple network clusters, elect clusterheads, and build a QoS routing overlay network. The network formation (initial discovery and cluster overlay formation) phase is illustrated in Figure 2 and Figure 3, respectively, and are further explained in the rest of this section.

Figure 2. A depiction of the HCPR-enabled ad hoc network after the network discovery phase

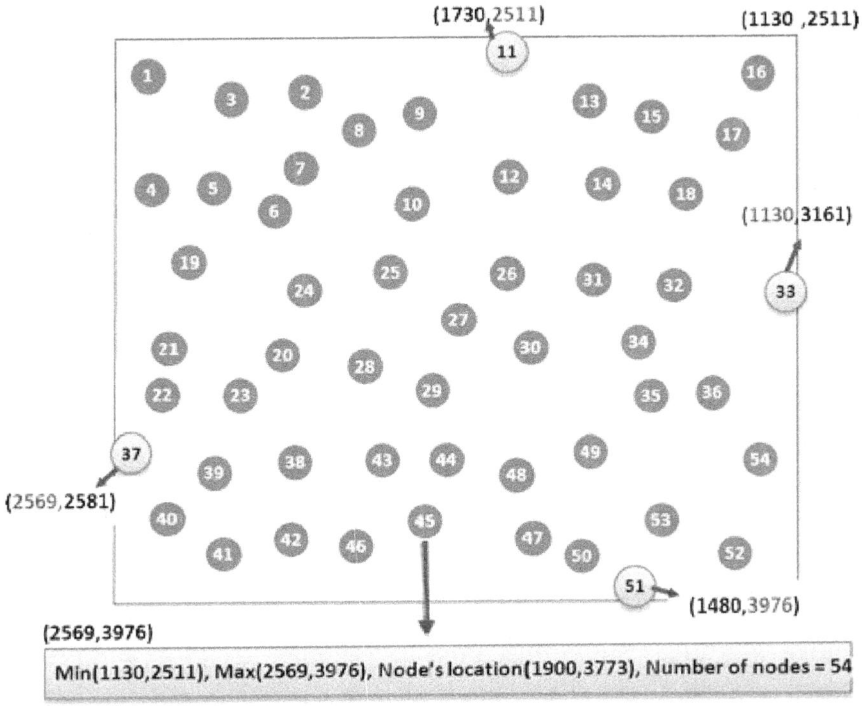

Figure 3. A depiction of the HCPR-enabled ad hoc network after the network formation phase

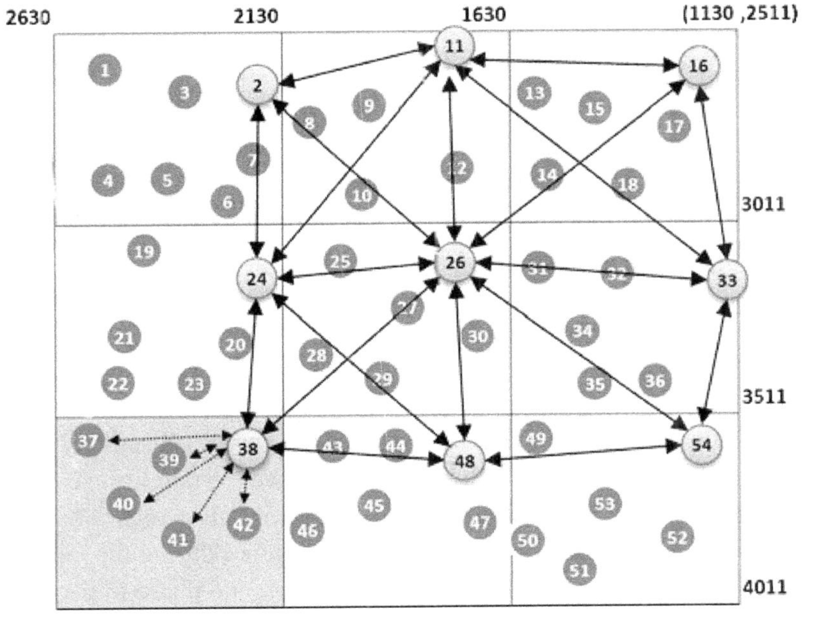

During the initial network discovery phase, each node broadcasts hello message containing its position to all other nodes in the network. As a node receives a hello message, it will use this message to build its routing table and will broadcast the message to other nodes provided the TTL (Time To Live) value is greater than zero. It is hoped that after a hello expiry period all the nodes in a network become aware of the geographic boundaries of the network as well as the position of all the nodes in the network. However, in a wireless environment, a single broadcast phase is inadequate to build a complete picture of the network. Therefore, the network discovery is achieved using a convergence process where the broadcast messages are iteratively sent in order to build a complete picture of the network. In each iteration of the convergence phase, each node in the network broadcasts the messages containing basic information of its routing table and iteratively builds a table containing information about the other nodes in the network, the nodes' respective geographic positions, and the geographic boundaries of the network. Each broadcast iteration of the convergence process within the initial network discovery phase is followed by an information (arithmetic) comparison process in order to test the initial network convergence. This process continues until the network convergence is achieved.

The network is said to have converged when all network nodes agree on the coordinates of the network geographic boundaries, nodes' positions and routing information. The comparison process is further explained as follows. It starts by sending a unicast message to all nodes to request an agreement on the information sent. When a node receives this agreement message request, it compares the information table sent with its own routing table. If the received information is matched with its table, an agreement message is sent back to the source node; otherwise, a disagreement will be sent to the source. All agreement and disagreement messages are saved in the states table of a node. The states table is a table used by a node to keep track of agreements and disagreements of other nodes. After agreement expiry, each node will check for all the acknowledgements received and will move to the cluster overlay formation phase if there is no disagreement. If even a single node disagrees, there will be a new iteration of convergence and comparison and it is repeated until all nodes are in agreement.

Note that the number of iterations in the network convergence phase can be configured at the start up and the network could be allowed to proceed to the next (overlay formation) phase without agreement. This is to reduce the network formation time and to allow network to be operational even in the case of disagreement on network boundaries. The disagreement could exist due to various problems associated the wireless environments such as unfair access to the network and hidden node problem. We have also experimented with some heuristics to automatically adjust node positions in case the network disagrees.

The OPNET software allows specifying the geographic location of the nodes using the latitude and longitude coordinate system. Figure 2 depicts the network after this initial phase where the location of nodes and the network geographic boundaries are shown in meters, in (X, Y) coordinates. The top right corner of the network is used as the reference point for the network and the location values increase towards the left (X) and the bottom (Y). For instance, as shown in the figure, the coordinates of the bottom-left-most geographic boundary of the network is (2569, 3976) and this is known to all nodes in the network including the node number 45. The X and Y coordinates of this bottom-leftmost boundary are due to the node 37 and 51, respectively.

The network discovery phase described above is followed by the cluster overlay formation phase. The information regarding the network boundaries acquired during the initial phase is then used in the second phase to partition the network into geographic regions. We used in this scheme a simple method for the network partitioning: given the knowledge of the total geographic area covered by the network and the total number of nodes; we partition the network into square geographic regions such that the number of nodes within a region and the geographic size of the network remains within a set of threshold values. This is depicted in Figure 3 where the network has been divided into *3 x 3* square regions. The regions do not have to be square; these regions can be of irregular shape and/or can be virtually independent of geographic boundaries. The partitioning algorithm is very simple because all nodes are aware of the network boundaries, their own geographic locations, and the partitioning criteria. Hence, each node can calculate the number of regions and the region where this node belongs.

Following the network partitioning, the nodes within each region form a cluster identity and select a clusterhead for themselves. Since the nodes are aware of the network and regional geographic boundaries, they could communicate and cooperate with each other to build a forwarding table for their own cluster. Similarly, the clusterhead for a particular region is the node which is the closest node to the top right reference point measured as Euclidean distance and it can be calculated individually by each node by a simple mathematical formula to compute the minimum Euclidean distance to the reference point. The reference point does not have to be the top right network boundary point; it can be any other point on the geographical boundaries. When a clusterhead discovers its leader status, it will firstly update its table and then send advertisement message to all nodes in the cluster informing them of its status. Ordinary nodes should by this stage have discovered the identity of their clusterhead and this should be in agreement with the message received from the clusterhead. The advertisement message is used to inform ordinary nodes to update their routing table for the destination outside the cluster. Local nodes, which are the ordinary nodes within the cluster, will update next address for all destinations outside cluster to be the clusterhead. The ordinary nodes send probe messages periodically to their clusterhead and maintain their tables proactively.

Once the clusterheads have been elected for each cluster, they discover each other, build routing tables to reach each other and form a network overlay. The network overlay uses a proactive routing strategy to other clusterhead nodes. However, it has the option to consider either a proactive or a reactive routing strategy to find ordinary destination nodes; this is discussed later in this section when we explain the network operational phase. Figure 3 depicts the state of an HCPR enabled ad hoc network after the network formation phase. The clusterheads are represented by red colour circles while the ordinary nodes are coloured blue. The figure illustrates the packet routing/forwarding between nodes. The ordinary nodes can talk to each other within a cluster (using multiple hops if necessary); however, an ordinary node must go through the clusterhead if it wishes to communicate with a node in another cluster. The figure also shows the network overlay formed by the clusterheads and the possible routes between them.

Note that there are many possible application scenarios one can think of where such a network formation phase, as described above, will be necessary, such as a sensor network dispatch to an area of land or sea, in an area of disaster, or a warzone. It is possible to use alternative strategies for the network formation phase where complete or partial information about the network, network boundaries, clusters, clusterheads, nodes, their position etc is available. In such cases, the discovery phase and/or the cluster formation phase information can be embedded into each node, or into a particular node (or nodes) known to everyone, allowing other nodes to acquire the information by appropriate protocols. In a mobile ad hoc networking environment, the network will continue to be in the network formation phase because it will have to reconfigure itself according to the changing network conditions.

Figure 4. A depiction of the HCPR-enabled ad hoc network in an operational phase (node 41 requesting a service connection to node 9)

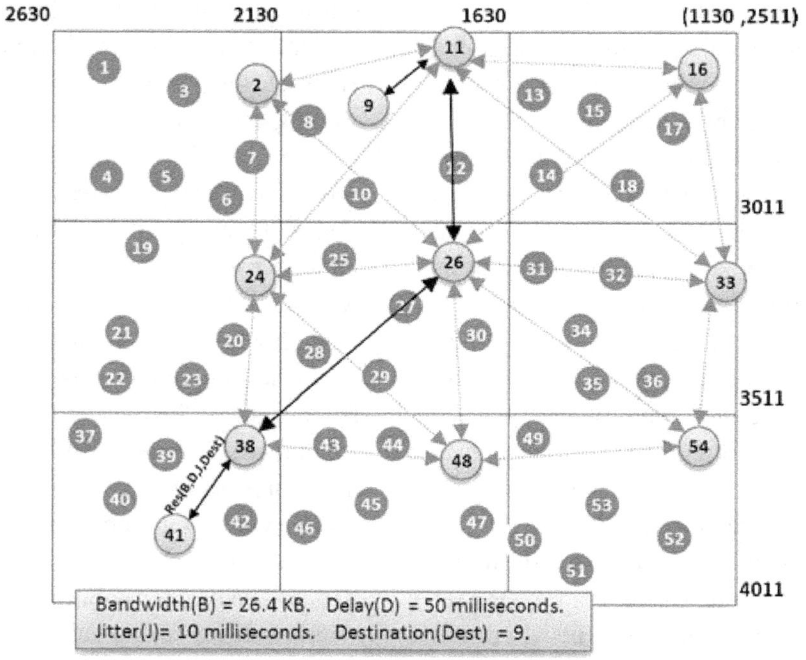

Network Operational Phase

Subsequent to the formation of the HCPR enabled network, a node in the network can make requests to connect to a destination node with some application. A node wishing to receive a service, however, does not talk to the destination node directly; rather it requests the clusterhead of its cluster to make a request to the destination on its behalf. This request is initiated from the application layer, which will call the corresponding function of HCPR (i.e. at the transport layer) that makes the reservation request to the clusterhead. Essentially, in an HCPR enabled network although the requests are made locally, these are propagated and provisioned by a hierarchical structure made up of clusterhead nodes. Consequently, the network provides high reliability and scalability.

Figure 4 depicts the operational phase of an HCPR enabled ad hoc network. This is illustrated by a connection request made by the source node 41 to the destination node 9 for an application requiring the following QoS parameters: 26.4 KB of bandwidth with maximum of 50 ms end-to-end delay and 10 ms jitter. The node can specify the time period for the connection or could send a request at the end to close down the connection. The clusterheads have their own routing tables and will contact other clusterheads accordingly to find the required route to the destination and whether the application requests could be accepted. The clusterheads use similar criteria to select their next hop to the destination node as used by the nodes of a region to select their clusterhead; i.e. the next hop is the node, which is the closest clusterhead node to the destination in terms of the Euclidean distance. If the available resources are insufficient to meet the application requirements, the application request will be rejected with a feedback to the source node (e.g. the maximum available resources are insufficient to meet the

Figure 5. Flow chart of the main functions of the HCPR process within the OPNET software

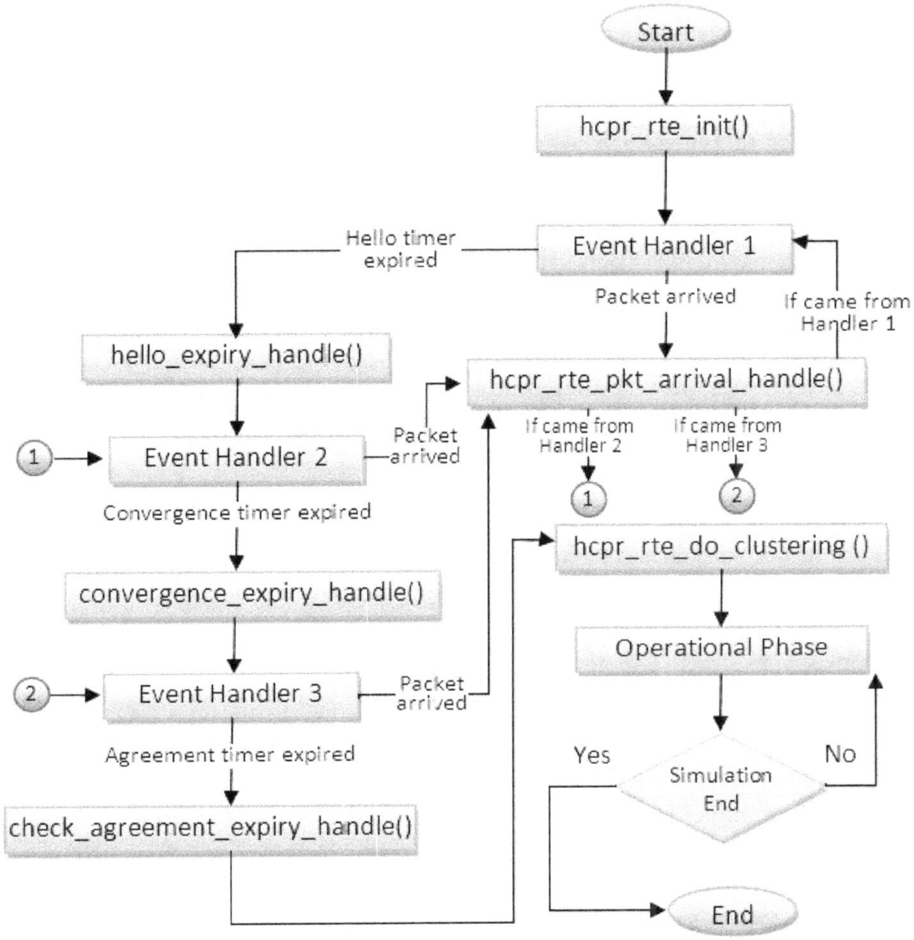

application request along with the detail of available resources).

As regards the process to find the destination node, two options have been considered. First options is that all the clusterheads proactively keep and maintain the list of the nodes associated with all clusterheads so they know in advance the clusterhead a particular destination node belongs to. The second option is that the clusterhead of the source node (reactively) broadcasts a request to find the clusterhead of the destination node. There are clear tradeoffs here in terms of overhead of maintaining tables and the delay of finding the node. In this paper, we have only implemented the proactive approach to keep record of the clusters and their associated destination nodes. The second option will be explored in the future.

The mobile environment where the topology of the network changes with time has not been explored for the HCPR scheme. However, HCPR can be enhanced to be used in such an environment as follows. Where nodes are mobile, the clusterheads are re-elected each time a clusterhead node moves away from the reference point (the top right hand coordinates), or move out from their cluster regions, or in any other case, such as the node failure. An ordinary node within a cluster initiates the election of a new clusterhead if it is

Figure 6. The location of the HCPR process within the WLAN node architecture

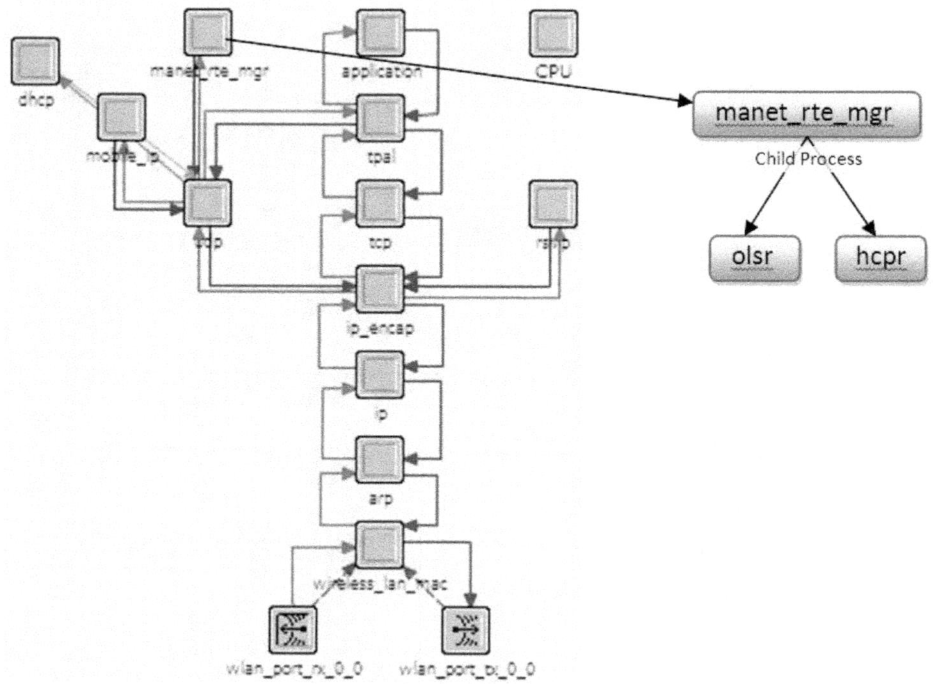

unable to reach its clusterhead using the criteria and protocols as described above. That is, the node closest to the top right reference point is elected as the new clusterhead and the tables are updated accordingly to reflect the network state following a node movement or a failure. The cluster maintains the cluster topology internally while the leader nodes maintain the network leader topology. In case of a leadership change, the topology state is transferred to the new respective leadership. Ordinary nodes are responsible to update their state if they move out of a cluster. The cluster leaders update their cluster state tables in case of any changes in their clusters and this updated information is propagated to other clusterheads as necessary.

Note that in this current work, we have only used two levels of hierarchy, the ordinary nodes and the clusterheads. In the future, we plan to propose and implement multiple hierarchies that are dynamically and recursively created according to the network size and other network parameters (for motivation, see our earlier work on network virtualization [Mehmood, Crowcroft, Hand, & Smith, 2005]).

Implementation

Figure 5 shows flow chart of the HCPR scheme, abstracted from its implementation in OPNET, with the focus on the six main functions mentioned before. Figure 6 shows the positioning and relationship of HCPR within a WLAN node architecture. The implementation details are given in (Alturki, 2011) in greater details and have been excluded here due to the space restrictions.

RESULTS AND ANALYSIS

We now analyse multimedia performance over wireless ad hoc networks using the proposed routing scheme HCPR and compare it with three well-known routing protocols AODV, OLSR, and

Table 2. The four traffic profiles and their respective intensities used for analysis

Traffic Profile	Voice		HTTP (Kbps)	Video (Kbps)	Total (Kbps)
	Codec	Kbps			
Low (L)	Low quality (G.723.1 5.3K)	21.312 (5.3K x 2 x 2)	10.2	68.488	100
Medium-Low (ML)	IP telephony (G.729 A)	32 (8K x 2 x 2)	20.4	947.6	1,000
Medium-High (MH)	GSM quality (GSM FR)	52.8 (13.2K x 2 x 2)	43.2	9,904	10,000
High (H)	PCM quality (G.711)	256 (64K x 2 x 2)	99.2	19,644.8	20,000

GRP. During the course of our study on multimedia networks, we have simulated a few hundred different networking scenarios by varying the routing protocols, traffic intensities, and numbers of nodes. Results show that different protocols act differently under different network conditions, and some protocols perform better at low traffic intensities while others give better performance for networks with higher numbers of nodes. In this section, we give a selection of scenarios and results from our simulation experiments. Our aim is to explore how the various protocols are affected by increasing the number of nodes, traffic intensities, and the mix of multimedia traffic applications.

The rest of the section is organised as follows. The next section describes the performance analysis methodology including the simulation scenarios, traffic profiles and network topologies. The following introduces and defines the performance measure used for the analysis in this chapter. Later sections present and discuss results for overall performance, HTTP, voice, and video performance respectively. Finally, we make some concluding remarks on the performance of HCPR and other protocols.

Performance Analysis Methodology

We describe here our performance analysis methodology that includes (a) the four traffic profiles that we have created using a mix of multimedia (Voice, Video, and HTTP) applications, (b) the variation in the network size and geographical structure, and (c) the network topology of multimedia applications.

Multimedia Traffic Profiles

In order to study the effect of how routing protocols are behaving with different traffic intensities, we have configured voice, video and HTTP applications into four traffic intensities. These are listed in Table 2. Rows 1 – 4 give details for each traffic intensity beginning with Low (L) traffic at 100Kbps up to the High (H) intensity at 20 Mbps. Columns 1 – 4 list bit rate for the voice, HTTP, and video including the codec name for voice. Column 5 lists the total bit rate for each traffic intensity. The bit rates for each application also give the number of connections. For example, in the third row, which represents the details of Medium-High traffic, the first and second columns show the voice codec used and the total voice bit rate. GSM FR codec is the codec used which has a bit rate of 12.3 kbps. The total voice bit rate is 52.3 which is a bit rate for 2 connections and each connection has two ways of voice. Columns 3 and 4 show the total HTTP and video traffics which are 43.2 Kbps and 9,900Kbps = 9.9 Mbps respectively. The last column is the total traffic populated in the network, including Voice, HTTP,

Figure 7. The network geographical structure and applications topology (5 nodes)

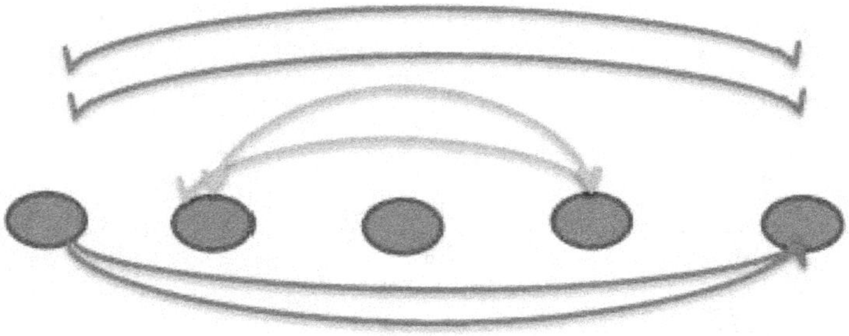

and Video, which is 10,000 kbps=10 Mbps. The last row shows the High intensity chosen at 20,000Kbps= 20 Mbps because this is usually the maximum achievable bandwidth for 802.11g at 54 Mbps. As can be seen from the table, traffic populated were a mix of HTTP, voice and video in all the four traffic intensities. Having a mix of data and multimedia applications in the same network was chosen to show how applications are performing with each other to simulate more realistic networking scenarios.

Variations in Network Size and Geographical Structure

Varying the network size against fixed traffic intensities was intricate and required careful configuration, as it is difficult to compare the performance of two scenarios unless if they are exactly the same, except one parameter, which is the one intended to study. In the case of this chapter, all parameters should be fixed and network sizes will be varied. The problem here is that increasing the number of nodes requires other parameters to change, which is the scenario of application sessions, that is which node is connected to which. For example, if a network has 5 nodes (A, B, C, D, and E), and there is a one video streaming from two nodes A and B. if number of nodes increases to 10, there will be 5 nodes added to the network. The challenge here is how to make use of the increase to ten nodes, to assess that increase, with keeping the same application scenario. Since the goal of this study is to see the effect of number of hops increase when increasing number of nodes, It has been decided in this chapter to have nodes fixed and distributed in a way that study the effect of the increase in the number of hops as following. The study will start by having five nodes organized in one row and having distance of 200 meters between each other. As number of nodes increases to 10, other row of five nodes will be added below and parallel to the first row. This is done consequently for 20, 30, 40, and 50 node scenarios. Figure 7 and Figure 8 illustrate the geographical structure of a network with 5 and 20 nodes respectively. The figures are explained further in the next section where we discuss applications topology.

The Network Topology of Multimedia Applications

Having applications deployed similarly across network sizes is very important to study the effects network size and hops increase. In all scenarios used in this study, there were two HTTP connections, two video streaming connections and two voice connections. In case of five nodes, the pairs of connections of HTTP and Video were initiated between nodes that are located in the edges of the network. The pair connections of Voice were between nodes that located just to the next of the edge nodes. Figure 7 shows applications'

Figure 8. The network geographical structure and applications topology (20 nodes)

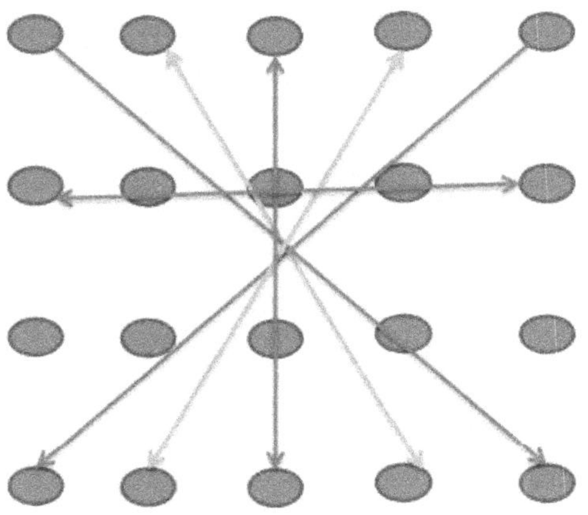

connections details for 5-nodes scenario. The red lines represent Video, purples for HTTP and green lines for Voice.

As number of nodes increases to 10, the Video connections will be between the two corners of the network. One Video connection is between two facing corners, which will be like a diagonal line. The Voice connection will be between nodes that are next to the two corners toward the middle of every row. The HTTP connections will be between the middle nodes in the first and second rows and between the two edges of the first row. The settings for 20, 30, 40, and 50 nodes scenarios will follow the same patterns. Figure 8 depicts the applications topology for a network with 20 nodes.

Performance Measures

We have used several key metrics to measure network performance. These are described below.

Traffic sent: is the traffic that has been successfully sent from the application layer to the next lower layer (transport layer).

Traffic intended to send: is the traffic that has been planned to send from application layer to lower layer. As it has been known that, not all traffic is intended to send is sent because some layers (e.g. physical) will not be able to handle the traffic for any reason.

Throughput: is the percentage of the average traffic received per second to the traffic that was intended to send per second in the network. In results presented in this Section; throughput is used be for every individual application Video, Voice, and HTTP, besides having throughout of total traffic in network.

Delivery Ratio (DR): is the percentage of the average traffic received per second to the traffic sent per second in the network. In this section, there will be Delivery Ratio (DR) for every individual application: Video, Voice, and HTTP, besides having the total Delivery Ratio (DR) of all traffic in network.

End-to-End Delay: is the average time spent in second for the packet to reach destination. The Video and Voice end-end-delay have slight different definitions by OPNET. The **Video End-to-End Delay** defined as the average time spent in second to transfer video packets from source's application layer to destination's application layer in the network. It is measured from the time of creating the packet until the time of receiving it. The **Voice End-to-End-Delay** is the average time in second that voice packet take to reach destination from time of encoding the analogue signals at source to the time the packet is decoded at destination. This is called mouth-to-mouth delay which include encoding delay, compression delay, network delay, decompression delay and decoding delay (OPNET Technologies Inc, 2008).

Delay Variation: is the average variance in packet end-to-end delay. In OPNET, user can collect Delay variation for Video and Voice packets besides having jitter for voice. **Voice Jitter** is the variation of delay in second in received voice packets (Cisco Systems, 2006). If two packets left the source voice application at t1 and t2 times consequently and arrived at destination's voice application at t3 and t4 consequently, the jitter is the result of this following equation: (t4-t3) – (t2-t1) (OPNET Technologies Inc, 2008).

Figure 9. The overall delivery ratio (DR) for low traffic across network sizes

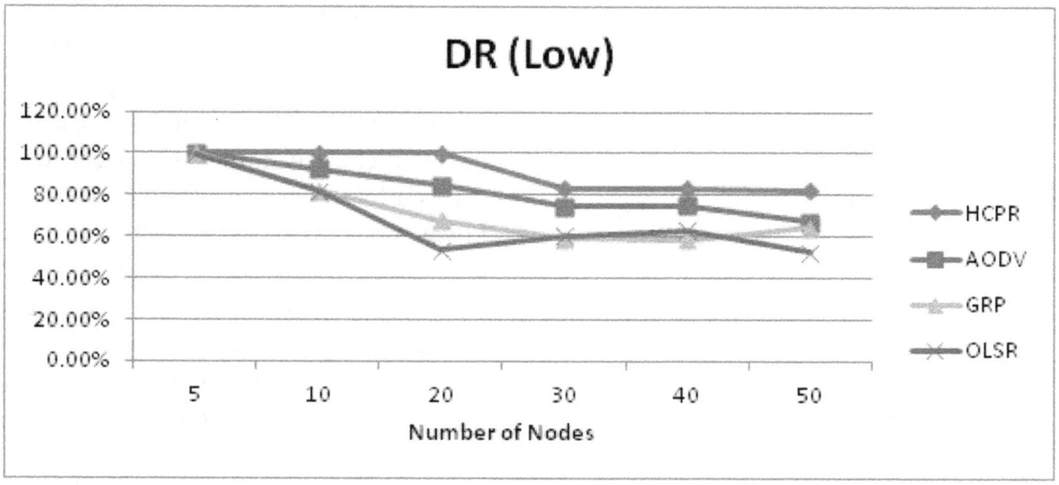

Voice Mean Opinion Score (MOS): "MOS is a subjective measurement representing the quality of digital multimedia including video, voice, or audio" (Rashid Mehmood, et al., 2011) and it represents the quality in a numerical format ranging from 1 as reference to bad quality to 5 as reference to Excellent quality. In the results presented in this chapter, the average of Voice MOS in the network per second was taken and used.

HTTP Page Response Time: is the average time in second required to retrieve the entire page with all objects (e.g. image) in the page.

The Overall Multimedia Performance for Traffic Profiles against Network Sizes

We will now focus on the analysis of overall multimedia performance measured in terms of Delivery Ratio (DR), comprising collective results of the three populated applications Video, Voice and HTTP traffic. We will study the overall performance for varying network sizes and a set of four traffic intensities; L, ML, MH, and H (see Table 2). The overall performance is plotted for each of these four traffic intensities in separately. The overall performance will be depicted against varying network sizes while keeping the traffic intensity constant.

Figure 9 plots the overall Delivery Ratio (DR) of low traffic intensity of 100kbps. As with five nodes, all routing protocols achieved around 100% DR and then started to decline afterword except HCPR. HCPR kept its 100% DR for 10 and 20-nodes network sizes, then declined when number of nodes reached 30 nodes, and continued on stable performance between 82.4-83.4% DR. AODV has shown the second best performance—after HCPR—with clear pattern of slight decrease in every step of increasing number of nodes until it reached 67.2 DR at 50-nodes network size. OLSR and GRP were the lowest when number of nodes was 10. OLSR then declined the most between 10 to 20 nodes network sizes to reach its second worst point of 53.8% DR. Then, OLSR improved considerably when number of nodes increased to 30, and continued the slight increase when number of nodes was 40. Then, it had a small decline when nodes reached 50. GRP continued in declining until it reached the 40 nodes network size, and then showed some improvement when number of nodes reached 50.

Figure 10 depicts the overall Delivery Ratios of Medium Low (ML) traffic intensity of 1Mbps.

Figure 10. The overall delivery ratio (DR) for medium low (ML) traffic across network sizes

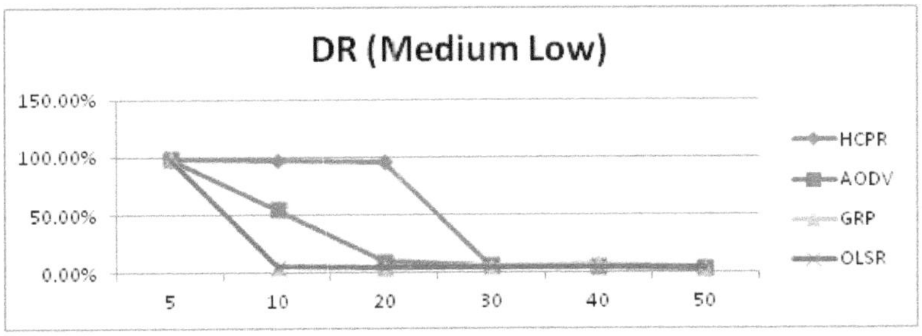

Figure 11. The overall delivery ratio (DR) for medium high (MH) traffic across network sizes

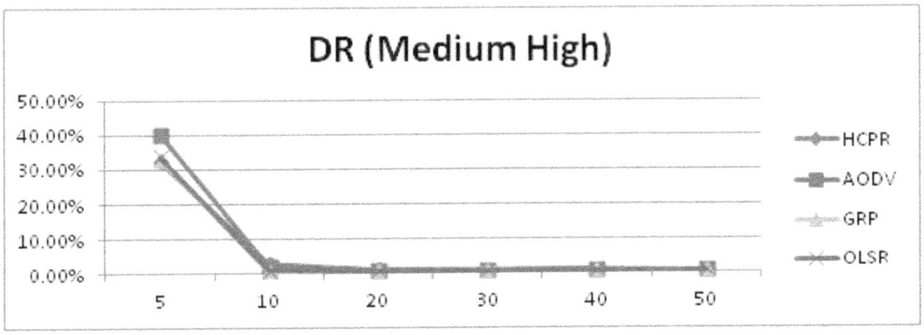

At five nodes network size, all routing protocols achieved above 97% DR and then started to decline sharply afterword except HCPR, which kept the DR above 95.9% for 10 and 20 node network sizes. GRP and OLSR made a big decline going from 5 to 10 nodes to reach DR below 5% and then kept the DR below 5% until 50 nodes except the slight increase of GRP when nodes are 40. On the other hand, AODV declined gradually until it reached below 10% DR when number of nodes is 20 and then decrease slightly until it achieved 3.4% DR at 50 nodes. There is no doubt that increasing the traffics from low to Medium-Low has affected the performance badly on AODV, GRP, and OLSR.

Figure 11 depicts the overall Delivery Ratio (DR) of Medium High (MH) traffic intensity of 10Mbps. In five node scenarios, routing protocols achieved DR between 32.3% and 40.2% and then started to decline sharply. As number of nodes increased to 10, the DR achieved for routing protocols were below 3.2% and then kept in slight declining until 50 nodes. There is a clear drop of the performance of MH traffic across network sizes in comparison to ML and L traffics.

Figure 12 shows the overall Delivery Ratio of High (H) traffic intensity of 20Mbps across network sizes. When number of nodes was five, all routing protocols achieved between 2.2% and 2.7% DR except AODV, which had DR of 8.9%. As number of nodes is increased, AODV has decreased gradually until 20 nodes, and then kept the same level for 30 and 40 nodes followed by slight decline at the end. OLSR has shown some decreases until it reached 10 nodes, and then kept the same level until 50 nodes. GRP was similar to OLSR trend at 10 and 20 nodes network sizes, and then showed some increases and then declined until it reached 50-nodes size. HCPR was the lowest performer for 10 and 20 nodes and then showed considerable increase in 30 nodes, and kept the level for 40 nodes and then small decline

Figure 12. The overall delivery ratio (DR) for high (H) traffic across network sizes

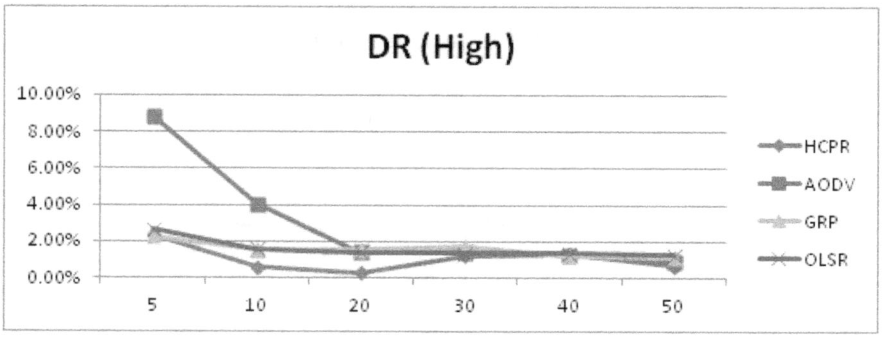

at 50 nodes. Overall performance of all routing protocols when populated with High traffic is not acceptable for any application.

Performance Overview for All Traffic Profiles

We present here an overview of the results presented earlier, where we analysed the overall multimedia performance against network sizes, separately for each traffic profile: L, ML, MH and H. It has been observed through the results presented that delivery ratio performance was around 100% at 5 nodes when network was populated with Low (L) traffic. HCPR was the only routing protocol that kept its high performance of 100% DR for 10 and 20-node networks at this traffic intensity. In Medium Low (ML) intensity, all routing protocols had over 98% DR at 5-node network, then went sharply for higher network sizes except HCPR, which showed 97.3% and 96% for 10 and 20-node networks respectively. At Medium High (MH) and High (H) traffic intensities (profiles), none of the routing protocols achieved acceptable DR at any network size.

HTTP Performance

Now, we focus on performance analysis of HTTP applications over ad hoc network for the four routing protocols. The HTTP applications share the network with Voice and Video traffic. We have studied HTTP performance for varying network sizes and, as before, a set of four traffic intensities; L, ML, MH, and H (see Alturki, 2011 for full set of results). However, in this chapter we only depict performance for the "L" traffic profile and discuss performance for three other traffic profiles in the summary subsection. Note that although the network was shared by Voice, Video, and HTTP traffic, we will only highlight HTTP performance perspective. We depict the HTTP performance against varying network sizes while keeping the traffic intensity constant. In each constant profile, we study delivery ratio, throughput, and page response time.

Figure 13 illustrate the performance of HTTP Delivery Ratio (DR) when the network is populated with Low (L) overall traffic of 100Kbps (HTTP traffic was 10.2 Kbps). As evident in Figure 13, HCPR did well with 100% DR across network sizes. GRP on the other hand did well at small and medium network sizes with 100% DR, and then decreased when number of nodes reached 40, and showed a slight increase at 50 nodes network size. AODV was excellent at 5 nodes network size by achieving 100%, and then went down gradually and slowly to achieve 95.2% at 50 nodes. OLSR showed some fluctuation by having 100% at five nodes network size, and then went down gradually to reach 92% when number of nodes was 20, and then slight increase at 30 nodes before the last

Figure 13. The HTTP delivery ratio (DR) for low (L) traffic across network sizes

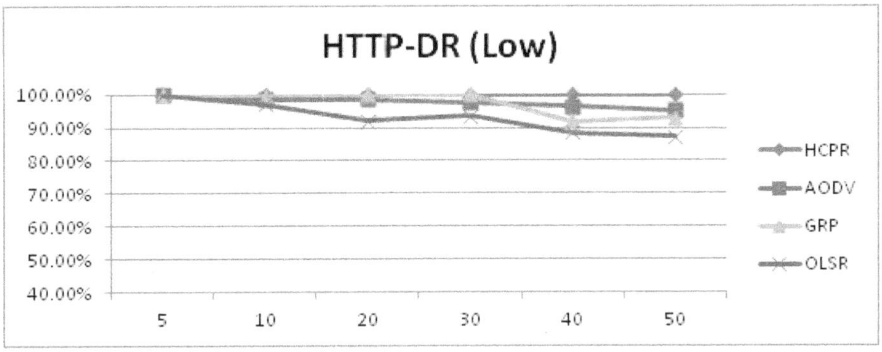

Figure 14. The HTTP page response time for low (L) traffic across network sizes

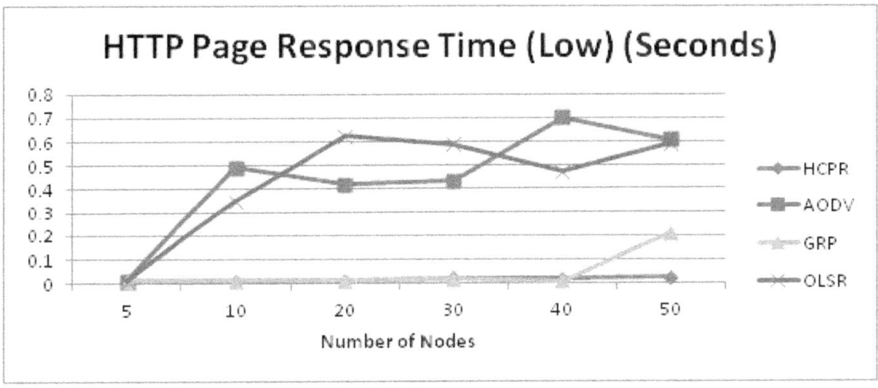

decrease to reach 87.3% at 50 nodes network size. OLSR was the worst DR performer in comparison to the other routing protocols.

Figure 14 illustrates the HTTP page response time for the Low (L) traffic across all network sizes. The general trend of the graph shows an increase in the values with the increase of network size with some exceptions for some cases. HCPR showed around zero response times across network sizes while GRP showed zero response times for small and medium size networks, and increased for large network sizes. AODV started by having zero response time at 5 nodes; an increase at 10 nodes; a decrease at 20 nodes; increase at 30 and 40 nodes and then small decrease at 50 nodes. OLSR increased gradually until 20 nodes network size, and then decreased at 30 and 40 nodes followed by an increase at 50-node network size. There is no clear reason for the fluctuation of values in this graph and by comparing these results to the results of overall, Video, Voice, and HTTP values, there is no clear connection. The relatively high page response time for AODV and OLSR compared to GRP and HCPR is because GRP and HCPR are benefitting by geographical routing. In addition, the AODV protocol is also affected by its reactive routing. Furthermore, as mentioned earlier, TCP is well known to have unstable performance in ad hoc networking environments, and, perhaps, it has created a little inconsistent behaviour by the non-geographical routing protocols, AODV and OLSR.

HTTP Performance Overview for All Traffic Profile

In this section, we give our conclusion about the performances analyses of HTTP applications. The HTTP application was sharing the network with Voice and Video traffic. We studied the HTTP performance for varying network sizes and a set of four traffic intensities; L, ML, MH, and H. However, we only depicted performance for the "L" traffic profile (Figure 13 and Figure 14 depicted the DR and page response time for "L" traffic profile). The performance for three other traffic profiles is summarised below.

When traffic profile intensity was Low (L), HCPR was the only routing protocol that achieved 100% across network sizes where others lost between 5 to 12% at 50-node network. HCPR also had 100% throughput across network sizes while others showed fluctuation. HCPR had constant Page Response Time below 0.03 second across network sizes while others showed higher than that. At Medium Low (ML) traffic intensity, HCPR achieved the best DR across network sizes with 100% at most sizes. HCPR was also the best throughput performer for all network sizes except 50-node where it was the worst. HCPR relatively had the best Page Response Time across network sizes followed by GRP. At Medium High (MH) traffic intensity, HCPR and GRP were the best DR performers; GRP for small networks and HCPR for large network size. Throughput trend for the routing protocols was fluctuating across network sizes; HCPR had the highest at 30 and 40-node networks. Page Response Time values were below 1 second for all routing protocols across network sizes. However, there were some variations of values where HCPR was the best for 5-nodes network size and GRP for the rest of the network. At High (H) traffic intensity, DR and throughput showed a fluctuation of values across network sizes where GRP had the best DR at 10 and 20-node networks and HCPR was the best for 30 and 40-node networks sizes. The general trend of Page Response Time was increasing where HCPR and GRP were collectively the best performers. We could conclude that HCPR was best HTTP performer for Low and Medium-Low traffic profiles. As intensity increased to Medium-High and High, HCPR performance was the best at 30 and 40-node networks.

Voice Performance

We focus now on performance analysis of Voice applications over ad hoc networks. As before, the Voice flows were sharing the network with Video and HTTP traffic. We study Voice performance for varying network sizes and a set of four traffic intensities; L, ML, MH, and H (see Table 2). As was the case for HTTP performance study in Section 5.4, we have studied voice performance for varying network sizes and a set of four traffic intensities; however, in this chapter we only depict performance for the "L" traffic profile and discuss performance for three other traffic profiles later in the summary subsection. Note that although the network is shared by Voice, Video, and HTTP traffic, we only present Voice performance metrics. In each traffic profile, we start by plotting DR and throughput followed by jitter and delay and finally by plotting the MOS.

Figure 15 depicts the performance of Voice DR when the network is populated with low overall traffic of 100kbps, and voice traffic of 21.3 Kbps (G.723 codec). As is evident from DR figure, all routing protocols achieved around 100% DR across networks sizes. Figure 16 depicts the performance of Voice Jitter when the network was populated with Low overall traffic of 100kbps. As can be seen from the figure, OLSR, GRP, and HCPR have similar Jitter of zero values across network sizes. On the other hand, AODV had negative values across networks sizes, although the values are so small that these can be considered zero.

Figure 17 depicts the Voice packet end-to-end delay when the network was populated with Low

Figure 15. The voice delivery ratio (DR) for low (L) traffic across network sizes

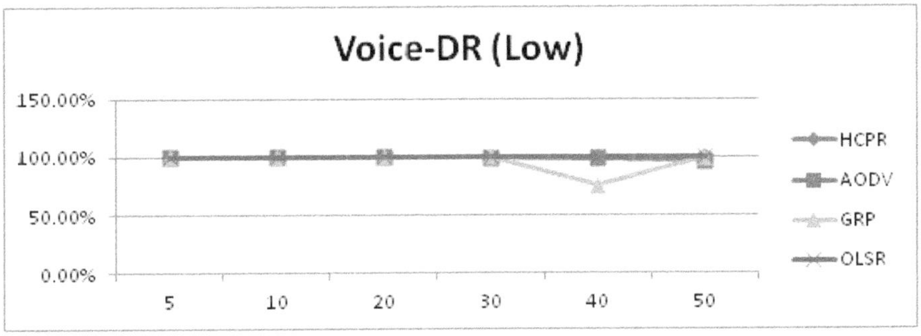

Figure 16. The voice jitter for low (L) traffic across network sizes

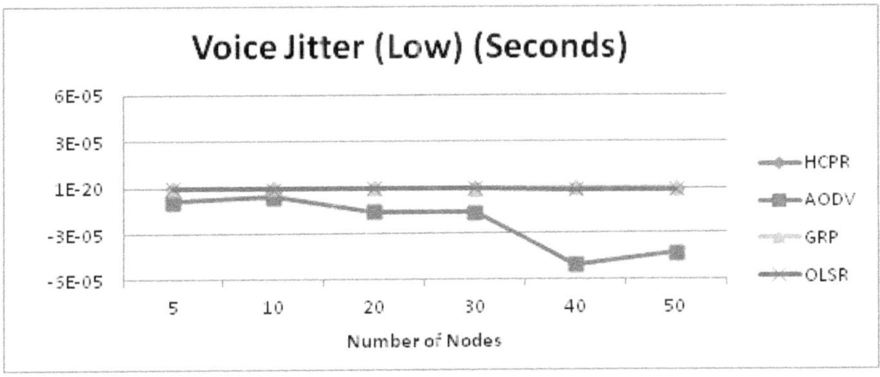

overall traffic of 100kbps. The figure shows that OLSR, HCPR, and GRP delays were stable across network sizes with delay of around 0.1 second. On the other hand, AODV had a delay of around 0.1 second at 5, 10, and 20 nodes, and then started to increase gradually to reach 0.15 second when number of nodes reached 50. This we believe is due to the reactive nature of AODV protocol. Figure 18 depicts the Voice MOS when the network was populated with low overall traffic of 100kbps. As can be observed from the figure, MOS values were 2.5 across network sizes for GRP, HCPR and OLSR. The Voice MOS performance for AODV was 2.5 at 5, 10, and 20 nodes and then went down gradually to reach around 1.3 at 50 nodes. The reason for AODV degradation with increasing number of nodes was most probably because of negative decrease of jitter and the increase of delay as network size was increasing.

Voice Performance Overview for all Traffic Profile

Here we summarise result of the performance analysis of Voice application. As before, the Voice flows were sharing the network with Video and HTTP traffic. We study the Voice performance for varying network sizes and a set of four traffic intensities; L, ML, MH and H. In each traffic profile, a conclusion of deliver ratio, throughput, jitter, delay MOS was collected.

At Low traffic intensity profile, Voice DR was over 99% for all routing protocols when network size was between 5 and 30. At 40 and 50-node networks, DR values were over 96% except in

Figure 17. The voice packet end-to-end delay for low (L) traffic across network sizes

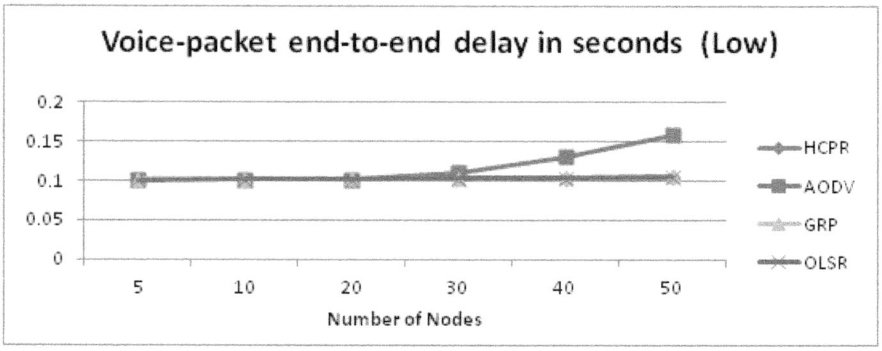

Figure 18. The voice MOS for low (L) traffic across network sizes

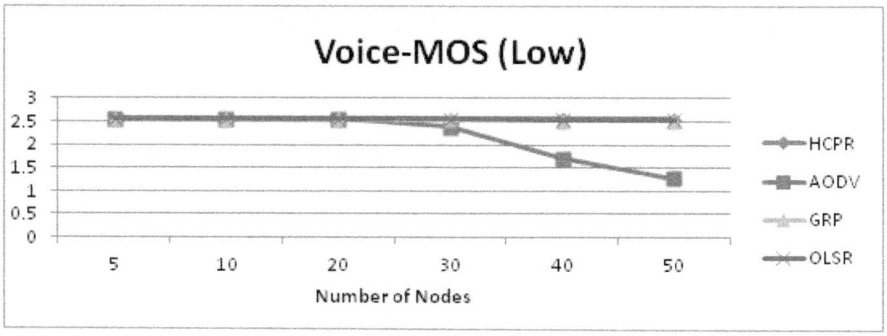

one case at 40-nodes network size where the GRP scored around 74% only. All routing protocols had zero jitter except AODV which showed fluctuation of jitter across network sizes. Delay was zero for all routing protocols across network size except for AODV at large network sizes. MOS values were around 2.5 across network sizes except AODV, which went down for larger network sizes. At Medium-Low intensity, all routing protocols achieved around 100% from 5 to 20 network sizes. OLSR is the only routing protocol that had over 96% until 50-node network while AODV and GRP did well at 30, and GRP did that again at 50-node network. Jitter and Delay were low at small network sizes and then increased for some routing protocols at larger network sizes. The highest MOS value achieved was 3 and HCPR was the best performer.

At Medium-High intensity, all routing protocols had just above 97% DR when network size was five except AODV which had around 84%. Afterward, all routing protocols showed some fluctuation across network sizes. OLSR and GRP were the best performers followed by AODV and then HCPR. Jitter and delay values were high for 5-node network then decreased for some routing protocols where HCPR was the worst. The highest MOS values did not go above 3.5, and the MOS trend was fluctuating, and GRP was the best MOS performer. At High intensity, OLSR had the best Delivery Ratio (DR) across network sizes while HCPR was the worst. At 5-node network size, jitter and delay were high for all routing protocols then decreased except HCPR, which was the worst. The highest MOS values were 3.6, and the general MOS trend showed some fluctuation. In conclusion, for ML profile, HCPR was the best

Figure 19. The video delivery ratio (DR) for low (L) traffic across network sizes

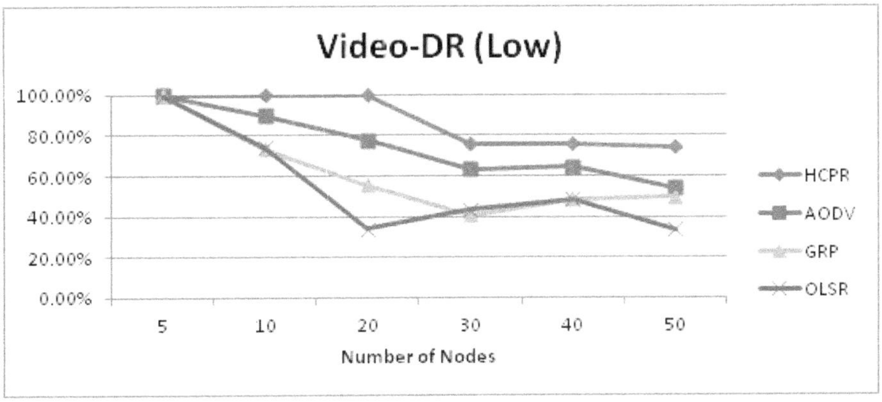

MOS performer among other routing protocols. However, HCPR was the worst at MH and H.

Video Performance

This section studies the performance of Video applications over ad hoc networks. As before, the Video application shared the network with Voice and HTTP traffics. The study of Video performance is for varying network sizes and a set of four traffic intensities; L, ML, MH, and H. We have studied video performance for varying network sizes and a set of four traffic intensities; however, in this chapter we only depict performance for the "L" traffic profile and discuss performance for other traffic profiles later in the summary subsection. We plot results for DR, delay and delay variation.

Figure 19 depicts the performance of Video Delivery Ratio (DR) when the network is populated with Low (L) overall traffic of 100kbps, and video traffic was around 68.5kbps of the total traffic. The Overall view of DR figure is a decrease with the increase of number of nodes except HCPR which showed excellent results at 10 and 20 nodes. That made HCPR to be the best achiever among routing protocols followed by AODV, which was in clear pattern and decreased gradually. OLSR dropped to around 34% when number of nodes was 20, rose at 30 and 40-node network sizes, and again dropped again at 50-node network size. GRP went down gradually until 30-node network size, and then started increasing slightly at 40 and 50 nodes scenarios.

Figure 20 shows the Video packet end-to-end delay across network sizes when the network populated with Low (L) overall traffic of 100kbps. The figure started by having Low delay of around 0 for all routing protocols and then started increasing at 30 nodes. AODV increased gradually from 30-node network size while HCPR increased sharply from 40-node network size. Although, HCPR was the worst at 50 nodes, it is has the least lost ratio than others and it was the same for AODV as well. That means sometimes having a bad performance for one metric does not necessarily means all metrics performances are bad but it could be a sign good performance of other metrics, in this case DR was achieving good. Figure 21 shows the Video delay variation across network sizes when network populated with Low (L) overall traffic of 100kbps. The variation was almost zero for all routing protocols across network sizes except for AODV and HCPR when number of nodes was 50 where the variation rose to just below 0.5 and 5 seconds consequently.

Figure 20. The video packet end-to-end delay for low (L) traffic across network sizes

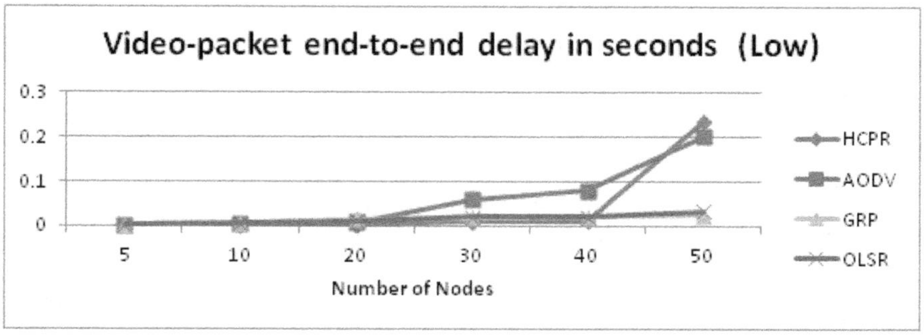

Figure 21. The video packet delay variation for low (L) traffic across network sizes

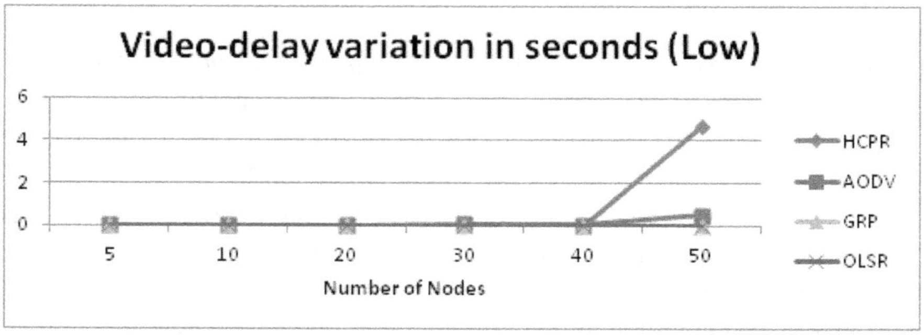

Video Performance Overview for All Traffic Profile

We summarise here the results video applications. A conclusion of each of the four traffic profiles is drawn using four metrics: Delivery Ratio, throughput, delay, and delay variation as follows. Generally, The performance of video DR was similar to overall DR performance because Video traffic intensity was over 60% of the total traffic of Low traffic and over 94% of ML, MH and H intensities. At Low traffic intensity profile, all routing protocols achieved around 100% DR at 5-node network. HCPR was the only routing protocol that kept its high performance of 100% DR at 10 and 20-node networks. Packet end-to-end delay was below 0.25 second for all network sizes and below 0.02 when network sizes were between 5 and 20 nodes. At Medium-Low traffic intensity profile, all nodes had over 98% DR at 5-node network, and then dropped to below 55% for larger networks except HCPR, which kept its high performance at 10 and 20-node networks to reach around 97.3% and 95.8% respectively. In all cases where DR was above 95%, we had a maximum of 0.07-second delay and zero delay variation. At Medium-High and High traffic intensities, none of the routing protocols achieved acceptable DR at any network sizes. In conclusion, HCPR had the best performance in Low and Medium-Low traffic intensities.

RESULTS AND ANALYSIS: AN OVERVIEW

In the previous section, we presented an extensive analysis of the HCPR performance for multimedia delivery over ad hoc networks. A number of scenarios were configured by varying different routing protocols, different intensities, and different numbers of nodes. It showed how protocols are affected by increasing the number of nodes and traffic intensities, and showed how HCPR was performing in comparison to other routing protocols (AODV, GRP, and OLSR) under different scenario settings. Network sizes were varied by the number of nodes from 5-node to 50-node networks. In addition, traffic intensities were set in four main intensity profiles, L, ML, MH, and H. Each intensity profile was a mix of three multimedia applications Video, Voice, and HTTP. We have studied the overall performance by fixing the network size and see how network was performing with various intensities. That was followed by a performance analyses of the overall and multimedia applications by fixing traffic intensity and varying network sizes. When studying the effect of varying traffic intensities across network sizes, HCPR was the only routing protocol that showed a good overall performance at Low (L) and Medium-Low (ML) traffic intensities for 10 and 20-node networks. When studying the effect of varying network sizes for each traffic intensity profile, the overall and video performance for HCPR outperformed other routing protocols at 10 and 20-node networks for L and ML traffic profiles. The high bitrate of Video and HCPR performer have resulted of having lower performance of Voice especially at MH and H traffic intensities. The results showed about HTTP, HCPR was the best performer at L and ML intensities, for higher traffic intensities, HCPR also outperformed other routing protocols at 30 and 40 network sizes.

CONCLUSION AND FUTURE WORK

Rapid developments in the advancement and convergence of the telecommunications industry are largely attributed to multimedia applications. People demand anytime anywhere mobility giving rise to heterogamous networks comprising a range of fixed, wireless, mobile and ad hoc networks. Multimedia applications pose stringent requirements on networks while ad hoc networks are inherently limited due to their autonomic and multi-hop nature and the lack or scarcity of infrastructure. Supporting multimedia applications over ad hoc networks is a challenge for the industry. The aim of our research is to address the challenges associated with the design and analysis of ad hoc networks supporting multimedia applications with scalable Quality of Service (QoS). We believe that we have been able to accomplish this aim to a large extent and have significantly advanced the state of the art in this area. We had approached these challenges from multiple perspectives and, accordingly, by adopting the cross-layer design approach, have been able to contribute in two major areas.

Firstly, we proposed the HCPR scheme, and implemented as an extension to the OPNET simulation software. The HCPR scheme was described, and it was analysed in detail for its QoS performance to deliver multimedia applications over ad hoc networks. It was compared with three well-known and widely used routing protocols; AODV, OLSR, and GRP. Around a hundred networking and applications scenarios were simulated and it was demonstrated that the HCPR enabled ad hoc network outperforms the well-known routing schemes, in particular for relatively large networks and high QoS network loads. These results show promising performance because many QoS schemes do work for small networks and low network loads but are unable to sustain performance for large networks and high QoS loads. A detailed account of the literature review, implementation details, and performance analysis is provided in the PhD thesis (Alturki, 2011).

Secondly, we developed and presented a novel analysis methodology for configuring multimedia applications and networking scenarios. This methodology was used in the analysis and comparison of the HCPR scheme for wireless ad hoc networks. The analysis methodology is to simulate and evaluate realistic multimedia applications scenarios for networks with a good number and type of applications - VoIP, Video, and HTTP, both elastic and non-elastic - so that the mutual affects of network applications can be taken into account. We had shown through a detailed literature review that earlier literature in this area lacks such an analysis approach and therefore our proposed analysis methodology would contribute to the advancements in the analysis of networks carrying multimedia applications.

The HCPR scheme proposed in this chapter can be enhanced in multiple ways, outlined as follows. First, various criteria to form clusters and clusterheads can be used to optimise performance for specific scenarios, for example by incorporating in the decision process the information regarding node's connectivity, battery lifetime and transmission range. Secondly, we have used here pro-active routing approaches for both intra- and inter-cluster routing in HCPR. The routing approaches used in HCPR can be configured such that the HCPR performance is optimised in different context based on for example the network and cluster sizes. Thirdly, we have proposed to integrate RSVP within HCPR in order to make QoS reservations. We would like to implement the reservation protocols and additional QoS metrics criteria within HCPR in order to investigate its impact on HCPR performance. Studying the effect of mobility, various mobility models, network topology on HCPR performance is another future direction of our research.

Finally, the work on analysis methodology can be extended and enhanced in the future by using additional types of multimedia traffic models (such as video codecs and models, additional VoIP models, etc.) and network configurations. We would like to formalise our analysis methodology by finding good and bad practices for networks and applications setup found in realistic environments.

ACKNOWLEDGMENT

We are thankful to all the reviewers for their time, very useful recommendations, and their comments for helping us to improve this chapter.

REFERENCES

Al Turki, R., & Mehmood, R. (2008). Multimedia ad hoc networks: Performance analysis. In *Proceedings of the Second UKSIM European Symposium on Computer Modeling and Simulation*, (pp. 561-566). UKSIM.

Alturki, R. (2011). *Multimedia ad hoc networks: Design, QoS, routing and analysis.* PhD Thesis. Swansea, UK: Swansea University.

An, C., Loyola, L., & Aad, I. (2007). *Analysis of UDP, TCP and voice performance in IEEE 802.11b multihop networks* Paper presented at the 13th European Wireless. Paris, France.

Bhatti, S. N., & Crowcroft, J. (2000). QoS-sensitive flows: Issues in IP packet handling. *IEEE Internet Computing*, *4*(4), 48–57. doi:10.1109/4236.865087

Blake, S., Black, D., Carlson, M., Davies, E., Wang, Z., & Weiss, W. (1998). An architecture for differentiated services. *RFC:2475*. Retrieved from http://www.ietf.org.

Braden, R., Clark, D., & Shenker, S. (1994). Integrated services in the internet architecture: An overview. *RFC 1633*. Retrieved from http://www.ietf.org.

Broch, J., Maltz, D. A., Johnson, D. B., Hu, Y.-C., & Jetcheva, J. (1998). *A performance comparison of multi-hop wireless ad hoc network routing protocols.* Paper presented at the Mobile Computing and Networking. New York, NY.

Cisco Systems. (2006). *Understanding jitter in packet voice networks*. Retrieved from http://www.cisco.com/en/US/tech/tk652/tk698/technologies_tech_note09186a00800945df.shtml.

Clausen, E., & Jacquet, E. (2003). Optimized link state routing protocol (OLSR). *RFC:3626*. Retrieved from http://www.ietf.org.

Conti, M., Maselli, G., & Turi, G. (2006). Design of a flexible cross-layer interface for ad hoc networks. In Al Agha, K., Guérin Lassous, I., & Pujolle, G. (Eds.), *Challenges in Ad Hoc Networking* (*Vol. 197*, pp. 189–198). Boston, MA: Springer. doi:10.1007/0-387-31173-4_21

Conti, M., Maselli, G., Turi, G., & Giordano, S. (2004). Cross-layering in mobile ad hoc network design. *Computer*, *37*(2), 48–51. doi:10.1109/MC.2004.1266295

Gharavi, H., & Ban, K. (2003). *Cross-layer feedback control for video communications via mobile ad-hoc networks*. Paper presented at the Vehicular Technology Conference, 2003. New York, NY.

Gottron, C., Konig, A., Hollick, M., Bergstrasser, S., Hildebrandt, T., & Steinmetz, R. (2009). *Quality of experience of voice communication in large-scale mobile ad hoc networks*. Paper presented at the Wireless Days (WD). Paris, France.

Hongqi, Z., Jiying, Z., & Yang, O. (2008). *Adaptive rate control for VoIP in wireless ad hoc networks*. Paper presented at the ICC 2008. Beijing, China.

IEEE. (2007). *Wireless LAN medium access control (MAC) and physical layer (PHY) specifications*. Retrieved from http://www.ieee.org.

Jeong, Y., Kakumanu, S., Tsao, C.-L., & Sivakumar, R. (2009). VoIP over Wi-Fi networks: Performance analysis and acceleration algorithms. *Mobile Networks and Applications*, *14*(4), 523–538. doi:10.1007/s11036-009-0157-6

Johnson, D., Hu, Y., & Maltz, D. (2007). The dynamic source routing protocol (DSR) for mobile ad hoc networks for IPv4. *RFC: 4728*. Retrieved from http://www.ietf.org.

Kitae, N., Ahmed, H., & Kuo, C. C. J. (2005). TCP over multihop 802.11 networks: Issues and performance enhancement. In *Proceedings of the 6th ACM International Symposium on Mobile Ad Hoc Networking and Computing*. Urbana-Champaign, IL: ACM Press.

Layuan, L., Chunlin, L., & Peiyan, Y. (2007). Performance evaluation and simulations of routing protocols in ad hoc networks. *Computer Communications*, *30*(8), 1890–1898. doi:10.1016/j.comcom.2007.02.015

Lee, G., & Song, H. (2010). Cross layer optimized video streaming based on IEEE 802.11 multi-rate over multi-hop mobile ad hoc networks. *Mobile Networks and Applications*, *15*(5), 652–663. doi:10.1007/s11036-009-0196-z

Li, J., Jannotti, J., Couto, D. S. J. D., Karger, D. R., & Morris, R. (2000). A scalable location service for geographic ad hoc routing. In *Proceedings of the 6th Annual International Conference on Mobile Computing and Networking*. Boston, MA: IEEE.

Li, X., & Cuthbert, L. (2005). Optimal QoS mechanism: Integrating multipath routing, DiffServ and distributed traffic control in mobile ad hoc networks. In Jia, X., Wu, J., & He, Y. (Eds.), *Mobile Ad-Hoc and Sensor Networks* (*Vol. 3794*, pp. 560–569). Berlin, Germany: Springer. doi:10.1007/11599463_55

Liu, C., & Kaiser, J. (2003). *A survey of mobile ad hoc network routing protocols*. Ulm, Germany: University of Ulm.

Mehmood, R., & Alturki, R. (2010). A scalable multimedia QoS architecture for ad hoc networks. *Multimedia Tools and Applications*, *54*, 1–18.

Mehmood, R., Alturki, R., & Zeadally, S. (2011). Multimedia applications over metropolitan area networks (MANs). *Journal of Network and Computer Applications*, *34*, 1518–1529. doi:10.1016/j.jnca.2010.08.002

Mehmood, R., Crowcroft, J., Hand, S., & Smith, S. (2005). Grid-level computing needs pervasive debugging. In *Proceedings of the 6th IEEE/ACM International Workshop on Grid Computing*. IEEE Press.

Melodia, T., & Akyildiz, I. F. (2010). Cross-layer QoS-aware communication for ultra wide band wireless multimedia sensor networks. *IEEE Journal on Selected Areas in Communications*, *28*(5), 653–663. doi:10.1109/JSAC.2010.100604

Mundarath, J. C., Ramanathan, P., & Veen, B. D. V. (2009). A quality of service aware cross-layer approach for wireless ad hoc networks with smart antennas. *Ad Hoc Networks*, *7*(5), 891–903. doi:10.1016/j.adhoc.2008.08.003

Murthy, C. S. R., & Manoj, B. S. (2004). *Ad hoc wireless networks: Architectures and protocols*. Upper Saddle River, NJ: Prentice Hall.

O'Hara, B., & Petrick, A. (2005). *The IEEE 802.11 handbook: A designer's companion* (2nd ed.). New York, NY: Standards Information Network IEEE Press.

OPNET Technologies Inc. (2008). *OPNET modeler-educational version (version 14.5)*. Retrieved from www.opnet.com.

Park, V., & Corson, S. (2001). Temporally-ordered routing algorithm (TORA) version 1 - Functional specification. *Internet Draft*. Retrieved from http://www.ietf.org.

Perkins, C., Belding-Royer, E., & Das, S. (2003). Ad hoc on-demand distance vector (AODV) routing. *RFC: 3561*. Retrieved from http://www.ietf.org.

Peterson, L., & Davie, B. (2011). *Computer networks: A systems approach* (5th ed.). New York, NY: Morgan Kaufmann.

Pompili, D., & Akyildiz, I. F. (2010). A multimedia cross-layer protocol for underwater acoustic sensor networks. *IEEE Transactions on Wireless Communications*, *9*(9), 2924–2933. doi:10.1109/TWC.2010.062910.100137

Ramanathan, R., & Redi, J. (2002). A brief overview of ad hoc networks: challenges and directions. *IEEE Communications Magazine*, *40*(5), 20–22. doi:10.1109/MCOM.2002.1006968

Rosen, E., Viswanathan, A., & Callon, R. (2001). Multiprotocol label switching architecture. *RFC 3031*. Retrieved from http://www.ietf.org.

Santos, N. P. (2009). *Voice traffic over mobile ad hoc networks: A performance analysis of the optimized link state routing protocol.* Master's Thesis. Wright-Patterson, OH: Air Force Institute of Technology.

Setton, E., Taesang, Y., Xiaoqing, Z., Goldsmith, A., & Girod, B. (2005). Cross-layer design of ad hoc networks for real-time video streaming. *IEEE Transactions on Wireless Communications*, *12*(4), 59–65. doi:10.1109/MWC.2005.1497859

Sondi, P., Gantsou, D., & Lecomte, S. (2010). *Performance evaluation of multimedia applications over an OLSR-based mobile ad hoc network using OPNET.* Paper presented at the Computer Modelling and Simulation (UKSim). London, UK.

Stallings, W. (2006). *Data and computer communications* (8th ed.). Upper Saddle River, NJ: Prentice-Hall, Inc.

Tanenbaum, A. S. (2010). *Computer networks* (5th ed.). Upper Saddle River, NJ: Prentice Hall.

Utsu, K., Chow, C., & Ishii, H. A. (2010). Study on video performance of multipoint-to-point video streaming with multiple description coding over ad hoc networks. *Electrical Engineering in Japan*, *170*(4), 43–50. doi:10.1002/eej.20999

Xue, Q., & Ganz, A. (2003). Ad hoc QoS on-demand routing (AQOR) in mobile ad hoc networks. *Journal of Parallel and Distributed Computing*, *63*(2), 154–165. doi:10.1016/S0743-7315(02)00061-8

Zhiguo, D., & Leung, K. K. (2011). Cross-layer routing using cooperative transmission in vehicular ad-hoc networks. *IEEE Journal on Selected Areas in Communications*, *29*(3), 571–581. doi:10.1109/JSAC.2011.110307

ENDNOTES

1 An ad hoc network is usually defined as a collection of nodes dynamically forming a network without the use of any existing network infrastructure or centralized administration.

2 the Open Systems Interconnection (OSI) layered model.

Chapter 20
Cross-Layer Optimization for Video Transmission over WLAN:
Cross-Layer Prioritization

Chih-Yu Wang
National Taiwan University, Taiwan

Yin-Cheng Huang
National Taiwan University, Taiwan

Cheng-Han Mai
National Taiwan University, Taiwan

Fu-Wang Chang
National Taiwan University, Taiwan

Hung-Yu Wei
National Taiwan University, Taiwan

ABSTRACT

As IEEE 802.11 wireless devices have become increasingly widespread, providing Quality of Service in the context of H.264/AVC, the video coding standard for future multimedia networking, has become an important issue in the fields of communication and networking. Cross-Layer Adaptive Video Prioritization (CAVP) is a cross-layer framework that prioritizes video frame transmission according to the application-layer information and the MAC layer transmission condition. In this chapter, a **Peak Signal-to-Noise Ratio** *(PSNR) estimation method is proposed to sort out different priorities of H.264/AVC (Advanced Video Coding) video frames at the application layer to provide user-centric media quality estimation. Compared to previous heuristic algorithms, the authors also investigate a theoretic access delay estimator to monitor the wireless medium access delay at the MAC layer. In addition, an admission control is employed to serve the delay-sensitive video application and to give higher priority to those critical video frames. Video packets are dynamically classified into different 802.11e* **access categories** *according to the level of wireless medium access delay and the priority of the video frames.*

DOI: 10.4018/978-1-4666-0960-0.ch020

The myths of naïvely prioritizing video packets based on I/P/B types as well as naïvely assign packets to high priority access categories in 802.11e are resolved. Rather than creating complex scheme that is unable to be implemented in practical scenarios, the authors design the proposed scheme with practical implementability in mind. The proposed scheme is implemented with Click kernel module and the MadWifi WLAN driver. The performance of proposed CAVP design is evaluated by both NS-2 simulations and real testbed experiments, and results show that it enhances receiving video quality in error-prone wireless networking environments.

INTRODUCTION

High-quality video streaming and video-on-demand services have increasingly burgeoned these years; however, the problems of video traffic being error-sensitive and network capacity consuming still remain. Therefore, IEEE 802.11e is introduced to achieve **Quality of Service** (QoS) in 802.11-based system; moreover, H.264/AVC (Advanced Video Coding), an emerging video coding standard is used to delivering videos over a wide variety of networks, of which the advanced video coding standard introduces several mechanisms, such as independent slice decoding and flexible NAL structures to improve content network delivery. Combining the advantages of both techniques, the delivery of H.264/AVC video content over IEEE 802.11e wireless networks can satisfy the rising demands of network multimedia applications.

BACKGROUND

IEEE 802.11 WLAN

The basic medium access mechanism of the IEEE 802.11 Distributed Coordination Function (DCF) is Carrier Sense Multiple Access with Collision Avoidance (CSMA/CA). An 802.11 wireless station senses the wireless medium before transmitting any packet. If the medium is busy, it applies a random backoff mechanism to defer the wireless medium access. The backoff counter is a random value between 0 and the current **Contention Window** (CW) value, which is initialized to the minimum window size. When the 802.11 station senses that the medium is free for at least the time duration DIFS (DCF Interframe Space), it begins to decrease the backoff counter. When the backoff counter value is 0 and the medium is still idle, the station can begin transmitting packets. After the packet transmission is done, the station which successfully received the data should send an Acknowledgement (ACK) frame back to the sending station. A Short Interframe Space (SIFS) waiting time is applied before replying the ACK message. The whole transmission procedure ends when the sending station successfully receives the ACK frame. If a data transmission is detected failed (no ACK is received), the backoff counter CW is doubled, until it reaches the maximum contention window size (CW_{max}). The CW value is reset to CW_{min} after successful data transmission.

In basic IEEE 802.11 operation, no quality of service is supported. All traffic contends for channel access with the same DIFS, CW, CW_{max}, and CW_{min}. Therefore, there is no difference between each packet. IEEE 802.11e (IEEE, 2005) is developed for QoS support in wireless LAN. The Enhanced Distributed Channel Access (EDCA) provides differentiated QoS support (IEEE, 2005; Mangold, et al., 2003). In EDCA, there are four Access Categories (AC): AC_VO for voice transmission, AC_VI for video transmission, AC_BE for best effort traffic, and AC_BK for background traffic. Each AC has different CW_{max}, CW_{min}, Arbitration Interframe Space (AIFS), which is similar to DIFS, and duration of Transmission Opportunity

(TxOP). For high priority AC, smaller contention window parameters and AIFS value are applied.

To date, IEEE 802.11e has become a fundamental part of 802.11 protocols. It has been merged with IEEE 802.11a,b,g as a new standard of 802.11 (IEEE, 2007). It is also widely supported by most commercial products. Most subsequent 802.11 family members, including the high-throughput 802.11n (IEEE, 2009), retain the support of EDCA for compatibility. Thus, a cross-layer design based on the popular EDCA mechanism can provide high compatibility to both existing and future 802.11 infrastructures and mobile devices.

H.264/AVC Video Codec

In recent years, the H.264/AVC video compression standard has been used for video delivery over various networking environments. Joint Video Team (JVT) in the VCEG of ITU-T and MPEG of ISO/IEC jointly developed H.264/AVC (Richardson, et al., 2003; Wiegand, et al., 2003). H.264/AVC consists of two major layers: The Video Coding Layer (VCL) and the Network Abstract Layer (NAL). VCL contains specifications of the video-encoding engine including motion compensation, transform coding of coefficients, and entropy coding. NAL is responsible for the encapsulation of the coded slices into transport entities of the network. The VCL data are mapped to NAL units prior to transmission or storage.

A macroblock is the fundamental coding block of H.264/AVC and can be encoded in intra or inter mode. A coded picture frame is partitioned into fixed-size macroblocks. A video frame is coded as one or more slices, each containing an integer number of macroblocks. A slice can be decoded independently and is the minimal self-decodable unit. I-slices may contain only **intra-mode** macroblocks. P-slices may contain **intra-mode** macroblocks and at least one **inter-mode** macroblock referencing another frame. In practice, most macroblocks in P-slices are inter-mode in order to have a better compression rate. B-slices may contain both intra-mode and inter-mode macroblocks referencing other frames. Typically, I-slices contain more video information. P-slice and B-slice have higher compression ratio by referencing other macroblocks.

Multimedia Networking with H.264/AVC

H.264/AVC is provisioned to become the mainstream video standard to be used in wireless networking environments. Superior performance can be achieved by H.264/AVC in error-prone network environments (Stockhammer, et al., 2005). In error-prone network transmission, H.264 video packet **prioritization** affects the video reception quality. An analytical model on packet-loss error distortion in video is proposed by Politis et al. (2007). The work of Liu et al. (2007) suggested to measure the quality of packet-loss video with both PSNR and error length to have a more accurate measurement. The work of Krause et al. (2008) on frame loss showed that the perception of viewer on frame loss depends on the content of video, and the quality degradation due to certain frame loss is tolerable.

Cross-layer design that integrates IEEE 802.11e and H.264/AVC is investigated. Based on centralized IEEE 802.11e HCCA (HCF Controlled Channel Access) protocol, van der Schaar et al. (2006) introduced the concept of subflow and admission control on channel access. POAC-QG, introduced by Lagkas et al. (2007), is based on HCCA and is to realize bandwidth allocation and admission control to maintain minimum quality requirements of multiple users. Shankar and van der Schaar (2007) focused on the airtime fairness and MAC/PHY parameters optimization in packet delivery. Xiao et al. introduced a frame-based approach (Xiao, 2008) to control the EDCA parameters based on the type of video packets, and drop all related P- and B-type frames when the I-type frame is lost. A popular method to enhance the delivery rate of high-priority packet is adjusting the retry limit, which was proposed by Hsu et al. (2009) and Chen et al. (2009). This increases the delivery rate of high-priority packets. In addition, it also provides another dimension of prioritization in transmissions. Finally, Chilamkurti et al. (2010) noticed that the current loading of ACs should also be considered since an overloaded AC will lead to unnecessary packet dropping. They proposed a heuristic method to choose ACs with concerns on the loading of ACs.

Ksentini et al. (2006) proposed the QoS Architecture where transmission depends on the type of video packets or non-video packets. They use the built-in IEEE 802.11e Access Categories to realize the prioritization delivery on packets. Chilamkurti et al. (2010) also prioritize the video packets according to naïve I/P/B types.They employ downward probabilities to map less significant packets to lower priority ACs when the queue length of high priority AC exceed some fixed threshold. Another heuristic method to prioritizing the packets is based on the theoretical rate distortion model (Mansour, et al., 2009; Milani, et al., 2009). This model relies on the fact that the distortion due to packet loss will propagate to subsequent frames. This is true in general, but the impact degree of each packet in the distortion propagation is still worth noting, especially when frames are packetizied into small packets and have different amount of contained information.

Following these cross-layer video transmission research, we address two myths of video transmission over IEEE 802.11e networks. The first myth is prioritizing packets with video frame type, and the second myth is delivering high-priority packets through high-priority AC. We find out these two myths are not generally true in the state-of-the-art wireless networks, especially when we consider a cross-layer design system. Therefore, we proposed a cross-layer mechanism, Cross-layer Adaptive Video Prioritization (CAVP), to resolve the two myths. The first part of CAVP is Video Frame Prioritization (VFP) on application layer. VFP prioritizes packets according to the impact of each packet on the video quality. Another part of CAVP is MAC-Layer Adaptive Prioritization (MAP). MAP uses the idea of load balancing to transmit packets with the least estimated waiting time. We conducted a mathematical model for estimating waiting time for each AC. We also provide simulation and real testbed experiments results in this chapter.

APPLICATION LAYER: VIDEO PACKET PRIORITIZATION

After H.264/AVC Codec encodes the video sequences, the video frames might be of either type I, P, or B. An I-type frame has all required data for the decoding process and could be decoded by itself. A P-type frame is predicted according to the previous I or P frames. A B-type frame is bi-directionally predicted according to both previous and future frames. If a whole I-type frame is lost, the following P-type or B-type frames cannot be directly decoded without error concealment. However, a myth emerges because of the structure of video frame.

Myth (1): I-type video packets are considered more important than P-type and P-type are more important than B-type; moreover, I-frame is mostly ranked higher than P-frame, and P-frame higher than B-frame.

The importance ranking of video frame type is correct in some condition. However, because of the limitation of network transmission, larger video frames must be packetized into several packets; thereby results in discrepancy in the importance ranking between the video packets and the original order of video frames. For instance, as shown in Figure 1, losing one network packet containing 11% of an I-frame might not cause video quality degradation as severe as one network packet containing 33% of an P-frame, especially when the error resilience techniques

Figure 1. Myth (1): after packetization, 1/9 of an I-frame may or may not be more important than 1/3 of a B-frame

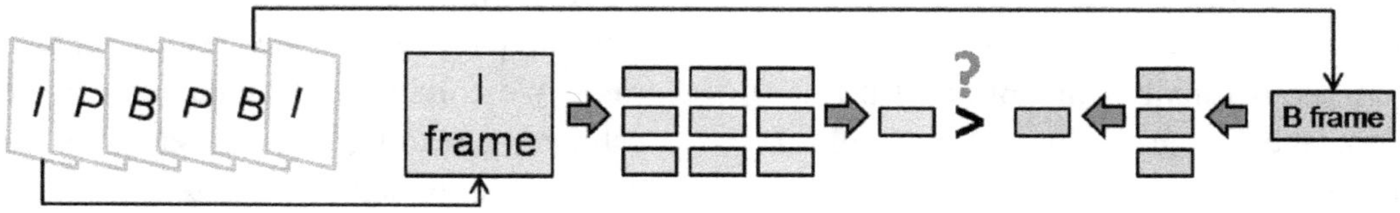

are commonly implemented in advanced video decoders. Here we show an example of this. We compressed the foreman QCIF video sequence with H.264/AVC and packetized with maximum packet size of 1000bytes. After packetization process, 722 packets are extracted. Then, one packet is removed and we decode the video using the rest of packets. The resulting video will have some distortions from the original one due to the loss of the removed packet. We repeat this experiment on all packets and get the corresponding distortion rates (measured in average PSNR of the affected frames). Finally, we sort the packets according to the distortion rate and we have the Table 1.

We find out that none of the top 10 most influential packets belong to I-frame. In fact, there is only one packet belonging to I-frame in the top 20 lists. This result roughly confirms our thinking about this myth. Thus, we will examine the priority of each video packet instead of naïvely classifying them based on the types[1].

Table 1. The most influential packets in Foreman QCIF clip

Packet #	Frame #	Frame Type	Distortion (PSNR)
488	284	P	16.07713
540	314	P	16.58123
492	286	B	17.42993
480	280	B	17.79678
476	278	P	17.83667
553	320	P	18.06566
128	74	P	18.272
464	272	P	18.66955
565	326	P	18.76652
498	290	P	19.12146
434	254	P	19.12482
502	292	B	19.48384
331	194	P	19.53234
468	274	B	19.54002
16	8	P	19.57196
493	287	I	19.64643
543	316	B	19.66341
578	332	P	19.68044
509	296	P	19.85425
424	248	P	19.97658

VFP Design

To determine the importance of the packetized video, PSNR is used as the utility function for quality assessment of the application layer. We proposed VFP to decide the importance of each application layer video packet.

To predict quality degradation in the video reception, we use a low-complexity first-order PSNR estimation to determine the importance of each packet. We remove one video packet from the video and calculate the PSNR of the Group Of Picture (GOP) as the PSNR estimation of this packet. Even though the first-order estimation method only takes the impact of single packet error into consideration, as shown in later sections, it performs fairly well in realistic wireless error environment where continuous error burst might occur. This PSNR computation for every video packet is denoted as PSNR Calculator in Figure

5. Since the P-type and B-type frames only refer to video frames within the same GOP, the quality degradation in one GOP does not propagate to next GOP in this single-error estimation process. Hence, the PSNR calculation can be limited to one GOP. This also leads to possible parallel implementation architecture for PSNR estimation in multi-core computing architecture.

PSNR Estimation

We classify video packets based on their importance with first-order PSNR estimation method. A video packet is more important when the corresponding PSNR estimation is lower. We know that a high PSNR value of a video means that the quality of video is good. If a video packet is lost and we get a higher PSNR value of that GOP, it implies that we can ignore that video packet and the quality of the video remains good. On the other hand, if we get a low PSNR value, meaning that the lost video packet is important, we should deliver that video packet in higher priority.

In VFP, we assign these packets into three priority types. We first defined the ratios of these three priorities as $r_1 < 1$, $r_2 < 1$, and $(1 - r_1 - r_2)$. We sort these packets with their PSNR estimations derived from the PSNR Calculator. If the PSNR belongs to the top $r_1\%$ of the sorted PSNR values, then the packet is tagged with priority 0 (high priority). If the PSNR belongs to medium $r_2\%$ of the sorted PSNR, then the packet is tagged with priority 1 (medium priority). If the PSNR belongs to the bottom of the sorted PSNR, then the packet is tagged with priority 2 (low priority). Typically, parameters r_1 and r_2 can be selected based on the MAC level throughput ratio between QoS categories. This process is denoted as PSNR Classifier in Figure 5.

MAC LAYER: PACKET TRANSMISSION PRIORITIZATION

In the original IEEE 802.11e, packets are sent into multiple independent access categories corresponding to the applications. The four access categories are AC_VO (voice), AC_VI (video), AC_BE (best effort), and AC_BK (background).

To clarify, we denote AC_VO, AC_VI, AC_BE, and AC_BK as AC_0, AC_1, AC_2, and AC_3 respectively. The smaller the AC number, the higher the priority. The MAC layer of higher priority AC is implemented with smaller values of AIFS, CW_{min}, and CW_{max} in IEEE 802.11e. In previous work, QoS Architecture Mechanism is proposed for transmitting different types of video packets with the corresponding ACs, AC_0 for I-type frames, AC_1 for P-type frames, AC_2 for B-type frames, and AC_3 for non-video traffic. Here comes the second myth.

Myth (2): High-priority packets are ought to transmit through the high-priority access categories.

In 802.11e systems, high-priority access category are expected to win most **contentions** because of their smaller AIFS, CW_{min}, and CW_{max}; however, packet delay has great impact on video transmission, which is affected not only by the successful transmission rate but also by the number of packets in the queue. Therefore, when the queue length increases, the packet delay time increases along with the longer waiting time Though naïve mechanism is able to assign high priority packets into high priority AC, and functions well when network load is low or evenly distributed, the load-aware mechanism presented here is deemed more comprehensive for low-delay video transmission for it considers both network load and AC priority. The most efficient way to send high-priority packets is not through the highest-priority AC but through the AC with

Figure 2. Myth (2): higher-priority AC must be with shorter latency?

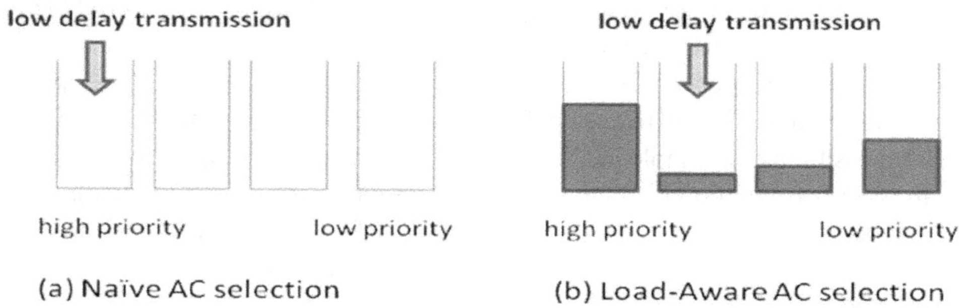

the lowest expected waiting time, as shown in Figure 2.

The expected waiting time in our design is only related to internal contention, because the total delay of packet transmission depends on the channel condition and operational environment, such as the number of contented stations, and the parameter sets of contented stations; therefore, the total expected delay of transmitting packet is difficult to compute. To solve this problem, we only consider the internal contentions when we compute the expected waiting time, which means that the expected waiting time of packet in our design is only connected to the parameter sets and the queue status of each AC of the transmitting station.

MAP Design

Typically, high-priority ACs are chosen for video packet transmission. Original IEEE 802.11e use only AC_1 while QoS architecture mechanism uses AC_0 for high priority packets, AC_1 for medium priority packets, and AC_2 for low priority packets. These methods may result in some serious packet lost or prolong the delay time because the traffic is not balanced according to current AC queuing status. For instance, QoS architecture chooses to deliver all I-type packets by AC_0, and the packets are jammed in the queue and suffered from long delay time. If some packets are delivered by lower-loaded AC_1, they can be transmitted with a shorter waiting time. To be able to adapt to current queuing status of different access categories, the MAC-Layer Adaptive Prioritization (MAP) design allows dynamic assignment of packets into multiple candidate access categories for load balancing. For example, a priority 0 packet can be sent into AC_0 or AC_1. A priority 1 packet can be sent into AC_0, AC_1, or AC_2. Figure 3 demonstrates how packets of different priority are sent into the ACs in MAP design.

Since 802.11e is a random access MAC, the packet delivery time depends on the randomly generated backoff counter. Hence, the number of packets waiting in queues of different access categories is affected not only by the packet assignments, but also by the random contention of each AC. The MAP scheme we proposed estimates the access delay of each AC candidate with Access Delay Estimator for different priority packets. It forwards the packet to the best AC candidate with the least access delay. This dynamic packet allocation enhances wireless networking performance in this changing environment.

In addition, because the video packet transmission is time-sensitive, the packets received after the decoding deadline is considered useless for video streaming application thereby be dropped. At the end of sender, if the estimated access delay of a packet is over the expected decoding deadline, we

Figure 3. Adaptive MAC layer video packet assignment to access categories

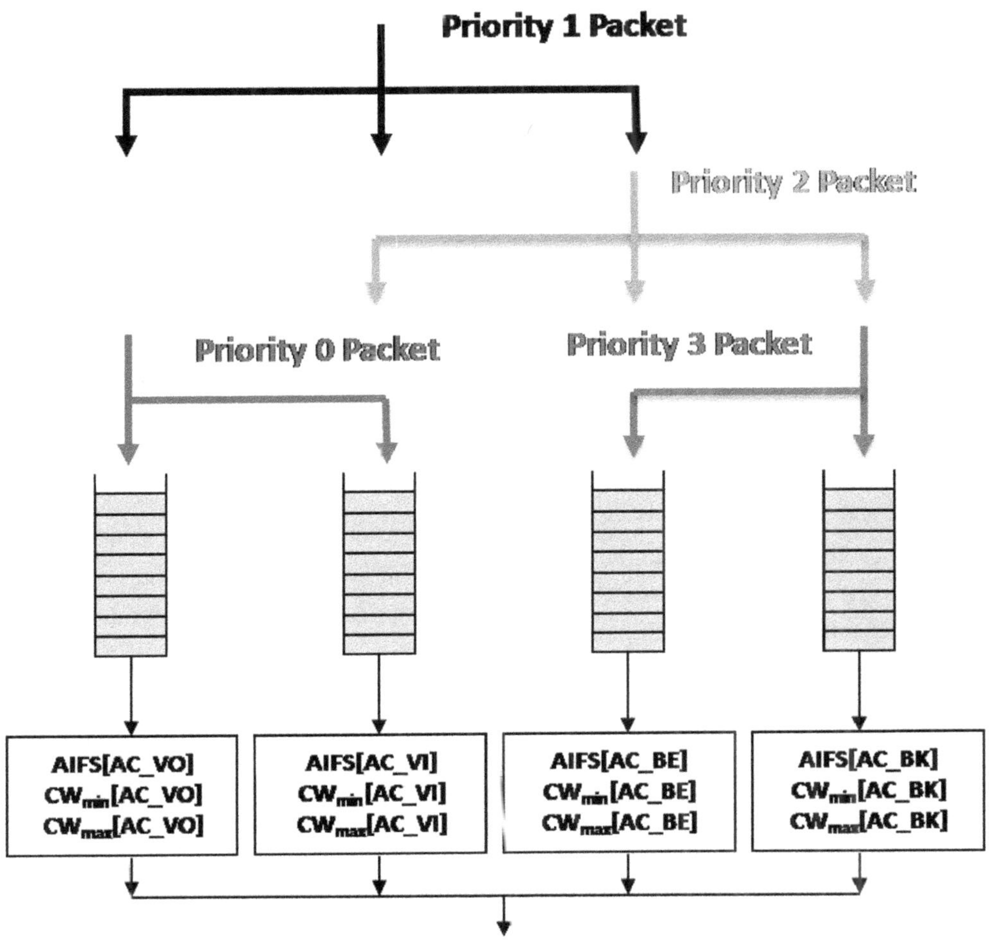

will drop it to preserve radio resource and transfer the transmission opportunity to other packets with later deadlines. This process is denoted as Admission Control block in Figure 5. The key issue here is how to correctly determine the choices of AC given the current loading of each ACs. Instead of just using some heuristic schemes (Chilamkurti, et al., 2010; Lin, et al., 2009), we would like to add some theoretical foundation on the estimation of expected waiting time in each ACs, which is given as follows. The inter-contention among 802.11 wireless devices has been investigated (Bianchi, 2000), while the analysis on the effect of EDCA on other wireless devices are also derived (Tinnirello, et al., 2010). Here we would like to provide a more straightforward analysis on the expected waiting time of each AC. The proposed estimation method is shown to be effectively in our simulations and experiments.

Expected Waiting Time Estimation

To compute expected waiting time E_i of AC_i, some variable must be denoted first. The sequence number of AC is denoted by i, $0 \leq i \leq 3$. The parameter set of the i-th AC includes $AIFS_i$, CW_{mini}, and CW_{maxi}. The current contention

Figure 4. Medium access procedure in IEEE 802.11e

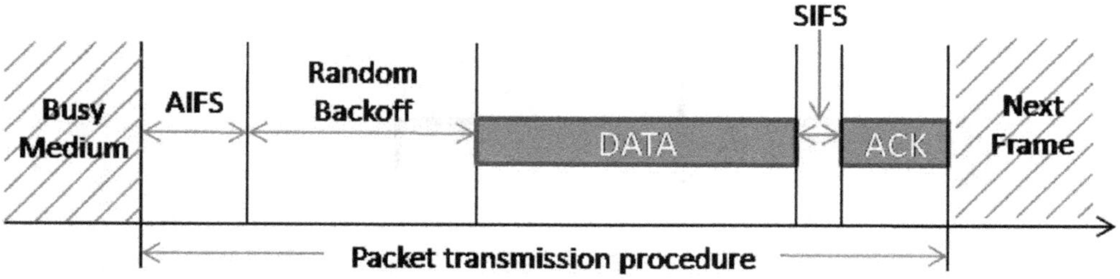

window size of the i-th AC is CW_i, the backoff counter of the i-th AC is τ_i, the transmission rate of the transmitter is r_t, the queue length of the i-th AC is l_i, the packet size of the k-th packet in the i-th AC is $s_{i,k}$, and the IEEE 802.11e slot time is t. According to IEEE 802.11 standard, the packet transmission procedure is shown in Figure 4. Thus, the packet time T_p of the k-th packet in the i-th AC is shown in the following equation.

$$T_p(i,k) \le \frac{s_{i,k}}{r_t} + (AIFS_i + CW_i + SIFS + ACK) * t \tag{1}$$

Next, we need to compute the access probability ap_i of AC_i. For AC_i, there are $(CW_i + 1)$ choices of backoff counter. Therefore, for internal contention, the number the combinations of backoff counters of the four ACs is $\prod_{j=0}^{3}(CW_j + 1)$.

When only the internal contention is taken into consideration, an AC wins the contention only if its backoff counter is smaller than other ACs.' For example, if AC_i wins the contention, the backoff counter of other ACs must be greater than the backoff counter τ_i of AC_i. For any other AC_k, $k \ne i$, the upper bound of the backoff counter is $(AIFS_k + CW_k)$. So there are $(AIFS_k + CW_k - \tau_i)$ possible choices for AC_k if AC_i wins the contention. Thus, the number of possible combinations when AC_i wins contention with backoff counter τ_i is $\prod_{k=0,k\ne i}^{3}(AIFS_k + CW_k - \tau_i)$.

On the other hand, since AC_0 has highest priority than any AC_i, AC_i can never win the contention if $\tau_i > CW_{min0} + AIFS_0$, which means the backoff counter of AC_0 is always smaller than τ_i. Therefore, the upper bound of τ_i is $(CW_{min0} + AIFS_0)$. We also know that the lower bound of the backoff counter choice is AIFS when random backoff time equals zero. To conclude, the access probability of AC_i can be described as the following equation.

$$ap_i = \frac{\sum_{AIFS_i}^{CW_{min0}+AIFS_0} \prod_{k=0,k\ne i}^{3} (AIFS_k + CW_k - \tau_i)}{\prod_{j=0}^{3}(CW_j + 1)}. \tag{2}$$

Next, to estimate the waiting time, we need to estimate the number of transmitted packets for any AC before the new packet is transmitted. If the new packet p_{i,l_i} is put in AC_i, the packet can only be transmitted if all the packets in the queue of AC_i are transmitted. Thus, for AC_i, the number of transmitted packets before p_{i,l_i+1} is l_i. For any other AC_j, the relative probability of successful channel access is ap_j / ap_i. Thus the estimated number of transmitted packets for AC_j will be $l_i \times ap_j / ap_i$. We can use the following expression to get the estimated number of transmitted packets $n_{i,j}$ for any AC_j when the new packet is put in AC_i.

$$n_{i,j} = \begin{cases} l_i \times ap_j / ap_i & \text{if } l_i \times ap_j / ap_i \le l_j, \\ l_j & \text{otherwise.} \end{cases} \quad (3)$$

Finally, the packet time of the new packet p_{i,l_i+1} can be expressed as the following equation.

$$T_p = \frac{s_{i,l_i+1}}{r_t} + (AIFS_i + CW_i) * t. \quad (4)$$

We do not take the ACK time into consideration because it is transmitted after p_{i,l_i+1}. From the derivation before, we can show as following equation the expected waiting time E_i if p_{i,l_i+1} is transmitted through AC_i.

$$E_i = T_p + \sum_{j=0}^{3} n_{i,j} * \frac{\sum_{k=0}^{l_k} T_p(j,k)}{l_k} \quad (5)$$

This process of estimating access delay of the packet is denoted as Access Delay Estimator. It is used for two purposed in MAP: one is to estimate the loading of each ACs by checking the corresponding expected waiting time. This helps MAP to determine the best choice of AC for the current packet. The other is to determine whether this packet can be transmitted and received before the decoding deadline. If the expected waiting time is longer than the decoding deadline, the packet should be dropped in order to preserve transmission resource.

CROSS-LAYER DESIGN IN WIRELESS VIDEO STREAMING SYSTEM

After addressing the two myths in video transmission prioritization, we observe that our findings potentially lead to a cross-layer framework in video transmission over wireless networks. We have confirmed that a better packet prioritization method on video packets can be applied in Application layer by the proposed VFP process. We also showed that the loading of ACs should be put into considerations in transmission prioritization in MAC layer by the proposed MAP process. The prioritization process in these two layers should be designed concurrently in order to extract most potential enhancements in addressing these two myths. Intuitively, the prioritization results in VFP should be the input of MAP. Thus, we proposed Cross-Layer Adaptive Video Prioritization (CAVP) to improve the quality of video streaming over IEEE 802.11. CAVP aims to demonstrate the potential enhancement in video transmission from getting rid of those two myths.

CAVP consists of two prioritization mechanisms in the application layer and MAC layer, respectively. Video Frame Prioritization (VFP), which determines the video frame priorities according to video application quality, is implemented in application layer. MAC-Layer Adaptive Prioritization (MAP), which determines adaptively the MAC layer Access Category (AC) for each packet, is implemented in MAC layer. In the process of CAVP, the video packets are prioritized by VFP first. It prioritizes all video packets into three levels by their importance on the video quality first. Then, these packets with their priority are sent to MAP, which determines the transmission priority according to the current AC loading and the importance of these packets.

When a new video packet arrives, MAP first gets the estimated access delay of all AC's with the help of Access Delay Estimator. Then, MAP determines whether this packet can be transmitted and received by the receiver before the decoding deadline in Admission Control block according to the results of Access Delay Estimator. If it reaches a negative conclusion, the packet will be dropped. Otherwise, MAP sends this packet to one of AC candidates. The choice of AC is determined by the expected waiting time of each AC and the AC candidates predefined in Figure 3. According to the packet's priority determined by VFP, MAP sends this packet to the AC with least expected waiting time in the AC candidates. Figure 5 shows the design architecture of CAVP.

To summarize, the CAVP design combines with utility-based application layer prioritization with the dynamic MAC layer prioritization optimizes adaptively the packet transmission according to the current MAC layer status. What is more, the adaptive priority assignment can balance the MAC layer load among ACs. We estimate the access delay of each packet and select the smallest estimated access delay of the candidate AC according to the frame types of the packets. The priority assignment avoids allocating high-priority video packets incorrectly into low-priority AC while keeping the flexibility of utilizing light loaded AC. Moreover, the MAP drops packets if the estimated access delay values exceed the expected decoding deadline for all AC candidates.

IMPLEMENTATION AND EVALUATION OF A CROSS-LAYER MULTIMEDIA SYSTEM

Packetized Video Analysis

We analyze the H.264/AVC video with the proposed VFP mechanism. First we examine the QCIF video test sequences. We collect video sequences from Video Traces Research Group (VTRG) for

Figure 5. System architecture

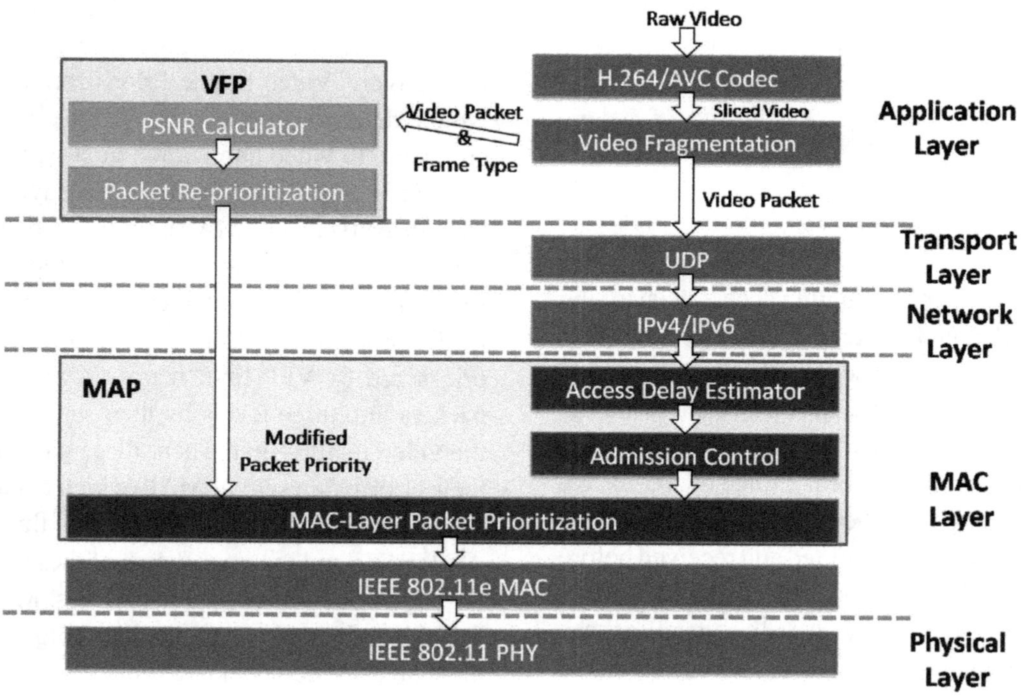

Figure 6. Video packet prioritization with VFP (QCIF)

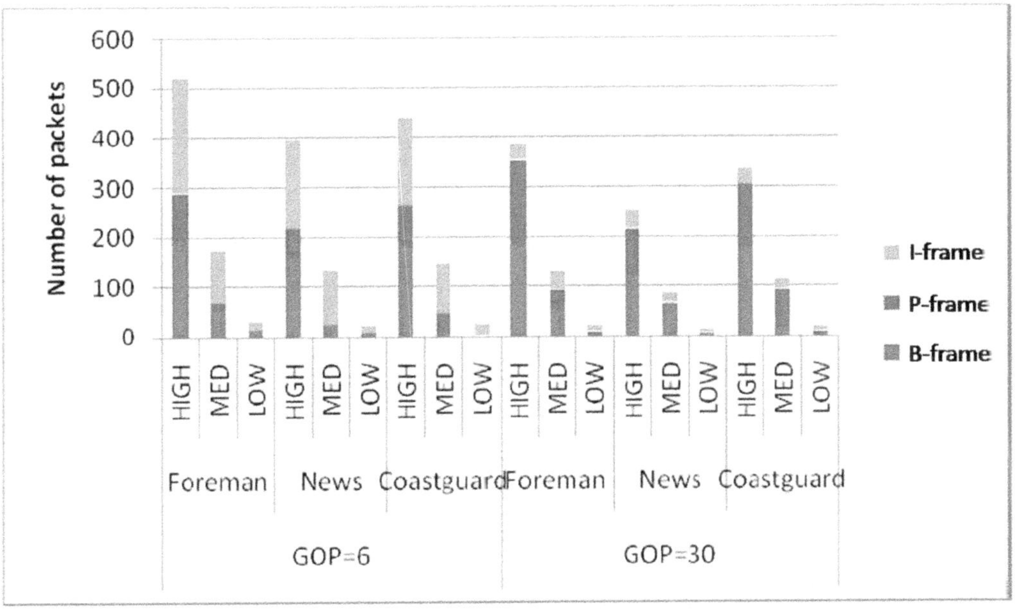

QCIF (176 x 144) YUV sequences. We choose Foreman, Coast Guard, and News, which are QCIF 30 fps format. These video sequences are compressed into CBR H.264/AVC video bit streams and packetized. Figure 6 shows the ratio of I/P/B type of video packets after prioritization. The numbers of packets assigned to different QoS priorities are proportional to the expected throughput in different 802.11e MAC QoS classes. On average, I-type is more important than P-type, and P-type is more important than B-type. However, some P-type or even B-type might be more important than I-type in terms of PSNR performance. Use the case of Foreman with GOP of 6 as example, only 43% of high-priority packets are I-type, 18% priority 0 packets are P-type, and 39% high-priority packets are B-type. In addition, 60% of medium-priority packets are I-type, and 5% are P-type. Even some I-type packets are classified as low-priority according to the previously described PSNR estimation method.

From intuition, we may think that I-type video packets are extremely important. However, the video analysis observations show that such impression does not reflect the reality. It is because the commonly used GOP structure in wireless network transmission is typically much smaller than the GOP structure used in low loss rate wired-line networking environments or stored video applications. A large I-frame might be packetized as several I-type packets. One I-type packet loss might just indicate a partial I-frame loss; hence, the PSNR degradation in single packet error scenario is less than expected. The built-in error concealment in decoder may help, too.

In another case with GOP of 30, although the ratio of P-type packets in low-priority increases, there are still some I-type packets with less impact on video quality than other P- and B-type packets. Some lost information of I-frame can be even partially recovered by the following P frames. The slice structure in H.264/AVC and the larger video frame, which leads to unavoidable packetization, give us a different view to video packet importance. All the analysis results of videos are presented in Figure 6.

As VFP is first-order estimation to video packet importance, we would like to investigate how it

Figure 7. PSNR of Foreman QCIF video clips in packet loss condition: comparing VFP, naïve I/P/B prioritization

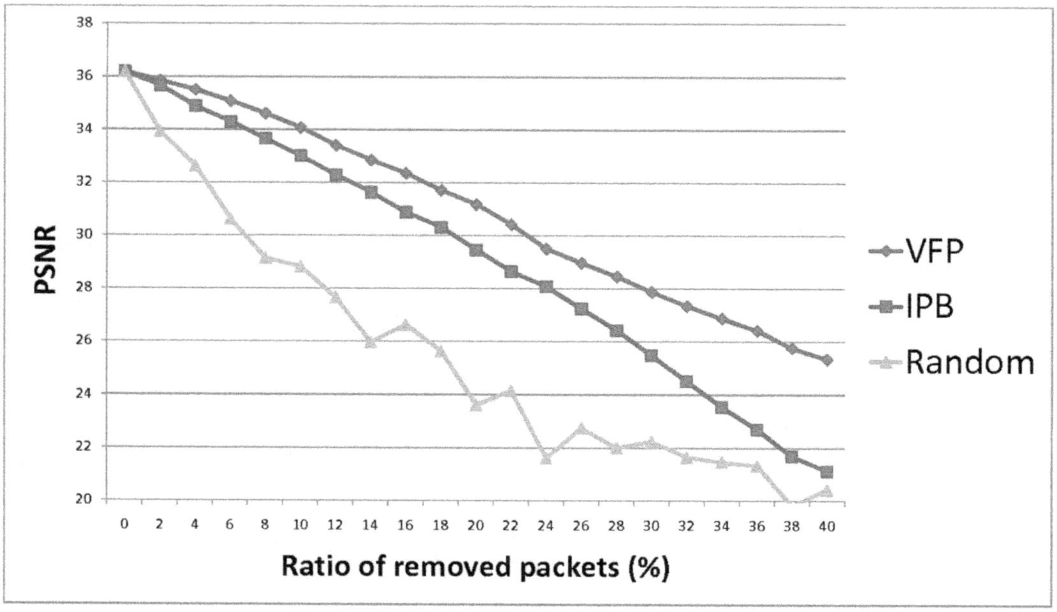

performs under more generalized packet error conditions. In the experiment, we drop x% of video packets and then calculate the PSNR. We compare three mechanisms: VFP, I/P/B prioritization, and Random mechanisms. Under VFP mechanism, we remove the packets with lowest priority first, which are those with highest PSNR estimations. In I/P/B mechanism, we randomly remove packets with the order of B-type, P-type, and I-type. In Random mechanism, we just randomly remove packets from the video without any prioritization. We control the ratio of removed packets, and measure the PSNR of the result videos. In the experiment, we used the Foreman QCIF clip with GOP of 6. The average video quality of modified Foreman clip is shown in Figure 7.

We observed that after removing the same ratio of low-priority packets, the VFP prioritization achieves better video quality than the naïve I/P/B mechanism in both QCIF and SDTV format. The prioritization process in VFP can more accurately rank those packets than the naïve I/P/B-type classification. In addition, both VFP and I/P/B mechanisms outperforms Random mechanism. In summary, the proposed VFP mechanism can effectively prioritize the video packets than I/P/B mechanism. The Myth (1) in 2 has been resolved.

Network Simulation

We first examine the performance of the proposed CAVP mechanism in different wireless network scenarios through simulations. We use the popular network simulator NS-2 ("NS-2") with TKN EDCA model (Wiethölter, et al., 2006) to implement and evaluate the proposed scheme. We add the video transmission module and implemented the proposed Video Frame Prioritization into NS-2. We also modified the TKN 802.11e EDCA model to implement the MAC-Layer Adaptive Prioritization scheme. The 802.11e parameters are the default ones in the IEEE 802.11e standard. We integrate the H.264/AVC reference software JM

12.4 ("H.264/AVC JM") with NS-2 to encode and decode the videos. We choose the QCIF Foreman clip as the test video. The parameters are listed in Table 2.

We setup a scenario for evaluating the proposed cross-layer CAVP mechanism and the MAC-layer MAP mechanism, as shown in Figure 8.

It is designed to examine the performance of different mechanisms under controlled traffic load in all ACs. We compare these two proposed mechanism with the original 802.11e mechanism and the QoS Architecture mechanism (Ksentini, et al., 2006). The original 802.11e mechanism allocates all video packets into AC_VO, which is AC_1. The QoS Architecture mechanism allocates video packets according to their type (i.e. AC_0 for I-type, AC_1 for P-type, and AC_2 for B-type). The proposed MAP mechanism is CAVP without VFP mechanism. The priority of the packet in MAP depends on its video type, which is the same as the QoS Architecture. The comparison between QoS and MAP can help us examine the performance enhancement from naïve AC selection to load-aware AC selection. In the proposed CAVP mechanism, the priority of the application layer is determined by previously described PSNR estimation while the priority of the MAC layer is adaptively adjusted using the MAP mechanism.

Table 2. Video test sequence settings

Parameter	Setting
Frame Rate	30fps
Maximum Packet Size	1000bytes
Buffer time	100ms
Group of Picture	6 (IBBPBB) [20] and 30 (IBB-PBB….PBB)

In the simulation scenario, there are two video streams and four Constant-Bit-Rate (CBR) data traffic flows, which are transmitted with each access category, respectively. We experiment video streaming in different CBR traffic load conditions. This scenario can help us simulate the scenario that all ACs are under background traffic load. The CBR traffic loads of all four access categories are the same, as denoted by the value of the x-axis in Figure 9. We replay the video stream five times in order to make sure the wireless network

Figure 8. Simulation scenario

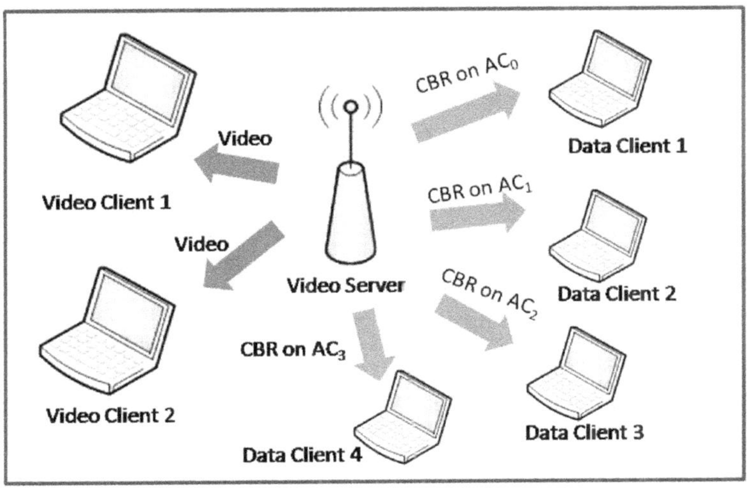

Scenario 2

Figure 9. The PSNR performance of video streaming with data traffic in all four access categories

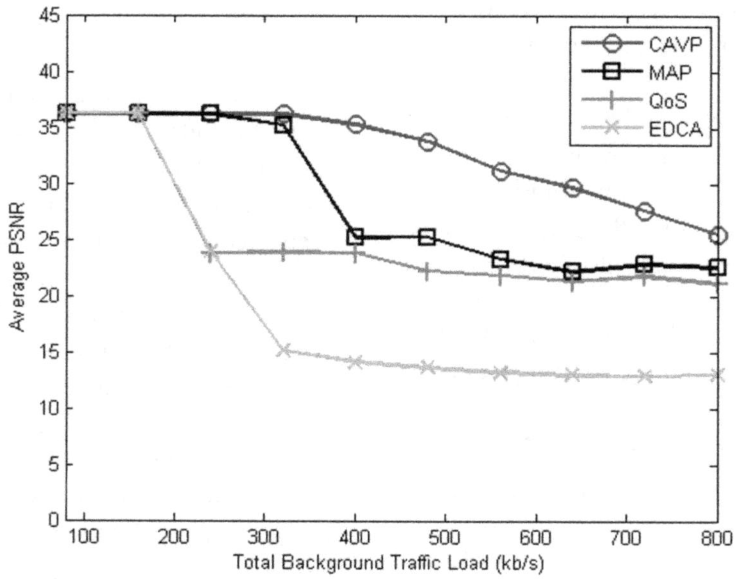

is stabilized. The simulation results are shown in Figure 9

We observe that as the CBR traffic load increases to 240kbps, the effective throughput of video streaming over the AC_2 decreases and leads to obvious video quality degradation. In the QoS Architecture mechanism, as observed in the turning point of the PSNR curve around 160kbps, most B-type video packets are lost or dropped. When the traffic load increases to 240kbps, the effective throughput of video streaming over AC_1 decreases and leads to more quality degradation. Therefore, decline in PSNR performance can be seen in IEEE 802.11e and QoS Architecture. In contrast, the load balancing between ACs is effective in MAP and CAVP. They preserve high video quality even when AC_1 is heavily loaded. In the original IEEE 802.11e, the video packets are forwarded to AC_1, so the degradation in PSNR performance is highly related to the traffic load of AC_1. Decline in PSNR also occurs in the QoS Architecture scheme. Since the percentage of packet transmission over AC_1 is less than that of the original 802.11e scheme, the decline is not as steep as that in the original 802.11e curve. When traffic load increases to 480kbps, the wireless network is overloaded, so the PSNR performances of all schemes deteriorate.

In this scenario, the proposed CAVP can precisely estimate access delay and forward adaptively packets to suitable ACs. When using the PSNR estimation, where the utility of a video packet is estimated in the case of losing only one packet, the experimental results demonstrate that this low-complexity PSNR estimation works effectively in realistic wireless networking environments where consecutive video packet loss may occur.

In the end of this section, we can answer the Myth (2) we addressed in Section 3 with our simulation results. The QoS Architecture mechanism can be treated as an example of naïve AC selection because it always chooses higher priority AC for higher priority packets. The proposed MAP mechanism is a realization of load-aware AC selection, which chooses AC based on both priority and expected waiting time. In the simulation results, we observed that MAP mechanism always outperforms QoS Architecture mechanism with better PSNR performance. Our experiment shows that load-aware AC selection outperforms naïve AC selection. We conclude our experiment results resolve the problem of Myth (2).

Testbed Implementation

We use Asus EeePC 900 and TOSHIBA Satellite M200 as our transceivers. EeePC 900 has built-in Atheros WLAN chip that can work smoothly with madwifi driver. We use D-Link AirPlus DWL-G650 cardbus adapter as the external wireless device on M200. DWL-650 can work well with M200 and madwifi driver. Ubuntu 8.04 is running on both EeePC 900 and M200. We use madwifi 0.9.4 version ("Madwifi") and modifies some of the codes for implementation of CAVP and VFP. Click version 1.7.0 ("Click") is used. The WLAN interfaces have their default value of IEEE 802.11e EDCA parameters.

In real testbed implementation, we choose madwifi as the wireless LAN interface driver on Linux kernel. Madwifi is one of the most advanced open-source WLAN drivers available for Linux. It is designed to cooperate with devices with Atheros chips. Although the driver itself is open source, it is highly dependent on Hardware Abstraction Layer (HAL), which is available only in binary form. Therefore, some of our design in CAVP cannot be directly realized through modifying madwifi source code. Therefore, we have some indirect methods of implementing the system.

Map with Madwifi

First, we modifies madwifi so the packet numbers in each AC will be recorded in system logs. We can use the kernel command "dmesg" to check the system message, which contains the information of the status of AC's. In MAP, we have to know the number of packets in each AC so that we can compute the estimated access delay of all AC's. As to other AC parameters (e.g. CW_{max}, CW_{min}, and AIFS), we can obtain from the wireless network interface by the Linux command "iwpriv." As a result, the information required for MAP can be gathered by madwifi and Linux system commands.

We also made another modification to madwifi to forcibly enable 802.11e EDCA. In original madwifi, we have to substantially change the header of packets to enable 802.11e EDCA function. With our modification, we only need to fill the tos field of packet header that the packets will go to the corresponding AC's. Therefore, the time spent on change headers could be reduced.

CAVP with Click

To complete the implementation of CAVP, we have to find a method to modify packet header. When we have an output packet with priority mark from VFP, the header of packet should be modified so that the packet can go to appropriate AC. However, the implementation of packet capsulation is deep in the kernel of operating system. It is difficult to reconstruct the kernel and make the kernel cooperate with our CAVP. Therefore, we utilize a tool named Click.

Click is a software architecture for building flexible and configurable modular routers. By arranging the elements (the modules in Click), one can construct a router with the desired function easily. Click can be thought as a intermediate language. The kernel of Click is written in C++, and is in charge of communications with network interfaces and Linux kernel. That is to say, it is like a black box that covers the complicated jobs. Using Click to arrange network interface behavior, we can concentrate on the system implementation design rather than on the complex detail to control the hardware.

By the aid of Click, we can easily modify the tos filed in MAC-layer header according to the priority mark in the packets. We also use Click to

Figure 10. Experiment scenarios

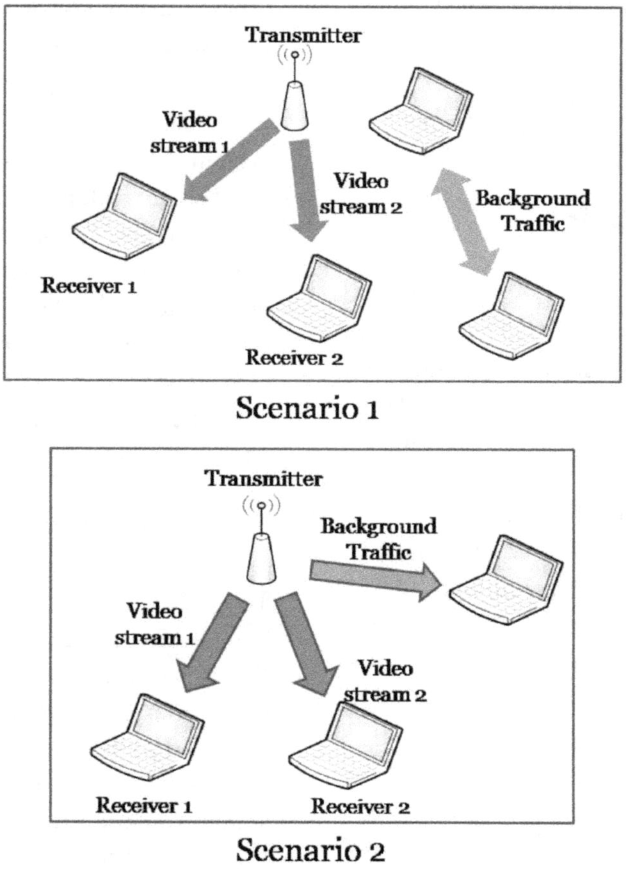

record the packet receiving information. We can know exactly the delay or loss of a packet from the record of Click.

Testbed Experiment

The video encoder and decoder used in testbed experiments are the same as which used in simulation. The video we choose in testbed experiments is Foreman, which is also used in simulation. The sizes of packets are no more than 1000 bytes. Video is encoded with Group Of Picture (GOP) length being 6, which is suggested by Lombardo et al. (2004) and Park et al. (2005). The video decoding deadline is set to 300 ms if not mentioned.

The experiment scenarios are shown in Figure 10. The settings of the scenarios are listed below.

Scenario 1: As shown in Figure 9, one transmitter sends two video streams at the same time. The two video streams are transmitted under the same QoS scheme (CAVP, QoS, or EDCA) in each experiment. Two wireless stations transmit background traffic. Background traffic is sent with AC_VI priority. All the stations are work in the same channel, with 1 Mbps transmission rate. There are inter-station contentions in this scenario. We can check if the system is resistant to congested channel.

Scenario 2: One transmitter sends two video streams and an AC_VI priority background traffic simultaneously, as shown in Figure 9. Just like

Figure 11. Scenario 1 testbed experiment result (CAVP, QoS, and EDCA)

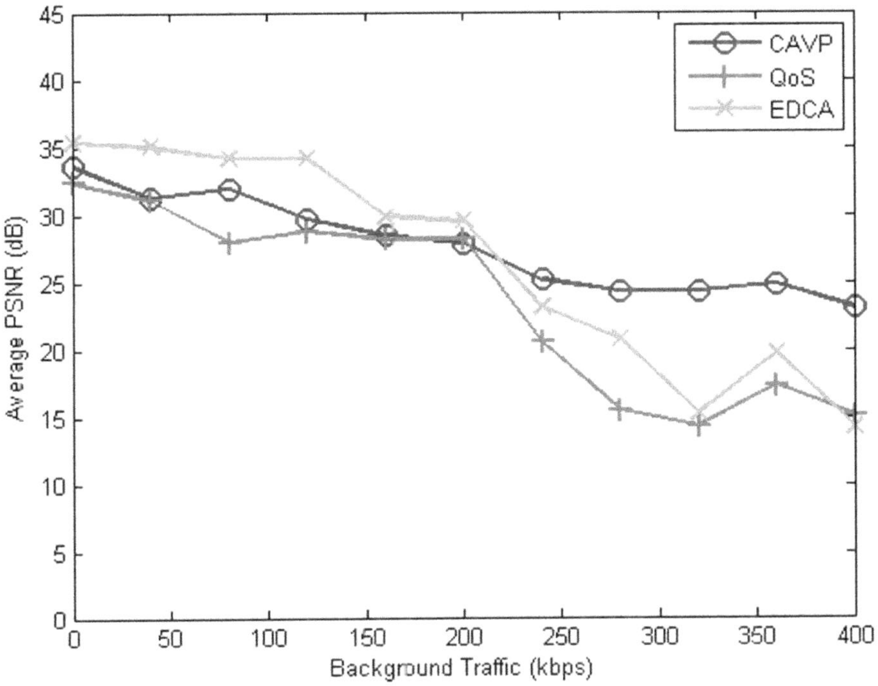

scenario 1, all the network traffic run in the same channel, and transmitted with 1 Mbps transmission rate. We use this scenario to check intra-station contentions. The ability of the system to face the saturated transmission queue can also be checked.

We compared the performance of several structures in the experiments: 1) Traditional IEEE 802.11e EDCA. All packets are sent to AC_VI. 2) QoS Architecture mechanism, allocates packets according to the frame types. I-type will be sent to AC_VO, P-type to AC_VI, and B-type to AC_BE. 3) VFP. 4) CAVP.

PSNR Results

We compare the performance of complete CAVP, QoS scheme, and IEEE 802.11e EDCA. We took experiments in both scenarios. For scenario 1, the result is shown in Figure 11. Unfortunately, CAVP cannot completely defeat EDCA from the result, although CAVP is still better than QoS. When the background traffic load is smaller than 200 kbps, EDCA outperforms the other two modified schemes. As the background traffic grows, the advantage of CAVP reveals. For EDCA and the QoS scheme, once the channel saturated, the performance of video stream dropped acutely. We can see that both EDCA and the QoS scheme show a deep drop immediately after the background traffic is larger than 200 kbps. In comparison, CAVP shows a gentler drop in PSNR. CAVP provides higher priority to the important packets, and only transmits the timely packets. Hence, once the channel is saturated, CAVP can provide higher Quality of Service than the other two schemes.

In our implementation of the QoS scheme, the computation effort on Application Layer of the QoS scheme is almost the same with EDCA implementation. The only difference between QoS and EDCA schemes is the use of AC's. QoS sends

Figure 12. Scenario 2 testbed experiment result (CAVP, QoS, and EDCA)

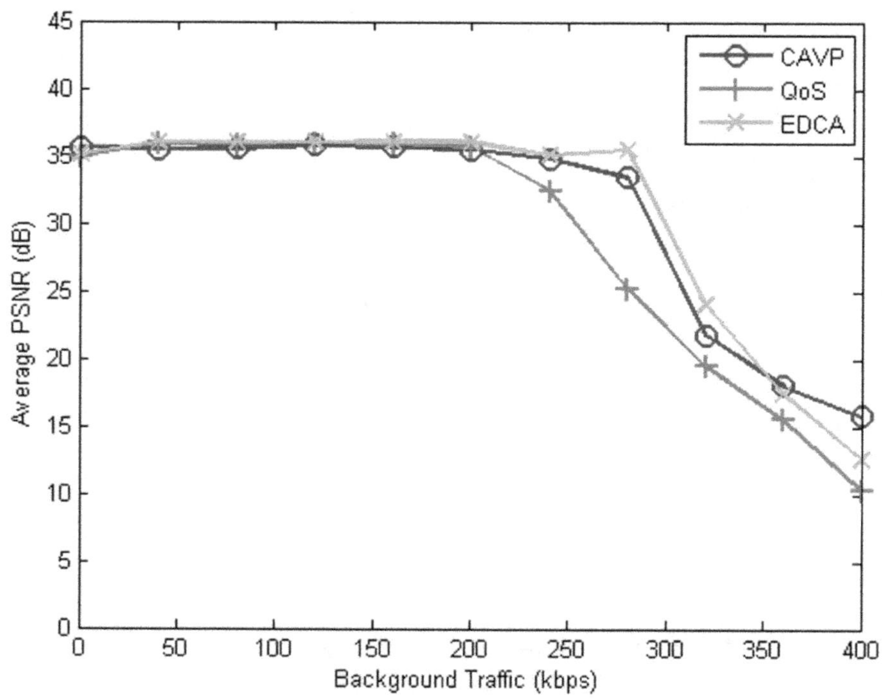

packets to different AC's while EDCA only sends packets to AC_VI. Thus, the performance drop may be by the implementation of EDCA on WLAN interfaces, which is related to the madwifi driver or hardware wireless chips. Using multiple ACs to transmit is likely to degrade the transmission efficiency in the experiments. This behavior also affects the performance of CAVP. Nevertheless, we can see that CAVP is more interference-resisted among the 3 schemes. CAVP works better when the load of background traffic in the air is heavy.

The result in scenario 2, as shown in Figure 12, shows a slightly different result. In scenario 1, the background traffic comes from different stations other than video transmitter, so the packets may collision with each other. On the other side, in scenario 2, all the packets come from the same station, so the packet collision will probably not happen. Therefore, in scenario 2, all the schemes perform well with little background traffic. This is because the transmission queues are not saturated in this condition. All the packets can be transmitted timely and successfully. When the traffic load increases, the transmission queues are filled up. The packets could be dropped by full queue or transmits with long latency. However, we can observe that CAVP still works better than the QoS scheme and almost the same with EDCA scheme.

Packet Loss Rate

Since the PSNR of video is highly related to network statistics, we look deeper into the packet loss rate and overdue packet rate in this section. Figure 13 shows loss rate of each scheme in scenario 1. We can observe that the loss rate of CAVP is much lower than the other schemes especially in heavy load environment. Different from other schemes, there are very few overdue packets in CAVP. This improvement can be traced to the feature of MAP,

Figure 13. Loss rate in scenario 1

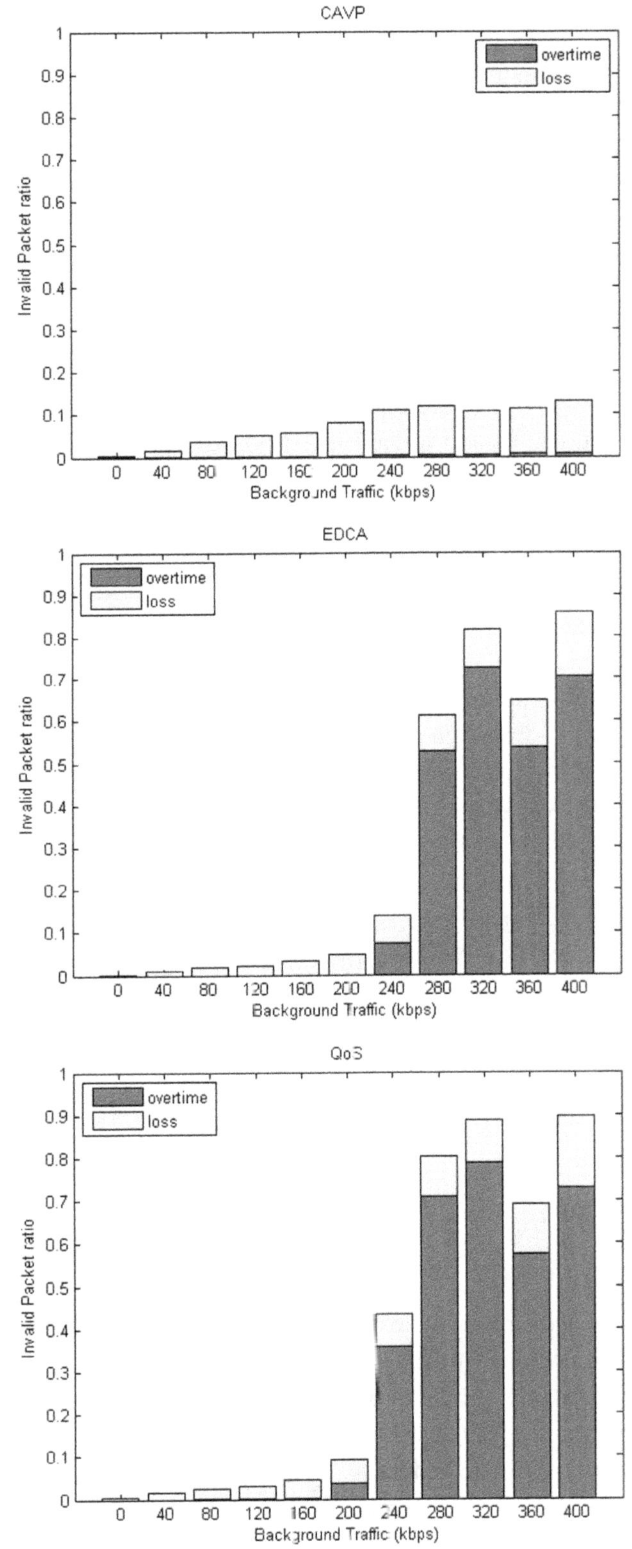

which chooses the one with the least estimated waiting time. Therefore, the packets transmitted are likely to be in time. However, comparing the loss rate results with the PSNR results, we find that PSNR results are not as inspiring as loss rate results. The relation between packet loss rate and PSNR is not simply linear. What is more, metrics of different layers show different faces of one experiment result.

Figure 14 shows the loss rate of three schemes under scenario 2. It can be seen that the ratio of overdue packets in scenario 2 is higher than in scenario 1. This is because all the traffic are sent from one station. When the transmission queue is full of packets, the waiting time of a packet can be long, leading to high overdue packet rate. Since the background traffic has AC_VI priority, so EDCA video stream is share the same AC with background traffic, results in the long latency of video stream. For QoS scheme, the background traffic shares the same AC with P-frame packets, and is more advantageous than B-frame packets. Therefore, the queuing delay increases and the overdue packets grow. As to CAVP, since AC_VI is likely to be filled, so the packets with the highest and the second highest priority will be sent to AC_VO or AC_BE. The waiting time of packets can be reduced because the packets are distributed into four AC's. Therefore, CAVP outperforms the QoS scheme. On the other side, EDCA is still over our expectation. This may be related to the hardware issue, which mentioned in the previous section.

The Effect of Non-Constant Background Traffic

In this section, we are going to examine the experimental result when the background traffic is not constant-bit-rate. We made some change in scenario-1 experiment. We used D-ITG (Distributed Internet Traffic Generator) as traffic generator instead of Iperf. D-ITG is able to generate traffic with different distribution on inter-packet time or packet size. We used D-ITG to generate background traffic with Poisson distribution inter-departure time. Other settings are the same with previous experiments. The result is shown in Figure 15.

Comparing the result in Figure 15 with the result in Figure 11, the performance of all three schemes degrade when the background traffic becomes non-constant. The traffic burst in the background makes the channel unstable. Even when the average load of background traffic is small, there are traffic bursts that makes channel congested. From the experimental result, we can find that the performance of CAVP degrades less than the other two schemes. CAVP provides higher priority for important and timely packets, and gives every packet minimum expected latency. Therefore, CAVP shows higher level of resistance to the unstable environment.

CONCLUSION

Cross-layer design is vital to H.264/AVC video streaming over IEEE 802.11e networks. In this chapter, we proposed CAVP design including application layer and MAC layer enhancements. The application layer Video Frame Prioritization scheme employs PSNR estimation method to identify critical video packets for error-prone wireless packet transmission. Compared with the conventional prioritization technique of which classification is based on the types of video frames involved, the PSNR-based application prioritization not only enhances video transmission quality in error-prone wireless networking environments, but also solves the myth of the naïve I/P/B-type classification. The MAC layer adaptive prioritization enhances delay-sensitive video streaming services by dynamically sorting packets into different access categories according to both estimated access delay and the importance of video packets. In addition, the myths of naïve AC selection are addressed. As shown in the NS-2

Figure 14. Loss rate in scenario 2

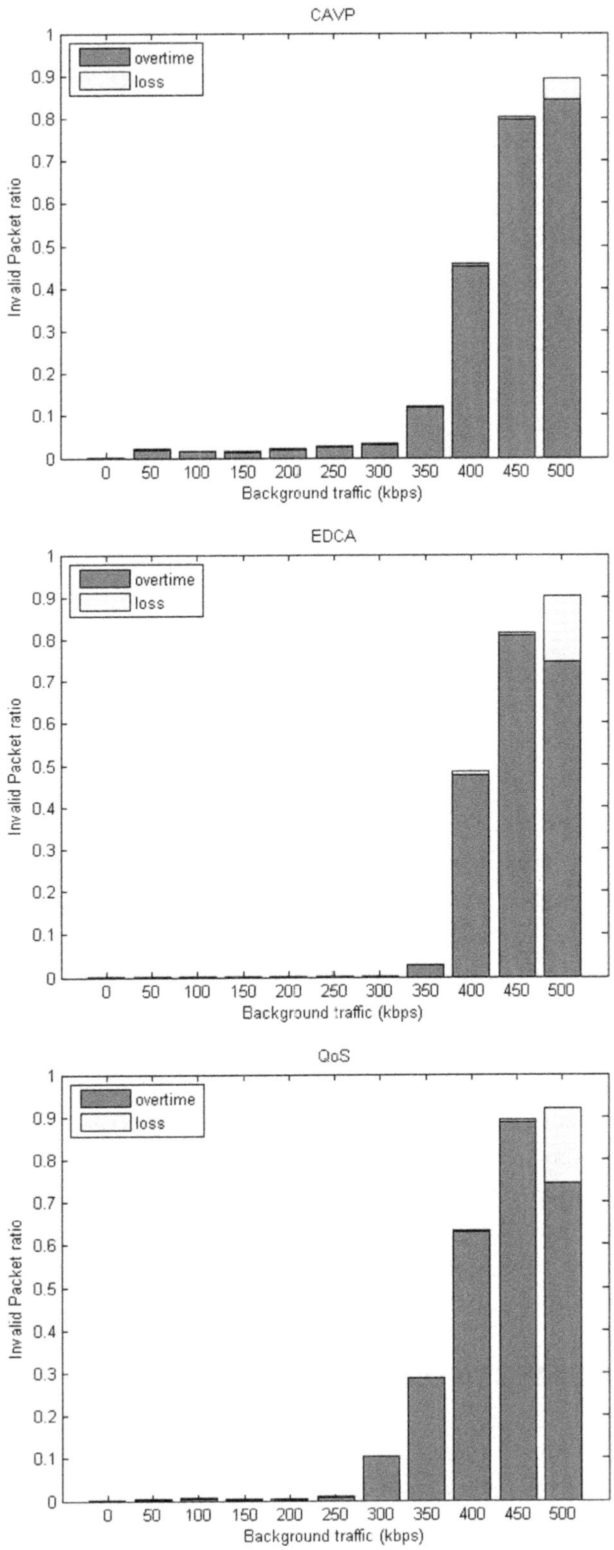

Figure 15. PSNR results: background traffic with Poisson distributed inter-packet time

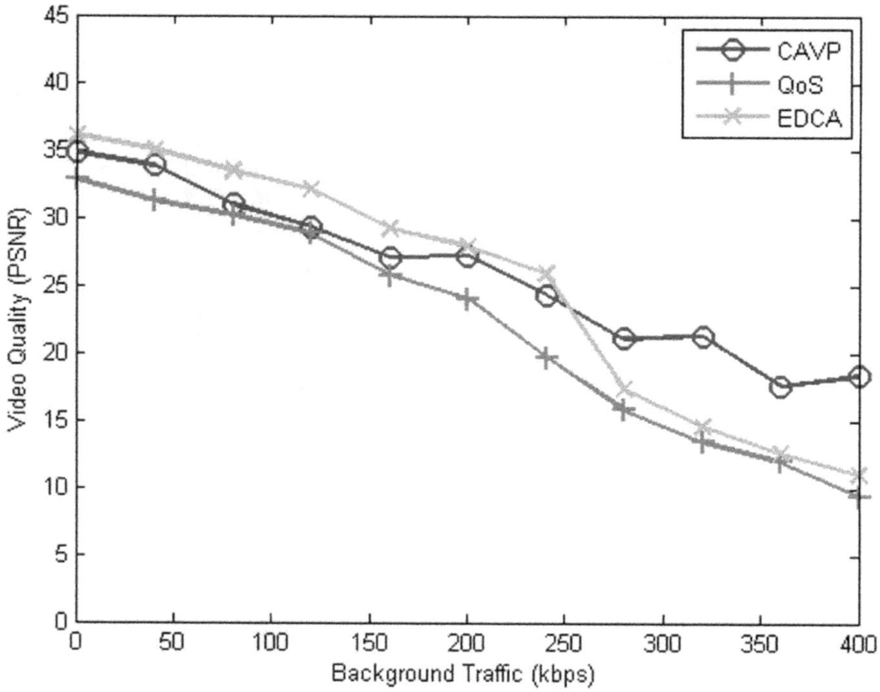

simulations, H.264/AVC video transmission quality has improved significantly with the proposed scheme. In addition, we also implement the proposed scheme with Click kernel module and the MadWifi WLAN driver. We have demonstrated the practicability and the enhanced performance in the testbed experiments. To accomplish the challenging task of delivering delay-sensitive video content over random access-based 802.11e wireless networks, the cross-layer prioritization scheme improves the system performance by considering both the application-layer H.264/AVC streaming quality and the MAC-level QoS.

FUTURE RESEARCH DIRECTIONS

Although we have shown the potential enhancements by addressing the two myths we proposed in this chapter, readers should consider this as the pioneer work in this direction. There are several possible extensions we can make in the proposed CAVP design. First, the complexity of PSNR estimation process in VFP can be reduce if we have a better understanding in the impact of packetization in distortion. A theoretical analysis on the structure of H.264/AVC video sequence combining with the existing distortion model may be helpful to achieve this. Second, in this chapter we have a simplified model for estimating the expected waiting time of each AC considering the internal contentions. An extension for concerning the inter-node contentions will be desirable for more accurate estimations. Finally, the implementation method we applied in this chapter unavoidably limits the performances of CAVP due to software and hardware constraints. A better wireless video streaming platform is required in order to show the full potentials of the proposed CAVP design.

ADDITIONAL READINGS

The readers are recommended to read the books of Petrick and O'Hara (2005) and Gast (2005) for better background on the design of 802.11 protocol. The readers are also suggested to read the book of Pratt (2001) for understanding H.264 AVC codec. For related works, Chen et al (2010) studied the optimal retry limit of packets under multimedia transmission with cross-layer concerns. Hsu and van der Schaar (2009) studied the effect of rety limits on the distortion of videos, and hereby provides a practical algoirthm for user utility maximization.. Chilamkurti et al. (2010) propose a similar idea to mapping the slices in H.264 AVC videos to appropriate ACs according to the information importance. Lin et al. (2009) consider the dynamic mapping with both the significance of the packet and the current loading of the ACs. Mansour et al. (2009) consider a more general wireless system with the capability to adjust data rates for different traffics. The cross-layer adjustment becomes more flexible in their scheme. Milani and Calvagno (2009) studied the possibility to adaptively tune the parameters of both video source encoder and wireless transmissions.

REFERENCES

Bianchi, G. (2000). Performance analysis of the IEEE 802.11 distributed coordination function. *IEEE Journal on Selected Areas in Communications*, *18*(3), 535–547. doi:10.1109/49.840210

Chen, C. M., Lin, C. W., & Chen, Y. C. (2010). Cross-layer packet retry limit adaptation for video transport over wireless LANs. *IEEE Transactions on Circuits and Systems for Video Technology*, *20*(11), 1448–1461. doi:10.1109/TCSVT.2010.2077475

Chilamkurti, N., Zeadally, S., Soni, R., & Giambene, G. (2010). Wireless multimedia delivery over 802.11e with cross-layer optimization techniques. *Multimedia Tools and Applications*, *47*(1), 189–205. doi:10.1007/s11042-009-0413-6

Click Modular Router. (2011). *Website*. Retrieved from http://read.cs.ucla.edu/click/.

D-ITG. (2011). *Distributed internet traffic generator*. Retrieved from http://www.grid.unina.it/software/ITG/index.php.

Gast, M. (2005). *802.11 wireless networks: The definitive guide*. New York, NY: O'Reilly Media.

Home, I. P. (2011). *H.264/AVC JM reference software*. Retrieved from http://iphome.hhi.de/suehring/tml.

Hsu, J. L., & van der Schaar, M. (2009). Cross layer design and analysis of multiuser wireless video streaming over 802.11e EDCA. *IEEE Signal Processing Letters*, *16*(4), 268–271. doi:10.1109/LSP.2008.2010821

IEEE Standard for Information Technology. (2005). Telecommunications and information exchange between systems - Local and metropolitan area networks - Specific requirements - Part 11: Wireless LAN medium access control (MAC) and physical layer (PHY) specifications - Amendment 8: Medium access method (MAC) quality of service enhancements. *IEEE Std 802.11e-2005*. Retrieved from http://www.ieee.org.

IEEE Standard for Information Technology. (2007). Telecommunications and information exchange between systems-Local and metropolitan area networks-Specific requirements - Part 11: Wireless LAN medium access control (MAC) and physical layer (PHY) specifications. *IEEE Std 802.11-2007.* Retrieved from http://www.ieee.org.

IEEE Standard for Information Technology. (2009). Telecommunications and information exchange between systems--Local and metropolitan area networks--Specific requirements: Part 11: Wireless LAN medium access control (MAC) and physical layer (PHY) specifications: Amendment 5: Enhancements for higher throughput. *IEEE Std 802.11n-2009.* Retrieved from http://www.ieee.org.

Krause, M., van Hartskamp, M., & Aarts, E. (2008). A quality metric for use with frame-rate based bandwidth adaptation algorithms. *Society of Photo-Optical Instrumentation Engineers (SPIE) Conference Series*. SPIE.

Ksentini, A., Naimi, M., & Gueroui, A. (2006). Toward an improvement of H. 264 video transmission over IEEE 802.11 e through a cross-layer architecture. *IEEE Communications Magazine, 44*(1), 107–114. doi:10.1109/MCOM.2006.1580940

Lagkas, T. D., Papadimitriou, G. I., Nicopolitidis, P., & Pomportsis, A. S. (2007). Priority-oriented adaptive control with QoS guarantee for wireless LANs. *IEEE Transactions on Vehicular Technology, 56*(4), 1761–1772. doi:10.1109/TVT.2007.897223

Lin, C. H., Shieh, C. K., Ke, C. H., Chilamkurti, N. K., & Zeadally, S. (2009). An adaptive cross-layer mapping algorithm for MPEG-4 video transmission over IEEE 802.11 e WLAN. *Telecommunication Systems, 42*(3), 223–234. doi:10.1007/s11235-009-9182-9

Liu, T., Wang, Y., Boyce, J., Wu, Z., & Yang, H. (2007). Subjective quality evaluation of decoded video in the presence of packet losses. In *Proceedings of the IEEE International Conference on Acoustics, Speech and Signal Processing (ICASSP),* (pp. 1125-1128). IEEE Press.

Lombardo, A., Morabito, G., & Schembra, G. (2004). Modeling intramedia and intermedia relationships in multimedia network analysis through multiple timescale statistics. *IEEE Transactions on Multimedia, 6*(1), 142–157. doi:10.1109/TMM.2003.819750

Madwifi Project. (2011). *Website.* Retrieved from http://madwifi-project.org/.

Mangold, S., Choi, S., Hiertz, G. R., Klein, O., Walke, B., Res, P., & Aachen, G. (2003). Analysis of IEEE 802.11 e for QoS support in wireless LANs. *IEEE Wireless Communications, 10*(6), 40–50. doi:10.1109/MWC.2003.1265851

Mansour, H., Fallah, Y. P., Nasiopoulos, P., & Krishnamurthy, V. (2009). Dynamic resource allocation for MGS H.264/AVC video transmission over link-adaptive networks. *IEEE Transactions on Multimedia, 11*(8), 1478–1491. doi:10.1109/TMM.2009.2032682

Milani, S., & Calvagno, G. (2009). A low-complexity cross-layer optimization algorithm for video communication over wireless networks. *IEEE Transactions on Multimedia, 11*(5), 810–821. doi:10.1109/TMM.2009.2021791

Network Simulator. (2011). *Ns-2.* Retrieved from http://www.isi.edu/nsnam/ns/.

Park, S. H., Lee, S., & Kim, J. W. (2005). Network-adaptive high definition MPEG-2 streaming over IEEE 802.11 a WLAN using frame-based prioritized packetization. In *Proceedings of the ACM International Workshop on Wireless Mobile Applications and Services on WLAN Hotspots,* (pp. 84-87). ACM Press.

Petrick, A., & O'Hara, B. (2005). *IEEE 802.11 handbook: A designer's companion*. New York, NY: IEEE Press.

Politis, I., Tsagkaropoulos, M., Pliakas, T., & Dagiuklas, T. (2007). Distortion optimized packet scheduling and prioritization of multiple video streams over 802.11 e networks. *Advances in Multimedia*, 1. doi:10.1155/2007/76846

Pratt, W. K. (2001). *Digital image processing* (3rd ed.). New York, NY: John Wiley & Sons Inc. doi:10.1002/0471221325

Richardson, I. E. G. (2003). *H. 264 and MPEG-4 video compression: Video coding for next-generation multimedia*. New York, NY: John Wiley & Sons Inc. doi:10.1002/0470869615

Sai Shankar, N., & van der Schaar, M. (2007). Performance analysis of video transmission over IEEE 802.11 a/e WLANs. *IEEE Transactions on Vehicular Technology, 56*(4), 2346–2362. doi:10.1109/TVT.2007.897646

Stockhammer, T., Hannuksela, M. M., Res, N., & Bergen, G. (2005). H. 264/AVC video for wireless transmission. *IEEE Wireless Communications, 12*(4), 6–13. doi:10.1109/MWC.2005.1497853

Tinnirello, I., & Bianchi, G. (2010). Rethinking the IEEE 802.11e EDCA performance modeling methodology. *IEEE/ACM Transactions on Networking, 18*(2), 540–553. doi:10.1109/TNET.2009.2029101

van der Schaar, M., Andreopoulos, Y., & Hu, Z. (2006). Optimized scalable video streaming over IEEE 802.11 a/e HCCA wireless networks under delay constraints. *IEEE Transactions on Mobile Computing, 5*(6), 755–678. doi:10.1109/TMC.2006.81

Wiegand, T., Sullivan, G. J., Bjontegaard, G., & Luthra, A. (2003). Overview of the H. 264/AVC video coding standard. *IEEE Transactions on Circuits and Systems for Video Technology, 13*(7), 560–576. doi:10.1109/TCSVT.2003.815165

Wiethölter, S., Emmelmann, M., Hoene, C., & Wolisz, A. (2006). *TKN EDCA model for ns-2*. Berlin, Germany: Technische Universität.

Xiao, Y., Du, X., Hu, F., & Zhang, J. (2008). A cross-layer approach for frame transmissions of MPEG-4 over the IEEE 802.11e wireless local area networks. In *Proceedings of the IEEE Wireless Communications and Networking Conference (WCNC)*, (pp. 1728-1733). IEEE Press.

KEY TERMS AND DEFINITIONS

Access Category (AC): A bundle of CWmin, CWmax, AIFS, TXOP parameters, an independent packet queue, and timers. The QoS support in IEEE 802.11e is provided by implementing a series of ACs.

Backoff: A timer triggered when a node senses an idle medium in a contention. The node can access the medium when the timer expires and the medium remains idle in the whole countdown.

Contention: A random access process for nodes in IEEE 802.11-based networks to compete for the opportunity to transmit data.

Contention Window (CW): The upper bound of the backoff counter in the current contention.

Inter-Mode: The frame/slice is expressed with one or several neighbor frame/slices, and cannot be decoded with its own contained information.

Intra-Mode: The frame/slice can be decoded only by the information it contained.

Peak Signal-to-Noise Ratio (PSNR): The ratio of maximum signal to noise, usually used in image and video processing as a benchmark.

Prioritization: A process to classify the importance of an object, demand, or service.

Quality of Service (QoS): A feature to support prioritization on multiple classes of demands or guarantee requirements on certain performance metrics.

ENDNOTE

1 We use the terminology "I-frame" to denote the video frame and "I-type" to denote the network packet that may contain only part of "I-frame."

Chapter 21
Cross-Layer Design in Wireless Sensor Networks

Weidong Yi
Graduate University of Chinese Academy of Sciences, China

Yongrui Chen
Graduate University of Chinese Academy of Sciences, China

ABSTRACT

New and diverse applications for Wireless Sensor Networks (WSNs) have led to new challenges. Cross-layer approaches have proven to be the most efficient optimization techniques for these problems, since they are able to take the behavior of the protocols at each layer into consideration to overcome the constraints for WSNs. Thus, this chapter focuses on identifying the core problems of WSNs and collects available cross layer solutions for them that have been proposed. The literature on cross-layer protocols, protocol improvements, and design methodologies for WSNs are reviewed, and a taxonomy is proposed in order to provide insights on the identification of cross-layer design in WSNs. The open issues are discussed in detail for future research, and precautionary guidelines for cross layer design to WSNs are indicated.

INTRODUCTION

Wireless sensor networks are an emerging technology consisting of small, low power, and low-cost devices that integrate limited computation, sensing, and radio communication capabilities. These sensor nodes can be deployed close to the detected area and used for continuous sensing, event detection, and local control of actuators. Therefore, wireless sensor networks can be widely applied into military, environment, healthcare, home, and commercial applications. Advance in low-power electronics design have made it possible to develop highly integrated, yet low cost, microsensor nodes, with the capabilities of sensing, processing, and wireless communications. These microsensor nodes can be deployed as Wireless Sensor Networks (WSNs) for close to

DOI: 10.4018/978-1-4666-0960-0.ch021

Figure 1. Traditional layered protocol stack for ad hoc network (from Goldsmith, et al., 2002)

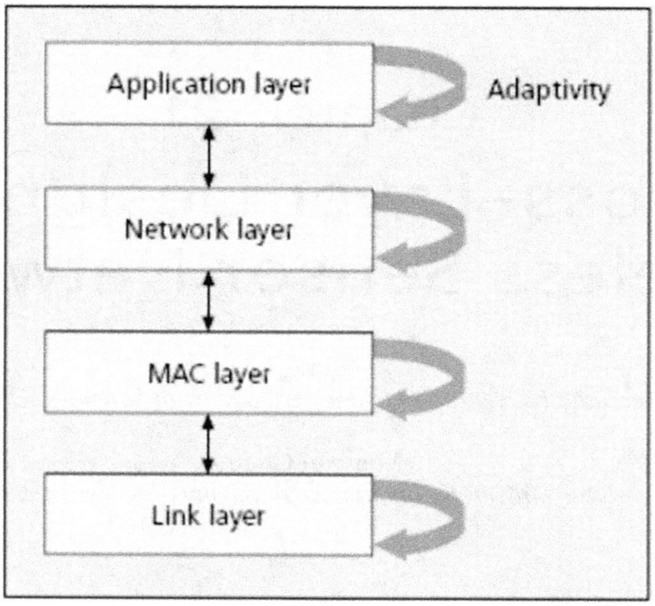

the detected area and used for continuous sensing, event detection, and local control of actuators. Once deployed, a network of thousands of these low-power micro-sensor nodes is expected to operate over years. Due to a large number of potential civil and military applications, a growing research interest has been directed in developing energy efficient self-organizing protocols for wireless sensor networks

Sensor nodes are usually battery operated and their operational lifetime should be maximized, hence energy consumption at sensor node is a crucial issue for applications of WSNs. A large amount of research aims to improve the energy efficiency of communication in Wireless Sensor Networks (WSNs). However, most of the existing solutions are based on the classical layered protocol architecture, which is mainly based on the layered stack as shown Figure 1. This layered model makes a significant contribution to simplifying network design. Consequently, the layer structure leads to robust and scalable protocols. However, the design and operation of each layer in the stack are isolated, and the interface between layers is static and independent of the individual network constraints and application. While these protocols may achieve very high performance related to each of these individual layers, they are not jointly optimized to maximize the overall network performance. Therefore, inheriting such a stack will lead to poor WSN performance in which resources, especially energy, bandwidth, memory size, and CPU speed are greatly constrained.

In addition, since wireless media are used to transmit data, packets are subject to interference from other transmission, resulting in errors. This issue is closely related to energy constraints, since packet errors will cause retransmission, requiring more energy from the sensor. Moreover, retransmission will affect delay and data throughput, also affecting QoS. The use of error control techniques can prevent retransmission, however at the cost of introducing some transmission overhead, and thus increasing the sensors energy consumption. Hence, it can be seen that transmission errors, QoS constraints, and energy consump-

Figure 2. Cross-layer protocol stack for WSN (from Goldsmith, et al., 2002)

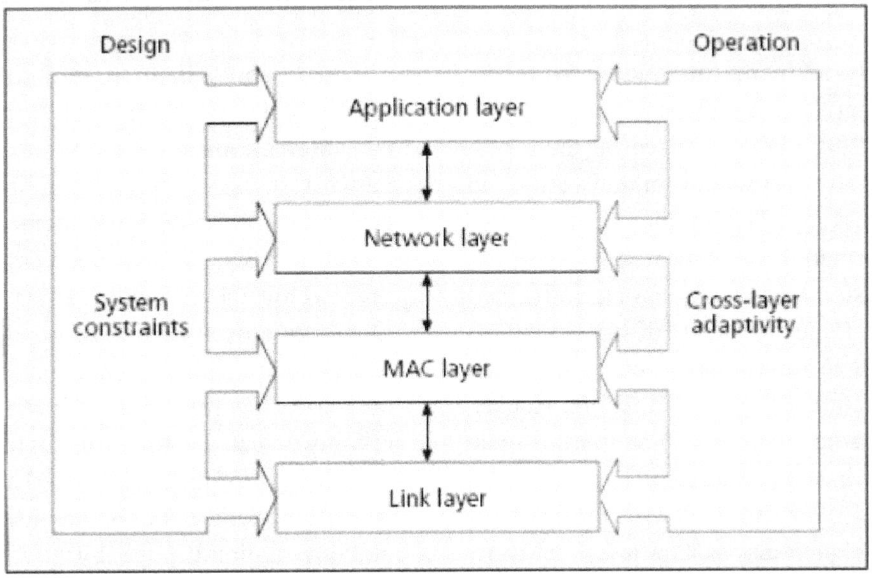

tion are closely related, requiring cross-layer solutions to address these problems simultaneously (Vuran, et al., 2009). Also, the effects of the use of Automatic Repeat Request (ARQ), Forward Error Correction (FEC), and their combination on the discussed parameters should be studied to find an optimal solution for WSNs.

Thus, new approaches are desirable to break the traditional border between the adjunct layers and create cross-layer paradigms. Possible cross-layer stack architecture is depicted in Figure 2.

Cross-layer design is a new technique which can be used to design and improve the performance in both wireless and wireline networks. The central idea of cross-layer design is to optimize the control and exchange of information over two or more layers to achieve significant performance improvements by exploiting the interactions between various protocol layers. A cross-layer design, on the other hand, stands as a very promising approach to achieve the greatest energy savings (Tommaso, et al., 2006).

Generally speaking, the following reasons necessitate the cross-layer design in WSNs:

1. All layers of the protocol stack contributes to energy consumption, therefore an overall energy consuming budget across all layers should be investigated and optimized (Lin, et al., 2006).
2. The significantly overhead of layered protocols results in high inefficiency. Hence, an efficient WSN requires a cross-layer design across all these layers.
3. Some empirical studies (Ganesan, et al., 2002; Zuniga, et al., 2004) reveal that the properties of low power radio transceivers and the wireless channel conditions in WSN are quite different with other networks. Thus, these physical layer features should be considered in the upper protocol design.
4. The dense deployment of sensor nodes and the continuous observation from a sensor node make the spatial and temporal correlation another significantly characteristic of sensor networks (Vuran, et al., 2004). Therefore, the data-centric approach of WSNs requires application-aware commu-

nication protocols design that is different with other networks.

Accordingly, an increasing number of recent papers (Fang, et al., 2004; Hoesel, et al., 2004; Vuran, et al., 2005) have focused on the cross-layer design for wireless sensor networks and reveal that cross-layer techniques result in significant performance improvement. Therefore, cross-layer design has become an attractive and active research topic in protocol design for WSNs in recent years. Although some effort has been made in the literature (Fang, et al., 2004; Hoesel, et al., 2004; Vuran, et al., 2005), numerous open issues, such as how to understand and apply this design principle, how to deal with problems of information exchange across stack layers, and how to realize a specific application requirement with global system constraints, remain. In this chapter, we will briefly review recent literatures that have tried to establish sound cross-layer design methodologies and protocols for WSNs.

The chapter is organized as follows. We overview the communication protocols devised for WSNs that focus on cross-layer design techniques. Then, cross-layer objective optimization techniques in WSN are discussed, which relates to apply cross-layer design to the resource allocation in WSNs. Next, we further review recent literatures that have tried to extend the cross-layer design methods horizontally to allow cooperation among multiple WSN nodes, and thus lead virtual Multiple Input Multiple Output (MIMO) systems to be created. Later, we discuss the cross-layer design for several specific applications of WSN such as WMSN (Wireless Multimedia Sensor Networks) and data fusion. Based on the experiences in the cross-layer design for WSNs, we list the potential open problems and precautions for WSNs in the next section. Finally, the conclusion of this chapter is drawn.

CROSS-LAYER PROTOCOLS FOR WIRELESS SENSOR NETWORKS

Existing works in both scientific studies and experimental work in WSNs revealed important interactions between different layers of the network stack. These interactions are especially important for the design of communication protocols in WSNs. WSNs are event-based systems that exploit the collective effort of densely deployed microsensor nodes, which continuously observe certain physical phenomenon. Generally, the main objective of any WSN application is to reliably detect or estimate event features from the collective information observed by sensor nodes. Nevertheless, the main challenge for achieving this objective is mainly posed by the severe energy and processing constraints of low-end wireless sensor nodes. Clearly, the collaborative sensing notion of the WSN achieved by the networked deployment of sensor nodes help to overcome the characteristic challenge of WSN, i.e., resource constraints. To this end, there has been significant amount of research effort that aims to develop networking protocols in order to achieve communication with maximum energy efficiency.

Differently from the conventional networks layers, WSNs consider only the following layers (Baronti, et al., 2007): Application (APP), Network (NWK or NET), Medium Access Control (MAC), and Physical (PHY). Although there is no transport layer, since it is complex and would waste sensors energy (Misra, et al., 2008), some WSN protocols have been designed for congestion control and reliable end-to-end communication (functions of the transport layer). Nevertheless, some authors defend the definition of a transport layer on WSNs, keeping the same layers stack as defined by TCP/IP. The considered stack can be seen in Figure 2 with two examples of cross-layer interactions. For instance, the QoS requirements at the application layer can be informed to the MAC layer in order to achieve better scheduling for the running application and the Channel State

Information (CSI) can be fed to the network layer so the routing protocol can avoid paths including channels in a bad state.

In addition to the collaborative sensing and networking in WSN, spatio-temporal correlation is another significant characteristic of sensor networks. Dense deployment of sensor nodes makes the sensor observations highly correlated in the space domain with the degree of correlation increasing with inter-node proximity. Similarly, some of WSN applications such as event tracking require sensor nodes to periodically sample and communicate the sensed event features, which yield temporal correlation between each consecutive observation of a sensor node. As follows, we overview representative communication protocols in WSN that are relevant to the cross-layer design. Moreover, we overview a single module solution for efficient communication in WSNs (Heinzelman, et al., 2002).

INTERACTIONS OF LAYER IN THE CROSS-LAYER DESIGN FOR WSN

In generally, a cross-layer design can be applied for communication protocol design in WSN with two ways. One is to aim at improving the performance of the communication protocol by taking into account parameters in other layers, while another is to merge relevant protocols into one component. Although the latter can allow much closer interaction among protocols, the functionality of the merged component can be very complex. We classify these studies in terms of interactions or modularity among Physical (PHY), Medium Access Control (MAC), routing, and Transport layers (TRAN).

Interactions: Combing MAC with PHY Layer

For physical layers, the power efficiency questions are similar to those addressed in wireless networks: how to transmit bits and frames in a power efficient way, respectively, to devices one-hop away. During the development of various protocols and services, the lowest layers of a WSN's system (e.g., the MAC layer) were increasingly being used to pass information to these higher layers. The overall result of these developments has led to the cross-layered approach to combine MAC with PHY layer. Apart from medium access control, these problems include identifying suitable modulation schemes, efficient FEC strategies, etc. (see (Akyildiz, et al., 2002; Karl, et al., 2006). The solutions of these problems are strongly affected by the sensor-device resources' constraints. The proposed solutions are generally independent from the applications; however, recently some authors (Vuran, et al., 2006) proposed to apply data-centric policies also at the MAC layer. The basic idea is to exploit spatial correlation among neighboring nodes to reduce the number of transmissions at the MAC layer.

The energy consumption analysis for physical and MAC layers is enabled to perform for three different MAC protocols (Haapola, et al., 2005). It concludes that single-hop communication can be more efficient if real radio models are used. Although this is an interesting result, the analysis is based on a linear network, which may not be practical in realistic scenarios.

A cross-layer solution among MAC layer, physical phenomenon, and the application layer for WSNs is proposed in Vuran et al. (2006). The spatial correlation in the observed physical phenomenon is exploited for MAC. Based on a theoretical framework, it is shown that a sensor node can act as a representative node for several other sensor nodes. Accordingly, a distributed, spatial Correlation-Based Collaborative Medium Access Control (CC-MAC) protocol is proposed. Simulation results show that exploiting spatial correlation for medium access results in high performance in terms of energy, packet drop rate, and latency.

Interactions: Combing MAC with Routing

Cross-layer designs that do not consider an optimization involving the medium access control layer are rare in the literatures. The MAC protocol consists of a fully distributed and self-organizing Time-Division Multiple Access (TDMA) scheme, in which each active node periodically listens to the channel and broadcasts a short control message. This control message is needed for medium access operation and is also used to piggyback various types of information at low energy costs. Information in the control message is used to create a maximal independent set of nodes. This set of nodes creates a connected network, and nodes in the set are active, while other nodes are passive and save energy by exploiting the infrastructure created by the connected network. The control message is also used by the routing protocol to establish and maintain efficient routes in a dynamic topology. The routing protocol uses local topology information gathered by the MAC protocol and is therefore efficient in re-establishing routes when they become disconnected.

As MAC and routing cross-layer modularity, the receiver-based routing is exploited in many works. Receiver-based routing has been independently proposed in Skraba et al. (2004), Zorzi et al. (2003), and Zorzi et al. (2004). In Akyildiz et al. (2006) and Seada et al. (2004), the authors discuss the energy efficiency, latency, and multi-hop performance of the algorithm. In Zorzi et al. (2004), the work in Zorzi et al. (2003) and Zorzi et al. (2004) is extended for a single radio node. In Skraba et al. (2003), the receiver based routing is also analyzed by a simple channel model and lossless links. The next hop of routing is chosen as a result of the contention in the neighborhood. Moreover, the latency performance of the protocol is presented based on different delay functions and collision rates. Although the authors provide insightful results for the receiver-based routing, the impact of physical layer is not considered in the protocol operation. Similarly in Ferrara et al. (2005), the routing decision is performed as a result of successive competitions at the medium access level. More specifically, the next hop is selected based on a weighted progress factor and the transmit power is increased successively until the most efficient node is found. Moreover, on-off schedules are used. The performance evaluations of all these propositions present the advantages of cross-layer approach at the routing and MAC layers. A joint scheduling and routing scheme is proposed in Sichitiu et al. (2004) for periodic traffic in WSNs. In this scheme, the nodes form distributed on-off schedules for each flow in the network while the routes are established such that the nodes are only awake when necessary. Since the traffic is periodic, the schedules are then maintained to favor maximum efficiency.

The trade-off between on-off schedules and the connectivity of the network are investigated in a cross-layer routing and MAC framework (Madan, et al., 2005). In this work, a TDMA-based MAC scheme is devised, where nodes distributively select their appropriate time slots based on local topology information. The routing protocol also exploits this information for route establishment. In Madan et al. (2005), the authors advocate the usage of cross-layer interaction through comparative simulations with a strict layered approach. WSNs are characterized by multiple flows from closely located nodes to a single sink. However, if this fact is not considered in route establishment, potential interfering routes can be established. In Yu et al. (2005), this effect of broadcast nature of MAC on routing is investigated. In this work, MAC interference between routes is minimized by constructing interference-aware routes. The routes are constructed using node code words that indicate the interference level of nodes and each packet contains a route indicator for route establishment. As a result, the routes are constructed to minimize the interference among them.

Hoesel et al. (2004) have proposed a cross-layer approach to combines medium access organization with routing. The authors show that a tightly integrated set of networking protocols is a good solution to reach the target of highly energy-efficient in WSNs. In the approach, the MAC is TDMA-based and the nodes in the network are capable of choosing their own time slot based on local information only. Nodes in the network can communicate with each other collision-free. Not every node needs to actively participate in communication in the network for global connectivity. Hence, the MAC protocol allows some nodes to be passive. These passive nodes save energy by not controlling a time slot, but make use of a backbone in the network formed by active nodes.

In this way, the MAC protocol overhead is greatly reduced for passive nodes. Passive nodes can communicate with active nodes, although this communication is not guaranteed to be collision-free. A simple yet effective algorithm is proposed for nodes to make the decision between the active and passive medium access protocol states. Consequently, the routing protocol benefits from local topology information already present in the MAC protocol. Only active nodes assist each other in forwarding messages to a destination that cannot be directly reached by the source node. An initial route is established by flooding of the active nodes of the network by the source. This is very energy consuming, yet inevitable. In this stage the benefit of the distinction between active and passive network participation of nodes already becomes clear. When a route gets disconnected (e.g., due to the highly dynamic topology of a WSN or energy depletion of nodes along the route), the protocol is able to efficiently reestablish a route between source and sink node.

In the static case, the difference is smaller, which is mainly due to the fact that routing protocols have to establish a route only once. The protocol has a small standard amount of data reserved for route updates; in the static case, this space is wasted. The approach clearly benefits from active and passive modes in the MAC protocol. In the mobile case, the roles are changed frequently, resulting in high network lifetimes. In the static case, nodes keep their MAC mode until their energy gets depleted. Therefore, roles should be reconsidered every now and then

Mitchell et al. (2010) have proposed the Aerial Platform Based Routing and Medium Access Control (APRMAC). In this protocol, each sensor node needs to discover their one-hop neighbor through sending hello packets. Then, all sensors send this information to an aerial node in LOS with all the sensors. The aerial platform is responsible for calculating the routes from all the sensors to their nearest sink nodes, and then it schedules medium access time for all the nodes. Thus, each node knows when it can transmit, when it is supposed to listen, and when it can enter sleep mode, saving at least 5% more sensor energy than TRAMA. Figure 3 shows an example of an APRMAC scenario.

Cheng et al. (2009) have argued that shortest path or minimum energy routing solutions have not considered the bandwidth constraints of WSNs. Thus, they have proposed a cross-layer routing protocol that is able to adjust the Time Division Multiple Access (TDMA) link rates according to their utilization. It has been proven in their work that the lifetime of the network is increased by 40% when compared to shortest path routing and the capacity of links is respected.

A new routing metric has been defined by Zhang and Zhang (2009) for their cross-layer Channel-Aware Geographic Informed Forwarding (CAGIF) protocol. The Efficient-Advancement Metric (EAM) determines the optimal relay node position to reach the next hop towards the destination. They have considered CDMA as medium access method, also taking interference into account. Through the use of the EAM and three positions (the current node, the source, and the destination), CAGIF provides routing with at least 50% less energy consumption and up to 4 times less packet control overhead when compared to GIF.

Figure 3. An example of an APRMAC (from Mitchell, et al., 2010)

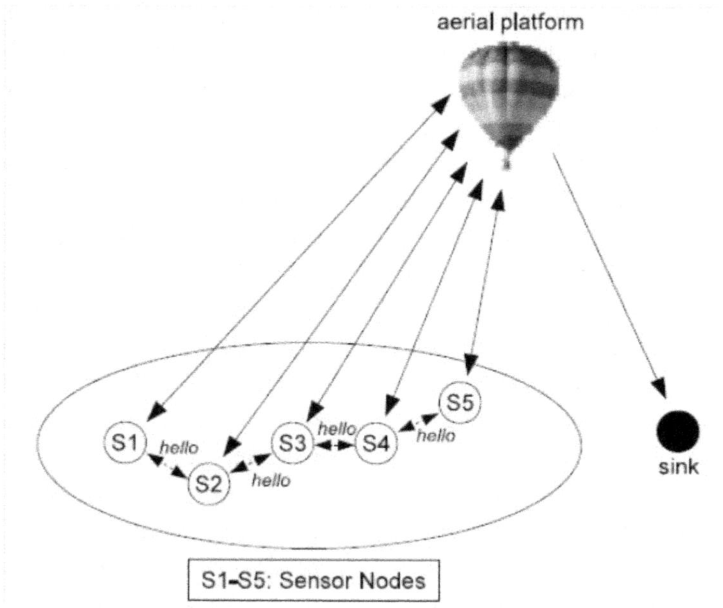

Interactions: Combing Routing with PHY

A cross-layer optimization of network throughput for multihop wireless networks is presented in Long et al. (2008). The authors split the throughput optimization problem into two sub-problems, i.e., multi-hop flow routing at the network layer and power allocation at the physical layer. The throughput is tied to the per-link data flow rates, which in turn depend on the link capacities and hence, the per-node radio power level. On the other hand, the power allocation problem is tied to interference as well as the link rate. Based on this solution, a CDMA/OFDM based solution is provided such that the power control and the routing are performed in a distributed manner. In Chiang et al. (2005), new forwarding strategies for geographic routing are proposed based on the results in Stann et al. (2003). The authors provide expressions for the optimal forwarding distance for networks with Automatic Repeat Request (ARQ) and without ARQ. Moreover, two forwarding strategies for these cases are provided. The forwarding algorithms require the packet reception rate of each neighbor for determination of the next hop and construct routes accordingly. Although the new forwarding metrics illustrate the advantages of cross-layer forwarding techniques in WSNs, the analysis for the distribution of optimal hop distance is based on a linear network structure.

Interactions: Combing TRAN with PHY Layer

In a wireless network with multihop transmissions and interference-limited link rates, it is important to balance power control in the physical layer and congestion control in the transport layer for enhancing the overall network performance while maintaining the architectural modularity between the layers. In Yuan et al. (2006), a cross-layer optimization solution for power control and congestion control is considered. The authors provide analytical analysis of power control and congestion control, and the trade-off between layered and cross-layer approach is presented. Based on this framework, a cross-layer communication protocol based on CDMA is proposed, where the transmission power and the transmission rate is controlled. However, the proposed solutions only apply to CDMA-based wireless multihop networks, which may not apply to WSNs that CDMA technology may not be feasible.

Interaction: 3-Layer Solutions

In addition to the proposed protocols that focus on pairwise cross-layer interaction, more general cross-layer approaches among three protocol layers exist. An optimization of transmission power, transmission rate, and link schedule for TDMA-based WSNs is proposed in Vuran et al. (2008). The optimization is performed to maximize the network lifetime, instead of minimizing the total average power consumption. In Messier et al. (2008), joint routing, MAC and link layer optimization is proposed. The authors consider a variable-length TDMA scheme and MQAM modulation. The optimization problem considers energy consumption that includes both transmission energy and circuit processing energy. Based on this analysis, it is shown that single-hop communication may be optimal in some cases where the circuit energy dominates the energy consumption instead of transmission energy. Although the optimization problems presented in Vuran et al. (2008) are insightful, no communication protocol for practical implementation is proposed. Moreover, the transport layer issues such as congestion and flow control are not considered.

Cui et al. (2007) have proposed a joint optimization of network, MAC, and physical layers. At the physical layer, M-ary Quadrature Amplitude Modulation (MQAM) is considered. The modulation adapts to the channel conditions by increasing its transmission rate (also increasing BER) when the channel is in a good state and decreasing it to be more robust when the channel is in a bad state. The MAC layer schedules more TDMA time slots to the links with less transmission rate to keep the throughput, and the network layer reroutes traffic to the links with better channel states. By solving the mathematical optimization problem considering these parameters, the proposed approach is able to reduce sensors energy consumption.

Cross-Layer Module (XLM) for Wireless Sensor Networks

In the cross-layer *interaction,* the traditional layered structure is preserved, while each layer is informed about the conditions of other layers However, the mechanisms of each layer still stay intact. Considering the mechanisms of network layers in a unified way, it provides a single communication module for efficient communication in WSNs. The cross-layer approach still necessitates a unified cross-layer communication protocol as a module design for efficient and reliable event communication that considers transport, routing, and medium access functionalities with physical layer effects for WSNs. So as to the cross-layer module design, the functionalities of multiple traditional layers are melted into a functional module.

Akydiz et al. (2006) have proposed a unified cross-layer protocol which combines the functionalities of the transport, network and medium access protocols into a single module. The unified Cross-Layer Module (XLM) achieves energy efficiency and reliable event communication. XLM is built upon the concept of letting a node decide whether it wants to participate in communication or not. A node makes its decision based an initiative determination procedure. The initiative determination procedure is a binary operation based on a set of four conditions. All four conditions must be satisfied for a node to participate. The first condition is to ensure that the link is reliable for communication. The received SNR of the packet must be above some specified threshold for communication. The second and third conditions ensure that there will be no congestion at the node. The node must be able to relay the data where the incoming traffic rate must not be greater than the outgoing traffic rate. In addition, the node must not experience any buffer overflow. The last condition ensures that the node has enough energy to participate. The remaining energy of a node must be above the minimum energy threshold. If all four conditions are satisfied, the node will participate in communication. Using this XLM concept, nodes will operate in duty-cycle fashion to save energy. After the initiative determination procedure, the node forwards the packet by performing the receiver contention operation. The receiver contention operation is based on the receiver-based routing (Skraba, et al., 2004; Zorzi, et al., 2003) which uses the routing level of each node to decide its priority in sending the packet. Simulation results show that XLM outperforms the tradition layered protocol in network performance and communication efficiency. An overview for the new communication paradigm, i.e., Cross-Layer Module (XLM) for WSN is followed as below.

The basis of communication in XLM is built on the initiative concept. The initiative concept constitutes the core of XLM and implicitly incorporates the intrinsic functionalities required for successful communication in WSN. A node in WSN initiates transmission by broadcasting an RTS packet to indicate its neighbors that it has a packet to send. Upon receiving an RTS packet, each neighbor of a node decides to participate in the communication through initiative determination. Denoting the initiative as I, which is determined as follows:

$$I = \begin{cases} 1, if \begin{cases} \xi_{RTS} \geq \xi_{Th} \\ \lambda_{relay} \leq \lambda_{relay}^{Th} \\ \beta \leq \beta^{\max} \\ E_{rem} \geq E_{rem}^{\min} \end{cases} \\ 0, \quad otherwise \end{cases} \quad (1)$$

where ξ_{RTS} is the received SNR value of the RTS packet, λ_{relay} is the rate of packets that are relayed by a node, β is the buffer occupancy of the node, and E_{rem} is the residual energy of the node, while the terms on the right side of the inequalities indicate the associated threshold values for these parameters, respectively. The initiative, I, is set to 1 if all four conditions in (1) are satisfied. The first condition ensures that reliable links be constructed for communication. The second and third conditions are used for local congestion control in XLM. The second condition prevents congestion by limiting the traffic that the transmission of a node can relay. The third condition ensures that the node does not experience any buffer overflow. The last condition ensures that the remaining energy of a node E_{rem} stays above a minimum value, $E_{rem}^{\min}$. The cross-layer functionalities of XLM are laid in these constraints that are defined as the initiative of a node to participate in communication. Using the initiative concept, XLM performs local congestion control, hop-by-hop reliability, and distributed operation. For a successful communication, the ðrst initiates transmission for a node by broadcasting a Request To Send(RTS)packet, which serves as a link quality indicator and also helps the potential destinations to perform receiver-based contention. Then, the nodes that hear this initiation perform initiative determination according to (1). The nodes that decide to participate in the communication contend for routing of the packet by transmitting the Clear To Send (CTS) packets. The waiting time for the CTS packet transmission is determined based on the advancement of a node for routing (Akyildiz, et al., 2006). Moreover, the local congestion control component of XLM ensures energy efficiency as well as reliable communication by a two-step congestion control. Analytical performance evaluation and simulation experiment results show that XLM significantly improves the communication performance and outperforms the traditional layered protocol architectures in terms of both network performance and implementation complexity (Akyildiz, et al., 2006).

Accordingly, the design principle of XLM is complete unified cross layering such that both the information and the functionalities of traditional communication layers are melted in a single module. The protocol operation of XLM is governed by the new concept of initiative determination. Based on this concept, XLM performs received based contention, local congestion control, and distributed duty cycle operation in order to realize efficient and reliable communication in WSN. The results from simulations are shown in Figure 3. XML is compared to GEO, PRR, PRR-SMAC, Flooding, and DD-RMST.

As results from simulation, the average goodput of the communication suites are shown in Figure 4a. Irrespective of the duty cycle value, δ, XLM provides very high reliability. The cross-layer communication paradigm of the XLM that is adaptive to the network topology enables such high performance even when the network operates at low duty cycle. Another important observation from Figure 4b is that the energy consumption per packet for XLM has a minimum at $\delta = 0.2$. Hence, we observe that the duty cycle value of $\delta = 0.2$ provides the most energy efficient performance for the operation of XLM. The advantages of using a separate routing layer in the layered protocol suites can be seen from Figure 4c, where the average hop count is shown. GEO (Seada, et al., 2003), PRR (Seada, et al., 2003), PRR-SMAC (Ye, et al., 2004), and DD-RMST (Stann, et al., 2003) result in less number of hops for the packets that reach the sink than XLM. Furthermore, as shown in Figure 4d, XLM results in end-to-end latency comparable to PRR. GEO results in smaller end-to-end delay since the routing is performed based only on geographical location. On the other hand, PRR-SMAC results in higher end-to-end latency due to the clustered scheduling of nodes. Figure 4d also clearly shows the trade off of DD-RMST in achieving high reliability. This results in significantly high latency values when compared to the other configurations.

Consequently, in a cross-layer simulation platform, the state-of-the-art layered protocol configurations have been implemented along with XLM to provide a complete evaluation. Analytical performance evaluation and simulation experiment results show that XLM significantly improves the communication performance and outperforms the traditional layered protocol architectures in terms of both network performance and implementation complexity.

CROSS-LAYER OBJECTIVE OPTIMIZATION TECHNIQUES IN WSN

Overview

In many cases, it is very challenging to design and implement a resource-efficient WSN. This is usually constrained by many factors and has several objectives to meet at same time. Often such factors and objectives are contradictory to each other. Optimization techniques can be more effective to solve these problems when taking into account the overall system and using all available knowledge. When this information has to be distributed to other sensor nodes, this effect is even larger. A solution in which such information is piggybacked to other messages can limit this extra message exchange. Thus, cross-layer objective optimization techniques are desirable to break the traditional border between the adjunct layer and create cross-layer paradigms in WSN, which are able to be interpreted as the distributed solution of resource allocation problems at different layers.

Resource allocation in the context of multi-hop wireless sensor networks has been extensively studied in the last few years, typically with the objectives of maximizing the network lifetime (Chang, et al., 2000), minimizing the energy consumption (Melodia, et al., 2005), and maximizing the network capacity (Radunovic, et al., 2004), which are conducted as an optimization

Figure 4. (a) average goodput, (b) average energy consumption per packet, (c) average hop count, and (d) average latency vs. duty cycle for layered protocol suites and XLM (from Akyildiz, et al., 2006)

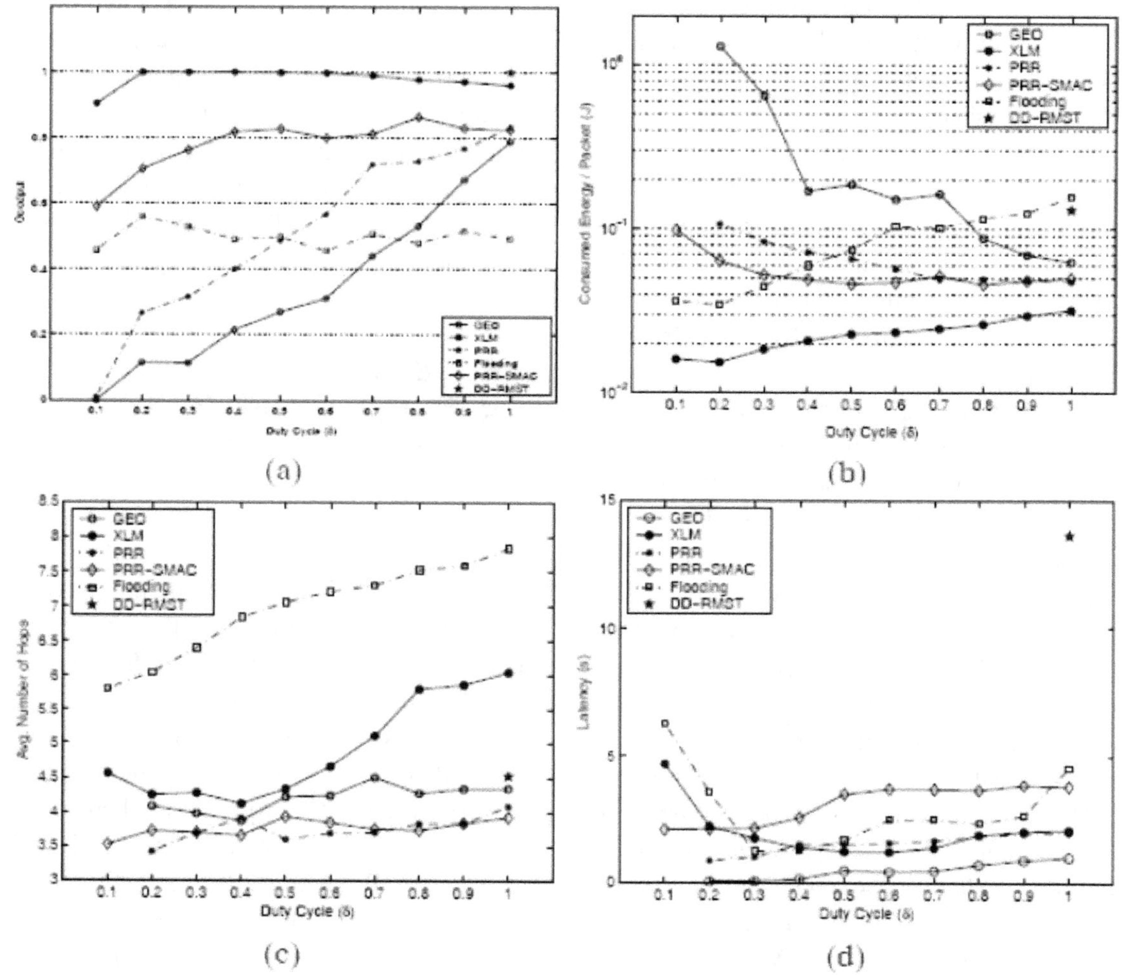

problem analytically. To solve these optimal problems, various cross-layer objective optimization techniques are applied for these existing studies. In these existing studies, some decompose the resource allocation problem at different layers, and consider allocation of the resources at each layer separately to achieve resource efficiency design. Some studies treat the resource allocation problems by considering interdependencies and pairwise interactions between isolated pairs of layers. Some other studies allocate resources in the way of tight coupling between functionalities handled at different layers. In this section, these mainly cross-layer objective optimization techniques for maximizing the network lifetime, the minimizing the energy consumption, and maximizing the network capacity are overviewed briefly.

Preliminaries

This section reviews definitions and concepts of optimization problems necessary for later discussion. It specifies how the optimization problem is formulated mathematically, and reviews the type of optimization problem and their solving for a corss layer design.

Optimization Problem

Usually, the computation of optimal transmission powers, rates, and link schedule in cross layer design for WSN can be solved through optimization programming techniques. The most of these problems are classified as a parametric optimization problem. A parametric optimization problem is used to find a set of design parameters, $x = \{x_1, x_2, x_3.x_n\}$, that can in some way be defined as optimal. In a simple case, this may be the minimization or maximization of some system characteristic that is dependent on *x*. In a more advanced formulation the objective function, *f(x)*, to be minimized or maximized, may be subject to constraints in the form of equality constraints, $G_i x = 0$ $i = 1,\ldots,m_e$ inequality constraints,

$G_i x > 0$ $i = m_e + 1,\ldots,m$, and/or parameter bounds, x_l, x_u. A General Problem (**GP**) description is stated as:

$$\begin{array}{lll} & \underset{x\in\Re^n}{\min\text{imize}}\ f(x) & \\ \text{subject to}: & G_i x = 0, & i = 1,\ldots,m_e \\ & G_i(x) \le 0, & i = m_e + 1,\ldots,m \\ & x_l \le x \le x_u & \end{array}$$

Where *x* is the vector of design parameters, $(x \in R^n)$, *f(x)* is the objective function that returns a scalar value ($f(x)$: $R^n \rightarrow, R$), and the vector function *G(x)* returns the values of the equality and inequality constraints evaluated at x($G(x)$: $R^n \rightarrow, R^m$).

An efficient and accurate solution to this problem is not only dependent on the size of the problem in terms of the number of constraints and design variables but also on characteristics of the objective function and constraints. When both the objective function and the constraints are linear functions of the design variable, the problem is known as a Linear Programming problem (LP). Quadratic Programming (QP) concerns the minimization or maximization of a quadratic objective function that is linearly constrained. For both the LP and QP problems, reliable solution procedures are readily available. More difficult to solve is the Nonlinear Programming (NP) problem in which the objective function and constraints may be nonlinear functions of the design variables. A solution of the NP problem generally requires an iterative procedure to establish a direction of search at each major iteration. This is usually achieved by the solution of an LP, a QP, or an unconstrained sub-problem.

The typical framework in joint optimal problem of cross layer design consists in defining an (possibly linear) optimization problem, in which some function of the network energy consumption has to be minimized (or, equivalently, the network lifetime has to be maximized). The constraints of the problem allow to model real constraints of the network. From a networking perspective, these formulations are cross-layer in nature, since the parameters of the objective function and the constraints usually depend on data that resides at different layers of the stack. Cross-layer design states that parameters of two or more layers can be retrieved and/or changed in order to achieve an optimization objective.

Convex and Non-Convex Optimization Problem

In an optimization problem, the types of mathematical relationships between the objective and constraints and the decision variables determine how hard it is to solve, the solution methods or algorithms that can be used for optimization, and the confidence we can have that the solution is truly optimal.

A key issue is whether the problem functions are convex or non-convex: A convex optimization problem is a problem where all of the constraints are convex functions, and the objective is a convex function if minimizing, or a concave function if maximizing. Linear functions are convex, so linear programming problems are convex problems. Conic optimization problems, the natural extension of linear programming problems, are also convex problems. In a convex optimization problem, the feasible region, the intersection of convex constraint functions, is a convex region.A non-convex optimization problem is any problem where the objective or any of the constraints are non-convex.Such a problem may have multiple feasible regions and multiple locally optimal points within each region. It can take time exponential in the number of variables and constraints to determine that a non-convex problem is infeasible, that the objective function is unbounded, or that an optimal solution is the "global optimum" across all feasible regions.

If the objective and all constraints are convex, we can be confident of determining whether there is a feasible solution, finding the globally optimal solution, and solving the problem up to very large size. If any functions are non-convex, the problem is much harder to solve and we cannot be certain of any of these things. All linear functions and some quadratic functions are convex, and the cone constraints in conic optimization problems are also convex functions. Some but not all smooth nonlinear functions are convex. Integer and constraint programming problems are inherently non-convex. Global optimization methods are designed to solve non-convex problems.

Cross-Layer Design for Lifetime Maximization of WSN

A network of wireless sensor nodes is distributed in a region. Each node has a limited energy supply and generates information that needs to be communicated to a sink node. It assumes that each node can vary its transmission power, modulation scheme, and duty cycle. The computation of optimal transmission powers, rates, and link schedule that maximize the network lifetime is desirable for the cross layer design in WSN efficiency.

The design challenges and the importance of cross-layer design for meeting application requirements in energy constrained networks were described in Tian et al. (2007). For wireless sensor networks, we may not always need to operate on the boundary of the achievable rates region. Hence, we have a choice among various transmission strategies (routing, power control, and scheduling) that we can exploit to increase the lifetimes of such networks based on cross-layer design. A brief overview of recent work on optimization of different layers of a wireless network for minimizing energy consumption and maximizing network lifetime is given below.

Maximize Lifetime in Interference-Limited WSN

In order to maximize the lifetime of interference-limited WSN, a joint optimal design is proposed in Madan et al. (2005). In this joint optimal design, it is on the computation of optimal transmission powers, rates, and link schedule that maximize the network lifetime. Through a joint scheduling and power control, it reduces energy consumption and increases single hop throughput. The problem of computing lifetime-optimal routing flow, link schedule, and link transmission powers for all active time slots is formulated as a non-linear optimization problem.

Cross-layer design based on computation of optimal power control, link schedule, and routing flow was described in Cruz et al. (2003). The authors assume a system with very low bandwidth efficiency The maximum achievable rate was assumed to be a linear function of Signal to Noise and Interference Ratio (SINR) and the goal was to minimize the total power consumption over an infinite horizon. The routing flow was computed in an incremental manner: it used the Lagrange multipliers obtained at each step by solving an optimization problem of possibly exponential complexity in the number of links. To improve it, a computational algorithm for cross-layer optimization in a very general is setup in Kozat et al. (2004). The algorithm computes competitive solutions that increase network lifetime for wireless networks with high bandwidth efficiency. The solutions computed by the algorithm are much better than commonly used heuristics. The algorithm alternates between link scheduling and computation of transmission powers and rates. The transmission scheme is computed during each iteration. Hence, the sub-optimality of the solution can be traded off with the required computational power. The main step of the algorithm involves the solution of a convex optimization problem for which the number of variables grows as $2LN$, where N is the number of time slots and L is the number of links in the network. Also, the number of constraints grows as $NL + 2|V|$, where $|V|$ is the number of nodes in the network.

As an optimization problem for computing the transmission powers, data rates, and link schedule to maximize the network lifetime, it is solved exactly if the MAC layer is restricted to only TDMA schemes (interference-free case). Without restriction on link schedules, mutually interfering links is allowed to be scheduled to transmit in the same time slot. Thus the problem formulation is not convex, it is difficult to solve. Hence, an alternate approach is taken to apply for it. For a fixed link schedule (i.e. fixed $Ln; n = 1,...,N$), the rate constraint is approximated as a convex constraint; This gives a convex set contained in the feasible set of the original problem. The resulting problem is a convex optimization problem that solves for optimal rates and powers for a given link schedule. During each iteration, both rates and powers are computed for a given link schedule, and then adapt the link schedule to the computed rates and powers.

Consequently, a convex approximation to the problem of computing an optimal data rate and transmission power, for each link and time slot, and a fixed link schedule is achieved by the algorithm that uses a heuristic to adapt the link schedule to the resulting optimal rates and powers. The results of numerical studies of the algorithm for different network topologies illustrate the advantages of frequency reuse, interference mitigation, load balancing, and multihop routing in energy-constrained wireless networks (Madan, et al., 2005).

Efficiency use of energy while providing an adequate level of connection to individual sessions is of paramount importance in multi-hop wireless networks. Energy efficiency and connection quality depend on mechanisms that span several communication layers due to the existing co-channel interference among competing flows that must reuse the limited radio spectrum. So as to do it, a framework for cross-layer design towards energy-efficiency communication is presented in Kozat et al. (2004). The approach is characterized by a synergy between the physical and the Medium Access Control (MAC) layers with a view towards inclusion of higher layers as well. More specifically, the joint problem of power control and scheduling is addressed with the objective of minimizing the total transmit power subject to the end-to-end Quality of Service (QoS) guarantees for sessions in terms of their bandwidth and bit error rate guarantees. Bearing to the NP-hardness of this combinatorial optimization problem, heuristic solutions are proposed that follow greedy approaches.

A general joint design of the physical, MAC, and routing layers to minimize network energy consumption is complex and hard to solve. In Cui et al. (2007), the authors focus on synchronous small-scale networks with interference-free link scheduling and practical MQAM link transmission schemes. It is shown that the cross-layer optimization problems can be closely approximated by convex optimization problems that can be efficiently solved. First, the total network energy consumption that includes both transmission and circuit energy is minimized, where the tradeoff between the two energy elements is explored.

Secondly, the routing flow, TDMA slot assignment, and MQAM modulation rate and power on each link is optimized. The results demonstrate that the minimum energy transmission scheme is a combination of multi-hop and single-hop transmissions for general networks; including circuit energy favors transmission schemes with fewer hops. Finally, based on the solved optimal transmission scheme, the best trade-off curve between delay and energy consumption is quantified, where a scheduling algorithm to minimize the worst-case packet delay is derived.

In Yuan et al. (2006), a distributed optimization framework for wireless multi-hop sensor networks is proposed and it shows that the cross-layer optimization problem can be decomposed into two sub-problems corresponding to two separate layers (the physical and the application layers) of the overall system. By modeling each sub-problem as a non-cooperative game, the non-convex application-layer rate-allocation and physical-layer power-allocation sub-problems can be solved in a distributive manner. Finally, it is shown that by using a set of dual variables as the market prices to coordinate the physical layer supply and the application layer demand, the overall optimization process strikes a right balance between the two layers in an overall cross-layer design.

In Vuran et al. (2008), a cross-layer solution for packet size optimization in WSN is proposed, which captures the effects of multi-hop routing, the broadcast nature of the physical wireless channel, and the effects of error control techniques. A significantly result of this paper is that contrary to the conventional wireless networks, in wireless sensor networks, longer packets reduce the collision probability. Consequently, an optimization solution is formalized by using three different objective functions, i.e., packet throughput, energy consumption, and resource utilization. Furthermore, the effects of end-to-end latency and reliability constraints are investigated that may be required by a particular application. As a result, a generic, cross-layer optimization framework is developed to determine the optimal packet size in WSN. This framework is further extended to determine the optimal packet size in underwater and underground sensor networks. From this framework, the optimal packet sizes under various network parameters are determined.

Maximize Lifetime under the Constraints of End-to-End Reliability

The problem of the lifetime maximization in WSN under the constraint of end-to-end transmission success probability is investigated by adopting a cross-layer strategy that considers physical layer (i.e., power control), MAC layer (i.e., retransmission control) and network layer (i.e., routing protocol) jointly (Kwon, et al., 2006). The problem is decoupled into separate sub-problem of each layer and the optimal algorithm is proposed as well as an alternative heuristic algorithm with lower complexity for each sub-problem.

The problem of minimizing the total energy consumption under the reliability constraint is formulated as follows:

Minimize

$$\sum_{i=1}^{H}\{E^{tx}(P_{l_i}) + E^{ack-rx}\} + (H-1)\cdot(E^{rx} + E^{ack-tx})$$

Subject to:

$$\prod_{i=1}^{H} P_s(G_i, l_i) \geq Q_s, l_i \in \{1, 2, ..., L\} \qquad (2)$$

Where, $E^{tx}(P_{l_i})$ is the energy consumption for each packet transmission at the *i-th* link of the path that the source node transmits its packets to a sink node; P_{l_i} is the transmission power consumption which is chosen by each sensor node from one of L values from [$P_{\min}$, $P_{\max}$], where the gap between two consecutive levels, i.e., ΔP (dBm), is constant. E^{rx} is the energy consumption for each packet reception, which is assumed to be constant. The energy consumptions for transmitting and receiving ACK packets are denoted by E^{ack-tx} and E^{ack-rx}, respectively. H is the hop of the path. $P_s(G_i, l_i)$ is the average success probability at the *i-th* link which is decided by the channel gain of the link G_i and the transmission power level l_i. The channel gain is given by the log-normal shadowing model (Zuniga et al. 2004), and the *per-hop* success probability function is based on the empirical study of Mica2 mote (Zuniga, et al., 2004) and determined by the SNR of the link. Q_s denotes the target end-to-end success probability.

For a given route, as l_i should be integer, the problem in (2) is a non-linear integer programming problem. Unfortunately, the integer programming problem is NP-complete (Gupta et al. 1998). To solve this problem, an algorithm is developed by using greedy approach and the incremental ratio ρ_i of the per-hop success probability contributed by the power level increment at the *i-th* hop to be:

$$\rho_i = \frac{P_s(G_i, \min(l_i + 1, L))}{P_s(G_i, l_i)} \qquad (3)$$

Then, considering a Greedy Power Allocation (GPA) algorithm that increases the power level of the link of which the increment is the highest as outlined below:

1. Allocate the lowest level to all links, i.e., $l_i \leftarrow 1$ for $i = 1, 2, \cdots, H$.
2. Increase, by one, the power level of the link with the highest value of ρ_i, i = 1, 2, · · ·, H, i.e., $l_i^* \leftarrow l_i^* + 1$ for $i^* = \arg\max_i \rho_i$.
3. Iterate the above process until the end-to-end success probability becomes equal to or larger than the target value.

This procedure is performed iteratively until the stability constraint is met. If a feasible solution does not exist, the power level is increased to the next level and then performs the retry limit allocation and the above routing algorithm again.

For performance comparisons, four different schemes are considered: (1) Non-optimal power allocation and CR (Cost-Based Route) (namely, NPA-CR), (2) GPA (Greedy Power Allocation) and CR (namely, GPA-CR), (3) GPA and optimal routing (namely, GPA-OR), and (4) Non-Optimal Power Allocation and Random Routing (namely, NPA-RR). In the case of NPA algorithm, the transmission power is allocated such that all links have the same per-hop success probability, which is set to the value that can guarantee the target end-to-end success probability for the path with the maximum number of hops. In the RR algorithm, we randomly choose a route for all sensor nodes.

It is demonstrated through simulations (Figure 5) that a trade-off relation exists between the network lifetime maximization and the reliability constraint, and the strategy that is designed by combining the proposed algorithm at each layer can significantly increase the network lifetime. The effect of the retransmission control on the energy efficiency is also investigated for different energy consumption models. Simulation results reveal that multiple retransmissions with low power yield little gain when the link distance is short and the power conversion efficiency of the amplifier increases with the transmission power.

Maximize Lifetime across Three Layers

Cross-layer design can include information exchange between different layers (not necessarily neighboring layers), adaptivity at each layer to this information, and diversity built into each layer to insure robustness (Goldsmith, et al., 2002; Cui, et al., 2005). Considering across the physical layer, MAC layer and routing layer to maximize lifetime of WSNs, the physical layer can deploy adaptive

Figure 5. Network lifetime of various schemes with respect to target end-to-end success probability (w = 50, update period = 200 frames) (from Kwon, et al., 2006)

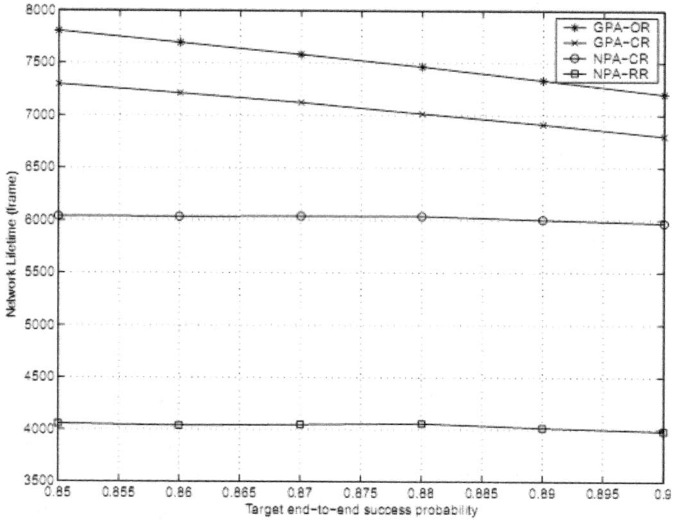

modulation and coding to exploit or compensate for the time-varying wireless channel. Then, this adaptivity at the physical layer can be used by higher layers to achieve better performance: The MAC layer can assign a longer channel usage time to links with low-rate modulation schemes to meet the throughput or energy constraint; the routing layer can reroute traffic to links supporting high-rate modulation schemes to minimize congestion; and the application layer can use multi-description codes to leverage the diversity of different routes.

Significant performance gain can be achieved by these interactions between different layers (Goldsmith, et al., 2002; Cui, et al., 2005). However, to maximize the achievable gain, the dynamics among different layers should be analytically modeled. Cui *et al.* (2007) have proposed a cross three layer optimization approach to minimize the total average power consumption across all nodes. In the approach, the problem formulation is restricted only in the sense with interference-free scheduling in a small-scale network where synchronization among nodes is achievable. Specifically, the approach focuses on the joint design of the routing, MAC, and physical layers to minimize the overall network energy consumption that includes both the transmission and the circuit processing energy consumption.

Instead of seeking heuristics or high complexity algorithms for cross layer design of arbitrary networks, authors focus on small-scale sensor networks that only involve several dozens of nodes, which are closely located. For such networks, interference free. TDMA is a good choice for the MAC layer because it is feasible in a relatively small network and is energy efficient given that nodes can be perfectly scheduled to be in sleep and active modes to save energy, where a central controller (the SINK node) is assumed and perfects synchronization among nodes. Due to the simplification at the MAC layer, the cross layer design problem can be relaxed to a convex one, for which it has known techniques to solve efficiently (Boyd, et al., 2003). The relative performance degradation caused by the relaxation is low. In addition, authors have proposed an algorithm that schedules transmissions to minimize the delay for a given routing, MAC, and physical layer transmission scheme, and then it is characterized to the delay-energy tradeoff using convex optimization techniques. In comparison to the model proposed in Yu et al. (2004), which only applies to tree topologies, the model proposed in Cu *et al.* (2007) applies to general topologies with one SINK node and no loops (for which the tree topology is a special case). In addition, the model proposed in this paper is based on convex optimization techniques, which approaches the solution from a different angle. A brief overview of this approach is given below.

To maximize the network lifetime, Cu *et al.* (2007) have formulated an optimization problem with the following constraints:

1. Flow Conservation: the flow conservation equations are satisfied over each frame of N time slots.
2. Rate Constraints: the maximum rate that can be sent over each link is *(1 + K log SINR)*.
3. Energy Conservation: the energy consumed by each node over time *Tnet* should be less than or equal to the initial energy at the node.
4. Range Constraints: the flow over a directed link can only be from the transmitter to the receiver. The transmission power at a node is less than or equal to the maximum transmission power at that node.

To describe the network, the following notation is defined:

1. For a wireless sensor network, let $G = (V, L)$ denotes the directed graph representing the network. V is the set of wireless nodes and L is the set of directed links. Let $A \in R^{|V| \times |L|}$ be the incidence matrix of the graph G. We have:

$$A(v.l) = \begin{cases} 1 & \text{if v is the transmitter of link l} \\ -1 & \text{if v is the receiver of link l} \\ 0 & \text{otherwise} \end{cases} \tag{4}$$

Let us write $A = A^{+} - A^{-}$, such that $A^{-}(v.l)$, $A^{-}(v.l) = 0$ if $A(v.l) = 0$, and A^{+}, A^{-} have only 0 and 1 entries.

2. Let N be the number of time slots in each frame of the periodic schedule. L_n denotes the set of links scheduled, i.e., allowed to transmit, during time slot n ∈ $\{1, ..., N\}$.
3. P_l^n and r_l^n denote the transmission power and rate per unit bandwidth, respectively, over link l and slot n. Also, r^n, $P^n \in R^{|L|}$ will be used to denote the corresponding vectors for time slot n. Let P_l^{max} be the maximum allowable transmission power of the transmitting node of link l. The corresponding vector is $P^{max} \in R^{|L|}$. Also, define the vectors $1_t(P^n)$, $1_r(P^n) \in R^{|V|}$ such that:

$$(1_t(P^n))_v = \begin{cases} 1 & \text{if } (a_v^{+})^T P^n > 0 \\ 0 & \text{otherwise} \end{cases} \tag{5}$$

$$(1_r(P^n))_v = \begin{cases} 1 & \text{if } (a_v^{-})^T P^n > 0 \\ 0 & \text{otherwise} \end{cases} \tag{6}$$

where $(a_v^{+})^T$ and $(a_v^{-})^T$ denote the v_{th} row of the matrices A^{+} and A^{-}, respectively. Thus, the vectors give the sets of nodes that transmit and receive data, respectively, in each time slot.

4. Let E ∈ $R^{|V|}$ be such that E_v denotes the initial amount of energy at node v. P_{ct} and P_{cr} give the power consumption of the transmitter and the receiver circuits at a node, respectively. These values are assumed to be the same across all nodes.
5. Let s_v denote the rate at which information is generated at node v. Let s ∈ $R^{|V|}$ be the vector whose entries are s_v. We note that $s_{\sin k} = -\sum_{v \in V, v \neq \sin k} s_v$.

6. $G \in R^{|L| \times |L|}$ is the link gain matrix of the network. G_{lk} denotes the power gain from the transmitter of link k to the receiver of link l. Let N_0 denote the total noise power over the bandwidth of operation.
7. Let T_v denote the lifetime of node v, i.e., the time at which it runs out of energy. Then the network lifetime is defined to be $T_{net} = \min_{v \in V, v \neq \sin k} T_v$.

Thus, the problem of maximizing the network lifetime can be written as the following optimization problem.

$$\max. \quad T_{net} \tag{7}$$

$$s.t. \quad \frac{1}{N} A(r^1 + ... + r^N) = s \tag{8}$$

$$\log(1 + K \frac{G_{ll} P_l^n}{\sum_{k \neq l} G_{lk} P_k^n + N_0}) \geq r_l^n \tag{9}$$

$$\frac{T_{net}}{N} \sum_{n=1}^{N} ((1+\alpha) A^{+} P^n + P_{ct} 1_t(P^n) + P_{cr} 1_t(P^n) + P_{cr} 1_r(P^n)) \preceq E \tag{10}$$

$$r^n \succeq 0, \; 0 \preceq P^n \preceq P^{\max} \tag{11}$$

Such as the network lifetime can be increased by using transmission schemes that have the following characteristics:

1. Multi-hop routing: the received power in wireless environments typically falls off as the *m-th* power of distance, with $2 \leq m \leq 6$. Hence, the transmission energy can be conserved by using multi-hop routing.
2. Load Balancing: a node will run out of energy very quickly if it is on the routes of many source destination pairs. Hence, load balancing is necessary to avoid the creation of hot spots where some nodes die out quickly and cause the network to fail.
3. Interference mitigation: links that strongly interfere with each other should be scheduled at different times to decrease the energy consumption on these links.
4. Frequency reuse: weakly interfering links should be scheduled simultaneously so that each link can transmit at a lower rate when active. This reduces the average transmission power on each link.

The procedure of the iterative approach to compute an approximate optimal strategy is summarized in Figure 6. It is shown that the cross-layer optimization problems can be closely approximated by convex optimization problems that can be efficiently solved. First, the total network energy consumption that includes both transmission and circuit energy is minimized, where the tradeoff between the two energy elements is explored. Secondly, the routing flow, TDMA slot assignment, and MQAM modulation rate and power on each link is optimized. The results demonstrate that the minimum energy transmission scheme is a combination of multi-hop and single-hop transmissions for general networks; including circuit energy favors transmission schemes with fewer hops. Finally, based on the solved optimal transmission scheme, the best trade-off curve between delay and energy consumption is quantified, where a scheduling algorithm to minimize the worst-case packet delay is derived.

The performance of the max-lifetime algorithm (namely *agl)* is compared with several transmission schemes with specific scheduling at the MAC layer, outlined below:

1. Uniform TDMA: Each link in the network is scheduled for an equal number of time slots.

Figure 6. Iterative approach to compute powers, rates and link schedule (from Cu, et al., 2007)

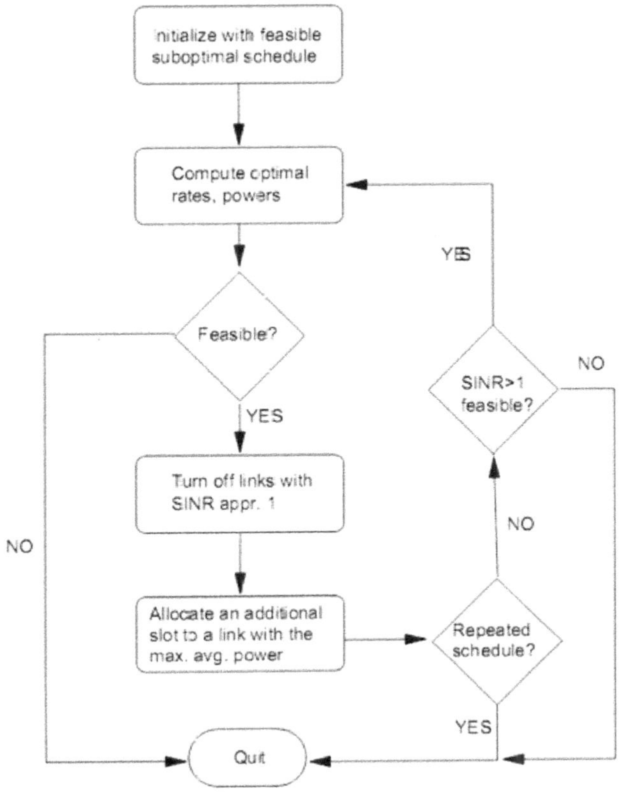

2. Optimal TDMA: This refers to an optimal TDMA schedule computed by the mixed integer-convex problem formulation.
3. Spatially periodic time sharing: This refers to a link layer scheduling scheme specific to one-dimensional (string and linear) topologies. A spatially periodic scheme with T time slots per frame is adopted. In each time slot every T *th* link is activated. Also, every link is activated once in every T slots.

The results shown in Figure 7 illustrate the advantages of *alg* with load balancing, multi-hop routing, frequency reuse, and interference mitigation in increasing the lifetime of energy-constrained networks to compare with uniform and optimal TDMA.

CROSS-LAYER DESIGN WITH COOPERATIVE COMMUNICATION FOR WIRELESS SENSOR NETWORKS

In wireless networks, the advantages of Multiple-Input Multiple-Output (MIMO) systems have been widely acknowledged, for their ability to achieve space diversity to combat the fading effects of wireless channel. Although MIMO technique is clearly advantageous on a cellular base station, it may not be practical for wireless sensor networks. Due to limitations in size, power, cost, or hardware, a node in a wireless sensor network may not be able to support multiple transmit antennas (Nosratina, et al., 2004).

Cooperative communication, which allows single-antenna sensor nodes to reap some of the

Figure 7. Linear topology-network lifetime (from Cu, et al., 2007)

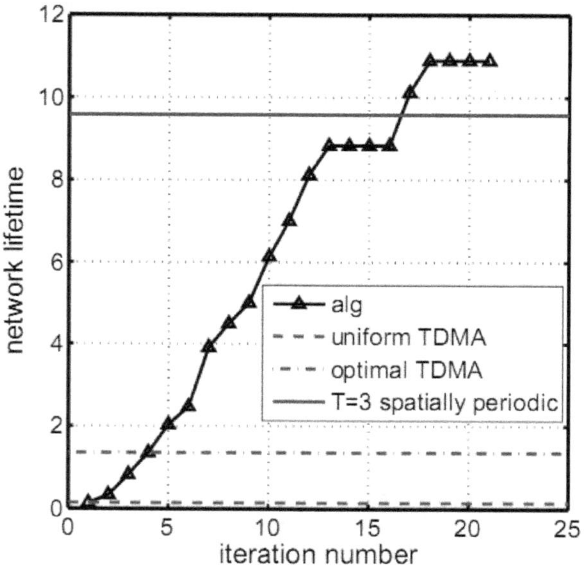

benefits of MIMO systems, is a powerful solution that can overcome the limitation of wireless sensor networks to improve the transmission reliability as well as the energy efficiency. The basic idea is that single-antenna nodes in a multi-user scenario can "share" their antennas in a manner that creates a virtual MIMO system. By forming a virtual antenna array, a node can achieve high diversity gains by using several partners to relay the transmitted signals, which is shown in Figure 8 (Nosratina, et al., 2004).

Figure 9 demonstrates a typical application of cooperative communication in clustered WSN (Zhou, et al., 2009). In the scenario, a WSN is deployed in a given region with many sensor nodes distributed randomly or artificially. The nodes are divided into several clusters for the sake of energy saving and lifetime improvement. Similar to LEACH (Heinzelman, et al., 2002), the behavior of network is broken up into round. In each round, the network has clusters. The data collected by the multiple source nodes will be transmitted to a remote sink node through multiple hops transmission. The data transfer is divided into two phases. In phase 1, the common nodes (also can be called cluster members) transmit their data packet to the cluster head one by one. In phase 2, the cluster head merges all the data and transfer the merged packet to the sink node through multi-hop transmission via a route constructed by several cluster heads. Since the distance between the cluster heads are much further than the distance between any cluster member and its cluster head, the data packet may be lost due to the unreliable wireless link. To enhance the reliability of transmission between cluster heads, cooperative communication techniques can be used. When the cluster head transmits the packet, several cooperative nodes can overhear the packet. Then they transmit the packet to the next cluster head (using space-time coding or other techniques). As the next cluster head collects several signal "samples," it can combine them and re-construct the data packet. In this way, the space diversity is achieved, and the data transmission reliability is improved, without additional energy consumption.

Figure 8. Cooperative communication (from Nosratina, et al., 2004)

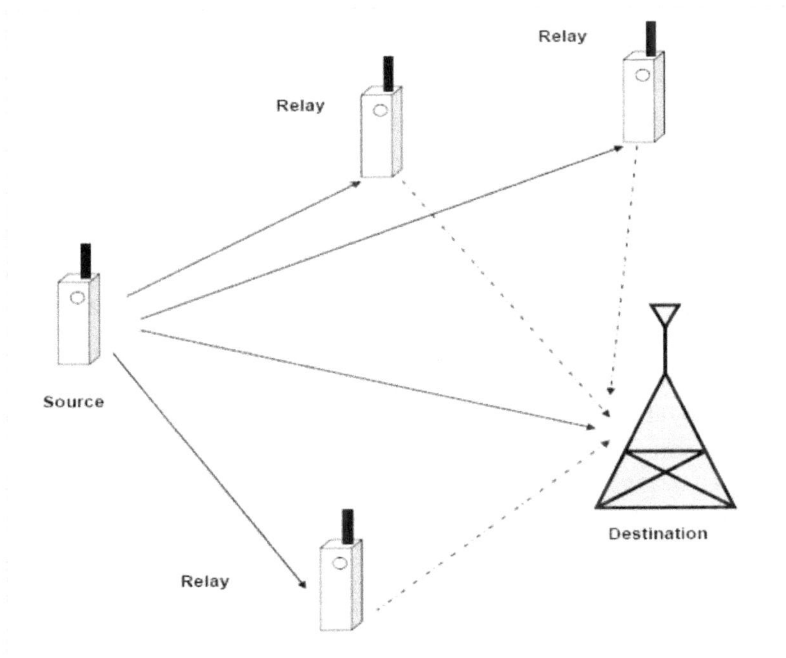

Figure 9. A system model of multihop cooperative MIMO transmission in WSN (from Zhou, et al., 2009)

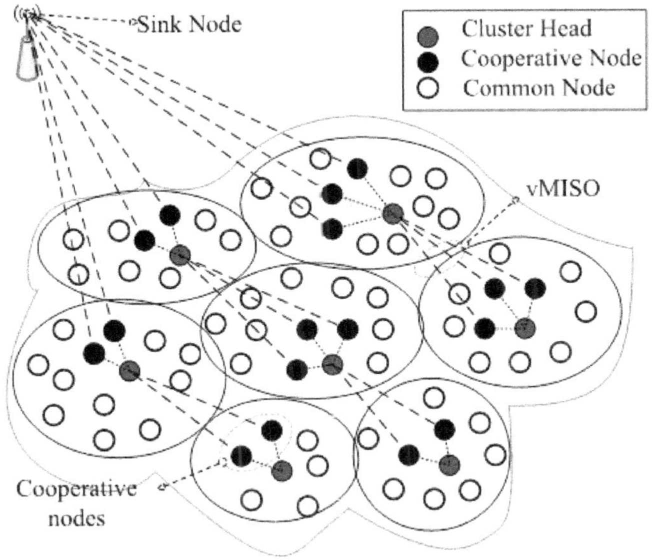

There has been a lot of work on improving the link performance in cooperative networks by using advanced signal processing or power allocation methods among a single source node and its relays. However, cooperative communication is inherently a network problem, as pointed out in Sendonaris et al. (2003), Laneman et al. (2003), and Laneman et al. (2004). It would be therefore fruitful to take into account additional higher layer network issues and use cross-layer design methods. There have been some efforts towards this such as combining node cooperation with ARQ in the link layer (Dai, et al., 2005), routing in the network layer (Luo, et al., 2004), or resource allocation in the MAC layer (Luo, et al., 2005).

In Chen et al. (2008), a unified cross-layer framework is presented for resource allocation in cooperative networks, which considers the physical and network layers jointly and can be applied for any cooperative transmission scheme. It is found that the fairness and energy constraint cannot be satisfied simultaneously if each node uses a fixed set of relays. To solve this problem, a multi-state cooperation methodology is proposed, where the energy is allocated among the nodes state-by-state via a geometric and network decomposition approach. Given the energy allocation, the duration of each state is then optimized so as to maximize the nodes utility. Numerical results will compare the performance of cooperative networks with and without resource allocation for cooperative beamforming and selection relaying. It is shown that without resource allocation, cooperation will result in a poor lifetime of the heavily used nodes. In contrast, the proposed framework will not only guarantee fairness, but will also provide significant throughput and diversity gain over conventional cooperation schemes.

The problem of energy consumption minimization while satisfying given throughput requirements is considered in Cui et al. (2006). For such networks, the cross-layer design coupled with the node cooperation can significantly reduces both energy consumption and delay. A cooperative Multiple-Input Multiple-Output (MIMO) technique is proposed where multiple nodes within a cluster cooperate in signal transmission and/or reception. In this scheme, local information exchange within the cluster is not necessary if Alamouti codes are used with appropriate transmission scheduling. A cross-layer design framework is then applied that optimizes routing to minimize the energy consumption and delay. For the cooperative MIMO scheme, routing is optimized based on an equivalent Single-Input Single-Out (SISO) system, where each cooperating cluster is treated as a super node. The best energy-delay tradeoff curve is derived under this optimization model and the cooperative MIMO approach is shown in Figures 10a and 10b, where we see that the optimal curve for the cooperative MIMO system (Coop MIMO) is strictly below that of the non-cooperative system (non-coop), which means cooperative MIMO schemes can reduce both energy and delay. For both the case where the circuit processing energy is included and the case where it is not included, we see that the two curves converge to the same point on the far right of the curves, which corresponds to the scenario where the source node transmits all the packets directly to the destination node. Consequently, the cooperative MIMO system can dramatically improve energy and delay performance, especially when link layer adaptation is used.

A cross-layer optimization framework for multihop wireless networks using cooperative diversity is proposed in Long et al. (2008), which provide solutions to fundamental relaying problems of determining who should be relays for whom and how to perform resource allocation for these relaying schemes jointly with routing and congestion control such that the system performance is optimized. A fully distributed algorithm is presented where the joint routing, relay selection, and power allocation problem to minimize network power consumption is solved by using convex optimization. By the dual decomposition, the master optimization problem is decomposed

Figure 10. The optimal energy-delay tradeoff curves for both the non-cooperative and cooperative MIMO systems: (a) transmission energy only (b) circuit energy included (from Cui, et al., 2006)

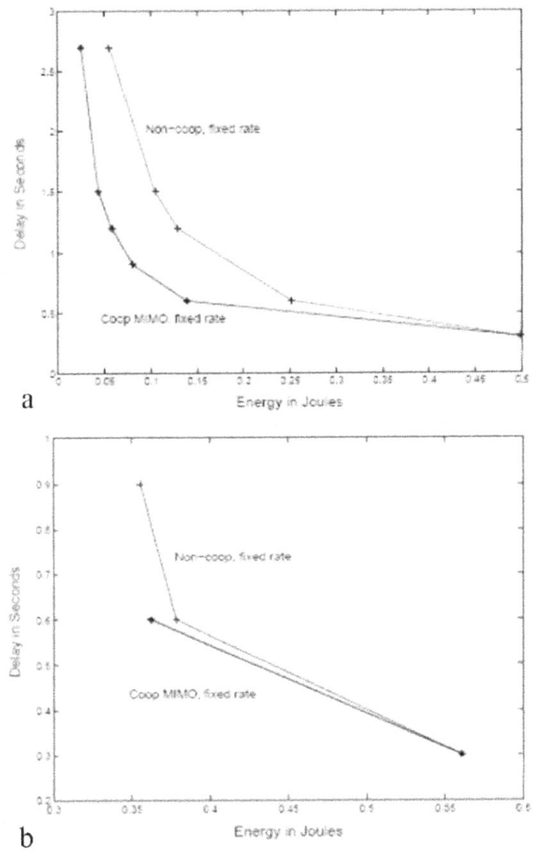

into a routing sub-problem in the network layer and a joint relay selection and power allocation sub-problem in the physical layer, which can be solved efficiently in a distributed manner. Then the framework is extended to incorporate congestion control and develop a framework for optimizing the sum rate utility and power tradeoff for wireless networks using cooperative diversity. The numerical results (Figure 11) show the convergence of the proposed algorithms and significant improvement in terms of power consumption and source rates due to cooperative diversity.

Considering a general framework for lifetime extension of WSN by exploiting cooperative diversity, it efficiently takes advantage of different locations and energy levels among distributed nodes (Himsoon, et al., 2007). In Himsoon et al. (2007), a lifetime maximization problem via cooperative nodes is considered and a performance analysis for M-ary PSK modulation is provided. With an objective to maximize the minimum device lifetime under a constraint on bit-error- rate performance, the optimization problem determines which nodes should cooperate and how much power should be allocated for cooperation. Since the formulated problem is NP hard, a closed-form solution for a two-node network is derived to obtain some insights. Based on the two-node solution, a fast suboptimal algorithm is developed for multi-node scenarios. Moreover, the device lifetime is further improved by a deployment of cooperative relays in order to help forward information of the distributed nodes in the network. Optimum location and power allocation for each cooperative relay are determined with an aim to maximize the minimum device lifetime. A sub-optimal algorithm is developed to solve the problem with multiple cooperative relays and cooperative nodes. Simulation results show that the minimum device lifetime of the network with cooperative nodes improves 2 times longer than the lifetime of the non-cooperative network. In addition, deploying a cooperative relay in a proper location leads up to 12 times longer lifetime than that of the non-cooperative network.

In Yuan et al. (2006), a novel multi-hop virtual MIMO communication protocol is proposed by the cross layer design to jointly improve the energy efficiency, reliability and End-To-End (ETE) QoS provisioning in Wireless Sensor Network (WSN). In the protocol, the traditional LEACH protocol is extended by incorporating the cooperative MIMO communication, multi- hop routing and hop-by-hop recovery schemes. Based on the protocol, the overall energy consumption per packet transmission is modeled and the optimal

Figure 11. Total power consumption versus source rate with and without cooperative diversity (from Long, et al., 2008)

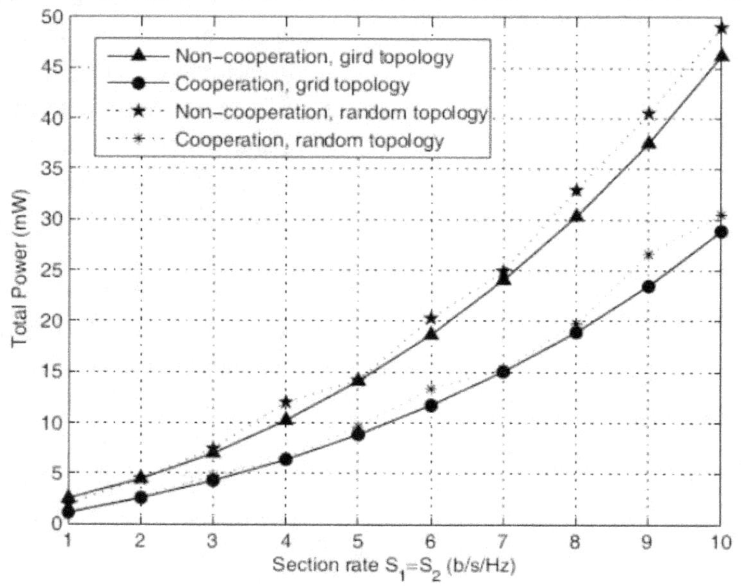

set of transmission parameters is found. Then, the issues of ETE QoS provisioning of the protocol are considered. The ETE latency and throughput of the protocol are modeled in terms of the BER performance of each link. Then, a nonlinear constrained programming model is developed to find the optimal BER performance of each link to meet the ETE QoS requirements with minimum energy consumption, and the Particle Swarm Optimization (PSO) algorithm is employed to solve the problem. Simulation results show the effectiveness of the proposed protocol in energy saving and QoS provisioning.

It is demonstrated through simulations that an energy efficiency trade-off exists in cooperation, and the trade-off depends on several parameters such as the average inter-node distance, the required QoS, the number of relays, the receiving power and so on (Yuan, et al., 2006; Chen, et al., 2010). Results also show that the proposed cross-layer strategy achieves significant energy savings compared to other recently proposed algorithms. The proposed strategies turned out to achieve better energy performance than other existing cooperative routing schemes. The impact of several parameters on the energy efficiency of the cooperative communication is also investigated in wireless sensor networks, thus can be used to provide guidelines to decide when and how to apply cooperation for a given setup. Simulation results (Figures 12-13) (Chen, et al., 2010) also show that the network lifetime of the cooperative network can be increased to 2-10 times when the BER constraint is 10^{-2} to 10^{-3}, and the improvement of network lifetime become more significant as the BER constraint is tighter. The reason not only lies on the reduced transmission power, but also on the more balanced traffic.

CROSS-LAYER SOLUTIONS FOR APPLICATIONS IN WIRELESS SENSOR NETWORKS

In this section, we briefly review how cross-layer design has been adopted on application of

Figure 12. Network lifetime comparison between different algorithms according to different BER targets (from Chen, et al., 2010)

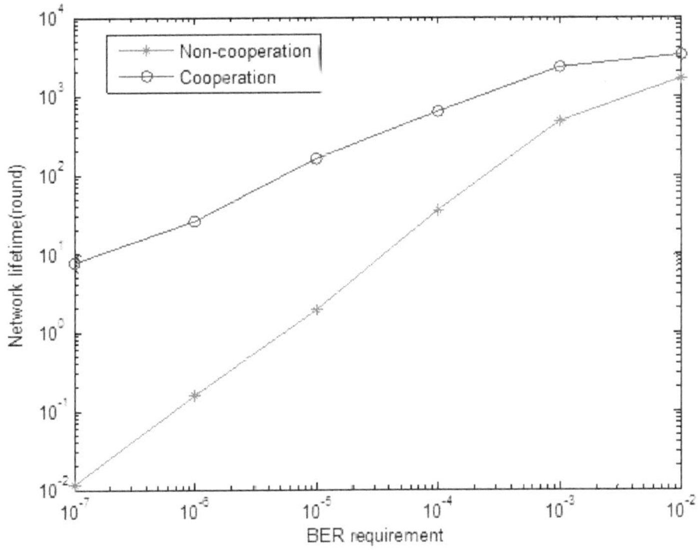

Figure 13. Network lifetime comparison between different algorithms according to different network density (from Chen, et al., 2010)

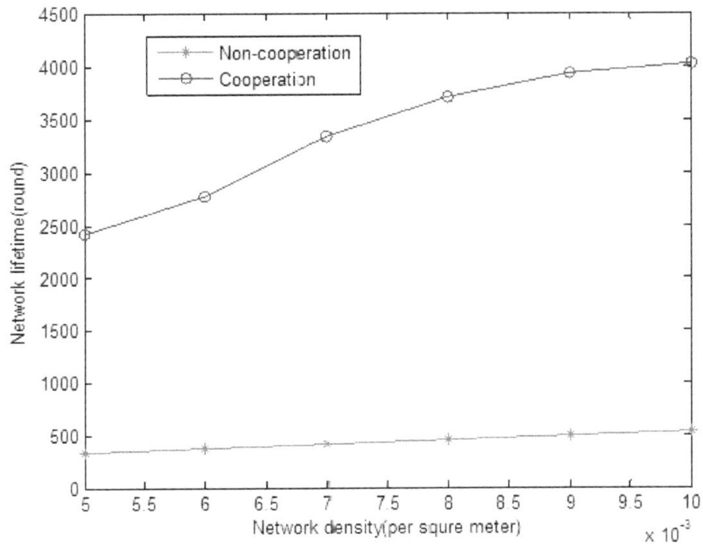

Wireless Sensor Networks (WSNs) to improve their performance by discussing the state-of-the-art literature on the subject such as multimedia streaming, real time and data fusion.

Multimedia Streaming Applications

The integration of low-power wireless networking technologies with inexpensive hardware such as Complementary Metal–Oxide Semiconductor (CMOS) cameras and microphones is now enabling the development of distributed, networked systems that we refer to as Wireless Multimedia Sensor Networks (WMSNs) (Akyildiz, et al., 2007). That is, networks of wireless, interconnected smart devices that enable retrieving video and audio streams, still images, and scalar sensor data. Most potential applications of a WMSN require the sensor network paradigm to be rethought to provide mechanisms to deliver multimedia content with a predetermined level of Quality of Service (QoS). Whereas minimizing energy consumption has been the main objective in the sensor network research, mechanisms to efficiently deliver application-level QoS and to map these requirements to network-layer metrics, such as latency and jitter, have not been primary concerns. Because of the shared nature of the wireless communication channel, in multihop wireless networks, there is a strict interdependence among functions handled at all layers of the communication stack. This interdependence must be explicitly considered when designing communication protocols aimed at QoS provisioning.

Joint capacity and flow assignment at the MAC and network layers is investigated in Adlakha et al. (2007), which include rate allocation at the application layer of each user. The proposed optimization aims to minimize the tradeoff between encoded video quality of all users versus overall network congestion. Compared to a scheme with oblivious layers, where capacity, flow, and video rates are assigned individually, simulation results show significant performance gain of the proposed cross-layer approach, in terms of maximum sustainable rate and quality of the video streams.

Congestion in wireless multimedia sensor networks not only causes packet loss, but also leads to excessive energy consumption. Therefore, congestion in WMSNs needs to be controlled in order to prolong system lifetime. In addition, this is also necessary to improve fairness and provide better Quality of Service (QoS), which is required by multimedia applications in wireless multimedia sensor networks. In Wang et al. (2007), a novel upstream congestion control protocol for WSNs is proposed, called Priority-Based Congestion Control Protocol (PCCP). Unlike existing work, PCCP innovatively measures congestion degree as the ratio of packet inter-arrival time along over packet service time. PCCP still introduced node priority index to reflect the importance of each sensor node. Based on the introduced congestion degree and node priority index, PCCP utilizes a cross-layer optimization and imposes a hop-by-hop approach to control congestion. It is demonstrated that PCCP achieves efficient congestion control and flexible weighted fairness for both single-path and multi-path routing. As a result, this leads to higher energy efficiency and better QoS in terms of both packet loss rate and delay.

Melodia and Akyildiz (2010) have proposed a Cross-Layer Control Unit (XLCU) to provide QoS for WMSNs. They have considered geographical forwarding to control the flows and hop-by-hop QoS requirements. Also, the XLCU controls the scheduling and rate assignment of the time hopping impulse radio UWB (TH-IR-UWB), preventing collisions and idle listening. Thus, throughput and reliability are increased and delay is reduced, according to the results.

MPEG and H.26 are the most used compression standards for video. The compression they provide allows more video to be stored and less bandwidth for video transmission is required in Chen et al. (2006). Chen et al. (2006) have proven that the IEEE 802.15.3 technology is not suitable for Variable Bitrate (VBR) video transmission.

Hence, they have proposed the Energy Diffuser Application-Aware Scheduling (EDAS). This cross-layer algorithm increases the quality of the received video by increasing the fraction of decodable frames to more than 85%, while the IEEE 802.15.3 achieves a worst case of 65%. Moreover, energy consumption is reduced up to 30% by compressing video with lower quality where QoS admits it, thus transmitting less data, and transmitting higher video quality where QoS constraints are stricter. Also, EDAS includes an admission control mechanism to provide the necessary QoS to all the video flows in transmission.

Cross-Layer Design for Real-Time QoS in WSNs

Supporting real-time QoS in sensor networks faces severe challenges due to the wireless nature, limited resource, low node reliability, distributed architecture, and dynamic network topology. There is trade-off between different application requirements including energy efficiency and delay performance. The following designs have provided real-time guarantee in cross-layer architecture.

Figure 14 illustrates a simplified cross-layer framework for real-time-WSNs (He, et al., 2006). Following the typical communication stack, parameters in lower layer are reported to adjacent higher layer. As shown in Figure 5, the cross layer design enables a higher layer to coordinate the behavior of lower layers. Parameters of distant layers can have interactions as well. For example, the application layer may have QoS requirement like end-to-end delay bound and service differentiation, which places a delay constraint and have priority scheduling in routing layer and MAC layer. Physical layer determines the amount of energy spent in transmission and circuit. It also provides the Link Quality Indication (LQI) to the MAC, routing and application layers. LQI can provide a reference for the scheduling at MAC layer, reflect the reliability and robustness of routing path, and also affect the source coding at the application layer. Besides, the active/idle/sleep action at MAC layer is able to optimize the performance of routing layer And there are impacts on one another significantly.

Real-Time Architecture and Protocols (RAP) (Lu, et al., 2002) is a cross-layer real-time communication architecture for large-scale WSNs. It is supported by a scalable and efficient protocol stack, which integrates a transport-layer Location-Addressed Protocol (LAP), a geographic routing protocol, Velocity Monotonic Scheduling (VMS) policy and a contention-based MAC with packet

Figure 14. Cross layer design for real-time WSNs (from He, et al., 2006)

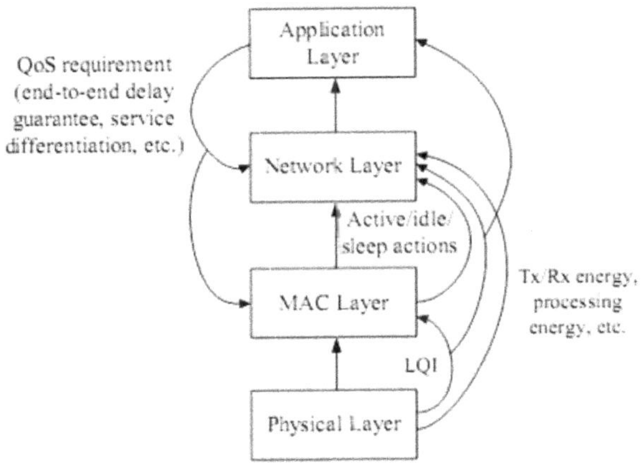

prioritization. The cornerstone of RAP is the VMS policy, which is based on packet requested velocity that reflects both distance and timing constraints. VMS reduces end-to-end deadline miss ratio by giving higher priority to packets with higher requested velocities.

MAC and Efficient Routing Integrated with Support for Localization (MERLIN) (Ruzzelli, et al., 2006) is a lightweight protocol that integrates routing and MAC protocols together to support energy efficiency and low latency in WSNs. Its MAC follows a hybrid TDMA/CSMA approach in which slot contention needs to be conducted at the beginning of each slot. For routing, the network is divided into time zones defined by the number of hops in a schedule table, by which nodes forward messages to the gateway over multiple hops. The localization procedure is based on triangulation and distance between pairs of nodes. The result shows MERLIN outperforms an integration of S-MAC and ESR (Eyes Source Routing) (Wu, et al., 2004) in both energy efficiency and latency. The delay performance of MERLIN should be investigated more specifically.

VigilNet (Wu, et al., 2004) is a large-scale real-time sensor system for target tracking, detection and classification, which aims at both timeliness and energy efficiencies. Its network infrastructure is a multi-path diffusion tree rooted at bases. B-MAC is the default MAC protocol. Deadline partition method is used to guarantee an end-to-end tracking deadline by satisfying a set of sub-deadlines. Multi-dimensional tradeoffs between real-time performance and other system properties are investigated.

Cross-Layer Solutions for Data Fusion Applications

Data fusion (also known as data aggregation) is a technique used to reduce measurement errors. In this case, data are not encoded to reduce the number of transmitted bits. Instead, data received from several sensors are compared to search for errors. Thus, a reliable approximation of the measured phenomena is achieved. Liang et al. (2007) have proposed a cross-layer solution to enhance data fusion and the network lifetime. They have used sensors Channel State Information (CSI) to schedule transmissions only to the sensors experiencing better channels, and they have also used this information to weight the reliability of a sensor measurement when performing data fusion. Hence, preventing sensors experiencing bad channels from transmitting, the bit error rate could be reduced up to 1000 times when compared to the maximum ratio combining fusion algorithm, thus wasting less energy. Data aggregation has also been used by the GRASS protocol, proposed by Al-Karaki et al. (2009). In the analyzed scenario, clusters are formed and the Cluster Heads (CHs) aggregate the data coming from the sensors in the cluster. Then, the CHs are responsible to find a route to the base station that maximizes the network lifetime. Results have shown a gain of at least 35% on the lifetime when compared to the directed diffusion protocol. Also, to avoid limiting the network lifetime to the CHs lifetime, new CHs are elected periodically inside each cluster.

OPEN ISSUES AND PRECAUTION ON CROSS-LAYER WSN

As explained in Sections 2 and 3, there exists some research on cross-layer interactions and design in developing new communication protocols for WSNs. However, there is still much to be gained by rethinking the protocol functions of network layers in a unified way so as to provide a single communication module for efficient communication in WSNs. In other words, the cross-layer approach emerged recently, still necessitates a unified cross-layer communication protocol for efficient and reliable event communication that considers transport, routing, and medium access functionalities with physical layer (wireless channel) effects for WSNs. There are several open research problems toward the development of systematic techniques for the cross-layer design of wireless sensor network protocols (Tommaso, et al., 2006):

1. **Identify Adequate Utility Functions:** A thorough study is needed to identify utility functions that: 1) represent the desired global design objectives of sensor networks, such as minimal energy consumption and maximum network lifetime; 2) exhibit particular properties, e.g., convexity, that allow to ðnd an unique global optimum with efficient methods and developing distributed implementations.
2. **Improved Understanding of Energy Consumption and Transmission Delay:** Existing studies on cross-layer optimization are mostly focused on jointly optimizing functionalities at different layers, usually with the overall objective of maximizing the network throughput. Conversely, in WSNs, the ultimate objective is usually to minimize the energy consumption and/or to maximize the network lifetime. Hence, further study is needed to develop models and methodologies suitable to solve energy-oriented problems. There is a need to develop sound models to include in the above framework an accurate description of the end-to-end delay as results from the interaction of the different layers. This is particularly important for the design of sensor network protocols for monitoring applications that require real-time delivery of event data, such as those encountered in Wireless Sensor and Actor Networks (WSAN) (Akyildiz, et al., 2004).
3. **Connectivity with Realistic Physical Layer:** Connectivity in wireless networks has been previously studied (Christian, et al., 2002), i.e., stochastic models have been developed to determine conditions under which a network is connected. These results, however, cannot be straightforwardly used, as they are based on the so-called unit disk graph communication model. However, recent experimental studies have demonstrated that the effects of the impairments of the wireless channel on higher-layer protocols are not negligible, as the availability of links further fluctuates because of channel fading phenomena that affect the wireless transmission medium. Furthermore, mobility of nodes is not considered. In fact, due to node mobility and node join and leave events, the network may be subject to frequent topological reconfigurations. Thus, links are continuously established and broken. For the above reasons, new analytical models are required to determine connectivity conditions that incorporate mobility and fading channels.
4. **Sensors Mobility**: Sensors mobility are rarely considered in cross layer design for WSN. Only a few researches consider sensors mobility so far. It has been argued that the researches of new routing protocols is not an easy task, and by considering mobility more complexity is expected. While low mobility is experimented by tracked assets in a warehouse, a vehicular traffic sensoring solution needs to cope with high mobility. This is certainly one of the larger challenges. The high complexity of the proposed routing protocols for WSNs is increased by considering mobility, demanding a great amount of effort from researchers to fill this gap.
5. **Cross-Layer Network Simulators and Testbed:** Cross layer network simulator architecture is based on a framework approach. This will allow for a maximal flexibility and adaptability at all levels. Although the advances in Cross layer network simulators have achieved such as OMNeT++ network simulator (Andras, et al., 2008) and Multi InteRfAce Cross Layer Extension for ns2 (MIRACLE) (Baldo, et al., 2007),. typical complex simulators have shown to spend

most of the computing time in event handling. Thus, this part of the simulator has to be carefully designed and optimized, but in the case of cross layer systems, the kind of designs and optimisations required depends on the part of the model being simulated. In order to allow several kinds of optimisation to be combined, cross layer simulator kernel will allow for transparent mixed-mode multi-level kernels. Current discrete-event network simulators such as OPNET (http://www.opnet.com/), NS-2 (http://www.isi.edu/nsnam/ns/), J-Sim (http://www.j-sim.org/), GloMoSim (http://pcl.cs.ucla.edu/projects/glomosim/) may be unsuitable to implement a cross-layer solution, since their inner structure is based on a layered architecture, and each implemented functionality run by the simulator engine is tightly tied to this architecture. Hence, implementing a cross-layer solution in one of these simulators may turn into a non-trivial task. For this reason, there is a need to develop new software simulators that are based on a new developing paradigm so as to ease the development and test of cross-layer algorithmic and protocol solutions. Although many researches on cross layer WSN have existed, none of the cross-layer solutions has been deployed on testbeds. Comparisons between the simulated or calculated WSN performances and results achieved in testbeds can be a source for further routing protocol research and development.

6. **Identify Efficient Routing Hierarchy:** The most efficient routing hierarchy of cross-layer WSNs has not been identified. Some solutions suppose it is the clustered hierarchy, others use single-hop approaches, and also flat hierarchies can be found. Thus, each hierarchy considered needs to be assessed in terms of the network objective in order to identify which one results in higher throughput, less end-to-end delay, and so on.

We described several open research problems toward the development of systematic techniques for cross-layer design of wireless sensor network protocols above. However, possible risks are raising when a cross-layer approach is followed, and propose precautionary guidelines and principles for cross-layer design beyond the open research issues are presented, where the pros and cons of cross layer design approach are evaluated in Goldsmith et al. (2002), In the following, some concerns and precautionary considerations, which need to be considered when a cross-layer design architecture is proposed, and suggest some possible research directions (Tommaso, et al., 2006):

1. **Modularity**: In the classical layered design approach, system architecture is broken down into modular components, and the interactions and dependencies between these components are systematically specified. This design philosophy allows to break complex problems into easier subproblems, which can then be solved in isolation, without considering all the details pertaining the overall system. This approach guarantees the inter-operability of subsystems in the overall system once each subsystem is tested and standardized, leading to quick proliferation of technology and mass production. Conversely, a cross-layer design approach may lose the decoupling between design and development process, which may impair both the design and the implementation development and slow the innovation down.
2. **System Enhancement**: Design improvements and innovations may become difficult in a cross-layer design, since it will be hard to assess how a new modification will interact with the already existing solutions. Furthermore, cross-layer architecture would be hard to upkeep, and the maintaining costs would be high. In the worst cases, rather than modifying just one subsystem, the entire system may need to be replaced. For

these reasons, we advocate keeping some degree of modularity in the design of cross-layer solutions. This could be achieved by relying on functional entities - as opposed to layers in the classical design philosophy - that implement particular functions. This would also have the positive consequence of limiting the duplication of functions that often characterizes a layered design. This functional redundancy is, in fact, one the cause for poor system performance.

3. **Instability:** In cross-layer design, the effect of any single design choice may affect the whole system, leading to various negative consequences such as instability. This is a non trivial problem to solve, since it is well known from control theory that stability is a paramount issue. Moreover, the fact that some interactions are not easily foreseen makes cross-layer design choices even trickier. Hence, great care should be paid to prevent design choices from negatively affecting the overall system performance. To this purpose, there is a need to integrate and further develop control theory techniques to study stability properties of system designed following a cross-layer approach. Dependency graphs, which may be used to capture the dependency relation between parameters, could be valuable means to prove stability, although hard to implement in some cases.

CONCLUSION

In this chapter, we reviewed and classified literatures on cross-layer protocols, improvements, and design methodologies for WSNs and focus on cross-layer design techniques to apply for the communication protocols devised for WSNs.

These techniques are based on the network layers they aim at replacing in the classical OSI network stack. Furthermore, we discussed systematic methodologies in the optimal design of cross-layer solution for WSNs as resource allocation problems in the framework of non-linear optimization and the cross-layer design with cooperative communication for WSNs. We outlined open research issues and precaution in the development of cross-layer methodologies for WSNs and discussed possible research directions. A cross-layer design methodology for energy-constrained WSNs is an appealing approach as long as cross-layer interactions are thoroughly studied and controlled. Consequently, we can come to a conclusion that cross layer techniques are extremely useful and are capable of providing high performance gain for WSNs.

REFERENCES

Adlakha, S., Zhu, X., Girod, B., & Goldsmith, A. (2007). Joint capacity, flow and rate allocation for multiuser video streaming over wireless ad-hoc networks. In *Proceedings of the IEEE International Conference on Communications (ICC 2007)*, (pp. 1747-1753). IEEE Press.

Akyildiz, I. F., & Kasimoglu, I. H. (2004). Wireless sensor and actor networks: Research challenges. *Ad Hoc Networks Journal*, *2*(4), 351–367. doi:10.1016/j.adhoc.2004.04.003

Akyildiz, I. F., Melodia, T., & Chowdury, K. R. (2007). Wireless multimedia sensor networks: A survey. *IEEE Wireless Communications*, *14*(6), 32–39. doi:10.1109/MWC.2007.4407225

Akyildiz, I. F., Su, W., Sankarasubramaniam, Y., & Cayirci, E. (2002). Wireless sensor networks: A survey. *Computer Networks*, *38*(4), 393–442. doi:10.1016/S1389-1286(01)00302-4

Akyildiz, I. F., Vuran, M. C., & Akan, O. B. (2006). A cross-layer protocol for wireless sensor networks. In *Proceedings of IEEE Conference on Information Sciences and Systems (CISS 2006)*, (pp. 1102-1107). IEEE Press.

Andras, V. Hornig, & Rudolf. (2008). An overview of the OMNeT++ simulation environment. In *Proceedings of the 1st International Conference on Simulation Tools and Techniques for Communications, Networks and Systems (SIMUTools 2008),* (pp. 209-214). IEEE.

Baldo, N., Maguolo, F., Miozzo, M., Rossi, M., & Zorzi, M. (2007). NS2-MIRACLE: A modular framework for multi-technology and cross-layer support in network simulator. In *Proceedings of the ACM International Workshop on Network Simulation Tools (NSTOOLS)*, (pp. 1131-1140). ACM Press.

Baronti, P., Pillai, P., Chook, V., Chessa, S., Gotta, A., & Hu, Y. F. (2007). Wireless sensor networks: A survey on the state of the art and the 802.15.4 and ZigBee standards. *Computer Communications*, *30*(7), 1655–1695. doi:10.1016/j.comcom.2006.12.020

Boyd, S., & Vandenberghe, L. (2003). *Convex optimization.* Cambridge, UK: Cambridge University Press.

Chang, J. H., & Tassiulas, L. (2000). Energy conserving routing in wireless ad hoc networks. In *Proceedings of IEEE International Conference on Computer Communications (INFOCOM 2000)*, (pp. 22–31). IEEE Press.

Chen, W., Dai, L., Letaief, B., & Cao, Z. (2008). A unified cross-layer framework for resource allocation in cooperative networks. *IEEE Transactions on Wireless Communications*, *7*(8), 3000–3012. doi:10.1109/TWC.2008.060831

Chen, Y., Yi, W., & Yang, Y. (2010). A cross-layer strategy for energy-efficient cooperative communication in wireless sensor network. In *Proceedings of The 1st IET International Conference on Wireless Sensor Network (IET-WSN 2010),* (pp. 203 – 208). IET Press.

Chen, Y., Yi, W., & Yang, Y. (2010). A cooperative routing algorithm for lifetime maximization in wireless sensor networks. In *Proceedings of The 1st IET International Conference on Wireless Sensor Network (IET-WSN 2010)*, (pp. 167-172). IET Press.

Cheng, M., Gong, X., & Cai, L. (2009). Joint routing and link rate allocation under bandwidth and energy constraints in sensor networks. *IEEE Transactions on Wireless Communications*, *8*(7), 3770–3779. doi:10.1109/TWC.2009.081134

Chiang, M. (2005). Geometric programming for communication systems. *Short Monograph in Foundations and Trends in Communications and Information Theory*, *2*, 1–154. doi:10.1561/0100000005

Christian, B. (2002). On the minimum node degree and connectivity of a wireless multihop network. In *Proceedings of ACM International Symposium on Mobile Ad Hoc Networking and Computing (MobiHoc)*, (pp. 80-91). ACM Press.

Cruz, R. L., & Santhanam, A. V. (2003). Optimal routing, link scheduling and power control in multi-hop wireless networks. In *Proceedings of IEEE International Conference on Computer Communications (INFOCOM),* (pp. 13-23). IEEE Press.

Cui, S., & Goldsmith, A. J. (2006). Cross-layer design of energy-constrained networks using cooperative MIMO techniques, *Eurasip's Signal Processing Journal.* Retrieved from http://www.ece.tamu.edu/~cui/publications/eurasip_05.pdf.

Cui, S., Madan, R., Goldsmith, A., & Lall, S. (2005). Joint routing, MAC, and link layer optimization in sensor networks with energy constraints. In *Proceedings of IEEE International Conference on Communications (ICC 2005),* (Vol. 2), (pp. 725 – 729). IEEE Press.

Cui, S., Madan, R., Goldsmith, A. J., & Lall, S. (2007). Cross-layer energy and delay optimization in small-scale sensor networks. *IEEE Transactions on Wireless Communications*, *6*(10), 3688–3699. doi:10.1109/TWC.2007.060072

Dai, L., & Letaief, K. B. (2005). Cross-layer design for combining cooperative diversity with truncated ARQ in ad-hoc networks. In *Proceedings of IEEE Global Telecommunications Conference (IEEE Globecom),* (pp. 3175–3179). IEEE Press.

Fang, Y., & McDonald, B. (2004). Dynamic codeword routing (DCR): A cross-layer approach for performance enhancement of general multi-hop wireless routing. In *Proceedings of IEEE Communications Society Conference on Sensor, Mesh and Ad Hoc Communications and Networks (SECON 2004),* (pp. 255 – 263). IEEE Press.

Ferrara, D., et al. (2005). MACRO: An integrated MAC/routing protocol for geographical forwarding in wireless sensor networks. In *Proceedings of IEEE International Conference on Computer Communications (Infocom 2005),* (Vol. 3), (pp. 1770 – 1781). IEEE Press.

Ganesan, D. (2002). *An empirical study of epidemic algorithms in large scale multihop wireless networks. Technical Report*. Palo Alto, CA: Intel Research.

Gill, P. E., Murray, W., & Wright, M. H. (1981). *Practical optimization*. New York, NY: Academic Press.

GloMoSim. (2011). *Website.* Retrieved Nov 23, 2011, from http://pcl.cs.ucla.edu/projects/glomosim/.

Goldsmith, A. J., & Wicker, S. B. (2002). Design challenges for energy constrained ad hoc wireless networks. *IEEE Wireless Communications Magazine*, *8*, 8–27. doi:10.1109/MWC.2002.1028874

Gupta, P., & Kumar, P. R. (1998). Critical power for asymptotic connectivity in wireless networks. In *Proceedings of Stochastic Analysis, Control, Optimization and Applications*, (Vol. 1), (pp. 1106-1110). IEEE.

Haapola, J., Shelby, Z. C., & Mahonen, P. (2005). Cross-layer energy analysis of multi-hop wireless sensor networks. In *Proceedings of European Workshop on Wireless Sensor Networks (EWSN 2005),* (pp. 33 – 44). EWSN.

He, T., Gu, L., Luo, L., Yan, T., Stankovic, J., Abdelzaher, T., & Son, S. (2006). An overview of data aggregation architecture for real-time tracking with sensor networks. In *Proceedings of IEEE Real-Time and Embedded Technology and Applications Symposium (RTAS),* (pp. 55–66). IEEE Press.

Heinzelman, W. B., Chandrakasan, A. P., & Balakrishnan, H. (2002). An application-specific protocol architecture for wireless microsensor networks. *IEEE Transactions on Wireless Communications*, *1*(4), 660–670. doi:10.1109/TWC.2002.804190

Himsoon, T., Pam, W. S., Han, Z., & Liu, K. J. R. (2007). Lifetime maximization via cooperative nodes and relay deployment in wireless networks. *IEEE Journal on Selected Areas in Communications*, *25*(2), 306–317. doi:10.1109/JSAC.2007.070207

Hoesel, L. V., Nieberg, T., Wu, J., & Havinga, P. J. M. (2004). Prolonging the lifetime of wireless sensor networks by cross-layer interaction. *IEEE Wireless Communications*, *11*(6), 78–86. doi:10.1109/MWC.2004.1368900

J-Sim. (2011). *Website.* Retrieved Nov 23, 2011, from http://www.j-sim.org/.

Karaki, J. N., Mustafa, R., & Kamal, A. E. (2009). Data aggregation and routing in wireless sensor networks: Optimal and heuristic algorithms. *Computer Networks, 53*(7), 945–960. doi:10.1016/j.comnet.2008.12.001

Karl, H., & Willig, A. (2006). *A short survey of wireless sensor networks*. TKN Technical Report: TKN-03-018. Retrieved from http://www.cs.uni-paderborn.de/fileadmin/Informatik/AG-Karl/Publications/Technical_Report/TechReport_03_018.pdf.

Kozat, U. C., Koutsopoulos, I., & Tassiulas, L. (2004). A framework for cross-layer design of energy-efficient communication with QoS provisioning in multi-hop wireless networks. In *Proceedings of IEEE International Conference on Computer Communications (INFOCOM 2004)*, (pp. 1-5). IEEE Press.

Kwon, H., Kim, T. H., Choi, S., & Lee, B. G. (2006). A cross-layer strategy for energy-efficient reliable delivery in wireless sensor networks. *IEEE Transactions on Wireless Communications, 5*(12), 3689–3699. doi:10.1109/TWC.2006.256992

Laneman, J. N., Tse, D. N. C., & Wornell, G. W. (2004). Cooperative diversity in wireless networks: Efficient protocols and outage behavior. *IEEE Transactions on Information Theory, 50*(12), 3062–3080. doi:10.1109/TIT.2004.838089

Laneman, J. N., & Wornell, G. W. (2003). Distributed space-time-coded protocols for exploiting cooperative diversity in wireless networks. *IEEE Transactions on Information Theory, 49*(10), 2415–2425. doi:10.1109/TIT.2003.817829

Liang, Q., Yuan, D., Wang, Y., & Chen, H.-H. (2007). A cross-layer transmission scheduling scheme for wireless sensor networks. *Computer Communications, 30*(15), 2987–2994. doi:10.1016/j.comcom.2007.05.025

Lin, X. J., Shroff, N. B., & Srikant, R. A. (2006). Tutorial on cross-layer optimization in wireless networks. *IEEE Journal on Selected Areas in Communications, 24*(8), 1452–1463. doi:10.1109/JSAC.2006.879351

Long, L., & Hossain, E. (2008). Cross-layer optimization frameworks for multihop wireless networks using cooperative diversity. *IEEE Transactions on Wireless Communications, 7*(7), 2592–2602. doi:10.1109/TWC.2008.060962

Lu, C., Blum, B. M., Abdelzaher, T. F., Stankovic, J. A., & He, T. (2002). RAP: A real-time communication architecture for large-scale wireless sensor networks. In *Proceedings of IEEE Real-Time and Embedded Technology and Applications Symposium (RTAS)*, (pp. 55–66). IEEE Press.

Luo, J., Blum, R., Cimini, L., Greenstein, L., & Haimovich, A. M. (2005). Power allocation in a transmit diversity system with mean channel gain information. *IEEE Communications Letters, 9*(7), 616–618. doi:10.1109/LCOMM.2005.1461682

Luo, J., Blum, R. S., Greenstein, L. J., Cimini, L. J., & Haimovich, A. M. (2004). New approaches for cooperative use of multiple antennas in ad-hoc wireless networks. In *Proceedings of IEEE Vehicular Technology Conference (VTC)*, (pp. 2769–2773). IEEE Press.

Madan, R., Cui, S., & Lall, S. (2005). Cross-layer design for lifetime maximization in interference-limited wireless sensor networks. In *Proceedings of IEEE International Conference on Computer Communications (INFOCOM 2005)*, (Vol. 3), (pp. 964 – 975). IEEE Press.

Melodia, T., & Akyildiz, I. F. (2010). Cross-layer QoS-aware communication for ultra wide band wireless multimedia sensor networks. *IEEE Journal on Selected Areas in Communications, 28*(5), 653–663. doi:10.1109/JSAC.2010.100604

Melodia, T., Pompili, D., & Akyildiz, I. F. (2005). On the interdependence of distributed topology control and geographical routing in ad hoc and sensor networks. *Journal of Selected Areas in Communications*, *23*(3), 520–532. doi:10.1109/JSAC.2004.842557

Messier, G. G., Hartwell, J. A., & Davies, R. J. (2008). A sensor network cross-layer power control algorithm that incorporates multiple-access interference. *IEEE Transactions on Wireless Communications*, *7*(8), 2877–2883. doi:10.1109/TWC.2008.070208

Misra, S., Reisslein, M., & Xue, G. (2008). A survey of multimedia streaming in wireless sensor networks. *IEEE Communications Surveys & Tutorials*, *10*(4), 18–39. doi:10.1109/SURV.2008.080404

Mitchell, P. D., Qiu, J., Li, H., & Grace, D. (2010). Use of aerial platforms for energy efficient medium access control in wireless sensor networks. *Computer Communications*, *33*(4), 500–512. doi:10.1016/j.comcom.2009.10.015

Nosratina, A., Hunter, T. E., & Hedayat, A. (2004). Cooperative communication in wireless networks. *IEEE Communications Magazine*, *10*(4), 74–80. doi:10.1109/MCOM.2004.1341264

NS-2. (2011). *Website.* Retrieved Nov 23, 2011, from http://www.isi.edu/nsnam/ns/.

OPNET. (2011). *Website.* Retrieved Nov 23, 2011, from http://www.opnet.com/.

Radunovic, B., Jean, Y., & Boudec, L. (2004). Rate performance objectives of multihop wireless networks. *IEEE Transactions on Mobile Computing*, *3*(4), 334–349. doi:10.1109/TMC.2004.45

Ruzzelli, A. G., O'Hare, G. M. P., Grady, M. J. O., & Tynan, R. (2006). MERLIN: A synergetic integration of MAC and routing protocol for distributed sensor networks. In *Proceedings of IEEE Communications Society Conference on Sensor, Mesh and Ad Hoc Communications and Networks (SECON)*, (pp. 266–275). IEEE Press.

Seada, K., Zuniga, M., Helmy, A., & Krishnamachari, B. (2004). Energy-efficient forwarding strategies for geographic routing in lossy wireless sensor networks. In *Proceedings of ACM International Conference on Embedded Networked Sensor Systems (Sensys 2004)*, (pp. 567-575). ACM Press.

Sendonaris, A., Erkip, E., & Aazhang, B. (2003). User cooperation diversity-Part I: System description. *IEEE Transactions on Communications*, *51*(11), 1927–1938. doi:10.1109/TCOMM.2003.818096

Sichitiu, M. L. (2004). Cross-layer scheduling for power efficiency in wireless sensor networks. In *Proceedings of IEEE International Conference on Computer Communications (Infocom 2004)*, (Vol. 3), (pp. 1740 – 1750). IEEE Press.

Skraba, P., Aghajan, H., & Bahai, A. (2004). Cross-layer optimization for high density sensor networks: Distributed passive routing decisions. In *Proceedings of International Conference for Ad Hoc Networks and Wireless (Ad-Hoc Now 2004)*, (pp. 266-279). IEEE.

Stann, F., & Heidemann, J. (2003). RMST: Reliable data transport in sensor networks. In *Proceedings of IEEE International Workshop on Sensor Network Protocols and Applications (SNPA 2003)*, (Vol. 1), (pp. 102-112). IEEE Press.

Tian, Y., & Ekici, E. (2007). Cross-layer collaborative in-network processing in multihop wireless sensor networks. *IEEE Transactions on Mobile Computing*, *6*(3), 297–310. doi:10.1109/TMC.2007.39

Tommaso, M., Vuran, M. C., & Pompili, D. (2006). The state of the art in cross-layer design for wireless sensor networks. In Cesana, M., & Fratta, L. (Eds.), *Wireless Systems/Network Architecture* (pp. 78–92). Berlin, Germany: Springer-Verlag.

Vuran, M. C., Akan, O. B., & Akyildiz, I. F. (2004). Spatio-temporal correlation: Theory and applications for wireless sensor networks. *Computer Networks Journal, 45*(3), 245–261. doi:10.1016/j.comnet.2004.03.007

Vuran, M. C., & Akyildiz, I. F. (2006). Correlation-based collaborative medium access in wireless sensor networks. *IEEE/ACM Transactions on Networking, 14*(2), 316–329. doi:10.1109/TNET.2006.872544

Vuran, M. C., & Akyildiz, I. F. (2008). Cross-layer packet size optimization for wireless terrestrial, underwater, and underground sensor networks. In *Proceedings of IEEE International Conference on Computer Communications (Infocom),* (pp. 226-230). IEEE Press.

Vuran, M. C., & Akyildiz, I. F. (2009). Error control in wireless sensor networks: a cross layer analysis. *IEEE/ACM Transactions on Networking, 17*(4), 1186–1199. doi:10.1109/TNET.2008.2009971

Vuran, M. C., Gungor, V. B., & Akan, O. B. (2005). On the interdependency of congestion and contention in wireless sensor networks. In *International Workshop on Measurement, Modeling, and Performance Analysis of Wireless Sensor Networks* (*SENMETRICS 2005*), (pp. 345-351). SENMETRICS.

Wang, C., Li, B., Sohraby, K., & Daneshmand, M. (2007). Upstream congestion control in wireless sensor networks through cross-layer optimization. *IEEE Journal on Selected Areas in Communications, 25*(4), 786–795. doi:10.1109/JSAC.2007.070514

Wu, J., Havinga, P., Dulman, S., & Nieberg, T. (2004). Eyes source routing protocol for wireless sensor networks. In *Proceedings of European Workshop on Wireless Sensor Networks (EWSN),* (pp. 334-341). EWSN.

Ye, W., Heidemann, J., & Estrin, D. (2004). Medium access control with coordinated adaptive sleeping for wireless sensor networks. *IEEE/ACM Transactions, 12*(3), 493–506.

Yu, W., & Jun, Y. (2005). Joint source coding, routing, and resource allocation for wireless sensor networks. In *Proceedings of IEEE International Conference on Communications (ICC)*, (Vol. 2), (pp. 737-741). IEEE Press.

Yu, Y., Krishnamachari, B., & Prasanna, V. K. (2004). Energy-latency tradeoffs for data gathering in wireless sensor networks. In *Proceedings of IEEE International Conference on Computer Communications (Infocom)*, (Vol. 1), (pp. 7-15). IEEE Press.

Yuan, J., & Yu, W. (2006). Distributed cross-layer optimization of wireless sensor networks: A game theoretic approach. In *Proceedings of IEEE Global Telecommunications Conference (Globecom),* (pp. 1-5). IEEE Press.

Yuan, Y., He, Z., & Chen, M. (2006). Virtual MIMO-based cross layer design for wireless sensor networks. *IEEE Transactions on Vehicular Technology, 55*(3), 856–864. doi:10.1109/TVT.2006.873837

Zhang, L., & Zhang, Y. (2009). Energy-efficient cross-layer protocol of channel-aware geographic- informed forwarding in wireless sensor networks. *IEEE Transactions on Vehicular Technology, 58*(6), 3041–3052. doi:10.1109/TVT.2008.2010046

Zhou, P., Liu, W., Yuan, W., Cheng, W., & Wang, S. (2009). An energy-efðcient cooperative MISO-based routing protocol for wireless sensor networks. In *Proceedings of IEEE Wireless Communications and Networking Conference (WCNC 2009),* (pp. 1-6). IEEE Press.

Zorzi, M. (2004). A new contention-based MAC protocol for geographic forwarding in ad hoc and sensor networks. In *Proceedings of IEEE International Conference on Communications*, (Vol. 6), (pp. 3481 – 3485). IEEE Press.

Zorzi, M., & Rao, R. (2003). Geographic random forwarding (GeRaF) for ad hoc and sensor networks: Multihop performance. *IEEE Transactions on Mobile Computing*, *2*(4), 337–348. doi:10.1109/TMC.2003.1255648

Zorzi, M., & Rao, R. (2004). Geographic random forwarding (GeRaF) for ad hoc and sensor networks: Energy and latency performance. *IEEE Transactions on Mobile Computing*, *2*(4), 349–365. doi:10.1109/TMC.2003.1255650

Zuniga, M., & Krishnamachari, B. (2004). Analyzing the transitional region in low power wireless links. In *Proceedings of IEEE Communications Society Conference on Sensor, Mesh and Ad Hoc Communications and Networks (SECON 2004)*, (Vol. 1), (pp. 517 – 526). IEEE Press.

KEY TERMS AND DEFINITIONS

Cooperative Communication: A communication technique which allows single-antenna nodes in a multi-user scenario can "share" their antennas in a manner that creates a virtual MIMO system to achieve high diversity gains.

Cross-Layer Design: A design approach for a communication system to optimize the control and exchange of information over two or more layers to achieve significant performance improvements by exploiting the interactions between various protocol layers.

Cross-Layer Module: A unified cross-layer protocol which combines the functionalities of the transport, network and medium access protocols into a single module.

Cross-Layer Optimization Design: A cross-layer design approach that solve the cross-layer problem through optimization programming techniques.

Cross-Layer Protocol Stack: A network stack which escape from the pure waterfall-like concept of the OSI communications model with virtually strict boundaries between layers.

Data Fusion: Techniques that combine data from multiple sources and gather that information into discrete, actionable items in order to achieve inferences, which will be more efficient and narrowly tailored than if they were achieved by means of disparate sources.

Energy Consuming Efficiency: The amount of energy required to provide certain products or services.

Lifetime Maximization: A goal of protocol design to maximize the time till the first node of the network "dies" because of energy depletion.

Wireless Sensor Networks: A self-organized wireless network consist of spatially distributed autonomous sensors to monitoring physical or environmental conditions, and to cooperatively send their data through the network to a main location.

Compilation of References

3 GPP TS 36.927 V2.0.0. (2011). *Potential solutions for energy saving for E-UTRAN (Release 10)*. Retrieved from http://www.3gpp.org.

3 GPP. (2006). Technical specification group services and system aspects: IP multimedia subsystem (IMS): Stage 2. *3GPP Std*. Retrieved from http://www.3gpp.org.

3 GPP. (2009). TS36.300: *Evolved universal terrestrial radio access (E-UTRA) and evolved universal terrestrial radio access network (E-UTRAN)*. Retrieved from http://www.3gpp.org.

3 GPP. (2011). *Mobile group*. Retrieved from http://www.3gpp.org/.

Aazami, A., Ghandeharizadeh, S., & Helmi, T. (2004). Near optimal number of replicas for continuous media in ad-hoc networks of wireless devices. In *Proceedings of the Multimedia Information Systems*, (pp. 40–49). College Park, MD: IEEE.

Accelops. (2011). *Accelops*. Retrieved from http://www.accelops.net/.

Adlakha, S., Zhu, X., Girod, B., & Goldsmith, A. (2007). Joint capacity, flow and rate allocation for multiuser video streaming over wireless ad-hoc networks. In *Proceedings of the IEEE International Conference on Communications (ICC 2007)*, (pp. 1747-1753). IEEE Press.

Agarwal, A. K., Wang, W., & McNair, J. Y. (2005). An experimental study of cross-layer security protocols in public access wireless networks. In *Proceedings of the IEEE Global Telecommunications Conference (GLOBECOM 2005)*, (vol. 3), (pp. 1747–1751). IEEE Press.

Agarwal, S., Krishnamurthy, S., Katz, R. H., & Dao, S. K. (2001). Distributed power control in ad-hoc wireless networks. In *Proceedings of PIMRC*. PIMRC.

Ahlswede, R., Cai, N., Li, S.-Y. R., & Yeung, R. W. (2000). Network information flow. *IEEE Transactions on Information Theory*, *46*(4), 1204–1216. doi:10.1109/18.850663

Ahn, K. S. (2008). Performance analysis of MIMO-MRC systems with channel estimation error in the presence of co-channel interferences. *IEEE Signal Processing Letters*, *15*, 445–448. doi:10.1109/LSP.2008.923788

Aïssa, S., & Aniba, G. (2008). BER analysis of M-QAM with packet combining over space-time block coded MIMO fading channels. *IEEE Transactions on Wireless Communications*, *7*(3), 799–805. doi:10.1109/TWC.2008.060745

Akbar, I. A., & Tranter, W. H. (2007). Dynamic spectrum allocation in cognitive radio using hidden Markov models: Poisson distributed case. In *Proceedings of IEEE Southeast Conference*, (pp. 196-201). IEEE Press.

Akbar, I. A., & Tranter, W. H. (2007). Dynamic Spectrum allocation in cognitive radio using hidden Markov models: Poisson distributed case. In *Proceedings of the IEEE Communications Society Conference on Sensor, Mesh and Ad Hoc Communications and Network*, (pp. 196–201). Richmond, VA: IEEE Press.

Akyildiz, I. F., Vuran, M. C., & Akan, O. B. (2006). A cross-layer protocol for wireless sensor networks. In *Proceedings of IEEE Conference on Information Sciences and Systems (CISS 2006)*, (pp. 1102-1107). IEEE Press.

Akyildiz, I. F., Vuran, M. C., & Akan, O. B. (2006). A cross-layer protocol for wireless sensor networks. In *Proceedings of the 40th Annual IEEE Conference on Information Sciences and Systems*, (pp. 1102 - 1107). Princeton, NJ: IEEE Press.

Akyildiz, I. F., & Kasimoglu, I. H. (2004). Wireless sensor and actor networks: Research challenges. *Ad Hoc Networks Journal*, *2*(4), 351–367. doi:10.1016/j.adhoc.2004.04.003

Akyildiz, I. F., Lee, W., Vuran, M. C., & Hohanty, S. (2008). A survey on spectrum management in cognitive radio networks. *IEEE Communications Magazine*, *46*(4), 40–48. doi:10.1109/MCOM.2008.4481339

Akyildiz, I. F., Melodia, T., & Chowdury, K. R. (2007). Wireless multimedia sensor networks: A survey. *IEEE Wireless Communications*, *14*(6), 32–39. doi:10.1109/MWC.2007.4407225

Akyildiz, I. F., Su, W., Sankarasubramaniam, Y., & Cayirci, E. (2002). Wireless sensor networks: A survey. *Computer Networks*, *38*(4), 393–442. doi:10.1016/S1389-1286(01)00302-4

Akyildiz, I. F., Su, W., Sankarasubramaniam, Y., & Cayirci, E. (2002a). A survey on sensor networks. *IEEE Communications Magazine*, *40*(8), 102–114. doi:10.1109/MCOM.2002.1024422

Akyildiz, I. F., & Wang, X. (2008). Cross-layer design in wireless mesh networks. *IEEE Transactions on Vehicular Technology*, *57*(2), 1061–1076. doi:10.1109/TVT.2007.911615

Akyildiz, I. F., Wang, X., & Wang, W. (2005). Wireless mesh networks: A survey. *Computer Networks*, *47*(4), 445–487. doi:10.1016/j.comnet.2004.12.001

Akyildiz, I. F., Xie, J., & Mohanty, S. (2004). A survey of mobility management in next-generation all-IP-based wireless systems. *IEEE Wireless Communications*, *11*(4), 16–28. doi:10.1109/MWC.2004.1325888

Akyildiz, I., Lee, W., Vuran, M., & Mohanty, S. (2006). Next generation/dynamic spectrum access/cognitive radio wireless networks: A survey. *Computer Networks*, *50*(13), 2127–2159. doi:10.1016/j.comnet.2006.05.001

Akyildiz, I., Melodia, T., & Chowdhury, K. (2008). Wireless multimedia sensor networks: Applications and testbeds. *Proceedings of the IEEE*, *96*(10), 1588–1605. doi:10.1109/JPROC.2008.928756

Al Mosawi, T., Shuaib, H., & Aghvami, A. H. (2009). A fast handover scheme based on smart triggers and SIP. In *Proceedings of the IEEE 70th Vehicular Technology Conference Fall (VTC 2009-Fall)*, (pp. 1-5). Anchorage, AK: IEEE Press.

Al Turki, R., & Mehmood, R. (2008). Multimedia ad hoc networks: Performance analysis. In *Proceedings of the Second UKSIM European Symposium on Computer Modeling and Simulation*, (pp. 561-566). UKSIM.

Alamouti, S. (1998). A simple transmit diversity technique for wireless communications. *IEEE Journal on Selected Areas in Communications*, *16*, 1451–1458. doi:10.1109/49.730453

Alfa, A. S. (2010). *Queueing theory for telecommunications - Discrete time modeling of a single node system*. Berlin, Germany: Springer.

Al-Fahoum, A., & Howitt, I. (1999). Combined wavelet transformation and radial basis neural networks for classifying life-threatening cardiac arrhythmias. *Journal of Medical and Biological Engineering and Computing*, *37*(5), 566–573. doi:10.1007/BF02513350

Ali, S., & Yu, F. R. (2009). Cross-layer QoS provisioning for multimedia transmissions in cognitive radio networks. In *Proceedings of the IEEE Wireless Communications and Networking Conference WCNC 2009*, (pp. 1-5). Budapest, Hungary: IEEE Press.

Al-Mualla, M. E., Canagrajah, C. N., & Bull, D. R. (2002). *Video coding for mobile communication*. San Diego, CA: Academic Press.

Alonso, L., & Agusti, R. (2004). Automatic rate adaptation and energy-saving mechanisms based on cross-layer information for packet-switched data networks. *IEEE Communications Magazine*, *42*(3), S15–S20. doi:10.1109/MCOM.2004.1273769

Alouini, M.-S., & Goldsmith, A. J. (2000). Adaptive modulation over Nakagami fading channels. *Wireless Personal Communications*, *13*(1-2), 119–143. doi:10.1023/A:1008979107539

Alouini, M.-S., Tang, X., & Goldsmith, A. J. (1999). An adaptive modulation scheme for simultaneous voice and data transmission over fading channels. *IEEE Journal on Selected Areas in Communications*, *17*(5), 337–350. doi:10.1109/49.768199

Alphand, O., Duda, A., Heusse, M., Ponsard, B., Rousseau, F., & Theoleyre, F. (2010). Towards the future internet of sensors. In *The Internet of Things* (pp. 309–318). New York, NY: Springer. doi:10.1007/978-1-4419-1674-7_30

Alshakhsi, S. A. A., & Hasbullah, H. (2010). Improving QoS of VoWLAN via cross-layer interaction approach. In *Proceedings of the 2010 International Symposium in Information Technology (ITSim),* (pp. 678–682). Kuala Lumpur, Malaysia: ITSim.

Al-Suhail, G., Fleury, M., & Al-Majeed, S. S. (2011). Mobile video streaming over heterogeneous networks. In Khalil, I. (Ed.), *Mobile Multimedia Communications and Applications: New Technologies*. Hershey, PA: IGI Global. doi:10.4018/978-1-60960-563-6.ch013

Altman, E. (1999). *Constrained Markov decision processes: Stochastic modeling*. London, UK: Chapman and Hall/CRC.

Alturki, R. (2011). *Multimedia ad hoc networks: Design, QoS, routing and analysis.* PhD Thesis. Swansea, UK: Swansea University.

Amjed, S. A. F. (2006). Quality assessment of ECG compression techniques using a wavelet-based diagnostic measure. *IEEE Transactions on Information Technology in Biomedicine*, *10*(1), 182–191. doi:10.1109/TITB.2005.855554

An, C., Loyola, L., & Aad, I. (2007). *Analysis of UDP, TCP and voice performance in IEEE 802.11b multihop networks* Paper presented at the 13th European Wireless. Paris, France.

Anastasi, G., Conti, M., Francesco, M. D., & Passarella, A. (2009). Energy conservation in wireless sensor networks: A survey. *Ad Hoc Networks*, *7*(3), 537–568. doi:10.1016/j.adhoc.2008.06.003

Andradottir, S. (1996). A global search method for discrete stochastic optimization. *SIAM Journal on Optimization*, *6*, 513–530. doi:10.1137/0806027

Andras, V. Hornig, & Rudolf. (2008). An overview of the OMNeT++ simulation environment. In *Proceedings of the 1st International Conference on Simulation Tools and Techniques for Communications, Networks and Systems (SIMUTools 2008),* (pp. 209-214). IEEE.

Andreopoulos, Y., Mastronarde, N., & Schaar, M. V. D. (2006). Cross-layer optimized video streaming over wireless multihop mesh networks. *IEEE Journal on Selected Areas in Communications*, *24*(11), 2104–2115. doi:10.1109/JSAC.2006.881614

Ansel, P., Ni, Q., & Turletti, T. (2004). An efficient scheduling scheme for IEEE 802.11e. In *Proceedings of the IEEE Workshop on Modeling and Optimization in Mobile, Ad Hoc and Wireless Networks*. Cambridge, UK: IEEE Press.

Anton-Haro, C. (2006). Cross-layer scheduling for multi-user MIMO systems. *IEEE Communications Magazine*, *44*(9), 39–45. doi:10.1109/MCOM.2006.1705977

Appadwedula, S., Veeravalli, V., & Jones, D. L. (2005). Energy efficient detection in sensor networks. *IEEE Selected Areas in Communications*, *23*(4), 693–702. doi:10.1109/JSAC.2005.843536

Argyriou, A. (2010). Cross-layer and cooperative opportunistic network coding in wireless ad hoc networks. *IEEE Transactions on Vehicular Technology*, *59*(2), 803–812. doi:10.1109/TVT.2009.2035811

Arienzo, L. (2010). An information-theoretic approach for energy-efficient collaborative tracking in wireless sensor networks. *EURASIP Journal on Wireless Communications and Networking*. Retrieved from http://downloads.hindawi.com/journals/wcn/2010/641632.pdf.

Armbrust, M., Fox, A., Griffith, R., Joseph, A., Katz, R., & Konwinski, A. (2010). A view of cloud computing. *Communications of the ACM*, *53*(4), 50–58. doi:10.1145/1721654.1721672

Arslan, H., & Yarkan, S. (2007). Cross-layer adaptation and optimization for cognitive radio. In Arslan, H. (Ed.), *Cognitive Radio, Software Defined Radio, and Adaptive Wireless Systems* (pp. 421–452). Springer. doi:10.1007/978-1-4020-5542-3_14

Arulampalam, M. S., Maskell, S., Gordon, N., & Clapp, T. (2002). A tutorial on particle filters for online nonlinear/non-gaussian bayesian tracking. *IEEE Transactions on Signal Processing*, *50*(2), 174–188. doi:10.1109/78.978374

Ashraf, I., Ho, L. T. W., & Claussen, H. (2010). Improving energy efficiency of femtocell base stations via user activity detection. In *Proceedings of the IEEE Wireless Communications and Networking Conference (WCNC 2010)*, (pp. 1-5). Sydney, Australia: IEEE Press.

Assunção, P. A. A., & Ghanbari, M. (1997). Congestion control of video traffic with transcoder. In *Proceedings of the International Conference on Image Processing*, (pp. 523-527). IEEE.

Astrom, K. J., & Wittenmark, B. (1995). *Adaptive control*. Boston, MA: Addison-Wesley Publishing Company.

Atkin, G., & Tam, C. (1992). Performance analysis of punctured convolutional codes for variable-bit-rate speech transmission. In *Proceedings of the IEEE 42nd Vehicular Technology Conference*, (pp. 183-186). IEEE Press.

Au, E. K. S., Jin, S., McKay, M. R., Mow, W. H., Gao, X., & Collings, I. B. (2008). Analytical performance of MIMO-SVD systems in Ricean fading channels with channel estimation error and feedback delay. *IEEE Transactions on Wireless Communications*, *7*(4), 1315–1325. doi:10.1109/TWC.2008.060912

Aune, F. (2004). *Cross-layer design tutorial*. Technical Report. Retrieved from http://www.iet.ntnu.no/projects/cuban/archive/1812041.pdf.

Azhari, V., & Todd, T. D. (2005). Fast connection recovery in WLAN-to-cellular vertical handoff. In *Proceedings of the IEEE International Conference on Mobile Adhoc and Sensor Systems*. Washington, DC: IEEE Press.

Babich, F., Kelly, O., & Lombardi, G. (2000). Generalized Markov modeling for flat fading. *IEEE Transactions on Communications*, *48*(4), 547–551. doi:10.1109/26.843121

Bachir, A., Barthel, D., Heusse, M., & Duda, A. (2006). Micro-frame preamble mac for multihop wireless sensor networks. In *Proceedings of the International Conference on Communications (ICC)*, (vol 7), (pp. 3365–3370). IEEE Press.

Bachir, A., Barthel, D., Heusse, M., & Duda, A. (2007). O(1)-reception routing for sensor networks. *Computer Communications*, *30*(13), 2603–2614. doi:10.1016/j.comcom.2007.05.054

Bai, Y., Liu, S., Sha, M., Lu, Y., & Xu, C. (2008). An energy optimization protocol based on cross-layer for wireless sensor networks. *The Journal of Communication*, *3*(6), 27–34.

Bajaj, M. G. L., Takai, M., Ahuja, R., & Bagrodia, R. (1999). *GloMoSim: A scalable network simulation environment*. Technical Report 990027. Los Angeles, CA: University of California.

Bajic, I. V. (2007). Efficient cross-layer error control for wireless video multicast. *IEEE Transactions on Broadcasting*, *53*(1), 276–285. doi:10.1109/TBC.2006.889211

Bajwa, W., Haupt, J., Sayeed, A., & Nowak, R. (2007). Joint source-channel communication for distributed estimation in sensor networks. *IEEE Transactions on Information Theory*, *53*(10), 3629–3653. doi:10.1109/TIT.2007.904835

Baldo, N., & Zorzi, M. (2007). Fuzzy logic for cross-layer optimization in cognitive radio networks. In *Proceedings of 4th IEEE Consumer Communications and Networking Conference*, (pp. 1128-1133). IEEE Press.

Baldo, N., Maguolo, F., Miozzo, M., Rossi, M., & Zorzi, M. (2007). NS2-MIRACLE: A modular framework for multi-technology and cross-layer support in network simulator. In *Proceedings of the ACM International Workshop on Network Simulation Tools (NSTOOLS)*, (pp. 1131-1140). ACM Press.

Baldo, N., & Zorzi, M. (2008). Fuzzy logic for cross-layer optimization in cognitive radio networks. *IEEE Communications Magazine*, *46*(4), 64–71. doi:10.1109/MCOM.2008.4481342

Balog, T. (2005). *Enterprise-wide network management with OpenNMS*. New York, NY: O'Really SysAdmin.

Bansal, D., & Sofat, S. (2010). Use of cross layer interactions for detecting denial of service attacks in WMN. In *Proceedings of the 2010 14th International Telecommunications Network Strategy and Planning Symposium (NETWORKS)*, (pp. 1-6). NETWORKS.

Bari, F., & Leung, V. C. (2009). Architectural aspects of automated network selection in heterogeneous wireless systems. *International Journal of Ad Hoc and Ubiquitous Computing, 45*, 282–291. doi:10.1504/IJAHUC.2009.027478

Barmada, B., Ghandi, M. M., Jones, E. V., & Ghanbari, M. (2006). Combined turbo coding and hierarchical QAM for unequal protection of H.264 coded video. *Signal Processing Image Communication, 21*(1), 390–395. doi:10.1016/j.image.2006.01.001

Baronti, P., Pillai, P., Chook, V., Chessa, S., Gotta, A., & Hu, Y. F. (2007). Wireless sensor networks: A survey on the state of the art and the 802.15.4 and ZigBee standards. *Computer Communications, 30*(7), 1655–1695. doi:10.1016/j.comcom.2006.12.020

Barrenetxea, G., Ingelrest, F., Schaefer, G., & Vetterli, M. (2008). The hitchhiker's guide to successful wireless sensor network deployments. In *Proceedings of the Conference on Embedded Network Sensor Systems (Sensys)*, (pp. 43–56). Raleigh, NC: ACM Press.

Barth, W. (2008). *Nagios: System and network monitoring*. New York, NY: No Starch Press.

Behzad, A., Rubin, I., & Mojibi-Yazdi, A. (2003). Distributed power controlled medium access control for ad-hoc wireless networks. In *Proceedings of IEEE 18th Annual Workshop on Computer Communications*, (pp. 47–53). IEEE Press.

Behzad, A., & Rubin, I. (2004). Multiple access protocol for power-controlled wireless access nets. *IEEE Transactions on Mobile Computing, 3*(4), 307–316. doi:10.1109/TMC.2004.44

Berenguer, I., Wang, X., & Krishnamurhty, V. (2005). Adaptive MIMO antenna selection via discrete stochastic optimization. *IEEE Transactions on Signal Processing, 53*, 4315–4329. doi:10.1109/TSP.2005.857056

Berry, R. A., & Gallager, R. G. (2002). Communication over fading channels with delay constraints. *IEEE Transactions on Information Theory, 48*(3), 1135–1149. doi:10.1109/18.995554

Bhatia, R., & Kodialam, M. (2001). On power efficient communication over multi-hop wireless networks: Joint routing, scheduling, and power control. In *Proceedings of the IEEE Annual Conference INFOCOM,* (pp. 1457–1466). IEEE Press.

Bhatnagar, A., Kumar, S., Janarthanan, A., Annamalai, A., Jr., Medley, M., & John Matyjas, J. (2009). A multilevel unequal error protection scheme for robust H.264/AVC video transmission over wireless channels. In *Proceedings of Military Communication Conference*. Military Communication Conference.

Bhatti, S., & Xu, J. (2009). Survey of target tracking protocols using wireless sensor network. In *Proceedings of the 5th International Conference on Wireless and Mobile Communications,* (pp. 110-115). Cannes, France: IEEE.

Bhatti, S. N., & Crowcroft, J. (2000). QoS-sensitive flows: Issues in IP packet handling. *IEEE Internet Computing, 4*(4), 48–57. doi:10.1109/4236.865087

Bianchi, G. (2000). Performance analysis of the IEEE 802.11 distributed coordination function. *IEEE Journal on Selected Areas in Communications, 18*(3), 535–547. doi:10.1109/49.840210

Biermann, T. (Ed.). (2009). D-5.2.0 description of generic path mechanism. *FP7-ICT-2007-1-216041-4WARD-"Architecture and Design for the Future Internet" Deliverables*. Retrieved January 2010, from http://www.4ward-project.eu/index.php?s=Deliverables.

Biglieri, E., Caire, G., & Taricco, G. (2001). Limiting performance of block fading channels with multiple antennas. *IEEE Transactions on Information Theory, 47*(4), 1273–1289. doi:10.1109/18.923715

Blake, S., Black, D., Carlson, M., Davies, E., Wang, Z., & Weiss, W. (1998). An architecture for differentiated services. *RFC:2475*. Retrieved from http://www.ietf.org.

Blakowski, G., & Steinmetz, R. (1996). A media synchronization survey: Reference model, specification, and case studies. *IEEE Journal on Selected Areas in Communications, 14*(1), 9–33. doi:10.1109/49.481691

Blondia, C. (1993). A discrete time batch Markovian arrival process as b-isdn traffic model. *Belgian Journal of Operations Research. Statistics and Computer Science, 32*(3/4), 3–23.

Bogineni, K., Ludwig, R., Mogensen, P., Nandlall, V., Vucetic, V., & Yi, B. (2009). LTE part I: Core network. *IEEE Communications Magazine*, *47*(2). doi:10.1109/MCOM.2009.4785378

Bogineni, K., Ludwig, R., Mogensen, P., Nandlall, V., Vucetic, V., & Yi, B. (2009). LTE part II: Radio access. *IEEE Communications Magazine*, *47*(4). doi:10.1109/MCOM.2009.4907405

Bonacorsi, D., Colling, D., Field, L., Fisher, S., Grandi, C., & Hobson, P. (2004). Scalability test of R-GMA based grid job monitoring system for CMS Monte Carlo data production. *IEEE Transactions on Nuclear Science*, *51*(6), 3026–3029. doi:10.1109/TNS.2004.839094

Bose, P., Morin, P., Stojmenovic, I., & Urrutia, J. (1999). Routing with guaranteed delivery in ad hoc wireless networks. In *Proceedings of the International Workshop on Discrete Algorithms and Methods for Mobile Computing and Communications (DIALM POMC), DIALM 1999*, (pp. 48–55). New York, NY: ACM Press.

Bose, S., & Kannan, A. (2008). Detecting denial of service attacks using cross layer based intrusion detection system in wireless ad hoc networks. In *Proceedings of the International Conference on Signal Processing, Communications and Networking (ICSCN 2008)*, (pp. 182-188). ICSCN.

Boskovic, A., & Despotovic, M. (2005). An efficient approach to ECG signal transmission via GPRS. In *Proceedings of the International Conference on Computer as a Tool*, (pp. 76-79). IEEE.

Bowling, J. (2008). Zenoss and the art of enterprise monitoring. *Linux Journal, 172*(6).

Boyd, S., & Vandenberghe, L. (2003). *Convex optimization*. Cambridge, UK: Cambridge University Press.

Braden, R., Clark, D., & Shenker, S. (1994). Integrated services in the internet architecture: An overview. *RFC 1633*. Retrieved from http://www.ietf.org.

Braden, R., Faber, T., & Handley, M. (2002). From protocol stack to protocol heap - Role-based architecture. In *Hot Topics in Networking (HOTNETS I)*. Princeton, NJ: Princeton University Press.

Brakmo, L. S., & Peterson, L. L. (1995). End to end congestion avoidance on a global internet. *IEEE Journal on Selected Areas in Communications*, *13*(8), 1465–1480. doi:10.1109/49.464716

Breslau, L., Cao, P., Fan, L., Phillips, G., & Shenker, S. (1999). Web caching and zipf-like distributions: Evidence and implications. In *Proceedings of IEEE Infocom* (pp. 126–134). New York, NY: IEEE Press. doi:10.1109/INFCOM.1999.749260

Brik, V., Rozner, E., & Banerjee, S. (2005). DSAP: A protocol for coordinated spectrum access. In *Proceedings of IEEE DySPAN,* (pp. 611-614). IEEE Press.

Broch, J., Maltz, D. A., Johnson, D. B., Hu, Y.-C., & Jetcheva, J. (1998). *A performance comparison of multi-hop wireless ad hoc network routing protocols.* Paper presented at the Mobile Computing and Networking. New York, NY.

Broustis, I. (Ed.). (2009). D4.2 first report on test-bed functionalities and implementation of network coding schemes. *FP7-ICT-215252 N-CRAVE Deliverables*. Retrieved December 2009, from http://www.n-crave.eu/deliverables.php.

Burks, A. W., Goldstine, H. H., & Neumann, J. V. (1963). *John von Neumann collected works* (Taub, A., Ed.). New York, NY: Pergamon Press.

Buyyaa, R., Shin Yeo, C., Venugopal, S., Broberg, J., & Brandic, I. (2009). Cloud computing and emerging IT platforms: Vision, hype, and reality for delivering computing as the 5th utility. *Future Generation Computer Systems*, *25*(6), 599–616. doi:10.1016/j.future.2008.12.001

Campbell, J. P., Campbell, W. M., Lewandowski, S. M., & Weinstein, C. J. (2006). Cognitive services for the user. In Fette, B. (Ed.), *Cognitive Radio Technology* (pp. 313–336). Burlington, MA: Newnes. doi:10.1016/B978-075067952-7/50011-8

Camp, J., & Knightly, E. (2010). Modulation rate adaptation in urban and vehicular environments: Cross-layer implementation and experimental evaluation. *IEEE/ACM Transactions on Networking*, *18*(6), 1949–1962. doi:10.1109/TNET.2010.2051454

Cao, G., Yang, D., An, R., Ye, X., Zheng, R., & Zhang, X. (2011). An adaptive sub-band allocation scheme for dense femtocell environment. In *Proceedings of the IEEE Wireless Communications and Networking Conference (WCNC 2011)*, (pp. 102-107). Quintana-roo, Mexico: IEEE Press.

Carneiro, G., Ruela, J., & Ricardo, M. (2004). Cross-layer design in 4G wireless terminals. *IEEE Wireless Communications*, *11*(2), 7–13. doi:10.1109/MWC.2004.1295732

Carrasco, L., Ramis, J., & Femenias, G. (2009). Multidimensional Markov models for the cross-layer design of multi-rate wireless systems using the effective capacity function. In *Proceedings of the IEEE 69th Vehicular Technology Conference VTC Spring 2009*, (pp.1-5). Barcelona, Spain: IEEE Press.

Cataldi, P., Grangetto, M., Tillo, T., Magli, E., & Olmo, G. (2010). Sliding-window raptor codes for efficient scalable wireless video broadcasting with unequal loss protection. *IEEE Transactions on Image Processing*, *19*(6), 1491–1503. doi:10.1109/TIP.2010.2042985

Ceriotti, M., Mottola, L., Picco, G. P., Murphy, A. L., Guna, S., & Corra, M. … Zanon, P. (2009). Monitoring heritage buildings with wireless sensor networks: The torre aquila deployment. In *Proceedings of the International Conference on Information Processing in Sensor Networks (IPSN)*, (pp. 277–288). Washington, DC: ACM Press.

Chakareski, J., Chou, P., & Girod, B. (2002). Rate-distortion optimized streaming from the edge of the network. In *Proceedings of IEEE Workshop on Multimedia Signal Processing*, (pp. 49–52). St. Thomas, US Virgin Islands: IEEE Press.

Chakareski, J., Apostolopoulos, J., Wee, S., Tan, W., & Girod, B. (2005). Rate-distortion hint tracks for adaptive video streaming. *IEEE Transactions on Circuits and Systems for Video Technology*, *15*(10), 1257–1269. doi:10.1109/TCSVT.2005.854227

Chakareski, J., & Chou, P. A. (2006). Radio edge: Rate-distortion optimized proxy-driven streaming from the network edge. *IEEE/ACM Transactions on Networking*, *14*(6), 1302–1312. doi:10.1109/TNET.2006.886299

Chakarvarthy, V., Li, X., Wu, Z., Temple, M., Garber, F., Kannan, R., & Vasilakos, A. (2009). Novel overlay/underlay cognitive radio waveforms using SD-SMSE framework to enhance spectrum efficiency – Part I: Theoretical framework and analysis in AWGN channel. *IEEE Transactions on Communications*, *57*(12), 3794–3804. doi:10.1109/TCOMM.2009.12.080400

Chakarvarthy, V., Li, X., Zhou, R., Wu, Z., & Temple, M. (2010). Novel overlay/underlay cognitive radio waveforms using SD-SMSE framework to enhance spectrum efficiency – Part II: Analysis in fading channel. *IEEE Transactions on Communications*, *58*(6), 1868–1976. doi:10.1109/TCOMM.2010.06.090176

Chakravarthi, R., Gomathy, C., Sebastian, S. K., Pushparaj, K., & Mon, V. B. (2010). A survey on congestion control in wireless sensor networks. *International Journal of Computer Science & Communication*, *1*(1), 161–164.

Chandrasekhar, V., Andrews, J. G., & Gatherer, A. (2008). Femtocell networks: A survey. *IEEE Communications Magazine*, *46*(9), 59–67. doi:10.1109/MCOM.2008.4623708

Chang, J. H., & Tassiulas, L. (2000). Energy conserving routing in wireless ad hoc networks. In *Proceedings of the IEEE Infocom*, (pp. 22-31). IEEE Press.

Chang, B.-J., & Chou, C.-M. (2007). Cross-layer based delay constraint adaptive polling for high density subscribers in IEEE 802.16 WiMAX networks. *Wireless Personal Communications*, *46*(3), 285–304. doi:10.1007/s11277-007-9434-5

Chang, R. I., Liu, C. T., Wang, T. C., & Wang, C. H. (2010). Improve QoS of mobile multimedia streaming via cross-layer predictive packet transmission over MIPv4. *Journal of Internet Technology*, *11*(4), 529–542.

Chao, H. C., Chang, C. Y., Chen, C. Y., & Chang, K. D. (2009). Survey of Cross-layer optimization techniques for wireless networks. In Adibi, S., Mobasher, A., & Tofigh, T. (Eds.), *The Fourth-Generation Wireless Networks: Applications and Innovations* (pp. 453–468). Hershey, PA: IGI Global. doi:10.4018/978-1-61520-674-2.ch021

Chao, H. C., & Huang, C. Y. (2003). Micro-mobility mechanism for smooth handoffs in an integrated ad-hoc and cellular IPv6 network under high-speed movement. *IEEE Transactions on Vehicular Technology*, *52*(6), 1576–1593. doi:10.1109/TVT.2003.819450

Chase, D. (1973). A combined coding and modulation approach for communications over dispersive channels. *IEEE Transactions on Communications*, *21*(3), 159–174. doi:10.1109/TCOM.1973.1091636

Chase, D. (1985). Code combining: A maximum-likelihood decoding approach for combining an arbitrary number of noisy packets. *IEEE Transactions on Communications*, *33*(5), 385–393. doi:10.1109/TCOM.1985.1096314

Chellappa Doss, R., & Pan, L. (2006). A case for cross layer design: The impact of physical layer properties on routing protocol performance in MANETs. In *Proceedings of the Australian Telecommunication Networks and Applications Conference*, (pp. 409-413). Melbourne, Australia: IEEE.

Chen, C. Y., & Chao, H. C. (2011). A survey of key distribution in wireless sensor networks. *Security and Communication Networks*. Retrieved from http://onlinelibrary.wiley.com/doi/10.1002/sec.354/abstract.

Chen, J., & Abedi, A. (2010). A hybrid framework for radio localization in broadband wireless systems. In *Proceedings of the IEEE Global Communications Conference*, (pp. 1-6). Miami, FL: IEEE Press.

Chen, J., Wang, C. C., & Lee, J. D. (2007). Pre-coordination mechanism for fast handover in WiMAX networks. In *Proceedings of the International Conference on Wireless Broadband and Ultra Wideband Communications AusWireless*, (pp. 15-21). AusWireless.

Chen, L., Cai, X., Sofia, R., & Huang, Z. (2007). A cross-layer fast handover scheme for mobile WiMAX. In *Proceedings of the IEEE 66th VTC- 2007 Fall Vehicular Technology Conference*, (pp. 1578-1582). IEEE Press.

Chen, L.-J., Kapoor, R., Sanadidi, M. Y., Lee, R., & Gerla, M. (2004). Audio streaming over Bluetooth: An adaptive ARQ timeout approach. In *Proceedings of the 24th International Conference on Distributed Computing Systems,* (pp. 196–201). IEEE.

Chen, S., & Zhang, Z. (2006). Localized algorithm for aggregate fairness in wireless sensor networks. In *Proceedings of the 12th Annual International Conference on Mobile Computing and Networking,* (pp. 274-285). Los Angeles, CA: IEEE Press.

Chen, Y. S., Chiu, K. L., & Hwang, R. H. (2007). SmSCTP: SIP-based MSCTP scheme for session mobility over WLAN/3G heterogeneous networks. In *Proceedings of the IEEE Wireless Communications and Networking Conference*, (pp. 3307-3312). IEEE Press.

Chen, Y. W., & Hsieh, F. Y. (2007). A cross layer design for handoff in 802.16e network with IPv6 mobility. In *Proceedings of the IEEE Wireless Communications and Networking Conference*, (pp. 3844–3849). IEEE Press.

Chen, Y., Yi, W., & Yang, Y. (2010). A cooperative routing algorithm for lifetime maximization in wireless sensor networks. In *Proceedings of The 1st IET International Conference on Wireless Sensor Network (IET-WSN 2010)*, (pp. 167-172). IET Press.

Chen, Y., Yi, W., & Yang, Y. (2010). A cross-layer strategy for energy-efficient cooperative communication in wireless sensor network. In *Proceedings of The 1st IET International Conference on Wireless Sensor Network (IET-WSN 2010),* (pp. 203 – 208). IET Press.

Chen, Y.-W., & Hsieh, F.-Y. (2007). A cross-layer design for handoff in 802.16e network with IPv6 mobility. In *Proceedings of the IEEE Wireless Communications and Networking Conference*, (pp. 3844-3849). IEEE Press.

Chen, Z., & Qiu, R. C. (2010). Prediction of channel state for cognitive radio using higher-order hidden Markov model. In *Proceedings of IEEE Southeast Conference,* (pp. 276-282). IEEE Press.

Chen, C.-M., Lin, C.-W., & Chen, Y.-C. (2010). Cross-layer packet retry limit adaptation for video transport over wireless LANs. *IEEE Transactions on Circuits and Systems for Video Technology*, *20*(11), 1448–1461. doi:10.1109/TCSVT.2010.2077475

Cheng, J.-F. T. (2006). Coding performance of hybrid ARQ schemes. *IEEE Transactions on Communications*, *54*(6), 1017–1029. doi:10.1109/TCOMM.2005.876846

Cheng, M., Gong, X., & Cai, L. (2009). Joint routing and link rate allocation under bandwidth and energy constraints in sensor networks. *IEEE Transactions on Wireless Communications*, *8*(7), 3770–3779. doi:10.1109/TWC.2009.081134

Chen, P., & Woods, J. W. (2004). Bidirectional MC-EZBC with lifting implementation. *IEEE Transactions on Circuits and Systems for Video Technology*, *14*(10), 1183–1194. doi:10.1109/TCSVT.2004.833165

Chen, W., Dai, L., Letaief, B., & Cao, Z. (2008). A unified cross-layer framework for resource allocation in cooperative networks. *IEEE Transactions on Wireless Communications*, *7*(8), 3000–3012. doi:10.1109/TWC.2008.060831

Chen, W., Letaief, K. B., & Cao, Z. (2008). A joint coding and scheduling method for delay optimal cognitive multiple access. [IEEE Press.]. *Proceedings of the IEEE, ICC*, 3558–3562.

Chen, Y. S., Chen, C. K., & Chuang, M. C. (2008). DeuceScan: Deuce-based fast handoff scheme in IEEE 802.11 wireless networks. *IEEE Transactions on Vehicular Technology*, *57*(2), 1126–1141. doi:10.1109/TVT.2007.907027

Chen, Y. S., Cho, C. H., Ilsun, Y., & Chao, H. C. (2011b). A cross-layer protocol of spectrum mobility and handover in cognitive LTE networks. *Simulation Modelling Practice and Theory*, *19*(8), 916–928. doi:10.1016/j.simpat.2010.09.007

Chen, Y. S., Hsiao, W. H., & Chiu, K. L. (2009). A cross-layer partner-based fast handoff mechanism for IEEE 802.11 wireless networks. *International Journal of Communication Systems*, *22*(12), 1515–1541. doi:10.1002/dac.1033

Chen, Y. S., & Wu, K. L. (2011a). A cross-layer partner-assisted handoff scheme for hierarchical mobile IPv6 in IEEE 802.16e systems. *Wireless Communications and Mobile Computing*, *11*(4), 522–541. doi:10.1002/wcm.844

Chen, Y., & Tellambura, C. (2005). Performance analysis of maximum ratio transmission with imperfect channel estimation. *IEEE Communications Letters*, *9*(4), 322–324. doi:10.1109/LCOMM.2005.1413621

Chiang, C., Wu, H., Liu, W., & Gerla, M. (1997). Routing in clustered multihop, mobile wireless networks with fading channel. In *Proceedings of the Singapore International Conference on Networks, IEEE SICON 1997*, (pp. 197-211). IEEE Press.

Chiang, J. T., & Hu, Y. C. (2011). Cross-layer jamming detection and mitigation in wireless broadcast networks. *IEEE/ACM Transactions on Networking*, *19*(1), 286–298. doi:10.1109/TNET.2010.2068576

Chiang, M. (2004). To layer or not to layer: Balancing transport and physical layers in wireless multihop networks. In *Proceedings of IEEE INFOCOM* (pp. 2525–2536). IEEE Press.

Chiang, M. (2005). Balancing transport and physical layers in wireless multihop networks: Jointly optimal congestion control and power control. *IEEE Journal on Selected Areas in Communications*, *23*(1), 104–116. doi:10.1109/JSAC.2004.837347

Chiang, M. (2005). Geometric programming for communication systems. *Short Monograph in Foundations and Trends in Communications and Information Theory*, *2*, 1–154. doi:10.1561/0100000005

Chiang, M., Low, S. H., Calderbank, A. C., & Doyle, J. C. (2007). Layering as optimization decomposition: A mathematical theory of network architectures. *Proceedings of the IEEE*, *95*(1), 255–312. doi:10.1109/JPROC.2006.887322

Chilamkurti, N., Zeadally, S., Vasilakos, A., & Sharma, V. (2009). Cross-layer support for energy efficient routing in wireless sensor networks. *Journal of Sensors*. Retrieved from downloads.hindawi.com/journals/js/2009/134165.pdf.

Chilamkurti, N., Zeadally, S., Soni, R., & Giambene, G. (2010). Wireless multimedia delivery over 802.11e with cross-layer optimization techniques. *Multimedia Tools and Applications*, *47*(1), 189–205. doi:10.1007/s11042-009-0413-6

Chittabrata, G., Cordeiro, C., Agrawal, D. P., & Rao, M. B. (2009). Markov chain existence and hidden Markov models in spectrum sensing. In *Proceedings of IEEE International Conference on Pervasive Computing and Communications*, (pp. 1-6). IEEE Press.

Chittabrata, G., Roy, S., Rao, M. B., & Agrawal, D. P. (2010). Spectrum occupancy validation and modeling using real-time measurements. In *Proceedings of ACM Workshop on Cognitive Radio Networks,* (pp. 25-30). ACM Press.

Chiu, K. L., Chen, Y. S., & Hwang, R. H. (2011a). Seamless session mobility scheme in heterogeneous wireless networks. *International Journal of Communication Systems*, *24*(6), 789–809. doi:10.1002/dac.1189

Chiu, K. L., Hwang, R. H., & Chen, Y. S. (2011b). Cross-layer design vehicle-aided handover scheme in VANETs. *Wireless Communications and Mobile Computing*, *11*(7), 916–928. doi:10.1002/wcm.861

Choi, J. D., Wasserman, K. M., & Stark, W. E. (1999). Effect of channel memory on retransmission protocols for low energy wireless data communications. *Proceedings IEEE International Conference on Communications, 3*, 1552-1556.

Choi, J. H., Yu, H. J., & Lee, Y. H. (2005). Adaptive MIMO decision feedback equalization for receivers with time-varying channels. *IEEE Transactions on Signal Processing*, *53*, 4295–4303. doi:10.1109/TSP.2005.857053

Choi, J.-Y., & Shin, J. (2007). Cross-layer error-control with low-overhead ARQ for H.264 video transmission over wireless LANs. *Computer Communications*, *30*(1), 1476–1486. doi:10.1016/j.comcom.2006.12.027

Chou, P. A. (2007). Streaming media on demand and live broadcast. In Chou, P. A., & van der Schaar, M. (Eds.), *Multimedia in IP and Wireless Networks* (pp. 453–501). Burlington, MA: Academic Press. doi:10.1016/B978-012088480-3/50015-1

Chou, P. A., & Miao, Z. (2006). Rate-distortion optimized streaming of packetized media. *IEEE Transactions on Multimedia*, *8*(2), 390–404. doi:10.1109/TMM.2005.864313

Chowdhury, K. R., Felice, M. D., & Akyildiz, I. F. (2009). TP-CRAHN: A transport protocol for cognitive radio adhoc networks. In *Proceedings of IEEE INFOCOM*, (pp. 2482-2490). IEEE Press.

Christian, B. (2002). On the minimum node degree and connectivity of a wireless multihop network. In *Proceedings of ACM International Symposium on Mobile Ad Hoc Networking and Computing (MobiHoc)*, (pp. 80-91). ACM Press.

Chung, S. T., & Goldsmith, A. J. (2001). Degrees of freedom in adaptive modulation: A unified view. *IEEE Transactions on Communications*, *49*, 1561–1571. doi:10.1109/26.950343

Cisco Systems. (2006). *Understanding jitter in packet voice networks*. Retrieved from http://www.cisco.com/en/US/tech/tk652/tk698/technologies_tech_note-09186a00800945df.shtml.

Clarke, P., & de Lamare, R. C. (2011). Joint transmit diversity optimization and relay selection for multi-relay cooperative MIMO systems using discrete stochastic algorithms. *IEEE Communications Letters*, *15*, 1035–1037. doi:10.1109/LCOMM.2011.082611.102262

Clarke, R. H. (1968). A statistical theory of mobile radio reception. *The Bell System Technical Journal*, *47*(6), 957–1000.

Clausen, E., & Jacquet, E. (2003). Optimized link state routing protocol (OLSR). *RFC:3626*. Retrieved from http://www.ietf.org.

Click Modular Router. (2011). *Website*. Retrieved from http://read.cs.ucla.edu/click/.

Cohen, E., & Shenker, S. (2002). Replication strategies in unstructured peer-to-peer networks. In *Proceedings of ACM SIGCOMM,* (pp. 177–190). Pittsburgh, PA. ACM Press.

Collins, B., & Cruz, R. (1999). Transmission policy for time varying channel with average delay constraints. In *Proceedings of Allerton Conference on Communications, Control and Computing*, (pp. 709-717). Allerton.

Committee, I. D. (2011). *IEEE 1900.6 working group on spectrum sensing interfaces and data structures for dynamic spectrum access and other advanced radio communication systems*. Retrieved March 2011 from http://www.ieee.org.

Conti, M., Maselli, G., & Turi, G. (2006). Design of a flexible cross-layer interface for ad hoc networks. In Al Agha, K., Guérin Lassous, I., & Pujolle, G. (Eds.), *Challenges in Ad Hoc Networking* (*Vol. 197*, pp. 189–198). Boston, MA: Springer. doi:10.1007/0-387-31173-4_21

Conti, M., Maselli, G., Turi, G., & Giordano, S. (2004). Cross-layering in mobile ad hoc network design. *IEEE Computer Society*, *37*(2), 48–51. doi:10.1109/MC.2004.1266295

Correia, L. M., Abramowicz, H., Johnsson, M., & Wünstel, K. (2011). *Architecture and design for the future internet*. Berlin, Germany: Springer Science.

Cox, J. R., Nolle, F. M., Fozzard, H. A., & Oliver, G. C. (1968). AZTEC: A preprocessing program for real-time ECG rhythm analysis. *IEEE Transactions on Bio-Medical Engineering*, *15*(2), 128–129. doi:10.1109/TBME.1968.4502549

Crow, B. P., Widjaja, I., Kim, J. G., & Sakai, P. T. (1997). IEEE 802.11 wireless local area networks. *IEEE Communications Magazine*, *35*, 116–126. doi:10.1109/35.620533

Cruz, R. L., & Santhanam, A. V. (2003). Optimal routing, link scheduling and power control in multi-hop wireless networks. In *Proceedings of IEEE International Conference on Computer Communications (INFOCOM)*, (pp. 13-23). IEEE Press.

Cui, S., & Goldsmith, A. J. (2006). Cross-layer design of energy-constrained networks using cooperative MIMO techniques, *Eurasip's Signal Processing Journal.* Retrieved from http://www.ece.tamu.edu/~cui/publications/eurasip_05.pdf.

Cui, S., Madan, R., Goldsmith, A., & Lall, S. (2005). Joint routing, MAC, and link layer optimization in sensor networks with energy constraints. In *Proceedings of IEEE International Conference on Communications (ICC 2005)*, (Vol. 2), (pp. 725 – 729). IEEE Press.

Cui, S., Madan, R., Goldsmith, A. J., & Lall, S. (2007). Cross-layer energy and delay optimization in small-scale sensor networks. *IEEE Transactions on Wireless Communications*, *6*(10), 3688–3699. doi:10.1109/TWC.2007.060072

Czylwik, A. (1997). Comparison between adaptive OFDM and single carrier modulation with frequency domain equalization. In *Proceedings of IEEE Vehicular Technology Conference,* (pp. 865-869). IEEE Press.

Dai, L., & Letaief, K. B. (2005). Cross-layer design for combining cooperative diversity with truncated ARQ in ad-hoc networks. In *Proceedings of IEEE Global Telecommunications Conference (IEEE Globecom)*, (pp. 3175–3179). IEEE Press.

Dam, T. V., & Langendoen, K. (2003). An adaptive energy-efficient MAC protocol for wireless sensor networks. In *Proceedings of the International Conference on Embedded Networked Sensor Systems,* (pp. 171-180). Los Angeles, CA: IEEE Press.

Das, S., & Zuniga, J. C. (2011). *IEEE 802.21*. Retrieved from http://www.ieee802.org/21/.

de Lamare, R. C. (2010a). Joint iterative power allocation and interference suppression algorithms for cooperative spread spectrum networks. In *Proceedings of the IEEE International Conference on Acoustics, Speech and Signal Processing*. Dallas, TX: IEEE Press.

de Lamare, R. C., & Alcaim, A. (2005c). Strategies to improve the performance of very low bit rate speech coders and application to a 1.2 kb/s codec. In *Proceedings of the IEEE Conference on Vision, Image and Signal Processing*, (p. 152). IEEE Press.

de Lamare, R. C., & Diniz, P. S. R. (2009b). Set-membership adaptive algorithms based on time-varying error bounds for CDMA interference suppression. *IEEE Transactions on Vehicular Technology*, *58*, 644–654. doi:10.1109/TVT.2008.926608

de Lamare, R. C., Haardt, M., & Sampaio-Neto, R. (2008). Blind adaptive constrained reduced-rank parameter estimation based on constant modulus design for CDMA interference suppression. *IEEE Transactions on Signal Processing*, 56.

de Lamare, R. C., & Sampaio-Neto, R. (2003). Adaptive MBER decision feedback multiuser receivers in frequency selective fading channels. *IEEE Communications Letters*, *7*(2), 73–75. doi:10.1109/LCOMM.2002.808373 de Lamare, R. C., & Sampaio-Neto, R. (2005a). Adaptive reduced-rank MMSE filtering with interpolated FIR filters and adaptive interpolators. *IEEE Signal Processing Letters*, 12.

de Lamare, R. C., & Sampaio-Neto, R. (2005b). Reduced-rank interference suppression for DS-CDMA based on interpolated FIR filters. *IEEE Communications Letters*, 9.

de Lamare, R. C., & Sampaio-Neto, R. (2005d). Blind adaptive code-constrained constant modulus algorithms for CDMA interference suppression in multipath channels. *IEEE Communications Letters*, *9*, 334–336.

de Lamare, R. C., & Sampaio-Neto, R. (2006). Low-complexity variable step-size mechanisms for stochastic gradient algorithms in minimum variance CDMA receivers. *IEEE Transactions on Signal Processing*, *54*, 2302–2317. doi:10.1109/TSP.2006.873651

de Lamare, R. C., & Sampaio-Neto, R. (2007b). Adaptive interference suppression for DS-CDMA systems based on interpolated FIR filters with adaptive interpolators in multipath channels. *IEEE Transactions on Vehicular Technology*, 56.

de Lamare, R. C., & Sampaio-Neto, R. (2007c). Reduced-rank adaptive filtering based on joint iterative optimization of adaptive filters. *IEEE Signal Processing Letters*, ▪▪▪, 14.

de Lamare, R. C., & Sampaio-Neto, R. (2008a). Minimum mean squared error iterative successive parallel arbitrated decision feedback detectors for DS-CDMA systems. *IEEE Transactions on Communications*, *56*, 778–789. doi:10.1109/TCOMM.2008.060209

de Lamare, R. C., & Sampaio-Neto, R. (2009a). Adaptive reduced-rank processing based on joint and iterative interpolation, decimation, and filtering. *IEEE Transactions on Signal Processing*, *57*, 2503–2514. doi:10.1109/TSP.2009.2018641

de Lamare, R. C., & Sampaio-Neto, R. (2010b). Reduced-rank space-time adaptive interference suppression with joint iterative least squares algorithms for spread-spectrum systems. *IEEE Transactions on Vehicular Technology*, *59*, 1217–1228. doi:10.1109/TVT.2009.2038391

de Lamare, R. C., & Sampaio-Neto, R. (2010c). Blind adaptive MIMO receivers for space-time block-coded DS-CDMA systems in multipath channels using the constant modulus criterion. *IEEE Transactions on Communications*, *58*, 21–27. doi:10.1109/TCOMM.2010.01.070549

de Lamare, R. C., Sampaio-Neto, R., & Haardt, M. (2011). Blind adaptive constrained constant-modulus reduced-rank interference suppression algorithms based on interpolation and switched decimation. *IEEE Transactions on Signal Processing*, *59*, 681–695. doi:10.1109/TSP.2010.2091274

de Lamare, R. C., Sampaio-Neto, R., & Hjorungnes, A. (2007a). Joint iterative interference cancellation and parameter estimation for CDMA systems. *IEEE Communications Letters*, *11*, 916–918. doi:10.1109/LCOMM.2007.070943

Derevenetz, O. (2011). *TclMon: Open source cross-platform network monitoring solution*. Retrieved from http://tclmon.sourceforge.net/.

DeVaul, R., Sung, M., Gips, J., & Pentland, A. (2003). MIThril 2003: Applications and architecture. In *Proceedings of the 7th IEEE International Symposium on Wearable Computers*, (pp. 4-11). IEEE Press.

Dhyan Infotech. (2011). *Dhyan network/element management system (DNMS)*. Retrieved from http://www.dhyanit.com/dnms_overview.html.

Dighe, P. A., Mallik, R. K., & Jamuar, S. S. (2003a). Analysis of transmit-receive diversity in Rayleigh fading. *IEEE Transactions on Communications*, *4*(51), 694–703. doi:10.1109/TCOMM.2003.810871

Dighe, P. A., Mallik, R. K., & Jamuar, S. S (2003b). Analysis of K transmit dual-receive diversity with cochannel interferers over a Rayleigh fading channel. *Wireless Personal Communications*, *25*(2), 87–100. doi:10.1023/A:1023699628908

Dimic, G., Sidiropoulos, N., & Zhang, R. (2004). Medium access control - Physical cross-layer design. *IEEE Signal Processing Magazine*, *21*(9), 40–50. doi:10.1109/MSP.2004.1328087

Ding, J.-W., Wang, W.-T., & Wang, C.-F. (2008). An efficient data replica scheme for peer-to-peer video streaming over wireless mesh community networks. In *Proceedings of the International Conference on Intelligent Information Hiding and Multimedia Signal Processing,* (pp. 767–770). Harbin Shi, China: IEEE.

Ding, M., & Cheng, X. (2009). Fault tolerant target tracking in sensor networks. In *Proceedings of the Tenth ACM International Symposium on Mobile Ad Hoc Networking and Computing,* (pp. 125-134). New Orleans, LA: ACM Press.

Ding, L., Melodia, T., Batalama, S. N., Matyjas, J. D., & Medley, M. J. (2010). Cross-layer routing and dynamic spectrum allocation in cognitive radio ad hoc networks. *IEEE Transactions on Vehicular Technology*, *59*(4), 1969–1979. doi:10.1109/TVT.2010.2045403

Ding, M., Liu, S., Luo, H., & Chen, W. (2010). MMSE based greedy antenna selection scheme for AF MIMO relay systems. *IEEE Signal Processing Letters*, *17*, 433–436. doi:10.1109/LSP.2010.2042531

Ding, Z., & Leung, K. K. (2011). Cross-layer routing using cooperative transmission in vehicular ad-hoc networks. *IEEE Journal on Selected Areas in Communications*, *29*(3), 571–581. doi:10.1109/JSAC.2011.110307

Dipersio, D. A. (1985). Evaluation of the fan method of adaptive sampling on human electrocardiograms. *Medical & Biological Engineering &. Computing*, *23*(5), 401–410.

Dirac Video Project. (2011). *Website.* Retrieved from http://diracvideo.org.

D-ITG. (2011). *Distributed internet traffic generator.* Retrieved from http://www.grid.unina.it/software/ITG/index.php.

Dixit, S., & Prasad, R. (Eds.). (2003). *Wireless IP and building the mobile internet.* London, UK: Artech House.

Djohan, A., Nguyen, T., & Tompkins, W. (1995). ECG compression using discrete symmetric wavelet transform. In *Proceedings of the IEEE 17th Annual Conference on Engineering in Medicine and Biology Society*, (pp. 167-168). IEEE Press.

Djonin, D. V., Karmokar, A. K., & Bhargava, V. K. (2008). Joint rate and power adaptation for type-I hybrid ARQ systems over correlated fading channels under different buffer-cost constraints. *IEEE Transactions on Vehicular Technology*, *57*, 421–435. doi:10.1109/TVT.2007.901876

Djonin, D. V., & Krishnamurthy, V. (2007). Q-learning algorithms for constrained Markov decision processes with randomized monotone policies: Application to MIMO transmission control. *IEEE Transactions on Signal Processing*, *55*(5), 2170–2181. doi:10.1109/TSP.2007.893228

Dobrota, V. (2011). Switching and routing systems. *Technical University of Cluj-Napoca.* Retrieved March 2011, from http://el.el.obs.utcluj.ro/scr/.

Dohler, M., & Li, Y. (2010). *Cooperative communications: Hardware, Channel and PHY.* New York, NY: John Wiley.

Dudzik, E., Abedi, A., Pereira da Cunha, M., & Hummels, D. (2008). Orthogonal code design for passive wireless sensors. In *Proceedings of the 24th Queen's Biennial Symposium on Communications,* (pp. 316 - 319). Kingston, Canada: IEEE Press.

Duong, T. Q., & Zepernick, H.-J. (2009). Cross-layer design for MRT Systems with channel estimation error and feedback delay. *Wireless Personal Communications*, *58*(4), 681–694. doi:10.1007/s11277-009-9901-2

Du, Q., & Zhang, X. (2009). A cross-layer framework for multi-layer-video multicast with QoS requirements in multirate wireless networks. *IEEE Communications Letters*, *13*(9), 658–660. doi:10.1109/LCOMM.2009.082211

Eastwood, L., Migaldi, S., Qiaobing, X., & Gupta, V. (2008). Mobility using IEEE 802.21 in a heterogeneous IEEE 802.16/802.11-based, IMT-advanced (4G) network. *IEEE Wireless Communications*, *15*(2), 26–34. doi:10.1109/MWC.2008.4492975

EDR. (2007). *Core specification of the Bluetooth system, version 2.1 + EDR*. Retrieved from http://www.bluetooth.com.

ElBatt, T., & Ephremides, A. (2004). Joint scheduling and power control for wireless ad hoc networks. *IEEE Transactions on Wireless Communications*, *3*, 74–85. doi:10.1109/TWC.2003.819032

El-Hoiydi, A., Decotignie, J.-D., & Hernandez, J. (2004). Low power mac protocols for infrastructure wireless sensor networks. In *Proceedings of the European Wireless Conference*, (pp. 563–569). Barcelona, Spain: European Wireless.

Elliott, E. O. (1963). Estimates of error rates for codes on burst-noise channels. *The Bell System Technical Journal*, *42*, 1977–1997.

ElRakabawy, S. M., & Lindemann, C. (2009). A clean-slate architecture for reliable data delivery in wireless mesh networks. In *Proceedings of the IEEE Wireless Communications and Networking Conference, WCNC 2009*, (pp. 1-6). Budapest, Hungary: IEEE Press.

Etkin, R., Parekh, A., & Tse, D. (2005). Spectrum sharing for unlicensed bands. In *Proceedings of IEEE DySPAN*, (pp. 251-258). IEEE Press.

Etoh, M., & Yoshimura, T. (2005). Advances in wireless video delivery. *Proceedings of the IEEE*, *93*(1), 111–122. doi:10.1109/JPROC.2004.839605

Ettefagh, A., Kuhn, M., Hammerstrom, I., & Wittneben, A. (2006). On the range performance of decode-andforward relay: IEEE 802.11 WLANs. In *Proceedings of the IEEE 17th International Symposium Personal, Indoor and Mobile Radio Communications*. IEEE Press.

Ettus Research LLC. (2011). *Website*. Retrieved from http://www.ettus.com/products.

Falconer, D., Ariyavisitakul, L. S., Benyamin-Seeyar, A., & Eidson, B. (2002). Frequency domain equalization for single-carrier broadband wireless systems. *IEEE Communications Magazine*, *40*(4), 58–66. doi:10.1109/35.995852

Fang, J. C., & Rao, R. R. (2003). An integrated and distributed scheduling and power control algorithm for maximizing network utility for wireless multihop networks. In *Proceedings of IEEE MILCOM*, (pp. 1011–1017). IEEE Press.

Fang, Y., & McDonald, B. (2004). Dynamic codeword routing (DCR): A cross-layer approach for performance enhancement of general multi-hop wireless routing. In *Proceedings of IEEE Communications Society Conference on Sensor, Mesh and Ad Hoc Communications and Networks (SECON 2004)*, (pp. 255 – 263). IEEE Press.

Fang, Z., Hua, Y., & Koshy, J. C. (2006). Joint source and relay optimzation for a non-regenerative MIMO relay. In *Proceedings of the Fourth IEEE Workshop on Sensor Array and Multichannel Processing*. Waltham, MA: IEEE Press.

Fan, Y., & Thompson, J. (2007). MIMO configurations for relay channels: Theory and practice. *IEEE Transactions on Wireless Communications*, *6*, 1774–1786. doi:10.1109/TWC.2007.360379

Fathi, M., Taheri, H., & Mehrjoo, M. (2010). Cross-layer joint rate control and scheduling for OFDMA wireless mesh networks. *IEEE Transactions on Vehicular Technology*, *59*(8), 3933–3941. doi:10.1109/TVT.2010.2064346

FCC Spectrum Policy Task Force. (2002). *Report of the spectrum efficiency working group*. Retrieved from http://www.fcc.gov/sptf/reports.html.

FCC. (2003). Notice of proposed rule making and order. *ET Docket No. 03-222*. Washington, DC: FCC.

Feeney, L., & Nilson, M. (2001). Investigating the energy consumption of a wireless network interface in an ad hoc networking environment. In *Proceedings of INFOCOM*, (pp. 1548–1557). Anchorage, AK: IEEE Press.

Felemban, E., Lee, C.-G., & Ekici, E. (2006). Mmspeed: Multipath multi-speed protocol for qos guarantee of reliability and. timeliness in wireless sensor networks. *IEEE Transactions on Mobile Computing*, *5*(6), 738–754. doi:10.1109/TMC.2006.79

Felice, M. (2008). *Cross-layer optimization in multi-hop ad-hoc networks*. Unpublished Doctoral Dissertation. Padova, Italy: Universita di Bologna.

Femenias, G., Ramis, J., & Carrasco, L. (2009). Using two dimensional Markov models and the effective-capacity approach for cross-layer design in AMC/ARQ-based wireless networks. *IEEE Transactions on Vehicular Technology*, *58*(8), 4193–4203. doi:10.1109/TVT.2009.2020723

Ferguson, T. J., & Ranowitz, J. H. (1984). Self-synchronizing Huffman codes. *IEEE Transactions on Information Theory*, *30*, 687–693. doi:10.1109/TIT.1984.1056931

Ferrara, D., et al. (2005). MACRO: An integrated MAC/routing protocol for geographical forwarding in wireless sensor networks. In *Proceedings of IEEE International Conference on Computer Communications (Infocom 2005),* (Vol. 3), (pp. 1770 – 1781). IEEE Press.

Fettweis, G. P., & Zimmermann, E. (2008). ICT energy consumption – Trends and challenges. In *Proceedings of the International Symposium on Wireless Personal Multimedia Communications*. Lapland, Finland: IEEE.

Fleury, M., Altaf, M., Moiron, M., Qadri, M., & Ghanbari, M. (2011). Source coding methods for robust wireless video streaming. In Farrugia, R., & Debono, C. J. (Eds.), *Multimedia Networking and Coding: From Capture to Display*. Hershey, PA: IGI Global.

Flierl, M., & Girod, B. (2003). Generalized B pictures and the draft H.264/AVC video compression standard. *IEEE Transactions on Circuits and Systems for Video Technology, 13*(7), 587–597. doi:10.1109/TCSVT.2003.814963

Foh, C. H., Zhang, Y., Ni, Z., Cai, J., & Ngan, K. N. (2007). Optimized cross-layer design for scalable transmission over IEEE 802.11e network. *IEEE Transactions on Circuits and Systems for Video Technology, 17*(12), 1665–1678. doi:10.1109/TCSVT.2007.903808

Foschini, G. J., & Gans, M. J. (1998). On limits of wireless communications in a fading environment when using multiple antennas. *Wireless Personal Communications, 6*, 311–335. doi:10.1023/A:1008889222784

Foukalas, F., Gazis, V., & Alonistioti, N. (2008). Cross-layer design proposals for wireless mobile networks: A survey and taxonomy. *IEEE Communications Surveys & Tutorials, 10*(1), 70–85. doi:10.1109/COMST.2008.4483671

Fowler, J. E., & Pesquet-Popescu, B. (2007). An overview on wavelets in source coding, communications, and networks. *EURASIP Journal on Image and Video Processing, 4*, 1–27. doi:10.1155/2007/60539

Frederiksen, F., & Kolding, T. (2002). Performance and modeling of WCDMA/HSDPA transmission/H-ARQ schemes. In *Proceedings of the VTC Fall,* (pp. 472-476). VTC.

Freeman, E., Freeman, E., Bates, B., & Kathy, S. (2004). *Headfirst design patterns*. New York, NY: O'Reilly Media.

Frommel, O. (2010). Monitoring: Aus observerNMS wird observium. *Linux Magazine Online*. Retrieved from http://www.linux-mag.com.

Fu, Y., Hu, R., Tian, G., & Wang, T. (1996). TCP-friendly rate control for streaming service over 3G network. In *Proceedings of the International Conference on Wireless Communications, Networking and Mobile Computing*, (pp. 1-4). IEEE.

Fu, A., Modiano, E., & Tsitsiklis, J. N. (2006). Optimal transmission scheduling over a fading channel with energy and deadline constraints. *IEEE Transactions on Wireless Communications, 5*, 630–641.

Fu, F., & vander Schaar, M. (2009). A new systematic framework for autonomous cross-layer optimization. *IEEE Transactions on Vehicular Technology, 58*(4), 1887–1903. doi:10.1109/TVT.2008.2007418

Fujitsu. (2010). *Fujitsu research reveals global consumer attitudes to data privacy crucial to realizing benefits of cloud computing*. Retrieved from http://solutions.us.fujitsu.com/www/content/news/newsdetail.php?nf=10678411.nitf.

Gandhi, S., Buragohain, C., Cao, L., Zheng, H., & Suri, S. (2007). A general framework for wireless spectrum auction. In *Proceedings of IEEE DySPAN*. IEEE Press.

Ganesan, D. (2002). *An empirical study of epidemic algorithms in large scale multihop wireless networks. Technical Report*. Palo Alto, CA: Intel Research.

Garcia-Frias, J., & Xiong, Z. (2005). Distributed source and joint source channel coding: From theory to practice. In *Proceedings of the 30th IEEE Acoustics, Speech, and Signal Processing Conference,* (pp. 1093-1096). IEEE Press.

Garcia, R., Llana, L., Malagon, C., & Pancorbo, J. (2010). Event prediction in network monitoring systems: Performing sequential pattern mining in osmius monitoring tool. *LNCS Advances in Data Mining: Applications and Theoretical Aspects, 6171*, 632–642. doi:10.1007/978-3-642-14400-4_49

Gast, M. (2005). *802.11 wireless networks: The definitive guide*. New York, NY: O'Reilly Media.

Gauvain, J. L. (1994). Maximum a posteriori estimation for multivariate Gaussian mixture observations of Markov chains. *IEEE Transactions on Speech and Audio Processing*, *2*(2), 291–298. doi:10.1109/89.279278

Gersho, A., & Gray, R. M. (1992). *Vector quantization and signal compression*. Dordrecht, The Netherlands: Kluwer Academic Press/Springer.

Gesbert, D., Hanly, S., Huang, H., Shitz, S., Simeone, O., & Yu, W. (2010). Multi-cell MIMO cooperative networks: A new look at interference. *IEEE Journal on Selected Areas in Communications*, *28*(9), 1380–1408. doi:10.1109/JSAC.2010.101202

Ghanbari, M. (2003). *Standard codecs: Image compression to advanced video coding*. London, UK: IEE. doi:10.1049/PBTE049E

Ghandeharizadeh, S., Krishnamachari, B., & Song, S. (2004). Placement of continuous media in wireless peer-to-peer networks. *IEEE Transactions on Multimedia*, *6*(2), 335–342. doi:10.1109/TMM.2003.822787

Ghandi, M. M., & Ghanbari, M. (2006). Layered H.264 video transmission with hierarchical QAM. *Journal of Visual Communication and Image Representation*, *17*(8), 451–466. doi:10.1016/j.jvcir.2005.05.005

Gharavi, H., & Ban, K. (2003). *Cross-layer feedback control for video communications via mobile ad-hoc networks*. Paper presented at the Vehicular Technology Conference, 2003. New York, NY.

Ghosh, C., Pagadarai, S., Agrawal, D., & Wyglinski, A. (2010). A framework for statistical wireless spectrum occupancy modeling. *IEEE Transactions on Wireless Communications*, *9*(1), 38–44. doi:10.1109/TWC.2010.01.081701

Gilbert, E. N. (1960). Capacity of a burst-noise channel. *The Bell System Technical Journal*, *39*, 1253–1266.

Gill, P. E., Murray, W., & Wright, M. H. (1981). *Practical optimization*. New York, NY: Academic Press.

Girod, B., & Farber, N. (2000). Wireless video. In Sun, M.-T., & Reibman, A. R. (Eds.), *Compressed Video over Networks* (pp. 1–38). New York, NY: Marcel Dekker.

GloMoSim. (1998). *Glomosim: A scalable network simulation environment*. Retrieved from http://pcl.cs.ucla.edu/projects/glomosim/.

GloMoSim. (2011). *Website*. Retrieved Nov 23, 2011, from http://pcl.cs.ucla.edu/projects/glomosim/.

Golden, G. D., Foschini, C. J., Valenzuela, R. A., & Wolniansky, P. W. (1999). Detection algorithm and initial laboratory results using V-BLAST space-time communication architecture. *Electronics Letters*, *35*(1). doi:10.1049/el:19990058

Goldsack, P., Guijarro, J., Loughran, S., Coles, A., Farrell, A., & Lain, A. (2009). The SmartFrog configuration management framework. *ACM SIGOPS Operating Systems Review*, *43*, 16–25. doi:10.1145/1496909.1496915

Goldsmith, A. (2005). *Wireless communications*. Cambridge, UK: Cambridge University Press.

Goldsmith, A. J., & Chua, S.-G. (1997). Variable-rate variable-power MQAM for fading channels. *IEEE Transactions on Communications*, *45*(10), 218–1230. doi:10.1109/26.634685

Goldsmith, A. J., & Chua, S.-G. (1998). Adaptive coded modulation for fading channels. *IEEE Transactions on Communications*, *46*(5), 595–602. doi:10.1109/26.668727

Goldsmith, A. J., & Varaiya, P. P. (1997). Capacity of fading channels with channel side information. *IEEE Transactions on Information Theory*, *43*(6), 1986–1992. doi:10.1109/18.641562

Goldsmith, A. J., & Wicker, S. B. (2002). Design challenges for energy constrained ad hoc wireless networks. *IEEE Wireless Communications Magazine*, *8*, 8–27. doi:10.1109/MWC.2002.1028874

Golwelkar, A., & Woods, J. W. (2007). Motion compensated temporal filtering and motion vector coding using biorthogonal filters. *IEEE Transactions on Circuits and Systems for Video Technology*, *17*(4), 417–428. doi:10.1109/TCSVT.2007.893834

Gomez, J., Campbell, A. T., Naghshineh, M., & Bisdikian, C. (2001). Conserving transmission power in wireless ad hoc networks. In *Proceedings of IEEE ICNP*. IEEE Press.

Gonzales, A. (Ed.). (2010). D-4.3 in-network management design. *FP7-ICT-2007-1-216041-4WARD-"Architecture and Design for the Future Internet" Deliverables*. Retrieved February 2010 from http://www.4ward-project.eu/index.php?s=Deliverables.

Görkemli, B., Sunay, M. O., & Tekalp, A. M. (2008). Video streaming over wireless DCCP. In *Proceedings of the IEEE International Conference on Image Processing*, (pp. 2028-2031). IEEE Press.

Gottron, C., Konig, A., Hollick, M., Bergstrasser, S., Hildebrandt, T., & Steinmetz, R. (2009). *Quality of experience of voice communication in large-scale mobile ad hoc networks*. Paper presented at the Wireless Days (WD). Paris, France.

Goyal, M., Kumar, A., & Sharma, V. (2008). Optimal cross-layer scheduling of transmissions over a fading multiaccess channel. *IEEE Transactions on Information Theory*, *54*(8), 3518–3537. doi:10.1109/TIT.2008.926335

Goyal, V. (2001). Theoretical foundations of transform coding. *IEEE Signal Processing Magazine*, *18*(9), 9–21. doi:10.1109/79.952802

Gradshteyn, I. S., & Ryzhik, I. M. (Eds.). (2000). *Table of integrals, series, and products* (6th ed.). San Diego, CA: Academic Press.

Granelli, F., & Kliazovich, D. (2005). Cross-layering for performance improvement in multi-hop wireless networks. In *Proceedings of the 8th International Symposium on Parallel Architectures, Algorithms and Networks ISPAN 2005*, (pp. 536-541). Trento, Italy: ISPAN.

Granelli, F., Costa, C. E., & Katsaggelos, A. K. (2007). A study on the usage of cross-layer power control and forward error correction for embedded video transmission over wireless links. *Advances in Multimedia*. Retrieved from http://www.hindawi.com/journals/am/2007/095807/abs/.

Gross, J. (2004). Cross-layer optimization of OFDM transmission systems for MPEG-4 video streaming. *Computer Communications*, *27*, 1044–1055. doi:10.1016/j.comcom.2004.01.010

Guillemin, F. (Ed.). (2009). D-5.1 architecture of a generic path. *FP7-ICT-2007-1-216041-4WARD-"Architecture and Design for the Future Internet" Deliverables*. Retrieved November 2009 from http://www.4ward-project.eu/index.php?s=Deliverables.

Gunasekaran, R., & Qi, H. (2008). Xlrp: Cross layer routing protocol for wireless sensor networks. In *Proceedings of the IEEE Wireless Communications and Networking Conference*, (pp. 2135-2140). Las Vegas, NV: IEEE Press.

Gundavelli, S., Leung, K., Devarapalli, V., Chowdhury, K., & Patil, B. (2008). Proxy mobile IPv6. *IETF RFC 5213*. Retrieved from http://www.ietf.org.

Gungor, V. C., Vuran, M. C., & Akan, O. B. (2007). On the cross-layer interactions between congestion and contention in wireless sensor and actor networks. *Ad Hoc Networks*, *5*(6), 897–909. doi:10.1016/j.adhoc.2007.02.007

Gungor, V., & Hancke, G. (2009). Industrial wireless sensor networks: Challenges, design principles, and technical approaches. *IEEE Transactions on Industrial Electronics*, *56*(10), 4258–4265. doi:10.1109/TIE.2009.2015754

Gupta, P., & Kumar, P. R. (1998). Critical power for asymptotic connectivity in wireless networks. In *Proceedings of Stochastic Analysis, Control, Optimization and Applications*, (Vol. 1), (pp. 1106-1110). IEEE.

Gyasi-Agyei, A., & Kim, S. L. (2006). Cross-layer multiservice opportunistic scheduling for wireless networks. *IEEE Communications Magazine*, *44*(6), 50–57. doi:10.1109/MCOM.2006.1668419

Haapola, J., Shelby, Z. C., & Mahonen, P. (2005). Cross-layer energy analysis of multi-hop wireless sensor networks. In *Proceedings of European Workshop on Wireless Sensor Networks (EWSN 2005)*, (pp. 33 – 44). EWSN.

Haartsen, J. (2000). The Bluetooth radio system. *IEEE Personal Communications*, *7*(1), 28–36. doi:10.1109/98.824570

Hagenauer, J. (1988). Rate-compatible punctured convolutional codes (RCPC codes) and their applications. *IEEE Transactions on Communications*, *36*(4), 389–400. doi:10.1109/26.2763

Hagenauer, J., & Stockhammer, T. (1999). Channel coding and transmission aspects for wireless multimedia. *Proceedings of the IEEE*, *87*(10), 1764–1777. doi:10.1109/5.790636

Hamzaoui, R., Stankovic, V., & Xiong, Z. (2007). Forward error control for packet loss and corruption. In van der Schaar, M., & Chou, P. (Eds.), *Multimedia over IP and Wireless Networks: Compression, Networking, and Systems* (pp. 271–291). California: Academic Press. doi:10.1016/B978-012088480-3/50010-2

Han, Z., Pandana, C., & Liu, K. J. R. (2007). Distributive opportunistic spectrum access for cognitive radio using correlated equilibrium and no-regret learning. In *Proceedings of IEEE WCNC*. IEEE Press.

Hara, T., & Madria, S. K. (2006). Data replica for improving data accessibility in ad hoc networks. *IEEE Transactions on Mobile Computing*, *5*(11), 1515–1532. doi:10.1109/TMC.2006.165

Haratcherev, I., Taal, J., Langendoen, K., Lagendijk, R., & Sips, H. (2006). Optimized video streaming over 802.11 by cross-layer signaling. *IEEE Communications Magazine*, *44*(1), 115–121. doi:10.1109/MCOM.2006.1580941

Hassibi, B., & Hochwald, B. M. (2003). How much training is needed in multiple-antenna wireless links? *IEEE Transactions on Information Theory*, *49*, 951–963. doi:10.1109/TIT.2003.809594

Hassibi, B., & Vikalo, H. (2005). On the sphere decoding algorithm: Part I, the expected complexity. *IEEE Transactions on Signal Processing*, *53*, 2806–2818. doi:10.1109/TSP.2005.850352

Hatfield, D., & Wiser, P. (2005). Property rights in spectrum: taking the next step. In *Proceedings of DySPAN* (pp. 43–55). IEEE Press. doi:10.1109/DYSPAN.2005.1542616

Haykin, S. (2002). *Adaptive filter theory* (4th ed.). Upper Saddle River, NJ: Prentice Hall.

Haykin, S. (2005). Cognitive radio: Brain-empowered wireless communications. *IEEE Journal on Selected Areas in Communications*, *23*(2), 201–220. doi:10.1109/JSAC.2004.839380

He, T., Gu, L., Luo, L., Yan, T., Stankovic, J., Abdelzaher, T., & Son, S. (2006). An overview of data aggregation architecture for real-time tracking with sensor networks. In *Proceedings of IEEE Real-Time and Embedded Technology and Applications Symposium (RTAS)*, (pp. 55–66). IEEE Press.

Heinzelman, W., Chandrakasan, A., & Balakrishnan, H. (2000). Energy-efficient communication protocol for wireless microsensor networks. In *Proceedings of the 33rd Hawaii International Conference on System Sciences*, (p. 10). IEEE.

Heinzelman, W. B., Chandrakasan, A. P., & Balakrishnan, H. (2002). An application-specific protocol architecture for wireless microsensor networks. *IEEE Transactions on Wireless Communications*, *1*(4), 660–670. doi:10.1109/TWC.2002.804190

Hilton, M. (1997). Wavelet and wavelet packet compression of electrocardiograms. *IEEE Transactions on Bio-Medical Engineering*, *44*(5), 394–402. doi:10.1109/10.568915

Himsoon, T., Pam, W. S., Han, Z., & Liu, K. J. R. (2007). Lifetime maximization via cooperative nodes and relay deployment in wireless networks. *IEEE Journal on Selected Areas in Communications*, *25*(2), 306–317. doi:10.1109/JSAC.2007.070207

Hoesel, L. V., Nieberg, T., Wu, J., & Havinga, P. J. M. (2004). Prolonging the lifetime of wireless sensor networks by cross-layer interaction. *IEEE Wireless Communications*, *11*(6), 78–86. doi:10.1109/MWC.2004.1368900

Holma, H., Kolding, T., Laselva, D., Pedersen, K., Rosa, C., & Viering, I. (2009). Radio resource management. In Holma, H. T. (Ed.), *LTE for UMTS - OFDMA and SC-FDMA Based Radio Access* (pp. 181–212). Chichester, UK: John Wiley & Sons.

Home, I. P. (2011). *H.264/AVC JM reference software*. Retrieved from http://iphome.hhi.de/suehring/tml.

Hongqi, Z., Jiying, Z., & Yang, O. (2008). *Adaptive rate control for VoIP in wireless ad hoc networks*. Paper presented at the ICC 2008. Beijing, China.

Honig, M. L., & Goldstein, J. S. (2002). Adaptive reduced-rank interference suppression based on the multistage Wiener filter. *IEEE Transactions on Communications*, *50*.

Honig, M. L., & Poor, H. V. (1998). Adaptive interference suppression. In Poor, H. V., & Wornell, G. W. (Eds.), *Wireless Communications: Signal Processing Perspectives*. Englewood Cliffs, NJ: Prentice-Hall.

Hoyhtya, M., Pollin, S., & Mammela, A. (2008). Performance improvement with predictive channel selection for cognitive radios. In *Proceedings of the Cognitive Radio and Advanced Spectrum Management*, (pp. 1–5). Aalborg, Denmark: IEEE.

Ho, Z. K. M., Lau, V. K. N., & Cheng, R. S.-K. (2009). Cross-layer design of FDD-OFDM systems based on ACK/NAK feedbacks. *IEEE Transactions on Information Theory, 55*(10), 4568–4584. doi:10.1109/TIT.2009.2025565

Hsu, C., & Hefeeda, M. (2007). Optimal bit allocation for fine-grained scalable video sequences in distributed streaming environments. In *Proceedings of ACM/SPIE Multimedia Computing and Networking Conference*, (pp. 1–12). San Jose, CA: ACM Press.

Hsu, Y. P., & Feng, K. T. (2008). Cross-layer routing for congestion control in wireless sensor networks. In *Proceedings of the IEEE Radio and Wireless Symposium*, (pp. 783-786). Orlando, FL: IEEE Press.

Hsu, J. L., & van der Schaar, M. (2009). Cross layer design and analysis of multiuser wireless video streaming over 802.11e EDCA. *IEEE Signal Processing Letters, 16*(4), 268–271. doi:10.1109/LSP.2008.2010821

Huang, C. M., Chao-Hsien Lee, C. H., & Tseng, P. H. (2009). Multihomed SIP-based network mobility using IEEE 802.21 media independent handover. In *Proceedings of the IEEE International Conference on Communications (IEEE ICC 2009)*. IEEE Press.

Huang, J., Berry, R., & Honig, M. (2005). Spectrum sharing with distributed interference compensation. In *Proceedings of IEEE DySPAN*, (pp.88-93). IEEE Press.

Huang, W. J., Hong, Y. W., & Kuo, C. C. J. (2007). Decode-and-forward cooperative relay with multi-user detection in uplink CDMA networks. In *Proceedings of the IEEE GLOBECOM*, (pp. 4397-4401). IEEE Press.

Hunter, G. W., Neudeck, P. G., Okojie, R. S., Beheim, G. M., & Powell, J. A. (2003). An overview of high-temperature electronics and sensor development at NASA Glenn research center. *Turbomachinery, 125*(4), 658–665. doi:10.1115/1.1579508

Hurwitz, J., Bloor, R., Kaufman, M., & Halper, F. (2009). *Cloud computing for dummies*. New York, NY: For Dummies.

IEEE Standard for Information Technology. (2005). Telecommunications and information exchange between systems - Local and metropolitan area networks - Specific requirements - Part 11: Wireless LAN medium access control (MAC) and physical layer (PHY) specifications - Amendment 8: Medium access method (MAC) quality of service enhancements. *IEEE Std 802.11e-2005*. Retrieved from http://www.ieee.org.

IEEE. (1999). *802.11 standard for wireless LAN medium access control and physical layer specifications*. New York, NY: IEEE Computer Society LAN MAN Standards Committee.

IEEE. (2003). *Draft supplement to part 11: Wireless LAN medium access control (MAC) and physical layer (PHY) specifications: Medium access control (MAC) enhancements for quality of service (QoS), IEEE 802.11e/d5.0*. Retrieved from http://www.ieee.org.

IEEE. (2004). *802.16 standard for local and metropolitan area networks, part 16: Air interface for fixed broadband wireless access systems*. New York, NY: IEEE Computer Society, IEEE Microwave Theory and Techniques Society.

IEEE. (2005). Standard for local and metropolitan area networks—Part 16: Air interface for fixed and mobile broadband wireless access systems—Amendment for physical and medium access control layers for combined fixed and mobile operation in licensed bands. *IEEE 802.16e/D8-2005*. Retrieved from http://www.ieee.org.

IEEE. (2006). Draft standard for local and metropolitan area networks: Media independent handover services. *IEEE P802.21/D03.00*. Retrieved from http://www.ieee.org.

IEEE. (2007). *Wireless LAN medium access control (MAC) and physical layer (PHY) specifications*. Retrieved from http://www.ieee.org.

IEEE. 802.16e-2005. (2005). *IEEE standard for local and metropolitan area networks: Part 16: Air interface for fixed and mobile broadband wireless access systems*. Retrieved from http://www.ieee.org.

IEEE. 802.16's Relay Task Group. (2011). *Website*. Retrieved from http://www.ieee802.org/16/relay/.

IEEE. 802.21 WG. (2008). *IEEE standard for local and metropolitan area networks, media independent handover services: IEEE 802.21-2008*. Retrieved from http://www.ieee.org.

IEEE. Computer Society. (2002). IEEE trial-use recommended practice for multi-vendor access point interoperability via an inter-access point protocol across distribution systems supporting IEEE 802.11 operation. *Internet Engineering Task Force (IETF) Internet-Draft*. Retrieved from http://www.ietf.org.

IEEE802. 11. (1997). *IEEE standard 802.11: Wireless LAN medium access control (MAC) and physical layer (PHY) specifications*. New York, NY: IEEE Computer Society

Ileri, O., Samardzija, D., & Manadayam, N. (2005). Demand responsive pricing and competitive spectrum allocation via a spectrum server. In *Proceedings of IEEE DySPAN,* (pp. 194-202). IEEE Press.

Inoiue, M., Mahmuhd, K., Murakaami, H., & Hasegawa, M. (2004). Novel out-of-band signaling for seamless interworking between heterogeneous networks. *IEEE Wireless Communications, 11*(2), 56–63. doi:10.1109/MWC.2004.1295740

Intanagonwiwat, C., Govindan, R., & Estrin, D. (2000). Directed diffusion: A scalable and robust communication paradigm for sensor networks. In *Proceedings of the Annual International Conference on Mobile Computing and Networking (MOBICOM)*, (pp. 56–67). Boston, MA: ACM Press.

Iskander, C.-D., & Mathiopoulos, P. T. (2003). Fast simulation of diversity Nakagami fading channels using finite-state Markov models. *IEEE Transactions on Broadcasting, 49*, 269–277. doi:10.1109/TBC.2003.817096

ISO. (2004). MPEG-4 visual reference software. *ISO/IEC 14496*. Retrieved from http://www.iso.org.

ITelSib, Ltd. (2011). *IPHost network monitor*. Retrieved from http://www.iphostmonitor.com/.

Jain, K., Padhye, J., Padmanabhan, V., & Qiu, L. (2003). Impact of interference on multi-hop wireless network performance. In *Proceedings of the ACM Mobicom*. San Diego, CA: ACM Press.

Jambli, M. N., & Tully, A. (2007). Cross-layer design for information dissemination in wireless sensor networks: State-of-the-art and research challenges. In *Proceedings of the 37th Annual IEEE/IFIP International Conference on Dependable Systems and Networks,* (pp 13-18). Edinburgh, UK: IEEE Press.

Jamieson, B. K., Balakrishnan, H., & Morris, R. (2001). An energy-efficient coordination algorithm for topology maintenance in ad hoc wireless networks. In *Proceedings of MOBICOM*. MOBICOM.

Jang, H. J., Jee, J. H., Han, Y. H., & Cha, J. (2007). Mobile IPv6 fast handovers over IEEE 802.16e networks. *Internet Engineering Task Force (IETF), Internet-Draft*. Retrieved from http://www.ietf.org.

Jang, J. (2007). Gaussian mixture model. In *Data Clustering and Pattern Recognition*. Retrieved from http://mirlab.org/jang.

Jeong, Y., Kakumanu, S., Tsao, C.-L., & Sivakumar, R. (2009). VoIP over Wi-Fi networks: Performance analysis and acceleration algorithms. *Mobile Networks and Applications, 14*(4), 523–538. doi:10.1007/s11036-009-0157-6

Jeruchim, M. C., Balaban, P., & Shanmugan, K. S. (2000). *Simulation of communication systems - Modeling, methodology, and techniques*. New York, NY: Kluver Academic/Plenum Publishers.

Jiang, C., Yuan, D., & Zhao, Y. (2009). Towards clustering algorithms in wireless sensor networks-A survey. In *Proceedings of the IEEE Wireless Communications and Networking Conference*, (pp. 1-6). IEEE Press.

Jin, S. (2004). Replication of partitioned media streams in wireless ad hoc networks. In *Proceedings of the ACM Multimedia,* (pp. 396–399). New York, NY: ACM Press.

Jin, S., & Wang, L. (2005). Content and service replica strategies in multi-hop wireless mesh networks. In *Proceedings of the International Symposium on Modeling Analysis and Simulation of Wireless and Mobile Systems,* (pp. 79–86). Montreal, Canada: IEEE.

Jing, Y., & Hassibi, B. (2006). Distributed space-time coding in wireless relay networks. *IEEE Transactions on Wireless Communications*, *5*(12), 3524–3536. doi:10.1109/TWC.2006.256975

Jo, O., & Cho, D. H. (2008). Efficient spectrum matching based on spectrum characteristics in cognitive radio systems. In *Proceedings of the IEEE Wireless Telecommunications Symposium*, (pp. 230–235). Pomona, CA: IEEE Press.

Johnson, D., Hu, Y., & Maltz, D. (2007). The dynamic source routing protocol (DSR) for mobile ad hoc networks for IPv4. *RFC: 4728*. Retrieved from http://www.ietf.org.

Johnson, D., Perkins, C., & Arkko, J. (2004). Mobility support in IPv6. *Internet Engineering Task Force (IETF), RFC-3775*. Retrieved from http://www.ietf.org.

Johnson, D. B., & Maltz, D. A. (1996). Dynamic source routing in ad hoc wireless networks. In *Mobile Computing* (pp. 153–181). Dordrecht, The Netherlands: Kluwer. doi:10.1007/978-0-585-29603-6_5

Joseph, J. F. C., Das, A., Seet, B. C., & Lee, B. S. (2007). Cross layer versus single layer approaches for intrusion detection in MANETs. In *Proceedings of the 15th IEEE International Conference on Networks (ICON 2007)*, (pp. 194-200). IEEE Press.

Joseph, J. F. C., Das, A., Seet, B. C., & Lee, B. S. (2008). CRADS: Integrated cross layer approach for detecting routing attacks in MANETs. In *Proceedings of the IEEE Wireless Communications and Networking Conference (WCNC 2008)*, (pp. 1525-1531). IEEE Press.

Joseph, J. F. C., Lee, B. S., Das, A., & Seet, B. C. (2011). Cross-layer detection of sinking behavior in wireless ad hoc networks using SVM and FDA. *IEEE Transactions on Dependable and Secure Computing*, *8*(2), 233–245. doi:10.1109/TDSC.2009.48

Joung, J., & Sayed, A. H. (2010). Multiuser two-way amplify-and-forward relay processing and power control methods for beamforming systems. *IEEE Transactions on Signal Processing*, *58*, 1833–1846. doi:10.1109/TSP.2009.2038668

J-Sim. (2001). *J-sim: An object-oriented library for discrete-time process-oriented simulation.* Retrieved from http://www.j-sim.zcu.cz/.

J-Sim. (2011). *Website.* Retrieved Nov 23, 2011, from http://www.j-sim.org/.

Juang, P., Oki, H., Wang, Y., Martonosi, M., Peh, L. S., & Rubenstein, D. (2002). Energy-efficient computing for wildlife tracking: Design tradeoffs and early experiences with zebranet. *SIGARCH Computer Architecture News*, *30*(5), 96–107. doi:10.1145/635506.605408

Juan, H.-H., Huang, H.-C., Huang, C., & Chiang, T. (2009). Cross-layer mobile WiMAX MAC designs for the H.264/SVC scalable video coding. *Wireless Networks*, *16*(1), 113–123. doi:10.1007/s11276-008-0118-1

Juels, A., & Waternberg, M. (1999). A fuzzy commitment scheme. In *Proceedings of the 6th ACM Conference on Computer and Communications Security*, (pp. 28-36). ACM Press.

Jung, E.-S., & Vaidya, N. H. (2002a). An energy efficient mac protocol for wireless LANs. In *Proceedings of INFOCOM*. INFOCOM.

Jung, E.-S., & Vaidya, N. H. (2002b). A power control MAC protocol for ad hoc networks. In *Proceedings of ACM MobiCom*, (pp. 36–47). ACM Press.

Jung, E.-S., & Vaidya, N. H. (2005). A power control MAC protocol for ad hoc networks. *Wireless Networks*, *11*(1–2), 55–66. doi:10.1007/s11276-004-4746-9

Jurdak, R. (2007). *Wireless ad hoc and sensor networks: A cross-layer design perspective*. Berlin, Germany: Springer Verlag.

Kalita, H. K., & Kar, A. (2009). Wireless sensor network security analysis. *International Journal of Next-Generation Networks*, *1*(1), 1–10.

Kalman, M., Steinbach, E., & Girod, B. (2002). Adaptive media playout for low delay video streaming over error-prone channels. In *Proceedings of the International Conference on Image Processing*, (pp. 841-851). IEEE.

Kanal, L., & Sastry, A. (1978). Models for channels with memory and their applications to error control. *Proceedings of the IEEE*, *66*(7), 724–744. doi:10.1109/PROC.1978.11013

Kang, K., Won, J., & Jeon, W. J. (2012). Differentiated protection of video layers to improve perceived quality. To appear in *IEEE Transactions on Mobile Computing, 11*(2), 292-304.

Kang, K., Cho, Y., & Shin, H. (2007). Energy-efficient MAC-layer error recovery for mobile multimedia applications in 3GPP2 BCMCS. *IEEE Transactions on Broadcasting, 53*(1), 338–349. doi:10.1109/TBC.2007.891699

Kang, K., & Kim, T. (2009). Improved error control for real-time video broadcasting over CDMA2000 networks. *IEEE Transactions on Vehicular Technology, 58*(1), 188–197. doi:10.1109/TVT.2008.926077

Kang, M., & Alouini, M.-S. (2004). A comparative study on the performance of MIMO MRC systems with and without cochannel interference. *IEEE Transactions on Communications, 52*(18), 1417–1425. doi:10.1109/TCOMM.2004.833027

Karaki, J. N., Mustafa, R., & Kamal, A. E. (2009). Data aggregation and routing in wireless sensor networks: Optimal and heuristic algorithms. *Computer Networks, 53*(7), 945–960. doi:10.1016/j.comnet.2008.12.001

Karl, H., & Willig, A. (2006). *A short survey of wireless sensor networks*. TKN Technical Report: TKN-03-018. Retrieved from http://www.cs.uni-paderborn.de/fileadmin/Informatik/AG-Karl/Publications/Technical_Report/TechReport_03_018.pdf.

Karl, H., & Willig, A. (2003). *A short survey of wireless sensor networks. TKN Technical Report*. Berlin, Germany: Technical University Berlin.

Karmokar, A. K., & Bhargava, V. K. (2009). Performance of cross-layer optimal adaptive transmission techniques over diversity Nakagami-m fading channels. *IEEE Transactions on Communications, 57*(12), 3640–3652. doi:10.1109/TCOMM.2009.12.080262

Karmokar, A. K., Djonin, D. V., & Bhargava, V. K. (2006). Optimal and suboptimal packet scheduling over time-varying flat fading channels. *IEEE Transactions on Wireless Communications, 5*, 446–457. doi:10.1109/TWC.2006.1611068

Karmokar, A. K., Djonin, D. V., & Bhargava, V. K. (2006). POMDP-based coding rate adaptation for type-i hybrid ARQ systems over fading channels with memory. *IEEE Transactions on Wireless Communications, 5*, 3512–3523. doi:10.1109/TWC.2006.256974

Karn, P. (1990). MACA-A new channel access method for packet radio. In *Proceedings of 9th ARRL Computer Networking Conference*. ARRL.

Kasiviswanath, N., Charles, B. S., & Reddy, P. C. (2007). A combined tuning of power control in wireless ad hoc networks using cross layer architecture. *Indian Journal of Computing Technology, 2*(2), 1–30.

Kasiviswanath, N., Charles, B. S., & Reddy, P. C. (2008). A new power conservative routing for wireless mobile ad hoc networks using cross layer design. *International Journal of Computer Science and Network Security, 8*(3).

Katsaggelos, A. K., Eisenberg, Y., Zhai, F., Berry, R., & Pappas, T. N. (2005). Advances in efficient resource allocation for packet-based real-time video transmission. *Proceedings of the IEEE, 93*(1), 135–147. doi:10.1109/JPROC.2004.839621

Kawadia, V., & Kumar, P. R. (2003). A cautionary perspective on cross layer design. *IEEE Wireless Communications Magazine, 12*(1), 3–11. doi:10.1109/MWC.2005.1404568

Kawadia, V., & Kumar, P. R. (2003). Power control and clustering in ad hoc networks. In *Proceedings of IEEE INFOCOM* (pp. 459–469). IEEE Press.

Kawadia, V., & Kumar, P. R. (2005). A cautionary perspective on cross-layer design. *IEEE Wireless Communications, 12*(1), 3–11. doi:10.1109/MWC.2005.1404568

Kawadia, V., & Kumar, P. R. (2005). Principles and protocols for power control in wireless ad hoc networks. *IEEE Journal on Selected Areas in Communications, 23*(1). doi:10.1109/JSAC.2004.837354

Khadra, L., Al-Fahoum, A., & Al-Nashash, H. (1997). Detection of lifethreatening cardiac arrhythmias using the wavelet transformation. *Medical & Biological Engineering & Computing, 35*(6), 626–632. doi:10.1007/BF02510970

Khan, A., Sun, L., & Ifeachor, E. (2009). Impact of video content on video quality for video over wireless networks. In *Proceedings of the International Conference on Autonomic and Autonomous Systems,* (pp. 277–282). Valencia, Spain: IEEE.

Khan, M. A., Moinuddin, A. A., Khan, E., & Ghanbari, M. (2010). Reliable transmission of wavelet-based scalable video over wireless networks using cross-layer approach. *IET Communications*, *4*(11), 1325–1336. doi:10.1049/iet-com.2009.0412

Khan, S., Peng, Y., Steinbach, E., Sgroi, M., & Kellerer, W. (2006). Application driven cross-layer optimization for video streaming over wireless networks. *IEEE Communications Magazine*, *44*(1), 122–130. doi:10.1109/MCOM.2006.1580942

Khatri, C. G. (1964). Distribution of the largest or the smallest characteristic root under null hyperthesis concerning complex multivariate normal populations. *Annals of Mathematical Statistics*, *35*, 1807–1810. doi:10.1214/aoms/1177700403

Kim, H., Kim, Y., & Yoo, H. J. (2008). A low cost quadratic level ECG compression algorithm and its hardware optimization for body sensor network system. In *Proceedings of the 30th Annual International Conference of the IEEE,* (pp. 5490-5493). IEEE Press.

Kim, D. P., & Koh, S. J. (2009). Performance enhancement of mSCTP for vertical handover across heterogeneous wireless networks. *International Journal of Communication Systems*, *22*(12), 1573–1591. doi:10.1002/dac.1038

Kim, J., & Park, K. H. (2009). An energy-efficient, transport-controlled MAC protocol for wireless sensor networks. *Computer Networks*, *53*(11), 1879–1902. doi:10.1016/j.comnet.2009.03.002

Kim, S. W., Kim, S.-Y., Kim, S., & Heo, J. (2008). Performance analysis of forward error correcting codes in IPTV. *IEEE Transactions on Consumer Electronics*, *54*(2), 376–380. doi:10.1109/TCE.2008.4560102

Kirschnick, J., Alcaraz Calero, J., Wilcock, L., & Edwards, N. (2010). Towards an architecture for the automated provisioning of cloud services. *IEEE Communications Magazine*, *48*(12). doi:10.1109/MCOM.2010.5673082

Kitae, N., Ahmed, H., & Kuo, C. C. J. (2005). TCP over multihop 802.11 networks: Issues and performance enhancement. In *Proceedings of the 6th ACM International Symposium on Mobile Ad Hoc Networking and Computing*. Urbana-Champaign, IL: ACM Press.

Kleinrock, L. (1975). *Queueing systems* (*Vol. 1*). New York, NY: Wiley.

Kliazovich, D., Gomes, P. H., Granelli, F., & da Fonseca, N. L. (2009). Receiver-driven queue management for achieving RTT-fairness in Wi-Fi networks. In *Proceedings from Globecom*. Honolulu, HI: IEEE Communications Society. doi:10.1109/GLOCOM.2009.5426120

Kliazovich, D., Granelli, F., & Gerla, M. (2007). Performance improvement in wireless networks using cross-layer ARQ. *Computer Networks*, *51*, 4396–4411. doi:10.1016/j.comnet.2007.06.002

Ko, J., Dawson-Haggerty, S., Gnawali, O., Culler, D., & Terzis, A. (2011a). Evaluating the performance of rpl and 6lowpan in tinyos. In *Proceedings of the International Symposium on Information Processing in Sensor Networks (IPSN), Workshop*. IEEE Press.

Kobayashi, H., Fukuhara, T., Yuan, H., & Takeuchi, Y. (2003). Proposal of single carrier OFDM technique with adaptive modulation method. In *Proceedings of IEEE Vehicular Technology Conference,* (pp. 1915-1919). IEEE Press.

Koetter, R., & Medard, M. (2003). An algebraic approach to network coding. *IEEE/ACM Transactions on Networking*, *11*(5), 782–795. doi:10.1109/TNET.2003.818197

Kohler, E., Handley, M., & Floyd, S. (2006). *Datagram congestion control protocol (DCCP), RFC 4340*. Retrieved from http://www.ietf.org.

Koh, S., Chang, M., & Lee, M. (2004). mSCTP for soft handover in transport layer. *IEEE Communications Letters*, *8*(3), 189–191. doi:10.1109/LCOMM.2004.823432

Ko, J., Terzis, A., Dawson-Haggerty, S., Culler, D., Hui, J., & Levis, P. (2011b). Connecting low-power and lossy networks to the internet. *IEEE Communications Magazine*, *49*(4), 96–101. doi:10.1109/MCOM.2011.5741163

Kolodzy, P. J. (2006). Interference temperature: A metric for dynamic spectrum utilization. *International Journal of Network Management*, *16*(12).

Koodli, R. (2005). Fast handovers for mobile IPv6. *Internet Engineering Task Force (IETF), RFC-4068*. Retrieved from http://www.ietf.org.

Kota, S., Giambene, G., & Candio, N. L. (2007). Cross-layer approach for an air interface of GEO satellite communication networks. *International Journal of Satellite Communications and Networking*, *25*(5), 481–499. doi:10.1002/sat.887

Koubâa, A., Cunha, A., Alves, M., & Tovar, E. (2008). Tdbs: A time division beacon scheduling mechanism for zigbee cluster-tree wireless sensor networks. *Real-Time Systems*, *40*(3), 321–354. doi:10.1007/s11241-008-9063-4

Kozat, U. C., Koutsopoulos, I., & Tassiulas, L. (2004). A framework for cross-layer design of energy-efficient communication with QoS provisioning in multi-hop wireless networks. In *Proceedings of IEEE International Conference on Computer Communications (INFOCOM 2004)*, (pp. 1-5). IEEE Press.

Kozat, U. C., Koutsopoulos, I., & Tassiulas, L. (2004). A framework for cross-layer design of energy-efficient communication with QoS provisioning in multi-hop wireless networks. In *Proceedings of IEEE INFOCOM* (pp. 1446–1456). IEEE Press. doi:10.1109/INFCOM.2004.1357029

Kramer, G., Maric, I., & Yates, R. D. (2006). Cooperative communications. *Foundations and Trends in Networking, 1*(3-4).

Kramer, G., Gastpar, M., & Gupta, P. (2005). Cooperative strategies and capacity theorems for relay networks. *IEEE Transactions on Information Theory*, *51*, 3037–3063. doi:10.1109/TIT.2005.853304

Kratos Network Solutions. (2011). *dopplerVUE*. Retrieved from http://www.kratosnetworks.com/products/dopplervue/.

Krause, M., van Hartskamp, M., & Aarts, E. (2008). A quality metric for use with frame-rate based bandwidth adaptation algorithms. *Society of Photo-Optical Instrumentation Engineers (SPIE) Conference Series*. SPIE.

Krishnamachari, S., van der Schaar, M., Choi, S., & Xu, X. (2003). Video streaming over wireless LANs: A cross-layer approach. In *Proceedings of the 13th International Packet Video Workshop*. IEEE.

Krishnamurthy, V., & Djonin, D. V. (2007). Structured threshold policies for dynamic sensor scheduling—A partially observed Markov decision process approach. *IEEE Transactions on Signal Processing*, *55*(10), 4938–4957. doi:10.1109/TSP.2007.897908

Ksentini, A., Naimi, M., & Gueroui, A. (2006). Toward an improvement of H.264 video transmission over IEEE 802.11e through a cross-layer architecture. *IEEE Communications Magazine*, *44*(1), 107–114. doi:10.1109/MCOM.2006.1580940

Kuhran, M. S., Gür, G., Togcu, T., & Alagöz, F. (2010). Applications of the cross-layer paradigm for improving the performance of WiMAX. *IEEE Wireless Communications*, *17*(3), 86–95. doi:10.1109/MWC.2010.5490983

Kulkarni, S. S., & Rosenberg, C. (2003). Opportunistic scheduling policies for wireless systems with short term fairness constraints. In *Proceedings of IEEE GLOBECOM 2003,* (Vol.1), (pp. 533–537). San Francisco, CA: IEEE Press.

Kulkarni, S., Iyer, A., & Rosenberg, C. (2006). An address-light, integrated MAC and routing protocol for wireless sensor networks. *IEEE/ACM Transactions on Networking*, *14*(4), 793–806. doi:10.1109/TNET.2006.880163

Kumar, P. R. (1985). A survey of some results in stochastic adaptive control. *Society for Industrial and Applied Mathematics Journal on Control and Optimization*, *23*(3), 329–380.

Kumwilaisak, W., Hou, Y. T., Zhang, Q., Zhu, W., Kuo, C.-C. J., & Zhang, Y.-Q. (2003). A cross-layer quality-of-service mapping architecture for video delivery in wireless network. *IEEE Journal on Selected Areas in Communications*, *21*(10), 1685–1698. doi:10.1109/JSAC.2003.816445

Kuran, M. S., Gur, G., Tugcu, T., & Alagoz, F. (2010). Applications of the cross-layer paradigm for improving the performance of WiMax. *IEEE Wireless Communications*, *17*(3), 86–95. doi:10.1109/MWC.2010.5490983

Kwon, H., Kim, T. H., Choi, S., & Lee, B. G. (2006). A cross-layer strategy for energy-efficient reliable delivery in wireless sensor networks. *IEEE Transactions on Wireless Communications*, *5*(12), 3689–3699. doi:10.1109/TWC.2006.256992

Kyritsi, P., Popovski, P., Eggers, P., Wang, Y., Khan, D. A., Bouaziz, A. E., et al. (2007). Cooperative transmission: A reality check using experimental data. In *Proceedings of the IEEE 65th Vehicular Technology Conference, 2007 (VTC 2007-Spring)*, (pp. 2281–2285). IEEE Press.

Lachenmann, A., Marron, P. J., Minder, D., & Rothermel, K. (2005). An analysis of cross-layer interactions in sensor network applications. In *Proceedings of the International Conference on Intelligent Sensors, Sensor Networks and Information Processing,* (pp. 121-126). Melbourne, Australia: IEEE Press.

Lagkas, T. D., Papadimitriou, G. I., Nicopolitidis, P., & Pomportsis, A. S. (2007). Priority-oriented adaptive control with QoS guarantee for wireless LANs. *IEEE Transactions on Vehicular Technology, 56*(4), 1761–1772. doi:10.1109/TVT.2007.897223

Lai, W. K., & Chiu, J. C. (2005). Improving handoff performance in wireless overlay networks by switching between two-layer IPv6 and one- layer IPv6 addressing. *IEEE Journal on Selected Areas in Communications, 23*(11), 621–628.

Lakshman, T., Ortega, A., & Reibman, A. (1998). VBR video: Tradeoffs and potentials. *Proceedings of the IEEE, 86*(5), 952–973. doi:10.1109/5.664282

Laneman, J. N., Tse, D. N. C., & Wornell, G. W. (2004). Cooperative diversity in wireless networks: Efficient protocols and outage behavior. *IEEE Transactions on Information Theory, 50*(12), 3062–3080. doi:10.1109/TIT.2004.838089

Laneman, J. N., & Wornell, G. W. (2003). Distributed space-time-coded protocols for exploiting cooperative diversity in wireless networks. *IEEE Transactions on Information Theory, 49*(10), 2415–2425. doi:10.1109/TIT.2003.817829

Lavlu, S. M. (2009). *Cacti 0.8 network monitoring*. New York, NY: Packt Publishing.

Layuan, L., Chunlin, L., & Peiyan, Y. (2007). Performance evaluation and simulations of routing protocols in ad hoc networks. *Computer Communications, 30*(8), 1890–1898. doi:10.1016/j.comcom.2007.02.015

Lazar, G., Dobrota, V., & Blaga, T. (2008). Practical evaluation of H.264 video streaming over IEEE 802.11e devices by cross-layer signaling. In *Proceedings of the 16th IEEE Workshop on Local and Metropolitan Area Networks LANMAN 2008,* (pp. 13-18). Cluj-Napoca, Romania: IEEE Press.

Lazos, L., & Poovendran, R. (2004). Cross-layer design for energy-efficient secure multicast communications in ad hoc networks. In *Proceedings of the 2004 IEEE International Conference on Communications (ICC)*, (Vol. 6), (pp. 3633-3639). IEEE Press.

Lazos, L., Salido, J., & Poovendran, R. (2004b). Vp3: Using vertex path and power proximity for energy efficient key distribution. In *Proceedings of the IEEE 60th Vehicular Technology Conference,* (Vol. 2), (pp. 1228-1232). Los Angeles, CA: IEEE Press.

Lee, C. W., Chen, L. M., Chen, M. C., & Sun, Y. S. (2005). A framework of handoffs in wireless overlay networks based on mobile IPv6. *IEEE Selected Areas in Communications, 23*(11), 2118–2128. doi:10.1109/JSAC.2005.856833

Lee, G., & Song, H. (2010). Cross layer optimized video streaming based on IEEE 802.11 multi-rate over multi-hop mobile ad hoc networks. *Mobile Networks and Applications, 15*(5), 652–663. doi:10.1007/s11036-009-0196-z

Lefol, D., Bull, D., & Canagarajah, N. (2006). Performance evaluation of transcoding algorithms for H.264. *IEEE Transactions on Consumer Electronics, 52*(1), 215–222. doi:10.1109/TCE.2006.1605050

Lehr, W., & Crowcroft, J. (2005). Managing shared access to a spectrum commons. In *Proceedings of IEEE DySPAN,* (pp. 110-118). IEEE Press.

Le, L. B., Hossain, E., & Alfa, A. S. (2006). Radio link level performance evaluation in wireless networks using multi-rate transmission with ARQ-based error control. *IEEE Transactions on Wireless Communications, 5*(10), 2647–2653. doi:10.1109/TWC.2006.04485

Leon-Garcia, A., & Widjaja, I. (2004). *Communication networks: Fundamental concepts and key architectures* (2nd ed.). New Delhi, India: McGraw-Hill.

Levis, P., et al. (2011). *The trickle algorithm*. draft 8, IETF. Retrieved from http://www.ietf.org.

Li, D., Cheung, G., Chuah, C.-N., & Yoo, S. B. (2004). Joint server/peer receiver-driven rate-distortion optimized video streaming using asynchronous clocks. In *Proceedings of IEEE International Conference on Image Processing,* (pp. 3149–3152). Singapore: IEEE Press.

Li, J., & Zhou, Y. (2010). *Wireless sensor networks: Application-centric design.* Retrieved from http://www.intechopen.com/articles/show/title/target-tracking-in-wireless-sensor-network.

Li, J., Jannotti, J., Couto, D. S. J. D., Karger, D. R., & Morris, R. (2000). A scalable location service for geographic ad hoc routing. In *Proceedings of the 6th Annual International Conference on Mobile Computing and Networking.* Boston, MA: IEEE.

Li, S., & Ephremides, A. (2006). Anonymous routing: A cross-layer coupling between application and network layer. In *Proceedings of the 2006 40th Annual Conference on Information Sciences and Systems,* (pp. 783-788). IEEE.

Li, X., Zhou, R., Chakravarthy, V., & Wu, Z. (2009). Intercarrier interference immune single carrier OFDM via magnitude shift keying modulation. In *Proceedings of IEEE Globecom,* (pp. 1-6). IEEE Press.

Liang, Q., Yuan, D., Wang, Y., & Chen, H.-H. (2007). A cross-layer transmission scheduling scheme for wireless sensor networks. *Computer Communications, 30*(15), 2987–2994. doi:10.1016/j.comcom.2007.05.025

Li, C., & Weihua, Z. (2002). Hybrid tdoa/aoa mobile user location for wideband CDMA cellular systems. *IEEE Transactions on Wireless Communications, 1*(3), 439–447. doi:10.1109/TWC.2002.800542

Lieberman, H. (2005). *Introduction to operations research.* New York, NY: McGraw Hill.

Li, F., Liu, G., & He, L. (2010). Cross-layer scheduling for multiuser H.264 video transmission over wireless networks. *IET Communications, 4*(8), 1012–1025. doi:10.1049/iet-com.2009.0618

Li, L. (2011). Overview of fine granularity scalability in MPEG-4 video standard. *IEEE Transactions on Circuits and Systems for Video Technology, 11*(3), 301–317.

Lin, Z., Sinclair, M., & Bittner, R. (2006). A phone-centered body sensor network platform cost, energy efficiency and user interface. In *Proceedings of the Wearable and Implantable Body Sensor Networks Workshop,* (pp. 179-182). IEEE.

Lin, C. H., Shieh, C. K., Ke, C. H., Chilamkurti, N. K., & Zeadally, S. (2009). An adaptive cross-layer mapping algorithm for MPEG-4 video transmission over IEEE 802.11 e WLAN. *Telecommunication Systems, 42*(3), 223–234. doi:10.1007/s11235-009-9182-9

Lin, C.-H., Ce-Kuen Shieh, N., Chilamkurti, K., Ke, C.-H., & Hwang, W.-S. (2008). A RED-FEC mechanism for video transmission over WLANs. *IEEE Transactions on Broadcasting, 54,* 517–524. doi:10.1109/TBC.2008.2001713

Lindeberg, M., Kristiansen, S., Plagemann, T., & Goebel, V. (2010). Challenges and techniques for video streaming over mobile ad-hoc networks. *Multimedia Systems, 17*(1), 51–82. doi:10.1007/s00530-010-0187-8

Lin, S., & Costello, D. J. (2004). *Error control coding: Fundamentals and applications* (2nd ed.). Englewood Cliffs, NJ: Prentice Hall.

Lin, S., Costello, D. J., & Miller, M. J. (1984). Automatic repeat error control schemes. *IEEE Communications Magazine, 22*(12), 5–17. doi:10.1109/MCOM.1984.1091865

Lin, T.-L., & Cosman, P. C. (2010). Efficient optimal RCPC code rate allocation with packet discarding for pre-encoded compressed video. *IEEE Signal Processing Letters, 17*(5), 505–508. doi:10.1109/LSP.2010.2044938

Lin, X. J., Shroff, N. B., & Srikant, R. A. (2006). Tutorial on cross-layer optimization in wireless networks. *IEEE Journal on Selected Areas in Communications, 24*(8), 1452–1463. doi:10.1109/JSAC.2006.879351

Li, P., Chang, Y., Feng, N., & Yang, F. (2010). A novel hierarchical QAM-based unequal error protection scheme for H.264/AVC video over frequency-selective fading channels. *IEEE Transactions on Consumer Electronics, 56*(4), 2741–2746. doi:10.1109/TCE.2010.5681164

Li, Q., & van der Schaar, M. (2004). Providing adaptive QoS to layered video over wireless local area networks through real-time retry limit adaptation. *IEEE Transactions on Multimedia, 6*(2), 278–290. doi:10.1109/TMM.2003.822792

Liu, K. R. (2011). Cognitive-radio games. *IEEE Spectrum*. Retrieved from http://spectrum.ieee.org/telecom/wireless/cognitive-radio-and-game-theory.

Liu, Q., Zhou, S., & Giannakis, G. B. (2004b). TCP performance in wireless access with adaptive modulation and coding. In *Proceedings of the IEEE International Communications Conference*, (pp. 3989-3993). Paris, France: IEEE Press.

Liu, T., Wang, Y., Boyce, J., Wu, Z., & Yang, H. (2007). Subjective quality evaluation of decoded video in the presence of packet losses. In *Proceedings of the IEEE International Conference on Acoustics, Speech and Signal Processing (ICASSP),* (pp. 1125-1128). IEEE Press.

Liu, Y., He, Y., Li, M., Wang, J., Liu, K., & Mo, L. … Li, X.-Y. (2011). Does wireless sensor network scale? A measurement study on greenorbs. In *Proceedings of INFOCOM*, (pp. 873–881). IEEE Press.

Liu, Y., Li, Y., & Man, H. (2005). Short paper: A distributed cross-layer intrusion detection system for ad hoc networks. In *Proceedings of the First International Conference on Security and Privacy for Emerging Areas in Communications Networks (SecureComm 2005)*, (pp. 418-420). SecureComm.

Liu, Z., Kwiatkowska, M., & Constantinou, C. (2005). A biologically inspired qos routing algorithm for mobile ad hoc networks. In *Proceedings of the International Conference on Advanced Information Networking and Applications (AINA)*, (vol 1), (pp. 426 – 431). IEEE Press.

Liu, B., Marins, P., & Bertin, P. (2010). Cross-layer design of the inter-RAT handover between UMTS and WiMAX. *EURASIP Journal on Wireless Communications and Networking*, 10.

Liu, C., & Kaiser, J. (2003). *A survey of mobile ad hoc network routing protocols*. Ulm, Germany: University of Ulm.

Liu, H., Ma, H., Zarki, M. E., & Gupta, S. (1997). Error control schemes for networks: An overview. *Mobile Networks and Applications*, *2*(2), 167–182. doi:10.1023/A:1013676531988

Liu, Q., Zhou, S., & Giannakis, G. (2005). Cross-layer scheduling with prescribed QoS guarantees in adaptive wireless networks. *IEEE Journal on Selected Areas in Communications*, *23*(5), 1056–1066. doi:10.1109/JSAC.2005.845430

Liu, Q., Zhou, S., & Giannakis, G. B. (2004). Cross-layer combining of adaptive modulation and coding with truncated ARQ over wireless links. *IEEE Transactions on Wireless Communications*, *3*(5), 1746–1755. doi:10.1109/TWC.2004.833474

Liu, Q., Zhou, S., & Giannakis, G. B. (2005). Cross-layer scheduling with prescribed QoS guarantees in adaptive wireless networks. *IEEE Journal on Selected Areas in Communications*, *23*(5), 1056–1066. doi:10.1109/JSAC.2005.845430

Liu, Q., Zhou, S., & Giannakis, G. B. (2005). Queueing with adaptive modulation and coding over wireless links: Cross-layer analysis and design. *IEEE Transactions on Wireless Communications*, *4*(3), 1142–1153. doi:10.1109/TWC.2005.847005

Li, W. (2001). Overview of fine granularity scalability in MPEG-4 video standard. *IEEE Transactions on Circuits and Systems for Video Technology*, *11*(3), 301–317. doi:10.1109/76.911157

Li, X., & Cuthbert, L. (2005). Optimal QoS mechanism: Integrating multipath routing, DiffServ and distributed traffic control in mobile ad hoc networks. In Jia, X., Wu, J., & He, Y. (Eds.), *Mobile Ad-Hoc and Sensor Networks* (*Vol. 3794*, pp. 560–569). Berlin, Germany: Springer. doi:10.1007/11599463_55

Li, X., Shu, W., Li, M., Huang, H. Y., Luo, P. E., & Wu, M. Y. (2009). Performance evaluation of vehicle-based mobile sensor networks for traffic monitoring. *IEEE Transactions on Vehicular Technology*, *58*(4), 1647–1653. doi:10.1109/TVT.2008.2005775

Li, Z. S. (2007). Joint source-channel-authentication resource allocation and unequal authenticity protection for multimedia over wireless networks. *IEEE Transactions on Multimedia*, *9*(4), 837–850. doi:10.1109/TMM.2007.893338

Lloyd, S. P. (1982). Least squares quantization in PCM. *IEEE Transactions on Information Theory*, *28*, 129–137. doi:10.1109/TIT.1982.1056489

Lombardo, A., Morabito, G., & Schembra, G. (2004). Modeling intramedia and intermedia relationships in multimedia network analysis through multiple timescale statistics. *IEEE Transactions on Multimedia*, *6*(1), 142–157. doi:10.1109/TMM.2003.819750

Long, L., & Hossain, E. (2008). Cross-layer optimization frameworks for multihop wireless networks using cooperative diversity. *IEEE Transactions on Wireless Communications*, *7*(7), 2592–2602. doi:10.1109/TWC.2008.060962

Lo, T. K. Y. (1990). Maximum ratio transmission. *IEEE Transactions on Communications*, *47*(10), 1458–1461. doi:10.1109/26.795811

Lu, C., Blum, B. M., Abdelzaher, T. F., Stankovic, J. A., & He, T. (2002). RAP: A real-time communication architecture for large-scale wireless sensor networks. In *Proceedings of IEEE Real-Time and Embedded Technology and Applications Symposium (RTAS)*, (pp. 55–66). IEEE Press.

Lu, M.-H., Steenkiste, P., & Chen, T. (2007). Time-aware opportunistic relay for video streaming over WLANs. In *Proceedings of the IEEE International Conference in Multimedia and Expo*, (vol 7), (pp. 1782–1785). IEEE Press.

Luby, M., Gasiba, T., Stockhammer, T., & Watson, M. (2007). Reliable multimedia download delivery in cellular broadcast networks. *IEEE Transactions on Broadcasting*, *53*(1), 235–246. doi:10.1109/TBC.2007.891703

Lu, J., Letaief, K. B., Chuang, J. C.-I., & Liou, M. L. (1999). M-PSK and M-QAM BER computation using signal-space concepts. *IEEE Transactions on Communications*, *47*(2), 181–184. doi:10.1109/26.752121

Lukic, M., Pavkovic, B., Mitton, N., & Stojmenovic, I. (2009). Greedy geographic routing algorithms in real environment. In *Proceedings of the International Conference on Mobile Ad-Hoc and Sensor Networks (MSN)*, (pp. 86 –93). IEEE Press.

Luo, J., Blum, R. S., Greenstein, L. J., Cimini, L. J., & Haimovich, A. M. (2004). New approaches for cooperative use of multiple antennas in ad-hoc wireless networks. In *Proceedings of IEEE Vehicular Technology Conference (VTC)*, (pp. 2769–2773). IEEE Press.

Luo, C., Yu, F. R., Ji, H., & Leung, V. C. M. (2010). Cross-layer design for TCP performance improvement in cognitive radio networks. *IEEE Transactions on Vehicular Technology*, *59*(5), 2485–2495. doi:10.1109/TVT.2010.2041802

Luo, J., Blum, R. S., Cimini, L. J., Greenstein, L. J., & Haimovich, A. M. (2007). Decode-and-forward cooperative diversity with power allocation in wireless networks. *IEEE Transactions on Wireless Communications*, 6.

Luo, J., Blum, R., Cimini, L., Greenstein, L., & Haimovich, A. M. (2005). Power allocation in a transmit diversity system with mean channel gain information. *IEEE Communications Letters*, *9*(7), 616–618. doi:10.1109/LCOMM.2005.1461682

Lu, Z., Kim, D. Y., & Pearlman, W. (2000). Wavelet compression of ECG signals by the set partitioning in hierarchical trees algorithm. *IEEE Transactions on Bio-Medical Engineering*, *47*(7), 849–856. doi:10.1109/10.846678

Ma, T., Hempel, M., Peng, D., & Sharif, H. (2010). Rate-switching unequal error protection for wireless electrocardiogram (ECG) transmission. In *Proceedings of the Military Communications Conference*, (pp. 1181-1186). IEEE.

Maani, E., & Katsaggelos, A. K. (2010). Unequal error protection for robust streaming of scalable video over packet lossy networks. *IEEE Transactions on Circuits and Systems for Video Technology*, *20*(3), 407–416. doi:10.1109/TCSVT.2009.2035846

Maaref, A., & Aïssa, S. (2004a). A cross-layer design for MIMO Rayleigh fading channels. In *Proceedings of the Electrical and Computer Engineering Canadian Conference*, (pp. 2247-2250). Ontario, Canada: IEEE.

Maaref, A., & Aïssa, S. (2004b). Combined adaptive modulation and truncated ARQ for packet data transmission in MIMO systems. In *Proceedings of the IEEE Global Commununications Conference*, (pp. 3818-3822). Dallas, TX: IEEE Press.

Maaref, A., & Aïssa, S. (2005, May). Cross-layer design for MIMO Nakagami fading channels in the presence of Gaussian channel estimation errors. In *Proceedings of the Electrical and Computer Engineering Canadian Conference*, (pp. 1380-1383). Saskatoon, Canada: IEEE.

Maaref, A., & Aïssa, S. (2005). Closed-form expressions for the outage and ergodic Shannon capacity of MIMO MRC systems. *IEEE Transactions on Communications, 53*(7), 1092–1095. doi:10.1109/TCOMM.2005.851564

Macfarlane, J. (2006). *Understanding IP routing in Cisco systems*. New York, NY: John Wiley & Sons Inc.

MacKay, D. (2003). *Information theory, inference, and learning algorithms*. Cambridge, UK: Cambridge University Press.

MacKay, D. J. C. (2005). Fountain codes. *IEEE Proceedings on Communications, 152*(6), 1062–1068. doi:10.1049/ip-com:20050237

Madan, R., Cui, S., & Lall, S. (2005). Cross-layer design for lifetime maximization in interference-limited wireless sensor networks. In *Proceedings of IEEE International Conference on Computer Communications (INFOCOM 2005),* (Vol. 3), (pp. 964 – 975). IEEE Press.

Madan, R., Cui, S., Lall, S., & Goldsmith, A. (2006). Layer design for lifetime maximization in interference-limited wireless sensor networks. *IEEE Transactions on Wireless Communications, 5*(11), 3142–3152. doi:10.1109/TWC.2006.04770

Madden, S. R., Franklin, M. J., Hellerstein, J. M., & Hong, W. (2005). Tinydb: An acquisitional query processing system for sensor networks. *ACM Transactions on Database Systems, 30*(1), 122–173. doi:10.1145/1061318.1061322

Madwifi Project. (2011). *Website.* Retrieved from http://madwifi-project.org/.

Magagula, L. A., & Chan, H. A. (2008). IEEE 802.21-assisted cross-layer design and PMIPv6 mobility management framework for next generation wireless networks. In *Proceedings of the IEEE International Conference on Wireless & Mobile Computing, Networking & Communication (IEEE WiMob)*. Avignon, France: IEEE Press.

Maham, B., & Nader-Esfahani, S. (2007). Performance analysis of distributed space-time codes in amplify-and-forward mode. In *Proceedings of the 8th IEEE Workshop on Signal Processing Advances in Wireless Communications*, (pp. 1-5). IEEE Press.

Mainwaring, A., Polastre, J., Szewczyk, R., Culler, D., & Anderson, J. (2002). Wireless sensor networks for habitat monitoring. In *Proceedings of the 1st ACM International Workshop on Wireless Sensor Networks and Applications,* (pp. 88-97). New York, NY: ACM Press.

Malkamaki, E., & Leib, H. (2000). Performance of truncated type-II hybrid ARQ schemes with noisy feedback over block fading channels. *IEEE Transactions on Communications, 48*(9), 1477–1487. doi:10.1109/26.870015

Mangold, S., Choi, S., Hiertz, G. R., Klein, O., Walke, B., Res, P., & Aachen, G. (2003). Analysis of IEEE 802.11 e for QoS support in wireless LANs. *IEEE Wireless Communications, 10*(6), 40–50. doi:10.1109/MWC.2003.1265851

Mansour, H., Fallah, Y. P., Nasiopoulos, P., & Krishnamurthy, V. (2009). Dynamic resource allocation for MGS H.264/AVC video transmission over link-adaptive networks. *IEEE Transactions on Multimedia, 11*(8), 1478–1491. doi:10.1109/TMM.2009.2032682

Marcus, M. (2005). Unlicensed cognitive sharing of TV spectrum: The controversy at the federal communications commission. *IEEE Communications Magazine*, •••, 24–25. doi:10.1109/MCOM.2005.1453413

Martucci, S. A., Sodagar, I., Chiang, T., & Zhang, Y.-Q. (1997). A zerotree wavelet video coder. *IEEE Transactions on Circuits and Systems for Video Technology, 7*(2), 109–118. doi:10.1109/76.554422

Ma, S., Gao, W., & Lu, Y., Y. (2005). Rate-distortion analysis for H.264/AVC video coding and its application to rate control. *IEEE Transactions on Circuits and Systems for Video Technology, 15*(12), 1533–1544. doi:10.1109/TCSVT.2005.857300

Massie, M. L., Chun, B., & Culler, D. (2004). The ganglia distributed monitoring system: Design, implementation, and experience. *Parallel Computing, 30*(7), 817–840. doi:10.1016/j.parco.2004.04.001

Mateos, R. (2011). *Pandora FMS: The free monitoring system*. Retrieved from http://pandora.sourceforge.net/en/index.php.

Max, J. (1960). Quantization for minimum distortion. *I.R.E. Transactions on Information Theory, 6*, 7–12. doi:10.1109/TIT.1960.1057548

Maxted, J. C., & Robinson, J. P. (1985). Error recovery for variable length codes. *IEEE Transactions on Information Theory*, *31*, 794–801. doi:10.1109/TIT.1985.1057110

McHenry, M. A. (2005). *NSF spectrum occupancy measurements project summary*. Shared Spectrum Company Report. Retrieved from http://www.sharedspectrum.com.

Medhi, D., & Ramasamy, K. (2007). *Network routing algorithms, protocols, and architectures*. New York, NY: Morgan Kauffman Publishers.

Mehmood, R., Crowcroft, J., Hand, S., & Smith, S. (2005). Grid-level computing needs pervasive debugging. In *Proceedings of the 6th IEEE/ACM International Workshop on Grid Computing*. IEEE Press.

Mehmood, R., & Alturki, R. (2010). A scalable multimedia QoS architecture for ad hoc networks. *Multimedia Tools and Applications*, *54*, 1–18.

Mehmood, R., Alturki, R., & Zeadally, S. (2011). Multimedia applications over metropolitan area networks (MANs). *Journal of Network and Computer Applications*, *34*, 1518–1529. doi:10.1016/j.jnca.2010.08.002

Melodia, T., & Akyildiz, I. F. (2010). Cross-layer QoS-aware communication for ultra wide band wireless multimedia sensor networks. *IEEE Journal on Selected Areas in Communications*, *28*(5), 653–663. doi:10.1109/JSAC.2010.100604

Melodia, T., Pompili, D., & Akyildiz, I. F. (2005). On the interdependence of distributed topology control and geographical routing in ad hoc and sensor networks. *Journal of Selected Areas in Communications*, *23*(3), 520–532. doi:10.1109/JSAC.2004.842557

Melodia, T., Vuran, M. C., & Pompili, D. (2006). *Wireless system/network architect*. Berlin, Germany: Springer-Verlag.

Memon, S., Lech, M., & Maddage, N. (2010). Information theoretic expectation maximization based gaussian mixture modeling for speaker verification. In *Proceedings of the 20th International Conference on Pattern Recognition*, (pp. 4536-4540). IEEE.

Mendes, L. D. P., & Rodrigues, J. J. P. C. (2011). A survey on cross-layer solutions for wireless sensor networks. *Journal of Network and Computer Applications*, *34*(2), 523–534. doi:10.1016/j.jnca.2010.11.009

Messier, G. G., Hartwell, J. A., & Davies, R. J. (2008). A sensor network cross-layer power control algorithm that incorporates multiple-access interference. *IEEE Transactions on Wireless Communications*, *7*(8), 2877–2883. doi:10.1109/TWC.2008.070208

Meyer, R. A., & Bagrodia, R. (1999). *PARSEC user manual for PARSEC release 1.1*. Retrieved from http://www.parsec.org.

Miao, X., Song, F., Qiu, F., Wan, M., Zhang, S., & Zhang, H. (2011). Redesigning transport layer architecture for future internet. In *Proceedings of the 7th International Conference on Wireless Communications, Networking and Mobile Computing (WiCOM 2011)*, (pp. 1-5). Wuhan, China: WiCOM.

Milani, S., & Calvagno, G. (2009). A low-complexity cross-layer optimization algorithm for video communication over wireless networks. *IEEE Transactions on Multimedia*, *11*(5), 810–821. doi:10.1109/TMM.2009.2021791

Millar, P. (2011). *MonAMI*. Retrieved from http://monami.sourceforge.net/.

Mirabbasi, S., & Martin, K. (2000). Hierarchical QAM: A spectrally efficient DC-free modulation scheme. *IEEE Communications Magazine*, *38*, 140–146. doi:10.1109/35.883503

Misra, S., Reisslein, M., & Xue, G. (2008). A survey of multimedia streaming in wireless sensor networks. *IEEE Communications Surveys & Tutorials*, *10*(4), 18–39. doi:10.1109/SURV.2008.080404

Mitchell, P. D., Qiu, J., Li, H., & Grace, D. (2010). Use of aerial platforms for energy efficient medium access control in wireless sensor networks. *Computer Communications*, *33*(4), 500–512. doi:10.1016/j.comcom.2009.10.015

Mitola, J. (1999). Cognitive radio for flexible mobile multimedia communications. In *Proceedings of IEEE International Workshop on Mobile Multimedia Communications*, (pp. 3-10). IEEE Press.

Mitola, J. (1999). Cognitive radio: Making software radio more personal. *IEEE Journal on Selected Areas in Communications*, *17*(4), 514–538. doi:10.1109/49.761033

Mitola, J. (2000). *Software radios: Wireless architecture for the 21st century*. New York, NY: Wiley. doi:10.1002/047121664X

Mitola, J. I. (2006). *Cognitive radio architecture - The engineering foundation of radio XML*. Hoboken, NJ: John Wiley & Sons, Inc.doi:10.1002/0471773735

Mitola, J. III, & Maguire, G. Q. Jr. (1999). Cognitive radio: Making software radios more personal. *IEEE Personal Communications*, *6*(4), 13–18. doi:10.1109/98.788210

Moinuddin, A. A., Khan, E., & Ghanbari, M. (2006). Wavelet-based low bit rate video coding using block-tree approach. In *Proceedings of IET International Conference on Visual Information Engineering* (*VIE 2006*), (pp. 287-291). Bangalore, India: IET.

Monks, J. P., Bharghavan, V., & Hwu, W.-M. W. (2001). A power controlled multiple access protocol for wireless packet networks. In *Proceedings of IEEE INFOCOM* (*Vol. 1*, pp. 219–228). INFOCOM. doi:10.1109/INFCOM.2001.916704

Moon, J. M., & Cho, D. H. (2010). Novel handoff decision algorithm in hierarchical macro/femto-cell networks. In *Proceedings of the IEEE Wireless Communications and Networking Conference (WCNC 2010)*, (pp. 1-6). Sydney, Australia: IEEE Press.

Morgan, D. P. (1991). *Surface-wave devices for signal processing*. London, UK: Elsevier Science Publishers.

Mpitziopoulos, A., Gavalas, D., Konstantopoulos, C., & Pantziou, G. (2009). A survey on jamming attacks and countermeasures in WSNS. *IEEE Communications Surveys & Tutorials*, *11*(4), 42–56. doi:10.1109/SURV.2009.090404

Mueller, W. C. (1978). Arrhythmia detection program for an ambulatory ECG monitor. *Biomedical Sciences Instrumentation*, *14*, 1–85.

Mukhopadhyay, S., Panigrahi, D., & Dey, S. (2004). Data aware, low cost error correction for wireless sensor networks. In *Proceedings of the IEEE Wireless Communications and Networking Conference*, (pp. 2492 - 2497). IEEE Press.

Mulligan, G., et al. (2011). *IPv6 over IEEE 802.15.4 BOF (6lowplan)*. Technical report, IETF Working group. Retrieved from http://www.ietf.org.

Mundarath, J. C., Ramanathan, P., & Veen, B. D. V. (2009). A quality of service aware cross-layer approach for wireless ad hoc networks with smart antennas. *Ad Hoc Networks*, *7*(5), 891–903. doi:10.1016/j.adhoc.2008.08.003

Muqattash, A., & Krunz, M. (2005). POWMAC: A single-channel power-control protocol for throughput enhancement in wireless ad hoc networks. *IEEE Journal on Selected Areas in Communications*, *23*, 1067–1084. doi:10.1109/JSAC.2005.845422

Muraleedharan, R., & Osadciw, L. A. (2006a). Cross layer denial of service attacks in wireless sensor network using swarm intelligence. In *Proceedings of the 2006 40th Annual Conference on Information Sciences and Systems*, (pp. 1653-1659). IEEE.

Muraleedharan, R., & Osadciw, L. A. (2006b). Security: Cross layer protocol in wireless sensor network. In *Proceedings of the 25th IEEE International Conference on Computer Communications (INFOCOM 2006)*, (pp. 1-2). IEEE Press.

Muramatsu, D. K. (2006). A markov chain monte carlo algorithm for bayesian dynamic signature verification. *IEEE Transaction on Information Forensics and Security*, *1*(1), 22–34. doi:10.1109/TIFS.2005.863507

Murthy, C. S. R., & Manoj, B. S. (2004). *Ad hoc wireless networks: Architectures and protocols*. Upper Saddle River, NJ: Prentice Hall.

Murthy, M., & Garcia-Luna-Aceves, J. J. (1996). An efficient routing protocol for wireless networks. *ACM Mobile Networks and Applications*, *1*(2), 183–197. doi:10.1007/BF01193336

Mussabbir, Q. B., Yao, W., Niu, Z., & Fu, X. (2007). Optimized FMIPv6 using IEEE 802.21 MIH services in vehicular networks. *IEEE Transactions on Vehicular Technology*, *56*(6), 3397–3407. doi:10.1109/TVT.2007.906987

Muthukumaran, P. S., de Paz, R., Spinar, R., & Pesch, D. (2009). Mesh-mac: Enabling mesh networking over IEEE 802.15.4 through distributed beacon scheduling. In *Proceedings of the International Conference on Ad Hoc Networks (AdHocNets)*. Niagara Falls, Canada: IEEE.

Nafaa, A., Taleb, T., & Murphy, L. (2008). Forward error correction strategies for media streaming over wireless networks. *IEEE Communications Magazine*, *46*(12), 72–79. doi:10.1109/MCOM.2008.4427233

Narayanaswamy, S., Kawadia, V., Sreenivas, R. S., & Kumar, P. R. (2002). Power control in ad hoc networks: Theory, architecture, algorithms and implementation of the COMPO protocol. In *Proceedings of European Wireless Conference*, (pp. 156–162). European Wireless Conference.

Nardi, S., & Pachter, M. (1998). GPS estimation algorithm using stochastic modeling. In *Proceedings of the 37th IEEE Conference on Decision and Control*, (pp. 4498–4502). IEEE Press.

Narten, T., & Nordmark, E. (1998). Neighbor discovery for IP version 6 (IPv6). *Internet Engineering Task Force (IETF), RFC-2461*. Retrieved from http://www.ietf.org.

Nave, G., & Cohen, A. (1993). ECG compression using long-term prediction. *IEEE Transactions on Bio-Medical Engineering*, *40*(9), 877–885. doi:10.1109/10.245608

Neely, M. J., Modiano, E., & Rohrs, C. E. (2005). Dynamic power allocation and routing for time-varying wireless networks. *IEEE Journal on Selected Areas in Communications*, *23*(1), 89–103. doi:10.1109/JSAC.2004.837349

NetCrunch Adrem. (2011). *NetCrunch 6*. Retrieved from http://www.adremsoft.com/netcrunch/.

Network Simulator. (2011). *Ns-2*. Retrieved from http://www.isi.edu/nsnam/ns/.

Neves, P., Sargneto, S., Pentikousis, K., & Fontes, F. (2009). WiMAX cross-layer system for next generation heterogeneous environments. In Dalal, U. D. D., & Kosta, Y. P. (Eds.), *WiMAX: New Developments* (pp. 357–376). Vukovar, Croatia: In-Tech.

Newman, H., Legrand, I., Galvez, P., & Voicu, C. (2003). MonALISA: A distributed monitoring service architecture. In *Proceedings of the Computing in High Energy and Nuclear Physics Conference*. La Jolla, CA: ACM Press.

Ngo, M. H., & Krishnamurthy, V. (2009). Optimality of threshold policies for transmission scheduling in correlated fading channels. *IEEE Transactions on Communications*, *57*, 2474–2483. doi:10.1109/TCOMM.2009.08.070350

Nilsson, M. (2001). Investigating the energy consumption of a wireless network interface in an ad hoc networking. In *Proceedings of IEEE INFOCOM*. IEEE Press.

Noh, K., Yoon, S., & Heo, J. (2009). Performance analysis of network coding with raptor codes for IPTV. *IEEE Transactions on Consumer Electronics*, *55*(1), 83–87. doi:10.1109/TCE.2009.4814418

Nokia. (2002). *Mobile internet technical architecture*, (Vol. 1). New York, NY: IT Press, Edita Publishing Inc.

Nosratina, A., Hunter, T. E., & Hedayat, A. (2004). Cooperative communication in wireless networks. *IEEE Communications Magazine*, *10*(4), 74–80. doi:10.1109/MCOM.2004.1341264

NS-2. (1997). *Ns-2: A discrete event network simulator* Retrieved from http://www.isi.edu/nsnam/ns/.

NS-2. (2011). *Website*. Retrieved Nov 23, 2011, from http://www.isi.edu/nsnam/ns/.

NS-3. (2006). *Ns-3: A discrete-event network simulator for internet systems*. Retrieved from http://www.nsnam.org/.

Nunzi, G., & Dudkowski, D. (Eds.). (2009). D-4.2 in-network management concept. *FP7-ICT-2007-1-216041-4WARD – "Architecture and Design for the Future Internet" Deliverables*. Retrieved November 2009, from http://www.4ward-project.eu/index.php?s=Deliverables.

O'Flynn, B., Bellis, S., Delaney, K., Barton, J., O'Mathuna, S., & Barroso, A. … Sreenan, C. (2005). The development of a novel minaturized modular platform for wireless sensor networks. In *Proceedings of the International Conference on Information Processing in Sensor Networks (IPSN)*, (pp. 370 – 375). IPSN.

OBD-II. (2010). *Website*. Retrieved from http://www.obdii.com/obdii.html.

O'Hara, B., & Petrick, A. (2005). *The IEEE 802.11 handbook: A designer's companion* (2nd ed.). New York, NY: Standards Information Network IEEE Press.

Oh, H., Yoo, K., Na, J., & Kim, C. (2009). A seamless handover scheme in IPv6-based mobile networks. *International Journal of Ad Hoc and Ubiquitous Computing*, *4*(1), 54–60. doi:10.1504/IJAHUC.2009.021914

Ohm, J.-R. (2004). *Multimedia communication technology*. Berlin, Germany: Springer-Verlag.

Ohm, J.-R. (2005). Advances in scalable video coding. *Proceedings of the IEEE, 93*(1), 42–56. doi:10.1109/JPROC.2004.839611

OPNET Technologies Inc. (2008). *OPNET modeler-educational version (version 14.5)*. Retrieved from www.opnet.com.

OPNET. (2000). *OPNET modeler: A software tool for network modeling and simulation.* Retrieved from http://www.opnet.com/.

OPNET. (2011). *Website.* Retrieved Nov 23, 2011, from http://www.opnet.com/.

Orchad, M. T., & Sullivan, G. J. (1994). Overlapped block motion compensation: an estimation-theoretic approach. *IEEE Transactions on Image Processing, 3*, 693–699. doi:10.1109/83.334974

Ouyang, Y. C., Jang, C. B., & Chen, H. T. (2007). A secure authentication policy for UMTS and WLAN interworking. In *Proceedings of the IEEE International Conference on Communications*, (pp. 1552-1557) Glasgow, UK: IEEE Press.

Pabst, R., Walke, B. H., Schultz, D. C., Herhold, P., Yanikomeroglu, H., & Mukherjee, S. (2004). Relay-based deployment concepts for wireless and mobile broadband radio. *IEEE Communications Magazine, 42*, 80–89. doi:10.1109/MCOM.2004.1336724

Padmavathi, G., & Shanmugapriya, D. (2009). A survey of attacks, security mechanisms, and challenges in wireless sensor networks. *International Journal of Computer Science and Information Security, 4*(1-2), 117–125.

Pados, D. A., & Batalama, S. N. (1997). Low complexity blind detection of DS/CDMA signals: Auxiliary-vector receivers. *IEEE Transactions on Communications, 45*, 1586–1594. doi:10.1109/26.650237

Palanki, R., & Yedidai, J. (2004). Rateless codes on noisy channels. In *Proceedings of the International Symposium on Information Theory,* (p. 37). IEEE.

Palat, S., & Godin, P. (2009). Network architecture. In Sesia, S., Toufik, I., & Baker, M. (Eds.), *LTE - The UMTS Long Term Evolution - From Theory to Practice* (pp. 23–50). Chichester, UK: John Wiley & Sons. doi:10.1002/9780470742891.ch2

Pan, J., & Tompkins, W. J. (1985). A real-time QRS detection algorithm. *IEEE Transactions on Bio-Medical Engineering, 32*(3), 230–236. doi:10.1109/TBME.1985.325532

Pan, M.-S., Tsai, C.-H., & Tseng, Y.-C. (2009). The orphan problem in zigbee wireless networks. *IEEE Transactions on Mobile Computing, 8*(11), 1573–1584. doi:10.1109/TMC.2009.60

Pantelidou, A., & Ephremides, A. (2010). A cross-layer view of optimal scheduling. *IEEE Transactions on Information Theory, 56*(11), 5568–5580. doi:10.1109/TIT.2010.2068970

Pantelidou, A., & Ephremides, A. (2011). Wireless multicast optimization: A cross-layer approach. *IEEE Transactions on Information Theory, 57*(7), 4333–4343. doi:10.1109/TIT.2011.2145530

Park, J. H., Han, K. Y., & Cho, D. H. (2007). Reducing inter-cell handover events based on cell ID information in multi-hop relay systems. In *Proceedings of the IEEE 65th VTC 2007-Spring Vehicular Technology Conference*, (pp.743–747). IEEE Press.

Park, J., Kwon, D. H., & Suh, Y. J. (2006). An integrated handover scheme for fast mobile IPv6 over IEEE 802.16e systems. In *Proceedings of the IEEE Vehicular Technology Conference,* (pp. 1-5). IEEE Press.

Park, S. H., Lee, S., & Kim, J. W. (2005). Network-adaptive high definition MPEG-2 streaming over IEEE 802.11 a WLAN using frame-based prioritized packetization. In *Proceedings of the ACM International Workshop on Wireless Mobile Applications and Services on WLAN Hotspots*, (pp. 84-87). ACM Press.

Park, V. D., & Corson, M. S. (1997). A highly adaptive distributed routing algorithm for mobile wireless networks. In *Proceedings of the 16th Annual Joint Conference of the IEEE Computer and Communications Societies, IEEE INFOCOM 1997,* (vol. 3), (pp. 1405-1413). IEEE Press.

Park, V., & Corson, S. (2001). Temporally-ordered routing algorithm (TORA) version 1 - Functional specification. *Internet Draft*. Retrieved from http://www.ietf.org.

Pavkovic, B., Theoleyre, F., Barthel, D., & Duda, A. (2010). Experimental analysis and characterization of a wireless sensor network environment. In *Proceedings of the International Symposium on Performance Evaluation of Wireless Ad Hoc, Sensor, and Ubiquitous Networks (PE-WASUN)*. Bodrum, Turkey: ACM Press.

Pei, Y., & Modestino, J. W. (2007). Cross-layer design for video transmission over wireless Rician slow-fading channels using an adaptive multiresolution modulation and coding scheme. *EURASIP Journal on Advances in Signal Processing*. Retrieved from http://downloads.hindawi.com/journals/asp/2007/086915.pdf.

Pelechrinis, K., Iliofotou, M., & Krishnamurthy, S. V. (2011). Denial of service attacks in wireless networks: The case of jammers. *IEEE Communications Surveys & Tutorials*, *13*(2), 245–257. doi:10.1109/SURV.2011.041110.00022

Peng, C., Zheng, H., & Zhao, B. Y. (2006). Utilization and fairness in spectrum assignment for opportunistic spectrum access. *ACM/Springer. Mobile Networks and Applications*, *11*(4), 555–576. doi:10.1007/s11036-006-7322-y

Perkins, C. E., & Royer, E. M. (1999). Ad-hoc on-demand distance vector routing. In *Proceedings of the 2nd IEEE Workshop on Mobile Computing Systems and Applications, IEEE WMCSA 1999,* (pp. 90-100). IEEE Press.

Perkins, C. E., Royer, E. M., & Chakeres, I. D. (2003). *Ad hoc on-demand distance vector (AODV) routing*. IETF Draft. Retrieved from http://www.ietf.org.

Perkins, C., & Bhagwat, P. (1994). Highly dynamic destination-sequenced distance-vector routing (DSDV) for mobile computers. In *Proceedings of the Conference on Communications Architectures, Protocols and Applications, ACM SIG-COMM 1994,* (vol. 24), (pp. 234-244). ACM Press.

Perkins, C., Belding-Royer, E., & Das, S. (2003). *Ad hoc on-demand distance vector routing*. Internet Engineering Task Force, RFC 3561. Retrieved from http://www.ietf.org.

Perkins, C., et al. (2002). IP mobility support for IPv4. *RFC 3344*. Retrieved from http://www.ietf.org.

Pesquet-Popescu, B., Li, S., & van der Schaar, M. (2007). Scalable video coding for adaptive streaming applications. In Chou, P. A., & van der Schaar, M. (Eds.), *Multimedia in IP and Wireless Networks* (pp. 117–158). Burlington, MA: Academic Press. doi:10.1016/B978-012088480-3/50006-0

Peters, S. W., Panah, A. Y., Truong, K. T., & Heath, R. W. (2009). Relay architextures for 3GPP LTE-advanced. *EURASIP Journal on Wireless Communications and Networking*. Retrieved from http://downloads.hindawi.com/journals/wcn/2009/618787.pdf.

Peterson, L. L., & Davie, B. S. (2003). *Computer networks: A systems approach* (3rd ed.). San Francisco, CA: Morgan Kaufmann Publishers.

Peterson, L., & Davie, B. (2011). *Computer networks: A systems approach* (5th ed.). New York, NY: Morgan Kaufmann.

Petrick, A., & O'Hara, B. (2005). *IEEE 802.11 handbook: A designer's companion*. New York, NY: IEEE Press.

Petrova, M., & Mahonen, P. (2007). Cognitive resource manager. In Fitzek, F. H., & Katz, M. D. (Eds.), *Cognitive Wireless Networks* (pp. 397–422). Springer. doi:10.1007/978-1-4020-5979-7_20

Poggioni, M., Rugini, L., & Banelli, P. (2010). QoS analysis of a scheduling policy for heterogeneous users employing AMC jointly with ARQ. *IEEE Transactions on Communications*, *58*(9), 2639–2652. doi:10.1109/TCOMM.2010.09.090276

Polastre, J., Hill, J., & Culler, D. (2004). Versatile low power media access for wireless sensor networks. In *Proceedings of the International Conference on Embedded Networked Sensor Systems (SENSYS)*, (pp. 95–107). Baltimore, MD: ACM Press.

Politis, I., Tsagkaropoulos, M., Pliakas, T., & Dagiuklas, T. (2007). Distortion optimized packet scheduling and prioritization of multiple video streams over 802.11 e networks. *Advances in Multimedia*, 1. doi:10.1155/2007/76846

Pompili, D., & Akyildiz, I. F. (2010). A multimedia cross-layer protocol for underwater acoustic sensor networks. *IEEE Transactions on Wireless Communications*, *9*(9), 2924–2933. doi:10.1109/TWC.2010.062910.100137

Pompili, D., Vuran, M. C., & Melodia, T. (2006). *Sensor network and configuration: Fundamentals, standards, platforms, and applications*. Berlin, Germany: Springer-Verlag.

Poon, C., Zhang, Y.-T., & Bao, S. D. (2006). A novel biometrics method to secure wireless body area sensor networks for telemedicine and m-health. *IEEE Communications Magazine*, *44*(4), 73–81. doi:10.1109/MCOM.2006.1632652

Pradas, D., Bouabdallah, A., Lacan, J., Vazquez Castro, M. A., & Michel Bousquet, M. (2009). Cross-layer optimization of unequal protected layered video over hierarchical Modulation. In *Proceedings of the IEEE Global Telecommunications Conference 2009 (GLOBECOM 2009)*. IEEE Press.

Pratt, W. K. (2001). *Digital image processing* (3rd ed.). New York, NY: John Wiley & Sons Inc. doi:10.1002/0471221325

Pudashine, K. (2009). Energy efficient cross layer routing for life time maximization in wireless sensor networks. In *Proceedings of the first Asian Himalayas International Conference on Internet,* (pp. 1-3.). Kathmandu, Nepal: IEEE.

Purple Pixie Systems. (2011). *FreeNATS network monitor*. Retrieved from http://www.purplepixie.org/freenats/.

Pursley, M. B., Russell, H. B., & Wysocarski, J. S. (2000). Energy-efficient transmission and routing protocols for wireless multiple-hop networks and spread-spectrum radios. In *Proceedings of EUROCOMM*, (pp. 1–5). EUROCOMM.

Puterman, M. L. (1994). *Markov decision processes: Discrete stochastic dynamic programming*. New York, NY: John Wiley.

Qi, J., & Aïssa, S. (2009). Cross-layer design for multiuser MIMO MRC systems with feedback constraints. *IEEE Transactions on Vehicular Technology*, *58*(7), 3347–3360. doi:10.1109/TVT.2009.2015668

Qi, J., Aïssa, S., & Maaref, A. (2010). Cross-layer design for MIMO over spatially correlated and keyhole Nakagami-*m* fading channels. *Wireless Communications and Mobile Computing*, *10*, 1055–1067.

Radha, H., Schaar, M. V. D., & Chen, Y. (2001). The MPEG-4 fine-grained scalable video coding method for multimedia streaming over IP. *IEEE Transactions on Multimedia*, *3*(1), 53–68. doi:10.1109/6046.909594

Radha, H., van der Schaar, M., & Chen, Y. (2001). The MPEG-4 fine_grained scalable video coding method for multimedia streaming over IP. *IEEE Transactions on Multimedia*, *3*(1), 53–68. doi:10.1109/6046.909594

Radio, G. N. U. (2011). *Website.* Retrieved from http://www.gnuradio.org.

Radunovic, B., Jean, Y., & Boudec, L. (2004). Rate performance objectives of multihop wireless networks. *IEEE Transactions on Mobile Computing*, *3*(4), 334–349. doi:10.1109/TMC.2004.45

Raghunathan, V., Schurghers, C., Park, S., & Srivastava, M. (2002). Energy-aware wireless microsensor networks. *IEEE Signal Processing Magazine*, *19*(2), 40–50. doi:10.1109/79.985679

Rahman, M. A., Saddik, A. E., & Gueaieb, W. (2008). *Sensors*. Berlin, Germany: Springer-Verlag.

Raisinghani, V., & Iyer, S. (2004). ECLAIR: An efficient cross layer architecture for wireless protocol stacks. In *Proceedings of the World Wireless Congress.* World Wireless Congress.

Raisinghani, V. T., & Iyer, S. (2004). Cross-layer design optimizations in wireless protocol stacks. *Computer Communications*, *27*(8), 720–724. doi:10.1016/j.comcom.2003.10.011

Raivio, Y. (2001). 4G-hype or reality. In *Proceedings of the International Conference on 3G Mobile Communication Technologies*, (Vol. 47), (pp. 346-350). London, UK: IEEE.

Rajan, D., Sabharwal, A., & Aazhang, B. (2004). Delay-bounded packet scheduling of bursty traffic over wireless channels. *IEEE Transactions on Information Theory*, *50*, 125–144. doi:10.1109/TIT.2003.821989

Rajbanshi, R., Chen, Q., Wyglinski, A., Minden, G., & Evans, J. (2007). Quantitative comparison of agile modulation technique for cognitive radio transceivers. In *Proceedings of CCNC*. CCNC.

Ramachandran, S., Fairhurs, G., Luglio, M., Roseti, C. & Provenzano, S. (2007). Network layer security: Design for a cross layer architecture. In *Proceedings of the International Workshop on Satellite and Space Communications (IWSSC 2007)*, (pp. 271-275). IEEE.

Ramakrishnan, K. K., Floyd, S., & Black, D. (2001). The addition of explicit congestion notification (ECN) to IP. *Internet Engineering Task Force RFC 3168*. Retrieved from http://www.ietf.org.

Ramanathan, R., & Redi, J. (2002). A brief overview of ad hoc networks: challenges and directions. *IEEE Communications Magazine*, *40*(5), 20–22. doi:10.1109/MCOM.2002.1006968

Raman, B., Chebrolu, K., Gokhale, D., & Sen, S. (2009). On the feasibility of the link abstraction in wireless mesh networks. *IEEE Transactions on Networking*, *17*(2), 528–541. doi:10.1109/TNET.2009.2013706

Ramis, J., Carrasco, L., & Femenias, G. (2008). A two-dimensional Markov model for cross-layer design in AMC/ARQ-based wireless networks. In *Proceedings of the IEEE GLOBECOM*, (pp. 4637-4642). IEEE Press.

Ramis, J., Femenias, G., & Carrasco, L. (2009). Cross-layer design of multi-rate wireless networks based on link layer truncated ARQ. In *Proceedings of IWCLD*. IWCLD.

Ramis, J., Femenias, G., Riera-Palou, F., & Carrasco, L. (2010). Cross-layer optimization of adaptive multirate wireless networks using truncated Chase combining HARQ. In *Proceedings of IEEE GLOBECOM*. IEEE Press.

Ramis, J., & Femenias, G. (2011). Cross-layer design of adaptive multirate wireless networks using truncated HARQ. *IEEE Transactions on Vehicular Technology*, *60*(3), 944–954. doi:10.1109/TVT.2011.2108324

Rappaport, T. S. (2002). *Wireless communications principles and practice* (2nd ed.). Upper Saddle River, NJ: Prentice Hall.

Ratnasamy, S., Karp, B., Yin, L., Yu, F., Estrin, D., Govindan, R., & Shenker, S. (2002). GHT: A geographic hash table for data-centric storage. In *Proceedings of the International Workshop on Wireless Sensor Networks and Applications (WSNA)*, (pp. 78–87). ACM Press.

Raychaudhuri, D. (2004). *4G network architectures: WLAN hot-spots, infostations and beyond*. London, UK: International 4G Forum.

Razavi, R., Fleury, M., & Ghanbari, M. (2008). Unequal protection of video streaming through adaptive modulation with a tri-zone buffer over Bluetooth enhanced data rate. *EURASIP Journal on Wireless Communications and Networking*. Retrieved from http://jwcn.eurasipjournals.com/content/2008/1/658794.

Razavi, R., Fleury, M., Jammeh, E., & Ghanbari, M. (2006). An efficient packetization scheme for Bluetooth video transmission. *Electronics Letters*, *42*(20), 1143–1145. doi:10.1049/el:20061638

Razi, A., Afghah, F., & Abedi, A. (2011). *Power optimized DSTBC assisted DMF relaying in wireless sensor networks with redundant super nodes*. Unpublished.

Razi, A., Yasami, K., & Abedi, A. (2011). On minimum number of wireless sensors required for reliable binary source estimation. In *Proceedings of the IEEE Wireless Communications and Networking Conference*, (pp. 1852 - 1857). Cancun, Mexico: IEEE Press.

Razzaque, M., Dobson, S., & Nixon, P. (2007). Cross-layer architectures for autonomic communications. *Journal of Network and Systems Management*, *15*(1), 13–27. doi:10.1007/s10922-006-9051-8

Reddy, Y. B., & Bullmaster, C. (2008). Cross-layer design in wireless cognitive networks. In *Proceedings of the International Conference on Parallel and Distributed Computing, Applications and Technologies*, (pp. 462-467). IEEE Computer Society.

Redmill, D. W., & Kingsbury, N. G. (1996). The EREC: An error resilient technique for coding variable length blocks of data. *IEEE Transactions on Image Processing*, *5*(4), 565–574. doi:10.1109/83.491333

Rexford, J., & Dovrolis, C. (2010). Future Internet architecture: Clean-slate versus evolutionary research. *Communications of the ACM, 53*(9), 36–40. doi:10.1145/1810891.1810906

Rhee, I., Warrier, A., Min, J., & Xu, L. (2006). Drand: Distributed randomized tdma scheduling for wireless ad-hoc networks. In *Proceedings of the International Symposium on Mobile Ad Hoc Networking & Computing (MOBIHOC)*. Florence, Italy: ACM Press.

Rhee, S., Seetharam, D., & Liu, S. (2004). Techniques for minimizing power consumption in low-data rate wireless sensor networks. In *Proceedings of the IEEE Wireless Communications and Networking Conference,* (pp. 1727 - 1731). IEEE Press.

Rhee, I., Warrier, A., Aia, M., Min, J., & Sichitiu, M. L. (2008). Z-mac: Hybrid mac for wireless sensor networks. *IEEE/ACM Transactions on Networking, 16*(3), 511–524. doi:10.1109/TNET.2007.900704

Richardson, I. E. G. (2003). *H. 264 and MPEG-4 video compression: Video coding for next-generation multimedia*. New York, NY: John Wiley & Sons Inc. doi:10.1002/0470869615

Rocket Software Inc. (2011). *Rocket OpenKBM*. Retrieved from http://www.rocketsoftware.com/networks/enterprise-management/openkbm/overview.

Rodrigues, J. J., & Chen, M. (2010). A cross-layer sleep and rate adaptation mechanism for slotted ALOHA wireless sensor networks. In *Proceedings of the International Conference on ICT Convergence*. Jeju Island, Korea: IEEE Communications Society.

Roman, R., Lopez, J., & Najera, P. (2011). A cross-layer approach for integrating security mechanisms in sensor networks architectures. *Wireless Communications and Mobile Computing, 11*, 267–276. doi:10.1002/wcm.1006

Rosen, E., Viswanathan, A., & Callon, R. (2001). Multiprotocol label switching architecture. *RFC 3031*. Retrieved from http://www.ietf.org.

Ross, S. (1996). *Stochastic processes*. Chichester, UK: Wiley & Sons.

Royer, E. M., & Keong Toh, C. (1999). A review of current routing protocols for ad hoc mobile wireless networks. *IEEE Personal Communications, 6*(2), 46–55. doi:10.1109/98.760423

Rui, X., Jin, R., & Geng, J. (2007). Performance analysis of MIMO-MRC systems in the presence of self-interference and co-channel interferences. *IEEE Signal Processing Letters, 14*(11), 801–803. doi:10.1109/LSP.2007.903254

Ruiz, J., Gallardo, J. R., Villasenor-Gonzalez, L., Makrakis, D., & Mouftah, H. T. (2009). *QUATTRO: QoS-capable cross-layer MAC protocol for wireless sensor networks*. Paper presented at the IEEE Global Telecommunications Conference GLOBECOM 2009. Honolulu, HI.

Rus, A. B. (2011). *Quality of service through cross-layer techniques for future internet*. Unpublished Ph.D. Thesis. Cluj-Napoca, Romania: Technical University of Cluj-Napoca.

Rus, A. B., Barabas, M., Boanea, G., Kiss, Z., Polgar, Z., & Dobrota, V. (2010). Cross-layer QoS and its application in congestion control. In *Proceedings of the 17th IEEE Workshop on Local and Metropolitan Area Networks LANMAN 2010,* (pp. 1-6). Long Branch, NJ: IEEE Press.

Rus, A. B., Dobrota, V., Vedinas, A., Boanea, G., & Barabas, M. (2010). Modified Dijkstra's algorithm with cross-layer QoS. *ACTA TECHNICA NAPOCENSIS. Electronics and Telecommunications, 51*(3), 75–80.

Ruzzelli, A. G., O'Hare, G. M. P., Grady, M. J. O., & Tynan, R. (2006). MERLIN: A synergetic integration of MAC and routing protocol for distributed sensor networks. In *Proceedings of IEEE Communications Society Conference on Sensor, Mesh and Ad Hoc Communications and Networks (SECON)*, (pp. 266–275). IEEE Press.

Sadeghi, P., Kennedy, R., Rapajic, P., & Shams, R. (2008). Finite-state Markov modeling of fading channels - A survey of principles and applications. *IEEE Signal Processing Magazine, 25*(5), 57–80. doi:10.1109/MSP.2008.926683

Sadka, A. H. (2002). *Compressed video communications*. London, UK: John Wiley & Sons, Ltd. doi:10.1002/0470846712

Safwat, A., Hassanein, H., & Mouftah, H. (2003). Optimal cross-layer designs for energy-efficient wireless ad hoc and sensor networks. In *Proceedings of the 22nd IEEE International Performance, Computing, and Communications Conference,* (pp. 123-128). Phoenix, AZ: IEEE Press.

Saglam, O., & Dalkilic, M. E. (2009). Cross layer implementation of key establishment protocols in wireless sensor networks. In *Proceedings of the 24th International Symposium on Computer and Information Sciences (ISCIS 2009)*, (pp. 346-351). ICIS.

Sai Shankar, N., & van der Schaar, M. (2007). Performance analysis of video transmission over IEEE 802.11 a/e WLANs. *IEEE Transactions on Vehicular Technology*, *56*(4), 2346–2362. doi:10.1109/TVT.2007.897646

Sajadieh, M., Kschischung, F., & Leon-Garcia, A. (1998). Modulation-assisted unequal error protection over the fading channel. *IEEE Transactions on Vehicular Technology*, *47*(3), 900–908. doi:10.1109/25.704843

Salsano, S., Veltri, L., Martiniello, G., & Polidoro, A. (2006). Seamless vertical handover of VoIP calls based on SIP session border controllers. In *Proceedings of the IEEE International Conference on Communications*, (Vol. 5), (pp. 2040-2047). IEEE Press.

Sanchez, J., Ruiz, P., & Marin-Perez, R. (2009). Beaconless geographic routing made practical: Challenges, design guidelines, and protocols. *IEEE Communications Magazine*, *47*(8), 85–91. doi:10.1109/MCOM.2009.5181897

Santos, N. P. (2009). *Voice traffic over mobile ad hoc networks: A performance analysis of the optimized link state routing protocol.* Master's Thesis. Wright-Patterson, OH: Air Force Institute of Technology.

Sarr, C., Chaudet, C., Chelius, G., & Guerin Lassous, I. (2008). Bandwidth estimation for IEEE 802.11-based ad hoc networks. *IEEE Transactions on Mobile Computing*, *7*(10), 1228–1241. doi:10.1109/TMC.2008.41

Savas, O., Alanyali, M., & Yener, B. (2004). Joint route and power assignment in asynchronous multi-hop wireless networks. In *Proceedings MedHocNet.* MedHocNet.

Scaglione, A., Goeckel, D. L., & Laneman, J. N. (2006) Cooperative communications in mobile ad hoc networks. *IEEE Signal Processing Magazine*, *23*(5), 18–29. doi:10.1109/MSP.2006.1708409

Schaar, M. V. D., & Shankar, N. S. (2005). Cross-layer wireless multimedia transmission: Challenges, principles, and new paradigms. *IEEE Transactions on Wireless Communications*, *12*(4), 50–58. doi:10.1109/MWC.2005.1497858

Scholl, G., Korden, C., Riha, E., Ruppel, C. W., Wolff, U., & Riha, G. (2003). SAW-based radio sensor systems for short-range applications. *IEEE Microwave Magazine*, *4*(4), 68–76. doi:10.1109/MMW.2003.1266068

Schopf, J. M., Pearlman, L., Miller, N., Kesselman, C., Foster, I., & D'Arcy, M. (2006). Monitoring the grid with the globus toolkit MDS4. *Journal of Physics: Conference Series*, *46*(46), 521. doi:10.1088/1742-6596/46/1/072

Schrader, K., Abedi, A., & Caccese, V. (2010). Impact localization and scaling for SHM of inflatable habitats. In *Proceedings of the 3rd IEEE/CANEUS Fly By Wireless Workshop*, (pp. 20-21). IEEE Press.

Schulzrinne, H., Rao, A., & Lampeter, R. (1998). *Real time streaming protocol (RTSP), proposed standard, RFC 2326*. Retrieved from http://www.ietf.org.

Schulzrinne, H., & Wedlund, E. (2000). Application layer mobility using SIP. *IEEE Service Portability and Virtual Customer Environments*, *4*(1), 29–36.

Schurgers, C., Aberthorne, O., & Srivastava, M. B. (2001) Modulation scaling for energy aware communication systems. In *Proceedings of the International Symposium on Low Power Electronics and Design*, (pp. 96 - 99). Huntington Beach, CA: IEEE.

Schwartz, M. (1988). *Telecommunication networks - Protocols, modeling and analysis*. Reading, MA: Addison-Wesley Publishing Company.

Schwarz, H., Marpe, D., & Wiegand, T. (2007). Overview of the scalable video coding extension of the H.264/AVC standard. *IEEE Transactions on Circuits and Systems for Video Technology*, *17*, 1103–1120. doi:10.1109/TCSVT.2007.905532

Seada, K., Zuniga, M., Helmy, A., & Krishnamachari, B. (2004). Energy-efficient forwarding strategies for geographic routing in lossy wireless sensor networks. In *Proceedings of ACM International Conference on Embedded Networked Sensor Systems (Sensys 2004)*, (pp. 567-575). ACM Press.

Sendonaris, A., Erkip, E., & Aazhang, B. (2003). User cooperation diversity - Parts I and II. *IEEE Transactions on Communications*, *51*, 2386–2391.

Sendonaris, A., Erkip, E., & Aazhang, B. (2003). User cooperation diversity-Part I: System description. *IEEE Transactions on Communications*, *51*(11), 1927–1938. doi:10.1109/TCOMM.2003.818096

Sen, J. (2009). A survey on wireless sensor network security. *International Journal of Communication Networks and Information Security*, *1*(2), 59–82.

Seo, J.-B., & Leung, V. C. M. (2011). Design and Analysis of cross-layer contention resolution algorithms for multi-packet reception slotted ALOHA systems. *IEEE Transactions on Wireless Communications*, *10*(3), 825–833. doi:10.1109/TWC.2011.122010.100032

Setton, E., Taesang, Y., Xiaoqing, Z., Goldsmith, A., & Girod, B. (2005). Cross-layer design of ad hoc networks for real-time video streaming. *IEEE Transactions on Wireless Communications*, *12*(4), 59–65. doi:10.1109/MWC.2005.1497859

Setton, E., Yoo, T., Xhu, X., Goldsmith, A., & Girod, B. (2005). Cross-layer design of ad-hoc networks for realtime video streaming. *IEEE Wireless Communications Magazine*, *12*(4), 59–65. doi:10.1109/MWC.2005.1497859

Shah, G. A., Weifa, L., & Xiaojun, S. (2010). Cross-layer design for QoS support in wireless multimedia sensor networks. In *Proceedings of the IEEE Global Telecommunications Conference GLOBECOM 2010,* (pp.1-5). Miami, FL: IEEE Press.

Shakkottai, S., Rappaport, T. S., & Karrisson, P. C. (2003). Cross-layer design for wireless networks. *IEEE Communications Magazine*, *41*(10), 74–80. doi:10.1109/MCOM.2003.1235598

Shannon, C., & Weaver, W. (1949). *The mathematical theory of communication*. Urbana, IL: University of Illinois Press.

Shao, M., Hu, W., Zhu, S., Cao, G., Krishnamurthy, S., & Porta, T. L. (2009). Cross-layer enhanced source location privacy in sensor networks. In *Proceedings of the 6th Annual IEEE Communications Society Conference on Sensor, Mesh and Ad Hoc Communications and Networks, (SECON 2009)*, (pp. 1-9). IEEE Press.

Shao, M., Zhu, S., Cao, G., Porta, T. L., & Mohapatra, P. (2008). A cross-layer dropping attack in video streaming over ad hoc networks. In *Proceedings of the 4th International Conference on Security and Privacy in Communication Networks, (SecureComm 2008)*. SecureComm.

Sharma, S., Zhu, N., & Chiueh, T. C. (2004). Low-latency mobile IP handoff for infrastructure-mode wireless LANs. *IEEE Journal on Selected Areas in Communications*, *22*(4), 643–652. doi:10.1109/JSAC.2004.825988

She, J., Yu, X., Po, P.-H., & Yang, E. H. (2009). A cross-layer design framework for robust IPTV services over IEEE 802.16 network. *IEEE Journal on Selected Areas in Communications*, *29*(2), 235–245. doi:10.1109/JSAC.2009.090214

Shih, E., Cho, S., Ickes, N., Min, R., Sinha, A., Wang, A., et al. (2001). Energy-efficient communication protocol for wireless microsensor networks physical layer driven protocol and algorithm design for energy-efficient wireless sensor networks. In *Proceedings of the 7th Annual International Conference on Mobile Computing and Networking*, (pp. 272--287). IEEE.

Shim, M., Kim, H., & Lee, S. (2006). A fast handover mechanism for IPv6 based WiBro system. In *Proceedings of the IEEE International Conference on Advanced Communication Technology*. IEEE Press.

Shi, Y., & Hou, Y. T. (2008). A distributed optimization algorithm for multi-hop cognitive radio networks. In *Proceedings of IEEE INFOCOM* (pp. 1966–1974). IEEE Press. doi:10.1109/INFOCOM.2008.186

Shi, Y., & Hou, Y. T. (2009). Per-node based optimal power control for multi-hop cognitive radio networks. *IEEE Transactions on Wireless Communications*, *8*(10), 5290–5299. doi:10.1109/TWC.2009.081702

Shrestha, R., Han, K. H., Choi, D. Y., & Han, S. J. (2010). A novel cross layer intrusion detection system in MANET. In *Proceedings of the 2010 24th IEEE International Conference on Advanced Information Networking and Applications (AINA)*, (pp. 647-654). IEEE Press.

Sichitiu, M. L. (2004). Cross-layer scheduling for power efficiency in wireless sensor networks. In *Proceedings of IEEE International Conference on Computer Communications (Infocom 2004)*, (Vol. 3), (pp. 1740 – 1750). IEEE Press.

Siddiqui, F., & Zeadally, S. (2006). Mobility management across hybrid wireless networks: Trends and challenges. *Computer Communications*, *29*(9), 1363–1385. doi:10.1016/j.comcom.2005.09.003

Silva, H., Holanda, R., Nogueira, M., & Santos, A. (2010). A cross-layer and adaptive scheme for balancing performance and security on WMN data routing. In *Proceedings of the 2010 IEEE Global Telecommunications Conference (GLOBECOM 2010)*, (pp. 1-5). IEEE Press.

Skraba, P., Aghajan, H., & Bahai, A. (2004). Cross-layer optimization for high density sensor networks: Distributed passive routing decisions. In *Proceedings of International Conference for Ad Hoc Networks and Wireless (Ad-Hoc Now 2004),* (pp. 266-279). IEEE.

Smadi, M., Todd, T. D., Zhao, D., & Kezys, V. (2006). Dynamically anchored conferencing handoff for dual-mode cellular/WLAN handsets. In *Proceedings of the IEEE International Conference on Communications*, (Vol. 5), (pp. 2028-2033). IEEE Press.

Smith, N., Reddy, Y. B., & Gajendar, N. (2009). Cross-layer design approach for wireless networks to improve the performance. In *Proceedings of the Sixth International Conference on Information Technology: New Generations,* (pp. 552-557). Las Vegas, NV: IEEE.

So, A., & Liang, B. (2005). Effect of relaying on capacity improvement in wireless local area networks. In *Proceedings of the IEEE Wireless Communications and Networking Conference*, (vol 3), (pp. 1539–1544). IEEE Press.

Soliman, H., Castelluccia, C., Malki, K., & Bellier, L. (2005). Hierarchical mobile IPv6 mobility management (HMIPv6). *Internet Engineering Task Force (IETF), RFC-4140*. Retrieved from http://www.ietf.org.

Sondi, P., Gantsou, D., & Lecomte, S. (2010). *Performance evaluation of multimedia applications over an OLSR-based mobile ad hoc network using OPNET.* Paper presented at the Computer Modelling and Simulation (UKSim). London, UK.

Song, L., & Hatzinakos, D. (2007). A cross-layer architecture of wireless sensor networks for target tracking. *IEEE/ACM Transactions on Networking*, *15*(1), 145–158. doi:10.1109/TNET.2006.890084

SOTEC Technical Solutions, Inc. (2011). *SOTEC technical solutions*. Retrieved from http://sotec-technical.com/.

Souryal, M. R., Vojcic, B. R., & Pickholtz, L. (2006). Adaptive modulation in ad hoc DS/CDMA packet radio networks. *IEEE Transactions on Communications*, *54*, 714–725. doi:10.1109/TCOMM.2006.873092

Spectrum Policy Task Force. (2002). *Spectrum policy task force report*. Washington, DC: Federal Communication Commission.

Spurgeon, C. (2000). *Ethernet - The definitive guide*. New York, NY: O'Reilly Media.

Srinivasan, K., Kazandjieva, M. A., Agarwal, S., & Levis, P. (2008). The β-factor: Measuring wireless link burstiness. In *Proceedings of the Conference on Embedded Network Sensor Systems (Sensys)*, (pp. 29–42). Raleigh, NC: ACM Press.

Srinivasan, K., & Levis, P. (2006). Rssi is under appreciated. In *EmNets*. Cambridge, MA: Harvard University Press.

Srivastava, V., & Motani, M. (2005). Cross-layer design: A survey and the road ahead. *IEEE Communications Magazine*, *43*(12), 112–119. doi:10.1109/MCOM.2005.1561928

Srivastava, V., & Motani, M. (2005). The road ahead for cross-layer design. *IEEE Communications Magazine*, *43*(12), 112–119. doi:10.1109/MCOM.2005.1561928

Srivastava, V., & Motani, M. (2007). Cross-layer design and optimization in wireless networks. In Mahmoud, Q. H. (Ed.), *Cognitive Networks* (pp. 121–146). Chichester, UK: John Wiley & Sons Ltd. doi:10.1002/9780470515143.ch6

Srivastav, E., & Motani, M. (2005). Cross-layer design: A survey and the road ahead. *IEEE Communications Magazine*, *43*(12), 712–721.

Stallings, W. (2006). *Data and computer communications* (8th ed.). Upper Saddle River, NJ: Prentice-Hall, Inc.

Stann, F., & Heidemann, J. (2003). RMST: Reliable data transport in sensor networks. In *Proceedings of IEEE International Workshop on Sensor Network Protocols and Applications (SNPA 2003),* (Vol. 1), (pp. 102-112). IEEE Press.

Steenstrup, M. E. (2005). Opportunistic use of radio-frequency spectrum: A network perspective. In *Proceedings of IEEE DySPAN,* (pp. 638-641). IEEE Press.

Stevens, W. R. (1994). *TCP/IP illustrated* (*Vol. 1*). Reading, MA: Addison-Wesley Publishing Company.

Stewart, R., et al. (2000). Stream control transmission protocol. *RFC 2960*. Retrieved from http://www.ietf.org.

Stockhammer, T. (2006). Robust system and cross-layer design for H.264/AVC-based wireless video applications. *EURASIP Journal on Applied Signal Processing*. Retrieved from http://asp.eurasipjournals.com/content/2006/1/089371.

Stockhammer, T., Hannuksela, M. M., Res, N., & Bergen, G. (2005). H. 264/AVC video for wireless transmission. *IEEE Wireless Communications*, *12*(4), 6–13. doi:10.1109/MWC.2005.1497853

Stuber, G. L., Barry, J. R., MacLaughlin, S. W., Li, Y., Ingram, M. A., & Pratt, T. G. (2004). Broadband MIMO-OFDM wireless communications. *Proceedings of the IEEE*, *92*.

Sufi, F., Mahmoud, S., & Khalil, I. (2008). A wavelet based secured ECG distribution technique for patient centric approach. In *Proceedings of the 5th International Summer School and Symposium on Medical Devices and Biosensors*, (pp. 301-304). IEEE.

Suh, C., Ko, Y. B., & Son, D. M. (2006). An energy efficient cross-layer MAC protocol for wireless sensor networks. In *Proceedings of the Eighth Asia Pacific Web Conference,* (pp. 410-419). Harbin, China: IEEE.

Su, H., & Zhang, X. (2008). Cross-layer based opportunistic MAC protocols for QoS provisioning over cognitive radio wireless networks. *IEEE Journal on Selected Areas in Communications*, *26*(1), 118–129. doi:10.1109/JSAC.2008.080111

Sullivan, G. J., & Wiegand, T. (2005). Video compression-from concepts to the H.264/AVC standard. *Proceedings of the IEEE*, *93*(1), 18–31. doi:10.1109/JPROC.2004.839617

Sun, H., Chen, X., & Chiang, T. (2005). *Digital video transcoding for transmission and storage*. Boca Raton, FL: CRC Press.

Sun, Y., Tripathi, V., & Honig, M. L. (2005). Adaptive, iterative, reduced-rank (turbo) equalization. *IEEE Transactions on Wireless Communications*, *4*, 2789–2800.

Sur, A., & Sicker, D. C. (2005). Multi layer rules-based framework for vertical handoff. In *Proceedings of the International Conference on Broadband Networks*, (Vol. 1), (pp. 71-580). IEEE.

Sutton, P. D., Jörg, L., Lahlou, H., Fahmy, S. A., Nolan, K. E., & Özgül, B. (2010). Iris: An architecture for cognitive radio networking testbeds. *IEEE Communications Magazine*, *48*(9), 114–122. doi:10.1109/MCOM.2010.5560595

Svensson, A. (2007). An introduction to adaptive QAM modulation schemes for known and predicted channels. *Proceedings of the IEEE*, *95*(12), 2322–2336. doi:10.1109/JPROC.2007.904442

Tan, C. C., & Beaulieu, N. C. (1997). Infinite series representations of the bivariate Rayleigh and Nakagami-m distributions. *IEEE Transactions on Communications*, *45*(10), 1159–1161. doi:10.1109/26.634675

Tan, C. C., & Beaulieu, N. C. (2000). On first-order Markov modeling for the Rayleigh fading channel. *IEEE Transactions on Communications*, *48*(12), 2032–2040. doi:10.1109/26.891214

Tanenbaum, A. S. (2010). *Computer networks* (5th ed.). Upper Saddle River, NJ: Prentice Hall.

Tarokh, V., Jafarkhani, H., & Calderbank, A. R. (1999). Space-time block codes from orthogonal designs. *IEEE Transactions on Information Theory*, *45*, 1456–1467. doi:10.1109/18.771146

Telatar, I. E. (1999). Capacity of multi-antenna gaussian channels. *European Transactions on Telecommunications*, *10*, 585–595. doi:10.1002/ett.4460100604

Teraoka, F. (2009). Redesigning layered network architecture for next generation networks. In [IEEE Press.]. *Proceedings of the IEEE GLOBECOM Workshops*, *2009*, 1–6. doi:10.1109/GLOCOMW.2009.5360742

Thamilarasu, G., & Sridhar, R. (2007a). Toward building a multi-level robust intrusion detection architecture for distributed mobile networks. In *Proceedings of the 27th International Conference on Distributed Computing Systems Workshops (ICDCSW 2007)*, (pp. 5-13). ICDCSW.

Thamilarasu, G., & Sridhar, R. (2007b). Exploring cross-layer techniques for security: Challenges and opportunities in wireless networks. In *Proceedings of the IEEE Military Communications Conference (MILCOM 2007)*, (pp. 1-6). MILCOM.

Thamilarasu, G., & Sridhar, R. (2009). XLSEC - A distributed cross-layer framework for security in wireless sensor networks. In *Proceedings of the 6th IEEE Consumer Communications and Networking Conference (CCNC 2009)*, (pp. 1-2). IEEE Press.

Thamilarasu, G., Balasubramanian, A., Mishra, S., & Sridhar, R. (2005). A cross-layer based intrusion detection approach for wireless ad hoc networks. In *Proceedings of the 2005 IEEE International Conference on Mobile Adhoc and Sensor Systems Conference (MASS) Workshop, WSNS 2005*, (pp. 861-868). IEEE Press.

Theoleyre, F. (2011). A route-aware mac for wireless multihop networks with a convergecast traffic pattern. *Computer Networks*, *55*(3), 822–837. doi:10.1016/j.comnet.2010.10.018

Thomson, S., & Narten, T. (2002). IPv6 stateless address autoconfiguration. *Internet Engineering Task Force (IETF) RFC-2462*. Retrieved from http://www.ietf.org.

Tian, L., Zhou, Y., Sun, Y., & Sun, S. (2010). Rate distortion model analysis for H.264/AVC video coding. *Journal on Advanced Materials Research*, *15*(2), 172–177. doi:10.4028/www.scientific.net/AMR.108-111.172

Tian, Y., & Ekici, E. (2007). Cross-layer collaborative in-network processing in multihop wireless sensor networks. *IEEE Transactions on Mobile Computing*, *6*(3), 297–310. doi:10.1109/TMC.2007.39

Tinnirello, I., & Bianchi, G. (2010). Rethinking the IEEE 802.11e EDCA performance modeling methodology. *IEEE/ACM Transactions on Networking*, *18*(2), 540–553. doi:10.1109/TNET.2009.2029101

Togneri, J. R. (2002). *Fundamentals of information theory and coding design*. Boca Raton, FL: CRC Press.

Tommaso, M., Vuran, M. C., & Pompili, D. (2006). The state of the art in cross-layer design for wireless sensor networks. In Cesana, M., & Fratta, L. (Eds.), *Wireless Systems/Network Architecture* (pp. 78–92). Berlin, Germany: Springer-Verlag.

Tompkins, J., & Abenstein, J. P. (1982). New data-reduction algorithm for real-time ECG analysis. *IEEE Transactions on Bio-Medical Engineering*, *29*(1), 43–48. doi:10.1109/TBME.1982.324962

Tumuluru, V. K., Wang, P., & Niyaoto, D. (2010). A neural network based spectrum prediction scheme for cognitive radio. [IEEE Press.]. *Proceedings of the IEEE, ICC*, 1–5.

Turnbull, J. (2007). *Pulling strings with puppet*. New York, NY: FristPress. doi:10.1007/978-1-4302-0622-4

UCLA. (2011). *Glomosim*. Retrieved from http://pcl.cs.ucla.edu.projects/glomosim.

Ukil, A., & Sen, J. (2009). QoS aware cross-layer optimized resource allocation in WiMAX systems. In *Proceedings of the First International Conference on Wireless Communication, Vehicular Technology, Information Theory and Aerospace & Electronic Systems Technology, Wireless VITAE 2009*, (pp.818-822). Aalborg, Denmark: VITAE.

Ulvan, A., Bestak, R., & Ulvan, M. (2010). The study of handover procedure in LTE-based femtocell network. In *Proceedings of the IEEE Wireless and Mobile Networking Conference (WMNC 2010)*, (pp. 1-6). Budapest, Hungary: IEEE Press.

Unnikrishnan, J., & Veeravalli, V. V. (2007). Cooperative spectrum sensing and detection for cognitive radio. In *Proceedings of IEEE Globecom*. IEEE Press.

Urbano Fullana, A., Ferrer Gomila, J. L., Payeras Capellá, M., Hinarejos Campos, M., & Huguet Rotger, L. (2011). Paper. In *Proceedings of the 2011 4th IFIP International Conference on New Technologies, Mobility and Security (NTMS)*, (pp. 1-5). IFIP.

Utsu, K., Chow, C., & Ishii, H. A. (2010). Study on video performance of multipoint-to-point video streaming with multiple description coding over ad hoc networks. *Electrical Engineering in Japan*, *170*(4), 43–50. doi:10.1002/eej.20999

Uysal-Biyikoglu, E., Gamal, A. E., & Prabhakar, B. (2002). Energy-eficient packet transmission over a wireless link. *IEEE/ACM Transactions on Networking*, *10*, 487–499. doi:10.1109/TNET.2002.801419

Valera, M., & Velastin, S. A. (2005). Intelligent distributed surveillance systems: A review. In *Proceedings of the IEEE Visual Image Signal Processing,* (Vol. 152), (pp. 192-204). IEEE Press.

Van der Auwera, G., & Reisslein, M. (2009). Implications of smoothing on statistical multiplexing of H.264/AVC and SVC video streams. *IEEE Transactions on Broadcasting, 55*(3), 541–558. doi:10.1109/TBC.2009.2027399

Van der Schaar, M. (2007). Cross-layer wireless multimedia. In van der Schaar, M., & Chou, P. A. (Eds.), *Multimedia over IP and Wireless Networks* (pp. 337–408). Burlington, MA: Academic Press. doi:10.1016/B978-012088480-3/50013-8

van der Schaar, M., Andreopoulos, Y., & Hu, Z. (2006). Optimized scalable video streaming over IEEE 802.11 a/e HCCA wireless networks under delay constraints. *IEEE Transactions on Mobile Computing, 5*(6), 755–678. doi:10.1109/TMC.2006.81

van der Schaar, M., & Chou, P. (2007). *Multimedia over IP and wireless networks: Compression, networking, and systems*. California: Academic Press.

van der Schaar, M., Krishnamachari, S., Choi, S., & Xu, X. (2003). Adaptive cross-layer protection strategies for robust scalable video transmission over 802.11 WLANs. *IEEE Journal on Selected Areas in Communications, 21*(10), 1752–1763. doi:10.1109/JSAC.2003.815231

van der Schaar, M., & Shankar, S. N. (2005). Cross-layer wireless multimedia transmission: Challenges, principles, and new paradigms. *IEEE Wireless Communication Magazine, 12*(4), 50–58. doi:10.1109/MWC.2005.1497858

Vardhe, K., Reynolds, D., & Valenti, M. C. (2008). The performance of multi-user cooperative diversity in an asynchronous CDMA uplink. *IEEE Transactions on Wireless Communications, 7*, 1930–1940. doi:10.1109/TWC.2008.070282

Vasseur, J., & Culler, D. (2011). Routing over low power and lossy networks (roll). *Working Group, IETF*. Retrieved from http://www.ietf.org.

Vaughan, R., & Bach Andersen, J. (2003). *Channels, propagation and antennas for mobile communications*. London, UK: The Institution of Electrical Engineers.

Vegesna, S. (2001). *IP quality of service*. New York, NY: Cisco Press.

Venturino, L., Wang, X., & Lops, M. (2006). Multiuser detection for cooperative networks and performance analysis. *IEEE Transactions on Signal Processing*, 54.

Verdu, S. (1998). *Multiuser detection*. Cambridge, UK: Cambridge University Press.

Villalón, J., Cuenca, P., Orozco-Barbosa, L., Seok, Y., & Turletti, T. (2007). Cross-layer architecture for adaptive video multicast streaming over multirate wireless LANs. *IEEE Journal on Selected Areas in Communications, 25*(4), 1–14. doi:10.1109/JSAC.2007.070507

Villaverde, B., De Paz Alberola, R., Rea, S., & Pesch, D. (2010). Experimental evaluation of beacon scheduling mechanisms for multihop IEEE 802.15.4 wireless sensor networks. In *Proceedings of the International Conference on Sensor Technologies and Applications (SENSORCOMM)*, (pp. 226–231). IARIA.

Vladishev, A. (2010). *Zabbix: An enterprise-class open source distributed monitoring solution*. Retrieved from http://www.zabbix.com/.

Vo, N.-S., Duong, T. Q., Shu, L., Du, X., Zepernick, H.-J., & Cheng, W. (2011). Cross-layer design for video replication strategy over multihop wireless networks. In *Proceedings of IEEE International Communications Conference*. Kyoto, Japan: IEEE Press.

Vucetic, B. (1991). An adaptive coding scheme for time-varying channels. *IEEE Transactions on Communications, 39*, 653–663. doi:10.1109/26.87156

Vuran, M. C., & Akyildiz, I. F. (2006). Cross-layer analysis of error control in wireless sensor networks. In *Proceedings of the 3rd Annual IEEE Communications Society on Sensor and Ad Hoc Communications and Networks,* (pp. 585-594). Reston, VA: IEEE Press.

Vuran, M. C., & Akyildiz, I. F. (2008). Cross-layer packet size optimization for wireless terrestrial, underwater, and underground sensor networks. In *Proceedings of IEEE International Conference on Computer Communications (Infocom),* (pp. 226-230). IEEE Press.

Vuran, M. C., Gungor, V. B., & Akan, O. B. (2005). On the interdependency of congestion and contention in wireless sensor networks. In *International Workshop on Measurement, Modeling, and Performance Analysis of Wireless Sensor Networks* (*SENMETRICS 2005*), (pp. 345-351). SENMETRICS.

Vuran, M. C., Akan, O. B., & Akyildiz, I. F. (2004). Spatio-temporal correlation: Theory and applications for wireless sensor networks. *Computer Networks Journal*, *45*(3), 245–261. doi:10.1016/j.comnet.2004.03.007

Vuran, M. C., & Akyildiz, I. F. (2006). Correlation-based collaborative medium access in wireless sensor networks. *IEEE/ACM Transactions on Networking*, *14*(2), 316–329. doi:10.1109/TNET.2006.872544

Vuran, M. C., & Akyildiz, I. F. (2009). Error control in wireless sensor networks: A cross layer analysis. *IEEE/ACM Transactions on Networking*, *17*(4), 1186–1199. doi:10.1109/TNET.2008.2009971

Vuran, M. C., & Akyildiz, I. F. (2010). XLP: A cross-layer protocol for efficient communication in wireless sensor networks. *IEEE Transactions on Mobile Computing*, *9*(11), 1578–1591. doi:10.1109/TMC.2010.125

Wada, W. (1989). Selective recovery of video packet loss using error concealment. *IEEE Journal on Selected Areas in Communications*, *7*(6), 807–814. doi:10.1109/49.32344

Walters, J. P., Liang, Z., Shi, W., & Chaudhary, V. (2006). Security in distributed, grid, and pervasive computing. In Xiao, Y. (Ed.), *Wireless Sensor Network Security: A Survey*. Boca Raton, FL: CRC Press.

Wang, G., Zhang, Q., Zhu, W., & Zhang, Y.-Q. (2001). Channel-adaptive unequal error protection for scalable video transmission over wireless channel. In *Proceedings of SPIE VCIP*, (pp. 648-655). San Jose, CA: SPIE.

Wang, Q., & Zheng, H. (2006). Route and spectrum selection in dynamic spectrum networks. In *Proceedings of CNNC*. CNNC.

Wang, Q., Chen, W. P., Zheng, R., Lee, K., & Sha, L. (2003). Acoustic target tracking using tiny wireless sensor devices. In *Proceedings of the 2nd International Conference on Information Processing in Sensor Network*, (pp. 642-657). IEEE.

Wang, W., Hua, K., Hempel, M., Peng, D., Sharif, H., & Chen, H.-H. (2010). A stochastic biometric authentication scheme using uniformed GMM in wireless body area sensor networks. In *Proceedings of the 2010 IEEE 21st International Symposium on Personal Indoor and Mobile Radio Communications*, (pp. 1620-1624). IEEE Press.

Wang, W., Sun, Y., Li, H., & Han, Z. (2010). Cross-layer attack and defense in cognitive radio networks. In *Proceedings of the 2010 IEEE Global Telecommunications Conference (GLOBECOM 2010)*, (pp. 1-6). IEEE Press.

Wang, X., Wong, J. S., Stanley, F., & Basu, S. (2009). Cross-layer based anomaly detection in wireless mesh networks. In *Proceedings of the Ninth Annual International Symposium on Applications and the Internet (SAINT 2009)*, (pp. 9-15). SAINT.

Wang, Y.-C., Lin, C.-H., Tsai, M.-F., & Shieh, C.-K. (2010) Cross-layer unequal error protection mechanism with an interleaved strategy for video streaming over wireless networks. In *Proceedings of the IEEE WCNC 2010*. IEEE Press.

Wang, Z., & Chen, Y. (2008). ECG signal compression based on MSPIHT algorithm. In *Proceedings of the International Conference on Information Technology and Applications in Biomedicine*, (pp. 415-419). IEEE.

Wang, C., Li, B., Sohraby, K., Daneshmand, M., & Hu, Y. (2007). Upstream congestion control in wireless sensor networks through cross-layer optimization. *IEEE Journal on Selected Areas in Communications*, *25*(4), 786–795. doi:10.1109/JSAC.2007.070514

Wang, H. S., & Moayeri, N. (1995). Finite-state Markov channel- A useful model for radio communication channels. *IEEE Transactions on Vehicular Technology*, *44*(1), 163–171. doi:10.1109/25.350282

Wang, H., Peng, D., Wang, W., Sharif, H., Chen, H. H., & Khoynezhad, A. (2010). Resource-aware secure ECG healthcare monitoring through body sensor networks. *IEEE Wireless Communications*, *17*(1), 12–19. doi:10.1109/MWC.2010.5416345

Wang, J. C., & X, W. (2010). On the sum rate of Gaussian multi-terminal source coding: New proofs and results. *IEEE Transactions on Information Theory*, *56*(8), 3946–3960. doi:10.1109/TIT.2010.2050960

Wang, X., Liu, Q., & Giannakis, G. B. (2007). Analyzing and optimizing adaptive modulation coding jointly with ARQ for QoS-guaranteed traffic. *IEEE Transactions on Vehicular Technology*, *56*(2), 710–720. doi:10.1109/TVT.2007.891465

Wang, X., & Poor, H. V. (1999). Iterative (turbo) soft interference cancellation and decoding for coded CDMA. *IEEE Transactions on Communications*, *47*, 1046–1061. doi:10.1109/26.774855

Wang, Y., Attebury, G., & Ramamurthy, B. (2006). A survey of security issues in wireless sensor networks. *IEEE Communications Surveys & Tutorials*, *8*(2), 2–23. doi:10.1109/COMST.2006.315852

Wang, Y., Ostermann, J., & Zhang, Y.-Q. (2002). *Video processing and communications*. Upper Saddle River, NJ: Prentice-Hall.

Wang, Y., Reibman, A. R., & Lee, S. (2005). Multiple description coding for video delivery. *Proceedings of the IEEE*, *93*(1), 57–70. doi:10.1109/JPROC.2004.839618

Wang, Y., & Zhu, Q.-F. (1998). Error control and concealment for video communication: A review. *Proceedings of the IEEE*, *86*(5), 974–997. doi:10.1109/5.664283

Wan, J., Xu, X., Feng, R., & Wu, Y. (2009). Cross-layer active predictive congestion control protocol for wireless sensor networks. *Sensors (Basel, Switzerland)*, *9*(10), 8278–8310. doi:10.3390/s91008278

Want, R., Farkas, K. I., & Narayanaswami, C. (2005). Guest editors' introduction: Energy harvesting and conservation. *IEEE Pervasive Computing / IEEE Computer Society [and] IEEE Communications Society*, *4*(1), 14–17. doi:10.1109/MPRV.2005.12

Webb, W. T., & Steele, R. (1995). Variable rate QAM for mobile radio. *IEEE Transactions on Communications*, *43*, 2223–2230. doi:10.1109/26.392965

Wen, H., Gong, G., & Ho, P. H. (2010). MIMO cross-layer secure communication architecture based on STBC. In *Proceedings of the 2010 IEEE Global Telecommunications Conference (GLOBECOM 2010)*, (pp. 1-6). IEEE Press.

Wen, Z., Luo, T., Xiang, W., Majhi, S., & Ma, Y. (2008). Autoregressive spectrum hole prediction model for cognitive radio systems. In *Proceedings of IEEE ICC Workshop on Communications*, (pp. 154-157). IEEE Press.

Wenger, S. (2003). H.264/AVC over IP. *IEEE Transactions on Circuits and Systems for Video Technology*, *13*(7), 645–655. doi:10.1109/TCSVT.2003.814966

Wiegand, T., Sullivan, G. J., Bjontegaard, G., & Luthra, A. (2003). Overview of the H. 264/AVC video coding standard. *IEEE Transactions on Circuits and Systems for Video Technology*, *13*(7), 560–576. doi:10.1109/TCSVT.2003.815165

Wiethölter, S., Emmelmann, M., Hoene, C., & Wolisz, A. (2006). *TKN EDCA model for ns-2*. Berlin, Germany: Technische Universität.

WiMAX Forum. (2011). *Website*. Retrieved from http://www.wimaxforum.org/home/.

Win, M. Z., & Scholtz, R. A. (2000). Ultra-wide bandwidth time-hopping spread-spectrum impulse radio for wireless multiple-acess communications. *IEEE Transactions on Communications*, *48*, 679–689. doi:10.1109/26.843135

Winter, R., Schiller, J. H., Nikaein, N., & Bonnet, C. (2006). CrossTalk: Cross-layer decision support based on global knowledge. *IEEE Communications Magazine*, *44*(1), 93–99. doi:10.1109/MCOM.2006.1580938

Woo, T. K. (2000). Orthogonal code design for quasi-synchronous CDMA. *IET Electronics Letters*, *36*(19), 1632–1633. doi:10.1049/el:20001136

Wu, D., & Ci, S. (2006). Cross-layer combination of hybrid ARQ and adaptive modulation and coding for QoS provisioning in wireless data networks. In *Proceedings of the 3rd International Conference on QoS in Heterogenoeous Wired/Wireless Networks*. Waterloo, Canada: IEEE.

Wu, J., Havinga, P., Dulman, S., & Nieberg, T. (2004). Eyes source routing protocol for wireless sensor networks. In *Proceedings of European Workshop on Wireless Sensor Networks (EWSN)*, (pp. 334-341). EWSN.

Wu, S., Zhang, X., Zheng, R., Yin, Z., Fang, Y., & Yang, D. (2009). Handover study concerning mobility in the two-hierarchy network. In *Proceedings of the IEEE Vehicular Technology Conference (VTC 2009-Spring)*, (pp. 1-5). Barcelona, Spain: IEEE Press.

Wu, Z., Ratazzi, P., Chakravarthy, V., & Hong, L. (2008). Performance evaluation of adaptive non-contiguous MC-CDMA and non-contiguous CI/MC-CDMA for dynamic spectrum access. In *Proceedings of 3rd International Conference on Cognitive Radio Oriented Wireless Networks and Communcations*. IEEE.

Wu, S.-L., Tseng, Y.-C., & Sheu, J.-P. (2000). Intelligent medium access for mobile ad hoc networks with busy tones and power control. *IEEE Journal on Selected Areas in Communications*, *18*(9), 1647–1657. doi:10.1109/49.872953

Wu, T. Y., Chen, C. Y., Chang, K. D., Chung, D., & Chao, H. C. (2007b). Pave the way to future smart living space - Cross-layer enhanced AAA for 4G core network. *International Journal of Smart Home*, *1*(2), 191–203.

Wu, T. Y., Chen, C. Y., Chen, C. H., Chang, K. D., Chung, J. H., & Chao, H. C. (2007a). Cross-layer AAA for various heterogeneous systems over the 4G core network environments. *Journal of Internet Technology*, *8*(2), 157–163.

Wu, W., Banerjee, N., Basu, K., & Dasu, S. K. (2005). SIP-based vertical handoff between WWANs and WLANs. *IEEE Wireless Communications*, *12*(3), 66–72. doi:10.1109/MWC.2005.1452856

Wyglinski, A. M., Nekovee, M., & Hou, Y. T. (2010). *Cognitive radio communications and networks: Principles and practice*. London, UK: Academic Press/Elsevier.

Xiao, M., Wang, X., & Yang, G. (2006). Cross-layer design for the security of wireless sensor networks. In *Proceedings of the Sixth World Congress on Intelligent Control and Automation,* (pp. 104-108). Dalian, China: IEEE.

Xiao, Y., Du, X., Hu, F., & Zhang, J. (2008). A cross-layer approach for frame transmissions of MPEG-4 over the IEEE 802.11e wireless local area networks. In *Proceedings of the IEEE Wireless Communications and Networking Conference (WCNC),* (pp. 1728-1733). IEEE Press.

Xie, P., Zhou, Z., Peng, Z., Yan, H., Hu, T., Cui, J.-H., et al. (2009). Aqua-sim: A ns-2 based simulator for underwater sensor networks. In *Proceedings of the IEEE/MTS Oceans.* IEEE Press.

Xie, L.-L., & Kumar, P. R. (2004). A network information theory for wireless communication: Scaling laws and optimal operation. *IEEE Transactions on Information Theory*, *50*(5), 748–767. doi:10.1109/TIT.2004.826631

Xin, C., Xie, B., & Shen, C.-C. (2005). A novel layered graph model for topology formation and routing in dynamic spectrum access networks. In *Proceedings of IEEE DySPAN,* (pp. 308-317). IEEE Press.

Xing, G., Lu, C., Zhang, Y., Huang, Q., & Pless, R. (2005). Minimun power configration in wireless sensor networks. In *Proceeding of the 6th ACM International Symposium on Mobile Ad Hoc Networking and Computing.* Urbana-Champaign, IL: ACM Press.

Xiong, Z. Y., Fan, X. P., Liu, S. Q., & Zhong, Z. (2010). Distributed image coding in wireless multimedia sensor networks: A survey. In *Proceedings of the Third International Workshop on Advanced Computational Intelligence,* (pp. 618 - 622). IEEE.

Xu, K., Gerla, M., & Bae, S. (2002). How effective is the IEEE 802.11 RTS/CTS handshake in ad hoc networks? In *Proceedings of Global Telecommunications Conference, IEEE GLOBECOM 2002,* (vol. 1), (pp. 72 – 76). IEEE Press.

Xu, L., Tonjes, R., Paila, T., Hansmann, W., Frank, M., & Albrecht, M. (2000). DRiVE-ing to the Internet: Dynamic radio for IP services in vehicular environments. In *Proceedings of 25th Annual IEEE Conference on Local Computer Networks,* (pp. 281-289). IEEE Press.

Xu, X., van der Schaar, M., Krishnamachari, S., Choi, S., & Wang, Y. (2004). Fine-granular-scalability video streaming over wireless LANs using cross-layer error control. In *Proceedings of the IEEE International Conference on Acoustic, Speech and Signal Processing* (*ICASSP 2004*), (pp. 989-992). IEEE Press.

Xu, C., Feng, G., & Kwak, K. S. (2001). A modified constrained constant modulus approach to blind adaptive multiuser detection. *IEEE Transactions on Communications*, 49.

Xue, Q., & Ganz, A. (2003). Ad hoc QoS on-demand routing (AQOR) in mobile ad hoc networks. *Journal of Parallel and Distributed Computing*, *63*(2), 154–165 doi:10.1016/S0743-7315(02)00061-8

Xu, Y., & Zhang, T. (2002). Variable shortened-and-punctured Reed-Solomon codes for packet loss protection. *IEEE Transactions on Broadcasting*, *48*(3), 237–245. doi:10.1109/TBC.2002.803706

Xu, Z., & Liu, P. (2002). Code-constrained blind detection of CDMA signals in multipath channels. *IEEE Signal Processing Letters*, 9.

Xu, Z., & Tsatsanis, M. K. (2006). Blind adaptive algorithms for minimum variance CDMA receivers. *IEEE Transactions on Communications*, 49.

Yang, Z., & Liu, Y. (2008). Quality of trilateration: Confidence based iterative localization. In *Proceedings of the 28th International Conference on Distributed Computing Systems*, (pp. 446–453). IEEE.

Yang, K., Ou, S., Chen, H. H., & He, J. (2007). A multihop peer-communication protocol with fairness guarantee for IEEE 802.16-based vehicular networks. *IEEE Transactions on Vehicular Technology*, *56*(6), 3358–3370. doi:10.1109/TVT.2007.906875

Yang, Y., Zhang, Q., Zhu, W., & Zhang, Y.-Q. (2007). End-to-end TCP-friendly streaming protocol and bit allocation for scalable video over wireless Internet. *IEEE Journal on Selected Areas in Communications*, *22*(4), 777–790. doi:10.1109/JSAC.2004.826008

Yarkan, S., & Arslan, H. (2007). Binary time series approach to spectrum prediction for cognitive radio. In *Proceedings of IEEE Vehicular Technology Conference*. IEEE Press.

Yasami, K., & Abedi, A. (2011). Effect of channel estimation error on physical layer network coding. In *Proceedings of the IEEE Consumer Communications and Networking Conference,* (pp. 973 - 974). Las Vegas, NV: IEEE Press.

Ye, W., & Heidemann, J. (2006). Ultra-low duty cycle mac with scheduled channel polling. In *Proceedings of the Conference on Embedded Networked Sensor Systems (SenSys)*, (pp. 321–334). Boulder, CO: ACM Press.

Ye, W., Heidemann, J., & Estrin, D. (2002). An energy-efficient MAC protocol for wireless sensor networks. In *Proceeding of the Twenty-First Annual Joint Conference of the IEEE Computer and Communications Societies,* (Vol. 3), (pp. 1567-1576). IEEE Press.

Ye, W., Heidemann, J., & Estrin, D. (2004). Medium access control with coordinated adaptive sleeping for wireless sensor networks. *IEEE/ACM Transactions, 12*(3), 493–506.

Ye, W., Heidemann, J., & Estrin, D. (2004). Medium access control with coordinated adaptive sleeping for wireless sensor networks. *IEEE/ACM Transactions on Networking*, *12*(3), 493–506. doi:10.1109/TNET.2004.828953

Yoo, T., & Goldsmith, A. (2006). Capacity and power allocation for fading MIMO channels with channel estimation error. *IEEE Transactions on Information Theory*, *52*(5), 2203–2214. doi:10.1109/TIT.2006.872984

Yost, G. A. (1998). Improvement in estimation of time of arrival (TOA) from timing advance (TA). In *Proceedings of the International Conference on Universal Personal Communications*, (pp. 1367–1372). IEEE.

Yu, H., Martin, P., & Hassanein, H. (2005). Cluster based replication for large-scale mobile ad-hoc networks. In *Proceedings of the International Conference on Wireless Networks, Communications and Mobile Computing,* (pp. 552–557). Hawaii, HI: IEEE.

Yu, W., & Jun, Y. (2005). Joint source coding, routing, and resource allocation for wireless sensor networks. In *Proceedings of IEEE International Conference on Communications (ICC)*, (Vol. 2), (pp. 737-741). IEEE Press.

Yu, Y., Krishnamachari, B., & Prasanna, V. K. (2004). Energy-latency tradeoffs for data gathering in wireless sensor networks. In *Proceedings of IEEE International Conference on Computer Communications (Infocom)*, (Vol. 1), (pp. 7-15). IEEE Press.

Yuan, J., & Yu, W. (2006). Distributed cross-layer optimization of wireless sensor networks: A game theoretic approach. In *Proceedings of IEEE Global Telecommunications Conference (Globecom),* (pp. 1-5). IEEE Press.

Yuan, Y., He, Z., & Chen, M. (2006). Virtual MIMO-based cross layer design for wireless sensor networks. *IEEE Transactions on Vehicular Technology*, *55*(3), 856–864. doi:10.1109/TVT.2006.873837

Yuksel, M., Ramakrishnan, K. K., & Doverspike, R. D. (2008). Cross-layer failure restoration techniques for a robust IPTV service. In *Proceedings of the 16th IEEE Workshop on Local and Metropolitan Area Networks LANMAN 2008,* (pp. 49-54). Cluj-Napoca, Romania: IEEE Press.

Yuliang Tang, Y., Lin, C. C., Kou, G., & Deng, D. J. (2010). Cross-layer handover scheme for multimedia communications in next generation wireless networks. *EURASIP Journal on Wireless Communications and Networking*, 10.

Zabin, F., Misra, S., Woungang, I., Rashvand, H., Ma, N.-W., & Ahsan Ali, M. (2008). Reep: Data-centric, energy-efficient and reliable routing protocol for wireless sensor networks. *IET Communications*, *2*(8), 995–1008. doi:10.1049/iet-com:20070424

Zafer, M., & Modiano, E. (2008). Optimal rate control for delay-constrained data transmission over a wireless channel. *IEEE Transactions on Information Theory*, *54*(9), 4020–4039. doi:10.1109/TIT.2008.928249

Zanikolas, S., & Sakellariou, R. (2005). A taxonomy of grid monitoring systems. *Future Generation Computer Systems*, *21*(1), 163–188. doi:10.1016/j.future.2004.07.002

Zeadally, S., & Siddiqui, F. (2007). An empirical analysis of handoff performance for SIP, mobile IP, and SCTP protocols. *Wireless Personal Communications*, *43*(2), 589–603. doi:10.1007/s11277-007-9251-x

Zeng, P., Hu, R., Yang, Y., & Song, J. (2010). A cross-layer key management scheme in ad hoc network. In *Proceedings of the 2010 2nd International Conference on Signal Processing Systems (ICSPS)*, (Vol. 1), (pp. 132-137). ICSPS.

Zeng, W., Arora, A., & Shroff, N. (2010). Maximizing energy efficiency for convergecast via joint duty cycle and route optimization. In *Proceedings of INFOCOM*, (pp. 1 –5). IEEE Press.

Zhang, H., Wen, X., Wang, B., Zheng, W., & Sun, Y. (2010). A novel handover mechanism between femtocell and macrocell for LTE based networks. In *Proceedings of the International Conference on Communication Software and Networks (ICCSN 2010)*, (pp. 228-231). Singapore: ICCSN.

Zhang, L., et al. (1997). *Resource ReSerVation protocol (RSVP): RFC 2205*. Retrieved from http://www.ietf.org.

Zhang, Y., & Lee, W. (2000). Intrusion detection in wireless ad-hoc networks. In *Proceedings of the 6th Annual International Conference on Mobile Computing and Networking (MobiCom)*, (pp. 275-283). MobiCom.

Zhang, H., Zheng, Y., Khojastepour, M., & Rangarajan, S. (2010). Cross-layer optimization for streaming scalable video over fading wireless networks. *IEEE Journal on Selected Areas in Communications*, *28*(3), 344–353. doi:10.1109/JSAC.2010.100406

Zhang, L., & Zhang, Y. (2009). Energy-efficient cross-layer protocol of channel-aware geographic- informed forwarding in wireless sensor networks. *IEEE Transactions on Vehicular Technology*, *58*(6), 3041–3052 doi:10.1109/TVT.2008.2010046

Zhang, Q., Yang, F., & Zhu, W. (2005). Cross-layer QoS support for multimedia delivery over wireless Internet. *EURASIP Journal on Applied Signal Processing*, *2*, 207–219. doi:10.1155/ASP.2005.207

Zhang, X., Jiang, H., Zhang, L., Zhang, C., Wang, Z., & Chen, X. (2010). An energy-efficient ASIC for wireless body sensor networks in medical applications. *IEEE Transactions on Biomedical Circuits and Systems*, *4*(1), 11–18. doi:10.1109/TBCAS.2009.2031627

Zhang, Y., Fu, F., & van der Schaar, M. (2010). On-line learning and optimization for wireless video transmission. *IEEE Transactions on Signal Processing*, *58*(6), 3108–3124. doi:10.1109/TSP.2010.2046040

Zhao, J., Wang, L., Li, S., Liu, X., Yuan, Z., & Gao, Z. (2010a). A survey of congestion control mechanisms in wireless sensor networks. In *Proceedings of the Sixth International Conference on Intelligent Information Hiding and Multimedia Signal Processing,* (pp. 719-722). Darmstadt, Germany: IEEE.

Zhao, J., Zheng, H., & Yang, G. (2005). Distributed coordination in dynamic spectrum allocation networks. In *Proceedings of IEEE DySPAN,* (pp. 259-268). IEEE Press.

Zhao, T., Guo, T. D., Yang, W. G., & Chen, G. (2010b). An optimal cross-layer model for maximum lifetime routing in wireless sensor networks. In *Proceedings of the International Conference on Communications and Mobile Computing,* (Vol. 3), (pp. 244-248). Shenzhen, China: IEEE.

Zhao, F., Shin, J., & Reich, J. (2002). Information-driven dynamic sensor collaboration for tracking applications. *IEEE Signal Processing Magazine*, *19*(2), 61–72. doi:10.1109/79.985685

Zhao, H., Garcia-Palacios, E., Wei, J., & Xi, Y. (2009). Accurate available bandwidth estimation in IEEE 802.11-based ad hoc networks. *Computer Communications*, *32*, 1050–1057. doi:10.1016/j.comcom.2008.12.031

Zhao, Q., & Sadler, M. (2007). A survey of dynamic spectrum access. *IEEE Signal Processing Magazine*, *24*(3), 79–89. doi:10.1109/MSP.2007.361604

Zhao, Q., Tong, L., Swami, A., & Chen, Y. (2007). Decentralized cognitive MAC for opportunistic spectrum access in ad hoc networks: A POMDP framework. *IEEE Journal on Selected Areas in Communications*, *25*(3), 589–600. doi:10.1109/JSAC.2007.070409

Zheng, H., & Cao, L. (2005). Device-centric spectrum management. In *Proceedings of IEEE DySPAN,* (pp. 56-65). IEEE Press.

Zheng, H., & Peng, C. (2005). Collaboration and fairness in opportunistic spectrum access. [IEEE Press.]. *Proceedings of the IEEE*, *ICC*, 3132–3136.

Zheng, L., & Tse, D. N. C. (2003). Diversity and multiplexing: A fundamental tradeoff in multiple-antenna channels. *IEEE Transactions on Information Theory*, *49*, 1073–1096. doi:10.1109/TIT.2003.810646

Zhiguo, D., & Leung, K. K. (2011). Cross-layer routing using cooperative transmission in vehicular ad-hoc networks. *IEEE Journal on Selected Areas in Communications*, *29*(3), 571–581. doi:10.1109/JSAC.2011.110307

Zhou, P., Liu, W., Yuan, W., Cheng, W., & Wang, S. (2009). An energy-efðcient cooperative MISO-based routing protocol for wireless sensor networks. In *Proceedings of IEEE Wireless Communications and Networking Conference (WCNC 2009),* (pp. 1-6). IEEE Press.

Zhou, Q., & Dai, H. (2006). Asymptotic analysis in MIMO MRT/MRC systems. *Eurasip Journal on Wireless Communications and Networking*. Retrieved from http://jwcn.eurasipjournals.com/content/2006/1/045831.

Zhou, R., Li, X., Wu, Z., Chakravarthy, V., & Li, H. (2010). The demonstration of SMSE based cognitive radio in mobile environment via software defined radio. In *Proceedings of the IEEE Globecom Demo Session.* IEEE Press.

Zhou, X., Gandi, S., Suri, S., & Zheng, H. (2008). eBay in the sky: Strategy-proof wireless spectrum auctions. In *Proceedings of ACM MobiCom*. ACM Press.

Zhou, Y., & Nettles, S. M. (2005). Balancing the hidden and exposed node problems with power control in CSMA/CA-based wireless networks. In *Proceedings of Wireless Communication and Networking Conference, IEEE WCNC 2005*. IEEE Press.

Zhou, L., & Chao, H. C. (2011). Multimedia traffic security architecture for internet of things. *IEEE Network*, *25*(3), 29–34. doi:10.1109/MNET.2011.5772059

Zhou, L., Geller, B., Wang, X., Wei, A., Zheng, B., & Chao, H. C. (2009). Multi-user video streaming over multiple heterogeneous wireless networks: A distributed, cross-layer design paradigm. *Journal of Internet Technology*, *10*(1), 1–11. doi:10.1145/1552291.1552295

Zhou, N., Zhu, X., Huang, Y., & Lin, H. (2010). Low complexity cross-layer design with packet dependent scheduling for heterogeneous traffic in multiuser OFDM systems. *IEEE Transactions on Wireless Communications*, *9*(6), 1912–1923. doi:10.1109/TWC.2010.06.081595

Zia, T., & Zomaya, A. (2006). Security issues in wireless sensor networks. In *Proceedings of the International Conference on Systems and Networks Communications,* (pp. 40-43). Tahiti, French Polynesia: IEEE.

Ziemer, R. E., Peterson, R. L., & Borth, D. E. (1995). *Introduction to spread spectrum communications*. Upper Saddle River, NJ: Prentice-Hall.

Zigel, Y., Cohen, A., & Katz, A. (2000). ECG signal compression using analysis by synthesis coding. *IEEE Transactions on Bio-Medical Engineering*, *47*(10), 1308–1316. doi:10.1109/10.871403

Zorzi, M. (2004). A new contention-based MAC protocol for geographic forwarding in ad hoc and sensor networks. In *Proceedings of IEEE International Conference on Communications*, (Vol. 6), (pp. 3481 – 3485). IEEE Press.

Zorzi, M., & Rao, R. (2003). Geographic random forwarding (GeRaF) for ad hoc and sensor networks: Multihop performance. *IEEE Transactions on Mobile Computing*, *2*(4), 337–348. doi:10.1109/TMC.2003.1255648

Zorzi, M., & Rao, R. (2004). Geographic random forwarding (GeRaF) for ad hoc and sensor networks: Energy and latency performance. *IEEE Transactions on Mobile Computing*, *2*(4), 349–365. doi:10.1109/TMC.2003.1255650

Zuniga, M., & Krishnamachari, B. (2004). Analyzing the transitional region in low power wireless links. In *Proceedings of IEEE Communications Society Conference on Sensor, Mesh and Ad Hoc Communications and Networks (SECON 2004)*, (Vol. 1), (pp. 517 – 526). IEEE Press.

Zyrion Inc. (2010). *Zyrion transverse*. Retrieved from http://www.zyrion.com/.

About the Contributors

Habib F. Rashvand received his B.Sc. in 1969 and post-graduate qualifications in 1970 from the University of Tehran. He was selected for a training mission as the head of division for development of a new Telecom Research Centre as under a new cooperation project between the Iranian PTT and Japanese Industries including the NTT, KTT following his Doctorate at the University of Kent in 1980. Since then he earned a rich blend of industrial research and development positions with industries in collaboration with many universities including University of Southampton, University of Reading, Portsmouth University, Warwick University, and Coventry University. His academic positions compile University of Tehran, University of Zambia, Coventry University, Magdeburg University and University of Warwick. His Professorship in Networks, Systems, and Protocols applied in 1998 to the German Ministry of Education succeeded in 2001. Since 2004, he headed a Special Academic Quality Research Operation under Directorship of Advanced Communication Systems, which involves ITU, CTO, WHO, IEEE/IEE/IET. He was the editor-in-chief, member of editorial board, and invited speaker for many research journals and conferences.

Yousef S. Kavian received the B.Sc. (Hons) degree in Electronic Engineering from the Shahid Beheshti University, Tehran, Iran, in 2001, the M.Sc. degree in Control Engineering from the Amkabir University, Tehran, Iran, in 2003, and the Ph.D. degree in Electronic Engineering from the Iran University of Science and Technology, Tehran, Iran, in 2007. After a one-year appointment at Shahid Beheshti University, in 2008, he joined the Shahid Chamran University as an Assistant Professor. He worked as a Postdoctoral Research fellow at Esslingen University and IAER, Germany, in 2010. His research interests include digital circuits and systems design, optical and wireless networking. Dr. Kavian has over 50 technical publications including journal and conference papers and book chapters in these fields. He is a Senior Industrial Engineer and Trainer with more than 10 years industrial experience.

* * *

Ali Abedi received his PhD in Electrical and Computer Engineering from University of Waterloo in 2004. He joined the University of Maine, Orono, in 2005, where he is currently Associate Professor of Electrical and Computer Engineering and Director of WiSe-Net Lab. His research works include analytical performance evaluation of block codes, new methods for performance and convergence analysis of Turbo-codes, and applications of error correction codes in wireless sensor networks for structural monitoring, space explorations, and biomedical applications. Dr. Abedi has received a number of awards and recognitions from Natural Sciences and Engineering Research Council of Canada (NSERC), Japan Society for the Promotion of Science (JSPS), and Canadian Space Agency (CSA), NASA, and IEEE. He is a Senior Member of IEEE and currently serves as IEEE Region-1 Technical Chapters Coordinator.

Laith Al-Jobouri received his BSc in 1995 from the University of Technology, Iraq and subsequently worked in different wireless and microwave projects. He obtained his MSc from the University of Mosul, Iraq, in 2001, and then again returned to work as a project manager and head of department for several national and international companies concerned with wireless and GSM. He is now undertaking PhD research for a degree at the University of Essex, UK. His research interests are in video over wireless, forward error correction, and wireless technologies.

Salah Al-Majeed holds a BSc from Baghdad University, Iraq, obtained in 1992. He went on to take an MSc in Mobile and Satellite Communication at Westminster University, London, UK. He went on to take a PhD on the subject of broadband wireless video streaming at the University of Essex, Colchester, UK. After receiving his PhD in 2011, Salah took up commercial work in video streaming applications. His research interests are cross-layer design of congestion control, quality-of-service issues, scheduling algorithms, bandwidth allocation mechanisms, and multi-connection techniques, especially in WiMAX broadband wireless.

Raad Alturki is an Assistant Professor of Computer Science at Imam Mohammed Bin Saud University in Riyadh, Saudi Arabia. Earlier he finished his PhD and MSc with distinction from Swansea University, UK. Also he holds a BSc in Computer Science from King Saud University, Riyadh. Mr. Alturki is a member of the Association for Computing Machinery (ACM) and British Computer Society (BCS). Raad has research interests in multimedia, ad-hoc networks, routing, security, and HCI.

Alagan Anpalagan directs a research group working on Radio Resource Management (RRM) and Radio Access and Networking (RAN) areas within the WINCORE Lab. His current research interests include cognitive radio resource allocation and management, wireless cross layer design and optimization, collaborative communication, green communications technologies and QoE-aware femtocells. Dr. Anpalagan serves as Associate Editor for the *IEEE Communications Letters* (2010-) and Springer *Wireless Personal Communications* (2009-), and past Editor for *EURASIP Journal of Wireless Communications and Networking* (2004-2009). He also served as EURASIP Guest Editor for two special issues in *Radio Resource Management in 3G+ Systems* (2006) and *Fairness in Radio Resource Management for Wireless Networks* (2008). Dr. Anpalagan served as TPC Co-Chair of: IEEE PIMRC 2011 Track on Cognitive Radio and Spectrum Management, IEEE IWCMC 2011 Workshop on Cooperative and Cognitive Networks, IEEE CCECE 2004/2008 and WirelessCom 2005 Symposium on Radio Resource Management. He received BASc (Hons), MASc, and PhD degrees from the University of Toronto.

Jose M. Alcaraz Calero holds a PhD in Computer Science from the Univeristy of Murcia, Spain, and is currently a Researcher at HP Labs in Bristol, UK. He is part of the Cloud and Security Lab focusing on developing technologies for highly automated, secure, and dynamic instantiation and management of cloud computing infrastructure and services.

Andrea F. Cattoni received his MSc in Telecommunications Engineering and his Ph.D. in ICT from the University of Genova, Italy, in 2004 and 2008, respectively. From November 2008 he is Post Doc at Aalborg University, and Research Consultant at Nokia Siemens Networks, Aalborg, Denmark. He is author of more than 30 scientific publications. He is Danish Substitute Member in the Management Committee of the COST Action IC0902. He serves as Technical Program Committee Member and as a reviewer in several International Conferences and Journals. His current research topics include cognitive radio, radio resource management for IMT-A systems, local area access methods.

Fu-Wang Chang graduated from National Taiwan University with an Electrical Engineering and Computer Science master degree in 2008. Now he works in Chunghwa Telecom Co., Ltd. as a System-Development Engineer and his main job is to maintain and ensure regular operation of systems is normal.

Han-Chieh Chao is a jointly appointed Professor of the Department of Electronic Engineering and Institute of Computer Science and Information Engineering, National Ilan University, I-Lan, Taiwan. He also holds a joint professorship of the Department of Electrical Engineering, National Dong Hwa University, Hualien, Taiwan. His research interests include high speed networks, wireless networks, and IPv6 based networks and applications. He received his MS and Ph.D. degrees in Electrical Engineering from Purdue University in 1989 and 1993, respectively. Dr. Chao is an IEEE Senior Member, IET and BCS Fellows.

Chi-Yuan Chen received the M.S. degree in Electrical Engineering from National Dong Hwa University, Hualien, Taiwan, in 2007. He is currently working toward the Ph.D. degree in Electrical Engineering at National Dong Hwa University. He is a student member of IEEE and ACM. His research interests include Next Generation Networks, Communication Services, and Network Security.

Hsiao-Hwa Chen received the B.Sc. and M.Sc. degrees from Zhejiang University, Hangzhou, China, and the Ph.D. degree from the University of Oulu, Oulu, Finland, in 1982, 1985, and 1991, respectively. He is currently a full Professor in the Department of Engineering Science, National Cheng Kung University, Taiwan. He has authored or coauthored over 250 technical papers in major international journals and conferences, five books, and three book chapters in the areas of communications. Dr. Chen served as general chair, TPC chair, and symposium chair for many international conferences. He served or is serving as an Editor or/and Guest Editor for numerous technical journals. He is the founding Editor-in-Chief of Wiley's *Security and Communication Networks Journal* (http://www.interscience.wiley.com/journal/security). He is the recipient of the Best Paper Award in IEEE WCNC 2008.

Yongrui Chen received the B.S. degree from Yanshan University, Qinghuangdao, China, in 2001, and the M.S. from Tsinghua University, Beijing, in 2007, and Ph.D. degrees from the Graduate University of Chinese Academic of Sciences (GUCAS) in 2011 respectively. He is currently a Post Doctoral at the Graduate School of Chinese Academic of Sciences. His research interests are in cross-layer design, reliable data transfer, MAC protocols, routing protocols, and cooperative communication in wireless sensor networks.

Yuh-Shyan Chen received the B.S. degree in Computer Science from Tamkang University, Taiwan, R. O. C., in June 1988 and the M.S. and Ph.D. degrees in Computer Science and Information Engineering from the National Central University, Taiwan, R. O. C., in June 1991 and January 1996, respectively. He joined the faculty of Department of Computer Science and Information Engineering at Chung-Hua University, Taiwan, R. O. C., as an Associate Professor in February 1996. He joined the Department of Statistic, National Taipei University in August 2000, and joined the Department of Computer Science and Information Engineering, National Chung Cheng University in August 2002. Since 2006, he has been a Professor at the Department of Computer Science and Information Engineering, National Taipei University, Taiwan. Prof. Chen now serves as Editor-in-Chief of *International Journal of Ad Hoc and Ubiquitous Computing* (SCIE), Regional Editor (Asia and Pacific) of *IET Commuication* (SCIE), and Editorial Board of *Telecommunication System Journal* (SCIE), *EURASIP Journal on Wireless Communications and Networking* (SCIE), *International Journal of Communication Systems* (SCIE), *Mobile Information Systems* (SCIE), and *Journal of Internet Technology* (SCIE). His paper wins the 2001 IEEE 15th ICOIN-15 Best Paper Award. Prof. Chen was a recipient of the 2005 Young Scholar Research Award, National Chung Cheng University, R.O.C. His recent research topics include wireless communications and mobile computing, and next-generation personal communication system. Dr. Chen is a senior member of the IEEE Communication Society and Phi Tau Phi Society.

Wenqing Cheng received the MS and Ph.D degree in Electronics and Information System from Huazhong University, China, in 1988 and 2005, respectively. Her research interests are Information System and Applications, Wireless Network Technology and Applications, and E-learning.

Patrick Clarke received the MEng and PhD degrees in Electronic Engineering from the University of York in 2008 and 2011, respectively. His areas of interest include wireless communications, adaptive signal processing, and cooperative communications.

Rodrigo C. de Lamare received the Diploma in Electronic Engineering from the Federal University of Rio de Janeiro (UFRJ) in 1998, and the M.Sc. and PhD degrees, both in Electrical Engineering, from the Pontifical Catholic University of Rio de Janeiro (PUC-Rio) in 2001 and 2004, respectively. Since January 2006, he has been with the Communications Research Group, Department of Electronics, University of York, where he is currently a senior lecturer in communications engineering. His research interests lie in communications and signal processing, areas in which he has published over 200 papers in refereed journals and conferences.

Virgil Dobrota holds Dipl.Eng. in Telecommunications (1987) and Ph.D. in Electronics and Telecommunications (1995) from Technical University of Cluj-Napoca. He is currently professor at Technical University of Cluj-Napoca and Head of Communications Department. Courses taught: Switching and Routing Systems, Internet Protocols, and Unified Communications. Professor Dobrota was involved in RACE-MAGIC, INCO-COPERNICUS 1529, ACTS AC235, COST 290, FP7-4WARD. Member of IEEE Communications Society, ACM-SIGCOMM, he served as TPC Co-Chair at IEEE LANMAN 2007 and General Co-Chair at IEEE LANMAN 2008.

Xu Du received the Ph.D degree in Electronics and Information System from Huazhong University, China, in 2002. His research interests are Distributed Network System and Applications, Streaming media distribution, Wireless Network Technology and Applications, and Broadband IP Network Technology and Applications.

Trung Q. Duong was born in HoiAn town, Quang Nam province, Vietnam, in 1979. He received the B.S. degree in Electrical Engineering from Ho Chi Minh City University of Technology, Vietnam in 2002, and the M.Sc. degree in Computer Engineering from Kyung Hee University, South Korea, in 2005. In December 2007, he joined the Blekinge Institute of Technology, Sweden, as a research staff working toward his Ph.D. degree in Radio Communications. His current research interests include cross-layer optimization, cooperative communications, and cognitive radio networks.

Guillem Femenias received the M. S. and Ph. D. degrees in Telecommunications Engineering from the Technical University of Catalonia (UPC), Spain, in 1987 and 1991, respectively. In the past, he was an Associate Professor with UPC and the University of the Balearic Islands (UIB), Spain. Currently, he is a Full Professor with UIB, and the Research Leader of the Mobile Communications Group. He has been the Project Manager of a number of projects funded by the Spanish and Balearic Islands Governments, and in the past, he was also involved with several European projects.

Martin Fleury worked as a Senior Lecturer at the University of Essex, where he is still a Visiting Fellow. He is now a free-lance consultant involved in business concerns. Martin has authored or co-authored two hundred and thirty articles and book chapters on topics such as document and image compression algorithms, performance prediction of parallel systems, software engineering, reconfigurable hardware, and vision systems. His current research interests are video communication over wireless networks. He has published a book on high-performance computing for low-level image- and signal-processing algorithms and has another book in print on peer-to-peer video streaming.

Mohammed Ghanbari is best known for his pioneering work on two-layer video coding for ATM networks (which earned him an IEEE Fellowship in 2001). He has served as an Associate Editor to *IEEE Transactions on Multimedia* and has registered for eleven international patents on various aspects of video networking. Mohammed is the author of *Video Coding: An Introduction to Standard Codecs* and *Standard Codecs: Image Compression to Advanced Video Coding*, both published by the IET press. Prof. Ghanbari has authored or co-authored at least 500 journal and conference papers, many of which have had a fundamental influence in this field.

Michael Hempel received his Ph.D. in computer engineering at the University of Nebraska-Lincoln, USA. He is currently working as a Research Assistant Professor at the Advanced Telecommunication Engineering Laboratory (TEL) at the University of Nebraska-Lincoln. His research interests include wireless communication protocol design and performance analysis, wireless multimedia services, and distributed computing. For his research in networking he has also been developing various network simulation and analysis tools for streaming media and WiFi/WiMAX technologies. He has authored or co-authored over 50 publications in major international journals and conferences, and served as a TPC member on numerous international conferences.

Yin-Cheng Huang received the M.S. degree in Electrical Engineering from National Taiwan University in 2009. He currently works in 4ipnet as Software Engineer. His research interests include multimedia streaming over wireless network and network mobility management.

Ashok Karmokar is a Ryerson Fellow in the Department of Electrical and Computer Engineering (ECE) at Ryerson University, Toronto, Canada. He received the Ph. D. degree in ECE from the University of British Columbia. He received his B.Sc. degree with distinction and M.Sc. degree in Electrical and Electronic Engineering from Bangladesh University of Engineering and Technology (BUET), Dhaka, Bangladesh. From 1998 to 2002, he was with BUET as a Lecturer/Assistant Professor. His research interests include energy-efficient cross-layer adaptive transmission and optimization, dynamic decision making on resource allocation and scheduling for cognitive and green radio wireless communications networks.

Ekram Khan obtained his Ph. D. (2003) degree from University of Essex, Colchester, UK, and is currently working as a Professor in the Department of Electronics Engineering. He was a recipient of the prestigious Commonwealth Scholarship to pursue his Ph. D. degree from University of Essex, UK. He has authored and co-authored over 85 papers in refereed academic journals and international conference proceedings. His areas of research are low complexity image/video coding, video transmission over wireless and IP networks, video streaming and biomedical image processing. Prof. Khan had spent summers of 2005 and 2006 as Academic Visitor to University of Essex (funded by Royal Society, UK). He is a member of IEEE.

Mohd Ayyub Khan is an Assistant Professor in Electronics Engineering, University Polytechnic, Aligarh Muslim University, Aligarh, India. He received the B.Tech (Hons) and M.Tech (Hons) from the Department of Electronics Engineering, Z.H College of Engineering and Technology, Aligarh Muslim University, Aligarh, India in 2002 and 2005, respectively. He is currently working towards the PhD degree in the Department of Electronics Engineering, AMU, Aligarh, India. His research interest includes Image and Video Coding and Transmission over wireless/IP channels. He is a member of IEEE.

Manas Khatua is presently pursuing his Ph.D. from the School of Information Technology, IIT Kharagpur, India, and is working as a Senior Research Fellow in an Underwater Sensor Network project funded by DIT, Government of India. He received the M.Tech. degree in IT from Bengal Engineering and Science University, Shibpur, India, in 2007, and B.Tech in Computer Science and Engineering from University of Kalyani, India, in 2003. He has teaching experience of 2 years and 1 month in an engineering college and was an Assistant System Engineer in TCS for 2 years and 6 months. His research interests include the application of biological, statistical, and economic model in wireless ad-hoc and sensor network security, cooperation, and cross-layer protocol design.

Johannes Kirschnick holds a degree in Computer Science from the Technical University of Munich and is currently a Researcher at HP Labs in Bristol, UK. He is part of the Cloud and Security Lab focusing on developing technologies for highly automated, secure, and dynamic instantiation and management of cloud computing infrastructure and services.

Benjamin Koenig studied Computer Science in Media at LMU Munich from 2005 to 2008. Graduated M. Sc. h. c. in Software Engineering from University of Augsburg, TU Munich, and LMU Munich in 2010. Currently work as a Software Developer / Integrator for AdNovum Informatik AG in Zurich, Switzerland.

Yun-Wei Lin received the B.S. degree in Computer and Information Science from the Aletheia University, Taiwan, ROC, in June 2003 and the M.S. and Ph.D. degrees in Computer Science and Information Engineering from National Chung Cheng University, Taiwan, ROC, in July 2005 and January 2011, respectively. His research interests include mobile ad hoc networks, wireless sensor network, and vehicular ad hoc networks.

Tao Ma received the B.Sc. and M.Sc. degrees from Xian Jiaotong University, China, in 2002 and 2005 respectively, both in electrical engineering. He is currently pursuing the Ph.D. degree and in Department of Computer and Electronics Engineering, University of Nebraska-Lincoln. His research areas are Cross-layer design for QoS provisioning in wireless data networks, and multi-media distribution.

Chang-Han Mai received the B.S. degree in Electrical Engineering from National Taiwan University in 2009. He is currently working toward the M.S. degree in the Graduate Institute of Communication Engineering, National Taiwan University. His research interests include IEEE 802.11 protocol design, wireless communication testbed, and multicast in wireless local network.

Rashid Mehmood holds a Lectureship in Computing and Communications at Swansea University. He leads a research group in High Performance Computational Engineering (HPCE) at Swansea, developing state-of-the-art technologies through cross disciplinary research. Rashid has a multidisciplinary background with qualifications and experience in electronics and communications engineering, business, and computing from institutions including Universities of Oxford, Birmingham, and Cambridge. His core expertise is in stochastic modelling, simulation/analysis of large-scale complex systems, and HPC, applied to telecommunications, intelligent transportation systems and healthcare. Rashid is a member of ACM, OSA, a Senior Member of IEEE, and the Vice-Chairman of the IET Wales South West Network.

Sudip Misra is an Assistant Professor in the School of Information Technology at the IIT Kharagpur, India. He is an Alexander von Humboldt Fellow (Germany) and a Senior Member of IEEE. He received his Ph.D. degree in Computer Science from Carleton University, Canada, and the masters and bachelors degrees, respectively, from the University of New Brunswick, Fredericton, Canada, and the Indian Institute of Technology, Kharagpur, India. Dr. Misra has several years of experience working in the academia and the private sectors in research, teaching, consulting, project management, software design and product engineering roles. Dr. Misra is the author/editor of over 120 scholarly research papers and books. Dr. Misra is the editor-in-chief of one journal and an associate editor and editorial board member of several others. He has won six research paper awards in different conferences. His research interests include algorithm design, wireless ad hoc networks, wireless sensor networks, wireless mesh networks, cellular networks, body sensor networks, underwater sensor networks, WiMAX, and WiFi.

Preben Mogensen received his MSc., EE., and Ph.D. degrees in 1988 and 1996, respectively, from Aalborg University (AAU), Denmark. From 1999 he has been part time professor at Department of Electronic Systems, AAU, where he is heading the Radio Access Technology (RATE) research section. Preben Mogensen is also Principal Engineer at Nokia Siemens Networks (NSN) – Aalborg, where he has been involved in LTE and LTE-Advanced standardisation research. He was nominated NSN fellow in 2009. He is author or co-author of more than 180 technical publications. His current research interest includes cognitive radio, heterogeneous networks, and network energy saving.

Athar Ali Moinuddin is an Associate Professor in the Department of Electronics Engineering, Aligarh Muslim University, Aligarh, India. In 2007, he received Commonwealth Split-Site Scholarship for one year to conduct research under Professor M. Ghanbari at Multimedia Networking Laboratory, University of Essex, UK. He has authored/co-authored over 25 papers in refereed journals and conference proceedings of international repute. His areas of research are low complexity image/video coding and image/video transmission over wireless networks. He is a member of IEEE.

Bogdan Pavkovic is PhD candidate in DRAKKAR team, LIG laboratory in Grenoble, France, working towards his thesis diploma. After Master diploma in Microcomputer Embedded Systems (2009) he became interested in domain of Wireless Sensor Networks. His research interests include routing protocols, topology control, realistic link quality metrics as well as emulation of large-scale networks and testbed implementations.

Dongming Peng received the B.A. and M.A. degrees in electrical engineering from Beijing University of Aeronautics and Astronautics, Beijing, China, in 1993 and 1996, respectively, and the Ph.D.degree in computer engineering from Texas A&M University, College Station, in 2003. From 1996 to 1997, he was a faculty member at Beijing University. In 2002, he joined the University of Nebraska-Lincoln, where he is currently an Associate Professor. His research interests include digital image processing, computer architectures, parallel and distributed computing, and sensor network. Dr. Peng is one of the recipients of the Best Paper Award in IEEE WCNC 2008. He has also served as a referee and program committee member for several conferences and journals.

Jaume Ramis was born in Sencelles, Spain, in 1972. He received the Engineer of Telecommunication degree from the Universitat Politècnica de Catalunya, Barcelona, Spain, in 1997. He is currently working toward the Ph.D. degree with the Department of Mathematics and Informatics, University of the Balearic Islands (UIB), Mallorca, Spain, where he is currently an Assistant Professor. He has been involved in projects concerned with wireless local area network design, funded by the Spanish and Balearic Islands governments (MARIMBA, XISPES, and COSMOS).

Fahimeh Rezaei received her B.Sc degree from Tehran Azad University, Iran 2006 in Computer Engineering- Hardware. She received her M.Sc degree from the University of Nebraska Lincoln in Telecommunications Engineering 2010. She is currently pursuing The PhD degree in Department of Computer and Electronics Engineering University of Nebraska Lincoln. Her research areas are wireless communications and protocols, computer Architecture, and image processing.

Rouzbeh Razavi graduated from the Kahjeh Nasir Toosi University of Technology in Tehran before taking an MSc at the University of Essex, UK, which he passed with distinction in 2005. He subsequently took a PhD at Essex and published 12 journal papers, 18 conference papers, and 3 book chapters on the subject of video over wireless networks, as well as publications on optical access networks. Having successfully completed his PhD in 2008, he took up a post as a research engineer with Bell Labs in Dublin, Ireland. His research interests are video streaming over wireless networks, computational intelligence, and autonomic computing systems.

Andrei Bogdan Rus received Dipl.Eng. (2007), M.Sc. (2009), and Ph.D. (2011) in Telecommunications at Technical University of Cluj-Napoca. Currently he is Teaching Assistant, being interested in cross-layer techniques for quality of service and Internet algorithms and protocols. Andrei Bogdan Rus is member of IEEE Communications Society and he was involved in COST 290 and FP7-4WARD.

Kale Schrader received his Bachelor of Science in Physics from Fort Hays State University, Hays, KS, in 2007. Kale matriculated to the University of Maine to pursue a Master of Science in Electrical Engineering, which he received in August 2010. His university work included high intensity laser research, wavelet theory, and impact localization using wireless sensor networks. After graduation, Kale joined Capstan Ag Systems in Topeka, KS, and focuses on hardware design, instrumentation, and embedded system development for precision agriculture applications.

Hamid Sharif received the B.Sc. degree from the University of Iowa, Iowa City, the M.Sc. degree from the University of Missouri, Columbia, and the Ph.D. degree from the University of Nebraska-Lincoln, all in electrical engineering. He is the Paul and Betty Henson Professor of Computer and Electronics Engineering Department at the University of Nebraska-Lincoln (UNL). He is also the Director of Advanced Telecommunications Engineering Laboratory (TEL) at UNL. His research areas include: wireless communication protocols, wireless sensor networks, communication system performance modeling, and communication and network security. He has authored/coauthored a large number of journal and conference papers. Dr. Sharif is serving as an editor for several journals including the Co-Editor- in-Chief of Wiley's *Security and Communication Networks* journal. He has contributed to the IEEE in many roles including Chair of the Nebraska Section, Chair of the Nebraska Computer and Communication Chapters, and currently, he is the Chapter Coordinator for the IEEE Region 4.

Pradhumna L Shrestha received his BE in Electronics and Communication from Tribhuvan University, Kathmandu, Nepal in 2007 and M.S. in Telecommunications Engineering from University of Nebraska-Lincoln in 2011. He is currently pursuing his Ph.D. in Computer and Electronics Engineering Department of University of Nebraska-Lincoln. His research interests include theoretical and mathematical modeling, wireless communications, signal processing, high speed networks, among others.

Lei Shu is currently a Specially Assigned Research Fellow in Department of Multimedia Engineering, Graduate School of Information Science and Technology, Osaka University, Japan. His research interests include semantic sensor networks, wireless sensor network, context aware and sensor network middleware.

Troels B. Sørensen graduated from Aalborg University in 1990 (M.Sc.E.E.) and received the Ph.D. degree from the same university in 2002. From 1991 to 1997 he worked in the R&D department of the former Danish national telecom operator. In 1997 he joined the Center for Person Kommunikation (CPK) at Aalborg University, and later the Radio Access Technology (RATE) section where currently he holds an Associate Professor position. Main research activities are related to physical layer and radio resource management for Long Term Evolution (LTE) and LTE-Advanced, a topic area within which he has been supervising several Ph.D. students.

Fabrice Theoleyre is a Research Scientist in the CNRS since 2007. After having spent two years in the Grenoble Informatics Laboratory, he is now affiliated with the University of Strasbourg. He received his PhD in Computer Science from INSA, Lyon (France) in 2005. He was a Visiting Researcher in INRIA Sophia-Antipolis in 2006 and a Visiting Scholar in University of Waterloo in 2006. He served in the TPC of various international conferences (including PIMRC, IWCMC, VTC) and is Associate Editor for *IEEE Communications Letters*. His research interests comprise wireless multihop networks, including distributed algorithms, testbeds, MAC layer, and mobility management.

N. Kasi Viswanath has completed B.E in Computer Science and Engineering from Marathwada University in the year 1992, M.S (Systems and Information) from Birla Institute of Technology and Science, Pilani, and obtained PhD in Computer Science and Engineering from Rayalaseema University, Kurnool. He has 18 years of teaching experience. He published 14 papers in National and International Journals, 4 International and 3 National conferences. He organized nearly 6 National level Technical Symposiums and 4 National Workshops. He is the author of a text book *Data Structures Using C++* and also he is a reviewer for the subject *Advanced Data Structures and Algorithms* for Pearson Education Publications Pvt. Ltd., India. Presently, he is working as Professor and Head of Computer Science and Engineering Department, G.Pulla Reddy Engineering College (Autonomous), Kurnool, A.P, India.

Nguyen-Son Vo was born in Hue City, Vietnam, in 1980. He received the B.S. degree in Electrical Engineering and M.S degree in Radio Engineering - Electronics from Ho Chi Minh City University of Technology in 2002 and 2005, respectively. He won the Chinese Government Scholarship to attend a three-year Ph.D program, China, in 2008. His current research focuses on cross-layer optimization and multimedia streaming over multihop wireless networks, especially interactive streaming media distribution over wireless mesh networks.

Bin Wang received the Ph.D. degree in Electrical Engineering from the Ohio State University in 2000. He joined the Wright State University in September 2000, where he is currently full Professor of Computer Science and Engineering. His research interests include optical networks, real-time computing, mobile and wireless networks, cognitive radio networks, trust and information security, and semantic web. He was a recipient of the US Department of Energy Career Award and the AFOSR Senior Research Fellowship. His research has been supported by US Department of Energy, National Science Foundation, Air Force Office of Scientific Research, Air Force Research Laboratories, Ohio Supercomputer Center, and the State of Ohio. He is a Senior Member of the IEEE.

Chih-Yu Wang received the B.S. degree in Electrical Engineering from National Taiwan University in 2007. He is currently working toward the Ph.D. degree in the Graduate Institute of Communication Engineering, National Taiwan University. His research interests include applications of game theory in wireless networks, femtocell, and cognitive radio.

Hung-Yu Wei received the B.S. degree in Electrical Engineering from National Taiwan University in 1999. He received the M.S. and the Ph.D. degree in Electrical Engineering from Columbia University in 2001 and 2005, respectively. He was a summer intern at Telcordia Applied Research in 2000 and 2001. He was with NEC Labs America from 2003 to 2005. He joined Department of Electrical Engineering at the National Taiwan University in July 2005. His research interests include wireless mesh networks, mobility management in mobile Internet, sensor networks, cross-layer design and optimization in wireless multimedia communications, and game theoretical models for communications networks.

Zhiqiang Wu received the Ph.D. degree in Electrical Engineering from Colorado State University in 2002. He joined the Department of Electrical Engineering, Wright State University in 2005, where he is currently an Associate Professor. Dr. Wu has been working in wireless communication industry for more than 15 years. He has worked for Motorola and Chinese Academy for Telecommunication Technologies. Dr. Wu has been actively conducting research in different areas of wireless communication, including RF spectrum processing, high performance/high security wireless network, space time coding and MIMO, and carrier interferometry technology, cognitive radio and cognitive network, UWB transmission, modulation detection, and signal classification. Recently, Dr. Wu's SDR based cognitive radio in mobile environment demonstration won the Best Demo Award of the prestigious IEEE Globecom Conference.

Zhongmei Yao received the BS degree in Engineering from Donghua University, Shanghai, China, in 1997, the MS degree in Computer Science from Louisiana Tech University, Ruston, in 2004, and the PhD degree in Computer Science from Texas A&M University, College Station, in 2009. Since 2009, she has been an Assistant Professor in the Department of Computer Science at The University of Dayton, Ohio. Her research interests include peer-to-peer systems, Internet measurement, and stochastic network modeling. She is a member of the IEEE.

Weidong Yi received the B.S degree from Peking University, China, in 1982, M.S from Brown University, USA, in 1991, and Ph.D degrees from the University of California at Berkeley in 1999. In 2004, he joined the Graduate University of Chinese Academic of Sciences (GUCAS), Beijing, China and is now a Professor at the College of Information Science and Technology of GUCAS. His research interests are in the design, modeling micorsensors and its applications in the internet of things, node platforms, data fusion, energy harvesting and consuming efficiency, cross-layer design, reliable data transfer and cooperative communication in wireless sensor networks system and its applications. Dr. Yi is the recipient of the 2008 One Hundreds Talents Program Award by Chinese Academy of Sciences for research in wireless sensor network and applications for hazard and industries.

Hans-Jürgen Zepernick received the Dipl.-Ing. degree in Electrical Engineering from the University of Siegen, Germany, in 1987 and the Dr.-Ing. degree from the University of Hagen, Germany, in 1994. From 1987 to 1989, he was with the Radio and Radar Department of Siemens AG, Germany. He held the positions of Professor at Curtin University of Technology, Australia; Deputy Director of the Australian Telecommunications Research Institute; and Associate Director of the Australian Telecommunications Cooperative Research Centre. He is currently a Professor at the School of Computing of the Blekinge Institute of Technology, Sweden. His research interests include radio channel characterization and modeling, radio transmission technologies, mobile multimedia communications.

Index

D

E

F

N

O

P

W

Z

www.ingramcontent.com/pod-product-compliance
Lightning Source LLC
LaVergne TN
LVHW062134200926
848484LV00007B/1813
9781466609600